THE MATHEMATICAL THEORY OF SYMMETRY IN SOLIDS

REPRESENTATION THEORY FOR POINT GROUPS AND SPACE GROUPS

THE MATHEMATICAL THEORY OF SYMMETRY IN SOLIDS

Representation theory for point groups and space groups

BY

C. J. BRADLEY
Jesus College, Oxford
AND

A. P. CRACKNELL
Carnegie Laboratory of Physics, University of Dundee

CLARENDON PRESS · OXFORD

OXFORD
UNIVERSITY PRESS

Great Clarendon Street, Oxford, OX2 6DP,
United Kingdom

Oxford University Press is a department of the University of Oxford.
It furthers the University's objective of excellence in research, scholarship,
and education by publishing worldwide. Oxford is a registered trade mark of
Oxford University Press in the UK and in certain other countries

First Edition published in 1972
First published in paperback with corrections 2010

Published in the United States of America by Oxford University Press
198 Madison Avenue, New York, NY 10016, United States of America

British Library Cataloguing in Publication Data
Data available

Library of Congress Cataloging in Publication Data
Data available

ISBN 978–0–19–958258–7

To L. A. W. from C. J. B.
to M. F. C. from A. P. C.
and to S. L. A. from both of us

Preface to the 2009 Reprint

At the time we wrote *The Mathematical Theory of Symmetry in Solids* we had both recently been involved in working on the band structure of metals and we were acutely conscious of the lack of a book of tables of the irreducible representations of space groups and of the magnetic groups and their corepresentations. Before we had completed our book, certain authors, named in the Bibliography, had published similar tables, but in very different notations. What is unique about our book is that (i) it is complete, (ii) it covers, in very great detail, the theory behind the tables of point group and space groups representations and (iii) it also gives a set of matrix representatives and not just the character tables, for each of the degenerate representations. *The Mathematical Theory of Symmetry in Solids* went out of print a long time ago and secondhand copies had become very difficult to obtain. So we are therefore very pleased, and indeed honoured, that Oxford University Press has undertaken a reprint to provide readier access to the material involved to the present generation.

There were a few errors that were found in the original version and the opportunity has now been taken to correct them. However the number of these errors is very small and that is a tribute to the very great care and attention to detail by the original typesetters and proofreaders at the Press.

C. J. Bradley
A. P. Cracknell

June 2009

Preface

As the sub-title suggests, this book is devoted to the theory of the deduction of the irreducible representations of point groups and space groups and to their tabulation, together with some discussion of the determination of symmetry-adapted functions that belong to these representations. Some consideration is also given to the co-representations of magnetic point groups and space groups. Most of the theory of the determination of space-group representations is, of course, available already in the literature, but it is very scattered and different authors have used various sets of notation. Two sets of tables of space-group representations have been published for some time (Faddeyev 1964, Kovalev 1965), but neither included any very comprehensive account either of the theory or of the properties of these representations; at a very late stage in the preparation of this manuscript the work of Miller and Love (1967) was published and at the proof stage the work of Zak, Casher, Glück, and Gur (1969) also became available. We have made use of all these works in checking many of our tables.

Complete tables are given of the single-valued and double-valued representations of the group of **k** at each point of symmetry and along each line of symmetry in the Brillouin zone of each of the 230 space groups. These tables include all the relevant abstract finite groups (of order $\leqslant$ 192) and we identify the group of each wave vector **k** in terms of the appropriate abstract group. In these tables both the character tables and the matrix representatives are given. Several of the tables have been derived or checked by computer, specifically Tables 2.6, 5.1, 5.7, and 6.8, but we shall not give any description of the computing techniques involved and refer the reader, for example, to the review article on computers and group theory by Cannon (1969). The compatibility of Table 6.13 with Table 5.7 has also been checked, by hand; see the footnote on p. 468. Comparison with existing sets of tables for some individual space groups and with the tables of Faddeyev, Kovalev, Miller and Love, and Zak, Casher, Glück, and Gur has been made, but a completely exhaustive comparison with all the existing work has proved impossible because of the many different notations and conventions that have been used by different authors. It would be very remarkable indeed if all the entries in these tables were correct, and we should be extremely grateful to receive details of any errors that readers may find. An arrangement has been made with the Institute of Physics for the publication of such errors that come to our notice in the form of Letters to the Editor of *Journal of Physics C: Solid State Physics.*

We have included as much as seemed to be necessary of the description of the mathematical crystallography of Bravais lattices, point groups, and space groups, but the reader may find it profitable on occasions to refer either to one of the various textbooks on crystallography or to the *International tables for X-ray crystallography* (Henry and Lonsdale 1965). In Chapter 3 we give a partially complete account of representation theory for space groups. The complete theory is presented in Chapter

4 in which we treat the theory in a general notation so that we cover not only the theory of space groups but also obtain a theory that is useful in other applications outside the theory of solids. It is hoped, therefore, that parts of Chapter 4 will be of use, as a preliminary study in the theory of induced and subduced representations, to workers in a variety of fields and not just to solid-state physicists. We have assumed that the reader has a basic working knowledge of the theory of groups and of group representations; however, we have summarized the relevant parts of this theory in sections 1.2 and 1.3.

Some omissions have necessarily been made. We have omitted any discussion of the symmetry properties of tensors in crystals partly because it makes little use of representation theory and partly because the subject of the symmetry properties of tensors is already well-documented both for non-magnetic crystals (Nye 1957) and for magnetic crystals (Birss 1964). Although we have given some careful consideration in Chapter 7 to the theory of the corepresentations of magnetic space groups and have given some examples, it has not been practicable, in the space available, to include tables of the irreducible corepresentations of all the 1191 black and white magnetic space groups; these are all tabulated by Miller and Love (1967). We should have liked to have included some detailed discussion of the fields in which the theory and the tables that we have given can be applied. However, neither space nor time was available to do this properly, since this would probably need a second large volume; therefore, at appropriate points in the text we have simply indicated possible physical applications and given references to suitable treatises or review articles. Finally, we have omitted any discussion of the non-crystallographic point groups on the grounds that they do not properly belong in a book that is concerned primarily with solids rather than with molecules.

We are particularly grateful to Dr. S. L. Altmann, who provided much of the original stimulus for the writing of this book and who has watched its progress with considerable interest, and to the various people with whom we have had helpful discussions or correspondence about either text or references at various stages; Dr. J. S. Rousseau and Dr. N. B. Backhouse for a careful reading of various chapters which led to the removal of a number of errors; Dr. B. L. Davies, for extending the tables of cubic lattice harmonics in Chapter 2 from an accuracy of 8 to 11 decimal places and for some general assistance with parts of Chapter 7 and some of the Russian references; Dr. R. J. Elliott; Dr. G. Harbeke; Dr. K. L. Jüngst, who carried out an independent check on the tables of lattice harmonics in Chapter 2 and subsequently provided a few corrections; Dr. D. Litvin; Prof. R. Loudon; Dr. W. Marzec, for supplying a list of corrections to the tables of Kovalev (1965); Dr. K Olbrychski; Dr. M. Schulz; Mr. J. Staněk, for some help with the indexing of chapter 7; Mr. D. E. Wallis, for writing computer programs to check Table 5.1; Mr. R. H. Whittaker and Prof. J. Zak. One of us (A.P.C.) would like to record the fact that most of his contributions to the writing of this book were made during his two previous appointments, in the Physics Departments of the University of Singapore and the University of Essex, and he is grateful to his former colleagues for their interest and encouragement during those times. We are grateful to the various authors, editors, and publishers who have granted permission to reproduce copyright figures and tables, the sources of which are indicated appropriately *in situ*. Finally, we are also grateful to the staff of the Clarendon Press for the care with which the production has been handled.

C. J. BRADLEY
A. P. CRACKNELL

August 1969

Contents

1

Symmetry and the solid state

1.1. Introduction

THE history of man's interest in symmetry goes back many centuries (Belov 1956*b*, Coxeter and Moser 1965, Steno 1669), but its study on a modern scientific basis can be considered to have been started by the Abbé Haüy. Haüy studied the behaviour of a specimen of calcite when it was cleaved and, by breaking it into smaller and smaller pieces and studying the angles between the faces of the fragments, he convinced himself that the crystal was made up by the repetition of a large number of identical units. Haüy (1815*a–d*) studied many other crystals as well and summarized his conclusions in his so-called *Loi de symmétrie*. The study of symmetry developed through the nineteenth century with the formulation of ideas about point groups, Bravais lattices, and space groups.

A *point group* is a set of symmetry operations acting at a point and obeying the requirements that they should form a group in the mathematical sense; the crystallographic point groups satisfy the extra requirement that they must be compatible with a space lattice. Only a finite number of different combinations of symmetry operations are observed to occur in real crystals. The derivation of these 32 point groups was published by Hessel (1830) but his work was neglected for over 30 years until they were derived again by Gadolin (1869). Since then the point groups have been studied extensively, both in their original crystallographic context and, more recently, in the context of group-theoretical studies of the physics and chemistry of molecules and solids. There are useful crystallographic texts, for example, by Buerger (1956) and Phillips (1963*a*). General discussions of the theory associated with the applications of the group-theoretical studies of the point groups are given by many authors (for example; Bhagavantam and Venkatarayudu 1962, Cracknell 1968*b*, Hamermesh 1962, Heine 1960, Tinkham 1964).

We can also consider another collection of groups, this time by considering translational symmetry operations. If we were to look at the internal structure of a crystal we would find that it is made up of a large number of atoms or molecules regularly arranged; it would be possible to find a set of points within the crystal which are similar. That is, the crystal looks exactly the same if viewed from any one of these points as it does if it is viewed from any other of them. If we consider such a set of identical points they make up what the mathematicians call a *lattice*. It is possible to show that there is only a small number of essentially different ways of arranging a set of identical points so that the environment of each one is the same. This was done by Bravais (1850) who showed that in a three-dimensional space there are only

14 different lattices possible; consequently these are now known as *Bravais lattices*, even though Frankenheim had deduced, incorrectly, 15 such lattices somewhat earlier.

A point group is concerned with the symmetry of a finite object and for natural crystals there are only 32 different point groups; a Bravais lattice is concerned with the arrangement in space of a collection of mathematical points. To study fully the internal structure of a crystal, that is, the exact detailed arrangement of the atoms within the unit cell of a crystal, one needs a further development of symmetry studies known as a *space group*. A space group takes into consideration the symmetry of an arrangement of a set of identical objects, each of which is now not a point but is a finite object or a collection of atoms having some symmetry of its own. The actual operations present in a space group may be operations of the type which are present in point groups, namely pure rotations, reflections, the inversion operation, and roto-inversion or roto-reflection operations. But other operations are possible as well in a space group: they are screw rotation—and glide reflection operations—of symmetry. These are symmetry operations in which either a rotation axis or an ordinary reflection plane has a bodily movement of the crystal combined with it. In the descriptions of the derivation of the 230 space groups it is usually indicated that we owe them to Fedorov and Schönflies and sometimes the name of Barlow is added. A review of the history of the derivation of the space groups, together with a list of the publications of Barlow, Fedorov, and Schönflies, is given in an article by Burckhardt (1967). The derivation of the space groups has its origins in the works of Jordan (1868, 1869) and of Sohncke (1879). Sohncke had derived those space groups, of which there are 65, that contain only proper rotations and he noted that Jordan had previously derived them mathematically but had not translated his results into the more graphic terms of geometry. Schönflies re-derived these 65 space groups and extended the theory to include the space groups containing reflection planes of symmetry (Schönflies, 1887*a*, *b*, 1889, 1891). Similar results were derived by Fedorov (1885, 1891*a*) but his work was written in Russian and has not become so well known in western Europe; an account of the life and work of E. S. Fedorov and a list of his publications is given (in Russian) in the book by Shafranovskii (1963). It is evident that these two scientists began their works independently, one (Fedorov) as the director of a mine in the Urals and the other (Schönflies) at the suggestion of F. Klein at Göttingen, but in the course of time they heard of each other's work and compared their results. Barlow (1883) was first concerned with spherical packings and then starting with Sohncke's 65 groups he, too, obtained the remaining space groups by including reflection operations of symmetry (Barlow 1894). Burckhardt (1967) concludes that although Schönflies was not actually the first to establish the existence of the 230 space groups his writings have been the means of making their enumeration and identification generally known to the scientific world. His work, which is but little later than that of Fedorov and is quite independent, culminates in the book

Krystallsysteme und Krystallstructur (Schönflies 1891). A letter from Schönflies to Fedorov, quoted by Burckhardt (1967), reads 'I express my great joy about the agreement with your own views; I am particularly pleased, because I am no longer alone with my theory; it will still take great efforts before we shall succeed in winning over the crystallographers. *I concede you the priority with pleasure*, it is of no primary importance to me.' A convenient detailed list of the space groups in modern notation can be found in Volume 1 of the *International tables for X-ray crystallography* (Henry and Lonsdale 1965). At the present time there are about 9000 compounds whose space groups have been identified (for recent lists see Donnay, Donnay, Cox, Kennard, and King (1963), Nowacki, Edenharter, and Matsumoto (1967), and Wyckoff (1963, 1964, 1965, 1966, 1968)). The discovery of the two-dimensional space groups, which are also listed in detail in Volume 1 of the *International tables for X-ray crystallography*, is lost in the mists of antiquity because they arose in practice, in many different civilizations, in the designs of wallpapers or tiled floors (see, for example, Coxeter and Moser (1965), p. 33).

Although studies of a vast number of crystal structures had been undertaken by X-ray methods and these crystals had been assigned to the appropriate space groups, the study of the theory of symmetry seemed not to advance very much, after the derivation of the 230 space groups in about 1890, until Shubnikov in 1951 published a book called *Symmetry and anti-symmetry of finite figures* (in Russian, though this work is now translated into English, together with a list of many references to other works of Shubnikov (Shubnikov and Belov 1964)). A review of the developments in the theory of symmetry over the last 50 years is given by Koptsik (1967*a*) and a brief biography of A. V. Shubnikov is given at the beginning of volume 2 of *Kristallografiya* (*Soviet Phys. Crystallogr.* (English transl.) (1957)). The new developments were connected with introducing an operation of *anti-symmetry*. The classical theory of symmetry, point groups, Bravais lattices, and space groups, was essentially a 3-dimensional study, that is, a point P would be specified by the vector $\mathbf{r}\{=(x, y, z)\}$, and we would consider the effect of symmetry operations on this point. Shubnikov's basic idea was to say that in addition to the ordinary coordinates x, y, and z of a point we now also give each point a fourth coordinate, s, which can only take one of two possible values. The coordinate s can be the spin of a particle and the two allowed values will then correspond to spin up and spin down. Or, in purely abstract terms, they may be two colours such as black and white. If we include the coordinate s and if the values of s for the various atoms are randomly specified then the symmetry of the lattice has been completely destroyed. But if the spins are all parallel to a particular direction or if they are arranged in some regular fashion it is possible for some fraction of the symmetry to survive. If we introduce a new operation, which we may call the *operation of anti-symmetry*, $\mathscr{R}$, and consider this in conjunction with all the ordinary point-group and space-group operations it is possible to obtain a whole collection of new point groups and space groups which are called *black and white groups*, or *magnetic groups*, or *Shubnikov groups*. The idea of black and white groups was actually introduced long before Shubnikov's work, by Heesch (1929*a*, *b*, 1930*a*, *b*) and also discussed by Woods (1935*a*–*c*), but at that time there seemed to be no very great use for these groups in the description of physical systems. It was only with the introduction of the use of neutron diffraction techniques that it became apparent that these groups could be used in the description of magnetically ordered structures. If we think of s as being the two allowed values of a magnet's direction, parallel and anti-parallel to a particular direction, then $\mathcal{R}$ is the operation that reverses a magnetic moment. $\mathcal{R}$ can then be thought of as being the operation of *time-inversion*.

The theory of finite groups dates from the time of Cauchy‡ who was responsible for noticing that a number of apparently disconnected facts could be explained simultaneously by introducing the concept of a group. Galois§ added to the theory a number of new concepts, including that of an invariant subgroup, and part of his work on the theory of equations was a first and most startling example of the power of group theory in its applications. However, it is to Serret (1866) that we owe the first connected account of group theory. Since then there has been an increasing flow of literature on the subject and today abstract group theory still flourishes as a major topic for research. Furthermore, the variety of applications of finite groups in a host of mathematical situations as diverse as the theory of permutations, the study of symmetry, and the theories of algebraic and differential equations, to mention just a few, means that a study of groups is essential for those engaged in many disciplines requiring mathematical techniques. The natural sciences are riddled with examples of problems requiring a knowledge of group theory and it is a safe assumption that the biological sciences and perhaps even the social sciences, as they become increasingly mathematical, will produce further interesting applications.

In a mathematical theory it is often possible to pick out a number of famous scholars who have been responsible for the major advances. Group theory is no exception. The only fear we have in mentioning certain names is that those of many others who have made great advances are likely to be omitted. However, it is surely no injustice to single out the names of Sylow, Frobenius, Burnside, Schur, Miller, and Mackey (apologizing immediately to Noether, Brauer, Ito, and many others who have made great contributions to the theory of abstract groups but whose work is not so directly related to the applications in this book).

Sylow (1872) made considerable progress in describing the structure of a finite group particularly in relation to its number of elements when this number is factorized as a product of primes. Frobenius (1896*a*, *b*, 1898) originated and was largely responsible for the theory of group representations and group characters, though Burnside (1903, 1911) made such significant simplifications and was responsible for so many original results that he also must be thought of as a group theoretician of great influence.

‡ 1789–1857. § 1811–32.

As a worker with a prodigious output (of approximately 800 papers between 1894 and 1946) Miller (1894, 1946) devoted considerable attention to the investigation of the structures and properties of various groups of finite order. He was responsible for determining the numbers of finite groups of various specific orders and studying the interrelationships between the structures of these groups, as exemplified by their generating relations.

The study of relations between representations of a group and those of an invariant subgroup leads inevitably to projective representations. Schur was the first to notice this and in an astounding sequence of definitive papers (1904, 1907, 1911) he not only laid the foundations of the general theory of projective representations but established most of the results that are regarded as being of particular significance. Again it was Frobenius (1898) who was responsible for the first construction of what is now called an induced representation. However, this particular notion, so important to applications in physics, was not developed significantly until after 1950 when Mackey in a series of papers (1951, 1952, 1953*a*, *b*, 1958) made extremely important advances that already find considerable application not only in the theory of space groups but throughout the whole realm of theoretical physics (see also Mackey (1968)).

We have described the importance of point groups, Bravais lattices, and space groups in specifying both the macroscopic symmetry of a crystal, as determined by goniometry, and the symmetry of the internal structure of a crystal, as determined by X-ray diffraction or neutron diffraction experiments. In classical physics there are some applications of group theory, such as, for instance, the investigation of the normal modes of vibration of a molecule or solid (Wigner 1930) or the determination, for a crystal belonging to a given point group, of relationships that may exist between the various components of a tensor describing some macroscopic property (see, for example, Nye (1957)). However, it was with the advent of quantum mechanics that all the powerful mathematics of group theory and representation theory really became most useful in helping to understand physical systems. Much of the pioneer work on the application of group theory in quantum mechanics was done by Weyl, Wigner, and von Neumann (see Weyl (1931), the translation of the classic book by Wigner (1959) and the collected works of von Neumann (1961, 1963)). In studying a crystal at the microscopic level one has to remember that each of the individual particles of which the crystal is composed obeys quantum mechanics rather than classical mechanics and therefore has to be described by an appropriate wave function ψ. The key to the application of group theory to quantum mechanics lies in the result that is expounded in Chapter 11 of Wigner's classic book (English translation, Wigner (1959)). If a quantum-mechanical system is described by the appropriate Schrödinger wave equation Wigner's theorem can be summarized as follows: '*the representation of the group of the Schrödinger equation which belongs to a particular eigenvalue is uniquely determined up to a similarity transformation.*' Apart from accidental degeneracies this representation will be irreducible. The irreducible representations are therefore important because they can be used to label unambiguously the energy levels of a quantum-mechanical system. The irreducible representations of the crystallographic point groups and double point groups were determined a long time ago (Bethe 1929) and have been used extensively in labelling the energy levels of molecules (reviews and treatises include those of Eyring, Walter, and Kimball (1944), Nussbaum (1968), Rosenthal and Murphy (1936), Slater (1963), and Wilson, Decius, and Cross (1955)) in labelling the energy levels of ions or molecules in a crystal (reviews and discussions include those of Herzfeld and Meijer (1961), Hutchings (1964), Judd (1963), and McClure (1959*a*, *b*)) and also in labelling excitons in a crystal (Overhauser 1956). A particularly useful summary of the important properties of the crystallographic point groups and their representations is given by Koster, Dimmock, Wheeler, and Statz (1963).

The theory that underlies the determination of the irreducible representations of a space group was studied by Seitz (1936*b*) and first applied to symmorphic space groups by Bouckaert, Smoluchowski, and Wigner (1936), to non-symmorphic space groups by Herring (1942), and to double space groups by Elliott (1954*b*). Subsequently, many authors have determined the irreducible representations of individual space groups and, in doing so, have employed many different sets of notation. A substantial review was written by Koster (1957) and there have recently been published some sets of complete tables of the irreducible representations of all the 230 space groups (Faddeyev 1964, Kovalev 1965, Miller and Love 1967, Zak, Casher, Glück and Gur 1969). The importance of the irreducible representations of the space groups lies in the fact that, as a result of Wigner's theorem, they can be used in labelling the energy levels of a particle or quasi-particle in a crystal; they can therefore be used in labelling the electronic energy band structure and the phonon dispersion curves in a crystalline solid (for reviews see Blount (1962), Jones (1960), Nussbaum (1966), Slater (1965*b*, 1967) on electronic band structure, and Maradudin and Vosko (1968), Warren (1968) on phonon dispersion curves). Similarly, the irreducible representations of a space group can also be assigned to the magnon dispersion curves in a magnetic crystal. However, there is an added complication because the black and white Shubnikov space groups possess *corepresentations* rather than ordinary representations (Dimmock and Wheeler 1962*b*, Karavaev, Kudryavtseva, and Chaldyshev 1962, Loudon 1968, Wigner 1959, 1960*a*, *b*).

It is doubtful whether all the effort that workers have expended on the determination of point-group and space-group irreducible representations would be considered worth while if the only result was a scheme for labelling energy levels. However, the irreducible representations also enable one to determine the exact way in which a wave function ψ_i will transform under the various operations of the Schrödinger group of a molecule or crystal. This often enables some simplifications to be made when an unknown wave function is expanded in terms of a set of known functions such as spherical harmonics (Altmann 1957, Altmann and Bradley 1963*b*, Bell 1954, Betts

1959, McIntosh 1963, von der Lage and Bethe 1947) or plane waves (Cornwell 1969, Luehrmann 1968, Schlosser 1962, Slater 1965*b*, 1967). By restricting the expansion of an unknown ψ_i for an energy level E_i to those functions that are known to belong to the representation of E_i considerable simplifications can very often be achieved in the actual process of solving Schrödinger's equation to determine ψ_i. The knowledge of the transformation properties of the wave functions ψ_i is also of importance when considering a transition of a system between two energy levels E_i and E_j. It is then possible to use the condition that the quantum-mechanical matrix element of the transition is a pure number in order to determine, for any given perturbing potential, whether a given transition is allowed or forbidden, that is, to determine *selection rules*. The group-theoretical determination of selection rules for transitions in isolated molecules and in ions or molecules in crystals involves the study of products of various point-group representations and this is discussed in the references we have already mentioned. To use the knowledge of the transformation properties of ψ_i to study selection rules for transitions involving non-localized states in crystals is more complicated and initial work has been done by several authors (Birman 1962*b*, 1963, Elliott and Loudon 1960, Lax and Hopfield 1961, Zak 1962).

1.2. Group theory

We begin the mathematical work of this book by giving a short account of the theory of groups and their representations. We do not give proofs of theorems as these appear in the first few chapters of many well-known books such as those by Hamermesh (1962), Lomont (1959), Lyubarskii (1960), and Wigner (1959). For the sake of clarity, however, we illustrate some of the definitions and theorems by means of an example; for this purpose we use a group containing six elements which, as an abstract group we call $\mathbf{G}_6^2$ (see Table 5.1) and which, in one of its realizations, is the symmetry group of an equilateral triangle.

There are two good reasons for starting with a preliminary account such as this. The first is that it makes clear what the background to the work is, and hence what it is recommended that the reader should be familiar with before proceeding with the rest of the book. The second reason is that it serves to introduce a large amount of notation; furthermore, when this is done on topics that are relatively familiar, then a reader can adjust himself more easily to the style and notation of the authors than if he is plunged immediately into new work.

The following set of definitions and theorems forms, therefore, the group-theoretical background to the work of this book. In later chapters some of them will be used as building blocks for further theorems that are either more advanced or more directly related to the study of solids. The groups that occur in the theory of solids have quite a complicated structure and, if the theorems needed for dealing with them are established rigorously and in complete detail, the proofs of such theorems require some advanced algebraic methods not commonly met in introductory courses on group theory. Some of these methods appear in Chapter 4 and they rest heavily for their appreciation on the material in this section and the next.

DEFINITION 1.2.1. *Group.* A *group* **G** is a set of elements together with a binary composition called a *product* such that

(i) the product of any two elements in the group is defined and is a member of the group: if $A, B \in \mathbf{G}$ then $AB \in \mathbf{G}$,
(ii) the product is associative: $A(BC) = (AB)C$ for all $A, B, C \in \mathbf{G}$,
(iii) there exists a unique identity‡ E in the group: $EA = AE = A$ for all $A \in \mathbf{G}$, and
(iv) every element has a unique inverse‡ element: given $A \in \mathbf{G}$ there exists a unique element A^{-1} such that $AA^{-1} = A^{-1}A = E$.

DEFINITION 1.2.2. *Order of a group.* The number of elements in a group, **G**, is called the *order* of the group.

In what follows we shall be dealing only with *groups of finite order.* The symbol $|\mathbf{G}|$ is often used to denote the order of **G**.

DEFINITION 1.2.3. *Order of an element.* The *order of an element* $A \in \mathbf{G}$ is the least positive integer s such that $A^s = E$.

From Definition 1.2.1 it follows that a group is completely defined by its multiplication table. In fact, it is sufficient to give a set of relations involving certain elements from which the whole multiplication table can be constructed. The elements $P_1, P_2, \ldots, P_m$ of a group **G** are called a set of *generators* (or sometimes *generating elements*) if every element of **G** is expressible as a finite product of powers (including negative powers) of $P_1, P_2, \ldots, P_m$. The set of relations $g_k(P_1, P_2, \ldots, P_m) = E$ $(k = 1, 2, \ldots, s)$ satisfied by the generators, which are sufficient to determine the whole multiplication table of **G** are called the *defining relations* (or sometimes *generating relations*) of **G**. For a lengthy discussion of generators and generating relations see, for example, the book by Coxeter and Moser (1965).

Example 1.2.1. In Table 1.1 we give the multiplication table for the group $\mathbf{G}_6^2$ of order 6 whose generators are P and Q and whose defining relations are $P^3 = E$, $Q^2 = E$, and $QP = P^2Q$. P is of order 3 and Q is of order 2.

Sets of generators and defining relations are not by any means unique; in fact it is often more straightforward to take more generators than are strictly necessary. There is always, of course, a minimum number of generators without which one

‡ It is not necessary to postulate the uniqueness and two-sidedness of the identity and of the inverse elements; for the existence of a right identity ($AE = A$) and right inverses ($AA^{-1} = E$) together with axioms (i) and (ii) is sufficient to establish uniqueness and two-sidedness. However these properties are so fundamental that many authors include them in their definition.

TABLE 1.1

The multiplication table for the group $\mathbf{G}_6^2$.

E	P	P^2	Q	PQ	P^2Q
P	P^2	E	PQ	P^2Q	Q
P^2	E	P	P^2Q	Q	PQ
Q	P^2Q	PQ	E	P^2	P
PQ	Q	P^2Q	P	E	P^2
P^2Q	PQ	Q	P^2	P	E

Notes to Table 1.1.

(i) The generating relations of this group are: $P^3 = E$; $Q^2 = E$; $QP = P^2Q$.

(ii) In order to obtain a product LM take the element in the row beginning with L and the column headed M. Thus, for example, $(PQ)(P^2Q) = P^2$.

cannot generate the group, but if one uses such a minimal set the defining relations can sometimes be extremely complicated and it is to avoid such complications that one often takes an over-determined set of generators. However in the simple example above we do have a minimal set.

A geometrical realization of the above group is the set of symmetry operations that carry an equilateral triangle, $\triangle ABC$, into itself. If the intersection of the medians of $\triangle ABC$ is denoted by O then the operation P may be thought of as the 120° *anti-clockwise* rotation about a line through O perpendicular to the plane ABC and Q may be thought of as the reflection in the line AO. Equally well we could have taken P to be a 120° *clockwise* rotation and Q to be the reflection in BO, or for that matter in CO. In each case we would obtain the same group but with its elements labelled differently. This is a simple example of the non-uniqueness of a set of generators. In this example all six of the above sets of generators lead to the same defining relations; as we shall see in the next few paragraphs this is no accident, but on the other hand it must not be thought that different sets of generators always lead to the same defining relations.

DEFINITION 1.2.4. *Homomorphism, isomorphism.* Given two groups $\mathbf{G}$ and $\mathbf{G}'$, a mapping θ of $\mathbf{G}$ onto $\mathbf{G}'$ which preserves multiplication is called a *homomorphism.* Thus for a homomorphism θ it follows that, for all $G_1, G_2 \in \mathbf{G}$,

$$(\theta G_1)(\theta G_2) = \theta(G_1 G_2). \tag{1.2.1}$$

If in addition θ is a one-to-one mapping it is called an *isomorphism*: $\mathbf{G}$ and $\mathbf{G}'$ are then said to be *isomorphic.* If θ is an isomorphism and $\mathbf{G} = \mathbf{G}'$ then θ is called an *automorphism.*

Example 1.2.2. Let $\mathbf{G} = \mathbf{G}_6^2$ and $\mathbf{G}' = \mathbf{G}_2^1$, the cyclic group of order 2 composed of elements E and P' with $P'^2 = E$ (E being the identity) (see Table 5.1). Then if θ is defined so that $\theta E = E, \theta P = E, \theta P^2 = E, \theta Q = P', \theta(PQ) = P'$, and $\theta(P^2Q) = P'$, then θ is a homomorphism of $\mathbf{G}_6^2$ onto $\mathbf{G}_2^1$.

On the other hand, if it is given that θ is a homomorphism of $\mathbf{G}_6^2$ onto $\mathbf{G}_2^1$ then, by virtue of eqn. (1.2.1), it is sufficient in order to define θ to specify its action only on the generators of $\mathbf{G}_6^2$. Thus, if θ is a homomorphism, $\theta P = E$ and $\theta Q = P'$ defines θ completely.

Example 1.2.3. Let $\mathbf{G} = \mathbf{G}' = \mathbf{G}_6^2$. Then if ϕ is such that $\phi(E) = E$, $\phi(P) = P^2$, $\phi(P^2) = P$, $\phi(Q) = Q$, $\phi(PQ) = P^2Q$, and $\phi(P^2Q) = PQ$, ϕ is an automorphism of $\mathbf{G}_6^2$ onto itself.

THEOREM 1.2.1. *If ϕ_1 and ϕ_2 are two automorphisms of a group* **G** *then the mapping product $\phi_1\phi_2$ (where $(\phi_1\phi_2)G = \phi_1(\phi_2 G)$ for all $G \in \mathbf{G}$) is also an automorphism. Further the set of all automorphisms of a group* **G** *is itself a group* **A(G)** *of which the binary composition is the mapping product just defined.*

Example 1.2.4. Consider $\mathbf{A}(\mathbf{G}_6^2)$. It is a group of six elements. Its generators may be taken to be ϕ, as defined in Example 1.2.3, and ρ, where $\rho(P) = P$ and $\rho(Q) = P^2Q$. If the identity automorphism is denoted by ε then $\phi^2 = \varepsilon$, $\rho^3 = \varepsilon$, and $\phi\rho = \rho^2\phi$.

Let $\mathbf{G} = \mathbf{G}_6^2$ and $\mathbf{G}' = \mathbf{A}(\mathbf{G}_6^2)$. Then **G** and **G**′ are isomorphic under the mapping α: $\alpha P = \rho$ and $\alpha Q = \phi$.

Example 1.2.5. $\mathbf{G}_6^2$ is isomorphic with $\mathbf{S}_3$, the group of permutations of three identical objects.

Example 1.2.6. Example 1.2.4 explains the six possible sets of generators for $\mathbf{G}_6^2$ mentioned at the end of Example 1.2.1, for each automorphism leads to a different labelling of the group.

DEFINITION 1.2.5. *Kernel.* If $\theta\mathbf{G} = \mathbf{G}'$ is a homomorphism of **G** onto **G**′ then the *kernel* of θ is the set of elements of **G** that is mapped onto the identity of **G**′.

Example 1.2.7. The kernel of an isomorphism consists of one element only, the identity of **G**.

Example 1.2.8. The kernel of the homomorphism θ defined in Example 1.2.2 consists of the elements E, P, and P^2.

DEFINITION 1.2.6. *Subgroup.* A subset **H** of a group **G** that is itself a group under the same binary composition as in **G** is called a *subgroup* of **G**.

Example 1.2.9. The following are subgroups of $\mathbf{G}_6^2$: (i) $\mathbf{G}_6^2$ itself, (ii) $\mathbf{G}_3^1$, consisting of E, P, and P^2, (iii) $\mathbf{G}_2^1$, consisting of E and Q, (iv) $\mathbf{G}_2^{1\prime}$, consisting of E and PQ (v) $\mathbf{G}_2^{1\prime\prime}$, consisting of E and P^2Q, and (vi) $\mathbf{G}_1^1$, consisting of the identity E alone.

A group always has at least two subgroups, namely the group itself and the group consisting of the identity alone. Such subgroups are called *improper subgroups*. Other subgroups besides these two are called *proper subgroups*. Thus $\mathbf{G}_6^2$ has 4 proper subgroups.

If $\mathbf{H}$ is a subset of $\mathbf{G}$ consisting of t elements $(H_1, H_2, \ldots, H_t)$ then by $A\mathbf{H}$, $A \in \mathbf{G}$, we mean the subset of t elements $(AH_1, AH_2, \ldots, AH_t)$.

If $\mathbf{K}$ is a subset of $\mathbf{G}$ consisting of s elements $(K_1, K_2, \ldots, K_s)$ then by $\mathbf{KH}$ we mean the set of st elements K_iH_j $(i = 1 \text{ to } s, j = 1 \text{ to } t)$; by $\mathbf{H} + \mathbf{K}$ we mean the set of $(t + s)$ elements $(H_1, H_2, \ldots, H_t, K_1, K_2, \ldots, K_s)$.

DEFINITION 1.2.7. *Conjugate elements.* Two elements $G_1, G_2 \in \mathbf{G}$ are said to be *conjugate* if there exists an element $G \in \mathbf{G}$ such that $G_2 = GG_1G^{-1}$.

Example 1.2.10. In the group $\mathbf{G}_6^2$, P and P^2 form a pair of conjugate elements. For it follows immediately from the defining relations that $P^2 = QPQ^{-1}$ and $P = QP^2Q^{-1}$.

DEFINITION 1.2.8. *Abelian group.* If $\mathbf{G}$ is a group and $G_1G_2 = G_2G_1$ for all $G_1, G_2 \in \mathbf{G}$ then $\mathbf{G}$ is called an *Abelian group* (or sometimes a *commutative group*).

Example 1.2.11. The group $\mathbf{G}_3^1$ mentioned in Example 1.2.9 is an Abelian group.

THEOREM 1.2.2. *A group* $\mathbf{G}$ *splits into 'conjugacy classes'* $C_1, C_2, \ldots, C_r$ *such that the following properties hold*:

(i) *every element of* $\mathbf{G}$ *is in some class, and no element of* $\mathbf{G}$ *is in more than one class, so that* $\mathbf{G} = C_1 + C_2 + \cdots + C_r$,

(ii) *all the elements in a given class are mutually conjugate and consequently have the same order (though, of course, not all elements of the same order necessarily belong to the same class)*,

(iii) *an element that commutes with all elements of the group is in a class by itself and is called a 'self-conjugate' element (the identity is always in a class by itself, and, further, if* $\mathbf{G}$ *is Abelian then every element of* $\mathbf{G}$ *is in a class by itself)*,

(iv) *the number of elements in a class is a divisor of the order of the group*,

(v)
$$C_iC_j = \sum_{k=1}^{r} h_{ij,k}C_k \tag{1.2.2}$$

where the coefficients $h_{ij,k}$ *are called the 'class multiplication coefficients' and are positive integers or zero. Also*

$$h_{ij,k} = h_{ji,k}. \tag{1.2.3}$$

Example 1.2.12. For the group $\mathbf{G}_6^2$ take $C_1 = E$; $C_2 = P, P^2$; and $C_3 = Q, PQ, P^2Q$. Then the conditions of Theorem 1.2.2 are satisfied with $r = 3$. Elements in C_2 are of order 3, and elements in C_3 are of order 2. The identity is the only element to be in a class by itself. 1, 2, and 3, the numbers of elements in the classes C_1, C_2, and C_3 are divisors of 6, the order of $\mathbf{G}_6^2$. The values of the $h_{ij,k}$ can easily be evaluated by using Table 1.1. They are $h_{11,1} = 1$, $h_{12,2} = 1$, $h_{13,3} = 1$, $h_{22,1} = 2$, $h_{22,2} = 1$, $h_{23,3} = 2$, $h_{33,1} = 3$, $h_{33,2} = 3$, and all others not derived from these by eqn. (1.2.3) are zero.

DEFINITION 1.2.9. *Coset.* If **H** is a subgroup of **G** and A is any element of **G** then the subset **H**A is called a *right coset* of **H**; similarly A**H** is called a *left coset.* A is called the *coset representative* (and is not in any way special because if $B \in \mathbf{H}A$ then the coset $\mathbf{H}B = \mathbf{H}A$, so that any element of a coset can serve as the coset representative).

THEOREM 1.2.3. *A group* **G** *with a subgroup* **H** *splits up into 'left cosets' of* **H**. *The following properties hold*:

(i) *every element of* **G** *is in some left coset and no element of* **G** *is in more than one left coset,*

(ii) *every left coset contains the same number of elements, this number being equal to the order of* **H**. *Thus the order of* **H** *is a divisor of the order of* **G**; *the quotient* $|\mathbf{G}|/|\mathbf{H}|$, *which is just the number, t, of left cosets of* **H**, *is called the 'index' of* **H** *in* **G**.

Example 1.2.13. Take $\mathbf{G} = \mathbf{G}_6^2$ and $\mathbf{H} = \mathbf{G}_3^1$, consisting of E, P, and P^2. If we write $\mathbf{G}_6^2 = (E\mathbf{G}_3^1 + Q\mathbf{G}_3^1)$ then this is a decomposition of $\mathbf{G}_6^2$ satisfying the conditions of Theorem 1.2.3 with $t = 2$.

Example 1.2.14. From Example 1.2.9 the orders of the subgroups of $\mathbf{G}_6^2$ are seen to be 1, 2, 3, and 6, and these numbers are all divisors of 6, the order of $\mathbf{G}_6^2$. This division property is called Lagrange's theorem.

DEFINITION 1.2.10. *Invariant subgroup.* If **H** is a subgroup of **G** such that, for all $G \in \mathbf{G}$ and all $H \in \mathbf{H}$, $GHG^{-1} \in \mathbf{H}$, then **H** is said to be an *invariant* subgroup (or sometimes a *normal* subgroup or a *self-conjugate* subgroup) of **G**.

THEOREM 1.2.4. *If* **H** *is a subgroup of* **G** *then* **H** *is invariant if and only if* $A\mathbf{H} = \mathbf{H}A$ *for all* $A \in \mathbf{G}$; *that is, if and only if all right and left cosets coincide.*

Example 1.2.15. An immediate result from Theorem 1.2.4 is that all subgroups of index 2 are invariant subgroups. In particular, we see from Example 1.2.13 that $\mathbf{G}_3^1$ is invariant in $\mathbf{G}_6^2$.

DEFINITION 1.2.11. *Inner automorphism.* The mapping β from **G** onto itself such that $\beta G = BGB^{-1}$ for all $G \in \mathbf{G}$ (and where B is a fixed element of **G**) is an automorphism, and an automorphism produced by conjugation in this way is called an *inner automorphism* (all others being called *outer automorphisms*).

Example 1.2.16. All automorphisms of $\mathbf{G}_6^2$ are inner. Indeed it can soon be verified from Example 1.2.4 and Table 1.1 that ρ corresponds to conjugation by P and ϕ corresponds to conjugation by Q.

THEOREM 1.2.5. *A subgroup* **H** *of* **G** *is an invariant subgroup if and only if* **H** *is invariant under all inner automorphisms of* **G**.

THEOREM 1.2.6. *A subgroup* **H** *of* **G** *is invariant if and only if* **H** *is composed of entire conjugacy classes of* **G**.

Example 1.2.17. From Theorem 1.2.6 and Examples 1.2.9 and 1.2.13 it follows at once that the subgroup $\mathbf{G}_3^1$ is the only proper invariant subgroup of $\mathbf{G}_6^2$.

DEFINITION 1.2.12. *Simple groups.* If a group $\mathbf{G}$ has no proper invariant subgroup then it is said to be *simple.* If a group $\mathbf{G}$ has no proper invariant Abelian subgroup it is said to be *semi-simple.*

Example 1.2.18. $\mathbf{G}_6^2$ is not simple. Nor is it even semi-simple, for its subgroup $\mathbf{G}_3^1$ is a proper invariant Abelian subgroup.

THEOREM 1.2.7.

(i) *If* $\mathbf{H}$ *is an invariant subgroup of* $\mathbf{G}$ *then the left cosets of* $\mathbf{H}$ *form a group with binary composition defined by the equation*

$$(G_i\mathbf{H})(G_j\mathbf{H}) = (G_iG_j)\mathbf{H}. \tag{1.2.4}$$

This group is called the quotient group $\mathbf{G}/\mathbf{H}$ *and its order is equal to the index of* $\mathbf{H}$ *in* $\mathbf{G}$. *The identity of* $\mathbf{G}/\mathbf{H}$ *is the subgroup* $\mathbf{H}$ *itself.*

(ii) *If* θ *is a homomorphism of* $\mathbf{G}$ *onto* $\mathbf{G}'$ *and* $\mathbf{H}$ *is the kernel of* θ *then* $\mathbf{H}$ *is an invariant subgroup of* $\mathbf{G}$ *and* $\mathbf{G}'$ *is isomorphic to* $\mathbf{G}/\mathbf{H}$.

(iii) *Conversely, if* $\mathbf{H}$ *is an invariant subgroup of* $\mathbf{G}$ *and* θ *is a mapping of* $\mathbf{G}$ *onto* $\mathbf{G}/\mathbf{H}$ *such that* $\theta G = G\mathbf{H}$ *for all* $G \in \mathbf{G}$, *then* θ *is a homomorphism and the kernel of* θ *is* $\mathbf{H}$.

Example 1.2.19. Let $\mathbf{G} = \mathbf{G}_6^2$ and $\mathbf{H} = \mathbf{G}_3^1$. From Example 1.2.15 $\mathbf{G}_3^1$ is invariant and so the quotient group $\mathbf{G}_6^2/\mathbf{G}_3^1$ is defined. From Example 1.2.13 it is seen to consist of the two elements $E\mathbf{G}_3^1$ and $Q\mathbf{G}_3^1$. From Example 1.2.2 and Theorem 1.2.7(ii) it follows that this quotient group is isomorphic with $\mathbf{G}' = \mathbf{G}_2^1$, the cyclic group of order 2 defined in Example 1.2.2.

DEFINITION 1.2.13. *Outer direct product.* Let $\mathbf{G}$ be a group with subgroups $\mathbf{H}$ and $\mathbf{K}$ such that

(i) if $H \in \mathbf{H}$, $K \in \mathbf{K}$ then $HK = KH$,

(ii) all $G \in \mathbf{G}$ can be expressed in the form $G = HK$ with $H \in \mathbf{H}$ and $K \in \mathbf{K}$, and

(iii) the intersection of $\mathbf{H}$ and $\mathbf{K}$, $\mathbf{H} \cap \mathbf{K} = \{E\}$, the set consisting of the identity element of $\mathbf{G}$.

Then $\mathbf{G}$ is called the *outer direct product* of $\mathbf{H}$ and $\mathbf{K}$ and the factorization in (ii) is unique. We write $\mathbf{G} = \mathbf{H} \otimes \mathbf{K} = \mathbf{K} \otimes \mathbf{H}$.

THEOREM 1.2.8. *If* $\mathbf{G} = \mathbf{H} \otimes \mathbf{K}$ *then*

(i) $\mathbf{H}$ *and* $\mathbf{K}$ *are both invariant subgroups of* $\mathbf{G}$, *and*

(ii) *the number of classes in* $\mathbf{G}$ *is the product of the number of classes in* $\mathbf{H}$ *and the number of classes in* $\mathbf{K}$.

If further, $\mathbf{H}$ *and* $\mathbf{K}$ *are Abelian, then* $\mathbf{G}$ *is also Abelian.*

Example 1.2.20. Given two groups $\mathbf{H}$ and $\mathbf{K}$ it is possible to form the outer direct product $\mathbf{H} \otimes \mathbf{K} = \mathbf{G}$ by writing the elements of $\mathbf{G}$ in the form (H, K) with multiplication

$$(H_1, K_1)(H_2, K_2) = (H_1 H_2, K_1 K_2). \tag{1.2.5}$$

Equation (1.2.5) ensures the validity of rules (i)–(iii) of Definition 1.2.13.‡

DEFINITION 1.2.14. *Inner direct product.* The subgroup of elements of the outer direct product $\mathbf{H} \otimes \mathbf{H}$ of the form (H, H) is a group $\mathbf{G}$ isomorphic to $\mathbf{H}$ called the *inner direct product* of $\mathbf{H}$ with itself. We write $\mathbf{G} = \mathbf{H} \boxtimes \mathbf{H}$. The notation is specially designed to distinguish between the outer direct product, $\mathbf{H} \otimes \mathbf{H}$, and the inner direct product, $\mathbf{H} \boxtimes \mathbf{H}$, of two isomorphic groups.

DEFINITION 1.2.15. *Normal series.* A *normal series* of a group $\mathbf{G} = \mathbf{H}_0$ is a series of subgroups $\mathbf{H}_0, \mathbf{H}_1, \ldots, \mathbf{H}_s, \mathbf{G}_1^1$ ($\mathbf{G}_1^1$ being the group consisting of the identity, E, alone) such that $\mathbf{H}_{k+1}$ is a proper invariant (normal) subgroup of $\mathbf{H}_k$, $k = 0$ to $(s - 1)$.

Example 1.2.21. From Example 1.2.9 it follows that $\mathbf{G}_6^2, \mathbf{G}_3^1, \mathbf{G}_1^1$ forms a normal series for $\mathbf{G}_6^2$.

DEFINITION 1.2.16. *Solvable groups.* A group $\mathbf{H}_0$ is said to be *solvable* if and only if it possesses a normal series $\mathbf{H}_0, \mathbf{H}_1, \ldots, \mathbf{H}_s, \mathbf{G}_1^1$ whose quotient groups $\mathbf{H}_0/\mathbf{H}_1, \mathbf{H}_1/\mathbf{H}_2, \ldots, \mathbf{H}_s/\mathbf{G}_1^1$ are Abelian.

Example 1.2.22. The group $\mathbf{G}_6^2$ is solvable, for $\mathbf{G}_6^2/\mathbf{G}_3^1$ is isomorphic with $\mathbf{G}_2^1$, and both $\mathbf{G}_2^1$ and $\mathbf{G}_3^1$ are Abelian.

DEFINITION 1.2.17. *Semi-direct product.* Let $\mathbf{G}$ be a group with subgroups $\mathbf{H}$ and $\mathbf{K}$ such that

(i) if $K \in \mathbf{K}$ then $K\mathbf{H} = \mathbf{H}K$,
(ii) all $G \in \mathbf{G}$ can be expressed in the form $G = HK$ with $H \in \mathbf{H}$ and $K \in \mathbf{K}$, and
(iii) the intersection of $\mathbf{H}$ and $\mathbf{K}$, $\mathbf{H} \cap \mathbf{K} = \{E\}$, the set consisting of the identity element of $\mathbf{G}$.

Then $\mathbf{G}$ is called the *semi-direct product* of $\mathbf{H}$ and $\mathbf{K}$ and the factorization in (ii) is unique. We write $\mathbf{G} = \mathbf{H} \wedge \mathbf{K}$. Note that in contrast to the symbol $\otimes$ the caret symbol $\wedge$ is not commutative. $\mathbf{H}$ is invariant in $\mathbf{G}$ but $\mathbf{K}$ is not necessarily invariant in $\mathbf{G}$. In an expression such as $\mathbf{H} \wedge \mathbf{K}$ *we always write the invariant subgroup first.*

Example 1.2.23. Using the notation of Example 1.2.9, $\mathbf{G}_6^2 = \mathbf{G}_3^1 \wedge \mathbf{G}_2^1$. Note that $\mathbf{G}_2^1$ is not invariant in $\mathbf{G}_6^2$, so that $\mathbf{G}_6^2$ is *not* the direct product of $\mathbf{G}_3^1$ and $\mathbf{G}_2^1$.

‡ The group $\mathbf{H}$ is isomorphic to the subgroup $\mathbf{H}'$ of $\mathbf{G}$ consisting of elements (H, E), all $H \in \mathbf{H}$. If $\mathbf{K}'$ is similarly defined then according to Definition 1.2.13 what we really have is $\mathbf{G} = \mathbf{H}' \otimes \mathbf{K}'$; however because of the isomorphism between $\mathbf{H}$ and $\mathbf{H}'$ and between $\mathbf{K}$ and $\mathbf{K}'$ it is customary to write $\mathbf{G} = \mathbf{H} \otimes \mathbf{K}$.

Let **G** be a group and **A(G)** its group of automorphisms. Then it is always possible to form the semi-direct product $\mathbf{G} \wedge \mathbf{A}(\mathbf{G})$. It consists of ordered pairs (G, α) with $G \in \mathbf{G}$ and $\alpha \in \mathbf{A}(\mathbf{G})$ and with the product defined by

$$(G_1, \alpha_1)(G_2, \alpha_2) = (G_1\alpha_1(G_2), \alpha_1\alpha_2). \tag{1.2.6}$$

DEFINITION 1.2.18. *Holomorph.* The group $\mathbf{G} \wedge \mathbf{A}(\mathbf{G})$ is called the *holomorph* of **G**.

Example 1.2.24. The holomorph of $\mathbf{G}_3^1$ is isomorphic with $\mathbf{G}_6^2$. (Indeed every semi-direct product is isomorphic with a subgroup of the holomorph of the invariant subgroup out of which the semi-direct product is formed.) This follows because $\mathbf{G}_6^2 = \mathbf{G}_3^1 \wedge \mathbf{G}_2^1$ and $\mathbf{G}_2^1$ is isomorphic with $\mathbf{A}(\mathbf{G}_3^1)$.

In some of the examples given above the reasoning is omitted or abbreviated. The reader is recommended to fill in the gaps and to convince himself where statements are made that seem to require further explanation.

1.3. Group representations

We now move on from the abstract theory of groups described in section 1.2 to the more specialized topic of group representations, for our main concern in this book is with the representations of groups that appear in the study of solids and not with the development of the abstract theory of groups. It is true that for pure mathematicians there are interesting topics from more advanced abstract group theory, such as the theory of central extensions and cohomology groups, which throw considerable light on the groups in which we are interested; but in a book for applied mathematicians and theoretical physicists a study of these would be a long and perhaps indigestible digression. In this section we continue to give some elementary definitions and theorems, illustrated again, where possible, by examples using the group $\mathbf{G}_6^2$.

DEFINITION 1.3.1. *Matrix group.* A *matrix group* $\boldsymbol{\Delta}$ is a group of non-singular matrices. If all the matrices of the group are unitary then it is said to be a *unitary matrix group*.

In what follows we shall be concerned with matrix groups of finite order and with matrices of finite dimension.

DEFINITION 1.3.2. *Equivalence.* Two matrices $\mathbf{D}_1$ and $\mathbf{D}_2$ are said to be conjugate if there exists a non-singular matrix **S** such that $\mathbf{D}_1 = \mathbf{S}\mathbf{D}_2\mathbf{S}^{-1}$.

Two matrix groups $\boldsymbol{\Delta}_1$ and $\boldsymbol{\Delta}_2$ are said to be *equivalent* if there exists a non-singular matrix **S** such that $\boldsymbol{\Delta}_1 = \mathbf{S}\boldsymbol{\Delta}_2\mathbf{S}^{-1}$. Note that the matrices of the two groups not only have to be conjugate in pairs but that the conjugation must be produced by using the *same* matrix **S** for all pairs. This is a very strong condition, so strong that it follows immediately that $\boldsymbol{\Delta}_1$ and $\boldsymbol{\Delta}_2$ must be isomorphic. The converse is not true because $\mathbf{\Delta}_1$ and $\mathbf{\Delta}_2$ can be isomorphic and yet have different dimensions and then they are obviously not equivalent.

THEOREM 1.3.1. *Every matrix group is equivalent to a unitary matrix group.*

The following symbols will be used in dealing with matrices:

$\mathbf{D}^T$ for the transpose of $\mathbf{D}$,
$\mathbf{D}^*$ for the complex conjugate of $\mathbf{D}$,
$\mathbf{D}^\dagger$ $[= (\mathbf{D}^*)^T]$ for the Hermitean conjugate of $\mathbf{D}$,
$\tilde{\mathbf{D}}$ $[= (\mathbf{D}^{-1})^T]$ for the contragredient of $\mathbf{D}$,
$\dim \mathbf{D}$ for the dimension of $\mathbf{D}$.

DEFINITION 1.3.3. *Trace, character.* The *trace* of a matrix $\mathbf{D}$ is the sum of its diagonal elements, written $\operatorname{tr} \mathbf{D}$.

The *character* of a matrix group $\mathbf{\Delta}$ is the function χ defined on all elements $\mathbf{D} \in \mathbf{\Delta}$ such that $\chi(\mathbf{D}) = \operatorname{tr} \mathbf{D}$.

THEOREM 1.3.2. *If $\mathbf{\Delta}$ is a matrix group with identity $\mathbf{E}$ then $\chi(\mathbf{E}) = \dim \mathbf{E}$; and further, if $\mathbf{C}$ and $\mathbf{D}$ are in the same conjugacy class in $\mathbf{\Delta}$, then $\chi(\mathbf{C}) = \chi(\mathbf{D})$.*

THEOREM 1.3.3. *Two matrix groups are isomorphic and have the same character if and only if they are equivalent.*

DEFINITION 1.3.4. *Representation of a group.* A *representation of a group* $\mathbf{G}$ is a homomorphism γ of $\mathbf{G}$ onto a group $\mathbf{T}$ of non-singular linear operators acting on a finite-dimensional vector space $\mathbf{V}$ over the complex field. We write $\gamma G = \mathbf{T}_G$, for all $G \in \mathbf{G}$.

From Definition 1.3.4 it follows that when γ is a representation then

(i) $\mathbf{T}_{G_1}(\mathbf{T}_{G_2}\mathbf{x}) = \mathbf{T}_{G_1G_2}\mathbf{x}$ for all $G_1, G_2 \in \mathbf{G}$ and for all $\mathbf{x} \in \mathbf{V}$,
(ii) $\mathbf{T}_E\mathbf{x} = \mathbf{x}$ for all $\mathbf{x} \in \mathbf{V}$; that is, $\mathbf{T}_E$ is the identity operator, and
(iii) $\mathbf{T}_G^{-1}\mathbf{x} = \mathbf{T}_{G^{-1}}\mathbf{x}$ for all $G \in \mathbf{G}$ and for all $\mathbf{x} \in \mathbf{V}$.

If γ is an isomorphism the representation is said to be *faithful*.

Suppose now that we choose a basis $\langle \mathbf{x}|$ consisting of linearly independent vectors $\mathbf{x}_1, \mathbf{x}_2, \ldots, \mathbf{x}_d$ spanning the space $\mathbf{V}$, and let us define matrices $\mathbf{\Gamma}_\mathbf{x}(G)$ by the equations

$$\mathbf{T}_G\mathbf{x}_i = \sum_{j=1}^{d} \mathbf{x}_j\mathbf{\Gamma}_\mathbf{x}(G)_{ji} \quad (i = 1 \text{ to } d), \tag{1.3.1}$$

then $\mathbf{\Gamma}_\mathbf{x}(G)$ is said to be the matrix representing G with respect to the basis $\langle \mathbf{x}|$ in the representation γ. The set of all distinct matrices $\mathbf{\Gamma}_\mathbf{x}(G)$ is a matrix group and it is the homomorphic image of $\mathbf{G}$ under the mapping $G \to \mathbf{\Gamma}_\mathbf{x}(G)$, the kernel of the homomorphism being the elements of $\mathbf{G}$ mapped onto the unit matrix.

Example 1.3.1. Let $\mathbf{G}$ be the group $\mathbf{G}_6^2$, the multiplication table of which is given in Table 1.1. A geometrical realization of this group was described in Example 1.2.1.

This involved an equilateral triangle ABC with centroid O. Let $\mathbf{OA} = \mathbf{a}$, $\mathbf{OB} = \mathbf{b}$, and $\mathbf{OC} = \mathbf{c}$; then $\mathbf{a} + \mathbf{b} + \mathbf{c} = 0$, and the plane of the triangle forms a vector space $\mathbf{V}$ of dimension 2. This we take to be the underlying vector space of the representation γ. Take as basis for this vector space $\mathbf{x}_1 = \mathbf{b}$ and $\mathbf{x}_2 = \mathbf{c}$. The representation γ maps P and Q onto elements $\mathbf{T}_P$ and $\mathbf{T}_Q$ which are respectively an anti-clockwise rotation of 120° about O and a reflection in the line AO. The operators $\mathbf{T}_P$ and $\mathbf{T}_Q$ are operators acting on $\mathbf{V}$ and from their definition $\mathbf{T}_P\mathbf{b} = \mathbf{c}$, $\mathbf{T}_P\mathbf{c} = \mathbf{a} = -\mathbf{b}-\mathbf{c}$, $\mathbf{T}_Q\mathbf{b} = \mathbf{c}$, and $\mathbf{T}_Q\mathbf{c} = \mathbf{b}$. From eqn. (1.3.1) it follows that

$$\mathbf{\Gamma}_{\mathbf{x}}(P) = \begin{pmatrix} 0 & -1 \\ 1 & -1 \end{pmatrix} \quad \text{and} \quad \mathbf{\Gamma}_{\mathbf{x}}(Q) = \begin{pmatrix} 0 & 1 \\ 1 & 0 \end{pmatrix}. \tag{1.3.2}$$

Since γ is a homomorphism the rest of the matrix group follows from multiplication. Thus, for example,

$$\mathbf{\Gamma}_{\mathbf{x}}(P^2) = \mathbf{\Gamma}_{\mathbf{x}}(P)\mathbf{\Gamma}_{\mathbf{x}}(P) = \begin{pmatrix} -1 & 1 \\ -1 & 0 \end{pmatrix}, \tag{1.3.3}$$

and so on. It can easily be checked that γ is a faithful representation. However it contains non-unitary matrices. γ is of dimension 2. Thus if χ_γ is the character of γ it follows from Theorem 1.3.2 and eqn. (1.3.2) that $\chi_\gamma(E) = 2$, $\chi_\gamma(P) = \chi_\gamma(P^2) = -1$, and $\chi_\gamma(Q) = \chi_\gamma(PQ) = \chi_\gamma(P^2Q) = 0$.

Example 1.3.2. Corresponding to the homomorphism θ defined in Example 1.2.2 there exists the 1-dimensional matrix representation B in which $\mathbf{B}(P) = 1$ and $\mathbf{B}(Q) = -1$. This is because of the isomorphism between $\mathbf{G}_2^1$ and the multiplicative group of order 2 consisting of the numbers 1 and -1.

Example 1.3.3. There always exists, for any group $\mathbf{G}$, the symmetric (or trivial) 1-dimensional representation A in which $\mathbf{A}(G) = 1$ for all $G \in \mathbf{G}$.

From Examples 1.3.2 and 1.3.3 it can be seen that we can often find a matrix representation of a group without specifying an underlying vector space and a choice of basis. This is because an abstract group has homomorphisms that are matrix groups, and are therefore matrix representations. However, as far as we are concerned, the way representations arise out of Definition 1.3.4, as illustrated by Example 1.3.1, seems to be more natural. This is because all the groups in which we are interested have geometrical realizations, and their representations are most easily constructed by considering the action of their elements within these realizations on appropriate vector spaces.

THEOREM 1.3.4. *Let* $\langle \mathbf{x}|$ *and* $\langle \mathbf{y}|$ *be two bases of* $\mathbf{V}$ *defined so that*

$$\mathbf{y}_k = \sum_{i=1}^{d} \mathbf{x}_i \mathbf{S}_{ik} \quad (k = 1 \text{ to } d) \tag{1.3.4}$$

where $\mathbf{S}$ *is non-singular, then*

$$\boldsymbol{\Gamma}_{\mathbf{y}}(G) = \mathbf{S}^{-1}\boldsymbol{\Gamma}_{\mathbf{x}}(G)\mathbf{S} \quad \textit{for all } G \in \mathbf{G}. \tag{1.3.5}$$

That is to say, a change of basis leads to matrix groups $\boldsymbol{\Gamma}_{\mathbf{x}}(G)$ and $\boldsymbol{\Gamma}_{\mathbf{y}}(G)$, which are equivalent. Hence, using Theorem 1.3.1, it is possible to choose a basis $\langle\mathbf{z}|$ in $\mathbf{V}$ such that $\boldsymbol{\Gamma}_{\mathbf{z}}(G)$ is a unitary matrix group.

Example 1.3.4. Using the same notation as in Example 1.3.1 we define the new basis $\langle\mathbf{z}|$ so that $\mathbf{z}_1 = (\frac{1}{3}\sqrt{3})(-\mathbf{b} + \mathbf{c})$ and $\mathbf{z}_2 = -(\mathbf{b} + \mathbf{c})$ then, since $\mathbf{x}_1 = \mathbf{b}$ and $\mathbf{x}_2 = \mathbf{c}$,

$$\langle\mathbf{z}_1, \mathbf{z}_2| = \langle\mathbf{x}_1, \mathbf{x}_2|\begin{pmatrix} -\frac{1}{3}\sqrt{3} & -1 \\ \frac{1}{3}\sqrt{3} & -1 \end{pmatrix}. \tag{1.3.6}$$

Hence, from Theorem 1.3.4,

$$\boldsymbol{\Gamma}_{\mathbf{z}}(G) = \mathbf{S}^{-1}\boldsymbol{\Gamma}_{\mathbf{z}}(G)\mathbf{S}, \tag{1.3.7}$$

where

$$\mathbf{S} = \begin{pmatrix} -\frac{1}{3}\sqrt{3} & -1 \\ \frac{1}{3}\sqrt{3} & -1 \end{pmatrix}. \tag{1.3.8}$$

It follows from eqns. (1.3.2), (1.3.7), and (1.3.8) that

$$\boldsymbol{\Gamma}_{\mathbf{z}}(P) = \begin{pmatrix} -\frac{1}{2} & -\frac{1}{2}\sqrt{3} \\ \frac{1}{2}\sqrt{3} & -\frac{1}{2} \end{pmatrix} \quad \text{and} \quad \boldsymbol{\Gamma}_{\mathbf{z}}(Q) = \begin{pmatrix} -1 & 0 \\ 0 & 1 \end{pmatrix}, \tag{1.3.9}$$

so that with the new basis $\langle\mathbf{z}|$ the matrix representation $\boldsymbol{\Gamma}_{\mathbf{z}}(G)$ is unitary. The geometrical reason for this is that $\mathbf{z}_1$ and $\mathbf{z}_2$ are orthogonal and equal in length, and that the operators $\mathbf{T}_P$ and $\mathbf{T}_Q$ are unitary, and therefore preserve angles and lengths. Note that in accordance with Theorem 1.3.3 $\boldsymbol{\Gamma}_{\mathbf{z}}(G)$ and $\boldsymbol{\Gamma}_{\mathbf{x}}(G)$ have the same characters.

We shall assume from now on that whenever we have a matrix representation $\boldsymbol{\Gamma}(G)$ *the basis in the vector space* $\mathbf{V}$ *has been chosen so that the matrices* $\boldsymbol{\Gamma}(G)$ *are unitary and that we can therefore use the following relationship:*

$$\boldsymbol{\Gamma}(G^{-1})_{ij} = \boldsymbol{\Gamma}^*(G)_{ji}. \tag{1.3.10}$$

Because of Theorems 1.3.1 and 1.3.3 there is no loss of generality in considering in this way only those matrix representations which are unitary.

DEFINITION 1.3.5. *Invariant subspace.* Let γ be a representation of $\mathbf{G}$ so that $\mathbf{T} = \gamma\mathbf{G}$ is a group of non-singular linear operators acting on a vector space $\mathbf{V}$. $\mathbf{U}$ is said to be an *invariant subspace* of $\mathbf{V}$ under $\mathbf{T}$ if

(i) $\mathbf{U}$ is a vector subspace of $\mathbf{V}$, and

(ii) $\mathbf{T}_G\mathbf{x} \in \mathbf{U}$ for all $\mathbf{T}_G \in \mathbf{T}$ and for all $\mathbf{x} \in \mathbf{U}$.

DEFINITION 1.3.6. *Irreducible representation.* If **V** has no proper invariant subspace under **T** (that is, no subspace invariant under **T** except **V** itself and the zero-vector) then γ is said to be an *irreducible representation.* If there exists a proper invariant subspace under **T** then γ is said to be *reducible.* If **V** can be split up into the direct sum of subspaces each of which is invariant under **T** and each of which is the carrier space for an irreducible representation of **G** then γ is said to be *completely reducible.*

THEOREM 1.3.5. *All representations of a finite group* **G** *are completely reducible.*

THEOREM 1.3.6. Schur's lemma. *Let* $\mathbf{\Gamma}(G)$ *and* $\mathbf{\Gamma}'(G)$ *be two irreducible representations of* **G** *such that* $\mathbf{\Gamma}(G)\mathbf{S} = \mathbf{S}\mathbf{\Gamma}'(G)$ *for all* $G \in \mathbf{G}$, *then either* $\mathbf{S} = 0$ *or* **S** *is a non-singular matrix and* $\mathbf{\Gamma}(G)$ *is equivalent to* $\mathbf{\Gamma}'(G)$.

The following criteria for irreducibility are useful.

THEOREM 1.3.7. *A representation is irreducible if, and only if, the only matrices which commute with all matrices of the representation are scalar multiples of the unit matrix.*

THEOREM 1.3.8. *Let* **G** *be a group of order* $|\mathbf{G}|$ *with elements* $G_1, G_2, \ldots, G_{|\mathbf{G}|}$. *Then* $\mathbf{\Gamma}(G)$ *is an irreducible representation if and only if*

$$\frac{1}{|\mathbf{G}|}\sum_{i=1}^{|\mathbf{G}|} |\chi_\gamma(G_i)|^2 = 1 \tag{1.3.11}$$

where $\chi_\gamma(G_i)$ *is the character of* $\mathbf{\Gamma}(G_i)$.

Example 1.3.5. The representations A, B, and $\mathbf{\Gamma}_z(G)$ of $\mathbf{G}_6^2$, as defined in Examples 1.3.2–1.3.4 are irreducible.

It should now be clear from the theorems and remarks made in this section that the task of finding all the representations of a finite group **G** can be limited to the task of finding all the non-equivalent unitary irreducible representations of **G**. *From now on we shall abbreviate the phrase 'unitary irreducible representation' by the word 'rep'.* In discussing the problem of the determination of the reps of a finite group, **G**, we shall use the following symbols:

G	a group,
$\lvert\mathbf{G}\rvert$	the order of **G**,
G_j	$j = 1$ to $\lvert\mathbf{G}\rvert$, the elements of **G**,
r	the number of classes of **G**,
C_t	$t = 1$ to r, the classes of **G**,
r_t	the number of elements in C_t,
Γ^i	the label for the ith rep of **G**,
d_i or dim Γ^i	the dimension of Γ^i
$\mathbf{V}^i$	the underlying vector space (or carrier space) of Γ^i

$\langle \mathbf{z}^i|$ a basis in $\mathbf{V}^i$ for the rep Γ^i,
$\mathbf{z}^i_p$ $p = 1$ to d_i, the elements of the basis $\langle \mathbf{z}^i|$,
$\chi^i(G_j)$ the character of G_j in Γ^i,
$\chi^i(C_t)$ the character of any element in the class C_t in Γ^i,
$\Gamma^i(G_j)_{pq}$ the (p, q) element of the matrix representative of G_j in the rep Γ^i

$$\delta_{pq} = \begin{cases} 1 & \text{if } p = q, \\ 0 & \text{if } p \neq q, \end{cases}$$

$$\delta^{ik} = \begin{cases} 1 & \text{if } \Gamma^i \text{ is \textit{identical} to } \Gamma^k, \\ 0 & \text{if } \Gamma^i \text{ is not equivalent to } \Gamma^k. \end{cases}$$

DEFINITION 1.3.7. *Classification of reps*
(i) A rep Γ is of the *first kind* if Γ is equivalent to a group of real matrices.
(ii) A rep Γ is of the *second kind* if Γ is equivalent to Γ^* but not to any group of real matrices.
(iii) A rep Γ is of the *third kind* if Γ is not equivalent to Γ^*.

THEOREM 1.3.9

$$\frac{1}{|\mathbf{G}|} \sum_{j=1}^{|\mathbf{G}|} \chi(G_j^2) = \begin{cases} 1 & \textit{if and only if } \Gamma \textit{ is of the first kind,} \\ -1 & \textit{if and only if } \Gamma \textit{ is of the second kind,} \\ 0 & \textit{if and only if } \Gamma \textit{ is of the third kind.} \end{cases} \quad (1.3.12)$$

Example 1.3.6. The reps A, B, and $\Gamma_z(G)$ of $\mathbf{G}_6^2$ are all of the first kind.

THEOREM 1.3.10. *The number of reps = the number of classes = r.*

Example 1.3.7. There are three classes in $\mathbf{G}_6^2$ (see Example 1.2.12) and so the three reps A, B, and $\Gamma_z(G)$ are the only reps that exist for this group, in the sense that any other rep must be equivalent to one or other of these three.

THEOREM 1.3.11

$$\sum_{i=1}^{r} d_i^2 = |\mathbf{G}|. \quad (1.3.13)$$

Example 1.3.8. As an example of Theorem 1.3.11 $\mathbf{G}_6^2$ is of order 6 and the dimensions of its reps are 1, 1, and 2.

THEOREM 1.3.12. Orthogonality relationships for matrix elements.

$$\sum_{j=1}^{|\mathbf{G}|} \Gamma^{i*}(G_j)_{pv} \Gamma^k(G_j)_{qw} = \frac{|\mathbf{G}|}{d_i} \delta^{ik} \delta_{pq} \delta_{vw}. \quad (1.3.14)$$

DEFINITION 1.3.8. *Character table.* The *character table* of a group is an $(r \times r)$ square array whose entries are $\chi^i(C_t)$ ($i = 1$ to r, $t = 1$ to r).

Note that as a result of Theorems 1.3.3 and 1.3.4 the character table of a group is invariant under changes of basis in the vector spaces $\mathbf{V}^i$. However if, instead of a character table, the full table of matrix elements $\mathbf{\Gamma}^i(G_j)_{pq}$ is given in a $(|\mathbf{G}| \times |\mathbf{G}|)$ square array ($i = 1$ to r, $p = 1$ to d_i, $q = 1$ to d_i, $j = 1$ to $|\mathbf{G}|$) then the bases $\langle \mathbf{z}^i|$ being used should be stated since this array is not invariant under a change of basis in any $\mathbf{V}^i$ for which $d_i > 1$.

If we consider as matrices $(|\mathbf{G}| \times |\mathbf{G}|)$ square arrays (with rows and columns labelled (ipq) and j respectively and with meaning for the labels as above) then Theorem 1.3.12 expresses the fact that the matrix whose entries are $\sqrt{(d_i/|\mathbf{G}|)}\mathbf{\Gamma}^i(G_j)_{pq}$ is a unitary matrix. Thus we have the immediate corollary:

$$\sum_{i=1}^{r} \sum_{p=1}^{d_i} \sum_{q=1}^{d_i} d_i \mathbf{\Gamma}^i(G_j)_{pq} \mathbf{\Gamma}^{i*}(G_m)_{pq} = |\mathbf{G}|\, \delta_{jm}. \tag{1.3.15}$$

Equation (1.3.14) provides the following orthogonality relationships for characters:

THEOREM 1.3.13. Orthogonality relationships for characters

$$\sum_{t=1}^{r} r_t \chi^{i*}(C_t)\chi^k(C_t) = |\mathbf{G}|\, \delta^{ik}. \tag{1.3.16}$$

Whereupon if we consider as matrices the $(r \times r)$ square arrays (with rows and columns labelled i and t respectively) then Theorem 1.3.13 expresses the fact that the matrix whose entries are $\sqrt{(r_t/|\mathbf{G}|)}\chi^i(C_t)$ is a unitary matrix. Thus we have the immediate corollary:

$$\sum_{i=1}^{r} \chi^{i*}(C_t)\chi^i(C_s) = \frac{|\mathbf{G}|}{r_t}\, \delta_{ts}. \tag{1.3.17}$$

Example 1.3.9. For $\mathbf{G}_6^2$ let $\Gamma^1 = A$, $\Gamma^2 = B$, and $\Gamma^3 = \mathbf{\Gamma}_{\mathbf{z}}(G)$. Also let $G_1 = E$, $G_2 = P$, $G_3 = P^2$, $G_4 = Q$, $G_5 = PQ$, and $G_6 = P^2Q$. Then $C_1 = \{G_1\}$, $C_2 = \{G_2, G_3\}$, and $C_3 = \{G_4, G_5, G_6\}$ so that $r_1 = 1$, $r_2 = 2$, and $r_3 = 3$. Further $d_1 = 1$, $d_2 = 1$, and $d_3 = 2$. In what follows we use matrices for Γ^3 given by eqn. (1.3.9) with basis in $\mathbf{V}^3$ as defined in Example 1.3.4. For the ordering of the composite labels (ipq) we use dictionary order: (111), (211), (311), (312), (321), (322). Then the following (3×3) unitary matrix with elements $\sqrt{(r_t/|\mathbf{G}|)}\chi^i(C_t)$ (with rows and columns labelled i and t respectively) demonstrates the orthogonality relationships for characters given by eqns. (1.3.16) and (1.3.17):

$$\begin{pmatrix} \frac{1}{\sqrt{6}} & \frac{1}{\sqrt{3}} & \frac{1}{\sqrt{2}} \\ \frac{1}{\sqrt{6}} & \frac{1}{\sqrt{3}} & -\frac{1}{\sqrt{2}} \\ \frac{2}{\sqrt{6}} & -\frac{1}{\sqrt{3}} & 0 \end{pmatrix},$$

while the following (6×6) unitary matrix with elements $\sqrt{(d_i/|\mathbf{G}|)}\boldsymbol{\Gamma}^i(G_j)_{pq}$ (with rows and columns labelled (ipq) and j respectively) demonstrates the orthogonality relationships for matrix elements given by eqns. (1.3.14) and (1.3.15):

$$\begin{pmatrix} \frac{1}{\sqrt{6}} & \frac{1}{\sqrt{6}} & \frac{1}{\sqrt{6}} & \frac{1}{\sqrt{6}} & \frac{1}{\sqrt{6}} & \frac{1}{\sqrt{6}} \\ \frac{1}{\sqrt{6}} & \frac{1}{\sqrt{6}} & \frac{1}{\sqrt{6}} & -\frac{1}{\sqrt{6}} & -\frac{1}{\sqrt{6}} & -\frac{1}{\sqrt{6}} \\ \frac{1}{\sqrt{3}} & -\frac{1}{2\sqrt{3}} & -\frac{1}{2\sqrt{3}} & -\frac{1}{\sqrt{3}} & \frac{1}{2\sqrt{3}} & \frac{1}{2\sqrt{3}} \\ 0 & -\frac{1}{2} & \frac{1}{2} & 0 & -\frac{1}{2} & \frac{1}{2} \\ 0 & \frac{1}{2} & -\frac{1}{2} & 0 & -\frac{1}{2} & \frac{1}{2} \\ \frac{1}{\sqrt{3}} & -\frac{1}{2\sqrt{3}} & -\frac{1}{2\sqrt{3}} & \frac{1}{\sqrt{3}} & -\frac{1}{2\sqrt{3}} & -\frac{1}{2\sqrt{3}} \end{pmatrix}$$

THEOREM 1.3.14. *Let Γ be an arbitrary matrix representation of* $\mathbf{G}$ *with character* χ *then when* Γ *has been completely reduced by suitable equivalence transformations to block-diagonal form it becomes a direct sum of reps* $\sum_{i=1}^{r} c_i\Gamma^i$, *where*

$$c_i = \frac{1}{|\mathbf{G}|}\sum_{t=1}^{r} r_t\chi(C_t)\chi^{i*}(C_t). \tag{1.3.18}$$

Theorems 1.3.11 and 1.3.13 help in the determination of character tables and together with the aid of the following theorem the task of calculating characters is made a completely determinate problem.

THEOREM 1.3.15

$$r_t r_s \chi^i(C_t)\chi^i(C_s) = d_i \sum_{w=1}^{r} h_{ts,w} r_w \chi^i(C_w), \tag{1.3.19}$$

where the $h_{ts,w}$ *are the class multiplication coefficients defined by eqn.* (1.2.2).

Example 1.3.10. If we write $\chi^i(C_t) = x_t$ and $d_i = d$, and if we substitute the values of the numbers r_t and $h_{ts,w}$ from Example 1.2.12 then for $\mathbf{G}_6^2$ eqn. (1.3.19) becomes $x_1^2 = dx_1$, $2x_1x_2 = 2dx_2$, $3x_1x_3 = 3dx_3$, $4x_2^2 = 2dx_1 + 2dx_2$, $6x_2x_3 = 6dx_3$, and $9x_3^2 = 3dx_1 + 6dx_2$. From Theorem 1.3.11 it follows that $d \leqslant 2$. The three solutions satisfying Theorem 1.3.13 as well are $d = 1, x_1 = 1, x_2 = 1, x_3 = 1$; $d = 1, x_1 = 1, x_2 = 1, x_3 = -1$; and $d = 2, x_1 = 2, x_2 = -1, x_3 = 0$. These correspond respectively to the reps A, B, and $\boldsymbol{\Gamma}_z(G)$.

THEOREM 1.3.16. Outer direct product. *Let* $\mathbf{G} = \mathbf{H} \otimes \mathbf{K}$; *if* Δ^i $(i = 1$ *to* $r)$ *and* Λ^j $(j = 1$ *to* $s)$ *are the reps of* $\mathbf{H}$ *and* $\mathbf{K}$ *respectively, then the reps of* $\mathbf{G}$ *can be constructed as follows.*

If $G = (H, K)$, *we define* $\mathbf{\Gamma}^{ij}(G) = \mathbf{\Delta}^i(H) \otimes \mathbf{\Lambda}^j(K)$, *that is, the Kronecker matrix product whose matrix elements are given by*

$$\mathbf{\Gamma}^{ij}(G)_{pq;\,rs} = \mathbf{\Delta}^i(H)_{pr}\mathbf{\Lambda}^j(K)_{qs}. \tag{1.3.20}$$

and then $\mathbf{\Gamma}^{ij}(G)$ *is a rep of* $\mathbf{G}$, *and each of the rs reps of* $\mathbf{G}$ *can be obtained in this way by running through all pairs of values of i and j.*

Also the basis for Γ^{ij} *is* $|\mathbf{z}^i\mathbf{t}^j\rangle$ *in the Cartesian product space* $\mathbf{V}^{ij} = \mathbf{U}^i \otimes \mathbf{W}^j$, *where* $|\mathbf{z}^i\rangle$ *is the basis for* Δ^i *in* $\mathbf{U}^i$, *and* $|\mathbf{t}^j\rangle$ *is the basis for* Λ^j *in* $\mathbf{W}^j$.

It follows also from eqn. (1.3.20) that the character of Γ^{ij} is given by the following rule:

$$\chi^{ij}(G) = \chi^i(H)\chi^j(K) \tag{1.3.21}$$

and in particular

$$\dim \Gamma^{ij} = (\dim \Delta^i)(\dim \Lambda^j). \tag{1.3.22}$$

$\mathbf{H} \boxtimes \mathbf{H}$, the inner direct product of $\mathbf{H}$ with itself, is the subgroup of elements of $\mathbf{H} \otimes \mathbf{H}$ of the form (H, H). Using the notation of Theorem 1.3.16 we obtain for the representatives of elements in this subgroup,

$$\mathbf{\Gamma}^{ij}(H, H) = \mathbf{\Delta}^i(H) \otimes \mathbf{\Delta}^j(H). \tag{1.3.23}$$

But on the inner direct product Γ^{ij} is reducible and is equal to the direct sum $\sum_{k=1}^{r} c_{ij,k}\,\Delta^k$ (for, since $\mathbf{H} \boxtimes \mathbf{H}$ is isomorphic to $\mathbf{H}$, its reps are just the same as the reps of $\mathbf{H}$). Also from Theorem 1.3.14 and eqn. (1.3.21) we obtain

$$c_{ij,k} = \frac{1}{|\mathbf{H}|}\sum_{p=1}^{|\mathbf{H}|} \chi^i(H_p)\chi^j(H_p)\chi^{k*}(H_p), \tag{1.3.24}$$

where H_p, $p = 1$ to $|\mathbf{H}|$, are the elements of $\mathbf{H}$.

DEFINITION 1.3.9. *Clebsch–Gordan decomposition.* The decomposition

$$(\Delta^i \boxtimes \Delta^j)(H) = \mathbf{\Delta}^i(H) \otimes \mathbf{\Delta}^j(H) \equiv \sum_{k=1}^{r} c_{ij,k}\,\mathbf{\Delta}^k(H), \tag{1.3.25}$$

in which the symbol $\equiv$ means 'equivalent to' in the technical sense given in Definition 1.3.2 and in which the $c_{ij,k}$ are given by eqn. (1.3.24), is called the *Clebsch–Gordan decomposition* of the *inner-direct product* $\Delta^i \boxtimes \Delta^j$ of the reps Δ^i and Δ^j, and the coefficients $c_{ij,k}$ are called the *Clebsch–Gordan coefficients*, or sometimes *Wigner coefficients*. Note that

$$c_{ij,k} = c_{ji,k}. \tag{1.3.26}$$

Example 1.3.11. Using the notation of Example 1.3.9, repeated use of eqn. (1.3.24) leads to the following values of the Clebsch–Gordan coefficients for $\mathbf{G}_6^2$: $c_{11,1} = 1$; $c_{12,2} = 1$; $c_{13,3} = 1$; $c_{22,1} = 1$; $c_{23,3} = 1$; $c_{33,1} = 1$; $c_{33,2} = 1$; $c_{33,3} = 1$; and all others not derived from these by eqn. (1.3.26) are zero.

1.4. Point groups

A 3-dimensional point group is a group of symmetry operators which act at a point O (and therefore keep O fixed) and which also leave invariant all distances and angles in a 3-dimensional Euclidean space, $E(3)$. The symmetry operators having these properties are the unitary operators acting at O, namely rotations about axes through O or products of such rotations and the inversion. Such products include reflections in planes through O.

If the group contains all possible rotations and no other elements it is called the *3-dimensional proper rotation group* and it is isomorphic with the group $SO(3)$ of all (3×3) orthogonal matrices having determinant $+1$. The reason for this is that under a rotation R a point P with position vector $\mathbf{r}$ will be moved to a point P' with position vector $\mathbf{r}'$ governed by an equation of the form

$$\mathbf{r}' = \mathbf{R}\mathbf{r}, \tag{1.4.1}$$

where $\mathbf{R}$ is a (3×3) matrix. Since all distances and angles are to be left invariant it follows that the equation

$$\mathbf{r}.\mathbf{s} = \mathbf{R}\mathbf{r}.\mathbf{R}\mathbf{s} = \mathbf{r}'.\mathbf{s}'. \tag{1.4.2}$$

must be true for all possible pairs of vectors $\mathbf{r}$, $\mathbf{s}$. This holds if and only if $\mathbf{R}$ is an orthogonal matrix. As is well known the determinant of any orthogonal matrix is either $+1$ or -1. In fact only those orthogonal matrices with determinant $+1$ correspond to rotations; this is because the rotation through zero angle is the identity and its determinant is $+1$ and from requirements of continuity it follows that the determinant of a matrix corresponding to any rotation is also $+1$.

If the group contains all possible rotations and their products with the inversion it is called the *3-dimensional rotation group* and it is isomorphic with the group $O(3)$ of all (3×3) orthogonal matrices. Operators that correspond to matrices with determinant -1 are called improper rotations and are products of a proper rotation and the inversion. The inversion commutes with all rotations; this follows at once from the fact that it is represented in $O(3)$ by minus the unit matrix.

It is important at this stage to make clear the convention we use when dealing with unitary operators. From what has so far been written the convention may already be clear, but, in case this is not so, it is worth emphasizing that we interpret unitary operators as acting on the points of space and that the coordinates of these points are always referred to a fixed set of axes. Operators that move the points of a space and leave the axes fixed are said to be *active*. Operators that move the axes and leave the points fixed are said to be *passive*. *In this book we shall always use the active convention*. The equation linking the two conventions is simply

$$\mathbf{R}_{\text{active}} = \mathbf{R}^{-1}_{\text{passive}}. \tag{1.4.3}$$

However, great care must be exercised in transferring from one convention to the other because the correspondence between the two conventions is not an automorphism but rather an anti-automorphism; that is to say, if **R**, **S**, and **T** are the matrices of three rotations such that in the active mode

$$\mathbf{R}_{\text{active}}\,\mathbf{S}_{\text{active}} = \mathbf{T}_{\text{active}} \tag{1.4.4}$$

then the corresponding equation for passive operators is

$$\mathbf{S}_{\text{passive}}\,\mathbf{R}_{\text{passive}} = \mathbf{T}_{\text{passive}}, \tag{1.4.5}$$

the order of multiplication of the operators on the left being reversed, rather than preserved as would have been the case for an automorphism.

Every subgroup of the 3-dimensional rotation group is a *point group*. Subgroups of the proper rotation group are called *proper point groups*. A proper point group contains only rotations about axes through O. In what follows we are concerned only with point groups of finite order.

We now classify the proper point groups of finite order. These are:

(i) *The cyclic groups*

The cyclic groups occur when there is only one axis of rotation, say Oz. Such groups are denoted by n in the international notation and by C_n in the Schönflies notation. For each positive integer n there is a well-defined cyclic group whose n elements are rotations about the given axis through angles $2\pi q/n$, where $q = 0, 1, 2, \ldots, (n-1)$. In the Schönflies notation the anti-clockwise rotation through an angle $2\pi/n$ about the axis Oz is denoted by C^{+}_{nz}; the corresponding clockwise rotation is denoted by C^{-}_{nz}. The superscript $\pm$ is omitted when $n = 2$.

(ii) *The dihedral groups*

Dihedral groups consist of all rotations that transform a regular n-sided prism into itself. Such groups are denoted in the international notation by $n22$, n even, and by $n2$, n odd, and by D_n in the Schönflies notation. For each positive integer $n \geqslant 2$ there is a well-defined dihedral group whose $2n$ elements are: n rotations about the axis of the prism parallel to its edges through angles $2\pi q/n$, where $q = 0, 1, 2, \ldots, (n-1)$, the axis of these rotations being called *the principal axis*; and n further rotations each of order 2 (that is, through an angle π)—each such rotation being about an axis called a *secondary axis*, which is perpendicular to the principal axis. There are n secondary axes and they are symmetrically placed within the prism so that the angle between any two adjacent secondary axes is π/n. When n is even half the secondary axes pass through the centres of the faces of the prism and half through the mid-points of the edges, and when n is odd each secondary axis passes both through the centre of a face and through the mid-point of the opposite edge. In the Schönflies notation the n rotations about the principal axis are labelled as in the cyclic group of order n, and a rotation about a secondary axis Op is denoted by C_{2p}.

(iii) *The tetrahedral group*
The tetrahedral group consists of all the rotations that transform a regular tetrahedron into itself. It is denoted by 23 in the international notation and by T in the Schönflies notation. This group has four axes of order 3 passing through the vertices of the tetrahedron, and three axes of order 2 that join the mid-points of its opposite edges. The three axes of order 2 form a right triad and are therefore conveniently labelled Ox, Oy, and Oz. If we denote the vertices of the tetrahedron by 1, 2, 3, and 4 then the four axes of order 3 are labelled 01, 02, 03, and 04 (see Fig. 1.3). Using the Schönflies notation, as described above, the 12 elements of 23 (T) are therefore E (the identity), $C_{3j}^{\pm}$ ($j = 1, 2, 3$, and 4) and C_{2m} ($m = x, y$, and z).

(iv) *The octahedral group*
The octahedral or cubic group consists of all the pure rotations that transform a cube into itself. It is denoted by 432 (sometimes abbreviated to 43) in the international notation and by O in the Schönflies notation. A regular tetrahedron 1234 can be inscribed in the cube so that its four vertices lie at four corners of the cube; in this way the elements of the cubic group can be related to the elements of the tetrahedral group—indeed 23 (T) is a subgroup of 432 (O). The axes 01, 02, 03, and 04 are still of order 3, but now the axes Ox, Oy, and Oz are of order 4. There are in addition six axes of order 2 which join the mid-points of opposite edges of the cube and which we denote, as in Fig. 1.3, by Oa, Ob, Oc, Od, Oe, and Of. Using the Schönflies notation the 24 elements of 432 (O) are therefore E, $C_{3j}^{\pm}$ ($j = 1, 2, 3$, and 4), $C_{4m}^{\pm}$ ($m = x, y$, and z), C_{2m} ($m = x, y$, and z), and C_{2p} ($p = a, b, c, d, e$, and f).

(v) *The icosahedral group*
The icosahedral group consists of all rotations that transform either a pentagonal dodecahedron or a regular cosahedron into itself. It has six axes of order 5, 10 axes of order 3, and 15 axes of order 2 and it consists therefore of 60 elements (Cohan 1958, Cotton 1963). This group is denoted by 532 in the international notation and by Y in the Schönflies notation.

If we denote by $\bar{1}$ (or C_i) the group consisting of the identity E and the inversion I, then given a proper point group $\mathbf{P}$ it is always possible to construct the outer direct product $\mathbf{P} \otimes \bar{1}$ (or $\mathbf{P} \otimes C_i$), as explained in Definition 1.2.13. This process leads to a new set of point groups that are subgroups of $O(3)$. These point groups, for which we give the international notation followed in brackets by its Schönflies equivalent, are as follows:

(i) From n (C_n), n odd, we derive $\bar{n}$ (S_{2n}); and from n (C_n), n even, we derive n/m (C_{nh}). (In the Schönflies notation S_2 is often replaced by C_i.)

(ii) From $n2$ (D_n), n odd, we derive $\bar{n}m$ (D_{nd}); and from $n22$ (D_n), n even, we derive n/mmm (D_{nh}). (When $n = 2$ the international symbol is usually given as mmm.)

(iii) From 23 (T) we derive $m3$ (T_h).

(iv) From 432 (O) we derive $m3m$ (O_h).

(v) From 532 (Y) we derive $53m$ (Y_h).

There is a second method of obtaining new point groups from a proper point group **P** which is fruitful whenever **P** has an invariant subgroup **Q** of index 2. In this case we can write

$$\mathbf{P} = \mathbf{Q} + R\mathbf{Q} \tag{1.4.6}$$

for some rotation R. Then the group

$$\mathbf{P}' = \mathbf{Q} + IR\mathbf{Q}, \tag{1.4.7}$$

where I is the inversion, is also a point group. The point groups derived in this way are as follows.

(i) From $2n$ (C_{2n}), n odd, and its invariant subgroup n (C_n) we derive $\overline{2n}$ (C_{nh}) (except that $\bar{2}$ (C_{1h}) is usually replaced by the symbols m (C_s)). From $2n$ (C_{2n}), n even, and its invariant subgroup n (C_n) we derive $\overline{2n}$ (S_{2n}).

(ii) From $n22$ or $n2$ (D_n) and its invariant subgroup n (C_n) we derive nmm or nm (C_{nv}); and from $(2n)2$ (D_{2n}) ($n \geqslant 2$) and its invariant subgroup $n22$ or $n2$ (D_n) we derive $(\overline{2n})2m$ (D_{nh}), n odd, and $(\overline{2n})2m$ (D_{nd}), n even.

(iii) 23 (T) has no invariant subgroup of index 2 and so no new group can be derived in this way from 23 (T).

(iv) From 432 (O) and its invariant subgroup 23 (T) we derive $\bar{4}3m$ (T_d).

(v) 532 (Y) has no invariant subgroups, and so no new group can be derived in this way from 532 (Y).

These two methods of obtaining further point groups from the proper point groups account for all the point groups of finite order which are subgroups of $O(3)$ but which do not consist entirely of proper rotations. This therefore completes the classification of the finite point groups. It will have been seen that there are at least two rival notations in use for the labelling of the point groups and their symmetry elements, the international notation and the Schönflies notation. For the labelling of the point groups and space groups in this book we use both the international notation and the Schönflies notation; the Schönflies label is always given in brackets following the international label.

We now elaborate slightly on the meaning of the notations used. In the international notation for the labelling of the point groups a number n implies the presence of a rotation axis of order n, a bar over a number, $\bar{n}$, implies the presence of a rotation-inversion axis of order n and the letter m implies the presence of a reflection plane.

Also, if the letter m is separated from a number by a solidus (/), then one plane of symmetry is perpendicular to the principal rotation axis. Finally, in the Schönflies notation for the point groups the subscripts h, v, and d stand for reflection planes that are respectively horizontal, vertical, and diagonal which is, of course, meaningless except when the point group concerned is in some standard orientation. In many cases we have to use settings of the point groups that are non-standard and this is why we prefer the international notation for the label of the point group itself. However no confusion can arise because in such non-standard settings all the elements will be given a definite label that the reader can identify by referring to Table 1.3.

We began this section by saying that a point group is a group of symmetry operations which act at a point O and also leave invariant all distances and angles in a 3-dimensional Euclidean space, $E(3)$. In this book we are concerned with the theory of the representations of the groups of symmetry operations of crystalline solids. We therefore shall be restricting our discussion to the *crystallographic point groups*. A crystallographic point group must satisfy the extra requirement that it be compatible with the translational symmetry of some crystalline solid. The possible point-group operations that may satisfy the extra requirement are the identity, the inversion, reflections in certain planes, and rotations about certain axes of orders 1, 2, 3, 4, or 6. It is possible to show that only a finite number of essentially different groups can be constructed from these operations. In fact there are 32 different crystallographic point groups and many books give discussions of various of their properties (for example, Buerger (1956), Koster, Dimmock, Wheeler, and Statz (1963), Phillips (1963*a*)). The 32 point groups can be classified into seven *crystal systems* (sometimes called *syngonies*) according to the order of the principal axis. There are five crystal systems for point groups with a single principal axis of order 1, 2, 3, 4, or 6, namely the triclinic, monoclinic, trigonal, tetragonal, and hexagonal crystal systems respectively. Two more systems also exist, to complete the seven, the orthorhombic system which has three mutually perpendicular rotation axes of order 2, and the cubic system which has four rotation axes directed towards the vertices of a regular tetrahedron, each of order 3. Most of the point groups that have been described in this chapter are crystallographic point groups. Examples of non-crystallographic point groups that we have described are the icosahedral group and the cyclic and dihedral groups of order n, where n takes any value except 1, 2, 3, 4, or 6.

Notation of point-group operations

There are, in fact, very many different notations to be found in the literature for the labelling of the symmetry elements of the point groups; none of them is entirely satisfactory (for a discussion of various notations see Warren (1968)). Some authors (for example, Kovalev (1965), Miller and Love (1967), Slater (1965*b*)), apparently disconcerted by this lack of uniformity, have taken, for example, to labelling the elements of the cubic group $m3m$ (O_h) by $R_1, R_2, \ldots, R_{48}$ and of the hexagonal

TABLE 1.2

Symmetry elements for the seven crystal systems

Triclinic	Monoclinic	Orthorhombic	Tetragonal	Trigonal	Hexagonal	Cubic
$\bar{1}$ (C_i)	$2/m$ (C_{2h})	mmm (D_{2h})	$4/mmm$ (D_{4h})	$\bar{3}m$ (D_{3d})	$6/mmm$ (D_{6h})	$m3m$ (O_h)
E, I	E, I C_{2z}, σ_z	E, I C_{2m}, σ_m	E, I $C_{4z}^{\pm}, S_{4z}^{\mp}$ C_{2m}, σ_m C_{2s}, σ_{ds}	E, I $C_3^{\pm}, S_6^{\mp}$ C'_{2i}, σ_{di}	E, I $C_6^{\pm}, S_3^{\mp}$ $C_3^{\pm}, S_6^{\mp}$ C_2, σ_h C'_{2i}, σ_{di} C''_{2i}, σ_{vi}	E, I C_{2m}, σ_m $C_{3j}^{\pm}, S_{6j}^{\mp}$ $C_{4m}^{\pm}, S_{4m}^{\mp}$ C_{2p}, σ_{dp}

Notes to Table 1.2

(i) There is an alternative setting for the trigonal system using C''_{2i} instead of C'_{2i} as a standard setting.

(ii) See Figs. 1.1, 1.2, and 1.3 for the positions of the following axes:
$m = x, y, z$; $s = a, b$; $i = 1, 2, 3$; $j = 1, 2, 3, 4$, and $p = a, b, c, d, e, f$.

group $6/mmm$ (D_{6h}) by $R_1, R_2, \ldots, R_{24}$; this scheme is very convenient when computers are being used. However, because the well-established notations, while far from perfect, do carry some meaning (which is not the case for an arbitrary labelling) we therefore use in this book the Schönflies notation for the actual symmetry operations of both point groups and space groups, elaborated so that each point-group

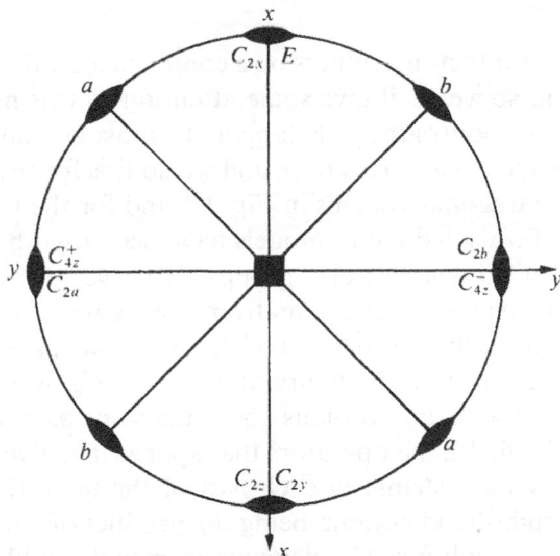

FIG. 1.1. Symmetry elements: triclinic, monoclinic, orthorhombic, and tetragonal systems. The point groups in these systems are subgroups of $m3m$ (O_h) and so the same notation is used. x, y, z form a right-handed set of axes. The labels of the symmetry operations are placed on the figure in the position to which the letter E is taken by that operation.

operator can be separately identified. The notation for the proper rotations has already been indicated in the classification of the proper point groups; the identification is made complete by reference to Tables 1.2 and 1.3 and to Figs. 1.1–1.3.

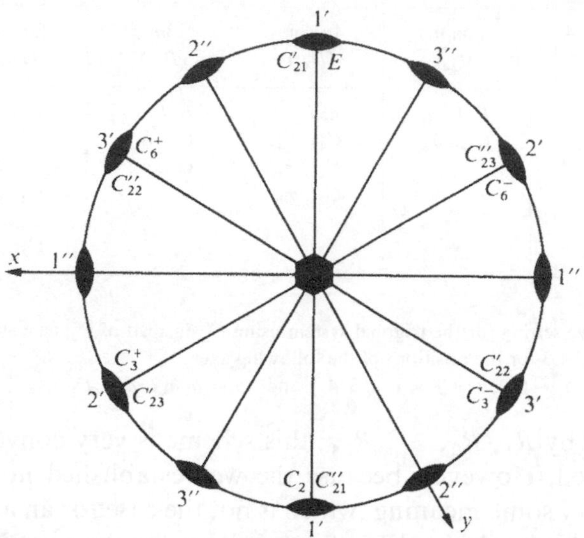

FIG. 1.2. Symmetry elements: trigonal and hexagonal systems. The z-axis is vertically upwards, out of the page. The labels of the symmetry operations are placed on the figure in the position to which the letter E is taken by that operation.

Many results depend for their usefulness on a complete identification of the symmetry elements present and so we shall give some attention to this matter. To identify the symmetry elements for point groups belonging to most crystal systems it is possible to use a plane figure such as a stereogram, and we do this for the triclinic, monoclinic, orthorhombic, and tetragonal systems in Fig. 1.1 and for the trigonal and hexagonal systems in Fig. 1.2. Two-sided paper models have been used by Schiff (1954) to help visualize the different non-cubic point groups. However, for the cubic system it is easier to identify the symmetry elements from the figure of a cube; this is done in Fig. 1.3. C_{nr}^+ is an anti-clockwise rotation of the points of space through $2\pi/n$ radians about the axis labelled by r on the figure in question. C_{nr}^- is a clockwise rotation of the points of space through $2\pi/n$ radians about the same axis. E is the identity and I is the inversion. In Table 1.2 the operators that appear in a given system are listed for each of the seven crystal systems; in each part of the table there are two columns, the elements in the right-hand column being the product of I with the corresponding elements in the left-hand column. The elements given in this table are those that appear when the appropriate point group is in the standard setting with respect to the x, y, and z axes of the figure. Non-standard settings of certain point groups will sometimes have to be used, but since each element is given a definite label this will be no cause for confusion. It should be noticed that reflection planes are always denoted by σ and other rotation reflections by S; it should also be noted that in the notation that we are using $IC_2 = \sigma$, $IC_3^{\pm} = S_6^{\mp}$, $IC_4^{\pm} = S_4^{\mp}$, $IC_6^{\pm} = S_3^{\mp}$ (that is, $IC_3^{+} = S_6^{-}$, and so on). This is because S_n^{+} denotes an anti-clockwise rotation by $2\pi/n$ followed by a reflection in the plane perpendicular to the rotation and hence $IC_n^{+} = \sigma C_2 C_n^{+} = \sigma(C_{2n}^{+})^{n+2} = (S_{2n}^{+})^{n+2}$.

A final and very important point, which we have mentioned earlier, must be stressed and that is that when dealing with a point-group operator it will be interpreted always as an *active* operator moving the points of space and leaving the axes fixed; similarly, when dealing with a space-group operator written in the form $\{R \mid \mathbf{v}\}$ (see section 1.5) it too will be interpreted as active.

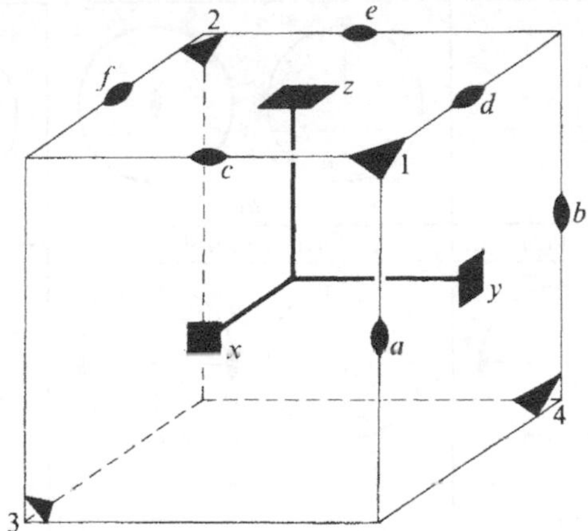

FIG. 1.3. Symmetry elements: cubic system (Altmann and Bradley 1963*b*).

In Table 1.3 we identify the symmetry operations that are present in each of the 32 crystallographic point groups. The point groups are arranged according to the seven crystal systems and all the elements of each point group are given. Wherever a point group has a principal axis this axis is set in the Oz direction and this is called the *standard setting*, but of course many other settings are possible. In Fig. 1.4 are shown stereograms for the 32 crystallographic point groups; for details of the theory of stereographic projection see, for example, Phillips (1963). In any given crystal system the point group that contains the largest number of symmetry operations is called the *holosymmetric* point group of that system.

The multiplication of point-group operations

The result of the application of two point-group operations in succession is easily obtained with the help of a stereogram for all point groups other than the cubic

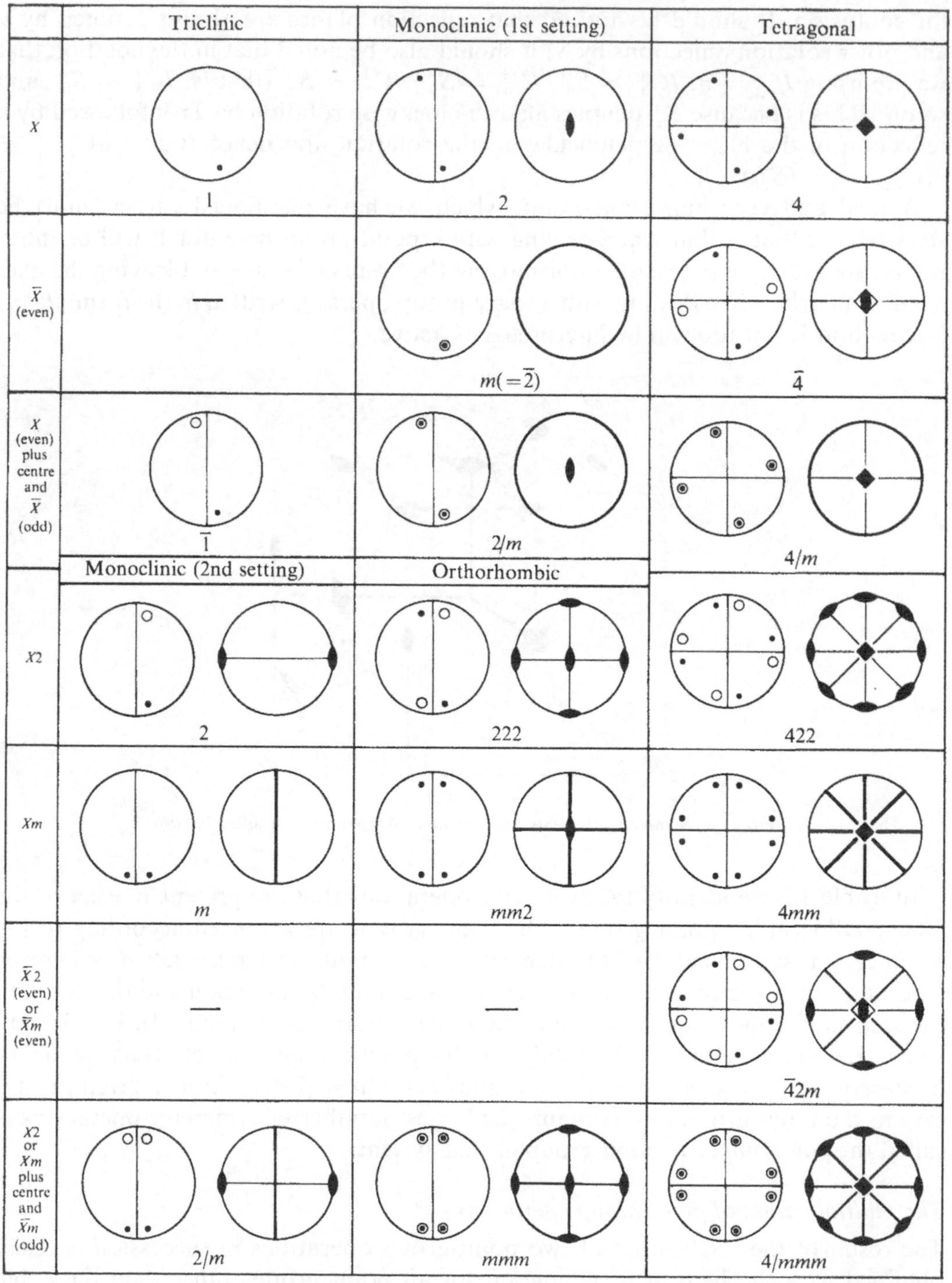

	Triclinic	Monoclinic (1st setting)	Tetragonal
X	1	2	4
$\bar{X}$ (even)	—	$m(=\bar{2})$	$\bar{4}$
X (even) plus centre and $\bar{X}$ (odd)	$\bar{1}$	$2/m$	$4/m$
	Monoclinic (2nd setting)	Orthorhombic	
$X2$	2	222	422
Xm	m	$mm2$	$4mm$
$\bar{X}2$ (even) or $\bar{X}m$ (even)	—	—	$\bar{4}2m$
$X2$ or Xm plus centre and $\bar{X}m$ (odd)	$2/m$	mmm	$4/mmm$

FIG. 1.4. Stereograms of poles of general equivalent directions and symmetry elements of each of the (Henry and

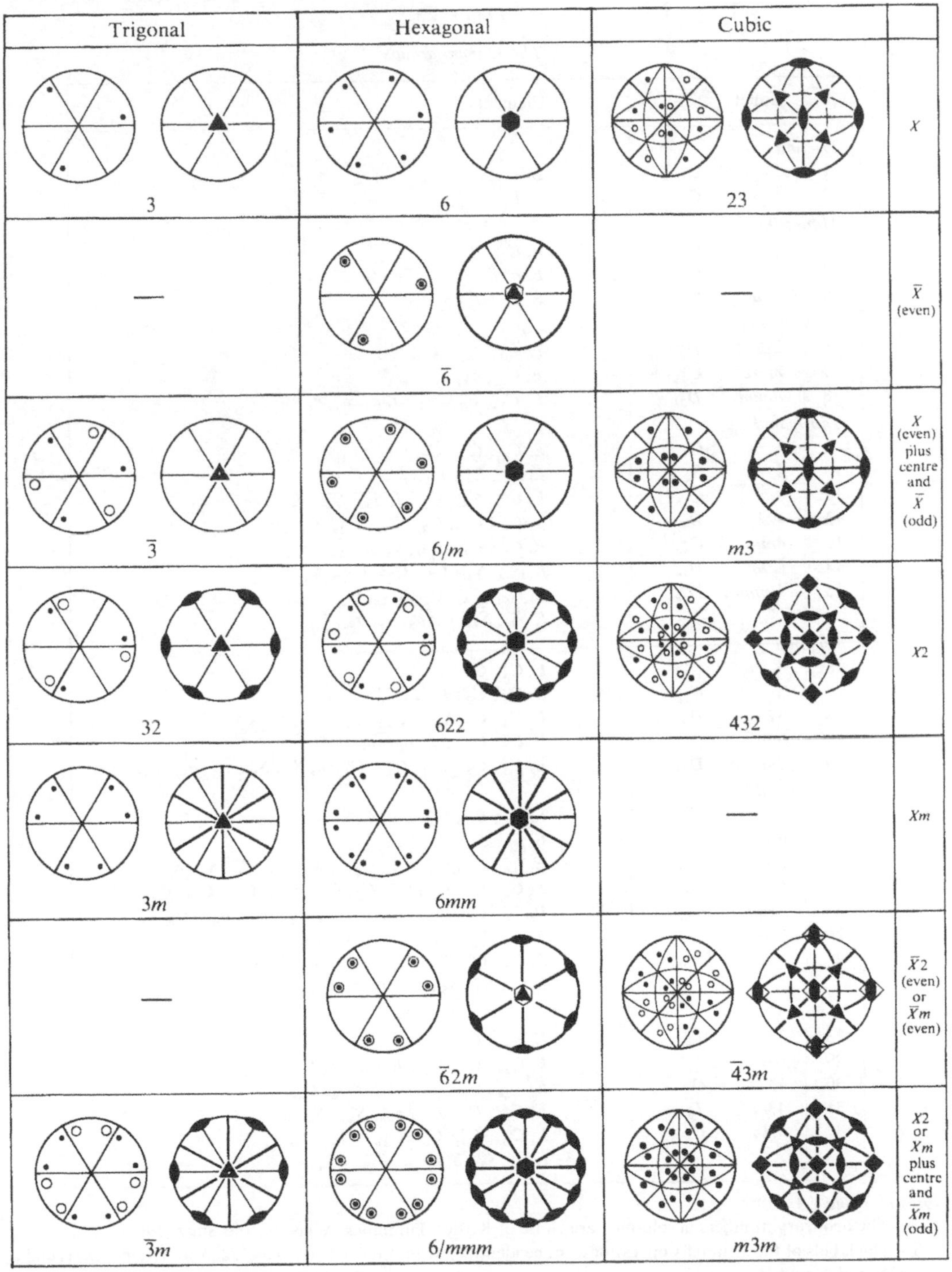

Trigonal	Hexagonal	Cubic	
3	6	23	*X*
—	$\bar{6}$	—	$\bar{X}$ (even)
$\bar{3}$	6/*m*	*m*3	*X* (even) plus centre and $\bar{X}$ (odd)
32	622	432	*X*2
3*m*	6*mm*	—	*Xm*
—	$\bar{6}2m$	$\bar{4}3m$	$\bar{X}2$ (even) or $\bar{X}m$ (even)
$\bar{3}m$	6/*mmm*	*m*3*m*	*X*2 or *Xm* plus centre and $\bar{X}m$ (odd)

32 point groups (*z*-axis normal to the paper in all drawings; for *m*3 and *m*3*m* only half the points are shown) Lonsdale 1965).

TABLE 1.3

The 32 point groups

No.	Label		Elements
Triclinic			
1	1	C_1	E
2	$\bar{1}$	C_i	E, I
Monoclinic			
3	2	C_2	E, C_{2z}
4	*m*	C_s, C_{1h}	E, σ_z
5	2/*m*	C_{2h}	E, C_{2z}, I, σ_z
Orthorhombic			
6	222	D_2	$E, C_{2x}, C_{2y}, C_{2z}$
7	*mm*2	C_{2v}	$E, C_{2z}, \sigma_x, \sigma_y$
8	*mmm*	D_{2h}	$E, C_{2x}, C_{2y}, C_{2z}, I, \sigma_x, \sigma_y, \sigma_z$
Tetragonal			
9	4	C_4	$E, C_{4z}^+, C_{4z}^-, C_{2z}$
10	$\bar{4}$	S_4	$E, S_{4z}^-, S_{4z}^+, C_{2z}$
11	4/*m*	C_{4h}	$E, C_{4z}^+, C_{4z}^-, C_{2z}, I, S_{4z}^-, S_{4z}^+, \sigma_z$
12	422	D_4	$E, C_{4z}^+, C_{4z}^-, C_{2z}, C_{2x}, C_{2y}, C_{2a}, C_{2b}$
13	4*mm*	C_{4v}	$E, C_{4z}^+, C_{4z}^-, C_{2z}, \sigma_x, \sigma_y, \sigma_{da}, \sigma_{db}$
14	$\bar{4}2m$	D_{2d}	$E, S_{4z}^+, S_{4z}^-, C_{2z}, C_{2x}, C_{2y}, \sigma_{da}, \sigma_{db}$
15	4/*mmm*	D_{4h}	$E, C_{4z}^+, C_{4z}^-, C_{2z}, C_{2x}, C_{2y}, C_{2a}, C_{2b}$
			$I, S_{4z}^-, S_{4z}^+, \sigma_z, \sigma_x, \sigma_y, \sigma_{da}, \sigma_{db}$
Trigonal			
16	3	C_3	E, C_3^+, C_3^-
17	$\bar{3}$	C_{3i}	$E, C_3^+, C_3^-, I, S_6^-, S_6^+$
18	32	D_3	$E, C_3^+, C_3^-, C'_{21}, C'_{22}, C'_{23}$
19	3*m*	C_{3v}	$E, C_3^+, C_3^-, \sigma_{d1}, \sigma_{d2}, \sigma_{d3}$
20	$\bar{3}m$	D_{3d}	$E, C_3^+, C_3^-, C'_{21}, C'_{22}, C'_{23}, I, S_6^-, S_6^+, \sigma_{d1}, \sigma_{d2}, \sigma_{d3}$
Hexagonal			
21	6	C_6	$E, C_6^+, C_6^-, C_3^+, C_3^-, C_2$
22	$\bar{6}$	C_{3h}	$E, S_3^-, S_3^+, C_3^+, C_3^-, \sigma_h$
23	6/*m*	C_{6h}	$E, C_6^+, C_6^-, C_3^+, C_3^-, C_2, I, S_3^-, S_3^+, S_6^-, S_6^+, \sigma_h$
24	622	D_6	$E, C_6^+, C_6^-, C_3^+, C_3^-, C_2, C'_{21}, C'_{22}, C'_{23}, C''_{21}, C''_{22}, C''_{23}$
25	6*mm*	C_{6v}	$E, C_6^+, C_6^-, C_3^+, C_3^-, C_2, \sigma_{d1}, \sigma_{d2}, \sigma_{d3}, \sigma_{v1}, \sigma_{v2}, \sigma_{v3}$
26	$\bar{6}2m$	D_{3h}	$E, S_3^-, S_3^+, C_3^+, C_3^-, \sigma_h, C'_{21}, C'_{22}, C'_{23}, \sigma_{v1}, \sigma_{v2}, \sigma_{v3}$
27	6/*mmm*	D_{6h}	$E, C_6^+, C_6^-, C_3^+, C_3^-, C_2, C'_{21}, C'_{22}, C'_{23}, C''_{21}, C''_{22}, C''_{23}$
			$I, S_3^-, S_3^+, S_6^-, S_6^+, \sigma_h, \sigma_{d1}, \sigma_{d2}, \sigma_{d3}, \sigma_{v1}, \sigma_{v2}, \sigma_{v3}$
Cubic			
28	23	T	$E, C_{2m}, C_{3j}^+, C_{3j}^-$
29	*m*3	T_h	$E, C_{2m}, C_{3j}^+, C_{3j}^-, I, \sigma_m, S_{6j}^-, S_{6j}^+$
30	432	O	$E, C_{2m}, C_{3j}^+, C_{3j}^-, C_{2p}, C_{4m}^+, C_{4m}^-$
31	$\bar{4}3m$	T_d	$E, C_{2m}, C_{3j}^+, C_{3j}^-, \sigma_{dp}, S_{4m}^-, S_{4m}^+$
32	*m*3*m*	O_h	$E, C_{2m}, C_{3j}^+, C_{3j}^-, C_{2p}, C_{4m}^+, C_{4m}^-$
			$I, \sigma_m, S_{6j}^-, S_{6j}^+, \sigma_{dp}, S_{4m}^-, S_{4m}^+$

Notes to Table 1.3

(i) The arbitrary numbers in column 1 are those of Koster, Dimmock, Wheeler, and Statz (1963).

(ii) The labels of the symmetry operations can be identified from Figs. 1.1–1.3; $j = 1, 2, 3$, and 4; $m = x, y$, and z; $p = a, b, c, d, e$, and f.

(iii) The principal axes have been set in the Oz direction but there are still possible alternative settings for some of the point groups, for example $\bar{4}2m$ (D_{2d}) may contain the elements, $E, S_{4z}^-, C_{2z}, S_{4z}^+, \sigma_x, \sigma_y, C_{2a}$, and C_{2b}; sometimes alternatives of this kind are important when one considers the space groups (see Chapter 3).

groups. For example, the multiplication $C_{2x}C_{4z}^{+} = C_{2b}$ is illustrated in Fig. 1.5, in which a dot represents a point above the plane of the drawing and a circle a point below the plane of the drawing. The point A is taken into the point B by the operation C_{4z}^{+}; B is in turn taken into the point C by the operation C_{2x}. The result is the operation that takes A directly into the point C, namely C_{2b}, as can be seen from Fig 1.1.

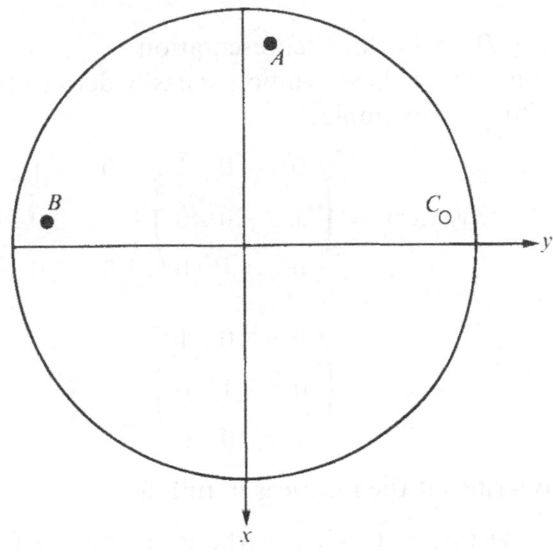

FIG. 1.5.

For a cubic point group the easiest way to obtain the group multiplication table is to use the Jones' faithful representation of the operators given in Table 1.4(a). The Jones symbol for the operator R is just the vector formed by the action of the operator R on the vector (x, y, z); these symbols may be recognized as the coordinates of equivalent positions in the unit cells of symmorphic cubic and hexagonal space groups in Table 4.3 of Volume 1 of the *International tables for X-ray crystallography* (Henry and Lonsdale 1965). If (x, y, z) is a vector referred to a basis of unit vectors $\mathbf{i}$, $\mathbf{j}$, and $\mathbf{k}$ along the coordinate axes, then, for example, $C_{4z}^{+}x\mathbf{i} = x\mathbf{j}$, $C_{4z}^{+}y\mathbf{j} = -y\mathbf{i}$, and $C_{4z}^{+}z\mathbf{k} = z\mathbf{k}$, so that we may write $C_{4z}^{+}(x, y, z) = (-y, x, z)$ or

$$C_{4z}^{+}(x\mathbf{i} + y\mathbf{j} + z\mathbf{k}) = (-y\mathbf{i} + x\mathbf{j} + z\mathbf{k}) \tag{1.4.8}$$

and the Jones' faithful representation of C_{4z}^{+} is then written as $(\bar{y}xz)$. The matrix representing C_{4z}^{+} with respect to the basis consisting of the row vector $\langle \mathbf{i}, \mathbf{j}, \mathbf{k}|$ is

$$\mathbf{D}(C_{4z}^{+}) = \begin{pmatrix} 0 & -1 & 0 \\ 1 & 0 & 0 \\ 0 & 0 & 1 \end{pmatrix} \tag{1.4.9}$$

and thus the Jones' symbol is an abbreviation for this matrix representative. Equation (1.4.9) follows from the fact that

$$C_{4z}^{+}\langle \mathbf{i}, \mathbf{j}, \mathbf{k}| = \langle \mathbf{i}, \mathbf{j}, \mathbf{k}| \begin{pmatrix} 0 & -1 & 0 \\ 1 & 0 & 0 \\ 0 & 0 & 1 \end{pmatrix}, \tag{1.4.10}$$

which illustrates why D is a faithful representation.

The multiplication rule for these symbols is easily derived from the multiplication rule for matrices. Thus, for example,

$$(zxy)(\bar{y}xz) = \begin{pmatrix} 0 & 0 & 1 \\ 1 & 0 & 0 \\ 0 & 1 & 0 \end{pmatrix} \begin{pmatrix} 0 & -1 & 0 \\ 1 & 0 & 0 \\ 0 & 0 & 1 \end{pmatrix}$$

$$= \begin{pmatrix} 0 & 0 & 1 \\ 0 & -1 & 0 \\ 1 & 0 & 0 \end{pmatrix} = (z\bar{y}x). \tag{1.4.11}$$

There is no need to write out the matrices in full, for

$$(zxy)(\bar{y}xz) = (z'x'y'), \quad \text{where} \quad (x'y'z') = (\bar{y}xz),$$

so that

$$(z'x'y') = (z\bar{y}x).$$

Equation (1.4.11) is a representation equation reading

$$\mathbf{D}(C_{31}^{+})\mathbf{D}(C_{4z}^{+}) = \mathbf{D}(C_{2c}) \tag{1.4.12}$$

as can be seen from Table 1.4(a) in which the Jones' faithful representation symbol is given for each of the 48 cubic point-group operations. Since D is a faithful representation it follows at once that $C_{31}^{+}C_{4z}^{+} = C_{2c}$. The Jones' symbols given in Table 1.4(a) can also be used for the elements of tetragonal or orthorhombic point groups and for monoclinic or triclinic point groups where the vector (x, y, z) is referred to a basis of unit vectors **i**, **j**, and **k** parallel to the axes of the crystal. A similar set of symbols in Table 1.4(b) gives the effect of the operations of a hexagonal or trigonal point group on a vector (x, y, z) where (x, y, z) is referred to a basis of unit vectors **i**, **j**, and **k**, where **i** and **j** are at 120° to each other and are in the plane normal to **k**.

In Tables 1.5 and 1.6 we give the group multiplication tables for the two point groups 432 (O) and 622 (D_6) (Belov 1957*b*, Belov and Tarkhova 1960, Hurley 1966, Koptsik 1966). These two point groups consist only of pure rotations; all those other

TABLE 1.4

Jones' faithful representation symbols

(a) *Cubic*								(b) *Hexagonal*			
E	xyz	I	$\bar{x}\bar{y}\bar{z}$	C_{4x}^{+}	$x\bar{z}y$	S_{4x}^{-}	$\bar{x}z\bar{y}$	E	x, y, z	I	$\bar{x}, \bar{y}, \bar{z}$
C_{2x}	$x\bar{y}\bar{z}$	σ_x	$\bar{x}yz$	C_{4y}^{+}	$zy\bar{x}$	S_{4y}^{-}	$\bar{z}\bar{y}x$	C_6^{+}	$x - y, x, z$	S_3^{-}	$-x + y, \bar{x}, \bar{z}$
C_{2y}	$\bar{x}y\bar{z}$	σ_y	$x\bar{y}z$	C_{4z}^{+}	$\bar{y}xz$	S_{4z}^{-}	$y\bar{x}\bar{z}$	C_3^{+}	$\bar{y}, x - y, z$	S_6^{-}	$y, -x + y, \bar{z}$
C_{2z}	$\bar{x}\bar{y}z$	σ_z	$xy\bar{z}$	C_{4x}^{-}	$xz\bar{y}$	S_{4x}^{+}	$\bar{x}\bar{z}y$	C_2	$\bar{x}, \bar{y}, z$	σ_h	$x, y, \bar{z}$
C_{31}^{+}	zxy	S_{61}^{-}	$\bar{z}\bar{x}\bar{y}$	C_{4y}^{-}	$\bar{z}yx$	S_{4y}^{+}	$z\bar{y}\bar{x}$	C_3^{-}	$-x + y, \bar{x}, z$	S_6^{+}	$x - y, x, \bar{z}$
C_{32}^{+}	$\bar{z}x\bar{y}$	S_{62}^{-}	$z\bar{x}y$	C_{4z}^{-}	$y\bar{x}z$	S_{4z}^{+}	$\bar{y}x\bar{z}$	C_6^{-}	$y, -x + y, z$	S_3^{+}	$\bar{y}, x - y, \bar{z}$
C_{33}^{+}	$\bar{z}\bar{x}y$	S_{63}^{-}	$zx\bar{y}$	C_{2a}	$yx\bar{z}$	σ_{da}	$\bar{y}\bar{x}z$	C'_{21}	$-x + y, y, \bar{z}$	σ_{d1}	$x - y, \bar{y}, z$
C_{34}^{+}	$z\bar{x}\bar{y}$	S_{64}^{-}	$\bar{z}xy$	C_{2b}	$\bar{y}\bar{x}\bar{z}$	σ_{db}	yxz	C'_{22}	$x, x - y, \bar{z}$	σ_{d2}	$\bar{x}, -x + y, z$
C_{31}^{-}	yzx	S_{61}^{+}	$\bar{y}\bar{z}\bar{x}$	C_{2c}	$z\bar{y}x$	σ_{dc}	$\bar{z}y\bar{x}$	C'_{23}	$\bar{y}, \bar{x}, \bar{z}$	σ_{d3}	y, x, z
C_{32}^{-}	$y\bar{z}\bar{x}$	S_{62}^{+}	$\bar{y}zx$	C_{2d}	$\bar{x}zy$	σ_{dd}	$x\bar{z}\bar{y}$	C''_{21}	$x - y, \bar{y}, \bar{z}$	σ_{v1}	$-x + y, y, z$
C_{33}^{-}	$\bar{y}z\bar{x}$	S_{63}^{+}	$y\bar{z}x$	C_{2e}	$\bar{z}\bar{y}\bar{x}$	σ_{de}	zyx	C''_{22}	$\bar{x}, -x + y, \bar{z}$	σ_{v2}	$x, x - y, z$
C_{34}^{-}	$\bar{y}\bar{z}x$	S_{64}^{+}	$yz\bar{x}$	C_{2f}	$\bar{x}\bar{z}\bar{y}$	σ_{df}	xzy	C''_{23}	$y, x, \bar{z}$	σ_{v3}	$\bar{y}, \bar{x}, z$

Notes to Table 1.4

(i) As mentioned in the text the symbol next to an (active) operator R is just the vector formed from the vector (x, y, z) by R.

(ii) In (a) the unit vectors **i**, **j**, and **k** are mutually orthogonal and are along the x, y, and z axes of Fig. 1.1. In (b) the unit vectors **i**, **j**, and **k** are along the x, y, and z axes of Fig. 1.2, therefore **i** and **j** are at 120° to each other and are in the plane normal to **k**.

(iii) Thus, for example, C_{4z}^{-} has the symbol $(y\bar{x}z)$ so that

$$C_{4z}^{-}(x\mathbf{i} + y\mathbf{j} + z\mathbf{k}) = (y\mathbf{i} - x\mathbf{j} + z\mathbf{k}).$$

(iv) (a) is also to be used for the tetragonal, orthorhombic, monoclinic, and triclinic systems; for the monoclinic and triclinic systems **i**, **j**, and **k** are parallel to the axes of the crystal. (b) is also to be used for the trigonal system.

(v) The symbols in (a) and (b) may be recognized as simply the coordinates of equivalent positions in the unit cells of symmorphic cubic and hexagonal space groups in Table 4.3 of Volume 1 of the *International tables for X-ray crystallography* (Henry and Lonsdale 1965).

point groups that consist only of pure rotations are subgroups of one or other of these two point groups. Each of those remaining point groups that contain operations that are not pure rotations is simply related to one of the point groups consisting of pure rotations only. Thus, from Tables 1.5 and 1.6 it is possible to obtain very quickly the group multiplication table of any of the 32 point groups.

1.5. Space groups

Before discussing the definition of a space group it is desirable to consider what is meant by a *Bravais lattice*. A lattice is a collection of mathematical points arranged in such a way that each lattice point has the same environment in the same orientation. With reference to crystalline solids, the atoms or molecules of which such a solid is composed are found to be arranged in a regular manner (Haüy 1815*a*–*d*). That is, within a crystalline solid there is a set of mathematical lattice points. If a Maxwell demon were to view such a crystal from any one of these points he would see an

TABLE 1.5

The group multiplication table for the point group 432 (O)

E	C_{2x}	C_{2y}	C_{2z}	C_{4x}^-	C_{4y}^-	C_{4z}^-	C_{4x}^+	C_{4y}^+	C_{4z}^+	C_{31}^-	C_{32}^-	C_{33}^-	C_{34}^-	C_{31}^+	C_{32}^+	C_{33}^+	C_{34}^+	C_{2a}	C_{2b}	C_{2c}	C_{2d}	C_{2e}	C_{2f}
C_{2x}	E	C_{2z}	C_{2y}	C_{4x}^+	C_{2e}	C_{2a}	C_{4x}^-	C_{2c}	C_{2b}	C_{32}^-	C_{31}^-	C_{34}^-	C_{33}^-	C_{34}^+	C_{33}^+	C_{32}^+	C_{31}^+	C_{4z}^-	C_{4z}^+	C_{4y}^+	C_{2f}	C_{4y}^-	C_{2d}
C_{2y}	C_{2z}	E	C_{2x}	C_{2d}	C_{4y}^+	C_{2b}	C_{2f}	C_{4y}^-	C_{2a}	C_{33}^-	C_{34}^-	C_{31}^-	C_{32}^-	C_{32}^+	C_{31}^+	C_{34}^+	C_{33}^+	C_{4z}^+	C_{4z}^-	C_{2e}	C_{4x}^-	C_{2c}	C_{4x}^+
C_{2z}	C_{2y}	C_{2x}	E	C_{2f}	C_{2c}	C_{4z}^+	C_{2d}	C_{2e}	C_{4z}^-	C_{34}^-	C_{33}^-	C_{32}^-	C_{31}^-	C_{33}^+	C_{34}^+	C_{31}^+	C_{32}^+	C_{2b}	C_{2a}	C_{4y}^-	C_{4x}^+	C_{4y}^+	C_{4x}^-
C_{4x}^-	C_{4x}^+	C_{2f}	C_{2d}	C_{2x}	C_{32}^+	C_{31}^-	E	C_{34}^+	C_{33}^-	C_{2a}	C_{4z}^-	C_{2b}	C_{4z}^+	C_{4y}^+	C_{2e}	C_{4y}^-	C_{2c}	C_{32}^-	C_{34}^-	C_{31}^-	C_{2y}	C_{33}^+	C_{2z}
C_{4y}^-	C_{2c}	C_{4y}^+	C_{2e}	C_{31}^-	C_{2y}	C_{33}^+	C_{34}^-	E	C_{32}^-	C_{2d}	C_{4x}^+	C_{4x}^-	C_{2f}	C_{4z}^+	C_{2a}	C_{2b}	C_{4z}^-	C_{31}^+	C_{34}^+	C_{2z}	C_{33}^-	C_{2x}	C_{32}^-
C_{4z}^-	C_{2b}	C_{2a}	C_{4z}^+	C_{34}^+	C_{31}^-	C_{2z}	C_{33}^+	C_{32}^-	E	C_{2c}	C_{2e}	C_{4y}^+	C_{4y}^-	C_{4x}^+	C_{4x}^-	C_{2d}	C_{2f}	C_{2x}	C_{2y}	C_{34}^-	C_{31}^-	C_{33}^-	C_{32}^+
C_{4x}^+	C_{4x}^-	C_{2d}	C_{2f}	E	C_{33}^+	C_{32}^-	C_{2x}	C_{31}^+	C_{34}^-	C_{4z}^-	C_{2a}	C_{4z}^+	C_{2b}	C_{2c}	C_{4y}^-	C_{2e}	C_{4y}^+	C_{31}^-	C_{33}^-	C_{34}^+	C_{2z}	C_{32}^+	C_{2y}
C_{4y}^+	C_{2e}	C_{4y}^-	C_{2c}	C_{33}^-	E	C_{34}^+	C_{32}^-	C_{2y}	C_{31}^+	C_{4x}^-	C_{2f}	C_{2d}	C_{4x}^+	C_{2a}	C_{4z}^+	C_{4z}^-	C_{2b}	C_{32}^+	C_{33}^+	C_{2x}	C_{31}^-	C_{2z}	C_{34}^-
C_{4z}^+	C_{2a}	C_{2b}	C_{4z}^-	C_{32}^+	C_{34}^-	E	C_{31}^+	C_{33}^-	C_{2z}	C_{4y}^-	C_{4y}^+	C_{2e}	C_{2c}	C_{2d}	C_{2f}	C_{4x}^+	C_{4x}^-	C_{2y}	C_{2x}	C_{31}^-	C_{33}^+	C_{32}^-	C_{34}^+
C_{31}^-	C_{34}^-	C_{32}^-	C_{33}^-	C_{2c}	C_{2a}	C_{2d}	C_{4y}^-	C_{4z}^-	C_{4x}^-	C_{31}^+	C_{33}^+	C_{34}^+	C_{32}^+	E	C_{2x}	C_{2y}	C_{2z}	C_{4x}^+	C_{2f}	C_{4z}^+	C_{4y}^+	C_{2b}	C_{2e}
C_{32}^-	C_{33}^-	C_{31}^-	C_{34}^-	C_{4y}^+	C_{4z}^-	C_{2f}	C_{2e}	C_{2a}	C_{4x}^+	C_{34}^+	C_{32}^+	C_{31}^+	C_{33}^+	C_{2x}	E	C_{2z}	C_{2y}	C_{4x}^-	C_{2d}	C_{2b}	C_{2c}	C_{4z}^+	C_{4y}^-
C_{33}^-	C_{32}^-	C_{34}^-	C_{31}^-	C_{2e}	C_{4z}^+	C_{4x}^-	C_{4y}^+	C_{2b}	C_{2d}	C_{32}^-	C_{34}^+	C_{33}^+	C_{31}^+	C_{2y}	C_{2z}	E	C_{2x}	C_{2f}	C_{4x}^+	C_{2a}	C_{4y}^-	C_{4z}^-	C_{2c}
C_{34}^-	C_{31}^-	C_{33}^-	C_{32}^-	C_{4y}^-	C_{2b}	C_{4x}^+	C_{2c}	C_{4z}^+	C_{2f}	C_{33}^+	C_{31}^+	C_{32}^+	C_{34}^+	C_{2z}	C_{2y}	C_{2x}	E	C_{2d}	C_{4x}^-	C_{4z}^-	C_{2e}	C_{2a}	C_{4y}^+
C_{31}^+	C_{32}^+	C_{33}^+	C_{34}^+	C_{4z}^+	C_{4x}^+	C_{4y}^+	C_{2a}	C_{2d}	C_{2c}	E	C_{2y}	C_{2z}	C_{2x}	C_{31}^-	C_{34}^-	C_{32}^-	C_{33}^-	C_{4y}^-	C_{2e}	C_{4x}^-	C_{4z}^-	C_{2f}	C_{2b}
C_{32}^+	C_{31}^+	C_{34}^+	C_{33}^+	C_{2a}	C_{2f}	C_{4y}^-	C_{4z}^+	C_{4x}^-	C_{2e}	C_{2y}	E	C_{2x}	C_{2z}	C_{33}^-	C_{32}^-	C_{34}^-	C_{31}^-	C_{4y}^+	C_{2c}	C_{2d}	C_{2b}	C_{4x}^+	C_{4z}^-
C_{33}^+	C_{34}^+	C_{31}^+	C_{32}^+	C_{4z}^-	C_{2d}	C_{2e}	C_{2b}	C_{4x}^+	C_{4y}^-	C_{2z}	C_{2x}	E	C_{2y}	C_{34}^-	C_{31}^-	C_{33}^-	C_{32}^-	C_{2c}	C_{4y}^+	C_{2f}	C_{4z}^+	C_{4x}^-	C_{2a}
C_{34}^+	C_{33}^+	C_{32}^+	C_{31}^+	C_{2b}	C_{4x}^-	C_{2c}	C_{4z}^-	C_{2f}	C_{4y}^+	C_{2x}	C_{2z}	C_{2y}	E	C_{32}^-	C_{33}^-	C_{31}^-	C_{34}^-	C_{2e}	C_{4y}^-	C_{4x}^+	C_{2a}	C_{2d}	C_{4z}^+
C_{2a}	C_{4z}^+	C_{4z}^-	C_{2b}	C_{31}^+	C_{32}^-	C_{2y}	C_{32}^+	C_{31}^-	C_{2x}	C_{4y}^+	C_{4y}^-	C_{2c}	C_{2e}	C_{4x}^-	C_{4x}^+	C_{2f}	C_{2d}	E	C_{2z}	C_{33}^-	C_{34}^+	C_{34}^-	C_{33}^+
C_{2b}	C_{4z}^-	C_{4z}^+	C_{2a}	C_{33}^+	C_{33}^-	C_{2x}	C_{34}^+	C_{34}^-	C_{2y}	C_{2e}	C_{2c}	C_{4y}^-	C_{4y}^+	C_{2f}	C_{2d}	C_{4x}^-	C_{4x}^-	C_{2z}	E	C_{32}^-	C_{32}^+	C_{31}^-	C_{31}^+
C_{2c}	C_{4y}^-	C_{2e}	C_{4y}^+	C_{34}^-	C_{2x}	C_{31}^+	C_{31}^-	C_{2z}	C_{34}^+	C_{4x}^+	C_{2d}	C_{2f}	C_{4x}^-	C_{4z}^-	C_{2b}	C_{2a}	C_{4z}^+	C_{33}^+	C_{32}^+	E	C_{32}^-	C_{2y}	C_{33}^-
C_{2d}	C_{2f}	C_{4x}^+	C_{4x}^-	C_{2z}	C_{31}^+	C_{33}^-	C_{2y}	C_{33}^+	C_{31}^-	C_{4z}^+	C_{2b}	C_{4z}^-	C_{2a}	C_{4y}^-	C_{2c}	C_{4y}^+	C_{2e}	C_{34}^-	C_{32}^-	C_{32}^+	E	C_{34}^+	C_{2x}
C_{2e}	C_{4y}^+	C_{2c}	C_{4y}^-	C_{32}^-	C_{2z}	C_{32}^+	C_{33}^-	C_{2x}	C_{33}^+	C_{2f}	C_{4x}^-	C_{4x}^+	C_{2d}	C_{2b}	C_{4z}^-	C_{4z}^+	C_{2a}	C_{34}^+	C_{31}^+	C_{2y}	C_{34}^-	E	C_{31}^-
C_{2f}	C_{2d}	C_{4x}^-	C_{4x}^+	C_{2y}	C_{34}^+	C_{34}^-	C_{2z}	C_{32}^+	C_{32}^-	C_{2b}	C_{4z}^+	C_{2a}	C_{4z}^-	C_{2e}	C_{4y}^+	C_{2c}	C_{4y}^-	C_{33}^-	C_{31}^-	C_{33}^+	C_{2x}	C_{31}^+	E

TABLE 1.6

The group multiplication table for the point group 622 (D_6)

E	C_6^+	C_3^+	C_2	C_3^-	C_6^-	C'_{21}	C'_{22}	C'_{23}	C''_{21}	C''_{22}	C''_{23}
C_6^+	C_3^+	C_2	C_3^-	C_6^-	E	C''_{22}	C''_{23}	C''_{21}	C'_{22}	C'_{23}	C'_{21}
C_3^+	C_2	C_3^-	C_6^-	E	C_6^+	C'_{23}	C'_{21}	C'_{22}	C''_{23}	C''_{21}	C''_{22}
C_2	C_3^-	C_6^-	E	C_6^+	C_3^+	C''_{21}	C''_{22}	C''_{23}	C'_{21}	C'_{22}	C'_{23}
C_3^-	C_6^-	E	C_6^+	C_3^+	C_2	C'_{22}	C'_{23}	C'_{21}	C''_{22}	C''_{23}	C''_{21}
C_6^-	E	C_6^+	C_3^+	C_2	C_3^-	C''_{23}	C''_{21}	C''_{22}	C'_{23}	C'_{21}	C'_{22}
C'_{21}	C''_{23}	C'_{22}	C''_{21}	C'_{23}	C''_{22}	E	C_3^+	C_3^-	C_2	C_6^-	C_6^+
C'_{22}	C''_{21}	C'_{23}	C''_{22}	C'_{21}	C''_{23}	C_3^-	E	C_3^+	C_6^+	C_2	C_6^-
C'_{23}	C''_{22}	C'_{21}	C''_{23}	C'_{22}	C''_{21}	C_3^+	C_3^-	E	C_6^-	C_6^+	C_2
C''_{21}	C'_{23}	C''_{22}	C'_{21}	C''_{23}	C'_{22}	C_2	C_6^-	C_6^+	E	C_3^+	C_3^-
C''_{22}	C'_{21}	C''_{23}	C'_{22}	C''_{21}	C'_{23}	C_6^+	C_2	C_6^-	C_3^-	E	C_3^+
C''_{23}	C'_{22}	C''_{21}	C'_{23}	C''_{22}	C'_{21}	C_6^-	C_6^+	C_2	C_3^+	C_3^-	E

exactly similar array of atoms or molecules to the array that he would see if he were to view the crystal from any other of these lattice points. Strictly speaking, in order to obtain complete similarity of the environment of each lattice point it is necessary that a mathematical lattice be of infinite extent. A real crystal clearly cannot contain such an infinite lattice but, remembering the actual sizes of atoms, it will be a close approximation to an infinite lattice. We may illustrate the idea of a lattice with a 2-dimensional example; if the set of points in Fig. 1.6, which are arranged at the

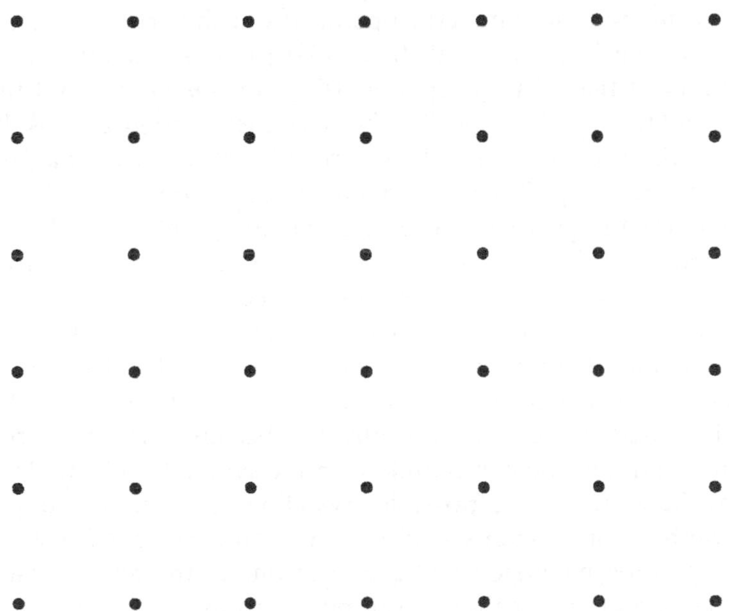

FIG. 1.6. The square 2-dimensional Bravais lattice, *p*.

corners of squares, is imagined to continue off the diagram to infinity, then these points constitute the simple square 2-dimensional Bravais lattice, *p*. There are five different 2-dimensional Bravais lattices (Alexander 1929, Alexander and Herrmann 1929, Belov 1959*b*) and they are described and illustrated in Volume 1 of the *International tables for X-ray crystallography* (Henry and Lonsdale 1965). In a 3-dimensional Euclidean space, $E(3)$, there are 14 essentially different lattices possible and these are called the (3-dimensional) *Bravais lattices* (Bravais 1849*a*, *b*, 1850). It should perhaps be emphasized that there is not necessarily a Bravais lattice point at the centre of every atom in a real crystal, nor is it necessary for Bravais lattice points to be occupied by atoms.

If we choose one particular lattice point to be the origin, O, then we may write $\mathbf{t}$, the position vector of any other lattice point, as

$$\mathbf{t} = n_1\mathbf{t}_1 + n_2\mathbf{t}_2 + n_3\mathbf{t}_3 \tag{1.5.1}$$

where n_1, n_2, and n_3 are integers and $\mathbf{t}_1$, $\mathbf{t}_2$, and $\mathbf{t}_3$ are the fundamental translations that join O to three of its (non-coplanar) nearest neighbours. The translation operations $\mathbf{t}$ defined by eqn. (1.5.1) form an infinite group which is called the *translation group* of the Bravais lattice. In addition to the symmetry of the group of translations, the Bravais lattice is also invariant under certain point-group operations acting at the lattice points and, sometimes at least, at other points too. The point-group operations at any lattice point form a point group $\mathbf{P}$, which is in fact the holosymmetric point group of one of the crystal systems or syngonies, i.e. $\mathbf{P}$ is the point group with the largest number of symmetry operations in the crystal system in question. In the example of the 2-dimensional Bravais lattice illustrated in Fig. 1.6 the lattice has the symmetry of the point group $4mm$ (C_{4v}), or $4/mmm$ (D_{4h}) if the dots are on both the back and front of the page. The holosymmetric point groups, $\mathbf{P}$, of the seven crystal systems, the triclinic, monoclinic, orthorhombic, tetragonal, trigonal, hexagonal, and cubic are $\bar{1}$ (C_i), $2/m$ (C_{2h}), mmm (D_{2h}), $4/mmm$ (D_{4h}), $\bar{3}m$ (D_{3d}), $6/mmm$ (D_{6h}), and $m3m$ (O_h) respectively. The fact that the point group $\mathbf{P}$ is the holosymmetric point group of the relevant crystal system arises from the following facts; (i) $\mathbf{P}$ must contain the space-inversion operation because if $\mathbf{t}$ is a vector of the Bravais lattice then so also is $-\mathbf{t}$, and (ii) together with axes of orders 3, 4, and 6, $\mathbf{P}$ also contains reflection planes that pass through these axes. It is therefore possible to assign each of the 14 Bravais lattices to one of the seven crystal systems and these Bravais lattices are illustrated in Fig. 1.7. Basically it is because the point-group symmetry of a crystalline solid must be compatible with the symmetry of the Bravais lattice of that solid that there are only 32 possible crystallographic point groups. Each of the 32 crystallographic point groups is either the point group of one of the Bravais lattices, i.e., the holosymmetric point group of one of the seven crystal systems, or else it is a subgroup of one of these seven point groups.

We have said that a Bravais lattice in three dimensions is an infinite array of points

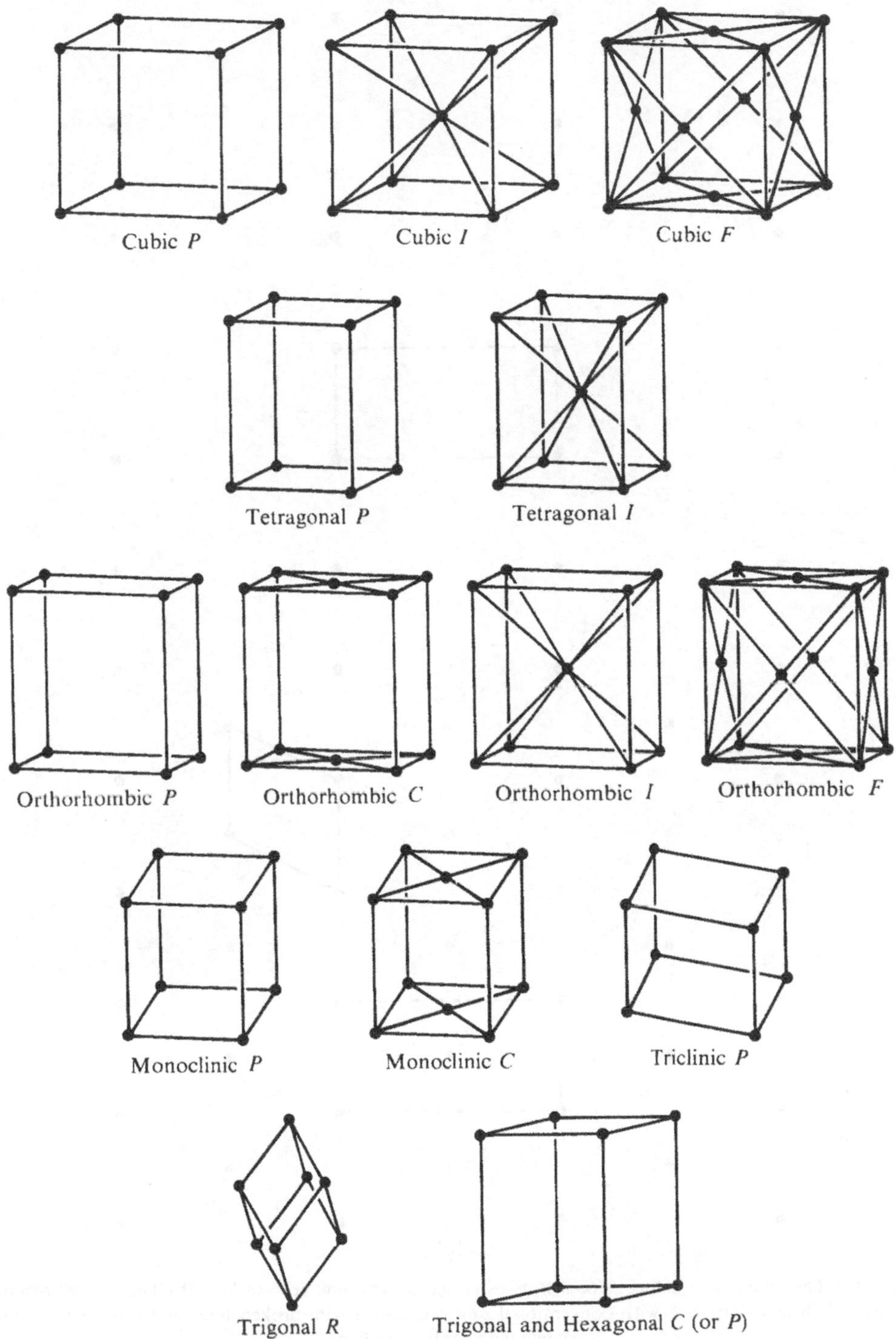

FIG. 1.7. The 14 Bravais lattices (Phillips 1963a).

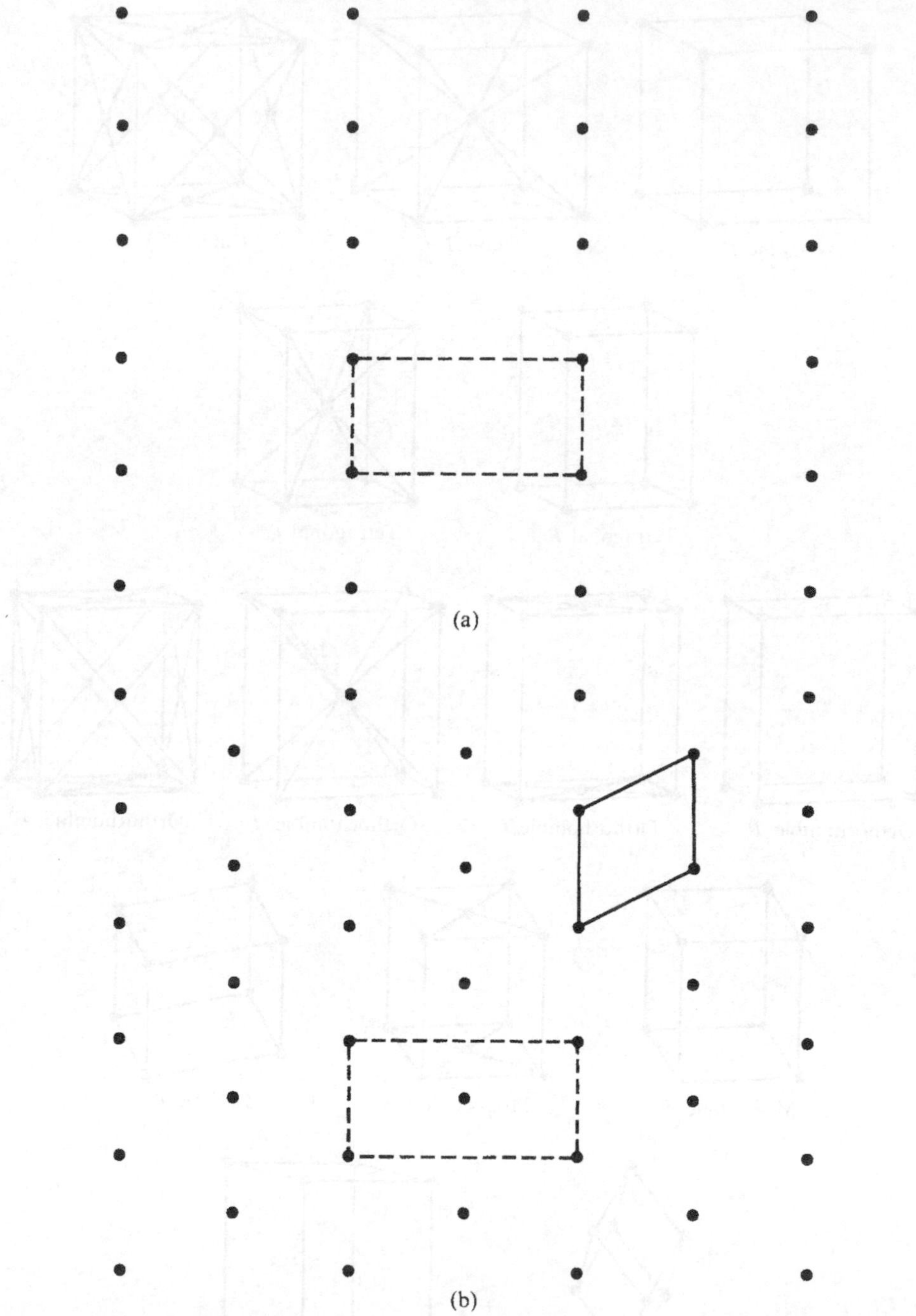

FIG. 1.8. (a) The rectangular 2-dimensional Bravais lattice, *p*, with unit cell marked. (b) The centred rectangular 2-dimensional Bravais lattice, *c*, with conventional unit cell shown with broken lines and fundamental unit cell shown with continuous lines.

such that each point has the same environment in the same orientation. A *unit cell* may be defined in various ways and it will be specified by three non-coplanar vectors which form the edges of a parallelepiped. The symbols **a**, **b**, and **c** are used to denote these three vectors. It is usually convenient to select the unit-cell vectors in such a way that the unit cell clearly exhibits the symmetry of the lattice. This conventional choice of unit cell may then not be primitive, that is, it may contain the equivalent of more than one lattice point. The parallelepiped constructed on the three basic vectors $\mathbf{t}_1$, $\mathbf{t}_2$, and $\mathbf{t}_3$ of the Bravais lattice (eqn. (1.5.1)) is called the *primitive unit cell*. In each Bravais lattice where the conventional unit cell and the primitive unit cell are not identical it is possible to establish the relationship between them. This is done for instance in Table 2.2.2 of Volume 1 of the *International tables for X-ray crystallography*. The unit cells illustrated in Fig. 1.7 are the conventional unit cells. The actual specification of $\mathbf{t}_1$, $\mathbf{t}_2$, and $\mathbf{t}_3$ for each of the 14 Bravais lattices will be given in Chapter 3 (Table 3.1). We can illustrate the distinction between conventional and primitive unit cells with a 2-dimensional example. The 2-dimensional primitive rectangular Bravais lattice, p, has its lattice points arranged on the corners of a set of rectangles and the conventional unit cell is one such rectangle, see Fig. 1.8(a); this is also a fundamental unit cell of this lattice. If a point is added to the centre of each of these cells a new Bravais lattice is obtained, the centred rectangular Bravais lattice, c, see Fig. 1.8(*b*). A conventional unit cell, which exhibits the rectangular symmetry, is shown with broken lines and a primitive unit cell is shown with continuous lines. The conventional unit cell contains effectively two lattice points while the primitive unit cell contains only one lattice point. Another unit cell that is sometimes used is the *Wigner–Seitz unit cell* (Wigner and Seitz 1933). The Wigner–Seitz unit cell of a Bravais lattice is obtained by choosing as origin, O, any one of the lattice points and drawing the planes that bisect perpendicularly the lines joining O to its nearest (and sometimes to its next-nearest) neighbours. The unit cell bounded by these planes is called the Wigner–Seitz unit cell; these unit cells are illustrated, for example, by Delaunay (1932).

A Bravais lattice, as we have described it, is an array of mathematical points that have position but no magnitude or shape. We now wish to include in our discussion the atoms or molecules of which the crystal is constituted. We are then effectively associating a similar collection of atoms with each Bravais lattice point and this set of atoms or molecules must itself possess symmetry that is compatible with the point-group symmetry of the Bravais lattice of the crystal. One type of symmetry operation of the crystal will be a compound operation consisting of some point-group operation followed by a translation $\mathbf{t}$ (see eqn. (1.5.1)) of the Bravais lattice. Thus, an especially simple type of *space group* arises by associating with each point of a Bravais lattice one of the point groups belonging to the same crystal system. In this way it is possible to obtain $\sum_{i=1}^{7} n_i p_i$ space groups where n_i is the number of point groups in a certain crystal system and p_i is the number of Bravais lattices in that system, the suffix i ranging over the seven crystal systems. For example, there are three cubic Bravais lattices and five cubic point groups, so that there are 15 cubic space groups that can be derived in this way. A space group of this form is said to be *symmorphic* and it is a semi-direct product group of the translation group of the Bravais lattice and a point group. The theory of semi-direct product groups will be discussed in Chapter 4. The above formula does not at once account for all the symmorphic space groups; there are two reasons for this. First, it is possible to associate a trigonal point group with the hexagonal Bravais lattice; so for the trigonal system we must take $p_i = 2$ even though there is only one trigonal Bravais lattice. Secondly, it does in fact happen that there can sometimes be two different and distinguishable orientations of a point group possible on a Bravais lattice; this leads to a few extra symmorphic space groups that occur irregularly as shown in column 6 of Table 1.7. There are in fact 73 different symmorphic space groups.

In a symmorphic space group the point-group operations and the operations of the translation group are essentially separable and the point-group operations that are present are, on their own, symmetry operations of the crystal. A second type of space group can arise by starting with a symmorphic space group and replacing some of the reflection planes by glide reflection planes and some of the rotation axes by screw rotation axes; these are commonly called glide planes and screw axes respectively. A glide reflection symmetry operation is a compound operation that consists of a reflection in a plane, m, together with a translation $\mathbf{v}$, which is often, but not necessarily, in the plane m itself. The final result of two successive performances of such a glide reflection operation is to produce a translation that must itself be a member of the group of translations of the Bravais lattice of the crystal. In a similar way we may have a screw axis of order n; a screw rotation symmetry operation consists of a rotation through $2\pi/n$ about an axis, followed by a translation, $\mathbf{v}$, which is often, but not necessarily, along the axis of rotation. If this screw rotation operation is performed n times in succession, the result is a pure translation that must again be a member of the group of translations of the Bravais lattice of the crystal. By including glide reflection planes and screw rotation axes it is possible to derive a further 157 space groups that are called *non-symmorphic space groups* or *asymmorphic space groups*. One can impose the condition that the translational part of each glide reflection operation must be in the reflection plane itself and that the translational part of each screw rotation operation should be along the axis of the rotation. If this condition is imposed then, when there are glide reflection planes or screw rotation axes present, the symmetry elements may not act all at one point in the crystal. When this is so the operators can always be transformed to act at one point but then the translations may lose their sense of being along the axis of the screw rotation or of being in the plane of the glide reflection and may be in some strange direction. The former viewpoint is visually more satisfying, but the latter viewpoint is essential for a systematic mathematical treatment, and we shall therefore adopt the convention that all space-group operations act at one point. The 157 non-symmorphic space groups together with the 73 symmorphic space groups make a total of 230 space groups in all. They are distributed among the crystal systems as shown in Table 1.7.

TABLE 1.7

The classification of the 230 space groups

Crystal system	s	n_i	p_i	$n_i p_i$	col. 6	col. 7	col. 8	col. 9
Triclinic	1	2	1	2	–	2	–	2
Monoclinic	2	3	2	6	–	6	7	13
Orthorhombic	–	3	4	12	1	13	46	59
Tetragonal	4	7	2	14	2	16	52	68
Trigonal	3	5	2	10	3	13	12	25
Hexagonal	6	7	1	7	1	8	19	27
Cubic	–	5	3	15	–	15	21	36
Total	–	32	15	66	7	73	157	230

Notes to Table 1.7

(i) s is the order of the principal axis in those cases where it exists. The operations about the principal axis can be proper or improper rotations.

(ii) n_i is the number of point groups of a given system. Note that $\bar{6}$ (C_{3h}) and $\bar{6}2m$ (D_{3h}) must be allocated to the hexagonal system. p_i is the number of Bravais lattices of a given system, except for the trigonal system where $p_i = 2$; for as explained in the text each trigonal point group can be associated with the hexagonal Bravais lattice. This lattice is therefore counted twice in the total.

(iii) Column 6 is the number of symmorphic space groups that occur because of alternative distinguishable settings of certain point groups on their Bravais lattices. Columns 7 and 8 are the total numbers of symmorphic and non-symmorphic space groups in a given system. Column 9 is the total number of space groups of either kind in a given system.

Seitz notation for space-group operators

A convenient notation to represent a general space-group operation is to use the symbol $\{R \mid \mathbf{v}\}$ with the meaning that R is a point-group operator and $\mathbf{v}$ is the translation vector to be associated with it; the interpretation of this operator is that it acts on the points of space and that these are always referred to a fixed set of axes, that is, it is an *active* operator (see section 1.4). The symbols $\{R \mid \mathbf{v}\}$ were introduced by Seitz (1936*b*) and they are called *Seitz space-group symbols*. Since these operators are active, $\{R_1 \mid \mathbf{v}_1\}$ operating on the vector $\mathbf{r}$ produces $\mathbf{r}'$, so that

$$\{R_1 \mid \mathbf{v}_1\}\mathbf{r} = \mathbf{r}' = R_1\mathbf{r} + \mathbf{v}_1, \tag{1.5.2}$$

where $R_1\mathbf{r}$ is the vector that is produced from $\mathbf{r}$ by the application of the active point-group operator R_1. There is a simple but important rule for the multiplication of these operators:

$$\{R_2 \mid \mathbf{v}_2\}\{R_1 \mid \mathbf{v}_1\} = \{R_2R_1 \mid \mathbf{v}_2 + R_2\mathbf{v}_1\} \tag{1.5.3}$$

as can easily be verified by applying $\{R_2 \mid \mathbf{v}_2\}$ to both sides of eqn. (1.5.2). From eqn. (1.5.3) it follows immediately that the inverse of $\{R_1 \mid \mathbf{v}_1\}$ is $\{R_1^{-1} \mid -R_1^{-1}\mathbf{v}_1\}$. The Seitz operators for all the 230 space groups are identified by various authors (Faddeyev 1964, Henry and Lonsdale 1965, Koptsik 1966, Kovalev 1965, Lyubarskii 1960, Miller and Love 1967) and they are listed in Chapter 3.

Notation for the labels of the space groups

A space group is classified first by the crystal system to which it belongs. Any secondary classification depends upon how much information one wants to convey. It is possible to label the point group to which the space group is *isogonal*, that is, the point group which is obtained by taking all the point-group operators present from amongst the operations of the space-group.‡ This is the position that is adopted when using the Schönflies notation (Hilton 1903). In this notation the space group is given the Schönflies label of its isogonal point group and different space groups having the same isogonal point group are given an arbitrary distinguishing superscript. This represented merely the order in which Schönflies (1891) presented the derivation of the space groups. There are, for example, 10 space groups isogonal to the cubic point group O_h ($m3m$) and these are distributed among the three cubic Bravais lattices, four primitive, four face-centred and two body-centred, and they are numbered O_h^1, O_h^2, ..., O_h^{10}. This notation has two disadvantages. First, it is not possible to tell at a glance on which Bravais lattice the space group is based; for example, O_h^8 is face-centred and O_h^9 is body-centred. Secondly, there is no indication of whether the space group is symmorphic or non-symmorphic. An even more arbitrary scheme in which each space group is given a number between 1 and 230 was developed by Astbury and Yardley (1924) and has been included in the *International tables for X-ray crystallography* (Henry and Lonsdale 1965).

The two disadvantages of the Schönflies notation are overcome by using the international notation, which not only determines the two features just mentioned but also gives some idea of the nature and orientation of the symmetry elements present in the space group (Hermann 1928*a*, Mauguin 1931, Schiebold 1929). Starting with the international point-group symbol, for example $m3m$, this is prefixed by the letter describing the Bravais lattice, for example P, F, I, for primitive, face-centred, and body-centred respectively, so that we obtain the symbols $Pm3m$, $Fm3m$, and $Im3m$. These symbols denote the three symmorphic space groups isogonal to the point group $m3m$ (O_h). When a point group can be situated on a Bravais lattice in more than one orientation, as for instance $\bar{4}2m$ (D_{2d}) in the tetragonal system, this is allowed for by attaching significance to the ordering of the characters in the international space-group symbol; thus $P\bar{4}2m$ (D_{2d}^1) and $P\bar{4}m2$ (D_{2d}^5) are distinct space groups. For non-symmorphic space groups the symbol is modified as follows. For any reflection plane

‡ Some writers use the word *isomorphous* where we use the word *isogonal*.

that has been replaced by a glide plane the appropriate part of the symbol is replaced by another letter denoting a glide plane; thus $Fd3m$ (O_h^7), the space group of the diamond structure, is face-centred and non-symmorphic, with glide planes in a diagonal direction. Other letters used include the crystal axis directions a, b, and c. For a screw axis the part of the space-group symbol denoting the order of such an axis, for example 4 of $P432$, is replaced by the same symbol with an added suffix denoting the size of the translation of the screw axis. Thus $P4_132$ (O^7), one of the cubic space groups isogonal to the point group 432 (O), is primitive with the fourfold rotation axis replaced by a fourfold screw axis so that a translation of one-quarter of a lattice vector in that direction is associated with each of the fourfold rotations. The international notation is not perfect, but is well known, carefully identified (Henry and Lonsdale 1965), and widely accepted. The original form of this notation is explained in the old *International tables for X-ray crystallography* (Bragg, von Laue, and Hermann 1935*a*, *b*) and the later modifications are explained in the new tables (Henry and Lonsdale 1965); the use of these tables in the identification of space groups will be discussed further in Chapter 3 (see section 3.5). In this book we always refer to a point group or space group by its international symbol followed by its Schönflies symbol in brackets.

Function space operators

If R is the rotational part of a Seitz space-group operator, that is, R is a point-group operation, then each operator R acts on the points of space $\mathbf{r}$ by means of an equation like eqn. (1.4.8),

$$R\mathbf{r} = \mathbf{r}'. \tag{1.5.4}$$

If a scalar function $f(\mathbf{r})$ is defined then the movement of the points $\mathbf{r}$ will induce a corresponding change in the function. This new function will be such that the new function at the transformed point is equal in value to the old function at the original point. That is to say,

$$g(\mathbf{r}') = f(\mathbf{r}). \tag{1.5.5}$$

It is customary to write $g = \bar{R}f$, so that to each operator R there is defined a corresponding *function space operator* $\bar{R}$. From eqns. (1.5.4) and (1.5.5) it follows that

$$\bar{R}f(\mathbf{r}) = f(R^{-1}\mathbf{r}). \tag{1.5.6}$$

It is perhaps unfortunate, but nevertheless a fact, that most writers do not use a separate symbol for $\bar{R}$ and indeed there is no need for a separate symbol provided one distinguishes carefully whether the operator is acting on a function or on a vector. Care must be taken to manipulate function space operators by means of eqn. (1.5.6). The following examples should act as sufficient warning.

Example 1.5.1. Let $\mathbf{r} = x\mathbf{i} + y\mathbf{j} + z\mathbf{k}$, and define $f(\mathbf{r}) = x$, $g(\mathbf{r}) = y$, and $h(\mathbf{r}) = z$, the projections of the vector $\mathbf{r}$ in the three coordinate directions. Then

$$\bar{C}_{4z}^{+} f(\mathbf{r}) = f(C_{4z}^{-}\mathbf{r}), \tag{1.5.7}$$

which is equal to $f(y\mathbf{i} - x\mathbf{j} + z\mathbf{k})$ (see Note (iii) to Table 1.4). The projection of the argument in the direction $\mathbf{i}$ is y so that

$$\bar{C}_{4z}^{+} f(\mathbf{r}) = y = g(\mathbf{r}). \tag{1.5.8}$$

Symbolically we have therefore both $C_{4z}^{+}\mathbf{i} = \mathbf{j}$ and $\bar{C}_{4z}^{+} f = g$. Indeed the matrix representing the operator C_{4z}^{+} with respect to the basis $\langle \mathbf{i}, \mathbf{j}, \mathbf{k}|$ is the same as the matrix representing the function space operator $\bar{C}_{4z}^{+}$ with respect to the basis $\langle f, g, h|$. After the warning to be careful this is a result that one would scarcely have dared to expect to be true.

Example 1.5.2. Let ϕ be the azimuthal angle (the angle between the x-axis and the projection of the position vector onto the xy-plane). Then

$$C_{4z}^{+}\phi = \phi + \tfrac{1}{2}\pi \tag{1.5.9}$$

but for $\bar{C}_{4z}^{+}$ as a function space operator we must write

$$\bar{C}_{4z}^{+} \cos\phi = \cos(C_{4z}^{-}\phi) = \cos(\phi - \tfrac{1}{2}\pi) = \sin\phi. \tag{1.5.10}$$

Equation (1.5.10) could have been deduced also from eqn. (1.5.8), because $f(\mathbf{r}) = r\sin\theta\cos\phi$ and $g(\mathbf{r}) = r\sin\theta\sin\phi$, and C_{4z}^{+} leaves r and θ unaltered. It should be noted that, since $\cos(\phi + \tfrac{1}{2}\pi) = -\sin\phi$, $\bar{C}_{4z}^{+}\cos\phi \neq \cos(C_{4z}^{+}\phi)$.

Active and passive space-group operators

As we have already pointed out an active operator is one that moves the points or position vectors of space, all vectors being referred to a fixed set of axes. The Seitz space-group operators that we shall use in this book are active and have been defined by eqn. (1.5.2). The definition is

$$\{R_a \mid \mathbf{v}\}\mathbf{r} = R_a\mathbf{r} + \mathbf{v}, \tag{1.5.11}$$

where we use the subscript a to emphasize that we mean an active operator. That is to say, an active rotation R_a is made first so that $\mathbf{r} \rightarrow R_a\mathbf{r}$ and then a translation $+\mathbf{v}$ is made to the new position vector, the result being $(R_a\mathbf{r} + \mathbf{v})$. We have already found the multiplication rule for the active Seitz space-group operators in eqn. (1.5.3), namely

$$\{S_a \mid \mathbf{w}\}\{R_a \mid \mathbf{v}\} = \{S_aR_a \mid \mathbf{w} + S_a\mathbf{v}\}. \tag{1.5.12}$$

The inverse of $\{R_a \mid \mathbf{v}\}$ is

$$\{R_a \mid \mathbf{v}\}^{-1} = \{R_a^{-1} \mid -R_a^{-1}\mathbf{v}\}. \tag{1.5.13}$$

A passive operator is one that moves the axes of space, all points of the space and hence all vector positions being left unmoved, so that after each operation a vector position is referred to a new set of axes. It is possible to define passive space-group operators (Johnston 1960). To distinguish them from active Seitz operators we use square brackets $[R_p \mid \mathbf{v}]$ and the rule of definition corresponding to eqn. (1.5.11) is

$$[R_p \mid \mathbf{v}]\mathbf{r} = R_p\mathbf{r} - R_p\mathbf{v}. \tag{1.5.14}$$

That is to say a translation of axes by $+\mathbf{v}$ is made first so that $\mathbf{r} \to (\mathbf{r} - \mathbf{v})$, and then a rotation of the new axes is made, the final result being $(R_p\mathbf{r} - R_p\mathbf{v})$, where R_p is now a passive rotation. The rule for the multiplication of these passive operators is

$$[S_p \mid \mathbf{w}][R_p \mid \mathbf{v}] = [S_pR_p \mid \mathbf{v} + R_p^{-1}\mathbf{w}] \tag{1.5.15}$$

and the inverse of $[R_p \mid \mathbf{v}]$ is

$$[R_p \mid \mathbf{v}]^{-1} = [R_p^{-1} \mid -R_p\mathbf{v}]. \tag{1.5.16}$$

If X is an operator, whether active or passive, then along with X there is another operator $\bar{X}$ where $\bar{X}$ is a function-space operator as defined in eqns. (1.5.4)–(1.5.6). If $f(\mathbf{r})$ is a scalar function of position then

$$\bar{X}f(\mathbf{r}) = f(X^{-1}\mathbf{r}). \tag{1.5.17}$$

It is important to realize that $(\bar{X}f)$ is the symbol for a function just as f is also the symbol for a function. Equation (1.5.17) is not arbitrary but states the fact that the value of the transformed function $(\bar{X}f)$ at the transformed point $X\mathbf{r}$ is equal in value to f at the point $\mathbf{r}$. Another point, stressed by Wigner (1959), must again be emphasized, and that is that if $Z = YX$ then we require $\bar{Z}f = \bar{Y}\bar{X}f$ because we want to preserve an isomorphism between groups of operators and the corresponding groups of function-space operators. The proof that this is true follows from repeated application of eqn. (1.5.17). If we write $\bar{X}f = g$ then

$$\begin{aligned}
\bar{Y}\bar{X}f(\mathbf{r}) &= \bar{Y}g(\mathbf{r}), \\
&= g(Y^{-1}\mathbf{r}) && \text{(from eqn. (1.5.17))}, \\
&= \bar{X}f(Y^{-1}\mathbf{r}) && \text{(since } \bar{X}f = g), \\
&= f(X^{-1}Y^{-1}\mathbf{r}) && \text{(from eqn. (1.5.17))}, \\
&= f(Z^{-1}\mathbf{r}) && \text{(since } Z = YX), \\
&= \bar{Z}f(\mathbf{r}) && \text{(from eqn. (1.5.17) again)}.
\end{aligned}$$

Thus $\bar{Y}\bar{X}f = \bar{Z}f$ as required. It is incorrect to operate directly on eqn. (1.5.17) with $\bar{Y}$ since it is a numerical relationship between two functions for different values of their arguments and operators cannot act on numbers (Altmann and Bradley 1965, Slater 1965*a*).

As a first example let us consider the effect of passive space-group operators on a function $\psi(\mathbf{r})$. From eqn. (1.5.17) it follows that

$$\begin{aligned}\overline{[S_p \mid \mathbf{w}]}\psi(\mathbf{r}) &= \psi([S_p \mid \mathbf{w}]^{-1}\mathbf{r}) \\ &= \psi([S_p^{-1} \mid -S_p\mathbf{w}]\mathbf{r}) \qquad \text{(from eqn. (1.5.16))} \qquad (1.5.18) \\ &= \psi(S_p^{-1}\mathbf{r} + \mathbf{w}) \qquad \text{(from eqn. (1.5.14))}. \qquad (1.5.19)\end{aligned}$$

From the rule for multiplying function-space operators and from eqn. (1.5.19)

$$\begin{aligned}\overline{[S_p \mid \mathbf{w}]}\,\overline{[R_p \mid \mathbf{v}]}\psi(\mathbf{r}) &= \overline{[R_p \mid \mathbf{v}]}\psi(S_p^{-1}\mathbf{r} + \mathbf{w}) \\ &= \psi(R_p^{-1}S_p^{-1}\mathbf{r} + R_p^{-1}\mathbf{w} + \mathbf{v}). \qquad (1.5.20)\end{aligned}$$

A derivation of eqn. (1.5.20) from eqn. (1.5.19) as follows would not be correct:

$$\begin{aligned}\overline{[S_p \mid \mathbf{w}]}\,\overline{[R_p \mid \mathbf{v}]}\psi(\mathbf{r}) &= \overline{[S_p \mid \mathbf{w}]}\psi(R_p^{-1}\mathbf{r} + \mathbf{v}) \qquad (1.5.21) \\ &= \psi(R_p^{-1}(S_p^{-1}\mathbf{r} + \mathbf{w}) + \mathbf{v}) \qquad (1.5.22) \\ &= \psi(R_p^{-1}S_p^{-1}\mathbf{r} + R_p^{-1}\mathbf{w} + \mathbf{v}). \qquad (1.5.20)\end{aligned}$$

Equation (1.5.21) is incorrect because, as mentioned above, $\overline{[S_p \mid \mathbf{w}]}$ cannot be applied directly to the numerical relationship $\overline{[R_p \mid \mathbf{v}]}\psi(\mathbf{r}) = \psi(R_p^{-1}\mathbf{r} + \mathbf{v})$; but just supposing it could, then eqn. (1.5.22) is also at fault, and is directly contrary to mathematical practice, because a function-space operator acts on the whole of the argument of a function and not just the part **r**.

The equations corresponding to (1.5.19) and (1.5.20) for active space-group operators are

$$\overline{\{S_a \mid \mathbf{w}\}}\psi(\mathbf{r}) = \psi(S_a^{-1}(\mathbf{r} - \mathbf{w})) \qquad (1.5.23)$$

and

$$\overline{\{S_a \mid \mathbf{w}\}}\,\overline{\{R_a \mid \mathbf{v}\}}\psi(\mathbf{r}) = \psi(R_a^{-1}S_a^{-1}\mathbf{r} - R_a^{-1}S_a^{-1}\mathbf{w} - R_a^{-1}\mathbf{v}). \qquad (1.5.24)$$

Equations (1.5.19) and (1.5.23) plus the fact that $S_p = S_a^{-1}$ imply that $\overline{[S_p \mid \mathbf{w}]} = \overline{\{S_a \mid \mathbf{w}\}}^{-1}$ a relationship that we should expect to hold between active and passive operators.

Example 1.5.3.Let $\{E \mid \mathbf{t}\}$ denote the translation **t**. Then

$$\{E \mid \mathbf{t}\}\mathbf{r} = \mathbf{r} + \mathbf{t}. \qquad (1.5.25)$$

But if $\overline{\{E \mid \mathbf{t}\}}$ now acts on the function $\psi(\mathbf{r})$ rather than on the vector **r** we have

$$\overline{\{E \mid \mathbf{t}\}}\psi(\mathbf{r}) = \psi(\{E \mid \mathbf{t}\}^{-1}\mathbf{r}) = \psi(\mathbf{r} - \mathbf{t}). \qquad (1.5.26)$$

As we have said before, we shall always use active point-group and space-group operators in this book, except that we shall occasionally illustrate how our theoretical discussions would have to be modified if one did choose to use passive operators.

2

Symmetry-adapted functions for the point groups

THE aim of this chapter is to derive the linear combinations of spherical harmonics that belong to the various rows of the matrix representations of the crystallographic point groups, and to give tables of these functions. Such functions as these have been given the name, by Melvin (1956), of *symmetry-adapted functions* because they have the correct properties required by the representations under transformation by the group elements. Work on the determination of symmetry-adapted functions has been done by Altmann (1956, 1957), Bell (1954), Bethe (1929), Betts (1959), Callen and Callen (1963), Cohan (1958), Cornwell (1969), Flodmark (1963), Flower, March, and Murray (1960), McIntosh (1960*a*, 1963), Melvin (1956), Meyer (1954), Nesbet (1961), Schiff (1955, 1956), and von der Lage and Bethe (1947). This chapter follows quite closely the treatment given by Altmann (1957), Altmann and Bradley (1963*a*, *b*, 1965), and Altmann and Cracknell (1965). Given a particular matrix representation of a group the problem is to find all possible bases that are symmetry-adapted to that representation. Then any function that can be expanded in terms of spherical harmonics and is to have the transformation properties of a given row of that representation, will include only those linear combinations of spherical harmonics that are symmetry-adapted to have the same transformation properties. We use the term *surface harmonic* for such a linear combination of spherical harmonics that is symmetry-adapted to one row of a point-group representation.

2.1. The matrix elements of the rotation group

During the course of determining symmetry-adapted functions it becomes necessary to determine how the spherical harmonics transform under various rotations of the 3-dimensional rotation group. In this section we give the formulae and definitions which are required for determining these transformation properties.

We use the following definition of the normalized spherical harmonics:

$$Y_l^m(\theta, \phi) = \sqrt{\left\{\frac{(2l+1)(l-|m|)!}{4\pi(l+|m|)!}\right\}}\, P_l^m(\cos\theta)\exp(im\phi) \tag{2.1.1}$$

where $P_l^m(\cos\theta)$ is the associated Legendre function:

$$P_l^m(\cos\theta) = \frac{1}{2^l l!}\sin^{|m|}\theta\,\frac{d^{\,l+|m|}}{(d\cos\theta)^{l+|m|}}\{(\cos^2\theta - 1)^l\}. \tag{2.1.2}$$

This is not the place for a lengthy discussion of the properties of the spherical harmonics, these are considered in a very large number of books (for example, Condon and Shortley (1935), Edmonds (1957), Gel'fand, Minlos, and Shapiro (1963), and Wigner (1959), to name only a few).

We specify a pure rotation R by its Euler angles α, β, and γ defined as follows: we perform first an active rotation by an angle $\alpha(0 \leqslant \alpha < 2\pi)$ about the z-axis, secondly an active rotation by an angle $\beta(0 \leqslant \beta \leqslant \pi)$ about the y-axis, and finally an active rotation by an angle $\gamma(0 \leqslant \gamma < 2\pi)$ about the z-axis. Therefore, the first rotation, through α, sends the point (r, θ, ϕ) to $(r, \theta, \phi + \alpha)$, the second rotation, through β, sends the point $(r, \theta, 0)$ to $(r, \theta + \beta, 0)$ and the third rotation, through γ, sends the point (r, θ, ϕ) to $(r, \theta, \phi + \gamma)$ (see Fig. 2.1). Then $R(\alpha, \beta, \gamma)$ means the product of these three rotations in succession. Note that, as always in this book, we use the active convention; our operators $R(\alpha, \beta, \gamma)$ move the points of space and leave the axes $Oxyz$ fixed. Also, all positive rotations are taken to be in the conventional anti-clockwise direction.

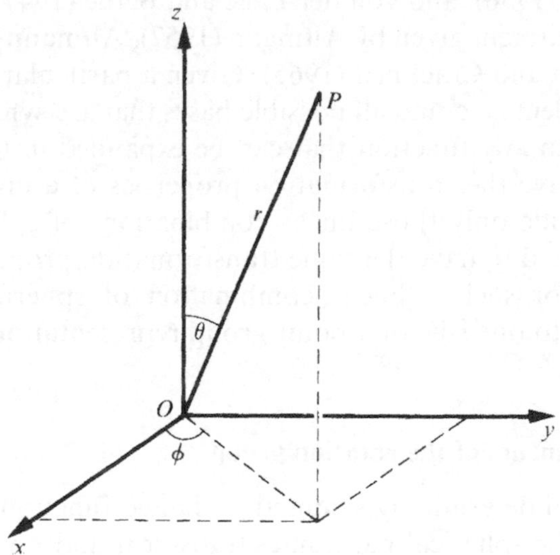

FIG. 2.1. The spherical polar coordinates (r, θ, ϕ).

With this definition for the Euler angles the law of transformation of the spherical harmonics can be derived. The result (Altmann 1957, Wigner 1959) is

$$R(\alpha, \beta, \gamma) Y_l^m(\theta, \phi) = \sum_{n=-l}^{l} Y_l^n(\theta, \phi)\mathscr{D}^l\{R(\alpha, \beta, \gamma)\}_{nm}, \tag{2.1.3}$$

where

$$\mathscr{D}^l\{R(\alpha, \beta, \gamma)\}_{nm} = C_{nm} \exp(-in\gamma) d^l(\beta)_{nm} \exp(-im\alpha). \tag{2.1.4}$$

In eqn. (2.1.4)

$$C_{nm} = (\mathrm{i}^{|n|+n})(\mathrm{i}^{-|m|-m}), \tag{2.1.5}$$

and

$$d^l(\beta)_{nm} = \sum_k \frac{(-1)^{k-m+n}\sqrt{\{(l+n)!(l+m)!(l-n)!(l-m)!\}}}{(l-n-k)!(l+m-k)!k!(k-m+n)!} \times \cos^{2l+m-n-2k}(\tfrac{1}{2}\beta)\sin^{2k+n-m}(\tfrac{1}{2}\beta), \tag{2.1.6}$$

where the summation over k goes from the larger of the numbers 0 and $(m-n)$ to the smaller of the numbers $(l-n)$ and $(l+m)$. To every rotation $R(\alpha, \beta, \gamma)$ of the 3-dimensional rotation group there is one and only one matrix $\mathscr{D}^l\{R(\alpha, \beta, \gamma)\}$ for each integer value of l. For any given l the matrices $\mathscr{D}^l\{R(\alpha, \beta, \gamma)\}$ form a $(2l+1)$-dimensional representation of the group of rotations $R(\alpha, \beta, \gamma)$ as defined in Definition 1.3.4. Equation (2.1.3) expresses the fact that the functions $Y_l^m(\theta, \phi)$ $(-l \leqslant m \leqslant l)$ form a basis for this representation. The representations $\mathscr{D}^j\{R(\alpha, \beta, \gamma)\}$ where j is half an odd integer will be discussed in Chapter 6.

The reduced matrix elements $d^l(\beta)_{nm}$ satisfy the following symmetry relations:

$$d^l(\beta)_{nm} = d^l(\beta)_{-m,-n} = (-1)^{m+n}\, d^l(\beta)_{mn}. \tag{2.1.7}$$

Because of the relations (2.1.7) it is only necessary to evaluate $d^l(\beta)_{nm}$ in the range $-l \leqslant m \leqslant l$, $|m| \leqslant n \leqslant l$. In this range Altmann and Bradley (1963*a*) have shown that $d^l(\beta)_{nm}$ is most easily calculated by means of the recurrence relation

$$\sqrt{\{(l+n)(l-n+1)\}}\, d^l(\beta)_{n-1,m} = (m-n)\cot(\tfrac{1}{2}\beta)\, d^l(\beta)_{nm} - \sqrt{\{(l+m)(l-m+1)\}}\, d^l(\beta)_{n,m-1} \tag{2.1.8}$$

with starting values

$$d^l(\beta)_{lm} = (-1)^{l-m}\sqrt{\left\{\frac{(2l)!}{(l+m)!(l-m)!}\right\}}\cos^{l+m}(\tfrac{1}{2}\beta)\sin^{l-m}(\tfrac{1}{2}\beta). \tag{2.1.9}$$

Equation (2.1.8) is only valid for integer values of l, m, and n.

The cases $\beta = 0$, $\beta = \pi$, and $\beta = \frac{1}{2}\pi$ deserve special attention for two reasons. First, as shown by Wigner (1959), the matrix representatives for arbitrary β can be obtained in terms of those for $\beta = \frac{1}{2}\pi$. Secondly, it is always possible (Altmann 1957) to choose axes so that the β angle for any rotation in a crystallographic point group takes one of the values 0, $\frac{1}{2}\pi$, and π. If $\beta = 0$ we have a rotation $(\alpha + \gamma)$ about the z-axis so that

$$\mathscr{D}^l\{R(\alpha, 0, \gamma)\}_{nm} = \exp(-\mathrm{i}m\alpha)\exp(-\mathrm{i}m\gamma)\,\delta_{nm}. \tag{2.1.10}$$

If $\beta = \pi$ then, as shown by Altmann (1957), the only non-zero elements are found when $n = -m$ and

$$\mathscr{D}^l\{R(\alpha, \pi, \gamma)\}_{nm} = (-1)^l \exp(-\mathrm{i}m\alpha)\exp(\mathrm{i}m\gamma)\,\delta_{n,-m}. \tag{2.1.11}$$

If $\beta = \frac{1}{2}\pi$ then no simple analytic formula exists for $d^l(\frac{1}{2}\pi)_{nm}$. However, tables of these functions have been compiled by Bradley (1961) for all values of l up to $l = 20$.

In order to complete the study of the transformation properties of spherical harmonics under point-group operations we include the extra relation for transformation under the inversion, I:

$$IY_l^m(\theta, \phi) = (-1)^l Y_l^m(\theta, \phi). \tag{2.1.12}$$

A table of the Euler angles for the crystallographic point-group operations introduced in Chapter 1 is given in Table 2.1.

TABLE 2.1.

Euler angles for the point groups 432 (O) *and* 622 (D_6).

(a) 432 (O)								(b) 622 (D_6)			
Element	α	β	γ	Element	α	β	γ	Element	α	β	γ
E	0	0	0	C_{4x}^+	$\frac{1}{2}\pi$	$\frac{1}{2}\pi$	$\frac{3}{2}\pi$	E	0	0	0
C_{2x}	π	π	0	C_{4y}^+	0	$\frac{1}{2}\pi$	0	C_6^+	$\frac{1}{3}\pi$	0	0
C_{2y}	0	π	0	C_{4z}^+	$\frac{1}{2}\pi$	0	0	C_3^+	$\frac{2}{3}\pi$	0	0
C_{2z}	π	0	0	C_{4x}^-	$\frac{3}{2}\pi$	$\frac{1}{2}\pi$	$\frac{1}{2}\pi$	C_2	π	0	0
C_{31}^+	$\frac{1}{2}\pi$	$\frac{1}{2}\pi$	0	C_{4y}^-	π	$\frac{1}{2}\pi$	π	C_3^-	$\frac{4}{3}\pi$	0	0
C_{32}^+	$\frac{3}{2}\pi$	$\frac{1}{2}\pi$	π	C_{4z}^-	$\frac{3}{2}\pi$	0	0	C_6^-	$\frac{5}{3}\pi$	0	0
C_{33}^+	$\frac{1}{2}\pi$	$\frac{1}{2}\pi$	π	C_{2a}	$\frac{1}{2}\pi$	π	0	C'_{21}	π	π	0
C_{34}^+	$\frac{3}{2}\pi$	$\frac{1}{2}\pi$	0	C_{2b}	$\frac{3}{2}\pi$	π	0	C'_{22}	$\frac{5}{3}\pi$	π	0
C_{31}^-	π	$\frac{1}{2}\pi$	$\frac{1}{2}\pi$	C_{2c}	π	$\frac{1}{2}\pi$	0	C'_{23}	$\frac{1}{3}\pi$	π	0
C_{32}^-	0	$\frac{1}{2}\pi$	$\frac{3}{2}\pi$	C_{2d}	$\frac{1}{2}\pi$	$\frac{1}{2}\pi$	$\frac{1}{2}\pi$	C''_{21}	0	π	0
C_{33}^-	0	$\frac{1}{2}\pi$	$\frac{1}{2}\pi$	C_{2e}	0	$\frac{1}{2}\pi$	π	C''_{22}	$\frac{2}{3}\pi$	π	0
C_{34}^-	π	$\frac{1}{2}\pi$	$\frac{3}{2}\pi$	C_{2f}	$\frac{3}{2}\pi$	$\frac{1}{2}\pi$	$\frac{3}{2}\pi$	C''_{23}	$\frac{4}{3}\pi$	π	0

Notes to Table 2.1.

(i) The symmetry operations were defined in Chapter 1 and are active.

(ii) The Euler angles are defined by a rotation α of the points of space about the z-axis (positive being defined so that a point on the x-axis moves towards the y-axis) followed by a rotation β about the y-axis and followed by a rotation γ about the z-axis. α, β, and γ are restricted so that $0 \leqslant \alpha < 2\pi$, $0 \leqslant \beta \leqslant \pi$, and $0 \leqslant \gamma < 2\pi$.

2.2. The generation of symmetry-adapted functions

In this section we give general rules for obtaining symmetry-adapted functions that belong to the irreducible representations of any finite group.

Suppose that $\mathbf{G}$ is a group of order $|\mathbf{G}|$ with elements $R, S, \ldots$, etc. Suppose further that its reps are $\Gamma^i\{R \to \mathbf{D}^i(R)\}$, and are known. Here $\mathbf{D}^i(R)$ is the matrix representative of R and is of dimension, say, d_i. The problem is to find a basis $\langle \phi_1^i, \phi_2^i, \ldots, \phi_{d_i}^i|$ such that

$$R\phi_s^i = \sum_{t=1}^{d_i} \phi_t^i \mathbf{D}^i(R)_{ts} \tag{2.2.1}$$

for all $R \in \mathbf{G}$. ϕ_s^i is then a symmetry-adapted function belonging to row s of the rep Γ^i. Actually the problem is not to find just one basis, but a complete set of bases such that the totality of functions ϕ_s^i (for all s and all i) forms a complete set in the linear space V in which the realization of the group operators act. The theory of the determination of such functions is a straightforward application of group theory.

We may define the elements W_{ts}^i of the group ring as follows:

$$W_{ts}^i = \frac{d_i}{|\mathbf{G}|} \sum_{R \in \mathbf{G}} \mathbf{D}^i(R)_{ts}^* R \tag{2.2.2}$$

where the sum is over all $R \in \mathbf{G}$.

Theorem 2.2.1. *If ϕ is an arbitrary function of V such that $W_{ss}^i \phi \neq 0$ (s is fixed and is a number in the range 1 to d_i) then the functions $W_{ts}^i \phi = \phi_t^i$, $t = 1$ to d_i, form a basis for the rep Γ^i.*

ϕ may then be regarded as a generating function of the symmetry-adapted function ϕ_t^i and W_{ss}^i is a *projection operator*. The proof of this theorem is to verify that eqn. (2.2.1) holds and it makes use of the unitary nature of the matrices $\mathbf{D}^i(R)$. The proof is as follows.

Let $S \in \mathbf{G}$. Then, by definition,

$$S\phi_p^i = SW_{ps}^i \phi = S \frac{d_i}{|\mathbf{G}|} \sum_{R \in \mathbf{G}} \mathbf{D}^i(R)_{ps}^* R\phi. \tag{2.2.3}$$

Take S under the summation sign and write $SR = T$, then as R runs over all the group elements so does T and $R = S^{-1}T$. Equation (2.2.3) then becomes

$$S\phi_p^i = \frac{d_i}{|\mathbf{G}|} \sum_{T \in \mathbf{G}} \mathbf{D}^i(S^{-1}T)_{ps}^* T\phi = \frac{d_i}{|\mathbf{G}|} \sum_T \sum_t \mathbf{D}^i(S)_{tp} \mathbf{D}^i(T)_{ts}^* T\phi \tag{2.2.4}$$

since Γ^i is both unitary and a representation. Thus

$$S\phi_p^i = \sum_t \mathbf{D}^i(S)_{tp} W_{ts}^i \phi = \sum_t \phi_t^i \mathbf{D}^i(S)_{tp} \tag{2.2.5}$$

and since S was an arbitrary element of $\mathbf{G}$ eqn. (2.2.1) is verified and the theorem is proved.

The power of this theorem for our purpose is immediate. All that we have to do is to vary ϕ in V and to apply the operators W_{ts}^i systematically on each ϕ until all required base functions ϕ_t^i have been obtained. The process is then repeated for each representation Γ^i and the problem is then solved in principle. In practice the process may well be lengthy and laborious. Indeed unless $R\phi$ can be evaluated analytically for all $\phi \in V$ the process is interminable.

In our case $\mathbf{G}$ is a point group, and the linear space V is the Hilbert space spanned by the spherical harmonics $Y_l^m(\theta, \phi)$. For the cyclic and dihedral groups the rotations

$R(\alpha, \beta, \gamma)$ are such that $\beta = 0$ or π; for such rotations $R(\alpha, \beta, \gamma)\, Y_l^m(\theta, \phi)$ can be evaluated analytically (see eqns. (2.1.10) and (2.1.11)). Hence for the cyclic and dihedral groups complete sets of base functions can be obtained. For some elements of the cubic groups, however, β is equal to $\frac{1}{2}\pi$ (see Table 2.1), and so we can calculate $R(\alpha, \beta, \gamma)\, Y_l^m(\theta, \phi)$ only for those l values for which $d^l(\frac{1}{2}\pi)$ is known. In fact, surface harmonics belonging to the representations of the cubic groups have been evaluated only for l values up to $l = 12$. But to have base functions up to $l = 12$ is sufficient for all practical purposes at the present time.

For the sake of completeness, we quote two other theorems concerning the operators W_{ts}^i.

THEOREM 2.2.2

$$W_{ts}^i W_{pq}^j = \delta^{ij}\, \delta_{sp} W_{tq}^i. \tag{2.2.6}$$

THEOREM 2.2.3

$$W_{ts}^{i\dagger} = W_{st}^i \tag{2.2.7}$$

where the dagger denotes the adjoint operator in the space V.

The proofs of these theorems are given, for example, by Altmann (1962).

2.3. Application to the point groups

Suppose that we are given a point group **G** with elements $R, S, \ldots$, and a matrix representation $\mathbf{D}^i(R), \mathbf{D}^i(S), \ldots$. For the generating functions we take the spherical harmonics $Y_l^m(\theta, \phi)$. We require to evaluate $W_{ts}^i Y_l^m(\theta, \phi)$. Using eqn. (2.2.2) we obtain

$$W_{ts}^i Y_l^m(\theta, \phi) = \frac{d_i}{|\mathbf{G}|} \sum_{R \in \mathbf{G}} \mathbf{D}^i(R)_{ts}^* R Y_l^m(\theta, \phi). \tag{2.3.1}$$

In order to proceed further we need an expression for $R Y_l^m(\theta, \phi)$. Now R is either a proper rotation or a product of the inversion, I, with a proper rotation, that is, an improper rotation. If R is a proper rotation then we find its Euler angles (α, β, γ) (see Table 2.1), and evaluate $R Y_l^m(\theta, \phi)$ directly by means of eqn. (2.1.3). On the other hand, if R is improper we express $R = IQ$ and use the Euler angles (α, β, γ) for the proper rotation Q, evaluating $Q Y_l^m(\theta, \phi)$ by means of eqn. (2.1.3); to complete the evaluation of $R Y_l^m(\theta, \phi)$ we use eqn. (2.1.12) for the transforms of $I Y_l^n(\theta, \phi)$. This simply adds an extra factor $(-1)^l$ if R is an improper rotation.

To summarize, eqn. (2.3.1) becomes

$$W_{ts}^i Y_l^m(\theta, \phi) = \frac{d_i}{|\mathbf{G}|} \sum_{R \in \mathbf{G}} P_R \mathbf{D}^i(R)_{ts}^* \exp(-im\alpha) \sum_n C_{nm} \exp(-in\gamma) \times d^l(\beta)_{nm}\, Y_l^n(\theta, \phi), \tag{2.3.2}$$

where we sum over all the operations R of the group and, when R is improper, we take in the right-hand side the quantities corresponding to the associated proper rotation Q. Also P_R is unity when R is proper and $(-1)^l$ when R is improper.

Before tabulating the results for all the point groups we deal with one further theoretical problem. This is that the surface harmonics generated by means of eqn. (2.3.2) for a given row of a given rep, that is for fixed i and t, are not necessarily orthogonal. For practical purposes it is desirable that any two bases for the same representation should consist of mutually orthogonal functions. All the expansions given in the tables that follow have been orthogonalized with the help of Theorems 2.2.2 and 2.2.3.

2.4. Symmetry-adapted functions for the crystallographic point groups

In Table 2.2 we give the character tables of the (single-valued) reps of the 32 crystallographic point groups. The reps are labelled in the notation of Mulliken (1933) which we shall follow in this book, but the Γ labels, given for example by Koster, Dimmock, Wheeler, and Statz (1963), are also included for reference. In Table 2.3 we give the matrices that we use for the degenerate point-group reps. The method of section 2.3 can be applied to the determination of the surface harmonics for the cyclic, dihedral, and cubic point groups and these functions are given in Tables 2.4–2.6, respectively (Altmann 1957, Altmann and Bradley 1963*b*, Altmann and Cracknell 1965). We give a few examples of the interpretation of these tables of surface harmonics.

TABLE 2.2

Character tables for the crystallographic point groups
($\omega = \exp(2\pi i/3)$)

$1(C_1)$		E
A	Γ_1	1

$\bar{1}\ (C_i)$		$2\ (C_2)$		$m\ (C_{1h})$		E / E / E	I / C_{2z} / σ_z
A_g	Γ_1^+	A	Γ_1	A'	Γ_1	1	1
A_u	Γ_1^-	B	Γ_2	A''	Γ_2	1	-1

$2/m = 2 \otimes \bar{1}\ (C_{2h} = C_2 \otimes C_i)$

$mm2$ (C_{2v})		222 (D_2)		E E	C_{2z} C_{2z}	σ_y C_{2y}	σ_x C_{2x}
A_1	Γ_1	A	Γ_1	1	1	1	1
B_2	Γ_4	B_3	Γ_4	1	−1	−1	1
A_2	Γ_3	B_1	Γ_3	1	1	−1	−1
B_1	Γ_2	B_2	Γ_2	1	−1	1	−1

$mmm = 222 \otimes \bar{1}$ ($D_{2h} = D_2 \otimes C_i$)

4 (C_4)		$\bar{4}$ (S_4)		E E	C_{2z} C_{2z}	C_{4z}^+ S_{4z}^-	C_{4z}^- S_{4z}^+
A	Γ_1	A	Γ_1	1	1	1	1
B	Γ_2	B	Γ_2	1	1	−1	−1
${}^1E$	Γ_4	${}^1E$	Γ_4	1	−1	−i	i
${}^2E$	Γ_3	${}^2E$	Γ_3	1	−1	i	−i

$4/m = 4 \otimes \bar{1}$ ($C_{4h} = C_4 \otimes C_i$)

3 (C_3)		E	C_3^+	C_3^-
A	Γ_1	1	1	1
${}^1E$	Γ_3	1	ω^*	ω
${}^2E$	Γ_2	1	ω	ω^*

$\bar{3} = 3 \otimes \bar{1}$ ($C_{3i} = C_3 \otimes C_i$)

32 (D_3)		$3m$ (C_{3v})		E E	$C_3^\pm$ $C_3^\pm$	C'_{2i} σ_{di}
A_1	Γ_1	A_1	Γ_1	1	1	1
A_2	Γ_2	A_2	Γ_2	1	1	−1
E	Γ_3	E	Γ_3	2	−1	0

$\bar{3}m = 32 \otimes \bar{1}$ ($D_{3d} = D_3 \otimes C_i$)

6 (C_6)		$\bar{6}$ (C_{3h})		E E	C_6^+ S_3^-	C_3^+ C_3^+	C_2 σ_h	C_3^- C_3^-	C_6^- S_3^+
A	Γ_1	A'	Γ_1	1	1	1	1	1	1
B	Γ_4	A''	Γ_4	1	−1	1	−1	1	−1
${}^1E_1$	Γ_6	${}^1E'$	Γ_3	1	ω	ω^*	1	ω	ω^*
${}^2E_1$	Γ_5	${}^2E'$	Γ_2	1	ω^*	ω	1	ω^*	ω
${}^1E_2$	Γ_3	${}^1E''$	Γ_6	1	$-\omega$	ω^*	−1	ω	$-\omega^*$
${}^2E_2$	Γ_2	${}^2E''$	Γ_5	1	$-\omega^*$	ω	−1	ω^*	$-\omega$

$6/m = 6 \otimes \bar{1}$ ($C_{6h} = C_6 \otimes C_i$)

422 (D_4)						E	C_{2z}	$C_{4z}^{\pm}$	C_{2x}, C_{2y}	C_{2a}, C_{2b}
		$4mm$ (C_{4v})				E	C_{2z}	$C_{4z}^{\pm}$	σ_x, σ_y	σ_{da}, σ_{db}
				$\bar{4}2m$ (D_{2d})		E	C_{2z}	$S_{4z}^{\pm}$	C_{2x}, C_{2y}	σ_{da}, σ_{db}
A_1	Γ_1	A_1	Γ_1	A_1	Γ_1	1	1	1	1	1
A_2	Γ_2	A_2	Γ_2	A_2	Γ_2	1	1	1	−1	−1
B_1	Γ_3	B_1	Γ_3	B_1	Γ_3	1	1	−1	1	−1
B_2	Γ_4	B_2	Γ_4	B_2	Γ_4	1	1	−1	−1	1
E	Γ_5	E	Γ_5	E	Γ_5	2	−2	0	0	0

$4/mmm = 422 \otimes \bar{1}$ ($D_{4h} = D_4 \otimes C_i$)

622 (D_6)						E	C_2	$C_3^{\pm}$	$C_6^{\pm}$	C'_{2i}	C''_{2i}
		$6mm$ (C_{6v})				E	C_2	$C_3^{\pm}$	$C_6^{\pm}$	σ_{di}	σ_{vi}
				$\bar{6}2m$ (D_{3h})		E	σ_h	$C_3^{\pm}$	$S_3^{\pm}$	C'_{2i}	σ_{vi}
A_1	Γ_1	A_1	Γ_1	A'_1	Γ_1	1	1	1	1	1	1
A_2	Γ_2	A_2	Γ_2	A'_2	Γ_2	1	1	1	1	−1	−1
B_1	Γ_3	B_2	Γ_3	A''_1	Γ_3	1	−1	1	−1	1	−1
B_2	Γ_4	B_1	Γ_4	A''_2	Γ_4	1	−1	1	−1	−1	1
E_2	Γ_6	E_2	Γ_6	E'	Γ_6	2	2	−1	−1	0	0
E_1	Γ_5	E_1	Γ_5	E''	Γ_5	2	−2	−1	1	0	0

$6/mmm = 622 \otimes \bar{1}$ ($D_{6h} = D_6 \otimes C_i$)

23 (T)		E	C_{2m}	C_{3j}^{+}	C_{3j}^{-}
A	Γ_1	1	1	1	1
${}^1E$	Γ_2	1	1	ω	ω^*
${}^2E$	Γ_3	1	1	ω^*	ω
T	Γ_4	3	−1	0	0

$m3 = 23 \otimes \bar{1}$ ($T_h = T \otimes C_i$)

432 (O)				E	$C_{3j}^{\pm}$	C_{2m}	C_{2p}	$C_{4m}^{\pm}$
		$\bar{4}3m$ (T_d)		E	$C_{3j}^{\pm}$	C_{2m}	σ_{dp}	$S_{4m}^{\pm}$
A_1	Γ_1	A_1	Γ_1	1	1	1	1	1
A_2	Γ_2	A_2	Γ_2	1	1	1	−1	−1
E	Γ_3	E	Γ_3	2	−1	2	0	0
T_2	Γ_5	T_2	Γ_5	3	0	−1	1	−1
T_1	Γ_4	T_1	Γ_4	3	0	−1	−1	1

$m3m = 432 \otimes \bar{1}$ ($O_h = O \otimes C_i$)

Notes to Table 2.2

(i) The names of the point groups are given in both the international and Schönflies notations. Sometimes two or three point groups have identical characters; such groups are tabulated together.

(ii) For each point group the names of the representations appear in the column headed by the name of the point group. The standard Mulliken notation for the representations is used (Margenau and Murphy 1956, Mulliken 1933). The Γ notation of Koster, Dimmock, Wheeler, and Statz (1963) is also included for reference, though we shall not actually use it in this book.

(iii) For each point group the names of the operators appear in the row begun by the name of that group. They are to be identified with respect to axes $Oxyz$ by means of Figs. 1.1–1.4 and Tables 1.2–1.6. Note that we have taken the first setting of the *International tables for X-ray crystallography* (Henry and Lonsdale 1965) for the point groups of the monoclinic system; the z-axis (being the polar axis) is more appropriate than the y-axis in the study of harmonic functions.

(iv) We have not given the character tables of those groups that are direct products of some other point group with $\bar{1}$ (C_i); the character tables of these direct product groups can be constructed as follows. If a group $\mathbf{G}'$ is given as a direct product of the form $\mathbf{G} \otimes \bar{1}$ then the reps of $\mathbf{G}'$ fall into pairs; each pair M_g and M_u arise out of a single rep M of $\mathbf{G}$ and the characters of M_g and M_u obey the following rules. If $R' = RI$ then for all $R \in \mathbf{G}$ the character of R in M_g and M_u is equal to the character of R in M; the character of R' in M_g is equal to the character of R in M, but the character of R' in M_u is minus the character of R in M. In the Γ notation of Koster, Dimmock, Wheeler, and Statz (1963), if $\Gamma \equiv M$ in $\mathbf{G}$, then $\Gamma^+ \equiv M_g$ and $\Gamma^- \equiv M_u$ in $\mathbf{G}'$.

(v) By using Theorem 1.3.9 one can easily show that all the point-group reps are of the first kind, except reps with complex characters and these are of the third kind.

(vi) The Kronecker products of the various reps of each point group and the compatibilities between the reps of a point group and those of its subgroups are given in the tables of Koster, Dimmock, Wheeler, and Statz (1963).

TABLE 2.3

Matrices for the degenerate representations of the crystallographic point groups

Tetragonal groups

Key: $\varepsilon = \begin{bmatrix} 1 & 0 \\ 0 & 1 \end{bmatrix}$ $\lambda = \begin{bmatrix} 1 & 0 \\ 0 & -1 \end{bmatrix}$ $\kappa = \begin{bmatrix} 0 & 1 \\ 1 & 0 \end{bmatrix}$ $\rho = \begin{bmatrix} 0 & -1 \\ 1 & 0 \end{bmatrix}$

Group Rep	422 (D_4) E	$4mm$ (C_{4v}) E	$\bar{4}2m$ (D_{2d}) E	Group Rep	422 (D_4) E	$4mm$ (C_{4v}) E	$\bar{4}2m$ (D_{2d}) E
E	ε	ε	ε	C_{2b}	$-\kappa$	.	.
C_{2z}	$-\varepsilon$	$-\varepsilon$	$-\varepsilon$	σ_y	.	λ	.
C_{4z}^+	ρ	ρ	.	σ_x	.	$-\lambda$	.
C_{4z}^-	$-\rho$	$-\rho$	.	σ_{db}	.	κ	κ
C_{2x}	λ	.	λ	σ_{da}	.	$-\kappa$	$-\kappa$
C_{2y}	$-\lambda$	.	$-\lambda$	S_{4z}^+	.	.	ρ
C_{2a}	κ	.	.	S_{4z}^-	.	.	$-\rho$

Trigonal and hexagonal groups

Key:

$$\varepsilon = \begin{bmatrix} 1 & 0 \\ 0 & 1 \end{bmatrix} \quad \alpha = \begin{bmatrix} -\frac{1}{2} & -\frac{1}{2}\sqrt{3} \\ \frac{1}{2}\sqrt{3} & -\frac{1}{2} \end{bmatrix} \quad \beta = \begin{bmatrix} -\frac{1}{2} & \frac{1}{2}\sqrt{3} \\ -\frac{1}{2}\sqrt{3} & -\frac{1}{2} \end{bmatrix} \quad \lambda = \begin{bmatrix} 1 & 0 \\ 0 & -1 \end{bmatrix} \quad \mu = \begin{bmatrix} -\frac{1}{2} & -\frac{1}{2}\sqrt{3} \\ -\frac{1}{2}\sqrt{3} & \frac{1}{2} \end{bmatrix} \quad \nu = \begin{bmatrix} -\frac{1}{2} & \frac{1}{2}\sqrt{3} \\ \frac{1}{2}\sqrt{3} & \frac{1}{2} \end{bmatrix}$$

Group	32 (D_3)	3m (C_{3v})	622 (D_6)		6mm (C_{6v})		$\bar{6}2m$ (D_{3h})	
Rep	E	E	E_1	E_2	E_1	E_2	E'	E''
E	ε	ε	ε	ε	ε	ε	ε	ε
C_6^+	.	.	$-\beta$	α	$-\beta$	α	.	.
C_6^-	.	.	$-\alpha$	β	$-\alpha$	β	.	.
C_3^+	α	α	α	β	α	β	α	α
C_3^-	β	β	β	α	β	α	β	β
C_2	.	.	$-\varepsilon$	ε	$-\varepsilon$	ε	.	.
C'_{21}	λ	.	λ	λ	.	.	λ	$-\lambda$
C'_{22}	μ	.	μ	ν	.	.	μ	$-\mu$
C'_{23}	ν	.	ν	μ	.	.	ν	$-\nu$
C''_{21}	.	.	$-\lambda$	λ	.	.	.	.
C''_{22}	.	.	$-\mu$	ν	.	.	.	.
C''_{23}	.	.	$-\nu$	μ	.	.	.	.
σ_{v1}	.	.	.	.	λ	λ	λ	λ
σ_{v2}	.	.	.	.	μ	ν	μ	μ
σ_{v3}	.	.	.	.	ν	μ	ν	ν
σ_{d1}	.	λ	.	.	$-\lambda$	λ	.	.
σ_{d2}	.	μ	.	.	$-\mu$	ν	.	.
σ_{d3}	.	ν	.	.	$-\nu$	μ	.	.
σ_h	.	.	.	.	.	.	ε	$-\varepsilon$
S_3^+	.	.	.	.	.	.	α	$-\alpha$
S_3^-	.	.	.	.	.	.	β	$-\beta$

The doubly-degenerate representations, E, of the cubic groups

432 (O), $\bar{4}3m$ (T_d):	E, C_{2x}, C_{2y}, C_{2z}	ε
432 (O), $\bar{4}3m$ (T_d):	C_{31}^+, C_{32}^+, C_{33}^+, C_{34}^+	α
432 (O), $\bar{4}3m$ (T_d);	C_{31}^-, C_{32}^-, C_{33}^-, C_{34}^-	β
432 (O): $\bar{4}3m$ (T_d):	C_{2a}, C_{4z}^+, C_{4z}^-, C_{2b} σ_{da}, S_{4z}^-, S_{4z}^+, σ_{db}	λ
432 (O): $\bar{4}3m$ (T_d):	C_{4x}^-, C_{4x}^+, C_{2f}, C_{2d} S_{4x}^+, S_{4x}^-, σ_{df}, σ_{dd}	μ
432 (O): $\bar{4}3m$ (T_d):	C_{4y}^+, C_{4y}^-, C_{2c}, C_{2e} S_{4y}^-, S_{4y}^+, σ_{dc}, σ_{de}	ν

See key to trigonal and hexagonal groups, above, for the identification of the matrices.

The threefold degenerate representations of the cubic groups

Given a representation of 432 (O) or $\bar{4}3m$ (T_d), the representatives for the operations of these groups that do not belong to 23 (T), which are listed under the headings 432 (O) and $\bar{4}3m$ (T_d) in the first part of the table, are obtained as follows: take the corresponding matrix from the first part of the table and post-multiply it with the matrix that appears under the representation chosen at the bottom of the table.

23 (T), 432 (O) $\bar{4}3m$ (T_d)	432 (O)	$\bar{4}3m$ (T_d)	
E	C_{2a}	σ_{da}	$\begin{bmatrix} 1 & 0 & 0 \\ 0 & 1 & 0 \\ 0 & 0 & 1 \end{bmatrix}$
C_{2x}	C_{4z}^-	S_{4z}^+	$\begin{bmatrix} 1 & 0 & 0 \\ 0 & -1 & 0 \\ 0 & 0 & -1 \end{bmatrix}$
C_{2y}	C_{4z}^+	S_{4z}^-	$\begin{bmatrix} -1 & 0 & 0 \\ 0 & 1 & 0 \\ 0 & 0 & -1 \end{bmatrix}$
C_{2z}	C_{2b}	σ_{db}	$\begin{bmatrix} -1 & 0 & 0 \\ 0 & -1 & 0 \\ 0 & 0 & 1 \end{bmatrix}$
C_{31}^-	C_{4x}^+	S_{4x}^-	$\begin{bmatrix} 0 & 1 & 0 \\ 0 & 0 & 1 \\ 1 & 0 & 0 \end{bmatrix}$
C_{32}^-	C_{4x}^-	S_{4x}^+	$\begin{bmatrix} 0 & 1 & 0 \\ 0 & 0 & -1 \\ -1 & 0 & 0 \end{bmatrix}$
C_{33}^-	C_{2f}	σ_{df}	$\begin{bmatrix} 0 & -1 & 0 \\ 0 & 0 & 1 \\ -1 & 0 & 0 \end{bmatrix}$
C_{34}^-	C_{2d}	σ_{dd}	$\begin{bmatrix} 0 & -1 & 0 \\ 0 & 0 & -1 \\ 1 & 0 & 0 \end{bmatrix}$
C_{31}^+	C_{4y}^-	S_{4y}^+	$\begin{bmatrix} 0 & 0 & 1 \\ 1 & 0 & 0 \\ 0 & 1 & 0 \end{bmatrix}$
C_{32}^+	C_{4y}^+	S_{4y}^-	$\begin{bmatrix} 0 & 0 & -1 \\ 1 & 0 & 0 \\ 0 & -1 & 0 \end{bmatrix}$
C_{33}^+	C_{2c}	σ_{dc}	$\begin{bmatrix} 0 & 0 & -1 \\ -1 & 0 & 0 \\ 0 & 1 & 0 \end{bmatrix}$
C_{34}^+	C_{2e}	σ_{de}	$\begin{bmatrix} 0 & 0 & 1 \\ -1 & 0 & 0 \\ 0 & -1 & 0 \end{bmatrix}$

432 (O)	$\bar{4}3m$ (T_d)	T_1	T_2
C_{2a}	σ_{da}	$\begin{bmatrix} 0 & 1 & 0 \\ 1 & 0 & 0 \\ 0 & 0 & -1 \end{bmatrix}$	$\begin{bmatrix} 0 & -1 & 0 \\ -1 & 0 & 0 \\ 0 & 0 & 1 \end{bmatrix}$

Note to Table 2.3

See Note (iv) to Table 2.2 concerning direct product groups. The note applies here with 'matrix representative' substituted for 'character' wherever the word 'character' appears.

TABLE 2.4

Surface harmonics for the cyclic groups

$\bar{1}$ (C_i)	l
A_g	0
A_u	1

2 (C_2)	m mod 2
A	0
B	1

m (C_{1h})	l	m mod 2
A'	0	0
	1	1
A''	2	1
	1	0

3 (C_3)	m mod 3
A	0
^1E	1
^2E	2

$\bar{6}$ (C_{3h})	l	m mod 6
A'	0	0
	3	3
A''	1	0
	4	3
$^1E'$	1	1
	2	−2
$^2E'$	1	−1
	2	2
$^1E''$	2	1
	3	−2
$^2E''$	2	−1
	3	2

4 (C_4)	m mod 4
A	0
B	2
^1E	1
^2E	3

$\bar{4}$ (S_4)	l	m mod 4
A	0	0
	3	2
B	1	0
	2	2
^1E	1	1
	2	−1
^2E	1	−1
	2	1

$6\,(C_6)$	m mod 6
A	0
B	3
$^1E_1$	4
$^2E_1$	2
$^1E_2$	1
$^2E_2$	5

Notes to Table 2.4

(i) The symmetry assignments for the surface harmonics that are given in this and the two following tables depend on the choice of axes. We always use right-handed axes and we have already indicated in Figs. 1.1–1.3 how they are oriented with respect to the symmetry elements of the groups.

(ii) The spherical harmonics $Y_l^m(\theta, \phi)$ are defined by eqns. (2.1.1) and (2.1.2). We also use the following two functions for $m \neq 0$:

$$Y_l^{m,c}(\theta, \phi) = (Y_l^m + Y_l^{-m})/\sqrt{2}, \tag{2.4.1}$$

$$Y_l^{m,s}(\theta, \phi) = -\mathrm{i}(Y_l^m - Y_l^{-m})/\sqrt{2}, \tag{2.4.2}$$

which are real and have ϕ-dependence cos $(m\phi)$ and sin $(m\phi)$ respectively.

(iii) The spherical harmonics that appear in Tables 2.4–2.6 (see Note (ii)) are normalized so that the integral

$$\int_{\theta=0}^{\pi}\int_{\phi=0}^{2\pi} |Y_l^m(\theta, \phi)|^2 \sin\theta \,\mathrm{d}\theta \,\mathrm{d}\phi = 1. \tag{2.4.3}$$

This implies that $Y_l^{m,c}(\theta, \phi)$ and $Y_l^{m,s}(\theta, \phi)$ are also normalized to unity.

(iv) In this table and Table 2.5 l is given mod $(+2)$, except that if no value of l is specified it means that there is no restriction on l; by l mod $(+2)$ we mean that multiples of 2 can be added to, but not subtracted from, the value of l given.

(v) Surface harmonics belonging to the reps M_g and M_u of a direct product group $\mathbf{G}' = \mathbf{G} \otimes \bar{1}$ (see Note (iv) to Table 2.2) are those of even and odd order in l, respectively, that belong to the rep M of $\mathbf{G}$. Because of this simple rule the harmonics for such groups are not separately tabulated.

(vi) By $m = 1$, mod 2 (e.g. rep B of 2 (C_2), Table 2.4), we mean the succession of values $m = 1, -1, 3, -3, 5, -5, \ldots$.

TABLE 2.5

Surface harmonics for the dihedral groups

$222\,(D_2)$	l	m mod $(+2)$	ϕ-dep
A_1	0	0	c
	3	2	s
B_1	2	2	s
	1	0	c
B_2	2	1	c
	1	1	s
B_3	2	1	s
	1	1	c

$mm2\,(C_{2v})$	m mod $(+2)$	ϕ-dep
A_1	0	c
A_2	2	s
B_1	1	c
B_2	1	s

32 (D_3)	l	m mod (+6)	ϕ-dep
A_1	0	0	c
	4	3	s
	3	3	c
	7	6	s
A_2	1	0	c
	3	3	s
	4	3	c
	6	6	s
E	1	1	(c, s)
	2	1	$(s, -c)$
	3	2	(s, c)
	2	2	$(c, -s)$
	5	4	$(s, -c)$
	4	4	(c, s)
	5	5	$(c, -s)$
	6	5	(s, c)

3m (C_{3v})	m mod (+3)	ϕ-dep
A_1	0	c
A_2	3	s
E	1	(c, s)
	2	$(c, -s)$

422 (D_4)	l	m mod (+4)	ϕ-dep
A_1	0	0	c
	5	4	s
A_2	4	4	s
	1	0	c
B_1	2	2	c
	3	2	s
B_2	2	2	s
	3	2	c
E	1	1	(c, s)
	2	1	$(s, -c)$
	3	3	$(c, -s)$
	4	3	(s, c)

4mm (C_{4v})	m mod (+4)	ϕ-dep
A_1	0	c
A_2	4	s
B_1	2	c
B_2	2	s
E	1	(c, s)
	3	$(c, -s)$

$\bar{4}2m$ (D_{2d})	l	m mod (+4)	ϕ-dep
A_1	0	0	c
	3	2	s
A_2	4	4	s
	3	2	c
B_1	2	2	c
	5	4	s
B_2	2	2	s
	1	0	c
E	1	1	(c, s)
	2	1	(s, c)
	3	3	$(c, -s)$
	4	3	$(s, -c)$.

622 (D_6)	l	m mod (+6)	ϕ-dep
A_1	0	0	c
	7	6	s
A_2	6	6	s
	1	0	c
B_1	4	3	s
	3	3	c
B_2	4	3	c
	3	3	s
E_1	1	1	(c, s)
	2	1	$(s, -c)$
	5	5	$(c, -s)$
	6	5	(s, c)
E_2	2	2	(c, s)
	3	2	$(s, -c)$
	4	4	$(c, -s)$
	5	4	(s, c)

6mm (C_{6v})	m mod (+6)	ϕ-dep
A_1	0	c
A_2	6	s
B_1	3	c
B_2	3	s
E_1	1	(c, s)
	5	$(c, -s)$
E_2	2	(c, s)
	4	$(c, -s)$

$\bar{6}2m$ (D_{3h})	l	m mod (+6)	ϕ-dep
A_1'	0	0	c
	3	3	c
A_2'	6	6	s
	3	3	s
A_1''	4	3	s
	7	6	s
A_2''	4	3	c
	1	0	c
E'	1	1	(c, s)
	2	2	$(c, -s)$
	4	4	(c, s)
	5	5	$(c, -s)$
E''	2	1	(c, s)
	3	2	$(c, -s)$
	5	4	(c, s)
	6	5	$(c, -s)$

Notes to Table 2.5

(i) See Notes (i)–(v) to Table 2.4 which also apply to Table 2.5. The harmonics are tabulated in the same way as for the cyclic groups, with the extra convention that the superscripts c and s of the allowed harmonics appear under the heading 'ϕ-dep'.

(ii) We use a slightly different notation for m from that used in Table 2.4. By $m = 1$, mod (+2) (e.g. rep B_2 of 222 (D_2)), we mean the succession of values 1, 3, 5, . . . , that is, multiples of 2 can be added to, but not subtracted from, the value of m given.

(iii) The symbol (c, s) denotes a degenerate basis $\{Y_l^{m,c}(\theta, \phi), Y_l^{m,s}(\theta, \phi)\}$. Note that $(c, -s)$, for instance, means $\{Y_l^{m,c}(\theta, \phi), -Y_l^{m,s}(\theta, \phi)\}$. The bases are given as *row* vectors abbreviated in the table with symbols of the form (c, s). Their transformation properties are given by post-multiplying them by the matrices listed in Table 2.3, the first function belonging to the first row of the representation and the second function to the second row.

TABLE 2.6

Surface harmonics for the cubic groups

One-dimensional real representations of the cubic groups

23 (T)	$m3$ (T_h)	$\bar{4}3m$ (T_d)	432 (O)	$m3m$ (O_h)	l	ϕ-dep	Surface harmonic
A	A_g	A_1	A_1	A_{1g}	0	–	1(0)
A	A_u	A_1	A_2	A_{2u}	3	s	1(2)
A	A_g	A_1	A_1	A_{1g}	4	c	0·76376261583(0) + 0·64549722437(4)
A	A_g	A_1	A_1	A_{1g}	6	c	0·35355339059(0) − 0·93541434669(4)
A	A_g	A_2	A_2	A_{2g}	6	c	0·82915619759(2) − 0·55901699438(6)
A	A_u	A_1	A_2	A_{2u}	7	s	0·73598007219(2) + 0·67700320038(6)
A	A_g	A_1	A_1	A_{1g}	8	c	0·71807033082(0) + 0·38188130791(4) + 0·58184333516(8)
A	A_u	A_1	A_2	A_{2u}	9	s	0·43301270189(2) − 0·90138781887(6)
A	A_u	A_2	A_1	A_{1u}	9	s	0·84162541153(4) − 0·54006172487(8)
A	A_g	A_1	A_1	A_{1g}	10	c	0·41142536788(0) − 0·58630196998(4) − 0·69783892602(8)
A	A_g	A_2	A_2	A_{2g}	10	c	0·80201568979(2) + 0·15728821740(6) − 0·57622152858(10)
A	A_u	A_1	A_2	A_{2u}	11	s	0·66536330928(2) + 0·45927932677(6) + 0·58851862049(10)
A	A_g	A_1	A_1	A_{1g}	12	c	0·69550266594(0) + 0·31412556680(4) + 0·34844953759(8) + 0·54422797585(12)
A	A_g	A_1	A_1	A_{1g}	12	c	0·55897937420(4) − 0·80626750818(8) + 0·19358399848(12)
A	A_g	A_2	A_2	A_{2g}	12	c	0·21040635288(2) − 0·82679728471(6) + 0·52166600107(10)

Complex representations of 23 (T) and $m3$ (T_h)

The harmonics of ^2E, ^2E_g, and ^2E_u are the complex conjugates of the expansions listed in this table.

23 (T)	$m3$ (T_h)	l	ϕ-dep	Surface harmonic
^1E	^1E_g	2	c	0·70710678119(0) − 0·70710678119(2)i
^1E	^1E_g	4	c	0·45643546459(0) − 0·54006172487(4) + 0·70710678119(2)i
^1E	^1E_u	5	s	0·70710678119(2) − 0·70710678119(4)i
^1E	^1E_g	6	c	0·66143782777(0) + 0·25(4) − {0·39528470752(2) + 0·58630196998(6)}i
^1E	^1E_u	7	s	0·47871355388(2) − 0·52041649987(6) + 0·70710678119(4)i
^1E	^1E_g	8	c	0·49212549213(0) − 0·27860539791(4) − 0·42448973163(8)
				+ {0·46010167179(2) + 0·53694175812(6)}i
^1E	^1E_g	8	c	0·59115341967(4) − 0·38799179683(8)
				+ {0·53694175812(2) − 0·46010167179(6)}i
^1E	^1E_u	9	s	0·38188130791(4) + 0·59511903571(8)
				+ {0·63737743920(2) + 0·30618621785(6)}i
^1E	^1E_g	10	c	0·64448784576(0) + 0·18714045988(4) + 0·22274170005(8)
				− {0·31464779160(2) + 0·34379897770(6) + 0·53178852016(10)}i
^1E	^1E_g	10	c	0·54139029200(4) − 0·45485882615(8)
				− {0·28174844083(2) − 0·60780956826(6) + 0·22624178400(10)}i
^1E	^1E_u	11	s	0·52786914414(2) − 0·28945394530(6) − 0·37090508249(10)
				+ {0·35023356693(4) + 0·61427717571(8)}i
^1E	^1E_u	11	s	0·55744745362(6) − 0·43503142007(10)
				+ {0·61427717571(4) − 0·35023356693(8)}i
^1E	^1E_g	12	c	0·50807284993(0) − 0·21500378261(4) − 0·23849688325(8) − 0·37249777088(12)
				+ {0·36487351840(2) + 0·38689303336(6) + 0·46602692660(10)}i
^1E	^1E_g	12	c	0·49820374066(4) + 0·23953506879(8) − 0·44092627911(12)
				+ {0·58713873906(2) − 0·09228708327(6) − 0·38308118638(10)}i

Two-dimensional representations of $\bar{4}3m$ (T_d), 432 (O) and $m3m$ (O_h)

For the representations marked with an asterisk the partners must be interchanged and the sign of one of them reversed.

$\bar{4}3m$ (T_d)	432 (O)	$m3m$ (O_h)	l	ϕ-dep	Surface harmonic
E	E	E_g	2	c	1(0)
				c	1(2)
E	E	E_g	4	c	0·64549722437(0) − 0·76376261583(4)
				c	−1(2)
E*	E	E_u	5	s	1(4)
				s	−1(2)
E	E	E_g	6	c	0·93541434670(0) + 0·35355339059(4)
				c	0·55901699438(2) + 0·82915619759(6)
E*	E	E_u	7	s	1(4)
				s	0·67700320039(2) − 0·73598007219(6)
E	E	E_g	8	c	0·69597054536(0) − 0·39400753227(4) − 0·60031913556(8)
				c	−0·65068202432(2) − 0·75935031654(6)
E	E	E_g	8	c	0·83601718355(4) − 0·54870326117(8)
				c	−0·75935031654(2) + 0·65068202432(6)
E*	E	E_u	9	s	0·54006172487(4) + 0·84162541153(8)
				s	−0·90138781887(2) − 0·43301270189(6)

Two-dimensional representations of $\bar{4}3m$ (T_d), 432 (O) and $m3m$ (O_h)—(continued)

$\bar{4}3m$ (T_d)	432 (O)	$m3m$ (O_h)	l	ϕ-dep	Surface harmonic
E	E	E_g	10	c	0·91144345226(0) + 0·26465657643(4) + 0·31500433312(8)
				c	0·44497917425(2) + 0·48620517700(6) + 0·75206253752(10)
E	E	E_g	10	c	0·76564149349(4) − 0·64326752090(8)
				c	0·39845246619(2) − 0·85957253477(6) + 0·31995419931(10)
E^*	E	E_u	11	s	0·49530506035(4) + 0·86871911295(8)
				s	0·74651970280(2) − 0·40934969512(6) − 0·52453899801(10)
E^*	E	E_u	11	s	0·86871911295(4) − 0·49530506035(8)
				s	0·78834974923(6) − 0·61522733432(10)
E	E	E_g	12	c	0·71852351504(0) − 0·30406126533(4) − 0·33728552687(8) − 0·52679139953(12)
				c	−0·51600907827(2) − 0·54714937497(6) − 0·65906160002(10)
E	E	E_g	12	c	0·70456648687(4) + 0·33875374295(8) − 0·62356392392(12)
				c	−0·83033956777(2) + 0·13051364479(6) + 0·54175860926(10)

Three-dimensional representations of the cubic groups

23 (T)	$m3$ (T_h)	$\bar{4}3m$ (T_d)	432 (O)	$m3m$ (O_h)	l	ϕ-dep	Surface harmonic
T	T_u	T_2	T_1	T_{1u}	1	c, s	1(1)
						c	1(0)
T	T_g	T_2	T_2	T_{2g}	2	s, c	1(1)
						s	1(2)
T	T_u	T_2	T_1	T_{1u}	3	c, s	0·61237243570(1) ∓ 0·79056941504(3)
						c	−1(0)
T	T_u	T_1	T_2	T_{2u}	3	c, s	∓0·79056941504(1) − 0·61237243570(3)
						c	1(2)
T	T_g	T_2	T_2	T_{2g}	4	s, c	0·35355339059(1) ∓ 0·93541434669(3)
						s	−1(2)
T	T_g	T_1	T_1	T_{1g}	4	s, c	∓0·93541434669(1) − 0·35355339059(3)
						s	1(4)
T	T_u	T_2	T_1	T_{1u}	5	c, s	0·48412291828(1) ∓ 0·52291251658(3) + 0·70156076002(5)
						c	1(0)
T	T_u	T_1	T_2	T_{2u}	5	c, s	±0·66143782777(1) − 0·30618621785(3) ∓ 0·68465319688(5)
						c	1(2)
T	T_u	T_2	T_1	T_{1u}	5	c, s	0·57282196187(1) ± 0·79549512883(3) + 0·19764235376(5)
						c	1(4)
T	T_g	T_2	T_2	T_{2g}	6	s, c	0·19764235376(1) ∓ 0·56250000000(3) + 0·80282703617(5)
						s	1(2)
T	T_g	T_1	T_1	T_{1g}	6	s, c	±0·43301270189(1) − 0·68465319688(3) ∓ 0·58630196998(5)
						s	1(4)
T	T_g	T_2	T_2	T_{2g}	6	s, c	0·87945295497(1) ± 0·46351240544(3) + 0·10825317547(5)
						s	1(6)
T	T_u	T_2	T_1	T_{1u}	7	c, s	0·41339864235(1) ∓ 0·42961647140(3) + 0·47495887980(5) ∓ 0·64725984929(7)
						c	−1(0)
T	T_u	T_1	T_2	T_{2u}	7	c, s	∓0·57409915846(1) + 0·41984465133(3) ∓ 0·07328774625(5) − 0·69912054129(7)
						c	1(2)

Three-dimensional representations of the cubic groups—(continued)

23 (T)	$m3$ (T_h)	$\bar{4}3m$ (T_d)	432 (O)	$m3m$ (O_h)	l	ϕ-dep	Surface harmonic
T	T_u	T_2	T_1	T_{1u}	7	c, s	0·53855274811(1) ± 0·10364452470(3) − 0·78125000000(5) ∓ 0·29810600044(7)
						c	−1(4)
T	T_u	T_1	T_2	T_{2u}	7	c, s	∓0·45768182862(1) − 0·79272818087(3) ∓ 0·39836089950(5) − 0·05846339667(7)
						c	1(6)
T	T_g	T_2	T_2	T_{2g}	8	s, c	0·13072812915(1) ∓ 0·38081430022(3) + 0·59086470004(5) ∓ 0·69912054129(7)
						s	−1(2)
T	T_g	T_1	T_1	T_{1g}	8	s, c	∓0·27421763711(1) + 0·60515364784(3) ∓ 0·33802043208(5) − 0·66658528149(7)
						s	1(4)
T	T_g	T_2	T_2	T_{2g}	8	s, c	0·45768182862(1) ∓ 0·47134697278(3) − 0·70883101389(5) ∓ 0·25674494883(7)
						s	−1(6)
T	T_g	T_1	T_1	T_{1g}	8	s, c	∓0·83560887232(1) − 0·51633473881(3) ∓ 0·18487749322(5) −0·03125000000(7)
						s	1(8)
T	T_u	T_2	T_1	T_{1u}	9	c, s	0·36685490256(1) ∓ 0·37548796377(3) + 0·39636409044(5) ∓ 0·44314852503(7) + 0·60904939218(9)
						c	1(0)
T	T_u	T_1	T_2	T_{2u}	9	c, s	±0·51301422373(1) − 0·42961647140(3) ± 0·25194555463(5) + 0·05633673868(7) ∓ 0·69684697253(9)
						c	1(2)
T	T_u	T_2	T_1	T_{1u}	9	c, s	0·49435287561(1) ∓ 0·13799626354(3) − 0·39218438744(5) ± 0·67232906169(7) + 0·36157613954(9)
						c	1(4)
T	T_u	T_1	T_2	T_{2u}	9	c, s	±0·45768182862(1) + 0·29810600044(3) ∓ 0·60515364784(5) − 0·56832917123(7) ∓ 0·11158481920(9)
						c	1(6)
T	T_u	T_2	T_1	T_{1u}	9	c, s	0·38519665736(1) ± 0·75268075591(3) + 0·50931268791(5) ± 0·15944009088(7) + 0·01657281518(9)
						c	1(8)
T	T_g	T_2	T_2	T_{2g}	10	s, c	0·09472152854(1) ∓ 0·27885262965(3) + 0·44538102543(5) ∓ 0·57486942301(7) + 0·62002413795(9)
						s	1(2)
T	T_g	T_1	T_1	T_{1g}	10	s, c	±0·19515618745(1) − 0·48613591207(3) ± 0·49410588440(5) − 0·09110862336(7) ∓ 0·68785502197(9)
						s	1(4)
T	T_g	T_2	T_2	T_{2g}	10	s, c	0·31049159296(1) ∓ 0·53906250000(3) − 0·01746928107(5) ± 0·69255289805(7) + 0·36479021288(9)
						s	1(6)
T	T_g	T_1	T_1	T_{1g}	10	s, c	±0·46456464835(1) − 0·31560952932(3) ∓ 0·70572436192(5) − 0·42100604954(7) ∓ 0·09631896880(9)
						s	1(8)
T	T_g	T_2	T_2	T_{2g}	10	s, c	0·80044772018(1) ± 0·54379714235(3) + 0·24319347525(5) ± 0·06594508991(7) + 0·00873464054(9)
						s	1(10)

Three-dimensional representations of the cubic groups—(continued)

23 (T)	$m3$ (T_h)	$\bar{4}3m$ (T_d)	432 (O)	$m3m$ (O_h)	l	ϕ-dep	Surface harmonic
T	T_u	T_2	T_1	T_{1u}	11	c, s	0·33321251269(1) ∓ 0·33846027668(3) + 0·35033967021(5)
							∓ 0·37296505975(7) + 0·41975832571(9) ∓ 0·57997947393(11)
						c	− 1(0)
T	T_u	T_1	T_2	T_{2u}	11	c, s	∓0·46765007670(1) + 0·41655170126(3) ∓ 0·31014124452(5)
							+ 0·13689999148(7) ± 0·13594928559(9) − 0·68875008419(11)
						c	1(2)
T	T_u	T_2	T_1	T_{1u}	11	c, s	0·45637974397(1) ∓ 0·23534953643(3) − 0·13435455877(5)
							± 0·49510851971(7) − 0·55722625444(9) ∓ 0·40329075444(11)
						c	− 1(4)
T	T_u	T_1	T_2	T_{2u}	11	c, s	∓0·43552935783(1) − 0·04764183952(3) ± 0·52272828294(5)
							− 0·32425698664(7) ∓ 0·63603688807(9) − 0·15847416019(11)
						c	1(6)
T	T_u	T_2	T_1	T_{1u}	11	c, s	0·40022386009(1) ± 0·39401846317(3) − 0·40784785677(5)
							∓ 0·65553753643(7) − 0·29501240333(9) ∓ 0·03832307983(11)
						c	− 1(8)
T	T_u	T_1	T_2	T_{2u}	11	c, s	∓0·33485130540(1) − 0·70641320968(3) ∓ 0·56871666444(5)
							− 0·24930093301(7) ∓ 0·05695963504(9) − 0·00458048414(11)
						c	1(10)
T	T_g	T_2	T_2	T_{2g}	12	s, c	0·07271293152(1) ∓ 0·21528718442(3) + 0·34870255999(5)
							∓ 0·46435521203(7) + 0·54718531932(7) ∓ 0·55833076742(11)
						s	− 1(2)
T	T_g	T_1	T_1	T_{1g}	12	s, c	∓0·14842464994(1) + 0·39257812500(3) ∓ 0·48401456158(5)
							+ 0·34122999866(7) ± 0·07446249039(9) − 0·68381274394(11)
						s	1(4)
T	T_g	T_2	T_2	T_{2g}	12	s, c	0·23130311098(1) ∓ 0·49003980203(3) + 0·27404717188(5)
							± 0·30237819269(7) − 0·58930833781(9) ∓ 0·43879508023(11)
						s	− 1(6)
T	T_g	T_1	T_1	T_{1g}	12	s, c	∓0·32928552212(1) + 0·45497333155(3) ± 0·23408184792(5)
							− 0·54296875000(7) ∓ 0·55492127556(9) − 0·16438769854(11)
						s	1(8)
T	T_g	T_2	T_2	T_{2g}	12	s, c	0·46435521203(1) ∓ 0·20164537722(3) − 0·65705604098(5)
							∓ 0·52110934256(7) − 0·19834078136(9) ∓ 0·03311684562(11)
						s	− 1(10)
T	T_g	T_1	T_1	T_{1g}	12	s, c	∓0·77144481702(1) − 0·55833076742(3) ∓ 0·28725880996(5)
							− 0·10066649535(7) ∓ 0·02196723023(9) − 0·00239207983(11)
						s	1(12)

Notes to Table 2.6

(i) See Notes (i)–(iii) to Table 2.4, which also apply to Table 2.6.

(ii) Unlike Tables 2.4 and 2.5 the value of l alongside a given harmonic is the only one that can be used and it cannot therefore be modified.

(iii) The bases are to be understood as *row* vectors. Their transformation properties are obtained by postmultiplying them with the matrix representatives listed in Table 2.3. The first function belongs to the first row of the representation, the second function to the second row, and so on.

(iv) A surface harmonic such as $aY_l^{m,c}(\theta, \phi) + bY_l^{n,c}(\theta, \phi) + cY_l^{p,c}(\theta, \phi)$ is given as follows: the values of l and the superscript c (or s) appear under the headings l and ϕ-dep, respectively. The rest of the expansion appears on the same line in the form $a(m) + b(n) + c(p)$. Degenerate representations are given in several lines, and they must be understood as row vectors, the successive lines corresponding to the successive columns of the vector. In the 3-dimensional representations the first two partners are given in one line: the first letter under 'ϕ-dep' and the upper sign in the expansion corresponding to the first partner. The expansions are given up to 11 decimal places and, as explained in the text, are given only up to $l = 12$. The expansions given are normalized to unity and for every representation they have been orthogonalized one to another.

(v) The surface harmonics for A_{1g} of $m3m$ (O_h) up to $l = 30$ are given by Mueller and Priestley (1966).

Example 2.4.1. We consider the rep A' of $\bar{6}$ (C_{3h}). From Table 2.4 the surface harmonics belonging to it are

$$Y_0^0(\theta, \phi),\ Y_2^0(\theta, \phi),\ Y_4^0(\theta, \phi),\ Y_6^0(\theta, \phi), \ldots$$
$$Y_6^6(\theta, \phi),\ Y_6^{-6}(\theta, \phi),\ Y_8^6(\theta, \phi),\ Y_8^{-6}(\theta, \phi), \ldots$$
$$Y_3^3(\theta, \phi),\ Y_3^{-3}(\theta, \phi),\ Y_5^3(\theta, \phi),\ Y_5^{-3}(\theta, \phi), \ldots$$

Example 2.4.2. For the representation E of $\bar{4}2m$ (D_{2d}) the bases $\{Y_1^{1,c}(\theta, \phi), Y_1^{1,s}(\theta, \phi)\}$, $\{Y_3^{1,c}(\theta, \phi), Y_3^{1,s}(\theta, \phi)\}$, $\{Y_5^{1,c}(\theta, \phi), Y_5^{1,s}(\theta, \phi)\}$, $\{Y_5^{5,c}(\theta, \phi), Y_5^{5,s}(\theta, \phi)\}$, $\ldots$, $\{Y_2^{1,s}(\theta, \phi), Y_2^{1,c}(\theta, \phi)\}$, $\{Y_4^{1,s}(\theta, \phi), Y_4^{1,c}(\theta, \phi)\}$, etc., span the representation E listed in Table 2.3.

It is instructive to consider the relation between crystal field theory and the functions given in Tables 2.4–2.6. Crystal field theory, as initiated by Bethe (1929), concerns itself with restricting the group of the Hamiltonian from the full rotation group, $O(3)$, in the case of an atomic electron to one of the crystallographic point groups for an electron in a crystal. This restriction, or *subduction* as it is sometimes called (Altmann and Cracknell 1965, Lomont 1959), often leads to a splitting of the atomic energy levels. An electron in an atom is commonly assumed to experience an electrostatic potential which possesses spherical symmetry and, consequently, the angular part of the electron's wave function, neglecting spin, belongs to one of the irreducible representations of $O(3)$, namely $\mathscr{D}^l\{R(\alpha, \beta, \gamma)\}$, where l is an integer. $\mathscr{D}^l\{R(\alpha, \beta, \gamma)\}$ is $(2l + 1)$-fold degenerate and the basis function of $\mathscr{D}^l\{R(\alpha, \beta, \gamma)\}$ is a vector whose $(2l + 1)$ components are the spherical harmonics $Y_l^m(\theta, \phi)$ $(-l \leqslant m \leqslant l)$. The energy level corresponding to these wave functions is thus $(2l + 1)$-fold degenerate. On reducing the symmetry of an electron's environment from that of $O(3)$ to that of one of the point groups, **G**, this $(2l + 1)$-fold degenerate energy level generally splits into several levels which now have to be labelled by the reps of **G**. Complete tables of these splittings are given by various authors (see, for example, Koster, Dimmock, Wheeler, and Statz (1963), Lomont (1959)), and are reproduced in Table 2.7.

TABLE 2.7

Compatibility tables for reps of $O(3)$ and point-group reps

l	$1\ (C_1)$
l	$(2l + 1)\ A$

l	$2\,(C_2)$	$m\,(C_{1h})$
0	A	A'
1	$A + 2B$	$2A' + A''$

$$\mathscr{D}^{2n+\lambda} = 2n \text{ reg} + \mathscr{D}^{\lambda} \quad (\lambda = 0, 1)$$

l	$222\,(D_2)$	$mm2\,(C_{2v})$
0	A	A_1
1	$B_1 + B_2 + B_3$	$A_1 + B_1 + B_2$

$$\mathscr{D}^{2n+\lambda} = n \text{ reg} + \mathscr{D}^{\lambda} \quad (\lambda = 0, 1)$$

l	$4\,(C_4)$	$\bar{4}\,(S_4)$
0	A	A
1	$A + {}^1E + {}^2E$	$B + {}^1E + {}^2E$
2	$A + 2B + {}^1E + {}^2E$	$A + 2B + {}^1E + {}^2E$
3	$A + 2B + 2{}^1E + 2{}^2E$	$2A + B + 2{}^1E + 2{}^2E$

$$\mathscr{D}^{4n+\lambda} = 2n \text{ reg} + \mathscr{D}^{\lambda} \quad (\lambda = 0, 1, 2, 3)$$

l	$422\,(D_4)$	$4mm\,(C_{4v})$	$\bar{4}2m\,(D_{2d})$
0	A_1	A_1	A_1
1	$A_2 + E$	$A_1 + E$	$B_2 + E$
2	$A_1 + B_1 + B_2 + E$	$A_1 + B_1 + B_2 + E$	$A_1 + B_1 + B_2 + E$
3	$A_2 + B_1 + B_2 + 2E$	$A_1 + B_1 + B_2 + 2E$	$A_1 + A_2 + B_2 + 2E$

$$\mathscr{D}^{4n+\lambda} = n \text{ reg} + \mathscr{D}^{\lambda} \quad (\lambda = 0, 1, 2, 3)$$

l	$3\,(C_3)$
0	A
1	$A + {}^1E + {}^2E$
2	$A + 2{}^1E + 2{}^2E$

$$\mathscr{D}^{3n+\lambda} = 2n \text{ reg} + \mathscr{D}^{\lambda} \quad (\lambda = 0, 1, 2)$$

l	32 (D_3)
0	A_1
1	$A_2 + E$
2	$A_1 + 2E$
3	$A_1 + 2A_2 + 2E$
4	$2A_1 + A_2 + 3E$
5	$A_1 + 2A_2 + 4E$

$$\mathscr{D}^{6n+\lambda} = 2n \text{ reg} + \mathscr{D}^{\lambda} \quad (\lambda = 0, 1, 2, 3, 4, 5)$$

l	$3m$ (C_{3v})
0	A_1
1	$A_1 + E$
2	$A_1 + 2E$

$$\mathscr{D}^{3n+\lambda} = n \text{ reg} + \mathscr{D}^{\lambda} \quad (\lambda = 0, 1, 2)$$

l	6 (C_6)	$\bar{6}$ (C_{3h})
0	A	A'
1	$A + {}^1E_1 + {}^2E_1$	$A'' + {}^1E' + {}^2E'$
2	$A + {}^1E_1 + {}^2E_1 + {}^1E_2 + {}^2E_2$	$A' + {}^1E'' + {}^2E'' + {}^1E' + {}^2E'$
3	$A + {}^1E_1 + {}^2E_1 + 2B + {}^1E_2 + {}^2E_2$	$A'' + {}^1E' + {}^2E' + 2A' + {}^1E'' + {}^2E''$
4	$A + {}^1E_1 + {}^2E_1 + 2B + 2{}^1E_2 + 2{}^2E_2$	$A' + {}^1E'' + {}^2E'' + 2A'' + 2{}^1E' + 2{}^2E'$
5	$A + 2{}^1E_1 + 2{}^2E_1 + 2B + 2{}^1E_2 + 2{}^2E_2$	$A'' + 2{}^1E' + 2{}^2E' + 2A' + 2{}^1E'' + 2{}^2E''$

$$\mathscr{D}^{6n+\lambda} = 2n \text{ reg} + \mathscr{D}^{\lambda} \quad (\lambda = 0, 1, 2, 3, 4, 5)$$

l	622 (D_6)	$6mm$ (C_{6v})	$\bar{6}2m$ (D_{3h})
0	A_1	A_1	A_1'
1	$A_2 + E_1$	$A_1 + E_1$	$A_2'' + E'$
2	$A_1 + E_1 + E_2$	$A_1 + E_1 + E_2$	$A_1' + E' + E''$
3	$A_2 + B_1 + B_2 + E_1 + E_2$	$A_1 + B_1 + B_2 + E_1 + E_2$	$A_1' + A_2' + A_2'' + E' + E''$
4	$A_1 + B_1 + B_2 + E_1 + 2E_2$	$A_1 + B_1 + B_2 + E_1 + 2E_2$	$A_1' + A_1'' + A_2'' + 2E' + E''$
5	$A_2 + B_1 + B_2 + 2E_1 + 2E_2$	$A_1 + B_1 + B_2 + 2E_1 + 2E_2$	$A_1' + A_2' + A_2'' + 2E' + 2E''$

$$\mathscr{D}^{6n+\lambda} = n \text{ reg} + \mathscr{D}^{\lambda} \quad (\lambda = 0, 1, 2, 3, 4, 5)$$

l	23 (T)
0	A
1	T
2	${}^1E + {}^2E + T$
3	$A + 2T$
4	$A + {}^1E + {}^2E + 2T$
5	${}^1E + {}^2E + 3T$

$\mathscr{D}^{6n+\lambda} = n \text{ reg} + \mathscr{D}^{\lambda} \quad (\lambda = 0, 1, 2, 3, 4, 5)$

l	432 (O)
0	A_1
1	T_1
2	$E + T_2$
3	$A_2 + T_1 + T_2$
4	$A_1 + E + T_1 + T_2$
5	$E + 2T_1 + T_2$

$\mathscr{D}^{\mu} = \text{reg} - \mathscr{D}^{11-\mu} (\mu = 6, 7, 8, 9, 10, 11)$

$\mathscr{D}^{12n+\lambda} = n \text{ reg} + \mathscr{D}^{\lambda} (\lambda = 0, 1, 2, \ldots, 11)$

$\bar{4}3m$ (T_d): as 432 (O), except that for l odd $A_1 \to A_2$, $A_2 \to A_1$, $T_1 \to T_2$ and $T_2 \to T_1$ (note if $l = 6, 7, 8, \ldots, 11$, $\mathscr{D}^l$ must first be found in 432 (O) and then this substitution made: the formula $\mathscr{D}^{\mu} + \mathscr{D}^{11-\mu} = \text{reg}$ is only valid for 432 (O) and not for $\bar{4}3m$ (T_d)).

Notes to Table 2.7

(i) The point-group reps may be identified from Table 2.2.

(ii) The character in $\mathscr{D}^l\{R(\alpha, \beta, \gamma)\}$ where $R(\alpha, \beta, \gamma)$ is a proper rotation through $2\pi/n$ is given by Bethe (1929),

$$\chi^l(2\pi/n) = \frac{\sin\{(l + \frac{1}{2})(2\pi/n)\}}{\sin(\pi/n)}.$$

For improper rotations there is an extra factor $(-1)^l$, see eqn. (2.1.12).

(iii) For those point groups that are direct products of some other point group **G** with $(E + I)$ one simply uses the table given above for **G** and adds a subscript g if l is even and a subscript u if l is odd. These point groups are

$$\begin{aligned}
\bar{1} &= 1 \otimes \bar{1}\ (C_i = C_1 \otimes C_i) \\
2/m &= 2 \otimes \bar{1}\ (C_{2h} = C_2 \otimes C_i) \\
mmm &= 222 \otimes \bar{1}\ (D_{2h} = D_2 \otimes C_i) \\
4/m &= 4 \otimes \bar{1}\ (C_{4h} = C_4 \otimes C_i) \\
\bar{3} &= 3 \otimes \bar{1}\ (C_{3i} = C_3 \otimes C_i) \\
\bar{3}m &= 32 \otimes \bar{1}\ (D_{3d} = D_3 \otimes C_i) \\
6/m &= 6 \otimes \bar{1}\ (C_{6h} = C_6 \otimes C_i) \\
4/mmm &= 422 \otimes \bar{1}\ (D_{4h} = D_4 \otimes C_i) \\
6/mmm &= 622 \otimes \bar{1}\ (D_{6h} = D_6 \otimes C_i) \\
m3 &= 23 \otimes \bar{1}\ (T_h = T \otimes C_i) \\
m3m &= 432 \otimes \bar{1}\ (O_h = O \otimes C_i).
\end{aligned}$$

(iv) The compatibilities for higher l values can be found by using the formula at the foot of the appropriate entry in the table. 'reg' denotes the regular representation which is the reducible representation which contains each of the reps Γ^i d_i times. Thus, for example, for 422 (D_4) for $l = 10$,

$$\text{reg} = A_1 + A_2 + B_1 + B_2 + 2E$$

and, using the table, we have

$$\begin{aligned}
\mathscr{D}^{10} &= 2 \text{ reg} + \mathscr{D}^2 \\
&= 2(A_1 + A_2 + B_1 + B_2 + 2E) + A_1 + B_1 + B_2 + E \\
&= 3A_1 + 2A_2 + 3B_1 + 3B_2 + 5E.
\end{aligned}$$

Example 2.4.3. We consider $l = 2$ (i.e. an atomic D term which is fivefold degenerate) and the point group $mm2$ (C_{2v}) as **G**. According to Table 2.7 this term splits in $mm2$ (C_{2v}) into five non-degenerate levels,

$$\mathscr{D}^2\{R(\alpha, \beta, \gamma)\} = 2A_1 + A_2 + B_1 + B_2. \tag{2.4.4}$$

The basis functions of $\mathscr{D}^2\{R(\alpha, \beta, \gamma)\}$ are $\{Y_2^{-2}(\theta, \phi), Y_2^{-1}(\theta, \phi), Y_2^0(\theta, \phi), Y_2^1(\theta, \phi), Y_2^2(\theta, \phi)\}$ and from Table 2.5 we see that these functions distribute themselves among the five reps of $mm2$ (C_{2v}) in eqn. (2.4.4) according to

A_1: $Y_2^0(\theta, \phi)$, $Y_2^{2,c}(\theta, \phi)$,
A_2: $Y_2^{2,s}(\theta, \phi)$,
B_1: $Y_2^{1,c}(\theta, \phi)$,
B_2: $Y_2^{1,s}(\theta, \phi)$,

see Fig. 2.2.

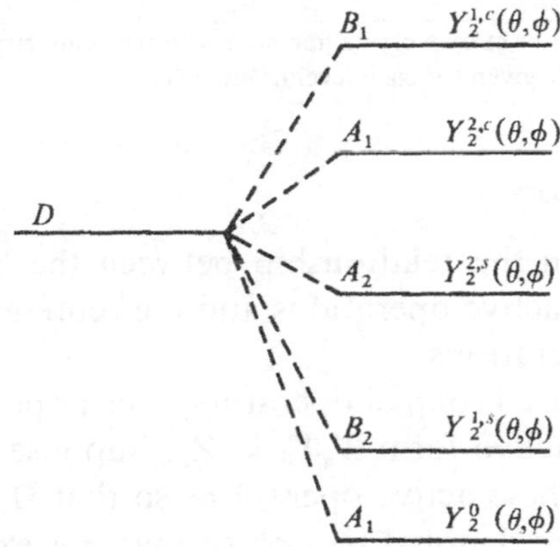

FIG. 2.2. Schematic splitting of atomic D term by a crystalline field with the symmetry of $mm2$ (C_{2v}).

Example 2.4.4. As a second example we consider $l = 3$, (i.e. an atomic F term which is sevenfold degenerate), with a cubic point group 432 (O) as **G**. According to Table 2.7 we have

$$\mathscr{D}^3\{R(\alpha, \beta, \gamma)\} = A_2 + T_1 + T_2. \tag{2.4.5}$$

A_2 is a non-degenerate rep of 432 (O) and T_1 and T_2 are threefold degenerate reps of 432 (O). The angular parts of the wave functions that correspond to these levels are linear combinations of $Y_3^0(\theta, \phi)$, $Y_3^{1,c}(\theta, \phi)$, $Y_3^{1,s}(\theta, \phi)$, $Y_3^{2,c}(\theta, \phi)$, $Y_3^{2,s}(\theta, \phi)$, $Y_3^{3,c}(\theta, \phi)$, and $Y_3^{3,s}(\theta, \phi)$. From Table 2.6 we see that the relevant surface harmonic for A_2 is $Y_3^{2,s}(\theta, \phi)$; for T_1 we have $[\{aY_3^{1,c}(\theta, \phi) - bY_3^{3,c}(\theta, \phi)\}, \{aY_3^{1,s}(\theta, \phi) + bY_3^{3,s}(\theta, \phi)\}, \{-Y_3^0(\theta, \phi)\}]$ and for T_2 we have $[\{-bY_3^{1,c}(\theta, \phi) - aY_3^{3,c}(\theta, \phi)\},$

$\{bY_3^{1,s}(\theta, \phi) - aY_3^{3,s}(\theta, \phi)\}$, $\{Y_3^{2,c}(\theta, \phi)\}]$ where $a = 0{\cdot}612\ 372\ 435\ 70$ and $b = 0{\cdot}790\ 569\ 415\ 04$. The splitting of the atomic F term in the crystalline field of 432 (O) is illustrated in Fig. 2.3.

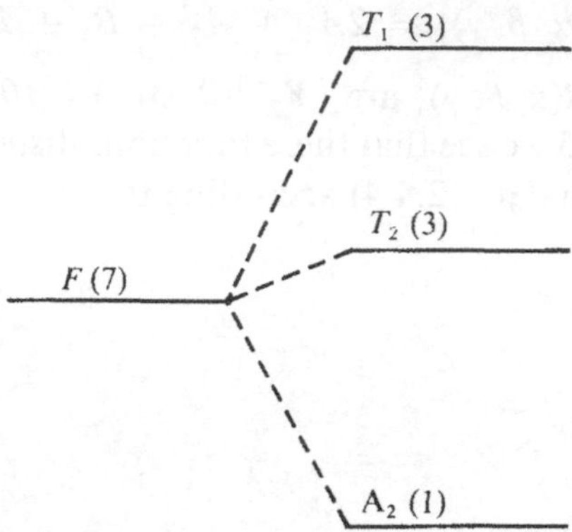

FIG. 2.3. Schematic splitting of atomic F term by a crystalline field with the symmetry of 432 (O). The degeneracy is given for each level in brackets.

2.5. Active and passive operators

For completeness we mention the relationship between the base functions of the representations of a group of active operations and the representations of the corresponding group of passive operations.

Suppose that $X, Y, Z, \ldots$ are a group of operators, such as point-group operations, and that considered in their active form $X_a Y_a = Z_a$. Suppose further that we have given a representation Γ_a^i of these active operations so that $\mathbf{D}_a^i(X_a)\mathbf{D}_a^i(Y_a) = \mathbf{D}_a^i(Z_a)$ and that the base functions are denoted by $\langle\phi_t^i|$ so that, for example,

$$\bar{X}_a \phi_t^i = \sum_j \phi_j^i \mathbf{D}_a^i(X_a)_{jt}, \tag{2.5.1}$$

where $\bar{X}_a$ is an active function space operator (see section 1.5). Suppose now that one wishes to use the operators in their passive form. First, $X_p = X_a^{-1}$, so that the multiplication table of the group is different. For example, we now have $Y_p X_p = Z_p$. The group is of course the same group; the multiplications differ because the mapping of an element on to its inverse is an anti-automorphism.

THEOREM 2.5.1. *If* $\langle\phi_t^i|$ *is the basis for a rep* Γ_a^i *of elements* $X_a, Y_a, Z_a, \ldots$ *of a group* **G** *then, if we set* $\mathbf{D}_p^i(X_p) = \mathbf{D}_a^i(X_a^{-1})$, $\langle\phi_t^i|$ *is also the basis for a rep* Γ_p^i *of elements* $X_p, Y_p, Z_p, \ldots$ *of* **G**, *where* $X_p = X_a^{-1}, Y_p = Y_a^{-1}, Z_p = Z_a^{-1}, \ldots$, etc.

This follows because

$$\bar{X}_p \phi_t^i = \bar{X}_a^{-1} \phi_t^i = \sum_j \phi_j^i \mathbf{D}_a^i(X_a^{-1})_{jt} = \sum_j \phi_j^i \mathbf{D}_p^i(X_p)_{jt} \tag{2.5.2}$$

and

$$\mathbf{D}_p^i(Y_p)\mathbf{D}_p^i(X_p) = \mathbf{D}_a^i(Y_a^{-1})\mathbf{D}_a^i(X_a^{-1}) = \mathbf{D}_a^i(Y_a^{-1}X_a^{-1})$$
$$= \mathbf{D}_a^i(Z_a^{-1}) = \mathbf{D}_p^i(Z_p). \quad (2.5.3)$$

As an example of the use of this theorem consider the tables that have been given in this chapter. In this chapter we use active operators and tabulate base functions for various matrix representations that are also tabulated in full. If one is using passive operators then one can use the same base functions provided one takes as matrix representatives the inverses of those tabulated. That is to say, one writes X (and means X_p) where we write X (and mean X_a). The multiplication table includes $YX = Z$ (meaning $Y_pX_p = Z_p$) in place of our multiplication table that includes $XY = Z$ (meaning $X_aY_a = Z_a$). $\langle\phi_t^i|$ is a basis for a representation Γ_p^i and $\mathbf{D}^i(Y)\mathbf{D}^i(X) = \mathbf{D}^i(Z)$ (meaning $\mathbf{D}_p^i(Y_p)\mathbf{D}_p^i(X_p) = \mathbf{D}_p^i(Z_p)$) and we will have written the same basis $\langle\phi_t^i|$ for a representation Γ_a^i and $\mathbf{D}^i(X)\mathbf{D}^i(Y) = \mathbf{D}^i(Z)$ (meaning $\mathbf{D}_a^i(X_a)\mathbf{D}_a^i(Y_a) = \mathbf{D}_a^i(Z_a)$); the relation between the matrices of Γ_a^i and Γ_p^i is $\mathbf{D}_p^i(X) = \mathbf{D}_a^i(X^{-1}) = \{\mathbf{D}_a^i(X)\}^{-1}$. We add one final word of warning: if $\mathbf{D}_a^i(X_a)$ and $\mathbf{D}_p^i(X_p)$ have different characters then the representations labelled by i, Γ_p^i and Γ_a^i, will have different labels in the character table although the base functions are the same.

2.6. Symmetrized and anti-symmetrized products of point-group reps

To conclude this chapter we give a brief account of the symmetrized and anti-symmetrized products of point-group reps. These products are of interest in their own right, for example, in connection with molecular vibrations (Jahn and Teller 1937) or with the Landau theory of second-order phase transitions (see, for example, Birman (1966*b*), Indenbom (1960*a*), Landau and Lifshitz (1958*b*), Lyubarskii (1960)). In addition to this, a study of these products for point-group reps forms a simple introduction to the work on symmetrized and anti-symmetrized products of space-group reps which will be discussed in section 4.8.

It is a well-known result in quantum mechanics that the wave function of a set of identical particles must be either symmetric or anti-symmetric with respect to the interchange of any pair of these identical particles (see any reputable book on quantum mechanics). The total wave function must be symmetric for bosons such as, for example, photons, α-particles, phonons, magnons, etc. and anti-symmetric for fermions such as, for example, electrons, protons, neutrons, etc.; the principle of anti-symmetry for fermions may be more commonly recognized as the *Pauli exclusion principle*. When there are two identical particles this leads to the concept of the symmetrized and anti-symmetrized Kronecker squares of a representation; the idea of a Kronecker product of two representations of a group has already been introduced in section 1.3 (see eqn. (1.3.23)).

Let **S** be a group and Δ a representation of **S** with basis $\langle\phi| = \langle\phi_1, \phi_2, \ldots, \phi_d|$ so that for all $s \in \mathbf{S}$

$$s\phi_i = \sum_{j=1}^{d} \phi_j \Delta(s)_{ji}. \tag{2.6.1}$$

Denoting the carrier space of the representation Δ by V we can consider the tensor product of V with itself, $V \otimes V$, to be the vector space spanned by linear combinations of ordered pairs of functions (ϕ_i, ϕ_j), $i = 1$ to d, $j = 1$ to d. The space $V \otimes V$ is of dimension d^2, and is invariant under the outer direct product $\mathbf{S} \otimes \mathbf{S}$. Since the inner direct product $\mathbf{S} \boxtimes \mathbf{S}$ is a subgroup of $\mathbf{S} \otimes \mathbf{S}$ isomorphic with **S** under the natural mapping $(s, s) \leftrightarrow s$ we may therefore regard $V \otimes V$ as being invariant under **S** itself and the representation of **S** so defined is called the inner Kronecker square of Δ. This is in keeping with the ideas and notation of section 1.3 and we denote this representation by $\Delta \boxtimes \Delta$. In order to obtain an explicit form for $\Delta \boxtimes \Delta$ we have

$$s(\phi_i, \phi_j) = (s\phi_i, s\phi_j) \tag{2.6.2}$$

$$= \sum_{k,l} (\phi_k, \phi_l)\Delta(s)_{ki}\Delta(s)_{lj} \tag{2.6.3}$$

$$= \sum_{k,l} (\phi_k, \phi_l)(\Delta \boxtimes \Delta)(s)_{kl,\,ij}. \tag{2.6.4}$$

Hence

$$(\Delta \boxtimes \Delta)(s) = \Delta(s) \otimes \Delta(s), \tag{2.6.5}$$

the Kronecker square of the matrix $\Delta(s)$; also from eqn. (2.6.3) the character of $\Delta \boxtimes \Delta$ is seen to be

$$\chi_{\Delta \boxtimes \Delta}(s) = \sum_{k,l} \Delta(s)_{kk}\Delta(s)_{ll} = \chi_\Delta^2(s). \tag{2.6.6}$$

In general the representation $\Delta \boxtimes \Delta$ will be reducible over **S** even when Δ is itself irreducible. When $d > 1$ a partial reduction can be achieved immediately separating $V \otimes V$ into its symmetric and anti-symmetric parts, see for example Lyubarskii (1960), Chapter 4. These subspaces of $V \otimes V$, which we denote by $[V \otimes V]$ and $\{V \otimes V\}$ respectively, are defined as follows: $[V \otimes V]$ is the vector space of dimension $\frac{1}{2}d(d + 1)$ spanned by linear combinations of the symmetrized pairs $(\phi_i, \phi_j) + (\phi_j, \phi_i)$, and $\{V \otimes V\}$ is the vector space of dimension $\frac{1}{2}d(d - 1)$ spanned by linear combinations of the anti-symmetrized pairs $(\phi_i, \phi_j) - (\phi_j, \phi_i)$. In order to see that these subspaces are invariant under **S** we consider the symmetrized space $[V \otimes V]$:

$$s[(\phi_i, \phi_j) + (\phi_j, \phi_i)] = \sum_{k,l} (\phi_k, \phi_l)\{\Delta(s)_{ki}\Delta(s)_{lj} + \Delta(s)_{kj}\Delta(s)_{li}\} \tag{2.6.7}$$

which on reversing the order of summation is seen to be equal to

$$\tfrac{1}{2} \sum_{k,l} \{(\phi_k, \phi_l) + (\phi_l, \phi_k)\}\{\Delta(s)_{ki}\Delta(s)_{lj} + \Delta(s)_{kj}\Delta(s)_{li}\} \tag{2.6.8}$$

and this sum clearly belongs to $[V \otimes V]$. The representation of $\mathbf{S}$ defined on $[V \otimes V]$ is denoted by $[\Delta \boxtimes \Delta]$. Similarly, $\{V \otimes V\}$ is invariant under $\mathbf{S}$ and defines a representation $\{\Delta \boxtimes \Delta\}$. These representations are called respectively the symmetrized and anti-symmetrized Kronecker squares of Δ and are sometimes denoted by $[\Delta]^2$ and $\{\Delta\}^2$, respectively.

In order to compute the character of the symmetrized square, $[\Delta \boxtimes \Delta]$ or $[\Delta]^2$, we have from expression (2.6.8) that

$$\begin{aligned} \chi_{[\Delta \boxtimes \Delta]}(s) &= \tfrac{1}{2} \sum_{k,l} \{\Delta(s)_{kk}\Delta(s)_{ll} + \Delta(s)_{kl}\Delta(s)_{lk}\} \\ &= \tfrac{1}{2}\{\chi_\Delta^2(s) + \chi_\Delta(s^2)\}. \end{aligned} \tag{2.6.9}$$

Similarly, for the anti-symmetrized square, $\{\Delta \boxtimes \Delta\}$ or $\{\Delta\}^2$

$$\chi_{\{\Delta \boxtimes \Delta\}}(s) = \tfrac{1}{2}\{\chi_\Delta^2(s) - \chi_\Delta(s^2)\}. \tag{2.6.10}$$

Incidentally, eqn. (2.6.10) follows also from eqns. (2.6.6) and (2.6.9) and the fact that

$$\Delta \boxtimes \Delta = [\Delta \boxtimes \Delta] + \{\Delta \boxtimes \Delta\}. \tag{2.6.11}$$

The reader who is slightly confused by the relevance of the notation of the ordered pair may like to think of the first member of the pair as referring to particle 1 and the second to particle 2, so that an expression such as $(\phi_i, \phi_j) - (\phi_j, \phi_i)$ then corresponds to the usual anti-symmetrized wave function $\phi_i(1)\phi_j(2) - \phi_j(1)\phi_i(2)$. The meaning of eqn. (2.6.2) in this context is also clear; the operator s acts simultaneously and in the same manner on the spaces of both particles 1 and 2. Clearly the idea of symmetrized and anti-symmetrized squares of representations can be extended to higher products of a representation of a group, see, for example, Lyubarskii (1960), Chapter 4.

We illustrate the use of eqns. (2.6.9) and (2.6.10) in identifying symmetrized and anti-symmetrized products of Δ by considering the examples provided by the crystallographic point groups. For a non-degenerate point-group rep eqns. (2.6.9) and (2.6.10) can be simplified; eqn. (2.6.9) becomes

$$\chi_{[\Delta]^2}(s) = \{\chi_\Delta(s)\}^2 \tag{2.6.12}$$

and eqn. (2.6.10) becomes

$$\chi_{\{\Delta\}^2}(s) = 0. \tag{2.6.13}$$

In other words, for a non-degenerate point-group rep the symmetrized square is simply the ordinary Kronecker square of Δ and the anti-symmetrized square vanishes.

For the degenerate point-group reps, $[\Delta]^2$ and $\{\Delta\}^2$ can be identified by using eqns. (2.6.9) and (2.6.10) and Table 2.2.; for example, for the rep E of 32 (D_3) we obtain

$$[E]^2 = A_1 + E \tag{2.6.14}$$

and

$$\{E\}^2 = A_2. \tag{2.6.15}$$

In Table 2.8 we identify $[\Delta]^2$ and $\{\Delta\}^2$ for the degenerate reps of each of the crystallographic point groups, while the symmetrized cubes of the degenerate point-group reps are given by Cracknell and Joshua (1968).

TABLE 2.8

$[\Delta]^2$ and $\{\Delta\}^2$ for the degenerate point-group reps

Point group	rep Δ	$[\Delta]^2$	$\{\Delta\}^2$
422 (D_4), 4*mm* (C_{4v}), $\bar{4}2m$ (D_{2d})			
	E	$A_1 + B_1 + B_2$	A_2
32 (D_3), 3*m* (C_{3v})			
	E	$A_1 + E$	A_2
622 (D_6), 6*mm* (C_{6v})			
	E_1	$A_1 + E_2$	A_2
	E_2	$A_1 + E_2$	A_2
$\bar{6}2m$ (C_{3h})			
	E'	$A_1' + E'$	A_2'
	E''	$A_1' + E'$	A_2'
23 (T)			
	T	$A + {}^1E + {}^2E + T$	T
432 (O), $\bar{4}3m$ (T_d)			
	E	$A_1 + E$	A_2
	T_1	$A_1 + E + T_2$	T_1
	T_2	$A_1 + E + T_2$	T_1

Note to Table 2.8

Groups that are direct products of a point group **G** with the point group $\bar{1}$ (C_i) are not included in Table 2.8 since their product representations can be found from those of **G** by adding g and u obeying the usual rules, see Note (iv) to Table 2.2.

3

Space groups

In section 1.5 we gave a descriptive account of the space groups and their labels and we introduced the Seitz notation for the symmetry operations that constitute the space group. In the present chapter we shall be concerned with the details of the identification of the 230 space groups, with the construction of reciprocal lattices and Brillouin zones and with an introduction to the problem of determining the irreducible representations of the space groups.

3.1. Bravais lattices

Every space group **G** has a set of pure translations, $\{E \mid \mathbf{t}\}$, which forms an invariant subgroup, **T**, of the space group, **G**. This is simply the group of the translational symmetry operations of the Bravais lattice on which the crystal is based. **t** is given by eqn. (1.5.1),

$$\mathbf{t} = n_1\mathbf{t}_1 + n_2\mathbf{t}_2 + n_3\mathbf{t}_3, \tag{1.5.1}$$

where n_1, n_2, and n_3 are integers and $\mathbf{t}_1$, $\mathbf{t}_2$, and $\mathbf{t}_3$ are the fundamental translations of the Bravais lattice. As we saw in section 1.5 the points determined by the vectors **t** are called the lattice points, and the parallelepiped constructed on the vectors $\mathbf{t}_1$, $\mathbf{t}_2$, and $\mathbf{t}_3$ is called the fundamental unit cell. The distinction between the fundamental unit cell and the conventional unit cell was mentioned in section 1.5 and illustrated in Fig. 1.8. The lattice as a whole is made up of a very large number of unit cells separated from each other by vectors **t** of the lattice. Strictly speaking, a Bravais lattice is of infinite extent. It would be possible to assume that a real crystal was of infinite extent and then to consider the theory of the representations of its space group and its translational subgroup. However, in practice one assumes that a crystal is of finite extent and then applies periodic boundary conditions. We suppose that there are N_i unit cells in direction $\mathbf{t}_i$ ($i = 1, 2$, or 3), where N_i is a very large number. In order that **G** shall still be a group we adopt the Born–von Kármán periodicity condition (Born and von Kármán 1913*a*, *b*) that if l_1, l_2, and l_3 are integers then

$$\{E \mid l_1N_1\mathbf{t}_1 + l_2N_2\mathbf{t}_2 + l_3N_3\mathbf{t}_3\} = \{E \mid \mathbf{0}\}. \tag{3.1.1}$$

The advantage of this is that the space group is then a finite group and so we can apply to it the theory of finite groups. It is customary then to let $N_i \to \infty$, the justification for this being that in this way we obtain all results of physical significance. There is no objection to dealing with infinite groups in the first place except that for infinite groups one has to reformulate many of the definitions and theorems on irreducible representations.

Before passing on to a complete description of the various possible lattices we make some general remarks. First, all primitive unit cells are of equal volume, V, given by

$$V = \mathbf{t}_1 \cdot (\mathbf{t}_2 \times \mathbf{t}_3). \tag{3.1.2}$$

Secondly, the volume of a parallelepiped constructed on any three independent lattice vectors of the Bravais lattice is not less than V. Thirdly, the basic vectors $\mathbf{t}_1$, $\mathbf{t}_2$, and $\mathbf{t}_3$ characterizing the lattice can often be chosen in many different ways. We shall make a definite choice (see Table 3.1) and from then on it may be assumed that we shall adhere to that choice.

In section 1.5 we said that there are 14 possible different Bravais lattices and they were illustrated in Fig. 1.7. Two Bravais lattices of the same crystal system are not different if one can be obtained from the other by a continuous deformation that does not involve the Bravais lattice passing through a crystal system of lower symmetry. For example, the face-centred cubic Bravais lattice, Γ_c^f, cannot be deformed into a body-centred cubic Bravais lattice, Γ_c^v, without passing through the trigonal Bravais lattice, Γ_{rh}. In Table 3.1 we give the list of all the 14 Bravais lattices together with the basic translations that we use, and the volumes of the corresponding unit cells. The *type* of a crystal structure, or of a space group, is the Bravais lattice on which it is based, denoted usually by the symbol in column 1 of Table 3.1. For the monoclinic system we use in Table 3.1 the first of the two alternative settings given in the

TABLE 3.1

The 14 Bravais lattices

Bravais lattice		**P**	Basic vectors, $\mathbf{t}_1$, $\mathbf{t}_2$, $\mathbf{t}_3$.	V
P Triclinic primitive	Γ_t	$\bar{1}$ (C_i)	arbitrary	$\mathbf{t}_1 \cdot (\mathbf{t}_2 \times \mathbf{t}_3)$
P Monoclinic primitive	Γ_m	$2/m$ (C_{2h})	$(0, -b, 0)$; $(a \sin\gamma, -a\cos\gamma, 0)$; $(0, 0, c)$	$abc \sin\gamma$
B Monoclinic base-centred	Γ_m^b	$2/m$ (C_{2h})	$(0, -b, 0)$; $\frac{1}{2}(a \sin\gamma, -a\cos\gamma, -c)$; $\frac{1}{2}(a\sin\gamma, -a\cos\gamma, c)$	$\frac{1}{2}abc \sin\gamma$
P Orthorhombic primitive	Γ_o	mmm (D_{2h})	$(0, -b, 0)$; $(a, 0, 0)$; $(0, 0, c)$	abc
C Orthorhombic base-centred	Γ_o^b	mmm (D_{2h})	$\frac{1}{2}(a, -b, 0)$; $\frac{1}{2}(a, b, 0)$; $(0, 0, c)$	$\frac{1}{2}abc$
I Orthorhombic body-centred	Γ_o^v	mmm (D_{2h})	$\frac{1}{2}(a, b, c)$; $\frac{1}{2}(-a, -b, c)$; $\frac{1}{2}(a, -b, -c)$	$\frac{1}{2}abc$
F Orthorhombic face-centred	Γ_o^f	mmm (D_{2h})	$\frac{1}{2}(a, 0, c)$; $\frac{1}{2}(0, -b, c)$; $\frac{1}{2}(a, -b, 0)$	$\frac{1}{4}abc$
P Tetragonal primitive	Γ_q	$4/mmm$ (D_{4h})	$(a, 0, 0)$; $(0, a, 0)$; $(0, 0, c)$	a^2c

I Tetragonal body-centred	Γ_q^v	$4/mmm$ (D_{4h})	$\frac{1}{2}(-a, a, c)$; $\frac{1}{2}(a, -a, c)$; $\frac{1}{2}(a, a, -c)$	$\frac{1}{2}a^2c$
R Trigonal primitive	Γ_{rh}	$\bar{3}m$ (D_{3d})	$(0, -a, c)$; $\frac{1}{2}(a\sqrt{(3)}, a, 2c)$; $\frac{1}{2}(-a\sqrt{(3)}, a, 2c)$	$\frac{1}{2}3\sqrt{(3)}a^2c$
P Hexagonal primitive	Γ_h	$6/mmm$ (D_{6h})	$(0, -a, 0)$; $\frac{1}{2}(a\sqrt{(3)}, a, 0)$; $(0, 0, c)$	$\frac{1}{2}\sqrt{(3)}a^2c$
P Cubic primitive	Γ_c	$m3m$ (O_h)	$(a, 0, 0)$; $(0, a, 0)$; $(0, 0, a)$	a^3
F Cubic face-centred	Γ_c^f	$m3m$ (O_h)	$\frac{1}{2}(0, a, a)$; $\frac{1}{2}(a, 0, a)$; $\frac{1}{2}(a, a, 0)$	$\frac{1}{4}a^3$
I Cubic body-centred	Γ_c^v	$m3m$ (O_h)	$\frac{1}{2}(-a, a, a)$; $\frac{1}{2}(a, -a, a)$; $\frac{1}{2}(a, a, -a)$	$\frac{1}{2}a^3$

Notes to Table 3.1

(i) Column 1 lists the Bravais lattice and the symbol used as its label; for diagrams of these lattices see Fig. 1.7, in which, however, it should be remembered that the diagrams show the conventional unit cell that is not necessarily the fundamental unit cell (see section 1.5).

(ii) Column 2 lists the holosymmetric point group of the crystal system to which the Bravais lattice belongs.

(iii) Column 3 lists the basic translations of the Bravais lattice referred to the right-handed orthogonal set of axes *Oxyz*. For example, $\mathbf{t}_1 = (p, q, r)$ means $\mathbf{t}_1 = p\mathbf{i} + q\mathbf{j} + r\mathbf{k}$, where **i**, **j**, and **k** are unit vectors in the *x*, *y*, and *z* directions.

(iv) Column 4 gives, in terms of *a*, *b*, and *c*, which are the lengths of the sides of the conventional unit cell, the volume of the fundamental unit cell (see eqn. (3.1.2)).

(v) An alternative orientation for the orthorhombic (*C*) base-centred Bravais lattice, Γ_o^b, is used in the *International tables for X-ray crystallography* (Henry and Lonsdale 1965) for certain space groups (see entries numbers 38–41 of Table 3.7). This is denoted by *A* and has basic vectors $\mathbf{t}_1 = \frac{1}{2}(0, -b, c)$, $\mathbf{t}_2 = (a, 0, 0)$ and $\mathbf{t}_3 = \frac{1}{2}(0, b, c)$ referred to axes *Oxyz*; we shall not use the *A* lattice in this book.

(vi) At this stage, in accordance with the *International tables for X-ray crystallography*, we impose no special restrictions on the relative lengths of *a*, *b*, and *c* beyond those actually imposed by symmetry. However, when we come to tabulate space-group representations we shall sometimes have occasion to impose restrictions on *a*, *b*, and *c*, see Notes to Figs. 3.2–3.15 and Notes to Table 3.6.

(vii) For the monoclinic system we use the first of the two settings given in the *International tables for X-ray crystallography*, that is, we take the twofold rotation axis of symmetry to be along the *z*-axis.

(viii) Alternative notations for the 14 Bravais lattices have been collected by Belov (1964) as follows.

		Schönflies	International		Wilson	Pearson
Triclinic	*P*	Γ_t	*P*	$\bar{1}$	*Z*	*AP*
Monoclinic	*P*	Γ_m	*P*	$2/m$	*M*	*MP*
	B	Γ_m^b	*B*, *C*	$2/m$	*X*	*MC*
Orthorhombic	*P*	Γ_o	*P*	*mmm*	*O*	*OP*
	C (or *A*)	Γ_o^b	*C*, *A*	*mmm*	*U*	*OC*
	I	Γ_o^v	*I*	*mmm*	*V*	*OI*
	F	Γ_o^f	*F*	*mmm*	*W*	*OF*
Tetragonal	*P*	Γ_q	*P*	$4/mmm$	*T*	*TP*(*C*)
	I	Γ_q^v	*I*	$4/mmm$	*Q*	*TI*(*F*)
Trigonal	*R*	Γ_{rh}	*R*	$\bar{3}m$	*R*	*HR*
Hexagonal	*P*	Γ_h	*P*	$6/mmm$	*H*	*HP*
Cubic	*P*	Γ_c	*P*	$m3m$	*K*	*CP*
	I	Γ_c^v	*I*	$m3m$	*B*	*CI*
	F	Γ_c^f	*F*	$m3m$	*F*	*CF*

International tables for X-ray crystallography (Henry and Lonsdale 1965). **a**, **b**, and **c** are the vectors along adjacent edges of the conventional unit cell; in the *International tables for X-ray crystallography* there are no restrictions on the relative magnitudes of **a**, **b**, and **c** other than those required by the symmetry of the Bravais lattice.

In spite of the obvious importance of the 14 Bravais lattices there is, as yet, no universally adopted notation that would describe each lattice succinctly and clearly. The generally accepted letters, *P*, *I*, *F*, *B*, *C*, and *R* given in Table 3.1, which are used in the *International tables for X-ray crystallography* merely establish the lattice as being primitive or centred in some particular way. The letter Γ used in the Schönflies symbol is superfluous and conveys no particularly useful information. The use of 14 different letters, one for each Bravais lattice, has been suggested by Wilson and the use of a two-letter symbol for each Bravais lattice has been suggested by Pearson. These alternative suggestions have been collated by Belov (1964) and are given in Note (viii) to Table 3.1.

If certain particular relationships exist between the lengths *a*, *b*, and *c* in a Bravais lattice of low symmetry they may make it into a Bravais lattice of higher symmetry. For example, if $a = c/\sqrt{2}$ then Γ_{rh} becomes Γ_c^f, if $a = \sqrt{(2)}c$ then Γ_{rh} becomes Γ_c and if $a = 2\sqrt{(2)}c$ then Γ_{rh} becomes Γ_c^v. Thus all the cubic lattices are particular cases of the trigonal lattice. Also, if $a = c$ then Γ_q becomes Γ_c and Γ_q^v becomes Γ_c^v; and if $c = \sqrt{(3)}b$ then Γ_o^b becomes Γ_h. Other such relationships can easily be deduced from Table 3.1.

The symmetry elements for the seven crystal systems are given in Table 1.2. It is useful to know what effect the point-group operations of the crystal systems have on the lattice translations. These are given in Table 3.2.

TABLE 3.2

Results of operators on the basic translations

Monoclinic

	Γ_m			Γ_m^b		
E	$\mathbf{t}_1$	$\mathbf{t}_2$	$\mathbf{t}_3$	$\mathbf{t}_1$	$\mathbf{t}_2$	$\mathbf{t}_3$
C_{2z}	$-\mathbf{t}_1$	$-\mathbf{t}_2$	$\mathbf{t}_3$	$-\mathbf{t}_1$	$-\mathbf{t}_3$	$-\mathbf{t}_2$

Orthorhombic

	Γ_o			Γ_o^b			. . .
E	$\mathbf{t}_1$	$\mathbf{t}_2$	$\mathbf{t}_3$	$\mathbf{t}_1$	$\mathbf{t}_2$	$\mathbf{t}_3$	
C_{2x}	$-\mathbf{t}_1$	$\mathbf{t}_2$	$-\mathbf{t}_3$	$\mathbf{t}_2$	$\mathbf{t}_1$	$-\mathbf{t}_3$	. . .
C_{2y}	$\mathbf{t}_1$	$-\mathbf{t}_2$	$-\mathbf{t}_3$	$-\mathbf{t}_2$	$-\mathbf{t}_1$	$-\mathbf{t}_3$	
C_{2z}	$-\mathbf{t}_1$	$-\mathbf{t}_2$	$\mathbf{t}_3$	$-\mathbf{t}_1$	$-\mathbf{t}_2$	$\mathbf{t}_3$	

. . .	Γ_o^v			Γ_o^f		
	$\mathbf{t}_1$	$\mathbf{t}_2$	$\mathbf{t}_3$	$\mathbf{t}_1$	$\mathbf{t}_2$	$\mathbf{t}_3$
. . .	$\mathbf{t}_3$	$-\mathbf{t}_1-\mathbf{t}_2-\mathbf{t}_3$	$\mathbf{t}_1$	$\mathbf{t}_3-\mathbf{t}_2$	$-\mathbf{t}_2$	$\mathbf{t}_1-\mathbf{t}_2$
	$-\mathbf{t}_1-\mathbf{t}_2-\mathbf{t}_3$	$\mathbf{t}_3$	$\mathbf{t}_2$	$-\mathbf{t}_1$	$\mathbf{t}_3-\mathbf{t}_1$	$\mathbf{t}_2-\mathbf{t}_1$
	$\mathbf{t}_2$	$\mathbf{t}_1$	$-\mathbf{t}_1-\mathbf{t}_2-\mathbf{t}_3$	$\mathbf{t}_2-\mathbf{t}_3$	$\mathbf{t}_1-\mathbf{t}_3$	$-\mathbf{t}_3$

Tetragonal

	Γ_q			Γ_q^v		
E	$\mathbf{t}_1$	$\mathbf{t}_2$	$\mathbf{t}_3$	$\mathbf{t}_1$	$\mathbf{t}_2$	$\mathbf{t}_3$
C_{4z}^+	$\mathbf{t}_2$	$-\mathbf{t}_1$	$\mathbf{t}_3$	$-\mathbf{t}_3$	$\mathbf{t}_1+\mathbf{t}_2+\mathbf{t}_3$	$-\mathbf{t}_2$
C_{2z}	$-\mathbf{t}_1$	$-\mathbf{t}_2$	$\mathbf{t}_3$	$\mathbf{t}_2$	$\mathbf{t}_1$	$-\mathbf{t}_1-\mathbf{t}_2-\mathbf{t}_3$
C_{4z}^-	$-\mathbf{t}_2$	$\mathbf{t}_1$	$\mathbf{t}_3$	$\mathbf{t}_1+\mathbf{t}_2+\mathbf{t}_3$	$-\mathbf{t}_3$	$-\mathbf{t}_1$
C_{2x}	$\mathbf{t}_1$	$-\mathbf{t}_2$	$-\mathbf{t}_3$	$-\mathbf{t}_1-\mathbf{t}_2-\mathbf{t}_3$	$\mathbf{t}_3$	$\mathbf{t}_2$
C_{2y}	$-\mathbf{t}_1$	$\mathbf{t}_2$	$-\mathbf{t}_3$	$\mathbf{t}_3$	$-\mathbf{t}_1-\mathbf{t}_2-\mathbf{t}_3$	$\mathbf{t}_1$
C_{2a}	$\mathbf{t}_2$	$\mathbf{t}_1$	$-\mathbf{t}_3$	$-\mathbf{t}_1$	$-\mathbf{t}_2$	$\mathbf{t}_1+\mathbf{t}_2+\mathbf{t}_3$
C_{2b}	$-\mathbf{t}_2$	$-\mathbf{t}_1$	$-\mathbf{t}_3$	$-\mathbf{t}_2$	$-\mathbf{t}_1$	$-\mathbf{t}_3$

Trigonal and hexagonal

	Γ_{rh}			Γ_h		
E	$\mathbf{t}_1$	$\mathbf{t}_2$	$\mathbf{t}_3$	$\mathbf{t}_1$	$\mathbf{t}_2$	$\mathbf{t}_3$
C_6^+	—	—	—	$\mathbf{t}_1+\mathbf{t}_2$	$-\mathbf{t}_1$	$\mathbf{t}_3$
C_3^+	$\mathbf{t}_2$	$\mathbf{t}_3$	$\mathbf{t}_1$	$\mathbf{t}_2$	$-\mathbf{t}_1-\mathbf{t}_2$	$\mathbf{t}_3$
C_2	—	—	—	$-\mathbf{t}_1$	$-\mathbf{t}_2$	$\mathbf{t}_3$
C_3^-	$\mathbf{t}_3$	$\mathbf{t}_1$	$\mathbf{t}_2$	$-\mathbf{t}_1-\mathbf{t}_2$	$\mathbf{t}_1$	$\mathbf{t}_3$
C_6^-	—	—	—	$-\mathbf{t}_2$	$\mathbf{t}_1+\mathbf{t}_2$	$\mathbf{t}_3$
C'_{21}	$-\mathbf{t}_1$	$-\mathbf{t}_3$	$-\mathbf{t}_2$	$-\mathbf{t}_1$	$\mathbf{t}_1+\mathbf{t}_2$	$-\mathbf{t}_3$
C'_{22}	$-\mathbf{t}_3$	$-\mathbf{t}_2$	$-\mathbf{t}_1$	$\mathbf{t}_1+\mathbf{t}_2$	$-\mathbf{t}_2$	$-\mathbf{t}_3$
C'_{23}	$-\mathbf{t}_2$	$-\mathbf{t}_1$	$-\mathbf{t}_3$	$-\mathbf{t}_2$	$-\mathbf{t}_1$	$-\mathbf{t}_3$
C''_{21}	—	—	—	$\mathbf{t}_1$	$-\mathbf{t}_1-\mathbf{t}_2$	$-\mathbf{t}_3$
C''_{22}	—	—	—	$-\mathbf{t}_1-\mathbf{t}_2$	$\mathbf{t}_2$	$-\mathbf{t}_3$
C''_{23}	—	—	—	$\mathbf{t}_2$	$\mathbf{t}_1$	$-\mathbf{t}_3$

Cubic

	Γ_c			Γ_c^f			Γ_c^v		
E	$\mathbf{t}_1$	$\mathbf{t}_2$	$\mathbf{t}_3$	$\mathbf{t}_1$	$\mathbf{t}_2$	$\mathbf{t}_3$	$\mathbf{t}_1$	$\mathbf{t}_2$	$\mathbf{t}_3$
C_{2x}	$\mathbf{t}_1$	$-\mathbf{t}_2$	$-\mathbf{t}_3$	$-\mathbf{t}_1$	$-\mathbf{t}_1+\mathbf{t}_3$	$-\mathbf{t}_1+\mathbf{t}_2$	$-\mathbf{t}_1-\mathbf{t}_2-\mathbf{t}_3$	$\mathbf{t}_3$	$\mathbf{t}_2$
C_{2y}	$-\mathbf{t}_1$	$\mathbf{t}_2$	$-\mathbf{t}_3$	$-\mathbf{t}_2+\mathbf{t}_3$	$-\mathbf{t}_2$	$\mathbf{t}_1-\mathbf{t}_2$	$\mathbf{t}_3$	$-\mathbf{t}_1-\mathbf{t}_2-\mathbf{t}_3$	$\mathbf{t}_1$
C_{2z}	$-\mathbf{t}_1$	$-\mathbf{t}_2$	$\mathbf{t}_3$	$\mathbf{t}_2-\mathbf{t}_3$	$\mathbf{t}_1-\mathbf{t}_3$	$-\mathbf{t}_3$	$\mathbf{t}_2$	$\mathbf{t}_1$	$-\mathbf{t}_1-\mathbf{t}_2-\mathbf{t}_3$
C_{31}^+	$\mathbf{t}_2$	$\mathbf{t}_3$	$\mathbf{t}_1$	$\mathbf{t}_2$	$\mathbf{t}_3$	$\mathbf{t}_1$	$\mathbf{t}_2$	$\mathbf{t}_3$	$\mathbf{t}_1$
C_{32}^+	$\mathbf{t}_2$	$-\mathbf{t}_3$	$-\mathbf{t}_1$	$-\mathbf{t}_2$	$\mathbf{t}_1-\mathbf{t}_2$	$-\mathbf{t}_2+\mathbf{t}_3$	$-\mathbf{t}_1-\mathbf{t}_2-\mathbf{t}_3$	$\mathbf{t}_1$	$\mathbf{t}_3$
C_{33}^+	$-\mathbf{t}_2$	$\mathbf{t}_3$	$-\mathbf{t}_1$	$\mathbf{t}_1-\mathbf{t}_3$	$-\mathbf{t}_3$	$\mathbf{t}_2-\mathbf{t}_3$	$\mathbf{t}_1$	$-\mathbf{t}_1-\mathbf{t}_2-\mathbf{t}_3$	$\mathbf{t}_2$

Cubic—(continued)

	Γ_c			Γ_c^f			Γ_c^v		
C_{34}^+	$-\mathbf{t}_2$	$-\mathbf{t}_3$	$\mathbf{t}_1$	$-\mathbf{t}_1+\mathbf{t}_3$	$-\mathbf{t}_1+\mathbf{t}_2$	$-\mathbf{t}_1$	$\mathbf{t}_3$	$\mathbf{t}_2$	$-\mathbf{t}_1-\mathbf{t}_2-\mathbf{t}_3$
C_{31}^-	$\mathbf{t}_3$	$\mathbf{t}_1$	$\mathbf{t}_2$	$\mathbf{t}_3$	$\mathbf{t}_1$	$\mathbf{t}_2$	$\mathbf{t}_3$	$\mathbf{t}_1$	$\mathbf{t}_2$
C_{32}^-	$-\mathbf{t}_3$	$\mathbf{t}_1$	$-\mathbf{t}_2$	$-\mathbf{t}_1+\mathbf{t}_2$	$-\mathbf{t}_1$	$-\mathbf{t}_1+\mathbf{t}_3$	$\mathbf{t}_2$	$-\mathbf{t}_1-\mathbf{t}_2-\mathbf{t}_3$	$\mathbf{t}_3$
C_{33}^-	$-\mathbf{t}_3$	$-\mathbf{t}_1$	$\mathbf{t}_2$	$\mathbf{t}_1-\mathbf{t}_2$	$-\mathbf{t}_2+\mathbf{t}_3$	$-\mathbf{t}_2$	$\mathbf{t}_1$	$\mathbf{t}_3$	$-\mathbf{t}_1-\mathbf{t}_2-\mathbf{t}_3$
C_{34}^-	$\mathbf{t}_3$	$-\mathbf{t}_1$	$-\mathbf{t}_2$	$-\mathbf{t}_3$	$\mathbf{t}_2-\mathbf{t}_3$	$\mathbf{t}_1-\mathbf{t}_3$	$-\mathbf{t}_1-\mathbf{t}_2-\mathbf{t}_3$	$\mathbf{t}_2$	$\mathbf{t}_1$
C_{4x}^+	$\mathbf{t}_1$	$\mathbf{t}_3$	$-\mathbf{t}_2$	$\mathbf{t}_2-\mathbf{t}_3$	$-\mathbf{t}_1+\mathbf{t}_2$	$\mathbf{t}_2$	$-\mathbf{t}_3$	$-\mathbf{t}_1$	$\mathbf{t}_1+\mathbf{t}_2+\mathbf{t}_3$
C_{4y}^+	$-\mathbf{t}_3$	$\mathbf{t}_2$	$\mathbf{t}_1$	$\mathbf{t}_3$	$-\mathbf{t}_1+\mathbf{t}_3$	$-\mathbf{t}_2+\mathbf{t}_3$	$\mathbf{t}_1+\mathbf{t}_2+\mathbf{t}_3$	$-\mathbf{t}_1$	$-\mathbf{t}_2$
C_{4z}^+	$\mathbf{t}_2$	$-\mathbf{t}_1$	$\mathbf{t}_3$	$\mathbf{t}_1-\mathbf{t}_3$	$\mathbf{t}_1$	$\mathbf{t}_1-\mathbf{t}_2$	$-\mathbf{t}_3$	$\mathbf{t}_1+\mathbf{t}_2+\mathbf{t}_3$	$-\mathbf{t}_2$
C_{4x}^-	$\mathbf{t}_1$	$-\mathbf{t}_3$	$\mathbf{t}_2$	$-\mathbf{t}_2+\mathbf{t}_3$	$\mathbf{t}_3$	$-\mathbf{t}_1+\mathbf{t}_3$	$-\mathbf{t}_2$	$\mathbf{t}_1+\mathbf{t}_2+\mathbf{t}_3$	$-\mathbf{t}_1$
C_{4y}^-	$\mathbf{t}_3$	$\mathbf{t}_2$	$-\mathbf{t}_1$	$\mathbf{t}_1-\mathbf{t}_2$	$\mathbf{t}_1-\mathbf{t}_3$	$\mathbf{t}_1$	$-\mathbf{t}_2$	$-\mathbf{t}_3$	$\mathbf{t}_1+\mathbf{t}_2+\mathbf{t}_3$
C_{4z}^-	$-\mathbf{t}_2$	$\mathbf{t}_1$	$\mathbf{t}_3$	$\mathbf{t}_2$	$\mathbf{t}_2-\mathbf{t}_3$	$-\mathbf{t}_1+\mathbf{t}_2$	$\mathbf{t}_1+\mathbf{t}_2+\mathbf{t}_3$	$-\mathbf{t}_3$	$-\mathbf{t}_1$
C_{2a}	$\mathbf{t}_2$	$\mathbf{t}_1$	$-\mathbf{t}_3$	$-\mathbf{t}_1+\mathbf{t}_3$	$-\mathbf{t}_2+\mathbf{t}_3$	$\mathbf{t}_3$	$-\mathbf{t}_1$	$-\mathbf{t}_2$	$\mathbf{t}_1+\mathbf{t}_2+\mathbf{t}_3$
C_{2b}	$-\mathbf{t}_2$	$-\mathbf{t}_1$	$-\mathbf{t}_3$	$-\mathbf{t}_2$	$-\mathbf{t}_1$	$-\mathbf{t}_3$	$-\mathbf{t}_2$	$-\mathbf{t}_1$	$-\mathbf{t}_3$
C_{2c}	$\mathbf{t}_3$	$-\mathbf{t}_2$	$\mathbf{t}_1$	$-\mathbf{t}_1+\mathbf{t}_2$	$\mathbf{t}_2$	$\mathbf{t}_2-\mathbf{t}_3$	$-\mathbf{t}_1$	$\mathbf{t}_1+\mathbf{t}_2+\mathbf{t}_3$	$-\mathbf{t}_3$
C_{2d}	$-\mathbf{t}_1$	$\mathbf{t}_3$	$\mathbf{t}_2$	$\mathbf{t}_1$	$\mathbf{t}_1-\mathbf{t}_2$	$\mathbf{t}_1-\mathbf{t}_3$	$\mathbf{t}_1+\mathbf{t}_2+\mathbf{t}_3$	$-\mathbf{t}_2$	$-\mathbf{t}_3$
C_{2e}	$-\mathbf{t}_3$	$-\mathbf{t}_2$	$-\mathbf{t}_1$	$-\mathbf{t}_3$	$-\mathbf{t}_2$	$-\mathbf{t}_1$	$-\mathbf{t}_3$	$-\mathbf{t}_2$	$-\mathbf{t}_1$
C_{2f}	$-\mathbf{t}_1$	$-\mathbf{t}_3$	$-\mathbf{t}_2$	$-\mathbf{t}_1$	$-\mathbf{t}_3$	$-\mathbf{t}_2$	$-\mathbf{t}_1$	$-\mathbf{t}_3$	$-\mathbf{t}_2$

Notes to Table 3.2

(i) To save space only the effects of the elements of the holosymmetric point groups of the crystal systems are tabulated. Since $I\mathbf{t} = -\mathbf{t}$ for all $\mathbf{t}$ the action of the inversion, I, and its product with each of the elements given, is also omitted.

(ii) The point-group operations are active and correspond to symmetry elements passing through the origin.

(iii) The translations $\mathbf{t}_1$, $\mathbf{t}_2$, and $\mathbf{t}_3$ are given in Table 3.1.

3.2. Reciprocal lattices and Brillouin zones

If the basic translation vectors of a Bravais lattice are $\mathbf{t}_1$, $\mathbf{t}_2$, and $\mathbf{t}_3$ it is possible to define a set of reciprocal lattice vectors $\mathbf{g}_1$, $\mathbf{g}_2$, and $\mathbf{g}_3$ where

$$\mathbf{g}_i \cdot \mathbf{t}_j = 2\pi\delta_{ij} \quad (i, j = 1, 2, 3). \tag{3.2.1}$$

The reciprocal lattice defined in this way is, of course, extensively used by X-ray crystallographers. Some authors define $\mathbf{g}_i$ without the factor 2π in this equation and consequently have to use the unwieldy phrase '2π times a reciprocal lattice vector' in later theory. From eqn. (3.2.1) $\mathbf{g}_1$, $\mathbf{g}_2$, and $\mathbf{g}_3$ can be written explicitly as

$$\left.\begin{aligned} \mathbf{g}_1 &= \frac{2\pi(\mathbf{t}_2 \times \mathbf{t}_3)}{\mathbf{t}_1\cdot(\mathbf{t}_2 \times \mathbf{t}_3)}, \\ \mathbf{g}_2 &= \frac{2\pi(\mathbf{t}_3 \times \mathbf{t}_1)}{\mathbf{t}_2\cdot(\mathbf{t}_3 \times \mathbf{t}_1)}, \\ \mathbf{g}_3 &= \frac{2\pi(\mathbf{t}_1 \times \mathbf{t}_2)}{\mathbf{t}_3\cdot(\mathbf{t}_1 \times \mathbf{t}_2)}. \end{aligned}\right\} \tag{3.2.2}$$

If $\mathbf{t}_1$, $\mathbf{t}_2$, and $\mathbf{t}_3$ are the basic vectors of a Bravais lattice then $\mathbf{g}_1$, $\mathbf{g}_2$, and $\mathbf{g}_3$ are usually basic vectors of the same Bravais lattice in reciprocal space. However, in four cases $\mathbf{g}_1$, $\mathbf{g}_2$, and $\mathbf{g}_3$ are basic vectors of another Bravais lattice in the same crystal system. For example, if we consider the face-centred cubic lattice its basic vectors are, from Table 3.1,

$$\left.\begin{aligned} \mathbf{t}_1 &= \tfrac{1}{2}a(0, 1, 1), \\ \mathbf{t}_2 &= \tfrac{1}{2}a(1, 0, 1), \\ \mathbf{t}_3 &= \tfrac{1}{2}a(1, 1, 0). \end{aligned}\right\} \tag{3.2.3}$$

If these vectors are substituted into eqn. (3.2.2) we find that

$$\left.\begin{aligned} \mathbf{g}_1 &= \frac{2\pi}{a}(-1, 1, 1), \\ \mathbf{g}_2 &= \frac{2\pi}{a}(1, -1, 1), \\ \mathbf{g}_3 &= \frac{2\pi}{a}(1, 1, -1), \end{aligned}\right\} \tag{3.2.4}$$

and by inspection of Table 3.1 these vectors can be seen to define a body-centred cubic lattice. Thus the reciprocal lattice for the cubic face-centred lattice is the cubic body-centred lattice, and vice versa. The reciprocal lattice for the orthorhombic face-centred lattice is also the orthorhombic body-centred lattice, and vice versa. The tetragonal body-centred lattice is its own reciprocal lattice because a tetragonal face-centred lattice is at the same time a tetragonal body-centred lattice with a different axial ratio. As with the Bravais lattice in direct space we may write the position vector, $\mathbf{g}$, of any reciprocal lattice point in terms of $\mathbf{g}_1$, $\mathbf{g}_2$, and $\mathbf{g}_3$, and by analogy with eqn. (1.5.1),

$$\mathbf{g} = n_1\mathbf{g}_1 + n_2\mathbf{g}_2 + n_3\mathbf{g}_3 \tag{3.2.5}$$

where n_1, n_2, and n_3 are integers.

We use x, y, and z for coordinates in direct space and k_x, k_y, and k_z for coordinates in reciprocal space. This is because a general vector in reciprocal space is customarily denoted by $\mathbf{k}$. It should be noted that if a diagram is drawn with the reciprocal lattice superimposed on the direct lattice then $\mathbf{g}_1$, being perpendicular to $\mathbf{t}_2$ and $\mathbf{t}_3$, is not necessarily in the direction of $\mathbf{t}_1$. The reason for the introduction of reciprocal space at all in connection with representation theory for space groups will become apparent in section 3.4, in which we discuss the representation theory of translation groups.

In Table 3.3 we give a list of the reciprocal lattice vectors for the 14 Bravais lattices. The transformations of the basis vectors $\mathbf{g}_1$, $\mathbf{g}_2$, and $\mathbf{g}_3$ of the reciprocal lattice under the point-group operations R of the appropriate crystal system can be seen from diagrams in reciprocal space or alternatively worked out analytically by applying R

TABLE 3.3

The reciprocal lattices

Bravais lattice		Reciprocal vectors, $\mathbf{g}_1, \mathbf{g}_2, \mathbf{g}_3$.	$8\pi^3/V$
Triclinic primitive	Γ_t	arbitrary	$\mathbf{g}_1 . (\mathbf{g}_2 \times \mathbf{g}_3)$
Monoclinic primitive	Γ_m	$2\pi/b(-\cot\gamma, -1, 0)$; $2\pi/a(\operatorname{cosec}\gamma, 0, 0)$; $2\pi/c(0, 0, 1)$	$8\pi^3/abc \sin\gamma$
Monoclinic base-centred	Γ_m^b	$2\pi/b(-\cot\gamma, -1, 0)$; $2\pi/ac(c \operatorname{cosec}\gamma, 0, -a)$; $2\pi/ac(c \operatorname{cosec}\gamma, 0, a)$	$16\pi^3/abc \sin\gamma$
Orthorhombic primitive	Γ_o	$2\pi/b(0, -1, 0)$; $2\pi/a(1, 0, 0)$; $2\pi/c(0, 0, 1)$	$8\pi^3/abc$
Orthorhombic base-centred	Γ_o^b	$2\pi/ba(b, -a, 0)$; $2\pi/ba(b, a, 0)$; $2\pi/c(0, 0, 1)$	$16\pi^3/abc$
Orthorhombic body-centred	Γ_o^v	$2\pi/ca(c, 0, a)$; $2\pi/cb(0, -c, b)$; $2\pi/ba(b, -a, 0)$	$16\pi^3/abc$
Orthorhombic face-centred	Γ_o^f	$2\pi(1/a, 1/b, 1/c)$; $2\pi(-1/a, -1/b, 1/c)$; $2\pi(1/a, -1/b, -1/c)$	$32\pi^3/abc$
Tetragonal primitive	Γ_q	$2\pi/a(1, 0, 0)$; $2\pi/a(0, 1, 0)$; $2\pi/c(0, 0, 1)$	$8\pi^3/a^2c$
Tetragonal body-centred	Γ_q^v	$2\pi/ca(0, c, a)$; $2\pi/ca(c, 0, a)$; $2\pi/a(1, 1, 0)$	$16\pi^3/a^2c$
Trigonal primitive	Γ_{rh}	$2\pi(0, -2/3a, 1/3c)$; $2\pi(1/\sqrt{(3)}a, 1/3a, 1/3c)$; $2\pi(-1/\sqrt{(3)}a, 1/3a, 1/3c)$	$16\pi^3/3\sqrt{(3)}a^2c$
Hexagonal primitive	Γ_h	$2\pi/a(1/\sqrt{3}, -1, 0)$; $2\pi/a(2/\sqrt{3}, 0, 0)$; $2\pi/c(0, 0, 1)$	$16\pi^3/\sqrt{(3)}a^2c$
Cubic primitive	Γ_c	$2\pi/a(1, 0, 0)$; $2\pi/a(0, 1, 0)$; $2\pi/a(0, 0, 1)$	$8\pi^3/a^3$
Cubic face-centred	Γ_c^f	$2\pi/a(-1, 1, 1)$; $2\pi/a(1, -1, 1)$; $2\pi/a(1, 1, -1)$	$32\pi^3/a^3$
Cubic body-centred	Γ_c^v	$2\pi/a(0, 1, 1)$; $2\pi/a(1, 0, 1)$; $2\pi/a(1, 1, 0)$	$16\pi^3/a^3$

Notes to Table 3.3

(i) Column 1 lists the Bravais lattice of the translation group, **T**, in direct space. The basic vectors of the Bravais lattices are given in Table 3.1.

(ii) Column 2 lists the coordinates of the translations $\mathbf{g}_1$, $\mathbf{g}_2$, and $\mathbf{g}_3$ of reciprocal space with respect to the k_x, k_y, k_z axes. For example, $\mathbf{g}_1 = (p, q, r)$ means that $\mathbf{g}_1 = p\mathbf{l} + q\mathbf{m} + r\mathbf{n}$, where **l**, **m**, and **n** are unit vectors in the k_x, k_y, and k_z directions.

(iii) Column 3 gives $(8\pi^3/V)$, the volume of the Brillouin zone.

TABLE 3.4

Results of operators on the reciprocal lattice vectors

Monoclinic

	Γ_m			Γ_m^b		
E	$\mathbf{g}_1$	$\mathbf{g}_2$	$\mathbf{g}_3$	$\mathbf{g}_1$	$\mathbf{g}_2$	$\mathbf{g}_3$
C_{2z}	$-\mathbf{g}_1$	$-\mathbf{g}_2$	$\mathbf{g}_3$	$-\mathbf{g}_1$	$-\mathbf{g}_3$	$-\mathbf{g}_2$

Orthorhombic

	Γ_o			Γ_o^b			. . .
E	$\mathbf{g}_1$	$\mathbf{g}_2$	$\mathbf{g}_3$	$\mathbf{g}_1$	$\mathbf{g}_2$	$\mathbf{g}_3$	
C_{2x}	$-\mathbf{g}_1$	$\mathbf{g}_2$	$-\mathbf{g}_3$	$\mathbf{g}_2$	$\mathbf{g}_1$	$-\mathbf{g}_3$	. . .
C_{2y}	$\mathbf{g}_1$	$-\mathbf{g}_2$	$-\mathbf{g}_3$	$-\mathbf{g}_2$	$-\mathbf{g}_1$	$-\mathbf{g}_3$	
C_{2z}	$-\mathbf{g}_1$	$-\mathbf{g}_2$	$\mathbf{g}_3$	$-\mathbf{g}_1$	$-\mathbf{g}_2$	$\mathbf{g}_3$	

. . .	Γ_o^v			Γ_o^f		
	$\mathbf{g}_1$	$\mathbf{g}_2$	$\mathbf{g}_3$	$\mathbf{g}_1$	$\mathbf{g}_2$	$\mathbf{g}_3$
. . .	$-\mathbf{g}_2+\mathbf{g}_3$	$-\mathbf{g}_2$	$\mathbf{g}_1-\mathbf{g}_2$	$\mathbf{g}_3$	$-\mathbf{g}_1-\mathbf{g}_2-\mathbf{g}_3$	$\mathbf{g}_1$
	$-\mathbf{g}_1$	$-\mathbf{g}_1+\mathbf{g}_3$	$-\mathbf{g}_1+\mathbf{g}_2$	$-\mathbf{g}_1-\mathbf{g}_2-\mathbf{g}_3$	$\mathbf{g}_3$	$\mathbf{g}_2$
	$\mathbf{g}_2-\mathbf{g}_3$	$\mathbf{g}_1-\mathbf{g}_3$	$-\mathbf{g}_3$	$\mathbf{g}_2$	$\mathbf{g}_1$	$-\mathbf{g}_1-\mathbf{g}_2-\mathbf{g}_3$

Tetragonal

	Γ_q			Γ_q^v		
E	$\mathbf{g}_1$	$\mathbf{g}_2$	$\mathbf{g}_3$	$\mathbf{g}_1$	$\mathbf{g}_2$	$\mathbf{g}_3$
C_{4z}^+	$\mathbf{g}_2$	$-\mathbf{g}_1$	$\mathbf{g}_3$	$\mathbf{g}_1-\mathbf{g}_3$	$\mathbf{g}_1$	$\mathbf{g}_1-\mathbf{g}_2$
C_{2z}	$-\mathbf{g}_1$	$-\mathbf{g}_2$	$\mathbf{g}_3$	$\mathbf{g}_2-\mathbf{g}_3$	$\mathbf{g}_1-\mathbf{g}_3$	$-\mathbf{g}_3$
C_{4z}^-	$-\mathbf{g}_2$	$\mathbf{g}_1$	$\mathbf{g}_3$	$\mathbf{g}_2$	$\mathbf{g}_2-\mathbf{g}_3$	$-\mathbf{g}_1+\mathbf{g}_2$
C_{2x}	$\mathbf{g}_1$	$-\mathbf{g}_2$	$-\mathbf{g}_3$	$-\mathbf{g}_1$	$-\mathbf{g}_1+\mathbf{g}_3$	$-\mathbf{g}_1+\mathbf{g}_2$
C_{2y}	$-\mathbf{g}_1$	$\mathbf{g}_2$	$-\mathbf{g}_3$	$-\mathbf{g}_2+\mathbf{g}_3$	$-\mathbf{g}_2$	$\mathbf{g}_1-\mathbf{g}_2$
C_{2a}	$\mathbf{g}_2$	$\mathbf{g}_1$	$-\mathbf{g}_3$	$-\mathbf{g}_1+\mathbf{g}_3$	$-\mathbf{g}_2+\mathbf{g}_3$	$\mathbf{g}_3$
C_{2b}	$-\mathbf{g}_2$	$-\mathbf{g}_1$	$-\mathbf{g}_3$	$-\mathbf{g}_2$	$-\mathbf{g}_1$	$-\mathbf{g}_3$

Trigonal and hexagonal

	Γ_{rh}			Γ_h		
E	$\mathbf{g}_1$	$\mathbf{g}_2$	$\mathbf{g}_3$	$\mathbf{g}_1$	$\mathbf{g}_2$	$\mathbf{g}_3$
C_6^+	—	—	—	$\mathbf{g}_2$	$-\mathbf{g}_1+\mathbf{g}_2$	$\mathbf{g}_3$
C_3^+	$\mathbf{g}_2$	$\mathbf{g}_3$	$\mathbf{g}_1$	$-\mathbf{g}_1+\mathbf{g}_2$	$-\mathbf{g}_1$	$\mathbf{g}_3$
C_2	—	—	—	$-\mathbf{g}_1$	$-\mathbf{g}_2$	$\mathbf{g}_3$
C_3^-	$\mathbf{g}_3$	$\mathbf{g}_1$	$\mathbf{g}_2$	$-\mathbf{g}_2$	$\mathbf{g}_1-\mathbf{g}_2$	$\mathbf{g}_3$
C_6^-	—	—	—	$\mathbf{g}_1-\mathbf{g}_2$	$\mathbf{g}_1$	$\mathbf{g}_3$
C'_{21}	$-\mathbf{g}_1$	$-\mathbf{g}_3$	$-\mathbf{g}_2$	$-\mathbf{g}_1+\mathbf{g}_2$	$\mathbf{g}_2$	$-\mathbf{g}_3$
C'_{22}	$-\mathbf{g}_3$	$-\mathbf{g}_2$	$-\mathbf{g}_1$	$\mathbf{g}_1$	$\mathbf{g}_1-\mathbf{g}_2$	$-\mathbf{g}_3$
C'_{23}	$-\mathbf{g}_2$	$-\mathbf{g}_1$	$-\mathbf{g}_3$	$-\mathbf{g}_2$	$-\mathbf{g}_1$	$-\mathbf{g}_3$
C''_{21}	—	—	—	$\mathbf{g}_1-\mathbf{g}_2$	$-\mathbf{g}_2$	$-\mathbf{g}_3$
C''_{22}	—	—	—	$-\mathbf{g}_1$	$-\mathbf{g}_1+\mathbf{g}_2$	$-\mathbf{g}_3$
C''_{23}	—	—	—	$\mathbf{g}_2$	$\mathbf{g}_1$	$-\mathbf{g}_3$

Cubic

	Γ_c			Γ_c^f			Γ_c^v		
E	$\mathbf{g}_1$	$\mathbf{g}_2$	$\mathbf{g}_3$	$\mathbf{g}_1$	$\mathbf{g}_2$	$\mathbf{g}_3$	$\mathbf{g}_1$	$\mathbf{g}_2$	$\mathbf{g}_3$
C_{2x}	$\mathbf{g}_1$	$-\mathbf{g}_2$	$-\mathbf{g}_3$	$-\mathbf{g}_1-\mathbf{g}_2-\mathbf{g}_3$	$\mathbf{g}_3$	$\mathbf{g}_2$	$-\mathbf{g}_1$	$-\mathbf{g}_1+\mathbf{g}_3$	$-\mathbf{g}_1+\mathbf{g}_2$
C_{2y}	$-\mathbf{g}_1$	$\mathbf{g}_2$	$-\mathbf{g}_3$	$\mathbf{g}_3$	$-\mathbf{g}_1-\mathbf{g}_2-\mathbf{g}_3$	$\mathbf{g}_1$	$-\mathbf{g}_2+\mathbf{g}_3$	$-\mathbf{g}_2$	$\mathbf{g}_1-\mathbf{g}_2$
C_{2z}	$-\mathbf{g}_1$	$-\mathbf{g}_2$	$\mathbf{g}_3$	$\mathbf{g}_2$	$\mathbf{g}_1$	$-\mathbf{g}_1-\mathbf{g}_2-\mathbf{g}_3$	$\mathbf{g}_2-\mathbf{g}_3$	$\mathbf{g}_1-\mathbf{g}_3$	$-\mathbf{g}_3$
C_{31}^+	$\mathbf{g}_2$	$\mathbf{g}_3$	$\mathbf{g}_1$	$\mathbf{g}_2$	$\mathbf{g}_3$	$\mathbf{g}_1$	$\mathbf{g}_2$	$\mathbf{g}_3$	$\mathbf{g}_1$
C_{32}^+	$\mathbf{g}_2$	$-\mathbf{g}_3$	$-\mathbf{g}_1$	$-\mathbf{g}_1-\mathbf{g}_2-\mathbf{g}_3$	$\mathbf{g}_1$	$\mathbf{g}_3$	$-\mathbf{g}_2$	$\mathbf{g}_1-\mathbf{g}_2$	$-\mathbf{g}_2+\mathbf{g}_3$
C_{33}^+	$-\mathbf{g}_2$	$\mathbf{g}_3$	$-\mathbf{g}_1$	$\mathbf{g}_1$	$-\mathbf{g}_1-\mathbf{g}_2-\mathbf{g}_3$	$\mathbf{g}_2$	$\mathbf{g}_1-\mathbf{g}_3$	$-\mathbf{g}_3$	$\mathbf{g}_2-\mathbf{g}_3$
C_{34}^+	$-\mathbf{g}_2$	$-\mathbf{g}_3$	$\mathbf{g}_1$	$\mathbf{g}_3$	$\mathbf{g}_2$	$-\mathbf{g}_1-\mathbf{g}_2-\mathbf{g}_3$	$-\mathbf{g}_1+\mathbf{g}_3$	$-\mathbf{g}_1+\mathbf{g}_2$	$-\mathbf{g}_1$
C_{31}^-	$\mathbf{g}_3$	$\mathbf{g}_1$	$\mathbf{g}_2$	$\mathbf{g}_3$	$\mathbf{g}_1$	$\mathbf{g}_2$	$\mathbf{g}_3$	$\mathbf{g}_1$	$\mathbf{g}_2$

Cubic—(continued)

	Γ_c			Γ_c^f			Γ_c^v		
C_{32}^-	$-\mathbf{g}_3$	$\mathbf{g}_1$	$-\mathbf{g}_2$	$\mathbf{g}_2$	$-\mathbf{g}_1-\mathbf{g}_2-\mathbf{g}_3$	$\mathbf{g}_3$	$-\mathbf{g}_1+\mathbf{g}_2$	$-\mathbf{g}_1$	$-\mathbf{g}_1+\mathbf{g}_3$
C_{33}^-	$-\mathbf{g}_3$	$-\mathbf{g}_1$	$\mathbf{g}_2$	$\mathbf{g}_1$	$\mathbf{g}_3$	$-\mathbf{g}_1-\mathbf{g}_2-\mathbf{g}_3$	$\mathbf{g}_1-\mathbf{g}_2$	$-\mathbf{g}_2+\mathbf{g}_3$	$-\mathbf{g}_2$
C_{34}^-	$\mathbf{g}_3$	$-\mathbf{g}_1$	$-\mathbf{g}_2$	$-\mathbf{g}_1-\mathbf{g}_2-\mathbf{g}_3$	$\mathbf{g}_2$	$\mathbf{g}_1$	$-\mathbf{g}_3$	$\mathbf{g}_2-\mathbf{g}_3$	$\mathbf{g}_1-\mathbf{g}_3$
C_{4x}^+	$\mathbf{g}_1$	$\mathbf{g}_3$	$-\mathbf{g}_2$	$-\mathbf{g}_3$	$-\mathbf{g}_1$	$\mathbf{g}_1+\mathbf{g}_2+\mathbf{g}_3$	$\mathbf{g}_2-\mathbf{g}_3$	$-\mathbf{g}_1+\mathbf{g}_2$	$\mathbf{g}_2$
C_{4y}^+	$-\mathbf{g}_3$	$\mathbf{g}_2$	$\mathbf{g}_1$	$\mathbf{g}_1+\mathbf{g}_2+\mathbf{g}_3$	$-\mathbf{g}_1$	$-\mathbf{g}_2$	$\mathbf{g}_3$	$-\mathbf{g}_1+\mathbf{g}_3$	$-\mathbf{g}_2+\mathbf{g}_3$
C_{4z}^+	$\mathbf{g}_2$	$-\mathbf{g}_1$	$\mathbf{g}_3$	$-\mathbf{g}_3$	$\mathbf{g}_1+\mathbf{g}_2+\mathbf{g}_3$	$-\mathbf{g}_2$	$\mathbf{g}_1-\mathbf{g}_3$	$\mathbf{g}_1$	$\mathbf{g}_1-\mathbf{g}_2$
C_{4x}^-	$\mathbf{g}_1$	$-\mathbf{g}_3$	$\mathbf{g}_2$	$-\mathbf{g}_2$	$\mathbf{g}_1+\mathbf{g}_2+\mathbf{g}_3$	$-\mathbf{g}_1$	$-\mathbf{g}_2+\mathbf{g}_3$	$\mathbf{g}_3$	$-\mathbf{g}_1+\mathbf{g}_3$
C_{4y}^-	$\mathbf{g}_3$	$\mathbf{g}_2$	$-\mathbf{g}_1$	$-\mathbf{g}_2$	$-\mathbf{g}_3$	$\mathbf{g}_1+\mathbf{g}_2+\mathbf{g}_3$	$\mathbf{g}_1-\mathbf{g}_2$	$\mathbf{g}_1-\mathbf{g}_3$	$\mathbf{g}_1$
C_{4z}^-	$-\mathbf{g}_2$	$\mathbf{g}_1$	$\mathbf{g}_3$	$\mathbf{g}_1+\mathbf{g}_2+\mathbf{g}_3$	$-\mathbf{g}_3$	$-\mathbf{g}_1$	$\mathbf{g}_2$	$\mathbf{g}_2-\mathbf{g}_3$	$-\mathbf{g}_1+\mathbf{g}_2$
C_{2a}	$\mathbf{g}_2$	$\mathbf{g}_1$	$-\mathbf{g}_3$	$-\mathbf{g}_1$	$-\mathbf{g}_2$	$\mathbf{g}_1+\mathbf{g}_2+\mathbf{g}_3$	$-\mathbf{g}_1+\mathbf{g}_3$	$-\mathbf{g}_2+\mathbf{g}_3$	$\mathbf{g}_3$
C_{2b}	$-\mathbf{g}_2$	$-\mathbf{g}_1$	$-\mathbf{g}_3$	$-\mathbf{g}_2$	$-\mathbf{g}_1$	$-\mathbf{g}_3$	$-\mathbf{g}_2$	$-\mathbf{g}_1$	$-\mathbf{g}_3$
C_{2c}	$\mathbf{g}_3$	$-\mathbf{g}_2$	$\mathbf{g}_1$	$-\mathbf{g}_1$	$\mathbf{g}_1+\mathbf{g}_2+\mathbf{g}_3$	$-\mathbf{g}_3$	$-\mathbf{g}_1+\mathbf{g}_2$	$\mathbf{g}_2$	$\mathbf{g}_2-\mathbf{g}_3$
C_{2d}	$-\mathbf{g}_1$	$\mathbf{g}_3$	$\mathbf{g}_2$	$\mathbf{g}_1+\mathbf{g}_2+\mathbf{g}_3$	$-\mathbf{g}_2$	$-\mathbf{g}_3$	$\mathbf{g}_1$	$\mathbf{g}_1-\mathbf{g}_2$	$\mathbf{g}_1-\mathbf{g}_3$
C_{2e}	$-\mathbf{g}_3$	$-\mathbf{g}_2$	$-\mathbf{g}_1$	$-\mathbf{g}_3$	$-\mathbf{g}_2$	$-\mathbf{g}_1$	$-\mathbf{g}_3$	$-\mathbf{g}_2$	$-\mathbf{g}_1$
C_{2f}	$-\mathbf{g}_1$	$-\mathbf{g}_3$	$-\mathbf{g}_2$	$-\mathbf{g}_1$	$-\mathbf{g}_3$	$-\mathbf{g}_2$	$-\mathbf{g}_1$	$-\mathbf{g}_3$	$-\mathbf{g}_2$

Notes to Table 3.4

(i) To save space only the effects of the elements of the holosymmetric point groups of the crystal systems are tabulated. Since $I\mathbf{g} = -\mathbf{g}$ for all $\mathbf{g}$ the action of the inversion is also omitted.

(ii) The point-group operations are active.

(iii) The basic translations $\mathbf{g}_1$, $\mathbf{g}_2$, and $\mathbf{g}_3$ of the reciprocal lattices are exactly as in Table 3.3.

to both sides of eqn. (3.2.2). The results are given in Table 3.4. The transformation of any reciprocal lattice vector $\mathbf{g}$ under R can then be found by applying R to eqn. (3.2.5) and using Table 3.4.

We now proceed to the consideration of Brillouin zones (Brillouin 1930*a–c*, 1931, 1953). The *first Brillouin zone* is essentially *a unit cell of the reciprocal lattice* of a crystal; this is a necessary condition if the Brillouin zone is to serve any useful purpose. The particular unit cell that is conventionally chosen is the Wigner–Seitz unit cell (see section 1.5) of the reciprocal lattice (Jones 1960, Slater 1965*b*). In this case a Brillouin zone is defined as follows:

DEFINITION 3.2.1. The set of all $\mathbf{k}$ vectors with one of the reciprocal lattice points as origin and having the property that no vector of shorter length can be reached from any of them by adding translation vectors of the reciprocal lattice forms what is called the *first Brillouin zone.*

In practice we shall not use this definition for triclinic and monoclinic space groups (see Definition 3.2.2 below).

To construct the first Brillouin zone using this definition we then choose one particular reciprocal-lattice point as origin and construct the vectors $\mathbf{g}$ joining it to the other reciprocal-lattice points. If the planes that are the perpendicular bisectors of the reciprocal-lattice vectors, $\mathbf{g}$, are drawn, then the first Brillouin zone is the smallest volume of space, surrounding the origin, that is enclosed by these planes. The equation of the plane that is the perpendicular bisector of $\mathbf{g}$ is

$$\mathbf{k}\cdot\mathbf{g} = \tfrac{1}{2}\,|\mathbf{g}|^2. \tag{3.2.6}$$

Each face of the first Brillouin zone is thus characterized by a reciprocal-lattice vector $\mathbf{g}$ and is, of course, the perpendicular bisector of $\mathbf{g}$. The reason for the choice of the Wigner–Seitz unit cell, rather than some other unit cell, of reciprocal space is because this unit cell exhibits the point-group symmetry of the reciprocal lattice (see, for example, Delaunay (1932)). However, for crystals belonging to crystal systems of low symmetry the application of eqn. (3.2.6) in the geometrical construction of the Brillouin zone is exceedingly tedious.

Another unit cell that is worthy of some consideration in this context is the primitive unit cell (see section 1.5) which is the parallelepiped that is centred at $\mathbf{k} = 0$ and has edges parallel to and equal in magnitude to $\mathbf{g}_1$, $\mathbf{g}_2$, and $\mathbf{g}_3$, where $\mathbf{g}_1$, $\mathbf{g}_2$, and $\mathbf{g}_3$ are the basic vectors of the reciprocal lattice. The volume of the first Brillouin zone is thus given by

$$\mathbf{g}_1\cdot(\mathbf{g}_2 \times \mathbf{g}_3) = \frac{8\pi^3}{V}, \tag{3.2.7}$$

where V is the volume of the primitive unit cell of the corresponding direct lattice; the value of $\mathbf{g}_1\cdot(\mathbf{g}_2 \times \mathbf{g}_3)$ is given for each reciprocal lattice in Table 3.3. For some reciprocal lattices (those of orthorhombic P, tetragonal P, and cubic P space groups) a Brillouin zone defined by using the primitive unit cell of reciprocal space would be identical with the conventional Brillouin zone defined by using the Wigner–Seitz unit cell of reciprocal space. For some other reciprocal lattices (those of orthorhombic I, tetragonal I, hexagonal P, cubic F, and cubic I space groups) the primitive unit cell does not exhibit the point-group symmetry of the Bravais lattice in the same

TABLE 3.5

In column 2 we indicate whether the Brillouin zone defined by the primitive unit cell of reciprocal space possesses, in the correct orientation, the point-group symmetry of the crystal. In column 3 we indicate those Bravais lattices that satisfy the previous condition *and* for which the Wigner–Seitz and primitive unit cells of reciprocal space lead to the same shape for the Brillouin zone.

Bravais lattice of crystal	2	3	Bravais lattice of crystal	2	3
Triclinic P	✓	×	Tetragonal P	✓	✓
Monoclinic P	✓	×	Tetragonal I	×	—
Monoclinic B	✓	×	Trigonal R	✓	×
Orthorhombic P	✓	✓	Hexagonal P	×	—
Orthorhombic C	✓	×	Cubic P	✓	✓
Orthorhombic I	×	—	Cubic F	×	—
Orthorhombic F	✓	×	Cubic I	×	—

orientation as in the real space lattice. In these cases the primitive unit cell would therefore not be acceptable as defining a Brillouin zone, see column 2 of Table 3.5. These eight Bravais lattices include nearly all the crystals whose Brillouin zones have been studied in practice and it is therefore not surprising that the choice of the Wigner–Seitz unit cell of reciprocal space has usually been taken to be a necessary part of the definition of the first Brillouin zone. However, for the remaining six Bravais lattices (those of triclinic *P*, monoclinic *P*, monoclinic *B*, orthorhombic *C*, orthorhombic *F*, and trigonal *R* space groups) one can establish by inspection that the primitive unit cell of the reciprocal lattice also possesses the point-group symmetry of the Bravais lattice of the crystal in the correct orientation. However, it is desirable, whenever possible, that surfaces of the Brillouin zone should be parallel to planes of symmetry in the crystal. This means that for orthorhombic *C* and *F* space groups it is desirable to retain the Wigner–Seitz cell in defining the Brillouin zone, see Figs. 3.6 and 3.8. Also, for the trigonal *R* lattice there seems to be no particularly strong argument in favour of replacing the conventional definition of the Brillouin zone by the fundamental unit cell of reciprocal space. However, for the three remaining lattices (triclinic *P*, monoclinic *P*, and monoclinic *B* space groups), as we have already mentioned, eqn. (3.2.6) is very difficult to apply in practice, both to draw the shape of the Brillouin zone and then, having drawn it, to visualize it. The details of the shape depend very much on the actual values of the parameters of the unit cell of the crystal in question. For monoclinic crystals some authors (Koster 1957, Sushkevich 1966) illustrate one particular example for each Bravais lattice, Luehrmann (1968) shows two examples for the monoclinic *B* lattice, while the comprehensive tables of Kovalev (1965) contain no diagrams at all of the Brillouin zones. The use of the fundamental unit cell in defining the Brillouin zone for a monoclinic or triclinic crystal would therefore have the advantage that the general shape, i.e. the numbers of edges and faces, would be independent of the axial ratios and inter-axial angles of the unit cell of the crystal.

We therefore suggest that, *while the Wigner–Seitz unit cell of reciprocal space is retained for defining the Brillouin zones of most crystals it should be replaced by the use of the primitive unit cell of reciprocal space for monoclinic and triclinic space groups* (Bradley and Cracknell 1970). We have followed this procedure in constructing Figs. 3.2–3.4.

DEFINITION 3.2.2. For a monoclinic or triclinic crystal the *first Brillouin zone* is the primitive unit cell of the reciprocal lattice, that is, the parallelepiped that is centred at $\mathbf{k} = 0$ and has edges parallel to and equal in magnitude to $\mathbf{g}_1$, $\mathbf{g}_2$, and $\mathbf{g}_3$, where $\mathbf{g}_1$, $\mathbf{g}_2$, and $\mathbf{g}_3$ are the basic vectors of the reciprocal lattice.

One final point should be mentioned. Equation (3.2.6) is sometimes said to be specially related to the appearance of discontinuities in the curves of the energy of an electron, phonon, or magnon, $E(\mathbf{k})$, against $\mathbf{k}$. This is, in fact, untrue because, both in the extended zone scheme and in the repeated zone scheme, $E(\mathbf{k})$ is a quasi-continuous function of $\mathbf{k}$ everywhere in reciprocal space (Koopmans' theorem). The discontinuities that appear when zone boundaries are defined are simply the separation of different curves $E(\mathbf{k})$. It is, of course, true that one can sometimes make a careful choice of zone boundary, so that $E(\mathbf{k})$ exhibits special degeneracies at the zone boundary; however, this property is a result of the requirement that the Brillouin zone should exhibit the correct point-group symmetry of the Bravais lattice, which we have not abandoned, rather than a result of the use of the Wigner–Seitz cell and eqn. (3.2.6).

In constructing the first Brillouin zone using eqn. (3.2.6) the reciprocal lattice vectors $\mathbf{g}$ will be the vectors joining the origin to its first nearest neighbours and, possibly, its second or higher nearest neighbours. It is possible to define the second Brillouin zone by using further planes defined by eqn. (3.2.6) to enclose a volume outside the first Brillouin zone that satisfies the following three conditions. (i) The volume of the second Brillouin zone is equal to the volume of the first Brillouin zone. (ii) For every $\mathbf{k}$ vector, $\mathbf{k}_1$, in the first Brillouin zone there is exactly one $\mathbf{k}$ vector, $\mathbf{k}_2$, in the second Brillouin zone such that

$$\mathbf{k}_2 - \mathbf{k}_1 = \mathbf{g}, \tag{3.2.8}$$

where $\mathbf{g}$ is a reciprocal lattice vector but is not necessarily the same for all $\mathbf{k}_1$. Every point in the second Brillouin zone is thus *equivalent*, in a sense to be defined rather more carefully in section 3.3, to some point in the first Brillouin zone. (iii) The $\mathbf{k}$ vectors in the second Brillouin zone must be those of shortest length that satisfy conditions (i) and (ii). It is possible similarly to define third, fourth, and higher Brillouin zones.

In defining the Brillouin zone of any given crystal, care must be taken to ensure that one uses the Bravais lattice describing the full translational symmetry of the crystal and not some lattice of larger unit cell (Barron and Fischer 1959). Once the full Bravais lattice has been correctly identified the reciprocal lattice and Brillouin zone structure are determined uniquely; they are then completely independent of the detailed configuration of the atoms within the unit cell. The Brillouin zones as defined above have the following properties:

(i) They contain one state for each primitive unit cell of the Bravais lattice in the crystal (for electrons the Pauli exclusion principle allows two electrons, one with spin $+\frac{1}{2}$ and the other with spin $-\frac{1}{2}$, to occupy each of these states; for phonons or magnons there is, of course, no such exclusion principle) (see eqn. (3.4.5)).

(ii) The energy of a particle or quasi-particle in the crystal is continuous throughout each zone.

(iii) The pieces of each higher zone can be transferred into the first zone by moving the separate regions by appropriate reciprocal lattice vectors, and the states of each zone then form a continuous energy band over the first zone; this is called the *reduced zone scheme*.

(iv) Cells of the shape of the first Brillouin zone can be fitted together to fill reciprocal space completely, i.e. leaving no empty spaces between neighbouring cells. This is called the *repeated* zone scheme and is often useful.

An alternative method of defining zones in reciprocal space (Jones 1934*a*) is based on saying that the only true zone boundaries are those planes that have a non-vanishing structure-factor associated with the corresponding Bragg reflection. These *Jones' zones* have been discussed in many standard textbooks (for example Jones (1960), Kittel (1956), Mott and Jones (1936), Wilson (1953)), but they appear to be used very little now.

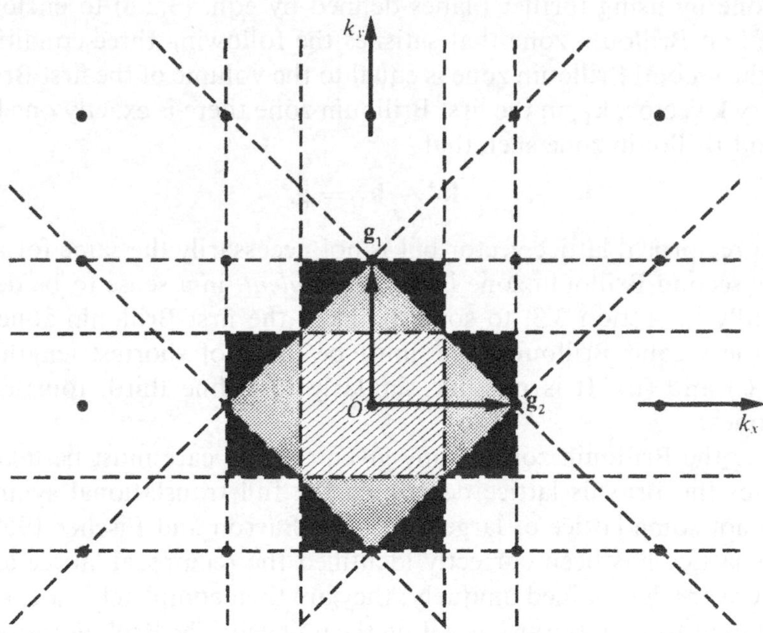

FIG. 3.1. The first three Brillouin zones for the square 2-dimensional Bravais lattice, *p*.

We can illustrate the idea of first, second, and higher order Brillouin zones by a 2-dimensional example. The basic vectors of the 2-dimensional square Bravais lattice, *p*, may be taken as

$$\left.\begin{aligned} \mathbf{t}_1 &= a\mathbf{i} \\ \mathbf{t}_2 &= a\mathbf{j} \end{aligned}\right\} \qquad (3.2.9)$$

where **i** and **j** are unit vectors in the x and y directions. The reciprocal lattice vectors $\mathbf{g}_1$ and $\mathbf{g}_2$ that satisfy equation (3.2.1) are thus

$$\left.\begin{aligned} \mathbf{g}_1 &= \frac{2\pi}{a}\mathbf{j} \\ \mathbf{g}_2 &= \frac{2\pi}{a}\mathbf{i}. \end{aligned}\right\} \tag{3.2.10}$$

The first Brillouin zone is the region between the lines $k_x = \pm\frac{1}{2}|\mathbf{g}_2| = \pm\pi/a$ and $k_y = \pm\frac{1}{2}|\mathbf{g}_1| = \pm\pi/a$, that is, the vectors **g** that go into eqn. (3.2.6) are $+\mathbf{g}_1$, $-\mathbf{g}_1$, $+\mathbf{g}_2$, and $-\mathbf{g}_2$. The first Brillouin zone is thus the square that is shaded in Fig. 3.1. The vectors that go into eqn. (3.2.6) for the second Brillouin zone are $\mathbf{g}_1 + \mathbf{g}_2$, $-\mathbf{g}_1 + \mathbf{g}_2$, $\mathbf{g}_1 - \mathbf{g}_2$, and $-\mathbf{g}_1 - \mathbf{g}_2$ and the second Brillouin zone is thus the region marked with dots in Fig. 3.1. In fact, however, we only use the first Brillouin zone so that from now on we shall omit the adjective 'first'.

According to Bouckaert, Smoluchowski, and Wigner (1936) the existence of these zones was noticed by various authors almost simultaneously (Bloch 1928, Morse 1930, Peierls 1930, Strutt 1928*a*, *b*) and it was Brillouin (1930*a*–*c* 1931) who pointed out their connection with X-ray reflection. A textbook on Brillouin zones has been written by Jones (1960).

3.3. The classification of points and lines of symmetry

For some lattices the shape of the Brillouin zone is unique, while for others there are two or more possible shapes depending on the relative sizes of the basic translations and the angles between them. The various Brillouin zones for the 14 Bravais lattices in direct space are shown in Figs. 3.2–3.15.

Two vectors $\mathbf{k}_1$ and $\mathbf{k}_2$ are said to be *equivalent* if $(\mathbf{k}_1 - \mathbf{k}_2) = a_1\mathbf{g}_1 + a_2\mathbf{g}_2 + a_3\mathbf{g}_3$ where a_1, a_2, and a_3 are integers. Because of its construction no two interior points of a Brillouin zone can be equivalent; but a point on the surface of the zone may be equivalent to one or more other points also on the surface of the zone. For example, the mid-point of a face is always equivalent to the mid-point of the opposite face, because opposite faces are always separated by a reciprocal lattice vector.

Given a Brillouin zone and a point **k** of the zone there exist certain elements of **P**, the holosymmetric point group of the corresponding crystal system, which transform **k** into itself or into some equivalent **k** vector. These elements form a subgroup of **P** that we denote by **P(k)** and name the *symmetry group of* **k**. We can now give the definitions of a point of symmetry, a line of symmetry and a plane of symmetry.

DEFINITION 3.3.1. **k** is a *point of symmetry* if there exists a neighbourhood N of **k** in which no point except **k** has the symmetry group **P(k)**.

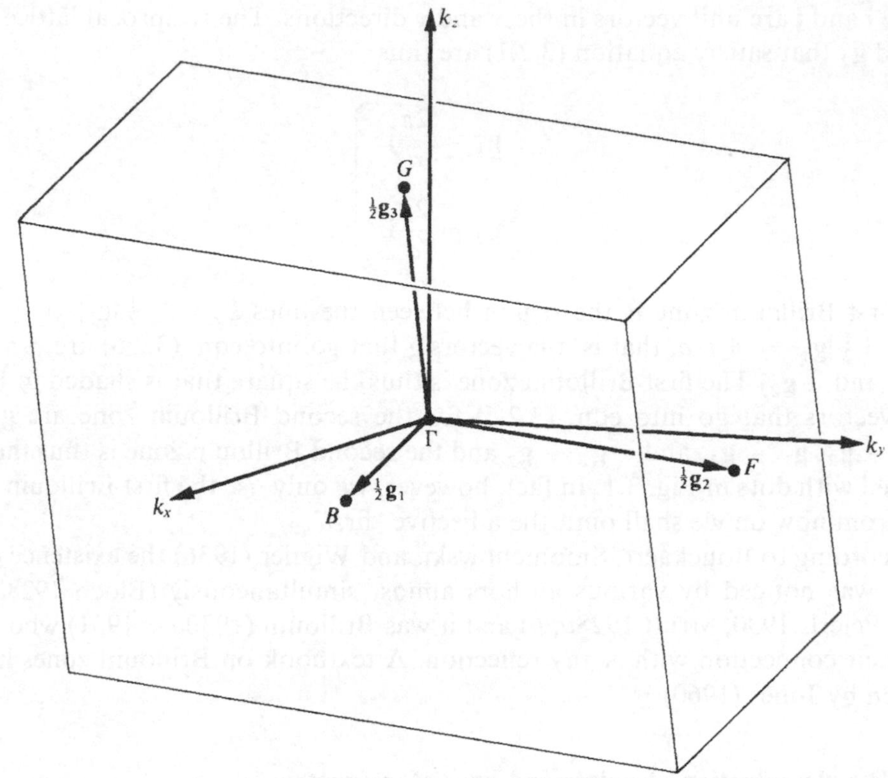

FIG. 3.2. The Brillouin zone for Γ_t. $\Gamma = (000)$; $B = (\frac{1}{2}00)$; $F = (0\frac{1}{2}0)$; $G = (00\frac{1}{2})$.

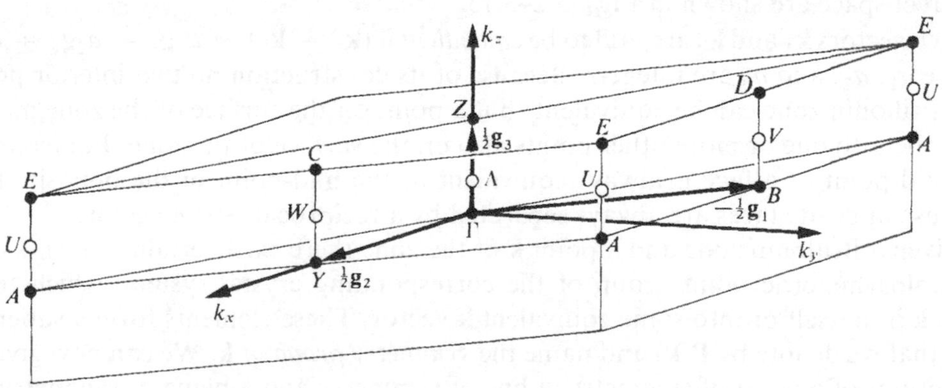

FIG. 3.3. The Brillouin zone for Γ_m. $\Gamma = (000)$; $B = (\bar{\frac{1}{2}}00)$; $Y = (0\frac{1}{2}0)$; $Z = (00\frac{1}{2})$; $C = (0\frac{1}{2}\frac{1}{2})$; $D = (\bar{\frac{1}{2}}0\frac{1}{2})$; $A = (\bar{\frac{1}{2}}\frac{1}{2}0)$, $(\bar{\frac{1}{2}}\bar{\frac{1}{2}}0)$, or $(\frac{1}{2}\frac{1}{2}0)$; $E = (\bar{\frac{1}{2}}\frac{1}{2}\frac{1}{2})$, $(\bar{\frac{1}{2}}\bar{\frac{1}{2}}\frac{1}{2})$, or $(\frac{1}{2}\frac{1}{2}\frac{1}{2})$.

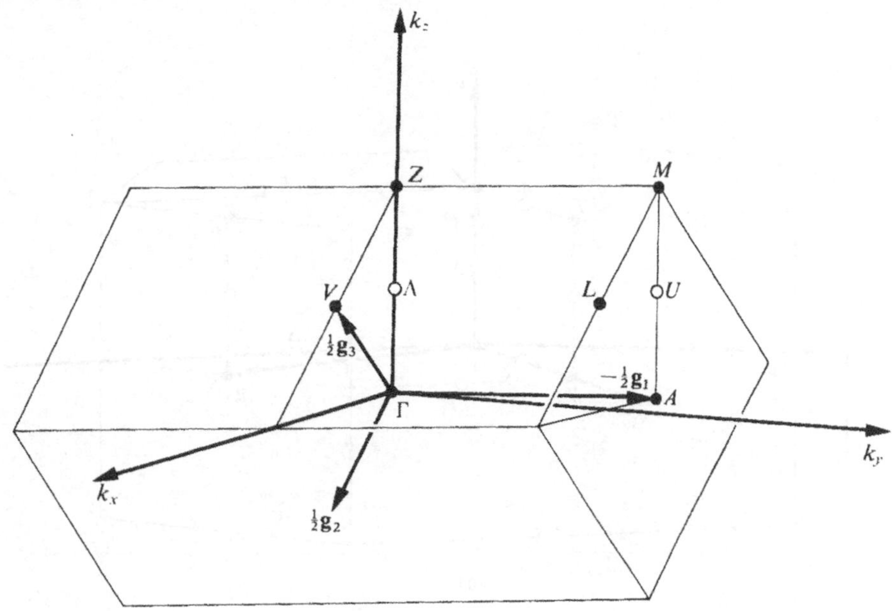

FIG. 3.4. The Brillouin zone for Γ_m^b. $\Gamma = (000)$; $A = (\bar{\tfrac{1}{2}}00)$; $Z = (0\bar{\tfrac{1}{2}}\tfrac{1}{2})$; $M = (\bar{\tfrac{1}{2}}\bar{\tfrac{1}{2}}\tfrac{1}{2})$; $L = (\bar{\tfrac{1}{2}}0\tfrac{1}{2})$; $V = (00\tfrac{1}{2})$.

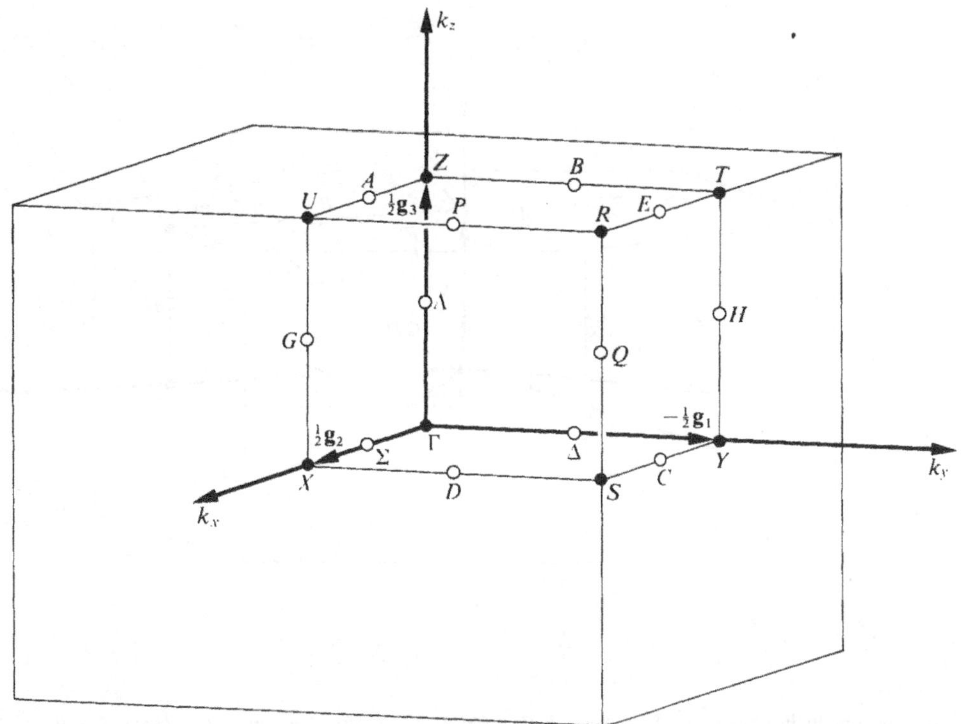

FIG. 3.5. The Brillouin zone for Γ_o. $\Gamma = (000)$; $Y = (\bar{\tfrac{1}{2}}00)$; $X = (0\tfrac{1}{2}0)$; $Z = (00\tfrac{1}{2})$; $U = (0\tfrac{1}{2}\tfrac{1}{2})$; $T = (\bar{\tfrac{1}{2}}0\tfrac{1}{2})$; $S = (\bar{\tfrac{1}{2}}\tfrac{1}{2}0)$; $R = (\bar{\tfrac{1}{2}}\tfrac{1}{2}\tfrac{1}{2})$.

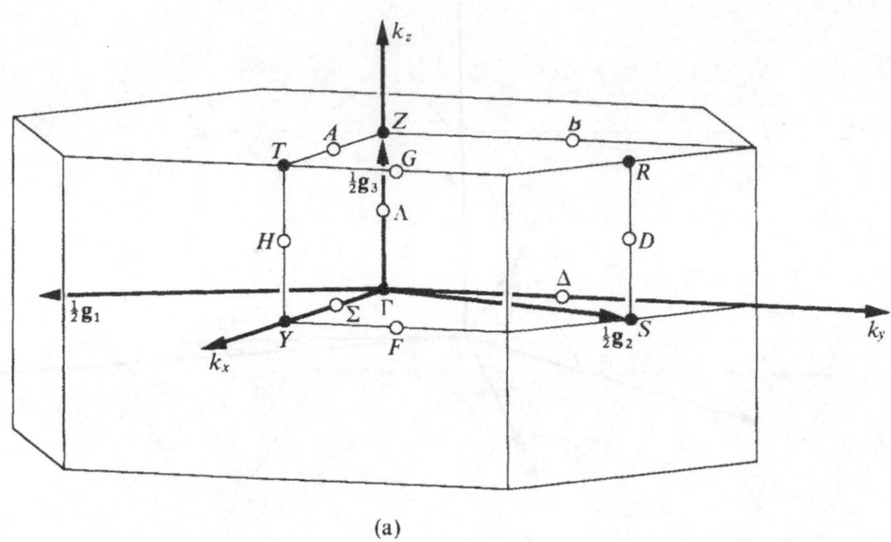

(a)

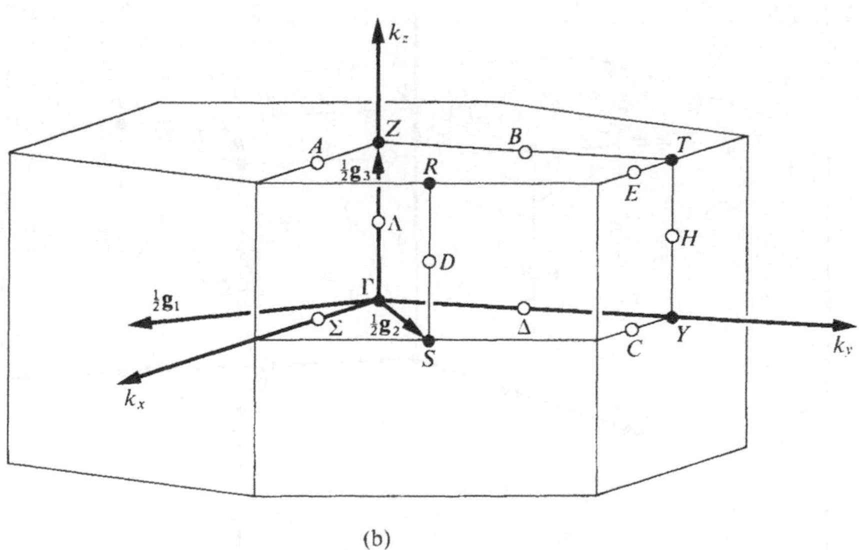

(b)

FIG. 3.6. The Brillouin zone for Γ_o^b. (a) $a > b$, $\Gamma = (000)$; $Y = (\frac{1}{2}\frac{1}{2}0)$; $Z = (00\frac{1}{2})$; $T = (\frac{1}{2}\frac{1}{2}\frac{1}{2})$; $S = (0\frac{1}{2}0)$; $R = (0\frac{1}{2}\frac{1}{2})$; (b) $b > a$, $\Gamma = (000)$; $Y = (\frac{\bar{1}}{2}\frac{1}{2}0)$; $Z = (00\frac{1}{2})$; $T = (\frac{\bar{1}}{2}\frac{1}{2}\frac{1}{2})$; $S = (0\frac{1}{2}0)$; $R = (0\frac{1}{2}\frac{1}{2})$.

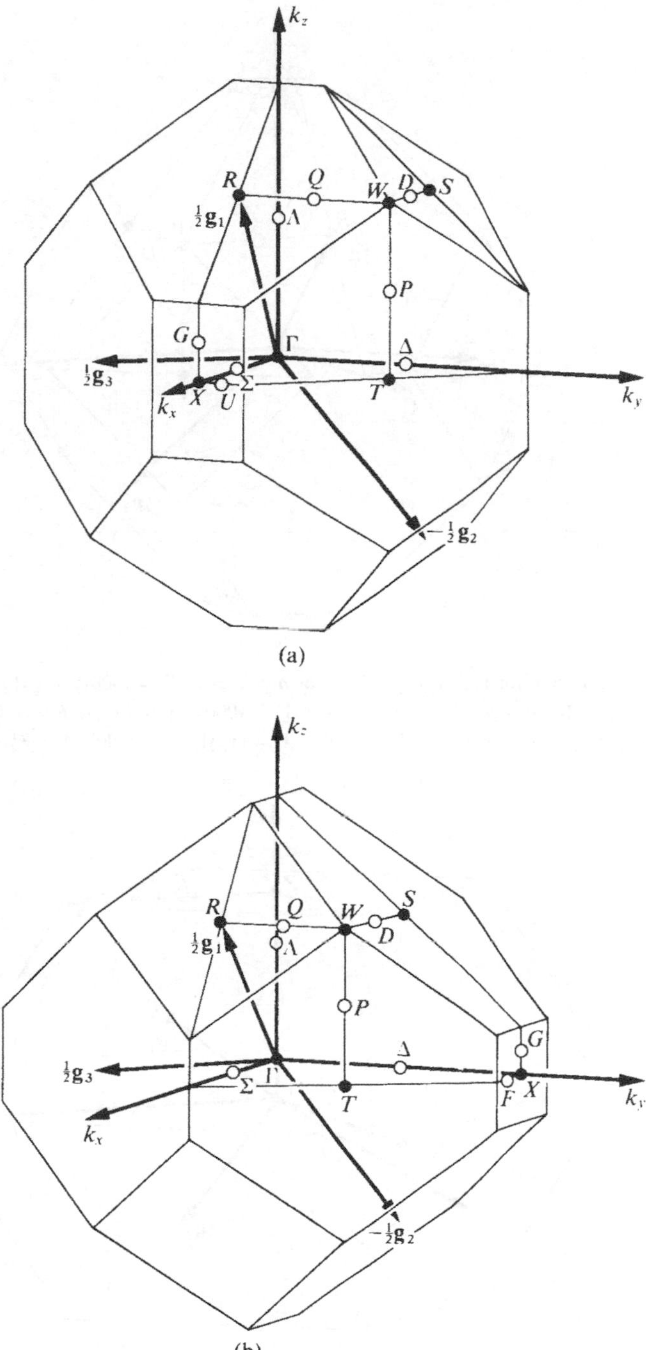

See page 100 for caption.

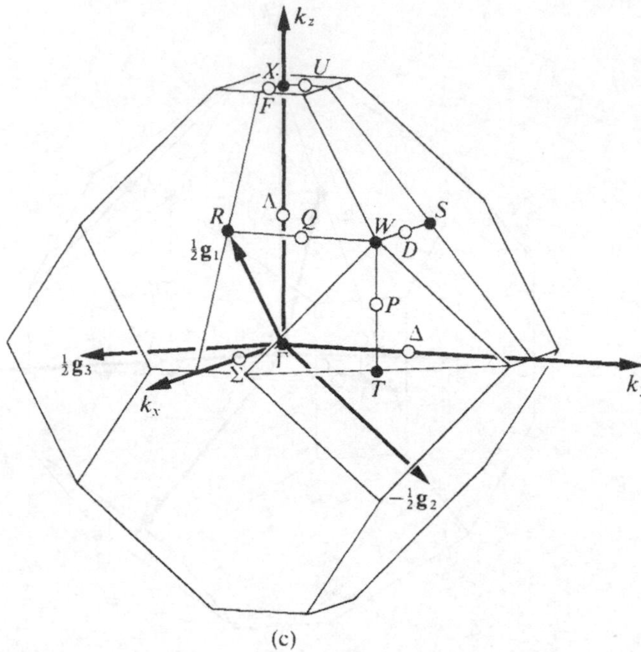

(c)

FIG. 3.7. The Brillouin zone for Γ_o^v. (a) $a > b > c$ or $a > c > b$, $\Gamma = (000)$; $X = (\frac{1}{2}\bar{\frac{1}{2}}\frac{1}{2})$; $R = (\frac{1}{2}00)$; $S = (\frac{1}{2}0\bar{\frac{1}{2}})$; $T = (\frac{1}{2}\bar{\frac{1}{2}}0)$; $W = (\frac{3}{4}\bar{\frac{1}{4}}\bar{\frac{1}{4}})$; (b) $b > a > c$ or $b > c > a$, $\Gamma = (000)$; $X = (\frac{1}{2}\bar{\frac{1}{2}}\bar{\frac{1}{2}})$; $R = (\frac{1}{2}00)$; $S = (\frac{1}{2}0\bar{\frac{1}{2}})$; $T = (\frac{1}{2}\bar{\frac{1}{2}}0)$; $W = (\frac{3}{4}\bar{\frac{1}{4}}\bar{\frac{1}{4}})$; (c) $c > b > a$ or $c > a > b$, $\Gamma = (000)$; $X = (\frac{1}{2}\frac{1}{2}\bar{\frac{1}{2}})$; $R = (\frac{1}{2}00)$; $S = (\frac{1}{2}0\bar{\frac{1}{2}})$; $T = (\frac{1}{2}\bar{\frac{1}{2}}0)$; $W = (\frac{3}{4}\bar{\frac{1}{4}}\bar{\frac{1}{4}})$.

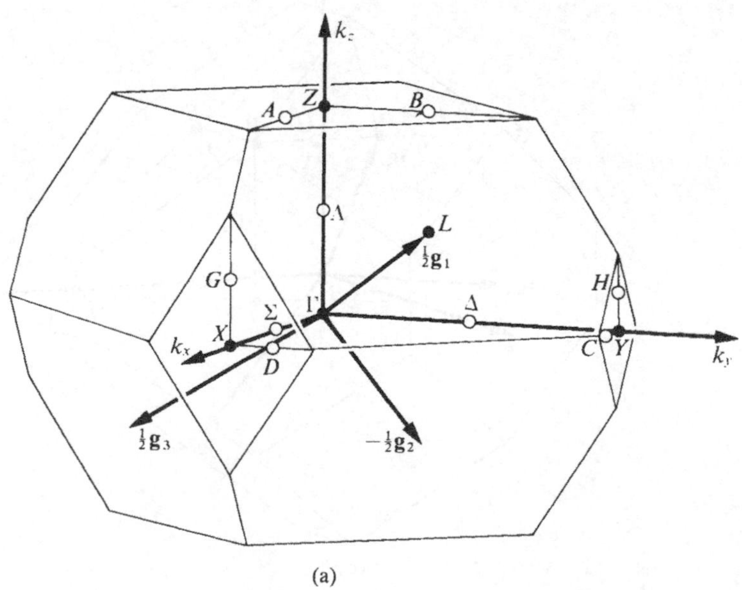

(a)

See page 103 for caption.

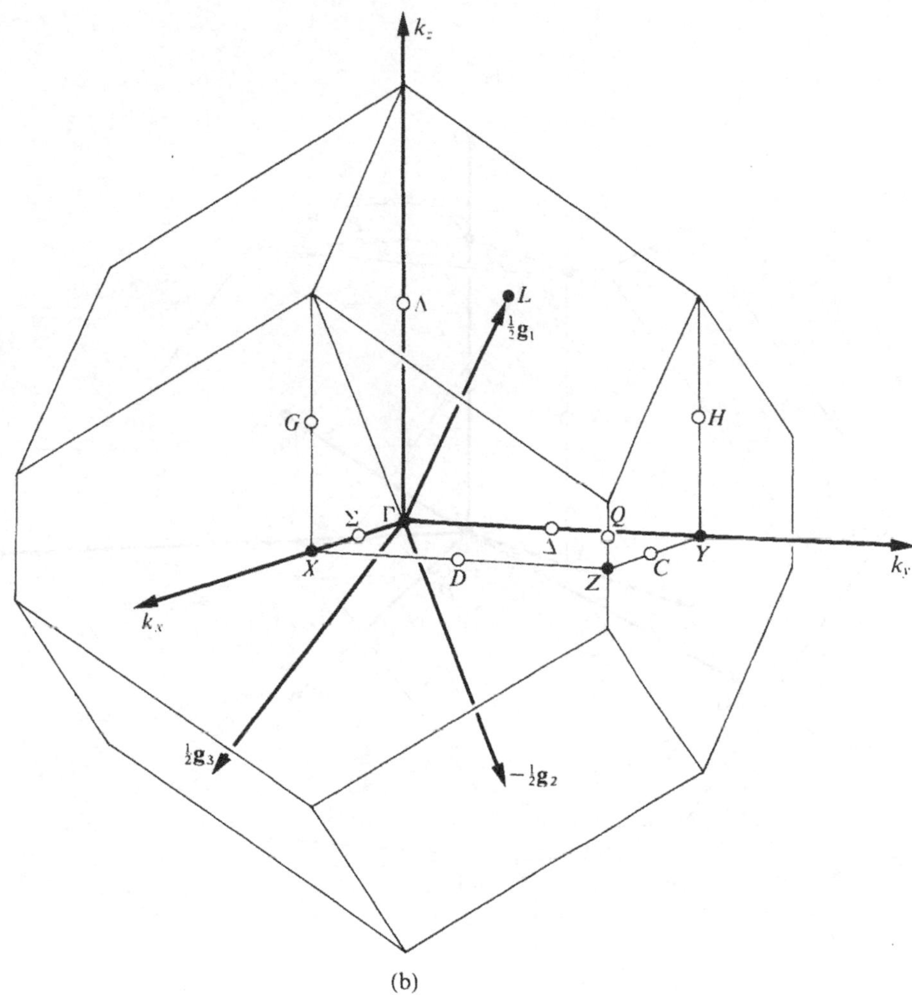

(b)

See page 103 for caption.

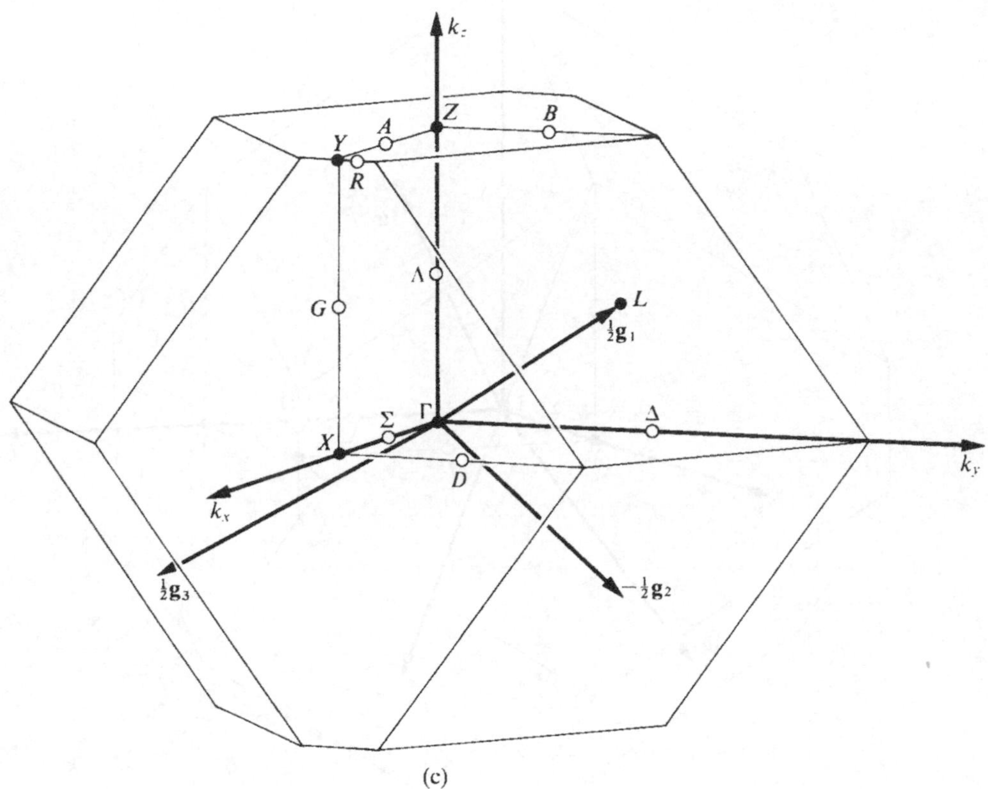

(c)

See page 103 for caption.

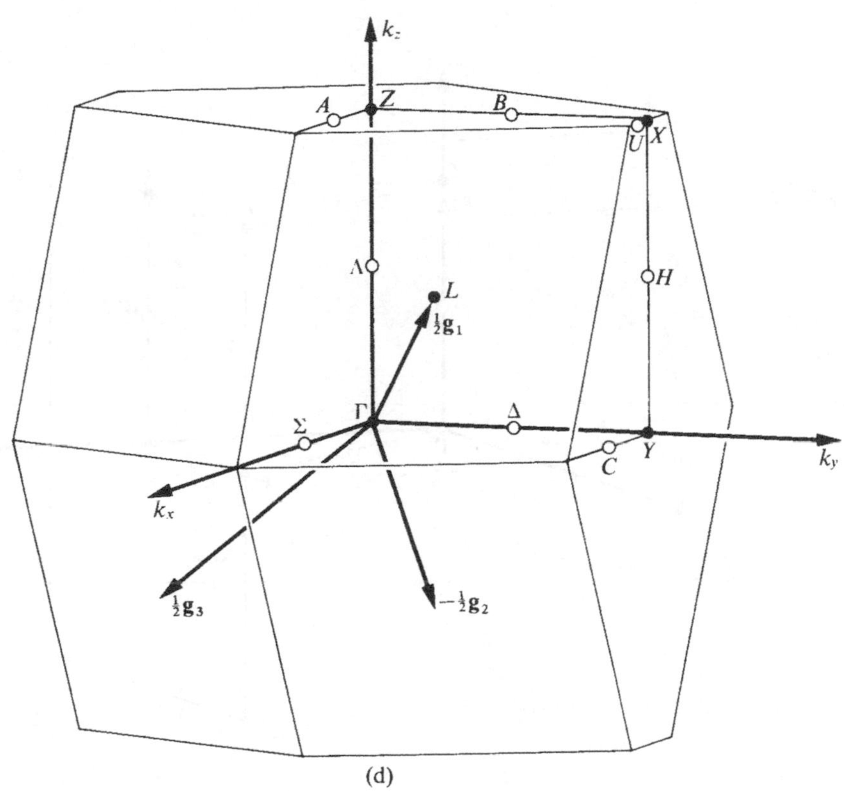

(d)

FIG. 3.8. The Brillouin zone for Γ_o^f. (a) $1/a^2 < 1/b^2 + 1/c^2$, $1/b^2 < 1/c^2 + 1/a^2$ and $1/c^2 < 1/a^2 + 1/b^2$, $\Gamma = (000)$; $Y = (0\overline{\tfrac{1}{2}\tfrac{1}{2}})$; $X = (\tfrac{1}{2}0\tfrac{1}{2})$; $Z = (\tfrac{1}{2}\tfrac{1}{2}0)$; $L = (\tfrac{1}{2}00)$; (b) $1/c^2 > 1/a^2 + 1/b^2$, $\Gamma = (000)$; $Y = (0\overline{\tfrac{1}{2}\tfrac{1}{2}})$; $X = (\tfrac{1}{2}0\tfrac{1}{2})$; $Z = (\tfrac{1}{2}\overline{\tfrac{1}{2}}0)$; $L = (\tfrac{1}{2}00)$; (c) $1/b^2 > 1/a^2 + 1/c^2$, $\Gamma = (000)$; $Y = (\overline{1}\tfrac{1}{2}\tfrac{1}{2})$; $X = (\tfrac{1}{2}0\tfrac{1}{2})$; $Z = (\tfrac{1}{2}\tfrac{1}{2}0)$; $L = (\tfrac{1}{2}00)$; (d) $1/a^2 > 1/b^2 + 1/c^2$, $\Gamma = (000)$; $Y = (0\overline{\tfrac{1}{2}\tfrac{1}{2}})$; $X = (\tfrac{1}{2}0\overline{\tfrac{1}{2}})$; $Z = (\tfrac{1}{2}\tfrac{1}{2}0)$; $L = (\tfrac{1}{2}00)$.

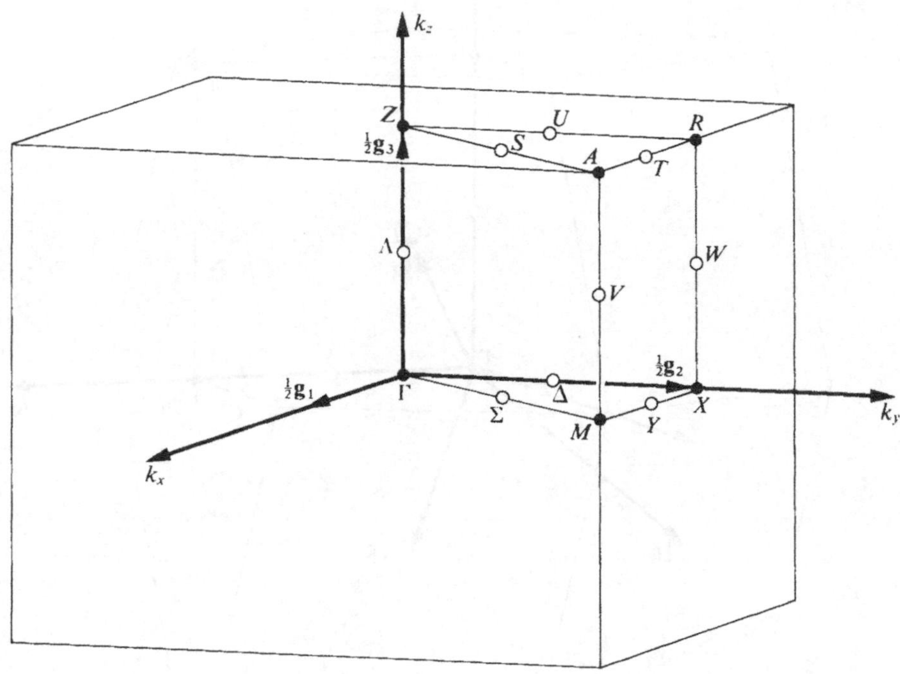

FIG. 3.9. The Brillouin zone for Γ_q. $\Gamma = (000)$; $M = (\frac{1}{2}\frac{1}{2}0)$; $Z = (00\frac{1}{2})$; $A = (\frac{1}{2}\frac{1}{2}\frac{1}{2})$; $R = (0\frac{1}{2}\frac{1}{2})$; $X = (0\frac{1}{2}0)$.

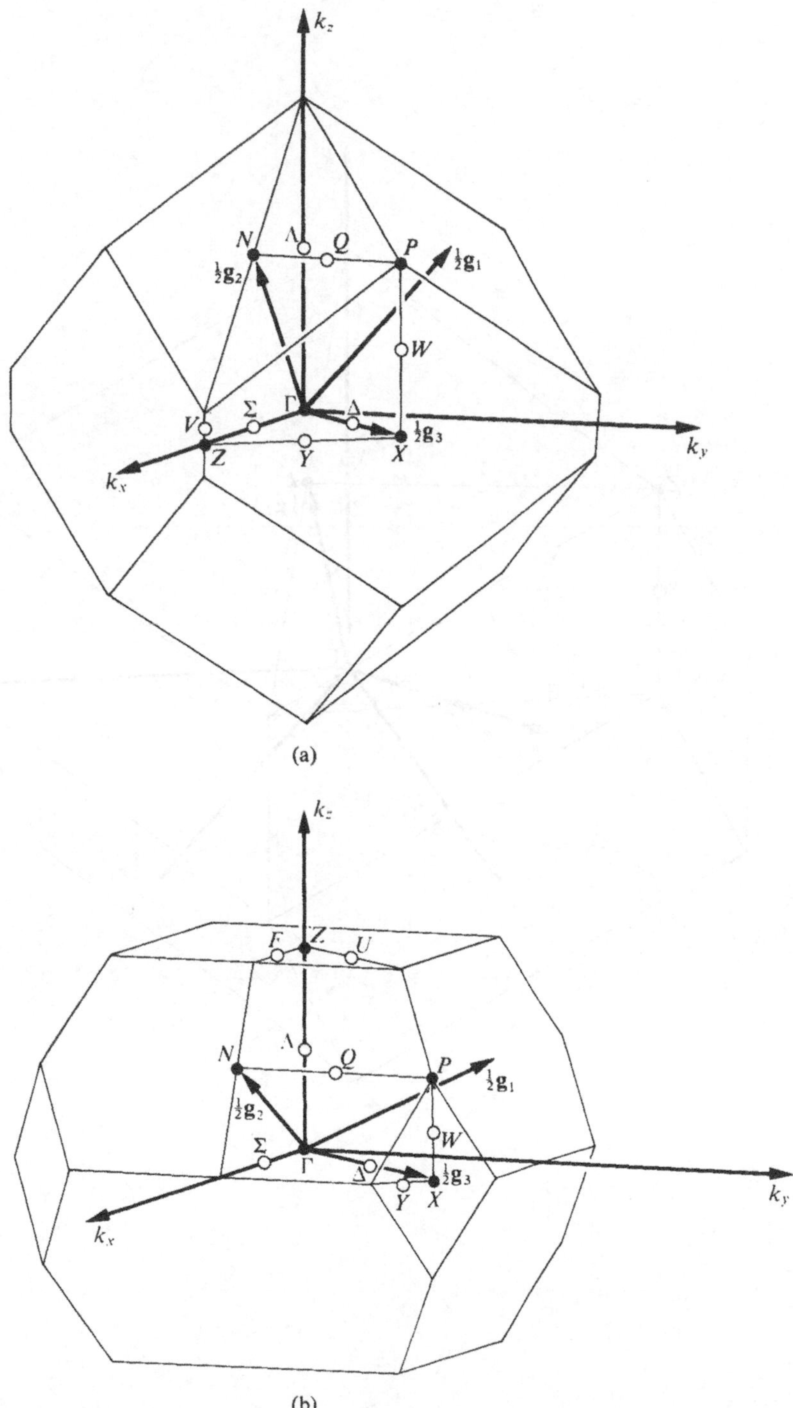

(b)

FIG. 3.10. The Brillouin zone for Γ_q^v. (a) $a > c$, $\Gamma = (000)$; $N = (0\frac{1}{2}0)$; $X = (00\frac{1}{2})$; $Z = (\bar{\frac{1}{2}}\frac{1}{2}\frac{1}{2})$; $P = (\frac{1}{4}\frac{1}{4}\frac{1}{4})$; (b) $c > a$, $\Gamma = (000)$; $N = (0\frac{1}{2}0)$; $X = (00\frac{1}{2})$; $Z = (\frac{1}{2}\frac{1}{2}\bar{\frac{1}{2}})$; $P = (\frac{1}{4}\frac{1}{4}\frac{1}{4})$.

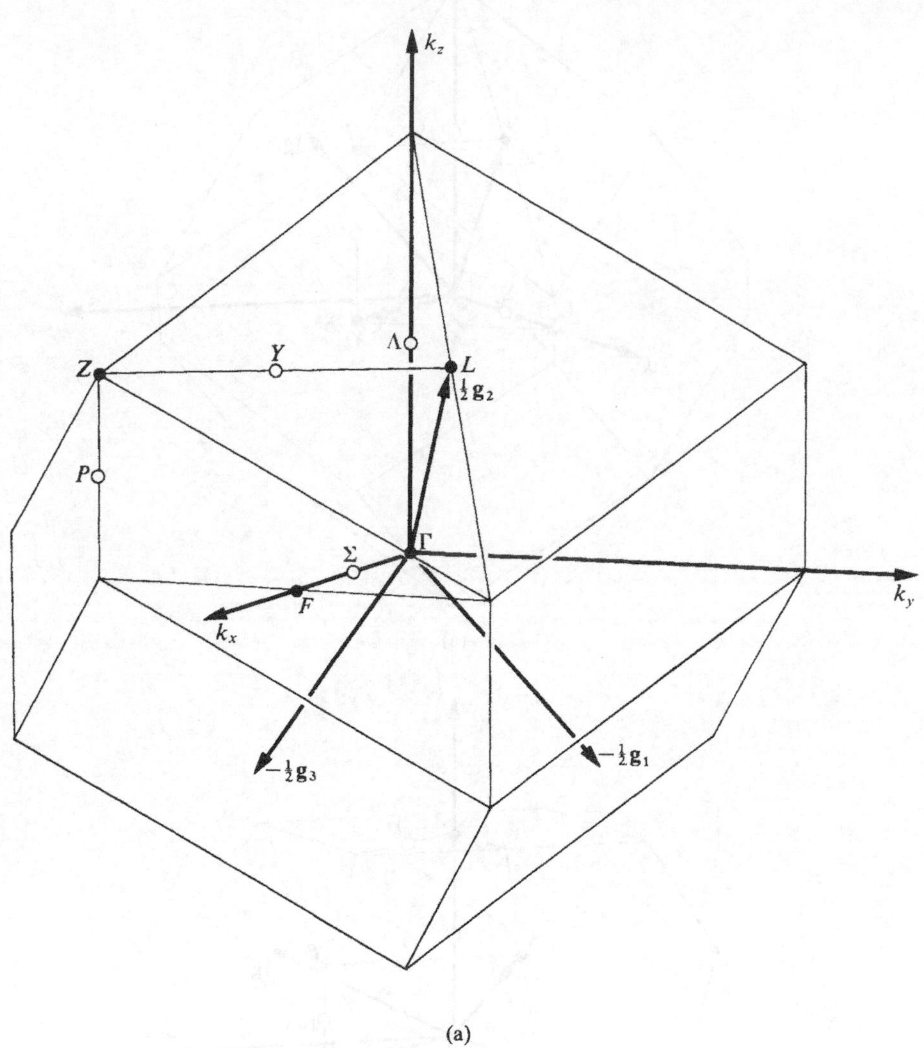
k_z
Λ
Z
Y
L
$\frac{1}{2}\mathbf{g}_2$
P
Γ
Σ
F
k_x
k_y
$-\frac{1}{2}\mathbf{g}_3$
$-\frac{1}{2}\mathbf{g}_1$

(a)

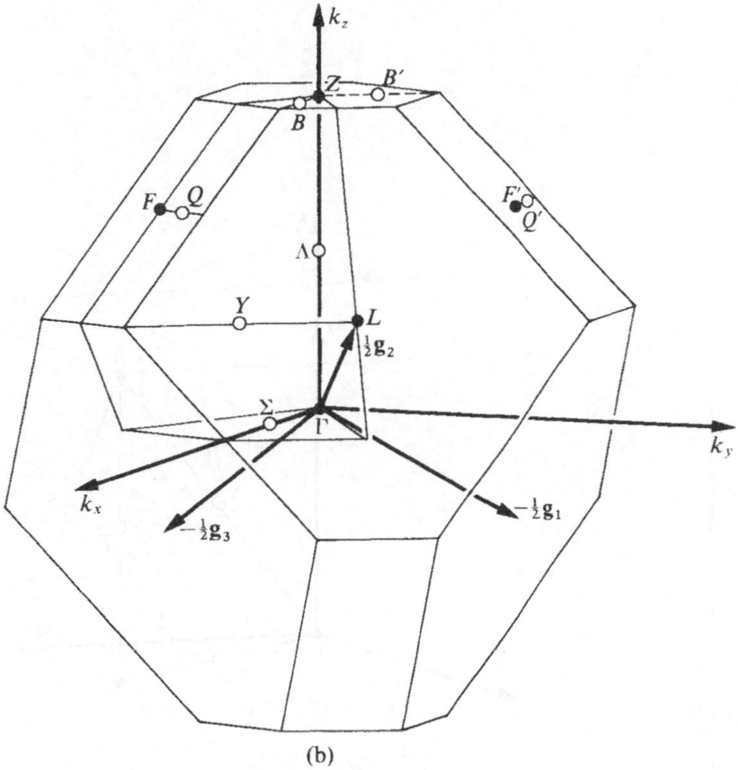

(b)

FIG. 3.11. The Brillouin zone for Γ_{rh}. (a) $a > \sqrt{(2)}c$, $\Gamma = (000)$: $Z = (\frac{1}{2}\frac{1}{2}\frac{1}{2})$; $L = (0\frac{1}{2}0)$; $F = (0\frac{1}{2}\frac{1}{2})$; (b) $\sqrt{(2)}c > a$, $\Gamma = (000)$; $Z = (\frac{1}{2}\frac{1}{2}\frac{1}{2})$; $L = (0\frac{1}{2}0)$; $F = (\frac{1}{2}\frac{1}{2}0)$. In (a) Σ and Y are two complete lines of symmetry; in (b) Q and B are simply related by a threefold rotation to Q' and B', respectively, which are continuations of the lines Σ and Y, respectively ($F' = (0\frac{1}{2}\frac{\bar{1}}{2})$ and is related to F by a threefold rotation; see also Table 3.6).

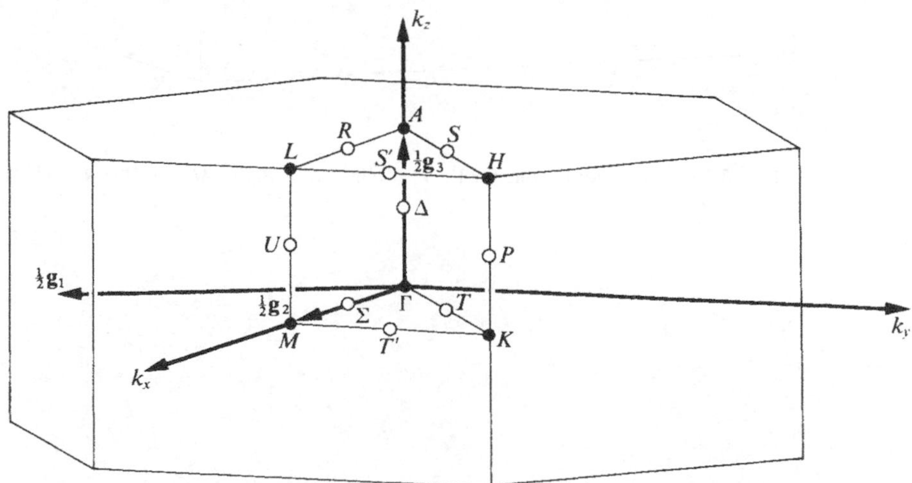

FIG. 3.12. The Brillouin zone for Γ_{h}. $\Gamma = (000)$; $M = (0\frac{1}{2}0)$; $A = (00\frac{1}{2})$; $L = (0\frac{1}{2}\frac{1}{2})$; $K = (\frac{\bar{1}}{3}\frac{2}{3}0)$; $H = (\frac{\bar{1}}{3}\frac{2}{3}\frac{1}{2})$.

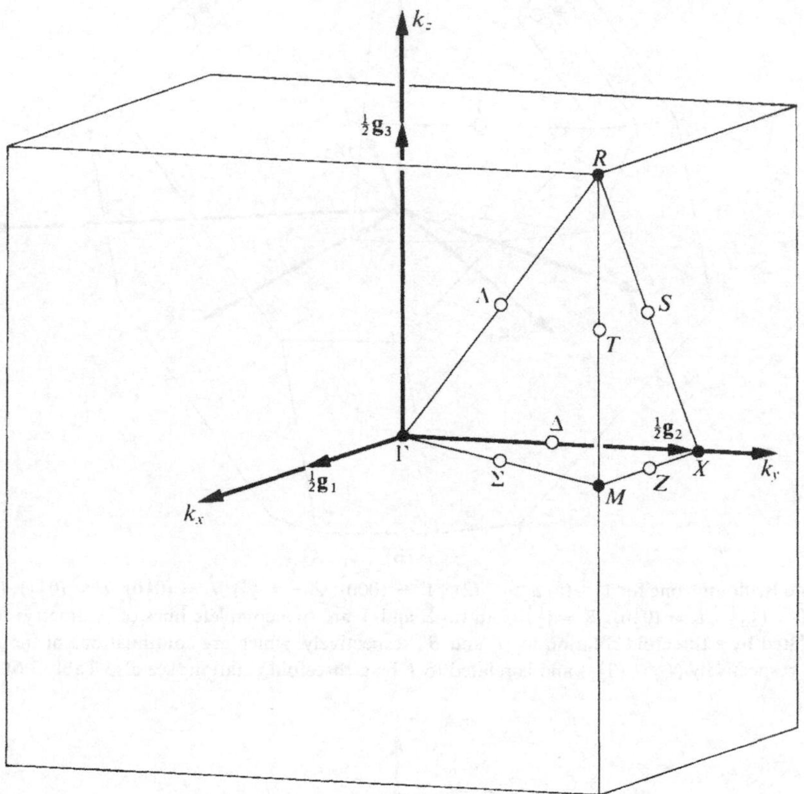

FIG. 3.13. The Brillouin zone for Γ_c. $\Gamma = (000)$; $X = (0\frac{1}{2}0)$; $M = (\frac{1}{2}\frac{1}{2}0)$; $R = (\frac{1}{2}\frac{1}{2}\frac{1}{2})$.

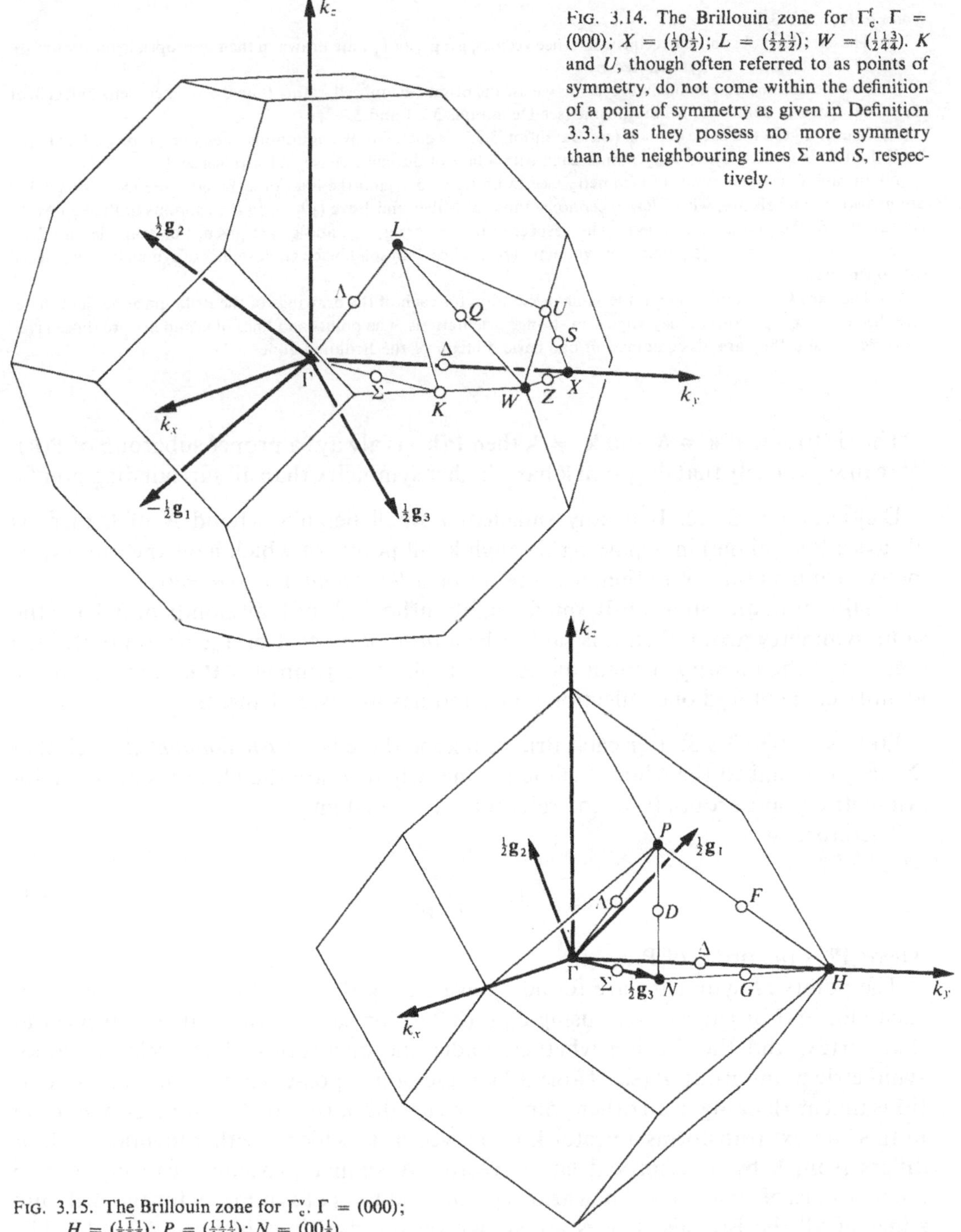

FIG. 3.14. The Brillouin zone for Γ_c^f. $\Gamma = (000)$; $X = (\frac{1}{2}0\frac{1}{2})$; $L = (\frac{1}{2}\frac{1}{2}\frac{1}{2})$; $W = (\frac{1}{2}\frac{1}{4}\frac{3}{4})$. K and U, though often referred to as points of symmetry, do not come within the definition of a point of symmetry as given in Definition 3.3.1, as they possess no more symmetry than the neighbouring lines Σ and S, respectively.

FIG. 3.15. The Brillouin zone for Γ_c^v. $\Gamma = (000)$; $H = (\frac{1}{2}\frac{\bar{1}}{2}\frac{1}{2})$; $P = (\frac{1}{4}\frac{1}{4}\frac{1}{4})$; $N = (00\frac{1}{2})$.

Notes to Figs. 3.2–3.15

(i) The axes $\Gamma k_x k_y k_z$ and the reciprocal lattice vectors, $\mathbf{g}_1$, $\mathbf{g}_2$, and $\mathbf{g}_3$ are drawn in their appropriate positions; see Table 3.3. The origin is always denoted by Γ.

(ii) For the triclinic and monoclinic systems we use the primitive unit cell, rather than the Wigner–Seitz unit cell, of reciprocal space to define the Brillouin zone (see Definitions 3.2.1 and 3.2.2).

(iii) In each figure the basic domain (see Definition 3.3.3) is enclosed by continuous lines, except for Γ_t. For Γ_t the basic domain is formed by three pyramids, each with a face of the cell as base and with vertex Γ.

(iv) In each figure the points of symmetry (see Definition 3.3.1) and the lines of symmetry (see Definition 3.3.2) are named; the labels are, with a few exceptions, those of Miller and Love (1967). In the captions to the figures the coordinates of the points of symmetry with respect to the vectors $\mathbf{g}_1$, $\mathbf{g}_2$, and $\mathbf{g}_3$ are given. For example, in Γ_m^b we have $M = -\frac{1}{2}\mathbf{g}_1 - \frac{1}{2}\mathbf{g}_2 + \frac{1}{2}\mathbf{g}_3$. Points of symmetry are marked with solid black circles; lines of symmetry are marked with open circles.

(v) The axes k_x, k_y, and k_z are in the same orientation for each of the drawings of the Brillouin zones and therefore the vectors $\mathbf{g}_1$, $\mathbf{g}_2$, and $\mathbf{g}_3$ may appear in strange orientations. The points and lines of symmetry are those given in Table 3.6 and they are all contained in one basic domain of the Brillouin zone.

That is to say, if $\mathbf{k}' \in N$ and $\mathbf{k}' \neq \mathbf{k}$ then $\mathbf{P}(\mathbf{k}')$ is always a proper subgroup of $\mathbf{P}(\mathbf{k})$. We can say loosely that the point $\mathbf{k}$ has a higher symmetry than all surrounding points.

DEFINITION 3.3.2. If in any sufficiently small neighbourhood N of $\mathbf{k}$ there is always a line (plane) in N passing through $\mathbf{k}$, all points of which have the same symmetry group as that of $\mathbf{k}$, then $\mathbf{k}$ is said to be a *line* (*plane*) *of symmetry*.

Finally, if in any sufficiently small neighbourhood N of $\mathbf{k}$ all points of N have the same symmetry group then $\mathbf{k}$ is said to be a *general point*. For a general point, $\mathbf{P}(\mathbf{k})$ consists of the identity element alone. For a plane of symmetry $\mathbf{P}(\mathbf{k})$ consists of the identity element and one reflection plane and has just two elements.

DEFINITION 3.3.3. For each Brillouin zone there is a *basic domain*, Ω, such that $(\Sigma_R R\Omega)$ is equal to the whole Brillouin zone, where R are the elements of the holosymmetric point group, $\mathbf{P}$, of the relevant crystal system.

Therefore,

$$\Omega = \frac{8\pi^3}{V\,|\mathbf{P}|} \tag{3.3.1}$$

where $|\mathbf{P}|$ is the order of $\mathbf{P}$.

The points of symmetry are found by evaluating the coordinates of the centre of each face and of each vertex, using eqn. (3.2.6) for each of the planes that meet at that vertex, and then seeing whether, under the operations of the relevant holosymmetric point group $\mathbf{P}$ (see Table 3.1), these points possess extra symmetry operations that are lacking in all other points in the neighbourhood. A symmetry operation in this context transforms a vector $\mathbf{k}$ into a vector $\mathbf{k}'$ which is either identical to $\mathbf{k}$ or differs from $\mathbf{k}$ by a reciprocal lattice vector. A similar procedure can be used to identify lines of symmetry. The various points and lines of symmetry for the Brillouin zones of all the Bravais lattices are shown on the drawings in Figs. 3.2–3.15. The letters that we have used in Figs. 3.2–3.15 for the points and lines of symmetry in the Brillouin zones are almost standard and agree, as far as possible, with the notation of previous authors (Bouckaert, Smoluchowski, and Wigner 1936, Herring 1942, Koster 1955, Luehrmann 1968, Meijer and Bauer 1962, Miller and Love 1967).

For some Bravais lattices the shape of the Brillouin zone is unique, for example for each of the three cubic Bravais lattices for which the Brillouin zones are illustrated in Figs. 3.13–3.15. But for some of the Bravais lattices there are various shapes possible depending on the axial ratios and the interaxial angles of the Bravais lattice. For the triclinic Bravais lattice (Fig. 3.2) and each of the monoclinic Bravais lattices (Figs. 3.3 and 3.4) we have followed Definition 3.2.2 and used the primitive unit cell of reciprocal space to define the Brillouin zone. For some of the other Bravais lattices there may be apparently different Brillouin zones (see, for example, Figs. 3.8 for Γ_o^f, 3.10 for Γ_q^v, and 3.11 for Γ_{rh}). However, when two apparently different Brillouin zones are possible there are only the same number of points of symmetry in each case. Thus in Fig. 3.11(a), for Γ_{rh} with $a > \sqrt{(2)}c$, the points of symmetry are Γ ($\mathbf{k} = 0$), L ($\mathbf{k} = \frac{1}{2}\mathbf{g}_2$), F ($\mathbf{k} = \frac{1}{2}\mathbf{g}_2 - \frac{1}{2}\mathbf{g}_3$), and Z ($\mathbf{k} = \frac{1}{2}\mathbf{g}_1 + \frac{1}{2}\mathbf{g}_2 - \frac{1}{2}\mathbf{g}_3$) and in the alternative Brillouin zone in Fig. 3.11(b), when $a < \sqrt{(2)}c$, the points of symmetry are Γ ($\mathbf{k} = 0$), L ($\mathbf{k} = \frac{1}{2}\mathbf{g}_2$), F ($\mathbf{k} = \frac{1}{2}\mathbf{g}_1 + \frac{1}{2}\mathbf{g}_2$), and Z ($\mathbf{k} = \frac{1}{2}\mathbf{g}_1 + \frac{1}{2}\mathbf{g}_2 + \frac{1}{2}\mathbf{g}_3$). Γ and L appear with the same $\mathbf{k}$ in both Brillouin zones while F and Z in Fig. 3.11(a) have different, but equivalent, $\mathbf{k}$ vectors in Fig. 3.11(b). For certain other Bravais lattices the Brillouin zone has the same appearance except that the orientation of the faces with respect to the k_x, k_y, and k_z axes is different, see Figs. 3.6, 3.7, and 3.8 for Γ_o^b, Γ_o^v, and Γ_o^f respectively. Once again, for a given Bravais lattice the same set of points of symmetry, at least up to equivalence, will be obtained however many different Brillouin zones actually appear in the relevant figure from Figs. 3.2–3.15. When more than one shape of Brillouin zone is possible for a given Bravais lattice, depending on the axial ratios, the relationship between lines of symmetry in the various cases is more complicated. What may sometimes happen with lines of symmetry may be illustrated by considering an example. Two shapes of Brillouin zone are possible for the orthorhombic base-centred lattice, Γ_o^b; the Brillouin zone if $a > b$ is illustrated in Fig. 3.6(a) and the Brillouin zone if $b > a$ is illustrated in Fig. 3.6(b). The equivalence between the points of symmetry in the two cases can be established by inspection of the captions to these two figures; thus for example T in Fig. 3.6(a) is ($\frac{1}{2}\mathbf{g}_1 + \frac{1}{2}\mathbf{g}_2 + \frac{1}{2}\mathbf{g}_3$) and T in Fig. 3.6(b) is ($-\frac{1}{2}\mathbf{g}_1 + \frac{1}{2}\mathbf{g}_2 + \frac{1}{2}\mathbf{g}_3$). If we consider the line of symmetry E in Fig. 3.6(b) its $\mathbf{k}$ vector is $\{(-\frac{1}{2} + \alpha)\mathbf{g}_1 + (\frac{1}{2} + \alpha)\mathbf{g}_2 + \frac{1}{2}\mathbf{g}_3\}$, i.e.

$$\mathbf{k}_E = \mathbf{k}_T + (\alpha\mathbf{g}_1 + \alpha\mathbf{g}_2). \tag{3.3.2}$$

The corresponding line of symmetry in Fig. 3.6(a) will then have $\mathbf{k}$ vector $\mathbf{k}_E$ given by eqn. (3.3.2), that is by the continuation of ZT; see Fig. 3.16. This line E is now outside the first Brillouin zone but it is equivalent to some line in the first Brillouin zone, namely E' where

$$\mathbf{k}_{E'} = \mathbf{k}_E - \mathbf{g}_1 - \mathbf{g}_2. \tag{3.3.3}$$

The line E' is simply related to the line A in Fig. 3.6(a) by symmetry and so does not need to be considered separately from A. Looked at from the other way this means that the line A in Fig. 3.6(a) becomes, in Fig. 3.6(b), separated into two pieces; one piece, for some values of α is A of Fig. 3.6(a), while the other piece, for the remaining values of α is E of Fig. 3.6(b). Similarly, the line Σ in Fig. 3.6(a) becomes divided in Fig. 3.6(b) into Σ and C. Conversely, the line Δ in Fig. 3.6(b) becomes divided into Δ and F in Fig. 3.6(a) and the line B in Fig. 3.6(b) becomes divided into B and G in Fig. 3.6(a).

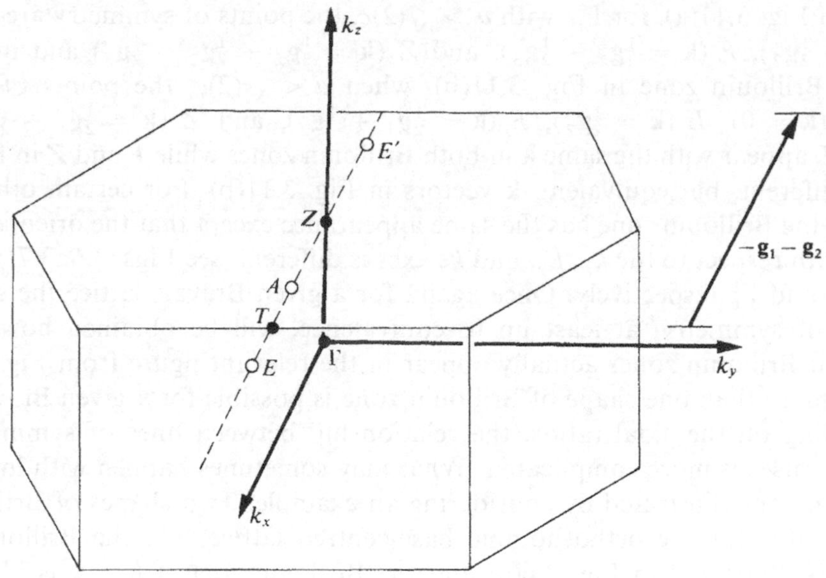

FIG. 3.16. The Brillouin zone for Γ_o^b.

In the tables that appear later in this book it will only be necessary, because of the equivalence that we have noted, to tabulate the points of symmetry for one Brillouin zone for each of the 14 Bravais lattices. Also, it will usually only be necessary to tabulate the lines of symmetry for one Brillouin zone for each of the 14 Bravais lattices. However, for lines of symmetry on the surface of the Brillouin zone there are a few exceptions. For example, although C in Fig. 3.6(b) is related to Σ in Fig. 3.6(a) there is a subtle difference between them that will become apparent later.

The group **P(k)**, the symmetry group of **k**, for each point and line of symmetry in the basic domain of each of the 14 Brillouin zones is given in Table 3.6. It will be noted that many of the edges of the basic domains in Figs. 3.2–3.15 are not lines of symmetry, but are just, in the sense of Definition 3.3.2, planes of symmetry. The reason for restricting ourselves to a single basic domain will become apparent later; see, for example, section 5.5.

TABLE 3.6

Symmetry groups **P(k)**

Bravais lattice	Point	Coordinates	**P(k)**	Elements of **P(k)**
Triclinic, P, Γ_t				
(Fig. 3.2)	Γ	(000)	$\bar{1}\ (C_i)$	E, I
	B	$(\frac{1}{2}00)$	$\bar{1}\ (C_i)$	E, I
	F	$(0\frac{1}{2}0)$	$\bar{1}\ (C_i)$	E, I
	G	$(00\frac{1}{2})$	$\bar{1}\ (C_i)$	E, I
Monoclinic, P, Γ_m				
(Fig. 3.3)	Γ	(000)	$2/m\ (C_{2h})$	E, C_{2z}, I, σ_z
	B	$(\frac{1}{2}00)$	$2/m\ (C_{2h})$	E, C_{2z}, I, σ_z
	Y	$(0\frac{1}{2}0)$	$2/m\ (C_{2h})$	E, C_{2z}, I, σ_z
	Z	$(00\frac{1}{2})$	$2/m\ (C_{2h})$	E, C_{2z}, I, σ_z
	C	$(0\frac{1}{2}\frac{1}{2})$	$2/m\ (C_{2h})$	E, C_{2z}, I, σ_z
	D	$(\frac{1}{2}0\frac{1}{2})$	$2/m\ (C_{2h})$	E, C_{2z}, I, σ_z
	A	$(\frac{1}{2}\frac{1}{2}0)$, $(\bar{\frac{1}{2}}\frac{1}{2}0)$, $(\bar{\frac{1}{2}}\bar{\frac{1}{2}}0)$	$2/m\ (C_{2h})$	E, C_{2z}, I, σ_z
	E	$(\frac{1}{2}\frac{1}{2}\frac{1}{2})$, $(\bar{\frac{1}{2}}\frac{1}{2}\frac{1}{2})$, $(\bar{\frac{1}{2}}\bar{\frac{1}{2}}\frac{1}{2})$	$2/m\ (C_{2h})$	E, C_{2z}, I, σ_z
	$\Lambda(\Gamma Z)$	(00α)	$2\ (C_2)$	E, C_{2z}
	$V(BD)$	$(\frac{1}{2}0\alpha)$	$2\ (C_2)$	E, C_{2z}
	$W(YC)$	$(0\frac{1}{2}\alpha)$	$2\ (C_2)$	E, C_{2z}
	$U(AE)$	$(\frac{1}{2}\frac{1}{2}\alpha)$, $(\bar{\frac{1}{2}}\frac{1}{2}\alpha)$, $(\bar{\frac{1}{2}}\bar{\frac{1}{2}}\alpha)$	$2\ (C_2)$	E, C_{2z}
Monoclinic, B, Γ_m^b				
(Fig. 3.4)	Γ	(000)	$2/m\ (C_{2h})$	E, C_{2z}, I, σ_z
	A	$(\frac{1}{2}00)$	$2/m\ (C_{2h})$	E, C_{2z}, I, σ_z
	Z	$(0\frac{1}{2}\frac{1}{2})$	$2/m\ (C_{2h})$	E, C_{2z}, I, σ_z
	M	$(\frac{1}{2}\frac{1}{2}\frac{1}{2})$	$2/m\ (C_{2h})$	E, C_{2z}, I, σ_z
	L	$(\frac{1}{2}0\frac{1}{2})$	$\bar{1}\ (C_i)$	E, I
	V	$(00\frac{1}{2})$	$\bar{1}\ (C_i)$	E, I
	$\Lambda(\Gamma Z)$	$(0\alpha\alpha)$	$2\ (C_2)$	E, C_{2z}
	$U(AM)$	$(\frac{1}{2}\alpha\alpha)$	$2\ (C_2)$	E, C_{2z}

Bravais lattice		Point	Coordinates	$\mathbf{P}(\mathbf{k})$	Elements of $\mathbf{P}(\mathbf{k})$
Orthorhombic, P, Γ_o					
(Fig. 3.5)		Γ	(000)	$mmm\ (D_{2h})$	$E, C_{2x}, C_{2y}, C_{2z}, I, \sigma_x, \sigma_y, \sigma_z$
		Y	$(\bar{\tfrac{1}{2}}00)$	$mmm\ (D_{2h})$	$E, C_{2x}, C_{2y}, C_{2z}, I, \sigma_x, \sigma_y, \sigma_z$
		X	$(0\tfrac{1}{2}0)$	$mmm\ (D_{2h})$	$E, C_{2x}, C_{2y}, C_{2z}, I, \sigma_x, \sigma_y, \sigma_z$
		Z	$(00\tfrac{1}{2})$	$mmm\ (D_{2h})$	$E, C_{2x}, C_{2y}, C_{2z}, I, \sigma_x, \sigma_y, \sigma_z$
		U	$(0\tfrac{1}{2}\tfrac{1}{2})$	$mmm\ (D_{2h})$	$E, C_{2x}, C_{2y}, C_{2z}, I, \sigma_x, \sigma_y, \sigma_z$
		T	$(\bar{\tfrac{1}{2}}0\tfrac{1}{2})$	$mmm\ (D_{2h})$	$E, C_{2x}, C_{2y}, C_{2z}, I, \sigma_x, \sigma_y, \sigma_z$
		S	$(\bar{\tfrac{1}{2}}\tfrac{1}{2}0)$	$mmm\ (D_{2h})$	$E, C_{2x}, C_{2y}, C_{2z}, I, \sigma_x, \sigma_y, \sigma_z$
		R	$(\bar{\tfrac{1}{2}}\tfrac{1}{2}\tfrac{1}{2})$	$mmm\ (D_{2h})$	$E, C_{2x}, C_{2y}, C_{2z}, I, \sigma_x, \sigma_y, \sigma_z$
		$\Delta(\Gamma Y)$	$(\bar{\alpha}00)$	$mm2\ (C_{2v})$	$E, C_{2y}, \sigma_z, \sigma_x$
		$D(XS)$	$(\bar{\alpha}\tfrac{1}{2}0)$	$mm2\ (C_{2v})$	$E, C_{2y}, \sigma_z, \sigma_x$
		$P(UR)$	$(\bar{\alpha}\tfrac{1}{2}\tfrac{1}{2})$	$mm2\ (C_{2v})$	$E, C_{2y}, \sigma_z, \sigma_x$
		$B(ZT)$	$(\bar{\alpha}0\tfrac{1}{2})$	$mm2\ (C_{2v})$	$E, C_{2y}, \sigma_z, \sigma_x$
		$\Sigma(\Gamma X)$	$(0\alpha0)$	$mm2\ (C_{2v})$	$E, C_{2x}, \sigma_y, \sigma_z$
		$C(YS)$	$(\bar{\tfrac{1}{2}}\alpha0)$	$mm2\ (C_{2v})$	$E, C_{2x}, \sigma_y, \sigma_z$
		$E(TR)$	$(\bar{\tfrac{1}{2}}\alpha\tfrac{1}{2})$	$mm2\ (C_{2v})$	$E, C_{2x}, \sigma_y, \sigma_z$
		$A(ZU)$	$(0\alpha\tfrac{1}{2})$	$mm2\ (C_{2v})$	$E, C_{2x}, \sigma_y, \sigma_z$
		$\Lambda(\Gamma Z)$	(00α)	$mm2\ (C_{2v})$	$E, C_{2z}, \sigma_x, \sigma_y$
		$H(YT)$	$(\bar{\tfrac{1}{2}}0\alpha)$	$mm2\ (C_{2v})$	$E, C_{2z}, \sigma_x, \sigma_y$
		$Q(SR)$	$(\bar{\tfrac{1}{2}}\tfrac{1}{2}\alpha)$	$mm2\ (C_{2v})$	$E, C_{2z}, \sigma_x, \sigma_y$
		$G(XU)$	$(0\tfrac{1}{2}\alpha)$	$mm2\ (C_{2v})$	$E, C_{2z}, \sigma_x, \sigma_y$
Orthorhombic, C, Γ_o^b					
(Fig. 3.6)		Γ	(000)	$mmm\ (D_{2h})$	$E, C_{2x}, C_{2y}, C_{2z}, I, \sigma_x, \sigma_y, \sigma_z$
	(a)	Y	$(\tfrac{1}{2}\tfrac{1}{2}0)$	$mmm\ (D_{2h})$	$E, C_{2x}, C_{2y}, C_{2z}, I, \sigma_x, \sigma_y, \sigma_z$
	(b)		$(\bar{\tfrac{1}{2}}\tfrac{1}{2}0)$		
		Z	$(00\tfrac{1}{2})$	$mmm\ (D_{2h})$	$E, C_{2x}, C_{2y}, C_{2z}, I, \sigma_x, \sigma_y, \sigma_z$
	(a)	T	$(\tfrac{1}{2}\tfrac{1}{2}\tfrac{1}{2})$	$mmm\ (D_{2h})$	$E, C_{2x}, C_{2y}, C_{2z}, I, \sigma_x, \sigma_y, \sigma_z$
	(b)		$(\bar{\tfrac{1}{2}}\tfrac{1}{2}\tfrac{1}{2})$		
		S	$(0\tfrac{1}{2}0)$	$2/m\ (C_{2h})$	E, C_{2z}, I, σ_z
		R	$(0\tfrac{1}{2}\tfrac{1}{2})$	$2/m\ (C_{2h})$	E, C_{2z}, I, σ_z
		$\Lambda(\Gamma Z)$	(00α)	$mm2\ (C_{2v})$	$E, C_{2z}, \sigma_x, \sigma_y$
	(a)	$H(YT)$	$(\tfrac{1}{2}\tfrac{1}{2}\alpha)$	$mm2\ (C_{2v})$	$E, C_{2z}, \sigma_x, \sigma_y$
	(b)		$(\bar{\tfrac{1}{2}}\tfrac{1}{2}\alpha)$		
		$D(SR)$	$(0\tfrac{1}{2}\alpha)$	$2\ (C_2)$	E, C_{2z}
		$A(ZT)$	$(\alpha\alpha\tfrac{1}{2})$	$mm2\ (C_{2v})$	$E, C_{2x}, \sigma_y, \sigma_z$
		$\Sigma(\Gamma Y)$	$(\alpha\alpha0)$	$mm2\ (C_{2v})$	$E, C_{2x}, \sigma_y, \sigma_z$
		$\Delta(\Gamma\Delta)$	$(\bar{\alpha}\alpha0)$	$mm2\ (C_{2v})$	$E, C_{2y}, \sigma_z, \sigma_x$
		$B(ZB)$	$(\bar{\alpha}\alpha\tfrac{1}{2})$	$mm2\ (C_{2v})$	$E, C_{2y}, \sigma_z, \sigma_x$
	(a)	$G(TG)$	$(\tfrac{1}{2} - \alpha, \tfrac{1}{2} + \alpha, \tfrac{1}{2})$	$mm2\ (C_{2v})$	$E, C_{2y}, \sigma_z, \sigma_x$
	(a)	$F(YF)$	$(\tfrac{1}{2} - \alpha, \tfrac{1}{2} + \alpha, 0)$	$mm2\ (C_{2v})$	$E, C_{2y}, \sigma_z, \sigma_x$
	(b)	$E(TE)$	$(-\tfrac{1}{2} + \alpha, \tfrac{1}{2} + \alpha, \tfrac{1}{2})$	$mm2\ (C_{2v})$	$E, C_{2x}, \sigma_y, \sigma_z$
	(b)	$C(YC)$	$(-\tfrac{1}{2} + \alpha, \tfrac{1}{2} + \alpha, 0)$	$mm2\ (C_{2v})$	$E, C_{2x}, \sigma_y, \sigma_z$

Bravais lattice	Point	Coordinates	$\mathbf{P}(\mathbf{k})$	Elements of $\mathbf{P}(\mathbf{k})$
Orthorhombic, I, Γ_o^v				
(Fig. 3.7)	Γ	(000)	$mmm\ (D_{2h})$	$E, C_{2x}, C_{2y}, C_{2z}, I, \sigma_x, \sigma_y, \sigma_z$
(a)	X	$(\tfrac{1}{2}\bar{\tfrac{1}{2}}\tfrac{1}{2})$	$mmm\ (D_{2h})$	$E, C_{2x}, C_{2y}, C_{2z}, I, \sigma_x, \sigma_y, \sigma_z$
(b)		$(\bar{\tfrac{1}{2}}\tfrac{1}{2}\tfrac{1}{2})$		
(c)		$(\tfrac{1}{2}\tfrac{1}{2}\bar{\tfrac{1}{2}})$		
	R	$(\tfrac{1}{2}00)$	$2/m\ (C_{2h})$	E, C_{2y}, I, σ_y
	S	$(\tfrac{1}{2}0\bar{\tfrac{1}{2}})$	$2/m\ (C_{2h})$	E, C_{2x}, I, σ_x
	T	$(\tfrac{1}{2}\bar{\tfrac{1}{2}}0)$	$2/m\ (C_{2h})$	E, C_{2z}, I, σ_z
	W	$(\tfrac{3}{4}\bar{\tfrac{1}{4}}\bar{\tfrac{1}{4}})$	$222\ (D_2)$	$E, C_{2x}, C_{2y}, C_{2z}$
(a), (b)	$\Lambda(\Gamma\Lambda)$	$(\alpha\alpha\bar{\alpha})$	$mm2\ (C_{2v})$	$E, C_{2z}, \sigma_x, \sigma_y$
(c)	$\Lambda(\Gamma X)$			
(a)	$G(XG)$	$(\tfrac{1}{2}+\alpha, -\tfrac{1}{2}+\alpha, \tfrac{1}{2}-\alpha)$	$mm2\ (C_{2v})$	$E, C_{2z}, \sigma_x, \sigma_y$
(b)		$(\tfrac{1}{2}+\alpha, -\tfrac{1}{2}+\alpha, -\tfrac{1}{2}-\alpha)$		
	$P(TW)$	$(\tfrac{1}{2}+\alpha, -\tfrac{1}{2}+\alpha, -\alpha)$	$2\ (C_2)$	E, C_{2z}
(b), (c)	$\Sigma(\Gamma\Sigma)$	$(\alpha\bar{\alpha}\alpha)$	$mm2\ (C_{2v})$	$E, C_{2x}, \sigma_y, \sigma_z$
(a)	$\Sigma(\Gamma X)$			
(b)	$F(XF)$	$(\tfrac{1}{2}+\alpha, -\tfrac{1}{2}-\alpha, -\tfrac{1}{2}+\alpha)$	$mm2\ (C_{2v})$	$E, C_{2x}, \sigma_y, \sigma_z$
(c)		$(\tfrac{1}{2}+\alpha, \tfrac{1}{2}-\alpha, -\tfrac{1}{2}+\alpha)$		
	$D(SW)$	$(\tfrac{1}{2}+\alpha, -\alpha, -\tfrac{1}{2}+\alpha)$	$2\ (C_2)$	E, C_{2x}
(a), (c)	$\Delta(\Gamma\Delta)$	$(\alpha\bar{\alpha}\bar{\alpha})$	$mm2\ (C_{2v})$	$E, C_{2y}, \sigma_z, \sigma_x$
(b)	$\Delta(\Gamma X)$			
(a)	$U(XU)$	$(\tfrac{1}{2}+\alpha, -\tfrac{1}{2}-\alpha, \tfrac{1}{2}-\alpha)$	$mm2\ (C_{2v})$	$E, C_{2y}, \sigma_z, \sigma_x$
(c)		$(\tfrac{1}{2}+\alpha, \tfrac{1}{2}-\alpha, -\tfrac{1}{2}-\alpha)$		
	$Q(RW)$	$(\tfrac{1}{2}+\alpha, -\alpha, -\alpha)$	$2\ (C_2)$	E, C_{2y}
Orthorhombic, F, Γ_o^f				
(Fig. 3.8)				
	Γ	(000)	$mmm\ (D_{2h})$	$E, C_{2x}, C_{2y}, C_{2z}, I, \sigma_x, \sigma_y, \sigma_z$
(a), (b), (d)	Y	$(0\bar{\tfrac{1}{2}}\bar{\tfrac{1}{2}})$	$mmm\ (D_{2h})$	$E, C_{2x}, C_{2y}, C_{2z}, I, \sigma_x, \sigma_y, \sigma_z$
(c)		$(1\tfrac{1}{2}\tfrac{1}{2})$		
(a), (b), (c)	X	$(\tfrac{1}{2}0\tfrac{1}{2})$	$mmm\ (D_{2h})$	$E, C_{2x}, C_{2y}, C_{2z}, I, \sigma_x, \sigma_y, \sigma_z$
(d)		$(\tfrac{1}{2}0\bar{\tfrac{1}{2}})$		
(a), (c), (d)	Z	$(\tfrac{1}{2}\tfrac{1}{2}0)$	$mmm\ (D_{2h})$	$E, C_{2x}, C_{2y}, C_{2z}, I, \sigma_x, \sigma_y, \sigma_z$
(b)		$(\tfrac{1}{2}\bar{\tfrac{1}{2}}0)$		
	L	$(\tfrac{1}{2}00)$	$\bar{1}\ (C_i)$	E, I
(a), (c), (d)	$\Lambda(\Gamma Z)$	$(\alpha\alpha0)$	$mm2\ (C_{2v})$	$E, C_{2z}, \sigma_x, \sigma_y$
(b)	$\Lambda(\Gamma\Lambda)$			
(a), (b)	$G(XG)$	$(\tfrac{1}{2}+\alpha, \alpha, \tfrac{1}{2})$	$mm2\ (C_{2v})$	$E, C_{2z}, \sigma_x, \sigma_y$
(c)	$G(XY)$			
(a), (b)	$H(YH)$	$(\alpha, -\tfrac{1}{2}+\alpha, -\tfrac{1}{2})$	$mm2\ (C_{2v})$	$E, C_{2z}, \sigma_x, \sigma_y$
(d)	$H(YX)$			
(b)	$Q(ZQ)$	$(\tfrac{1}{2}+\alpha, -\tfrac{1}{2}+\alpha, 0)$	$mm2\ (C_{2v})$	$E, C_{2z}, \sigma_x, \sigma_y$
(a), (b), (c)	$\Sigma(\Gamma X)$	$(\alpha0\alpha)$	$mm2\ (C_{2v})$	$E, C_{2x}, \sigma_y, \sigma_z$
(d)	$\Sigma(\Gamma\Sigma)$			
(a), (d)	$C(YC)$	$(\alpha, -\tfrac{1}{2}, -\tfrac{1}{2}+\alpha)$	$mm2\ (C_{2v})$	$E, C_{2x}, \sigma_y, \sigma_z$
(b)	$C(YZ)$			
(a), (d)	$A(ZA)$	$(\tfrac{1}{2}+\alpha, \tfrac{1}{2}, \alpha)$	$mm2\ (C_{2v})$	$E, C_{2x}, \sigma_y, \sigma_z$
(c)	$A(ZY)$			
(d)	$U(XU)$	$(\tfrac{1}{2}+\alpha, 0, -\tfrac{1}{2}+\alpha)$	$mm2\ (C_{2v})$	$E, C_{2x}, \sigma_y, \sigma_z$
(a), (b), (d)	$\Delta(\Gamma Y)$	$(0\bar{\alpha}\bar{\alpha})$	$mm2\ (C_{2v})$	$E, C_{2y}, \sigma_z, \sigma_x$
(c)	$\Delta(\Gamma\Delta)$			
(a), (c)	$D(XD)$	$(\tfrac{1}{2}, -\alpha, \tfrac{1}{2}-\alpha)$	$mm2\ (C_{2v})$	$E, C_{2y}, \sigma_z, \sigma_x$
(b)	$D(XZ)$			
(a), (c)	$B(ZB)$	$(\tfrac{1}{2}, \tfrac{1}{2}-\alpha, -\alpha)$	$mm2\ (C_{2v})$	$E, C_{2y}, \sigma_z, \sigma_x$
(d)	$B(ZX)$			
(c)	$R(YR)$	$(1, \tfrac{1}{2}-\alpha, \tfrac{1}{2}-\alpha)$	$mm2\ (C_{2v})$	$E, C_{2y}, \sigma_z, \sigma_x$

Bravais lattice	Point	Coordinates	**P(k)**	Elements of **P(k)**
Tetragonal, P, Γ_q (Fig. 3.9)	Γ	(000)	$4/mmm\ (D_{4h})$	$E, C_{4z}^{\pm}, C_{2z}, C_{2x}, C_{2y}, C_{2a}, C_{2b}, I, S_{4z}^{\mp}, \sigma_z, \sigma_x, \sigma_y, \sigma_{da}, \sigma_{db}$
	M	$(\frac{1}{2}\frac{1}{2}0)$	$4/mmm\ (D_{4h})$	$E, C_{4z}^{\pm}, C_{2z}, C_{2x}, C_{2y}, C_{2a}, C_{2b}, I, S_{4z}^{\mp}, \sigma_z, \sigma_x, \sigma_y, \sigma_{da}, \sigma_{db}$
	Z	$(00\frac{1}{2})$	$4/mmm\ (D_{4h})$	$E, C_{4z}^{\pm}, C_{2z}, C_{2x}, C_{2y}, C_{2a}, C_{2b}, I, S_{4z}^{\mp}, \sigma_z, \sigma_x, \sigma_y, \sigma_{da}, \sigma_{db}$
	A	$(\frac{1}{2}\frac{1}{2}\frac{1}{2})$	$4/mmm\ (D_{4h})$	$E, C_{4z}^{\pm}, C_{2z}, C_{2x}, C_{2y}, C_{2a}, C_{2b}, I, S_{4z}^{\mp}, \sigma_z, \sigma_x, \sigma_y, \sigma_{da}, \sigma_{db}$
	R	$(0\frac{1}{2}\frac{1}{2})$	$mmm\ (D_{2h})$	$E, C_{2x}, C_{2y}, C_{2z}, I, \sigma_x, \sigma_y, \sigma_z$
	X	$(0\frac{1}{2}0)$	$mmm\ (D_{2h})$	$E, C_{2x}, C_{2y}, C_{2z}, I, \sigma_x, \sigma_y, \sigma_z$
	$\Delta(\Gamma X)$	$(0\alpha 0)$	$mm2\ (C_{2v})$	$E, C_{2y}, \sigma_z, \sigma_x$
	$U(ZR)$	$(0\alpha\frac{1}{2})$	$mm2\ (C_{2v})$	$E, C_{2y}, \sigma_z, \sigma_x$
	$\Lambda(\Gamma Z)$	(00α)	$4mm\ (C_{4v})$	$E, C_{4z}^{\pm}, C_{2z}, \sigma_x, \sigma_y, \sigma_{da}, \sigma_{db}$
	$V(MA)$	$(\frac{1}{2}\frac{1}{2}\alpha)$	$4mm\ (C_{4v})$	$E, C_{4z}^{\pm}, C_{2z}, \sigma_x, \sigma_y, \sigma_{da}, \sigma_{db}$
	$\Sigma(\Gamma M)$	$(\alpha\alpha 0)$	$mm2\ (C_{2v})$	$E, C_{2a}, \sigma_z, \sigma_{db}$
	$S(ZA)$	$(\alpha\alpha\frac{1}{2})$	$mm2\ (C_{2v})$	$E, C_{2a}, \sigma_z, \sigma_{db}$
	$Y(XM)$	$(\alpha\frac{1}{2}0)$	$mm2\ (C_{2v})$	$E, C_{2x}, \sigma_y, \sigma_z$
	$T(RA)$	$(\alpha\frac{1}{2}\frac{1}{2})$	$mm2\ (C_{2v})$	$E, C_{2x}, \sigma_y, \sigma_z$
	$W(XR)$	$(0\frac{1}{2}\alpha)$	$mm2\ (C_{2v})$	$E, C_{2z}, \sigma_x, \sigma_y$
Tetragonal, I, Γ_q^v (Fig. 3.10)	Γ	(000)	$4/mmm\ (D_{4h})$	$E, C_{4z}^{\pm}, C_{2z}, C_{2x}, C_{2y}, C_{2a}, C_{2b}, I, S_{4z}^{\mp}, \sigma_z, \sigma_x, \sigma_y, \sigma_{da}, \sigma_{db}$
	N	$(0\frac{1}{2}0)$	$2/m\ (C_{2h})$	E, C_{2y}, I, σ_y
	X	$(00\frac{1}{2})$	$mmm\ (D_{2h})$	$E, C_{2z}, C_{2a}, C_{2b}, I, \sigma_z, \sigma_{da}, \sigma_{db}$
(a) (b)	Z	$(\bar{\frac{1}{2}}\frac{1}{2}\frac{1}{2})$ $(\frac{1}{2}\frac{1}{2}\bar{\frac{1}{2}})$	$4/mmm\ (D_{4h})$	$E, C_{4z}^{\pm}, C_{2z}, C_{2x}, C_{2y}, C_{2a}, C_{2b}, I, S_{4z}^{\mp}, \sigma_z, \sigma_x, \sigma_y, \sigma_{da}, \sigma_{db}$
	P	$(\frac{1}{4}\frac{1}{4}\frac{1}{4})$	$\bar{4}2m\ (D_{2d})$	$E, C_{2x}, C_{2y}, C_{2z}, \sigma_{da}, \sigma_{db}, S_{4z}^{\mp}$
(a) (b)	$\Lambda(\Gamma\Lambda)$ $\Lambda(\Gamma Z)$	$(\alpha\alpha\bar{\alpha})$	$4mm\ (C_{4v})$	$E, C_{4z}^{\pm}, C_{2z}, \sigma_x, \sigma_y, \sigma_{da}, \sigma_{db}$
(a)	$V(ZV)$	$(-\frac{1}{2}+\alpha, \frac{1}{2}+\alpha, \frac{1}{2}-\alpha)$	$4mm\ (C_{4v})$	$E, C_{4z}^{\pm}, C_{2z}, \sigma_x, \sigma_y, \sigma_{da}, \sigma_{db}$
	$W(XP)$	$(\alpha, \alpha, \frac{1}{2}-\alpha)$	$mm2\ (C_{2v})$	$E, C_{2z}, \sigma_{da}, \sigma_{db}$
(a) *(b)*	$\Sigma(\Gamma Z)$ $\Sigma(\Gamma\Sigma)$	$(\bar{\alpha}\alpha\alpha)$	$mm2\ (C_{2v})$	$E, C_{2x}, \sigma_y, \sigma_z$
(b)	$F(ZF)$	$(\frac{1}{2}-\alpha, \frac{1}{2}+\alpha, -\frac{1}{2}+\alpha)$	$mm2\ (C_{2v})$	$E, C_{2x}, \sigma_y, \sigma_z$
	$Q(NP)$	$(\alpha, \frac{1}{2}-\alpha, \alpha)$	$2\ (C_2)$	E, C_{2y}
	$\Delta(\Gamma X)$	(00α)	$mm2\ (C_{2v})$	$E, C_{2a}, \sigma_z, \sigma_{db}$
(b)	$U(ZU)$	$(\frac{1}{2}, \frac{1}{2}, -\frac{1}{2}+\alpha)$	$mm2\ (C_{2v})$	$E, C_{2a}, \sigma_z, \sigma_{db}$
(a) *(b)*	$Y(XZ)$ $Y(XY)$	$(\bar{\alpha}, \alpha, \frac{1}{2})$	$mm2\ (C_{2v})$	$E, C_{2b}, \sigma_z, \sigma_{da}$

Bravais lattice	Point	Coordinates	**P(k)**	Elements of **P(k)**
Trigonal, *R*, Γ_{rh} (Fig. 3.11)	Γ	(000)	$\bar{3}m\ (D_{3d})$	$E, C_3^{\pm}, C'_{21}, C'_{22}, C'_{23}, I, S_6^{\mp}, \sigma_{d1}, \sigma_{d2}, \sigma_{d3}$
{(a), (b)}	Z	{$(\frac{1}{2}\frac{1}{2}\bar{\frac{1}{2}})$, $(\frac{1}{2}\frac{1}{2}\frac{1}{2})$}	$\bar{3}m\ (D_{3d})$	$E, C_3^{\pm}, C'_{21}, C'_{22}, C'_{23}, I, S_6^{\mp}, \sigma_{d1}, \sigma_{d2}, \sigma_{d3}$
	L	$(0\frac{1}{2}0)$	$2/m\ (C_{2h})$	$E, C'_{22}, I, \sigma_{d2}$
(a)	F	$(0\frac{1}{2}\bar{\frac{1}{2}})$	$2/m\ (C_{2h})$	$E, C'_{21}, I, \sigma_{d1}$ ‡
(b)	F	$(\frac{1}{2}\frac{1}{2}0)$	$2/m\ (C_{2h})$	$E, C'_{23}, I, \sigma_{d3}$ ‡
{(a), (b)}	$\Lambda(\Gamma\Lambda)$, $\Lambda(\Gamma Z)$	$(\alpha\alpha\alpha)$	$3m\ (C_{3v})$	$E, C_3^{\pm}, \sigma_{d1}, \sigma_{d2}, \sigma_{d3}$
(a)	$P(ZP)$	$(\frac{1}{2}-\alpha, \frac{1}{2}-\alpha, -\frac{1}{2}-\alpha)$	$3m\ (C_{3v})$	$E, C_3^{\pm}, \sigma_{d1}, \sigma_{d2}, \sigma_{d3}$
(b)	$B(ZB)$	$(\frac{1}{2}, \frac{1}{2}+\alpha, \frac{1}{2}-\alpha)$	$2\ (C_2)$	E, C'_{21}
{(a), (b)}	$\Sigma(\Gamma F)$, $\Sigma(\Gamma\Sigma)$	$(0\alpha\bar{\alpha})$	$2\ (C_2)$	E, C'_{21}
(b)	$Q(FQ)$	$(\frac{1}{2}-\alpha, \frac{1}{2}+\alpha, 0)$	$2\ (C_2)$	E, C'_{23}
{(a), (b)}	$Y(LZ)$, $Y(LY)$	$(\alpha, \frac{1}{2}, \bar{\alpha})$	$2\ (C_2)$	E, C'_{22}
Hexagonal, *P*, Γ_h (Fig. 3.12)	Γ	(000)	$6/mmm\ (D_{6h})$	$E, C_6^{\pm}, C_3^{\pm}, C_2, C'_{21}, C'_{22}, C'_{23}, C''_{21}, C''_{22}, C''_{23}, I, S_3^{\mp}, S_6^{\mp}, \sigma_h, \sigma_{d1}, \sigma_{d2}, \sigma_{d3}, \sigma_{v1}, \sigma_{v2}, \sigma_{v3}$
	M	$(0\frac{1}{2}0)$	$mmm\ (D_{2h})$	$E, C_2, C'_{21}, C''_{21}, I, \sigma_h, \sigma_{d1}, \sigma_{v1}$
	A	$(00\frac{1}{2})$	$6/mmm\ (D_{6h})$	$E, C_6^{\pm}, C_3^{\pm}, C_2, C'_{21}, C'_{22}, C'_{23}, C''_{21}, C''_{22}, C''_{23}, I, S_3^{\mp}, S_6^{\mp}, \sigma_h, \sigma_{d1}, \sigma_{d2}, \sigma_{d3}, \sigma_{v1}, \sigma_{v2}, \sigma_{v3}$
	L	$(0\frac{1}{2}\frac{1}{2})$	$mmm\ (D_{2h})$	$E, C_2, C'_{21}, C''_{21}, I, \sigma_h, \sigma_{d1}, \sigma_{v1}$
	K	$(\bar{\frac{1}{3}}\frac{2}{3}0)$	$\bar{6}2m\ (D_{3h})$	$E, C_3^{\pm}, C''_{21}, C''_{22}, C''_{23}, \sigma_h, S_3^{\pm}, \sigma_{d1}, \sigma_{d2}, \sigma_{d3}$
	H	$(\bar{\frac{1}{3}}\frac{2}{3}\frac{1}{2})$	$\bar{6}2m\ (D_{3h})$	$E, C_3^{\pm}, C''_{21}, C''_{22}, C''_{23}, \sigma_h, S_3^{\pm}, \sigma_{d1}, \sigma_{d2}, \sigma_{d3}$
	$\Delta(\Gamma A)$	(00α)	$6mm\ (C_{6v})$	$E, C_6^{\pm}, C_3^{\pm}, C_2, \sigma_{v1}, \sigma_{v2}, \sigma_{v3}, \sigma_{d1}, \sigma_{d2}, \sigma_{d3}$
	$U(ML)$	$(0\frac{1}{2}\alpha)$	$mm2\ (C_{2v})$	$E, C_2, \sigma_{v1}, \sigma_{d1}$
	$P(KH)$	$(\bar{\frac{1}{3}}\frac{2}{3}\alpha)$	$3m\ (C_{3v})$	$E, C_3^{\pm}, \sigma_{d1}, \sigma_{d2}, \sigma_{d3}$
	$T(\Gamma K)$	$(\bar{\alpha}, 2\alpha, 0)$	$mm2\ (C_{2v})$	$E, C''_{22}, \sigma_h, \sigma_{d2}$
	$S(AH)$	$(\bar{\alpha}, 2\alpha, \frac{1}{2})$	$mm2\ (C_{2v})$	$E, C''_{22}, \sigma_h, \sigma_{d2}$
	$T'(MK)$	$(2\bar{\alpha}, \frac{1}{2}+\alpha, 0)$	$mm2\ (C_{2v})$	$E, C''_{21}, \sigma_h, \sigma_{d1}$
	$S'(LH)$	$(2\bar{\alpha}, \frac{1}{2}+\alpha, \frac{1}{2})$	$mm2\ (C_{2v})$	$E, C''_{21}, \sigma_h, \sigma_{d1}$
	$\Sigma(\Gamma M)$	$(0\alpha 0)$	$mm2\ (C_{2v})$	$E, C'_{21}, \sigma_h, \sigma_{v1}$
	$R(AL)$	$(0\alpha\frac{1}{2})$	$mm2\ (C_{2v})$	$E, C'_{21}, \sigma_h, \sigma_{v1}$

‡ In order to keep F within the chosen basic domain we have to allow the elements of **P(k)** in (b) to be different from those of **P(k)** at F in (a), although the points are clearly related. We could preserve **P(k)** $= E, C'_{21}, I, \sigma_{d1}$ for (b) by allowing F to be outside the chosen basic domain (i.e. at F' in Fig. 3.11(b)) but we prefer not to do this.

Bravais lattice	Point	Coordinates	P(k)	Elements of P(k)
Cubic, *P*, Γ_c				
(Fig. 3.13)	Γ	(000)	$m3m\ (O_h)$	$E, C_{3j}^{\pm}, C_{2m}, C_{4m}^{\pm}, C_{2p},$
				$I, S_{6j}^{\mp}, \sigma_m, S_{4m}^{\mp}, \sigma_{dp}$
	X	$(0\frac{1}{2}0)$	$4/mmm\ (D_{4h})$	$E, C_{2y}, C_{4y}^{\pm}, C_{2x}, C_{2z}, C_{2c}, C_{2e},$
				$I, \sigma_y, S_{4y}^{\mp}, \sigma_x, \sigma_z, \sigma_{dc}, \sigma_{de}$
	M	$(\frac{1}{2}\frac{1}{2}0)$	$4/mmm\ (D_{4h})$	$E, C_{2z}, C_{4z}^{\pm}, C_{2x}, C_{2y}, C_{2a}, C_{2b},$
				$I, \sigma_z, S_{4z}^{\mp}, \sigma_x, \sigma_y, \sigma_{da}, \sigma_{db}$
	R	$(\frac{1}{2}\frac{1}{2}\frac{1}{2})$	$m3m\ (O_h)$	$E, C_{3j}^{\pm}, C_{2m}, C_{4m}^{\pm}, C_{2p},$
				$I, S_{6j}^{\mp}, \sigma_m, S_{4m}^{\mp}, \sigma_{dp}$
	$\Delta(\Gamma X)$	$(0\alpha 0)$	$4mm\ (C_{4v})$	$E, C_{2y}, C_{4y}^{\pm}, \sigma_x, \sigma_z, \sigma_{dc}, \sigma_{de}$
	$\Sigma(\Gamma M)$	$(\alpha\alpha 0)$	$mm2\ (C_{2v})$	$E, C_{2a}, \sigma_z, \sigma_{db}$
	$\Lambda(\Gamma R)$	$(\alpha\alpha\alpha)$	$3m\ (C_{3v})$	$E, C_{31}^{\pm}, \sigma_{db}, \sigma_{de}, \sigma_{df}$
	$S(XR)$	$(\alpha\frac{1}{2}\alpha)$	$mm2\ (C_{2v})$	$E, C_{2c}, \sigma_{de}, \sigma_y$
	$Z(XM)$	$(\alpha\frac{1}{2}0)$	$mm2\ (C_{2v})$	$E, C_{2x}, \sigma_y, \sigma_z$
	$T(MR)$	$(\frac{1}{2}\frac{1}{2}\alpha)$	$4mm\ (C_{4v})$	$E, C_{2z}, C_{4z}^{\pm}, \sigma_x, \sigma_y, \sigma_{da}, \sigma_{db}$
Cubic, *F*, Γ_c^f				
(Fig. 3.14)	Γ	(000)	$m3m\ (O_h)$	$E, C_{3j}^{\pm}, C_{2m}, C_{4m}^{\pm}, C_{2p},$
				$I, S_{6j}^{\mp}, \sigma_m, S_{4m}^{\mp}, \sigma_{dp}$
	X	$(\frac{1}{2}0\frac{1}{2})$	$4/mmm\ (D_{4h})$	$E, C_{2y}, C_{4y}^{\pm}, C_{2x}, C_{2z}, C_{2c}, C_{2e},$
				$I, \sigma_y, S_{4y}^{\mp}, \sigma_x, \sigma_{dc}, \sigma_{de}$
	L	$(\frac{1}{2}\frac{1}{2}\frac{1}{2})$	$\bar{3}m\ (D_{3d})$	$E, C_{31}^{\pm}, C_{2b}, C_{2e}, C_{2f},$
				$I, S_{61}^{\mp}, \sigma_{db}, \sigma_{de}, \sigma_{df}$
	W	$(\frac{1}{2}\frac{1}{4}\frac{3}{4})$	$\bar{4}2m\ (D_{2d})$	$E, C_{2x}, C_{2d}, C_{2f}, \sigma_y, \sigma_z, S_{4x}^{\pm}$
	$\Delta(\Gamma X)$	$(\alpha 0\alpha)$	$4mm\ (C_{4v})$	$E, C_{2y}, C_{4y}^{\pm}, \sigma_x, \sigma_z, \sigma_{dc}, \sigma_{de}$
	$\Lambda(\Gamma L)$	$(\alpha\alpha\alpha)$	$3m\ (C_{3v})$	$E, C_{31}^{\pm}, \sigma_{db}, \sigma_{de}, \sigma_{df}$
	$\Sigma(\Gamma\Sigma)$	$(\alpha, \alpha, 2\alpha)$	$mm2\ (C_{2v})$	$E, C_{2a}, \sigma_z, \sigma_{db}$
	$S(XS)$	$(\frac{1}{2} + \alpha, 2\alpha, \frac{1}{2} + \alpha)$	$mm2\ (C_{2v})$	$E, C_{2c}, \sigma_{de}, \sigma_y$
	$Z(XW)$	$(\frac{1}{2}, \alpha, \frac{1}{2} + \alpha)$	$mm2\ (C_{2v})$	$E, C_{2x}, \sigma_y, \sigma_z$
	$Q(LW)$	$(\frac{1}{2}, \frac{1}{2} - \alpha, \frac{1}{2} + \alpha)$	$2\ (C_2)$	E, C_{2f}
Cubic, *I*, Γ_c^v				
(Fig. 3.15)	Γ	(000)	$m3m\ (O_h)$	$E, C_{3j}^{\pm}, C_{2m}, C_{4m}^{\pm}, C_{2p},$
				$I, S_{6j}^{\mp}, \sigma_m, S_{4m}^{\mp}, \sigma_{dp}$
	H	$(\frac{1}{2}\bar{\frac{1}{2}}\frac{1}{2})$	$m3m\ (O_h)$	$E, C_{3j}^{\pm}, C_{2m}, C_{4m}^{\pm}, C_{2p},$
				$I, S_{6j}^{\mp}, \sigma_m, S_{4m}^{\mp}, \sigma_{dp}$
	P	$(\frac{1}{4}\frac{1}{4}\frac{1}{4})$	$\bar{4}3m\ (T_d)$	$E, C_{3j}^{\pm}, C_{2m}, S_{4m}^{\pm}, \sigma_{dp}$
	N	$(00\frac{1}{2})$	$mmm\ (D_{2h})$	$E, C_{2a}, C_{2z}, C_{2b}, I, \sigma_{da}, \sigma_z, \sigma_{db}$
	$\Sigma(\Gamma N)$	(00α)	$mm2\ (C_{2v})$	$E, C_{2a}, \sigma_z, \sigma_{db}$
	$\Delta(\Gamma H)$	$(\alpha\bar{\alpha}\alpha)$	$4mm\ (C_{4v})$	$E, C_{2y}, C_{4y}^{\pm}, \sigma_x, \sigma_z, \sigma_{dc}, \sigma_{de}$
	$\Lambda(\Gamma P)$	$(\alpha\alpha\alpha)$	$3m\ (C_{3v})$	$E, C_{31}^{\pm}, \sigma_{db}, \sigma_{de}, \sigma_{df}$
	$D(NP)$	$(\alpha, \alpha, \frac{1}{2} - \alpha)$	$mm2\ (C_{2v})$	$E, C_{2z}, \sigma_{da}, \sigma_{db}$
	$G(HN)$	$(\alpha\bar{\alpha}\frac{1}{2})$	$mm2\ (C_{2v})$	$E, C_{2b}, \sigma_z, \sigma_{da}$
	$F(PH)$	$(\frac{1}{4} + \alpha, \frac{1}{4} - 3\alpha, \frac{1}{4} + \alpha)$	$3m\ (C_{3v})$	$E, C_{34}^{\pm}, \sigma_{da}, \sigma_{dd}, \sigma_{de}$

Notes to Table 3.6

(i) Column 1 lists the Bravais lattice in direct space, and indicates the appropriate part of Figs. 3.2–3.15 in which the positions of the points and lines of symmetry are illustrated.

(ii) Column 2 lists the points and lines of symmetry. These are to be identified from Figs. 3.2–3.15 or alternatively from column 3, which lists the coordinates of the points and lines with respect to $\mathbf{g}_1$, $\mathbf{g}_2$, and $\mathbf{g}_3$. For example, W in Γ_m is a point on the line YC and its $\mathbf{k}$ vector is $(\frac{1}{2}\mathbf{g}_2 + \alpha\mathbf{g}_3)$, $0 < \alpha < \frac{1}{2}$. The labels used for the points and lines of symmetry are, with a few exceptions, those of Miller and Love (1967).

(iii) Column 4 lists the name of the symmetry group, $\mathbf{P}(\mathbf{k})$, of each of the points and lines. For example, W in Γ_m has the symmetry group 2 (C_2). For the definition of symmetry group see the text of section 3.3. For the relative orientation of the vectors $\mathbf{g}_1$, $\mathbf{g}_2$, and $\mathbf{g}_3$ with respect to the Cartesian axes $\Gamma k_x k_y k_z$ see Table 3.3.

(iv) Column 5 lists the elements of the symmetry groups, $\mathbf{P}(\mathbf{k})$. These are to be identified by means of Figs. 1.1–1.3 with axes $Oxyz$ coincident with axes $\Gamma k_x k_y k_z$ of Figs. 3.2–3.15 in every case.

3.4. The irreducible representations of the translation groups

Every space group has an invariant subgroup, $\mathbf{T}$, consisting of the translational symmetry operations of the Bravais lattice on which the space group is based. In the present section we consider the determination of the irreducible representations of the group of the translational symmetry operations of a Bravais lattice.

$\mathbf{T}$ consists of elements $\{E \mid n_1\mathbf{t}_1 + n_2\mathbf{t}_2 + n_3\mathbf{t}_3\}$, where n_i is an integer that satisfies $0 \leqslant n_i < N_i$, $i = 1, 2$, or 3 and where we apply Born–von Kármán cyclic boundary conditions (Born and von Kármán 1913*a*, *b*). Because the multiplication rule is of the form $\{E \mid \mathbf{t}\}\{E \mid \mathbf{s}\} = \{E \mid \mathbf{t} + \mathbf{s}\}$ and because of the periodicity imposed by eqn. (3.1.1), $\mathbf{T}$ is the outer direct product of three cyclic groups of orders N_1, N_2, and N_3 so that in an obvious notation $\mathbf{T} = \mathbf{T}_1 \otimes \mathbf{T}_2 \otimes \mathbf{T}_3$. The reps of cyclic groups are well known. For example, the reps Δ of $\mathbf{T}_1$ are N_1 in number and each is labelled by an integer p_1 $(0 \leqslant p_1 < N_1)$ so that

$$\Delta^{p_1}(\{E \mid n_1\mathbf{t}_1\}) = \exp(-2\pi i n_1 p_1/N_1). \tag{3.4.1}$$

The reps of $\mathbf{T}$ can now easily be deduced from Theorem (1.3.16) on direct product groups. They are 1-dimensional and are $N_1N_2N_3$ in number. Each rep is labelled by three integers p_1, p_2, and p_3 $(0 \leqslant p_i < N_i)$. If we write $k_i = p_i/N_i$ then we can take $\mathbf{k} = (k_1, k_2, k_3)$ to be a vector in reciprocal space, that is,

$$\mathbf{k} = k_1\mathbf{g}_1 + k_2\mathbf{g}_2 + k_3\mathbf{g}_3. \tag{3.4.2}$$

$\mathbf{k}$ can then be taken over as the label for the reps of $\mathbf{T}$ and these reps are

$$\Delta^{\mathbf{k}}[\{E \mid n_1\mathbf{t}_1 + n_2\mathbf{t}_2 + n_3\mathbf{t}_3\}] = \exp\{-i\mathbf{k}.(n_1\mathbf{t}_1 + n_2\mathbf{t}_2 + n_3\mathbf{t}_3)\}. \tag{3.4.3}$$

We can now see the reason for defining two $\mathbf{k}$ vectors to be equivalent if they differ by a reciprocal lattice vector. For if $\mathbf{k}' = \mathbf{k} + \mathbf{g}$ then, since $\exp(-i\mathbf{g}.\mathbf{t}) = 1$,

$$\Delta^{\mathbf{k}'}(\mathbf{t}) = \exp\{-i(\mathbf{k} + \mathbf{g}).\mathbf{t}\} = \Delta^{\mathbf{k}}(\mathbf{t}). \tag{3.4.4}$$

Hence we can take as allowed values of $\mathbf{k}$ any set of $N_1N_2N_3$ non-equivalent $\mathbf{k}$ vectors. One such set is defined above and they terminate either within or on the surface of the primitive unit cell of reciprocal space (except for a change of origin) which for triclinic and monoclinic Bravais lattices we took to define the Brillouin zone (see Definition 3.2.2). For all other Bravais lattices we may take as the set of $N_1N_2N_3$ non-equivalent **k** vectors the set of vectors which terminate either within or on the surface of the Wigner–Seitz unit cell of reciprocal space (see Definition 3.2.2). Thus in every case we may therefore think of the reps of **T** as being defined by points **k** of the Brillouin zone.

It will be useful to have a formula for the density of allowed **k** vectors. There are $N_1N_2N_3$ in a volume $8\pi^3/V$ where V is the volume of a unit cell of the crystal. The density $n(\mathbf{k})$ is therefore $N_1N_2N_3V/8\pi^3 = W/8\pi^3$ where W is the volume of the crystal. Thus

$$n(\mathbf{k}) = \frac{W}{8\pi^3}. \tag{3.4.5}$$

This implies that in the limit of an infinite crystal the **k** values become arbitrarily close together and so fill densely the whole of the Brillouin zone.

The basis functions for $\Delta^{\mathbf{k}}$ can be taken to be exp (i**k**.**r**) because, see Example 1.5.3,

$$\begin{aligned}\{E \mid \mathbf{t}\} \exp(i\mathbf{k}.\mathbf{r}) &= \exp\{i\mathbf{k}.(\mathbf{r}-\mathbf{t})\}\\ &= \exp(-i\mathbf{k}.\mathbf{t})\exp(i\mathbf{k}.\mathbf{r}) = \Delta^{\mathbf{k}}\{E \mid \mathbf{t}\}\exp(i\mathbf{k}.\mathbf{r}).\end{aligned} \tag{3.4.6}$$

Instead of exp (i**k**.**r**) we can use any function of the form

$$\Psi_{\mathbf{k}}(\mathbf{r}) = \exp(i\mathbf{k}.\mathbf{r})u_{\mathbf{k}}(\mathbf{r}) \tag{3.4.7}$$

where

$$u_{\mathbf{k}}(\mathbf{r}) = u_{\mathbf{k}}(\mathbf{r}+\mathbf{t}) \tag{3.4.8}$$

for all **t**. That the basis functions of the translation group **T** are of the form defined by eqns. (3.4.7) and (3.4.8) is commonly known as *Bloch's theorem*.

THEOREM 3.4.1. Bloch's theorem (Bloch 1928). *The wave function of a particle or quasi-particle that moves in a periodic potential,* $V(\mathbf{r})$, *with periodicity defined by* **T**, *can be written in the form*

$$\Psi_{\mathbf{k}}(\mathbf{r}) = \exp(i\mathbf{k}.\mathbf{r})u_{\mathbf{k}}(\mathbf{r}) \tag{3.4.7}$$

where $u_{\mathbf{k}}(\mathbf{r})$ *has the same periodicity as* $V(\mathbf{r})$ *and where* **k** *is determined by eqn.* (3.4.2).

We must emphasize, however, that the label **k** on the Bloch function $\Psi_{\mathbf{k}}(\mathbf{r})$, just as with the reps themselves, is determined only up to equivalence and can be altered by vectors **g** of the reciprocal lattice. This follows from the fact that exp (i**g**.**r**) satisfies eqn. (3.4.8), so that any such factor can be absorbed into the $u_{\mathbf{k}}(\mathbf{r})$ part of the Bloch function. Modifications to the theory are necessary if the crystal under consideration is placed in a uniform external electric or magnetic field (see, for example, Ashby and Miller (1965)).

3.5. The classification of the 230 3-dimensional space groups

So far in this chapter we have discussed the translational symmetry of space groups. A space group **G** has a 3-dimensional subgroup of pure translations **T**; this determines the Bravais lattice. The Bravais lattice has the symmetry of **T**, but in addition it is invariant under operations of a point group **P**. In fact, if **Γ** is the full symmetry group of the Bravais lattice, then $\mathbf{\Gamma} = \mathbf{T} \wedge \mathbf{P}$, the semi-direct product of **T** and **P** (see Definition (1.2.17) for the meaning of semi-direct product). **P** determines the crystal system of **G**. **Γ** is itself a space group. If **G** is symmorphic then **G** either coincides with **Γ** or is a subgroup of **Γ**. This accounts for 73 space groups, see section 1.5. It is clear that in this case $\mathbf{G} = \mathbf{T} \wedge \mathbf{Q}$ where **Q** is a point group in the crystal system of **P**. **Q** determines the point group of the crystal of which the space group is **G**.

We can account for the remaining 157 non-symmorphic space groups by replacing some of the elements of **Q** by screw-axis rotations or glide-plane reflections. **Q** is then no longer a group and one can no longer write $\mathbf{G} = \mathbf{T} \wedge \mathbf{Q}$; this is because if the elements of the set **Q** could be chosen to form a group then it would be isomorphic to a point group and **G** would in turn be isomorphic to one of the symmorphic space groups.

Let us now consider the restrictions that must be placed on the elements $\{R \mid \mathbf{v}\}$ of **Q** so that **G** will be a space group. The following discussion applies trivially to symmorphic space groups and may be taken therefore as a complete characterization. First **G** must be a group. Hence, if $\{R \mid \mathbf{v}\}$ is an element of **G**, its inverse $\{R^{-1} \mid -R^{-1}\mathbf{v}\}$ must also be an element of **G**, see eqn. (1.5.3). Therefore, if $\{R \mid \mathbf{w}\}$ is in **G** as well, then so also is

$$\{R^{-1} \mid -R^{-1}\mathbf{v}\}\{R \mid \mathbf{w}\} = \{E \mid R^{-1}\mathbf{w} - R^{-1}\mathbf{v}\}. \tag{3.5.1}$$

Indeed, this must be an element of **T**, the translational subgroup, since its rotational part is the identity. Secondly, **G** has a Bravais lattice that belongs to a crystal system whose holosymmetric point group, **P**, must contain the operation R. So that if, as eqn. (3.5.1) implies, $R^{-1}(\mathbf{w} - \mathbf{v})$ is a translation of **Γ** then so is $(\mathbf{w} - \mathbf{v})$. This means that the translational parts associated with any point-group operation are separated by elements of **T**. It is therefore possible to write **G** in terms of left coset representatives of **T** (see Definition 1.2.9),

$$\mathbf{G} = \{R_1 \mid \mathbf{v}_1\}\mathbf{T} + \{R_2 \mid \mathbf{v}_2\}\mathbf{T} + \cdots + \{R_h \mid \mathbf{v}_h\}\mathbf{T}, \tag{3.5.2}$$

where by choice we can set $R_1 = E$ and $\mathbf{v}_i$ to be a vector associated with R_i ($i = 1$ to h) of least possible magnitude. For a symmorphic space group all $\mathbf{v}_i$ are therefore equal to zero, but for a non-symmorphic space group there must be one or more $\mathbf{v}_i$ non-zero. **Q** consists of the coset representatives $\{R_i \mid \mathbf{v}_i\}$. The h elements $R_1, R_2, \ldots, R_h$ form a point group **F** which is the isogonal point group of **G** (see section 1.5). **F** is a subgroup of **P**. For a symmorphic space group **F** coincides with **Q**. h is said to be the macroscopic order of **G**; it is the index of **T** in **G**.

In the above specification the non-zero vectors **v** are not arbitrary because of the group requirements of **G**. For example, if R is of order n then $\{R \mid \mathbf{v}\}^n$ must belong to **T**. That is, $(\mathbf{v} + R\mathbf{v} + \cdots + R^{n-1}\mathbf{v})$ must be a translation. This and other such requirements limit the number of non-symmorphic space groups to 157 making a total of 230 space groups in all.

THEOREM 3.5.1. **T** *is an invariant subgroup of* **G**. For if $\{R \mid \mathbf{v}\}$ is an element of **G** and if $\{E \mid \mathbf{t}\}$ is an element of **T** then

$$\{R \mid \mathbf{v}\}\{E \mid \mathbf{t}\}\{R \mid \mathbf{v}\}^{-1} = \{R \mid \mathbf{v} + R\mathbf{t}\}\{R^{-1} \mid -R^{-1}\mathbf{v}\} = \{E \mid R\mathbf{t}\} \quad (3.5.3)$$

which belongs to **T** since $R\mathbf{t}$ is a translation of **Γ**.

The Seitz space-group symbols for the elements of each space group are given by various authors (Faddeyev 1964, Koptsik 1966, Kovalev 1965, Lyubarskii 1960, Miller and Love 1967) or they can be written down using the *International tables for X-ray crystallography* (Henry and Lonsdale 1965). In section 4.3 of those tables there is given, for each space group, a set of coordinates of equivalent positions in the unit cell; for each of the non-cubic space groups there is also a diagram illustrating the symmetry elements present. It is possible to use these tables not only to obtain the elements of the space group in the form of the Seitz symbols, $\{R_i \mid \mathbf{v}_i\}$, but also to interpret these symmetry operations in their geometrical aspect; for a screw rotation axis, for example, one can determine the direction of the axis and the magnitudes of the rotation and the translation along the axis (Wondratschek and Neubüser 1967). We shall illustrate both these uses of the equivalent positions given in the *International Tables*.

We show how to use the equivalent positions given in the *International Tables* to determine the appropriate Seitz space-group symbols by considering one example, that of the space group *Pmmn* (D_{2h}^{13}) which is based on the primitive orthorhombic Bravais lattice. These equivalent positions are given both for a completely general point in the unit cell and also for points lying on the various symmetry elements. In Fig. 3.17 we reproduce the two pages of section 4.3 of the first volume of the *International Tables* relevant to the space group *Pmmn* (D_{2h}^{13}). In Fig. 3.17(a) the first set of equivalent positions corresponds to a general point (x, y, z) in the unit cell; the remaining sets of equivalent positions correspond to points (x, y, z) that are situated on some axis or plane of symmetry and these do not concern us here. Starting with the point $\mathbf{r}_1\{=(x, y, z)\}$ and performing the space-group operation denoted by $\{R_1 \mid \mathbf{v}_1\}$ we obtain from eqn. (1.5.2)

$$\mathbf{r}'_1 = R_1\mathbf{r}_1 + \mathbf{v}_1 \quad (3.5.4)$$

and the resulting vectors $\mathbf{r}'_1$ are given by

$$x, y, z; \quad \bar{x}, \bar{y}, z; \quad \bar{x}, y, z; \quad x, \bar{y}, z;$$

$$\tfrac{1}{2} - x, \tfrac{1}{2} - y, \bar{z}; \quad \tfrac{1}{2} - x, \tfrac{1}{2} + y, \bar{z}; \quad \tfrac{1}{2} + x, \tfrac{1}{2} + y, \bar{z}; \quad \tfrac{1}{2} + x, \tfrac{1}{2} - y, \bar{z}.$$

The fractions refer to fractions of the lengths of the edges, in the x, y, and z directions, of the conventional unit cell that is drawn for each space group, except the cubic space

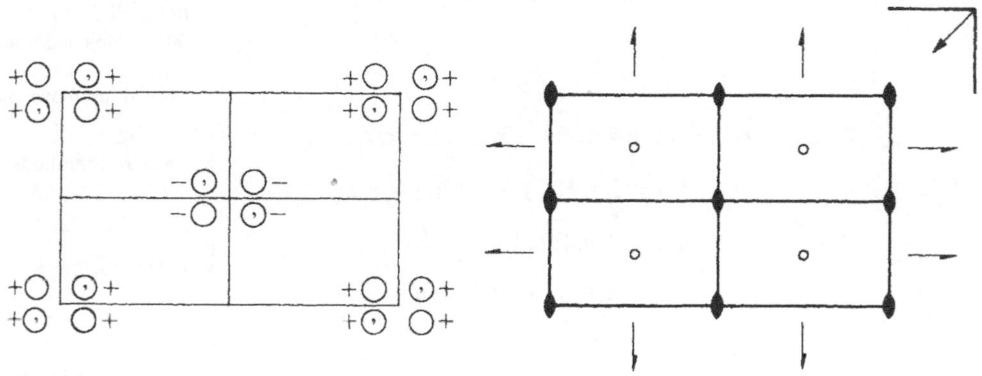

Number of positions	Wyckoff notation	Point symmetry	Coordinates of equivalent positions	Conditions limiting possible reflections
(a)			Origin at *mmn*, at $\frac{1}{4}, \frac{1}{4}, 0$ from $\bar{1}$	
				General:
8	*g*	1	$x, y, z; \bar{x}, \bar{y}, z; \frac{1}{2} - x, \frac{1}{2} - y, \bar{z}; \frac{1}{2} - x, \frac{1}{2} + y, \bar{z};$ $\bar{x}, y, z; x, \bar{y}, z; \frac{1}{2} + x, \frac{1}{2} + y, \bar{z}; \frac{1}{2} + x, \frac{1}{2} - y, \bar{z}.$	hkl:, $0kl$:, $h0l$: No conditions $hk0$: $h + k = 2n$ $h00$: $(h = 2n)$ $0k0$: $(k = 2n)$ $00l$: No conditions
				Special: as above, plus
4	*f*	*m*	$x, 0, z; \bar{x}, 0, z; \frac{1}{2} - x, \frac{1}{2}, \bar{z}; \frac{1}{2} + x, \frac{1}{2}, \bar{z}$	no extra conditions
4	*e*	*m*	$0, y, z; 0, \bar{y}, z; \frac{1}{2}, \frac{1}{2} - y, z; \frac{1}{2}, \frac{1}{2} + y, z.$	
4	*d*	$\bar{1}$	$\frac{1}{4}, \frac{1}{4}, \frac{1}{2}; \frac{3}{4}, \frac{3}{4}, \frac{1}{2}; \frac{1}{4}, \frac{3}{4}, \frac{1}{2}; \frac{3}{4}, \frac{1}{4}, \frac{1}{2}.$	hkl: $h = 2n; k = 2n$
4	*c*	$\bar{1}$	$\frac{1}{4}, \frac{1}{4}, 0; \frac{3}{4}, \frac{3}{4}, 0; \frac{1}{4}, \frac{3}{4}, 0; \frac{3}{4}, \frac{1}{4}, 0.$	
2	*b*	*mm*	$0, \frac{1}{2}, z; \frac{1}{2}, 0, \bar{z}.$	no extra conditions
2	*a*	*mm*	$0, 0, z; \frac{1}{2}, \frac{1}{2}, \bar{z}.$	

(b)			Origin at $\bar{1}$, at $\bar{\tfrac{1}{4}}, \bar{\tfrac{1}{4}}, 0$ from *mmn*	
				General:
8	g	1	$x, y, z; \tfrac{1}{2} - x, y, z; x, \tfrac{1}{2} - y, z; \tfrac{1}{2} - x, \tfrac{1}{2} - y, z;$ $\bar{x}, \bar{y}, \bar{z}; \tfrac{1}{2} + x, \bar{y}, \bar{z}; \bar{x}, \tfrac{1}{2} + y, \bar{z}; \tfrac{1}{2} + x, \tfrac{1}{2} + y, \bar{z}.$	hkl, $0kl$:, $h0l$: } No conditions $hk0$: $h + k = 2n$ $h00$: $(h = 2\text{n})$ $0k0$: $(k = 2n)$ $00l$: No conditions
				Special: as above, plus
4	f	m	$x, \tfrac{1}{4}, z; \bar{x}, \tfrac{3}{4}, \bar{z}; \tfrac{1}{2} - x, \tfrac{1}{4}, z; \tfrac{1}{2} + x, \tfrac{3}{4}, \bar{z}.$	} no extra conditions
4	e	m	$\tfrac{1}{4}, y, z; \tfrac{3}{4}, \bar{y}, \bar{z}; \tfrac{1}{4}, \tfrac{1}{2} - y, z; \tfrac{3}{4}, \tfrac{1}{2} + y, \bar{z}.$	
4	d	$\bar{1}$	$0, 0, \tfrac{1}{2}; \tfrac{1}{2}, 0, \tfrac{1}{2}; 0, \tfrac{1}{2}, \tfrac{1}{2}; \tfrac{1}{2}, \tfrac{1}{2}, \tfrac{1}{2}.$	} hkl: $h = 2n; k = 2n$
4	c	$\bar{1}$	$0, 0, 0; \tfrac{1}{2}, 0, 0; 0, \tfrac{1}{2}, 0; \tfrac{1}{2}, \tfrac{1}{2}, 0.$	
2	b	mm	$\tfrac{1}{4}, \tfrac{3}{4}, z; \tfrac{3}{4}, \tfrac{1}{4}, \bar{z}.$	} no extra conditions
2	a	mm	$\tfrac{1}{4}, \tfrac{1}{4}, z; \tfrac{3}{4}, \tfrac{3}{4}, \bar{z}.$	

FIG. 3.17. Excerpt from the *International Tables* (Henry and Lonsdale 1965) illustrating alternative settings for the group

Pmmn (D_{2h}^{13}) (orthorhombic, *mmm*, $P2_1/m2_1/m2/n$, No. 59).

groups, in the *International Tables*. We can identify R_1 for each of the eight vectors $\mathbf{r}_1'$ by making use of Table 1.4 from which we find

$$\begin{array}{cccccccc} xyz; & \bar{x}\bar{y}z; & \bar{x}yz; & x\bar{y}z; & \bar{x}\bar{y}\bar{z}; & \bar{x}y\bar{z}; & xy\bar{z}; & x\bar{y}\bar{z}; \\ E & C_{2z} & \sigma_x & \sigma_y & I & C_{2y} & \sigma_z & C_{2x} \end{array}$$

which gives the rotational parts of the space-group symbols. The translation vector $\mathbf{v}_1$ associated with each of the first four of these is clearly zero. Since $\mathbf{t}_1 = (0, -b, 0)$ and $\mathbf{t}_2 = (a, 0, 0)$ for the primitive orthorhombic Bravais lattice (see Table 3.1), the translation vector associated with each of the last four of these symbols is $-\tfrac{1}{2}\mathbf{t}_1 + \tfrac{1}{2}\mathbf{t}_2$. We can add an integer multiple of $\mathbf{t}_1$, $\mathbf{t}_2$, or $\mathbf{t}_3$ to the translation vector $\mathbf{v}_1$ so that we choose $\mathbf{v}_1 = +\tfrac{1}{2}\mathbf{t}_1 + \tfrac{1}{2}\mathbf{t}_2$ and the elements of the space group *Pmmn* (D_{2h}^{13}) are therefore

$$\{E \mid 000\}, \{C_{2z} \mid 000\}, \{\sigma_x \mid 000\}, \{\sigma_y \mid 000\},$$

$$\{I \mid \tfrac{1}{2}\tfrac{1}{2}0\}, \{C_{2y} \mid \tfrac{1}{2}\tfrac{1}{2}0\}, \{\sigma_z \mid \tfrac{1}{2}\tfrac{1}{2}0\}, \text{ and } \{C_{2x} \mid \tfrac{1}{2}\tfrac{1}{2}0\}.$$

It is not necessary to list all these and in Table 3.7 we have listed only the generating elements of each space group; under *Pmmn* (D_{2h}^{13}) one can see that generating elements $\{C_{2x} \mid \frac{1}{2}\frac{1}{2}0\}$, $\{C_{2y} \mid \frac{1}{2}\frac{1}{2}0\}$, and $\{I \mid \frac{1}{2}\frac{1}{2}0\}$ have been chosen.

It is important to be able to see how the form of any given Seitz space-group symbol $\{R_1 \mid \mathbf{v}_1\}$ is affected if the origin of the coordinate system used to describe the crystal is moved to a different point in the unit cell. Suppose that $\mathbf{r}_1$ and $\mathbf{r}_1'$ are the positions of a point before and after the performance of an active space-group symmetry operation; this operation is denoted by $\{R_1 \mid \mathbf{v}_1\}$ in the first choice of coordinate axes $O_1x_1y_1z_1$. Suppose now that a new choice of coordinate axes

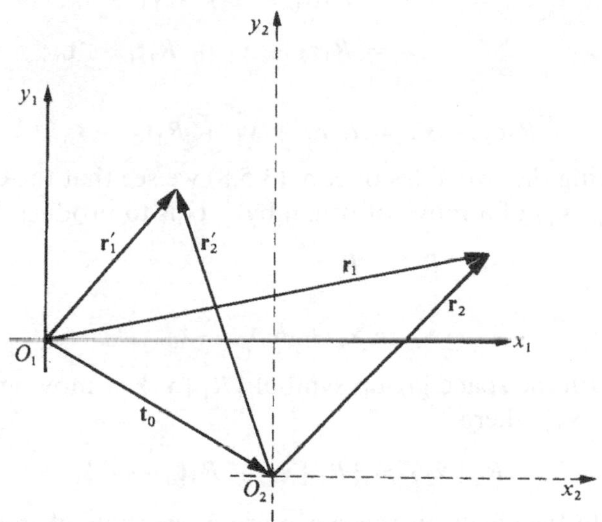

FIG. 3.18. Change of origin by a translation.

$O_2x_2y_2z_2$ is made where the origin is displaced, without any rotation of the axes, by a vector $+\mathbf{t}_0$ from the origin of the axes $O_1x_1y_1z_1$, see Fig. 3.18. From this figure we can see that

$$\mathbf{r}_1 = \mathbf{t}_0 + \mathbf{r}_2 \tag{3.5.5}$$

and

$$\mathbf{r}_1' = \mathbf{t}_0 + \mathbf{r}_2' \tag{3.5.6}$$

and, from the definition of the space-group symbol $\{R_1 \mid \mathbf{v}_1\}$ which acts in the system $O_1x_1y_1z_1$, we have

$$\mathbf{r}_1' = R_1\mathbf{r}_1 + \mathbf{v}_1. \tag{3.5.4}$$

If this space-group operation is denoted by $\{R_2 \mid \mathbf{v}_2\}$ in the coordinate system $O_2x_2y_2z_2$ then $\mathbf{r}_2'$ and $\mathbf{r}_2$ are related by

$$\mathbf{r}_2' = R_2\mathbf{r}_2 + \mathbf{v}_2 \tag{3.5.7}$$

and we can use all these equations to determine $\{R_2 \mid \mathbf{v}_2\}$ in terms of $\{R_1 \mid \mathbf{v}_1\}$. From eqn. (3.5.7)

$$\begin{aligned} R_2\mathbf{r}_2 + \mathbf{v}_2 &= \mathbf{r}_2', \\ &= \mathbf{r}_1' - \mathbf{t}_0, && \text{from (3.5.6)}, \\ &= R_1\mathbf{r}_1 + \mathbf{v}_1 - \mathbf{t}_0, && \text{from (3.5.4)}, \\ &= R_1(\mathbf{t}_0 + \mathbf{r}_2) + \mathbf{v}_1 - \mathbf{t}_0, && \text{from (3.5.5)}, \\ &= R_1\mathbf{r}_2 + \mathbf{v}_1 + R_1\mathbf{t}_0 - \mathbf{t}_0; \end{aligned}$$

that is

$$R_2\mathbf{r}_2 + \mathbf{v}_2 = R_1\mathbf{r}_2 + \mathbf{v}_1 + R_1\mathbf{t}_0 - \mathbf{t}_0. \tag{3.5.8}$$

Therefore comparing the two sides of eqn. (3.5.8) we see that the effect on the space-group symbol $\{R_1 \mid \mathbf{v}_1\}$ of a move of origin by $+\mathbf{t}_0$ is to produce $\{R_2 \mid \mathbf{v}_2\}$ where

$$R_2 = R_1 \tag{3.5.9}$$

and

$$\mathbf{v}_2 = \mathbf{v}_1 + R_1\mathbf{t}_0 - \mathbf{t}_0. \tag{3.5.10}$$

That is, the effect on the space-group symbol $\{R_1 \mid \mathbf{v}_1\}$ of moving the origin by $+\mathbf{t}_0$ is to produce $\{R_2 \mid \mathbf{v}_2\}$ where

$$\{R_2 \mid \mathbf{v}_2\} = \{R_1 \mid \mathbf{v}_1 + R_1\mathbf{t}_0 - \mathbf{t}_0\}. \tag{3.5.11}$$

We can illustrate the effect on the space-group symbols of moving the origin by considering the space group *Pmmn* (D_{2h}^{13}) again. The *International Tables* give two different positions for this space group, see Fig. 3.17. We have already used the first setting (Fig. 3.17(a)) to write down the space-group elements. In the second setting, see Fig. 3.17(b), the origin has been moved by $\{(-\frac{1}{4} \times$ the repeat distance along the x-axis) + $(-\frac{1}{4} \times$ the repeat distance along the y-axis)$\}$, that is, from Table 3.1, by

$$\mathbf{t}_0 = \tfrac{1}{4}\mathbf{t}_1 - \tfrac{1}{4}\mathbf{t}_2. \tag{3.5.12}$$

Therefore, if we make use of eqns. (3.5.9) and (3.5.10) we can convert the space-group elements we obtained before to the second set of axes, for example, $\{C_{2z} \mid 000\}$ becomes replaced by $\{R_2 \mid \mathbf{v}_2\}$ where, from eqn. (3.5.9), $R_2 = C_{2z}$ and

$$\begin{aligned} \mathbf{v}_2 &= \mathbf{v}_1 + R_1\mathbf{t}_0 - \mathbf{t}_0 && (3.5.10) \\ &= \mathbf{0} + C_{2z}(\tfrac{1}{4}\mathbf{t}_1 - \tfrac{1}{4}\mathbf{t}_2) - (\tfrac{1}{4}\mathbf{t}_1 - \tfrac{1}{4}\mathbf{t}_2) \\ &= (-\tfrac{1}{4}\mathbf{t}_1 + \tfrac{1}{4}\mathbf{t}_2) - (\tfrac{1}{4}\mathbf{t}_1 - \tfrac{1}{4}\mathbf{t}_2) \quad \text{from Table 3.2} \\ &= -\tfrac{1}{2}\mathbf{t}_1 + \tfrac{1}{2}\mathbf{t}_2, && (3.5.13) \end{aligned}$$

so that the element in the second coordinate system is $\{C_{2z} \mid -\frac{1}{2}\frac{1}{2}0\}$ to which we may add $\mathbf{t}_1$ to obtain $\{C_{2z} \mid \frac{1}{2}\frac{1}{2}0\}$ as an element of the space group referred to the new origin. This agrees with the equivalent positions shown in Fig. 3.17(b) where $\{C_{2z} \mid -\frac{1}{2}\frac{1}{2}0\}$ acting on (x, y, z) produces $(\frac{1}{2} - x, \frac{1}{2} - y, z)$ which is one of the eight general equivalent positions given in Fig. 3.17(b). Repeating this use of eqns. (3.5.9) and (3.5.10) with the same $\mathbf{t}_0 = \frac{1}{4}\mathbf{t}_1 - \frac{1}{4}\mathbf{t}_2$ we obtain all the elements of the space group *Pmmn* (D_{2h}^{13}) referred to the new origin:

$$\{E \mid 000\}, \{C_{2z} \mid \tfrac{1}{2}\tfrac{1}{2}0\}, \{\sigma_x \mid 0\tfrac{1}{2}0\}, \{\sigma_y \mid \tfrac{1}{2}00\},$$
$$\{I \mid 000\}, \{\sigma_z \mid \tfrac{1}{2}\tfrac{1}{2}0\}, \{C_{2x} \mid 0\tfrac{1}{2}0\}, \{C_{2y} \mid \tfrac{1}{2}00\}.$$

In this list we have added or subtracted vectors of the translation group of the primitive orthorhombic Bravais lattice in certain cases to simplify the form of the space-group elements.

The generating elements of each of the 230 space groups are listed in Table 3.7. In the first three columns we identify the space groups using the international and Schönflies symbols and in column 4 we give the Seitz space-group symbols of the

TABLE 3.7

The 230 space groups

International number	International symbol	Schönflies symbol	Generating elements	$\mathbf{t}_0$
1	*P*1	$\Gamma_t C_1^1$	$\{E \mid 000\}$	0
2	$P\bar{1}$	$\Gamma_t C_i^1$	$\{I \mid 000\}$	0
3	*P*2	$\Gamma_m C_2^1$	$\{C_{2z} \mid 000\}$	0
4	$P2_1$	$\Gamma_m C_2^2$	$\{C_{2z} \mid 00\frac{1}{2}\}$	0
5	*B*2	$\Gamma_m^b C_2^3$	$\{C_{2z} \mid 000\}$	0
6	*Pm*	$\Gamma_m C_{1h}^1$	$\{\sigma_z \mid 000\}$	0
7	*Pb*	$\Gamma_m C_{1h}^2$	$\{\sigma_z \mid \frac{1}{2}00\}$	0
8	*Bm*	$\Gamma_m^b C_{1h}^3$	$\{\sigma_z \mid 000\}$	0
9	*Bb*	$\Gamma_m^b C_{1h}^4$	$\{\sigma_z \mid \frac{1}{2}00\}$	0
10	*P*2/*m*	$\Gamma_m C_{2h}^1$	$\{C_{2z} \mid 000\}, \{I \mid 000\}$	0
11	$P2_1/m$	$\Gamma_m C_{2h}^2$	$\{C_{2z} \mid 00\frac{1}{2}\}, \{I \mid 00\frac{1}{2}\}$, $\{C_{2z} \mid 00\frac{1}{2}\}, \{I \mid 000\}$	$\frac{1}{4}\mathbf{t}_3$
12	*B*2/*m*	$\Gamma_m^b C_{2h}^3$	$\{C_{2z} \mid 000\}, \{I \mid 000\}$	0
13	*P*2/*b*	$\Gamma_m C_{2h}^4$	$\{C_{2z} \mid 000\}, \{I \mid \frac{1}{2}00\}$, $\{C_{2z} \mid \frac{1}{2}00\}, \{I \mid 000\}$	$\frac{1}{4}\mathbf{t}_1$
14	$P2_1/b$	$\Gamma_m C_{2h}^5$	$\{C_{2z} \mid 00\frac{1}{2}\}, \{I \mid \frac{1}{2}0\frac{1}{2}\}$, $\{C_{2z} \mid \frac{1}{2}0\frac{1}{2}\}, \{I \mid 000\}$	$-\frac{1}{4}\mathbf{t}_1 + \frac{1}{4}\mathbf{t}_3$
15	*B*2/*b*	$\Gamma_m^b C_{2h}^6$	$\{C_{2z} \mid 000\}, \{I \mid \frac{1}{2}00\}$, $\{C_{2z} \mid \frac{1}{2}00\}, \{I \mid 000\}$	$\frac{1}{4}\mathbf{t}_1$
16	*P*222	$\Gamma_o D_2^1$	$\{C_{2x} \mid 000\}, \{C_{2y} \mid 000\}$	0
17	$P222_1$	$\Gamma_o D_2^2$	$\{C_{2x} \mid 00\frac{1}{2}\}, \{C_{2y} \mid 000\}$, $\{C_{2x} \mid 000\}, \{C_{2y} \mid 00\frac{1}{2}\}$	*

International number	International symbol	Schönflies symbol	Generating elements	$\mathbf{t}_0$
18	$P2_12_12$	$\Gamma_oD_2^3$	$\{C_{2x} \mid \tfrac{1}{2}\tfrac{1}{2}0\}$, $\{C_{2y} \mid \tfrac{1}{2}\tfrac{1}{2}0\}$	0
19	$P2_12_12_1$	$\Gamma_oD_2^4$	$\{C_{2x} \mid 0\tfrac{1}{2}\tfrac{1}{2}\}$, $\{C_{2y} \mid \tfrac{1}{2}\tfrac{1}{2}0\}$, $\{C_{2x} \mid \tfrac{1}{2}\tfrac{1}{2}0\}$, $\{C_{2y} \mid \tfrac{1}{2}0\tfrac{1}{2}\}$	*
20	$C222_1$	$\Gamma_o^bD_2^5$	$\{C_{2x} \mid 000\}$, $\{C_{2y} \mid 00\tfrac{1}{2}\}$	0
21	$C222$	$\Gamma_o^bD_2^6$	$\{C_{2x} \mid 000\}$, $\{C_{2y} \mid 000\}$	0
22	$F222$	$\Gamma_o^fD_2^7$	$\{C_{2x} \mid 000\}$, $\{C_{2y} \mid 000\}$	0
23	$I222$	$\Gamma_o^vD_2^8$	$\{C_{2x} \mid 000\}$, $\{C_{2y} \mid 000\}$	0
24	$I2_12_12_1$	$\Gamma_o^vD_2^9$	$\{C_{2x} \mid 0\tfrac{1}{2}\tfrac{1}{2}\}$, $\{C_{2y} \mid \tfrac{1}{2}\tfrac{1}{2}0\}$, $\{C_{2x} \mid \tfrac{1}{2}\tfrac{1}{2}0\}$, $\{C_{2y} \mid \tfrac{1}{2}0\tfrac{1}{2}\}$	$\tfrac{1}{2}\mathbf{t}_1$
25	$Pmm2$	$\Gamma_oC_{2v}^1$	$\{\sigma_x \mid 000\}$, $\{\sigma_y \mid 000\}$	0
26	$Pmc2_1$	$\Gamma_oC_{2v}^2$	$\{\sigma_x \mid 000\}$, $\{\sigma_y \mid 00\tfrac{1}{2}\}$	0
27	$Pcc2$	$\Gamma_oC_{2v}^3$	$\{\sigma_x \mid 00\tfrac{1}{2}\}$, $\{\sigma_y \mid 00\tfrac{1}{2}\}$	0
28	$Pma2$	$\Gamma_oC_{2v}^4$	$\{\sigma_x \mid \tfrac{1}{2}00\}$, $\{\sigma_y \mid 000\}$, $\{\sigma_x \mid 0\tfrac{1}{2}0\}$, $\{\sigma_y \mid 0\tfrac{1}{2}0\}$	$-\tfrac{1}{4}\mathbf{t}_1$*
29	$Pca2_1$	$\Gamma_oC_{2v}^5$	$\{\sigma_x \mid \tfrac{1}{2}00\}$, $\{\sigma_y \mid 00\tfrac{1}{2}\}$, $\{\sigma_x \mid 0\tfrac{1}{2}\tfrac{1}{2}\}$, $\{\sigma_y \mid 0\tfrac{1}{2}0\}$	$-\tfrac{1}{4}\mathbf{t}_1$*
30	$Pnc2$	$\Gamma_oC_{2v}^6$	$\{\sigma_x \mid \tfrac{1}{2}0\tfrac{1}{2}\}$, $\{\sigma_y \mid 00\tfrac{1}{2}\}$, $\{\sigma_x \mid \tfrac{1}{2}0\tfrac{1}{2}\}$, $\{\sigma_y \mid \tfrac{1}{2}0\tfrac{1}{2}\}$	$-\tfrac{1}{4}\mathbf{t}_1$
31	$Pmn2_1$	$\Gamma_oC_{2v}^7$	$\{\sigma_x \mid \tfrac{1}{2}0\tfrac{1}{2}\}$, $\{\sigma_y \mid 000\}$, $\{\sigma_x \mid 000\}$, $\{\sigma_y \mid 0\tfrac{1}{2}\tfrac{1}{2}\}$	*
32	$Pba2$	$\Gamma_oC_{2v}^8$	$\{\sigma_x \mid \tfrac{1}{2}00\}$, $\{\sigma_y \mid 0\tfrac{1}{2}0\}$, $\{\sigma_x \mid \tfrac{1}{2}\tfrac{1}{2}0\}$, $\{\sigma_y \mid \tfrac{1}{2}\tfrac{1}{2}0\}$	$-\tfrac{1}{4}\mathbf{t}_1 + \tfrac{1}{4}\mathbf{t}_2$
33	$Pna2_1$	$\Gamma_oC_{2v}^9$	$\{\sigma_x \mid \tfrac{1}{2}00\}$, $\{\sigma_y \mid 0\tfrac{1}{2}\tfrac{1}{2}\}$, $\{\sigma_x \mid \tfrac{1}{2}\tfrac{1}{2}\tfrac{1}{2}\}$, $\{\sigma_y \mid \tfrac{1}{2}\tfrac{1}{2}0\}$	$-\tfrac{1}{4}\mathbf{t}_1 + \tfrac{1}{4}\mathbf{t}_2$*
34	$Pnn2$	$\Gamma_oC_{2v}^{10}$	$\{\sigma_x \mid \tfrac{1}{2}0\tfrac{1}{2}\}$, $\{\sigma_y \mid 0\tfrac{1}{2}\tfrac{1}{2}\}$, $\{\sigma_x \mid \tfrac{1}{2}\tfrac{1}{2}\tfrac{1}{2}\}$, $\{\sigma_y \mid \tfrac{1}{2}\tfrac{1}{2}\tfrac{1}{2}\}$	$-\tfrac{1}{4}\mathbf{t}_1 + \tfrac{1}{4}\mathbf{t}_2$
35	$Cmm2$	$\Gamma_o^bC_{2v}^{11}$	$\{\sigma_x \mid 000\}$, $\{\sigma_y \mid 000\}$	0
36	$Cmc2_1$	$\Gamma_o^bC_{2v}^{12}$	$\{\sigma_x \mid 00\tfrac{1}{2}\}$, $\{\sigma_y \mid 000\}$, $\{\sigma_x \mid 000\}$, $\{\sigma_y \mid 00\tfrac{1}{2}\}$	*
37	$Ccc2$	$\Gamma_o^bC_{2v}^{13}$	$\{\sigma_x \mid 00\tfrac{1}{2}\}$, $\{\sigma_y \mid 00\tfrac{1}{2}\}$	0
38	$Amm2$‡	$\Gamma_o^bC_{2v}^{14}$	$\{\sigma_z \mid 000\}$, $\{\sigma_x \mid 000\}$, $\{\sigma_x \mid 000\}$, $\{\sigma_y \mid 000\}$†	*
39	$Abm2$‡	$\Gamma_o^bC_{2v}^{15}$	$\{\sigma_z \mid \tfrac{1}{2}\tfrac{1}{2}0\}$, $\{\sigma_x \mid \tfrac{1}{2}\tfrac{1}{2}0\}$, $\{\sigma_x \mid \tfrac{1}{2}0\tfrac{1}{2}\}$, $\{\sigma_y \mid \tfrac{1}{2}0\tfrac{1}{2}\}$†	*
40	$Ama2$‡	$\Gamma_o^bC_{2v}^{16}$	$\{\sigma_z \mid 00\tfrac{1}{2}\}$, $\{\sigma_x \mid 00\tfrac{1}{2}\}$, $\{\sigma_x \mid 0\tfrac{1}{2}0\}$, $\{\sigma_y \mid 0\tfrac{1}{2}0\}$†	*
41	$Aba2$‡	$\Gamma_o^bC_{2v}^{17}$	$\{\sigma_z \mid \tfrac{1}{2}\tfrac{1}{2}\tfrac{1}{2}\}$, $\{\sigma_x \mid \tfrac{1}{2}\tfrac{1}{2}\tfrac{1}{2}\}$, $\{\sigma_x \mid \tfrac{1}{2}\tfrac{1}{2}\tfrac{1}{2}\}$, $\{\sigma_y \mid \tfrac{1}{2}\tfrac{1}{2}\tfrac{1}{2}\}$†	*
42	$Fmm2$	$\Gamma_o^fC_{2v}^{18}$	$\{\sigma_x \mid 000\}$, $\{\sigma_y \mid 000\}$	0
43	$Fdd2$	$\Gamma_o^fC_{2v}^{19}$	$\{\sigma_x \mid 0\tfrac{1}{2}0\}$, $\{\sigma_y \mid \tfrac{1}{2}00\}$, $\{\sigma_x \mid \tfrac{3}{4}\tfrac{3}{4}\tfrac{3}{4}\}$, $\{\sigma_y \mid \tfrac{3}{4}\tfrac{3}{4}\tfrac{3}{4}\}$	$\tfrac{1}{4}\mathbf{t}_3$
44	$Imm2$	$\Gamma_o^vC_{2v}^{20}$	$\{\sigma_x \mid 000\}$, $\{\sigma_y \mid 000\}$	0

‡ In this book we do not use the *A* lattice for these space groups, see Note (v) to Table 3.1 and Note (iv) to the present table.

† Referred to $\mathbf{t}_1$, $\mathbf{t}_2$, $\mathbf{t}_3$ in Note (v) to Table 3.1.

International number	International symbol	Schönflies symbol	Generating elements	$\mathbf{t}_0$
45	*Iba*2	$\Gamma_o^v C_{2v}^{21}$	$\{\sigma_x \mid \frac{1}{2}\frac{1}{2}0\}, \{\sigma_y \mid \frac{1}{2}\frac{1}{2}0\}$	0
46	*Ima*2	$\Gamma_o^v C_{2v}^{22}$	$\{\sigma_x \mid 0\frac{1}{2}\frac{1}{2}\}, \{\sigma_y \mid 0\frac{1}{2}\frac{1}{2}\}$,	*
			$\{\sigma_x \mid \frac{1}{2}0\frac{1}{2}\}, \{\sigma_y \mid \frac{1}{2}0\frac{1}{2}\}$	
47	*Pmmm*	$\Gamma_o D_{2h}^1$	$\{C_{2x} \mid 000\}, \{C_{2y} \mid 000\}, \{I \mid 000\}$	0
48	*Pnnn*	$\Gamma_o D_{2h}^2$	$\{C_{2x} \mid 000\}, \{C_{2y} \mid 000\}, \{I \mid \frac{1}{2}\frac{1}{2}\frac{1}{2}\}$	0
49	*Pccm*	$\Gamma_o D_{2h}^3$	$\{C_{2x} \mid 000\}, \{C_{2y} \mid 000\}, \{I \mid 00\frac{1}{2}\}$,	$\frac{1}{4}\mathbf{t}_3$
			$\{C_{2x} \mid 00\frac{1}{2}\}, \{C_{2y} \mid 00\frac{1}{2}\}, \{I \mid 000\}$	
50	*Pban*	$\Gamma_o D_{2h}^4$	$\{C_{2x} \mid 000\}, \{C_{2y} \mid 000\}, \{I \mid \frac{1}{2}\frac{1}{2}0\}$	0
51	*Pmma*	$\Gamma_o D_{2h}^5$	$\{C_{2x} \mid 00\frac{1}{2}\}, \{C_{2y} \mid 000\}, \{I \mid 000\}$,	*
			$\{C_{2x} \mid 0\frac{1}{2}0\}, \{C_{2y} \mid 000\}, \{I \mid 000\}$	
52	*Pnna*	$\Gamma_o D_{2h}^6$	$\{C_{2x} \mid 00\frac{1}{2}\}, \{C_{2y} \mid 000\}, \{I \mid \frac{1}{2}\frac{1}{2}0\}$,	$\frac{1}{4}\mathbf{t}_1 + \frac{1}{4}\mathbf{t}_2$*
			$\{C_{2x} \mid \frac{1}{2}0\frac{1}{2}\}, \{C_{2y} \mid \frac{1}{2}\frac{1}{2}\frac{1}{2}\}, \{I \mid 000\}$	
53	*Pmna*	$\Gamma_o D_{2h}^7$	$\{C_{2x} \mid 00\frac{1}{2}\}, \{C_{2y} \mid 000\}, \{I \mid \frac{1}{2}00\}$,	$-\frac{1}{4}\mathbf{t}_1$*
			$\{C_{2x} \mid 000\}, \{C_{2y} \mid 0\frac{1}{2}\frac{1}{2}\}, \{I \mid 000\}$	
54	*Pcca*	$\Gamma_o D_{2h}^8$	$\{C_{2x} \mid 00\frac{1}{2}\}, \{C_{2y} \mid 000\}, \{I \mid 0\frac{1}{2}0\}$,	$\frac{1}{4}\mathbf{t}_2$*
			$\{C_{2x} \mid 0\frac{1}{2}\frac{1}{2}\}, \{C_{2y} \mid 00\frac{1}{2}\}, \{I \mid 000\}$	
55	*Pbam*	$\Gamma_o D_{2h}^9$	$\{C_{2x} \mid \frac{1}{2}\frac{1}{2}0\}, \{C_{2y} \mid \frac{1}{2}\frac{1}{2}0\}, \{I \mid 000\}$	0
56	*Pccn*	$\Gamma_o D_{2h}^{10}$	$\{C_{2x} \mid \frac{1}{2}\frac{1}{2}0\}, \{C_{2y} \mid \frac{1}{2}\frac{1}{2}0\}, \{I \mid \frac{1}{2}\frac{1}{2}\frac{1}{2}\}$,	$-\frac{1}{4}\mathbf{t}_1 + \frac{1}{4}\mathbf{t}_2 + \frac{1}{4}\mathbf{t}_3$
			$\{C_{2x} \mid 0\frac{1}{2}\frac{1}{2}\}, \{C_{2y} \mid \frac{1}{2}0\frac{1}{2}\}, \{I \mid 000\}$	
57	*Pbcm*	$\Gamma_o D_{2h}^{11}$	$\{C_{2x} \mid \frac{1}{2}\frac{1}{2}0\}, \{C_{2y} \mid \frac{1}{2}\frac{1}{2}0\}, \{I \mid 0\frac{1}{2}0\}$,	$\frac{1}{4}\mathbf{t}_2$*
			$\{C_{2x} \mid \frac{1}{2}00\}, \{C_{2y} \mid \frac{1}{2}0\frac{1}{2}\}, \{I \mid 000\}$	
58	*Pnnm*	$\Gamma_o D_{2h}^{12}$	$\{C_{2x} \mid \frac{1}{2}\frac{1}{2}0\}, \{C_{2y} \mid \frac{1}{2}\frac{1}{2}0\}, \{I \mid 00\frac{1}{2}\}$,	$-\frac{1}{4}\mathbf{t}_3$
			$\{C_{2x} \mid \frac{1}{2}\frac{1}{2}\frac{1}{2}\}, \{C_{2y} \mid \frac{1}{2}\frac{1}{2}\frac{1}{2}\}, \{I \mid 000\}$	
59	*Pmmn*	$\Gamma_o D_{2h}^{13}$	$\{C_{2x} \mid \frac{1}{2}\frac{1}{2}0\}, \{C_{2y} \mid \frac{1}{2}\frac{1}{2}0\}, \{I \mid \frac{1}{2}\frac{1}{2}0\}$	0
60	*Pbcn*	$\Gamma_o D_{2h}^{14}$	$\{C_{2x} \mid \frac{1}{2}\frac{1}{2}0\}, \{C_{2y} \mid \frac{1}{2}\frac{1}{2}0\}, \{I \mid \frac{1}{2}0\frac{1}{2}\}$,	$-\frac{1}{4}\mathbf{t}_1 + \frac{1}{4}\mathbf{t}_3$*
			$\{C_{2x} \mid \frac{1}{2}\frac{1}{2}0\}, \{C_{2y} \mid 00\frac{1}{2}\}, \{I \mid 000\}$	
61	*Pbca*	$\Gamma_o D_{2h}^{15}$	$\{C_{2x} \mid 0\frac{1}{2}\frac{1}{2}\}, \{C_{2y} \mid \frac{1}{2}\frac{1}{2}0\}, \{I \mid 000\}$,	*
			$\{C_{2x} \mid \frac{1}{2}\frac{1}{2}0\}, \{C_{2y} \mid \frac{1}{2}0\frac{1}{2}\}, \{I \mid 000\}$	
62	*Pnma*	$\Gamma_o D_{2h}^{16}$	$\{C_{2x} \mid 0\frac{1}{2}\frac{1}{2}\}, \{C_{2y} \mid \frac{1}{2}\frac{1}{2}0\}, \{I \mid \frac{1}{2}\frac{1}{2}0\}$,	$\frac{1}{4}\mathbf{t}_1 + \frac{1}{4}\mathbf{t}_2$
			$\{C_{2x} \mid \frac{1}{2}\frac{1}{2}\frac{1}{2}\}, \{C_{2y} \mid \frac{1}{2}00\}, \{I \mid 000\}$	
63	*Cmcm*	$\Gamma_o^b D_{2h}^{17}$	$\{C_{2x} \mid 000\}, \{C_{2y} \mid 00\frac{1}{2}\}, \{I \mid 000\}$	0
64	*Cmca*	$\Gamma_o^b D_{2h}^{18}$	$\{C_{2x} \mid 000\}, \{C_{2y} \mid 00\frac{1}{2}\}, \{I \mid \frac{1}{2}\frac{1}{2}0\}$,	$\frac{1}{4}\mathbf{t}_1 + \frac{1}{4}\mathbf{t}_2$
			$\{C_{2x} \mid 000\}, \{C_{2y} \mid \frac{1}{2}\frac{1}{2}\frac{1}{2}\}, \{I \mid 000\}$	
65	*Cmmm*	$\Gamma_o^b D_{2h}^{19}$	$\{C_{2x} \mid 000\}, \{C_{2y} \mid 000\}, \{I \mid 000\}$	0
66	*Cccm*	$\Gamma_o^b D_{2h}^{20}$	$\{C_{2x} \mid 000\}, \{C_{2y} \mid 000\}, \{I \mid 00\frac{1}{2}\}$,	$\frac{1}{4}\mathbf{t}_3$
			$\{C_{2x} \mid 00\frac{1}{2}\}, \{C_{2y} \mid 00\frac{1}{2}\}, \{I \mid 000\}$	
67	*Cmma*	$\Gamma_o^b D_{2h}^{21}$	$\{C_{2x} \mid 000\}, \{C_{2y} \mid 000\}, \{I \mid \frac{1}{2}\frac{1}{2}0\}$,	$\frac{1}{4}\mathbf{t}_1 + \frac{1}{4}\mathbf{t}_2$
			$\{C_{2x} \mid 000\}, \{C_{2y} \mid \frac{1}{2}\frac{1}{2}0\}, \{I \mid 000\}$	
68	*Ccca*	$\Gamma_o^b D_{2h}^{22}$	$\{C_{2x} \mid 000\}, \{C_{2y} \mid 000\}, \{I \mid \frac{1}{2}\frac{1}{2}\frac{1}{2}\}$,	$-\frac{1}{4}\mathbf{t}_1 + \frac{1}{4}\mathbf{t}_2 + \frac{1}{4}\mathbf{t}_3$
			$\{C_{2x} \mid \frac{1}{2}\frac{1}{2}\frac{1}{2}\}, \{C_{2y} \mid 00\frac{1}{2}\}, \{I \mid 000\}$	
69	*Fmmm*	$\Gamma_o^f D_{2h}^{23}$	$\{C_{2x} \mid 000\}, \{C_{2y} \mid 000\}, \{I \mid 000\}$	0
70	*Fddd*	$\Gamma_o^f D_{2h}^{24}$	$\{C_{2x} \mid 000\}, \{C_{2y} \mid 000\}, \{I \mid \frac{1}{4}\frac{1}{4}\frac{1}{4}\}$,	*
			$\{C_{2x} \mid 000\}, \{C_{2y} \mid 000\}, \{I \mid \frac{3}{4}\frac{3}{4}\frac{3}{4}\}$	
71	*Immm*	$\Gamma_o^v D_{2h}^{25}$	$\{C_{2x} \mid 000\}, \{C_{2y} \mid 000\}, \{I \mid 000\}$	0
72	*Ibam*	$\Gamma_o^v D_{2h}^{26}$	$\{C_{2x} \mid 000\}, \{C_{2y} \mid 000\}, \{I \mid \frac{1}{2}\frac{1}{2}0\}$,	$\frac{1}{4}\mathbf{t}_1 + \frac{1}{4}\mathbf{t}_2$
			$\{C_{2x} \mid \frac{1}{2}\frac{1}{2}0\}, \{C_{2y} \mid \frac{1}{2}\frac{1}{2}0\}, \{I \mid 000\}$	
73	*Ibca*	$\Gamma_o^v D_{2h}^{27}$	$\{C_{2x} \mid 0\frac{1}{2}\frac{1}{2}\}, \{C_{2y} \mid \frac{1}{2}\frac{1}{2}0\}, \{I \mid 000\}$,	$\frac{1}{2}\mathbf{t}_1$
			$\{C_{2x} \mid \frac{1}{2}\frac{1}{2}0\}, \{C_{2y} \mid \frac{1}{2}0\frac{1}{2}\}, \{I \mid 000\}$	

International number	International symbol	Schönflies symbol	Generating elements	$\mathbf{t}_0$
74	*Imma*	$\Gamma_o^v D_{2h}^{28}$	$\{C_{2x} \mid 0\frac{1}{2}\frac{1}{2}\}, \{C_{2y} \mid \frac{1}{2}\frac{1}{2}0\}, \{I \mid \frac{1}{2}\frac{1}{2}0\},$ $\{C_{2x} \mid 000\}, \{C_{2y} \mid 0\frac{1}{2}\frac{1}{2}\}, \{I \mid 000\}$	$\frac{3}{4}\mathbf{t}_1 + \frac{1}{4}\mathbf{t}_2$
75	$P4$	$\Gamma_q C_4^1$	$\{C_{4z}^+ \mid 000\}$	0
76	$P4_1$	$\Gamma_q C_4^2$	$\{C_{4z}^+ \mid 00\frac{1}{4}\}$	0
77	$P4_2$	$\Gamma_q C_4^3$	$\{C_{4z}^+ \mid 00\frac{1}{2}\}$	0
78	$P4_3$	$\Gamma_q C_4^4$	$\{C_{4z}^+ \mid 00\frac{3}{4}\}$	0
79	$I4$	$\Gamma_q^v C_4^5$	$\{C_{4z}^+ \mid 000\}$	0
80	$I4_1$	$\Gamma_q^v C_4^6$	$\{C_{4z}^+ \mid \frac{3}{4}\frac{1}{4}\frac{1}{2}\}$	0
81	$P\bar{4}$	$\Gamma_q S_4^1$	$\{S_{4z}^+ \mid 000\}$	0
82	$I\bar{4}$	$\Gamma_q^v S_4^2$	$\{S_{4z}^+ \mid 000\}$	0
83	$P4/m$	$\Gamma_q C_{4h}^1$	$\{C_{4z}^+ \mid 000\}, \{I \mid 000\}$	0
84	$P4_2/m$	$\Gamma_q C_{4h}^2$	$\{C_{4z}^+ \mid 00\frac{1}{2}\}, \{I \mid 00\frac{1}{2}\},$ $\{C_{4z}^+ \mid 00\frac{1}{2}\}, \{I \mid 000\}$	$\frac{1}{4}\mathbf{t}_3$
85	$P4/n$	$\Gamma_q C_{4h}^3$	$\{C_{4z}^+ \mid \frac{1}{2}\frac{1}{2}0\}, \{I \mid \frac{1}{2}\frac{1}{2}0\}$	0
86	$P4_2/n$	$\Gamma_q C_{4h}^4$	$\{C_{4z}^+ \mid \frac{1}{2}\frac{1}{2}\frac{1}{2}\}, \{I \mid \frac{1}{2}\frac{1}{2}\frac{1}{2}\}$	0
87	$I4/m$	$\Gamma_q^v C_{4h}^5$	$\{C_{4z}^+ \mid 000\}, \{I \mid 000\}$	0
88	$I4_1/a$	$\Gamma_q^v C_{4h}^6$	$\{C_{4z}^+ \mid \frac{3}{4}\frac{1}{4}\frac{1}{2}\}, \{I \mid \frac{3}{4}\frac{1}{4}\frac{1}{2}\}$	0
89	$P422$	$\Gamma_q D_4^1$	$\{C_{4z}^+ \mid 000\}, \{C_{2x} \mid 000\}$	0
90	$P42_12$	$\Gamma_q D_4^2$	$\{C_{4z}^+ \mid 000\}, \{C_{2x} \mid \frac{1}{2}\frac{1}{2}0\},$ $\{C_{4z}^+ \mid \frac{1}{2}\frac{1}{2}0\}, \{C_{2x} \mid \frac{1}{2}\frac{1}{2}0\}$	$\frac{1}{2}\mathbf{t}_1$
91	$P4_122$	$\Gamma_q D_4^3$	$\{C_{4z}^+ \mid 00\frac{1}{4}\}, \{C_{2x} \mid 000\},$ $\{C_{4z}^+ \mid 00\frac{1}{4}\}, \{C_{2x} \mid 00\frac{1}{2}\}$	$\frac{1}{4}\mathbf{t}_3$
92	$P4_12_12$	$\Gamma_q D_4^4$	$\{C_{4z}^+ \mid 00\frac{1}{4}\}, \{C_{2x} \mid \frac{1}{2}\frac{1}{2}0\},$ $\{C_{4z}^+ \mid \frac{1}{2}\frac{1}{2}\frac{1}{4}\}, \{C_{2x} \mid \frac{1}{2}\frac{1}{2}\frac{3}{4}\}$	$\frac{1}{2}\mathbf{t}_1 - \frac{3}{8}\mathbf{t}_3$
93	$P4_222$	$\Gamma_q D_4^5$	$\{C_{4z}^+ \mid 00\frac{1}{2}\}, \{C_{2x} \mid 000\}$	0
94	$P4_22_12$	$\Gamma_q D_4^6$	$\{C_{4z}^+ \mid 00\frac{1}{2}\}, \{C_{2x} \mid \frac{1}{2}\frac{1}{2}0\},$ $\{C_{4z}^+ \mid \frac{1}{2}\frac{1}{2}\frac{1}{2}\}, \{C_{2x} \mid \frac{1}{2}\frac{1}{2}\frac{1}{2}\}$	$\frac{1}{2}\mathbf{t}_1 + \frac{1}{4}\mathbf{t}_3$
95	$P4_322$	$\Gamma_q D_4^7$	$\{C_{4z}^+ \mid 00\frac{3}{4}\}, \{C_{2x} \mid 000\},$ $\{C_{4z}^+ \mid 00\frac{3}{4}\}, \{C_{2x} \mid 00\frac{1}{2}\}$	$\frac{1}{4}\mathbf{t}_3$
96	$P4_32_12$	$\Gamma_q D_4^8$	$\{C_{4z}^+ \mid 00\frac{3}{4}\}, \{C_{2x} \mid \frac{1}{2}\frac{1}{2}0\},$ $\{C_{4z}^+ \mid \frac{1}{2}\frac{1}{2}\frac{3}{4}\}, \{C_{2x} \mid \frac{1}{2}\frac{1}{2}\frac{1}{4}\}$	$\frac{1}{2}\mathbf{t}_1 - \frac{1}{8}\mathbf{t}_3$
97	$I422$	$\Gamma_q^v D_4^9$	$\{C_{4z}^+ \mid 000\}, \{C_{2x} \mid 000\}$	0
98	$I4_122$	$\Gamma_q^v D_4^{10}$	$\{C_{4z}^+ \mid \frac{3}{4}\frac{1}{4}\frac{1}{2}\}, \{C_{2x} \mid 0\frac{1}{2}\frac{1}{2}\},$ $\{C_{4z}^+ \mid \frac{3}{4}\frac{1}{4}\frac{1}{2}\}, \{C_{2x} \mid \frac{3}{4}\frac{1}{4}\frac{1}{2}\}$	$\frac{1}{8}\mathbf{t}_1 + \frac{1}{8}\mathbf{t}_2$
99	$P4mm$	$\Gamma_q C_{4v}^1$	$\{C_{4z}^+ \mid 000\}, \{\sigma_x \mid 000\}$	0
100	$P4bm$	$\Gamma_q C_{4v}^2$	$\{C_{4z}^+ \mid 000\}, \{\sigma_x \mid \frac{1}{2}\frac{1}{2}0\}$	0
101	$P4_2cm$	$\Gamma_q C_{4v}^3$	$\{C_{4z}^+ \mid 00\frac{1}{2}\}, \{\sigma_x \mid 00\frac{1}{2}\}$	0
102	$P4_2nm$	$\Gamma_q C_{4v}^4$	$\{C_{4z}^+ \mid 00\frac{1}{2}\}, \{\sigma_x \mid \frac{1}{2}\frac{1}{2}\frac{1}{2}\},$ $\{C_{4z}^+ \mid \frac{1}{2}\frac{1}{2}\frac{1}{2}\}, \{\sigma_x \mid \frac{1}{2}\frac{1}{2}\frac{1}{2}\}$	$\frac{1}{2}\mathbf{t}_1$
103	$P4cc$	$\Gamma_q C_{4v}^5$	$\{C_{4z}^+ \mid 000\}, \{\sigma_x \mid 00\frac{1}{2}\}$	0
104	$P4nc$	$\Gamma_q C_{4v}^6$	$\{C_{4z}^+ \mid 000\}, \{\sigma_x \mid \frac{1}{2}\frac{1}{2}\frac{1}{2}\}$	0
105	$P4_2mc$	$\Gamma_q C_{4v}^7$	$\{C_{4z}^+ \mid 00\frac{1}{2}\}, \{\sigma_x \mid 000\}$	0
106	$P4_2bc$	$\Gamma_q C_{4v}^8$	$\{C_{4z}^+ \mid 00\frac{1}{2}\}, \{\sigma_x \mid \frac{1}{2}\frac{1}{2}0\}$	0
107	$I4mm$	$\Gamma_q^v C_{4v}^9$	$\{C_{4z}^+ \mid 000\}, \{\sigma_x \mid 000\}$	0
108	$I4cm$	$\Gamma_q^v C_{4v}^{10}$	$\{C_{4z}^+ \mid 000\}, \{\sigma_x \mid \frac{1}{2}\frac{1}{2}0\}$	0
109	$I4_1md$	$\Gamma_q^v C_{4v}^{11}$	$\{C_{4z}^+ \mid \frac{3}{4}\frac{1}{4}\frac{1}{2}\}, \{\sigma_x \mid 000\}$	0
110	$I4_1cd$	$\Gamma_q^v C_{4v}^{12}$	$\{C_{4z}^+ \mid \frac{3}{4}\frac{1}{4}\frac{1}{2}\}, \{\sigma_x \mid \frac{1}{2}\frac{1}{2}0\}$	0
111	$P\bar{4}2m$	$\Gamma_q D_{2d}^1$	$\{S_{4z}^+ \mid 000\}, \{C_{2x} \mid 000\}$	0

International number	International symbol	Schönflies symbol	Generating elements	$\mathbf{t}_0$
112	$P\bar{4}2c$	$\Gamma_q D_{2d}^2$	$\{S_{4z}^+ \mid 000\}, \{C_{2x} \mid 00\frac{1}{2}\}$	0
113	$P\bar{4}2_1m$	$\Gamma_q D_{2d}^3$	$\{S_{4z}^+ \mid 000\}, \{C_{2x} \mid \frac{1}{2}\frac{1}{2}0\}$	0
114	$P\bar{4}2_1c$	$\Gamma_q D_{2d}^4$	$\{S_{4z}^+ \mid 000\}, \{C_{2x} \mid \frac{1}{2}\frac{1}{2}\frac{1}{2}\}$	0
115	$P\bar{4}m2$	$\Gamma_q D_{2d}^5$	$\{S_{4z}^+ \mid 000\}, \{C_{2a} \mid 000\}$	0
116	$P\bar{4}c2$	$\Gamma_q D_{2d}^6$	$\{S_{4z}^+ \mid 000\}, \{C_{2a} \mid 00\frac{1}{2}\}$	0
117	$P\bar{4}b2$	$\Gamma_q D_{2d}^7$	$\{S_{4z}^+ \mid 000\}, \{C_{2a} \mid \frac{1}{2}\frac{1}{2}0\}$	0
118	$P\bar{4}n2$	$\Gamma_q D_{2d}^8$	$\{S_{4z}^+ \mid 000\}, \{C_{2a} \mid \frac{1}{2}\frac{1}{2}\frac{1}{2}\}$	0
119	$I\bar{4}m2$	$\Gamma_q^v D_{2d}^9$	$\{S_{4z}^+ \mid 000\}, \{C_{2a} \mid 000\}$	0
120	$I\bar{4}c2$	$\Gamma_q^v D_{2d}^{10}$	$\{S_{4z}^+ \mid 000\}, \{C_{2a} \mid \frac{1}{2}\frac{1}{2}0\}$	0
121	$I\bar{4}2m$	$\Gamma_q^v D_{2d}^{11}$	$\{S_{4z}^+ \mid 000\}, \{C_{2x} \mid 000\}$	0
122	$I\bar{4}2d$	$\Gamma_q^v D_{2d}^{12}$	$\{S_{4z}^+ \mid 000\}, \{C_{2x} \mid \frac{1}{4}\frac{3}{4}\frac{1}{2}\}$,	*
			$\{S_{4z}^+ \mid 000\}, \{C_{2x} \mid \frac{3}{4}\frac{1}{4}\frac{1}{2}\}$	
123	$P4/mmm$	$\Gamma_q D_{4h}^1$	$\{C_{4z}^+ \mid 000\}, \{C_{2x} \mid 000\}, \{I \mid 000\}$	0
124	$P4/mcc$	$\Gamma_q D_{4h}^2$	$\{C_{4z}^+ \mid 000\}, \{C_{2x} \mid 000\}, \{I \mid 00\frac{1}{2}\}$,	$\frac{1}{4}\mathbf{t}_3$
			$\{C_{4z}^+ \mid 000\}, \{C_{2x} \mid 00\frac{1}{2}\}, \{I \mid 000\}$	
125	$P4/nbm$	$\Gamma_q D_{4h}^3$	$\{C_{4z}^+ \mid \frac{1}{2}\frac{1}{2}0\}, \{C_{2x} \mid 000\}, \{I \mid \frac{1}{2}\frac{1}{2}0\}$,	$\frac{1}{4}\mathbf{t}_1 - \frac{1}{4}\mathbf{t}_2$
			$\{C_{4z}^+ \mid \frac{1}{2}00\}, \{C_{2x} \mid 0\frac{1}{2}0\}, \{I \mid 000\}$	
126	$P4/nnc$	$\Gamma_q D_{4h}^4$	$\{C_{4z}^+ \mid \frac{1}{2}\frac{1}{2}0\}, \{C_{2x} \mid 000\}, \{I \mid \frac{1}{2}\frac{1}{2}\frac{1}{2}\}$,	$\frac{1}{2}\mathbf{t}_1$
			$\{C_{4z}^+ \mid 000\}, \{C_{2x} \mid 000\}, \{I \mid \frac{1}{2}\frac{1}{2}\frac{1}{2}\}$	
127	$P4/mbm$	$\Gamma_q D_{4h}^5$	$\{C_{4z}^+ \mid \frac{1}{2}\frac{1}{2}0\}, \{C_{2x} \mid \frac{1}{2}\frac{1}{2}0\}, \{I \mid 000\}$,	$\frac{1}{2}\mathbf{t}_1$
			$\{C_{4z}^+ \mid 000\}, \{C_{2x} \mid \frac{1}{2}\frac{1}{2}0\}, \{I \mid 000\}$	
128	$P4/mnc$	$\Gamma_q D_{4h}^6$	$\{C_{4z}^+ \mid \frac{1}{2}\frac{1}{2}0\}, \{C_{2x} \mid \frac{1}{2}\frac{1}{2}0\}, \{I \mid 00\frac{1}{2}\}$,	$\frac{1}{2}\mathbf{t}_1 + \frac{1}{4}\mathbf{t}_3$
			$\{C_{4z}^+ \mid 000\}, \{C_{2x} \mid \frac{1}{2}\frac{1}{2}\frac{1}{2}\}, \{I \mid 000\}$	
129	$P4/nmm$	$\Gamma_q D_{4h}^7$	$\{C_{4z}^+ \mid 000\}, \{C_{2x} \mid \frac{1}{2}\frac{1}{2}0\}, \{I \mid \frac{1}{2}\frac{1}{2}0\}$,	$\frac{1}{2}\mathbf{t}_1$
			$\{C_{4z}^+ \mid \frac{1}{2}\frac{1}{2}0\}, \{C_{2x} \mid \frac{1}{2}\frac{1}{2}0\}, \{I \mid \frac{1}{2}\frac{1}{2}0\}$	
130	$P4/ncc$	$\Gamma_q D_{4h}^8$	$\{C_{4z}^+ \mid 000\}, \{C_{2x} \mid \frac{1}{2}\frac{1}{2}0\}, \{I \mid \frac{1}{2}\frac{1}{2}\frac{1}{2}\}$,	$\frac{1}{2}\mathbf{t}_1 + \frac{1}{4}\mathbf{t}_3$
			$\{C_{4z}^+ \mid \frac{1}{2}\frac{1}{2}0\}, \{C_{2x} \mid \frac{1}{2}\frac{1}{2}\frac{1}{2}\}, \{I \mid \frac{1}{2}\frac{1}{2}0\}$	
131	$P4_2/mmc$	$\Gamma_q D_{4h}^9$	$\{C_{4z}^+ \mid 00\frac{1}{2}\}, \{C_{2x} \mid 000\}, \{I \mid 000\}$	0
132	$P4_2/mcm$	$\Gamma_q D_{4h}^{10}$	$\{C_{4z}^+ \mid 00\frac{1}{2}\}, \{C_{2x} \mid 000\}, \{I \mid 00\frac{1}{2}\}$,	$\frac{1}{4}\mathbf{t}_3$
			$\{C_{4z}^+ \mid 00\frac{1}{2}\}, \{C_{2x} \mid 00\frac{1}{2}\}, \{I \mid 000\}$	
133	$P4_2/nbc$	$\Gamma_q D_{4h}^{11}$	$\{C_{4z}^+ \mid \frac{1}{2}\frac{1}{2}\frac{1}{2}\}, \{C_{2x} \mid 000\}, \{I \mid \frac{1}{2}\frac{1}{2}0\}$,	$\frac{1}{4}\mathbf{t}_3$
			$\{C_{4z}^+ \mid \frac{1}{2}\frac{1}{2}\frac{1}{2}\}, \{C_{2x} \mid 00\frac{1}{2}\}, \{I \mid \frac{1}{2}\frac{1}{2}\frac{1}{2}\}$	
134	$P4_2/nnm$	$\Gamma_q D_{4h}^{12}$	$\{C_{4z}^+ \mid \frac{1}{2}\frac{1}{2}\frac{1}{2}\}, \{C_{2x} \mid 000\}, \{I \mid \frac{1}{2}\frac{1}{2}\frac{1}{2}\}$	0
135	$P4_2/mbc$	$\Gamma_q D_{4h}^{13}$	$\{C_{4z}^+ \mid \frac{1}{2}\frac{1}{2}\frac{1}{2}\}, \{C_{2x} \mid \frac{1}{2}\frac{1}{2}0\}, \{I \mid 000\}$,	$\frac{1}{2}\mathbf{t}_1$
			$\{C_{4z}^+ \mid 00\frac{1}{2}\}, \{C_{2x} \mid \frac{1}{2}\frac{1}{2}0\}, \{I \mid 000\}$	
136	$P4_2/mnm$	$\Gamma_q D_{4h}^{14}$	$\{C_{4z}^+ \mid \frac{1}{2}\frac{1}{2}\frac{1}{2}\}, \{C_{2x} \mid \frac{1}{2}\frac{1}{2}0\}, \{I \mid 00\frac{1}{2}\}$,	$\frac{1}{4}\mathbf{t}_3$
			$\{C_{4z}^+ \mid \frac{1}{2}\frac{1}{2}\frac{1}{2}\}, \{C_{2x} \mid \frac{1}{2}\frac{1}{2}\frac{1}{2}\}, \{I \mid 000\}$	
137	$P4_2/nmc$	$\Gamma_q D_{4h}^{15}$	$\{C_{4z}^+ \mid 00\frac{1}{2}\}, \{C_{2x} \mid \frac{1}{2}\frac{1}{2}0\}, \{I \mid \frac{1}{2}\frac{1}{2}0\}$,	$\frac{1}{2}\mathbf{t}_1 + \frac{1}{4}\mathbf{t}_3$
			$\{C_{4z}^+ \mid \frac{1}{2}\frac{1}{2}\frac{1}{2}\}, \{C_{2x} \mid \frac{1}{2}\frac{1}{2}\frac{1}{2}\}, \{I \mid \frac{1}{2}\frac{1}{2}\frac{1}{2}\}$	
138	$P4_2/ncm$	$\Gamma_q D_{4h}^{16}$	$\{C_{4z}^+ \mid 00\frac{1}{2}\}, \{C_{2x} \mid \frac{1}{2}\frac{1}{2}0\}, \{I \mid \frac{1}{2}\frac{1}{2}\frac{1}{2}\}$,	$\frac{1}{2}\mathbf{t}_1$
			$\{C_{4z}^+ \mid \frac{1}{2}\frac{1}{2}\frac{1}{2}\}, \{C_{2x} \mid \frac{1}{2}\frac{1}{2}0\}, \{I \mid \frac{1}{2}\frac{1}{2}\frac{1}{2}\}$	
139	$I4/mmm$	$\Gamma_q^v D_{4h}^{17}$	$\{C_{4z}^+ \mid 000\}, \{C_{2x} \mid 000\}, \{I \mid 000\}$	0
140	$I4/mcm$	$\Gamma_q^v D_{4h}^{18}$	$\{C_{4z}^+ \mid \frac{1}{2}\frac{1}{2}0\}, \{C_{2x} \mid 000\}, \{I \mid \frac{1}{2}\frac{1}{2}0\}$,	$\frac{3}{4}\mathbf{t}_1 + \frac{1}{4}\mathbf{t}_2 + \frac{1}{2}\mathbf{t}_3$
			$\{C_{4z}^+ \mid 000\}, \{C_{2x} \mid \frac{1}{2}\frac{1}{2}0\}, \{I \mid 000\}$	
141	$I4_1/amd$	$\Gamma_q^v D_{4h}^{19}$	$\{C_{4z}^+ \mid 0\frac{1}{2}0\}, \{C_{2x} \mid \frac{1}{2}\frac{1}{2}0\}, \{I \mid \frac{1}{2}\frac{1}{2}0\}$,	$\frac{1}{4}\mathbf{t}_1 + \frac{1}{4}\mathbf{t}_2$
			$\{C_{4z}^+ \mid 0\frac{1}{2}0\}, \{C_{2x} \mid 000\}, \{I \mid 000\}$	
142	$I4_1/acd$	$\Gamma_q^v D_{4h}^{20}$	$\{C_{4z}^+ \mid \frac{1}{2}00\}, \{C_{2x} \mid \frac{1}{2}\frac{1}{2}0\}, \{I \mid 000\}$,	*
			$\{C_{4z}^+ \mid 0\frac{1}{2}0\}, \{C_{2x} \mid \frac{1}{2}\frac{1}{2}0\}, \{I \mid 000\}$	

International number	International symbol	Schönflies symbol	Generating elements	$\mathbf{t}_0$
143	$P3$	$\Gamma_h C_3^1$	$\{C_3^+ \mid 000\}$	0
144	$P3_1$	$\Gamma_h C_3^2$	$\{C_3^+ \mid 00\frac{1}{3}\}$	0
145	$P3_2$	$\Gamma_h C_3^3$	$\{C_3^+ \mid 00\frac{2}{3}\}$	0
146	$R3$	$\Gamma_{rh} C_3^4$	$\{C_3^+ \mid 000\}$	0
147	$P\bar{3}$	$\Gamma_h C_{3i}^1$	$\{S_6^+ \mid 000\}$	0
148	$R\bar{3}$	$\Gamma_{rh} C_{3i}^2$	$\{S_6^+ \mid 000\}$	0
149	$P312$	$\Gamma_h D_3^1$	$\{C_3^+ \mid 000\}, \{C'_{21} \mid 000\}$	0
150	$P321$	$\Gamma_h D_3^2$	$\{C_3^+ \mid 000\}, \{C''_{21} \mid 000\}$	0
151	$P3_112$	$\Gamma_h D_3^3$	$\{C_3^+ \mid 00\frac{1}{3}\}, \{C'_{21} \mid 00\frac{2}{3}\}$, $\{C_3^+ \mid 00\frac{1}{3}\}, \{C'_{21} \mid 00\frac{1}{3}\}$	$\frac{1}{6}\mathbf{t}_3$
152	$P3_121$	$\Gamma_h D_3^4$	$\{C_3^+ \mid 00\frac{1}{3}\}, \{C''_{21} \mid 00\frac{2}{3}\}$	0
153	$P3_212$	$\Gamma_h D_3^5$	$\{C_3^+ \mid 00\frac{2}{3}\}, \{C'_{21} \mid 00\frac{1}{3}\}$, $\{C_3^+ \mid 00\frac{2}{3}\}, \{C'_{21} \mid 00\frac{2}{3}\}$	$-\frac{1}{6}\mathbf{t}_3$
154	$P3_221$	$\Gamma_h D_3^6$	$\{C_3^+ \mid 00\frac{2}{3}\}, \{C''_{21} \mid 00\frac{1}{3}\}$	0
155	$R32$	$\Gamma_{rh} D_3^7$	$\{C_3^+ \mid 000\}, \{C'_{21} \mid 000\}$, $\{C_3^+ \mid 000\}, \{C''_{21} \mid 000\}$	*
156	$P3m1$	$\Gamma_h C_{3v}^1$	$\{C_3^+ \mid 000\}, \{\sigma_{v1} \mid 000\}$	0
157	$P31m$	$\Gamma_h C_{3v}^2$	$\{C_3^+ \mid 000\}, \{\sigma_{d1} \mid 000\}$	0
158	$P3c1$	$\Gamma_h C_{3v}^3$	$\{C_3^+ \mid 000\}, \{\sigma_{v1} \mid 00\frac{1}{2}\}$	0
159	$P31c$	$\Gamma_h C_{3v}^4$	$\{C_3^+ \mid 000\}, \{\sigma_{d1} \mid 00\frac{1}{2}\}$	0
160	$R3m$	$\Gamma_{rh} C_{3v}^5$	$\{C_3^+ \mid 000\}, \{\sigma_{d1} \mid 000\}$, $\{C_3^+ \mid 000\}, \{\sigma_{v1} \mid 000\}$	*
161	$R3c$	$\Gamma_{rh} C_{3v}^6$	$\{C_3^+ \mid 000\}, \{\sigma_{d1} \mid \frac{1}{2}\frac{1}{2}\frac{1}{2}\}$, $\{C_3^+ \mid 000\}, \{\sigma_{v1} \mid \frac{1}{2}\frac{1}{2}\frac{1}{2}\}$	*
162	$P\bar{3}1m$	$\Gamma_h D_{3d}^1$	$\{S_6^+ \mid 000\}, \{\sigma_{d1} \mid 000\}$	0
163	$P\bar{3}1c$	$\Gamma_h D_{3d}^2$	$\{S_6^+ \mid 000\}, \{\sigma_{d1} \mid 00\frac{1}{2}\}$	0
164	$P\bar{3}m1$	$\Gamma_h D_{3d}^3$	$\{S_6^+ \mid 000\}, \{\sigma_{v1} \mid 000\}$	0
165	$P\bar{3}c1$	$\Gamma_h D_{3d}^4$	$\{S_6^+ \mid 000\}, \{\sigma_{v1} \mid 00\frac{1}{2}\}$	0
166	$R\bar{3}m$	$\Gamma_{rh} D_{3d}^5$	$\{S_6^+ \mid 000\}, \{\sigma_{d1} \mid 000\}$, $\{S_6^+ \mid 000\}, \{\sigma_{v1} \mid 000\}$	*
167	$R\bar{3}c$	$\Gamma_{rh} D_{3d}^6$	$\{S_6^+ \mid 000\}, \{\sigma_{d1} \mid \frac{1}{2}\frac{1}{2}\frac{1}{2}\}$, $\{S_6^+ \mid 000\}, \{\sigma_{v1} \mid \frac{1}{2}\frac{1}{2}\frac{1}{2}\}$	*
168	$P6$	$\Gamma_h C_6^1$	$\{C_6^+ \mid 000\}$	0
169	$P6_1$	$\Gamma_h C_6^2$	$\{C_6^+ \mid 00\frac{1}{6}\}$	0
170	$P6_5$	$\Gamma_h C_6^3$	$\{C_6^+ \mid 00\frac{5}{6}\}$	0
171	$P6_2$	$\Gamma_h C_6^4$	$\{C_6^+ \mid 00\frac{1}{3}\}$	0
172	$P6_4$	$\Gamma_h C_6^5$	$\{C_6^+ \mid 00\frac{2}{3}\}$	0
173	$P6_3$	$\Gamma_h C_6^6$	$\{C_6^+ \mid 00\frac{1}{2}\}$	0
174	$P\bar{6}$	$\Gamma_h C_{3h}^1$	$\{S_3^+ \mid 000\}$	0
175	$P6/m$	$\Gamma_h C_{6h}^1$	$\{C_6^+ \mid 000\}, \{\sigma_h \mid 000\}$	0
176	$P6_3/m$	$\Gamma_h C_{6h}^2$	$\{C_6^+ \mid 00\frac{1}{2}\}, \{\sigma_h \mid 000\}$, $\{C_6^+ \mid 00\frac{1}{2}\}, \{\sigma_h \mid 00\frac{1}{2}\}$	$\frac{1}{4}\mathbf{t}_3$
177	$P622$	$\Gamma_h D_6^1$	$\{C_6^+ \mid 000\}, \{C'_{21} \mid 000\}$	0
178	$P6_122$	$\Gamma_h D_6^2$	$\{C_6^+ \mid 00\frac{1}{6}\}, \{C'_{21} \mid 000\}$, $\{C_6^+ \mid 00\frac{1}{6}\}, \{C''_{21} \mid 000\}$	*
179	$P6_522$	$\Gamma_h D_6^3$	$\{C_6^+ \mid 00\frac{5}{6}\}, \{C'_{21} \mid 000\}$, $\{C_6^+ \mid 00\frac{5}{6}\}, \{C''_{21} \mid 000\}$	*
180	$P6_222$	$\Gamma_h D_6^4$	$\{C_6^+ \mid 00\frac{1}{3}\}, \{C'_{21} \mid 000\}$	0

International number	International symbol	Schönflies symbol	Generating elements	$\mathbf{t}_0$
181	$P6_422$	$\Gamma_h D_6^5$	$\{C_6^+ \mid 00\tfrac{2}{3}\}, \{C'_{21} \mid 000\}$	0
182	$P6_322$	$\Gamma_h D_6^6$	$\{C_6^+ \mid 00\tfrac{1}{2}\}, \{C'_{21} \mid 000\},$ $\{C_6^+ \mid 00\tfrac{1}{2}\}, \{C''_{21} \mid 000\}$	*
183	$P6mm$	$\Gamma_h C_{6v}^1$	$\{C_6^+ \mid 000\}, \{\sigma_{v1} \mid 000\}$	0
184	$P6cc$	$\Gamma_h C_{6v}^2$	$\{C_6^+ \mid 000\}, \{\sigma_{v1} \mid 00\tfrac{1}{2}\}$	0
185	$P6_3cm$	$\Gamma_h C_{6v}^3$	$\{C_6^+ \mid 00\tfrac{1}{2}\}, \{\sigma_{v1} \mid 00\tfrac{1}{2}\}$	0
186	$P6_3mc$	$\Gamma_h C_{6v}^4$	$\{C_6^+ \mid 00\tfrac{1}{2}\}, \{\sigma_{v1} \mid 000\}$	0
187	$P\bar{6}m2$	$\Gamma_h D_{3h}^1$	$\{S_3^+ \mid 000\}, \{\sigma_{v1} \mid 000\}$	0
188	$P\bar{6}c2$	$\Gamma_h D_{3h}^2$	$\{S_3^+ \mid 000\}, \{\sigma_{v1} \mid 00\tfrac{1}{2}\},$ $\{S_3^+ \mid 00\tfrac{1}{2}\}, \{\sigma_{v1} \mid 00\tfrac{1}{2}\}$	$\tfrac{1}{4}\mathbf{t}_3$
189	$P\bar{6}2m$	$\Gamma_h D_{3h}^3$	$\{S_3^+ \mid 000\}, \{\sigma_{d1} \mid 000\}$	0
190	$P\bar{6}2c$	$\Gamma_h D_{3h}^4$	$\{S_3^+ \mid 000\}, \{\sigma_{d1} \mid 00\tfrac{1}{2}\}$ $\{S_3^+ \mid 00\tfrac{1}{2}\}, \{\sigma_{d1} \mid 00\tfrac{1}{2}\}$	$\tfrac{1}{4}\mathbf{t}_3$
191	$P6/mmm$	$\Gamma_h D_{6h}^1$	$\{C_6^+ \mid 000\}, \{C'_{21} \mid 000\}, \{I \mid 000\}$	0
192	$P6/mcc$	$\Gamma_h D_{6h}^2$	$\{C_6^+ \mid 000\}, \{C'_{21} \mid 00\tfrac{1}{2}\}, \{I \mid 000\}$	0
193	$P6_3/mcm$	$\Gamma_h D_{6h}^3$	$\{C_6^+ \mid 00\tfrac{1}{2}\}, \{C'_{21} \mid 000\}, \{I \mid 000\}$	0
194	$P6_3/mmc$	$\Gamma_h D_{6h}^4$	$\{C_6^+ \mid 00\tfrac{1}{2}\}, \{C'_{21} \mid 00\tfrac{1}{2}\}, \{I \mid 000\}$	0
195	$P23$	$\Gamma_c T^1$	$\{C_{2z} \mid 000\}, \{C_{2x} \mid 000\}, \{C_{31}^+ \mid 000\}$	0
196	$F23$	$\Gamma_c^f T^2$	$\{C_{2z} \mid 000\}, \{C_{2x} \mid 000\}, \{C_{31}^+ \mid 000\}$	0
197	$I23$	$\Gamma_c^v T^3$	$\{C_{2z} \mid 000\}, \{C_{2x} \mid 000\}, \{C_{31}^+ \mid 000\}$	0
198	$P2_13$	$\Gamma_c T^4$	$\{C_{2z} \mid \tfrac{1}{2}0\tfrac{1}{2}\}, \{C_{2x} \mid \tfrac{1}{2}\tfrac{1}{2}0\}, \{C_{31}^+ \mid 000\}$	0
199	$I2_13$	$\Gamma_c^v T^5$	$\{C_{2z} \mid \tfrac{1}{2}0\tfrac{1}{2}\}, \{C_{2x} \mid \tfrac{1}{2}\tfrac{1}{2}0\}, \{C_{31}^+ \mid 000\}$	0
200	$Pm3$	$\Gamma_c T_h^1$	$\{C_{2z} \mid 000\}, \{C_{2x} \mid 000\}, \{C_{31}^+ \mid 000\},$ $\{I \mid 000\}$	0
201	$Pn3$	$\Gamma_c T_h^2$	$\{C_{2z} \mid 000\}, \{C_{2x} \mid 000\}, \{C_{31}^+ \mid 000\},$ $\{I \mid \tfrac{1}{2}\tfrac{1}{2}\tfrac{1}{2}\}$	0
202	$Fm3$	$\Gamma_c^f T_h^3$	$\{C_{2z} \mid 000\}, \{C_{2x} \mid 000\}, \{C_{31}^+ \mid 000\},$ $\{I \mid 000\}$	0
203	$Fd3$	$\Gamma_c^f T_h^4$	$\{C_{2z} \mid 000\}, \{C_{2x} \mid 000\}, \{C_{31}^+ \mid 000\},$ $\{I \mid \tfrac{1}{4}\tfrac{1}{4}\tfrac{1}{4}\}$	0
204	$Im3$	$\Gamma_c^v T_h^5$	$\{C_{2z} \mid 000\}, \{C_{2x} \mid 000\}, \{C_{31}^+ \mid 000\},$ $\{I \mid 000\}$	0
205	$Pa3$	$\Gamma_c T_h^6$	$\{C_{2z} \mid \tfrac{1}{2}0\tfrac{1}{2}\}, \{C_{2x} \mid \tfrac{1}{2}\tfrac{1}{2}0\}, \{C_{31}^+ \mid 000\},$ $\{I \mid 000\}$	0
206	$Ia3$	$\Gamma_c^v T_h^7$	$\{C_{2z} \mid \tfrac{1}{2}0\tfrac{1}{2}\}, \{C_{2x} \mid \tfrac{1}{2}\tfrac{1}{2}0\}, \{C_{31}^+ \mid 000\},$ $\{I \mid 000\}$	0
207	$P432$	$\Gamma_c O^1$	$\{C_{2z} \mid 000\}, \{C_{2x} \mid 000\}, \{C_{2a} \mid 000\},$ $\{C_{31}^+ \mid 000\}$	0
208	$P4_232$	$\Gamma_c O^2$	$\{C_{2z} \mid 000\}, \{C_{2x} \mid 000\}, \{C_{2a} \mid \tfrac{1}{2}\tfrac{1}{2}\tfrac{1}{2}\},$ $\{C_{31}^+ \mid 000\}$	0
209	$F432$	$\Gamma_c^f O^3$	$\{C_{2z} \mid 000\}, \{C_{2x} \mid 000\}, \{C_{2a} \mid 000\},$ $\{C_{31}^+ \mid 000\}$	0
210	$F4_132$	$\Gamma_c^f O^4$	$\{C_{2z} \mid 000\}, \{C_{2x} \mid 000\}, \{C_{2a} \mid \tfrac{1}{4}\tfrac{1}{4}\tfrac{1}{4}\},$ $\{C_{31}^+ \mid 000\}$	0
211	$I432$	$\Gamma_c^v O^5$	$\{C_{2z} \mid 000\}, \{C_{2x} \mid 000\}, \{C_{2a} \mid 000\},$ $\{C_{31}^+ \mid 000\}$	0
212	$P4_332$	$\Gamma_c O^6$	$\{C_{2z} \mid \tfrac{1}{2}0\tfrac{1}{2}\}, \{C_{2x} \mid \tfrac{1}{2}\tfrac{1}{2}0\}, \{C_{2a} \mid \tfrac{1}{4}\tfrac{3}{4}\tfrac{3}{4}\},$ $\{C_{31}^+ \mid 000\}$	0

International number	International symbol	Schönflies symbol	Generating elements	$\mathbf{t}_0$
213	$P4_132$	$\Gamma_c O^7$	$\{C_{2z} \mid \tfrac{1}{2}0\tfrac{1}{2}\}, \{C_{2x} \mid \tfrac{1}{2}\tfrac{1}{2}0\}, \{C_{2a} \mid \tfrac{3}{4}\tfrac{1}{4}\tfrac{1}{4}\},$ $\{C_{31}^+ \mid 000\}$	0
214	$I4_132$	$\Gamma_c^v O^8$	$\{C_{2z} \mid \tfrac{1}{2}0\tfrac{1}{2}\}, \{C_{2x} \mid \tfrac{1}{2}\tfrac{1}{2}0\}, \{C_{2a} \mid \tfrac{1}{2}00\},$ $\{C_{31}^+ \mid 000\}$	0
215	$P\bar{4}3m$	$\Gamma_c T_d^1$	$\{C_{2z} \mid 000\}, \{C_{2x} \mid 000\}, \{\sigma_{du} \mid 000\},$ $\{C_{31}^+ \mid 000\}$	0
216	$F\bar{4}3m$	$\Gamma_c^f T_d^2$	$\{C_{2z} \mid 000\}, \{C_{2x} \mid 000\}, \{\sigma_{da} \mid 000\},$ $\{C_{31}^+ \mid 000\}$	0
217	$I\bar{4}3m$	$\Gamma_c^v T_d^3$	$\{C_{2z} \mid 000\}, \{C_{2x} \mid 000\}, \{\sigma_{da} \mid 000\},$ $\{C_{31}^+ \mid 000\}$	0
218	$P\bar{4}3n$	$\Gamma_c T_d^4$	$\{C_{2z} \mid 000\}, \{C_{2x} \mid 000\}, \{\sigma_{da} \mid \tfrac{1}{2}\tfrac{1}{2}\tfrac{1}{2}\},$ $\{C_{31}^+ \mid 000\}$	0
219	$F\bar{4}3c$	$\Gamma_c^f T_d^5$	$\{C_{2z} \mid 000\}, \{C_{2x} \mid 000\}, \{\sigma_{da} \mid \tfrac{1}{2}\tfrac{1}{2}\tfrac{1}{2}\},$ $\{C_{31}^+ \mid 000\}$	0
220	$I\bar{4}3d$	$\Gamma_c^v T_d^6$	$\{C_{2z} \mid \tfrac{1}{2}0\tfrac{1}{2}\}, \{C_{2x} \mid \tfrac{1}{2}\tfrac{1}{2}0\}, \{\sigma_{da} \mid \tfrac{1}{2}00\},$ $\{C_{31}^+ \mid 000\}$	0
221	$Pm3m$	$\Gamma_c O_h^1$	$\{C_{2z} \mid 000\}, \{C_{2x} \mid 000\}, \{C_{2a} \mid 000\},$ $\{C_{31}^+ \mid 000\}, \{I \mid 000\}$	0
222	$Pn3n$	$\Gamma_c O_h^2$	$\{C_{2z} \mid 000\}, \{C_{2x} \mid 000\}, \{C_{2a} \mid 000\},$ $\{C_{31}^+ \mid 000\}, \{I \mid \tfrac{1}{2}\tfrac{1}{2}\tfrac{1}{2}\}$	0
223	$Pm3n$	$\Gamma_c O_h^3$	$\{C_{2z} \mid 000\}, \{C_{2x} \mid 000\}, \{C_{2a} \mid \tfrac{1}{2}\tfrac{1}{2}\tfrac{1}{2}\},$ $\{C_{31}^+ \mid 000\}, \{I \mid 000\}$	0
224	$Pn3m$	$\Gamma_c O_h^4$	$\{C_{2z} \mid 000\}, \{C_{2x} \mid 000\}, \{C_{2a} \mid \tfrac{1}{2}\tfrac{1}{2}\tfrac{1}{2}\},$ $\{C_{31}^+ \mid 000\}, \{I \mid \tfrac{1}{2}\tfrac{1}{2}\tfrac{1}{2}\}$	0
225	$Fm3m$	$\Gamma_c^f O_h^5$	$\{C_{2z} \mid 000\}, \{C_{2x} \mid 000\}, \{C_{2a} \mid 000\},$ $\{C_{31}^+ \mid 000\}, \{I \mid 000\}$	0
226	$Fm3c$	$\Gamma_c^f O_h^6$	$\{C_{2z} \mid 000\}, \{C_{2x} \mid 000\}, \{C_{2a} \mid 000\},$ $\{C_{31}^+ \mid 000\}, \{I \mid \tfrac{1}{2}\tfrac{1}{2}\tfrac{1}{2}\}$	$\tfrac{1}{4}\mathbf{t}_1 + \tfrac{1}{4}\mathbf{t}_2 + \tfrac{1}{4}\mathbf{t}_3$
			$\{C_{2z} \mid 000\}, \{C_{2x} \mid 000\}, \{C_{2a} \mid \tfrac{1}{2}\tfrac{1}{2}\tfrac{1}{2}\},$ $\{C_{31}^+ \mid 000\}, \{I \mid 000\}$	
227	$Fd3m$	$\Gamma_c^f O_h^7$	$\{C_{2z} \mid 000\}, \{C_{2x} \mid 000\}, \{C_{2a} \mid \tfrac{1}{4}\tfrac{1}{4}\tfrac{1}{4}\},$ $\{C_{31}^+ \mid 000\}, \{I \mid \tfrac{1}{4}\tfrac{1}{4}\tfrac{1}{4}\}$	0
228	$Fd3c$	$\Gamma_c^f O_h^8$	$\{C_{2z} \mid 000\}, \{C_{2x} \mid 000\}, \{C_{2a} \mid \tfrac{1}{4}\tfrac{1}{4}\tfrac{1}{4}\},$ $\{C_{31}^- \mid 000\}, \{I \mid \tfrac{3}{4}\tfrac{3}{4}\tfrac{3}{4}\}$	0
229	$Im3m$	$\Gamma_c^v O_h^9$	$\{C_{2z} \mid 000\}, \{C_{2x} \mid 000\}, \{C_{2a} \mid 000\},$ $\{C_{31}^+ \mid 000\}, \{I \mid 000\}$	0
230	$Ia3d$	$\Gamma_c^v O_h^{10}$	$\{C_{2z} \mid \tfrac{1}{2}0\tfrac{1}{2}\}, \{C_{2x} \mid \tfrac{1}{2}\tfrac{1}{2}0\}, \{C_{2a} \mid \tfrac{1}{2}00\},$ $\{C_{31}^+ \mid 000\}, \{I \mid 000\}$	0

Notes to Table 3.7

(i) We give, for each space group, its international number, its international symbol and its Schönflies symbol in columns 1, 2, and 3 respectively. The Schönflies symbol in column 3 is prefixed by the symbol for the appropriate Bravais lattice (see Table 3.1).

(ii) The name of a space group, as given in columns 1, 2, and 3 implies (a) the Bravais lattice on which it is situated and (b) the isogonal point group. It is therefore unnecessary to tabulate separately the elements of the translation group **T** (these are given in Table 3.1), or the elements of the isogonal point group **F** (these are given in Table 2.2). It remains to give a number of coset representatives (elements of **Q** as described in section 3.5). We give enough to determine all the elements of the space group **G**; these are given in column 4 in which $\{R \mid pqr\}$ means $\{R \mid p\mathbf{t}_1 + q\mathbf{t}_2 + r\mathbf{t}_3\}$ with $\mathbf{t}_1$, $\mathbf{t}_2$, and $\mathbf{t}_3$ as defined in Table 3.1. These operations are given as usual in the Seitz notation that was explained in section 1.5 (see eqns. (1.5.2) and (1.5.3)).

(iii) For some space groups the elements that we have used differ in some way from those given in the *International Tables* (Henry and Lonsdale 1965). In these cases we give, in the first row of column 4, the Seitz space-group symbols that we have used for the elements of **Q**; in the second row we then give the elements used in the *International Tables*.

(iv) Although we retain the international labels for the space groups numbers 38–41, *Amm*2 (C_{2v}^{14}), *Abm*2 (C_{2v}^{15}), *Ama*2 (C_{2v}^{16}) and *Aba*2 (C_{2v}^{17}), we do not use the *A* lattice because this would require an extra block in Table 3.6, under Γ_o^b, and an extra figure to add to Fig. 3.6 (see Note (v) to Table 3.1). Instead, we use the *C* lattice throughout for all monoclinic base-centred space groups and change the orientation of the point-group operations for space groups 38–41 so that the twofold rotation axis is along the *y*-axis, whereas in the *International Tables* it is along the *z*-axis.

(v) If a translation $\mathbf{t}_0$ is given in column 5, this is the translation of the origin that, substituted in eqn. (3.5.11), changes the elements $\{R_1 \mid \mathbf{v}_1\}$ that we use into the elements $\{R_2 \mid \mathbf{v}_2\}$ of the *International Tables*. An asterisk in column 5 indicates that our origin coincides with that of the *International Tables* but that the orientations of the axes are different. Where there is both a vector $\mathbf{t}_0$ and an asterisk in column 5 this means that the vector $\mathbf{t}_0$ changes the elements $\{R_1 \mid \mathbf{v}_1\}$ that we use into the elements $\{R_2 \mid \mathbf{v}_2\}$, according to eqn. (3.5.11), where the $\{R_2 \mid \mathbf{v}_2\}$ differ only in orientation from the elements in the *International Tables*.

(vi) Modifications to the international symbols of certain space groups have been suggested by Buerger (1967). A reversion is urged to the practice used in the old *International Tables* of always indicating the relative orientation of lattice and symmetry axes entirely through the Bravais lattice part of the space-group symbol and not at all through the point-group (or its substituent) part of the symbol. If the symbol *E* is used for an end-centred lattice without implying orientation, then the permissible orientations of *E* for orthorhombic space groups are *A*, *B*, and *C*. If *S* and *Z* are the symbols used for simple and body-centred lattices without implying orientation the corresponding symbols indicating orientation are as follows.

	Orientation symbol	Orientation of axis of point-group symbol First symbol	Second symbol
S	*P*	[011]	in (100)
	D	[001]	[110] orthogonal; [210] hexagonal
	R	[111]	[101]
Z	*I*	[001]	[100]
	J	[001]	[110]

For tetragonal space groups we need to use alternative symbols for two permissible orientations of *S* (*P* and *D*) and for two possible orientations of *Z* (*I* and *J*). For trigonal crystals we need to use alternative symbols for three permissible orientations of *S* (*P*, *D*, and *R*). Buerger urges the adoption of these lattice symbols together with the outlawing of the use of the symbol '1' except in the point groups 1 (C_1) and $\bar{1}$ (C_i) and the space groups *P*1 (C_1^1) and $P\bar{1}$ (C_i^1). Only the following space groups would be affected by such changes:

Point group	Number	Schönflies symbol	Bragg, von Laue, and Hermann (1935*a*, *b*)	Henry and Lonsdale (1965)	Buerger (1967)
32 (D_3)	149	D_3^1	*H*32	*P*312	*D*32
	150	D_3^2	*C*32	*P*321	*P*32
	151	D_3^3	$H3_12$	$P3_112$	$D3_12$
	152	D_3^4	$C3_12$	$P3_121$	$P3_12$

Point group	Number	Schönflies symbol	Bragg, von Laue, and Hermann (1935*a*, *b*)	Henry and Lonsdale (1965)	Buerger (1967)
	153	D_3^5	$H3_22$	$P3_212$	$D3_22$
	154	D_3^6	$C3_22$	$P3_221$	$P3_22$
$3m$ (C_{3v})	156	C_{3v}^1	$C3m$	$P3m1$	$P3m$
	157	C_{3v}^2	$H3m$	$P31m$	$D3m$
	158	C_{3v}^3	$C3c$	$P3c1$	$P3c$
	159	C_{3v}^4	$H3c$	$P31c$	$D3c$
$\bar{3}m$ (D_{3d})	162	D_{3d}^1	$H\bar{3}m$	$P\bar{3}1m$	$D\bar{3}m$
	163	D_{3d}^2	$H\bar{3}c$	$P\bar{3}1c$	$D\bar{3}c$
	164	D_{3d}^3	$C\bar{3}m$	$P\bar{3}m1$	$P\bar{3}m$
	165	D_{3d}^4	$C\bar{3}c$	$P\bar{3}c1$	$P\bar{3}c$
$\bar{6}2m$ (D_{3h})	187	D_{3h}^1	$C\bar{6}m2$	$P\bar{6}m2$	$D\bar{6}2m$
	188	D_{3h}^2	$C\bar{6}c2$	$P\bar{6}c2$	$D\bar{6}2c$
$\bar{4}2m$ (D_{2d})	115	D_{2d}^5	$C\bar{4}2m$	$P\bar{4}m2$	$D\bar{4}2m$
	116	D_{2d}^6	$C\bar{4}2c$	$P\bar{4}c2$	$D\bar{4}2c$
	117	D_{2d}^7	$C\bar{4}2b$	$P\bar{4}b2$	$D\bar{4}2b$
	118	D_{2d}^8	$C\bar{4}2n$	$P\bar{4}n2$	$D\bar{4}2n$
	119	D_{2d}^9	$F\bar{4}2m$	$I\bar{4}m2$	$J\bar{4}2m$
	120	D_{2d}^{10}	$F\bar{4}2c$	$I\bar{4}c2$	$J\bar{4}2c$

With these modifications introduced by Buerger (1967) the first symbol in any space-group label indicates the general Bravais lattice type and its mutual orientation with respect to the symmetry axes. The remaining symbols in the label are then either the symbols of a point group or substituents thereof. One result of these modifications would be that for every space group these symbols would be in exactly the same sequence as in the standard sequence for the corresponding point group.

generating elements. For some space groups we have used a unit cell that does not coincide with the unit cell given in Volume 1 of the *International tables for X-ray crystallography* (Henry and Lonsdale 1965); in these cases the vector $\mathbf{t}_0$ relating the two origins is given in column 5. For a few space groups the orientation of the symmetry elements with respect to the Bravais lattice that we have used is different from the orientation used in the *International Tables*; these cases are indicated by an asterisk in column 5 of Table 3.7. Although the space groups themselves are officially defined in the *International Tables* the point that one chooses as the origin and the orientation that one chooses for the coordinate axes are still open to some choice; the *International Tables* themselves give two alternative settings for many space groups, as we have seen for the example of *Pmmn* (D_{2h}^{13}). To quote from p. 52 of the *International Tables* 'No official importance is attached to the particular setting of the space group adopted as standard . . .'.

A brief description of the logic of the international space-group symbols was given in section 1.5. These symbols were used in the first edition of the *International Tables*

(Bragg, von Laue, and Hermann 1935*a*, *b*) and then came into common use. The symbols used then were not entirely satisfactory in the Bravais lattice part of the symbol and some changes were made when the new *International Tables* were compiled (Henry and Lonsdale 1965); the new version of the symbols is used in Table 3.7. For example, the tetragonal point group $\bar{4}2m$ can be placed in the unit cell of the primitive tetragonal Bravais lattice with the pure rotation axis parallel either to the x-axis or to the line $x = y$. In the old *International Tables* these two orientations were indicated by using the two different Bravais lattice symbols P and C. For the two orientations on the body-centred tetragonal lattice the two different symbols I and F were used. In the new *International Tables* any given Bravais lattice is denoted by the same symbol in each space group based on that Bravais lattice (with the exception of space groups numbers 38–41, $Amm2$ (C_{2v}^{14}), $Abm2$ (C_{2v}^{15}), $Ama2$ (C_{2v}^{16}), and $Aba2$ (C_{2v}^{17})). The problem of distinguishing between different orientations of a point group on a Bravais lattice is then overcome either by reversing the order of the last two symbols in the point-group symbols $\bar{4}2m$ and $\bar{6}2m$ (to give $\bar{4}m2$ and $\bar{6}m2$ respectively) or, for trigonal space groups, by adding the symbol '1' in different places in the space-group symbol. As a consequence, the symmetry symbols are now often interpreted, in part, as signifying not only which symmetry elements are present but also their orientations, see Table 3.3.2 of Henry and Lonsdale (1965). This question of the identification of the various parts of the international point-group labels will concern us again in Chapter 7. For certain space groups the space-group symbols used in the new *International Tables* (Henry and Lonsdale 1965) have been criticized by Buerger (1967). The modifications suggested by Buerger are given in Note (vi) to Table 3.7 and, if adopted, would only affect a small number of space groups.

To show how to use the equivalent positions given in the *International Tables* to determine the geometrical description of the symmetry elements we follow the method of Wondratschek and Neubüser (1967). The distinction between a *symmetry element*, such as a rotation axis or reflection plane of symmetry which is, loosely speaking, a physical property of a crystal, and a *symmetry operation*, which is a covering operation of the crystal and an element of the abstract mathematical group of the point group or space group, should be apparent from sections 1.4 and 1.5. The effect of one of the space-group symmetry operations $\{R_i \mid \mathbf{r}_i\}$ on the point x, y, z is to produce one of the points in an equivalent position which may be specified in the general form

$$x', y', z' = r_1 + a_{11}x + a_{12}y + a_{13}z,\quad r_2 + a_{21}x + a_{22}y + a_{23}z,\ r_3 + a_{31}x + a_{32}y + a_{33}z \qquad (3.5.14)$$

where the r_i are rational numbers and the a_{ij} are $+1$, -1 or 0.

Example 3.5.1. In $Pmmn$ (D_{2h}^{13}) one finds, among other equivalent positions,

$\frac{1}{2} + x, \frac{1}{2} - y, \bar{z}$; this means that $a_{11} = +1, a_{22} = a_{33} = -1, r_1 = r_2 = \frac{1}{2}, r_3 = 0$, and all other $a_{ij} = 0$.

Equation (3.5.14) can conveniently be written in matrix form,

$$\mathbf{M}(\{R_i \mid \mathbf{r}_i\}) = \left(\begin{array}{ccc|c} a_{11} & a_{12} & a_{13} & r_1 \\ a_{21} & a_{22} & a_{23} & r_2 \\ a_{31} & a_{32} & a_{33} & r_3 \\ \hline 0 & 0 & 0 & 1 \end{array}\right) = (\mathbf{A}, \mathbf{r}_i) \tag{3.5.15}$$

where $\mathbf{M}(\{R_i \mid \mathbf{r}_i\})$ acts on the vector $\begin{pmatrix} x \\ y \\ z \\ 1 \end{pmatrix}$. Clearly the matrix a_{ij} has already been given, in contracted form, in the Jones' symbols in Table 1.4. The determination of the geometrical significance of the R_i part of $\{R_i \mid \mathbf{r}_i\}$ can of course be made by inspection of Table 1.4. Alternatively, the character of $\mathbf{A}$, $\chi(\mathbf{A})$, and the determinant of $\mathbf{A}$, det $(\mathbf{A})$, can be evaluated quite simply and these two quantities characterize the nature of R_i completely (Bethe 1929) according to Table 3.8, where C_n indicates a rotation through $2\pi/n$.

TABLE 3.8

The determination of the form of $\mathbf{R}_i$ *from* $\chi(\mathbf{A})$ *and* det $(\mathbf{A})$

R_i	C_1	C_2	C_3	C_4	C_6	IC_1 I	IC_2 σ	IC_3 S_6	IC_4 S_4	IC_6 S_3
$\chi(\mathbf{A})$	3	-1	0	1	2	-3	1	0	-1	-2
det $(\mathbf{A})$	1	1	1	1	1	-1	-1	-1	-1	-1

If det $(\mathbf{A}) = +1$ the symmetry element is either a pure rotation axis or a screw rotation axis. We therefore seek to determine the direction of this axis and, for a screw rotation axis, the screw vector, $\mathbf{w}$, i.e. the component of the translation $\mathbf{r}_i$, resolved along the axis. If $\mathbf{q}$ is a vector that passes through the origin and is parallel to the screw axis then $\mathbf{Aq} = \mathbf{q}$ so that $\mathbf{q}$, which gives the *direction* of the screw axis, can be found by solving

$$(\mathbf{A} - \mathbf{E})\mathbf{q} = 0. \tag{3.5.16}$$

To determine $\mathbf{w}$, for a screw rotation, we note that

$$\{R_i \mid \mathbf{r}_i\}\mathbf{q} = \mathbf{Aq} + \mathbf{r}_i = \mathbf{Eq} + \mathbf{r}_i = \mathbf{q} + \mathbf{r}_i \tag{3.5.17}$$

therefore

$$\{R_i \mid \mathbf{r}_i\}^2\mathbf{q} = \mathbf{q} + \mathbf{r}_i + \mathbf{A}\mathbf{r}_i \tag{3.5.18}$$

and

$$\begin{aligned}\{R_i \mid \mathbf{r}_i\}^n\mathbf{q} &= \mathbf{q} + \mathbf{r}_i + \mathbf{A}\mathbf{r}_i + \mathbf{A}^2\mathbf{r}_i + \cdots + \mathbf{A}^{n-1}\mathbf{r}_i \\ &= \mathbf{q} + n\mathbf{w}\end{aligned} \tag{3.5.19}$$

since n applications of the screw rotation operation are equivalent to the pure translation $n\mathbf{w}$. By rearranging eqn. (3.5.19) we obtain an explicit expression for the *screw vector* $\mathbf{w}$,

$$\mathbf{w} = (1/n)(\mathbf{E} + \mathbf{A} + \mathbf{A}^2 + \cdots + \mathbf{A}^{n-1})\mathbf{r}_i. \tag{3.5.20}$$

The direction of the screw axis is of course given by the direction of $\mathbf{w}$, but this direction can be found more easily from eqn. (3.5.16). The *localization* of the screw axis may be determined from the fact that a point $\mathbf{x}$ on the axis satisfies $\mathbf{A}\mathbf{x} + \mathbf{r}_i = \mathbf{x} + \mathbf{w}$, i.e.

$$(\mathbf{A} - \mathbf{E})\mathbf{x} = \left(\frac{1}{n}\mathbf{B} - \mathbf{E}\right)\mathbf{r}_i \tag{3.5.21}$$

where

$$\mathbf{B} = \mathbf{E} + \mathbf{A} + \mathbf{A}^2 + \cdots + \mathbf{A}^{n-1}. \tag{3.5.22}$$

If $n = 2$, eqn. (3.5.21) has a particularly simple solution, namely $\mathbf{x} = \frac{1}{2}\mathbf{r}_i$.

If det $(\mathbf{A}) = -1$ there are three possible cases to be considered: (a) I $(=IC_1)$; (b) σ $(=IC_2)$; and (c) S_6 $(=IC_3)$; S_4 $(=IC_4)$, and S_3 $(=IC_6)$.

(a) $I(=IC_1)$. The inversion centre $\mathbf{x}$ can be found from the fact that $\mathbf{x} = \{I \mid \mathbf{r}_i\}\mathbf{x} = -\mathbf{x} + \mathbf{r}_i$ and therefore $\mathbf{x} = \frac{1}{2}\mathbf{r}_i$.

(b) σ $(=IC_2)$. The orientation of the plane can be specified by a vector $\mathbf{q}$ that passes through the origin and is normal to the plane; $\mathbf{q}$ then satisfies $\mathbf{A}\mathbf{q} = -\mathbf{q}$, i.e.

$$(\mathbf{A} + \mathbf{E})\mathbf{q} = 0. \tag{3.5.23}$$

For a glide plane the glide vector, $\mathbf{w}$, can be found from the fact that two successive applications of $\{R_i \mid \mathbf{r}_i\}$ to a point $\mathbf{x}$ in the plane moves it to $\mathbf{x} + 2\mathbf{w}$ so that, from eqn. (3.5.18),

$$\{R_i \mid \mathbf{r}_i\}^2\mathbf{x} = \mathbf{x} + \mathbf{r}_i + \mathbf{A}\mathbf{r}_i = \mathbf{x} + 2\mathbf{w}. \tag{3.5.24}$$

Therefore

$$\mathbf{w} = \tfrac{1}{2}(\mathbf{A} + \mathbf{E})\mathbf{r}_i. \tag{3.5.25}$$

Any point $\mathbf{x}$ on the plane then obeys $\{R_i \mid \mathbf{r}_i\}\mathbf{x} = \mathbf{A}\mathbf{x} + \mathbf{r}_i = \mathbf{x} + \mathbf{w}$ so that $\mathbf{x}$ satisfies

$$(\mathbf{A} - \mathbf{E})\mathbf{x} = \tfrac{1}{2}(\mathbf{A} - \mathbf{E})\mathbf{r}_i, \tag{3.5.26}$$

which obviously has the special solution $\mathbf{x} = \frac{1}{2}\mathbf{r}_i$.

(c) S_6 $(=IC_3)$; S_4 $(=IC_4)$; S_3 $(=IC_6)$. If $\mathbf{q}$ is a vector through the origin and parallel to one of these axes then $\mathbf{Aq} = -\mathbf{q}$ and $\mathbf{q}$ is again given by solving

$$(\mathbf{A}+\mathbf{E})\mathbf{q} = 0. \tag{3.5.23}$$

There are no screw vectors $\mathbf{w}$. The inversion centre can be found from $\{R_i \mid \mathbf{r}_i\}\mathbf{x} = \mathbf{Ax} + \mathbf{r}_i = \mathbf{x}$, i.e.

$$(\mathbf{E}-\mathbf{A})\mathbf{x} = \mathbf{r}_i. \tag{3.5.27}$$

In the *International Tables* every matrix $\mathbf{A}$ belongs to one of four different types that are shown in Table 3.9. Having determined the form of R_i and the type of $\mathbf{A}$

TABLE 3.9

The types of matrix $\mathbf{A}$ *that arise from the International Tables for X-ray crystallography*

(i) $\mathbf{A} = \begin{bmatrix} \pm1 & 0 & 0 \\ 0 & \pm1 & 0 \\ 0 & 0 & \pm1 \end{bmatrix}$; (iii) (a) $\mathbf{A} = \begin{bmatrix} 0 & 0 & \pm1 \\ \pm1 & 0 & 0 \\ 0 & \pm1 & 0 \end{bmatrix}$,

(ii) (a) $\mathbf{A} = \begin{bmatrix} \pm1 & 0 & 0 \\ 0 & 0 & \pm1 \\ 0 & \pm1 & 0 \end{bmatrix}$, (b) $\mathbf{A} = \begin{bmatrix} 0 & \pm1 & 0 \\ 0 & 0 & \pm1 \\ \pm1 & 0 & 0 \end{bmatrix}$;

(b) $\mathbf{A} = \begin{bmatrix} 0 & 0 & \pm1 \\ 0 & \pm1 & 0 \\ \pm1 & 0 & 0 \end{bmatrix}$, (iv) $\mathbf{A} = \begin{bmatrix} a_{11} & a_{12} & 0 \\ a_{21} & a_{22} & 0 \\ 0 & 0 & \pm1 \end{bmatrix}$ with one of the $a_{ik} = 0$, the other three $a_{ik} = \pm1$; $i, k = 1, 2$.

(c) $\mathbf{A} = \begin{bmatrix} 0 & \pm1 & 0 \\ \pm1 & 0 & 0 \\ 0 & 0 & \pm1 \end{bmatrix}$;

Note to Table 3.9

(i) It is assumed that a setting given by Henry and Lonsdale (1965) is being used; for other settings the appropriate equations from eqns. (3.5.16)–(3.5.27) will have to be solved.

one can then use Table 3.10, which makes use of eqns. (3.5.16)–(3.5.27), to determine the orientation and position of the symmetry element corresponding to each of the equivalent positions for the space group in the *International Tables*.

Example 3.5.2. To illustrate the use of Table 3.10 we use the matrix $(\mathbf{A}, \mathbf{r})$ that was determined in Example 3.5.1, namely

$$(\mathbf{A}, \mathbf{r}) = \left(\begin{array}{ccc|c} 1 & 0 & 0 & \frac{1}{2} \\ 0 & -1 & 0 & \frac{1}{2} \\ 0 & 0 & -1 & 0 \\ \hline 0 & 0 & 0 & 1 \end{array}\right) \tag{3.5.28}$$

TABLE 3.10

Determination of the symmetry elements
(*After Wondratschek and Neubüser* (1967))

(i) C_1	Translation with components (r_1, r_2, r_3)		
(i) I	Inversion centre in $(\frac{1}{2}r_1, \frac{1}{2}r_2, \frac{1}{2}r_3)$		
(i) C_2	One $a_{ii} = +1$, the other two $a_{kk} = -1, k \neq i$		
Direction of axis given by	$\mathbf{q} = \mathbf{a}$ for $i = 1$; $\mathbf{q} = \mathbf{b}$ for $i = 2$; $\mathbf{q} = \mathbf{c}$ for $i = 3$		
Screw vector	$r_i\mathbf{q}$		
Coordinates of a point on the axis	$(\frac{1}{2}r_1, \frac{1}{2}r_2, \frac{1}{2}r_3)$		
(i) σ	Two $a_{ii} = +1$, the third $a_{kk} = -1, k \neq i$		
Normal to symmetry plane	$\mathbf{a}$ for $k = 1$, $\mathbf{b}$ for $k = 2$, $\mathbf{c}$ for $k = 3$		
Glide vector	$\frac{1}{2}\{(1 + a_{11})r_1\mathbf{a} + (1 + a_{22})r_2\mathbf{b} + (1 + a_{33})r_3\mathbf{c}\}$		
Coordinates of a point on symmetry plane	$(\frac{1}{2}r_1, \frac{1}{2}r_2, \frac{1}{2}r_3)$		
(ii) C_2	(a)	(b)	(c)
Direction of axis given by	$\mathbf{q} = [\mathbf{b} + a_{23}\mathbf{c}]$	$\mathbf{q} = [a_{31}\mathbf{a} + \mathbf{c}]$	$\mathbf{q} = [\mathbf{a} + a_{12}\mathbf{b}]$
Screw vector	$\frac{1}{2}(r_2 + a_{23}r_3)\mathbf{q}$	$\frac{1}{2}(r_3 + a_{31}r_1)\mathbf{q}$	$\frac{1}{2}(r_1 + a_{12}r_2)\mathbf{q}$
Coordinates of a point on the axis	$(\frac{1}{2}(r_1, 0, -a_{23}r_2 + r_3))$	$(\frac{1}{2}(-a_{31}r_3 + r_1, r_2, 0))$	$(\frac{1}{2}(0, -a_{12}r_1 + r_2, r_3))$
(ii) σ	(a)	(b)	(c)
Normal to symmetry plane	$[0\ 1\ -a_{23}]$	$[-a_{31}\ 0\ 1]$	$[1\ -a_{12}\ 0]$
Glide vector	$\frac{1}{2}(r_2 + a_{23}r_3)(\mathbf{b} + a_{23}\mathbf{c}) + r_1\mathbf{a}$	$\frac{1}{2}(r_3 + a_{31}r_1)(\mathbf{c} + a_{31}\mathbf{a}) + r_2\mathbf{b}$	$\frac{1}{2}(r_1 + a_{12}r_2)(\mathbf{a} + a_{12}\mathbf{b}) + r_3\mathbf{c}$
Coordinates of a point on symmetry plane	$(\frac{1}{2}(0, 0, -a_{23}r_2 + r_3))$	$(\frac{1}{2}(-a_{31}r_3 + r_1, 0, 0))$	$(\frac{1}{2}(0, -a_{12}r_1 + r_2, 0))$
(ii) C_4	(a)	(b)	(c)
Direction of axis	[100]	[010]	[001]
Screw vector	$r_1\mathbf{a}$	$r_2\mathbf{b}$	$r_3\mathbf{c}$
Coordinates of a point on the axis	$(\frac{1}{2}(0, r_2 + a_{23}r_3, a_{32}r_2 + r_3))$	$(\frac{1}{2}(a_{13}r_3 + r_1, 0, r_3 + a_{31}r_1))$	$(\frac{1}{2}(r_1 + a_{12}r_2, a_{21}r_1 + r_2, 0))$

(Remark: If the product (a) $a_{32}r_1$ or (b) $a_{13}r_2$ or (c) $a_{21}r_3$ has the value $(4n + 1)/4$, n integer, a 4_1-axis is obtained; for $(4n + 3)/4$ one obtains a 4_3-axis, for $(4n + 2)/4$ a 4_2-axis.)

(ii) S_4	(a)	(b)	(c)
Direction of inversion axis	[100]	[010]	[001]
Coordinates of inversion point	$(\frac{1}{2}(r_1, r_2 + a_{23}r_3, a_{32}r_2 + r_3))$	$(\frac{1}{2}(a_{13}r_3 + r_1, r_2, r_3 + a_{31}r_1))$	$(\frac{1}{2}(r_1 + a_{12}r_2, a_{21}r_1 + r_2, r_3))$

(Remark: no translation components occur.)

(iii) C_3	(a)	(b)
Direction of axis given by	$\mathbf{q} = a_{32}\mathbf{a} + a_{13}\mathbf{b} + a_{21}\mathbf{c}$	$\mathbf{q} = a_{23}\mathbf{a} + a_{31}\mathbf{b} + a_{12}\mathbf{c}$
Screw vector	$\frac{1}{3}(a_{32}r_1 + a_{13}r_2 + a_{21}r_3)\mathbf{q}$	$\frac{1}{3}(a_{23}r_1 + a_{31}r_2 + a_{12}r_3)\mathbf{q}$
Coordinates of point on the axis	$(\frac{1}{3}(r_1 - a_{21}r_2, r_2 - a_{32}r_3, r_3 - a_{13}r_1))$	$(\frac{1}{3}(r_1 - a_{31}r_3, r_2 - a_{12}r_1, r_3 - a_{23}r_2))$

[Remark: The symmetry operation belongs to a 3_1-axis if in (a) the screw vector is $[(3n + 1)/3]\mathbf{q}$, in (b) $[(3n + 2)/3]\mathbf{q}$, n integer. Similarly a screw vector $[(3n + 2)/3]\mathbf{q}$ in (a) and $[(3n + 1)/3]\mathbf{q}$ in (b) belongs to a 3_2-axis.)

(iii) S_6 $(=IC_3)$	(a)	(b)
Direction of inversion axis	$-[a_{32}, a_{13}, a_{21}]$	$-[a_{23}, a_{31}, a_{12}]$
Coordinates of inversion centre	$(\frac{1}{2}(r_1 - a_{21}r_2 + a_{13}r_3, a_{21}r_1 + r_2 - a_{32}r_3, -a_{13}r_1 + a_{32}r_2 + r_3))$	$(\frac{1}{2}(r_1 + a_{12}r_2 - a_{31}r_3, -a_{12}r_1 + r_2 + a_{23}r_3, a_{31}r_1 - a_{23}r_2 + r_3))$

(Remark: no translation components occur.)

(iv) C_2	
Direction of axis given by	$\mathbf{q} = (1 + a_{11} + a_{12})\mathbf{a} + (1 + a_{22} + a_{21})\mathbf{b}$
Screw vector	$\frac{1}{2}[(1 + a_{11} - a_{21}{}^2)r_1 + (1 + a_{22} - a_{12}{}^2)r_2]\mathbf{q}$
Coordinates of a point on the axis	$(\frac{1}{2}r_1, \frac{1}{2}r_2, \frac{1}{2}r_3)$

(iv) σ	
Direction of normal to symmetry plane	$[1 + a_{22} - a_{12}, 1 + a_{11} - a_{21}, 0]$
Glide vector	$\frac{1}{2}\{[(1 + a_{11})r_1 + a_{12}r_2]\mathbf{a} + [a_{21}r_1 + (1 + a_{22})r_2]\mathbf{b} + 2r_3\mathbf{c}\}$
Coordinates of a point on the symmetry plane	$(\frac{1}{2}r_1, \frac{1}{2}r_2, 0)$

(iv) C_3	
Direction of axis	[001]
Screw vector	$r_3\mathbf{c}$

Coordinates of a point on the axis	$(\frac{1}{3}[(1 - a_{22})r_1 + a_{12}r_2, a_{21}r_1 + (1 - a_{11})r_2, 0])$

(Remark: If $a_{21}r_3 = (3n + 1)/3$, n integer, a 3_1-axis is obtained; for $a_{21}r_3 = (3n + 2)/3$ one obtains a 3_2-axis.)

(iv) C_6

Direction of axis	[001]
Screw vector	$r_3\mathbf{c}$
Coordinates of a point on the axis	$(a_{11}r_1 + a_{12}r_2, a_{21}r_1 + a_{22}r_2, 0)$

(Remark: The symmetry operation belongs to a 6_1-, 6_2-, 6_3-, 6_4-, 6_5-axis, if $a_{21}r_3 = (6n + 1)/6; (6n + 2)/6; (6n + 3)/6; (6n + 4)/6; (6n + 5)/6$ respectively.)

(iv) S_6 $(=IC_3)$

Direction of inversion axis	[001]
Coordinates of inversion centre	$(a_{11}r_1 + a_{12}r_2, a_{21}r_1 + a_{22}r_2, r_3/2)$

(Remark: no translation components occur.)

(iv) S_3 $(=IC_6)$

Direction of inversion axis	[001]
Coordinates of inversion point	$(\frac{1}{3}\{(1 - a_{22})r_1 + a_{12}r_2, a_{21}r_1 + (1 - a_{11})r_2, \frac{3}{2}r_3\})$

(Remark: no translation components occur.)

Notes to Table 3.10

(i) As in the *International Tables* a, b, and c are the lengths of three edges of the conventional unit cell that meet at one vertex; **a**, **b**, and **c** are the vectors along these edges.

(ii) The matrix **A** (a_{ij}) and the vector $\mathbf{r}_i$ are defined in the text; see eqns. (3.5.14) and (3.5.15).

(iii) The type (i), (ii), (iii), or (iv) of **A** is to be determined by using Table 3.9.

(iv) C_n is a pure rotation through an angle of $2\pi/n$.

(v) This table is for settings of the unit cell given by Henry and Lonsdale (1965); for other settings the appropriate equations from eqns. (3.5.16)–(3.5.27) will have to be solved.

so that

$$\mathbf{A} = \begin{pmatrix} 1 & 0 & 0 \\ 0 & -1 & 0 \\ 0 & 0 & -1 \end{pmatrix} \tag{3.5.29}$$

and therefore

$$\chi(\mathbf{A}) = -1 \tag{3.5.30}$$

and

$$\det(\mathbf{A}) = +1, \tag{3.5.31}$$

From Table 3.8 $\{R_i \mid \mathbf{r}_i\}$ is therefore a C_2 type rotation and from Table 3.10 we see that, since $i = 1$, it is along the x direction, i.e. is C_{2x}. The direction of the screw vector is thus $r_1\mathbf{q} = \frac{1}{2}\mathbf{a}$ and the axis is located by the fact that it passes through the point $(\frac{1}{2}r_1, \frac{1}{2}r_2, \frac{1}{2}r_3)$, i.e. through $(\frac{1}{4}a, \frac{1}{4}b, 0)$.

3.6. The action of space-group operations on Bloch functions

We now discuss the effect of the application of space-group operators on Bloch functions. Since we are using active operations in this book the discussion of function-space operators in section 1.5 gives

$$\{S \mid \mathbf{w}\}\psi(\mathbf{r}) = \psi(S^{-1}(\mathbf{r} - \mathbf{w})) \tag{3.6.1}$$

where we have dropped the subscript a and the bar from the function-space operator.

First we prove that

$$\exp(\mathrm{i}\mathbf{k}.S\mathbf{r}) = \exp(\mathrm{i}S^{-1}\mathbf{k}.\mathbf{r}). \tag{3.6.2}$$

This follows because

$$\mathbf{k}.S\mathbf{r} = \sum_i \sum_j k_i S_{ij} r_j = \sum_i \sum_j S^T_{ji} k_i r_j = \sum_i \sum_j S^{-1}_{ji} k_i r_j, \tag{3.6.3}$$

since S_{ij} is an orthogonal matrix. This means that from eqns. (3.6.1) and (3.6.2)

$$\begin{aligned} \{S \mid \mathbf{w}\} \exp(\mathrm{i}\mathbf{k}.\mathbf{r}) &= \exp\{\mathrm{i}\mathbf{k}.S^{-1}(\mathbf{r} - \mathbf{w})\} \\ &= \exp\{\mathrm{i}S\mathbf{k}.(\mathbf{r} - \mathbf{w})\}, \end{aligned} \tag{3.6.4}$$

which is almost the same function except that it is labelled with reciprocal vector $S\mathbf{k}$ and is centred at $\mathbf{r} = \mathbf{w}$.

Similarly we may define

$$u_{S\mathbf{k}}(\mathbf{r} - \mathbf{w}) = u_{\mathbf{k}}\{S^{-1}(\mathbf{r} - \mathbf{w})\} = \{S \mid \mathbf{w}\}\, u_{\mathbf{k}}(\mathbf{r}), \tag{3.6.5}$$

where $u_{\mathbf{k}}(\mathbf{r})$ satisfies eqn. (3.4.8); that is, it has the periodicity of the lattice. Using eqns. (3.6.4) and (3.6.5) together with the definition of a Bloch function,

$$\Psi_{\mathbf{k}}(\mathbf{r}) = \exp(\mathrm{i}\mathbf{k}.\mathbf{r})\, u_{\mathbf{k}}(\mathbf{r}), \tag{3.4.7}$$

we obtain

$$\begin{aligned}\{S \mid \mathbf{w}\}\psi_{\mathbf{k}}(\mathbf{r}) &= \exp\{\mathrm{i}S\mathbf{k}.(\mathbf{r}-\mathbf{w})\}\,u_{S\mathbf{k}}(\mathbf{r}-\mathbf{w})\\ &= \psi_{S\mathbf{k}}(\mathbf{r}-\mathbf{w}),\end{aligned} \tag{3.6.6}$$

say. Also from eqn. (1.5.24) it follows that

$$\{S \mid \mathbf{w}\}\{R \mid \mathbf{v}\}\psi_{\mathbf{k}}(\mathbf{r}) = \psi_{SR\mathbf{k}}(\mathbf{r}-\mathbf{w}-S\mathbf{v}). \tag{3.6.7}$$

It should be noticed that $u_{S\mathbf{k}}$ also satisfies eqn. (3.4.8), because if $\mathbf{t}$ is a translation so is $\mathbf{t}' = S^{-1}\mathbf{t}$; therefore

$$\begin{aligned}u_{S\mathbf{k}}(\mathbf{r}-\mathbf{w}+\mathbf{t}) &= u_{\mathbf{k}}\{S^{-1}(\mathbf{r}-\mathbf{w}+\mathbf{t})\}\\ &= u_{\mathbf{k}}\{S^{-1}(\mathbf{r}-\mathbf{w})+\mathbf{t}'\}\\ &= u_{\mathbf{k}}(S^{-1}(\mathbf{r}-\mathbf{w})\} \quad \text{(from eqn. (3.4.8))}\\ &= u_{S\mathbf{k}}(\mathbf{r}-\mathbf{w}).\end{aligned} \tag{3.6.8}$$

This means that the right-hand side of eqn. (3.6.6) is a Bloch function. Thus the action of $\{S \mid \mathbf{w}\}$ on a Bloch function labelled with the wave vector $\mathbf{k}$ and centred at $\mathbf{r} = 0$ is to transform it into a Bloch function labelled with the wave vector $S\mathbf{k}$ and centred at $\mathbf{r} = \mathbf{w}$. In spite of the similarity between eqns. (3.6.4) and (3.6.6) there is one important difference. If $S\mathbf{k} = \mathbf{k}$ then $\{S \mid \mathbf{w}\}$ exp $(\mathrm{i}\mathbf{k}.\mathbf{r})$ = exp $\{\mathrm{i}\mathbf{k}.(\mathbf{r}-\mathbf{w})\}$ which is exactly the same function but centred at $\mathbf{r} = \mathbf{w}$. Whereas if $S\mathbf{k} \equiv \mathbf{k}$ it is not necessarily true that $\{S \mid \mathbf{w}\}\psi_{\mathbf{k}}(\mathbf{r}) = \psi_{\mathbf{k}}(\mathbf{r}-\mathbf{w})$; this is because the label $\mathbf{k}$ on $\psi_{\mathbf{k}}(\mathbf{r})$ is determined only up to equivalence. An example should help to make this clear. Suppose $\mathbf{k} = 0$ so that $S\mathbf{k} = \mathbf{k}$ trivially and let $\psi_0(\mathbf{r}) = 1 + \exp(\mathrm{i}\mathbf{g}.\mathbf{r})$; then from eqn. (3.6.5) we have

$$\begin{aligned}\{S \mid \mathbf{w}\}\psi_0(\mathbf{r}) &= \psi_0(S^{-1}\mathbf{r} - S^{-1}\mathbf{w})\\ &= 1 + \exp\{\mathrm{i}\mathbf{g}.S^{-1}(\mathbf{r}-\mathbf{w})\}\\ &= 1 + \exp\{\mathrm{i}S\mathbf{g}.(\mathbf{r}-\mathbf{w})\}.\end{aligned} \tag{3.6.9}$$

But the right-hand side of eqn. (3.6.9) is equal to $\psi_0(\mathbf{r}-\mathbf{w})$ only if $S\mathbf{g} = \mathbf{g}$. What is true, however, is that because $S\mathbf{g}$ is equivalent to $\mathbf{g}$, which in turn is equivalent to zero, the wave vector associated with the right-hand side of eqn. (3.6.9) is still the zero vector.

THEOREM 3.6.1. *If* $\psi_{\mathbf{k}}(\mathbf{r})$ *is a Bloch function with wave vector* $\mathbf{k}$ *so that* $\{E \mid \mathbf{t})\psi_{\mathbf{k}}(\mathbf{r}) = \exp(-\mathrm{i}\mathbf{k}.\mathbf{t})\psi_{\mathbf{k}}(\mathbf{r})$, *then* $\{S \mid \mathbf{w}\}\psi_{\mathbf{k}}(\mathbf{r})$ *is a Bloch function with the wave vector* $S\mathbf{k}$.

The reader should appreciate that this does not follow from applying $\{E \mid \mathbf{t}\}$ to both sides of eqn. (3.6.6) and the already known fact that $\psi_{S\mathbf{k}}(\mathbf{r}-\mathbf{w})$ is a Bloch function with the wave vector $S\mathbf{k}$, because eqn. (3.6.6) is a numerical rather than a functional relation. To prove the theorem analytically it is necessary to go through the following series of equations:

$$\begin{aligned}\{E \mid \mathbf{t}\}\{S \mid \mathbf{w}\}\psi_{\mathbf{k}}(\mathbf{r}) &= \{S \mid \mathbf{w}\}\psi_{\mathbf{k}}(\mathbf{r} - \mathbf{t}) \\ &= \psi_{\mathbf{k}}\{S^{-1}(\mathbf{r} - \mathbf{t} - \mathbf{w})\} = \psi_{S\mathbf{k}}(\mathbf{r} - \mathbf{t} - \mathbf{w}) \\ &= \{E \mid \mathbf{t}\}\psi_{S\mathbf{k}}(\mathbf{r} - \mathbf{w}) = \exp(-\mathrm{i}S\mathbf{k}.\mathbf{t})\psi_{S\mathbf{k}}(\mathbf{r} - \mathbf{w}) \\ &= \exp(-\mathrm{i}S\mathbf{k}.\mathbf{t})\{S \mid \mathbf{w}\}\psi_{\mathbf{k}}(\mathbf{r}). \qquad (3.6.10)\end{aligned}$$

However, the theorem can of course be argued through by starting from eqn. (3.6.6) and by using a geometrical notion of what is meant by Bloch periodicity. As a corollary to the theorem, if $S\mathbf{k} \equiv \mathbf{k}$ then $\{S \mid \mathbf{w}\}\psi_{\mathbf{k}}(\mathbf{r})$ is a Bloch function with wave vector $\mathbf{k}$. We use this theorem in the next section.

3.7. A descriptive account of the representation theory of space groups

In this section we describe how to obtain all the reps of a given space group. The description given is not entirely rigorous; however the main gaps in the arguments are mentioned in the text. These gaps will be filled in Chapter 4 in which a rigorous and complete account is given of the representation theory of groups having a proper invariant Abelian subgroup, of which space groups are a particular example. Many readers will probably prefer to concentrate their attention on this section for their basic understanding of the representation theory of space groups, and even omit altogether or leave until much later the rather more abstract and mathematically difficult work involved in the next chapter. For this reason the account given here is made almost complete and is lacking only in that proofs of certain readily acceptable results are postponed; however, the basic concepts of a star, and the little group (or group of $\mathbf{k}$) and its small representations are properly introduced, the need for the introduction of projective representations is explained and their theory is described in some detail. In section 3.8 two complete examples will be given, those of the face-centred cubic structure and the diamond structure.

We consider then a space group $\mathbf{G}$ with a translational subgroup $\mathbf{T}$. The reps of $\mathbf{T}$ are given in section 3.4 and are characterized by the vectors $\mathbf{k}$ of the Brillouin zone of the Bravais lattice Γ on which $\mathbf{G}$ is based. These reps $\Delta^{\mathbf{k}}$ are such that (see eqn. (3.4.3))

$$\Delta^{\mathbf{k}}(\{E \mid \mathbf{t}\}) = \exp(-\mathrm{i}\mathbf{k}.\mathbf{t}). \qquad (3.7.1)$$

We can write $\mathbf{G}$ in terms of left coset representatives of $\mathbf{T}$

$$\mathbf{G} = \{R_1 \mid \mathbf{v}_1\}\mathbf{T} + \{R_2 \mid \mathbf{v}_2\}\mathbf{T} + \cdots + \{R_h \mid \mathbf{v}_h\}\mathbf{T} \qquad (3.5.2)$$

in which, as always, we can choose $R_1 = E$, $\mathbf{v}_1 = 0$ and where $\mathbf{v}_i$ is a vector associated with R_i. The coset representatives form a set $\mathbf{Q}$ which is not necessarily a group. The factor group $\mathbf{G}/\mathbf{T}$ is, however, a group and it is isomorphic with the point group

F consisting of the operators $R_1, R_2, \ldots, R_h$. **F** is the isogonal point group of **G**. Its order h is the macroscopic order of **G**.

DEFINITION 3.7.1. *Representation domain.* For any space group **G** there is a representation domain Φ, of the appropriate Brillouin zone, such that $(\sum_R R\Phi)$ is equal to the whole Brillouin zone, where the sum over R runs through the elements of the group **F**, the isogonal point group of **G**.

It is helpful to compare this definition of the representation domain with that given for the basic domain in Definition 3.3.3. If the space group **G** has as its isogonal point group the holosymmetric point group of the relevant crystal system, that is, if **F** coincides with **P**, then the representation domain Φ is of the same volume as the basic domain Ω and can be chosen to coincide with it. In general, however, **F** is a subgroup of **P**; and if $|\mathbf{P}|/|\mathbf{F}| = i$, then the representation domain Φ will have a volume equal to i times that of the basic domain Ω. Note also that whereas Ω is fixed for each Brillouin zone, Φ depends on the space group under consideration. The reason for introducing Φ for a given space group **G** is that it is of critical importance in the prescription for deriving a complete set of reps of **G**.

As an abstract mathematical problem it is interesting to determine the reps of a space group, **G**. However, much of the motivation behind the study of space-group reps is connected with their usefulness in investigating the eigenfunctions and eigenvalues of the Hamiltonian of a quantum-mechanical particle or quasi-particle in a regular crystalline solid. It is a well-known, and possibly trivial, statement to say that *if a system has a certain group of symmetry operations*, **G**, *then any physical observable of that system must also possess these symmetry operations*; this statement is sometimes known as *Neumann's principle* and is used, for instance, in simplifying the form of a tensor that describes some physical property of a crystal (Birss 1964, Nye 1957). However, a wave function ψ is not a physical observable of a system and therefore one cannot necessarily expect that it will exhibit all the symmetry of **G**. It is, nevertheless, still possible to obtain some information about the transformation properties of ψ under the operations of the elements of **G**. The basic principle behind the application of group theory to the study of a quantum-mechanical system lies in Wigner's theorem which we mentioned in section 1.1; there are various ways of stating the theorem, of which we give just one.

THEOREM 3.7.1. (Wigner's theorem). *If R is a symmetry operation of the Hamiltonian operator*, **H**, *which describes a quantum-mechanical system and if ψ is an eigenfunction of* **H**, *then $R\psi$ is also an eigenfunction of* **H** *which has the same eigenvalue E as ψ.*

For a discussion of the proof of this theorem we refer the reader to the classic book on group theory and quantum mechanics by Wigner (English translation, Wigner (1959)). An alternative statement of the result would be to say that the wave function ψ of a particle or quasi-particle must be one component of a basis of one of the irreducible representations of **G**. In our present problem **G** is a space group so that the wave function ψ of a particle or quasi-particle in a crystal that is described by a space group **G** must be one component of a basis of one of the reps of **G**.

The method we shall now follow is to suppose that we are given an arbitrary rep of **G** so that we can discuss its properties. Then we shall assume that the properties we find are not only necessary but sufficient to characterize the reps of **G**; this assumption is one point that prevents this treatment from being entirely rigorous. Then, having done this, we shall show how to build up all structures that have those required properties. The assumption then implies that all such structures are in fact reps of **G**. Another point that prevents the present treatment from being entirely complete is that we shall not show how to write down the matrix elements of the reps in full; this is left for the next chapter. We are more concerned here with the properties of the reps and we shall not go all the way in describing how to write them down in detail.

Suppose then that Γ is a rep of **G** acting in a vector space V of dimension d. If we consider only those elements of Γ that represent translations then such elements form a representation of **T** that in general is reducible. Suppose it reduces into reps of **T** characterized by the vectors $\mathbf{k}_1, \mathbf{k}_2, \ldots, \mathbf{k}_d$. These reps of **T** are 1-dimensional and, as a result of the discussion in section 3.4, it follows that a basis can be found in V which consists of Bloch functions $\psi_{\mathbf{k}_1}(\mathbf{r}), \psi_{\mathbf{k}_2}(\mathbf{r}), \ldots, \psi_{\mathbf{k}_d}(\mathbf{r})$, and that with respect to this basis the elements of **T** are represented by diagonal matrices; and from eqn. (3.7.1) we see that the diagonal element $\boldsymbol{\Gamma}(\{E \mid \mathbf{t}\})_{pp}$ is $\exp(-i\mathbf{k}_p.\mathbf{t})$. It may be that the vectors $\mathbf{k}_1, \mathbf{k}_2, \ldots, \mathbf{k}_d$ are not all different; in that case we shall need a second index on the functions $\psi_{\mathbf{k}_p}(\mathbf{r})$ to distinguish between two functions labelled with the same **k** vector. To see whether this is possible let us consider closely how the vectors $\mathbf{k}_1, \mathbf{k}_2, \ldots, \mathbf{k}_d$ arise in the first place. We make use of two facts: that V is irreducible under **G**, and Theorem 3.6.1 that if $\psi_{\mathbf{k}}(\mathbf{r})$ is a Bloch function with wave vector **k** then $\{S \mid \mathbf{w}\}\psi_{\mathbf{k}}(\mathbf{r})$ is a Bloch function with a wave vector $S\mathbf{k}$. The fact that Γ is a representation implies that if we start with $\psi_{\mathbf{k}_1}(\mathbf{r})$ and generate the Bloch functions $\{S \mid \mathbf{w}\}\psi_{\mathbf{k}_1}(\mathbf{r})$ where $\{S \mid \mathbf{w}\}$ is any member of **G** then we obtain some linear combination of the d Bloch functions in the basis; that is to say, using Theorem 3.6.1, $S\mathbf{k}_1$ is one of the vectors $\mathbf{k}_1, \mathbf{k}_2, \ldots, \mathbf{k}_d$. Moreover the irreducibility of V under **G** implies that as we run over all elements of **G** operating on the $\psi_{\mathbf{k}_1}(\mathbf{r})$ we shall generate the entire vector space V. That is, each of the vectors $\mathbf{k}_1, \mathbf{k}_2, \ldots, \mathbf{k}_d$ appears as the transform of $\mathbf{k}_1$ under some element of **F**.

DEFINITION 3.7.2. *Star.* A set of *distinct* **k** vectors chosen from the set $\mathbf{k}_i$ ($i = 1$ to d) is called the *star* of Γ. (In this context remember that two **k** vectors are not distinct if they are equivalent—see eqn. (3.4.4).)

Since the above discussion holds irrespective of whether we start with $\mathbf{k}_1$ or with some other member of the star we have proved that a star can be generated from any one of its members by operating on that member by elements of **F**.

DEFINITION 3.7.3. *Little co-group.* The set of elements of **F** which leave $\mathbf{k}_1$ invariant, that is elements $R \in \mathbf{F}$ such that $R\mathbf{k}_1 \equiv \mathbf{k}_1$, forms a subgroup of **F** called the little co-group of $\mathbf{k}_1$ and which we denote by $\overline{\mathbf{G}}^{\mathbf{k}_1}$. (We are using here the symbol $\equiv$ for equivalence, as defined in section 3.3.)

From Theorem 1.2.3(ii) the order of $\overline{\mathbf{G}}^{\mathbf{k}_1}$ is a divisor of h, the order of **F**. Suppose now that $R\mathbf{k}_1 \equiv \mathbf{k}_1$ and S is any fixed element such that $S\mathbf{k}_1 \equiv \mathbf{k}_2$, then $SRS^{-1}\mathbf{k}_2 \equiv SR\mathbf{k}_1 \equiv S\mathbf{k}_1 \equiv \mathbf{k}_2$. This implies that each element of $S\overline{\mathbf{G}}^{\mathbf{k}_1}S^{-1}$ is contained in $\overline{\mathbf{G}}^{\mathbf{k}_2}$. Therefore the order of $\overline{\mathbf{G}}^{\mathbf{k}_2}$ is greater than or equal to the order of $\overline{\mathbf{G}}^{\mathbf{k}_1}$. But $\mathbf{k}_1 \equiv S^{-1}\mathbf{k}_2$ so that each element of $S^{-1}\overline{\mathbf{G}}^{\mathbf{k}_2}S$ is contained in $\overline{\mathbf{G}}^{\mathbf{k}_1}$. Thus the orders of the little co-groups of the members of a star are equal and because of relations of the form $\overline{\mathbf{G}}^{\mathbf{k}_2} = S\overline{\mathbf{G}}^{\mathbf{k}_1}S^{-1}$ which must hold between all the little co-groups we find that the little co-groups of the members of a star form a set of *conjugate* subgroups.

We determine next the number of elements $P \in \mathbf{F}$ such that $P\mathbf{k}_1 \equiv \mathbf{k}_2$. Let T be any fixed element of **F** such that $T\mathbf{k}_2 \equiv \mathbf{k}_1$. Then $TP\mathbf{k}_1 \equiv T\mathbf{k}_2 \equiv \mathbf{k}_1$. Hence $TP \in \overline{\mathbf{G}}^{\mathbf{k}_1}$. Therefore P belongs to the left coset $T^{-1}\overline{\mathbf{G}}^{\mathbf{k}_1}$ and so the number of such elements P is equal to the order of the little co-group. We have proved that if we write **F** as the sum of left cosets of the little co-group of any member of the star then these cosets are in one-to-one correspondence with the members of the star. And as a corollary, the product of the order of the little co-group and the number of members of the star is equal to the order of **F**. The geometrical interpretation of these theorems in terms of the symmetry properties of the Brillouin zone is made very clear by Bouckaert, Smoluchowski, and Wigner (1936). The important geometrical fact from our point of view is that *without loss of generality the vector* $\mathbf{k}_1$ *out of which the star is generated can always be chosen to lie within or on the surface of the representation domain of the Brillouin zone.* Indeed, careful consideration of Definitions 3.7.1 and 3.7.2 shows that the representation domain is defined specifically in order that it should contain at least one **k** vector from each star.

When $\mathbf{k}_1$ is in the *basic domain* the elements $\overline{\mathbf{G}}^{\mathbf{k}_1}$ for all lines and points of symmetry for each of the space groups can now be read off from Table 3.6. *In each case the elements of* $\overline{\mathbf{G}}^{\mathbf{k}_1}$ *are the elements of* $\mathbf{P}(\mathbf{k}_1)$ *that are also members of* **F**, *the isogonal point group of* **G**.

In the case in which $\mathbf{k}_1$ is in a part of the representation domain other than in the basic domain, the elements of $\overline{\mathbf{G}}^{\mathbf{k}_1}$ are very simply related to the elements of $\overline{\mathbf{G}}^{\mathbf{k}}$, where **k** is a point that lies in the basic domain and can be obtained from $\mathbf{k}_1$ by operating on it with some element R of the holosymmetric point group, **P**. Indeed suppose $R \in \mathbf{P}$ is such that $R\mathbf{k}_1 \equiv \mathbf{k}$, then it follows immediately that $\mathbf{P}(\mathbf{k}_1) = R^{-1}\mathbf{P}(\mathbf{k})R$. $\mathbf{P}(\mathbf{k})$ can be obtained from Table 3.6, and then $\mathbf{P}(\mathbf{k}_1)$ follows by conjugation with R. As before, $\overline{\mathbf{G}}^{\mathbf{k}_1}$ is the intersection of $\mathbf{P}(\mathbf{k}_1)$ and **F**. We shall return to this whole question again in Chapter 5, for it is one that affects the completeness of the tabulations of the reps of the space groups; indeed, we tabulate in detail only those reps which arise when $\mathbf{k}_1$ lies in the basic domain, but we shall demonstrate by example how to proceed when this is not the case.

DEFINITION 3.7.4. *The little group.* Suppose now that the little co-group $\overline{\mathbf{G}}^{\mathbf{k}_1}$ consists of the b elements $S_1, S_2, \ldots, S_b$. Then if we look at the factorization in eqn. (3.5.2) of $\mathbf{G}$ into left cosets with respect to $\mathbf{T}$, and form the set theoretical sum of the b left cosets whose coset representatives have the rotational parts $S_1, S_2, \ldots, S_b$ then this sum is a subgroup of $\mathbf{G}$ called the *little group* of $\mathbf{k}_1$ and which we denote by $\mathbf{G}^{\mathbf{k}_1}$. Just as $\overline{\mathbf{G}}^{\mathbf{k}_1}$ is not necessarily an invariant subgroup of $\mathbf{F}$ so also $\mathbf{G}^{\mathbf{k}_1}$ is not necessarily an invariant subgroup of $\mathbf{G}$.

However, just as we can write $\mathbf{F}$ as a sum of left cosets with respect to $\overline{\mathbf{G}}^{\mathbf{k}_1}$:

$$\mathbf{F} = T_1\overline{\mathbf{G}}^{\mathbf{k}_1} + T_2\overline{\mathbf{G}}^{\mathbf{k}_1} + \cdots + T_q\overline{\mathbf{G}}^{\mathbf{k}_1}, \tag{3.7.2}$$

where $q = h/b$ is the number of elements in the star of Γ and where, if $T_i\mathbf{k}_1 \equiv \mathbf{k}_i$, then all elements of the left coset $T_i\overline{\mathbf{G}}^{\mathbf{k}_1}$ transform $\mathbf{k}_1$ into $\mathbf{k}_i$ so that the left cosets are in one-to-one correspondence with the vectors $\mathbf{k}_1, \mathbf{k}_2, \ldots, \mathbf{k}_q$ of the star, then we can also write $\mathbf{G}$ as a sum of left cosets with respect to $\mathbf{G}^{\mathbf{k}_1}$:

$$\mathbf{G} = \{T_1 \mid \mathbf{x}_1\}\mathbf{G}^{\mathbf{k}_1} + \{T_2 \mid \mathbf{x}_2\}\mathbf{G}^{\mathbf{k}_1} + \cdots + \{T_q \mid \mathbf{x}_q\}\mathbf{G}^{\mathbf{k}_1} \tag{3.7.3}$$

and maintain the same one-to-one correspondence. This follows from the fact that any element of the coset $\{T_i \mid \mathbf{x}_i\}\mathbf{G}^{\mathbf{k}_1}$ transforms a Bloch function with wave vector $\mathbf{k}_1$ into a Bloch function with wave vector $\mathbf{k}_i$. The proof of this is that an element of $\{T_i \mid \mathbf{x}_i\}\mathbf{G}^{\mathbf{k}_1}$ can be written in the form $\{T_i \mid \mathbf{x}_i\}\{S \mid \mathbf{w}\}$ where $\{S \mid \mathbf{w}\} \in \mathbf{G}^{\mathbf{k}_1}$ and from eqn. (3.6.7)

$$\{T_i \mid \mathbf{x}_i\}\{S \mid \mathbf{w}\}\psi_{\mathbf{k}_1}(\mathbf{r}) = \psi_{T_iS\mathbf{k}_1}(\mathbf{r} - \mathbf{x}_i - T_i\mathbf{w})$$

which, on account of the fact that $T_iS\mathbf{k}_1 \equiv T_i\mathbf{k}_1 \equiv \mathbf{k}_i$ is a Bloch function with wave vector $\mathbf{k}_i$. See now Theorem 3.6.1, which proves that the left-hand side has the same property. The characteristic property of the little group $\mathbf{G}^{\mathbf{k}_1}$ is that each of its elements transforms a Bloch function with wave vector $\mathbf{k}_1$ into another Bloch function also with wave vector $\mathbf{k}_1$.

We can now come back to our original question: is it possible that the vectors $\mathbf{k}_1, \mathbf{k}_2, \ldots, \mathbf{k}_d$ characterizing the rep Γ are not all distinct or is d necessarily equal to q? The answer, which we shall now obtain, is that d must be a fixed multiple of q: $d = tq$ where t is an integer, and the star $\mathbf{k}_1, \mathbf{k}_2, \ldots, \mathbf{k}_q$ is repeated exactly t times in the set $\mathbf{k}_1, \mathbf{k}_2, \ldots, \mathbf{k}_d$.

The proof of this assertion is as follows. From the discussion on little groups the only elements of $\mathbf{G}$ that can possibly generate out of $\psi_{\mathbf{k}_1}(\mathbf{r})$ a new Bloch function with wave vector $\mathbf{k}_1$ are the elements of $\mathbf{G}^{\mathbf{k}_1}$, and since $\{E \mid \mathbf{t}\}\{S_i \mid \mathbf{w}_i\}$ generates the same

Bloch function out of $\psi_{\mathbf{k}_1}(\mathbf{r})$ as $\{S_i \mid \mathbf{w}_i\}$ (apart from a phase factor that can be ignored) it follows that the number of Bloch functions with wave vector $\mathbf{k}_1$ in the basis $\psi_{\mathbf{k}_1}(\mathbf{r}), \psi_{\mathbf{k}_2}(\mathbf{r}), \ldots, \psi_{\mathbf{k}_d}(\mathbf{r})$ is equal to the maximum number of linearly independent functions that can be chosen from the b functions $\{S_1 \mid \mathbf{w}_1\}\psi_{\mathbf{k}_1}(\mathbf{r})$, $\{S_2 \mid \mathbf{w}_2\}\psi_{\mathbf{k}_1}(\mathbf{r}), \ldots, \{S_b \mid \mathbf{w}_b\}\psi_{\mathbf{k}_1}(\mathbf{r})$. Suppose in fact that we find a maximum of t independent functions $\psi_{\mathbf{k}_1,1}(\mathbf{r}), \psi_{\mathbf{k}_1,2}(\mathbf{r}), \ldots, \psi_{\mathbf{k}_1,t}(\mathbf{r})$, then we know that because of the independence of these functions

$$\sum_{i=1}^{t} \lambda_i \psi_{\mathbf{k}_1,i}(\mathbf{r}) = 0 \tag{3.7.4}$$

implies $\lambda_i = 0$, $i = 1$ to t. Suppose now that

$$\sum_{i=1}^{t} \lambda_i \{T_j \mid \mathbf{x}_j\}\psi_{\mathbf{k}_1,i}(\mathbf{r}) = 0,$$

then premultiplying by $\{T_j \mid \mathbf{x}_j\}^{-1}$ and using the relation (3.7.4) we see that $\lambda_i = 0$, $i = 1$ to t. Hence the functions $\{T_j \mid \mathbf{x}_j\}\psi_{\mathbf{k}_1,i}(\mathbf{r})$ ($i = 1$ to t) are linearly independent Bloch functions with wave vectors $\mathbf{k}_j$. Also there are no others that can be generated out of $\psi_{\mathbf{k}_1}(\mathbf{r})$ because if there were then by reversing the argument we could find other linearly independent functions with wave vector $\mathbf{k}_1$ contrary to the assertion that we have chosen already as many as possible.

Another important fact is that the t wave functions $\psi_{\mathbf{k}_1,1}(\mathbf{r}), \psi_{\mathbf{k}_1,2}(\mathbf{r}), \ldots, \psi_{\mathbf{k}_1,t}(\mathbf{r})$ form a basis for an irreducible representation of $\mathbf{G}^{\mathbf{k}_1}$. That they form a representation follows from the way they were generated. The irreducibility of the subspace $V^{\mathbf{k}_1}$ spanned by these t vectors under $\mathbf{G}^{\mathbf{k}_1}$ follows from the irreducibility of the whole space V under $\mathbf{G}$: symbolically we can write $V = \sum_j \{T_j \mid \mathbf{x}_j\}V^{\mathbf{k}_1}$ and if $V^{\mathbf{k}_1}$ reduces so that $V^{\mathbf{k}_1} = U^{\mathbf{k}_1} \oplus W^{\mathbf{k}_1}$ where $U^{\mathbf{k}_1}$ and $W^{\mathbf{k}_1}$ are proper subspaces of $V^{\mathbf{k}_1}$ then V also reduces so that $V = U \oplus W$ where

$$U = \sum_j \{T_j \mid \mathbf{x}_j\}U^{\mathbf{k}_1} \quad \text{and} \quad W = \sum_j \{T_j \mid \mathbf{x}_j\}W^{\mathbf{k}_1}$$

and are proper subspaces of V. The details of this symbolic proof are left to the reader.

We have now characterized the basis of the rep Γ completely. Assuming the sufficiency of this characterization we can now outline a scheme for obtaining all the reps of $\mathbf{G}$. Collecting all the results together the scheme is as follows.

(i) Choose a vector $\mathbf{k}_1$ out of the representation domain of the Brillouin zone.

(ii) Determine the little co-group of $\mathbf{k}_1$, $\overline{\mathbf{G}}^{\mathbf{k}_1}$ which consists of b elements S_j such that $S_j\mathbf{k}_1 \equiv \mathbf{k}_1$ and from $\overline{\mathbf{G}}^{\mathbf{k}_1}$ construct the little group of $\mathbf{k}_1$ in $\mathbf{G}$, $\mathbf{G}^{\mathbf{k}_1}$.

(iii) Determine the star containing $\mathbf{k}_1$, or equivalently, write $\mathbf{G}$ in terms of its left cosets with respect to $\mathbf{G}^{\mathbf{k}_1}$; for we have shown that the vectors of the star are in one-to-one correspondence with these left cosets. The factorization into left cosets is given by eqn. (3.7.3). There are q elements in the star where $q = h/b$, and h is the macroscopic order of the crystal.

(iv) Determine an irreducible representation $\Gamma_p^{\mathbf{k}_1}$ of $\mathbf{G}^{\mathbf{k}_1}$ whose basis has the additional property that its elements are Bloch functions with wave vector $\mathbf{k}_1$, (eigenfunctions of the translation group $\mathbf{T}$ with eigenvalues for $\{E \mid \mathbf{t}\}$ equal to $\exp(-i\mathbf{k}_1 . \mathbf{t})$). Let this basis consist of the functions $\psi_{\mathbf{k}_1,1}(\mathbf{r}), \psi_{\mathbf{k}_1,2}(\mathbf{r}), \ldots, \psi_{\mathbf{k}_1,t}(\mathbf{r})$. $\Gamma_p^{\mathbf{k}_1}$ is called a *small representation* of the little group of $\mathbf{k}_1$.

(v) Form the vector space V consisting of the linear closure of all the functions $\{T_j \mid \mathbf{x}_j\}\psi_{\mathbf{k}_1,i}(\mathbf{r}), j = 1$ to $q, i = 1$ to t. ($\{T_j \mid \mathbf{x}_j\}$ are any set of left coset representatives of $\mathbf{G}^{\mathbf{k}_1}$ in $\mathbf{G}$). Then the result is that the vector space V is irreducible under $\mathbf{G}$ and the basis of a rep Γ may be chosen to consist of the d functions $\{T_j \mid \mathbf{x}_j\}\psi_{\mathbf{k}_1,i}(\mathbf{r})$, with $d = qt = ht/b$.

(vi) Repeat the process from stage (iv) for all possible small representations $\Gamma_p^{\mathbf{k}_1}$ that have the extra property mentioned in (iv), thereby obtaining all the reps of $\mathbf{G}$ having the star characterized by $\mathbf{k}_1$.

(vii) Repeat the whole process for each vector $\mathbf{k}_1$ in the representation domain. Then the result is that in this way we obtain all the reps of $\mathbf{G}$.

From this scheme we see that each rep of $\mathbf{G}$ is characterized by a star containing a vector $\mathbf{k}_1$ of the representation domain of the Brillouin zone, and the label of one of the small representations of the little group of $\mathbf{k}_1$.

There appears, at first sight, to be a conflict between the mathematical approach that we have outlined above and the approach that is usually adopted in textbooks of solid-state physics. In the approach that we have adopted we have used Wigner's theorem which ascribes a wave function ψ to one of the reps of the complete space group, $\mathbf{G}$; such a rep needs two labels to characterize it completely, $\mathbf{k}_1$ and a label, p, of one of the small reps of $\mathbf{G}^{\mathbf{k}_1}$. That is, ψ belongs to one of the reps induced in $\mathbf{G}$ by $\Gamma_p^{\mathbf{k}_1}$, one of the small reps of the little group $\mathbf{G}^{\mathbf{k}_1}$; the basis that is constructed in (v) above is a basis of such an induced representation of $\mathbf{G}$. (The theory of induced representations will be discussed in Chapter 4.) The approach adopted in textbooks of solid-state physics is usually to say that the wave function of a particle with wave vector $\mathbf{k}_1$ simply belongs to one of the components of a basis of the rep $\Gamma_p^{\mathbf{k}_1}$ of the little group $\mathbf{G}^{\mathbf{k}_1}$ of the wave vector $\mathbf{k}_1$. The point that we wish to emphasize at this stage is that, strictly speaking mathematically, the wave function ψ must belong to a rep of $\mathbf{G}$ as a result of Wigner's theorem; it is only because these reps can in fact all be constructed as the reps induced in $\mathbf{G}$ from the (small) reps of the little groups $\mathbf{G}^{\mathbf{k}_1}$ that it is permissible for practical purposes to regard ψ as belonging to one of the reps $\Gamma_p^{\mathbf{k}_1}$ of the little group $\mathbf{G}^{\mathbf{k}_1}$ of the wave vector $\mathbf{k}_1$ (Bouckaert, Smoluchowski, and Wigner 1936, Herring 1942, Seitz 1936*b*). One does have to be careful when considering the problem of determining selection rules in crystals but, even then, so long as one is prepared to consider all wave vectors in the star of $\mathbf{k}_1$ and not just $\mathbf{k}_1$ it is still possible in practice to avoid the explicit use of induced representations if one so wishes.

Our reason for expressing the theory in terms of induced representations of **G**, rather than only in terms of representations of little groups $\mathbf{G}^{\mathbf{k}_1}$, is that it is part of our aim to expound the mathematical justification for the physical approach that is commonly used.

We can perhaps illustrate the significance of these remarks by a more pictorial approach. Suppose that $\mathbf{k}_1$ is some wave vector in the representation domain of the Brillouin zone shown in Fig. 3.19. In both the mathematical and the physical

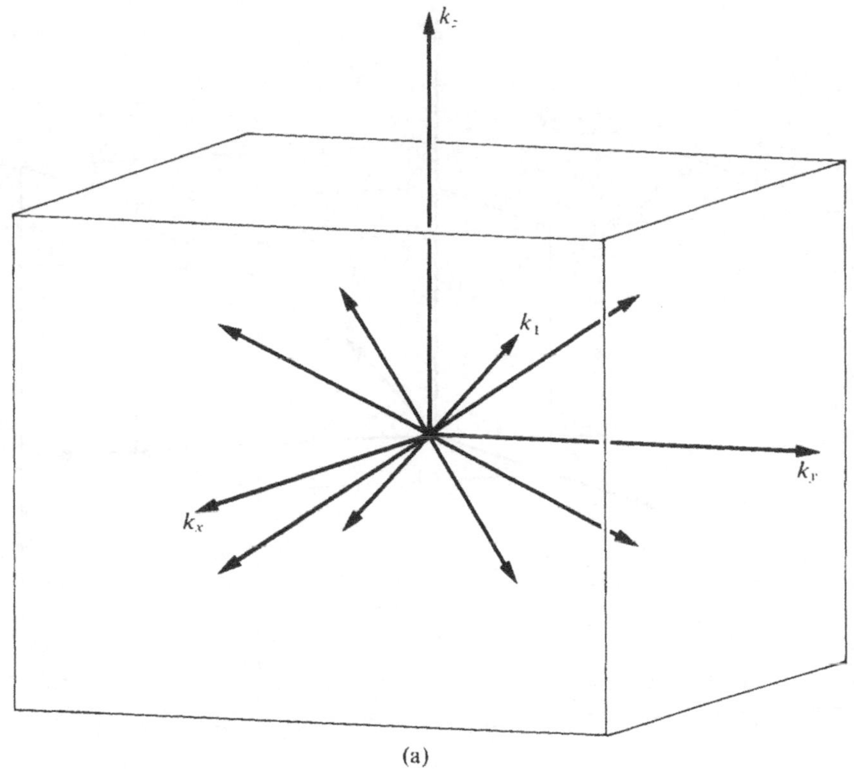

FIG. 3.19(a) The Brillouin zone and the vectors of a star determining a space group representation.

approaches that we have described one identifies $\mathbf{G}^{\mathbf{k}_1}$ and determines the small reps $\Gamma_p^{\mathbf{k}_1}$ of $\mathbf{G}^{\mathbf{k}_1}$. We suppose that the order of $\bar{\mathbf{G}}^{\mathbf{k}_1}$ is b, that the number of elements in the star of $\mathbf{k}_1$ is q and that $\Gamma_p^{\mathbf{k}_1}$ is of dimension t. In the strict mathematical approach we study the rep of **G** induced by $\Gamma_p^{\mathbf{k}_1}$. However, as will be shown in Chapter 4, this induced rep $D_p^{\mathbf{k}_1}$ of **G** has a rather special structure; it has dimension equal to q times the dimension of $\Gamma_p^{\mathbf{k}_1}$ and it is formed by the 'sticking together' of q reps, one from each member of the star of $\mathbf{k}_1$, where each of these reps has the same dimension as $\Gamma_p^{\mathbf{k}_1}$ and is very simply related to $\Gamma_p^{\mathbf{k}_1}$ (see section 5.5).

In the physical approach one would say that there was a set of t degenerate eigenfunctions, with eigenvalue λ_1, at $\mathbf{k}_1$ belonging to $\Gamma_p^{\mathbf{k}_1}$ and that there was a similar set of t degenerate eigenfunctions (with the *same* eigenvalue λ_1) at each of the other wave vectors $S\mathbf{k}_1$ in the star of $\mathbf{k}_1$. However, since each of these other members of the star of $\mathbf{k}_1$ is outside the representation domain, and since one is usually only interested in the numerical values of the eigenvalues, which are the same in each representation domain, we only consider the t eigenfunctions and t-fold degenerate

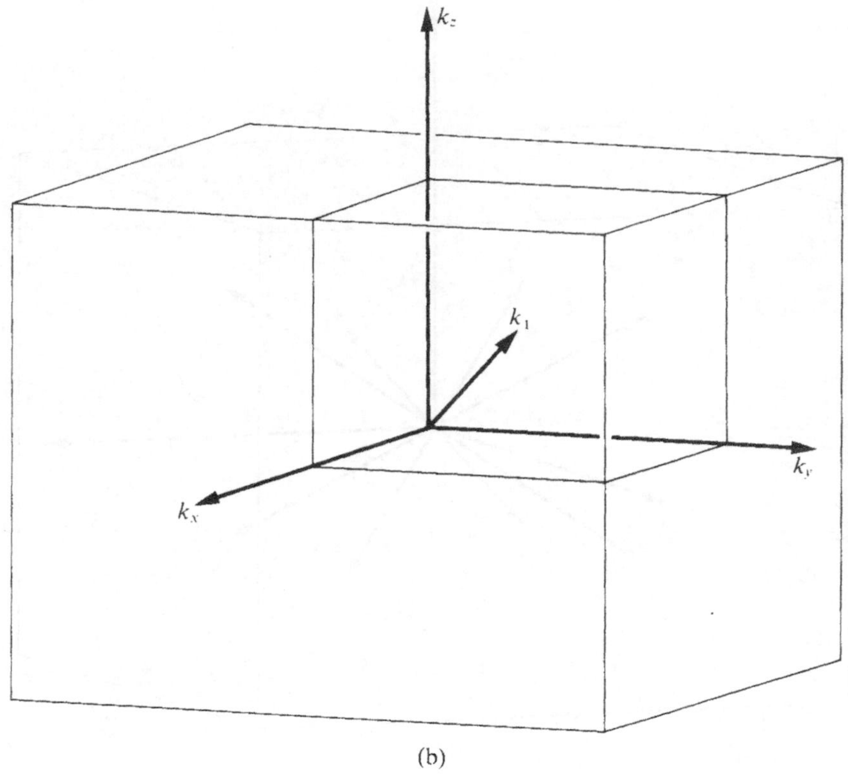

(b)

FIG. 3.19(b) The representation domain and the vector determining a small representation.

eigenvalues at $\mathbf{k}_1$. In the stricter mathematical approach there is a set of (qt) eigenfunctions with the eigenvalue λ_1 and there is a set of t of these functions associated with each wave vector of the star of $\mathbf{k}_1$. That is, a representation of $\mathbf{G}$ involves all the wave vectors in the star of $\mathbf{k}_1$ and therefore the use of reps of $\mathbf{G}$ involves using the whole Brillouin zone, see Fig. 3.19(a). A (small) representation of the little group $\mathbf{G}^{\mathbf{k}_1}$ is only associated with the single wave vector $\mathbf{k}_1$ which will be in the representation domain, see Fig. 3.19(b).

Having produced a scheme for finding the reps of $\mathbf{G}$ we now consider how it works in practice. The crucial step is the determination of the small representations; the rest of the scheme is straightforward. Here we encounter a basic difficulty because the group $\mathbf{G}^{\mathbf{k}_1}$ is itself a space group. It would seem that we are almost back where we started! Fortunately this is not the case for, whereas in the reps of $\mathbf{G}$ $\{E \mid \mathbf{t}\}$ is represented by a *diagonal* matrix, in the reps of $\mathbf{G}^{\mathbf{k}_1}$ that we are looking for $\{E \mid \mathbf{t}\}$ is represented by a *scalar* matrix, $\exp(-i\mathbf{k}_1 . \mathbf{t})\mathbf{1}$. This implies as we shall see that the matrices of $\Gamma_p^{\mathbf{k}_1}$ obey certain very simple rules. To see this we factorize $\mathbf{G}^{\mathbf{k}_1}$ into left cosets with respect to $\mathbf{T}$:

$$\mathbf{G}^{\mathbf{k}_1} = \{S_1 \mid \mathbf{w}_1\}\mathbf{T} + \{S_2 \mid \mathbf{w}_2\}\mathbf{T} + \cdots + \{S_b \mid \mathbf{w}_b\}\mathbf{T}, \tag{3.7.5}$$

and suppose that

$$\{S_i \mid \mathbf{w}_i\}\{S_j \mid \mathbf{w}_j\} = \{S_iS_j \mid \mathbf{w}_i + S_i\mathbf{w}_j\} = \{E \mid \mathbf{t}_{ij}\}\{S_k \mid \mathbf{w}_k\}, \tag{3.7.6}$$

where $\{E \mid \mathbf{t}_{ij}\} \in \mathbf{T}$ and, from eqn. (3.7.6),

$$\mathbf{t}_{ij} = \mathbf{w}_i + S_i\mathbf{w}_j - \mathbf{w}_k. \tag{3.7.7}$$

Because $\mathbf{t}_{ij}$ is a translation it follows that

$$\Gamma_p^{\mathbf{k}_1}(\{S_i \mid \mathbf{w}_i\})\Gamma_p^{\mathbf{k}_1}(\{S_j \mid \mathbf{w}_j\}) = \exp(-i\mathbf{k}_1 .(\mathbf{w}_i + S_i\mathbf{w}_j - \mathbf{w}_k))\Gamma_p^{\mathbf{k}_1}(\{S_k \mid \mathbf{w}_k\}). \tag{3.7.8}$$

If we now set, for all $\{R \mid \mathbf{v}\} \in \mathbf{G}^{\mathbf{k}_1}$,

$$\Gamma_p^{\mathbf{k}_1}(\{R \mid \mathbf{v}\}) = \exp(-i\mathbf{k}_1 . \mathbf{v})\mathbf{D}_p^{\mathbf{k}_1}(\{R \mid \mathbf{v}\}) \tag{3.7.9}$$

then from eqn. (3.7.8) we obtain

$$\mathbf{D}_p^{\mathbf{k}_1}(\{S_i \mid \mathbf{w}_i\})\mathbf{D}_p^{\mathbf{k}_1}(\{S_j \mid \mathbf{w}_j\}) = \exp(-i\mathbf{g}_i . \mathbf{w}_j)\mathbf{D}_p^{\mathbf{k}_1}(\{S_k \mid \mathbf{w}_k\}) \tag{3.7.10}$$

where $\mathbf{g}_i$, remembering that $S_i\mathbf{k}_1 \equiv \mathbf{k}_1$, is a reciprocal lattice vector defined by the equation

$$S_i^{-1}\mathbf{k}_1 = \mathbf{k}_1 + \mathbf{g}_i. \tag{3.7.11}$$

Also, if $\{E \mid \mathbf{t}\} \in \mathbf{T}$, then from eqn. (3.7.9)

$$\mathbf{D}_p^{\mathbf{k}_1}(\{E \mid \mathbf{t}\}) = \mathbf{1}. \tag{3.7.12}$$

It is also true that if $\{E \mid \mathbf{t}\} \in \mathbf{T}$ then

$$\exp\{-i\mathbf{g}_i .(\mathbf{w}_j + \mathbf{t})\} = \exp(-i\mathbf{g}_i . \mathbf{w}_j). \tag{3.7.13}$$

Equations (3.7.12), (3.7.13), and (3.7.10) together imply that the matrices of $D_p^{\mathbf{k}_1}$ are the same for all members of any fixed coset in eqn. (3.7.5). That is to say, $D_p^{\mathbf{k}_1}$ is a matrix-valued function on the elements of the factor group $\mathbf{G}^{\mathbf{k}_1}/\mathbf{T}$. But this factor group is isomorphic with the little co-group $\overline{\mathbf{G}}^{\mathbf{k}_1}$. Therefore $D_p^{\mathbf{k}_1}$ is a matrix-valued function on the elements of the little co-group. Instead of studying the little group which is a space group it is only necessary to study the properties of the little co-group.

We can therefore rewrite eqn. (3.7.10) in the form

$$\mathbf{D}_p^{\mathbf{k}_1}(S_i)\mathbf{D}_p^{\mathbf{k}_1}(S_j) = \exp(-i\mathbf{g}_i \,.\, \mathbf{w}_j)\mathbf{D}_p^{\mathbf{k}_1}(S_k), \tag{3.7.14}$$

where $S_iS_j = S_k$ and $S_i^{-1}\mathbf{k}_1 = \mathbf{k}_1 + \mathbf{g}_i$.

From (3.7.14) we see that $D_p^{\mathbf{k}_1}$ is a homomorphism of $\overline{\mathbf{G}}^{\mathbf{k}_1}$ onto a set of matrices in two important cases:

(i) If $\mathbf{w}_j = 0$ for all $j = 1$ to b. This is the case when the space group $\mathbf{G}$ is symmorphic.

(ii) If $S_i^{-1}\mathbf{k}_1 = \mathbf{k}_1$ (i.e. $\mathbf{g}_i = 0$) for all $i = 1$ to b. This is the case when $\mathbf{k}_1$ is an interior point of the Brillouin zone, because then the only point of the Brillouin zone equivalent to $\mathbf{k}_1$ is $\mathbf{k}_1$ itself.

When $D_p^{\mathbf{k}_1}$ is a homomorphism it is a representation of the little co-group and moreover the irreducibility of $\Gamma_p^{\mathbf{k}_1}$ implies that $D_p^{\mathbf{k}_1}$ must also be irreducible. Therefore, in the two cases listed above, all we need to do is to find the irreducible representations of the little co-group. Since this is a known point group we can pick up the required reps from Table 2.2. The matrices for $\Gamma_p^{\mathbf{k}_1}$ can then be deduced immediately from eqn. (3.7.9).

The only remaining case to discuss is when $\mathbf{k}_1$ is on the surface of the Brillouin zone and the space group is non-symmorphic.

If $\mathbf{k}_1$ is a general point then the little co-group consists of the identity element alone and we are back in case (ii). The only difficult cases then are for non-symmorphic space groups when $\mathbf{k}_1$ is a point of symmetry or when $\mathbf{k}_1$ is on a line or plane of symmetry on the surface of the Brillouin zone.

To cover the difficult cases we need to introduce the concept of the projective representations of a finite group.

DEFINITION 3.7.5. *Projective representations.* A non-singular matrix function Δ on a group $\mathbf{H}$ consisting of elements H_i, $i = 1$ to $|\mathbf{H}|$ is a projective representation of $\mathbf{H}$ if it satisfies the following rule.

If for each group product $H_iH_j = H_k$ there exists a scalar function $\mu(H_i, H_j)$ on the ordered pair of group elements H_i, H_j such that

$$\Delta(H_i)\Delta(H_j) = \mu(H_i, H_j)\Delta(H_k) \tag{3.7.15}$$

and if for all i, j, l

$$\mu(H_i, H_jH_l)\mu(H_j, H_l) = \mu(H_iH_j, H_l)\mu(H_i, H_j). \tag{3.7.16}$$

Equation (3.7.16) has to be imposed because of required associativity of matrix multiplication.

The function $\mu(H_i, H_j)$ forms what is called a *factor system* for the projective representation Δ. If $\mu(H_i, H_j) = 1$ for all i and j the projective representation Δ

becomes an ordinary (vector) representation. If $|\mu(H_i, H_j)| = 1$ for all i and j it is always possible to perform an equivalence transformation on the matrices Δ so that they become unitary. In the same way that equivalence transformations are used to reduce ordinary representations, so also they may be used to reduce projective representations. We therefore have the concept of an irreducible projective representation. The equation corresponding to eqn. (1.3.14) for irreducible unitary projective representations having the same factor system is

$$\sum_{j=1}^{|\mathbf{H}|} \boldsymbol{\Delta}^i(H_j)^*_{pv}\boldsymbol{\Delta}^k(H_j)_{qw} = \frac{|\mathbf{H}|}{d_i}\,\delta^{ik}\,\delta_{pq}\,\delta_{vw}, \tag{3.7.17}$$

with the corresponding relations for characters,

$$\sum_{j=1}^{|\mathbf{H}|} \chi^{i*}(H_j)\chi^k(H_j) = |\mathbf{H}|\,\delta^{ik}. \tag{3.7.18}$$

Given a projective representation Δ with factor system μ then if we set

$$\boldsymbol{\Delta}'(H_i) = C_i\boldsymbol{\Delta}(H_i) \quad \text{for all } i, \tag{3.7.19}$$

where C_i is a complex number, then the equation corresponding to eqn. (3.7.15) is

$$\boldsymbol{\Delta}'(H_i)\boldsymbol{\Delta}'(H_j) = \frac{C_iC_j}{C_k}\,\mu(H_i, H_j)\boldsymbol{\Delta}'(H_k). \tag{3.7.20}$$

If we now write

$$\nu(H_i, H_j) = \frac{C_iC_j}{C_k}\mu(H_i, H_j) \tag{3.7.21}$$

we see that Δ' is a projective representation of $\mathbf{H}$ with factor system ν. Two factor systems that are related by an equation of the form (3.7.21) are said to belong to the same *class*. Because of the simple relationship (3.7.19) we need study only one factor system from each class. To choose an appropriate factor system from a particular class suppose we are given a projective representation Δ with factor system μ. Then $\boldsymbol{\Delta}(E)\boldsymbol{\Delta}(E) = \mu(E, E)\boldsymbol{\Delta}(E)$ so that

$$\boldsymbol{\Delta}(E) = \mu(E, E)\mathbf{1} \tag{3.7.22}$$

If we now set $\boldsymbol{\Delta}'(H_i) = \boldsymbol{\Delta}(H_i)/\mu(E, E)$ for all i then

$$\boldsymbol{\Delta}'(E) = \mathbf{1}. \tag{3.7.23}$$

Let a new factor system ν be defined by eqn. (3.7.21) with $1/C_i = \mu(E, E)$ for all i then because

$$\boldsymbol{\Delta}'(H_i)\boldsymbol{\Delta}'(E) = \nu(H_i, E)\boldsymbol{\Delta}'(H_i)$$

we must have

$$\nu(H_i, E) = 1 \quad \text{for all } H_i. \tag{3.7.24}$$

Similarly,

$$\nu(E, H_i) = 1 \quad \text{for all } H_i. \tag{3.7.25}$$

(It is of course true for any factor system μ that $\mu(H_i, E) = \mu(E, H_i) = \mu(E, E)$ and that $\mu(H_i, H_i^{-1}) = \mu(H_i^{-1}, H_i)$.)

We now quote without proof certain theorems due to Schur on projective representations.

THEOREM 3.7.1. *The number m of classes of factor systems of a finite group* **H** *is finite; and the classes can be put in one-to-one correspondence with the elements of an Abelian group* **M** *of order m, called the multiplier of* **H**.

The one-to-one correspondence is such that if we denote the elements of **M** by $M_1, M_2, \ldots, M_m$ and if μ_p and μ_q are any factor systems chosen from the classes corresponding to M_p and M_q then the factor system μ_r such that for all i and j

$$\mu_r(H_i, H_j) = \mu_p(H_i, H_j)\mu_q(H_i, H_j) \tag{3.7.26}$$

always belongs to the class corresponding to M_r where $M_r = M_pM_q$. Moreover, the order of every element of **M** is a factor of the order of **H**. In particular, if M_p has order α_p then a representative factor system can be chosen from the class corresponding to M_p whose values are all α_pth roots of unity. As we observed above, an equivalence transformation can then be found so that the projective representation is unitary. Therefore every projective representation is equivalent to a unitary one with a factor system in the same class. The identity element M_1 of **M** corresponds to the class containing a factor system whose values are all equal to unity. Thus the reps of **H** are the irreducible unitary projective representations of the class corresponding to M_1.

The theory developed by Schur then goes on to show that for every finite group **H** of order $|\mathbf{H}|$, whose multiplier **M** is of order m, a group $\mathbf{H}_M$ of order $|\mathbf{H}|m$ can be constructed so that **M** is isomorphic with a subgroup of the centre of $\mathbf{H}_M$. (The centre of a group is the invariant subgroup of elements which commute with all elements of the group, that is the set of all elements each of which is in a class by itself.) And identifying **M** with its image in $\mathbf{H}_M$ we also have $\mathbf{H}_M/\mathbf{M}$ isomorphic with **H**. $\mathbf{H}_M$ is said to be a *central extension of* **H** *with kernel* **M**. We can now quote the most important theorem of Schur on projective representations.

THEOREM 3.7.2. *By running through all the reps of* $\mathbf{H}_M$ *we obtain on restriction to elements of* **H**‡ *(up to equivalence) all possible unitary irreducible projective representations of* **H**.

‡ It is important to realize in connection with this theorem that although the elements of **H** can be identified as the coset representatives of $\mathbf{H}_M/\mathbf{M}$ the construction of $\mathbf{H}_M$ is such that, except in the trivial case $m = 1$, the multiplication rule in $\mathbf{H}_M$ for elements identified as elements of **H** is not the same as for those elements considered as part of the group **H**. So that although **H** is a group with respect to its own multiplication it is not a subgroup of $\mathbf{H}_M$.

Since this theorem is rather stronger than we shall need we shall not show how to construct $\mathbf{H}_M$. It will be remembered that we only require projective representations belonging to a given factor system. Instead of constructing $\mathbf{H}_M$ we shall show how to construct another group $\mathbf{H}^*$ that will give us not all the projective representations of $\mathbf{H}$ but certainly those that we require.

Suppose then that we have a group $\mathbf{H}$ and that we want to find its unitary irreducible projective representations belonging to a given factor system μ. From the discussion so far we know that we can choose μ so that for all j, k

$$\mu(H_j, H_k) = \exp\{2\pi i a(H_j, H_k)/g\}, \tag{3.7.27}$$

where g and $a(H_j, H_k)$ are integers fixed by our knowledge of $\mu(H_j, H_k)$ and the condition $0 \leqslant a(H_j, H_k) \leqslant g - 1$ and (see eqns. (3.7.24) and (3.7.25))

$$a(H_j, E) = a(E, H_j) = 0. \tag{3.7.28}$$

Also, since $\mu(H_j, H_j^{-1}) = \mu(H_j^{-1}, H_j)$, then

$$a(H_j, H_j^{-1}) = a(H_j^{-1}, H_j) \tag{3.7.29}$$

and, from eqns. (3.7.16) and (3.7.27),

$$a(H_i, H_jH_l) + a(H_j, H_l) = a(H_iH_j, H_l) + a(H_i, H_j), \bmod g. \tag{3.7.30}$$

Let $\mathbf{Z}_g$ be the cyclic group of integers $0, 1, \ldots, (g - 1)$ with group product defined as addition modulo g.

Let $\mathbf{H}^*$ consist of the $g\,|\mathbf{H}|$ elements that are pairs of elements (H_j, α) one from $\mathbf{H}$ and one from $\mathbf{Z}_g$. Define a multiplication rule by the equation

$$(H_j, \alpha)(H_k, \beta) = (H_jH_k, \alpha + \beta + a(H_j, H_k)). \tag{3.7.31}$$

With respect to this multiplication rule $\mathbf{H}^*$ is a group: closure is obvious; associativity follows from associativity in $\mathbf{H}$ and $\mathbf{Z}_g$, and eqn. (3.7.30); the identity element of $\mathbf{H}^*$ is $(E, 0)$; and the inverse of (H_j, α) is $(H_j^{-1}, -\alpha - a(H_j, H_j^{-1}))$. Note that the set of g elements (E, α) forms a subgroup of $\mathbf{H}^*$ isomorphic with $\mathbf{Z}_g$, but that the set of elements $(H_i, 0)$, though in one-to-one correspondence with elements of $\mathbf{H}$, is not a group (closure is not satisfied, as can be seen at once from eqn. (3.7.31)), and $\mathbf{H}$ is therefore certainly not a subgroup of $\mathbf{H}^*$. (See also the footnote to Theorem 3.7.2.)

From eqns. (3.7.31) and (3.7.28) it follows that for all α, β, and k

$$(E, \alpha)(H_k, \beta) = (H_k, \beta)(E, \alpha) = (H_k, \alpha + \beta). \tag{3.7.32}$$

Equation (3.7.32) implies that the subgroup of g elements (E, α) lies in the centre of $\mathbf{H}^*$. Since (E, α) commutes with all the elements of $\mathbf{H}^*$ it follows from Schur's lemma (see Theorem 1.3.6) that if Γ is a rep of $\mathbf{H}^*$ then $\mathbf{\Gamma}(E, \alpha)$ is a scalar multiple of the unit matrix.

Suppose now that there exists a rep of $\mathbf{H}^*$ such that for all $\alpha \in \mathbf{Z}_g$

$$\boldsymbol{\Gamma}(E, \alpha) = \exp(2\pi i\alpha/g)\mathbf{1}. \tag{3.7.33}$$

Then

$$\boldsymbol{\Gamma}(H_k, \beta) = \boldsymbol{\Gamma}(H_k, 0)\boldsymbol{\Gamma}(E, \beta) = \boldsymbol{\Gamma}(H_k, 0)\exp(2\pi i\beta/g) \tag{3.7.34}$$

and writing $\boldsymbol{\Gamma}(H_k, 0) = \boldsymbol{\Delta}(H_k)$ we have

$$\begin{aligned}\boldsymbol{\Delta}(H_j)\boldsymbol{\Delta}(H_k) &= \boldsymbol{\Gamma}(H_j, \alpha)\boldsymbol{\Gamma}(H_k, \beta)\exp\{-2\pi i(\alpha + \beta)/g\} \\ &= \boldsymbol{\Gamma}(H_jH_k, \alpha + \beta + a(H_j, H_k))\exp\{-2\pi i(\alpha + \beta)/g\} \\ &= \boldsymbol{\Delta}(H_jH_k)\exp\{2\pi i a(H_j, H_k)/g\}.\end{aligned} \tag{3.7.35}$$

From eqns. (3.7.35) and (3.7.27) it follows that Δ is a projective representation of **H** with factor system μ. Moreover it is unitary (because Γ is unitary) and it is irreducible (because if Δ is reducible then since (E, α) is represented by a scalar matrix for all α then all the matrices of Γ are in the same reduced block-diagonal form as the matrices of Δ, contrary to the hypothesis that Γ is irreducible).

Conversely, if Δ is a unitary irreducible projective representation of **H** and if we set

$$\boldsymbol{\Gamma}(H_k, \beta) = \boldsymbol{\Delta}(H_k)\exp(2\pi i\beta/g) \tag{3.7.36}$$

for all k and β then Γ is a rep of $\mathbf{H}^*$. Thus in the language of the pure mathematician all the irreducible projective representations of **H** can be *lifted* into $\mathbf{H}^*$, that is they can be found from the ordinary vector reps of $\mathbf{H}^*$.

We can now go back to the little co-group and look at the factor system we are interested in. We identify $\overline{\mathbf{G}}^{\mathbf{k}_1}$ with **H** and $D_p^{\mathbf{k}_1}$ with Δ. The factor system is given by

$$\mu(S_i, S_j) = \exp(-i\mathbf{g}_i \cdot \mathbf{w}_j) \tag{3.7.37}$$

where

$$S_i^{-1}\mathbf{k}_1 = \mathbf{k}_1 + \mathbf{g}_i \tag{3.7.11}$$

and $\mathbf{w}_j$ is the translational part associated with S_j in the decomposition (3.7.5) of $\mathbf{G}^{\mathbf{k}_1}$ into left cosets.

From eqn. (3.7.11) we see that $\mathbf{g}_i = 0$ when $S_i = E$, so that

$$\mu(E, S_j) = 1 \quad \text{for all } j.$$

And furthermore, because $\mathbf{w}_j$ is always a fractional part of a translation, then $\mu(S_j, S_k)$ is always of the form of (3.7.27) and what is more important, g is bound to be a small number (in fact, either 2, 3, 4, or 6). Thus the factor system defined by eqn. (3.7.37) is already in a form amenable to the treatment we have just given. This means that to obtain all the reps of a space group it is only going to be necessary to find the reps of a few groups of comparatively small order in addition to those of the point groups that are already known.

As a last point in this section, by analyzing the dimension numbers of the reps of $\mathbf{H}_M$, and noting which of them become irreducible unitary projective representations of a given factor system, one can prove a theorem that is the analogue of Theorem 1.3.11, namely that if the dimensions of the projective reps of a given factor system are d_i then $\sum_i d_i^2 = |\mathbf{H}|$. This allows eqn. (3.7.17) to be inverted to produce another orthogonality relation for the matrix elements. Note, however, that the number of values of i is not necessarily equal to the number of classes of $\mathbf{H}$, as for the ordinary vector representations.

3.8. Examples: cubic close-packed and diamond structures

In this section we give two examples to illustrate the theory of the preceding sections; one example is a symmorphic space group, the cubic close-packed, $Fm3m$ (O_h^5) and the other is a non-symmorphic space group, the space group of the diamond structure, $Fd3m$ (O_h^7). These space groups were first treated by Bouckaert, Smoluchowski, and Wigner (1936) and Herring (1942), respectively.

They are based on the face-centred cubic, F, Bravais lattice Γ_c^f so that from Table 3.1 the lattice translations are in both cases

$$\left.\begin{aligned} \mathbf{t}_1 &= \tfrac{1}{2}(0, a, a), \\ \mathbf{t}_2 &= \tfrac{1}{2}(a, 0, a), \\ \mathbf{t}_3 &= \tfrac{1}{2}(a, a, 0). \end{aligned}\right\} \qquad (3.8.1)$$

In both cases the isogonal point group $\mathbf{F}$ is $m3m$ (O_h), the full cubic group.

The reciprocal lattice vectors are seen from Table 3.3 to be

$$\left.\begin{aligned} \mathbf{g}_1 &= (2\pi/a)(-1, 1, 1), \\ \mathbf{g}_2 &= (2\pi/a)(1, -1, 1), \\ \mathbf{g}_3 &= (2\pi/a)(1, 1, -1), \end{aligned}\right\} \qquad (3.8.2)$$

and the Brillouin zone is shown in Fig. 3.14.

From Table 3.7 we see that for $Fm3m$ (O_h^5) each rotation operator of $m3m$ (O_h) is associated with pure lattice translations only, so that an appropriate decomposition into left cosets with respect to the translation group $\mathbf{T}$ is

$$Fm3m\ (O_h^5) = \sum_R \{R \mid 000\}\mathbf{T} \qquad (3.8.3)$$

where the sum over R runs over all elements of $m3m$ (O_h). But for $Fd3m$ (O_h^7) only the operators belonging to the subgroup $\bar{4}3m$ (T_d) are associated with pure lattice translations and the corresponding decomposition into left cosets is

$$Fd3m\ (O_h^7) = \sum_R \{R \mid 000\}\mathbf{T} + \sum_R \{RI \mid \tfrac{1}{4}\tfrac{1}{4}\tfrac{1}{4}\}\mathbf{T} \qquad (3.8.4)$$

where now the sums over R run over all elements of $\bar{4}3m$ (T_d) and where, as usual, I is the inversion. The coset representatives for those cosets corresponding to elements of $m3m$ (O_h) not in $\bar{4}3m$ (T_d) can always be taken to be equal to $\mathbf{v} = \frac{1}{4}(\mathbf{t}_1 + \mathbf{t}_2 + \mathbf{t}_3) = \frac{1}{4}(a, a, a)$ when referred to axes $Oxyz$. The action of the operators R on a non-zero translation (p, q, r) referred to axes $Oxyz$ can be read off from Table 1.4. For example, $C_{4z}^{-}(p, q, r) = (q, -p, r)$. The action of the operators R on the basic vectors $\mathbf{t}_1$, $\mathbf{t}_2$, and $\mathbf{t}_3$ of the Bravais lattice can be read off from Table 3.2. For example, for a cubic F lattice $C_{4z}^{-}\mathbf{t}_1 = \mathbf{t}_2$, $C_{4z}^{-}\mathbf{t}_2 = \mathbf{t}_2 - \mathbf{t}_3$ and $C_{4z}^{-}\mathbf{t}_3 = -\mathbf{t}_1 + \mathbf{t}_2$.

In Table 3.11 we list the points, lines, and planes of symmetry, together with a general point, of the basic domain of the Brillouin zone which we shall need in the

TABLE 3.11

The basic domain of the Brillouin zone of Γ_c^f

$\mathbf{k}_1$	Coordinates $\Gamma\mathbf{g}_1\mathbf{g}_2\mathbf{g}_3$	$\Gamma k_x k_y k_z$	Star	Little co-group $\bar{\mathbf{G}}^{\mathbf{k}_1}$
Γ	(000)	(000)	$1I$	$E, C_{3j}^{\pm}, C_{2m}, C_{4m}^{\pm}, C_{2p}$, $I, S_{6j}^{\mp}, \sigma_m, S_{4m}^{\mp}, \sigma_{dp}$
X	$(\frac{1}{2}0\frac{1}{2})$	(010)	$3S$	$E, C_{2y}, C_{4y}^{\pm}, C_{2x}, C_{2z}, C_{2c}, C_{2e}$ $I, \sigma_y, S_{4y}^{\mp}, \sigma_x, \sigma_z, \sigma_{dc}, \sigma_{de}$
L	$(\frac{1}{2}\frac{1}{2}\frac{1}{2})$	$(\frac{1}{2}\frac{1}{2}\frac{1}{2})$	$4S$	$E, C_{31}^{\pm}, C_{2b}, C_{2e}, C_{2f}$, $I, S_{61}^{\mp}, \sigma_{db}, \sigma_{de}, \sigma_{df}$
W	$(\frac{1}{2}\frac{1}{4}\frac{3}{4})$	$(\frac{1}{2}10)$	$6S$	$E, C_{2x}, C_{2d}, C_{2f}$, $\sigma_y, \sigma_z, S_{4x}^{\mp}$
Δ	$(\alpha 0 \alpha)$	$(0, 2\alpha, 0)$	$6I$	$E, C_{2y}, C_{4y}^{\pm}, \sigma_x, \sigma_z, \sigma_{dc}, \sigma_{de}$
Λ	$(\alpha\alpha\alpha)$	$(\alpha\alpha\alpha)$	$8I$	$E, C_{31}^{\pm}, \sigma_{db}, \sigma_{de}, \sigma_{df}$
Σ	$(\alpha, \alpha, 2\alpha)$	$(2\alpha, 2\alpha, 0)$	$12I$	$E, C_{2a}, \sigma_z, \sigma_{db}$
S	$(\frac{1}{2} + \alpha, 2\alpha, \frac{1}{2} + \alpha)$	$(2\alpha, 1, 2\alpha)$	$12S$	$E, C_{2c}, \sigma_{de}, \sigma_y$
Z	$(\frac{1}{2}, \alpha, \frac{1}{2} + \alpha)$	$(2\alpha, 1, 0)$	$12S$	$E, C_{2x}, \sigma_z, \sigma_y$
Q	$(\frac{1}{2}, \frac{1}{2} - \alpha, \frac{1}{2} + \alpha)$	$(\frac{1}{2}, \frac{1}{2} + 2\alpha, \frac{1}{2} - 2\alpha)$	$24S$	E, C_{2f}
C	$(\alpha + \beta, \alpha + \beta, 2\alpha)$	$(2\alpha, 2\alpha, 2\beta)$	$24I$	E, σ_{db}
O	$(\beta, \alpha, \alpha + \beta)$	$(2\alpha, 2\beta, 0)$	$24I$	E, σ_z
J	$(\alpha + \beta, 2\alpha, \alpha + \beta)$	$(2\alpha, 2\beta, 2\alpha)$	$24I$	E, σ_{de}
B	$(\frac{1}{2} + \beta, \alpha + \beta, \frac{1}{2} + \alpha)$	$(2\alpha, 1, 2\beta)$	$24S$	E, σ_y
A	$(\beta + \gamma, \gamma + \alpha, \alpha + \beta)$	$(2\alpha, 2\beta, 2\gamma)$	$48I$ or S	E

Notes to Table 3.11

(i) In columns 1, 2, and 3 we list respectively in a given row of the table a point $\mathbf{k}_1$, its coordinates with respect to axes $\Gamma\mathbf{g}_1\mathbf{g}_2\mathbf{g}_3$, and its coordinates in units of $2\pi/a$ with respect to axes $\Gamma k_x k_y k_z$. The points $\mathbf{k}_1$ are chosen to be the generators of the stars of the reps.

(ii) In column 4 we list in the row corresponding to $\mathbf{k}_1$ the number, q, of elements in the star of $\mathbf{k}_1$ and a letter I or S depending on whether the point $\mathbf{k}_1$ is an internal or surface point of the Brillouin zone.

(iii) In column 5 we list, in the row corresponding to $\mathbf{k}_1$, the b elements of the little co-group $\bar{\mathbf{G}}^{\mathbf{k}_1}$. In all cases $bq = 48$, the order of $m3m$ (O_h).

(iv) This table is common to both space groups and indeed to all space groups based on Γ_c^f and with isogonal point group $m3m$ (O_h); see also the appropriate section of Table 3.6. The extra points in Table 3.11 not in Table 3.6 are the points on planes of symmetry and a general point. Except in this example we shall not tabulate the planes because as we shall see they work out very easily.

determination of the reps of $Fm3m$ (O_h^5) and $Fd3m$ (O_h^7). These are the points $\mathbf{k}_1$ of section 3.7. In both cases, since $\mathbf{F}$ is the holosymmetric point group of the Bravais lattice, the representation domain coincides with the basic domain and is therefore the region indicated in Fig. 3.14.

We now consider the factorization of the little group $\mathbf{G}^{\mathbf{k}_1}$ into left cosets with respect to the translation group $\mathbf{T}$. For $Fm3m$ we have

$$\mathbf{G}^{\mathbf{k}_1} = \sum_S \{S \mid 000\}\mathbf{T} \tag{3.8.5}$$

where the sum over S runs over all the elements of $\bar{\mathbf{G}}^{\mathbf{k}_1}$. (See Table 3.11 for these groups.) But for $Fd3m$ we have

$$\mathbf{G}^{\mathbf{k}_1} = \sum_P \{P \mid 000\}\mathbf{T} + \sum_Q \{Q \mid \tfrac{1}{4}\tfrac{1}{4}\tfrac{1}{4}\}\mathbf{T}, \tag{3.8.6}$$

where the sum over P runs over all the elements common to $\bar{\mathbf{G}}^{\mathbf{k}_1}$ and $\bar{4}3m$ (T_d), and where the sum over Q runs over the remaining elements of $\bar{\mathbf{G}}^{\mathbf{k}_1}$.

Fm3m

We can now dispose with $Fm3m$ very quickly. Given any point $\mathbf{k}_1$ we have to determine the projective reps of $\bar{\mathbf{G}}^{\mathbf{k}_1}$ with factor system $\exp(-i\mathbf{g}_i . \mathbf{w}_j)$ where $\mathbf{g}_i$ is given by eqn. (3.7.11) and $\mathbf{w}_j$ is the translation associated with S_j in the decomposition (3.8.5). We see at once that $\mathbf{w}_j = 0$ for all j. Therefore the factor system consists entirely of units and belongs to the class corresponding to the identity element of the multiplier of $\bar{\mathbf{G}}^{\mathbf{k}_1}$. Therefore the appropriate representations $D_p^{\mathbf{k}_1}$ of $\bar{\mathbf{G}}^{\mathbf{k}_1}$ are just the ordinary vector representations. And since $\bar{\mathbf{G}}^{\mathbf{k}_1}$ is a point group (admittedly in some cases in a non-standard setting) we can obtain the representations $D_p^{\mathbf{k}_1}$ from Table 2.2.

The small representations $\Gamma_p^{\mathbf{k}_1}$ of $\mathbf{G}^{\mathbf{k}_1}$ now follow immediately from eqn. (3.7.9). The matrices of $\Gamma_p^{\mathbf{k}_1}$ are in all cases related to those of $D_p^{\mathbf{k}_1}$ by the equation

$$\mathbf{\Gamma}_p^{\mathbf{k}_1}(\{S \mid \mathbf{t}\}) = \exp(-i\mathbf{k}_1 . \mathbf{t})\mathbf{D}_p^{\mathbf{k}_1}(S). \tag{3.8.7}$$

It is possible therefore to dismiss $Fm3m$ by naming the point groups $\bar{\mathbf{G}}^{\mathbf{k}_1}$. These are as follows: for Γ, $m3m$ (O_h); for X, $4/mmm$ (D_{4h}); for L, $\bar{3}m$ (D_{3d}); for W, $\bar{4}2m$ (D_{2d}); for Δ, $4mm$ (C_{4v}); for Λ, $3m$ (C_{3v}); for Σ, $mm2$ (C_{2v}); for S, $mm2$ (C_{2v}); for Z, $mm2$ (C_{2v}); for Q, 2 (C_2); for C, m (C_{1h}); for O, m (C_{1h}); for J, m (C_{1h}) for B, m (C_{1h}); and for A, 1 (C_1).

Fd3m

The case of diamond is not quite so easily dealt with. The first thing to notice is that if in the decomposition (3.8.6) the rotational part P of the coset representative is one of the elements E, $C_{3j}^{\pm}$, C_{2m}, σ_{dp} or $S_{4m}^{\pm}$ its translational part $\mathbf{w}_j = 0$, but if its rotational part Q is one of the elements I, $S_{6j}^{\mp}$, σ_m, C_{2p}, or $C_{4m}^{\pm}$ its translational part

$\mathbf{w}_j = \mathbf{v} = \frac{1}{4}(a, a, a)$. Therefore the factor system $\exp(-i\mathbf{g}_i . \mathbf{w}_j)$ consists entirely of units if $\mathbf{g}_i = 0$ for all i, or if the little group $\overline{\mathbf{G}}^{\mathbf{k}_1}$ has all its elements of the form P and none of the form Q. The former is the case if $\mathbf{k}_1$ is an internal point of the Brillouin zone $\{\Gamma, \Delta, \Lambda, \Sigma, C, O, J, A(I)\}$, and the latter is the case if $\overline{\mathbf{G}}^{\mathbf{k}_1}$ is composed entirely of elements in $\bar{4}3m\,(T_d)$ $\{\Lambda, C, J, A(S)\}$. In these cases the required projective representations $D_p^{\mathbf{k}_1}$ are just the ordinary reps of the point group $\overline{\mathbf{G}}^{\mathbf{k}_1}$.

Thus in $Fd3m$, if $\mathbf{k}_1 = \Gamma, \Delta, \Lambda, \Sigma, C, O, J$, or A all we have to do to find $D_p^{\mathbf{k}_1}$ is to look up the reps of $\overline{\mathbf{G}}^{\mathbf{k}_1}$ in Table 2.2. The names of the point groups $\overline{\mathbf{G}}^{\mathbf{k}_1}$ corresponding to these points are to be found above in the treatment of $Fm3m$. However, even for these points there is an essential difference between $Fd3m$ and $Fm3m$ and that is, having obtained $D_p^{\mathbf{k}_1}$, the small representations $\Gamma_p^{\mathbf{k}_1}$ do not follow quite in the same way as for $Fm3m$. In this case we see from eqn. (3.7.9) that

$$\Gamma_p^{\mathbf{k}_1}(\{P \mid \mathbf{t}\}) = \exp(-i\mathbf{k}_1 . \mathbf{t})\, \mathbf{D}_p^{\mathbf{k}_1}(P), \tag{3.8.8}$$

but that

$$\Gamma_p^{\mathbf{k}_1}(\{Q \mid \mathbf{t} + \mathbf{v}\}) = \exp(-i\mathbf{k}_1 . (\mathbf{t} + \mathbf{v}))\, \mathbf{D}_p^{\mathbf{k}_1}(Q), \tag{3.8.9}$$

where P and Q have the same meaning as in the decomposition (3.8.6). Since it is customary to tabulate the small representations $\Gamma_p^{\mathbf{k}_1}$ rather than $D_p^{\mathbf{k}_1}$, certain of the elements appearing in the tables carry an extra factor $\exp(-i\mathbf{k}_1 . \mathbf{v})$ that does not appear in the character table for the corresponding point group. (It is worth while noting that this factor is always unity for Γ so that the reps of a non-symmorphic space group belonging to Γ are the same as for the reps of the corresponding symmorphic space group.)

We now consider in detail each of the surface points of the Brillouin zone B, Q, Z, S, L, W, and X. We shall use the notation of the later paragraphs of section 3.7 without further explanation.

B

Coset representatives of $\mathbf{G}^B$: $\{E \mid 000\}$, $\{\sigma_y \mid \frac{1}{4}\frac{1}{4}\frac{1}{4}\}$.
$EB = B$, so that $\mathbf{g}_E = 0$.
$\sigma_y B = B - \mathbf{g}_1 - \mathbf{g}_3$, so that $\mathbf{g}_{\sigma_y} = -\mathbf{g}_1 - \mathbf{g}_3 = (-1, 0, -1)$.
Factor system: $\mu(E, E) = 1$, $\mu(E, \sigma_y) = 1$, $\mu(\sigma_y, E) = 1$, and $\mu(\sigma_y, \sigma_y) = \exp(2\pi i(\frac{1}{4} + 0 + \frac{1}{4})) = -1$.

The central extension $\overline{\mathbf{G}}^{B*}$ consists of the four elements $(E, 0), (E, 1), (\sigma_y, 0), (\sigma_y, 1)$, with multiplication table:

$(E, 0)$	$(E, 1)$	$(\sigma_y, 0)$	$(\sigma_y, 1)$
$(E, 1)$	$(E, 0)$	$(\sigma_y, 1)$	$(\sigma_y, 0)$
$(\sigma_y, 0)$	$(\sigma_y, 1)$	$(E, 1)$	$(E, 0)$
$(\sigma_y, 1)$	$(\sigma_y, 0)$	$(E, 0)$	$(E, 1)$

from which we see that $\overline{\mathbf{G}}^{B*}$ is isomorphic with C_4. We are interested in those representations of $\overline{\mathbf{G}}^{B*}$ for which $D(E, 1) = -\mathbf{1}$ (see eqn. (3.7.33) with $\alpha = 1$ and $g = 2$). From Table 2.2 we see that these are the two 1-dimensional representations:

	$(E, 0)$	$(\sigma_y, 0)$	$(E, 1)$	$(\sigma_y, 1)$
${}^1E$	1	$-i$	-1	i
${}^2E$	1	i	-1	$-i$

From eqns. (3.8.8) and (3.8.9) we see that the corresponding entries in the character table for the small representations of $\mathbf{G}^B$ are

B	$\{E \mid 000\}$	$\{\sigma_y \mid \frac{1}{4}\frac{1}{4}\frac{1}{4}\}$
${}^1E$	1	$-i\zeta$
${}^2E$	1	$i\zeta$

where $\zeta = \exp(-i\mathbf{k}_1 . \mathbf{v})$.

Q

Coset representatives of $\mathbf{G}^Q$: $\{E \mid 000\}$, $\{C_{2f} \mid \frac{1}{4}\frac{1}{4}\frac{1}{4}\}$.

$EQ = Q$, so that $\mathbf{g}_E = 0$.

$C_{2f}Q = Q - \mathbf{g}_1 - \mathbf{g}_2 - \mathbf{g}_3$ so that $\mathbf{g}_Q = -(\mathbf{g}_1 + \mathbf{g}_2 + \mathbf{g}_3) = (-1, -1, -1)$.

Factor system: $\mu(E, E) = 1$, $\mu(E, C_{2f}) = 1$, $\mu(C_{2f}, E) = 1$ and $\mu(C_{2f}, C_{2f}) = -i$.

The central extension $\overline{\mathbf{G}}^{Q*}$ consists of eight elements and is isomorphic with C_8 in such a way that $(C_{2f}, 0)^8 = (E, 0)$ and $(C_{2f}, 0)^6 = (E, 1)$. We are interested in those representations of $\overline{\mathbf{G}}^{Q*}$ such that $D(E, 1) = i\mathbf{1}$ (see eqn. (3.7.33) with $\alpha = 1$ and $g = 4$). These are two 1-dimensional representations and in them the appropriate entries in the character table are

	$(E, 0)$	$(C_{2f}, 0)$
${}^1E_3$	1	$\exp(3\pi i/4)$
${}^2E_1$	1	$\exp(7\pi i/4)$

Since $\exp(-i\mathbf{k}_1 . \mathbf{v}) = \exp(-3\pi i/4)$ we see that the corresponding entries in the character table for the small representations of $\mathbf{G}_Q$ are

Q	$\{E\,|\,000\}$	$\{C_{2f}\,|\,\frac{1}{4}\frac{1}{4}\frac{1}{4}\}$
${}^1E_3$	1	+1
${}^2E_1$	1	−1

Z

Coset representatives of $\mathbf{G}^Z$: $\{E\,|\,000\}$, $\{C_{2x}\,|\,000\}$, $\{\sigma_y\,|\,\frac{1}{4}\frac{1}{4}\frac{1}{4}\}$, $\{\sigma_z\,|\,\frac{1}{4}\frac{1}{4}\frac{1}{4}\}$.
$\mathbf{g}_E = 0$, $\mathbf{g}_{C_{2x}} = (-1, 0, -1)$, $\mathbf{g}_{\sigma_z} = 0$, $\mathbf{g}_{\sigma_y} = (-1, 0, -1)$.
The elements of the factor system are all equal to 1 except $\mu(C_{2x}, \sigma_z) = \mu(C_{2x}, \sigma_y) = \mu(\sigma_y, \sigma_z) = \mu(\sigma_y, \sigma_y) = -1$.

The central extension $\overline{\mathbf{G}}^{Z*}$ consists of eight elements and is isomorphic with 422 (D_4) in such a way that $(\sigma_y, 0)^4 = (\sigma_z, 0)^2 = (E, 0)$ and $(\sigma_z, 0)(\sigma_y, 0) = (\sigma_y, 0)^3 (\sigma_z, 0) = (C_{2x}, 0)$. We are interested in the representations of $\overline{\mathbf{G}}^{Z*}$ for which $D(E, 1) = -\mathbf{1}$ and from Table 2.2 we see that this is the representation E whose characters are

	$(E, 0)$	$(E, 1)$	$(\sigma_y, 0)(\sigma_y, 1)$	$(\sigma_z, 0)(\sigma_z, 1)$	$(C_{2x}, 0)(C_{2x}, 1)$
E	2	−2	0	0	0

with possible matrices:

$$D_E(E, 0) = \begin{pmatrix} 1 & 0 \\ 0 & 1 \end{pmatrix}, \quad D_E(\sigma_y, 0) = \begin{pmatrix} 0 & 1 \\ -1 & 0 \end{pmatrix},$$

$$D_E(\sigma_z, 0) = \begin{pmatrix} 1 & 0 \\ 0 & -1 \end{pmatrix}, \quad D_E(C_{2x}, 0) = \begin{pmatrix} 0 & 1 \\ 1 & 0 \end{pmatrix}.$$

Writing $\zeta = \exp(-i\mathbf{k}_1 . \mathbf{v})$, we obtain the corresponding entries in the matrix representation table for the small representation of $\mathbf{G}^Z$.

Z	$\{E\,|\,000\}$	$\{\sigma_y\,|\,\frac{1}{4}\frac{1}{4}\frac{1}{4}\}$	$\{\sigma_z\,|\,\frac{1}{4}\frac{1}{4}\frac{1}{4}\}$	$\{C_{2x}\,|\,000\}$
E	$\begin{pmatrix} 1 & 0 \\ 0 & 1 \end{pmatrix}$	$\begin{pmatrix} 0 & \zeta \\ -\zeta & 0 \end{pmatrix}$	$\begin{pmatrix} \zeta & 0 \\ 0 & -\zeta \end{pmatrix}$	$\begin{pmatrix} 0 & 1 \\ 1 & 0 \end{pmatrix}$

S

Coset representatives of $\mathbf{G}^S$: $\{E \mid 000\}$, $\{C_{2c} \mid \frac{1}{4}\frac{1}{4}\frac{1}{4}\}$, $\{\sigma_{de} \mid 000\}$, $\{\sigma_y \mid \frac{1}{4}\frac{1}{4}\frac{1}{4}\}$.
$\mathbf{g}_E = 0$, $\mathbf{g}_{C_{2c}} = (-1, 0, -1)$, $\mathbf{g}_{\sigma_{de}} = 0$, $\mathbf{g}_{\sigma_y} = (-1, 0, -1)$.
The elements of the factor system are all equal to 1 except $\mu(C_{2c}, C_{2c}) = \mu(C_{2c}, \sigma_y) = \mu(\sigma_y, C_{2c}) = \mu(\sigma_y, \sigma_y) = -1$.

The central extension $\bar{\mathbf{G}}^{S*}$ consists of eight elements and is isomorphic with $4/m$ (C_{4h}) in such a way that $(\sigma_y, 0)^4 = (\sigma_{de}, 0)^2 = (E, 0)$ and $(\sigma_y, 0)(\sigma_{de}, 0) = (\sigma_{de}, 0)(\sigma_y, 0) = (C_{2c}, 0)$. We are interested in those representations of $\bar{\mathbf{G}}^{S*}$ for which $\mathbf{D}(E, 1) = -\mathbf{1}$ and from Table 2.2 we see that these are four 1-dimensional representations whose appropriate characters are

	$(E, 0)$	$(\sigma_{de}, 0)$	$(\sigma_y, 0)$	$(C_{2c}, 0)$
${}^1E_1$	1	1	$-i$	$-i$
${}^2E_1$	1	1	i	i
${}^2E_2$	1	-1	i	$-i$
${}^1E_2$	1	-1	$-i$	i

and writing $\zeta = \exp(-i\mathbf{k}_1 . \mathbf{v})$ we obtain for the corresponding entries in the character table of the small representations of $\mathbf{G}^S$

S	$\{E \mid 000\}$	$\{\sigma_{de} \mid 000\}$	$\{\sigma_y \mid \frac{1}{4}\frac{1}{4}\frac{1}{4}\}$	$\{C_{2c} \mid \frac{1}{4}\frac{1}{4}\frac{1}{4}\}$
${}^1E_1$	1	1	$-i\zeta$	$-i\zeta$
${}^2E_1$	1	1	$i\zeta$	$i\zeta$
${}^2E_2$	1	-1	$i\zeta$	$-i\zeta$
${}^1E_2$	1	-1	$-i\zeta$	$i\zeta$

The methods we have used to obtain the small representations for the wave vectors, $\mathbf{k}$, on the surface of the Brillouin zone for non-symmorphic space groups can of course be used for the points of symmetry also. However there is an alternative method due to Herring (1942) which avoids the use of projective representations for such points of symmetry and since it is rather more direct in its approach (it uses a factor group of the little group rather than the little co-group) we shall describe it and apply it as an example to the points W, X, and L for $Fd3m$ (O_h^7). There is very little to choose between the two methods because they both involve finding the characters of a group of large order.

Herring's method is as follows. In any small representation

$$\mathbf{\Gamma}_p^{\mathbf{k}_1}(\{E \mid \mathbf{t}\}) = \exp(-i\mathbf{k}_1 . \mathbf{t})\mathbf{1}. \qquad (3.8.10)$$

If $\mathbf{k}_1$ is a point of symmetry then the phase factor $\exp(-\mathrm{i}\mathbf{k}_1 . \mathbf{t}) = 1$ for a large number of translations $\mathbf{t}$. Denote the subgroup of $\mathbf{T}$ of elements that have this property by $\mathbf{T}^{\mathbf{k}_1}$. Then $\mathbf{T}^{\mathbf{k}_1}$ is an invariant subgroup of $\mathbf{G}^{\mathbf{k}_1}$ and the factor group $\mathbf{G}^{\mathbf{k}_1}/\mathbf{T}^{\mathbf{k}_1}$, which we denote by ${}^{H}\mathbf{G}^{\mathbf{k}_1}$, is of small order compared with that of $\mathbf{G}^{\mathbf{k}_1}$. (The maximum possible order of ${}^{H}\mathbf{G}^{\mathbf{k}_1}$ that occurs in determining the single-valued reps of the space groups is 96.) Now because $\boldsymbol{\Gamma}_p^{\mathbf{k}_1}(\{E \mid \mathbf{t}\}) = \mathbf{1}$ for all $\mathbf{t} \in \mathbf{T}^{\mathbf{k}_1}$ it follows that each element of $\mathbf{G}^{\mathbf{k}_1}$ in a given coset of $\mathbf{T}^{\mathbf{k}_1}$ is represented by the same matrix. Thus, since the cosets of $\mathbf{T}^{\mathbf{k}_1}$ are the elements of ${}^{H}\mathbf{G}^{\mathbf{k}_1}$, it follows that $\Gamma_p^{\mathbf{k}_1}$ can be taken over as a representation of the factor group. Loosely speaking, we identify all the elements of a given coset. Conversely, using this identification, we can also take the reps of the factor group to be the reps of the little group. Of course only some of them will be small reps because of the extra requirement imposed by eqn. (3.8.10); but certainly all the small reps can be deduced in this way by lifting them from ${}^{H}\mathbf{G}^{\mathbf{k}_1}$ into $\mathbf{G}^{\mathbf{k}_1}$ because a small rep always has the property that $\boldsymbol{\Gamma}_p^{\mathbf{k}_1}(\{E \mid \mathbf{t}\}) = \mathbf{1}$ for $\mathbf{t} \in \mathbf{T}^{\mathbf{k}_1}$. In order to make the working seem natural it is customary to use the coset representatives as the symbols for the cosets. These symbols then form a group isomorphic with the factor group provided, when working with them, we identify any two symbols that only differ from each other by translations $\mathbf{t} \in \mathbf{T}^{\mathbf{k}_1}$. Since this can be done at sight, multiplications within the factor group can be deduced immediately from the ordinary space-group multiplication rule.

Returning to the case of diamond we shall now show how Herring's method works for the points W, X, and L.

W

Since $W = \frac{1}{2}\mathbf{g}_1 + \frac{1}{4}\mathbf{g}_2 + \frac{3}{4}\mathbf{g}_3$ the group $\mathbf{G}^W/\mathbf{T}^W$ contains four translations $\{E \mid \mathbf{0}\}$, $\{E \mid \mathbf{t}_1\}$, $\{E \mid \mathbf{t}_2\}$, $\{E \mid \mathbf{t}_3\}$, represented respectively by $\mathbf{1}$, $-\mathbf{1}$, $-\mathrm{i}\mathbf{1}$, and $\mathrm{i}\mathbf{1}$. There are therefore 32 elements in ${}^{H}\mathbf{G}^W$, the products in Herring's sense of the eight elements $\{E \mid \mathbf{0}\}$, $\{C_{2x} \mid \mathbf{0}\}$, $\{C_{2d} \mid \mathbf{v}\}$, $\{C_{2f} \mid \mathbf{v}\}$, $\{\sigma_y \mid \mathbf{v}\}$, $\{\sigma_z \mid \mathbf{v}\}$, $\{S_{4x}^+ \mid \mathbf{0}\}$, $\{S_{4x}^- \mid \mathbf{0}\}$ with the four translations. This group of order 32 can be identified as G_{32}^4 in Table 5.1 with $\{S_{4x}^+ \mid \mathbf{0}\} = P$, $\{C_{2f} \mid \mathbf{v}\} = R$, and $\{E \mid \mathbf{t}_3\} = Q$. (All one needs to do is to check that the generating relations $P^4 = R^2 = Q^4 = E$, $PQ = QP$, $RQ = QR$, and $RP = Q^3P^3R$ are satisfied.) We require the reps in which $\{E \mid 001\} = Q$ is represented by $\mathrm{i}\mathbf{1}$. These are two 2-dimensional representations. The relevant section of the character table of ${}^{H}\mathbf{G}^W$ is

W	$\{E \mid 000\}$	$\{E \mid 001\}$	$\{E \mid 100\}$	$\{E \mid 010\}$	$\{C_{2x} \mid 000\}$ $\{C_{2x} \mid 100\}$	$\{C_{2x} \mid 001\}$ $\}C_{2x} \mid 010\}$	$\{S_{4x}^+ \mid 000\}$ $\{S_{4x}^- \mid 010\}$
${}^1F_1$	2	2i	-2	-2i	0	0	$1 - \mathrm{i}$
${}^1F_2$	2	2i	-2	-2i	0	0	$-1 + \mathrm{i}$

W	$\{S_{4x}^+\|001\}$ $\{S_{4x}^-\|000\}$	$\{S_{4x}^+\|100\}$ $\{S_{4x}^-\|001\}$	$\{S_{4x}^+\|010\}$ $\{S_{4x}^-\|100\}$	$\{\sigma_y\|\frac{1}{4}\frac{1}{4}\frac{5}{4}\}$ $\{\sigma_y\|\frac{1}{4}\frac{5}{4}\frac{1}{4}\}$ $\{\sigma_z\|\frac{1}{4}\frac{1}{4}\frac{1}{4}\}$ $\{\sigma_z\|\frac{5}{4}\frac{1}{4}\frac{1}{4}\}$	$\{\sigma_y\|\frac{1}{4}\frac{1}{4}\frac{1}{4}\}$ $\{\sigma_y\|\frac{5}{4}\frac{1}{4}\frac{1}{4}\}$ $\{\sigma_z\|\frac{1}{4}\frac{1}{4}\frac{5}{4}\}$ $\{\sigma_z\|\frac{1}{4}\frac{5}{4}\frac{1}{4}\}$	$\{C_{2f}\|\frac{1}{4}\frac{1}{4}\frac{5}{4}\}$ $\{C_{2f}\|\frac{1}{4}\frac{5}{4}\frac{1}{4}\}$ $\{C_{2d}\|\frac{1}{4}\frac{1}{4}\frac{1}{4}\}$ $\{C_{2d}\|\frac{5}{4}\frac{1}{4}\frac{1}{4}\}$	$\{C_{2f}\|\frac{1}{4}\frac{1}{4}\frac{1}{4}\}$ $\{C_{2f}\|\frac{5}{4}\frac{1}{4}\frac{1}{4}\}$ $\{C_{2d}\|\frac{1}{4}\frac{1}{4}\frac{5}{4}\}$ $\{C_{2d}\|\frac{1}{4}\frac{5}{4}\frac{1}{4}\}$
${}^1F_1$	$1+\mathrm{i}$	$-1+\mathrm{i}$	$-1-\mathrm{i}$	0	0	0	0
${}^1F_2$	$-1-\mathrm{i}$	$1-\mathrm{i}$	$1+\mathrm{i}$	0	0	0	0

Suitable matrices generating these representations are

	$\{C_{2f}\|\frac{1}{4}\frac{1}{4}\frac{1}{4}\}$	$\{S_{4x}^+\|000\}$
${}^1F_1$	$\begin{pmatrix}0 & 1\\ 1 & 0\end{pmatrix}$	$\begin{pmatrix}1 & 0\\ 0 & -\mathrm{i}\end{pmatrix}$
${}^1F_2$	$\begin{pmatrix}0 & 1\\ 1 & 0\end{pmatrix}$	$\begin{pmatrix}\mathrm{i} & 0\\ 0 & -1\end{pmatrix}$

X

Since $X = (\frac{1}{2}0\frac{1}{2})$, the group ${}^H\mathbf{G}^X$ contains two translations $\{E\,|\,\mathbf{0}\}$ and $\{E\,|\,\mathbf{t}_1\}$ represented respectively by $\mathbf{1}$ and $-\mathbf{1}$. There are therefore 32 elements in ${}^H\mathbf{G}^X$, the products in Herring's sense of the 16 elements $\{E\,|\,\mathbf{0}\}$, $\{C_{2y}\,|\,\mathbf{0}\}$, $\{C_{4y}^{\pm}\,|\,\mathbf{v}\}$, $\{C_{2x}\,|\,\mathbf{0}\}$, $\{C_{2z}\,|\,\mathbf{0}\}$, $\{C_{2c}\,|\,\mathbf{v}\}$, $\{C_{2e}\,|\,\mathbf{v}\}$, $\{I\,|\,\mathbf{v}\}$, $\{\sigma_y\,|\,\mathbf{v}\}$, $\{S_{4y}^{\mp}\,|\,\mathbf{0}\}$, $\{\sigma_x\,|\,\mathbf{v}\}$, $\{\sigma_z\,|\,\mathbf{v}\}$, $\{\sigma_{dc}\,|\,\mathbf{0})$ and $\{\sigma_{de}\,|\,\mathbf{0}\}$ with these two translations. This group of order 32 can be identified as G_{32}^2 in Table 5.1 with $\{\sigma_x\,|\,\mathbf{v}\} = P$, $\{S_{4y}^+\,|\,\mathbf{0}\} = Q$, and $\{C_{2x}\,|\,\mathbf{0}\} = R$. We require the reps in which $\{E\,|\,\mathbf{t}_1\} = P^2$ is represented by $-\mathbf{1}$. These are four 2-dimensional representations. The relevant section of the character table of ${}^H\mathbf{G}^X$ is

X	$\{E\|000\}$	$\{E\|100\}$	$\{C_{2y}\|000\}$	$\{C_{2y}\|100\}$	$\{I\|\frac{1}{4}\frac{1}{4}\frac{1}{4}\}$ $\{I\|\frac{5}{4}\frac{1}{4}\frac{1}{4}\}$	$\{\sigma_y\|\frac{1}{4}\frac{1}{4}\frac{1}{4}\}$ $\{\sigma_y\|\frac{5}{4}\frac{1}{4}\frac{1}{4}\}$	$\{\sigma_{dc}\|000\}$ $\{\sigma_{de}\|000\}$
E_1	2	-2	2	-2	0	0	2
E_2	2	-2	-2	2	0	0	0
E_3	2	-2	2	-2	0	0	-2
E_4	2	-2	-2	2	0	0	0

X	$\{\sigma_{dc}\|100\}$ $\{\sigma_{de}\|100\}$	$\{C_{2e}\|\frac{1}{4}\frac{1}{4}\frac{1}{4}\}$ $\{C_{2c}\|\frac{5}{4}\frac{1}{4}\frac{1}{4}\}$	$\{C_{2e}\|\frac{5}{4}\frac{1}{4}\frac{1}{4}\}$ $\{C_{2c}\|\frac{1}{4}\frac{1}{4}\frac{1}{4}\}$	$\{C_{4y}^{\pm}\|\frac{1}{4}\frac{1}{4}\frac{1}{4}\}$ $\{C_{4y}^{\pm}\|\frac{5}{4}\frac{1}{4}\frac{1}{4}\}$	$\{S_{4y}^{\mp}\|000\}$ $\{S_{4y}^{\mp}\|100\}$	$\{C_{2z}\|000\}$ $\{C_{2z}\|100\}$ $\{C_{2x}\|000\}$ $\{C_{2x}\|100\}$	$\{\sigma_z\|\frac{1}{4}\frac{1}{4}\frac{1}{4}\}$ $\{\sigma_z\|\frac{5}{4}\frac{1}{4}\frac{1}{4}\}$ $\{\sigma_x\|\frac{1}{4}\frac{1}{4}\frac{1}{4}\}$ $\{\sigma_x\|\frac{5}{4}\frac{1}{4}\frac{1}{4}\}$
E_1	-2	0	0	0	0	0	0
E_2	0	2	-2	0	0	0	0
E_3	2	0	0	0	0	0	0
E_4	0	-2	2	0	0	0	0

Suitable matrices generating these representations are

	$\{\sigma_x \mid \frac{1}{4}\frac{1}{4}\frac{1}{4}\}$	$\{S_{4y}^{+} \mid 000\}$	$\{C_{2x} \mid 000\}$
E_1	$\begin{pmatrix} 0 & -1 \\ 1 & 0 \end{pmatrix}$	$\begin{pmatrix} 0 & 1 \\ 1 & 0 \end{pmatrix}$	$\begin{pmatrix} 0 & 1 \\ 1 & 0 \end{pmatrix}$
E_2	$\begin{pmatrix} 0 & -1 \\ 1 & 0 \end{pmatrix}$	$\begin{pmatrix} 0 & 1 \\ -1 & 0 \end{pmatrix}$	$\begin{pmatrix} 0 & 1 \\ 1 & 0 \end{pmatrix}$
E_3	$\begin{pmatrix} 0 & -1 \\ 1 & 0 \end{pmatrix}$	$\begin{pmatrix} 0 & -1 \\ -1 & 0 \end{pmatrix}$	$\begin{pmatrix} 0 & 1 \\ 1 & 0 \end{pmatrix}$
E_4	$\begin{pmatrix} 0 & -1 \\ 1 & 0 \end{pmatrix}$	$\begin{pmatrix} 0 & -1 \\ 1 & 0 \end{pmatrix}$	$\begin{pmatrix} 0 & 1 \\ 1 & 0 \end{pmatrix}$

L

Since $L = (\frac{1}{2}\frac{1}{2}\frac{1}{2})$, the group ${}^{H}\mathbf{G}^{L}$ contains two translations $\{E \mid \mathbf{0}\}$ and $\{E \mid \mathbf{t}_1\}$ represented respectively by $\mathbf{1}$ and $-\mathbf{1}$. The group is of order 24 and is isomorphic with $3m \otimes \mathbf{J} \otimes \mathbf{T}_1$ where $3m$ (C_{3v}) contains the six elements $\{\mathbf{E} \mid \mathbf{0}\}$, $\{C_{31}^{\pm} \mid \mathbf{0}\}$, $\{\sigma_{db} \mid \mathbf{0}\}$, $\{\sigma_{de} \mid \mathbf{0}\}$, $\{\sigma_{df} \mid \mathbf{0}\}$, where $\mathbf{J}$ contains $\{E \mid \mathbf{0}\}$ and $\{I \mid \mathbf{v}\}$ and where $\mathbf{T}_1$ contains the two translations $\{E \mid \mathbf{0}\}$ and $\{E \mid \mathbf{t}_1\}$. There are therefore six small representations and a section of the character table of ${}^{H}\mathbf{G}^{L}$ can be seen from Table 2.2 to be as follows.

L	$\{E \mid 000\}$	$\{C_{31}^{\pm} \mid 000\}$	$\{\sigma_{db} \mid 000\}$ $\{\sigma_{de} \mid 000\}$ $\{\sigma_{df} \mid 000\}$	$\{I \mid \frac{1}{4}\frac{1}{4}\frac{1}{4}\}$	$\{S_{61}^{\mp} \mid \frac{1}{4}\frac{1}{4}\frac{1}{4}\}$	$\{C_{2b} \mid \frac{1}{4}\frac{1}{4}\frac{1}{4}\}$ $\{C_{2e} \mid \frac{1}{4}\frac{1}{4}\frac{1}{4}\}$ $\{C_{2f} \mid \frac{1}{4}\frac{1}{4}\frac{1}{4}\}$
A_{1g}	1	1	1	1	1	1
A_{2g}	1	1	−1	1	1	−1
E_g	2	−1	0	2	−1	0
A_{1u}	1	1	1	−1	−1	−1
A_{2u}	1	1	−1	−1	−1	1
E_u	2	−1	0	−2	1	0

Generating matrices for the 2-dimensional representations can be obtained from Table 2.3. It will be seen that by chance the form of the reps belonging to Q and L in $Fd3m$ are just like those in $Fm3m$. The rules we gave for when this certainly happens are therefore sufficient but not necessary, and from this example we see that even if $\mathbf{k}_1$ is on the surface of the Brillouin zone and the space group is non-symmorphic then the small reps of $\mathbf{G}^{\mathbf{k}_1}$ occasionally have the same form as the reps of $\overline{\mathbf{G}}^{\mathbf{k}_1}$.

4

The representations of a group in terms of the representations of an invariant subgroup

In the last two sections of Chapter 3 we gave only a partially complete account of the representation theory of space groups. A number of arguments were presented only in outline, while some statements were left unproved. There were several reasons for presenting the work in this way. First, it was felt that for some readers the account given in Chapter 3 would be sufficient. Indeed, the details of an entirely rigorous account may often distract a reader from a central theme and thereby lose for him in obscurity as much as it gains for him in accuracy. Secondly, the completely rigorous scheme will be found to be valid in a wider group-theoretical context and is of use not only for the classical space groups but in other applications also. It seemed sensible, therefore, to present the complete group theory in a general notation that is not restricted to space groups and is set apart from those particular considerations involved in the last chapter.

The purpose of this chapter is, therefore, to treat in complete generality the representation theory of groups with an invariant subgroup, of which space groups are a particular example where the invariant subgroup is the 3-dimensional translation group and hence is Abelian. The extra theory will enable us to fill in the gaps of the last chapter, primarily to show that the scheme outlined for obtaining the reps of space groups does in fact produce all of them, neither too few nor any spurious ones that are equivalent to others, but also to show how to obtain in precise matrix form complete representations from the tabulated small representations of little groups. It also happens that this extra theory will enable us to discuss more easily further points of interest in the theory of solids, such as the reality of the reps and their inner direct products with each other. The reality is important because of its usefulness in studying the degeneracy of individual particle or quasi-particle states in a crystal as a result of time-reversal symmetry (Herring 1937*a*, Wigner 1932). The inner direct products are important because of their usefulness in determining selection rules for various processes involving electrons, phonons, magnons, etc. in solids (Elliott and Loudon 1960).

4.1 Induced representations

Suppose that $\mathbf{G}$ is a group and that $\mathbf{K}_1$ is a subgroup of $\mathbf{G}$ which is not necessarily invariant. For example, $\mathbf{G}$ might be a space group and $\mathbf{K}_1$ the little group of a vector

$\mathbf{k}_1$ of the Brillouin zone. We denote the elements of $\mathbf{K}_1$ by k_a ($a = 1$ to $|\mathbf{K}_1|$). Let us factorize $\mathbf{G}$ into left cosets with respect to $\mathbf{K}_1$ so that (compare with eqns. (3.7.2)–(3.7.4))

$$\mathbf{G} = \sum_\alpha p_\alpha \mathbf{K}_1, \tag{4.1.1}$$

where p_α are left coset representatives, which we shall suppose chosen once for all;‡ for example we always take $p_1 = e$, the identity, and whenever we write p_λ we mean the particular coset representative of the coset $p_\lambda \mathbf{K}_1$ that occurs in the expansion (4.1.1), and not just any element of the coset. In eqn. (4.1.1) the sum over α runs from 1 to $|\mathbf{G}|/|\mathbf{K}_1|$, where $|\mathbf{G}|$ is the order of $\mathbf{G}$; this equation means that each element $g \in \mathbf{G}$ is expressible in the form $p_\lambda k_s$ where p_λ is one of the chosen coset representatives and where both p_λ and $k_s \in \mathbf{K}_1$ are *uniquely* determined by g.

THEOREM 4.1.1. *Every element $g \in \mathbf{G}$ is expressible in the form $g = k_b p_\lambda^{-1}$ where b and λ are uniquely determined by g.*

For, by what we have just inferred from eqn. (4.1.1) $g^{-1} = p_\lambda k_s$ where λ and s are uniquely determined by g^{-1}. Hence $g = k_s^{-1} p_\lambda^{-1} = k_b p_\lambda^{-1}$, where since $\mathbf{K}_1$ is a group $k_b = k_s^{-1} \in \mathbf{K}_1$ and where, moreover, b is uniquely determined. Notice that p_λ^{-1} is not necessarily one of the chosen coset representatives. This is because the $|\mathbf{G}|/|\mathbf{K}_1|$ coset representatives p_α do not necessarily form a group, even when $\mathbf{K}_1$ is invariant.

Now let Ω_1 be a vector space of dimension d_j, irreducible under $\mathbf{K}_1$, and choose a basis $\langle \phi_r|$, $r = 1$ to d_j, so that, for all $k_a \in \mathbf{K}_1$,

$$k_a \phi_r = \sum_{p=1}^{d_j} \phi_p \Gamma^j(k_a)_{pr}. \tag{4.1.2}$$

Denote the character of k_a in the rep j by $\chi^j(k_a)$. Furthermore, let Ω denote the vector space spanned by the $d_j|\mathbf{G}|/|\mathbf{K}_1|$ functions $\phi_{\alpha r}$ where for all α and r

$$\phi_{\alpha r} = p_\alpha \phi_r. \tag{4.1.3}$$

Equation (4.1.3) is a functional identity, so that for equal values of their arguments $\phi_{\alpha r}$ and $p_\alpha \phi_r$ have the same value. We met the same procedure in Chapter 3, section 3.7, when we defined functions of the form $\{T_i \mid \mathbf{x}_i\}\psi_{\mathbf{k}}$.

THEOREM 4.1.2. *Ω is invariant under* $\mathbf{G}$.

For writing $g = p_\lambda k_s$, then $g\phi_{\tau r} = p_\lambda k_s p_\tau \phi_r = p_\gamma k_t \phi_r$, where γ and t are uniquely determined by g and τ. Thus, using eqns. (4.1.2) and (4.1.3) we obtain

$$\begin{aligned} g\phi_{\tau r} &= p_\gamma \sum_p \phi_p \Gamma^j(k_t)_{pr} \\ &= \sum_p \phi_{\gamma p} \Gamma^j(k_t)_{pr}. \end{aligned} \tag{4.1.4}$$

‡ The results of the theory in no way depend upon the particular choice made. By this we mean that all choices are equivalent in the sense that induced representations arising from different choices of coset representatives are equivalent; however, once a choice is made it should not be altered because the mathematical details, particularly in the proofs of theorems, are affected by the choice.

Equation (4.1.4) defines what is called the induced representation of Γ^j in $\mathbf{G}$, written conveniently as $\Gamma^j \uparrow \mathbf{G}$. We denote its character by $\chi(g)$. It is of dimension $d_j|\mathbf{G}|/|\mathbf{K}_1|$. What eqn. (4.1.4) means is that if $gp_\tau = p_\gamma k_t$ (so that $k_t = p_\gamma^{-1}gp_\tau$) then for g the (γ, τ) block matrix (of dimension d_j) in $\Gamma^j \uparrow \mathbf{G}$ is $\Gamma^j(p_\gamma^{-1}gp_\tau)$ and furthermore, since γ is uniquely determined by g and τ, it is the only non-zero block matrix in the column labelled τ. Similarly, it is the only non-zero block matrix in the row labelled γ. It follows at once that if Γ^j is a unitary representation then $\Gamma^j \uparrow \mathbf{G}$ is also unitary. However, although Γ^j is irreducible it does not follow that $\Gamma^j \uparrow \mathbf{G}$ is also irreducible. We shall discuss conditions for irreducibility in due course.

It should now be clear that the reps of space groups are in fact the induced representations of the small reps of little groups. So we have already completed part of our task, that of showing how to obtain the detailed matrix form of the reps of space groups from tabulated small reps of little groups. The steps involved are covered by eqns. (4.1.1)–(4.1.4) with $\mathbf{G}$ identified as the space group, $\mathbf{K}_1$ as the little group $\mathbf{G}^{\mathbf{k}_1}$ (eqn. (3.7.3)), and Γ^j as the small representation $\Gamma_p^{\mathbf{k}_1}$ (see item (iv) of the scheme in section 3.7 for obtaining the reps of $\mathbf{G}$). One important fact is that in a space group $|\mathbf{G}| = N_1N_2N_3h$ and $|\mathbf{K}_1| = N_1N_2N_3b$, so that given a small rep of dimension t the dimension of the corresponding rep of the space group, $d = th/b = qt =$ (number of vectors in star) $\times$ (dimension of small rep), a formula in keeping with that derived in Chapter 3 (summarized in item (v) of the scheme in section 3.7 for obtaining the reps of $\mathbf{G}$).

THEOREM 4.1.3. *Define Ω_α to be the vector space spanned by the d_j functions $\phi_{\alpha r}$, $r = 1$ to d_j. Then, for any fixed α, Ω_α is irreducible under the subgroup $\mathbf{K}_\alpha = (p_\alpha k_a p_\alpha^{-1};\, k_a \in \mathbf{K}_1)$.*

For, using eqns. (4.1.2) and (4.1.3), we have

$$p_\alpha k_a p_\alpha^{-1}\phi_{\alpha r} = p_\alpha k_a \phi_r = \sum_s \phi_{\alpha s}\Gamma^j(k_a)_{sr}. \tag{4.1.5}$$

What eqn. (4.1.5) means is that the representative of $p_\alpha k_a p_\alpha^{-1}$ with respect to the basis $\langle\phi_{\alpha r}|$ is $\Gamma^j(k_a)$. But the group of matrices $\Gamma^j(k_a)$ is irreducible (by definition) and so the theorem is established.

The subgroups $\mathbf{K}_\alpha$ may be written in the form $p_\alpha\mathbf{K}_1p_\alpha^{-1}$. Therefore each one of the subgroups $\mathbf{K}_\alpha$, $\alpha = 1$ to $|\mathbf{G}|/|\mathbf{K}_1|$, has the following properties.

(i) It has the same order as $\mathbf{K}_1$.

(ii) It is conjugate to $\mathbf{K}_1$ under $\mathbf{G}$, and therefore to every other member of the set $\mathbf{K}_\alpha$ (because conjugacy is an equivalence relation).

(iii) Provided the subgroups are distinct each one is in direct correspondence with one and only one of the cosets $p_\alpha\mathbf{K}_1$ in the decomposition (4.1.1).

We recognize that we have met this situation before. For if we identify $\mathbf{G}$ with the isogonal point group $\mathbf{F}$ of a space group and $\mathbf{K}_1$ with the little co-group of $\mathbf{k}_1$, then we can identify $\mathbf{K}_\alpha$ as the little co-group of $\mathbf{k}_\alpha$, where $\mathbf{k}_\alpha$ is the member of the star defined by $\mathbf{k}_1$ such that $p_\alpha \mathbf{k}_1 = \mathbf{k}_\alpha$ (see section 3.7, following Definitions 3.7.2 and 3.7.3). We leave it as an exercise for the reader to show that there are no subgroups of $\mathbf{G}$ conjugate to $\mathbf{K}_1$ under $\mathbf{G}$ other than those in the set $\mathbf{K}_\alpha$.

THEOREM 4.1.4. Johnston's irreducibility criterion. *Denote the character of g in Ω_α by $\chi_\alpha(g)$ so that $\chi(g)$, the character of g in $\Gamma^j \uparrow \mathbf{G}$, is equal to $\sum_\alpha \chi_\alpha(g)$, then Ω is irreducible under $\mathbf{G}$ provided that*

$$\sum_t \chi_\alpha^*(t)\chi_\beta(t) = 0 \tag{4.1.6}$$

for all pairs α, β such that $\alpha \neq \beta$, where for each choice α, β the sum over t runs over all the elements of $\mathbf{K}_\alpha \cap \mathbf{K}_\beta$.

The proof of this useful theorem as given by Johnston (1960) is as follows.
From Theorem 1.3.8 Ω is irreducible under $\mathbf{G}$ if and only if

$$\sum_g \chi^*(g)\chi(g) = |\mathbf{G}|, \tag{4.1.7}$$

where the sum over g runs over all elements of $\mathbf{G}$. That is, if

$$\sum_\alpha \sum_g |\chi_\alpha(g)|^2 + \sum_{\alpha \neq \beta} \sum_g \{\chi_\alpha^*(g)\chi_\beta(g)\} = |\mathbf{G}|. \tag{4.1.8}$$

Now from eqn. (4.1.4) it is clear that if g is expressible in the form $p_\tau k_t p_\tau^{-1}$ then

$$\chi_\tau(g) = \chi^j(k_t), \tag{4.1.9a}$$

but that if not then

$$\chi_\tau(g) = 0. \tag{4.1.9b}$$

Thus $\chi_\tau(g) = \chi^j(k_t)$ if $g \in \mathbf{K}_\tau$ and $g = p_\tau k_t p_\tau^{-1}$, but is zero for all g not in $\mathbf{K}_\tau$. It follows that

$$\sum_\alpha \sum_g |\chi_\alpha(g)|^2 = \sum_\alpha \sum_a |\chi^j(k_a)|^2 = \sum_\alpha |\mathbf{K}_1| = |\mathbf{G}|.$$

(Since Γ^j is irreducible under $\mathbf{K}_1$ Theorem 1.3.8 holds and it has been applied in evaluating the sum over a.)

$\Gamma^j \uparrow \mathbf{G}$ is therefore irreducible if and only if

$$\sum_{\alpha \neq \beta} \sum_g \chi_\alpha^*(g)\chi_\beta(g) = 0. \tag{4.1.10}$$

In particular, if for all α, β such that $\alpha \neq \beta$

$$\sum_g \chi_\alpha^*(g)\chi_\beta(g) = 0 \tag{4.1.11}$$

then $\Gamma^j \uparrow \mathbf{G}$ is irreducible.

By virtue of eqn. (4.1.9) the only possibility of a non-zero term in the left-hand side of eqn. (4.1.11) is for elements of **G** in both $\mathbf{K}_\alpha$ and $\mathbf{K}_\beta$. This yields condition (4.1.6) and the theorem is proved.

COROLLARIES

(i) If $\mathbf{K}_1$ is invariant in **G** then $\mathbf{K}_\alpha = \mathbf{K}_1$ for all α, and condition (4.1.6) becomes

$$\sum_a \chi_\alpha^*(k_a)\chi_\beta(k_a) = 0 \tag{4.1.12}$$

for all α, β such that $\alpha \neq \beta$. This is satisfied if the irreducible representations of $\mathbf{K}_1$ spanned by the bases $\langle\phi_{\alpha r}|$, $\alpha = 1$ to $|\mathbf{G}|/|\mathbf{K}_1|$ are mutually inequivalent.

(ii) The condition (4.1.6) involves for each pair α, β such that $\alpha \neq \beta$ a sum over all the elements of the group $\mathbf{K}_\alpha \cap \mathbf{K}_\beta$. The condition imposed is that this sum must vanish. The sum will vanish as we shall prove shortly if for *any* subgroup $\mathbf{H}_{\alpha\beta}$ of $\mathbf{K}_\alpha \cap \mathbf{K}_\beta$ it can be established that

$$\sum_h \chi_\alpha^*(h)\chi_\beta(h) = 0, \tag{4.1.13}$$

where the sum over h runs over all elements of $\mathbf{H}_{\alpha\beta}$. This is particularly useful if $\mathbf{H}_{\alpha\beta}$ can be chosen to be the same subgroup **H** of **G** for all pairs α, β.

In order to prove that (4.1.13) is sufficient to imply (4.1.6) for a particular choice of the pair α, β let us consider more closely the meaning of eqn. (4.1.13). What it means is that the two representations of $\mathbf{H}_{\alpha\beta}$ one with basis $\langle\phi_{\alpha r}|$ and the other with basis $\langle\phi_{\beta r}|$ have, when reduced, no irreducible components in common; this is because the orthogonality relationships for characters of the reps of $\mathbf{H}_{\alpha\beta}$ ensures that the sum (4.1.13) contains only positive or zero terms and that non-zero terms occur when and only when there are irreducible subspaces of Ω_α and Ω_β one in Ω_α and one in Ω_β which generate the same rep of $\mathbf{H}_{\alpha\beta}$. The sum (4.1.13) is therefore equal to $n\,|\mathbf{H}_{\alpha\beta}|$ where n is the number of pairs of such irreducible subspaces. n is sometimes called the *intertwining number* of Ω_α and Ω_β under $\mathbf{H}_{\alpha\beta}$. In this case $n = 0$. But $\mathbf{H}_{\alpha\beta}$ is a subgroup of $\mathbf{K}_\alpha \cap \mathbf{K}_\beta$ so if Ω_α and Ω_β have no pairs of irreducible subspaces under $\mathbf{H}_{\alpha\beta}$ that generate the same rep, there certainly cannot be any under $\mathbf{K}_\alpha \cap \mathbf{K}_\beta$. Hence the intertwining number of Ω_α and Ω_β under $\mathbf{K}_\alpha \cap \mathbf{K}_\beta$ is also zero and so if (4.1.13) holds then so does (4.1.6). The important application of this corollary is when $\mathbf{K}_\alpha$, $\alpha = 1$ to $|\mathbf{G}|/|\mathbf{K}_1|$, are the little groups of a star and when $\mathbf{H}_{\alpha\beta}$ for all pairs α, β is taken to be the translation group **T** of the space group which, being a subgroup of all $\mathbf{K}_\alpha$, is a subgroup of all $\mathbf{K}_\alpha \cap \mathbf{K}_\beta$. Now different members of a star generate mutually inequivalent representations of **T** (the whole point of a star being that no two $\mathbf{k}_\alpha$ are equivalent) so that in this case all sums (4.1.6) vanish. This ensures that $\Gamma_p^{\mathbf{k}} \uparrow \mathbf{G}$, the space group rep induced out of a small rep of the little group, is indeed irreducible.

THEOREM 4.1.5. The Frobenius reciprocity theorem. *Let* Γ *be a rep of* **G**, *character* $\chi^\Gamma(g)$; *then the number of times that* Γ *appears in the decomposition of* $\Gamma^j \uparrow \mathbf{G}$

into irreducible representations of **G** *is equal to the number of times the rep* Γ^j *appears in* $\Gamma \downarrow \mathbf{K}_1$. *Here* $\Gamma \downarrow \mathbf{K}_1$ *denotes the restriction of* Γ *to elements of* $\mathbf{K}_1$ *and is commonly called the representation of* $\mathbf{K}_1$ subduced *by* Γ; *since* $\mathbf{K}_1$ *is a subgroup of* **G** *it is clear that* $\Gamma \downarrow \mathbf{K}_1$ *is a representation of* $\mathbf{K}_1$ *of the same dimension as* Γ.

The proof of this theorem is as follows. The number of times Γ appears in $\Gamma^j \uparrow \mathbf{G}$ is equal to the intertwining number $(1/|\mathbf{G}|) \sum_g \chi(g)\chi^\Gamma(g^{-1})$

$$= \frac{1}{|\mathbf{G}|} \sum_g \sum_\alpha \chi_\alpha(g)\chi^\Gamma(g^{-1}) \tag{4.1.14}$$

$$= \frac{1}{|\mathbf{G}|} \sum_a \sum_\alpha \chi^j(k_a)\chi^\Gamma(p_\alpha k_a^{-1} p_\alpha^{-1}) \tag{4.1.15}$$

$$= \frac{1}{|\mathbf{K}_1|} \sum_a \chi^j(k_a)\chi^\Gamma(k_a^{-1}), \tag{4.1.16}$$

which is an intertwining number representing the number of times Γ^j appears in $\Gamma \downarrow \mathbf{K}_1$. In going from (4.1.14) to (4.1.15) we have used (4.1.9), and in going from (4.1.15) to (4.1.16) we have used the fact that k_a^{-1} and $p_\alpha k_a^{-1} p_\alpha^{-1}$ are in the same conjugacy class in **G** and so have the same character $\chi^\Gamma(k_a^{-1})$ for all $\alpha = 1$ to $|\mathbf{G}|/|\mathbf{K}_1|$. The importance of this theorem should be emphasized because it provides the answer to many counting problems. A word of warning is necessary however; the conditions of the theorem that Γ should be irreducible under **G** and Γ^j should be irreducible under $\mathbf{K}_1$ must not be forgotten.

4.2. Groups with an invariant subgroup

In this section we shall consider certain properties of a group **G** with a proper invariant subgroup **T**. (If **T** is Abelian then **G** has a structure similar to that of a space group; see for example equation (3.5.2).)

We expand **G** in terms of left cosets with respect to **T**:

$$\mathbf{G} = \sum_\alpha r_\alpha \mathbf{T}, \tag{4.2.1}$$

where, as usual, the left coset representatives r_α are assumed chosen once for all. We always take $r_1 = e$, the identity of **G**, and we denote the identity of **T** by $t_1 = e$.

If it is possible to choose the left coset representatives to form a group **F** we shall assume that such a choice has been made.‡

‡ It is often possible that of all the choices of left coset representatives r_α, some will form a group and some will not. Thus the point group 432 (O) can be expanded for example as

$$E\mathbf{T} + C_{31}^+\mathbf{T} + C_{31}^-\mathbf{T} + C_{2b}\mathbf{T} + C_{2e}\mathbf{T} + C_{2f}\mathbf{T}$$

where **T** is the point group 222 (D_2) ($= E + C_{2x} + C_{2y} + C_{2z}$); it can also be expanded as

$$E\mathbf{T} + C_{31}^+\mathbf{T} + C_{31}^-\mathbf{T} + C_{2a}\mathbf{T} + C_{2c}\mathbf{T} + C_{2d}\mathbf{T}.$$

In the first case the r_α form the point group 32 (D_3) (admittedly in a non-standard orientation) so that

DEFINITION 4.2.1. If the left coset representatives r_α of a group **G** with respect to an invariant subgroup **T** can be chosen to form a subgroup **F** then **G** is said to be the *semi-direct product* of **T** and **F** and we write $\mathbf{G} = \mathbf{T} \wedge \mathbf{F}$.
(See Definition 1.2.17 for an equivalent definition of semi-direct product.) The contents of this section are true, however, whether **G** is a semi-direct product or not.

We denote an element of **G** by $(r_\alpha t_a)$. (These symbols must not be confused with the Seitz space-group operators.) Their multiplication rule is

$$(r_\alpha t_a)(r_\beta t_b) = (r_{\alpha\beta} t_{\alpha\beta} [t_a]_\beta t_b), \tag{4.2.2}$$

where $[t_a]_\beta = r_\beta^{-1} t_a r_\beta \in \mathbf{T}$ since **T** is invariant, and where

$$r_\alpha r_\beta = r_{\alpha\beta} t_{\alpha\beta}, \tag{4.2.3}$$

the right-hand side of eqn. (4.2.3) being the unique decomposition of the element $r_\alpha r_\beta \in \mathbf{G}$ as the product of one of the chosen left coset representatives $r_{\alpha\beta}$ and an element $t_{\alpha\beta} \in \mathbf{T}$. For a semi-direct product group $t_{\alpha\beta} = t_1$ for all α, β.

The identity of **G** is $(r_1 t_1)$ and from eqns. (4.2.2) and (4.2.3) we obtain the following relations:

$$[t_a]_1 = t_a, \quad [t_1]_\alpha = t_1, \quad r_{\alpha 1} = r_{1\alpha} = r_\alpha, \quad \text{and} \quad t_{\alpha 1} = t_{1\alpha} = t_1; \tag{4.2.4}$$

$$[t_a]_\alpha [t_b]_\alpha = [t_a t_b]_\alpha \tag{4.2.5}$$

and

$$[t_a^{-1}]_\alpha = ([t_a]_\alpha)^{-1}. \tag{4.2.6}$$

Associativity of the group **G** then implies that for all a, α, β, γ (using eqns. (4.2.2) and (4.2.5)),

$$r_{\alpha\beta,\gamma} = r_{\alpha,\beta\gamma} \tag{4.2.7}$$

and

$$t_{\alpha\beta,\gamma} [t_{\alpha\beta}]_\gamma [[t_a]_\beta]_\gamma = t_{\alpha,\beta\gamma} [t_a]_{\beta\gamma} t_{\beta\gamma}. \tag{4.2.8}$$

Note that from the above sets of equations

$$r_{\alpha,\beta 1} = r_{\alpha\beta,1} = r_{\alpha\beta} = r_{1,\alpha\beta} = r_{1\alpha,\beta} = r_{\alpha,1\beta} = r_{\alpha 1,\beta} \tag{4.2.9}$$

but that, although

$$t_{\alpha,\beta 1} = t_{1\alpha,\beta} = t_{\alpha,1\beta} = t_{\alpha 1,\beta} = t_{\alpha\beta}, \qquad t_{\alpha\beta,1} = t_{1,\alpha\beta} = t_1. \tag{4.2.10}$$

By putting $a = 1$ in eqn. (4.2.8) we obtain an important relation for all α, β, γ that

$$t_{\alpha\beta,\gamma} [t_{\alpha\beta}]_\gamma = t_{\alpha,\beta\gamma} t_{\beta\gamma}. \tag{4.2.11}$$

$432 = 32 \wedge 222$ ($O = D_3 \wedge D_2$), but in the second case the r_α do not form a group, since for example $C_{31}^- C_{2a} = C_{4x}^+$ and the closure property is not satisfied. The point here is that if the coset representatives *can* be chosen to form a group then they *must* be so chosen.

The following points are worth noticing. The inverse of $(r_1 t_a)$ is $(r_1 t_a^{-1})$. Given r_α there exists a unique coset representative $r_{\bar{\alpha}}$ such that $r_{\alpha\bar{\alpha}} = r_{\bar{\alpha}\alpha} = r_1$; the inverse of $(r_\alpha t_1)$ is $(r_{\bar{\alpha}} t_{\alpha\bar{\alpha}}^{-1})$,

$$t_{\bar{\alpha}\alpha} = [t_{\alpha\bar{\alpha}}]_\alpha, \tag{4.2.12}$$

and

$$[[t_a]_\alpha]_{\bar{\alpha}} = t_{\alpha\bar{\alpha}}^{-1} t_a t_{\alpha\bar{\alpha}}. \tag{4.2.13}$$

And finally, it is easy to check that

$$r_\alpha t_a r_\alpha^{-1} = t_{\alpha\bar{\alpha}} [t_a]_{\bar{\alpha}} t_{\alpha\bar{\alpha}}^{-1}. \tag{4.2.14}$$

The right-hand side of eqn. (4.2.14) we take to be the definition of $[t_a]_{\alpha^{-1}}$, which is not otherwise defined since r_α^{-1} is not necessarily one of the chosen coset representatives.

THEOREM 4.2.1. *The inverse of* $(r_\alpha t_a)$ *is* $(r_{\bar{\alpha}}[t_a^{-1}]_{\bar{\alpha}} t_{\alpha\bar{\alpha}}^{-1})$.

For

$$(r_\alpha t_a)(r_{\bar{\alpha}}[t_a^{-1}]_{\bar{\alpha}} t_{\alpha\bar{\alpha}}^{-1}) = (r_{\alpha\bar{\alpha}} t_{\alpha\bar{\alpha}} [t_a]_{\bar{\alpha}} [t_a^{-1}]_{\bar{\alpha}} t_{\alpha\bar{\alpha}}^{-1}),$$

which as a result of eqn. (4.2.5) and the fact that $r_{\alpha\bar{\alpha}} = r_1$ is equal to the identity. We leave it as an exercise to the reader to verify by using eqns. (4.2.12) and (4.2.13) that $(r_{\bar{\alpha}}[t_a^{-1}]_{\bar{\alpha}} t_{\alpha\bar{\alpha}}^{-1})$ is also the left inverse of $(r_\alpha t_a)$.

THEOREM 4.2.2. *The suffices of the coset representatives form a group of order* $|\mathbf{G}|/|\mathbf{T}|$, *which we denote by* **R**. *It is, of course, isomorphic to the factor group* **G**/**T**.

The multiplication rule in **R** is $\alpha\beta = \gamma$ if $r_\alpha r_\beta \in r_\gamma \mathbf{T}$. Closure follows by construction, associativity from eqn. (4.2.7), the identity is 1 (see eqns. (4.2.4)) and the inverse of α is $\bar{\alpha}$. The theorem is in fact nothing but another demonstration of the existence of the factor group. Note that if $\mathbf{G} = \mathbf{T} \wedge \mathbf{F}$ then **F**, **R**, and **G**/**T** are mutually isomorphic, the correspondence being $r_\alpha \leftrightarrow \alpha \leftrightarrow r_\alpha \mathbf{T}$.

We now particularize the results of section 4.1 to the present case in which the subgroup is invariant. To do this we identify $\mathbf{K}_1$, p_α, and Γ^j of section 4.1 with the symbols **T**, r_α, and D^i in this section.

D^i is a rep of **T** of dimension d_i with basis $\langle\phi_r|$ so that (see eqn. (4.1.2))

$$t_a \phi_r = \sum_p \phi_p \mathbf{D}^i(t_a)_{pr}. \tag{4.2.15}$$

The functions $\phi_{\alpha p} = r_\alpha \phi_p$ ($\alpha = 1$ to $|\mathbf{G}|/|\mathbf{T}|$, $p = 1$ to d_i) span a vector space Ω which is invariant under **G** (see Theorem 4.1.2). The subgroup $\mathbf{T}_\alpha = (r_\alpha t_a r_\alpha^{-1};\ t_a \in \mathbf{T})$ is in this case identical with **T**, since **T** is invariant. Theorem (4.1.3) then gives us the following important result: Ω_α is irreducible under **T**. Let us see how the rep D^i_α of **T** generated by Ω_α is related to D^i. We have

$$t_a \phi_{\alpha r} = t_a r_\alpha \phi_r = r_\alpha [t_a]_\alpha \phi_r = \sum_p \phi_{\alpha p} \mathbf{D}^i([t_a]_\alpha)_{pr}. \tag{4.2.16}$$

Equation (4.2.16) implies that

$$\mathbf{D}_\alpha^i(t_a) = \mathbf{D}^i([t_a]_\alpha). \tag{4.2.17}$$

Since this equation holds for all $t_a \in \mathbf{T}$, D_α^i is said to be *conjugate* to D^i under $\mathbf{G}$. The set of representations D_α^i for $\alpha = 1$ to $|\mathbf{G}|/|\mathbf{T}|$ forms a *set of conjugate representations.* Two conjugate representations may be equivalent. This does not, however, follow in any way from eqn. (4.2.17) for although t_a and $[t_a]_\alpha$ are in the same class in $\mathbf{G}$ they are not necessarily in the same class in $\mathbf{T}$.

A maximal set of mutually inequivalent representations chosen from a set of conjugate representations is called a *star* (or by some authors an *orbit*). It is quite easy to see that this fits in with the definition already applied to space groups. If we identify t_a with a translation $\{E \mid \mathbf{t}\}$ and r_α with an operation $\{R \mid \mathbf{v}\} \in \mathbf{G}$, then

$$[t_a]_\alpha \to \{R \mid \mathbf{v}\}^{-1}\{E \mid \mathbf{t}\}\{R \mid \mathbf{v}\} = \{E \mid R^{-1}\mathbf{t}\}. \tag{4.2.18}$$

If we identify D^i with $\Delta^{\mathbf{k}}$ then from eqns. (4.2.17), (4.2.18), (3.4.3), and (3.6.2)

$$\begin{aligned} \mathbf{D}_\alpha^i(t_a) &= \mathbf{D}^i([t_a]_\alpha) \to \boldsymbol{\Delta}^{\mathbf{k}}(\{E \mid R^{-1}\mathbf{t}\}) \\ &= \exp(-i\mathbf{k}.R^{-1}\mathbf{t})\mathbf{1} = \exp(-iR\mathbf{k}.\mathbf{t})\mathbf{1} \\ &= \boldsymbol{\Delta}^{R\mathbf{k}}(\{E \mid \mathbf{t}\}). \end{aligned} \tag{4.2.19}$$

The equivalence between the two definitions is now brought about by the fact that if $R\mathbf{k} \equiv \mathbf{k}$ then $\Delta^{R\mathbf{k}} \equiv \Delta^{\mathbf{k}}$, so that to choose a maximal set of mutually inequivalent representations from a set of conjugate representations of the translation group $\mathbf{T}$ is to form a star; whether we think of the star as consisting of a set of mutually inequivalent conjugate reps $\Delta^{\mathbf{k}}$ or as the corresponding set of vectors $\mathbf{k}$ does not matter, provided the reason for the correspondence is not obscured.

The set of suffices $\gamma \in \mathbf{R}$ for which $D_\gamma^i \equiv D^i$ forms a group $\overline{\mathbf{K}}^i$ called the *little co-group* of D^i in $\mathbf{G}$. The correspondence with space group terminology (see Definition 3.7.3) is immediate; there the little co-group $\overline{\mathbf{G}}^{\mathbf{k}}$ of $\mathbf{k}$ in $\mathbf{G}$ is defined to consist of all $S \in \mathbf{F}$ such that $S\mathbf{k} \equiv \mathbf{k}$, that is $\Delta^{S\mathbf{k}} \equiv \Delta^{\mathbf{k}}$. To make the correspondence entirely complete note that the group $\overline{\mathbf{K}}_\alpha^i = \{\alpha\gamma\bar{\alpha};\ \gamma \in \overline{\mathbf{K}}^i\}$ is the little co-group of D_α^i in $\mathbf{G}$. This follows because, if we denote $\mathbf{D}_\alpha^i(t_a) = \mathbf{D}^i([t_a]_\alpha)$ by $\boldsymbol{\Delta}^i(t_a)$ then, for all γ,

$$\begin{aligned} \boldsymbol{\Delta}_{\alpha\gamma\bar{\alpha}}^i(t_a) &= \boldsymbol{\Delta}^i([t_a]_{\alpha\gamma\bar{\alpha}}) = \mathbf{D}^i([[t_a]_{\alpha\gamma\bar{\alpha}}]_\alpha) \\ &= \mathbf{D}^i([t_a]_{\alpha\gamma}) = \mathbf{D}_\gamma^i([t_a]_\alpha) = \mathbf{D}^i([t_a]_\alpha) = \mathbf{D}_\alpha^i(t_a) = \boldsymbol{\Delta}^i(t_a). \end{aligned}$$

In space group theory the corresponding result is that $\overline{\mathbf{G}}^{R\mathbf{k}} = \{RSR^{-1};\ S \in \overline{\mathbf{G}}^{\mathbf{k}}\}$.

If all the members of a set of conjugate reps are inequivalent then the little co-group consists of 1 alone. This corresponds to $\mathbf{k}$ being a general point of the Brillouin zone.

The group $\mathbf{K}^i$ defined by the equation

$$\mathbf{K}^i = \sum_{\gamma \in \overline{\mathbf{K}}^i} r_\gamma \mathbf{T} \tag{4.2.20}$$

where γ runs over the little co-group, is called the *little group* of i in **G**. The little co-group is isomorphic with the factor group $\mathbf{K}^i/\mathbf{T}$.

We now investigate the induced representation $D^i \uparrow \mathbf{G}$ determined by the space Ω. The equation corresponding to (4.1.4) is

$$(r_\alpha t_a)\phi_{\tau r} = \sum_p \phi_{\sigma p}\mathbf{D}^i(t_{\alpha\tau}[t_a]_\tau)_{pr} \tag{4.2.21}$$

where $\sigma = \alpha\tau$.

The equation corresponding to (4.1.9) is

$$\chi(r_\alpha t_a) = \delta_{\alpha 1}\sum_\tau \chi_\tau^i(t_a). \tag{4.2.22}$$

That is, only the elements of **T** have non-zero character in $D^i \uparrow \mathbf{G}$ and in **T** each element has a character that is a sum of its characters taken over a complete set of reps conjugate to D^i.

Since **T** is invariant, Corollary (i) of Johnston's irreducibility criterion (Theorem 4.1.4) applies and asserts that if no two members of the set of reps conjugate to D^i are equivalent then $D^i \uparrow \mathbf{G}$ is irreducible. And so for a space group **G** if **k** is a general point of the Brillouin zone $\Delta^{\mathbf{k}} \uparrow \mathbf{G}$ is irreducible.

From eqn. (4.2.22) we see that

$$(D^i \uparrow \mathbf{G}) \downarrow \mathbf{T} = \sum_\tau D_\tau^i. \tag{4.2.23}$$

In the case when $D^i \uparrow \mathbf{G}$ is irreducible all D_τ^i, $\tau = 1$ to $|\mathbf{G}|/|\mathbf{T}|$, are inequivalent and the right-hand side of eqn. (4.2.23) is equal to the star containing D^i (the star of i, for short). And by the Frobenius reciprocity Theorem 4.1.5 it follows that $D_\tau^i \uparrow \mathbf{G}$ is equivalent to $D^i \uparrow \mathbf{G}$, for by the theorem $D_\tau^i \uparrow \mathbf{G}$ contains $D^i \uparrow \mathbf{G}$ once and since both are irreducible and are of the same dimension they must be equivalent.

In the general case when $D^i \uparrow \mathbf{G}$ is reducible the little co-group consists of elements other than 1. From eqn. (4.2.20) we see that it consists of $|\mathbf{K}^i|/|\mathbf{T}|$ elements; this means that in the sum $\sum_\tau D_\tau^i$ the star of i will appear $|\mathbf{K}^i|/|\mathbf{T}|$ times.

So far we have shown that when the little co-group consists of only one element the representation $D^i \uparrow \mathbf{G}$ is irreducible and that it can be induced not only from D^i but from any member of the star of i. Such an irreducible representation when restricted to **T** contains each member of the star once and once only. Not every rep of **G** can be obtained in this way because there exist sets of conjugate representations of **T** which contain equivalent reps. This latter case holds when $\overline{\mathbf{K}}^i$ consists of more than one element. In space-group theory this corresponds to the cases in which **k** is a point of symmetry or lies on a line or plane of symmetry. In the following section we give the theory to cover such cases and which, by the way, embraces the present case as a particularly simple example.

4.3 The theory of little groups

As in section 4.2 **G** is a group with an invariant subgroup **T**. The aim of this section is to establish rigorously the scheme for obtaining all the reps of **G**, the same scheme that was outlined for space groups in Chapter 3, section 3.7. The scheme is as follows.

(i) Distribute the reps of **T** into stars i.
(ii) Select arbitrarily one rep D^i from the star i.
(iii) Determine the little group $\mathbf{K}^i$ of i in **G**; we remember that $\mathbf{K}^i = \sum_\gamma r_\gamma \mathbf{T}$ where the sum over γ ranges over the little co-group $\bar{\mathbf{K}}^i$ of elements such that $D^i_\gamma \equiv D^i$.
(iv) Expand **G** in terms of left cosets with respect to $\mathbf{K}^i$; as in section 4.1 we write $\mathbf{G} = \sum_\alpha p_\alpha \mathbf{K}^i$ where the p_α are supposed chosen once for all, and by construction of the little group they have the property that D^i_α, $\alpha = 1$ to $|\mathbf{G}|/|\mathbf{K}^i|$, are mutually inequivalent and form the star of D^i.
(v) Find the reps of $\mathbf{K}^i$ and denote those by Γ^i_j which, when restricted to **T**, yield multiples of D^i. These are called the small reps of $\mathbf{K}^i$.
(vi) Determine $\Gamma^i_j \uparrow \mathbf{G}$ for each j. This is to be done by the method established in section 4.1 identifying $\mathbf{K}^i$ here with the group $\mathbf{K}_1$ of that section and Γ^i_j here with the rep Γ^j of that section.
(vii) Repeat (ii) to (vi) for each star i.

Steps (i) to (iv) of the above scheme have already been discussed in detail. In order to justify step (v) we prove the following lemma.

LEMMA 4.3.1. *If Γ^i_j is a rep of $\mathbf{K}^i$ such that $\Gamma^i_j \downarrow \mathbf{T}$ contains D^i then it contains only D^i but perhaps more than once.*

First note, from the Frobenius reciprocity Theorem 4.1.5 that $\Gamma^i_j \downarrow \mathbf{T}$ contains D^i if and only if $D^i \uparrow \mathbf{K}^i$ contains Γ^i_j.

Suppose then that $D^i \uparrow \mathbf{K}^i = \sum_j \lambda^i_j \Gamma^i_j = \Gamma$. The theorem then states that $\Gamma^i_j \downarrow \mathbf{T}$ contains D^i just λ^i_j times. To prove that it contains nothing else it is sufficient to remember that $\mathbf{K}^i = \sum_\gamma r_\gamma \mathbf{T}$ where $\gamma \in \bar{\mathbf{K}}^i$. Now **T** is invariant in $\mathbf{K}^i$ so we may use the result expressed by eqn. (4.2.23) that $\Gamma \downarrow \mathbf{T} = \sum_\gamma D^i_\gamma$. But $\gamma \in \bar{\mathbf{K}}^i$ so that $D^i_\gamma \equiv D^i$. Hence $\Gamma \downarrow \mathbf{T} = (|\mathbf{K}^i|/|\mathbf{T}|)D^i$. In particular, $\Gamma^i_j \downarrow \mathbf{T} = \lambda^i_j D^i$, which establishes the lemma.

Incidentally, we have proved further that $(|\mathbf{K}^i|/|\mathbf{T}|)D^i = \Gamma \downarrow \mathbf{T} = (\sum_j \lambda^i_j \Gamma^i_j) \downarrow \mathbf{T} = \sum_j (\lambda^i_j)^2 D^i$ so that $\sum_j (\lambda^i_j)^2 = |\mathbf{K}^i|/|\mathbf{T}| = |\bar{\mathbf{K}}^i|$, the order of the little co-group of i in **G**. This very important fact means that the sum of the squares of the dimensions of the small reps of $\mathbf{K}^i$ is $\sum_j (\lambda^i_j d_i)^2 = d_i^2\, |\bar{\mathbf{K}}^i|$, the square of the dimension of D^i times the order of the little co-group.

THEOREM 4.3.1. *The representations $\Gamma^i_j \uparrow \mathbf{G}$ (all i, all j) are all irreducible, and all the reps of **G** are obtained in this way once and once only.*

Note, of course, that for those stars i for which $\bar{\mathbf{K}}^i$ consists of the identity alone, $\mathbf{K}^i = \mathbf{T}$; the only $\Gamma^i_j = D^i$ and $D^i \uparrow \mathbf{G}$ is irreducible so that Theorem 4.3.1 will include the final result of section 4.2 as a trivial case.

We first prove that $\Gamma^i_j \uparrow \mathbf{G}$ is irreducible. We remember that $\mathbf{G} = \sum_\alpha p_\alpha \mathbf{K}^i$ where the characteristic property of the p_α is that the set D^i_α forms the star of D^i, so that the members of the set D^i_α are mutually inequivalent.

The irreducibility of $\Gamma^i_j \uparrow \mathbf{G}$ follows at once from Corollary (ii) to Theorem 4.1.4. We take $\mathbf{H}_{\alpha\beta} = \mathbf{T}$ to be the common subgroup of $\mathbf{K}^i_\alpha \cap \mathbf{K}^i_\beta$ for all $\alpha \neq \beta$ and the corollary asserts that if

$$\sum_{t_a \in \mathbf{T}} \chi^{i*}_\alpha(t_a)\chi^i_\beta(t_a) = 0, \quad \text{all } \alpha \neq \beta \tag{4.3.1}$$

then $\Gamma^i_j \uparrow \mathbf{G}$ is irreducible. Here $\chi^i_\alpha(t_a)$ is the character of t_a in the representation spanned by the space Ω_α, that is the (reducible) representation $\lambda^i_j D^i_\alpha$. Similarly, $\chi^i_\beta(t_a)$ is the character of t_a in $\lambda^i_j D^i_\beta$. Since D^i_α and D^i_β are inequivalent the sum on the left-hand side of eqn. (4.3.1) vanishes and $\Gamma^i_j \uparrow \mathbf{G}$ is therefore irreducible.

It remains to prove that by repeating the induction process for each star i and all small reps j we obtain all reps of $\mathbf{G}$ once and once only. There are two things to take into account; first, the possibility that by doing this we miss some rep of $\mathbf{G}$ altogether, and secondly the possibility that we count some of the reps more than once. To counter the second possibility we must prove that the reps $\Gamma^i_j \uparrow \mathbf{G}$ are mutually inequivalent. To counter the first possibility we must then show that the sums of the squares of their dimensions is equal to $|\mathbf{G}|$, which will prove that we have obtained all the reps of $\mathbf{G}$. In order to prove these facts it is necessary to establish certain lemmas. In the following we shall denote $\Gamma^i_j \uparrow \mathbf{G}$ by Δ^i_j.

LEMMA 4.3.2. Transitivity of subduction

$$(\Delta^i_j \downarrow \mathbf{K}^i) \downarrow \mathbf{T} = \Delta^i_j \downarrow \mathbf{T}. \tag{4.3.2}$$

This follows because both sides consist of an identical set of matrices.

LEMMA 4.3.3. Transitivity of induction

$$(D^i \uparrow \mathbf{K}^i) \uparrow \mathbf{G} \equiv D^i \uparrow \mathbf{G} \tag{4.3.3}$$

where, since $D^i \uparrow \mathbf{K}^i = \sum_j \lambda^i_j \Gamma^i_j$, *the left-hand side of eqn.* (4.3.3) *is taken to mean* $\sum_j \lambda^i_j(\Gamma^i_j \uparrow \mathbf{G}) = \sum_j \lambda^i_j \Delta^i_j$.

From the text following eqn. (4.2.15) we recall that the basis spanning the representation $D^i \uparrow \mathbf{G}$ consists of the functions $r_\alpha \phi_p = \phi_{\alpha p}$, $\alpha = 1$ to $|\mathbf{G}|/|\mathbf{T}|$, $p = 1$ to d_i, where $\langle \phi_p|$ is the basis for D^i and the r_α are the left coset representatives in the decomposition (4.2.1). Since $\mathbf{K}^i = \sum_\gamma r_\gamma \mathbf{T}$, where γ ranges over $\overline{\mathbf{K}}^i$, the basis spanning $D^i \uparrow \mathbf{K}^i$ consists of the functions $r_\gamma \phi_p = \phi_{\gamma p}$, $\gamma = 1$ to $|\mathbf{K}^i|/|\mathbf{T}|$, $p = 1$ to d_i. Further, since $\mathbf{G} = \sum_\beta p_\beta \mathbf{K}^i$, $\beta = 1$ to $|\mathbf{G}|/|\mathbf{K}^i|$, the basis of the representation $(D^i \uparrow \mathbf{K}^i) \uparrow \mathbf{G}$ consists of a linear combination of the $d_i|\mathbf{G}|/|\mathbf{T}|$ functions $p_\beta \phi_{\gamma p}$. Now the various p_β can be chosen so that the set $\{p_\beta r_\gamma\} = \{r_\alpha\}$. In this way the basis $\langle p_\beta \phi_{\gamma p}|$ is identical with the basis $\langle \phi_{\alpha p}|$. The essential point is that however the coset representatives are chosen the representations of **G** on the two sides of eqn. (4.3.3.) both derive from the space Ω that is spanned by the $d_i|\mathbf{G}|/|\mathbf{T}|$ functions $\phi_{\alpha p}$. A different choice of coset representatives will in neither case take us out of the space Ω. The two representations are therefore equivalent. An immediate corollary is that

$$D^i \uparrow \mathbf{G} \equiv \sum_j \lambda^i_j \Delta^i_j. \tag{4.3.4}$$

But Δ^i_j is irreducible in **G**. Therefore, from Frobenius' reciprocity theorem $\Delta^i_j \downarrow \mathbf{T}$ contains D^i just λ^i_j times.

LEMMA 4.3.4

$$\Delta^i_j \downarrow \mathbf{T} = \sum_\alpha \lambda^i_j D^i_\alpha \tag{4.3.5}$$

in which the set D^i_α forms the star of D^i.

We have shown above that $\Delta^i_j \downarrow \mathbf{T}$ contains D^i just λ^i_j times and since Δ^i_j is an induced representation from $\mathbf{K}^i$ into **G** it follows that it must contain all members of the star of D^i an equal number of times.

We can now prove very easily that $\Gamma^i_j \uparrow \mathbf{G} = \Delta^i_j$, all i, all j are mutually inequivalent. First, suppose we are given $\Delta^{i_1}_{j_1}$ and $\Delta^{i_2}_{j_2}$ with $i_1 \neq i_2$ then from Lemma 4.3.4 and from the fact that pairs of reps taken from two distinct stars are mutually inequivalent it follows that $\Delta^{i_1}_{j_1} \downarrow \mathbf{T}$ and $\Delta^{i_2}_{j_2} \downarrow \mathbf{T}$ have no irreducible components in common. The same argument that was used in Corollary (ii) to Theorem 4.1.4 can now be applied. **T** is a subgroup of **G**: hence $\Delta^{i_1}_{j_1}$ and $\Delta^{i_2}_{j_2}$ have no irreducible components in common and they are therefore inequivalent. Secondly suppose that $i_1 = i_2 = i$ but that $j_1 \neq j_2$, then since $\Gamma^i_{j_1} \uparrow \mathbf{G} = \Delta^i_{j_1}$ is irreducible it follows from Frobenius' reciprocity theorem that $\Delta^i_{j_1} \downarrow \mathbf{K}^i$ contains $\Gamma^i_{j_1}$ just once. Similarly $\Delta^i_{j_2} \downarrow \mathbf{K}^i$ contains $\Gamma^i_{j_2}$ just once. Now if $\Delta^i_{j_1} \equiv \Delta^i_{j_2}$ then $\Delta^i_{j_1} \downarrow \mathbf{K}^i$ must contain both $\Gamma^i_{j_1}$ and $\Gamma^i_{j_2}$ which by hypothesis are inequivalent. Using Lemma 4.3.2 we see that $\Delta^i_{j_1} \downarrow \mathbf{T}$ must contain both $\Gamma^i_{j_1} \downarrow \mathbf{T}$ and $\Gamma^i_{j_2} \downarrow \mathbf{T}$ and must therefore contain D^i at least $(\lambda^i_{j_1} + \lambda^i_{j_2})$ times. This is a contradiction to Lemma 4.3.4. And so the inequivalence of the Δ^i_j is finally established. To see now that we have got all the reps of **G** we shall evaluate the sums of the squares of the dimensions of the Δ^i_j. From Lemma 4.3.4 we see that dim $\Delta^i_j = q_i \lambda^i_j d_i$ where q_i is the number of elements in the star of i. But $q_i = |\mathbf{G}|/|\mathbf{K}^i|$. Therefore

$$\sum_i \sum_j (\dim \Delta^i_j)^2 = \sum_i \sum_j (|\mathbf{G}|/|\mathbf{K}^i|)^2 (\lambda^i_j)^2 d_i^2 = \sum_i \frac{|\mathbf{G}|^2}{|\mathbf{K}^i|\,|\mathbf{T}|} d_i^2$$

$$= \frac{|\mathbf{G}|}{|\mathbf{T}|} \sum_i q_i d_i^2 = \frac{|\mathbf{G}|}{|\mathbf{T}|} |\mathbf{T}| = |\mathbf{G}|.$$

This completes the proof of Theorem 4.3.1.

4.4 The small representations of little groups

We have established that, in the notation of the previous section, each rep of **G** is to be obtained from some rep Γ^i_j of $\mathbf{K}^i$ by the method of induction. Indeed each rep of **G** is characterized by two labels: i, the label of the star, and j, the label of the small rep. The method described in section 4.3 forms a practical scheme provided it is possible to obtain the reps of $\mathbf{K}^i$. However we often meet the same difficulty, as exemplified by space-group theory, that $\mathbf{K}^i$ is not necessarily easy to analyse. Indeed for some i we know that $\mathbf{K}^i = \mathbf{G}$, the whole group. The purpose of this section is to discuss a method of obtaining the small reps Γ^i_j of $\mathbf{K}^i$ without necessarily obtaining all reps of $\mathbf{K}^i$: for it must be realized that we are interested only in those reps Γ for which $\Gamma = \Gamma^i_j$ where $\Gamma^i_j \downarrow \mathbf{T} \equiv \lambda^i_j D^i$, some multiple of D^i.

With a suitable choice of basis we can therefore write

$$\mathbf{\Gamma}^i_j(t_a) = \begin{pmatrix} \mathbf{D}^i(t_a) & 0 & \cdots \\ 0 & \mathbf{D}^i(t_a) & \cdots \\ \vdots & \vdots & \end{pmatrix}, \tag{4.4.1}$$

the direct sum of λ^i_j matrices $\mathbf{D}^i(t_a)$. It is sufficient therefore to obtain matrix representatives $\mathbf{\Gamma}^i_j(r_\gamma)$ where $\gamma \in \bar{\mathbf{K}}^i$. For then matrix representatives of all elements of $\mathbf{K}^i$ can be obtained by forming all products of the form $\mathbf{\Gamma}^i_j(r_\gamma)\mathbf{\Gamma}^i_j(t_a)$.

Case (i). **T** *Abelian*

This covers the case of space groups and indeed the case of all finite groups with a proper invariant subgroup that is a direct product of cyclic groups. Then D^i is 1-dimensional and the dimension of Γ^i_j is equal to λ^i_j. Furthermore $\mathbf{\Gamma}^i_j(t_a)$ is equal to the λ^i_j by λ^i_j matrix $\mathbf{D}^i(t_a)\mathbf{1}$, where $\mathbf{D}^i(t_a)$ is a known complex number of unit modulus. Now Γ^i_j is to be irreducible and the matrices for **T** are *scalar* matrices. This means that the set of matrices $\mathbf{\Gamma}^i_j(r_\gamma)$ must be irreducible.

We now recall that the following relations must hold for all $\beta, \gamma, \delta \in \bar{\mathbf{K}}^i$:

$$r_\gamma r_\delta = r_{\gamma\delta} t_{\gamma\delta}, \tag{4.4.2}$$

$$t_{\beta\gamma,\delta}[t_{\beta\gamma}]_\delta = t_{\beta,\gamma\delta} t_{\gamma\delta} \tag{4.4.3}$$

(see eqns. (4.2.3) and (4.2.11)). In the rep Γ^i_j it follows therefore that we must have

$$\mathbf{\Gamma}^i_j(r_\gamma)\mathbf{\Gamma}^i_j(r_\delta) = \mathbf{\Gamma}^i_j(r_{\gamma\delta})\mathbf{\Gamma}^i_j(t_{\gamma\delta}) \tag{4.4.4}$$

$$= \mathbf{D}^i(t_{\gamma\delta})\mathbf{\Gamma}^i_j(r_{\gamma\delta}) \tag{4.4.5}$$

and

$$\mathbf{D}^i(t_{\beta\gamma,\delta})\mathbf{D}^i([t_{\beta\gamma}]_\delta) = \mathbf{D}^i(t_{\beta,\gamma\delta})\mathbf{D}^i(t_{\gamma\delta}). \tag{4.4.6}$$

But $D^i = D^i_\delta$, since D^i is 1-dimensional and $\delta \in \bar{\mathbf{K}}^i$ (equivalent 1-dimensional matrices being equal). Hence eqn. (4.4.6) becomes

$$\mathbf{D}^i(t_{\beta\gamma,\delta})\mathbf{D}^i(t_{\beta\gamma}) = \mathbf{D}^i(t_{\beta,\gamma\delta})\mathbf{D}^i(t_{\gamma\delta}). \tag{4.4.7}$$

Consider now the group $\bar{\mathbf{K}}^i$. Set $\mathbf{\Gamma}^i_j(r_\gamma) = \mathbf{\Lambda}^i_j(\gamma)$ and $\mathbf{D}^i(t_{\beta\gamma}) = \mu^i(\beta, \gamma)$, then eqns. (4.4.5) and (4.4.7) become

$$\mathbf{\Lambda}^i_j(\gamma)\mathbf{\Lambda}^i_j(\delta) = \mu^i(\gamma, \delta)\mathbf{\Lambda}^i_j(\gamma\delta) \tag{4.4.8}$$

and

$$\mu^i(\beta\gamma, \delta)\mu^i(\beta, \gamma) = \mu^i(\beta, \gamma\delta)\mu^i(\gamma, \delta). \tag{4.4.9}$$

Equations (4.4.8) and (4.4.9) imply that the complex numbers μ^i are a factor system for a projective representation of $\bar{\mathbf{K}}^i$. Moreover this projective representation is of dimension λ^i_j and is irreducible because $\mathbf{\Gamma}^i_j(r_\gamma) = \mathbf{\Lambda}^i_j(\gamma)$ is to form an irreducible set of matrices. Conversely if we have a projective rep Λ^i_j with factor system μ^i we obtain a small rep Γ^i_j by setting $\mathbf{\Gamma}^i_j(r_\gamma) = \mathbf{\Lambda}^i_j(\gamma)$. We thus obtain all allowed Γ^i_j of $\mathbf{K}^i$ by determining all non-equivalent projective reps of the little co-group $\bar{\mathbf{K}}^i$ with a known factor system $\mu^i(\beta, \gamma) = \mathbf{D}^i(t_{\beta\gamma})$.

Incidentally we have verified that, since $\sum_j (\lambda^i_j)^2 = |\bar{\mathbf{K}}^i|$, the sum of the squares of the dimensions of these projective reps of $\bar{\mathbf{K}}^i$ is equal to the order of $\bar{\mathbf{K}}^i$.

A simplification occurs when the elements r_γ can be chosen to form a group. For then $t_{\gamma\delta} = t_1 = e$ for all $\gamma, \delta \in \bar{\mathbf{K}}^i$. The factor system then consists of numbers all equal to one and the projective reps become the ordinary vector reps. This always happens for semi-direct product groups but (as we saw in the previous chapter) it also happens for certain i when $\mathbf{G}$ is not expressible as a semi-direct product group.

Case (ii). **T** *non-Abelian*

If D^i is a 1-dimensional rep then the theory is precisely as in case (i). Indeed the only place in case (i), in which the fact that **T** was Abelian was used, was at the beginning where we remarked that D^i in that case was necessarily 1-dimensional. If D^i is of dimension $d_i > 1$ then the following facts hold.

(i) Γ^i_j is of dimension $d_i\lambda^i_j$ where $\Gamma^i_j \downarrow \mathbf{T} = \lambda^i_j D^i$.

(ii) The matrix form (4.4.1) for the representative $\mathbf{\Gamma}^i_j(t_a)$ still holds, but $\mathbf{D}^i(t_a)$ is now a d_i by d_i matrix and not a complex number.

(iii) Equations (4.4.5) and (4.4.7) are replaced by

$$\mathbf{\Gamma}^i_j(r_\gamma)\mathbf{\Gamma}^i_j(r_\delta) = \mathbf{\Gamma}^i_j(r_{\gamma\delta})\mathbf{\Gamma}^i_j(t_{\gamma\delta}) \tag{4.4.4}$$

and

$$\mathbf{D}^i(t_{\beta\gamma,\delta})\mathbf{D}^i_\delta(t_{\beta\gamma}) = \mathbf{D}^i(t_{\beta,\gamma\delta})\mathbf{D}^i(t_{\gamma\delta}), \tag{4.4.10}$$

where $\mathbf{\Gamma}^i_j(t_{\gamma\delta})$ is given by eqn. (4.4.1), and where $\mathbf{D}^i_\delta(t_{\beta\gamma}) = \mathbf{D}^i(r_\delta^{-1}t_{\beta\gamma}r_\delta)$. Note now that although $D^i_\delta \equiv D^i$ the equivalence cannot now be sharpened to an equality.

The problem from here does not have an easy solution. If **G** is a semi-direct product group then $t_{\gamma\delta} = t_1$ for all γ, δ and eqn. (4.4.4) reduces to

$$\mathbf{\Gamma}_j^i(r_\gamma)\mathbf{\Gamma}_j^i(r_\delta) = \mathbf{\Gamma}_j^i(r_{\gamma\delta}). \tag{4.4.11}$$

That is, Γ_j^i is a vector representation of the little co-group of dimension $d_i\lambda_j^i$. Since $\sum_j d_i^2\, \lambda_j^{i2} = d_i^2|\mathbf{\bar{K}}^i| > |\mathbf{\bar{K}}^i|$ these representations are in general reducible under $\mathbf{\bar{K}}^i$. A procedure is given by Mackey (1958) for determining Γ_j^i in this case. This procedure is rather complicated and since the theory of space groups rests entirely on case (i) we omit the details. Case (ii), when **G** is a semi-direct product, is also discussed by Jansen and Boon (1967).

4.5 The point groups as semi-direct products

The method described in the previous sections of this chapter is very general and is suitable for the classification and reduction of groups other than space groups. In particular it brings out clearly the relations, which might appropriately be called genealogical, between the symmetries of two point groups, one of which is a subgroup of the other. These relationships may be illustrated diagrammatically as shown in Fig. 4.1. The purpose of this section is therefore twofold. First we shall discuss the representations of point groups in terms of those of their invariant subgroups and we shall see how in this case it is possible to make certain simplifications of the general theory of sections 4.1–4.4. Secondly, by doing this for these relatively small finite groups, we shall be giving some simple examples of the use of the methods of sections 4.1–4.4. When the reader has practised on one or two simple examples of this sort the scheme outlined at the beginning of section 4.3 should become quite clear and the harder examples provided by the space groups should be easier to appreciate.

It is a fact that the point groups are *solvable*; that is, they admit of a *composition series*

$$\mathbf{G} = \mathbf{G}_1, \mathbf{G}_2, \ldots, \mathbf{G}_m\ (= E),$$

where $\mathbf{G}_{i+1}$ is a maximal invariant subgroup of $\mathbf{G}_i$ and $\mathbf{G}_i/\mathbf{G}_{i+1}$ is a cyclic group of prime order. (An invariant subgroup is maximal if it is not a subgroup of another proper invariant subgroup of larger order.) Note that this definition of solvability might at first sight seem to be more restrictive than that given by Definition 1.2.16, but the two definitions are easily shown to be equivalent. All groups of order less than 60 are solvable, the icosahedral group being the group of least order that is not solvable. This means that if we take a point group **G** and a maximal invariant subgroup **N** then **G**/**N** is cyclic. It also happens, and this is a matter for direct verification, that the coset representatives of **N** can in all cases be chosen to form a cyclic group **C**

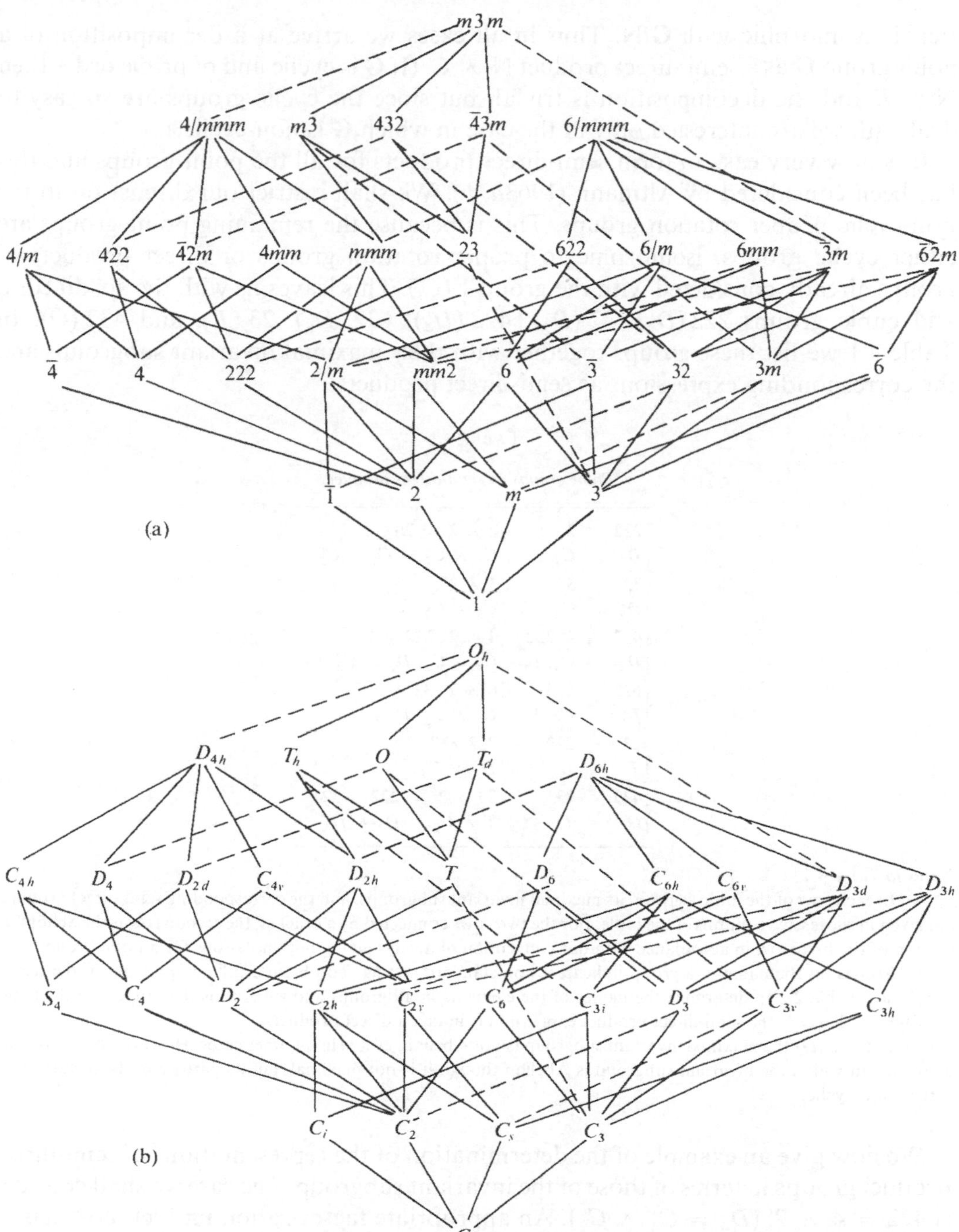

FIG. 4.1. The genealogical relations between the point groups. A continuous line indicates that a subgroup is invariant. (a) international notation. (b) Schönflies notation.

that is isomorphic with **G**/**N**. Thus in all cases we arrive at a decomposition of a point group **G** as a semi-direct product **N** $\wedge$ **C**. (If **G** is cyclic and of prime order then **N** $\equiv E$ and the decomposition is trivial, but since the cyclic groups are so easy to deal with we are interested only in the case in which **G** is non-cyclic.)

It is now very easy to form semi-direct products for all the point groups and this has been considered by Altmann (1963*a*, *b*). We shall restrict our discussion to the non-cyclic proper rotation groups. This is because the remaining point groups are either cyclic groups, isomorphic to proper rotation groups or direct products of groups already considered with the group $\bar{1}$ (C_i). This leaves us with the six dihedral and cubic groups 222 (D_2), 32 (D_3), 422 (D_4), 622 (D_6), 23 (T), and 432 (O). In Table 4.1 we list these groups together with their maximal invariant subgroups and the corresponding expressions as semi-direct products.

TABLE 4.1

Point groups as semi-direct products

222	2	$2 \wedge 2' = 2 \otimes 2'$
D_2	C_2	$C_2 \wedge C_2' = C_2 \otimes C_2'$
32	3	$3 \wedge 2'$
D_3	C_3	$C_3 \wedge C_2'$
422	4, 222	$4 \wedge 2'$, $222 \wedge 2''$
D_4	C_4, D_2	$C_4 \wedge C_2'$, $D_2 \wedge C_2''$
622	6, 32	$6 \wedge 2'$, $32 \wedge 2$
D_6	C_6, D_3	$C_6 \wedge C_2'$, $D_3 \wedge C_2$
23	222	$222 \wedge 3'$
T	D_2	$D_2 \wedge C_3'$
432	23	$23 \wedge 2'' = 222 \wedge 32'$
O	T	$T \wedge C_2'' = D_2 \wedge D_3'$

Notes to Table 4.1

(i) In each row of the table a group, its maximal invariant subgroups, and the corresponding semi-direct products are given in successive columns of the table. For the two rows connected by a bracket, the second row is the Schönflies version of the first row, in accordance with our convention of always giving both notations for a point group.

(ii) Primes on the name of a group indicate a non-standard setting. To obtain a full comprehension the reader should use Table 2.2 to determine the names of the elements in all groups and subgroups that appear in the table.

(iii) For 222 (D_2), the semi-direct product expression is in fact a direct product.

(iv) For 432 (O) the maximal invariant subgroup is not Abelian, as it is in all other cases. However, an expression does exist in which the invariant subgroup is Abelian though it is not maximal, but its partner in the factorization is then non-cyclic.

We now give an example of the determination of the representations of semi-direct product groups in terms of those of the invariant subgroup. The case we shall consider is $422 = 4 \wedge 2'$ ($D_4 = C_4 \wedge C_2'$). An appropriate factorization into left cosets is

$$\begin{gathered} 422 = E.4 + C_{2x}.4 \\ (D_4 = E.C_4 + C_{2x}.C_4). \end{gathered} \tag{4.5.1}$$

The reps of 4 (C_4) are seen from Table 2.2 to be four 1-dimensional reps A, B, ${}^1E$, and ${}^2E$ in which C_{4z}^+ is represented by, 1, -1, $-$i, and i respectively. The conjugacy relations which establish 4 (C_4) as an invariant subgroup are

$$C_{2x}^{-1}EC_{2x} = E,\ C_{2x}^{-1}C_{4z}^+C_{2x} = C_{4z}^-,\ C_{2x}^{-1}C_{2z}C_{2x} = C_{2z},\ \text{and}\ C_{2x}^{-1}C_{4z}^-C_{2x} = C_{4z}^+.$$

The reps of 4 (C_4) separate into three stars containing A alone, B alone, and ${}^1E$ and ${}^2E$ together. The fact that ${}^1E$ and ${}^2E$ come together follows from equations of the form ${}^1\mathbf{E}(C_{2x}^{-1}C_{4z}^+C_{2x}) = {}^1\mathbf{E}(C_{4z}^-) = \mathrm{i} = {}^2\mathbf{E}(C_{4z}^+)$, which imply that ${}^1E$ and ${}^2E$ are conjugate (whereas A and B are self-conjugate).

The little co-groups in the three cases are for stars A and B the group 2′ (C_2') containing E and C_{2x}, and for the remaining star, E, the group 1 (C_1). Since we have a semi-direct product the required reps of the little co-groups are their ordinary reps and these are well known to be, for 2′ (C_2') the 1-dimensional reps Λ_1 and Λ_2 with $\mathbf{\Lambda}_1(C_{2x}) = 1$ and $\mathbf{\Lambda}_2(C_{2x}) = -1$, and for 1 ($C_1$) the trivial 1-dimensional rep Λ with $\mathbf{\Lambda}(E) = 1$. Using Definition 3.7.4 or eqn. (4.2.20) and the factorization of 422 (D_4) in eqn. (4.5.1), it can be seen that the sum of the elements in those cosets that have E and C_{2x} as their coset representatives just makes up the group 422 (D_4). Thus the little group of A is the whole point group 422 (D_4), and the small reps are two in number, Λ_1^A and Λ_2^A. In these reps $\mathbf{\Lambda}_1^A(C_{4z}^+) = \mathbf{\Lambda}_2^A(C_{4z}^+) = \mathbf{A}(C_{4z}^+) = 1$, and $\mathbf{\Lambda}_1^A(C_{2x}) = \mathbf{\Lambda}_1(C_{2x}) = 1$ and $\mathbf{\Lambda}_2^A(C_{2x}) = \mathbf{\Lambda}_2(C_{2x}) = -1$. Since the little group is the whole group the corresponding reps of 422 (D_4) coincide with the small reps. Similarly for the star B there are 2 reps of 422 (D_4) which we denote by Λ_1^B and Λ_2^B. In these reps $\mathbf{\Lambda}_1^B(C_{4z}^+) = \mathbf{\Lambda}_2^B(C_{4z}^+) = \mathbf{B}(C_{4z}^+) = -1$, and $\mathbf{\Lambda}_1^B(C_{2x}) = \mathbf{\Lambda}_1(C_{2x}) = 1$ and $\mathbf{\Lambda}_2^B(C_{2x}) = \mathbf{\Lambda}_2(C_{2x}) = -1$. Again using the factorization of 422 (D_4) in eqn. (4.5.1) the little group of E can be seen to be 4 (C_4). From the star E we choose arbitrarily one of the reps, say ${}^1E$. Then the only small rep is $\Lambda^{{}^1E}$ with $\mathbf{\Lambda}^{{}^1E}(C_{4z}^+) = {}^1\mathbf{E}(C_{4z}^+) = -\mathrm{i}$. The corresponding rep of 422 (D_4) is to be found by forming the induced rep $\Lambda^{{}^1E} \uparrow 422\ (D_4)$. Let us denote the basis function of $\Lambda^{{}^1E}$ by ϕ so that

$$C_{4z}^+\phi = \phi\,{}^1\mathbf{E}(C_{4z}^+) = -\mathrm{i}\phi. \tag{4.5.2}$$

Let us write $C_{2x}\phi = \psi$. Then $\langle\phi, \psi|$ is a basis for the rep $\Lambda^{{}^1E} \uparrow 422\ (D_4) = \Delta$. We calculate the induced matrix reps of C_{4z}^+ and C_{2x}. Now

$$C_{4z}^+\psi = C_{4z}^+C_{2x}\phi = C_{2x}C_{4z}^-\phi = C_{2x}\phi\,{}^1\mathbf{E}(C_{4z}^-) = \mathrm{i}\psi. \tag{4.5.3}$$

From eqns. (4.5.2) and (4.5.3) we see that

$$C_{4z}^+\langle\phi, \psi| = \langle -\mathrm{i}\phi, \mathrm{i}\psi| = \langle\phi, \psi|\begin{pmatrix} -\mathrm{i} & 0 \\ 0 & \mathrm{i} \end{pmatrix}$$

and therefore

$$\mathbf{\Delta}(C_{4z}^+) = \begin{pmatrix} -\mathrm{i} & 0 \\ 0 & \mathrm{i} \end{pmatrix}. \tag{4.5.4}$$

Also

$$C_{2x}\phi = \psi \tag{4.5.5}$$

by definition, and so

$$C_{2x}\psi = C_{2x}C_{2x}\phi = \phi, \tag{4.5.6}$$

and therefore

$$\Delta(C_{2x}) = \begin{pmatrix} 0 & 1 \\ 1 & 0 \end{pmatrix}. \tag{4.5.7}$$

The above results are summarized in Table 4.2, in which to conform with Table 2.2 we make the identification $\Lambda_1^A = A_1$, $\Lambda_2^A = A_2$, $\Lambda_1^B = B_1$, $\Lambda_2^B = B_2$, $\Delta = E$.

TABLE 4.2

The reps of 422 (D_4) in relation to those of 4 (C_4) (Altmann 1963*a*)

4 (C_4)	Star	Little co-group	E	C_{4z}^+	C_{2z}	C_{4z}^-	422 (D_4)	E	$C_{4z}^{\pm}$	C_{2z}	C_{2x} C_{2y}	C_{2a} C_{2b}
A	A	$2'$ (C_2')	1	1	1	1	A_1	1	1	1	1	1
							A_2	1	1	1	−1	−1
B	B	$2'$ (C_2')	1	−1	1	−1	B_1	1	−1	1	1	−1
							B_2	1	−1	1	−1	1
${}^1E$	E	1 (C_1)	1	−i	−1	i	E	2	0	−2	0	0
${}^2E$			1	i	−1	−i						

In Table 4.3 we reproduce a summary of similar relations for the remaining semi-direct products appearing in Table 4.1.

TABLE 4.3

The relations between the reps of point groups and those of their invariant subgroups (Altmann 1963*a*)

$\omega = \exp(2\pi i/3)$

2 (C_2)	Star	Little co-group	E	C_{2z}	222 (D_2)	E	C_{2z}	C_{2x}	C_{2y}
A	A	$2'$ (C_2')	1	1	A_1	1	1	1	1
					B_1	1	1	−1	−1
B	B	$2'$ (C_2')	1	−1	B_3	1	−1	1	−1
					B_2	1	−1	−1	1

$3\,(C_3)$	Star	Little co-group	E	C_3^+	C_3^-	$32\,(D_3)$	E	$C_3^\pm$	C'_{2i}
A	A	$2'\,(C'_2)$	1	1	1	A_1	1	1	1
						A_2	1	1	−1
^1E	E	$1\,(C_1)$	1	ω^*	ω	E	2	−1	0
^2E			1	ω	ω^*				

$222\,(D_2)$	Star	Little co-group	E	C_{2z}	C_{2x}	C_{2y}	$422\,(D_4)$	E	C_{2z}	C_{2x}, C_{2y}	$C_{4z}^\pm$	C_{2a}, C_{2b}
A_1	A	$2''\,(C''_2)$	1	1	1	1	A_1	1	1	1	1	1
							B_1	1	1	1	−1	−1
B_1	B	$2''\,(C''_2)$	1	1	−1	−1	A_2	1	1	−1	1	−1
							B_2	1	1	−1	−1	1
B_3	E	$1\,(C_1)$	1	−1	1	−1	E	2	−2	0	0	0
B_2			1	−1	−1	1						

$6\,(C_6)$	Star	Little co-group	E	C_6^+	C_3^+	C_2	C_3^-	C_6^-	$622\,(D_6)$	E	C_2	$C_3^\pm$	$C_6^\pm$	C'_{2s}	C''_{2s}
A	A	$2'\,(C'_2)$	1	1	1	1	1	1	A_1	1	1	1	1	1	1
									A_2	1	1	1	1	−1	−1
B	B	$2'\,(C'_2)$	1	−1	1	−1	1	−1	B_1	1	−1	1	−1	1	−1
									B_2	1	−1	1	−1	−1	1
$^1E_1$	E_1	$1\,(C_1)$	1	ω	ω^*	1	ω	ω^*	E_2	2	2	−1	−1	0	0
$^2E_1$			1	ω^*	ω	1	ω^*	ω							
$^1E_2$	E_2	$1\,(C_1)$	1	$-\omega$	ω^*	−1	ω	$-\omega^*$	E_1	2	−2	−1	1	0	0
$^2E_2$			1	$-\omega^*$	ω	−1	ω^*	$-\omega$							

$32\,(D_3)$	Star	Little co-group	E	$C_3^\pm$	C'_{2s}	$622\,(D_6)$	E	$C_3^\pm$	C'_{2s}	C_2	$C_6^\pm$	C''_{2s}
A_1	A_1	$2\,(C_2)$	1	1	1	A_1	1	1	1	1	1	1
						B_1	1	1	1	−1	−1	−1
A_2	A_2	$2\,(C_2)$	1	1	−1	A_2	1	1	−1	1	1	−1
						B_2	1	1	−1	−1	−1	1
E	E	$2\,(C_2)$	2	−1	0	E_2	2	−1	0	2	−1	0
						E_1	2	−1	0	−2	1	0

$222\,(D_2)$	Star	Little co-group	E	C_{2z}	C_{2x}	C_{2y}	$23\,(T)$	E	C_{2m}	C_{3j}^+	C_{3j}^-
A_1	A	$3'\,(C'_3)$	1	1	1	1	A	1	1	1	1
							^1E	1	1	ω	ω^*
							^2E	1	1	ω^*	ω
B_1	B	$1\,(C_1)$	1	1	−1	−1	T	3	−1	0	0
B_3			1	−1	1	−1					
B_2			1	−1	−1	1					

222 (D_2)	Star	Little co-group	E	C_{2z}	C_{2x}	C_{2y}	432 (O)	E	C_{2m}	$C_{3j}^{\pm}$	C_{2p}	$C_{4m}^{\pm}$
							A_1	1	1	1	1	1
A_1	A	32′ (D'_3)	1	1	1	1	A_2	1	1	1	−1	−1
							E	2	2	−1	0	0
B_1			1	1	−1	−1						
B_3	B	2′ (C'_2)	1	−1	1	−1	T_2	3	−1	0	1	−1
B_2			1	−1	−1	1	T_1	3	−1	0	−1	1

Of the examples in Table 4.3 only the factorization 432 = 222 ∧ 32′ ($O = D_2 \wedge D'_3$) is of any real difficulty. This is because it provides a first example of a case in which the little group is neither the whole group nor the invariant subgroup; in the star B the little group is 422′ (D'_4) (the orientation of the little group depends on whether one starts from B_1, B_2, or B_3) and the small reps are just two of the 1-dimensional reps of 422′ (D'_4), and this is why one obtains two 3-dimensional reps of 432 (O), the order of 432 (O) being three times the order of 422′ (D'_4). The corresponding case in space-group theory is when the point in **k**-space being considered has more symmetry than just the translation group but does not have the whole symmetry of the space group.

It is interesting to contrast the formation of the reps of the quaternion group **Q** (the group $\mathbf{G}_8^5$ in Table 5.1) out of the cyclic group of order 4 (which we call $\mathbf{G}_4^1$ as we are not considering it as a point group) with the formation given above of the reps of 422 (D_4) from the same cyclic group. If the elements of $\mathbf{G}_4^1$, the cyclic group of order 4, are E, P, P^2, and P^3, (P corresponds to C_{4z}^+ when considering 422 (D_4)), then the quaternion group has 8 elements which form 5 classes, E; P^2; P, P^3; Q, P^2Q; and PQ, P^3Q. Alternatively we can just say that the generating relations are $P^4 = E$, $Q^4 = E$, $P^2 = Q^2$, and $QP = P^3Q$. The factorization into left cosets is then

$$\mathbf{Q} = E\mathbf{G}_4^1 + Q\mathbf{G}_4^1. \tag{4.5.8}$$

All elements of $Q\mathbf{G}_4^1$ have order 4 so the group **Q** is not a semi-direct product. It is analogous in space-group theory to a non-symmorphic space group. Just as before, there are three stars, A, B, and E and the little groups are **Q**, **Q**, and $\mathbf{G}_4^1$ respectively. The star E once again contains the two reps ${}^1E$ and ${}^2E$. The little co-groups are, for the stars A and B the cyclic group of order 2 which we call $\mathbf{G}_2^1$, and for the star E the trivial group $\mathbf{G}_1^1$ consisting of just the identity element. Since in this case we no longer have a semi-direct product the required reps of the little co-groups require further investigation. For the little co-group $\mathbf{G}_2^1$ consisting, say, of elements e and q we require reps (possibly projective) so that

$$\Lambda^A(q)\Lambda^A(q) = \mathrm{A}(P^2)\Lambda^A(e), \tag{4.5.9}$$

and

$$\Lambda^B(q)\Lambda^B(q) = \mathbf{B}(P^2)\Lambda^B(e). \tag{4.5.10}$$

However $\mathbf{A}(P^2) = \mathbf{B}(P^2) = 1$ so once again the small reps are the ordinary vector reps. Hence the four 1-dimensional reps Λ_1^A, Λ_2^A, Λ_1^B, and Λ_2^B have for $\mathbf{Q}$ exactly the same structure as they do for 422 (D_4). For the star E we form out of the rep ${}^1E$ the rep $\Gamma = {}^1E \uparrow \mathbf{Q}$ and this is the only rep of $\mathbf{Q}$ which derives from the star E. Denote the basis function of ${}^1E$ by ϕ so that

$$P\phi = -\mathrm{i}\phi, \tag{4.5.11}$$

and let us write $Q\phi = \psi$. Then $\langle\phi, \psi|$ is a basis for Γ. Now

$$P\psi = PQ\phi = QP^3\phi = \mathrm{i}Q\phi = \mathrm{i}\psi \tag{4.5.12}$$

and so from eqns. (4.5.11) and (4.5.12) we see that

$$\mathbf{\Gamma}(P) = \begin{pmatrix} -\mathrm{i} & 0 \\ 0 & \mathrm{i} \end{pmatrix}. \tag{4.5.13}$$

This is exactly as for 422 (D_4) (see eqn. (4.5.4)). Also $Q\phi = \psi$, by definition, and so

$$Q\psi = Q^2\phi = P^2\phi = -\phi. \tag{4.5.14}$$

Here at last is the only difference between the reps of $\mathbf{Q}$ and 422 (D_4) for now

$$\mathbf{\Gamma}(Q) = \begin{pmatrix} 0 & -1 \\ 1 & 0 \end{pmatrix} \tag{4.5.15}$$

and this is to be contrasted with eqn. (4.5.7) for $\mathbf{\Delta}(C_{2x})$. Note, however, that in spite of this the character tables of the two groups are the same. This demonstrates the remarkable fact that two non-isomorphic groups can have identical character tables.

It has already been mentioned that every finite group $\mathbf{G}$ which is solvable has a maximal invariant subgroup $\mathbf{N}$ of prime index. In particular every crystallographic point group contains a maximal invariant subgroup of index either 2 or 3. As a result of this it is possible to show that every space group also contains a maximal invariant subgroup of index either 2 or 3 (Zak 1960). It is therefore appropriate to investigate as a particular case of the theory in sections 4.1–4.4 how the representations of $\mathbf{G}$ derive from $\mathbf{N}$ in the case in which $\mathbf{G}/\mathbf{N}$ is a cyclic group of prime order p. This theory formed the basis of the derivation originally given by Seitz (1936*b*) in showing how to reduce the space groups; more recently certain methods employed by Zak (1960) have depended upon this theory. Of course the theory of section 4.4 covers perfectly well what happens when an invariant subgroup is Abelian (see case (i) of that section); and this is quite sufficient for dealing with the space groups, for which the invariant subgroup can be chosen to be the invariant subgroup of translations, $\mathbf{T}$. However, the following theory is included not just as a mathematical exercise, nor merely because of its historical interest, but rather because it throws further light on the structure of the groups with which we deal. For example, it shows how to make use of a factorization such as $432 = 23 \wedge 2''$ ($O = T \wedge C_2''$) (see Table 4.1) which, on account of the fact that 23 (T) is not Abelian, does not fall within the scope of case (i) of section 4.4. Incidentally the theory shows that when $\mathbf{G}/\mathbf{N}$ is of prime order case (ii) of section 4.4 is not as intractable as it might seem at first sight.

Suppose then that $\mathbf{G}$ is a group and that $\mathbf{N}$ is a maximal invariant subgroup of $\mathbf{G}$ of prime index p. In this case we can factorize $\mathbf{G}$ into left cosets in the form

$$\mathbf{G} = \mathbf{N} + r\mathbf{N} + r^2\mathbf{N} + \cdots + r^{p-1}\mathbf{N} \tag{4.5.16}$$

where $r^p \in \mathbf{N}$. When r^p is equal to E, the identity, $\mathbf{G}$ is the semi-direct product of $\mathbf{N}$ and the cyclic group of order p. More generally, however, suppose that $r^p = \bar{n} \in \mathbf{N}$ and that $\bar{n}$ is of order l. Then $E = \bar{n}^l = r^{lp}$. If $1 \leqslant l \leqslant p$ it follows, since p is prime, that we can choose $s = r^l$ instead of r and form the alternative coset decomposition

$$\mathbf{G} = \mathbf{N} + s\mathbf{N} + s^2\mathbf{N} + \cdots + s^{p-1}\mathbf{N} \tag{4.5.17}$$

where now $s^p = E$. Hence with a value of l in the above range we can make a fresh choice of r so that $l = 1$. On the other hand, if $l = p$ no such alternative choice is possible. Further arguments along these lines show that a choice of r can always be made so that l has a value that is a power of p; that is, $l = 1, p, p^2, \ldots$. So in the decomposition (4.5.16) we may assume without loss of generality that $r^{(p^m)} = E$ where $m \geqslant 1$.

We next consider the class structure of $\mathbf{G}$ relative to that of $\mathbf{N}$. Suppose that $C_1, C_2, \ldots, C_n$ are the classes of $\mathbf{N}$. We can show that conjugation by r produces a permutation of these classes. Suppose that $n_i \in C_i$ and $rn_ir^{-1} = n_j \in C_j$. Let n_i' be any other element of C_i. Then there exists $n \in \mathbf{N}$ such that $n_i' = nn_in^{-1}$ and $rn_i'r^{-1} = rnn_in^{-1}r^{-1}$. Now since $\mathbf{N}$ is invariant in $\mathbf{G}$, $rn = n'r$ for some $n' \in \mathbf{N}$. Hence, $rn_i'r^{-1} = n'rn_ir^{-1}n'^{-1} = n'n_jn'^{-1} = n_j' \in C_j$. This shows that $rC_ir^{-1} \subset C_j$ and consequently that the number of elements s_i in C_i is less than or equal to the number of elements s_j in C_j. A similar argument using conjugation by r^{-1} shows that $r^{-1}C_jr \subset C_i$ and $s_j \leqslant s_i$. It follows that $rC_ir^{-1} = C_j$ and $s_i = s_j$. Let C_k be another class such that $rC_kr^{-1} = C_j$ then $C_i = r^{-1}C_jr = C_k$. Hence conjugation by r produces a permutation $\pi(r)$ on the classes of $\mathbf{N}$. Similarly, it can be shown that conjugation by r^m produces a permutation $\pi(r^m)$ on the classes of $\mathbf{N}$ for $m = 1, 2, \ldots p$. Since $r^p \in \mathbf{N}$ it follows that $\pi(r^p)$ is the identity permutation ε. Also $\{\pi(r)\}^2\, C_i = \pi(r)rC_ir^{-1} = r^2C_ir^{-2} = \pi(r^2)C_i$, so that $\{\pi(r)\}^2 = \pi(r^2)$. Similarly $\{\pi(r)\}^m = \pi(r^m)$. Hence $\{\pi(r)\}^p = \pi(r^p) = \varepsilon$. The order of $\pi(r)$ therefore divides p. Since p is prime it follows that $\pi(r)$ is of order either 1 or p. This means that if $\pi(r)$ is decomposed into a product of disjoint cycles then the length of any cycle is either 1 or p. Therefore the classes of $\mathbf{N}$ are of two kinds: those which are self-conjugate under $r(rC_lr^{-1} = C_l)$ and those which occur in cycles of length p

such that C_i, rC_ir^{-1}, $r^2C_ir^{-2}, \ldots, r^{p-1}C_ir^{-(p-1)}$ form p distinct classes of **N**. The class structure of **G** relative to **N** is therefore determined as follows. *The self-conjugate classes of* **N** *under r are also classes of* **G**. *Those which are not self-conjugate under r become linked with* $(p-1)$ *other classes in the manner described above to form a single class in* **G**.

We next show how to determine the reps of **G** in terms of those of **N**. Suppose that Δ_0 is a rep of **N** of dimension d and suppose that ϕ_{0i}, $i = 1$ to d, is a basis for Δ_0 so that, for all $n \in \mathbf{N}$,

$$n\phi_{0i} = \sum_{j=1}^{d} \phi_{0j}\Delta_0(n)_{ji} \tag{4.5.18}$$

$i = 1, 2, \ldots, d$. In the same manner as in section 4.1 we define the functions

$$\phi_{mi} = r^m\phi_{0i} \tag{4.5.19}$$

$m = 0, 1, \ldots, (p-1)$, $i = 1, 2, \ldots, d$, then by analogy with eqn. (4.2.16) it follows that for fixed m

$$n\phi_{mi} = \sum_{j=1}^{d} \phi_{mj}\Delta_0(r^{-m}nr^m)_{ji} \tag{4.5.20}$$

$i = 1, 2, \ldots, d$. From the theory of section 4.2 we see that in this way we define p reps Δ_m of **N** such that

$$\Delta_m(n) = \Delta_0(r^{-m}nr^m) \tag{4.5.21}$$

$m = 0, 1, \ldots (p-1)$. Now in the present case the group **G**/**N** is isomorphic with C_p, the cyclic group of order p. Since p is prime only two cases can occur. The order of the little co-group of Δ_0, being a factor of the order of **G**/**N**, must be either 1 or p and the star of Δ_0 contains respectively either p reps or 1 rep. Thus we have a result analogous to that for the classes. *Either* Δ_0 *is self-conjugate under r, the* Δ_m *are mutually equivalent, and the little group is the whole group* **G**. *Or the* Δ_m *are mutually inequivalent and the little group is just the invariant subgroup* **N**.

Case (i)

Suppose the Δ_m, $m = 0, 1, \ldots (p-1)$, are mutually inequivalent. Then it follows immediately from the theory in section 4.2 that $\Delta_0 \uparrow \mathbf{G}$ is irreducible. The character of $\Delta_0 \uparrow \mathbf{G}$ is found at once from eqn. (4.2.22)

$$\chi_{\Delta_0 \uparrow \mathbf{G}}(n) = \sum_{m=0}^{p-1} \chi_{\Delta_0}(r^{-m}nr^m) = \sum_{m=0}^{p-1} \chi_{\Delta_m}(n) \tag{4.5.22}$$

and

$$\chi_{\Delta_0 \uparrow \mathbf{G}}(r^in) = 0 \tag{4.5.23}$$

$i = 1, 2, \ldots, (p-1)$.

Case (ii)

Suppose the Δ_m, $m = 0, 1, \ldots (p-1)$ are mutually equivalent. Then $\Delta_0 \uparrow \mathbf{G}$ is reducible. Since the Δ_m have identical characters the character of $\Delta_0 \uparrow \mathbf{G}$ is given by

$$\chi_{\Delta_0 \uparrow \mathbf{G}}(n) = p\chi_{\Delta_0}(n) \tag{4.5.24}$$

and

$$\chi_{\Delta_0 \uparrow \mathbf{G}}(r^i n) = 0 \tag{4.5.25}$$

$i = 1, 2, \ldots (p-1)$. Thus the intertwining number of $\Delta_0 \uparrow \mathbf{G}$ with itself is

$$\begin{aligned}\frac{1}{|\mathbf{G}|}\sum_{g\in\mathbf{G}} |\chi_{\Delta_0 \uparrow \mathbf{G}}(g)|^2 &= \frac{1}{|\mathbf{G}|} p^2 \sum_{n\in\mathbf{N}} |\chi_{\Delta_0}(n)|^2 \\ &= p^2\,|\mathbf{N}|/|\mathbf{G}| = p.\end{aligned} \tag{4.5.26}$$

It is plausible therefore that in this case $\Delta_0 \uparrow \mathbf{G}$ is reducible into p mutually inequivalent reps of $\mathbf{G}$ each of dimension d. We next show that this is indeed the case and, in particular, we show how to effect the reduction and how to obtain the characters of the component reps. Since every element of $\mathbf{G}$ can be written in the form $r^i n$ for some i in the range 0 to $(p-1)$ and for some $n \in \mathbf{N}$, it is sufficient, in order to reduce $\Delta_0 \uparrow \mathbf{G}$, to find a change of basis which simultaneously reduces the representatives of n and r to the same block-diagonal form. Now with respect to the basis $\langle \phi_{mi}| = \langle \phi_{0i}, \phi_{1i}, \ldots, \phi_{(p-1)i}|$ the representative of n is clearly

$$(\Delta_0 \uparrow \mathbf{G})(n) = \begin{pmatrix} \Delta_0(n) & 0 & 0 & \ldots & 0 \\ 0 & \Delta_0(r^{-1}nr) & 0 & \ldots & 0 \\ 0 & 0 & \Delta_0(r^{-2}nr^2) & \ldots & 0 \\ \vdots & \vdots & \vdots & & \vdots \\ 0 & 0 & 0 & \ldots & \Delta_0(r^{-p+1}nr^{p-1}) \end{pmatrix}. \tag{4.5.27}$$

Also, since $\phi_{mi} = r^m \phi_{0i}$, it follows that

$$\begin{aligned} r\langle \phi_{0i}, \phi_{1i}, \ldots \phi_{(p-1)i}| &= \langle r\phi_{0i}, r^2\phi_{0i}, \ldots, r^p\phi_{0i}| \\ &= \langle \phi_{1i}, \phi_{2i}, \ldots, \phi_{(p-1)i}, \sum_j \phi_{0j}\Delta_0(r^p)_{ji}| \end{aligned} \tag{4.5.28}$$

where, in the last component, we have used the fact that $r^p \in \mathbf{N}$. Therefore

$$(\Delta_0 \uparrow \mathbf{G})(r) = \begin{pmatrix} 0 & 0 & 0 & \ldots & 0 & \Delta_0(r^p) \\ 1 & 0 & 0 & \ldots & 0 & 0 \\ 0 & 1 & 0 & \ldots & 0 & 0 \\ \vdots & \vdots & \vdots & & \vdots & \vdots \\ 0 & 0 & 0 & \ldots & 1 & 0 \end{pmatrix}. \tag{4.5.29}$$

Now we are given that Δ_1 and Δ_0 are equivalent. Hence there exists a unitary matrix $\mathbf{P}$ such that, for all $n \in \mathbf{N}$,

$$\Delta_1(n) = \mathbf{P}\Delta_0(n)\mathbf{P}^{-1}. \tag{4.5.30}$$

Thus $\Delta_2(n) = \Delta_0(r^{-2}nr^2) = \Delta_0(r^{-1}(r^{-1}nr)r) = \Delta_0(r^{-1}n'r) = \Delta_1(n') = \mathbf{P}\Delta_0(n')\mathbf{P}^{-1} = \mathbf{P}\Delta_0(r^{-1}nr)\mathbf{P}^{-1} = \mathbf{P}^2\Delta_0(n)\mathbf{P}^{-2}$ and, more generally, by induction that

$$\Delta_m(n) = \mathbf{P}^m\Delta_0(n)\mathbf{P}^{-m} \tag{4.5.31}$$

$m = 0, 1, 2, \ldots, (p-1)$. Let us therefore define a new basis

$$\psi_{mi} = \sum_{j=1}^{d} \phi_{mj}(\mathbf{P}^m)_{ji} \tag{4.5.32}$$

$m = 0, 1, \ldots, (p-1)$ and $i = 1, 2, \ldots, d$, then a short calculation shows that with respect to this new basis

$$(\Delta_0 \uparrow \mathbf{G})(n) = \begin{pmatrix} \Delta_0(n) & 0 & 0 & \ldots & 0 \\ 0 & \Delta_0(n) & 0 & \ldots & 0 \\ 0 & 0 & \Delta_0(n) & \ldots & 0 \\ \vdots & \vdots & \vdots & & \vdots \\ 0 & 0 & 0 & \ldots & \Delta_0(n) \end{pmatrix} \tag{4.5.33}$$

and

$$(\Delta_0 \uparrow \mathbf{G})(r) = \begin{pmatrix} 0 & 0 & 0 & \ldots & 0 & \Delta_0(r^p)\mathbf{P}^{p-1} \\ \mathbf{P}^{-1} & 0 & 0 & \ldots & 0 & 0 \\ 0 & \mathbf{P}^{-1} & 0 & \ldots & 0 & 0 \\ \vdots & \vdots & \vdots & & \vdots & \vdots \\ 0 & 0 & 0 & \ldots & \mathbf{P}^{-1} & 0 \end{pmatrix}. \tag{4.5.34}$$

The next step is to show that $\Delta_0(r^p)\mathbf{P}^p$ is a scalar multiple of the unit matrix. Let n be any element of $\mathbf{N}$ then since $r^p \in \mathbf{N}$ we have

$$\begin{aligned}\Delta_0(n)\Delta_0(r^p)\mathbf{P}^p &= \Delta_0(nr^p)\mathbf{P}^p = \Delta_0(r^p)\Delta_0(r^{-p}nr^p)\mathbf{P}^p \\ &= \Delta_0(r^p)\Delta_1(r^{-p+1}nr^{p-1})\mathbf{P}^p = \Delta_0(r^p)\mathbf{P}\Delta_0(r^{-p+1}nr^{p-1})\mathbf{P}^{p-1} \\ &= \Delta_0(r^p)\mathbf{P}^p\Delta_0(n),\end{aligned}$$

by repetition of the argument. Hence $\Delta_0(r^p)\mathbf{P}^p$ commutes with the rep Δ_0 of $\mathbf{N}$ and so by Schur's lemma

$$\Delta_0(r^p)\mathbf{P}^p = \lambda\mathbf{1} \tag{4.5.35}$$

where λ is a phase factor. From eqn. (4.5.30) one can see that the matrix $\mathbf{P}$ can be modified by a phase factor and in particular it is possible to choose $\mathbf{P}$ so that $\lambda = 1$. Assuming this choice is made and writing $\mathbf{P}^{-1} = \mathbf{R}$ the following equations then hold:

$$\mathbf{\Delta}_1(n) = \mathbf{R}^{-1}\mathbf{\Delta}_0(n)\mathbf{R} \tag{4.5.36}$$

$$\mathbf{\Delta}_0(r^p) = \mathbf{R}^p \tag{4.5.37}$$

$$(\Delta_0 \uparrow \mathbf{G})(r) = \begin{pmatrix} 0 & 0 & 0 & \dots & 0 & \mathbf{R} \\ \mathbf{R} & 0 & 0 & \dots & 0 & 0 \\ 0 & \mathbf{R} & 0 & \dots & 0 & 0 \\ \vdots & \vdots & \vdots & & \vdots & \vdots \\ 0 & 0 & 0 & \dots & \mathbf{R} & 0 \end{pmatrix}, \tag{4.5.38}$$

with eqn. (4.5.33) still valid for $(\Delta_0 \uparrow \mathbf{G})(n)$. The basis spanning this representation may be denoted by $\langle \boldsymbol{\chi}_0, \boldsymbol{\chi}_1, \dots, \boldsymbol{\chi}_{p-1}|$, where $\boldsymbol{\chi}_m$ stands for the row vector $(\chi_{m1}, \chi_{m2}, \dots, \chi_{md})$. We now look for a further change of basis in an attempt to block-diagonalize $(\Delta_0 \uparrow \mathbf{G})(r)$ whilst keeping the same form for $(\Delta_0 \uparrow \mathbf{G})(n)$. To see that this is indeed possible we try

$$\boldsymbol{\eta} = \sum_{m=0}^{p-1} a_m \boldsymbol{\chi}_m. \tag{4.5.39}$$

Clearly, for any choice of the coefficients a_m,

$$n\boldsymbol{\eta} = \sum_{m=0}^{p-1} a_m n\boldsymbol{\chi}_m = \sum_{m=0}^{p-1} a_m \boldsymbol{\chi}_m \mathbf{\Delta}_0(n) = \boldsymbol{\eta}\mathbf{\Delta}_0(n). \tag{4.5.40}$$

Also, from eqn. (4.5.38),

$$\begin{aligned} r\boldsymbol{\eta} &= r(a_0\boldsymbol{\chi}_0 + a_1\boldsymbol{\chi}_1 + \cdots + a_{p-1}\boldsymbol{\chi}_{p-1}) \\ &= (a_0\boldsymbol{\chi}_1 + a_1\boldsymbol{\chi}_2 + \cdots + a_{p-1}\boldsymbol{\chi}_0)\mathbf{R} \end{aligned} \tag{4.5.41}$$

The question arises now as to whether it is possible to choose the a_m and a scalar ε so that

$$r\boldsymbol{\eta} = \boldsymbol{\eta}\varepsilon\mathbf{R} \tag{4.5.42}$$

Comparing the right-hand sides of eqns. (4.5.41) and (4.5.42) we need to satisfy simultaneously

$$\left.\begin{aligned} \varepsilon a_0 &= a_{p-1} \\ \varepsilon a_1 &= a_0 \\ &\vdots \\ \varepsilon a_{p-1} &= a_{p-2} \end{aligned}\right\} \tag{4.5.43}$$

which allows p possible solutions $\varepsilon = \varepsilon_q = \exp(2\pi q i/p)$ for $q = 0, 1, \ldots, (p-1)$, the pth roots of unity. This means that the vector space spanning $\Delta_0 \uparrow \mathbf{G}$ can be reduced into p invariant subspaces each of dimension d, one for each of the pth roots of unity. Also in the vector space determined by $\varepsilon = \varepsilon_q$ we have a vector $\boldsymbol{\eta}_q$ defined by values of the a_m corresponding to $\varepsilon = \varepsilon_q$ found from eqns. (4.5.43) for which

$$\left.\begin{aligned} n\boldsymbol{\eta}_q &= \boldsymbol{\eta}_q\boldsymbol{\Delta}_0(n) \\ r\boldsymbol{\eta}_q &= \boldsymbol{\eta}_q\,\varepsilon_q\mathbf{R} \end{aligned}\right\}. \tag{4.5.44}$$

Hence, we determine p representations of $\mathbf{G}$ which we may label Δ_{0q} such that

$$\left.\begin{aligned} \boldsymbol{\Delta}_{0q}(n) &= \boldsymbol{\Delta}_0(n) \\ \boldsymbol{\Delta}_{0q}(r) &= \varepsilon_q\mathbf{R} \end{aligned}\right\} \tag{4.5.45}$$

$q = 0, 1, \ldots, (p-1)$. Finally since Δ_0 is irreducible under $\mathbf{N}$ no further reductions are possible and hence the Δ_{0q} are irreducible under $\mathbf{G}$. Also, as we have seen above from eqn. (4.5.26), the intertwining number of $\Delta_0 \uparrow \mathbf{G}$ with itself is p. Since Δ_{0q} is equivalent to itself for $q = 0$ to $(p-1)$ this exhausts the p intertwinings and does not allow for any further equivalences of the form $\Delta_{0q_1} \equiv \Delta_{0q_2}$, $q_1 \neq q_2$; thus it is not possible for any two reps Δ_{0q_1} and Δ_{0q_2} to be equivalent if $q_1 \neq q_2$.

As an example of the use of the above theory we derive the reps of the point group 432 (O) in terms of those of 23 (T). We use the character table and matrix representatives of the group 23 (T) given in Tables 2.2 and 2.3. Here $\mathbf{G}$ is the group 432 (O) and $\mathbf{N}$ is the group 23 (T) so that $p = 2$. The element r can be chosen to be C_{2a}, so that $r^p = C_{2a}^2 = E$, the identity.

First take $\Delta_0 = A$ of 23 (T). It is clearly self-conjugate since $A(n) = 1$ for all $n \in 23$ (T). Since A is 1-dimensional and we require $\mathbf{R}^2 = 1$ we can choose $\mathbf{R} = +\mathbf{1}$ and so $A \uparrow 432$ (O) decomposes into the two 1-dimensional reps of 432 (O) labelled A_1 and A_2 in Table 2.2. Furthermore, since the square roots of unity are ± 1 we have $A_1(C_{2a}) = 1$ and $A_2(C_{2a}) = -1$. Since $A_1(n) = A_2(n) = A(n)$ for all $n \in 23$ (T) these equations determine entirely the characters of A_1 and A_2.

Next take $\Delta_0 = {}^1E$ of 23 (T). Since $C_{2a}C_{31}^+C_{2a} = C_{32}^-$ (see Table 1.5) we have $\Delta_1(C_{31}^+) = \Delta_0(C_{2a}C_{31}^+C_{2a}) = \Delta_0(C_{32}^-) = {}^1\mathrm{E}(C_{32}^-) = \omega^* = {}^2\mathrm{E}(C_{31}^+)$. Hence $\Delta_1 = {}^2E$. Since ${}^1E$ and ${}^2E$ are inequivalent it follows that ${}^1E \uparrow 432$ (O) is irreducible. Thus ${}^1E$ and ${}^2E$ come together to form the rep of 432 (O) labelled E in Table 2.2. The matrices for E given in Table 2.3 differ from those obtained immediately from the induction ${}^1E \uparrow 432$ (O) by an equivalence transformation. It can easily be shown that the induced matrices are generated by the following three matrices:

$$\left.\begin{aligned}({}^1E\uparrow 432)(C_{31}^{+}) &= \begin{pmatrix}\omega & 0\\ 0 & \omega^*\end{pmatrix}\\ ({}^1E\uparrow 432)(C_{2x}) &= \begin{pmatrix}1 & 0\\ 0 & 1\end{pmatrix}\\ ({}^1E\uparrow 432)(C_{2a}) &= \begin{pmatrix}0 & 1\\ 1 & 0\end{pmatrix}\end{aligned}\right\} \tag{4.5.46}$$

Finally take $\Delta_0 = T$ of 23 (T). Typical matrices for Δ_0 are

$$\begin{aligned}\Delta_0(C_{2x}) &= \begin{pmatrix}1 & 0 & 0\\ 0 & -1 & 0\\ 0 & 0 & -1\end{pmatrix}\\ \Delta_0(C_{31}^{+}) &= \begin{pmatrix}0 & 0 & 1\\ 1 & 0 & 0\\ 0 & 1 & 0\end{pmatrix}.\end{aligned} \tag{4.5.47}$$

Since $C_{2a}C_{2x}C_{2a} = C_{2y}$ and $C_{2a}C_{31}^{+}C_{2a} = C_{32}^{-}$ it follows that

$$\begin{aligned}\Delta_1(C_{2x}) = \Delta_0(C_{2y}) &= \begin{pmatrix}-1 & 0 & 0\\ 0 & 1 & 0\\ 0 & 0 & -1\end{pmatrix}\\ \Delta_1(C_{31}^{+}) = \Delta_0(C_{32}^{-}) &= \begin{pmatrix}0 & 1 & 0\\ 0 & 0 & -1\\ -1 & 0 & 0\end{pmatrix}.\end{aligned} \tag{4.5.48}$$

From this we see that the character of Δ_1 is equal to that of Δ_0 and hence that T is self-conjugate. In order to reduce $T\uparrow 432$ (O) it is necessary to find a matrix $\mathbf{R}$ (see eqns. (4.5.36) and (4.5.37)) such that

$$\left.\begin{aligned}\mathbf{R}\Delta_1(C_{2x}) &= \Delta_0(C_{2x})\mathbf{R},\\ \mathbf{R}\Delta_1(C_{31}^{+}) &= \Delta_0(C_{31}^{+})\mathbf{R},\\ \mathbf{R}^2 = \Delta_0(C_{2a}^2) &= \mathbf{1}.\end{aligned}\right\} \tag{4.5.49}$$

Using eqns. (4.5.47)–(4.5.49) it can be shown that a matrix **R** satisfying these conditions is

$$\mathbf{R} = \begin{pmatrix} 0 & 1 & 0 \\ 1 & 0 & 0 \\ 0 & 0 & -1 \end{pmatrix}. \tag{4.5.50}$$

Hence $T \uparrow 432\ (O)$ decomposes into two 3-dimensional reps, Δ_{00} and Δ_{01}, of 432 (O), the matrices of which can be found by using eqn. (4.5.45). The matrix representatives of C_{2a} are then $\Delta_{00}(C_{2a}) = +\mathbf{R}$ and $\Delta_{01}(C_{2a}) = -\mathbf{R}$. Δ_{00} and Δ_{01} can then be identified with the reps labelled T_1 and T_2 respectively in Table 2.2 and eqn. (4.5.45) gives just the matrix representatives that were given in Table 2.3.

If the matrix **R** is not readily obtained by an *ad hoc* method it can be shown that a matrix **P** that satisfies eqn. (4.5.30) is given by

$$\mathbf{P} = (1/|\mathbf{N}|) \sum_{n \in \mathbf{N}} \mathbf{\Delta}_1(n)\mathbf{X}\mathbf{\Delta}_0^{-1}(n) \tag{4.5.51}$$

where **X** is a matrix conveniently chosen so that **P** is unitary; $\mathbf{X} = \mathbf{1}$ is not always a suitable choice since this choice for **X** may cause the right-hand side of eqn. (4.5.51) to vanish.

4.6 The reality of representations induced from little groups

We refer to Definition 1.3.7 and to Theorem 1.3.9 in which reps of **G** are seen to be of three kinds according as the sum

$$\frac{1}{|\mathbf{G}|} \sum_{j=1}^{|\mathbf{G}|} \chi(G_j^2) = \begin{cases} 1 \\ -1. \\ 0 \end{cases} \tag{4.6.1}$$

We shall now apply this test to the rep $\Delta_j^i = \Gamma_j^i \uparrow \mathbf{G}$ induced in **G** from a small rep Γ_j^i of the little group $\mathbf{K}^i$. The notation is the same as in section 4.3. The elements of $\mathbf{K}^i$ are of the form $(r_\gamma t_a)$ where $\gamma \in \overline{\mathbf{K}}^i$ and $t_a \in \mathbf{T}$. We denote the character of the small rep of $\mathbf{K}^i$ by χ_j^i. The aim is to reduce the condition (4.6.1) from a sum over all the elements of **G** to a sum over a relatively small number of elements. To be able to do this we must clearly make use of the formula for the character χ of Δ_j^i in terms of the the character χ_j^i of Γ_j^i.

We recall the decomposition of **G** into left cosets with respect to $\mathbf{K}^i$ (eqn. (4.1.1)),

$$\mathbf{G} = \sum_{\tau} p_\tau \mathbf{K}^i. \tag{4.6.2}$$

We also recall the content of eqn. (4.1.9), that

$$\chi_\tau(g^2) = \begin{cases} \chi^i_j(r_\gamma t_a) & \text{if } g^2 = p_\tau r_\gamma t_a p_\tau^{-1} \in \mathbf{K}^i_\tau, \\ 0 & \text{otherwise,} \end{cases} \tag{4.6.3}$$

and that

$$\chi(g^2) = \sum_\tau \chi_\tau(g^2). \tag{4.6.4}$$

Then the sum (4.6.1) becomes

$$\frac{1}{|\mathbf{G}|}\sum_{g\in\mathbf{G}}\sum_\tau \chi_\tau(g^2) = \frac{1}{|\mathbf{G}|}\sum_\tau\sum_{g\in\mathbf{G}} \chi^i_j(r_\gamma t_a), \tag{4.6.5}$$

in which the sum over g is now restricted to elements such that $g^2 = p_\tau r_\gamma t_a p_\tau^{-1} \in \mathbf{K}^i_\tau$. Now if $h^2 \in \mathbf{K}^i$, that is, if $h^2 = r_\gamma t_a$ for some γ, a then $g = p_\tau h p_\tau^{-1}$ is such that $g^2 = p_\tau r_\gamma t_a p_\tau^{-1}$ with the same γ, a. This means that the sum (4.6.5) splits into $q^i = |\mathbf{G}|/|\mathbf{K}^i|$ equal parts, one part for each member of the star of i. Thus performing the sum over τ, (4.6.5) becomes

$$\frac{q_i}{|\mathbf{G}|}\sum_{h\in\mathbf{G}} \chi^i_j(h^2) \tag{4.6.6}$$

in which the sum over h is restricted to elements such that $h^2 \in \mathbf{K}^i$.

Now if $h = r_\alpha t_b$ then $h^2 = r_{\alpha\alpha} t_{\alpha\alpha}[t_b]_\alpha t_b$ so that $h^2 \in \mathbf{K}^i$ if $\alpha^2 \in \overline{\mathbf{K}}^i$. The sum (4.6.6) now becomes

$$\frac{q_i}{|\mathbf{G}|}\sum_{\alpha^2\in\mathbf{K}^i}\sum_b \chi^i_j(r_{\alpha\alpha}t_{\alpha\alpha}[t_b]_\alpha t_b). \tag{4.6.7}$$

We now remember that in Γ^i_j the matrix of the element t is the scalar matrix $D^i(t)\mathbf{1}$ so the sum (4.6.7) reduces to

$$\frac{q_i}{|\mathbf{G}|}\sum_{\alpha^2\in\mathbf{K}^i}\sum_b \chi^i_j(r_{\alpha\alpha})\mathbf{D}^i(t_{\alpha\alpha})\mathbf{D}^i([t_b]_\alpha)\mathbf{D}^i(t_b). \tag{4.6.8}$$

Now from eqn. (4.2.17) $\mathbf{D}^i([t_b]_\alpha) = \mathbf{D}^i_\alpha(t_b)$ and the sum over b, $\sum_b \mathbf{D}^i_\alpha(t_b)\mathbf{D}^i(t_b)$, will vanish by virtue of the orthogonality relations unless α is such that $\mathbf{D}^i_\alpha(t_a) = \mathbf{D}^i(t_a)^{-1}$. Note that all such α satisfy $\alpha^2 \in \overline{\mathbf{K}}^i$. When the sum over b does not vanish its value is $|\mathbf{T}|$. Writing $|\mathbf{G}|/|\mathbf{T}| = h$ the sum (4.6.8) becomes

$$\frac{q_i}{h}\sum_\alpha \chi^i_j(r^2_\alpha) \tag{4.6.9}$$

where in eqn. (4.6.9) the sum over α is restricted to those elements $\alpha \in \mathbf{R}$ (the group of suffices of the coset representatives r_α of $\mathbf{G}$ with respect to $\mathbf{T}$; see Theorem 4.2.2) such that $D^i_\alpha = (D^i)^{-1}$. The classification of Δ^i_j into kinds now depends upon whether this sum is $+1$, -1, or 0. We may summarize this result in

THEOREM 4.6.1. *The rep $\Delta^i_j = \Gamma^i_j \uparrow \mathbf{G}$ induced in $\mathbf{G}$ from a small rep Γ^i_j of the little group $\mathbf{K}^i$ is of the first, second, or third kind according as*

$$\frac{q_i}{h}\sum_{\alpha}\chi^i_j(r^2_\alpha) = 1,\ -1,\ \text{or } 0, \tag{4.6.10}$$

where the sum over α is restricted so that $D^i_\alpha = (D^i)^{-1}$.

The advantage of using induced representations is that expression (4.6.10) involves a summation over a much smaller number of elements than would be involved in the direct use of eqn. (4.6.1). In section 2.4 we have already noted that eqns. (1.3.12) or (4.6.1) can be applied directly to the reps of the crystallographic point groups which were given in Table 2.2 (see Note (v) to Table 2.2); as a result it is found that all the reps are of the first kind, except reps with complex characters and these are of the third kind. This result is, of course, obtained without any use of the theory of induced representations. However, they do provide convenient simple examples by which we can illustrate the use of induced representations in determining the reality of a representation. As an example of this use of Theorem 4.6.1 in determining the reality of reps we consider the 2-dimensional reps of the point group 422 (D_4). We determine the reality of E by using the fact that 422 (D_4) can be written as

$$\begin{aligned} 422 &= E.4 \ + C_{2x}.4 \\ (D_4 &= E.C_4 + C_{2x}.C_4) \end{aligned} \tag{4.5.1}$$

and E can be regarded as the representation induced by the rep ${}^1E$ of 4 (C_4) in 422 (D_4), i.e. as ${}^1E \uparrow 422$; therefore $h = 2$. The star of E has two members ${}^1E$ and ${}^2E$, see Table 4.2 and the text of section 4.5, so that $q_i = 2$. The sum in expression (4.6.10) is over those coset representatives r_α that satisfy the condition $D^i_\alpha = (D^i)^{-1}$; in words this condition means that the result of conjugating the representation ${}^1E$ with r_α must be to produce the representation in which each element of 4 (C_4) is represented by the inverse of its representative in ${}^1E$. Thus for ${}^1E$ we have

		E	C^+_{4z}	C_{2z}	C^-_{4z}
	D^i	1	$-i$	-1	i
	$(D^i)^{-1}$	1	i	-1	$-i$
$r_\alpha = E$	D^i_α	1	$-i$	-1	i
$r_\alpha = C_{2x}$	D^i_α	1	i	-1	$-i$

and the condition is only satisfied for $r_\alpha = C_{2x}$. The sum in expression (4.6.10) therefore becomes $(q_i/h)\chi^{{}^1E}(C^2_{2x}) = (2/2)\chi^{{}^1E}(E) = +1$. The rep E of 422 (D_4) is, therefore, of the first kind, as we found earlier by the more direct method in section 2.4.

A similar example is provided by the quaternion group which was discussed in section 4.5, see eqns. (4.5.8)–(4.5.15). For **Q** the star of $\Gamma = {}^1E \uparrow \mathbf{Q}$ has two members ${}^1E$ and ${}^2E$, $h = 2$, and the only r_α that satisfies the condition $D^i_\alpha = (D^i)^{-1}$ is Q. Now $Q^2 = P^2$ and in the small rep $\chi^{{}^1E}(P^2) = -1$. Hence the sum (4.6.10) becomes $(2/2).(-1) = -1$ and so the rep Γ of **Q** is of the second kind.

The simplification given in eqn. (4.6.10) of the condition to determine the reality of a rep of **G** can be applied to space groups. If Γ^i_j is a small rep $\Gamma^{\mathbf{k}}_j$ of a little group $\mathbf{G}^{\mathbf{k}}$ then eqn. (4.6.10) can be used to determine the reality of the rep $\Delta^{\mathbf{k}}_j$ induced by $\Gamma^{\mathbf{k}}_j$ in **G**. Thus if **G** is a space group h is the macroscopic order of symmetry of **G**, $q_i \equiv q_{\mathbf{k}}$ is the number of elements in the star containing **k**, and the r_α are elements $\{R_\alpha \mid \mathbf{v}_\alpha\}$ in the decomposition of **G** into left cosets with respect to **T**, see eqn. (3.7.2). The condition that the sum over α in eqn. (4.6.10) is restricted to those α for which $D^i_\alpha = (D^i)^{-1}$ becomes, in this context, the condition on α that $R_\alpha\mathbf{k} \equiv -\mathbf{k}$. We should perhaps clarify one point of notation, namely that although $\Delta^{\mathbf{k}}_j$ has a superscript **k** it is a rep of the whole space group **G** and not a rep of the little group $\mathbf{G}^{\mathbf{k}}$; the **k** superscript in $\Delta^{\mathbf{k}}_j$ merely indicates that it is induced from a rep of $\mathbf{G}^{\mathbf{k}}$. Therefore, making use of eqn. (4.6.10) we have

THEOREM 4.6.2. $\Delta^{\mathbf{k}}_j = \Gamma^{\mathbf{k}}_j \uparrow \mathbf{G}$ *is of the first, second, or third kind according as*

$$\frac{q_{\mathbf{k}}}{h} \sum_{R_\alpha} \chi^{\mathbf{k}}_j(\{R_\alpha \mid \mathbf{v}_\alpha\}^2) = +1, -1, \text{ or } 0, \tag{4.6.11}$$

where $\chi^{\mathbf{k}}_j$ *is the character of the small rep j of the little group of* **k** *in* **G** *and where the sum is restricted to coset representatives* $\{R_\alpha \mid \mathbf{v}_\alpha\}$ of **G** *with respect to* **T** *whose rotational parts send* **k** *into a vector equivalent to* $-\mathbf{k}$.

As one might expect, the sum (4.6.11) is independent of which of the possible coset representatives $\{R_\alpha \mid \mathbf{v}_\alpha\}$ is chosen for fixed α, because if **t** is a translation of the Bravais lattice and $R_\alpha\mathbf{k} \equiv -\mathbf{k}$ it can be shown that $\chi^{\mathbf{k}}_j(\{R_\alpha \mid \mathbf{v}_\alpha + \mathbf{t}\}^2) = \chi^{\mathbf{k}}_j(\{R_\alpha \mid \mathbf{v}_\alpha\}^2)$. We consider this problem again in Chapter 5 and an example of the application of Theorem 4.6.1 to space groups is given in section 5.4 where we derive the reality of the reps of the space group $F\bar{4}3c$ (T^5_d).

The physical importance of determining the reality of the reps of a point group or space group lies in the fact that it can be used to determine whether or not extra degeneracies of energy levels are produced by the addition of θ, the operation of time reversal symmetry, to the operations of the point group or space group (Herring 1937*a*; see also sections 5.4, 7.5, and 7.6).

4.7 Direct products of induced representations

We have seen in previous sections that if **G** is a group and **T** is an invariant Abelian subgroup of **G** then the reps of **G** can be determined from those of **T**. How this is done is outlined at the beginning of section 4.3. In this section we shall show how to determine the inner direct product of two reps of **G**. That is to say, if $(\Gamma^i_p \uparrow \mathbf{G})$ and $(\Gamma^j_q \uparrow \mathbf{G})$ are any two reps of **G**, we shall show how to determine the coefficients $C^{ij,l}_{pq,r}$ in the Clebsch–Gordan decomposition

$$(\Gamma^i_p \uparrow \mathbf{G}) \boxtimes (\Gamma^j_q \uparrow \mathbf{G}) \equiv \sum_l \sum_r C^{ij,l}_{pq,r}(\Gamma^l_r \uparrow \mathbf{G}). \tag{4.7.1}$$

To make any progress with this problem it is necessary to reformulate some of the results in section 1.3 and section 4.1, and to introduce some additional concepts, notably that of double cosets.

DEFINITION 4.7.1. *Complex conjugate representation.* Let **G** be a group and Γ a unitary representation of **G**. Then the representation Γ^* defined by

$$\Gamma^*(g) = \{\Gamma(g)\}^* \tag{4.7.2}$$

is called the *complex conjugate representation* of Γ.

THEOREM 4.7.1. *Let* **G** *be a group with reps* Γ^i *and suppose* Γ *is equivalent to the direct sum* $\sum_{i=1}^r c_i\Gamma^i$ *then* c_i *is the frequency of* $A(\mathbf{G})$ (*the totally symmetric rep of* **G**) *in* $\Gamma^* \boxtimes \Gamma^i$ (*or* $\Gamma \boxtimes \Gamma^{i*}$).

This is almost immediate. From eqn. (1.3.18)

$$c_i = \frac{1}{|\mathbf{G}|}\sum_g \chi^*(g)\chi^i(g) \tag{4.7.3}$$

$$= \frac{1}{|\mathbf{G}|}\sum_g \chi^{\Gamma^*\boxtimes\Gamma^i}(g) \tag{4.7.4}$$

(see eqn. (1.3.21) for the character of a direct product of reps). But $\chi^A(g) = 1$ for all $g \in \mathbf{G}$ so that eqn. (4.7.4) becomes

$$c_i = \frac{1}{|\mathbf{G}|}\sum_g \chi^{A*}(g)\chi^{\Gamma^*\boxtimes\Gamma^i}(g), \tag{4.7.5}$$

which on using eqn. (1.3.18) in reverse is the frequency of $A(\mathbf{G})$ in $\Gamma^* \boxtimes \Gamma^i$.

From Theorem 4.7.1 it follows that $C^{ij,l}_{pq,r}$ is the frequency of the totally symmetric rep in the triple direct product

$$(\Gamma^i_p \uparrow \mathbf{G}) \boxtimes (\Gamma^j_q \uparrow \mathbf{G}) \boxtimes (\Gamma^{l*}_r \uparrow \mathbf{G}).$$

We shall therefore study this equivalent and, as it turns out, easier problem.

We next reformulate a result of section 4.1. If **G** is a group and **K** is any subgroup with a representation Λ then the induced rep $\Lambda \uparrow \mathbf{G}$ is defined by eqn. (4.1.4):

$$g\phi_{\tau r} = \sum_t \phi_{\gamma t}\Lambda(k_m)_{tr}. \tag{4.7.6}$$

Here g is any element $p_\lambda k_s$ in the left coset decomposition $\mathbf{G} = \sum_\sigma p_\sigma \mathbf{K}$ and $k_m = p_\gamma^{-1} g p_\tau$. If we define $p_{g\tau} = p_\gamma$ by virtue of the fact that $gp_\tau \in p_\gamma \mathbf{K}$ then we can rewrite eqn. (4.7.6) in the block matrix form

$$(\Lambda \uparrow \mathbf{G})(g)_{\gamma\tau} = \Lambda(p_\gamma^{-1} g p_\tau)\delta_{\gamma,(g\tau)}. \tag{4.7.7}$$

Let the character of Λ be ψ and that of $\Lambda \uparrow \mathbf{G}$ be χ then

$$\chi(g) = \sum_\sigma \mathrm{Tr}\,\{(\Lambda \uparrow \mathbf{G})(g)_{\sigma\sigma}\} \tag{4.7.8}$$

$$= \sum_\sigma \mathrm{Tr}\,\{\Lambda(p_\sigma^{-1} g p_\sigma)\}\delta_{\sigma,(g\sigma)} \tag{4.7.9}$$

$$= \sum_\sigma \psi(p_\sigma^{-1} g p_\sigma)\delta_{\sigma,(g\sigma)}. \tag{4.7.10}$$

Note that the sum over σ is over those cosets only for which $p_\sigma^{-1} g p_\sigma \in \mathbf{K}$, that is those σ for which $p_\sigma \mathbf{K} p_\sigma^{-1} = \mathbf{K}_\sigma$ contains g.

Incidentally the formula (4.7.10) justifies the footnote to the comment following eqn. (4.1.1). For suppose we had chosen a different set of coset representatives $q_\sigma = p_\sigma k_\sigma$ where the $k_\sigma \in \mathbf{K}$, then the formula for the character of the induced rep would be

$$\sum_\sigma \psi(q_\sigma^{-1} g q_\sigma)\delta_{\sigma,(g\sigma)}. \tag{4.7.11}$$

But $q_\sigma \mathbf{K} q_\sigma^{-1} = p_\sigma k_\sigma \mathbf{K} k_\sigma^{-1} p_\sigma^{-1} = p_\sigma \mathbf{K} p_\sigma^{-1} = \mathbf{K}_\sigma$ and $q_\sigma^{-1} g q_\sigma$ and $p_\sigma^{-1} g p_\sigma$ are in the same class in $\mathbf{K}$ and so have the same character. Expression (4.7.11) has therefore the same value as the right-hand side of eqn. (4.7.10). This implies that two reps induced using two choices of coset representatives are equivalent. It is in this sense that the choice of coset representatives is immaterial.

Note that formula (4.7.10) holds whether or not Λ is irreducible under $\mathbf{K}$. This is important because we shall be concerned not only with the case in which $\mathbf{K}$ is a little group and Λ is a small rep and therefore irreducible but also with cases in which Λ is in general reducible and $\mathbf{K}$ is not necessarily a little group.

THEOREM 4.7.2. Frequency theorem. *Using the above notation, the frequency of $A(\mathbf{G})$ in $\Lambda \uparrow \mathbf{G}$ is equal to the frequency of $A(\mathbf{K})$ in Λ.*

We write $f(\Gamma^i) \mid \Gamma$ for the frequency of Γ^i in Γ. Then, from eqns. (4.7.3) and (4.7.10),

$$f\{A(\mathbf{G})\} \mid (\Lambda \uparrow \mathbf{G}) = \frac{1}{|\mathbf{G}|} \sum_{g \in \mathbf{G}} \sum_\sigma \psi(p_\rho^{-1} g p_\sigma)\delta_{\sigma,(g\sigma)}. \tag{4.7.12}$$

Reversing the order of summation this becomes

$$f\{A(\mathbf{G})\} \mid (\Lambda \uparrow \mathbf{G}) = \frac{1}{|\mathbf{G}|} \sum_\sigma \sum_{g \in \mathbf{K}_\sigma} \psi(p_\sigma^{-1} g p_\sigma) \tag{4.7.13}$$

$$= \frac{1}{|\mathbf{G}|} \sum_\sigma \sum_{k \in \mathbf{K}} \psi(k). \tag{4.7.14}$$

There are $|\mathbf{G}|/|\mathbf{K}|$ cosets σ so that this

$$= \frac{1}{|\mathbf{K}|} \sum_{k \in \mathbf{K}} \psi(k) \tag{4.7.15}$$

$$= f\{A(\mathbf{K})\} \mid \Lambda, \tag{4.7.16}$$

which proves the theorem.

THEOREM 4.7.3. *Let Λ be a representation of* $\mathbf{K}$ *with character* ψ, *and* Γ *a representation of* $\mathbf{G}$ *with character* χ, *where again* $\mathbf{K}$ *is a subgroup of* $\mathbf{G}$, *then*

$$\Gamma \boxtimes (\Lambda \uparrow \mathbf{G}) \equiv [(\Gamma \downarrow \mathbf{K}) \boxtimes \Lambda] \uparrow \mathbf{G}. \tag{4.7.17}$$

We establish the equivalence by showing that the characters of the two representations coincide. The character of g in $\Gamma \boxtimes (\Lambda \uparrow \mathbf{G})$ is

$$\chi(g) \sum_{\sigma} \psi(p_\sigma^{-1} g p_\sigma) \delta_{\sigma, (g\sigma)}$$

and the character of g in $[(\Gamma \downarrow \mathbf{K}) \boxtimes \Lambda] \uparrow \mathbf{G}$ is

$$\sum_{\sigma} \chi(p_\sigma^{-1} g p_\sigma) \psi(p_\sigma^{-1} g p_\sigma) \delta_{\sigma,(g\sigma)}.$$

But g and $p_\sigma^{-1} g p_\sigma$ are in the same class in $\mathbf{G}$ so $\chi(g) = \chi(p_\sigma^{-1} g p_\sigma)$ for all σ. The result follows.

We now prove the theorem of Transitivity of induction under more general circumstances than in Lemma 4.3.3.

THEOREM 4.7.4. Transitivity of induction. *Let* $\mathbf{K}$ *be a subgroup of* $\mathbf{H}$ *and* $\mathbf{H}$ *a subgroup of* $\mathbf{G}$ *and let* Λ *be a representation of* $\mathbf{K}$. *Then,*

$$(\Lambda \uparrow \mathbf{H}) \uparrow \mathbf{G} \equiv \Lambda \uparrow \mathbf{G}. \tag{4.7.18}$$

We establish the equivalence by showing that if Γ is a rep of $\mathbf{G}$ then its frequency in $(\Lambda \uparrow \mathbf{H}) \uparrow \mathbf{G}$ is the same as its frequency in $\Lambda \uparrow \mathbf{G}$.

$$\begin{aligned} & f(\Gamma) \mid (\Lambda \uparrow \mathbf{G}), \\ &= f\{A(\mathbf{G})\} \mid \Gamma^* \boxtimes (\Lambda \uparrow \mathbf{G}), && \text{by Theorem 4.7.1,} \\ &= f\{A(\mathbf{G})\} \mid [(\Gamma^* \downarrow \mathbf{K}) \boxtimes \Lambda] \uparrow \mathbf{G}, && \text{by Theorem 4.7.3,} \\ &= f\{A(\mathbf{K})\} \mid (\Gamma^* \downarrow \mathbf{K}) \boxtimes \Lambda. && \text{by Theorem 4.7.2.} \end{aligned}$$

But $f(\Gamma) \mid (\Lambda \uparrow \mathbf{H}) \uparrow \mathbf{G}$

$$\begin{aligned} &= f\{A(\mathbf{G})\} \mid \Gamma^* \boxtimes [(\Lambda \uparrow \mathbf{H}) \uparrow \mathbf{G}], && \text{by Theorem 4.7.1,} \\ &= f\{A(\mathbf{G})\} \mid [(\Gamma^* \downarrow \mathbf{H}) \boxtimes (\Lambda \uparrow \mathbf{H})] \uparrow \mathbf{G}, && \text{by Theorem 4.7.3,} \\ &= f\{A(\mathbf{H})\} \mid (\Gamma^* \downarrow \mathbf{H}) \boxtimes (\Lambda \uparrow \mathbf{H}), && \text{by Theorem 4.7.2,} \\ &= f\{A(\mathbf{H})\} \mid [(\Gamma^* \downarrow \mathbf{H}) \downarrow \mathbf{K} \boxtimes \Lambda] \uparrow \mathbf{H}, && \text{by Theorem 4.7.3,} \\ &= f\{A(\mathbf{K})\} \mid (\Gamma^* \downarrow \mathbf{K}) \boxtimes \Lambda, && \text{by Theorem 4.7.2,} \end{aligned}$$

and because of the transitivity of subduction $(\Gamma^* \downarrow \mathbf{H}) \downarrow \mathbf{K} \equiv (\Gamma^* \downarrow \mathbf{K})$, which is obvious, both sides consisting of the same matrices. This proves the theorem.

DEFINITION 4.7.2. Let **H** and **K** be any two subgroups of **G**. Then

$$\mathbf{G} = \sum_{\alpha} \mathbf{H}d_\alpha\mathbf{K}, \tag{4.7.19}$$

where in the complex $\mathbf{H}d_\alpha\mathbf{K}$ we count an element once only however many times it may appear, is called the *double coset* decomposition of **G** with respect to **H** and **K**. The d_α are called *double coset representatives*.

The expansion (4.7.19) is unique in much the same way as is the ordinary single coset decomposition. The double coset representatives are not unique: any element of the double coset serves equally well as its representative. Suppose that $g \in \mathbf{H}d_\alpha\mathbf{K}$ and $\mathbf{H}d_\beta\mathbf{K}$, two different cosets. Then $\exists\ h_1, h_2 \in \mathbf{H}$ and $k_1, k_2 \in \mathbf{K}$ such that $g = h_1 d_\alpha k_1 = h_2 d_\beta k_2$. This implies $h_1 d_\alpha k_1 \in \mathbf{H}d_\beta\mathbf{K}$. Hence $\mathbf{H}d_\alpha\mathbf{K} \subset \mathbf{H}\mathbf{H}d_\beta\mathbf{K}\mathbf{K} = \mathbf{H}d_\beta\mathbf{K}$. Similarly $\mathbf{H}d_\beta\mathbf{K} \subset \mathbf{H}d_\alpha\mathbf{K}$. Therefore $\mathbf{H}d_\alpha\mathbf{K} = \mathbf{H}d_\beta\mathbf{K}$. Hence two double cosets are either entirely distinct or their elements coincide (in which case only one would appear in the expansion (4.7.19)). To see how many elements appear in a given double coset we now prove a theorem of Frobenius.

THEOREM 4.7.5. *The double coset* $\mathbf{H}d_\alpha\mathbf{K}$ *contains* $|\mathbf{H}|/|\mathbf{L}_\alpha|$ *left cosets of* **K** *(and therefore* $|\mathbf{H}|\,|\mathbf{K}|/|\mathbf{L}_\alpha|$ *elements), where* $\mathbf{L}_\alpha = \mathbf{H} \cap \mathbf{K}_\alpha$ *(i.e. the intersection of* **H** *and* $\mathbf{K}_\alpha$ *and therefore a subgroup of both* **H** *and* $\mathbf{K}_\alpha$*) and* $\mathbf{K}_\alpha = d_\alpha\mathbf{K}d_\alpha^{-1}$ *is a subgroup conjugate to* **K**.

The proof of this is as follows.

Let
$$\mathbf{H} = \sum_{\beta} q_{\alpha\beta}\mathbf{L}_\alpha. \tag{4.7.20}$$

Suppose h_1 and h_2 belong to $\mathbf{L}_\alpha$. Then for any β $q_{\alpha\beta}h_1d_\alpha\mathbf{K} = q_{\alpha\beta}h_2d_\alpha\mathbf{K}$. This follows because we may write $h_1 = d_\alpha k_1 d_\alpha^{-1}$ and $h_2 = d_\alpha k_2 d_\alpha^{-1}$, where $k_1, k_2 \in \mathbf{K}$, so that $q_{\alpha\beta}h_1d_\alpha\mathbf{K} = q_{\alpha\beta}d_\alpha k_1 d_\alpha d_\alpha^{-1}\mathbf{K} = q_{\alpha\beta}d_\alpha k_1\mathbf{K} = q_{\alpha\beta}d_\alpha\mathbf{K} = q_{\alpha\beta}d_\alpha k_2\mathbf{K} = q_{\alpha\beta}h_2d_\alpha\mathbf{K}$. Also the $q_{\alpha\beta}d_\alpha\mathbf{K}$ for different $q_{\alpha\beta}$ are distinct left cosets of **K**, for if not we would have, say, $q_{\alpha1}d_\alpha\mathbf{K} = q_{\alpha2}d_\alpha\mathbf{K}$ and so $q_{\alpha1}d_\alpha \in q_{\alpha2}d_\alpha\mathbf{K}$; that is $q_{\alpha1} \in q_{\alpha2}\mathbf{K}_\alpha$, but also $q_{\alpha1}$ and $q_{\alpha2} \in \mathbf{H}$, which in turn implies $q_{\alpha1} \in q_{\alpha2}\mathbf{L}_\alpha$, contrary to the hypothesis that $q_{\alpha1}$ and $q_{\alpha2}$ are distinct coset representatives in eqn. (4.7.20). What we have proved is that $\mathbf{H}d_\alpha\mathbf{K}$ contains an integral number of left cosets of **K** and that these are in one to one correspondence with the left cosets of $\mathbf{L}_\alpha$ in **H**. The number of such left cosets is therefore $|\mathbf{H}|/|\mathbf{L}_\alpha|$. We note that if D is a representation of **K** and if $k_\alpha \in \mathbf{K}_\alpha$ (so that $\exists\ k \in \mathbf{K}$ such that $k_\alpha = d_\alpha k d_\alpha^{-1}$) then $\mathbf{D}_\alpha(k_\alpha) = \mathbf{D}(k)$ defines a representation D_α of $\mathbf{K}_\alpha$. Using the notation just defined we now prove an important theorem.

THEOREM 4.7.6. Mackey's subgroup theorem

$$(D \uparrow \mathbf{G}) \downarrow \mathbf{H} \equiv \sum_{\alpha} (D_\alpha \downarrow \mathbf{L}_\alpha) \uparrow \mathbf{H}. \tag{4.7.21}$$

We recall the decompositions $\mathbf{G} = \sum_\sigma p_\sigma \mathbf{K}$, $\mathbf{G} = \sum_\alpha \mathbf{H} d_\alpha \mathbf{K}$, and $\mathbf{H} = \sum_\beta q_{\alpha\beta} \mathbf{L}_\alpha$ and we note that because of Theorem 4.7.5 we may choose $p_\sigma = q_{\alpha\beta} d_\alpha$ and that as α and β run over all possible values so does σ. To prove Theorem 4.7.6 we show that the characters of the two representations coincide. From eqn. (4.7.10) the character of h in $(D \uparrow \mathbf{G}) \downarrow \mathbf{H}$ is

$$\chi(h) = \sum_\sigma \psi(p_\sigma^{-1} h p_\sigma) \delta_{\sigma,(h\sigma)}. \tag{4.7.22}$$

That is, the sum over σ is restricted to those σ for which $h \in \mathbf{K}_\sigma$. Now $\mathbf{K}_\sigma = p_\sigma \mathbf{K} p_\sigma^{-1} = q_{\alpha\beta} d_\alpha \mathbf{K} d_\alpha^{-1} q_{\alpha\beta}^{-1} = q_{\alpha\beta} \mathbf{K}_\alpha q_{\alpha\beta}^{-1}$. Hence

$$\chi(h) = \sum_{\alpha,\beta} \psi(d_\alpha^{-1} q_{\alpha\beta}^{-1} h q_{\alpha\beta} d_\alpha) \tag{4.7.23}$$

where the sum over α, β is restricted to those α, β for which $q_{\alpha\beta} \mathbf{K}_\alpha q_{\alpha\beta}^{-1} \supset h$.

Now
$$\mathbf{D}(d_\alpha^{-1} q_{\alpha\beta}^{-1} h q_{\alpha\beta} d_\alpha) = \mathbf{D}_\alpha(q_{\alpha\beta}^{-1} h q_{\alpha\beta}).$$

Therefore

$$\chi(h) = \sum_\alpha \{ \sum_\beta \operatorname{Tr} \mathbf{D}_\alpha(q_{\alpha\beta}^{-1} h q_{\alpha\beta})\}, \tag{4.7.24}$$

where for a given α the sum over β is restricted to those β for which $q_{\alpha\beta}^{-1} h q_{\alpha\beta} \in \mathbf{K}_\alpha$. But $q_{\alpha\beta} \in \mathbf{H}$ so $q_{\alpha\beta}^{-1} h q_{\alpha\beta} \in \mathbf{L}_\alpha$ is in fact the restriction. From eqn. (4.7.10) with h replacing g, $q_{\alpha\beta}$ replacing p_σ, and $\mathbf{L}_\alpha$ replacing $\mathbf{K}$ we see that the right-hand side of eqn. (4.7.24) is just the character of $\sum_\alpha (D_\alpha \downarrow \mathbf{L}_\alpha) \uparrow \mathbf{H}$, which proves the theorem.

We are now in a position to prove a theorem on the direct product of two induced representations.

THEOREM 4.7.7. *Let D be a representation of* $\mathbf{K}$ *and C a representation of* $\mathbf{H}$, *then*

$$(D \uparrow \mathbf{G}) \boxtimes (C \uparrow \mathbf{G}) \equiv \sum_\alpha \{(D_\alpha \boxtimes C) \downarrow \mathbf{L}_\alpha\} \uparrow \mathbf{G}. \tag{4.7.25}$$

From Theorem 4.7.3

$$(D \uparrow \mathbf{G}) \boxtimes (C \uparrow \mathbf{G}) \equiv [\{(D \uparrow \mathbf{G}) \downarrow \mathbf{H}\} \boxtimes C] \uparrow \mathbf{G}$$

which from Theorem 4.7.6 is equivalent to

$$\sum_\alpha \{[(D_\alpha \downarrow \mathbf{L}_\alpha) \uparrow \mathbf{H}] \boxtimes C\} \uparrow \mathbf{G}.$$

Now C is a representation of $\mathbf{H}$ and $(D_\alpha \downarrow \mathbf{L}_\alpha) \uparrow \mathbf{H}$ is a representation induced from $\mathbf{L}_\alpha$ into $\mathbf{H}$ so we may use Theorem 4.7.3 again to yield a representation equivalent to

$$\sum_\alpha \{[(C \downarrow \mathbf{L}_\alpha) \boxtimes (D_\alpha \downarrow \mathbf{L}_\alpha)] \uparrow \mathbf{H}\} \uparrow \mathbf{G}.$$

Now subduction is distributive with respect to the inner direct product and induction is transitive (Theorem 4.7.4) and so we obtain finally

$$\sum_{\alpha} \{(D_\alpha \boxtimes C) \downarrow \mathbf{L}_\alpha\} \uparrow \mathbf{G}, \text{ which proves the theorem.}$$

We shall write this in the form $\sum_\alpha E_\alpha \uparrow \mathbf{G}$ where E_α is a representation of $\mathbf{L}_\alpha = \mathbf{H} \cap \mathbf{K}_\alpha$. Suppose now that $\mathbf{M}$ is a subgroup of $\mathbf{G}$ and that B is a representation of $\mathbf{M}$. If we write $\mathbf{G} = \sum_\beta \mathbf{L}_\alpha b_\beta \mathbf{M}$ and let $\mathbf{N}_{\alpha\beta} = \mathbf{L}_\alpha \cap \mathbf{M}_\beta$ where $\mathbf{M}_\beta = b_\beta \mathbf{M} b_\beta^{-1}$ and define a representation B_β on $\mathbf{M}_\beta$ by the equation $\mathbf{B}_\beta(m_\beta) = \mathbf{B}(m)$ where $m_\beta = b_\beta m b_\beta^{-1}$ ($m_\beta \in \mathbf{M}_\beta$, $m \in \mathbf{M}$) then from Theorem 4.7.7

$$\begin{aligned}(B \uparrow \mathbf{G}) \boxtimes (E_\alpha \uparrow \mathbf{G}) &\equiv \sum_\beta \{(B_\beta \boxtimes E_\alpha) \downarrow \mathbf{N}_{\alpha\beta}\} \uparrow \mathbf{G} \\ &\equiv \sum_\beta \{(B_\beta \boxtimes D_\alpha \boxtimes C) \downarrow (\mathbf{M}_\beta \cap \mathbf{K}_\alpha \cap \mathbf{H})\} \uparrow \mathbf{G},\end{aligned}$$

from which we conclude

$$\begin{aligned}&(B \uparrow \mathbf{G}) \boxtimes (D \uparrow \mathbf{G}) \boxtimes (C \uparrow \mathbf{G}) \\ &\qquad \equiv \sum_\alpha \sum_\beta \{(B_\beta \boxtimes D_\alpha \boxtimes C) \downarrow (\mathbf{M}_\beta \cap \mathbf{K}_\alpha \cap \mathbf{H}) \uparrow \mathbf{G}\}. \qquad (4.7.26)\end{aligned}$$

Equation (4.7.26) can obviously be generalized by repeated application of Theorem 4.7.7 to give an expression for a direct product of as many induced representations as we want. However, we are interested in the particular case in which $\mathbf{M} = \mathbf{K}^i$, $\mathbf{K} = \mathbf{K}^j$ and $\mathbf{H} = \mathbf{K}^l$ (i.e. little groups in $\mathbf{G}$) and in which $B = \Gamma_p^i$, $D = \Gamma_q^j$ and $C = \Gamma_r^{l*}$ (i.e. small reps). What we want to know is the frequency of the rep A of $\mathbf{G}$ in $(\Gamma_p^i \uparrow \mathbf{G}) \boxtimes (\Gamma_q^j \uparrow \mathbf{G}) \boxtimes (\Gamma_r^{l*} \uparrow \mathbf{G})$. Using eqn. (4.7.26) and the frequency Theorem 4.7.2 this is the double sum over α and β of the frequencies of the identity rep in $(\Gamma_p^i)_\beta \boxtimes (\Gamma_q^j)_\alpha \boxtimes \Gamma_r^{l*}$ on the intersection $(\mathbf{K}_\beta^i \cap \mathbf{K}_\alpha^j \cap \mathbf{K}^l)$.

In the case of a space group, $\mathbf{G}$ is the whole space group itself and $\mathbf{K}^i$, $\mathbf{K}^j$, and $\mathbf{K}^l$ are little groups $\mathbf{G}^{\mathbf{k}_i}$, $\mathbf{G}^{\mathbf{k}_j}$, and $\mathbf{G}^{\mathbf{k}_l}$. Γ_p^i, Γ_q^j, and Γ_r^l are reps of the little groups $\mathbf{G}^{\mathbf{k}_i}$, $\mathbf{G}^{\mathbf{k}_j}$, and $\mathbf{G}^{\mathbf{k}_l}$ and $(\Gamma_p^i \uparrow \mathbf{G})$, $(\Gamma_q^j \uparrow \mathbf{G})$, and $(\Gamma_r^l \uparrow \mathbf{G})$ are representations induced by Γ_p^i, Γ_q^j, and Γ_r^l respectively, in $\mathbf{G}$. We therefore see that the d_α are the double coset representatives of $\mathbf{G}$ with respect to $\mathbf{G}^{\mathbf{k}_l}$ $(= \mathbf{K}^l = \mathbf{H})$ and $\mathbf{G}^{\mathbf{k}_j}$ $(= \mathbf{K}^j = \mathbf{K})$ (see eqn. (4.7.19)) and the b_β are the double coset representatives of $\mathbf{G}$ with respect to $\mathbf{G}^{\mathbf{k}_l} \cap \mathbf{G}_\alpha^{\mathbf{k}_j}$ $(= \mathbf{L}_\alpha = \mathbf{H} \cap \mathbf{K}_\alpha)$ and $\mathbf{G}^{\mathbf{k}_i}$ $(= \mathbf{K}^i = \mathbf{M})$. If $\mathbf{K}^i$ is the little group $\mathbf{G}^{\mathbf{k}_i}$ and b_β is identified with $\{\beta \mid \mathbf{v}\}$ then $\mathbf{K}_\beta^i$ is the little group of $\beta\mathbf{k}_i$ and $(\Gamma_p^i)_\beta$ is a small rep of $\mathbf{K}_\beta^i$. Remember that if $\{\gamma \mid \mathbf{w}\} \in \mathbf{K}_\beta^i$ then $(\boldsymbol{\Gamma}_p^i)_\beta(\{\gamma \mid \mathbf{w}\}) = \boldsymbol{\Gamma}_p^i(\{\beta \mid \mathbf{v}\}^{-1}\{\gamma \mid \mathbf{w}\}\{\beta \mid \mathbf{v}\})$, so that in evaluating the required frequency we shall need only the characters of the tabulated small reps of $\mathbf{G}^{\mathbf{k}_i}$ (as given in the tables in Chapter 5). Let the order of $\mathbf{K}_\beta^i \cap \mathbf{K}_\alpha^j \cap \mathbf{K}^l$ be $|\mathbf{N}_{\alpha\beta}|$ then putting our various results together and writing χ_p^i for the character of Γ_p^i we obtain the following formulae for $C^{ij,l}_{pq,r}$:

$$C^{ij,l}_{pq,r} = \sum_{\alpha}\sum_{\beta} \frac{1}{|\mathbf{N}_{\alpha\beta}|} \sum_{\{\gamma|\mathbf{w}\}\in\mathbf{N}_{\alpha\beta}} \chi^i_p(\{\beta|\mathbf{v}\}^{-1}\{\gamma|\mathbf{w}\}\{\beta|\mathbf{v}\}).$$

$$\chi^j_q(\{\alpha|\mathbf{u}\}^{-1}\{\gamma|\mathbf{w}\}\{\alpha|\mathbf{u}\}).\chi^{l*}_r(\{\gamma|\mathbf{w}\}) \quad (4.7.27)$$

where in eqn. (4.7.27) we identify $\{\alpha|\mathbf{u}\}$ with d_α. Performing the summation over translations $\{E|\mathbf{t}\}$ it is clear that $C^{ij,l}_{pq,r}$ vanishes unless $\exists\ \alpha, \beta$ such that

$$\beta\mathbf{k}_i + \alpha\mathbf{k}_j \equiv \mathbf{k}_l. \quad (4.7.28)$$

Denoting this restriction by a prime on the summations and writing $\{\gamma|\mathbf{w}\}$ as typical coset representatives of $\mathbf{T}$ with respect to $\mathbf{N}_{\alpha\beta}$ we obtain

$$C^{ij,l}_{pq,r} = \sum_{\alpha}{}' \sum_{\beta}{}' \frac{|\mathbf{T}|}{|\mathbf{N}_{\alpha\beta}|} \sum_{\{\gamma|\mathbf{w}\}\in\mathbf{N}_{\alpha\beta}/\mathbf{T}} \chi^i_p(\{\beta|\mathbf{v}\}^{-1}\{\gamma|\mathbf{w}\}\{\beta|\mathbf{v}\}).$$

$$\chi^j_q(\{\alpha|\mathbf{u}\}^{-1}\{\gamma|\mathbf{w}\}\{\alpha|\mathbf{u}\}).\chi^{l*}_r(\{\gamma|\mathbf{w}\}). \quad (4.7.29)$$

Since, in general, there are very few elements in $\mathbf{N}_{\alpha\beta}/\mathbf{T}$ and the restrictions on α and β are so severe that there is rarely more than one appropriate α and β to survive the triple summation the expression (4.7.29) is easy to calculate.

Therefore, summarizing, to evaluate $C^{ij,l}_{pq,r}$ *using eqn.* (4.7.29) *we must determine the co-star of* $\mathbf{k}_j$ *with respect to* $\mathbf{G}^{\mathbf{k}_l}$ *(i.e. all possible* $\alpha\mathbf{k}_j$*) and the co-star of* $\mathbf{k}_i$ *with respect to* $\mathbf{G}^{\mathbf{k}_l} \cap \mathbf{G}^{\mathbf{k}_j}_{\alpha}$ *(i.e. all possible* $\beta\mathbf{k}_i$*), and find all triples of vectors, one vector from each co-star, which together with* $\mathbf{k}_l$ *satisfy eqn.* (4.7.28).

To illustrate the results of this section we give an example using the space group $\mathbf{G} = P23\ (T^1)$ (Bradley 1966). This example is chosen to produce cases in which more than one term survives in the double summation in eqn. (4.7.29). To find examples in which this summation over α and β is not trivial it appears to be necessary to consider a space group for which the representation domain of the Brillouin zone (see Definition 3.7.1) is larger than the basic domain (see Definition 3.3.3), that is, where the order of the isogonal point group is lower than the order of the holosymmetric point group of the appropriate crystal system. The space group $P23\ (T^1)$ is based on the simple cubic Bravais lattice. Details of this lattice and its Brillouin zone appear in Chapter 3, the notation used for the operators being exactly the same as defined in Chapter 1. The isogonal point group of $P23\ (T^1)$ is the tetrahedral group $23\ (T)$ which contains the 12 elements $E, C_{2m}, C^{\pm}_{3j}$ $(m = x, y, z; j = 1, 2, 3, 4)$. We consider only two points of the representation domain of the Brillouin zone (see Fig. 3.13): $\Gamma = (0, 0, 0)$ and $M = (\frac{1}{2}, \frac{1}{2}, 0)$. Coordinates here are given as usual in units of the reciprocal lattice vectors $\mathbf{g}_1, \mathbf{g}_2$, and $\mathbf{g}_3$ which are in the k_x, k_y, and k_z directions, see Table 3.3. We define the points $M^+ = C^+_{31}M = (0, \frac{1}{2}, \frac{1}{2})$ and $M^- = C^-_{31}M =$

$(\frac{1}{2}, 0, \frac{1}{2})$. The star of Γ is just the point Γ by itself and the star of M consists of the three points M, M^+, and M^-.

TABLE 4.4

The small reps of $\mathbf{G}^\Gamma$ *and* $\mathbf{G}^M$

$\mathbf{G}^\Gamma$	E	$3C_{2m}$	$4C_{3j}^+$	$4C_{3j}^-$	$\mathbf{G}^M$	E	C_{2x}	C_{2y}	C_{2z}
A	1	1	1	1	A_1	1	1	1	1
${}^1E$	1	1	ω^*	ω	B_1	1	-1	-1	1
${}^2E$	1	1	ω	ω^*	B_2	1	-1	1	-1
T	3	-1	0	0	B_3	1	1	-1	-1

Notes to Table 4.4

(i) $\omega = \exp(2\pi i/3)$.

(ii) All translations in $\mathbf{G}^\Gamma$ are represented by the identity.

(iii) In $\mathbf{G}^M$, $\{E \mid \mathbf{t}_1\}$ and $\{E \mid \mathbf{t}_2\}$ are represented by -1 and $\{E \mid \mathbf{t}_3\}$ is represented by $+1$.

In Table 4.4 we list the characters of the small reps of the little groups $\mathbf{G}^\Gamma$ and $\mathbf{G}^M$. The little group of Γ is the whole space group, but the little group of M contains multiples of the translations with just four rotation operators: E, C_{2x}, C_{2y}, and C_{2z}.

For our example we consider the inner Kronecker products of pairs of reps induced in $\mathbf{G}$ from $\mathbf{G}^M$. We write MA_1 for $A_1 \uparrow \mathbf{G}$, etc. In the notation of the earlier part of this section $\mathbf{k}_j = M = (\frac{1}{2}, \frac{1}{2}, 0)$ and $\mathbf{k}_i = M = (\frac{1}{2}, \frac{1}{2}, 0)$. It is soon verified that with this choice of $\mathbf{k}_j$ and $\mathbf{k}_i$ the only possible values of $\mathbf{k}_l$ that can appear on the right-hand side of eqn. (4.7.1) are $\mathbf{k}_l = \Gamma = (0, 0, 0)$ and $\mathbf{k}_l = M = (\frac{1}{2}, \frac{1}{2}, 0)$.

Consider first the case $\mathbf{k}_l = \Gamma$. From eqn. (4.7.19) with $\mathbf{H} = \mathbf{G}^\Gamma$ and $\mathbf{K} = \mathbf{G}^M$ it is soon verified that there is only one term in the first double coset decomposition and the single representative $d_\alpha = \{\alpha \mid \mathbf{0}\}$ may be chosen to be the identity $\{E \mid \mathbf{0}\}$. Hence $\mathbf{K}_\alpha = \mathbf{G}^M$ and the group intersection $\mathbf{L}_\alpha = \mathbf{G}^\Gamma \cap \mathbf{G}^M = \mathbf{G}^M$. Next we form the second double coset decomposition: $\mathbf{G} = \sum_\beta \mathbf{G}^M b_\beta \mathbf{G}^M$, and this time it can be verified that three terms survive, and the three representatives $b_\beta = \{\beta \mid \mathbf{0}\}$ may be chosen to be $\{E \mid \mathbf{0}\}$, $\{C_{31}^+ \mid \mathbf{0}\}$, and $\{C_{31}^- \mid \mathbf{0}\}$. Thus the set $(\alpha\mathbf{k}_j)$ consists of the point M alone and the set $(\beta\mathbf{k}_i)$ consists of the three points M, M^+, and M^-. The only pair to survive the restriction of eqn. (4.7.28)

$$\beta\mathbf{k}_i + \alpha\mathbf{k}_j \equiv \mathbf{k}_l \tag{4.7.28}$$

has $\alpha = \beta = E$ (i.e. $M + M \sim \Gamma$). The triple intersection group $\mathbf{N}_{\alpha\beta} = \mathbf{G}^M \cap \mathbf{G}^M = \mathbf{G}^M$.

Consider now the case $\mathbf{k}_l = M$. This time the first double coset decomposition consists of three terms and the corresponding values of α are E, C_{31}^+, and C_{31}^-. Since $\mathbf{G}^M$ is invariant under these three operations the groups $\mathbf{G}^M$, $\mathbf{G}^{M^+}$, and $\mathbf{G}^{M^-}$ coincide so that in each case $\mathbf{L}_\alpha = \mathbf{G}^M$. In all three cases the second coset decomposition contains three terms and the corresponding values of β are again E, C_{31}^{+}, and C_{31}^{-}. Of the nine possible pairs of α, β values, we now have two pairs that are compatible with the restriction of eqn. (4.7.28),

$$\beta\mathbf{k}_i + \alpha\mathbf{k}_j \equiv \mathbf{k}_l \tag{4.7.28}$$

when $\beta\mathbf{k}_i = M^+$ and $\alpha\mathbf{k}_j = M^-$, and when $\beta\mathbf{k}_i = M^-$ and $\alpha\mathbf{k}_j = M^+$. In both cases $\mathbf{N}_{\alpha\beta} = \mathbf{G}^M$.

Consider now in detail the product $MB_2 \boxtimes MB_3$. There are three cases to consider:

(i) $\alpha\mathbf{k}_j = M, \quad \beta\mathbf{k}_i = M, \quad \mathbf{k}_l = \Gamma.$

From Table 4.4 the appropriate part of the sum (4.7.29) becomes

$$\tfrac{1}{4}\{\chi_r^\Gamma(E) - \chi_r^\Gamma(C_{2x}) - \chi_r^\Gamma(C_{2y}) + \chi_r^\Gamma(C_{2z})\},$$

which is zero when $r = A$, ${}^1E$, or ${}^2E$ and unity when $r = T$.

(ii) $\alpha\mathbf{k}_j = M^+, \quad \beta\mathbf{k}_i = M^-, \quad \mathbf{k}_l = M.$

Now $C_{31}^- C_{2x} C_{31}^+ = C_{2z}$, $C_{31}^- C_{2y} C_{31}^+ = C_{2x}$, $C_{31}^- C_{2z} C_{31}^+ = C_{2y}$, and on using these relations and Table 4.4 the appropriate part of the sum (4.7.29) is now $\tfrac{1}{4}\{\chi_r^M(E) + \chi_r^M(C_{2x}) + \chi_r^M(C_{2y}) + \chi_r^M(C_{2z})\}$, which is zero when $r = B_1, B_2, B_3$ and unity when $r = A_1$.

TABLE 4.5

The inner Kronecker products of representations belonging to M for P23

$MA_1 \boxtimes MA_1 \equiv \Gamma A + \Gamma^1E + \Gamma^2E + 2MA_1$	$MB_1 \boxtimes MB_2 \equiv \Gamma T + MB_3 + MA_1$
$MA_1 \boxtimes MB_1 \equiv \Gamma T + MB_2 + MB_3$	$MB_1 \boxtimes MB_3 \equiv \Gamma T + MB_2 + MA_1$
$MA_1 \boxtimes MB_2 \equiv \Gamma T + MB_3 + MB_1$	$MB_2 \boxtimes MB_2 \equiv \Gamma A + \Gamma^1E + \Gamma^2E + 2MB_2$
$MA_1 \boxtimes MB_3 \equiv \Gamma T + MB_1 + MB_2$	$MB_2 \boxtimes MB_3 \equiv \Gamma T + MB_1 + MA_1$
$MB_1 \boxtimes MB_1 \equiv \Gamma A + \Gamma^1E + \Gamma^2E + 2MB_1$	$MB_3 \boxtimes MB_3 \equiv \Gamma A + \Gamma^1E + \Gamma^2E + 2MB_3$

Notes to Table 4.5

(i) Each line of the table is an equation like eqn. (4.7.30).

(ii) To obtain a Clebsch–Gordan coefficient $C_{pq,r}^{ij,l}$, select the number multiplying the symbol corresponding to r on the right-hand side of the equation whose left-hand side contains the symbols p and q. For example, $C_{A_1A_1,A_1}^{MM,M} = 2$.

(iii) $\alpha\mathbf{k}_j = M^-, \quad \beta\mathbf{k}_i = M^+, \quad \mathbf{k}_l = M.$

This time the appropriate part of the sum (4.7.29) is $\tfrac{1}{4}\{\chi_r^M(E) - \chi_r^M(C_{2x}) - \chi_r^M(C_{2y}) + \chi_r^M(C_{2z})\}$ which is zero when $r = A_1$, B_2, or B_3 and unity when $r = B_1$.

Collecting these results together, the equation corresponding to (4.7.1) is

$$MB_2 \boxtimes MB_3 \equiv \Gamma T + MA_1 + MB_1. \tag{4.7.30}$$

As a check on eqn. (4.7.30), note that the dimension of each side is 9, it being remembered that reps induced from $\mathbf{G}^M$ in $P23$ are of dimension 3. In Table 4.5 we list all 10 such products for the reps induced from $\mathbf{G}^M$.

Space-group selection rules in crystals

The importance, in physical terms, of the theory described in this section lies in the determination of selection rules for various quantum-mechanical processes in crystalline ionic, metallic, or semiconducting solids. One example of such a process might be, for instance, the absorption of a photon by a specimen of a solid and the production within the solid of two phonons (with wave vectors $\mathbf{k}$ and $-\mathbf{k}$ to conserve momentum). Other processes might involve the absorption of a photon and the production of two magnons (one at $\mathbf{k}$ and one at $-\mathbf{k}$), or of one phonon (at $\mathbf{k}$) and one magnon (at $-\mathbf{k}$) or of just one phonon (at $\mathbf{k} = 0$) or of just one magnon (at $\mathbf{k} = 0$). Selection rules will also arise in studying the interactions between quantum-mechanical particles or quasi-particles in solids, for example, electron–phonon interactions or magnon–phonon interactions. The determination of the selection rules that govern whether such a process is allowed or not involves the reduction of Kronecker products of space-group reps (or, possibly, the reduction of symmetrized Kronecker products of space-group reps, see section 4.8). Suppose that Ψ^i_r is a wave function belonging to row r of the rep Γ^i and that W is a self-adjoint operator belonging to the rep Γ^W. The transition between an initial state Ψ^k_t and a final state Ψ^i_r under the operator W is governed by the value of the matrix element

$$W^{ik}_{rt} = \langle \Psi^i_r, W\Psi^k_t \rangle. \tag{4.7.31}$$

The transition is said to be forbidden by symmetry if the triple Kronecker product $\Gamma^{i*} \boxtimes \Gamma^k \boxtimes \Gamma^W$ does not contain the identity representation of $\mathbf{G}$, for then W^{ik}_{rt} vanishes identically (for proof of this result see Hamermesh (1962), Chapter 6). Various authors have studied the problem of determining these selection rules in crystalline solids (Balkanski and Nusimovici 1964, Birman 1962*b*, 1963, 1966*a*, Birman, Lax, and Loudon 1966, Bradley 1966, Burstein, Johnson, and Loudon 1965, Chen and Hsieh 1965, Cornwell 1966, Elliott and Loudon 1960, Elliott and Thorpe 1967, Gorzkowski 1964*b*, Hsieh and Chen 1964, Hsü and Hsieh 1965, Lax 1965, Lax and Hopfield 1961, Loudon 1965, Winston and Halford 1949, Zak 1962).

In determining these selection rules the essential problem is to see whether it is possible to show group-theoretically that the matrix element W^{ik}_{rt} must vanish, where W is the quantum-mechanical operator of the physical influence that causes the transition. It is possible to assign Ψ^i_r, Ψ^k_t, and W to the various reps of the complete space group $\mathbf{G}$ and to determine whether any particular process is forbidden by showing that $\langle \Psi^i_r, W\Psi^k_t \rangle$, which is then a product of three space-group reps, does not belong to the totally symmetrical rep of $\mathbf{G}$; this is called the *full-group method*. The analysis of such a triple product is best done in two stages, first by forming the Kronecker product $\Gamma^k \boxtimes \Gamma^W$ of the reps to which Ψ^k_t and W belong. Since the Kronecker product of the rep Γ^{i*} of $\mathbf{G}$, to which Ψ^{i*}_r belongs, with the representation $\Gamma^k \boxtimes \Gamma^W$ must contain the totally symmetrical representation of $\mathbf{G}$ if the process is to be allowed, it follows that we have to determine whether or not $\Gamma^k \boxtimes \Gamma^W$ contains

Γ^i. The use of eqn. (4.7.29) will in fact determine exactly how many times $\Gamma^k \boxtimes \Gamma^W$ contains Γ^i. In practice one often wants to try several different operators W (for example, electric dipole, magnetic dipole, electric quadrupole, etc.) between two given states to see which physical interaction leads to an allowed transition between these two states; in this situation one would form $\Gamma^k \boxtimes \Gamma^{i*}$ first, rather than $\Gamma^k \boxtimes \Gamma^W$. Since the full-group method uses the complete space-group reps, that is the induced reps $(\Gamma_p^{\mathbf{k}} \uparrow \mathbf{G})$, the following question will have been answered: 'Is the transition allowed between Ψ_t^k and Ψ_r^i, where Ψ_t^k is a Bloch function of given space-group symmetry belonging to *any vector* from the star of $\mathbf{k}_k$, and Ψ_r^i belongs to *any vector* from the star of $\mathbf{k}_i$?'

In an alternative method, the *subgroup method*, Ψ_t^k, Ψ_r^i, and W are assigned directly to small reps of little groups $\mathbf{G}^{\mathbf{k}_k}$, $\mathbf{G}^{\mathbf{k}_i}$, and $\mathbf{G}^{\mathbf{k}_W}$ (instead of to the reps induced in $\mathbf{G}$ from these small reps) and the product of these reps is tested in a similar way (Elliott and Loudon 1960). Since the subgroup method uses the reps of little groups the following question will have been answered (after $\mathbf{k}_W$ has been allowed to vary over all vectors in its star): 'Is the transition allowed between Ψ_t^k and Ψ_r^i, where Ψ_t^k is a Bloch function of given space-group symmetry of *precisely* the wave vector $\mathbf{k}_k$ and with a similar restriction for Ψ_r^i?' The method to use will depend on which of the two questions one wishes to answer. If one wished to use the subgroup method to answer the full-group question it would be necessary to allow $\mathbf{k}_k$ and $\mathbf{k}_i$ to vary over their respective stars, unless, of course, one has recourse to eqn. (4.7.28) and its implications. Indeed this equation can be thought of as a device for reducing the excessive labour that one would otherwise have in answering the full-group question by recourse to subgroup methods. It turns out that the restrictions on α and β over which the summations of eqn. (4.7.29) are performed are so severe that in very many cases only one set of α and β survives, in which case one application of the subgroup method is sufficient to answer the full-group question. This is not the case if eqn. (4.7.28) is satisfied for more than one set of values of α and β. It is in this more complicated situation that a mistake can easily be made if one is proceeding by trial and error rather than by strict use of eqn. (4.7.28). To see how this could happen, note that the vector equations corresponding to the two allowed α, β pairs in the above example are $M^+ + M^- \equiv M$ and $M^- + M^+ \equiv M$. Someone relying only on the subgroup method, having discovered one of these equations, might well overlook the possibility of a contribution from the other. If the precise value of the Clebsch–Gordan coefficient is required it is imperative to use either the method of this section, which in effect makes the subgroup technique precise by deriving it from first principles, or an entirely full-group method such as that employed by Birman (1962*b*, 1963).

4.8. Symmetrized and anti-symmetrized squares of induced representations

In section 2.6 we introduced the ideas of symmetrized and anti-symmetrized powers of a rep Δ; we illustrated the general theory by considering the symmetrized and anti-symmetrized squares of the reps of the crystallographic point groups. In this section we shall show how to decompose the space of the Kronecker square of an induced representation into its symmetric and anti-symmetric parts. We shall then apply the theory to space-group representations by extending the example of the previous section to demonstrate the results. The symmetrized squares and cubes of space-group reps for some points in the Brillouin zones of the diamond structure ($Fd3m$, O_h^7) and the zinc blende structure ($F\bar{4}3m$, T_d^2) have been tabulated by Birman (1962*b*), while the symmetrized squares for the hexagonal close-packed ($P6_3/mmc$, D_{6h}^4) and wurtzite ($P6_3mc$, C_{6v}^4) structures have been tabulated by Chen and Hsieh (1965). The theory presented in this section is an amplification of the work of Mackey (1952) presented from a somewhat different point of view (Bradley and Davies 1970); in particular, our proofs are constructive in nature and lead to a definite prescription for carrying out the decomposition.

We illustrate, briefly, the physical application of the ideas of symmetrized and anti-symmetrized products to the study of quantum-mechanical selection rules. If we suppose that the group of the unitary symmetry operations of the Hamiltonian of a system is the space group **G**, then we have seen in the previous section how to determine selection rules for various processes that may occur in the system. If now the group **G** is augmented by the addition of θ, the operation of time-reversal symmetry, the product $\Gamma^i \boxtimes \Gamma^k$ of the initial and final states in the usual triple Kronecker product used in section 4.7 may have to be replaced by a symmetrized or anti-symmetrized product. This happens when the final state is related to the time-reverse of the initial state and the effect is that extra selection rules may be provided. The original papers demonstrating this were on the subject of the Jahn–Teller effect (Jahn 1938, Jahn and Teller 1937) but the following piece of theory due to Lax (1962) seems to be all that is really necessary. If the wave function Ψ_r^i of the final state is related by time-reversal symmetry to the initial state, that is, $\Psi_r^i = \theta\Psi_r^k$, then we consider the matrix element

$$W_{rt}^k = \langle\theta\Psi_r^k, W\Psi_t^k\rangle \tag{4.8.1}$$

$$= \langle\theta W\Psi_t^k, \theta^2\Psi_r^k\rangle \tag{4.8.2}$$

since θ is anti-unitary (see Chapter 7). Now $\Delta(\theta^2) = \omega = \pm 1$, and we make the not very restrictive assumption that $(\theta W\theta^{-1})^\dagger = \alpha W$, where the dagger denotes the adjoint operator, and where $\alpha = \pm 1$. Substituting these relations into eqn. (4.8.2) we find that

$$W_{rt}^k = \alpha\omega W_{tr}^k = \pm W_{tr}^k. \tag{4.8.3}$$

Since both Ψ_t^k and $\Phi_r^k = (\theta\Psi_r^k)^*$ belong to Γ^k we conclude from eqn. (4.8.3) that the existence of a selection rule depends in this case on the behaviour of $\int(\Phi_r^k W\Psi_t^k \pm \Phi_t^k W\Psi_r^k)\, d\tau$ under the operations of **G**. The invariance of the subspaces spanned by the functions $(\Phi_r^k\Psi_t^k \pm \Phi_t^k\Psi_r^k)$ taken together with the usual theory on selection rules shows that in this case the transition is forbidden by symmetry if the product $([\Gamma^k \boxtimes \Gamma^k] \boxtimes \Gamma^W)$ in the case of the plus sign holding in eqn. (4.8.3) (or if the product $(\{\Gamma^k \boxtimes \Gamma^k\} \boxtimes \Gamma^W)$ in the case of the minus sign holding in eqn. (4.8.3)), does not contain the identity representation of **G**. An equivalent criterion is whether the symmetrized (or anti-symmetrized) square of Γ^k contains any of the irreducible components of Γ^{W*}. The above theory can be extended from the case of $\Psi_r^i = \theta\Psi_r^k$ to the case of $\Psi_r^i = \{R \mid \mathbf{v}\}\theta\Psi_r^k$ where $\{R \mid \mathbf{v}\}$ is any element of the space group **G** (see Lax (1962)). Another physical application which requires for its analysis the symmetrized and anti-symmetrized powers of a given rep is the Landau theory of second-order phase transitions, which was mentioned at the beginning of section 2.6 (for references see section 2.6).

We now wish to study the symmetrized and anti-symmetrized squares of *induced* representations of a group **G**. That is, we wish to consider what happens when **S** (in the notation of section 2.6) is identified with a group **G**, **H** is a subgroup of **G** and Δ (in the notation of section 2.6) is identified with the representation of **G** induced from a representation D of **H**. The decomposition of the Kronecker square of $D \uparrow \mathbf{G}$ is covered by the theory of the last section, the main result being that of eqn. (4.7.25)

$$(D \uparrow \mathbf{G}) \boxtimes (D \uparrow \mathbf{G}) \equiv \sum_{\alpha} [(D_\alpha \boxtimes D) \downarrow \mathbf{L}_\alpha] \uparrow \mathbf{G}. \tag{4.8.4}$$

Here the terms in the sum over α are in one-to-one correspondence with the double coset decomposition

$$\mathbf{G} = \sum_{\alpha} \mathbf{H} d_\alpha \mathbf{H}, \tag{4.8.5}$$

and $\mathbf{L}_\alpha = \mathbf{H} \cap \mathbf{H}_\alpha$, where $\mathbf{H}_\alpha = d_\alpha \mathbf{H} d_\alpha^{-1}$. Also D_α is the representation of $\mathbf{H}_\alpha$ such that $\mathbf{D}_\alpha(d_\alpha h d_\alpha^{-1}) = \mathbf{D}(h)$ for all $h \in \mathbf{H}$. We require to rearrange or to modify the right-hand side of eqn. (4.8.4) so that one part can be identified with the symmetrized square $[(D \uparrow \mathbf{G}) \boxtimes (D \uparrow \mathbf{G})]$ and the other part with the anti-symmetrized square $\{(D \uparrow \mathbf{G}) \boxtimes (D \uparrow \mathbf{G})\}$. We use the same notation as in the previous section: D is a representation of **H** of dimension d with basis $\langle\phi| = \langle\phi_1, \phi_2, \ldots, \phi_d|$, and D_α is the representation of $\mathbf{H}_\alpha$ with basis $d_\alpha\langle\phi| = \langle\phi_\alpha| = \langle\phi_{\alpha 1}, \phi_{\alpha 2}, \ldots, \phi_{\alpha d}|$. Indeed

$$\begin{aligned} d_\alpha h d_\alpha^{-1}\langle\phi_\alpha| &= d_\alpha h\langle\phi| \\ &= d_\alpha\langle\phi|\, \mathbf{D}(h) \\ &= \langle\phi_\alpha|\, \mathbf{D}(h), \end{aligned} \tag{4.8.6}$$

thereby checking the relation $\mathbf{D}_\alpha(d_\alpha h d_\alpha^{-1}) = \mathbf{D}(h)$. The representation $(D \uparrow \mathbf{G})$ has basis $p_\sigma\langle\phi|$ for $\sigma = 1$ to $|\mathbf{G}|/|\mathbf{H}|$ where the p_σ are the coset representatives in the decomposition $\mathbf{G} = \sum_\sigma p_\sigma \mathbf{H}$. Also, from the last section, we remember the coset decomposition $\mathbf{H} = \sum_\gamma q_{\alpha\gamma} \mathbf{L}_\alpha$ and hence the basis for $[(D_\alpha \boxtimes D) \downarrow \mathbf{L}_\alpha] \uparrow \mathbf{G}$ for fixed α

is the set of functions $p_\sigma q_{\alpha\gamma}(\phi_{\alpha i}, \phi_j)$, $i = 1$ to d, $j = 1$ to d, $\gamma = 1$ to $|\mathbf{H}|/|\mathbf{L}_\alpha|$, $\sigma = 1$ to $|\mathbf{G}|/|\mathbf{H}|$. By virtue of Theorem 4.7.7 the vector space V_α spanned by these $d^2|\mathbf{G}|/|\mathbf{L}_\alpha|$ functions is invariant under $(D \uparrow \mathbf{G}) \boxtimes (D \uparrow \mathbf{G})$. We may write symbolically

$$V_\alpha = \sum_{\sigma, \gamma, i, j} p_\sigma q_{\alpha\gamma}(\phi_{\alpha i}, \phi_j), \tag{4.8.7}$$

where the right-hand side means the set of all linear combinations of the functions $p_\sigma q_{\alpha\gamma}(\phi_{\alpha i}, \phi_j)$. The space V_α is independent of the particular double coset representative chosen, in the sense that if instead of d_α we use $d_\beta = h_a d_\alpha h_b$, h_a, $h_b \in \mathbf{H}$, then V_β coincides with V_α. This was implicit in section 4.7 but can be seen as follows:

$$\begin{aligned} \mathbf{L}_\beta = \mathbf{H} \cap \mathbf{H}_\beta &= \mathbf{H} \cap d_\beta \mathbf{H} d_\beta^{-1} = \mathbf{H} \cap h_a d_\alpha h_b \mathbf{H} h_b^{-1} d_\alpha^{-1} h_a^{-1} \\ &= \mathbf{H} \cap h_a d_\alpha \mathbf{H} d_\alpha^{-1} h_a^{-1} = \mathbf{H} \cap h_a \mathbf{H}_\alpha h_a^{-1}, \quad \text{and} \end{aligned}$$

hence $h_a^{-1}\mathbf{L}_\beta h_a = h_a^{-1}\mathbf{H}h_a \cap \mathbf{H}_\alpha = \mathbf{H} \cap \mathbf{H}_\alpha = \mathbf{L}_\alpha$. Thus if $\mathbf{H} = \sum_\gamma q_{\alpha\gamma}\mathbf{L}_\alpha$ then we also have $\mathbf{H} = \mathbf{H}h_a^{-1} = \sum_\gamma q_{\alpha\gamma}h_a^{-1}\mathbf{L}_\beta$ and so for each γ we may choose $q_{\beta\gamma} = q_{\alpha\gamma}h_a^{-1}$ and $\mathbf{H} = \sum_\gamma q_{\beta\gamma}\mathbf{L}_\beta$. Then

$$\begin{aligned} V_\beta &= \sum_{\sigma, \gamma, i, j} p_\sigma q_{\beta\gamma}(\phi_{\beta i}, \phi_j) \\ &= \sum_{\sigma, \gamma, i, j} p_\sigma q_{\alpha\gamma} h_a^{-1}(d_\beta \phi_i, \phi_j) \\ &= \sum_{\sigma, \gamma, i, j} p_\sigma q_{\alpha\gamma}(d_\alpha h_b \phi_i, h_a^{-1}\phi_j) \\ &= \sum_{\sigma, \gamma, k, l} p_\sigma q_{\alpha\gamma}(\phi_{\alpha k}, \phi_l) \\ &= V_\alpha. \end{aligned} \tag{4.8.8}$$

The vector space V, which is the carrier space for the entire representation $(D \uparrow \mathbf{G}) \boxtimes (D \uparrow \mathbf{G})$ is, of course, already proved to be the direct sum $\sum_\alpha V_\alpha$, where the sum over α is taken over the distinct double cosets, as in eqn. (4.8.5). Thus no two vector spaces V_α and V_β coincide unless d_α and d_β belong to the same double coset.

For reasons which will soon emerge, given a particular double coset representative d_α it is useful to define $d_\alpha^{-1}\phi_i \equiv \phi_{\bar{\alpha}i}$ and to write $\mathbf{L}_{\bar{\alpha}} = \mathbf{H} \cap \mathbf{H}_{\bar{\alpha}} = \mathbf{H} \cap d_\alpha^{-1}\mathbf{H}d_\alpha$. Note that $\mathbf{L}_{\bar{\alpha}} = d_\alpha^{-1}\mathbf{L}_\alpha d_\alpha$ and that because $\mathbf{G} = \sum_\sigma p_\sigma \mathbf{H}$ and $\mathbf{H} = \sum_\gamma q_{\alpha\gamma}\mathbf{L}_\alpha$ then $\mathbf{G} = \mathbf{G}d_\alpha = \sum_{\sigma, \gamma} p_\sigma q_{\alpha\gamma}\mathbf{L}_\alpha d_\alpha = \sum_{\sigma, \gamma} p_\sigma q_{\alpha\gamma} d_\alpha \mathbf{L}_{\bar{\alpha}}$. Hence as σ and γ run over all possible values the set $p_\sigma q_{\alpha\gamma} d_\alpha$ forms a complete set of coset representatives of $\mathbf{L}_{\bar{\alpha}}$ in $\mathbf{G}$. Thus if we define the vector space $W_{\bar{\alpha}}$ as the carrier space of $[(D \boxtimes D_{\bar{\alpha}}) \downarrow \mathbf{L}_{\bar{\alpha}}] \uparrow \mathbf{G}$ we find

$$\begin{aligned} W_{\bar{\alpha}} &= \sum_{\sigma, \gamma, i, j} p_\sigma q_{\alpha\gamma} d_\alpha(\phi_i, \phi_{\bar{\alpha}j}) \\ &= \sum_{\sigma, \gamma, i, j} p_\sigma q_{\alpha\gamma}(\phi_{\alpha i}, \phi_j) = V_\alpha. \end{aligned} \tag{4.8.9}$$

Hence the vector spaces $W_{\bar{\alpha}}$ and V_α coincide. Similarly W_α and $V_{\bar{\alpha}}$ coincide. Now the character of $[(D \boxtimes D_\alpha) \downarrow \mathbf{L}_\alpha] \uparrow \mathbf{G}$ is clearly the same as that of $[(D_\alpha \boxtimes D) \downarrow \mathbf{L}_\alpha] \uparrow \mathbf{G}$

and since the first of these is defined on $W_\alpha = V_{\bar{\alpha}}$ and the second is defined on V_α it follows that the representations defined on V_α and $V_{\bar{\alpha}}$ are equivalent. There are now two cases which can occur: either V_α and $V_{\bar{\alpha}}$ coincide (and trivially define equivalent representations by some equivalence transformation) or V_α and $V_{\bar{\alpha}}$ are distinct (but nevertheless define equivalent representations). Now from what has gone previously we know that V_α and $V_{\bar{\alpha}}$ coincide if and only if $\mathbf{H}d_\alpha\mathbf{H} = \mathbf{H}d_\alpha^{-1}\mathbf{H}$, so that which of the two cases occurs depends critically upon whether the double coset $\mathbf{H}d_\alpha\mathbf{H}$ is self-inverse or not (for, of course, $(\mathbf{H}d_\alpha\mathbf{H})^{-1} = \mathbf{H}d_\alpha^{-1}\mathbf{H}$, since $\mathbf{H} = \mathbf{H}^{-1}$).

Consider first the case in which V_α and $V_{\bar{\alpha}}$ are distinct. Now $V_\alpha = \sum_{\sigma,\gamma,i,j} p_\sigma q_{\alpha\gamma}(\phi_{\alpha i}, \phi_j)$ and $V_{\bar{\alpha}} = W_\alpha = \sum_{\sigma,\gamma,i,j} p_\sigma q_{\alpha\gamma}(\phi_i, \phi_{\alpha j})$. These are distinct spaces and define equivalent representations. Clearly it is possible to form the direct sum $(V_\alpha + V_{\bar{\alpha}})$ and to decompose this direct sum into an alternative one $(V_\alpha^+ + V_\alpha^-)$ in which $V_\alpha^\pm$ are given by

$$V_\alpha^\pm = \sum_{\sigma,\gamma,i,j} p_\sigma q_{\alpha\gamma}\{(\phi_{\alpha i}, \phi_j) \pm (\phi_j, \phi_{\alpha i})\} \tag{4.8.10}$$

with upper and lower signs corresponding. From the form of the right-hand side of eqn. (4.8.10) it is also clear that V_α^+ is a subspace of V^+ and V_α^- is a subspace of V^-, where V^+ and V^- are the symmetrized and anti-symmetrized subspaces of V itself. Moreover, the representations on V_α^+, V_α^-, V_α, $V_{\bar{\alpha}}$ are all of them equivalent, the first two being derived from the last two by the simple equivalence transformation expressed by eqn. (4.8.10). Thus if V_α and $V_{\bar{\alpha}}$ are distinct, that is, if d_α and d_α^{-1} belong to different double cosets, $[(D\uparrow\mathbf{G}) \boxtimes (D\uparrow\mathbf{G})]$ and $\{(D\uparrow\mathbf{G}) \boxtimes (D\uparrow\mathbf{G})\}$ will both contain a representation equivalent to $(D_\alpha \boxtimes D)\downarrow\mathbf{L}_\alpha)\uparrow\mathbf{G}$.

Now consider the case in which $V_\alpha = V_{\bar{\alpha}}$. There are two possibilities which require separate treatment, as the decomposition depends upon whether the carrier spaces of D_α and D are identical or distinct. These carrier spaces are identical if and only if $d_\alpha \in \mathbf{H}$, in which case d_α may be chosen to be the identity d_ε. The $\mathbf{H}_\varepsilon = \mathbf{H}$ and $\mathbf{L}_\varepsilon = \mathbf{H} \cap \mathbf{H}_\varepsilon = \mathbf{H}$. The term now under consideration in eqn. (4.8.4) is just $(D \boxtimes D)\uparrow\mathbf{G}$ and $V_\varepsilon = \sum_{\sigma,i,j} p_\sigma(\phi_i, \phi_j)$. It is immediately clear that V_ε decomposes into the direct sum $(V_\varepsilon^+ + V_\varepsilon^-)$ where

$$V_\varepsilon^\pm = \sum_{\sigma,i,j} p_\sigma\{(\phi_i, \phi_j) \pm (\phi_j, \phi_i)\} \tag{4.8.11}$$

and that V_ε^+ is a subspace of V^+, and V_ε^- is a subspace of V^-. From eqn. (4.8.11) we see that V_ε^+ defines the representation $[D \boxtimes D]\uparrow\mathbf{G}$ and V_ε^- defines the representation $\{D \boxtimes D\}\uparrow\mathbf{G}$, the dimensions of these representations being $\frac{1}{2}d(d+1)k$ and $\frac{1}{2}d(d-1)k$ respectively with $k = |\mathbf{G}|/|\mathbf{H}|$. The difference in dimension between these two representations is therefore equal to dk. Since $[(D\uparrow\mathbf{G}) \boxtimes (D\uparrow\mathbf{G})]$ and $\{(D\uparrow\mathbf{G}) \boxtimes (D\uparrow\mathbf{G})\}$ are of dimension $\frac{1}{2}dk(dk+1)$ and $\frac{1}{2}dk(dk-1)$, and also have a difference in dimension equal to dk, it follows that this difference is entirely accounted for by the decomposition of V_ε. This makes it reasonable that any other space V_α for which $V_\alpha = V_{\bar{\alpha}}$ will somehow yield spaces of equal dimensionality in the final decomposition of V. We now embark upon some fairly elaborate analysis to show how this in fact occurs.

We are now considering the case in which $\mathbf{H}d_\alpha\mathbf{H} = \mathbf{H}d_\alpha^{-1}\mathbf{H}$ and d_α cannot be chosen to be equal to the identity d_ε. Therefore there exist $h_i, h_j, h_k, h_l \in \mathbf{H}$ such that $h_i d_\alpha h_j = h_k d_\alpha^{-1} h_l$; that is $d_\alpha h_j h_l^{-1} = h_i^{-1} h_k d_\alpha^{-1}$. Thus $d_\alpha\mathbf{H} \cap \mathbf{H}d_\alpha^{-1}$ is non empty. Let z be any element of this intersection; then z is not in $\mathbf{H}$ and we may write $z = d_\alpha h = h' d_\alpha^{-1}$ where $h, h' \in \mathbf{H}$. Now

$$z^2 = d_\alpha h h' d_\alpha^{-1} = h'h \in d_\alpha \mathbf{H} d_\alpha^{-1} \cap \mathbf{H} = \mathbf{L}_\alpha$$

and furthermore

$$\begin{aligned} z\mathbf{H}z^{-1} &= d_\alpha h \mathbf{H} h^{-1} d_\alpha^{-1} = \mathbf{H}_\alpha, \\ z\mathbf{H}_\alpha z^{-1} &= h' d_\alpha^{-1} d_\alpha \mathbf{H} d_\alpha^{-1} d_\alpha h'^{-1} = \mathbf{H}. \end{aligned}$$

Thus we may construct the group

$$\mathbf{M}_\alpha = \mathbf{L}_\alpha + z\mathbf{L}_\alpha \tag{4.8.12}$$

and $\mathbf{L}_\alpha$ is an invariant subgroup of $\mathbf{M}_\alpha$ of index 2. Starting from the representation $(D_\alpha \boxtimes D)$ on $\mathbf{L}_\alpha$ with basis $\langle\phi_\alpha| \langle\phi|$ we can construct the induced representation $C_\alpha = (D_\alpha \boxtimes D) \uparrow \mathbf{M}_\alpha$ with basis $(\langle\phi_\alpha| \langle\phi|, z\langle\phi_\alpha| \langle\phi|)$. The analysis is analogous to that given in section 4.5 with $p = 2$ and $r = z$ so we shall leave out the details, merely quoting the result. If $l \in \mathbf{L}_\alpha$, then the induced representation is given in block-matrix form by the expressions

$$\mathbf{C}_\alpha(l) = \begin{pmatrix} \mathbf{D}(d_\alpha^{-1} l d_\alpha) \otimes \mathbf{D}(l) & \mathbf{0} \\ \mathbf{0} & \mathbf{D}(d_\alpha^{-1} z^{-1} l z d_\alpha) \otimes \mathbf{D}(z^{-1} l z) \end{pmatrix}, \tag{4.8.13}$$

$$\mathbf{C}_\alpha(z) = \begin{pmatrix} \mathbf{0} & \mathbf{D}(d_\alpha^{-1} z^2 d_\alpha) \otimes \mathbf{D}(z^2) \\ \mathbf{1} & \mathbf{0} \end{pmatrix}. \tag{4.8.14}$$

For convenience we shall write

$$\mathbf{N}_\alpha(l) = \mathbf{D}(d_\alpha^{-1} l d_\alpha) \otimes \mathbf{D}(l) \tag{4.8.15}$$

and

$${}_z\mathbf{N}_\alpha(l) = \mathbf{N}_\alpha(z^{-1} l z), \tag{4.8.16}$$

so that

$$\mathbf{C}_\alpha(l) = \begin{pmatrix} \mathbf{N}_\alpha(l) & \mathbf{0} \\ \mathbf{0} & {}_z\mathbf{N}_\alpha(l) \end{pmatrix}, \tag{4.8.17}$$

$$\mathbf{C}_\alpha(z) = \begin{pmatrix} \mathbf{0} & \mathbf{N}_\alpha(z^2) \\ \mathbf{1} & \mathbf{0} \end{pmatrix}. \tag{4.8.18}$$

Now from the analysis in section 4.5 we know that if we can find a matrix **P** such that

$$\mathbf{P}^2 = \mathbf{N}_\alpha(z^2) \tag{4.8.19}$$

and

$$\mathbf{P}\,{}_z\mathbf{N}_\alpha(l)\mathbf{P}^{-1} = \mathbf{N}_\alpha(l) \tag{4.8.20}$$

for all $l \in \mathbf{L}_\alpha$, then ${}_zN_\alpha$ and N_α are equivalent; and with respect to a transformed basis $(\langle\phi_\alpha|\,\langle\phi|,\ z\langle\phi_\alpha|\,\langle\phi|\,\mathbf{P}^{-1})$ the representation C_α becomes transformed into the equivalent representation C'_α given by

$$\mathbf{C}'_\alpha(l) = \begin{pmatrix}\mathbf{N}_\alpha(l) & \mathbf{0}\\ \mathbf{0} & \mathbf{N}_\alpha(l)\end{pmatrix}, \tag{4.8.21}$$

$$\mathbf{C}'_\alpha(z) = \begin{pmatrix}\mathbf{0} & \mathbf{P}\\ \mathbf{P} & \mathbf{0}\end{pmatrix}. \tag{4.8.22}$$

A matrix **P** that satisfies eqns. (4.8.19) and (4.8.20) can be found and is given by

$$\mathbf{P}_{ts,uv} = \mathbf{D}(h')_{su}\mathbf{D}(h)_{tv} = \mathbf{D}(h') \otimes \mathbf{D}(h)_{st,uv}. \tag{4.8.23}$$

For example with this choice of **P** we can soon check that eqn. (4.8.19) holds

$$\begin{aligned}
\mathbf{P}^2{}_{ab,cd} &= \sum_{e,f} \mathbf{P}_{ab,ef}\mathbf{P}_{ef,cd}\\
&= \sum_{e,f} \mathbf{D}(h')_{be}\mathbf{D}(h)_{af}\mathbf{D}(h')_{fc}\mathbf{D}(h)_{ed}\\
&= \mathbf{D}(h'h)_{bd}\mathbf{D}(hh')_{ac}\\
&= \mathbf{D}(z^2)_{bd}\mathbf{D}(d_\alpha^{-1}z^2d_\alpha)_{ac}\\
&= \mathbf{D}(d_\alpha^{-1}z^2d_\alpha) \otimes \mathbf{D}(z^2)_{ab,cd}\\
&= \mathbf{N}_\alpha(z^2)_{ab,cd}.
\end{aligned}$$

In order to verify eqn. (4.8.20) we compute $\mathbf{P}\,{}_z\mathbf{N}_\alpha(l)$ and show it to be equal to $\mathbf{N}_\alpha(l)\mathbf{P}$:

$$\begin{aligned}
(\mathbf{P}\,{}_z\mathbf{N}_\alpha(l))_{ab,cd} &= \sum_{e,f} \mathbf{P}_{ab,ef}\,{}_z\mathbf{N}_\alpha(l)_{ef,cd}\\
&= \sum_{e,f} \mathbf{D}(h')_{be}\mathbf{D}(h)_{af}\,\mathbf{D}(d_\alpha^{-1}z^{-1}lzd_\alpha)_{ec}\mathbf{D}(z^{-1}lz)_{fd}\\
&= \mathbf{D}(lzd_\alpha)_{bc}\mathbf{D}(d_\alpha^{-1}lz)_{ad} &&(4.8.24)\\
&= \sum_{e,f} \mathbf{D}(l)_{bf}\mathbf{D}(h')_{fc}\mathbf{D}(d_\alpha^{-1}ld_\alpha)_{ae}\mathbf{D}(h)_{ed} &&(4.8.25)\\
&= \sum_{e,f} \mathbf{N}_\alpha(l)_{ab,ef}\mathbf{P}_{ef,cd}\\
&= (\mathbf{N}_\alpha(l)\mathbf{P})_{ab,cd}.
\end{aligned}$$

In deriving eqns. (4.8.24) and (4.8.25) we have used the expressions $z = d_\alpha h = h'd_\alpha^{-1}$

and the fact that D is a homomorphism on the elements of $\mathbf{H}$. All this analysis would be fruitless if it were not for the fact that the basis for C'_α now takes on a particularly simple form. To see this let us compute the (kl) element of the second member of the basis:

$$\begin{aligned}(z\langle\phi_\alpha|\langle\phi|\,\mathbf{P}^{-1})_{kl} &= \sum_{i,j} z(\phi_{\alpha i}, \phi_j)\mathbf{P}^{-1}_{ij,kl} \\ &= \sum_{i,j} (h'd_\alpha^{-1}\phi_{\alpha i}, d_\alpha h\phi_j)\mathbf{P}^{-1}_{ij,kl} \\ &= \sum_{i,j,t} (h'\phi_i, d_\alpha\phi_t\mathbf{D}(h)_{tj})\mathbf{P}^{-1}_{ij,kl} \\ &= \sum_{i,j,t,s} (\phi_s, \phi_{\alpha t})\mathbf{D}(h')_{si}\mathbf{D}(h)_{tj}\mathbf{P}^{-1}_{ij,kl} \\ &= \sum_{i,j,t,s} (\phi_s, \phi_{\alpha t})\mathbf{P}_{ts,ij}\mathbf{P}^{-1}_{ij,kl} \\ &= (\phi_l, \phi_{\alpha k}).\end{aligned} \tag{4.8.26}$$

Also

$$\langle\phi_\alpha|\langle\phi|_{kl} = (\phi_{\alpha k}, \phi_l). \tag{4.8.27}$$

Hence

$$(\langle\phi_\alpha|\langle\phi| \pm z\langle\phi|\langle\phi_\alpha|\mathbf{P}^{-1})_{kl} = ((\phi_{\alpha k}, \phi_l) \pm (\phi_l, \phi_{\alpha k})). \tag{4.8.28}$$

Using as a new basis the two functions on the left-hand side of eqn. (4.8.28) the representation C'_α becomes transformed into the equivalent representation C''_α given by

$$\mathbf{C}''_\alpha(l) = \begin{pmatrix}\mathbf{N}_\alpha(l) & \mathbf{0} \\ \mathbf{0} & \mathbf{N}_\alpha(l)\end{pmatrix}, \tag{4.8.29}$$

$$\mathbf{C}''_\alpha(z) = \begin{pmatrix}\mathbf{P} & \mathbf{0} \\ \mathbf{0} & -\mathbf{P}\end{pmatrix}. \tag{4.8.30}$$

Thus, the symmetrized basis $(\phi_{\alpha k}, \phi_l) + (\phi_l, \phi_{\alpha k})$ yields a representation N_α^+ of $\mathbf{M}_\alpha$ given by

$$\left.\begin{aligned}\mathbf{N}_\alpha^+(l) &= \mathbf{N}_\alpha(l) \\ \mathbf{N}_\alpha^+(z) &= \mathbf{P}\end{aligned}\right\} \tag{4.8.31}$$

and the anti-symmetrized basis $(\phi_{\alpha k}, \phi_l) - (\phi_l, \phi_{\alpha k})$ yields a representation N_α^- of $\mathbf{M}_\alpha$ given by

$$\left.\begin{aligned}\mathbf{N}_\alpha^-(l) &= \mathbf{N}_\alpha(l) \\ \mathbf{N}_\alpha^-(z) &= -\mathbf{P}\end{aligned}\right\}. \tag{4.8.32}$$

Furthermore when the representations N_α^+ and N_α^- are induced from $\mathbf{M}_\alpha$ into $\mathbf{G}$ they are clearly defined respectively on the spaces V_α^+ and V_α^-, where $V_\alpha = (V_\alpha^+ + V_\alpha^-)$ is a direct sum decomposition of V_α into its symmetrized and anti-symmetrized parts. Thus $[(D\uparrow\mathbf{G}) \boxtimes (D\uparrow\mathbf{G})]$ contains $N_\alpha^+\uparrow\mathbf{G}$, and $\{(D\uparrow\mathbf{G}) \boxtimes (D\uparrow\mathbf{G})\}$ contains $N_\alpha^-\uparrow\mathbf{G}$, and these two representations are of equal dimension, as expected.

Also, from eqns. (4.8.23), (4.8.31), and (4.8.32) the characters of N_α^+ and N_α^- are easily calculated and are, for all $l \in \mathbf{L}_\alpha$,

$$\chi_{N_\alpha^\pm}(l) = \chi_D(d_\alpha^{-1}ld_\alpha)\chi_D(l) \tag{4.8.33}$$

$$\chi_{N_\alpha^\pm}(zl) = \pm\chi_D(zlzl). \tag{4.8.34}$$

For example

$$\begin{aligned}\sum_{i,j} \mathbf{N}_\alpha^+(zl)_{ij,ij} &= \sum_{i,j,k,l} \mathbf{P}_{ij,kl}\mathbf{N}_\alpha(l)_{kl,ij}\\ &= \sum_{i,j,k,l} \{\mathbf{D}(h') \otimes \mathbf{D}(h)\}_{ji,kl}\{\mathbf{D}(d_\alpha^{-1}ld_\alpha) \otimes \mathbf{D}(l)\}_{kl,ij}\\ &= \sum_{i,j} \{\mathbf{D}(h'd_\alpha^{-1}ld_\alpha) \otimes \mathbf{D}(hl)\}_{ji,ij}\\ &= \sum_{i,j} \mathbf{D}(h'd_\alpha^{-1}ld_\alpha)_{ji}\mathbf{D}(hl)_{ij}\\ &= \sum_{j} \mathbf{D}(h'd_\alpha^{-1}ld_\alpha hl)_{jj}\\ &= \chi_D(zlzl).\end{aligned} \tag{4.8.35}$$

This completes the decomposition. On collecting together the various different cases we are provided with the following theorem.

THEOREM 4.8.1. *Let D be a representation of the subgroup $\mathbf{H}$ of the finite group $\mathbf{G}$. Let A be the set of all self-inverse double cosets $\mathbf{H}d_\alpha\mathbf{H} = \mathbf{H}d_\alpha^{-1}\mathbf{H}$ except $\mathbf{H}$ itself. Let B be the set of all sets of the form $\mathbf{H}d_\beta\mathbf{H} \cup \mathbf{H}d_\beta^{-1}\mathbf{H}$, where $\mathbf{H}d_\beta\mathbf{H}$ is a non self-inverse double coset. For each $\beta \in B$ let $\mathbf{N}_\beta(l)$ be the representation $\mathbf{D}(d_\beta^{-1}ld_\beta) \otimes \mathbf{D}(l)$ of $\mathbf{L}_\beta = \mathbf{H} \cap d_\beta\mathbf{H}d_\beta^{-1}$. For each $\alpha \in A$, $d_\alpha\mathbf{H} \cap \mathbf{H}d_\alpha^{-1}$ is non-empty. Let z be any one of its members, say $z = d_\alpha h = h'd_\alpha^{-1}$. Let $\mathbf{L}_\alpha = \mathbf{H} \cap d_\alpha\mathbf{H}d_\alpha^{-1}$ and let $\mathbf{M}_\alpha$ be the subgroup generated by $\mathbf{L}_\alpha$ and z. Then $\mathbf{L}_\alpha$ is an invariant subgroup of $\mathbf{M}_\alpha$ of index 2. Let $\mathbf{N}_\alpha(l)$ be the representation $\mathbf{D}(d_\alpha^{-1}ld_\alpha) \otimes \mathbf{D}(l)$ of $\mathbf{L}_\alpha$ and let $\mathbf{P}$ be the matrix given by $\mathbf{P}_{ab,cd} = \mathbf{D}(h')_{bc}\mathbf{D}(h)_{ad}$. Then there exist extensions N_α^+ and N_α^- of N_α into the group $\mathbf{M}_\alpha$ such that $\mathbf{N}_\alpha^+(z) = \mathbf{P}$ and $\mathbf{N}_\alpha^-(z) = -\mathbf{P}$. Finally we have*

$$[(D\uparrow\mathbf{G}) \boxtimes (D\uparrow\mathbf{G})] \equiv [D \boxtimes D]\uparrow\mathbf{G} + \sum_{\alpha\in A} N_\alpha^+\uparrow\mathbf{G} + \sum_{\beta\in B} N_\beta\uparrow\mathbf{G} \tag{4.8.36}$$

$$\{(D\uparrow\mathbf{G}) \boxtimes (D\uparrow\mathbf{G})\} \equiv \{D \boxtimes D\}\uparrow\mathbf{G} + \sum_{\alpha\in A} N_\alpha^-\uparrow\mathbf{G} + \sum_{\beta\in B} N_\beta\uparrow\mathbf{G}. \tag{4.8.37}$$

As an example consider the application of the above theorem to the space group $\mathbf{G} = P23(T^1)$, the example of section 4.7. $\mathbf{H}$ is now the little group $\mathbf{G}^M$. The double coset decomposition $\mathbf{G} = \sum_\lambda \mathbf{G}^M d_\lambda \mathbf{G}^M$ contains three terms with $d_\lambda = \{E\,|\,\mathbf{0}\}$, $\{C_{31}^+\,|\,\mathbf{0}\}$ and $\{C_{31}^-\,|\,\mathbf{0}\}$. Hence the set A is empty and the set B contains the single term

$\mathbf{G}^M\{C_{31}^+ \mid \mathbf{0}\}\mathbf{G}^M \cup \mathbf{G}^M\{C_{31}^- \mid \mathbf{0}\}\mathbf{G}^M$. If we take $D = A_1$ then in the expansion (4.8.4) the term with $d_\lambda = d_\varepsilon = \{E \mid \mathbf{0}\}$ yields on induction to $\mathbf{G}$ the three reps ΓA, $\Gamma^1 E$, and $\Gamma^2 E$, the term with $d_\lambda = \{C_{31}^+ \mid \mathbf{0}\}$ yields MA_1, and likewise the term with $d_\lambda = \{C_{31}^- \mid \mathbf{0}\}$ yields MA_1 (illustrating the fact that the reps coming from inverse double cosets are equivalent)—see Table 4.5 to check these results. Also since A_1 is 1-dimensional it follows that $\{A_1 \boxtimes A_1\}$ is empty. The expansions (4.8.36) and (4.8.37) can now be written down and are

$$[MA_1 \boxtimes MA_1] \equiv \Gamma A + \Gamma^1 E + \Gamma^2 E + MA_1, \tag{4.8.38}$$

$$\{MA_1 \boxtimes MA_1\} \equiv MA_1. \tag{4.8.39}$$

Note particularly how little extra work is needed in practice to obtain the symmetrized and anti-symmetrized Kronecker square decompositions when the decomposition for the Kronecker square is already known; although it must be admitted that the above case is very favourable in that there is no self-inverse double coset and in that the rep A_1 is 1-dimensional so that $\{A_1 \boxtimes A_1\}$ vanishes.

5

The single-valued representations of the 230 space groups

In this chapter we tabulate the reps of each of the 230 space groups at each point of symmetry and along each line of symmetry in the basic domain of the appropriate Brillouin zone. The theory that has been used in the derivation of these reps was discussed in the later sections of Chapter 3 and, in more abstract terms, in Chapter 4. In addition to tabulating the results we shall include, in section 5.4, some examples to illustrate the extraction of information from the tables which are, of necessity, in a relatively condensed form.

5.1 Abstract groups

For any wave vector **k** in the Brillouin zone of a space group, the little group $\mathbf{G}^{\mathbf{k}}$ is an infinite group. However, this infinite group can be related to a finite group ${}^{H}\mathbf{G}^{\mathbf{k}} = \mathbf{G}^{\mathbf{k}}/\mathbf{T}^{\mathbf{k}}$ (see section 3.8) for a point of symmetry, or $\overline{\mathbf{G}}^{\mathbf{k}*}$, the central extension of the little co-group (see Theorem 3.7.2) otherwise. One often finds that there are several groups ${}^{H}\mathbf{G}^{\mathbf{k}}$ or $\overline{\mathbf{G}}^{\mathbf{k}*}$, perhaps for different wave vectors in the Brillouin zone of one space group, or for the same wave vector in the Brillouin zones of different space groups, or for different wave vectors in the Brillouin zones of different space groups, which are all isomorphic to one abstract group. Because of the frequent recurrence of the same abstract group for many different space groups we identify completely in Table 5.1 all the irreducible representations of all the abstract groups that occur among the (single-valued and double-valued) representations of the space groups.

Each abstract group is characterized by a set of generating elements P, Q, R, $S, \ldots$, and each element of the group can be written in the form $P^{\alpha}Q^{\beta}R^{\gamma}S^{\delta}\ldots$. Each group is completely defined if its group multiplication table is given; to determine this it is sufficient to have the set of *defining relations* or *generating relations* for the elements P, Q, R, S, etc., and they can be used to determine the group multiplication table (Coxeter and Moser 1965, Hall and Senior 1964, Miller 1894, 1946). This was illustrated for a very simple group in Example 1.2.1. In Table 5.1 we give the generating relations, the division of the elements among the classes of the group and the character table of the group. The representations of each abstract group are labelled R_1, $R_2, \ldots, R_r$ and for each of the degenerate representations we give the matrix representatives for the generating elements. The complete set of matrix representatives for all the elements of the group can then be determined for any given degenerate representation.

TABLE 5.1

Defining relations, classes, character tables, and matrix representatives for the abstract groups needed in the tabulation of the reps of the 230 space groups

$\mathbf{G}_1^1$

$C_1 = E.$

	C_1
R_1	1

$\mathbf{G}_2^1$

$P^2 = E.$

$C_1 = E;\ C_2 = P.$

	C_1	C_2
R_1	1	1
R_2	1	−1

$\mathbf{G}_3^1$

$P^3 = E.$

$C_1 = E;\ C_2 = P;\ C_3 = P^2.$

	C_1	C_2	C_3
R_1	1	1	1
R_2	1	ω	ω^*
R_3	1	ω^*	ω

$\mathbf{G}_4^1$

$P^4 = E.$

$C_1 = E;\ C_2 = P;\ C_3 = P^2;\ C_4 = P^3.$

	C_1	C_2	C_3	C_4
R_1	1	1	1	1
R_2	1	i	−1	−i
R_3	1	−1	1	−1
R_4	1	−i	−1	i

$\mathbf{G}_4^2$

$P^2 = E;\ Q^2 = E;\ QP = PQ.$

$C_1 = E;\ C_2 = P;\ C_3 = Q;\ C_4 = PQ.$

	C_1	C_2	C_3	C_4
R_1	1	1	1	1
R_2	1	1	−1	−1
R_3	1	−1	1	−1
R_4	1	−1	−1	1

$\mathbf{G}_6^1$

$P^6 = E.$

$C_1 = E;\ C_2 = P;\ C_3 = P^2;\ C_4 = P^3;\ C_5 = P^4;\ C_6 = P^5.$

	C_1	C_2	C_3	C_4	C_5	C_6
R_1	1	1	1	1	1	1
R_2	1	$-\omega^*$	ω	−1	ω^*	$-\omega$
R_3	1	ω	ω^*	1	ω	ω^*
R_4	1	−1	1	−1	1	−1
R_5	1	ω^*	ω	1	ω^*	ω
R_6	1	$-\omega$	ω^*	−1	ω	$-\omega^*$

G_6^2

$P^3 = E$; $Q^2 = E$; $QP = P^2Q$.
$C_1 = E$; $C_2 = P, P^2$; $C_3 = Q, PQ, P^2Q$.

	C_1	C_2	C_3
R_1	1	1	1
R_2	1	1	−1
R_3	2	−1	0

R_3: $P = \alpha$; $Q = \lambda$.

G_8^1

$P^8 = E$.
$C_1 = E$; $C_2 = P$; $C_3 = P^2$; $C_4 = P^3$; $C_5 = P^4$; $C_6 = P^5$; $C_7 = P^6$; $C_8 = P^7$.

	C_1	C_2	C_3	C_4	C_5	C_6	C_7	C_8
R_1	1	1	1	1	1	1	1	1
R_2	1	θ	i	$-\theta^*$	−1	$-\theta$	−i	θ^*
R_3	1	i	−1	−i	1	i	−1	−i
R_4	1	$-\theta^*$	−i	θ	−1	θ^*	i	$-\theta$
R_5	1	−1	1	−1	1	−1	1	−1
R_6	1	$-\theta$	i	θ^*	−1	θ	−i	$-\theta^*$
R_7	1	−i	−1	i	1	−i	−1	i
R_8	1	θ^*	−i	$-\theta$	−1	$-\theta^*$	i	θ

G_8^2

$P^4 = E$; $Q^2 = E$; $QP = PQ$.
$C_1 = E$; $C_2 = P$; $C_3 = P^2$; $C_4 = P^3$; $C_5 = Q$; $C_6 = PQ$; $C_7 = P^2Q$; $C_8 = P^3Q$.

	C_1	C_2	C_3	C_4	C_5	C_6	C_7	C_8
R_1	1	1	1	1	1	1	1	1
R_2	1	i	−1	−i	1	i	−1	−i
R_3	1	−1	1	−1	1	−1	1	−1
R_4	1	−i	−1	i	1	−i	−1	i
R_5	1	1	1	1	−1	−1	−1	−1
R_6	1	i	−1	−i	−1	−i	1	i
R_7	1	−1	1	−1	−1	1	−1	1
R_8	1	−i	−1	i	−1	i	1	−i

G_8^3

$P^2 = E;\ Q^2 = E;\ R^2 = E;\ QP = PQ;\ RP = PR;\ RQ = QR.$
$C_1 = E;\ C_2 = P;\ C_3 = Q;\ C_4 = PQ;\ C_5 = R;\ C_6 = PR;\ C_7 = QR;\ C_8 = PQR.$

	C_1	C_2	C_3	C_4	C_5	C_6	C_7	C_8
R_1	1	1	1	1	1	1	1	1
R_2	1	−1	1	−1	1	−1	1	−1
R_3	1	1	−1	−1	1	1	−1	−1
R_4	1	−1	−1	1	1	−1	−1	1
R_5	1	1	1	1	−1	−1	−1	−1
R_6	1	−1	1	−1	−1	1	−1	1
R_7	1	1	−1	−1	−1	−1	1	1
R_8	1	−1	−1	1	−1	1	1	−1

G_8^4

$P^4 = E;\ Q^2 = E;\ QP = P^3Q.$
$C_1 = E;\ C_2 = P^2;\ C_3 = P, P^3;\ C_4 = Q, P^2Q;\ C_5 = PQ, P^3Q.$

	C_1	C_2	C_3	C_4	C_5
R_1	1	1	1	1	1
R_2	1	1	1	−1	−1
R_3	1	1	−1	1	−1
R_4	1	1	−1	−1	1
R_5	2	−2	0	0	0

R_5: $P = \kappa;\ Q = \lambda.$

G_8^5 (The quaternion group.)

$P^4 = E;\ Q^4 = E;\ QP = P^3Q;\ Q^2 = P^2.$
$C_1 = E;\ C_2 = P^2;\ C_3 = P, P^3;\ C_4 = Q, P^2Q;\ C_5 = PQ, P^3Q.$

	C_1	C_2	C_3	C_4	C_5
R_1	1	1	1	1	1
R_2	1	1	1	−1	−1
R_3	1	1	−1	1	−1
R_4	1	1	−1	−1	1
R_5	2	−2	0	0	0

R_5: $P = \kappa;\ Q = i\phi.$

G_{12}^{1}

$P^{12} = E$.

$C_1 = E$; $C_2 = P$; $C_3 = P^2$; $C_4 = P^3$; $C_5 = P^4$; $C_6 = P^5$; $C_7 = P^6$; $C_8 = P^7$; $C_9 = P^8$; $C_{10} = P^9$; $C_{11} = P^{10}$; $C_{12} = P^{11}$.

	C_1	C_2	C_3	C_4	C_5	C_6	C_7	C_8	C_9	C_{10}	C_{11}	C_{12}
R_1	1	1	1	1	1	1	1	1	1	1	1	1
R_2	1	$-i\omega$	$-\omega^*$	i	ω	$-i\omega^*$	−1	$i\omega$	ω^*	−i	$-\omega$	$i\omega^*$
R_3	1	$-\omega^*$	ω	−1	ω^*	$-\omega$	1	$-\omega^*$	ω	−1	ω^*	$-\omega$
R_4	1	i	−1	−i	1	i	−1	−i	1	i	−1	−i
R_5	1	ω	ω^*	1	ω	ω^*	1	ω	ω^*	1	ω	ω^*
R_6	1	$-i\omega^*$	$-\omega$	i	ω^*	$-i\omega$	−1	$i\omega^*$	ω	−i	$-\omega^*$	$i\omega$
R_7	1	−1	1	−1	1	−1	1	−1	1	−1	1	−1
R_8	1	$i\omega$	$-\omega^*$	−i	ω	$i\omega^*$	−1	$-i\omega$	ω^*	i	$-\omega$	$-i\omega^*$
R_9	1	ω^*	ω	1	ω^*	ω	1	ω^*	ω	1	ω^*	ω
R_{10}	1	−i	−1	i	1	−i	−1	i	1	−i	−1	i
R_{11}	1	$-\omega$	ω^*	−1	ω	$-\omega^*$	1	$-\omega$	ω^*	−1	ω	$-\omega^*$
R_{12}	1	$i\omega^*$	$-\omega$	−i	ω^*	$i\omega$	−1	$-i\omega^*$	ω	i	$-\omega^*$	$-i\omega$

G_{12}^{2}

$P^3 = E$; $Q^2 = E$; $R^2 = E$; $QP = PQ$; $RQ = QR$; $RP = PR$.

$C_1 = E$; $C_2 = P$; $C_3 = P^2$; $C_4 = Q$; $C_5 = PQ$; $C_6 = P^2Q$; $C_7 = R$; $C_8 = PR$; $C_9 = P^2R$; $C_{10} = QR$; $C_{11} = PQR$; $C_{12} = P^2QR$.

	C_1	C_2	C_3	C_4	C_5	C_6	C_7	C_8	C_9	C_{10}	C_{11}	C_{12}
R_1	1	1	1	1	1	1	1	1	1	1	1	1
R_2	1	ω	ω^*	1	ω	ω^*	1	ω	ω^*	1	ω	ω^*
R_3	1	ω^*	ω	1	ω^*	ω	1	ω^*	ω	1	ω^*	ω
R_4	1	1	1	−1	−1	−1	1	1	1	−1	−1	−1
R_5	1	ω	ω^*	−1	$-\omega$	$-\omega^*$	1	ω	ω^*	−1	$-\omega$	$-\omega^*$
R_6	1	ω^*	ω	−1	$-\omega^*$	$-\omega$	1	ω^*	ω	−1	$-\omega^*$	$-\omega$
R_7	1	1	1	1	1	1	−1	−1	−1	−1	−1	−1
R_8	1	ω	ω^*	1	ω	ω^*	−1	$-\omega$	$-\omega^*$	−1	$-\omega$	$-\omega^*$
R_9	1	ω^*	ω	1	ω^*	ω	−1	$-\omega^*$	$-\omega$	−1	$-\omega^*$	$-\omega$
R_{10}	1	1	1	−1	−1	−1	−1	−1	−1	1	1	1
R_{11}	1	ω	ω^*	−1	$-\omega$	$-\omega^*$	−1	$-\omega$	$-\omega^*$	1	ω	ω^*
R_{12}	1	ω^*	ω	−1	$-\omega^*$	$-\omega$	−1	$-\omega^*$	$-\omega$	1	ω^*	ω

G_{12}^3

$P^6 = E$; $Q^2 = E$; $QP = P^5Q$.
$C_1 = E$; $C_2 = P^3$; $C_3 = P, P^5$; $C_4 = P^2, P^4$; $C_5 = Q, P^2Q, P^4Q$; $C_6 = PQ, P^3Q, P^5Q$.

	C_1	C_2	C_3	C_4	C_5	C_6
R_1	1	1	1	1	1	1
R_2	1	1	1	1	−1	−1
R_3	1	−1	−1	1	1	−1
R_4	1	−1	−1	1	−1	1
R_5	2	2	−1	−1	0	0
R_6	2	−2	1	−1	0	0

R_5: $P = \alpha$; $Q = \lambda$. R_6: $P = \beta$; $Q = \lambda$.

G_{12}^4

$P^6 = E$; $Q^4 = E$; $Q^2 = P^3$; $QP = P^5Q$.
$C_1 = E$; $C_2 = P^3$; $C_3 = P, P^5$; $C_4 = P^2, P^4$; $C_5 = Q, P^2Q, P^4Q$; $C_6 = PQ, P^3Q, P^5Q$.

	C_1	C_2	C_3	C_4	C_5	C_6
R_1	1	1	1	1	1	1
R_2	1	1	1	1	−1	−1
R_3	1	−1	−1	1	i	−i
R_4	1	−1	−1	1	−i	i
R_5	2	2	−1	−1	0	0
R_6	2	−2	1	−1	0	0

R_5: $P = \alpha$; $Q = \lambda$. R_6: $P = \beta$; $Q = \mathrm{i}\lambda$.

G_{12}^5

$P^3 = E$; $Q^2 = E$; $R^2 = E$; $QP = PR$; $RQ = QR$; $RP = PQR$.
$C_1 = E$; $C_2 = Q, R, QR$; $C_3 = P, PQ, PR, PQR$; $C_4 = P^2, P^2Q, P^2R, P^2QR$.

	C_1	C_2	C_3	C_4
R_1	1	1	1	1
R_2	1	1	ω	ω^*
R_3	1	1	ω^*	ω
R_4	3	−1	0	0

R_4: $P = A$; $Q = B$; $R = C$.
Note that $R = P^2QP$.

$\mathbf{G}_{12}^{6}$

$P^6 = E;\ Q^2 = E;\ QP = PQ.$

$C_1 = E;\ C_2 = P;\ C_3 = P^2;\ C_4 = P^3;\ C_5 = P^4;\ C_6 = P^5;\ C_7 = Q;\ C_8 = PQ;\ C_9 = P^2Q;\ C_{10} = P^3Q;$
$C_{11} = P^4Q;\ C_{12} = P^5Q.$

	C_1	C_2	C_3	C_4	C_5	C_6	C_7	C_8	C_9	C_{10}	C_{11}	C_{12}
R_1	1	1	1	1	1	1	1	1	1	1	1	1
R_2	1	$-\omega^*$	ω	-1	ω^*	$-\omega$	1	$-\omega^*$	ω	-1	ω^*	$-\omega$
R_3	1	ω	ω^*	1	ω	ω^*	1	ω	ω^*	1	ω	ω^*
R_4	1	-1	1	-1	1	-1	1	-1	1	-1	1	-1
R_5	1	ω^*	ω	1	ω^*	ω	1	ω^*	ω	1	ω^*	ω
R_6	1	$-\omega$	ω^*	-1	ω	$-\omega^*$	1	$-\omega$	ω^*	-1	ω	$-\omega^*$
R_7	1	1	1	1	1	1	-1	-1	-1	-1	-1	-1
R_8	1	$-\omega^*$	ω	-1	ω^*	$-\omega$	-1	ω^*	$-\omega$	1	$-\omega^*$	ω
R_9	1	ω	ω^*	1	ω	ω^*	-1	$-\omega$	$-\omega^*$	-1	$-\omega$	$-\omega^*$
R_{10}	1	-1	1	-1	1	-1	-1	1	-1	1	-1	1
R_{11}	1	ω^*	ω	1	ω^*	ω	-1	$-\omega^*$	$-\omega$	-1	$-\omega^*$	$-\omega$
R_{12}	1	$-\omega$	ω^*	-1	ω	$-\omega^*$	-1	ω	$-\omega^*$	1	$-\omega$	ω^*

$\mathbf{G}_{16}^{1}$

$P^{16} = E.$

$C_s = P^{s-1},\ s = 1 \text{ to } 16.$

There are 16 reps $R_1, R_2, \ldots, R_{16}$ such that the character in the rep R_t of the element in C_s is given by

$$\chi_t(C_s) = \sigma^{(t-1)(s-1)},\ t = 1 \text{ to } 16 \text{ and } s = 1 \text{ to } 16.$$

Note that information about *all* cyclic groups can be abbreviated in this way. As an example the reader should form for himself a similar abbreviation for $\mathbf{G}_{12}^{1}$ and then check with the version tabulated above.

$\mathbf{G}_{16}^{2}$

$P^8 = E;\ Q^2 = E;\ QP = PQ.$

$C_1 = E;\ C_2 = P;\ C_3 = P^2;\ C_4 = P^3;\ C_5 = P^4;\ C_6 = P^5;\ C_7 = P^6;\ C_8 = P^7;\ C_9 = Q;\ C_{10} = PQ;$
$C_{11} = P^2Q;\ C_{12} = P^3Q;\ C_{13} = P^4Q;\ C_{14} = P^5Q;\ C_{15} = P^6Q;\ C_{16} = P^7Q.$

	C_1	C_2	C_3	C_4	C_5	C_6	C_7	C_8	C_9	C_{10}	C_{11}	C_{12}	C_{13}	C_{14}	C_{15}	C_{16}
R_1	1	1	1	1	1	1	1	1	1	1	1	1	1	1	1	1
R_2	1	θ	i	$-\theta^*$	-1	$-\theta$	$-$i	θ^*	1	θ	i	$-\theta^*$	-1	$-\theta$	$-$i	θ^*
R_3	1	i	-1	$-$i	1	i	-1	$-$i	1	i	-1	$-$i	1	i	-1	$-$i
R_4	1	$-\theta^*$	$-$i	θ	-1	θ^*	i	$-\theta$	1	$-\theta^*$	$-$i	θ	-1	θ^*	i	$-\theta$
R_5	1	-1	1	-1	1	-1	1	-1	1	-1	1	-1	1	-1	1	-1
R_6	1	$-\theta$	i	θ^*	-1	θ	$-$i	$-\theta^*$	1	$-\theta$	i	θ^*	-1	θ	$-$i	$-\theta^*$
R_7	1	$-$i	-1	i	1	$-$i	-1	i	1	$-$i	-1	i	1	$-$i	-1	i
R_8	1	θ^*	$-$i	$-\theta$	-1	$-\theta^*$	i	θ	1	θ^*	$-$i	$-\theta$	-1	$-\theta^*$	i	θ
R_9	1	1	1	1	1	1	1	1	-1	-1	-1	-1	-1	-1	-1	-1
R_{10}	1	θ	i	$-\theta^*$	-1	$-\theta$	$-$i	θ^*	-1	$-\theta$	$-$i	θ^*	1	θ	i	$-\theta^*$
R_{11}	1	i	-1	$-$i	1	i	-1	$-$i	-1	$-$i	1	i	-1	$-$i	1	i
R_{12}	1	$-\theta^*$	$-$i	θ	-1	θ^*	i	$-\theta$	-1	θ^*	i	$-\theta$	1	$-\theta^*$	$-$i	θ
R_{13}	1	-1	1	-1	1	-1	1	-1	-1	1	-1	1	-1	1	-1	1
R_{14}	1	$-\theta$	i	θ^*	-1	θ	$-$i	$-\theta^*$	-1	θ	$-$i	$-\theta^*$	1	$-\theta$	i	θ^*
R_{15}	1	$-$i	-1	i	1	$-$i	-1	i	-1	i	1	$-$i	-1	i	1	$-$i
R_{16}	1	θ^*	$-$i	$-\theta$	-1	$-\theta^*$	i	θ	-1	$-\theta^*$	i	θ	1	θ^*	$-$i	$-\theta$

Note that this can be abbreviated as follows:

	C_1 to C_8	C_9 to C_{16}
R_1 to R_8	$(\mathbf{G}_8^1)$	$(\mathbf{G}_8^1)$
R_9 to R_{16}	$(\mathbf{G}_8^1)$	$-(\mathbf{G}_8^1)$

Where $(\mathbf{G}_8^1)$ stands for the array of complex numbers forming the character table of $\mathbf{G}_8^1$; this is because $\mathbf{G}_{16}^2 = \mathbf{G}_8^1 \otimes \mathbf{G}_2^1$. In the remainder of this table we shall use similar abbreviations for certain direct product groups. In connection with such abbreviations it should be understood that any multiplying factor that precedes the symbol of an array implies that every element in that array is to be multiplied by that factor.

$\mathbf{G}_{16}^3$

$P^4 = E;\ Q^4 = E;\ QP = PQ.$
$C_1 = E;\ C_2 = P;\ C_3 = P^2;\ C_4 = P^3;\ C_5 = Q;\ C_6 = PQ;\ C_7 = P^2Q;\ C_8 = P^3Q;\ C_9 = Q^2;\ C_{10} = PQ^2;$
$C_{11} = P^2Q^2;\ C_{12} = P^3Q^2;\ C_{13} = Q^3;\ C_{14} = PQ^3;\ C_{15} = P^2Q^3;\ C_{16} = P^3Q^3.$

	C_1 to C_4	C_5 to C_8	C_9 to C_{12}	C_{13} to C_{16}
R_1 to R_4	$(\mathbf{G}_4^1)$	$(\mathbf{G}_4^1)$	$(\mathbf{G}_4^1)$	$(\mathbf{G}_4^1)$
R_5 to R_8	$(\mathbf{G}_4^1)$	$i(\mathbf{G}_4^1)$	$-(\mathbf{G}_4^1)$	$-i(\mathbf{G}_4^1)$
R_9 to R_{12}	$(\mathbf{G}_4^1)$	$-(\mathbf{G}_4^1)$	$(\mathbf{G}_4^1)$	$-(\mathbf{G}_4^1)$
R_{13} to R_{16}	$(\mathbf{G}_4^1)$	$-i(\mathbf{G}_4^1)$	$-(\mathbf{G}_4^1)$	$i(\mathbf{G}_4^1)$

$\mathbf{G}_{16}^4$

$P^4 = E;\ Q^2 = E;\ R^2 = E;\ QP = PQ;\ RP = PR;\ RQ = QR.$
$C_1 = E;\ C_2 = P;\ C_3 = P^2;\ C_4 = P^3;\ C_5 = Q;\ C_6 = PQ;\ C_7 = P^2Q;\ C_8 = P^3Q;\ C_9 = R;\ C_{10} = PR;$
$C_{11} = P^2R;\ C_{12} = P^3R;\ C_{13} = QR;\ C_{14} = PQR;\ C_{15} = P^2QR;\ C_{16} = P^3QR.$

	C_1 to C_8	C_9 to C_{16}
R_1 to R_8	$(\mathbf{G}_8^2)$	$(\mathbf{G}_8^2)$
R_9 to R_{16}	$(\mathbf{G}_8^2)$	$-(\mathbf{G}_8^2)$

$\mathbf{G}_{16}^5$

$P^2 = E;\ Q^2 = E;\ R^2 = E;\ S^2 = E;\ QP = PQ;\ RP = PR;\ RQ = QR;\ SP = PS;\ SQ = QS;\ SR = RS.$
$C_1 = E;\ C_2 = P;\ C_3 = Q;\ C_4 = PQ;\ C_5 = R;\ C_6 = PR;\ C_7 = QR;\ C_8 = PQR;\ C_9 = S;\ C_{10} = PS;$
$C_{11} = QS;\ C_{12} = PQS;\ C_{13} = RS;\ C_{14} = PRS;\ C_{15} = QRS;\ C_{16} = PQRS.$

	C_1 to C_8	C_9 to C_{16}
R_1 to R_8	$(\mathbf{G}_8^3)$	$(\mathbf{G}_8^3)$
R_9 to R_{16}	$(\mathbf{G}_8^3)$	$-(\mathbf{G}_8^3)$

G_{16}^{6}

$P^8 = E; Q^2 = E; QP = P^5Q$

$C_1 = E; C_2 = P^2; C_3 = P^4; C_4 = P^6; C_5 = P, P^5; C_6 = P^3, P^7;$
$C_7 = Q, P^4Q; C_8 = P^2Q, P^6Q; C_9 = PQ, P^5Q; C_{10} = P^3Q, P^7Q.$

	C_1	C_2	C_3	C_4	C_5	C_6	C_7	C_8	C_9	C_{10}
R_1	1	1	1	1	1	1	1	1	1	1
R_2	1	1	1	1	1	1	−1	−1	−1	−1
R_3	1	1	1	1	−1	−1	1	1	−1	−1
R_4	1	1	1	1	−1	−1	−1	−1	1	1
R_5	1	−1	1	−1	i	−i	1	−1	i	−i
R_6	1	−1	1	−1	i	−i	−1	1	−i	i
R_7	1	−1	1	−1	−i	i	1	−1	−i	i
R_8	1	−1	1	−1	−i	i	−1	1	i	−i
R_9	2	2i	−2	−2i	0	0	0	0	0	0
R_{10}	2	−2i	−2	2i	0	0	0	0	0	0

R_9: $P = \theta\lambda$; $Q = -i\kappa$. R_{10}: $P = \theta^*\lambda$; $Q = -i\kappa$.

G_{16}^{7}

$P^4 = E;\ Q^2 = E;\ R^2 = E;\ QP = PQ;\ RP = PR;\ RQ = P^2QR.$

$C_1 = E;\ C_2 = P;\ C_3 = P^2;\ C_4 = P^3;\ C_5 = Q, P^2Q;\ C_6 = PQ, P^3Q;\ C_7 = R, P^2R;\ C_8 = PR, P^3R;$
$C_9 = QR, P^2QR;\ C_{10} = PQR, P^3QR.$

	C_1	C_2	C_3	C_4	C_5	C_6	C_7	C_8	C_9	C_{10}
R_1	1	1	1	1	1	1	1	1	1	1
R_2	1	1	1	1	1	1	−1	−1	−1	−1
R_3	1	1	1	1	−1	−1	1	1	−1	−1
R_4	1	1	1	1	−1	−1	−1	−1	1	1
R_5	1	−1	1	−1	1	−1	1	−1	1	−1
R_6	1	−1	1	−1	1	−1	−1	1	−1	1
R_7	1	−1	1	−1	−1	1	1	−1	−1	1
R_8	1	−1	1	−1	−1	1	−1	1	1	−1
R_9	2	2i	−2	−2i	0	0	0	0	0	0
R_{10}	2	−2i	−2	2i	0	0	0	0	0	0

R_9: $P = i\varepsilon$; $Q = \phi$; $R = \lambda$. R_{10}: $P = -i\varepsilon$; $Q = \phi$; $R = \lambda$.

G_{16}^{8}

$P^4 = E;\ Q^4 = E;\ QP = P^3Q.$

$C_1 = E;\ C_2 = P^2;\ C_3 = Q^2;\ C_4 = P^2Q^2;\ C_5 = P, P^3;\ C_6 = PQ^2, P^3Q^2;\ C_7 = Q, P^2Q;\ C_8 = PQ, P^3Q;$
$C_9 = Q^3, P^2Q^3;\ C_{10} = PQ^3, P^3Q^3.$

	C_1	C_2	C_3	C_4	C_5	C_6	C_7	C_8	C_9	C_{10}
R_1	1	1	1	1	1	1	1	1	1	1
R_2	1	1	1	1	1	1	−1	−1	−1	−1
R_3	1	1	1	1	−1	−1	1	−1	1	−1
R_4	1	1	1	1	−1	−1	−1	1	−1	1
R_5	1	1	−1	−1	−1	1	i	−i	−i	i
R_6	1	1	−1	−1	−1	1	−i	i	i	−i
R_7	1	1	−1	−1	1	−1	−i	−i	i	i
R_8	1	1	−1	−1	1	−1	i	i	−i	−i
R_9	2	−2	2	−2	0	0	0	0	0	0
R_{10}	2	−2	−2	2	0	0	0	0	0	0

R_9: $P = i\lambda$; $Q = \phi$. R_{10}: $P = i\lambda$; $Q = \kappa$.

$\mathbf{G}_{16}^{9}$

$P^4 = E$; $Q^2 = E$; $R^2 = E$; $QP = P^3Q$; $RP = PR$; $RQ = QR$.
$C_1 = E$; $C_2 = P^2$; $C_3 = P, P^3$; $C_4 = Q, P^2Q$; $C_5 = PQ, P^3Q$; $C_6 = R$; $C_7 = P^2R$; $C_8 = PR, P^3R$; $C_9 = QR, P^2QR$; $C_{10} = PQR, P^3QR$.

	C_1 to C_5	C_6 to C_{10}
R_1 to R_5	$(\mathbf{G}_8^4)$	$(\mathbf{G}_8^4)$
R_6 to R_{10}	$(\mathbf{G}_8^4)$	$-(\mathbf{G}_8^4)$

R_5: $P = \kappa$; $Q = \lambda$; $R = \varepsilon$.　　R_{10}: $P = \kappa$; $Q = \lambda$; $R = -\varepsilon$.

$\mathbf{G}_{16}^{10}$

$P^4 = E$; $Q^2 = E$; $R^2 = E$; $QP = PQ$; $RP = PQR$; $RQ = QR$.
$C_1 = E$; $C_2 = Q$; $C_3 = P^2$; $C_4 = P^2Q$; $C_5 = R, QR$; $C_6 = P^2R, P^2QR$; $C_7 = P, PQ$; $C_8 = P^3, P^3Q$; $C_9 = PR, PQR$; $C_{10} = P^3R, P^3QR$.

	C_1	C_2	C_3	C_4	C_5	C_6	C_7	C_8	C_9	C_{10}
R_1	1	1	1	1	1	1	1	1	1	1
R_2	1	1	1	1	−1	−1	−1	−1	1	1
R_3	1	1	1	1	1	1	−1	−1	−1	−1
R_4	1	1	1	1	−1	−1	1	1	−1	−1
R_5	1	1	−1	−1	1	−1	i	−i	i	−i
R_6	1	1	−1	−1	1	−1	−i	i	−i	i
R_7	1	1	−1	−1	−1	1	−i	i	i	−i
R_8	1	1	−1	−1	−1	1	i	−i	−i	i
R_9	2	−2	2	−2	0	0	0	0	0	0
R_{10}	2	−2	−2	2	0	0	0	0	0	0

R_9: $P = \phi$; $Q = -\varepsilon$; $R = \lambda$.　　R_{10}: $P = \kappa$; $Q = -\varepsilon$; $R = \lambda$.

$\mathbf{G}_{16}^{11}$

$P^4 = E$; $Q^4 = E$; $Q^2 = P^2$; $R^2 = E$; $QP = P^3Q$; $RP = PR$; $RQ = QR$.
$C_1 = E$; $C_2 = P^2$; $C_3 = P, P^3$; $C_4 = Q, P^2Q$; $C_5 = PQ, P^3Q$; $C_6 = R$; $C_7 = P^2R$; $C_8 = PR, P^3R$; $C_9 = QR, P^2QR$; $C_{10} = PQR, P^3QR$.

	C_1 to C_5	C_6 to C_{10}
R_1 to R_5	$(\mathbf{G}_8^5)$	$(\mathbf{G}_8^5)$
R_6 to R_{10}	$(\mathbf{G}_8^5)$	$-(\mathbf{G}_8^5)$

R_5: $P = \kappa$; $Q = \mathrm{i}\phi$; $R = \varepsilon$.　　R_{10}: $P = \kappa$; $Q = \mathrm{i}\phi$; $R = -\varepsilon$.

$\mathbf{G}_{16}^{12}$

$P^8 = E;\ Q^2 = E;\ QP = P^7Q.$
$C_1 = E;\ C_2 = P^4;\ C_3 = P^2, P^6;\ C_4 = P, P^7;\ C_5 = P^3, P^5;\ C_6 = Q, P^2Q, P^4Q, P^6Q;$
$C_7 = PQ, P^3Q, P^5Q, P^7Q.$

	C_1	C_2	C_3	C_4	C_5	C_6	C_7
R_1	1	1	1	1	1	1	1
R_2	1	1	1	1	1	−1	−1
R_3	1	1	1	−1	−1	1	−1
R_4	1	1	1	−1	−1	−1	1
R_5	2	2	−2	0	0	0	0
R_6	2	−2	0	$\sqrt{2}$	$-\sqrt{2}$	0	0
R_7	2	−2	0	$-\sqrt{2}$	$\sqrt{2}$	0	0

R_5: $P = \kappa;\ Q = \lambda.$ R_6: $P = \delta;\ Q = \lambda.$
R_7: $P = \eta;\ Q = \lambda.$

$\mathbf{G}_{16}^{13}$

$P^8 = E;\ Q^2 = E;\ QP = P^3Q.$
$C_1 = E;\ C_2 = P^4;\ C_3 = P^2, P^6;\ C_4 = P, P^3;\ C_5 = P^5, P^7;\ C_6 = Q, P^2Q, P^4Q, P^6Q;$
$C_7 = PQ, P^3Q, P^5Q, P^7Q.$

	C_1	C_2	C_3	C_4	C_5	C_6	C_7
R_1	1	1	1	1	1	1	1
R_2	1	1	1	1	1	−1	−1
R_3	1	1	1	−1	−1	1	−1
R_4	1	1	1	−1	−1	−1	1
R_5	2	2	−2	0	0	0	0
R_6	2	−2	0	$i\sqrt{2}$	$-i\sqrt{2}$	0	0
R_7	2	−2	0	$-i\sqrt{2}$	$i\sqrt{2}$	0	0

R_5: $P = \kappa;\ Q = \lambda.$ R_6: $P = \xi;\ Q = \lambda.$
R_7: $P = \xi^*;\ Q = \lambda.$

$\mathbf{G}_{16}^{14}$

$P^8 = E;\ Q^4 = E;\ Q^2 = P^4;\ QP = P^7Q.$
$C_1 = E;\ C_2 = P^4;\ C_3 = P^2, P^6;\ C_4 = P, P^7;\ C_5 = P^3, P^5;\ C_6 = Q, P^2Q, P^4Q, P^6Q;$
$C_7 = PQ, P^3Q, P^5Q, P^7Q.$

	C_1	C_2	C_3	C_4	C_5	C_6	C_7
R_1	1	1	1	1	1	1	1
R_2	1	1	1	1	1	−1	−1
R_3	1	1	1	−1	−1	1	−1
R_4	1	1	1	−1	−1	−1	1
R_5	2	2	−2	0	0	0	0
R_6	2	−2	0	$\sqrt{2}$	$-\sqrt{2}$	0	0
R_7	2	−2	0	$-\sqrt{2}$	$\sqrt{2}$	0	0

R_5: $P = \kappa;\ Q = \lambda.$ R_6: $P = \delta;\ Q = i\phi.$
R_7: $P = \eta;\ Q = i\phi.$

$\mathbf{G}_{24}^{1}$

$P^4 = E$; $Q^3 = E$; $R^2 = E$; $QP = PQ^2$; $RP = P^3R$; $RQ = QR$.
$C_1 = E$; $C_2 = P^2$; $C_3 = Q, Q^2$; $C_4 = P^2Q, P^2Q^2$; $C_5 = R, P^2R$; $C_6 = Q^2R, P^2QR$; $C_7 = QR, P^2Q^2R$;
$C_8 = P, PQ, PQ^2, P^3, P^3Q, P^3Q^2$; $C_9 = PR, PQR, PQ^2R, P^3R, P^3QR, P^3Q^2R$.

	C_1	C_2	C_3	C_4	C_5	C_6	C_7	C_8	C_9
R_1	1	1	1	1	1	1	1	1	1
R_2	1	1	1	1	1	1	1	-1	-1
R_3	1	1	1	1	-1	-1	-1	1	-1
R_4	1	1	1	1	-1	-1	-1	-1	1
R_5	2	2	-1	-1	-2	1	1	0	0
R_6	2	2	-1	-1	2	-1	-1	0	0
R_7	2	-2	-1	1	0	$i\sqrt{3}$	$-i\sqrt{3}$	0	0
R_8	2	-2	-1	1	0	$-i\sqrt{3}$	$i\sqrt{3}$	0	0
R_9	2	-2	2	-2	0	0	0	0	0

R_5: $P = \phi$; $Q = \pi^*$; $R = -\varepsilon$.
R_6: $P = \phi$; $Q = \pi^*$; $R = \varepsilon$.
R_7: $P = -i\phi$; $Q = \pi^*$; $R = \lambda$.
R_8: $P = i\phi$; $Q = \pi$; $R = \lambda$.
R_9: $P = i\phi$; $Q = \varepsilon$; $R = \lambda$.

$\mathbf{G}_{24}^{2}$

$P^{12} = E$; $Q^2 = E$; $QP = P^{11}Q$.
$C_1 = E$; $C_2 = P^6$; $C_3 = P^4, P^8$; $C_4 = P^2, P^{10}$; $C_5 = P^3, P^9$; $C_6 = P, P^{11}$; $C_7 = P^5, P^7$;
$C_8 = Q, P^2Q, P^4Q, P^6Q, P^8Q, P^{10}Q$; $C_9 = PQ, P^3Q, P^5Q, P^7Q, P^9Q, P^{11}Q$.

	C_1	C_2	C_3	C_4	C_5	C_6	C_7	C_8	C_9
R_1	1	1	1	1	1	1	1	1	1
R_2	1	1	1	1	1	1	1	-1	-1
R_3	1	1	1	1	-1	-1	-1	1	-1
R_4	1	1	1	1	-1	-1	-1	-1	1
R_5	2	2	-1	-1	-2	1	1	0	0
R_6	2	2	-1	-1	2	-1	-1	0	0
R_7	2	-2	-1	1	0	$\sqrt{3}$	$-\sqrt{3}$	0	0
R_8	2	-2	-1	1	0	$-\sqrt{3}$	$\sqrt{3}$	0	0
R_9	2	-2	2	-2	0	0	0	0	0

R_5: $P = -\pi^*$; $Q = \phi$.
R_6: $P = \pi$; $Q = \phi$.
R_7: $P = \rho$; $Q = \phi$.
R_8: $P = -\rho^*$; $Q = \phi$.
R_9: $P = i\lambda$; $Q = \phi$.

$\mathbf{G}_{24}^{3}$

$P^6 = E$; $Q^4 = E$; $Q^2 = P^3$; $R^2 = E$; $QP = P^5Q$; $RP = PR$; $RQ = QR$.
$C_1 = E$; $C_2 = P^3$; $C_3 = P, P^5$; $C_4 = P^2, P^4$; $C_5 = Q, P^2Q, P^4Q$; $C_6 = PQ, P^3Q, P^5Q$; $C_7 = R$;
$C_8 = P^3R$; $C_9 = PR, P^5R$; $C_{10} = P^2R, P^4R$; $C_{11} = QR, P^2QR, P^4QR$; $C_{12} = PQR, P^3QR, P^5QR$.

	C_1 to C_6	C_7 to C_{12}
R_1 to R_6	$(\mathbf{G}_{12}^{4})$	$(\mathbf{G}_{12}^{4})$
R_7 to R_{12}	$(\mathbf{G}_{12}^{4})$	$-(\mathbf{G}_{12}^{4})$

R_5: $P = \alpha$; $Q = \lambda$; $R = \varepsilon$.
R_6: $P = \beta$; $Q = i\lambda$; $R = \varepsilon$.
R_{11}: $P = \alpha$; $Q = \lambda$; $R = -\varepsilon$.
R_{12}: $P = \beta$; $Q = i\lambda$; $R = -\varepsilon$.

$\mathbf{G}_{24}^{4}$

$P^3 = E;\ Q^2 = E;\ R^4 = E;\ QP = P^2Q;\ RP = PR;\ RQ = QR.$
$C_1 = E;\ C_2 = P, P^2;\ C_3 = Q, PQ, P^2Q;\ C_4 = R;\ C_5 = PR, P^2R;\ C_6 = QR, PQR, P^2QR;\ C_7 = R^2;$
$C_8 = PR^2, P^2R^2;\ C_9 = QR^2, PQR^2, P^2QR^2;\ C_{10} = R^3;\ C_{11} = PR^3, P^2R^3;\ C_{12} = QR^3, PQR^3, P^2QR^3$

	C_1 to C_3	C_4 to C_6	C_7 to C_9	C_{10} to C_{12}
R_1 to R_3	$(\mathbf{G}_6^2)$	$(\mathbf{G}_6^2)$	$(\mathbf{G}_6^2)$	$(\mathbf{G}_6^2)$
R_4 to R_6	$(\mathbf{G}_6^2)$	$\mathrm{i}(\mathbf{G}_6^2)$	$-(\mathbf{G}_6^2)$	$-\mathrm{i}(\mathbf{G}_6^2)$
R_7 to R_9	$(\mathbf{G}_6^2)$	$-(\mathbf{G}_6^2)$	$(\mathbf{G}_6^2)$	$-(\mathbf{G}_6^2)$
R_{10} to R_{12}	$(\mathbf{G}_6^2)$	$-\mathrm{i}(\mathbf{G}_6^2)$	$-(\mathbf{G}_6^2)$	$\mathrm{i}(\mathbf{G}_6^2)$

R_3: $P = \alpha;\ Q = \lambda;\ R = \varepsilon.$ R_6: $P = \alpha;\ Q = \lambda;\ R = \mathrm{i}\varepsilon.$
R_9: $P = \alpha;\ Q = \lambda;\ R = -\varepsilon.$ R_{12}: $P = \alpha;\ Q = \lambda;\ R = -\mathrm{i}\varepsilon.$

$\mathbf{G}_{24}^{5}$

$P^3 = E;\ Q^2 = E;\ R^2 = E;\ S^2 = E;\ QP = P^2Q;\ RP = PR;\ RQ = QR;\ SP = PS;\ SQ = QS;\ SR = RS.$
$C_1 = E;\ C_2 = P, P^2;\ C_3 = Q, PQ, P^2Q;\ C_4 = R;\ C_5 = PR, P^2R;\ C_6 = QR, PQR, P^2QR;\ C_7 = S;$
$C_8 = PS, P^2S;\ C_9 = QS, PQS, P^2QS;\ C_{10} = RS;\ C_{11} = PRS, P^2RS;\ C_{12} = QRS, PQRS, P^2QRS.$

	C_1 to C_3	C_4 to C_6	C_7 to C_9	C_{10} to C_{12}
R_1 to R_3	$(\mathbf{G}_6^2)$	$(\mathbf{G}_6^2)$	$(\mathbf{G}_6^2)$	$(\mathbf{G}_6^2)$
R_4 to R_6	$(\mathbf{G}_6^2)$	$-(\mathbf{G}_6^2)$	$(\mathbf{G}_6^2)$	$-(\mathbf{G}_6^2)$
R_7 to R_9	$(\mathbf{G}_6^2)$	$(\mathbf{G}_6^2)$	$-(\mathbf{G}_6^2)$	$-(\mathbf{G}_6^2)$
R_{10} to R_{12}	$(\mathbf{G}_6^2)$	$-(\mathbf{G}_6^2)$	$-(\mathbf{G}_6^2)$	$(\mathbf{G}_6^2)$

R_3: $P = \alpha;\ Q = \lambda;\ R = \varepsilon;\ S = \varepsilon.$ R_6: $P = \alpha;\ Q = \lambda;\ R = -\varepsilon;\ S = \varepsilon.$
R_9: $P = \alpha;\ Q = \lambda;\ R = \varepsilon;\ S = -\varepsilon.$ R_{12}: $P = \alpha;\ Q = \lambda;\ R = -\varepsilon;\ S = -\varepsilon.$

$\mathbf{G}_{24}^{6}$

$P^{12} = E;\ Q^2 = E;\ QP = P^7Q.$
$C_1 = E;\ C_2 = P, P^7;\ C_3 = P^2;\ C_4 = P^3, P^9;\ C_5 = P^4;\ C_6 = P^5, P^{11};\ C_7 = P^6;\ C_8 = P^8;\ C_9 = P^{10};$
$C_{10} = Q, P^6Q;\ C_{11} = PQ, P^7Q;\ C_{12} = P^2Q, P^5Q;\ C_{13} = P^3Q, P^9Q;\ C_{14} = P^4Q, P^{10}Q;\ C_{15} = P^5Q, P^{11}Q.$

	C_1	C_2	C_3	C_4	C_5	C_6	C_7	C_8	C_9	C_{10}	C_{11}	C_{12}	C_{13}	C_{14}	C_{15}
R_1	1	1	1	1	1	1	1	1	1	1	1	1	1	1	1
R_2	1	1	1	1	1	1	1	1	1	-1	-1	-1	-1	-1	-1
R_3	1	$-\omega^*$	ω	-1	ω^*	$-\omega$	1	ω	ω^*	1	$-\omega^*$	ω	-1	ω^*	$-\omega$
R_4	1	$-\omega^*$	ω	-1	ω^*	$-\omega$	1	ω	ω^*	-1	ω^*	$-\omega$	1	$-\omega^*$	ω
R_5	1	ω	ω^*	1	ω	ω^*	1	ω^*	ω	1	ω	ω^*	1	ω	ω^*
R_6	1	ω	ω^*	1	ω	ω^*	1	ω^*	ω	-1	$-\omega$	$-\omega^*$	-1	$-\omega$	$-\omega^*$
R_7	1	-1	1	-1	1	-1	1	1	1	1	-1	1	-1	1	-1
R_8	1	-1	1	-1	1	-1	1	1	1	-1	1	-1	1	-1	1
R_9	1	ω^*	ω	1	ω^*	ω	1	ω	ω^*	1	ω^*	ω	1	ω^*	ω
R_{10}	1	ω^*	ω	1	ω^*	ω	1	ω	ω^*	-1	$-\omega^*$	$-\omega$	-1	$-\omega^*$	$-\omega$
R_{11}	1	$-\omega$	ω^*	-1	ω	$-\omega^*$	1	ω^*	ω	1	$-\omega$	ω^*	-1	ω	$-\omega^*$
R_{12}	1	$-\omega$	ω^*	-1	ω	$-\omega^*$	1	ω^*	ω	-1	ω	$-\omega^*$	1	$-\omega$	ω^*
R_{13}	2	0	$-2\omega^*$	0	2ω	0	-2	$2\omega^*$	-2ω	0	0	0	0	0	0
R_{14}	2	0	-2	0	2	0	-2	2	-2	0	0	0	0	0	0
R_{15}	2	0	-2ω	0	$2\omega^*$	0	-2	2ω	$-2\omega^*$	0	0	0	0	0	0

R_{13}: $P = -\mathrm{i}\omega\lambda;\ Q = \phi.$ R_{14}: $P = \mathrm{i}\lambda;\ Q = \phi.$
R_{15}: $P = \mathrm{i}\omega^*\lambda;\ Q = \phi.$

G_{24}^{7}

$P^3 = E$; $Q^2 = E$; $R^2 = E$; $S^2 = E$; $QP = PR$; $RP = PQR$; $RQ = QR$; $SP = P^2RS$; $SQ = QS$; $SR = QRS$.
$C_1 = E$; $C_2 = R, Q, QR$; $C_3 = PS, RS, P^2S, PQRS, QRS, P^2RS$; $C_4 = S, QS, PQS, PRS, P^2QRS, P^2QS$; $C_5 = P, P^2, PQ, PR, PQR, P^2QR, P^2R, P^2Q$.

	C_1	C_2	C_3	C_4	C_5
R_1	1	1	1	1	1
R_2	1	1	-1	-1	1
R_3	2	2	0	0	-1
R_4	3	-1	1	-1	0
R_5	3	-1	-1	1	0

R_3: $P = -\beta$; $Q = \varepsilon$; $R = \varepsilon$; $S = \lambda$.
R_4: $P = D$; $Q = B$; $R = G$; $S = F$.
R_5: $P = D$; $Q = B$; $R = G$; $S = H$.

G_{24}^{8}

$P^6 = E$; $Q^2 = E$; $R^2 = E$; $QP = PR$; $QP^3 = P^3Q$; $RP = P^4QR$; $RQ = QR$.
$C_1 = E$; $C_2 = P^3$; $C_3 = Q, R, P^3QR$; $C_4 = P^3Q, P^3R, QR$; $C_5 = PQ, PR, P^4, P^4QR$; $C_6 = P^4Q, P^4R, P, PQR$; $C_7 = P^2, P^2QR, P^5Q, P^5R$; $C_8 = P^5, P^5QR, P^2Q, P^2R$.

	C_1	C_2	C_3	C_4	C_5	C_6	C_7	C_8
R_1	1	1	1	1	1	1	1	1
R_2	1	1	1	1	ω	ω	ω^*	ω^*
R_3	1	1	1	1	ω^*	ω^*	ω	ω
R_4	1	-1	-1	1	1	-1	1	-1
R_5	1	-1	-1	1	ω	$-\omega$	ω^*	$-\omega^*$
R_6	1	-1	-1	1	ω^*	$-\omega^*$	ω	$-\omega$
R_7	3	3	-1	-1	0	0	0	0
R_8	3	-3	1	-1	0	0	0	0

R_7: $P = A$; $Q = G$; $R = B$.
R_8: $P = L$; $Q = M$; $R = N$.

G_{24}^{9}

$P^6 = E$; $Q^4 = E$; $R^4 = E$; $P^3 = Q^2 = R^2$; $QP = PR$; $RP = PQR$; $RQ = P^3QR$.
$C_1 = E$; $C_2 = P^3$; $C_3 = Q, R, P^3Q, P^3R, QR, P^3QR$; $C_4 = P, PQ, PQR, PR$; $C_5 = P^4, P^4Q, P^4QR, P^4R$; $C_6 = P^2, P^5R, P^5QR, P^5Q$; $C_7 = P^5, P^2R, P^2QR, P^2Q$.

	C_1	C_2	C_3	C_4	C_5	C_6	C_7
R_1	1	1	1	1	1	1	1
R_2	1	1	1	ω	ω	ω^*	ω^*
R_3	1	1	1	ω^*	ω^*	ω	ω
R_4	2	-2	0	1	-1	-1	1
R_5	2	-2	0	ω	$-\omega$	$-\omega^*$	ω^*
R_6	2	-2	0	ω^*	$-\omega^*$	$-\omega$	ω
R_7	3	3	-1	0	0	0	0

R_4: $P = -\pi'^*$; $Q = -i\phi$.
R_5: $P = -\omega\pi'^*$; $Q = -i\phi$.
R_6: $P = -\omega^*\pi'^*$; $Q = -i\phi$.
R_7: $P = A$; $Q = B$.
Note that $R = P^5QP$.

$\mathbf{G}_{24}^{10}$

$P^6 = E$; $Q^2 = E$; $R^2 = E$; $QP = PQR$; $RP = PQ$; $RQ = QR$.
$C_1 = E$; $C_2 = Q, R, QR$; $C_3 = P^2, P^2Q, P^2R, P^2QR$; $C_4 = P^4, P^4Q, P^4R, P^4QR$; $C_5 = P^3$;
$C_6 = P^3Q, P^3R, P^3QR$; $C_7 = P^5, P^5Q, P^5R, P^5QR$; $C_8 = P, PQ, PR, PQR$.

	C_1	C_2	C_3	C_4	C_5	C_6	C_7	C_8
R_1	1	1	1	1	1	1	1	1
R_2	1	1	ω	ω^*	1	1	ω	ω^*
R_3	1	1	ω^*	ω	1	1	ω^*	ω
R_4	3	-1	0	0	3	-1	0	0
R_5	1	1	1	1	-1	-1	-1	-1
R_6	1	1	ω	ω^*	-1	-1	$-\omega$	$-\omega^*$
R_7	1	1	ω^*	ω	-1	-1	$-\omega^*$	$-\omega$
R_8	3	-1	0	0	-3	1	0	0

R_4: $P = D$; $Q = B$; $R = C$. R_8: $P = J$; $Q = B$; $R = C$.

$\mathbf{G}_{24}^{11}$

$P^{12} = E$; $Q^4 = E$; $Q^2 = P^6$; $QP = P^{11}Q$.
$C_1 = E$; $C_2 = P^6$; $C_3 = P, P^{11}$; $C_4 = P^5, P^7$; $C_5 = P^2, P^{10}$; $C_6 = P^4, P^8$; $C_7 = P^3, P^9$;
$C_8 = Q, P^2Q, P^4Q, P^6Q, P^8Q, P^{10}Q$; $C_9 = PQ, P^3Q, P^5Q, P^7Q, P^9Q, P^{11}Q$.

	C_1	C_2	C_3	C_4	C_5	C_6	C_7	C_8	C_9
R_1	1	1	1	1	1	1	1	1	1
R_2	1	1	1	1	1	1	1	-1	-1
R_3	1	1	-1	-1	1	1	-1	1	-1
R_4	1	1	-1	-1	1	1	-1	-1	1
R_5	2	2	-1	-1	-1	-1	2	0	0
R_6	2	2	1	1	-1	-1	-2	0	0
R_7	2	-2	0	0	-2	2	0	0	0
R_8	2	-2	$\sqrt{3}$	$-\sqrt{3}$	1	-1	0	0	0
R_9	2	-2	$-\sqrt{3}$	$\sqrt{3}$	1	-1	0	0	0

R_5: $P = -\pi^*$; $Q = \phi$. R_6: $P = \pi$; $Q = \phi$.
R_7: $P = \rho$; $Q = \mathrm{i}\phi$. R_8: $P = -\rho^*$; $Q = \mathrm{i}\phi$.
R_9: $P = \mathrm{i}\lambda$; $Q = \mathrm{i}\phi$.

$\mathbf{G}_{24}^{12}$

$P^{12} = E$; $Q^2 = E$; $QP = PQ$.
$C_1 = E$; $C_2 = P$; $C_3 = P^2$; $C_4 = P^3$; $C_5 = P^4$; $C_6 = P^5$; $C_7 = P^6$; $C_8 = P^7$; $C_9 = P^8$; $C_{10} = P^9$;
$C_{11} = P^{10}$; $C_{12} = P^{11}$; $C_{13} = Q$; $C_{14} = PQ$; $C_{15} = P^2Q$; $C_{16} = P^3Q$; $C_{17} = P^4Q$; $C_{18} = P^5Q$;
$C_{19} = P^6Q$; $C_{20} = P^7Q$; $C_{21} = P^8Q$; $C_{22} = P^9Q$; $C_{23} = P^{10}Q$; $C_{24} = P^{11}Q$.

	C_1 to C_{12}	C_{13} to C_{24}
R_1 to R_{12}	$(\mathbf{G}_{12}^1)$	$(\mathbf{G}_{12}^1)$
R_{13} to R_{24}	$(\mathbf{G}_{12}^1)$	$-(\mathbf{G}_{12}^1)$

G_{32}^{1}

$P^2 = E$; $Q^4 = E$; $R^4 = E$; $QP = PQ^3$; $RP = PR$; $RQ = QR^3$.
$C_1 = E$; $C_2 = Q^2$; $C_3 = R^2$; $C_4 = Q^2R^2$; $C_5 = R, R^3$; $C_6 = Q^2R, Q^2R^3$; $C_7 = PR, PQ^2R^3$;
$C_8 = PQ^2R, PR^3$; $C_9 = P, PQ^2$; $C_{10} = PR^2, PQ^2R^2$; $C_{11} = PQ, PQ^3, PQR^2, PQ^3R^2$;
$C_{12} = PQR, PQR^3, PQ^3R, PQ^3R^3$; $C_{13} = Q, Q^3, QR^2, Q^3R^2$; $C_{14} = Q^3R, QR, Q^3R^3, QR^3$.

	C_1	C_2	C_3	C_4	C_5	C_6	C_7	C_8	C_9	C_{10}	C_{11}	C_{12}	C_{13}	C_{14}
R_1	1	1	1	1	1	1	1	1	1	1	1	1	1	1
R_2	1	1	1	1	1	1	1	1	1	1	−1	−1	−1	−1
R_3	1	1	1	1	−1	−1	−1	−1	1	1	1	−1	1	−1
R_4	1	1	1	1	−1	−1	−1	−1	1	1	−1	1	−1	1
R_5	1	1	1	1	1	1	−1	−1	−1	−1	1	1	−1	−1
R_6	1	1	1	1	1	1	−1	−1	−1	−1	−1	−1	1	1
R_7	1	1	1	1	−1	−1	1	1	−1	−1	1	−1	−1	1
R_8	1	1	1	1	−1	−1	1	1	−1	−1	−1	1	1	−1
R_9	2	2	−2	−2	0	0	0	0	2	−2	0	0	0	0
R_{10}	2	−2	2	−2	2	−2	0	0	0	0	0	0	0	0
R_{11}	2	−2	2	−2	−2	2	0	0	0	0	0	0	0	0
R_{12}	2	−2	−2	2	0	0	2i	−2i	0	0	0	0	0	0
R_{13}	2	−2	−2	2	0	0	−2i	2i	0	0	0	0	0	0
R_{14}	2	2	−2	−2	0	0	0	0	−2	2	0	0	0	0

R_9: $P = \varepsilon$; $Q = \phi$; $R = -i\lambda$.
R_{10}: $P = -\lambda$; $Q = \kappa$; $R = \varepsilon$.
R_{11}: $P = -\lambda$; $Q = \kappa$; $R = -\varepsilon$.
R_{12}: $P = -\lambda$; $Q = \kappa$; $R = -i\lambda$.
R_{13}: $P = -\lambda$; $Q = \kappa$; $R = i\lambda$.
R_{14}: $P = -\varepsilon$; $Q = \phi$; $R = -i\lambda$.

G_{32}^{2}

$P^4 = E$; $Q^4 = E$; $R^2 = E$; $QP = P^3Q^3$; $QP^2 = P^2Q$; $RP = P^3R$; $RQ = Q^3R$.
$C_1 = E$; $C_2 = Q^2$; $C_3 = P^2$; $C_4 = P^2Q^2$; $C_5 = P^3R, PR$; $C_6 = P^3Q^2R, PQ^2R$; $C_7 = QR, Q^3R$;
$C_8 = P^2Q^3R, P^2QR$; $C_9 = PQ, P^3Q^3$; $C_{10} = PQ^3, P^3Q$; $C_{11} = R, P^2R, Q^2R, P^2Q^2R$;
$C_{12} = P, P^3, PQ^2, P^3Q^2$; $C_{13} = Q, P^2Q, Q^3, P^2Q^3$; $C_{14} = PQR, P^3QR, PQ^3R, P^3Q^3R$.

	C_1	C_2	C_3	C_4	C_5	C_6	C_7	C_8	C_9	C_{10}	C_{11}	C_{12}	C_{13}	C_{14}
R_1	1	1	1	1	1	1	1	1	1	1	1	1	1	1
R_2	1	1	1	1	1	1	1	1	1	1	−1	−1	−1	−1
R_3	1	1	1	1	−1	−1	1	1	−1	−1	1	−1	1	−1
R_4	1	1	1	1	−1	−1	1	1	−1	−1	−1	1	−1	1
R_5	1	1	1	1	1	1	−1	−1	−1	−1	1	1	−1	−1
R_6	1	1	1	1	1	1	−1	−1	−1	−1	−1	−1	1	1
R_7	1	1	1	1	−1	−1	−1	−1	1	1	1	−1	−1	1
R_8	1	1	1	1	−1	−1	−1	−1	1	1	−1	1	1	−1
R_9	2	−2	2	−2	2	−2	0	0	0	0	0	0	0	0
R_{10}	2	2	−2	−2	0	0	2	−2	0	0	0	0	0	0
R_{11}	2	−2	−2	2	0	0	0	0	2	−2	0	0	0	0
R_{12}	2	−2	2	−2	−2	2	0	0	0	0	0	0	0	0
R_{13}	2	2	−2	−2	0	0	−2	2	0	0	0	0	0	0
R_{14}	2	−2	−2	2	0	0	0	0	−2	2	0	0	0	0

R_9: $P = \phi$; $Q = \kappa$; $R = \phi$.
R_{10}: $P = -\kappa$; $Q = \phi$; $R = \phi$.
R_{11}: $P = -\kappa$; $Q = \kappa$; $R = \phi$.
R_{12}: $P = -\phi$; $Q = \kappa$; $R = \phi$.
R_{13}: $P = -\kappa$; $Q = -\phi$; $R = \phi$.
R_{14}: $P = -\kappa$; $Q = -\kappa$; $R = \phi$.

$\mathbf{G}_{32}^{3}$

$P^4 = E$; $Q^2 = E$; $R^2 = E$; $S^2 = E$; $QP = P^3Q$; $RP = PR$; $RQ = QR$; $SP = PS$; $SQ = QS$; $SR = RS$.
$C_1 = E$; $C_2 = P^2$; $C_3 = P, P^3$; $C_4 = Q, P^2Q$; $C_5 = PQ, P^3Q$; $C_6 = R$; $C_7 = P^2R$; $C_8 = PR, P^3R$;
$C_9 = QR, P^2QR$; $C_{10} = PQR, P^3QR$; $C_{11} = S$; $C_{12} = P^2S$; $C_{13} = PS, P^3S$; $C_{14} = QS, P^2QS$;
$C_{15} = PQS, P^3QS$; $C_{16} = RS$; $C_{17} = P^2RS$; $C_{18} = PRS, P^3RS$; $C_{19} = QRS, P^2QRS$;
$C_{20} = PQRS, P^3QRS$.

	C_1 to C_5	C_6 to C_{10}	C_{11} to C_{15}	C_{16} to C_{20}
R_1 to R_5	$(\mathbf{G}_8^4)$	$(\mathbf{G}_8^4)$	$(\mathbf{G}_8^4)$	$(\mathbf{G}_8^4)$
R_6 to R_{10}	$(\mathbf{G}_8^4)$	$-(\mathbf{G}_8^4)$	$(\mathbf{G}_8^4)$	$-(\mathbf{G}_8^4)$
R_{11} to R_{15}	$(\mathbf{G}_8^4)$	$(\mathbf{G}_8^4)$	$-(\mathbf{G}_8^4)$	$-(\mathbf{G}_8^4)$
R_{16} to R_{20}	$(\mathbf{G}_8^4)$	$-(\mathbf{G}_8^4)$	$-(\mathbf{G}_8^4)$	$(\mathbf{G}_8^4)$

R_5: $P = \kappa$; $Q = \lambda$; $R = \varepsilon$; $S = \varepsilon$.
R_{10}: $P = \kappa$; $Q = \lambda$; $R = -\varepsilon$; $S = \varepsilon$.
R_{15}: $P = \kappa$; $Q = \lambda$; $R = \varepsilon$; $S = -\varepsilon$.
R_{20}: $P = \kappa$; $Q = \lambda$; $R = -\varepsilon$; $S = -\varepsilon$.

$\mathbf{G}_{32}^{4}$

$P^4 = E$; $Q^4 = E$; $R^2 = E$; $QP = PQ$; $RP = P^3Q^3R$; $RP^2 = P^2Q^2R$; $RP^3 = PQR$; $RQ = QR$.
$C_1 = E$; $C_2 = Q$; $C_3 = Q^2$; $C_4 = Q^3$; $C_5 = P^2, P^2Q^2$; $C_6 = P^2Q, P^2Q^3$; $C_7 = P, P^3Q^3$; $C_8 = PQ, P^3$;
$C_9 = PQ^2, P^3Q$; $C_{10} = PQ^3, P^3Q^2$; $C_{11} = PQR, P^3R, PQ^3R, P^3Q^2R$; $C_{12} = PR, PQ^2R, P^3QR, P^3Q^3R$;
$C_{13} = QR, Q^3R, P^2R, P^2Q^2R$; $C_{14} = R, Q^2R, P^2QR, P^2Q^3R$.

	C_1	C_2	C_3	C_4	C_5	C_6	C_7	C_8	C_9	C_{10}	C_{11}	C_{12}	C_{13}	C_{14}
R_1	1	1	1	1	1	1	1	1	1	1	1	1	1	1
R_2	1	1	1	1	1	1	1	1	1	1	−1	−1	−1	−1
R_3	1	−1	1	−1	−1	1	i	−i	i	−i	−i	i	−1	1
R_4	1	−1	1	−1	−1	1	i	−i	i	−i	i	−i	1	−1
R_5	1	1	1	1	1	1	−1	−1	−1	−1	−1	−1	1	1
R_6	1	1	1	1	1	1	−1	−1	−1	−1	1	1	−1	−1
R_7	1	−1	1	−1	−1	1	−i	i	−i	i	i	−i	−1	1
R_8	1	−1	1	−1	−1	1	−i	i	−i	i	−i	i	1	−1
R_9	2	2	2	2	−2	−2	0	0	0	0	0	0	0	0
R_{10}	2	−2	2	−2	2	−2	0	0	0	0	0	0	0	0
R_{11}	2	2i	−2	−2i	0	0	1 − i	1 + i	−1 + i	−1 − i	0	0	0	0
R_{12}	2	2i	−2	−2i	0	0	−1 + i	−1 − i	1 − i	1 + i	0	0	0	0
R_{13}	2	−2i	−2	2i	0	0	1 + i	1 − i	−1 − i	−1 + i	0	0	0	0
R_{14}	2	−2i	−2	2i	0	0	−1 − i	−1 + i	1 + i	1 − i	0	0	0	0

R_9: $P = \mathrm{i}\lambda$; $Q = \varepsilon$; $R = \phi$.
R_{10}: $P = \lambda$; $Q = -\varepsilon$; $R = \phi$.
R_{11}: $P = \nu^*$; $Q = \mathrm{i}\varepsilon$; $R = \phi$.
R_{12}: $P = \mathrm{i}\nu$; $Q = \mathrm{i}\varepsilon$; $R = \phi$.
R_{13}: $P = \nu$; $Q = -\mathrm{i}\varepsilon$; $R = \phi$.
R_{14}: $P = -\mathrm{i}\nu^*$; $Q = -\mathrm{i}\varepsilon$; $R = \phi$.

G_{32}^{5}

$P^4 = E$; $Q^2 = E$; $R^2 = E$; $S^2 = E$; $QP = PQ$; $RP = PQR$; $RQ = QR$; $SP = PS$; $SQ = QS$; $SR = RS$.
$C_1 = E$; $C_2 = Q$; $C_3 = P^2$; $C_4 = P^2Q$; $C_5 = R, QR$; $C_6 = P^2R, P^2QR$; $C_7 = P, PQ$; $C_8 = P^3, P^3Q$; $C_9 = PR, PQR$; $C_{10} = P^3R, P^3QR$; $C_{11} = S$; $C_{12} = QS$; $C_{13} = P^2S$; $C_{14} = P^2QS$; $C_{15} = RS, QRS$; $C_{16} = P^2RS, P^2QRS$; $C_{17} = PS, PQS$; $C_{18} = P^3S, P^3QS$; $C_{19} = PRS, PQRS$; $C_{20} = P^3RS, P^3QRS$.

	C_1 to C_{10}	C_{11} to C_{20}
R_1 to R_{10}	(G_{16}^{10})	(G_{16}^{10})
R_{11} to R_{20}	(G_{16}^{10})	$-(G_{16}^{10})$

R_9: $P = \phi$; $Q = -\varepsilon$; $R = \lambda$; $S = \varepsilon$.
R_{10}: $P = \kappa$; $Q = -\varepsilon$; $R = \lambda$; $S = \varepsilon$.
R_{11}: $P = \phi$; $Q = -\varepsilon$; $R = \lambda$; $S = -\varepsilon$.
R_{12}: $P = \kappa$; $Q = -\varepsilon$; $R = \lambda$; $S = -\varepsilon$.

G_{32}^{6}

$P^4 = E$; $Q^4 = E$; $R^2 = E$; $QP = P^3Q^3$; $QP^2 = P^2Q$; $RP = PR$; $RQ = P^2QR$.
$C_1 = E$; $C_2 = P^2$; $C_3 = P, P^3Q^2$; $C_4 = P^3, PQ^2$; $C_5 = Q^2$; $C_6 = P^2Q^2$; $C_7 = R, P^2R$; $C_8 = PR, PQ^2R$; $C_9 = P^3R, P^3Q^2R$; $C_{10} = Q^2R, P^2Q^2R$; $C_{11} = PQ^3, P^3Q^3, PQ, P^3Q$; $C_{12} = Q, Q^3, P^2Q, P^2Q^3$; $C_{13} = PQ^3R, P^3Q^3R, PQR, P^3QR$; $C_{14} = QR, Q^3R, P^2QR, P^2Q^3R$.

	C_1	C_2	C_3	C_4	C_5	C_6	C_7	C_8	C_9	C_{10}	C_{11}	C_{12}	C_{13}	C_{14}
R_1	1	1	1	1	1	1	1	1	1	1	1	1	1	1
R_2	1	1	1	1	1	1	1	1	1	1	−1	−1	−1	−1
R_3	1	1	−1	−1	1	1	1	−1	−1	1	1	−1	1	−1
R_4	1	1	−1	−1	1	1	1	−1	−1	1	−1	1	−1	1
R_5	1	1	1	1	1	1	−1	−1	−1	−1	1	1	−1	−1
R_6	1	1	1	1	1	1	−1	−1	−1	−1	−1	−1	1	1
R_7	1	1	−1	−1	1	1	−1	1	1	−1	1	−1	−1	1
R_8	1	1	−1	−1	1	1	−1	1	1	−1	−1	1	1	−1
R_9	2	−2	0	0	2	−2	0	2i	−2i	0	0	0	0	0
R_{10}	2	−2	0	0	2	−2	0	−2i	2i	0	0	0	0	0
R_{11}	2	2	0	0	−2	−2	2	0	0	−2	0	0	0	0
R_{12}	2	2	0	0	−2	−2	−2	0	0	2	0	0	0	0
R_{13}	2	−2	2i	−2i	−2	2	0	0	0	0	0	0	0	0
R_{14}	2	−2	−2i	2i	−2	2	0	0	0	0	0	0	0	0

R_9: $P = i\lambda$; $Q = \phi$; $R = \lambda$.
R_{10}: $P = -i\lambda$; $Q = \phi$; $R = \lambda$.
R_{11}: $P = \lambda$; $Q = \kappa$; $R = \varepsilon$.
R_{12}: $P = \lambda$; $Q = \kappa$; $R = -\varepsilon$.
R_{13}: $P = i\varepsilon$; $Q = \kappa$; $R = \lambda$.
R_{14}: $P = -i\varepsilon$; $Q = \kappa$; $R = \lambda$.

$\mathbf{G}_{32}^{7}$

$P^4 = E$; $Q^4 = E$; $R^2 = E$; $QP = P^3Q$; $RP = P^3Q^2R$; $RQ = QR$.
$C_1 = E$; $C_2 = P^2$; $C_3 = Q^2$; $C_4 = P^2Q^2$; $C_5 = P, P^3, PQ^2, P^3Q^2$; $C_6 = Q, P^2Q$; $C_7 = Q^3, P^2Q^3$; $C_8 = PQ, P^3Q, PQ^3, P^3Q^3$; $C_9 = R, P^2Q^2R$; $C_{10} = P^2R, Q^2R$; $C_{11} = PR, P^3R, PQ^2R, P^3Q^2R$; $C_{12} = QR, Q^3R$; $C_{13} = P^2QR, P^2Q^3R$; $C_{14} = PQR, P^3QR, PQ^3R, P^3Q^3R$.

	C_1	C_2	C_3	C_4	C_5	C_6	C_7	C_8	C_9	C_{10}	C_{11}	C_{12}	C_{13}	C_{14}
R_1	1	1	1	1	1	1	1	1	1	1	1	1	1	1
R_2	1	1	1	1	1	1	1	1	−1	−1	−1	−1	−1	−1
R_3	1	1	1	1	1	−1	−1	−1	1	1	1	−1	−1	−1
R_4	1	1	1	1	1	−1	−1	−1	−1	−1	−1	1	1	1
R_5	1	1	1	1	−1	1	1	−1	1	1	−1	1	1	−1
R_6	1	1	1	1	−1	1	1	−1	−1	−1	1	−1	−1	1
R_7	1	1	1	1	−1	−1	−1	1	1	1	−1	−1	−1	1
R_8	1	1	1	1	−1	−1	−1	1	−1	−1	1	1	1	−1
R_9	2	−2	2	−2	0	0	0	0	0	0	0	2	−2	0
R_{10}	2	−2	2	−2	0	0	0	0	0	0	0	−2	2	0
R_{11}	2	−2	−2	2	0	0	0	0	2	−2	0	0	0	0
R_{12}	2	−2	−2	2	0	0	0	0	−2	2	0	0	0	0
R_{13}	2	2	−2	−2	0	2i	−2i	0	0	0	0	0	0	0
R_{14}	2	2	−2	−2	0	−2i	2i	0	0	0	0	0	0	0

R_9: $P = \kappa$; $Q = \lambda$; $R = \lambda$.
R_{10}: $P = \kappa$; $Q = \lambda$; $R = -\lambda$.
R_{11}: $P = \mathrm{i}\lambda$; $Q = \kappa$; $R = \varepsilon$.
R_{12}: $P = \mathrm{i}\lambda$; $Q = \kappa$; $R = -\varepsilon$.
R_{13}: $P = \phi$; $Q = \mathrm{i}\varepsilon$; $R = \lambda$.
R_{14}: $P = \phi$; $Q = -\mathrm{i}\varepsilon$; $R = \lambda$.

$\mathbf{G}_{32}^{8}$

$P^8 = E$; $Q^4 = E$; $QP = P^5Q^3$; $QP^2 = P^2Q$.
$C_1 = E$; $C_2 = P^6Q^2$; $C_3 = P^4$; $C_4 = P^2Q^2$; $C_5 = Q, P^4Q^3$; $C_6 = P^6Q^3, P^2Q$; $C_7 = P, P^5Q^2$; $C_8 = P^7Q^2, P^3$; $C_9 = PQ, P^5Q^3$; $C_{10} = P^7Q^3, P^3Q$; $C_{11} = Q^2$; $C_{12} = P^6$; $C_{13} = P^4Q^2$; $C_{14} = P^2$; $C_{15} = Q^3, P^4Q$; $C_{16} = P^6Q, P^2Q^3$; $C_{17} = PQ^2, P^5$; $C_{18} = P^7, P^3Q^2$; $C_{19} = PQ^3, P^5Q$; $C_{20} = P^7Q, P^3Q^3$.

	C_1	C_2	C_3	C_4	C_5	C_6	C_7	C_8	C_9	C_{10}
R_1	1	1	1	1	1	1	1	1	1	1
R_2	1	1	1	1	−1	−1	1	1	−1	−1
R_3	1	1	1	1	1	1	−1	−1	−1	−1
R_4	1	1	1	1	−1	−1	−1	−1	1	1
R_5	1	−1	1	−1	1	−1	i	−i	i	−i
R_6	1	−1	1	−1	−1	1	i	−i	−i	i
R_7	1	−1	1	−1	1	−1	−i	i	−i	i
R_8	1	−1	1	−1	−1	1	−i	i	i	−i
R_9	2	−2i	−2	2i	0	0	0	0	0	0
R_{10}	2	2i	−2	−2i	0	0	0	0	0	0
R_{11}	1	i	−1	−i	i	−1	θ	$-\theta^*$	$-\theta^*$	$-\theta$
R_{12}	1	i	−1	−i	i	−1	$-\theta$	θ^*	θ^*	θ
R_{13}	1	−i	−1	i	i	1	θ^*	$-\theta$	θ	θ^*
R_{14}	1	−i	−1	i	i	1	$-\theta^*$	θ	$-\theta$	$-\theta^*$
R_{15}	1	i	−1	−i	−i	1	θ	$-\theta^*$	θ^*	θ
R_{16}	1	i	−1	−i	−i	1	$-\theta$	θ^*	$-\theta^*$	$-\theta$
R_{17}	1	−i	−1	i	−i	−1	θ^*	$-\theta$	$-\theta$	$-\theta^*$
R_{18}	1	−i	−1	i	−i	−1	$-\theta^*$	θ	θ	θ^*
R_{19}	2	−2	2	−2	0	0	0	0	0	0
R_{20}	2	2	2	2	0	0	0	0	0	0

	C_{11}	C_{12}	C_{13}	C_{14}	C_{15}	C_{16}	C_{17}	C_{18}	C_{19}	C_{20}
R_1	1	1	1	1	1	1	1	1	1	1
R_2	1	1	1	1	-1	-1	1	1	-1	-1
R_3	1	1	1	1	1	1	-1	-1	-1	-1
R_4	1	1	1	1	-1	-1	-1	-1	1	1
R_5	1	-1	1	-1	1	-1	i	$-$i	i	$-$i
R_6	1	-1	1	-1	-1	1	i	$-$i	$-$i	i
R_7	1	-1	1	-1	1	-1	$-$i	i	$-$i	i
R_8	1	-1	1	-1	-1	1	$-$i	i	i	$-$i
R_9	2	-2i	-2	2i	0	0	0	0	0	0
R_{10}	2	2i	-2	-2i	0	0	0	0	0	0
R_{11}	-1	$-$i	1	i	$-$i	1	$-\theta$	θ^*	θ^*	θ
R_{12}	-1	$-$i	1	i	$-$i	1	θ	$-\theta^*$	$-\theta^*$	$-\theta$
R_{13}	-1	i	1	$-$i	$-$i	-1	$-\theta^*$	θ	$-\theta$	$-\theta^*$
R_{14}	-1	i	1	$-$i	$-$i	-1	θ^*	$-\theta$	θ	θ^*
R_{15}	-1	$-$i	1	i	i	-1	$-\theta$	θ^*	$-\theta^*$	$-\theta$
R_{16}	-1	$-$i	1	i	i	-1	θ	$-\theta^*$	θ^*	θ
R_{17}	-1	i	1	$-$i	i	1	$-\theta^*$	θ	θ	θ^*
R_{18}	-1	i	1	$-$i	i	1	θ^*	$-\theta$	$-\theta$	$-\theta^*$
R_{19}	-2	2	-2	2	0	0	0	0	0	0
R_{20}	-2	-2	-2	-2	0	0	0	0	0	0

R_9: $P = \theta\lambda$; $Q = -\mathrm{i}\kappa$.
R_{10}: $P = \theta^*\lambda$; $Q = -\mathrm{i}\kappa$.
R_{19}: $P = \phi$; $Q = \mathrm{i}\lambda$.
R_{20}: $P = \kappa$; $Q = \mathrm{i}\lambda$.

$\mathbf{G}_{32}^{9}$

$P^8 = E$; $Q^4 = E$; $R^2 = E$; $Q^2 = P^4$; $QP = P^7Q$; $RP = PR$; $RQ = QR$.
$C_1 = E$; $C_2 = P^4$; $C_3 = P^2, P^6$; $C_4 = P, P^7$; $C_5 = P^3, P^5$; $C_6 = Q, P^2Q, P^4Q, P^6Q$;
$C_7 = PQ, P^3Q, P^5Q, P^7Q$; $C_8 = R$; $C_9 = P^4R$; $C_{10} = P^2R, P^6R$; $C_{11} = PR, P^7R$; $C_{12} = P^3R, P^5R$;
$C_{13} = QR, P^2QR, P^4QR, P^6QR$; $C_{14} = PQR, P^3QR, P^5QR, P^7QR$.

	C_1 to C_7	C_8 to C_{14}
R_1 to R_7	$(\mathbf{G}_{16}^{14})$	$(\mathbf{G}_{16}^{14})$
R_8 to R_{14}	$(\mathbf{G}_{16}^{14})$	$-(\mathbf{G}_{16}^{14})$

R_5: $P = \kappa$; $Q = \lambda$; $R = \varepsilon$.
R_6: $P = \delta$; $Q = \mathrm{i}\phi$; $R = \varepsilon$.
R_7: $P = \eta$; $Q = \mathrm{i}\phi$; $R = \varepsilon$.
R_{12}: $P = \kappa$; $Q = \lambda$; $R = -\varepsilon$.
R_{13}: $P = \delta$; $Q = \mathrm{i}\phi$; $R = -\varepsilon$.
R_{14}: $P = \eta$; $Q = \mathrm{i}\phi$; $R = -\varepsilon$.

$\mathbf{G}_{32}^{10}$

$P^8 = E$; $Q^4 = E$; $QP = P^7Q$.
$C_1 = E$; $C_2 = P^4$; $C_3 = P^2, P^6$; $C_4 = P, P^7$; $C_5 = P^3, P^5$; $C_6 = Q, P^2Q, P^4Q, P^6Q$;
$C_7 = PQ, P^3Q, P^5Q, P^7Q$; $C_8 = Q^2$; $C_9 = P^4Q^2$; $C_{10} = P^2Q^2, P^6Q^2$; $C_{11} = PQ^2, P^7Q^2$;
$C_{12} = P^3Q^2, P^5Q^2$; $C_{13} = Q^3, P^2Q^3, P^4Q^3, P^6Q^3$; $C_{14} = PQ^3, P^3Q^3, P^5Q^3, P^7Q^3$.

	C_1	C_2	C_3	C_4	C_5	C_6	C_7	C_8	C_9	C_{10}	C_{11}	C_{12}	C_{13}	C_{14}
R_1	1	1	1	1	1	1	1	1	1	1	1	1	1	1
R_2	1	1	1	1	1	−1	−1	1	1	1	1	1	−1	−1
R_3	1	1	1	−1	−1	1	−1	1	1	1	−1	−1	1	−1
R_4	1	1	1	−1	−1	−1	1	1	1	1	−1	−1	−1	1
R_5	2	2	−2	0	0	0	0	2	2	−2	0	0	0	0
R_6	2	−2	0	$\sqrt{2}$	$-\sqrt{2}$	0	0	2	−2	0	$\sqrt{2}$	$-\sqrt{2}$	0	0
R_7	2	−2	0	$-\sqrt{2}$	$\sqrt{2}$	0	0	2	−2	0	$-\sqrt{2}$	$\sqrt{2}$	0	0
R_8	1	1	1	1	1	i	i	−1	−1	−1	−1	−1	−i	−i
R_9	1	1	1	−1	−1	i	−i	−1	−1	−1	1	1	−i	i
R_{10}	1	1	1	1	1	−i	−i	−1	−1	−1	−1	−1	i	i
R_{11}	1	1	1	−1	−1	−i	i	−1	−1	−1	1	1	i	−i
R_{12}	2	−2	0	$\sqrt{2}$	$-\sqrt{2}$	0	0	−2	2	0	$-\sqrt{2}$	$\sqrt{2}$	0	0
R_{13}	2	−2	0	$-\sqrt{2}$	$\sqrt{2}$	0	0	−2	2	0	$\sqrt{2}$	$-\sqrt{2}$	0	0
R_{14}	2	2	−2	0	0	0	0	−2	−2	2	0	0	0	0

R_5: $P = \kappa$; $Q = \lambda$.
R_7: $P = \eta$; $Q = \lambda$.
R_{13}: $P = \eta$; $Q = \mathrm{i}\lambda$.
R_6: $P = \delta$; $Q = \lambda$.
R_{12}: $P = \delta$; $Q = \mathrm{i}\lambda$.
R_{14}: $P = \mathrm{i}\lambda$; $Q = \kappa$.

$\mathbf{G}_{32}^{11}$

$P^8 = E$; $Q^4 = E$; $QP = P^3Q^3$; $Q^2P = PQ^2$.
$C_1 = E$; $C_2 = P^4$; $C_3 = P^2, P^6$; $C_4 = Q^2$; $C_5 = P^4Q^2$; $C_6 = P^2Q^2, P^6Q^2$; $C_7 = P, P^3Q^2$;
$C_8 = PQ^2, P^3$; $C_9 = P^5, P^7Q^2$; $C_{10} = P^5Q^2, P^7$; $C_{11} = PQ, P^3Q^3, P^5Q, P^7Q^3$;
$C_{12} = PQ^3, P^3Q, P^5Q^3, P^7Q$; $C_{13} = Q, P^2Q^3, P^4Q, P^6Q^3$; $C_{14} = Q^3, P^2Q, P^4Q^3, P^6Q$.

	C_1	C_2	C_3	C_4	C_5	C_6	C_7	C_8	C_9	C_{10}	C_{11}	C_{12}	C_{13}	C_{14}
R_1	1	1	1	1	1	1	1	1	1	1	1	1	1	1
R_2	1	1	1	1	1	1	1	1	1	1	−1	−1	−1	−1
R_3	1	1	1	1	1	1	−1	−1	−1	−1	−1	−1	1	1
R_4	1	1	1	1	1	1	−1	−1	−1	−1	1	1	−1	−1
R_5	2	2	−2	2	2	−2	0	0	0	0	0	0	0	0
R_6	2	−2	0	2	−2	0	$\mathrm{i}\sqrt{2}$	$\mathrm{i}\sqrt{2}$	$-\mathrm{i}\sqrt{2}$	$-\mathrm{i}\sqrt{2}$	0	0	0	0
R_7	2	−2	0	2	−2	0	$-\mathrm{i}\sqrt{2}$	$-\mathrm{i}\sqrt{2}$	$\mathrm{i}\sqrt{2}$	$\mathrm{i}\sqrt{2}$	0	0	0	0
R_8	1	1	−1	−1	−1	1	i	−i	i	−i	−1	1	i	−i
R_9	1	1	−1	−1	−1	1	i	−i	i	−i	1	−1	−i	i
R_{10}	1	1	−1	−1	−1	1	−i	i	−i	i	1	−1	i	−i
R_{11}	1	1	−1	−1	−1	1	−i	i	−i	i	−1	1	−i	i
R_{12}	2	2	2	−2	−2	−2	0	0	0	0	0	0	0	0
R_{13}	2	−2	0	−2	2	0	$-\sqrt{2}$	$\sqrt{2}$	$\sqrt{2}$	$-\sqrt{2}$	0	0	0	0
R_{14}	2	−2	0	−2	2	0	$\sqrt{2}$	$-\sqrt{2}$	$-\sqrt{2}$	$\sqrt{2}$	0	0	0	0

R_5: $P = \kappa$; $Q = \lambda$.
R_7: $P = \xi^*$; $Q = \lambda$.
R_{13}: $P = \mathrm{i}\xi$; $Q = \mathrm{i}\lambda$.
R_6: $P = \xi$; $Q = \lambda$.
R_{12}: $P = \mathrm{i}\kappa$; $Q = \mathrm{i}\lambda$.
R_{14}: $P = \mathrm{i}\xi^*$; $Q = \mathrm{i}\lambda$.

$\mathbf{G}_{32}^{12}$

$P^8 = E$; $Q^2 = E$; $R^2 = E$; $QP = PQ$; $RP = PQR$; $RQ = QR$.
$C_1 = E$; $C_2 = Q$; $C_3 = P^2$; $C_4 = P^2Q$; $C_5 = P^4$; $C_6 = P^4Q$; $C_7 = P^6$; $C_8 = P^6Q$; $C_9 = R, QR$; $C_{10} = P^2R, P^2QR$; $C_{11} = P^4R, P^4QR$; $C_{12} = P^6R, P^6QR$; $C_{13} = P, PQ$; $C_{14} = P^3, P^3Q$; $C_{15} = PR, PQR$; $C_{16} = P^3R, P^3QR$; $C_{17} = P^5, P^5Q$; $C_{18} = P^7, P^7Q$; $C_{19} = P^5R, P^5QR$; $C_{20} = P^7R, P^7QR$.

	C_1	C_2	C_3	C_4	C_5	C_6	C_7	C_8	C_9	C_{10}
R_1	1	1	1	1	1	1	1	1	1	1
R_2	1	1	1	1	1	1	1	1	−1	−1
R_3	1	1	1	1	1	1	1	1	1	1
R_4	1	1	1	1	1	1	1	1	−1	−1
R_5	1	1	−1	−1	1	1	−1	−1	1	−1
R_6	1	1	−1	−1	1	1	−1	−1	1	−1
R_7	1	1	−1	−1	1	1	−1	−1	−1	1
R_8	1	1	−1	−1	1	1	−1	−1	−1	1
R_9	2	−2	2	−2	2	−2	2	−2	0	0
R_{10}	2	−2	−2	2	2	−2	−2	2	0	0
R_{11}	1	1	i	i	−1	−1	−i	−i	1	i
R_{12}	1	1	i	i	−1	−1	−i	−i	−1	−i
R_{13}	1	1	i	i	−1	−1	−i	−i	1	i
R_{14}	1	1	i	i	−1	−1	−i	−i	−1	−i
R_{15}	1	1	−i	−i	−1	−1	i	i	1	−i
R_{16}	1	1	−i	−i	−1	−1	i	i	1	−i
R_{17}	1	1	−i	−i	−1	−1	i	i	−1	i
R_{18}	1	1	−i	−i	−1	−1	i	i	−1	i
R_{19}	2	−2	2i	−2i	−2	2	−2i	2i	0	0
R_{20}	2	−2	−2i	2i	−2	2	2i	−2i	0	0

	C_{11}	C_{12}	C_{13}	C_{14}	C_{15}	C_{16}	C_{17}	C_{18}	C_{19}	C_{20}
R_1	1	1	1	1	1	1	1	1	1	1
R_2	−1	−1	−1	−1	1	1	−1	−1	1	1
R_3	1	1	−1	−1	−1	−1	−1	−1	−1	−1
R_4	−1	−1	1	1	−1	−1	1	1	−1	−1
R_5	1	−1	i	−i	i	−i	i	−i	i	−i
R_6	1	−1	−i	i	−i	i	−i	i	−i	i
R_7	−1	1	−i	i	i	−i	−i	i	i	−i
R_8	−1	1	i	−i	−i	i	i	−i	−i	i
R_9	0	0	0	0	0	0	0	0	0	0
R_{10}	0	0	0	0	0	0	0	0	0	0
R_{11}	−1	−i	θ	$-\theta^*$	θ	$-\theta^*$	$-\theta$	θ^*	$-\theta$	θ^*
R_{12}	1	i	$-\theta$	θ^*	θ	$-\theta^*$	θ	$-\theta^*$	$-\theta$	θ^*
R_{13}	−1	−i	$-\theta$	θ^*	$-\theta$	θ^*	θ	$-\theta^*$	θ	$-\theta^*$
R_{14}	1	i	θ	$-\theta^*$	$-\theta$	θ^*	$-\theta$	θ^*	θ	$-\theta^*$
R_{15}	−1	i	$-\theta^*$	θ	$-\theta^*$	θ	θ^*	$-\theta$	θ^*	$-\theta$
R_{16}	−1	i	θ^*	$-\theta$	θ^*	$-\theta$	$-\theta^*$	θ	$-\theta^*$	θ
R_{17}	1	−i	θ^*	$-\theta$	$-\theta^*$	θ	$-\theta^*$	θ	θ^*	$-\theta$
R_{18}	1	−i	$-\theta^*$	θ	θ^*	$-\theta$	θ^*	$-\theta$	$-\theta^*$	θ
R_{19}	0	0	0	0	0	0	0	0	0	0
R_{20}	0	0	0	0	0	0	0	0	0	0

R_9: $P = \phi$; $Q = -\varepsilon$; $R = \lambda$.
R_{10}: $P = \kappa$; $Q = -\varepsilon$; $R = \lambda$.
R_{19}: $P = \theta\phi$; $Q = -\varepsilon$; $R = \lambda$.
R_{20}: $P = \theta\kappa$; $Q = -\varepsilon$; $R = \lambda$.

G_{32}^{13}

$P^4 = E$; $Q^4 = E$; $R^2 = E$; $QP = PQ$; $RP = P^3Q^2R$; $RQ = QR$.
$C_1 = E$; $C_2 = P, P^3Q^2$; $C_3 = P^2$; $C_4 = P^3, PQ^2$; $C_5 = Q$; $C_6 = PQ, P^3Q^3$; $C_7 = P^2Q$; $C_8 = P^3Q, PQ^3$; $C_9 = Q^2$; $C_{10} = P^2Q^2$; $C_{11} = Q^3$; $C_{12} = P^2Q^3$; $C_{13} = R, P^2Q^2R$; $C_{14} = PR, P^3Q^2R$; $C_{15} = P^2R, Q^2R$; $C_{16} = P^3R, PQ^2R$; $C_{17} = QR, P^2Q^3R$; $C_{18} = PQR, P^3Q^3R$; $C_{19} = P^2QR, Q^3R$; $C_{20} = P^3QR, PQ^3R$.

	C_1	C_2	C_3	C_4	C_5	C_6	C_7	C_8	C_9	C_{10}
R_1	1	1	1	1	1	1	1	1	1	1
R_2	1	1	1	1	1	1	1	1	1	1
R_3	1	−1	1	−1	1	−1	1	−1	1	1
R_4	1	−1	1	−1	1	−1	1	−1	1	1
R_5	1	i	−1	−i	i	−1	−i	1	−1	1
R_6	1	i	−1	−i	i	−1	−i	1	−1	1
R_7	1	−i	−1	i	i	1	−i	−1	−1	1
R_8	1	−i	−1	i	i	1	−i	−1	−1	1
R_9	1	1	1	1	−1	−1	−1	−1	1	1
R_{10}	1	1	1	1	−1	−1	−1	−1	1	1
R_{11}	1	−1	1	−1	−1	1	−1	1	1	1
R_{12}	1	−1	1	−1	−1	1	−1	1	1	1
R_{13}	1	i	−1	−i	−i	1	i	−1	−1	1
R_{14}	1	i	−1	−i	−i	1	i	−1	−1	1
R_{15}	1	−i	−1	i	−i	−1	i	1	−1	1
R_{16}	1	−i	−1	i	−i	−1	i	1	−1	1
R_{17}	2	0	−2	0	2	0	−2	0	2	−2
R_{18}	2	0	2	0	2i	0	2i	0	−2	−2
R_{19}	2	0	−2	0	−2	0	2	0	2	−2
R_{20}	2	0	2	0	−2i	0	−2i	0	−2	−2

	C_{11}	C_{12}	C_{13}	C_{14}	C_{15}	C_{16}	C_{17}	C_{18}	C_{19}	C_{20}
R_1	1	1	1	1	1	1	1	1	1	1
R_2	1	1	−1	−1	−1	−1	−1	−1	−1	−1
R_3	1	1	1	−1	1	−1	1	−1	1	−1
R_4	1	1	−1	1	−1	1	−1	1	−1	1
R_5	−i	i	1	i	−1	−i	i	−1	−i	1
R_6	−i	i	−1	−i	1	i	−i	1	i	−1
R_7	−i	i	1	−i	−1	i	i	1	−i	−1
R_8	−i	i	−1	i	1	−i	−i	−1	i	1
R_9	−1	−1	1	1	1	1	−1	−1	−1	−1
R_{10}	−1	−1	−1	−1	−1	−1	1	1	1	1
R_{11}	−1	−1	1	−1	1	−1	1	−1	1	−1
R_{12}	−1	−1	−1	1	−1	1	−1	1	−1	1
R_{13}	i	−i	1	i	−1	−i	−i	1	i	1
R_{14}	i	−i	−1	−i	1	i	i	−1	−i	−1
R_{15}	i	−i	1	−i	−1	i	−i	−1	i	1
R_{16}	i	−i	−1	i	1	−i	i	1	−i	−1
R_{17}	2	−2	0	0	0	0	0	0	0	0
R_{18}	−2i	−2i	0	0	0	0	0	0	0	0
R_{19}	−2	2	0	0	0	0	0	0	0	0
R_{20}	2i	2i	0	0	0	0	0	0	0	0

R_{17}: $P = i\lambda$; $Q = \varepsilon$; $R = \phi$.
R_{18}: $P = \lambda$; $Q = i\varepsilon$; $R = \phi$.
R_{19}: $P = i\lambda$; $Q = -\varepsilon$; $R = \phi$.
R_{20}: $P = \lambda$; $Q = -i\varepsilon$; $R = \phi$.

$\mathbf{G}_{32}^{14}$

$P^4 = E$; $Q^4 = E$; $R^2 = E$; $QP = P^3Q$; $RP = PR$; $RQ = QR$.
$C_1 = E$; $C_2 = P^2$; $C_3 = Q^2$; $C_4 = P^2Q^2$; $C_5 = P, P^3$; $C_6 = PQ^2, P^3Q^2$; $C_7 = Q, P^2Q$; $C_8 = PQ, P^3Q$; $C_9 = Q^3, P^2Q^3$; $C_{10} = PQ^3, P^3Q^3$; $C_{11} = R$; $C_{12} = P^2R$; $C_{13} = Q^2R$; $C_{14} = P^2Q^2R$; $C_{15} = PR, P^3R$; $C_{16} = PQ^2R, P^3Q^2R$; $C_{17} = QR, P^2QR$; $C_{18} = PQR, P^3QR$; $C_{19} = Q^3R, P^2Q^3R$; $C_{20} = PQ^3R, P^3Q^3R$.

	C_1 to C_{10}	C_{11} to C_{20}
R_1 to R_{10}	$(\mathbf{G}_{16}^{8})$	$(\mathbf{G}_{16}^{8})$
R_{11} to R_{20}	$(\mathbf{G}_{16}^{8})$	$-(\mathbf{G}_{16}^{8})$

R_9: $P = i\lambda$; $Q = \phi$; $R = \varepsilon$.
R_{10}: $P = i\lambda$; $Q = \kappa$; $R = \varepsilon$.
R_{19}: $P = i\lambda$; $Q = \phi$; $R = -\varepsilon$.
R_{20}: $P = i\lambda$; $Q = \kappa$; $R = -\varepsilon$.

$\mathbf{G}_{32}^{15}$

$P^4 = E$; $Q^4 = E$; $Q^2 = P^2$; $R^2 = E$; $S^2 = E$; $QP = P^3Q$; $RP = PR$; $RQ = QR$; $SP = PS$; $SQ = QS$; $SR = RS$.
$C_1 = E$; $C_2 = P^2$; $C_3 = P, P^3$; $C_4 = Q, P^2Q$; $C_5 = PQ, P^3Q$; $C_6 = R$; $C_7 = P^2R$; $C_8 = PR, P^3R$; $C_9 = QR, P^2QR$; $C_{10} = PQR, P^3QR$; $C_{11} = S$; $C_{12} = P^2S$; $C_{13} = PS, P^3S$; $C_{14} = QS, P^2QS$; $C_{15} = PQS, P^3QS$; $C_{16} = RS$; $C_{17} = P^2RS$; $C_{18} = PRS, P^3RS$; $C_{19} = QRS, P^2QRS$; $C_{20} = PQRS, P^3QRS$.

	C_1 to C_{10}	C_{11} to C_{20}
R_1 to R_{10}	$(\mathbf{G}_{16}^{11})$	$(\mathbf{G}_{16}^{11})$
R_{11} to R_{20}	$(\mathbf{G}_{16}^{11})$	$-(\mathbf{G}_{16}^{11})$

R_5: $P = \kappa$; $Q = i\phi$; $R = \varepsilon$; $S = \varepsilon$.
R_{10}: $P = \kappa$; $Q = i\phi$; $R = -\varepsilon$; $S = \varepsilon$.
R_{15}: $P = \kappa$; $Q = i\phi$; $R = \varepsilon$; $S = -\varepsilon$.
R_{20}: $P = \kappa$; $Q = i\phi$; $R = -\varepsilon$; $S = -\varepsilon$.

$\mathbf{G}_{32}^{16}$

$P^4 = E$; $Q^4 = E$; $R^4 = E$; $QP = P^3Q$; $Q^2 = P^2$; $RP = PR$; $RQ = QR$.
$C_1 = E$; $C_2 = P^2$; $C_3 = P, P^3$; $C_4 = Q, P^2Q$; $C_5 = PQ, P^3Q$; $C_6 = R$; $C_7 = P^2R$; $C_8 = PR, P^3R$; $C_9 = QR, P^2QR$; $C_{10} = PQR, P^3QR$; $C_{11} = R^2$; $C_{12} = P^2R^2$; $C_{13} = PR^2, P^3R^2$; $C_{14} = QR^2, P^2QR^2$; $C_{15} = PQR^2, P^3QR^2$; $C_{16} = R^3$; $C_{17} = P^2R^3$; $C_{18} = PR^3, P^3R^3$; $C_{19} = QR^3, P^2QR^3$; $C_{20} = PQR^3, P^3QR^3$.

	C_1 to C_5	C_6 to C_{10}	C_{11} to C_{15}	C_{16} to C_{20}
R_1 to R_5	$(\mathbf{G}_8^5)$	$(\mathbf{G}_8^5)$	$(\mathbf{G}_8^5)$	$(\mathbf{G}_8^5)$
R_6 to R_{10}	$(\mathbf{G}_8^5)$	$i(\mathbf{G}_8^5)$	$-(\mathbf{G}_8^5)$	$-i(\mathbf{G}_8^5)$
R_{11} to R_{15}	$(\mathbf{G}_8^5)$	$-(\mathbf{G}_8^5)$	$(\mathbf{G}_8^5)$	$-(\mathbf{G}_8^5)$
R_{16} to R_{20}	$(\mathbf{G}_8^5)$	$-i(\mathbf{G}_8^5)$	$-(\mathbf{G}_8^5)$	$i(\mathbf{G}_8^5)$

R_5: $P = \kappa$; $Q = i\phi$; $R = \varepsilon$.
R_{10}: $P = \kappa$; $Q = i\phi$; $R = i\varepsilon$.
R_{15}: $P = \kappa$; $Q = i\phi$; $R = -\varepsilon$.
R_{20}: $P = \kappa$; $Q = i\phi$; $R = -i\varepsilon$.

G_{32}^{17}

$P^8 = E;\ Q^4 = E;\ QP = PQ.$
$C_1 = E;\ C_2 = P;\ C_3 = P^2;\ C_4 = P^3;\ C_5 = P^4;\ C_6 = P^5;\ C_7 = P^6;\ C_8 = P^7;\ C_9 = Q;\ C_{10} = PQ;$
$C_{11} = P^2Q;\ C_{12} = P^3Q;\ C_{13} = P^4Q;\ C_{14} = P^5Q;\ C_{15} = P^6Q;\ C_{16} = P^7Q;\ C_{17} = Q^2;\ C_{18} = PQ^2;$
$C_{19} = P^2Q^2;\ C_{20} = P^3Q^2;\ C_{21} = P^4Q^2;\ C_{22} = P^5Q^2;\ C_{23} = P^6Q^2;\ C_{24} = P^7Q^2;\ C_{25} = Q^3;\ C_{26} = PQ^3;$
$C_{27} = P^2Q^3;\ C_{28} = P^3Q^3;\ C_{29} = P^4Q^3;\ C_{30} = P^5Q^3;\ C_{31} = P^6Q^3;\ C_{32} = P^7Q^3.$

	C_1 to C_8	C_9 to C_{16}	C_{17} to C_{24}	C_{25} to C_{32}
R_1 to R_8	(G_8^1)	(G_8^1)	(G_8^1)	(G_8^1)
R_9 to R_{16}	(G_8^1)	$i(G_8^1)$	$-(G_8^1)$	$-i(G_8^1)$
R_{17} to R_{24}	(G_8^1)	$-(G_8^1)$	(G_8^1)	$-(G_8^1)$
R_{25} to R_{32}	(G_8^1)	$-i(G_8^1)$	$-(G_8^1)$	$i(G_8^1)$

G_{48}^1

$P^{12} = E;\ Q^2 = E;\ R^2 = E;\ QP = P^{11}Q;\ RP = P^7R;\ RQ = QR.$
$C_1 = E;\ C_2 = P^6;\ C_3 = P^3R, P^9R;\ C_4 = P^4, P^8;\ C_5 = P^2, P^{10};\ C_6 = PR, P^5R, P^7R, P^{11}R;$
$C_7 = PQ, P^3Q, P^5Q, P^7Q, P^9Q, P^{11}Q;\ C_8 = QR, P^4QR, P^8QR;\ C_9 = P^2QR, P^6QR, P^{10}QR;\ C_{10} = R, P^6R;$
$C_{11} = P^3, P^9;\ C_{12} = P^2R, P^4R, P^8R, P^{10}R;\ C_{13} = P, P^5, P^7, P^{11};\ C_{14} = Q, P^2Q, P^4Q, P^6Q, P^8Q, P^{10}Q;$
$C_{15} = PQR, P^3QR, P^5QR, P^7QR, P^9QR, P^{11}QR.$

	C_1	C_2	C_3	C_4	C_5	C_6	C_7	C_8	C_9	C_{10}	C_{11}	C_{12}	C_{13}	C_{14}	C_{15}
R_1	1	1	1	1	1	1	1	1	1	1	1	1	1	1	1
R_2	1	1	−1	1	1	−1	1	−1	−1	−1	1	−1	1	1	−1
R_3	1	1	−1	1	1	−1	−1	1	1	−1	1	−1	1	−1	1
R_4	1	1	1	1	1	1	−1	−1	−1	1	1	1	1	−1	−1
R_5	1	1	−1	1	1	−1	−1	1	1	1	−1	1	−1	1	−1
R_6	1	1	1	1	1	1	−1	−1	−1	−1	−1	−1	−1	1	1
R_7	1	1	−1	1	1	−1	1	−1	−1	1	−1	1	−1	−1	1
R_8	1	1	1	1	1	1	1	1	1	−1	−1	−1	−1	−1	−1
R_9	2	2	2	−1	−1	−1	0	0	0	2	2	−1	−1	0	0
R_{10}	2	2	−2	−1	−1	1	0	0	0	−2	2	1	−1	0	0
R_{11}	2	2	−2	−1	−1	1	0	0	0	2	−2	−1	1	0	0
R_{12}	2	2	2	−1	−1	−1	0	0	0	−2	−2	1	1	0	0
R_{13}	2	−2	0	2	−2	0	0	2	−2	0	0	0	0	0	0
R_{14}	2	−2	0	2	−2	0	0	−2	2	0	0	0	0	0	0
R_{15}	4	−4	0	−2	2	0	0	0	0	0	0	0	0	0	0

R_9: $P = \pi;\ Q = \phi;\ R = \varepsilon.$ R_{10}: $P = \pi;\ Q = \phi;\ R = -\varepsilon.$
R_{11}: $P = -\pi^*;\ Q = \phi;\ R = \varepsilon.$ R_{12}: $P = -\pi^*;\ Q = \phi;\ R = -\varepsilon.$
R_{13}: $P = i\lambda;\ Q = \phi;\ R = \phi.$ R_{14}: $P = i\lambda;\ Q = \phi;\ R = -\phi.$

$$R_{15}:\ P = \left(\begin{array}{c|c} \rho & 0 \\ \hline 0 & -\rho^* \end{array}\right);\quad Q = \left(\begin{array}{c|c} \phi & 0 \\ \hline 0 & \phi \end{array}\right);\quad R = \left(\begin{array}{c|c} 0 & \phi \\ \hline \phi & 0 \end{array}\right).$$

$\mathbf{G}_{48}^{2}$

$P^4 = E$; $Q^3 = E$; $R^2 = E$; $S^2 = E$; $QP = PQ^2$; $RP = P^3R$; $RQ = QR$; $SP = PS$; $SQ = QS$; $SR = RS$.
$C_1 = E$; $C_2 = P^2$; $C_3 = Q, Q^2$; $C_4 = P^2Q, P^2Q^2$; $C_5 = R, P^2R$; $C_6 = Q^2R, P^2QR$; $C_7 = QR, P^2Q^2R$;
$C_8 = P, PQ, PQ^2, P^3, P^3Q, P^3Q^2$; $C_9 = PR, PQR, PQ^2R, P^3R, P^3QR, P^3Q^2R$; $C_{10} = S$; $C_{11} = P^2S$;
$C_{12} = QS, Q^2S$; $C_{13} = P^2QS, P^2Q^2S$; $C_{14} = RS, P^2RS$; $C_{15} = Q^2RS, P^2QRS$; $C_{16} = QRS, P^2Q^2RS$;
$C_{17} = PS, PQS, PQ^2S, P^3S, P^3QS, P^3Q^2S$; $C_{18} = PRS, PQRS, PQ^2RS, P^3RS, P^3QRS, P^3Q^2RS$.

	C_1 to C_9	C_{10} to C_{18}
R_1 to R_9	$(\mathbf{G}_{24}^{1})$	$(\mathbf{G}_{24}^{1})$
R_{10} to R_{18}	$(\mathbf{G}_{24}^{1})$	$-(\mathbf{G}_{24}^{1})$

R_5: $P = \phi$; $Q = \pi^*$; $R = -\varepsilon$; $S = \varepsilon$.
R_6: $P = \phi$; $Q = \pi^*$; $R = \varepsilon$; $S = \varepsilon$.
R_7: $P = -i\phi$; $Q = \pi^*$; $R = \lambda$; $S = \varepsilon$.
R_8: $P = i\phi$; $Q = \pi$; $R = \lambda$; $S = \varepsilon$.
R_9: $P = i\phi$; $Q = \varepsilon$; $R = \lambda$; $S = \varepsilon$.
R_{14}: $P = \phi$; $Q = \pi^*$; $R = -\varepsilon$; $S = -\varepsilon$.
R_{15}: $P = \phi$; $Q = \pi^*$; $R = \varepsilon$; $S = -\varepsilon$.
R_{16}: $P = -i\phi$; $Q = \pi^*$; $R = \lambda$; $S = -\varepsilon$.
R_{17}: $P = i\phi$; $Q = \pi$; $R = \lambda$; $S = -\varepsilon$.
R_{18}: $P = i\phi$; $Q = \varepsilon$; $R = \lambda$; $S = -\varepsilon$.

$\mathbf{G}_{48}^{3}$

$P^{12} = E$; $Q^2 = E$; $R^2 = E$; $QP = P^7R$; $QP^3 = P^3Q$; $RP = P^{10}QR$; $RQ = P^6QR$.
$C_1 = E$; $C_2 = P^3$; $C_3 = P^6$; $C_4 = P^9$; $C_5 = Q, R, P^6Q, P^6R, P^3QR, P^9QR$;
$C_6 = QR, P^3Q, P^3R, P^6QR, P^9Q, P^9R$; $C_7 = P^4, PQ, P^7R, P^4QR$; $C_8 = P^7, P^4Q, P^{10}R, P^7QR$;
$C_9 = P^{10}, P^7Q, PR, P^{10}QR$; $C_{10} = P, P^{10}Q, P^4R, PQR$; $C_{11} = P^8, P^{11}Q, P^5R, P^2QR$;
$C_{12} = P^{11}, P^2Q, P^8R, P^5QR$; $C_{13} = P^2, P^5Q, P^{11}R, P^8QR$; $C_{14} = P^5, P^8Q, P^2R, P^{11}QR$.

	C_1	C_2	C_3	C_4	C_5	C_6	C_7	C_8	C_9	C_{10}	C_{11}	C_{12}	C_{13}	C_{14}
R_1	1	1	1	1	1	1	1	1	1	1	1	1	1	1
R_2	1	1	1	1	1	1	ω	ω	ω	ω	ω^*	ω^*	ω^*	ω^*
R_3	1	1	1	1	1	1	ω^*	ω^*	ω^*	ω^*	ω	ω	ω	ω
R_4	1	-1	1	-1	-1	1	1	-1	1	-1	1	-1	1	-1
R_5	1	-1	1	-1	-1	1	ω	$-\omega$	ω	$-\omega$	ω^*	$-\omega^*$	ω^*	$-\omega^*$
R_6	1	-1	1	-1	-1	1	ω^*	$-\omega^*$	ω^*	$-\omega^*$	ω	$-\omega$	ω	$-\omega$
R_7	2	2i	-2	-2i	0	0	-1	$-$i	1	i	-1	$-$i	1	i
R_8	2	2i	-2	-2i	0	0	$-\omega$	$-i\omega$	ω	$i\omega$	$-\omega^*$	$-i\omega^*$	ω^*	$i\omega^*$
R_9	2	2i	-2	-2i	0	0	$-\omega^*$	$-i\omega^*$	ω^*	$i\omega^*$	$-\omega$	$-i\omega$	ω	$i\omega$
R_{10}	2	-2i	-2	2i	0	0	-1	i	1	$-$i	-1	i	1	$-$i
R_{11}	2	-2i	-2	2i	0	0	$-\omega^*$	$i\omega^*$	ω^*	$-i\omega^*$	$-\omega$	$i\omega$	ω	$-i\omega$
R_{12}	2	-2i	-2	2i	0	0	$-\omega$	$i\omega$	ω	$-i\omega$	$-\omega^*$	$i\omega^*$	ω^*	$-i\omega^*$
R_{13}	3	3	3	3	-1	-1	0	0	0	0	0	0	0	0
R_{14}	3	-3	3	-3	1	-1	0	0	0	0	0	0	0	0

R_7: $P = -i\pi$; $Q = \chi$.
R_8: $P = -i\omega^*\mu$; $Q = \chi$.
R_9: $P = -i\mu$; $Q = \chi$.
R_{10}: $P = i\pi^*$; $Q = \chi^*$.
R_{11}: $P = i\omega\mu^*$; $Q = \chi^*$.
R_{12}: $P = i\mu^*$; $Q = \chi^*$.
R_{13}: $P = A$; $Q = B$.
R_{14}: $P = L$; $Q = N$.
Note that $R = P^5QP$.

$\mathbf{G}_{48}^{4}$

$P^6 = E$; $Q^4 = E$; $R^4 = E$; $P^3 = Q^2 = R^2$; $S^2 = E$; $QP = PR$; $RP = PQR$; $RQ = P^3QR$; $SP = PS$; $SQ = QS$; $SR = RS$.
$C_1 = E$; $C_2 = P^3$; $C_3 = Q, R, P^3Q, P^3R, QR, P^3QR$; $C_4 = P, PQ, PQR, PR$; $C_5 = P^4, P^4Q, P^4QR, P^4R$; $C_6 = P^2, P^5R, P^5QR, P^5Q$; $C_7 = P^5, P^2R, P^2QR, P^2Q$; $C_8 = S$; $C_9 = P^3S$;
$C_{10} = QS, RS, P^3QS, P^3RS, QRS, P^3QRS$; $C_{11} = PS, PQS, PQRS, PRS$; $C_{12} = P^4S, P^4QS, P^4QRS, P^4RS$; $C_{13} = P^2S, P^5RS, P^5QRS, P^5QS$; $C_{14} = P^5S, P^2RS, P^2QRS, P^2QS$.

	C_1 to C_7	C_8 to C_{14}
R_1 to R_7	$(\mathbf{G}_{24}^{9})$	$(\mathbf{G}_{24}^{9})$
R_8 to R_{14}	$(\mathbf{G}_{24}^{9})$	$-(\mathbf{G}_{24}^{9})$

R_4: $P = -\pi$; $Q = \psi$; $S = \varepsilon$.
R_5: $P = -\omega^*\mu$; $Q = \psi$; $S = \varepsilon$.
R_6: $P = -\mu$; $Q = \psi$; $S = \varepsilon$.
R_7: $P = A$; $Q = B$; $S = E$.
R_{11}: $P = -\pi$; $Q = \psi$; $S = -\varepsilon$.
R_{12}: $P = -\omega^*\mu$; $Q = \psi$; $S = -\varepsilon$.
R_{13}: $P = -\mu$; $Q = \psi$; $S = -\varepsilon$.
R_{14}: $P = A$; $Q = B$; $S = I$.
Note that $R = P^5QP$.

$\mathbf{G}_{48}^{5}$

$P^6 = E$; $Q^2 = E$; $R^2 = E$; $S^2 = E$; $QP = PR$; $QP^3 = P^3Q$; $RP = P^4QR$; $RQ = QR$; $SP = PS$; $SQ = QS$; $SR = RS$.
$C_1 = E$; $C_2 = P^3$; $C_3 = Q, R, P^3QR$; $C_4 = P^3Q, P^3R, QR$; $C_5 = PQ, PR, P^4, P^4QR$; $C_6 = P^4Q, P^4R, P, PQR$; $C_7 = P^2, P^2QR, P^5Q, P^5R$; $C_8 = P^5, P^5QR, P^2Q, P^2R$; $C_9 = S$; $C_{10} = P^3S$; $C_{11} = QS, RS, P^3QRS$; $C_{12} = P^3QS, P^3RS, QRS$; $C_{13} = PQS, PRS, P^4S, P^4QRS$; $C_{14} = P^4QS, P^4RS, PS, PQRS$; $C_{15} = P^2S, P^2QRS, P^5QS, P^5RS$; $C_{16} = P^5S, P^5QRS, P^2QS, P^2RS$.

	C_1 to C_8	C_9 to C_{16}
R_1 to R_8	$(\mathbf{G}_{24}^{8})$	$(\mathbf{G}_{24}^{8})$
R_9 to R_{16}	$(\mathbf{G}_{24}^{8})$	$-(\mathbf{G}_{24}^{8})$

R_7: $P = A$; $Q = G$; $R = B$; $S = E$.
R_8: $P = L$; $Q = M$; $R = N$; $S = E$.
R_{15}: $P = A$; $Q = G$; $R = B$; $S = I$.
R_{16}: $P = L$; $Q = M$; $R = N$; $S = I$.

G_{48}^{6}

$P^4 = E$; $Q^4 = E$; $Q^2 = P^2$; $R^3 = E$; $S^2 = E$; $QP = P^3Q$; $RP = P^3QR$; $RQ = PR$; $SP = P^2QS$; $SQ = P^3S$; $SR = R^2S$.
$C_1 = E$; $C_2 = P^2$; $C_3 = P, Q, PQ, P^3, P^2Q, P^3Q$; $C_4 = R, R^2, P^3R, P^2QR, PQR, QR^2\ PR^2, P^3QR^2$;
$C_5 = P^2R, P^2R^2, PR, QR, P^3QR, P^2QR^2, P^3R^2, PQR^2$; $C_6 = QS, P^3S, P^2QRS, P^3QRS, PR^2S, PQR^2S$;
$C_7 = P^2QS, PS, QRS, PQRS, P^3R^2S, P^3QR^2S$;
$C_8 = PQS, P^3QS, RS, P^2RS, R^2S, P^2R^2S, S, P^2S, P^3RS, PRS, QR^2S, P^2QR^2S$.

	C_1	C_2	C_3	C_4	C_5	C_6	C_7	C_8
R_1	1	1	1	1	1	1	1	1
R_2	1	1	1	1	1	−1	−1	−1
R_3	2	2	2	−1	−1	0	0	0
R_4	2	−2	0	−1	1	$-i\sqrt{2}$	$i\sqrt{2}$	0
R_5	2	−2	0	−1	1	$i\sqrt{2}$	$-i\sqrt{2}$	0
R_6	3	3	−1	0	0	1	1	−1
R_7	3	3	−1	0	0	−1	−1	1
R_8	4	−4	0	1	−1	0	0	0

R_3: $P = \varepsilon$; $Q = \varepsilon$; $R = \pi$; $S = \phi$.
R_4: $P = \zeta$; $Q = \gamma$; $R = \pi$; $S = \phi$.
R_5: $P = \zeta^*$; $Q = \gamma^*$; $R = \pi^*$; $S = \phi$.
R_6: $P = C$; $Q = G$; $R = K$; $S = F$.
R_7: $P = C$; $Q = G$; $R = K$; $S = H$.

$$R_8\colon\ P = \left(\begin{array}{c|c}\alpha' & 0\\ \hline 0 & \beta'\end{array}\right);\quad Q = \left(\begin{array}{c|c}\beta' & 0\\ \hline 0 & \alpha'\end{array}\right);\quad R = \left(\begin{array}{c|c}\lambda' & \nu'\\ \hline \rho' & \mu'\end{array}\right);\quad S = \left(\begin{array}{c|c}\lambda & 0\\ \hline 0 & \lambda\end{array}\right).$$

G_{48}^{7}

$P^6 = E$; $Q^2 = E$; $R^2 = E$; $S^2 = E$; $QP = PR$; $QP^3 = P^3Q$; $RP = P^4QR$; $RQ = QR$; $SP = P^2QS$; $SQ = RS$; $SR = QS$.
$C_1 = E$; $C_2 = P^3$; $C_3 = Q, R, P^3QR$; $C_4 = P^3Q, P^3R, QR$; $C_5 = PR, P^4, PQ, P^4QR, P^5Q, P^2, P^2QR, P^5R$;
$C_6 = P^4R, P, P^4Q, PQR, P^2Q, P^5, P^5QR, P^2R$; $C_7 = P^4S, P^5QS, P^3QS, PRS, P^2S, P^3RS$;
$C_8 = PS, P^2QS, QS, P^4RS, P^5S, RS$; $C_9 = S, QRS, P^2QRS, P^4QRS, P^5RS, PQS$;
$C_{10} = P^3S, P^3QRS, P^5QRS, PQRS, P^2RS, P^4QS$.

	C_1	C_2	C_3	C_4	C_5	C_6	C_7	C_8	C_9	C_{10}
R_1	1	1	1	1	1	1	1	1	1	1
R_2	1	1	1	1	1	1	−1	−1	−1	−1
R_3	1	−1	−1	1	1	−1	−1	1	−1	1
R_4	1	−1	−1	1	1	−1	1	−1	1	−1
R_5	2	2	2	2	−1	−1	0	0	0	0
R_6	2	−2	−2	2	−1	1	0	0	0	0
R_7	3	3	−1	−1	0	0	1	1	−1	−1
R_8	3	3	−1	−1	0	0	−1	−1	1	1
R_9	3	−3	1	−1	0	0	1	−1	−1	1
R_{10}	3	−3	1	−1	0	0	−1	1	1	−1

R_5: $P = \alpha$; $Q = \varepsilon$; $R = \varepsilon$; $S = \iota$.
R_6: $P = -\alpha$; $Q = -\varepsilon$; $R = -\varepsilon$; $S = \iota$.
R_7: $P = A$; $Q = G$; $R = B$; $S = U$.
R_8: $P = A$; $Q = G$; $R = B$; $S = V$.
R_9: $P = L$; $Q = M$; $R = N$; $S = U$.
R_{10}: $P = L$; $Q = M$; $R = N$; $S = V$.
Note that $R = P^5QP$.

$\mathbf{G}_{48}^{8}$

$P^4 = E$; $Q^4 = E$; $R^3 = E$; $QP = PQR$; $QP^2 = P^3R$; $RP = P^3Q^2R^2$; $RP^2 = PQ$; $RQ = QR^2$.
$C_1 = E$; $C_2 = Q^2$; $C_3 = P^3Q, PQ^3, R, P^2Q^2R, PQ^3R, P^2Q^2R^2, R^2, P^3QR$;
$C_4 = P^3Q^3, PQ, Q^2R, P^2R, PQR, P^2R^2, Q^2P^2, P^3Q^3R$; $C_5 = P^2Q^2, P^3QR^2, PQ^3R^2$;
$C_6 = P^2, P^3Q^3R^2, PQR^2$; $C_7 = Q, PR, QR, P^2Q^3R^2, P^3Q^2R^2, QR^2$;
$C_8 = Q^3, PQ^2R, Q^3R, P^2QR^2, P^3R^2, Q^3R^2$; $C_9 = P, PR^2, P^2Q^3, P^3Q^2, P^2Q^3R, P^3Q^2R$;
$C_{10} = PQ^2, PQ^2R^2, P^2Q, P^3, P^2QR, P^3R$.

	C_1	C_2	C_3	C_4	C_5	C_6	C_7	C_8	C_9	C_{10}
R_1	1	1	1	1	1	1	1	1	1	1
R_2	1	1	1	1	1	1	−1	−1	−1	−1
R_3	1	−1	1	−1	1	−1	i	−i	i	−i
R_4	1	−1	1	−1	1	−1	−i	i	−i	i
R_5	2	2	−1	−1	2	2	0	0	0	0
R_6	2	−2	−1	1	2	−2	0	0	0	0
R_7	3	3	0	0	−1	−1	1	1	−1	−1
R_8	3	3	0	0	−1	−1	−1	−1	1	1
R_9	3	−3	0	0	−1	1	i	−i	−i	i
R_{10}	3	−3	0	0	−1	1	−i	i	i	−i

R_5: $P = \gamma'$; $Q = \phi$; $R = \pi$.
R_6: $P = \delta'$; $Q = \kappa$; $R = \pi$.
R_7: $P = W$; $Q = X$; $R = A$.
R_8: $P = -W$; $Q = -X$; $R = A$.
R_9: $P = \mathrm{i}W$; $Q = \mathrm{i}X$; $R = A$.
R_{10}: $P = -\mathrm{i}W$; $Q = -\mathrm{i}X$; $R = A$.

$\mathbf{G}_{48}^{9}$

$P^6 = E$; $Q^4 = E$; $R^2 = E$; $QP = P^5Q$; $RP = PR$; $RQ = P^3QR$.
$C_1 = E$; $C_2 = P^3$; $C_3 = P^2, P^4$; $C_4 = P, P^5$; $C_5 = Q^2$; $C_6 = P^3Q^2$; $C_7 = P^2Q^2, P^4Q^2$;
$C_8 = PQ^2, P^5Q^2$; $C_9 = R, P^3R$; $C_{10} = Q^2R, P^3Q^2R$; $C_{11} = P^2R, PR$; $C_{12} = P^2Q^2R, PQ^2R$;
$C_{13} = P^5R, P^4R$; $C_{14} = P^5Q^2R, P^4Q^2R$; $C_{15} = Q, PQ, P^2Q, P^3Q, P^4Q, P^5Q$;
$C_{16} = Q^3, PQ^3, P^2Q^3, P^3Q^3, P^4Q^3, P^5Q^3$; $C_{17} = QR, PQR, P^2QR, P^3QR, P^4QR, P^5QR$;
$C_{18} = Q^3R, PQ^3R, P^2Q^3R, P^3Q^3R, P^4Q^3R, P^5Q^3R$.

	C_1	C_2	C_3	C_4	C_5	C_6	C_7	C_8	C_9
R_1	1	1	1	1	1	1	1	1	1
R_2	1	1	1	1	1	1	1	1	1
R_3	1	1	1	1	1	1	1	1	−1
R_4	1	1	1	1	1	1	1	1	−1
R_5	1	1	1	1	−1	−1	−1	−1	1
R_6	1	1	1	1	−1	−1	−1	−1	1
R_7	1	1	1	1	−1	−1	−1	−1	−1
R_8	1	1	1	1	−1	−1	−1	−1	−1
R_9	2	−2	2	−2	2	−2	2	−2	0
R_{10}	2	−2	2	−2	−2	2	−2	2	0
R_{11}	2	2	−1	−1	2	2	−1	−1	2
R_{12}	2	2	−1	−1	2	2	−1	−1	−2
R_{13}	2	−2	−1	1	2	−2	−1	1	0
R_{14}	2	−2	−1	1	2	−2	−1	1	0
R_{15}	2	2	−1	−1	−2	−2	1	1	2
R_{16}	2	2	−1	−1	−2	−2	1	1	−2
R_{17}	2	−2	−1	1	−2	2	1	−1	0
R_{18}	2	−2	−1	1	−2	2	1	−1	0

[*continued on p. 254*]

	C_{10}	C_{11}	C_{12}	C_{13}	C_{14}	C_{15}	C_{16}	C_{17}	C_{18}
R_1	1	1	1	1	1	1	1	1	1
R_2	1	1	1	1	1	−1	−1	−1	−1
R_3	−1	−1	−1	−1	−1	1	1	−1	−1
R_4	−1	−1	−1	−1	−1	−1	−1	1	1
R_5	−1	1	−1	1	−1	i	−i	i	−i
R_6	−1	1	−1	1	−1	−i	i	−i	i
R_7	1	−1	1	−1	1	i	−i	−i	i
R_8	1	−1	1	−1	1	−i	i	i	−i
R_9	0	0	0	0	0	0	0	0	0
R_{10}	0	0	0	0	0	0	0	0	0
R_{11}	2	−1	−1	−1	−1	0	0	0	0
R_{12}	−2	1	1	1	1	0	0	0	0
R_{13}	0	$-i\sqrt{3}$	$-i\sqrt{3}$	$i\sqrt{3}$	$i\sqrt{3}$	0	0	0	0
R_{14}	0	$i\sqrt{3}$	$i\sqrt{3}$	$-i\sqrt{3}$	$-i\sqrt{3}$	0	0	0	0
R_{15}	−2	−1	1	−1	1	0	0	0	0
R_{16}	2	1	−1	1	−1	0	0	0	0
R_{17}	0	$-i\sqrt{3}$	$i\sqrt{3}$	$i\sqrt{3}$	$-i\sqrt{3}$	0	0	0	0
R_{18}	0	$i\sqrt{3}$	$-i\sqrt{3}$	$-i\sqrt{3}$	$i\sqrt{3}$	0	0	0	0

R_9: $P = -\varepsilon$; $Q = \lambda$; $R = \phi$.
R_{10}: $P = -\varepsilon$; $Q = i\lambda$; $R = \phi$.
R_{11}: $P = \pi$; $Q = \phi$; $R = \varepsilon$.
R_{12}: $P = \pi$; $Q = \phi$; $R = -\varepsilon$.
R_{13}: $P = -\pi$; $Q = \phi$; $R = \lambda$.
R_{14}: $P = -\pi$; $Q = \phi$; $R = -\lambda$.
R_{15}: $P = \pi$; $Q = i\phi$; $R = \varepsilon$.
R_{16}: $P = \pi$; $Q = i\phi$; $R = -\varepsilon$.
R_{17}: $P = -\pi$; $Q = i\phi$; $R = \lambda$.
R_{18}: $P = -\pi$; $Q = i\phi$; $R = -\lambda$.

$\mathbf{G}_{48}^{10}$

$P^8 = E$; $Q^3 = E$; $R^4 = E$; $R^2 = P^4$; $QP = P^6Q^2R$; $QP^2 = P^3R$; $QP^4 = P^4Q$; $Q^2P = P^3Q$; $RP = P^6Q^2$; $RP^2 = PQ$; $RQ = Q^2R$.
$C_1 = E$; $C_2 = P^4$; $C_3 = P^2, P^5QR, P^7QR, P^6, PQR, P^3QR$;
$C_4 = P^4Q, P^6Q, P^5R, P^7R, P^4Q^2, PQ^2R, P^3Q^2R, P^2Q^2$; $C_5 = Q, P^2Q, PR, P^3R, Q^2, P^5Q^2R, P^7Q^2R, P^6Q^2$;
$C_6 = P, P^3Q^2, P^6R, P^7, P^2Q^2R, P^5Q$; $C_7 = P^5, P^7Q^2, P^2R, P^3, P^6Q^2R, PQ$;
$C_8 = P^3Q, R, PQ^2, P^6QR, P^4Q^2R, P^4QR, P^7Q, P^4R, P^5Q^2, P^2QR, Q^2R, QR$.

	C_1	C_2	C_3	C_4	C_5	C_6	C_7	C_8
R_1	1	1	1	1	1	1	1	1
R_2	1	1	1	1	1	−1	−1	−1
R_3	2	2	2	−1	−1	0	0	0
R_4	2	−2	0	1	−1	$\sqrt{2}$	$-\sqrt{2}$	0
R_5	2	−2	0	1	−1	$-\sqrt{2}$	$\sqrt{2}$	0
R_6	3	3	−1	0	0	1	1	−1
R_7	3	3	−1	0	0	−1	−1	1
R_8	4	−4	0	−1	1	0	0	0

R_3: $P = \lambda\alpha$; $Q = -\beta$; $R = \lambda$.
R_4: $P = -\xi'$; $Q = \pi'^*$; $R = \xi'\pi'$.
R_5: $P = \xi'$; $Q = \pi'^*$; $R = -\xi'\pi'$.
R_6: $P = Z$; $Q = D$; $R = -A'$.
R_7: $P = -Z$; $Q = D$; $R = A'$.

$$R_8:\ P = \frac{1}{\sqrt{2}}\begin{pmatrix} \lambda\alpha & -i\lambda\alpha \\ -i\lambda\alpha & \lambda\alpha \end{pmatrix};\quad Q = \frac{1}{2}\begin{pmatrix} (-1+i)\alpha^2 & (1-i)\alpha^2 \\ (-1-i)\alpha^2 & (-1-i)\alpha^2 \end{pmatrix};\quad R = \frac{1}{\sqrt{2}}\begin{pmatrix} 0 & (-1-i)\lambda \\ (1-i)\lambda & 0 \end{pmatrix}$$

Note that $R = P^5QP^2$.

$\mathbf{G}_{48}^{11}$

$P^{12} = E;\ Q^2 = E;\ R^2 = E;\ QP = P^7QR;\ RP = PR;\ RQ = QR.$
$C_1 = E;\ C_2 = P, P^7R;\ C_3 = P^2;\ C_4 = P^3, P^9R;\ C_5 = P^4;\ C_6 = P^5, P^{11}R;\ C_7 = P^6;\ C_8 = P^8;\ C_9 = P^{10};$
$C_{10} = Q, P^6QR;\ C_{11} = PQ, P^7QR;\ C_{12} = P^2Q, P^8QR;\ C_{13} = P^3Q, P^9QR;\ C_{14} = P^4Q, P^{10}QR;$
$C_{15} = P^5Q, P^{11}QR;\ C_{16} = R;\ C_{17} = PR, P^7;\ C_{18} = P^2R;\ C_{19} = P^3R, P^9;\ C_{20} = P^4R;\ C_{21} = P^5R, P^{11};$
$C_{22} = P^6R;\ C_{23} = P^8R;\ C_{24} = P^{10}R;\ C_{25} = QR, P^6Q;\ C_{26} = PQR, P^7Q;\ C_{27} = P^2QR, P^8Q;$
$C_{28} = P^3QR, P^9Q;\ C_{29} = P^4QR, P^{10}Q;\ C_{30} = P^5QR, P^{11}Q.$

	C_1	C_2	C_3	C_4	C_5	C_6	C_7	C_8	C_9	C_{10}	C_{11}	C_{12}	C_{13}	C_{14}	C_{15}
R_1	1	1	1	1	1	1	1	1	1	1	1	1	1	1	1
R_2	1	1	1	1	1	1	1	1	1	−1	−1	−1	−1	−1	−1
R_3	1	$-\omega^*$	ω	−1	ω^*	$-\omega$	1	ω	ω^*	1	$-\omega^*$	ω	−1	ω^*	$-\omega$
R_4	1	$-\omega^*$	ω	−1	ω^*	$-\omega$	1	ω	ω^*	−1	ω^*	$-\omega$	1	$-\omega^*$	ω
R_5	1	ω	ω^*	1	ω	ω^*	1	ω^*	ω	1	ω	ω^*	1	ω	ω^*
R_6	1	ω	ω^*	1	ω	ω^*	1	ω^*	ω	−1	$-\omega$	$-\omega^*$	−1	$-\omega$	$-\omega^*$
R_7	1	−1	1	−1	1	−1	1	1	1	1	−1	1	−1	1	−1
R_8	1	−1	1	−1	1	−1	1	1	1	−1	1	−1	1	−1	1
R_9	1	ω^*	ω	1	ω^*	ω	1	ω	ω^*	1	ω^*	ω	1	ω^*	ω
R_{10}	1	ω^*	ω	1	ω^*	ω	1	ω	ω^*	−1	$-\omega^*$	$-\omega$	−1	$-\omega^*$	$-\omega$
R_{11}	1	$-\omega$	ω^*	−1	ω	$-\omega^*$	1	ω^*	ω	1	$-\omega$	ω^*	−1	ω	$-\omega^*$
R_{12}	1	$-\omega$	ω^*	−1	ω	$-\omega^*$	1	ω^*	ω	−1	ω	$-\omega^*$	1	$-\omega$	ω^*
R_{13}	1	−i	−1	i	1	−i	−1	1	−1	1	−i	−1	i	1	−i
R_{14}	1	−i	−1	i	1	−i	−1	1	−1	−1	i	1	−i	−1	i
R_{15}	1	$i\omega^*$	$-\omega$	−i	ω^*	$i\omega$	−1	ω	$-\omega^*$	1	$i\omega^*$	$-\omega$	−i	ω^*	$i\omega$
R_{16}	1	$i\omega^*$	$-\omega$	−i	ω^*	$i\omega$	−1	ω	$-\omega^*$	−1	$-i\omega^*$	ω	i	$-\omega^*$	$-i\omega$
R_{17}	1	$-i\omega$	$-\omega^*$	i	ω	$-i\omega^*$	−1	ω^*	$-\omega$	1	$-i\omega$	$-\omega^*$	i	ω	$-i\omega^*$
R_{18}	1	$-i\omega$	$-\omega^*$	i	ω	$-i\omega^*$	−1	ω^*	$-\omega$	−1	$i\omega$	ω^*	−i	$-\omega$	$i\omega^*$
R_{19}	1	i	−1	−i	1	i	−1	1	−1	1	i	−1	−i	1	i
R_{20}	1	i	−1	−i	1	i	−1	1	−1	−1	−i	1	i	−1	−i
R_{21}	1	$-i\omega^*$	$-\omega$	i	ω^*	$-i\omega$	−1	ω	$-\omega^*$	1	$-i\omega^*$	$-\omega$	i	ω^*	$-i\omega$
R_{22}	1	$-i\omega^*$	$-\omega$	i	ω^*	$-i\omega$	−1	ω	$-\omega^*$	−1	$i\omega^*$	ω	−i	$-\omega^*$	$i\omega$
R_{23}	1	$i\omega$	$-\omega^*$	−i	ω	$i\omega^*$	−1	ω^*	$-\omega$	1	$i\omega$	$-\omega^*$	−i	ω	$i\omega^*$
R_{24}	1	$i\omega$	$-\omega^*$	−i	ω	$i\omega^*$	−1	ω^*	$-\omega$	−1	$-i\omega$	ω^*	i	$-\omega$	$-i\omega^*$
R_{25}	2	0	$-2\omega^*$	0	2ω	0	−2	$2\omega^*$	-2ω	0	0	0	0	0	0
R_{26}	2	0	−2	0	2	0	−2	2	−2	0	0	0	0	0	0
R_{27}	2	0	-2ω	0	$2\omega^*$	0	−2	2ω	$-2\omega^*$	0	0	0	0	0	0
R_{28}	2	0	$2\omega^*$	0	2ω	0	2	$2\omega^*$	2ω	0	0	0	0	0	0
R_{29}	2	0	2	0	2	0	2	2	2	0	0	0	0	0	0
R_{30}	2	0	2ω	0	$2\omega^*$	0	2	2ω	$2\omega^*$	0	0	0	0	0	0

[*continued on p. 256*]

	C_{16}	C_{17}	C_{18}	C_{19}	C_{20}	C_{21}	C_{22}	C_{23}	C_{24}	C_{25}	C_{26}	C_{27}	C_{28}	C_{29}	C_{30}
R_1	1	1	1	1	1	1	1	1	1	1	1	1	1	1	1
R_2	1	1	1	1	1	1	1	1	1	-1	-1	-1	-1	-1	-1
R_3	1	$-\omega^*$	ω	-1	ω^*	$-\omega$	1	ω	ω^*	1	$-\omega^*$	ω	-1	ω^*	$-\omega$
R_4	1	$-\omega^*$	ω	-1	ω^*	$-\omega$	1	ω	ω^*	-1	ω^*	$-\omega$	1	$-\omega^*$	ω
R_5	1	ω	ω^*	1	ω	ω^*	1	ω^*	ω	1	ω	ω^*	1	ω	ω^*
R_6	1	ω	ω^*	1	ω	ω^*	1	ω^*	ω	-1	$-\omega$	$-\omega^*$	-1	$-\omega$	$-\omega^*$
R_7	1	-1	1	-1	1	-1	1	1	1	1	-1	1	-1	1	-1
R_8	1	-1	1	-1	1	-1	1	1	1	-1	1	-1	1	-1	1
R_9	1	ω^*	ω	1	ω^*	ω	1	ω	ω^*	1	ω^*	ω	1	ω^*	ω
R_{10}	1	ω^*	ω	1	ω^*	ω	1	ω	ω^*	-1	$-\omega^*$	$-\omega$	-1	$-\omega^*$	$-\omega$
R_{11}	1	$-\omega$	ω^*	-1	ω	$-\omega^*$	1	ω^*	ω	1	$-\omega$	ω^*	-1	ω	$-\omega^*$
R_{12}	1	$-\omega$	ω^*	-1	ω	$-\omega^*$	1	ω^*	ω	-1	ω	$-\omega^*$	1	$-\omega$	ω^*
R_{13}	-1	i	1	$-$i	-1	i	1	-1	1	-1	i	1	$-$i	-1	i
R_{14}	-1	i	1	$-$i	-1	i	1	-1	1	1	$-$i	-1	i	1	$-$i
R_{15}	-1	$-\mathrm{i}\omega^*$	ω	i	$-\omega^*$	$-\mathrm{i}\omega$	1	$-\omega$	ω^*	-1	$-\mathrm{i}\omega^*$	ω	i	$-\omega^*$	$-\mathrm{i}\omega$
R_{16}	-1	$-\mathrm{i}\omega^*$	ω	i	$-\omega^*$	$-\mathrm{i}\omega$	1	$-\omega$	ω^*	1	$\mathrm{i}\omega^*$	$-\omega$	$-$i	ω^*	$\mathrm{i}\omega$
R_{17}	-1	$\mathrm{i}\omega$	ω^*	$-$i	$-\omega$	$\mathrm{i}\omega^*$	1	$-\omega^*$	ω	-1	$\mathrm{i}\omega$	ω^*	$-$i	$-\omega$	$\mathrm{i}\omega^*$
R_{18}	-1	$\mathrm{i}\omega$	ω^*	$-$i	$-\omega$	$\mathrm{i}\omega^*$	1	$-\omega^*$	ω	1	$-\mathrm{i}\omega$	$-\omega^*$	i	ω	$-\mathrm{i}\omega^*$
R_{19}	-1	$-$i	1	i	-1	$-$i	1	-1	1	-1	$-$i	1	i	-1	$-$i
R_{20}	-1	$-$i	1	i	-1	$-$i	1	-1	1	1	i	-1	$-$i	1	i
R_{21}	-1	$\mathrm{i}\omega^*$	ω	$-$i	$-\omega^*$	$\mathrm{i}\omega$	1	$-\omega$	ω^*	-1	$\mathrm{i}\omega^*$	ω	$-$i	$-\omega^*$	$\mathrm{i}\omega$
R_{22}	-1	$\mathrm{i}\omega^*$	ω	$-$i	$-\omega^*$	$\mathrm{i}\omega$	1	$-\omega$	ω^*	1	$-\mathrm{i}\omega^*$	$-\omega$	i	ω^*	$-\mathrm{i}\omega$
R_{23}	-1	$-\mathrm{i}\omega$	ω^*	i	$-\omega$	$-\mathrm{i}\omega^*$	1	$-\omega^*$	ω	-1	$-\mathrm{i}\omega$	ω^*	i	$-\omega$	$-\mathrm{i}\omega^*$
R_{24}	-1	$-\mathrm{i}\omega$	ω^*	i	$-\omega$	$-\mathrm{i}\omega^*$	1	$-\omega^*$	ω	1	$\mathrm{i}\omega$	$-\omega^*$	$-$i	ω	$\mathrm{i}\omega^*$
R_{25}	2	0	$-2\omega^*$	0	2ω	0	-2	$2\omega^*$	-2ω	0	0	0	0	0	0
R_{26}	2	0	-2	0	2	0	-2	2	-2	0	0	0	0	0	0
R_{27}	2	0	-2ω	0	$2\omega^*$	0	-2	2ω	$-2\omega^*$	0	0	0	0	0	0
R_{28}	-2	0	$-2\omega^*$	0	-2ω	0	-2	$-2\omega^*$	-2ω	0	0	0	0	0	0
R_{29}	-2	0	-2	0	-2	0	-2	-2	-2	0	0	0	0	0	0
R_{30}	-2	0	-2ω	0	$-2\omega^*$	0	-2	-2ω	$-2\omega^*$	0	0	0	0	0	0

R_{25}: $P = -\mathrm{i}\omega\lambda$; $Q = \phi$; $R = \varepsilon$.
R_{26}: $P = \mathrm{i}\lambda$; $Q = \phi$; $R = \varepsilon$.
R_{27}: $P = \mathrm{i}\omega^*\lambda$; $Q = \phi$; $R = \varepsilon$.
R_{28}: $P = \omega\lambda$; $Q = \phi$; $R = -\varepsilon$.
R_{29}: $P = -\lambda$; $Q = \phi$; $R = -\varepsilon$.
R_{30}: $P = -\omega^*\lambda$; $Q = \phi$; $R = -\varepsilon$.

$\mathbf{G}_{48}^{12}$

$P^{12} = E;\ Q^4 = E;\ QP = P^{11}Q.$

$C_1 = E;\ C_2 = P^6;\ C_3 = P^4, P^8;\ C_4 = P^2, P^{10};\ C_5 = P^3, P^9;\ C_6 = P, P^{11};\ C_7 = P^5, P^7;$
$C_8 = Q, P^2Q, P^4Q, P^6Q, P^8Q, P^{10}Q;\ C_9 = PQ, P^3Q, P^5Q, P^7Q, P^9Q, P^{11}Q;\ C_{10} = Q^2;\ C_{11} = P^6Q^2;$
$C_{12} = P^4Q^2, P^8Q^2;\ C_{13} = P^2Q^2, P^{10}Q^2;\ C_{14} = P^3Q^2, P^9Q^2;\ C_{15} = PQ^2, P^{11}Q^2;\ C_{16} = P^5Q^2, P^7Q^2;$
$C_{17} = Q^3, P^2Q^3, P^4Q^3, P^6Q^3, P^8Q^3, P^{10}Q^3;\ C_{18} = PQ^3, P^3Q^3, P^5Q^3, P^7Q^3, P^9Q^3, P^{11}Q^3.$

	C_1	C_2	C_3	C_4	C_5	C_6	C_7	C_8	C_9
R_1	1	1	1	1	1	1	1	1	1
R_2	1	1	1	1	1	1	1	−1	−1
R_3	1	1	1	1	−1	−1	−1	1	−1
R_4	1	1	1	1	−1	−1	−1	−1	1
R_5	2	2	−1	−1	−2	1	1	0	0
R_6	2	2	−1	−1	2	−1	−1	0	0
R_7	2	−2	−1	1	0	$\sqrt{3}$	$-\sqrt{3}$	0	0
R_8	2	−2	−1	1	0	$-\sqrt{3}$	$\sqrt{3}$	0	0
R_9	2	−2	2	−2	0	0	0	0	0
R_{10}	1	1	1	1	1	1	1	i	i
R_{11}	1	1	1	1	1	1	1	−i	−i
R_{12}	1	1	1	1	−1	−1	−1	i	−i
R_{13}	1	1	1	1	−1	−1	−1	−i	i
R_{14}	2	2	−1	−1	−2	1	1	0	0
R_{15}	2	2	−1	−1	2	−1	−1	0	0
R_{16}	2	−2	−1	1	0	$\sqrt{3}$	$-\sqrt{3}$	0	0
R_{17}	2	−2	−1	1	0	$-\sqrt{3}$	$\sqrt{3}$	0	0
R_{18}	2	2	2	−2	0	0	0	0	0

	C_{10}	C_{11}	C_{12}	C_{13}	C_{14}	C_{15}	C_{16}	C_{17}	C_{18}
R_1	1	1	1	1	1	1	1	1	1
R_2	1	1	1	1	1	1	1	−1	−1
R_3	1	1	1	1	−1	−1	−1	1	−1
R_4	1	1	1	1	−1	−1	−1	−1	1
R_5	2	2	−1	−1	−2	1	1	0	0
R_6	2	2	−1	−1	2	−1	−1	0	0
R_7	2	−2	−1	1	0	$\sqrt{3}$	$-\sqrt{3}$	0	0
R_8	2	−2	−1	1	0	$-\sqrt{3}$	$\sqrt{3}$	0	0
R_9	2	−2	2	−2	0	0	0	0	0
R_{10}	−1	−1	−1	−1	−1	−1	−1	−i	−i
R_{11}	−1	−1	−1	−1	−1	−1	−1	i	i
R_{12}	−1	−1	−1	−1	1	1	1	−i	i
R_{13}	−1	−1	−1	−1	1	1	1	i	−i
R_{14}	−2	−2	1	1	2	−1	−1	0	0
R_{15}	−2	−2	1	1	−2	1	1	0	0
R_{16}	−2	2	1	−1	0	$-\sqrt{3}$	$\sqrt{3}$	0	0
R_{17}	−2	2	1	−1	0	$\sqrt{3}$	$-\sqrt{3}$	0	0
R_{18}	−2	2	−2	2	0	0	0	0	0

R_5: $P = -\pi^*;\ Q = \phi.$
R_6: $P = \pi;\ Q = \phi.$
R_7: $P = \rho;\ Q = \phi.$
R_8: $P = -\rho^*;\ Q = \phi.$
R_9: $P = \mathrm{i}\lambda;\ Q = \phi.$
R_{14}: $P = -\pi^*;\ Q = \mathrm{i}\phi.$
R_{15}: $P = \pi;\ Q = \mathrm{i}\phi.$
R_{16}: $P = \rho;\ Q = \mathrm{i}\phi.$
R_{17}: $P = -\rho^*;\ Q = \mathrm{i}\phi.$
R_{18}: $P = \mathrm{i}\lambda;\ Q = \mathrm{i}\phi.$

$\mathbf{G}_{48}^{13}$

$P^6 = E$; $Q^4 = E$; $R^4 = E$; $Q^2 = P^3$; $QP = P^5Q$; $RP = PR$; $RQ = P^3QR$.
$C_1 = E$; $C_2 = P^3$; $C_3 = R^2$; $C_4 = P^3R^2$; $C_5 = P, P^5$; $C_6 = P^2, P^4$; $C_7 = PR^2, P^5R^2$;
$C_8 = P^2R^2, P^4R^2$; $C_9 = R, P^3R$; $C_{10} = R^3, P^3R^3$; $C_{11} = PR, P^2R$; $C_{12} = P^4R, P^5R$; $C_{13} = PR^3, P^2R^3$;
$C_{14} = P^4R^3, P^5R^3$; $C_{15} = Q, PQ, P^2Q, P^3Q, P^4Q, P^5Q$; $C_{16} = QR^2, PQR^2, P^2QR^2, P^3QR^2, P^4QR^2, P^5QR^2$;
$C_{17} = QR, PQR, P^2QR, P^3QR, P^4QR, P^5QR$; $C_{18} = QR^3, PQR^3, P^2QR^3, P^3QR^3, P^4QR^3, P^5QR^3$.

	C_1	C_2	C_3	C_4	C_5	C_6	C_7	C_8	C_9
R_1	1	1	1	1	1	1	1	1	1
R_2	1	1	1	1	1	1	1	1	1
R_3	1	1	1	1	1	1	1	1	−1
R_4	1	1	1	1	1	1	1	1	−1
R_5	1	1	−1	−1	1	1	−1	−1	i
R_6	1	1	−1	−1	1	1	−1	−1	i
R_7	1	1	−1	−1	1	1	−1	−1	−i
R_8	1	1	−1	−1	1	1	−1	−1	−i
R_9	2	2	2	2	−1	−1	−1	−1	2
R_{10}	2	2	2	2	−1	−1	−1	−1	−2
R_{11}	2	2	−2	−2	−1	−1	1	1	2i
R_{12}	2	2	−2	−2	−1	−1	1	1	−2i
R_{13}	2	−2	2	−2	−2	2	−2	2	0
R_{14}	2	−2	2	−2	1	−1	1	−1	0
R_{15}	2	−2	2	−2	1	−1	1	−1	0
R_{16}	2	−2	−2	2	−2	2	2	−2	0
R_{17}	2	−2	−2	2	1	−1	−1	1	0
R_{18}	2	−2	−2	2	1	−1	−1	1	0

	C_{10}	C_{11}	C_{12}	C_{13}	C_{14}	C_{15}	C_{16}	C_{17}	C_{18}
R_1	1	1	1	1	1	1	1	1	1
R_2	1	1	1	1	1	−1	−1	−1	−1
R_3	−1	−1	−1	−1	−1	1	1	−1	−1
R_4	−1	−1	−1	−1	−1	−1	−1	1	1
R_5	−i	i	i	−i	−i	1	−1	i	−i
R_6	−i	i	i	−i	−i	−1	1	−i	i
R_7	i	−i	−i	i	i	1	−1	−i	i
R_8	i	−i	−i	i	i	−1	1	i	−i
R_9	2	−1	−1	−1	−1	0	0	0	0
R_{10}	−2	1	1	1	1	0	0	0	0
R_{11}	−2i	−i	−i	i	i	0	0	0	0
R_{12}	2i	i	i	−i	−i	0	0	0	0
R_{13}	0	0	0	0	0	0	0	0	0
R_{14}	0	$i\sqrt{3}$	$-i\sqrt{3}$	$i\sqrt{3}$	$-i\sqrt{3}$	0	0	0	0
R_{15}	0	$-i\sqrt{3}$	$i\sqrt{3}$	$-i\sqrt{3}$	$i\sqrt{3}$	0	0	0	0
R_{16}	0	0	0	0	0	0	0	0	0
R_{17}	0	$\sqrt{3}$	$-\sqrt{3}$	$-\sqrt{3}$	$\sqrt{3}$	0	0	0	0
R_{18}	0	$-\sqrt{3}$	$\sqrt{3}$	$\sqrt{3}$	$-\sqrt{3}$	0	0	0	0

R_9: $P = \alpha$; $Q = \lambda$; $R = \varepsilon$.
R_{10}: $P = \alpha$; $Q = \lambda$; $R = -\varepsilon$.
R_{11}: $P = \alpha$; $Q = \lambda$; $R = i\varepsilon$.
R_{12}: $P = \alpha$; $Q = \lambda$; $R = -i\varepsilon$.
R_{13}: $P = -\varepsilon$; $Q = i\phi$; $R = \lambda$.
R_{14}: $P = -\alpha$; $Q = i\lambda$; $R = -i\kappa$.
R_{15}: $P = -\alpha$; $Q = i\lambda$; $R = i\kappa$.
R_{16}: $P = -\varepsilon$; $Q = i\phi$; $R = i\lambda$.
R_{17}: $P = -\alpha$; $Q = i\lambda$; $R = -\kappa$.
R_{18}: $P = -\alpha$; $Q = i\lambda$; $R = \kappa$.

$\mathbf{G}_{48}^{14}$

$P^4 = E$; $Q^6 = E$; $R^4 = E$; $R^2 = Q^3$; $QP = PQ^5$; $QP^2 = P^2Q$; $RP = P^3R$; $RQ = QR$.
$C_1 = E$; $C_2 = P^2$; $C_3 = Q, Q^5$; $C_4 = P^2Q, P^2Q^5$; $C_5 = R, P^2R$; $C_6 = Q^2R, P^2Q^4R$; $C_7 = QR, P^2Q^5R$;
$C_8 = P, PQ^4, PQ^2, P^3, P^3Q^4, P^3Q^2$; $C_9 = PR, PQ^4R, PQ^2R, P^3R, P^3Q^4R, P^3Q^2R$; $C_{10} = Q^3$; $C_{11} = P^2Q^3$;
$C_{12} = Q^4, Q^2$; $C_{13} = P^2Q^4, P^2Q^2$; $C_{14} = Q^3R, P^2Q^3R$; $C_{15} = Q^5R, P^2QR$; $C_{16} = Q^4R, P^2Q^2R$;
$C_{17} = PQ^3, PQ, PQ^5, P^3Q^3, P^3Q, P^3Q^5$; $C_{18} = PQ^3R, PQR, PQ^5R, P^3Q^3R, P^3QR, P^3Q^5R$.

	C_1	C_2	C_3	C_4	C_5	C_6	C_7	C_8	C_9
R_1	1	1	1	1	1	1	1	1	1
R_2	1	1	1	1	1	1	1	−1	−1
R_3	1	1	1	1	−1	−1	−1	1	−1
R_4	1	1	1	1	−1	−1	−1	−1	1
R_5	1	1	−1	−1	i	i	−i	1	i
R_6	1	1	−1	−1	i	i	−i	−1	−i
R_7	1	1	−1	−1	−i	−i	i	1	−i
R_8	1	1	−1	−1	−i	−i	i	−1	i
R_9	2	2	−1	−1	2	−1	−1	0	0
R_{10}	2	2	−1	−1	−2	1	1	0	0
R_{11}	2	2	1	1	2i	−i	i	0	0
R_{12}	2	2	1	1	−2i	i	−i	0	0
R_{13}	2	−2	−1	1	0	$i\sqrt{3}$	$-i\sqrt{3}$	0	0
R_{14}	2	−2	−1	1	0	$-i\sqrt{3}$	$i\sqrt{3}$	0	0
R_{15}	2	−2	1	−1	0	$\sqrt{3}$	$\sqrt{3}$	0	0
R_{16}	2	−2	1	−1	0	$-\sqrt{3}$	$-\sqrt{3}$	0	0
R_{17}	2	−2	2	−2	0	0	0	0	0
R_{18}	2	−2	−2	2	0	0	0	0	0

	C_{10}	C_{11}	C_{12}	C_{13}	C_{14}	C_{15}	C_{16}	C_{17}	C_{18}
R_1	1	1	1	1	1	1	1	1	1
R_2	1	1	1	1	1	1	1	−1	−1
R_3	1	1	1	1	−1	−1	−1	1	−1
R_4	1	1	1	1	−1	−1	−1	−1	1
R_5	−1	−1	1	1	−i	−i	i	−1	−i
R_6	−1	−1	1	1	−i	−i	i	1	i
R_7	−1	−1	1	1	i	i	−i	−1	i
R_8	−1	−1	1	1	i	i	−i	1	−i
R_9	2	2	−1	−1	2	−1	−1	0	0
R_{10}	2	2	−1	−1	−2	1	1	0	0
R_{11}	−2	−2	−1	−1	−2i	i	−i	0	0
R_{12}	−2	−2	−1	−1	2i	−i	i	0	0
R_{13}	2	−2	−1	1	0	$i\sqrt{3}$	$-i\sqrt{3}$	0	0
R_{14}	2	−2	−1	1	0	$-i\sqrt{3}$	$i\sqrt{3}$	0	0
R_{15}	−2	2	−1	1	0	$-\sqrt{3}$	$-\sqrt{3}$	0	0
R_{16}	−2	2	−1	1	0	$\sqrt{3}$	$\sqrt{3}$	0	0
R_{17}	2	−2	2	−2	0	0	0	0	0
R_{18}	−2	2	2	−2	0	0	0	0	0

R_9: $P = \phi$; $Q = \pi^*$; $R = \varepsilon$.
R_{10}: $P = \phi$; $Q = \pi^*$; $R = -\varepsilon$.
R_{11}: $P = \phi$; $Q = -\pi^*$; $R = i\varepsilon$.
R_{12}: $P = \phi$; $Q = -\pi^*$; $R = -i\varepsilon$.
R_{13}: $P = -i\phi$; $Q = \pi^*$; $R = \lambda$.
R_{14}: $P = i\phi$; $Q = \pi$; $R = \lambda$.
R_{15}: $P = -i\phi$; $Q = -\pi^*$; $R = -i\lambda$.
R_{16}: $P = i\phi$; $Q = -\pi$; $R = -i\lambda$.
R_{17}: $P = i\phi$; $Q = \varepsilon$; $R = \lambda$.
R_{18}: $P = i\phi$; $Q = -\varepsilon$; $R = i\lambda$.

$\mathbf{G}_{48}^{15}$

$P^{12} = E;\ Q^4 = E;\ R^2 = E;\ Q^2 = P^6;\ QP = P^{11}Q;\ RP = PR;\ RQ = QR.$
$C_1 = E;\ C_2 = P^6;\ C_3 = P, P^{11};\ C_4 = P^5, P^7;\ C_5 = P^2, P^{10};\ C_6 = P^4, P^8;\ C_7 = P^3, P^9;$
$C_8 = Q, P^2Q, P^4Q, P^6Q, P^8Q, P^{10}Q;\ C_9 = PQ, P^3Q, P^5Q, P^7Q, P^9Q, P^{11}Q;\ C_{10} = R;\ C_{11} = P^6R;$
$C_{12} = PR, P^{11}R;\ C_{13} = P^5R, P^7R;\ C_{14} = P^2R, P^{10}R;\ C_{15} = P^4R, P^8R;\ C_{16} = P^3R, P^9R;$
$C_{17} = QR, P^2QR, P^4QR, P^6QR, P^8QR, P^{10}QR;\ C_{18} = PQR, P^3QR, P^5QR, P^7QR, P^9QR, P^{11}QR.$

	C_1 to C_9	C_{10} to C_{18}
R_1 to R_9	$(\mathbf{G}_{24}^{11})$	$(\mathbf{G}_{24}^{11})$
R_{10} to R_{18}	$(\mathbf{G}_{24}^{11})$	$-(\mathbf{G}_{24}^{11})$

R_5: $P = -\pi^*;\ Q = \phi;\ R = \varepsilon.$
R_6: $P = \pi;\ Q = \phi;\ R = \varepsilon.$
R_7: $P = \rho;\ Q = i\phi;\ R = \varepsilon.$
R_8: $P = -\rho^*;\ Q = i\phi;\ R = \varepsilon.$
R_9: $P = i\lambda;\ Q = i\phi;\ R = \varepsilon.$
R_{14}: $P = -\pi^*;\ Q = \phi;\ R = -\varepsilon.$
R_{15}: $P = \pi;\ Q = \phi;\ R = -\varepsilon.$
R_{16}: $P = \rho;\ Q = i\phi;\ R = -\varepsilon.$
R_{17}: $P = -\rho^*;\ Q = i\phi;\ R = -\varepsilon.$
R_{18}: $P = i\lambda;\ Q = i\phi;\ R = -\varepsilon.$

$\mathbf{G}_{64}^{1}$

$P^2 = E;\ Q^4 = E;\ R^8 = E;\ QP = PQ^3R^4;\ RP = PR;\ RQ = QR^7.$
$C_1 = E;\ C_2 = R^4;\ C_3 = Q^2;\ C_4 = Q^2R^4;\ C_5 = R^2, R^6;\ C_6 = Q^2R^2, Q^2R^6;\ C_7 = R, R^7;\ C_8 = R^5, R^3;$
$C_9 = Q^2R, Q^2R^7;\ C_{10} = Q^2R^5, Q^2R^3;\ C_{11} = PR, PQ^2R^3;\ C_{12} = PR^5, PQ^2R^7;\ C_{13} = PQ^2R, PR^3;$
$C_{14} = PQ^2R^5, PR^7;\ C_{15} = P, PQ^2R^4;\ C_{16} = PR^4, PQ^2;\ C_{17} = PR^2, PQ^2R^2;\ C_{18} = PR^6, PQ^2R^6;$
$C_{19} = PQ, PQR^2, PQ^3R^4, PQ^3R^6, PQR^4, PQR^6, PQ^3, PQ^3R^2;$
$C_{20} = PQR, PQR^3, PQ^3R^5, PQ^3R^7, PQR^5, PQR^7, PQ^3R, PQ^3R^3;$
$C_{21} = Q, QR^2, Q^3R^4, Q^3R^6, QR^4, QR^6, Q^3, Q^3R^2;\ C_{22} = QR^3, QR, Q^3R^7, Q^3R^5, QR^7, QR^5, Q^3R^3, Q^3R.$

	C_1	C_2	C_3	C_4	C_5	C_6	C_7	C_8	C_9	C_{10}	C_{11}
R_1	1	1	1	1	1	1	1	1	1	1	1
R_2	1	1	1	1	1	1	1	1	1	1	1
R_3	1	1	1	1	1	1	−1	−1	−1	−1	−1
R_4	1	1	1	1	1	1	−1	−1	−1	−1	−1
R_5	1	1	1	1	1	1	1	1	1	1	−1
R_6	1	1	1	1	1	1	1	1	1	1	−1
R_7	1	1	1	1	1	1	−1	−1	−1	−1	1
R_8	1	1	1	1	1	1	−1	−1	−1	−1	1
R_9	2	2	2	2	−2	−2	0	0	0	0	0
R_{10}	2	2	−2	−2	2	−2	2	2	−2	−2	0
R_{11}	2	2	−2	−2	2	−2	−2	−2	2	2	0
R_{12}	2	2	−2	−2	−2	2	0	0	0	0	2i
R_{13}	2	2	−2	−2	−2	2	0	0	0	0	−2i
R_{14}	2	2	2	2	−2	−2	0	0	0	0	0
R_{15}	2	−2	−2	2	0	0	$\sqrt{2}$	$-\sqrt{2}$	$-\sqrt{2}$	$\sqrt{2}$	$\sqrt{2}$
R_{16}	2	−2	−2	2	0	0	$\sqrt{2}$	$-\sqrt{2}$	$-\sqrt{2}$	$\sqrt{2}$	$-\sqrt{2}$
R_{17}	2	−2	−2	2	0	0	$-\sqrt{2}$	$\sqrt{2}$	$\sqrt{2}$	$-\sqrt{2}$	$-\sqrt{2}$
R_{18}	2	−2	−2	2	0	0	$-\sqrt{2}$	$\sqrt{2}$	$\sqrt{2}$	$-\sqrt{2}$	$\sqrt{2}$
R_{19}	2	−2	2	−2	0	0	$\sqrt{2}$	$-\sqrt{2}$	$\sqrt{2}$	$-\sqrt{2}$	$i\sqrt{2}$
R_{20}	2	−2	2	−2	0	0	$\sqrt{2}$	$-\sqrt{2}$	$\sqrt{2}$	$-\sqrt{2}$	$-i\sqrt{2}$
R_{21}	2	−2	2	−2	0	0	$-\sqrt{2}$	$\sqrt{2}$	$-\sqrt{2}$	$\sqrt{2}$	$-i\sqrt{2}$
R_{22}	2	−2	2	−2	0	0	$-\sqrt{2}$	$\sqrt{2}$	$-\sqrt{2}$	$\sqrt{2}$	$i\sqrt{2}$

[*continued on p. 261*]

	C_{12}	C_{13}	C_{14}	C_{15}	C_{16}	C_{17}	C_{18}	C_{19}	C_{20}	C_{21}	C_{22}
R_1	1	1	1	1	1	1	1	1	1	1	1
R_2	1	1	1	1	1	1	1	-1	-1	-1	-1
R_3	-1	-1	-1	1	1	1	1	1	-1	1	-1
R_4	-1	-1	-1	1	1	1	1	-1	1	-1	1
R_5	-1	-1	-1	-1	-1	-1	-1	1	1	-1	-1
R_6	-1	-1	-1	-1	-1	-1	-1	-1	-1	1	1
R_7	1	1	1	-1	-1	-1	-1	1	-1	-1	1
R_8	1	1	1	-1	-1	-1	-1	-1	1	1	-1
R_9	0	0	0	2	2	-2	-2	0	0	0	0
R_{10}	0	0	0	0	0	0	0	0	0	0	0
R_{11}	0	0	0	0	0	0	0	0	0	0	0
R_{12}	2i	-2i	-2i	0	0	0	0	0	0	0	0
R_{13}	-2i	2i	2i	0	0	0	0	0	0	0	0
R_{14}	0	0	0	-2	-2	2	2	0	0	0	0
R_{15}	$-\sqrt{2}$	$-\sqrt{2}$	$\sqrt{2}$	2	-2	0	0	0	0	0	0
R_{16}	$\sqrt{2}$	$\sqrt{2}$	$-\sqrt{2}$	-2	2	0	0	0	0	0	0
R_{17}	$\sqrt{2}$	$\sqrt{2}$	$-\sqrt{2}$	2	-2	0	0	0	0	0	0
R_{18}	$-\sqrt{2}$	$-\sqrt{2}$	$\sqrt{2}$	-2	2	0	0	0	0	0	0
R_{19}	$-i\sqrt{2}$	$i\sqrt{2}$	$-i\sqrt{2}$	0	0	2i	-2i	0	0	0	0
R_{20}	$i\sqrt{2}$	$-i\sqrt{2}$	$i\sqrt{2}$	0	0	-2i	2i	0	0	0	0
R_{21}	$i\sqrt{2}$	$-i\sqrt{2}$	$i\sqrt{2}$	0	0	2i	-2i	0	0	0	0
R_{22}	$-i\sqrt{2}$	$i\sqrt{2}$	$-i\sqrt{2}$	0	0	-2i	2i	0	0	0	0

R_9: $P = \varepsilon$; $Q = \phi$; $R = -i\lambda$.
R_{10}: $P = -\lambda$; $Q = \kappa$; $R = \varepsilon$.
R_{11}: $P = -\lambda$; $Q = \kappa$; $R = -\varepsilon$.
R_{12}: $P = -\lambda$; $Q = \kappa$; $R = -i\lambda$.
R_{13}: $P = -\lambda$; $Q = \kappa$; $R = i\lambda$.
R_{14}: $P = -\varepsilon$; $Q = \phi$; $R = -i\lambda$.
R_{15}: $P = \varepsilon$; $Q = i\lambda$; $R = \delta$.
R_{16}: $P = -\varepsilon$; $Q = i\lambda$; $R = \delta$.
R_{17}: $P = \varepsilon$; $Q = i\lambda$; $R = -\delta$.
R_{18}: $P = -\varepsilon$; $Q = i\lambda$; $R = -\delta$.
R_{19}: $P = -i\kappa$; $Q = \lambda$; $R = \delta$.
R_{20}: $P = i\kappa$; $Q = \lambda$; $R = \delta$.
R_{21}: $P = -i\kappa$; $Q = \lambda$; $R = -\delta$.
R_{22}: $P = i\kappa$; $Q = \lambda$; $R = -\delta$.

G_{64}^{2}

$P^8 = E;\ Q^4 = E;\ R^4 = E;\ R^2 = P^4;\ QP = P^3Q^3;\ QP^2 = P^6Q;\ RP = P^7R;\ RQ = P^4Q^3R.$
$C_1 = E;\ C_2 = P^4;\ C_3 = P^4Q^2;\ C_4 = Q^2;\ C_5 = P^2, P^6;\ C_6 = P^2Q^2, P^6Q^2;\ C_7 = P^3R, PR, P^7R, P^5R;$
$C_8 = P^3Q^2R, PQ^2R, P^7Q^2R, P^5Q^2R;\ C_9 = QR, P^4Q^3R;\ C_{10} = P^4QR, Q^3R;$
$C_{11} = P^2Q^3R, P^6QR, P^6Q^3R, P^2QR;\ C_{12} = PQ, P^7Q^3, P^5Q, P^3Q^3;\ C_{13} = PQ^3, P^7Q, P^5Q^3, P^3Q;$
$C_{14} = R, P^2R, P^4Q^2R, P^6Q^2R, P^4R, P^6R, Q^2R, P^2Q^2R;\ C_{15} = P, P^7, P^5Q^2, P^3Q^2;$
$C_{16} = P^5, P^3, PQ^2, P^7Q^2;\ C_{17} = Q, P^2Q, P^4Q^3, P^6Q^3, P^4Q, P^6Q, Q^3, P^2Q^3;$
$C_{18} = PQR, P^7QR, P^5Q^3R, P^3Q^3R;\ C_{19} = P^5QR, P^3QR, PQ^3R, P^7Q^3R.$

	C_1	C_2	C_3	C_4	C_5	C_6	C_7	C_8	C_9	C_{10}
R_1	1	1	1	1	1	1	1	1	1	1
R_2	1	1	1	1	1	1	1	1	1	1
R_3	1	1	1	1	1	1	−1	−1	1	1
R_4	1	1	1	1	1	1	−1	−1	1	1
R_5	1	1	1	1	1	1	1	1	−1	−1
R_6	1	1	1	1	1	1	1	1	−1	−1
R_7	1	1	1	1	1	1	−1	−1	−1	−1
R_8	1	1	1	1	1	1	−1	−1	−1	−1
R_9	2	2	−2	−2	2	−2	2	−2	0	0
R_{10}	2	2	2	2	−2	−2	0	0	2	2
R_{11}	2	2	−2	−2	−2	2	0	0	0	0
R_{12}	2	2	−2	−2	2	−2	−2	2	0	0
R_{13}	2	2	2	2	−2	−2	0	0	−2	−2
R_{14}	2	2	−2	−2	−2	2	0	0	0	0
R_{15}	2	−2	2	−2	0	0	0	0	−2	2
R_{16}	2	−2	2	−2	0	0	0	0	2	−2
R_{17}	2	−2	2	−2	0	0	0	0	−2	2
R_{18}	2	−2	2	−2	0	0	0	0	2	−2
R_{19}	4	−4	−4	4	0	0	0	0	0	0

[*continued on p. 263*]

	C_{11}	C_{12}	C_{13}	C_{14}	C_{15}	C_{16}	C_{17}	C_{18}	C_{19}
R_1	1	1	1	1	1	1	1	1	1
R_2	1	1	1	−1	−1	−1	−1	−1	−1
R_3	1	−1	−1	1	−1	−1	1	−1	−1
R_4	1	−1	−1	−1	1	1	−1	1	1
R_5	−1	−1	−1	1	1	1	−1	−1	−1
R_6	−1	−1	−1	−1	−1	−1	1	1	1
R_7	−1	1	1	1	−1	−1	−1	1	1
R_8	−1	1	1	−1	1	1	1	−1	−1
R_9	0	0	0	0	0	0	0	0	0
R_{10}	−2	0	0	0	0	0	0	0	0
R_{11}	0	2	−2	0	0	0	0	0	0
R_{12}	0	0	0	0	0	0	0	0	0
R_{13}	2	0	0	0	0	0	0	0	0
R_{14}	0	−2	2	0	0	0	0	0	0
R_{15}	0	0	0	0	$\sqrt{2}$	$-\sqrt{2}$	0	$-\sqrt{2}$	$\sqrt{2}$
R_{16}	0	0	0	0	$\sqrt{2}$	$-\sqrt{2}$	0	$\sqrt{2}$	$-\sqrt{2}$
R_{17}	0	0	0	0	$-\sqrt{2}$	$\sqrt{2}$	0	$\sqrt{2}$	$-\sqrt{2}$
R_{18}	0	0	0	0	$-\sqrt{2}$	$\sqrt{2}$	0	$-\sqrt{2}$	$\sqrt{2}$
R_{19}	0	0	0	0	0	0	0	0	0

R_9: $P = \phi$; $Q = \kappa$; $R = \phi$.
R_{10}: $P = -\kappa$; $Q = \phi$; $R = \phi$.
R_{11}: $P = -\kappa$; $Q = \kappa$; $R = \phi$.
R_{12}: $P = -\phi$; $Q = \kappa$; $R = \phi$.
R_{13}: $P = -\kappa$; $Q = -\phi$; $R = \phi$.
R_{14}: $P = -\kappa$; $Q = -\kappa$; $R = \phi$.
R_{15}: $P = \delta$; $Q = i\phi$; $R = i\phi$.
R_{16}: $P = \delta$; $Q = -i\phi$; $R = i\phi$.
R_{17}: $P = -\delta$; $Q = i\phi$; $R = i\phi$.
R_{18}: $P = -\delta$; $Q = -i\phi$; $R = i\phi$.

$$R_{19}:\ P = \left(\begin{array}{c|c} \xi' & 0 \\ \hline 0 & -\xi' \end{array}\right); \qquad Q = \left(\begin{array}{c|c} 0 & -i\kappa \\ \hline -i\kappa & 0 \end{array}\right); \qquad R = \left(\begin{array}{c|c} i\lambda & 0 \\ \hline 0 & i\lambda \end{array}\right).$$

$\mathbf{G}_{64}^{3}$

$P^8 = E$; $Q^8 = E$; $QP = P^3Q^7$; $QP^4 = P^4Q$; $Q^2P = PQ^2$.
$C_1 = E$; $C_2 = P^4$; $C_3 = P^4Q^2$; $C_4 = Q^2$; $C_5 = Q^4$; $C_6 = P^4Q^4$; $C_7 = P^4Q^6$; $C_8 = Q^6$;
$C_9 = P^2, P^6Q^4$; $C_{10} = P^6, P^2Q^4$; $C_{11} = P^6Q^2, P^2Q^6$; $C_{12} = P^2Q^2, P^6Q^6$; $C_{13} = P, P^3Q^6$;
$C_{14} = P^5, P^7Q^6$; $C_{15} = P^5Q^2, P^7$; $C_{16} = PQ^2, P^3$; $C_{17} = PQ^4, P^3Q^2$; $C_{18} = P^5Q^4, P^7Q^2$;
$C_{19} = P^5Q^6, P^7Q^4$; $C_{20} = PQ^6, P^3Q^4$; $C_{21} = Q, P^4Q^5, P^2Q^7, P^6Q^3$; $C_{22} = P^4Q, Q^5, P^6Q^7, P^2Q^3$;
$C_{23} = P^4Q^3, Q^7, P^6Q, P^2Q^5$; $C_{24} = Q^3, P^4Q^7, P^2Q, P^6Q^5$; $C_{25} = P^5Q^3, PQ^7, P^7Q, P^3Q^5$;
$C_{26} = PQ^3, P^5Q^7, P^3Q, P^7Q^5$; $C_{27} = PQ^5, P^5Q, P^3Q^3, P^7Q^7$; $C_{28} = P^5Q^5, PQ, P^7Q^3, P^3Q^7$.

	C_1	C_2	C_3	C_4	C_5	C_6	C_7	C_8	C_9	C_{10}	C_{11}	C_{12}	C_{13}	C_{14}
R_1	1	1	1	1	1	1	1	1	1	1	1	1	1	1
R_2	1	1	1	1	1	1	1	1	1	1	1	1	1	1
R_3	1	1	−1	−1	1	1	−1	−1	−1	−1	1	1	−i	−i
R_4	1	1	−1	−1	1	1	−1	−1	−1	−1	1	1	−i	−i
R_5	1	1	1	1	1	1	1	1	1	1	1	1	−1	−1
R_6	1	1	1	1	1	1	1	1	1	1	1	1	−1	−1
R_7	1	1	−1	−1	1	1	−1	−1	−1	−1	1	1	i	i
R_8	1	1	−1	−1	1	1	−1	−1	−1	−1	1	1	i	i
R_9	1	−1	i	−i	−1	1	−i	i	−i	i	1	−1	iθ	−iθ
R_{10}	1	−1	i	−i	−1	1	−i	i	−i	i	1	−1	iθ	−iθ
R_{11}	1	−1	i	−i	−1	1	−i	i	−i	i	1	−1	−iθ	iθ
R_{12}	1	−1	i	−i	−1	1	−i	i	−i	i	1	−1	−iθ	iθ
R_{13}	1	−1	−i	i	−1	1	i	−i	i	−i	1	−1	θ	−θ
R_{14}	1	−1	−i	i	−1	1	i	−i	i	−i	1	−1	θ	−θ
R_{15}	1	−1	−i	i	−1	1	i	−i	i	−i	1	−1	−θ	θ
R_{16}	1	−1	−i	i	−1	1	i	−i	i	−i	1	−1	−θ	θ
R_{17}	2	2	2	2	2	2	2	2	−2	−2	−2	−2	0	0
R_{18}	2	2	−2	−2	2	2	−2	−2	2	2	−2	−2	0	0
R_{19}	2	−2	2i	−2i	−2	2	−2i	2i	2i	−2i	−2	2	0	0
R_{20}	2	−2	−2i	2i	−2	2	2i	−2i	−2i	2i	−2	2	0	0
R_{21}	2	2	−2i	−2i	−2	−2	2i	2i	0	0	0	0	1 + i	1 + i
R_{22}	2	2	−2i	−2i	−2	−2	2i	2i	0	0	0	0	−1 − i	−1 − i
R_{23}	2	2	2i	2i	−2	−2	−2i	−2i	0	0	0	0	1 − i	1 − i
R_{24}	2	2	2i	2i	−2	−2	−2i	−2i	0	0	0	0	−1 + i	−1 + i
R_{25}	2	−2	2	−2	2	−2	2	−2	0	0	0	0	$\sqrt{2}$	$-\sqrt{2}$
R_{26}	2	−2	2	−2	2	−2	2	−2	0	0	0	0	$-\sqrt{2}$	$\sqrt{2}$
R_{27}	2	−2	−2	2	2	−2	−2	2	0	0	0	0	$i\sqrt{2}$	$-i\sqrt{2}$
R_{28}	2	−2	−2	2	2	−2	−2	2	0	0	0	0	$-i\sqrt{2}$	$i\sqrt{2}$

[*continued on p. 265*]

	C_{15}	C_{16}	C_{17}	C_{18}	C_{19}	C_{20}	C_{21}	C_{22}	C_{23}	C_{24}	C_{25}	C_{26}	C_{27}	C_{28}
R_1	1	1	1	1	1	1	1	1	1	1	1	1	1	1
R_2	1	1	1	1	1	1	-1	-1	-1	-1	-1	-1	-1	-1
R_3	i	i	$-i$	$-i$	i	i	i	i	$-i$	$-i$	-1	-1	1	1
R_4	i	i	$-i$	$-i$	i	i	$-i$	$-i$	i	i	1	1	-1	-1
R_5	-1	-1	-1	-1	-1	-1	-1	-1	-1	-1	1	1	1	1
R_6	-1	-1	-1	-1	-1	-1	1	1	1	1	-1	-1	-1	-1
R_7	$-i$	$-i$	i	i	$-i$	$-i$	$-i$	$-i$	i	i	-1	-1	1	1
R_8	$-i$	$-i$	i	i	$-i$	$-i$	i	i	$-i$	$-i$	1	1	-1	-1
R_9	$-\theta$	θ	$-i\theta$	$i\theta$	θ	$-\theta$	$i\theta$	$-i\theta$	$-\theta$	θ	1	-1	i	$-i$
R_{10}	$-\theta$	θ	$-i\theta$	$i\theta$	θ	$-\theta$	$-i\theta$	$i\theta$	θ	$-\theta$	-1	1	$-i$	i
R_{11}	θ	$-\theta$	$i\theta$	$-i\theta$	$-\theta$	θ	$i\theta$	$-i\theta$	$-\theta$	θ	-1	1	$-i$	i
R_{12}	θ	$-\theta$	$i\theta$	$-i\theta$	$-\theta$	θ	$-i\theta$	$i\theta$	θ	$-\theta$	1	-1	i	$-i$
R_{13}	$-i\theta$	$i\theta$	$-\theta$	θ	$i\theta$	$-i\theta$	θ	$-\theta$	$-i\theta$	$i\theta$	1	-1	$-i$	i
R_{14}	$-i\theta$	$i\theta$	$-\theta$	θ	$i\theta$	$-i\theta$	$-\theta$	θ	$i\theta$	$-i\theta$	-1	1	i	$-i$
R_{15}	$i\theta$	$-i\theta$	θ	$-\theta$	$-i\theta$	$i\theta$	θ	$-\theta$	$-i\theta$	$i\theta$	-1	1	i	$-i$
R_{16}	$i\theta$	$-i\theta$	θ	$-\theta$	$-i\theta$	$i\theta$	$-\theta$	θ	$i\theta$	$-i\theta$	1	-1	$-i$	i
R_{17}	0	0	0	0	0	0	0	0	0	0	0	0	0	0
R_{18}	0	0	0	0	0	0	0	0	0	0	0	0	0	0
R_{19}	0	0	0	0	0	0	0	0	0	0	0	0	0	0
R_{20}	0	0	0	0	0	0	0	0	0	0	0	0	0	0
R_{21}	$1-i$	$1-i$	$-1-i$	$-1-i$	$-1+i$	$-1+i$	0	0	0	0	0	0	0	0
R_{22}	$-1+i$	$-1+i$	$1+i$	$1+i$	$1-i$	$1-i$	0	0	0	0	0	0	0	0
R_{23}	$1+i$	$1+i$	$-1+i$	$-1+i$	$-1-i$	$-1-i$	0	0	0	0	0	0	0	0
R_{24}	$-1-i$	$-1-i$	$1-i$	$1-i$	$1+i$	$1+i$	0	0	0	0	0	0	0	0
R_{25}	$\sqrt{2}$	$-\sqrt{2}$	$\sqrt{2}$	$-\sqrt{2}$	$\sqrt{2}$	$-\sqrt{2}$	0	0	0	0	0	0	0	0
R_{26}	$-\sqrt{2}$	$\sqrt{2}$	$-\sqrt{2}$	$\sqrt{2}$	$-\sqrt{2}$	$\sqrt{2}$	0	0	0	0	0	0	0	0
R_{27}	$-i\sqrt{2}$	$i\sqrt{2}$	$i\sqrt{2}$	$-i\sqrt{2}$	$-i\sqrt{2}$	$i\sqrt{2}$	0	0	0	0	0	0	0	0
R_{28}	$i\sqrt{2}$	$-i\sqrt{2}$	$-i\sqrt{2}$	$i\sqrt{2}$	$i\sqrt{2}$	$-i\sqrt{2}$	0	0	0	0	0	0	0	0

R_{17}: $P = \kappa$; $Q = \lambda$.
R_{18}: $P = \lambda$; $Q = \kappa$.
R_{19}: $P = \theta\lambda$; $Q = \theta\kappa$.
R_{20}: $P = \theta\kappa$; $Q = \theta\lambda$.
R_{21}: $P = i\nu^*$; $Q = i\phi\nu$.
R_{22}: $P = -i\nu^*$; $Q = i\phi\nu$.
R_{23}: $P = \nu^*$; $Q = \phi\nu$.
R_{24}: $P = -\nu^*$; $Q = \phi\nu$.
R_{25}: $P = \delta$; $Q = i\delta\phi$.
R_{26}: $P = -\delta$; $Q = i\delta\phi$.
R_{27}: $P = i\delta$; $Q = -\delta\phi$.
R_{28}: $P = -i\delta$; $Q = -\delta\phi$.

$\mathbf{G}_{64}^{4}$

$P^8 = E$; $Q^4 = E$; $R^2 = E$; $QP = P^3Q^3$; $RP = PR$; $RQ = QR$.
$C_1 = E$; $C_2 = P^4$; $C_3 = P^2, P^6$; $C_4 = Q^2$; $C_5 = P^4Q^2$; $C_6 = P^2Q^2, P^6Q^2$; $C_7 = P, P^3Q^2$;
$C_8 = PQ^2, P^3$; $C_9 = P^5, P^7Q^2$; $C_{10} = P^5Q^2, P^7$; $C_{11} = PQ, P^3Q^3, P^5Q, P^7Q^3$;
$C_{12} = PQ^3, P^3Q, P^5Q^3, P^7Q$; $C_{13} = Q, P^2Q^3, P^4Q, P^6Q^3$; $C_{14} = Q^3, P^2Q, P^4Q^3, P^6Q$; $C_{15} = R$;
$C_{16} = P^4R$; $C_{17} = P^2R, P^6R$; $C_{18} = Q^2R$; $C_{19} = P^4Q^2R$; $C_{20} = P^2Q^2R, P^6Q^2R$; $C_{21} = PR, P^3Q^2R$;
$C_{22} = PQ^2R, P^3R$; $C_{23} = P^5R, P^7Q^2R$; $C_{24} = P^5Q^2R, P^7R$; $C_{25} = PQR, P^3Q^3R, P^5Q, P^7Q^3R$;
$C_{26} = PQ^3R, P^3QR, P^5Q^3R, P^7QR$; $C_{27} = QR, P^2Q^3R, P^4QR, P^6Q^3R$; $C_{28} = Q^3R, P^2QR, P^4Q^3R, P^6QR$.

	C_1 to C_{14}	C_{15} to C_{28}
R_1 to R_{14}	$(\mathbf{G}_{32}^{11})$	$(\mathbf{G}_{32}^{11})$
R_{15} to R_{28}	$(\mathbf{G}_{32}^{11})$	$-(\mathbf{G}_{32}^{11})$

R_5: $P = \kappa$; $Q = \lambda$; $R = \varepsilon$.
R_6: $P = \xi$; $Q = \lambda$; $R = \varepsilon$.
R_7: $P = \xi^*$; $Q = \lambda$; $R = \varepsilon$.
R_{12}: $P = \mathrm{i}\kappa$; $Q = \mathrm{i}\lambda$; $R = \varepsilon$.
R_{13}: $P = \mathrm{i}\xi$; $Q = \mathrm{i}\lambda$; $R = \varepsilon$.
R_{14}: $P = \mathrm{i}\xi^*$; $Q = \mathrm{i}\lambda$; $R = \varepsilon$.
R_{19}: $P = \kappa$; $Q = \lambda$; $R = -\varepsilon$.
R_{20}: $P = \xi$; $Q = \lambda$; $R = -\varepsilon$.
R_{21}: $P = \xi^*$; $Q = \lambda$; $R = -\varepsilon$.
R_{26}: $P = \mathrm{i}\kappa$; $Q = \mathrm{i}\lambda$; $R = -\varepsilon$.
R_{27}: $P = \mathrm{i}\xi$; $Q = \mathrm{i}\lambda$; $R = -\varepsilon$.
R_{28}: $P = \mathrm{i}\xi^*$; $Q = \mathrm{i}\lambda$; $R = -\varepsilon$.

$\mathbf{G}_{64}^{5}$

$P^8 = E$; $Q^4 = E$; $R^2 = E$; $QP = P^7Q$; $QP^2 = P^6Q$; $RP = P^5Q^2R$; $RQ = P^4Q^3R$.
$C_1 = E$; $C_2 = P^4$; $C_3 = P^4Q^2$; $C_4 = Q^2$; $C_5 = P^7QR, P^5Q^3R, P^3QR, PQ^3R$;
$C_6 = P^3Q^3R, PQR, P^7Q^3R, P^5QR$; $C_7 = P^2, P^6$; $C_8 = P^6Q^2, P^2Q^2$;
$C_9 = P^4Q^3R, P^6Q^3R, QR, P^2QR, Q^3R, P^2Q^3R, P^4QR, P^6QR$; $C_{10} = P, P^7, P^5Q^2, P^3Q^2$;
$C_{11} = P^5, P^3, PQ^2, P^7Q^2$; $C_{12} = R, P^4Q^2R$; $C_{13} = P^4R, Q^2R$; $C_{14} = P^7Q, PQ, P^3Q, P^5Q$;
$C_{15} = P^3Q^3, P^5Q^3, P^7Q^3, PQ^3$; $C_{16} = P^2R, P^6R, P^6Q^2R, P^2Q^2R$;
$C_{17} = Q, P^2Q, P^4Q^3, P^6Q^3, P^4Q, P^6Q, Q^3, P^2Q^3$; $C_{18} = P^5Q^2R, P^3Q^2R, PR, P^7R$;
$C_{19} = PQ^2R, P^7Q^2R, P^5R, P^3R$.

	C_1	C_2	C_3	C_4	C_5	C_6	C_7	C_8	C_9	C_{10}
R_1	1	1	1	1	1	1	1	1	1	1
R_2	1	1	1	1	1	1	1	1	−1	−1
R_3	1	1	1	1	−1	−1	1	1	1	−1
R_4	1	1	1	1	−1	−1	1	1	−1	1
R_5	1	1	1	1	1	1	1	1	1	1
R_6	1	1	1	1	1	1	1	1	−1	−1
R_7	1	1	1	1	−1	−1	1	1	1	−1
R_8	1	1	1	1	−1	−1	1	1	−1	1
R_9	2	2	−2	−2	0	0	2	−2	0	0
R_{10}	2	2	−2	−2	0	0	2	−2	0	0
R_{11}	2	2	2	2	0	0	−2	−2	0	0
R_{12}	2	2	2	2	0	0	−2	−2	0	0
R_{13}	2	2	−2	−2	2i	−2i	−2	2	0	0
R_{14}	2	2	−2	−2	−2i	2i	−2	2	0	0
R_{15}	2	−2	2	−2	0	0	0	0	0	$\sqrt{2}$
R_{16}	2	−2	2	−2	0	0	0	0	0	$\sqrt{2}$
R_{17}	2	−2	2	−2	0	0	0	0	0	$-\sqrt{2}$
R_{18}	2	−2	2	−2	0	0	0	0	0	$-\sqrt{2}$
R_{19}	4	−4	−4	4	0	0	0	0	0	0

[*continued on p. 267*]

	C_{11}	C_{12}	C_{13}	C_{14}	C_{15}	C_{16}	C_{17}	C_{18}	C_{19}
R_1	1	1	1	1	1	1	1	1	1
R_2	−1	1	1	1	1	1	−1	−1	−1
R_3	−1	1	1	−1	−1	1	1	−1	−1
R_4	1	1	1	−1	−1	1	−1	1	1
R_5	1	−1	−1	−1	−1	−1	−1	−1	−1
R_6	−1	−1	−1	−1	−1	−1	1	1	1
R_7	−1	−1	−1	1	1	−1	−1	1	1
R_8	1	−1	−1	1	1	−1	1	−1	−1
R_9	0	0	0	2i	−2i	0	0	0	0
R_{10}	0	0	0	−2i	2i	0	0	0	0
R_{11}	0	2	2	0	0	−2	0	0	0
R_{12}	0	−2	−2	0	0	2	0	0	0
R_{13}	0	0	0	0	0	0	0	0	0
R_{14}	0	0	0	0	0	0	0	0	0
R_{15}	$-\sqrt{2}$	2	−2	0	0	0	0	$\sqrt{2}$	$-\sqrt{2}$
R_{16}	$-\sqrt{2}$	−2	2	0	0	0	0	$-\sqrt{2}$	$\sqrt{2}$
R_{17}	$\sqrt{2}$	2	−2	0	0	0	0	$-\sqrt{2}$	$\sqrt{2}$
R_{18}	$\sqrt{2}$	−2	2	0	0	0	0	$\sqrt{2}$	$-\sqrt{2}$
R_{19}	0	0	0	0	0	0	0	0	0

R_9: $P = i\kappa$; $Q = -\kappa$; $R = \lambda$.
R_{10}: $P = i\kappa$; $Q = \kappa$; $R = \lambda$.
R_{11}: $P = \kappa$; $Q = \lambda$; $R = \varepsilon$.
R_{12}: $P = \kappa$; $Q = \lambda$; $R = -\varepsilon$.
R_{13}: $P = \kappa$; $Q = i\lambda$; $R = \phi$.
R_{14}: $P = \kappa$; $Q = i\lambda$; $R = -\phi$.
R_{15}: $P = \delta$; $Q = i\lambda$; $R = \varepsilon$.
R_{16}: $P = \delta$; $Q = i\lambda$; $R = -\varepsilon$.
R_{17}: $P = -\delta$; $Q = i\lambda$; $R = \varepsilon$.
R_{18}: $P = -\delta$; $Q = i\lambda$; $R = -\varepsilon$.

R_{19}: $P = \begin{pmatrix} -i\delta & 0 \\ 0 & i\delta \end{pmatrix}$; $\quad Q = \begin{pmatrix} 0 & i\delta\lambda \\ -i\delta\lambda & 0 \end{pmatrix}$; $\quad R = \begin{pmatrix} 0 & \varepsilon \\ \varepsilon & 0 \end{pmatrix}$.

$\mathbf{G}^1_{96}$

$P^{12} = E$; $Q^2 = E$; $R^2 = E$; $S^4 = E$; $S^2 = P^6R$; $QP = P^7R$; $QP^3 = P^3Q$; $RP = P^{10}QR$;
$RQ = P^6QR$; $SP = P^2RS$; $SQ = P^3QRS$; $SR = RS$.
$C_1 = E$; $C_2 = P^3$; $C_3 = P^6$; $C_4 = P^9$; $C_5 = Q, R, P^9QR, P^6Q, P^6R, P^3QR$;
$C_6 = P^3Q, P^9Q, P^3R, P^9R, QR, P^6QR$; $C_7 = P^7R, P^4, PQ, P^4QR, P^8, P^{11}Q, P^2QR, P^5R$;
$C_8 = P^{10}R, P^7, P^4Q, P^7QR, P^{11}, P^2Q, P^5QR, P^8R$; $C_9 = PR, P^{10}, P^7Q, P^{10}QR, P^2, P^5Q, P^8QR, P^{11}R$;
$C_{10} = P^4R, P, P^{10}Q, PQR, P^5, P^8Q, P^{11}QR, P^2R$; $C_{11} = S, P^8QRS, P^{10}QRS, P^3RS, P^2S, PRS$;
$C_{12} = P^3S, P^{11}QRS, PQRS, P^6RS, P^5S, P^4RS$; $C_{13} = P^6S, P^2QRS, P^4QRS, P^9RS, P^8S, P^7RS$;
$C_{14} = P^9S, P^5QRS, P^7QRS, RS, P^{11}S, P^{10}RS$;
$C_{15} = P^4S, P^{10}S, P^7QS, PQS, P^{11}QS, P^5QS, P^9QS, P^3QS, P^5RS, P^{11}RS, P^6QRS, QRS$;
$C_{16} = P^7S, PS, P^{10}QS, P^4QS, P^2QS, P^8QS, QS, P^6QS, P^8RS, P^2RS, P^9QRS, P^3QRS$.

	C_1	C_2	C_3	C_4	C_5	C_6	C_7	C_8	C_9	C_{10}	C_{11}	C_{12}	C_{13}	C_{14}	C_{15}	C_{16}
R_1	1	1	1	1	1	1	1	1	1	1	1	1	1	1	1	1
R_2	1	1	1	1	1	1	1	1	1	1	−1	−1	−1	−1	−1	−1
R_3	1	−1	1	−1	−1	1	1	−1	1	−1	i	−i	i	−i	i	−i
R_4	1	−1	1	−1	−1	1	1	−1	1	−1	−i	i	−i	i	−i	i
R_5	2	2	2	2	2	2	−1	−1	−1	−1	0	0	0	0	0	0
R_6	2	−2	2	−2	−2	2	−1	1	−1	1	0	0	0	0	0	0
R_7	2	2i	−2	−2i	0	0	−1	−i	1	i	1 − i	1 + i	−1 + i	−1 − i	0	0
R_8	2	2i	−2	−2i	0	0	−1	−i	1	i	−1 + i	−1 − i	1 − i	1 + i	0	0
R_9	2	−2i	−2	2i	0	0	−1	i	1	−i	1 + i	1 − i	−1 − i	−1 + i	0	0
R_{10}	2	−2i	−2	2i	0	0	−1	i	1	−i	−1 − i	−1 + i	1 + i	1 − i	0	0
R_{11}	3	3	3	3	−1	−1	0	0	0	0	1	1	1	1	−1	−1
R_{12}	3	3	3	3	−1	−1	0	0	0	0	−1	−1	−1	−1	1	1
R_{13}	3	−3	3	−3	1	−1	0	0	0	0	i	−i	i	−i	−i	i
R_{14}	3	−3	3	−3	1	−1	0	0	0	0	−i	i	−i	i	i	−i
R_{15}	4	4i	−4	−4i	0	0	1	i	−1	−i	0	0	0	0	0	0
R_{16}	4	−4i	−4	4i	0	0	1	−i	−1	i	0	0	0	0	0	0

R_5: $P = \pi$; $Q = \varepsilon$; $S = \phi$.
R_6: $P = -\pi$; $Q = -\varepsilon$; $S = \kappa$.
R_7: $P = -\mathrm{i}\pi$; $Q = \chi$; $S = \sigma'$.
R_8: $P = -\mathrm{i}\pi$; $Q = \chi$; $S = -\sigma'$.
R_9: $P = \mathrm{i}\pi^*$; $Q = \chi^*$; $S = \tau'$.
R_{10}: $P = \mathrm{i}\pi^*$; $Q = \chi^*$; $S = -\tau'$.
R_{11}: $P = A$; $Q = B$; $S = Y$.
R_{12}: $P = A$; $Q = B$; $S = -Y$.
R_{13}: $P = L$; $Q = N$; $S = \mathrm{i}Y$.
R_{14}: $P = L$; $Q = N$; $S = -\mathrm{i}Y$.

$$R_{15}:\ P = \left(\begin{array}{c|c} \frac{1-\mathrm{i}}{4}\psi' & \sqrt{\left(\frac{3}{8}\right)}\psi' \\ \hline \mathrm{i}\sqrt{\left(\frac{3}{8}\right)}\psi' & \frac{1-\mathrm{i}}{4}\psi' \end{array}\right); \quad Q = \left(\begin{array}{c|c} \phi & 0 \\ \hline 0 & \phi \end{array}\right); \quad S = \left(\begin{array}{c|c} \nu^* & 0 \\ \hline 0 & -\nu^* \end{array}\right).$$

$$R_{16}:\ P = \left(\begin{array}{c|c} \frac{1+\mathrm{i}}{4}\chi' & \sqrt{\left(\frac{3}{8}\right)}\chi' \\ \hline -\mathrm{i}\sqrt{\left(\frac{3}{8}\right)}\chi' & \frac{1+\mathrm{i}}{4}\chi' \end{array}\right); \quad Q = \left(\begin{array}{c|c} \phi & 0 \\ \hline 0 & \phi \end{array}\right); \quad S = \left(\begin{array}{c|c} \nu & 0 \\ \hline 0 & -\nu \end{array}\right).$$

Note that $R = P^5QP$.

$\mathbf{G}_{96}^{2}$

$P^6 = E$; $Q^2 = E$; $R^2 = E$; $S^2 = E$; $T^2 = E$; $QP = PR$; $RP = PQR$; $RQ = QR$; $SP = P^5RS$; $SQ = QS$; $SR = QRS$; $TP = PT$; $TQ = QT$; $TR = RT$; $TS = P^3ST$.
$C_1 = E$; $C_2 = P^3$; $C_3 = R, Q, QR$; $C_4 = P^3R, P^3Q, P^3QR$; $C_5 = T, P^3T$;
$C_6 = RT, QT, QRT, P^3RT, P^3QT, P^3QRT$; $C_7 = P^2, P^4, P^2Q, P^2R, P^2QR, P^4QR, P^4R, P^4Q$;
$C_8 = P, P^5, P^5Q, P^5R, P^5QR, PQR, PR, PQ$;
$C_9 = PS, RS, P^2S, PQRS, QRS, P^2RS, P^4S, P^3RS, P^5S, P^4QRS, P^3QRS, P^5RS$;
$C_{10} = S, QS, PQS, PRS, P^2QRS, P^2QS, P^3S, P^3QS, P^4QS, P^4RS, P^5QRS, P^5QS$;
$C_{11} = P^5T, P^5QT, P^5RT, P^5QRT, P^4T, P^4QT, P^4RT, P^4QRT$;
$C_{12} = PT, PQT, PRT, PQRT, P^2T, P^2QT, P^2RT, P^2QRT$;
$C_{13} = PST, RST, P^2ST, PQRST, QRST, P^2RST, P^4ST, P^3RST, P^5ST, P^4QRST, P^3QRST, P^5RST$;
$C_{14} = ST, QST, PQST, PRST, P^2QRST, P^2QST, P^3ST, P^3QST, P^4QST, P^4RST, P^5QRST, P^5QST$.

	C_1	C_2	C_3	C_4	C_5	C_6	C_7	C_8	C_9	C_{10}	C_{11}	C_{12}	C_{13}	C_{14}
R_1	1	1	1	1	1	1	1	1	1	1	1	1	1	1
R_2	1	1	1	1	−1	−1	1	1	1	1	−1	−1	−1	−1
R_3	1	1	1	1	1	1	1	1	−1	−1	1	1	−1	−1
R_4	1	1	1	1	−1	−1	1	1	−1	−1	−1	−1	1	1
R_5	2	2	2	2	2	2	−1	−1	0	0	−1	−1	0	0
R_6	2	2	2	2	−2	−2	−1	−1	0	0	1	1	0	0
R_7	2	−2	2	−2	0	0	2	−2	0	0	0	0	0	0
R_8	2	−2	2	−2	0	0	−1	1	0	0	$i\sqrt{3}$	$-i\sqrt{3}$	0	0
R_9	2	−2	2	−2	0	0	−1	1	0	0	$-i\sqrt{3}$	$i\sqrt{3}$	0	0
R_{10}	3	3	−1	−1	3	−1	0	0	1	−1	0	0	1	−1
R_{11}	3	3	−1	−1	−3	1	0	0	1	−1	0	0	−1	1
R_{12}	3	3	−1	−1	3	−1	0	0	−1	1	0	0	−1	1
R_{13}	3	3	−1	−1	−3	1	0	0	−1	1	0	0	1	−1
R_{14}	6	−6	−2	2	0	0	0	0	0	0	0	0	0	0

R_5: $P = \pi$; $Q = \varepsilon$; $R = \varepsilon$; $S = \phi$; $T = \varepsilon$.
R_6: $P = \pi$; $Q = \varepsilon$; $R = \varepsilon$; $S = \phi$; $T = -\varepsilon$.
R_7: $P = -\varepsilon$; $Q = \varepsilon$; $R = \varepsilon$; $S = \phi$; $T = \lambda$.
R_8: $P = -\pi$; $Q = \varepsilon$; $R = \varepsilon$; $S = \phi$; $T = \lambda$.
R_9: $P = -\pi$; $Q = \varepsilon$; $R = \varepsilon$; $S = \phi$; $T = -\lambda$.
R_{10}: $P = D$; $Q = B$; $R = G$; $S = F$; $T = E$.
R_{11}: $P = D$; $Q = B$; $R = G$; $S = F$; $T = I$.
R_{12}: $P = D$; $Q = B$; $R = G$; $S = H$; $T = E$.
R_{13}: $P = D$; $Q = B$; $R = G$; $S = H$; $T = I$.
R_{14}: $P = \mathscr{A}$; $Q = \mathscr{B}$; $R = \mathscr{C}$; $S = \mathscr{D}$; $T = \mathscr{F}$.

$\mathbf{G}_{96}^{3}$

$P^6 = E$; $Q^2 = E$; $R^2 = E$; $S^4 = E$; $S^2 = P^3$; $T^2 = E$; $QP = PR$; $RP = PQR$; $RQ = QR$; $SP = P^5RS$; $SQ = QS$; $SR = QRS$; $TP = PT$; $TQ = QT$; $TR = RT$; $TS = P^3ST$.

$C_1 = E$; $C_2 = P^3$; $C_3 = R, Q, QR$; $C_4 = P^3R, P^3Q, P^3QR$; $C_5 = T, P^3T$;

$C_6 = RT, QT, QRT, P^3RT, P^3QT, P^3QRT$; $C_7 = P^2, P^4, P^2Q, P^2R, P^2QR, P^4QR, P^4R, P^4Q$;

$C_8 = P, P^5, P^5Q, P^5R, P^5QR, PQR, PR, PQ$;

$C_9 = PS, RS, P^2S, PQRS, QRS, P^2RS, P^4S, P^3RS, P^5S, P^4QRS, P^3QRS, P^5RS$;

$C_{10} = S, QS, PQS, PRS, P^2QRS, P^2QS, P^3S, P^3QS, P^4QS, P^4RS, P^5QRS, P^5QS$;

$C_{11} = P^5T, P^5QT, P^5RT, P^5QRT, P^4T, P^4QT, P^4RT, P^4QRT$;

$C_{12} = PT, PQT, PRT, PQRT, P^2T, P^2QT, P^2RT, P^2QRT$;

$C_{13} = PST, RST, P^2ST, PQRST, QRST, P^2RST, P^4ST, P^3RST, P^5ST, P^4QRST, P^3QRST, P^5RST$;

$C_{14} = ST, QST, PQST, PRST, P^2QRST, P^2QST, P^3ST, P^3QRST, P^4QST, P^4RST, P^5QRST, P^5QST$.

	C_1	C_2	C_3	C_4	C_5	C_6	C_7	C_8	C_9	C_{10}	C_{11}	C_{12}	C_{13}	C_{14}
R_1	1	1	1	1	1	1	1	1	1	1	1	1	1	1
R_2	1	1	1	1	−1	−1	1	1	1	1	−1	−1	−1	−1
R_3	1	1	1	1	1	1	1	1	−1	−1	1	1	−1	−1
R_4	1	1	1	1	−1	−1	1	1	−1	−1	−1	−1	1	1
R_5	2	2	2	2	2	2	−1	−1	0	0	−1	−1	0	0
R_6	2	2	2	2	−2	−2	−1	−1	0	0	1	1	0	0
R_7	2	−2	2	−2	0	0	2	−2	0	0	0	0	0	0
R_8	2	−2	2	−2	0	0	−1	1	0	0	$i\sqrt{3}$	$-i\sqrt{3}$	0	0
R_9	2	−2	2	−2	0	0	−1	1	0	0	$-i\sqrt{3}$	$i\sqrt{3}$	0	0
R_{10}	3	3	−1	−1	3	−1	0	0	1	−1	0	0	1	−1
R_{11}	3	3	−1	−1	−3	1	0	0	1	−1	0	0	−1	1
R_{12}	3	3	−1	−1	3	−1	0	0	−1	1	0	0	−1	1
R_{13}	3	3	−1	−1	−3	1	0	0	−1	1	0	0	1	−1
R_{14}	6	−6	−2	2	0	0	0	0	0	0	0	0	0	0

R_5: $P = \pi$; $Q = \varepsilon$; $R = \varepsilon$; $S = \phi$; $T = \varepsilon$.

R_6: $P = \pi$; $Q = \varepsilon$; $R = \varepsilon$; $S = \phi$; $T = -\varepsilon$.

R_7: $P = -\varepsilon$; $Q = \varepsilon$; $R = \varepsilon$; $S = i\phi$; $T = \lambda$.

R_8: $P = -\pi$; $Q = \varepsilon$; $R = \varepsilon$; $S = i\phi$; $T = \lambda$.

R_9: $P = -\pi$; $Q = \varepsilon$; $R = \varepsilon$; $S = i\phi$; $T = -\lambda$.

R_{10}: $P = D$; $Q = B$; $R = G$; $S = F$; $T = E$.

R_{11}: $P = D$; $Q = B$; $R = G$; $S = F$; $T = I$.

R_{12}: $P = D$; $Q = B$; $R = G$; $S = H$; $T = E$.

R_{13}: $P = D$; $Q = B$; $R = G$; $S = H$; $T = I$.

R_{14}: $P = \mathscr{A}$; $Q = \mathscr{B}$; $R = \mathscr{C}$; $S = i\mathscr{D}$; $T = \mathscr{F}$.

$\mathbf{G}_{96}^{4}$

$P^6 = E$; $Q^2 = E$; $R^2 = E$; $S^4 = E$; $S^2 = P^3$; $T^2 = E$; $QP = PR$; $QP^3 = P^3Q$; $RP = P^4QR$; $RQ = QR$; $SP = P^2QS$; $SQ = RS$; $SR = QS$; $TP = PT$; $TQ = QT$; $TR = RT$; $TS = P^3ST$.

$C_1 = E$; $C_2 = P^3$; $C_3 = P^3Q, P^3R, QR$; $C_4 = Q, R, P^3QR$; $C_5 = T, P^3T$;
$C_6 = QT, RT, P^3QRT, P^3QT, P^3RT, QRT$; $C_7 = P^2, P^4, P^5Q, P^5R, P^2QR, P^4QR, PQ, PR$;
$C_8 = P, P^5, P^2Q, P^2R, P^5QR, PQR, P^4Q, P^4R$;
$C_9 = PS, RS, P^2S, P^4S, P^3RS, P^5S, QS, P^3QS, PRS, P^4RS, P^2QS, P^5QS$;
$C_{10} = PQRS, QRS, P^2RS, P^4QRS, P^3QRS, P^5RS, S, P^3S, PQS, P^4QS, P^2QRS, P^5QRS$;
$C_{11} = P^5T, P^4T, P^2QT, P^2RT, P^5QRT, P^4QRT, PQT, PRT$;
$C_{12} = P^2T, PT, P^5QT, P^5RT, P^2QRT, PQRT, P^4QT, P^4RT$;
$C_{13} = PST, RST, P^2ST, P^4ST, P^3RST, P^5ST, QST, P^3QST, PRST, P^4RST, P^2QST, P^5QST$;
$C_{14} = PQRST, QRST, P^2RST, P^4QRST, P^3QRST, P^5RST, ST, P^3ST, P^4QST, PQST, P^2QRST, P^5QRST$.

	C_1	C_2	C_3	C_4	C_5	C_6	C_7	C_8	C_9	C_{10}	C_{11}	C_{12}	C_{13}	C_{14}
R_1	1	1	1	1	1	1	1	1	1	1	1	1	1	1
R_2	1	1	1	1	−1	−1	1	1	1	1	−1	−1	−1	−1
R_3	1	1	1	1	1	1	1	1	−1	−1	1	1	−1	−1
R_4	1	1	1	1	−1	−1	1	1	−1	−1	−1	−1	1	1
R_5	2	2	2	2	2	2	−1	−1	0	0	−1	−1	0	0
R_6	2	2	2	2	−2	−2	−1	−1	0	0	1	1	0	0
R_7	2	−2	2	−2	0	0	2	−2	0	0	0	0	0	0
R_8	2	−2	2	−2	0	0	−1	1	0	0	$i\sqrt{3}$	$-i\sqrt{3}$	0	0
R_9	2	−2	2	−2	0	0	−1	1	0	0	$-i\sqrt{3}$	$i\sqrt{3}$	0	0
R_{10}	3	3	−1	−1	3	−1	0	0	1	−1	0	0	1	−1
R_{11}	3	3	−1	−1	−3	1	0	0	1	−1	0	0	−1	1
R_{12}	3	3	−1	−1	3	−1	0	0	−1	1	0	0	−1	1
R_{13}	3	3	−1	−1	−3	1	0	0	−1	1	0	0	1	−1
R_{14}	6	−6	−2	2	0	0	0	0	0	0	0	0	0	0

R_5: $P = \pi$; $Q = \varepsilon$; $R = \varepsilon$; $S = \phi$; $T = \varepsilon$.
R_6: $P = \pi$; $Q = \varepsilon$; $R = \varepsilon$; $S = \phi$; $T = -\varepsilon$.
R_7: $P = -\varepsilon$; $Q = -\varepsilon$; $R = -\varepsilon$; $S = -\lambda$; $T = \phi$.
R_8: $P = -\pi$; $Q = -\varepsilon$; $R = -\varepsilon$; $S = \phi$; $T = \lambda$.
R_9: $P = -\pi$; $Q = -\varepsilon$; $R = -\varepsilon$; $S = \phi$; $T = -\lambda$.
R_{10}: $P = A$; $Q = G$; $R = B$; $S = U$; $T = E$.
R_{11}: $P = A$; $Q = G$; $R = B$; $S = U$; $T = I$.
R_{12}: $P = A$; $Q = G$; $R = B$; $S = V$; $T = E$.
R_{13}: $P = A$; $Q = G$; $R = B$; $S = V$; $T = I$.
R_{14}: $P = \mathscr{G}$; $Q = \mathscr{H}$; $R = \mathscr{L}$; $S = i\mathscr{M}$; $T = \mathscr{N}$.

$\mathbf{G}_{96}^{5}$

$P^6 = E$; $Q^4 = E$; $R^4 = E$; $P^3 = Q^2 = R^2$; $S^4 = E$; $QP = PR$; $RP = PQR$; $RQ = P^3QR$; $SP = PS$; $SQ = QS$; $SR = RS$.

$C_1 = E$; $C_2 = P^3$; $C_3 = Q, R, P^3Q, P^3R, QR, P^3QR$; $C_4 = P, PQ, PQR, PR$; $C_5 = P^4, P^4Q, P^4QR, P^4R$; $C_6 = P^2, P^5R, P^5QR, P^5Q$; $C_7 = P^5, P^2R, P^2QR, P^2Q$; $C_8 = S$; $C_9 = P^3S$; $C_{10} = QS, RS, P^3QS, P^3RS, QRS, P^3QRS$; $C_{11} = PS, PQS, PQRS, PRS$; $C_{12} = P^4S, P^4QS, P^4QRS, P^4RS$; $C_{13} = P^2S, P^5RS, P^5QRS, P^5QS$; $C_{14} = P^5S, P^2RS, P^2QRS, P^2QS$; $C_{15} = S^2$; $C_{16} = P^3S^2$; $C_{17} = QS^2, RS^2, P^3QS^2, P^3RS^2, QRS^2, P^3QRS^2$; $C_{18} = PS^2, PQS^2, PQRS^2, PRS^2$; $C_{19} = P^4S^2, P^4QS^2, P^4QRS^2, P^4RS^2$; $C_{20} = P^2S^2, P^5RS^2, P^5QRS^2, P^5QS^2$; $C_{21} = P^5S^2, P^2RS^2, P^2QRS^2, P^2QS^2$; $C_{22} = S^3$; $C_{23} = P^3S^3$; $C_{24} = QS^3, RS^3, P^3QS^3, P^3RS^3, QRS^3, P^3QRS^3$; $C_{25} = PS^3, PQS^3, PQRS^3, PRS^3$; $C_{26} = P^4S^3, P^4QS^3, P^4QRS^3, P^4RS^3$; $C_{27} = P^2S^3, P^5RS^3, P^5QRS^3, P^5QS^3$; $C_{28} = P^5S^3, P^2RS^3, P^2QRS^3, P^2QS^3$.

	C_1 to C_7	C_8 to C_{14}	C_{15} to C_{21}	C_{22} to C_{28}
R_1 to R_7	$(\mathbf{G}_{24}^{9})$	$(\mathbf{G}_{24}^{9})$	$(\mathbf{G}_{24}^{9})$	$(\mathbf{G}_{24}^{9})$
R_8 to R_{14}	$(\mathbf{G}_{24}^{9})$	$\mathrm{i}(\mathbf{G}_{24}^{9})$	$-(\mathbf{G}_{24}^{9})$	$-\mathrm{i}(\mathbf{G}_{24}^{9})$
R_{15} to R_{21}	$(\mathbf{G}_{24}^{9})$	$-(\mathbf{G}_{24}^{9})$	$(\mathbf{G}_{24}^{9})$	$-(\mathbf{G}_{24}^{9})$
R_{22} to R_{28}	$(\mathbf{G}_{24}^{9})$	$-\mathrm{i}(\mathbf{G}_{24}^{9})$	$-(\mathbf{G}_{24}^{9})$	$\mathrm{i}(\mathbf{G}_{24}^{9})$

R_4: $P = -\pi$; $Q = \psi$; $R = \varepsilon$.
R_5: $P = -\omega^*\mu$; $Q = \psi$; $R = \varepsilon$.
R_6: $P = -\mu$; $Q = \psi$; $R = \varepsilon$.
R_7: $P = A$; $Q = B$; $R = E$.
R_{11}: $P = -\pi$; $Q = \psi$; $R = \mathrm{i}\varepsilon$.
R_{12}: $P = -\omega^*\mu$; $Q = \psi$; $R = \mathrm{i}\varepsilon$.
R_{13}: $P = -\mu$; $Q = \psi$; $R = \mathrm{i}\varepsilon$.
R_{14}: $P = A$; $Q = B$; $R = \mathrm{i}E$.
R_{18}: $P = -\pi$; $Q = \psi$; $R = -\varepsilon$.
R_{19}: $P = -\omega^*\mu$; $Q = \psi$; $R = -\varepsilon$.
R_{20}: $P = -\mu$; $Q = \psi$; $R = -\varepsilon$.
R_{21}: $P = A$; $Q = B$; $R = -E$.
R_{25}: $P = -\pi$; $Q = \psi$; $R = -\mathrm{i}\varepsilon$.
R_{26}: $P = -\omega^*\mu$; $Q = \psi$; $R = -\mathrm{i}\varepsilon$.
R_{27}: $P = -\mu$; $Q = \psi$; $R = -\mathrm{i}\varepsilon$.
R_{28}: $P = A$; $Q = B$; $R = -\mathrm{i}E$.

$\mathbf{G}_{96}^{6}$

$P^6 = E$; $Q^4 = E$; $R^4 = E$; $P^3 = Q^2 = R^2$; $S^2 = E$; $T^2 = E$; $QP = PR$; $RP = PQR$; $RQ = P^3QR$; $SP = PS$; $SQ = QS$; $SR = RS$; $TP = PT$; $TQ = QT$; $TR = RT$; $TS = ST$.

$C_1 = E$; $C_2 = P^3$; $C_3 = Q, R, P^3Q, P^3R, QR, P^3QR$; $C_4 = P, PQ, PQR, PR$; $C_5 = P^4, P^4Q, P^4QR, P^4R$; $C_6 = P^2, P^5R, P^5QR, P^5Q$; $C_7 = P^5, P^2R, P^2QR, P^2Q$; $C_8 = S$; $C_9 = P^3S$; $C_{10} = QS, RS, P^3QS, P^3RS, QRS, P^3QRS$; $C_{11} = PS, PQS, PQRS, PRS$; $C_{12} = P^4S, P^4QS, P^4QRS, P^4RS$; $C_{13} = P^2S, P^5RS, P^5QRS, P^5QS$; $C_{14} = P^5S, P^2RS, P^2QRS, P^2QS$; $C_{15} = T$; $C_{16} = P^3T$; $C_{17} = QT, RT, P^3QT, P^3RT, QRT, P^3QRT$; $C_{18} = PT, PQT, PQRT, PRT$; $C_{19} = P^4T, P^4QT, P^4QRT, P^4RT$; $C_{20} = P^2T, P^5RT, P^5QRT, P^5QT$; $C_{21} = P^5T, P^2RT, P^2QRT, P^2QT$; $C_{22} = ST$; $C_{23} = P^3ST$; $C_{24} = QST, RST, P^3QST, P^3RST, QRST, P^3QRST$; $C_{25} = PST, PQST, PQRST, PRST$; $C_{26} = P^4ST, P^4QST, P^4QRST, P^4RST$; $C_{27} = P^2ST, P^5RST, P^5QRST, P^5QST$; $C_{28} = P^5ST, P^2RST, P^2QRST, P^2QST$.

	C_1 to C_{14}	C_{15} to C_{28}
R_1 to R_{14}	$(\mathbf{G}_{48}^{4})$	$(\mathbf{G}_{48}^{4})$
R_{15} to R_{28}	$(\mathbf{G}_{48}^{4})$	$-(\mathbf{G}_{48}^{4})$

R_4: $P = -\pi$; $Q = \psi$; $S = \varepsilon$; $T = \varepsilon$.
R_5: $P = -\omega^*\mu$; $Q = \psi$; $S = \varepsilon$; $T = \varepsilon$.
R_6: $P = -\mu$; $Q = \psi$; $S = \varepsilon$; $T = \varepsilon$.
R_7: $P = A$; $Q = B$; $S = E$; $T = E$.
R_{11}: $P = -\pi$; $Q = \psi$; $S = -\varepsilon$; $T = \varepsilon$.
R_{12}: $P = -\omega^*\mu$; $Q = \psi$; $S = -\varepsilon$; $T = \varepsilon$.
R_{13}: $P = -\mu$; $Q = \psi$; $S = -\varepsilon$; $T = \varepsilon$.
R_{14}: $P = A$; $Q = B$; $S = I$; $T = E$.
R_{18}; $P = -\pi$; $Q = \psi$; $S = \varepsilon$; $T = -\varepsilon$.
R_{19}: $P = -\omega^*\mu$; $Q = \psi$; $S = \varepsilon$; $T = -\varepsilon$.
R_{20}: $P = -\mu$; $Q = \psi$; $S = \varepsilon$; $T = -\varepsilon$.
R_{21}: $P = A$; $Q = B$; $S = E$; $T = I$.
R_{25}: $P = -\pi$; $Q = \psi$; $S = -\varepsilon$; $T = -\varepsilon$.
R_{26}: $P = -\omega^*\mu$; $Q = \psi$; $S = -\varepsilon$; $T = -\varepsilon$.
R_{27}: $P = -\mu$; $Q = \psi$; $S = -\varepsilon$; $T = -\varepsilon$.
R_{28}: $P = A$; $Q = B$; $S = I$; $T = I$.

Note that $R = P^5QP$.

$\mathbf{G}_{96}^{7}$

$P^4 = E;\ Q^4 = E;\ R^6 = E;\ S^4 = E;\ Q^2 = P^2;\ S^2 = R^3;\ QP = P^3Q;\ RP = P^3QR;\ RQ = PR^4;$
$SP = P^2QR^3S;\ SQ = P^3R^3S;\ SR = R^5S.$
$C_1 = E;\ C_2 = R^3;\ C_3 = P^2R^3;\ C_4 = P^2;\ C_5 = P, P^3, QR^3, P^2QR^3, P^3Q, PQ;$
$C_6 = PR^3, P^3R^3, Q, P^2Q, P^3QR^3, PQR^3;\ C_7 = R, P^3R^4, P^2QR, PQR^4, R^5, QR^5, P^3QR^2, PR^2;$
$C_8 = R^4, P^3R, P^2QR^4, PQR, R^2, QR^2, P^3QR^5, PR^5;\ C_9 = P^2R^4, PR, QR^4, P^3QR, P^2R^2, P^2QR^2, PQR^5, P^3R^5;$
$C_{10} = P^2R, PR^4, QR, P^3QR^4, P^2R^5, P^2QR^5, PQR^2, P^3R^2;$
$C_{11} = P^2QS, PR^3S, QR^4S, PQRS, P^3R^5S, P^3QR^5S;\ C_{12} = P^2QR^3S, PS, QRS, PQR^4S, P^3R^2S, P^3QR^2S;$
$C_{13} = QR^3S, P^3S, P^2QRS, P^3QR^4S, PR^2S, PQR^2S;\ C_{14} = QS, P^3R^3S, P^2QR^4S, P^3QRS, PR^5S, PQR^5S;$
$C_{15} = P^2R^3S, P^3QS, P^2QR^5S, P^2R^5S, P^3R^4S, P^2RS, R^3S, PQS, QR^5S, R^5S, PR^4S, RS;$
$C_{16} = P^2S, P^3QR^3S, P^2QR^2S, P^2R^2S, P^3RS, P^2R^4S, S, PQR^3S, QR^2S, R^2S, PRS, R^4S.$

	C_1	C_2	C_3	C_4	C_5	C_6	C_7	C_8	C_9	C_{10}	C_{11}	C_{12}	C_{13}	C_{14}	C_{15}	C_{16}
R_1	1	1	1	1	1	1	1	1	1	1	1	1	1	1	1	1
R_2	1	1	1	1	1	1	1	1	1	1	−1	−1	−1	−1	−1	−1
R_3	1	−1	−1	1	−1	1	−1	1	1	−1	i	−i	−i	i	−i	i
R_4	1	−1	−1	1	−1	1	−1	1	1	−1	−i	i	i	−i	i	−i
R_5	2	2	2	2	2	2	−1	−1	−1	−1	0	0	0	0	0	0
R_6	2	2	−2	−2	0	0	−1	−1	1	1	$-i\sqrt{2}$	$-i\sqrt{2}$	$i\sqrt{2}$	$i\sqrt{2}$	0	0
R_7	2	2	−2	−2	0	0	−1	−1	1	1	$i\sqrt{2}$	$i\sqrt{2}$	$-i\sqrt{2}$	$-i\sqrt{2}$	0	0
R_8	2	−2	−2	2	−2	2	1	−1	−1	1	0	0	0	0	0	0
R_9	2	−2	2	−2	0	0	1	−1	1	−1	$\sqrt{2}$	$-\sqrt{2}$	$\sqrt{2}$	$-\sqrt{2}$	0	0
R_{10}	2	−2	2	−2	0	0	1	−1	1	−1	$-\sqrt{2}$	$\sqrt{2}$	$-\sqrt{2}$	$\sqrt{2}$	0	0
R_{11}	3	3	3	3	−1	−1	0	0	0	0	1	1	1	1	−1	−1
R_{12}	3	3	3	3	−1	−1	0	0	0	0	−1	−1	−1	−1	1	1
R_{13}	3	−3	−3	3	1	−1	0	0	0	0	i	−i	−i	i	i	−i
R_{14}	3	−3	−3	3	1	−1	0	0	0	0	−i	i	i	−i	−i	i
R_{15}	4	4	−4	−4	0	0	1	1	−1	−1	0	0	0	0	0	0
R_{16}	4	−4	4	−4	0	0	−1	1	−1	1	0	0	0	0	0	0

R_5: $P = \varepsilon;\ Q = \varepsilon;\ R = \pi;\ S = \phi.$

R_6: $P = \zeta^*;\ Q = \gamma^*;\ R = \pi^*;\ S = \phi.$

R_7: $P = \zeta;\ Q = \gamma;\ R = \pi;\ S = \phi.$

R_8: $P = -\varepsilon;\ Q = \varepsilon;\ R = -\pi;\ S = i\phi.$

R_9: $P = -\zeta^*;\ Q = \gamma^*;\ R = -\pi^*;\ S = i\phi.$

R_{10}: $P = -\zeta;\ Q = \gamma;\ R = -\pi;\ S = i\phi.$

R_{11}: $P = C;\ Q = G;\ R = K;\ S = F.$

R_{12}: $P = C;\ Q = G;\ R = K;\ S = H.$

R_{13}: $P = -C;\ Q = G;\ R = -K;\ S = iF.$

R_{14}: $P = -C;\ Q = G;\ R = -K;\ S = iH.$

$$R_{15}:\ P = \left(\begin{array}{c|c} \alpha' & 0 \\ \hline 0 & \beta' \end{array}\right);\quad Q = \left(\begin{array}{c|c} \beta' & 0 \\ \hline 0 & \alpha' \end{array}\right);\quad R = \left(\begin{array}{c|c} \lambda' & \nu' \\ \hline \rho' & \mu' \end{array}\right);\quad S = \left(\begin{array}{c|c} \lambda & 0 \\ \hline 0 & \lambda \end{array}\right).$$

$$R_{16}:\ P = \left(\begin{array}{c|c} -\alpha' & 0 \\ \hline 0 & -\beta' \end{array}\right);\quad Q = \left(\begin{array}{c|c} \beta' & 0 \\ \hline 0 & \alpha' \end{array}\right);\quad R = \left(\begin{array}{c|c} -\lambda' & -\nu' \\ \hline -\rho' & -\mu' \end{array}\right);\quad S = \left(\begin{array}{c|c} i\lambda & 0 \\ \hline 0 & i\lambda \end{array}\right).$$

$\mathbf{G}_{96}^{8}$

$P^8 = E$; $Q^3 = E$; $R^4 = E$; $S^2 = E$; $R^2 = P^4$; $QP = P^6Q^2R$; $QP^2 = P^3R$; $QP^4 = P^4Q$; $Q^2P = P^3Q$; $RP = P^6Q^2$; $RP^2 = PQ$; $RQ = Q^2R$; $SP = PS$; $SQ = QS$; $SR = RS$.
$C_1 = E$; $C_2 = P^4$; $C_3 = P^2, P^5QR, P^7QR, P^6, PQR, P^3QR$;
$C_4 = P^4Q, P^6Q, P^5R, P^7R, P^4Q^2, PQ^2R, P^3Q^2R, P^2Q^2$; $C_5 = Q, P^2Q, PR, P^3R, Q^2, P^5Q^2R, P^7Q^2R, P^6Q^2$;
$C_6 = P, P^3Q^2, P^6R, P^7, P^2Q^2R, P^5Q$; $C_7 = P^5, P^7Q^2, P^2R, P^3, P^6Q^2R, PQ$;
$C_8 = P^3Q, R, PQ^2, P^6QR, P^4Q^2R, P^4QR, P^7Q, P^4R, P^5Q^2, P^2QR, Q^2R, QR$; $C_9 = S$; $C_{10} = P^4S$;
$C_{11} = P^2S, P^5QRS, P^7QRS, P^6S, PQRS, P^3QRS$;
$C_{12} = P^4QS, P^6QS, P^5RS, P^7RS, P^4Q^2S, PQ^2RS, P^3Q^2RS, P^2Q^2S$;
$C_{13} = QS, P^2QS, PRS, P^3RS, Q^2S, P^5Q^2RS, P^7Q^2RS, P^6Q^2S$;
$C_{14} = PS, P^3Q^2S, P^6RS, P^7S, P^2Q^2RS, P^5QS$; $C_{15} = P^5S, P^7Q^2S, P^2RS, P^3S, P^6Q^2RS, PQS$;
$C_{16} = P^3QS, RS, PQ^2S, P^6QRS, P^4Q^2RS, P^4QRS, P^7QS, P^4RS, P^5Q^2S, P^2QRS, Q^2RS, QRS$.

	C_1 to C_8	C_9 to C_{16}
R_1 to R_8	$(\mathbf{G}_{48}^{10})$	$(\mathbf{G}_{48}^{10})$
R_9 to R_{16}	$(\mathbf{G}_{48}^{10})$	$-(\mathbf{G}_{48}^{10})$

R_3: $P = \lambda\alpha$; $Q = -\beta$; $R = \lambda$; $S = \varepsilon$.
R_4: $P = \delta$; $Q = \kappa'$; $R = \phi'$; $S = \varepsilon$.
R_5: $P = -\delta$; $Q = \kappa'$; $R = -\phi'$; $S = \varepsilon$.
R_6: $P = Z$; $Q = D$; $R = -A'$; $S = E$.
R_7: $P = -Z$; $Q = D$; $R = A'$; $S = E$.

$$R_8\text{: } P = \left(\begin{array}{c|c} \kappa\alpha' & 0 \\ \hline 0 & -\alpha'\kappa \end{array}\right); \quad Q = \left(\begin{array}{c|c} \lambda' & \nu' \\ \hline \rho' & \mu' \end{array}\right); \quad R = \mathrm{i}\left(\begin{array}{c|c} \lambda\lambda' & \lambda\nu' \\ \hline \lambda\rho' & \lambda\mu' \end{array}\right); \quad S = \left(\begin{array}{c|c} \varepsilon & 0 \\ \hline 0 & \varepsilon \end{array}\right).$$

R_{11}: $P = \lambda\alpha$; $Q = -\beta$; $R = \lambda$; $S = -\varepsilon$.
R_{12}: $P = \delta$; $Q = \kappa'$; $R = \phi'$; $S = -\varepsilon$.
R_{13}: $P = -\delta$; $Q = \kappa'$; $R = -\phi'$; $S = -\varepsilon$.
R_{14}: $P = Z$; $Q = D$; $R = -A'$; $S = -E$.
R_{15}: $P = -Z$; $Q = D$; $R = A'$; $S = -E$.

$$R_{16}\text{: } P = \left(\begin{array}{c|c} \kappa\alpha' & 0 \\ \hline 0 & -\alpha'\kappa \end{array}\right); \quad Q = \left(\begin{array}{c|c} \lambda' & \nu' \\ \hline \rho' & \mu' \end{array}\right); \quad R = \mathrm{i}\left(\begin{array}{c|c} \lambda\lambda' & \lambda\nu' \\ \hline \lambda\rho' & \lambda\mu' \end{array}\right); \quad S = \left(\begin{array}{c|c} -\varepsilon & 0 \\ \hline 0 & -\varepsilon \end{array}\right).$$

Note that $R = P^5QP^2$.

$\mathbf{G}_{96}^{9}$

$P^4 = E$; $Q^6 = E$; $R^4 = E$; $S^4 = E$; $Q^3 = R^2 = S^2$; $QP = PQ^5$; $RP = P^3R$; $RQ = QR$; $SP = PQ^3S$; $SQ = QS$; $SR = RS$.

$C_1 = E$; $C_2 = Q^3$; $C_3 = P^2$; $C_4 = P^2Q^3$; $C_5 = Q, Q^5$; $C_6 = Q^4, Q^2$; $C_7 = P^2Q, P^2Q^5$; $C_8 = P^2Q^4, P^2Q^2$; $C_9 = R, P^2R$; $C_{10} = Q^3R, P^2Q^3R$; $C_{11} = Q^2R, P^2Q^4R$; $C_{12} = Q^5R, P^2QR$; $C_{13} = QR, P^2Q^5R$; $C_{14} = Q^4R, P^2Q^2R$;
$C_{15} = P, PQ, PQ^2, P^3, P^3Q, P^3Q^2, PQ^3, PQ^4, PQ^5, P^3Q^3, P^3Q^4, P^3Q^5$;
$C_{16} = PR, PQR, PQ^2R, P^3R, P^3QR, P^3Q^2R, PQ^3R, PQ^4R, PQ^5R, P^3Q^3R, P^3Q^4R, P^3Q^5R$; $C_{17} = S, Q^3S$;
$C_{18} = P^2S, P^2Q^3S$; $C_{19} = QS, Q^2S$; $C_{20} = Q^4S, Q^5S$; $C_{21} = P^2QS, P^2Q^2S$; $C_{22} = P^2Q^4S, P^2Q^5S$;
$C_{23} = RS, P^2Q^3RS$; $C_{24} = Q^3RS, P^2RS$; $C_{25} = Q^2RS, P^2QRS$; $C_{26} = Q^5RS, P^2Q^4RS$;
$C_{27} = QRS, P^2Q^2RS$; $C_{28} = Q^4RS, P^2Q^5RS$;
$C_{29} = PS, PQS, PQ^2S, P^3S, P^3QS, P^3Q^2S, PQ^3S, PQ^4S, PQ^5S, P^3Q^3S, P^3Q^4S, P^3Q^5S$;
$C_{30} = PRS, PQRS, PQ^2RS, P^3RS, P^3QRS, P^3Q^2RS, PQ^3RS, PQ^4RS, PQ^5RS, P^3Q^3RS, P^3Q^4RS, P^3Q^5RS$.

	C_1	C_2	C_3	C_4	C_5	C_6	C_7	C_8	C_9	C_{10}	C_{11}	C_{12}	C_{13}	C_{14}	C_{15}
R_1	1	1	1	1	1	1	1	1	1	1	1	1	1	1	1
R_2	1	1	1	1	1	1	1	1	1	1	1	1	1	1	1
R_3	1	1	1	1	1	1	1	1	1	1	1	1	1	1	-1
R_4	1	1	1	1	1	1	1	1	1	1	1	1	1	1	-1
R_5	1	1	1	1	1	1	1	1	-1	-1	-1	-1	-1	-1	1
R_6	1	1	1	1	1	1	1	1	-1	-1	-1	-1	-1	-1	1
R_7	1	1	1	1	1	1	1	1	-1	-1	-1	-1	-1	-1	-1
R_8	1	1	1	1	1	1	1	1	-1	-1	-1	-1	-1	-1	-1
R_9	2	2	2	2	-1	-1	-1	-1	2	2	-1	-1	-1	-1	0
R_{10}	2	2	2	2	-1	-1	-1	-1	2	2	-1	-1	-1	-1	0
R_{11}	2	2	2	2	-1	-1	-1	-1	-2	-2	1	1	1	1	0
R_{12}	2	2	2	2	-1	-1	-1	-1	-2	-2	1	1	1	1	0
R_{13}	2	2	-2	-2	2	2	-2	-2	0	0	0	0	0	0	0
R_{14}	2	2	-2	-2	2	2	-2	-2	0	0	0	0	0	0	0
R_{15}	2	2	-2	-2	-1	-1	1	1	0	0	$i\sqrt{3}$	$i\sqrt{3}$	$-i\sqrt{3}$	$-i\sqrt{3}$	0
R_{16}	2	2	-2	-2	-1	-1	1	1	0	0	$i\sqrt{3}$	$i\sqrt{3}$	$-i\sqrt{3}$	$-i\sqrt{3}$	0
R_{17}	2	2	-2	-2	-1	-1	1	1	0	0	$-i\sqrt{3}$	$-i\sqrt{3}$	$i\sqrt{3}$	$i\sqrt{3}$	0
R_{18}	2	2	-2	-2	-1	-1	1	1	0	0	$-i\sqrt{3}$	$-i\sqrt{3}$	$i\sqrt{3}$	$i\sqrt{3}$	0
R_{19}	2	-2	2	-2	-2	2	-2	2	$2i$	$-2i$	$2i$	$-2i$	$-2i$	$2i$	0
R_{20}	2	-2	2	-2	-2	2	-2	2	$-2i$	$2i$	$-2i$	$2i$	$2i$	$-2i$	0
R_{21}	2	-2	2	-2	1	-1	1	-1	$2i$	$-2i$	$-i$	i	i	$-i$	0
R_{22}	2	-2	2	-2	1	-1	1	-1	$2i$	$-2i$	$-i$	i	i	$-i$	0
R_{23}	2	-2	2	-2	1	-1	1	-1	$-2i$	$2i$	i	$-i$	$-i$	i	0
R_{24}	2	-2	2	-2	1	-1	1	-1	$-2i$	$2i$	i	$-i$	$-i$	i	0
R_{25}	2	-2	-2	2	-2	2	2	-2	0	0	0	0	0	0	0
R_{26}	2	-2	-2	2	-2	2	2	-2	0	0	0	0	0	0	0
R_{27}	2	-2	-2	2	1	-1	-1	1	0	0	$\sqrt{3}$	$-\sqrt{3}$	$\sqrt{3}$	$-\sqrt{3}$	0
R_{28}	2	-2	-2	2	1	-1	-1	1	0	0	$\sqrt{3}$	$-\sqrt{3}$	$\sqrt{3}$	$-\sqrt{3}$	0
R_{29}	2	-2	-2	2	1	-1	-1	1	0	0	$-\sqrt{3}$	$\sqrt{3}$	$-\sqrt{3}$	$\sqrt{3}$	0
R_{30}	2	-2	-2	2	1	-1	-1	1	0	0	$-\sqrt{3}$	$\sqrt{3}$	$-\sqrt{3}$	$\sqrt{3}$	0

[*continued on p. 276*]

	C_{16}	C_{17}	C_{18}	C_{19}	C_{20}	C_{21}	C_{22}	C_{23}	C_{24}	C_{25}	C_{26}	C_{27}	C_{28}	C_{29}	C_{30}
R_1	1	1	1	1	1	1	1	1	1	1	1	1	1	1	1
R_2	1	−1	−1	−1	−1	−1	−1	−1	−1	−1	−1	−1	−1	−1	−1
R_3	−1	1	1	1	1	1	1	1	1	1	1	1	1	−1	−1
R_4	−1	−1	−1	−1	−1	−1	−1	−1	−1	−1	−1	−1	−1	1	1
R_5	−1	1	1	1	1	1	1	−1	−1	−1	−1	−1	−1	1	−1
R_6	−1	−1	−1	−1	−1	−1	−1	1	1	1	1	1	1	−1	1
R_7	1	1	1	1	1	1	1	−1	−1	−1	−1	−1	−1	−1	1
R_8	1	−1	−1	−1	−1	−1	−1	1	1	1	1	1	1	1	−1
R_9	0	2	2	−1	−1	−1	−1	2	2	−1	−1	−1	−1	0	0
R_{10}	0	−2	−2	1	1	1	1	−2	−2	1	1	1	1	0	0
R_{11}	0	2	2	−1	−1	−1	−1	−2	−2	1	1	1	1	0	0
R_{12}	0	−2	−2	1	1	1	1	2	2	−1	−1	−1	−1	0	0
R_{13}	0	2	−2	2	2	−2	−2	0	0	0	0	0	0	0	0
R_{14}	0	−2	2	−2	−2	2	2	0	0	0	0	0	0	0	0
R_{15}	0	2	−2	−1	−1	1	1	0	0	$i\sqrt{3}$	$i\sqrt{3}$	$-i\sqrt{3}$	$-i\sqrt{3}$	0	0
R_{16}	0	−2	2	1	1	−1	−1	0	0	$-i\sqrt{3}$	$-i\sqrt{3}$	$i\sqrt{3}$	$i\sqrt{3}$	0	0
R_{17}	0	2	−2	−1	−1	1	1	0	0	$-i\sqrt{3}$	$-i\sqrt{3}$	$i\sqrt{3}$	$i\sqrt{3}$	0	0
R_{18}	0	−2	2	1	1	−1	−1	0	0	$i\sqrt{3}$	$i\sqrt{3}$	$-i\sqrt{3}$	$-i\sqrt{3}$	0	0
R_{19}	0	0	0	0	0	0	0	0	0	0	0	0	0	0	0
R_{20}	0	0	0	0	0	0	0	0	0	0	0	0	0	0	0
R_{21}	0	0	0	$\sqrt{3}$	$-\sqrt{3}$	$\sqrt{3}$	$-\sqrt{3}$	0	0	$i\sqrt{3}$	$-i\sqrt{3}$	$i\sqrt{3}$	$-i\sqrt{3}$	0	0
R_{22}	0	0	0	$-\sqrt{3}$	$\sqrt{3}$	$-\sqrt{3}$	$\sqrt{3}$	0	0	$-i\sqrt{3}$	$i\sqrt{3}$	$-i\sqrt{3}$	$i\sqrt{3}$	0	0
R_{23}	0	0	0	$\sqrt{3}$	$-\sqrt{3}$	$\sqrt{3}$	$-\sqrt{3}$	0	0	$-i\sqrt{3}$	$i\sqrt{3}$	$-i\sqrt{3}$	$i\sqrt{3}$	0	0
R_{24}	0	0	0	$-\sqrt{3}$	$\sqrt{3}$	$-\sqrt{3}$	$\sqrt{3}$	0	0	$i\sqrt{3}$	$-i\sqrt{3}$	$i\sqrt{3}$	$-i\sqrt{3}$	0	0
R_{25}	0	0	0	0	0	0	0	2	−2	2	−2	−2	2	0	0
R_{26}	0	0	0	0	0	0	0	−2	2	−2	2	2	−2	0	0
R_{27}	0	0	0	$-\sqrt{3}$	$\sqrt{3}$	$\sqrt{3}$	$-\sqrt{3}$	2	−2	−1	1	1	−1	0	0
R_{28}	0	0	0	$\sqrt{3}$	$-\sqrt{3}$	$-\sqrt{3}$	$\sqrt{3}$	−2	2	1	−1	−1	1	0	0
R_{29}	0	0	0	$\sqrt{3}$	$-\sqrt{3}$	$-\sqrt{3}$	$\sqrt{3}$	2	−2	−1	1	1	−1	0	0
R_{30}	0	0	0	$-\sqrt{3}$	$\sqrt{3}$	$\sqrt{3}$	$-\sqrt{3}$	−2	2	1	−1	−1	1	0	0

R_9: $P = \phi$; $Q = \pi^*$; $R = \varepsilon$; $S = \varepsilon$.
R_{10}: $P = \phi$; $Q = \pi^*$; $R = \varepsilon$; $S = -\varepsilon$.
R_{11}: $P = \phi$; $Q = \pi^*$; $R = -\varepsilon$; $S = \varepsilon$.
R_{12}: $P = \phi$; $Q = \pi^*$; $R = -\varepsilon$; $S = -\varepsilon$.
R_{13}: $P = i\phi$; $Q = \varepsilon$; $R = \lambda$; $S = \varepsilon$.
R_{14}: $P = i\phi$; $Q = \varepsilon$; $R = \lambda$; $S = -\varepsilon$.
R_{15}: $P = -i\phi$; $Q = \pi^*$; $R = \lambda$; $S = \varepsilon$.
R_{16}: $P = -i\phi$; $Q = \pi^*$; $R = \lambda$; $S = -\varepsilon$.
R_{17}: $P = i\phi$; $Q = \pi$; $R = \lambda$; $S = \varepsilon$.
R_{18}: $P = i\phi$; $Q = \pi$; $R = \lambda$; $S = -\varepsilon$.
R_{19}: $P = \phi$; $Q = -\varepsilon$; $R = i\varepsilon$; $S = i\lambda$.
R_{20}: $P = \phi$; $Q = -\varepsilon$; $R = -i\varepsilon$; $S = -i\lambda$.
R_{21}: $P = \phi$; $Q = -\pi^*$; $R = i\varepsilon$; $S = -i\lambda$.
R_{22}: $P = \phi$; $Q = -\pi^*$; $R = i\varepsilon$; $S = i\lambda$.
R_{23}: $P = \phi$; $Q = -\pi$; $R = -i\varepsilon$; $S = i\lambda$.
R_{24}: $P = \phi$; $Q = -\pi$; $R = -i\varepsilon$; $S = -i\lambda$.
R_{25}: $P = i\phi$; $Q = -\varepsilon$; $R = i\lambda$; $S = -i\lambda$.
R_{26}: $P = i\phi$; $Q = -\varepsilon$; $R = i\lambda$; $S = i\lambda$.
R_{27}: $P = i\phi$; $Q = -\pi$; $R = i\lambda$; $S = -i\lambda$.
R_{28}: $P = i\phi$; $Q = -\pi$; $R = i\lambda$; $S = i\lambda$.
R_{29}: $P = i\phi$; $Q = -\pi$; $R = -i\lambda$; $S = i\lambda$.
R_{30}: $P = i\phi$; $Q = -\pi$; $R = -i\lambda$; $S = -i\lambda$.

G_{96}^{10}

$P^{12} = E$; $Q^4 = E$; $R^2 = E$; $QP = P^{11}Q$; $RP = P^7Q^2R$; $RQ = QR$.
$C_1 = E$; $C_2 = P^6$; $C_3 = Q^2$; $C_4 = P^6Q^2$; $C_5 = P^3R, P^9R, P^3Q^2R, P^9Q^2R$; $C_6 = P^4, P^8$;
$C_7 = P^4Q^2, P^8Q^2$; $C_8 = P^2, P^{10}$; $C_9 = P^2Q^2, P^{10}Q^2$; $C_{10} = PR, P^5Q^2R, P^7Q^2R, P^{11}R$;
$C_{11} = PQ^2R, P^5R, P^7R, P^{11}Q^2R$;
$C_{12} = PQ, P^3Q, P^5Q, P^7Q, P^9Q, P^{11}Q, PQ^3, P^3Q^3, P^5Q^3, P^7Q^3, P^9Q^3, P^{11}Q^3$;
$C_{13} = QR, P^4QR, P^8QR, Q^3R, P^4Q^3R, P^8Q^3R$; $C_{14} = P^2QR, P^6QR, P^{10}QR, P^2Q^3R, P^6Q^3R, P^{10}Q^3R$;
$C_{15} = R, P^6Q^2R$; $C_{16} = Q^2R, P^6R$; $C_{17} = P^3, P^9, P^3Q^2, P^9Q^2$; $C_{18} = P^2R, P^4Q^2R, P^8Q^2R, P^{10}R$;
$C_{19} = P^2Q^2R, P^4R, P^8R, P^{10}Q^2R$; $C_{20} = P, P^5Q^2, P^7Q^2, P^{11}$; $C_{21} = PQ^2, P^5, P^7, P^{11}Q^2$;
$C_{22} = Q, P^2Q, P^4Q, P^6Q, P^8Q, P^{10}Q$; $C_{23} = Q^3, P^2Q^3, P^4Q^3, P^6Q^3, P^8Q^3, P^{10}Q^3$;
$C_{24} = PQR, P^3QR, P^5QR, P^7QR, P^9QR, P^{11}QR, PQ^3R, P^3Q^3R, P^5Q^3R, P^7Q^3R, P^9Q^3R, P^{11}Q^3R$.

	C_1	C_2	C_3	C_4	C_5	C_6	C_7	C_8	C_9	C_{10}	C_{11}	C_{12}
R_1	1	1	1	1	1	1	1	1	1	1	1	1
R_2	1	1	1	1	1	1	1	1	1	1	1	1
R_3	1	1	1	1	1	1	1	1	1	1	1	−1
R_4	1	1	1	1	1	1	1	1	1	1	1	−1
R_5	1	1	1	1	−1	1	1	1	1	−1	−1	1
R_6	1	1	1	1	−1	1	1	1	1	−1	−1	1
R_7	1	1	1	1	−1	1	1	1	1	−1	−1	−1
R_8	1	1	1	1	−1	1	1	1	1	−1	−1	−1
R_9	2	2	2	2	2	−1	−1	−1	−1	−1	−1	0
R_{10}	2	2	2	2	2	−1	−1	−1	−1	−1	−1	0
R_{11}	2	2	2	2	−2	−1	−1	−1	−1	1	1	0
R_{12}	2	2	2	2	−2	−1	−1	−1	−1	1	1	0
R_{13}	2	−2	2	−2	0	2	2	−2	−2	0	0	0
R_{14}	2	−2	2	−2	0	2	2	−2	−2	0	0	0
R_{15}	2	2	−2	−2	0	2	−2	2	−2	0	0	0
R_{16}	2	2	−2	−2	0	2	−2	2	−2	0	0	0
R_{17}	2	−2	−2	2	0	2	−2	−2	2	0	0	0
R_{18}	2	−2	−2	2	0	2	−2	−2	2	0	0	0
R_{19}	2	−2	−2	2	0	−1	1	1	−1	$\sqrt{3}$	$-\sqrt{3}$	0
R_{20}	2	−2	−2	2	0	−1	1	1	−1	$\sqrt{3}$	$-\sqrt{3}$	0
R_{21}	2	−2	−2	2	0	−1	1	1	−1	$-\sqrt{3}$	$\sqrt{3}$	0
R_{22}	2	−2	−2	2	0	−1	1	1	−1	$-\sqrt{3}$	$\sqrt{3}$	0
R_{23}	4	−4	4	−4	0	−2	−2	2	2	0	0	0
R_{24}	4	4	−4	−4	0	−2	2	−2	2	0	0	0

[*continued on p. 278*]

	C_{13}	C_{14}	C_{15}	C_{16}	C_{17}	C_{18}	C_{19}	C_{20}	C_{21}	C_{22}	C_{23}	C_{24}
R_1	1	1	1	1	1	1	1	1	1	1	1	1
R_2	1	1	−1	−1	−1	−1	−1	−1	−1	−1	−1	−1
R_3	−1	−1	1	1	1	1	1	1	1	−1	−1	−1
R_4	−1	−1	−1	−1	−1	−1	−1	−1	−1	1	1	1
R_5	−1	−1	1	1	−1	1	1	−1	−1	−1	−1	1
R_6	−1	−1	−1	−1	1	−1	−1	1	1	1	1	−1
R_7	1	1	1	1	−1	1	1	−1	−1	1	1	−1
R_8	1	1	−1	−1	1	−1	−1	1	1	−1	−1	1
R_9	0	0	2	2	2	−1	−1	−1	−1	0	0	0
R_{10}	0	0	−2	−2	−2	1	1	1	1	0	0	0
R_{11}	0	0	2	2	−2	−1	−1	1	1	0	0	0
R_{12}	0	0	−2	−2	2	1	1	−1	−1	0	0	0
R_{13}	2	−2	0	0	0	0	0	0	0	0	0	0
R_{14}	−2	2	0	0	0	0	0	0	0	0	0	0
R_{15}	0	0	0	0	0	0	0	0	0	2i	−2i	0
R_{16}	0	0	0	0	0	0	0	0	0	−2i	2i	0
R_{17}	0	0	2	−2	0	−2	2	0	0	0	0	0
R_{18}	0	0	−2	2	0	2	−2	0	0	0	0	0
R_{19}	0	0	2	−2	0	1	−1	$\sqrt{3}$	$-\sqrt{3}$	0	0	0
R_{20}	0	0	−2	2	0	−1	1	$-\sqrt{3}$	$\sqrt{3}$	0	0	0
R_{21}	0	0	2	−2	0	1	−1	$-\sqrt{3}$	$\sqrt{3}$	0	0	0
R_{22}	0	0	−2	2	0	−1	1	$\sqrt{3}$	$-\sqrt{3}$	0	0	0
R_{23}	0	0	0	0	0	0	0	0	0	0	0	0
R_{24}	0	0	0	0	0	0	0	0	0	0	0	0

R_9: $P = \pi$; $Q = \phi$; $R = \varepsilon$.
R_{10}: $P = -\pi^*$; $Q = \phi$; $R = -\varepsilon$.
R_{11}: $P = -\pi^*$; $Q = \phi$; $R = \varepsilon$.
R_{12}: $P = \pi$; $Q = \phi$; $R = -\varepsilon$.
R_{13}: $P = \mathrm{i}\lambda$; $Q = \phi$; $R = \phi$.
R_{14}: $P = \mathrm{i}\lambda$; $Q = \phi$; $R = -\phi$.
R_{15}: $P = \lambda$; $Q = \mathrm{i}\varepsilon$; $R = \phi$.
R_{16}: $P = \lambda$; $Q = -\mathrm{i}\varepsilon$; $R = \phi$.
R_{17}: $P = \mathrm{i}\lambda$; $Q = \mathrm{i}\phi$; $R = \varepsilon$.
R_{18}: $P = \mathrm{i}\lambda$; $Q = \mathrm{i}\phi$; $R = -\varepsilon$.
R_{19}: $P = -\mathrm{i}\lambda\pi$; $Q = \mathrm{i}\phi$; $R = \varepsilon$.
R_{20}: $P = \mathrm{i}\lambda\pi$; $Q = \mathrm{i}\phi$; $R = -\varepsilon$.
R_{21}: $P = \mathrm{i}\lambda\pi$; $Q = \mathrm{i}\phi$; $R = \varepsilon$.
R_{22}: $P = -\mathrm{i}\lambda\pi$; $Q = \mathrm{i}\phi$; $R = -\varepsilon$.

$$R_{23}:\ P = \left(\begin{array}{c|c} \rho & 0 \\ \hline 0 & -\rho^* \end{array}\right); \qquad Q = \left(\begin{array}{c|c} \phi & 0 \\ \hline 0 & \phi \end{array}\right); \qquad R = \left(\begin{array}{c|c} 0 & \phi \\ \hline \phi & 0 \end{array}\right).$$

$$R_{24}:\ P = \left(\begin{array}{c|c} \pi & 0 \\ \hline 0 & -\pi \end{array}\right); \qquad Q = \left(\begin{array}{c|c} \mathrm{i}\phi & 0 \\ \hline 0 & \mathrm{i}\phi \end{array}\right); \qquad R = \left(\begin{array}{c|c} 0 & \lambda \\ \hline \lambda & 0 \end{array}\right).$$

$\mathbf{G}^1_{192}$

$P^{12} = E$; $Q^4 = E$; $R^4 = E$; $S^8 = E$; $Q^2 = R^2$; $S^2 = P^6Q^2R$; $QP = P^7Q^2R$; $QP^3 = P^3Q$;
$RP = P^{10}QR$; $RQ = P^6Q^3R$; $SP = P^2RS$; $SQ = P^3Q^3RS$; $SR = RS$.
$C_1 = E$; $C_2 = P^3Q^2$; $C_3 = P^6$; $C_4 = P^9Q^2$; $C_5 = P^6R, Q^3, P^9Q^3R, Q^2R, P^6Q, P^3QR$;
$C_6 = P^9Q^2R, P^3Q, QR, P^3R, P^9Q^3, P^6Q^3R$; $C_7 = P^{11}Q, P^8Q^2, P^5Q^2R, P^2Q^3R, P^7R, P^4Q^2, P^4QR, PQ^3$;
$C_8 = P^2Q^3, P^{11}, P^8R, P^5QR, P^{10}Q^2R, P^7, P^7Q^3R, P^4Q$;
$C_9 = P^5Q, P^2Q^2, P^{11}Q^2R, P^8Q^3R, PR, P^{10}Q^2, P^{10}QR, P^7Q^3$;
$C_{10} = P^8Q^3, P^5, P^2R, P^{11}QR, P^4Q^2R, P, PQ^3R, P^{10}Q$; $C_{11} = S, P^8Q^3RS, P^{10}QRS, P^3Q^2RS, P^2Q^2S, PRS$;
$C_{12} = P^3Q^2S, P^{11}QRS, PQ^3RS, P^6RS, P^5S, P^4Q^2RS$; $C_{13} = P^6S, P^2Q^3RS, P^4QRS, P^9Q^2RS, P^8Q^2S, P^7RS$;
$C_{14} = P^9Q^2S, P^5QRS, P^7Q^3RS, RS, P^{11}S, P^{10}Q^2RS$;
$C_{15} = P^4Q^2S, P^7QS, P^{11}QS, P^9Q^3S, P^5Q^2RS, P^6Q^3RS, P^{10}S, PQ^3S, P^5Q^3S, P^3QS, P^{11}RS, QRS$;
$C_{16} = P^7S, P^{10}Q^3S, P^2Q^3S, QS, P^8RS, P^9QRS, PQ^2S, P^4QS, P^8QS, P^6Q^3S, P^2Q^2RS, P^3Q^3RS$;
$C_{17} = Q^2$; $C_{18} = P^3$; $C_{19} = P^6Q^2$; $C_{20} = P^9$; $C_{21} = P^6Q^2R, Q, P^9QR, R, P^6Q^3, P^3Q^3R$;
$C_{22} = P^9R, P^3Q^3, Q^3R, P^3Q^2R, P^9Q, P^6QR$; $C_{23} = P^{11}Q^3, P^8, P^5R, P^2QR, P^7Q^2R, P^4, P^4Q^3R, PQ$;
$C_{24} = P^2Q, P^{11}Q^2, P^8Q^2R, P^5Q^3R, P^{10}R, P^7Q^2, P^7QR, P^4Q^3$;
$C_{25} = P^5Q^3, P^2, P^{11}R, P^8QR, PQ^2R, P^{10}, P^{10}Q^3R, P^7Q$;
$C_{26} = P^8Q, P^5Q^2, P^2Q^2R, P^{11}Q^3R, P^4R, PQ^2, PQR, P^{10}Q^3$;
$C_{27} = Q^2S, P^8QRS, P^{10}Q^3RS, P^3RS, P^2S, PQ^2RS$; $C_{28} = P^3S, P^{11}Q^3RS, PQRS, P^6Q^2RS, P^5Q^2S, P^4RS$;
$C_{29} = P^6Q^2S, P^2QRS, P^4Q^3RS, P^9RS, P^8S, P^7Q^2RS$; $C_{30} = P^9S, P^5Q^3RS, P^7QRS, Q^2RS, P^{11}Q^2S, P^{10}RS$;
$C_{31} = P^4S, P^7Q^3S, P^{11}Q^3S, P^9QS, P^5RS, P^6QRS, P^{10}Q^2S, PQS, P^5QS, P^3Q^3S, P^{11}Q^2RS, Q^3RS$;
$C_{32} = P^7Q^2S, P^{10}QS, P^2QS, Q^3S, P^8Q^2RS, P^9Q^3RS, PS, P^4Q^3S, P^8Q^3S, P^6QS, P^2RS, P^3QRS$.

	C_1	C_2	C_3	C_4	C_5	C_6	C_7	C_8	C_9	C_{10}	C_{11}	C_{12}	C_{13}	C_{14}	C_{15}	C_{16}
R_1	1	1	1	1	1	1	1	1	1	1	1	1	1	1	1	1
R_2	1	1	1	1	1	1	1	1	1	1	−1	−1	−1	−1	−1	−1
R_3	1	−1	1	−1	−1	1	1	−1	1	−1	i	−i	i	−i	i	−i
R_4	1	−1	1	−1	−1	1	1	−1	1	−1	−i	i	−i	i	−i	i
R_5	2	2	2	2	2	2	−1	−1	−1	−1	0	0	0	0	0	0
R_6	2	−2	2	−2	−2	2	−1	1	−1	1	0	0	0	0	0	0
R_7	2	2i	−2	−2i	0	0	−1	−i	1	i	$\theta^*\sqrt{2}$	$\theta\sqrt{2}$	$-\theta^*\sqrt{2}$	$-\theta\sqrt{2}$	0	0
R_8	2	2i	−2	−2i	0	0	−1	−i	1	i	$-\theta^*\sqrt{2}$	$-\theta\sqrt{2}$	$\theta^*\sqrt{2}$	$\theta\sqrt{2}$	0	0
R_9	2	−2i	−2	2i	0	0	−1	i	1	−i	$\theta\sqrt{2}$	$\theta^*\sqrt{2}$	$-\theta\sqrt{2}$	$-\theta^*\sqrt{2}$	0	0
R_{10}	2	−2i	−2	2i	0	0	−1	i	1	−i	$-\theta\sqrt{2}$	$-\theta^*\sqrt{2}$	$\theta\sqrt{2}$	$\theta^*\sqrt{2}$	0	0
R_{11}	3	3	3	3	−1	−1	0	0	0	0	1	1	1	1	−1	−1
R_{12}	3	3	3	3	−1	−1	0	0	0	0	−1	−1	−1	−1	1	1
R_{13}	3	−3	3	−3	1	−1	0	0	0	0	i	−i	i	−i	−i	i
R_{14}	3	−3	3	−3	1	−1	0	0	0	0	−i	i	−i	i	i	−i
R_{15}	4	4i	−4	−4i	0	0	1	i	−1	−i	0	0	0	0	0	0
R_{16}	4	−4i	−4	4i	0	0	1	−i	−1	i	0	0	0	0	0	0
R_{17}	1	−i	−1	i	−i	−1	−1	i	1	−i	$-\theta$	$-\theta^*$	θ	θ^*	θ	θ^*
R_{18}	1	−i	−1	i	−i	−1	−1	i	1	−i	θ	θ^*	$-\theta$	$-\theta^*$	$-\theta$	$-\theta^*$
R_{19}	1	i	−1	−i	i	−1	−1	−i	1	i	$-\theta^*$	$-\theta$	θ^*	θ	θ^*	θ
R_{20}	1	i	−1	−i	i	−1	−1	−i	1	i	θ^*	θ	$-\theta^*$	$-\theta$	$-\theta^*$	$-\theta$
R_{21}	2	−2i	−2	2i	−2i	−2	1	−i	−1	i	0	0	0	0	0	0
R_{22}	2	2i	−2	−2i	2i	−2	1	i	−1	−i	0	0	0	0	0	0

[*continued on p. 280*]

	C_1	C_2	C_3	C_4	C_5	C_6	C_7	C_8	C_9	C_{10}	C_{11}	C_{12}	C_{13}	C_{14}	C_{15}	C_{16}
R_{23}	2	-2	2	-2	0	0	1	-1	1	-1	$-i\sqrt{2}$	$i\sqrt{2}$	$-i\sqrt{2}$	$i\sqrt{2}$	0	0
R_{24}	2	-2	2	-2	0	0	1	-1	1	-1	$i\sqrt{2}$	$-i\sqrt{2}$	$i\sqrt{2}$	$-i\sqrt{2}$	0	0
R_{25}	2	2	2	2	0	0	1	1	1	1	$-\sqrt{2}$	$-\sqrt{2}$	$-\sqrt{2}$	$-\sqrt{2}$	0	0
R_{26}	2	2	2	2	0	0	1	1	1	1	$\sqrt{2}$	$\sqrt{2}$	$\sqrt{2}$	$\sqrt{2}$	0	0
R_{27}	3	$-3i$	-3	$3i$	i	1	0	0	0	0	$-\theta$	$-\theta^*$	θ	θ^*	$-\theta$	$-\theta^*$
R_{28}	3	$-3i$	-3	$3i$	i	1	0	0	0	0	θ	θ^*	$-\theta$	$-\theta^*$	θ	θ^*
R_{29}	3	$3i$	-3	$-3i$	$-i$	1	0	0	0	0	$-\theta^*$	$-\theta$	θ^*	θ	$-\theta^*$	$-\theta$
R_{30}	3	$3i$	-3	$-3i$	$-i$	1	0	0	0	0	θ^*	θ	$-\theta^*$	$-\theta$	θ^*	θ
R_{31}	4	-4	4	-4	0	0	-1	1	-1	1	0	0	0	0	0	0
R_{32}	4	4	4	4	0	0	-1	-1	-1	-1	0	0	0	0	0	0

Note that for R_1 to R_{16} $\chi(C_{n+16}) = \chi(C_n)$ and for R_{17} to R_{32} $\chi(C_{n+16}) = -\chi(C_n)$.

R_5: $P = \pi$; $Q = \varepsilon$; $S = \phi$.
R_6: $P = -\pi$; $Q = -\varepsilon$; $S = \kappa$.
R_7: $P = -i\pi$; $Q = \chi$; $S = \sigma'$.
R_8: $P = -i\pi$; $Q = \chi$; $S = -\sigma'$.
R_9: $P = i\pi^*$; $Q = \chi^*$; $S = \tau'$.
R_{10}: $P = i\pi^*$; $Q = \chi^*$; $S = -\tau'$.
R_{11}: $P = A$; $Q = B$; $S = Y$.
R_{12}: $P = A$; $Q = B$; $S = -Y$.
R_{13}: $P = L$; $Q = N$; $S = iY$.
R_{14}: $P = L$; $Q = N$; $S = -iY$.

$$R_{15}\colon P = \left(\begin{array}{c|c} \frac{1-i}{4}\psi' & \sqrt{\left(\frac{3}{8}\right)}\psi' \\ \hline i\sqrt{\left(\frac{3}{8}\right)}\psi' & \frac{1-i}{4}\psi' \end{array}\right); \quad Q = \left(\begin{array}{c|c} \phi & 0 \\ \hline 0 & \phi \end{array}\right); \quad S = \left(\begin{array}{c|c} \nu^* & 0 \\ \hline 0 & -\nu^* \end{array}\right).$$

$$R_{16}\colon P = \left(\begin{array}{c|c} \frac{1+i}{4}\chi' & \sqrt{\left(\frac{3}{8}\right)}\chi' \\ \hline -i\sqrt{\left(\frac{3}{8}\right)}\chi' & \frac{1+i}{4}\chi' \end{array}\right); \quad Q = \left(\begin{array}{c|c} \phi & 0 \\ \hline 0 & \phi \end{array}\right); \quad S = \left(\begin{array}{c|c} \nu & 0 \\ \hline 0 & -\nu \end{array}\right).$$

R_{21}: $P = -i\pi$; $Q = i\varepsilon$; $S = \theta\phi$.
R_{22}: $P = i\pi$; $Q = -i\varepsilon$; $S = \theta\kappa$.
R_{23}: $P = \pi$; $Q = -i\chi$; $S = -i\theta\sigma'$.
R_{24}: $P = \pi$; $Q = -i\chi$; $S = i\theta\sigma'$.
R_{25}: $P = -\pi^*$; $Q = -i\chi^*$; $S = i\theta\tau'$.
R_{26}: $P = -\pi^*$; $Q = -i\chi^*$; $S = -i\theta\tau'$.
R_{27}: $P = -iA$; $Q = iB$; $S = -\theta Y$.
R_{28}: $P = -iA$; $Q = iB$; $S = \theta Y$.
R_{29}: $P = -iL$; $Q = iN$; $S = i\theta Y$.
R_{30}: $P = -iL$; $Q = iN$; $S = -i\theta Y$.

$$R_{31}\colon P = i\left(\begin{array}{c|c} \frac{1-i}{4}\psi' & \sqrt{\left(\frac{3}{8}\right)}\psi' \\ \hline i\sqrt{\left(\frac{3}{8}\right)}\psi' & \frac{1-i}{4}\psi' \end{array}\right); \quad Q = -i\left(\begin{array}{c|c} \phi & 0 \\ \hline 0 & \phi \end{array}\right); \quad S = \theta^*\left(\begin{array}{c|c} \nu^* & 0 \\ \hline 0 & -\nu^* \end{array}\right).$$

$$R_{32}\colon P = i\left(\begin{array}{c|c} \frac{1+i}{4}\chi' & \sqrt{\left(\frac{3}{8}\right)}\chi' \\ \hline -i\sqrt{\left(\frac{3}{8}\right)}\chi' & \frac{1+i}{4}\chi' \end{array}\right); \quad S = -i\left(\begin{array}{c|c} \phi & 0 \\ \hline 0 & \phi \end{array}\right); \quad S = \theta^*\left(\begin{array}{c|c} \nu & 0 \\ \hline 0 & -\nu \end{array}\right).$$

Note that $R = P^5Q^3P$.

$\mathbf{G}_{192}^{2}$

$P^8 = E$; $Q^4 = E$; $R^6 = E$; $S^2 = E$; $Q^2 = P^4$; $QP = P^7Q$; $RP = P^3R^5$; $RQ = P^6R$; $SP = PR^3S$; $SQ = QS$; $SR = RS$.

$C_1 = E$; $C_2 = P^4$; $C_3 = R^3$; $C_4 = P^4R^3$; $C_5 = Q, P^2Q, P^2, P^4Q, P^6Q, P^6$;
$C_6 = QR^3, P^2QR^3, P^2R^3, P^4QR^3, P^6QR^3, P^6R^3$; $C_7 = S, R^3S$; $C_8 = P^4S, P^4R^3S$;
$C_9 = QS, P^2QS, P^2S, P^4QS, P^6QS, P^6S, QR^3S, P^2QR^3S, P^2R^3S, P^4QR^3S, P^6QR^3S, P^6R^3S$;
$C_{10} = P^4R^4, P^4QR^4, P^2QR^4, P^2R^4, P^4R^2, P^6QR^2, P^6R^2, QR^2$;
$C_{11} = R^4, QR^4, P^6QR^4, P^6R^4, R^2, P^2QR^2, P^2R^2, P^4QR^2$;
$C_{12} = P^4R, P^4QR, P^2QR, P^2R, P^4R^5, P^6QR^5, P^6R^5, QR^5$;
$C_{13} = R, QR, P^6QR, P^6R, R^5, P^2QR^5, P^2R^5, P^4QR^5$;
$C_{14} = P, P^7, P^3R^4, P^3QR^4, P^7QR^2, P^5R^2, PR^3, P^7R^3, P^3R, P^3QR, P^7QR^5, P^5R^5$;
$C_{15} = P^5, P^3, P^7R^4, P^7QR^4, P^3QR^2, PR^2, P^5R^3, P^3R^3, P^7R, P^7QR, P^3QR^5, PR^5$;
$C_{16} = PQ, P^3Q, P^5R^4, P^5QR^4, P^7R^2, P^5QR^2, PQR^3, P^3QR^3, P^5R, P^5QR, P^7R^5, P^5QR^5, P^5Q, P^7Q, PR^4, PQR^4, P^3R^2, PQR^2, P^5QR^3, P^7QR^3, PR, PQR, P^3R^5, PQR^5$;
$C_{17} = P^4R^4S, P^4QR^4S, P^2QR^4S, P^2R^4S, P^4R^5S, P^6QR^5S, P^6R^5S, QR^5S$;
$C_{18} = R^4S, QR^4S, P^6QR^4S, P^6R^4S, R^5S, P^2QR^5S, P^2R^5S, P^4QR^5S$;
$C_{19} = P^4RS, P^4QRS, P^2QRS, P^2RS, P^4R^2S, P^6QR^2S, P^6R^2S, QR^2S$;
$C_{20} = RS, QRS, P^6QRS, P^6RS, R^2S, P^2QR^2S, P^2R^2S, P^4QR^2S$;
$C_{21} = PR^3S, P^7R^3S, P^3RS, P^3QRS, P^5R^5S, P^7QR^5S, PR^5S, P^7S, P^3R^4S, P^3QR^4S, P^5R^2S, P^7QR^2S$;
$C_{22} = P^5R^3S, P^3R^3S, P^7RS, P^7QRS, PR^5S, P^3QR^5S, P^5R^5S, P^3S, P^7R^4S, P^7QR^4S, PR^2S, P^3QR^2S$;
$C_{23} = PQR^3S, P^3QR^3S, P^5RS, P^5QRS, P^7R^5S, P^5QR^5S, PQS, P^3QS, P^5R^4S, P^5QR^4S, P^7R^2S, P^5QR^2S, P^5QR^3S, P^7QR^3S, PRS, PQRS, P^3R^5S, PQR^5S, P^5QS, P^7QS, PR^4S, P^2QR^4S, P^3R^2S, PQR^2S$.

	C_1	C_2	C_3	C_4	C_5	C_6	C_7	C_8	C_9	C_{10}	C_{11}	C_{12}
R_1	1	1	1	1	1	1	1	1	1	1	1	1
R_2	1	1	1	1	1	1	−1	−1	−1	1	1	1
R_3	1	1	1	1	1	1	1	1	1	1	1	1
R_4	1	1	1	1	1	1	−1	−1	−1	1	1	1
R_5	2	2	2	2	2	2	2	2	2	−1	−1	−1
R_6	2	2	2	2	2	2	−2	−2	−2	−1	−1	−1
R_7	2	2	−2	−2	2	−2	0	0	0	2	2	−2
R_8	2	2	−2	−2	2	−2	0	0	0	−1	−1	1
R_9	2	2	−2	−2	2	−2	0	0	0	−1	−1	1
R_{10}	2	−2	2	−2	0	0	2	−2	0	1	−1	1
R_{11}	2	−2	2	−2	0	0	−2	2	0	1	−1	1
R_{12}	2	−2	2	−2	0	0	2	−2	0	1	−1	1
R_{13}	2	−2	2	−2	0	0	−2	2	0	1	−1	1
R_{14}	3	3	3	3	−1	−1	3	3	−1	0	0	0
R_{15}	3	3	3	3	−1	−1	−3	−3	1	0	0	0
R_{16}	3	3	3	3	−1	−1	3	3	−1	0	0	0
R_{17}	3	3	3	3	−1	−1	−3	−3	1	0	0	0
R_{18}	4	−4	−4	4	0	0	0	0	0	2	−2	−2
R_{19}	4	−4	4	−4	0	0	4	−4	0	−1	1	−1
R_{20}	4	−4	4	−4	0	0	−4	4	0	−1	1	−1
R_{21}	4	−4	−4	4	0	0	0	0	0	−1	1	1
R_{22}	4	−4	−4	4	0	0	0	0	0	−1	1	1
R_{23}	6	6	−6	−6	−2	2	0	0	0	0	0	0

[*continued on p. 282*]

	C_{13}	C_{14}	C_{15}	C_{16}	C_{17}	C_{18}	C_{19}	C_{20}	C_{21}	C_{22}	C_{23}
R_1	1	1	1	1	1	1	1	1	1	1	1
R_2	1	1	1	1	−1	−1	−1	−1	−1	−1	−1
R_3	1	−1	−1	−1	1	1	1	1	−1	−1	−1
R_4	1	−1	−1	−1	−1	−1	−1	−1	1	1	1
R_5	−1	0	0	0	−1	−1	−1	−1	0	0	0
R_6	−1	0	0	0	1	1	1	1	0	0	0
R_7	−2	0	0	0	0	0	0	0	0	0	0
R_8	1	0	0	0	$i\sqrt{3}$	$i\sqrt{3}$	$-i\sqrt{3}$	$-i\sqrt{3}$	0	0	0
R_9	1	0	0	0	$-i\sqrt{3}$	$-i\sqrt{3}$	$i\sqrt{3}$	$i\sqrt{3}$	0	0	0
R_{10}	−1	$\sqrt{2}$	$-\sqrt{2}$	0	1	−1	1	−1	$\sqrt{2}$	$-\sqrt{2}$	0
R_{11}	−1	$\sqrt{2}$	$-\sqrt{2}$	0	−1	1	−1	1	$-\sqrt{2}$	$\sqrt{2}$	0
R_{12}	−1	$-\sqrt{2}$	$\sqrt{2}$	0	1	−1	1	−1	$-\sqrt{2}$	$\sqrt{2}$	0
R_{13}	−1	$-\sqrt{2}$	$\sqrt{2}$	0	−1	1	−1	1	$\sqrt{2}$	$-\sqrt{2}$	0
R_{14}	0	1	1	−1	0	0	0	0	1	1	−1
R_{15}	0	1	1	−1	0	0	0	0	−1	−1	1
R_{16}	0	−1	−1	1	0	0	0	0	−1	−1	1
R_{17}	0	−1	−1	1	0	0	0	0	1	1	−1
R_{18}	2	0	0	0	0	0	0	0	0	0	0
R_{19}	1	0	0	0	−1	1	−1	1	0	0	0
R_{20}	1	0	0	0	1	−1	1	−1	0	0	0
R_{21}	−1	0	0	0	$i\sqrt{3}$	$-i\sqrt{3}$	$-i\sqrt{3}$	$i\sqrt{3}$	0	0	0
R_{22}	−1	0	0	0	$-i\sqrt{3}$	$i\sqrt{3}$	$i\sqrt{3}$	$-i\sqrt{3}$	0	0	0
R_{23}	0	0	0	0	0	0	0	0	0	0	0

R_5: $P = \gamma'^*$; $Q = \varepsilon$; $R = \pi$; $S = \varepsilon$.
R_6: $P = \gamma'^*$; $Q = \varepsilon$; $R = \pi$; $S = -\varepsilon$.
R_7: $P = \phi$; $Q = \varepsilon$; $R = -\varepsilon$; $S = \lambda$.
R_8: $P = \gamma'^*$; $Q = \varepsilon$; $R = -\pi$; $S = \lambda$.
R_9: $P = \gamma'^*$; $Q = \varepsilon$; $R = -\pi$; $S = -\lambda$.
R_{10}: $P = -\xi'$; $Q = \kappa$; $R = \pi'$; $S = \varepsilon$.
R_{11}: $P = -\xi'$; $Q = \kappa$; $R = \pi'$; $S = -\varepsilon$.
R_{12}: $P = \xi'$; $Q = \kappa$; $R = \pi'$; $S = \varepsilon$.
R_{13}: $P = \xi'$; $Q = \kappa$; $R = \pi'$; $S = -\varepsilon$.
R_{14}: $P = B'$; $Q = B$; $R = C'$; $S = E$.
R_{15}: $P = B'$; $Q = B$; $R = C'$; $S = -E$.
R_{16}: $P = -B'$; $Q = B$; $R = C'$; $S = E$.
R_{17}: $P = -B'$; $Q = B$; $R = C'$; $S = -E$.

$$R_{18}\colon P = \left(\begin{array}{c|c} \zeta' & 0 \\ \hline 0 & \eta' \end{array}\right); \quad Q = \left(\begin{array}{c|c} i\phi & 0 \\ \hline 0 & i\phi \end{array}\right); \quad R = \left(\begin{array}{c|c} -\alpha'' & 0 \\ \hline 0 & -\alpha'' \end{array}\right); \quad S = \left(\begin{array}{c|c} 0 & \varepsilon \\ \hline -\varepsilon & 0 \end{array}\right).$$

$$R_{19}\colon P = \left(\begin{array}{c|c} 0 & \zeta' \\ \hline \zeta' & 0 \end{array}\right); \quad Q = \left(\begin{array}{c|c} i\phi & 0 \\ \hline 0 & i\phi \end{array}\right); \quad R = \left(\begin{array}{c|c} \beta'' & 0 \\ \hline 0 & \gamma'' \end{array}\right); \quad S = \left(\begin{array}{c|c} \varepsilon & 0 \\ \hline 0 & \varepsilon \end{array}\right).$$

$$R_{20}\colon P = \left(\begin{array}{c|c} 0 & \zeta' \\ \hline \zeta' & 0 \end{array}\right); \quad Q = \left(\begin{array}{c|c} i\phi & 0 \\ \hline 0 & i\phi \end{array}\right); \quad R = \left(\begin{array}{c|c} \beta'' & 0 \\ \hline 0 & \gamma'' \end{array}\right); \quad S = \left(\begin{array}{c|c} -\varepsilon & 0 \\ \hline 0 & -\varepsilon \end{array}\right).$$

$$R_{21}\colon P = \left(\begin{array}{c|c} 0 & \zeta' \\ \hline \zeta' & 0 \end{array}\right); \quad Q = \left(\begin{array}{c|c} i\phi & 0 \\ \hline 0 & i\phi \end{array}\right); \quad R = \left(\begin{array}{c|c} -\beta'' & 0 \\ \hline 0 & -\gamma'' \end{array}\right); \quad S = \left(\begin{array}{c|c} \varepsilon & 0 \\ \hline 0 & -\varepsilon \end{array}\right).$$

$$R_{22}\colon P = \left(\begin{array}{c|c} 0 & \zeta' \\ \hline \zeta' & 0 \end{array}\right); \quad Q = \left(\begin{array}{c|c} i\phi & 0 \\ \hline 0 & i\phi \end{array}\right); \quad R = \left(\begin{array}{c|c} -\beta'' & 0 \\ \hline 0 & -\gamma'' \end{array}\right); \quad S = \left(\begin{array}{c|c} -\varepsilon & 0 \\ \hline 0 & \varepsilon \end{array}\right).$$

R_{23}: $P = \mathscr{P}$; $Q = \mathscr{B}$; $R = \mathscr{Q}$; $S = \mathscr{F}$.

Notes to Table 5.1

(i) The name of the group has the following significance. $\mathbf{G}^n_{|\mathbf{G}|}$ is the nth group of order $|\mathbf{G}|$ in the table.

(ii) The names of the elements of the groups have no special significance except that E is always the identity; $P, Q, R, \ldots$, etc. are used for other elements. No concerted attempt is made to link a group with any of its subgroups so that, for example, an element labelled P in one group bears in general no relation to an element labelled P in another.

(iii) The reps and the classes are numbered and are put in such an order as to make the character tables of the groups look tidy. For example rep R_1 is always the identity rep, and class C_1 always contains the identity element alone.

(iv) Each group is completely characterized by certain generating relations put immediately under the name of the group and they serve amongst other things to distinguish groups of the same order one from another.

(v) The elements of a given class are tabulated underneath the defining relations in terms of the generators. Note that it has not always been convenient to use a minimum set of such generators; indeed, for some groups the algebraic manipulations involved in the use of a minimum set of generators are very laborious. We have always used sufficient generators to make the algebra relatively easy.

(vi) The character tables are given, as usual, as square arrays of complex numbers. Each row refers to a given rep R and each column to a given class C.

(vii) In order to save space the matrix representatives of the degenerate reps are given for generating elements only. These are listed just below the character table concerned. A key to the letters used for various matrices and for all complex numbers is given below.

(viii) An asterisk always denotes the complex conjugate.

(ix) The character tables of certain direct product groups are given in an abbreviated form explained in the table under group $\mathbf{G}^2_{16}$, where the method of abbreviation is first used.

Key to Table 5.1

(i) *Complex numbers*

'i' is, as usual, the square root of minus one, with argument equal to $\frac{1}{2}\pi$.

$$\omega = \exp(2\pi i/3) = -\tfrac{1}{2} + i\sqrt{(3)}/2.$$
$$\omega^* = \exp(-2\pi i/3) = -\tfrac{1}{2} - i\sqrt{(3)}/2.$$
$$\theta = \exp(2\pi i/8) = (1 + i)/\sqrt{2}.$$
$$\theta^* = \exp(-2\pi i/8) = (1 - i)/\sqrt{2}.$$
$$\sigma = \exp(2\pi i/16) = (1/2^{\frac{3}{4}})[\{\sqrt{(2)} + 1\}^{\frac{1}{2}} + i\{\sqrt{(2)} - 1\}^{\frac{1}{2}}].$$
$$\vartheta = \exp(2\pi i/24).$$
$$\tilde{\omega} = \sqrt{(2)} + i.$$
$$\tilde{\omega}^* = \sqrt{(2)} - i.$$
$$\tau = \sqrt{(3)} + i.$$
$$\tau^* = \sqrt{(3)} - i.$$

(ii) *Two-dimensional matrices*

$$\alpha = \begin{pmatrix} -\frac{1}{2} & -\frac{1}{2}\sqrt{3} \\ \frac{1}{2}\sqrt{3} & -\frac{1}{2} \end{pmatrix}; \quad \beta = \begin{pmatrix} \frac{1}{2} & -\frac{1}{2}\sqrt{3} \\ \frac{1}{2}\sqrt{3} & \frac{1}{2} \end{pmatrix}; \quad \gamma = \frac{-i}{\sqrt{6}}\begin{pmatrix} \sqrt{2} & \tau^* \\ \tau & -\sqrt{2} \end{pmatrix}; \quad \delta = \frac{1}{\sqrt{2}}\begin{pmatrix} 1 & 1 \\ -1 & 1 \end{pmatrix};$$

$$\varepsilon = \begin{pmatrix} 1 & 0 \\ 0 & 1 \end{pmatrix}; \quad \zeta = \frac{i}{\sqrt{6}}\begin{pmatrix} -\sqrt{2} & \tau \\ \tau^* & \sqrt{2} \end{pmatrix}; \quad \eta = \frac{1}{\sqrt{2}}\begin{pmatrix} -1 & 1 \\ -1 & -1 \end{pmatrix}; \quad \iota = \begin{pmatrix} -\frac{1}{2} & \frac{1}{2}\sqrt{3} \\ \frac{1}{2}\sqrt{3} & \frac{1}{2} \end{pmatrix};$$

$$\kappa = \begin{pmatrix} 0 & 1 \\ -1 & 0 \end{pmatrix}; \quad \lambda = \begin{pmatrix} 1 & 0 \\ 0 & -1 \end{pmatrix}; \quad \mu = \begin{pmatrix} 1 & 0 \\ 0 & \omega \end{pmatrix}; \quad \nu = \begin{pmatrix} 1 & 0 \\ 0 & i \end{pmatrix}; \quad \xi = \frac{1}{\sqrt{2}}\begin{pmatrix} i & 1 \\ 1 & i \end{pmatrix}; \quad \pi = \begin{pmatrix} \omega & 0 \\ 0 & \omega^* \end{pmatrix};$$

$$\rho = \begin{pmatrix} -i\omega & 0 \\ 0 & i\omega^* \end{pmatrix}; \quad \phi = \begin{pmatrix} 0 & 1 \\ 1 & 0 \end{pmatrix}; \quad \chi = \frac{1}{\sqrt{3}}\begin{pmatrix} -1 & -1+i \\ -1-i & 1 \end{pmatrix}; \quad \psi = \frac{1}{\sqrt{3}}\begin{pmatrix} i & -1+i \\ 1+i & -i \end{pmatrix};$$

$$\alpha' = \frac{1}{\sqrt{2}}\begin{pmatrix} -i & -1 \\ 1 & i \end{pmatrix}; \quad \beta' = \frac{1}{\sqrt{2}}\begin{pmatrix} i & -1 \\ 1 & -i \end{pmatrix}; \quad \gamma' = \begin{pmatrix} 0 & \omega \\ \omega^* & 0 \end{pmatrix}; \quad \delta' = \begin{pmatrix} 0 & \omega \\ -\omega^* & 0 \end{pmatrix}; \quad \zeta' = \begin{pmatrix} \theta^7 & 0 \\ 0 & \theta \end{pmatrix};$$

$$\eta' = \begin{pmatrix} \theta^3 & 0 \\ 0 & \theta^5 \end{pmatrix}; \quad \kappa' = \frac{1}{2}\begin{pmatrix} -1 & i\tilde{\omega}^* \\ i\tilde{\omega} & -1 \end{pmatrix}; \quad \lambda' = \frac{1}{4}\begin{pmatrix} 1 & -\tilde{\omega} \\ \tilde{\omega}^* & 1 \end{pmatrix}; \quad \mu' = \frac{1}{4}\begin{pmatrix} 1 & -\tilde{\omega}^* \\ \tilde{\omega} & 1 \end{pmatrix};$$

$$\nu' = \frac{\sqrt{3}}{4}\begin{pmatrix} \tilde{\omega} & 1 \\ -1 & \tilde{\omega}^* \end{pmatrix}; \quad \xi' = \frac{1}{\sqrt{2}}\begin{pmatrix} -1 & i \\ i & -1 \end{pmatrix}; \quad \pi' = \frac{1}{2}\begin{pmatrix} -1-i & 1+i \\ -1+i & -1+i \end{pmatrix}; \quad \rho' = \frac{\sqrt{3}}{4}\begin{pmatrix} \tilde{\omega}^* & 1 \\ -1 & \tilde{\omega} \end{pmatrix};$$

$$\sigma' = \frac{1}{3-\sqrt{3}}\begin{pmatrix} (2-\sqrt{3})-i & \omega(1-\sqrt{3}) \\ i\omega^*(1-\sqrt{3}) & 1-i(2-\sqrt{3}) \end{pmatrix}; \quad \tau' = \frac{1}{3-\sqrt{3}}\begin{pmatrix} (2-\sqrt{3})+i & \omega^*(1-\sqrt{3}) \\ -i\omega(1-\sqrt{3}) & 1+i(2-\sqrt{3}) \end{pmatrix};$$

$$\phi' = \frac{1}{2}\begin{pmatrix} i & \tilde{\omega}^* \\ -\tilde{\omega} & -i \end{pmatrix}; \quad \chi' = \begin{pmatrix} 1 & -i \\ 1 & i \end{pmatrix}; \quad \psi' = \begin{pmatrix} 1 & i \\ 1 & -i \end{pmatrix}; \quad \alpha'' = \frac{1}{\sqrt{2}}\begin{pmatrix} \theta^3 & \theta^3 \\ \theta & \theta^5 \end{pmatrix};$$

$$\beta'' = \frac{1}{\sqrt{2}}\begin{pmatrix} \vartheta^{17} & \vartheta^{17} \\ \vartheta^{11} & \vartheta^{23} \end{pmatrix}; \quad \gamma'' = \frac{1}{\sqrt{2}}\begin{pmatrix} \vartheta & \vartheta \\ \vartheta^{19} & \vartheta^{7} \end{pmatrix}$$

(iii) *Three-dimensional matrices*

Abbreviated symbols are used: for example $K = (\bar{c}\bar{a}b)$ stands for the matrix $\begin{pmatrix} 0 & 0 & -1 \\ -1 & 0 & 0 \\ 0 & 1 & 0 \end{pmatrix}$, the first letter of the symbol referring to the first row and so on.

$A = (cab)$; $B = (\bar{a}\bar{b}c)$; $C = (\bar{a}b\bar{c})$; $D = (bca)$; $E = (abc)$; $F = (ba\bar{c})$; $G = (a\bar{b}\bar{c})$; $H = (\bar{b}\bar{a}c)$; $I = (\bar{a}\bar{b}\bar{c})$;
$J = (\bar{b}\bar{c}\bar{a})$; $K = (\bar{c}\bar{a}b)$; $L = (\bar{c}\bar{a}\bar{b})$; $M = (\bar{a}bc)$; $N = (ab\bar{c})$; $U = (c\bar{b}a)$; $V = (\bar{c}b\bar{a})$; $W = (b\bar{a}\bar{c})$;
$X = (acb)$; $Y = (cb\bar{a})$; $Z = (a\bar{c}b)$; $A' = (bac)$; $B' = (\bar{c}ba)$; $C' = (\bar{b}ca)$.

(iv) *Four-dimensional matrices*

These are listed as 2-by-2 block matrices, each entry being a 2-by-2 matrix to be found in the key, (ii), above.

(v) *Six-dimensional matrices*

Abbreviated symbols are used: for example $\mathscr{A} = (ced\bar{a}\bar{f}b)$ stands for the matrix

$$\begin{pmatrix} 0 & 0 & 1 & 0 & 0 & 0 \\ 0 & 0 & 0 & 0 & 1 & 0 \\ 0 & 0 & 0 & 1 & 0 & 0 \\ -1 & 0 & 0 & 0 & 0 & 0 \\ 0 & 0 & 0 & 0 & 0 & -1 \\ 0 & 1 & 0 & 0 & 0 & 0 \end{pmatrix}$$

$\mathscr{A} = (ced\bar{a}\bar{f}b)$; $\mathscr{B} = (ab\bar{c}\bar{d}\bar{e}\bar{f})$; $\mathscr{C} = (\bar{a}\bar{b}cde\bar{f})$; $\mathscr{D} = (\bar{a}\bar{b}\bar{d}\bar{c}\bar{f}\bar{e})$; $\mathscr{F} = (bae\bar{f}c\bar{d})$; $\mathscr{G} = (\bar{c}\bar{a}\bar{b}\bar{f}d\bar{e})$; $\mathscr{H} = (\bar{a}bc\bar{d}ef)$;
$\mathscr{L} = (ab\bar{c}de\bar{f})$; $\mathscr{M} = (c\bar{b}af\bar{e}\bar{d})$; $\mathscr{N} = (defabc)$; $\mathscr{P} = (\bar{c}ead\bar{b}\bar{f})$; $\mathscr{Q} = (ced\bar{a}f\bar{b})$.

It is possible to identify each of the crystallographic point groups, whose character tables were given in Table 2.2, with one of these abstract groups (Belova, Belov, and Shubnikov 1948). For example, the abstract group $\mathbf{G}_8^4$ is isomorphic with the point group 422 (D_4). By inspection it is fairly easy to see which elements of the point group 422 (D_4) correspond to the generating elements of $\mathbf{G}_8^4$, particularly as the choice of P and Q is restricted by the matrices of the representation E, which are, from the key to Table 5.1,

$$P = \begin{pmatrix} 0 & 1 \\ -1 & 0 \end{pmatrix}, \quad Q = \begin{pmatrix} 1 & 0 \\ 0 & -1 \end{pmatrix}.$$

From Table 2.3 P must therefore be C_{4z}^- and Q must be C_{2x}, and the complete correspondence between the two groups is

	422 (D_4)	$\mathbf{G}_8^4$
C_1	E	E
C_2	C_{2z}	P^2
C_3	C_{4z}^-	P
	C_{4z}^+	P^3
C_4	C_{2x}	Q
	C_{2y}	P^2Q
C_5	C_{2b}	PQ
	C_{2a}	P^3Q

The identification of the elements has of course to be made with proper regard to the isomorphism between the abstract group and its realization as a point group, for example, $PQ = C_{4z}^-C_{2x} = C_{2b}$ (by using the Jones' symbols or Table 1.5).

The identification of each of the crystallographic point groups as a manifestation of one of the abstract groups of Table 5.1 is given in Table 5.2; the identification is not always uniquely determined and in some cases the matrix representatives differ from those given in Chapter 2 by a unitary transformation between the two groups.

5.2. The single-valued representations of the 230 space groups

In compiling tables of all the space-group reps we have used the theory that was described in detail in section 3.7 and illustrated in section 3.8 for the examples of the cubic close-packed structure ($Fm3m$, O_h^5) and the diamond structure ($Fd3m$, O_h^7). We wish to determine the reps of $\mathbf{G}^{\mathbf{k}}$, the little group of the wave vector $\mathbf{k}$ for all wave vectors in the basic domain of the appropriate Brillouin zone. $\mathbf{G}^{\mathbf{k}}$ is, of course, an infinite group. For a point of symmetry the problem simplifies to the determination of the reps of Herring's little group ${}^H\mathbf{G}^{\mathbf{k}} = \mathbf{G}^{\mathbf{k}}/\mathbf{T}^{\mathbf{k}}$, where $\mathbf{T}^{\mathbf{k}}$ is the group of all those symmetry operations, $\{E \mid \mathbf{t}_i\}$, of the Bravais lattice for which $\exp(-i\mathbf{k}.\mathbf{t}_i) = +1$.

TABLE 5.2

The identification of the point groups in terms of the abstract groups

Point group	Abstract group	Identification of generating elements
$1\ (C_1)$	$\mathbf{G}_1^1$	$P = E$
$\bar{1}\ (C_i)$	$\mathbf{G}_2^1$	$P = I$
$2\ (C_2)$	$\mathbf{G}_2^1$	$P = C_{2z}$
$m\ (C_{1h})$	$\mathbf{G}_2^1$	$P = \sigma_z$
$2/m\ (C_{2h})$	$\mathbf{G}_4^2$	$P = C_{2z}, Q = I$
$222\ (D_2)$	$\mathbf{G}_4^2$	$P = C_{2z}, Q = C_{2y}$
$mm2\ (C_{2v})$	$\mathbf{G}_4^2$	$P = C_{2z}, Q = \sigma_y$
$mmm\ (D_{2h})$	$\mathbf{G}_8^3$	$P = C_{2z}, Q = C_{2y}, R = I$
$4\ (C_4)$	$\mathbf{G}_4^1$	$P = C_{4z}^+$
$\bar{4}\ (S_4)$	$\mathbf{G}_4^1$	$P = S_{4z}^-$
$4/m\ (C_{4h})$	$\mathbf{G}_8^2$	$P = C_{4z}^+, Q = I$
$422\ (D_4)$	$\mathbf{G}_8^4$	$P = C_{4z}^+, Q = C_{2x}$
$4mm\ (C_{4v})$	$\mathbf{G}_8^4$	$P = C_{4z}^+, Q = \sigma_x$
$\bar{4}2m\ (D_{2d})$	$\mathbf{G}_8^4$	$P = S_{4z}^-, Q = C_{2x}$
$4/mmm\ (D_{4h})$	$\mathbf{G}_{16}^9$	$P = C_{4z}^+, Q = C_{2x}, R = I$
$3\ (C_3)$	$\mathbf{G}_3^1$	$P = C_3^+$
$\bar{3}\ (C_{3i})$	$\mathbf{G}_6^1$	$P = S_6^-$
$32\ (D_3)$	$\mathbf{G}_6^2$	$P = C_3^+, Q = C'_{21}$
$3m\ (C_{3v})$	$\mathbf{G}_6^2$	$P = C_3^+, Q = \sigma_{d1}$
$\bar{3}m\ (D_{3d})$	$\mathbf{G}_{12}^3$	$P = S_6^-, Q = C'_{21}$
$6\ (C_6)$	$\mathbf{G}_6^1$	$P = C_6^+$
$\bar{6}\ (C_{3h})$	$\mathbf{G}_6^1$	$P = S_3^-$
$6/m\ (C_{6h})$	$\mathbf{G}_{12}^2$	$P = C_3^+, Q = C_2, R = I$
$622\ (D_6)$	$\mathbf{G}_{12}^3$	$P = C_6^+, Q = C'_{21}$
$6mm\ (C_{6v})$	$\mathbf{G}_{12}^3$	$P = C_6^+, Q = \sigma_{d1}$
$\bar{6}2m\ (D_{3h})$	$\mathbf{G}_{12}^3$	$P = S_3^-, Q = C'_{21}$
$6/mmm\ (D_{6h})$	$\mathbf{G}_{24}^5$	$P = C_3^+, Q = C'_{21}, R = C_2, S = I$
$23\ (T)$	$\mathbf{G}_{12}^5$	$P = C_{31}^+, Q = C_{2x}, R = C_{2z}$
$m3\ (T_h)$	$\mathbf{G}_{24}^{10}$	$P = S_{61}^+, Q = C_{2x}, R = C_{2z}$
$432\ (O)$	$\mathbf{G}_{24}^7$	$P = C_{31}^+, Q = C_{2x}, R = C_{2z}, S = C_{2d}$
$\bar{4}3m\ (T_d)$	$\mathbf{G}_{24}^7$	$P = C_{31}^+, Q = C_{2x}, R = C_{2z}, S = \sigma_{dd}$
$m3m\ (O_h)$	$\mathbf{G}_{48}^7$	$P = S_{61}^+, Q = \sigma_x, R = \sigma_z, S = C_{2c}$

$\mathbf{T}^{\mathbf{k}}$ will be a subgroup of $\mathbf{T}$, the group of all the translational symmetry operations of the Bravais lattice. The elements of the group ${}^{H}\mathbf{G}^{\mathbf{k}}$ can be determined from Tables 3.6 and 3.7. The group multiplication table of ${}^{H}\mathbf{G}^{\mathbf{k}}$ can then be determined by direct multiplication of the Seitz space-group symbols, replacing $\{E \mid \mathbf{t}_i\}$ by $\{E \mid \mathbf{0}\}$ whenever $\exp(-i\mathbf{k}.\mathbf{t}_i) = +1$. We call ${}^{H}\mathbf{G}^{\mathbf{k}}$ Herring's little group since this method of dealing with the points of symmetry was first applied by Herring (1942) to determine the character tables for the space groups of the hexagonal close-packed and diamond structures. This group can then be identified with one of the abstract groups $\mathbf{G}_{|\mathbf{G}|}^n$ in Table 5.1 and the appropriate reps of $\mathbf{G}_{|\mathbf{G}|}^n$ that form small reps (that is, reps for which the elements of the translation group of the Bravais lattice satisfy eqn. (3.4.3)) can easily be identified. For lines of symmetry we have to obtain the reps of $\mathbf{G^k}$ by finding projective reps of $\bar{\mathbf{G}}^{\mathbf{k}*}$, the central extension. For such $\mathbf{k}$, $\mathbf{T/T^k}$ does not contain a finite number of elements, so that in an attempt to determine the small reps of $\mathbf{G^k}$ from the reps of a finite group we cannot profitably use $\mathbf{G^k/T^k}$, since this group is now infinite. Instead, to find a suitable finite group we have to use the central extension $\bar{\mathbf{G}}^{\mathbf{k}*}$. The group multiplication table of the elements of $\bar{\mathbf{G}}^{\mathbf{k}*}$ is determined by using the multiplication rule

$$(H_j, \alpha)(H_k, \beta) = (H_jH_k, \alpha + \beta + a(H_j, H_k)) \qquad (3.7.31)$$

with $a(H_j, H_k)$ given by eqns. (3.7.27) and (3.7.37). $\bar{\mathbf{G}}^{\mathbf{k}*}$ can then also be identified with one of the abstract groups $\mathbf{G}^n_{|\mathbf{G}|}$ in Table 5.1 and the reps of $\mathbf{G}^n_{|\mathbf{G}|}$ tested with eqn. (3.7.33) to determine which of them can be lifted by means of eqn. (3.7.9) into small reps of $\mathbf{G^k}$.

The above method has been applied separately *ab initio* for each point of symmetry and each line of symmetry in the basic domain of the Brillouin zone of each of the 230 space groups. This procedure was applied to both symmorphic and non-symmorphic space groups in the manner illustrated in section 3.8 for the two space groups $Fm3m$ (O_h^5) and $Fd3m$ (O_h^7). That is, the group-multiplication table of Herring's little group ${}^H\mathbf{G^k}$, or of the central extension $\bar{\mathbf{G}}^{\mathbf{k}*}$, has been determined separately each time and then identified with one of the abstract groups whose representations are given in Table 5.1. To go from the allowed reps of ${}^H\mathbf{G^k}$ or $\bar{\mathbf{G}}^{\mathbf{k}*}$ to the small reps of $\mathbf{G^k}$ is very straightforward; the small reps of ${}^H\mathbf{G^k}$ immediately become small reps of $\mathbf{G^k}$ by representing all elements of $\mathbf{T^k}$ by $\mathbf{1}$, and the allowed reps of $\bar{\mathbf{G}}^{\mathbf{k}*}$ become small reps of $\mathbf{G^k}$ once they have been lifted into $\mathbf{G^k}$ by means of eqn. (3.7.9). While the details of the algebra are often more complicated for non-symmorphic space groups, there is no essential difference between the basic theory for symmorphic and non-symmorphic space groups. Other authors have sometimes used slightly different methods to deduce these reps. For instance Zak (1960) was able to show that it was only necessary to determine *ab initio* the reps of $\mathbf{G^k}$ for the symmorphic space groups (see also Klauder and Gay 1968). This is a fairly easy process because ${}^H\mathbf{G^k}$ or $\bar{\mathbf{G}}^{\mathbf{k}*}$ for a symmorphic space group is isomorphic either with one of the 32 crystallographic point groups or with a direct product of $\mathbf{T}_n$ (an Abelian group of small order, n) with one of these point groups. The reps of ${}^H\mathbf{G^k}$ or $\bar{\mathbf{G}}^{\mathbf{k}*}$ for any non-symmorphic space group can then be found, from those of ${}^H\mathbf{G^k}$ or $\bar{\mathbf{G}}^{\mathbf{k}*}$ in some suitably chosen symmorphic space group, by making use of the theory of induced representations which we have described in Chapter 4. In section 4.5 we illustrated the use of the theory of induced representations to determine the reps of one of the point groups, 432 (O), from those of one of its invariant subgroups, 23 (T), see p. 199. This theory can also be applied to space groups to determine the reps of larger space groups from those of smaller space groups. It can be verified fairly easily, for example by inspection of Fig. 4.1, that every point group contains an invariant subgroup of index either 2 or 3. As a result, it is possible to show that every space group also contains an invariant subgroup of index either 2 or 3. As we saw in section 4.5, when a finite group G contains an invariant subgroup of prime index the theory of induced representations is quite easy to apply.

Any space group can be written in terms of left coset representatives of the translation group, **T**, of its Bravais lattice

$$\mathbf{G} = \{R_1 \mid \mathbf{v}_1\}\mathbf{T} + \{R_2 \mid \mathbf{v}_2\}\mathbf{T} + \cdots + \{R_h \mid \mathbf{v}_h\}\mathbf{T}. \tag{3.5.2}$$

In a symmorphic group we are able to choose the coset representatives $\{R_1 \mid \mathbf{v}_1\}$, $\{R_2 \mid \mathbf{v}_2\}, \ldots, \{R_h \mid \mathbf{v}_h\}$ in such a way that they are the elements of a point group, i.e. $\mathbf{v}_1, \mathbf{v}_2, \ldots, \mathbf{v}_h$ are all zero; however, for a non-symmorphic space group at least one of the $\mathbf{v}_j$ will have to be non-zero. The space-group reps of a symmorphic space group can be determined very easily; this was illustrated for $Fm3m$ (O_h^5) as the first example in section 3.8. Zak's technique for a given non-symmorphic space group, **G**, begins by finding an invariant subgroup **H**, of **G**, of index 2 or 3. Let us suppose that we are fortunate and that **H** is a symmorphic space group. This means that for a given ${}^H\mathbf{G}^{\mathbf{k}}$, for a point of symmetry, we obtain an invariant subgroup ${}^H\mathbf{H}^{\mathbf{k}}$ consisting only of space-group elements with translation vectors equal to zero. For a given $\overline{\mathbf{G}}^{\mathbf{k}*}$, for a line of symmetry, we obtain a group $\overline{\mathbf{H}}^{\mathbf{k}*}$ of elements (R_i, α_i) such that all the α_i are zero. The reps of ${}^H\mathbf{H}^{\mathbf{k}}$ or of $\overline{\mathbf{H}}^{\mathbf{k}*}$ can be determined very easily as was illustrated for $Fm3m$ (O_h^5) in section 3.8. Zak's method then consists of applying the method of induced representations, which we have described in Chapter 4, to determining the reps of ${}^H\mathbf{G}^{\mathbf{k}}$ or $\overline{\mathbf{G}}^{\mathbf{k}*}$ from the reps of ${}^H\mathbf{H}^{\mathbf{k}}$ or $\overline{\mathbf{H}}^{\mathbf{k}*}$ respectively. The character tables of symmorphic space groups can be obtained very simply, so that if a non-symmorphic space group, **G**, contains an invariant subgroup, **H**, of index 2 or 3 the problem of determining the reps of **G** can be solved immediately by this method. A very large number of non-symmorphic space groups do in fact contain an invariant symmorphic subgroup of index 2 or 3. To show how to use this method for the remaining non-symmorphic space groups when **H** is non-symmorphic, we note that all non-trivial space groups contain some invariant subgroup of index 2 or 3, even if they do not contain an invariant symmorphic subgroup of index 2 or 3. Therefore, in using Zak's method, if an invariant subgroup, **H**, of a non-symmorphic space group, **G**, is also non-symmorphic we have first to find the character tables of the non-symmorphic subgroup, **H**. This can easily be done if the non-symmorphic subgroup **H** has an invariant symmorphic subgroup of index 2 or 3 itself. If not, this process is repeated until an invariant subgroup of index 2 or 3 which is symmorphic is obtained.

We illustrate this method of determining the space-group reps by considering the example of $Pn3n$ (O_h^2) (Zak 1960). The space group $Pn3n$ (O_h^2) has an invariant subgroup of index 2, namely the space group $P432$ (O^1), which is a symmorphic space group (see Table 3.7). If P is the space-group element $\{I \mid \frac{1}{2}\frac{1}{2}\frac{1}{2}\}$ then the left coset representatives of $Pn3n$ (O_h^2) in an expansion of the form of eqn. (3.5.2) consist of all the elements of the point group 432 (O), associated with zero translations, together with the products of these elements with P. The Brillouin zone for the space group $Pn3n$ (O_h^2) was illustrated in Fig. 3.13 and it is re-drawn in Fig. 5.1. The elements of the little group of **k** at the points of symmetry can be found from Table 3.6 and they are given again in Table 5.3. The space-group representations at the points of symmetry can now be found. At Γ $^H\mathbf{G}^\mathbf{k}$ is isomorphic with the point group

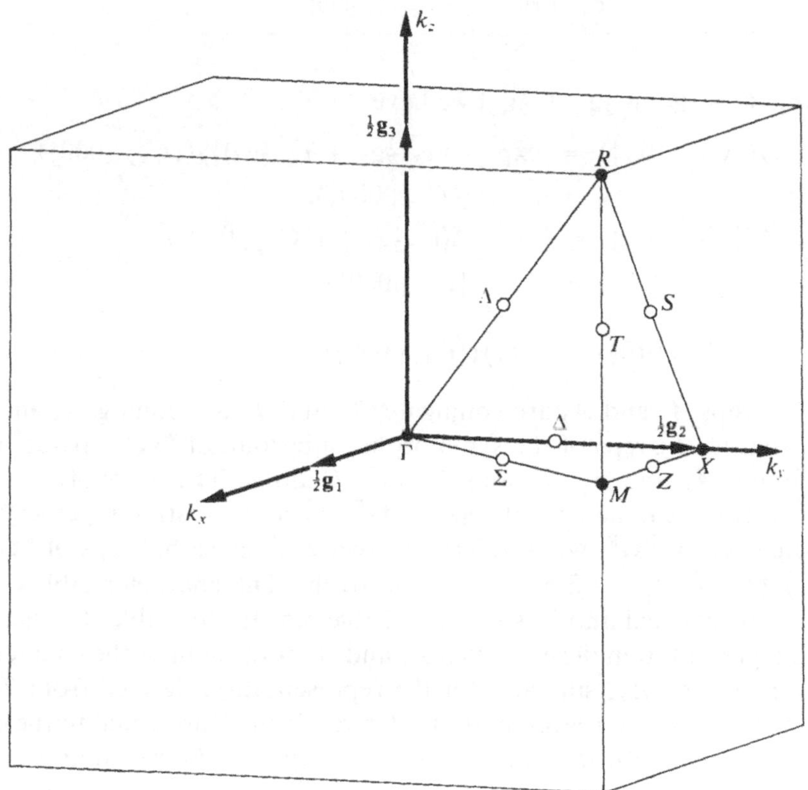

FIG. 5.1. The Brillouin zone for Γ_c. $\Gamma = (000)$; $X = (0\frac{1}{2}0)$; $M = (\frac{1}{2}\frac{1}{2}0)$; $R = (\frac{1}{2}\frac{1}{2}\frac{1}{2})$.

$m3m$ (O_h) and its representations can therefore be found directly from Chapter 2. At R the coset representatives of the symmorphic subgroup are of the form $\{R_i \mid \mathbf{0}\}$ where the rotational parts R_i make up the point group 432 (O) whose character table is given in Table 2.2; this is reproduced in Table 5.4. In order to determine which of the representations are conjugate and which are self-conjugate we proceed in a similar way to the examples considered in section 4.5. We examine the characters of

$\{R_i \mid 000\}$ and of $P^{-1}\{R_i \mid 000\}P$, i.e. of $\{I \mid \frac{1}{2}\frac{1}{2}\frac{1}{2}\}\{R_i \mid 000\}\{I \mid \frac{1}{2}\frac{1}{2}\frac{1}{2}\}$; for specimen elements of each class one obtains

$\{R_i \mid 000\}$	$\{I \mid \frac{1}{2}\frac{1}{2}\frac{1}{2}\}\{R_i \mid 000\}\{I \mid \frac{1}{2}\frac{1}{2}\frac{1}{2}\}$
$\{E \mid 000\}$	$\{E \mid 000\}$
$\{C_{31}^+ \mid 000\}$	$\{C_{31}^+ \mid 000\}$
$\{C_{2x} \mid 000\}$	$\{C_{2x} \mid 011\}$
$\{C_{2a} \mid 000\}$	$\{C_{2a} \mid 001\}$
$\{C_{4x}^+ \mid 000\}$	$\{C_{4x}^+ \mid 010\}$

At the point R $(=\frac{1}{2}\mathbf{g}_1 + \frac{1}{2}\mathbf{g}_2 + \frac{1}{2}\mathbf{g}_3)$ we have

$$\begin{aligned}\chi(\{C_{2x} \mid 011\}) &= \{\exp - \mathrm{i}(\mathbf{t}_2 . \tfrac{1}{2}\mathbf{g}_2 + \mathbf{t}_3 . \tfrac{1}{2}\mathbf{g}_3)\}\chi(\{C_{2x} \mid 000\}) \\ &= +1 . \chi(\{C_{2x} \mid 000\}) \end{aligned} \tag{5.2.1}$$

$$\begin{aligned}\chi(\{C_{2a} \mid 001\}) &= \{\exp - \mathrm{i}(\mathbf{t}_3 . \tfrac{1}{2}\mathbf{g}_3)\}\chi(\{C_{2a} \mid 000\}) \\ &= -1 . \chi(\{C_{2a} \mid 000\}) \end{aligned} \tag{5.2.2}$$

and

$$\chi(\{C_{4x}^+ \mid 010\}) = -1 . \chi(\{C_{4x}^+ \mid 000\}). \tag{5.2.3}$$

Therefore the reps A_1 and A_2 are conjugate, T_1 and T_2 are conjugate, and E is self-conjugate; that is, the reps of the symmorphic subgroup of ${}^H\mathbf{G}^R$ separate into three stars containing A_1 and A_2, T_1 and T_2, and E alone. Therefore A_1 and A_2 stick together to form a 2-dimensional rep of ${}^H\mathbf{G}^R$, T_1 and T_2 stick together to form a 6-dimensional rep of ${}^H\mathbf{G}^R$, while E leads to two 2-dimensional reps of ${}^H\mathbf{G}^R$ characterized by $\chi(\{I \mid \frac{1}{2}\frac{1}{2}\frac{1}{2}\}) = +2$ and -2 respectively. The character table of ${}^H\mathbf{G}^R$ can therefore be constructed and it is given in Table 5.5. In this table the character of a class of the representation derived from A_1 and A_2 is the sum of the characters of the same class in A_1 and A_2; similarly for the representation derived from T_1 and T_2. The characters of the two representations derived from E are equal to the characters of E for the elements of the symmorphic subgroup; for the remaining elements we have, from eqn. (4.5.45),

$$\chi(P\{R_i \mid \mathbf{0}\}) = \pm\chi(\{R_i \mid \mathbf{0}\}). \tag{5.2.4}$$

The character tables for the reps of the other special points of symmetry, X and M, can be found in a similar way; the results are given in Table 5.6.

Another example, that of the space group $P2_13$ (T^4), has also been considered by Zak (1960). This example is slightly more complicated because at some points of symmetry the method has to be used two, or even three, times to relate ${}^H\mathbf{G}^\mathbf{k}$ to a symmorphic subgroup. We refer the reader to Zak's original work for the details of this example.

TABLE 5.3

The elements of $\overline{\mathbf{G}}^{\mathbf{k}}$ at the points of symmetry in Pn3n (O_h^2)

Point	$\mathbf{k}$	Elements of $P(\mathbf{k})$
Γ	(000)	$E, 8C_{3j}^{\pm}, 3C_{2m}, 6C_{4m}^{\pm}, 6C_{2p}, I, 8S_{6j}^{\mp}, 3\sigma_m, 6S_{4m}^{\mp}, 6\sigma_{dp}$
X	$(0\frac{1}{2}0)$	$E, C_{2y}, C_{4y}^{\pm}, C_{2x}, C_{2z}, C_{2c}, C_{2e}, I, \sigma_y, S_{4y}^{\mp}, \sigma_x, \sigma_z, \sigma_{dc}, \sigma_{de}$
M	$(\frac{1}{2}\frac{1}{2}0)$	$E, C_{2z}, C_{4z}^{\pm}, C_{2x}, C_{2y}, C_{2a}, C_{2b}, I, \sigma_z, S_{4z}^{\mp}, \sigma_x, \sigma_y, \sigma_{da}, \sigma_{db}$
R	$(\frac{1}{2}\frac{1}{2}\frac{1}{2})$	$E, 8C_{3j}^{\pm}, 3C_{2m}, 6C_{4m}^{\pm}, 6C_{2p}, I, 8S_{6j}^{\mp}, 3\sigma_m, 6S_{4m}^{\mp}, 6\sigma_{dp}$

Notes to Table 5.3

(i) The points Γ, X, M, and R are shown in Fig. 5.1.

(ii) Column 2 gives the $\mathbf{k}$ vector in terms of $\mathbf{g}_1$, $\mathbf{g}_2$, and $\mathbf{g}_3$.

(iii) Column 3 identifies the elements in $\overline{\mathbf{G}}^{\mathbf{k}}$.

TABLE 5.4

Character table of ${}^H\mathbf{H}^R$ for P432 (O^1), the invariant symmorphic subgroup of Pn3n (O_h^2)

	$\{E\mid\mathbf{0}\}$	$\{C_{3j}^{\pm}\mid\mathbf{0}\}$	$\{C_{2m}\mid\mathbf{0}\}$	$\{C_{2p}\mid\mathbf{0}\}$	$\{C_{4m}^{\pm}\mid\mathbf{0}\}$	$\{E\mid\mathbf{t}\}$	$\{C_{3j}^{\pm}\mid\mathbf{t}\}$	$\{C_{2m}\mid\mathbf{t}\}$	$\{C_{2p}\mid\mathbf{t}\}$	$\{C_{4m}^{\pm}\mid\mathbf{t}\}$
A_1	1	1	1	1	1	−1	−1	−1	−1	−1
A_2	1	1	1	−1	−1	−1	−1	−1	1	1
E	2	−1	2	0	0	−2	1	−2	0	0
T_2	3	0	−1	1	−1	−3	0	1	−1	1
T_1	3	0	−1	−1	1	−3	0	1	1	−1

Note to Table 5.4

(i) $m = x, y, z; j = 1, 2, 3, 4; p = a, b, c, d, e, f; \mathbf{t} = \mathbf{t}_1$ or $\mathbf{t}_2$ or $\mathbf{t}_3$.

TABLE 5.5

Character table of ${}^H\mathbf{G}^R$ for Pn3n (O_h^2)

$\{E\mid\mathbf{0}\}$	$\{E\mid\mathbf{t}\}$	$\{C_{2m}\mid\mathbf{0}\}$	$\{C_{2m}\mid\mathbf{t}\}$	$\{I\mid\boldsymbol{\tau}\}$ $\{I\mid\mathbf{t}+\boldsymbol{\tau}\}$	$\{\sigma_m\mid\boldsymbol{\tau}\}$ $\{\sigma_m\mid\mathbf{t}+\boldsymbol{\tau}\}$	$\{C_{3j}^{\pm}\mid\mathbf{0}\}$	
2	−2	2	−2	0	0	2	R_7
2	−2	2	−2	0	0	−1	R_8
2	−2	2	−2	0	0	−1	R_9
6	−6	−2	2	0	0	0	R_{14}
C_1	C_2	C_3	C_4	C_5	C_6	C_7	

$\{C_{3j}^{\pm}\mid\mathbf{t}\}$	$\{C_{4m}^{\pm}\mid\mathbf{0}\}$ $\{C_{4m}^{\pm}\mid\mathbf{t}\}$	$\{C_{2p}\mid\mathbf{0}\}$ $\{C_{2p}\mid\mathbf{t}\}$	$\{S_{6j}^{+}\mid\boldsymbol{\tau}\}$ $\{S_{6j}^{-}\mid\mathbf{t}+\boldsymbol{\tau}\}$	$\{S_{6j}^{-}\mid\boldsymbol{\tau}\}$ $\{S_{6j}^{+}\mid\mathbf{t}+\boldsymbol{\tau}\}$	$\{S_{4m}^{\pm}\mid\boldsymbol{\tau}\}$ $\{S_{4m}^{\pm}\mid\mathbf{t}+\boldsymbol{\tau}\}$	$\{\sigma_{dp}\mid\boldsymbol{\tau}\}$ $\{\sigma_{dp}\mid\mathbf{t}+\boldsymbol{\tau}\}$	
−2	0	0	0	0	0	0	R_7
1	0	0	$i\sqrt{3}$	$-i\sqrt{3}$	0	0	R_8
1	0	0	$-i\sqrt{3}$	$i\sqrt{3}$	0	0	R_9
0	0	0	0	0	0	0	R_{14}
C_8	C_9	C_{10}	C_{11}	C_{12}	C_{13}	C_{14}	

Notes to Table 5.5

(i) $m = x, y, z; j = 1, 2, 3, 4; p = a, b, c, d, e, f; \boldsymbol{\tau} = \frac{1}{2}\mathbf{t}_1 + \frac{1}{2}\mathbf{t}_2 + \frac{1}{2}\mathbf{t}_3; \mathbf{t} = \mathbf{t}_1$ or $\mathbf{t}_2$ or $\mathbf{t}_3$.

(ii) The last row of the table identifies the classes in terms of those of the abstract group $\mathbf{G}_{96}^2$ and the last column identifies the reps in terms of those of $\mathbf{G}_{96}^2$ (see Tables 5.1 and 5.7).

TABLE 5.6

The character tables for ${}^{H}\mathbf{G}^{X}$ and ${}^{H}\mathbf{G}^{M}$ of Pn3n (O_h^2)

(a) X

C_1	C_2	C_3	C_4	C_5	C_6	C_7	C_8	C_9	C_{10}	C_{11}	C_{12}	C_{13}	C_{14}	
2	−2	2	−2	2	−2	0	0	0	0	0	0	0	0	R_{10}
2	−2	2	−2	−2	2	0	0	0	0	0	0	0	0	R_{11}
2	−2	−2	2	0	0	2i	−2i	0	0	0	0	0	0	R_{12}
2	−2	−2	2	0	0	−2i	2i	0	0	0	0	0	0	R_{13}

C_1: $\{E \mid 000\}$

C_2: $\{E \mid 010\}$

C_3: $\{C_{2y} \mid 000\}$

C_4: $\{C_{2y} \mid 010\}$

C_5: $\{C_{4y}^- \mid 000\}, \{C_{4y}^+ \mid 000\}$

C_6: $\{C_{4y}^- \mid 010\}, \{C_{4y}^+ \mid 010\}$

C_7: $\{S_{4y}^+ \mid \tfrac{1}{2}\tfrac{1}{2}\tfrac{1}{2}\}, \{S_{4y}^- \mid \tfrac{1}{2}\tfrac{\bar{1}}{2}\tfrac{1}{2}\}$

C_8: $\{S_{4y}^+ \mid \tfrac{1}{2}\tfrac{\bar{1}}{2}\tfrac{1}{2}\}, \{S_{4y}^- \mid \tfrac{1}{2}\tfrac{1}{2}\tfrac{1}{2}\}$

C_9: $\{I \mid \tfrac{1}{2}\tfrac{1}{2}\tfrac{1}{2}\}, \{I \mid \tfrac{1}{2}\tfrac{\bar{1}}{2}\tfrac{1}{2}\}$

C_{10}: $\{\sigma_y \mid \tfrac{1}{2}\tfrac{1}{2}\tfrac{1}{2}\}, \{\sigma_y \mid \tfrac{1}{2}\tfrac{\bar{1}}{2}\tfrac{1}{2}\}$

C_{11}: $\{C_{2x} \mid 000\}, \{C_{2x} \mid 010\}, \{C_{2z} \mid 000\}, \{C_{2z} \mid 010\}$

C_{12}: $\{C_{2e} \mid 000\}, \{C_{2c} \mid 000\}, \{C_{2e} \mid 010\}, \{C_{2c} \mid 010\}$

C_{13}: $\{\sigma_x \mid \tfrac{1}{2}\tfrac{1}{2}\tfrac{1}{2}\}, \{\sigma_x \mid \tfrac{1}{2}\tfrac{\bar{1}}{2}\tfrac{1}{2}\}, \{\sigma_z \mid \tfrac{1}{2}\tfrac{1}{2}\tfrac{1}{2}\}, \{\sigma_z \mid \tfrac{1}{2}\tfrac{\bar{1}}{2}\tfrac{1}{2}\}$

C_{14}: $\{\sigma_{de} \mid \tfrac{1}{2}\tfrac{\bar{1}}{2}\tfrac{1}{2}\}, \{\sigma_{de} \mid \tfrac{1}{2}\tfrac{1}{2}\tfrac{1}{2}\}, \{\sigma_{dc} \mid \tfrac{1}{2}\tfrac{\bar{1}}{2}\tfrac{1}{2}\}, \{\sigma_{dc} \mid \tfrac{1}{2}\tfrac{1}{2}\tfrac{1}{2}\}$

(b) M

C_1	C_2	C_3	C_4	C_5	C_6	C_7	C_8	C_9	C_{10}	C_{11}	C_{12}	C_{13}	C_{14}	
2	−2	2	−2	2	−2	0	0	0	0	0	0	0	0	R_9
2	−2	−2	2	0	0	0	0	2	−2	0	0	0	0	R_{11}
2	−2	2	−2	−2	2	0	0	0	0	0	0	0	0	R_{12}
2	−2	−2	2	0	0	0	0	−2	2	0	0	0	0	R_{14}

C_1: $\{E \mid 000\}$

C_2: $\{E \mid 010\}$

C_3: $\{C_{2z} \mid 000\}$

C_4: $\{C_{2z} \mid 010\}$

C_5: $\{C_{2a} \mid 000\}, \{C_{2b} \mid 000\}$

C_6: $\{C_{2a} \mid 010\}, \{C_{2b} \mid 010\}$

C_7: $\{I \mid \tfrac{1}{2}\tfrac{1}{2}\tfrac{1}{2}\}, \{I \mid \tfrac{1}{2}\tfrac{\bar{1}}{2}\tfrac{1}{2}\}$

C_8: $\{\sigma_z \mid \tfrac{1}{2}\tfrac{\bar{1}}{2}\tfrac{1}{2}\}, \{\sigma_z \mid \tfrac{1}{2}\tfrac{1}{2}\tfrac{1}{2}\}$

C_9: $\{\sigma_{db} \mid \tfrac{1}{2}\tfrac{\bar{1}}{2}\tfrac{1}{2}\}, \{\sigma_{da} \mid \tfrac{1}{2}\tfrac{1}{2}\tfrac{1}{2}\}$

C_{10}: $\{\sigma_{db} \mid \tfrac{1}{2}\tfrac{1}{2}\tfrac{1}{2}\}, \{\sigma_{da} \mid \tfrac{1}{2}\tfrac{\bar{1}}{2}\tfrac{1}{2}\}$

C_{11}: $\{C_{2x} \mid 000\}, \{C_{2y} \mid 000\}, \{C_{2x} \mid 010\}, \{C_{2y} \mid 010\}$

C_{12}: $\{C_{4z}^- \mid 000\}, \{C_{4z}^+ \mid 000\}, \{C_{4z}^- \mid 010\}, \{C_{4z}^+ \mid 010\}$

C_{13}: $\{\sigma_x \mid \tfrac{1}{2}\tfrac{1}{2}\tfrac{1}{2}\}, \{\sigma_y \mid \tfrac{1}{2}\tfrac{1}{2}\tfrac{1}{2}\}, \{\sigma_x \mid \tfrac{1}{2}\tfrac{\bar{1}}{2}\tfrac{1}{2}\}, \{\sigma_y \mid \tfrac{1}{2}\tfrac{\bar{1}}{2}\tfrac{1}{2}\}$

C_{14}: $\{S_{4z}^+ \mid \tfrac{1}{2}\tfrac{\bar{1}}{2}\tfrac{1}{2}\}, \{S_{4z}^- \mid \tfrac{1}{2}\tfrac{\bar{1}}{2}\tfrac{1}{2}\}, \{S_{4z}^+ \mid \tfrac{1}{2}\tfrac{1}{2}\tfrac{1}{2}\}, \{S_{4z}^- \mid \tfrac{1}{2}\tfrac{1}{2}\tfrac{1}{2}\}$

Notes to Table 5.6

(i) The translational parts of the space-group elements are given in terms of $\mathbf{t}_1$, $\mathbf{t}_2$, and $\mathbf{t}_3$ (see Table 3.1).

(ii) In (a) the numbering of the classes of ${}^{H}\mathbf{G}^{X}$ is that of the abstract group $\mathbf{G}_{32}^{1}$ and the last column identifies the reps in terms of those of $\mathbf{G}_{32}^{1}$ (see Tables 5.1 and 5.7).

(iii) In (b) the numbering of the classes of ${}^{H}\mathbf{G}^{M}$ is that of the abstract group $\mathbf{G}_{32}^{2}$ and the last column identifies the reps in terms of those of $\mathbf{G}_{32}^{2}$ (see Tables 5.1 and 5.7).

In Table 5.7 we list for each point of symmetry and line of symmetry in the basic domain of each of the 230 space groups the identification of ${}^{H}\mathbf{G}^{\mathbf{k}}$ or $\overline{\mathbf{G}}^{\mathbf{k}*}$ in terms of one of the abstract groups $\mathbf{G}^{n}_{|G|}$ of Table 5.1, together with the elements that are to be taken as the generating elements of $\mathbf{G}^{n}_{|G|}$. The rules for constructing the small reps of $\mathbf{G}^{\mathbf{k}}$ for those wave vectors, $\mathbf{k}$, that are in the representation domain but not in the basic domain will be given in section 5.5. The space groups are listed in the order given in Volume 1 of the *International tables for X-ray crystallography* (Henry and Lonsdale 1965) which was also the order we used in Table 3.7. For each space group we also give references to the work of other authors on the determination of the reps of that space group. In some cases ${}^{H}\mathbf{G}^{\mathbf{k}}$ is a direct product of a smaller group with an Abelian group $\mathbf{T}_s$, of order s, whose elements are pure translation operations such as $\{E \mid \mathbf{t}_i\}$, $\{E \mid \mathbf{t}_i\}^2, \ldots, \{E \mid \mathbf{t}_i\}^{s-1}$, where $\{E \mid \mathbf{t}_i\}^s$ is equivalent to $\{E \mid \mathbf{0}\}$, i.e., $\exp\{-i\mathbf{k}.(s\mathbf{t}_i)\} = +1$. Not all the reps of the abstract group $\mathbf{G}^{n}_{|G|}$ are necessarily allowable as small reps of $\mathbf{G}^{\mathbf{k}}$. This is because the representations of ${}^{H}\mathbf{G}^{\mathbf{k}}, = \mathbf{G}^{\mathbf{k}}/\mathbf{T}^{\mathbf{k}}$, must be compatible with the representations of the translational subgroup $\mathbf{T}$. Therefore, if ${}^{H}\mathbf{G}^{\mathbf{k}}$ contains any elements that are pure translations of the Bravais lattice of the space group only those reps of $\mathbf{G}^{n}_{|G|}$ which satisfy eqn. (3.4.3)

$$\Delta^{\mathbf{k}}[\{E \mid n_1\mathbf{t}_1 + n_2\mathbf{t}_2 + n_3\mathbf{t}_3\}] = \exp\{-i\mathbf{k}.(n_1\mathbf{t}_1 + n_2\mathbf{t}_2 + n_3\mathbf{t}_3)\}. \quad (3.4.3)$$

are allowed as small reps. This means that, sometimes, in different manifestations of the same abstract group a different set of its irreducible representations may actually be allowed as small reps. For example, when $\mathbf{G}^{1}_{4}$ appears at Γ of $P4$ (C^{1}_{4}) all four reps R_1, R_2, R_3, and R_4 are allowable, but when $\mathbf{G}^{1}_{4}$ appears at Z of $P2_1$ (C^{2}_{2}) only R_2 and R_4 are allowable. This is because at Z the group ${}^{H}\mathbf{G}^{\mathbf{k}}$ contains the element $P^2 = \{C_{2z} \mid \frac{1}{2}\mathbf{t}_3\}^2 = \{E \mid \mathbf{t}_3\}$ (from Table 3.2) which must be represented by $\exp\{-i\mathbf{k}.\mathbf{t}_3\} = -1$ (from Table 3.6) and therefore only the representations in which $\chi(P^2) = -1$ are compatible with the representations of the translational subgroup of the Bravais lattice of the crystal. In Table 5.7 we list only those reps of $\mathbf{G}^{n}_{|G|}$ that do satisfy eqn. (3.4.3).

TABLE 5.7

The single-valued reps of the 230 space groups

1 $P1$ C^{1}_{1}

($F1$; $K7$; $M5$; $Z1$.)

Γ $\mathbf{G}^{1}_{1}$: $\{E \mid 000\}$: 1, 1: a.
B $\mathbf{G}^{1}_{1} \otimes \mathbf{T}_2$: $\{E \mid 000\}$; $\mathbf{t}_1$: 1, 1: a.
F $\mathbf{G}^{1}_{1} \otimes \mathbf{T}_2$: $\{E \mid 000\}$; $\mathbf{t}_2$: 1, 1: a.
G $\mathbf{G}^{1}_{1} \otimes \mathbf{T}_2$: $\{E \mid 000\}$; $\mathbf{t}_3$: 1, 1: a.

2 $P\bar{1}$ C_i^1

($F1$; $K7$; $M5$; $Z1$.)

Γ $\mathbf{G}_2^1$: $\{I\,|\,000\}$: 1, 1; 2, 1: a.
B $\mathbf{G}_2^1 \otimes \mathbf{T}_2$: $\{I\,|\,000\}$; $\mathbf{t}_1$: 1, 1; 2, 1: a.
F $\mathbf{G}_2^1 \otimes \mathbf{T}_2$: $\{I\,|\,000\}$; $\mathbf{t}_2$: 1, 1; 2, 1: a.
G $\mathbf{G}_2^1 \otimes \mathbf{T}_2$: $\{I\,|\,000\}$; $\mathbf{t}_3$: 1, 1; 2, 1: a.

3 $P2$ C_2^1

($F1$; $K7$; $M5$; $S15$; $Z1$.)

Γ $\mathbf{G}_2^1$: $\{C_{2z}\,|\,000\}$: 1, 1; 2, 1: b.
B $\mathbf{G}_2^1 \otimes \mathbf{T}_2$: $\{C_{2z}\,|\,000\}$; $\mathbf{t}_1$: 1, 1; 2, 1: b.
Y $\mathbf{G}_2^1 \otimes \mathbf{T}_2$: $\{C_{2z}\,|\,000\}$; $\mathbf{t}_2$: 1, 1; 2, 1: b.
Z $\mathbf{G}_2^1 \otimes \mathbf{T}_2$: $\{C_{2z}\,|\,000\}$; $\mathbf{t}_3$: 1, 1; 2, 1: b.
C $\mathbf{G}_2^1 \otimes \mathbf{T}_2$: $\{C_{2z}\,|\,000\}$; $\mathbf{t}_2$ or $\mathbf{t}_3$: 1, 1; 2, 1: b.
D $\mathbf{G}_2^1 \otimes \mathbf{T}_2$: $\{C_{2z}\,|\,000\}$; $\mathbf{t}_1$ or $\mathbf{t}_3$: 1, 1; 2, 1: b.
A $\mathbf{G}_2^1 \otimes \mathbf{T}_2$: $\{C_{2z}\,|\,000\}$; $\mathbf{t}_1$ or $\mathbf{t}_2$: 1, 1; 2, 1: b.
E $\mathbf{G}_2^1 \otimes \mathbf{T}_2$: $\{C_{2z}\,|\,000\}$; $\mathbf{t}_1$ or $\mathbf{t}_2$ or $\mathbf{t}_3$: 1, 1; 2, 1: b.

Λ^x $\mathbf{G}_2^1$: $(C_{2z}, 0)$: 1, x; 2, x: b.
V^x $\mathbf{G}_2^1$: $(C_{2z}, 0)$: 1, x; 2, x: b.
W^x $\mathbf{G}_2^1$: $(C_{2z}, 0)$: 1, x; 2, x: b.
U^x $\mathbf{G}_2^1$: $(C_{2z}, 0)$: 1, x; 2, x: b.

4 $P2_1$ C_2^2

($F1$; $K7$; $M5$; $S15$; $Z1$.)

Γ $\mathbf{G}_2^1$: $\{C_{2z}\,|\,00\tfrac{1}{2}\}$: 1, 1; 2, 1: b.
B $\mathbf{G}_2^1 \otimes \mathbf{T}_2$: $\{C_{2z}\,|\,00\tfrac{1}{2}\}$; $\mathbf{t}_1$: 1, 1; 2, 1: b.
Y $\mathbf{G}_2^1 \otimes \mathbf{T}_2$: $\{C_{2z}\,|\,00\tfrac{1}{2}\}$; $\mathbf{t}_2$: 1, 1; 2, 1: b.
Z $\mathbf{G}_4^1$: $\{C_{2z}\,|\,00\tfrac{1}{2}\}$: 2, 3; 4, 3: a.
C $\mathbf{G}_4^1$: $\{C_{2z}\,|\,00\tfrac{1}{2}\}$: 2, 3; 4, 3: a.
D $\mathbf{G}_4^1$: $\{C_{2z}\,|\,00\tfrac{1}{2}\}$: 2, 3; 4, 3: a.
A $\mathbf{G}_2^1 \otimes \mathbf{T}_2$: $\{C_{2z}\,|\,00\tfrac{1}{2}\}$; $\mathbf{t}_1$ or $\mathbf{t}_2$: 1, 1; 2, 1: b.
E $\mathbf{G}_4^1$: $\{C_{2z}\,|\,00\tfrac{1}{2}\}$: 2, 3; 4, 3: a.

Λ^x $\mathbf{G}_2^1$: $(C_{2z}, 0)$: 1, x; 2, x: b.
V^x $\mathbf{G}_2^1$: $(C_{2z}, 0)$: 1, x; 2, x: b.
W^x $\mathbf{G}_2^1$: $(C_{2z}, 0)$: 1, x; 2, x: b.
U^x $\mathbf{G}_2^1$: $(C_{2z}, 0)$: 1, x; 2, x: b.

5 *B2* C_2^3

(*F*1; *K*7; *M*5; *S*15; *Z*1.)

Γ $\mathbf{G}_2^1$: $\{C_{2z} | 000\}$: 1, 1; 2, 1: *b*.
A $\mathbf{G}_2^1 \otimes \mathbf{T}_2$: $\{C_{2z} | 000\}$; $\mathbf{t}_1$: 1, 1; 2, 1: *b*.
Z $\mathbf{G}_2^1 \otimes \mathbf{T}_2$: $\{C_{2z} | 000\}$; $\mathbf{t}_2$ or $\mathbf{t}_3$: 1, 1; 2, 1: *b*.
M $\mathbf{G}_2^1 \otimes \mathbf{T}_2$: $\{C_{2z} | 000\}$; $\mathbf{t}_1$ or $\mathbf{t}_2$ or $\mathbf{t}_3$: 1, 1; 2, 1: *b*.
L $\mathbf{G}_1^1 \otimes \mathbf{T}_2$: $\{E | 000\}$; $\mathbf{t}_1$ or $\mathbf{t}_3$: 1, 1: *a*.
V $\mathbf{G}_1^1 \otimes \mathbf{T}_2$: $\{E | 000\}$; $\mathbf{t}_3$: 1, 1: *a*.

Λ^x $\mathbf{G}_2^1$: $(C_{2z}, 0)$: 1, *x*; 2, *x*: *b*.
U^x $\mathbf{G}_2^1$: $(C_{2z}, 0)$: 1, *x*; 2, *x*: *b*.

6 *Pm* C_{1h}^1

(*F*1; *K*7; *M*5; *S*15; *Z*1.)

Γ $\mathbf{G}_2^1$: $\{\sigma_z | 000\}$: 1, 1; 2, 1: *c*.
B $\mathbf{G}_2^1 \otimes \mathbf{T}_2$: $\{\sigma_z | 000\}$; $\mathbf{t}_1$: 1, 1; 2, 1: *c*.
Y $\mathbf{G}_2^1 \otimes \mathbf{T}_2$: $\{\sigma_z | 000\}$; $\mathbf{t}_2$: 1, 1; 2, 1: *c*.
Z $\mathbf{G}_2^1 \otimes \mathbf{T}_2$: $\{\sigma_z | 000\}$; $\mathbf{t}_3$: 1, 1; 2, 1: *c*.
C $\mathbf{G}_2^1 \otimes \mathbf{T}_2$: $\{\sigma_z | 000\}$; $\mathbf{t}_2$ or $\mathbf{t}_3$: 1, 1; 2, 1: *c*.
D $\mathbf{G}_2^1 \otimes \mathbf{T}_2$: $\{\sigma_z | 000\}$; $\mathbf{t}_1$ or $\mathbf{t}_3$: 1, 1; 2, 1: *c*.
A $\mathbf{G}_2^1 \otimes \mathbf{T}_2$: $\{\sigma_z | 000\}$; $\mathbf{t}_1$ or $\mathbf{t}_2$: 1, 1; 2, 1: *c*.
E $\mathbf{G}_2^1 \otimes \mathbf{T}_2$: $\{\sigma_z | 000\}$; $\mathbf{t}_1$ or $\mathbf{t}_2$ or $\mathbf{t}_3$: 1, 1; 2, 1: *c*.

Λ^x $\mathbf{G}_1^1$: $(E, 0)$: 1, 1: *a*.
V^x $\mathbf{G}_1^1$: $(E, 0)$: 1, 1: *a*.
W^x $\mathbf{G}_1^1$: $(E, 0)$: 1, 1: *a*.
U^x $\mathbf{G}_1^1$: $(E, 0)$: 1, 1: *a*.

7 *Pb* C_{1h}^2

(*F*1; *K*7; *M*5; *S*15; *Z*1.)

Γ $\mathbf{G}_2^1$: $\{\sigma_z | \tfrac{1}{2}00\}$: 1, 1; 2, 1: *c*.
B $\mathbf{G}_4^1$: $\{\sigma_z | \tfrac{1}{2}00\}$: 2, 3; 4, 3: *a*.
Y $\mathbf{G}_2^1 \otimes \mathbf{T}_2$: $\{\sigma_z | \tfrac{1}{2}00\}$; $\mathbf{t}_2$: 1, 1; 2, 1: *c*.
Z $\mathbf{G}_2^1 \otimes \mathbf{T}_2$: $\{\sigma_z | \tfrac{1}{2}00\}$; $\mathbf{t}_3$: 1, 1; 2, 1: *c*.
C $\mathbf{G}_2^1 \otimes \mathbf{T}_2$: $\{\sigma_z | \tfrac{1}{2}00\}$; $\mathbf{t}_2$ or $\mathbf{t}_3$: 1, 1; 2, 1: *c*.
D $\mathbf{G}_4^1$: $\{\sigma_z | \tfrac{1}{2}00\}$: 2, 3; 4, 3: *a*.
A $\mathbf{G}_4^1$: $\{\sigma_z | \tfrac{1}{2}00\}$: 2, 3; 4, 3: *a*.
E $\mathbf{G}_4^1$: $\{\sigma_z | \tfrac{1}{2}00\}$: 2, 3; 4, 3: *a*.

Λ^x $\mathbf{G}_1^1$: $(E, 0)$: 1, 1: *a*.
V^x $\mathbf{G}_1^1$: $(E, 0)$: 1, 2: *a*.
W^x $\mathbf{G}_1^1$: $(E, 0)$: 1, 1: *a*.
U^x $\mathbf{G}_1^1$: $(E, 0)$: 1, 2: *a*.

8 *Bm* C_{1h}^3

(*F*1; *K*7; *M*5; *S*15; *Z*1.)

Γ $\mathbf{G}_2^1$: $\{\sigma_z \mid 000\}$: 1, 1; 2, 1: *c*.
A $\mathbf{G}_2^1 \otimes \mathbf{T}_2$: $\{\sigma_z \mid 000\}$; $\mathbf{t}_1$: 1, 1; 2, 1: *c*.
Z $\mathbf{G}_2^1 \otimes \mathbf{T}_2$: $\{\sigma_z \mid 000\}$; $\mathbf{t}_2$ or $\mathbf{t}_3$: 1, 1; 2, 1: *c*.
M $\mathbf{G}_2^1 \otimes \mathbf{T}_2$: $\{\sigma_z \mid 000\}$; $\mathbf{t}_1$ or $\mathbf{t}_2$ or $\mathbf{t}_3$: 1, 1; 2, 1: *c*.
L $\mathbf{G}_1^1 \otimes \mathbf{T}_2$: $\{E \mid 000\}$; $\mathbf{t}_1$ or $\mathbf{t}_3$: 1, 1: *a*.
V $\mathbf{G}_1^1 \otimes \mathbf{T}_2$: $\{E \mid 000\}$; $\mathbf{t}_3$: 1, 1: *a*.

Λ^x $\mathbf{G}_1^1$: $(E, 0)$: 1, 1: *a*.
U^x $\mathbf{G}_1^1$: $(E, 0)$: 1, 1: *a*.

9 *Bb* C_{1h}^4

(*F*1; *K*7; *M*5; *S*15; *Z*1.)

Γ $\mathbf{G}_2^1$: $\{\sigma_z \mid \frac{1}{2}00\}$: 1, 1; 2, 1: *c*.
A $\mathbf{G}_4^1$: $\{\sigma_z \mid \frac{1}{2}00\}$: 2, 3; 4, 3: *a*.
Z $\mathbf{G}_2^1 \otimes \mathbf{T}_2$: $\{\sigma_z \mid \frac{1}{2}00\}$; $\mathbf{t}_2$ or $\mathbf{t}_3$: 1, 1; 2, 1: *c*.
M $\mathbf{G}_4^1$: $\{\sigma_z \mid \frac{1}{2}00\}$: 2, 3; 4, 3: *a*.
L $\mathbf{G}_1^1 \otimes \mathbf{T}_2$: $\{E \mid 000\}$ $\mathbf{t}_1$ or $\mathbf{t}_3$: 1, 1: *a*.
V $\mathbf{G}_1^1 \otimes \mathbf{T}_2$: $\{E \mid 000\}$ $\mathbf{t}_3$: 1, 1: *a*.

Λ^x $\mathbf{G}_1^1$: $(E, 0)$: 1, 1: *a*.
U^x $\mathbf{G}_1^1$: $(E, 0)$: 1, 2: *a*.

10 *P*2/*m* C_{2h}^1

(*F*1; *K*7; *M*5; *S*15; *Z*1.)

Γ $\mathbf{G}_4^2$: $\{C_{2z} \mid 000\}$, $\{I \mid 000\}$: 1, 1; 2, 1; 3, 1; 4, 1: *a*.
B $\mathbf{G}_4^2 \otimes \mathbf{T}_2$: $\{C_{2z} \mid 000\}$, $\{I \mid 000\}$; $\mathbf{t}_1$: 1, 1; 2, 1; 3, 1; 4, 1: *a*.
Y $\mathbf{G}_4^2 \otimes \mathbf{T}_2$: $\{C_{2z} \mid 000\}$, $\{I \mid 000\}$; $\mathbf{t}_2$: 1, 1; 2, 1; 3, 1; 4, 1: *a*.
Z $\mathbf{G}_4^2 \otimes \mathbf{T}_2$: $\{C_{2z} \mid 000\}$, $\{I \mid 000\}$; $\mathbf{t}_3$: 1, 1; 2, 1; 3, 1; 4, 1: *a*.
C $\mathbf{G}_4^2 \otimes \mathbf{T}_2$: $\{C_{2z} \mid 000\}$, $\{I \mid 000\}$; $\mathbf{t}_2$ or $\mathbf{t}_3$: 1, 1; 2, 1; 3, 1; 4, 1: *a*.
D $\mathbf{G}_4^2 \otimes \mathbf{T}_2$: $\{C_{2z} \mid 000\}$, $\{I \mid 000\}$; $\mathbf{t}_1$ or $\mathbf{t}_3$: 1, 1; 2, 1; 3, 1; 4, 1: *a*.
A $\mathbf{G}_4^2 \otimes \mathbf{T}_2$: $\{C_{2z} \mid 000\}$, $\{I \mid 000\}$; $\mathbf{t}_1$ or $\mathbf{t}_2$: 1, 1; 2, 1; 3, 1; 4, 1: *a*.
E $\mathbf{G}_4^2 \otimes \mathbf{T}_2$: $\{C_{2z} \mid 000\}$, $\{I \mid 000\}$; $\mathbf{t}_1$ or $\mathbf{t}_2$ or $\mathbf{t}_3$: 1, 1; 2, 1; 3, 1; 4, 1: *a*.

Λ^x $\mathbf{G}_2^1$: $(C_{2z}, 0)$: 1, 1; 2, 1: *b*.
V^x $\mathbf{G}_2^1$: $(C_{2z}, 0)$: 1, 1; 2, 1: *b*.
W^x $\mathbf{G}_2^1$: $(C_{2z}, 0)$: 1, 1; 2, 1: *b*.
U^x $\mathbf{G}_2^1$: $(C_{2z}, 0)$: 1, 1; 2, 1: *b*.

11 $P2_1/m$ C_{2h}^2

($F1$; $K7$; $M5$; $S15$; $Z1$.)

Γ $\mathbf{G}_4^2$: $\{C_{2z}\,|\,00\frac{1}{2}\}$, $\{I\,|\,00\frac{1}{2}\}$: 1, 1; 2, 1; 3, 1; 4, 1: a.
B $\mathbf{G}_4^2 \otimes \mathbf{T}_2$: $\{C_{2z}\,|\,00\frac{1}{2}\}$, $\{I\,|\,00\frac{1}{2}\}$; $\mathbf{t}_1$: 1, 1; 2, 1; 3, 1; 4, 1: a.
Y $\mathbf{G}_4^2 \otimes \mathbf{T}_2$: $\{C_{2z}\,|\,00\frac{1}{2}\}$, $\{I\,|\,00\frac{1}{2}\}$; $\mathbf{t}_2$: 1, 1; 2, 1; 3, 1; 4, 1: a.
Z $\mathbf{G}_8^4$: $\{C_{2z}\,|\,00\frac{1}{2}\}$, $\{I\,|\,00\frac{1}{2}\}$: 5, 1: a.
C $\mathbf{G}_8^4$: $\{C_{2z}\,|\,00\frac{1}{2}\}$, $\{I\,|\,00\frac{1}{2}\}$: 5, 1: a.
D $\mathbf{G}_8^4$: $\{C_{2z}\,|\,00\frac{1}{2}\}$, $\{I\,|\,00\frac{1}{2}\}$: 5, 1: a.
A $\mathbf{G}_4^2 \otimes \mathbf{T}_2$: $\{C_{2z}\,|\,00\frac{1}{2}\}$, $\{I\,|\,00\frac{1}{2}\}$; $\mathbf{t}_1$ or $\mathbf{t}_2$: 1, 1; 2, 1; 3, 1; 4, 1: a.
E $\mathbf{G}_8^4$: $\{C_{2z}\,|\,00\frac{1}{2}\}$, $\{I\,|\,00\frac{1}{2}\}$: 5, 1: a.

Λ^x $\mathbf{G}_2^1$: $(C_{2z}, 0)$: 1, 1; 2, 1: b.
V^x $\mathbf{G}_2^1$: $(C_{2z}, 0)$: 1, 1; 2, 1: b.
W^x $\mathbf{G}_2^1$: $(C_{2z}, 0)$: 1, 1; 2, 1: b.
U^x $\mathbf{G}_2^1$: $(C_{2z}, 0)$: 1, 1; 2, 1: b.

12 $B2/m$ C_{2h}^3

($F1$; $K7$; $M5$; $S15$; $Z1$.)

Γ $\mathbf{G}_4^2$: $\{C_{2z}\,|\,000\}$, $\{I\,|\,000\}$: 1, 1; 2, 1; 3, 1; 4, 1: a.
A $\mathbf{G}_4^2 \otimes \mathbf{T}_2$: $\{C_{2z}\,|\,000\}$, $\{I\,|\,000\}$; $\mathbf{t}_1$: 1, 1; 2, 1; 3, 1; 4, 1: a.
Z $\mathbf{G}_4^2 \otimes \mathbf{T}_2$: $\{C_{2z}\,|\,000\}$, $\{I\,|\,000\}$; $\mathbf{t}_2$ or $\mathbf{t}_3$: 1, 1; 2, 1; 3, 1; 4, 1: a.
M $\mathbf{G}_4^2 \otimes \mathbf{T}_2$: $\{C_{2z}\,|\,000\}$, $\{I\,|\,000\}$; $\mathbf{t}_1$ or $\mathbf{t}_2$ or $\mathbf{t}_3$: 1, 1; 2, 1; 3, 1; 4, 1: a.
L $\mathbf{G}_2^1 \otimes \mathbf{T}_2$: $\{I\,|\,000\}$; $\mathbf{t}_1$ or $\mathbf{t}_3$: 1, 1; 2, 1: a.
V $\mathbf{G}_2^1 \otimes \mathbf{T}_2$: $\{I\,|\,000\}$; $\mathbf{t}_3$: 1, 1; 2, 1: a.

Λ^x $\mathbf{G}_2^1$: $(C_{2z}, 0)$: 1, 1; 2, 1: b.
U^x $\mathbf{G}_2^1$: $(C_{2z}, 0)$: 1, 1; 2, 1: b.

13 $P2/b$ C_{2h}^4

($F1$; $K7$; $M5$; $S15$; $Z1$.)

Γ $\mathbf{G}_4^2$: $\{C_{2z}\,|\,000\}$, $\{I\,|\,\frac{1}{2}00\}$: 1, 1; 2, 1; 3, 1; 4, 1: a.
B $\mathbf{G}_8^4$: $\{\sigma_z\,|\,\frac{1}{2}00\}$, $\{I\,|\,\frac{1}{2}00\}$: 5, 1: a.
Y $\mathbf{G}_4^2 \otimes \mathbf{T}_2$: $\{C_{2z}\,|\,000\}$, $\{I\,|\,\frac{1}{2}00\}$; $\mathbf{t}_2$: 1, 1; 2, 1; 3, 1; 4, 1: a.
Z $\mathbf{G}_4^2 \otimes \mathbf{T}_2$: $\{C_{2z}\,|\,000\}$, $\{I\,|\,\frac{1}{2}00\}$; $\mathbf{t}_3$: 1, 1; 2, 1; 3, 1; 4, 1: a.
C $\mathbf{G}_4^2 \otimes \mathbf{T}_2$: $\{C_{2z}\,|\,000\}$, $\{I\,|\,\frac{1}{2}00\}$; $\mathbf{t}_2$ or $\mathbf{t}_3$: 1, 1; 2, 1; 3, 1; 4, 1: a.
D $\mathbf{G}_8^4$: $\{\sigma_z\,|\,\frac{1}{2}00\}$, $\{I\,|\,\frac{1}{2}00\}$: 5, 1: a.
A $\mathbf{G}_8^4$: $\{\sigma_z\,|\,\frac{1}{2}00\}$, $\{I\,|\,\frac{1}{2}00\}$: 5, 1: a.
E $\mathbf{G}_8^4$: $\{\sigma_z\,|\,\frac{1}{2}00\}$, $\{I\,|\,\frac{1}{2}00\}$: 5, 1: a.

Λ^x $\mathbf{G}_2^1$: $(C_{2z}, 0)$: 1, 1; 2, 1: b.
V^x $\mathbf{G}_2^1$: $(C_{2z}, 0)$: 1, 3; 2, 3: b.
W^x $\mathbf{G}_2^1$: $(C_{2z}, 0)$: 1, 1; 2, 1: b.
U^x $\mathbf{G}_2^1$: $(C_{2z}, 0)$: 1, 3; 2, 3: b.

14 $P2_1/b$ C_{2h}^5

($F1$; $J2$; $K7$; $M5$; $S15$; $Z1$.)

Γ $\mathbf{G}_4^2$: $\{C_{2z} | 00\frac{1}{2}\}, \{I | \frac{1}{2}0\frac{1}{2}\}$: 1, 1; 2, 1; 3, 1; 4, 1: *a*.
B $\mathbf{G}_8^4$: $\{\sigma_z | \frac{1}{2}00\}, \{I | \frac{1}{2}0\frac{1}{2}\}$: 5, 1: *a*.
Y $\mathbf{G}_4^2 \otimes \mathbf{T}_2$: $\{C_{2z} | 00\frac{1}{2}\}, \{I | \frac{1}{2}0\frac{1}{2}\}$; $\mathbf{t}_2$: 1, 1; 2, 1; 3, 1; 4, 1: *a*.
Z $\mathbf{G}_8^4$: $\{C_{2z} | 00\frac{1}{2}\}, \{I | \frac{1}{2}0\frac{1}{2}\}$: 5, 1: *a*.
C $\mathbf{G}_8^4$: $\{C_{2z} | 00\frac{1}{2}\}, \{I | \frac{1}{2}0\frac{1}{2}\}$: 5, 1: *a*.
D $\mathbf{G}_8^2$: $\{C_{2z} | 00\frac{1}{2}\}, \{I | \frac{1}{2}0\frac{1}{2}\}$: 2, 3; 4, 3; 6, 3; 8, 3: *a*.
A $\mathbf{G}_8^4$: $\{\sigma_z | \frac{1}{2}00\}, \{I | \frac{1}{2}0\frac{1}{2}\}$: 5, 1: *a*.
E $\mathbf{G}_8^2$: $\{C_{2z} | 00\frac{1}{2}\}, \{I | \frac{1}{2}0\frac{1}{2}\}$: 2, 3; 4, 3; 6, 3; 8, 3: *a*.

Λ^x $\mathbf{G}_2^1$: $(C_{2z}, 0)$: 1, 1; 2, 1: *b*.
V^x $\mathbf{G}_2^1$: $(C_{2z}, 0)$: 1, 3; 2, 3: *b*.
W^x $\mathbf{G}_2^1$: $(C_{2z}, 0)$: 1, 1; 2, 1: *b*.
U^x $\mathbf{G}_2^1$: $(C_{2z}, 0)$: 1, 3; 2, 3: *b*.

15 $B2/b$ C_{2h}^6

($C8$; $D1$; $F1$; $K7$; $M5$; $S15$; $Z1$.)

Γ $\mathbf{G}_4^2$: $\{C_{2z} | 000\}, \{I | \frac{1}{2}00\}$: 1, 1; 2, 1; 3, 1; 4, 1: *a*.
A $\mathbf{G}_8^4$: $\{\sigma_z | \frac{1}{2}00\}, \{I | \frac{1}{2}00\}$: 5, 1: *a*.
Z $\mathbf{G}_4^2 \otimes \mathbf{T}_2$: $\{C_{2z} | 000\}, \{I | \frac{1}{2}00\}$; $\mathbf{t}_2$ or $\mathbf{t}_3$: 1, 1; 2, 1; 3, 1; 4, 1: *a*.
M $\mathbf{G}_8^4$: $\{\sigma_z | \frac{1}{2}00\}, \{I | \frac{1}{2}00\}$: 5, 1: *a*.
L $\mathbf{G}_2^1 \otimes \mathbf{T}_2$: $\{I | \frac{1}{2}00\}$; $\mathbf{t}_1$ or $\mathbf{t}_3$: 1, 1; 2, 1: *a*.
V $\mathbf{G}_2^1 \otimes \mathbf{T}_2$: $\{I | \frac{1}{2}00\}$; $\mathbf{t}_3$: 1, 1; 2, 1: *a*.

Λ^x $\mathbf{G}_2^1$: $(C_{2z}, 0)$: 1, 1; 2, 1: *b*.
U^x $\mathbf{G}_2^1$: $(C_{2z}, 0)$: 1, 3; 2, 3: *b*.

16 $P222$ D_2^1

($F1$; $K7$; $M5$; $T1$; $Z1$.)

Γ $\mathbf{G}_4^2$: $\{C_{2z} | 000\}, \{C_{2y} | 000\}$: 1, 1; 2, 1; 3, 1; 4, 1: *b*.
Y $\mathbf{G}_4^2 \otimes \mathbf{T}_2$: $\{C_{2z} | 000\}, \{C_{2y} | 000\}$; $\mathbf{t}_1$: 1, 1; 2, 1; 3, 1; 4, 1: *b*.
X $\mathbf{G}_4^2 \otimes \mathbf{T}_2$: $\{C_{2z} | 000\}, \{C_{2y} | 000\}$; $\mathbf{t}_2$: 1, 1; 2, 1; 3, 1; 4, 1: *b*.
Z $\mathbf{G}_4^2 \otimes \mathbf{T}_2$: $\{C_{2z} | 000\}, \{C_{2y} | 000\}$; $\mathbf{t}_3$: 1, 1; 2, 1; 3, 1; 4, 1: *b*.
U $\mathbf{G}_4^2 \otimes \mathbf{T}_2$: $\{C_{2z} | 000\}, \{C_{2y} | 000\}$; $\mathbf{t}_2$ or $\mathbf{t}_3$: 1, 1; 2, 1; 3, 1; 4, 1: *b*.
T $\mathbf{G}_4^2 \otimes \mathbf{T}_2$: $\{C_{2z} | 000\}, \{C_{2y} | 000\}$; $\mathbf{t}_1$ or $\mathbf{t}_3$: 1, 1; 2, 1; 3, 1; 4, 1: *b*.
S $\mathbf{G}_4^2 \otimes \mathbf{T}_2$: $\{C_{2z} | 000\}, \{C_{2y} | 000\}$; $\mathbf{t}_1$ or $\mathbf{t}_2$: 1, 1; 2, 1; 3, 1; 4, 1: *b*.
R $\mathbf{G}_4^2 \otimes \mathbf{T}_2$: $\{C_{2z} | 000\}, \{C_{2y} | 000\}$; $\mathbf{t}_1$ or $\mathbf{t}_2$ or $\mathbf{t}_3$: 1, 1; 2, 1; 3, 1; 4, 1: *b*.

Δ^x $\mathbf{G}_2^1$: $(C_{2y}, 0)$: 1, 1; 2, 1: *b*.
D^x $\mathbf{G}_2^1$: $(C_{2y}, 0)$: 1, 1; 2, 1: *b*.
P^x $\mathbf{G}_2^1$: $(C_{2y}, 0)$: 1, 1; 2, 1: *b*.
B^x $\mathbf{G}_2^1$: $(C_{2y}, 0)$: 1, 1; 2, 1: *b*.
Σ^x $\mathbf{G}_2^1$: $(C_{2x}, 0)$: 1, 1; 2, 1: *b*.
C^x $\mathbf{G}_2^1$: $(C_{2x}, 0)$: 1, 1; 2, 1: *b*.
E^x $\mathbf{G}_2^1$: $(C_{2x}, 0)$: 1, 1; 2, 1: *b*.
A^x $\mathbf{G}_2^1$: $(C_{2x}, 0)$: 1, 1; 2, 1: *b*.
Λ^x $\mathbf{G}_2^1$: $(C_{2z}, 0)$: 1, 1; 2, 1: *b*.
H^x $\mathbf{G}_2^1$: $(C_{2z}, 0)$: 1, 1; 2, 1: *b*.
Q^x $\mathbf{G}_2^1$: $(C_{2z}, 0)$: 1, 1; 2, 1: *b*.
G^x $\mathbf{G}_2^1$: $(C_{2z}, 0)$: 1, 1; 2, 1: *b*.

17 $P222_1$ D_2^2

($F1$; $K7$; $M5$; $T1$; $Z1$.)

Γ $\mathbf{G}_4^2$: $\{C_{2z}|00\frac{1}{2}\}$, $\{C_{2y}|000\}$: 1, 1; 2, 1; 3, 1; 4, 1: b.
Y $\mathbf{G}_4^2 \otimes \mathbf{T}_2$: $\{C_{2z}|00\frac{1}{2}\}$, $\{C_{2y}|000\}$; $\mathbf{t}_1$: 1, 1; 2, 1; 3, 1; 4, 1: b.
X $\mathbf{G}_4^2 \otimes \mathbf{T}_2$: $\{C_{2z}|00\frac{1}{2}\}$, $\{C_{2y}|000\}$; $\mathbf{t}_2$: 1, 1; 2, 1; 3, 1; 4, 1: b.
Z $\mathbf{G}_8^4$: $\{C_{2z}|00\frac{1}{2}\}$, $\{C_{2y}|000\}$: 5, 1: a.
U $\mathbf{G}_8^4$: $\{C_{2z}|00\frac{1}{2}\}$, $\{C_{2y}|000\}$: 5, 1: a.
T $\mathbf{G}_8^4$: $\{C_{2z}|00\frac{1}{2}\}$, $\{C_{2y}|000\}$: 5, 1: a.
S $\mathbf{G}_4^2 \otimes \mathbf{T}_2$: $\{C_{2z}|00\frac{1}{2}\}$, $\{C_{2y}|000\}$; $\mathbf{t}_1$ or $\mathbf{t}_2$: 1, 1; 2, 1; 3, 1; 4, 1: b.
R $\mathbf{G}_8^4$: $\{C_{2z}|00\frac{1}{2}\}$, $\{C_{2y}|000\}$: 5, 1: a.

Δ^x $\mathbf{G}_2^1$: $(C_{2y}, 0)$: 1, 1; 2, 1: b.
D^x $\mathbf{G}_2^1$: $(C_{2y}, 0)$: 1, 1; 2, 1: b.
P^x $\mathbf{G}_2^1$: $(C_{2y}, 0)$: 1, 3; 2, 3: b.
B^x $\mathbf{G}_2^1$: $(C_{2y}, 0)$: 1, 3; 2, 3: b.
Σ^x $\mathbf{G}_2^1$: $(C_{2x}, 0)$: 1, 1; 2, 1: b.
C^x $\mathbf{G}_2^1$: $(C_{2x}, 0)$: 1, 1; 2, 1: b.
E^x $\mathbf{G}_4^1$: $(C_{2x}, 0)$: 2, 3; 4, 3: a.
A^x $\mathbf{G}_4^1$: $(C_{2x}, 0)$: 2, 3; 4, 3: a.
Λ^x $\mathbf{G}_2^1$: $(C_{2z}, 0)$: 1, 1; 2, 1: b.
H^x $\mathbf{G}_2^1$: $(C_{2z}, 0)$: 1, 1; 2, 1: b.
Q^x $\mathbf{G}_2^1$: $(C_{2z}, 0)$: 1, 1; 2, 1: b.
G^x $\mathbf{G}_2^1$: $(C_{2z}, 0)$: 1, 1; 2, 1: b.

18 $P2_12_12$ D_2^3

($F1$; $K7$; $M5$; $T1$; $Z1$.)

Γ $\mathbf{G}_4^2$: $\{C_{2z}|000\}$, $\{C_{2y}|\frac{1}{2}\frac{1}{2}0\}$: 1, 1; 2, 1; 3, 1; 4, 1: b.
Y $\mathbf{G}_8^4$: $\{C_{2y}|\frac{1}{2}\frac{1}{2}0\}$, $\{C_{2z}|000\}$: 5, 1: a.
X $\mathbf{G}_8^4$: $\{C_{2x}|\frac{1}{2}\frac{1}{2}0\}$, $\{C_{2z}|000\}$: 5, 1: a.
Z $\mathbf{G}_4^2 \otimes \mathbf{T}_2$: $\{C_{2z}|000\}$, $\{C_{2y}|\frac{1}{2}\frac{1}{2}0\}$; $\mathbf{t}_3$: 1, 1; 2, 1; 3, 1; 4, 1: b.
U $\mathbf{G}_8^4$: $\{C_{2x}|\frac{1}{2}\frac{1}{2}0\}$, $\{C_{2z}|000\}$: 5, 1: a.
T $\mathbf{G}_8^4$: $\{C_{2y}|\frac{1}{2}\frac{1}{2}0\}$, $\{C_{2z}|000\}$: 5, 1: a.
S $\mathbf{G}_8^2$: $\{C_{2y}|\frac{1}{2}\frac{1}{2}0\}$, $\{C_{2z}|000\}$: 2, 3; 4, 3; 6, 3; 8, 3: d.
R $\mathbf{G}_8^2$: $\{C_{2y}|\frac{1}{2}\frac{1}{2}0\}$, $\{C_{2z}|000\}$: 2, 3; 4, 3; 6, 3; 8, 3: d.

Δ^x $\mathbf{G}_2^1$: $(C_{2y}, 0)$: 1, 1; 2, 1: b.
D^x $\mathbf{G}_4^1$: $(C_{2y}, 0)$: 2, 3; 4, 3: a.
P^x $\mathbf{G}_4^1$: $(C_{2y}, 0)$: 2, 3; 4, 3: a.
B^x $\mathbf{G}_2^1$: $(C_{2y}, 0)$: 1, 1; 2, 1: b.
Σ^x $\mathbf{G}_2^1$: $(C_{2x}, 0)$: 1, 1; 2, 1: b.
C^x $\mathbf{G}_4^1$: $(C_{2x}, 0)$: 2, 3; 4, 3: a.
E^x $\mathbf{G}_4^1$: $(C_{2x}, 0)$: 2, 3; 4, 3: a.
A^x $\mathbf{G}_2^1$: $(C_{2x}, 0)$: 1, 1; 2, 1: b.
Λ^x $\mathbf{G}_2^1$: $(C_{2z}, 0)$: 1, 1; 2, 1: b.
H^x $\mathbf{G}_2^1$: $(C_{2z}, 0)$: 1, 3; 2, 3: b.
Q^x $\mathbf{G}_2^1$: $(C_{2z}, 0)$: 1, 2; 2, 2: b.
G^x $\mathbf{G}_2^1$: $(C_{2z}, 0)$: 1, 3; 2, 3: b.

19 $P2_12_12_1$ D_2^4

($F1$; $K7$; $M5$; $T1$; $Z1$.)

Γ $\mathbf{G}_4^2$: $\{C_{2z}\,|\,\frac{1}{2}0\frac{1}{2}\}$, $\{C_{2y}\,|\,\frac{1}{2}\frac{1}{2}0\}$: 1, 1; 2, 1; 3, 1; 4, 1: b.
Y $\mathbf{G}_8^4$: $\{C_{2y}\,|\,\frac{1}{2}\frac{1}{2}0\}$, $\{C_{2z}\,|\,\frac{1}{2}0\frac{1}{2}\}$: 5, 1: a.
X $\mathbf{G}_8^4$: $\{C_{2x}\,|\,0\frac{1}{2}\frac{1}{2}\}$, $\{C_{2z}\,|\,\frac{1}{2}0\frac{1}{2}\}$: 5, 1: a.
Z $\mathbf{G}_8^4$: $\{C_{2z}\,|\,\frac{1}{2}0\frac{1}{2}\}$, $\{C_{2x}\,|\,0\frac{1}{2}\frac{1}{2}\}$: 5, 1: a.
U $\mathbf{G}_8^2$: $\{C_{2z}\,|\,\frac{1}{2}0\frac{1}{2}\}$, $\{C_{2y}\,|\,\frac{1}{2}\frac{1}{2}0\}$: 2, 3; 4, 3; 6, 3; 8, 3: d.
T $\mathbf{G}_8^2$: $\{C_{2z}\,|\,\frac{1}{2}0\frac{1}{2}\}$, $\{C_{2x}\,|\,0\frac{1}{2}\frac{1}{2}\}$: 2, 3; 4, 3; 6, 3; 8, 3: d.
S $\mathbf{G}_8^2$: $\{C_{2x}\,|\,0\frac{1}{2}\frac{1}{2}\}$, $\{C_{2z}\,|\,\frac{1}{2}0\frac{1}{2}\}$: 2, 3; 4, 3; 6, 3; 8, 3: d.
R $\mathbf{G}_8^5$: $\{C_{2z}\,|\,\frac{1}{2}0\frac{1}{2}\}$, $\{C_{2y}\,|\,\frac{1}{2}\frac{1}{2}0\}$: 5, 2: a.

Δ^x $\mathbf{G}_2^1$: $(C_{2y}, 0)$: 1, 1; 2, 1: b.
D^x $\mathbf{G}_4^1$: $(C_{2y}, 0)$: 2, 3; 4, 3: a.
P^x $\mathbf{G}_4^1$: $(C_{2y}, 0)$: 2, 2; 4, 2: a.
B^x $\mathbf{G}_2^1$: $(C_{2y}, 0)$: 1, 3; 2, 3: b.
Σ^x $\mathbf{G}_2^1$: $(C_{2x}, 0)$: 1, 1; 2, 1: b.
C^x $\mathbf{G}_2^1$: $(C_{2x}, 0)$: 1, 3; 2, 3: b.
E^x $\mathbf{G}_4^1$: $(C_{2x}, 0)$: 2, 2; 4, 2: a.
A^x $\mathbf{G}_4^1$: $(C_{2x}, 0)$: 2, 3; 4, 3: a.
Λ^x $\mathbf{G}_2^1$: $(C_{2z}, 0)$: 1, 1; 2, 1: b.
H^x $\mathbf{G}_4^1$: $(C_{2z}, 0)$: 2, 3; 4, 3: a.
Q^x $\mathbf{G}_4^1$: $(C_{2z}, 0)$: 2, 2; 4, 2: a.
G^x $\mathbf{G}_2^1$: $(C_{2z}, 0)$: 1, 3; 2, 3: b.

20 $C222_1$ D_2^5

($F1$; $K7$; $M5$; $T3$; $Z1$.)

Γ $\mathbf{G}_4^2$: $\{C_{2z}\,|\,00\frac{1}{2}\}$, $\{C_{2y}\,|\,00\frac{1}{2}\}$: 1, 1; 2, 1; 3, 1; 4, 1: b.
Y $\mathbf{G}_4^2 \otimes \mathbf{T}_2$: $\{C_{2z}\,|\,00\frac{1}{2}\}$, $\{C_{2y}\,|\,00\frac{1}{2}\}$; $\mathbf{t}_1$ or $\mathbf{t}_2$: 1, 1; 2, 1; 3, 1; 4, 1: b.
Z $\mathbf{G}_8^4$: $\{C_{2z}\,|\,00\frac{1}{2}\}$, $\{C_{2x}\,|\,000\}$: 5, 1: a.
T $\mathbf{G}_8^4$: $\{C_{2z}\,|\,00\frac{1}{2}\}$, $\{C_{2x}\,|\,000\}$: 5, 1: a.
S $\mathbf{G}_2^1 \otimes \mathbf{T}_2$: $\{C_{2z}\,|\,00\frac{1}{2}\}$; $\mathbf{t}_2$: 1, 1; 2, 1: b.
R $\mathbf{G}_4^1$: $\{C_{2z}\,|\,00\frac{1}{2}\}$: 2, 3; 4, 3: a.

Λ^x $\mathbf{G}_2^1$: $(C_{2z}, 0)$: 1, 1; 2, 1: b.
H^x $\mathbf{G}_2^1$: $(C_{2z}, 0)$: 1, 1; 2, 1: b.
D^x $\mathbf{G}_2^1$: $(C_{2z}, 0)$: 1, x; 2, x: b.
A^x $\mathbf{G}_2^1$: $(C_{2x}, 0)$: 1, 3; 2, 3: b.
Σ^x $\mathbf{G}_2^1$: $(C_{2x}, 0)$: 1, 1; 2, 1: b.
Δ^x $\mathbf{G}_2^1$: $(C_{2y}, 0)$: 1, 1; 2, 1: b.
B^x $\mathbf{G}_4^1$: $(C_{2y}, 0)$: 2, 3; 4, 3: a.
G^x $\mathbf{G}_4^1$: $(C_{2y}, 0)$: 2, 3; 4, 3: a.
F^x $\mathbf{G}_2^1$: $(C_{2y}, 0)$: 1, 1; 2, 1: b.
E^x $\mathbf{G}_2^1$: $(C_{2x}, 0)$: 1, 3; 2, 3: b.
C^x $\mathbf{G}_2^1$: $(C_{2x}, 0)$: 1, 1; 2, 1: b.

21 *C*222 D_2^6

(*F*1; *K*7; *M*5; *T*3; *Z*1.)

Γ $\mathbf{G}_4^2$: $\{C_{2z}|000\}, \{C_{2y}|000\}$: 1, 1; 2, 1; 3, 1; 4, 1: *b*.
Y $\mathbf{G}_4^2 \otimes \mathbf{T}_2$: $\{C_{2z}|000\}, \{C_{2y}|000\}$; $\mathbf{t}_1$ or $\mathbf{t}_2$: 1, 1; 2, 1; 3, 1; 4, 1: *b*.
Z $\mathbf{G}_4^2 \otimes \mathbf{T}_2$: $\{C_{2z}|000\}, \{C_{2y}|000\}$; $\mathbf{t}_3$: 1, 1; 2, 1; 3, 1; 4, 1: *b*.
T $\mathbf{G}_4^2 \otimes \mathbf{T}_2$: $\{C_{2z}|000\}, \{C_{2y}|000\}$; $\mathbf{t}_1$ or $\mathbf{t}_2$ or $\mathbf{t}_3$: 1, 1; 2, 1; 3, 1; 4, 1: *b*.
S $\mathbf{G}_2^1 \otimes \mathbf{T}_2$: $\{C_{2z}|000\}$; $\mathbf{t}_2$: 1, 1; 2, 1: *b*.
R $\mathbf{G}_2^1 \otimes \mathbf{T}_2$: $\{C_{2z}|000\}$; $\mathbf{t}_2$ or $\mathbf{t}_3$: 1, 1; 2, 1: *b*.

Λ^x $\mathbf{G}_2^1$: $(C_{2z}, 0)$: 1, 1; 2, 1: *b*.
H^x $\mathbf{G}_2^1$: $(C_{2z}, 0)$: 1, 1; 2, 1: *b*.
D^x $\mathbf{G}_2^1$: $(C_{2z}, 0)$: 1, *x*; 2, *x*: *b*.
A^x $\mathbf{G}_2^1$: $(C_{2x}, 0)$: 1, 1; 2, 1: *b*.
Σ^x $\mathbf{G}_2^1$: $(C_{2x}, 0)$: 1, 1; 2, 1: *b*.
Δ^x $\mathbf{G}_2^1$: $(C_{2y}, 0)$: 1, 1; 2, 1: *b*.
B^x $\mathbf{G}_2^1$: $(C_{2y}, 0)$: 1, 1; 2, 1: *b*.
G^x $\mathbf{G}_2^1$: $(C_{2y}, 0)$: 1, 1; 2, 1: *b*.
F^x $\mathbf{G}_2^1$: $(C_{2y}, 0)$: 1, 1; 2, 1: *b*.
E^x $\mathbf{G}_2^1$: $(C_{2x}, 0)$: 1, 1; 2, 1: *b*.
C^x $\mathbf{G}_2^1$: $(C_{2x}, 0)$: 1, 1; 2, 1: *b*.

22 *F*222 D_2^7

(*F*1; *K*7; *M*5; *S*14; *Z*1.)

Γ $\mathbf{G}_4^2$: $\{C_{2z}|000\}, \{C_{2y}|000\}$: 1, 1; 2, 1; 3, 1; 4, 1: *b*.
Y $\mathbf{G}_4^2 \otimes \mathbf{T}_2$: $\{C_{2z}|000\}, \{C_{2y}|000\}$; $\mathbf{t}_2$ or $\mathbf{t}_3$: 1, 1; 2, 1; 3, 1; 4, 1: *b*.
X $\mathbf{G}_4^2 \otimes \mathbf{T}_2$: $\{C_{2z}|000\}, \{C_{2y}|000\}$; $\mathbf{t}_1$ or $\mathbf{t}_3$: 1, 1; 2, 1; 3, 1; 4, 1: *b*.
Z $\mathbf{G}_4^2 \otimes \mathbf{T}_2$: $\{C_{2z}|000\}, \{C_{2y}|000\}$; $\mathbf{t}_1$ or $\mathbf{t}_2$: 1, 1; 2, 1; 3, 1; 4, 1: *b*.
L $\mathbf{G}_1^1 \otimes \mathbf{T}_2$: $\{E|000\}$; $\mathbf{t}_1$: 1, 1: *a*.

Λ^x $\mathbf{G}_2^1$: $(C_{2z}, 0)$: 1, 1; 2, 1: *b*.
G^x $\mathbf{G}_2^1$: $(C_{2z}, 0)$: 1, 1; 2, 1: *b*.
H^x $\mathbf{G}_2^1$: $(C_{2z}, 0)$: 1, 1; 2, 1: *b*.
Q^x $\mathbf{G}_2^1$: $(C_{2z}, 0)$: 1, 1; 2, 1: *b*.
Σ^x $\mathbf{G}_2^1$: $(C_{2x}, 0)$: 1, 1; 2, 1: *b*.
C^x $\mathbf{G}_2^1$: $(C_{2x}, 0)$: 1, 1; 2, 1: *b*.
A^x $\mathbf{G}_2^1$: $(C_{2x}, 0)$: 1, 1; 2, 1: *b*.
U^x $\mathbf{G}_2^1$: $(C_{2x}, 0)$: 1, 1; 2, 1: *b*.
Δ^x $\mathbf{G}_2^1$: $(C_{2y}, 0)$: 1, 1; 2, 1: *b*.
D^x $\mathbf{G}_2^1$: $(C_{2y}, 0)$: 1, 1; 2, 1: *b*.
B^x $\mathbf{G}_2^1$: $(C_{2y}, 0)$: 1, 1; 2, 1: *b*.
R^x $\mathbf{G}_2^1$: $(C_{2y}, 0)$: 1, 1; 2, 1: *b*.

23 $I222$ D_2^8

($F1$; $K7$; $M5$; $Z1$.)

Γ $\mathbf{G}_4^2$: $\{C_{2z}|000\}$, $\{C_{2y}|000\}$: 1, 1; 2, 1; 3, 1; 4, 1: *b*.
X $\mathbf{G}_4^2 \otimes \mathbf{T}_2$: $\{C_{2z}|000\}$, $\{C_{2y}|000\}$; $\mathbf{t}_1$ or $\mathbf{t}_2$ or $\mathbf{t}_3$: 1, 1; 2, 1; 3, 1; 4, 1: *b*.
R $\mathbf{G}_2^1 \otimes \mathbf{T}_2$: $\{C_{2y}|000\}$; $\mathbf{t}_1$: 1, 1; 2, 1: *b*.
S $\mathbf{G}_2^1 \otimes \mathbf{T}_2$: $\{C_{2x}|000\}$; $\mathbf{t}_1$ or $\mathbf{t}_3$: 1, 1; 2, 1: *b*.
T $\mathbf{G}_2^1 \otimes \mathbf{T}_2$: $\{C_{2z}|000\}$; $\mathbf{t}_1$ or $\mathbf{t}_2$: 1, 1; 2, 1: *b*.
W $\mathbf{G}_4^2 \otimes \mathbf{T}_4$: $\{C_{2z}|000\}$, $\{C_{2y}|000\}$; $\mathbf{t}_1$ or $\mathbf{t}_2$ or $\mathbf{t}_3$; 1, x; 2, x; 3, x; 4, x: *b*.

Λ^x $\mathbf{G}_2^1$: $(C_{2z}, 0)$: 1, 1; 2, 1: *b*.
G^x $\mathbf{G}_2^1$: $(C_{2z}, 0)$: 1, 1; 2, 1: *b*.
P^x $\mathbf{G}_2^1$: $(C_{2z}, 0)$: 1, x; 2, x: *b*.
Σ^x $\mathbf{G}_2^1$: $(C_{2x}, 0)$: 1, 1; 2, 1: *b*.
F^x $\mathbf{G}_2^1$: $(C_{2x}, 0)$: 1, 1; 2, 1: *b*.
D^x $\mathbf{G}_2^1$: $(C_{2x}, 0)$: 1, x; 2, x: *b*.
Δ^x $\mathbf{G}_2^1$: $(C_{2y}, 0)$: 1, 1; 2, 1: *b*.
U^x $\mathbf{G}_2^1$: $(C_{2y}, 0)$: 1, 1; 2, 1: *b*.
Q^x $\mathbf{G}_2^1$: $(C_{2y}, 0)$: 1, x; 2, x: *b*.

24 $I2_12_12_1$ D_2^9

($F1$; $K7$; $M5$; $Z1$.)

Γ $\mathbf{G}_4^2$: $\{C_{2z}|\frac{1}{2}0\frac{1}{2}\}$, $\{C_{2y}|\frac{1}{2}\frac{1}{2}0\}$: 1, 1; 2, 1; 3, 1; 4, 1: *b*.
X $\mathbf{G}_4^2 \otimes \mathbf{T}_2$: $\{C_{2z}|\frac{1}{2}0\frac{1}{2}\}$, $\{C_{2y}|\frac{1}{2}\frac{1}{2}0\}$; $\mathbf{t}_1$ or $\mathbf{t}_2$ or $\mathbf{t}_3$: 1, 1; 2, 1; 3, 1; 4, 1: *b*.
R $\mathbf{G}_2^1 \otimes \mathbf{T}_2$: $\{C_{2y}|\frac{1}{2}\frac{1}{2}0\}$; $\mathbf{t}_1$: 1, 1; 2, 1: *b*.
S $\mathbf{G}_2^1 \otimes \mathbf{T}_2$: $\{C_{2x}|0\frac{1}{2}\frac{1}{2}\}$; $\mathbf{t}_1$ or $\mathbf{t}_3$: 1, 1; 2, 1: *b*.
T $\mathbf{G}_2^1 \otimes \mathbf{T}_2$: $\{C_{2z}|\frac{1}{2}0\frac{1}{2}\}$; $\mathbf{t}_1$ or $\mathbf{t}_2$: 1, 1; 2, 1: *b*.
W $\mathbf{G}_{16}^7$: $\{E|100\}$, $\{C_{2z}|\frac{1}{2}0\frac{1}{2}\}$, $\{C_{2y}|\frac{1}{2}\frac{1}{2}0\}$: 9, x: *f*.

Λ^x $\mathbf{G}_2^1$: $(C_{2z}, 0)$: 1, 1; 2, 1: *b*.
G^x $\mathbf{G}_2^1$: $(C_{2z}, 0)$: 1, 1; 2, 1: *b*.
P^x $\mathbf{G}_4^1$: $(C_{2z}, 0)$: 2, x; 4, x: *a*.
Σ^x $\mathbf{G}_2^1$: $(C_{2x}, 0)$: 1, 1; 2, 1: *b*.
F^x $\mathbf{G}_2^1$: $(C_{2x}, 0)$: 1, 1; 2, 1: *b*.
D^x $\mathbf{G}_4^1$: $(C_{2x}, 0)$: 2, x; 4, x: *a*.
Δ^x $\mathbf{G}_2^1$: $(C_{2y}, 0)$: 1, 1; 2, 1: *b*.
U^x $\mathbf{G}_2^1$: $(C_{2y}, 0)$: 1, 1; 2, 1: *b*.
Q^x $\mathbf{G}_4^1$: $(C_{2y}, 0)$: 2, x; 4, x: *a*.

25 *Pmm2* C_{2v}^1

($F1$; $K7$; $M5$; $T1$; $Z1$.)

Γ $\mathbf{G}_4^2$: $\{C_{2z} \mid 000\}$, $\{\sigma_y \mid 000\}$: 1, 1; 2, 1; 3, 1; 4, 1: *c*.
Y $\mathbf{G}_4^2 \otimes \mathbf{T}_2$: $\{C_{2z} \mid 000\}$, $\{\sigma_y \mid 000\}$; $\mathbf{t}_1$: 1, 1; 2, 1; 3, 1; 4, 1: *c*.
X $\mathbf{G}_4^2 \otimes \mathbf{T}_2$: $\{C_{2z} \mid 000\}$, $\{\sigma_y \mid 000\}$; $\mathbf{t}_2$: 1, 1; 2, 1; 3, 1; 4, 1: *c*.
Z $\mathbf{G}_4^2 \otimes \mathbf{T}_2$: $\{C_{2z} \mid 000\}$, $\{\sigma_y \mid 000\}$; $\mathbf{t}_3$: 1, 1; 2, 1; 3, 1; 4, 1: *c*.
U $\mathbf{G}_4^2 \otimes \mathbf{T}_2$: $\{C_{2z} \mid 000\}$, $\{\sigma_y \mid 000\}$; $\mathbf{t}_2$ or $\mathbf{t}_3$: 1, 1; 2, 1; 3, 1; 4, 1: *c*.
T $\mathbf{G}_4^2 \otimes \mathbf{T}_2$: $\{C_{2z} \mid 000\}$, $\{\sigma_y \mid 000\}$; $\mathbf{t}_1$ or $\mathbf{t}_3$: 1, 1; 2, 1; 3, 1; 4, 1: *c*.
S $\mathbf{G}_4^2 \otimes \mathbf{T}_2$: $\{C_{2z} \mid 000\}$, $\{\sigma_y \mid 000\}$; $\mathbf{t}_1$ or $\mathbf{t}_2$: 1, 1; 2, 1; 3, 1; 4, 1: *c*.
R $\mathbf{G}_4^2 \otimes \mathbf{T}_2$: $\{C_{2z} \mid 000\}$, $\{\sigma_y \mid 000\}$; $\mathbf{t}_1$ or $\mathbf{t}_2$ or $\mathbf{t}_3$: 1, 1; 2, 1; 3, 1; 4, 1: *c*.

Δ^x $\mathbf{G}_2^1$: $(\sigma_x, 0)$: 1, 1; 2, 1: *c*.
D^x $\mathbf{G}_2^1$: $(\sigma_x, 0)$: 1, 1; 2, 1: *c*.
P^x $\mathbf{G}_2^1$: $(\sigma_x, 0)$: 1, 1; 2, 1: *c*.
B^x $\mathbf{G}_2^1$: $(\sigma_x, 0)$: 1, 1; 2, 1: *c*.
Σ^x $\mathbf{G}_2^1$: $(\sigma_y, 0)$: 1, 1; 2, 1: *c*.
C^x $\mathbf{G}_2^1$: $(\sigma_y, 0)$: 1, 1; 2, 1: *c*.
E^x $\mathbf{G}_2^1$: $(\sigma_y, 0)$: 1, 1; 2, 1: *c*.
A^x $\mathbf{G}_2^1$: $(\sigma_y, 0)$: 1, 1; 2, 1: *c*.
Λ^x $\mathbf{G}_4^2$: $(C_{2z}, 0)$, $(\sigma_y, 0)$: 1, *x*; 2, *x*; 3, *x*; 4, *x*: *c*.
H^x $\mathbf{G}_4^2$: $(C_{2z}, 0)$, $(\sigma_y, 0)$: 1, *x*; 2, *x*; 3, *x*; 4, *x*: *c*.
Q^x $\mathbf{G}_4^2$: $(C_{2z}, 0)$, $(\sigma_y, 0)$: 1, *x*; 2, *x*; 3, *x*; 4, *x*: *c*.
G^x $\mathbf{G}_4^2$: $(C_{2z}, 0)$, $(\sigma_y, 0)$: 1, *x*; 2, *x*; 3, *x*; 4, *x*: *c*.

26 *Pmc*2_1 C_{2v}^2

($F1$; $K7$; $M5$; $T1$; $Z1$.)

Γ $\mathbf{G}_4^2$: $\{C_{2z} \mid 00\frac{1}{2}\}$, $\{\sigma_y \mid 00\frac{1}{2}\}$: 1, 1; 2, 1; 3, 1; 4, 1: *c*.
Y $\mathbf{G}_4^2 \otimes \mathbf{T}_2$: $\{C_{2z} \mid 00\frac{1}{2}\}$, $\{\sigma_y \mid 00\frac{1}{2}\}$; $\mathbf{t}_1$: 1, 1; 2, 1; 3, 1; 4, 1: *c*.
X $\mathbf{G}_4^2 \otimes \mathbf{T}_2$: $\{C_{2z} \mid 00\frac{1}{2}\}$, $\{\sigma_y \mid 00\frac{1}{2}\}$; $\mathbf{t}_2$: 1, 1; 2, 1; 3, 1; 4, 1: *c*.
Z $\mathbf{G}_8^2$: $\{C_{2z} \mid 00\frac{1}{2}\}$, $\{\sigma_x \mid 000\}$: 2, 3; 4, 3; 6, 3; 8, 3: *c*.
U $\mathbf{G}_8^2$: $\{C_{2z} \mid 00\frac{1}{2}\}$, $\{\sigma_x \mid 000\}$: 2, 3; 4, 3; 6, 3; 8, 3: *c*.
T $\mathbf{G}_8^2$: $\{C_{2z} \mid 00\frac{1}{2}\}$, $\{\sigma_x \mid 000\}$: 2, 3; 4, 3; 6, 3; 8, 3: *c*.
S $\mathbf{G}_4^2 \otimes \mathbf{T}_2$: $\{C_{2z} \mid 00\frac{1}{2}\}$, $\{\sigma_y \mid 00\frac{1}{2}\}$; $\mathbf{t}_1$ or $\mathbf{t}_2$: 1, 1; 2, 1; 3, 1; 4, 1: *c*.
R $\mathbf{G}_8^2$: $\{C_{2z} \mid 00\frac{1}{2}\}$, $\{\sigma_x \mid 000\}$: 2, 3; 4, 3; 6, 3; 8, 3: *c*.

Δ^x $\mathbf{G}_2^1$: $(\sigma_x, 0)$: 1, 1; 2, 1: *c*.
D^x $\mathbf{G}_2^1$: $(\sigma_x, 0)$: 1, 1; 2, 1: *c*.
P^x $\mathbf{G}_2^1$: $(\sigma_x, 0)$: 1, 2; 2, 2: *c*.
B^x $\mathbf{G}_2^1$: $(\sigma_x, 0)$: 1, 2; 2, 2: *c*.
Σ^x $\mathbf{G}_2^1$: $(\sigma_y, 0)$: 1, 1; 2, 1: *c*.
C^x $\mathbf{G}_2^1$: $(\sigma_y, 0)$: 1, 1; 2, 1: *c*.
E^x $\mathbf{G}_2^1$: $(\sigma_y, 0)$: 1, 3; 2, 3: *c*.
A^x $\mathbf{G}_2^1$: $(\sigma_y, 0)$: 1, 3; 2, 3: *c*.
Λ^x $\mathbf{G}_4^2$: $(C_{2z}, 0)$, $(\sigma_y, 0)$: 1, *x*; 2, *x*; 3, *x*; 4, *x*: *c*.
H^x $\mathbf{G}_4^2$: $(C_{2z}, 0)$, $(\sigma_y, 0)$: 1, *x*; 2, *x*; 3, *x*; 4, *x*: *c*.
Q^x $\mathbf{G}_4^2$: $(C_{2z}, 0)$, $(\sigma_y, 0)$: 1, *x*; 2, *x*; 3, *x*; 4, *x*: *c*.
G^x $\mathbf{G}_4^2$: $(C_{2z}, 0)$, $(\sigma_y, 0)$: 1, *x*; 2, *x*; 3, *x*; 4, *x*: *c*.

27 *Pcc*2 C_{2v}^{3}

($F1$; $K7$; $M5$; $T1$; $Z1$.)

Γ $\mathbf{G}_4^2$: $\{C_{2z}|000\}$, $\{\sigma_y|00\frac{1}{2}\}$: 1, 1; 2, 1; 3, 1; 4, 1: *c*.
Y $\mathbf{G}_4^2 \otimes \mathbf{T}_2$: $\{C_{2z}|000\}$, $\{\sigma_y|00\frac{1}{2}\}$; $\mathbf{t}_1$: 1, 1; 2, 1; 3, 1; 4, 1: *c*.
X $\mathbf{G}_4^2 \otimes \mathbf{T}_2$: $\{C_{2z}|000\}$, $\{\sigma_y|00\frac{1}{2}\}$; $\mathbf{t}_2$: 1, 1; 2, 1; 3, 1; 4, 1: *c*.
Z $\mathbf{G}_8^2$: $\{\sigma_y|00\frac{1}{2}\}$, $\{C_{2z}|000\}$: 2, 3; 4, 3; 6, 3; 8, 3: *d*.
U $\mathbf{G}_8^2$: $\{\sigma_y|00\frac{1}{2}\}$, $\{C_{2z}|000\}$: 2, 3; 4, 3; 6, 3; 8, 3: *d*.
T $\mathbf{G}_8^2$: $\{\sigma_y|00\frac{1}{2}\}$, $\{C_{2z}|000\}$: 2, 3; 4, 3; 6, 3; 8, 3: *d*.
S $\mathbf{G}_4^2 \otimes \mathbf{T}_2$: $\{C_{2z}|000\}$, $\{\sigma_y|00\frac{1}{2}\}$; $\mathbf{t}_1$ or $\mathbf{t}_2$: 1, 1; 2, 1; 3, 1; 4, 1: *c*.
R $\mathbf{G}_8^2$: $\{\sigma_y|00\frac{1}{2}\}$, $\{C_{2z}|000\}$: 2, 3; 4, 3; 6, 3; 8, 3: *d*.

Δ^x $\mathbf{G}_2^1$: $(\sigma_x, 0)$: 1, 1; 2, 1: *c*.
D^x $\mathbf{G}_2^1$: $(\sigma_x, 0)$: 1, 1; 2, 1: *c*.
P^x $\mathbf{G}_2^1$: $(\sigma_x, 0)$: 1, 3; 2, 3: *c*.
B^x $\mathbf{G}_2^1$: $(\sigma_x, 0)$: 1, 3; 2, 3: *c*.
Σ^x $\mathbf{G}_2^1$: $(\sigma_y, 0)$: 1, 1; 2, 1: *c*.
C^x $\mathbf{G}_2^1$: $(\sigma_y, 0)$: 1, 1; 2, 1: *c*.
E^x $\mathbf{G}_2^1$: $(\sigma_y, 0)$: 1, 3; 2, 3: *c*.
A^x $\mathbf{G}_2^1$: $(\sigma_y, 0)$: 1, 3; 2, 3: *c*.
Λ^x $\mathbf{G}_4^2$: $(C_{2z}, 0)$, $(\sigma_y, 0)$: 1, *x*; 2, *x*; 3, *x*; 4, *x*: *c*.
H^x $\mathbf{G}_4^2$: $(C_{2z}, 0)$, $(\sigma_y, 0)$: 1, *x*; 2, *x*; 3, *x*; 4, *x*: *c*.
Q^x $\mathbf{G}_4^2$: $(C_{2z}, 0)$, $(\sigma_y, 0)$: 1, *x*; 2, *x*; 3, *x*; 4, *x*: *c*.
G^x $\mathbf{G}_4^2$: $(C_{2z}, 0)$, $(\sigma_y, 0)$: 1, *x*; 2, *x*; 3, *x*; 4, *x*: *c*.

28 *Pma*2 C_{2v}^{4}

($F1$; $K7$; $M5$; $T1$; $Z1$.)

Γ $\mathbf{G}_4^2$: $\{C_{2z}|\frac{1}{2}00\}$, $\{\sigma_y|000\}$: 1, 1; 2, 1; 3, 1; 4, 1: *c*.
Y $\mathbf{G}_8^4$: $\{\sigma_x|\frac{1}{2}00\}$, $\{\sigma_y|000\}$: 5, 1: *a*.
X $\mathbf{G}_4^2 \otimes \mathbf{T}_2$: $\{C_{2z}|\frac{1}{2}00\}$, $\{\sigma_y|000\}$; $\mathbf{t}_2$: 1, 1; 2, 1; 3, 1; 4, 1: *c*.
Z $\mathbf{G}_4^2 \otimes \mathbf{T}_2$: $\{C_{2z}|\frac{1}{2}00\}$, $\{\sigma_y|000\}$; $\mathbf{t}_3$: 1, 1; 2, 1; 3, 1; 4, 1: *c*.
U $\mathbf{G}_4^2 \otimes \mathbf{T}_2$: $\{C_{2z}|\frac{1}{2}00\}$, $\{\sigma_y|000\}$; $\mathbf{t}_2$ or $\mathbf{t}_3$: 1, 1; 2, 1; 3, 1; 4, 1: *c*.
T $\mathbf{G}_8^4$: $\{\sigma_x|\frac{1}{2}00\}$, $\{\sigma_y|000\}$: 5, 1: *a*.
S $\mathbf{G}_8^4$: $\{\sigma_x|\frac{1}{2}00\}$, $\{\sigma_y|000\}$: 5, 1: *a*.
R $\mathbf{G}_8^4$: $\{\sigma_x|\frac{1}{2}00\}$, $\{\sigma_y|000\}$: 5, 1: *a*.

Δ^x $\mathbf{G}_2^1$: $(\sigma_x, 0)$: 1, 1; 2, 1: *c*.
D^x $\mathbf{G}_2^1$: $(\sigma_x, 0)$: 1, 1; 2, 1: *c*.
P^x $\mathbf{G}_2^1$: $(\sigma_x, 0)$: 1, 1; 2, 1: *c*.
B^x $\mathbf{G}_2^1$: $(\sigma_x, 0)$: 1, 1; 2, 1: *c*.
Σ^x $\mathbf{G}_2^1$: $(\sigma_y, 0)$: 1, 1; 2, 1: *c*.
C^x $\mathbf{G}_2^1$: $(\sigma_y, 0)$: 1, 3; 2, 3; *c*.
E^x $\mathbf{G}_2^1$: $(\sigma_y, 0)$: 1, 3; 2, 3: *c*.
A^x $\mathbf{G}_2^1$: $(\sigma_y, 0)$: 1, 1; 2, 1: *c*.
Λ^x $\mathbf{G}_4^2$: $(C_{2z}, 0)$, $(\sigma_y, 0)$: 1, *x*; 2, *x*; 3, *x*; 4, *x*: *c*.
H^x $\mathbf{G}_8^4$: $(C_{2z}, 0)$, $(\sigma_y, 0)$: 5, *x*: *a*.
Q^x $\mathbf{G}_8^4$: $(C_{2z}, 0)$, $(\sigma_y, 0)$: 5, *x*: *a*.
G^x $\mathbf{G}_4^2$: $(C_{2z}, 0)$, $(\sigma_y, 0)$: 1, *x*; 2, *x*; 3, *x*; 4, *x*: *c*.

29 $Pca2_1$ C_{2v}^5

($F1$; $K7$; $M5$; $T1$; $Z1$.)

Γ $\mathbf{G}_4^2$: $\{C_{2z}|\frac{1}{2}0\frac{1}{2}\}, \{\sigma_y|00\frac{1}{2}\}$: 1, 1; 2, 1; 3, 1; 4, 1: *c*.
Y $\mathbf{G}_8^4$: $\{\sigma_x|\frac{1}{2}00\}, \{C_{2z}|\frac{1}{2}0\frac{1}{2}\}$: 5, 1: *a*.
X $\mathbf{G}_4^2 \otimes \mathbf{T}_2$: $\{C_{2z}|\frac{1}{2}0\frac{1}{2}\}, \{\sigma_y|00\frac{1}{2}\}$; $\mathbf{t}_2$: 1, 1; 2, 1; 3, 1; 4, 1: *c*.
Z $\mathbf{G}_8^2$: $\{C_{2z}|\frac{1}{2}0\frac{1}{2}\}, \{\sigma_x|\frac{1}{2}00\}$: 2, 3; 4, 3; 6, 3; 8, 3: *c*.
U $\mathbf{G}_8^2$: $\{C_{2z}|\frac{1}{2}0\frac{1}{2}\}, \{\sigma_x|\frac{1}{2}00\}$: 2, 3; 4, 3; 6, 3; 8, 3: *c*.
T $\mathbf{G}_8^5$: $\{C_{2z}|\frac{1}{2}0\frac{1}{2}\}, \{\sigma_y|00\frac{1}{2}\}$: 5, 2: *a*.
S $\mathbf{G}_8^4$: $\{\sigma_x|\frac{1}{2}00\}, \{C_{2z}|\frac{1}{2}0\frac{1}{2}\}$: 5, 1: *a*.
R $\mathbf{G}_8^5$: $\{C_{2z}|\frac{1}{2}0\frac{1}{2}\}, \{\sigma_y|00\frac{1}{2}\}$: 5, 2: *a*.

Δ^x $\mathbf{G}_2^1$: $(\sigma_x, 0)$: 1, 1; 2, 1: *c*.
D^x $\mathbf{G}_2^1$: $(\sigma_x, 0)$: 1, 1; 2, 1: *c*.
P^x $\mathbf{G}_2^1$: $(\sigma_x, 0)$: 1, 2; 2, 2: *c*.
B^x $\mathbf{G}_2^1$: $(\sigma_x, 0)$: 1, 2; 2, 2: *c*.
Σ^x $\mathbf{G}_2^1$: $(\sigma_y, 0)$: 1, 1; 2, 1: *c*.
C^x $\mathbf{G}_2^1$: $(\sigma_y, 0)$: 1, 3; 2, 3: *c*.
E^x $\mathbf{G}_2^1$: $(\sigma_y, 0)$: 1, 2; 2, 2: *c*.
A^x $\mathbf{G}_2^1$: $(\sigma_y, 0)$: 1, 3; 2, 3: *c*.
Λ^x $\mathbf{G}_4^2$: $(C_{2z}, 0), (\sigma_y, 0)$: 1, x; 2, x; 3, x; 4, x: *c*.
H^x $\mathbf{G}_8^4$: $(C_{2z}, 0), (\sigma_y, 0)$: 5, x: *a*.
Q^x $\mathbf{G}_8^4$: $(C_{2z}, 0), (\sigma_y, 0)$: 5, x: *a*.
G^x $\mathbf{G}_4^2$: $(C_{2z}, 0), (\sigma_y, 0)$: 1, x; 2, x; 3, x; 4, x: *c*.

30 $Pnc2$ C_{2v}^6

($F1$; $K7$; $M5$; $T1$; $Z1$.)

Γ $\mathbf{G}_4^2$: $\{C_{2z}|\frac{1}{2}00\}, \{\sigma_y|00\frac{1}{2}\}$: 1, 1; 2, 1; 3, 1; 4, 1: *c*.
Y $\mathbf{G}_8^4$: $\{\sigma_x|\frac{1}{2}0\frac{1}{2}\}, \{\sigma_y|00\frac{1}{2}\}$: 5, 1: *a*.
X $\mathbf{G}_4^2 \otimes \mathbf{T}_2$: $\{C_{2z}|\frac{1}{2}00\}, \{\sigma_y|00\frac{1}{2}\}$; $\mathbf{t}_2$: 1, 1; 2, 1; 3, 1; 4, 1: *c*.
Z $\mathbf{G}_8^2$: $\{\sigma_x|\frac{1}{2}0\frac{1}{2}\}, \{C_{2z}|\frac{1}{2}00\}$: 2, 3; 4, 3; 6, 3; 8, 3: *d*.
U $\mathbf{G}_8^2$: $\{\sigma_x|\frac{1}{2}0\frac{1}{2}\}, \{C_{2z}|\frac{1}{2}00\}$: 2, 3; 4, 3; 6, 3; 8, 3: *d*.
T $\mathbf{G}_8^4$: $\{\sigma_y|00\frac{1}{2}\}, \{C_{2z}|\frac{1}{2}00\}$: 5, 1: *a*.
S $\mathbf{G}_8^4$: $\{\sigma_x|\frac{1}{2}0\frac{1}{2}\}, \{\sigma_y|00\frac{1}{2}\}$: 5, 1: *a*.
R $\mathbf{G}_8^4$: $\{\sigma_y|00\frac{1}{2}\}, \{C_{2z}|\frac{1}{2}00\}$: 5, 1: *a*.

Δ^x $\mathbf{G}_2^1$: $(\sigma_x, 0)$: 1, 1; 2, 1: *c*.
D^x $\mathbf{G}_2^1$: $(\sigma_x, 0)$: 1, 1; 2, 1: *c*.
P^x $\mathbf{G}_2^1$: $(\sigma_x, 0)$: 1, 3; 2, 3: *c*.
B^x $\mathbf{G}_2^1$: $(\sigma_x, 0)$: 1, 3; 2, 3: *c*.
Σ^x $\mathbf{G}_2^1$: $(\sigma_y, 0)$: 1, 1; 2, 1: *c*.
C^x $\mathbf{G}_2^1$: $(\sigma_y, 0)$: 1, 3; 2, 3: *c*.
E^x $\mathbf{G}_2^1$: $(\sigma_y, 0)$: 1, 1; 2, 1: *c*.
A^x $\mathbf{G}_2^1$: $(\sigma_y, 0)$: 1, 3; 2, 3: *c*.
Λ^x $\mathbf{G}_4^2$: $(C_{2z}, 0), (\sigma_y, 0)$: 1, x; 2, x; 3, x; 4, x: *c*.
H^x $\mathbf{G}_8^4$: $(C_{2z}, 0), (\sigma_y, 0)$: 5, x: *a*.
Q^x $\mathbf{G}_8^4$: $(C_{2z}, 0), (\sigma_y, 0)$: 5, x: *a*.
G^x $\mathbf{G}_4^2$: $(C_{2z}, 0), (\sigma_y, 0)$: 1, x; 2, x; 3, x; 4, x: *c*.

31 *Pmn*2_1 C_{2v}^7

($F1$; $K7$; $M5$; $T1$; $Z1$.)

Γ $\mathbf{G}_4^2$: $\{C_{2z}|\frac{1}{2}0\frac{1}{2}\}$, $\{\sigma_y|000\}$: 1, 1; 2, 1; 3, 1; 4, 1: *c*.
Y $\mathbf{G}_8^4$: $\{\sigma_x|\frac{1}{2}0\frac{1}{2}\}$, $\{\sigma_y|000\}$: 5, 1: *a*.
X $\mathbf{G}_4^2 \otimes \mathbf{T}_2$: $\{C_{2z}|\frac{1}{2}0\frac{1}{2}\}$, $\{\sigma_y|000\}$; $\mathbf{t}_2$: 1, 1; 2, 1; 3, 1; 4, 1: *c*.
Z $\mathbf{G}_8^2$: $\{C_{2z}|\frac{1}{2}0\frac{1}{2}\}$, $\{\sigma_y|000\}$: 2, 3; 4, 3; 6, 3; 8, 3: *c*.
U $\mathbf{G}_8^2$: $\{C_{2z}|\frac{1}{2}0\frac{1}{2}\}$, $\{\sigma_y|000\}$: 2, 3; 4, 3; 6, 3; 8, 3: *c*.
T $\mathbf{G}_8^4$: $\{C_{2z}|\frac{1}{2}0\frac{1}{2}\}$, $\{\sigma_y|000\}$: 5, 1: *a*.
S $\mathbf{G}_8^4$: $\{\sigma_x|\frac{1}{2}0\frac{1}{2}\}$, $\{\sigma_y|000\}$: 5, 1: *a*.
R $\mathbf{G}_8^4$: $\{C_{2z}|\frac{1}{2}0\frac{1}{2}\}$, $\{\sigma_y|000\}$: 5, 1: *a*.

Δ^x $\mathbf{G}_2^1$: $(\sigma_x, 0)$: 1, 1; 2, 1: *c*.
D^x $\mathbf{G}_2^1$: $(\sigma_x, 0)$: 1, 1; 2, 1: *c*.
P^x $\mathbf{G}_2^1$: $(\sigma_x, 0)$: 1, 3; 2, 3: *c*.
B^x $\mathbf{G}_2^1$: $(\sigma_x, 0)$: 1, 3; 2, 3: *c*.
Σ^x $\mathbf{G}_2^1$: $(\sigma_y, 0)$: 1, 1; 2, 1: *c*.
C^x $\mathbf{G}_2^1$: $(\sigma_y, 0)$: 1, 3; 2, 3: *c*.
E^x $\mathbf{G}_2^1$: $(\sigma_y, 0)$: 1, 3; 2, 3: *c*.
A^x $\mathbf{G}_2^1$: $(\sigma_y, 0)$: 1, 2; 2, 2: *c*.
Λ^x $\mathbf{G}_4^2$: $(C_{2z}, 0)$, $(\sigma_y, 0)$: 1, *x*; 2, *x*; 3, *x*; 4, *x*: *c*.
H^x $\mathbf{G}_8^4$: $(C_{2z}, 0)$, $(\sigma_y, 0)$: 5, *x*: *a*.
Q^x $\mathbf{G}_8^4$: $(C_{2z}, 0)$, $(\sigma_y, 0)$: 5, *x*: *a*.
G^x $\mathbf{G}_4^2$: $(C_{2z}, 0)$, $(\sigma_y, 0)$: 1, *x*; 2, *x*; 3, *x*; 4, *x*: *c*.

32 *Pba*2 C_{2v}^8

($F1$; $K7$; $M5$; $T1$; $Z1$.)

Γ $\mathbf{G}_4^2$: $\{C_{2z}|\frac{1}{2}\frac{1}{2}0\}$, $\{\sigma_y|0\frac{1}{2}0\}$: 1, 1; 2, 1; 3, 1; 4, 1: *c*.
Y $\mathbf{G}_8^4$: $\{\sigma_x|\frac{1}{2}00\}$, $\{C_{2z}|\frac{1}{2}\frac{1}{2}0\}$: 5, 1: *a*.
X $\mathbf{G}_8^4$: $\{\sigma_y|0\frac{1}{2}0\}$, $\{C_{2z}|\frac{1}{2}\frac{1}{2}0\}$: 5, 1: *a*.
Z $\mathbf{G}_4^2 \otimes \mathbf{T}_2$: $\{C_{2z}|\frac{1}{2}\frac{1}{2}0\}$, $\{\sigma_y|0\frac{1}{2}0\}$; $\mathbf{t}_3$: 1, 1; 2, 1; 3, 1; 4, 1: *c*.
U $\mathbf{G}_8^4$: $\{\sigma_y|0\frac{1}{2}0\}$, $\{C_{2z}|\frac{1}{2}\frac{1}{2}0\}$: 5, 1: *a*.
T $\mathbf{G}_8^4$: $\{\sigma_x|\frac{1}{2}00\}$, $\{C_{2z}|\frac{1}{2}\frac{1}{2}0\}$: 5, 1: *a*.
S $\mathbf{G}_8^2$: $\{\sigma_x|\frac{1}{2}00\}$, $\{C_{2z}|\frac{1}{2}\frac{1}{2}0\}$: 2, 3; 4, 3; 6, 3; 8, 3: *d*.
R $\mathbf{G}_8^2$: $\{\sigma_x|\frac{1}{2}00\}$, $\{C_{2z}|\frac{1}{2}\frac{1}{2}0\}$: 2, 3; 4, 3; 6, 3; 8, 3: *d*.

Δ^x $\mathbf{G}_2^1$: $(\sigma_x, 0)$: 1, 1; 2, 1: *c*.
D^x $\mathbf{G}_2^1$: $(\sigma_x, 0)$: 1, 3; 2, 3: *c*.
P^x $\mathbf{G}_2^1$: $(\sigma_x, 0)$: 1, 3; 2, 3: *c*.
B^x $\mathbf{G}_2^1$: $(\sigma_x, 0)$: 1, 1; 2, 1: *c*.
Σ^x $\mathbf{G}_2^1$: $(\sigma_y, 0)$: 1, 1; 2, 1: *c*.
C^x $\mathbf{G}_2^1$: $(\sigma_y, 0)$: 1, 3; 2, 3: *c*.
E^x $\mathbf{G}_2^1$: $(\sigma_y, 0)$: 1, 3; 2, 3: *c*.
A^x $\mathbf{G}_2^1$: $(\sigma_y, 0)$: 1, 1; 2, 1: *c*.
Λ^x $\mathbf{G}_4^2$: $(C_{2z}, 0)$, $(\sigma_y, 0)$: 1, *x*; 2, *x*; 3, *x*; 4, *x*: *c*.
H^x $\mathbf{G}_8^4$: $(C_{2z}, 0)$, $(\sigma_y, 0)$: 5, *x*: *a*.
Q^x $\mathbf{G}_8^3$: $(C_{2z}, 0)$, $(\sigma_y, 0)$, $(E, 1)$: 5, *x*; 6, *x*; 7, *x*; 8, *x*: *a*.
G^x $\mathbf{G}_8^4$: $(C_{2z}, 0)$, $(\sigma_y, 0)$: 5, *x*: *a*.

33 *Pna*2_1 C_{2v}^9

(*F*1; *K*7; *M*5; *T*1; *Z*1.)

Γ $\mathbf{G}_4^2$: $\{C_{2z}|\frac{1}{2}\frac{1}{2}\frac{1}{2}\}$, $\{\sigma_y|0\frac{1}{2}\frac{1}{2}\}$: 1, 1; 2, 1; 3, 1; 4, 1: *c*.
Y $\mathbf{G}_8^4$: $\{\sigma_x|\frac{1}{2}00\}$, $\{C_{2z}|\frac{1}{2}\frac{1}{2}\frac{1}{2}\}$: 5, 1: *a*.
X $\mathbf{G}_8^4$: $\{\sigma_y|0\frac{1}{2}\frac{1}{2}\}$, $\{C_{2z}|\frac{1}{2}\frac{1}{2}\frac{1}{2}\}$: 5, 1: *a*.
Z $\mathbf{G}_8^2$: $\{C_{2z}|\frac{1}{2}\frac{1}{2}\frac{1}{2}\}$, $\{\sigma_x|\frac{1}{2}00\}$: 2, 3; 4, 3; 6, 3; 8, 3: *c*.
U $\mathbf{G}_8^4$: $\{C_{2z}|\frac{1}{2}\frac{1}{2}\frac{1}{2}\}$, $\{\sigma_y|0\frac{1}{2}\frac{1}{2}\}$: 5, 1: *a*.
T $\mathbf{G}_8^5$: $\{C_{2z}|\frac{1}{2}\frac{1}{2}\frac{1}{2}\}$, $\{\sigma_y|0\frac{1}{2}\frac{1}{2}\}$: 5, 2: *a*.
S $\mathbf{G}_8^2$: $\{\sigma_y|0\frac{1}{2}\frac{1}{2}\}$, $\{C_{2z}|\frac{1}{2}\frac{1}{2}\frac{1}{2}\}$: 2, 3; 4, 3; 6, 3; 8, 3: *d*.
R $\mathbf{G}_8^2$: $\{C_{2z}|\frac{1}{2}\frac{1}{2}\frac{1}{2}\}$, $\{\sigma_y|0\frac{1}{2}\frac{1}{2}\}$: 2, 3; 4, 3; 6, 3; 8, 3: *c*.

Δ^x $\mathbf{G}_2^1$: $(\sigma_x, 0)$: 1, 1; 2, 1: *c*.
D^x $\mathbf{G}_2^1$: $(\sigma_x, 0)$: 1, 3; 2, 3: *c*.
P^x $\mathbf{G}_2^1$: $(\sigma_x, 0)$: 1, 3; 2, 3: *c*.
B^x $\mathbf{G}_2^1$: $(\sigma_x, 0)$: 1, 2; 2, 2: *c*.
Σ^x $\mathbf{G}_2^1$: $(\sigma_y, 0)$: 1, 1; 2, 1: *c*.
C^x $\mathbf{G}_2^1$: $(\sigma_y, 0)$: 1, 3; 2, 3: *c*.
E^x $\mathbf{G}_2^1$: $(\sigma_y, 0)$: 1, 2; 2, 2: *c*.
A^x $\mathbf{G}_2^1$: $(\sigma_y, 0)$: 1, 3; 2, 3: *c*.
Λ^x $\mathbf{G}_4^2$: $(C_{2z}, 0)$, $(\sigma_y, 0)$: 1, *x*; 2, *x*; 3, *x*; 4, *x*: *c*.
H^x $\mathbf{G}_8^4$: $(C_{2z}, 0)$, $(\sigma_y, 0)$: 5, *x*: *a*.
Q^x $\mathbf{G}_8^3$: $(C_{2z}, 0)$, $(\sigma_y, 0)$, $(E, 1)$: 5, *x*; 6, *x*; 7, *x*; 8, *x*: *a*.
G^x $\mathbf{G}_8^4$: $(C_{2z}, 0)$, $(\sigma_y, 0)$: 5, *x*: *a*.

34 *Pnn*2 C_{2v}^{10}

(*F*1; *K*7; *M*5; *T*1; *Z*1.)

Γ $\mathbf{G}_4^2$: $\{C_{2z}|\frac{1}{2}\frac{1}{2}0\}$, $\{\sigma_y|0\frac{1}{2}\frac{1}{2}\}$: 1, 1; 2, 1; 3, 1; 4, 1: *c*.
Y $\mathbf{G}_8^4$: $\{\sigma_x|\frac{1}{2}0\frac{1}{2}\}$, $\{C_{2z}|\frac{1}{2}\frac{1}{2}0\}$: 5, 1: *a*.
X $\mathbf{G}_8^4$: $\{\sigma_y|0\frac{1}{2}\frac{1}{2}\}$, $\{C_{2z}|\frac{1}{2}\frac{1}{2}0\}$: 5, 1: *a*.
Z $\mathbf{G}_8^2$: $\{\sigma_y|0\frac{1}{2}\frac{1}{2}\}$, $\{C_{2z}|\frac{1}{2}\frac{1}{2}0\}$: 2, 3; 4, 3; 6, 3; 8, 3: *d*.
U $\mathbf{G}_8^4$: $\{\sigma_x|\frac{1}{2}0\frac{1}{2}\}$, $\{C_{2z}|\frac{1}{2}\frac{1}{2}0\}$: 5, 1: *a*.
T $\mathbf{G}_8^4$: $\{\sigma_y|0\frac{1}{2}\frac{1}{2}\}$, $\{C_{2z}|\frac{1}{2}\frac{1}{2}0\}$: 5, 1: *a*.
S $\mathbf{G}_8^2$: $\{\sigma_y|0\frac{1}{2}\frac{1}{2}\}$, $\{C_{2z}|\frac{1}{2}\frac{1}{2}0\}$: 2, 3; 4, 3; 6, 3; 8, 3: *d*.
R $\mathbf{G}_4^2 \otimes \mathbf{T}_2$: $\{C_{2z}|\frac{1}{2}\frac{1}{2}0\}$, $\{\sigma_y|0\frac{1}{2}\frac{1}{2}\}$; $\mathbf{t}_1$ or $\mathbf{t}_2$ or $\mathbf{t}_3$: 1, 1; 2, 1; 3, 1; 4, 1: *c*.

Δ^x $\mathbf{G}_2^1$: $(\sigma_x, 0)$: 1, 1; 2, 1: *c*.
D^x $\mathbf{G}_2^1$: $(\sigma_x, 0)$: 1, 3; 2, 3: *c*.
P^x $\mathbf{G}_2^1$: $(\sigma_x, 0)$: 1, 1; 2, 1: *c*.
B^x $\mathbf{G}_2^1$: $(\sigma_x, 0)$: 1, 3; 2, 3: *c*.
Σ^x $\mathbf{G}_2^1$: $(\sigma_y, 0)$: 1, 1; 2, 1: *c*.
C^x $\mathbf{G}_2^1$: $(\sigma_y, 0)$: 1, 3; 2, 3: *c*.
E^x $\mathbf{G}_2^1$: $(\sigma_y, 0)$: 1, 1; 2, 1: *c*.
A^x $\mathbf{G}_2^1$: $(\sigma_y, 0)$: 1, 3; 2, 3: *c*.
Λ^x $\mathbf{G}_4^2$: $(C_{2z}, 0)$, $(\sigma_y, 0)$: 1, *x*; 2, *x*; 3, *x*; 4, *x*: *c*.
H^x $\mathbf{G}_8^4$: $(C_{2z}, 0)$, $(\sigma_y, 0)$: 5, *x*: *a*.
Q^x $\mathbf{G}_8^3$: $(C_{2z}, 0)$, $(\sigma_y, 0)$, $(E, 1)$: 5, *x*; 6, *x*; 7, *x*; 8, *x*: *a*.
G^x $\mathbf{G}_8^4$: $(C_{2z}, 0)$, $(\sigma_y, 0)$: 5, *x*: *a*.

35 *Cmm*2 C_{2v}^{11}

($F1$; $K7$; $M5$; $T3$; $Z1$.)

Γ $\mathbf{G}_4^2$: $\{C_{2z}|000\}$, $\{\sigma_y|000\}$: 1, 1; 2, 1; 3, 1; 4, 1: *c*.
Y $\mathbf{G}_4^2 \otimes \mathbf{T}_2$: $\{C_{2z}|000\}$, $\{\sigma_y|000\}$; $\mathbf{t}_1$ or $\mathbf{t}_2$: 1, 1; 2, 1; 3, 1; 4, 1: *c*.
Z $\mathbf{G}_4^2 \otimes \mathbf{T}_2$: $\{C_{2z}|000\}$, $\{\sigma_y|000\}$; $\mathbf{t}_3$: 1, 1; 2, 1; 3, 1; 4, 1: *c*.
T $\mathbf{G}_4^2 \otimes \mathbf{T}_2$: $\{C_{2z}|000\}$, $\{\sigma_y|000\}$; $\mathbf{t}_1$ or $\mathbf{t}_2$ or $\mathbf{t}_3$: 1, 1; 2, 1; 3, 1; 4, 1: *c*.
S $\mathbf{G}_2^1 \otimes \mathbf{T}_2$: $\{C_{2z}|000\}$; $\mathbf{t}_2$: 1, 1; 2, 1: *b*.
R $\mathbf{G}_2^1 \otimes \mathbf{T}_2$: $\{C_{2z}|000\}$; $\mathbf{t}_2$ or $\mathbf{t}_3$: 1, 1; 2, 1: *b*.

Λ^x $\mathbf{G}_4^2$: $(C_{2z}, 0)$, $(\sigma_y, 0)$: 1, x; 2, x; 3, x; 4, x: *c*.
H^x $\mathbf{G}_4^2$: $(C_{2z}, 0)$, $(\sigma_y, 0)$: 1, x; 2, x; 3, x; 4, x: *c*.
D^x $\mathbf{G}_2^1$: $(C_{2z}, 0)$: 1, x; 2, x: *b*.
A^x $\mathbf{G}_2^1$: $(\sigma_y, 0)$: 1, 1; 2, 1: *c*.
Σ^x $\mathbf{G}_2^1$: $(\sigma_y, 0)$: 1, 1; 2, 1: *c*.
Δ^x $\mathbf{G}_2^1$: $(\sigma_x, 0)$: 1, 1; 2, 1: *c*.
B^x $\mathbf{G}_2^1$: $(\sigma_x, 0)$: 1, 1; 2, 1: *c*.
G^x $\mathbf{G}_2^1$: $(\sigma_x, 0)$: 1, 1; 2, 1: *c*.
F^x $\mathbf{G}_2^1$: $(\sigma_x, 0)$: 1, 1; 2, 1: *c*.
E^x $\mathbf{G}_2^1$: $(\sigma_y, 0)$: 1, 1; 2, 1: *c*.
C^x $\mathbf{G}_2^1$: $(\sigma_y, 0)$: 1, 1; 2, 1: *c*.

36 *Cmc*2_1 C_{2v}^{12}

($F1$; $K7$; $M5$; $T3$; $Z1$.)

Γ $\mathbf{G}_4^2$: $\{C_{2z}|00\frac{1}{2}\}$, $\{\sigma_y|000\}$: 1, 1; 2, 1; 3, 1; 4, 1: *c*.
Y $\mathbf{G}_4^2 \otimes \mathbf{T}_2$: $\{C_{2z}|00\frac{1}{2}\}$, $\{\sigma_y|000\}$; $\mathbf{t}_1$ or $\mathbf{t}_2$: 1, 1; 2, 1; 3, 1; 4, 1: *c*
Z $\mathbf{G}_8^2$: $\{C_{2z}|00\frac{1}{2}\}$, $\{\sigma_y|000\}$: 2, 3; 4, 3; 6, 3; 8, 3: *c*.
T $\mathbf{G}_8^2$: $\{C_{2z}|00\frac{1}{2}\}$, $\{\sigma_y|000\}$: 2, 3; 4, 3; 6, 3; 8, 3: *c*.
S $\mathbf{G}_2^1 \otimes \mathbf{T}_2$: $\{C_{2z}|00\frac{1}{2}\}$; $\mathbf{t}_2$: 1, 1; 2, 1: *b*.
R $\mathbf{G}_4^1$: $\{C_{2z}|00\frac{1}{2}\}$: 2, 3; 4, 3: *a*.

Λ^x $\mathbf{G}_4^2$: $(C_{2z}, 0)$, $(\sigma_y, 0)$: 1, x; 2, x; 3, x; 4, x: *c*.
H^x $\mathbf{G}_4^2$: $(C_{2z}, 0)$, $(\sigma_y, 0)$: 1, x; 2, x; 3, x; 4, x: *c*.
D^x $\mathbf{G}_2^1$: $(C_{2z}, 0)$: 1, x; 2, x: *b*.
A^x $\mathbf{G}_2^1$: $(\sigma_y, 0)$: 1, 2; 2, 2: *c*.
Σ^x $\mathbf{G}_2^1$: $(\sigma_y, 0)$: 1, 1; 2, 1: *c*.
Δ^x $\mathbf{G}_2^1$: $(\sigma_x, 0)$: 1, 1; 2, 1: *c*.
B^x $\mathbf{G}_2^1$: $(\sigma_x, 0)$: 1, 3; 2, 3: *c*.
G^x $\mathbf{G}_2^1$: $(\sigma_x, 0)$: 1, 3; 2, 3: *c*.
F^x $\mathbf{G}_2^1$: $(\sigma_x, 0)$: 1, 1; 2, 1: *c*.
E^x $\mathbf{G}_2^1$: $(\sigma_y, 0)$: 1, 2; 2, 2: *c*.
C^x $\mathbf{G}_2^1$: $(\sigma_y, 0)$: 1, 1; 2, 1: *c*.

37 *Ccc*2 C_{2v}^{13}

(*F*1; *K*7; *M*5; *T*3; *Z*1.)

Γ $\mathbf{G}_4^2$: $\{C_{2z}\,|\,000\}$, $\{\sigma_y\,|\,00\frac{1}{2}\}$: 1, 1; 2, 1; 3, 1; 4, 1: *c*.
Y $\mathbf{G}_4^2 \otimes \mathbf{T}_2$: $\{C_{2z}\,|\,000\}$, $\{\sigma_y\,|\,00\frac{1}{2}\}$; $\mathbf{t}_1$ or $\mathbf{t}_2$: 1, 1; 2, 1; 3, 1; 4, 1: *c*.
Z $\mathbf{G}_8^2$: $\{\sigma_x\,|\,00\frac{1}{2}\}$, $\{C_{2z}\,|\,000\}$: 2, 3; 4, 3; 6, 3; 8, 3: *d*.
T $\mathbf{G}_8^2$: $\{\sigma_x\,|\,00\frac{1}{2}\}$, $\{C_{2z}\,|\,000\}$: 2, 3; 4, 3; 6, 3; 8, 3: *d*.
S $\mathbf{G}_2^1 \otimes \mathbf{T}_2$: $\{C_{2z}\,|\,000\}$; $\mathbf{t}_2$: 1, 1; 2, 1: *b*.
R $\mathbf{G}_2^1 \otimes \mathbf{T}_2$: $\{C_{2z}\,|\,000\}$; $\mathbf{t}_2$ or $\mathbf{t}_3$: 1, 1; 2, 1: *b*.

Λ^x $\mathbf{G}_4^2$: $(C_{2z}, 0)$, $(\sigma_y, 0)$: 1, *x*; 2, *x*; 3, *x*; 4, *x*: *c*.
H^x $\mathbf{G}_4^2$: $(C_{2z}, 0)$, $(\sigma_y, 0)$: 1, *x*; 2, *x*; 3, *x*; 4, *x*: *c*.
D^x $\mathbf{G}_2^1$: $(C_{2z}, 0)$: 1, *x*; 2, *x*: *b*.
A^x $\mathbf{G}_2^1$: $(\sigma_y, 0)$: 1, 3; 2, 3: *c*.
Σ^x $\mathbf{G}_2^1$: $(\sigma_y, 0)$: 1, 1; 2, 1: *c*.
Δ^x $\mathbf{G}_2^1$: $(\sigma_x, 0)$: 1, 1; 2, 1: *c*.
B^x $\mathbf{G}_2^1$: $(\sigma_x, 0)$: 1, 3; 2, 3: *c*.
G^x $\mathbf{G}_2^1$: $(\sigma_x, 0)$: 1, 3; 2, 3: *c*.
F^x $\mathbf{G}_2^1$: $(\sigma_x, 0)$: 1, 1; 2, 1: *c*.
E^x $\mathbf{G}_2^1$: $(\sigma_y, 0)$: 1, 3; 2, 3: *c*.
C^x $\mathbf{G}_2^1$: $(\sigma_y, 0)$: 1, 1; 2, 1: *c*.

38 *Amm*2 C_{2v}^{14}

(*F*1; *K*7; *M*5; *T*3; *Z*1.)

Γ $\mathbf{G}_4^2$: $\{C_{2y}\,|\,000\}$, $\{\sigma_x\,|\,000\}$: 1, 1; 2, 1; 3, 1; 4, 1: *c*.
Y $\mathbf{G}_4^2 \otimes \mathbf{T}_2$: $\{C_{2y}\,|\,000\}$, $\{\sigma_x\,|\,000\}$; $\mathbf{t}_1$ or $\mathbf{t}_2$: 1, 1; 2, 1; 3, 1; 4, 1: *c*.
Z $\mathbf{G}_4^2 \otimes \mathbf{T}_2$: $\{C_{2y}\,|\,000\}$, $\{\sigma_x\,|\,000\}$; $\mathbf{t}_3$: 1, 1; 2, 1; 3, 1; 4, 1: *c*.
T $\mathbf{G}_4^2 \otimes \mathbf{T}_2$: $\{C_{2y}\,|\,000\}$, $\{\sigma_x\,|\,000\}$; $\mathbf{t}_1$ or $\mathbf{t}_2$ or $\mathbf{t}_3$: 1, 1; 2, 1; 3, 1; 4, 1: *c*.
S $\mathbf{G}_2^1 \otimes \mathbf{T}_2$: $\{\sigma_z\,|\,000\}$; $\mathbf{t}_2$: 1, 1; 2, 1: *c*.
R $\mathbf{G}_2^1 \otimes \mathbf{T}_2$: $\{\sigma_z\,|\,000\}$; $\mathbf{t}_2$ or $\mathbf{t}_3$: 1, 1; 2, 1: *c*.

Λ^x $\mathbf{G}_2^1$: $(\sigma_x, 0)$: 1, 1; 2, 1: *c*.
H^x $\mathbf{G}_2^1$: $(\sigma_x, 0)$: 1, 1; 2, 1: *c*.
D^x $\mathbf{G}_1^1$: $(E, 0)$: 1, 1: *a*.
A^x $\mathbf{G}_2^1$: $(\sigma_z, 0)$: 1, 1; 2, 1: *c*.
Σ^x $\mathbf{G}_2^1$: $(\sigma_z, 0)$: 1, 1; 2, 1: *c*.
Δ^x $\mathbf{G}_4^2$: $(C_{2y}, 0)$, $(\sigma_x, 0)$: 1, *x*; 2, *x*; 3, *x*; 4, *x*: *c*.
B^x $\mathbf{G}_4^2$: $(C_{2y}, 0)$, $(\sigma_x, 0)$: 1, *x*; 2, *x*; 3, *x*; 4, *x*: *c*.
G^x $\mathbf{G}_4^2$: $(C_{2y}, 0)$, $(\sigma_x, 0)$: 1, *x*; 2, *x*; 3, *x*; 4, *x*: *c*.
F^x $\mathbf{G}_4^2$: $(C_{2y}, 0)$, $(\sigma_x, 0)$: 1, *x*; 2, *x*; 3, *x*; 4, *x*: *c*.
E^x $\mathbf{G}_2^1$: $(\sigma_z, 0)$: 1, 1; 2, 1: *c*.
C^x $\mathbf{G}_2^1$: $(\sigma_z, 0)$: 1, 1; 2, 1: *c*.

39 *Abm2* C_{2v}^{15}

($F1$; $K7$; $M5$; $T3$; $Z1$.)

Γ $\mathbf{G}_4^2$: $\{C_{2y}|000\}$, $\{\sigma_x|\frac{1}{2}\frac{1}{2}0\}$: 1, 1; 2, 1; 3, 1; 4, 1: *c*.
Y $\mathbf{G}_4^2 \otimes \mathbf{T}_2$: $\{C_{2y}|000\}$, $\{\sigma_x|\frac{1}{2}\frac{1}{2}0\}$; $\mathbf{t}_1$ or $\mathbf{t}_2$: 1, 1; 2, 1; 3, 1; 4, 1: *c*.
Z $\mathbf{G}_4^2 \otimes \mathbf{T}_2$: $\{C_{2y}|000\}$, $\{\sigma_x|\frac{1}{2}\frac{1}{2}0\}$; $\mathbf{t}_3$: 1, 1; 2, 1; 3, 1; 4, 1: *c*.
T $\mathbf{G}_4^2 \otimes \mathbf{T}_2$: $\{C_{2y}|000\}$, $\{\sigma_x|\frac{1}{2}\frac{1}{2}0\}$; $\mathbf{t}_1$ or $\mathbf{t}_2$ or $\mathbf{t}_3$: 1, 1; 2, 1; 3, 1; 4, 1: *c*.
S $\mathbf{G}_4^1$: $\{\sigma_z|\frac{1}{2}\frac{1}{2}0\}$: 2, 3; 4, 3: *a*.
R $\mathbf{G}_4^1$: $\{\sigma_z|\frac{1}{2}\frac{1}{2}0\}$: 2, 3; 4, 3: *a*.

Λ^x $\mathbf{G}_2^1$: $(\sigma_x, 0)$: 1, 1; 2, 1: *c*.
H^x $\mathbf{G}_2^1$: $(\sigma_x, 0)$: 1, 1; 2, 1: *c*.
D^x $\mathbf{G}_1^1$: $(E, 0)$: 1, 2: *a*.
A^x $\mathbf{G}_2^1$: $(\sigma_z, 0)$: 1, 1; 2, 1: *c*.
Σ^x $\mathbf{G}_2^1$: $(\sigma_z, 0)$: 1, 1; 2, 1: *c*.
Δ^x $\mathbf{G}_4^2$: $(C_{2y}, 0)$, $(\sigma_x, 0)$: 1, x; 2, x; 3, x; 4, x: *c*.
B^x $\mathbf{G}_4^2$: $(C_{2y}, 0)$, $(\sigma_x, 0)$: 1, x; 2, x; 3, x; 4, x: *c*.
G^x $\mathbf{G}_4^2$: $(C_{2y}, 0)$, $(\sigma_x, 0)$: 1, x; 2, x; 3, x; 4, x: *c*.
F^x $\mathbf{G}_4^2$: $(C_{2y}, 0)$, $(\sigma_x, 0)$: 1, x; 2, x; 3, x; 4, x: *c*.
E^x $\mathbf{G}_2^1$: $(\sigma_z, 0)$: 1, 1; 2, 1: *c*.
C^x $\mathbf{G}_2^1$: $(\sigma_z, 0)$: 1, 1; 2, 1: *c*.

40 *Ama2* C_{2v}^{16}

($F1$; $K7$; $M5$; $T3$; $Z1$.)

Γ $\mathbf{G}_4^2$: $\{C_{2y}|000\}$, $\{\sigma_x|00\frac{1}{2}\}$: 1, 1; 2, 1; 3, 1; 4, 1: *c*.
Y $\mathbf{G}_4^2 \otimes \mathbf{T}_2$: $\{C_{2y}|000\}$, $\{\sigma_x|00\frac{1}{2}\}$; $\mathbf{t}_1$ or $\mathbf{t}_2$: 1, 1; 2, 1; 3, 1; 4, 1: *c*.
Z $\mathbf{G}_8^4$: $\{\sigma_x|00\frac{1}{2}\}$, $\{C_{2y}|000\}$: 5, 1: *a*.
T $\mathbf{G}_8^4$: $\{\sigma_x|00\frac{1}{2}\}$, $\{C_{2y}|000\}$: 5, 1: *a*.
S $\mathbf{G}_2^1 \otimes \mathbf{T}_2$: $\{\sigma_z|00\frac{1}{2}\}$; $\mathbf{t}_2$: 1, 1; 2, 1; *c*.
R $\mathbf{G}_2^1 \otimes \mathbf{T}_2$: $\{\sigma_z|00\frac{1}{2}\}$; $\mathbf{t}_2$ or $\mathbf{t}_3$: 1, 1; 2, 1: *c*.

Λ^x $\mathbf{G}_2^1$: $(\sigma_x, 0)$: 1, 1; 2, 1: *c*.
H^x $\mathbf{G}_2^1$: $(\sigma_x, 0)$: 1, 1; 2, 1: *c*.
D^x $\mathbf{G}_1^1$: $(E, 0)$: 1, 1: *a*.
A^x $\mathbf{G}_4^1$: $(\sigma_z, 0)$: 2, 3; 4, 3: *a*.
Σ^x $\mathbf{G}_2^1$: $(\sigma_z, 0)$: 1, 1; 2, 1: *c*.
Δ^x $\mathbf{G}_4^2$: $(C_{2y}, 0)$, $(\sigma_x, 0)$: 1, x; 2, x; 3, x; 4, x: *c*.
B^x $\mathbf{G}_8^4$: $(\sigma_z, 0)$, $(C_{2y}, 0)$: 5, x: *a*.
G^x $\mathbf{G}_8^4$: $(\sigma_z, 0)$, $(C_{2y}, 0)$: 5, x: *a*.
F^x $\mathbf{G}_4^2$: $(C_{2y}, 0)$, $(\sigma_x, 0)$: 1, x; 2, x; 3, x; 4, x: *c*.
E^x $\mathbf{G}_4^1$: $(\sigma_z, 0)$: 2, 3; 4, 3: *a*.
C^x $\mathbf{G}_2^1$: $(\sigma_z, 0)$: 1, 1; 2, 1: *c*.

41 *Aba*2 C_{2v}^{17}

(*F*1; *K*7; *M*5; *T*3; *Z*1.)

Γ $\mathbf{G}_4^2$: $\{C_{2y}|000\}$, $\{\sigma_x|\frac{1}{2}\frac{1}{2}\frac{1}{2}\}$: 1, 1; 2, 1; 3, 1; 4, 1: *c*.
Y $\mathbf{G}_4^2 \otimes \mathbf{T}_2$: $\{C_{2y}|000\}$, $\{\sigma_x|\frac{1}{2}\frac{1}{2}\frac{1}{2}\}$; $\mathbf{t}_1$ or $\mathbf{t}_2$: 1, 1; 2, 1; 3, 1; 4, 1: *c*.
Z $\mathbf{G}_8^4$: $\{\sigma_x|\frac{1}{2}\frac{1}{2}\frac{1}{2}\}$, $\{C_{2y}|000\}$: 5, 1: *a*.
T $\mathbf{G}_8^4$: $\{\sigma_x|\frac{1}{2}\frac{1}{2}\frac{1}{2}\}$, $\{C_{2y}|000\}$: 5, 1: *a*.
S $\mathbf{G}_4^1$: $\{\sigma_z|\frac{1}{2}\frac{1}{2}\frac{1}{2}\}$: 2, 3; 4, 3: *a*.
R $\mathbf{G}_4^1$: $\{\sigma_z|\frac{1}{2}\frac{1}{2}\frac{1}{2}\}$: 2, 3; 4, 3: *a*.

Λ^x $\mathbf{G}_2^1$: $(\sigma_x, 0)$: 1, 1; 2, 1: *c*.
H^x $\mathbf{G}_2^1$: $(\sigma_x, 0)$: 1, 1; 2, 1: *c*.
D^x $\mathbf{G}_1^1$: $(E, 0)$: 1, 2: *a*.
A^x $\mathbf{G}_4^1$: $(\sigma_z, 0)$: 2, 3; 4, 3: *a*.
Σ^x $\mathbf{G}_2^1$: $(\sigma_z, 0)$: 1, 1; 2, 1: *c*.
Δ^x $\mathbf{G}_4^2$: $(C_{2y}, 0)$, $(\sigma_x, 0)$: 1, *x*; 2, *x*; 3, *x*; 4, *x*: *c*.
B^x $\mathbf{G}_8^4$: $(\sigma_z, 0)$, $(C_{2y}, 0)$: 5, *x*: *a*.
G^x $\mathbf{G}_8^4$: $(\sigma_z, 0)$, $(C_{2y}, 0)$: 5, *x*: *a*.
F^x $\mathbf{G}_4^2$: $(C_{2y}, 0)$, $(\sigma_x, 0)$: 1, *x*; 2, *x*; 3, *x*; 4, *x*: *c*.
E^x $\mathbf{G}_4^1$: $(\sigma_z, 0)$: 2, 3; 4, 3: *a*.
C^x $\mathbf{G}_2^1$: $(\sigma_z, 0)$: 1, 1; 2, 1: *c*.

42 *Fmm*2 C_{2v}^{18}

(*F*1; *K*7; *M*5; *S*14; *Z*1.)

Γ $\mathbf{G}_4^2$: $\{C_{2z}|000\}$, $\{\sigma_y|000\}$: 1, 1; 2, 1; 3, 1; 4, 1: *c*.
Y $\mathbf{G}_4^2 \otimes \mathbf{T}_2$: $\{C_{2z}|000\}$, $\{\sigma_y|000\}$; $\mathbf{t}_2$ or $\mathbf{t}_3$: 1, 1; 2, 1; 3, 1; 4, 1: *c*.
X $\mathbf{G}_4^2 \otimes \mathbf{T}_2$: $\{C_{2z}|000\}$, $\{\sigma_y|000\}$; $\mathbf{t}_1$ or $\mathbf{t}_3$: 1, 1; 2, 1; 3, 1; 4, 1: *c*.
Z $\mathbf{G}_4^2 \otimes \mathbf{T}_2$: $\{C_{2z}|000\}$, $\{\sigma_y|000\}$; $\mathbf{t}_1$ or $\mathbf{t}_2$: 1, 1; 2, 1; 3, 1; 4, 1: *c*.
L $\mathbf{G}_1^1 \otimes \mathbf{T}_2$: $\{E|000\}$; $\mathbf{t}_1$: 1, 1: *a*.

Λ^x $\mathbf{G}_4^2$: $(C_{2z}, 0)$, $(\sigma_y, 0)$: 1, *x*; 2, *x*; 3, *x*; 4, *x*: *c*.
G^x $\mathbf{G}_4^2$: $(C_{2z}, 0)$, $(\sigma_y, 0)$: 1, *x*; 2, *x*; 3, *x*; 4, *x*: *c*.
H^x $\mathbf{G}_4^2$: $(C_{2z}, 0)$, $(\sigma_y, 0)$: 1, *x*; 2, *x*; 3, *x*; 4, *x*: *c*.
Q^x $\mathbf{G}_4^2$: $(C_{2z}, 0)$, $(\sigma_y, 0)$: 1, *x*; 2, *x*; 3, *x*; 4, *x*: *c*.
Σ^x $\mathbf{G}_2^1$: $(\sigma_y, 0)$: 1, 1; 2, 1: *c*.
C^x $\mathbf{G}_2^1$: $(\sigma_y, 0)$: 1, 1; 2, 1: *c*.
A^x $\mathbf{G}_2^1$: $(\sigma_y, 0)$: 1, 1; 2, 1: *c*.
U^x $\mathbf{G}_2^1$: $(\sigma_y, 0)$: 1, 1; 2, 1: *c*.
Δ^x $\mathbf{G}_2^1$: $(\sigma_x, 0)$: 1, 1; 2, 1: *c*.
D^x $\mathbf{G}_2^1$: $(\sigma_x, 0)$: 1, 1; 2, 1: *c*.
B^x $\mathbf{G}_2^1$: $(\sigma_x, 0)$: 1, 1; 2, 1: *c*.
R^x $\mathbf{G}_2^1$: $(\sigma_x, 0)$: 1, 1; 2, 1: *c*.

43 *Fdd2* C_{2v}^{19}

($F1$; $K7$; $M5$; $S14$; $Z1$.)

Γ $\mathbf{G}_4^2$: $\{C_{2z}|00\tfrac{1}{2}\}$, $\{\sigma_y|\tfrac{1}{2}00\}$: 1, 1; 2, 1; 3, 1; 4, 1: *c*.
Y $\mathbf{G}_8^4$: $\{\sigma_x|0\tfrac{1}{2}0\}$, $\{C_{2z}|00\tfrac{1}{2}\}$: 5, 1: *a*.
X $\mathbf{G}_8^4$: $\{\sigma_y|\tfrac{1}{2}00\}$, $\{C_{2z}|00\tfrac{1}{2}\}$: 5, 1: *a*.
Z $\mathbf{G}_8^2$: $\{\sigma_x|0\tfrac{1}{2}0\}$, $\{C_{2z}|00\tfrac{1}{2}\}$: 2, 3; 4, 3; 6, 3; 8, 3: *d*.
L $\mathbf{G}_1^1 \otimes \mathbf{T}_2$: $\{E|000\}$; $\mathbf{t}_1$: 1, 1: *a*.

Λ^x $\mathbf{G}_4^2$: $(C_{2z}, 0)$, $(\sigma_y, 0)$: 1, *x*; 2, *x*; 3, *x*; 4, *x*: *c*.
G^x $\mathbf{G}_8^4$: $(C_{2z}, 0)$, $(\sigma_y, 0)$: 5, *x*: *a*.
H^x $\mathbf{G}_8^4$: $(C_{2z}, 0)$, $(\sigma_y, 0)$: 5, *x*: *a*.
Q^x $\mathbf{G}_8^3$: $(C_{2z}, 0)$, $(\sigma_y, 0)$, $(E, 1)$: 5, *x*; 6, *x*; 7, *x*; 8, *x*: *a*.
Σ^x $\mathbf{G}_2^1$: $(\sigma_y, 0)$: 1, 1; 2, 1: *c*.
C^x $\mathbf{G}_2^1$: $(\sigma_y, 0)$: 1, 3; 2, 3: *c*.
A^x $\mathbf{G}_2^1$: $(\sigma_y, 0)$: 1, 3; 2, 3: *c*.
U^x $\mathbf{G}_2^1$: $(\sigma_y, 0)$: 1, 1; 2, 1: *c*.
Δ^x $\mathbf{G}_2^1$: $(\sigma_x, 0)$: 1, 1; 2, 1: *c*.
D^x $\mathbf{G}_2^1$: $(\sigma_x, 0)$: 1, 3; 2, 3: *c*.
B^x $\mathbf{G}_2^1$: $(\sigma_x, 0)$: 1, 3; 2, 3: *c*.
R^x $\mathbf{G}_2^1$: $(\sigma_x, 0)$: 1, 1; 2, 1: *c*.

44 *Imm2* C_{2v}^{20}

($F1$; $K7$; $M5$; $Z1$.)

Γ $\mathbf{G}_4^2$: $\{C_{2z}|000\}$, $\{\sigma_y|000\}$: 1, 1; 2, 1; 3, 1; 4, 1: *c*.
X $\mathbf{G}_4^2 \otimes \mathbf{T}_2$: $\{C_{2z}|000\}$, $\{\sigma_y|000\}$; $\mathbf{t}_1$ or $\mathbf{t}_2$ or $\mathbf{t}_3$: 1, 1; 2, 1; 3, 1; 4, 1: *c*.
R $\mathbf{G}_2^1 \otimes \mathbf{T}_2$: $\{\sigma_y|000\}$; $\mathbf{t}_1$: 1, 1; 2, 1: *c*.
S $\mathbf{G}_2^1 \otimes \mathbf{T}_2$: $\{\sigma_x|000\}$; $\mathbf{t}_1$ or $\mathbf{t}_3$: 1, 1; 2, 1: *c*.
T $\mathbf{G}_2^1 \otimes \mathbf{T}_2$: $\{C_{2z}|000\}$; $\mathbf{t}_1$ or $\mathbf{t}_2$: 1, 1; 2, 1: *b*.
W $\mathbf{G}_2^1 \otimes \mathbf{T}_4$: $\{C_{2z}|000\}$; $\mathbf{t}_1$ or $\mathbf{t}_2$ or $\mathbf{t}_3$: 1, 1; 2, 1: *b*.

Λ^x $\mathbf{G}_4^2$: $(C_{2z}, 0)$, $(\sigma_y, 0)$: 1, *x*; 2, *x*; 3, *x*; 4, *x*; *c*.
G^x $\mathbf{G}_4^2$: $(C_{2z}, 0)$, $(\sigma_y, 0)$: 1, *x*; 2, *x*; 3, *x*; 4, *x*: *c*.
P^x $\mathbf{G}_2^1$: $(C_{2z}, 0)$: 1, *x*; 2, *x*: *b*.
Σ^x $\mathbf{G}_2^1$: $(\sigma_y, 0)$: 1, 1; 2, 1: *c*.
F^x $\mathbf{G}_2^1$: $(\sigma_y, 0)$: 1, 1; 2, 1: *c*.
D^x $\mathbf{G}_1^1$: $(E, 0)$: 1, 1: *a*.
Δ^x $\mathbf{G}_2^1$: $(\sigma_x, 0)$: 1, 1; 2, 1: *c*.
U^x $\mathbf{G}_2^1$: $(\sigma_x, 0)$: 1, 1; 2, 1: *c*.
Q^x $\mathbf{G}_1^1$: $(E, 0)$: 1, 1: *a*.

45 *Iba*2 C_{2v}^{21}

(*F*1; *K*7; *M*5; *Z*1.)

Γ $\mathbf{G}_4^2$: $\{C_{2z}|000\}$, $\{\sigma_y|\frac{1}{2}\frac{1}{2}0\}$: 1, 1; 2, 1; 3, 1; 4, 1: *c*.
X $\mathbf{G}_4^2 \otimes \mathbf{T}_2$: $\{C_{2z}|000\}$, $\{\sigma_y|\frac{1}{2}\frac{1}{2}0\}$; $\mathbf{t}_1$ or $\mathbf{t}_2$ or $\mathbf{t}_3$: 1, 1; 2, 1; 3, 1; 4, 1: *c*.
R $\mathbf{G}_4^1$: $\{\sigma_y|\frac{1}{2}\frac{1}{2}0\}$: 2, 3; 4, 3: *a*.
S $\mathbf{G}_4^1$: $\{\sigma_x|\frac{1}{2}\frac{1}{2}0\}$: 2, 3; 4, 3: *a*.
T $\mathbf{G}_2^1 \otimes \mathbf{T}_2$: $\{C_{2z}|000\}$; $\mathbf{t}_1$ or $\mathbf{t}_2$: 1, 1; 2, 1: *b*.
W $\mathbf{G}_2^1 \otimes \mathbf{T}_4$: $\{C_{2z}|000\}$; $\mathbf{t}_1$ or $\mathbf{t}_2$ or $\mathbf{t}_3$: 1, 2; 2, 2: *b*.

Λ^x $\mathbf{G}_4^2$: $(C_{2z}, 0)$, $(\sigma_y, 0)$: 1, *x*; 2, *x*; 3, *x*; 4, *x*: *c*.
G^x $\mathbf{G}_4^2$: $(C_{2z}, 0)$, $(\sigma_y, 0)$: 1, *x*; 2, *x*; 3, *x*; 4, *x*: *c*.
P^x $\mathbf{G}_2^1$: $(C_{2z}, 0)$: 1, *x*; 2, *x*: *b*.
Σ^x $\mathbf{G}_2^1$: $(\sigma_y, 0)$: 1, 1; 2, 1: *c*.
F^x $\mathbf{G}_2^1$: $(\sigma_y, 0)$: 1, 1; 2, 1: *c*.
D^x $\mathbf{G}_1^1$: $(E, 0)$: 1, 2: *a*.
Δ^x $\mathbf{G}_2^1$: $(\sigma_x, 0)$: 1, 1; 2, 1: *c*.
U^x $\mathbf{G}_2^1$: $(\sigma_x, 0)$: 1, 1; 2, 1: *c*.
Q^x $\mathbf{G}_1^1$: $(E, 0)$: 1, 2: *a*.

46 *Ima*2 C_{2v}^{22}

(*F*1; *K*7; *M*5; *Z*1.)

Γ $\mathbf{G}_4^2$: $\{C_{2z}|000\}$, $\{\sigma_y|0\frac{1}{2}\frac{1}{2}\}$: 1, 1; 2, 1; 3, 1; 4, 1: *c*.
X $\mathbf{G}_4^2 \otimes \mathbf{T}_2$: $\{C_{2z}|000\}$, $\{\sigma_y|0\frac{1}{2}\frac{1}{2}\}$; $\mathbf{t}_1$ or $\mathbf{t}_2$ or $\mathbf{t}_3$: 1, 1; 2, 1; 3, 1; 4, 1: *c*.
R $\mathbf{G}_2^1 \otimes \mathbf{T}_2$: $\{\sigma_y|0\frac{1}{2}\frac{1}{2}\}$; $\mathbf{t}_1$: 1, 1; 2, 1: *c*.
S $\mathbf{G}_4^1$: $\{\sigma_x|0\frac{1}{2}\frac{1}{2}\}$: 2, 3; 4, 3: *a*.
T $\mathbf{G}_2^1 \otimes \mathbf{T}_2$: $\{C_{2z}|000\}$; $\mathbf{t}_1$ or $\mathbf{t}_2$: 1, 1; 2, 1: *b*.
W $\mathbf{G}_2^1 \otimes \mathbf{T}_4$: $\{C_{2z}|000\}$; $\mathbf{t}_1$ or $\mathbf{t}_2$ or $\mathbf{t}_3$: 1, 3; 2, 3: *b*.

Λ^x $\mathbf{G}_4^2$: $(C_{2z}, 0)$, $(\sigma_y, 0)$: 1, *x*; 2, *x*; 3, *x*; 4, *x*: *c*.
G^x $\mathbf{G}_4^2$: $(C_{2z}, 0)$, $(\sigma_y, 0)$: 1, *x*; 2, *x*; 3, *x*; 4, *x*: *c*.
P^x $\mathbf{G}_2^1$: $(C_{2z}, 0)$: 1, *x*; 2, *x*: *b*.
Σ^x $\mathbf{G}_2^1$: $(\sigma_y, 0)$: 1, 1; 2, 1: *c*.
F^x $\mathbf{G}_2^1$: $(\sigma_y, 0)$: 1, 1; 2, 1: *c*.
D^x $\mathbf{G}_1^1$: $(E, 0)$: 1, 2: *a*.
Δ^x $\mathbf{G}_2^1$: $(\sigma_x, 0)$: 1, 1; 2, 1: *c*.
U^x $\mathbf{G}_2^1$: $(\sigma_x, 0)$: 1, 1; 2, 1: *c*.
Q^x $\mathbf{G}_1^1$: $(E, 0)$: 1, 1: *a*.

47 *Pmmm* D_{2h}^1

($F1$; $K7$; $M5$; $T1$; $Z1$.)

Γ $\mathbf{G}_8^3$: $\{C_{2z}|000\}$, $\{C_{2y}|000\}$, $\{I|000\}$: 1, 1; 2, 1; 3, 1; 4, 1; 5, 1; 6, 1; 7, 1; 8, 1: *b*.
Y $\mathbf{G}_8^3 \otimes \mathbf{T}_2$: $\{C_{2z}|000\}$, $\{C_{2y}|000\}$, $\{I|000\}$; $\mathbf{t}_1$: 1, 1; 2, 1; 3, 1; 4, 1; 5, 1; 6, 1; 7, 1; 8, 1: *b*.
X $\mathbf{G}_8^3 \otimes \mathbf{T}_2$: $\{C_{2z}|000\}$, $\{C_{2y}|000\}$, $\{I|000\}$; $\mathbf{t}_2$: 1, 1; 2, 1; 3, 1; 4, 1; 5, 1; 6, 1; 7, 1; 8, 1: *b*.
Z $\mathbf{G}_8^3 \otimes \mathbf{T}_2$: $\{C_{2z}|000\}$, $\{C_{2y}|000\}$, $\{I|000\}$; $\mathbf{t}_3$: 1, 1; 2, 1; 3, 1; 4, 1; 5, 1; 6, 1; 7, 1; 8, 1: *b*.
U $\mathbf{G}_8^3 \otimes \mathbf{T}_2$: $\{C_{2z}|000\}$, $\{C_{2y}|000\}$, $\{I|000\}$; $\mathbf{t}_2$ or $\mathbf{t}_3$: 1, 1; 2, 1; 3, 1; 4, 1; 5, 1; 6, 1; 7, 1; 8, 1: *b*.
T $\mathbf{G}_8^3 \otimes \mathbf{T}_2$: $\{C_{2z}|000\}$, $\{C_{2y}|000\}$, $\{I|000\}$; $\mathbf{t}_1$ or $\mathbf{t}_3$: 1, 1; 2, 1; 3, 1; 4, 1; 5, 1; 6, 1; 7, 1; 8, 1: *b*.
S $\mathbf{G}_8^3 \otimes \mathbf{T}_2$: $\{C_{2z}|000\}$, $\{C_{2y}|000\}$, $\{I|000\}$; $\mathbf{t}_1$ or $\mathbf{t}_2$: 1, 1; 2, 1; 3, 1; 4, 1; 5, 1; 6, 1; 7, 1; 8, 1: *b*.
R $\mathbf{G}_8^3 \otimes \mathbf{T}_2$: $\{C_{2z}|000\}$, $\{C_{2y}|000\}$, $\{I|000\}$; $\mathbf{t}_1$ or $\mathbf{t}_2$ or $\mathbf{t}_3$: 1, 1; 2, 1; 3, 1; 4, 1; 5, 1; 6, 1; 7, 1; 8, 1: *b*.

Δ^x $\mathbf{G}_4^2$: $(C_{2y}, 0)$, $(\sigma_x, 0)$: 1, 1; 2, 1; 3, 1; 4, 1: *c*.
D^x $\mathbf{G}_4^2$: $(C_{2y}, 0)$, $(\sigma_x, 0)$: 1, 1; 2, 1; 3, 1; 4, 1: *c*.
P^x $\mathbf{G}_4^2$: $(C_{2y}, 0)$, $(\sigma_x, 0)$: 1, 1; 2, 1; 3, 1; 4, 1: *c*.
B^x $\mathbf{G}_4^2$: $(C_{2y}, 0)$, $(\sigma_x, 0)$: 1, 1; 2, 1; 3, 1; 4, 1: *c*.
Σ^x $\mathbf{G}_4^2$: $(C_{2x}, 0)$, $(\sigma_z, 0)$: 1, 1; 2, 1; 3, 1; 4, 1: *c*.
C^x $\mathbf{G}_4^2$: $(C_{2x}, 0)$, $(\sigma_z, 0)$: 1, 1; 2, 1; 3, 1; 4, 1: *c*.
E^x $\mathbf{G}_4^2$: $(C_{2x}, 0)$, $(\sigma_z, 0)$: 1, 1; 2, 1; 3, 1; 4, 1: *c*.
A^x $\mathbf{G}_4^2$: $(C_{2x}, 0)$, $(\sigma_z, 0)$: 1, 1; 2, 1; 3, 1; 4, 1: *c*.
Λ^x $\mathbf{G}_4^2$: $(C_{2z}, 0)$, $(\sigma_y, 0)$: 1, 1; 2, 1; 3, 1; 4, 1: *c*.
H^x $\mathbf{G}_4^2$: $(C_{2z}, 0)$, $(\sigma_y, 0)$: 1, 1; 2, 1; 3, 1; 4, 1: *c*.
Q^x $\mathbf{G}_4^2$: $(C_{2z}, 0)$, $(\sigma_y, 0)$: 1, 1; 2, 1; 3, 1; 4, 1: *c*.
G^x $\mathbf{G}_4^2$: $(C_{2z}, 0)$, $(\sigma_y, 0)$: 1, 1; 2, 1; 3, 1; 4, 1: *c*.

48 *Pnnn* D_{2h}^2

($F1$; $K7$; $M5$; $T1$; $Z1$.)

Γ $\mathbf{G}_8^3$: $\{C_{2z}|000\}$, $\{C_{2y}|000\}$, $\{I|\frac{1}{2}\frac{1}{2}\frac{1}{2}\}$: 1, 1; 2, 1; 3, 1; 4, 1; 5, 1; 6, 1; 7, 1; 8, 1: *b*.
Y $\mathbf{G}_{16}^9$: $\{\sigma_z|\frac{1}{2}\frac{1}{2}\frac{1}{2}\}$, $\{I|\frac{1}{2}\frac{1}{2}\frac{1}{2}\}$, $\{C_{2y}|000\}$: 5, 1; 10, 1: *a*.
X $\mathbf{G}_{16}^9$: $\{\sigma_y|\frac{1}{2}\frac{1}{2}\frac{1}{2}\}$, $\{I|\frac{1}{2}\frac{1}{2}\frac{1}{2}\}$, $\{C_{2x}|000\}$: 5, 1; 10, 1: *a*.
Z $\mathbf{G}_{16}^9$: $\{\sigma_x|\frac{1}{2}\frac{1}{2}\frac{1}{2}\}$, $\{I|\frac{1}{2}\frac{1}{2}\frac{1}{2}\}$, $\{C_{2z}|000\}$: 5, 1; 10, 1: *a*.
U $\mathbf{G}_{16}^9$: $\{\sigma_z|\frac{1}{2}\frac{1}{2}\frac{1}{2}\}$, $\{I|\frac{1}{2}\frac{1}{2}\frac{1}{2}\}$, $\{C_{2y}|000\}$: 5, 1; 10, 1: *a*.
T $\mathbf{G}_{16}^9$: $\{\sigma_y|\frac{1}{2}\frac{1}{2}\frac{1}{2}\}$, $\{I|\frac{1}{2}\frac{1}{2}\frac{1}{2}\}$, $\{C_{2x}|000\}$: 5, 1; 10, 1: *a*.
S $\mathbf{G}_{16}^9$: $\{\sigma_x|\frac{1}{2}\frac{1}{2}\frac{1}{2}\}$, $\{I|\frac{1}{2}\frac{1}{2}\frac{1}{2}\}$, $\{C_{2z}|000\}$: 5, 1; 10, 1: *a*.
R $\mathbf{G}_8^3 \otimes \mathbf{T}_2$: $\{C_{2z}|000\}$, $\{C_{2y}|000\}$, $\{I|\frac{1}{2}\frac{1}{2}\frac{1}{2}\}$; $\mathbf{t}_1$ or $\mathbf{t}_2$ or $\mathbf{t}_3$: 1, 1; 2, 1; 3, 1; 4, 1; 5, 1; 6, 1; 7, 1; 8, 1: *b*.

Δ^x $\mathbf{G}_4^2$: $(C_{2y}, 0)$, $(\sigma_x, 0)$: 1, 1; 2, 1; 3, 1; 4, 1: *c*.
D^x $\mathbf{G}_8^4$: $(\sigma_x, 0)$, $(C_{2y}, 0)$: 5, 1: *a*.
P^x $\mathbf{G}_8^2$: $(\sigma_z, 0)$, $(C_{2y}, 0)$: 2, 1; 4, 1; 6, 1; 8, 1: *d*.
B^x $\mathbf{G}_8^4$: $(\sigma_z, 0)$, $(C_{2y}, 0)$: 5, 1: *a*.
Σ^x $\mathbf{G}_4^2$: $(C_{2x}, 0)$, $(\sigma_y, 0)$: 1, 1; 2, 1; 3, 1; 4, 1: *c*.
C^x $\mathbf{G}_8^4$: $(\sigma_y, 0)$, $(C_{2x}, 0)$: 5, 1: *a*.
E^x $\mathbf{G}_8^2$: $(\sigma_z, 0)$, $(C_{2x}, 0)$: 2, 1; 4, 1; 6, 1; 8, 1: *d*.
A^x $\mathbf{G}_8^4$: $(\sigma_z, 0)$, $(C_{2x}, 0)$: 5, 1: *a*.
Λ^x $\mathbf{G}_4^2$: $(C_{2z}, 0)$, $(\sigma_y, 0)$: 1, 1; 2, 1; 3, 1; 4, 1: *c*.
H^x $\mathbf{G}_8^4$: $(\sigma_y, 0)$, $(C_{2z}, 0)$: 5, 1: *a*.
Q^x $\mathbf{G}_8^2$: $(\sigma_x, 0)$, $(C_{2z}, 0)$: 2, 1; 4, 1; 6, 1; 8, 1: *d*.
G^x $\mathbf{G}_8^4$: $(\sigma_x, 0)$, $(C_{2z}, 0)$: 5, 1: *a*.

49 *Pccm* D_{2h}^3

($F1$; $K7$; $M5$; $T1$; $Z1$.)

Γ $\mathbf{G}_8^3$: $\{C_{2z}|000\}, \{C_{2y}|000\}, \{I|00\tfrac{1}{2}\}$: 1, 1; 2, 1; 3, 1; 4, 1; 5, 1; 6, 1; 7, 1; 8, 1: *b*.
Y $\mathbf{G}_8^3 \otimes \mathbf{T}_2$: $\{C_{2z}|000\}, \{C_{2y}|000\}, \{I|00\tfrac{1}{2}\}$; $\mathbf{t}_1$: 1, 1; 2, 1; 3, 1; 4, 1; 5, 1; 6, 1; 7, 1; 8, 1: *b*.
X $\mathbf{G}_8^3 \otimes \mathbf{T}_2$: $\{C_{2z}|000\}, \{C_{2y}|000\}, \{I|00\tfrac{1}{2}\}$; $\mathbf{t}_2$: 1, 1; 2, 1; 3, 1; 4, 1; 5, 1; 6, 1; 7, 1; 8, 1: *b*.
Z $\mathbf{G}_{16}^9$: $\{\sigma_x|00\tfrac{1}{2}\}, \{I|00\tfrac{1}{2}\}, \{C_{2z}|000\}$: 5, 1; 10, 1: *a*.
U $\mathbf{G}_{16}^9$: $\{\sigma_x|00\tfrac{1}{2}\}, \{I|00\tfrac{1}{2}\}, \{C_{2z}|000\}$: 5, 1; 10, 1: *a*.
T $\mathbf{G}_{16}^9$: $\{\sigma_x|00\tfrac{1}{2}\}, \{I|00\tfrac{1}{2}\}, \{C_{2z}|000\}$: 5, 1; 10, 1: *a*.
S $\mathbf{G}_8^3 \otimes \mathbf{T}_2$: $\{C_{2z}|000\}, \{C_{2y}|000\}, \{I|00\tfrac{1}{2}\}$; $\mathbf{t}_1$ or $\mathbf{t}_2$: 1, 1; 2, 1; 3, 1; 4, 1; 5, 1; 6, 1; 7, 1; 8, 1: *b*.
R $\mathbf{G}_{16}^9$: $\{\sigma_x|00\tfrac{1}{2}\}, \{I|00\tfrac{1}{2}\}, \{C_{2z}|000\}$: 5, 1; 10, 1: *a*.

Δ^x $\mathbf{G}_4^2$: $(C_{2y}, 0), (\sigma_x, 0)$: 1, 1; 2, 1; 3, 1; 4, 1: *c*.
D^x $\mathbf{G}_4^2$: $(C_{2y}, 0), (\sigma_x, 0)$: 1, 1; 2, 1; 3, 1; 4, 1: *c*.
P^x $\mathbf{G}_8^4$: $(\sigma_z, 0), (C_{2y}, 0)$: 5, 1: *a*.
B^x $\mathbf{G}_8^4$: $(\sigma_z, 0), (C_{2y}, 0)$: 5, 1: *a*.
Σ^x $\mathbf{G}_4^2$: $(C_{2x}, 0), (\sigma_z, 0)$: 1, 1; 2, 1; 3, 1; 4, 1: *c*.
C^x $\mathbf{G}_4^2$: $(C_{2x}, 0), (\sigma_z, 0)$: 1, 1; 2, 1; 3, 1; 4, 1: *c*.
E^x $\mathbf{G}_8^4$: $(\sigma_z, 0), (C_{2x}, 0)$: 5, 1: *a*.
A^x $\mathbf{G}_8^4$: $(\sigma_z, 0), (C_{2x}, 0)$: 5, 1: *a*.
Λ^x $\mathbf{G}_4^2$: $(C_{2z}, 0), (\sigma_y, 0)$: 1, 1; 2, 1; 3, 1; 4, 1: *c*.
H^x $\mathbf{G}_4^2$: $(C_{2z}, 0), (\sigma_y, 0)$: 1, 1; 2, 1; 3, 1; 4, 1: *c*.
Q^x $\mathbf{G}_4^2$: $(C_{2z}, 0), (\sigma_y, 0)$: 1, 1; 2, 1; 3, 1; 4, 1: *c*.
G^x $\mathbf{G}_4^2$: $(C_{2z}, 0), (\sigma_y, 0)$: 1, 1; 2, 1; 3, 1; 4, 1: *c*.

50 *Pban* D_{2h}^4

($F1$; $K7$; $M5$; $T1$; $Z1$.)

Γ $\mathbf{G}_8^3$: $\{C_{2z}|000\}, \{C_{2y}|000\}, \{I|\tfrac{1}{2}\tfrac{1}{2}0\}$: 1, 1; 2, 1; 3, 1; 4, 1; 5, 1; 6, 1; 7, 1; 8, 1: *b*.
Y $\mathbf{G}_{16}^9$: $\{\sigma_z|\tfrac{1}{2}\tfrac{1}{2}0\}, \{I|\tfrac{1}{2}\tfrac{1}{2}0\}, \{C_{2y}|000\}$: 5, 1; 10, 1: *a*.
X $\mathbf{G}_{16}^9$: $\{\sigma_z|\tfrac{1}{2}\tfrac{1}{2}0\}, \{I|\tfrac{1}{2}\tfrac{1}{2}0\}, \{C_{2x}|000\}$: 5, 1; 10, 1: *a*.
Z $\mathbf{G}_8^3 \otimes \mathbf{T}_2$: $\{C_{2z}|000\}, \{C_{2y}|000\}, \{I|\tfrac{1}{2}\tfrac{1}{2}0\}$; $\mathbf{t}_3$: 1, 1; 2, 1; 3, 1; 4, 1; 5, 1; 6, 1; 7, 1; 8, 1: *b*.
U $\mathbf{G}_{16}^9$: $\{\sigma_z|\tfrac{1}{2}\tfrac{1}{2}0\}, \{I|\tfrac{1}{2}\tfrac{1}{2}0\}, \{C_{2x}|000\}$: 5, 1; 10, 1: *a*.
T $\mathbf{G}_{16}^9$: $\{\sigma_z|\tfrac{1}{2}\tfrac{1}{2}0\}, \{I|\tfrac{1}{2}\tfrac{1}{2}0\}, \{C_{2y}|000\}$: 5, 1; 10, 1: *a*.
S $\mathbf{G}_{16}^9$: $\{\sigma_x|\tfrac{1}{2}\tfrac{1}{2}0\}, \{I|\tfrac{1}{2}\tfrac{1}{2}0\}, \{C_{2z}|000\}$: 5, 1; 10, 1: *a*.
R $\mathbf{G}_{16}^9$: $\{\sigma_x|\tfrac{1}{2}\tfrac{1}{2}0\}, \{I|\tfrac{1}{2}\tfrac{1}{2}0\}, \{C_{2z}|000\}$: 5, 1; 10, 1: *a*.

Δ^x $\mathbf{G}_4^2$: $(C_{2y}, 0), (\sigma_x, 0)$: 1, 1; 2, 1; 3, 1; 4, 1: *c*.
D^x $\mathbf{G}_8^4$: $(\sigma_x, 0), (C_{2y}, 0)$: 5, 1: *a*.
P^x $\mathbf{G}_8^4$: $(\sigma_x, 0), (C_{2y}, 0)$: 5, 1: *a*.
B^x $\mathbf{G}_4^2$: $(C_{2y}, 0), (\sigma_x, 0)$: 1, 1; 2, 1; 3, 1; 4, 1: *c*.
Σ^x $\mathbf{G}_4^2$: $(C_{2x}, 0), (\sigma_z, 0)$: 1, 1; 2, 1; 3, 1; 4, 1: *c*.
C^x $\mathbf{G}_8^4$: $(\sigma_y, 0), (C_{2x}, 0)$: 5, 1: *a*.
E^x $\mathbf{G}_8^4$: $(\sigma_y, 0), (C_{2x}, 0)$: 5, 1: *a*.
A^x $\mathbf{G}_4^2$: $(C_{2x}, 0), (\sigma_z, 0)$: 1, 1; 2, 1; 3, 1; 4, 1: *c*.
Λ^x $\mathbf{G}_4^2$: $(C_{2z}, 0), (\sigma_y, 0)$: 1, 1; 2, 1; 3, 1; 4, 1: *c*.
H^x $\mathbf{G}_8^4$: $(\sigma_y, 0), (C_{2z}, 0)$: 5, 1: *a*.
Q^x $\mathbf{G}_8^2$: $(\sigma_y, 0), (C_{2z}, 0)$: 2, 1; 4, 1; 6, 1; 8, 1: *d*.
G^x $\mathbf{G}_8^4$: $(\sigma_x, 0), (C_{2z}, 0)$: 5, 1: *a*.

51 *Pmma* D_{2h}^5

($F1$; $K7$; $M5$; $T1$; $Z1$.)

Γ $\mathbf{G}_8^3$: $\{C_{2z}|00\frac{1}{2}\}$, $\{C_{2y}|000\}$, $\{I|000\}$: 1, 1; 2, 1; 3, 1; 4, 1; 5, 1; 6, 1; 7, 1; 8, 1: *b*.

Y $\mathbf{G}_8^3 \otimes \mathbf{T}_2$: $\{C_{2z}|00\frac{1}{2}\}$, $\{C_{2y}|000\}$, $\{I|000\}$; $\mathbf{t}_1$: 1, 1; 2, 1; 3, 1; 4, 1; 5, 1; 6, 1; 7, 1; 8, 1: *b*.

X $\mathbf{G}_8^3 \otimes \mathbf{T}_2$: $\{C_{2z}|00\frac{1}{2}\}$, $\{C_{2y}|000\}$, $\{I|000\}$; $\mathbf{t}_2$: 1, 1; 2, 1; 3, 1; 4, 1; 5, 1; 6, 1; 7, 1; 8, 1: *b*.

Z $\mathbf{G}_{16}^9$: $\{C_{2z}|00\frac{1}{2}\}$, $\{I|000\}$, $\{\sigma_y|000\}$: 5, 1; 10, 1: *b*.

U $\mathbf{G}_{16}^9$: $\{C_{2z}|00\frac{1}{2}\}$, $\{I|000\}$, $\{\sigma_y|000\}$: 5, 1; 10, 1: *b*.

T $\mathbf{G}_{16}^9$: $\{C_{2z}|00\frac{1}{2}\}$, $\{I|000\}$, $\{\sigma_y|000\}$: 5, 1; 10, 1: *b*.

S $\mathbf{G}_8^3 \otimes \mathbf{T}_2$: $\{C_{2z}|00\frac{1}{2}\}$, $\{C_{2y}|000\}$, $\{I|000\}$; $\mathbf{t}_1$ or $\mathbf{t}_2$: 1, 1; 2, 1; 3, 1; 4, 1; 5, 1; 6, 1; 7, 1; 8, 1: *b*.

R $\mathbf{G}_{16}^9$: $\{C_{2z}|00\frac{1}{2}\}$, $\{I|000\}$, $\{\sigma_y|000\}$: 5, 1; 10, 1: *b*.

Δ^x $\mathbf{G}_4^2$: $(C_{2y}, 0)$, $(\sigma_x, 0)$: 1, 1; 2, 1; 3, 1; 4, 1: *c*.

D^x $\mathbf{G}_4^2$: $(C_{2y}, 0)$, $(\sigma_x, 0)$: 1, 1; 2, 1; 3, 1; 4, 1: *c*.

P^x $\mathbf{G}_8^4$: $(\sigma_z, 0)$, $(C_{2y}, 0)$: 5, 1: *a*.

B^x $\mathbf{G}_8^4$: $(\sigma_z, 0)$, $(C_{2y}, 0)$: 5, 1: *a*.

Σ^x $\mathbf{G}_4^2$: $(C_{2x}, 0)$, $(\sigma_z, 0)$: 1, 1; 2, 1; 3, 1; 4, 1: *c*.

C^x $\mathbf{G}_4^2$: $(C_{2x}, 0)$, $(\sigma_z, 0)$: 1, 1; 2, 1; 3, 1; 4, 1: *c*.

E^x $\mathbf{G}_8^2$: $(C_{2x}, 0)$, $(\sigma_y, 0)$: 2, 3; 4, 3; 6, 3; 8, 3: *c*.

A^x $\mathbf{G}_8^2$: $(C_{2x}, 0)$, $(\sigma_y, 0)$: 2, 3; 4, 3; 6, 3; 8, 3: *c*.

Λ^x $\mathbf{G}_4^2$: $(C_{2z}, 0)$, $(\sigma_y, 0)$: 1, 1; 2, 1; 3, 1; 4, 1: *c*.

H^x $\mathbf{G}_4^2$: $(C_{2z}, 0)$, $(\sigma_y, 0)$: 1, 1; 2, 1; 3, 1; 4, 1: *c*.

Q^x $\mathbf{G}_4^2$: $(C_{2z}, 0)$, $(\sigma_y, 0)$: 1, 1; 2, 1; 3, 1; 4, 1: *c*.

G^x $\mathbf{G}_4^2$: $(C_{2z}, 0)$, $(\sigma_y, 0)$: 1, 1; 2, 1; 3, 1; 4, 1: *c*.

52 *Pnna* D_{2h}^6

($F1$; $K7$; $M5$; $T1$; $Z1$.)

Γ $\mathbf{G}_8^3$: $\{C_{2z}|00\frac{1}{2}\}$, $\{C_{2y}|000\}$, $\{I|\frac{1}{2}\frac{1}{2}0\}$: 1, 1; 2, 1; 3, 1; 4, 1; 5, 1; 6, 1; 7, 1; 8, 1: *b*.

Y $\mathbf{G}_{16}^9$: $\{\sigma_z|\frac{1}{2}\frac{1}{2}\frac{1}{2}\}$, $\{I|\frac{1}{2}\frac{1}{2}0\}$, $\{C_{2y}|000\}$: 5, 1; 10, 1: *a*.

X $\mathbf{G}_{16}^9$: $\{\sigma_y|\frac{1}{2}\frac{1}{2}0\}$, $\{I|\frac{1}{2}\frac{1}{2}0\}$, $\{C_{2x}|00\frac{1}{2}\}$: 5, 1; 10, 1: *a*.

Z $\mathbf{G}_{16}^9$: $\{\sigma_x|\frac{1}{2}\frac{1}{2}\frac{1}{2}\}$, $\{I|\frac{1}{2}\frac{1}{2}0\}$, $\{\sigma_y|\frac{1}{2}\frac{1}{2}0\}$: 5, 1; 10, 1: *b*.

U $\mathbf{G}_{16}^7$: $\{\sigma_z|\frac{1}{2}\frac{1}{2}\frac{1}{2}\}$, $\{C_{2y}|000\}$, $\{I|\frac{1}{2}\frac{1}{2}0\}$: 9, 3; 10, 3: *a*.

T $\mathbf{G}_{16}^9$: $\{\sigma_z|\frac{1}{2}\frac{1}{2}\frac{1}{2}\}$, $\{\sigma_y|\frac{1}{2}\frac{1}{2}0\}$, $\{I|\frac{1}{2}\frac{1}{2}0\}$: 5, 1; 10, 1: *c*.

S $\mathbf{G}_{16}^9$: $\{\sigma_x|\frac{1}{2}\frac{1}{2}\frac{1}{2}\}$, $\{I|\frac{1}{2}\frac{1}{2}0\}$, $\{C_{2z}|00\frac{1}{2}\}$: 5, 1; 10, 1: *a*.

R $\mathbf{G}_{16}^9$: $\{\sigma_y|\frac{1}{2}\frac{1}{2}0\}$, $\{I|\frac{1}{2}\frac{1}{2}0\}$, $\{\sigma_x|\frac{1}{2}\frac{1}{2}\frac{1}{2}\}$: 5, 1; 10, 1: *b*.

Δ^x $\mathbf{G}_4^2$: $(C_{2y}, 0)$, $(\sigma_x, 0)$: 1, 1; 2, 1; 3, 1; 4, 1: *c*.

D^x $\mathbf{G}_8^4$: $(\sigma_x, 0)$, $(C_{2y}, 0)$: 5, 1: *a*.

P^x $\mathbf{G}_8^2$: $(\sigma_z, 0)$, $(C_{2y}, 0)$: 2, 3; 4, 3; 6, 3; 8, 3: *d*.

B^x $\mathbf{G}_8^4$: $(\sigma_z, 0)$, $(C_{2y}, 0)$: 5, 1: *a*.

Σ^x $\mathbf{G}_4^2$: $(C_{2x}, 0)$, $(\sigma_z, 0)$: 1, 1; 2, 1; 3, 1; 4, 1: *c*.

C^x $\mathbf{G}_8^4$: $(\sigma_y, 0)$, $(C_{2x}, 0)$: 5, 1: *a*.

E^x $\mathbf{G}_8^5$: $(\sigma_y, 0)$, $(\sigma_z, 0)$: 5, 1: *a*.

A^x $\mathbf{G}_8^2$: $(C_{2x}, 0)$, $(\sigma_y, 0)$: 2, 3; 4, 3; 6, 3; 8, 3: *c*.

Λ^x $\mathbf{G}_4^2$: $(C_{2z}, 0)$, $(\sigma_y, 0)$: 1, 1; 2, 1; 3, 1; 4, 1: *c*.

H^x $\mathbf{G}_8^4$: $(\sigma_y, 0)$, $(C_{2z}, 0)$: 5, 1: *a*.

Q^x $\mathbf{G}_8^2$: $(\sigma_y, 0)$, $(C_{2z}, 0)$: 2, 1; 4, 1; 6, 1; 8, 1: *d*.

G^x $\mathbf{G}_8^4$: $(\sigma_x, 0)$, $(C_{2z}, 0)$: 5, 1: *a*.

53 *Pmna* D_{2h}^{7}

($F1$; $K7$; $M5$; $T1$; $Z1$.)

Γ $\mathbf{G}_8^3$: $\{C_{2z}|00\frac{1}{2}\}$, $\{C_{2y}|000\}$, $\{I|\frac{1}{2}00\}$: 1, 1; 2, 1; 3, 1; 4, 1; 5, 1; 6, 1; 7, 1; 8, 1: *b*.
Y $\mathbf{G}_{16}^9$: $\{\sigma_z|\frac{1}{2}0\frac{1}{2}\}$, $\{I|\frac{1}{2}00\}$, $\{C_{2y}|000\}$: 5, 1; 10, 1: *a*.
X $\mathbf{G}_8^3 \otimes \mathbf{T}_2$: $\{C_{2z}|00\frac{1}{2}\}$, $\{C_{2y}|000\}$, $\{I|\frac{1}{2}00\}$; $\mathbf{t}_2$: 1, 1; 2, 1; 3, 1; 4, 1; 5, 1; 6, 1; 7, 1; 8, 1: *b*.
Z $\mathbf{G}_{16}^9$: $\{C_{2z}|00\frac{1}{2}\}$, $\{I|\frac{1}{2}00\}$, $\{\sigma_y|\frac{1}{2}00\}$: 5, 1; 10, 1: *b*.
U $\mathbf{G}_{16}^9$: $\{C_{2z}|00\frac{1}{2}\}$, $\{I|\frac{1}{2}00\}$, $\{\sigma_y|\frac{1}{2}00\}$: 5, 1; 10, 1: *b*.
T $\mathbf{G}_{16}^9$: $\{C_{2z}|00\frac{1}{2}\}$, $\{\sigma_y|\frac{1}{2}00\}$, $\{I|\frac{1}{2}00\}$: 5, 1; 10, 1: *c*.
S $\mathbf{G}_{16}^9$: $\{\sigma_z|\frac{1}{2}0\frac{1}{2}\}$, $\{I|\frac{1}{2}00\}$, $\{C_{2y}|000\}$: 5, 1; 10, 1: *a*.
R $\mathbf{G}_{16}^9$: $\{C_{2z}|00\frac{1}{2}\}$, $\{\sigma_y|\frac{1}{2}00\}$, $\{I|\frac{1}{2}00\}$: 5, 1; 10, 1: *c*.

Δ^x $\mathbf{G}_4^2$: $(C_{2y}, 0)$, $(\sigma_x, 0)$: 1, 1; 2, 1; 3, 1; 4, 1: *c*.
D^x $\mathbf{G}_4^2$: $(C_{2y}, 0)$, $(\sigma_x, 0)$: 1, 1; 2, 1; 3, 1; 4, 1: *c*.
P^x $\mathbf{G}_8^4$: $(\sigma_z, 0)$, $(C_{2y}, 0)$: 5, 1: *a*.
B^x $\mathbf{G}_8^4$: $(\sigma_z, 0)$, $(C_{2y}, 0)$: 5, 1: *a*.
Σ^x $\mathbf{G}_4^2$: $(C_{2x}, 0)$, $(\sigma_z, 0)$: 1, 1; 2, 1; 3, 1; 4, 1: *c*.
C^x $\mathbf{G}_8^4$: $(\sigma_y, 0)$, $(C_{2x}, 0)$: 5, 1: *a*.
E^x $\mathbf{G}_8^5$: $(\sigma_y, 0)$, $(C_{2x}, 0)$: 5, 1: *a*.
A^x $\mathbf{G}_8^2$: $(C_{2x}, 0)$, $(\sigma_y, 0)$: 2, 3; 4, 3; 6, 3; 8, 3: *c*.
Λ^x $\mathbf{G}_4^2$: $(C_{2z}, 0)$, $(\sigma_y, 0)$: 1, 1; 2, 1; 3, 1; 4, 1: *c*.
H^x $\mathbf{G}_8^4$: $(\sigma_y, 0)$, $(C_{2z}, 0)$: 5, 1: *a*.
Q^x $\mathbf{G}_8^4$: $(\sigma_y, 0)$, $(C_{2z}, 0)$: 5, 1: *a*.
G^x $\mathbf{G}_4^2$: $(C_{2z}, 0)$, $(\sigma_y, 0)$: 1, 1; 2, 1; 3, 1; 4, 1: *c*.

54 *Pcca* D_{2h}^{8}

($F1$; $K7$; $M5$; $T1$; $Z1$.)

Γ $\mathbf{G}_8^3$: $\{C_{2z}|00\frac{1}{2}\}$, $\{C_{2y}|000\}$, $\{I|0\frac{1}{2}0\}$: 1, 1; 2, 1; 3, 1; 4, 1; 5, 1; 6, 1; 7, 1; 8, 1: *b*.
Y $\mathbf{G}_8^3 \otimes \mathbf{T}_2$: $\{C_{2z}|00\frac{1}{2}\}$, $\{C_{2y}|000\}$, $\{I|0\frac{1}{2}0\}$; $\mathbf{t}_1$: 1, 1; 2, 1; 3, 1; 4, 1; 5, 1; 6, 1; 7, 1; 8, 1: *b*.
X $\mathbf{G}_{16}^9$: $\{\sigma_y|0\frac{1}{2}0\}$, $\{I|0\frac{1}{2}0\}$, $\{C_{2x}|00\frac{1}{2}\}$: 5, 1; 10, 1: *a*.
Z $\mathbf{G}_{16}^9$: $\{C_{2z}|00\frac{1}{2}\}$, $\{I|0\frac{1}{2}0\}$, $\{\sigma_y|0\frac{1}{2}0\}$: 5, 1; 10, 1: *b*.
U $\mathbf{G}_{16}^7$: $\{\sigma_z|0\frac{1}{2}\frac{1}{2}\}$; $\{C_{2y}|000\}$; $\{I|0\frac{1}{2}0\}$: 9, 3; 10, 3: *a*.
T $\mathbf{G}_{16}^9$: $\{C_{2z}|00\frac{1}{2}\}$; $\{I|0\frac{1}{2}0\}$; $\{\sigma_y|0\frac{1}{2}0\}$: 5, 1; 10, 1: *b*.
S $\mathbf{G}_{16}^9$: $\{\sigma_y|0\frac{1}{2}0\}$; $\{I|0\frac{1}{2}0\}$; $\{C_{2x}|00\frac{1}{2}\}$: 5, 1; 10, 1: *a*.
R $\mathbf{G}_{16}^7$: $\{\sigma_z|0\frac{1}{2}\frac{1}{2}\}$; $\{C_{2y}|000\}$; $\{I|0\frac{1}{2}0\}$: 9, 3; 10, 3: *a*.

Δ^x $\mathbf{G}_4^2$: $(C_{2y}, 0)$, $(\sigma_x, 0)$: 1, 1; 2, 1; 3, 1; 4, 1: *c*.
D^x $\mathbf{G}_8^4$: $(\sigma_x, 0)$, $(C_{2y}, 0)$: 5, 1: *a*.
P^x $\mathbf{G}_8^2$: $(\sigma_x, 0)$, $(C_{2y}, 0)$: 2, 3; 4, 3; 6, 3; 8, 3: *d*.
B^x $\mathbf{G}_8^4$: $(\sigma_z, 0)$, $(C_{2y}, 0)$: 5, 1: *a*.
Σ^x $\mathbf{G}_4^2$: $(C_{2x}, 0)$, $(\sigma_z, 0)$: 1, 1; 2, 1; 3, 1; 4, 1: *c*.
C^x $\mathbf{G}_4^2$: $(C_{2x}, 0)$, $(\sigma_z, 0)$: 1, 1; 2, 1; 3, 1; 4, 1: *c*.
E^x $\mathbf{G}_8^2$: $(C_{2x}, 0)$, $(\sigma_y, 0)$: 2, 3; 4, 3; 6, 3; 8, 3: *c*.
A^x $\mathbf{G}_8^2$: $(C_{2x}, 0)$, $(\sigma_y, 0)$: 2, 3; 4, 3; 6, 3; 8, 3: *c*.
Λ^x $\mathbf{G}_4^2$: $(C_{2z}, 0)$, $(\sigma_y, 0)$: 1, 1; 2, 1; 3, 1; 4, 1: *c*.
H^x $\mathbf{G}_4^2$: $(C_{2z}, 0)$, $(\sigma_y, 0)$: 1, 1; 2, 1; 3, 1; 4, 1: *c*.
Q^x $\mathbf{G}_8^4$: $(\sigma_x, 0)$, $(C_{2z}, 0)$: 5, 1: *a*.
G^x $\mathbf{G}_8^4$: $(\sigma_x, 0)$, $(C_{2z}, 0)$: 5, 1: *a*.

55 *Pbam* D_{2h}^{9}

($F1$; $K7$; $M5$; $T1$; $Z1$.)

Γ $\mathbf{G}_8^3$: $\{C_{2z}|000\}$, $\{C_{2y}|\frac{1}{2}\frac{1}{2}0\}$, $\{I|000\}$: 1, 1; 2, 1; 3, 1; 4, 1; 5, 1; 6, 1; 7, 1; 8, 1: *b*.

Y $\mathbf{G}_{16}^9$: $\{\sigma_x|\frac{1}{2}\frac{1}{2}0\}$, $\{I|000\}$, $\{\sigma_z|000\}$: 5, 1; 10, 1: *b*.

X $\mathbf{G}_{16}^9$: $\{\sigma_y|\frac{1}{2}\frac{1}{2}0\}$, $\{I|000\}$, $\{\sigma_z|000\}$: 5, 1; 10, 1: *b*.

Z $\mathbf{G}_8^3 \otimes \mathbf{T}_2$: $\{C_{2z}|000\}$, $\{C_{2y}|\frac{1}{2}\frac{1}{2}0\}$, $\{I|000\}$; $\mathbf{t}_3$: 1, 1; 2, 1; 3, 1; 4, 1; 5, 1; 6, 1; 7, 1; 8, 1: *b*.

U $\mathbf{G}_{16}^9$: $\{\sigma_y|\frac{1}{2}\frac{1}{2}0\}$, $\{I|000\}$, $\{\sigma_z|000\}$: 5, 1; 10, 1: *b*.

T $\mathbf{G}_{16}^9$: $\{\sigma_x|\frac{1}{2}\frac{1}{2}0\}$, $\{I|000\}$, $\{\sigma_z|000\}$: 5, 1; 10, 1: *b*.

S $\mathbf{G}_{16}^4$: $\{C_{2x}|\frac{1}{2}\frac{1}{2}0\}$, $\{C_{2z}|000\}$, $\{I|000\}$: 2, 3; 4, 3; 6, 3; 8, 3; 10, 3; 12, 3; 14, 3; 16, 3: *a*.

R $\mathbf{G}_{16}^4$: $\{C_{2x}|\frac{1}{2}\frac{1}{2}0\}$, $\{C_{2z}|000\}$, $\{I|000\}$: 2, 3; 4, 3; 6, 3; 8, 3; 10, 3; 12, 3; 14, 3; 16, 3: *a*.

Δ^x $\mathbf{G}_4^2$: $(C_{2y}, 0)$, $(\sigma_x, 0)$: 1, 1; 2, 1; 3, 1; 4, 1: *c*.

D^x $\mathbf{G}_8^2$: $(C_{2y}, 0)$, $(\sigma_z, 0)$: 2, 3; 4, 3; 6, 3; 8, 3: *c*.

P^x $\mathbf{G}_8^2$: $(C_{2y}, 0)$, $(\sigma_z, 0)$: 2, 3; 4, 3; 6, 3; 8, 3: *c*.

B^x $\mathbf{G}_4^2$: $(C_{2y}, 0)$, $(\sigma_x, 0)$: 1, 1; 2, 1; 3, 1; 4, 1: *c*.

Σ^x $\mathbf{G}_4^2$: $(C_{2x}, 0)$, $(\sigma_z, 0)$: 1, 1; 2, 1; 3, 1; 4, 1: *c*.

C^x $\mathbf{G}_8^2$: $(C_{2x}, 0)$, $(\sigma_z, 0)$: 2, 3; 4, 3; 6, 3; 8, 3: *c*.

E^x $\mathbf{G}_8^2$: $(C_{2x}, 0)$, $(\sigma_z, 0)$: 2, 3; 4, 3; 6, 3; 8, 3: *c*.

A^x $\mathbf{G}_4^2$: $(C_{2x}, 0)$, $(\sigma_z, 0)$: 1, 1; 2, 1; 3, 1; 4, 1: *c*.

Λ^x $\mathbf{G}_4^2$: $(C_{2z}, 0)$, $(\sigma_y, 0)$: 1, 1; 2, 1; 3, 1; 4, 1: *c*.

H^x $\mathbf{G}_8^4$: $(\sigma_y, 0)$, $(C_{2z}, 0)$: 5, 1: *a*.

Q^x $\mathbf{G}_8^2$: $(\sigma_y, 0)$, $(C_{2z}, 0)$: 2, 3; 4, 3; 6, 3; 8, 3: *d*.

G^x $\mathbf{G}_8^4$: $(\sigma_x, 0)$, $(C_{2z}, 0)$: 5, 1: *a*.

56 *Pccn* D_{2h}^{10}

($F1$; $K7$; $M5$; $T1$; $Z1$.)

Γ $\mathbf{G}_8^3$: $\{C_{2z}|000\}$, $\{C_{2y}|\frac{1}{2}\frac{1}{2}0\}$, $\{I|\frac{1}{2}\frac{1}{2}\frac{1}{2}\}$: 1, 1; 2, 1; 3, 1; 4, 1; 5, 1; 6, 1; 7, 1; 8, 1: *b*.

Y $\mathbf{G}_{16}^9$: $\{\sigma_z|\frac{1}{2}\frac{1}{2}\frac{1}{2}\}$, $\{I|\frac{1}{2}\frac{1}{2}\frac{1}{2}\}$, $\{\sigma_x|00\frac{1}{2}\}$: 5, 1; 10, 1: *b*.

X $\mathbf{G}_{16}^9$: $\{\sigma_z|\frac{1}{2}\frac{1}{2}\frac{1}{2}\}$, $\{I|\frac{1}{2}\frac{1}{2}\frac{1}{2}\}$, $\{\sigma_y|00\frac{1}{2}\}$: 5, 1; 10, 1: *b*.

Z $\mathbf{G}_{16}^9$: $\{\sigma_x|00\frac{1}{2}\}$, $\{I|\frac{1}{2}\frac{1}{2}\frac{1}{2}\}$, $\{C_{2z}|000\}$: 5, 1; 10, 1: *a*.

U $\mathbf{G}_{16}^7$: $\{\sigma_x|00\frac{1}{2}\}$, $\{C_{2z}|000\}$, $\{I|\frac{1}{2}\frac{1}{2}\frac{1}{2}\}$: 9, 3; 10, 3: *a*.

T $\mathbf{G}_{16}^7$: $\{\sigma_y|00\frac{1}{2}\}$, $\{C_{2z}|000\}$, $\{I|\frac{1}{2}\frac{1}{2}\frac{1}{2}\}$: 9, 3; 10, 3: *a*.

S $\mathbf{G}_{16}^9$: $\{C_{2y}|\frac{1}{2}\frac{1}{2}0\}$, $\{I|\frac{1}{2}\frac{1}{2}\frac{1}{2}\}$, $\{C_{2z}|000\}$: 5, 1; 10, 1: *a*.

R $\mathbf{G}_{16}^4$: $\{C_{2x}|\frac{1}{2}\frac{1}{2}0\}$, $\{C_{2z}|000\}$, $\{I|\frac{1}{2}\frac{1}{2}\frac{1}{2}\}$: 2, 3; 4, 3; 6, 3; 8, 3; 10, 3; 12, 3; 14, 3; 16, 3: *a*.

Δ^x $\mathbf{G}_4^2$: $(C_{2y}, 0)$, $(\sigma_x, 0)$: 1, 1; 2, 1; 3, 1; 4, 1: *c*.

D^x $\mathbf{G}_8^4$: $(C_{2y}, 0)$, $(\sigma_x, 0)$: 5, 1: *a*.

P^x $\mathbf{G}_8^2$: $(C_{2y}, 0)$, $(\sigma_x, 0)$: 2, 3; 4, 3; 6, 3; 8, 3: *c*.

B^x $\mathbf{G}_8^4$: $(\sigma_z, 0)$, $(C_{2y}, 0)$: 5, 1: *a*.

Σ^x $\mathbf{G}_4^2$: $(C_{2x}, 0)$, $(\sigma_z, 0)$: 1, 1; 2, 1; 3, 1; 4, 1: *c*.

C^x $\mathbf{G}_8^4$: $(C_{2x}, 0)$, $(\sigma_z, 0)$: 5, 1: *a*.

E^x $\mathbf{G}_8^2$: $(C_{2x}, 0)$, $(\sigma_y, 0)$: 2, 3; 4, 3; 6, 3; 8, 3: *c*.

A^x $\mathbf{G}_8^4$: $(\sigma_z, 0)$, $(C_{2x}, 0)$: 5, 1: *a*.

Λ^x $\mathbf{G}_4^2$: $(C_{2z}, 0)$, $(\sigma_y, 0)$: 1, 1; 2, 1; 3, 1; 4, 1: *c*.

H^x $\mathbf{G}_4^2$: $(C_{2z}, 0)$, $(\sigma_y, 0)$: 1, 3; 2, 3; 3, 3; 4, 3: *c*.

Q^x $\mathbf{G}_4^2$: $(C_{2z}, 0)$, $(\sigma_y, 0)$: 1, 3; 2, 3; 3, 3; 4, 3: *c*.

G^x $\mathbf{G}_4^2$: $(C_{2z}, 0)$, $(\sigma_y, 0)$: 1, 3; 2, 3; 3, 3; 4, 3: *c*.

57 *Pbcm* D_{2h}^{11}

($F1$; $K7$; $M5$; $T1$; $Z1$.)

Γ $\mathbf{G}_8^3$: $\{C_{2z}|000\}, \{C_{2y}|\frac{1}{2}\frac{1}{2}0\}, \{I|0\frac{1}{2}0\}$: 1, 1; 2, 1; 3, 1; 4, 1; 5, 1; 6, 1; 7, 1; 8, 1: *b*.

Y $\mathbf{G}_{16}^9$: $\{\sigma_x|\frac{1}{2}00\}, \{I|0\frac{1}{2}0\}, \{\sigma_z|0\frac{1}{2}0\}$: 5, 1; 10, 1: *b*.

X $\mathbf{G}_{16}^9$: $\{\sigma_z|0\frac{1}{2}0\}, \{I|0\frac{1}{2}0\}, \{\sigma_y|\frac{1}{2}00\}$: 5, 1; 10, 1: *b*.

Z $\mathbf{G}_8^3 \otimes \mathbf{T}_2$: $\{C_{2z}|000\}, \{C_{2y}|\frac{1}{2}\frac{1}{2}0\}, \{I|0\frac{1}{2}0\}$; $\mathbf{t}_3$: 1, 1; 2, 1; 3, 1; 4, 1; 5, 1; 6, 1; 7, 1; 8, 1: *b*.

U $\mathbf{G}_{16}^9$: $\{\sigma_z|0\frac{1}{2}0\}, \{I|0\frac{1}{2}0\}, \{\sigma_y|\frac{1}{2}00\}$: 5, 1; 10, 1: *b*.

T $\mathbf{G}_{16}^9$: $\{\sigma_x|\frac{1}{2}00\}, \{I|0\frac{1}{2}0\}, \{\sigma_z|0\frac{1}{2}0\}$: 5, 1; 10, 1: *b*.

S $\mathbf{G}_{16}^7$: $\{C_{2x}|\frac{1}{2}\frac{1}{2}0\}, \{C_{2z}|000\}, \{I|0\frac{1}{2}0\}$: 9, 3; 10, 3: *b*.

R $\mathbf{G}_{16}^7$: $\{C_{2x}|\frac{1}{2}\frac{1}{2}0\}, \{C_{2z}|000\}, \{I|0\frac{1}{2}0\}$: 9, 3; 10, 3: *b*.

Δ^x $\mathbf{G}_4^2$: $(C_{2y}, 0), (\sigma_x, 0)$: 1, 1; 2, 1; 3, 1; 4, 1: *c*.

D^x $\mathbf{G}_8^4$: $(C_{2y}, 0), (\sigma_x, 0)$: 5, 1: *a*.

P^x $\mathbf{G}_8^4$: $(C_{2y}, 0), (\sigma_x, 0)$: 5, 1: *a*.

B^x $\mathbf{G}_4^2$: $(C_{2y}, 0), (\sigma_x, 0)$: 1, 1; 2, 1; 3, 1; 4, 1: *c*.

Σ^x $\mathbf{G}_4^2$: $(C_{2x}, 0), (\sigma_z, 0)$: 1, 1; 2, 1; 3, 1; 4, 1: *c*.

C^x $\mathbf{G}_8^2$: $(C_{2x}, 0), (\sigma_z, 0)$: 2, 3; 4, 3; 6, 3; 8, 3: *c*.

E^x $\mathbf{G}_8^2$: $(C_{2x}, 0), (\sigma_z, 0)$: 2, 3; 4, 3; 6, 3; 8, 3: *c*.

A^x $\mathbf{G}_4^2$: $(C_{2x}, 0), (\sigma_z, 0)$: 1, 1; 2, 1; 3, 1; 4, 1: *c*.

Λ^x $\mathbf{G}_4^2$: $(C_{2z}, 0), (\sigma_y, 0)$: 1, 1; 2, 1; 3, 1; 4, 1: *c*.

H^x $\mathbf{G}_8^4$: $(\sigma_y, 0), (C_{2z}, 0)$: 5, 1: *a*.

Q^x $\mathbf{G}_8^4$: $(\sigma_y, 0), (C_{2z}, 0)$: 5, 2: *a*.

G^x $\mathbf{G}_4^2$: $(C_{2z}, 0), (\sigma_y, 0)$: 1, 3; 2, 3; 3, 3; 4, 3: *c*.

58 *Pnnm* D_{2h}^{12}

($C7$; $D2$; $F1$; $G7$; $K7$; $M5$; $T1$; $Z1$.)

Γ $\mathbf{G}_8^3$: $\{C_{2z}|000\}, \{C_{2y}|\frac{1}{2}\frac{1}{2}0\}, \{I|00\frac{1}{2}\}$: 1, 1; 2, 1; 3, 1; 4, 1; 5, 1; 6, 1; 7, 1; 8, 1: *b*.

Y $\mathbf{G}_{16}^9$: $\{\sigma_x|\frac{1}{2}\frac{1}{2}\frac{1}{2}\}, \{I|00\frac{1}{2}\}, \{\sigma_z|00\frac{1}{2}\}$: 5, 1; 10, 1: *b*.

X $\mathbf{G}_{16}^9$: $\{\sigma_y|\frac{1}{2}\frac{1}{2}\frac{1}{2}\}, \{I|00\frac{1}{2}\}, \{\sigma_z|00\frac{1}{2}\}$: 5, 1; 10, 1: *b*.

Z $\mathbf{G}_{16}^9$: $\{\sigma_x|\frac{1}{2}\frac{1}{2}\frac{1}{2}\}, \{I|00\frac{1}{2}\}, \{C_{2z}|000\}$: 5, 1; 10, 1: *a*.

U $\mathbf{G}_{16}^9$: $\{\sigma_x|\frac{1}{2}\frac{1}{2}\frac{1}{2}\}, \{C_{2z}|000\}, \{I|00\frac{1}{2}\}$: 5, 1; 10, 1: *c*.

T $\mathbf{G}_{16}^9$: $\{\sigma_y|\frac{1}{2}\frac{1}{2}\frac{1}{2}\}, \{C_{2z}|000\}, \{I|00\frac{1}{2}\}$: 5, 1; 10, 1: *c*.

S $\mathbf{G}_{16}^4$: $\{\sigma_x|\frac{1}{2}\frac{1}{2}\frac{1}{2}\}, \{C_{2z}|000\}, \{I|00\frac{1}{2}\}$: 2, 3; 4, 3; 6, 3; 8, 3; 10, 3; 12, 3; 14, 3; 16, 3: *a*.

R $\mathbf{G}_{16}^9$: $\{C_{2y}|\frac{1}{2}\frac{1}{2}0\}, \{I|00\frac{1}{2}\}, \{C_{2z}|000\}$: 5, 1; 10, 1: *a*.

Δ^x $\mathbf{G}_4^2$: $(C_{2y}, 0), (\sigma_x, 0)$: 1, 1; 2, 1; 3, 1; 4, 1: *c*.

D^x $\mathbf{G}_8^2$: $(C_{2y}, 0), (\sigma_z, 0)$: 2, 3; 4, 3; 6, 3; 8, 3: *c*.

P^x $\mathbf{G}_8^5$: $(C_{2y}, 0), (\sigma_z, 0)$: 5, 1: *a*.

B^x $\mathbf{G}_8^4$: $(\sigma_z, 0), (C_{2y}, 0)$: 5, 1: *a*.

Σ^x $\mathbf{G}_4^2$: $(C_{2x}, 0), (\sigma_z, 0)$: 1, 1; 2, 1; 3, 1; 4, 1: *c*.

C^x $\mathbf{G}_8^2$: $(C_{2x}, 0), (\sigma_z, 0)$: 2, 3; 4, 3; 6, 3; 8, 3: *c*.

E^x $\mathbf{G}_8^5$: $(C_{2x}, 0), (\sigma_z, 0)$: 5, 1: *a*.

A^x $\mathbf{G}_8^4$: $(\sigma_z, 0), (C_{2x}, 0)$: 5, 1: *a*.

Λ^x $\mathbf{G}_4^2$: $(C_{2z}, 0), (\sigma_y, 0)$: 1, 1; 2, 1; 3, 1; 4, 1: *c*.

H^x $\mathbf{G}_8^4$: $(\sigma_y, 0), (C_{2z}, 0)$: 5, 1: *a*.

Q^x $\mathbf{G}_8^2$: $(\sigma_y, 0), (C_{2z}, 0)$: 2, 3; 4, 3; 6, 3; 8, 3: *d*.

G^x $\mathbf{G}_8^4$: $(\sigma_x, 0), (C_{2z}, 0)$: 5, 1: *a*.

59 *Pmmn* D_{2h}^{13}

($F1$; $K7$; $M5$; $T1$; $Z1$.)

Γ $\mathbf{G}_8^3$: $\{C_{2z}|000\}, \{C_{2y}|\tfrac{1}{2}\tfrac{1}{2}0\}, \{I|\tfrac{1}{2}\tfrac{1}{2}0\}$: 1, 1; 2, 1; 3, 1; 4, 1; 5, 1; 6, 1; 7, 1; 8, 1: *b*.

Y $\mathbf{G}_{16}^9$: $\{\sigma_z|\tfrac{1}{2}\tfrac{1}{2}0\}, \{I|\tfrac{1}{2}\tfrac{1}{2}0\}, \{\sigma_x|000\}$: 5, 1; 10, 1: *b*.

X $\mathbf{G}_{16}^9$: $\{\sigma_z|\tfrac{1}{2}\tfrac{1}{2}0\}, \{I|\tfrac{1}{2}\tfrac{1}{2}0\}, \{\sigma_y|000\}$: 5, 1; 10, 1: *b*.

Z $\mathbf{G}_8^3 \otimes \mathbf{T}_2$: $\{C_{2z}|000\}, \{C_{2y}|\tfrac{1}{2}\tfrac{1}{2}0\}, \{I|\tfrac{1}{2}\tfrac{1}{2}0\}$; $\mathbf{t}_3$: 1, 1; 2, 1; 3, 1; 4, 1; 5, 1; 6, 1; 7, 1; 8, 1: *b*.

U $\mathbf{G}_{16}^9$: $\{\sigma_z|\tfrac{1}{2}\tfrac{1}{2}0\}, \{I|\tfrac{1}{2}\tfrac{1}{2}0\}, \{\sigma_y|000\}$: 5, 1; 10, 1: *b*.

T $\mathbf{G}_{16}^9$: $\{\sigma_z|\tfrac{1}{2}\tfrac{1}{2}0\}, \{I|\tfrac{1}{2}\tfrac{1}{2}0\}, \{\sigma_x|000\}$: 5, 1; 10, 1: *b*.

S $\mathbf{G}_{16}^9$: $\{C_{2x}|\tfrac{1}{2}\tfrac{1}{2}0\}, \{I|\tfrac{1}{2}\tfrac{1}{2}0\}, \{C_{2z}|000\}$: 5, 1; 10, 1: *a*.

R $\mathbf{G}_{16}^9$: $\{C_{2x}|\tfrac{1}{2}\tfrac{1}{2}0\}, \{I|\tfrac{1}{2}\tfrac{1}{2}0\}, \{C_{2z}|000\}$: 5, 1; 10, 1: *a*.

Δ^x $\mathbf{G}_4^2$: $(C_{2y}, 0), (\sigma_x, 0)$: 1, 1; 2, 1; 3, 1; 4, 1: *c*.

D^x $\mathbf{G}_8^4$: $(C_{2y}, 0), (\sigma_x, 0)$: 5, 1: *a*.

P^x $\mathbf{G}_8^4$: $(C_{2y}, 0), (\sigma_x, 0)$: 5, 1: *a*.

B^x $\mathbf{G}_4^2$: $(C_{2y}, 0), (\sigma_x, 0)$: 1, 1; 2, 1; 3, 1; 4, 1: *c*.

Σ^x $\mathbf{G}_4^2$: $(C_{2x}, 0), (\sigma_z, 0)$: 1, 1; 2, 1; 3, 1; 4, 1: *c*.

C^x $\mathbf{G}_8^4$: $(C_{2x}, 0), (\sigma_z, 0)$: 5, 1: *a*.

E^x $\mathbf{G}_8^4$: $(C_{2x}, 0), (\sigma_z, 0)$: 5, 1: *a*.

A^x $\mathbf{G}_4^2$: $(C_{2x}, 0), (\sigma_z, 0)$: 1, 1; 2, 1; 3, 1; 4, 1: *c*.

Λ^x $\mathbf{G}_4^2$: $(C_{2z}, 0), (\sigma_y, 0)$: 1, 1; 2, 1; 3, 1; 4, 1: *c*.

H^x $\mathbf{G}_4^2$: $(C_{2z}, 0), (\sigma_y, 0)$: 1, 3; 2, 3; 3, 3; 4, 3: *c*.

Q^x $\mathbf{G}_4^2$: $(C_{2z}, 0), (\sigma_y, 0)$: 1, 3; 2, 3; 3, 3; 4, 3: *c*.

G^x $\mathbf{G}_4^2$: $(C_{2z}, 0), (\sigma_y, 0)$: 1, 3; 2, 3; 3, 3; 4, 3: *c*.

60 *Pbcn* D_{2h}^{14}

($F1$; $K7$; $M5$; $T1$; $Z1$.)

Γ $\mathbf{G}_8^3$: $\{C_{2z}|000\}, \{C_{2y}|\tfrac{1}{2}\tfrac{1}{2}0\}, \{I|\tfrac{1}{2}0\tfrac{1}{2}\}$: 1, 1; 2, 1; 3, 1; 4, 1; 5, 1; 6, 1; 7, 1; 8, 1: *b*.

Y $\mathbf{G}_{16}^9$: $\{\sigma_z|\tfrac{1}{2}0\tfrac{1}{2}\}, \{I|\tfrac{1}{2}0\tfrac{1}{2}\}, \{\sigma_x|0\tfrac{1}{2}\tfrac{1}{2}\}$: 5, 1; 10, 1: *b*.

X $\mathbf{G}_{16}^9$: $\{\sigma_y|0\tfrac{1}{2}\tfrac{1}{2}\}, \{I|\tfrac{1}{2}0\tfrac{1}{2}\}, \{\sigma_z|\tfrac{1}{2}0\tfrac{1}{2}\}$: 5, 1; 10, 1: *b*.

Z $\mathbf{G}_{16}^9$: $\{\sigma_x|0\tfrac{1}{2}\tfrac{1}{2}\}, \{I|\tfrac{1}{2}0\tfrac{1}{2}\}, \{C_{2z}|000\}$: 5, 1; 10, 1: *a*.

U $\mathbf{G}_{16}^9$: $\{\sigma_x|0\tfrac{1}{2}\tfrac{1}{2}\}, \{C_{2z}|000\}, \{I|\tfrac{1}{2}0\tfrac{1}{2}\}$: 5, 1; 10, 1: *c*.

T $\mathbf{G}_{16}^7$: $\{\sigma_y|0\tfrac{1}{2}\tfrac{1}{2}\}, \{C_{2z}|000\}, \{I|\tfrac{1}{2}0\tfrac{1}{2}\}$: 9, 3; 10, 3: *a*.

S $\mathbf{G}_{16}^7$: $\{C_{2y}|\tfrac{1}{2}\tfrac{1}{2}0\}, \{C_{2z}|000\}, \{I|\tfrac{1}{2}0\tfrac{1}{2}\}$: 9, 3; 10, 3: *b*.

R $\mathbf{G}_{16}^7$: $\{C_{2x}|\tfrac{1}{2}\tfrac{1}{2}0\}, \{C_{2z}|000\}, \{I|\tfrac{1}{2}0\tfrac{1}{2}\}$: 9, 3; 10, 3: *b*.

Δ^x $\mathbf{G}_4^2$: $(C_{2y}, 0), (\sigma_x, 0)$: 1, 1; 2, 1; 3, 1; 4, 1: *c*.

D^x $\mathbf{G}_8^2$: $(C_{2y}, 0), (\sigma_z, 0)$: 2, 3; 4, 3; 6, 3; 8, 3: *c*.

P^x $\mathbf{G}_8^5$: $(C_{2y}, 0), (\sigma_x, 0)$: 5, 1: *a*.

B^x $\mathbf{G}_8^4$: $(\sigma_z, 0), (C_{2y}, 0)$: 5, 1: *a*.

Σ^x $\mathbf{G}_4^2$: $(C_{2x}, 0), (\sigma_z, 0)$: 1, 1; 2, 1; 3, 1; 4, 1: *c*.

C^x $\mathbf{G}_8^4$: $(C_{2x}, 0), (\sigma_z, 0)$: 5, 1: *a*.

E^x $\mathbf{G}_8^2$: $(C_{2x}, 0), (\sigma_y, 0)$: 2, 3; 4, 3; 6, 3; 8, 3: *c*.

A^x $\mathbf{G}_8^4$: $(\sigma_z, 0), (C_{2x}, 0)$: 5, 1: *a*.

Λ^x $\mathbf{G}_4^2$: $(C_{2z}, 0), (\sigma_y, 0)$: 1, 1; 2, 1; 3, 1; 4, 1: *c*.

H^x $\mathbf{G}_4^2$: $(C_{2z}, 0), (\sigma_y, 0)$: 1, 3; 2, 3; 3, 3; 4, 3: *c*.

Q^x $\mathbf{G}_8^4$: $(\sigma_x, 0), (C_{2z}, 0)$: 5, 2: *a*.

G^x $\mathbf{G}_8^4$: $(\sigma_x, 0), (C_{2z}, 0)$: 5, 1: *a*.

61 *Pbca* D_{2h}^{15}

($F1$; $K3$; $K7$; $M5$; $T1$; $T2$; $Z1$.)

Γ $\mathbf{G}_8^3$: $\{C_{2z}|\tfrac{1}{2}0\tfrac{1}{2}\}, \{C_{2y}|\tfrac{1}{2}\tfrac{1}{2}0\}, \{I|000\}$: 1, 1; 2, 1; 3, 1; 4, 1; 5, 1; 6, 1; 7, 1; 8, 1: *b*.
Y $\mathbf{G}_{16}^9$: $\{\sigma_z|\tfrac{1}{2}0\tfrac{1}{2}\}, \{I|000\}, \{\sigma_x|0\tfrac{1}{2}\tfrac{1}{2}\}$: 5, 1; 10, 1: *b*.
X $\mathbf{G}_{16}^9$: $\{\sigma_y|\tfrac{1}{2}\tfrac{1}{2}0\}, \{I|000\}, \{\sigma_z|\tfrac{1}{2}0\tfrac{1}{2}\}$: 5, 1; 10, 1: *b*.
Z $\mathbf{G}_{16}^9$: $\{\sigma_x|0\tfrac{1}{2}\tfrac{1}{2}\}, \{I|000\}, \{\sigma_y|\tfrac{1}{2}\tfrac{1}{2}0\}$: 5, 1; 10, 1: *b*.
U $\mathbf{G}_{16}^7$: $\{C_{2x}|0\tfrac{1}{2}\tfrac{1}{2}\}, \{C_{2y}|\tfrac{1}{2}\tfrac{1}{2}0\}, \{I|000\}$: 9, 3; 10, 3: *b*.
T $\mathbf{G}_{16}^7$: $\{C_{2z}|\tfrac{1}{2}0\tfrac{1}{2}\}, \{C_{2x}|0\tfrac{1}{2}\tfrac{1}{2}\}, \{I|000\}$: 9, 3; 10, 3: *b*.
S $\mathbf{G}_{16}^7$: $\{C_{2y}|\tfrac{1}{2}\tfrac{1}{2}0\}, \{C_{2z}|\tfrac{1}{2}0\tfrac{1}{2}\}, \{I|000\}$: 9, 3; 10, 3: *b*.
R $\mathbf{G}_{16}^{11}$: $\{C_{2x}|0\tfrac{1}{2}\tfrac{1}{2}\}, \{C_{2y}|\tfrac{1}{2}\tfrac{1}{2}0\}, \{I|000\}$: 5, 2; 10, 2: *a*.

Δ^x $\mathbf{G}_4^2$: $(C_{2y}, 0), (\sigma_x, 0)$: 1, 1; 2, 1; 3, 1; 4, 1: *c*.
D^x $\mathbf{G}_8^2$: $(C_{2y}, 0), (\sigma_z, 0)$: 2, 3; 4, 3; 6, 3; 8, 3: *c*.
P^x $\mathbf{G}_8^5$: $(C_{2y}, 0), (\sigma_x, 0)$: 5, 2: *a*.
B^x $\mathbf{G}_8^4$: $(\sigma_z, 0), (C_{2y}, 0)$: 5, 1: *a*.
Σ^x $\mathbf{G}_4^2$: $(C_{2x}, 0), (\sigma_z, 0)$: 1, 1; 2, 1; 3, 1; 4, 1: *c*.
C^x $\mathbf{G}_8^4$: $(\sigma_y, 0), (C_{2x}, 0)$: 5, 1: *a*.
E^x $\mathbf{G}_8^5$: $(C_{2x}, 0), (\sigma_z, 0)$: 5, 2: *a*.
A^x $\mathbf{G}_8^2$: $(C_{2x}, 0), (\sigma_y, 0)$: 2, 3; 4, 3; 6, 3; 8, 3: *c*.
Λ^x $\mathbf{G}_4^2$: $(C_{2z}, 0), (\sigma_y, 0)$: 1, 1; 2, 1; 3, 1; 4, 1: *c*.
H^x $\mathbf{G}_8^2$: $(C_{2z}, 0), (\sigma_x, 0)$: 2, 3; 4, 3; 6, 3; 8, 3: *c*.
Q^x $\mathbf{G}_8^5$: $(C_{2z}, 0), (\sigma_y, 0)$: 5, 2: *a*.
G^x $\mathbf{G}_8^4$: $(\sigma_x, 0), (C_{2z}, 0)$: 5, 1: *a*.

62 *Pnma* D_{2h}^{16}

($F1$; $G1$; $K2$; $K7$; $M5$; $T1$; $S8$; $Z1$.)

Γ $\mathbf{G}_8^3$: $\{C_{2z}|\tfrac{1}{2}0\tfrac{1}{2}\}, \{C_{2y}|\tfrac{1}{2}\tfrac{1}{2}0\}, \{I|\tfrac{1}{2}\tfrac{1}{2}0\}$: 1, 1; 2, 1; 3, 1; 4, 1; 5, 1; 6, 1; 7, 1; 8, 1: *b*.
Y $\mathbf{G}_{16}^9$: $\{\sigma_x|\tfrac{1}{2}0\tfrac{1}{2}\}, \{I|\tfrac{1}{2}\tfrac{1}{2}0\}, \{\sigma_z|0\tfrac{1}{2}\tfrac{1}{2}\}$: 5, 1; 10, 1: *b*.
X $\mathbf{G}_{16}^9$: $\{\sigma_z|0\tfrac{1}{2}\tfrac{1}{2}\}, \{I|\tfrac{1}{2}\tfrac{1}{2}0\}, \{\sigma_y|000\}$: 5, 1; 10, 1: *b*.
Z $\mathbf{G}_{16}^9$: $\{\sigma_x|\tfrac{1}{2}0\tfrac{1}{2}\}, \{I|\tfrac{1}{2}\tfrac{1}{2}0\}, \{\sigma_y|000\}$: 5, 1; 10, 1: *b*.
U $\mathbf{G}_{16}^4$: $\{C_{2x}|0\tfrac{1}{2}\tfrac{1}{2}\}, \{C_{2y}|\tfrac{1}{2}\tfrac{1}{2}0\}, \{I|\tfrac{1}{2}\tfrac{1}{2}0\}$: 2, 3; 4, 3; 6, 3; 8, 3; 10, 3; 12, 3; 14, 3; 16, 3: *a*.
T $\mathbf{G}_{16}^9$: $\{C_{2y}|\tfrac{1}{2}\tfrac{1}{2}0\}, \{I|\tfrac{1}{2}\tfrac{1}{2}0\}, \{C_{2x}|0\tfrac{1}{2}\tfrac{1}{2}\}$: 5, 1; 10, 1: *a*.
S $\mathbf{G}_{16}^7$: $\{C_{2x}|0\tfrac{1}{2}\tfrac{1}{2}\}, \{\sigma_y|000\}, \{I|\tfrac{1}{2}\tfrac{1}{2}0\}$: 9, 3; 10, 3: *b*.
R $\mathbf{G}_{16}^7$: $\{\sigma_z|0\tfrac{1}{2}\tfrac{1}{2}\}, \{\sigma_y|000\}, \{I|\tfrac{1}{2}\tfrac{1}{2}0\}$: 9, 3; 10, 3: *a*.

Δ^x $\mathbf{G}_4^2$: $(C_{2y}, 0), (\sigma_x, 0)$: 1, 1; 2, 1; 3, 1; 4, 1: *c*.
D^x $\mathbf{G}_8^4$: $(C_{2y}, 0), (\sigma_x, 0)$: 5, 1: *a*.
P^x $\mathbf{G}_8^2$: $(C_{2y}, 0), (\sigma_x, 0)$: 2, 3; 4, 3; 6, 3; 8, 3: *c*.
B^x $\mathbf{G}_8^4$: $(\sigma_z, 0), (C_{2y}, 0)$: 5, 1: *a*.
Σ^x $\mathbf{G}_4^2$: $(C_{2x}, 0), (\sigma_z, 0)$: 1, 1; 2, 1; 3, 1; 4, 1: *c*.
C^x $\mathbf{G}_4^2$: $(C_{2x}, 0), (\sigma_z, 0)$: 1, 3; 2, 3; 3, 3; 4, 3: *c*.
E^x $\mathbf{G}_8^2$: $(C_{2x}, 0), (\sigma_y, 0)$: 2, 3; 4, 3; 6, 3; 8, 3: *c*.
A^x $\mathbf{G}_8^2$: $(C_{2x}, 0), (\sigma_y, 0)$: 2, 3; 4, 3; 6, 3; 8, 3: *c*.
Λ^x $\mathbf{G}_4^2$: $(C_{2z}, 0), (\sigma_y, 0)$: 1, 1; 2, 1; 3, 1; 4, 1: *c*.
H^x $\mathbf{G}_8^4$: $(C_{2z}, 0), (\sigma_y, 0)$: 5, 1: *a*.
Q^x $\mathbf{G}_8^4$: $(C_{2z}, 0), (\sigma_y, 0)$: 5, 2: *a*.
G^x $\mathbf{G}_4^2$: $(C_{2z}, 0), (\sigma_y, 0)$: 1, 3; 2, 3; 3, 3; 4, 3: *c*.

63 *Cmcm* D_{2h}^{17}

($F1$; $J1$; $K7$; $M5$; $S12$; $T3$; $Z1$.)

Γ $\mathbf{G}_8^3$: $\{C_{2z}|00\frac{1}{2}\}$, $\{C_{2y}|00\frac{1}{2}\}$, $\{I|000\}$: 1, 1; 2, 1; 3, 1; 4, 1; 5, 1; 6, 1; 7, 1; 8, 1: *b*.
Y $\mathbf{G}_8^3 \otimes \mathbf{T}_2$: $\{C_{2z}|00\frac{1}{2}\}$, $\{C_{2y}|00\frac{1}{2}\}$, $\{I|000\}$; $\mathbf{t}_1$ or $\mathbf{t}_2$: 1, 1; 2, 1; 3, 1; 4, 1; 5, 1; 6, 1; 7, 1; 8, 1: *b*.
Z $\mathbf{G}_{16}^9$: $\{C_{2z}|00\frac{1}{2}\}$, $\{I|000\}$, $\{\sigma_x|000\}$: 5, 1; 10, 1: *b*.
T $\mathbf{G}_{16}^9$: $\{C_{2z}|00\frac{1}{2}\}$, $\{I|000\}$, $\{\sigma_x|000\}$: 5, 1; 10, 1: *b*.
S $\mathbf{G}_4^2 \otimes \mathbf{T}_2$: $\{C_{2z}|00\frac{1}{2}\}$, $\{I|000\}$; $\mathbf{t}_2$: 1, 1; 2, 1; 3, 1; 4, 1: *a*.
R $\mathbf{G}_8^4$: $\{C_{2z}|00\frac{1}{2}\}$, $\{I|000\}$: 5, 1: *a*.

Λ^x $\mathbf{G}_4^2$: $(C_{2z}, 0)$, $(\sigma_y, 0)$: 1, 1; 2, 1; 3, 1; 4, 1: *c*.
H^x $\mathbf{G}_4^2$: $(C_{2z}, 0)$, $(\sigma_y, 0)$: 1, 1; 2, 1; 3, 1; 4, 1: *c*.
D^x $\mathbf{G}_2^1$: $(C_{2z}, 0)$: 1, 1; 2, 1: *b*.
A^x $\mathbf{G}_8^4$: $(\sigma_z, 0)$, $(C_{2x}, 0)$: 5, 1: *a*.
Σ^x $\mathbf{G}_4^2$: $(C_{2x}, 0)$, $(\sigma_z, 0)$: 1, 1; 2, 1; 3, 1; 4, 1: *c*.
Δ^x $\mathbf{G}_4^2$: $(C_{2y}, 0)$, $(\sigma_x, 0)$: 1, 1; 2, 1; 3, 1; 4, 1: *c*.
B^x $\mathbf{G}_8^2$: $(C_{2y}, 0)$, $(\sigma_x, 0)$: 2, 3; 4, 3; 6, 3; 8, 3: *c*.
G^x $\mathbf{G}_8^2$: $(C_{2y}, 0)$, $(\sigma_x, 0)$: 2, 3; 4, 3; 6, 3; 8, 3: *c*.
F^x $\mathbf{G}_4^2$: $(C_{2y}, 0)$, $(\sigma_x, 0)$: 1, 1; 2, 1; 3, 1; 4, 1: *c*.
E^x $\mathbf{G}_8^4$: $(\sigma_z, 0)$, $(C_{2x}, 0)$: 5, 1: *a*.
C^x $\mathbf{G}_4^2$: $(C_{2x}, 0)$, $(\sigma_z, 0)$: 1, 1; 2, 1; 3, 1; 4, 1: *c*.

64 *Cmca* D_{2h}^{18}

($F1$; $K7$; $M5$; $S8$; $S9$; $T3$; $W1$; $Z1$.)

Γ $\mathbf{G}_8^3$: $\{C_{2z}|00\frac{1}{2}\}$, $\{C_{2y}|00\frac{1}{2}\}$, $\{I|\frac{1}{2}\frac{1}{2}0\}$: 1, 1; 2, 1; 3, 1; 4, 1; 5, 1; 6, 1; 7, 1; 8, 1: *b*.
Y $\mathbf{G}_8^3 \otimes \mathbf{T}_2$: $\{C_{2z}|00\frac{1}{2}\}$, $\{C_{2y}|00\frac{1}{2}\}$, $\{I|\frac{1}{2}\frac{1}{2}0\}$; $\mathbf{t}_1$ or $\mathbf{t}_2$: 1, 1; 2, 1; 3, 1; 4, 1; 5, 1; 6, 1; 7, 1; 8, 1: *b*.
Z $\mathbf{G}_{16}^9$: $\{C_{2z}|00\frac{1}{2}\}$, $\{I|\frac{1}{2}\frac{1}{2}0\}$, $\{\sigma_x|\frac{1}{2}\frac{1}{2}0\}$: 5, 1; 10, 1: *b*.
T $\mathbf{G}_{16}^9$: $\{C_{2z}|00\frac{1}{2}\}$, $\{I|\frac{1}{2}\frac{1}{2}0\}$, $\{\sigma_x|\frac{1}{2}\frac{1}{2}0\}$: 5, 1; 10, 1: *b*.
S $\mathbf{G}_8^4$: $\{\sigma_z|\frac{1}{2}\frac{1}{2}\frac{1}{2}\}$, $\{I|\frac{1}{2}\frac{1}{2}0\}$: 5, 1: *a*.
R $\mathbf{G}_8^2$: $\{\sigma_z|\frac{1}{2}\frac{1}{2}\frac{1}{2}\}$, $\{I|\frac{1}{2}\frac{1}{2}0\}$: 2, 3; 4, 3; 6, 3; 8, 3: *a*.

Λ^x $\mathbf{G}_4^2$: $(C_{2z}, 0)$, $(\sigma_y, 0)$: 1, 1; 2, 1; 3, 1; 4, 1: *c*.
H^x $\mathbf{G}_4^2$: $(C_{2z}, 0)$, $(\sigma_y, 0)$: 1, 1; 2, 1; 3, 1; 4, 1: *c*.
D^x $\mathbf{G}_2^1$: $(C_{2z}, 0)$: 1, 3; 2, 3: *b*.
A^x $\mathbf{G}_8^4$: $(\sigma_z, 0)$, $(C_{2x}, 0)$: 5, 1: *a*.
Σ^x $\mathbf{G}_4^2$: $(C_{2x}, 0)$, $(\sigma_z, 0)$: 1, 1; 2, 1; 3, 1; 4, 1: *c*.
Δ^x $\mathbf{G}_4^2$: $(C_{2y}, 0)$, $(\sigma_x, 0)$: 1, 1; 2, 1; 3, 1; 4, 1: *c*.
B^x $\mathbf{G}_8^2$: $(C_{2y}, 0)$, $(\sigma_x, 0)$: 2, 3; 4, 3; 6, 3; 8, 3: *c*.
G^x $\mathbf{G}_8^2$: $(C_{2y}, 0)$, $(\sigma_x, 0)$: 2, 3; 4, 3; 6, 3; 8, 3: *c*.
F^x $\mathbf{G}_4^2$: $(C_{2y}, 0)$, $(\sigma_x, 0)$: 1, 1; 2, 1; 3, 1; 4, 1: *c*.
E^x $\mathbf{G}_8^4$: $(\sigma_z, 0)$, $(C_{2x}, 0)$: 5, 1: *a*.
C^x $\mathbf{G}_4^2$: $(C_{2x}, 0)$, $(\sigma_z, 0)$: 1, 1; 2, 1; 3, 1; 4, 1: *c*.

65 *Cmmm* D_{2h}^{19}

($F1$; $K7$; $M5$; $T3$; $Z1$.)

Γ $\mathbf{G}_8^3$: $\{C_{2z}|000\}$, $\{C_{2y}|000\}$, $\{I|000\}$: 1, 1; 2, 1; 3, 1; 4, 1; 5, 1; 6, 1; 7, 1; 8, 1: *b*.
Y $\mathbf{G}_8^3 \otimes \mathbf{T}_2$: $\{C_{2z}|000\}$, $\{C_{2y}|000\}$, $\{I|000\}$; $\mathbf{t}_1$ or $\mathbf{t}_2$: 1, 1; 2, 1; 3, 1; 4, 1; 5, 1; 6, 1; 7, 1; 8, 1: *b*.
Z $\mathbf{G}_8^3 \otimes \mathbf{T}_2$: $\{C_{2z}|000\}$, $\{C_{2y}|000\}$, $\{I|000\}$; $\mathbf{t}_3$: 1, 1; 2, 1; 3, 1; 4, 1; 5, 1; 6, 1; 7, 1; 8, 1: *b*.
T $\mathbf{G}_8^3 \otimes \mathbf{T}_2$: $\{C_{2z}|000\}$, $\{C_{2y}|000\}$, $\{I|000\}$; $\mathbf{t}_1$ or $\mathbf{t}_2$ or $\mathbf{t}_3$: 1, 1; 2, 1; 3, 1; 4, 1; 5, 1; 6, 1; 7, 1; 8, 1: *b*.
S $\mathbf{G}_4^2 \otimes \mathbf{T}_2$: $\{C_{2z}|000\}$, $\{I|000\}$; $\mathbf{t}_2$: 1, 1; 2, 1; 3, 1; 4, 1: *a*.
R $\mathbf{G}_4^2 \otimes \mathbf{T}_2$: $\{C_{2z}|000\}$, $\{I|000\}$; $\mathbf{t}_2$ or $\mathbf{t}_3$: 1, 1; 2, 1; 3, 1; 4, 1: *a*.

Λ^x $\mathbf{G}_4^2$: $(C_{2z}, 0)$, $(\sigma_y, 0)$: 1, 1; 2, 1; 3, 1; 4, 1: *c*.
H^x $\mathbf{G}_4^2$: $(C_{2z}, 0)$, $(\sigma_y, 0)$: 1, 1; 2, 1; 3, 1; 4, 1: *c*.
D^x $\mathbf{G}_2^1$: $(C_{2z}, 0)$: 1, 1; 2, 1: *b*.
A^x $\mathbf{G}_4^2$: $(C_{2x}, 0)$, $(\sigma_z, 0)$: 1, 1; 2, 1; 3, 1; 4, 1: *c*.
Σ^x $\mathbf{G}_4^2$: $(C_{2x}, 0)$, $(\sigma_z, 0)$: 1, 1; 2, 1; 3, 1; 4, 1: *c*.
Δ^x $\mathbf{G}_4^2$: $(C_{2y}, 0)$, $(\sigma_x, 0)$: 1, 1; 2, 1; 3, 1; 4, 1: *c*.
B^x $\mathbf{G}_4^2$: $(C_{2y}, 0)$, $(\sigma_x, 0)$: 1, 1; 2, 1; 3, 1; 4, 1: *c*.
G^x $\mathbf{G}_4^2$: $(C_{2y}, 0)$, $(\sigma_x, 0)$: 1, 1; 2, 1; 3, 1; 4, 1: *c*.
F^x $\mathbf{G}_4^2$: $(C_{2y}, 0)$, $(\sigma_x, 0)$: 1, 1; 2, 1; 3, 1; 4, 1: *c*.
E^x $\mathbf{G}_4^2$: $(C_{2x}, 0)$, $(\sigma_z, 0)$: 1, 1; 2, 1; 3, 1; 4, 1: *c*.
C^x $\mathbf{G}_4^2$: $(C_{2x}, 0)$, $(\sigma_z, 0)$: 1, 1; 2, 1; 3, 1; 4, 1: *c*.

66 *Cccm* D_{2h}^{20}

($F1$; $K7$; $M5$; $T3$; $Z1$.)

Γ $\mathbf{G}_8^3$: $\{C_{2z}|000\}$, $\{C_{2y}|000\}$, $\{I|00\tfrac{1}{2}\}$: 1, 1; 2, 1; 3, 1; 4, 1; 5, 1; 6, 1; 7, 1; 8, 1: *b*.
Y $\mathbf{G}_8^3 \otimes \mathbf{T}_2$: $\{C_{2z}|000\}$, $\{C_{2y}|000\}$, $\{I|00\tfrac{1}{2}\}$; $\mathbf{t}_1$ or $\mathbf{t}_2$: 1, 1; 2, 1; 3, 1; 4, 1; 5, 1; 6, 1; 7, 1; 8, 1: *b*.
Z $\mathbf{G}_{16}^9$: $\{\sigma_x|00\tfrac{1}{2}\}$, $\{I|00\tfrac{1}{2}\}$, $\{C_{2z}|000\}$: 5, 1; 10, 1: *a*.
T $\mathbf{G}_{16}^9$: $\{\sigma_x|00\tfrac{1}{2}\}$, $\{I|00\tfrac{1}{2}\}$, $\{C_{2z}|000\}$: 5, 1; 10, 1: *a*.
S $\mathbf{G}_4^2 \otimes \mathbf{T}_2$: $\{C_{2z}|000\}$, $\{I|00\tfrac{1}{2}\}$; $\mathbf{t}_2$: 1, 1; 2, 1; 3, 1; 4, 1: *a*.
R $\mathbf{G}_4^2 \otimes \mathbf{T}_2$: $\{C_{2z}|000\}$, $\{I|00\tfrac{1}{2}\}$; $\mathbf{t}_2$ or $\mathbf{t}_3$: 1, 1; 2, 1; 3, 1; 4, 1: *a*.

Λ^x $\mathbf{G}_4^2$: $(C_{2z}, 0)$, $(\sigma_y, 0)$: 1, 1; 2, 1; 3, 1; 4, 1: *c*.
H^x $\mathbf{G}_4^2$: $(C_{2z}, 0)$, $(\sigma_y, 0)$: 1, 1; 2, 1; 3, 1; 4, 1: *c*.
D^x $\mathbf{G}_2^1$: $(C_{2z}, 0)$: 1, 1; 2, 1: *b*.
A^x $\mathbf{G}_8^4$: $(\sigma_z, 0)$, $(C_{2x}, 0)$: 5, 1: *a*.
Σ^x $\mathbf{G}_4^2$: $(C_{2x}, 0)$, $(\sigma_z, 0)$: 1, 1; 2, 1; 3, 1; 4, 1: *c*.
Δ^x $\mathbf{G}_4^2$: $(C_{2y}, 0)$, $(\sigma_x, 0)$: 1, 1; 2, 1; 3, 1; 4, 1: *c*.
B^x $\mathbf{G}_8^4$: $(\sigma_z, 0)$, $(C_{2y}, 0)$: 5, 1: *a*.
G^x $\mathbf{G}_8^4$: $(\sigma_z, 0)$, $(C_{2y}, 0)$: 5, 1: *a*.
F^x $\mathbf{G}_4^2$: $(C_{2y}, 0)$, $(\sigma_x, 0)$: 1, 1; 2, 1; 3, 1; 4, 1: *c*.
E^x $\mathbf{G}_8^4$: $(\sigma_z, 0)$, $(C_{2x}, 0)$: 5, 1: *a*.
C^x $\mathbf{G}_4^2$: $(C_{2x}, 0)$, $(\sigma_z, 0)$: 1, 1; 2, 1; 3, 1; 4, 1: *c*.

67 *Cmma* D_{2h}^{21}

($F1$; $K7$; $M5$; $T3$; $Z1$.)

Γ $\mathbf{G}_8^3$: $\{C_{2z} \mid 000\}$, $\{C_{2y} \mid 000\}$, $\{I \mid \frac{1}{2}\frac{1}{2}0\}$: 1, 1; 2, 1; 3, 1; 4, 1; 5, 1; 6, 1; 7, 1; 8, 1: *b*.
Y $\mathbf{G}_8^3 \otimes \mathbf{T}_2$: $\{C_{2z} \mid 000\}$, $\{C_{2y} \mid 000\}$, $\{I \mid \frac{1}{2}\frac{1}{2}0\}$; $\mathbf{t}_1$ or $\mathbf{t}_2$: 1, 1; 2, 1; 3, 1; 4, 1; 5, 1; 6, 1; 7, 1; 8, 1: *b*.
Z $\mathbf{G}_8^3 \otimes \mathbf{T}_2$: $\{C_{2z} \mid 000\}$, $\{C_{2y} \mid 000\}$, $\{I \mid \frac{1}{2}\frac{1}{2}0\}$; $\mathbf{t}_3$: 1, 1; 2, 1; 3, 1; 4, 1; 5, 1; 6, 1; 7, 1; 8, 1: *b*.
T $\mathbf{G}_8^3 \otimes \mathbf{T}_2$: $\{C_{2z} \mid 000\}$, $\{C_{2y} \mid 000\}$, $\{I \mid \frac{1}{2}\frac{1}{2}0\}$; $\mathbf{t}_1$ or $\mathbf{t}_2$ or $\mathbf{t}_3$: 1, 1; 2, 1; 3, 1; 4, 1; 5, 1; 6, 1; 7, 1; 8, 1: *b*.
S $\mathbf{G}_8^4$: $\{\sigma_z \mid \frac{1}{2}\frac{1}{2}0\}$, $\{I \mid \frac{1}{2}\frac{1}{2}0\}$: 5, 1: *a*.
R $\mathbf{G}_8^4$: $\{\sigma_z \mid \frac{1}{2}\frac{1}{2}0\}$, $\{I \mid \frac{1}{2}\frac{1}{2}0\}$: 5, 1: *a*.

Λ^x $\mathbf{G}_4^2$: $(C_{2z}, 0)$, $(\sigma_y, 0)$: 1, 1; 2, 1; 3, 1; 4, 1: *c*.
H^x $\mathbf{G}_4^2$: $(C_{2z}, 0)$, $(\sigma_y, 0)$: 1, 1; 2, 1; 3, 1; 4, 1: *c*.
D^x $\mathbf{G}_2^1$: $(C_{2z}, 0)$: 1, 3; 2, 3: *b*.
A^x $\mathbf{G}_4^2$: $(C_{2x}, 0)$, $(\sigma_z, 0)$: 1, 1; 2, 1; 3, 1; 4, 1: *c*.
Σ^x $\mathbf{G}_4^2$: $(C_{2x}, 0)$, $(\sigma_z, 0)$: 1, 1; 2, 1; 3, 1; 4, 1: *c*.
Δ^x $\mathbf{G}_4^2$: $(C_{2y}, 0)$, $(\sigma_x, 0)$: 1, 1; 2, 1; 3, 1; 4, 1: *c*.
B^x $\mathbf{G}_4^2$: $(C_{2y}, 0)$, $(\sigma_x, 0)$: 1, 1; 2, 1; 3, 1; 4, 1: *c*.
G^x $\mathbf{G}_4^2$: $(C_{2y}, 0)$, $(\sigma_x, 0)$: 1, 1; 2, 1; 3, 1; 4, 1: *c*.
F^x $\mathbf{G}_4^2$: $(C_{2y}, 0)$, $(\sigma_x, 0)$: 1, 1; 2, 1; 3, 1; 4, 1: *c*.
E^x $\mathbf{G}_4^2$: $(C_{2x}, 0)$, $(\sigma_z, 0)$: 1, 1; 2, 1; 3, 1; 4, 1: *c*.
C^x $\mathbf{G}_4^2$: $(C_{2x}, 0)$, $(\sigma_z, 0)$: 1, 1; 2, 1; 3, 1; 4, 1: *c*.

68 *Ccca* D_{2h}^{22}

($F1$; $K7$; $M5$; $T3$; $Z1$.)

Γ $\mathbf{G}_8^3$: $\{C_{2z} \mid 000\}$, $\{C_{2y} \mid 000\}$, $\{I \mid \frac{1}{2}\frac{1}{2}\frac{1}{2}\}$: 1, 1; 2, 1; 3, 1; 4, 1; 5, 1; 6, 1; 7, 1; 8, 1: *b*.
Y $\mathbf{G}_8^3 \otimes \mathbf{T}_2$: $\{C_{2z} \mid 000\}$, $\{C_{2y} \mid 000\}$, $\{I \mid \frac{1}{2}\frac{1}{2}\frac{1}{2}\}$; $\mathbf{t}_1$ or $\mathbf{t}_2$: 1, 1; 2, 1; 3, 1; 4, 1; 5, 1; 6, 1; 7, 1; 8, 1: *b*.
Z $\mathbf{G}_{16}^9$: $\{\sigma_x \mid \frac{1}{2}\frac{1}{2}\frac{1}{2}\}$, $\{I \mid \frac{1}{2}\frac{1}{2}\frac{1}{2}\}$, $\{C_{2z} \mid 000\}$: 5, 1; 10, 1: *a*.
T $\mathbf{G}_{16}^9$: $\{\sigma_x \mid \frac{1}{2}\frac{1}{2}\frac{1}{2}\}$, $\{I \mid \frac{1}{2}\frac{1}{2}\frac{1}{2}\}$, $\{C_{2z} \mid 000\}$: 5, 1; 10, 1: *a*.
S $\mathbf{G}_8^4$: $\{\sigma_z \mid \frac{1}{2}\frac{1}{2}\frac{1}{2}\}$, $\{I \mid \frac{1}{2}\frac{1}{2}\frac{1}{2}\}$: 5, 1: *a*.
R $\mathbf{G}_8^4$: $\{\sigma_z \mid \frac{1}{2}\frac{1}{2}\frac{1}{2}\}$, $\{I \mid \frac{1}{2}\frac{1}{2}\frac{1}{2}\}$: 5, 1: *a*.

Λ^x $\mathbf{G}_4^2$: $(C_{2z}, 0)$, $(\sigma_y, 0)$: 1, 1; 2, 1; 3, 1; 4, 1: *c*.
H^x $\mathbf{G}_4^2$: $(C_{2z}, 0)$, $(\sigma_y, 0)$: 1, 1; 2, 1; 3, 1; 4, 1: *c*.
D^x $\mathbf{G}_2^1$: $(C_{2z}, 0)$: 1, 3; 2, 3: *b*.
A^x $\mathbf{G}_8^4$: $(\sigma_z, 0)$, $(C_{2x}, 0)$: 5, 1: *a*.
Σ^x $\mathbf{G}_4^2$: $(C_{2x}, 0)$, $(\sigma_z, 0)$: 1, 1; 2, 1; 3, 1; 4, 1: *c*.
Δ^x $\mathbf{G}_4^2$: $(C_{2y}, 0)$, $(\sigma_x, 0)$: 1, 1; 2, 1; 3, 1; 4, 1: *c*.
B^x $\mathbf{G}_8^4$: $(\sigma_z, 0)$, $(C_{2y}, 0)$: 5, 1: *a*.
G^x $\mathbf{G}_8^4$: $(\sigma_z, 0)$, $(C_{2y}, 0)$: 5, 1: *a*.
F^x $\mathbf{G}_4^2$: $(C_{2y}, 0)$, $(\sigma_x, 0)$: 1, 1; 2, 1; 3, 1; 4, 1: *c*.
E^x $\mathbf{G}_8^4$: $(\sigma_z, 0)$, $(C_{2x}, 0)$: 5, 1: *a*.
C^x $\mathbf{G}_4^2$: $(C_{2x}, 0)$, $(\sigma_z, 0)$: 1, 1; 2, 1; 3, 1; 4, 1: *c*.

69 *Fmmm* D_{2h}^{23}

(F1; K7; M5; S14; Z1.)

Γ $\mathbf{G}_8^3$: $\{C_{2z}|000\}, \{C_{2y}|000\}, \{I|000\}$: 1, 1; 2, 1; 3, 1; 4, 1; 5, 1; 6, 1; 7, 1; 8, 1: *b*.
Y $\mathbf{G}_8^3 \otimes \mathbf{T}_2$: $\{C_{2z}|000\}, \{C_{2y}|000\}, \{I|000\}$; $\mathbf{t}_2$ or $\mathbf{t}_3$: 1, 1; 2, 1; 3, 1; 4, 1; 5, 1; 6, 1; 7, 1; 8, 1: *b*.
X $\mathbf{G}_8^3 \otimes \mathbf{T}_2$: $\{C_{2z}|000\}, \{C_{2y}|000\}, \{I|000\}$; $\mathbf{t}_1$ or $\mathbf{t}_3$: 1, 1; 2, 1; 3, 1; 4, 1; 5, 1; 6, 1; 7, 1; 8, 1: *b*.
Z $\mathbf{G}_8^3 \otimes \mathbf{T}_2$: $\{C_{2z}|000\}, \{C_{2y}|000\}, \{I|000\}$; $\mathbf{t}_1$ or $\mathbf{t}_2$: 1, 1; 2, 1; 3, 1; 4, 1; 5, 1; 6, 1; 7, 1; 8, 1: *b*.
L $\mathbf{G}_2^1 \otimes \mathbf{T}_2$: $\{I|000\}$; $\mathbf{t}_1$: 1, 1; 2, 1: *a*.

Λ^x $\mathbf{G}_4^2$: $(C_{2z}, 0), (\sigma_y, 0)$: 1, 1; 2, 1; 3, 1; 4, 1: *c*.
G^x $\mathbf{G}_4^2$: $(C_{2z}, 0), (\sigma_y, 0)$: 1, 1; 2, 1; 3, 1; 4, 1: *c*.
H^x $\mathbf{G}_4^2$: $(C_{2z}, 0), (\sigma_y, 0)$: 1, 1; 2, 1; 3, 1; 4, 1: *c*.
Q^x $\mathbf{G}_4^2$: $(C_{2z}, 0), (\sigma_y, 0)$: 1, 1; 2, 1; 3, 1; 4, 1: *c*.
Σ^x $\mathbf{G}_4^2$: $(C_{2x}, 0), (\sigma_z, 0)$: 1, 1; 2, 1; 3, 1; 4, 1: *c*.
C^x $\mathbf{G}_4^2$: $(C_{2x}, 0), (\sigma_z, 0)$: 1, 1; 2, 1; 3, 1; 4, 1: *c*.
A^x $\mathbf{G}_4^2$: $(C_{2x}, 0), (\sigma_z, 0)$: 1, 1; 2, 1; 3, 1; 4, 1: *c*.
U^x $\mathbf{G}_4^2$: $(C_{2x}, 0), (\sigma_z, 0)$: 1, 1; 2, 1; 3, 1; 4, 1: *c*.
Δ^x $\mathbf{G}_4^2$: $(C_{2y}, 0), (\sigma_x, 0)$: 1, 1; 2, 1; 3, 1; 4, 1: *c*.
D^x $\mathbf{G}_4^2$: $(C_{2y}, 0), (\sigma_x, 0)$: 1, 1; 2, 1; 3, 1; 4, 1: *c*.
B^x $\mathbf{G}_4^2$: $(C_{2y}, 0), (\sigma_x, 0)$: 1, 1; 2, 1; 3, 1; 4, 1: *c*.
R^x $\mathbf{G}_4^2$: $(C_{2y}, 0), (\sigma_x, 0)$: 1, 1; 2, 1; 3, 1; 4, 1: *c*.

70 *Fddd* D_{2h}^{24}

(F1; K7; M5; S14; Z1.)

Γ $\mathbf{G}_8^3$: $\{C_{2z}|000\}, \{C_{2y}|000\}, \{I|\frac{1}{4}\frac{1}{4}\frac{1}{4}\}$: 1, 1; 2, 1; 3, 1; 4, 1; 5, 1; 6, 1; 7, 1; 8, 1: *b*.
Y $\mathbf{G}_{16}^9$: $\{\sigma_x|\frac{1}{4}\frac{1}{4}\frac{1}{4}\}, \{I|\frac{1}{4}\frac{1}{4}\frac{1}{4}\}, \{C_{2y}|000\}$: 5, 1; 10, 1: *a*.
X $\mathbf{G}_{16}^9$: $\{\sigma_z|\frac{1}{4}\frac{1}{4}\frac{1}{4}\}, \{I|\frac{1}{4}\frac{1}{4}\frac{1}{4}\}, \{C_{2x}|000\}$: 5, 1; 10, 1: *a*.
Z $\mathbf{G}_{16}^9$: $\{\sigma_y|\frac{1}{4}\frac{1}{4}\frac{1}{4}\}, \{I|\frac{1}{4}\frac{1}{4}\frac{1}{4}\}, \{C_{2z}|000\}$: 5, 1; 10, 1: *a*.
L $\mathbf{G}_2^1 \otimes \mathbf{T}_2$: $\{I|\frac{1}{4}\frac{1}{4}\frac{1}{4}\}$; $\mathbf{t}_1$: 1, 1; 2, 1: *a*.

Λ^x $\mathbf{G}_4^2$: $(C_{2z}, 0), (\sigma_y, 0)$: 1, 1; 2, 1; 3, 1; 4, 1: *c*.
G^x $\mathbf{G}_8^4$: $(\sigma_x, 0), (C_{2z}, 0)$: 5, 1: *a*.
H^x $\mathbf{G}_8^4$: $(\sigma_y, 0), (C_{2z}, 0)$: 5, 1: *a*.
Q^x $\mathbf{G}_8^2$: $(\sigma_x, 0), (C_{2z}, 0)$: 2, 1; 4, 1; 6, 1; 8, 1: *d*.
Σ^x $\mathbf{G}_4^2$: $(C_{2x}, 0), (\sigma_z, 0)$: 1, 1; 2, 1; 3, 1; 4, 1: *c*.
C^x $\mathbf{G}_8^4$: $(\sigma_y, 0), (C_{2x}, 0)$: 5, 1: *a*.
A^x $\mathbf{G}_8^4$: $(\sigma_z, 0), (C_{2x}, 0)$: 5, 1: *a*.
U^x $\mathbf{G}_8^2$: $(\sigma_y, 0), (C_{2x}, 0)$: 2, 1; 4, 1; 6, 1; 8, 1: *d*.
Δ^x $\mathbf{G}_4^2$: $(C_{2y}, 0), (\sigma_x, 0)$: 1, 1; 2, 1; 3, 1; 4, 1: *c*.
D^x $\mathbf{G}_8^4$: $(\sigma_x, 0), (C_{2y}, 0)$: 5, 1: *a*.
B^x $\mathbf{G}_8^4$: $(\sigma_z, 0), (C_{2y}, 0)$: 5, 1: *a*.
R^x $\mathbf{G}_8^2$: $(\sigma_x, 0), (C_{2y}, 0)$: 2, 1; 4, 1; 6, 1; 8, 1: *d*.

71 *Immm* D_{2h}^{25}

($F1$; $K7$; $M5$; $Z1$.)

Γ $\mathbf{G}_8^3$: $\{C_{2z}|000\}, \{C_{2y}|000\}, \{I|000\}$: 1, 1; 2, 1; 3, 1; 4, 1; 5, 1; 6, 1; 7, 1; 8, 1: *b*.
X $\mathbf{G}_8^3 \otimes \mathbf{T}_2$: $\{C_{2z}|000\}, \{C_{2y}|000\}, \{I|000\}$; $\mathbf{t}_1$ or $\mathbf{t}_2$ or $\mathbf{t}_3$: 1, 1; 2, 1; 3, 1; 4, 1; 5, 1; 6, 1; 7, 1; 8, 1: *b*.
R $\mathbf{G}_4^2 \otimes \mathbf{T}_2$: $\{C_{2y}|000\}, \{I|000\}$; $\mathbf{t}_1$: 1, 1; 2, 1; 3, 1; 4, 1: *a*.
S $\mathbf{G}_4^2 \otimes \mathbf{T}_2$: $\{C_{2x}|000\}, \{I|000\}$; $\mathbf{t}_1$ or $\mathbf{t}_3$: 1, 1; 2, 1; 3, 1; 4, 1: *a*.
T $\mathbf{G}_4^2 \otimes \mathbf{T}_2$: $\{C_{2z}|000\}, \{I|000\}$; $\mathbf{t}_1$ or $\mathbf{t}_2$: 1, 1; 2, 1; 3, 1; 4, 1: *a*.
W $\mathbf{G}_4^2 \otimes \mathbf{T}_4$: $\{C_{2z}|000\}, \{C_{2y}|000\}$; $\mathbf{t}_1$ or $\mathbf{t}_2$ or $\mathbf{t}_3$: 1, 1; 2, 1; 3, 1; 4, 1: *b*.

Λ^x $\mathbf{G}_4^2$: $(C_{2z}, 0), (\sigma_y, 0)$: 1, 1; 2, 1; 3, 1; 4, 1: *c*.
G^x $\mathbf{G}_4^2$: $(C_{2z}, 0), (\sigma_y, 0)$: 1, 1; 2, 1; 3, 1; 4, 1: *c*.
P^x $\mathbf{G}_2^1$: $(C_{2z}, 0)$: 1, 1; 2, 1: *b*.
Σ^x $\mathbf{G}_4^2$: $(C_{2x}, 0), (\sigma_z, 0)$: 1, 1; 2, 1; 3, 1; 4, 1: *c*.
F^x $\mathbf{G}_4^2$: $(C_{2x}, 0), (\sigma_z, 0)$: 1, 1; 2, 1; 3, 1; 4, 1: *c*.
D^x $\mathbf{G}_2^1$: $(C_{2x}, 0)$: 1, 1; 2, 1: *b*.
Δ^x $\mathbf{G}_4^2$: $(C_{2y}, 0), (\sigma_x, 0)$: 1, 1; 2, 1; 3, 1; 4, 1: *c*.
U^x $\mathbf{G}_4^2$: $(C_{2y}, 0), (\sigma_x, 0)$: 1, 1; 2, 1; 3, 1; 4, 1: *c*.
Q^x $\mathbf{G}_2^1$: $(C_{2y}, 0)$: 1, 1; 2, 1: *b*.

72 *Ibam* D_{2h}^{26}

($F1$; $K7$; $M5$; $Z1$.)

Γ $\mathbf{G}_8^3$: $\{C_{2z}|000\}, \{C_{2y}|000\}, \{I|\tfrac{1}{2}\tfrac{1}{2}0\}$: 1, 1; 2, 1; 3, 1; 4, 1; 5, 1; 6, 1; 7, 1; 8, 1: *b*.
X $\mathbf{G}_8^3 \otimes \mathbf{T}_2$: $\{C_{2z}|000\}, \{C_{2y}|000\}, \{I|\tfrac{1}{2}\tfrac{1}{2}0\}$; $\mathbf{t}_1$ or $\mathbf{t}_2$ or $\mathbf{t}_3$: 1, 1; 2, 1; 3, 1; 4, 1; 5, 1; 6, 1; 7, 1; 8, 1: *b*.
R $\mathbf{G}_8^4$: $\{\sigma_y|\tfrac{1}{2}\tfrac{1}{2}0\}, \{I|\tfrac{1}{2}\tfrac{1}{2}0\}$: 5, 1: *a*.
S $\mathbf{G}_8^4$: $\{\sigma_x|\tfrac{1}{2}\tfrac{1}{2}0\}, \{I|\tfrac{1}{2}\tfrac{1}{2}0\}$: 5, 1: *a*.
T $\mathbf{G}_4^2 \otimes \mathbf{T}_2$: $\{C_{2z}|000\}, \{I|\tfrac{1}{2}\tfrac{1}{2}0\}$; $\mathbf{t}_1$ or $\mathbf{t}_2$: 1, 1; 2, 1; 3, 1; 4, 1: *a*.
W $\mathbf{G}_4^2 \otimes \mathbf{T}_4$: $\{C_{2z}|000\}, \{C_{2y}|000\}$; $\mathbf{t}_1$ or $\mathbf{t}_2$ or $\mathbf{t}_3$: 1, 3; 2, 3; 3, 3; 4, 3: *b*.

Λ^x $\mathbf{G}_4^2$: $(C_{2z}, 0), (\sigma_y, 0)$: 1, 1; 2, 1; 3, 1; 4, 1: *c*.
G^x $\mathbf{G}_4^2$: $(C_{2z}, 0), (\sigma_y, 0)$: 1, 1; 2, 1; 3, 1; 4, 1: *c*.
P^x $\mathbf{G}_2^1$: $(C_{2z}, 0)$: 1, 1; 2, 1: *b*.
Σ^x $\mathbf{G}_4^2$: $(C_{2x}, 0), (\sigma_z, 0)$: 1, 1; 2, 1; 3, 1; 4, 1: *c*.
F^x $\mathbf{G}_4^2$: $(C_{2x}, 0), (\sigma_z, 0)$: 1, 1; 2, 1; 3, 1; 4, 1: *c*.
D^x $\mathbf{G}_2^1$: $(C_{2x}, 0)$: 1, 3; 2, 3: *b*.
Δ^x $\mathbf{G}_4^2$: $(C_{2y}, 0), (\sigma_x, 0)$: 1, 1; 2, 1; 3, 1; 4, 1: *c*.
U^x $\mathbf{G}_4^2$: $(C_{2y}, 0), (\sigma_x, 0)$: 1, 1; 2, 1; 3, 1; 4, 1: *c*.
Q^x $\mathbf{G}_2^1$: $(C_{2y}, 0)$: 1, 3; 2, 3: *b*.

73 *Ibca* D_{2h}^{27}

($F1$; $K7$; $M5$; $Z1$.)

Γ $\mathbf{G}_8^3$: $\{C_{2z}|\frac{1}{2}0\frac{1}{2}\}$, $\{C_{2y}|\frac{1}{2}\frac{1}{2}0\}$, $\{I|000\}$: 1, 1; 2, 1; 3, 1; 4, 1; 5, 1; 6, 1; 7, 1; 8, 1: *b*.
X $\mathbf{G}_8^3 \otimes \mathbf{T}_2$: $\{C_{2z}|\frac{1}{2}0\frac{1}{2}\}$, $\{C_{2y}|\frac{1}{2}\frac{1}{2}0\}$, $\{I|000\}$; $\mathbf{t}_1$ or $\mathbf{t}_2$ or $\mathbf{t}_3$: 1, 1; 2, 1; 3, 1; 4, 1; 5, 1; 6, 1; 7, 1; 8, 1: *b*.
R $\mathbf{G}_8^4$: $\{\sigma_y|\frac{1}{2}\frac{1}{2}0\}$, $\{I|000\}$: 5, 1: *a*.
S $\mathbf{G}_8^4$: $\{\sigma_x|0\frac{1}{2}\frac{1}{2}\}$, $\{I|000\}$: 5, 1: *a*.
T $\mathbf{G}_8^4$: $\{\sigma_z|\frac{1}{2}0\frac{1}{2}\}$, $\{I|000\}$: 5, 1: *a*.
W $\mathbf{G}_{16}^7$: $\{E|100\}$, $\{C_{2z}|\frac{1}{2}0\frac{1}{2}\}$, $\{C_{2y}|\frac{1}{2}\frac{1}{2}0\}$: 9, 2: *f*.

Λ^x $\mathbf{G}_4^2$: $(C_{2z}, 0)$, $(\sigma_y, 0)$: 1, 1; 2, 1; 3, 1; 4, 1: *c*.
G^x $\mathbf{G}_4^2$: $(C_{2z}, 0)$, $(\sigma_y, 0)$: 1, 1; 2, 1; 3, 1; 4, 1: *c*.
P^x $\mathbf{G}_4^1$: $(C_{2z}, 0)$: 2, 3; 4, 3: *a*.
Σ^x $\mathbf{G}_4^2$: $(C_{2x}, 0)$, $(\sigma_z, 0)$: 1, 1; 2, 1; 3, 1; 4, 1: *c*.
F^x $\mathbf{G}_4^2$: $(C_{2x}, 0)$, $(\sigma_z, 0)$: 1, 1; 2, 1; 3, 1; 4, 1: *c*.
D^x $\mathbf{G}_4^1$: $(C_{2x}, 0)$: 2, 3; 4, 3: *a*.
Δ^x $\mathbf{G}_4^2$: $(C_{2y}, 0)$, $(\sigma_x, 0)$: 1, 1; 2, 1; 3, 1; 4, 1: *c*.
U^x $\mathbf{G}_4^2$: $(C_{2y}, 0)$, $(\sigma_x, 0)$: 1, 1; 2, 1; 3, 1; 4, 1: *c*.
Q^x $\mathbf{G}_4^1$: $(C_{2y}, 0)$: 2, 3; 4, 3: *a*.

74 *Imma* D_{2h}^{28}

($F1$; $K7$; $M5$; $Z1$.)

Γ $\mathbf{G}_8^3$: $\{C_{2z}|\frac{1}{2}0\frac{1}{2}\}$, $\{C_{2y}|\frac{1}{2}\frac{1}{2}0\}$, $\{I|\frac{1}{2}\frac{1}{2}0\}$: 1, 1; 2, 1; 3, 1; 4, 1; 5, 1; 6, 1; 7, 1; 8, 1: *b*.
X $\mathbf{G}_8^3 \otimes \mathbf{T}_2$: $\{C_{2z}|\frac{1}{2}0\frac{1}{2}\}$, $\{C_{2y}|\frac{1}{2}\frac{1}{2}0\}$, $\{I|\frac{1}{2}\frac{1}{2}0\}$; $\mathbf{t}_1$ or $\mathbf{t}_2$ or $\mathbf{t}_3$: 1, 1; 2, 1; 3, 1; 4, 1; 5, 1; 6, 1; 7, 1; 8, 1: *b*.
R $\mathbf{G}_4^2 \otimes \mathbf{T}_2$: $\{C_{2y}|\frac{1}{2}\frac{1}{2}0\}$, $\{I|\frac{1}{2}\frac{1}{2}0\}$; $\mathbf{t}_1$: 1, 1; 2, 1; 3, 1; 4, 1: *a*.
S $\mathbf{G}_4^2 \otimes \mathbf{T}_2$: $\{C_{2x}|0\frac{1}{2}\frac{1}{2}\}$, $\{I|\frac{1}{2}\frac{1}{2}0\}$; $\mathbf{t}_1$ or $\mathbf{t}_3$: 1, 1; 2, 1; 3, 1; 4, 1: *a*.
T $\mathbf{G}_8^4$: $\{\sigma_z|0\frac{1}{2}\frac{1}{2}\}$, $\{I|\frac{1}{2}\frac{1}{2}0\}$: 5, 1: *a*.
W $\mathbf{G}_{16}^7$: $\{E|100\}$, $\{C_{2z}|\frac{1}{2}0\frac{1}{2}\}$, $\{C_{2y}|\frac{1}{2}\frac{1}{2}0\}$: 9, 1: *f*.

Λ^x $\mathbf{G}_4^2$: $(C_{2z}, 0)$, $(\sigma_y, 0)$: 1, 1; 2, 1; 3, 1; 4, 1: *c*.
G^x $\mathbf{G}_4^2$: $(C_{2z}, 0)$, $(\sigma_y, 0)$: 1, 1; 2, 1; 3, 1; 4, 1: *c*.
P^x $\mathbf{G}_4^1$: $(C_{2z}, 0)$: 2, 3; 4, 3: *a*.
Σ^x $\mathbf{G}_4^2$: $(C_{2x}, 0)$, $(\sigma_z, 0)$: 1, 1; 2, 1; 3, 1; 4, 1: *c*.
F^x $\mathbf{G}_4^2$: $(C_{2x}, 0)$, $(\sigma_z, 0)$: 1, 1; 2, 1; 3, 1; 4, 1: *c*.
D^x $\mathbf{G}_4^1$: $(C_{2x}, 0)$: 2, 1; 4, 1: *a*.
Δ^x $\mathbf{G}_4^2$: $(C_{2y}, 0)$, $(\sigma_x, 0)$: 1, 1; 2, 1; 3, 1; 4, 1: *c*.
U^x $\mathbf{G}_4^2$: $(C_{2y}, 0)$, $(\sigma_x, 0)$: 1, 1; 2, 1; 3, 1; 4, 1: *c*.
Q^x $\mathbf{G}_4^1$: $(C_{2y}, 0)$: 2, 1; 4, 1: *a*.

75 $P4$ C_4^1

($F1$; $K7$; $M5$; $Z1$.)

Γ $\mathbf{G}_4^1$: $\{C_{4z}^+ \mid 000\}$: 1, 1; 2, 3; 3, 1; 4, 3: b.
M $\mathbf{G}_4^1 \otimes \mathbf{T}_2$: $\{C_{4z}^+ \mid 000\}$; $\mathbf{t}_1$ or $\mathbf{t}_2$: 1, 1; 2, 3; 3, 1; 4, 3: b.
Z $\mathbf{G}_4^1 \otimes \mathbf{T}_2$: $\{C_{4z}^+ \mid 000\}$; $\mathbf{t}_3$: 1, 1; 2, 3; 3, 1; 4, 3: b.
A $\mathbf{G}_4^1 \otimes \mathbf{T}_2$: $\{C_{4z}^+ \mid 000\}$; $\mathbf{t}_1$ or $\mathbf{t}_2$ or $\mathbf{t}_3$: 1, 1; 2, 3; 3, 1; 4, 3: b.
R $\mathbf{G}_2^1 \otimes \mathbf{T}_2$: $\{C_{2z} \mid 000\}$; $\mathbf{t}_2$ or $\mathbf{t}_3$: 1, 1; 2, 1: b.
X $\mathbf{G}_2^1 \otimes \mathbf{T}_2$: $\{C_{2z} \mid 000\}$; $\mathbf{t}_2$: 1, 1; 2, 1: b.

Δ^x $\mathbf{G}_1^1$: $(E, 0)$: 1, 1: a.
U^x $\mathbf{G}_1^1$: $(E, 0)$: 1, 1: a.
Λ^x $\mathbf{G}_4^1$: $(C_{4z}^+, 0)$: 1, x; 2, x; 3, x; 4, x: b.
V^x $\mathbf{G}_4^1$: $(C_{4z}^+, 0)$: 1, x; 2, x; 3, x; 4, x: b.
Σ^x $\mathbf{G}_1^1$: $(E, 0)$: 1, 1: a.
S^x $\mathbf{G}_1^1$: $(E, 0)$: 1, 1: a.
Y^x $\mathbf{G}_1^1$: $(E, 0)$: 1, 1: a.
T^x $\mathbf{G}_1^1$: $(E, 0)$: 1, 1: a.
W^x $\mathbf{G}_2^1$: $(C_{2z}, 0)$: 1, x; 2, x: b.

76 $P4_1$ C_4^2

($F1$; $K7$; $M5$; $Z1$.)

Γ $\mathbf{G}_4^1$: $\{C_{4z}^+ \mid 00\frac{1}{4}\}$: 1, 1; 2, 3; 3, 1; 4, 3: b.
M $\mathbf{G}_4^1 \otimes \mathbf{T}_2$: $\{C_{4z}^+ \mid 00\frac{1}{4}\}$; $\mathbf{t}_1$ or $\mathbf{t}_2$: 1, 1; 2, 3; 3, 1; 4, 3: b.
Z $\mathbf{G}_8^1$: $\{C_{4z}^+ \mid 00\frac{1}{4}\}$: 2, 3; 4, 3; 6, 3; 8, 3: a.
A $\mathbf{G}_8^1$: $\{C_{4z}^+ \mid 00\frac{1}{4}\}$: 2, 3; 4, 3; 6, 3; 8, 3: a.
R $\mathbf{G}_4^1$: $\{C_{2z} \mid 00\frac{1}{2}\}$: 2, 3; 4, 3: a.
X $\mathbf{G}_2^1 \otimes \mathbf{T}_2$: $\{C_{2z} \mid 00\frac{1}{2}\}$; $\mathbf{t}_2$: 1, 1; 2, 1: b.

Δ^x $\mathbf{G}_1^1$: $(E, 0)$: 1, 1: a.
U^x $\mathbf{G}_1^1$: $(E, 0)$: 1, 2: a.
Λ^x $\mathbf{G}_4^1$: $(C_{4z}^+, 0)$: 1, x; 2, x; 3, x; 4, x: b.
V^x $\mathbf{G}_4^1$: $(C_{4z}^+, 0)$: 1, x; 2, x; 3, x; 4, x: b.
Σ^x $\mathbf{G}_1^1$: $(E, 0)$: 1, 1: a.
S^x $\mathbf{G}_1^1$: $(E, 0)$: 1, 2: a.
Y^x $\mathbf{G}_1^1$: $(E, 0)$: 1, 1: a.
T^x $\mathbf{G}_1^1$: $(E, 0)$: 1, 2: a.
W^x $\mathbf{G}_2^1$: $(C_{2z}, 0)$: 1, x; 2, x: b.

77 $P4_2$ C_4^3

($F1$; $K7$; $M5$; $Z1$.)

Γ $\mathbf{G}_4^1$: $\{C_{4z}^+ | 00\frac{1}{2}\}$: 1, 1; 2, 3; 3, 1; 4, 3: *b*.
M $\mathbf{G}_4^1 \otimes \mathbf{T}_2$: $\{C_{4z}^+ | 00\frac{1}{2}\}$; $\mathbf{t}_1$ or $\mathbf{t}_2$: 1, 1; 2, 3; 3, 1; 4, 3: *b*.
Z $\mathbf{G}_4^1 \otimes \mathbf{T}_2$: $\{C_{4z}^+ | 00\frac{1}{2}\}$; $\mathbf{t}_3$: 1, 1; 2, 3; 3, 1; 4, 3: *b*.
A $\mathbf{G}_4^1 \otimes \mathbf{T}_2$: $\{C_{4z}^+ | 00\frac{1}{2}\}$; $\mathbf{t}_1$ or $\mathbf{t}_2$ or $\mathbf{t}_3$: 1, 1; 2, 3; 3, 1; 4, 3: *b*.
R $\mathbf{G}_2^1 \otimes \mathbf{T}_2$: $\{C_{2z} | 000\}$; $\mathbf{t}_2$ or $\mathbf{t}_3$: 1, 1; 2, 1: *b*.
X $\mathbf{G}_2^1 \otimes \mathbf{T}_2$: $\{C_{2z} | 000\}$; $\mathbf{t}_2$: 1, 1; 2, 1: *b*.

Δ^x $\mathbf{G}_1^1$: $(E, 0)$: 1, 1: *a*.
U^x $\mathbf{G}_1^1$: $(E, 0)$: 1, 1: *a*.
Λ^x $\mathbf{G}_4^1$: $(C_{4z}^+, 0)$: 1, *x*; 2, *x*; 3, *x*; 4, *x*: *b*.
V^x $\mathbf{G}_4^1$: $(C_{4z}^+, 0)$: 1, *x*; 2, *x*; 3, *x*; 4, *x*: *b*.
Σ^x $\mathbf{G}_1^1$: $(E, 0)$: 1, 1: *a*.
S^x $\mathbf{G}_1^1$: $(E, 0)$: 1, 1: *a*.
Y^x $\mathbf{G}_1^1$: $(E, 0)$: 1, 1: *a*.
T^x $\mathbf{G}_1^1$: $(E, 0)$: 1, 1: *a*.
W^x $\mathbf{G}_2^1$: $(C_{2z}, 0)$: 1, *x*; 2, *x*: *b*.

78 $P4_3$ C_4^4

($F1$; $K7$; $M5$; $Z1$.)

Γ $\mathbf{G}_4^1$: $\{C_{4z}^+ | 00\frac{3}{4}\}$: 1, 1; 2, 3; 3, 1; 4, 3: *b*.
M $\mathbf{G}_4^1 \otimes \mathbf{T}_2$: $\{C_{4z}^+ | 00\frac{3}{4}\}$; $\mathbf{t}_1$ or $\mathbf{t}_2$: 1, 1; 2, 3; 3, 1; 4, 3: *b*.
Z $\mathbf{G}_8^1$: $\{C_{4z}^+ | 00\frac{3}{4}\}$: 2, 3; 4, 3; 6, 3; 8, 3: *a*.
A $\mathbf{G}_8^1$: $\{C_{4z}^+ | 00\frac{3}{4}\}$: 2, 3; 4, 3; 6, 3; 8, 3: *a*.
R $\mathbf{G}_4^1$: $\{C_{2z} | 00\frac{1}{2}\}$: 2, 3; 4, 3: *a*.
X $\mathbf{G}_2^1 \otimes \mathbf{T}_2$: $\{C_{2z} | 00\frac{1}{2}\}$; $\mathbf{t}_2$: 1, 1; 2, 1: *b*.

Δ^x $\mathbf{G}_1^1$: $(E, 0)$: 1, 1: *a*.
U^x $\mathbf{G}_1^1$: $(E, 0)$: 1, 2: *a*.
Λ^x $\mathbf{G}_4^1$: $(C_{4z}^+, 0)$: 1, *x*; 2, *x*; 3, *x*; 4, *x*: *b*.
V^x $\mathbf{G}_4^1$: $(C_{4z}^+, 0)$: 1, *x*; 2, *x*; 3, *x*; 4, *x*: *b*.
Σ^x $\mathbf{G}_1^1$: $(E, 0)$: 1, 1: *a*.
S^x $\mathbf{G}_1^1$: $(E, 0)$: 1, 2: *a*.
Y^x $\mathbf{G}_1^1$: $(E, 0)$: 1, 1: *a*.
T^x $\mathbf{G}_1^1$: $(E, 0)$: 1, 2: *a*.
W^x $\mathbf{G}_2^1$: $(C_{2z}, 0)$: 1, *x*; 2, *x*: *b*.

79 $I4$ C_4^5

($F1$; $K7$; $M5$; $Z1$.)

Γ $\mathbf{G}_4^1$: $\{C_{4z}^+ | 000\}$: 1, 1; 2, 3; 3, 1; 4, 3: b.
N $\mathbf{G}_1^1 \otimes \mathbf{T}_2$: $\{E | 000\}$; $\mathbf{t}_2$: 1, 1: a.
X $\mathbf{G}_2^1 \otimes \mathbf{T}_2$: $\{C_{2z} | 000\}$; $\mathbf{t}_3$: 1, 1; 2, 1: b.
Z $\mathbf{G}_4^1 \otimes \mathbf{T}_2$: $\{C_{4z}^+ | 000\}$; $\mathbf{t}_1$ or $\mathbf{t}_2$ or $\mathbf{t}_3$: 1, 1; 2, 3; 3, 1; 4, 3: b.
P $\mathbf{G}_2^1 \otimes \mathbf{T}_4$: $\{C_{2z} | 000\}$; $\mathbf{t}_1$ or $\mathbf{t}_2$ or $\mathbf{t}_3$: 1, 1; 2, 2: b.

Λ^x $\mathbf{G}_4^1$: $(C_{4z}^+, 0)$: 1, x; 2, x; 3, x; 4, x: b.
V^x $\mathbf{G}_4^1$: $(C_{4z}^+, 0)$: 1, x; 2, x; 3, x; 4, x: b.
W^x $\mathbf{G}_2^1$: $(C_{2z}, 0)$: 1, x; 2, x: b.
Σ^x $\mathbf{G}_1^1$: $(E, 0)$: 1, 1: a.
F^x $\mathbf{G}_1^1$: $(E, 0)$: 1, 1: a.
Q^x $\mathbf{G}_1^1$: $(E, 0)$: 1, x: a.
Δ^x $\mathbf{G}_1^1$: $(E, 0)$: 1, 1: a.
U^x $\mathbf{G}_1^1$: $(E, 0)$: 1, 1: a.
Y^x $\mathbf{G}_1^1$: $(E, 0)$: 1, 1: a.

80 $I4_1$ C_4^6

($F1$; $K7$; $M5$; $Z1$.)

Γ $\mathbf{G}_4^1$: $\{C_{4z}^+ | \frac{3}{4}\frac{1}{4}\frac{1}{2}\}$: 1, 1; 2, 3; 3, 1; 4, 3: b.
N $\mathbf{G}_1^1 \otimes \mathbf{T}_2$: $\{E | 000\}$; $\mathbf{t}_2$: 1, 1: a.
X $\mathbf{G}_2^1 \otimes \mathbf{T}_2$: $\{C_{2z} | 000\}$; $\mathbf{t}_3$: 1, 1; 2, 1: b.
Z $\mathbf{G}_4^1 \otimes \mathbf{T}_2$: $\{C_{4z}^+ | \frac{3}{4}\frac{1}{4}\frac{1}{2}\}$; $\mathbf{t}_1$ or $\mathbf{t}_2$ or $\mathbf{t}_3$: 1, 1; 2, 3; 3, 1; 4, 3: b.
P $\mathbf{G}_2^1 \otimes \mathbf{T}_4$: $\{C_{2z} | 000\}$; $\mathbf{t}_1$ or $\mathbf{t}_2$ or $\mathbf{t}_3$: 1, 3; 2, 3: b.

Λ^x $\mathbf{G}_4^1$: $(C_{4z}^+, 0)$: 1, x; 2, x; 3, x; 4, x: b.
V^x $\mathbf{G}_8^2$: $(C_{4z}^+, 0)$, $(E, 1)$: 5, x; 6, x; 7, x; 8, x: b.
W^x $\mathbf{G}_2^1$: $(C_{2z}, 0)$: 1, x; 2, x: b.
Σ^x $\mathbf{G}_1^1$: $(E, 0)$: 1, 1: a.
F^x $\mathbf{G}_1^1$: $(E, 0)$: 1, 1: a.
Q^x $\mathbf{G}_1^1$: $(E, 0)$: 1, x: a.
Δ^x $\mathbf{G}_1^1$: $(E, 0)$: 1, 1: a.
U^x $\mathbf{G}_1^1$: $(E, 0)$: 1, 1: a.
Y^x $\mathbf{G}_1^1$: $(E, 0)$: 1, 1: a.

81 $P\bar{4}$ S_4^1

($F1$; $K7$; $M5$; $Z1$.)

Γ $\mathbf{G}_4^1$: $\{S_{4z}^+ | 000\}$: 1, 1; 2, 3; 3, 1; 4, 3: b.
M $\mathbf{G}_4^1 \otimes \mathbf{T}_2$: $\{S_{4z}^+ | 000\}$; $\mathbf{t}_1$ or $\mathbf{t}_2$: 1, 1; 2, 3; 3, 1; 4, 3: b.
Z $\mathbf{G}_4^1 \otimes \mathbf{T}_2$: $\{S_{4z}^+ | 000\}$; $\mathbf{t}_3$: 1, 1; 2, 3; 3, 1; 4, 3: b.
A $\mathbf{G}_4^1 \otimes \mathbf{T}_2$: $\{S_{4z}^+ | 000\}$; $\mathbf{t}_1$ or $\mathbf{t}_2$ or $\mathbf{t}_3$: 1, 1; 2, 3; 3, 1; 4, 3: b.
R $\mathbf{G}_2^1 \otimes \mathbf{T}_2$: $\{C_{2z} | 000\}$; $\mathbf{t}_2$ or $\mathbf{t}_3$: 1, 1; 2, 1: b.
X $\mathbf{G}_2^1 \otimes \mathbf{T}_2$: $\{C_{2z} | 000\}$; $\mathbf{t}_2$: 1, 1; 2, 1: b.

Δ^x $\mathbf{G}_1^1$: $(E, 0)$: 1, 1: a.
U^x $\mathbf{G}_1^1$: $(E, 0)$: 1, 1: a.
Λ^x $\mathbf{G}_2^1$: $(C_{2z}, 0)$: 1, 1; 2, 2: b.
V^x $\mathbf{G}_2^1$: $(C_{2z}, 0)$: 1, 1; 2, 2: b.
Σ^x $\mathbf{G}_1^1$: $(E, 0)$: 1, 1: a.
S^x $\mathbf{G}_1^1$: $(E, 0)$: 1, 1: a.
Y^x $\mathbf{G}_1^1$: $(E, 0)$: 1, 1: a.
T^x $\mathbf{G}_1^1$: $(E, 0)$: 1, 1: a.
W^x $\mathbf{G}_2^1$: $(C_{2z}, 0)$: 1, x; 2, x: b.

82 $I\bar{4}$ S_4^2

($F1$; $K7$; $M5$; $Z1$.)

Γ $\mathbf{G}_4^1$: $\{S_{4z}^+ \mid 000\}$: 1, 1; 2, 3; 3, 1; 4, 3: b.
N $\mathbf{G}_1^1 \otimes \mathbf{T}_2$: $\{E \mid 000\}$; $\mathbf{t}_2$: 1, 1: a.
X $\mathbf{G}_2^1 \otimes \mathbf{T}_2$: $\{C_{2z} \mid 000\}$; $\mathbf{t}_3$: 1, 1; 2, 1: b.
Z $\mathbf{G}_4^1 \otimes \mathbf{T}_2$: $\{S_{4z}^+ \mid 000\}$; $\mathbf{t}_1$ or $\mathbf{t}_2$ or $\mathbf{t}_3$: 1, 1; 2, 3; 3, 1; 4, 3: b.
P $\mathbf{G}_4^1 \otimes \mathbf{T}_4$: $\{S_{4z}^+ \mid 000\}$; $\mathbf{t}_1$ or $\mathbf{t}_2$ or $\mathbf{t}_3$: 1, x; 2, x; 3, x; 4, x: b.

Λ^x $\mathbf{G}_2^1$: $(C_{2z}, 0)$: 1, 1; 2, 2: b.
V^x $\mathbf{G}_2^1$: $(C_{2z}, 0)$: 1, 1; 2, 2: b.
W^x $\mathbf{G}_2^1$: $(C_{2z}, 0)$: 1, x; 2, x: b.
Σ^x $\mathbf{G}_1^1$: $(E, 0)$: 1, 1: a.
F^x $\mathbf{G}_1^1$: $(E, 0)$: 1, 1: a.
Q^x $\mathbf{G}_1^1$: $(E, 0)$: 1, x: a.
Δ^x $\mathbf{G}_1^1$: $(E, 0)$: 1, 1: a.
U^x $\mathbf{G}_1^1$: $(E, 0)$: 1, 1: a.
Y^x $\mathbf{G}_1^1$: $(E, 0)$: 1, 1: a.

83 $P4/m$ C_{4h}^1

($F1$; $K7$; $M5$; $Z1$.)

Γ $\mathbf{G}_8^2$: $\{C_{4z}^+ \mid 000\}$, $\{I \mid 000\}$: 1, 1; 2, 3; 3, 1; 4, 3; 5, 1; 6, 3; 7, 1; 8, 3: e.
M $\mathbf{G}_8^2 \otimes \mathbf{T}_2$: $\{C_{4z}^+ \mid 000\}$, $\{I \mid 000\}$; $\mathbf{t}_1$ or $\mathbf{t}_2$: 1, 1; 2, 3; 3, 1; 4, 3; 5, 1; 6, 3; 7, 1; 8, 3: e.
Z $\mathbf{G}_8^2 \otimes \mathbf{T}_2$: $\{C_{4z}^+ \mid 000\}$, $\{I \mid 000\}$; $\mathbf{t}_3$: 1, 1; 2, 3; 3, 1; 4, 3; 5, 1; 6, 3; 7, 1; 8, 3: e.
A $\mathbf{G}_8^2 \otimes \mathbf{T}_2$: $\{C_{4z}^+ \mid 000\}$, $\{I \mid 000\}$; $\mathbf{t}_1$ or $\mathbf{t}_2$ or $\mathbf{t}_3$: 1, 1; 2, 3; 3, 1; 4, 3; 5, 1; 6, 3; 7, 1; 8, 3: e.
R $\mathbf{G}_4^2 \otimes \mathbf{T}_2$: $\{C_{2z} \mid 000\}$, $\{I \mid 000\}$; $\mathbf{t}_2$ or $\mathbf{t}_3$: 1, 1; 2, 1; 3, 1; 4, 1: a.
X $\mathbf{G}_4^2 \otimes \mathbf{T}_2$: $\{C_{2z} \mid 000\}$, $\{I \mid 000\}$; $\mathbf{t}_2$: 1, 1; 2, 1; 3, 1; 4, 1: a.

Δ^x $\mathbf{G}_2^1$: $(\sigma_z, 0)$: 1, 1; 2, 1: c.
U^x $\mathbf{G}_2^1$: $(\sigma_z, 0)$: 1, 1; 2, 1: c.
Λ^x $\mathbf{G}_4^1$: $(C_{4z}^+, 0)$: 1, 1; 2, 3; 3, 1; 4, 3: b.
V^x $\mathbf{G}_4^1$: $(C_{4z}^+, 0)$: 1, 1; 2, 3; 3, 1; 4, 3: b.
Σ^x $\mathbf{G}_2^1$: $(\sigma_z, 0)$: 1, 1; 2, 1: c.
S^x $\mathbf{G}_2^1$: $(\sigma_z, 0)$: 1, 1; 2, 1: c.
Y^x $\mathbf{G}_2^1$: $(\sigma_z, 0)$: 1, 1; 2, 1: c.
T^x $\mathbf{G}_2^1$: $(\sigma_z, 0)$: 1, 1; 2, 1: c.
W^x $\mathbf{G}_2^1$: $(C_{2z}, 0)$: 1, 1; 2, 1: b.

84 $P4_2/m$ C_{4h}^2

($F1$; $K7$; $M5$; $Z1$.)

Γ $\mathbf{G}_8^2$: $\{C_{4z}^+ \mid 00\tfrac{1}{2}\}$, $\{I \mid 00\tfrac{1}{2}\}$: 1, 1; 2, 3; 3, 1; 4, 3; 5, 1; 6, 3; 7, 1; 8, 3: e.
M $\mathbf{G}_8^2 \otimes \mathbf{T}_2$: $\{C_{4z}^+ \mid 00\tfrac{1}{2}\}$, $\{I \mid 00\tfrac{1}{2}\}$; $\mathbf{t}_1$ or $\mathbf{t}_2$: 1, 1; 2, 3; 3, 1; 4, 3; 5, 1; 6, 3; 7, 1; 8, 3: e.
Z $\mathbf{G}_{16}^{10}$: $\{C_{4z}^+ \mid 00\tfrac{1}{2}\}$, $\{E \mid 001\}$, $\{I \mid 00\tfrac{1}{2}\}$: 9, 1; 10, 1: d.
A $\mathbf{G}_{16}^{10}$: $\{C_{4z}^+ \mid 00\tfrac{1}{2}\}$, $\{E \mid 001\}$, $\{I \mid 00\tfrac{1}{2}\}$: 9, 1; 10, 1: d.
R $\mathbf{G}_4^2 \otimes \mathbf{T}_2$: $\{C_{2z} \mid 000\}$, $\{I \mid 00\tfrac{1}{2}\}$; $\mathbf{t}_2$ or $\mathbf{t}_3$: 1, 1; 2, 1; 3, 1; 4, 1: a.
X $\mathbf{G}_4^2 \otimes \mathbf{T}_2$: $\{C_{2z} \mid 000\}$, $\{I \mid 00\tfrac{1}{2}\}$; $\mathbf{t}_2$: 1, 1; 2, 1; 3, 1; 4, 1: a.

Δ^x $\mathbf{G}_2^1$: $(\sigma_z, 0)$: 1, 1; 2, 1: c.
U^x $\mathbf{G}_4^1$: $(\sigma_z, 0)$: 2, 1; 4, 1: a.
Λ^x $\mathbf{G}_4^1$: $(C_{4z}^+, 0)$: 1, 1; 2, 3; 3, 1; 4, 3: b.
V^x $\mathbf{G}_4^1$: $(C_{4z}^+, 0)$: 1, 1; 2, 3; 3, 1; 4, 3: b.
Σ^x $\mathbf{G}_2^1$: $(\sigma_z, 0)$: 1, 1; 2, 1: c.
S^x $\mathbf{G}_4^1$: $(\sigma_z, 0)$: 2, 1; 4, 1: a.
Y^x $\mathbf{G}_2^1$: $(\sigma_z, 0)$: 1, 1; 2, 1: c.
T^x $\mathbf{G}_4^1$: $(\sigma_z, 0)$: 2, 1; 4, 1: a.
W^x $\mathbf{G}_2^1$: $(C_{2z}, 0)$: 1, 1; 2, 1: b.

85 $P4/n$ C_{4h}^3

($F1$; $K7$; $M5$; $Z1$.)

Γ $\mathbf{G}_8^2$: $\{C_{4z}^+ \mid \frac{1}{2}\frac{1}{2}0\}$, $\{I \mid \frac{1}{2}\frac{1}{2}0\}$: 1, 1; 2, 3; 3, 1; 4, 3; 5, 1; 6, 3; 7, 1; 8, 3: *e*.
M $\mathbf{G}_{16}^{10}$: $\{C_{4z}^+ \mid \frac{1}{2}\frac{1}{2}0\}$, $\{E \mid 010\}$, $\{I \mid \frac{1}{2}\frac{1}{2}0\}$: 9, 1; 10, 1: *d*.
Z $\mathbf{G}_8^2 \otimes \mathbf{T}_2$: $\{C_{4z}^+ \mid \frac{1}{2}\frac{1}{2}0\}$, $\{I \mid \frac{1}{2}\frac{1}{2}0\}$; $\mathbf{t}_3$: 1, 1; 2, 3; 3, 1; 4, 3; 5, 1; 6, 3; 7, 1; 8, 3: *e*.
A $\mathbf{G}_{16}^{10}$: $\{C_{4z}^+ \mid \frac{1}{2}\frac{1}{2}0\}$, $\{E \mid 010\}$, $\{I \mid \frac{1}{2}\frac{1}{2}0\}$: 9, 1; 10, 1: *d*.
R $\mathbf{G}_8^4$: $\{\sigma_z \mid \frac{1}{2}\frac{1}{2}0\}$, $\{I \mid \frac{1}{2}\frac{1}{2}0\}$: 5, 1: *a*.
X $\mathbf{G}_8^4$: $\{\sigma_z \mid \frac{1}{2}\frac{1}{2}0\}$, $\{I \mid \frac{1}{2}\frac{1}{2}0\}$: 5, 1: *a*.

Δ^x $\mathbf{G}_2^1$: $(\sigma_z, 0)$: 1, 1; 2, 1: *c*.
U^x $\mathbf{G}_2^1$: $(\sigma_z, 0)$: 1, 1; 2, 1: *c*.
Λ^x $\mathbf{G}_4^1$: $(C_{4z}^+, 0)$: 1, 1; 2, 3; 3, 1; 4, 3: *b*.
V^x $\mathbf{G}_8^2$: $(C_{4z}^+, 0)$, $(E, 1)$: 5, 3; 6, 1; 7, 3; 8, 1: *b*.
Σ^x $\mathbf{G}_2^1$: $(\sigma_z, 0)$: 1, 1; 2, 1: *c*.
S^x $\mathbf{G}_2^1$: $(\sigma_z, 0)$: 1, 1; 2, 1: *c*.
Y^x $\mathbf{G}_2^1$: $(\sigma_z, 0)$: 1, 1; 2, 1: *c*.
T^x $\mathbf{G}_2^1$: $(\sigma_z, 0)$: 1, 1; 2, 1: *c*.
W^x $\mathbf{G}_2^1$: $(C_{2z}, 0)$: 1, 3; 2, 3: *b*.

86 $P4_2/n$ C_{4h}^4

($F1$; $K7$; $M5$; $Z1$.)

Γ $\mathbf{G}_8^2$: $\{C_{4z}^+ \mid \frac{1}{2}\frac{1}{2}\frac{1}{2}\}$, $\{I \mid \frac{1}{2}\frac{1}{2}\frac{1}{2}\}$: 1, 1; 2, 3; 3, 1; 4, 3; 5, 1; 6, 3; 7, 1; 8, 3: *e*.
M $\mathbf{G}_{16}^{10}$: $\{C_{4z}^+ \mid \frac{1}{2}\frac{1}{2}\frac{1}{2}\}$, $\{E \mid 010\}$, $\{I \mid \frac{1}{2}\frac{1}{2}\frac{1}{2}\}$: 9, 1; 10, 1: *d*.
Z $\mathbf{G}_{16}^{10}$: $\{C_{4z}^+ \mid \frac{1}{2}\frac{1}{2}\frac{1}{2}\}$, $\{E \mid 001\}$, $\{I \mid \frac{1}{2}\frac{1}{2}\frac{1}{2}\}$: 9, 1; 10, 1: *d*.
A $\mathbf{G}_8^2 \otimes \mathbf{T}_2$: $\{C_{4z}^+ \mid \frac{1}{2}\frac{1}{2}\frac{1}{2}\}$, $\{I \mid \frac{1}{2}\frac{1}{2}\frac{1}{2}\}$; $\mathbf{t}_1$ or $\mathbf{t}_2$ or $\mathbf{t}_3$: 1, 1; 2, 3; 3, 1; 4, 3; 5, 1; 6, 3; 7, 1; 8, 3: *e*.
R $\mathbf{G}_8^4$: $\{\sigma_z \mid \frac{1}{2}\frac{1}{2}\frac{1}{2}\}$, $\{I \mid \frac{1}{2}\frac{1}{2}\frac{1}{2}\}$: 5, 1: *a*.
X $\mathbf{G}_8^4$: $\{\sigma_z \mid \frac{1}{2}\frac{1}{2}\frac{1}{2}\}$, $\{I \mid \frac{1}{2}\frac{1}{2}\frac{1}{2}\}$: 5, 1: *a*.

Δ^x $\mathbf{G}_2^1$: $(\sigma_z, 0)$: 1, 1; 2, 1: *a*.
U^x $\mathbf{G}_4^1$: $(\sigma_z, 0)$: 2, 1; 4, 1: *a*.
Λ^x $\mathbf{G}_4^1$: $(C_{4z}^+, 0)$: 1, 1; 2, 3; 3, 1; 4, 3: *b*.
V^x $\mathbf{G}_8^2$: $(C_{4z}^+, 0)$, $(E, 1)$: 5, 3; 6, 1; 7, 3; 8, 1: *b*.
Σ^x $\mathbf{G}_2^1$: $(\sigma_z, 0)$: 1, 1; 2, 1: *c*.
S^x $\mathbf{G}_4^1$: $(\sigma_z, 0)$: 2, 1; 4, 1: *a*.
Y^x $\mathbf{G}_2^1$: $(\sigma_z, 0)$: 1, 1; 2, 1: *c*.
T^x $\mathbf{G}_4^1$: $(\sigma_z, 0)$: 2, 1; 4, 1: *a*.
W^x $\mathbf{G}_2^1$: $(C_{2z}, 0)$: 1, 3; 2, 3: *b*.

87 $I4/m$ C_{4h}^5

($F1$; $K7$; $M5$; $Z1$.)

Γ $\mathbf{G}_8^2$: $\{C_{4z}^+ \mid 000\}$, $\{I \mid 000\}$: 1, 1; 2, 3; 3, 1; 4, 3; 5, 1; 6, 3; 7, 1; 8, 3: *e*.
N $\mathbf{G}_2^1 \otimes \mathbf{T}_2$: $\{I \mid 000\}$; $\mathbf{t}_2$: 1, 1; 2, 1: *a*.
X $\mathbf{G}_4^2 \otimes \mathbf{T}_2$: $\{C_{2z} \mid 000\}$, $\{I \mid 000\}$; $\mathbf{t}_3$: 1, 1; 2, 1; 3, 1; 4, 1: *a*.
Z $\mathbf{G}_8^2 \otimes \mathbf{T}_2$: $\{C_{4z}^+ \mid 000\}$, $\{I \mid 000\}$; $\mathbf{t}_1$ or $\mathbf{t}_2$ or $\mathbf{t}_3$: 1, 1; 2, 3; 3, 1; 4, 3; 5, 1; 6, 3; 7, 1; 8, 3: *e*.
P $\mathbf{G}_4^1 \otimes \mathbf{T}_4$: $\{S_{4z}^+ \mid 000\}$; $\mathbf{t}_1$ or $\mathbf{t}_2$ or $\mathbf{t}_3$: 1, 1; 2, 3; 3, 1; 4, 3: *b*.

Λ^x $\mathbf{G}_4^1$: $(C_{4z}^+, 0)$: 1, 1; 2, 3; 3, 1; 4, 3: *b*.
V^x $\mathbf{G}_4^1$: $(C_{4z}^+, 0)$: 1, 1; 2, 3; 3, 1; 4, 3: *b*.
W^x $\mathbf{G}_2^1$: $(C_{2z}, 0)$: 1, 1; 2, 1: *b*.
Σ^x $\mathbf{G}_2^1$: $(\sigma_z, 0)$: 1, 1; 2, 1: *c*.
F^x $\mathbf{G}_2^1$: $(\sigma_z, 0)$: 1, 1; 2, 1: *c*.
Q^x $\mathbf{G}_1^1$: $(E, 0)$: 1, 1: *a*.
Δ^x $\mathbf{G}_2^1$: $(\sigma_z, 0)$: 1, 1; 2, 1: *c*.
U^x $\mathbf{G}_2^1$: $(\sigma_z, 0)$: 1, 1; 2, 1: *c*.
Y^x $\mathbf{G}_2^1$: $(\sigma_z, 0)$: 1, 1; 2, 1: *c*.

88 $I4_1/a$ C_{4h}^6

($F1$; $K7$; $M5$; $Z1$.)

Γ $\mathbf{G}_8^2$: $\{C_{4z}^+ \mid \frac{3}{4}\frac{1}{4}\frac{1}{2}\}$, $\{I \mid \frac{3}{4}\frac{1}{4}\frac{1}{2}\}$: 1, 1; 2, 3; 3, 1; 4, 3; 5, 1; 6, 3; 7, 1; 8, 3: *e*.
N $\mathbf{G}_2^1 \otimes \mathbf{T}_2$: $\{I \mid \frac{3}{4}\frac{1}{4}\frac{1}{2}\}$; $\mathbf{t}_2$: 1, 1; 2, 1: *a*.
X $\mathbf{G}_8^4$: $\{\sigma_z \mid \frac{3}{4}\frac{1}{4}\frac{1}{2}\}$, $\{I \mid \frac{3}{4}\frac{1}{4}\frac{1}{2}\}$: 5, 1: *a*.
Z $\mathbf{G}_{16}^{10}$: $\{C_{4z}^+ \mid \frac{3}{4}\frac{1}{4}\frac{1}{2}\}$, $\{E \mid 001\}$, $\{I \mid \frac{3}{4}\frac{1}{4}\frac{1}{2}\}$: 9, 1; 10, 1: *d*.
P $\mathbf{G}_4^1 \otimes \mathbf{T}_4$: $\{S_{4z}^+ \mid 000\}$; $\mathbf{t}_1$ or $\mathbf{t}_2$ or $\mathbf{t}_3$: 1, 3; 2, 3; 3, 3; 4, 3: *b*.

Λ^x $\mathbf{G}_4^1$: $(C_{4z}^+, 0)$: 1, 1; 2, 3; 3, 1; 4, 3: *b*.
V^x $\mathbf{G}_8^2$: $(C_{4z}^+, 0)$, $(E, 1)$: 5, 3; 6, 1; 7, 3; 8, 1: *b*.
W^x $\mathbf{G}_2^1$: $(C_{2z}, 0)$: 1, 3; 2, 3: *b*.
Σ^x $\mathbf{G}_2^1$: $(\sigma_z, 0)$: 1, 1; 2, 1: *c*.
F^x $\mathbf{G}_4^1$: $(\sigma_z, 0)$: 2, 1; 4, 1: *a*.
Q^x $\mathbf{G}_1^1$: $(E, 0)$: 1, 1: *a*.
Δ^x $\mathbf{G}_2^1$: $(\sigma_z, 0)$: 1, 1; 2, 1: *c*.
U^x $\mathbf{G}_4^1$: $(\sigma_z, 0)$: 2, 1; 4, 1: *a*.
Y^x $\mathbf{G}_2^1$: $(\sigma_z, 0)$: 1, 1; 2, 1: *c*.

89 $P422$ D_4^1

($F1$; $K7$; $M5$; $Z1$.)

Γ $\mathbf{G}_8^4$: $\{C_{4z}^+ \mid 000\}$, $\{C_{2x} \mid 000\}$: 1, 1; 2, 1; 3, 1; 4, 1; 5, 1: *b*.
M $\mathbf{G}_8^4 \otimes \mathbf{T}_2$: $\{C_{4z}^+ \mid 000\}$, $\{C_{2x} \mid 000\}$; $\mathbf{t}_1$ or $\mathbf{t}_2$: 1, 1; 2, 1; 3, 1; 4, 1; 5, 1: *b*.
Z $\mathbf{G}_8^4 \otimes \mathbf{T}_2$: $\{C_{4z}^+ \mid 000\}$, $\{C_{2x} \mid 000\}$; $\mathbf{t}_3$: 1, 1; 2, 1; 3, 1; 4, 1; 5, 1: *b*.
A $\mathbf{G}_8^4 \otimes \mathbf{T}_2$: $\{C_{4z}^+ \mid 000\}$, $\{C_{2x} \mid 000\}$; $\mathbf{t}_1$ or $\mathbf{t}_2$ or $\mathbf{t}_3$: 1, 1; 2, 1; 3, 1; 4, 1; 5, 1: *b*.
R $\mathbf{G}_4^2 \otimes \mathbf{T}_2$: $\{C_{2z} \mid 000\}$, $\{C_{2y} \mid 000\}$; $\mathbf{t}_2$ or $\mathbf{t}_3$: 1, 1; 2, 1; 3, 1; 4, 1: *b*.
X $\mathbf{G}_4^2 \otimes \mathbf{T}_2$: $\{C_{2z} \mid 000\}$, $\{C_{2y} \mid 000\}$; $\mathbf{t}_2$: 1, 1; 2, 1; 3, 1; 4, 1: *b*.

Δ^x $\mathbf{G}_2^1$: $(C_{2y}, 0)$: 1, 1; 2, 1: *b*.
U^x $\mathbf{G}_2^1$: $(C_{2y}, 0)$: 1, 1; 2, 1: *b*.
Λ^x $\mathbf{G}_4^1$: $(C_{4z}^+, 0)$: 1, 1; 2, 1; 3, 1; 4, 1: *b*.
V^x $\mathbf{G}_4^1$: $(C_{4z}^+, 0)$: 1, 1; 2, 1; 3, 1; 4, 1: *b*.
Σ^x $\mathbf{G}_2^1$: $(C_{2a}, 0)$: 1, 1; 2, 1: *b*.
S^x $\mathbf{G}_2^1$: $(C_{2a}, 0)$: 1, 1; 2, 1: *b*.
Y^x $\mathbf{G}_2^1$: $(C_{2x}, 0)$: 1, 1; 2, 1: *b*.
T^x $\mathbf{G}_2^1$: $(C_{2x}, 0)$: 1, 1; 2, 1: *b*.
W^x $\mathbf{G}_2^1$: $(C_{2z}, 0)$: 1, 1; 2, 1: *b*.

90 $P42_12$ D_4^2

($F1$; $K7$; $M5$; $Z1$.)

Γ $\mathbf{G}_8^4$: $\{C_{4z}^+ \mid 000\}$, $(C_{2x} \mid \frac{1}{2}\frac{1}{2}0)$: 1, 1; 2, 1; 3, 1; 4, 1; 5, 1: *b*.
M $\mathbf{G}_{16}^{10}$: $\{C_{4z}^+ \mid 000\}$, $\{C_{2z} \mid 010\}$, $\{C_{2b} \mid \frac{1}{2}\frac{1}{2}0\}$: 5, 3; 6, 3; 7, 3; 8, 3; 9, 1: *a*.
Z $\mathbf{G}_8^4 \otimes \mathbf{T}_2$: $\{C_{4z}^+ \mid 000\}$, $(C_{2x} \mid \frac{1}{2}\frac{1}{2}0)$; $\mathbf{t}_3$: 1, 1; 2, 1; 3, 1; 4, 1; 5, 1: *b*.
A $\mathbf{G}_{16}^{10}$: $\{C_{4z}^+ \mid 000\}$, $\{C_{2z} \mid 010\}$, $\{C_{2b} \mid \frac{1}{2}\frac{1}{2}0\}$: 5, 3; 6, 3; 7, 3; 8, 3; 9, 1: *a*.
R $\mathbf{G}_8^4$: $\{C_{2y} \mid \frac{1}{2}\frac{1}{2}0\}$, $\{C_{2x} \mid \frac{1}{2}\frac{1}{2}0\}$: 5, 1: *a*.
X $\mathbf{G}_8^4$: $\{C_{2y} \mid \frac{1}{2}\frac{1}{2}0\}$, $\{C_{2x} \mid \frac{1}{2}\frac{1}{2}0\}$: 5, 1: *a*.

Δ^x $\mathbf{G}_2^1$: $(C_{2y}, 0)$: 1, 1; 2, 1: *b*.
U^x $\mathbf{G}_2^1$: $(C_{2y}, 0)$: 1, 1; 2, 1: *b*.
Λ^x $\mathbf{G}_4^1$: $(C_{4z}^+, 0)$: 1, 1; 2, 1; 3, 1; 4, 1: *b*.
V^x $\mathbf{G}_4^1$: $(C_{4z}^+, 0)$: 1, 3; 2, 3; 3, 3; 4, 3: *b*.
Σ^x $\mathbf{G}_2^1$: $(C_{2a}, 0)$: 1, 1; 2, 1: *b*.
S^x $\mathbf{G}_2^1$: $(C_{2a}, 0)$: 1, 1; 2, 1: *b*.
Y^x $\mathbf{G}_4^1$: $(C_{2x}, 0)$: 2, 3; 4, 3: *a*.
T^x $\mathbf{G}_4^1$: $(C_{2x}, 0)$: 2, 3; 4, 3: *a*.
W^x $\mathbf{G}_2^1$: $(C_{2z}, 0)$: 1, 3; 2, 3: *b*.

91 $P4_122$ D_4^3

($F1$; $K7$; $M5$; $Z1$.)

Γ $\mathbf{G}_8^4$: $\{C_{4z}^+ | 00\frac{1}{4}\}, \{C_{2x} | 000\}$: 1, 1; 2, 1; 3, 1; 4, 1; 5, 1: *b*.
M $\mathbf{G}_8^4 \otimes \mathbf{T}_2$: $\{C_{4z}^+ | 00\frac{1}{4}\}, \{C_{2x} | 000\}$; $\mathbf{t}_1$ or $\mathbf{t}_2$: 1, 1; 2, 1; 3, 1; 4, 1; 5, 1: *b*.
Z $\mathbf{G}_{16}^{12}$: $\{C_{4z}^+ | 00\frac{1}{4}\}, \{C_{2x} | 000\}$: 6, 1; 7, 1: *a*.
A $\mathbf{G}_{16}^{12}$: $\{C_{4z}^+ | 00\frac{1}{4}\}, \{C_{2x} | 000\}$: 6, 1; 7, 1: *a*.
R $\mathbf{G}_8^4$: $\{C_{2z} | 00\frac{1}{2}\}, \{C_{2x} | 000\}$: 5, 1: *a*.
X $\mathbf{G}_4^2 \otimes \mathbf{T}_2$: $\{C_{2z} | 00\frac{1}{2}\}, \{C_{2y} | 00\frac{1}{2}\}$; $\mathbf{t}_2$: 1, 1; 2, 1; 3, 1; 4, 1: *b*.

Δ^x $\mathbf{G}_2^1$: $(C_{2y}, 0)$: 1, 1; 2, 1: *b*.
U^x $\mathbf{G}_4^1$: $(C_{2y}, 0)$: 2, 3; 4, 3: *a*.
Λ^x $\mathbf{G}_4^1$: $(C_{4z}^+, 0)$: 1, 1; 2, 1; 3, 1; 4, 1: *b*.
V^x $\mathbf{G}_4^1$: $(C_{4z}^+, 0)$: 1, 1; 2, 1; 3, 1; 4, 1: *b*.
Σ^x $\mathbf{G}_2^1$: $(C_{2a}, 0)$: 1, 1; 2, 1: *b*.
S^x $\mathbf{G}_8^1$: $(C_{2a}, 0)$: 2, 3; 6, 3: *b*.
Y^x $\mathbf{G}_2^1$: $(C_{2x}, 0)$: 1, 1; 2, 1: *b*.
T^x $\mathbf{G}_2^1$: $(C_{2x}, 0)$: 1, 3; 2, 3: *b*.
W^x $\mathbf{G}_2^1$: $(C_{2z}, 0)$: 1, 1; 2, 1: *b*.

92 $P4_12_12$ D_4^4

($F1$; $K7$; $M5$; $Z1$.)

Γ $\mathbf{G}_8^4$: $\{C_{4z}^+ | 00\frac{1}{4}\}, \{C_{2x} | \frac{1}{2}\frac{1}{2}0\}$: 1, 1; 2, 1; 3, 1; 4, 1; 5, 1: *b*.
M $\mathbf{G}_{16}^{10}$: $\{C_{4z}^+ | 00\frac{1}{4}\}, \{C_{2z} | 01\frac{1}{2}\}, \{C_{2b} | \frac{1}{2}\frac{1}{2}\frac{3}{4}\}$: 5, 3; 6, 3; 7, 3; 8, 3; 9, 1: *a*.
Z $\mathbf{G}_{16}^{12}$: $\{C_{4z}^+ | 00\frac{1}{4}\}, \{C_{2x} | \frac{1}{2}\frac{1}{2}0\}$: 6, 1; 7, 1: *a*.
A $\mathbf{G}_{16}^{13}$: $\{C_{4z}^+ | 00\frac{1}{4}\}, \{C_{2a} | \frac{1}{2}\frac{1}{2}\frac{1}{4}\}$: 6, 3; 7, 3: *a*.
R $\mathbf{G}_8^2$: $\{C_{2z} | 00\frac{1}{2}\}, \{C_{2x} | \frac{1}{2}\frac{1}{2}0\}$: 2, 3; 4, 3; 6, 3; 8, 3: *d*.
X $\mathbf{G}_8^4$: $\{C_{2y} | \frac{1}{2}\frac{1}{2}\frac{1}{2}\}, \{C_{2x} | \frac{1}{2}\frac{1}{2}0\}$: 5, 1: *a*.

Δ^x $\mathbf{G}_2^1$: $(C_{2y}, 0)$: 1, 1; 2, 1: *b*.
U^x $\mathbf{G}_4^1$: $(C_{2y}, 0)$: 2, 3; 4, 3: *a*.
Λ^x $\mathbf{G}_4^1$: $(C_{4z}^+, 0)$: 1, 1; 2, 1; 3, 1; 4, 1: *b*.
V^x $\mathbf{G}_4^1$: $(C_{4z}^+, 0)$: 1, 3; 2, 3; 3, 3; 4, 3: *b*.
Σ^x $\mathbf{G}_2^1$: $(C_{2a}, 0)$: 1, 1; 2, 1: *b*.
S^x $\mathbf{G}_8^1$: $(C_{2a}, 0)$: 2, 3; 6, 3: *b*.
Y^x $\mathbf{G}_4^1$: $(C_{2x}, 0)$: 2, 3; 4, 3: *a*.
T^x $\mathbf{G}_4^1$: $(C_{2x}, 0)$: 2, 2; 4, 2: *a*.
W^x $\mathbf{G}_2^1$: $(C_{2z}, 0)$: 1, 3; 2, 3: *b*.

93 $P4_222$ D_4^5

($F1$; $K7$; $M5$; $Z1$.)

Γ $\mathbf{G}_8^4$: $\{C_{4z}^+ | 00\frac{1}{2}\}, \{C_{2x} | 000\}$: 1, 1; 2, 1; 3, 1; 4, 1; 5, 1: *b*.
M $\mathbf{G}_8^4 \otimes \mathbf{T}_2$: $\{C_{4z}^+ | 00\frac{1}{2}\}, \{C_{2x} | 000\}$; $\mathbf{t}_1$ or $\mathbf{t}_2$: 1, 1; 2, 1; 3, 1; 4, 1; 5, 1: *b*.
Z $\mathbf{G}_{16}^9$: $\{C_{4z}^+ | 00\frac{1}{2}\}, \{C_{2x} | 000\}, \{C_{2z} | 000\}$: 5, 1; 6, 1; 7, 1; 8, 1; 9, 1: *f*.
A $\mathbf{G}_{16}^9$: $\{C_{4z}^+ | 00\frac{1}{2}\}, \{C_{2x} | 000\}, \{C_{2z} | 000\}$: 5, 1; 6, 1; 7, 1; 8, 1; 9, 1: *f*.
R $\mathbf{G}_4^2 \otimes \mathbf{T}_2$: $\{C_{2z} | 000\}, \{C_{2y} | 000\}$; $\mathbf{t}_2$ or $\mathbf{t}_3$: 1, 1; 2, 1; 3, 1; 4, 1: *b*.
X $\mathbf{G}_4^2 \otimes \mathbf{T}_2$: $\{C_{2z} | 000\}, \{C_{2y} | 000\}$; $\mathbf{t}_2$: 1, 1; 2, 1; 3, 1; 4, 1: *b*.

Δ^x $\mathbf{G}_2^1$: $(C_{2y}, 0)$: 1, 1; 2, 1: *b*.
U^x $\mathbf{G}_2^1$: $(C_{2y}, 0)$: 1, 1; 2, 1: *b*.
Λ^x $\mathbf{G}_4^1$: $(C_{4z}^+, 0)$: 1, 1; 2, 1; 3, 1; 4, 1: *b*.
V^x $\mathbf{G}_4^1$: $(C_{4z}^+, 0)$: 1, 1; 2, 1; 3, 1; 4, 1: *b*.
Σ^x $\mathbf{G}_2^1$: $(C_{2a}, 0)$: 1, 1; 2, 1: *b*.
S^x $\mathbf{G}_4^1$: $(C_{2a}, 0)$: 2, 1; 4, 1: *a*.
Y^x $\mathbf{G}_2^1$: $(C_{2x}, 0)$: 1, 1; 2, 1: *b*.
T^x $\mathbf{G}_2^1$: $(C_{2x}, 0)$: 1, 1; 2, 1: *b*.
W^x $\mathbf{G}_2^1$: $(C_{2z}, 0)$: 1, 1; 2, 1: *b*.

94 $P4_22_12$ D_4^6

($F1$; $K7$; $M5$; $Z1$.)

Γ $\mathbf{G}_8^4$: $\{C_{4z}^+ \mid 00\tfrac{1}{2}\}, \{C_{2x} \mid \tfrac{1}{2}\tfrac{1}{2}0\}$: 1, 1; 2, 1; 3, 1; 4, 1; 5, 1: *b*.

M $\mathbf{G}_{16}^{10}$: $\{C_{4z}^+ \mid 00\tfrac{1}{2}\}, \{C_{2z} \mid 010\}, \{C_{2b} \mid \tfrac{1}{2}\tfrac{1}{2}\tfrac{1}{2}\}$: 5, 3; 6, 3; 7, 3; 8, 3; 9, 1: *a*.

Z $\mathbf{G}_{16}^{9}$: $\{C_{4z}^+ \mid 00\tfrac{1}{2}\}, \{C_{2x} \mid \tfrac{1}{2}\tfrac{1}{2}0\}, \{C_{2z} \mid 000\}$: 5, 1; 6, 1; 7, 1; 8, 1; 9, 1: *f*.

A $\mathbf{G}_{16}^{10}$: $\{C_{4z}^+ \mid 00\tfrac{1}{2}\}, \{C_{2z} \mid 000\}, \{C_{2a} \mid \tfrac{1}{2}\tfrac{1}{2}\tfrac{1}{2}\}$: 5, 3; 6, 3; 7, 3; 8, 3; 9, 1: *a*.

R $\mathbf{G}_8^4$: $\{C_{2y} \mid \tfrac{1}{2}\tfrac{1}{2}0\}, \{C_{2z} \mid 000\}$: 5, 1: *a*.

X $\mathbf{G}_8^4$: $\{C_{2y} \mid \tfrac{1}{2}\tfrac{1}{2}0\}, \{C_{2z} \mid 000\}$: 5, 1: *a*.

Δ^x $\mathbf{G}_2^1$: $(C_{2y}, 0)$: 1, 1; 2, 1: *b*.

U^x $\mathbf{G}_2^1$: $(C_{2y}, 0)$: 1, 1; 2, 1: *b*.

Λ^x $\mathbf{G}_4^1$: $(C_{4z}^+, 0)$: 1, 1; 2, 1; 3, 1; 4, 1: *b*.

V^x $\mathbf{G}_4^1$: $(C_{4z}^+, 0)$: 1, 3; 2, 3; 3, 3; 4, 3: *b*.

Σ^x $\mathbf{G}_2^1$: $(C_{2a}, 0)$: 1, 1; 2, 1: *b*.

S^x $\mathbf{G}_4^1$: $(C_{2a}, 0)$: 2, 1; 4, 1: *a*.

Y^x $\mathbf{G}_4^1$: $(C_{2x}, 0)$: 2, 3; 4, 3: *a*.

T^x $\mathbf{G}_4^1$: $(C_{2x}, 0)$: 2, 3; 4, 3: *a*.

W^x $\mathbf{G}_2^1$: $(C_{2z}, 0)$: 1, 3; 2, 3: *b*.

95 $P4_322$ D_4^7

($F1$; $K7$; $M5$; $Z1$.)

Γ $\mathbf{G}_8^4$: $\{C_{4z}^+ \mid 00\tfrac{3}{4}\}, \{C_{2x} \mid 000\}$: 1, 1; 2, 1; 3, 1; 4, 1; 5, 1: *b*.

M $\mathbf{G}_8^4 \otimes \mathbf{T}_2$: $\{C_{4z}^+ \mid 00\tfrac{3}{4}\}, \{C_{2x} \mid 000\}$; $\mathbf{t}_1$ or $\mathbf{t}_2$: 1, 1; 2, 1; 3, 1; 4, 1; 5, 1: *b*.

Z $\mathbf{G}_{16}^{12}$: $\{C_{4z}^+ \mid 00\tfrac{3}{4}\}, \{C_{2x} \mid 000\}$: 6, 1; 7, 1: *a*.

A $\mathbf{G}_{16}^{12}$: $\{C_{4z}^+ \mid 00\tfrac{3}{4}\}, \{C_{2x} \mid 000\}$: 6, 1; 7, 1: *a*.

R $\mathbf{G}_8^4$: $\{C_{2z} \mid 00\tfrac{1}{2}\}, \{C_{2x} \mid 000\}$: 5, 1: *a*.

X $\mathbf{G}_4^2 \otimes \mathbf{T}_2$: $\{C_{2z} \mid 00\tfrac{1}{2}\}, \{C_{2y} \mid 00\tfrac{1}{2}\}$; $\mathbf{t}_2$: 1, 1; 2, 1; 3, 1; 4, 1: *b*.

Δ^x $\mathbf{G}_2^1$: $(C_{2y}, 0)$: 1, 1; 2, 1: *b*.

U^x $\mathbf{G}_4^1$: $(C_{2y}, 0)$: 2, 3; 4, 3: *a*.

Λ^x $\mathbf{G}_4^1$: $(C_{4z}^+, 0)$: 1, 1; 2, 1; 3, 1; 4, 1: *b*.

V^x $\mathbf{G}_4^1$: $(C_{4z}^+, 0)$: 1, 1; 2, 1; 3, 1; 4, 1: *b*.

Σ^x $\mathbf{G}_2^1$: $(C_{2a}, 0)$: 1, 1; 2, 1: *b*.

S^x $\mathbf{G}_8^1$: $(C_{2a}, 0)$: 4, 3; 8, 3: *c*.

Y^x $\mathbf{G}_2^1$: $(C_{2x}, 0)$: 1, 1; 2, 1: *b*.

T^x $\mathbf{G}_2^1$: $(C_{2x}, 0)$: 1, 3; 2, 3: *b*.

W^x $\mathbf{G}_2^1$: $(C_{2z}, 0)$: 1, 1; 2, 1: *b*.

96 $P4_32_12$ D_4^8

($F1$; $K7$; $M5$; $Z1$.)

Γ $\mathbf{G}_8^4$: $\{C_{4z}^+ \mid 00\tfrac{3}{4}\}, \{C_{2x} \mid \tfrac{1}{2}\tfrac{1}{2}0\}$: 1, 1; 2, 1; 3, 1; 4, 1; 5, 1: *b*.

M $\mathbf{G}_{16}^{10}$: $\{C_{4z}^+ \mid 00\tfrac{3}{4}\}, \{C_{2z} \mid 01\tfrac{1}{2}\}, \{C_{2b} \mid \tfrac{1}{2}\tfrac{1}{2}\tfrac{1}{4}\}$: 5, 3; 6, 3; 7, 3; 8, 3; 9, 1: *a*.

Z $\mathbf{G}_{16}^{12}$: $\{C_{4z}^+ \mid 00\tfrac{3}{4}\}, \{C_{2x} \mid \tfrac{1}{2}\tfrac{1}{2}0\}$: 6, 1; 7, 1: *a*.

A $\mathbf{G}_{16}^{13}$: $\{C_{4z}^+ \mid 00\tfrac{3}{4}\}, \{C_{2a} \mid \tfrac{1}{2}\tfrac{1}{2}\tfrac{3}{4}\}$: 6, 3; 7, 3: *a*.

R $\mathbf{G}_8^2$: $\{C_{2z} \mid 00\tfrac{1}{2}\}, \{C_{2x} \mid \tfrac{1}{2}\tfrac{1}{2}0\}$: 2, 3; 4, 3; 6, 3; 8, 3: *d*.

X $\mathbf{G}_8^4$: $\{C_{2y} \mid \tfrac{1}{2}\tfrac{1}{2}\tfrac{1}{2}\}, \{C_{2x} \mid \tfrac{1}{2}\tfrac{1}{2}0\}$: 5, 1: *a*.

Δ^x $\mathbf{G}_2^1$: $(C_{2y}, 0)$: 1, 1; 2, 1: *b*.

U^x $\mathbf{G}_4^1$: $(C_{2y}, 0)$: 2, 3; 4, 3: *a*.

Λ^x $\mathbf{G}_4^1$: $(C_{4z}^+, 0)$: 1, 1; 2, 1; 3, 1; 4, 1: *b*.

V^x $\mathbf{G}_4^1$: $(C_{4z}^+, 0)$: 1, 3; 2, 3; 3, 3; 4, 3: *b*.

Σ^x $\mathbf{G}_2^1$: $(C_{2a}, 0)$: 1, 1; 2, 1: *b*.

S^x $\mathbf{G}_8^1$: $(C_{2a}, 0)$: 4, 3; 8, 3: *c*.

Y^x $\mathbf{G}_4^1$: $(C_{2x}, 0)$: 2, 3; 4, 3: *a*.

T^x $\mathbf{G}_4^1$: $(C_{2x}, 0)$: 2, 2; 4, 2: *a*.

W^x $\mathbf{G}_2^1$: $(C_{2z}, 0)$: 1, 3; 2, 3: *b*.

97 $I422$ D_4^9

($F1$; $K7$; $M5$; $Z1$.)

Γ $\mathbf{G}_8^4$: $\{C_{4z}^+|000\}$, $\{C_{2x}|000\}$: 1, 1; 2, 1; 3, 1; 4, 1; 5, 1: *b*.
N $\mathbf{G}_2^1 \otimes \mathbf{T}_2$: $\{C_{2y}|000\}$; $\mathbf{t}_2$: 1, 1; 2, 1: *b*.
X $\mathbf{G}_4^2 \otimes \mathbf{T}_2$: $\{C_{2z}|000\}$, $\{C_{2b}|000\}$; $\mathbf{t}_3$: 1, 1; 2, 1; 3, 1; 4, 1: *b*.
Z $\mathbf{G}_8^4 \otimes \mathbf{T}_2$: $\{C_{4z}^+|000\}$, $\{C_{2x}|000\}$; $\mathbf{t}_1$ or $\mathbf{t}_2$ or $\mathbf{t}_3$: 1, 1; 2, 1; 3, 1; 4, 1; 5, 1: *b*.
P $\mathbf{G}_4^2 \otimes \mathbf{T}_4$: $\{C_{2z}|000\}$, $\{C_{2y}|000\}$; $\mathbf{t}_1$ or $\mathbf{t}_2$ or $\mathbf{t}_3$: 1, 1; 2, 1; 3, 3; 4, 3: *b*.

Λ^x $\mathbf{G}_4^1$: $(C_{4z}^+, 0)$: 1, 1; 2, 1; 3, 1; 4, 1: *b*.
V^x $\mathbf{G}_4^1$: $(C_{4z}^+, 0)$: 1, 1; 2, 1; 3, 1; 4, 1: *b*.
W^x $\mathbf{G}_2^1$: $(C_{2z}, 0)$: 1, 1; 2, 1: *b*.
Σ^x $\mathbf{G}_2^1$: $(C_{2x}, 0)$: 1, 1; 2, 1: *b*.
F^x $\mathbf{G}_2^1$: $(C_{2x}, 0)$: 1, 1; 2, 1: *b*.
Q^x $\mathbf{G}_2^1$: $(C_{2y}, 0)$: 1, *x*; 2, *x*: *b*.
Δ^x $\mathbf{G}_2^1$: $(C_{2a}, 0)$: 1, 1; 2, 1: *b*.
U^x $\mathbf{G}_2^1$: $(C_{2a}, 0)$: 1, 1; 2, 1: *b*.
Y^x $\mathbf{G}_2^1$: $(C_{2b}, 0)$: 1, 1; 2, 1: *b*.

98 $I4_122$ D_4^{10}

($F1$; $K7$; $M5$; $Z1$.)

Γ $\mathbf{G}_8^4$: $\{C_{4z}^+|\frac{3}{4}\frac{1}{4}\frac{1}{2}\}$, $\{C_{2x}|0\frac{1}{2}\frac{1}{2}\}$: 1, 1; 2, 1; 3, 1; 4, 1; 5, 1: *b*.
N $\mathbf{G}_2^1 \otimes \mathbf{T}_2$: $\{C_{2y}|0\frac{1}{2}\frac{1}{2}\}$; $\mathbf{t}_2$: 1, 1; 2, 1: *b*.
X $\mathbf{G}_4^2 \otimes \mathbf{T}_2$: $\{C_{2z}|000\}$, $\{C_{2b}|\frac{1}{4}\frac{1}{4}0\}$; $\mathbf{t}_3$: 1, 1; 2, 1; 3, 1; 4, 1: *b*.
Z $\mathbf{G}_8^4 \otimes \mathbf{T}_2$: $\{C_{4z}^+|\frac{3}{4}\frac{1}{4}\frac{1}{2}\}$, $\{C_{2x}|0\frac{1}{2}\frac{1}{2}\}$; $\mathbf{t}_1$ or $\mathbf{t}_2$ or $\mathbf{t}_3$: 1, 1; 2, 1; 3, 1; 4, 1; 5, 1: *b*.
P $\mathbf{G}_{16}^7$: $\{E|001\}$, $\{C_{2z}|000\}$, $\{C_{2y}|0\frac{1}{2}\frac{1}{2}\}$: 10, 1: *e*.

Λ^x $\mathbf{G}_4^1$: $(C_{4z}^+, 0)$: 1, 1; 2, 1; 3, 1; 4, 1: *b*.
V^x $\mathbf{G}_8^2$: $(C_{4z}^+, 0)$, $(E, 1)$: 5, 1; 6, 1; 7, 1; 8, 1: *b*.
W^x $\mathbf{G}_2^1$: $(C_{2z}, 0)$: 1, 1; 2, 1: *b*.
Σ^x $\mathbf{G}_2^1$: $(C_{2x}, 0)$: 1, 1; 2, 1: *b*.
F^x $\mathbf{G}_2^1$: $(C_{2x}, 0)$: 1, 1; 2, 1: *b*.
Q^x $\mathbf{G}_4^1$: $(C_{2y}, 0)$: 2, *x*; 4, *x*: *a*.
Δ^x $\mathbf{G}_2^1$: $(C_{2a}, 0)$: 1, 1; 2, 1: *b*.
U^x $\mathbf{G}_4^1$: $(C_{2a}, 0)$: 2, 1; 4, 1: *a*.
Y^x $\mathbf{G}_2^1$: $(C_{2b}, 0)$: 1, 1; 2, 1: *b*.

99 $P4mm$ C_{4v}^1

($F1$; $K7$; $M5$; $Z1$.)

Γ $\mathbf{G}_8^4$: $\{C_{4z}^+|000\}$, $\{\sigma_y|000\}$: 1, 1; 2, 1; 3, 1; 4, 1; 5, 1: *b*.
M $\mathbf{G}_8^4 \otimes \mathbf{T}_2$: $\{C_{4z}^+|000\}$, $\{\sigma_y|000\}$; $\mathbf{t}_1$ or $\mathbf{t}_2$: 1, 1; 2, 1; 3, 1; 4, 1; 5, 1: *b*.
Z $\mathbf{G}_8^4 \otimes \mathbf{T}_2$: $\{C_{4z}^+|000\}$, $\{\sigma_y|000\}$; $\mathbf{t}_3$: 1, 1; 2, 1; 3, 1; 4, 1; 5, 1: *b*.
A $\mathbf{G}_8^4 \otimes \mathbf{T}_2$: $\{C_{4z}^+|000\}$, $\{\sigma_y|000\}$; $\mathbf{t}_1$ or $\mathbf{t}_2$ or $\mathbf{t}_3$: 1, 1; 2, 1; 3, 1; 4, 1; 5, 1: *b*.
R $\mathbf{G}_4^2 \otimes \mathbf{T}_2$: $\{C_{2z}|000\}$, $\{\sigma_y|000\}$; $\mathbf{t}_2$ or $\mathbf{t}_3$: 1, 1; 2, 1; 3, 1; 4, 1: *c*.
X $\mathbf{G}_4^2 \otimes \mathbf{T}_2$: $\{C_{2z}|000\}$, $\{\sigma_y|000\}$; $\mathbf{t}_2$: 1, 1; 2, 1; 3, 1; 4, 1: *c*.

Δ^x $\mathbf{G}_2^1$: $(\sigma_x, 0)$: 1, 1; 2, 1: *c*.
U^x $\mathbf{G}_2^1$: $(\sigma_x, 0)$: 1, 1; 2, 1: *c*.
Λ^x $\mathbf{G}_8^4$: $(C_{4z}^+, 0)$, $(\sigma_y, 0)$: 1, *x*; 2, *x*; 3, *x*; 4, *x*; 5, *x*: *b*.
V^x $\mathbf{G}_8^4$: $(C_{4z}^+, 0)$, $(\sigma_y, 0)$: 1, *x*; 2, *x*; 3, *x*; 4, *x*; 5, *x*: *b*.
Σ^x $\mathbf{G}_2^1$: $(\sigma_{db}, 0)$: 1, 1; 2, 1: *c*.
S^x $\mathbf{G}_2^1$: $(\sigma_{db}, 0)$: 1, 1; 2, 1: *c*.
Y^x $\mathbf{G}_2^1$: $(\sigma_y, 0)$: 1, 1; 2, 1: *c*.
T^x $\mathbf{G}_2^1$: $(\sigma_y, 0)$: 1, 1; 2, 1: *c*.
W^x $\mathbf{G}_4^2$: $(C_{2z}, 0)$, $(\sigma_y, 0)$: 1, *x*; 2, *x*; 3, *x*; 4, *x*: *c*.

100 $P4bm$ C_{4v}^2

($F1$; $K7$; $M5$; $Z1$.)

Γ $\mathbf{G}_8^4$: $\{C_{4z}^+|000\}$, $\{\sigma_y|\frac{1}{2}\frac{1}{2}0\}$: 1, 1; 2, 1; 3, 1; 4, 1; 5, 1: b.
M $\mathbf{G}_{16}^{10}$: $\{C_{4z}^+|000\}$, $\{C_{2z}|010\}$, $\{\sigma_{db}|\frac{1}{2}\frac{1}{2}0\}$: 5, 3; 6, 3; 7, 3; 8, 3; 9, 1: b.
Z $\mathbf{G}_8^4\otimes\mathbf{T}_2$: $\{C_{4z}^+|000\}$, $\{\sigma_y|\frac{1}{2}\frac{1}{2}0\}$; $\mathbf{t}_3$: 1, 1; 2, 1; 3, 1; 4, 1; 5, 1: b.
A $\mathbf{G}_{16}^{10}$: $\{C_{4z}^+|000\}$, $\{C_{2z}|010\}$, $\{\sigma_{db}|\frac{1}{2}\frac{1}{2}0\}$: 5, 3; 6, 3; 7, 3; 8, 3; 9, 1: b.
R $\mathbf{G}_8^4$: $\{\sigma_x|\frac{1}{2}\frac{1}{2}0\}$, $\{C_{2z}|000\}$: 5, 1: a.
X $\mathbf{G}_8^4$: $\{\sigma_x|\frac{1}{2}\frac{1}{2}0\}$, $\{C_{2z}|000\}$: 5, 1: a.

Δ^x $\mathbf{G}_2^1$: $(\sigma_x, 0)$: 1, 1; 2, 1: c.
U^x $\mathbf{G}_2^1$: $(\sigma_x, 0)$: 1, 1; 2, 1: c.
Λ^x $\mathbf{G}_8^4$: $(C_{4z}^+, 0)$, $(\sigma_y, 0)$: 1, x; 2, x; 3, x; 4, x; 5, x: b.
V^x $\mathbf{G}_{16}^{10}$: $(C_{4z}^+, 0)$, $(C_{2z}, 1)$, $(\sigma_{da}, 0)$: 5, x; 6, x; 7, x; 8, x; 9, x: b.
Σ^x $\mathbf{G}_2^1$: $(\sigma_{db}, 0)$: 1, 1; 2, 1: c.
S^x $\mathbf{G}_2^1$: $(\sigma_{db}, 0)$: 1, 1; 2, 1: c.
Y^x $\mathbf{G}_4^1$: $(\sigma_y, 0)$: 2, 3; 4, 3: a.
T^x $\mathbf{G}_4^1$: $(\sigma_y, 0)$: 2, 3; 4, 3: a.
W^x $\mathbf{G}_8^4$: $(\sigma_y, 0)$, $(C_{2z}, 0)$: 5, x: a.

101 $P4_2cm$ C_{4v}^3

($F1$; $K7$; $M5$; $Z1$.)

Γ $\mathbf{G}_8^4$: $\{C_{4z}^+|00\frac{1}{2}\}$, $\{\sigma_y|00\frac{1}{2}\}$: 1, 1; 2, 1; 3, 1; 4, 1; 5, 1: b.
M $\mathbf{G}_8^4\otimes\mathbf{T}_2$: $\{C_{4z}^+|00\frac{1}{2}\}$, $\{\sigma_y|00\frac{1}{2}\}$; $\mathbf{t}_1$ or $\mathbf{t}_2$: 1, 1; 2, 1; 3, 1; 4, 1; 5, 1: b.
Z $\mathbf{G}_{16}^{10}$: $\{C_{4z}^+|00\frac{1}{2}\}$, $\{C_{2z}|000\}$, $\{\sigma_{db}|000\}$: 5, 3; 6, 3; 7, 3; 8, 3; 9, 1: b.
A $\mathbf{G}_{16}^{10}$: $\{C_{4z}^+|00\frac{1}{2}\}$, $\{C_{2z}|000\}$, $\{\sigma_{db}|000\}$: 5, 3; 6, 3; 7, 3; 8, 3; 9, 1: b.
R $\mathbf{G}_8^2$: $\{\sigma_x|00\frac{1}{2}\}$, $\{C_{2z}|000\}$: 2, 3; 4, 3; 6, 3; 8, 3: d.
X $\mathbf{G}_4^2\otimes\mathbf{T}_2$: $\{C_{2z}|000\}$, $\{\sigma_y|00\frac{1}{2}\}$; $\mathbf{t}_2$: 1, 1; 2, 1; 3, 1; 4, 1: c.

Δ^x $\mathbf{G}_2^1$: $(\sigma_x, 0)$: 1, 1; 2, 1: c.
U^x $\mathbf{G}_2^1$: $(\sigma_x, 0)$: 1, 3; 2, 3: c.
Λ^x $\mathbf{G}_8^4$: $(C_{4z}^+, 0)$, $(\sigma_y, 0)$: 1, x; 2, x; 3, x; 4, x; 5, x: b.
V^x $\mathbf{G}_8^4$: $(C_{4z}^+, 0)$, $(\sigma_y, 0)$: 1, x; 2, x; 3, x; 4, x; 5, x: b.
Σ^x $\mathbf{G}_2^1$: $(\sigma_{db}, 0)$: 1, 1; 2, 1: c.
S^x $\mathbf{G}_2^1$: $(\sigma_{db}, 0)$: 1, 1; 2, 1: c.
Y^x $\mathbf{G}_2^1$: $(\sigma_y, 0)$: 1, 1; 2, 1: c.
T^x $\mathbf{G}_2^1$: $(\sigma_y, 0)$: 1, 3; 2, 3: c.
W^x $\mathbf{G}_4^2$: $(C_{2z}, 0)$, $(\sigma_y, 0)$: 1, x; 2, x; 3, x; 4, x: c.

102 $P4_2nm$ C_{4v}^4

($F1$; $K7$; $M5$; $Z1$.)

Γ $\mathbf{G}_8^4$: $\{C_{4z}^+|00\frac{1}{2}\}$, $\{\sigma_y|\frac{1}{2}\frac{1}{2}\frac{1}{2}\}$: 1, 1; 2, 1; 3, 1; 4, 1; 5, 1: b.
M $\mathbf{G}_{16}^{10}$: $\{C_{4z}^+|00\frac{1}{2}\}$, $\{C_{2z}|010\}$, $\{\sigma_{db}|\frac{1}{2}\frac{1}{2}0\}$: 5, 3; 6, 3; 7, 3; 8, 3; 9, 1: b.
Z $\mathbf{G}_{16}^{10}$: $\{C_{4z}^+|00\frac{1}{2}\}$, $\{C_{2z}|000\}$, $\{\sigma_{db}|\frac{1}{2}\frac{1}{2}0\}$: 5, 3; 6, 3; 7, 3; 8, 3; 9, 1: b.
A $\mathbf{G}_8^4\otimes\mathbf{T}_2$: $\{C_{4z}^+|00\frac{1}{2}\}$, $\{\sigma_{db}|\frac{1}{2}\frac{1}{2}0\}$; $\mathbf{t}_1$ or $\mathbf{t}_2$ or $\mathbf{t}_3$: 1, 1; 2, 1; 3, 1; 4, 1; 5, 1: b.
R $\mathbf{G}_8^4$: $\{\sigma_y|\frac{1}{2}\frac{1}{2}\frac{1}{2}\}$, $\{C_{2z}|000\}$: 5, 1: a.
X $\mathbf{G}_8^4$: $\{\sigma_x|\frac{1}{2}\frac{1}{2}\frac{1}{2}\}$, $\{C_{2z}|000\}$: 5, 1: a.

Δ^x $\mathbf{G}_2^1$: $(\sigma_x, 0)$: 1, 1; 2, 1: c.
U^x $\mathbf{G}_2^1$: $(\sigma_x, 0)$: 1, 3; 2, 3: c.
Λ^x $\mathbf{G}_8^4$: $(C_{4z}^+, 0)$, $(\sigma_y, 0)$: 1, x; 2, x; 3, x; 4, x; 5, x: b.
V^x $\mathbf{G}_{16}^{10}$: $(C_{4z}^+, 0)$, $(C_{2z}, 1)$, $(\sigma_{da}, 0)$: 5, x; 6, x; 7, x; 8, x; 9, x: b.
Σ^x $\mathbf{G}_2^1$: $(\sigma_{db}, 0)$: 1, 1; 2, 1: c.
S^x $\mathbf{G}_2^1$: $(\sigma_{db}, 0)$: 1, 1; 2, 1: c.
Y^x $\mathbf{G}_4^1$: $(\sigma_y, 0)$: 2, 3; 4, 3: a.
T^x $\mathbf{G}_4^1$: $(\sigma_y, 0)$: 2, 1; 4, 1: a.
W^x $\mathbf{G}_8^4$: $(\sigma_y, 0)$, $(C_{2z}, 0)$: 5, x: a.

103 *P*4*cc* C_{4v}^5

(*F*1; *K*7; *M*5; *Z*1.)

Γ $\mathbf{G}_8^4$: $\{C_{4z}^+|000\}, \{\sigma_y|00\tfrac{1}{2}\}$: 1, 1; 2, 1; 3, 1; 4, 1; 5, 1: *b*.
M $\mathbf{G}_8^4 \otimes \mathbf{T}_2$: $\{C_{4z}^+|000\}, \{\sigma_y|00\tfrac{1}{2}\}$; $\mathbf{t}_1$ or $\mathbf{t}_2$: 1, 1; 2, 1; 3, 1; 4, 1; 5, 1: *b*.
Z $\mathbf{G}_{16}^8$: $\{C_{4z}^+|000\}, \{\sigma_{db}|00\tfrac{1}{2}\}$: 5, 3; 6, 3; 7, 3; 8, 3; 10, 2: *a*.
A $\mathbf{G}_{16}^8$: $\{C_{4z}^+|000\}, \{\sigma_{db}|00\tfrac{1}{2}\}$: 5, 3; 6, 3; 7, 3; 8, 3; 10, 2: *a*.
R $\mathbf{G}_8^2$: $\{\sigma_x|00\tfrac{1}{2}\}, \{C_{2z}|000\}$: 2, 3; 4, 3; 6, 3; 8, 3: *d*.
X $\mathbf{G}_4^2 \otimes \mathbf{T}_2$: $\{C_{2z}|000\}, \{\sigma_y|00\tfrac{1}{2}\}$; $\mathbf{t}_2$: 1, 1; 2, 1; 3, 1; 4, 1: *c*.

Δ^x $\mathbf{G}_2^1$: $(\sigma_x, 0)$: 1, 1; 2, 1: *c*.
U^x $\mathbf{G}_2^1$: $(\sigma_x, 0)$: 1, 3; 2, 3: *c*.
Λ^x $\mathbf{G}_8^4$: $(C_{4z}^+, 0), (\sigma_y, 0)$: 1, *x*; 2, *x*; 3, *x*; 4, *x*; 5, *x*: *b*.
V^x $\mathbf{G}_8^4$: $(C_{4z}^+, 0), (\sigma_y, 0)$: 1, *x*; 2, *x*; 3, *x*; 4, *x*; 5, *x*: *b*.
Σ^x $\mathbf{G}_2^1$: $(\sigma_{db}, 0)$: 1, 1; 2, 1: *c*.
S^x $\mathbf{G}_2^1$: $(\sigma_{db}, 0)$: 1, 3; 2, 3: *c*.
Y^x $\mathbf{G}_2^1$: $(\sigma_y, 0)$: 1, 1; 2, 1: *c*.
T^x $\mathbf{G}_2^1$: $(\sigma_y, 0)$: 1, 3; 2, 3: *c*.
W^x $\mathbf{G}_4^2$: $(C_{2z}, 0), (\sigma_y, 0)$: 1, *x*; 2, *x*; 3, *x*; 4, *x*: *c*.

104 *P*4*nc* C_{4v}^6

(*F*1; *K*7; *M*5; *Z*1.)

Γ $\mathbf{G}_8^4$: $\{C_{4z}^+|000\}, \{\sigma_y|\tfrac{1}{2}\tfrac{1}{2}\tfrac{1}{2}\}$: 1, 1; 2, 1; 3, 1; 4, 1; 5, 1: *b*.
M $\mathbf{G}_{16}^{10}$: $\{C_{4z}^+|000\}, \{C_{2z}|010\}, \{\sigma_{db}|\tfrac{1}{2}\tfrac{1}{2}\tfrac{1}{2}\}$: 5, 3; 6, 3; 7, 3; 8, 3; 9, 1: *b*.
Z $\mathbf{G}_{16}^8$: $\{C_{4z}^+|000\}, \{\sigma_{db}|\tfrac{1}{2}\tfrac{1}{2}\tfrac{1}{2}\}$: 5, 3; 6, 3; 7, 3; 8, 3; 10, 2: *a*.
A $\mathbf{G}_{16}^{10}$: $\{C_{4z}^+|000\}, \{C_{2z}|010\}, \{\sigma_x|\tfrac{1}{2}\tfrac{1}{2}\tfrac{1}{2}\}$: 5, 3; 6, 3; 7, 3; 8, 3; 9, 1: *b*.
R $\mathbf{G}_8^4$: $\{\sigma_y|\tfrac{1}{2}\tfrac{1}{2}\tfrac{1}{2}\}, \{C_{2z}|000\}$: 5, 1: *a*.
X $\mathbf{G}_8^4$: $\{\sigma_x|\tfrac{1}{2}\tfrac{1}{2}\tfrac{1}{2}\}, \{C_{2z}|000\}$: 5, 1: *a*.

Δ^x $\mathbf{G}_2^1$: $(\sigma_x, 0)$: 1, 1; 2, 1: *c*.
U^x $\mathbf{G}_2^1$: $(\sigma_x, 0)$: 1, 3; 2, 3: *c*.
Λ^x $\mathbf{G}_8^4$: $(C_{4z}^+, 0), (\sigma_y, 0)$: 1, *x*; 2, *x*; 3, *x*; 4, *x*; 5, *x*: *b*.
V^x $\mathbf{G}_{16}^{10}$: $(C_{4z}^+, 0), (C_{2z}, 1), (\sigma_{da}, 0)$: 5, *x*; 6, *x*; 7, *x*; 8, *x*; 9, *x*: *b*.
Σ^x $\mathbf{G}_2^1$: $(\sigma_{db}, 0)$: 1, 1; 2, 1: *c*.
S^x $\mathbf{G}_2^1$: $(\sigma_{db}, 0)$: 1, 3; 2, 3: *c*.
Y^x $\mathbf{G}_4^1$: $(\sigma_y, 0)$: 2, 3; 4, 3: *a*.
T^x $\mathbf{G}_4^1$: $(\sigma_y, 0)$: 2, 1; 4, 1: *a*.
W^x $\mathbf{G}_8^4$: $(\sigma_y, 0), (C_{2z}, 0)$: 5, *x*: *a*.

105 *P*4$_2$*mc* C_{4v}^7

(*F*1; *K*7; *M*5; *Z*1.)

Γ $\mathbf{G}_8^4$: $\{C_{4z}^+|00\tfrac{1}{2}\}, \{\sigma_y|000\}$: 1, 1; 2, 1; 3, 1; 4, 1; 5, 1: *b*.
M $\mathbf{G}_8^4 \otimes \mathbf{T}_2$: $\{C_{4z}^+|00\tfrac{1}{2}\}, \{\sigma_y|000\}$; $\mathbf{t}_1$ or $\mathbf{t}_2$: 1, 1; 2, 1; 3, 1; 4, 1; 5, 1: *b*.
Z $\mathbf{G}_{16}^{10}$: $\{C_{4z}^+|00\tfrac{1}{2}\}, \{C_{2z}|000\}, \{\sigma_x|000\}$: 5, 3; 6, 3; 7, 3; 8, 3; 9, 1: *b*.
A $\mathbf{G}_{16}^{10}$: $\{C_{4z}^+|00\tfrac{1}{2}\}, \{C_{2z}|000\}, \{\sigma_x|000\}$: 5, 3; 6, 3; 7, 3; 8, 3; 9, 1: *b*.
R $\mathbf{G}_4^2 \otimes \mathbf{T}_2$: $\{C_{2z}|000\}, \{\sigma_y|000\}$; $\mathbf{t}_2$ or $\mathbf{t}_3$: 1, 1; 2, 1; 3, 1; 4, 1: *c*.
X $\mathbf{G}_4^2 \otimes \mathbf{T}_2$: $\{C_{2z}|000\}, \{\sigma_y|000\}$; $\mathbf{t}_2$: 1, 1; 2, 1; 3, 1; 4, 1: *c*.

Δ^x $\mathbf{G}_2^1$: $(\sigma_x, 0)$: 1, 1; 2, 1: *c*.
U^x $\mathbf{G}_2^1$: $(\sigma_x, 0)$: 1, 1; 2, 1: *c*.
Λ^x $\mathbf{G}_8^4$: $(C_{4z}^+, 0), (\sigma_y, 0)$: 1, *x*; 2, *x*; 3, *x*; 4, *x*; 5, *x*: *b*.
V^x $\mathbf{G}_8^4$: $(C_{4z}^+, 0), (\sigma_y, 0)$: 1, *x*; 2, *x*; 3, *x*; 4, *x*; 5, *x*: *b*.
Σ^x $\mathbf{G}_2^1$: $(\sigma_{db}, 0)$: 1, 1; 2, 1: *c*.
S^x $\mathbf{G}_2^1$: $(\sigma_{db}, 0)$: 1, 3; 2, 3: *c*.
Y^x $\mathbf{G}_2^1$: $(\sigma_y, 0)$: 1, 1; 2, 1: *c*.
T^x $\mathbf{G}_2^1$: $(\sigma_y, 0)$: 1, 1; 2, 1: *c*.
W^x $\mathbf{G}_4^2$: $(C_{2z}, 0), (\sigma_y, 0)$: 1, *x*; 2, *x*; 3, *x*; 4, *x*: *c*.

106 $P4_2bc$ C_{4v}^8

($F1$; $K7$; $M5$; $Z1$.)

Γ $\mathbf{G}_8^4$: $\{C_{4z}^+ \mid 00\frac{1}{2}\}, \{\sigma_y \mid \frac{1}{2}\frac{1}{2}0\}$: 1, 1; 2, 1; 3, 1; 4, 1; 5, 1: *b*.
M $\mathbf{G}_{16}^{10}$: $\{C_{4z}^+ \mid 00\frac{1}{2}\}, \{C_{2z} \mid 010\}, \{\sigma_{db} \mid \frac{1}{2}\frac{1}{2}\frac{1}{2}\}$: 5, 3; 6, 3; 7, 3; 8, 3; 9, 1: *b*.
Z $\mathbf{G}_{16}^{10}$: $\{C_{4z}^+ \mid 00\frac{1}{2}\}, \{C_{2z} \mid 000\}, \{\sigma_y \mid \frac{1}{2}\frac{1}{2}0\}$: 5, 3; 6, 3; 7, 3; 8, 3; 9, 1: *b*.
A $\mathbf{G}_{16}^{8}$: $\{C_{4z}^+ \mid 00\frac{1}{2}\}, \{\sigma_{db} \mid \frac{1}{2}\frac{1}{2}\frac{1}{2}\}$: 5, 3; 6, 3; 7, 3; 8, 3; 10, 2: *a*.
R $\mathbf{G}_8^4$: $\{\sigma_x \mid \frac{1}{2}\frac{1}{2}0\}, \{C_{2z} \mid 000\}$: 5, 1: *a*.
X $\mathbf{G}_8^4$: $\{\sigma_x \mid \frac{1}{2}\frac{1}{2}0\}, \{C_{2z} \mid 000\}$: 5, 1: *a*.

Δ^x $\mathbf{G}_2^1$: $(\sigma_x, 0)$: 1, 1; 2, 1: *c*.
U^x $\mathbf{G}_2^1$: $(\sigma_x, 0)$: 1, 1; 2, 1: *c*.
Λ^x $\mathbf{G}_8^4$: $(C_{4z}^+, 0), (\sigma_y, 0)$: 1, x; 2, x; 3, x; 4, x; 5, x: *b*.
V^x $\mathbf{G}_{16}^{10}$: $(C_{4z}^+, 0), (C_{2z}, 1), (\sigma_{da}, 0)$: 5, x; 6, x; 7, x; 8, x; 9, x: *b*.
Σ^x $\mathbf{G}_2^1$: $(\sigma_{db}, 0)$: 1, 1; 2, 1: *c*.
S^x $\mathbf{G}_2^1$: $(\sigma_{db}, 0)$: 1, 3; 2, 3: *c*.
Y^x $\mathbf{G}_4^1$: $(\sigma_y, 0)$: 2, 3; 4, 3: *a*.
T^x $\mathbf{G}_4^1$: $(\sigma_y, 0)$: 2, 3; 4, 3: *a*.
W^x $\mathbf{G}_8^4$: $(\sigma_y, 0), (C_{2z}, 0)$: 5, x: *a*.

107 $I4mm$ C_{4v}^9

($F1$; $K7$; $M5$; $Z1$.)

Γ $\mathbf{G}_8^4$: $\{C_{4z}^+ \mid 000\}, \{\sigma_y \mid 000\}$: 1, 1; 2, 1; 3, 1; 4, 1; 5, 1: *b*.
N $\mathbf{G}_2^1 \otimes \mathbf{T}_2$: $\{\sigma_y \mid 000\}$; $\mathbf{t}_2$: 1, 1; 2, 1: *c*.
X $\mathbf{G}_4^2 \otimes \mathbf{T}_2$: $\{C_{2z} \mid 000\}, \{\sigma_{db} \mid 000\}$; $\mathbf{t}_3$: 1, 1; 2, 1; 3, 1; 4, 1: *c*.
Z $\mathbf{G}_8^4 \otimes \mathbf{T}_2$: $\{C_{4z}^+ \mid 000\}, \{\sigma_y \mid 000\}$; $\mathbf{t}_1$ or $\mathbf{t}_2$ or $\mathbf{t}_3$: 1, 1; 2, 1; 3, 1; 4, 1; 5, 1: *b*.
P $\mathbf{G}_4^2 \otimes \mathbf{T}_4$: $\{C_{2z} \mid 000\}, \{\sigma_{db} \mid 000\}$; $\mathbf{t}_1$ or $\mathbf{t}_2$ or $\mathbf{t}_3$: 1, 1; 2, 1; 3, 3; 4, 3: *c*.

Λ^x $\mathbf{G}_8^4$: $(C_{4z}^+, 0), (\sigma_y, 0)$: 1, x; 2, x; 3, x; 4, x; 5, x: *b*.
V^x $\mathbf{G}_8^4$: $(C_{4z}^+, 0), (\sigma_y, 0)$: 1, x; 2, x; 3, x; 4, x; 5, x: *b*.
W^x $\mathbf{G}_4^2$: $(C_{2z}, 0), (\sigma_{db}, 0)$: 1, x; 2, x; 3, x; 4, x: *c*.
Σ^x $\mathbf{G}_2^1$: $(\sigma_y, 0)$: 1, 1; 2, 1: *c*.
F^x $\mathbf{G}_2^1$: $(\sigma_y, 0)$: 1, 1; 2, 1: *c*.
Q^x $\mathbf{G}_1^1$: $(E, 0)$: 1, 1: *a*.
Δ^x $\mathbf{G}_2^1$: $(\sigma_{db}, 0)$: 1, 1; 2, 1: *c*.
U^x $\mathbf{G}_2^1$: $(\sigma_{db}, 0)$: 1, 1; 2, 1: *c*.
Y^x $\mathbf{G}_2^1$: $(\sigma_{da}, 0)$: 1, 1; 2, 1: *c*.

108 $I4cm$ C_{4v}^{10}

($F1$; $K7$; $M5$; $Z1$.)

Γ $\mathbf{G}_8^4$: $\{C_{4z}^+ \mid 000\}, \{\sigma_y \mid \frac{1}{2}\frac{1}{2}0\}$: 1, 1; 2, 1; 3, 1; 4, 1; 5, 1: *b*.
N $\mathbf{G}_4^1$: $\{\sigma_y \mid \frac{1}{2}\frac{1}{2}0\}$: 2, 3; 4, 3: *a*.
X $\mathbf{G}_4^2 \otimes \mathbf{T}_2$: $\{C_{2z} \mid 000\}, \{\sigma_{db} \mid \frac{1}{2}\frac{1}{2}0\}$; $\mathbf{t}_3$: 1, 1; 2, 1; 3, 1; 4, 1: *c*.
Z $\mathbf{G}_8^4 \otimes \mathbf{T}_2$: $\{C_{4z}^+ \mid 000\}, \{\sigma_y \mid \frac{1}{2}\frac{1}{2}0\}$; $\mathbf{t}_1$ or $\mathbf{t}_2$ or $\mathbf{t}_3$: 1, 1; 2, 1; 3, 1; 4, 1; 5, 1: *b*.
P $\mathbf{G}_4^2 \otimes \mathbf{T}_4$: $\{\sigma_{db} \mid \frac{3}{2}\frac{1}{2}0\}, \{C_{2z} \mid 000\}$; $\mathbf{t}_1$ or $\mathbf{t}_2$ or $\mathbf{t}_3$: 1, 3; 2, 2; 3, 3; 4, 2: *c*.

Λ^x $\mathbf{G}_8^4$: $(C_{4z}^+, 0), (\sigma_y, 0)$: 1, x; 2, x; 3, x; 4, x; 5, x: *b*.
V^x $\mathbf{G}_8^4$: $(C_{4z}^+, 0), (\sigma_y, 0)$: 1, x; 2, x; 3, x; 4, x; 5, x: *b*.
W^x $\mathbf{G}_4^2$: $(C_{2z}, 0), (\sigma_{db}, 0)$: 1, x; 2, x; 3, x; 4, x: *c*.
Σ^x $\mathbf{G}_2^1$: $(\sigma_y, 0)$: 1, 1; 2, 1: *c*.
F^x $\mathbf{G}_2^1$: $(\sigma_y, 0)$: 1, 1; 2, 1: *c*.
Q^x $\mathbf{G}_1^1$: $(E, 0)$: 1, 2: *a*.
Δ^x $\mathbf{G}_2^1$: $(\sigma_{db}, 0)$: 1, 1; 2, 1: *c*.
U^x $\mathbf{G}_2^1$: $(\sigma_{db}, 0)$: 1, 1; 2, 1: *c*.
Y^x $\mathbf{G}_2^1$: $(\sigma_{da}, 0)$: 1, 1; 2, 1: *c*.

109 $I4_1md$ C_{4v}^{11}

($F1$; $K7$; $M5$; $Z1$.)

Γ $\mathbf{G}_8^4$: $\{C_{4z}^+|\frac{3}{4}\frac{1}{4}\frac{1}{2}\}, \{\sigma_y|000\}$: 1, 1; 2, 1; 3, 1; 4, 1; 5, 1: b.
N $\mathbf{G}_2^1 \otimes \mathbf{T}_2$: $\{\sigma_y|000\}$; $\mathbf{t}_2$: 1, 1; 2, 1: c.
X $\mathbf{G}_8^4$: $\{\sigma_{db}|\frac{3}{4}\frac{1}{4}\frac{1}{2}\}, \{C_{2z}|000\}$: 5, 1: a.
Z $\mathbf{G}_{16}^{10}$: $\{C_{4z}^+|\frac{3}{4}\frac{1}{4}\frac{1}{2}\}, \{C_{2z}|000\}, \{\sigma_x|000\}$: 5, 3; 6, 3; 7, 3; 8, 3; 9, 1: b.
P $\mathbf{G}_{16}^6$: $\{\sigma_{da}|\frac{3}{4}\frac{1}{4}\frac{1}{2}\}, \{C_{2z}|000\}$: 10, 1: a.

Λ^x $\mathbf{G}_8^4$: $(C_{4z}^+, 0), (\sigma_y, 0)$: 1, x; 2, x; 3, x; 4, x; 5, x: b.
V^x $\mathbf{G}_{16}^{10}$: $(C_{4z}^+, 0), (C_{2z}, 0), (\sigma_y, 0)$: 5, x; 6, x; 7, x; 8, x; 9, x: b.
W^x $\mathbf{G}_8^4$: $(\sigma_{da}, 0), (C_{2z}, 0)$: 5, x: a.
Σ^x $\mathbf{G}_2^1$: $(\sigma_y, 0)$: 1, 1; 2, 1: c.
F^x $\mathbf{G}_2^1$: $(\sigma_y, 0)$: 1, 1; 2, 1: c.
Q^x $\mathbf{G}_1^1$: $(E, 0)$: 1, 1: a.
Δ^x $\mathbf{G}_2^1$: $(\sigma_{db}, 0)$: 1, 1; 2, 1: c.
U^x $\mathbf{G}_2^1$: $(\sigma_{db}, 0)$: 1, 3; 2, 3: c.
Y^x $\mathbf{G}_4^1$: $(\sigma_{da}, 0)$: 2, 3; 4, 3: a.

110 $I4_1cd$ C_{4v}^{12}

($F1$; $K7$; $M5$; $Z1$.)

Γ $\mathbf{G}_8^4$: $\{C_{4z}^+|\frac{3}{4}\frac{1}{4}\frac{1}{2}\}, \{\sigma_y|\frac{1}{2}\frac{1}{2}0\}$: 1, 1; 2, 1; 3, 1; 4, 1; 5, 1: b.
N $\mathbf{G}_4^1$: $\{\sigma_y|\frac{1}{2}\frac{1}{2}0\}$: 2, 3; 4, 3: a.
X $\mathbf{G}_8^4$: $\{\sigma_{db}|\frac{1}{4}\frac{3}{4}\frac{1}{2}\}, \{C_{2z}|000\}$: 5, 1: a.
Z $\mathbf{G}_{16}^{10}$: $\{C_{4z}^+|\frac{3}{4}\frac{1}{4}\frac{1}{2}\}, \{C_{2z}|000\}, \{\sigma_x|\frac{1}{2}\frac{1}{2}0\}$: 5, 3; 6, 3; 7, 3; 8, 3; 9, 1: b.
P $\mathbf{G}_{16}^5$: $\{\sigma_{da}|\frac{1}{4}\frac{3}{4}\frac{1}{2}\}, \{C_{2z}|000\}$: 10, 2: a.

Λ^x $\mathbf{G}_8^4$: $(C_{4z}^+, 0), (\sigma_y, 0)$: 1, x; 2, x; 3, x; 4, x; 5, x: b.
V^x $\mathbf{G}_{16}^{10}$: $(C_{4z}^+, 0), (C_{2z}, 0), (\sigma_y, 0)$: 5, x; 6, x; 7, x; 8, x; 9, x: b.
W^x $\mathbf{G}_8^4$: $(\sigma_{da}, 0), (C_{2z}, 0)$: 5, x: a.
Σ^x $\mathbf{G}_2^1$: $(\sigma_y, 0)$: 1, 1; 2, 1: c.
F^x $\mathbf{G}_2^1$: $(\sigma_y, 0)$: 1, 1; 2, 1: c.
Q^x $\mathbf{G}_1^1$: $(E, 0)$: 1, 2: a.
Δ^x $\mathbf{G}_2^1$: $(\sigma_{db}, 0)$: 1, 1; 2, 1: c.
U^x $\mathbf{G}_2^1$: $(\sigma_{db}, 0)$: 1, 3; 2, 3: c.
Y^x $\mathbf{G}_4^1$: $(\sigma_{da}, 0)$: 2, 3; 4, 3: a.

111 $P\bar{4}2m$ D_{2d}^1

($F1$; $G8$; $K7$; $M5$; $Z1$.)

Γ $\mathbf{G}_8^4$: $\{S_{4z}^+|000\}, \{C_{2x}|000\}$: 1, 1; 2, 1; 3, 1; 4, 1; 5, 1: b.
M $\mathbf{G}_8^4 \otimes \mathbf{T}_2$: $\{S_{4z}^+|000\}, \{C_{2x}|000\}$; $\mathbf{t}_1$ or $\mathbf{t}_2$: 1, 1; 2, 1; 3, 1; 4, 1; 5, 1: b.
Z $\mathbf{G}_8^4 \otimes \mathbf{T}_2$: $\{S_{4z}^+|000\}, \{C_{2x}|000\}$; $\mathbf{t}_3$: 1, 1; 2, 1; 3, 1; 4, 1; 5, 1: b.
A $\mathbf{G}_8^4 \otimes \mathbf{T}_2$: $\{S_{4z}^+|000\}, \{C_{2x}|000\}$; $\mathbf{t}_1$ or $\mathbf{t}_2$ or $\mathbf{t}_3$: 1, 1; 2, 1; 3, 1; 4, 1; 5, 1: b.
R $\mathbf{G}_4^2 \otimes \mathbf{T}_2$: $\{C_{2z}|000\}, \{C_{2y}|000\}$; $\mathbf{t}_2$ or $\mathbf{t}_3$: 1, 1; 2, 1; 3, 1; 4, 1: b.
X $\mathbf{G}_4^2 \otimes \mathbf{T}_2$: $\{C_{2z}|000\}, \{C_{2y}|000\}$; $\mathbf{t}_2$: 1, 1; 2, 1; 3, 1; 4, 1: b.

Δ^x $\mathbf{G}_2^1$: $(C_{2y}, 0)$: 1, 1; 2, 1: b.
U^x $\mathbf{G}_2^1$: $(C_{2y}, 0)$: 1, 1; 2, 1: b.
Λ^x $\mathbf{G}_4^2$: $(C_{2z}, 0), (\sigma_{db}, 0)$: 1, 1; 2, 1; 3, 3; 4, 3: c.
V^x $\mathbf{G}_4^2$: $(C_{2z}, 0), (\sigma_{db}, 0)$: 1, 1; 2, 1; 3, 3; 4, 3: c.
Σ^x $\mathbf{G}_2^1$: $(\sigma_{db}, 0)$: 1, 1; 2, 1: c.
S^x $\mathbf{G}_2^1$: $(\sigma_{db}, 0)$: 1, 1; 2, 1: c.
Y^x $\mathbf{G}_2^1$: $(C_{2x}, 0)$: 1, 1; 2, 1: b.
T^x $\mathbf{G}_2^1$: $(C_{2x}, 0)$: 1, 1; 2, 1: b.
W^x $\mathbf{G}_2^1$: $(C_{2z}, 0)$: 1, 1; 2, 1: b.

112 $P\bar{4}2c$ D_{2d}^2

($F1$; $K7$; $M5$; $Z1$.)

Γ $\mathbf{G}_8^4$: $\{S_{4z}^+ | 000\}$, $\{C_{2x} | 00\frac{1}{2}\}$: 1, 1; 2, 1; 3, 1; 4, 1; 5, 1: b.
M $\mathbf{G}_8^4 \otimes \mathbf{T}_2$: $\{S_{4z}^+ | 000\}$, $\{C_{2x} | 00\frac{1}{2}\}$; $\mathbf{t}_1$ or $\mathbf{t}_2$: 1, 1; 2, 1; 3, 1; 4, 1; 5, 1: b.
Z $\mathbf{G}_{16}^{10}$: $\{S_{4z}^+ | 000\}$, $\{C_{2z} | 001\}$, $\{C_{2x} | 00\frac{1}{2}\}$: 5, 3; 6, 3; 7, 3; 8, 3; 9, 1: a.
A $\mathbf{G}_{16}^{10}$: $\{S_{4z}^+ | 000\}$, $\{C_{2z} | 001\}$, $\{C_{2x} | 00\frac{1}{2}\}$: 5, 3; 6, 3; 7, 3; 8, 3; 9, 1: a.
R $\mathbf{G}_4^2 \otimes \mathbf{T}_2$: $\{C_{2z} | 000\}$, $\{C_{2y} | 00\frac{1}{2}\}$; $\mathbf{t}_2$ or $\mathbf{t}_3$: 1, 1; 2, 1; 3, 1; 4, 1: b.
X $\mathbf{G}_4^2 \otimes \mathbf{T}_2$: $\{C_{2z} | 000\}$, $\{C_{2y} | 00\frac{1}{2}\}$; $\mathbf{t}_2$: 1, 1; 2, 1; 3, 1; 4, 1: b.

Δ^x $\mathbf{G}_2^1$: $(C_{2y}, 0)$: 1, 1; 2, 1: b.
U^x $\mathbf{G}_4^1$: $(C_{2y}, 0)$: 2, 1; 4, 1: a.
Λ^x $\mathbf{G}_4^2$: $(C_{2z}, 0)$, $(\sigma_{db}, 0)$: 1, 1; 2, 1; 3, 3; 4, 3: c.
V^x $\mathbf{G}_4^2$: $(C_{2z}, 0)$, $(\sigma_{db}, 0)$: 1, 1; 2, 1; 3, 3; 4, 3: c.
Σ^x $\mathbf{G}_2^1$: $(\sigma_{db}, 0)$: 1, 1; 2, 1: c.
S^x $\mathbf{G}_2^1$: $(\sigma_{db}, 0)$: 1, 3; 2, 3: c.
Y^x $\mathbf{G}_2^1$: $(C_{2x}, 0)$: 1, 1; 2, 1: b.
T^x $\mathbf{G}_4^1$: $(C_{2x}, 0)$: 2, 1; 4, 1: a.
W^x $\mathbf{G}_2^1$: $(C_{2z}, 0)$: 1, 1; 2, 1: b.

113 $P\bar{4}2_1m$ D_{2d}^3

($F1$; $K7$; $M5$; $Z1$.)

Γ $\mathbf{G}_8^4$: $\{S_{4z}^+ | 000\}$, $\{C_{2x} | \frac{1}{2}\frac{1}{2}0\}$: 1, 1; 2, 1; 3, 1; 4, 1; 5, 1: b.
M $\mathbf{G}_{16}^{10}$: $\{S_{4z}^+ | 000\}$, $\{C_{2z} | 010\}$, $\{\sigma_{db} | \frac{1}{2}\frac{1}{2}0\}$: 5, 3; 6, 3; 7, 3; 8, 3; 9, 1: b.
Z $\mathbf{G}_8^4 \otimes \mathbf{T}_2$: $\{S_{4z}^+ | 000\}$, $\{C_{2x} | \frac{1}{2}\frac{1}{2}0\}$; $\mathbf{t}_3$: 1, 1; 2, 1; 3, 1; 4, 1; 5, 1: b.
A $\mathbf{G}_{16}^{10}$: $\{S_{4z}^+ | 000\}$, $\{C_{2z} | 010\}$, $\{\sigma_{db} | \frac{1}{2}\frac{1}{2}0\}$: 5, 3; 6, 3; 7, 3; 8, 3; 9, 1: b.
R $\mathbf{G}_8^4$: $\{C_{2y} | \frac{1}{2}\frac{1}{2}0\}$, $\{C_{2z} | 000\}$: 5, 1: a.
X $\mathbf{G}_8^4$: $\{C_{2y} | \frac{1}{2}\frac{1}{2}0\}$, $\{C_{2z} | 000\}$: 5, 1: a.

Δ^x $\mathbf{G}_2^1$: $(C_{2y}, 0)$: 1, 1; 2, 1: b.
U^x $\mathbf{G}_2^1$: $(C_{2y}, 0)$: 1, 1; 2, 1: b.
Λ^x $\mathbf{G}_4^2$: $(C_{2z}, 0)$, $(\sigma_{db}, 0)$: 1, 1; 2, 1; 3, 3; 4, 3: c.
V^x $\mathbf{G}_4^2$: $(C_{2z}, 0)$, $(\sigma_{db}, 0)$: 1, 3; 2, 3; 3, 2; 4, 2: c.
Σ^x $\mathbf{G}_2^1$: $(\sigma_{db}, 0)$: 1, 1; 2, 1: c.
S^x $\mathbf{G}_2^1$: $(\sigma_{db}, 0)$: 1, 1; 2, 1: c.
Y^x $\mathbf{G}_4^1$: $(C_{2x}, 0)$: 2, 3; 4, 3: a.
T^x $\mathbf{G}_4^1$: $(C_{2x}, 0)$: 2, 3; 4, 3: a.
W^x $\mathbf{G}_2^1$: $(C_{2z}, 0)$: 1, 3; 2, 3: b.

114 $P\bar{4}2_1c$ D_{2d}^4

($F1$; $K7$; $M5$; $Z1$.)

Γ $\mathbf{G}_8^4$: $\{S_{4z}^+ | 000\}$, $\{C_{2x} | \frac{1}{2}\frac{1}{2}\frac{1}{2}\}$: 1, 1; 2, 1; 3, 1; 4, 1; 5, 1: b.
M $\mathbf{G}_{16}^{10}$: $\{S_{4z}^+ | 000\}$, $\{C_{2z} | 010\}$, $\{\sigma_{db} | \frac{1}{2}\frac{1}{2}\frac{1}{2}\}$: 5, 3; 6, 3; 7, 3; 8, 3; 9, 1: b.
Z $\mathbf{G}_{16}^{10}$: $\{S_{4z}^+ | 000\}$, $\{C_{2z} | 001\}$, $\{C_{2x} | \frac{1}{2}\frac{1}{2}\frac{1}{2}\}$: 5, 3; 6, 3; 7, 3; 8, 3; 9, 1: a.
A $\mathbf{G}_{16}^{8}$: $\{S_{4z}^+ | 000\}$, $\{\sigma_{db} | \frac{1}{2}\frac{1}{2}\frac{1}{2}\}$: 5, 3; 6, 3; 7, 3; 8, 3; 10, 2: a.
R $\mathbf{G}_8^4$: $\{C_{2y} | \frac{1}{2}\frac{1}{2}\frac{1}{2}\}$, $\{C_{2z} | 000\}$: 5, 1: a.
X $\mathbf{G}_8^4$: $\{C_{2y} | \frac{1}{2}\frac{1}{2}\frac{1}{2}\}$, $\{C_{2z} | 000\}$: 5, 1: a.

Δ^x $\mathbf{G}_2^1$: $(C_{2y}, 0)$: 1, 1; 2, 1: b.
U^x $\mathbf{G}_4^1$: $(C_{2y}, 0)$: 2, 1; 4, 1: a.
Λ^x $\mathbf{G}_4^2$: $(C_{2z}, 0)$, $(\sigma_{db}, 0)$: 1, 1; 2, 1; 3, 3; 4, 3: c.
V^x $\mathbf{G}_4^2$: $(C_{2z}, 0)$, $(\sigma_{db}, 0)$: 1, 3; 2, 3; 3, 2; 4, 2: c.
Σ^x $\mathbf{G}_2^1$: $(\sigma_{db}, 0)$: 1, 1; 2, 1: c.
S^x $\mathbf{G}_2^1$: $(\sigma_{db}, 0)$: 1, 3; 2, 3: c.
Y^x $\mathbf{G}_4^1$: $(C_{2x}, 0)$: 2, 3; 4, 3: a.
T^x $\mathbf{G}_2^1$: $(C_{2x}, 0)$: 1, 3; 2, 3: b.
W^x $\mathbf{G}_2^1$: $(C_{2z}, 0)$: 1, 3; 2, 3: b.

115 $P\bar{4}m2$ D_{2d}^5

($F1$; $K7$; $M5$; $Z1$.)

Γ $\mathbf{G}_8^4$: $\{S_{4z}^+|000\}$, $\{C_{2a}|000\}$: 1, 1; 2, 1; 3, 1; 4, 1; 5, 1: b.
M $\mathbf{G}_8^4 \otimes \mathbf{T}_2$: $\{S_{4z}^+|000\}$, $\{C_{2a}|000\}$; $\mathbf{t}_1$ or $\mathbf{t}_2$: 1, 1; 2, 1; 3, 1; 4, 1; 5, 1: b.
Z $\mathbf{G}_8^4 \otimes \mathbf{T}_2$: $\{S_{4z}^+|000\}$, $\{C_{2a}|000\}$; $\mathbf{t}_3$: 1, 1; 2, 1; 3, 1; 4, 1; 5, 1: b.
A $\mathbf{G}_8^4 \otimes \mathbf{T}_2$: $\{S_{4z}^+|000\}$, $\{C_{2a}|000\}$; $\mathbf{t}_1$ or $\mathbf{t}_2$ or $\mathbf{t}_3$: 1, 1; 2, 1; 3, 1; 4, 1; 5, 1: b.
R $\mathbf{G}_4^2 \otimes \mathbf{T}_2$: $\{C_{2z}|000\}$, $\{\sigma_y|000\}$; $\mathbf{t}_2$ or $\mathbf{t}_3$: 1, 1; 2, 1; 3, 1; 4, 1: c.
X $\mathbf{G}_4^2 \otimes \mathbf{T}_2$: $\{C_{2z}|000\}$, $\{\sigma_y|000\}$; $\mathbf{t}_2$: 1, 1; 2, 1; 3, 1; 4, 1: c.

Δ^x $\mathbf{G}_2^1$: $(\sigma_x, 0)$: 1, 1; 2, 1: c.
U^x $\mathbf{G}_2^1$: $(\sigma_x, 0)$: 1, 1; 2, 1: c.
Λ^x $\mathbf{G}_4^2$: $(C_{2z}, 0)$, $(\sigma_y, 0)$: 1, 1; 2, 1; 3, 3; 4, 3: c.
V^x $\mathbf{G}_4^2$: $(C_{2z}, 0)$, $(\sigma_y, 0)$: 1, 1; 2, 1; 3, 3; 4, 3: c.
Σ^x $\mathbf{G}_2^1$: $(C_{2a}, 0)$: 1, 1; 2, 1: b.
S^x $\mathbf{G}_2^1$: $(C_{2a}, 0)$: 1, 1; 2, 1: b.
Y^x $\mathbf{G}_2^1$: $(\sigma_y, 0)$: 1, 1; 2, 1: c.
T^x $\mathbf{G}_2^1$: $(\sigma_y, 0)$: 1, 1; 2, 1: c.
W^x $\mathbf{G}_4^2$: $(C_{2z}, 0)$, $(\sigma_y, 0)$: 1, x; 2, x; 3, x; 4, x: c.

116 $P\bar{4}c2$ D_{2d}^6

($F1$; $K7$; $M5$; $Z1$.)

Γ $\mathbf{G}_8^4$: $\{S_{4z}^+|000\}$, $\{C_{2a}|00\frac{1}{2}\}$: 1, 1; 2, 1; 3, 1; 4, 1; 5, 1: b.
M $\mathbf{G}_8^4 \otimes \mathbf{T}_2$: $\{S_{4z}^+|000\}$, $\{C_{2a}|00\frac{1}{2}\}$; $\mathbf{t}_1$ or $\mathbf{t}_2$: 1, 1; 2, 1; 3, 1; 4, 1; 5, 1: b.
Z $\mathbf{G}_{16}^{10}$: $\{S_{4z}^+|000\}$, $\{C_{2z}|001\}$, $\{C_{2a}|00\frac{1}{2}\}$: 5, 3; 6, 3; 7, 3; 8, 3; 9, 1: a.
A $\mathbf{G}_{16}^{10}$: $\{S_{4z}^+|000\}$, $\{C_{2z}|001\}$, $\{C_{2a}|00\frac{1}{2}\}$: 5, 3; 6, 3; 7, 3; 8, 3; 9, 1: a.
R $\mathbf{G}_8^2$: $\{\sigma_x|00\frac{1}{2}\}$, $\{C_{2z}|000\}$: 2, 3; 4, 3; 6, 3; 8, 3: d.
X $\mathbf{G}_4^2 \otimes \mathbf{T}_2$: $\{C_{2z}|000\}$, $\{\sigma_y|00\frac{1}{2}\}$; $\mathbf{t}_2$: 1, 1; 2, 1; 3, 1; 4, 1: c.

Δ^x $\mathbf{G}_2^1$: $(\sigma_x, 0)$: 1, 1; 2, 1: c.
U^x $\mathbf{G}_2^1$: $(\sigma_x, 0)$: 1, 3; 2, 3: c.
Λ^x $\mathbf{G}_4^2$: $(C_{2z}, 0)$, $(\sigma_y, 0)$: 1, 1; 2, 1; 3, 3; 4, 3: c.
V^x $\mathbf{G}_4^2$: $(C_{2z}, 0)$, $(\sigma_y, 0)$: 1, 1; 2, 1; 3, 3; 4, 3: c.
Σ^x $\mathbf{G}_2^1$: $(C_{2a}, 0)$: 1, 1; 2, 1: b.
S^x $\mathbf{G}_4^1$: $(C_{2a}, 0)$: 2, 1; 4, 1: a.
Y^x $\mathbf{G}_2^1$: $(\sigma_y, 0)$: 1, 1; 2, 1: c.
T^x $\mathbf{G}_2^1$: $(\sigma_y, 0)$: 1, 3; 2, 3: c.
W^x $\mathbf{G}_4^2$: $(C_{2z}, 0)$, $(\sigma_y, 0)$: 1, x; 2, x; 3, x; 4, x: c.

117 $P\bar{4}b2$ D_{2d}^7

($F1$; $K7$; $M5$; $Z1$.)

Γ $\mathbf{G}_8^4$: $\{S_{4z}^+|000\}$, $\{C_{2a}|\frac{1}{2}\frac{1}{2}0\}$: 1, 1; 2, 1; 3, 1; 4, 1; 5, 1: b.
M $\mathbf{G}_{16}^{10}$: $\{S_{4z}^+|000\}$, $\{C_{2z}|010\}$, $\{C_{2a}|\frac{1}{2}\frac{1}{2}0\}$: 5, 3; 6, 3; 7, 3; 8, 3; 9, 1: a.
Z $\mathbf{G}_8^4 \otimes \mathbf{T}_2$: $\{S_{4z}^+|000\}$, $\{C_{2a}|\frac{1}{2}\frac{1}{2}0\}$; $\mathbf{t}_3$: 1, 1; 2, 1; 3, 1; 4, 1; 5, 1: b.
A $\mathbf{G}_{16}^{10}$: $\{S_{4z}^+|000\}$, $\{C_{2z}|010\}$, $\{C_{2a}|\frac{1}{2}\frac{1}{2}0\}$: 5, 3; 6, 3; 7, 3; 8, 3; 9, 1: a.
R $\mathbf{G}_8^4$: $\{\sigma_x|\frac{1}{2}\frac{1}{2}0\}$, $\{C_{2z}|000\}$: 5, 1: a.
X $\mathbf{G}_8^4$: $\{\sigma_x|\frac{1}{2}\frac{1}{2}0\}$, $\{C_{2z}|000\}$: 5, 1: a.

Δ^x $\mathbf{G}_2^1$: $(\sigma_x, 0)$: 1, 1; 2, 1: c.
U^x $\mathbf{G}_2^1$: $(\sigma_x, 0)$: 1, 1; 2, 1: c.
Λ^x $\mathbf{G}_4^2$: $(C_{2z}, 0)$, $(\sigma_y, 0)$: 1, 1; 2, 1; 3, 3; 4, 3: c.
V^x $\mathbf{G}_8^2$: $(\sigma_x, 0)$, $(C_{2z}, 0)$: 2, 1; 4, 1; 6, 3; 8, 3: d.
Σ^x $\mathbf{G}_2^1$: $(C_{2a}, 0)$: 1, 1; 2, 1: b.
S^x $\mathbf{G}_2^1$: $(C_{2a}, 0)$: 1, 1; 2, 1: b.
Y^x $\mathbf{G}_4^1$: $(\sigma_y, 0)$: 2, 3; 4, 3: a.
T^x $\mathbf{G}_4^1$: $(\sigma_y, 0)$: 2, 3; 4, 3: a.
W^x $\mathbf{G}_8^4$: $(\sigma_y, 0)$, $(C_{2z}, 0)$: 5, x: a.

118 $P\bar{4}n2$ D_{2d}^{8}

($F1$; $K7$; $M5$; $Z1$.)

Γ $\mathbf{G}_8^4$: $\{S_{4z}^+ | 000\}$, $\{C_{2a} | \frac{1}{2}\frac{1}{2}\frac{1}{2}\}$: 1, 1; 2, 1; 3, 1; 4, 1; 5, 1: b.
M $\mathbf{G}_{16}^{10}$: $\{S_{4z}^+ | 000\}$, $\{C_{2z} | 010\}$, $\{C_{2a} | \frac{1}{2}\frac{1}{2}\frac{1}{2}\}$: 5, 3; 6, 3; 7, 3; 8, 3; 9, 1: a.
Z $\mathbf{G}_{16}^{10}$: $\{S_{4z}^+ | 000\}$, $\{C_{2z} | 001\}$, $\{C_{2a} | \frac{1}{2}\frac{1}{2}\frac{1}{2}\}$: 5, 3; 6, 3; 7, 3; 8, 3; 9, 1: a.
A $\mathbf{G}_8^4 \otimes \mathbf{T}_2$: $\{S_{4z}^+ | 000\}$, $\{C_{2a} | \frac{1}{2}\frac{1}{2}\frac{1}{2}\}$; $\mathbf{t}_1$ or $\mathbf{t}_2$ or $\mathbf{t}_3$: 1, 1; 2, 1; 3, 1; 4, 1; 5, 1: b.
R $\mathbf{G}_8^4$: $\{\sigma_y | \frac{1}{2}\frac{1}{2}\frac{1}{2}\}$, $\{C_{2z} | 000\}$: 5, 1: a.
X $\mathbf{G}_8^4$: $\{\sigma_x | \frac{1}{2}\frac{1}{2}\frac{1}{2}\}$, $\{C_{2z} | 000\}$: 5, 1: a.

Δ^x $\mathbf{G}_2^1$: $(\sigma_x, 0)$: 1, 1; 2, 1: c.
U^x $\mathbf{G}_2^1$: $(\sigma_x, 0)$: 1, 3; 2, 3: c.
Λ^x $\mathbf{G}_4^2$: $(C_{2z}, 0)$, $(\sigma_y, 0)$: 1, 1; 2, 1; 3, 3; 4, 3: c.
V^x $\mathbf{G}_8^2$: $(\sigma_x, 0)$, $(C_{2z}, 0)$: 2, 1; 4, 1; 6, 3; 8, 3: d.
Σ^x $\mathbf{G}_2^1$: $(C_{2a}, 0)$: 1, 1; 2, 1: b.
S^x $\mathbf{G}_4^1$: $(C_{2a}, 0)$: 2, 1; 4, 1: a.
Y^x $\mathbf{G}_4^1$: $(\sigma_y, 0)$: 2, 3; 4, 3: a.
T^x $\mathbf{G}_4^1$: $(\sigma_y, 0)$: 2, 1; 4, 1: a.
W^x $\mathbf{G}_8^4$: $(\sigma_y, 0)$, $(C_{2z}, 0)$: 5, x: a.

119 $I\bar{4}m2$ D_{2d}^{9}

($F1$; $K7$; $M5$; $Z1$.)

Γ $\mathbf{G}_8^4$: $\{S_{4z}^+ | 000\}$, $\{C_{2a} | 000\}$: 1, 1; 2, 1; 3, 1; 4, 1; 5, 1: b.
N $\mathbf{G}_2^1 \otimes \mathbf{T}_2$: $\{\sigma_y | 000\}$; $\mathbf{t}_2$: 1, 1; 2, 1: c.
X $\mathbf{G}_4^2 \otimes \mathbf{T}_2$: $\{C_{2z} | 000\}$, $\{C_{2b} | 000\}$; $\mathbf{t}_3$: 1, 1; 2, 1; 3, 1; 4, 1: b.
Z $\mathbf{G}_8^4 \otimes \mathbf{T}_2$: $\{S_{4z}^+ | 000\}$, $\{C_{2a} | 000\}$; $\mathbf{t}_1$ or $\mathbf{t}_2$ or $\mathbf{t}_3$: 1, 1; 2, 1; 3, 1; 4, 1; 5, 1: b.
P $\mathbf{G}_4^1 \otimes \mathbf{T}_4$: $\{S_{4z}^+ | 000\}$; $\mathbf{t}_1$ or $\mathbf{t}_2$ or $\mathbf{t}_3$: 1, 1; 2, 1; 3, 1; 4, 1: b.

Λ^x $\mathbf{G}_4^2$: $(C_{2z}, 0)$, $(\sigma_y, 0)$: 1, 1; 2, 1; 3, 3; 4, 3: c.
V^x $\mathbf{G}_4^2$: $(C_{2z}, 0)$, $(\sigma_y, 0)$: 1, 1; 2, 1; 3, 3; 4, 3: c.
W^x $\mathbf{G}_2^1$: $(C_{2z}, 0)$: 1, 1; 2, 1: b.
Σ^x $\mathbf{G}_2^1$: $(\sigma_y, 0)$: 1, 1; 2, 1: c.
F^x $\mathbf{G}_2^1$: $(\sigma_y, 0)$: 1, 1; 2, 1: c.
Q^x $\mathbf{G}_1^1$: $(E, 0)$: 1, 1: a.
Δ^x $\mathbf{G}_2^1$: $(C_{2a}, 0)$: 1, 1; 2, 1: b.
U^x $\mathbf{G}_2^1$: $(C_{2a}, 0)$: 1, 1; 2, 1: b.
Y^x $\mathbf{G}_2^1$: $(C_{2b}, 0)$: 1, 1; 2, 1: b.

120 $I\bar{4}c2$ D_{2d}^{10}

($F1$; $K7$; $M5$; $Z1$.)

Γ $\mathbf{G}_8^4$: $\{S_{4z}^+ | 000\}$, $\{C_{2a} | \frac{1}{2}\frac{1}{2}0\}$: 1, 1; 2, 1; 3, 1; 4, 1; 5, 1: b.
N $\mathbf{G}_4^1$: $\{\sigma_y | \frac{1}{2}\frac{1}{2}0\}$: 2, 3; 4, 3: a.
X $\mathbf{G}_4^2 \otimes \mathbf{T}_2$: $\{C_{2z} | 000\}$, $\{C_{2b} | \frac{1}{2}\frac{1}{2}0\}$; $\mathbf{t}_3$: 1, 1; 2, 1; 3, 1; 4, 1: b.
Z $\mathbf{G}_8^4 \otimes \mathbf{T}_2$: $\{S_{4z}^+ | 000\}$, $\{C_{2a} | \frac{1}{2}\frac{1}{2}0\}$; $\mathbf{t}_1$ or $\mathbf{t}_2$ or $\mathbf{t}_3$: 1, 1; 2, 1; 3, 1; 4, 1; 5, 1: b.
P $\mathbf{G}_4^1 \otimes \mathbf{T}_4$: $\{S_{4z}^+ | 000\}$; $\mathbf{t}_1$ or $\mathbf{t}_2$ or $\mathbf{t}_3$: 1, 3; 2, 3; 3, 3; 4, 3: b.

Λ^x $\mathbf{G}_4^2$: $(C_{2z}, 0)$, $(\sigma_y, 0)$: 1, 1; 2, 1; 3, 3; 4, 3: c.
V^x $\mathbf{G}_4^2$: $(C_{2z}, 0)$, $(\sigma_y, 0)$: 1, 1; 2, 1; 3, 3; 4, 3: c.
W^x $\mathbf{G}_2^1$: $(C_{2z}, 0)$: 1, 1; 2, 1: b.
Σ^x $\mathbf{G}_2^1$: $(\sigma_y, 0)$: 1, 1; 2, 1: c.
F^x $\mathbf{G}_2^1$: $(\sigma_y, 0)$: 1, 1; 2, 1: c.
Q^x $\mathbf{G}_1^1$: $(E, 0)$: 1, 2: a.
Δ^x $\mathbf{G}_2^1$: $(C_{2a}, 0)$: 1, 1; 2, 1: b.
U^x $\mathbf{G}_2^1$: $(C_{2a}, 0)$: 1, 1; 2, 1: b.
Y^x $\mathbf{G}_2^1$: $(C_{2b}, 0)$: 1, 1; 2, 1: b.

121 $I\bar{4}2m$ D_{2d}^{11}

($F1$; $K7$; $M5$; $Z1$.)

Γ $\mathbf{G}_8^4$: $\{S_{4z}^+|000\}$, $\{C_{2x}|000\}$: 1, 1; 2, 1; 3, 1; 4, 1; 5, 1: *b*.
N $\mathbf{G}_2^1 \otimes \mathbf{T}_2$: $\{C_{2y}|000\}$; $\mathbf{t}_2$: 1, 1; 2, 1: *b*.
X $\mathbf{G}_4^2 \otimes \mathbf{T}_2$: $\{C_{2z}|000\}$, $\{\sigma_{db}|000\}$; $\mathbf{t}_3$: 1, 1; 2, 1; 3, 1; 4, 1: *c*.
Z $\mathbf{G}_8^4 \otimes \mathbf{T}_2$: $\{S_{4z}^+|000\}$, $\{C_{2x}|000\}$; $\mathbf{t}_1$ or $\mathbf{t}_2$ or $\mathbf{t}_3$: 1, 1; 2, 1; 3, 1; 4, 1; 5, 1: *b*.
P $\mathbf{G}_8^4 \otimes \mathbf{T}_4$: $\{S_{4z}^+|000\}$, $\{C_{2x}|000\}$; $\mathbf{t}_1$ or $\mathbf{t}_2$ or $\mathbf{t}_3$: 1, x; 2, x; 3, x; 4, x; 5, x: *b*.

Λ^x $\mathbf{G}_4^2$: $(C_{2z}, 0)$, $(\sigma_{db}, 0)$: 1, 1; 2, 1; 3, 3; 4, 3: *c*.
V^x $\mathbf{G}_4^2$: $(C_{2z}, 0)$, $(\sigma_{db}, 0)$: 1, 1; 2, 1; 3, 3; 4, 3: *c*.
W^x $\mathbf{G}_4^2$: $(C_{2z}, 0)$, $(\sigma_{db}, 0)$: 1, x; 2, x; 3, x; 4, x: *c*.
Σ^x $\mathbf{G}_2^1$: $(C_{2x}, 0)$: 1, 1; 2, 1: *b*.
F^x $\mathbf{G}_2^1$: $(C_{2x}, 0)$: 1, 1; 2, 1: *b*.
Q^x $\mathbf{G}_2^1$: $(C_{2y}, 0)$: 1, x; 2, x: *b*.
Δ^x $\mathbf{G}_2^1$: $(\sigma_{db}, 0)$: 1, 1; 2, 1: *c*.
U^x $\mathbf{G}_2^1$: $(\sigma_{db}, 0)$: 1, 1; 2, 1: *c*.
Y^x $\mathbf{G}_2^1$: $(\sigma_{da}, 0)$: 1, 1; 2, 1: *c*.

122 $I\bar{4}2d$ D_{2d}^{12}

($C2$; $F1$; $K7$; $M5$; $S1$; $S7$; $Z1$.)

Γ $\mathbf{G}_8^4$: $\{S_{4z}^+|000\}$, $\{C_{2x}|\frac{1}{4}\frac{3}{4}\frac{1}{2}\}$: 1, 1; 2, 1; 3, 1; 4, 1; 5, 1: *b*.
N $\mathbf{G}_2^1 \otimes \mathbf{T}_2$: $\{C_{2y}|\frac{1}{4}\frac{3}{4}\frac{1}{2}\}$; $\mathbf{t}_2$: 1, 1; 2, 1: *b*.
X $\mathbf{G}_8^4$: $\{\sigma_{db}|\frac{1}{4}\frac{3}{4}\frac{1}{2}\}$, $\{C_{2z}|000\}$: 5, 1: *a*.
Z $\mathbf{G}_{16}^{10}$: $\{S_{4z}^+|000\}$, $\{C_{2z}|100\}$, $\{C_{2x}|\frac{1}{4}\frac{3}{4}\frac{1}{2}\}$: 5, 3; 6, 3; 7, 3; 8, 3; 9, 1: *a*.
P $\mathbf{G}_{32}^4$: $\{S_{4z}^+|000\}$, $\{E|100\}$, $\{C_{2y}|\frac{1}{4}\frac{3}{4}\frac{1}{2}\}$: 13, x; 14, x: *a*.

Λ^x $\mathbf{G}_4^2$: $(C_{2z}, 0)$, $(\sigma_{db}, 0)$: 1, 1; 2, 1; 3, 3; 4, 3: *c*.
V^x $\mathbf{G}_8^2$: $(\sigma_{da}, 0)$, $(C_{2z}, 0)$: 2, 1; 4, 1; 6, 3; 8, 3: *d*.
W^x $\mathbf{G}_8^4$: $(\sigma_{da}, 0)$, $(C_{2z}, 0)$: 5, x: *a*.
Σ^x $\mathbf{G}_2^1$: $(C_{2x}, 0)$: 1, 1; 2, 1: *b*.
F^x $\mathbf{G}_4^1$: $(C_{2x}, 0)$: 2, 1; 4, 1: *a*.
Q^x $\mathbf{G}_8^1$: $(C_{2y}, 0)$: 4, x; 8, x: *c*.
Δ^x $\mathbf{G}_2^1$: $(\sigma_{db}, 0)$: 1, 1; 2, 1: *c*.
U^x $\mathbf{G}_2^1$: $(\sigma_{db}, 0)$: 1, 3; 2, 3: *c*.
Y^x $\mathbf{G}_4^1$: $(\sigma_{da}, 0)$: 2, 3; 4, 3: *a*.

123 *P4/mmm* D_{4h}^1

(*F*1; *K*7; *M*5; *O*2; *Z*1.)

Γ $\mathbf{G}_{16}^9$: $\{C_{4z}^+|000\}$, $\{C_{2x}|000\}$, $\{I|000\}$: 1, 1; 2, 1; 3, 1; 4, 1; 5, 1; 6, 1; 7, 1; 8, 1; 9, 1; 10, 1: *e*.
M $\mathbf{G}_{16}^9 \otimes \mathbf{T}_2$: $\{C_{4z}^+|000\}$, $\{C_{2x}|000\}$, $\{I|000\}$; $\mathbf{t}_1$ or $\mathbf{t}_2$: 1, 1; 2, 1; 3, 1; 4, 1; 5, 1; 6, 1; 7, 1; 8, 1; 9, 1; 10, 1: *e*.
Z $\mathbf{G}_{16}^9 \otimes \mathbf{T}_2$: $\{C_{4z}^+|000\}$, $\{C_{2x}|000\}$, $\{I|000\}$; $\mathbf{t}_3$: 1, 1; 2, 1; 3, 1; 4, 1; 5, 1; 6, 1; 7, 1; 8, 1; 9, 1; 10, 1: *e*.
A $\mathbf{G}_{16}^9 \otimes \mathbf{T}_2$: $\{C_{4z}^+|000\}$, $\{C_{2x}|000\}$, $\{I|000\}$; $\mathbf{t}_1$ or $\mathbf{t}_2$ or $\mathbf{t}_3$: 1, 1; 2, 1; 3, 1; 4, 1; 5, 1; 6, 1; 7, 1; 8, 1; 9, 1; 10, 1: *e*.
R $\mathbf{G}_8^3 \otimes \mathbf{T}_2$: $\{C_{2z}|000\}$, $\{C_{2y}|000\}$, $\{I|000\}$; $\mathbf{t}_2$ or $\mathbf{t}_3$: 1, 1; 2, 1; 3, 1; 4, 1; 5, 1; 6, 1; 7, 1; 8, 1: *b*.
X $\mathbf{G}_8^3 \otimes \mathbf{T}_2$: $\{C_{2z}|000\}$, $\{C_{2y}|000\}$, $\{I|000\}$; $\mathbf{t}_2$: 1, 1; 2, 1; 3, 1; 4, 1; 5, 1; 6, 1; 7, 1; 8, 1: *b*.

Δ^x $\mathbf{G}_4^2$: $(C_{2y}, 0)$, $(\sigma_x, 0)$: 1, 1; 2, 1; 3, 1; 4, 1: *c*.
U^x $\mathbf{G}_4^2$: $(C_{2y}, 0)$, $(\sigma_x, 0)$: 1, 1; 2, 1; 3, 1; 4, 1: *c*.
Λ^x $\mathbf{G}_8^4$: $(C_{4z}^+, 0)$, $(\sigma_y, 0)$: 1, 1; 2, 1; 3, 1; 4, 1; 5, 1: *b*.
V^x $\mathbf{G}_8^4$: $(C_{4z}^+, 0)$, $(\sigma_y, 0)$: 1, 1; 2, 1; 3, 1; 4, 1; 5, 1: *b*.
Σ^x $\mathbf{G}_4^2$: $(C_{2a}, 0)$, $(\sigma_z, 0)$: 1, 1; 2, 1; 3, 1; 4, 1: *c*.
S^x $\mathbf{G}_4^2$: $(C_{2a}, 0)$, $(\sigma_z, 0)$: 1, 1; 2, 1; 3, 1; 4, 1: *c*.
Y^x $\mathbf{G}_4^2$: $(C_{2x}, 0)$, $(\sigma_z, 0)$: 1, 1; 2, 1; 3, 1; 4, 1: *c*.
T^x $\mathbf{G}_4^2$: $(C_{2x}, 0)$, $(\sigma_z, 0)$: 1, 1; 2, 1; 3, 1; 4, 1: *c*.
W^x $\mathbf{G}_4^2$: $(C_{2z}, 0)$, $(\sigma_y, 0)$: 1, 1; 2, 1; 3, 1; 4, 1: *c*.

124 *P4/mcc* D_{4h}^2

(*F*1; *K*7; *M*5; *O*2; *Z*1.)

Γ $\mathbf{G}_{16}^9$: $\{C_{4z}^+|000\}$, $\{C_{2x}|000\}$, $\{I|00\tfrac{1}{2}\}$: 1, 1; 2, 1; 3, 1; 4, 1; 5, 1; 6, 1; 7, 1; 8, 1; 9, 1; 10, 1: *e*.
M $\mathbf{G}_{16}^9 \otimes \mathbf{T}_2$: $\{C_{4z}^+|000\}$, $\{C_{2x}|000\}$, $\{I|00\tfrac{1}{2}\}$; $\mathbf{t}_1$ or $\mathbf{t}_2$: 1, 1; 2, 1; 3, 1; 4, 1; 5, 1; 6, 1; 7, 1; 8, 1; 9, 1; 10, 1: *e*.
Z $\mathbf{G}_{32}^1$: $\{I|00\tfrac{1}{2}\}$, $\{\sigma_{db}|00\tfrac{1}{2}\}$, $\{C_{4z}^+|000\}$: 10, 1; 11, 1; 12, 3; 13, 3: *a*.
A $\mathbf{G}_{32}^1$: $\{I|00\tfrac{1}{2}\}$, $\{\sigma_{db}|00\tfrac{1}{2}\}$, $\{C_{4z}^+|000\}$: 10, 1; 11, 1; 12, 3; 13, 3: *a*.
R $\mathbf{G}_{16}^9$: $\{\sigma_x|00\tfrac{1}{2}\}$, $\{C_{2y}|000\}$, $\{C_{2z}|000\}$: 5, 1; 10, 1: *a*.
X $\mathbf{G}_8^3 \otimes \mathbf{T}_2$: $\{C_{2z}|000\}$, $\{C_{2y}|000\}$, $\{I|00\tfrac{1}{2}\}$; $\mathbf{t}_2$: 1, 1; 2, 1; 3, 1; 4, 1; 5, 1; 6, 1; 7, 1; 8, 1: *b*.

Δ^x $\mathbf{G}_4^2$: $(C_{2y}, 0)$, $(\sigma_x, 0)$: 1, 1; 2, 1; 3, 1; 4, 1: *c*.
U^x $\mathbf{G}_8^4$: $(\sigma_z, 0)$, $(C_{2y}, 0)$: 5, 1: *a*.
Λ^x $\mathbf{G}_8^4$: $(C_{4z}^+, 0)$, $(\sigma_y, 0)$: 1, 1; 2, 1; 3, 1; 4, 1; 5, 1: *b*.
V^x $\mathbf{G}_8^4$: $(C_{4z}^+, 0)$, $(\sigma_y, 0)$: 1, 1; 2, 1; 3, 1; 4, 1; 5, 1: *b*.
Σ^x $\mathbf{G}_4^2$: $(C_{2a}, 0)$, $(\sigma_z, 0)$: 1, 1; 2, 1; 3, 1; 4, 1: *c*.
S^x $\mathbf{G}_8^4$: $(\sigma_z, 0)$, $(C_{2a}, 0)$: 5, 1: *a*.
Y^x $\mathbf{G}_4^2$: $(C_{2x}, 0)$, $(\sigma_z, 0)$: 1, 1; 2, 1; 3, 1; 4, 1: *c*.
T^x $\mathbf{G}_8^4$: $(\sigma_z, 0)$, $(C_{2x}, 0)$: 5, 1: *a*.
W^x $\mathbf{G}_4^2$: $(C_{2z}, 0)$, $(\sigma_y, 0)$: 1, 1; 2, 1; 3, 1; 4, 1: *c*.

125 *P4/nbm* D_{4h}^3

(*F*1; *K*7; *M*5; *O*2; *Z*1.)

Γ $\mathbf{G}_{16}^9$: $\{C_{4z}^+|\frac{1}{2}\frac{1}{2}0\}$, $\{C_{2x}|000\}$, $\{I|\frac{1}{2}\frac{1}{2}0\}$: 1, 1; 2, 1; 3, 1; 4, 1; 5, 1; 6, 1; 7, 1; 8, 1; 9, 1; 10, 1: *e*.
M $\mathbf{G}_{32}^2$: $\{C_{4z}^+|\frac{1}{2}\frac{1}{2}0\}$, $\{\sigma_x|\frac{1}{2}\frac{1}{2}0\}$, $\{C_{2x}|000\}$: 9, 1; 11, 1; 12, 1; 14, 1: *a*.
Z $\mathbf{G}_{16}^9 \otimes \mathbf{T}_2$: $\{C_{4z}^+|\frac{1}{2}\frac{1}{2}0\}$, $\{C_{2x}|000\}$, $\{I|\frac{1}{2}\frac{1}{2}0\}$; $\mathbf{t}_3$: 1, 1; 2, 1; 3, 1; 4, 1; 5, 1; 6, 1; 7, 1; 8, 1; 9, 1; 10, 1: *e*.
A $\mathbf{G}_{32}^2$: $\{C_{4z}^+|\frac{1}{2}\frac{1}{2}0\}$, $\{\sigma_x|\frac{1}{2}\frac{1}{2}0\}$, $\{C_{2x}|000\}$: 9, 1; 11, 1; 12, 1; 14, 1: *a*.
R $\mathbf{G}_{16}^9$: $\{\sigma_x|\frac{1}{2}\frac{1}{2}0\}$, $\{\sigma_y|\frac{1}{2}\frac{1}{2}0\}$, $\{C_{2y}|000\}$: 5, 1; 10, 1: *a*.
X $\mathbf{G}_{16}^9$: $\{\sigma_x|\frac{1}{2}\frac{1}{2}0\}$, $\{\sigma_y|\frac{1}{2}\frac{1}{2}0\}$, $\{C_{2y}|000\}$: 5, 1; 10, 1: *a*.

Δ^x $\mathbf{G}_4^2$: $(C_{2y}, 0)$, $(\sigma_x, 0)$: 1, 1; 2, 1; 3, 1; 4, 1: *c*.
U^x $\mathbf{G}_4^2$: $(C_{2y}, 0)$, $(\sigma_x, 0)$: 1, 1; 2, 1; 3, 1; 4, 1: *c*.
Λ^x $\mathbf{G}_8^4$: $(C_{4z}^+, 0)$, $(\sigma_y, 0)$: 1, 1; 2, 1; 3, 1; 4, 1; 5, 1: *b*.
V^x $\mathbf{G}_{16}^{10}$: $(C_{4z}^+, 0)$, $(C_{2z}, 0)$, $(\sigma_{da}, 0)$: 5, 1; 6, 1; 7, 1; 8, 1; 9, 1: *b*.
Σ^x $\mathbf{G}_4^2$: $(C_{2a}, 0)$, $(\sigma_z, 0)$: 1, 1; 2, 1; 3, 1; 4, 1: *c*.
S^x $\mathbf{G}_4^2$: $(C_{2a}, 0)$, $(\sigma_z, 0)$: 1, 1; 2, 1; 3, 1; 4, 1: *c*.
Y^x $\mathbf{G}_8^4$: $(\sigma_y, 0)$, $(C_{2x}, 0)$: 5, 1: *a*.
T^x $\mathbf{G}_8^4$: $(\sigma_y, 0)$, $(C_{2x}, 0)$: 5, 1: *a*.
W^x $\mathbf{G}_8^4$: $(\sigma_y, 0)$, $(C_{2z}, 0)$: 5, 1: *a*.

126 *P4/nnc* D_{4h}^4

(*F*1; *K*7; *M*5; *O*2; *Z*1.)

Γ $\mathbf{G}_{16}^9$: $\{C_{4z}^+|\frac{1}{2}\frac{1}{2}0\}$, $\{C_{2x}|000\}$, $\{I|\frac{1}{2}\frac{1}{2}\frac{1}{2}\}$: 1, 1; 2, 1; 3, 1; 4, 1; 5, 1; 6, 1; 7, 1; 8, 1; 9, 1; 10, 1: *e*.
M $\mathbf{G}_{32}^2$: $\{C_{4z}^+|\frac{1}{2}\frac{1}{2}0\}$, $\{\sigma_x|\frac{1}{2}\frac{1}{2}\frac{1}{2}\}$, $\{C_{2x}|000\}$: 9, 1; 11, 1; 12, 1; 14, 1: *a*.
Z $\mathbf{G}_{32}^1$: $\{I|\frac{1}{2}\frac{1}{2}\frac{1}{2}\}$, $\{\sigma_{db}|00\frac{1}{2}\}$, $\{C_{4z}^+|\frac{1}{2}\frac{1}{2}0\}$: 10, 1; 11, 1; 12, 3; 13, 3: *a*.
A $\mathbf{G}_{32}^2$: $\{C_{4z}^+|\frac{1}{2}\frac{1}{2}1\}$, $\{\sigma_{db}|00\frac{1}{2}\}$, $\{C_{2b}|\frac{1}{2}\frac{1}{2}1\}$: 9, 1; 11, 1; 12, 1; 14, 1: *a*.
R $\mathbf{G}_{16}^9$: $\{\sigma_y|\frac{1}{2}\frac{1}{2}\frac{1}{2}\}$, $\{C_{2y}|000\}$, $\{C_{2x}|000\}$: 5, 1; 10, 1: *a*.
X $\mathbf{G}_{16}^9$: $\{\sigma_x|\frac{1}{2}\frac{1}{2}\frac{1}{2}\}$, $\{\sigma_y|\frac{1}{2}\frac{1}{2}\frac{1}{2}\}$, $\{C_{2y}|000\}$: 5, 1; 10, 1: *a*.

Δ^x $\mathbf{G}_4^2$: $(C_{2y}, 0)$, $(\sigma_x, 0)$: 1, 1; 2, 1; 3, 1; 4, 1: *c*.
U^x $\mathbf{G}_8^4$: $(\sigma_z, 0)$, $(C_{2y}, 0)$: 5, 1: *a*.
Λ^x $\mathbf{G}_8^4$: $(C_{4z}^+, 0)$, $(\sigma_y, 0)$: 1, 1; 2, 1; 3, 1; 4, 1; 5, 1: *b*.
V^x $\mathbf{G}_{16}^{10}$: $(C_{4z}^+, 0)$, $(C_{2z}, 0)$, $(\sigma_{da}, 0)$: 5, 1; 6, 1; 7, 1; 8, 1; 9, 1: *b*.
Σ^x $\mathbf{G}_4^2$: $(C_{2a}, 0)$, $(\sigma_z, 0)$: 1, 1; 2, 1; 3, 1; 4, 1: *c*.
S^x $\mathbf{G}_8^4$: $(\sigma_z, 0)$, $(C_{2a}, 0)$: 5, 1: *a*.
Y^x $\mathbf{G}_8^4$: $(\sigma_y, 0)$, $(C_{2x}, 0)$: 5, 1: *a*.
T^x $\mathbf{G}_8^2$: $(\sigma_y, 0)$, $(C_{2x}, 0)$: 2, 1; 4, 1; 6, 1; 8, 1: *d*.
W^x $\mathbf{G}_8^4$: $(\sigma_y, 0)$, $(C_{2z}, 0)$: 5, 1: *a*.

127 $P4/mbm$ D_{4h}^5

($F1$; $K4$; $K7$; $M5$; $O2$; $Z1$.)

Γ $\mathbf{G}_{16}^9$: $\{C_{4z}^+|\frac{1}{2}\frac{1}{2}0\}$, $\{C_{2x}|\frac{1}{2}\frac{1}{2}0\}$, $\{I|000\}$: 1, 1; 2, 1; 3, 1; 4, 1; 5, 1; 6, 1; 7, 1; 8, 1; 9, 1; 10, 1: *e*.
M $\mathbf{G}_{32}^5$: $\{C_{2x}|\frac{1}{2}\frac{1}{2}0\}$, $\{C_{2z}|000\}$, $\{\sigma_{db}|000\}$, $\{I|000\}$: 5, 3; 6, 3; 7, 3; 8, 3; 10, 1; 15, 3; 16, 3; 17, 3; 18, 3; 20, 1: *a*.
Z $\mathbf{G}_{16}^9 \otimes \mathbf{T}_2$: $\{C_{4z}^+|\frac{1}{2}\frac{1}{2}0\}$, $\{C_{2x}|\frac{1}{2}\frac{1}{2}0\}$, $\{I|000\}$; $\mathbf{t}_3$: 1, 1; 2, 1; 3, 1; 4, 1; 5, 1; 6, 1; 7, 1; 8, 1; 9, 1; 10, 1: *e*.
A $\mathbf{G}_{32}^5$: $\{C_{2x}|\frac{1}{2}\frac{1}{2}0\}$, $\{C_{2z}|000\}$, $\{\sigma_{db}|000\}$, $\{I|000\}$: 5, 3; 6, 3; 7, 3; 8, 3; 10, 1; 15, 3; 16, 3; 17, 3; 18, 3; 20, 1: *a*.
R $\mathbf{G}_{16}^9$: $\{\sigma_x|\frac{1}{2}\frac{1}{2}0\}$, $\{\sigma_y|\frac{1}{2}\frac{1}{2}0\}$, $\{\sigma_z|000\}$: 5, 1; 10, 1: *b*.
X $\mathbf{G}_{16}^9$: $\{\sigma_x|\frac{1}{2}\frac{1}{2}0\}$, $\{\sigma_y|\frac{1}{2}\frac{1}{2}0\}$, $\{\sigma_z|000\}$: 5, 1; 10, 1: *b*.

Δ^x $\mathbf{G}_4^2$: $(C_{2y}, 0)$, $(\sigma_x, 0)$: 1, 1; 2, 1; 3, 1; 4, 1: *c*.
U^x $\mathbf{G}_4^2$: $(C_{2y}, 0)$, $(\sigma_x, 0)$: 1, 1; 2, 1; 3, 1; 4, 1: *c*.
Λ^x $\mathbf{G}_8^4$: $(C_{4z}^+, 0)$, $(\sigma_y, 0)$: 1, 1; 2, 1; 3, 1; 4, 1; 5, 1: *b*.
V^x $\mathbf{G}_{16}^{10}$: $(\sigma_x, 0)$, $(C_{2z}, 0)$, $(\sigma_{da}, 0)$: 5, 3; 6, 3; 7, 3; 8, 3; 10, 1: *c*.
Σ^x $\mathbf{G}_4^2$: $(C_{2a}, 0)$, $(\sigma_z, 0)$: 1, 1; 2, 1; 3, 1; 4, 1: *c*.
S^x $\mathbf{G}_4^2$: $(C_{2a}, 0)$, $(\sigma_z, 0)$: 1, 1; 2, 1; 3, 1; 4, 1: *c*.
Y^x $\mathbf{G}_8^2$: $(\sigma_y, 0)$, $(\sigma_z, 0)$: 2, 3; 4, 3; 6, 3; 8, 3: *c*.
T^x $\mathbf{G}_8^2$: $(\sigma_y, 0)$, $(\sigma_z, 0)$: 2, 3; 4, 3; 6, 3; 8, 3: *c*.
W^x $\mathbf{G}_8^4$: $(\sigma_y, 0)$, $(C_{2z}, 0)$: 5, 1: *a*.

128 $P4/mnc$ D_{4h}^6

($F1$; $K7$; $M5$; $O2$; $Z1$.)

Γ $\mathbf{G}_{16}^9$: $\{C_{4z}^+|\frac{1}{2}\frac{1}{2}0\}$, $\{C_{2x}|\frac{1}{2}\frac{1}{2}0\}$, $\{I|00\frac{1}{2}\}$: 1, 1; 2, 1; 3, 1; 4, 1; 5, 1; 6, 1; 7, 1; 8, 1; 9, 1; 10, 1: *e*.
M $\mathbf{G}_{32}^5$: $\{C_{2x}|\frac{1}{2}\frac{1}{2}0\}$, $\{C_{2z}|000\}$, $\{\sigma_{db}|00\frac{1}{2}\}$, $\{I|00\frac{1}{2}\}$: 5, 3; 6, 3; 7, 3; 8, 3; 10, 1; 15, 3; 16, 3; 17, 3; 18, 3; 20, 1: *a*.
Z $\mathbf{G}_{32}^1$: $\{I|00\frac{1}{2}\}$, $\{\sigma_{db}|00\frac{1}{2}\}$, $\{C_{4z}^+|\frac{1}{2}\frac{1}{2}0\}$: 10, 1; 11, 1; 12, 3; 13, 3: *a*.
A $\mathbf{G}_{32}^1$: $\{I|00\frac{1}{2}\}$, $\{\sigma_{db}|00\frac{1}{2}\}$, $\{S_{4z}^+|\frac{1}{2}\frac{1}{2}\frac{1}{2}\}$: 10, 1; 11, 1; 12, 3; 13, 3: *a*.
R $\mathbf{G}_{16}^9$: $\{C_{2y}|\frac{1}{2}\frac{1}{2}0\}$, $\{C_{2z}|000\}$, $\{I|00\frac{1}{2}\}$: 5, 1; 10, 1: *c*.
X $\mathbf{G}_{16}^9$: $\{\sigma_x|\frac{1}{2}\frac{1}{2}\frac{1}{2}\}$, $\{\sigma_y|\frac{1}{2}\frac{1}{2}\frac{1}{2}\}$, $\{\sigma_z|00\frac{1}{2}\}$: 5, 1; 10, 1: *b*.

Δ^x $\mathbf{G}_4^2$: $(C_{2y}, 0)$, $(\sigma_x, 0)$: 1, 1; 2, 1; 3, 1; 4, 1: *c*.
U^x $\mathbf{G}_8^4$: $(\sigma_z, 0)$, $(C_{2y}, 0)$: 5, 1: *a*.
Λ^x $\mathbf{G}_8^4$: $(C_{4z}^+, 0)$, $(\sigma_y, 0)$: 1, 1; 2, 1; 3, 1; 4, 1; 5, 1: *b*.
V^x $\mathbf{G}_{16}^{10}$: $(\sigma_x, 0)$, $(C_{2z}, 0)$, $(\sigma_{da}, 0)$: 5, 3; 6, 3; 7, 3; 8, 3; 10, 1: *c*.
Σ^x $\mathbf{G}_4^2$: $(C_{2a}, 0)$, $(\sigma_z, 0)$: 1, 1; 2, 1; 3, 1; 4, 1: *c*.
S^x $\mathbf{G}_8^4$: $(\sigma_z, 0)$, $(C_{2a}, 0)$: 5, 1: *a*.
Y^x $\mathbf{G}_8^2$: $(\sigma_y, 0)$, $(\sigma_z, 0)$: 2, 3; 4, 3; 6, 3; 8, 3: *c*.
T^x $\mathbf{G}_8^5$: $(C_{2x}, 0)$, $(\sigma_y, 0)$: 5, 1: *a*.
W^x $\mathbf{G}_8^4$: $(\sigma_y, 0)$, $(C_{2z}, 0)$: 5, 1: *a*.

129 *P4/nmm* D_{4h}^7

($F1$; $K7$; $M5$; $O2$; $Z1$.)

Γ $\mathbf{G}_{16}^9$: $\{C_{4z}^+|000\}$, $\{C_{2x}|\frac{1}{2}\frac{1}{2}0\}$, $\{I|\frac{1}{2}\frac{1}{2}0\}$: 1, 1; 2, 1; 3, 1; 4, 1; 5, 1; 6, 1; 7, 1; 8, 1; 9, 1; 10, 1: *e*.
M $\mathbf{G}_{32}^2$: $\{C_{4z}^+|000\}$, $\{C_{2x}|\frac{1}{2}\frac{1}{2}0\}$, $\{\sigma_x|000\}$: 9, 1; 11, 1; 12, 1; 14, 1: *b*.
Z $\mathbf{G}_{16}^9 \otimes \mathbf{T}_2$: $\{C_{4z}^+|000\}$, $\{C_{2x}|\frac{1}{2}\frac{1}{2}0\}$, $\{I|\frac{1}{2}\frac{1}{2}0\}$; $\mathbf{t}_3$: 1, 1; 2, 1; 3, 1; 4, 1; 5, 1; 6, 1; 7, 1; 8, 1; 9, 1; 10, 1: *e*.
A $\mathbf{G}_{32}^2$: $\{C_{4z}^+|000\}$, $\{C_{2x}|\frac{1}{2}\frac{1}{2}0\}$, $\{\sigma_x|000\}$: 9, 1; 11, 1; 12, 1; 14, 1: *b*.
R $\mathbf{G}_{16}^9$: $\{C_{2y}|\frac{1}{2}\frac{1}{2}0\}$, $\{C_{2x}|\frac{1}{2}\frac{1}{2}0\}$, $\{\sigma_x|000\}$: 5, 1; 10, 1: *b*.
X $\mathbf{G}_{16}^9$: $\{C_{2y}|\frac{1}{2}\frac{1}{2}0\}$, $\{C_{2x}|\frac{1}{2}\frac{1}{2}0\}$, $\{\sigma_x|000\}$: 5, 1; 10, 1: *b*.

Δ^x $\mathbf{G}_4^2$: $(C_{2y}, 0)$, $(\sigma_x, 0)$: 1, 1; 2, 1; 3, 1; 4, 1: *c*.
U^x $\mathbf{G}_4^2$: $(C_{2y}, 0)$, $(\sigma_x, 0)$: 1, 1; 2, 1; 3, 1; 4, 1: *c*.
Λ^x $\mathbf{G}_8^4$: $(C_{4z}^+, 0)$, $(\sigma_y, 0)$: 1, 1; 2, 1; 3, 1; 4, 1; 5, 1: *b*.
V^x $\mathbf{G}_8^4$: $(C_{4z}^+, 0)$, $(\sigma_y, 0)$: 1, 3; 2, 3; 3, 3; 4, 3; 5, 1: *b*.
Σ^x $\mathbf{G}_4^2$: $(C_{2a}, 0)$, $(\sigma_z, 0)$: 1, 1; 2, 1; 3, 1; 4, 1: *c*.
S^x $\mathbf{G}_4^2$: $(C_{2a}, 0)$, $(\sigma_z, 0)$: 1, 1; 2, 1; 3, 1; 4, 1: *c*.
Y^x $\mathbf{G}_8^4$: $(C_{2x}, 0)$, $(\sigma_y, 0)$: 5, 1: *a*.
T^x $\mathbf{G}_8^4$: $(C_{2x}, 0)$, $(\sigma_y, 0)$: 5, 1: *a*.
W^x $\mathbf{G}_4^2$: $(C_{2z}, 0)$, $(\sigma_y, 0)$: 1, 3; 2, 3; 3, 3; 4, 3: *c*.

130 *P4/ncc* D_{4h}^8

($F1$; $K7$; $M5$; $O2$; $Z1$.)

Γ $\mathbf{G}_{16}^9$: $\{C_{4z}^+|000\}$, $\{C_{2x}|\frac{1}{2}\frac{1}{2}0\}$, $\{I|\frac{1}{2}\frac{1}{2}\frac{1}{2}\}$: 1, 1; 2, 1; 3, 1; 4, 1; 5, 1; 6, 1; 7, 1; 8, 1; 9, 1; 10, 1: *e*.
M $\mathbf{G}_{32}^2$: $\{C_{4z}^+|000\}$, $\{C_{2x}|\frac{1}{2}\frac{1}{2}0\}$, $\{\sigma_x|00\frac{1}{2}\}$: 9, 1; 11, 1; 12, 1; 14, 1: *b*.
Z $\mathbf{G}_{32}^1$: $\{I|\frac{1}{2}\frac{1}{2}\frac{1}{2}\}$, $\{\sigma_{db}|00\frac{1}{2}\}$, $\{C_{4z}^+|000\}$: 10, 1; 11, 1; 12, 3; 13, 3: *a*.
A $\mathbf{G}_{32}^6$: $\{C_{2x}|\frac{1}{2}\frac{1}{2}0\}$, $\{C_{4z}^+|000\}$, $\{I|\frac{1}{2}\frac{1}{2}\frac{1}{2}\}$: 9, 3; 10, 3; 13, 3; 14, 3: *a*.
R $\mathbf{G}_{16}^7$: $\{\sigma_y|00\frac{1}{2}\}$, $\{C_{2x}|\frac{1}{2}\frac{1}{2}0\}$, $\{I|\frac{1}{2}\frac{1}{2}\frac{1}{2}\}$: 9, 3; 10, 3: *d*.
X $\mathbf{G}_{16}^9$: $\{C_{2y}|\frac{1}{2}\frac{1}{2}0\}$, $\{C_{2x}|\frac{1}{2}\frac{1}{2}0\}$, $\{\sigma_x|00\frac{1}{2}\}$: 5, 1; 10, 1: *b*.

Δ^x $\mathbf{G}_4^2$: $(C_{2y}, 0)$, $(\sigma_x, 0)$: 1, 1; 2, 1; 3, 1; 4, 1: *c*.
U^x $\mathbf{G}_8^4$: $(\sigma_z, 0)$, $(C_{2y}, 0)$: 5, 1: *a*.
Λ^x $\mathbf{G}_8^4$: $(C_{4z}^+, 0)$, $(\sigma_y, 0)$: 1, 1; 2, 1; 3, 1; 4, 1; 5, 1: *b*.
V^x $\mathbf{G}_8^4$: $(C_{4z}^+, 0)$, $(\sigma_y, 0)$: 1, 3; 2, 3; 3, 3; 4, 3; 5, 1: *b*.
Σ^x $\mathbf{G}_4^2$: $(C_{2a}, 0)$, $(\sigma_z, 0)$: 1, 1; 2, 1; 3, 1; 4, 1: *c*.
S^x $\mathbf{G}_8^4$: $(\sigma_z, 0)$, $(C_{2a}, 0)$: 5, 1: *a*.
Y^x $\mathbf{G}_8^4$: $(C_{2x}, 0)$, $(\sigma_y, 0)$: 5, 1: *a*.
T^x $\mathbf{G}_8^2$: $(C_{2x}, 0)$, $(\sigma_y, 0)$: 2, 3; 4, 3; 6, 3; 8, 3: *c*.
W^x $\mathbf{G}_4^2$: $(C_{2z}, 0)$, $(\sigma_y, 0)$: 1, 3; 2, 3; 3, 3; 4, 3: *c*.

131 $P4_2/mmc$ D_{4h}^9

($F1$; $K7$; $M5$; $O2$; $S8$‡; $Z1$.)

Γ G_{16}^9: $\{C_{4z}^+|00\frac{1}{2}\}, \{C_{2x}|000\}, \{I|000\}$: 1, 1; 2, 1; 3, 1; 4, 1; 5, 1; 6, 1; 7, 1; 8, 1; 9, 1; 10, 1: e.
M $G_{16}^9 \otimes T_2$: $\{C_{4z}^+|00\frac{1}{2}\}, \{C_{2x}|000\}, \{I|000\}$; t_1 or t_2: 1, 1; 2, 1; 3, 1; 4, 1; 5, 1; 6, 1; 7, 1; 8, 1; 9, 1; 10, 1: e.
Z G_{32}^2: $\{C_{4z}^+|00\frac{1}{2}\}, \{\sigma_{db}|00\frac{1}{2}\}, \{C_{2b}|00\frac{1}{2}\}$: 9, 1; 11, 1; 12, 1; 14, 1: a.
A G_{32}^2: $\{C_{4z}^+|00\frac{1}{2}\}, \{\sigma_{db}|00\frac{1}{2}\}, \{C_{2b}|00\frac{1}{2}\}$: 9, 1; 11, 1; 12, 1; 14, 1: a.
R $G_8^3 \otimes T_2$: $\{C_{2z}|000\}, \{C_{2y}|000\}, \{I|000\}$; t_2 or t_3: 1, 1; 2, 1; 3, 1; 4, 1; 5, 1; 6, 1; 7, 1; 8, 1: b.
X $G_8^3 \otimes T_2$: $\{C_{2z}|000\}, \{C_{2y}|000\}, \{I|000\}$; t_2: 1, 1; 2, 1; 3, 1; 4, 1; 5, 1; 6, 1; 7, 1; 8, 1: b.

Δ^x G_4^2: $(C_{2y}, 0), (\sigma_x, 0)$: 1, 1; 2, 1; 3, 1; 4, 1: c.
U^x G_4^2: $(C_{2y}, 0), (\sigma_x, 0)$: 1, 1; 2, 1; 3, 1; 4, 1: c.
Λ^x G_8^4: $(C_{4z}^+, 0), (\sigma_y, 0)$: 1, 1; 2, 1; 3, 1; 4, 1; 5, 1: b.
V^x G_8^4: $(C_{4z}^+, 0), (\sigma_y, 0)$: 1, 1; 2, 1; 3, 1; 4, 1; 5, 1: b.
Σ^x G_4^2: $(C_{2a}, 0), (\sigma_z, 0)$: 1, 1; 2, 1; 3, 1; 4, 1: c.
S^x G_8^4: $(C_{2a}, 0), (\sigma_z, 0)$: 5, 1: a.
Y^x G_4^2: $(C_{2x}, 0), (\sigma_z, 0)$: 1, 1; 2, 1; 3, 1; 4, 1: c.
T^x G_4^2: $(C_{2x}, 0), (\sigma_z, 0)$: 1, 1; 2, 1; 3, 1; 4, 1: c.
W^x G_4^2: $(C_{2z}, 0), (\sigma_y, 0)$: 1, 1; 2, 1; 3, 1; 4, 1: c.

132 $P4_2/mcm$ D_{4h}^{10}

($F1$; $K7$; $M5$; $O2$; $Z1$.)

Γ G_{16}^9: $\{C_{4z}^+|00\frac{1}{2}\}, \{C_{2x}|000\}, \{I|00\frac{1}{2}\}$: 1, 1; 2, 1; 3, 1; 4, 1; 5, 1; 6, 1; 7, 1; 8, 1; 9, 1; 10, 1: e.
M $G_{16}^9 \otimes T_2$: $\{C_{4z}^+|00\frac{1}{2}\}, \{C_{2x}|000\}, \{I|00\frac{1}{2}\}$; t_1 or t_2: 1, 1; 2, 1; 3, 1; 4, 1; 5, 1; 6, 1; 7, 1; 8, 1; 9, 1; 10, 1: e.
Z G_{32}^2: $\{C_{4z}^+|00\frac{1}{2}\}, \{\sigma_x|00\frac{1}{2}\}, \{C_{2x}|000\}$: 9, 1; 11, 1; 12, 1; 14, 1: a.
A G_{32}^2: $\{C_{4z}^+|00\frac{1}{2}\}, \{\sigma_x|00\frac{1}{2}\}, \{C_{2x}|000\}$: 9, 1; 11, 1; 12, 1; 14, 1: a.
R G_{16}^9: $\{\sigma_x|00\frac{1}{2}\}, \{C_{2y}|000\}, \{C_{2z}|000\}$: 5, 1; 10, 1: a.
X $G_8^3 \otimes T_2$: $\{C_{2z}|000\}, \{C_{2y}|000\}, \{I|00\frac{1}{2}\}$; t_2: 1, 1; 2, 1; 3, 1; 4, 1; 5, 1; 6, 1; 7, 1; 8, 1: b.

Δ^x G_4^2: $(C_{2y}, 0), (\sigma_x, 0)$: 1, 1; 2, 1; 3, 1; 4, 1: c.
U^x G_8^4: $(\sigma_z, 0), (C_{2y}, 0)$: 5, 1: a.
Λ^x G_8^4: $(C_{4z}^+, 0), (\sigma_y, 0)$: 1, 1; 2, 1; 3, 1; 4, 1; 5, 1: b.
V^x G_8^4: $(C_{4z}^+, 0), (\sigma_y, 0)$: 1, 1; 2, 1; 3, 1; 4, 1; 5, 1: b.
Σ^x G_4^2: $(C_{2a}, 0), (\sigma_z, 0)$: 1, 1; 2, 1; 3, 1; 4, 1: c.
S^x G_8^2: $(C_{2a}, 0), (\sigma_{db}, 0)$: 2, 1; 4, 1; 6, 1; 8, 1: c.
Y^x G_4^2: $(C_{2x}, 0), (\sigma_z, 0)$: 1, 1; 2, 1; 3, 1; 4, 1: c.
T^x G_8^4: $(\sigma_z, 0), (C_{2x}, 0)$: 5, 1: a.
W^x G_4^2: $(C_{2z}, 0), (\sigma_y, 0)$: 1, 1; 2, 1; 3, 1; 4, 1: c.

The tables given by Slater (1965b) for $P4_2/mnm$ (D_{4h}^{14}) in fact apply to $P4_2/mmc$ (D_{4h}^9) (see Gay, Albers, and Arlinghaus (1968)).

133 $P4_2/nbc$ D_{4h}^{11}

($F1$; $K7$; $M5$; $O2$; $Z1$.)

Γ $\mathbf{G}_{16}^{9}$: $\{C_{4z}^{+}|\frac{1}{2}\frac{1}{2}\frac{1}{2}\}$, $\{C_{2x}|000\}$, $\{I|\frac{1}{2}\frac{1}{2}0\}$: 1, 1; 2, 1; 3, 1; 4, 1; 5, 1; 6, 1; 7, 1; 8, 1; 9, 1; 10, 1: *e*.
M $\mathbf{G}_{32}^{2}$: $\{C_{4z}^{+}|\frac{1}{2}\frac{1}{2}\frac{1}{2}\}$, $\{\sigma_x|\frac{1}{2}\frac{1}{2}0\}$, $\{C_{2x}|000\}$: 9, 1; 11, 1; 12, 1; 14, 1: *a*.
Z $\mathbf{G}_{32}^{2}$: $\{C_{4z}^{+}|\frac{1}{2}\frac{1}{2}\frac{1}{2}\}$, $\{\sigma_{db}|00\frac{1}{2}\}$, $\{C_{2b}|\frac{1}{2}\frac{1}{2}\frac{1}{2}\}$: 9, 1; 11, 1; 12, 1; 14, 1: *a*.
A $\mathbf{G}_{32}^{1}$: $\{I|\frac{1}{2}\frac{1}{2}0\}$, $\{\sigma_x|\frac{1}{2}\frac{1}{2}0\}$, $\{C_{4z}^{+}|\frac{1}{2}\frac{1}{2}\frac{1}{2}\}$: 10, 1; 11, 1; 12, 3; 13, 3: *a*.
R $\mathbf{G}_{16}^{9}$: $\{\sigma_x|\frac{1}{2}\frac{1}{2}0\}$, $\{\sigma_y|\frac{1}{2}\frac{1}{2}0\}$, $\{C_{2y}|000\}$: 5, 1; 10, 1: *a*
X $\mathbf{G}_{16}^{9}$: $\{\sigma_x|\frac{1}{2}\frac{1}{2}0\}$, $\{\sigma_y|\frac{1}{2}\frac{1}{2}0\}$, $\{C_{2y}|000\}$: 5, 1; 10, 1: *a*.

Δ^x $\mathbf{G}_{4}^{2}$: $(C_{2y}, 0)$, $(\sigma_x, 0)$: 1, 1; 2, 1; 3, 1; 4, 1: *c*.
U^x $\mathbf{G}_{4}^{2}$: $(C_{2y}, 0)$, $(\sigma_x, 0)$: 1, 1; 2, 1; 3, 1; 4, 1: *c*.
Λ^x $\mathbf{G}_{8}^{4}$: $(C_{4z}^{+}, 0)$, $(\sigma_y, 0)$: 1, 1; 2, 1; 3, 1; 4, 1; 5, 1: *b*.
V^x $\mathbf{G}_{16}^{10}$: $(C_{4z}^{+}, 0)$, $(C_{2z}, 0)$, $(\sigma_{da}, 0)$: 5, 1; 6, 1; 7, 1; 8, 1; 9, 1: *b*.
Σ^x $\mathbf{G}_{4}^{2}$: $(C_{2a}, 0)$, $(\sigma_z, 0)$: 1, 1; 2, 1; 3, 1; 4, 1: *c*.
S^x $\mathbf{G}_{8}^{4}$: $(C_{2a}, 0)$, $(\sigma_z, 0)$: 5, 1: *a*.
Y^x $\mathbf{G}_{8}^{4}$: $(\sigma_y, 0)$, $(C_{2x}, 0)$: 5, 1: *a*.
T^x $\mathbf{G}_{8}^{4}$: $(\sigma_y, 0)$, $(C_{2x}, 0)$: 5, 1: *a*.
W^x $\mathbf{G}_{8}^{4}$: $(\sigma_y, 0)$, $(C_{2z}, 0)$: 5, 1: *a*.

134 $P4_2/nnm$ D_{4h}^{12}

($F1$; $K7$; $M5$; $O2$; $Z1$.)

Γ $\mathbf{G}_{16}^{9}$: $\{C_{4z}^{+}|\frac{1}{2}\frac{1}{2}\frac{1}{2}\}$, $\{C_{2x}|000\}$, $\{I|\frac{1}{2}\frac{1}{2}\frac{1}{2}\}$: 1, 1; 2, 1; 3, 1; 4, 1; 5, 1; 6, 1; 7, 1; 8, 1; 9, 1; 10, 1: *e*.
M $\mathbf{G}_{32}^{2}$: $\{C_{4z}^{+}|\frac{1}{2}\frac{1}{2}\frac{1}{2}\}$, $\{\sigma_x|\frac{1}{2}\frac{1}{2}\frac{1}{2}\}$, $\{C_{2x}|000\}$: 9, 1; 11, 1; 12, 1; 14, 1: *a*.
Z $\mathbf{G}_{32}^{2}$: $\{C_{4z}^{+}|\frac{1}{2}\frac{1}{2}\frac{1}{2}\}$, $\{\sigma_x|\frac{1}{2}\frac{1}{2}\frac{1}{2}\}$, $\{C_{2x}|000\}$: 9, 1; 11, 1; 12, 1; 14, 1: *a*.
A $\mathbf{G}_{16}^{9} \otimes \mathbf{T}_2$: $\{C_{4z}^{+}|\frac{1}{2}\frac{1}{2}\frac{1}{2}\}$, $\{C_{2x}|000\}$, $\{I|\frac{1}{2}\frac{1}{2}\frac{1}{2}\}$; $\mathbf{t}_1$ or $\mathbf{t}_2$ or $\mathbf{t}_3$: 1, 1; 2, 1; 3, 1; 4, 1; 5, 1; 6, 1; 7, 1; 8, 1; 9, 1; 10, 1: *e*.
R $\mathbf{G}_{16}^{9}$: $\{\sigma_y|\frac{1}{2}\frac{1}{2}\frac{1}{2}\}$, $\{C_{2y}|000\}$, $\{C_{2x}|000\}$: 5, 1; 10, 1: *a*.
X $\mathbf{G}_{16}^{9}$: $\{\sigma_x|\frac{1}{2}\frac{1}{2}\frac{1}{2}\}$, $\{\sigma_y|\frac{1}{2}\frac{1}{2}\frac{1}{2}\}$, $\{C_{2y}|000\}$: 5, 1; 10, 1: *a*.

Δ^x $\mathbf{G}_{4}^{2}$: $(C_{2y}, 0)$, $(\sigma_x, 0)$: 1, 1; 2, 1; 3, 1; 4, 1: *c*.
U^x $\mathbf{G}_{8}^{4}$: $(\sigma_z, 0)$, $(C_{2y}, 0)$: 5, 1: *a*.
Λ^x $\mathbf{G}_{8}^{4}$: $(C_{4z}^{+}, 0)$, $(\sigma_y, 0)$: 1, 1; 2, 1; 3, 1; 4, 1; 5, 1: *b*.
V^x $\mathbf{G}_{16}^{10}$: $(C_{4z}^{+}, 0)$, $(C_{2z}, 0)$, $(\sigma_{da}, 0)$: 5, 1; 6, 1; 7, 1; 8, 1; 9, 1: *b*.
Σ^x $\mathbf{G}_{4}^{2}$: $(C_{2a}, 0)$, $(\sigma_z, 0)$: 1, 1; 2, 1; 3, 1; 4, 1: *c*.
S^x $\mathbf{G}_{8}^{2}$: $(C_{2a}, 0)$, $(\sigma_{db}, 0)$: 2, 1; 4, 1; 6, 1; 8, 1: *c*.
Y^x $\mathbf{G}_{8}^{4}$: $(\sigma_y, 0)$, $(C_{2x}, 0)$: 5, 1: *a*.
T^x $\mathbf{G}_{8}^{2}$: $(\sigma_y, 0)$, $(C_{2x}, 0)$: 2, 1; 4, 1; 6, 1; 8, 1: *d*.
W^x $\mathbf{G}_{8}^{4}$: $(\sigma_y, 0)$, $(C_{2z}, 0)$: 5, 1: *a*.

135 $P4_2/mbc$ D_{4h}^{13}

($F1$; $K7$; $M5$; $O2$; $Z1$.)

Γ $\mathbf{G}_{16}^{9}$: $\{C_{4z}^{+}|\tfrac{1}{2}\tfrac{1}{2}\tfrac{1}{2}\}$, $\{C_{2x}|\tfrac{1}{2}\tfrac{1}{2}0\}$, $\{I|000\}$: 1, 1; 2, 1; 3, 1; 4, 1; 5, 1; 6, 1; 7, 1; 8, 1; 9, 1; 10, 1: *e*.
M $\mathbf{G}_{32}^{5}$: $\{C_{2x}|\tfrac{1}{2}\tfrac{1}{2}0\}$, $\{C_{2z}|000\}$, $\{\sigma_{db}|00\tfrac{1}{2}\}$, $\{I|000\}$: 5, 3; 6, 3; 7, 3; 8, 3; 10, 1; 15, 3; 16, 3; 17, 3; 18, 3; 20, 1: *a*.
Z $\mathbf{G}_{32}^{2}$: $\{C_{4z}^{+}|\tfrac{1}{2}\tfrac{1}{2}\tfrac{1}{2}\}$, $\{\sigma_{db}|00\tfrac{1}{2}\}$, $\{C_{2b}|00\tfrac{1}{2}\}$: 9, 1; 11, 1; 12, 1; 14, 1: *a*.
A $\mathbf{G}_{32}^{6}$: $\{C_{2x}|\tfrac{1}{2}\tfrac{1}{2}0\}$, $\{C_{4z}^{+}|\tfrac{1}{2}\tfrac{1}{2}\tfrac{1}{2}\}$, $\{I|000\}$: 9, 3; 10, 3; 13, 3; 14, 3: *a*.
R $\mathbf{G}_{16}^{9}$: $\{\sigma_x|\tfrac{1}{2}\tfrac{1}{2}0\}$, $\{\sigma_y|\tfrac{1}{2}\tfrac{1}{2}0\}$, $\{\sigma_z|000\}$: 5, 1; 10, 1: *b*.
X $\mathbf{G}_{16}^{9}$: $\{\sigma_x|\tfrac{1}{2}\tfrac{1}{2}0\}$, $\{\sigma_y|\tfrac{1}{2}\tfrac{1}{2}0\}$, $\{\sigma_z|000\}$: 5, 1; 10, 1: *b*.

Δ^x $\mathbf{G}_{4}^{2}$: $(C_{2y}, 0)$, $(\sigma_x, 0)$: 1, 1; 2, 1; 3, 1; 4, 1: *c*.
U^x $\mathbf{G}_{4}^{2}$: $(C_{2y}, 0)$, $(\sigma_x, 0)$: 1, 1; 2, 1; 3, 1; 4, 1: *c*.
Λ^x $\mathbf{G}_{8}^{4}$: $(C_{4z}^{+}, 0)$, $(\sigma_y, 0)$: 1, 1; 2, 1; 3, 1; 4, 1; 5, 1: *b*.
V^x $\mathbf{G}_{16}^{10}$: $(\sigma_x, 0)$, $(C_{2z}, 0)$, $(\sigma_{da}, 0)$: 5, 3; 6, 3; 7, 3; 8, 3; 10, 1: *c*.
Σ^x $\mathbf{G}_{4}^{2}$: $(C_{2a}, 0)$, $(\sigma_z, 0)$: 1, 1; 2, 1; 3, 1; 4, 1: *c*.
S^x $\mathbf{G}_{8}^{4}$: $(C_{2a}, 0)$, $(\sigma_z, 0)$: 5, 1: *a*.
Y^x $\mathbf{G}_{8}^{2}$: $(C_{2x}, 0)$, $(\sigma_z, 0)$: 2, 3; 4, 3; 6, 3; 8, 3: *c*.
T^x $\mathbf{G}_{8}^{2}$: $(C_{2x}, 0)$, $(\sigma_z, 0)$: 2, 3; 4, 3; 6, 3; 8, 3: *c*.
W^x $\mathbf{G}_{8}^{4}$: $(\sigma_y, 0)$, $(C_{2z}, 0)$: 5, 1: *a*.

136 $P4_2/mnm$ D_{4h}^{14}

($D2$; $F1$; $G3$; $K7$; $M5$; $O1$; $O2$; $S8$‡; $Z1$.)

Γ $\mathbf{G}_{16}^{9}$: $\{C_{4z}^{+}|\tfrac{1}{2}\tfrac{1}{2}\tfrac{1}{2}\}$, $\{C_{2x}|\tfrac{1}{2}\tfrac{1}{2}0\}$, $\{I|00\tfrac{1}{2}\}$: 1, 1; 2, 1; 3, 1; 4, 1; 5, 1; 6, 1; 7, 1; 8, 1; 9, 1; 10, 1: *e*.
M $\mathbf{G}_{32}^{5}$: $\{C_{2x}|\tfrac{1}{2}\tfrac{1}{2}0\}$, $\{C_{2z}|000\}$, $\{\sigma_{db}|000\}$, $\{I|00\tfrac{1}{2}\}$: 5, 3; 6, 3; 7, 3; 8, 3; 10, 1; 15, 3; 16, 3; 17, 3; 18, 3; 20, 1: *a*.
Z $\mathbf{G}_{32}^{2}$: $\{C_{4z}^{+}|\tfrac{1}{2}\tfrac{1}{2}\tfrac{1}{2}\}$, $\{\sigma_x|\tfrac{1}{2}\tfrac{1}{2}\tfrac{1}{2}\}$, $\{C_{2x}|\tfrac{1}{2}\tfrac{1}{2}0\}$: 9, 1; 11, 1; 12, 1; 14, 1: *a*.
A $\mathbf{G}_{32}^{2}$: $\{C_{4z}^{+}|\tfrac{1}{2}\tfrac{1}{2}\tfrac{1}{2}\}$, $\{C_{2x}|\tfrac{1}{2}\tfrac{1}{2}0\}$, $\{\sigma_x|\tfrac{1}{2}\tfrac{1}{2}\tfrac{1}{2}\}$: 9, 1; 11, 1; 12, 1; 14, 1: *b*.
R $\mathbf{G}_{16}^{9}$: $\{C_{2y}|\tfrac{1}{2}\tfrac{1}{2}0\}$, $\{C_{2z}|000\}$, $\{I|00\tfrac{1}{2}\}$: 5, 1; 10, 1: *c*.
X $\mathbf{G}_{16}^{9}$: $\{\sigma_x|\tfrac{1}{2}\tfrac{1}{2}\tfrac{1}{2}\}$, $\{\sigma_y|\tfrac{1}{2}\tfrac{1}{2}\tfrac{1}{2}\}$, $\{\sigma_z|00\tfrac{1}{2}\}$: 5, 1; 10, 1: *b*.

Δ^x $\mathbf{G}_{4}^{2}$: $(C_{2y}, 0)$, $(\sigma_x, 0)$: 1, 1; 2, 1; 3, 1; 4, 1: *c*.
U^x $\mathbf{G}_{8}^{4}$: $(\sigma_z, 0)$, $(C_{2y}, 0)$: 5, 1: *a*.
Λ^x $\mathbf{G}_{8}^{4}$: $(C_{4z}^{+}, 0)$, $(\sigma_y, 0)$: 1, 1; 2, 1; 3, 1; 4, 1; 5, 1: *b*.
V^x $\mathbf{G}_{16}^{10}$: $(\sigma_x, 0)$, $(C_{2z}, 0)$, $(\sigma_{da}, 0)$: 5, 3; 6, 3; 7, 3; 8, 3; 10, 1: *c*.
Σ^x $\mathbf{G}_{4}^{2}$: $(C_{2a}, 0)$, $(\sigma_z, 0)$: 1, 1; 2, 1; 3, 1; 4, 1: *c*.
S^x $\mathbf{G}_{8}^{2}$: $(C_{2a}, 0)$, $(\sigma_{db}, 0)$: 2, 1; 4, 1; 6, 1; 8, 1: *c*.
Y^x $\mathbf{G}_{8}^{2}$: $(C_{2x}, 0)$, $(\sigma_z, 0)$: 2, 3; 4, 3; 6, 3; 8, 3: *c*.
T^x $\mathbf{G}_{8}^{5}$: $(C_{2x}, 0)$, $(\sigma_z, 0)$: 5, 1: *a*.
W^x $\mathbf{G}_{8}^{4}$: $(\sigma_y, 0)$, $(C_{2z}, 0)$: 5, 1: *a*.

‡ The tables given by Slater (1965*b*) for $P4_2/mnm$ (D_{4h}^{14}) do not in fact apply to this space group but to $P4_2/mmc$ (D_{4h}^{9}) (see Gay, Albers, and Arlinghaus (1968)).

137 $P4_2/nmc$ D_{4h}^{15}

($F1$; $K7$; $M5$; $O2$; $Z1$.)

Γ $\mathbf{G}_{16}^{9}$: $\{C_{4z}^{+}|00\frac{1}{2}\}$, $\{C_{2x}|\frac{1}{2}\frac{1}{2}0\}$, $\{I|\frac{1}{2}\frac{1}{2}0\}$: 1, 1; 2, 1; 3, 1; 4, 1; 5, 1; 6, 1; 7, 1; 8, 1; 9, 1; 10, 1: *e*.
M $\mathbf{G}_{32}^{2}$: $\{C_{4z}^{+}|00\frac{1}{2}\}$, $\{C_{2x}|\frac{1}{2}\frac{1}{2}0\}$, $\{\sigma_x|000\}$: 9, 1; 11, 1; 12, 1; 14, 1: *b*.
Z $\mathbf{G}_{32}^{2}$: $\{C_{4z}^{+}|00\frac{1}{2}\}$, $\{\sigma_{db}|00\frac{1}{2}\}$, $\{C_{2b}|\frac{1}{2}\frac{1}{2}\frac{1}{2}\}$: 9, 1; 11, 1; 12, 1; 14, 1: *a*.
A $\mathbf{G}_{32}^{1}$: $\{I|\frac{1}{2}\frac{1}{2}0\}$, $\{\sigma_{db}|00\frac{1}{2}\}$, $\{S_{4z}^{+}|\frac{1}{2}\frac{1}{2}\frac{1}{2}\}$: 10, 1; 11, 1; 12, 3; 13, 3: *a*.
R $\mathbf{G}_{16}^{9}$: $\{C_{2y}|\frac{1}{2}\frac{1}{2}0\}$, $\{C_{2x}|\frac{1}{2}\frac{1}{2}0\}$, $\{\sigma_x|000\}$: 5, 1; 10, 1: *b*.
X $\mathbf{G}_{16}^{9}$: $\{C_{2y}|\frac{1}{2}\frac{1}{2}0\}$, $\{C_{2x}|\frac{1}{2}\frac{1}{2}0\}$, $\{\sigma_x|000\}$: 5, 1; 10, 1: *b*.

Δ^x $\mathbf{G}_{4}^{2}$: $(C_{2y}, 0)$, $(\sigma_x, 0)$: 1, 1; 2, 1; 3, 1; 4, 1: *c*.
U^x $\mathbf{G}_{4}^{2}$: $(C_{2y}, 0)$, $(\sigma_x, 0)$: 1, 1; 2, 1; 3, 1; 4, 1: *c*.
Λ^x $\mathbf{G}_{8}^{4}$: $(C_{4z}^{+}, 0)$, $(\sigma_y, 0)$: 1, 1; 2, 1; 3, 1; 4, 1; 5, 1: *b*.
V^x $\mathbf{G}_{8}^{4}$: $(C_{4z}^{+}, 0)$, $(\sigma_y, 0)$: 1, 3; 2, 3; 3, 3; 4, 3; 5, 1: *b*.
Σ^x $\mathbf{G}_{4}^{2}$: $(C_{2a}, 0)$, $(\sigma_z, 0)$: 1, 1; 2, 1; 3, 1; 4, 1: *c*.
S^x $\mathbf{G}_{8}^{4}$: $(C_{2a}, 0)$, $(\sigma_z, 0)$: 5, 1: *a*.
Y^x $\mathbf{G}_{8}^{4}$: $(C_{2x}, 0)$, $(\sigma_z, 0)$: 5, 1: *a*.
T^x $\mathbf{G}_{8}^{4}$: $(C_{2x}, 0)$, $(\sigma_z, 0)$: 5, 1: *a*.
W^x $\mathbf{G}_{4}^{2}$: $(C_{2z}, 0)$, $(\sigma_y, 0)$: 1, 3; 2, 3; 3, 3; 4, 3: *c*.

138 $P4_2/ncm$ D_{4h}^{16}

($F1$; $K7$; $M5$; $O2$; $Z1$.)

Γ $\mathbf{G}_{16}^{9}$: $\{C_{4z}^{+}|00\frac{1}{2}\}$, $\{C_{2x}|\frac{1}{2}\frac{1}{2}0\}$, $\{I|\frac{1}{2}\frac{1}{2}\frac{1}{2}\}$: 1, 1; 2, 1; 3, 1; 4, 1; 5, 1; 6, 1; 7, 1; 8, 1; 9, 1; 10, 1: *e*.
M $\mathbf{G}_{32}^{2}$: $\{C_{4z}^{+}|00\frac{1}{2}\}$, $\{C_{2x}|\frac{1}{2}\frac{1}{2}0\}$, $\{\sigma_x|00\frac{1}{2}\}$: 9, 1; 11, 1; 12, 1; 14, 1: *b*.
Z $\mathbf{G}_{32}^{2}$: $\{C_{4z}^{+}|00\frac{1}{2}\}$, $\{\sigma_x|00\frac{1}{2}\}$, $\{C_{2x}|\frac{1}{2}\frac{1}{2}0\}$: 9, 1; 11, 1; 12, 1; 14, 1: *a*.
A $\mathbf{G}_{32}^{5}$: $\{C_{2x}|\frac{1}{2}\frac{1}{2}0\}$, $\{C_{2z}|000\}$, $\{\sigma_{db}|000\}$, $\{I|\frac{1}{2}\frac{1}{2}\frac{1}{2}\}$: 5, 3; 6, 3; 7, 3; 8, 3; 10, 1; 15, 3; 16, 3; 17, 3;
18, 3; 20, 1: *a*.
R $\mathbf{G}_{16}^{7}$: $\{\sigma_y|00\frac{1}{2}\}$, $\{C_{2x}|\frac{1}{2}\frac{1}{2}0\}$, $\{I|\frac{1}{2}\frac{1}{2}\frac{1}{2}\}$: 9, 3; 10, 3: *d*.
X $\mathbf{G}_{16}^{9}$: $\{C_{2y}|\frac{1}{2}\frac{1}{2}0\}$, $\{C_{2x}|\frac{1}{2}\frac{1}{2}0\}$, $\{\sigma_x|00\frac{1}{2}\}$: 5, 1; 10, 1: *b*.

Δ^x $\mathbf{G}_{4}^{2}$: $(C_{2y}, 0)$, $(\sigma_x, 0)$: 1, 1; 2, 1; 3, 1; 4, 1: *c*.
U^x $\mathbf{G}_{8}^{4}$: $(\sigma_z, 0)$, $(C_{2y}, 0)$: 5, 1: *a*.
Λ^x $\mathbf{G}_{8}^{4}$: $(C_{4z}^{+}, 0)$, $(\sigma_y, 0)$: 1, 1; 2, 1; 3, 1; 4, 1; 5, 1: *b*.
V^x $\mathbf{G}_{8}^{4}$: $(C_{4z}^{+}, 0)$, $(\sigma_y, 0)$: 1, 3; 2, 3; 3, 3; 4, 3; 5, 1: *b*.
Σ^x $\mathbf{G}_{4}^{2}$: $(C_{2a}, 0)$, $(\sigma_z, 0)$: 1, 1; 2, 1; 3, 1; 4, 1: *c*.
S^x $\mathbf{G}_{8}^{2}$: $(C_{2a}, 0)$, $(\sigma_{db}, 0)$: 2, 1; 4, 1; 6, 1; 8, 1: *c*.
Y^x $\mathbf{G}_{8}^{4}$: $(C_{2x}, 0)$, $(\sigma_z, 0)$: 5, 1: *a*.
T^x $\mathbf{G}_{8}^{2}$: $(C_{2x}, 0)$, $(\sigma_y, 0)$: 2, 3; 4, 3; 6, 3; 8, 3: *c*.
W^x $\mathbf{G}_{4}^{2}$: $(C_{2z}, 0)$, $(\sigma_y, 0)$: 1, 3; 2, 3; 3, 3; 4, 3: *c*.

139 $I4/mmm$ D_{4h}^{17}

($F1$; $K7$; $M5$; $S5$; $V1$; $Z1$.)

Γ G_{16}^{9}: $\{C_{4z}^{+}|000\}$, $\{C_{2x}|000\}$, $\{I|000\}$: 1, 1; 2, 1; 3, 1; 4, 1; 5, 1; 6, 1; 7, 1; 8, 1; 9, 1; 10, 1: *e*.
N $G_{4}^{2}\otimes T_{2}$: $\{C_{2y}|000\}$, $\{I|000\}$; t_2: 1, 1; 2, 1; 3, 1; 4, 1: *a*.
X $G_{8}^{3}\otimes T_{2}$: $\{C_{2z}|000\}$, $\{C_{2a}|000\}$, $\{I|000\}$; t_3: 1, 1; 2, 1; 3, 1; 4, 1; 5, 1; 6, 1; 7, 1; 8, 1: *b*.
Z $G_{16}^{9}\otimes T_{2}$: $\{C_{4z}^{+}|000\}$, $\{C_{2x}|000\}$, $\{I|000\}$; t_1 or t_2 or t_3: 1, 1; 2, 1; 3, 1; 4, 1; 5, 1; 6, 1; 7, 1; 8, 1; 9, 1; 10, 1: *e*.
P $G_{8}^{4}\otimes T_{4}$: $\{S_{4z}^{+}|000\}$, $\{C_{2x}|000\}$; t_1 or t_2 or t_3: 1, 1; 2, 1; 3, 1; 4, 1; 5, 1: *b*.

Λ^x G_{8}^{4}: $(C_{4z}^{+}, 0)$, $(\sigma_y, 0)$: 1, 1; 2, 1; 3, 1; 4, 1; 5, 1: *b*.
V^x G_{8}^{4}: $(C_{4z}^{+}, 0)$, $(\sigma_y, 0)$: 1, 1; 2, 1; 3, 1; 4, 1; 5, 1: *b*.
W^x G_{4}^{2}: $(C_{2z}, 0)$, $(\sigma_{db}, 0)$: 1, 1; 2, 1; 3, 1; 4, 1: *c*.
Σ^x G_{4}^{2}: $(C_{2x}, 0)$, $(\sigma_z, 0)$: 1, 1; 2, 1; 3, 1; 4, 1: *c*.
F^x G_{4}^{2}: $(C_{2x}, 0)$, $(\sigma_z, 0)$: 1, 1; 2, 1; 3, 1; 4, 1: *c*.
Q^x G_{2}^{1}: $(C_{2y}, 0)$: 1, 1; 2, 1: *b*.
Δ^x G_{4}^{2}: $(C_{2a}, 0)$, $(\sigma_z, 0)$: 1, 1; 2, 1; 3, 1; 4, 1: *c*.
U^x G_{4}^{2}: $(C_{2a}, 0)$, $(\sigma_z, 0)$: 1, 1; 2, 1; 3, 1; 4, 1: *c*.
Y^x G_{4}^{2}: $(C_{2b}, 0)$, $(\sigma_{da}, 0)$: 1, 1; 2, 1; 3, 1; 4, 1: *c*.

140 $I4/mcm$ D_{4h}^{18}

($F1$; $G2$; $K7$; $M5$; $S5$; $V1$; $Z1$.)

Γ G_{16}^{9}: $\{C_{4z}^{+}|\frac{1}{2}\frac{1}{2}0\}$, $\{C_{2x}|000\}$, $\{I|\frac{1}{2}\frac{1}{2}0\}$: 1, 1; 2, 1; 3, 1; 4, 1; 5, 1; 6, 1; 7, 1; 8, 1; 9, 1; 10, 1: *e*.
N G_{8}^{4}: $\{\sigma_y|\frac{1}{2}\frac{1}{2}0\}$, $\{C_{2y}|000\}$: 5, 1: *a*.
X $G_{8}^{3}\otimes T_{2}$: $\{C_{2z}|000\}$, $\{C_{2a}|\frac{1}{2}\frac{1}{2}0\}$, $\{I|\frac{1}{2}\frac{1}{2}0\}$; t_3: 1, 1; 2, 1; 3, 1; 4, 1; 5, 1; 6, 1; 7, 1; 8, 1: *b*.
Z $G_{16}^{9}\otimes T_{2}$: $\{C_{4z}^{+}|\frac{1}{2}\frac{1}{2}0\}$, $\{C_{2x}|000\}$, $\{I|\frac{1}{2}\frac{1}{2}0\}$; t_1 or t_2 or t_3: 1, 1; 2, 1; 3, 1; 4, 1; 5, 1; 6, 1; 7, 1; 8, 1; 9, 1; 10, 1: *e*.
P $G_{8}^{4}\otimes T_{4}$: $\{S_{4z}^{+}|000\}$, $\{C_{2x}|000\}$; t_1 or t_2 or t_3: 1, 3; 2, 3; 3, 3; 4, 3; 5, 1: *b*.

Λ^x G_{8}^{4}: $(C_{4z}^{+}, 0)$, $(\sigma_y, 0)$: 1, 1; 2, 1; 3, 1; 4, 1; 5, 1: *b*.
V^x G_{8}^{4}: $(C_{4z}^{+}, 0)$, $(\sigma_y, 0)$: 1, 1; 2, 1; 3, 1; 4, 1; 5, 1: *b*.
W^x G_{4}^{2}: $(C_{2z}, 0)$, $(\sigma_{db}, 0)$: 1, 1; 2, 1; 3, 1; 4, 1: *c*.
Σ^x G_{4}^{2}: $(C_{2x}, 0)$, $(\sigma_z, 0)$: 1, 1; 2, 1; 3, 1; 4, 1: *c*.
F^x G_{4}^{2}: $(C_{2x}, 0)$, $(\sigma_z, 0)$: 1, 1; 2, 1; 3, 1; 4, 1: *c*.
Q^x G_{2}^{1}: $(C_{2y}, 0)$: 1, 3; 2, 3: *b*.
Δ^x G_{4}^{2}: $(C_{2a}, 0)$, $(\sigma_z, 0)$: 1, 1; 2, 1; 3, 1; 4, 1: *c*.
U^x G_{4}^{2}: $(C_{2a}, 0)$, $(\sigma_z, 0)$: 1, 1; 2, 1; 3, 1; 4, 1: *c*.
Y^x G_{4}^{2}: $(C_{2b}, 0)$, $(\sigma_{da}, 0)$: 1, 1; 2, 1; 3, 1; 4, 1: *c*.

141 $I4_1/amd$ D_{4h}^{19}

($C3$; $F1$; $K7$; $M2$; $M4$; $M5$; $O3$; $S5$; $S13$; $V1$; $Z1$.)

Γ $\mathbf{G}_{16}^{9}$: $\{C_{4z}^{+}\mid 0\tfrac{1}{2}0\}$, $\{C_{2x}\mid \tfrac{1}{2}\tfrac{1}{2}0\}$, $\{I\mid \tfrac{1}{2}\tfrac{1}{2}0\}$: 1, 1; 2, 1; 3, 1; 4, 1; 5, 1; 6, 1; 7, 1; 8, 1; 9, 1; 10, 1: *e*.
N $\mathbf{G}_{4}^{2}\otimes\mathbf{T}_2$: $\{C_{2y}\mid 0\tfrac{1}{2}\tfrac{1}{2}\}$, $\{I\mid \tfrac{1}{2}\tfrac{1}{2}0\}$; $\mathbf{t}_2$: 1, 1; 2, 1; 3, 1; 4, 1: *a*.
X $\mathbf{G}_{16}^{9}$: $\{\sigma_z\mid 0\tfrac{1}{2}\tfrac{1}{2}\}$, $\{\sigma_{da}\mid 0\tfrac{1}{2}0\}$, $\{C_{2a}\mid \tfrac{1}{2}00\}$: 5, 1; 10, 1: *a*.
Z $\mathbf{G}_{32}^{2}$: $\{C_{4z}^{+}\mid 0\tfrac{1}{2}0\}$, $\{\sigma_{db}\mid 00\tfrac{1}{2}\}$, $\{C_{2a}\mid \tfrac{1}{2}00\}$: 9, 1; 11, 1; 12, 1; 14, 1: *a*.
P $\mathbf{G}_{32}^{4}$: $\{S_{4z}^{-}\mid \tfrac{1}{2}00\}$, $\{E\mid 100\}$, $\{C_{2y}\mid 0\tfrac{1}{2}\tfrac{1}{2}\}$: 13, 1; 14, 1: *a*.

Λ^x $\mathbf{G}_{8}^{4}$: $(C_{4z}^{+}, 0)$, $(\sigma_y, 0)$: 1, 1; 2, 1; 3, 1; 4, 1; 5, 1: *b*.
V^x $\mathbf{G}_{16}^{9}$: $(C_{4z}^{+}, 0)$, $(\sigma_x, 0)$, $(E, 1)$: 6, 1; 7, 1; 8, 1; 9, 1; 10, 1: *d*.
W^x $\mathbf{G}_{8}^{4}$: $(C_{2z}, 0)$, $(\sigma_{db}, 0)$: 5, 1: *a*.
Σ^x $\mathbf{G}_{4}^{2}$: $(C_{2x}, 0)$, $(\sigma_z, 0)$: 1, 1; 2, 1; 3, 1; 4, 1: *c*.
F^x $\mathbf{G}_{4}^{2}$: $(C_{2x}, 0)$, $(\sigma_z, 0)$: 1, 1; 2, 1; 3, 1; 4, 1: *c*.
Q^x $\mathbf{G}_{4}^{1}$: $(C_{2y}, 0)$: 2, 1; 4, 1: *a*.
Δ^x $\mathbf{G}_{4}^{2}$: $(C_{2a}, 0)$, $(\sigma_z, 0)$: 1, 1; 2, 1; 3, 1; 4, 1: *c*.
U^x $\mathbf{G}_{8}^{4}$: $(C_{2a}, 0)$, $(\sigma_z, 0)$: 5, 1: *a*.
Y^x $\mathbf{G}_{8}^{4}$: $(C_{2b}, 0)$, $(\sigma_{da}, 0)$: 5, 1: *a*.

142 $I4_1/acd$ D_{4h}^{20}

($F1$; $K7$; $M5$; $S5$; $V1$; $Z1$.)

Γ $\mathbf{G}_{16}^{9}$: $\{C_{4z}^{+}\mid \tfrac{1}{2}00\}$, $\{C_{2x}\mid \tfrac{1}{2}\tfrac{1}{2}0\}$, $\{I\mid 000\}$: 1, 1; 2, 1; 3, 1; 4, 1; 5, 1; 6, 1; 7, 1; 8, 1; 9, 1; 10, 1: *e*.
N $\mathbf{G}_{8}^{4}$: $\{\sigma_y\mid 0\tfrac{1}{2}\tfrac{1}{2}\}$, $\{I\mid 000\}$: 5, 1: *a*.
X $\mathbf{G}_{16}^{9}$: $\{\sigma_z\mid \tfrac{1}{2}0\tfrac{1}{2}\}$, $\{C_{2z}\mid \tfrac{1}{2}0\tfrac{1}{2}\}$, $\{C_{2a}\mid 0\tfrac{1}{2}0\}$: 5, 1; 10, 1: *a*.
Z $\mathbf{G}_{32}^{2}$: $\{C_{4z}^{+}\mid \tfrac{1}{2}00\}$, $\{\sigma_{db}\mid 00\tfrac{1}{2}\}$, $\{C_{2b}\mid 00\tfrac{1}{2}\}$: 9, 1; 11, 1; 12, 1; 14, 1: *a*.
P $\mathbf{G}_{32}^{4}$: $\{S_{4z}^{-}\mid \tfrac{1}{2}00\}$, $\{E\mid 100\}$, $\{C_{2y}\mid 0\tfrac{1}{2}\tfrac{1}{2}\}$: 13, 3; 14, 3: *a*.

Λ^x $\mathbf{G}_{8}^{4}$: $(C_{4z}^{+}, 0)$, $(\sigma_y, 0)$: 1, 1; 2, 1; 3, 1; 4, 1; 5, 1: *b*.
V^x $\mathbf{G}_{16}^{9}$: $(C_{4z}^{+}, 0)$, $(\sigma_x, 0)$, $(E, 1)$: 6, 1; 7, 1; 8, 1; 9, 1; 10, 1: *d*.
W^x $\mathbf{G}_{8}^{4}$: $(C_{2z}, 0)$, $(\sigma_{da}, 0)$: 5, 1: *a*.
Σ^x $\mathbf{G}_{4}^{2}$: $(C_{2x}, 0)$, $(\sigma_z, 0)$: 1, 1; 2, 1; 3, 1; 4, 1: *c*.
F^x $\mathbf{G}_{4}^{2}$: $(C_{2x}, 0)$, $(\sigma_z, 0)$: 1, 1; 2, 1; 3, 1; 4, 1: *c*.
Q^x $\mathbf{G}_{4}^{1}$: $(C_{2y}, 0)$: 2, 3; 4, 3: *a*.
Δ^x $\mathbf{G}_{4}^{2}$: $(C_{2a}, 0)$, $(\sigma_z, 0)$: 1, 1; 2, 1; 3, 1; 4, 1: *c*.
U^x $\mathbf{G}_{8}^{4}$: $(C_{2a}, 0)$, $(\sigma_z, 0)$: 5, 1: *a*.
Y^x $\mathbf{G}_{8}^{4}$: $(C_{2b}, 0)$, $(\sigma_{da}, 0)$: 5, 1: *a*.

143 $P3$ C_3^1

($F1$; $K6$; $K7$; $M5$; $Z1$.)

Γ $\mathbf{G}_{3}^{1}$: $\{C_3^{+}\mid 000\}$: 1, 1; 2, 3; 3, 3: *a*.
M $\mathbf{G}_{1}^{1}\otimes\mathbf{T}_2$: $\{E\mid 000\}$; $\mathbf{t}_2$: 1, 1: *a*.
A $\mathbf{G}_{3}^{1}\otimes\mathbf{T}_2$: $\{C_3^{+}\mid 000\}$; $\mathbf{t}_3$: 1, 1; 2, 3; 3, 3: *a*.
L $\mathbf{G}_{1}^{1}\otimes\mathbf{T}_2$: $\{E\mid 000\}$; $\mathbf{t}_2$ or $\mathbf{t}_3$: 1, 1: *a*.
K $\mathbf{G}_{3}^{1}\otimes\mathbf{T}_3$: $\{C_3^{+}\mid 000\}$; $\mathbf{t}_1$ or $\mathbf{t}_2$: 1, *x*; 2, *x*; 3, *x*: *a*.
H $\mathbf{G}_{3}^{1}\otimes\mathbf{T}_3\otimes\mathbf{T}_2$: $\{C_3^{+}\mid 000\}$; $\mathbf{t}_1$ or $\mathbf{t}_2$; $\mathbf{t}_3$: 1, *x*; 2, *x*; 3, *x*: *a*.

Δ^x $\mathbf{G}_{3}^{1}$: $(C_3^{+}, 0)$: 1, *x*; 2, *x*; 3, *x*: *a*.
U^x $\mathbf{G}_{1}^{1}$: $(E, 0)$: 1, *x*: *a*.
P^x $\mathbf{G}_{3}^{1}$: $(C_3^{+}, 0)$: 1, *x*; 2, *x*; 3, *x*: *a*.
T^x $\mathbf{G}_{1}^{1}$: $(E, 0)$: 1, *x*: *a*.
S^x $\mathbf{G}_{1}^{1}$: $(E, 0)$: 1, *x*: *a*.
T'^x $\mathbf{G}_{1}^{1}$: $(E, 0)$: 1, *x*: *a*.
S'^x $\mathbf{G}_{1}^{1}$: $(E, 0)$: 1, *x*: *a*.
Σ^x $\mathbf{G}_{1}^{1}$: $(E, 0)$: 1, *x*: *a*.
R^x $\mathbf{G}_{1}^{1}$: $(E, 0)$: 1, *x*: *a*.

144 $P3_1$ C_3^2

($F1$; $K6$; $K7$; $M5$; $Z1$.)

Γ $\mathbf{G}_3^1$: $\{C_3^+ \mid 00\frac{1}{3}\}$: 1, 1; 2, 3; 3, 3: a.
M $\mathbf{G}_1^1 \otimes \mathbf{T}_2$: $\{E \mid 000\}$; $\mathbf{t}_2$: 1, 1: a.
A $\mathbf{G}_3^1 \otimes \mathbf{T}_2$: $\{C_3^- \mid 00\frac{2}{3}\}$; $\mathbf{t}_3$: 1, 1; 2, 3; 3, 3: a.
L $\mathbf{G}_1^1 \otimes \mathbf{T}_2$: $\{E \mid 000\}$; $\mathbf{t}_2$ or $\mathbf{t}_3$: 1, 1: a.
K $\mathbf{G}_3^1 \otimes \mathbf{T}_3$: $\{C_3^+ \mid 00\frac{1}{3}\}$; $\mathbf{t}_1$ or $\mathbf{t}_2$: 1, x; 2, x; 3, x: a.
H $\mathbf{G}_3^1 \otimes \mathbf{T}_3 \otimes \mathbf{T}_2$: $\{C_3^- \mid 00\frac{2}{3}\}$; $\mathbf{t}_1$ or $\mathbf{t}_2$; $\mathbf{t}_3$: 1, x; 2, x; 3, x: a.

Δ^x $\mathbf{G}_3^1$: $(C_3^+, 0)$: 1, x; 2, x; 3, x: a.
U^x $\mathbf{G}_1^1$: $(E, 0)$: 1, x: a.
P^x $\mathbf{G}_3^1$: $(C_3^+, 0)$: 1, x; 2, x; 3, x: a.
T^x $\mathbf{G}_1^1$: $(E, 0)$: 1, x: a.
S^x $\mathbf{G}_1^1$: $(E, 0)$: 1, x: a.
T'^x $\mathbf{G}_1^1$: $(E, 0)$: 1, x: a.
S'^x $\mathbf{G}_1^1$: $(E, 0)$: 1, x: a.
Σ^x $\mathbf{G}_1^1$: $(E, 0)$: 1, x: a.
R^x $\mathbf{G}_1^1$: $(E, 0)$: 1, x: a.

145 $P3_2$ C_3^3

($F1$; $K6$; $K7$; $M5$; $Z1$.)

Γ $\mathbf{G}_3^1$: $\{C_3^+ \mid 00\frac{2}{3}\}$: 1, 1; 2, 3; 3, 3: a.
M $\mathbf{G}_1^1 \otimes \mathbf{T}_2$: $\{E \mid 000\}$; $\mathbf{t}_2$: 1, 1: a.
A $\mathbf{G}_3^1 \otimes \mathbf{T}_2$: $\{C_3^+ \mid 00\frac{2}{3}\}$; $\mathbf{t}_3$: 1, 1; 2, 3; 3, 3: a.
L $\mathbf{G}_1^1 \otimes \mathbf{T}_2$: $\{E \mid 000\}$; $\mathbf{t}_2$ or $\mathbf{t}_3$: 1, 1: a.
K $\mathbf{G}_3^1 \otimes \mathbf{T}_3$: $\{C_3^+ \mid 00\frac{2}{3}\}$; $\mathbf{t}_1$ or $\mathbf{t}_2$: 1, x; 2, x; 3, x: a.
H $\mathbf{G}_3^1 \otimes \mathbf{T}_3 \otimes \mathbf{T}_2$: $\{C_3^+ \mid 00\frac{2}{3}\}$; $\mathbf{t}_1$ or $\mathbf{t}_2$; $\mathbf{t}_3$: 1, x; 2, x; 3, x: a.

Δ^x $\mathbf{G}_3^1$: $(C_3^+, 0)$: 1, x; 2, x; 3, x: a.
U^x $\mathbf{G}_1^1$: $(E, 0)$: 1, x: a.
P^x $\mathbf{G}_3^1$: $(C_3^+, 0)$: 1, x; 2, x; 3, x: a.
T^x $\mathbf{G}_1^1$: $(E, 0)$: 1, x: a.
S^x $\mathbf{G}_1^1$: $(E, 0)$: 1, x: a.
T'^x $\mathbf{G}_1^1$: $(E, 0)$: 1, x: a.
S'^x $\mathbf{G}_1^1$: $(E, 0)$: 1, x: a.
Σ^x $\mathbf{G}_1^1$: $(E, 0)$: 1, x: a.
R^x $\mathbf{G}_1^1$: $(E, 0)$: 1, x: a.

146 $R3$ C_3^4

($F1$; $K6$; $K7$; $M5$; $Z1$.)

Γ $\mathbf{G}_3^1$: $\{C_3^+ \mid 000\}$: 1, 1; 2, 3; 3, 3: a.
Z $\mathbf{G}_3^1 \otimes \mathbf{T}_2$: $\{C_3^+ \mid 000\}$; $\mathbf{t}_1$ or $\mathbf{t}_2$ or $\mathbf{t}_3$: 1, 1; 2, 3; 3, 3: a.
L $\mathbf{G}_1^1 \otimes \mathbf{T}_2$: $\{E \mid 000\}$; $\mathbf{t}_2$: 1, 1: a.
{(a) F $\mathbf{G}_1^1 \otimes \mathbf{T}_2$: $\{E \mid 000\}$; $\mathbf{t}_2$ or $\mathbf{t}_3$: 1, 1: a.}
{(b) F $\mathbf{G}_1^1 \otimes \mathbf{T}_2$: $\{E \mid 000\}$; $\mathbf{t}_1$ or $\mathbf{t}_2$: 1, 1: a.}

Λ^x $\mathbf{G}_3^1$: $(C_3^+, 0)$: 1, x; 2, x; 3, x: a.
P^x $\mathbf{G}_3^1$: $(C_3^+, 0)$: 1, x; 2, x; 3, x: a.
B^x $\mathbf{G}_1^1$: $(E, 0)$: 1, x: a.
Σ^x $\mathbf{G}_1^1$: $(E, 0)$: 1, x: a.
Q^x $\mathbf{G}_1^1$: $(E, 0)$: 1, x: a.
Y^x $\mathbf{G}_1^1$: $(E, 0)$: 1, x: a.

147 $P\bar{3}$ C_{3i}^1

($F1$; $K6$; $K7$; $M5$; $Z1$.)

Γ G_6^1: $\{S_6^+ \mid 000\}$: 1, 1; 2, 3; 3, 3; 4, 1; 5, 3; 6, 3: a.
M $G_2^1 \otimes T_2$: $\{I \mid 000\}$; t_2: 1, 1; 2, 1: a.
A $G_6^1 \otimes T_2$: $\{S_6^+ \mid 000\}$; t_3: 1, 1; 2, 3; 3, 3; 4, 1; 5, 3; 6, 3: a.
L $G_2^1 \otimes T_2$: $\{I \mid 000\}$; t_2 or t_3: 1, 1; 2, 1: a.
K $G_3^1 \otimes T_3$: $\{C_3^+ \mid 000\}$; t_1 or t_2: 1, 1; 2, 3; 3, 3: a.
H $G_3^1 \otimes T_3 \otimes T_2$: $\{C_3^+ \mid 000\}$; t_1 or t_2; t_3: 1, 1; 2, 3; 3, 3: a.

Δ^x G_3^1: $(C_3^+, 0)$: 1, 1; 2, 3; 3, 3: a.
U^x G_1^1: $(E, 0)$: 1, 1: a.
P^x G_3^1: $(C_3^+, 0)$: 1, 1; 2, 3; 3, 3: a.
T^x G_1^1: $(E, 0)$: 1, 1: a.
S^x G_1^1: $(E, 0)$: 1, 1: a.
T'^x G_1^1: $(E, 0)$: 1, 1: a.
S'^x G_1^1: $(E, 0)$: 1, 1: a.
Σ^x G_1^1: $(E, 0)$: 1, 1: a.
R^x G_1^1: $(E, 0)$: 1, 1: a.

148 $R\bar{3}$ C_{3i}^2

($F1$; $K6$; $K7$; $M5$; $S8$; $Z1$.)

Γ G_6^1: $\{S_6^+ \mid 000\}$: 1, 1; 2, 3; 3, 3; 4, 1; 5, 3; 6, 3: a.
Z $G_6^1 \otimes T_2$: $\{S_6^+ \mid 000\}$; t_1 or t_2 or t_3: 1, 1; 2, 3; 3, 3; 4, 1; 5, 3; 6, 3: a.
L $G_2^1 \otimes T_2$: $\{I \mid 000\}$; t_2: 1, 1; 2, 1: a.
{(a) F $G_2^1 \otimes T_2$: $\{I \mid 000\}$; t_2 or t_3: 1, 1; 2, 1: a.}
{(b) F $G_2^1 \otimes T_2$: $\{I \mid 000\}$; t_1 or t_2: 1, 1; 2, 1: a.}

Λ^x G_3^1: $(C_3^+, 0)$: 1, 1; 2, 3; 3, 3: a.
P^x G_3^1: $(C_3^+, 0)$: 1, 1; 2, 3; 3, 3: a.
B^x G_1^1: $(E, 0)$: 1, 1: a.
Σ^x G_1^1: $(E, 0)$: 1, 1: a.
Q^x G_1^1: $(E, 0)$: 1, 1: a.
Y^x G_1^1: $(E, 0)$: 1, 1: a.

149 $P312$ D_3^1

($F1$; $K6$; $K7$; $M5$; $Z1$.)

Γ G_6^2: $\{C_3^+ \mid 000\}$, $\{C'_{21} \mid 000\}$: 1, 1; 2, 1; 3, 1: a.
M $G_2^1 \otimes T_2$: $\{C'_{21} \mid 000\}$; t_2: 1, 1; 2, 1: b.
A $G_6^2 \otimes T_2$: $\{C_3^+ \mid 000\}$, $\{C'_{21} \mid 000\}$; t_3: 1, 1; 2, 1; 3, 1: a.
L $G_2^1 \otimes T_2$: $\{C'_{21} \mid 000\}$; t_2 or t_3: 1, 1; 2, 1: b.
K $G_3^1 \otimes T_3$: $\{C_3^+ \mid 000\}$; t_1 or t_2: 1, 1; 2, 1; 3, 1: a.
H $G_3^1 \otimes T_3 \otimes T_2$: $\{C_3^+ \mid 000\}$; t_1 or t_2; t_3: 1, 1; 2, 1; 3, 1: a.

Δ^x G_3^1: $(C_3^+, 0)$: 1, 1; 2, 1; 3, 1: a.
U^x G_1^1: $(E, 0)$: 1, 1: a.
P^x G_3^1: $(C_3^+, 0)$: 1, 1; 2, 1; 3, 1: a.
T^x G_1^1: $(E, 0)$: 1, 1: a.
S^x G_1^1: $(E, 0)$: 1, 1: a.
T'^x G_1^1: $(E, 0)$: 1, 1: a.
S'^x G_1^1: $(E, 0)$: 1, 1: a.
Σ^x G_2^1: $(C'_{21}, 0)$: 1, x; 2, x: b.
R^x G_2^1: $(C'_{21}, 0)$: 1, x; 2, x: b.

150 $P321$ D_3^2

($F1$; $K6$; $K7$; $M5$; $Z1$.)

Γ $\mathbf{G}_6^2$: $\{C_3^+ \mid 000\}$, $\{C_{21}'' \mid 000\}$: 1, 1; 2, 1; 3, 1: a.
M $\mathbf{G}_2^1 \otimes \mathbf{T}_2$: $\{C_{21}'' \mid 000\}$; $\mathbf{t}_2$: 1, 1; 2, 1: b.
A $\mathbf{G}_6^2 \otimes \mathbf{T}_2$: $\{C_3^+ \mid 000\}$, $\{C_{21}'' \mid 000\}$; $\mathbf{t}_3$: 1, 1; 2, 1; 3, 1: a.
L $\mathbf{G}_2^1 \otimes \mathbf{T}_2$: $\{C_{21}'' \mid 000\}$; $\mathbf{t}_2$ or $\mathbf{t}_3$: 1, 1; 2, 1: b.
K $\mathbf{G}_6^2 \otimes \mathbf{T}_3$: $\{C_3^+ \mid 000\}$, $\{C_{21}'' \mid 000\}$; $\mathbf{t}_1$ or $\mathbf{t}_2$: 1, x; 2, x; 3, x: a.
H $\mathbf{G}_6^2 \otimes \mathbf{T}_3 \otimes \mathbf{T}_2$: $\{C_3^+ \mid 000\}$, $\{C_{21}'' \mid 000\}$; $\mathbf{t}_1$ or $\mathbf{t}_2$; $\mathbf{t}_3$: 1, x; 2, x; 3, x: a.

Δ^x $\mathbf{G}_3^1$: $(C_3^+, 0)$: 1, 1; 2, 1; 3, 1: a.
U^x $\mathbf{G}_1^1$: $(E, 0)$: 1, 1: a.
P^x $\mathbf{G}_3^1$: $(C_3^+, 0)$: 1, x; 2, x; 3, x: a.
T^x $\mathbf{G}_2^1$: $(C_{22}'', 0)$: 1, x; 2, x: b.
S^x $\mathbf{G}_2^1$: $(C_{22}'', 0)$: 1, x; 2, x: b.
T'^x $\mathbf{G}_2^1$: $(C_{21}'', 0)$: 1, x; 2, x: b.
S'^x $\mathbf{G}_2^1$: $(C_{21}'', 0)$: 1, x; 2, x: b.
Σ^x $\mathbf{G}_1^1$: $(E, 0)$: 1, 1: a.
R^x $\mathbf{G}_1^1$: $(E, 0)$: 1, 1: a.

151 $P3_112$ D_3^3

($A5$; $F1$; $F4$; $F5$; $K6$; $K7$; $M5$; $Z1$.)

Γ $\mathbf{G}_6^2$: $\{C_3^+ \mid 00\frac{1}{3}\}$, $\{C_{21}' \mid 00\frac{2}{3}\}$: 1, 1; 2, 1; 3, 1: a.
M $\mathbf{G}_2^1 \otimes \mathbf{T}_2$: $\{C_{21}' \mid 00\frac{2}{3}\}$; $\mathbf{t}_2$: 1, 1; 2, 1: b.
A $\mathbf{G}_6^2 \otimes \mathbf{T}_2$: $\{C_3^- \mid 00\frac{2}{3}\}$, $\{C_{21}' \mid 00\frac{2}{3}\}$; $\mathbf{t}_3$: 1, 1; 2, 1; 3, 1: a.
L $\mathbf{G}_2^1 \otimes \mathbf{T}_2$: $\{C_{21}' \mid 00\frac{2}{3}\}$; $\mathbf{t}_2$ or $\mathbf{t}_3$: 1, 1; 2, 1: b.
K $\mathbf{G}_3^1 \otimes \mathbf{T}_3$: $\{C_3^+ \mid 00\frac{1}{3}\}$; $\mathbf{t}_1$ or $\mathbf{t}_2$: 1, 1; 2, 1; 3, 1: a.
H $\mathbf{G}_3^1 \otimes \mathbf{T}_3 \otimes \mathbf{T}_2$: $\{C_3^- \mid 00\frac{2}{3}\}$; $\mathbf{t}_1$ or $\mathbf{t}_2$; $\mathbf{t}_3$: 1, 1; 2, 1; 3, 1: a.

Δ^x $\mathbf{G}_3^1$: $(C_3^+, 0)$: 1, 1; 2, 1; 3, 1: a.
U^x $\mathbf{G}_1^1$: $(E, 0)$: 1, 1: a.
P^x $\mathbf{G}_3^1$: $(C_3^+, 0)$: 1, 1; 2, 1; 3, 1: a.
T^x $\mathbf{G}_1^1$: $(E, 0)$: 1, 1: a.
S^x $\mathbf{G}_1^1$: $(E, 0)$: 1, 1: a.
T'^x $\mathbf{G}_1^1$: $(E, 0)$: 1, 1: a.
S'^x $\mathbf{G}_1^1$: $(E, 0)$: 1, 1: a.
Σ^x $\mathbf{G}_2^1$: $(C_{21}', 0)$: 1, x; 2, x: b.
R^x $\mathbf{G}_6^1$: $(C_{21}', 0)$: 3, x; 6, x: b.

152 $P3_121$ D_3^4

($A6$; $F1$; $F5$; $F7$; $F8$; $K6$; $K7$; $M5$; $N2$; $N3$; $P2$; $R1$; $R4$; $S8$; $T6$; $Z1$.)

Γ $\mathbf{G}_6^2$: $\{C_3^+ \mid 00\frac{1}{3}\}$, $\{C_{21}'' \mid 00\frac{2}{3}\}$: 1, 1; 2, 1; 3, 1: a.
M $\mathbf{G}_2^1 \otimes \mathbf{T}_2$: $\{C_{21}'' \mid 00\frac{2}{3}\}$; $\mathbf{t}_2$: 1, 1; 2, 1: b.
A $\mathbf{G}_6^2 \otimes \mathbf{T}_2$: $\{C_3^- \mid 00\frac{2}{3}\}$, $\{C_{21}'' \mid 00\frac{2}{3}\}$; $\mathbf{t}_3$: 1, 1; 2, 1; 3, 1: a.
L $\mathbf{G}_2^1 \otimes \mathbf{T}_2$: $\{C_{21}'' \mid 00\frac{2}{3}\}$; $\mathbf{t}_2$ or $\mathbf{t}_3$: 1, 1; 2, 1: b.
K $\mathbf{G}_6^2 \otimes \mathbf{T}_3$: $\{C_3^+ \mid 00\frac{1}{3}\}$, $\{C_{21}'' \mid 00\frac{2}{3}\}$; $\mathbf{t}_1$ or $\mathbf{t}_2$: 1, x; 2, x; 3, x: a.
H $\mathbf{G}_6^2 \otimes \mathbf{T}_3 \otimes \mathbf{T}_2$: $\{C_3^- \mid 00\frac{2}{3}\}$, $\{C_{21}'' \mid 00\frac{2}{3}\}$; $\mathbf{t}_1$ or $\mathbf{t}_2$; $\mathbf{t}_3$: 1, x; 2, x; 3, x: a.

Δ^x $\mathbf{G}_3^1$: $(C_3^+, 0)$: 1, 1; 2, 1; 3, 1: a.
U^x $\mathbf{G}_1^1$: $(E, 0)$: 1, 1: a.
P^x $\mathbf{G}_3^1$: $(C_3^+, 0)$: 1, x; 2, x; 3, x: a.
T^x $\mathbf{G}_2^1$: $(C_{22}'', 0)$: 1, x; 2, x: b.
S^x $\mathbf{G}_6^1$: $(C_{22}'', 0)$: 2, x; 5, x: c.
T'^x $\mathbf{G}_2^1$: $(C_{21}'', 0)$: 1, x; 2, x: b.
S'^x $\mathbf{G}_6^1$: $(C_{21}'', 0)$: 3, x; 6, x: b.
Σ^x $\mathbf{G}_1^1$: $(E, 0)$: 1, 1: a.
R^x $\mathbf{G}_1^1$: $(E, 0)$: 1, 1: a.

153 $P3_212$ D_3^5

($F1$; $K6$; $K7$; $M5$; $Z1$.)

Γ $\mathbf{G}_6^2$: $\{C_3^+ \mid 00\frac{2}{3}\}$, $\{C'_{21} \mid 00\frac{1}{3}\}$: 1, 1; 2, 1; 3, 1: a.
M $\mathbf{G}_2^1 \otimes \mathbf{T}_2$: $\{C'_{21} \mid 00\frac{1}{3}\}$; $\mathbf{t}_2$: 1, 1; 2, 1: b.
A $\mathbf{G}_6^2 \otimes \mathbf{T}_2$: $\{C_3^+ \mid 00\frac{2}{3}\}$, $\{C'_{21} \mid 00\frac{1}{3}\}$; $\mathbf{t}_3$: 1, 1; 2, 1; 3, 1: a.
L $\mathbf{G}_2^1 \otimes \mathbf{T}_2$: $\{C'_{21} \mid 00\frac{1}{3}\}$; $\mathbf{t}_2$ or $\mathbf{t}_3$: 1, 1; 2, 1: b.
K $\mathbf{G}_3^1 \otimes \mathbf{T}_3$: $\{C_3^+ \mid 00\frac{2}{3}\}$; $\mathbf{t}_1$ or $\mathbf{t}_2$: 1, 1; 2, 1; 3, 1: a.
H $\mathbf{G}_3^1 \otimes \mathbf{T}_3 \otimes \mathbf{T}_2$: $\{C_3^+ \mid 00\frac{2}{3}\}$; $\mathbf{t}_1$ or $\mathbf{t}_2$; $\mathbf{t}_3$: 1, 1; 2, 1; 3, 1: a.

Δ^x $\mathbf{G}_3^1$: $(C_3^+, 0)$: 1, 1; 2, 1; 3, 1: a.
U^x $\mathbf{G}_1^1$: $(E, 0)$: 1, 1: a.
P^x $\mathbf{G}_3^1$: $(C_3^+, 0)$: 1, 1; 2, 1; 3, 1: a.
T^x $\mathbf{G}_1^1$: $(E, 0)$: 1, 1: a.
S^x $\mathbf{G}_1^1$: $(E, 0)$: 1, 1: a.
T'^x $\mathbf{G}_1^1$: $(E, 0)$: 1, 1: a.
S'^x $\mathbf{G}_1^1$: $(E, 0)$: 1, 1: a.
Σ^x $\mathbf{G}_2^1$: $(C'_{21}, 0)$: 1, x; 2, x: b.
R^x $\mathbf{G}_6^1$: $(C'_{21}, 0)$: 2, x; 5, x: c.

154 $P3_221$ D_3^6

($A6$; $F1$; $F5$; $F7$; $F8$; $K6$; $K7$; $M5$; $N2$; $N3$; $P2$; $R1$; $Z1$.)

Γ $\mathbf{G}_6^2$: $\{C_3^+ \mid 00\frac{2}{3}\}$, $\{C''_{21} \mid 00\frac{1}{3}\}$: 1, 1; 2, 1; 3, 1: a.
M $\mathbf{G}_2^1 \otimes \mathbf{T}_2$: $\{C''_{21} \mid 00\frac{1}{3}\}$; $\mathbf{t}_2$: 1, 1; 2, 1: b.
A $\mathbf{G}_6^2 \otimes \mathbf{T}_2$: $\{C_3^+ \mid 00\frac{2}{3}\}$, $\{C''_{21} \mid 00\frac{1}{3}\}$; $\mathbf{t}_3$: 1, 1; 2, 1; 3, 1: a.
L $\mathbf{G}_2^1 \otimes \mathbf{T}_2$: $\{C''_{21} \mid 00\frac{1}{3}\}$; $\mathbf{t}_2$ or $\mathbf{t}_3$: 1, 1; 2, 1: b.
K $\mathbf{G}_6^2 \otimes \mathbf{T}_3$: $\{C_3^+ \mid 00\frac{2}{3}\}$, $\{C''_{21} \mid 00\frac{1}{3}\}$; $\mathbf{t}_1$ or $\mathbf{t}_2$: 1, x; 2, x; 3, x: a.
H $\mathbf{G}_6^2 \otimes \mathbf{T}_3 \otimes \mathbf{T}_2$: $\{C_3^+ \mid 00\frac{2}{3}\}$, $\{C''_{21} \mid 00\frac{1}{3}\}$; $\mathbf{t}_1$ or $\mathbf{t}_2$; $\mathbf{t}_3$: 1, x; 2, x; 3, x: a.

Δ^x $\mathbf{G}_3^1$: $(C_3^+, 0)$: 1, 1; 2, 1; 3, 1: a.
U^x $\mathbf{G}_1^1$: $(E, 0)$: 1, 1: a.
P^x $\mathbf{G}_3^1$: $(C_3^+, 0)$: 1, x; 2, x; 3, x: a.
T^x $\mathbf{G}_2^1$: $(C''_{22}, 0)$: 1, x; 2, x: b.
S^x $\mathbf{G}_6^1$: $(C''_{22}, 0)$: 3, x; 6, x: b.
T'^x $\mathbf{G}_2^1$: $(C''_{21}, 0)$: 1, x; 2, x: b.
S'^x $\mathbf{G}_6^1$: $(C''_{21}, 0)$: 2, x; 5, x: c.
Σ^x $\mathbf{G}_1^1$: $(E, 0)$: 1, 1: a.
R^x $\mathbf{G}_1^1$: $(E, 0)$: 1, 1: a.

155 $R32$ D_3^7

($F1$; $K6$; $K7$; $M5$; $S8$; $Z1$.)

Γ $\mathbf{G}_6^2$: $\{C_3^+ \mid 000\}$, $\{C'_{21} \mid 000\}$: 1, 1; 2, 1; 3, 1: a.
Z $\mathbf{G}_6^2 \otimes \mathbf{T}_2$: $\{C_3^+ \mid 000\}$, $\{C'_{21} \mid 000\}$; $\mathbf{t}_1$ or $\mathbf{t}_2$ or $\mathbf{t}_3$: 1, 1; 2, 1; 3, 1: a.
L $\mathbf{G}_2^1 \otimes \mathbf{T}_2$: $\{C'_{22} \mid 000\}$; $\mathbf{t}_2$: 1, 1; 2, 1: b.
{(a) F $\mathbf{G}_2^1 \otimes \mathbf{T}_2$: $\{C'_{21} \mid 000\}$; $\mathbf{t}_2$ or $\mathbf{t}_3$: 1, 1; 2, 1: b.}
{(b) F $\mathbf{G}_2^1 \otimes \mathbf{T}_2$: $\{C'_{23} \mid 000\}$; $\mathbf{t}_1$ or $\mathbf{t}_2$: 1, 1; 2, 1: b.}

Λ^x $\mathbf{G}_3^1$: $(C_3^+, 0)$: 1, 1; 2, 1; 3, 1: a.
P^x $\mathbf{G}_3^1$: $(C_3^+, 0)$: 1, 1; 2, 1; 3, 1: a.
B^x $\mathbf{G}_2^1$: $(C'_{21}, 0)$: 1, x; 2, x: b.
Σ^x $\mathbf{G}_2^1$: $(C'_{21}, 0)$: 1, x; 2, x: b.
Q^x $\mathbf{G}_2^1$: $(C'_{23}, 0)$: 1, x; 2, x: b.
Y^x $\mathbf{G}_2^1$: $(C'_{22}, 0)$: 1, x; 2, x: b.

156 *P3m*1 C_{3v}^1

($F1$; $K6$; $K7$; $M5$; $Z1$.)

Γ $\mathbf{G}_6^2$: $\{C_3^+|000\}$, $\{\sigma_{v1}|000\}$: 1, 1; 2, 1; 3, 1: *a*.
M $\mathbf{G}_2^1 \otimes \mathbf{T}_2$: $\{\sigma_{v1}|000\}$; $\mathbf{t}_2$: 1, 1; 2, 1: *c*.
A $\mathbf{G}_6^2 \otimes \mathbf{T}_2$: $\{C_3^+|000\}$, $\{\sigma_{v1}|000\}$; $\mathbf{t}_3$: 1, 1; 2, 1; 3, 1: *a*.
L $\mathbf{G}_2^1 \otimes \mathbf{T}_2$: $\{\sigma_{v1}|000\}$; $\mathbf{t}_2$ or $\mathbf{t}_3$: 1, 1; 2, 1: *c*.
K $\mathbf{G}_3^1 \otimes \mathbf{T}_3$: $\{C_3^+|000\}$; $\mathbf{t}_1$ or $\mathbf{t}_2$: 1, 1; 2, 1; 3, 1: *a*.
H $\mathbf{G}_3^1 \otimes \mathbf{T}_3 \otimes \mathbf{T}_2$: $\{C_3^+|000\}$; $\mathbf{t}_1$ or $\mathbf{t}_2$; $\mathbf{t}_3$: 1, 1; 2, 1; 3, 1: *a*.

Δ^x $\mathbf{G}_6^2$: $(C_3^+, 0)$, $(\sigma_{v1}, 0)$: 1, *x*; 2, *x*; 3, *x*: *a*.
U^x $\mathbf{G}_2^1$: $(\sigma_{v1}, 0)$: 1, *x*; 2, *x*: *c*.
P^x $\mathbf{G}_3^1$: $(C_3^+, 0)$: 1, *x*; 2, *x*; 3, *x*: *a*.
T^x $\mathbf{G}_1^1$: $(E, 0)$: 1, 1: *a*.
S^x $\mathbf{G}_1^1$: $(E, 0)$: 1, 1: *a*.
T'^x $\mathbf{G}_1^1$: $(E, 0)$: 1, 1: *a*.
S'^x $\mathbf{G}_1^1$: $(E, 0)$: 1, 1: *a*.
Σ^x $\mathbf{G}_2^1$: $(\sigma_{v1}, 0)$: 1, *x*; 2, *x*: *c*.
R^x $\mathbf{G}_2^1$: $(\sigma_{v1}, 0)$: 1, *x*; 2, *x*: *c*.

157 *P*31*m* C_{3v}^2

($F1$; $K6$; $K7$; $M5$; $Z1$.)

Γ $\mathbf{G}_6^2$: $\{C_3^+|000\}$, $\{\sigma_{d1}|000\}$: 1, 1; 2, 1; 3, 1: *a*.
M $\mathbf{G}_2^1 \otimes \mathbf{T}_2$: $\{\sigma_{d1}|000\}$; $\mathbf{t}_2$: 1, 1; 2, 1: *c*.
A $\mathbf{G}_6^2 \otimes \mathbf{T}_2$: $\{C_3^+|000\}$, $\{\sigma_{d1}|000\}$; $\mathbf{t}_3$: 1, 1; 2, 1; 3, 1: *a*.
L $\mathbf{G}_2^1 \otimes \mathbf{T}_2$: $\{\sigma_{d1}|000\}$; $\mathbf{t}_2$ or $\mathbf{t}_3$: 1, 1; 2, 1: *c*.
K $\mathbf{G}_6^2 \otimes \mathbf{T}_3$: $\{C_3^+|000\}$, $\{\sigma_{d1}|000\}$; $\mathbf{t}_1$ or $\mathbf{t}_2$: 1, *x*; 2, *x*; 3, *x*: *a*.
H $\mathbf{G}_6^2 \otimes \mathbf{T}_3 \otimes \mathbf{T}_2$: $\{C_3^+|000\}$, $\{\sigma_{d1}|000\}$; $\mathbf{t}_1$ or $\mathbf{t}_2$; $\mathbf{t}_3$: 1, *x*; 2, *x*; 3, *x*: *a*.

Δ^x $\mathbf{G}_6^2$: $(C_3^+, 0)$, $(\sigma_{d1}, 0)$: 1, *x*; 2, *x*; 3, *x*: *a*.
U^x $\mathbf{G}_2^1$: $(\sigma_{d1}, 0)$: 1, *x*; 2, *x*: *c*.
P^x $\mathbf{G}_6^2$: $(C_3^+, 0)$, $(\sigma_{d1}, 0)$: 1, *x*; 2, *x*; 3, *x*: *a*.
T^x $\mathbf{G}_2^1$: $(\sigma_{d2}, 0)$: 1, *x*; 2, *x*: *c*.
S^x $\mathbf{G}_2^1$: $(\sigma_{d2}, 0)$: 1, *x*; 2, *x*: *c*.
T'^x $\mathbf{G}_2^1$: $(\sigma_{d1}, 0)$: 1, *x*; 2, *x*: *c*.
S'^x $\mathbf{G}_2^1$: $(\sigma_{d1}, 0)$: 1, *x*; 2, *x*: *c*.
Σ^x $\mathbf{G}_1^1$: $(E, 0)$: 1, 1: *a*.
R^x $\mathbf{G}_1^1$: $(E, 0)$: 1, 1: *a*.

158 *P*3*c*1 C_{3v}^3

($F1$; $K6$; $K7$; $M5$; $Z1$.)

Γ $\mathbf{G}_6^2$: $\{C_3^+|000\}$, $\{\sigma_{v1}|00\tfrac{1}{2}\}$: 1, 1; 2, 1; 3, 1: *a*.
M $\mathbf{G}_2^1 \otimes \mathbf{T}_2$: $\{\sigma_{v1}|00\tfrac{1}{2}\}$; $\mathbf{t}_2$: 1, 1; 2, 1: *c*.
A $\mathbf{G}_{12}^4$: $\{C_3^+|001\}$, $\{\sigma_{v1}|00\tfrac{1}{2}\}$: 3, 3; 4, 3; 6, 2: *a*.
L $\mathbf{G}_4^1$: $\{\sigma_{v1}|00\tfrac{1}{2}\}$: 2, 3; 4, 3: *a*.
K $\mathbf{G}_3^1 \otimes \mathbf{T}_3$: $\{C_3^+|000\}$; $\mathbf{t}_1$ or $\mathbf{t}_2$: 1, 1; 2, 1; 3, 1: *a*.
H $\mathbf{G}_3^1 \otimes \mathbf{T}_3 \otimes \mathbf{T}_2$: $\{C_3^+|000\}$; $\mathbf{t}_1$ or $\mathbf{t}_2$; $\mathbf{t}_3$: 1, 2; 2, 2; 3, 2: *a*.

Δ^x $\mathbf{G}_6^2$: $(C_3^+, 0)$, $(\sigma_{v1}, 0)$: 1, *x*; 2, *x*; 3, *x*: *a*.
U^x $\mathbf{G}_2^1$: $(\sigma_{v1}, 0)$: 1, *x*; 2, *x*: *c*.
P^x $\mathbf{G}_3^1$: $(C_3^+, 0)$: 1, *x*; 2, *x*; 3, *x*: *a*.
T^x $\mathbf{G}_1^1$: $(E, 0)$: 1, 1: *a*.
S^x $\mathbf{G}_1^1$: $(E, 0)$: 1, 2: *a*.
T'^x $\mathbf{G}_1^1$: $(E, 0)$: 1, 1: *a*.
S'^x $\mathbf{G}_1^1$: $(E, 0)$: 1, 2: *a*.
Σ^x $\mathbf{G}_2^1$: $(\sigma_{v1}, 0)$: 1, *x*; 2, *x*: *c*.
R^x $\mathbf{G}_2^1$: $(\sigma_{v1}, 0)$: 1, *x*; 2, *x*: *c*.

159 $P31c$ C_{3v}^4

($F1$; $K6$; $K7$; $M5$; $Z1$.)

Γ $\mathbf{G}_6^2$: $\{C_3^+ \mid 000\}$, $\{\sigma_{d1} \mid 00\frac{1}{2}\}$: 1, 1; 2, 1; 3, 1: *a*.
M $\mathbf{G}_2^1 \otimes \mathbf{T}_2$: $\{\sigma_{d1} \mid 00\frac{1}{2}\}$; $\mathbf{t}_2$: 1, 1; 2, 1: *c*.
A $\mathbf{G}_{12}^4$: $\{C_3^+ \mid 001\}$, $\{\sigma_{d1} \mid 00\frac{1}{2}\}$: 3, 3; 4, 3; 6, 2: *a*.
L $\mathbf{G}_4^1$: $\{\sigma_{d1} \mid 00\frac{1}{2}\}$: 2, 3; 4, 3: *a*.
K $\mathbf{G}_6^2 \otimes \mathbf{T}_3$: $\{C_3^+ \mid 000\}$, $\{\sigma_{d1} \mid 00\frac{1}{2}\}$; $\mathbf{t}_1$ or $\mathbf{t}_2$: 1, x; 2, x; 3, x: *a*.
H $\mathbf{G}_{12}^4 \otimes \mathbf{T}_3$: $\{C_3^+ \mid 001\}$, $\{\sigma_{d1} \mid 00\frac{1}{2}\}$; $\mathbf{t}_1$ or $\mathbf{t}_2$: 3, x; 4, x; 6, x: *a*.

Δ^x $\mathbf{G}_6^2$: $(C_3^+, 0)$, $(\sigma_{d1}, 0)$: 1, x; 2, x; 3, x: *a*.
U^x $\mathbf{G}_2^1$: $(\sigma_{d1}, 0)$: 1, x; 2, x: *c*.
P^x $\mathbf{G}_6^2$: $(C_3^+, 0)$, $(\sigma_{d1}, 0)$: 1, x; 2, x; 3, x: *a*.
T^x $\mathbf{G}_2^1$: $(\sigma_{d2}, 0)$: 1, x; 2, x: *c*.
S^x $\mathbf{G}_2^1$: $(\sigma_{d2}, 0)$: 1, x; 2, x: *c*.
T'^x $\mathbf{G}_2^1$: $(\sigma_{d1}, 0)$: 1, x; 2, x: *c*.
S'^x $\mathbf{G}_2^1$: $(\sigma_{d1}, 0)$: 1, x; 2, x: *c*.
Σ^x $\mathbf{G}_1^1$: $(E, 0)$: 1, 1: *a*.
R^x $\mathbf{G}_1^1$: $(E, 0)$: 1, 2: *a*.

160 $R3m$ C_{3v}^5

($F1$; $K6$; $K7$; $M5$; $Z1$.)

Γ $\mathbf{G}_6^2$: $\{C_3^+ \mid 000\}$, $\{\sigma_{d1} \mid 000\}$: 1, 1; 2, 1; 3, 1: *a*.
Z $\mathbf{G}_6^2 \otimes \mathbf{T}_2$: $\{C_3^+ \mid 000\}$, $\{\sigma_{d1} \mid 000\}$; $\mathbf{t}_1$ or $\mathbf{t}_2$ or $\mathbf{t}_3$: 1, 1; 2, 1; 3, 1: *a*.
L $\mathbf{G}_2^1 \otimes \mathbf{T}_2$: $\{\sigma_{d2} \mid 000\}$; $\mathbf{t}_2$: 1, 1; 2, 1: *c*.
(a) F $\mathbf{G}_2^1 \otimes \mathbf{T}_2$: $\{\sigma_{d1} \mid 000\}$; $\mathbf{t}_2$ or $\mathbf{t}_3$: 1, 1; 2, 1: *c*.
(b) F $\mathbf{G}_2^1 \otimes \mathbf{T}_2$: $\{\sigma_{d3} \mid 000\}$; $\mathbf{t}_1$ or $\mathbf{t}_2$: 1, 1; 2, 1: *c*.

Λ^x $\mathbf{G}_6^2$: $(C_3^+, 0)$, $(\sigma_{d1}, 0)$: 1, x; 2, x; 3, x: *a*.
P^x $\mathbf{G}_6^2$: $(C_3^+, 0)$, $(\sigma_{d1}, 0)$: 1, x; 2, x; 3, x: *a*.
B^x $\mathbf{G}_1^1$: $(E, 0)$: 1, 1: *a*.
Σ^x $\mathbf{G}_1^1$: $(E, 0)$: 1, 1: *a*.
Q^x $\mathbf{G}_1^1$: $(E, 0)$: 1, 1: *a*.
Y^x $\mathbf{G}_1^1$: $(E, 0)$: 1, 1: *a*.

161 $R3c$ C_{3v}^6

($F1$; $K6$; $K7$; $M5$; $Z1$.)

Γ $\mathbf{G}_6^2$: $\{C_3^+ \mid 000\}$, $\{\sigma_{d1} \mid \frac{1}{2}\frac{1}{2}\frac{1}{2}\}$: 1, 1; 2, 1; 3, 1: *a*.
Z $\mathbf{G}_{12}^4$: $\{C_3^+ \mid 100\}$, $\{\sigma_{d1} \mid \frac{1}{2}\frac{1}{2}\frac{1}{2}\}$: 3, 3; 4, 3; 6, 2: *a*.
L $\mathbf{G}_4^1$: $\{\sigma_{d2} \mid \frac{1}{2}\frac{1}{2}\frac{1}{2}\}$: 2, 3; 4, 3: *a*.
(a) F $\mathbf{G}_2^1 \otimes \mathbf{T}_2$: $\{\sigma_{d1} \mid \frac{1}{2}\frac{1}{2}\frac{1}{2}\}$; $\mathbf{t}_2$ or $\mathbf{t}_3$: 1, 1; 2, 1: *c*.
(b) F $\mathbf{G}_2^1 \otimes \mathbf{T}_2$: $\{\sigma_{d3} \mid \frac{1}{2}\frac{1}{2}\frac{1}{2}\}$; $\mathbf{t}_1$ or $\mathbf{t}_2$: 1, 1; 2, 1: *c*.

Λ^x $\mathbf{G}_6^2$: $(C_3^+, 0)$, $(\sigma_{d1}, 0)$: 1, x; 2, x; 3, x: *a*.
P^x $\mathbf{G}_6^2$: $(C_3^+, 0)$, $(\sigma_{d1}, 0)$: 1, x; 2, x; 3, x: *a*.
B^x $\mathbf{G}_1^1$: $(E, 0)$: 1, 2: *a*.
Σ^x $\mathbf{G}_1^1$: $(E, 0)$: 1, 1: *a*.
Q^x $\mathbf{G}_1^1$: $(E, 0)$: 1, 2: *a*.
Y^x $\mathbf{G}_1^1$: $(E, 0)$: 1, 2: *a*.

162 $P\bar{3}1m$ D_{3d}^1

($F1$; $K6$; $K7$; $M5$; $Z1$.)

Γ $\mathbf{G}_{12}^3$: $\{S_6^+ \mid 000\}, \{C'_{21} \mid 000\}$: 1, 1; 2, 1; 3, 1; 4, 1; 5, 1; 6, 1: a.
M $\mathbf{G}_4^2 \otimes \mathbf{T}_2$: $\{C'_{21} \mid 000\}, \{I \mid 000\}$; $\mathbf{t}_2$: 1, 1; 2, 1; 3, 1; 4, 1: a.
A $\mathbf{G}_{12}^3 \otimes \mathbf{T}_2$: $\{S_6^+ \mid 000\}, \{C'_{21} \mid 000\}$; $\mathbf{t}_3$: 1, 1; 2, 1; 3, 1; 4, 1; 5, 1; 6, 1: a.
L $\mathbf{G}_4^2 \otimes \mathbf{T}_2$: $\{C'_{21} \mid 000\}, \{I \mid 000\}$; $\mathbf{t}_2$ or $\mathbf{t}_3$: 1, 1; 2, 1; 3, 1; 4, 1: a.
K $\mathbf{G}_6^2 \otimes \mathbf{T}_3$: $\{C_3^+ \mid 000\}, \{\sigma_{d1} \mid 000\}$; $\mathbf{t}_1$ or $\mathbf{t}_2$: 1, 1; 2, 1; 3, 1: a.
H $\mathbf{G}_6^2 \otimes \mathbf{T}_3 \otimes \mathbf{T}_2$: $\{C_3^+ \mid 000\}, \{\sigma_{d1} \mid 000\}$; $\mathbf{t}_1$ or $\mathbf{t}_2$; $\mathbf{t}_3$: 1, 1; 2, 1; 3, 1: a.

Δ^x $\mathbf{G}_6^2$: $(C_3^+, 0), (\sigma_{d1}, 0)$: 1, 1; 2, 1; 3, 1: a.
U^x $\mathbf{G}_2^1$: $(\sigma_{d1}, 0)$: 1, 1; 2, 1: c.
P^x $\mathbf{G}_6^2$: $(C_3^+, 0), (\sigma_{d1}, 0)$: 1, 1; 2, 1; 3, 1: a.
T^x $\mathbf{G}_2^1$: $(\sigma_{d2}, 0)$: 1, 1; 2, 1: c.
S^x $\mathbf{G}_2^1$: $(\sigma_{d2}, 0)$: 1, 1; 2, 1: c.
T'^x $\mathbf{G}_2^1$: $(\sigma_{d1}, 0)$: 1, 1; 2, 1: c.
S'^x $\mathbf{G}_2^1$: $(\sigma_{d1}, 0)$: 1, 1; 2, 1: c.
Σ^x $\mathbf{G}_2^1$: $(C'_{21}, 0)$: 1, 1; 2, 1: b.
R^x $\mathbf{G}_2^1$: $(C'_{21}, 0)$: 1, 1; 2, 1: b.

163 $P\bar{3}1c$ D_{3d}^2

($F1$; $K6$; $K7$; $M5$; $Z1$.)

Γ $\mathbf{G}_{12}^3$: $\{S_6^+ \mid 000\}, \{C'_{21} \mid 00\tfrac{1}{2}\}$: 1, 1; 2, 1; 3, 1; 4, 1; 5, 1; 6, 1: a.
M $\mathbf{G}_4^2 \otimes \mathbf{T}_2$: $\{C'_{21} \mid 00\tfrac{1}{2}\}, \{I \mid 000\}$; $\mathbf{t}_2$: 1, 1; 2, 1; 3, 1; 4, 1: a.
A $\mathbf{G}_{24}^1$: $\{\sigma_{d1} \mid 00\tfrac{1}{2}\}, \{C_3^+ \mid 000\}, \{I \mid 000\}$: 7, 3; 8, 3; 9, 1: a.
L $\mathbf{G}_8^4$: $\{\sigma_{d1} \mid 00\tfrac{1}{2}\}, \{I \mid 000\}$: 5, 1: a.
K $\mathbf{G}_6^2 \otimes \mathbf{T}_3$: $\{C_3^+ \mid 000\}, \{\sigma_{d1} \mid 00\tfrac{1}{2}\}$; $\mathbf{t}_1$ or $\mathbf{t}_2$: 1, 1; 2, 1; 3, 1: a.
H $\mathbf{G}_{12}^4 \otimes \mathbf{T}_3$: $\{C_3^+ \mid 001\}, \{\sigma_{d1} \mid 00\tfrac{1}{2}\}$; $\mathbf{t}_1$ or $\mathbf{t}_2$: 3, 1; 4, 1; 6, 1: a.

Δ^x $\mathbf{G}_6^2$: $(C_3^+, 0), (\sigma_{d1}, 0)$: 1, 1; 2, 1; 3, 1: a.
U^x $\mathbf{G}_2^1$: $(\sigma_{d1}, 0)$: 1, 1; 2, 1: c.
P^x $\mathbf{G}_6^2$: $(C_3^+, 0), (\sigma_{d1}, 0)$: 1, 1; 2, 1; 3, 1: a.
T^x $\mathbf{G}_2^1$: $(\sigma_{d2}, 0)$: 1, 1; 2, 1: c.
S^x $\mathbf{G}_2^1$: $(\sigma_{d2}, 0)$: 1, 1; 2, 1: c.
T'^x $\mathbf{G}_2^1$: $(\sigma_{d1}, 0)$: 1, 1; 2, 1: c.
S'^x $\mathbf{G}_2^1$: $(\sigma_{d1}, 0)$: 1, 1; 2, 1: c.
Σ^x $\mathbf{G}_2^1$: $(C'_{21}, 0)$: 1, 1; 2, 1: b.
R^x $\mathbf{G}_4^1$: $(C'_{21}, 0)$: 2, 3; 4, 3: a.

164 $P\bar{3}m1$ D_{3d}^3

($F1$; $K6$; $K7$; $M5$; $S8$; $Z1$.)

Γ $\mathbf{G}_{12}^3$: $\{S_6^+ \mid 000\}, \{C''_{21} \mid 000\}$: 1, 1; 2, 1; 3, 1; 4, 1; 5, 1; 6, 1: a.
M $\mathbf{G}_4^2 \otimes \mathbf{T}_2$: $\{C''_{21} \mid 000\}, \{I \mid 000\}$; $\mathbf{t}_2$: 1, 1; 2, 1; 3, 1; 4, 1: a.
A $\mathbf{G}_{12}^3 \otimes \mathbf{T}_2$: $\{S_6^+ \mid 000\}, \{C''_{21} \mid 000\}$; $\mathbf{t}_3$: 1, 1; 2, 1; 3, 1; 4, 1; 5, 1; 6, 1: a.
L $\mathbf{G}_4^2 \otimes \mathbf{T}_2$: $\{C''_{21} \mid 000\}, \{I \mid 000\}$; $\mathbf{t}_2$ or $\mathbf{t}_3$: 1, 1; 2, 1; 3, 1; 4, 1: a.
K $\mathbf{G}_6^2 \otimes \mathbf{T}_3$: $\{C_3^+ \mid 000\}, \{C''_{21} \mid 000\}$; $\mathbf{t}_1$ or $\mathbf{t}_2$: 1, 1; 2, 1; 3, 1: a.
H $\mathbf{G}_6^2 \otimes \mathbf{T}_3 \otimes \mathbf{T}_2$: $\{C_3^+ \mid 000\}, \{C''_{21} \mid 000\}$; $\mathbf{t}_1$ or $\mathbf{t}_2$; $\mathbf{t}_3$: 1, 1; 2, 1; 3, 1: a.

Δ^x $\mathbf{G}_6^2$: $(C_3^+, 0), (\sigma_{v1}, 0)$: 1, 1; 2, 1; 3, 1: a.
U^x $\mathbf{G}_2^1$: $(\sigma_{v1}, 0)$: 1, 1; 2, 1: c.
P^x $\mathbf{G}_3^1$: $(C_3^+, 0)$: 1, 1; 2, 3; 3, 3: a.
T^x $\mathbf{G}_2^1$: $(C''_{22}, 0)$: 1, 1; 2, 1: b.
S^x $\mathbf{G}_2^1$: $(C''_{22}, 0)$: 1, 1; 2, 1: b.
T'^x $\mathbf{G}_2^1$: $(C''_{21}, 0)$: 1, 1; 2, 1: b.
S'^x $\mathbf{G}_2^1$: $(C''_{21}, 0)$: 1, 1; 2, 1: b.
Σ^x $\mathbf{G}_2^1$: $(\sigma_{v1}, 0)$: 1, 1; 2, 1: c.
R^x $\mathbf{G}_2^1$: $(\sigma_{v1}, 0)$: 1, 1; 2, 1: c.

165 $P\bar{3}c1$ D_{3d}^4

($F1$; $K6$; $K7$; $M5$; $Z1$.)

Γ $\mathbf{G}_{12}^3$: $\{S_6^+ \mid 000\}, \{C_{21}'' \mid 00\frac{1}{2}\}$: 1, 1; 2, 1; 3, 1; 4, 1; 5, 1; 6, 1: *a*.
M $\mathbf{G}_4^2 \otimes \mathbf{T}_2$: $\{C_{21}'' \mid 00\frac{1}{2}\}, \{I \mid 000\}$; $\mathbf{t}_2$: 1, 1; 2, 1; 3, 1; 4, 1: *a*.
A $\mathbf{G}_{24}^1$: $\{\sigma_{v1} \mid 00\frac{1}{2}\}, \{C_3^+ \mid 000\}, \{I \mid 000\}$: 7, 3; 8, 3; 9, 1: *a*.
L $\mathbf{G}_8^4$: $\{\sigma_{v1} \mid 00\frac{1}{2}\}, \{I \mid 000\}$: 5, 1: *a*.
K $\mathbf{G}_6^2 \otimes \mathbf{T}_3$: $\{C_3^+ \mid 000\}, \{C_{21}'' \mid 00\frac{1}{2}\}$; $\mathbf{t}_1$ or $\mathbf{t}_2$: 1, 1; 2, 1; 3, 1: *a*.
H $\mathbf{G}_{12}^3 \otimes \mathbf{T}_3$: $\{C_3^+ \mid 001\}, \{C_{21}'' \mid 00\frac{1}{2}\}$; $\mathbf{t}_1$ or $\mathbf{t}_2$: 3, 3; 4, 3; 6, 2: *b*.

Δ^x $\mathbf{G}_6^2$: $(C_3^+, 0), (\sigma_{v1}, 0)$: 1, 1; 2, 1; 3, 1: *a*.
U^x $\mathbf{G}_2^1$: $(\sigma_{v1}, 0)$: 1, 1; 2, 1: *c*.
P^x $\mathbf{G}_3^1$: $(C_3^+, 0)$: 1, 1; 2, 3; 3, 3: *a*.
T^x $\mathbf{G}_2^1$: $(C_{22}'', 0)$: 1, 1; 2, 1: *b*.
S^x $\mathbf{G}_4^1$: $(C_{22}'', 0)$: 2, 3; 4, 3: *a*.
T'^x $\mathbf{G}_2^1$: $(C_{21}'', 0)$: 1, 1; 2, 1: *b*.
S'^x $\mathbf{G}_4^1$: $(C_{21}'', 0)$: 2, 3; 4, 3: *a*.
Σ^x $\mathbf{G}_2^1$: $(\sigma_{v1}, 0)$: 1, 1; 2, 1: *c*.
R^x $\mathbf{G}_2^1$: $(\sigma_{v1}, 0)$: 1, 1; 2, 1: *c*.

166 $R\bar{3}m$ D_{3d}^5

($F1$; $F3$; $K6$; $K7$; $L2$; $M1$; $M5$; $S8$; $Y2$; $Z1$.)

Γ $\mathbf{G}_{12}^3$: $\{S_6^+ \mid 000\}, \{C_{21}' \mid 000\}$: 1, 1; 2, 1; 3, 1; 4, 1; 5, 1; 6, 1: *a*.
Z $\mathbf{G}_{12}^3 \otimes \mathbf{T}_2$: $\{S_6^+ \mid 000\}, \{C_{21}' \mid 000\}$; $\mathbf{t}_1$ or $\mathbf{t}_2$ or $\mathbf{t}_3$: 1, 1; 2, 1; 3, 1; 4, 1; 5, 1; 6, 1: *a*.
L $\mathbf{G}_4^2 \otimes \mathbf{T}_2$: $\{C_{22}' \mid 000\}, \{I \mid 000\}$; $\mathbf{t}_2$: 1, 1; 2, 1; 3, 1; 4, 1: *a*.
{(a) F $\mathbf{G}_4^2 \otimes \mathbf{T}_2$: $\{C_{21}' \mid 000\}, \{I \mid 000\}$; $\mathbf{t}_2$ or $\mathbf{t}_3$: 1, 1; 2, 1; 3, 1; 4, 1: *a*.}
{(b) F $\mathbf{G}_4^2 \otimes \mathbf{T}_2$: $\{C_{23}' \mid 000\}, \{I \mid 000\}$; $\mathbf{t}_1$ or $\mathbf{t}_2$: 1, 1; 2, 1; 3, 1; 4, 1: *a*.}

Λ^x $\mathbf{G}_6^2$: $(C_3^+, 0), (\sigma_{d1}, 0)$: 1, 1; 2, 1; 3, 1: *a*.
P^x $\mathbf{G}_6^2$: $(C_3^+, 0), (\sigma_{d1}, 0)$: 1, 1; 2, 1; 3, 1: *a*.
B^x $\mathbf{G}_2^1$: $(C_{21}', 0)$: 1, 1; 2, 1: *b*.
Σ^x $\mathbf{G}_2^1$: $(C_{21}', 0)$: 1, 1; 2, 1: *b*.
Q^x $\mathbf{G}_2^1$: $(C_{23}', 0)$: 1, 1; 2, 1: *b*.
Y^x $\mathbf{G}_2^1$: $(C_{22}', 0)$: 1, 1; 2, 1: *b*.

167 $R\bar{3}c$ D_{3d}^6

($F1$; $K6$; $K7$; $M5$; $S8$; $Z1$.)

Γ $\mathbf{G}_{12}^3$: $\{S_6^+ \mid 000\}, \{C_{21}' \mid \frac{1}{2}\frac{1}{2}\frac{1}{2}\}$: 1, 1; 2, 1; 3, 1; 4, 1; 5, 1; 6, 1: *a*.
Z $\mathbf{G}_{24}^1$: $\{\sigma_{d1} \mid \frac{1}{2}\frac{1}{2}\frac{1}{2}\}, \{C_3^+ \mid 000\}, \{I \mid 000\}$: 7, 3; 8, 3; 9, 1: *a*.
L $\mathbf{G}_8^4$: $\{\sigma_{d2} \mid \frac{1}{2}\frac{1}{2}\frac{1}{2}\}, \{I \mid 000\}$: 5, 1: *a*.
{(a) F $\mathbf{G}_4^2 \otimes \mathbf{T}_2$: $\{C_{21}' \mid \frac{1}{2}\frac{1}{2}\frac{1}{2}\}, \{I \mid 000\}$; $\mathbf{t}_2$ or $\mathbf{t}_3$: 1, 1; 2, 1; 3, 1; 4, 1: *a*.}
{(b) F $\mathbf{G}_4^2 \otimes \mathbf{T}_2$: $\{C_{23}' \mid \frac{1}{2}\frac{1}{2}\frac{1}{2}\}, \{I \mid 000\}$; $\mathbf{t}_1$ or $\mathbf{t}_2$: 1, 1; 2, 1; 3, 1; 4, 1: *a*.}

Λ^x $\mathbf{G}_6^2$: $(C_3^+, 0), (\sigma_{d1}, 0)$: 1, 1; 2, 1; 3, 1: *a*.
P^x $\mathbf{G}_6^2$: $(C_3^+, 0), (\sigma_{d1}, 0)$: 1, 1; 2, 1; 3, 1: *a*.
B^x $\mathbf{G}_4^1$: $(C_{21}', 0)$: 2, 3; 4, 3: *a*.
Σ^x $\mathbf{G}_2^1$: $(C_{21}', 0)$: 1, 1; 2, 1: *b*.
Q^x $\mathbf{G}_2^1$: $(C_{23}', 0)$: 1, 1; 2, 1: *b*.
Y^x $\mathbf{G}_4^1$: $(C_{22}', 0)$: 2, 3; 4, 3: *a*.

168 $P6$ C_6^1

($F1$; $K6$; $K7$; $M5$; $Z1$.)

Γ $\mathbf{G}_6^1$: $\{C_6^+ \mid 000\}$: 1, 1; 2, 3; 3, 3; 4, 1; 5, 3; 6, 3: *d*.
M $\mathbf{G}_2^1 \otimes \mathbf{T}_2$: $\{C_2 \mid 000\}$; $\mathbf{t}_2$: 1, 1; 2, 1: *b*.
A $\mathbf{G}_6^1 \otimes \mathbf{T}_2$: $\{C_6^+ \mid 000\}$; $\mathbf{t}_3$: 1, 1; 2, 3; 3, 3; 4, 1; 5, 3; 6, 3: *d*.
L $\mathbf{G}_2^1 \otimes \mathbf{T}_2$: $\{C_2 \mid 000\}$; $\mathbf{t}_2$ or $\mathbf{t}_3$: 1, 1; 2, 1: *b*.
K $\mathbf{G}_3^1 \otimes \mathbf{T}_3$: $\{C_3^+ \mid 000\}$; $\mathbf{t}_1$ or $\mathbf{t}_2$: 1, 1; 2, 3; 3, 3: *a*.
H $\mathbf{G}_3^1 \otimes \mathbf{T}_3 \otimes \mathbf{T}_2$: $\{C_3^+ \mid 000\}$; $\mathbf{t}_1$ or $\mathbf{t}_2$; $\mathbf{t}_3$: 1, 1; 2, 3; 3, 3: *a*.

Δ^x $\mathbf{G}_6^1$: $(C_6^+, 0)$: 1, x; 2, x; 3, x; 4, x; 5, x; 6, x: *d*.
U^x $\mathbf{G}_2^1$: $(C_2, 0)$: 1, x; 2, x: *b*.
P^x $\mathbf{G}_3^1$: $(C_3^+, 0)$: 1, x; 2, x; 3, x: *a*.
T^x $\mathbf{G}_1^1$: $(E, 0)$: 1, 1: *a*.
S^x $\mathbf{G}_1^1$: $(E, 0)$: 1, 1: *a*.
T'^x $\mathbf{G}_1^1$: $(E, 0)$: 1, 1: *a*.
S'^x $\mathbf{G}_1^1$: $(E, 0)$: 1, 1: *a*.
Σ^x $\mathbf{G}_1^1$: $(E, 0)$: 1, 1: *a*.
R^x $\mathbf{G}_1^1$: $(E, 0)$: 1, 1: *a*.

169 $P6_1$ C_6^2

($F1$; $K6$; $K7$; $M5$; $Z1$.)

Γ $\mathbf{G}_6^1$: $\{C_6^+ \mid 00\frac{1}{6}\}$: 1, 1; 2, 3; 3, 3; 4, 1; 5, 3; 6, 3: *d*.
M $\mathbf{G}_2^1 \otimes \mathbf{T}_2$: $\{C_2 \mid 00\frac{1}{2}\}$; $\mathbf{t}_2$: 1, 1; 2, 1: *b*.
A $\mathbf{G}_{12}^1$: $\{C_6^+ \mid 00\frac{1}{6}\}$: 2, 3; 4, 3; 6, 3; 8, 3; 10, 3; 12, 3: *a*.
L $\mathbf{G}_4^1$: $\{C_2 \mid 00\frac{1}{2}\}$: 2, 3; 4, 3: *a*.
K $\mathbf{G}_3^1 \otimes \mathbf{T}_3$: $\{C_3^+ \mid 00\frac{1}{3}\}$; $\mathbf{t}_1$ or $\mathbf{t}_2$: 1, 1; 2, 3; 3, 3: *a*.
H $\mathbf{G}_3^1 \otimes \mathbf{T}_3 \otimes \mathbf{T}_2$: $\{C_3^- \mid 00\frac{2}{3}\}$; $\mathbf{t}_1$ or $\mathbf{t}_2$; $\mathbf{t}_3$: 1, 2; 2, 3; 3, 3: *a*.

Δ^x $\mathbf{G}_6^1$: $(C_6^+, 0)$: 1, x; 2, x; 3, x; 4, x; 5, x; 6, x: *d*.
U^x $\mathbf{G}_2^1$: $(C_2, 0)$: 1, x; 2, x: *b*.
P^x $\mathbf{G}_3^1$: $(C_3^+, 0)$: 1, x; 2, x; 3, x: *a*.
T^x $\mathbf{G}_1^1$: $(E, 0)$: 1, 1: *a*.
S^x $\mathbf{G}_1^1$: $(E, 0)$: 1, 2: *a*.
T'^x $\mathbf{G}_1^1$: $(E, 0)$: 1, 1: *a*.
S'^x $\mathbf{G}_1^1$: $(E, 0)$: 1, 2: *a*.
Σ^x $\mathbf{G}_1^1$: $(E, 0)$: 1, 1: *a*.
R^x $\mathbf{G}_1^1$: $(E, 0)$: 1, 2: *a*.

170 $P6_5$ C_6^3

($F1$; $K6$; $K7$; $M5$; $Z1$.)

Γ $\mathbf{G}_6^1$: $\{C_6^+ \mid 00\frac{5}{6}\}$: 1, 1; 2, 3; 3, 3; 4, 1; 5, 3; 6, 3: *d*.
M $\mathbf{G}_2^1 \otimes \mathbf{T}_2$: $\{C_2 \mid 00\frac{1}{2}\}$; $\mathbf{t}_2$: 1, 1; 2, 1: *b*.
A $\mathbf{G}_{12}^1$: $\{C_6^+ \mid 00\frac{5}{6}\}$: 2, 3; 4, 3; 6, 3; 8, 3; 10, 3; 12, 3: *a*.
L $\mathbf{G}_4^1$: $\{C_2 \mid 00\frac{1}{2}\}$: 2, 3; 4, 3: *a*.
K $\mathbf{G}_3^1 \otimes \mathbf{T}_3$: $\{C_3^+ \mid 00\frac{2}{3}\}$; $\mathbf{t}_1$ or $\mathbf{t}_2$: 1, 1; 2, 3; 3, 3: *a*.
H $\mathbf{G}_3^1 \otimes \mathbf{T}_3 \otimes \mathbf{T}_2$: $\{C_3^+ \mid 00\frac{2}{3}\}$; $\mathbf{t}_1$ or $\mathbf{t}_2$; $\mathbf{t}_3$: 1, 2; 2, 3; 3, 3: *a*.

Δ^x $\mathbf{G}_6^1$: $(C_6^+, 0)$: 1, x; 2, x; 3, x; 4, x; 5, x; 6, x: *d*.
U^x $\mathbf{G}_2^1$: $(C_2, 0)$: 1, x; 2, x: *b*.
P^x $\mathbf{G}_3^1$: $(C_3^+, 0)$: 1, x; 2, x; 3, x: *a*.
T^x $\mathbf{G}_1^1$: $(E, 0)$: 1, 1: *a*.
S^x $\mathbf{G}_1^1$: $(E, 0)$: 1, 2: *a*.
T'^x $\mathbf{G}_1^1$: $(E, 0)$: 1, 1: *a*.
S'^x $\mathbf{G}_1^1$: $(E, 0)$: 1, 2: *a*.
Σ^x $\mathbf{G}_1^1$: $(E, 0)$: 1, 1: *a*.
R^x $\mathbf{G}_1^1$: $(E, 0)$: 1, 2: *a*.

171 $P6_2$ C_6^4

($F1$; $K6$; $K7$; $M5$; $Z1$.)

Γ $\mathbf{G}_6^1$: $\{C_6^+ \mid 00\frac{1}{3}\}$: 1, 1; 2, 3; 3, 3; 4, 1; 5, 3; 6, 3: *d*.
M $\mathbf{G}_2^1 \otimes \mathbf{T}_2$: $\{C_2 \mid 000\}$; $\mathbf{t}_2$: 1, 1; 2, 1: *b*.
A $\mathbf{G}_6^1 \otimes \mathbf{T}_2$: $\{C_6^+ \mid 00\frac{1}{3}\}$; $\mathbf{t}_3$: 1, 1; 2, 3; 3, 3; 4, 1; 5, 3; 6, 3: *d*.
L $\mathbf{G}_2^1 \otimes \mathbf{T}_2$: $\{C_2 \mid 000\}$; $\mathbf{t}_2$ or $\mathbf{t}_3$: 1, 1; 2, 1: *b*.
K $\mathbf{G}_3^1 \otimes \mathbf{T}_3$: $\{C_3^+ \mid 00\frac{2}{3}\}$; $\mathbf{t}_1$ or $\mathbf{t}_2$: 1, 1; 2, 3; 3, 3: *a*.
H $\mathbf{G}_3^1 \otimes \mathbf{T}_3 \otimes \mathbf{T}_2$: $\{C_3^+ \mid 00\frac{2}{3}\}$; $\mathbf{t}_1$ or $\mathbf{t}_2$; $\mathbf{t}_3$: 1, 1; 2, 3; 3, 3: *a*.

Δ^x $\mathbf{G}_6^1$: $(C_6^+, 0)$: 1, x; 2, x; 3, x; 4, x; 5, x; 6, x: *d*.
U^x $\mathbf{G}_2^1$: $(C_2, 0)$: 1, x; 2, x: *b*.
P^x $\mathbf{G}_3^1$: $(C_3^+, 0)$: 1, x; 2, x; 3, x: *a*.
T^x $\mathbf{G}_1^1$: $(E, 0)$: 1, 1: *a*.
S^x $\mathbf{G}_1^1$: $(E, 0)$: 1, 1: *a*.
T'^x $\mathbf{G}_1^1$: $(E, 0)$: 1, 1: *a*.
S'^x $\mathbf{G}_1^1$: $(E, 0)$: 1, 1: *a*.
Σ^x $\mathbf{G}_1^1$: $(E, 0)$: 1, 1: *a*.
R^x $\mathbf{G}_1^1$: $(E, 0)$: 1, 1: *a*.

172 $P6_4$ C_6^5

($F1$; $K6$; $K7$; $M5$; $Z1$.)

Γ $\mathbf{G}_6^1$: $\{C_6^+ \mid 00\frac{2}{3}\}$: 1, 1; 2, 3; 3, 3; 4, 1; 5, 3; 6, 3: *d*.
M $\mathbf{G}_2^1 \otimes \mathbf{T}_2$: $\{C_2 \mid 000\}$; $\mathbf{t}_2$: 1, 1; 2, 1: *b*.
A $\mathbf{G}_6^1 \otimes \mathbf{T}_2$: $\{C_6^+ \mid 00\frac{2}{3}\}$; $\mathbf{t}_3$: 1, 1; 2, 3; 3, 3; 4, 1; 5, 3; 6, 3: *d*.
L $\mathbf{G}_2^1 \otimes \mathbf{T}_2$: $\{C_2 \mid 000\}$; $\mathbf{t}_2$ or $\mathbf{t}_3$: 1, 1; 2, 1: *b*.
K $\mathbf{G}_3^1 \otimes \mathbf{T}_3$: $\{C_3^+ \mid 00\frac{1}{3}\}$; $\mathbf{t}_1$ or $\mathbf{t}_2$: 1, 1; 2, 3; 3, 3: *a*.
H $\mathbf{G}_3^1 \otimes \mathbf{T}_3 \otimes \mathbf{T}_2$: $\{C_3^- \mid 00\frac{2}{3}\}$; $\mathbf{t}_1$ or $\mathbf{t}_2$; $\mathbf{t}_3$: 1, 1; 2, 3; 3, 3: *a*.

Δ^x $\mathbf{G}_6^1$: $(C_6^+, 0)$: 1, x; 2, x; 3, x; 4, x; 5, x; 6, x: *d*.
U^x $\mathbf{G}_2^1$: $(C_2, 0)$: 1, x; 2, x: *b*.
P^x $\mathbf{G}_3^1$: $(C_3^+, 0)$: 1, x; 2, x; 3, x: *a*.
T^x $\mathbf{G}_1^1$: $(E, 0)$: 1, 1: *a*.
S^x $\mathbf{G}_1^1$: $(E, 0)$: 1, 1: *a*.
T'^x $\mathbf{G}_1^1$: $(E, 0)$: 1, 1: *a*.
S'^x $\mathbf{G}_1^1$: $(E, 0)$: 1, 1: *a*.
Σ^x $\mathbf{G}_1^1$: $(E, 0)$: 1, 1: *a*.
R^x $\mathbf{G}_1^1$: $(E, 0)$: 1, 1: *a*.

173 $P6_3$ C_6^6

($F1$; $K6$; $K7$; $M5$; $Z1$.)

Γ $\mathbf{G}_6^1$: $\{C_6^+ \mid 00\tfrac{1}{2}\}$: 1, 1; 2, 3; 3, 3; 4, 1; 5, 3; 6, 3: *d*.
M $\mathbf{G}_2^1 \otimes \mathbf{T}_2$: $\{C_2 \mid 00\tfrac{1}{2}\}$; $\mathbf{t}_2$: 1, 1; 2, 1: *b*.
A $\mathbf{G}_{12}^1$: $\{C_6^+ \mid 00\tfrac{1}{2}\}$: 2, 3; 4, 3; 6, 3; 8, 3; 10, 3; 12, 3: *a*.
L $\mathbf{G}_4^1$: $\{C_2 \mid 00\tfrac{1}{2}\}$: 2, 3; 4, 3: *a*.
K $\mathbf{G}_3^1 \otimes \mathbf{T}_3$: $\{C_3^+ \mid 000\}$; $\mathbf{t}_1$ or $\mathbf{t}_2$: 1, 1; 2, 3; 3, 3: *a*.
H $\mathbf{G}_3^1 \otimes \mathbf{T}_3 \otimes \mathbf{T}_2$: $\{C_3^+ \mid 000\}$; $\mathbf{t}_1$ or $\mathbf{t}_2$; $\mathbf{t}_3$: 1, 2; 2, 3; 3, 3: *a*.

Δ^x $\mathbf{G}_6^1$: $(C_6^+, 0)$: 1, *x*; 2, *x*; 3, *x*; 4, *x*; 5, *x*; 6, *x*: *d*.
U^x $\mathbf{G}_2^1$: $(C_2, 0)$: 1, *x*; 2, *x*: *b*.
P^x $\mathbf{G}_3^1$: $(C_3^+, 0)$: 1, *x*; 2, *x*; 3, *x*: *a*.
T^x $\mathbf{G}_1^1$: $(E, 0)$: 1, 1: *a*.
S^x $\mathbf{G}_1^1$: $(E, 0)$: 1, 2: *a*.
T'^x $\mathbf{G}_1^1$: $(E, 0)$: 1, 1: *a*.
S'^x $\mathbf{G}_1^1$: $(E, 0)$: 1, 2: *a*.
Σ^x $\mathbf{G}_1^1$: $(E, 0)$: 1, 1: *a*.
R^x $\mathbf{G}_1^1$: $(E, 0)$: 1, 2: *a*.

174 $P\bar{6}$ C_{3h}^1

($F1$; $K6$; $K7$; $M5$; $Z1$.)

Γ $\mathbf{G}_6^1$: $\{S_3^+ \mid 000\}$: 1, 1; 2, 3; 3, 3; 4, 1; 5, 3; 6, 3: *e*.
M $\mathbf{G}_2^1 \otimes \mathbf{T}_2$: $\{\sigma_h \mid 000\}$; $\mathbf{t}_2$: 1, 1; 2, 1: *c*.
A $\mathbf{G}_6^1 \otimes \mathbf{T}_2$: $\{S_3^+ \mid 000\}$; $\mathbf{t}_3$: 1, 1; 2, 3; 3, 3; 4, 1; 5, 3; 6, 3: *e*.
L $\mathbf{G}_2^1 \otimes \mathbf{T}_2$: $\{\sigma_h \mid 000\}$; $\mathbf{t}_2$ or $\mathbf{t}_3$: 1, 1; 2, 1: *c*.
K $\mathbf{G}_6^1 \otimes \mathbf{T}_3$: $\{S_3^+ \mid 000\}$; $\mathbf{t}_1$ or $\mathbf{t}_2$: 1, *x*; 2, *x*; 3, *x*; 4, *x*; 5, *x*; 6, *x*: *e*.
H $\mathbf{G}_6^1 \otimes \mathbf{T}_3 \otimes \mathbf{T}_2$: $\{S_3^+ \mid 000\}$; $\mathbf{t}_1$ or $\mathbf{t}_2$; $\mathbf{t}_3$: 1, *x*; 2, *x*; 3, *x*; 4, *x*; 5, *x*; 6, *x*: *e*.

Δ^x $\mathbf{G}_3^1$: $(C_3^+, 0)$: 1, 1; 2, 3; 3, 3: *a*.
U^x $\mathbf{G}_1^1$: $(E, 0)$: 1, 1: *a*.
P^x $\mathbf{G}_3^1$: $(C_3^+, 0)$: 1, *x*; 2, *x*; 3, *x*: *a*.
T^x $\mathbf{G}_2^1$: $(\sigma_h, 0)$: 1, *x*; 2, *x*: *c*.
S^x $\mathbf{G}_2^1$: $(\sigma_h, 0)$: 1, *x*; 2, *x*: *c*.
T'^x $\mathbf{G}_2^1$: $(\sigma_h, 0)$: 1, *x*; 2, *x*: *c*.
S'^x $\mathbf{G}_2^1$: $(\sigma_h, 0)$: 1, *x*; 2, *x*: *c*.
Σ^x $\mathbf{G}_2^1$: $(\sigma_h, 0)$: 1, *x*; 2, *x*: *c*.
R^x $\mathbf{G}_2^1$: $(\sigma_h, 0)$: 1, *x*; 2, *x*: *c*.

175 $P6/m$ C_{6h}^1

($F1$; $K6$; $K7$; $M5$; $Z1$.)

Γ $\mathbf{G}_{12}^2$: $\{C_3^+ \mid 000\}$, $\{C_2 \mid 000\}$, $\{I \mid 000\}$: 1, 1; 2, 3; 3, 3; 4, 1; 5, 3; 6, 3; 7, 1; 8, 3; 9, 3; 10, 1; 11, 3; 12, 3: *a*.
M $\mathbf{G}_4^2 \otimes \mathbf{T}_2$: $\{C_2 \mid 000\}$, $\{I \mid 000\}$; $\mathbf{t}_2$: 1, 1; 2, 1; 3, 1; 4, 1: *a*.
A $\mathbf{G}_{12}^2 \otimes \mathbf{T}_2$: $\{C_3^+ \mid 000\}$, $\{C_2 \mid 000\}$, $\{I \mid 000\}$; $\mathbf{t}_3$: 1, 1; 2, 3; 3, 3; 4, 1; 5, 3; 6, 3; 7, 1; 8, 3; 9, 3; 10, 1; 11, 3; 12, 3: *a*.
L $\mathbf{G}_4^2 \otimes \mathbf{T}_2$: $\{C_2 \mid 000\}$, $\{I \mid 000\}$; $\mathbf{t}_2$ or $\mathbf{t}_3$: 1, 1; 2, 1; 3, 1; 4, 1: *a*.
K $\mathbf{G}_6^1 \otimes \mathbf{T}_3$: $\{S_3^+ \mid 000\}$; $\mathbf{t}_1$ or $\mathbf{t}_2$: 1, 1; 2, 3; 3, 3; 4, 1; 5, 3; 6, 3: *e*.
H $\mathbf{G}_6^1 \otimes \mathbf{T}_3 \otimes \mathbf{T}_2$: $\{S_3^+ \mid 000\}$; $\mathbf{t}_1$ or $\mathbf{t}_2$; $\mathbf{t}_3$: 1, 1; 2, 3; 3, 3; 4, 1; 5, 3; 6, 3: *e*.

Δ^x $\mathbf{G}_6^1$: $(C_6^+, 0)$: 1, 1; 2, 3; 3, 3; 4, 1; 5, 3; 6, 3: *d*.
U^x $\mathbf{G}_2^1$: $(C_2, 0)$: 1, 1; 2, 1: *b*.
P^x $\mathbf{G}_3^1$: $(C_3^+, 0)$: 1, 1; 2, 3; 3, 3: *a*.
T^x $\mathbf{G}_2^1$: $(\sigma_h, 0)$: 1, 1; 2, 1: *c*.
S^x $\mathbf{G}_2^1$: $(\sigma_h, 0)$: 1, 1; 2, 1: *c*.
T'^x $\mathbf{G}_2^1$: $(\sigma_h, 0)$: 1, 1; 2, 1: *c*.
S'^x $\mathbf{G}_2^1$: $(\sigma_h, 0)$: 1, 1; 2, 1: *c*.
Σ^x $\mathbf{G}_2^1$: $(\sigma_h, 0)$: 1, 1; 2, 1: *c*.
R^x $\mathbf{G}_2^1$: $(\sigma_h, 0)$: 1, 1; 2, 1: *c*.

176 $P6_3/m$ C_{6h}^2

($F1$; $K6$; $K7$; $M5$; $M8$; $Z1$.)

Γ $\mathbf{G}_{12}^2$: $\{C_3^+ \mid 000\}$, $\{C_2 \mid 00\frac{1}{2}\}$, $\{I \mid 00\frac{1}{2}\}$: 1, 1; 2, 3; 3, 3; 4, 1; 5, 3; 6, 3; 7, 1; 8, 3; 9, 3; 10, 1; 11, 3; 12, 3: *a*.
M $\mathbf{G}_4^2 \otimes \mathbf{T}_2$: $\{C_2 \mid 00\frac{1}{2}\}$, $\{I \mid 00\frac{1}{2}\}$; $\mathbf{t}_2$: 1, 1; 2, 1; 3, 1; 4, 1: *a*.
A $\mathbf{G}_{24}^6$: $\{C_6^+ \mid 00\frac{1}{2}\}$, $\{I \mid 00\frac{1}{2}\}$: 13, 3; 14, 1; 15, 3: *a*.
L $\mathbf{G}_8^4$: $\{C_2 \mid 00\frac{1}{2}\}$, $\{I \mid 00\frac{1}{2}\}$: 5, 1: *a*.
K $\mathbf{G}_6^1 \otimes \mathbf{T}_3$: $\{S_3^+ \mid 000\}$; $\mathbf{t}_1$ or $\mathbf{t}_2$: 1, 1; 2, 3; 3, 3; 4, 1; 5, 3; 6, 3: *e*.
H $\mathbf{G}_6^1 \otimes \mathbf{T}_3 \otimes \mathbf{T}_2$: $\{S_3^+ \mid 000\}$; $\mathbf{t}_1$ or $\mathbf{t}_2$; $\mathbf{t}_3$: 1, 3; 2, 3; 3, 3; 4, 3; 5, 3; 6, 3: *e*.

Δ^x $\mathbf{G}_6^1$: $(C_6^+, 0)$: 1, 1; 2, 3; 3, 3; 4, 1; 5, 3; 6, 3: *d*.
U^x $\mathbf{G}_2^1$: $(C_2, 0)$: 1, 1; 2, 1: *b*.
P^x $\mathbf{G}_3^1$: $(C_3^+, 0)$: 1, 1; 2, 3; 3, 3: *a*.
T^x $\mathbf{G}_2^1$: $(\sigma_h, 0)$: 1, 1; 2, 1: *c*.
S^x $\mathbf{G}_2^1$: $(\sigma_h, 0)$: 1, 3; 2, 3: *c*.
T'^x $\mathbf{G}_2^1$: $(\sigma_h, 0)$: 1, 1; 2, 1: *c*.
S'^x $\mathbf{G}_2^1$: $(\sigma_h, 0)$: 1, 3; 2, 3: *c*.
Σ^x $\mathbf{G}_2^1$: $(\sigma_h, 0)$: 1, 1; 2, 1: *c*.
R^x $\mathbf{G}_2^1$: $(\sigma_h, 0)$: 1, 3; 2, 3: *c*.

177 $P622$ D_6^1

($F1$; $K6$; $K7$; $M5$; $Z1$.)

Γ $\mathbf{G}_{12}^3$: $\{C_6^+ \mid 000\}$, $\{C'_{21} \mid 000\}$: 1, 1; 2, 1; 3, 1; 4, 1; 5, 1; 6, 1: *c*.
M $\mathbf{G}_4^2 \otimes \mathbf{T}_2$: $\{C_2 \mid 000\}$, $\{C''_{21} \mid 000\}$; $\mathbf{t}_2$: 1, 1; 2, 1; 3, 1; 4, 1: *b*.
A $\mathbf{G}_{12}^3 \otimes \mathbf{T}_2$: $\{C_6^+ \mid 000\}$, $\{C'_{21} \mid 000\}$; $\mathbf{t}_3$: 1, 1; 2, 1; 3, 1; 4, 1; 5, 1; 6, 1: *c*.
L $\mathbf{G}_4^2 \otimes \mathbf{T}_2$: $\{C_2 \mid 000\}$, $\{C''_{21} \mid 000\}$; $\mathbf{t}_2$ or $\mathbf{t}_3$: 1, 1; 2, 1; 3, 1; 4, 1: *b*.
K $\mathbf{G}_6^2 \otimes \mathbf{T}_3$: $\{C_3^+ \mid 000\}$, $\{C''_{21} \mid 000\}$; $\mathbf{t}_1$ or $\mathbf{t}_2$: 1, 1; 2, 1; 3, 1: *a*.
H $\mathbf{G}_6^2 \otimes \mathbf{T}_3 \otimes \mathbf{T}_2$: $\{C_3^+ \mid 000\}$, $\{C''_{21} \mid 000\}$; $\mathbf{t}_1$ or $\mathbf{t}_2$; $\mathbf{t}_3$: 1, 1; 2, 1; 3, 1: *a*.

Δ^x $\mathbf{G}_6^1$: $(C_6^+, 0)$: 1, 1; 2, 1; 3, 1; 4, 1; 5, 1; 6, 1: *d*.
U^x $\mathbf{G}_2^1$: $(C_2, 0)$: 1, 1; 2, 1: *b*.
P^x $\mathbf{G}_3^1$: $(C_3^+, 0)$: 1, 1; 2, 1; 3, 1: *a*.
T^x $\mathbf{G}_2^1$: $(C''_{22}, 0)$: 1, 1; 2, 1: *b*.
S^x $\mathbf{G}_2^1$: $(C''_{22}, 0)$: 1, 1; 2, 1: *b*.
T'^x $\mathbf{G}_2^1$: $(C''_{21}, 0)$: 1, 1; 2, 1: *b*.
S'^x $\mathbf{G}_2^1$: $(C''_{21}, 0)$: 1, 1; 2, 1: *b*.
Σ^x $\mathbf{G}_2^1$: $(C'_{21}, 0)$: 1, 1; 2, 1: *b*.
R^x $\mathbf{G}_2^1$: $(C'_{21}, 0)$: 1, 1; 2, 1: *b*.

178 $P6_122$ D_6^2

($F1$; $K6$; $K7$; $M5$; $Z1$.)

Γ $\mathbf{G}_{12}^3$: $\{C_6^+ \mid 00\frac{1}{6}\}$, $\{C'_{21} \mid 000\}$: 1, 1; 2, 1; 3, 1; 4, 1; 5, 1; 6, 1: *c*.
M $\mathbf{G}_4^2 \otimes \mathbf{T}_2$: $\{C_2 \mid 00\frac{1}{2}\}$, $\{C''_{21} \mid 00\frac{1}{2}\}$; $\mathbf{t}_2$: 1, 1; 2, 1; 3, 1; 4, 1: *b*.
A $\mathbf{G}_{24}^2$: $\{C_6^+ \mid 00\frac{1}{6}\}$, $\{C'_{21} \mid 000\}$: 7, 1; 8, 1; 9, 1: *a*.
L $\mathbf{G}_8^4$: $\{C_2 \mid 00\frac{1}{2}\}$, $\{C''_{21} \mid 00\frac{1}{2}\}$: 5, 1: *a*.
K $\mathbf{G}_6^2 \otimes \mathbf{T}_3$: $\{C_3^+ \mid 00\frac{1}{3}\}$, $\{C''_{21} \mid 00\frac{1}{2}\}$; $\mathbf{t}_1$ or $\mathbf{t}_2$: 1, 1; 2, 1; 3, 1: *a*.
H $\mathbf{G}_6^2 \otimes \mathbf{T}_3 \otimes \mathbf{T}_2$: $\{C_3^- \mid 00\frac{2}{3}\}$, $\{C''_{21} \mid 00\frac{1}{2}\}$; $\mathbf{t}_1$ or $\mathbf{t}_2$; $\mathbf{t}_3$: 1, 3; 2, 3; 3, 1: *a*.

Δ^x $\mathbf{G}_6^1$: $(C_6^+, 0)$: 1, 1; 2, 1; 3, 1; 4, 1; 5, 1; 6, 1: *d*.
U^x $\mathbf{G}_2^1$: $(C_2, 0)$: 1, 1; 2, 1: *b*.
P^x $\mathbf{G}_3^1$: $(C_3^+, 0)$: 1, 1; 2, 1; 3, 1: *a*.
T^x $\mathbf{G}_2^1$: $(C''_{22}, 0)$: 1, 1; 2, 1: *b*.
S^x $\mathbf{G}_{12}^1$: $(C''_{22}, 0)$: 2, 3; 8, 3: *c*.
T'^x $\mathbf{G}_2^1$: $(C''_{21}, 0)$: 1, 1; 2, 1: *b*.
S'^x $\mathbf{G}_4^1$: $(C''_{21}, 0)$: 2, 3; 4, 3: *a*.
Σ^x $\mathbf{G}_2^1$: $(C'_{21}, 0)$: 1, 1; 2, 1: *b*.
R^x $\mathbf{G}_2^1$: $(C'_{21}, 0)$: 1, 3; 2, 3: *b*.

179 $P6_522$ D_6^3

($F1$; $K6$; $K7$; $M5$; $Z1$.)

Γ $\mathbf{G}_{12}^3$: $\{C_6^+ \mid 00\tfrac{5}{6}\}, \{C_{21}' \mid 000\}$: 1, 1; 2, 1; 3, 1; 4, 1; 5, 1; 6, 1: *c*.
M $\mathbf{G}_4^2 \otimes \mathbf{T}_2$: $\{C_2 \mid 00\tfrac{1}{2}\}, \{C_{21}'' \mid 00\tfrac{1}{2}\}$; $\mathbf{t}_2$: 1, 1; 2, 1; 3, 1; 4, 1: *b*.
A $\mathbf{G}_{24}^2$: $\{C_6^+ \mid 00\tfrac{5}{6}\}, \{C_{21}' \mid 000\}$: 7, 1; 8, 1; 9, 1: *a*.
L $\mathbf{G}_8^4$: $\{C_2 \mid 00\tfrac{1}{2}\}, \{C_{21}'' \mid 00\tfrac{1}{2}\}$: 5, 1: *a*.
K $\mathbf{G}_6^2 \otimes \mathbf{T}_3$: $\{C_3^+ \mid 00\tfrac{2}{3}\}, \{C_{21}'' \mid 00\tfrac{1}{2}\}$; $\mathbf{t}_1$ or $\mathbf{t}_2$: 1, 1; 2, 1; 3, 1: *a*.
H $\mathbf{G}_6^2 \otimes \mathbf{T}_3 \otimes \mathbf{T}_2$: $\{C_3^+ \mid 00\tfrac{2}{3}\}, \{C_{21}'' \mid 00\tfrac{1}{2}\}$; $\mathbf{t}_1$ or $\mathbf{t}_2$; $\mathbf{t}_3$: 1, 3; 2, 3; 3, 1: *a*.

Δ^x $\mathbf{G}_6^1$: $(C_6^+, 0)$: 1, 1; 2, 1; 3, 1; 4, 1; 5, 1; 6, 1: *d*.
U^x $\mathbf{G}_2^1$: $(C_2, 0)$: 1, 1; 2, 1: *b*.
P^x $\mathbf{G}_3^1$: $(C_3^+, 0)$: 1, 1; 2, 1; 3, 1: *a*.
T^x $\mathbf{G}_2^1$: $(C_{22}'', 0)$: 1, 1; 2, 1: *b*.
S^x $\mathbf{G}_{12}^1$: $(C_{22}'', 0)$: 6, 3; 12, 3: *b*.
T'^x $\mathbf{G}_2^1$: $(C_{21}'', 0)$: 1, 1; 2, 1: *b*.
S'^x $\mathbf{G}_4^1$: $(C_{21}'', 0)$: 2, 3; 4, 3: *a*.
Σ^x $\mathbf{G}_2^1$: $(C_{21}', 0)$: 1, 1; 2, 1: *b*.
R^x $\mathbf{G}_2^1$: $(C_{21}', 0)$: 1, 3; 2, 3: *b*.

180 $P6_222$ D_6^4

($F1$; $F7$; $F8$; $K6$; $K7$; $M5$; $Z1$.)

Γ $\mathbf{G}_{12}^3$: $\{C_6^+ \mid 00\tfrac{1}{3}\}, \{C_{21}' \mid 000\}$: 1, 1; 2, 1; 3, 1; 4, 1; 5, 1; 6, 1: *c*.
M $\mathbf{G}_4^2 \otimes \mathbf{T}_2$: $\{C_2 \mid 000\}, \{C_{21}'' \mid 000\}$; $\mathbf{t}_2$: 1, 1; 2, 1; 3, 1; 4, 1: *b*.
A $\mathbf{G}_{12}^3 \otimes \mathbf{T}_2$: $\{C_6^+ \mid 00\tfrac{1}{3}\}, \{C_{21}' \mid 000\}$; $\mathbf{t}_3$: 1, 1; 2, 1; 3, 1; 4, 1; 5, 1; 6, 1: *c*.
L $\mathbf{G}_4^2 \otimes \mathbf{T}_2$: $\{C_2 \mid 000\}, \{C_{21}'' \mid 000\}$; $\mathbf{t}_2$ or $\mathbf{t}_3$: 1, 1; 2, 1; 3, 1; 4, 1: *b*.
K $\mathbf{G}_6^2 \otimes \mathbf{T}_3$: $\{C_3^+ \mid 00\tfrac{2}{3}\}, \{C_{21}'' \mid 000\}$; $\mathbf{t}_1$ or $\mathbf{t}_2$: 1, 1; 2, 1; 3, 1: *a*.
H $\mathbf{G}_6^2 \otimes \mathbf{T}_3 \otimes \mathbf{T}_2$: $\{C_3^+ \mid 00\tfrac{2}{3}\}, \{C_{21}'' \mid 000\}$; $\mathbf{t}_1$ or $\mathbf{t}_2$; $\mathbf{t}_3$: 1, 1; 2, 1; 3, 1: *a*.

Δ^x $\mathbf{G}_6^1$: $(C_6^+, 0)$: 1, 1; 2, 1; 3, 1; 4, 1; 5, 1; 6, 1: *d*.
U^x $\mathbf{G}_2^1$: $(C_2, 0)$: 1, 1; 2, 1: *b*.
P^x $\mathbf{G}_3^1$: $(C_3^+, 0)$: 1, 1; 2, 1; 3, 1: *a*.
T^x $\mathbf{G}_2^1$: $(C_{22}'', 0)$: 1, 1; 2, 1: *b*.
S^x $\mathbf{G}_6^1$: $(C_{22}'', 0)$: 2, 1; 5, 1: *c*.
T'^x $\mathbf{G}_2^1$: $(C_{21}'', 0)$: 1, 1; 2, 1: *b*.
S'^x $\mathbf{G}_2^1$: $(C_{21}'', 0)$: 1, 1; 2, 1: *b*.
Σ^x $\mathbf{G}_2^1$: $(C_{21}', 0)$: 1, 1; 2, 1: *b*.
R^x $\mathbf{G}_2^1$: $(C_{21}', 0)$: 1, 1; 2, 1: *b*.

181 $P6_422$ D_6^5

($F1$; $F7$; $F8$; $K6$; $K7$; $M5$; $Z1$.)

Γ $\mathbf{G}_{12}^3$: $\{C_6^+ \mid 00\tfrac{2}{3}\}, \{C_{21}' \mid 000\}$: 1, 1; 2, 1; 3, 1; 4, 1; 5, 1; 6, 1: *c*.
M $\mathbf{G}_4^2 \otimes \mathbf{T}_2$: $\{C_2 \mid 000\}, \{C_{21}'' \mid 000\}$; $\mathbf{t}_2$: 1, 1; 2, 1; 3, 1; 4, 1: *b*.
A $\mathbf{G}_{12}^3 \otimes \mathbf{T}_2$: $\{C_6^+ \mid 00\tfrac{2}{3}\}, \{C_{21}' \mid 000\}$; $\mathbf{t}_3$: 1, 1; 2, 1; 3, 1; 4, 1; 5, 1; 6, 1: *c*.
L $\mathbf{G}_4^2 \otimes \mathbf{T}_2$: $\{C_2 \mid 000\}, \{C_{21}'' \mid 000\}$; $\mathbf{t}_2$ or $\mathbf{t}_3$: 1, 1; 2, 1; 3, 1; 4, 1: *b*.
K $\mathbf{G}_6^2 \otimes \mathbf{T}_3$: $\{C_3^- \mid 00\tfrac{2}{3}\}, \{C_{21}'' \mid 000\}$; $\mathbf{t}_1$ or $\mathbf{t}_2$: 1, 1; 2, 1; 3, 1: *a*.
H $\mathbf{G}_6^2 \otimes \mathbf{T}_3 \otimes \mathbf{T}_2$: $\{C_3^- \mid 00\tfrac{2}{3}\}, \{C_{21}'' \mid 000\}$; $\mathbf{t}_1$ or $\mathbf{t}_2$; $\mathbf{t}_3$: 1, 1; 2, 1; 3, 1: *a*.

Δ^x $\mathbf{G}_6^1$: $(C_6^+, 0)$: 1, 1; 2, 1; 3, 1; 4, 1; 5, 1; 6, 1: *d*.
U^x $\mathbf{G}_2^1$: $(C_2, 0)$: 1, 1; 2, 1: *b*.
P^x $\mathbf{G}_3^1$: $(C_3^+, 0)$: 1, 1; 2, 1; 3, 1: *a*.
T^x $\mathbf{G}_2^1$: $(C_{22}'', 0)$: 1, 1; 2, 1: *b*.
S^x $\mathbf{G}_6^1$: $(C_{22}'', 0)$: 3, 1; 6, 1: *b*.
T'^x $\mathbf{G}_2^1$: $(C_{21}'', 0)$: 1, 1; 2, 1: *b*.
S'^x $\mathbf{G}_2^1$: $(C_{21}'', 0)$: 1, 1; 2, 1: *b*.
Σ^x $\mathbf{G}_2^1$: $(C_{21}', 0)$: 1, 1; 2, 1: *b*.
R^x $\mathbf{G}_2^1$: $(C_{21}', 0)$: 1, 1; 2, 1: *b*.

182 $P6_322$ D_6^6

($F1$; $K6$; $K7$; $M5$; $Z1$.)

Γ $\mathbf{G}_{12}^3$: $\{C_6^+ \mid 00\frac{1}{2}\}$, $\{C'_{21} \mid 000\}$: 1, 1; 2, 1; 3, 1; 4, 1; 5, 1; 6, 1: *c*.
M $\mathbf{G}_4^2 \otimes \mathbf{T}_2$: $\{C_2 \mid 00\frac{1}{2}\}$, $\{C''_{21} \mid 00\frac{1}{2}\}$; $\mathbf{t}_2$: 1, 1; 2, 1; 3, 1; 4, 1: *b*.
A $\mathbf{G}_{24}^2$: $\{C_6^+ \mid 00\frac{1}{2}\}$, $\{C'_{21} \mid 000\}$: 7, 1; 8, 1; 9, 1: *a*.
L $\mathbf{G}_8^4$: $\{C_2 \mid 00\frac{1}{2}\}$, $\{C''_{21} \mid 00\frac{1}{2}\}$: 5, 1: *a*.
K $\mathbf{G}_6^2 \otimes \mathbf{T}_3$: $\{C_3^+ \mid 000\}$, $\{C''_{21} \mid 00\frac{1}{2}\}$; $\mathbf{t}_1$ or $\mathbf{t}_2$: 1, 1; 2, 1; 3, 1: *a*.
H $\mathbf{G}_6^2 \otimes \mathbf{T}_3 \otimes \mathbf{T}_2$: $\{C_3^+ \mid 000\}$, $\{C''_{21} \mid 00\frac{1}{2}\}$; $\mathbf{t}_1$ or $\mathbf{t}_2$; $\mathbf{t}_3$: 1, 3; 2, 3; 3, 1: *a*.

Δ^x $\mathbf{G}_6^1$: $(C_6^+, 0)$: 1, 1; 2, 1; 3, 1; 4, 1; 5, 1; 6, 1: *d*.
U^x $\mathbf{G}_2^1$: $(C_2, 0)$: 1, 1; 2, 1: *b*.
P^x $\mathbf{G}_3^1$: $(C_3^+, 0)$: 1, 1; 2, 1; 3, 1: *a*.
T^x $\mathbf{G}_2^1$: $(C''_{22}, 0)$: 1, 1; 2, 1: *b*.
S^x $\mathbf{G}_4^1$: $(C''_{22}, 0)$: 2, 3; 4, 3: *a*.
T'^x $\mathbf{G}_2^1$: $(C''_{21}, 0)$: 1, 1; 2, 1: *b*.
S'^x $\mathbf{G}_4^1$: $(C''_{21}, 0)$: 2, 3; 4, 3: *a*.
Σ^x $\mathbf{G}_2^1$: $(C'_{21}, 0)$: 1, 1; 2, 1: *b*.
R^x $\mathbf{G}_2^1$: $(C'_{21}, 0)$: 1, 3; 2, 3: *b*.

183 $P6mm$ C_{6v}^1

($F1$; $K6$; $K7$; $M5$; $Z1$.)

Γ $\mathbf{G}_{12}^3$: $\{C_6^+ \mid 000\}$, $\{\sigma_{v1} \mid 000\}$: 1, 1; 2, 1; 3, 1; 4, 1; 5, 1; 6, 1: *d*.
M $\mathbf{G}_4^2 \otimes \mathbf{T}_2$: $\{C_2 \mid 000\}$, $\{\sigma_{v1} \mid 000\}$; $\mathbf{t}_2$: 1, 1; 2, 1; 3, 1; 4, 1: *c*.
A $\mathbf{G}_{12}^3 \otimes \mathbf{T}_2$: $\{C_6^+ \mid 000\}$, $\{\sigma_{v1} \mid 000\}$; $\mathbf{t}_3$: 1, 1; 2, 1; 3, 1; 4, 1; 5, 1; 6, 1: *d*.
L $\mathbf{G}_4^2 \otimes \mathbf{T}_2$: $\{C_2 \mid 000\}$, $\{\sigma_{v1} \mid 000\}$; $\mathbf{t}_2$ or $\mathbf{t}_3$: 1, 1; 2, 1; 3, 1; 4, 1: *c*.
K $\mathbf{G}_6^2 \otimes \mathbf{T}_3$: $\{C_3^+ \mid 000\}$, $\{\sigma_{d1} \mid 000\}$; $\mathbf{t}_1$ or $\mathbf{t}_2$: 1, 1; 2, 1; 3, 1: *a*.
H $\mathbf{G}_6^2 \otimes \mathbf{T}_3 \otimes \mathbf{T}_2$: $\{C_3^+ \mid 000\}$, $\{\sigma_{d1} \mid 000\}$; $\mathbf{t}_1$ or $\mathbf{t}_2$; $\mathbf{t}_3$: 1, 1; 2, 1; 3, 1: *a*.

Δ^x $\mathbf{G}_{12}^3$: $(C_6^+, 0)$, $(\sigma_{v1}, 0)$: 1, x; 2, x; 3, x; 4, x; 5, x; 6, x: *d*.
U^x $\mathbf{G}_4^2$: $(C_2, 0)$, $(\sigma_{v1}, 0)$: 1, x; 2, x; 3, x; 4, x: *c*.
P^x $\mathbf{G}_6^2$: $(C_3^+, 0)$, $(\sigma_{d1}, 0)$: 1, x; 2, x; 3, x: *a*.
T^x $\mathbf{G}_2^1$: $(\sigma_{d2}, 0)$: 1, 1; 2, 1: *c*.
S^x $\mathbf{G}_2^1$: $(\sigma_{d2}, 0)$: 1, 1; 2, 1: *c*.
T'^x $\mathbf{G}_2^1$: $(\sigma_{d1}, 0)$: 1, 1; 2, 1: *c*.
S'^x $\mathbf{G}_2^1$: $(\sigma_{d1}, 0)$: 1, 1; 2, 1: *c*.
Σ^x $\mathbf{G}_2^1$: $(\sigma_{v1}, 0)$: 1, 1; 2, 1: *c*.
R^x $\mathbf{G}_2^1$: $(\sigma_{v1}, 0)$: 1, 1; 2, 1: *c*.

184 $P6cc$ C_{6v}^2

($F1$; $K6$; $K7$; $M5$; $Z1$.)

Γ $\mathbf{G}_{12}^3$: $\{C_6^+ \mid 000\}$, $\{\sigma_{v1} \mid 00\frac{1}{2}\}$: 1, 1; 2, 1; 3, 1; 4, 1; 5, 1; 6, 1: *d*.
M $\mathbf{G}_4^2 \otimes \mathbf{T}_2$: $\{C_2 \mid 000\}$, $\{\sigma_{v1} \mid 00\frac{1}{2}\}$; $\mathbf{t}_2$: 1, 1; 2, 1; 3, 1; 4, 1: *c*.
A $\mathbf{G}_{24}^3$: $\{C_3^+ \mid 001\}$, $\{\sigma_{v1} \mid 00\frac{1}{2}\}$, $\{C_2 \mid 000\}$: 3, 3; 4, 3; 6, 2; 9, 3; 10, 3; 12, 2: *a*.
L $\mathbf{G}_8^2$: $\{\sigma_{v1} \mid 00\frac{1}{2}\}$, $\{C_2 \mid 000\}$: 2, 3; 4, 3; 6, 3; 8, 3: *d*.
K $\mathbf{G}_6^2 \otimes \mathbf{T}_3$: $\{C_3^+ \mid 000\}$, $\{\sigma_{d1} \mid 00\frac{1}{2}\}$; $\mathbf{t}_1$ or $\mathbf{t}_2$: 1, 1; 2, 1; 3, 1: *a*.
H $\mathbf{G}_{12}^4 \otimes \mathbf{T}_3$: $\{C_3^+ \mid 001\}$, $\{\sigma_{d1} \mid 00\frac{1}{2}\}$; $\mathbf{t}_1$ or $\mathbf{t}_2$: 3, 3; 4, 3; 6, 2: *a*.

Δ^x $\mathbf{G}_{12}^3$: $(C_6^+, 0)$, $(\sigma_{v1}\ 0)$: 1, x; 2, x; 3, x; 4, x; 5, x; 6, x: *d*.
U^x $\mathbf{G}_4^2$: $(C_2, 0)$, $(\sigma_{v1}, 0)$: 1, x; 2, x; 3, x; 4, x: *c*.
P^x $\mathbf{G}_6^2$: $(C_3^+, 0)$, $(\sigma_{d1}, 0)$: 1, x; 2, x; 3, x: *a*.
T^x $\mathbf{G}_2^1$: $(\sigma_{d2}, 0)$: 1, 1; 2, 1: *c*.
S^x $\mathbf{G}_2^1$: $(\sigma_{d2}, 0)$: 1, 3; 2, 3: *c*.
T'^x $\mathbf{G}_2^1$: $(\sigma_{d1}, 0)$: 1, 1; 2, 1: *c*.
S'^x $\mathbf{G}_2^1$: $(\sigma_{d1}, 0)$: 1, 3; 2, 3: *c*.
Σ^x $\mathbf{G}_2^1$: $(\sigma_{v1}, 0)$: 1, 1; 2, 1: *c*.
R^x $\mathbf{G}_2^1$: $(\sigma_{v1}, 0)$: 1, 3; 2, 3: *c*.

185 $P6_3cm$ C_{6v}^3

($F1$; $K6$; $K7$; $M5$; $Z1$.)

Γ $\mathbf{G}_{12}^3$: $\{C_6^+ \mid 00\frac{1}{2}\}$, $\{\sigma_{v1} \mid 00\frac{1}{2}\}$: 1, 1; 2, 1; 3, 1; 4, 1; 5, 1; 6, 1: d.
M $\mathbf{G}_4^2 \otimes \mathbf{T}_2$: $\{C_2 \mid 00\frac{1}{2}\}$, $\{\sigma_{v1} \mid 00\frac{1}{2}\}$; $\mathbf{t}_2$: 1, 1; 2, 1; 3, 1; 4, 1: c.
A $\mathbf{G}_{24}^4$: $\{C_3^+ \mid 000\}$, $\{\sigma_{d1} \mid 000\}$, $\{C_2 \mid 00\frac{1}{2}\}$: 4, 3; 5, 3; 6, 3; 10, 3; 11, 3; 12, 3: a.
L $\mathbf{G}_8^2$: $\{\sigma_{v1} \mid 00\frac{1}{2}\}$, $\{\sigma_{d1} \mid 000\}$: 2, 3; 4, 3; 6, 3; 8, 3: c.
K $\mathbf{G}_6^2 \otimes \mathbf{T}_3$: $\{C_3^+ \mid 000\}$, $\{\sigma_{d1} \mid 000\}$; $\mathbf{t}_1$ or $\mathbf{t}_2$: 1, 1; 2, 1; 3, 1: a.
H $\mathbf{G}_6^2 \otimes \mathbf{T}_3 \otimes \mathbf{T}_2$: $\{C_3^+ \mid 000\}$, $\{\sigma_{d1} \mid 000\}$; $\mathbf{t}_1$ or $\mathbf{t}_2$; $\mathbf{t}_3$: 1, 2; 2, 2; 3, 2: a.

Δ^x $\mathbf{G}_{12}^3$: $(C_6^+, 0)$, $(\sigma_{v1}, 0)$: 1, x; 2, x; 3, x; 4, x; 5, x; 6, x: d.
U^x $\mathbf{G}_4^2$: $(C_2, 0)$, $(\sigma_{v1}, 0)$: 1, x; 2, x; 3, x; 4, x: c.
P^x $\mathbf{G}_6^2$: $(C_3^+, 0)$, $(\sigma_{d1}, 0)$: 1, x; 2, x; 3, x: a.
T^x $\mathbf{G}_2^1$: $(\sigma_{d2}, 0)$: 1, 1; 2, 1: c.
S^x $\mathbf{G}_2^1$: $(\sigma_{d2}, 0)$: 1, 2; 2, 2: c.
T'^x $\mathbf{G}_2^1$: $(\sigma_{d1}, 0)$: 1, 1; 2, 1: c.
S'^x $\mathbf{G}_2^1$: $(\sigma_{d1}, 0)$: 1, 2; 2, 2: c.
Σ^x $\mathbf{G}_2^1$: $(\sigma_{v1}, 0)$: 1, 1; 2, 1: c.
R^x $\mathbf{G}_2^1$: $(\sigma_{v1}, 0)$: 1, 3; 2, 3: c.

186 $P6_3mc$ C_{6v}^4

($A1$; $B3$; $B4$; $C1$; $F1$; $G4$; $H2$; $K6$; $K7$; $M5$; $N1$; $P3$; $P4$; $R2$; $R3$; $S8$; $S10$; $Z1$.)

Γ $\mathbf{G}_{12}^3$: $\{C_6^+ \mid 00\frac{1}{2}\}$, $\{\sigma_{v1} \mid 000\}$: 1, 1; 2, 1; 3, 1; 4, 1; 5, 1; 6, 1: d.
M $\mathbf{G}_4^2 \otimes \mathbf{T}_2$: $\{C_2 \mid 00\frac{1}{2}\}$, $\{\sigma_{v1} \mid 000\}$; $\mathbf{t}_2$: 1, 1; 2, 1; 3, 1; 4, 1: c.
A $\mathbf{G}_{24}^4$: $\{C_3^+ \mid 000\}$, $\{\sigma_{v1} \mid 000\}$, $\{C_2 \mid 00\frac{1}{2}\}$: 4, 3; 5, 3; 6, 3; 10, 3; 11, 3; 12, 3: a.
L $\mathbf{G}_8^2$: $\{\sigma_{d1} \mid 00\frac{1}{2}\}$, $\{\sigma_{v1} \mid 000\}$: 2, 3; 4, 3; 6, 3; 8, 3: c.
K $\mathbf{G}_6^2 \otimes \mathbf{T}_3$: $\{C_3^+ \mid 000\}$, $\{\sigma_{d1} \mid 00\frac{1}{2}\}$; $\mathbf{t}_1$ or $\mathbf{t}_2$: 1, 1; 2, 1; 3, 1: a.
H $\mathbf{G}_{12}^4 \otimes \mathbf{T}_3$: $\{C_3^+ \mid 001\}$, $\{\sigma_{d1} \mid 00\frac{1}{2}\}$; $\mathbf{t}_1$ or $\mathbf{t}_2$: 3, 3; 4, 3; 6, 1: a.

Δ^x $\mathbf{G}_{12}^3$: $(C_6^+, 0)$, $(\sigma_{v1}, 0)$: 1, x; 2, x; 3, x; 4, x; 5, x; 6, x: d.
U^x $\mathbf{G}_4^2$: $(C_2, 0)$, $(\sigma_{v1}, 0)$: 1, x; 2, x; 3, x; 4, x: c.
P^x $\mathbf{G}_6^2$: $(C_3^+, 0)$, $(\sigma_{d1}, 0)$: 1, x; 2, x; 3, x: a.
T^x $\mathbf{G}_2^1$: $(\sigma_{d2}, 0)$: 1, 1; 2, 1: c.
S^x $\mathbf{G}_2^1$: $(\sigma_{d2}, 0)$: 1, 3; 2, 3: c.
T'^x $\mathbf{G}_2^1$: $(\sigma_{d1}, 0)$: 1, 1; 2, 1: c.
S'^x $\mathbf{G}_2^1$: $(\sigma_{d1}, 0)$: 1, 3; 2, 3: c.
Σ^x $\mathbf{G}_2^1$: $(\sigma_{v1}, 0)$: 1, 1; 2, 1: c.
R^x $\mathbf{G}_2^1$: $(\sigma_{v1}, 0)$: 1, 2; 2, 2: c.

187 $P\bar{6}m2$ D_{3h}^1

($F1$; $K6$; $K7$; $M5$; $Z1$.)

Γ $\mathbf{G}_{12}^3$: $\{S_3^+ \mid 000\}$, $\{C'_{21} \mid 000\}$: 1, 1; 2, 1; 3, 1; 4, 1; 5, 1; 6, 1: e.
M $\mathbf{G}_4^2 \otimes \mathbf{T}_2$: $\{C'_{21} \mid 000\}$, $\{\sigma_{v1} \mid 000\}$; $\mathbf{t}_2$: 1, 1; 2, 1; 3, 1; 4, 1: c.
A $\mathbf{G}_{12}^3 \otimes \mathbf{T}_2$: $\{S_3^+ \mid 000\}$, $\{C'_{21} \mid 000\}$; $\mathbf{t}_3$: 1, 1; 2, 1; 3, 1; 4, 1; 5, 1; 6, 1: e.
L $\mathbf{G}_4^2 \otimes \mathbf{T}_2$: $\{C'_{21} \mid 000\}$, $\{\sigma_{v1} \mid 000\}$; $\mathbf{t}_2$ or $\mathbf{t}_3$: 1, 1; 2, 1; 3, 1; 4, 1: c.
K $\mathbf{G}_6^1 \otimes \mathbf{T}_3$: $\{S_3^+ \mid 000\}$; $\mathbf{t}_1$ or $\mathbf{t}_2$: 1, 1; 2, 1; 3, 1; 4, 1; 5, 1; 6, 1: e.
H $\mathbf{G}_6^1 \otimes \mathbf{T}_3 \otimes \mathbf{T}_2$: $\{S_3^+ \mid 000\}$; $\mathbf{t}_1$ or $\mathbf{t}_2$; $\mathbf{t}_3$: 1, 1; 2, 1; 3, 1; 4, 1; 5, 1; 6, 1: e.

Δ^x $\mathbf{G}_6^2$: $(C_3^+, 0)$, $(\sigma_{v1}, 0)$: 1, 1; 2, 1; 3, 1: a.
U^x $\mathbf{G}_2^1$: $(\sigma_{v1}, 0)$: 1, 1; 2, 1: c.
P^x $\mathbf{G}_3^1$: $(C_3^+, 0)$: 1, 1; 2, 1; 3, 1: a.
T^x $\mathbf{G}_2^1$: $(\sigma_h, 0)$: 1, 1; 2, 1: c.
S^x $\mathbf{G}_2^1$: $(\sigma_h, 0)$: 1, 1; 2, 1: c.
T'^x $\mathbf{G}_2^1$: $(\sigma_h, 0)$: 1, 1; 2, 1: c.
S'^x $\mathbf{G}_2^1$: $(\sigma_h, 0)$: 1, 1; 2, 1: c.
Σ^x $\mathbf{G}_4^2$: $(C'_{21}, 0)$, $(\sigma_{v1}, 0)$: 1, x; 2, x; 3, x; 4, x: c.
R^x $\mathbf{G}_4^2$: $(C'_{21}, 0)$, $(\sigma_{v1}, 0)$: 1, x; 2, x; 3, x; 4, x: c.

188 $P\bar{6}c2$ D^2_{3h}

($F1$; $K6$; $K7$; $M5$; $Z1$.)

Γ $\mathbf{G}^3_{12}$: $\{S_3^+ | 000\}$, $\{C'_{21} | 00\tfrac{1}{2}\}$: 1, 1; 2, 1; 3, 1; 4, 1; 5, 1; 6, 1: e.
M $\mathbf{G}^2_4 \otimes \mathbf{T}_2$: $\{C'_{21} | 00\tfrac{1}{2}\}$, $\{\sigma_{v1} | 00\tfrac{1}{2}\}$; $\mathbf{t}_2$: 1, 1; 2, 1; 3, 1; 4, 1: c.
A $\mathbf{G}^1_{24}$: $\{\sigma_{v1} | 00\tfrac{1}{2}\}$, $\{C_3^+ | 000\}$, $\{\sigma_h | 000\}$: 7, 3; 8, 3; 9, 1: b.
L $\mathbf{G}^4_8$: $\{\sigma_{v1} | 00\tfrac{1}{2}\}$, $\{\sigma_h | 000\}$: 5, 1: a.
K $\mathbf{G}^1_6 \otimes \mathbf{T}_3$: $\{S_3^+ | 000\}$; $\mathbf{t}_1$ or $\mathbf{t}_2$: 1, 1; 2, 1; 3, 1; 4, 1; 5, 1; 6, 1: e.
H $\mathbf{G}^1_6 \otimes \mathbf{T}_3 \otimes \mathbf{T}_2$: $\{S_3^+ | 000\}$; $\mathbf{t}_1$ or $\mathbf{t}_2$; $\mathbf{t}_3$: 1, 3; 2, 3; 3, 3; 4, 3; 5, 3; 6, 3: e.

Δ^x $\mathbf{G}^2_6$: $(C_3^+, 0)$, $(\sigma_{v1}, 0)$: 1, 1; 2, 1; 3, 1: a.
U^x $\mathbf{G}^1_2$: $(\sigma_{v1}, 0)$: 1, 1; 2, 1: c.
P^x $\mathbf{G}^1_3$: $(C_3^+, 0)$: 1, 1; 2, 1; 3, 1: a.
T^x $\mathbf{G}^1_2$: $(\sigma_h, 0)$: 1, 1; 2, 1: c.
S^x $\mathbf{G}^1_2$: $(\sigma_h, 0)$: 1, 3; 2, 3: c.
T'^x $\mathbf{G}^1_2$: $(\sigma_h, 0)$: 1, 1; 2, 1: c.
S'^x $\mathbf{G}^1_2$: $(\sigma_h, 0)$: 1, 3; 2, 3: c.
Σ^x $\mathbf{G}^2_4$: $(C'_{21}, 0)$, $(\sigma_{v1}, 0)$: 1, x; 2, x; 3, x; 4, x: c.
R^x $\mathbf{G}^4_8$: $(C'_{21}, 0)$, $(\sigma_h, 0)$: 5, x: a.

189 $P\bar{6}2m$ D^3_{3h}

($F1$; $K6$; $K7$; $M5$; $Z1$.)

Γ $\mathbf{G}^3_{12}$: $\{S_3^+ | 000\}$, $\{C''_{21} | 000\}$: 1, 1; 2, 1; 3, 1; 4, 1; 5, 1; 6, 1: e.
M $\mathbf{G}^2_4 \otimes \mathbf{T}_2$: $\{C''_{21} | 000\}$, $\{\sigma_{d1} | 000\}$; $\mathbf{t}_2$: 1, 1; 2, 1; 3, 1; 4, 1: c.
A $\mathbf{G}^3_{12} \otimes \mathbf{T}_2$: $\{S_3^+ | 000\}$, $\{C''_{21} | 000\}$; $\mathbf{t}_3$: 1, 1; 2, 1; 3, 1; 4, 1; 5, 1; 6, 1: e.
L $\mathbf{G}^2_4 \otimes \mathbf{T}_2$: $\{C''_{21} | 000\}$, $\{\sigma_{d1} | 000\}$; $\mathbf{t}_2$ or $\mathbf{t}_3$: 1, 1; 2, 1; 3, 1; 4, 1: c.
K $\mathbf{G}^3_{12} \otimes \mathbf{T}_3$: $\{S_3^+ | 000\}$, $\{C''_{21} | 000\}$; $\mathbf{t}_1$ or $\mathbf{t}_2$: 1, x; 2, x; 3, x; 4, x; 5, x; 6, x: e.
H $\mathbf{G}^3_{12} \otimes \mathbf{T}_3 \otimes \mathbf{T}_2$: $\{S_3^+ | 000\}$, $\{C''_{21} | 000\}$; $\mathbf{t}_1$ or $\mathbf{t}_2$; $\mathbf{t}_3$: 1, x; 2, x; 3, x; 4, x; 5, x; 6, x: e.

Δ^x $\mathbf{G}^2_6$: $(C_3^+, 0)$, $(\sigma_{d1}, 0)$: 1, 1; 2, 1; 3, 1: a.
U^x $\mathbf{G}^1_2$: $(\sigma_{d1}, 0)$: 1, 1; 2, 1: c.
P^x $\mathbf{G}^2_6$: $(C_3^+, 0)$, $(\sigma_{d1}, 0)$: 1, x; 2, x; 3, x: a.
T^x $\mathbf{G}^2_4$: $(C''_{22}, 0)$, $(\sigma_{d2}, 0)$: 1, x; 2, x; 3, x; 4, x: c.
S^x $\mathbf{G}^2_4$: $(C''_{22}, 0)$, $(\sigma_{d2}, 0)$: 1, x; 2, x; 3, x; 4, x: c.
T'^x $\mathbf{G}^2_4$: $(C''_{21}, 0)$, $(\sigma_{d1}, 0)$: 1, x; 2, x; 3, x; 4, x: c.
S'^x $\mathbf{G}^2_4$: $(C''_{21}, 0)$, $(\sigma_{d1}, 0)$: 1, x; 2, x; 3, x; 4, x: c.
Σ^x $\mathbf{G}^1_2$: $(\sigma_h, 0)$: 1, 1; 2, 1: c.
R^x $\mathbf{G}^1_2$: $(\sigma_h, 0)$: 1, 1; 2, 1: c.

190 $P\bar{6}2c$ D^4_{3h}

($F1$; $K6$; $K7$; $M5$; $Z1$.)

Γ $\mathbf{G}^3_{12}$: $\{S_3^+ | 000\}$, $\{C''_{21} | 00\tfrac{1}{2}\}$: 1, 1; 2, 1; 3, 1; 4, 1; 5, 1; 6, 1: e.
M $\mathbf{G}^2_4 \otimes \mathbf{T}_2$: $\{C''_{21} | 00\tfrac{1}{2}\}$, $\{\sigma_{d1} | 00\tfrac{1}{2}\}$; $\mathbf{t}_2$: 1, 1; 2, 1; 3, 1; 4, 1: c.
A $\mathbf{G}^1_{24}$: $\{\sigma_{d1} | 00\tfrac{1}{2}\}$, $\{C_3^+ | 000\}$, $\{\sigma_h | 000\}$: 7, 3; 8, 3; 9, 1: a.
L $\mathbf{G}^4_8$: $\{\sigma_{d1} | 00\tfrac{1}{2}\}$, $\{\sigma_h | 000\}$: 5, 1: a.
K $\mathbf{G}^3_{12} \otimes \mathbf{T}_3$: $\{S_3^+ | 000\}$, $\{C''_{21} | 00\tfrac{1}{2}\}$; $\mathbf{t}_1$ or $\mathbf{t}_2$: 1, x; 2, x; 3, x; 4, x; 5, x; 6, x: e.
H $\mathbf{G}^1_{24} \otimes \mathbf{T}_3$: $\{\sigma_{d1} | 00\tfrac{1}{2}\}$, $\{C_3^+ | 000\}$, $\{\sigma_h | 000\}$; $\mathbf{t}_1$ or $\mathbf{t}_2$: 7, x; 8, x; 9, x: a.

Δ^x $\mathbf{G}^2_6$: $(C_3^+, 0)$, $(\sigma_{d1}, 0)$: 1, 1; 2, 1; 3, 1: a.
U^x $\mathbf{G}^1_2$: $(\sigma_{d1}, 0)$: 1, 1; 2, 1: c.
P^x $\mathbf{G}^2_6$: $(C_3^+, 0)$, $(\sigma_{d1}, 0)$: 1, x; 2, x; 3, x: a.
T^x $\mathbf{G}^2_4$: $(C''_{22}, 0)$, $(\sigma_{d2}, 0)$: 1, x; 2, x; 3, x; 4, x: c.
S^x $\mathbf{G}^4_8$: $(C''_{22}, 0)$, $(\sigma_h, 0)$: 5, x: a.
T'^x $\mathbf{G}^2_4$: $(C''_{21}, 0)$, $(\sigma_{d1}, 0)$: 1, x; 2, x; 3, x; 4, x: c.
S'^x $\mathbf{G}^4_8$: $(C''_{21}, 0)$, $(\sigma_h, 0)$: 5, x: a.
Σ^x $\mathbf{G}^1_2$: $(\sigma_h, 0)$: 1, 1; 2, 1: c.
R^x $\mathbf{G}^1_2$: $(\sigma_h, 0)$: 1, 3; 2, 3: c.

191 *P6/mmm* D_{6h}^1

($F1$; $K6$; $K7$; $M5$; $Z1$.)

Γ $\mathbf{G}_{24}^5$: $\{C_3^+ \mid 000\}$, $\{C'_{21} \mid 000\}$, $\{C_2 \mid 000\}$, $\{I \mid 000\}$: 1, 1; 2, 1; 3, 1; 4, 1; 5, 1; 6, 1; 7, 1; 8, 1; 9, 1; 10, 1; 11, 1; 12, 1: *a*.

M $\mathbf{G}_8^3 \otimes \mathbf{T}_2$: $\{C_2 \mid 000\}$, $\{C''_{21} \mid 000\}$, $\{I \mid 000\}$; $\mathbf{t}_2$: 1, 1; 2, 1; 3, 1; 4, 1; 5, 1; 6, 1; 7, 1; 8, 1: *b*.

A $\mathbf{G}_{24}^5 \otimes \mathbf{T}_2$: $\{C_3^+ \mid 000\}$, $\{C'_{21} \mid 000\}$, $\{C_2 \mid 000\}$, $\{I \mid 000\}$; $\mathbf{t}_3$: 1, 1; 2, 1; 3, 1; 4, 1; 5, 1; 6, 1; 7, 1; 8, 1; 9, 1; 10, 1; 11, 1; 12, 1: *a*.

L $\mathbf{G}_8^3 \otimes \mathbf{T}_2$: $\{C_2 \mid 000\}$, $\{C''_{21} \mid 000\}$, $\{I \mid 000\}$; $\mathbf{t}_2$ or $\mathbf{t}_3$: 1, 1; 2, 1; 3, 1; 4, 1; 5, 1; 6, 1; 7, 1; 8, 1: *b*.

K $\mathbf{G}_{12}^3 \otimes \mathbf{T}_3$: $\{S_3^+ \mid 000\}$, $\{C''_{21} \mid 000\}$; $\mathbf{t}_1$ or $\mathbf{t}_2$: 1, 1; 2, 1; 3, 1; 4, 1; 5, 1; 6, 1: *e*.

H $\mathbf{G}_{12}^3 \otimes \mathbf{T}_3 \otimes \mathbf{T}_2$: $\{S_3^+ \mid 000\}$, $\{C''_{21} \mid 000\}$; $\mathbf{t}_1$ or $\mathbf{t}_2$; $\mathbf{t}_3$: 1, 1; 2, 1; 3, 1; 4, 1; 5, 1; 6, 1: *e*.

Δ^x $\mathbf{G}_{12}^3$: $(C_6^+, 0)$, $(\sigma_{v1}, 0)$: 1, 1; 2, 1; 3, 1; 4, 1; 5, 1; 6, 1: *d*.

U^x $\mathbf{G}_4^2$: $(C_2, 0)$, $(\sigma_{v1}, 0)$: 1, 1; 2, 1; 3, 1; 4, 1: *c*.

P^x $\mathbf{G}_6^2$: $(C_3^+, 0)$, $(\sigma_{d1}, 0)$: 1, 1; 2, 1; 3, 1: *a*.

T^x $\mathbf{G}_4^2$: $(C''_{22}, 0)$, $(\sigma_{d2}, 0)$: 1, 1; 2, 1; 3, 1; 4, 1: *c*.

S^x $\mathbf{G}_4^2$: $(C''_{22}, 0)$, $(\sigma_{d2}, 0)$: 1, 1; 2, 1; 3, 1; 4, 1: *c*.

T'^x $\mathbf{G}_4^2$: $(C''_{21}, 0)$, $(\sigma_{d1}, 0)$: 1, 1; 2, 1; 3, 1; 4, 1: *c*.

S'^x $\mathbf{G}_4^2$: $(C''_{21}, 0)$, $(\sigma_{d1}, 0)$: 1, 1; 2, 1; 3, 1; 4, 1: *c*.

Σ^x $\mathbf{G}_4^2$: $(C'_{21}, 0)$, $(\sigma_{v1}, 0)$: 1, 1; 2, 1; 3, 1; 4, 1: *c*.

R^x $\mathbf{G}_4^2$: $(C'_{21}, 0)$, $(\sigma_{v1}, 0)$: 1, 1; 2, 1; 3, 1; 4, 1: *c*.

192 *P6/mcc* D_{6h}^2

($F1$; $K6$; $K7$; $M5$; $Z1$.)

Γ $\mathbf{G}_{24}^5$: $\{C_3^+ \mid 000\}$, $\{C'_{21} \mid 00\tfrac{1}{2}\}$, $\{C_2 \mid 000\}$, $\{I \mid 000\}$: 1, 1; 2, 1; 3, 1; 4, 1; 5, 1; 6, 1; 7, 1; 8, 1; 9, 1; 10, 1; 11, 1; 12, 1: *a*.

M $\mathbf{G}_8^3 \otimes \mathbf{T}_2$: $\{C_2 \mid 000\}$, $\{C''_{21} \mid 00\tfrac{1}{2}\}$, $\{I \mid 000\}$; $\mathbf{t}_2$: 1, 1; 2, 1; 3, 1; 4, 1; 5, 1; 6, 1; 7, 1; 8, 1: *b*.

A $\mathbf{G}_{48}^2$: $\{\sigma_{d1} \mid 00\tfrac{1}{2}\}$, $\{C_3^+ \mid 000\}$, $\{\sigma_h \mid 000\}$, $\{C_2 \mid 000\}$: 7, 3; 8, 3; 9, 1; 16, 3; 17, 3; 18, 1: *a*.

L $\mathbf{G}_{16}^9$: $\{\sigma_{v1} \mid 00\tfrac{1}{2}\}$, $\{I \mid 000\}$, $\{C_2 \mid 000\}$: 5, 1; 10, 1: *a*.

K $\mathbf{G}_{12}^3 \otimes \mathbf{T}_3$: $\{S_3^+ \mid 000\}$, $\{C''_{21} \mid 00\tfrac{1}{2}\}$; $\mathbf{t}_1$ or $\mathbf{t}_2$: 1, 1; 2, 1; 3, 1; 4, 1; 5, 1; 6, 1: *e*.

H $\mathbf{G}_{24}^1 \otimes \mathbf{T}_3$: $\{\sigma_{d1} \mid 00\tfrac{1}{2}\}$, $\{C_3^+ \mid 000\}$, $\{\sigma_h \mid 000\}$; $\mathbf{t}_1$ or $\mathbf{t}_2$: 7, 3; 8, 3; 9, 1: *a*.

Δ^x $\mathbf{G}_{12}^3$: $(C_6^+, 0)$, $(\sigma_{v1}, 0)$: 1, 1; 2, 1; 3, 1; 4, 1; 5, 1; 6, 1: *d*.

U^x $\mathbf{G}_4^2$: $(C_2, 0)$, $(\sigma_{v1}, 0)$: 1, 1; 2, 1; 3, 1; 4, 1: *c*.

P^x $\mathbf{G}_6^2$: $(C_3^+, 0)$, $(\sigma_{d1}, 0)$: 1, 1; 2, 1; 3, 1: *a*.

T^x $\mathbf{G}_4^2$: $(C''_{22}, 0)$, $(\sigma_{d2}, 0)$: 1, 1; 2, 1; 3, 1; 4, 1: *c*.

S^x $\mathbf{G}_8^4$: $(C''_{22}, 0)$, $(\sigma_h, 0)$: 5, 1: *a*.

T'^x $\mathbf{G}_4^2$: $(C''_{21}, 0)$, $(\sigma_{d1}, 0)$: 1, 1; 2, 1; 3, 1; 4, 1: *c*.

S'^x $\mathbf{G}_8^4$: $(C''_{21}, 0)$, $(\sigma_h, 0)$: 5, 1: *a*.

Σ^x $\mathbf{G}_4^2$: $(C'_{21}, 0)$, $(\sigma_{v1}, 0)$: 1, 1; 2, 1; 3, 1; 4, 1: *c*.

R^x $\mathbf{G}_8^4$: $(C'_{21}, 0)$, $(\sigma_h, 0)$: 5, 1: *a*.

193 $P6_3/mcm$ D_{6h}^3

($F1$; $K6$; $K7$; $M5$; $Z1$.)

Γ $\mathbf{G}_{24}^5$: $\{C_3^+ \mid 000\}$, $\{C'_{21} \mid 000\}$, $\{C_2 \mid 00\frac{1}{2}\}$, $\{I \mid 000\}$: 1, 1; 2, 1; 3, 1; 4, 1; 5, 1; 6, 1; 7, 1; 8, 1; 9, 1; 10, 1; 11, 1; 12, 1: *a*.
M $\mathbf{G}_8^3 \otimes \mathbf{T}_2$: $\{C_2 \mid 00\frac{1}{2}\}$, $\{C''_{21} \mid 00\frac{1}{2}\}$, $\{I \mid 000\}$; $\mathbf{t}_2$: 1, 1; 2, 1; 3, 1; 4, 1; 5, 1; 6, 1; 7, 1; 8, 1: *b*.
A $\mathbf{G}_{48}^1$: $\{C_6^+ \mid 00\frac{1}{2}\}$, $\{C'_{21} \mid 000\}$, $\{I \mid 000\}$: 13, 1; 14, 1; 15, 1: *a*.
L $\mathbf{G}_{16}^9$: $\{C_2 \mid 00\frac{1}{2}\}$, $\{I \mid 000\}$, $\{\sigma_{d1} \mid 000\}$: 5, 1; 10, 1: *b*.
K $\mathbf{G}_{12}^3 \otimes \mathbf{T}_3$: $\{S_3^+ \mid 00\frac{1}{2}\}$, $\{C''_{21} \mid 00\frac{1}{2}\}$; $\mathbf{t}_1$ or $\mathbf{t}_2$: 1, 1; 2, 1; 3, 1; 4, 1; 5, 1; 6, 1: *e*.
H $\mathbf{G}_{12}^3 \otimes \mathbf{T}_3 \otimes \mathbf{T}_2$: $\{S_3^+ \mid 00\frac{1}{2}\}$, $\{C''_{21} \mid 00\frac{1}{2}\}$; $\mathbf{t}_1$ or $\mathbf{t}_2$; $\mathbf{t}_3$: 1, 3; 2, 3; 3, 3; 4, 3; 5, 3; 6, 3: *e*.

Δ^x $\mathbf{G}_{12}^3$: $(C_6^+, 0)$, $(\sigma_{v1}, 0)$: 1, 1; 2, 1; 3, 1; 4, 1; 5, 1; 6, 1: *d*.
U^x $\mathbf{G}_4^2$: $(C_2, 0)$, $(\sigma_{v1}, 0)$: 1, 1; 2, 1; 3, 1; 4, 1: *c*.
P^x $\mathbf{G}_6^2$: $(C_3^+, 0)$, $(\sigma_{d1}, 0)$: 1, 1; 2, 1; 3, 1: *a*.
T^x $\mathbf{G}_4^2$: $(C''_{22}, 0)$, $(\sigma_{d2}, 0)$: 1, 1; 2, 1; 3, 1; 4, 1: *c*.
S^x $\mathbf{G}_8^2$: $(C''_{22}, 0)$, $(\sigma_{d2}, 0)$: 2, 3; 4, 3; 6, 3; 8, 3: *c*.
T'^x $\mathbf{G}_4^2$: $(C''_{21}, 0)$, $(\sigma_{d1}, 0)$: 1, 1; 2, 1; 3, 1; 4, 1: *c*.
S'^x $\mathbf{G}_8^2$: $(C''_{21}, 0)$, $(\sigma_{d1}, 0)$: 2, 3; 4, 3; 6, 3; 8, 3: *c*.
Σ^x $\mathbf{G}_4^2$: $(C'_{21}, 0)$, $(\sigma_{v1}, 0)$: 1, 1; 2, 1; 3, 1; 4, 1: *c*.
R^x $\mathbf{G}_8^4$: $(\sigma_h, 0)$, $(C'_{21}, 0)$: 5, 1: *a*.

194 $P6_3/mmc$ D_{6h}^4

($A2$; $A3$; $A5$; $B2$; $C5$; $E1$; $E3$; $F1$; $F2$; $F4$; $H1$; $H2$; $J1$; $K5$; $K6$; $K7$; $M3$; $M5$; $S2$; $S3$; $S8$; $Z1$.)

Γ $\mathbf{G}_{24}^5$: $\{C_3^+ \mid 000\}$, $\{C'_{21} \mid 00\frac{1}{2}\}$, $\{C_2 \mid 00\frac{1}{2}\}$, $\{I \mid 000\}$: 1, 1; 2, 1; 3, 1; 4, 1; 5, 1; 6, 1; 7, 1; 8, 1; 9, 1; 10, 1; 11, 1; 12, 1: *a*.
M $\mathbf{G}_8^3 \otimes \mathbf{T}_2$: $\{C_2 \mid 00\frac{1}{2}\}$, $\{C''_{21} \mid 000\}$, $\{I \mid 000\}$; $\mathbf{t}_2$: 1, 1; 2, 1; 3, 1; 4, 1; 5, 1; 6, 1; 7, 1; 8, 1: *b*.
A $\mathbf{G}_{48}^1$: $\{C_6^+ \mid 00\frac{1}{2}\}$, $\{C''_{21} \mid 000\}$, $\{I \mid 000\}$: 13, 1; 14, 1; 15, 1: *a*.
L $\mathbf{G}_{16}^9$: $\{C_2 \mid 00\frac{1}{2}\}$, $\{I \mid 000\}$, $\{\sigma_{v1} \mid 000\}$: 5, 1; 10, 1: *b*.
K $\mathbf{G}_{12}^3 \otimes \mathbf{T}_3$: $\{S_3^+ \mid 00\frac{1}{2}\}$, $\{C''_{21} \mid 000\}$; $\mathbf{t}_1$ or $\mathbf{t}_2$: 1, 1; 2, 1; 3, 1; 4, 1; 5, 1; 6, 1: *e*.
H $\mathbf{G}_{24}^1 \otimes \mathbf{T}_3$: $\{\sigma_{d1} \mid 00\frac{1}{2}\}$, $\{C_3^+ \mid 000\}$, $\{\sigma_h \mid 00\frac{1}{2}\}$; $\mathbf{t}_1$ or $\mathbf{t}_2$: 7, 1; 8, 1; 9, 1: *a*.

Δ^x $\mathbf{G}_{12}^3$: $(C_6^+, 0)$, $(\sigma_{v1}, 0)$: 1, 1; 2, 1; 3, 1; 4, 1; 5, 1; 6, 1: *d*.
U^x $\mathbf{G}_4^2$: $(C_2, 0)$, $(\sigma_{v1}, 0)$: 1, 1; 2, 1; 3, 1; 4, 1: *c*.
P^x $\mathbf{G}_6^2$: $(C_3^+, 0)$, $(\sigma_{d1}, 0)$: 1, 1; 2, 1; 3, 1: *a*.
T^x $\mathbf{G}_4^2$: $(C''_{22}, 0)$, $(\sigma_{d2}, 0)$: 1, 1; 2, 1; 3, 1; 4, 1: *c*.
S^x $\mathbf{G}_8^4$: $(\sigma_h, 0)$, $(C''_{22}, 0)$: 5, 1: *a*.
T'^x $\mathbf{G}_4^2$: $(C''_{21}, 0)$, $(\sigma_{d1}, 0)$: 1, 1; 2, 1; 3, 1; 4, 1: *c*.
S'^x $\mathbf{G}_8^4$: $(\sigma_h, 0)$, $(C''_{21}, 0)$: 5, 1: *a*.
Σ^x $\mathbf{G}_4^2$: $(C'_{21}, 0)$, $(\sigma_{v1}, 0)$: 1, 1; 2, 1; 3, 1; 4, 1: *c*.
R^x $\mathbf{G}_8^2$: $(\sigma_h, 0)$, $(\sigma_{v1}, 0)$: 2, 3; 4, 3; 6, 3; 8, 3: *c*.

195 *P*23 T^1

(*F*1; *K*7; *M*5; *T*5; *Z*1.)

Γ $\mathbf{G}_{12}^5$: $\{C_{31}^+|000\}, \{C_{2z}|000\}, \{C_{2y}|000\}$: 1, 1; 2, 3; 3, 3; 4, 1: *a*.
X $\mathbf{G}_4^2 \otimes \mathbf{T}_2$: $\{C_{2z}|000\}, \{C_{2y}|000\}$; $\mathbf{t}_2$: 1, 1; 2, 1; 3, 1; 4, 1: *b*.
M $\mathbf{G}_4^2 \otimes \mathbf{T}_2$: $\{C_{2z}|000\}, \{C_{2y}|000\}$; $\mathbf{t}_1$ or $\mathbf{t}_2$: 1, 1; 2, 1; 3, 1; 4, 1: *b*.
R $\mathbf{G}_{12}^5 \otimes \mathbf{T}_2$: $\{C_{31}^+|000\}, \{C_{2z}|000\}, \{C_{2y}|000\}$; $\mathbf{t}_1$ or $\mathbf{t}_2$ or $\mathbf{t}_3$: 1, 1; 2, 3; 3, 3; 4, 1: *a*.

Δ^x $\mathbf{G}_2^1$: $(C_{2y}, 0)$: 1, 1; 2, 1: *b*.
Σ^x $\mathbf{G}_1^1$: $(E, 0)$: 1, 1: *a*.
Λ^x $\mathbf{G}_3^1$: $(C_{31}^+, 0)$: 1, *x*; 2, *x*; 3, *x*: *a*.
S^x $\mathbf{G}_1^1$: $(E, 0)$: 1, 1: *a*.
Z^x $\mathbf{G}_2^1$: $(C_{2x}, 0)$: 1, 1; 2, 1: *b*.
T^x $\mathbf{G}_2^1$: $(C_{2z}, 0)$: 1, 1; 2, 1: *b*.

196 *F*23 T^2

(*F*1; *K*7; *M*5; *T*5; *Z*1.)

Γ $\mathbf{G}_{12}^5$: $\{C_{31}^+|000\}, \{C_{2z}|000\}, \{C_{2y}|000\}$: 1, 1; 2, 3; 3, 3; 4, 1: *a*.
X $\mathbf{G}_4^2 \otimes \mathbf{T}_2$: $\{C_{2z}|000\}, \{C_{2y}|000\}$; $\mathbf{t}_1$ or $\mathbf{t}_3$: 1, 1; 2, 1; 3, 1; 4, 1: *b*.
L $\mathbf{G}_3^1 \otimes \mathbf{T}_2$: $\{C_{31}^+|000\}$; $\mathbf{t}_1$ or $\mathbf{t}_2$ or $\mathbf{t}_3$: 1, 1; 2, 3; 3, 3: *a*.
W $\mathbf{G}_2^1 \otimes \mathbf{T}_4$: $\{C_{2x}|000\}$; $\mathbf{t}_2$ or $\mathbf{t}_3$: 1, 1; 2, 1: *b*.

Δ^x $\mathbf{G}_2^1$: $(C_{2y}, 0)$: 1, 1; 2, 1: *b*.
Λ^x $\mathbf{G}_3^1$: $(C_{31}^+, 0)$: 1, *x*; 2, *x*; 3, *x*: *a*.
Σ^x $\mathbf{G}_1^1$: $(E, 0)$: 1, 1: *a*.
S^x $\mathbf{G}_1^1$: $(E, 0)$: 1, 1: *a*.
Z^x $\mathbf{G}_2^1$: $(C_{2x}, 0)$: 1, 1; 2, 1: *b*.
Q^x $\mathbf{G}_1^1$: $(E, 0)$: 1, *x*: *a*.

197 *I*23 T^3

(*F*1; *K*7; *M*5; *T*5; *Z*1.)

Γ $\mathbf{G}_{12}^5$: $\{C_{31}^+|000\}, \{C_{2z}|000\}, \{C_{2y}|000\}$: 1, 1; 2, 3; 3, 3; 4, 1: *a*.
H $\mathbf{G}_{12}^5 \otimes \mathbf{T}_2$: $\{C_{31}^+|000\}, \{C_{2z}|000\}, \{C_{2y}|000\}$; $\mathbf{t}_1$ or $\mathbf{t}_2$ or $\mathbf{t}_3$: 1, 1; 2, 3; 3, 3; 4, 1: *a*.
P $\mathbf{G}_{12}^5 \otimes \mathbf{T}_4$: $\{C_{31}^+|000\}, \{C_{2z}|000\}, \{C_{2y}|000\}$; $\mathbf{t}_1$ or $\mathbf{t}_2$ or $\mathbf{t}_3$: 1, *x*; 2, *x*; 3, *x*; 4, *x*: *a*.
N $\mathbf{G}_2^1 \otimes \mathbf{T}_2$: $\{C_{2z}|000\}$; $\mathbf{t}_3$: 1, 1; 2, 1: *b*.

Σ^x $\mathbf{G}_1^1$: $(E, 0)$: 1, 1: *a*.
Δ^x $\mathbf{G}_2^1$: $(C_{2y}, 0)$: 1, 1; 2, 1: *b*.
Λ^x $\mathbf{G}_3^1$: $(C_{31}^+, 0)$: 1, *x*; 2, *x*; 3, *x*: *a*.
D^x $\mathbf{G}_2^1$: $(C_{2z}, 0)$: 1, *x*; 2, *x*: *b*.
G^x $\mathbf{G}_1^1$: $(E, 0)$: 1, 1: *a*.
F^x $\mathbf{G}_3^1$: $(C_{34}^+, 0)$: 1, *x*; 2, *x*; 3, *x*: *a*.

198 $P2_13$ T^4

($F1$; $K7$; $M5$; $T5$; $Z1$.)

Γ $\mathbf{G}_{12}^5$: $\{C_{31}^+ \mid 000\}$, $\{C_{2z} \mid \frac{1}{2}0\frac{1}{2}\}$, $\{C_{2y} \mid 0\frac{1}{2}\frac{1}{2}\}$: 1, 1; 2, 3; 3, 3; 4, 1: *a*.
X $\mathbf{G}_8^4$: $\{C_{2y} \mid 0\frac{1}{2}\frac{1}{2}\}$, $\{C_{2z} \mid \frac{1}{2}0\frac{1}{2}\}$: 5, 1: *a*.
M $\mathbf{G}_8^2$: $\{C_{2y} \mid 0\frac{1}{2}\frac{1}{2}\}$, $\{C_{2z} \mid \frac{1}{2}0\frac{1}{2}\}$: 2, 3; 4, 3; 6, 3; 8, 3: *d*.
R $\mathbf{G}_{24}^9$: $\{C_{31}^- \mid 010\}$, $\{C_{2x} \mid \frac{1}{2}\frac{3}{2}0\}$, $\{C_{2y} \mid 0\frac{3}{2}\frac{1}{2}\}$: 4, 2; 5, 3; 6, 3: *a*.

Δ^x $\mathbf{G}_2^1$: $(C_{2y}, 0)$: 1, 1; 2, 1: *b*.
Σ^x $\mathbf{G}_1^1$: $(E, 0)$: 1, 1: *a*.
Λ^x $\mathbf{G}_3^1$: $(C_{31}^+, 0)$: 1, x; 2, x; 3, x: *a*.
S^x $\mathbf{G}_1^1$: $(E, 0)$: 1, 2: *a*.
Z^x $\mathbf{G}_4^1$: $(C_{2x}, 0)$: 2, 3; 4, 3: *a*.
T^x $\mathbf{G}_4^1$: $(C_{2z}, 0)$: 2, 2; 4, 2: *a*.

199 $I2_13$ T^5

($C6$; $F1$; $K7$; $M5$; $T5$; $Z1$.)

Γ $\mathbf{G}_{12}^5$: $\{C_{31}^+ \mid 000\}$, $\{C_{2z} \mid \frac{1}{2}0\frac{1}{2}\}$, $\{C_{2y} \mid 0\frac{1}{2}\frac{1}{2}\}$: 1, 1; 2, 3; 3, 3; 4, 1: *a*.
H $\mathbf{G}_{24}^8$: $\{C_{31}^- \mid 111\}$, $\{C_{2x} \mid \frac{1}{2}\frac{1}{2}0\}$, $\{C_{2y} \mid 0\frac{1}{2}\frac{1}{2}\}$: 4, 1; 5, 3; 6, 3; 8, 1: *a*.
P $\mathbf{G}_{48}^3$: $\{C_{31}^+ \mid 100\}$, $\{C_{2x} \mid \frac{1}{2}\frac{1}{2}0\}$, $\{C_{2z} \mid \frac{3}{2}1\frac{1}{2}\}$: 7, x; 8, x; 9, x: *a*.
N $\mathbf{G}_2^1 \otimes \mathbf{T}_2$: $\{C_{2z} \mid \frac{1}{2}0\frac{1}{2}\}$; $\mathbf{t}_3$: 1, 1; 2, 1: *b*.

Σ^x $\mathbf{G}_1^1$: $(E, 0)$: 1, 1: *a*.
Δ^x $\mathbf{G}_2^1$: $(C_{2y}, 0)$: 1, 1; 2, 1: *b*.
Λ^x $\mathbf{G}_3^1$: $(C_{31}^+, 0)$: 1, x; 2, x; 3, x: *a*.
D^x $\mathbf{G}_4^1$: $(C_{2z}, 0)$: 2, x; 4, x: *a*.
G^x $\mathbf{G}_1^1$: $(E, 0)$: 1, 1: *a*.
F^x $\mathbf{G}_6^1$: $(C_{34}^+, 0)$: 2, x; 4, x; 6, x: *f*.

200 $Pm3$ T_h^1

($F1$; $K7$; $M5$; $T5$; $Z1$.)

Γ $\mathbf{G}_{24}^{10}$: $\{S_{61}^+ \mid 000\}$, $\{C_{2z} \mid 000\}$, $\{C_{2y} \mid 000\}$: 1, 1; 2, 3; 3, 3; 4, 1; 5, 1; 6, 3; 7, 3; 8, 1: *a*.
X $\mathbf{G}_8^3 \otimes \mathbf{T}_2$: $\{C_{2z} \mid 000\}$, $\{C_{2y} \mid 000\}$, $\{I \mid 000\}$; $\mathbf{t}_2$: 1, 1; 2, 1; 3, 1; 4, 1; 5, 1; 6, 1; 7, 1; 8, 1: *b*.
M $\mathbf{G}_8^3 \otimes \mathbf{T}_2$: $\{C_{2z} \mid 000\}$, $\{C_{2y} \mid 000\}$, $\{I \mid 000\}$; $\mathbf{t}_1$ or $\mathbf{t}_2$: 1, 1; 2, 1; 3, 1; 4, 1; 5, 1; 6, 1; 7, 1; 8, 1: *b*.
R $\mathbf{G}_{24}^{10} \otimes \mathbf{T}_2$: $\{S_{61}^+ \mid 000\}$, $\{C_{2z} \mid 000\}$, $\{C_{2y} \mid 000\}$; $\mathbf{t}_1$ or $\mathbf{t}_2$ or $\mathbf{t}_3$: 1, 1; 2, 3; 3, 3; 4, 1; 5, 1; 6, 3; 7, 3; 8, 1: *a*.

Δ^x $\mathbf{G}_4^2$: $(C_{2y}, 0)$, $(\sigma_x, 0)$: 1, 1; 2, 1; 3, 1; 4, 1: *c*.
Σ^x $\mathbf{G}_2^1$: $(\sigma_z, 0)$: 1, 1; 2, 1: *c*.
Λ^x $\mathbf{G}_3^1$: $(C_{31}^+, 0)$: 1, 1; 2, 3; 3, 3: *a*.
S^x $\mathbf{G}_2^1$: $(\sigma_y, 0)$: 1, 1; 2, 1: *c*.
Z^x $\mathbf{G}_4^2$: $(C_{2x}, 0)$, $(\sigma_z, 0)$: 1, 1; 2, 1; 3, 1; 4, 1: *c*.
T^x $\mathbf{G}_4^2$: $(C_{2z}, 0)$, $(\sigma_y, 0)$: 1, 1; 2, 1; 3, 1; 4, 1: *c*.

201 $Pn3$ T_h^2

($F1$; $K7$; $M5$; $T5$; $Z1$.)

Γ $\mathbf{G}_{24}^{10}$: $\{S_{61}^+ | \frac{1}{2}\frac{1}{2}\frac{1}{2}\}$, $\{C_{2z} | 000\}$, $\{C_{2y} | 000\}$: 1, 1; 2, 3; 3, 3; 4, 1; 5, 1; 6, 3; 7, 3; 8, 1: *a*.
X $\mathbf{G}_{16}^{9}$: $\{\sigma_z | \frac{1}{2}\frac{1}{2}\frac{1}{2}\}$, $\{\sigma_y | \frac{1}{2}\frac{1}{2}\frac{1}{2}\}$, $\{C_{2y} | 000\}$: 5, 1; 10, 1: *a*.
M $\mathbf{G}_{16}^{9}$: $\{\sigma_x | \frac{1}{2}\frac{1}{2}\frac{1}{2}\}$, $\{\sigma_z | \frac{1}{2}\frac{1}{2}\frac{1}{2}\}$, $\{C_{2z} | 000\}$: 5, 1; 10, 1: *a*.
R $\mathbf{G}_{24}^{10} \otimes \mathbf{T}_2$: $\{S_{61}^+ | \frac{1}{2}\frac{1}{2}\frac{1}{2}\}$, $\{C_{2z} | 000\}$, $\{C_{2y} | 000\}$; $\mathbf{t}_1$ or $\mathbf{t}_2$ or $\mathbf{t}_3$: 1, 1; 2, 3; 3, 3; 4, 1; 5, 1; 6, 3; 7, 3; 8, 1: *a*.

Δ^x $\mathbf{G}_4^2$: $(C_{2y}, 0)$, $(\sigma_x, 0)$: 1, 1; 2, 1; 3, 1; 4, 1: *c*.
Σ^x $\mathbf{G}_2^1$: $(\sigma_z, 0)$: 1, 1; 2, 1: *c*.
Λ^x $\mathbf{G}_3^1$: $(C_{31}^+, 0)$: 1, 1; 2, 3; 3, 3: *a*.
S^x $\mathbf{G}_4^1$: $(\sigma_y, 0)$: 2, 1; 4, 1: *a*.
Z^x $\mathbf{G}_8^4$: $(\sigma_y, 0)$, $(C_{2x}, 0)$: 5, 1: *a*.
T^x $\mathbf{G}_8^2$: $(\sigma_y, 0)$, $(C_{2z}, 0)$: 2, 1; 4, 1; 6, 1; 8, 1: *d*.

202 $Fm3$ T_h^3

($F1$; $K7$; $M5$; $T5$; $Z1$.)

Γ $\mathbf{G}_{24}^{10}$: $\{S_{61}^+ | 000\}$, $\{C_{2z} | 000\}$, $\{C_{2y} | 000\}$: 1, 1; 2, 3; 3, 3; 4, 1; 5, 1; 6, 3; 7, 3; 8, 1: *a*.
X $\mathbf{G}_8^3 \otimes \mathbf{T}_2$: $\{C_{2z} | 000\}$, $\{C_{2y} | 000\}$, $\{I | 000\}$; $\mathbf{t}_1$ or $\mathbf{t}_3$: 1, 1; 2, 1: 3, 1; 4, 1; 5, 1; 6, 1; 7, 1; 8, 1: *b*.
L $\mathbf{G}_6^1 \otimes \mathbf{T}_2$: $\{S_{61}^+ | 000\}$; $\mathbf{t}_1$ or $\mathbf{t}_2$ or $\mathbf{t}_3$: 1, 1; 2, 3; 3, 3; 4, 1; 5, 3; 6, 3: *a*.
W $\mathbf{G}_4^2 \otimes \mathbf{T}_4$: $\{C_{2x} | 000\}$, $\{\sigma_z | 000\}$; $\mathbf{t}_2$ or $\mathbf{t}_3$: 1, 1; 2, 1; 3, 1; 4, 1: *c*.

Δ^x $\mathbf{G}_4^2$: $(C_{2y}, 0)$, $(\sigma_x, 0)$: 1, 1; 2, 1; 3, 1; 4, 1: *c*.
Λ^x $\mathbf{G}_3^1$: $(C_{31}^+, 0)$: 1, 1; 2, 3; 3, 3: *a*.
Σ^x $\mathbf{G}_2^1$: $(\sigma_z, 0)$: 1, 1; 2, 1: *c*.
S^x $\mathbf{G}_2^1$: $(\sigma_y, 0)$: 1, 1; 2, 1: *c*.
Z^x $\mathbf{G}_4^2$: $(C_{2x}, 0)$, $(\sigma_z, 0)$: 1, 1; 2, 1; 3, 1; 4, 1: *c*.
Q^x $\mathbf{G}_1^1$: $(E, 0)$: 1, 1: *a*.

203 $Fd3$ T_h^4

($F1$; $K7$; $M5$; $T5$; $Z1$.)

Γ $\mathbf{G}_{24}^{10}$: $\{S_{61}^+ | \frac{1}{4}\frac{1}{4}\frac{1}{4}\}$, $\{C_{2z} | 000\}$, $\{C_{2y} | 000\}$: 1, 1; 2, 3; 3, 3; 4, 1; 5, 1; 6, 3; 7, 3; 8, 1: *a*.
X $\mathbf{G}_{16}^{9}$: $\{\sigma_z | \frac{1}{4}\frac{1}{4}\frac{1}{4}\}$, $\{\sigma_y | \frac{1}{4}\frac{1}{4}\frac{1}{4}\}$, $\{C_{2y} | 000\}$: 5, 1; 10, 1: *a*.
L $\mathbf{G}_6^1 \otimes \mathbf{T}_2$: $\{S_{61}^+ | \frac{1}{4}\frac{1}{4}\frac{1}{4}\}$; $\mathbf{t}_1$ or $\mathbf{t}_2$ or $\mathbf{t}_3$: 1, 1; 2, 3; 3, 3; 4, 1; 5, 3; 6, 3: *a*.
W $\mathbf{G}_{16}^{6}$: $\{\sigma_z | \frac{1}{4}\frac{1}{4}\frac{1}{4}\}$, $\{C_{2x} | 000\}$: 9, 1: *b*.

Δ^x $\mathbf{G}_4^2$: $(C_{2y}, 0)$, $(\sigma_x, 0)$: 1, 1; 2, 1; 3, 1; 4, 1: *c*.
Λ^x $\mathbf{G}_3^1$: $(C_{31}^+, 0)$: 1, 1; 2, 3; 3, 3: *a*.
Σ^x $\mathbf{G}_2^1$: $(\sigma_z, 0)$: 1, 1; 2, 1: *c*.
S^x $\mathbf{G}_4^1$: $(\sigma_y, 0)$: 2, 1; 4, 1: *a*.
Z^x $\mathbf{G}_8^4$: $(\sigma_y, 0)$, $(C_{2x}, 0)$: 5, 1: *a*.
Q^x $\mathbf{G}_1^1$: $(E, 0)$: 1, 1: *a*.

204 $Im3$ T_h^5

($F1$; $K7$; $M5$; $T5$; $Z1$.)

Γ $\mathbf{G}_{24}^{10}$: $\{S_{61}^+ \mid 000\}$, $\{C_{2z} \mid 000\}$, $\{C_{2y} \mid 000\}$: 1, 1; 2, 3; 3, 3; 4, 1; 5, 1; 6, 3; 7, 3; 8, 1: a.
H $\mathbf{G}_{24}^{10} \otimes \mathbf{T}_2$: $\{S_{61}^+ \mid 000\}$, $\{C_{2z} \mid 000\}$, $\{C_{2y} \mid 000\}$; $\mathbf{t}_1$ or $\mathbf{t}_2$ or $\mathbf{t}_3$: 1, 1; 2, 3; 3, 3; 4, 1; 5, 1; 6, 3; 7, 3; 8, 1: a.
P $\mathbf{G}_{12}^{5} \otimes \mathbf{T}_4$: $\{C_{31}^+ \mid 000\}$, $\{C_{2z} \mid 000\}$, $\{C_{2y} \mid 000\}$; $\mathbf{t}_1$ or $\mathbf{t}_2$ or $\mathbf{t}_3$: 1, 1; 2, 3; 3, 3; 4, 1: a.
N $\mathbf{G}_{4}^{2} \otimes \mathbf{T}_2$: $\{C_{2z} \mid 000\}$, $\{I \mid 000\}$; $\mathbf{t}_3$: 1, 1; 2, 1; 3, 1; 4, 1: a.

Σ^x $\mathbf{G}_2^1$: $(\sigma_z, 0)$: 1, 1; 2, 1: c.
Δ^x $\mathbf{G}_4^2$: $(C_{2y}, 0)$, $(\sigma_x, 0)$: 1, 1; 2, 1; 3, 1; 4, 1: c.
Λ^x $\mathbf{G}_3^1$: $(C_{31}^+, 0)$: 1, 1; 2, 3; 3, 3: a.
D^x $\mathbf{G}_2^1$: $(C_{2z}, 0)$: 1, 1; 2, 1: b.
G^x $\mathbf{G}_2^1$: $(\sigma_z, 0)$: 1, 1; 2, 1: c.
F^x $\mathbf{G}_3^1$: $(C_{34}^+, 0)$: 1, 1; 2, 3; 3, 3: a.

205 $Pa3$ T_h^6

($F1$; $G5$; $K7$; $M5$; $S8$; $T5$; $Z1$; see also Table 5.11 and text of section 5.5.)

Γ $\mathbf{G}_{24}^{10}$: $\{S_{61}^+ \mid 000\}$, $\{C_{2z} \mid \frac{1}{2}0\frac{1}{2}\}$, $\{C_{2y} \mid 0\frac{1}{2}\frac{1}{2}\}$: 1, 1; 2, 3; 3, 3; 4, 1; 5, 1; 6, 3; 7, 3; 8, 1: a.
X $\mathbf{G}_{16}^{9}$: $\{C_{2y} \mid 0\frac{1}{2}\frac{1}{2}\}$, $\{C_{2z} \mid \frac{1}{2}0\frac{1}{2}\}$, $\{\sigma_z \mid \frac{1}{2}0\frac{1}{2}\}$: 5, 1; 10, 1: b.
M $\mathbf{G}_{16}^{7}$: $\{C_{2x} \mid \frac{1}{2}\frac{1}{2}0\}$, $\{C_{2z} \mid \frac{1}{2}0\frac{1}{2}\}$, $\{I \mid 000\}$: 9, 3; 10, 3: d.
R $\mathbf{G}_{48}^{4}$: $\{C_{31}^- \mid 010\}$, $\{C_{2x} \mid \frac{1}{2}\frac{3}{2}0\}$, $\{C_{2y} \mid 0\frac{3}{2}\frac{1}{2}\}$, $\{I \mid 000\}$: 4, 2; 5, 3; 6, 3; 11, 2; 12, 3; 13, 3: a.

Δ^x $\mathbf{G}_4^2$: $(C_{2y}, 0)$, $(\sigma_x, 0)$: 1, 1; 2, 1; 3, 1; 4, 1: c.
Σ^x $\mathbf{G}_2^1$: $(\sigma_z, 0)$: 1, 1; 2, 1: c.
Λ^x $\mathbf{G}_3^1$: $(C_{31}^+, 0)$: 1, 1; 2, 3; 3, 3: a.
S^x $\mathbf{G}_4^1$: $(\sigma_y, 0)$: 2, 3; 4, 3: a.
Z^x $\mathbf{G}_8^2$: $(C_{2x}, 0)$, $(\sigma_z, 0)$: 2, 3; 4, 3; 6, 3; 8, 3: c.
T^x $\mathbf{G}_8^5$: $(C_{2z}, 0)$, $(\sigma_x, 0)$: 5, 2: a.

206 $Ia3$ T_h^7

($C6$; $F1$; $K7$; $M5$; $S8$; $T5$; $Z1$.)

Γ $\mathbf{G}_{24}^{10}$: $\{S_{61}^+ \mid 000\}$, $\{C_{2z} \mid \frac{1}{2}0\frac{1}{2}\}$, $\{C_{2y} \mid 0\frac{1}{2}\frac{1}{2}\}$: 1, 1; 2, 3; 3, 3; 4, 1; 5, 1; 6, 3; 7, 3; 8, 1: a.
H $\mathbf{G}_{48}^{5}$: $\{C_{31}^- \mid 111\}$, $\{C_{2x} \mid \frac{1}{2}\frac{1}{2}0\}$, $\{C_{2y} \mid 0\frac{1}{2}\frac{1}{2}\}$, $\{I \mid 000\}$: 4, 1; 5, 3; 6, 3; 8, 1; 12, 1; 13, 3; 14, 3; 16, 1: a.
P $\mathbf{G}_{48}^{3}$: $\{C_{31}^+ \mid 100\}$, $\{C_{2x} \mid \frac{1}{2}\frac{1}{2}0\}$, $\{C_{2z} \mid \frac{3}{2}1\frac{1}{2}\}$: 7, 2; 8, 3; 9, 3: a.
N $\mathbf{G}_{8}^{4}$: $\{\sigma_z \mid \frac{1}{2}0\frac{1}{2}\}$, $\{I \mid 000\}$: 5, 1: a.

Σ^x $\mathbf{G}_2^1$: $(\sigma_z, 0)$: 1, 1; 2, 1: c.
Δ^x $\mathbf{G}_4^2$: $(C_{2y}, 0)$, $(\sigma_x, 0)$: 1, 1; 2, 1; 3, 1; 4, 1: c.
Λ^x $\mathbf{G}_3^1$: $(C_{31}^+, 0)$: 1, 1; 2, 3; 3, 3: a.
D^x $\mathbf{G}_4^1$: $(C_{2z}, 0)$: 2, 3; 4, 3: a.
G^x $\mathbf{G}_2^1$: $(\sigma_z, 0)$: 1, 1; 2, 1: c.
F^x $\mathbf{G}_6^1$: $(C_{34}^+, 0)$: 2, 3; 4, 1; 6, 3: f.

207 *P*432 O^1

(*F*1; *K*7; *M*5; *T*5; *Z*1.)

Γ $\mathbf{G}_{24}^{7}$: $\{C_{31}^{-}\mid 000\}$, $\{C_{2z}\mid 000\}$, $\{C_{2x}\mid 000\}$, $\{C_{2a}\mid 000\}$: 1, 1; 2, 1; 3, 1; 4, 1; 5, 1: *a*.
X $\mathbf{G}_{8}^{4}\otimes\mathbf{T}_2$: $\{C_{4y}^{+}\mid 000\}$, $\{C_{2z}\mid 000\}$; $\mathbf{t}_2$: 1, 1; 2, 1; 3, 1; 4, 1; 5, 1: *b*.
M $\mathbf{G}_{8}^{4}\otimes\mathbf{T}_2$: $\{C_{4z}^{+}\mid 000\}$, $\{C_{2x}\mid 000\}$; $\mathbf{t}_1$ or $\mathbf{t}_2$: 1, 1; 2, 1; 3, 1; 4, 1; 5, 1: *b*.
R $\mathbf{G}_{24}^{7}\otimes\mathbf{T}_2$: $\{C_{31}^{-}\mid 000\}$, $\{C_{2z}\mid 000\}$, $\{C_{2x}\mid 000\}$, $\{C_{2a}\mid 000\}$; $\mathbf{t}_1$ or $\mathbf{t}_2$ or $\mathbf{t}_3$: 1, 1; 2, 1; 3, 1; 4, 1; 5, 1: *a*.

Δ^x $\mathbf{G}_{4}^{1}$: $(C_{4y}^{+}, 0)$: 1, 1; 2, 1; 3, 1; 4, 1: *b*.
Σ^x $\mathbf{G}_{2}^{1}$: $(C_{2a}, 0)$: 1, 1; 2, 1: *b*.
Λ^x $\mathbf{G}_{3}^{1}$: $(C_{31}^{+}, 0)$: 1, 1; 2, 1; 3, 1: *a*.
S^x $\mathbf{G}_{2}^{1}$: $(C_{2c}, 0)$: 1, 1; 2, 1: *b*.
Z^x $\mathbf{G}_{2}^{1}$: $(C_{2x}, 0)$: 1, 1; 2, 1: *b*.
T^x $\mathbf{G}_{4}^{1}$: $(C_{4z}^{+}, 0)$: 1, 1; 2, 1; 3, 1; 4, 1: *b*.

208 *P*4$_2$32 O^2

(*F*1; *K*7; *M*5; *T*5; *Z*1.)

Γ $\mathbf{G}_{24}^{7}$: $\{C_{31}^{-}\mid 000\}$, $\{C_{2z}\mid 000\}$, $\{C_{2x}\mid 000\}$, $\{C_{2a}\mid \frac{1}{2}\frac{1}{2}\frac{1}{2}\}$: 1, 1; 2, 1; 3, 1; 4, 1; 5, 1: *a*.
X $\mathbf{G}_{8}^{4}\otimes\mathbf{T}_2$: $\{C_{4y}^{+}\mid \frac{1}{2}\frac{1}{2}\frac{1}{2}\}$, $\{C_{2z}\mid 000\}$; $\mathbf{t}_2$: 1, 1; 2, 1; 3, 1; 4, 1; 5, 1: *b*.
M $\mathbf{G}_{8}^{4}\otimes\mathbf{T}_2$: $\{C_{4z}^{+}\mid \frac{1}{2}\frac{1}{2}\frac{1}{2}\}$, $\{C_{2x}\mid 000\}$; $\mathbf{t}_1$ or $\mathbf{t}_2$: 1, 1; 2, 1; 3, 1; 4, 1; 5, 1: *b*.
R $\mathbf{G}_{24}^{7}\otimes\mathbf{T}_2$: $\{C_{31}^{-}\mid 000\}$, $\{C_{2z}\mid 000\}$, $\{C_{2x}\mid 000\}$, $\{C_{2a}\mid \frac{1}{2}\frac{1}{2}\frac{1}{2}\}$; $\mathbf{t}_1$ or $\mathbf{t}_2$ or $\mathbf{t}_3$: 1, 1; 2, 1; 3, 1; 4, 1; 5, 1: *a*.

Δ^x $\mathbf{G}_{4}^{1}$: $(C_{4y}^{+}, 0)$: 1, 1; 2, 1; 3, 1; 4, 1: *b*.
Σ^x $\mathbf{G}_{2}^{1}$: $(C_{2a}, 0)$: 1, 1; 2, 1: *b*.
Λ^x $\mathbf{G}_{3}^{1}$: $(C_{31}^{+}, 0)$: 1, 1; 2, 1; 3, 1: *a*.
S^x $\mathbf{G}_{4}^{1}$: $(C_{2c}, 0)$: 2, 1; 4, 1: *a*.
Z^x $\mathbf{G}_{2}^{1}$: $(C_{2x}, 0)$: 1, 1; 2, 1: *b*.
T^x $\mathbf{G}_{8}^{2}$: $(C_{4z}^{+}, 0), (E, 1)$: 5, 1; 6, 1; 7, 1; 8, 1: *b*.

209 *F*432 O^3

(*F*1; *K*7; *M*5; *T*5; *Z*1.)

Γ $\mathbf{G}_{24}^{7}$: $\{C_{31}^{-}\mid 000\}$, $\{C_{2z}\mid 000\}$, $\{C_{2x}\mid 000\}$, $\{C_{2a}\mid 000\}$: 1, 1; 2, 1; 3, 1; 4, 1; 5, 1: *a*.
X $\mathbf{G}_{8}^{4}\otimes\mathbf{T}_2$: $\{C_{4y}^{+}\mid 000\}$, $\{C_{2z}\mid 000\}$; $\mathbf{t}_1$ or $\mathbf{t}_3$: 1, 1; 2, 1; 3, 1; 4, 1; 5, 1: *b*.
L $\mathbf{G}_{6}^{2}\otimes\mathbf{T}_2$: $\{C_{31}^{+}\mid 000\}$, $\{C_{2b}\mid 000\}$; $\mathbf{t}_1$ or $\mathbf{t}_2$ or $\mathbf{t}_3$: 1, 1; 2, 1; 3, 1: *a*.
W $\mathbf{G}_{4}^{2}\otimes\mathbf{T}_4$: $\{C_{2x}\mid 000\}$, $\{C_{2d}\mid 000\}$; $\mathbf{t}_2$ or $\mathbf{t}_3$: 1, 1; 2, 1; 3, 3; 4, 3: *b*.

Δ^x $\mathbf{G}_{4}^{1}$: $(C_{4y}^{+}, 0)$: 1, 1; 2, 1; 3, 1; 4, 1: *b*.
Λ^x $\mathbf{G}_{3}^{1}$: $(C_{31}^{+}, 0)$: 1, 1; 2, 1; 3, 1: *a*.
Σ^x $\mathbf{G}_{2}^{1}$: $(C_{2a}, 0)$: 1, 1; 2, 1: *b*.
S^x $\mathbf{G}_{2}^{1}$: $(C_{2c}, 0)$: 1, 1; 2, 1: *b*.
Z^x $\mathbf{G}_{2}^{1}$: $(C_{2x}, 0)$: 1, 1; 2, 1: *b*.
Q^x $\mathbf{G}_{2}^{1}$: $(C_{2f}, 0)$: 1, *x*; 2, *x*: *b*.

210 $F4_132$ O^4

($F1$; $K7$; $M5$; $T5$; $Z1$.)

Γ $\mathbf{G}_{24}^{7}$: $\{C_{31}^{-}\mid 000\}$, $\{C_{2z}\mid 000\}$, $\{C_{2x}\mid 000\}$, $\{C_{2a}\mid \frac{1}{4}\frac{1}{4}\frac{1}{4}\}$: 1, 1; 2, 1; 3, 1; 4, 1; 5, 1: *a*.
X $\mathbf{G}_{8}^{4}\otimes\mathbf{T}_2$: $\{C_{4y}^{-}\mid \frac{1}{4}\frac{1}{4}\frac{1}{4}\}$, $\{C_{2z}\mid 000\}$; $\mathbf{t}_1$ or $\mathbf{t}_3$: 1, 1; 2, 1; 3, 1; 4, 1; 5, 1: *b*.
L $\mathbf{G}_{6}^{2}\otimes\mathbf{T}_2$: $\{C_{31}^{+}\mid 000\}$, $\{C_{2b}\mid \frac{1}{4}\frac{1}{4}\frac{1}{4}\}$; $\mathbf{t}_1$ or $\mathbf{t}_2$ or $\mathbf{t}_3$: 1, 1; 2, 1; 3, 1: *a*.
W $\mathbf{G}_{16}^{7}$: $\{E\mid 010\}$, $\{C_{2x}\mid 000\}$, $\{C_{2f}\mid \frac{1}{4}\frac{1}{4}\frac{1}{4}\}$: 10, 1: *c*.

Δ^x $\mathbf{G}_{4}^{1}$: $(C_{4y}^{+}, 0)$: 1, 1; 2, 1; 3, 1; 4, 1: *b*.
Λ^x $\mathbf{G}_{3}^{1}$: $(C_{31}^{+}, 0)$: 1, 1; 2, 1; 3, 1: *a*.
Σ^x $\mathbf{G}_{2}^{1}$: $(C_{2a}, 0)$: 1, 1; 2, 1: *b*.
S^x $\mathbf{G}_{4}^{1}$: $(C_{2c}, 0)$: 2, 1; 4, 1: *a*.
Z^x $\mathbf{G}_{2}^{1}$: $(C_{2x}, 0)$: 1, 1; 2, 1: *b*.
Q^x $\mathbf{G}_{8}^{1}$: $(C_{2f}, 0)$: 4, *x*; 8, *x*: *c*.

211 $I432$ O^5

($F1$; $K7$; $M5$; $T5$; $Z1$.)

Γ $\mathbf{G}_{24}^{7}$: $\{C_{31}^{-}\mid 000\}$, $\{C_{2z}\mid 000\}$, $\{C_{2x}\mid 000\}$, $\{C_{2a}\mid 000\}$: 1, 1; 2, 1; 3, 1; 4, 1; 5, 1: *a*.
H $\mathbf{G}_{24}^{7}\otimes\mathbf{T}_2$: $\{C_{31}^{-}\mid 000\}$, $\{C_{2z}\mid 000\}$, $\{C_{2x}\mid 000\}$, $\{C_{2a}\mid 000\}$; $\mathbf{t}_1$ or $\mathbf{t}_2$ or $\mathbf{t}_3$: 1, 1; 2, 1; 3, 1; 4, 1; 5, 1: *a*.
P $\mathbf{G}_{12}^{5}\otimes\mathbf{T}_4$: $\{C_{31}^{+}\mid 000\}$, $\{C_{2z}\mid 000\}$, $\{C_{2y}\mid 000\}$; $\mathbf{t}_1$ or $\mathbf{t}_2$ or $\mathbf{t}_3$: 1, 1; 2, 3; 3, 3; 4, 1: *a*.
N $\mathbf{G}_{4}^{2}\otimes\mathbf{T}_2$: $\{C_{2z}\mid 000\}$, $\{C_{2b}\mid 000\}$; $\mathbf{t}_3$: 1, 1; 2, 1; 3, 1; 4, 1: *b*.

Σ^x $\mathbf{G}_{2}^{1}$: $(C_{2a}, 0)$: 1, 1; 2, 1: *b*.
Δ^x $\mathbf{G}_{4}^{1}$: $(C_{4y}^{+}, 0)$: 1, 1; 2, 1; 3, 1; 4, 1: *b*.
Λ^x $\mathbf{G}_{3}^{1}$: $(C_{31}^{+}, 0)$: 1, 1; 2, 1; 3, 1: *a*.
D^x $\mathbf{G}_{2}^{1}$: $(C_{2z}, 0)$: 1, 1; 2, 1: *b*.
G^x $\mathbf{G}_{2}^{1}$: $(C_{2b}, 0)$: 1, 1; 2, 1: *b*.
F^x $\mathbf{G}_{3}^{1}$: $(C_{34}^{+}, 0)$: 1, 1; 2, 1; 3, 1: *a*.

212 $P4_332$ O^6

($F1$; $K7$; $M5$; $T5$; $Z1$.)

Γ $\mathbf{G}_{24}^{7}$: $\{C_{31}^{-}\mid 000\}$, $\{C_{2z}\mid \frac{1}{2}0\frac{1}{2}\}$, $\{C_{2x}\mid \frac{1}{2}\frac{1}{2}0\}$, $\{C_{2a}\mid \frac{1}{4}\frac{3}{4}\frac{3}{4}\}$: 1, 1; 2, 1; 3, 1; 4, 1; 5, 1: *a*.
X $\mathbf{G}_{16}^{12}$: $\{C_{4y}^{-}\mid \frac{3}{4}\frac{1}{4}\frac{3}{4}\}$, $\{C_{2x}\mid \frac{1}{2}\frac{1}{2}0\}$: 6, 1; 7, 1: *a*.
M $\mathbf{G}_{16}^{10}$: $\{C_{4z}^{-}\mid \frac{3}{4}\frac{3}{4}\frac{1}{4}\}$, $\{C_{2z}\mid \frac{1}{2}0\frac{1}{2}\}$, $\{C_{2a}\mid \frac{1}{4}\frac{3}{4}\frac{3}{4}\}$: 5, 3; 6, 3; 7, 3; 8, 3; 9, 1: *a*.
R $\mathbf{G}_{48}^{6}$: $\{C_{2x}\mid \frac{1}{2}\frac{1}{2}0\}$, $\{C_{2y}\mid 0\frac{1}{2}\frac{1}{2}\}$, $\{C_{31}^{-}\mid 000\}$, $\{C_{2b}\mid \frac{1}{4}\frac{1}{4}\frac{1}{4}\}$: 4, 3; 5, 3; 8, 1: *a*.

Δ^x $\mathbf{G}_{4}^{1}$: $(C_{4y}^{+}, 0)$: 1, 1; 2, 1; 3, 1; 4, 1: *b*.
Σ^x $\mathbf{G}_{2}^{1}$: $(C_{2a}, 0)$: 1, 1; 2, 1: *b*.
Λ^x $\mathbf{G}_{3}^{1}$: $(C_{31}^{+}, 0)$: 1, 1; 2, 1; 3, 1: *a*.
S^x $\mathbf{G}_{8}^{1}$: $(C_{2c}, 0)$: 4, 3; 8, 3: *c*.
Z^x $\mathbf{G}_{4}^{1}$: $(C_{2x}, 0)$: 2, 3; 4, 3: *a*.
T^x $\mathbf{G}_{16}^{3}$: $(C_{4z}^{+}, 0)$, $(E, 1)$: 5, 3; 6, 3; 7, 3; 8, 3: *a*.

213 $P4_132$ O^7

($F1$; $K7$; $M5$; $T5$; $Z1$.)

Γ $\mathbf{G}_{24}^{7}$: $\{C_{31}^{-}\,|\,000\}$, $\{C_{2z}\,|\,\frac{1}{2}0\frac{1}{2}\}$, $\{C_{2x}\,|\,\frac{1}{2}\frac{1}{2}0\}$, $\{C_{2a}\,|\,\frac{3}{4}\frac{1}{4}\frac{1}{4}\}$: 1, 1; 2, 1; 3, 1; 4, 1; 5, 1: *a*.

X $\mathbf{G}_{16}^{12}$: $\{C_{4y}^{-}\,|\,\frac{1}{4}\frac{3}{4}\frac{1}{4}\}$, $\{C_{2x}\,|\,\frac{1}{2}\frac{1}{2}0\}$: 6, 1; 7, 1: *a*.

M $\mathbf{G}_{16}^{10}$: $\{C_{4z}^{-}\,|\,\frac{1}{4}\frac{1}{4}\frac{3}{4}\}$, $\{C_{2z}\,|\,\frac{1}{2}1\frac{1}{2}\}$, $\{C_{2b}\,|\,\frac{3}{4}\frac{3}{4}\frac{3}{4}\}$: 5, 3; 6, 3; 7, 3; 8, 3; 9, 1: *a*.

R $\mathbf{G}_{48}^{6}$: $\{C_{2x}\,|\,\frac{1}{2}\frac{1}{2}0\}$, $\{C_{2y}\,|\,0\frac{1}{2}\frac{1}{2}\}$, $\{C_{31}^{-}\,|\,000\}$, $\{C_{2b}\,|\,\frac{3}{4}\frac{3}{4}\frac{3}{4}\}$: 4, 3; 5, 3; 8, 1: *a*.

Δ^x $\mathbf{G}_{4}^{1}$: $(C_{4y}^{+}, 0)$: 1, 1; 2, 1; 3, 1; 4, 1: *b*.

Σ^x $\mathbf{G}_{2}^{1}$: $(C_{2a}, 0)$: 1, 1; 2, 1: *b*.

Λ^x $\mathbf{G}_{3}^{1}$: $(C_{31}^{+}, 0)$: 1, 1; 2, 1; 3, 1: *a*.

S^x $\mathbf{G}_{8}^{1}$: $(C_{2c}, 0)$: 2, 3; 6, 3: *b*.

Z^x $\mathbf{G}_{4}^{1}$: $(C_{2x}, 0)$: 2, 3; 4, 3: *a*.

T^x $\mathbf{G}_{16}^{3}$: $(C_{4z}^{+}, 0)$, $(E, 1)$: 5, 3; 6, 3; 7, 3; 8, 3: *a*.

214 $I4_132$ O^8

($F1$; $K7$; $M5$; $T5$; $Z1$.)

Γ $\mathbf{G}_{24}^{7}$: $\{C_{31}^{-}\,|\,000\}$, $\{C_{2z}\,|\,\frac{1}{2}0\frac{1}{2}\}$, $\{C_{2x}\,|\,\frac{1}{2}\frac{1}{2}0\}$, $\{C_{2a}\,|\,\frac{1}{2}00\}$: 1, 1; 2, 1; 3, 1; 4, 1; 5, 1: *a*.

H $\mathbf{G}_{48}^{7}$: $\{C_{32}^{-}\,|\,\frac{1}{2}\frac{1}{2}0\}$, $\{C_{2x}\,|\,\frac{1}{2}\frac{1}{2}0\}$, $\{C_{2y}\,|\,0\frac{1}{2}\frac{1}{2}\}$, $\{C_{2a}\,|\,\frac{1}{2}00\}$: 2, 1; 3, 1; 6, 1; 9, 1; 10, 1: *b*.

P $\mathbf{G}_{48}^{3}$: $\{C_{31}^{+}\,|\,100\}$, $\{C_{2x}\,|\,\frac{1}{2}\frac{1}{2}0\}$, $\{C_{2z}\,|\,\frac{3}{2}1\frac{1}{2}\}$: 7, 1; 8, 1; 9, 1: *a*.

N $\mathbf{G}_{4}^{2} \otimes \mathbf{T}_{2}$: $\{C_{2a}\,|\,\frac{1}{2}00\}$, $\{C_{2b}\,|\,\frac{1}{2}\frac{1}{2}\frac{1}{2}\}$; $\mathbf{t}_3$: 1, 1; 2, 1; 3, 1; 4, 1: *b*.

Σ^x $\mathbf{G}_{2}^{1}$: $(C_{2a}, 0)$: 1, 1; 2, 1: *b*.

Δ^x $\mathbf{G}_{4}^{1}$: $(C_{4y}^{+}, 0)$: 1, 1; 2, 1; 3, 1; 4, 1: *b*.

Λ^x $\mathbf{G}_{3}^{1}$: $(C_{31}^{+}, 0)$: 1, 1; 2, 1; 3, 1: *a*.

D^x $\mathbf{G}_{4}^{1}$: $(C_{2z}, 0)$: 2, 1; 4, 1: *a*.

G^x $\mathbf{G}_{4}^{1}$: $(C_{2b}, 0)$: 2, 1; 4, 1: *a*.

F^x $\mathbf{G}_{6}^{1}$: $(C_{34}^{+}, 0)$: 2, 1; 4, 1; 6, 1: *f*.

215 $P\bar{4}3m$ T_d^1

($F1$; $K7$; $M5$; $T5$; $Z1$.)

Γ $\mathbf{G}_{24}^{7}$: $\{C_{31}^{-}\,|\,000\}$, $\{C_{2z}\,|\,000\}$, $\{C_{2x}\,|\,000\}$, $\{\sigma_{da}\,|\,000\}$: 1, 1; 2, 1; 3, 1; 4, 1; 5, 1: *a*.

X $\mathbf{G}_{8}^{4} \otimes \mathbf{T}_{2}$: $\{S_{4y}^{+}\,|\,000\}$, $\{C_{2z}\,|\,000\}$; $\mathbf{t}_2$: 1, 1; 2, 1; 3, 1; 4, 1; 5, 1: *b*.

M $\mathbf{G}_{8}^{4} \otimes \mathbf{T}_{2}$: $\{S_{4z}^{+}\,|\,000\}$, $\{C_{2x}\,|\,000\}$; $\mathbf{t}_1$ or $\mathbf{t}_2$: 1, 1; 2, 1; 3, 1; 4, 1; 5, 1: *b*.

R $\mathbf{G}_{24}^{7} \otimes \mathbf{T}_{2}$: $\{C_{31}^{-}\,|\,000\}$, $\{C_{2z}\,|\,000\}$, $\{C_{2x}\,|\,000\}$, $\{\sigma_{da}\,|\,000\}$; $\mathbf{t}_1$ or $\mathbf{t}_2$ or $\mathbf{t}_3$: 1, 1; 2, 1; 3, 1; 4, 1; 5, 1: *a*.

Δ^x $\mathbf{G}_{4}^{2}$: $(C_{2y}, 0)$, $(\sigma_{dc}, 0)$: 1, 1; 2, 1; 3, 3; 4, 3: *c*.

Σ^x $\mathbf{G}_{2}^{1}$: $(\sigma_{db}, 0)$: 1, 1; 2, 1: *c*.

Λ^x $\mathbf{G}_{6}^{2}$: $(C_{31}^{+}, 0)$, $(\sigma_{db}, 0)$: 1, *x*; 2, *x*; 3, *x*: *a*.

S^x $\mathbf{G}_{2}^{1}$: $(\sigma_{de}, 0)$: 1, 1; 2, 1: *c*.

Z^x $\mathbf{G}_{2}^{1}$: $(C_{2x}, 0)$: 1, 1; 2, 1: *b*.

T^x $\mathbf{G}_{4}^{2}$: $(C_{2z}, 0)$, $(\sigma_{da}, 0)$: 1, 1; 2, 1; 3, 3; 4, 3: *c*.

216 $F\bar{4}3m$ T_d^2

($A1$; $B3$; $B4$; $B5$; $B6$; $C2$; $D4$; $F1$; $K5$; $K7$; $M5$; $M6$; $P1$; $S1$; $S6$; $S8$; $T5$; $Z1$.)

Γ $\mathbf{G}_{24}^7$: $\{C_{31}^- | 000\}$, $\{C_{2z} | 000\}$, $\{C_{2x} | 000\}$, $\{\sigma_{da} | 000\}$: 1, 1; 2, 1; 3, 1; 4, 1; 5, 1: *a*.
X $\mathbf{G}_8^4 \otimes \mathbf{T}_2$: $\{S_{4y}^+ | 000\}$, $\{C_{2x} | 000\}$; $\mathbf{t}_1$ or $\mathbf{t}_3$: 1, 1; 2, 1; 3, 1; 4, 1; 5, 1: *b*.
L $\mathbf{G}_6^2 \otimes \mathbf{T}_2$: $\{C_{31}^+ | 000\}$, $\{\sigma_{db} | 000\}$; $\mathbf{t}_1$ or $\mathbf{t}_2$ or $\mathbf{t}_3$: 1, 1; 2, 1; 3, 1: *a*.
W $\mathbf{G}_4^1 \otimes \mathbf{T}_4$: $\{S_{4x}^+ | 000\}$; $\mathbf{t}_2$ or $\mathbf{t}_3$: 1, 1; 2, 1; 3, 1; 4, 1: *b*.

Δ^x $\mathbf{G}_4^2$: $(C_{2y}, 0)$, $(\sigma_{dc}, 0)$: 1, 1; 2, 1; 3, 3; 4, 3: *c*.
Λ^x $\mathbf{G}_6^2$: $(C_{31}^+, 0)$, $(\sigma_{db}, 0)$: 1, x; 2, x; 3, x: *a*.
Σ^x $\mathbf{G}_2^1$: $(\sigma_{db}, 0)$: 1, 1; 2, 1: *c*.
S^x $\mathbf{G}_2^1$: $(\sigma_{de}, 0)$: 1, 1; 2, 1: *c*.
Z^x $\mathbf{G}_2^1$: $(C_{2x}, 0)$: 1, 1; 2, 1: *b*.
Q^x $\mathbf{G}_1^1$: $(E, 0)$: 1, 1: *a*.

217 $I\bar{4}3m$ T_d^3

($F1$; $J1$; $K7$; $M5$; $T5$; $Z1$.)

Γ $\mathbf{G}_{24}^7$: $\{C_{31}^- | 000\}$, $\{C_{2z} | 000\}$, $\{C_{2x} | 000\}$, $\{\sigma_{da} | 000\}$: 1, 1; 2, 1; 3, 1; 4, 1; 5, 1: *a*.
H $\mathbf{G}_{24}^7 \otimes \mathbf{T}_2$: $\{C_{31}^- | 000\}$, $\{C_{2z} | 000\}$, $\{C_{2x} | 000\}$, $\{\sigma_{da} | 000\}$; $\mathbf{t}_1$ or $\mathbf{t}_2$ or $\mathbf{t}_3$: 1, 1; 2, 1; 3, 1; 4, 1; 5, 1: *a*.
P $\mathbf{G}_{24}^7 \otimes \mathbf{T}_4$: $\{C_{31}^- | 000\}$, $\{C_{2z} | 000\}$, $\{C_{2x} | 000\}$, $\{\sigma_{da} | 000\}$; $\mathbf{t}_1$ or $\mathbf{t}_2$ or $\mathbf{t}_3$: 1, x; 2, x; 3, x; 4, x; 5, x: *a*.
N $\mathbf{G}_4^2 \otimes \mathbf{T}_2$: $\{C_{2z} | 000\}$, $\{\sigma_{db} | 000\}$; $\mathbf{t}_3$: 1, 1; 2, 1; 3, 1; 4, 1: *c*.

Σ^x $\mathbf{G}_2^1$: $(\sigma_{db}, 0)$: 1, 1; 2, 1: *c*.
Δ^x $\mathbf{G}_4^2$: $(C_{2y}, 0)$, $(\sigma_{de}, 0)$: 1, 1; 2, 1; 3, 3; 4, 3: *c*.
Λ^x $\mathbf{G}_6^2$: $(C_{31}^+, 0)$, $(\sigma_{db}, 0)$: 1, x; 2, x; 3, x: *a*.
D^x $\mathbf{G}_4^2$: $(C_{2z}, 0)$, $(\sigma_{db}, 0)$: 1, x; 2, x; 3, x; 4, x: *c*.
G^x $\mathbf{G}_2^1$: $(\sigma_{da}, 0)$: 1, 1; 2, 1: *c*.
F^x $\mathbf{G}_6^2$: $(C_{34}^+, 0)$, $(\sigma_{da}, 0)$: 1, x; 2, x; 3, x: *a*.

218 $P\bar{4}3n$ T_d^4

($F1$; $K7$; $M5$; $T5$; $Z1$.)

Γ $\mathbf{G}_{24}^7$: $\{C_{31}^- | 000\}$, $\{C_{2z} | 000\}$, $\{C_{2x} | 000\}$, $\{\sigma_{da} | \frac{1}{2}\frac{1}{2}\frac{1}{2}\}$: 1, 1; 2, 1; 3, 1; 4, 1; 5, 1: *a*.
X $\mathbf{G}_{16}^{10}$: $\{S_{4y}^+ | \frac{1}{2}\frac{1}{2}\frac{1}{2}\}$, $\{C_{2y} | 010\}$, $\{C_{2z} | 000\}$: 5, 3; 6, 3; 7, 3; 8, 3; 9, 1: *a*.
M $\mathbf{G}_8^4 \otimes \mathbf{T}_2$: $\{S_{4z}^+ | \frac{1}{2}\frac{1}{2}\frac{1}{2}\}$, $\{C_{2x} | 000\}$: $\mathbf{t}_1$ or $\mathbf{t}_2$: 1, 1; 2, 1; 3, 1; 4, 1; 5, 1: *b*.
R $\mathbf{G}_{48}^8$: $\{S_{4x}^+ | \frac{1}{2}\frac{1}{2}\frac{1}{2}\}$, $\{\sigma_{da} | \frac{1}{2}\frac{1}{2}\frac{1}{2}\}$, $\{C_{33}^- | 000\}$: 3, 3; 4, 3; 6, 2; 9, 3; 10, 3: *a*.

Δ^x $\mathbf{G}_4^2$: $(C_{2y}, 0)$, $(\sigma_{dc}, 0)$: 1, 1; 2, 1; 3, 3; 4, 3: *c*.
Σ^x $\mathbf{G}_2^1$: $(\sigma_{db}, 0)$: 1, 1; 2, 1: *c*.
Λ^x $\mathbf{G}_6^2$: $(C_{31}^+, 0)$, $(\sigma_{db}, 0)$: 1, x; 2, x; 3, x: *a*.
S^x $\mathbf{G}_2^1$: $(\sigma_{de}, 0)$: 1, 3; 2, 3: *c*.
Z^x $\mathbf{G}_2^1$: $(C_{2x}, 0)$: 1, 1; 2, 1: *b*.
T^x $\mathbf{G}_4^2$: $(C_{2z}, 0)$, $(\sigma_{da}, 0)$: 1, 3; 2, 3; 3, 1; 4, 1: *c*.

219 $F\bar{4}3c$ T_d^5

($F1$; $K7$; $M5$; $T5$; $Z1$; see also section 5.4.)

Γ $\mathbf{G}_{24}^{7}$: $\{C_{31}^{-}\,|\,000\}$, $\{C_{2z}\,|\,000\}$, $\{C_{2x}\,|\,000\}$, $\{\sigma_{da}\,|\,\frac{1}{2}\frac{1}{2}\frac{1}{2}\}$: 1, 1; 2, 1; 3, 1; 4, 1; 5, 1: *a*.

X $\mathbf{G}_{8}^{4}\otimes\mathbf{T}_2$: $\{S_{4y}^{+}\,|\,\frac{1}{2}\frac{1}{2}\frac{1}{2}\}$, $\{C_{2x}\,|\,000\}$; $\mathbf{t}_1$ or $\mathbf{t}_3$: 1, 1; 2, 1; 3, 1; 4, 1; 5, 1: *b*.

L $\mathbf{G}_{12}^{4}$: $\{C_{31}^{-}\,|\,001\}$, $\{\sigma_{db}\,|\,\frac{1}{2}\frac{1}{2}\frac{1}{2}\}$: 3, 3; 4, 3; 6, 2: *a*.

W $\mathbf{G}_{4}^{1}\otimes\mathbf{T}_4$: $\{S_{4x}^{+}\,|\,\frac{1}{2}\frac{1}{2}\frac{1}{2}\}$; $\mathbf{t}_2$ or $\mathbf{t}_3$: 1, 3; 2, 3; 3, 3; 4, 3: *b*.

Δ^x $\mathbf{G}_{4}^{2}$: $(C_{2y}, 0)$, $(\sigma_{dc}, 0)$: 1, 1; 2, 1; 3, 3; 4, 3: *c*.

Λ^x $\mathbf{G}_{6}^{2}$: $(C_{31}^{+}, 0)$, $(\sigma_{db}, 0)$: 1, x; 2, x; 3, x: *a*.

Σ^x $\mathbf{G}_{2}^{1}$: $(\sigma_{db}, 0)$: 1, 1; 2, 1: *c*.

S^x $\mathbf{G}_{2}^{1}$: $(\sigma_{de}, 0)$: 1, 1; 2, 1: *c*.

Z^x $\mathbf{G}_{2}^{1}$: $(C_{2x}, 0)$: 1, 1; 2, 1: *b*.

Q^x $\mathbf{G}_{1}^{1}$: $(E, 0)$: 1, 2: *a*.

220 $I\bar{4}3d$ T_d^6

($F1$; $K1$; $K7$; $M5$; $T5$; $Z1$.)

Γ $\mathbf{G}_{24}^{7}$: $\{C_{31}^{-}\,|\,000\}$, $\{C_{2z}\,|\,\frac{1}{2}0\frac{1}{2}\}$, $\{C_{2x}\,|\,\frac{1}{2}\frac{1}{2}0\}$, $\{\sigma_{da}\,|\,\frac{1}{2}00\}$: 1, 1; 2, 1; 3, 1; 4, 1; 5, 1: *a*.

H $\mathbf{G}_{48}^{8}$: $\{S_{4x}^{+}\,|\,\frac{1}{2}00\}$, $\{\sigma_{da}\,|\,\frac{1}{2}00\}$, $\{C_{33}^{-}\,|\,1\frac{1}{2}\frac{1}{2}\}$: 3, 3; 4, 3; 6, 2; 9, 3; 10, 3: *a*.

P $\mathbf{G}_{96}^{1}$: $\{C_{32}^{+}\,|\,0\frac{1}{2}\frac{1}{2}\}$, $\{C_{2y}\,|\,0\frac{1}{2}\frac{1}{2}\}$, $\{C_{2x}\,|\,\frac{3}{2}\frac{3}{2}0\}$, $\{S_{4x}^{+}\,|\,\frac{1}{2}11\}$: 9, x; 10, x; 16, x: *a*.

N $\mathbf{G}_{8}^{4}$: $\{\sigma_{db}\,|\,\frac{1}{2}\frac{1}{2}\frac{1}{2}\}$, $\{C_{2z}\,|\,\frac{1}{2}0\frac{1}{2}\}$: 5, 1: *a*.

Σ^x $\mathbf{G}_{2}^{1}$: $(\sigma_{db}, 0)$: 1, 1; 2, 1: *c*.

Δ^x $\mathbf{G}_{4}^{2}$: $(C_{2y}, 0)$, $(\sigma_{de}, 0)$: 1, 1; 2, 1; 3, 3; 4, 3: *c*.

Λ^x $\mathbf{G}_{6}^{2}$: $(C_{31}^{+}, 0)$, $(\sigma_{db}, 0)$: 1, x; 2, x; 3, x: *a*.

D^x $\mathbf{G}_{8}^{4}$: $(C_{2z}, 0)$, $(\sigma_{db}, 0)$: 5, x: *a*.

G^x $\mathbf{G}_{2}^{1}$: $(\sigma_{da}, 0)$: 1, 3; 2, 3: *c*.

F^x $\mathbf{G}_{12}^{3}$: $(C_{34}^{+}, 0)$, $(\sigma_{da}, 0)$: 3, x; 4, x; 6, x: *f*.

221 $Pm3m$ O_h^1

($A4$; $B2$; $B8$; $E1$; $F1$; $H3$; $J1$; $K5$; $K7$; $M5$; $N2$; $N3$; $S4$; $S8$; $T4$; $T5$; $Y1$; $Z1$.)

Γ $\mathbf{G}_{48}^{7}$: $\{S_{61}^{-}\,|\,000\}$, $\{\sigma_x\,|\,000\}$, $\{\sigma_z\,|\,000\}$, $\{C_{2c}\,|\,000\}$: 1, 1; 2, 1; 3, 1; 4, 1; 5, 1; 6, 1; 7, 1; 8, 1; 9, 1; 10, 1: *a*.

X $\mathbf{G}_{16}^{9}\otimes\mathbf{T}_2$: $\{C_{4y}^{+}\,|\,000\}$, $\{C_{2z}\,|\,000\}$, $\{I\,|\,000\}$; $\mathbf{t}_2$: 1, 1; 2, 1; 3, 1; 4, 1; 5, 1; 6, 1; 7, 1; 8, 1; 9, 1; 10, 1: *e*.

M $\mathbf{G}_{16}^{9}\otimes\mathbf{T}_2$: $\{C_{4z}^{+}\,|\,000\}$, $\{C_{2x}\,|\,000\}$, $\{I\,|\,000\}$; $\mathbf{t}_1$ or $\mathbf{t}_2$: 1, 1; 2, 1; 3, 1; 4, 1; 5, 1; 6, 1; 7, 1; 8, 1; 9, 1; 10, 1: *e*.

R $\mathbf{G}_{48}^{7}\otimes\mathbf{T}_2$: $\{S_{61}^{-}\,|\,000\}$, $\{\sigma_x\,|\,000\}$, $\{\sigma_z\,|\,000\}$, $\{C_{2c}\,|\,000\}$; $\mathbf{t}_1$ or $\mathbf{t}_2$ or $\mathbf{t}_3$: 1, 1; 2, 1; 3, 1; 4, 1; 5, 1; 6, 1; 7, 1; 8, 1; 9, 1; 10, 1: *a*.

Δ^x $\mathbf{G}_{8}^{4}$: $(C_{4y}^{+}, 0)$, $(\sigma_x, 0)$: 1, 1; 2, 1; 3, 1; 4, 1; 5, 1: *b*.

Σ^x $\mathbf{G}_{4}^{2}$: $(C_{2a}, 0)$, $(\sigma_z, 0)$: 1, 1; 2, 1; 3, 1; 4, 1: *c*.

Λ^x $\mathbf{G}_{6}^{2}$: $(C_{31}^{+}, 0)$, $(\sigma_{db}, 0)$: 1, 1; 2, 1; 3, 1: *a*.

S^x $\mathbf{G}_{4}^{2}$: $(C_{2c}, 0)$, $(\sigma_y, 0)$: 1, 1; 2, 1; 3, 1; 4, 1: *c*.

Z^x $\mathbf{G}_{4}^{2}$: $(C_{2x}, 0)$, $(\sigma_z, 0)$: 1, 1; 2, 1; 3, 1; 4, 1: *c*.

T^x $\mathbf{G}_{8}^{4}$: $(C_{4z}^{+}, 0)$, $(\sigma_y, 0)$: 1, 1; 2, 1; 3, 1; 4, 1; 5, 1: *b*.

222 *Pn3n* O_h^2

($F1$; $H3$; $K7$; $M5$; $T4$; $T5$; $Z1$.)

Γ $\mathbf{G}_{48}^7$: $\{S_{61}^- \mid \frac{1}{2}\frac{1}{2}\frac{1}{2}\}$, $\{\sigma_x \mid \frac{1}{2}\frac{1}{2}\frac{1}{2}\}$, $\{\sigma_z \mid \frac{1}{2}\frac{1}{2}\frac{1}{2}\}$, $\{C_{2c} \mid 000\}$: 1, 1; 2, 1; 3, 1; 4, 1; 5, 1; 6, 1; 7, 1; 8, 1; 9, 1; 10, 1: *a*.
X $\mathbf{G}_{32}^1$: $\{I \mid \frac{1}{2}\frac{1}{2}\frac{1}{2}\}$, $\{\sigma_x \mid \frac{1}{2}\frac{1}{2}\frac{1}{2}\}$, $\{C_{4y}^- \mid 000\}$: 10, 1; 11, 1; 12, 3; 13, 3: *a*.
M $\mathbf{G}_{32}^2$: $\{C_{4z}^- \mid 000\}$, $\{\sigma_x \mid \frac{1}{2}\frac{1}{2}\frac{1}{2}\}$, $\{C_{2x} \mid 000\}$: 9, 1; 11, 1; 12, 1; 14, 1: *a*.
R $\mathbf{G}_{96}^2$: $\{C_{32}^- \mid 010\}$, $\{C_{2x} \mid 000\}$, $\{C_{2y} \mid 000\}$, $\{C_{2f} \mid 000\}$, $\{I \mid \frac{1}{2}\frac{1}{2}\frac{1}{2}\}$: 7, 1; 8, 3; 9, 3; 14, 1: *a*.

Δ^x $\mathbf{G}_8^4$: $(C_{4y}^+, 0)$, $(\sigma_x, 0)$: 1, 1; 2, 1; 3, 1; 4, 1; 5, 1: *b*.
Σ^x $\mathbf{G}_4^2$: $(C_{2a}, 0)$, $(\sigma_z, 0)$: 1, 1; 2, 1; 3, 1; 4, 1: *c*.
Λ^x $\mathbf{G}_6^2$: $(C_{31}^+, 0)$, $(\sigma_{db}, 0)$: 1, 1; 2, 1; 3, 1: *a*.
S^x $\mathbf{G}_8^4$: $(\sigma_y, 0)$, $(C_{2c}, 0)$: 5, 1: *a*.
Z^x $\mathbf{G}_8^4$: $(\sigma_y, 0)$, $(C_{2x}, 0)$: 5, 1: *a*.
T^x $\mathbf{G}_{16}^{10}$: $(C_{4z}^+, 0)$, $(C_{2z}, 1)$, $(\sigma_{da}, 0)$: 5, 1; 6, 1; 7, 1; 8, 1; 9, 1: *b*.

223 *Pm3n* O_h^3

($F1$; $G6$; $H3$; $K7$; $M5$; $T4$; $T5$; $Z1$.)

Γ $\mathbf{G}_{48}^7$: $\{S_{61}^- \mid 000\}$, $\{\sigma_x \mid 000\}$, $\{\sigma_z \mid 000\}$, $\{C_{2c} \mid \frac{1}{2}\frac{1}{2}\frac{1}{2}\}$: 1, 1; 2, 1; 3, 1; 4, 1; 5, 1; 6, 1; 7, 1; 8, 1; 9, 1; 10, 1: *a*.
X $\mathbf{G}_{32}^2$: $\{S_{4y}^+ \mid \frac{1}{2}\frac{1}{2}\frac{1}{2}\}$, $\{\sigma_{dc} \mid \frac{1}{2}\frac{1}{2}\frac{1}{2}\}$, $\{C_{2c} \mid \frac{1}{2}\frac{1}{2}\frac{1}{2}\}$: 9, 1; 11, 1; 12, 1; 14, 1: *b*.
M $\mathbf{G}_{16}^9 \otimes \mathbf{T}_2$: $\{C_{4z}^- \mid \frac{1}{2}\frac{1}{2}\frac{1}{2}\}$, $\{C_{2x} \mid 000\}$, $\{I \mid 000\}$; $\mathbf{t}_1$ or $\mathbf{t}_2$: 1, 1; 2, 1; 3, 1; 4, 1; 5, 1; 6, 1; 7, 1; 8, 1; 9, 1; 10, 1: *e*.
R $\mathbf{G}_{96}^2$: $\{C_{32}^- \mid 010\}$, $\{C_{2x} \mid 000\}$, $\{C_{2y} \mid 000\}$, $\{C_{2f} \mid \frac{1}{2}\frac{1}{2}\frac{1}{2}\}$, $\{I \mid 000\}$: 7, 1; 8, 3; 9, 3; 14, 1: *a*.

Δ^x $\mathbf{G}_8^4$: $(C_{4y}^+, 0)$, $(\sigma_x, 0)$: 1, 1; 2, 1; 3, 1; 4, 1; 5, 1: *b*.
Σ^x $\mathbf{G}_4^2$: $(C_{2a}, 0)$, $(\sigma_z, 0)$: 1, 1; 2, 1; 3, 1; 4, 1: *c*.
Λ^x $\mathbf{G}_6^2$: $(C_{31}^+, 0)$, $(\sigma_{db}, 0)$: 1, 1; 2, 1; 3, 1: *a*.
S^x $\mathbf{G}_8^4$: $(C_{2c}, 0)$, $(\sigma_y, 0)$: 5, 1: *a*.
Z^x $\mathbf{G}_4^2$: $(C_{2x}, 0)$, $(\sigma_z, 0)$: 1, 1; 2, 1; 3, 1; 4, 1: *c*.
T^x $\mathbf{G}_{16}^9$: $(C_{4z}^+, 0)$, $(\sigma_x, 0)$, $(E, 1)$: 6, 1; 7, 1; 8, 1; 9, 1; 10, 1: *d*.

224 *Pn3m* O_h^4

($B7$; $F1$; $H3$; $K7$; $M5$; $M7$; $T4$; $T5$; $Z1$.)

Γ $\mathbf{G}_{48}^7$: $\{S_{61}^- \mid \frac{1}{2}\frac{1}{2}\frac{1}{2}\}$, $\{\sigma_x \mid \frac{1}{2}\frac{1}{2}\frac{1}{2}\}$, $\{\sigma_z \mid \frac{1}{2}\frac{1}{2}\frac{1}{2}\}$, $\{C_{2c} \mid \frac{1}{2}\frac{1}{2}\frac{1}{2}\}$: 1, 1; 2, 1; 3, 1; 4, 1; 5, 1; 6, 1; 7, 1; 8, 1; 9, 1; 10, 1: *a*.
X $\mathbf{G}_{32}^2$: $\{C_{4y}^- \mid \frac{1}{2}\frac{1}{2}\frac{1}{2}\}$, $\{\sigma_x \mid \frac{1}{2}\frac{1}{2}\frac{1}{2}\}$, $\{C_{2x} \mid 000\}$: 9, 1; 11, 1; 12, 1; 14, 1: *a*.
M $\mathbf{G}_{32}^2$: $\{C_{4z}^- \mid \frac{1}{2}\frac{1}{2}\frac{1}{2}\}$, $\{\sigma_y \mid \frac{1}{2}\frac{1}{2}\frac{1}{2}\}$, $\{C_{2y} \mid 000\}$: 9, 1; 11, 1; 12, 1; 14, 1: *a*.
R $\mathbf{G}_{48}^7 \otimes \mathbf{T}_2$: $\{S_{61}^- \mid \frac{1}{2}\frac{1}{2}\frac{1}{2}\}$, $\{\sigma_x \mid \frac{1}{2}\frac{1}{2}\frac{1}{2}\}$, $\{\sigma_z \mid \frac{1}{2}\frac{1}{2}\frac{1}{2}\}$, $\{C_{2c} \mid \frac{1}{2}\frac{1}{2}\frac{1}{2}\}$; $\mathbf{t}_1$ or $\mathbf{t}_2$ or $\mathbf{t}_3$: 1, 1; 2, 1; 3, 1; 4, 1; 5, 1; 6, 1; 7, 1; 8, 1; 9, 1; 10, 1: *a*.

Δ^x $\mathbf{G}_8^4$: $(C_{4y}^+, 0)$, $(\sigma_x, 0)$: 1, 1; 2, 1; 3, 1; 4, 1; 5, 1; *b*.
Σ^x $\mathbf{G}_4^2$: $(C_{2a}, 0)$, $(\sigma_z, 0)$: 1, 1; 2, 1; 3, 1; 4, 1: *c*.
Λ^x $\mathbf{G}_6^2$: $(C_{31}^+, 0)$, $(\sigma_{db}, 0)$: 1, 1; 2, 1; 3, 1: *a*.
S^x $\mathbf{G}_8^2$: $(C_{2c}, 0)$, $(\sigma_{de}, 0)$: 2, 1; 4, 1; 6, 1; 8, 1: *c*.
Z^x $\mathbf{G}_8^4$: $(\sigma_y, 0)$, $(C_{2x}, 0)$: 5, 1: *a*.
T^x $\mathbf{G}_{16}^{10}$: $(\sigma_x, 0)$, $(C_{2z}, 0)$, $(\sigma_{db}, 0)$: 5, 1; 6, 1; 7, 1; 8, 1; 10, 1: *c*.

225 *Fm3m* O_h^5

(See section 3.8; *A*4; *B*2; *B*8; *C*4; *C*5; *E*1; *F*1; *F*4; *F*6; *H*3; *J*1; *K*5; *K*7; *M*5; *P*1; *S*4; *S*8; *T*4; *T*5; *Y*3; *Z*1.)

Γ $\mathbf{G}_{48}^7$: $\{S_{61}^- \mid 000\}, \{\sigma_x \mid 000\}, \{\sigma_z \mid 000\}, \{C_{2c} \mid 000\}$: 1, 1; 2, 1; 3, 1; 4, 1; 5, 1; 6, 1; 7, 1; 8, 1; 9, 1; 10, 1: *a*.
X $\mathbf{G}_{16}^9 \otimes \mathbf{T}_2$: $\{C_{4y}^+ \mid 000\}, \{C_{2x} \mid 000\}, \{I \mid 000\}$; $\mathbf{t}_1$ or $\mathbf{t}_3$: 1, 1; 2, 1; 3, 1; 4, 1; 5, 1; 6, 1; 7, 1; 8, 1; 9, 1; 10, 1: *e*.
L $\mathbf{G}_{12}^3 \otimes \mathbf{T}_2$: $\{S_{61}^+ \mid 000\}, \{C_{2b} \mid 000\}$; $\mathbf{t}_1$ or $\mathbf{t}_2$ or $\mathbf{t}_3$: 1, 1; 2, 1; 3, 1; 4, 1; 5, 1; 6, 1: *a*.
W $\mathbf{G}_8^4 \otimes \mathbf{T}_4$: $\{S_{4x}^+ \mid 000\}, \{C_{2d} \mid 000\}$; $\mathbf{t}_2$ or $\mathbf{t}_3$: 1, 1; 2, 1; 3, 1; 4, 1; 5, 1: *b*.

Δ^x $\mathbf{G}_8^4$: $(C_{4y}^+, 0), (\sigma_x, 0)$: 1, 1; 2, 1; 3, 1; 4, 1; 5, 1: *b*.
Λ^x $\mathbf{G}_6^2$: $(C_{31}^+, 0), (\sigma_{db}, 0)$: 1, 1; 2, 1; 3, 1: *a*.
Σ^x $\mathbf{G}_4^2$: $(C_{2a}, 0), (\sigma_z, 0)$: 1, 1; 2, 1; 3, 1; 4, 1: *c*.
S^x $\mathbf{G}_4^2$: $(C_{2c}, 0), (\sigma_y, 0)$: 1, 1; 2, 1; 3, 1; 4, 1: *c*.
Z^x $\mathbf{G}_4^2$: $(C_{2x}, 0), (\sigma_z, 0)$: 1, 1; 2, 1; 3, 1; 4, 1: *c*.
Q^x $\mathbf{G}_2^1$: $(C_{2f}, 0)$: 1, 1; 2, 1: *b*.

226 *Fm3c* O_h^6

(*F*1; *H*3; *K*7; *M*5; *T*4; *T*5; *Z*1.)

Γ $\mathbf{G}_{48}^7$: $\{S_{61}^- \mid \frac{1}{2}\frac{1}{2}\frac{1}{2}\}, \{\sigma_x \mid \frac{1}{2}\frac{1}{2}\frac{1}{2}\}, \{\sigma_z \mid \frac{1}{2}\frac{1}{2}\frac{1}{2}\}, \{C_{2c} \mid 000\}$: 1, 1; 2, 1; 3, 1; 4, 1; 5, 1; 6, 1; 7, 1; 8, 1; 9, 1; 10, 1: *a*.
X $\mathbf{G}_{16}^9 \otimes \mathbf{T}_2$: $\{C_{4y}^+ \mid 000\}, \{C_{2x} \mid 000\}, \{I \mid \frac{1}{2}\frac{1}{2}\frac{1}{2}\}$; $\mathbf{t}_1$ or $\mathbf{t}_3$: 1, 1; 2, 1; 3, 1; 4, 1; 5, 1; 6, 1; 7, 1; 8, 1; 9, 1; 10, 1: *e*.
L $\mathbf{G}_{24}^1$: $\{\sigma_{db} \mid \frac{1}{2}\frac{1}{2}\frac{1}{2}\}, \{C_{31}^- \mid 000\}, \{I \mid \frac{1}{2}\frac{1}{2}\frac{1}{2}\}$: 7, 3; 8, 3; 9, 1: *a*.
W $\mathbf{G}_{32}^{13}$: $\{S_{4x}^- \mid \frac{1}{2}\frac{1}{2}\frac{1}{2}\}, \{E \mid 010\}, \{C_{2d} \mid 000\}$: 13, 3; 14, 3; 15, 3; 16, 3; 20, 1: *a*.

Δ^x $\mathbf{G}_8^4$: $(C_{4y}^+, 0), (\sigma_x, 0)$: 1, 1; 2, 1; 3, 1; 4, 1; 5, 1; *b*.
Λ^x $\mathbf{G}_8^4$: $(C_{31}^+, 0), (\sigma_{db}, 0)$: 1, 1; 2, 1; 3, 1: *a*.
Σ^x $\mathbf{G}_4^2$: $(C_{2a}, 0), (\sigma_z, 0)$: 1, 1; 2, 1; 3, 1; 4, 1: *c*.
S^x $\mathbf{G}_4^2$: $(C_{2c}, 0), (\sigma_y, 0)$: 1, 1; 2, 1; 3, 1; 4, 1: *c*.
Z^x $\mathbf{G}_4^2$: $(C_{2x}, 0), (\sigma_z, 0)$: 1, 1; 2, 1; 3, 1; 4, 1: *c*.
Q^x $\mathbf{G}_2^1$: $(C_{2f}, 0)$: 1, 3; 2, 3: *b*.

227 *Fd3m* O_h^7

(See section 3.8; *B*5; *B*6; *D*3; *E*1; *E*2; *F*1; *H*1; *H*3; *J*1; *K*3; *K*5; *K*7; *L*1; *M*2; *M*5; *M*6; *N*3; *P*1; *P*4; *S*8; *S*11; *T*4; *T*5; *Y*3; *Z*1.)

Γ $\mathbf{G}_{48}^7$: $\{S_{61}^- \mid \frac{1}{4}\frac{1}{4}\frac{1}{4}\}, \{\sigma_x \mid \frac{1}{4}\frac{1}{4}\frac{1}{4}\}, \{\sigma_z \mid \frac{1}{4}\frac{1}{4}\frac{1}{4}\}, \{C_{2c} \mid \frac{1}{4}\frac{1}{4}\frac{1}{4}\}$: 1, 1; 2, 1; 3, 1; 4, 1; 5, 1; 6, 1; 7, 1; 8, 1; 9, 1; 10, 1: *a*.
X $\mathbf{G}_{32}^2$: $\{\sigma_x \mid \frac{1}{4}\frac{1}{4}\frac{1}{4}\}, \{S_{4y}^+ \mid 000\}, \{C_{2x} \mid 000\}$: 10, 1; 11, 1; 13, 1; 14, 1: *c*.
L $\mathbf{G}_{12}^3 \otimes \mathbf{T}_2$: $\{S_{61}^+ \mid \frac{1}{4}\frac{1}{4}\frac{1}{4}\}, \{C_{2b} \mid \frac{1}{4}\frac{1}{4}\frac{1}{4}\}$; $\mathbf{t}_1$ or $\mathbf{t}_2$ or $\mathbf{t}_3$: 1, 1; 2, 1; 3, 1; 4, 1; 5, 1; 6, 1: *a*.
W $\mathbf{G}_{32}^4$: $\{S_{4x}^+ \mid 000\}, \{E \mid 001\}, \{C_{2f} \mid \frac{1}{4}\frac{1}{4}\frac{1}{4}\}$: 11, 1; 12, 1: *b*.

Δ^x $\mathbf{G}_8^4$: $(C_{4y}^+, 0), (\sigma_x, 0)$: 1, 1; 2, 1; 3, 1; 4, 1; 5, 1: *b*.
Λ^x $\mathbf{G}_6^2$: $(C_{31}^+, 0), (\sigma_{db}, 0)$: 1, 1; 2, 1; 3, 1: *a*.
Σ^x $\mathbf{G}_4^2$: $(C_{2a}, 0), (\sigma_z, 0)$: 1, 1; 2, 1; 3, 1; 4, 1: *c*.
S^x $\mathbf{G}_8^2$: $(\sigma_y, 0), (\sigma_{de}, 0)$: 2, 1; 4, 1; 6, 1; 8, 1: *c*.
Z^x $\mathbf{G}_8^4$: $(\sigma_y, 0), (\sigma_z, 0)$: 5, 1: *a*.
Q^x $\mathbf{G}_8^1$: $(C_{2f}, 0)$: 4, 1; 8, 1: *c*.

228 *Fd3c* O_h^8

(*F*1; *H*3; *K*7; *M*5; *T*4; *T*5; *Z*1.)

Γ $\mathbf{G}_{48}^{7}$: $\{S_{61}^{-}\,|\,\tfrac{3}{4}\tfrac{3}{4}\tfrac{3}{4}\}$, $\{\sigma_x\,|\,\tfrac{3}{4}\tfrac{3}{4}\tfrac{3}{4}\}$, $\{\sigma_z\,|\,\tfrac{3}{4}\tfrac{3}{4}\tfrac{3}{4}\}$, $\{C_{2c}\,|\,\tfrac{1}{4}\tfrac{1}{4}\tfrac{1}{4}\}$: 1, 1; 2, 1; 3, 1; 4, 1; 5, 1; 6, 1; 7, 1; 8, 1; 9, 1; 10, 1: *a*.
X $\mathbf{G}_{32}^{2}$: $\{C_{4y}^{-}\,|\,\tfrac{1}{4}\tfrac{1}{4}\tfrac{1}{4}\}$, $\{\sigma_x\,|\,\tfrac{3}{4}\tfrac{3}{4}\tfrac{3}{4}\}$, $\{C_{2x}\,|\,000\}$: 9, 1; 11, 1; 12, 1; 14, 1: *a*.
L $\mathbf{G}_{24}^{1}$: $\{\sigma_{db}\,|\,\tfrac{1}{2}\tfrac{1}{2}\tfrac{1}{2}\}$, $\{C_{31}^{-}\,|\,000\}$, $\{I\,|\,\tfrac{3}{4}\tfrac{3}{4}\tfrac{3}{4}\}$: 7, 3; 8, 3; 9, 1: *a*.
W $\mathbf{G}_{32}^{4}$: $\{S_{4x}^{+}\,|\,\tfrac{1}{2}\tfrac{1}{2}\tfrac{1}{2}\}$, $\{E\,|\,010\}$, $\{C_{2d}\,|\,\tfrac{1}{4}\tfrac{5}{4}\tfrac{1}{4}\}$: 13, 3; 14, 3: *a*.

Δ^x $\mathbf{G}_{8}^{4}$: $(C_{4y}^{+}, 0)$, $(\sigma_x, 0)$: 1, 1; 2, 1; 3, 1; 4, 1; 5, 1: *b*.
Λ^x $\mathbf{G}_{6}^{2}$: $(C_{31}^{+}, 0)$, $(\sigma_{db}, 0)$: 1, 1; 2, 1; 3, 1: *a*.
Σ^x $\mathbf{G}_{4}^{2}$: $(C_{2a}, 0)$, $(\sigma_z, 0)$: 1, 1; 2, 1; 3, 1; 4, 1: *c*.
S^x $\mathbf{G}_{8}^{2}$: $(\sigma_y, 0)$, $(\sigma_{de}, 0)$: 2, 1; 4, 1; 6, 1; 8, 1: *d*.
Z^x $\mathbf{G}_{8}^{4}$: $(\sigma_y, 0)$, $(\sigma_z, 0)$: 5, 1: *a*.
Q^x $\mathbf{G}_{8}^{1}$: $(C_{2f}, 0)$: 4, 3; 8, 3: *c*.

229 *Im3m* O_h^9

(*A*4; *B*1; *B*2; *B*8; *C*5; *E*1; *F*1; *F*4; *H*3; *J*1; *K*5; *K*7; *M*5; *S*4; *S*8; *T*4; *T*5; *Z*1.)

Γ $\mathbf{G}_{48}^{7}$: $\{S_{61}^{-}\,|\,000\}$, $\{\sigma_x\,|\,000\}$, $\{\sigma_z\,|\,000\}$, $\{C_{2c}\,|\,000\}$: 1, 1; 2, 1; 3, 1; 4, 1; 5, 1; 6, 1; 7, 1; 8, 1; 9, 1; 10, 1: *a*.
H $\mathbf{G}_{48}^{7}\otimes\mathbf{T}_2$: $\{S_{61}^{-}\,|\,000\}$, $\{\sigma_x\,|\,000\}$, $\{\sigma_z\,|\,000\}$, $\{C_{2c}\,|\,000\}$; $\mathbf{t}_1$ or $\mathbf{t}_2$ or $\mathbf{t}_3$: 1, 1; 2, 1; 3, 1; 4, 1; 5, 1; 6, 1; 7, 1; 8, 1; 9, 1; 10, 1: *a*.
P $\mathbf{G}_{24}^{7}\otimes\mathbf{T}_4$: $\{C_{31}^{-}\,|\,000\}$, $\{C_{2z}\,|\,000\}$, $\{C_{2x}\,|\,000\}$, $\{\sigma_{da}\,|\,000\}$; $\mathbf{t}_1$ or $\mathbf{t}_2$ or $\mathbf{t}_3$: 1, 1; 2, 1; 3, 1; 4, 1; 5, 1: *a*.
N $\mathbf{G}_{8}^{3}\otimes\mathbf{T}_2$: $\{C_{2z}\,|\,000\}$, $\{C_{2b}\,|\,000\}$, $\{I\,|\,000\}$; $\mathbf{t}_3$: 1, 1; 2, 1; 3, 1; 4, 1; 5, 1; 6, 1; 7, 1; 8, 1: *b*.

Σ^x $\mathbf{G}_{4}^{2}$: $(C_{2a}, 0)$, $(\sigma_z, 0)$: 1, 1; 2, 1; 3, 1; 4, 1: *c*.
Δ^x $\mathbf{G}_{8}^{4}$: $(C_{4y}^{+}, 0)$, $(\sigma_x, 0)$: 1, 1; 2, 1; 3, 1; 4, 1; 5, 1: *b*.
Λ^x $\mathbf{G}_{6}^{2}$: $(C_{31}^{+}, 0)$, $(\sigma_{db}, 0)$: 1, 1; 2, 1; 3, 1: *a*.
D^x $\mathbf{G}_{4}^{2}$: $(C_{2z}, 0)$, $(\sigma_{db}, 0)$: 1, 1; 2, 1; 3, 1; 4, 1: *c*.
G^x $\mathbf{G}_{4}^{2}$: $(C_{2b}, 0)$, $(\sigma_{da}, 0)$: 1, 1; 2, 1; 3, 1; 4, 1: *c*.
F^x $\mathbf{G}_{6}^{2}$: $(C_{34}^{+}, 0)$, $(\sigma_{da}, 0)$: 1, 1; 2, 1; 3, 1: *a*.

230 *Ia3d* O_h^{10}

(*F*1; *H*3; *K*7; *M*5; *S*8; *T*4; *T*5; *Z*1.)

Γ $\mathbf{G}_{48}^{7}$: $\{S_{61}^{-}\,|\,000\}$, $\{\sigma_x\,|\,\tfrac{1}{2}\tfrac{1}{2}0\}$, $\{\sigma_z\,|\,\tfrac{1}{2}0\tfrac{1}{2}\}$, $\{C_{2c}\,|\,00\tfrac{1}{2}\}$: 1, 1; 2, 1; 3, 1; 4, 1; 5, 1; 6, 1; 7, 1; 8, 1; 9, 1; 10, 1: *a*.
H $\mathbf{G}_{96}^{4}$: $\{C_{32}^{-}\,|\,\tfrac{1}{2}\tfrac{1}{2}0\}$, $\{C_{2x}\,|\,\tfrac{1}{2}\tfrac{1}{2}0\}$, $\{C_{2y}\,|\,0\tfrac{1}{2}\tfrac{1}{2}\}$, $\{\sigma_{da}\,|\,\tfrac{1}{2}00\}$, $\{I\,|\,000\}$: 7, 1; 8, 3; 9, 3; 14, 1: *a*.
P $\mathbf{G}_{96}^{1}$: $\{C_{32}^{+}\,|\,0\tfrac{1}{2}\tfrac{1}{2}\}$, $\{C_{2y}\,|\,0\tfrac{1}{2}\tfrac{1}{2}\}$, $\{C_{2x}\,|\,\tfrac{3}{2}\tfrac{3}{2}0\}$, $\{S_{4x}^{+}\,|\,\tfrac{1}{2}11\}$: 9, 3; 10, 3; 16, 1: *a*.
N $\mathbf{G}_{16}^{9}$: $\{\sigma_{db}\,|\,\tfrac{1}{2}\tfrac{1}{2}\tfrac{1}{2}\}$, $\{C_{2b}\,|\,\tfrac{1}{2}\tfrac{1}{2}\tfrac{1}{2}\}$, $\{C_{2a}\,|\,\tfrac{1}{2}00\}$: 5, 1; 10, 1: *a*.

Σ^x $\mathbf{G}_{4}^{2}$: $(C_{2a}, 0)$, $(\sigma_z, 0)$: 1, 1; 2, 1; 3, 1; 4, 1: *c*.
Δ^x $\mathbf{G}_{8}^{4}$: $(C_{4y}^{+}, 0)$, $(\sigma_x, 0)$: 1, 1; 2, 1; 3, 1; 4, 1; 5, 1: *b*.
Λ^x $\mathbf{G}_{6}^{2}$: $(C_{31}^{+}, 0)$, $(\sigma_{db}, 0)$: 1, 1; 2, 1; 3, 1: *a*.
D^x $\mathbf{G}_{8}^{4}$: $(C_{2z}, 0)$, $(\sigma_{db}, 0)$: 5, 1: *a*.
G^x $\mathbf{G}_{8}^{4}$: $(C_{2b}, 0)$, $(\sigma_{da}, 0)$: 5, 1: *a*.
F^x $\mathbf{G}_{12}^{3}$: $(C_{34}^{+}, 0)$, $(\sigma_{da}, 0)$: 3, 1; 4, 1; 6, 1: *f*.

Notes to Table 5.7

(i) The space groups are arranged in the order and notation used in Volume 1 of the *International tables for X-ray crystallography* (Henry and Lonsdale 1965) and in Table 3.7 of this book.

(ii) The Brillouin zone for any space group can be identified from Table 3.7 and Figs. 3.2–3.15.

(iii) The first column lists the special points of symmetry and the lines of symmetry in the basic domain of the Brillouin zone (see section 3.3); their positions in the Brillouin zone can be identified from Table 3.6. A superscript x is used to indicate a line of symmetry, for example, in the case of $F\bar{4}3c$ (T_d^5) Δ, Λ, Σ, S, Z, and Q are lines of symmetry while Γ, X, L, and W are points of symmetry.

Points of symmetry

(iv) Following the label of a point of symmetry the little group ${}^H\mathbf{G}^{\mathbf{k}}$ (in the sense of Herring (1942), see sections 3.8 and 5.2) is identified. This is done as follows: the abstract group $\mathbf{G}^n_{|\mathbf{G}|}$ (see Table 5.1) is given and the space-group elements following $\mathbf{G}^n_{|\mathbf{G}|}$ are to be identified with $P, Q, R, \ldots$, the generators of $\mathbf{G}^n_{|\mathbf{G}|}$ in the order in which they occur. For example, for $F\bar{4}3c$ (T_d^5) at the point Γ, ${}^H\mathbf{G}^{\mathbf{k}}$ is $\mathbf{G}^7_{24}$ and the generators of ${}^H\mathbf{G}^{\mathbf{k}}$ are, in terms of the generators of $\mathbf{G}^7_{24}$,

$$\{C_{31}^- \mid 000\} = P,$$

$$\{C_{2z} \mid 000\} = Q,$$

$$\{C_{2x} \mid 000\} = R,$$

and

$$\{\sigma_{da} \mid \tfrac{1}{2}\tfrac{1}{2}\tfrac{1}{2}\} = S.$$

The rest of the elements of ${}^H\mathbf{G}^{\mathbf{k}}$ can then be constructed and assigned to the five classes of $\mathbf{G}^7_{24}$ using Tables 1.5, 1.6, 3.2, and 5.1. The recognition of an element, e.g. P^2RS, the last element in the class C_3,

$$= \{C_{31}^- \mid 000\}^2\{C_{2x} \mid 000\}\{\sigma_{da} \mid \tfrac{1}{2}\tfrac{1}{2}\tfrac{1}{2}\},$$

and its identification as a single space-group element can be made by using the multiplication rule for the Seitz space-group symbols (see sections 1.4 and 1.5). In this case

$$P^2RS = \{C_{31}^- \mid 000\}^2\{C_{2x} \mid 000\}\{\sigma_{da} \mid \tfrac{1}{2}\tfrac{1}{2}\tfrac{1}{2}\} = \{C_{31}^+ \mid 000\}\{S_{4z}^+ \mid -\tfrac{3}{2}\tfrac{1}{2}\tfrac{1}{2}\} = \{S_{4y}^- \mid \tfrac{1}{2} -\tfrac{3}{2}\tfrac{1}{2}\}.$$

The multiplication of the rotational parts of the Seitz space-group symbols can be done with the help of Tables 1.5 and 1.6.

(v) In some cases for points of symmetry ${}^H\mathbf{G}^{\mathbf{k}}$ is written as $\mathbf{G}^n_{|\mathbf{G}|} \otimes \mathbf{T}_m$, a direct product of some group $\mathbf{G}^n_{|\mathbf{G}|}$ with $\mathbf{T}_m$, an Abelian group of small order, m, consisting only of translation operations of the form $\{E \mid \mathbf{t}\}$. In these cases a suitable generating element of the Abelian group is indicated after the generators of $\mathbf{G}^n_{|\mathbf{G}|}$ by giving the translation $\mathbf{t}$ of $\{E \mid \mathbf{t}\}$. For example, in the case of $F\bar{4}3c$ (T_d^5) at the point W, ${}^H\mathbf{G}^{\mathbf{k}}$ is the direct product of the groups $\mathbf{G}^1_4$ (with $P = \{S_{4x}^+ \mid \tfrac{1}{2}\tfrac{1}{2}\tfrac{1}{2}\}$) and $\mathbf{T}_4$ (with $\{E \mid \mathbf{t}\} = \{E \mid \mathbf{t}_2\}$ or $\{E \mid \mathbf{t}_3\}$). $\mathbf{T}_4$ is the Abelian group of order 4 whose elements are thus $\{E \mid 0\}$, $\{E \mid \mathbf{t}_i\}$, $\{E \mid 2\mathbf{t}_i\}$, and $\{E \mid 3\mathbf{t}_i\}$, with $i = 2$ or 3 and with the actual form of $\mathbf{t}_i$ being identified from Table 3.1. In a few cases ${}^H\mathbf{G}^{\mathbf{k}}$ takes the form $\mathbf{G}^n_{|\mathbf{G}|} \otimes \mathbf{T}_m \otimes \mathbf{T}_s$ where $\mathbf{T}_m$ and $\mathbf{T}_s$ are two Abelian groups with generating elements $\{E \mid \mathbf{t}_m\}$ and $\{E \mid \mathbf{t}_s\}$ respectively; $\mathbf{t}_m$ and $\mathbf{t}_s$ are then listed following the generators of $\mathbf{G}^n_{|\mathbf{G}|}$ (see, for example, $P6_3/mcm$ (D_{6h}^3) at the point H).

(vi) After the above information, and separated from it by a colon, is the identification of the labelling assigned to the reps of ${}^H\mathbf{G}^{\mathbf{k}}$ and of the reality of the induced space-group reps ($\Gamma_p^{\mathbf{k}} \uparrow \mathbf{G}$) (see sections 4.6, 5.2, and 7.6). Between one pair of semicolons is the code 'p, q' which is to be understood as meaning that the rep R_p of $\mathbf{G}^n_{|\mathbf{G}|}$ corresponds to a small (allowed) rep, $\Gamma_p^{\mathbf{k}}$, of $\mathbf{G}^{\mathbf{k}}$, and that ($\Gamma_p^{\mathbf{k}} \uparrow \mathbf{G}$) is of the qth kind ($q = 1$, 2, or 3). The usefulness of the reality of ($\Gamma_p^{\mathbf{k}} \uparrow \mathbf{G}$) lies in determining whether or not extra degeneracies occur if the crystal described by the space group $\mathbf{G}$ also possesses time-inversion symmetry. The relation between the reality of ($\Gamma_p^{\mathbf{k}} \uparrow \mathbf{G}$) and these extra degeneracies described in the text of section 5.2 (see p. 388) is

Reality	Degeneracy
1	a
2	b
3	c

When there is no element in the space group that transforms $\mathbf{k}$ into $-\mathbf{k}$ the addition of time-inversion symmetry produces a type (x) degeneracy (see p. 389 and p. 652) and in this case the symbol x is used in the table in place of the reality 1, 2, or 3.

(vii) In the cases where ${}^H\mathbf{G}^{\mathbf{k}}$ contains one or more elements of the form $\{E \mid \mathbf{t}\}$ only those reps of ${}^H\mathbf{G}^{\mathbf{k}}$ are included that are compatible with the requirement of eqn (3.4.3) $\Delta^{\mathbf{k}}(\{E \mid \mathbf{t}\}) = \exp(-i\mathbf{k}.\mathbf{t})\mathbf{1}$ for a representation of the translation group $\mathbf{T}$ of which $\{E \mid \mathbf{t}\}$ is a member; that is, only reps of ${}^H\mathbf{G}^{\mathbf{k}}$ which lead to small reps of $\mathbf{G}^{\mathbf{k}}$ are included.

(viii) The labels used for the space-group reps derived from $R_1, R_2, R_3, \ldots$, etc., the reps of the abstract group $\mathbf{G}^n_{|G|}$ are indicated by the small letter $a, b, c, \ldots$, etc., at the end of the entry for the point of symmetry in question. The letters $a, b, c, \ldots$, etc., refer the reader to the relevant part of Table 5.8 (see the notes to Table 5.8).

Lines of symmetry

(ix) Following the label of the line of symmetry the central extension $\overline{\mathbf{G}}^{\mathbf{k}*}$ (see sections 3.7 and 3.8) is identified; the elements of $\overline{\mathbf{G}}^{\mathbf{k}*}$ are of the form (H_j, α) where H_j is the point-group operation part of a space-group element and α is a member of the cyclic group $\mathbf{Z}_g$ (see eqn (3.7.31)). The (H_j, α) form the abstract group which is specified and the elements that are given are to be identified with $P, Q, R, \ldots$, the generators of $\mathbf{G}^n_{|G|}$ in the order in which they occur. Where alternative values of $\mathbf{k}$ for a line of symmetry were given in Table 3.6 we have used the first entry in deriving this table.

(x) After the generators of $\overline{\mathbf{G}}^{\mathbf{k}*}$ and separated from them by a colon is the labelling of those reps of $\overline{\mathbf{G}}^{\mathbf{k}*}$ that satisfy eqn (3.7.33) and therefore lead to small reps of $\mathbf{G}^{\mathbf{k}}$; the reality of the induced space-group reps $(\Gamma^{\mathbf{k}}_p \uparrow \mathbf{G})$ is also indicated. Between one pair of semicolons is the code 'p, q' which is to be understood as meaning that the rep $(\Gamma^{\mathbf{k}}_p \uparrow \mathbf{G})$ derived from R_p of $\mathbf{G}^n_{|G|}$ is of the qth kind ($q = 1, 2$, or 3); see Note (vi) above. Reps that do not satisfy eqn (3.7.33) are not included. The letters $a, b, c, \ldots$, etc. refer the reader to the relevant part of Table 5.8 for the letters used to label the space-group reps (see the notes to Table 5.8).

(xi) References to work on the representations of individual space groups are given in the table, in coded form, under the appropriate space group according to the following key:

*A*1 Adler 1962.
*A*2 Altmann 1956.
*A*3 Altmann and Bradley 1965.
*A*4 Altmann and Cracknell 1965.
*A*5 Antončik and Trlifaj 1952.
*A*6 Asendorf 1957.
*B*1 Bates and Stevens 1961.
*B*2 Bell 1954.
*B*3 Birman 1959*a*
*B*4 Birman 1959*b*.
*B*5 Birman 1962*a*.
*B*6 Birman 1962*b*, 1963.
*B*7 Bordure, Lecoy, and Savelli 1963*a*, *b*.
*B*8 Bouckaert, Smoluchowski, and Wigner 1936.
*C*1 Casella 1959.
*C*2 Chaldyshev and Karavaev 1963.
*C*3 Chen 1967.
*C*4 Chen, Berenson and Birman 1968.
*C*5 Cornwell 1969.
*C*6 Cracknell 1965.
*C*7 Cracknell 1967*c*.
*C*8 Cracknell and Joshua 1969*a*.
*D*1 Daniel and Cracknell 1969.
*D*2 Dimmock and Wheeler 1962*b*.
*D*3 Döring and Zehler 1953.
*D*4 Dresselhaus 1955.
*E*1 Elliott 1954*b*.
*E*2 Elliott and Loudon 1960.
*E*3 Erdman 1960.
*F*1 Faddeyev 1964.
*F*2 Falicov and Cohen 1963.
*F*3 Falicov and Golin 1965.
*F*4 Falicov and Ruvalds 1968.
*F*5 Firsov 1957.
*F*6 Flower, March, and Murray 1960.
*F*7 Frei 1967*a*.
*F*8 Frei 1967*b*.
*G*1 Gashimzade 1960*a*.
*G*2 Gashimzade 1960*b*.
*G*3 Gay, Albers, and Arlinghaus 1968.
*G*4 Glasser 1959.
*G*5 Gorzkowski 1963*a*.
*G*6 Gorzkowski 1963*b*, 1964*b*.
*G*7 Gorzkowski 1964*a*.
*G*8 Gubanov and Gashimzade 1959.
*H*1 Herring 1942.
*H*2 Hsieh and Chen 1964.
*H*3 Hurley 1966*a*.
*J*1 Jones 1960.
*J*2 Joshua and Cracknell 1969.
*K*1 Karavaev, Kudryavtseva, and Chaldyshev 1962.
*K*2 Karpus and Batarūnas 1961.
*K*3 Khartsiev 1962.
*K*4 Kitz 1965*a*.
*K*5 Koster 1957.
*K*6 Kovalev 1961*a*, *b*.
*K*7 Kovalev 1965.
*L*1 Lax and Hopfield 1961.
*L*2 Lee and Pincherle 1963.
*M*1 Mase 1958, 1959*a*.
*M*2 Mase 1959*b*; Miąsek 1960, 1963, 1966.

*M*3 Miąsek 1957*a*, *b*, 1958.
*M*4 Miąsek and Suffczyński 1961*a*, *b*, *c*.
*M*5 Miller and Love 1967.
*M*6 Montgomery 1969.
*M*7 Moskalenko 1960.
*M*8 Murphy, Caspers, and Buchanan 1964.
*N*1 Nusimovici 1965.
*N*2 Nussbaum 1962.
*N*3 Nussbaum 1966.
*O*1 Olbrychski 1961.
*O*2 Olbrychski 1963*b*.
*O*3 Olbrychski and Van Huong 1970.
*P*1 Parmenter 1955.
*P*2 Pikus 1961*a*.
*P*3 Pikus 1961*b*.
*P*4 Pikus 1961*c*.
*R*1 Rabotnikov 1960.
*R*2 Rashba 1959.
*R*3 Rashba and Sheka 1959.
*R*4 Rudra 1965*b*.
*S*1 Sandrock and Treusch 1964.
*S*2 Schiff 1955.
*S*3 Schiff 1956.
*S*4 Schlosser 1962.
*S*5 Sek 1963.
*S*6 Sheka 1960.
*S*7 Shur 1966.
*S*8 Slater 1965*b*.
*S*9 Slater, Koster, and Wood 1962; Koster 1962.
*S*10 Šlechta 1966.
*S*11 Streitwolf 1964.
*S*12 Suffczyński 1960.
*S*13 Suffczyński 1961.
*S*14 Sushkevich 1965.
*S*15 Sushkevich 1966.
*T*1 Tovstyuk and Bercha 1964.
*T*2 Tovstyuk and Gemus 1963.
*T*3 Tovstyuk and Sushkevich 1964.
*T*4 Tovstyuk and Tarnavskaya 1963.
*T*5 Tovstyuk and Tarnavskaya 1964.
*T*6 Treusch and Sandrock 1966.
*V*1 Van Huong and Olbrychski 1964.
*W*1 Waeber 1969.
*Y*1 Yamazaki 1957*a*.
*Y*2 Yamazaki 1957*b*.
*Y*3 Yanagawa 1953.
*Z*1 Zak, Casher, Glück, and Gur 1969.

In Table 5.7 we give, effectively, the reps of $\mathbf{G}^{\mathbf{k}}$, say $\Gamma_j^{\mathbf{k}}$, and hence the degeneracies of energy levels $E_j^{\mathbf{k}}$ of particles or quasi-particles with wave vector $\mathbf{k}$ are available; they are just the degeneracies of the $\Gamma_j^{\mathbf{k}}$. However, if we are considering a crystal in which no magnetic ordering exists the crystal will possess time-reversal symmetry in addition to all the spatial symmetry operations of $\mathbf{G}$. The complete group of the crystal is then the direct product $\mathbf{G} \otimes (E + \theta)$ where θ is the operation of time-inversion. The addition of the extra symmetry operation, θ, may cause some extra degeneracies in the energy levels of a particle or quasi-particle with wave vector $\mathbf{k}$. If $E_j^{\mathbf{k}}$ is an energy level belonging to the rep $\Gamma_j^{\mathbf{k}}$ of $\mathbf{G}^{\mathbf{k}}$ then there are three possibilities,

(a) there is no change in the degeneracy of $E_j^{\mathbf{k}}$,
(b) the degeneracy of $E_j^{\mathbf{k}}$ becomes doubled, that is two different energy levels, both described by the same rep $\Gamma_j^{\mathbf{k}}$, become degenerate, and
(c) the degeneracy of $E_j^{\mathbf{k}}$ becomes doubled but, unlike (b), two different (i.e. inequivalent) reps $\Gamma_j^{\mathbf{k}}$ and $\Gamma_{j'}^{\mathbf{k}}$ of $\mathbf{G}^{\mathbf{k}}$ become degenerate because of the addition of θ to the space group of the crystal.

In Table 5.7 we indicate for each small rep whether it belongs to case (a), case (b), or case (c) and hence it is possible to see whether or not the addition of time-reversal symmetry causes any extra degeneracy in the spectrum of the energy eigenvalues at $\mathbf{k}$. However, it is convenient to postpone until section 7.6 the discussion of the theory of the determination, for any given rep $\Gamma_j^{\mathbf{k}}$ whether it belongs to (a), (b), or (c); this makes use of the theory of the reality of the induced space-group representations $\Delta_j^{\mathbf{k}} = \Gamma_j^{\mathbf{k}} \uparrow \mathbf{G}$ (see section 4.6).

In addition to the possibility of causing some extra degeneracies in the spectrum of the energy eigenvalues at **k**, the inclusion of time-reversal symmetry will always cause degeneracies between **k** and $-\mathbf{k}$ when these two vectors do not appear in the same star. Of course, if **k** and $-\mathbf{k}$ do appear in the same star, as for example when the crystal has a centre of inversion symmetry so that $\{I \mid \mathbf{v}\}$ is in **G**, the spectrum of the energy eigenvalues at **k** is the same as at $-\mathbf{k}$ even in the absence of time-reversal symmetry. This extra degeneracy is very easy to notice as it occurs *when time-reversal symmetry is present and there are no space-group elements that transform* **k** *into* $-\mathbf{k}$; we call this a type (x) degeneracy. This degeneracy is rather different from the types (a), (b), and (c) mentioned above, which govern what happens at the point **k**, due to time-reversal symmetry, when **k** and $-\mathbf{k}$ are in the same star although, as we shall see in section 7.6, the mathematical origins of case (c) and case (x) degeneracies are very similar. Degeneracies of type (b) or (c) are extra degeneracies at the point **k**, the spectra at **k** and $-\mathbf{k}$ already being identical. A degeneracy of type (x) has the property of making the spectra at **k** and $-\mathbf{k}$ identical, when they are not already required by the space-group symmetry to be identical. Proofs of all these assertions will be given in Chapter 7.

Finally, in Table 5.7 we indicate for each point or line of symmetry the part of Table 5.8 that contains the labels that we use for the reps of $\mathbf{G}^{n}_{|\mathbf{G}|}$ in their particular manifestation as space-group reps of ${}^{H}\mathbf{G}^{\mathbf{k}}$ or $\overline{\mathbf{G}}^{\mathbf{k}*}$. The assignment of labels to the space-group reps is a complicated problem and we leave this discussion to the next section.

5.3. The labels of space-group representations

In Table 5.8 we give the labels used for the reps of ${}^{H}\mathbf{G}^{\mathbf{k}}$ or $\overline{\mathbf{G}}^{\mathbf{k}*}$. We use an extension of the two notations used for point-group representations in Table 2.2, namely the Mulliken (1933) notation with *A*, *B*, *E*, *T*, etc., and the Γ_1, Γ_2, Γ_3, etc. notation of Koster, Dimmock, Wheeler, and Statz (1963). For many space groups ${}^{H}\mathbf{G}^{\mathbf{k}}$ or $\overline{\mathbf{G}}^{\mathbf{k}*}$ is isomorphic to one of the 32 point groups and in this case we simply use the labels used for that point group in Table 2.2. Where ${}^{H}\mathbf{G}^{\mathbf{k}}$ or $\overline{\mathbf{G}}^{\mathbf{k}*}$ is not isomorphic to a point group the notation has been extended as follows, where by 'real' we mean that all the characters are real and by 'complex' we mean that some of the characters are complex.

Rep	Dimension	Label
real	1	*A* or *B*
complex	1	${}^{1}E$, ${}^{2}E$ (conjugate pair)
real	2	*E*
complex	2	${}^{1}F$, ${}^{2}F$ (conjugate pair)
real	3	*T*
complex	3	${}^{1}H$, ${}^{2}H$ (conjugate pair)
real	4	*F*
complex	4	${}^{1}J$, ${}^{2}J$ (conjugate pair)
real	6	*H*

Subscripts 1, 2, 3, etc., subscripts g and u, and superscript primes and double primes are used to distinguish between reps of the same dimension in a similar way to that used in the point-group lavels. If ${}^{H}\mathbf{G}^{\mathbf{k}}$ is a direct product of $(\{E\,|\,\mathbf{0}\} + \{I\,|\,\mathbf{0}\})$ with some group $\mathbf{G}$ then R_g and R_u are used to denote the two reps derived from the rep R of $\mathbf{G}$; in R_g $\chi(I\,|\,\mathbf{0}\}$ is $+\chi\{E\,|\,\mathbf{0}\}$ and in R_u $\chi(I\,|\,\mathbf{0}\}$ is $-\chi\{E\,|\,\mathbf{0}\}$. Primes and double primes are used to distinguish between two reps R_i' and R_i'' which differ only in the sign of the character of an important reflection plane. In the Γ notation the reps are labelled Γ_1, Γ_2, Γ_3, etc.; superscripts + and − are usually used for the reps of direct product groups in the same way as g and u above (except by Miller and Love (1967)). At any point of symmetry other than Γ or along any line of symmetry Γ is replaced by the letter denoting that point or line of symmetry; therefore, for example, at M of $P4_332$ (O^6) the space-group reps that are identified with the reps R_5, R_6, R_7, R_8, and R_9 of the abstract group $\mathbf{G}_{16}^{10}$ are labelled either as ${}^2E_1$, ${}^1E_1$, ${}^1E_2$, ${}^2E_2$, and E or as M_1, M_2, M_3, M_4, and M_5 respectively, from entry 'a' under $\mathbf{G}_{16}^{10}$ of Table 5.8, where Γ_i has been replaced by M_i.

It is impossible in either of these two notations to be completely logical in defining a set of labels. It is also not practical to adopt a synthesis of all those labels used by the various authors who have studied individual space groups or small collections of space groups.

TABLE 5.8

The labelling of the space-group reps

$\mathbf{G}_1^1$	R_1
a	A Γ_1

$\mathbf{G}_2^1$	R_1	R_2
a	A_g Γ_1^+	A_u Γ_1^-
b	A Γ_1	B Γ_2
c	A' Γ_1	A'' Γ_2

$\mathbf{G}_3^1$	R_1	R_2	R_3
a	A	${}^2E$	${}^1E$
b	A	${}^1E$	${}^2E$
	Γ_1	Γ_2	Γ_3

$\mathbf{G}_4^1$	R_1	R_2	R_3	R_4
a	— —	${}^2E$ Γ_1	— —	${}^1E$ Γ_2
b	A Γ_1	${}^2E$ Γ_2	B Γ_3	${}^1E$ Γ_4

$\mathbf{G}_4^2$	R_1	R_2	R_3	R_4
a	A_g	A_u	B_g	B_u
	Γ_1^+	Γ_1^-	Γ_2^+	Γ_2^-
b	A	B_1	B_2	B_3
c	A_1	A_2	B_1	B_2
	Γ_1	Γ_3	Γ_2	Γ_4

$\mathbf{G}_6^1$	R_1	R_2	R_3	R_4	R_5	R_6
a	A_g	${}^1E_u$	${}^2E_g$	A_u	${}^1E_g$	${}^2E_u$
	Γ_1^+	Γ_3^-	Γ_2^+	Γ_1^-	Γ_3^+	Γ_2^-
b	—	—	${}^1E_1$	—	—	${}^1E_2$
	—	—	Γ_1	—	—	Γ_2
c	—	${}^1E_1$	—	—	${}^1E_2$	—
	—	Γ_1	—	—	Γ_2	—
d	A	${}^2E_2$	${}^1E_1$	B	${}^2E_1$	${}^1E_2$
	Γ_1	Γ_3	Γ_5	Γ_4	Γ_6	Γ_2
e	A'	${}^1E''$	${}^2E'$	A''	${}^1E'$	${}^2E''$
	Γ_1	Γ_5	Γ_3	Γ_4	Γ_2	Γ_6
f	—	${}^2E$	—	A	—	${}^1E$
	—	Γ_3	—	Γ_1	—	Γ_2

$\mathbf{G}_6^2$	R_1	R_2	R_3
a	A_1	A_2	E
	Γ_1	Γ_2	Γ_3

$\mathbf{G}_8^1$	R_2	R_4	R_6	R_8
a	${}^2E_1$	${}^1E_2$	${}^2E_2$	${}^1E_1$
	Γ_1	Γ_2	Γ_3	Γ_4
b	${}^1E_1$	—	${}^1E_2$	—
	Γ_1	—	Γ_2	—
c	—	${}^1E_1$	—	${}^1E_2$
	—	Γ_1	—	Γ_2

$\mathbf{G}_8^2$	R_1	R_2	R_3	R_4	R_5	R_6	R_7	R_8
a	—	${}^2E_g$	—	${}^1E_g$	—	${}^2E_u$	—	${}^1E_u$
	—	Γ_1^+	—	Γ_2^+	—	Γ_1^-	—	Γ_2^-
b	—	—	—	—	A	${}^2E$	B	${}^1E$
	—	—	—	—	Γ_1	Γ_4	Γ_2	Γ_3
c	—	${}^2E'$	—	${}^1E'$	—	${}^2E''$	—	${}^1E''$
d	—	${}^2E_1$	—	${}^1E_1$	—	${}^2E_2$	—	${}^1E_2$
	—	Γ_1	—	Γ_2	—	Γ_3	—	Γ_4
e	A_g	${}^2E_g$	B_g	${}^1E_g$	A_u	${}^2E_u$	B_u	${}^1E_u$
	Γ_1^+	Γ_4^+	Γ_2^+	Γ_3^+	Γ_1^-	Γ_4^-	Γ_2^-	Γ_3^-

$\mathbf{G}_8^3$	R_1	R_2	R_3	R_4	R_5	R_6	R_7	R_8
a	—	—	—	—	A_1	B_1	A_2	B_2
	—	—	—	—	Γ_1	Γ_2	Γ_3	Γ_4
b	A_g	B_{2g}	B_{1g}	B_{3g}	A_u	B_{2u}	B_{1u}	B_{3u}
	Γ_1^+	Γ_2^+	Γ_3^+	Γ_4^+	Γ_1^-	Γ_2^-	Γ_3^-	Γ_4^-

$\mathbf{G}_8^4$	R_1	R_2	R_3	R_4	R_5
a	—	—	—	—	E
	—	—	—	—	Γ_1
b	A_1	A_2	B_1	B_2	E
	Γ_1	Γ_2	Γ_3	Γ_4	Γ_5

$\mathbf{G}_8^5$	R_5
a	E
	Γ_1

$\mathbf{G}_{12}^1$	R_2	R_4	R_6	R_8	R_{10}	R_{12}
a	${}^2E_2$	A_1	${}^1E_2$	${}^2E_1$	A_2	${}^1E_1$
	Γ_6	Γ_1	Γ_5	Γ_4	Γ_2	Γ_3
b	—	—	${}^1E_1$	—	—	${}^1E_2$
	—	—	Γ_1	—	—	Γ_2
c	${}^1E_1$	—	—	${}^1E_2$	—	—
	Γ_1	—	—	Γ_2	—	—

$\mathbf{G}_{12}^2$	R_1	R_2	R_3	R_4	R_5	R_6	R_7	R_8	R_9	R_{10}	R_{11}	R_{12}
a	A_g	${}^2E_{1g}$	${}^1E_{1g}$	B_g	${}^2E_{2g}$	${}^1E_{2g}$	A_u	${}^2E_{1u}$	${}^1E_{1u}$	B_u	${}^2E_{2u}$	${}^1E_{2u}$
	Γ_1^+	Γ_6^+	Γ_5^+	Γ_4^+	Γ_3^+	Γ_2^+	Γ_1^-	Γ_6^-	Γ_5^-	Γ_4^-	Γ_2^-	Γ_3^-

$\mathbf{G}_{12}^3$	R_1	R_2	R_3	R_4	R_5	R_6
a	A_{1g}	A_{2g}	A_{1u}	A_{2u}	E_g	E_u
	Γ_1^+	Γ_2^+	Γ_1^-	Γ_2^-	Γ_3^+	Γ_3^-
b	—	—	A_1	A_2	—	E
	—	—	Γ_1	Γ_2	—	Γ_3
c	A_1	A_2	B_1	B_2	E_2	E_1
d	A_1	A_2	B_2	B_1	E_2	E_1
e	A_1'	A_2'	A_1''	A_2''	E'	E''
	Γ_1	Γ_2	Γ_3	Γ_4	Γ_5	Γ_6
f	—	—	A'	A''	—	E
	—	—	Γ_1	Γ_2	—	Γ_3

$\mathbf{G}_{12}^4$	R_3	R_4	R_6
a	${}^1E_1$	${}^2E_1$	E_2
	Γ_1	Γ_2	Γ_3

$\mathbf{G}_{12}^{5}$	R_1	R_2	R_3	R_4
a	A	^1E	^2E	T
	Γ_1	Γ_3	Γ_2	Γ_4

$\mathbf{G}_{16}^{3}$	R_5	R_6	R_7	R_8
a	A	^2E	B	^1E
	Γ_1	Γ_2	Γ_3	Γ_4

$\mathbf{G}_{16}^{4}$	R_2	R_4	R_6	R_8	R_{10}	R_{12}	R_{14}	R_{16}
a	$^2E_{1g}$	$^1E_{1g}$	$^2E_{2g}$	$^1E_{2g}$	$^2E_{1u}$	$^1E_{1u}$	$^2E_{2u}$	$^1E_{2u}$
	Γ_1^+	Γ_2^+	Γ_3^+	Γ_4^+	Γ_1^-	Γ_2^-	Γ_3^-	Γ_4^-
b	—	E_1'	—	E_1''	—	E_2'	—	E_2''
	—	Γ_1	—	Γ_2	—	Γ_3	—	Γ_4

$\mathbf{G}_{16}^{6}$	R_9	R_{10}
a	—	E
	—	Γ_1
b	E	—
	Γ_1	—

$\mathbf{G}_{16}^{7}$	R_9	R_{10}
a	E'	E''
b	E_1	E_2
	Γ_1	Γ_2
c	—	E
	—	Γ_1
d	^2F	^1F
	Γ_1	Γ_2
e	—	^1F
	—	Γ_1
f	^1F	—
	Γ_1	—

$\mathbf{G}_{16}^{8}$	R_5	R_6	R_7	R_8	R_{10}
a	$^2E_2$	$^1E_2$	$^1E_1$	$^2E_1$	E
	Γ_1	Γ_2	Γ_3	Γ_4	Γ_5

$\mathbf{G}_{16}^{9}$	R_1	R_2	R_3	R_4	R_5	R_6	R_7	R_8	R_9	R_{10}
a	—	—	—	—	E_1	—	—	—	—	E_2
b	—	—	—	—	E'	—	—	—	—	E''
	—	—	—	—	Γ_1	—	—	—	—	Γ_2
c	—	—	—	—	E_g	—	—	—	—	E_u
	—	—	—	—	Γ_1^+	—	—	—	—	Γ_1^-
d	—	—	—	—	—	A_1	A_2	B_1	B_2	E
	—	—	—	—	—	Γ_1	Γ_2	Γ_3	Γ_4	Γ_5
e	A_{1g}	A_{2g}	B_{1g}	B_{2g}	E_g	A_{1u}	A_{2u}	B_{1u}	B_{2u}	E_u
	Γ_1^+	Γ_2^+	Γ_3^+	Γ_4^+	Γ_5^+	Γ_1^-	Γ_2^-	Γ_3^-	Γ_4^-	Γ_5^-
f	—	—	—	—	E	A_1	A_2	B_1	B_2	—
	—	—	—	—	Γ_5	Γ_1	Γ_2	Γ_3	Γ_4	—

G_{16}^{10}	R_5	R_6	R_7	R_8	R_9	R_{10}
a	${}^2E_1$	${}^1E_1$	${}^1E_2$	${}^2E_2$	E	—
b	${}^2E'$	${}^1E'$	${}^1E''$	${}^2E''$	E	—
	Γ_1	Γ_2	Γ_3	Γ_4	Γ_5	—
c	${}^2E'$	${}^1E'$	${}^1E''$	${}^2E''$	—	E
	Γ_1	Γ_2	Γ_3	Γ_4	—	Γ_5
d	—	—	—	—	E_2	E_1
	—	—	—	—	Γ_1	Γ_2

G_{16}^{11}	R_5	R_{10}
a	E_g	E_u
	Γ_1^+	Γ_1^-

G_{16}^{12}	R_6	R_7
a	E_1	E_2
	Γ_1	Γ_2

G_{16}^{13}	R_6	R_7
a	${}^1F$	${}^2F$
	Γ_1	Γ_2

G_{24}^{1}	R_7	R_8	R_9
a	${}^1F$	${}^2F$	E
	Γ_1	Γ_2	Γ_3

G_{24}^{2}	R_7	R_8	R_9
a	${}^1F$	${}^2F$	E
	Γ_1	Γ_2	Γ_3

G_{24}^{3}	R_3	R_4	R_6	R_9	R_{10}	R_{12}
a	${}^1E_1$	${}^2E_1$	E_3	${}^1E_2$	${}^2E_2$	E_4
	Γ_1	Γ_2	Γ_3	Γ_4	Γ_5	Γ_6

G_{24}^{4}	R_4	R_5	R_6	R_{10}	R_{11}	R_{12}
a	${}^1E_1$	${}^1E_2$	${}^1F$	${}^2E_1$	${}^2E_2$	${}^2F$
	Γ_1	Γ_2	Γ_3	Γ_4	Γ_5	Γ_6

G_{24}^{5}	R_1	R_2	R_3	R_4	R_5	R_6	R_7	R_8	R_9	R_{10}	R_{11}	R_{12}
a	A_{1g}	A_{2g}	E_{2g}	B_{1g}	B_{2g}	E_{1g}	A_{1u}	A_{2u}	E_{2u}	B_{1u}	B_{2u}	E_{1u}
	Γ_1^+	Γ_2^+	Γ_6^+	Γ_3^+	Γ_4^+	Γ_5^+	Γ_1^-	Γ_2^-	Γ_6^-	Γ_3^-	Γ_4^-	Γ_5^-

G_{24}^{6}	R_{13}	R_{14}	R_{15}
a	^2F	E	^1F
	Γ_1	Γ_2	Γ_3

G_{24}^{7}	R_1	R_2	R_3	R_4	R_5
a	A_1	A_2	E	T_1	T_2
	Γ_1	Γ_2	Γ_3	Γ_4	Γ_5

G_{24}^{8}	R_4	R_5	R_6	R_8
a	A	^1E	^2E	T
	Γ_1	Γ_2	Γ_3	Γ_4

G_{24}^{9}	R_4	R_5	R_6
a	E	^2F	^1F
	Γ_1	Γ_2	Γ_3

G_{24}^{10}	R_1	R_2	R_3	R_4	R_5	R_6	R_7	R_8
a	A_g	^1E_g	^2E_g	T_g	A_u	^1E_u	^2E_u	T_u
	Γ_1^+	Γ_2^+	Γ_3^+	Γ_4^+	Γ_1^-	Γ_2^-	Γ_3^-	Γ_4^-

G_{32}^{1}	R_{10}	R_{11}	R_{12}	R_{13}
a	E_1	E_2	^1F	^2F
	Γ_1	Γ_2	Γ_3	Γ_4

G_{32}^{2}	R_9	R_{10}	R_{11}	R_{12}	R_{13}	R_{14}
a	E_1	—	E'	E_2	—	E''
b	E'	—	E_1	E''	—	E_2
	Γ_1	—	Γ_2	Γ_3	—	Γ_4
c	—	E_1	E_2	—	E_3	E_4
	—	Γ_1	Γ_2	—	Γ_3	Γ_4
d	E_1	—	E_2	E_3	E_4	—
	Γ_1	—	Γ_2	Γ_3	Γ_4	—

G_{32}^{4}	R_{11}	R_{12}	R_{13}	R_{14}
a	—	—	^1F	^2F
	—	—	Γ_1	Γ_2
b	$^1F_1$	$^1F_2$	—	—
	Γ_1	Γ_2	—	—

G_{32}^{5}	R_5	R_6	R_7	R_8	R_{10}	R_{15}	R_{16}	R_{17}	R_{18}	R_{20}
a	$^2E_g'$	$^1E_g'$	$^1E_g''$	$^2E_g''$	E_g	$^2E_u'$	$^1E_u'$	$^1E_u''$	$^2E_u''$	E_u
b	$^2E_{1g}$	$^1E_{1g}$	$^1E_{2g}$	$^2E_{2g}$	E_g	$^2E_{1u}$	$^1E_{1u}$	$^1E_{2u}$	$^2E_{2u}$	E_u
	Γ_1^+	Γ_2^+	Γ_3^+	Γ_4^+	Γ_5^+	Γ_1^-	Γ_2^-	Γ_3^-	Γ_4^-	Γ_5^-

$\mathbf{G}_{32}^{6}$	R_9	R_{10}	R_{13}	R_{14}
a	${}^2F_1$ Γ_1	${}^1F_1$ Γ_2	${}^2F_2$ Γ_3	${}^1F_2$ Γ_4

$\mathbf{G}_{32}^{13}$	R_{13}	R_{14}	R_{15}	R_{16}	R_{20}
a	${}^1E_1$ Γ_1	${}^1E_2$ Γ_2	${}^1E_3$ Γ_3	${}^1E_4$ Γ_4	${}^1F_1$ Γ_5

$\mathbf{G}_{48}^{1}$	R_{13}	R_{14}	R_{15}
a	E' Γ_1	E'' Γ_2	F Γ_3

$\mathbf{G}_{48}^{2}$	R_7	R_8	R_9	R_{16}	R_{17}	R_{18}
a	${}^1F_1$ Γ_1	${}^2F_1$ Γ_2	E_1 Γ_3	${}^1F_2$ Γ_4	${}^2F_2$ Γ_5	E_2 Γ_6

$\mathbf{G}_{48}^{3}$	R_7	R_8	R_9
a	${}^1F_1$ Γ_1	${}^1F_2$ Γ_2	${}^1F_3$ Γ_3

$\mathbf{G}_{48}^{4}$	R_4	R_5	R_6	R_{11}	R_{12}	R_{13}
a	E_g Γ_1^+	${}^2F_g$ Γ_2^+	${}^1F_g$ Γ_3^+	E_u Γ_1^-	${}^2F_u$ Γ_2^-	${}^1F_u$ Γ_3^-

$\mathbf{G}_{48}^{5}$	R_4	R_5	R_6	R_8	R_{12}	R_{13}	R_{14}	R_{16}
a	A_g Γ_1^+	${}^1E_g$ Γ_2^+	${}^2E_g$ Γ_3^+	T_g Γ_4^+	A_u Γ_1^-	${}^1E_u$ Γ_2^-	${}^2E_u$ Γ_3^-	T_u Γ_4^-

$\mathbf{G}_{48}^{6}$	R_4	R_5	R_8
a	${}^1F_1$ Γ_1	${}^2F_1$ Γ_2	F_2 Γ_3

$\mathbf{G}_{48}^{7}$	R_1	R_2	R_3	R_4	R_5	R_6	R_7	R_8	R_9	R_{10}
a	A_{1g} Γ_1^+	A_{2g} Γ_2^+	A_{2u} Γ_2^-	A_{1u} Γ_1^-	E_g Γ_3^+	E_u Γ_3^-	T_{1g} Γ_4^+	T_{2g} Γ_5^+	T_{1u} Γ_4^-	T_{2u} Γ_5^-
b	— —	A_2 Γ_2	A_1 Γ_1	— —	— —	E Γ_3	— —	— —	T_1 Γ_4	T_2 Γ_5

$\mathbf{G}_{48}^{8}$	R_3	R_4	R_6	R_9	R_{10}
a	${}^2E_1$ Γ_1	${}^1E_1$ Γ_2	E_2 Γ_3	${}^2H$ Γ_4	${}^1H$ Γ_5

$\mathbf{G}_{96}^{1}$	R_9	R_{10}	R_{16}
a	${}^1F_1$ Γ_1	${}^1F_2$ Γ_2	${}^1J$ Γ_3

$\mathbf{G}_{96}^{2}$	R_7	R_8	R_9	R_{14}
a	E Γ_1	${}^1F$ Γ_2	${}^2F$ Γ_3	H Γ_4

$\mathbf{G}_{96}^{4}$	R_7	R_8	R_9	R_{14}
a	E Γ_1	${}^1F$ Γ_2	${}^2F$ Γ_3	H Γ_4

Notes to Table 5.8

(i) For a point of symmetry $^{H}\mathbf{G}^{\mathbf{k}}$ was identified in Table 5.7 with one of the abstract groups $\mathbf{G}^{n}_{|G|}$ and the labelling of the reps indicated by a small arabic letter *a, b, c*, . . ., etc. The labels given to the space-group reps can then be found by consulting the entry under $\mathbf{G}^{n}_{|G|}$ in Table 5.8. The entry corresponding to this small arabic letter gives the labels assigned to the reps $R_1, R_2, R_3, \ldots$, etc. of $\mathbf{G}^{n}_{|G|}$ as the reps of $^{H}\mathbf{G}^{\mathbf{k}}$ for this space group.

(ii) For a line of symmetry the interpretation of the labelling is similar except that the reps $R_1, R_2, R_3, \ldots$, etc. of the abstract group now correspond to the space-group reps derived from the reps of $\bar{\mathbf{G}}^{\mathbf{k}*}$.

(iii) The space-group reps are labelled, as explained in the text of section 5.3, in an extended version both of the notation of Mulliken (1933) and of the Γ notation introduced by Bouckaert, Smoluchowski, and Wigner (1936). In the latter notation Γ is replaced by the letter denoting the relevant point or line of symmetry in the Brillouin zone. For example, at M of $P4_332$ (O^6) the space-group reps that are identified with the reps R_5, R_6, R_7, R_8, and R_9 of the abstract group $\mathbf{G}^{10}_{16}$ are labelled as M_1, M_2, M_3, M_4, and M_5 respectively, where M replaces Γ under entry 'a' for $\mathbf{G}^{10}_{16}$ in Table 5.8.

5.4 Example of the use of the tables of the space-group representations

The space-group reps for the cubic close-packed and diamond structures ($Fm3m$ (O_h^5) and $Fd3m$ (O_h^7)) were deduced in section 3.8 in illustration of the theory of the

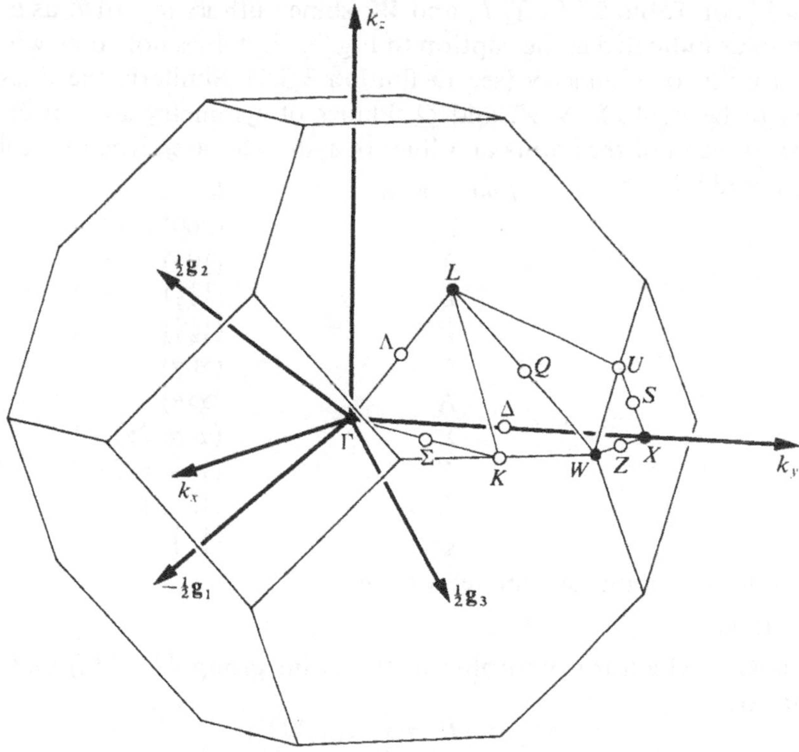

FIG. 5.2. The Brillouin zone for Γ_c^f. $\Gamma = (000)$; $X = (\frac{1}{2}0\frac{1}{2})$; $L = (\frac{1}{2}\frac{1}{2}\frac{1}{2})$; $W = (\frac{1}{2}\frac{1}{4}\frac{3}{4})$. K and U though often referred to as points of symmetry, do not come within the definition of a point of symmetry as given in Definition 3.3.1, as they possess no more symmetry than the neighbouring lines Σ and S, respectively.

deduction of the space-group reps. We now give an example to show how to use Tables 5.7 and 5.8 to retrieve similar information for any of the 230 space groups. We leave it to the reader as an additional example to see how the space-group reps of $Fm3m$ (O_h^5) and $Fd3m$ (O_h^7) considered in section 3.8 can be obtained from Tables 5.7 and 5.8.

The space group $F\bar{4}3c$ (T_d^5) was used on one or two occasions by way of illustration in the notes to Table 5.7, and it is a suitable choice for further illustrating the use of that table. $F\bar{4}3c$ (T_d^5) is a cubic space group related to the point group $\bar{4}3m$ (T_d) and based on the face-centred cubic Bravais lattice Γ_c^f whose basic vectors are given in Table 3.1 as

$$\left.\begin{aligned} \mathbf{t}_1 &= \tfrac{1}{2}(0, a, a), \\ \mathbf{t}_2 &= \tfrac{1}{2}(a, 0, a), \\ \text{and} \quad \mathbf{t}_3 &= \tfrac{1}{2}(a, a, 0). \end{aligned}\right\} \tag{5.4.1}$$

The reciprocal lattice is therefore a body-centred cubic lattice (see Table 3.3), and its Brillouin zone is illustrated in Fig. 3.14 and reproduced in Fig. 5.2. The special points of symmetry in this Brillouin zone are, from either Table 3.6, under the heading Γ_c^f, or Table 5.7 Γ, X, L, and W; some authors regard K as a point of symmetry but, as indicated in the caption to Fig. 3.14, it does not come within our definition of a point of symmetry (see Definition 3.3.1). Similarly the lines of symmetry are seen to be Δ, Λ, Σ, S, Z, and Q. Planes of symmetry are not considered. The **k** vectors of each of the points and lines of symmetry are given in Table 3.6 in terms of $\mathbf{g}_1$, $\mathbf{g}_2$, and $\mathbf{g}_3$:

Point or line	**k**
Γ	(000)
X	$(\frac{1}{2}0\frac{1}{2})$
L	$(\frac{1}{2}\frac{1}{2}\frac{1}{2})$
W	$(\frac{1}{2}\frac{1}{4}\frac{3}{4})$
Δ	$(\alpha 0 \alpha)$
Λ	$(\alpha\alpha\alpha)$
Σ	$(\alpha, \alpha, 2\alpha)$
S	$(\frac{1}{2}+\alpha, 2\alpha, \frac{1}{2}+\alpha)$
Z	$(\frac{1}{2}, \alpha, \frac{1}{2}+\alpha)$
Q	$(\frac{1}{2}, \frac{1}{2}-\alpha, \frac{1}{2}+\alpha)$.

We consider the points of symmetry in turn.

Γ: $\mathbf{k} = (000)$.

${}^H\mathbf{G}^\Gamma$ is $\mathbf{G}_{24}^7$ which is isomorphic to the point group $\bar{4}3m$ (T_d) and its generating elements are

$$\begin{aligned} P &= \{C_{31}^- \mid 000\}, \\ Q &= \{C_{2z} \mid 000\}, \\ R &= \{C_{2x} \mid 000\}, \\ \text{and} \quad S &= \{\sigma_{da} \mid \tfrac{1}{2}\tfrac{1}{2}\tfrac{1}{2}\}. \end{aligned}$$

The five classes (see Table 5.1) are

$$C_1 = \{E\,|\,000\};$$
$$C_2 = \{C_{2x}\,|\,000\}, \{C_{2y}\,|\,000\}, \{C_{2z}\,|\,000\};$$
$$C_3 = \{S_{4x}^-\,|\,\tfrac{1}{2}\tfrac{1}{2}\tfrac{1}{2}\}, \{S_{4y}^-\,|\,\tfrac{1}{2}\tfrac{1}{2}\tfrac{1}{2}\}, \{S_{4z}^-\,|\,\tfrac{1}{2}\tfrac{1}{2}\tfrac{1}{2}\}, \{S_{4x}^+\,|\,\tfrac{1}{2}\tfrac{1}{2}\tfrac{1}{2}\}, \{S_{4y}^+\,|\,\tfrac{1}{2}\tfrac{1}{2}\tfrac{1}{2}\}, \{S_{4z}^+\,|\,\tfrac{1}{2}\tfrac{1}{2}\tfrac{1}{2}\};$$
$$C_4 = \{\sigma_{da}\,|\,\tfrac{1}{2}\tfrac{1}{2}\tfrac{1}{2}\}, \{\sigma_{db}\,|\,\tfrac{1}{2}\tfrac{1}{2}\tfrac{1}{2}\}, \{\sigma_{dc}\,|\,\tfrac{1}{2}\tfrac{1}{2}\tfrac{1}{2}\}, \{\sigma_{dd}\,|\,\tfrac{1}{2}\tfrac{1}{2}\tfrac{1}{2}\}, \{\sigma_{de}\,|\,\tfrac{1}{2}\tfrac{1}{2}\tfrac{1}{2}\}, \{\sigma_{df}\,|\,\tfrac{1}{2}\tfrac{1}{2}\tfrac{1}{2}\};$$
$$C_5 = \{C_{31}^-\,|\,000\}, \{C_{32}^-\,|\,000\}, \{C_{33}^-\,|\,000\}, \{C_{34}^-\,|\,000\}, \{C_{31}^+\,|\,000\}, \{C_{32}^+\,|\,000\}, \{C_{33}^+\,|\,000\}, \{C_{34}^+\,|\,000\}.$$

To identify which of the elements of $\mathbf{G}_{24}^7$ corresponds to a particular space-group element is rather laborious. It has to be done by direct multiplication, see Note (iv) to Table 5.7 where it is shown that P^2RS of $\mathbf{G}_{24}^7$ corresponds to $\{S_{4y}^-\,|\,\tfrac{1}{2}\,-\tfrac{3}{2}\tfrac{1}{2}\}$ or, at Γ, to $\{S_{4y}^-\,|\,\tfrac{1}{2}\tfrac{1}{2}\tfrac{1}{2}\}$. This identification must of course be made if one wishes to find the actual matrix representatives and not just the characters in the cases of degenerate representations. The character table for ${}^H\mathbf{G}^\Gamma$ can thus be found from Table 5.1 under $\mathbf{G}_{24}^7$, but rather than labelling the reps as R_1, R_2, R_3, R_4, and R_5 they are labelled according to the entry 'a' under $\mathbf{G}_{24}^7$ in Table 5.8

	R_1	R_2	R_3	R_4	R_5
a	A_1 Γ_1	A_2 Γ_2	E Γ_3	T_1 Γ_4	T_2 Γ_5

The space-group reps can be seen from Table 5.7 to be all of the first kind so that the character table for ${}^H\mathbf{G}^\Gamma$ is (with the reality of the corresponding space-group reps shown in the right-hand column)

	C_1	C_2	C_3	C_4	C_5	
Γ_1 A_1	1	1	1	1	1	1
Γ_2 A_2	1	1	−1	−1	1	1
Γ_3 E	2	2	0	0	−1	1
Γ_4 T_1	3	−1	1	−1	0	1
Γ_5 T_2	3	−1	−1	1	0	1

and there are no extra degeneracies if time-reversal symmetry is included.

X: $\mathbf{k} = (\frac{1}{2}0\frac{1}{2})$

${}^H\mathbf{G}^X$ is the direct product of $\mathbf{G}_8^4$, with generating elements

$$P = \{S_{4y}^+ \mid \tfrac{1}{2}\tfrac{1}{2}\tfrac{1}{2}\},$$
$$Q = \{C_{2x} \mid 000\},$$

and $\mathbf{T}_2$, with elements $\{E \mid \mathbf{0}\}$ and $\{E \mid \mathbf{t}_s\}$ where $s = 1$ or 3. The elements of $\mathbf{G}_8^4$ are therefore

E	$\{E \mid 000\}$	Q	$\{C_{2x} \mid 000\}$
P	$\{S_{4y}^+ \mid \frac{1}{2}\frac{1}{2}\frac{1}{2}\}$	PQ	$\{C_{2c} \mid \frac{1}{2}\frac{1}{2}\frac{1}{2}\}$
P^2	$\{C_{2y} \mid 000\}$	P^2Q	$\{C_{2z} \mid 000\}$
P^3	$\{S_{4y}^- \mid \frac{1}{2}\frac{1}{2}\frac{1}{2}\}$	P^3Q	$\{C_{2e} \mid \frac{1}{2}\frac{1}{2}\frac{1}{2}\}$

and from Table 5.1 these can be assigned to the classes $C_1, C_2, \ldots, C_5$. The five reps are labelled, from Table 5.8, as A_1, A_2, B_1, B_2, and E (or X_1, X_2, X_3, X_4, and X_5) respectively. Thus the character table for ${}^H\mathbf{G}^X$ is (with the reality of the corresponding space-group reps shown in the right-hand column)

		$\{E \mid 000\}$	$\{C_{2y} \mid 000\}$	$\{S_{4y}^+ \mid \frac{1}{2}\frac{1}{2}\frac{1}{2}\}$ $\{S_{4y}^- \mid \frac{1}{2}\frac{1}{2}\frac{1}{2}\}$	$\{C_{2x} \mid 000\}$ $\{C_{2z} \mid 000\}$	$\{C_{2c} \mid \frac{1}{2}\frac{1}{2}\frac{1}{2}\}$ $\{C_{2e} \mid \frac{1}{2}\frac{1}{2}\frac{1}{2}\}$	
X_1	A_1	1	1	1	1	1	1
X_2	A_2	1	1	1	−1	−1	1
X_3	B_1	1	1	−1	1	−1	1
X_4	B_2	1	1	−1	−1	1	1
X_5	E	2	−2	0	0	0	1

and if time-reversal symmetry is included there will be no extra degeneracies. If the matrices for the representation $E(X_5)$ are required they are also available from R_5 of $\mathbf{G}_8^4$ in Table 5.1.

L: $\mathbf{k} = (\frac{1}{2}\frac{1}{2}\frac{1}{2})$

${}^H\mathbf{G}^L$ is $\mathbf{G}_{12}^4$ with

$$P = \{C_{31}^- \mid 001\}$$

and

$$Q = \{\sigma_{db} \mid \tfrac{1}{2}\tfrac{1}{2}\tfrac{1}{2}\}$$

and the twelve members of ${}^H\mathbf{G}^L$ are

E	$\{E \mid 000\}$	Q	$\{\sigma_{db} \mid \frac{1}{2}\frac{1}{2}\frac{1}{2}\}$
P	$\{C_{31}^- \mid 001\}$	PQ	$\{\sigma_{df} \mid \frac{1}{2}\frac{1}{2}-\frac{1}{2}\}$
P^2	$\{C_{31}^+ \mid 000\}$	P^2Q	$\{\sigma_{de} \mid \frac{1}{2}\frac{1}{2}\frac{1}{2}\}$

$$\begin{array}{llll} P^3 & \{E\,|\,001\} & P^3Q & \{\sigma_{db}\,|\,\tfrac{1}{2}\tfrac{1}{2}-\tfrac{1}{2}\} \\ P^4 & \{C_{31}^-\,|\,000\} & P^4Q & \{\sigma_{df}\,|\,\tfrac{1}{2}\tfrac{1}{2}\tfrac{1}{2}\} \\ P^5 & \{C_{31}^+\,|\,001\} & P^5Q & \{\sigma_{de}\,|\,\tfrac{1}{2}\tfrac{1}{2}-\tfrac{1}{2}\}. \end{array}$$

These can be assigned to the classes of $\mathbf{G}_{12}^4$ and the character table of ${}^H\mathbf{G}^L$ constructed:

	$\{E\|000\}$	$\{E\|001\}$	$\{C_{31}^-\|001\}$ $\{C_{31}^+\|001\}$	$\{C_{31}^+\|000\}$ $\{C_{31}^-\|000\}$	$\{\sigma_{db}\|\frac{1}{2}\frac{1}{2}\frac{1}{2}\}$ $\{\sigma_{de}\|\frac{1}{2}\frac{1}{2}\frac{1}{2}\}$ $\{\sigma_{df}\|\frac{1}{2}\frac{1}{2}\frac{1}{2}\}$	$\{\sigma_{db}\|\frac{1}{2}\frac{1}{2}-\frac{1}{2}\}$ $\{\sigma_{de}\|\frac{1}{2}\frac{1}{2}-\frac{1}{2}\}$ $\{\sigma_{df}\|\frac{1}{2}\frac{1}{2}-\frac{1}{2}\}$	
L_1 ${}^2E$	1	−1	−1	1	i	−i	3
L_2 ${}^1E$	1	−1	−1	1	−i	i	3
L_3 E	2	−2	1	−1	0	0	2

The matrices for $E(L_3)$ are available from R_6 of $\mathbf{G}_{12}^4$ in Table 5.1. The reality included in the extreme right-hand column indicates that if time-reversal symmetry is included there is an extra degeneracy between the reps ${}^1E$ and ${}^2E$ (L_2 and L_1) and also an extra degeneracy between two different sets of basis functions belonging to the rep $E(L_3)$.

W: $\mathbf{k} = (\frac{1}{2}\frac{1}{4}\frac{3}{4})$

${}^H\mathbf{G}^W$ is the direct product of $\mathbf{G}_4^1$ and $\mathbf{T}_4$, where $\mathbf{T}_4$ consists of $\{E\,|\,\mathbf{0}\}$, $\{E\,|\,\mathbf{t}_2\}$, $\{E\,|\,2\mathbf{t}_2\}$, and $\{E\,|\,3\mathbf{t}_2\}$ and $\mathbf{G}_4^1$ has as its generating element $P = \{S_{4x}^+\,|\,\frac{1}{2}\frac{1}{2}\frac{1}{2}\}$. Each of the elements of $\mathbf{G}_4^1$ is in a class by itself and the character table of ${}^H\mathbf{G}^W$ is thus

	$\{E\|000\}$	$\{S_{4x}^+\|\frac{1}{2}\frac{1}{2}\frac{1}{2}\}$	$\{C_{2x}\|000\}$	$\{S_{4x}^-\|\frac{1}{2}\frac{1}{2}\frac{1}{2}\}$	
W_1 A	1	1	1	1	3
W_2 ${}^2E$	1	i	−1	−i	3
W_3 B	1	−1	1	−1	3
W_4 ${}^1E$	1	−i	−1	i	3

The reality included in the extreme right-hand column indicates that if time-reversal symmetry is included there is an extra degeneracy between pairs of eigenfunctions belonging to different reps; in fact reps $A(W_1)$ and $B(W_3)$ become degenerate and reps ${}^1E(W_4)$ and ${}^2E(W_2)$ become degenerate (see sections 7.6 and 7.7).

This completes the consideration of the points of symmetry for this space group and we now proceed to the consideration of the lines of symmetry which involves the theory of projective representations which was described in section 3.7. For each line of symmetry we first identify the central extension $\overline{\mathbf{G}}^{\mathbf{k}*}$ and then from this obtain the representations of the little group $\mathbf{G}^{\mathbf{k}}$. The elements of $\overline{\mathbf{G}}^{\mathbf{k}*}$ take the form (H_j, α) where H_j is the point-group operation part of a space-group element and α is a small whole number which is a member of the cyclic group $\mathbf{Z}_g$ defined in eqn. (3.7.27). The group $\overline{\mathbf{G}}^{\mathbf{k}*}$ is specified in Table 5.7 and its generating elements $P, Q, \ldots,$ in that order, are given in the form (H_j, α). Thus the character table and matrix representatives of $\overline{\mathbf{G}}^{\mathbf{k}*}$ can be identified from Table 5.1 and the reps of $\mathbf{G}^{\mathbf{k}}$ can be found for each element $\{R \mid \mathbf{v}\}$ by multiplying the matrix representative of $(R, 0)$ by the factor $\exp(-i\mathbf{k}.\mathbf{v})$ (see eqns. (3.8.8) and (3.8.9)). The labels of the reps can be found by consulting the part of Table 5.8 indicated in the relevant line of Table 5.7. It will be noted that reps of $\overline{\mathbf{G}}^{\mathbf{k}*}$ are omitted when they are not compatible with the allowed representations of the cyclic group $\mathbf{Z}_g$ generated by $(E, 1)$ (see eqn. (3.7.33)). It should be noted that, in general, the class structure of $\mathbf{G}^{\mathbf{k}}$ may differ from that of $\overline{\mathbf{G}}^{\mathbf{k}*}$ and that, since $\mathbf{G}^{\mathbf{k}}$ is infinite, it may be quite a complicated process to determine the class structure of $\mathbf{G}^{\mathbf{k}}$. The realities of the corresponding space-group reps, and therefore the extra degeneracies due to the inclusion of time-reversal symmetry, are also available from Table 5.7. We now illustrate this for the lines of symmetry in our chosen space group $F\bar{4}3c$ (T_d^5).

Δ: $\mathbf{k} = (\alpha 0 \alpha)$

The central extension $\overline{\mathbf{G}}^{\Delta*}$ is the group $\mathbf{G}_4^2$ and its generating elements are

$$P = (C_{2y}, 0)$$
$$Q = (\sigma_{dc}, 0).$$

Since $\mathbf{G}_4^2$ is of order 4 (which is the same as the order of the point group containing E, C_{2y}, σ_{dc}, and σ_{de}) it is clear that g must be equal to 1 so that the four elements of $\overline{\mathbf{G}}^{\Delta*}$ are, in their classes,

C_1	E	$(E, 0)$
C_2	P	$(C_{2y}, 0)$
C_3	Q	$(\sigma_{dc}, 0)$
C_4	PQ	$(\sigma_{de}, 0)$.

The character table of $\overline{\mathbf{G}}^{\Delta*}$ is thus, from Table 5.1,

	$(E, 0)$	$(C_{2y}, 0)$	$(\sigma_{dc}, 0)$	$(\sigma_{de}, 0)$
Δ_1 A_1	1	1	1	1
Δ_3 A_2	1	1	-1	-1
Δ_2 B_1	1	-1	1	-1
Δ_4 B_2	1	-1	-1	1

using Table 5.8 to give the labels of the reps. The small reps of $\mathbf{G}^{\Delta}$ are the above but with each entry multiplied by the factor $\exp(-i\mathbf{k}.\mathbf{v})$:

	$\{E \mid 000\}$	$\{C_{2y} \mid 000\}$	$\{\sigma_{dc} \mid \frac{1}{2}\frac{1}{2}\frac{1}{2}\}$	$\{\sigma_{de} \mid \frac{1}{2}\frac{1}{2}\frac{1}{2}\}$	
Δ_1 A_1	1	1	A	A	1
Δ_3 A_2	1	1	$-A$	$-A$	1
Δ_2 B_1	1	-1	A	$-A$	3
Δ_4 B_2	1	-1	$-A$	A	3

where $A = \exp(-2\pi i\alpha)$. The inclusion of time-reversal symmetry leads to no extra degeneracy in the reps $A_1(\Delta_1)$ and $A_2(\Delta_3)$ but causes the reps $B_1(\Delta_2)$ and $B_2(\Delta_4)$ to become degenerate.

Λ: $\mathbf{k} = (\alpha\alpha\alpha)$

$\overline{\mathbf{G}}^{\Lambda*}$ is $\mathbf{G}_6^2$ which is of order 6, $g = 1$, and its generating elements are

$$P = (C_{31}^+, 0)$$

and

$$Q = (\sigma_{db}, 0)$$

and its elements are, in classes,

C_1	E	$(E, 0)$,
C_2	P	$(C_{31}^+, 0)$,
	P^2	$(C_{31}^-, 0)$,
C_3	Q	$(\sigma_{db}, 0)$,
	PQ	$(\sigma_{de}, 0)$,
	P^2Q	$(\sigma_{df}, 0)$.

The character table of $\overline{\mathbf{G}}^{\Lambda*}$ is thus

	$(E, 0)$	$(C_{31}^+, 0)$ $(C_{31}^-, 0)$	$(\sigma_{db}, 0)$ $(\sigma_{de}, 0)$ $(\sigma_{df}, 0)$
Λ_1 A_1	1	1	1
Λ_2 A_2	1	1	-1
Λ_3 E	2	-1	0

and the matrices for the representation E can also be found from Table 5.1. The small reps of $\mathbf{G}^{\Lambda}$ are therefore

	$\{E \mid 000\}$	$\{C_{31}^+ \mid 000\}$ $\{C_{31}^- \mid 000\}$	$\{\sigma_{db} \mid \frac{1}{2}\frac{1}{2}\frac{1}{2}\}$ $\{\sigma_{de} \mid \frac{1}{2}\frac{1}{2}\frac{1}{2}\}$ $\{\sigma_{df} \mid \frac{1}{2}\frac{1}{2}\frac{1}{2}\}$	
Λ_1 A_1	1	1	B	x
Λ_2 A_2	1	1	$-B$	x
Λ_3 E	2	-1	0	x

where $B = \exp(-3\pi i\alpha)$. The addition of time-reversal symmetry leads to a case (x) degeneracy, that is, the spectrum of the eigenvalues of a particle or quasi-particle is the same at $\mathbf{k}$ and $-\mathbf{k}$.

Σ: $\mathbf{k} = (\alpha, \alpha, 2\alpha)$
$\overline{\mathbf{G}}^{\Sigma *}$ is $\mathbf{G}_2^1$ which is of order 2, $g = 1$, and its character table is

	$(E, 0)$	$(\sigma_{db}, 0)$
Σ_1 A'	1	1
Σ_2 A''	1	-1

so that the small reps of $\mathbf{G}^{\Sigma}$ are

	$\{E \mid 000\}$	$\{\sigma_{db} \mid \frac{1}{2}\frac{1}{2}\frac{1}{2}\}$	
Σ_1 A'	1	C	1
Σ_2 A''	1	$-C$	1

where $C = \exp(-4\pi i\alpha)$. From the reality of the reps we see that there is no extra degeneracy at Σ if time-reversal symmetry is included.

S: $\mathbf{k} = (\frac{1}{2} + \alpha, 2\alpha, \frac{1}{2} + \alpha)$
$\overline{\mathbf{G}}^{S*}$ is $\mathbf{G}_2^1$, which is of order 2, $g = 1$, and its character table is

	$(E, 0)$	$(\sigma_{de}, 0)$
S_1 A'	1	1
S_2 A''	1	-1

so that the small reps of $\mathbf{G}^S$ are

	$\{E \mid 000\}$	$\{\sigma_{de} \mid \frac{1}{2}\frac{1}{2}\frac{1}{2}\}$	
S_1 A'	1	D	1
S_2 A''	1	$-D$	1

where $D = \exp(-\pi i(1 + 4\alpha))$. From the reality we see that there is no extra degeneracy at S if time-reversal symmetry is included.

Z: $\mathbf{k} = (\frac{1}{2}, \alpha, \frac{1}{2} + \alpha)$
$\overline{\mathbf{G}}^{Z*}$ is $\mathbf{G}_2^1$, $g = 1$, and its character table is

	$(E, 0)$	$(C_{2x}, 0)$
Z_1 A	1	1
Z_2 B	1	-1

and since $\mathbf{v} = 0$ for both these elements $\{R \mid \mathbf{v}\}$ then the small reps of $\mathbf{G}^Z$ are

	$\{E \mid 000\}$	$\{C_{2x} \mid 000\}$	
Z_1 A	1	1	1
Z_2 B	1	-1	1

From the reality we see that there is no extra degeneracy at Z if time-reversal symmetry is included.
Q: $\mathbf{k} = (\frac{1}{2}, \frac{1}{2} - \alpha, \frac{1}{2} + \alpha)$
$\overline{\mathbf{G}}^{Q*}$ is $\mathbf{G}_1^1$ so that the rep of $\mathbf{G}^Q$ is

	$\{E \mid 000\}$	
Q_1 A	1	2

and from the reality we see that the addition of time-reversal symmetry causes the eigenfunctions belonging to $A(Q_1)$ to become degenerate in pairs.

This example of $F\bar{4}3c$ (T_d^5), together with those in section 3.8 (which include examples in which $g \neq 1$) should make it clear how to use Table 5.7 and the associated Tables 5.1 and 5.8 to construct the reps of any of the 230 space groups, for any of the points of symmetry or lines of symmetry in the basic domain of the appropriate Brillouin zone, and to determine the extra degeneracies that will arise if time-reversal symmetry is added to the space group.

The physical significance of space-group reps

The importance of determining the reps of the space groups lies in the application of Wigner's theorem that we have already mentioned in sections 1.1 and 3.7 (see Theorem 3.7.1). The result of this theorem may be expressed in physical terms by saying that if ψ is the wave function of a particle or quasi-particle within a crystal, whose symmetry is described by a space group $\mathbf{G}$, then ψ must transform under the operations of $\mathbf{G}$ like one component of a basis of one of the reps of $\mathbf{G}$. These reps are uniquely identified by the small reps of the various little groups $\mathbf{G}^{\mathbf{k}}$, so that ψ may be regarded as belonging to a basis of one of the small reps of $\mathbf{G}^{\mathbf{k}}$. In other words, we may regard the eigenvalues of the Hamiltonian, $\mathbf{H}$, of the crystal, together with their associated eigenfunctions, as being distributed among all the allowed wave vectors $\mathbf{k}$

(defined by eqn. (3.4.2)) in the Brillouin zone. The space-group reps can therefore be used in providing a scheme for classifying and labelling the eigenfunctions of **H**; the degeneracies of the space-group reps will specify the essential degeneracies of the corresponding eigenvalues of **H**. This is similar to the use of the point-group reps in labelling the electronic or vibrational energy levels of a molecule.

The use of the space-group reps to label the eigenvalues and eigenfunctions of the electronic band structure of a crystal was introduced by Bouckaert, Smoluchowski, and Wigner (1936); these labels have been used extensively ever since then. Let us suppose that $\psi^{\mathbf{k}}_{p,q}$, the wave function of an electron with wave vector **k**, belongs to some particular small rep $\Gamma^{\mathbf{k}}_p$ and that $\psi^{\mathbf{k}}_{p,q}$ has been expanded in terms of some standard functions, such as spherical harmonics or plane waves. It is then possible to determine linear combinations of the spherical harmonics, or of the plane waves, which are symmetry-adapted to the rep $\Gamma^{\mathbf{k}}_p$. These symmetry-adapted functions can be determined by the use of the operator W^i_{ts} which was described in section 2.2. For any wave vector **k** for which $\bar{\mathbf{G}}^{\mathbf{k}}$ is non-trivial $\psi^{\mathbf{k}}_{p,q}$ can be expanded in terms of these symmetry-adapted functions and there will then be fewer arbitrary coefficients (for any given order of accuracy) than if the expansion were performed directly in terms of spherical harmonics or plane waves. The use of such symmetry-adapted functions has led to a considerable simplification of band-structure calculations, which was very valuable in the days when computer technology did not allow large expansions for general wave vectors, **k**, to be used (Altmann and Bradley 1965; Altmann and Cracknell 1965, Bell 1954, Cornwell 1969, Howarth and Jones 1952, Jones 1960, Mase 1958, 1959a, von der Lage and Bethe 1947). A list of references on the determination of symmetry-adapted functions for some particular space groups is given in Appendix 3 of the book by Cornwell (1969). In relativistic band-structure calculations it is necessary to use the double-valued space-group reps which will be discussed in Chapter 6.

The use of space-group reps has now been extended to the labelling of the normal modes of vibration of a crystal lattice, that is, of the phonon dispersion curves (Johnson and Loudon 1964; for reviews see Maradudin and Vosko (1968) and Warren (1968)) and also to the labelling of the spin–wave dispersion relations in a magnetic crystal (Dimmock and Wheeler 1962*b*, Joshua and Cracknell 1969, Loudon 1968). Originally the experimental measurement of phonon dispersion curves had been confined to directions of high symmetry in elemental cubic materials; for such situations the normal modes could be classified and labelled quite adequately as pure longitudinal acoustic (LA), transverse acoustic (TA), longitudinal optic (LO), or transverse optic (TO). However, this simple scheme is inadequate for directions of low symmetry or for more complicated crystals so that the normal modes are now usually labelled by the use of the space-group reps. In the same way that the normal modes of vibration of a molecule can be investigated group-theoretically and assigned to the various reps of the point group of that molecule (see, for example,

Cotton (1963), Cracknell (1968*b*), Leech and Newman (1969)) so also the normal modes of vibration of a crystal with any given wave vector **k** can be assigned to the various reps of the space group **G** of the crystal. This can be done by studying the transformation properties, under the operations of the space group **G**, of a general displacement $\mathbf{x}_i$ of the atoms in the unit cell of the crystal. In this way a representation Γ will be obtained of which $\mathbf{x}_i$ is a basis and, in general, this representation will be reducible. The small reps of $\mathbf{G}^{\mathbf{k}}$ to which the normal modes at **k** belong can then be obtained from the reduction of Γ and the labels of the small reps are then used to label the phonon dispersion relations for the **k** vectors in the Brillouin zone. Alternatively, the symmetry properties of a normal mode with wave vector **k** can be determined by studying the transformation properties of the creation and annihilation operators $a_{\mathbf{k}}^{\dagger}$ and $a_{\mathbf{k}}$ under the operations of the space group **G** and assigning them to the appropriate small reps of $\mathbf{G}^{\mathbf{k}}$; $a_{\mathbf{k}}^{\dagger}$ and $a_{\mathbf{k}}$ are the operators which create and annihilate, respectively, a phonon with wave vector **k** and their form is given by, for example, Kittel (1963). The labelling of the spin–wave dispersion relations for a magnetic crystal in terms of space-group reps can be determined by studying the transformation properties, under the operations of **G**, of the appropriate operators, similar to $a_{\mathbf{k}}^{\dagger}$ and $a_{\mathbf{k}}$, that create and annihilate magnons (Loudon 1968).

5.5. Representation domain and basic domain

In Table 5.7 (and also in the tables of Miller and Love (1967)) the space-group reps are only tabulated for the special points of symmetry and lines of symmetry in just one basic domain (see Definition 3.3.3) of the Brillouin zone of each of the 230 space groups. It will be recalled that the *basic domain*, Ω, has the property that $(\sum_R R\Omega)$ is equal to the whole Brillouin zone, where the operators R are the elements of the holosymmetric point group, **P**, of the appropriate crystal system. Thus, for example, for a cubic crystal the basic domain is only one-fortyeighth of the Brillouin zone. Since they may be required in the discussion of physical problems, it is necessary for completeness to consider how to determine the small reps associated with other points or lines of symmetry in the Brillouin zone, given those in just one basic domain. It will be remembered that the *representation domain*, Φ (see Definition 3.7.1), has the property that $(\Sigma_S S\Phi)$ is equal to the whole Brillouin zone, where the operators S are the elements of the isogonal point group, **F**, of the space group, **G**, under consideration. Thus, for example, for the space group $Pa3$ (T_h^6), the isogonal point group is $m3$ (T_h), which contains 24 symmetry operations, and Φ is therefore one-twentyfourth of the Brillouin zone; the basic domain, Ω, is determined by the holosymmetric cubic point group $m3m$ (O_h), which contains 48 symmetry operations, and Ω is one-fortyeighth of the Brillouin zone. So in this case the size of the representation domain is twice that of the basic domain. In general, since **F** is either equal to **P** or is a proper subgroup of **P**, the representation domain Φ is either the same as Ω or is the sum of an integral number of volumes congruent to Ω.

There are thus two situations to consider, the first when the representation domain is the same size as the basic domain, and the second when it is larger. If the representation domain, Φ, is the same size as the basic domain, Ω, that is, $\mathbf{F} = \mathbf{P}$, then any wave vector $\mathbf{k}_B$ that is outside the basic domain is related to some wave vector $\mathbf{k}_A$ within the basic domain by an expression

$$\mathbf{k}_B \equiv R\mathbf{k}_A \tag{5.5.1}$$

where $R \in \mathbf{P}$. Since $\mathbf{P} = \mathbf{F}$, this means that in this case there is always some wave vector $\mathbf{k}_A$ in the star of $\mathbf{k}_B$, which is in the basic domain of the Brillouin zone. This means that any physical observable associated with a particle or quasi-particle with wave vector $\mathbf{k}_B$ will have the same values as those possessed by a particle or quasi-particle with wave vector $\mathbf{k}_A$. Therefore, *if the representation domain, Φ, is the same size as the basic domain, Ω, no new physical insight is obtained by studying any wave vector outside the basic domain.* It is, probably, because most of the space groups that have actually been used in band structure calculations or studies of phonon dispersion relations, come into this classification, that there can arise the mistaken assumption that it is always adequate to consider only wave vectors within the basic domain of the Brillouin zone.

In the second situation the volume of the representation domain, Φ, is some small integral multiple of the volume of the basic domain, Ω. But now, for a vector $\mathbf{k}_B$ in Φ but not in Ω, there is no operation $R \in \mathbf{F}$ that relates $\mathbf{k}_B$ by means of eqn. (5.5.1) to any vector $\mathbf{k}_A$ within the basic domain. In this case the operations R that satisfy eqn. (5.5.1) are in $\mathbf{P}$ but not in $\mathbf{F}$. What remains true however is that *all significantly different wave vectors, from a physical point of view, will be obtained by considering all the wave vectors in the representation domain, Φ.* For example, in all cases, the energy eigenvalue surfaces $E(\mathbf{k}) = E$ of a band structure calculation will possess the point-group symmetry of the isogonal point group, $\mathbf{F}$. Wave vectors that are outside the representation domain still do not need to be considered, but it is necessary to be able to determine the space-group reps for wave vectors $\mathbf{k}_B$ which are within Φ but outside Ω and which are therefore not immediately available from Table 5.7. In this connection we note first the following theorem.

THEOREM 5.5.1. *If* **F** *is a point group of a given crystal system and* **P** *is the holosymmetric point group of that crystal system then* **F** *is an invariant subgroup of* **P**.

The proof is most simply performed by enumeration and inspection. We have noted previously that every non-trivial space group possesses an invariant subgroup of index either 2 or 3 (Zak (1960), see section 5.2). A related result in which we shall be interested in the present connection is that *nearly* every space group, **G**, whose isogonal point group, **F**, is of lower order than the order of **P** (the holosymmetric point group of that crystal system), is an invariant subgroup of some *parent space group* $\mathbf{G}_o$, where $\mathbf{G}_o$ is based on the same Bravais lattice as **G** and the isogonal point group of $\mathbf{G}_o$ is $\mathbf{P}$. This too can be verified by enumeration and inspection of the various space groups and it is found that there are only 24 exceptions to this rule:

$$\left.\begin{array}{c} P4_1\ (C_4^2),\ P4_3\ (C_4^4), \\ P4_122\ (D_4^3),\ P4_322\ (D_4^7),\ P4_12_12\ (D_4^4),\ P4_32_12\ (D_4^8) \\ P3_1\ (C_3^2),\ P3_2\ (C_3^3), \\ P3_112\ (D_3^3),\ P3_212\ (D_3^5),\ P3_121\ (D_3^4),\ P3_221\ (D_3^6) \\ P6_1\ (C_6^2),\ P6_5\ (C_6^3),\ P6_2\ (C_6^4),\ P6_4\ (C_6^5) \\ P6_122\ (D_6^2),\ P6_522\ (D_6^3),\ P6_222\ (D_6^4),\ P6_422\ (D_6^5), \\ P2_13\ (T^4), \\ Pa3\ (T_h^6), \\ P4_332\ (O^6) \text{ and } P4_132\ (O^7). \end{array}\right\} \qquad (5.5.2)$$

These 24 space groups comprise the 11 isomorphic but crystallographically distinguishable pairs together with two exceptional space groups $P2_13$ (T^4) and $Pa3$ (T_h^6).

Suppose that $\mathbf{k}_B$ is a wave vector in Φ but not in Ω. Then there exists a wave vector $\mathbf{k}_A$ in the basic domain, Ω, and a point-group operation R which is not in $\mathbf{F}$, the isogonal point group of $\mathbf{G}$, but which is in $\mathbf{P}$, the holosymmetric point group of the appropriate crystal system, such that

$$\mathbf{k}_B \equiv R\mathbf{k}_A. \qquad (5.5.1)$$

Since $\mathbf{k}_A$ is in Ω, the little group $\mathbf{G}^{\mathbf{k}_A}$ and the small reps of $\mathbf{G}^{\mathbf{k}_A}$ are available and can be identified from Table 5.7.

THEOREM 5.5.2. *If $\overline{\mathbf{G}}^{\mathbf{k}_A}$ is the little co-group of $\mathbf{k}_A$, then the little co-group of $\mathbf{k}_B$ is given by $\overline{\mathbf{G}}^{\mathbf{k}_B} = R\overline{\mathbf{G}}^{\mathbf{k}_A}R^{-1}$.*

This theorem can be proved as follows. Let $S \in \overline{\mathbf{G}}^{\mathbf{k}_A}$, then

$$S\mathbf{k}_A \equiv \mathbf{k}_A, \qquad (5.5.3)$$

which, from eqn. (5.5.1), may be written as

$$SR^{-1}\mathbf{k}_B \equiv R^{-1}\mathbf{k}_B \qquad (5.5.4)$$

and therefore

$$(RSR^{-1})\mathbf{k}_B \equiv \mathbf{k}_B. \qquad (5.5.5)$$

Now $S \in \overline{\mathbf{G}}^{\mathbf{k}_A} \subset \mathbf{F}$ and $R \in \mathbf{P}$. Since $\mathbf{F}$ is invariant in $\mathbf{P}$ (see Theorem 5.5.1) it follows that $RSR^{-1} \in \mathbf{F}$. This fact and eqn. (5.5.5) together imply that $RSR^{-1} \in \overline{\mathbf{G}}^{\mathbf{k}_B}$. Since this is true for all $S \in \overline{\mathbf{G}}^{\mathbf{k}_A}$, it follows that $R\overline{\mathbf{G}}^{\mathbf{k}_A}R^{-1} \subset \overline{\mathbf{G}}^{\mathbf{k}_B}$. Starting from an element $T \in \overline{\mathbf{G}}^{\mathbf{k}_B}$, it can be proved similarly that $R^{-1}TR \in \overline{\mathbf{G}}^{\mathbf{k}_A}$, and hence that $R^{-1}\overline{\mathbf{G}}^{\mathbf{k}_B}R \subset \overline{\mathbf{G}}^{\mathbf{k}_A}$, or $\overline{\mathbf{G}}^{\mathbf{k}_B} \subset R\overline{\mathbf{G}}^{\mathbf{k}_A}R^{-1}$. Thus the reverse inclusion also holds, and therefore

$$\overline{\mathbf{G}}^{\mathbf{k}_B} = R\overline{\mathbf{G}}^{\mathbf{k}_A}R^{-1}. \qquad (5.5.6)$$

The little co-group $\overline{\mathbf{G}}^{\mathbf{k}_B}$ is therefore conjugate to the little co-group $\overline{\mathbf{G}}^{\mathbf{k}_A}$ with respect to **P**. There is clearly an isomorphism between $\overline{\mathbf{G}}^{\mathbf{k}_B}$ and $\overline{\mathbf{G}}^{\mathbf{k}_A}$, because, if $S_i, S_j \in \overline{\mathbf{G}}^{\mathbf{k}_A}$ and

$$S_i S_j = S_k, \tag{5.5.7}$$

then, writing $T_i = RS_iR^{-1}$, etc., it follows that

$$\begin{aligned} T_i T_j &= RS_iR^{-1}RS_jR^{-1} = RS_iS_jR^{-1} \\ &= RS_kR^{-1} = T_k. \end{aligned} \tag{5.5.8}$$

Very often the little co-group $\overline{\mathbf{G}}^{\mathbf{k}_B}$ will coincide with $\overline{\mathbf{G}}^{\mathbf{k}_A}$.

The question now arises as to whether the little group $\mathbf{G}^{\mathbf{k}_B}$ is conjugate to the little group $\mathbf{G}^{\mathbf{k}_A}$ with respect to the parent space group $\mathbf{G}_o$ (whose isogonal point group is **P**, the holosymmetric point group of the crystal system of **G**), where the space group **G** is an invariant subgroup of $\mathbf{G}_o$.

THEOREM 5.5.3. $\mathbf{G}^{\mathbf{k}_B}$ *is conjugate to* $\mathbf{G}^{\mathbf{k}_A}$ *with respect to the parent space group* $\mathbf{G}_o$ *(provided* $\mathbf{G}_o$ *exists).*

The proof of this theorem is trivial. If $\mathbf{G}_o$ exists it will contain an operation $\{R \mid \mathbf{v}\}$ whose rotational part R satisfies eqn. (5.5.1). Then, since **G** is invariant in $\mathbf{G}_o$, it follows exactly as in the proof of Theorem 5.5.2 that

$$\mathbf{G}^{\mathbf{k}_B} = \{R \mid \mathbf{v}\}\mathbf{G}^{\mathbf{k}_A}\{R \mid \mathbf{v}\}^{-1}. \tag{5.5.9}$$

A word of warning is perhaps necessary at this stage, and that is that the setting of $\mathbf{G}_o$ may have to be adapted from its form in the tables before the selection of $\{R \mid \mathbf{v}\}$ is made. This is because in any set of tables a space group stands for a whole class of crystallographically indistinguishable groups, and naturally only one setting will have been chosen for the purposes of tabulation. Thus the setting of $\mathbf{G}_o$ may have to be altered so that it contains **G** properly, and not just an isomorphic image of **G**.

Now let $D_p^{\mathbf{k}_A}[\{S \mid \mathbf{w}\}]$ be a small rep of $\mathbf{G}^{\mathbf{k}_A}$, then $D_p^{\mathbf{k}_A}$ is available from Table 5.7. It will be remembered that a small rep has to satisfy two properties: (i) it must be an irreducible representation of the little group, and (ii) it must be compatible with the representations of the translational symmetry operations $\{E \mid \mathbf{t}\}$ of the Bravais lattice. Thus

$$D_p^{\mathbf{k}_A}[\{E \mid \mathbf{t}\}] = \exp(-i\mathbf{k}_A \cdot \mathbf{t})\mathbf{1} \tag{5.5.10}$$

for all **t**.

THEOREM 5.5.4. *The representation* $D_p^{\mathbf{k}_B}$ *of* $\mathbf{G}^{\mathbf{k}_B}$ *defined by*

$$D_p^{\mathbf{k}_B}[\{R \mid \mathbf{v}\}\{S \mid \mathbf{w}\}\{R \mid \mathbf{v}\}^{-1}] = D_p^{\mathbf{k}_A}[\{S \mid \mathbf{w}\}] \tag{5.5.11}$$

is a small rep of $\mathbf{G}^{\mathbf{k}_B}$.

That property (i) is satisfied is clear enough and follows immediately from the fact that $D_p^{\mathbf{k}_A}$ is an irreducible representation. To verify property (ii) we evaluate $D_p^{\mathbf{k}_B}[\{E \mid \mathbf{t}\}]$. From eqn. (5.5.11)

$$\begin{aligned} D_p^{\mathbf{k}_B}[\{E \mid \mathbf{t}\}] &= D_p^{\mathbf{k}_A}[\{R \mid \mathbf{v}\}^{-1}\{E \mid \mathbf{t}\}\{R \mid \mathbf{v}\}] \\ &= D_p^{\mathbf{k}_A}[\{E \mid R^{-1}\mathbf{t}\}] \\ &= \exp(-i\mathbf{k}_A . R^{-1}\mathbf{t})\mathbf{1} \end{aligned}$$

from eqn. (5.5.10). Hence from eqn. (5.5.1) it follows that

$$D_p^{\mathbf{k}_B}[\{E \mid \mathbf{t}\}] = \exp(-iR\mathbf{k}_A . \mathbf{t})\mathbf{1} = \exp(-i\mathbf{k}_B . \mathbf{t})\mathbf{1}, \tag{5.5.12}$$

which demonstrates that $D_p^{\mathbf{k}_B}$ is a small rep as required. We therefore conclude that *if the small reps $D_p^{\mathbf{k}_A}$ of $\mathbf{G}^{\mathbf{k}_A}$ are known then the small reps $D_p^{\mathbf{k}_B}$ of $\mathbf{G}^{\mathbf{k}_B}$ can be constructed by using eqn.* (5.5.11) *where $\{R \mid \mathbf{v}\}$ is in $\mathbf{G}_o$ but not in* $\mathbf{G}$ *(provided $\mathbf{G}_o$ exists).*

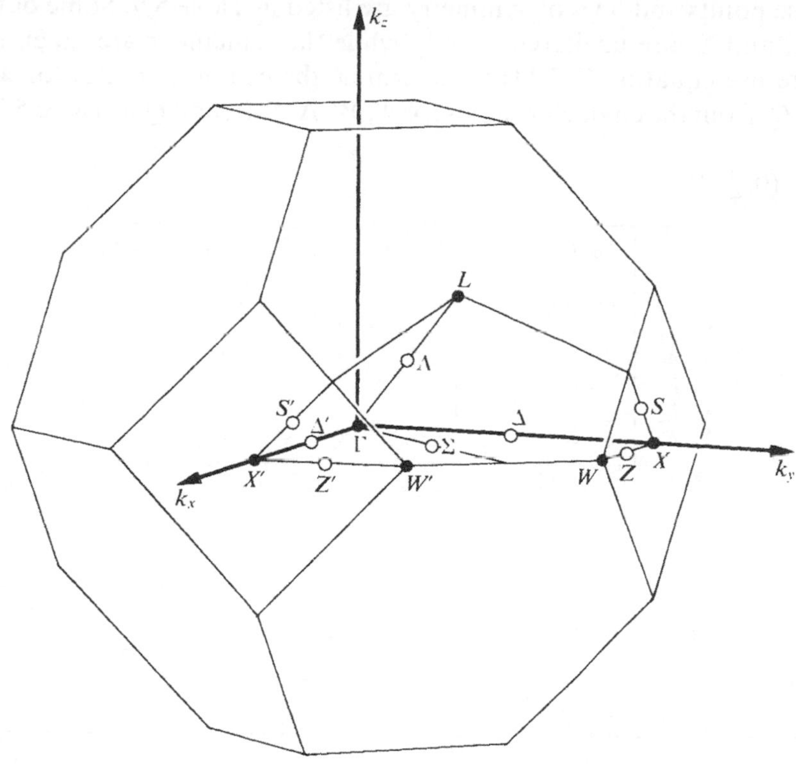

FIG. 5.3. The representation domain, Φ, of the Brillouin zone Γ_c^f of $F4_132$ (O^4).

It should be noted that if more than one $\mathbf{G}_o$ exists there is no contradiction. Any $\mathbf{G}_o$ that contains $\mathbf{G}$ as an invariant subgroup is satisfactory for the purpose. We thus have a rule for constructing all the reps of a space group for which the order of **F** is less than the order of **P**, except for the 24 space groups in (5.5.2) for which no $\mathbf{G}_o$ exist. We illustrate this now with an example.

An example of a space group in which **F** is of lower order than **P**, and therefore Φ is larger than Ω, is the space group $F4_132$ (O^4). Since **F** is the point group 432 (O) which is of order 24 and **P** is the point group $m3m$ (O_h) which is of order 48, the volume of Φ is $|\mathbf{P}|/|\mathbf{F}|$ times the volume of Ω, i.e. is twice the volume of Ω. Either of the space groups $Fd3m$ (O_h^7) and $Fd3c$ (O_h^8) could be chosen as the parent space group $\mathbf{G}_o$; we choose $Fd3m$ (O_h^7) as $\mathbf{G}_o$ and we choose $\{\sigma_{db} \mid 000\}$ as $\{R \mid \mathbf{v}\}$. The Brillouin zone of $F4_132$ (O^4) is illustrated in Fig. 3.14 and the basic domain is shown. The representation domain Φ is illustrated in Fig. 5.3 and consists of Ω, the basic domain used in Fig. 3.14, together with a region which we may denote by $\sigma_{db}\Omega$. The special points and lines of symmetry in $\sigma_{db}\Omega$ can be obtained from those in Ω which are given in Table 3.6 by using the fact that, from Table 3.4, $\sigma_{db}\mathbf{g}_1 = \mathbf{g}_2$, $\sigma_{db}\mathbf{g}_2 = \mathbf{g}_1$, and $\sigma_{db}\mathbf{g}_3 = \mathbf{g}_3$. These points and lines of symmetry are listed in Table 5.9. Some of them, namely Γ, L, Λ, and Σ, are unaltered by σ_{db}, while the remainder are taken out of Ω. We therefore use equation (5.5.11) to determine the character tables for X', W', Δ', S', Z', and Q' from the character tables for X, W, Δ, S, Z, and Q in Table 5.7 and we find:

X': $\mathbf{k} = (0, \frac{1}{2}, \frac{1}{2})$

	$\{E \mid 000\}$	$\{C_{2x} \mid 010\}$	$\{C_{4x}^+ \mid \frac{1}{4}\frac{1}{4}\frac{1}{4}\}$ $\{C_{4x}^- \mid \frac{1}{4}\frac{5}{4}\frac{1}{4}\}$	$\{C_{2z} \mid 000\}$ $\{C_{2y} \mid 010\}$	$\{C_{2f} \mid \frac{1}{4}\frac{1}{4}\frac{1}{4}\}$ $\{C_{2d} \mid \frac{1}{4}\frac{5}{4}\frac{1}{4}\}$
X_1' A_1	1	1	1	1	1
X_2' A_2	1	1	1	−1	−1
X_3' B_1	1	1	−1	1	−1
X_4' B_2	1	1	−1	−1	1
X_5' E	2	−2	0	0	0

W': $\mathbf{k} = (\frac{1}{4}, \frac{1}{2}, \frac{3}{4})$

	$\{E \mid 000\}$	$\{E \mid 100\}$	$\{E \mid 200\}$	$\{E \mid 300\}$	$\{C_{2y} \mid 000\}$ $\{C_{2y} \mid 200\}$	$\{C_{2y} \mid 100\}$ $\{C_{2y} \mid 300\}$	...
W_1' E	2	−2i	−2	2i	0	0	...

...	$\{C_{2e} \mid \frac{1}{4}\frac{1}{4}\frac{1}{4}\}$ $\{C_{2e} \mid \frac{\bar{7}}{4}\frac{1}{4}\frac{1}{4}\}$	$\{C_{2e} \mid \frac{5}{4}\frac{1}{4}\frac{1}{4}\}$ $\{C_{2e} \mid \frac{\bar{3}}{4}\frac{1}{4}\frac{1}{4}\}$	$\{C_{2c} \mid \frac{1}{4}\frac{5}{4}\frac{1}{4}\}$ $\{C_{2c} \mid \frac{1}{4}\frac{1}{4}\frac{1}{4}\}$	$\{C_{2c} \mid \frac{5}{4}\frac{5}{4}\frac{1}{4}\}$ $\{C_{2c} \mid \frac{5}{4}\frac{1}{4}\frac{1}{4}\}$
...	0	0	0	0

Δ': $\mathbf{k} = (0, \alpha, \alpha)$

	$\{E\mid 000\}$	$\{C_{4x}^{+}\mid \frac{1}{4}\frac{1}{4}\frac{1}{4}\}$	$\{C_{2x}\mid 000\}$	$\{C_{4x}^{-}\mid \frac{1}{4}\frac{1}{4}\frac{1}{4}\}$
Δ_1' A	1	A	1	A
Δ_2' ${}^2E$	1	$\mathrm{i}A$	-1	$-\mathrm{i}A$
Δ_3' B	1	$-A$	1	$-A$
Δ_4' ${}^1E$	1	$-\mathrm{i}A$	-1	$\mathrm{i}A$

where $A = \exp(-\pi\mathrm{i}\alpha)$

S': $\mathbf{k} = (2\alpha, \frac{1}{2} + \alpha, \frac{1}{2} + \alpha)$

	$\{E\mid 000\}$	$\{C_{2d}\mid \frac{1}{4}\frac{1}{4}\frac{1}{4}\}$
S_1' ${}^2E$	1	$\mathrm{i}B$
S_2' ${}^1E$	1	$-\mathrm{i}B$

where $B = \exp(-\frac{1}{2}\pi\mathrm{i}(1 + 4\alpha))$

Z': $\mathbf{k} = (\alpha, \frac{1}{2}, \frac{1}{2} + \alpha)$

	$\{E\mid 000\}$	$\{C_{2y}\mid 000\}$
Z_1' A	1	1
Z_2' B	1	-1

Q': $\mathbf{k} = (\frac{1}{2}, \frac{1}{2} + \alpha, \frac{1}{2} - \alpha)$

	$\{E\mid 000\}$	$\{C_{2e}\mid \frac{1}{4}\frac{1}{4}\frac{1}{4}\}$
Q_1' ${}^1E_1$	1	1
Q_2' ${}^1E_2$	1	-1

Although we have given the small reps for all those points and lines of symmetry that occur in $\sigma_{db}\Omega$ but not in Ω, they do not necessarily all lead to new induced reps $\Delta_p^{\mathbf{k}_B}$ ($= D_p^{\mathbf{k}_B} \uparrow \mathbf{G}$) of the space group $\mathbf{G}$. New reps $\Delta_p^{\mathbf{k}_B}$ of $\mathbf{G}$ will only be obtained from those points and lines of symmetry $\mathbf{k}_B$ for which $\mathbf{k}_B$ is not in the star of the corresponding wave vector $\mathbf{k}_A$ in Ω. Therefore in this example of $F4_132$ (O^4) only Q' leads to new induced reps $\Delta_p^{\mathbf{k}_B}$ that could not be obtained by using only the basic domain Ω.

TABLE 5.9

The points and lines of symmetry in $\sigma_{db}\Omega$ of $F4_132$ (O^4)

	$\mathbf{k}_A$		$\mathbf{k}_B = \sigma_{db}\mathbf{k}_A$
Γ	$(0, 0, 0)$	Γ	$(0, 0, 0)$
X	$(\frac{1}{2}, 0, \frac{1}{2})$	X'	$(0, \frac{1}{2}, \frac{1}{2})$
L	$(\frac{1}{2}, \frac{1}{2}, \frac{1}{2})$	L	$(\frac{1}{2}, \frac{1}{2}, \frac{1}{2})$
W	$(\frac{1}{2}, \frac{1}{4}, \frac{3}{4})$	W'	$(\frac{1}{4}, \frac{1}{2}, \frac{3}{4})$
Δ	$(\alpha, 0, \alpha)$	Δ'	$(0, \alpha, \alpha)$
Λ	(α, α, α)	Λ	(α, α, α)
Σ	$(\alpha, \alpha, 2\alpha)$	Σ	$(\alpha, \alpha, 2\alpha)$
$S(XS)$	$(\frac{1}{2} + \alpha, 2\alpha, \frac{1}{2} + \alpha)$	$S'(X'S')$	$(2\alpha, \frac{1}{2} + \alpha, \frac{1}{2} + \alpha)$
$Z(XW)$	$(\frac{1}{2}, \alpha, \frac{1}{2} + \alpha)$	$Z'(X'W')$	$(\alpha, \frac{1}{2}, \frac{1}{2} + \alpha)$
$Q(LQ)$	$(\frac{1}{2}, \frac{1}{2} - \alpha, \frac{1}{2} + \alpha)$	$Q'(LQ')$	$(\frac{1}{2} - \alpha, \frac{1}{2}, \frac{1}{2} + \alpha)$

We have to consider separately the derivation of the small reps of $\mathbf{G}^{\mathbf{k}_B}$ for the 24 space groups listed in (5.5.2). We divide these into four sets:

(a) $P2_13\ (T^4)$;
(b) $\{P4_1\ (C_4^2), P4_3\ (C_4^4)\}$,
$\{P3_1\ (C_3^2), P3_2\ (C_3^3)\}$,
$\{P3_112\ (D_3^3), P3_212\ (D_3^5)\}$, $\{P3_121\ (D_3^4), P3_221\ (D_3^6)\}$,
$\{P6_1\ (C_6^2), P6_5\ (C_6^3)\}$, $\{P6_2\ (C_6^4), P6_4\ (C_6^5)\}$;
(c) $\{P4_122\ (D_4^3), P4_322\ (D_4^7)\}$, $\{P4_12_12\ (D_4^4), P4_32_12\ (D_4^8)\}$,
$\{P6_122\ (D_6^2), P6_522\ (D_6^3)\}$, $\{P6_222\ (D_6^4), P6_422\ (D_6^5)\}$,
$\{P4_332\ (O^6), P4_132\ (O^7)\}$;
(d) $Pa3\ (T_h^6)$;

where isomorphic pairs are enclosed in brackets $\{\ \}$. None of the groups listed under (c) and (d) is an invariant subgroup of any larger space group. The groups listed under (a) and (b), though not invariant subgroups of a parent group $\mathbf{G}_o$, with isogonal point group $\mathbf{P}$, are, however, invariant subgroups of some other space group (or space groups) $\mathbf{G}'$, where the order of the isogonal point group of $\mathbf{G}'$ is intermediate between the order of $\mathbf{F}$ and the order of $\mathbf{P}$. The groups $\mathbf{G}'$ are indicated in Table 5.10.

TABLE 5.10

Invariance properties of 13 space groups

G	G′
$P2_13\ (T^4)$	$P4_332\ (O^6)$, $P4_132\ (O^7)$, $Pa3\ (T_h^6)$
$P4_1\ (C_4^2)$	$P4_122\ (D_4^3)$, $P4_12_12\ (D_4^4)$
$P4_3\ (C_4^4)$	$P4_322\ (D_4^7)$, $P4_32_12\ (D_4^8)$
$P6_1\ (C_6^2)$, $P3_112\ (D_3^3)$, $P3_121\ (D_3^4)$, $P3_1\ (C_3^2)$	$P6_122\ (D_6^2)$
$P6_5\ (C_6^3)$, $P3_212\ (D_3^5)$, $P3_221\ (D_3^6)$, $P3_2\ (C_3^3)$	$P6_522\ (D_6^3)$
$P6_2\ (C_6^4)$, $P3_212\ (D_3^5)$, $P3_221\ (D_3^6)$, $P3_2\ (C_3^3)$	$P6_222\ (D_6^4)$
$P6_4\ (C_6^5)$, $P3_112\ (D_3^3)$, $P3_121\ (D_3^4)$, $P3_1\ (C_3^2)$	$P6_422\ (D_6^5)$

Note to Table 5.10

The space groups listed in a given row and in the first column of the table are invariant subgroups of the space groups listed in the same row and in the second column.

Before making use of these invariance properties we discuss the eleven isomorphic pairs. We assume that these are being treated in a standard setting, so that if $\mathbf{G}_1$ and $\mathbf{G}_2$ are an isomorphic pair then given $\{S \mid \mathbf{w}\} \in \mathbf{G}_1$ its image in $\mathbf{G}_2$ under the isomorphism may be written as $\{S \mid -\mathbf{w}\}$. Now suppose that $\mathbf{k}$ is a vector of the basic domain, Ω. In no case with the eleven pairs under consideration is there a centre of inversion, so that we may choose the representation domain, Φ, so that it contains the vector $-\mathbf{k}$. From the tables of space group reps (Table 5.7) we already have at our disposal the little groups $\mathbf{G}_1^{\mathbf{k}}$ and $\mathbf{G}_2^{\mathbf{k}}$, and the small reps $D_{1p}^{\mathbf{k}}$ and $D_{2p}^{\mathbf{k}}$, for the point $\mathbf{k} \in \Omega$.

THEOREM 5.5.5. *The little groups and small reps for* $-\mathbf{k}$ *are given by*

$$\mathbf{G}_i^{-\mathbf{k}} = \mathbf{G}_i^{\mathbf{k}}, \quad i = 1, 2, \tag{5.5.13}$$

and

$$D_{ip}^{-\mathbf{k}}[\{S \mid \mathbf{w}\}] = D_{jp}^{\mathbf{k}}[\{S \mid -\mathbf{w}\}], \quad i = 1, j = 2; \quad i = 2, j = 1. \tag{5.5.14}$$

Equation (5.5.13) follows immediately from the fact that if a rotation operator S leaves $\mathbf{k}$ invariant then it also leaves $-\mathbf{k}$ invariant, and conversely. That $D_{ip}^{-\mathbf{k}}$ defined by eqn. (5.5.14) is an irreducible representation, follows immediately from the two facts: (i) that $D_{jp}^{\mathbf{k}}$ is an irreducible representation and (ii) that $\mathbf{G}_1$ and $\mathbf{G}_2$ are isomorphic. That $D_{ip}^{-\mathbf{k}}$ is a *small* rep is also easily checked. From eqn. (5.5.14)

$$D_{ip}^{-\mathbf{k}}[\{E \mid \mathbf{t}\}] = D_{jp}^{\mathbf{k}}[\{E \mid -\mathbf{t}\}] = \exp[i(\mathbf{k}.\mathbf{t})]\mathbf{1} \tag{5.5.15}$$

as required.

We now have sufficient information to give rules for obtaining the small reps for $\mathbf{k}$ vectors in Φ but not in Ω for the space groups listed under (a), (b), and (c). For groups listed under (b) and (c) we first use eqn. (5.5.14). Groups in list (c) are thereby disposed of, and for groups in list (b) we shall have obtained all small reps appropriate to the volume $(\Omega + I\Omega)$, where I is the inversion. The groups in lists (a) and (b) can now be completed by using eqn. (5.5.11) and the invariance properties listed in Table 5.10. In connection with this second step two explanatory comments are necessary. First, for the groups in list (b) the vector $\mathbf{k}_A$ is in $(\Omega + I\Omega)$ rather than Ω; but this is acceptable in view of eqn. (5.5.14). Secondly, for the group of list (a), $P2_13$ (T^4), it must be pointed out that its invariance in *both* $P4_332$ (O^6) or $P4_132$ (O^7) *and* $Pa3$ (T_h^6) compensates for the lack of a parent space group $\mathbf{G}_0$ in the sense described before. Its representation domain Φ is four times the size of the basic domain Ω and for any vector $\mathbf{k}_B$ within Φ but outside Ω there will be an element R belonging *either* to 432 (O) *or* $m3$ (T_h) such that eqn. (5.5.1) holds with $\mathbf{k}_A$ in Ω. The invariance properties in Table 5.10 and the analysis given previously is then sufficient to dispose of this group.

It appears that the group in list (d), $Pa3$ (T_h^6), the pyrite structure, is exceptional; neither is it an invariant subgroup of any larger space group based on the same Bravais lattice, nor is it a member of a pair of isomorphic space groups. Therefore, neither of the methods described so far in this section can be used to construct the small reps of $\mathbf{G}^{\mathbf{k}_B}$ given those of $\mathbf{G}^{\mathbf{k}_A}$. It seems that for completeness of our prescription for finding the small reps of $\mathbf{G}^{\mathbf{k}}$ for all vectors $\mathbf{k}$ throughout the representation domain of any space group, it is necessary, just for this one space group $Pa3$ (T_h^6), to consider $\mathbf{k}$ vectors throughout the whole representation domain. Restriction to a single basic domain, reasonable as this is for 229 out of the 230 space groups, is in the one remaining case invalid. Indeed our results show that the remaining case provides a rather interesting group-theoretical anomaly that should be capable of physical observation.

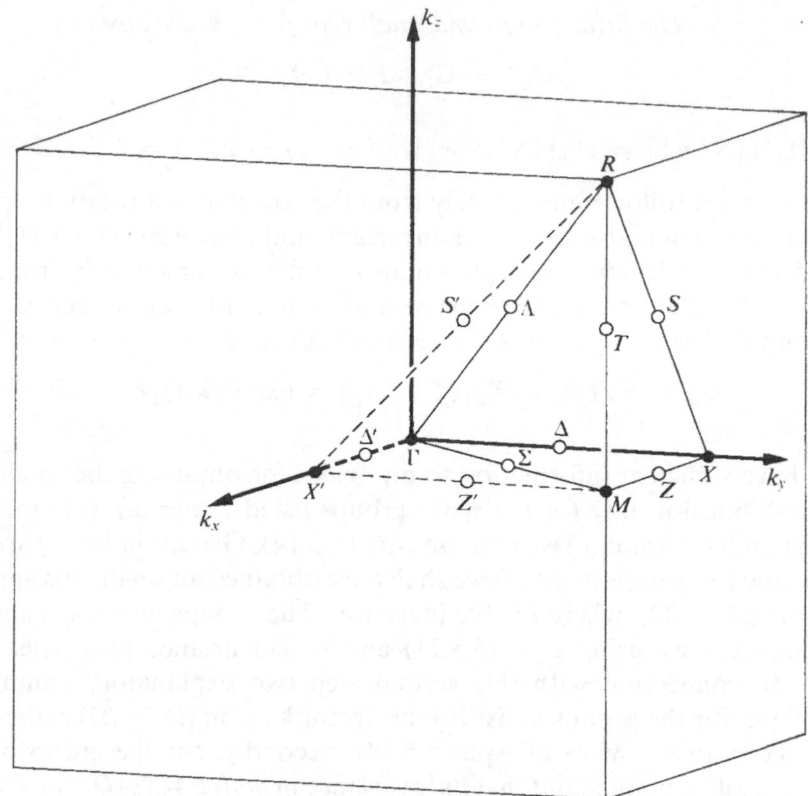

FIG. 5.4. The representation domain, $\Omega + \sigma_{db}\Omega$, of $Pa3$ (T_h^6).

$Pa3$ (T_h^6) is based on the simple cubic lattice and its Brillouin zone is shown in Fig. 3.13 and reproduced in Fig. 5.1. Figure 5.1 illustrates the basic domain that we have used in Tables 3.6 and 5.7 and Fig. 5.4 illustrates the representation domain Φ. We choose $\Phi = (\Omega + \sigma_{db}\Omega)$, where σ_{db} is the reflection in the plane ΓMR. The points and lines of symmetry Γ, M, R, Σ, Λ, and T are left invariant by the operation σ_{db}, so that we only need to identify the small reps of the little group of **k** for X', Δ′, S', and Z' which are generated by the action of σ_{db} on X, Δ, S, and Z respectively. The **k** vectors of these additional points or lines of symmetry can easily be determined using Tables 3.4 and 3.6 and are

$$X': \ (\tfrac{1}{2}, 0, 0)$$
$$\Delta': \ (\alpha, 0, 0)$$
$$S': \ (\tfrac{1}{2}, \alpha, \alpha)$$
$$Z': \ (\tfrac{1}{2}, \alpha, 0),$$

where $\mathbf{g}_1$, $\mathbf{g}_2$, and $\mathbf{g}_3$ are defined in Table 3.3. X', Δ′, and S' are in the stars of X, Δ, and S, respectively, being related to those points by the threefold rotation axis

TABLE 5.11

*The additional space-group rep for the representation domain of Pa*3 (T_h^6)

Z'^x G_8^4: $(\sigma_x, O), (C_{2y}, O)$: 5, 1: *a*.

Note to Table 5.11

The notation is that of Table 5.7 (see Notes to Table 5.7).

through ΓR which is an operation of $m3$ (T_h). There is therefore no difference, in any physically observable sense, between the small reps at X', Δ', and S' and those at X, Δ, and S, respectively. If we deal only with points and lines of symmetry it remains to compare the points Z' and Z, see Table 5.11. While Miller and Love (1967) tabulate Z and not Z', Slater (1965) in his treatment of this space group tabulates Z' and not Z (through using an alternative setting of the basic domain). The interesting point is that the character tables for Z and Z' are very dissimilar; while at Z' there is one twofold degenerate rep there are four non-degenerate reps at Z. This difference between the degeneracies at Z and Z' is a useful demonstration of the importance of considering the representation domain and this difference should manifest itself in the energy eigenvalues of electrons or phonons at Z' and Z. The inclusion of time-reversal symmetry does cause the non-degenerate reps at Z to stick together in pairs (see Table 5.7), but still does not remove the anomaly completely (see also section 6.5).

6

The double-valued representations of the 32 point groups and the 230 space groups

6.1. The double-valued representations of the point groups

IN section 2.1 the representation $\mathscr{D}^l\{R(\alpha, \beta, \gamma)\}$ of the 3-dimensional rotation group, $O(3)$, was considered, where l was assumed to be an integer. The basis of this representation is the vector whose components are the spherical harmonics $Y_l^m(\theta, \phi)$, $(-l \leqslant m \leqslant l)$. The irreducible representations of the point groups which were listed in Chapter 2, see Tables 2.2 and 2.3, can be thought of as being obtained from these representations by subduction, that is, by restricting the representation $\mathscr{D}^l\{R(\alpha, \beta, \gamma)\}$ to the elements of a point group, see under Theorem 4.1.5. This was done, for instance, by Bethe (1929). These representations of the point groups are called the single-valued representations of the point groups since they are derived from the representations $\mathscr{D}^l\{R(\alpha, \beta, \gamma)\}$ of the 3-dimensional rotation group, where l is an integer, which are themselves single-valued (that is, with one matrix representative for each operator).

The definition of Euler's angles α, β, and γ that we use was given in section 2.1. To every rotation $R(\alpha, \beta, \gamma)$ of the 3-dimensional rotation group there is one and only one matrix $\mathscr{D}^l\{R(\alpha, \beta, \gamma)\}$ for each integer value of l and for any given l the matrices $\mathscr{D}^l\{R(\alpha, \beta, \gamma)\}$ form a representation of the group of rotations $R(\alpha, \beta, \gamma)$ as defined in Definition 1.3.4. The spherical harmonics $Y_l^m(\theta, \phi)$ are well known to be the eigenfunctions of the quantum-mechanical angular momentum operators so that, for instance, the angular part of the wave function of an electron in an atom contains a factor $Y_l^m(\theta, \phi)$ and the magnitude and z component of the orbital angular momentum of such an electron are $\sqrt{\{l(l+1)\}}\,\hbar$ and $m\hbar$.

However, it is often necessary to take into account the existence of the spin of an electron due to the introduction of relativistic effects into band structure calculations or to the inclusion of spin-orbit coupling in crystal field theory. Then, instead of considering the representations $\mathscr{D}^l\{R(\alpha, \beta, \gamma)\}$ where l is an integer it is necessary to consider a second set of representations $\mathscr{D}^j\{R(\alpha, \beta, \gamma)\}$ where j is half an odd integer. The basis of such a representation is then a spinor rather than a vector. The simplest of these representations is $\mathscr{D}^{\frac{1}{2}}\{R(\alpha, \beta, \gamma)\}$ which is given by

$$\begin{aligned}\mathscr{D}^{\frac{1}{2}}\{R(\alpha, \beta, \gamma)\} &= \pm\mathbf{u}\{R(\alpha, \beta, \gamma)\}\\ &= \pm\begin{pmatrix} e^{\frac{1}{2}i\alpha}\cos\frac{1}{2}\beta\, e^{\frac{1}{2}i\gamma} & e^{-\frac{1}{2}i\alpha}\sin\frac{1}{2}\beta\, e^{\frac{1}{2}i\gamma} \\ -e^{\frac{1}{2}i\alpha}\sin\frac{1}{2}\beta\, e^{-\frac{1}{2}i\gamma} & e^{-\frac{1}{2}i\alpha}\cos\frac{1}{2}\beta\, e^{-\frac{1}{2}i\gamma}\end{pmatrix}\end{aligned} \tag{6.1.1}$$

as we shall see in section 6.2 (eqn (6.2.13)). The $\pm$ signs indicate that there are two matrices corresponding to each rotation $R(\alpha, \beta, \gamma)$ and the homomorphism of the matrices $\mathscr{D}^j\{R(\alpha, \beta, \gamma)\}$ onto the rotations $R(\alpha, \beta, \gamma)$ is a two to one homomorphism; the matrices $\mathscr{D}^j\{R(\alpha, \beta, \gamma)\}$ only form a group at all if both matrices are included for each set of α, β, and γ. The matrix elements of $\mathscr{D}^j\{R(\alpha, \beta, \gamma)\}$ can be written as

$$\pm\mathscr{D}^j\{R(\alpha, \beta, \gamma)\}_{nm} = \exp(-im\alpha)\exp(-in\gamma)d^j(\beta)_{nm} \tag{6.1.2}$$

where $d^j(\beta)_{nm}$ is given by replacing l by j in eqn (2.1.6). The factor C_{nm} which was used in Chapter 2 is not necessary here because for the spinors we are using the phase factors used by Condon and Shortley (1935) although for the spherical harmonics we used different phase factors from those of Condon and Shortley. If we allow α, β, and γ to take all the allowed values so that the operations $R(\alpha, \beta, \gamma)$ include all the members of the group of pure rotations in three dimensions then the matrices $\pm\mathbf{u}\{R(\alpha, \beta, \gamma)\}$ will include all the unitary unimodular matrices in two dimensions, that is all the members of the group $SU(2)$. These representations are thus often called *double-valued representations*. From these double-valued representations of the rotation group it is then also possible to obtain double-valued representations of the point groups by the process of subduction; these double-valued representations of the point groups were considered by Bethe (1929) and Opechowski (1940).

Any member of a point group is either a pure rotation operation $R(\alpha, \beta, \gamma)$ or it can be expressed as the product of some such pure rotation operation with the inversion operation. A point group may thus be made up entirely of pure rotations or it may be such that half of its elements are expressible as a product of the inversion operation and a pure rotation. In the second case the point group is either isomorphic to a point group consisting only of pure rotations or else it is a direct product of a point group consisting only of pure rotations with the group $\bar{1}(C_i)$ which contains the identity and the inversion operations. Thus it should be clear that it is really only necessary to obtain the reps of those point groups that consist only of pure rotations and the reps of the remaining point groups follow immediately.

Suppose that $\mathbf{G}$ is a point group, of order $|\mathbf{G}|$, which consists of pure rotations $R(\alpha, \beta, \gamma)$, then to every rotation $R(\alpha, \beta, \gamma)$ in the point group there correspond two matrices $\mathscr{D}^{\frac{1}{2}}\{R(\alpha, \beta, \gamma)\}$ given by eqn. (6.1.1). These matrices form a group of order $2|\mathbf{G}|$, called the '*double group*' of $\mathbf{G}$ and denoted by $\mathbf{G}^\dagger$.

DEFINITION 6.1.1 (Opechowski 1940). The *double group* $\mathbf{G}^\dagger$ of a group $\mathbf{G}$ of order $|\mathbf{G}|$, which is a subgroup of $O(3)$, the 3-dimensional rotation group, is the abstract group of order $2|\mathbf{G}|$ having the same group multiplication table as the $2|\mathbf{G}|$ matrices of $SU(2)$ which correspond to the elements of the group $\mathbf{G}$.

Although the double group is defined in terms of $\mathscr{D}^{\frac{1}{2}}\{R(\alpha, \beta, \gamma)\}$ exactly the same abstract group would be obtained by using $\mathscr{D}^j\{R(\alpha, \beta, \gamma)\}$ where j has any other half-odd-integer value. It is possible to find the irreducible representations of this group by the ordinary methods using Theorems 1.3.8, 1.3.10, 1.3.11, 1.3.13, and 1.3.15. **G** is a subgroup of the group of rotations in three dimensions while $\mathbf{G}^\dagger$ is a subgroup of $SU(2)$, the group of all two by two unitary unimodular matrices; it will be remembered that there is a two to one homomorphism between the group $SU(2)$ and the group of pure rotations in three dimensions.

There is a set of rules derived by Opechowski (1940) which make it possible to simplify the separation of the group $\mathbf{G}^\dagger$ into classes and the evaluation of the character table of $\mathbf{G}^\dagger$. However these rules are of somewhat limited application; they are useful in finding the character table of the double group but they are of little use if one wishes to go beyond that and deduce the actual matrix representatives themselves and not just their characters. We shall simply quote the more important of these results which are sometimes referred to as '*Opechowski's rules*', before proceeding.

THEOREM 6.1.1. *To each class C_i of the point group* **G** *which is not a class of rotations through π there correspond exactly two classes of $\mathbf{G}^\dagger$, C_i', and C_i''.*

THEOREM 6.1.2. *To each class C_π of the point group* **G** *which is a class of rotations through π there correspond either one or two classes of $\mathbf{G}^\dagger$. If in* **G** *there is no rotation through π about an axis at right angles to the axis of one of the rotations of C_π then there will be two classes in $\mathbf{G}^\dagger$ corresponding to C_π. But if in* **G** *there does exist a rotation through π about an axis at right angles to the axis of one of the rotations of C_π there will only be one class in $\mathbf{G}^\dagger$ corresponding to C_π.*

THEOREM 6.1.3. *Each rep of* **G** *is also a rep of $\mathbf{G}^\dagger$, where $\chi(C_i'') = +\chi(C_i')$; it is called a single-valued rep of* **G**.

THEOREM 6.1.4. *The remaining reps of $\mathbf{G}^\dagger$ are called double-valued reps of* **G** *and are such that $\chi(C_i'') = -\chi(C_i')$.*

It is customary to think of the representations of the group $\mathbf{G}^\dagger$ as representations of the point group **G** itself as in Theorems 6.1.3 and 6.1.4; that is, they are called single-valued representations of **G** if the characters of the matrix representatives of the two elements of $\mathbf{G}^\dagger$ derived from an element $R(\alpha, \beta, \gamma)$ are always the same and they are called double-valued representations of **G** if the characters of the matrix representatives of these two elements of $\mathbf{G}^\dagger$ always differ by a factor of -1. But it should be noted that strictly in the sense in which we defined the representation of a group in section 1.3 and in which we have used it so far these are not really representations of **G** but of $\mathbf{G}^\dagger$.

We now consider the deduction of the double groups of the point groups, and then by finding their reps we thereby determine the double-valued representations of the point groups. The procedure that is adopted can be outlined as follows. The Euler angles α, β, and γ are first determined for each of the elements of the point group **G**; from these the matrices $\mathscr{D}^{\frac{1}{2}}\{R(\alpha, \beta, \gamma)\}$ can be found and hence $\mathbf{G}^\dagger$ can be constructed. The irreducible representations of $\mathbf{G}^{\dagger}$ can be found in the usual way and sorted into single-valued and double-valued representations of the point group **G**. In deducing the Euler angles of all the operations that may occur in any of these point groups consisting of pure rotations we need consider the elements of only two point groups 432 (O) and 622 (D_6); the triclinic, monoclinic, orthorhombic, and tetragonal point groups are subgroups of the cubic point group 432 (O) while the trigonal point groups are subgroups of the hexagonal point group 622 (D_6). The Euler angles of the elements of the point groups were given in Table 2.1. The values of α, β, and γ given in Table 2.1 can be substituted into the right-hand side of eqn (6.1.1). The matrices $\mathscr{D}^{\frac{1}{2}}\{R(\alpha, \beta, \gamma)\}$ obtained in this way are listed for the two point groups 432 (O) and 622 (D_6) in Table 6.1. There is, of course, the choice of + or − sign in eqn (6.1.1) and we have only given one matrix for each point-group element. In addition to all the matrices given in Table 6.1 there is an equal number of matrices that are just the negatives of those listed. It is not possible to obtain a group by using just the matrices in Table 6.1(a) or 6.1(b). The matrices in Table 6.1(a) or 6.1(b) together with their negatives form $G^{\dagger}$, the double group of the point group 432 (O) or 622 (D_6). One-quarter of the group multiplication table for each of these double groups is given in Table 6.2. It is easy enough to form the other three-quarters of each of the group multiplication tables by inspection. The theory of section 6.2 which leads to eqn (6.1.1) ensures that the elements appearing in the matrix $\mathscr{D}^{\frac{1}{2}}\{R(\alpha, \beta, \gamma)\}$ are such that Tables 6.2(a) and 6.2(b) coincide exactly with Tables 1.5 and 1.6 when barred and unbarred operators in Table 6.2 are identified. That is, the two to one homomorphism between $SU(2)$ and $SO(3)$ automatically ensures that the symbols here have the same geometrical significance in 3-dimensional Euclidean space as the corresponding symbols in previous chapters.

We can illustrate the assertion which we made that the double group $\mathbf{G}^{\dagger}$ of any point group consisting only of pure rotation operations could be obtained easily from Table 6.2. We consider, for example, the point group 422 (D_4) which, from Table 2.2, consists of the following elements:

$$\begin{array}{ll} C_1 & E \\ C_2 & C_{2z} \\ C_3 & C_{4z}^{+}, C_{4z}^{-} \\ C_4 & C_{2x}, C_{2y} \\ C_5 & C_{2a}, C_{2b}. \end{array}$$

The elements in the double group of 422 (D_4) are

$$\begin{array}{l} E, C_{2z}, C_{4z}^{+}, C_{4z}^{-}, C_{2x}, C_{2y}, C_{2a}, C_{2b}, \\ \bar{E}, \bar{C}_{2z}, \bar{C}_{4z}^{+}, \bar{C}_{4z}^{-}, \bar{C}_{2x}, \bar{C}_{2y}, \bar{C}_{2a}, \bar{C}_{2b}. \end{array}$$

Since this double group is a subgroup of the double group of 432 (O) its multiplication table can easily be found by selecting the products of these elements from the large

TABLE 6.1

The matrices of SU(2) corresponding to the point groups 432 (O) and 622 (D_6)

($\varepsilon = \exp(2\pi i/12)$)

(a) 432 (O)

E	$\begin{pmatrix}1 & 0\\ 0 & 1\end{pmatrix}$	C_{4x}^+	$\frac{1}{\sqrt{2}}\begin{pmatrix}1 & -i\\ -i & 1\end{pmatrix}$
C_{2x}	$\begin{pmatrix}0 & -i\\ -i & 0\end{pmatrix}$	C_{4y}^+	$\frac{1}{\sqrt{2}}\begin{pmatrix}1 & 1\\ -1 & 1\end{pmatrix}$
C_{2y}	$\begin{pmatrix}0 & -1\\ 1 & 0\end{pmatrix}$	C_{4z}^+	$\frac{1}{\sqrt{2}}\begin{pmatrix}1+i & 0\\ 0 & 1-i\end{pmatrix}$
C_{2z}	$\begin{pmatrix}-i & 0\\ 0 & i\end{pmatrix}$	C_{4x}^-	$\frac{1}{\sqrt{2}}\begin{pmatrix}1 & i\\ i & 1\end{pmatrix}$
C_{31}^-	$\frac{1}{2}\begin{pmatrix}1-i & -1+i\\ 1+i & 1+i\end{pmatrix}$	C_{4y}^-	$\frac{1}{\sqrt{2}}\begin{pmatrix}1 & -1\\ 1 & 1\end{pmatrix}$
C_{32}^-	$\frac{1}{2}\begin{pmatrix}1-i & 1-i\\ -1+i & 1+i\end{pmatrix}$	C_{4z}^-	$\frac{1}{\sqrt{2}}\begin{pmatrix}1-i & 0\\ 0 & 1+i\end{pmatrix}$
C_{33}^-	$\frac{1}{2}\begin{pmatrix}1+i & 1+i\\ -1+i & 1-i\end{pmatrix}$	C_{2a}	$\frac{1}{\sqrt{2}}\begin{pmatrix}0 & -1+i\\ 1+i & 0\end{pmatrix}$
C_{34}^-	$\frac{1}{2}\begin{pmatrix}1+i & -1-i\\ 1-i & 1-i\end{pmatrix}$	C_{2b}	$\frac{1}{\sqrt{2}}\begin{pmatrix}0 & -1-i\\ 1-i & 0\end{pmatrix}$
C_{31}^+	$\frac{1}{2}\begin{pmatrix}1+i & 1-i\\ -1-i & 1-i\end{pmatrix}$	C_{2c}	$\frac{1}{\sqrt{2}}\begin{pmatrix}-i & i\\ i & i\end{pmatrix}$
C_{32}^+	$\frac{1}{2}\begin{pmatrix}1+i & -1+i\\ 1+i & 1-i\end{pmatrix}$	C_{2d}	$\frac{1}{\sqrt{2}}\begin{pmatrix}-i & -1\\ 1 & i\end{pmatrix}$
C_{33}^+	$\frac{1}{2}\begin{pmatrix}1-i & -1-i\\ 1-i & 1+i\end{pmatrix}$	C_{2e}	$\frac{1}{\sqrt{2}}\begin{pmatrix}-i & -i\\ -i & i\end{pmatrix}$
C_{34}^+	$\frac{1}{2}\begin{pmatrix}1-i & 1+i\\ -1-i & 1+i\end{pmatrix}$	C_{2f}	$\frac{1}{\sqrt{2}}\begin{pmatrix}i & -1\\ 1 & -i\end{pmatrix}$

(b) 622 (D_6)

E	$\begin{pmatrix}1 & 0\\ 0 & 1\end{pmatrix}$	C'_{21}	$\begin{pmatrix}0 & -i\\ -i & 0\end{pmatrix}$
C_6^+	$\begin{pmatrix}\varepsilon & 0\\ 0 & \varepsilon^*\end{pmatrix}$	C'_{22}	$\begin{pmatrix}0 & \varepsilon\\ -\varepsilon^* & 0\end{pmatrix}$
C_6^-	$\begin{pmatrix}\varepsilon^* & 0\\ 0 & \varepsilon\end{pmatrix}$	C'_{23}	$\begin{pmatrix}0 & -\varepsilon^*\\ \varepsilon & 0\end{pmatrix}$
C_3^+	$\begin{pmatrix}\varepsilon^2 & 0\\ 0 & \varepsilon^{*2}\end{pmatrix}$	C''_{21}	$\begin{pmatrix}0 & -1\\ 1 & 0\end{pmatrix}$
C_3^-	$\begin{pmatrix}\varepsilon^{*2} & 0\\ 0 & \varepsilon^2\end{pmatrix}$	C''_{22}	$\begin{pmatrix}0 & -\varepsilon^{*2}\\ \varepsilon^2 & 0\end{pmatrix}$
C_2	$\begin{pmatrix}-i & 0\\ 0 & i\end{pmatrix}$	C''_{23}	$\begin{pmatrix}0 & -\varepsilon^2\\ \varepsilon^{*2} & 0\end{pmatrix}$

The double groups of 432 (*O*) *and* 622 (D_6)

(a) 432 (*O*)

E	C_{2x}	C_{2y}	C_{2z}	C_{4x}^{-}	C_{4y}^{-}	C_{4z}^{-}	C_{4x}^{+}	C_{4y}^{+}	C_{4z}^{+}	C_{31}^{-}	C_{32}^{-}	C_{33}^{-}	C_{34}^{-}	C_{31}^{+}	C_{32}^{+}	C_{33}^{+}	C_{34}^{+}	C_{2a}	C_{2b}	C_{2c}	C_{2d}	C_{2e}	C_{2f}
C_{2x}	$\bar{E}$	C_{2z}	$\bar{C}_{2y}$	C_{4x}^{+}	C_{2e}	$\bar{C}_{2a}$	$\bar{C}_{4x}^{-}$	$\bar{C}_{2c}$	C_{2b}	C_{32}^{-}	$\bar{C}_{31}^{-}$	C_{34}^{-}	$\bar{C}_{33}^{-}$	$\bar{C}_{34}^{+}$	C_{33}^{+}	$\bar{C}_{32}^{+}$	C_{31}^{+}	C_{4z}^{-}	$\bar{C}_{4z}^{+}$	C_{4y}^{+}	$\bar{C}_{2f}$	$\bar{C}_{4y}^{-}$	C_{2d}
C_{2y}	$\bar{C}_{2z}$	$\bar{E}$	C_{2x}	C_{2d}	$\bar{C}_{4y}^{+}$	C_{2b}	C_{2f}	C_{4y}^{-}	C_{2a}	$\bar{C}_{33}^{-}$	C_{34}^{-}	C_{31}^{-}	$\bar{C}_{32}^{-}$	C_{32}^{+}	$\bar{C}_{31}^{+}$	$\bar{C}_{34}^{+}$	C_{33}^{+}	$\bar{C}_{4z}^{+}$	$\bar{C}_{4z}^{-}$	C_{2e}	$\bar{C}_{4x}^{-}$	$\bar{C}_{2c}$	$\bar{C}_{4x}^{+}$
C_{2z}	C_{2y}	$\bar{C}_{2x}$	$\bar{E}$	$\bar{C}_{2f}$	C_{2c}	$\bar{C}_{4z}^{+}$	C_{2d}	C_{2e}	C_{4z}^{-}	$\bar{C}_{34}^{-}$	$\bar{C}_{33}^{-}$	C_{32}^{-}	C_{31}^{-}	C_{33}^{+}	C_{34}^{+}	$\bar{C}_{31}^{+}$	$\bar{C}_{32}^{+}$	$\bar{C}_{2b}$	C_{2a}	$\bar{C}_{4y}^{-}$	$\bar{C}_{4x}^{+}$	$\bar{C}_{4y}^{+}$	C_{4x}^{-}
C_{4x}^{-}	C_{4x}^{+}	C_{2f}	C_{2d}	$\bar{C}_{2x}$	C_{32}^{+}	C_{31}^{-}	E	C_{34}^{+}	C_{33}^{-}	C_{2a}	C_{4z}^{-}	$\bar{C}_{2b}$	C_{4z}^{+}	C_{4y}^{+}	$\bar{C}_{2e}$	C_{4y}^{-}	C_{2c}	$\bar{C}_{32}^{-}$	C_{34}^{-}	$\bar{C}_{31}^{+}$	C_{2y}	C_{33}^{+}	$\bar{C}_{2z}$
C_{4y}^{-}	$\bar{C}_{2c}$	$\bar{C}_{4y}^{+}$	C_{2e}	C_{31}^{-}	C_{2y}	C_{33}^{+}	C_{34}^{-}	E	C_{32}^{+}	C_{2d}	C_{4x}^{+}	C_{4x}^{-}	C_{2f}	C_{4z}^{+}	C_{2a}	C_{2b}	C_{4z}^{-}	$\bar{C}_{31}^{+}$	$\bar{C}_{34}^{+}$	C_{2z}	$\bar{C}_{33}^{-}$	C_{2x}	$\bar{C}_{32}^{-}$
C_{4z}^{-}	C_{2b}	C_{2a}	$\bar{C}_{4z}^{+}$	C_{34}^{+}	C_{31}^{-}	C_{2z}	C_{33}^{+}	C_{32}^{-}	E	C_{2c}	C_{2e}	C_{4y}^{+}	C_{4y}^{-}	C_{4x}^{+}	C_{4x}^{-}	C_{2d}	$\bar{C}_{2f}$	$\bar{C}_{2x}$	C_{2y}	$\bar{C}_{34}^{-}$	$\bar{C}_{31}^{+}$	$\bar{C}_{33}^{-}$	C_{32}^{+}
C_{4x}^{+}	$\bar{C}_{4x}^{-}$	C_{2d}	$\bar{C}_{2f}$	E	C_{33}^{+}	C_{32}^{-}	C_{2x}	C_{31}^{+}	C_{34}^{-}	C_{4z}^{-}	$\bar{C}_{2a}$	C_{4z}^{+}	C_{2b}	$\bar{C}_{2c}$	C_{4y}^{-}	C_{2e}	C_{4y}^{+}	C_{31}^{-}	$\bar{C}_{33}^{-}$	C_{34}^{+}	C_{2z}	$\bar{C}_{32}^{+}$	C_{2y}
C_{4y}^{+}	C_{2e}	C_{4y}^{-}	C_{2c}	C_{33}^{-}	E	C_{34}^{+}	C_{32}^{-}	$\bar{C}_{2y}$	C_{31}^{+}	C_{4x}^{-}	$\bar{C}_{2f}$	$\bar{C}_{2d}$	C_{4x}^{+}	$\bar{C}_{2a}$	C_{4z}^{+}	C_{4z}^{-}	$\bar{C}_{2b}$	C_{32}^{+}	C_{33}^{+}	$\bar{C}_{2x}$	C_{31}^{-}	C_{2z}	C_{34}^{-}
C_{4z}^{+}	$\bar{C}_{2a}$	C_{2b}	C_{4z}^{-}	C_{32}^{+}	C_{34}^{-}	E	C_{31}^{+}	C_{33}^{-}	$\bar{C}_{2z}$	C_{4y}^{-}	C_{4y}^{+}	$\bar{C}_{2e}$	$\bar{C}_{2c}$	$\bar{C}_{2d}$	C_{2f}	C_{4x}^{+}	C_{4x}^{-}	C_{2y}	C_{2x}	C_{31}^{-}	C_{33}^{+}	C_{32}^{-}	$\bar{C}_{34}^{+}$
C_{31}^{-}	C_{34}^{-}	$\bar{C}_{32}^{-}$	$\bar{C}_{33}^{-}$	C_{2c}	C_{2a}	C_{2d}	C_{4y}^{-}	C_{4z}^{-}	C_{4x}^{-}	$\bar{C}_{31}^{+}$	C_{33}^{+}	C_{34}^{+}	C_{32}^{+}	E	$\bar{C}_{2x}$	C_{2y}	C_{2z}	$\bar{C}_{4x}^{+}$	C_{2f}	$\bar{C}_{4z}^{+}$	$\bar{C}_{4y}^{+}$	C_{2b}	$\bar{C}_{2e}$
C_{32}^{-}	$\bar{C}_{33}^{-}$	C_{31}^{-}	$\bar{C}_{34}^{-}$	C_{4y}^{+}	C_{4z}^{-}	$\bar{C}_{2f}$	C_{2e}	$\bar{C}_{2a}$	C_{4x}^{+}	C_{34}^{+}	$\bar{C}_{32}^{+}$	C_{31}^{+}	C_{33}^{+}	C_{2x}	E	C_{2z}	$\bar{C}_{2y}$	C_{4x}^{-}	C_{2d}	$\bar{C}_{2b}$	C_{2c}	$\bar{C}_{4z}^{+}$	C_{4y}^{-}
C_{33}^{-}	C_{32}^{-}	C_{34}^{-}	C_{31}^{-}	$\bar{C}_{2e}$	C_{4z}^{+}	C_{4x}^{-}	C_{4y}^{+}	$\bar{C}_{2b}$	$\bar{C}_{2d}$	C_{32}^{+}	C_{34}^{+}	$\bar{C}_{33}^{+}$	C_{31}^{+}	$\bar{C}_{2y}$	$\bar{C}_{2z}$	E	$\bar{C}_{2x}$	C_{2f}	C_{4x}^{+}	C_{2a}	C_{4y}^{-}	C_{4z}^{-}	$\bar{C}_{2c}$
C_{34}^{-}	$\bar{C}_{31}^{-}$	$\bar{C}_{33}^{-}$	C_{32}^{-}	C_{4y}^{-}	C_{2b}	C_{4x}^{+}	$\bar{C}_{2c}$	C_{4z}^{+}	C_{2f}	C_{33}^{+}	C_{31}^{+}	C_{32}^{+}	$\bar{C}_{34}^{+}$	$\bar{C}_{2z}$	C_{2y}	C_{2x}	E	C_{2d}	$\bar{C}_{4x}^{-}$	C_{4z}^{-}	C_{2e}	$\bar{C}_{2a}$	$\bar{C}_{4y}^{+}$
C_{31}^{+}	$\bar{C}_{32}^{+}$	C_{33}^{+}	C_{34}^{+}	C_{4z}^{+}	C_{4x}^{+}	C_{4y}^{+}	$\bar{C}_{2a}$	$\bar{C}_{2d}$	$\bar{C}_{2c}$	E	$\bar{C}_{2y}$	$\bar{C}_{2z}$	C_{2x}	$\bar{C}_{31}^{-}$	C_{34}^{-}	$\bar{C}_{32}^{-}$	C_{33}^{-}	C_{4y}^{-}	C_{2e}	C_{4x}^{-}	C_{4z}^{-}	$\bar{C}_{2f}$	C_{2b}
C_{32}^{+}	C_{31}^{+}	$\bar{C}_{34}^{+}$	C_{33}^{+}	C_{2a}	C_{2f}	C_{4y}^{-}	C_{4z}^{+}	C_{4x}^{-}	$\bar{C}_{2e}$	C_{2y}	E	$\bar{C}_{2x}$	$\bar{C}_{2z}$	C_{33}^{-}	$\bar{C}_{32}^{-}$	C_{34}^{-}	C_{31}^{-}	$\bar{C}_{4y}^{+}$	$\bar{C}_{2c}$	C_{2d}	C_{2b}	C_{4x}^{+}	$\bar{C}_{4z}^{-}$
C_{33}^{+}	$\bar{C}_{34}^{+}$	$\bar{C}_{31}^{+}$	$\bar{C}_{32}^{+}$	C_{4z}^{-}	C_{2d}	C_{2e}	C_{2b}	C_{4x}^{+}	C_{4y}^{-}	C_{2z}	C_{2x}	E	C_{2y}	C_{34}^{-}	C_{31}^{-}	$\bar{C}_{33}^{-}$	C_{32}^{-}	C_{2c}	$\bar{C}_{4y}^{+}$	$\bar{C}_{2f}$	$\bar{C}_{4z}^{+}$	$\bar{C}_{4x}^{-}$	C_{2a}
C_{34}^{+}	C_{33}^{+}	C_{32}^{+}	$\bar{C}_{31}^{+}$	$\bar{C}_{2b}$	C_{4x}^{-}	C_{2c}	C_{4z}^{-}	$\bar{C}_{2f}$	C_{4y}^{+}	$\bar{C}_{2x}$	C_{2z}	$\bar{C}_{2y}$	E	C_{32}^{-}	C_{33}^{-}	C_{31}^{-}	$\bar{C}_{34}^{-}$	$\bar{C}_{2e}$	C_{4y}^{-}	$\bar{C}_{4x}^{+}$	C_{2a}	C_{2d}	C_{4z}^{+}
C_{2a}	C_{4z}^{+}	$\bar{C}_{4z}^{-}$	C_{2b}	$\bar{C}_{31}^{+}$	$\bar{C}_{32}^{-}$	C_{2y}	C_{32}^{+}	C_{31}^{-}	$\bar{C}_{2x}$	$\bar{C}_{4y}^{+}$	C_{4y}^{-}	C_{2c}	$\bar{C}_{2e}$	C_{4x}^{-}	$\bar{C}_{4x}^{+}$	C_{2f}	C_{2d}	$\bar{E}$	$\bar{C}_{2z}$	$\bar{C}_{33}^{-}$	$\bar{C}_{34}^{+}$	C_{34}^{-}	$\bar{C}_{33}^{+}$
C_{2b}	$\bar{C}_{4z}^{-}$	$\bar{C}_{4z}^{+}$	$\bar{C}_{2a}$	C_{33}^{+}	$\bar{C}_{33}^{-}$	C_{2x}	$\bar{C}_{34}^{+}$	C_{34}^{-}	C_{2y}	C_{2e}	$\bar{C}_{2c}$	C_{4y}^{-}	$\bar{C}_{4y}^{+}$	C_{2f}	C_{2d}	$\bar{C}_{4x}^{-}$	C_{4x}^{+}	C_{2z}	$\bar{E}$	C_{32}^{-}	$\bar{C}_{32}^{+}$	$\bar{C}_{31}^{-}$	$\bar{C}_{31}^{+}$
C_{2c}	C_{4y}^{-}	$\bar{C}_{2e}$	$\bar{C}_{4y}^{+}$	$\bar{C}_{34}^{-}$	$\bar{C}_{2x}$	$\bar{C}_{31}^{+}$	C_{31}^{-}	C_{2z}	C_{34}^{+}	$\bar{C}_{4x}^{+}$	C_{2d}	$\bar{C}_{2f}$	C_{4x}^{-}	C_{4z}^{-}	$\bar{C}_{2b}$	C_{2a}	$\bar{C}_{4z}^{+}$	$\bar{C}_{33}^{+}$	C_{32}^{+}	$\bar{E}$	$\bar{C}_{32}^{-}$	C_{2y}	C_{33}^{-}
C_{2d}	C_{2f}	$\bar{C}_{4x}^{+}$	$\bar{C}_{4x}^{-}$	C_{2z}	$\bar{C}_{31}^{+}$	$\bar{C}_{33}^{-}$	C_{2y}	C_{33}^{+}	C_{31}^{-}	$\bar{C}_{4z}^{+}$	C_{2b}	C_{4z}^{-}	C_{2a}	C_{4y}^{-}	C_{2c}	$\bar{C}_{4y}^{+}$	C_{2e}	$\bar{C}_{34}^{-}$	$\bar{C}_{32}^{-}$	$\bar{C}_{32}^{+}$	$\bar{E}$	$\bar{C}_{34}^{+}$	$\bar{C}_{2x}$
C_{2e}	$\bar{C}_{4y}^{+}$	C_{2c}	$\bar{C}_{4y}^{-}$	C_{32}^{-}	C_{2z}	$\bar{C}_{32}^{+}$	$\bar{C}_{33}^{-}$	C_{2x}	C_{33}^{+}	$\bar{C}_{2f}$	$\bar{C}_{4x}^{-}$	C_{4x}^{+}	C_{2d}	C_{2b}	C_{4z}^{-}	$\bar{C}_{4z}^{+}$	$\bar{C}_{2a}$	C_{34}^{+}	$\bar{C}_{31}^{+}$	$\bar{C}_{2y}$	$\bar{C}_{34}^{-}$	$\bar{E}$	C_{31}^{-}
C_{2f}	$\bar{C}_{2d}$	$\bar{C}_{4x}^{-}$	C_{4x}^{+}	C_{2y}	$\bar{C}_{34}^{+}$	C_{34}^{-}	$\bar{C}_{2z}$	C_{32}^{+}	$\bar{C}_{32}^{-}$	C_{2b}	C_{4z}^{+}	C_{2a}	$\bar{C}_{4z}^{-}$	$\bar{C}_{2e}$	$\bar{C}_{4y}^{+}$	$\bar{C}_{2c}$	C_{4y}^{-}	$\bar{C}_{33}^{-}$	$\bar{C}_{31}^{-}$	C_{33}^{+}	C_{2x}	C_{31}^{+}	$\bar{E}$

(b) 622 (D_6)

E	C_6^{+}	C_3^{+}	C_2	C_3^{-}	C_6^{-}	C'_{21}	C'_{22}	C'_{23}	C''_{21}	C''_{22}	C''_{23}	C'_{21}	C''_{23}	$\bar{C}'_{22}$	$\bar{C}''_{21}$	$\bar{C}'_{23}$	$\bar{C}''_{22}$	$\bar{E}$	C_3^{+}	C_3^{-}	C_2	C_6^{-}	$\bar{C}_6^{+}$
C_6^{+}	C_3^{+}	$\bar{C}_2$	C_3^{-}	C_6^{-}	E	$\bar{C}''_{22}$	$\bar{C}''_{23}$	C''_{21}	$\bar{C}'_{22}$	C'_{23}	C'_{21}	C'_{22}	$\bar{C}''_{21}$	$\bar{C}'_{23}$	C''_{22}	$\bar{C}'_{21}$	$\bar{C}''_{23}$	C_3^{-}	$\bar{E}$	C_3^{+}	C_6^{+}	$\bar{C}_2$	C_6^{-}
C_3^{+}	$\bar{C}_2$	$\bar{C}_3^{-}$	C_6^{-}	E	C_6^{+}	$\bar{C}'_{23}$	$\bar{C}'_{21}$	$\bar{C}'_{22}$	C''_{23}	C''_{21}	$\bar{C}''_{22}$	C'_{23}	C''_{22}	$\bar{C}'_{21}$	C''_{23}	$\bar{C}'_{22}$	C''_{21}	C_3^{+}	C_3^{-}	$\bar{E}$	$\bar{C}_6^{-}$	$\bar{C}_6^{+}$	$\bar{C}_2$
C_2	C_3^{-}	C_6^{-}	$\bar{E}$	$\bar{C}_6^{+}$	$\bar{C}_3^{+}$	C''_{21}	$\bar{C}''_{22}$	$\bar{C}''_{23}$	$\bar{C}'_{21}$	C'_{22}	C'_{23}	C''_{21}	C'_{23}	C''_{22}	C'_{21}	C''_{23}	$\bar{C}'_{22}$	$\bar{C}_2$	C_6^{-}	$\bar{C}_6^{+}$	$\bar{E}$	$\bar{C}_3^{+}$	$\bar{C}_3^{-}$
C_3^{-}	C_6^{-}	E	$\bar{C}_6^{+}$	$\bar{C}_3^{+}$	C_2	$\bar{C}'_{22}$	$\bar{C}'_{23}$	$\bar{C}'_{21}$	C''_{22}	$\bar{C}''_{23}$	C''_{21}	C''_{22}	$\bar{C}'_{21}$	$\bar{C}''_{23}$	$\bar{C}'_{22}$	C''_{21}	C'_{23}	C_6^{+}	C_2	$\bar{C}_6^{-}$	$\bar{C}_3^{-}$	$\bar{E}$	C_3^{+}
C_6^{-}	E	C_6^{+}	$\bar{C}_3^{+}$	C_2	C_3^{-}	C''_{23}	$\bar{C}''_{21}$	C''_{22}	C'_{23}	$\bar{C}'_{21}$	$\bar{C}'_{22}$	C''_{23}	$\bar{C}'_{22}$	C''_{21}	$\bar{C}'_{23}$	$\bar{C}''_{22}$	C'_{21}	$\bar{C}_6^{-}$	C_6^{+}	C_2	$\bar{C}_3^{+}$	C_3^{-}	$\bar{E}$

Notes to Table 6.2

(i) The elements in the table are such that Tables 6.2(a) and 6.2(b) coincide exactly with Tables 1.5 and 1.6, respectively, when barred and unbarred operators are identified. A bar above an entry indicates that the matrix is -1 times the matrix indicated in Table 6.1, e.g. for the group 432 (*O*) C_{2d} denotes

$$\frac{1}{\sqrt{2}}\begin{pmatrix} -\mathrm{i} & -1 \\ 1 & \mathrm{i} \end{pmatrix} \text{ and } \bar{C}_{2d} \text{ denotes } \frac{1}{\sqrt{2}}\begin{pmatrix} \mathrm{i} & 1 \\ -1 & -\mathrm{i} \end{pmatrix}.$$

(ii) Only one-quarter of the group multiplication table is included; the other three-quarters can be determined by inspection.

(iii) Column 1 of Table 6.2(b) is continued in column 13, and so on.

TABLE 6.3

The double group of 422 (D_4)

E	C_{4z}^+	C_{2z}	C_{4z}^-	C_{2x}	C_{2y}	C_{2a}	C_{2b}
C_{4z}^+	$\bar{C}_{2z}$	C_{4z}^-	E	$\bar{C}_{2a}$	C_{2b}	C_{2y}	C_{2x}
C_{2z}	C_{4z}^-	$\bar{E}$	$\bar{C}_{4z}^+$	C_{2y}	$\bar{C}_{2x}$	$\bar{C}_{2b}$	C_{2a}
C_{4z}^-	E	$\bar{C}_{4z}^+$	C_{2z}	C_{2b}	C_{2a}	$\bar{C}_{2x}$	C_{2y}
C_{2x}	C_{2b}	$\bar{C}_{2y}$	$\bar{C}_{2a}$	$\bar{E}$	C_{2z}	C_{4z}^-	$\bar{C}_{4z}^+$
C_{2y}	C_{2a}	C_{2x}	C_{2b}	$\bar{C}_{2z}$	$\bar{E}$	$\bar{C}_{4z}^+$	$\bar{C}_{4z}^-$
C_{2a}	$\bar{C}_{2x}$	C_{2b}	C_{2y}	C_{4z}^+	$\bar{C}_{4z}^-$	$\bar{E}$	$\bar{C}_{2z}$
C_{2b}	C_{2y}	$\bar{C}_{2a}$	C_{2x}	$\bar{C}_{4z}^-$	$\bar{C}_{4z}^+$	C_{2z}	$\bar{E}$

Note to Table 6.3

Only one-quarter of the group multiplication table is given, the remainder follows from the fact that $\bar{E}$ commutes with every element of the group.

group multiplication table for 432 (O) in Table 6.2; the result is shown in Table 6.3.

It is possible to use Opechowski's rules, Theorems 6.1.1–6.1.4, to determine the number of classes in the double group of 422 (D_4). Using Theorem 6.1.1 it is clear that classes C_1 and C_3 each lead to two classes in the double group. From Theorem 6.1.2 each of the classes C_2, C_4, and C_5 is a class of rotations through π and such that there are twofold rotations at right angles to them and therefore each of these classes leads to only one class in the double group. Therefore this double group has seven classes which contain the following numbers of elements:

$$\begin{aligned}
C_1 &\to \begin{cases} C_1' & 1 \\ C_1'' & 1, \end{cases} \\
C_2 &\to \;\; C_2' \quad\;\; 2, \\
C_3 &\to \begin{cases} C_3' & 2 \\ C_3'' & 2, \end{cases} \\
C_4 &\to \;\; C_4' \quad\;\; 4, \\
C_5 &\to \;\; C_5' \quad\;\; 4.
\end{aligned}$$

It is obvious, from inspection, to which classes most of the elements of the double group belong, that is C_2' consists of C_{2z} and $\bar{C}_{2z}$, C_4' consists of C_{2x}, C_{2y}, $\bar{C}_{2x}$ and $\bar{C}_{2y}$ and C_5' consists of C_{2a}, C_{2b}, $\bar{C}_{2a}$ and $\bar{C}_{2b}$. By definition C_1' consists of E and C_1'' of $\bar{E}$, but it is not immediately obvious from Opechowski's rules which two of the four elements C_{4z}^+, C_{4z}^-, $\bar{C}_{4z}^+$ and $\bar{C}_{4z}^-$ belong to C_3' and which two belong to C_3''. One is not justified in assuming that C_3' contains C_{4z}^+ and C_{4z}^- while C_3'' contains $\bar{C}_{4z}^+$ and $\bar{C}_{4z}^-$. To determine this, one has to use the ordinary methods for finding the classes of a group, that is to form the products $X^{-1}C_{4z}^+X$ where X ranges over the elements of the group. If one does this it does so happen that C_{4z}^+ and C_{4z}^- are in one class, which we can call C_3' and therefore $\bar{C}_{4z}^+$ and $\bar{C}_{4z}^-$ are in the other class which we can call C_3''. We repeat that this need not have happened because for each entry in Table 6.1 an arbitrary choice was made as to whether to take the + or − sign of eqn (6.1.1); in other cases one could, in principle, have a class C_i leading to two classes C_i' and

C_i'' in which C_i' contains a mixture of elements such as A, $\bar{B}$, C, and D so that C_i'' therefore contains a similar mixture $\bar{A}$, B, $\bar{C}$, and $\bar{D}$. Thus the classes in the double group of 422 (D_4) are

$$\begin{array}{llll}
C_1' & E & C_1'' & \bar{E} \\
C_2' & C_{2z}, \bar{C}_{2z} & & \\
C_3' & C_{4z}^+, C_{4z}^- & C_3'' & \bar{C}_{4z}^+, \bar{C}_{4z}^- \\
C_4' & C_{2x}, C_{2y}, \bar{C}_{2x}, \bar{C}_{2y} & & \\
C_5' & C_{2a}, C_{2b}, \bar{C}_{2a}, \bar{C}_{2b}. & &
\end{array}$$

Since there are seven classes there will be seven reps of the double group $\mathbf{G}^\dagger$. Of these seven, five will form single-valued representations of the point group 422 (D_4) and two will form double-valued representations of 422 (D_4). The five that form single-valued representations of 422 (D_4) can easily be determined from the character table of that group in Table 2.2, and in these, see Theorem 6.1.3, $\chi(C_i') = +\chi(C_i'')$.

		C_1'	C_1''	C_2'	C_3'	C_3''	C_4'	C_5'
A_1	Γ_1	1	1	1	1	1	1	1
A_2	Γ_2	1	1	1	1	1	−1	−1
B_1	Γ_3	1	1	1	−1	−1	1	−1
B_2	Γ_4	1	1	1	−1	−1	−1	1
E	Γ_5	2	2	−2	0	0	0	0

For the other two representations, those that form double-valued representations of 422 (D_4), $\chi(C_i') = -\chi(C_i'')$ from Theorem 6.1.4; thus the characters of the classes C_2', C_4', and C_5' are all zero. These two representations must be of dimension equal to two, from Theorem 1.3.11. Therefore $\chi(C_1') = 2$ and $\chi(C_1'') = -2$, and then also, using eqn (1.3.11), it follows that $\chi(C_3') = \pm\sqrt{2}$ and $\chi(C_3'') = \mp\sqrt{2}$. Therefore the rest of the character table is

		C_1'	C_1''	C_2'	C_3'	C_3''	C_4'	C_5'
$\bar{E}_1$	Γ_6	2	−2	0	$\sqrt{2}$	$-\sqrt{2}$	0	0
$\bar{E}_2$	Γ_7	2	−2	0	$-\sqrt{2}$	$\sqrt{2}$	0	0

There are other ways of finding the reps of the double group $\mathbf{G}^\dagger$ than by using Opechowski's rules, i.e. Theorems 6.1.1–6.1.4. It is possible, having separated the double group $\mathbf{G}^\dagger$ into classes to work out the class-multiplication coefficients $h_{ij,k}$, see eqn (1.2.2), and use them in eqn (1.3.19) to work out the character table of $\mathbf{G}^\dagger$ and hence to find the double-valued reps of $\mathbf{G}$. Yet another way to find the reps of $\mathbf{G}^\dagger$ is to establish an isomorphism between $\mathbf{G}^\dagger$ and one of the abstract groups $\mathbf{G}_{|\mathbf{G}|}^n$ in Table 5.1. To establish this isomorphism is not as difficult as may appear at first sight. First $|\mathbf{G}|$, the order of the required abstract group $\mathbf{G}_{|\mathbf{G}|}^n$, is the order of $\mathbf{G}^\dagger$ which, in any particular case is known. Then if there is to be an isomorphism between

$\mathbf{G}^n_{|G|}$ and $\mathbf{G}^\dagger$ the number of classes in $\mathbf{G}^n_{|G|}$ the abstract group must be the same as the number of classes in $\mathbf{G}^\dagger$ and this is also known; this enables most of the abstract groups of the correct order to be rejected. By this stage there is probably only one abstract group remaining that can possibly be isomorphic to $\mathbf{G}^\dagger$ and at most there will only be a small number of groups left as possibilities. To distinguish between such possibilities and to identify the generating elements of the correct abstract group $\mathbf{G}^n_{|G|}$ with suitable elements of the double group $\mathbf{G}^\dagger$ it is best to proceed by inspection. Having made this identification the isomorphism is completely determined and the reps of $\mathbf{G}^\dagger$ are just the reps of $\mathbf{G}^n_{|G|}$ whose character table and matrix representatives, in the case of degenerate representations, are given in Table 5.1. Although Table 5.1 does not contain all the abstract groups of all the various orders it does contain all the abstract groups that are encountered in the double groups of the point groups and space groups.

We can illustrate this procedure for the double group of 422 (D_4) which we have already considered and whose multiplication table is given in Table 6.3. This group is of order 16 and has seven classes, and therefore of all the groups of order 16 only $\mathbf{G}^{12}_{16}$, $\mathbf{G}^{13}_{16}$, and $\mathbf{G}^{14}_{16}$ can possibly be isomorphic to the double group of 422 (D_4). Each of these three abstract groups has the required class structure, i.e. two classes of one element each, three classes of two elements each and two classes of four elements each. If we examine Table 6.3 we can determine the order of each of the elements in the double group, for example C_{2x}, C_{2y}, and C_{2z} are of order 4 while C^+_{4z} and C^-_{4z} are of order 8, and the total numbers of elements of each order are

Order	1	2	4	8
Number of elements	1	1	10	4

and doing the same for $\mathbf{G}^{12}_{16}$, $\mathbf{G}^{13}_{16}$, and $\mathbf{G}^{14}_{16}$ we obtain

$\mathbf{G}^{12}_{16}$	Order	1	2	4	8
	Number of elements	1	9	2	4
$\mathbf{G}^{13}_{16}$	Order	1	2	4	8
	Number of elements	1	5	6	4
$\mathbf{G}^{14}_{16}$	Order	1	2	4	8
	Number of elements	1	1	10	4

It is thus clear that $\mathbf{G}^{12}_{16}$ and $\mathbf{G}^{13}_{16}$ cannot be isomorphic to the double group of 422 (D_4), and therefore it is $\mathbf{G}^{14}_{16}$ that is required. It then remains to identify the generating elements P and Q of $\mathbf{G}^{14}_{16}$ with two suitable elements of the double group of 422 (D_4).

P can be identified with any of the elements C_{4z}^+, C_{4z}^-, $\bar{C}_{4z}^+$, and $\bar{C}_{4z}^-$ of the double group, and we choose C_{4z}^+. Q will then be one of the elements of order 4 which satisfies the generating relations, in this case just $Q^2 = P^4$, $QP = P^7Q$, and a possible choice is $Q = C_{2x}$ as can easily be verified. Thus we arrive at the conclusion that the double group of 422 (D_4) is $\mathbf{G}_{16}^{14}$, with its generating elements, P and Q being

$$P = C_{4z}^+$$

and

$$Q = C_{2x}.$$

The character table of the double group can be constructed immediately from that of $\mathbf{G}_{16}^{14}$ and this is shown below where the labels of the classes used in the previous treatment are also given.

	C_1	C_2	C_3	C_4	C_5	C_6	C_7
R_1	1	1	1	1	1	1	1
R_2	1	1	1	1	1	−1	−1
R_3	1	1	1	−1	−1	1	−1
R_4	1	1	1	−1	−1	−1	1
R_5	2	2	−2	0	0	0	0
R_6	2	−2	0	$\sqrt{2}$	$-\sqrt{2}$	0	0
R_7	2	−2	0	$-\sqrt{2}$	$\sqrt{2}$	0	0

$$\begin{aligned}
C_1 &= C_1' = E \\
C_2 &= C_1'' = \bar{E} \\
C_3 &= C_2' = C_{2z}, \bar{C}_{2z} \\
C_4 &= C_3' = C_{4z}^+, C_{4z}^- \\
C_5 &= C_3'' = \bar{C}_{4z}^+, \bar{C}_{4z}^- \\
C_6 &= C_4' = C_{2x}, C_{2y}, \bar{C}_{2x}, \bar{C}_{2y} \\
C_7 &= C_5' = C_{2a}, C_{2b}, \bar{C}_{2a}, \bar{C}_{2b}.
\end{aligned}$$

It is clear that this is the same as was obtained previously by using Opechowski's rules, Theorems 6.1.1–6.1.4. The single-valued reps of **G**, i.e. 422 (D_4), are clearly R_1, R_2, R_3, R_4, and R_5 for which $\chi(C_2) = \chi(C_1)$ and $\chi(C_5) = \chi(C_4)$ while the double-valued reps are R_6 and R_7 for which $\chi(C_2) = -\chi(C_1)$ and $\chi(C_5) = -\chi(C_4)$.

Nothing has been said explicitly about the double-valued reps of those point groups that contain improper rotations, for each of these point groups is bound to be simply related to one of those point groups that consists only of pure rotations. Suppose that **G** is a point group consisting only of pure rotations then there are two ways in which a point group **G**′ containing improper rotations can be derived from it, either

$$\mathbf{G}' = \mathbf{G} + I\mathbf{G} \tag{6.1.3}$$

where I is the inversion operation, so that **G**′ is a direct product group, or

$$\mathbf{G}' = \mathbf{H} + I(\mathbf{G} - \mathbf{H}) \tag{6.1.4}$$

where **H** is a halving subgroup of **G**, that is a subgroup of **G** of index 2. In each of these two cases the reps of **G**′ are simply related to the reps of **G**. Therefore, since we have already shown what methods are available for the deduction of the double-valued representations of those point groups consisting only of pure rotations it is straightforward to deduce the double-valued representations of all the crystallographic point groups. In Table 6.4 we identify each of the double point groups as a manifestation of one of the abstract groups that were given in Table 5.1. The character

TABLE 6.4

The identification of the double point groups in terms of the abstract groups

	Point group	Abstract group, see Table 5.1	Identification of generating elements
1	$1\ (C_1)$	$\mathbf{G}_2^1$	$P = \bar{E}$
2	$\bar{1}\ (C_i)$	$\mathbf{G}_4^2$	$P = I,\ Q = \bar{E}$
3	$2\ (C_2)$	$\mathbf{G}_4^1$	$P = C_{2z}$
4	$m\ (C_{1h})$	$\mathbf{G}_4^1$	$P = \sigma_z$
5	$2/m\ (C_{2h})$	$\mathbf{G}_8^2$	$P = C_{2z},\ Q = I$
6	$222\ (D_2)$	$\mathbf{G}_8^5$	$P = C_{2z},\ Q = C_{2y}$
7	$mm2\ (C_{2v})$	$\mathbf{G}_8^5$	$P = C_{2z},\ Q = \sigma_y$
8	$mmm\ (D_{2h})$	$\mathbf{G}_{16}^{11}$	$P = C_{2z},\ Q = C_{2y},\ R = I$
9	$4\ (C_4)$	$\mathbf{G}_8^1$	$P = C_{4z}^+$
10	$\bar{4}\ (S_4)$	$\mathbf{G}_8^1$	$P = S_{4z}^-$
11	$4/m\ (C_{4h})$	$\mathbf{G}_{16}^2$	$P = C_{4z}^+,\ Q = I$
12	$422\ (D_4)$	$\mathbf{G}_{16}^{14}$	$P = C_{4z}^+,\ Q = C_{2x}$
13	$4mm\ (C_{4v})$	$\mathbf{G}_{16}^{14}$	$P = C_{4z}^+,\ Q = \sigma_x$
14	$\bar{4}2m\ (D_{2d})$	$\mathbf{G}_{16}^{14}$	$P = S_{4z}^-,\ Q = C_{2x}$
15	$4/mmm\ (D_{4h})$	$\mathbf{G}_{32}^9$	$P = C_{4z}^+,\ Q = C_{2x},\ R = I$
16	$3\ (C_3)$	$\mathbf{G}_6^1$	$P = C_3^+$
17	$\bar{3}\ (C_{3i})$	$\mathbf{G}_{12}^6$	$P = S_6^+,\ Q = I$
18	$32\ (D_3)$	$\mathbf{G}_{12}^4$	$P = C_3^+,\ Q = C'_{21}$
19	$3m\ (C_{3v})$	$\mathbf{G}_{12}^4$	$P = C_3^+,\ Q = \sigma_{d1}$
20	$\bar{3}m\ (D_{3d})$	$\mathbf{G}_{24}^3$	$P = C_3^+,\ Q = C'_{21},\ R = I$
21	$6\ (C_6)$	$\mathbf{G}_{12}^1$	$P = C_6^+$
22	$\bar{6}\ (C_{3h})$	$\mathbf{G}_{12}^1$	$P = S_3^-$
23	$6/m\ (C_{6h})$	$\mathbf{G}_{24}^{12}$	$P = C_6^+,\ Q = I$
24	$622\ (D_6)$	$\mathbf{G}_{24}^{11}$	$P = C_6^+,\ Q = C'_{21}$
25	$6mm\ (C_{6v})$	$\mathbf{G}_{24}^{11}$	$P = C_6^+,\ Q = \sigma_{d1}$
26	$\bar{6}2m\ (D_{3h})$	$\mathbf{G}_{24}^{11}$	$P = S_3^-,\ Q = C'_{21}$
27	$6/mmm\ (D_{6h})$	$\mathbf{G}_{48}^{15}$	$P = C_6^+,\ Q = C'_{21},\ R = I$
28	$23\ (T)$	$\mathbf{G}_{24}^9$	$P = C_{31}^-,\ Q = C_{2x},\ R = \bar{C}_{2y}$
29	$m3\ (T_h)$	$\mathbf{G}_{48}^4$	$P = C_{31}^-,\ Q = C_{2x},\ R = \bar{C}_{2y},\ S = I$
30	$432\ (O)$	$\mathbf{G}_{48}^{10}$	$P = C_{4x}^+,\ Q = \bar{C}_{31}^-,\ R = C_{2b}$
31	$\bar{4}3m\ (T_d)$	$\mathbf{G}_{48}^{10}$	$P = S_{4x}^-,\ Q = \bar{C}_{31}^-,\ R = \sigma_{db}$
32	$m3m\ (O_h)$	$\mathbf{G}_{96}^8$	$P = C_{4x}^+,\ Q = \bar{C}_{31}^-,\ R = C_{2b},\ S = I$

tables of the double point groups are given in Table 6.5 (Cracknell and Wong 1967, Fick 1957, Hamermesh 1962, Kitz 1965*a*, Koster, Dimmock, Wheeler and Statz 1963, Opechowski 1940).

TABLE 6.5

The double-valued representations of the crystallographic point groups

($\omega = \exp(2\pi i/3)$; $\tilde{\omega} = \sqrt{2} + i$; $\theta = (1 + i)/\sqrt{2}$)

$1\ (C_1)$	E	$\bar{E}$
$A\ \Gamma_1$	1	1
$\bar{A}\ \Gamma_2$	1	−1

$\bar{1}\ (C_i)$	E	I	$\bar{E}$	$\bar{I}$
$A_g\ \Gamma_1^+$	1	1	1	1
$A_u\ \Gamma_1^-$	1	−1	1	−1
$\bar{A}_g\ \Gamma_2^+$	1	1	−1	−1
$\bar{A}_u\ \Gamma_2^-$	1	−1	−1	1

$2\ (C_2)$	$m\ (C_{1h})$	E E	C_{2z} σ_z	$\bar{E}$ $\bar{E}$	$\bar{C}_{2z}$ $\bar{\sigma}_z$
$A\ \Gamma_1$	$A'\ \Gamma_1$	1	1	1	1
$^1\bar{E}\ \Gamma_3$	$^1\bar{E}\ \Gamma_3$	1	i	−1	−i
$B\ \Gamma_2$	$A''\ \Gamma_2$	1	−1	1	−1
$^2\bar{E}\ \Gamma_4$	$^2\bar{E}\ \Gamma_4$	1	−i	−1	i

$2/m = 2 \otimes \bar{1}$ $(C_{2h} = C_2 \otimes C_i)$.

$mm2\ (C_{2v})$	$222\ (D_2)$	E E	$\bar{E}$ $\bar{E}$	$\sigma_x, \bar{\sigma}_x$ $C_{2x}, \bar{C}_{2x}$	$\sigma_y, \bar{\sigma}_y$ $C_{2y}, \bar{C}_{2y}$	$C_{2z}, \bar{C}_{2z}$ $C_{2z}, \bar{C}_{2z}$
$A_1\ \Gamma_1$	$A\ \Gamma_1$	1	1	1	1	1
$B_2\ \Gamma_4$	$B_3\ \Gamma_4$	1	1	1	−1	−1
$B_1\ \Gamma_2$	$B_2\ \Gamma_2$	1	1	−1	1	−1
$A_2\ \Gamma_3$	$B_1\ \Gamma_3$	1	1	−1	−1	1
$\bar{E}\ \Gamma_5$	$\bar{E}\ \Gamma_5$	2	−2	0	0	0

$mm2\ (C_{2v})$	$222\ (D_2)$	σ_x C_{2x}	σ_y C_{2y}
$\bar{E}\ \Gamma_5$	$\bar{E}\ \Gamma_5$	$\begin{pmatrix} 0 & 1 \\ -1 & 0 \end{pmatrix}$	$\begin{pmatrix} 0 & i \\ i & 0 \end{pmatrix}$

$mmm = 222 \otimes \bar{1}$ $(D_{2h} = D_2 \otimes C_i)$.

$4\ (C_4)$		E	C_{4z}^-	C_{2z}	$\bar{C}_{4z}^+$	$\bar{E}$	$\bar{C}_{4z}^-$	$\bar{C}_{2z}$	C_{4z}^+
	$\bar{4}\ (S_4)$	E	S_{4z}^+	C_{2z}	$\bar{S}_{4z}^-$	$\bar{E}$	$\bar{S}_{4z}^+$	$\bar{C}_{2z}$	S_{4z}^-
A Γ_1	A Γ_1	1	1	1	1	1	1	1	1
${}^2\bar{E}_1$ Γ_5	${}^2\bar{E}_1$ Γ_5	1	θ	i	$-\theta^*$	-1	$-\theta$	$-$i	θ^*
${}^1E$ Γ_4	${}^1E$ Γ_4	1	i	-1	$-$i	1	i	-1	$-$i
${}^1\bar{E}_2$ Γ_8	${}^1\bar{E}_2$ Γ_8	1	$-\theta^*$	$-$i	θ	-1	θ^*	i	$-\theta$
B Γ_2	B Γ_2	1	-1	1	-1	1	-1	1	-1
${}^2\bar{E}_2$ Γ_7	${}^2\bar{E}_2$ Γ_7	1	$-\theta$	i	θ^*	-1	θ	$-$i	$-\theta^*$
${}^2E$ Γ_3	${}^2E$ Γ_3	1	$-$i	-1	i	1	$-$i	-1	i
${}^1\bar{E}_1$ Γ_6	${}^1\bar{E}_1$ Γ_6	1	θ^*	$-$i	$-\theta$	-1	$-\theta^*$	i	θ

$4/m = 4 \otimes \bar{1}\ (C_{4h} = C_4 \otimes C_i)$.

$3\ (C_3)$	E	C_3^+	$\bar{C}_3^-$	$\bar{E}$	$\bar{C}_3^+$	C_3^-
A Γ_1	1	1	1	1	1	1
${}^2\bar{E}$ Γ_5	1	$-\omega^*$	ω	-1	ω^*	$-\omega$
${}^2E$ Γ_2	1	ω	ω^*	1	ω	ω^*
$\bar{A}$ Γ_6	1	-1	1	-1	1	-1
${}^1E$ Γ_3	1	ω^*	ω	1	ω^*	ω
${}^1\bar{E}$ Γ_4	1	$-\omega$	ω^*	-1	ω	$-\omega^*$

$\bar{3} = 3 \otimes \bar{1}\ (C_{3i} = C_3 \otimes C_i)$.

$32\ (D_3)$		E	$\bar{E}$	C_3^+, C_3^-	$\bar{C}_3^-, \bar{C}_3^+$	$C'_{21}, C'_{22}, C'_{23}$	$\bar{C}'_{21}, \bar{C}'_{22}, \bar{C}'_{23}$
	$3m\ (C_{3v})$	E	$\bar{E}$	C_3^+, C_3^-	$\bar{C}_3^-, \bar{C}_3^+$	$\sigma_{d1}, \sigma_{d2}, \sigma_{d3}$	$\bar{\sigma}_{d1}, \bar{\sigma}_{d2}, \bar{\sigma}_{d3}$
A_1 Γ_1	A_1 Γ_1	1	1	1	1	1	1
A_2 Γ_2	A_2 Γ_2	1	1	1	1	-1	-1
${}^1\bar{E}$ Γ_5	${}^1\bar{E}$ Γ_5	1	-1	-1	1	i	$-$i
${}^2\bar{E}$ Γ_6	${}^2\bar{E}$ Γ_6	1	-1	-1	1	$-$i	i
E Γ_3	E Γ_3	2	2	-1	-1	0	0
$\bar{E}_1$ Γ_4	$\bar{E}_1$ Γ_4	2	-2	1	-1	0	0

$32\ (D_3)$		C_3^+	C'_{21}
	$3m\ (C_{3v})$	C_3^+	σ_{d1}
$\bar{E}_1$ Γ_4	$\bar{E}_1$ Γ_4	$\begin{pmatrix} \frac{1}{2} & -\frac{1}{2}\sqrt{3} \\ \frac{1}{2}\sqrt{3} & \frac{1}{2} \end{pmatrix}$	$\begin{pmatrix} \mathrm{i} & 0 \\ 0 & -\mathrm{i} \end{pmatrix}$

$\bar{3}m = 32 \otimes \bar{1}\ (D_{3d} = D_3 \otimes C_i)$.

6 (C_6)		E	C_6^+	C_3^+	$\bar{C}_2$	$\bar{C}_3^-$	$\bar{C}_6^-$	$\bar{E}$	$\bar{C}_6^+$	$\bar{C}_3^+$	C_2	C_3^-	C_6^-
	$\bar{6}$ (C_{3h})	E	S_3^-	C_3^+	$\bar{\sigma}_h$	$\bar{C}_3^-$	$\bar{S}_3^+$	$\bar{E}$	$\bar{S}_3^-$	$\bar{C}_3^+$	σ_h	C_3^-	S_3^+
A Γ_1	A' Γ_1	1	1	1	1	1	1	1	1	1	1	1	1
$^2\bar{E}_3$ Γ_8	$^2\bar{E}_3$ Γ_8	1	$-\mathrm{i}\omega$	$-\omega^*$	i	ω	$-\mathrm{i}\omega^*$	-1	$\mathrm{i}\omega$	ω^*	$-\mathrm{i}$	$-\omega$	$\mathrm{i}\omega^*$
$^2E_2$ Γ_2	$^2E''$ Γ_5	1	$-\omega^*$	ω	-1	ω^*	$-\omega$	1	$-\omega^*$	ω	-1	ω^*	$-\omega$
$^1\bar{E}_1$ Γ_{11}	$^1\bar{E}_1$ Γ_{11}	1	i	-1	$-\mathrm{i}$	1	i	-1	$-\mathrm{i}$	1	i	-1	$-\mathrm{i}$
$^1E_1$ Γ_6	$^1E'$ Γ_3	1	ω	ω^*	1	ω	ω^*	1	ω	ω^*	1	ω	ω^*
$^2\bar{E}_2$ Γ_9	$^2\bar{E}_2$ Γ_9	1	$-\mathrm{i}\omega^*$	$-\omega$	i	ω^*	$-\mathrm{i}\omega$	-1	$\mathrm{i}\omega^*$	ω	$-\mathrm{i}$	$-\omega^*$	$\mathrm{i}\omega$
B Γ_4	A'' Γ_4	1	-1	1	-1	1	-1	1	-1	1	-1	1	-1
$^1\bar{E}_2$ Γ_{10}	$^1\bar{E}_2$ Γ_{10}	1	$\mathrm{i}\omega$	$-\omega^*$	$-\mathrm{i}$	ω	$\mathrm{i}\omega^*$	-1	$-\mathrm{i}\omega$	ω^*	i	$-\omega$	$-\mathrm{i}\omega^*$
$^2E_1$ Γ_5	$^2E'$ Γ_2	1	ω^*	ω	1	ω^*	ω	1	ω^*	ω	1	ω^*	ω
$^2\bar{E}_1$ Γ_{12}	$^2\bar{E}_1$ Γ_{12}	1	$-\mathrm{i}$	-1	i	1	$-\mathrm{i}$	-1	i	1	$-\mathrm{i}$	-1	i
$^1E_2$ Γ_3	$^1E''$ Γ_6	1	$-\omega$	ω^*	-1	ω	$-\omega^*$	1	$-\omega$	ω^*	-1	ω	$-\omega^*$
$^1\bar{E}_3$ Γ_7	$^1\bar{E}_3$ Γ_7	1	$\mathrm{i}\omega^*$	$-\omega$	$-\mathrm{i}$	ω^*	$\mathrm{i}\omega$	-1	$-\mathrm{i}\omega^*$	ω	i	$-\omega^*$	$-\mathrm{i}\omega$

$6/m = 6 \otimes \bar{1}$ ($C_{6h} = C_6 \otimes C_i$).

422 (D_4)			E	$\bar{E}$	$C_{2z}, \bar{C}_{2z}$	C_{4z}^-, C_{4z}^+	$\bar{C}_{4z}^+, \bar{C}_{4z}^-$	C_{2x}, C_{2y} $\bar{C}_{2x}, \bar{C}_{2y}$	C_{2a}, C_{2b} $\bar{C}_{2a}, \bar{C}_{2b}$
	4mm (C_{4v})		E	$\bar{E}$	$C_{2z}, \bar{C}_{2z}$	C_{4z}^-, C_{4z}^+	$\bar{C}_{4z}^+, \bar{C}_{4z}^-$	σ_x, σ_y $\bar{\sigma}_x, \bar{\sigma}_y$	σ_{da}, σ_{db} $\bar{\sigma}_{da}, \bar{\sigma}_{db}$
		$\bar{4}2m$ (D_{2d})	E	$\bar{E}$	$C_{2z}, \bar{C}_{2z}$	S_{4z}^+, S_{4z}^-	$\bar{S}_{4z}^-, \bar{S}_{4z}^+$	C_{2x}, C_{2y} $\bar{C}_{2x}, \bar{C}_{2y}$	σ_{da}, σ_{db} $\bar{\sigma}_{da}, \bar{\sigma}_{db}$
A_1 Γ_1	A_1 Γ_1	A_1 Γ_1	1	1	1	1	1	1	1
A_2 Γ_2	A_2 Γ_2	A_2 Γ_2	1	1	1	1	1	-1	-1
B_1 Γ_3	B_1 Γ_3	B_1 Γ_3	1	1	1	-1	-1	1	-1
B_2 Γ_4	B_2 Γ_4	B_2 Γ_4	1	1	1	-1	-1	-1	1
E Γ_5	E Γ_5	E Γ_5	2	2	-2	0	0	0	0
$\bar{E}_1$ Γ_6	$\bar{E}_1$ Γ_6	$\bar{E}_1$ Γ_6	2	-2	0	$\sqrt{2}$	$-\sqrt{2}$	0	0
$\bar{E}_2$ Γ_7	$\bar{E}_2$ Γ_7	$\bar{E}_2$ Γ_7	2	-2	0	$-\sqrt{2}$	$\sqrt{2}$	0	0

422 (D_4)			C_{4z}^- C_{4z}^- S_{4z}^+	C_{2x} σ_x C_{2x}
	4mm (C_{4v})			
		$\bar{4}2m$ (D_{2d})		
$\bar{E}_1$ Γ_6	$\bar{E}_1$ Γ_6	$\bar{E}_1$ Γ_6	$\frac{1}{\sqrt{2}}\begin{pmatrix} 1 & 1 \\ -1 & 1 \end{pmatrix}$	$\begin{pmatrix} 0 & \mathrm{i} \\ \mathrm{i} & 0 \end{pmatrix}$
$\bar{E}_2$ Γ_7	$\bar{E}_2$ Γ_7	$\bar{E}_2$ Γ_7	$\frac{1}{\sqrt{2}}\begin{pmatrix} -1 & 1 \\ -1 & -1 \end{pmatrix}$	$\begin{pmatrix} 0 & \mathrm{i} \\ \mathrm{i} & 0 \end{pmatrix}$

$4/mmm = 422 \otimes \bar{1}$ ($D_{4h} = D_4 \otimes C_i$).

622 (D_6)		6mm (C_{6v})		$\bar{6}2m$ (D_{3h})										
622 (D_6)						E	$\bar{E}$	$\bar{C}_3^+, \bar{C}_3^-$	C_3^+, C_3^-	$\bar{C}_2, C_2$	C_6^+, C_6^-	$\bar{C}_6^-, \bar{C}_6^+$	$C'_{21}, \bar{C}'_{21}$ $C'_{22}, \bar{C}'_{22}$ $C'_{23}, \bar{C}'_{23}$	$C''_{21}, \bar{C}''_{21}$ $C''_{22}, \bar{C}''_{22}$ $C''_{23}, \bar{C}''_{23}$
		6mm (C_{6v})				E	$\bar{E}$	$\bar{C}_3^+, \bar{C}_3^-$	C_3^+, C_3^-	$\bar{C}_2, C_2$	C_6^+, C_6^-	$\bar{C}_6^-, \bar{C}_6^+$	$\sigma_{d1}, \bar{\sigma}_{d1}$ $\sigma_{d2}, \bar{\sigma}_{d2}$ $\sigma_{d3}, \bar{\sigma}_{d3}$	$\sigma_{v1}, \bar{\sigma}_{v1}$ $\sigma_{v2}, \bar{\sigma}_{v2}$ $\sigma_{v3}, \bar{\sigma}_{v3}$
				$\bar{6}2m$ (D_{3h})		E	$\bar{E}$	$\bar{C}_3^+, \bar{C}_3^-$	C_3^+, C_3^-	$\bar{\sigma}_h, \sigma_h$	S_3^-, S_3^+	$\bar{S}_3^+, \bar{S}_3^-$	$C'_{21}, \bar{C}'_{21}$ $C'_{22}, \bar{C}'_{22}$ $C'_{23}, \bar{C}'_{23}$	$\sigma_{v1}, \bar{\sigma}_{v1}$ $\sigma_{v2}, \bar{\sigma}_{v2}$ $\sigma_{v3}, \bar{\sigma}_{v3}$
A_1	Γ_1	A_1	Γ_1	A'_1	Γ_1	1	1	1	1	1	1	1	1	1
A_2	Γ_2	A_2	Γ_2	A'_2	Γ_2	1	1	1	1	1	1	1	−1	−1
B_1	Γ_3	B_2	Γ_3	A''_1	Γ_3	1	1	1	1	−1	−1	−1	1	−1
B_2	Γ_4	B_1	Γ_4	A''_2	Γ_4	1	1	1	1	−1	−1	−1	−1	1
E_1	Γ_5	E_1	Γ_5	E''	Γ_5	2	2	−1	−1	−2	1	1	0	0
E_2	Γ_6	E_2	Γ_6	E'	Γ_6	2	2	−1	−1	2	−1	−1	0	0
$\bar{E}_1$	Γ_7	$\bar{E}_1$	Γ_7	$\bar{E}_1$	Γ_7	2	−2	−1	1	0	$\sqrt{3}$	$-\sqrt{3}$	0	0
$\bar{E}_2$	Γ_8	$\bar{E}_2$	Γ_8	$\bar{E}_2$	Γ_8	2	−2	−1	1	0	$-\sqrt{3}$	$\sqrt{3}$	0	0
$\bar{E}_3$	Γ_9	$\bar{E}_3$	Γ_9	$\bar{E}_3$	Γ_9	2	−2	2	−2	0	0	0	0	0

622 (D_6)		6mm (C_{6v})		$\bar{6}2m$ (D_{3h})		C_6^+ C_6^+ S_3^-	C'_{21} σ_{d1} C'_{21}
$\bar{E}_1$	Γ_7	$\bar{E}_1$	Γ_7	$\bar{E}_1$	Γ_7	$\begin{pmatrix} -i\omega & 0 \\ 0 & i\omega^* \end{pmatrix}$	$\begin{pmatrix} 0 & i \\ i & 0 \end{pmatrix}$
$\bar{E}_2$	Γ_8	$\bar{E}_2$	Γ_8	$\bar{E}_2$	Γ_8	$\begin{pmatrix} -i\omega^* & 0 \\ 0 & i\omega \end{pmatrix}$	$\begin{pmatrix} 0 & i \\ i & 0 \end{pmatrix}$
$\bar{E}_3$	Γ_9	$\bar{E}_3$	Γ_9	$\bar{E}_3$	Γ_9	$\begin{pmatrix} i & 0 \\ 0 & -i \end{pmatrix}$	$\begin{pmatrix} 0 & i \\ i & 0 \end{pmatrix}$

$6/mmm = 622 \otimes \bar{1}$ $(D_{6h} = D_6 \otimes C_i)$.

23 (T)		E	$\bar{E}$	C_{2x}, C_{2y}, C_{2z} $\bar{C}_{2x}, \bar{C}_{2y}, \bar{C}_{2z}$	C_{31}^-, C_{32}^- C_{33}^-, C_{34}^-	$\bar{C}_{31}^-, \bar{C}_{32}^-$ $\bar{C}_{33}^-, \bar{C}_{34}^-$	$\bar{C}_{31}^+, \bar{C}_{32}^+$ $\bar{C}_{33}^+, \bar{C}_{34}^+$	C_{31}^+, C_{32}^+ C_{33}^+, C_{34}^+
A	Γ_1	1	1	1	1	1	1	1
${}^2E$	Γ_3	1	1	1	ω	ω	ω^*	ω^*
${}^1E$	Γ_2	1	1	1	ω^*	ω^*	ω	ω
$\bar{E}$	Γ_5	2	−2	0	1	−1	−1	1
${}^1\bar{F}$	Γ_6	2	−2	0	ω	$-\omega$	$-\omega^*$	ω^*
${}^2\bar{F}$	Γ_7	2	−2	0	ω^*	$-\omega^*$	$-\omega$	ω
T	Γ_4	3	3	−1	0	0	0	0

23 (T)	C_{31}^-	C_{2x}	$\bar{C}_{2y}$
$\bar{E}$ Γ_5	$\frac{1}{2}\begin{pmatrix} 1-i & -1+i \\ 1+i & 1+i \end{pmatrix}$	$\begin{pmatrix} 0 & -i \\ -i & 0 \end{pmatrix}$	$\begin{pmatrix} 0 & 1 \\ -1 & 0 \end{pmatrix}$
${}^1\bar{F}$ Γ_6	$\frac{\omega}{2}\begin{pmatrix} 1-i & -1+i \\ 1+i & 1+i \end{pmatrix}$	$\begin{pmatrix} 0 & -i \\ -i & 0 \end{pmatrix}$	$\begin{pmatrix} 0 & 1 \\ -1 & 0 \end{pmatrix}$
${}^2\bar{F}$ Γ_7	$\frac{\omega^*}{2}\begin{pmatrix} 1-i & -1+i \\ 1+i & 1+i \end{pmatrix}$	$\begin{pmatrix} 0 & -i \\ -i & 0 \end{pmatrix}$	$\begin{pmatrix} 0 & 1 \\ -1 & 0 \end{pmatrix}$

$m3 = 23 \otimes \bar{1}$ $\quad (T_h = T \otimes C_i)$.

432 (O)	$\bar{4}3m$ (T_d)	C_1	C_2	C_3	C_4	C_5	C_6	C_7	C_8
A_1 Γ_1	A_1 Γ_1	1	1	1	1	1	1	1	1
A_2 Γ_2	A_2 Γ_2	1	1	1	1	1	−1	−1	−1
E Γ_3	E Γ_3	2	2	2	−1	−1	0	0	0
T_1 Γ_4	T_1 Γ_4	3	3	−1	0	0	1	1	−1
T_2 Γ_5	T_2 Γ_5	3	3	−1	0	0	−1	−1	1
$\bar{E}_1$ Γ_6	$\bar{E}_1$ Γ_6	2	−2	0	1	−1	$\sqrt{2}$	$-\sqrt{2}$	0
$\bar{E}_2$ Γ_7	$\bar{E}_2$ Γ_7	2	−2	0	1	−1	$-\sqrt{2}$	$\sqrt{2}$	0
$\bar{F}$ Γ_8	$\bar{F}$ Γ_8	4	−4	0	−1	1	0	0	0

432 (O)

C_1 $\quad E$
C_2 $\quad \bar{E}$
C_3 $\quad C_{2x}, C_{2y}, C_{2z}, \bar{C}_{2x}, \bar{C}_{2y}, \bar{C}_{2z}$
C_4 $\quad C_{31}^-, C_{32}^-, C_{33}^-, C_{34}^-, C_{31}^+, C_{32}^+, C_{33}^+, C_{34}^+$
C_5 $\quad \bar{C}_{31}^-, \bar{C}_{32}^-, \bar{C}_{33}^-, \bar{C}_{34}^-, \bar{C}_{31}^+, \bar{C}_{32}^+, \bar{C}_{33}^+, \bar{C}_{34}^+$
C_6 $\quad C_{4x}^+, C_{4y}^+, C_{4z}^+, C_{4x}^-, C_{4y}^-, C_{4z}^-$
C_7 $\quad \bar{C}_{4x}^+, \bar{C}_{4y}^+, \bar{C}_{4z}^+, \bar{C}_{4x}^-, \bar{C}_{4y}^-, \bar{C}_{4z}^-$
C_8 $\quad C_{2a}, C_{2b}, C_{2c}, C_{2d}, C_{2e}, C_{2f}, \bar{C}_{2a}, \bar{C}_{2b}, \bar{C}_{2c}, \bar{C}_{2d}, \bar{C}_{2e}, \bar{C}_{2f}$

$\bar{4}3m$ (T_d)

C_1 $\quad E$
C_2 $\quad \bar{E}$
C_3 $\quad C_{2x}, C_{2y}, C_{2z}, \bar{C}_{2x}, \bar{C}_{2y}, \bar{C}_{2z}$
C_4 $\quad C_{31}^-, C_{32}^-, C_{33}^-, C_{34}^-, C_{31}^+, C_{32}^+, C_{33}^+, C_{34}^+$
C_5 $\quad \bar{C}_{31}^-, \bar{C}_{32}^-, \bar{C}_{33}^-, \bar{C}_{34}^-, \bar{C}_{31}^+, \bar{C}_{32}^+, \bar{C}_{33}^+, \bar{C}_{34}^+$
C_6 $\quad S_{4x}^-, S_{4y}^-, S_{4z}^-, S_{4x}^+, S_{4y}^+, S_{4z}^+$
C_7 $\quad \bar{S}_{4x}^-, \bar{S}_{4y}^-, \bar{S}_{4z}^-, \bar{S}_{4x}^+, \bar{S}_{4y}^+, \bar{S}_{4z}^+$
C_8 $\quad \sigma_{da}, \sigma_{db}, \sigma_{dc}, \sigma_{dd}, \sigma_{de}, \sigma_{df}, \bar{\sigma}_{da}, \bar{\sigma}_{db}, \bar{\sigma}_{dc}, \bar{\sigma}_{dd}, \bar{\sigma}_{de}, \bar{\sigma}_{df}$

432 (O)		C_{4x}^{+}	$\bar{C}_{31}^{-}$	C_{2b}
	$\bar{4}3m$ (T_d)	S_{4x}^{-}	$\bar{C}_{31}^{-}$	σ_{db}
$\bar{E}_1$ Γ_6	$\bar{E}_1$ Γ_6	$\frac{1}{\sqrt{2}}\begin{pmatrix} 1 & -\mathrm{i} \\ -\mathrm{i} & 1 \end{pmatrix}$	$\frac{1}{2}\begin{pmatrix} -1+\mathrm{i} & 1-\mathrm{i} \\ -1-\mathrm{i} & -1-\mathrm{i} \end{pmatrix}$	$\frac{1}{\sqrt{2}}\begin{pmatrix} 0 & -1-\mathrm{i} \\ 1-\mathrm{i} & 0 \end{pmatrix}$
$\bar{E}_2$ Γ_7	$\bar{E}_2$ Γ_7	$\frac{1}{\sqrt{2}}\begin{pmatrix} -1 & \mathrm{i} \\ \mathrm{i} & -1 \end{pmatrix}$	$\frac{1}{2}\begin{pmatrix} -1+\mathrm{i} & 1-\mathrm{i} \\ -1-\mathrm{i} & -1-\mathrm{i} \end{pmatrix}$	$\frac{1}{\sqrt{2}}\begin{pmatrix} 0 & 1+\mathrm{i} \\ -1+\mathrm{i} & 0 \end{pmatrix}$
$\bar{F}$ Γ_8	$\bar{F}$ Γ_8	$\frac{1}{\sqrt{2}}\begin{pmatrix} \mu & -\mathrm{i}\mu \\ -\mathrm{i}\mu & \mu \end{pmatrix}$	$\frac{1}{2}\begin{pmatrix} (-1+\mathrm{i})\beta & (1-\mathrm{i})\beta \\ (-1-\mathrm{i})\beta & (-1-\mathrm{i})\beta \end{pmatrix}$	$\frac{1}{\sqrt{2}}\begin{pmatrix} 0 & (-1-\mathrm{i})\lambda \\ (1-\mathrm{i})\lambda & 0 \end{pmatrix}$

$$\beta = \begin{pmatrix} -\frac{1}{2} & \frac{1}{2}\sqrt{3} \\ -\frac{1}{2}\sqrt{3} & -\frac{1}{2} \end{pmatrix}; \quad \lambda = \begin{pmatrix} 1 & 0 \\ 0 & -1 \end{pmatrix}; \quad \mu = \begin{pmatrix} -\frac{1}{2} & -\frac{1}{2}\sqrt{3} \\ -\frac{1}{2}\sqrt{3} & \frac{1}{2} \end{pmatrix}.$$

$m3m = 432 \otimes \bar{1}$ $(O_h = O \otimes C_i)$.

Notes to Table 6.5

(i) The names of the point groups are given in the top left-hand blocks of the tables and, where appropriate, several point groups are tabulated together.

(ii) The characters of the single-valued reps, which were given in Table 2.2 are included for reference.

(iii) The matrix representatives for the generating elements are given for the degenerate double-valued reps. The matrices of the other elements can be found by using the group multiplication tables in Table 6.2.

(iv) The elements of the point groups can be identified by reference to Figs. 1.1–1.4 and Tables 1.2–1.6.

(v) The scheme used in Table 2.2 for labelling the single-valued reps of the point groups has been extended to cover the double-valued reps as well. Each double-valued point-group rep is labelled either by an extension of the Mulliken (1933) notation or by the Γ notation of Koster, Dimmock, Wheeler, and Statz (1963). In the extended Mulliken notation the symbols labelling the double-valued reps have a bar placed over them.

(vi) We have not given the character tables of those groups that are direct products of some other point group with $\bar{1}$ (C_i); the character tables of these direct product groups can be constructed as follows. If a group $\mathbf{G}'$ is given as a direct product of the form $\mathbf{G} \otimes \bar{1}$ then the reps of $\mathbf{G}'$ fall into pairs; each pair M_g and M_u arise out of a single rep M of $\mathbf{G}$ and the characters of M_g and M_u obey the following rules. If $R' = RI$ then for all $R \in \mathbf{G}$ the character of R in M_g and M_u is equal to the character of R in M; the character of R' in M_g is equal to the character of R in M, but the character of R' in M_u is minus the character of R in M. In the Γ notation of Koster, Dimmock, Wheeler, and Statz (1963), if $\Gamma \equiv M$ in $\mathbf{G}$, then $\Gamma^+ \equiv M_g$ and $\Gamma^- \equiv M_u$ in $\mathbf{G}'$.

(vii) The reality (see Definition 1.3.7) of the double-valued point-group reps is as follows:

non-degenerate, all characters real	first kind
degenerate, all characters real	second kind
non-degenerate, some characters complex degenerate, some characters complex	third kind

In Table 6.6 we give the compatibility tables between the double-valued representations $\mathscr{D}^j\{R(\alpha, \beta, \gamma)\}$ (j = half odd integer) of the 3-dimensional rotation group, $O(3)$, and the double-valued point-group reps. This table, like Table 2.7, is relevant to the study of the splitting of an energy level of a free atom, characterized by a half-odd-integer total angular momentum quantum number j, in the presence of an electrostatic field with the symmetry of any one of the crystallographic point groups

TABLE 6.6

Compatibility tables for double-valued reps of O(3) and point-group reps

j	$1\,(C_1)$
j	$(2j+1)\bar{A}$

j	$2\,(C_2)$ $m\,(C_{1h})$
j	$(j+\frac{1}{2})^1\bar{E} + (j+\frac{1}{2})^2\bar{E}$

j	$222\,(D_2)$ $mm2\,(C_{2v})$
j	$(j+\frac{1}{2})\bar{E}$

j	$4\,(C_4)$ $\bar{4}\,(S_4)$
$\frac{1}{2}$	${}^1\bar{E}_1 + {}^2\bar{E}_1$
$\frac{3}{2}$	${}^1\bar{E}_1 + {}^2\bar{E}_1 + {}^1\bar{E}_2 + {}^2\bar{E}_2$
$\frac{5}{2}$	${}^1\bar{E}_1 + {}^2\bar{E}_1 + 2{}^1\bar{E}_2 + 2{}^2\bar{E}_2$
$\frac{7}{2}$	$2{}^1\bar{E}_1 + 2{}^2\bar{E}_1 + 2{}^1\bar{E}_2 + 2{}^2\bar{E}_2$

$\mathscr{D}^{4n+\lambda} = 2n({}^1\bar{E}_1 + {}^2\bar{E}_1 + {}^1\bar{E}_2 + {}^2\bar{E}_2) + \mathscr{D}^\lambda$ $(\lambda = \frac{1}{2}, \frac{3}{2}, \frac{5}{2}, \frac{7}{2})$.

j	$422\,(D_4)$ $4mm\,(C_{4v})$ $\bar{4}2m\,(D_{2d})$
$\frac{1}{2}$	$\bar{E}_1$
$\frac{3}{2}$	$\bar{E}_1 + \bar{E}_2$
$\frac{5}{2}$	$\bar{E}_1 + 2\bar{E}_2$
$\frac{7}{2}$	$2\bar{E}_1 + 2\bar{E}_2$

$\mathscr{D}^{4n+\lambda} = 2n(\bar{E}_1 + \bar{E}_2) + \mathscr{D}^\lambda$ $(\lambda = \frac{1}{2}, \frac{3}{2}, \frac{5}{2}, \frac{7}{2})$.

j	$3\,(C_3)$
$\frac{1}{2}$	${}^1\bar{E} + {}^2\bar{E}$
$\frac{3}{2}$	$2\bar{A} + {}^1\bar{E} + {}^2\bar{E}$
$\frac{5}{2}$	$2\bar{A} + 2{}^1\bar{E} + 2{}^2\bar{E}$

$\mathscr{D}^{3n+\lambda} = 2n(\bar{A} + {}^1\bar{E} + {}^2\bar{E}) + \mathscr{D}^\lambda$ $(\lambda = \frac{1}{2}, \frac{3}{2}, \frac{5}{2})$.

j	$32\,(D_3)$ $3m\,(C_{3v})$
$\frac{1}{2}$	$\bar{E}_1$
$\frac{3}{2}$	${}^1\bar{E} + {}^2\bar{E} + \bar{E}_1$
$\frac{5}{2}$	${}^1\bar{E} + {}^2\bar{E} + 2\bar{E}_1$

$\mathscr{D}^{3n+\lambda} = n({}^1\bar{E} + {}^2\bar{E} + 2\bar{E}_1) + \mathscr{D}^\lambda$ $(\lambda = \frac{1}{2}, \frac{3}{2}, \frac{5}{2})$.

j	$6\,(C_6)$ $\bar{6}\,(C_{3h})$
$\frac{1}{2}$	${}^1\bar{E}_3 + {}^2\bar{E}_3$
$\frac{3}{2}$	${}^1\bar{E}_1 + {}^2\bar{E}_1 + {}^1\bar{E}_3 + {}^2\bar{E}_3$
$\frac{5}{2}$	${}^1\bar{E}_1 + {}^2\bar{E}_1 + {}^1\bar{E}_2 + {}^2\bar{E}_2 + {}^1\bar{E}_3 + {}^2\bar{E}_3$
$\frac{7}{2}$	${}^1\bar{E}_1 + {}^2\bar{E}_1 + 2{}^1\bar{E}_2 + 2{}^2\bar{E}_2 + {}^1\bar{E}_3 + {}^2\bar{E}_3$
$\frac{9}{2}$	$2{}^1\bar{E}_1 + 2{}^2\bar{E}_1 + 2{}^1\bar{E}_2 + 2{}^2\bar{E}_2 + {}^1\bar{E}_3 + {}^2\bar{E}_3$
$\frac{11}{2}$	$2{}^1\bar{E}_1 + 2{}^2\bar{E}_1 + 2{}^1\bar{E}_2 + 2{}^2\bar{E}_2 + 2{}^1\bar{E}_3 + 2{}^2\bar{E}_3$

$\mathscr{D}^{6n+\lambda} = 2n({}^1\bar{E}_1 + {}^2\bar{E}_1 + {}^1\bar{E}_2 + {}^2\bar{E}_2 + {}^1\bar{E}_3 + {}^2\bar{E}_3) + \mathscr{D}^\lambda$ $(\lambda = \frac{1}{2}, \frac{3}{2}, \ldots, \frac{11}{2})$.

j	$622\,(D_6)$ $6mm\,(C_{6v})$ $\bar{6}2m\,(D_{3h})$
$\frac{1}{2}$	$\bar{E}_1$
$\frac{3}{2}$	$\bar{E}_1 + \bar{E}_3$
$\frac{5}{2}$	$\bar{E}_1 + \bar{E}_2 + \bar{E}_3$
$\frac{7}{2}$	$\bar{E}_1 + 2\bar{E}_2 + \bar{E}_3$
$\frac{9}{2}$	$\bar{E}_1 + 2\bar{E}_2 + 2\bar{E}_3$
$\frac{11}{2}$	$2\bar{E}_1 + 2\bar{E}_2 + 2\bar{E}_3$

$\mathscr{D}^{6n+\lambda} = 2n(\bar{E}_1 + \bar{E}_2 + \bar{E}_3) + \mathscr{D}^\lambda$ $(\lambda = \frac{1}{2}, \frac{3}{2}, \ldots, \frac{11}{2})$.

j	23 (T)
$\frac{1}{2}$	$\bar{E}$
$\frac{3}{2}$	${}^1\bar{F} + {}^2\bar{F}$
$\frac{5}{2}$	$\bar{E} + {}^1\bar{F} + {}^2\bar{F}$

$\mathscr{D}^{3n+\lambda} = n(\bar{E} + {}^1\bar{F} + {}^2\bar{F}) + \mathscr{D}^{\lambda} \quad (\lambda = \frac{1}{2}, \frac{3}{2}, \frac{5}{2})$.

j	432 (O) $\bar{4}3m$ (T_d)
$\frac{1}{2}$	$\bar{E}_1$
$\frac{3}{2}$	$\bar{F}$
$\frac{5}{2}$	$\bar{E}_2 + \bar{F}$
$\frac{7}{2}$	$\bar{E}_1 + \bar{E}_2 + \bar{F}$
$\frac{9}{2}$	$\bar{E}_1 + 2\bar{F}$
$\frac{11}{2}$	$\bar{E}_1 + \bar{E}_2 + 2\bar{F}$
$\frac{13}{2}$	$\bar{E}_1 + 2\bar{E}_2 + 2\bar{F}$
$\frac{15}{2}$	$\bar{E}_1 + \bar{E}_2 + 3\bar{F}$
$\frac{17}{2}$	$2\bar{E}_1 + \bar{E}_2 + 3\bar{F}$
$\frac{19}{2}$	$2\bar{E}_1 + 2\bar{E}_2 + 3\bar{F}$
$\frac{21}{2}$	$\bar{E}_1 + 2\bar{E}_2 + 4\bar{F}$
$\frac{23}{2}$	$2\bar{E}_1 + 2\bar{E}_2 + 4\bar{F}$

$\mathscr{D}^{12n+\lambda} = 2n(\bar{E}_1 + \bar{E}_2 + 2\bar{F}) + \mathscr{D}^{\lambda} \quad (\lambda = \frac{1}{2}, \frac{3}{2}, \ldots, \frac{23}{2})$.

Notes to Table 6.6

(i) The double-valued point-group reps may be identified from Table 6.5.

(ii) The character in $\mathscr{D}^j\{R(\alpha, \beta, \gamma)\}$ where $R(\alpha, \beta, \gamma)$ is a proper rotation through $2\pi/n$ is given by Bethe (1929),

$$\chi^j(2\pi/n) = \frac{\sin\{(j + \frac{1}{2})(2\pi/n)\}}{\sin(\pi/n)}.$$

The parity of the spin functions can be taken to be $+1$ so that $\mathscr{D}^j(I) = +1$.

(iii) For those point groups that are direct products of some other point group **G** with $(E + I)$ one simply uses the table given above for **G** and adds a subscript g, since we have assumed that the parity of the spin functions is $+1$. These point groups are:

$\bar{1} = 1 \otimes \bar{1} \quad (C_i = C_1 \otimes C_i)$
$2/m = 2 \otimes \bar{1} \quad (C_{2h} = C_2 \otimes C_i)$
$mmm = 222 \otimes \bar{1} \quad (D_{2h} = D_2 \otimes C_i)$
$4/m = 4 \otimes \bar{1} \quad (C_{4h} = C_4 \otimes C_i)$
$\bar{3} = 3 \otimes \bar{1} \quad (C_{3i} = C_3 \otimes C_i)$
$\bar{3}m = 32 \otimes \bar{1} \quad (D_{3d} = D_3 \otimes C_i)$
$6/m = 6 \otimes \bar{1} \quad (C_{6h} = C_6 \otimes C_i)$
$4/mmm = 422 \otimes \bar{1} \quad (D_{4h} = D_4 \otimes C_i)$
$6/mmm = 622 \otimes \bar{1} \quad (D_{6h} = D_6 \otimes C_i)$
$m3 = 23 \otimes \bar{1} \quad (T_h = T \otimes C_i)$
$m3m = 432 \otimes \bar{1} \quad (O_h = O \otimes C_i)$

(iv) The compatibilities for higher j values can be found by using the formula at the foot of the appropriate entry in the table, see also Table 2.7.

G. In crystal field theory one can study various different situations and the relations between them; they are

(1) free atom, neglecting spin–orbit coupling,
(2) free atom, with spin–orbit coupling,
(3) strong spin–orbit coupling and weak crystal field,
(4) weak spin–orbit coupling and strong crystal field,
(5) crystal field (no spin–orbit coupling).

The procedure for determining the splitting of any given term for various relative strengths of spin–orbit coupling and of the crystalline field is well known and a particularly clear description is given in Chapter 4 of the book by Tinkham (1964). (1) and (2) are not confined to crystal field theory but belong to the general theory of fine structure. The sequence in which crystal field theory proceeds is usually either from (1) to (5) and then back to (4), or from (1) to (2) and then on to (3). Each of these two sequences involves two steps, one is the subduction of an irreducible representation of one group onto a smaller group and the other is the reduction of the inner Kronecker product of two representations. The inner Kronecker product will either involve two single-valued representations, if S in an integer, or one single-valued and one double-valued representation if S is half an odd integer. Tables of these Kronecker products are given for all the thirty-two crystallographic point groups by Koster, Dimmock, Wheeler, and Statz (1963) and Cracknell (1968*a*). In the first sequence the subduction is performed first.

A: (1) → (5), subduction of the rep $\mathscr{D}^L$ of $O(3)$ onto the point group **G**,

B: (5) → (4), reduction of the Kronecker product of the rep $\mathscr{D}^S$ and the reps of **G** obtained in A.

In the second sequence the reduction of the Kronecker product is performed first:

C: (1) → (2), reduction of the Kronecker product of the reps $\mathscr{D}^L$ and $\mathscr{D}^S$ of $O(3)$, corresponding to the orbital and spin angular momenta, to give the reps $\mathscr{D}^J$ of $O(3)$,

D: (2) → (3), subduction of the rep $\mathscr{D}^J$ of $O(3)$ onto the point group **G** (these reps will be single-valued or double-valued depending on the value of J).

The continuity of (3) and (4), from strong spin–orbit coupling and weak crystal field to weak spin–orbit coupling and strong crystal field is established by inspection of the splitting in the two cases and making the two fit together as the relative strengths of the spin–orbit coupling and the crystal field are varied.

6.2. Symmetry-adapted functions for double point groups

In Chapter 2 we considered the derivation of symmetry-adapted functions for the point groups; these functions were listed explicitly in Tables 2.4–2.6. Such a function is either an individual spherical harmonic $Y_l^m(\theta, \phi)$ or else some linear combination of spherical harmonics with various m values but with a fixed l value. These symmetry-adapted functions are of use in calculations on molecules and solids in the approximation in which spin is neglected. If one wishes to include a proper consideration of spin it is necessary, as we have already mentioned earlier in this chapter, to consider the double-valued reps of the point groups. The problem of determining the basis functions of these double-valued reps has been considered by various authors (Cracknell 1969*a*, Cracknell and Joshua 1970, Onodera and Okazaki 1966, Teleman and Glodeanu 1967).

The spherical harmonics $Y_l^m(\theta, \phi)$ defined in eqn (2.1.1) are eigenfunctions of the quantum-mechanical operators L^2, the square of the orbital angular momentum and L_z, the z component of the angular momentum. These functions correspond to eigenvalues $l(l+1)\hbar^2$ and $m\hbar$ for L^2 and L_z. The set of spherical harmonics $Y_l^m(\theta, \phi)$ $(-l \leqslant m \leqslant +l)$ for a given integer value of l form a vector which is a basis of the rep $\mathscr{D}^l\{R(\alpha, \beta, \gamma)\}$ of the 3-dimensional rotation group. In considering these $\mathscr{D}^l\{R(\alpha, \beta, \gamma)\}$, l is an integer. In this section we shall be concerned with reps $\mathscr{D}^j\{R(\alpha, \beta, \gamma)\}$ of the 3-dimensional rotation group where j is half an odd integer.

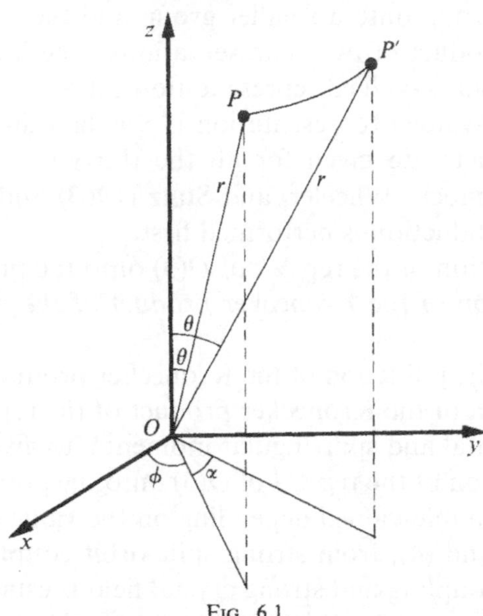

FIG. 6.1.

One example, that with $j = \frac{1}{2}$, was given in eqn. (6.1.1). The basis of one of these double-valued reps is not a vector whose components are the spherical harmonics $Y_l^m(\theta, \phi)$ but is a spinor. The theory of the deduction of the bases of the double-valued representations of the 3-dimensional rotation group and therefore also of the double-valued representations of the point groups is given by various authors (see, for example, Hamermesh (1962), Heine (1960), Tinkham (1964), Wigner (1959) (Chapters 15 and 20), while an extensive set of further references on the quantum mechanics of angular momentum is given by Biedenharn and van Dam (1965)).

A rotation in 3-dimensional space can be specified by three Euler angles α, β, and γ which we have already defined (see section 2.1) by first a rotation through α about the z-axis, secondly, a rotation through β about the y-axis and finally a rotation through γ about the z-axis. The restrictions on these angles are $0 \leqslant \alpha < 2\pi$, $0 \leqslant \beta \leqslant \pi$ and $0 \leqslant \gamma < 2\pi$ and all rotations are active, i.e. act on the points of space rather than the axes, and anticlockwise rotations are positive. It is therefore clear that the rotation $(\alpha, 0, 0)$ moves the point P, with spherical polar coordinates r, θ, and ϕ to P' with coordinates r, θ, and $(\phi + \alpha)$ (see Fig. 6.1) and the rotation $(0, 0, \gamma)$ moves

r, θ, ϕ to $r, \theta, (\phi + \gamma)$, where $(\phi + \alpha)$ and $(\phi + \gamma)$ are both mod 2π. Similarly, the rotation $(0, \beta, 0)$ moves the point $(r, \theta, 0)$ to $(r, \theta + \beta, 0)$ if $\theta + \beta \leqslant \pi$ and to $(r, 2\pi - \theta - \beta, \pi)$ if $\theta + \beta > \pi$.

In section 2.1 we saw that the fact that the spherical harmonics $Y_l^m(\theta, \phi)$ form a basis of the representation of dimension $(2l + 1)$ of the rotation group $O(3)$ means that

$$R(\alpha, \beta, \gamma) Y_l^m(\theta, \phi) = \sum_{n=-l}^{l} Y_l^n(\theta, \phi)\mathscr{D}^l\{R(\alpha, \beta, \gamma)\}_{nm}. \tag{6.2.1}$$

This equation gives the effect of $R(\alpha, \beta, \gamma)$ on the spherical harmonic $Y_l^m(\theta, \phi)$ and was much used in the production of symmetry-adapted functions by the application of projection operators. If $\beta = 0$ or π, $\mathscr{D}^l\{R(\alpha, \beta, \gamma)\}_{nm}$ takes the particularly simple forms

$$\mathscr{D}^l\{R(\alpha, 0, \gamma)\}_{nm} = \exp(-im\alpha)\exp(-im\gamma)\delta_{nm} \tag{2.1.10}$$

and

$$\mathscr{D}^l\{R(\alpha, \pi, \gamma)\}_{nm} = (-1)^l \exp(-im\alpha)\exp(im\gamma)\delta_{n,-m}. \tag{2.1.11}$$

The 2-dimensional representation $\mathscr{D}^{\frac{1}{2}}\{R(\alpha, \beta, \gamma)\}$ of the 3-dimensional rotation group may be determined as follows (Wigner (1959), Chapter 15). Suppose that $\mathbf{u}$ is any unitary unimodular 2×2 matrix, then it can be written as

$$\mathbf{u} = \begin{pmatrix} a & b \\ -b^* & a^* \end{pmatrix} \tag{6.2.2}$$

where $|a|^2 + |b|^2 = 1$. Given the Pauli matrices

$$\mathbf{s}_x = \begin{pmatrix} 0 & 1 \\ 1 & 0 \end{pmatrix}, \quad \mathbf{s}_y = \begin{pmatrix} 0 & \mathrm{i} \\ -\mathrm{i} & 0 \end{pmatrix} \quad \text{and} \quad \mathbf{s}_z = \begin{pmatrix} -1 & 0 \\ 0 & 1 \end{pmatrix}$$

we can express *any* Hermitean matrix of trace zero, say

$$\mathbf{h} = \begin{pmatrix} -z & x + \mathrm{i}y \\ x - \mathrm{i}y & z \end{pmatrix} \tag{6.2.3}$$

in the form $\mathbf{h} = x\mathbf{s}_x + y\mathbf{s}_y + z\mathbf{s}_z$ (x, y, z real). If $\mathbf{h}$ is now transformed by the unitary transformation $\mathbf{u}$, we obtain a matrix $\bar{\mathbf{h}} = \mathbf{uhu}^{-1}$, which is also Hermitean and of trace zero. Thus there will exist real numbers x', y', and z' such that $\bar{\mathbf{h}} = x'\mathbf{s}_x + y'\mathbf{s}_y + z'\mathbf{s}_z$ and

$$\bar{\mathbf{h}} = \begin{pmatrix} -z' & x' + \mathrm{i}y' \\ x' - \mathrm{i}y' & z' \end{pmatrix} = \begin{pmatrix} a & b \\ -b^* & a^* \end{pmatrix}\begin{pmatrix} -z & x + \mathrm{i}y \\ x - \mathrm{i}y & z \end{pmatrix}\begin{pmatrix} a^* & -b \\ b^* & a \end{pmatrix}. \tag{6.2.4}$$

It is therefore possible to express x', y', and z' in terms of x, y, and z,

$$\left.\begin{aligned} x' &= \tfrac{1}{2}(a^2 + a^{*2} - b^2 - b^{*2})x + \tfrac{1}{2}\mathrm{i}(a^2 - a^{*2} + b^2 - b^{*2})y \\ &\quad + (a^*b^* + ab)z, \\ y' &= \tfrac{1}{2}\mathrm{i}(a^{*2} - a^2 + b^2 - b^{*2})x + \tfrac{1}{2}(a^2 + a^{*2} + b^2 + b^{*2})y \\ &\quad + \mathrm{i}(a^*b^* - ab)z, \\ z' &= -(a^*b + ab^*)x + \mathrm{i}(a^*b - ab^*)y + (aa^* - bb^*)z, \end{aligned}\right\} \tag{6.2.5}$$

and, because det $\mathbf{h}$ = det $\bar{\mathbf{h}}$ we also have

$$x'^2 + y'^2 + z'^2 = x^2 + y^2 + z^2. \tag{6.2.6}$$

As a consequence of eqns. (6.2.5) and (6.2.6) it follows that the unitary transformation in eqn. (5.2.4) is equivalent to a rotation, R_u, of the point (x, y, z) to the position (x', y', z'). Since the determinant of the orthogonal transformation (6.2.5) is $+1$ it follows that every 2-dimensional unitary unimodular matrix $\mathbf{u}$ corresponds to a 3-dimensional rotation R_u. The relationship between $\mathbf{u}$ and R_u is given by eqn (6.2.4) or by eqn (6.2.5). As we have noted in the previous section of this chapter the correspondence between $\mathbf{u}$ and R_u is such that the matrices $\mathbf{u}$ form a double-valued representation of $SO(3)$, the group of pure rotations in three dimensions.

We now seek to determine explicit expressions for the matrix elements a and b of $\mathbf{u}$ explicitly in terms of α, β, and γ, the Euler angles of the rotation R_u to which $\mathbf{u}$ corresponds. We write

$$\mathscr{D}^{\frac{1}{2}}\{R(\alpha, \beta, \gamma)\} = \mathbf{u}_1(0, 0, \gamma)\mathbf{u}_2(0, \beta, 0)\mathbf{u}_1(\alpha, 0, 0). \tag{6.2.7}$$

The expression for the components of the vector OP' $(=(x', y', z'))$ that is produced from OP $(=(x, y, z))$ by the performance of the three rotations α, β, and γ can be shown to be

$$\begin{pmatrix} x' \\ y' \\ z' \end{pmatrix} = \begin{pmatrix} \cos\gamma\cos\beta\cos\alpha - \sin\gamma\sin\alpha & -\cos\gamma\cos\beta\sin\alpha - \sin\gamma\cos\alpha & \cos\gamma\sin\beta \\ \sin\gamma\cos\beta\cos\alpha + \cos\gamma\sin\alpha & -\sin\gamma\cos\beta\sin\alpha + \cos\gamma\cos\alpha & \sin\gamma\sin\beta \\ -\sin\beta\cos\alpha & \sin\beta\sin\alpha & \cos\beta \end{pmatrix} \begin{pmatrix} x \\ y \\ z \end{pmatrix}. \tag{6.2.8}$$

We therefore have to choose values of a and b which when substituted in eqn (6.2.5) produce eqn (6.2.8). By substituting in eqn (6.2.5) one can check that the values of a and b are

$$\left.\begin{aligned} a &= \mathrm{e}^{\frac{1}{2}\mathrm{i}\gamma}\cos\tfrac{1}{2}\beta\, \mathrm{e}^{\frac{1}{2}\mathrm{i}\alpha} \\ b &= \mathrm{e}^{\frac{1}{2}\mathrm{i}\gamma}\sin\tfrac{1}{2}\beta\, \mathrm{e}^{-\frac{1}{2}\mathrm{i}\alpha} \end{aligned}\right\} \tag{6.2.9}$$

so that

$$\mathbf{u}_1(\alpha, 0, 0) = \begin{pmatrix} e^{\frac{1}{2}i\alpha} & 0 \\ 0 & e^{-\frac{1}{2}i\alpha} \end{pmatrix} \tag{6.2.10}$$

$$\mathbf{u}_1(0, 0, \gamma) = \begin{pmatrix} e^{\frac{1}{2}i\gamma} & 0 \\ 0 & e^{-\frac{1}{2}i\gamma} \end{pmatrix} \tag{6.2.11}$$

and

$$\mathbf{u}_2(0, \beta, 0) = \begin{pmatrix} \cos\frac{1}{2}\beta & \sin\frac{1}{2}\beta \\ -\sin\frac{1}{2}\beta & \cos\frac{1}{2}\beta \end{pmatrix}. \tag{6.2.12}$$

Therefore, substituting in eqn (6.2.2) or (6.2.7) we have

$$\mathscr{D}^{\frac{1}{2}}\{R(\alpha, \beta, \gamma)\} = \pm\mathbf{u}\{R(\alpha, \beta, \gamma)\} = \pm\begin{pmatrix} e^{\frac{1}{2}i\gamma}\cos\frac{1}{2}\beta\, e^{\frac{1}{2}i\alpha} & e^{\frac{1}{2}i\gamma}\sin\frac{1}{2}\beta\, e^{-\frac{1}{2}i\alpha} \\ -e^{-\frac{1}{2}i\gamma}\sin\frac{1}{2}\beta\, e^{\frac{1}{2}i\alpha} & e^{-\frac{1}{2}i\gamma}\cos\frac{1}{2}\beta\, e^{-\frac{1}{2}i\alpha} \end{pmatrix}, \tag{6.2.13}$$

the $\pm$ sign appearing because if the signs of both a and b are changed simultaneously eqn (6.2.5) remains unaltered. From here it is not difficult to show that the mapping $\mathbf{u} \to R_u$ defined by eqns (6.2.2) and (6.2.5) is a 2 to 1 homomorphism of $SU(2)$ onto $SO(3)$ (Wigner 1959 (Chapter 15)). That is, there are two matrices $\mathbf{u}$ and $-\mathbf{u}$ corresponding to any given rotation R_u and the matrices $\mathbf{u}$ form a double-valued representation of $SO(3)$. This means that if $\mathbf{u}_1\mathbf{u}_2 = \mathbf{u}$ then $R_{u_1}R_{u_2} = R_u$ but that if $R_{u_1}R_{u_2} = R_u$ then one may have either

$$\mathbf{u}_1\mathbf{u}_2 = +\mathbf{u} \tag{6.2.14}$$

or

$$\mathbf{u}_1\mathbf{u}_2 = -\mathbf{u}. \tag{6.2.15}$$

Which one of eqns (6.2.14) and (6.2.15) in fact holds can only be determined by actually multiplying together the matrices $\mathbf{u}_1$ and $\mathbf{u}_2$. If we take all the sets of values α, β, and γ of the elements of $SO(3)$ and always take the $+$ sign in eqn. (6.2.13) we shall find that sooner or later a product of $\mathbf{u}_1$ and $\mathbf{u}_2$ that follows eqn (6.2.15) will arise and consequently it is found that it is only possible to satisfy the condition of closure by including for each (α, β, γ) both the matrices given in eqn (6.2.13).

To determine symmetry adapted functions for the double-valued reps of the point groups we make use of group operators. In section 2.2 we described the use of these operators for determining basis functions, or symmetry-adapted functions, for the single-valued reps of the point groups. Suppose that R is an element of a point group $\mathbf{G}$ and that $\phi(x, y, z, s)$ is some function of the space and spin coordinates. Then if the rep D^i has dimension equal to d_i we defined in eqn (2.2.2) the operator

$$W^i_{ts} = \frac{d_i}{|\mathbf{G}|}\sum_{R\in\mathbf{G}} \mathbf{D}^i(R)^*_{ts}R. \tag{2.2.2}$$

The operator W^i_{ts} can then be used as a result of Theorem 2.2.1 to determine symmetry-adapted functions for the rep D^i. This theorem states that if ϕ is an arbitrary function such that $W^i_{ss}\phi \neq 0$ (where s is any given number in the range 1 to d_i) then the functions $W^i_{ts}\phi = \phi^i_t(t = 1, 2, \ldots, d_i)$ form a basis for the rep D^i. If $\phi(x, y, z, s)$ is assumed to be of the form of a product of a space function and a spin function, i.e.

$$\phi(x, y, z, s) = \phi_1(x, y, z)\phi_2(s) \tag{6.2.16}$$

then the operator R can also be written as a product of two operators

$$R = R_r \,.\, R_s, \tag{6.2.17}$$

where R_r acts only on the space function $\phi_1(x, y, z)$ and R_s acts only on the spin function $\phi_2(s)$. Therefore

$$R\phi(x, y, z, s) = \{R_r\phi_1(x, y, z)\}\{R_s\phi_2(s)\}. \tag{6.2.18}$$

The space function $\phi_1(x, y, z)$ is, as in Chapter 2, either an individual spherical harmonic or a linear combination of spherical harmonics while the spin function $\phi_2(s)$ is $u_- = \begin{bmatrix}1\\0\end{bmatrix}$ if the spin coordinate s is $-\frac{1}{2}$ and is $u_+ = \begin{bmatrix}0\\1\end{bmatrix}$ if s is $+\frac{1}{2}$. The effect of the operator R_r, one of the point-group operations, on the spherical harmonic $Y_l^m(\theta, \phi)$ was considered in Chapter 2,

$$R_r Y_l^m(\theta, \phi) = \sum_{m'=-l}^{l} Y_l^{m'}(\theta, \phi)\mathscr{D}^l\{R(\alpha, \beta, \gamma)\}_{m'm} \tag{6.2.19}$$

from eqn (2.1.3), it being remembered of course that R_r is a function-space operator in this equation. It therefore remains, if we are to use the operator W^i_{tu}, to determine the effect of the operator R_s on $\phi_2(s)$. To do this we need to study the theory of spinors themselves in a little more detail (see, for example, Bade and Jehle (1953), Temple (1960), Wigner (1959)). If R is the matrix corresponding to a rotation in 3-dimensional space then R has only one real eigenvalue and one real eigenvector; this eigenvector is directed along the axis of rotation. If one tries to determine the eigenvalues and eigenvectors of R by simply solving $R\mathbf{x} = \lambda\mathbf{x}$ one can find two other, complex, eigenvalues and eigenvectors. The complex eigenvectors form a complex conjugate pair and they are called *isotropic vectors*. The absolute magnitude of an isotropic vector is zero. For example, using as R the matrix corresponding to $(\alpha, 0, 0)$ in eqn (6.2.8) we have

$$\begin{pmatrix}\cos\alpha & -\sin\alpha & 0\\ \sin\alpha & \cos\alpha & 0\\ 0 & 0 & 1\end{pmatrix}\begin{pmatrix}x\\y\\z\end{pmatrix} = \lambda\begin{pmatrix}x\\y\\z\end{pmatrix} \tag{6.2.20}$$

which on solution gives eigenvalues $\lambda = 1$, $e^{i\alpha}$ and $e^{-i\alpha}$ and the corresponding eigenvectors are (0, 0, 1), (1, −i, 0), and (1, i, 0) respectively. The magnitude over the real numbers of each of the pair of complex conjugate eigenvectors of R is clearly equal to zero. If we write an isotropic vector $\mathbf{v}$ $(=(v_x, v_y, v_z))$ then

$$v_x^2 + v_y^2 + v_z^2 = 0 \tag{6.2.21}$$

and we write the components of $\mathbf{v}$ in terms of a parameter ξ, where

$$\left.\begin{aligned} \xi &= \frac{v_x + iv_y}{v_z} \\ \xi^{-1} &= \frac{-v_x + iv_y}{v_z} \end{aligned}\right\} \tag{6.2.22}$$

and ξ is called the *isotropic parameter* (ξ is a complex number so that there is no difficulty about $v_z = 0$, which corresponds in the complex plane to the point $\xi = \infty$). If we perform a rotation operation R on the isotropic vector $\mathbf{v}$ specified by ξ it will be turned into an isotropic vector $\mathbf{v}'$ specified by ξ'. To obtain an expression for ξ' in terms of ξ it is convenient to regard the rotation R as the successive reflection in two planes M and N (Temple 1960). For the reflection of $\mathbf{v}$ in the plane M the vector $\bar{\mathbf{v}}$ is produced,

$$\bar{\mathbf{v}} = \mathbf{v} - 2\mathbf{m}(\mathbf{v} \, . \, \mathbf{m}) \tag{6.2.23}$$

where $\mathbf{m}$ is the (real) unit vector normal to M and from this we find

$$\bar{\xi} = \frac{-m_z\xi + (m_x + im_y)}{m_z + (m_x - im_y)\xi} \tag{6.2.24}$$

$$= \frac{\alpha\xi + \beta}{\beta^*\xi - \alpha^*} \tag{6.2.25}$$

where $\alpha = -m_z$ and $\beta = m_x + im_y$. For the second reflection, in the plane N, $\bar{\mathbf{v}}$ is reflected to $\mathbf{v}'$ and similarly

$$\xi' = \frac{\gamma\bar{\xi} + \delta}{\delta^*\bar{\xi} - \gamma^*} \tag{6.2.26}$$

where $\gamma = -n_z$ and $\delta = n_x + in_y$ and $\mathbf{n}$ is the unit vector normal to N. We may rewrite ξ' in terms of ξ,

$$\xi' = \frac{\rho\xi + \sigma}{-\sigma^*\xi + \rho^*} \tag{6.2.27}$$

where

$$\left.\begin{aligned} \rho &= \gamma\alpha + \delta\beta^* \\ \sigma &= \gamma\beta - \delta\alpha^* \end{aligned}\right\}. \tag{6.2.28}$$

If we write the isotropic parameter ξ as a quotient

$$\xi = \frac{\xi_1}{\xi_0} \tag{6.2.29}$$

eqn (6.2.27), which gives the effect on ξ of the rotation R, can be replaced by

$$\left.\begin{aligned} \xi_1' &= \rho\xi_1 + \sigma\xi_0, \\ \xi_0' &= -\sigma^*\xi_1 + \rho^*\xi_0 \end{aligned}\right\}. \tag{6.2.30}$$

A spinor is then defined as an ordered pair of numbers $\begin{bmatrix}\xi_1\\ \xi_0\end{bmatrix}$ which have the property that they transform under R according to eqn. (6.2.30), or under a reflection according to eqn (6.2.25). By writing the space-inversion operation I as $\sigma_x\sigma_y\sigma_z$ and using eqn (6.2.25) three times one can prove that $I\xi = \pm\xi$.

The 2×2 matrix obtained when eqn. (6.2.30) is written in matrix form, i.e.

$$\begin{bmatrix}\xi_1'\\ \xi_0'\end{bmatrix} = \begin{pmatrix}\rho & \sigma\\ -\sigma^* & \rho^*\end{pmatrix}\begin{bmatrix}\xi_1\\ \xi_0\end{bmatrix} \tag{6.2.31}$$

is in fact the same as the matrix $\mathscr{D}^{\frac{1}{2}}\{R(\alpha, \beta, \gamma)\}$ given in eqn. (6.2.13). In other words if R_s acts on the spinor $\begin{bmatrix}\xi_1\\ \xi_0\end{bmatrix}$ the effect is to produce $\begin{bmatrix}\xi_1'\\ \xi_0'\end{bmatrix}$ which we may write as $R_s\begin{bmatrix}\xi_1\\ \xi_0\end{bmatrix}$,

$$R_s\begin{bmatrix}\xi_1\\ \xi_0\end{bmatrix} = \mathscr{D}^{\frac{1}{2}}\{R(\alpha, \beta, \gamma)\}\begin{bmatrix}\xi_1\\ \xi_0\end{bmatrix} \tag{6.2.32}$$

where $\mathscr{D}^{\frac{1}{2}}\{R(\alpha, \beta, \gamma)\}$ is given in Table 6.1. We can illustrate that the matrix in eqn (6.2.31) is $\mathscr{D}^{\frac{1}{2}}\{R(\alpha, \beta, \gamma)\}$ with an example.

The rotation C_{4z}^+ can be written as the product of two reflections,

$$C_{4z}^+ = \sigma_{db}\sigma_y, \tag{6.2.33}$$

see Table 1.5. Thus $\mathbf{m} = (0, 1, 0)$ and $\mathbf{n} = (-1, 1, 0)/\sqrt{2}$ and therefore

$$\left.\begin{aligned} \alpha &= 0, & \beta &= \mathrm{i} \\ \gamma &= 0, & \delta &= (-1 + \mathrm{i})/\sqrt{2} \end{aligned}\right\} \tag{6.2.34}$$

and from eqn (6.2.28)

$$\left.\begin{aligned} \rho &= (1 + \mathrm{i})/\sqrt{2} \\ \sigma &= 0 \end{aligned}\right\} \tag{6.2.35}$$

Thus the matrix in eqn (6.2.31) is $\frac{1}{\sqrt{2}}\begin{pmatrix}1 + \mathrm{i} & 0\\ 0 & 1 - \mathrm{i}\end{pmatrix}$ which is, from Table 6.1, just the matrix $\mathscr{D}^{\frac{1}{2}}(C_{4z}^+)$. The double-valued properties arise by noting that ρ and σ

can be replaced by $-\rho$ and $-\sigma$, by choosing, for example, $\mathbf{m} = (0, -1, 0)$ and $\mathbf{n} = (-1, 1, 0)/\sqrt{2}$. We can therefore simply use eqn (6.2.32) to determine the effect of R_s on the two spinors $u_- = \begin{bmatrix}1\\0\end{bmatrix}$ and $u_+ = \begin{bmatrix}0\\1\end{bmatrix}$. If we write

$$\mathscr{D}^{\frac{1}{2}}\{R(\alpha, \beta, \gamma)\} = \begin{pmatrix} a & b \\ -b^* & a^* \end{pmatrix} \tag{6.2.36}$$

then

$$R_s u_- = \begin{pmatrix} a & b \\ -b^* & a^* \end{pmatrix}\begin{bmatrix}1\\0\end{bmatrix} = \begin{bmatrix}a\\-b^*\end{bmatrix} = au_- - b^* u_+ \tag{6.2.37}$$

and

$$R_s u_+ = \begin{pmatrix} a & b \\ -b^* & a^* \end{pmatrix}\begin{bmatrix}0\\1\end{bmatrix} = \begin{bmatrix}b\\a^*\end{bmatrix} = bu_- + a^* u_+. \tag{6.2.38}$$

We can think of the two spinors u_- and u_+ as the components of a column vector and therefore

$$\begin{aligned} R_s\begin{pmatrix}u_-\\u_+\end{pmatrix} &= \begin{pmatrix}au_- - b^*u_+\\ bu_- + a^*u_+\end{pmatrix} = \begin{pmatrix} a & -b^* \\ b & a^* \end{pmatrix}\begin{pmatrix}u_-\\u_+\end{pmatrix} \\ &= \mathscr{D}^{\frac{1}{2}}\{R(\alpha, \beta, \gamma)\}^{\mathrm{T}}\begin{pmatrix}u_-\\u_+\end{pmatrix}. \end{aligned} \tag{6.2.39}$$

Or if we have a row vector $\langle u_-, u_+|$ then

$$\begin{aligned} R_s\langle u_-, u_+| &= \langle au_- - b^*u_+, bu_- + a^*u_+| \\ &= \langle u_-, u_+| \begin{pmatrix} a & b \\ -b^* & a^* \end{pmatrix} \\ &= \langle u_-, u_+|\, \mathscr{D}^{\frac{1}{2}}\{R(\alpha, \beta, \gamma)\}. \end{aligned} \tag{6.2.40}$$

Equations (6.2.37) and (6.2.38) or (6.2.39) or (6.2.40) can be used to give the effect of the point-group operation R on the spin function $\phi_2(s)$. In Table 6.7 we give the effect of the point-group operations on the two spin functions $u_-(= \phi_2(-\frac{1}{2}))$ and $u_+(= \phi_2(+\frac{1}{2}))$.

Thus with the knowledge of the transformation properties of the spherical harmonics $Y_l^m(\theta, \phi)$ which were studied in detail in Chapter 2, together with the transformation properties of the spin functions u_- and u_+ given in Table 6.7 it is possible to use Theorem 2.2.1 to determine symmetry-adapted functions for the double point groups. For the non-cubic point groups the use of this theorem and the group operator W^i_{tu} is quite straightforward and the results are given in Table 6.8; the actual

TABLE 6.7

The effect of the point-group operations on the spin functions

($\varepsilon = \exp(\pi i/6)$)

	u_-	u_+
E, I	u_-	u_+
C_{2x}, σ_x	$-iu_+$	$-iu_-$
C_{2y}, σ_y	u_+	$-u_-$
C_{2z}, σ_z	$-iu_-$	iu_+
C_{31}^+, S_{61}^-	$\frac{1}{2}((1+i)u_- + (-1-i)u_+)$	$\frac{1}{2}((1-i)u_- + (1-i)u_+)$
C_{32}^+, S_{62}^-	$\frac{1}{2}((1+i)u_- + (1+i)u_+)$	$\frac{1}{2}((-1+i)u_- + (1-i)u_+)$
C_{33}^+, S_{63}^-	$\frac{1}{2}((1-i)u_- + (1-i)u_+)$	$\frac{1}{2}((-1-i)u_- + (1+i)u_+)$
C_{34}^+, S_{64}^-	$\frac{1}{2}((1-i)u_- + (-1+i)u_+)$	$\frac{1}{2}((1+i)u_- + (1+i)u_+)$
C_{31}^-, S_{61}^+	$\frac{1}{2}((1-i)u_- + (1+i)u_+)$	$\frac{1}{2}((-1+i)u_- + (1+i)u_+)$
C_{32}^-, S_{62}^+	$\frac{1}{2}((1-i)u_- + (-1-i)u_+)$	$\frac{1}{2}((1-i)u_- - (1+i)u_+)$
C_{33}^-, S_{63}^+	$\frac{1}{2}((1+i)u_- + (-1+i)u_+)$	$\frac{1}{2}((1+i)u_- + (1-i)u_+)$
C_{34}^-, S_{64}^+	$\frac{1}{2}((1+i)u_- + (1-1)u_+)$	$\frac{1}{2}((-1-i)u_- + (1-i)u_+)$

	u_-	u_+		u_-	u_+
C_{4x}^+, S_{4x}^-	$(u_- - iu_+)/\sqrt{2}$	$(-iu_- + u_+)/\sqrt{2}$	E, I	u_-	u_+
C_{4y}^+, S_{4y}^-	$(u_- - u_+)/\sqrt{2}$	$(u_- + u_+)/\sqrt{2}$	C_6^+, S_3^-	εu_-	$\varepsilon^* u_+$
C_{4z}^+, S_{4z}^-	$(1+i)u_-/\sqrt{2}$	$(1-i)u_+/\sqrt{2}$	C_3^+, S_6^-	$\varepsilon^2 u_-$	$\varepsilon^{*2} u_+$
C_{4x}^-, S_{4x}^+	$(u_- + iu_+)/\sqrt{2}$	$(iu_- + u_+)/\sqrt{2}$	C_2, σ_h	$-iu_-$	iu_+
C_{4y}^-, S_{4y}^+	$(u_- + u_+)/\sqrt{2}$	$(-u_- + u_+)/\sqrt{2}$	C_3^-, S_6^+	$\varepsilon^{*2} u_-$	$\varepsilon^2 u_+$
C_{4z}^-, S_{4z}^+	$(1-i)u_-/\sqrt{2}$	$(1+i)u_+/\sqrt{2}$	C_6^-, S_3^+	$\varepsilon^* u_-$	εu_+
C_{2a}, σ_{da}	$(1+i)u_+/\sqrt{2}$	$(-1+i)u_-/\sqrt{2}$	C_{21}', σ_{d1}	$-iu_+$	$-iu_-$
C_{2b}, σ_{db}	$(1-i)u_+/\sqrt{2}$	$(-1-i)u_-/\sqrt{2}$	C_{22}', σ_{d2}	$-\varepsilon^* u_+$	εu_-
C_{2c}, σ_{dc}	$(-iu_- + iu_+)/\sqrt{2}$	$(iu_- + iu_+)/\sqrt{2}$	C_{23}', σ_{d3}	εu_+	$-\varepsilon^* u_-$
C_{2d}, σ_{dd}	$(-iu_- + u_+)/\sqrt{2}$	$(-u_- + iu_+)/\sqrt{2}$	C_{21}'', σ_{v1}	u_+	$-u_-$
C_{2e}, σ_{de}	$(-iu_- - iu_+)/\sqrt{2}$	$(-iu_- + iu_+)/\sqrt{2}$	C_{22}'', σ_{v2}	$\varepsilon^2 u_+$	$-\varepsilon^{*2} u_-$
C_{2f}, σ_{df}	$(iu_- + u_+)/\sqrt{2}$	$(-u_- - iu_+)/\sqrt{2}$	C_{23}'', σ_{v3}	$\varepsilon^{*2} u_+$	$-\varepsilon^2 u_-$

matrix representatives can be identified from Tables 6.2 and 6.5. This table may be compared with Tables 2.4 and 2.5, which give the symmetry-adapted functions for the single-valued reps of the non-cubic point groups.

For the cubic groups, however, the direct use of the group operators is cumbersome. This is because some of the members of a cubic point group have their second Euler angle β not equal to 0 or π but equal to $\frac{1}{2}\pi$, see Table 2.1, and the matrix representatives $d^l_{m'm}(\frac{1}{2}\pi)$ are very complicated (Bradley 1961). For the single-valued representations of the cubic point groups this difficulty manifests itself in the tables of symmetry-adapted functions in Table 2.6 which are relatively more complicated than Tables 2.4 and 2.5 for non-cubic point groups. Since tables of symmetry-adapted

TABLE 6.8
Symmetry-adapted functions for the non-cubic double point groups

Point group			Rep		Symmetry-adapted function	m	m mod
1	1	C_1	$\bar{A}$	Γ_2	$u_- Y_l^m$	0	1
					$u_+ Y_l^m$	0	1
2	$\bar{1}$	C_i	$\bar{A}_g$	Γ_2^+	$u_- Y_l^m$	0	1
					$u_+ Y_l^m$	0	1
			$\bar{A}_u$	Γ_2^-	$u_- Y_l^m$	0	1
					$u_+ Y_l^m$	0	1
3	2	C_2	${}^1\bar{E}$	Γ_3	$u_- Y_l^m$	1	2
					$u_+ Y_l^m$	0	2
			${}^2\bar{E}$	Γ_4	$u_- Y_l^m$	0	2
					$u_+ Y_l^m$	1	2
4	m	C_{1h}	${}^1\bar{E}$	Γ_3	$u_- Y_l^m$ $(l + m)$ odd		2
					$u_+ Y_l^m$ $(l + m)$ even		2
			${}^2\bar{E}$	Γ_4	$u_- Y_l^m$ $(l + m)$ even		2
					$u_+ Y_l^m$ $(l + m)$ odd		2
5	$2/m$	C_{2h}			direct product $= 2 \otimes \bar{1}$		
6	222	D_2	$\bar{E}$	Γ_5	l even:		
					$(u_- Y_l^m, +iu_+ Y_l^{-m})$	0	2
					$(u_- Y_l^m, -iu_+ Y_l^{-m})$	1	2
					l odd:		
					replace u_+ by $-u_+$ above		
7	$mm2$	C_{2v}			l even and l odd of $mm2$ are same as l even of 222 (D_2)		
8	mmm	D_{2h}			direct product $= 222 \otimes \bar{1}$		
9	4	C_4	${}^1\bar{E}_1$	Γ_6	$u_- Y_l^m$	0	4
					$u_+ Y_l^m$	3	4
			${}^2\bar{E}_1$	Γ_5	$u_- Y_l^m$	1	4
					$u_+ Y_l^m$	0	4
			${}^1\bar{E}_2$	Γ_8	$u_- Y_l^m$	2	4
					$u_+ Y_l^m$	1	4
			${}^2\bar{E}_2$	Γ_7	$u_- Y_l^m$	3	4
					$u_+ Y_l^m$	2	4
10	$\bar{4}$	S_4			l even:		
					exactly as $4(C_4)$		
					l odd:		
			${}^1\bar{E}_1$	Γ_6	$u_- Y_l^m$	2	4
					$u_+ Y_l^m$	1	4
			${}^2\bar{E}_1$	Γ_5	$u_- Y_l^m$	3	4
					$u_+ Y_l^m$	2	4
			${}^1\bar{E}_2$	Γ_8	$u_- Y_l^m$	0	4
					$u_+ Y_l^m$	3	4
			${}^2\bar{E}_2$	Γ_7	$u_- Y_l^m$	1	4
					$u_+ Y_l^m$	0	4
11	$4/m$	C_{4h}			direct product $= 4 \otimes \bar{1}$		
12	422	D_4			l even:		
			$\bar{E}_1$	Γ_6	$\{(u_- Y_l^m - iu_+ Y_l^{-m}), +i(u_- Y_l^m + iu_+ Y_l^{-m})\}$	0	4
					$\{(u_- Y_l^m - iu_+ Y_l^{-m}), -i(u_- Y_l^m + iu_+ Y_l^{-m})\}$	1	4
			$\bar{E}_2$	Γ_7	$\{(u_- Y_l^m + iu_+ Y_l^{-m}), -i(u_- Y_l^m - iu_+ Y_l^{-m})\}$	2	4
					$\{(u_- Y_l^m + iu_+ Y_l^{-m}), +i(u_- Y_l^m - iu_+ Y_l^{-m})\}$	3	4
					l odd:		
					replace u_+ by $-u_+$ above		

Point group			Rep		Symmetry-adapted function	m	m mod
13	$4mm$	C_{4v}			l even and l odd of $4mm$ are same as l even of 422 (D_4)		
14	$\bar{4}2m$	D_{2d}			l even:		
					same as 422 (D_4)		
					l odd:		
			$\bar{E}_1$	Γ_6	$\{(u_- Y_l^m + \mathrm{i}u_+ Y_l^{-m}), +\mathrm{i}(u_- Y_l^m - \mathrm{i}u_+ Y_l^{-m})\}$	2	4
					$\{(u_- Y_l^m + \mathrm{i}u_+ Y_l^{-m}), -\mathrm{i}(u_- Y_l^m - \mathrm{i}u_+ Y_l^{-m})\}$	3	4
			$\bar{E}_2$	Γ_7	$\{(u_- Y_l^m - \mathrm{i}u_+ Y_l^{-m}), -\mathrm{i}(u_- Y_l^m + \mathrm{i}u_+ Y_l^{-m})\}$	0	4
					$\{(u_- Y_l^m - \mathrm{i}u_+ Y_l^{-m}), +\mathrm{i}(u_- Y_l^m + \mathrm{i}u_+ Y_l^{-m})\}$	1	4
15	$4/mmm$	D_{4h}			direct product $= 422 \otimes \bar{1}$		
16	3	C_3	$\bar{A}$	Γ_6	$u_- Y_l^m$	2	3
					$u_+ Y_l^m$	1	3
			$^1\bar{E}$	Γ_4	$u_- Y_l^m$	1	3
					$u_+ Y_l^m$	0	3
			$^2\bar{E}$	Γ_5	$u_- Y_l^m$	0	3
					$u_+ Y_l^m$	2	3
17	$\bar{3}$	C_{3i}			direct product $= 3 \otimes \bar{1}$		
18	32	D_3	$^1\bar{E}$	Γ_5	$u_- Y_l^m - (-1)^{l+m} u_+ Y_l^{-m}$	2	3
			$^2\bar{E}$	Γ_6	$u_- Y_l^m + (-1)^{l+m} u_+ Y_l^{-m}$	2	3
			$\bar{E}_1$	Γ_4	$\{(u_- Y_l^m - (-1)^{l+m} u_+ Y_l^{-m}), +\mathrm{i}(u_- Y_l^m + (-1)^{l+m} u_+ Y_l^{-m})\}$	0	3
					$\{(u_- Y_l^m - (-1)^{l+m} u_+ Y_l^{-m}), -\mathrm{i}(u_- Y_l^m + (-1)^{l+m} u_+ Y_l^{-m})\}$	1	3
19	$3m$	C_{3v}			as 32, but replace $(-1)^{l+m}$ by $(-1)^m$		
20	$\bar{3}m$	D_{3d}			direct product $= 32 \otimes \bar{1}$		
21	6	C_6	$^1\bar{E}_1$	Γ_{11}	$u_- Y_l^m$	5	6
					$u_+ Y_l^m$	4	6
			$^2\bar{E}_1$	Γ_{12}	$u_- Y_l^m$	2	6
					$u_+ Y_l^m$	1	6
			$^1\bar{E}_2$	Γ_{10}	$u_- Y_l^m$	3	6
					$u_+ Y_l^m$	2	6
			$^2\bar{E}_2$	Γ_9	$u_- Y_l^m$	4	6
					$u_+ Y_l^m$	3	6
			$^1\bar{E}_3$	Γ_7	$u_- Y_l^m$	1	6
					$u_+ Y_l^m$	0	6
			$^2\bar{E}_3$	Γ_8	$u_- Y_l^m$	0	6
					$u_+ Y_l^m$	5	6
22	$\bar{6}$	C_{3h}			l even:		
					exactly as 6 (C_6)		
					l odd:		
			$^1\bar{E}_1$	Γ_{11}	$u_- Y_l^m$	2	6
					$u_+ Y_l^m$	1	6
			$^2\bar{E}_1$	Γ_{12}	$u_- Y_l^m$	5	6
					$u_+ Y_l^m$	4	6
			$^1\bar{E}_2$	Γ_{10}	$u_- Y_l^m$	0	6
					$u_+ Y_l^m$	5	6
			$^2\bar{E}_2$	Γ_9	$u_- Y_l^m$	1	6
					$u_+ Y_l^m$	0	6
			$^1\bar{E}_3$	Γ_7	$u_- Y_l^m$	4	6
					$u_+ Y_l^m$	3	6
			$^2\bar{E}_3$	Γ_8	$u_- Y_l^m$	3	6
					$u_+ Y_l^m$	2	6
23	$6/m$	C_{6h}			direct product $= 6 \otimes \bar{1}$		

Point group			Rep		Symmetry-adapted function	m	m mod
24	622	D_6	$\bar{E}_1$	Γ_7	$\{u_- Y_l^m, (-1)^{l+1} u_+ Y_l^{-m}\}$	0	6
					$\{u_- Y_l^m, (-1)^l u_+ Y_l^{-m}\}$	5	6
			$\bar{E}_2$	Γ_8	$\{u_- Y_l^m, (-1)^{l+1} u_+ Y_l^{-m}\}$	4	6
					$\{u_- Y_l^m, (-1)^l u_+ Y_l^{-m}\}$	3	6
			$\bar{E}_3$	Γ_9	$\{u_- Y_l^m, (-1)^{l+1} u_+ Y_l^m\}$	2	6
					$\{u_- Y_l^{-m}, (-1)^l u_+ Y_l^m\}$	1	6
25	6*mm*	C_{6v}			as 622 (D_6) except that factor $(-1)^l$ is to be removed throughout		
26	$\bar{6}2m$	D_{3h}	$\bar{E}_1$	Γ_7	$\{(-1)^{l+1} u_+ Y_l^{-m}, u_- Y_l^m\}$	4	6
					$\{(-1)^l u_- Y_l^{-m}, u_+ Y_l^m\}$	3	6
			$\bar{E}_2$	Γ_8	$\{(-1)^{l+1} u_+ Y_l^{-m}, u_- Y_l^m)$	0	6
					$\{(-1)^l u_- Y_l^{-m}, u_+ Y_l^m\}$	5	6
			$\bar{E}_3$	Γ_9	$\{(-1)^l u_+ Y_l^m, u_- Y_l^{-m}\}$	1	6
					$\{(-1)^{l+1} u_- Y_l^m, u_+ Y_l^{-m}\}$	2	6
27	6/*mmm*	D_{6h}			direct product $= 622 \otimes \bar{1}$		

Notes to Table 6.8

(i) In the first three columns we identify a non-cubic point group **G** and in the next two columns we identify a double-valued rep Γ of **G**. In column 6 we identify functions that belong to Γ and in columns 7 and 8 we give the restrictions that apply to the values of m for the rep Γ.

(ii) For degenerate reps the basis functions are given as row vectors and the matrix representatives are available from Table 6.5.

functions for the single-valued reps of the cubic point groups have been given already it is easier to obtain the bases for the double-valued reps by the reduction of Kronecker products of $\mathscr{D}^{\frac{1}{2}}\{R(\alpha, \beta, \gamma)\}$ with the single-valued reps rather than to apply the group operator W^i_{tu} to the functions $u_- Y_l^m(\theta, \phi)$ and $u_+ Y_l^m(\theta, \phi)$ *ab initio*.

Suppose that Γ^i *is a single-valued rep of one of the point groups, then the bases of* Γ^i are available from Tables 2.4–2.6; we write $\langle \phi^i_p| = \langle \phi^i_1, \phi^i_2, \ldots, \phi^i_{d_i}|$ for a basis of Γ^i. The bases used in Chapter 2 were for row representations so that

$$R\phi^i_p = \sum_q \phi^i_q \mathbf{D}^i(R)_{qp} \tag{6.2.41}$$

and the matrix representatives $\mathbf{D}^i(R)_{qp}$ were given in Chapter 2 for each point-group rep Γ^i. From eqn. (6.2.40) we see that $\langle u_-, u_+|$ is a basis of $\mathscr{D}^{\frac{1}{2}}\{R(\alpha, \beta, \gamma)\}$ and the transformation properties of the spin functions u_- and u_+ were given in eqn. (6.2.40),

$$R_s\langle u_-, u_+| = \langle u_-, u_+|\, \mathscr{D}^{\frac{1}{2}}(R). \tag{6.2.42}$$

When $\mathscr{D}^{\frac{1}{2}}\{R(\alpha, \beta, \gamma)\}$ is restricted from the 3-dimensional rotation group, $O(3)$, to one of the point groups, **G**, i.e. is subduced onto **G**, it forms a representation of **G** that may be reducible or irreducible. If we write down the function

$$\langle \psi^k_p| = \langle u_-\phi^i_1, u_-\phi^i_2, \ldots, u_-\phi^i_{d_i}, u_+\phi^i_1, u_+\phi^i_2, \ldots, u_+\phi^i_{d_i}| \tag{6.2.43}$$

this forms a basis of the Kronecker product representation Γ^k where

$$\Gamma^k = \mathscr{D}^{\frac{1}{2}} \boxtimes \Gamma^i, \tag{6.2.44}$$

which is the Kronecker product of the single-valued rep Γ^i of **G** with the representation $\mathscr{D}^{\frac{1}{2}}$ of **G** to which the spin functions belong. The matrix representative $\mathbf{D}^k(R)$ can then be constructed by using eqns (6.2.41)–(6.2.43),

$$\mathbf{D}^k(R) = \begin{pmatrix} \mathscr{D}^{\frac{1}{2}}(R)_{11}\mathbf{D}^i(R) & \mathscr{D}^{\frac{1}{2}}(R)_{12}\mathbf{D}^i(R) \\ \mathscr{D}^{\frac{1}{2}}(R)_{21}\mathbf{D}^i(R) & \mathscr{D}^{\frac{1}{2}}(R)_{22}\mathbf{D}^i(R) \end{pmatrix}. \tag{6.2.45}$$

Γ^k is a double-valued representation of **G** that may be reducible or irreducible. If Γ^k is irreducible then it can be identified with one of the double-valued irreducible representations of **G** in Table 6.5 and, because we know $\langle\phi^i_p|$ from Tables 2.4–2.6, then eqn (6.2.43) immediately gives functions that are bases of Γ^k. On the other hand, if Γ^k is a reducible representation of **G** it can be transformed by some unitary transformation into a sum of irreducible double-valued representations of **G**. If R is an element of **G** then we may apply a unitary transformation to reduce Γ^k,

$$\mathbf{D}^{k'}(R) = \mathbf{M}^{-1}\mathbf{D}^k(R)\mathbf{M} \tag{6.2.46}$$

where

$$R\langle\psi^k_p| = \langle\psi^k_p|\,\mathbf{D}^k(R) \tag{6.2.47}$$

and some direct sum of irreducible double-valued representations of **G** will be produced. By inspection of the sequence in which these irreducible representations of **G** appear we can assign the various components of the basis of $\Gamma^{k'}$ to the appropriate representation. If $\langle\psi^k_p|$ is a basis of Γ^k then the transformed basis of $\Gamma^{k'}$ is given by

$$\langle\psi^{k'}_p| = \langle\psi^k_p|\,\mathbf{M} \tag{6.2.48}$$

where **M** is given by eqn (6.2.46).

When $\mathscr{D}^{\frac{1}{2}}$ of $O(3)$ is subduced onto one of the point groups it can be identified from Tables 6.1 and 6.5. The compatibilities between $\mathscr{D}^{\frac{1}{2}}$ of $O(3)$ and the double-valued reps of the cubic point groups are given in Table 6.9. For each of the cubic point groups we therefore form the Kronecker product of the appropriate rep in Table 6.9 with each of the single-valued reps Γ^i of that point group. The reduction of each of these Kronecker products is given in Table 6.10. Therefore if $\langle\phi^i_p|$ is a function from Table 2.6 that belongs to A, ${}^1E$, or ${}^2E$ of 23 (T) or A_1, A_2, or E of 432 (O) we can use eqn (6.2.43) to write down

$$\langle\psi^k_p| = \langle u_-\phi^i_p,\, u_+\phi^i_p|, \tag{6.2.49}$$

which forms a basis of the corresponding double-valued representation Γ^k listed on the right in Table 6.10 and which is irreducible. The only cases for which Γ^k is reducible are then $T \boxtimes \bar{E}$ for 23 (T) and $T_1 \boxtimes \bar{E}_1$ and $T_2 \boxtimes \bar{E}_1$ for 432 (O). For these representations it is necessary to find a matrix **M** that can be used in eqn (6.2.46) to reduce Γ^k. **M** can be determined by writing down the matrix representatives in Γ^k using eqn (6.2.45) and Tables 2.3 and 6.1 and by finding the matrix whose columns are the normalized eigenvectors of the matrix which is the sum of the matrices $\mathbf{D}^k(R)$ for all the elements in one class of $\mathbf{G}$. For $T \boxtimes \bar{E}$ of 23 (T) the matrix $\mathbf{M}$ is

$$\mathbf{M}_1 = \frac{1}{\sqrt{3}} \begin{pmatrix} 0 & 1 & 0 & -\omega^* & 0 & \omega \\ 0 & \mathrm{i} & 0 & -\mathrm{i} & 0 & \mathrm{i} \\ -1 & 0 & \omega & 0 & -\omega^* & 0 \\ 1 & 0 & -\omega^* & 0 & \omega & 0 \\ -\mathrm{i} & 0 & \mathrm{i} & 0 & -\mathrm{i} & 0 \\ 0 & 1 & 0 & -\omega & 0 & \omega^* \end{pmatrix} \tag{6.2.50}$$

where $\omega = \exp(2\pi\mathrm{i}/3)$; for $T_1 \boxtimes \bar{E}_1$ of 432 (O) the matrix $\mathbf{M}$ is

$$\mathbf{M}_2 = \frac{1}{\sqrt{6}} \begin{pmatrix} 0 & \sqrt{2} & 0 & 0 & -1 & \sqrt{3} \\ 0 & \mathrm{i}\sqrt{2} & 0 & 0 & -\mathrm{i} & -\mathrm{i}\sqrt{3} \\ -\sqrt{2} & 0 & -2 & 0 & 0 & 0 \\ \sqrt{2} & 0 & -1 & \sqrt{3} & 0 & 0 \\ -\mathrm{i}\sqrt{2} & 0 & \mathrm{i} & \mathrm{i}\sqrt{3} & 0 & 0 \\ 0 & \sqrt{2} & 0 & 0 & 2 & 0 \end{pmatrix} \tag{6.2.51}$$

and for $T_2 \boxtimes \bar{E}_1$ of 432 (O) the matrix $\mathbf{M}$ is

$$\mathbf{M}_3 = \frac{1}{\sqrt{6}} \begin{pmatrix} 0 & \sqrt{2} & 0 & 0 & \sqrt{3} & 1 \\ 0 & \mathrm{i}\sqrt{2} & 0 & 0 & -\mathrm{i}\sqrt{3} & \mathrm{i} \\ -\sqrt{2} & 0 & 0 & 2 & 0 & 0 \\ \sqrt{2} & 0 & \sqrt{3} & 1 & 0 & 0 \\ -\mathrm{i}\sqrt{2} & 0 & \mathrm{i}\sqrt{3} & -\mathrm{i} & 0 & 0 \\ 0 & \sqrt{2} & 0 & 0 & 0 & -2 \end{pmatrix}. \tag{6.2.52}$$

By applying the matrix $\mathbf{M}_1$ in eqn. (6.2.48) to the function $\langle\psi_p^k| = \langle u_-\phi_p^T, u_+\phi_p^T|$, where $\langle\phi_p^T|$ is a basis of the rep T of 23 (T) determined from Table 2.6, one obtains a transformed basis $\langle\psi_p^{k'}|$; the components of $\langle\psi_p^{k'}|$ can then be assigned to the double-valued reps $\bar{E}$, ${}^1\bar{F}$, and ${}^2\bar{F}$ of 23 (T). When $\mathbf{M}_1$ is applied to the matrices $\mathbf{D}^k(R)$ for $T \boxtimes \bar{E}$ one finds that the reps appear down the leading diagonal *in the order* $\bar{E}$, ${}^1\bar{F}$, and ${}^2\bar{F}$ respectively so that the first two components of $\langle\psi_p^{k'}|$ belong to

$\bar{E}$, the second two components belong to ${}^1\bar{F}$ and the third two components belong to ${}^2\bar{F}$. When $\mathbf{M}_2$ is applied to $T_1 \boxtimes E_1$ of 432 (O) one finds that $\bar{E}_1$ appears in the top left-hand corner of $\mathbf{D}^{k'}(R)$ and $\bar{F}$ in the lower right-hand corner, so that the first two components of $\langle\psi_p^{k'}|$ given by eqn (6.2.48) belong to $\bar{E}_1$ and the last four components belong to $\bar{F}$. Similarly, when $\mathbf{M}_3$ is applied to $T_2 \boxtimes \bar{E}_1$ of 432 (O) the first two components of $\langle\psi_p^{k'}|$ given by eqn (6.2.48) belong to $\bar{E}_2$ and the last four components belong to $\bar{F}$. The matrix representatives obtained in the ways just described are given in Table 6.11.

TABLE 6.9

The compatibilities of $\mathscr{D}^{\frac{1}{2}}$ of O(3) and the reps of the double cubic point groups

Point group			$\mathscr{D}^{\frac{1}{2}}$	
28	23	T	$\bar{E}$	Γ_5
29	$m3$	T_h	$\bar{E}_g$	Γ_5^+
30	432	O	$\bar{E}_1$	Γ_6
31	$\bar{4}3m$	T_d	$\bar{E}_1$	Γ_6
32	$m3m$	O_h	$\bar{E}_{1g}$	Γ_6^+

TABLE 6.10

Γ^k for the cubic point groups

Point group			Kronecker product Γ^k
28	23	T	$A \boxtimes \bar{E} = \bar{E}$
			${}^1E \boxtimes \bar{E} = {}^1\bar{F}$
			${}^2E \boxtimes \bar{E} = {}^2\bar{F}$
			$T \boxtimes \bar{E} = \bar{E} + {}^1\bar{F} + {}^2\bar{F}$
29	$m3$	T_h	$= 23 \otimes \bar{1}$ $(T \otimes C_i)$
30	432	O	$A_1 \boxtimes \bar{E}_1 = \bar{E}_1$
			$A_2 \boxtimes \bar{E}_1 = \bar{E}_2$
			$E \boxtimes \bar{E}_1 = \bar{F}$
			$T_1 \boxtimes \bar{E}_1 = \bar{E}_1 + \bar{F}$
			$T_2 \boxtimes \bar{E}_1 = \bar{E}_2 + \bar{F}$
31	$\bar{4}3m$	T_d	as 432 (O)
32	$m3m$	O_h	$= 432 \otimes \bar{1}$ $(T \otimes C_i)$

TABLE 6.11

Matrix representatives for bases of cubic double point groups

23 (T)

	C_{31}^-	C_{2x}	$\bar{C}_{2y}$
$\bar{E}$	$\frac{1}{2}\begin{pmatrix}1-\mathrm{i} & -1+\mathrm{i}\\ 1+\mathrm{i} & 1+\mathrm{i}\end{pmatrix}$	$\begin{pmatrix}0 & -\mathrm{i}\\ -\mathrm{i} & 0\end{pmatrix}$	$\begin{pmatrix}0 & 1\\ -1 & 0\end{pmatrix}$
${}^1\bar{F}$	$\frac{\omega}{2}\begin{pmatrix}1-\mathrm{i} & -1+\mathrm{i}\\ 1+\mathrm{i} & 1+\mathrm{i}\end{pmatrix}$	$\begin{pmatrix}0 & -\mathrm{i}\\ -\mathrm{i} & 0\end{pmatrix}$	$\begin{pmatrix}0 & 1\\ -1 & 0\end{pmatrix}$
${}^2\bar{F}$	$\frac{\omega^*}{2}\begin{pmatrix}1-\mathrm{i} & -1+\mathrm{i}\\ 1+\mathrm{i} & 1+\mathrm{i}\end{pmatrix}$	$\begin{pmatrix}0 & -\mathrm{i}\\ -\mathrm{i} & 0\end{pmatrix}$	$\begin{pmatrix}0 & 1\\ -1 & 0\end{pmatrix}$

432 (O)

	C_{4x}^+	C_{31}^-	C_{2b}
$\bar{E}_1$	$\frac{1}{\sqrt{2}}\begin{pmatrix}1 & -\mathrm{i}\\ -\mathrm{i} & 1\end{pmatrix}$	$\frac{1}{2}\begin{pmatrix}-1+\mathrm{i} & 1-\mathrm{i}\\ -1-\mathrm{i} & -1-\mathrm{i}\end{pmatrix}$	$\frac{1}{\sqrt{2}}\begin{pmatrix}0 & -1-\mathrm{i}\\ 1-\mathrm{i} & 0\end{pmatrix}$
$\bar{E}_2$	$\frac{1}{\sqrt{2}}\begin{pmatrix}-1 & \mathrm{i}\\ \mathrm{i} & -1\end{pmatrix}$	$\frac{1}{2}\begin{pmatrix}-1+\mathrm{i} & 1-\mathrm{i}\\ -1-\mathrm{i} & -1-\mathrm{i}\end{pmatrix}$	$\frac{1}{\sqrt{2}}\begin{pmatrix}0 & 1+\mathrm{i}\\ -1+\mathrm{i} & 0\end{pmatrix}$
$\bar{F}$	$\frac{1}{\sqrt{2}}\begin{pmatrix}\mu & -\mathrm{i}\mu\\ -\mathrm{i}\mu & \mu\end{pmatrix}$	$\frac{1}{2}\begin{pmatrix}(-1+\mathrm{i})\beta & (1-\mathrm{i})\beta\\ (-1-\mathrm{i})\beta & (-1-\mathrm{i})\beta\end{pmatrix}$	$\frac{1}{\sqrt{2}}\begin{pmatrix}0 & (-1-\mathrm{i})\lambda\\ (1-\mathrm{i})\lambda & 0\end{pmatrix}$

Notes to Table 6.11

(i) $\beta = \begin{pmatrix}-\frac{1}{2} & \frac{1}{2}\sqrt{3}\\ -\frac{1}{2}\sqrt{3} & -\frac{1}{2}\end{pmatrix};\quad \lambda = \begin{pmatrix}1 & 0\\ 0 & -1\end{pmatrix};\quad \mu = \begin{pmatrix}-\frac{1}{2} & -\frac{1}{2}\sqrt{3}\\ -\frac{1}{2}\sqrt{3} & \frac{1}{2}\end{pmatrix}.$

(ii) Only the matrices for the generating elements are given; the matrices for the other elements can be found by using Table 6.2.

(iii) In the reductions of $T \boxtimes \bar{E}$ of 23 (T) and of $T_1 \boxtimes \bar{E}_1$ and $T_2 \boxtimes \bar{E}_1$ of 432 (O) the representations appear down the leading diagonal in the following order:

$$T \boxtimes \bar{E}:\ \bar{E},\ {}^1\bar{F},\ {}^2\bar{F}$$
$$T_1 \boxtimes \bar{E}_1:\ \bar{E}_1,\ \bar{F}$$
$$T_2 \boxtimes \bar{E}_1:\ \bar{E}_2,\ \bar{F}$$

In Table 6.12 we give the expressions for the symmetry-adapted functions for the cubic point groups in terms of $\langle\phi_p^i|$ where $\langle\phi_p^i|$ is a basis of one of the single-valued reps Γ^i of that point group and can be found from Table 2.6.

TABLE 6.12

Symmetry-adapted functions for the cubic double point groups

($\omega = \exp(2\pi i/3)$)

Point group	Rep Γ	Γ^i	Basis of Γ
23 (T)	$\bar{E}$	A	$\langle u_-\phi_1^i;\ u_+\phi_1^i|$
		T	$\langle -u_-\phi_3^i + u_+\phi_1^i - iu_+\phi_2^i;\ u_-\phi_1^i + iu_-\phi_2^i + u_+\phi_3^i|$
	$^1\bar{F}$	^1E	$\langle u_-\phi_1^i;\ u_+\phi_1^i|$
		T	$\langle \omega u_-\phi_3^i - \omega^* u_+\phi_1^i + iu_+\phi_2^i;\ -\omega^* u_-\phi_1^i - iu_-\phi_2^i - \omega u_+\phi_3^i|$
	$^2\bar{F}$	^2E	$\langle u_-\phi_1^i;\ u_+\phi_1^i|$
		T	$\langle -\omega^* u_-\phi_3^i + \omega u_+\phi_1^i - iu_+\phi_2^i;\ \omega u_-\phi_1^i + iu_-\phi_2^i + \omega^* u_+\phi_3^i|$
432 (O)	$\bar{E}_1$	A_1	$\langle u_-\phi_1^i;\ u_+\phi_1^i|$
		T_1	$\langle -u_-\phi_3^i + u_+\phi_1^i - iu_+\phi_2^i;\ u_-\phi_1^i + iu_-\phi_2^i + u_+\phi_3^i|$
	$\bar{E}_2$	A_2	$\langle u_-\phi_1^i;\ u_+\phi_1^i|$
		T_2	$\langle -u_-\phi_3^i + u_+\phi_1^i - iu_+\phi_2^i;\ u_-\phi_1^i + iu_-\phi_2^i + u_+\phi_3^i|$
	$\bar{F}$	E	$\langle u_-\phi_1^i;\ u_-\phi_2^i;\ u_+\phi_1^i;\ u_+\phi_2^i|$
		T_1	$\langle -2u_-\phi_3^i - u_+\phi_1^i + iu_+\phi_2^i;\ \sqrt{(3)}u_+\phi_1^i + i\sqrt{(3)}u_+\phi_2^i;\ -u_-\phi_1^i - iu_-\phi_2^i + 2u_+\phi_3^i;\ \sqrt{(3)}u_-\phi_1^i - i\sqrt{(3)}u_-\phi_2^i|$
		T_2	$\langle \sqrt{(3)}u_+\phi_1^i + i\sqrt{(3)}u_+\phi_2^i;\ 2u_-\phi_3^i + u_+\phi_1^i - iu_+\phi_2^i;\ \sqrt{(3)}u_-\phi_1^i - i\sqrt{(3)}u_-\phi_2^i;\ u_-\phi_1^i + iu_-\phi_2^i - 2u_+\phi_3^i|$

Notes to Table 6.12

(i) $\langle\phi_p^i|$ is a basis of the single-valued representation listed in column 3 and can be found from Table 2.6.

(ii) Table 6.11 gives, for the generating elements of the group, the matrices according to which the *row vector* in column 4 transforms.

(iii) Only the point groups 23 (T) and 432 (O) are included in the table.

(iv) For the point group $\bar{4}3m$ (T_d) the set of rules for constructing the bases of the double-valued reps is exactly the same as the set of rules for 432 (O) which is given in the table. This does not, of course, imply that the actual bases of the double-valued reps are the same for the two point groups; this is because the bases $\langle\phi_p^i|$ of the single-valued rep Γ^i of 432 (O) are not necessarily the same as those of the rep Γ^i of $\bar{4}3m$ (T_d).

(v) $m3$ (T_h) and $m3m$ (O_h) are direct product groups and the rules for these groups can very easily be found since the parity of the spin functions is taken to be +1. Therefore the rules for these groups can be found for $m3$ (T_h) by adding the same subscript, either g or u, to the labels in both column 2 and column 3 of the table for 23 (T), and for $m3m$ (O_h) by adding the same subscript, again either g or u, to the labels in both column 2 and column 3 of the table for 432 (O).

(vi) Some of the functions in the table have been denormalized; to normalize them again each basis should be divided by the square root of the sum of the squares of the magnitudes of the coefficients of the contributions to any one component, e.g. for $\bar{E}$ of 23 (T) the basis derived from A is already normalized but the basis shown in the table derived from T should be divided by $\sqrt{3}$ to normalize it.

6.3. The double-valued representations of the space groups

We have seen in the previous two sections how an entity with half-odd-integral spin and in a system with the symmetry of one of the point groups, **G**, requires the use of the double group $\mathbf{G}^\dagger$ or the double-valued representations of **G**. In a similar way if we study an entity with half-odd-integral spin and in a system with the symmetry of one of the 230 space groups it is necessary to study the double-valued representations of the space groups. The double-valued space-group reps are, therefore, important when spin-dependent terms are included in the Hamiltonian used in the determination of the electronic band structure of a crystalline solid (Elliott 1954*b*). It is necessary to extend the scope of Definition 6.1.1 to cover space groups. A space group $\mathbf{G}$ is made up of elements $\{R \mid \mathbf{v}\}$ which obey the multiplication rule

$$\{R_2 \mid \mathbf{v}_2\}\{R_1 \mid \mathbf{v}_1\} = \{R_2R_1 \mid \mathbf{v}_2 + R_2\mathbf{v}_1\} \tag{1.5.3}$$

and form a group; $\mathbf{G}$ can be expressed in terms of left coset representatives of $\mathbf{T}$, the translation group of one of the Bravais lattices

$$\mathbf{G} = \{R_1 \mid \mathbf{v}_1\}\mathbf{T} + \{R_2 \mid \mathbf{v}_2\}\mathbf{T} + \cdots + \{R_h \mid \mathbf{v}_h\}\mathbf{T}. \tag{3.5.2}$$

The rotational parts $R_1, R_2, \ldots, R_h$ form one of the 32 crystallographic point groups and the rule for constructing its double group was given in Definition 6.1.1. Corresponding to every element R_i of this point group there are two elements, say R_i and $\bar{R}_i$, in the double point group. In considering a double space group the R_i and $\bar{R}_i$ are both to be regarded as having the same effect on a vector $\mathbf{v}_j$. That is,

$$\bar{R}_i\mathbf{v}_j = R_i\mathbf{v}_j. \tag{6.3.1}$$

When forming group products R_i and $\bar{R}_i$ multiply according to the multiplication rule for the double point group, that is R_i and $\bar{R}_i$ correspond to $+\mathbf{u}$ and $-\mathbf{u}$ respectively in the isomorphism of the double point group with the subgroup of $SU(2)$. It is then possible to define a double space group as follows,

DEFINITION 6.3.1. The *double group* $\mathbf{G}^\dagger$ of a space group $\mathbf{G}$, defined by eqn. (3.5.2) is given by

$$\mathbf{G} = \{R_1 \mid \mathbf{v}_1\}\mathbf{T} + \{\bar{R}_1 \mid \mathbf{v}_1\}\mathbf{T} + \{R_2 \mid \mathbf{v}_2\}\mathbf{T} + \{\bar{R}_2 \mid \mathbf{v}_2\}\mathbf{T} + \cdots + \{R_h \mid \mathbf{v}_h\}\mathbf{T} + \{\bar{R}_h \mid \mathbf{v}_h\}\mathbf{T}, \tag{6.3.2}$$

where R_i and $\bar{R}_i$ are the elements of the double point group corresponding to the element R_i in the point group of $R_1, R_2, \ldots, R_h$, and $\mathbf{T}$ is the translation group of the Bravais lattice of the space group $\mathbf{G}$.

The multiplication rule for the members of the double space group $\mathbf{G}^\dagger$ is, by analogy with eqn (1.5.3)

$$\{R_2 \mid \mathbf{v}_2\}\{R_1 \mid \mathbf{v}_1\} = \{R_2R_1 \mid \mathbf{v}_2 + R_2\mathbf{v}_1\}. \tag{6.3.3}$$

Similar products involving the barred elements can also be written down,

$$\{\bar{R}_2 \mid \mathbf{v}_2\}\{R_1 \mid \mathbf{v}_1\} = \{\bar{R}_2R_1 \mid \mathbf{v}_2 + R_2\mathbf{v}_1\}, \tag{6.3.4}$$

$$\{R_2 \mid \mathbf{v}_2\}\{\bar{R}_1 \mid \mathbf{v}_1\} = \{R_2\bar{R}_1 \mid \mathbf{v}_2 + R_2\mathbf{v}_1\}, \tag{6.3.5}$$

$$\{\bar{R}_2 \mid \mathbf{v}_2\}\{\bar{R}_1 \mid \mathbf{v}_1\} = \{\bar{R}_2\bar{R}_1 \mid \mathbf{v}_2 + R_2\mathbf{v}_1\}. \tag{6.3.6}$$

It is possible to construct the double space group $\mathbf{G}^\dagger$ for each of the 230 space groups by making use of the list of the generating elements of the space groups in Table 3.7 together with the group multiplication tables for the double point groups in Table 6.2. Any desired part of the group multiplication table can thus be constructed, but the mere construction of the group multiplication table on its own is not very illuminating. It is with the irreducible representations of $\mathbf{G}^\dagger$ that we are concerned, and in particular those which lead to the double-valued reps of $\mathbf{G}$. The theory given in section 3.7 and in Chapter 4 of the representation of a space group $\mathbf{G}$ is immediately applicable to finding the representations of $\mathbf{G}^\dagger$, the double space group. We emphasize once again that one can either talk about the single-valued and double-valued representations of a space group $\mathbf{G}$, in which case the single-valued representations have already been given in Chapter 5, or alternatively one can simply talk about the reps of the double group $\mathbf{G}^\dagger$. In the second case they are ordinary single-valued reps of $\mathbf{G}^\dagger$ which therefore possess all those properties of irreducible representations which were given in section 1.3. We can apply the theory of section 3.7 and Chapter 4 to the double space groups $\mathbf{G}^\dagger$ in a way exactly similar to what was done in Chapter 5 for the space groups $\mathbf{G}$. When this has been done the reps of $\mathbf{G}^\dagger$ can be divided into two sets, first those for which $\chi(\{\bar{E}\mid 000\}) = +\chi(\{E\mid 000\})$ and which therefore make up single-valued reps of $\mathbf{G}$ and secondly those for which $\chi(\{\bar{E}\mid 000\}) = -\chi(\{E\mid 000\})$ and which therefore make up double-valued reps of $\mathbf{G}$.

The task of finding the reps of the complete double group $\mathbf{G}^\dagger$ involves determining the small reps of the little group $\mathbf{G}^{\dagger\mathbf{k}}$, see Definition 3.7.4, for each wave vector $\mathbf{k}$ in the Brillouin zone of that space group. The problem of identifying the elements which are present in $\mathbf{G}^{\dagger\mathbf{k}}$ presents no difficulty because if we choose a particular space group and a particular point in its Brillouin zone we can read off from the appropriate part of Table 5.7 the generating elements of the space group $\mathbf{G}$ and then identify ${}^H\mathbf{G}^{\mathbf{k}}$ or $\overline{\mathbf{G}}^{\mathbf{k}*}$ in terms of one of the abstract groups in Table 5.1. When we attempt to construct the double space group $\mathbf{G}^\dagger$ and the corresponding little group $\mathbf{G}^{\dagger\mathbf{k}}$ this is less difficult than might appear at first sight. Suppose that in $\mathbf{G}$

$$\{R_2\mid \mathbf{v}_2\}\{R_1\mid \mathbf{v}_1\} = \{R_3\mid \mathbf{v}_3\} \tag{6.3.7}$$

then from eqn (6.3.3) in the double group $\mathbf{G}^\dagger$

$$\{R_2\mid \mathbf{v}_2\}\{R_1\mid \mathbf{v}_1\} = \{X\mid \mathbf{v}_3\} \tag{6.3.8}$$

where $\mathbf{v}_3$ is exactly the same as before. X may be either R_3 or $\bar{R}_3$ and which one it is can be determined by inspection of Table 6.2.

Having constructed ${}^H\mathbf{G}^{\dagger\mathbf{k}}$ or $\overline{\mathbf{G}}^{\dagger\mathbf{k}*}$, depending on whether it is a point of symmetry or a line of symmetry that is being studied, it then remains to identify this group as one of the abstract groups in Table 5.1 in a similar way to that described for point groups in section 6.1 and illustrated for the double group of 422 (D_4). To simplify this identification it is advisable to separate ${}^H\mathbf{G}^{\dagger\mathbf{k}}$ or $\overline{\mathbf{G}}^{\dagger\mathbf{k}*}$ into classes and it is not unreasonable to suppose that the classes of a double space group are simply related to the classes of the ordinary space group from which it has been derived. For a symmorphic space group, $^{H}\mathbf{G}^{\dagger\mathbf{k}}$ or $\bar{\mathbf{G}}^{\dagger\mathbf{k}*}$ is always one of the 32 double point groups so that Opechowski's rules, Theorems 6.1.1–6.1.4 still apply to them. But for non-symmorphic space groups some adaptation of Opechowski's rules is necessary (Elliott 1954*b*, Glück, Gur, and Zak 1967). Theorem 6.1.1 is still valid for $^{H}\mathbf{G}^{\dagger\mathbf{k}}$ for double space groups although it needs to be formulated slightly differently:

THEOREM 6.3.1. *To each class C_i of elements, $\{R_i \mid \mathbf{v}_i\}$, of the group $^{H}\mathbf{G}^{\mathbf{k}}$ in which the R_i are not rotations through π there correspond exactly two classes of $^{H}\mathbf{G}^{\dagger\mathbf{k}}$, C_i', and C_i''.*

There is no simple theorem which really corresponds to Theorem 6.1.2 for space groups although the following theorem is sometimes quoted; however it is little more than a re-statement of the definition of a class.

THEOREM 6.3.2. *To each class C_π of the group $^{H}\mathbf{G}^{\mathbf{k}}$ the rotational parts of whose elements are rotations through π there correspond either one or two classes of $^{H}\mathbf{G}^{\dagger\mathbf{k}}$. If there is no rotation through π about an axis at right angles to the axis of one of the rotations of C_π then there will be two classes in $^{H}\mathbf{G}^{\dagger\mathbf{k}}$ corresponding to C_π. But if $\{R_i \mid \mathbf{v}_i\}$ is an element of C_π and there exists an element $\{R_j \mid \mathbf{v}_j\}$ where R_j is a rotation about an axis at right angles to the axis of R_i then there will only be one class in $^{H}\mathbf{G}^{\dagger\mathbf{k}}$ corresponding to C_π if*

$$\{R_j \mid \mathbf{v}_j\}^{-1}\{R_i \mid \mathbf{v}_i\}\{R_j \mid \mathbf{v}_j\} \equiv \{\bar{R}_i \mid \mathbf{v}_i\}, \tag{6.3.9}$$

that is, if

$$\{R_i \mid \mathbf{v}_i\}\{R_j \mid \mathbf{v}_j\} \equiv \{R_j \mid \mathbf{v}_j\}\{\bar{R}_i \mid \mathbf{v}_i\}. \tag{6.3.10}$$

Theorems 6.1.1 and 6.1.2 could also be adapted to $\bar{\mathbf{G}}^{\dagger\mathbf{k}*}$ instead of to $^{H}\mathbf{G}^{\dagger\mathbf{k}}$. The remaining two theorems, Theorems 6.1.3 and 6.1.4, simply constitute the definition of single-valued and double-valued representations of a point group $\mathbf{G}$ and are equally valid for a space group $\mathbf{G}$.

We have discussed earlier, in section 5.2, the effect on the energy levels $E_j^{\mathbf{k}}$ if a crystal possesses time-reversal symmetry in addition to all the spatial symmetry operations of $\mathbf{G}$. In a similar way it is possible to consider the effect on the small reps of $\mathbf{G}^{\dagger\mathbf{k}}$ if θ, the operation of time inversion, is present as a symmetry operation of a crystal. For the present we simply note that the presence of θ may cause some extra degeneracies in the energy levels $E_j^{\mathbf{k}}$ that belong to the rep $\Gamma_j^{\mathbf{k}}$ of $\mathbf{G}^{\dagger\mathbf{k}}$ (see section 5.2, p. 388):

(a) there is no change in the degeneracy of $E_j^{\mathbf{k}}$,

(b) the degeneracy of $E_j^{\mathbf{k}}$ becomes doubled, that is two different energy levels, both described by the same rep $\Gamma_j^{\mathbf{k}}$, become degenerate,

(c) the degeneracy of $E_j^{\mathbf{k}}$ becomes doubled but, unlike (b), two different (i.e., inequivalent) reps $\Gamma_j^{\mathbf{k}}$ and $\Gamma_{j'}^{\mathbf{k}}$ of $\mathbf{G}^{\dagger\mathbf{k}}$ become degenerate, and

(d) when **k** and $-$**k** are not in the same star, the spectrum of the eigenvalues at $-$**k** becomes identical with the spectrum of the eigenvalues at $+$**k**.

Once again it is convenient to postpone until Chapter 7 the theory of the determination, for any given rep $\Gamma_j^{\mathbf{k}}$ which of the above situations applies (see section 7.6); this theory involves the reality of the induced rep $(\Gamma_j^{\mathbf{k}} \uparrow \mathbf{G}^{\dagger})$, see Note (vii) to Table 6.13.

6.4. An example of the deduction of the representations of a double space group

In section 5.4 we illustrated the use of Table 5.7 by showing how to obtain from it the single-valued reps of the space group $F\bar{4}3c$ (T_d^5). We now show how it is possible to use that table to deduce the double-valued reps of the same space group $F\bar{4}3c$ (T_d^5).

The Brillouin zone of the space group $F\bar{4}3c$ (T_d^5) is illustrated in Fig. 3.14 and in Fig. 5.2. The **k** vectors of each of the points and lines of symmetry are, from Table 3.6, in terms of $\mathbf{g}_1$, $\mathbf{g}_2$, and $\mathbf{g}_3$:

Point or line	**k**
Γ	$(0, 0, 0)$
X	$(\frac{1}{2}, 0, \frac{1}{2})$
L	$(\frac{1}{2}, \frac{1}{2}, \frac{1}{2})$
W	$(\frac{1}{2}, \frac{1}{4}, \frac{3}{4})$
Δ	$(\alpha, 0, \alpha)$
Λ	(α, α, α)
Σ	$(\alpha, \alpha, 2\alpha)$
S	$(\frac{1}{2} + \alpha, 2\alpha, \frac{1}{2} + \alpha)$
Z	$(\frac{1}{2}, \alpha, \frac{1}{2} + \alpha)$
Q	$(\frac{1}{2}, \frac{1}{2} - \alpha, \frac{1}{2} + \alpha)$.

We consider first the points of symmetry in turn.

Γ: $\mathbf{k} = (0, 0, 0)$

${}^H\mathbf{G}^{\Gamma}$ is $\mathbf{G}_{24}^{7}$ which is just the point group $\bar{4}3m$ (T_d), see section 5.4, and therefore ${}^H\mathbf{G}^{\dagger\Gamma}$ is the double group of $\bar{4}3m$ (T_d) which is given in Table 6.5. This group can also be identified as the abstract group $\mathbf{G}_{48}^{10}$ and its generating elements identified as

$$P = \{S_{4x}^{-} \mid \tfrac{1}{2}\tfrac{1}{2}\tfrac{1}{2}\}, \qquad Q = \{\bar{C}_{31}^{-} \mid 000\}, \qquad R = \{\sigma_{db} \mid \tfrac{1}{2}\tfrac{1}{2}\tfrac{1}{2}\}.$$

The eight classes of this group are thus

C_1 $\{E \mid 000\}$
C_2 $\{\bar{E} \mid 000\}$
C_3 $\{C_{2x} \mid 000\}$, $\{C_{2y} \mid 000\}$, $\{C_{2z} \mid 000\}$, $\{\bar{C}_{2x} \mid 000\}$, $\{\bar{C}_{2y} \mid 000\}$, $\{\bar{C}_{2z} \mid 000\}$
C_4 $\{C_{31}^{-} \mid 000\}$, $\{C_{32}^{-} \mid 000\}$, $\{C_{33}^{-} \mid 000\}$, $\{C_{34}^{-} \mid 000\}$, $\{C_{31}^{+} \mid 000\}$, $\{C_{32}^{+} \mid 000\}$, $\{C_{33}^{+} \mid 000\}$, $\{C_{34}^{+} \mid 000\}$

C_5 $\{\bar{C}_{31}^{-} \mid 000\}$, $\{\bar{C}_{32}^{-} \mid 000\}$, $\{\bar{C}_{33}^{-} \mid 000\}$, $\{\bar{C}_{34}^{-} \mid 000\}$, $\{\bar{C}_{31}^{+} \mid 000\}$, $\{\bar{C}_{32}^{+} \mid 000\}$, $\{\bar{C}_{33}^{+} \mid 000\}$, $\{\bar{C}_{34}^{+} \mid 000\}$

C_6 $\{S_{4x}^{-} \mid \tfrac{1}{2}\tfrac{1}{2}\tfrac{1}{2}\}$, $\{S_{4y}^{-} \mid \tfrac{1}{2}\tfrac{1}{2}\tfrac{1}{2}\}$, $\{S_{4z}^{-} \mid \tfrac{1}{2}\tfrac{1}{2}\tfrac{1}{2}\}$, $\{S_{4x}^{+} \mid \tfrac{1}{2}\tfrac{1}{2}\tfrac{1}{2}\}$, $\{S_{4y}^{+} \mid \tfrac{1}{2}\tfrac{1}{2}\tfrac{1}{2}\}$, $\{S_{4z}^{+} \mid \tfrac{1}{2}\tfrac{1}{2}\tfrac{1}{2}\}$

C_7 $\{\bar{S}_{4x}^{-} \mid \tfrac{1}{2}\tfrac{1}{2}\tfrac{1}{2}\}$, $\{\bar{S}_{4y}^{-} \mid \tfrac{1}{2}\tfrac{1}{2}\tfrac{1}{2}\}$, $\{\bar{S}_{4z}^{-} \mid \tfrac{1}{2}\tfrac{1}{2}\tfrac{1}{2}\}$, $\{\bar{S}_{4x}^{+} \mid \tfrac{1}{2}\tfrac{1}{2}\tfrac{1}{2}\}$, $\{\bar{S}_{4y}^{+} \mid \tfrac{1}{2}\tfrac{1}{2}\tfrac{1}{2}\}$, $\{\bar{S}_{4z}^{+} \mid \tfrac{1}{2}\tfrac{1}{2}\tfrac{1}{2}\}$

C_8 $\{\sigma_{da} \mid \tfrac{1}{2}\tfrac{1}{2}\tfrac{1}{2}\}$, $\{\sigma_{db} \mid \tfrac{1}{2}\tfrac{1}{2}\tfrac{1}{2}\}$, $\{\sigma_{dc} \mid \tfrac{1}{2}\tfrac{1}{2}\tfrac{1}{2}\}$, $\{\sigma_{dd} \mid \tfrac{1}{2}\tfrac{1}{2}\tfrac{1}{2}\}$, $\{\sigma_{de} \mid \tfrac{1}{2}\tfrac{1}{2}\tfrac{1}{2}\}$, $\{\sigma_{df} \mid \tfrac{1}{2}\tfrac{1}{2}\tfrac{1}{2}\}$, $\{\bar{\sigma}_{da} \mid \tfrac{1}{2}\tfrac{1}{2}\tfrac{1}{2}\}$, $\{\bar{\sigma}_{db} \mid \tfrac{1}{2}\tfrac{1}{2}\tfrac{1}{2}\}$, $\{\bar{\sigma}_{dc} \mid \tfrac{1}{2}\tfrac{1}{2}\tfrac{1}{2}\}$, $\{\bar{\sigma}_{dd} \mid \tfrac{1}{2}\tfrac{1}{2}\tfrac{1}{2}\}$, $\{\bar{\sigma}_{de} \mid \tfrac{1}{2}\tfrac{1}{2}\tfrac{1}{2}\}$, $\{\bar{\sigma}_{df} \mid \tfrac{1}{2}\tfrac{1}{2}\tfrac{1}{2}\}$

and the three representations which correspond to double-valued reps of ${}^H\mathbf{G}^\Gamma$ are

	C_1	C_2	C_3	C_4	C_5	C_6	C_7	C_8	
Γ_6 $\bar{E}_1$	2	−2	0	1	−1	$\sqrt{2}$	$-\sqrt{2}$	0	2
Γ_7 $\bar{E}_2$	2	−2	0	1	−1	$-\sqrt{2}$	$\sqrt{2}$	0	2
Γ_8 $\bar{F}$	4	−4	0	−1	1	0	0	0	2

The matrices for these representations can be found from Table 5.1 or Table 6.5. From the realities in the extreme right-hand column it follows that the addition of time-reversal symmetry causes no extra degeneracy of the small reps of $\mathbf{G}^\Gamma$.

X: $\mathbf{k} = (\tfrac{1}{2}, 0, \tfrac{1}{2})$

${}^H\mathbf{G}^X$ was identified in detail with the direct product $\mathbf{G}_8^4 \otimes \mathbf{T}_2$ in section 5.4:

E	$\{E \mid 000\}$	E	R	$\{E \mid 001\}$	E'
P	$\{S_{4y}^{+} \mid \tfrac{1}{2}\tfrac{1}{2}\tfrac{1}{2}\}$	W	PR	$\{S_{4y}^{+} \mid \tfrac{1}{2}\tfrac{1}{2}-\tfrac{1}{2}\}$	W'
P^2	$\{C_{2y} \mid 000\}$	C	P^2R	$\{C_{2y} \mid 001\}$	C'
P^3	$\{S_{4y}^{-} \mid \tfrac{1}{2}\tfrac{1}{2}\tfrac{1}{2}\}$	V	P^3R	$\{S_{4y}^{-} \mid \tfrac{1}{2}\tfrac{1}{2}-\tfrac{1}{2}\}$	V'
Q	$\{C_{2x} \mid 000\}$	B	QR	$\{C_{2x} \mid 001\}$	B'
PQ	$\{\sigma_{dc} \mid \tfrac{1}{2}\tfrac{1}{2}\tfrac{1}{2}\}$	T	PQR	$\{\sigma_{dc} \mid \tfrac{1}{2}\tfrac{1}{2}-\tfrac{1}{2}\}$	T'
P^2Q	$\{C_{2z} \mid 000\}$	D	P^2QR	$\{C_{2z} \mid 001\}$	D'
P^3Q	$\{\sigma_{de} \mid \tfrac{1}{2}\tfrac{1}{2}\tfrac{1}{2}\}$	U	P^3QR	$\{\sigma_{de} \mid \tfrac{1}{2}\tfrac{1}{2}-\tfrac{1}{2}\}$	U'

where the symbols in the last columns are used as a shorthand notation, the prime being used to distinguish between $\{R_i \mid \mathbf{v}_i\}$ with the same R_i and different $\mathbf{v}_i$.

There are, therefore, 32 elements in the Herring little group ${}^H\mathbf{G}^{\dagger X}$ and using Table 6.2 and the generating relations of the group $\mathbf{G}_8^4$, that is

$$P^4 = E; \quad Q^2 = E; \quad QP = P^3Q,$$

we can easily construct the group multiplication table. Part of it is shown here:

E	B	C	D	T	U	V	W
B	$\bar{E}$	D	$\bar{C}$	V	$\bar{W}$	$\bar{T}$	U
C	$\bar{D}$	$\bar{E}$	B	U	$\bar{T}$	W	$\bar{V}$
D	C	$\bar{B}$	$\bar{E}$	$\bar{W}$	$\bar{V}$	U	T
T	W	$\bar{U}$	$\bar{V}$	$\bar{E}$	C	D	$\bar{B}$
U	$\bar{V}$	T	$\bar{W}$	$\bar{C}$	$\bar{E}$	B	D
V	U	W	T	$\bar{B}$	D	$\bar{C}$	E
W	$\bar{T}$	$\bar{V}$	U	D	B	E	C

and the rest can readily be constructed from this since E', $\bar{E}$, and $\bar{E}'$ commute with all the elements of the group. We can then identify this with one of the groups of order 32 in Table 5.1; the number of elements of the various orders are

Order	1	2	4	8
Number of elements	1	3	20	8

and by separating the elements into classes we find that there are 14 classes. By inspection, using information such as the order of each element, the number of classes and the fact that the representations of ${}^H\mathbf{G}^{\dagger X}$ must be related to those of ${}^H\mathbf{G}^X$ we can identify ${}^H\mathbf{G}^{\dagger X}$ with $\mathbf{G}_{32}^9$ where

$$P = W = \{S_{4y}^+ \mid \tfrac{1}{2}\tfrac{1}{2}\tfrac{1}{2}\}$$
$$Q = B = \{C_{2x} \mid 000\}$$

and
$$R = E' = \{E \mid 001\}$$

which is also a direct product group, $\mathbf{G}_{16}^{14} \otimes \mathbf{T}_2$. Identifying the classes of $\mathbf{G}_{16}^{14}$ we find

C_1	E
C_2	$\bar{E}$
C_3	$C, \bar{C}$
C_4	W, V
C_5	$\bar{V}, \bar{W}$
C_6	$B, \bar{D}, \bar{B}, D$
C_7	$\bar{T}, \bar{U}, T, U.$

The double-valued reps of ${}^H\mathbf{G}^X$ are then those reps of $\mathbf{G}_{16}^{14}$ for which $\chi(C_2) = -\chi(C_1)$, i.e. R_6 and R_7:

		C_1	C_2	C_3	C_4	C_5	C_6	C_7	
X_6	$\bar{E}_1$	2	-2	0	$\sqrt{2}$	$-\sqrt{2}$	0	0	2
X_7	$\bar{E}_2$	2	-2	0	$-\sqrt{2}$	$\sqrt{2}$	0	0	2

The matrices for these reps can be identified from Table 5.1 and from the realities included in the right-hand column we see that if the crystal possesses time-reversal symmetry the degeneracies of the reps $\bar{E}_1$ and $\bar{E}_2$ (X_6 and X_7) are not changed.

L: $\mathbf{k} = (\tfrac{1}{2}, \tfrac{1}{2}, \tfrac{1}{2})$

${}^H\mathbf{G}^L$ for this point was identified in detail with the abstract group $\mathbf{G}_{12}^4$ in section 5.4:

E	$\{E\mid 000\}$	E	Q	$\{\sigma_{db}\mid \frac{1}{2}\frac{1}{2}\frac{1}{2}\}$	Q
P	$\{C_{31}^{-}\mid 001\}$	F'	PQ	$\{\sigma_{df}\mid \frac{1}{2}\frac{1}{2}-\frac{1}{2}\}$	I'
P^2	$\{C_{31}^{+}\mid 000\}$	K	P^2Q	$\{\sigma_{de}\mid \frac{1}{2}\frac{1}{2}\frac{1}{2}\}$	U
P^3	$\{E\mid 001\}$	E'	P^3Q	$\{\sigma_{db}\mid \frac{1}{2}\frac{1}{2}-\frac{1}{2}\}$	Q'
P^4	$\{C_{31}^{-}\mid 000\}$	F	P^4Q	$\{\sigma_{df}\mid \frac{1}{2}\frac{1}{2}\frac{1}{2}\}$	I
P^5	$\{C_{31}^{+}\mid 001\}$	K'	P^5Q	$\{\sigma_{de}\mid \frac{1}{2}\frac{1}{2}-\frac{1}{2}\}$	U'

where again the symbols in the last columns are used as a shorthand notation. It is then easy to construct the group multiplication table for ${}^H\mathbf{G}^{\dagger L}$:

E	F	K	Q	U	I
F	$\bar{K}$	E	I	Q	$\bar{U}$
K	E	$\bar{F}$	U	$\bar{I}$	Q
Q	U	I	$\bar{E}'$	$\bar{F}'$	$\bar{K}'$
U	$\bar{I}$	Q	$\bar{K}'$	$\bar{E}'$	F'
I	Q	$\bar{U}$	$\bar{F}'$	K'	$\bar{E}'$

The rest of this group multiplication table can readily be constructed since E', $\bar{E}$, and $\bar{E}'$ commute with all the members of the group. Separating the 24 elements of the group ${}^H\mathbf{G}^{\dagger L}$ into classes we obtain 12 classes and by inspection this group can be identified with the abstract group $\mathbf{G}_{24}^{3}$ with

$$P = F' = \{C_{31}^{-}\mid 001\},$$
$$Q = Q = \{\sigma_{db}\mid \tfrac{1}{2}\tfrac{1}{2}\tfrac{1}{2}\},$$
and
$$R = E' = \{E\mid 001\},$$

which is a direct product group $\mathbf{G}_{12}^{4} \otimes \mathbf{T}_2$. Identifying the classes of $\mathbf{G}_{12}^{4}$ we find

C_1	E
C_2	$\bar{E}'$
C_3	F', K'
C_4	$\bar{K}, \bar{F}$
C_5	$Q, \bar{U}, \bar{I}$
C_6	$I', \bar{Q}', U'$

and the reps that lead to double-valued reps of ${}^H\mathbf{G}^{L}$ are those of $\mathbf{G}_{12}^{4}$ for which $\chi(C_1) = +\chi(C_2)$; this means that only R_1, R_2, and R_5 are acceptable:

		C_1	C_2	C_3	C_4	C_5	C_6	
L_4	$\bar{A}_1$	1	1	1	1	1	1	1
L_5	$\bar{A}_2$	1	1	1	1	−1	−1	1
L_6	$\bar{E}$	2	2	−1	−1	0	0	1

The matrices for $\bar{E}$ (L_6) can be found from Table 5.1. From the realities included in the extreme right-hand column we can see that the addition of time-reversal symmetry causes the degeneracy of each of the reps $\bar{A}_1$ (L_4), $\bar{A}_2$ (L_5) and $\bar{E}$ (L_6) to be doubled.

W: $\mathbf{k} = (\frac{1}{2}, \frac{1}{4}, \frac{3}{4})$

${}^H\mathbf{G}^W$ is the direct product of $\mathbf{G}_4^1$ and $\mathbf{T}_4$, that is, of the point group $\bar{4}$ (S_4) and $\mathbf{T}_4$; ${}^H\mathbf{G}^{\dagger W}$ is therefore also a direct product, that of the double group of $\bar{4}$ (S_4) and the group $\mathbf{T}_4$. The double group of $\bar{4}$ (S_4) in this case is actually different from the $\bar{4}$ (S_4) of Table 6.5 because the axis of symmetry is along the x-axis rather than the z-axis; the elements of $\bar{4}$ (S_4) are

$$\begin{array}{lll} E & \{E \mid 000\} & E \\ P & \{S_{4x}^{+} \mid \frac{1}{2}\frac{1}{2}\frac{1}{2}\} & Y \\ P^2 & \{C_{2x} \mid 000\} & B \\ P^3 & \{S_{4x}^{-} \mid \frac{1}{2}\frac{1}{2}\frac{1}{2}\} & X \end{array}$$

and the group multiplication table for the corresponding double group ${}^H\mathbf{G}^{\dagger W}$ is

$$\begin{array}{cccc} E & B & X & Y \\ B & \bar{E} & \bar{Y} & X \\ X & \bar{Y} & B & E \\ Y & X & E & \bar{B} \end{array}$$

which is isomorphic with $\mathbf{G}_8^1$ where

$$P = X = \{S_{4x}^{-} \mid \tfrac{1}{2}\tfrac{1}{2}\tfrac{1}{2}\}$$

and the classes are thus

$$\begin{array}{llll} C_1 & \{E \mid 000\}, & C_5 & \{\bar{E} \mid 000\}, \\ C_2 & \{S_{4x}^{-} \mid \frac{1}{2}\frac{1}{2}\frac{1}{2}\}, & C_6 & \{\bar{S}_{4x}^{-} \mid \frac{1}{2}\frac{1}{2}\frac{1}{2}\}, \\ C_3 & \{C_{2x} \mid 000\}, & C_7 & \{\bar{C}_{2x} \mid 000\}, \\ C_4 & \{\bar{S}_{4x}^{+} \mid \frac{1}{2}\frac{1}{2}\frac{1}{2}\}, & C_8 & \{S_{4x}^{+} \mid \frac{1}{2}\frac{1}{2}\frac{1}{2}\}. \end{array}$$

The reps which lead to double-valued small reps of $\mathbf{G}^W$ are those for which $\chi(\{\bar{E} \mid 000\}) = -\chi(\{E \mid 000\})$, i.e.

		C_1	C_2	C_3	C_4	C_5	C_6	C_7	C_8	
W_5	${}^1\bar{E}_1$	1	θ	i	$-\theta^*$	-1	$-\theta$	$-$i	θ^*	3
W_6	${}^1\bar{E}_2$	1	$-\theta^*$	$-$i	θ	-1	θ^*	i	$-\theta$	3
W_7	${}^2\bar{E}_2$	1	$-\theta$	i	θ^*	-1	θ	$-$i	$-\theta^*$	3
W_8	${}^2\bar{E}_1$	1	θ^*	$-$i	$-\theta$	-1	$-\theta^*$	i	θ	3

The realities included in the extreme right-hand column indicate that the addition of time-reversal symmetry causes the double-valued small reps of $\mathbf{G}^W$ to stick together in pairs, each pair consisting of two complex conjugate reps of $\mathbf{G}^W$.

We now consider the lines of symmetry and as in the case of the study of the single-valued reps in section 5.4 this involves the theory of projective representations. We have to identify the central extension $\bar{\mathbf{G}}^{\dagger\mathbf{k}*}$ and its representations as the major step towards finding the reps of $\mathbf{G}^{\dagger\mathbf{k}}$ and hence the double-valued small reps of $\mathbf{G}^{\mathbf{k}}$. $\bar{\mathbf{G}}^{\dagger\mathbf{k}}$ will, of course, contain twice as many elements as $\bar{\mathbf{G}}^{\mathbf{k}}$. The multiplication rule of $\bar{\mathbf{G}}^{\mathbf{k}*}$ is

$$(H_j, \alpha)(H_k, \beta) = (H_jH_k, \alpha + \beta + a(H_j, H_k)) \tag{3.7.31}$$

and if we proceed to the double group then the value of $a(H_j, H_k)$ is unaltered. This is because $a(H_j, H_k)$ is determined by

$$\mu(H_j, H_k) = \exp(2\pi i a(H_j, H_k)/g) \tag{3.7.27}$$

where $\mu(H_j, H_k)$, see eqn. (3.7.37), depends only on the spatial properties of the space-group elements $\{S_j \mid \mathbf{w}_j\}$ and $\{S_k \mid \mathbf{w}_k\}$. This means that

$$a(H_j, H_k) = a(H_j, \bar{H}_k) = a(\bar{H}_j, H_k) = a(\bar{H}_j, \bar{H}_k) \tag{6.4.1}$$

where H_j, $\bar{H}_j$, H_k, and $\bar{H}_k$ derive from the elements $\{S_j \mid \mathbf{w}_j\}$, $\{\bar{S}_j \mid \mathbf{w}_j\}$, $\{S_k \mid \mathbf{w}_k\}$ and $\{\bar{S}_k \mid \mathbf{w}_k\}$ respectively of the double space group $\mathbf{G}^\dagger$. The multiplication rule of eqn (3.7.31) therefore becomes modified in $\bar{\mathbf{G}}^{\dagger\mathbf{k}*}$ to

$$(H_j, \alpha)(H_k, \beta) = (H_jH_k, \alpha + \beta + a(H_j, H_k)) \tag{6.4.2}$$

$$(\bar{H}_j, \alpha)(H_k, \beta) = (\bar{H}_jH_k, \alpha + \beta + a(H_j, H_k)) \tag{6.4.3}$$

$$(H_j, \alpha)(\bar{H}_k, \beta) = (H_j\bar{H}_k, \alpha + \beta + a(H_j, H_k)) \tag{6.4.4}$$

$$(\bar{H}_j, \alpha)(\bar{H}_k, \beta) = (\bar{H}_j\bar{H}_k, \alpha + \beta + a(H_j, H_k)). \tag{6.4.5}$$

The evaluation of the products H_jH_k, $\bar{H}_jH_k$, etc. has to be done using the known multiplication properties of the elements of the space group together with the multiplication tables for the double groups given in Table 6.2.

We can now proceed to the consideration of the lines of symmetry for the example that we have chosen, namely the double group of $F\bar{4}3c$ (T_d^5).

Δ: $\mathbf{k} = (\alpha, 0, \alpha)$

From section 5.4 we can see that $\bar{\mathbf{G}}^{\Delta*}$ is isomorphic to $\mathbf{G}_4^2$ with elements

E	$(E, 0)$	E
P	$(C_{2y}, 0)$	C
Q	$(\sigma_{dc}, 0)$	T
PQ	$(\sigma_{de}, 0)$	U

so that $\bar{\mathbf{G}}^{\dagger\Delta*}$ contains the eight elements

$(E, 0)$	E	$(\bar{E}, 0)$	$\bar{E}$
$(C_{2y}, 0)$	C	$(\bar{C}_{2y}, 0)$	$\bar{C}$
$(\sigma_{dc}, 0)$	T	$(\bar{\sigma}_{dc}, 0)$	$\bar{T}$
$(\sigma_{de}, 0)$	U	$(\bar{\sigma}_{de}, 0)$	$\bar{U}$.

From Table 6.2 its group multiplication table is

$$\begin{array}{cccc} E & C & T & U \\ C & \bar{E} & U & \bar{T} \\ T & \bar{U} & E & C \\ U & T & \bar{C} & \bar{E}. \end{array}$$

The rest of the table can readily be constructed from this since $\bar{E}$ commutes with all the elements of the group. This group is therefore isomorphic to $\mathbf{G}_8^5$ with

$$P = C = (C_{2y}, 0)$$

and
$$Q = T = (\sigma_{dc}, 0)$$

so that the classes of $\overline{\mathbf{G}}^{\dagger\Delta*}$ are

$$\begin{array}{ll} C_1 & (E, 0) \\ C_2 & (\bar{E}, 0) \\ C_3 & (C_{2y}, 0), (\bar{C}_{2y}, 0) \\ C_4 & (\sigma_{dc}, 0), (\bar{\sigma}_{dc}, 0) \\ C_5 & (\sigma_{de}, 0), (\bar{\sigma}_{de}, 0) \end{array}$$

There are five irreducible representations of this group of which only one leads to a double-valued representation of $\overline{\mathbf{G}}^{\Delta*}$; this is

	C_1	C_2	C_3	C_4	C_5
Δ_5 $\bar{E}$	2	−2	0	0	0

the matrices of which are given in Table 5.1. The double-valued small reps of $\mathbf{G}^{\Delta}$ can thus be found by multiplying each matrix representative by a factor $\exp(-i\mathbf{k}\,.\,\mathbf{v})$ which gives

	$\{E \mid 000\}$	$\{\bar{E} \mid 000\}$	$\{C_{2y} \mid 000\}$ $\{\bar{C}_{2y} \mid 000\}$	$\{\sigma_{dc} \mid \frac{1}{2}\frac{1}{2}\frac{1}{2}\}$ $\{\bar{\sigma}_{dc} \mid \frac{1}{2}\frac{1}{2}\frac{1}{2}\}$	$\{\sigma_{de} \mid \frac{1}{2}\frac{1}{2}\frac{1}{2}\}$ $\{\bar{\sigma}_{de} \mid \frac{1}{2}\frac{1}{2}\frac{1}{2}\}$	
Δ_5 $\bar{E}$	2	−2	0	0	0	2

as before, but where the matrices for $\{\sigma_{dc} \mid \frac{1}{2}\frac{1}{2}\frac{1}{2}\}$, $\{\bar{\sigma}_{dc} \mid \frac{1}{2}\frac{1}{2}\frac{1}{2}\}$, $\{\sigma_{de} \mid \frac{1}{2}\frac{1}{2}\frac{1}{2}\}$, and $\{\bar{\sigma}_{de} \mid \frac{1}{2}\frac{1}{2}\frac{1}{2}\}$ must all be multiplied by a factor $A = \exp(-2\pi i\alpha)$. From the reality in the extreme right-hand column the addition of time-reversal symmetry causes no extra degeneracy of the double-valued small reps of $\mathbf{G}^{\Delta}$.

Λ: $\mathbf{k} = (\alpha, \alpha, \alpha)$

In section 5.4 $\overline{\mathbf{G}}^{\Lambda*}$ was shown to be isomorphic to $\mathbf{G}_6^2$ and the complete identification of the elements was given there. The elements in $\overline{\mathbf{G}}^{\dagger\Lambda^*}$ are thus:

$$
\begin{array}{llcll}
(E, 0) & E & \quad & (\bar{E}, 0) & \bar{E} \\
(C_{31}^{+}, 0) & K & & (\bar{C}_{31}^{+}, 0) & \bar{K} \\
(C_{31}^{-}, 0) & F & & (\bar{C}_{31}^{-}, 0) & \bar{F} \\
(\sigma_{db}, 0) & Q & & (\bar{\sigma}_{db}, 0) & \bar{Q} \\
(\sigma_{de}, 0) & U & & (\bar{\sigma}_{de}, 0) & \bar{U} \\
(\sigma_{df}, 0) & I & & (\bar{\sigma}_{df}, 0) & \bar{I}.
\end{array}
$$

The group multiplication table for $\overline{\mathbf{G}}^{\dagger\Lambda^*}$ is thus

$$
\begin{array}{cccccc}
E & F & K & Q & U & I \\
F & \bar{K} & E & I & Q & \bar{U} \\
K & E & \bar{F} & U & \bar{I} & Q \\
Q & U & I & \bar{E} & \bar{F} & \bar{K} \\
U & \bar{I} & Q & \bar{K} & \bar{E} & F \\
I & Q & \bar{U} & \bar{F} & K & \bar{E}
\end{array}
$$

where the other three-quarters of the table can readily be constructed since $\bar{E}$ commutes with all the other elements. This group can be seen by inspection to be isomorphic to $\mathbf{G}_{12}^{4}$ with

$$P = F = (C_{31}^{-}, 0)$$

and

$$Q = Q = (\sigma_{db}, 0)$$

and the classes are

$$
\begin{array}{ll}
C_1 & (E, 0) \\
C_2 & (\bar{E}, 0) \\
C_3 & (C_{31}^{+}, 0), (C_{31}^{-}, 0) \\
C_4 & (\bar{C}_{31}^{+}, 0), (\bar{C}_{31}^{-}, 0) \\
C_5 & (\sigma_{db}, 0), (\bar{\sigma}_{de}, 0), (\bar{\sigma}_{df}, 0) \\
C_6 & (\bar{\sigma}_{db}, 0), (\sigma_{de}, 0), (\sigma_{df}, 0).
\end{array}
$$

The reps of $\mathbf{G}_{12}^{4}$ that lead to double-valued reps of $\mathbf{G}^{\Lambda}$ are those for which $\chi(C_2) = -\chi(C_1)$, i.e. R_3, R_4, and R_6.

		C_1	C_2	C_3	C_4	C_5	C_6
Λ_5	$^1\bar{E}$	1	−1	−1	1	i	−i
Λ_6	$^2\bar{E}$	1	−1	−1	1	−i	i
Λ_4	$\bar{E}_1$	2	−2	1	−1	0	0

The matrices for the 2-dimensional representation can be found from Table 5.1. The double-valued small reps of $\mathbf{G}^{\Lambda}$ can thus be found by multiplying each matrix representative by a factor exp $(-\mathrm{i}\mathbf{k}.\mathbf{v})$ which is $+1$ for classes C_1 to C_4 and is $B = \exp(-3\pi\mathrm{i}\alpha)$ for classes C_5 and C_6:

	$\{E\mid 000\}$	$\{\bar{E}\mid 000\}$	$\{C_{31}^+\mid 000\}$ $\{C_{31}^-\mid 000\}$	$\{\bar{C}_{31}^+\mid 000\}$ $\{\bar{C}_{31}^-\mid 000\}$	$\{\sigma_{db}\mid\frac{1}{2}\frac{1}{2}\frac{1}{2}\}$ $\{\bar{\sigma}_{de}\mid\frac{1}{2}\frac{1}{2}\frac{1}{2}\}$ $\{\bar{\sigma}_{df}\mid\frac{1}{2}\frac{1}{2}\frac{1}{2}\}$	$\{\bar{\sigma}_{db}\mid\frac{1}{2}\frac{1}{2}\frac{1}{2}\}$ $\{\sigma_{de}\mid\frac{1}{2}\frac{1}{2}\frac{1}{2}\}$ $\{\sigma_{df}\mid\frac{1}{2}\frac{1}{2}\frac{1}{2}\}$	
Λ_5 $^1\bar{E}$	1	-1	-1	1	$\mathrm{i}B$	$-\mathrm{i}B$	x
Λ_6 $^2\bar{E}$	1	-1	-1	1	$-\mathrm{i}B$	$\mathrm{i}B$	x
Λ_4 $\bar{E}_1$	2	-2	1	-1	0	0	x

The addition of time-reversal symmetry leads to a case (x) degeneracy, that is, the spectrum of the eigenvalues of a particle or quasi-particle is the same at $\mathbf{k}$ and $-\mathbf{k}$.

Σ: $\mathbf{k} = (\alpha, \alpha, 2\alpha)$

From section 5.4 we can see that $\bar{\mathbf{G}}^{\Sigma*}$ is isomorphic to $\mathbf{G}_2^1$ so that $\bar{\mathbf{G}}^{\dagger\Sigma*}$ contains the four elements

$$\begin{array}{llll} (E, 0) & E & (\bar{E}, 0) & \bar{E} \\ (\sigma_{db}, 0) & Q & (\bar{\sigma}_{db}, 0) & \bar{Q}. \end{array}$$

The group multiplication table is

$$\begin{array}{cccc} E & Q & \bar{E} & \bar{Q} \\ Q & \bar{E} & \bar{Q} & E \\ \bar{E} & \bar{Q} & E & Q \\ \bar{Q} & E & Q & \bar{E} \end{array}$$

which is clearly isomorphic to $\mathbf{G}_4^1$ with $P = Q = (\sigma_{db}, 0)$ and the representations that lead to double-valued small reps of $\mathbf{G}^{\Sigma}$ are thus R_2 and R_4, that is

	$(E, 0)$	$(\sigma_{db}, 0)$	$(\bar{E}, 0)$	$(\bar{\sigma}_{db}, 0)$
Σ_3 $^2\bar{E}$	1	i	-1	$-$i
Σ_4 $^1\bar{E}$	1	$-$i	-1	i

The double-valued small reps of $\mathbf{G}^{\Sigma}$ are therefore

	$\{E\mid 000\}$	$\{\sigma_{db}\mid\frac{1}{2}\frac{1}{2}\frac{1}{2}\}$	$\{\bar{E}\mid 000\}$	$\{\bar{\sigma}_{db}\mid\frac{1}{2}\frac{1}{2}\frac{1}{2}\}$	
Σ_3 $^2\bar{E}$	1	$\mathrm{i}C$	-1	$-\mathrm{i}C$	2
Σ_4 $^1\bar{E}$	1	$-\mathrm{i}C$	-1	$\mathrm{i}C$	2

where $C = \exp(-4\pi\mathrm{i}\alpha)$. The addition of time-reversal symmetry leads to no extra degeneracy of the double-valued small reps of $\mathbf{G}^{\Sigma}$.

S: $\mathbf{k} = (\frac{1}{2} + \alpha, 2\alpha, \frac{1}{2} + \alpha)$

This is similar to the case of Σ with $\{\sigma_{de}\mid\frac{1}{2}\frac{1}{2}\frac{1}{2}\}$ replacing $\{\sigma_{db}\mid\frac{1}{2}\frac{1}{2}\frac{1}{2}\}$ so that the double-valued small reps of $\mathbf{G}^S$ are

	$\{E \mid 000\}$	$\{\sigma_{de} \mid \frac{1}{2}\frac{1}{2}\frac{1}{2}\}$	$\{\bar{E} \mid 000\}$	$\{\bar{\sigma}_{de} \mid \frac{1}{2}\frac{1}{2}\frac{1}{2}\}$	
S_3 $^2\bar{E}$	1	$\mathrm{i}D$	-1	$-\mathrm{i}D$	2
S_4 $^1\bar{E}$	1	$-\mathrm{i}D$	-1	$\mathrm{i}D$	2

where $D = \exp(-\pi\mathrm{i}(1 + 4\alpha))$. The addition of time-reversal symmetry causes no extra degeneracy of the double-valued small reps of $\mathbf{G}^S$.

Z: $\mathbf{k} = (\frac{1}{2}, \alpha, \frac{1}{2} + \alpha)$

This again is similar to the case of Σ with $\{C_{2x} \mid 000\}$ replacing $\{\sigma_{db} \mid \frac{1}{2}\frac{1}{2}\frac{1}{2}\}$ so that the double-valued small reps of $\mathbf{G}^Z$ are

	$\{E \mid 000\}$	$\{C_{2x} \mid 000\}$	$\{\bar{E} \mid 000\}$	$\{\bar{C}_{2x} \mid 000\}$	
Z_3 $^2\bar{E}$	1	i	-1	$-$i	2
Z_4 $^1\bar{E}$	1	$-$i	-1	i	2

The addition of time-reversal symmetry causes no extra degeneracy of the double-valued small reps of $\mathbf{G}^Z$.

Q: $\mathbf{k} = (\frac{1}{2}, \frac{1}{2} - \alpha, \frac{1}{2} + \alpha)$

$\bar{\mathbf{G}}^{\dagger Q^*}$ is $\mathbf{G}_2^1$ so that the allowed double-valued small rep of $\mathbf{G}^Q$ is

	$\{E \mid 000\}$	$\{\bar{E} \mid 000\}$	
Q_2 $\bar{A}$	1	-1	1

and the addition of time-reversal symmetry causes the eigenfunctions belonging to $\bar{A}$ (Q_2) to become degenerate in pairs.

This completes the discussion of the example of the derivation of the double-valued reps of the space group $F\bar{4}3c$ (T_d^5). It is possible by working in a similar fashion to obtain the double-valued reps of any of the 230 space groups.

6.5. The double-valued representations of the 230 space groups

*This section is devoted to the tabulation of the double-valued reps of the 3-dimen*sional space groups and their classification into kinds as given in Definition 1.3.7 and for which the test in section 4.6 has been used. It is analogous to section 5.2 which presented similar tables for the single-valued reps. The tabulation of the double-valued reps and their labels is presented in Tables 6.13 and 6.14 in a similar way to that of the single-valued reps. Table 6.13 is analogous to Table 5.7 and

Table 6.14 is analogous to Table 5.8, that is, Table 6.13 identifies the double-valued reps in terms of one of the abstract groups of Table 5.1 and Table 6.14 identifies the labels used for the various double-valued reps of the space groups. With the notes to these tables, together with the example of the use of Tables 5.7 and 5.8 considered in section 5.4 and the example already considered in section 6.4 it should be fairly easy for the reader to obtain the double-valued small reps of $\mathbf{G}^{\mathbf{k}}$ for any wave vector $\mathbf{k}$ in the basic domain of any one of the 230 space groups. We give again in Table 6.13, in coded form, references to the work of other authors on each space group. The space-group reps are labelled both in an extension of the Mulliken (1933) notation and an extension of the Γ notation used by various authors. The extra degeneracies that may arise if time-reversal symmetry is present in a crystal can also be identified from Table 6.13 (see Note (vii) to Table 6.13).

In deriving the double-valued reps of ${}^{H}\mathbf{G}^{\mathbf{k}}$ or $\overline{\mathbf{G}}^{\mathbf{k}*}$ for a point or line of symmetry, respectively, we have used the reps of some abstract group $\mathbf{G}_2$ whose order is twice the order of the abstract group $\mathbf{G}_1$ that was used in Chapter 5 for finding the single-valued reps of ${}^{H}\mathbf{G}^{\mathbf{k}}$ or $\overline{\mathbf{G}}^{\mathbf{k}*}$. We have already noted that the single-valued reps of ${}^{H}\mathbf{G}^{\mathbf{k}}$ or $\overline{\mathbf{G}}^{\mathbf{k}*}$ are also contained among the reps of $\mathbf{G}_2$. We have not, however, in Table 6.13 identified explicitly the reps of $\mathbf{G}_2$ that lead to single-valued small reps of $\mathbf{G}^{\mathbf{k}}$, because this would duplicate the results already given in Table 5.7 and would considerably enlarge Table 6.13.‡ On the other hand, we chose to identify the single-valued small reps of $\mathbf{G}^{\mathbf{k}}$ separately in Chapter 5 so that those readers who only require the single-valued small reps of $\mathbf{G}^{\mathbf{k}}$ would not have to handle abstract groups $\mathbf{G}_2$ of twice the necessary order. Thus the reader who needs only the single-valued space-group reps profits at the slight expense of the reader who needs both the single-valued and double-valued space-group reps.

To obtain all physically significant double-valued space-group reps it is necessary to consider all $\mathbf{k}$ vectors in the representation domain rather than just the basic domain. The rules for constructing the small reps of $\mathbf{G}^{\mathbf{k}_B}$, where $\mathbf{k}_B$ is within the representation domain but outside the basic domain have already been discussed completely in section 5.5; they can be applied directly to double-valued space-group reps. It is only necessary to provide additional discussion here of the one exceptional space group $Pa3$ (T_h^6). The representation domain of the space group $Pa3$ (T_h^6) was identified in Fig. 5.4. In Table 6.15 we present, in a table similar to Table 5.11, the identification of the double-valued small reps of $\mathbf{G}^{\mathbf{k}}$ for the additional special $\mathbf{k}$ vector Z' which is in the representation domain but lies outside the basic domain and is not in the star of any $\mathbf{k}$ vector in the basic domain. As was found for the single-valued reps considered in Chapter 5, we also find for the double-valued reps that there is a difference between the degeneracies at Z and Z'. Whilst at Z' there are four non-degenerate double-valued reps there is one twofold degenerate double-

‡ We have, however, used this compatibility to check the derivation of table 6.13.

valued rep at Z. Once again this difference between the degeneracies at Z and at Z' is a useful demonstration of the importance of considering the whole representation domain, Φ, and not just the basic domain, Ω. This example appears to be the only case where the small reps of $\mathbf{G}^{\mathbf{k}_A}$ and $\mathbf{G}^{\mathbf{k}_B}$ have different dimensionalities. It can be seen from Tables 5.7, 5.11, 6.13, and 6.15 that the anomaly is not removed when time-reversal symmetry is added to the space group symmetry. When time-reversal symmetry is included the degeneracy of the twofold degenerate double-valued rep at Z is doubled again (see Table 6.13) and at Z' the addition of time-reversal symmetry causes the four non-degenerate reps to stick together in pairs (see Table 6.15).

TABLE 6.13

The double-valued reps of the 230 space groups

1 $P1$ C_1^1

($F1$; $K7$; $M5$; $Z1$.)

Γ $\mathbf{G}_2^1$: $\{\bar{E}\,|\,000\}$: 2, 1: *a*.
B $\mathbf{G}_2^1 \otimes \mathbf{T}_2$: $\{\bar{E}\,|\,000\}$; $\mathbf{t}_1$: 2, 1: *a*.
F $\mathbf{G}_2^1 \otimes \mathbf{T}_2$: $\{\bar{E}\,|\,000\}$; $\mathbf{t}_2$: 2, 1: *a*.
G $\mathbf{G}_2^1 \otimes \mathbf{T}_2$: $\{\bar{E}\,|\,000\}$; $\mathbf{t}_3$: 2, 1: *a*.

2 $P\bar{1}$ C_i^1

($F1$; $K7$; $M5$; $Z1$.)

Γ $\mathbf{G}_4^2$: $\{I\,|\,000\}$, $\{\bar{E}\,|\,000\}$: 2, 1; 4, 1: *a*.
B $\mathbf{G}_4^2 \otimes \mathbf{T}_2$: $\{I\,|\,000\}$, $\{\bar{E}\,|\,000\}$; $\mathbf{t}_1$: 2, 1; 4, 1: *a*.
F $\mathbf{G}_4^2 \otimes \mathbf{T}_2$: $\{I\,|\,000\}$, $\{\bar{E}\,|\,000\}$; $\mathbf{t}_2$: 2, 1; 4, 1: *a*.
G $\mathbf{G}_4^2 \otimes \mathbf{T}_2$: $\{I\,|\,000\}$, $\{\bar{E}\,|\,000\}$; $\mathbf{t}_3$: 2, 1; 4, 1: *a*.

3 $P2$ C_2^1

($F1$; $K7$; $M5$; $S15$; $Z1$.)

Γ $\mathbf{G}_4^1$: $\{C_{2z}\,|\,000\}$: 2, 3; 4, 3: *b*.
B $\mathbf{G}_4^1 \otimes \mathbf{T}_2$: $\{C_{2z}\,|\,000\}$; $\mathbf{t}_1$: 2, 3; 4, 3: *b*.
Y $\mathbf{G}_4^1 \otimes \mathbf{T}_2$: $\{C_{2z}\,|\,000\}$; $\mathbf{t}_2$: 2, 3; 4, 3: *b*.
Z $\mathbf{G}_4^1 \otimes \mathbf{T}_2$: $\{C_{2z}\,|\,000\}$; $\mathbf{t}_3$: 2, 3; 4, 3: *b*.
C $\mathbf{G}_4^1 \otimes \mathbf{T}_2$: $\{C_{2z}\,|\,000\}$; $\mathbf{t}_2$ or $\mathbf{t}_3$: 2, 3; 4, 3: *b*.
D $\mathbf{G}_4^1 \otimes \mathbf{T}_2$: $\{C_{2z}\,|\,000\}$; $\mathbf{t}_1$ or $\mathbf{t}_3$: 2, 3; 4, 3: *b*.
A $\mathbf{G}_4^1 \otimes \mathbf{T}_2$: $\{C_{2z}\,|\,000\}$; $\mathbf{t}_1$ or $\mathbf{t}_2$: 2, 3; 4, 3: *b*.
E $\mathbf{G}_4^1 \otimes \mathbf{T}_2$: $\{C_{2z}\,|\,000\}$; $\mathbf{t}_1$ or $\mathbf{t}_2$ or $\mathbf{t}_3$: 2, 3; 4, 3: *b*.

Λ^x $\mathbf{G}_4^1$: $(C_{2z}, 0)$: 2, x; 4, x: *b*.
V^x $\mathbf{G}_4^1$: $(C_{2z}, 0)$: 2, x; 4, x: *b*.
W^x $\mathbf{G}_4^1$: $(C_{2z}, 0)$: 2, x; 4, x: *b*.
U^x $\mathbf{G}_4^1$: $(C_{2z}, 0)$: 2, x; 4, x: *b*.

4 *P*2_1 C_2^2

(*F*1; *K*7; *M*5; *S*15; *Z*1.)

Γ $\mathbf{G}_4^1$: $\{C_{2z}\,|\,00\frac{1}{2}\}$: 2, 3; 4, 3: *b*.
B $\mathbf{G}_4^1 \otimes \mathbf{T}_2$: $\{C_{2z}\,|\,00\frac{1}{2}\}$; $\mathbf{t}_1$: 2, 3; 4, 3: *b*.
Y $\mathbf{G}_4^1 \otimes \mathbf{T}_2$: $\{C_{2z}\,|\,00\frac{1}{2}\}$; $\mathbf{t}_2$: 2, 3; 4, 3: *b*.
Z $\mathbf{G}_8^2$: $\{C_{2z}\,|\,00\frac{1}{2}\}$, $\{\bar{E}\,|\,000\}$: 5, 1; 7, 1: *a*.
C $\mathbf{G}_8^2$: $\{C_{2z}\,|\,00\frac{1}{2}\}$, $\{\bar{E}\,|\,000\}$: 5, 1; 7, 1: *a*.
D $\mathbf{G}_8^2$: $\{C_{2z}\,|\,00\frac{1}{2}\}$, $\{\bar{E}\,|\,000\}$: 5, 1; 7, 1: *a*.
A $\mathbf{G}_4^1 \otimes \mathbf{T}_2$: $\{C_{2z}\,|\,00\frac{1}{2}\}$; $\mathbf{t}_1$ or $\mathbf{t}_2$: 2, 3; 4, 3: *b*.
E $\mathbf{G}_8^2$: $\{C_{2z}\,|\,00\frac{1}{2}\}$, $\{\bar{E}\,|\,000\}$: 5, 1; 7, 1: *a*.

Λ^x $\mathbf{G}_4^1$: $(C_{2z}, 0)$: 2, *x*; 4, *x*: *b*.
V^x $\mathbf{G}_4^1$: $(C_{2z}, 0)$: 2, *x*; 4, *x*: *b*.
W^x $\mathbf{G}_4^1$: $(C_{2z}, 0)$: 2, *x*; 4, *x*: *b*.
U^x $\mathbf{G}_4^1$: $(C_{2z}, 0)$: 2, *x*; 4, *x*: *b*.

5 *B*2 C_2^3

(*F*1; *K*7; *M*5; *S*15; *Z*1.)

Γ $\mathbf{G}_4^1$: $\{C_{2z}\,|\,000\}$: 2, 3; 4, 3: *b*.
A $\mathbf{G}_4^1 \otimes \mathbf{T}_2$: $\{C_{2z}\,|\,000\}$; $\mathbf{t}_1$: 2, 3; 4, 3: *b*.
Z $\mathbf{G}_4^1 \otimes \mathbf{T}_2$: $\{C_{2z}\,|\,000\}$; $\mathbf{t}_2$ or $\mathbf{t}_3$: 2, 3; 4, 3: *b*.
M $\mathbf{G}_4^1 \otimes \mathbf{T}_2$: $\{C_{2z}\,|\,000\}$; $\mathbf{t}_1$ or $\mathbf{t}_2$ or $\mathbf{t}_3$: 2, 3; 4, 3: *b*.
L $\mathbf{G}_2^1 \otimes \mathbf{T}_2$: $\{\bar{E}\,|\,000\}$; $\mathbf{t}_1$ or $\mathbf{t}_3$: 2, 1: *a*.
V $\mathbf{G}_2^1 \otimes \mathbf{T}_2$: $\{\bar{E}\,|\,000\}$; $\mathbf{t}_3$: 2, 1: *a*.

Λ^x $\mathbf{G}_4^1$: $(C_{2z}, 0)$: 2, *x*; 4, *x*: *b*.
U^x $\mathbf{G}_4^1$: $(C_{2z}, 0)$: 2, *x*; 4, *x*: *b*.

6 *Pm* C_{1h}^1

(*F*1; *K*7; *M*5; *S*15; *Z*1.)

Γ $\mathbf{G}_4^1$: $\{\sigma_z\,|\,000\}$: 2, 3; 4, 3: *b*.
B $\mathbf{G}_4^1 \otimes \mathbf{T}_2$: $\{\sigma_z\,|\,000\}$; $\mathbf{t}_1$: 2, 3; 4, 3: *b*.
Y $\mathbf{G}_4^1 \otimes \mathbf{T}_2$: $\{\sigma_z\,|\,000\}$; $\mathbf{t}_2$: 2, 3; 4, 3: *b*.
Z $\mathbf{G}_4^1 \otimes \mathbf{T}_2$: $\{\sigma_z\,|\,000\}$; $\mathbf{t}_3$: 2, 3; 4, 3: *b*.
C $\mathbf{G}_4^1 \otimes \mathbf{T}_2$: $\{\sigma_z\,|\,000\}$; $\mathbf{t}_2$ or $\mathbf{t}_3$: 2, 3; 4, 3: *b*.
D $\mathbf{G}_4^1 \otimes \mathbf{T}_2$: $\{\sigma_z\,|\,000\}$; $\mathbf{t}_1$ or $\mathbf{t}_3$: 2, 3; 4, 3: *b*.
A $\mathbf{G}_4^1 \otimes \mathbf{T}_2$: $\{\sigma_z\,|\,000\}$; $\mathbf{t}_1$ or $\mathbf{t}_2$: 2, 3; 4, 3: *b*.
E $\mathbf{G}_4^1 \otimes \mathbf{T}_2$: $\{\sigma_z\,|\,000\}$; $\mathbf{t}_1$ or $\mathbf{t}_2$ or $\mathbf{t}_3$: 2, 3; 4, 3: *b*.

Λ^x $\mathbf{G}_2^1$: $(\bar{E}, 0)$: 2, 2: *a*.
V^x $\mathbf{G}_2^1$: $(\bar{E}, 0)$: 2, 2: *a*.
W^x $\mathbf{G}_2^1$: $(\bar{E}, 0)$: 2, 2: *a*.
U^x $\mathbf{G}_2^1$: $(\bar{E}, 0)$: 2, 2: *a*.

7 *Pb* C_{1h}^2

(*F*1; *K*7; *M*5; *S*15; *Z*1.)

Γ $\mathbf{G}_4^1$: $\{\sigma_z \mid \frac{1}{2}00\}$: 2, 3; 4, 3: *b*.
B $\mathbf{G}_8^2$: $\{\sigma_z \mid \frac{1}{2}00\}$, $\{\bar{E} \mid 000\}$: 5, 1; 7, 1: *a*.
Y $\mathbf{G}_4^1 \otimes \mathbf{T}_2$: $\{\sigma_z \mid \frac{1}{2}00\}$; $\mathbf{t}_2$: 2, 3; 4, 3: *b*.
Z $\mathbf{G}_4^1 \otimes \mathbf{T}_2$: $\{\sigma_z \mid \frac{1}{2}00\}$; $\mathbf{t}_3$: 2, 3; 4, 3: *b*.
C $\mathbf{G}_4^1 \otimes \mathbf{T}_2$: $\{\sigma_z \mid \frac{1}{2}00\}$; $\mathbf{t}_2$ or $\mathbf{t}_3$: 2, 3; 4, 3: *b*.
D $\mathbf{G}_8^2$: $\{\sigma_z \mid \frac{1}{2}00\}$, $\{\bar{E} \mid 000\}$: 5, 1; 7, 1: *a*.
A $\mathbf{G}_8^2$: $\{\sigma_z \mid \frac{1}{2}00\}$, $\{\bar{E} \mid 000\}$: 5, 1; 7, 1: *a*.
E $\mathbf{G}_8^2$: $\{\sigma_z \mid \frac{1}{2}00\}$, $\{\bar{E} \mid 000\}$: 5, 1; 7, 1: *a*.

Λ^x $\mathbf{G}_2^1$: $(\bar{E}, 0)$: 2, 2: *a*.
V^x $\mathbf{G}_2^1$: $(\bar{E}, 0)$: 2, 1: *a*.
W^x $\mathbf{G}_2^1$: $(\bar{E}, 0)$: 2, 2: *a*.
U^x $\mathbf{G}_2^1$: $(\bar{E}, 0)$: 2, 1: *a*.

8 *Bm* C_{1h}^3

(*F*1; *K*7; *M*5; *S*15; *Z*1.)

Γ $\mathbf{G}_4^1$: $\{\sigma_z \mid 000\}$: 2, 3; 4, 3: *b*.
A $\mathbf{G}_4^1 \otimes \mathbf{T}_2$: $\{\sigma_z \mid 000\}$; $\mathbf{t}_1$: 2, 3; 4, 3: *b*.
Z $\mathbf{G}_4^1 \otimes \mathbf{T}_2$: $\{\sigma_z \mid 000\}$; $\mathbf{t}_2$ or $\mathbf{t}_3$: 2, 3; 4, 3: *b*.
M $\mathbf{G}_4^1 \otimes \mathbf{T}_2$: $\{\sigma_z \mid 000\}$; $\mathbf{t}_1$ or $\mathbf{t}_2$ or $\mathbf{t}_3$: 2, 3; 4, 3: *b*.
L $\mathbf{G}_2^1 \otimes \mathbf{T}_2$: $\{\bar{E} \mid 000\}$; $\mathbf{t}_1$ or $\mathbf{t}_3$: 2, 1: *a*.
V $\mathbf{G}_2^1 \otimes \mathbf{T}_2$: $\{\bar{E} \mid 000\}$; $\mathbf{t}_3$: 2, 1: *a*.

Λ^x $\mathbf{G}_2^1$: $(\bar{E}, 0)$: 2, 2: *a*.
U^x $\mathbf{G}_2^1$: $(\bar{E}, 0)$: 2, 2: *a*.

9 *Bb* C_{1h}^4

(*F*1; *K*7; *M*5; *S*15; *Z*1.)

Γ $\mathbf{G}_4^1$: $\{\sigma_z \mid \frac{1}{2}00\}$: 2, 3; 4, 3: *b*.
A $\mathbf{G}_4^1 \otimes \mathbf{T}_2$: $\{\sigma_z \mid \frac{1}{2}00\}$; $\mathbf{t}_1$: 1, 1; 3, 1: *c*.
Z $\mathbf{G}_4^1 \otimes \mathbf{T}_2$: $\{\sigma_z \mid \frac{1}{2}00\}$; $\mathbf{t}_2$ or $\mathbf{t}_3$: 2, 3; 4, 3: *b*.
M $\mathbf{G}_4^1 \otimes \mathbf{T}_2$: $\{\sigma_z \mid \frac{1}{2}00\}$; $\mathbf{t}_1$ or $\mathbf{t}_2$ or $\mathbf{t}_3$: 1, 1; 3, 1: *c*.
L $\mathbf{G}_2^1 \otimes \mathbf{T}_2$: $\{\bar{E} \mid 000\}$; $\mathbf{t}_1$ or $\mathbf{t}_3$: 2, 1: *a*.
V $\mathbf{G}_2^1 \otimes \mathbf{T}_2$: $\{\bar{E} \mid 000\}$; $\mathbf{t}_3$: 2, 1: *a*.

Λ^x $\mathbf{G}_2^1$: $(\bar{E}, 0)$: 2, 2: *a*.
U^x $\mathbf{G}_2^1$: $(\bar{E}, 0)$: 2, 1: *a*.

10 $P2/m$ C^1_{2h}

($F1$; $K7$; $M5$; $S15$; $Z1$.)

Γ $\mathbf{G}^2_8$: $\{C_{2z}|000\}$, $\{I|000\}$: 2, 3; 4, 3; 6, 3; 8, 3: b.
B $\mathbf{G}^2_8 \otimes \mathbf{T}_2$: $\{C_{2z}|000\}$, $\{I|000\}$; $\mathbf{t}_1$: 2, 3; 4, 3; 6, 3; 8, 3: b.
Y $\mathbf{G}^2_8 \otimes \mathbf{T}_2$: $\{C_{2z}|000\}$, $\{I|000\}$; $\mathbf{t}_2$: 2, 3; 4, 3; 6, 3; 8, 3: b.
Z $\mathbf{G}^2_8 \otimes \mathbf{T}_2$: $\{C_{2z}|000\}$, $\{I|000\}$; $\mathbf{t}_3$: 2, 3; 4, 3; 6, 3; 8, 3: b.
C $\mathbf{G}^2_8 \otimes \mathbf{T}_2$: $\{C_{2z}|000\}$, $\{I|000\}$; $\mathbf{t}_2$ or $\mathbf{t}_3$: 2, 3; 4, 3; 6, 3; 8, 3: b.
D $\mathbf{G}^2_8 \otimes \mathbf{T}_2$: $\{C_{2z}|000\}$, $\{I|000\}$; $\mathbf{t}_1$ or $\mathbf{t}_3$: 2, 3; 4, 3; 6, 3; 8, 3: b.
A $\mathbf{G}^2_8 \otimes \mathbf{T}_2$: $\{C_{2z}|000\}$, $\{I|000\}$; $\mathbf{t}_1$ or $\mathbf{t}_2$: 2, 3; 4, 3; 6, 3; 8, 3: b.
E $\mathbf{G}^2_8 \otimes \mathbf{T}_2$: $\{C_{2z}|000\}$, $\{I|000\}$; $\mathbf{t}_1$ or $\mathbf{t}_2$ or $\mathbf{t}_3$: 2, 3; 4, 3; 6, 3; 8, 3: b.

Λ^x $\mathbf{G}^1_4$: $(C_{2z}, 0)$: 2, 3; 4, 3: b.
V^x $\mathbf{G}^1_4$: $(C_{2z}, 0)$: 2, 3; 4, 3: b.
W^x $\mathbf{G}^1_4$: $(C_{2z}, 0)$: 2, 3; 4, 3: b.
U^x $\mathbf{G}^1_4$: $(C_{2z}, 0)$: 2, 3; 4, 3: b.

11 $P2_1/m$ C^2_{2h}

($F1$; $K7$; $M5$; $S15$; $Z1$.)

Γ $\mathbf{G}^2_8$: $\{C_{2z}|00\frac{1}{2}\}$, $\{I|00\frac{1}{2}\}$: 2, 3; 4, 3; 6, 3; 8, 3: b.
B $\mathbf{G}^2_8 \otimes \mathbf{T}_2$: $\{C_{2z}|00\frac{1}{2}\}$, $\{I|00\frac{1}{2}\}$; $\mathbf{t}_1$: 2, 3; 4, 3; 6, 3; 8, 3: b.
Y $\mathbf{G}^2_8 \otimes \mathbf{T}_2$: $\{C_{2z}|00\frac{1}{2}\}$, $\{I|00\frac{1}{2}\}$; $\mathbf{t}_2$: 2, 3; 4, 3; 6, 3; 8, 3: b.
Z $\mathbf{G}^{10}_{16}$: $\{C_{2z}|00\frac{1}{2}\}$, $\{E|001\}$, $\{I|00\frac{1}{2}\}$: 9, 1: a.
C $\mathbf{G}^{10}_{16}$: $\{C_{2z}|00\frac{1}{2}\}$, $\{E|001\}$, $\{I|00\frac{1}{2}\}$: 9, 1: a.
D $\mathbf{G}^{10}_{16}$: $\{C_{2z}|00\frac{1}{2}\}$, $\{E|001\}$, $\{I|00\frac{1}{2}\}$: 9, 1: a.
A $\mathbf{G}^2_8 \otimes \mathbf{T}_2$: $\{C_{2z}|00\frac{1}{2}\}$, $\{I|00\frac{1}{2}\}$; $\mathbf{t}_1$ or $\mathbf{t}_2$: 2, 3; 4, 3; 6, 3; 8, 3: b.
E $\mathbf{G}^{10}_{16}$: $\{C_{2z}|00\frac{1}{2}\}$, $\{E|001\}$, $\{I|00\frac{1}{2}\}$: 9, 1: a.

Λ^x $\mathbf{G}^1_4$: $(C_{2z}, 0)$: 2, 3; 4, 3: b.
V^x $\mathbf{G}^1_4$: $(C_{2z}, 0)$: 2, 3; 4, 3: b.
W^x $\mathbf{G}^1_4$: $(C_{2z}, 0)$: 2, 3; 4, 3: b.
U^x $\mathbf{G}^1_4$: $(C_{2z}, 0)$: 2, 3; 4, 3: b.

12 $B2/m$ C^3_{2h}

($F1$; $K7$; $M5$; $S15$; $Z1$.)

Γ $\mathbf{G}^2_8$: $\{C_{2z}|000\}$, $\{I|000\}$: 2, 3; 4, 3; 6, 3; 8, 3: b.
A $\mathbf{G}^2_8 \otimes \mathbf{T}_2$: $\{C_{2z}|000\}$, $\{I|000\}$; $\mathbf{t}_1$: 2, 3; 4, 3; 6, 3; 8, 3: b.
Z $\mathbf{G}^2_8 \otimes \mathbf{T}_2$: $\{C_{2z}|000\}$, $\{I|000\}$; $\mathbf{t}_2$ or $\mathbf{t}_3$: 2, 3; 4, 3; 6, 3; 8, 3: b.
M $\mathbf{G}^2_8 \otimes \mathbf{T}_2$: $\{C_{2z}|000\}$, $\{I|000\}$; $\mathbf{t}_1$ or $\mathbf{t}_2$ or $\mathbf{t}_3$: 2, 3; 4, 3; 6, 3; 8, 3: b.
L $\mathbf{G}^2_4 \otimes \mathbf{T}_2$: $\{I|000\}$, $\{\bar{E}|000\}$; $\mathbf{t}_1$ or $\mathbf{t}_3$: 2, 1; 4, 1: a.
V $\mathbf{G}^2_4 \otimes \mathbf{T}_2$: $\{I|000\}$, $\{\bar{E}|000\}$; $\mathbf{t}_3$: 2, 1; 4, 1: a.

Λ^x $\mathbf{G}^1_4$: $(C_{2z}, 0)$: 2, 3; 4, 3: b.
U^x $\mathbf{G}^1_4$: $(C_{2z}, 0)$: 2, 3; 4, 3: b.

13 $P2/b$ C_{2h}^4

($F1$; $K7$; $M5$; $S15$; $Z1$.)

Γ $\mathbf{G}_8^2$: $\{C_{2z}|000\}$, $\{I|\frac{1}{2}00\}$: 2, 3; 4, 3; 6, 3; 8, 3: b.
B $\mathbf{G}_{16}^{10}$: $\{\sigma_z|\frac{1}{2}00\}$, $\{E|100\}$, $\{I|\frac{1}{2}00\}$: 9, 1: a.
Y $\mathbf{G}_8^2 \otimes \mathbf{T}_2$: $\{C_{2z}|000\}$, $\{I|\frac{1}{2}00\}$; $\mathbf{t}_2$: 2, 3; 4, 3; 6, 3; 8, 3: b.
Z $\mathbf{G}_8^2 \otimes \mathbf{T}_2$: $\{C_{2z}|000\}$, $\{I|\frac{1}{2}00\}$; $\mathbf{t}_3$: 2, 3; 4, 3; 6, 3; 8, 3: b.
C $\mathbf{G}_8^2 \otimes \mathbf{T}_2$: $\{C_{2z}|000\}$, $\{I|\frac{1}{2}00\}$; $\mathbf{t}_2$ or $\mathbf{t}_3$: 2, 3; 4, 3; 6, 3; 8, 3: b.
D $\mathbf{G}_{16}^{10}$: $\{\sigma_z|\frac{1}{2}00\}$, $\{E|100\}$, $\{I|\frac{1}{2}00\}$: 9, 1: a.
A $\mathbf{G}_{16}^{10}$: $\{\sigma_z|\frac{1}{2}00\}$, $\{E|100\}$, $\{I|\frac{1}{2}00\}$: 9, 1: a.
E $\mathbf{G}_{16}^{10}$: $\{\sigma_z|\frac{1}{2}00\}$, $\{E|100\}$, $\{I|\frac{1}{2}00\}$: 9, 1: a.

Λ^x $\mathbf{G}_4^1$: $(C_{2z}, 0)$: 2, 3; 4, 3: b.
V^x $\mathbf{G}_4^1$: $(C_{2z}, 0)$: 2, 1; 4, 1: b.
W^x $\mathbf{G}_4^1$: $(C_{2z}, 0)$: 2, 3; 4, 3: b.
U^x $\mathbf{G}_4^1$: $(C_{2z}, 0)$: 2, 1; 4, 1: b.

14 $P2_1/b$ C_{2h}^5

($F1$; $J2$; $K7$; $M5$; $S15$; $Z1$.)

Γ $\mathbf{G}_8^2$: $\{C_{2z}|00\frac{1}{2}\}$, $\{I|\frac{1}{2}0\frac{1}{2}\}$: 2, 3; 4, 3; 6, 3; 8, 3: b.
B $\mathbf{G}_{16}^{10}$: $\{\sigma_z|\frac{1}{2}00\}$, $\{E|100\}$, $\{I|\frac{1}{2}0\frac{1}{2}\}$: 9, 1: a.
Y $\mathbf{G}_8^2 \otimes \mathbf{T}_2$: $\{C_{2z}|00\frac{1}{2}\}$, $\{I|\frac{1}{2}0\frac{1}{2}\}$; $\mathbf{t}_2$: 2, 3; 4, 3; 6, 3; 8, 3: b.
Z $\mathbf{G}_{16}^{10}$: $\{C_{2z}|00\frac{1}{2}\}$, $\{E|001\}$, $\{I|\frac{1}{2}0\frac{1}{2}\}$: 9, 1: a.
C $\mathbf{G}_{16}^{10}$: $\{C_{2z}|00\frac{1}{2}\}$, $\{E|001\}$, $\{I|\frac{1}{2}0\frac{1}{2}\}$: 9, 1: a.
D $\mathbf{G}_8^2 \otimes \mathbf{T}_2$: $\{C_{2z}|00\frac{1}{2}\}$, $\{I|\frac{1}{2}0\frac{1}{2}\}$; $\mathbf{t}_1$ or $\mathbf{t}_3$: 1, 1; 3, 1; 5, 1; 7, 1: c.
A $\mathbf{G}_{16}^{10}$: $\{\sigma_z|\frac{1}{2}00\}$, $\{E|100\}$, $\{I|\frac{1}{2}0\frac{1}{2}\}$: 9, 1: a.
E $\mathbf{G}_8^2 \otimes \mathbf{T}_2$: $\{C_{2z}|00\frac{1}{2}\}$, $\{I|\frac{1}{2}0\frac{1}{2}\}$; $\mathbf{t}_1$ or $\mathbf{t}_2$ or $\mathbf{t}_3$: 1, 1; 3, 1; 5, 1; 7, 1: c.

Λ^x $\mathbf{G}_4^1$: $(C_{2z}, 0)$: 2, 3; 4, 3: b.
V^x $\mathbf{G}_4^1$: $(C_{2z}, 0)$: 2, 1; 4, 1: b.
W^x $\mathbf{G}_4^1$: $(C_{2z}, 0)$: 2, 3; 4, 3: b.
U^x $\mathbf{G}_4^1$: $(C_{2z}, 0)$: 2, 1; 4, 1: b.

15 $B2/b$ C_{2h}^6

($C8$; $D1$; $F1$; $K7$; $M5$; $S15$; $Z1$.)

Γ $\mathbf{G}_8^2$: $\{C_{2z}|000\}$, $\{I|\frac{1}{2}00\}$: 2, 3; 4, 3; 6, 3; 8, 3: b.
A $\mathbf{G}_{16}^{10}$: $\{\sigma_z|\frac{1}{2}00\}$, $\{E|100\}$, $\{I|\frac{1}{2}00\}$: 9, 1: a.
Z $\mathbf{G}_8^2 \otimes \mathbf{T}_2$: $\{C_{2z}|000\}$, $\{I|\frac{1}{2}00\}$; $\mathbf{t}_2$ or $\mathbf{t}_3$: 2, 3; 4, 3; 6, 3; 8, 3: b.
M $\mathbf{G}_{16}^{10}$: $\{\sigma_z|\frac{1}{2}00\}$, $\{E|100\}$, $\{I|\frac{1}{2}00\}$: 9, 1: a.
L $\mathbf{G}_4^2 \otimes \mathbf{T}_2$: $\{I|\frac{1}{2}00\}$, $\{\bar{E}|000\}$; $\mathbf{t}_1$ or $\mathbf{t}_3$: 2, 1; 4, 1: a.
V $\mathbf{G}_4^2 \otimes \mathbf{T}_2$: $\{I|\frac{1}{2}00\}$, $\{\bar{E}|000\}$; $\mathbf{t}_3$: 2, 1; 4, 1: a.

Λ^x $\mathbf{G}_4^1$: $(C_{2z}, 0)$: 2, 3; 4, 3: b.
U^x $\mathbf{G}_4^1$: $(C_{2z}, 0)$: 2, 1; 4, 1: b.

16 $P222$ D_2^1

($F1$; $K7$; $M5$; $T1$; $Z1$.)

Γ $\mathbf{G}_8^5$: $\{C_{2z}|000\}, \{C_{2y}|000\}$: 5, 2: a.
Y $\mathbf{G}_8^5 \otimes \mathbf{T}_2$: $\{C_{2z}|000\}, \{C_{2y}|000\}$; $\mathbf{t}_1$: 5, 2: a.
X $\mathbf{G}_8^5 \otimes \mathbf{T}_2$: $\{C_{2z}|000\}, \{C_{2y}|000\}$; $\mathbf{t}_2$: 5, 2: a.
Z $\mathbf{G}_8^5 \otimes \mathbf{T}_2$: $\{C_{2z}|000\}, \{C_{2y}|000\}$; $\mathbf{t}_3$: 5, 2: a.
U $\mathbf{G}_8^5 \otimes \mathbf{T}_2$: $\{C_{2z}|000\}, \{C_{2y}|000\}$; $\mathbf{t}_2$ or $\mathbf{t}_3$: 5, 2: a.
T $\mathbf{G}_8^5 \otimes \mathbf{T}_2$: $\{C_{2z}|000\}, \{C_{2y}|000\}$; $\mathbf{t}_1$ or $\mathbf{t}_3$: 5, 2: a.
S $\mathbf{G}_8^5 \otimes \mathbf{T}_2$: $\{C_{2z}|000\}, \{C_{2y}|000\}$; $\mathbf{t}_1$ or $\mathbf{t}_2$: 5, 2: a.
R $\mathbf{G}_8^5 \otimes \mathbf{T}_2$: $\{C_{2z}|000\}, \{C_{2y}|000\}$; $\mathbf{t}_1$ or $\mathbf{t}_2$ or $\mathbf{t}_3$: 5, 2: a.

Δ^x $\mathbf{G}_4^1$: $(C_{2y}, 0)$: 2, 2; 4, 2: b.
D^x $\mathbf{G}_4^1$: $(C_{2y}, 0)$: 2, 2; 4, 2: b.
P^x $\mathbf{G}_4^1$: $(C_{2y}, 0)$: 2, 2; 4, 2: b.
B^x $\mathbf{G}_4^1$: $(C_{2y}, 0)$: 2, 2; 4, 2: b.
Σ^x $\mathbf{G}_4^1$: $(C_{2x}, 0)$: 2, 2; 4, 2: b.
C^x $\mathbf{G}_4^1$: $(C_{2x}, 0)$: 2, 2; 4, 2: b.
E^x $\mathbf{G}_4^1$: $(C_{2x}, 0)$: 2, 2; 4, 2: b.
A^x $\mathbf{G}_4^1$: $(C_{2x}, 0)$: 2, 2; 4, 2: b.
Λ^x $\mathbf{G}_4^1$: $(C_{2z}, 0)$: 2, 2; 4, 2: b.
H^x $\mathbf{G}_4^1$: $(C_{2z}, 0)$: 2, 2; 4, 2: b.
Q^x $\mathbf{G}_4^1$: $(C_{2z}, 0)$: 2, 2; 4, 2: b.
G^x $\mathbf{G}_4^1$: $(C_{2z}, 0)$: 2, 2; 4, 2: b.

17 $P222_1$ D_2^2

($F1$; $K7$; $M5$; $T1$; $Z1$.)

Γ $\mathbf{G}_8^5$: $\{C_{2z}|00\frac{1}{2}\}, \{C_{2y}|000\}$: 5, 2: a.
Y $\mathbf{G}_8^5 \otimes \mathbf{T}_2$: $\{C_{2z}|00\frac{1}{2}\}, \{C_{2y}|000\}$; $\mathbf{t}_1$: 5, 2: a.
X $\mathbf{G}_8^5 \otimes \mathbf{T}_2$: $\{C_{2z}|00\frac{1}{2}\}, \{C_{2y}|000\}$; $\mathbf{t}_2$: 5, 2: a.
Z $\mathbf{G}_{16}^8$: $\{C_{2z}|00\frac{1}{2}\}, \{C_{2y}|000\}$: 5, 3; 6, 3; 7, 3; 8, 3: a.
U $\mathbf{G}_{16}^8$: $\{C_{2z}|00\frac{1}{2}\}, \{C_{2y}|000\}$: 5, 3; 6, 3; 7, 3; 8, 3: a.
T $\mathbf{G}_{16}^8$: $\{C_{2z}|00\frac{1}{2}\}, \{C_{2y}|000\}$: 5, 3; 6, 3; 7, 3; 8, 3: a.
S $\mathbf{G}_8^5 \otimes \mathbf{T}_2$: $\{C_{2z}|00\frac{1}{2}\}, \{C_{2y}|000\}$; $\mathbf{t}_1$ or $\mathbf{t}_2$: 5, 2: a.
R $\mathbf{G}_{16}^8$: $\{C_{2z}|00\frac{1}{2}\}, \{C_{2y}|000\}$: 5, 3; 6, 3; 7, 3; 8, 3: a.

Δ^x $\mathbf{G}_4^1$: $(C_{2y}, 0)$: 2, 2; 4, 2: b.
D^x $\mathbf{G}_4^1$: $(C_{2y}, 0)$: 2, 2; 4, 2: b.
P^x $\mathbf{G}_4^1$: $(C_{2y}, 0)$: 2, 3; 4, 3: b.
B^x $\mathbf{G}_4^1$: $(C_{2y}, 0)$: 2, 3; 4, 3: b.
Σ^x $\mathbf{G}_4^1$: $(C_{2x}, 0)$: 2, 2; 4, 2: b.
C^x $\mathbf{G}_4^1$: $(C_{2x}, 0)$: 2, 2; 4, 2: b.
E^x $\mathbf{G}_8^2$: $(C_{2x}, 0), (\bar{E}, 0)$: 5, 3; 7, 3: a.
A^x $\mathbf{G}_8^2$: $(C_{2x}, 0), (\bar{E}, 0)$: 5, 3; 7, 3: a.
Λ^x $\mathbf{G}_4^1$: $(C_{2z}, 0)$: 2, 2; 4, 2: b.
H^x $\mathbf{G}_4^1$: $(C_{2z}, 0)$: 2, 2; 4, 2: b.
Q^x $\mathbf{G}_4^1$: $(C_{2z}, 0)$: 2, 2; 4, 2: b.
G^x $\mathbf{G}_4^1$: $(C_{2z}, 0)$: 2, 2; 4, 2: b.

18 $P2_12_12$ D_2^3

($F1$; $K7$; $M5$; $T1$; $Z1$.)

Γ $\mathbf{G}_8^5$: $\{C_{2z}\,|\,000\}, \{C_{2y}\,|\,\tfrac{1}{2}\tfrac{1}{2}0\}$: 5, 2: *a*.

Y $\mathbf{G}_{16}^8$: $\{C_{2y}\,|\,\tfrac{1}{2}\tfrac{1}{2}0\}, \{C_{2z}\,|\,000\}$: 5, 3; 6, 3; 7, 3; 8, 3: *a*.

X $\mathbf{G}_{16}^8$: $\{C_{2x}\,|\,\tfrac{1}{2}\tfrac{1}{2}0\}, \{C_{2z}\,|\,000\}$: 5, 3; 6, 3; 7, 3; 8, 3: *a*.

Z $\mathbf{G}_8^5 \otimes \mathbf{T}_2$: $\{C_{2z}\,|\,000\}, \{C_{2y}\,|\,\tfrac{1}{2}\tfrac{1}{2}0\}$; $\mathbf{t}_3$: 5, 2: *a*.

U $\mathbf{G}_{16}^8$: $\{C_{2x}\,|\,\tfrac{1}{2}\tfrac{1}{2}0\}, \{C_{2z}\,|\,000\}$: 5, 3; 6, 3; 7, 3; 8, 3: *a*.

T $\mathbf{G}_{16}^8$: $\{C_{2y}\,|\,\tfrac{1}{2}\tfrac{1}{2}0\}, \{C_{2z}\,|\,000\}$: 5, 3; 6, 3; 7, 3; 8, 3: *a*.

S $\mathbf{G}_{16}^8$: $\{C_{2z}\,|\,000\}, \{C_{2x}\,|\,\tfrac{1}{2}\tfrac{1}{2}0\}$: 9, 1: *b*.

R $\mathbf{G}_{16}^8$: $\{C_{2z}\,|\,000\}, \{C_{2x}\,|\,\tfrac{1}{2}\tfrac{1}{2}0\}$: 9, 1: *b*.

Δ^x $\mathbf{G}_4^1$: $(C_{2y}, 0)$: 2, 2; 4, 2: *b*.

D^x $\mathbf{G}_8^2$: $(C_{2y}, 0), (\bar{E}, 0)$: 5, 3; 7, 3: *a*.

P^x $\mathbf{G}_8^2$: $(C_{2y}, 0), (\bar{E}, 0)$: 5, 3; 7, 3: *a*.

B^x $\mathbf{G}_4^1$: $(C_{2y}, 0)$: 2, 2; 4, 2: *b*.

Σ^x $\mathbf{G}_4^1$: $(C_{2x}, 0)$: 2, 2; 4, 2: *b*.

C^x $\mathbf{G}_8^2$: $(C_{2x}, 0), (\bar{E}, 0)$: 5, 3; 7, 3: *a*.

E^x $\mathbf{G}_8^2$: $(C_{2x}, 0), (\bar{E}, 0)$: 5, 3; 7, 3: *a*.

A^x $\mathbf{G}_4^1$: $(C_{2x}, 0)$: 2, 2; 4, 2: *b*.

Λ^x $\mathbf{G}_4^1$: $(C_{2z}, 0)$: 2, 2; 4, 2: *b*.

H^x $\mathbf{G}_4^1$: $(C_{2z}, 0)$: 2, 3; 4, 3: *b*.

Q^x $\mathbf{G}_4^1$: $(C_{2z}, 0)$: 2, 1; 4, 1: *b*.

G^x $\mathbf{G}_4^1$: $(C_{2z}, 0)$: 2, 3; 4, 3: *b*.

19 $P2_12_12_1$ D_2^4

($F1$; $K7$; $M5$; $T1$; $Z1$.)

Γ $\mathbf{G}_8^5$: $\{C_{2z}\,|\,\tfrac{1}{2}0\tfrac{1}{2}\}, \{C_{2y}\,|\,\tfrac{1}{2}\tfrac{1}{2}0\}$: 5, 2: *a*.

Y $\mathbf{G}_{16}^8$: $\{C_{2y}\,|\,\tfrac{1}{2}\tfrac{1}{2}0\}, \{C_{2z}\,|\,\tfrac{1}{2}0\tfrac{1}{2}\}$: 5, 3; 6, 3; 7, 3; 8, 3: *a*.

X $\mathbf{G}_{16}^8$: $\{C_{2x}\,|\,0\tfrac{1}{2}\tfrac{1}{2}\}, \{C_{2z}\,|\,\tfrac{1}{2}0\tfrac{1}{2}\}$: 5, 3; 6, 3; 7, 3; 8, 3: *a*.

Z $\mathbf{G}_{16}^8$: $\{C_{2z}\,|\,\tfrac{1}{2}0\tfrac{1}{2}\}, \{C_{2y}\,|\,\tfrac{1}{2}\tfrac{1}{2}0\}$: 5, 3; 6, 3; 7, 3; 8, 3: *a*.

U $\mathbf{G}_{16}^8$: $\{C_{2y}\,|\,\tfrac{1}{2}\tfrac{1}{2}0\}, \{C_{2z}\,|\,\tfrac{1}{2}0\tfrac{1}{2}\}$: 9, 1: *b*.

T $\mathbf{G}_{16}^8$: $\{C_{2x}\,|\,0\tfrac{1}{2}\tfrac{1}{2}\}, \{C_{2y}\,|\,\tfrac{1}{2}\tfrac{1}{2}0\}$: 9, 1: *b*.

S $\mathbf{G}_{16}^8$: $\{C_{2z}\,|\,\tfrac{1}{2}0\tfrac{1}{2}\}, \{C_{2x}\,|\,0\tfrac{1}{2}\tfrac{1}{2}\}$: 9, 1: *b*.

R $\mathbf{G}_{16}^{11}$: $\{C_{2x}\,|\,0\tfrac{1}{2}\tfrac{1}{2}\}, \{C_{2y}\,|\,\tfrac{1}{2}\tfrac{1}{2}0\}, \{\bar{E}\,|\,000\}$: 6, 1; 7, 1; 8, 1; 9, 1: *a*.

Δ^x $\mathbf{G}_4^1$: $(C_{2y}, 0)$: 2, 2; 4, 2: *b*.

D^x $\mathbf{G}_8^2$: $(C_{2y}, 0), (\bar{E}, 0)$: 5, 3; 7, 3: *a*.

P^x $\mathbf{G}_8^2$: $(C_{2y}, 0), (\bar{E}, 0)$: 5, 1; 7, 1: *a*.

B^x $\mathbf{G}_4^1$: $(C_{2y}, 0)$: 2, 3; 4, 3: *b*.

Σ^x $\mathbf{G}_4^1$: $(C_{2x}, 0)$: 2, 2; 4, 2: *b*.

C^x $\mathbf{G}_4^1$: $(C_{2x}, 0)$: 2, 3; 4, 3: *b*.

E^x $\mathbf{G}_8^2$: $(C_{2x}, 0), (\bar{E}, 0)$: 5, 1; 7, 1: *a*.

A^x $\mathbf{G}_8^2$: $(C_{2x}, 0), (\bar{E}, 0)$: 5, 3; 7, 3: *a*.

Λ^x $\mathbf{G}_4^1$: $(C_{2z}, 0)$: 2, 2; 4, 2: *b*.

H^x $\mathbf{G}_8^2$: $(C_{2z}, 0), (\bar{E}, 0)$: 5, 3; 7, 3: *a*.

Q^x $\mathbf{G}_8^2$: $(C_{2z}, 0), (\bar{E}, 0)$: 5, 1; 7, 1: *a*.

G^x $\mathbf{G}_4^1$: $(C_{2z}, 0)$: 2, 3; 4, 3: *b*.

20 $C222_1$ D_2^5

($F1$; $K7$; $M5$; $T3$; $Z1$.)

Γ $\mathbf{G}_8^5$: $\{C_{2z}\,|\,00\tfrac{1}{2}\}$, $\{C_{2y}\,|\,00\tfrac{1}{2}\}$: 5, 2: a.
Y $\mathbf{G}_8^5 \otimes \mathbf{T}_2$: $\{C_{2z}\,|\,00\tfrac{1}{2}\}$, $\{C_{2y}\,|\,00\tfrac{1}{2}\}$; $\mathbf{t}_1$ or $\mathbf{t}_2$: 5, 2: a.
Z $\mathbf{G}_{16}^8$: $\{C_{2z}\,|\,00\tfrac{1}{2}\}$, $\{C_{2y}\,|\,00\tfrac{1}{2}\}$: 5, 3; 6, 3; 7, 3; 8, 3: a.
T $\mathbf{G}_{16}^8$: $\{C_{2z}\,|\,00\tfrac{1}{2}\}$, $\{C_{2y}\,|\,00\tfrac{1}{2}\}$: 5, 3; 6, 3; 7, 3; 8, 3: a.
S $\mathbf{G}_4^1 \otimes \mathbf{T}_2$: $\{C_{2z}\,|\,00\tfrac{1}{2}\}$; $\mathbf{t}_2$: 2, 3; 4, 3: b.
R $\mathbf{G}_8^2$: $\{C_{2z}\,|\,00\tfrac{1}{2}\}$, $\{\bar{E}\,|\,000\}$: 5, 1; 7, 1: a.

Λ^x $\mathbf{G}_4^1$: $(C_{2z}, 0)$: 2, 2; 4, 2: b.
H^x $\mathbf{G}_4^1$: $(C_{2z}, 0)$: 2, 2; 4, 2: b.
D^x $\mathbf{G}_4^1$: $(C_{2z}, 0)$: 2, x; 4, x: b.
A^x $\mathbf{G}_4^1$: $(C_{2x}, 0)$: 2, 3; 4, 3: b.
Σ^x $\mathbf{G}_4^1$: $(C_{2x}, 0)$: 2, 2; 4, 2: b.
Δ^x $\mathbf{G}_4^1$: $(C_{2y}, 0)$: 2, 2; 4, 2: b.
B^x $\mathbf{G}_8^2$: $(C_{2y}, 0)$, $(\bar{E}, 0)$: 5, 3; 7, 3: a.
G^x $\mathbf{G}_8^2$: $(C_{2y}, 0)$, $(\bar{E}, 0)$: 5, 3; 7, 3: a.
F^x $\mathbf{G}_4^1$: $(C_{2y}, 0)$: 2, 2; 4, 2: b.
E^x $\mathbf{G}_4^1$: $(C_{2x}, 0)$: 2, 3; 4, 3: b.
C^x $\mathbf{G}_4^1$: $(C_{2x}, 0)$: 2, 2; 4, 2: b.

21 $C222$ D_2^6

($F1$; $K7$; $M5$; $T3$; $Z1$.)

Γ $\mathbf{G}_8^5$: $\{C_{2z}\,|\,000\}$, $\{C_{2y}\,|\,000\}$: 5, 2: a.
Y $\mathbf{G}_8^5 \otimes \mathbf{T}_2$: $\{C_{2z}\,|\,000\}$, $\{C_{2y}\,|\,000\}$; $\mathbf{t}_1$ or $\mathbf{t}_2$: 5, 2: a.
Z $\mathbf{G}_8^5 \otimes \mathbf{T}_2$: $\{C_{2z}\,|\,000\}$, $\{C_{2y}\,|\,000\}$; $\mathbf{t}_3$: 5, 2: a.
T $\mathbf{G}_8^5 \otimes \mathbf{T}_2$: $\{C_{2z}\,|\,000\}$, $\{C_{2y}\,|\,000\}$; $\mathbf{t}_1$ or $\mathbf{t}_2$ or $\mathbf{t}_3$: 5, 2: a.
S $\mathbf{G}_4^1 \otimes \mathbf{T}_2$: $\{C_{2z}\,|\,000\}$; $\mathbf{t}_2$: 2, 3; 4, 3: b.
R $\mathbf{G}_4^1 \otimes \mathbf{T}_2$: $\{C_{2z}\,|\,000\}$; $\mathbf{t}_2$ or $\mathbf{t}_3$: 2, 3; 4, 3: b.

Λ^x $\mathbf{G}_4^1$: $(C_{2z}, 0)$: 2, 2; 4, 2: b.
H^x $\mathbf{G}_4^1$: $(C_{2z}, 0)$: 2, 2; 4, 2: b.
D^x $\mathbf{G}_4^1$: $(C_{2z}, 0)$: 2, x; 4, x: b.
A^x $\mathbf{G}_4^1$: $(C_{2x}, 0)$: 2, 2; 4, 2: b.
Σ^x $\mathbf{G}_4^1$: $(C_{2x}, 0)$: 2, 2; 4, 2: b.
Δ^x $\mathbf{G}_4^1$: $(C_{2y}, 0)$: 2, 2; 4, 2: b.
B^x $\mathbf{G}_4^1$: $(C_{2y}, 0)$: 2, 2; 4, 2: b.
G^x $\mathbf{G}_4^1$: $(C_{2y}, 0)$: 2, 2; 4, 2: b.
F^x $\mathbf{G}_4^1$: $(C_{2y}, 0)$: 2, 2; 4, 2: b.
E^x $\mathbf{G}_4^1$: $(C_{2x}, 0)$: 2, 2; 4, 2: b.
C^x $\mathbf{G}_4^1$: $(C_{2x}, 0)$: 2, 2; 4, 2: b.

22 *F*222 D_2^7

(*F*1; *K*7; *M*5; *S*14; *Z*1.)

Γ $\mathbf{G}_8^5$: $\{C_{2z}|000\}$, $\{C_{2y}|000\}$: 5, 2: *a*.
Y $\mathbf{G}_8^5 \otimes \mathbf{T}_2$: $\{C_{2z}|000\}$, $\{C_{2y}|000\}$; $\mathbf{t}_2$ or $\mathbf{t}_3$: 5, 2: *a*.
X $\mathbf{G}_8^5 \otimes \mathbf{T}_2$: $\{C_{2z}|000\}$, $\{C_{2y}|000\}$; $\mathbf{t}_1$ or $\mathbf{t}_3$: 5, 2: *a*.
Z $\mathbf{G}_8^5 \otimes \mathbf{T}_2$: $\{C_{2z}|000\}$, $\{C_{2y}|000\}$; $\mathbf{t}_1$ or $\mathbf{t}_2$: 5, 2: *a*.
L $\mathbf{G}_2^1 \otimes \mathbf{T}_2$: $\{\bar{E}|000\}$; $\mathbf{t}_1$: 2, 1: *a*.

Λ^x $\mathbf{G}_4^1$: $(C_{2z}, 0)$: 2, 2; 4, 2: *b*.
G^x $\mathbf{G}_4^1$: $(C_{2z}, 0)$: 2, 2; 4, 2: *b*.
H^x $\mathbf{G}_4^1$: $(C_{2z}, 0)$: 2, 2; 4, 2: *b*.
Q^x $\mathbf{G}_4^1$: $(C_{2z}, 0)$: 2, 2; 4, 2: *b*.
Σ^x $\mathbf{G}_4^1$: $(C_{2x}, 0)$: 2, 2; 4, 2: *b*.
C^x $\mathbf{G}_4^1$: $(C_{2x}, 0)$: 2, 2; 4, 2: *b*.
A^x $\mathbf{G}_4^1$: $(C_{2x}, 0)$: 2, 2; 4, 2: *b*.
U^x $\mathbf{G}_4^1$: $(C_{2x}, 0)$: 2, 2; 4, 2: *b*.
Δ^x $\mathbf{G}_4^1$: $(C_{2y}, 0)$: 2, 2; 4, 2: *b*.
D^x $\mathbf{G}_4^1$: $(C_{2y}, 0)$: 2, 2; 4, 2: *b*.
B^x $\mathbf{G}_4^1$: $(C_{2y}, 0)$: 2, 2; 4, 2: *b*.
R^x $\mathbf{G}_4^1$: $(C_{2y}, 0)$: 2, 2; 4, 2: *b*.

23 *I*222 D_2^8

(*F*1; *K*7; *M*5; *Z*1.)

Γ $\mathbf{G}_8^5$: $\{C_{2z}|000\}$, $\{C_{2y}|000\}$: 5, 2: *a*.
X $\mathbf{G}_8^5 \otimes \mathbf{T}_2$: $\{C_{2z}|000\}$, $\{C_{2y}|000\}$; $\mathbf{t}_1$ or $\mathbf{t}_2$ or $\mathbf{t}_3$: 5, 2: *a*.
R $\mathbf{G}_4^1 \otimes \mathbf{T}_2$: $\{C_{2y}|000\}$; $\mathbf{t}_1$: 2, 3; 4, 3: *b*.
S $\mathbf{G}_4^1 \otimes \mathbf{T}_2$: $\{C_{2x}|000\}$; $\mathbf{t}_1$ or $\mathbf{t}_3$: 2, 3; 4, 3: *b*.
T $\mathbf{G}_4^1 \otimes \mathbf{T}_2$: $\{C_{2z}|000\}$; $\mathbf{t}_1$ or $\mathbf{t}_2$: 2, 3; 4, 3: *b*.
W $\mathbf{G}_8^5 \otimes \mathbf{T}_4$: $\{C_{2z}|000\}$, $\{C_{2y}|000\}$; $\mathbf{t}_1$ or $\mathbf{t}_2$ or $\mathbf{t}_3$: 5, *x*: *a*.

Λ^x $\mathbf{G}_4^1$: $(C_{2z}, 0)$: 2, 2; 4, 2: *b*.
G^x $\mathbf{G}_4^1$: $(C_{2z}, 0)$: 2, 2; 4, 2: *b*.
P^x $\mathbf{G}_4^1$: $(C_{2z}, 0)$: 2, *x*; 4, *x*: *b*.
Σ^x $\mathbf{G}_4^1$: $(C_{2x}, 0)$: 2, 2; 4, 2: *b*.
F^x $\mathbf{G}_4^1$: $(C_{2x}, 0)$: 2, 2; 4, 2: *b*.
D^x $\mathbf{G}_4^1$: $(C_{2x}, 0)$: 2, *x*; 4, *x*: *b*.
Δ^x $\mathbf{G}_4^1$: $(C_{2y}, 0)$: 2, 2; 4, 2: *b*.
U^x $\mathbf{G}_4^1$: $(C_{2y}, 0)$: 2, 2; 4, 2: *b*.
Q^x $\mathbf{G}_4^1$: $(C_{2y}, 0)$: 2, *x*; 4, *x*: *b*.

24 $I2_12_12_1$ D_2^9

($F1$; $K7$; $M5$; $Z1$.)

Γ $\mathbf{G}_8^5$: $\{C_{2z}|\frac{1}{2}0\frac{1}{2}\}, \{C_{2y}|\frac{1}{2}\frac{1}{2}0\}$: 5, 2: a.
X $\mathbf{G}_8^5 \otimes \mathbf{T}_2$: $\{C_{2z}|\frac{1}{2}0\frac{1}{2}\}, \{C_{2y}|\frac{1}{2}\frac{1}{2}0\}$; $\mathbf{t}_1$ or $\mathbf{t}_2$ or $\mathbf{t}_3$: 5, 2: a.
R $\mathbf{G}_4^1 \otimes \mathbf{T}_2$: $\{C_{2y}|\frac{1}{2}\frac{1}{2}0\}$; $\mathbf{t}_1$: 2, 3; 4, 3: b.
S $\mathbf{G}_4^1 \otimes \mathbf{T}_2$: $\{C_{2x}|0\frac{1}{2}\frac{1}{2}\}$; $\mathbf{t}_1$ or $\mathbf{t}_3$: 2, 3; 4, 3: b.
T $\mathbf{G}_4^1 \otimes \mathbf{T}_2$: $\{C_{2z}|\frac{1}{2}0\frac{1}{2}\}$; $\mathbf{t}_1$ or $\mathbf{t}_2$: 2, 3; 4, 3: b.
W $\mathbf{G}_8^5 \otimes \mathbf{T}_4$: $\{C_{2z}|-\frac{1}{2}0\frac{1}{2}\}, \{C_{2y}|-\frac{1}{2}\frac{1}{2}0\}$; $\mathbf{t}_1$ or $\mathbf{t}_2$ or $\mathbf{t}_3$: 1, x; 2, x; 3, x; 4, x: b.

Λ^x $\mathbf{G}_4^1$: $(C_{2z}, 0)$: 2, 2; 4, 2: b.
G^x $\mathbf{G}_4^1$: $(C_{2z}, 0)$: 2, 2; 4, 2: b.
P^x $\mathbf{G}_8^2$: $(C_{2z}, 0), (\bar{E}, 0)$: 5, x; 7, x: a.
Σ^x $\mathbf{G}_4^1$: $(C_{2x}, 0)$: 2, 2; 4, 2: b.
F^x $\mathbf{G}_4^1$: $(C_{2x}, 0)$: 2, 2; 4, 2: b.
D^x $\mathbf{G}_8^2$: $(C_{2x}, 0), (\bar{E}, 0)$: 5, x; 7, x: a.
Δ^x $\mathbf{G}_4^1$: $(C_{2y}, 0)$: 2, 2; 4, 2: b.
U^x $\mathbf{G}_4^1$: $(C_{2y}, 0)$: 2, 2; 4, 2: b.
Q^x $\mathbf{G}_8^2$: $(C_{2y}, 0), (\bar{E}, 0)$: 5, x; 7, x: a.

25 $Pmm2$ C_{2v}^1

($F1$; $K7$; $M5$; $T1$; $Z1$.)

Γ $\mathbf{G}_8^5$: $\{C_{2z}|000\}, \{\sigma_y|000\}$: 5, 2: a.
Y $\mathbf{G}_8^5 \otimes \mathbf{T}_2$: $\{C_{2z}|000\}, \{\sigma_y|000\}$; $\mathbf{t}_1$: 5, 2: a.
X $\mathbf{G}_8^5 \otimes \mathbf{T}_2$: $\{C_{2z}|000\}, \{\sigma_y|000\}$; $\mathbf{t}_2$: 5, 2: a.
Z $\mathbf{G}_8^5 \otimes \mathbf{T}_2$: $\{C_{2z}|000\}, \{\sigma_y|000\}$; $\mathbf{t}_3$: 5, 2: a.
U $\mathbf{G}_8^5 \otimes \mathbf{T}_2$: $\{C_{2z}|000\}, \{\sigma_y|000\}$; $\mathbf{t}_2$ or $\mathbf{t}_3$: 5, 2: a.
T $\mathbf{G}_8^5 \otimes \mathbf{T}_2$: $\{C_{2z}|000\}, \{\sigma_y|000\}$; $\mathbf{t}_1$ or $\mathbf{t}_3$: 5, 2: a.
S $\mathbf{G}_8^5 \otimes \mathbf{T}_2$: $\{C_{2z}|000\}, \{\sigma_y|000\}$; $\mathbf{t}_1$ or $\mathbf{t}_2$: 5, 2: a.
R $\mathbf{G}_8^5 \otimes \mathbf{T}_2$: $\{C_{2z}|000\}, \{\sigma_y|000\}$; $\mathbf{t}_1$ or $\mathbf{t}_2$ or $\mathbf{t}_3$: 5, 2: a.

Δ^x $\mathbf{G}_4^1$: $(\sigma_x, 0)$: 2, 2; 4, 2: b.
D^x $\mathbf{G}_4^1$: $(\sigma_x, 0)$: 2, 2; 4, 2: b.
P^x $\mathbf{G}_4^1$: $(\sigma_x, 0)$: 2, 2; 4, 2: b.
B^x $\mathbf{G}_4^1$: $(\sigma_x, 0)$: 2, 2; 4, 2: b.
Σ^x $\mathbf{G}_4^1$: $(\sigma_y, 0)$: 2, 2; 4, 2: b.
C^x $\mathbf{G}_4^1$: $(\sigma_y, 0)$: 2, 2; 4, 2: b.
E^x $\mathbf{G}_4^1$: $(\sigma_y, 0)$: 2, 2; 4, 2: b.
A^x $\mathbf{G}_4^1$: $(\sigma_y, 0)$: 2, 2; 4, 2: b.
Λ^x $\mathbf{G}_8^5$: $(C_{2z}, 0), (\sigma_y, 0)$: 5, x: a.
H^x $\mathbf{G}_8^5$: $(C_{2z}, 0), (\sigma_y, 0)$: 5, x: a.
Q^x $\mathbf{G}_8^5$: $(C_{2z}, 0), (\sigma_y, 0)$: 5, x: a.
G^x $\mathbf{G}_8^5$: $(C_{2z}, 0), (\sigma_y, 0)$: 5, x: a.

26 *Pmc*2_1 C_{2v}^2

(*F*1; *K*7; *M*5; *T*1; *Z*1.)

Γ $\mathbf{G}_8^5$: $\{C_{2z}|00\frac{1}{2}\}$, $\{\sigma_y|00\frac{1}{2}\}$: 5, 2: *a*.
Y $\mathbf{G}_8^5 \otimes \mathbf{T}_2$: $\{C_{2z}|00\frac{1}{2}\}$, $\{\sigma_y|00\frac{1}{2}\}$; $\mathbf{t}_1$: 5, 2: *a*.
X $\mathbf{G}_8^5 \otimes \mathbf{T}_2$: $\{C_{2z}|00\frac{1}{2}\}$, $\{\sigma_y|00\frac{1}{2}\}$; $\mathbf{t}_2$: 5, 2: *a*.
Z $\mathbf{G}_{16}^8$: $\{\sigma_x|000\}$, $\{\sigma_y|00\frac{1}{2}\}$: 9, 1: *b*.
U $\mathbf{G}_{16}^8$: $\{\sigma_x|000\}$, $\{\sigma_y|00\frac{1}{2}\}$: 9, 1: *b*.
T $\mathbf{G}_{16}^8$: $\{\sigma_x|000\}$, $\{\sigma_y|00\frac{1}{2}\}$: 9, 1: *b*.
S $\mathbf{G}_8^5 \otimes \mathbf{T}_2$: $\{C_{2z}|00\frac{1}{2}\}$, $\{\sigma_y|00\frac{1}{2}\}$; $\mathbf{t}_1$ or $\mathbf{t}_2$: 5, 2: *a*.
R $\mathbf{G}_{16}^8$: $\{\sigma_x|000\}$, $\{\sigma_y|00\frac{1}{2}\}$: 9, 1: *b*.

Δ^x $\mathbf{G}_4^1$: $(\sigma_x, 0)$: 2, 2; 4, 2: *b*.
D^x $\mathbf{G}_4^1$: $(\sigma_x, 0)$: 2, 2; 4, 2: *b*.
P^x $\mathbf{G}_4^1$: $(\sigma_x, 0)$: 2, 1; 4, 1: *b*.
B^x $\mathbf{G}_4^1$: $(\sigma_x, 0)$: 2, 1; 4, 1: *b*.
Σ^x $\mathbf{G}_4^1$: $(\sigma_y, 0)$: 2, 2; 4, 2: *b*.
C^x $\mathbf{G}_4^1$: $(\sigma_y, 0)$: 2, 2; 4, 2: *b*.
E^x $\mathbf{G}_4^1$: $(\sigma_y, 0)$: 2, 3; 4, 3: *b*.
A^x $\mathbf{G}_4^1$: $(\sigma_y, 0)$: 2, 3; 4, 3: *b*.
Λ^x $\mathbf{G}_8^5$: $(C_{2z}, 0)$, $(\sigma_y, 0)$: 5, *x*: *a*.
H^x $\mathbf{G}_8^5$: $(C_{2z}, 0)$, $(\sigma_y, 0)$: 5, *x*: *a*.
Q^x $\mathbf{G}_8^5$: $(C_{2z}, 0)$, $(\sigma_y, 0)$: 5, *x*: *a*.
G^x $\mathbf{G}_8^5$: $(C_{2z}, 0)$, $(\sigma_y, 0)$: 5, *x*: *a*.

27 *Pcc*2 C_{2v}^3

(*F*1; *K*7; *M*5; *T*1; *Z*1.)

Γ $\mathbf{G}_8^5$: $\{C_{2z}|000\}$, $\{\sigma_y|00\frac{1}{2}\}$: 5, 2: *a*.
Y $\mathbf{G}_8^5 \otimes \mathbf{T}_2$: $\{C_{2z}|000\}$, $\{\sigma_y|00\frac{1}{2}\}$; $\mathbf{t}_1$: 5, 2: *a*.
X $\mathbf{G}_8^5 \otimes \mathbf{T}_2$: $\{C_{2z}|000\}$, $\{\sigma_y|00\frac{1}{2}\}$; $\mathbf{t}_2$: 5, 2: *a*.
Z $\mathbf{G}_{16}^8$: $\{C_{2z}|000\}$, $\{\sigma_x|00\frac{1}{2}\}$: 9, 1: *b*.
U $\mathbf{G}_{16}^8$: $\{C_{2z}|000\}$, $\{\sigma_x|00\frac{1}{2}\}$: 9, 1: *b*.
T $\mathbf{G}_{16}^8$: $\{C_{2z}|000\}$, $\{\sigma_x|00\frac{1}{2}\}$: 9, 1: *b*.
S $\mathbf{G}_8^5 \otimes \mathbf{T}_2$: $\{C_{2z}|000\}$, $\{\sigma_y|00\frac{1}{2}\}$; $\mathbf{t}_1$ or $\mathbf{t}_2$: 5, 2: *a*.
R $\mathbf{G}_{16}^8$: $\{C_{2z}|000\}$, $\{\sigma_x|00\frac{1}{2}\}$: 9, 1: *b*.

Δ^x $\mathbf{G}_4^1$: $(\sigma_x, 0)$: 2, 2; 4, 2: *b*.
D^x $\mathbf{G}_4^1$: $(\sigma_x, 0)$: 2, 2; 4, 2: *b*.
P^x $\mathbf{G}_4^1$: $(\sigma_x, 0)$: 2, 3; 4, 3: *b*.
B^x $\mathbf{G}_4^1$: $(\sigma_x, 0)$: 2, 3; 4, 3: *b*.
Σ^x $\mathbf{G}_4^1$: $(\sigma_y, 0)$: 2, 2; 4, 2: *b*.
C^x $\mathbf{G}_4^1$: $(\sigma_y, 0)$: 2, 2; 4, 2: *b*.
E^x $\mathbf{G}_4^1$: $(\sigma_y, 0)$: 2, 3; 4, 3: *b*.
A^x $\mathbf{G}_4^1$: $(\sigma_y, 0)$: 2, 3; 4, 3: *b*.
Λ^x $\mathbf{G}_8^5$: $(C_{2z}, 0)$, $(\sigma_y, 0)$: 5, *x*: *a*.
H^x $\mathbf{G}_8^5$: $(C_{2z}, 0)$, $(\sigma_y, 0)$: 5, *x*: *a*.
Q^x $\mathbf{G}_8^5$: $(C_{2z}, 0)$, $(\sigma_y, 0)$: 5, *x*: *a*.
G^x $\mathbf{G}_8^5$: $(C_{2z}, 0)$, $(\sigma_y, 0)$: 5, *x*: *a*.

28 *Pma*2 C_{2v}^4

(*F*1; *K*7; *M*5; *T*1; *Z*1.)

Γ $\mathbf{G}_8^5$: $\{C_{2z}|\frac{1}{2}00\}, \{\sigma_y|000\}$: 5, 2: *a*.
Y $\mathbf{G}_{16}^8$: $\{\sigma_x|\frac{1}{2}00\}, \{\sigma_y|000\}$: 5, 3; 6, 3; 7, 3; 8, 3: *a*.
X $\mathbf{G}_8^5 \otimes \mathbf{T}_2$: $\{C_{2z}|\frac{1}{2}00\}, \{\sigma_y|000\}$; $\mathbf{t}_2$: 5, 2: *a*.
Z $\mathbf{G}_8^5 \otimes \mathbf{T}_2$: $\{C_{2z}|\frac{1}{2}00\}, \{\sigma_y|000\}$; $\mathbf{t}_3$: 5, 2: *a*.
U $\mathbf{G}_8^5 \otimes \mathbf{T}_2$: $\{C_{2z}|\frac{1}{2}00\}, \{\sigma_y|000\}$; $\mathbf{t}_2$ or $\mathbf{t}_3$: 5, 2: *a*.
T $\mathbf{G}_{16}^8$: $\{\sigma_x|\frac{1}{2}00\}, \{\sigma_y|000\}$: 5, 3; 6, 3; 7, 3; 8, 3: *a*.
S $\mathbf{G}_{16}^8$: $\{\sigma_x|\frac{1}{2}00\}, \{\sigma_y|000\}$: 5, 3; 6, 3; 7, 3; 8, 3: *a*.
R $\mathbf{G}_{16}^8$: $\{\sigma_x|\frac{1}{2}00\}, \{\sigma_y|000\}$: 5, 3; 6, 3; 7, 3; 8, 3: *a*.

Δ^x $\mathbf{G}_4^1$: $(\sigma_x, 0)$: 2, 2; 4, 2: *b*.
D^x $\mathbf{G}_4^1$: $(\sigma_x, 0)$: 2, 2; 4, 2: *b*.
P^x $\mathbf{G}_4^1$: $(\sigma_x, 0)$: 2, 2; 4, 2: *b*.
B^x $\mathbf{G}_4^1$: $(\sigma_x, 0)$: 2, 2; 4, 2: *b*.
Σ^x $\mathbf{G}_4^1$: $(\sigma_y, 0)$: 2, 2; 4, 2: *b*.
C^x $\mathbf{G}_4^1$: $(\sigma_y, 0)$: 2, 3; 4, 3: *b*.
E^x $\mathbf{G}_4^1$: $(\sigma_y, 0)$: 2, 3; 4, 3: *b*.
A^x $\mathbf{G}_4^1$: $(\sigma_y, 0)$: 2, 2; 4, 2: *b*.
Λ^x $\mathbf{G}_8^5$: $(C_{2z}, 0), (\sigma_y, 0)$: 5, *x*: *a*.
H^x $\mathbf{G}_{16}^8$: $(C_{2z}, 0), (\sigma_y, 0)$: 5, *x*; 6, *x*; 7, *x*; 8, *x*: *a*.
Q^x $\mathbf{G}_{16}^8$: $(C_{2z}, 0), (\sigma_y, 0)$: 5, *x*; 6, *x*; 7, *x*; 8, *x*: *a*.
G^x $\mathbf{G}_8^5$: $(C_{2z}, 0), (\sigma_y, 0)$: 5, *x*: *a*.

29 *Pca*2$_1$ C_{2v}^5

(*F*1; *K*7; *M*5; *T*1; *Z*1.)

Γ $\mathbf{G}_8^5$: $\{C_{2z}|\frac{1}{2}0\frac{1}{2}\}, \{\sigma_y|00\frac{1}{2}\}$: 5, 2: *a*.
Y $\mathbf{G}_{16}^8$: $\{\sigma_x|\frac{1}{2}00\}, (C_{2z}|\frac{1}{2}0\frac{1}{2}\}$: 5, 3; 6, 3; 7, 3; 8, 3: *a*.
X $\mathbf{G}_8^5 \otimes \mathbf{T}_2$: $\{C_{2z}|\frac{1}{2}0\frac{1}{2}\}, \{\sigma_y|00\frac{1}{2}\}$; $\mathbf{t}_2$: 5, 2: *a*.
Z $\mathbf{G}_{16}^8$: $\{\sigma_x|\frac{1}{2}00\}, \{\sigma_y|00\frac{1}{2}\}$: 9, 1: *b*.
U $\mathbf{G}_{16}^8$: $\{\sigma_x|\frac{1}{2}00\}, \{\sigma_y|00\frac{1}{2}\}$: 9, 1: *b*.
T $\mathbf{G}_{16}^{11}$: $\{\sigma_x|\frac{1}{2}00\}, \{\sigma_y|00\frac{1}{2}\}, \{\bar{E}|000\}$: 6, 1; 7, 1; 8, 1; 9, 1: *b*.
S $\mathbf{G}_{16}^8$: $\{\sigma_x|\frac{1}{2}00\}, \{C_{2z}|\frac{1}{2}0\frac{1}{2}\}$: 5, 3; 6, 3; 7, 3; 8, 3: *a*.
R $\mathbf{G}_{16}^{11}$: $\{\sigma_x|\frac{1}{2}00\}, \{\sigma_y|00\frac{1}{2}\}, \{\bar{E}|000\}$: 6, 1; 7, 1; 8, 1; 9, 1: *b*.

Δ^x $\mathbf{G}_4^1$: $(\sigma_x, 0)$: 2, 2; 4, 2: *b*.
D^x $\mathbf{G}_4^1$: $(\sigma_x, 0)$: 2, 2; 4, 2: *b*.
P^x $\mathbf{G}_4^1$: $(\sigma_x, 0)$: 2, 1; 4, 1: *b*.
B^x $\mathbf{G}_4^1$: $(\sigma_x, 0)$: 2, 1; 4, 1: *b*.
Σ^x $\mathbf{G}_4^1$: $(\sigma_y, 0)$: 2, 2; 4, 2: *b*.
C^x $\mathbf{G}_4^1$: $(\sigma_y, 0)$: 2, 3; 4, 3: *b*.
E^x $\mathbf{G}_4^1$: $(\sigma_y, 0)$: 2, 1; 4, 1: *b*.
A^x $\mathbf{G}_4^1$: $(\sigma_y, 0)$: 2, 3; 4, 3: *b*.
Λ^x $\mathbf{G}_8^5$: $(C_{2z}, 0), (\sigma_y, 0)$: 5, *x*: *a*.
H^x $\mathbf{G}_{16}^8$: $(C_{2z}, 0), (\sigma_y, 0)$: 5, *x*; 6, *x*; 7, *x*; 8, *x*: *a*.
Q^x $\mathbf{G}_{16}^8$: $(C_{2z}, 0), (\sigma_y, 0)$: 5, *x*; 6, *x*; 7, *x*; 8, *x*: *a*.
G^x $\mathbf{G}_8^5$: $(C_{2z}, 0), (\sigma_y, 0)$: 5, *x*: *a*.

30 $Pnc2$ C_{2v}^6

($F1$; $K7$; $M5$; $T1$; $Z1$.)

Γ $\mathbf{G}_8^5$: $\{C_{2z} \mid \frac{1}{2}00\}$, $\{\sigma_y \mid 00\frac{1}{2}\}$: 5, 2: a.
Y $\mathbf{G}_{16}^8$: $\{\sigma_x \mid \frac{1}{2}0\frac{1}{2}\}$, $\{\sigma_y \mid 00\frac{1}{2}\}$: 5, 3; 6, 3; 7, 3; 8, 3: a.
X $\mathbf{G}_8^5 \otimes \mathbf{T}_2$: $\{C_{2z} \mid \frac{1}{2}00\}$, $\{\sigma_y \mid 00\frac{1}{2}\}$; $\mathbf{t}_2$: 5, 2: a.
Z $\mathbf{G}_{16}^8$: $\{C_{2z} \mid \frac{1}{2}00\}$, $\{\sigma_x \mid \frac{1}{2}0\frac{1}{2}\}$: 9, 1: b.
U $\mathbf{G}_{16}^8$: $\{C_{2z} \mid \frac{1}{2}00\}$, $\{\sigma_x \mid \frac{1}{2}0\frac{1}{2}\}$: 9, 1: b.
T $\mathbf{G}_{16}^8$: $\{\sigma_y \mid 00\frac{1}{2}\}$, $\{C_{2z} \mid \frac{1}{2}00\}$: 5, 3; 6, 3; 7, 3; 8, 3: a.
S $\mathbf{G}_{16}^8$: $\{\sigma_x \mid \frac{1}{2}0\frac{1}{2}\}$, $\{\sigma_y \mid 00\frac{1}{2}\}$: 5, 3; 6, 3; 7, 3; 8, 3: a.
R $\mathbf{G}_{16}^8$: $\{\sigma_y \mid 00\frac{1}{2}\}$, $\{C_{2z} \mid \frac{1}{2}00\}$: 5, 3; 6, 3; 7, 3; 8, 3: a.

Δ^x $\mathbf{G}_4^1$: $(\sigma_x, 0)$: 2, 2; 4, 2: b.
D^x $\mathbf{G}_4^1$: $(\sigma_x, 0)$: 2, 2; 4, 2: b.
P^x $\mathbf{G}_4^1$: $(\sigma_x, 0)$: 2, 3; 4, 3: b.
B^x $\mathbf{G}_4^1$: $(\sigma_x, 0)$: 2, 3; 4, 3: b.
Σ^x $\mathbf{G}_4^1$: $(\sigma_y, 0)$: 2, 2; 4, 2: b.
C^x $\mathbf{G}_4^1$: $(\sigma_y, 0)$: 2, 3; 4, 3: b.
E^x $\mathbf{G}_4^1$: $(\sigma_y, 0)$: 2, 2; 4, 2: b.
A^x $\mathbf{G}_4^1$: $(\sigma_y, 0)$: 2, 3; 4, 3: b.
Λ^x $\mathbf{G}_8^5$: $(C_{2z}, 0)$, $(\sigma_y, 0)$: 5, x: a.
H^x $\mathbf{G}_{16}^8$: $(C_{2z}, 0)$, $(\sigma_y, 0)$: 5, x; 6, x; 7, x; 8, x: a.
Q^x $\mathbf{G}_{16}^8$: $(C_{2z}, 0)$, $(\sigma_y, 0)$: 5, x; 6, x; 7, x; 8, x: a.
G^x $\mathbf{G}_8^5$: $(C_{2z}, 0)$, $(\sigma_y, 0)$: 5, x: a.

31 $Pmn2_1$ C_{2v}^7

($F1$; $K7$; $M5$; $T1$; $Z1$.)

Γ $\mathbf{G}_8^5$: $\{C_{2z} \mid \frac{1}{2}0\frac{1}{2}\}$, $\{\sigma_y \mid 000\}$: 5, 2: a.
Y $\mathbf{G}_{16}^8$: $\{\sigma_x \mid \frac{1}{2}0\frac{1}{2}\}$, $\{\sigma_y \mid 000\}$: 5, 3; 6, 3; 7, 3; 8, 3: a.
X $\mathbf{G}_8^5 \otimes \mathbf{T}_2$: $\{C_{2z} \mid \frac{1}{2}0\frac{1}{2}\}$, $\{\sigma_y \mid 000\}$; $\mathbf{t}_2$: 5, 2: a.
Z $\mathbf{G}_{16}^8$: $\{\sigma_y \mid 000\}$, $\{C_{2z} \mid \frac{1}{2}0\frac{1}{2}\}$: 9, 1: b.
U $\mathbf{G}_{16}^8$: $\{\sigma_y \mid 000\}$, $\{C_{2z} \mid \frac{1}{2}0\frac{1}{2}\}$: 9, 1: b.
T $\mathbf{G}_{16}^8$: $\{C_{2z} \mid \frac{1}{2}0\frac{1}{2}\}$, $\{\sigma_y \mid 000\}$: 5, 3; 6, 3; 7, 3; 8, 3: a.
S $\mathbf{G}_{16}^8$: $\{\sigma_x \mid \frac{1}{2}0\frac{1}{2}\}$, $\{\sigma_y \mid 000\}$: 5, 3; 6, 3; 7, 3; 8, 3: a.
R $\mathbf{G}_{16}^8$: $\{C_{2z} \mid \frac{1}{2}0\frac{1}{2}\}$, $\{\sigma_y \mid 000\}$: 5, 3; 6, 3; 7, 3; 8, 3: a.

Δ^x $\mathbf{G}_4^1$: $(\sigma_x, 0)$: 2, 2; 4, 2: b.
D^x $\mathbf{G}_4^1$: $(\sigma_x, 0)$: 2, 2; 4, 2: b.
P^x $\mathbf{G}_4^1$: $(\sigma_x, 0)$: 2, 3; 4, 3: b.
B^x $\mathbf{G}_4^1$: $(\sigma_x, 0)$: 2, 3; 4, 3: b.
Σ^x $\mathbf{G}_4^1$: $(\sigma_y, 0)$: 2, 2; 4, 2: b.
C^x $\mathbf{G}_4^1$: $(\sigma_y, 0)$: 2, 3; 4, 3: b.
E^x $\mathbf{G}_4^1$: $(\sigma_y, 0)$: 2, 3; 4, 3: b.
A^x $\mathbf{G}_4^1$: $(\sigma_y, 0)$: 2, 1; 4, 1: b.
Λ^x $\mathbf{G}_8^5$: $(C_{2z}, 0)$, $(\sigma_y, 0)$: 5, x: a.
H^x $\mathbf{G}_{16}^8$: $(C_{2z}, 0)$, $(\sigma_y, 0)$: 5, x; 6, x; 7, x; 8, x: a.
Q^x $\mathbf{G}_{16}^8$: $(C_{2z}, 0)$, $(\sigma_y, 0)$: 5, x; 6, x; 7, x; 8, x: a.
G^x $\mathbf{G}_8^5$: $(C_{2z}, 0)$, $(\sigma_y, 0)$: 5, x: a.

32 *Pba*2 C_{2v}^{8}

($F1$; $K7$; $M5$; $T1$; $Z1$.)

Γ $\mathbf{G}_{8}^{5}$: $\{C_{2z}|\frac{1}{2}\frac{1}{2}0\}$, $\{\sigma_y|0\frac{1}{2}0\}$: 5, 2: *a*.
Y $\mathbf{G}_{16}^{8}$: $\{\sigma_x|\frac{1}{2}00\}$, $\{C_{2z}|\frac{1}{2}\frac{1}{2}0\}$: 5, 3; 6, 3; 7, 3; 8, 3: *a*.
X $\mathbf{G}_{16}^{8}$: $\{\sigma_y|0\frac{1}{2}0\}$, $\{C_{2z}|\frac{1}{2}\frac{1}{2}0\}$: 5, 3; 6, 3; 7, 3; 8, 3: *a*.
Z $\mathbf{G}_{8}^{5}\otimes\mathbf{T}_2$: $\{C_{2z}|\frac{1}{2}\frac{1}{2}0\}$, $\{\sigma_y|0\frac{1}{2}0\}$; $\mathbf{t}_3$: 5, 2: *a*.
U $\mathbf{G}_{16}^{8}$: $\{\sigma_y|0\frac{1}{2}0\}$, $\{C_{2z}|\frac{1}{2}\frac{1}{2}0\}$: 5, 3; 6, 3; 7, 3; 8, 3: *a*.
T $\mathbf{G}_{16}^{8}$: $\{\sigma_x|\frac{1}{2}00\}$, $\{C_{2z}|\frac{1}{2}\frac{1}{2}0\}$: 5, 3; 6, 3; 7, 3; 8, 3: *a*.
S $\mathbf{G}_{16}^{8}$: $\{C_{2z}|\frac{1}{2}\frac{1}{2}0\}$, $\{\sigma_x|\frac{1}{2}00\}$: 9, 1: *b*.
R $\mathbf{G}_{16}^{8}$: $\{C_{2z}|\frac{1}{2}\frac{1}{2}0\}$, $\{\sigma_x|\frac{1}{2}00\}$: 9, 1: *b*.

Δ^x $\mathbf{G}_{4}^{1}$: $(\sigma_x, 0)$: 2, 2; 4, 2: *b*.
D^x $\mathbf{G}_{4}^{1}$: $(\sigma_x, 0)$: 2, 3; 4, 3: *b*.
P^x $\mathbf{G}_{4}^{1}$: $(\sigma_x, 0)$: 2, 3; 4, 3: *b*.
B^x $\mathbf{G}_{4}^{1}$: $(\sigma_x, 0)$: 2, 2; 4, 2: *b*.
Σ^x $\mathbf{G}_{4}^{1}$: $(\sigma_y, 0)$: 2, 2; 4, 2: *b*.
C^x $\mathbf{G}_{4}^{1}$: $(\sigma_y, 0)$: 2, 3; 4, 3: *b*.
E^x $\mathbf{G}_{4}^{1}$: $(\sigma_y, 0)$: 2, 3; 4, 3: *b*.
A^x $\mathbf{G}_{4}^{1}$: $(\sigma_y, 0)$: 2, 2; 4, 2: *b*.
Λ^x $\mathbf{G}_{8}^{5}$: $(C_{2z}, 0)$, $(\sigma_y, 0)$: 5, x: *a*.
H^x $\mathbf{G}_{16}^{8}$: $(C_{2z}, 0)$, $(\sigma_y, 0)$: 5, x; 6, x; 7, x; 8, x: *a*.
Q^x $\mathbf{G}_{16}^{11}$: $(\sigma_x, 0)$, $(\sigma_y, 0)$, $(E, 1)$: 10, x: *d*.
G^x $\mathbf{G}_{16}^{8}$: $(C_{2z}, 0)$, $(\sigma_y, 0)$: 5, x; 6, x; 7, x; 8, x: *a*.

33 *Pna*2$_1$ C_{2v}^{9}

($F1$; $K7$; $M5$; $T1$; $Z1$.)

Γ $\mathbf{G}_{8}^{5}$: $\{C_{2z}|\frac{1}{2}\frac{1}{2}\frac{1}{2}\}$, $\{\sigma_y|0\frac{1}{2}\frac{1}{2}\}$: 5, 2: *a*.
Y $\mathbf{G}_{16}^{8}$: $\{\sigma_x|\frac{1}{2}00\}$, $\{C_{2z}|\frac{1}{2}\frac{1}{2}\frac{1}{2}\}$: 5, 3; 6, 3; 7, 3; 8, 3: *a*.
X $\mathbf{G}_{16}^{8}$: $\{\sigma_y|0\frac{1}{2}\frac{1}{2}\}$, $\{C_{2z}|\frac{1}{2}\frac{1}{2}\frac{1}{2}\}$: 5, 3; 6, 3; 7, 3; 8, 3: *a*.
Z $\mathbf{G}_{16}^{8}$: $\{\sigma_x|\frac{1}{2}00\}$, $\{\sigma_y|0\frac{1}{2}\frac{1}{2}\}$: 9, 1: *b*.
U $\mathbf{G}_{16}^{8}$: $\{C_{2z}|\frac{1}{2}\frac{1}{2}\frac{1}{2}\}$, $\{\sigma_y|0\frac{1}{2}\frac{1}{2}\}$: 5, 3; 6, 3; 7, 3; 8, 3: *a*.
T $\mathbf{G}_{16}^{11}$: $\{\sigma_x|\frac{1}{2}00\}$, $\{\sigma_y|0\frac{1}{2}\frac{1}{2}\}$, $\{\bar{E}|000\}$: 6, 1; 7, 1; 8, 1; 9, 1: *b*.
S $\mathbf{G}_{16}^{8}$: $\{C_{2z}|\frac{1}{2}\frac{1}{2}\frac{1}{2}\}$, $\{\sigma_x|\frac{1}{2}00\}$: 9, 1: *b*.
R $\mathbf{G}_{16}^{8}$: $\{\sigma_y|0\frac{1}{2}\frac{1}{2}\}$, $\{C_{2z}|\frac{1}{2}\frac{1}{2}\frac{1}{2}\}$: 9, 1: *b*.

Δ^x $\mathbf{G}_{4}^{1}$: $(\sigma_x, 0)$: 2, 2; 4, 2: *b*.
D^x $\mathbf{G}_{4}^{1}$: $(\sigma_x, 0)$: 2, 3; 4, 3: *b*.
P^x $\mathbf{G}_{4}^{1}$: $(\sigma_x, 0)$: 2, 3; 4, 3: *b*.
B^x $\mathbf{G}_{4}^{1}$: $(\sigma_x, 0)$: 2, 1; 4, 1: *b*.
Σ^x $\mathbf{G}_{4}^{1}$: $(\sigma_y, 0)$: 2, 2; 4, 2: *b*.
C^x $\mathbf{G}_{4}^{1}$: $(\sigma_y, 0)$: 2, 3; 4, 3: *b*.
E^x $\mathbf{G}_{4}^{1}$: $(\sigma_y, 0)$: 2, 1; 4, 1: *b*.
A^x $\mathbf{G}_{4}^{1}$: $(\sigma_y, 0)$: 2, 3; 4, 3: *b*.
Λ^x $\mathbf{G}_{8}^{5}$: $(C_{2z}, 0)$, $(\sigma_y, 0)$: 5, x: *a*.
H^x $\mathbf{G}_{16}^{8}$: $(C_{2z}, 0)$, $(\sigma_y, 0)$: 5, x; 6, x; 7, x; 8, x: *a*.
Q^x $\mathbf{G}_{16}^{11}$: $(\sigma_x, 0)$, $(\sigma_y, 0)$, $(E, 1)$: 10, x: *d*.
G^x $\mathbf{G}_{16}^{8}$: $(C_{2z}, 0)$, $(\sigma_y, 0)$: 5, x; 6, x; 7, x; 8, x: *a*.

34 *Pnn2* C_{2v}^{10}

($F1$; $K7$; $M5$; $T1$; $Z1$.)

Γ $\mathbf{G}_8^5$: $\{C_{2z}|\frac{1}{2}\frac{1}{2}0\}, \{\sigma_y|0\frac{1}{2}\frac{1}{2}\}$: 5, 2: *a*.
Y $\mathbf{G}_{16}^8$: $\{\sigma_x|\frac{1}{2}0\frac{1}{2}\}, \{C_{2z}|\frac{1}{2}\frac{1}{2}0\}$: 5, 3; 6, 3; 7, 3; 8, 3: *a*.
X $\mathbf{G}_{16}^8$: $\{\sigma_y|0\frac{1}{2}\frac{1}{2}\}, \{C_{2z}|\frac{1}{2}\frac{1}{2}0\}$: 5, 3; 6, 3; 7, 3; 8, 3: *a*.
Z $\mathbf{G}_{16}^8$: $\{C_{2z}|\frac{1}{2}\frac{1}{2}0\}, \{\sigma_x|\frac{1}{2}0\frac{1}{2}\}$: 9, 1: *b*.
U $\mathbf{G}_{16}^8$: $\{\sigma_x|\frac{1}{2}0\frac{1}{2}\}, \{C_{2z}|\frac{1}{2}\frac{1}{2}0\}$: 5, 3; 6, 3; 7, 3; 8, 3: *a*.
T $\mathbf{G}_{16}^8$: $\{\sigma_y|0\frac{1}{2}\frac{1}{2}\}, \{C_{2z}|\frac{1}{2}\frac{1}{2}0\}$: 5, 3; 6, 3; 7, 3; 8, 3: *a*.
S $\mathbf{G}_{16}^8$: $\{C_{2z}|\frac{1}{2}\frac{1}{2}0\}, \{\sigma_x|\frac{1}{2}0\frac{1}{2}\}$: 9, 1: *b*.
R $\mathbf{G}_8^5 \otimes \mathbf{T}_2$: $\{C_{2z}|\frac{1}{2}\frac{1}{2}0\}, \{\sigma_y|0\frac{1}{2}\frac{1}{2}\}$; $\mathbf{t}_1$ or $\mathbf{t}_2$ or $\mathbf{t}_3$: 5, 2: *a*.

Δ^x $\mathbf{G}_4^1$: $(\sigma_x, 0)$: 2, 2; 4, 2: *b*.
D^x $\mathbf{G}_4^1$: $(\sigma_x, 0)$: 2, 3; 4, 3: *b*.
P^x $\mathbf{G}_4^1$: $(\sigma_x, 0)$: 2, 2; 4, 2: *b*.
B^x $\mathbf{G}_4^1$: $(\sigma_x, 0)$: 2, 3; 4, 3: *b*.
Σ^x $\mathbf{G}_4^1$: $(\sigma_y, 0)$: 2, 2; 4, 2: *b*.
C^x $\mathbf{G}_4^1$: $(\sigma_y, 0)$: 2, 3; 4, 3: *b*.
E^x $\mathbf{G}_4^1$: $(\sigma_y, 0)$: 2, 2; 4, 2: *b*.
A^x $\mathbf{G}_4^1$: $(\sigma_y, 0)$: 2, 3; 4, 3: *b*.
Λ^x $\mathbf{G}_8^5$: $(C_{2z}, 0), (\sigma_y, 0)$: 5, x: *a*.
H^x $\mathbf{G}_{16}^8$: $(C_{2z}, 0), (\sigma_y, 0)$: 5, x; 6, x; 7, x; 8, x: *a*.
Q^x $\mathbf{G}_{16}^{11}$: $(\sigma_x, 0), (\sigma_y, 0), (E, 1)$: 10, x: *d*.
G^x $\mathbf{G}_{16}^8$: $(C_{2z}, 0), (\sigma_y, 0)$: 5, x; 6, x; 7, x; 8, x: *a*.

35 *Cmm2* C_{2v}^{11}

($F1$; $K7$; $M5$; $T3$; $Z1$.)

Γ $\mathbf{G}_8^5$: $\{C_{2z}|000\}, \{\sigma_y|000\}$: 5, 2: *a*.
Y $\mathbf{G}_8^5 \otimes \mathbf{T}_2$: $\{C_{2z}|000\}, \{\sigma_y|000\}$; $\mathbf{t}_1$ or $\mathbf{t}_2$: 5, 2: *a*.
Z $\mathbf{G}_8^5 \otimes \mathbf{T}_2$: $\{C_{2z}|000\}, \{\sigma_y|000\}$; $\mathbf{t}_3$: 5, 2: *a*.
T $\mathbf{G}_8^5 \otimes \mathbf{T}_2$: $\{C_{2z}|000\}, \{\sigma_y|000\}$; $\mathbf{t}_1$ or $\mathbf{t}_2$ or $\mathbf{t}_3$: 5, 2: *a*.
S $\mathbf{G}_4^1 \otimes \mathbf{T}_2$: $\{C_{2z}|000\}$; $\mathbf{t}_2$: 2, 3; 4, 3: *b*.
R $\mathbf{G}_4^1 \otimes \mathbf{T}_2$: $\{C_{2z}|000\}$; $\mathbf{t}_2$ or $\mathbf{t}_3$: 2, 3; 4, 3: *b*.

Λ^x $\mathbf{G}_8^5$: $(C_{2z}, 0), (\sigma_y, 0)$: 5, x: *a*.
H^x $\mathbf{G}_8^5$: $(C_{2z}, 0), (\sigma_y, 0)$: 5, x: *a*.
D^x $\mathbf{G}_4^1$: $(C_{2z}, 0)$: 2, x; 4, x: *b*.
A^x $\mathbf{G}_4^1$: $(\sigma_y, 0)$: 2, 2; 4, 2: *b*.
Σ^x $\mathbf{G}_4^1$: $(\sigma_y, 0)$: 2, 2; 4, 2: *b*.
Δ^x $\mathbf{G}_4^1$: $(\sigma_x, 0)$: 2, 2; 4, 2: *b*.
B^x $\mathbf{G}_4^1$: $(\sigma_x, 0)$: 2, 2; 4, 2: *b*.
G^x $\mathbf{G}_4^1$: $(\sigma_x, 0)$: 2, 2; 4, 2: *b*.
F^x $\mathbf{G}_4^1$: $(\sigma_x, 0)$: 2, 2; 4, 2: *b*.
E^x $\mathbf{G}_4^1$: $(\sigma_y, 0)$: 2, 2; 4, 2: *b*.
C^x $\mathbf{G}_4^1$: $(\sigma_y, 0)$: 2, 2; 4, 2: *b*.

36 *Cmc*2_1 C_{2v}^{12}

(*F*1; *K*7; *M*5; *T*3; *Z*1.)

Γ $\mathbf{G}_8^5$: $\{C_{2z}\,|\,00\tfrac{1}{2}\}$, $\{\sigma_y\,|\,000\}$: 5, 2: *a*.
Y $\mathbf{G}_8^5 \otimes \mathbf{T}_2$: $\{C_{2z}\,|\,00\tfrac{1}{2}\}$, $\{\sigma_y\,|\,000\}$; $\mathbf{t}_1$ or $\mathbf{t}_2$: 5, 2: *a*.
Z $\mathbf{G}_{16}^8$: $\{\sigma_y\,|\,000\}$, $\{C_{2z}\,|\,00\tfrac{1}{2}\}$: 9, 1: *b*.
T $\mathbf{G}_{16}^8$: $\{\sigma_y\,|\,000\}$, $\{C_{2z}\,|\,00\tfrac{1}{2}\}$: 9, 1: *b*.
S $\mathbf{G}_4^1 \otimes \mathbf{T}_2$: $\{C_{2z}\,|\,00\tfrac{1}{2}\}$; $\mathbf{t}_2$: 2, 3; 4, 3: *b*.
R $\mathbf{G}_8^2$: $\{C_{2z}\,|\,00\tfrac{1}{2}\}$, $\{\bar{E}\,|\,000\}$: 5, 1; 7, 1: *a*.

Λ^x $\mathbf{G}_8^5$: $(C_{2z}, 0)$, $(\sigma_y, 0)$: 5, *x*: *a*.
H^x $\mathbf{G}_8^5$: $(C_{2z}, 0)$, $(\sigma_y, 0)$: 5, *x*: *a*.
D^x $\mathbf{G}_4^1$: $(C_{2z}, 0)$: 2, *x*; 4, *x*: *b*.
A^x $\mathbf{G}_4^1$: $(\sigma_y, 0)$: 2, 1; 4, 1: *b*.
Σ^x $\mathbf{G}_4^1$: $(\sigma_y, 0)$: 2, 2; 4, 2: *b*.
Δ^x $\mathbf{G}_4^1$: $(\sigma_x, 0)$: 2, 2; 4, 2: *b*.
B^x $\mathbf{G}_4^1$: $(\sigma_x, 0)$: 2, 3; 4, 3: *b*.
G^x $\mathbf{G}_4^1$: $(\sigma_x, 0)$: 2, 3; 4, 3: *b*.
F^x $\mathbf{G}_4^1$: $(\sigma_x, 0)$: 2, 2; 4, 2: *b*.
E^x $\mathbf{G}_4^1$: $(\sigma_y, 0)$: 2, 1; 4, 1: *b*.
C^x $\mathbf{G}_4^1$: $(\sigma_y, 0)$: 2, 2; 4, 2: *b*.

37 *Ccc*2 C_{2v}^{13}

(*F*1; *K*7; *M*5; *T*3; *Z*1.)

Γ $\mathbf{G}_8^5$: $\{C_{2z}\,|\,000\}$, $\{\sigma_y\,|\,00\tfrac{1}{2}\}$: 5, 2: *a*.
Y $\mathbf{G}_8^5 \otimes \mathbf{T}_2$: $\{C_{2z}\,|\,000\}$, $\{\sigma_y\,|\,00\tfrac{1}{2}\}$; $\mathbf{t}_1$ or $\mathbf{t}_2$: 5, 2: *a*.
Z $\mathbf{G}_{16}^8$: $\{C_{2z}\,|\,000\}$, $\{\sigma_x\,|\,00\tfrac{1}{2}\}$: 9, 1: *b*.
T $\mathbf{G}_{16}^8$: $\{C_{2z}\,|\,000\}$, $\{\sigma_x\,|\,00\tfrac{1}{2}\}$: 9, 1: *b*.
S $\mathbf{G}_4^1 \otimes \mathbf{T}_2$: $\{C_{2z}\,|\,000\}$; $\mathbf{t}_2$: 2, 3; 4, 3: *b*.
R $\mathbf{G}_4^1 \otimes \mathbf{T}_2$: $\{C_{2z}\,|\,000\}$; $\mathbf{t}_2$ or $\mathbf{t}_3$: 2, 3; 4, 3: *b*.

Λ^x $\mathbf{G}_8^5$: $(C_{2z}, 0)$, $(\sigma_y, 0)$: 5, *x*: *a*.
H^x $\mathbf{G}_8^5$: $(C_{2z}, 0)$, $(\sigma_y, 0)$: 5, *x*: *a*.
D^x $\mathbf{G}_4^1$: $(C_{2z}, 0)$: 2, *x*; 4, *x*: *b*.
A^x $\mathbf{G}_4^1$: $(\sigma_y, 0)$: 2, 3; 4, 3: *b*.
Σ^x $\mathbf{G}_4^1$: $(\sigma_y, 0)$: 2, 2; 4, 2: *b*.
Δ^x $\mathbf{G}_4^1$: $(\sigma_x, 0)$: 2, 2; 4, 2: *b*.
B^x $\mathbf{G}_4^1$: $(\sigma_x, 0)$: 2, 3; 4, 3: *b*.
G^x $\mathbf{G}_4^1$: $(\sigma_x, 0)$: 2, 3; 4, 3: *b*.
F^x $\mathbf{G}_4^1$: $(\sigma_x, 0)$: 2, 2; 4, 2: *b*.
E^x $\mathbf{G}_4^1$: $(\sigma_y, 0)$: 2, 3; 4, 3: *b*.
C^x $\mathbf{G}_4^1$: $(\sigma_y, 0)$: 2, 2; 4, 2: *b*.

38 *Amm2* C_{2v}^{14}

(*F*1; *K*7; *M*5; *T*3; *Z*1.)

Γ $\mathbf{G}_8^5$: $\{C_{2y} | 000\}, \{\sigma_x | 000\}$: 5, 2: *a*.
Y $\mathbf{G}_8^5 \otimes \mathbf{T}_2$: $\{C_{2y} | 000\}, \{\sigma_x | 000\}$; $\mathbf{t}_1$ or $\mathbf{t}_2$: 5, 2: *a*.
Z $\mathbf{G}_8^5 \otimes \mathbf{T}_2$: $\{C_{2y} | 000\}, \{\sigma_x | 000\}$; $\mathbf{t}_3$: 5, 2: *a*.
T $\mathbf{G}_8^5 \otimes \mathbf{T}_2$: $\{C_{2y} | 000\}, \{\sigma_x | 000\}$; $\mathbf{t}_1$ or $\mathbf{t}_2$ or $\mathbf{t}_3$: 5, 2: *a*.
S $\mathbf{G}_4^1 \otimes \mathbf{T}_2$: $\{\sigma_z | 000\}$; $\mathbf{t}_2$: 2, 3; 4, 3: *b*.
R $\mathbf{G}_4^1 \otimes \mathbf{T}_2$: $\{\sigma_z | 000\}$; $\mathbf{t}_2$ or $\mathbf{t}_3$: 2, 3; 4, 3: *b*.

Λ^x $\mathbf{G}_4^1$: $(\sigma_x, 0)$: 2, 2; 4, 2: *b*.
H^x $\mathbf{G}_4^1$: $(\sigma_x, 0)$: 2, 2; 4, 2: *b*.
D^x $\mathbf{G}_2^1$: $(\bar{E}, 0)$: 2, 2: *a*.
A^x $\mathbf{G}_4^1$: $(\sigma_z, 0)$: 2, 2; 4, 2: *b*.
Σ^x $\mathbf{G}_4^1$: $(\sigma_z, 0)$: 2, 2; 4, 2: *b*.
Δ^x $\mathbf{G}_8^5$: $(C_{2y}, 0), (\sigma_x, 0)$: 5, *x*: *a*.
B^x $\mathbf{G}_8^5$: $(C_{2y}, 0), (\sigma_x, 0)$: 5, *x*: *a*.
G^x $\mathbf{G}_8^5$: $(C_{2y}, 0), (\sigma_x, 0)$: 5, *x*: *a*.
F^x $\mathbf{G}_8^5$: $(C_{2y}, 0), (\sigma_x, 0)$: 5, *x*: *a*.
E^x $\mathbf{G}_4^1$: $(\sigma_z, 0)$: 2, 2; 4, 2: *b*.
C^x $\mathbf{G}_4^1$: $(\sigma_z, 0)$: 2, 2; 4, 2: *b*.

39 *Abm2* C_{2v}^{15}

(*F*1; *K*7; *M*5; *T*3; *Z*1.)

Γ $\mathbf{G}_8^5$: $\{C_{2y} | 000\}, \{\sigma_x | \frac{1}{2}\frac{1}{2}0\}$: 5, 2: *a*.
Y $\mathbf{G}_8^5 \otimes \mathbf{T}_2$: $\{C_{2y} | 000\}, \{\sigma_x | \frac{1}{2}\frac{1}{2}0\}$; $\mathbf{t}_1$ or $\mathbf{t}_2$: 5, 2: *a*.
Z $\mathbf{G}_8^5 \otimes \mathbf{T}_2$: $\{C_{2y} | 000\}, \{\sigma_x | \frac{1}{2}\frac{1}{2}0\}$; $\mathbf{t}_3$: 5, 2: *a*.
T $\mathbf{G}_8^5 \otimes \mathbf{T}_2$: $\{C_{2y} | 000\}, \{\sigma_x | \frac{1}{2}\frac{1}{2}0\}$; $\mathbf{t}_1$ or $\mathbf{t}_2$ or $\mathbf{t}_3$: 5, 2: *a*.
S $\mathbf{G}_8^2$: $\{\sigma_z | \frac{1}{2}\frac{1}{2}0\}, \{\bar{E} | 000\}$: 5, 1; 7, 1: *a*.
R $\mathbf{G}_8^2$: $\{\sigma_z | \frac{1}{2}\frac{1}{2}0\}, \{\bar{E} | 000\}$: 5, 1; 7, 1: *a*.

Λ^x $\mathbf{G}_4^1$: $(\sigma_x, 0)$: 2, 2; 4, 2: *b*.
H^x $\mathbf{G}_4^1$: $(\sigma_x, 0)$: 2, 2; 4, 2: *b*.
D^x $\mathbf{G}_2^1$: $(\bar{E}, 0)$: 2, 1: *a*.
A^x $\mathbf{G}_4^1$: $(\sigma_z, 0)$: 2, 2; 4, 2: *b*.
Σ^x $\mathbf{G}_4^1$: $(\sigma_z, 0)$: 2, 2; 4, 2: *b*.
Δ^x $\mathbf{G}_8^5$: $(C_{2y}, 0), (\sigma_x, 0)$: 5, *x*: *a*.
B^x $\mathbf{G}_8^5$: $(C_{2y}, 0), (\sigma_x, 0)$: 5, *x*: *a*.
G^x $\mathbf{G}_8^5$: $(C_{2y}, 0), (\sigma_x, 0)$: 5, *x*: *a*.
F^x $\mathbf{G}_8^5$: $(C_{2y}, 0), (\sigma_x, 0)$: 5, *x*: *a*.
E^x $\mathbf{G}_4^1$: $(\sigma_z, 0)$: 2, 2; 4, 2: *b*.
C^x $\mathbf{G}_4^1$: $(\sigma_z, 0)$: 2, 2; 4, 2: *b*.

40 *Ama*2 C_{2v}^{16}

($F1$; $K7$; $M5$; $T3$; $Z1$.)

Γ $\mathbf{G}_8^5$: $\{C_{2y}|000\}$, $\{\sigma_x|00\frac{1}{2}\}$: 5, 2: *a*.
Y $\mathbf{G}_8^5 \otimes \mathbf{T}_2$: $\{C_{2y}|000\}$, $\{\sigma_x|00\frac{1}{2}\}$; $\mathbf{t}_1$ or $\mathbf{t}_2$: 5, 2: *a*.
Z $\mathbf{G}_{16}^8$: $\{\sigma_x|00\frac{1}{2}\}$, $\{C_{2y}|000\}$: 5, 3; 6, 3; 7, 3; 8, 3: *a*.
T $\mathbf{G}_{16}^8$: $\{\sigma_x|00\frac{1}{2}\}$, $\{C_{2y}|000\}$: 5, 3; 6, 3; 7, 3; 8, 3: *a*.
S $\mathbf{G}_4^1 \otimes \mathbf{T}_2$: $\{\sigma_z|00\frac{1}{2}\}$; $\mathbf{t}_1$: 2, 3; 4, 3: *b*.
R $\mathbf{G}_4^1 \otimes \mathbf{T}_2$: $\{\sigma_z|00\frac{1}{2}\}$; $\mathbf{t}_2$ or $\mathbf{t}_3$: 2, 3; 4, 3: *b*.

Λ^x $\mathbf{G}_4^1$: $(\sigma_x, 0)$: 2, 2; 4, 2: *b*.
H^x $\mathbf{G}_4^1$: $(\sigma_x, 0)$: 2, 2; 4, 2: *b*.
D^x $\mathbf{G}_2^1$: $(\bar{E}, 0)$: 2, 2: *a*.
A^x $\mathbf{G}_8^2$: $(\sigma_z, 0)$, $(\bar{E}, 0)$: 5, 3; 7, 3: *a*.
Σ^x $\mathbf{G}_4^1$: $(\sigma_z, 0)$: 2, 2; 4, 2: *b*.
Δ^x $\mathbf{G}_8^5$: $(C_{2y}, 0)$, $(\sigma_x, 0)$: 5, x: *a*.
B^x $\mathbf{G}_{16}^8$: $(\sigma_z, 0)$, $(C_{2y}, 0)$: 5, x; 6, x; 7, x; 8, x: *a*.
G^x $\mathbf{G}_{16}^8$: $(\sigma_z, 0)$, $(C_{2y}, 0)$: 5, x; 6, x; 7, x; 8, x: *a*.
F^x $\mathbf{G}_8^5$: $(C_{2y}, 0)$, $(\sigma_x, 0)$: 5, x: *a*.
E^x $\mathbf{G}_8^2$: $(\sigma_z, 0)$, $(\bar{E}, 0)$: 5, 3; 7, 3: *a*.
C^x $\mathbf{G}_4^1$: $(\sigma_z, 0)$: 2, 2; 4, 2: *b*.

41 *Aba*2 C_{2v}^{17}

($F1$; $K7$; $M5$; $T3$; $Z1$.)

Γ $\mathbf{G}_8^5$: $\{C_{2y}|000\}$, $\{\sigma_x|\frac{1}{2}\frac{1}{2}\frac{1}{2}\}$: 5, 2: *a*.
Y $\mathbf{G}_8^5 \otimes \mathbf{T}_2$: $\{C_{2y}|000\}$, $\{\sigma_x|\frac{1}{2}\frac{1}{2}\frac{1}{2}\}$; $\mathbf{t}_1$ or $\mathbf{t}_2$: 5, 2: *a*.
Z $\mathbf{G}_{16}^8$: $\{\sigma_x|\frac{1}{2}\frac{1}{2}\frac{1}{2}\}$, $\{C_{2y}|000\}$: 5, 3; 6, 3; 7, 3; 8, 3: *a*.
T $\mathbf{G}_{16}^8$: $\{\sigma_x|\frac{1}{2}\frac{1}{2}\frac{1}{2}\}$, $\{C_{2y}|000\}$: 5, 3; 6, 3; 7, 3; 8, 3: *a*.
S $\mathbf{G}_8^2$: $\{\sigma_z|\frac{1}{2}\frac{1}{2}\frac{1}{2}\}$, $\{\bar{E}|000\}$: 5, 1; 7, 1: *a*.
R $\mathbf{G}_8^2$: $\{\sigma_z|\frac{1}{2}\frac{1}{2}\frac{1}{2}\}$, $\{\bar{E}|000\}$: 5, 1; 7, 1: *a*.

Λ^x $\mathbf{G}_4^1$: $(\sigma_x, 0)$: 2, 2; 4, 2: *b*.
H^x $\mathbf{G}_4^1$: $(\sigma_x, 0)$: 2, 2; 4, 2: *b*.
D^x $\mathbf{G}_2^1$: $(\bar{E}, 0)$: 2, 1: *a*.
A^x $\mathbf{G}_8^2$: $(\sigma_z, 0)$, $(\bar{E}, 0)$: 5, 3; 7, 3: *a*.
Σ^x $\mathbf{G}_4^1$: $(\sigma_z, 0)$: 2, 2; 4, 2: *b*.
Δ^x $\mathbf{G}_8^5$: $(C_{2y}, 0)$, $(\sigma_x, 0)$: 5, x: *a*.
B^x $\mathbf{G}_{16}^8$: $(\sigma_z, 0)$, $(C_{2y}, 0)$: 5, x; 6, x; 7, x; 8, x: *a*.
G^x $\mathbf{G}_{16}^8$: $(\sigma_z, 0)$, $(C_{2y}, 0)$: 5, x; 6, x; 7, x; 8, x: *a*.
F^x $\mathbf{G}_8^5$: $(C_{2y}, 0)$, $(\sigma_x, 0)$: 5, x: *a*.
E^x $\mathbf{G}_8^2$: $(\sigma_z, 0)$, $(\bar{E}, 0)$: 5, 3; 7, 3: *a*.
C^x $\mathbf{G}_4^1$: $(\sigma_z, 0)$: 2, 2; 4, 2: *b*.

42 *Fmm2* C_{2v}^{18}

(*F*1; *K*7; *M*5; *S*14; *Z*1.)

Γ $\mathbf{G}_8^5$: $\{C_{2z}|000\}, \{\sigma_y|000\}$: 5, 2: *a*.
Y $\mathbf{G}_8^5 \otimes \mathbf{T}_2$: $\{C_{2z}|000\}, \{\sigma_y|000\}$; $\mathbf{t}_2$ or $\mathbf{t}_3$: 5, 2: *a*.
X $\mathbf{G}_8^5 \otimes \mathbf{T}_2$: $\{C_{2z}|000\}, \{\sigma_y|000\}$; $\mathbf{t}_1$ or $\mathbf{t}_3$: 5, 2: *a*.
Z $\mathbf{G}_8^5 \otimes \mathbf{T}_2$: $\{C_{2z}|000\}, \{\sigma_y|000\}$; $\mathbf{t}_1$ or $\mathbf{t}_2$: 5, 2: *a*.
L $\mathbf{G}_2^1 \otimes \mathbf{T}_2$: $\{\bar{E}|000\}$; $\mathbf{t}_1$: 2, 1: *a*.

Λ^x $\mathbf{G}_8^5$: $(C_{2z}, 0), (\sigma_y, 0)$: 5, *x*: *a*.
G^x $\mathbf{G}_8^5$: $(C_{2z}, 0), (\sigma_y, 0)$: 5, *x*: *a*.
H^x $\mathbf{G}_8^5$: $(C_{2z}, 0), (\sigma_y, 0)$: 5, *x*: *a*.
Q^x $\mathbf{G}_8^5$: $(C_{2z}, 0), (\sigma_y, 0)$: 5, *x*: *a*.
Σ^x $\mathbf{G}_4^1$: $(\sigma_y, 0)$: 2, 2; 4, 2: *b*.
C^x $\mathbf{G}_4^1$: $(\sigma_y, 0)$: 2, 2; 4, 2: *b*.
A^x $\mathbf{G}_4^1$: $(\sigma_y, 0)$: 2, 2; 4, 2: *b*.
U^x $\mathbf{G}_4^1$: $(\sigma_y, 0)$: 2, 2; 4, 2: *b*.
Δ^x $\mathbf{G}_4^1$: $(\sigma_x, 0)$: 2, 2; 4, 2: *b*.
D^x $\mathbf{G}_4^1$: $(\sigma_x, 0)$: 2, 2; 4, 2: *b*.
B^x $\mathbf{G}_4^1$: $(\sigma_x, 0)$: 2, 2; 4, 2: *b*.
R^x $\mathbf{G}_4^1$: $(\sigma_x, 0)$: 2, 2; 4, 2: *b*.

43 *Fdd2* C_{2v}^{19}

(*F*1; *K*7; *M*5; *S*14; *Z*1.)

Γ $\mathbf{G}_8^5$: $\{C_{2z}|00\tfrac{1}{2}\}, \{\sigma_y|\tfrac{1}{2}00\}$: 5, 2: *a*.
Y $\mathbf{G}_{16}^8$: $\{\sigma_x|0\tfrac{1}{2}0\}, \{C_{2z}|00\tfrac{1}{2}\}$: 5, 3; 6, 3; 7, 3; 8, 3: *a*.
X $\mathbf{G}_{16}^8$: $\{\sigma_y|\tfrac{1}{2}00\}, \{C_{2z}|00\tfrac{1}{2}\}$: 5, 3; 6, 3; 7, 3; 8, 3: *a*.
Z $\mathbf{G}_{16}^8$: $\{C_{2z}|00\tfrac{1}{2}\}, \{\sigma_x|0\tfrac{1}{2}0\}$: 9, 1: *b*.
L $\mathbf{G}_2^1 \otimes \mathbf{T}_2$: $\{\bar{E}|000\}$; $\mathbf{t}_1$: 2, 1: *a*.

Λ^x $\mathbf{G}_8^5$: $(C_{2z}, 0), (\sigma_y, 0)$: 5, *x*: *a*.
G^x $\mathbf{G}_{16}^8$: $(C_{2z}, 0), (\sigma_y, 0)$: 5, *x*; 6, *x*; 7, *x*; 8, *x*: *a*.
H^x $\mathbf{G}_{16}^8$: $(C_{2z}, 0), (\sigma_y, 0)$: 5, *x*; 6, *x*; 7, *x*; 8, *x*: *a*.
Q^x $\mathbf{G}_{16}^{11}$: $(C_{2z}, 0), (\sigma_y, 0), (\bar{E}, 1)$: 5, *x*: *e*.
Σ^x $\mathbf{G}_4^1$: $(\sigma_y, 0)$: 2, 2; 4, 2: *b*.
C^x $\mathbf{G}_4^1$: $(\sigma_y, 0)$: 2, 3; 4, 3: *b*.
A^x $\mathbf{G}_4^1$: $(\sigma_y, 0)$: 2, 3; 4, 3: *b*.
U^x $\mathbf{G}_4^1$: $(\sigma_y, 0)$: 2, 2; 4, 2: *b*.
Δ^x $\mathbf{G}_4^1$: $(\sigma_x, 0)$: 2, 2; 4, 2: *b*.
D^x $\mathbf{G}_4^1$: $(\sigma_x, 0)$: 2, 3; 4, 3: *b*.
B^x $\mathbf{G}_4^1$: $(\sigma_x, 0)$: 2, 3; 4, 3: *b*.
R^x $\mathbf{G}_4^1$: $(\sigma_x, 0)$: 2, 2; 4, 2: *b*.

44 *Imm*2 C_{2v}^{20}

(*F*1; *K*7; *M*5; *Z*1.)

Γ $\mathbf{G}_8^5$: $\{C_{2z}\,|\,000\}$, $\{\sigma_y\,|\,000\}$: 5, 2: *a*.
X $\mathbf{G}_8^5\otimes\mathbf{T}_2$: $\{C_{2z}\,|\,000\}$, $\{\sigma_y\,|\,000\}$; $\mathbf{t}_1$ or $\mathbf{t}_2$ or $\mathbf{t}_3$: 5, 2: *a*.
R $\mathbf{G}_4^1\otimes\mathbf{T}_2$: $\{\sigma_y\,|\,000\}$; $\mathbf{t}_1$: 2, 2; 4, 2: *b*.
S $\mathbf{G}_4^1\otimes\mathbf{T}_2$: $\{\sigma_x\,|\,000\}$; $\mathbf{t}_1$ or $\mathbf{t}_3$: 2, 2; 4, 2: *b*.
T $\mathbf{G}_4^1\otimes\mathbf{T}_2$: $\{C_{2z}\,|\,000\}$; $\mathbf{t}_1$ or $\mathbf{t}_2$: 2, 2; 4, 2: *b*.
W $\mathbf{G}_4^1\otimes\mathbf{T}_4$: $\{C_{2z}\,|\,000\}$; $\mathbf{t}_1$ or $\mathbf{t}_2$ or $\mathbf{t}_3$: 2, 2; 4, 2: *b*.

Λ^x $\mathbf{G}_8^5$: $(C_{2z}, 0)$, $(\sigma_y, 0)$: 5, *x*: *a*.
G^x $\mathbf{G}_8^5$: $(C_{2z}, 0)$, $(\sigma_y, 0)$: 5, *x*: *a*.
P^x $\mathbf{G}_4^1$: $(C_{2z}, 0)$: 2, *x*; 4, *x*: *b*.
Σ^x $\mathbf{G}_4^1$: $(\sigma_y, 0)$: 2, 2; 4, 2: *b*.
F^x $\mathbf{G}_4^1$: $(\sigma_y, 0)$: 2, 2; 4, 2: *b*.
D^x $\mathbf{G}_2^1$: $(\bar{E}, 0)$: 2, 2: *a*.
Δ^x $\mathbf{G}_4^1$: $(\sigma_x, 0)$: 2, 2; 4, 2: *b*.
U^x $\mathbf{G}_4^1$: $(\sigma_x, 0)$: 2, 2; 4, 2: *b*.
Q^x $\mathbf{G}_2^1$: $(\bar{E}, 0)$: 2, 2: *a*.

45 *Iba*2 C_{2v}^{21}

(*F*1; *K*7; *M*5; *Z*1.)

Γ $\mathbf{G}_8^5$: $\{C_{2z}\,|\,000\}$, $\{\sigma_y\,|\,\frac{1}{2}\frac{1}{2}0\}$: 5, 2: *a*.
X $\mathbf{G}_8^5\otimes\mathbf{T}_2$: $\{C_{2z}\,|\,000\}$, $\{\sigma_y\,|\,\frac{1}{2}\frac{1}{2}0\}$; $\mathbf{t}_1$ or $\mathbf{t}_2$ or $\mathbf{t}_3$: 5, 2: *a*.
R $\mathbf{G}_8^2$: $\{\sigma_y\,|\,\frac{1}{2}\frac{1}{2}0\}$, $\{\bar{E}\,|\,000\}$: 5, 1; 7, 1: *a*.
S $\mathbf{G}_8^2$: $\{\sigma_x\,|\,\frac{1}{2}\frac{1}{2}0\}$, $\{\bar{E}\,|\,000\}$: 5, 1; 7, 1: *a*.
T $\mathbf{G}_4^1\otimes\mathbf{T}_2$: $\{C_{2z}\,|\,000\}$; $\mathbf{t}_1$ or $\mathbf{t}_2$: 2, 3; 4, 3: *b*.
W $\mathbf{G}_4^1\otimes\mathbf{T}_4$: $\{C_{2z}\,|\,000\}$; $\mathbf{t}_1$ or $\mathbf{t}_2$ or $\mathbf{t}_3$: 2, 1; 4, 1: *b*.

Λ^x $\mathbf{G}_8^5$: $(C_{2z}, 0)$, $(\sigma_y, 0)$: 5, *x*: *a*.
G^x $\mathbf{G}_8^5$: $(C_{2z}, 0)$, $(\sigma_y, 0)$: 5, *x*: *a*.
P^x $\mathbf{G}_4^1$: $(C_{2z}, 0)$: 2, *x*; 4, *x*: *b*.
Σ^x $\mathbf{G}_4^1$: $(\sigma_y, 0)$: 2, 2; 4, 2: *b*.
F^x $\mathbf{G}_4^1$: $(\sigma_y, 0)$: 2, 2; 4, 2: *b*.
D^x $\mathbf{G}_2^1$: $(\bar{E}, 0)$: 2, 1: *a*.
Δ^x $\mathbf{G}_4^1$: $(\sigma_x, 0)$: 2, 2; 4, 2: *b*.
U^x $\mathbf{G}_4^1$: $(\sigma_x, 0)$: 2, 2; 4, 2: *b*.
Q^x $\mathbf{G}_2^1$: $(\bar{E}, 0)$: 2, 1: *a*.

46 *Ima*2 C_{2v}^{22}

($F1$; $K7$; $M5$; $Z1$.)

Γ $\mathbf{G}_8^5$: $\{C_{2z}|000\}$, $\{\sigma_y|0\frac{1}{2}\frac{1}{2}\}$: 5, 2: *a*.
X $\mathbf{G}_8^5 \otimes \mathbf{T}_2$: $\{C_{2z}|000\}$, $\{\sigma_y|0\frac{1}{2}\frac{1}{2}\}$; $\mathbf{t}_1$ or $\mathbf{t}_2$ or $\mathbf{t}_3$: 5, 2: *b*.
R $\mathbf{G}_4^1 \otimes \mathbf{T}_2$: $\{\sigma_y|0\frac{1}{2}\frac{1}{2}\}$; $\mathbf{t}_1$: 2, 2; 4, 2: *b*.
S $\mathbf{G}_8^2$: $\{\sigma_x|0\frac{1}{2}\frac{1}{2}\}$, $\{\bar{E}|000\}$: 5, 1; 7, 1: *a*.
T $\mathbf{G}_4^1 \otimes \mathbf{T}_2$: $\{C_{2z}|000\}$; $\mathbf{t}_1$ or $\mathbf{t}_2$: 2, 3; 4, 3: *b*.
W $\mathbf{G}_4^1 \otimes \mathbf{T}_4$: $\{C_{2z}|000\}$; $\mathbf{t}_1$ or $\mathbf{t}_2$ or $\mathbf{t}_3$: 2, 3; 4, 3: *b*.

Λ^x $\mathbf{G}_8^5$: $(C_{2z}, 0)$, $(\sigma_y, 0)$: 5, *x*: *a*.
G^x $\mathbf{G}_8^5$: $(C_{2z}, 0)$, $(\sigma_y, 0)$: 5, *x*: *a*.
P^x $\mathbf{G}_4^1$: $(C_{2z}, 0)$: 2, *x*; 4, *x*: *b*.
Σ^x $\mathbf{G}_4^1$: $(\sigma_y, 0)$: 2, 2; 4, 2: *b*.
F^x $\mathbf{G}_4^1$: $(\sigma_y, 0)$: 2, 2; 4, 2: *b*.
D^x $\mathbf{G}_2^1$: $(\bar{E}, 0)$: 2, 1: *a*.
Δ^x $\mathbf{G}_4^1$: $(\sigma_x, 0)$: 2, 2; 4, 2: *b*.
U^x $\mathbf{G}_4^1$: $(\sigma_x, 0)$: 2, 2; 4, 2: *b*.
Q^x $\mathbf{G}_2^1$: $(\bar{E}, 0)$: 2, 2: *a*.

47 *Pmmm* D_{2h}^1

($F1$; $K7$; $M5$; $T1$; $Z1$.)

Γ $\mathbf{G}_{16}^{11}$: $\{C_{2z}|000\}$, $\{C_{2y}|000\}$, $\{I|000\}$: 5, 2; 10, 2: *c*.
Y $\mathbf{G}_{16}^{11} \otimes \mathbf{T}_2$: $\{C_{2z}|000\}$, $\{C_{2y}|000\}$, $\{I|000\}$; $\mathbf{t}_1$: 5, 2; 10, 2: *c*.
X $\mathbf{G}_{16}^{11} \otimes \mathbf{T}_2$: $\{C_{2z}|000\}$, $\{C_{2y}|000\}$, $\{I|000\}$; $\mathbf{t}_2$: 5, 2; 10, 2: *c*.
Z $\mathbf{G}_{16}^{11} \otimes \mathbf{T}_2$: $\{C_{2z}|000\}$, $\{C_{2y}|000\}$, $\{I|000\}$; $\mathbf{t}_3$: 5, 2; 10, 2: *c*.
U $\mathbf{G}_{16}^{11} \otimes \mathbf{T}_2$: $\{C_{2z}|000\}$, $\{C_{2y}|000\}$, $\{I|000\}$; $\mathbf{t}_2$ or $\mathbf{t}_3$: 5, 2; 10, 2: *c*.
T $\mathbf{G}_{16}^{11} \otimes \mathbf{T}_2$: $\{C_{2z}|000\}$, $\{C_{2y}|000\}$, $\{I|000\}$; $\mathbf{t}_1$ or $\mathbf{t}_3$: 5, 2; 10, 2: *c*.
S $\mathbf{G}_{16}^{11} \otimes \mathbf{T}_2$: $\{C_{2z}|000\}$, $\{C_{2y}|000\}$, $\{I|000\}$; $\mathbf{t}_1$ or $\mathbf{t}_2$: 5, 2; 10, 2: *c*.
R $\mathbf{G}_{16}^{11} \otimes \mathbf{T}_2$: $\{C_{2z}|000\}$, $\{C_{2y}|000\}$, $\{I|000\}$; $\mathbf{t}_1$ or $\mathbf{t}_2$ or $\mathbf{t}_3$: 5, 2; 10, 2: *c*.

Δ^x $\mathbf{G}_8^5$: $(C_{2y}, 0)$, $(\sigma_x, 0)$: 5, 2: *a*.
D^x $\mathbf{G}_8^5$: $(C_{2y}, 0)$, $(\sigma_x, 0)$: 5, 2: *a*.
P^x $\mathbf{G}_8^5$: $(C_{2y}, 0)$, $(\sigma_x, 0)$: 5, 2: *a*.
B^x $\mathbf{G}_8^5$: $(C_{2y}, 0)$, $(\sigma_x, 0)$: 5, 2: *a*.
Σ^x $\mathbf{G}_8^5$: $(C_{2x}, 0)$, $(\sigma_z, 0)$: 5, 2: *a*.
C^x $\mathbf{G}_8^5$: $(C_{2x}, 0)$, $(\sigma_z, 0)$: 5, 2: *a*.
E^x $\mathbf{G}_8^5$: $(C_{2x}, 0)$, $(\sigma_z, 0)$: 5, 2: *a*.
A^x $\mathbf{G}_8^5$: $(C_{2x}, 0)$, $(\sigma_z, 0)$: 5, 2: *a*.
Λ^x $\mathbf{G}_8^5$: $(C_{2z}, 0)$, $(\sigma_y, 0)$: 5, 2: *a*.
H^x $\mathbf{G}_8^5$: $(C_{2z}, 0)$, $(\sigma_y, 0)$: 5, 2: *a*.
Q^x $\mathbf{G}_8^5$: $(C_{2z}, 0)$, $(\sigma_y, 0)$: 5, 2: *a*.
G^x $\mathbf{G}_8^5$: $(C_{2z}, 0)$, $(\sigma_y, 0)$: 5, 2: *a*.

48 *Pnnn* D_{2h}^{2}

($F1$; $K7$; $M5$; $T1$; $Z1$.)

Γ $\mathbf{G}_{16}^{11}$: $\{C_{2z}|000\}$, $\{C_{2y}|000\}$, $\{I|\frac{1}{2}\frac{1}{2}\frac{1}{2}\}$: 5, 2; 10, 2: *c*.
Y $\mathbf{G}_{32}^{7}$: $\{\sigma_z|\frac{1}{2}\frac{1}{2}\frac{1}{2}\}$, $\{\sigma_y|\frac{1}{2}\frac{1}{2}\frac{1}{2}\}$, $\{I|\frac{1}{2}\frac{1}{2}\frac{1}{2}\}$: 13, 3; 14, 3: *a*.
X $\mathbf{G}_{32}^{7}$: $\{\sigma_y|\frac{1}{2}\frac{1}{2}\frac{1}{2}\}$, $\{\sigma_x|\frac{1}{2}\frac{1}{2}\frac{1}{2}\}$, $\{I|\frac{1}{2}\frac{1}{2}\frac{1}{2}\}$: 13, 3; 14, 3: *a*.
Z $\mathbf{G}_{32}^{7}$: $\{\sigma_x|\frac{1}{2}\frac{1}{2}\frac{1}{2}\}$, $\{\sigma_z|\frac{1}{2}\frac{1}{2}\frac{1}{2}\}$, $\{I|\frac{1}{2}\frac{1}{2}\frac{1}{2}\}$: 13, 3; 14, 3: *a*.
U $\mathbf{G}_{32}^{7}$: $\{\sigma_z|\frac{1}{2}\frac{1}{2}\frac{1}{2}\}$, $\{\sigma_y|\frac{1}{2}\frac{1}{2}\frac{1}{2}\}$, $\{I|\frac{1}{2}\frac{1}{2}\frac{1}{2}\}$: 13, 3; 14, 3: *a*.
T $\mathbf{G}_{32}^{7}$: $\{\sigma_y|\frac{1}{2}\frac{1}{2}\frac{1}{2}\}$, $\{\sigma_x|\frac{1}{2}\frac{1}{2}\frac{1}{2}\}$, $\{I|\frac{1}{2}\frac{1}{2}\frac{1}{2}\}$: 13, 3; 14, 3: *a*.
S $\mathbf{G}_{32}^{7}$: $\{\sigma_x|\frac{1}{2}\frac{1}{2}\frac{1}{2}\}$, $\{\sigma_z|\frac{1}{2}\frac{1}{2}\frac{1}{2}\}$, $\{I|\frac{1}{2}\frac{1}{2}\frac{1}{2}\}$: 13, 3; 14, 3: *a*.
R $\mathbf{G}_{16}^{11}\otimes\mathbf{T}_2$: $\{C_{2z}|000\}$, $\{C_{2y}|000\}$, $\{I|\frac{1}{2}\frac{1}{2}\frac{1}{2}\}$; $\mathbf{t}_1$ or $\mathbf{t}_2$ or $\mathbf{t}_3$: 5, 2; 10, 2: *c*.

Δ^x $\mathbf{G}_{8}^{5}$: $(C_{2y}, 0)$, $(\sigma_x, 0)$: 5, 2: *a*.
D^x $\mathbf{G}_{16}^{8}$: $(\sigma_x, 0)$, $(C_{2y}, 0)$: 5, 3; 6, 3; 7, 3; 8, 3: *a*.
P^x $\mathbf{G}_{16}^{8}$: $(C_{2y}, 0)$, $(\sigma_z, 0)$: 9, 2: *b*.
B^x $\mathbf{G}_{16}^{8}$: $(\sigma_z, 0)$, $(C_{2y}, 0)$: 5, 3; 6, 3; 7, 3; 8, 3: *a*.
Σ^x $\mathbf{G}_{8}^{5}$: $(C_{2x}, 0)$, $(\sigma_y, 0)$: 5, 2: *a*.
C^x $\mathbf{G}_{16}^{8}$: $(\sigma_y, 0)$, $(C_{2x}, 0)$: 5, 3; 6, 3; 7, 3; 8, 3: *a*.
E^x $\mathbf{G}_{16}^{8}$: $(C_{2x}, 0)$, $(\sigma_y, 0)$: 9, 2: *b*.
A^x $\mathbf{G}_{16}^{8}$: $(\sigma_z, 0)$, $(\sigma_y, 0)$: 5, 3; 6, 3; 7, 3; 8, 3: *a*.
Λ^x $\mathbf{G}_{8}^{5}$: $(C_{2z}, 0)$, $(\sigma_y, 0)$: 5, 2: *a*.
H^x $\mathbf{G}_{16}^{8}$: $(\sigma_y, 0)$, $(C_{2z}, 0)$: 5, 3; 6, 3; 7, 3; 8, 3: *a*.
Q^x $\mathbf{G}_{16}^{8}$: $(C_{2z}, 0)$, $(\sigma_x, 0)$: 9, 2: *b*.
G^x $\mathbf{G}_{16}^{8}$: $(\sigma_x, 0)$, $(C_{2z}, 0)$: 5, 3; 6, 3; 7, 3; 8, 3: *a*.

49 *Pccm* D_{2h}^{3}

($F1$; $K7$; $M5$; $T1$; $Z1$.)

Γ $\mathbf{G}_{16}^{11}$: $\{C_{2z}|000\}$, $\{C_{2y}|000\}$, $\{I|00\frac{1}{2}\}$: 5, 2; 10, 2: *c*.
Y $\mathbf{G}_{16}^{11}\otimes\mathbf{T}_2$: $\{C_{2z}|000\}$, $\{C_{2y}|000\}$, $\{I|00\frac{1}{2}\}$; $\mathbf{t}_1$: 5, 2; 10, 2: *c*.
X $\mathbf{G}_{16}^{11}\otimes\mathbf{T}_2$: $\{C_{2z}|000\}$, $\{C_{2y}|000\}$, $\{I|00\frac{1}{2}\}$; $\mathbf{t}_2$: 5, 2; 10, 2: *c*.
Z $\mathbf{G}_{32}^{7}$: $\{\sigma_x|00\frac{1}{2}\}$, $\{\sigma_z|00\frac{1}{2}\}$, $\{I|00\frac{1}{2}\}$: 13, 3; 14, 3: *a*.
U $\mathbf{G}_{32}^{7}$: $\{\sigma_x|00\frac{1}{2}\}$, $\{\sigma_z|00\frac{1}{2}\}$, $\{I|00\frac{1}{2}\}$: 13, 3; 14, 3: *a*.
T $\mathbf{G}_{32}^{7}$: $\{\sigma_x|00\frac{1}{2}\}$, $\{\sigma_z|00\frac{1}{2}\}$, $\{I|00\frac{1}{2}\}$: 13, 3; 14, 3: *a*.
S $\mathbf{G}_{16}^{11}\otimes\mathbf{T}_2$: $\{C_{2z}|000\}$, $\{C_{2y}|000\}$, $\{I|00\frac{1}{2}\}$; $\mathbf{t}_1$ or $\mathbf{t}_2$: 5, 2; 10, 2: *c*.
R $\mathbf{G}_{32}^{7}$: $\{\sigma_x|00\frac{1}{2}\}$, $\{\sigma_z|00\frac{1}{2}\}$, $\{I|00\frac{1}{2}\}$: 13, 3; 14, 3: *a*.

Δ^x $\mathbf{G}_{8}^{5}$: $(C_{2y}, 0)$, $(\sigma_x, 0)$: 5, 2: *a*.
D^x $\mathbf{G}_{8}^{5}$: $(C_{2y}, 0)$, $(\sigma_x, 0)$: 5, 2: *a*.
P^x $\mathbf{G}_{16}^{8}$: $(\sigma_z, 0)$, $(C_{2y}, 0)$: 5, 3; 6, 3; 7, 3; 8, 3: *a*.
B^x $\mathbf{G}_{16}^{8}$: $(\sigma_z, 0)$, $(C_{2y}, 0)$: 5, 3; 6, 3; 7, 3; 8, 3: *a*.
Σ^x $\mathbf{G}_{8}^{5}$: $(C_{2x}, 0)$, $(\sigma_y, 0)$: 5, 2: *a*.
C^x $\mathbf{G}_{8}^{5}$: $(C_{2x}, 0)$, $(\sigma_y, 0)$: 5, 2: *a*.
E^x $\mathbf{G}_{16}^{8}$: $(\sigma_z, 0)$, $(\sigma_y, 0)$: 5, 3; 6, 3; 7, 3; 8, 3: *a*.
A^x $\mathbf{G}_{16}^{8}$: $(\sigma_z, 0)$, $(\sigma_y, 0)$: 5, 3; 6, 3; 7, 3; 8, 3: *a*.
Λ^x $\mathbf{G}_{8}^{5}$: $(C_{2z}, 0)$, $(\sigma_y, 0)$: 5, 2: *a*.
H^x $\mathbf{G}_{8}^{5}$: $(C_{2z}, 0)$, $(\sigma_y, 0)$: 5, 2: *a*.
Q^x $\mathbf{G}_{8}^{5}$: $(C_{2z}, 0)$, $(\sigma_y, 0)$: 5, 2: *a*.
G^x $\mathbf{G}_{8}^{5}$: $(C_{2z}, 0)$, $(\sigma_y, 0)$: 5, 2: *a*.

50 *Pban* D_{2h}^{4}

($F1$; $K7$; $M5$; $T1$; $Z1$.)

Γ $\mathbf{G}_{16}^{11}$: $\{C_{2z}\,|\,000\}, \{C_{2y}\,|\,000\}, \{I\,|\,\frac{1}{2}\frac{1}{2}0\}$: 5, 2; 10, 2: *c*.

Y $\mathbf{G}_{32}^{7}$: $\{\sigma_z\,|\,\frac{1}{2}\frac{1}{2}0\}, \{\sigma_y\,|\,\frac{1}{2}\frac{1}{2}0\}, \{I\,|\,\frac{1}{2}\frac{1}{2}0\}$: 13, 3; 14, 3: *a*.

X $\mathbf{G}_{32}^{7}$: $\{\sigma_z\,|\,\frac{1}{2}\frac{1}{2}0\}, \{\sigma_x\,|\,\frac{1}{2}\frac{1}{2}0\}, \{I\,|\,\frac{1}{2}\frac{1}{2}0\}$: 13, 3; 14, 3: *a*.

Z $\mathbf{G}_{16}^{11} \otimes \mathbf{T}_2$: $\{C_{2z}\,|\,000\}, \{C_{2y}\,|\,000\}, \{I\,|\,\frac{1}{2}\frac{1}{2}0\}$; $\mathbf{t}_3$: 5, 2; 10, 2: *c*.

U $\mathbf{G}_{32}^{7}$: $\{\sigma_z\,|\,\frac{1}{2}\frac{1}{2}0\}, \{\sigma_x\,|\,\frac{1}{2}\frac{1}{2}0\}, \{I\,|\,\frac{1}{2}\frac{1}{2}0\}$: 13, 3; 14, 3: *a*.

T $\mathbf{G}_{32}^{7}$: $\{\sigma_z\,|\,\frac{1}{2}\frac{1}{2}0\}, \{\sigma_y\,|\,\frac{1}{2}\frac{1}{2}0\}, \{I\,|\,\frac{1}{2}\frac{1}{2}0\}$: 13, 3; 14, 3: *a*.

S $\mathbf{G}_{32}^{7}$: $\{\sigma_x\,|\,\frac{1}{2}\frac{1}{2}0\}, \{\sigma_z\,|\,\frac{1}{2}\frac{1}{2}0\}, \{I\,|\,\frac{1}{2}\frac{1}{2}0\}$: 13, 3; 14, 3: *a*.

R $\mathbf{G}_{32}^{7}$: $\{\sigma_x\,|\,\frac{1}{2}\frac{1}{2}0\}, \{\sigma_z\,|\,\frac{1}{2}\frac{1}{2}0\}, \{I\,|\,\frac{1}{2}\frac{1}{2}0\}$: 13, 3; 14, 3: *a*.

Δ^x $\mathbf{G}_{8}^{5}$: $(C_{2y}, 0), (\sigma_x, 0)$: 5, 2: *a*.

D^x $\mathbf{G}_{16}^{8}$: $(\sigma_x, 0), (C_{2y}, 0)$: 5, 3; 6, 3; 7, 3; 8, 3: *a*.

P^x $\mathbf{G}_{16}^{8}$: $(\sigma_x, 0), (C_{2y}, 0)$: 5, 3; 6, 3; 7, 3; 8, 3: *a*.

B^x $\mathbf{G}_{8}^{5}$: $(C_{2y}, 0), (\sigma_x, 0)$: 5, 2: *a*.

Σ^x $\mathbf{G}_{8}^{5}$: $(C_{2x}, 0), (\sigma_y, 0)$: 5, 2: *a*.

C^x $\mathbf{G}_{16}^{8}$: $(\sigma_y, 0), (C_{2x}, 0)$: 5, 3; 6, 3; 7, 3; 8, 3: *a*.

E^x $\mathbf{G}_{16}^{8}$: $(\sigma_y, 0), (C_{2x}, 0)$: 5, 3; 6, 3; 7, 3; 8, 3: *a*.

A^x $\mathbf{G}_{8}^{5}$: $(C_{2x}, 0), (\sigma_y, 0)$: 5, 2: *a*.

Λ^x $\mathbf{G}_{8}^{5}$: $(C_{2z}, 0), (\sigma_y, 0)$: 5, 2: *a*.

H^x $\mathbf{G}_{16}^{8}$: $(\sigma_y, 0), (C_{2z}, 0)$: 5, 3; 6, 3; 7, 3; 8, 3: *a*.

Q^x $\mathbf{G}_{16}^{8}$: $(C_{2z}, 0), (\sigma_x, 0)$: 9, 2: *b*.

G^x $\mathbf{G}_{16}^{8}$: $(\sigma_x, 0), (C_{2z}, 0)$: 5, 3; 6, 3; 7, 3; 8, 3: *a*.

51 *Pmma* D_{2h}^{5}

($F1$; $K7$; $M5$; $T1$; $Z1$.)

Γ $\mathbf{G}_{16}^{11}$: $\{C_{2z}\,|\,00\frac{1}{2}\}, \{C_{2y}\,|\,000\}, \{I\,|\,000\}$: 5, 2; 10, 2: *c*.

Y $\mathbf{G}_{16}^{11} \otimes \mathbf{T}_2$: $\{C_{2z}\,|\,00\frac{1}{2}\}, \{C_{2y}\,|\,000\}, \{I\,|\,000\}$; $\mathbf{t}_1$: 5, 2; 10, 2: *c*.

X $\mathbf{G}_{16}^{11} \otimes \mathbf{T}_2$: $\{C_{2z}\,|\,00\frac{1}{2}\}, \{C_{2y}\,|\,000\}, \{I\,|\,000\}$; $\mathbf{t}_2$: 5, 2; 10, 2: *c*.

Z $\mathbf{G}_{32}^{7}$: $\{C_{2z}\,|\,00\frac{1}{2}\}, \{C_{2y}\,|\,000\}, \{I\,|\,000\}$: 13, 3; 14, 3: *a*.

U $\mathbf{G}_{32}^{7}$: $\{C_{2z}\,|\,00\frac{1}{2}\}, \{C_{2y}\,|\,000\}, \{I\,|\,000\}$: 13, 3; 14, 3: *a*.

T $\mathbf{G}_{32}^{7}$: $\{C_{2z}\,|\,00\frac{1}{2}\}, \{C_{2y}\,|\,000\}, \{I\,|\,000\}$: 13, 3; 14, 3: *a*.

S $\mathbf{G}_{16}^{11} \otimes \mathbf{T}_2$: $\{C_{2z}\,|\,00\frac{1}{2}\}, \{C_{2y}\,|\,000\}, \{I\,|\,000\}$; $\mathbf{t}_1$ or $\mathbf{t}_2$: 5, 2; 10, 2: *c*.

R $\mathbf{G}_{32}^{7}$: $\{C_{2z}\,|\,00\frac{1}{2}\}, \{C_{2y}\,|\,000\}, \{I\,|\,000\}$: 13, 3; 14, 3: *a*.

Δ^x $\mathbf{G}_{8}^{5}$: $(C_{2y}, 0), (\sigma_x, 0)$: 5, 2: *a*.

D^x $\mathbf{G}_{8}^{5}$: $(C_{2y}, 0), (\sigma_x, 0)$: 5, 2: *a*.

P^x $\mathbf{G}_{16}^{8}$: $(\sigma_z, 0), (C_{2y}, 0)$: 5, 3; 6, 3; 7, 3; 8, 3: *a*.

B^x $\mathbf{G}_{16}^{8}$: $(\sigma_z, 0), (C_{2y}, 0)$: 5, 3; 6, 3; 7, 3; 8, 3: *a*.

Σ^x $\mathbf{G}_{8}^{5}$: $(C_{2x}, 0), (\sigma_y, 0)$: 5, 2: *a*.

C^x $\mathbf{G}_{8}^{5}$: $(C_{2x}, 0), (\sigma_y, 0)$: 5, 2: *a*.

E^x $\mathbf{G}_{16}^{8}$: $(\sigma_y, 0), (\sigma_z, 0)$: 9, 1: *b*.

A^x $\mathbf{G}_{16}^{8}$: $(\sigma_y, 0), (\sigma_z, 0)$: 9, 1: *b*.

Λ^x $\mathbf{G}_{8}^{5}$: $(C_{2z}, 0), (\sigma_y, 0)$: 5, 2: *a*.

H^x $\mathbf{G}_{8}^{5}$: $(C_{2z}, 0), (\sigma_y, 0)$: 5, 2: *a*.

Q^x $\mathbf{G}_{8}^{5}$: $(C_{2z}, 0), (\sigma_y, 0)$: 5, 2: *a*.

G^x $\mathbf{G}_{8}^{5}$: $(C_{2z}, 0), (\sigma_y, 0)$: 5, 2: *a*.

52 *Pnna* D_{2h}^{6}

($F1$; $K7$; $M5$; $T1$; $Z1$.)

Γ $\mathbf{G}_{16}^{11}$: $\{C_{2z}|00\frac{1}{2}\}, \{C_{2y}|000\}, \{I|\frac{1}{2}\frac{1}{2}0\}$: 5, 2; 10, 2: *c*.
Y $\mathbf{G}_{32}^{7}$: $\{\sigma_z|\frac{1}{2}\frac{1}{2}\frac{1}{2}\}, \{\sigma_y|\frac{1}{2}\frac{1}{2}0\}, \{I|\frac{1}{2}\frac{1}{2}0\}$: 13, 3; 14, 3: *a*.
X $\mathbf{G}_{32}^{7}$: $\{\sigma_y|\frac{1}{2}\frac{1}{2}0\}, \{\sigma_x|\frac{1}{2}\frac{1}{2}\frac{1}{2}\}, \{I|\frac{1}{2}\frac{1}{2}0\}$: 13, 3; 14, 3: *a*.
Z $\mathbf{G}_{32}^{7}$: $\{\sigma_x|\frac{1}{2}\frac{1}{2}\frac{1}{2}\}, \{C_{2y}|000\}, \{I|\frac{1}{2}\frac{1}{2}0\}$: 13, 3; 14, 3: *a*.
U $\mathbf{G}_{32}^{7}$: $\{C_{2y}|000\}, \{\sigma_z|\frac{1}{2}\frac{1}{2}\frac{1}{2}\}, \{I|\frac{1}{2}\frac{1}{2}0\}$: 9, 1; 10, 1: *b*.
T $\mathbf{G}_{32}^{14}$: $\{\sigma_z|\frac{1}{2}\frac{1}{2}\frac{1}{2}\}, \{\sigma_y|\frac{1}{2}\frac{1}{2}0\}, \{I|\frac{1}{2}\frac{1}{2}0\}$: 5, 3; 6, 3; 7, 3; 8, 3; 15, 3; 16, 3; 17, 3; 18, 3: *a*.
S $\mathbf{G}_{32}^{7}$: $\{\sigma_x|\frac{1}{2}\frac{1}{2}\frac{1}{2}\}, \{\sigma_z|\frac{1}{2}\frac{1}{2}\frac{1}{2}\}, \{I|\frac{1}{2}\frac{1}{2}0\}$: 13, 3; 14, 3: *a*.
R $\mathbf{G}_{32}^{7}$: $\{\sigma_y|\frac{1}{2}\frac{1}{2}0\}, \{C_{2x}|00\frac{1}{2}\}, \{I|\frac{1}{2}\frac{1}{2}0\}$: 13, 3; 14, 3: *a*.

Δ^x $\mathbf{G}_{8}^{5}$: $(C_{2y}, 0), (\sigma_x, 0)$: 5, 2: *a*.
D^x $\mathbf{G}_{16}^{8}$: $(\sigma_x, 0), (C_{2y}, 0)$: 5, 3; 6, 3; 7, 3; 8, 3: *a*.
P^x $\mathbf{G}_{16}^{8}$: $(C_{2y}, 0), (\sigma_z, 0)$: 9, 1: *b*.
B^x $\mathbf{G}_{16}^{8}$: $(\sigma_z, 0), (C_{2y}, 0)$: 5, 3; 6, 3; 7, 3; 8, 3: *a*.
Σ^x $\mathbf{G}_{8}^{5}$: $(C_{2x}, 0), (\sigma_y, 0)$: 5, 2: *a*.
C^x $\mathbf{G}_{16}^{8}$: $(\sigma_y, 0), (C_{2x}, 0)$: 5, 3; 6, 3: 7, 3; 8, 3: *a*.
E^x $\mathbf{G}_{16}^{11}$: $(\sigma_y, 0), (\sigma_z, 0), (\bar{E}, 0)$: 6, 3; 7, 3; 8, 3; 9, 3: *b*.
A^x $\mathbf{G}_{16}^{8}$: $(\sigma_y, 0), (\sigma_z, 0)$: 9, 1: *b*.
Λ^x $\mathbf{G}_{8}^{5}$: $(C_{2z}, 0), (\sigma_y, 0)$: 5, 2: *a*.
H^x $\mathbf{G}_{16}^{8}$: $(\sigma_y, 0), (C_{2z}, 0)$: 5, 3; 6, 3; 7, 3; 8, 3: *a*.
Q^x $\mathbf{G}_{16}^{8}$: $(C_{2z}, 0), (\sigma_x, 0)$: 9, 2: *b*.
G^x $\mathbf{G}_{16}^{8}$: $(\sigma_x, 0), (C_{2z}, 0)$: 5, 3; 6, 3; 7, 3; 8, 3: *a*.

53 *Pmna* D_{2h}^{7}

($F1$; $K7$; $M5$; $T1$; $Z1$.)

Γ $\mathbf{G}_{16}^{11}$: $\{C_{2z}|00\frac{1}{2}\}, \{C_{2y}|000\}, \{I|\frac{1}{2}00\}$: 5, 2; 10, 2: *c*.
Y $\mathbf{G}_{32}^{7}$: $\{\sigma_z|\frac{1}{2}0\frac{1}{2}\}, \{\sigma_y|\frac{1}{2}00\}, \{I|\frac{1}{2}00\}$: 13, 3; 14, 3: *a*.
X $\mathbf{G}_{16}^{11} \otimes \mathbf{T}_2$: $\{C_{2z}|00\frac{1}{2}\}, \{C_{2y}|000\}, \{I|\frac{1}{2}00\}$; $\mathbf{t}_2$: 5, 2; 10, 2: *c*.
Z $\mathbf{G}_{32}^{7}$: $\{C_{2z}|00\frac{1}{2}\}, \{C_{2y}|000\}, \{I|\frac{1}{2}00\}$: 13, 3; 14, 3: *a*.
U $\mathbf{G}_{32}^{7}$: $\{C_{2z}|00\frac{1}{2}\}, \{C_{2y}|000\}, \{I|\frac{1}{2}00\}$: 13, 3; 14, 3: *a*.
T $\mathbf{G}_{32}^{14}$: $\{C_{2z}|00\frac{1}{2}\}, \{\sigma_y|\frac{1}{2}00\}, \{I|\frac{1}{2}00\}$: 5, 3; 6, 3; 7, 3; 8, 3; 15, 3; 16, 3; 17, 3; 18, 3: *a*.
S $\mathbf{G}_{32}^{7}$: $\{\sigma_z|\frac{1}{2}0\frac{1}{2}\}, \{\sigma_y|\frac{1}{2}00\}, \{I|\frac{1}{2}00\}$: 13, 3; 14, 3: *a*.
R $\mathbf{G}_{32}^{14}$: $\{C_{2z}|00\frac{1}{2}\}, \{\sigma_y|\frac{1}{2}00\}, \{I|\frac{1}{2}00\}$: 5, 3; 6, 3; 7, 3; 8, 3; 15, 3; 16, 3; 17, 3; 18, 3: *a*.

Δ^x $\mathbf{G}_{8}^{5}$: $(C_{2y}, 0), (\sigma_x, 0)$: 5, 2: *a*.
D^x $\mathbf{G}_{8}^{5}$: $(C_{2y}, 0), (\sigma_x, 0)$: 5, 2: *a*.
P^x $\mathbf{G}_{16}^{8}$: $(\sigma_z, 0), (C_{2y}, 0)$: 5, 3; 6, 3; 7, 3; 8, 3: *a*.
B^x $\mathbf{G}_{16}^{8}$: $(\sigma_z, 0), (C_{2y}, 0)$: 5, 3; 6, 3; 7, 3; 8, 3: *a*.
Σ^x $\mathbf{G}_{8}^{5}$: $(C_{2x}, 0), (\sigma_z, 0)$: 5, 2: *a*.
C^x $\mathbf{G}_{16}^{8}$: $(\sigma_y, 0), (C_{2x}, 0)$: 5, 3; 6, 3; 7, 3; 8, 3: *a*.
E^x $\mathbf{G}_{16}^{11}$: $(\sigma_y, 0), (C_{2x}, 0), (\bar{E}, 0)$: 6, 3; 7, 3; 8, 3; 9, 3: *b*.
A^x $\mathbf{G}_{16}^{8}$: $(\sigma_y, 0), (\sigma_z, 0)$: 9, 1: *b*.
Λ^x $\mathbf{G}_{8}^{5}$: $(C_{2z}, 0), (\sigma_y, 0)$: 5, 2: *a*.
H^x $\mathbf{G}_{16}^{8}$: $(\sigma_y, 0), (C_{2z}, 0)$: 5, 3; 6, 3; 7, 3; 8, 3: *a*.
Q^x $\mathbf{G}_{16}^{8}$: $(\sigma_y, 0), (C_{2z}, 0)$: 5, 3; 6, 3; 7, 3; 8, 3: *a*.
G^x $\mathbf{G}_{8}^{5}$: $(C_{2z}, 0), (\sigma_y, 0)$: 5, 2: *a*.

54 *Pcca* D_{2h}^{8}

($F1$; $K7$; $M5$; $T1$; $Z1$.)

Γ $\mathbf{G}_{16}^{11}$: $\{C_{2z}|00\frac{1}{2}\}$, $\{C_{2y}|000\}$, $\{I|0\frac{1}{2}0\}$: 5, 2; 10, 2: *c*.
Y $\mathbf{G}_{16}^{11} \otimes \mathbf{T}_2$: $\{C_{2z}|00\frac{1}{2}\}$, $\{C_{2y}|000\}$, $\{I|0\frac{1}{2}0\}$; $\mathbf{t}_1$: 5, 2; 10, 2: *c*.
X $\mathbf{G}_{32}^{7}$: $\{\sigma_y|0\frac{1}{2}0\}$, $\{\sigma_x|0\frac{1}{2}\frac{1}{2}\}$, $\{I|0\frac{1}{2}0\}$: 13, 3; 14, 3: *a*.
Z $\mathbf{G}_{32}^{7}$: $\{C_{2z}|00\frac{1}{2}\}$, $\{C_{2y}|000\}$, $\{I|0\frac{1}{2}0\}$: 13, 3; 14, 3: *a*.
U $\mathbf{G}_{32}^{7}$: $\{C_{2y}|000\}$, $\{\sigma_z|0\frac{1}{2}\frac{1}{2}\}$, $\{I|0\frac{1}{2}0\}$: 9, 1; 10, 1: *b*.
T $\mathbf{G}_{32}^{7}$: $\{C_{2z}|00\frac{1}{2}\}$, $\{C_{2y}|000\}$, $\{I|0\frac{1}{2}0\}$: 13, 3; 14, 3: *a*.
S $\mathbf{G}_{32}^{7}$: $\{\sigma_y|0\frac{1}{2}0\}$, $\{\sigma_x|0\frac{1}{2}\frac{1}{2}\}$, $\{I|0\frac{1}{2}0\}$: 13, 3; 14, 3: *a*.
R $\mathbf{G}_{32}^{7}$: $\{C_{2y}|000\}$, $\{\sigma_z|0\frac{1}{2}\frac{1}{2}\}$, $\{I|0\frac{1}{2}0\}$: 9, 1; 10, 1: *b*.

Δ^x $\mathbf{G}_{8}^{5}$: $(C_{2y}, 0)$, $(\sigma_x, 0)$: 5, 2: *a*.
D^x $\mathbf{G}_{16}^{8}$: $(\sigma_x, 0)$, $(C_{2y}, 0)$: 5, 3; 6, 3; 7, 3; 8, 3: *a*.
P^x $\mathbf{G}_{16}^{8}$: $(C_{2y}, 0)$, $(\sigma_z, 0)$: 9, 1: *b*.
B^x $\mathbf{G}_{16}^{8}$: $(\sigma_z, 0)$, $(C_{2y}, 0)$: 5, 3; 6, 3; 7, 3; 8, 3: *a*.
Σ^x $\mathbf{G}_{8}^{5}$: $(C_{2x}, 0)$, $(\sigma_z, 0)$: 5, 2: *a*.
C^x $\mathbf{G}_{8}^{5}$: $(C_{2x}, 0)$, $(\sigma_y, 0)$: 5, 2: *a*.
E^x $\mathbf{G}_{16}^{8}$: $(\sigma_y, 0)$, $(\sigma_z, 0)$: 9, 1: *b*.
A^x $\mathbf{G}_{16}^{8}$: $(\sigma_y, 0)$, $(\sigma_z, 0)$: 9, 1: *b*.
Λ^x $\mathbf{G}_{8}^{5}$: $(C_{2z}, 0)$, $(\sigma_y, 0)$: 5, 2: *a*.
H^x $\mathbf{G}_{8}^{5}$: $(C_{2z}, 0)$, $(\sigma_y, 0)$: 5, 2: *a*.
Q^x $\mathbf{G}_{16}^{8}$: $(\sigma_x, 0)$, $(C_{2z}, 0)$: 5, 3; 6, 3; 7, 3; 8, 3: *a*.
G^x $\mathbf{G}_{16}^{8}$: $(\sigma_x, 0)$, $(C_{2z}, 0)$: 5, 3; 6, 3; 7, 3; 8, 3: *a*.

55 *Pbam* D_{2h}^{9}

($F1$; $K7$; $M5$; $T1$; $Z1$.)

Γ $\mathbf{G}_{16}^{11}$: $\{C_{2z}|000\}$, $\{C_{2y}|\frac{1}{2}\frac{1}{2}0\}$, $\{I|000\}$: 5, 2; 10, 2: *c*.
Y $\mathbf{G}_{32}^{7}$: $\{\sigma_x|\frac{1}{2}\frac{1}{2}0\}$, $\{C_{2z}|000\}$, $\{I|000\}$: 13, 3; 14, 3: *a*.
X $\mathbf{G}_{32}^{7}$: $\{\sigma_y|\frac{1}{2}\frac{1}{2}0\}$, $\{C_{2z}|000\}$, $\{I|000\}$: 13, 3; 14, 3: *a*.
Z $\mathbf{G}_{16}^{11} \otimes \mathbf{T}_2$: $\{C_{2z}|000\}$, $\{C_{2y}|\frac{1}{2}\frac{1}{2}0\}$, $\{I|000\}$; $\mathbf{t}_3$: 5, 2; 10, 2: *c*.
U $\mathbf{G}_{32}^{7}$: $\{\sigma_y|\frac{1}{2}\frac{1}{2}0\}$, $\{C_{2z}|000\}$, $\{I|000\}$: 13, 3; 14, 3: *a*.
T $\mathbf{G}_{32}^{7}$: $\{\sigma_x|\frac{1}{2}\frac{1}{2}0\}$, $\{C_{2z}|000\}$, $\{I|000\}$: 13, 3; 14, 3: *a*.
S $\mathbf{G}_{32}^{14}$: $\{C_{2z}|000\}$, $\{C_{2x}|\frac{1}{2}\frac{1}{2}0\}$, $\{I|000\}$: 9, 1; 19, 1: *b*.
R $\mathbf{G}_{32}^{14}$: $\{C_{2z}|000\}$, $\{C_{2x}|\frac{1}{2}\frac{1}{2}0\}$, $\{I|000\}$: 9, 1; 19, 1: *b*.

Δ^x $\mathbf{G}_{8}^{5}$: $(C_{2y}, 0)$, $(\sigma_x, 0)$: 5, 2: *a*.
D^x $\mathbf{G}_{16}^{8}$: $(\sigma_z, 0)$, $(\sigma_x, 0)$: 9, 1: *b*.
P^x $\mathbf{G}_{16}^{8}$: $(\sigma_z, 0)$, $(\sigma_x, 0)$: 9, 1: *b*.
B^x $\mathbf{G}_{8}^{5}$: $(C_{2y}, 0)$, $(\sigma_x, 0)$: 5, 2: *a*.
Σ^x $\mathbf{G}_{8}^{5}$: $(C_{2x}, 0)$, $(\sigma_y, 0)$: 5, 2: *a*.
C^x $\mathbf{G}_{16}^{8}$: $(\sigma_z, 0)$, $(C_{2x}, 0)$: 9, 1: *b*.
E^x $\mathbf{G}_{16}^{8}$: $(\sigma_z, 0)$, $(C_{2x}, 0)$: 9, 1: *b*.
A^x $\mathbf{G}_{8}^{5}$: $(C_{2x}, 0)$, $(\sigma_y, 0)$: 5, 2: *a*.
Λ^x $\mathbf{G}_{8}^{5}$: $(C_{2z}, 0)$, $(\sigma_y, 0)$: 5, 2: *a*.
H^x $\mathbf{G}_{16}^{8}$: $(\sigma_y, 0)$, $(C_{2z}, 0)$: 5, 3; 6, 3; 7, 3; 8, 3: *a*.
Q^x $\mathbf{G}_{16}^{8}$: $(C_{2z}, 0)$, $(\sigma_x, 0)$: 9, 1: *b*.
G^x $\mathbf{G}_{16}^{8}$: $(\sigma_x, 0)$, $(C_{2z}, 0)$: 5, 3; 6, 3; 7, 3; 8, 3: *a*.

56 *Pccn* D_{2h}^{10}

($F1$; $K7$; $M5$; $T1$; $Z1$.)

Γ $\mathbf{G}_{16}^{11}$: $\{C_{2z}|000\}, \{C_{2y}|\frac{1}{2}\frac{1}{2}0\}, \{I|\frac{1}{2}\frac{1}{2}\frac{1}{2}\}$: 5, 2; 10, 2: *c*.

Y $\mathbf{G}_{32}^{7}$: $\{\sigma_z|\frac{1}{2}\frac{1}{2}\frac{1}{2}\}, \{C_{2x}|\frac{1}{2}\frac{1}{2}0\}, \{I|\frac{1}{2}\frac{1}{2}\frac{1}{2}\}$: 13, 3; 14, 3: *a*.

X $\mathbf{G}_{32}^{7}$: $\{\sigma_z|\frac{1}{2}\frac{1}{2}\frac{1}{2}\}, \{C_{2y}|\frac{1}{2}\frac{1}{2}0\}, \{I|\frac{1}{2}\frac{1}{2}\frac{1}{2}\}$: 13, 3; 14, 3: *a*.

Z $\mathbf{G}_{32}^{7}$: $\{\sigma_x|00\frac{1}{2}\}, \{\sigma_z|\frac{1}{2}\frac{1}{2}\frac{1}{2}\}, \{I|\frac{1}{2}\frac{1}{2}\frac{1}{2}\}$: 13, 3; 14, 3: *a*.

U $\mathbf{G}_{32}^{7}$: $\{C_{2z}|000\}, \{\sigma_x|00\frac{1}{2}\}, \{I|\frac{1}{2}\frac{1}{2}\frac{1}{2}\}$: 9, 1; 10, 1: *b*.

T $\mathbf{G}_{32}^{7}$: $\{C_{2z}|000\}, \{\sigma_y|00\frac{1}{2}\}, \{I|\frac{1}{2}\frac{1}{2}\frac{1}{2}\}$: 9, 1; 10, 1: *b*.

S $\mathbf{G}_{32}^{7}$: $\{C_{2y}|\frac{1}{2}\frac{1}{2}0\}, \{\sigma_z|\frac{1}{2}\frac{1}{2}\frac{1}{2}\}, \{I|\frac{1}{2}\frac{1}{2}\frac{1}{2}\}$: 13, 3; 14, 3: *a*.

R $\mathbf{G}_{32}^{14}$: $\{C_{2z}|000\}, \{C_{2x}|\frac{1}{2}\frac{1}{2}0\}, \{I|\frac{1}{2}\frac{1}{2}\frac{1}{2}\}$: 9, 1; 19, 1: *b*.

Δ^x $\mathbf{G}_{8}^{5}$: $(C_{2y}, 0), (\sigma_x, 0)$: 5, 2: *a*.

D^x $\mathbf{G}_{16}^{8}$: $(C_{2y}, 0), (\sigma_x, 0)$: 5, 3; 6, 3; 7, 3; 8, 3: *a*.

P^x $\mathbf{G}_{16}^{8}$: $(\sigma_x, 0), (C_{2y}, 0)$: 9, 1: *b*.

B^x $\mathbf{G}_{16}^{8}$: $(\sigma_z, 0), (C_{2y}, 0)$: 5, 3; 6, 3; 7, 3; 8, 3: *a*.

Σ^x $\mathbf{G}_{8}^{5}$: $(C_{2x}, 0), (\sigma_y, 0)$: 5, 2: *a*.

C^x $\mathbf{G}_{16}^{8}$: $(C_{2x}, 0), (\sigma_z, 0)$: 5, 3; 6, 3; 7, 3; 8, 3: *a*.

E^x $\mathbf{G}_{16}^{8}$: $(\sigma_y, 0), (\sigma_z, 0)$: 9, 1: *b*.

A^x $\mathbf{G}_{16}^{8}$: $(\sigma_z, 0), (\sigma_y, 0)$: 5, 3; 6, 3; 7, 3; 8, 3: *a*.

Λ^x $\mathbf{G}_{8}^{5}$: $(C_{2z}, 0), (\sigma_y, 0)$: 5, 2: *a*.

H^x $\mathbf{G}_{8}^{5}$: $(C_{2z}, 0), (\sigma_y, 0)$: 5, 1: *a*.

Q^x $\mathbf{G}_{8}^{5}$: $(C_{2z}, 0), (\sigma_y, 0)$: 5, 1: *a*.

G^x $\mathbf{G}_{8}^{5}$: $(C_{2z}, 0), (\sigma_y, 0)$: 5, 1: *a*.

57 *Pbcm* D_{2h}^{11}

($F1$; $K7$; $M5$; $T1$; $Z1$.)

Γ $\mathbf{G}_{16}^{11}$: $\{C_{2z}|000\}, \{C_{2y}|\frac{1}{2}\frac{1}{2}0\}, \{I|0\frac{1}{2}0\}$: 5, 2; 10, 2: *c*.

Y $\mathbf{G}_{32}^{7}$: $\{\sigma_x|\frac{1}{2}00\}, \{C_{2z}|000\}, \{I|0\frac{1}{2}0\}$: 13, 3; 14, 3: *a*.

X $\mathbf{G}_{32}^{7}$: $\{\sigma_z|0\frac{1}{2}0\}, \{C_{2y}|\frac{1}{2}\frac{1}{2}0\}, \{I|0\frac{1}{2}0\}$: 13, 3; 14, 3: *a*.

Z $\mathbf{G}_{16}^{11} \otimes \mathbf{T}_2$: $\{C_{2z}|000\}, \{C_{2y}|\frac{1}{2}\frac{1}{2}0\}, \{I|0\frac{1}{2}0\}; \mathbf{t}_3$: 5, 2; 10, 2: *c*.

U $\mathbf{G}_{32}^{7}$: $\{\sigma_z|0\frac{1}{2}0\}, \{C_{2y}|\frac{1}{2}\frac{1}{2}0\}, \{I|0\frac{1}{2}0\}$: 13, 3; 14, 3: *a*.

T $\mathbf{G}_{32}^{7}$: $\{\sigma_x|\frac{1}{2}00\}, \{C_{2z}|000\}, \{I|0\frac{1}{2}0\}$: 13, 3; 14, 3: *a*.

S $\mathbf{G}_{32}^{7}$: $\{C_{2z}|000\}, \{C_{2x}|\frac{1}{2}\frac{1}{2}0\}, \{I|0\frac{1}{2}0\}$: 9, 1; 10, 1: *b*.

R $\mathbf{G}_{32}^{7}$: $\{C_{2z}|000\}, \{C_{2x}|\frac{1}{2}\frac{1}{2}0\}, \{I|0\frac{1}{2}0\}$: 9, 1; 10, 1: *b*.

Δ^x $\mathbf{G}_{8}^{5}$: $(C_{2y}, 0), (\sigma_x, 0)$: 5, 2: *a*.

D^x $\mathbf{G}_{16}^{8}$: $(C_{2y}, 0), (\sigma_x, 0)$: 5, 3; 6, 3; 7, 3; 8, 3: *a*.

P^x $\mathbf{G}_{16}^{8}$: $(C_{2y}, 0), (\sigma_x, 0)$: 5, 3; 6, 3; 7, 3; 8, 3: *a*.

B^x $\mathbf{G}_{8}^{5}$: $(C_{2y}, 0), (\sigma_x, 0)$: 5, 2: *a*.

Σ^x $\mathbf{G}_{8}^{5}$: $(C_{2x}, 0), (\sigma_y, 0)$: 5, 2: *a*.

C^x $\mathbf{G}_{16}^{8}$: $(\sigma_z, 0), (C_{2x}, 0)$: 9, 1: *b*.

E^x $\mathbf{G}_{16}^{8}$: $(\sigma_z, 0), (C_{2x}, 0)$: 9, 1: *b*.

A^x $\mathbf{G}_{8}^{5}$: $(C_{2x}, 0), (\sigma_y, 0)$: 5, 2: *a*.

Λ^x $\mathbf{G}_{8}^{5}$: $(C_{2z}, 0), (\sigma_y, 0)$: 5, 2: *a*.

H^x $\mathbf{G}_{16}^{8}$: $(\sigma_y, 0), (C_{2z}, 0)$: 5, 3; 6, 3; 7, 3; 8, 3: *a*.

Q^x $\mathbf{G}_{16}^{8}$: $(\sigma_y, 0), (C_{2z}, 0)$: 5, 1; 6, 1; 7, 1; 8, 1: *a*.

G^x $\mathbf{G}_{8}^{5}$: $(C_{2z}, 0), (\sigma_y, 0)$: 5, 1: *a*.

58 *Pnnm* D_{2h}^{12}

(*C*7; *D*2; *F*1; *G*7; *K*7; *M*5; *T*1; *Z*1.)

Γ $\mathbf{G}_{16}^{11}$: $\{C_{2z}\,|\,000\}$, $\{C_{2y}\,|\,\frac{1}{2}\frac{1}{2}0\}$, $\{I\,|\,00\frac{1}{2}\}$: 5, 2; 10, 2: *c*.
Y $\mathbf{G}_{32}^{7}$: $\{\sigma_x\,|\,\frac{1}{2}\frac{1}{2}\frac{1}{2}\}$, $\{C_{2z}\,|\,000\}$, $\{I\,|\,00\frac{1}{2}\}$: 13, 3; 14, 3: *a*.
X $\mathbf{G}_{32}^{7}$: $\{\sigma_y\,|\,\frac{1}{2}\frac{1}{2}\frac{1}{2}\}$, $\{C_{2z}\,|\,000\}$, $\{I\,|\,00\frac{1}{2}\}$: 13, 3; 14, 3: *a*.
Z $\mathbf{G}_{32}^{7}$: $\{\sigma_x\,|\,\frac{1}{2}\frac{1}{2}\frac{1}{2}\}$, $\{\sigma_z\,|\,00\frac{1}{2}\}$, $\{I\,|\,00\frac{1}{2}\}$: 13, 3; 14, 3: *a*.
U $\mathbf{G}_{32}^{14}$: $\{\sigma_x\,|\,\frac{1}{2}\frac{1}{2}\frac{1}{2}\}$, $\{C_{2z}\,|\,000\}$, $\{I\,|\,00\frac{1}{2}\}$: 5, 3; 6, 3; 7, 3; 8, 3; 15, 3; 16, 3; 17, 3; 18, 3: *a*.
T $\mathbf{G}_{32}^{14}$: $\{\sigma_y\,|\,\frac{1}{2}\frac{1}{2}\frac{1}{2}\}$, $\{C_{2z}\,|\,000\}$, $\{I\,|\,00\frac{1}{2}\}$: 5, 3; 6, 3; 7, 3; 8, 3; 15, 3; 16, 3; 17, 3; 18, 3: *a*.
S $\mathbf{G}_{32}^{14}$: $\{C_{2z}\,|\,000\}$, $\{\sigma_x\,|\,\frac{1}{2}\frac{1}{2}\frac{1}{2}\}$, $\{I\,|\,00\frac{1}{2}\}$: 9, 1; 19, 1: *b*.
R $\mathbf{G}_{32}^{7}$: $\{C_{2y}\,|\,\frac{1}{2}\frac{1}{2}0\}$, $\{\sigma_z\,|\,00\frac{1}{2}\}$, $\{I\,|\,00\frac{1}{2}\}$: 13, 3; 14, 3: *a*.

Δ^x $\mathbf{G}_{8}^{5}$: $(C_{2y}, 0)$, $(\sigma_x, 0)$: 5, 2: *a*.
D^x $\mathbf{G}_{16}^{8}$: $(\sigma_z, 0)$, $(\sigma_x, 0)$: 9, 1: *b*.
P^x $\mathbf{G}_{16}^{11}$: $(C_{2y}, 0)$, $(\sigma_x, 0)$, $(\bar{E}, 0)$: 6, 3; 7, 3; 8, 3; 9, 3: *b*.
B^x $\mathbf{G}_{16}^{8}$: $(\sigma_z, 0)$, $(C_{2y}, 0)$: 5, 3; 6, 3; 7, 3; 8, 3: *a*.
Σ^x $\mathbf{G}_{8}^{5}$: $(C_{2x}, 0)$, $(\sigma_y, 0)$: 5, 2: *a*.
C^x $\mathbf{G}_{16}^{8}$: $(\sigma_z, 0)$, $(C_{2x}, 0)$: 9, 1: *b*.
E^x $\mathbf{G}_{16}^{11}$: $(C_{2x}, 0)$, $(\sigma_z, 0)$, $(\bar{E}, 0)$: 6, 3; 7, 3; 8, 3; 9, 3: *b*.
A^x $\mathbf{G}_{16}^{8}$: $(\sigma_z, 0)$, $(\sigma_y, 0)$: 5, 3; 6, 3; 7, 3; 8, 3: *a*.
Λ^x $\mathbf{G}_{8}^{5}$: $(C_{2z}, 0)$, $(\sigma_y, 0)$: 5, 2: *a*.
H^x $\mathbf{G}_{16}^{8}$: $(\sigma_y, 0)$, $(C_{2z}, 0)$: 5, 3; 6, 3; 7, 3; 8, 3: *a*.
Q^x $\mathbf{G}_{16}^{8}$: $(C_{2z}, 0)$, $(\sigma_x, 0)$: 9, 1: *b*.
G^x $\mathbf{G}_{16}^{8}$: $(\sigma_x, 0)$, $(C_{2z}, 0)$: 5, 3; 6, 3; 7, 3; 8, 3: *a*.

59 *Pmmn* D_{2h}^{13}

(*F*1; *K*7; *M*5; *T*1; *Z*1.)

Γ $\mathbf{G}_{16}^{11}$: $\{C_{2z}\,|\,000\}$, $\{C_{2y}\,|\,\frac{1}{2}\frac{1}{2}0\}$, $\{I\,|\,\frac{1}{2}\frac{1}{2}0\}$: 5, 2; 10, 2: *c*.
Y $\mathbf{G}_{32}^{7}$: $\{\sigma_z\,|\,\frac{1}{2}\frac{1}{2}0\}$, $\{C_{2x}\,|\,\frac{1}{2}\frac{1}{2}0\}$, $\{I\,|\,\frac{1}{2}\frac{1}{2}0\}$: 13, 3; 14, 3: *a*.
X $\mathbf{G}_{32}^{7}$: $\{\sigma_z\,|\,\frac{1}{2}\frac{1}{2}0\}$, $\{C_{2y}\,|\,\frac{1}{2}\frac{1}{2}0\}$, $\{I\,|\,\frac{1}{2}\frac{1}{2}0\}$: 13, 3; 14, 3: *a*.
Z $\mathbf{G}_{16}^{11} \otimes \mathbf{T}_2$: $\{C_{2z}\,|\,000\}$, $\{C_{2y}\,|\,\frac{1}{2}\frac{1}{2}0\}$, $\{I\,|\,\frac{1}{2}\frac{1}{2}0\}$; $\mathbf{t}_3$: 5, 2; 10, 2: *c*.
U $\mathbf{G}_{32}^{7}$: $\{\sigma_z\,|\,\frac{1}{2}\frac{1}{2}0\}$, $\{C_{2y}\,|\,\frac{1}{2}\frac{1}{2}0\}$, $\{I\,|\,\frac{1}{2}\frac{1}{2}0\}$: 13, 3; 14, 3: *a*.
T $\mathbf{G}_{32}^{7}$: $\{\sigma_z\,|\,\frac{1}{2}\frac{1}{2}0\}$, $\{C_{2x}\,|\,\frac{1}{2}\frac{1}{2}0\}$, $\{I\,|\,\frac{1}{2}\frac{1}{2}0\}$: 13, 3; 14, 3: *a*.
S $\mathbf{G}_{32}^{7}$: $\{C_{2x}\,|\,\frac{1}{2}\frac{1}{2}0\}$, $\{\sigma_z\,|\,\frac{1}{2}\frac{1}{2}0\}$, $\{I\,|\,\frac{1}{2}\frac{1}{2}0\}$: 13, 3; 14, 3: *a*.
R $\mathbf{G}_{32}^{7}$: $\{C_{2x}\,|\,\frac{1}{2}\frac{1}{2}0\}$, $\{\sigma_z\,|\,\frac{1}{2}\frac{1}{2}0\}$, $\{I\,|\,\frac{1}{2}\frac{1}{2}0\}$: 13, 3; 14, 3: *a*.

Δ^x $\mathbf{G}_{8}^{5}$: $(C_{2y}, 0)$, $(\sigma_x, 0)$: 5, 2: *a*.
D^x $\mathbf{G}_{16}^{8}$: $(C_{2y}, 0)$, $(\sigma_x, 0)$: 5, 3; 6, 3; 7, 3; 8, 3: *a*.
P^x $\mathbf{G}_{16}^{8}$: $(C_{2y}, 0)$, $(\sigma_x, 0)$: 5, 3; 6, 3; 7, 3; 8, 3: *a*.
B^x $\mathbf{G}_{8}^{5}$: $(C_{2y}, 0)$, $(\sigma_x, 0)$: 5, 2: *a*.
Σ^x $\mathbf{G}_{8}^{5}$: $(C_{2x}, 0)$, $(\sigma_y, 0)$: 5, 2: *a*.
C^x $\mathbf{G}_{16}^{8}$: $(C_{2x}, 0)$, $(\sigma_z, 0)$: 5, 3; 6, 3; 7, 3; 8, 3: *a*.
E^x $\mathbf{G}_{16}^{8}$: $(C_{2x}, 0)$, $(\sigma_z, 0)$: 5, 3; 6, 3; 7, 3; 8, 3: *a*.
A^x $\mathbf{G}_{8}^{5}$: $(C_{2x}, 0)$, $(\sigma_y, 0)$: 5, 2: *a*.
Λ^x $\mathbf{G}_{8}^{5}$: $(C_{2z}, 0)$, $(\sigma_y, 0)$: 5, 2: *a*.
H^x $\mathbf{G}_{8}^{5}$: $(C_{2z}, 0)$, $(\sigma_y, 0)$: 5, 1: *a*.
Q^x $\mathbf{G}_{8}^{5}$: $(C_{2z}, 0)$, $(\sigma_y, 0)$: 5, 1: *a*.
G^x $\mathbf{G}_{8}^{5}$: $(C_{2z}, 0)$, $(\sigma_y, 0)$: 5, 1: *a*.

60 *Pbcn* D_{2h}^{14}

($F1$; $K7$; $M5$; $T1$; $Z1$.)

Γ $\mathbf{G}_{16}^{11}$: $\{C_{2z}\,|\,000\}$, $\{C_{2y}\,|\,\frac{1}{2}\frac{1}{2}0\}$, $\{I\,|\,\frac{1}{2}0\frac{1}{2}\}$: 5, 2; 10, 2: *c*.
Y $\mathbf{G}_{32}^{7}$: $\{\sigma_z\,|\,\frac{1}{2}0\frac{1}{2}\}$, $\{C_{2x}\,|\,\frac{1}{2}\frac{1}{2}0\}$, $\{I\,|\,\frac{1}{2}0\frac{1}{2}\}$: 13, 3; 14, 3: *a*.
X $\mathbf{G}_{32}^{7}$: $\{\sigma_y\,|\,0\frac{1}{2}\frac{1}{2}\}$, $\{C_{2z}\,|\,000\}$, $\{I\,|\,\frac{1}{2}0\frac{1}{2}\}$: 13, 3; 14, 3: *a*.
Z $\mathbf{G}_{32}^{7}$: $\{\sigma_x\,|\,0\frac{1}{2}\frac{1}{2}\}$, $\{\sigma_z\,|\,\frac{1}{2}0\frac{1}{2}\}$, $\{I\,|\,\frac{1}{2}0\frac{1}{2}\}$: 13, 3; 14, 3: *a*.
U $\mathbf{G}_{32}^{14}$: $\{\sigma_x\,|\,0\frac{1}{2}\frac{1}{2}\}$, $\{C_{2z}\,|\,000\}$, $\{I\,|\,\frac{1}{2}0\frac{1}{2}\}$: 5, 3; 6, 3; 7, 3; 8, 3; 15, 3; 16, 3; 17, 3; 18, 3: *a*.
T $\mathbf{G}_{32}^{7}$: $\{C_{2z}\,|\,000\}$, $\{\sigma_y\,|\,0\frac{1}{2}\frac{1}{2}\}$, $\{I\,|\,\frac{1}{2}0\frac{1}{2}\}$: 9, 1; 10, 1: *b*.
S $\mathbf{G}_{32}^{7}$: $\{C_{2z}\,|\,000\}$, $\{C_{2y}\,|\,\frac{1}{2}\frac{1}{2}0\}$, $\{I\,|\,\frac{1}{2}0\frac{1}{2}\}$: 9, 1; 10, 1: *b*.
R $\mathbf{G}_{32}^{7}$: $\{C_{2z}\,|\,000\}$, $\{C_{2x}\,|\,\frac{1}{2}\frac{1}{2}0\}$, $\{I\,|\,\frac{1}{2}0\frac{1}{2}\}$: 9, 1; 10, 1: *b*.

Δ^x $\mathbf{G}_{8}^{5}$: $(C_{2y}, 0)$, $(\sigma_x, 0)$: 5, 2: *a*.
D^x $\mathbf{G}_{16}^{8}$: $(\sigma_z, 0)$, $(\sigma_x, 0)$: 9, 1: *b*.
P^x $\mathbf{G}_{16}^{11}$: $(C_{2y}, 0)$, $(\sigma_x, 0)$, $(\bar{E}, 0)$: 6, 3; 7, 3; 8, 3; 9, 3: *b*.
B^x $\mathbf{G}_{16}^{8}$: $(\sigma_z, 0)$, $(C_{2y}, 0)$: 5, 3; 6, 3; 7, 3; 8, 3: *a*.
Σ^x $\mathbf{G}_{8}^{5}$: $(C_{2x}, 0)$, $(\sigma_y, 0)$: 5, 2: *a*.
C^x $\mathbf{G}_{16}^{8}$: $(C_{2x}, 0)$, $(\sigma_z, 0)$: 5, 3; 6, 3; 7, 3; 8, 3: *a*.
E^x $\mathbf{G}_{16}^{8}$: $(\sigma_y, 0)$, $(\sigma_z, 0)$: 9, 1: *b*.
A^x $\mathbf{G}_{16}^{8}$: $(\sigma_z, 0)$, $(\sigma_y, 0)$: 5, 3; 6, 3; 7, 3; 8, 3: *a*.
Λ^x $\mathbf{G}_{8}^{5}$: $(C_{2z}, 0)$, $(\sigma_y, 0)$: 5, 2: *a*.
H^x $\mathbf{G}_{8}^{5}$: $(C_{2z}, 0)$, $(\sigma_y, 0)$: 5, 1: *a*.
Q^x $\mathbf{G}_{16}^{8}$: $(\sigma_x, 0)$, $(C_{2z}, 0)$: 5, 1; 6, 1; 7, 1; 8, 1: *a*.
G^x $\mathbf{G}_{16}^{8}$: $(\sigma_x, 0)$, $(C_{2z}, 0)$: 5, 3; 6, 3; 7, 3; 8, 3: *a*.

61 *Pbca* D_{2h}^{15}

($F1$; $K3$; $K7$; $M5$; $T1$; $T2$; $Z1$.)

Γ $\mathbf{G}_{16}^{11}$: $\{C_{2z}\,|\,\frac{1}{2}0\frac{1}{2}\}$, $\{C_{2y}\,|\,\frac{1}{2}\frac{1}{2}0\}$, $\{I\,|\,000\}$: 5, 2; 10, 2: *c*.
Y $\mathbf{G}_{32}^{7}$: $\{\sigma_z\,|\,\frac{1}{2}0\frac{1}{2}\}$, $\{C_{2x}\,|\,0\frac{1}{2}\frac{1}{2}\}$, $\{I\,|\,000\}$: 13, 3; 14, 3: *a*.
X $\mathbf{G}_{32}^{7}$: $\{\sigma_y\,|\,\frac{1}{2}\frac{1}{2}0\}$, $\{C_{2z}\,|\,\frac{1}{2}0\frac{1}{2}\}$, $\{I\,|\,000\}$: 13, 3; 14, 3: *a*.
Z $\mathbf{G}_{32}^{7}$: $\{\sigma_x\,|\,0\frac{1}{2}\frac{1}{2}\}$, $\{C_{2y}\,|\,\frac{1}{2}\frac{1}{2}0\}$, $\{I\,|\,000\}$: 13, 3; 14, 3: *a*.
U $\mathbf{G}_{32}^{7}$: $\{C_{2y}\,|\,\frac{1}{2}\frac{1}{2}0\}$, $\{C_{2x}\,|\,0\frac{1}{2}\frac{1}{2}\}$, $\{I\,|\,000\}$: 9, 1; 10, 1: *b*.
T $\mathbf{G}_{32}^{7}$: $\{C_{2x}\,|\,0\frac{1}{2}\frac{1}{2}\}$, $\{C_{2z}\,|\,\frac{1}{2}0\frac{1}{2}\}$, $\{I\,|\,000\}$: 9, 1; 10, 1: *b*.
S $\mathbf{G}_{32}^{7}$: $\{C_{2z}\,|\,\frac{1}{2}0\frac{1}{2}\}$, $\{C_{2y}\,|\,\frac{1}{2}\frac{1}{2}0\}$, $\{I\,|\,000\}$: 9, 1; 10, 1: *b*.
R $\mathbf{G}_{32}^{15}$: $\{C_{2x}\,|\,0\frac{1}{2}\frac{1}{2}\}$, $\{C_{2y}\,|\,\frac{1}{2}\frac{1}{2}0\}$, $\{\bar{E}\,|\,000\}$, $\{I\,|\,000\}$: 6, 1; 7, 1; 8, 1; 9, 1; 16, 1; 17, 1; 18, 1; 19, 1: *a*.

Δ^x $\mathbf{G}_{8}^{5}$: $(C_{2y}, 0)$, $(\sigma_x, 0)$: 5, 2: *a*.
D^x $\mathbf{G}_{16}^{8}$: $(\sigma_z, 0)$, $(\sigma_x, 0)$: 9, 1: *b*.
P^x $\mathbf{G}_{16}^{11}$: $(C_{2y}, 0)$, $(\sigma_x, 0)$, $(\bar{E}, 0)$: 6, 1; 7, 1; 8, 1; 9, 1: *b*.
B^x $\mathbf{G}_{16}^{8}$: $(\sigma_z, 0)$, $(C_{2y}, 0)$: 5, 3; 6, 3; 7, 3; 8, 3: *a*.
Σ^x $\mathbf{G}_{8}^{5}$: $(C_{2x}, 0)$, $(\sigma_y, 0)$: 5, 2: *a*.
C^x $\mathbf{G}_{16}^{8}$: $(\sigma_y, 0)$, $(C_{2x}, 0)$: 5, 3; 6, 3; 7, 3; 8, 3: *a*.
E^x $\mathbf{G}_{16}^{11}$: $(C_{2x}, 0)$, $(\sigma_z, 0)$, $(\bar{E}, 0)$: 6, 1; 7, 1; 8, 1; 9, 1: *b*.
A^x $\mathbf{G}_{16}^{8}$: $(\sigma_y, 0)$, $(\sigma_z, 0)$: 9, 1: *b*.
Λ^x $\mathbf{G}_{8}^{5}$: $(C_{2z}, 0)$, $(\sigma_y, 0)$: 5, 2: *a*.
H^x $\mathbf{G}_{16}^{8}$: $(\sigma_x, 0)$, $(\sigma_y, 0)$: 9, 1: *b*.
Q^x $\mathbf{G}_{16}^{11}$: $(C_{2z}, 0)$, $(\sigma_y, 0)$, $(\bar{E}, 0)$: 6, 1; 7, 1; 8, 1; 9, 1: *b*.
G^x $\mathbf{G}_{16}^{8}$: $(\sigma_x, 0)$, $(C_{2z}, 0)$: 5, 3; 6, 3; 7, 3; 8, 3: *a*.

62 *Pnma* D_{2h}^{16}

(*F*1; *G*1; *K*2; *K*7; *M*5; *T*1; *S*8; *Z*1.)

Γ $\mathbf{G}_{16}^{11}$: $\{C_{2z}|\frac{1}{2}0\frac{1}{2}\}$, $\{C_{2y}|\frac{1}{2}\frac{1}{2}0\}$, $\{I|\frac{1}{2}\frac{1}{2}0\}$: 5, 2; 10, 2: *c*.
Y $\mathbf{G}_{32}^{7}$: $\{\sigma_x|\frac{1}{2}0\frac{1}{2}\}$, $\{C_{2z}|\frac{1}{2}0\frac{1}{2}\}$, $\{I|\frac{1}{2}\frac{1}{2}0\}$: 13, 3; 14, 3: *a*.
X $\mathbf{G}_{32}^{7}$: $\{\sigma_z|0\frac{1}{2}\frac{1}{2}\}$, $\{C_{2y}|\frac{1}{2}\frac{1}{2}0\}$, $\{I|\frac{1}{2}\frac{1}{2}0\}$: 13, 3; 14, 3: *a*.
Z $\mathbf{G}_{32}^{7}$: $\{\sigma_x|\frac{1}{2}0\frac{1}{2}\}$, $\{C_{2y}|\frac{1}{2}\frac{1}{2}0\}$, $\{I|\frac{1}{2}\frac{1}{2}0\}$: 13, 3; 14, 3: *a*.
U $\mathbf{G}_{32}^{14}$: $\{C_{2y}|\frac{1}{2}\frac{1}{2}0\}$, $\{C_{2z}|\frac{1}{2}0\frac{1}{2}\}$, $\{I|\frac{1}{2}\frac{1}{2}0\}$: 9, 1; 19, 1: *b*.
T $\mathbf{G}_{32}^{7}$: $\{C_{2y}|\frac{1}{2}\frac{1}{2}0\}$, $\{\sigma_x|\frac{1}{2}0\frac{1}{2}\}$, $\{I|\frac{1}{2}\frac{1}{2}0\}$: 13, 3; 14, 3: *a*.
S $\mathbf{G}_{32}^{7}$: $\{\sigma_y|000\}$, $\{C_{2x}|0\frac{1}{2}\frac{1}{2}\}$, $\{I|\frac{1}{2}\frac{1}{2}0\}$: 9, 1; 10, 1: *b*.
R $\mathbf{G}_{32}^{7}$: $\{\sigma_y|000\}$, $\{\sigma_z|0\frac{1}{2}\frac{1}{2}\}$, $\{I|\frac{1}{2}\frac{1}{2}0\}$: 9, 1; 10, 1: *b*.

Δ^x $\mathbf{G}_{8}^{5}$: $(C_{2y}, 0)$, $(\sigma_x, 0)$: 5, 2: *a*.
D^x $\mathbf{G}_{16}^{8}$: $(C_{2y}, 0)$, $(\sigma_x, 0)$: 5, 3; 6, 3; 7, 3; 8, 3: *a*.
P^x $\mathbf{G}_{16}^{8}$: $(\sigma_x, 0)$, $(\sigma_z, 0)$: 9, 1: *b*.
B^x $\mathbf{G}_{16}^{8}$: $(\sigma_z, 0)$, $(C_{2y}, 0)$: 5, 3; 6, 3; 7, 3; 8, 3: *a*.
Σ^x $\mathbf{G}_{8}^{5}$: $(C_{2x}, 0)$, $(\sigma_y, 0)$: 5, 2: *a*.
C^x $\mathbf{G}_{8}^{5}$: $(C_{2x}, 0)$, $(\sigma_y, 0)$: 5, 1: *a*.
E^x $\mathbf{G}_{16}^{8}$: $(\sigma_y, 0)$, $(\sigma_z, 0)$: 9, 1: *b*.
A^x $\mathbf{G}_{16}^{8}$: $(\sigma_y, 0)$, $(\sigma_z, 0)$: 9, 1: *b*.
Λ^x $\mathbf{G}_{8}^{5}$: $(C_{2z}, 0)$, $(\sigma_y, 0)$: 5, 2: *a*.
H^x $\mathbf{G}_{16}^{8}$: $(C_{2z}, 0)$, $(\sigma_y, 0)$: 5, 3; 6, 3; 7, 3; 8, 3: *a*.
Q^x $\mathbf{G}_{16}^{8}$: $(C_{2z}, 0)$, $(\sigma_y, 0)$: 5, 1; 6, 1; 7, 1; 8, 1: *a*.
G^x $\mathbf{G}_{8}^{5}$: $(C_{2z}, 0)$, $(\sigma_y, 0)$: 5, 1: *a*.

63 *Cmcm* D_{2h}^{17}

(*F*1; *J*1; *K*7; *M*5; *S*12; *T*3; *Z*1.)

Γ $\mathbf{G}_{16}^{11}$: $\{C_{2z}|00\frac{1}{2}\}$, $\{C_{2y}|00\frac{1}{2}\}$, $\{I|000\}$: 5, 2; 10, 2: *c*.
Y $\mathbf{G}_{16}^{11} \otimes \mathbf{T}_2$: $\{C_{2z}|00\frac{1}{2}\}$, $\{C_{2y}|00\frac{1}{2}\}$, $\{I|000\}$; $\mathbf{t}_1$ or $\mathbf{t}_2$: 5, 2; 10, 2: *c*.
Z $\mathbf{G}_{32}^{7}$: $\{C_{2z}|00\frac{1}{2}\}$, $\{C_{2x}|000\}$, $\{I|000\}$: 13, 3; 14, 3: *a*.
T $\mathbf{G}_{32}^{7}$: $\{C_{2z}|00\frac{1}{2}\}$, $\{C_{2x}|000\}$, $\{I|000\}$: 13, 3; 14, 3: *a*.
S $\mathbf{G}_{8}^{2} \otimes \mathbf{T}_2$: $\{C_{2z}|00\frac{1}{2}\}$, $\{I|000\}$; $\mathbf{t}_2$: 2, 3; 4, 3; 6, 3; 8, 3: *b*.
R $\mathbf{G}_{16}^{10}$: $\{C_{2z}|00\frac{1}{2}\}$, $\{E|001\}$, $\{I|000\}$: 9, 1: *a*.

Λ^x $\mathbf{G}_{8}^{5}$: $(C_{2z}, 0)$, $(\sigma_y, 0)$: 5, 2: *a*.
H^x $\mathbf{G}_{8}^{5}$: $(C_{2z}, 0)$, $(\sigma_y, 0)$: 5, 2: *a*.
D^x $\mathbf{G}_{4}^{1}$: $(C_{2z}, 0)$: 2, 3; 4, 3: *b*.
A^x $\mathbf{G}_{16}^{8}$: $(\sigma_z, 0)$, $(\sigma_y, 0)$: 5, 3; 6, 3; 7, 3; 8, 3: *a*.
Σ^x $\mathbf{G}_{8}^{5}$: $(C_{2x}, 0)$, $(\sigma_y, 0)$: 5, 2: *a*.
Δ^x $\mathbf{G}_{8}^{5}$: $(C_{2y}, 0)$, $(\sigma_x, 0)$: 5, 2: *a*.
B^x $\mathbf{G}_{16}^{8}$: $(\sigma_x, 0)$, $(\sigma_z, 0)$: 9, 1: *b*.
G^x $\mathbf{G}_{16}^{8}$: $(\sigma_x, 0)$, $(\sigma_z, 0)$: 9, 1: *b*.
F^x $\mathbf{G}_{8}^{5}$: $(C_{2y}, 0)$, $(\sigma_x, 0)$: 5, 2: *a*.
E^x $\mathbf{G}_{16}^{8}$: $(\sigma_z, 0)$, $(\sigma_y, 0)$: 5, 3; 6, 3; 7, 3; 8, 3: *a*.
C^x $\mathbf{G}_{8}^{5}$: $(C_{2x}, 0)$, $(\sigma_y, 0)$: 5, 2: *a*.

64 *Cmca* D_{2h}^{18}

($F1$; $K7$; $M5$; $S8$; $S9$; $T3$; $W1$; $Z1$.)

Γ $\mathbf{G}_{16}^{11}$: $\{C_{2z}\,|\,00\tfrac{1}{2}\}$, $\{C_{2y}\,|\,00\tfrac{1}{2}\}$, $\{I\,|\,\tfrac{1}{2}\tfrac{1}{2}0\}$: 5, 2; 10, 2: *c*.

Y $\mathbf{G}_{16}^{11} \otimes \mathbf{T}_2$: $\{C_{2z}\,|\,00\tfrac{1}{2}\}$, $\{C_{2y}\,|\,00\tfrac{1}{2}\}$, $\{I\,|\,\tfrac{1}{2}\tfrac{1}{2}0\}$; $\mathbf{t}_1$ or $\mathbf{t}_2$: 5, 2; 10, 2: *c*.

Z $\mathbf{G}_{32}^{7}$: $\{C_{2z}\,|\,00\tfrac{1}{2}\}$, $\{C_{2x}\,|\,000\}$, $\{I\,|\,\tfrac{1}{2}\tfrac{1}{2}0\}$: 13, 3; 14, 3: *a*.

T $\mathbf{G}_{32}^{7}$: $\{C_{2z}\,|\,00\tfrac{1}{2}\}$, $\{C_{2x}\,|\,000\}$, $\{I\,|\,\tfrac{1}{2}\tfrac{1}{2}0\}$: 13, 3; 14, 3: *a*.

S $\mathbf{G}_{16}^{10}$: $\{\sigma_z\,|\,\tfrac{1}{2}\tfrac{1}{2}\tfrac{1}{2}\}$, $\{E\,|\,010\}$, $\{I\,|\,\tfrac{1}{2}\tfrac{1}{2}0)$: 9, 1: *a*.

R $\mathbf{G}_{8}^{2} \otimes \mathbf{T}_2$: $\{\sigma_z\,|\,\tfrac{1}{2}\tfrac{1}{2}\tfrac{1}{2}\}$, $\{I\,|\,\tfrac{1}{2}\tfrac{1}{2}0\}$; $\mathbf{t}_2$ or $\mathbf{t}_3$: 1, 1; 3, 1; 5, 1; 7, 1: *c*.

Λ^x $\mathbf{G}_{8}^{5}$: $(C_{2z}, 0)$, $(\sigma_y, 0)$: 5, 2: *a*.

H^x $\mathbf{G}_{8}^{5}$: $(C_{2z}, 0)$, $(\sigma_y, 0)$: 5, 2: *a*.

D^x $\mathbf{G}_{4}^{1}$: $(C_{2z}, 0)$: 2, 1; 4, 1: *b*.

A^x $\mathbf{G}_{16}^{8}$: $(\sigma_z, 0)$, $(\sigma_y, 0)$: 5, 3; 6, 3; 7, 3; 8, 3: *a*.

Σ^x $\mathbf{G}_{8}^{5}$: $(C_{2x}, 0)$, $(\sigma_y, 0)$: 5, 2: *a*.

Δ^x $\mathbf{G}_{8}^{5}$: $(C_{2y}, 0)$, $(\sigma_x, 0)$: 5, 2: *a*.

B^x $\mathbf{G}_{16}^{8}$: $(\sigma_x, 0)$, $(\sigma_z, 0)$: 9, 1: *b*.

G^x $\mathbf{G}_{16}^{8}$: $(\sigma_x, 0)$, $(\sigma_z, 0)$: 9, 1: *b*.

F^x $\mathbf{G}_{8}^{5}$: $(C_{2y}, 0)$, $(\sigma_x, 0)$: 5, 2: *a*.

E^x $\mathbf{G}_{16}^{8}$: $(\sigma_z, 0)$, $(\sigma_y, 0)$: 5, 3; 6, 3; 7, 3; 8, 3: *a*.

C^x $\mathbf{G}_{8}^{5}$: $(C_{2x}, 0)$, $(\sigma_y, 0)$: 5, 2: *a*.

65 *Cmmm* D_{2h}^{19}

($F1$; $K7$; $M5$; $T3$; $Z1$.)

Γ $\mathbf{G}_{16}^{11}$: $\{C_{2z}\,|\,000\}$, $\{C_{2y}\,|\,000\}$, $\{I\,|\,000\}$: 5, 2; 10, 2: *c*.

Y $\mathbf{G}_{16}^{11} \otimes \mathbf{T}_2$: $\{C_{2z}\,|\,000\}$, $\{C_{2y}\,|\,000\}$, $\{I\,|\,000\}$; $\mathbf{t}_1$ or $\mathbf{t}_2$: 5, 2; 10, 2: *c*.

Z $\mathbf{G}_{16}^{11} \otimes \mathbf{T}_2$: $\{C_{2z}\,|\,000\}$, $\{C_{2y}\,|\,000\}$, $\{I\,|\,000\}$; $\mathbf{t}_3$: 5, 2; 10, 2: *c*.

T $\mathbf{G}_{16}^{11} \otimes \mathbf{T}_2$: $\{C_{2z}\,|\,000\}$, $\{C_{2y}\,|\,000\}$, $\{I\,|\,000\}$; $\mathbf{t}_1$ or $\mathbf{t}_2$ or $\mathbf{t}_3$: 5, 2; 10, 2: *c*.

S $\mathbf{G}_{8}^{2} \otimes \mathbf{T}_2$: $\{C_{2z}\,|\,000\}$, $\{I\,|\,000\}$; $\mathbf{t}_2$: 2, 3; 4, 3; 6, 3; 8, 3: *b*.

R $\mathbf{G}_{8}^{2} \otimes \mathbf{T}_2$: $\{C_{2z}\,|\,000\}$,$\{I\,|\,000\}$; $\mathbf{t}_2$ or $\mathbf{t}_3$: 2, 3; 4, 3; 6, 3; 8, 3: *b*.

Λ^x $\mathbf{G}_{8}^{5}$: $(C_{2z}, 0)$, $(\sigma_y, 0)$: 5, 2: *a*.

H^x $\mathbf{G}_{8}^{5}$: $(C_{2z}, 0)$, $(\sigma_y, 0)$: 5, 2: *a*.

D^x $\mathbf{G}_{4}^{1}$: $(C_{2z}, 0)$: 2, 3; 4, 3: *b*.

A^x $\mathbf{G}_{8}^{5}$: $(C_{2x}, 0)$, $(\sigma_y, 0)$: 5, 2: *a*.

Σ^x $\mathbf{G}_{8}^{5}$: $(C_{2x}, 0)$, $(\sigma_y, 0)$: 5, 2: *a*.

Δ^x $\mathbf{G}_{8}^{5}$: $(C_{2y}, 0)$, $(\sigma_x, 0)$: 5, 2: *a*.

B^x $\mathbf{G}_{8}^{5}$: $(C_{2y}, 0)$, $(\sigma_x, 0)$: 5, 2: *a*.

G^x $\mathbf{G}_{8}^{5}$: $(C_{2y}, 0)$, $(\sigma_x, 0)$: 5, 2: *a*.

F^x $\mathbf{G}_{8}^{5}$: $(C_{2y}, 0)$, $(\sigma_x, 0)$: 5, 2: *a*.

E^x $\mathbf{G}_{8}^{5}$: $(C_{2x}, 0)$, $(\sigma_y, 0)$: 5, 2: *a*.

C^x $\mathbf{G}_{8}^{5}$: $(C_{2x}, 0)$, $(\sigma_y, 0)$: 5, 2: *a*.

66 *Cccm* D_{2h}^{20}

($F1$; $K7$; $M5$; $T3$; $Z1$.)

Γ $\mathbf{G}_{16}^{11}$: $\{C_{2z}\,|\,000\}$, $\{C_{2y}\,|\,000\}$, $\{I\,|\,00\frac{1}{2}\}$: 5, 2; 10, 2: *c*.
Y $\mathbf{G}_{16}^{11}\otimes\mathbf{T}_2$: $\{C_{2z}\,|\,000\}$, $\{C_{2y}\,|\,000\}$, $\{I\,|\,00\frac{1}{2}\}$; $\mathbf{t}_1$ or $\mathbf{t}_2$: 5, 2; 10, 2: *c*.
Z $\mathbf{G}_{32}^{7}$: $\{\sigma_x\,|\,00\frac{1}{2}\}$, $\{\sigma_z\,|\,00\frac{1}{2}\}$, $\{I\,|\,00\frac{1}{2}\}$: 13, 3; 14, 3: *a*.
T $\mathbf{G}_{32}^{7}$: $\{\sigma_x\,|\,00\frac{1}{2}\}$, $\{\sigma_z\,|\,00\frac{1}{2}\}$, $\{I\,|\,00\frac{1}{2}\}$: 13, 3; 14, 3: *a*.
S $\mathbf{G}_{8}^{2}\otimes\mathbf{T}_2$: $\{C_{2z}\,|\,000\}$, $\{I\,|\,00\frac{1}{2}\}$; $\mathbf{t}_2$: 2, 3; 4, 3; 6, 3; 8, 3: *b*.
R $\mathbf{G}_{8}^{2}\otimes\mathbf{T}_2$: $\{C_{2z}\,|\,000\}$, $\{I\,|\,00\frac{1}{2}\}$; $\mathbf{t}_2$ or $\mathbf{t}_3$: 2, 3; 4, 3; 6, 3; 8, 3: *b*.

Λ^x $\mathbf{G}_{8}^{5}$: $(C_{2z}, 0)$, $(\sigma_y, 0)$: 5, 2: *a*.
H^x $\mathbf{G}_{8}^{5}$: $(C_{2z}, 0)$, $(\sigma_y, 0)$: 5, 2: *a*.
D^x $\mathbf{G}_{4}^{1}$: $(C_{2z}, 0)$: 2, 3; 4, 3: *b*.
A^x $\mathbf{G}_{16}^{8}$: $(\sigma_z, 0)$, $(\sigma_y, 0)$: 5, 3; 6, 3; 7, 3; 8, 3: *a*.
Σ^x $\mathbf{G}_{8}^{5}$: $(C_{2x}, 0)$, $(\sigma_y, 0)$: 5, 2: *a*.
Δ^x $\mathbf{G}_{8}^{5}$: $(C_{2y}, 0)$, $(\sigma_x, 0)$: 5, 2: *a*.
B^x $\mathbf{G}_{16}^{8}$: $(\sigma_z, 0)$, $(C_{2y}, 0)$: 5, 3; 6, 3; 7, 3; 8, 3: *a*.
G^x $\mathbf{G}_{16}^{8}$: $(\sigma_z, 0)$, $(C_{2y}, 0)$: 5, 3; 6, 3; 7, 3; 8, 3: *a*.
F^x $\mathbf{G}_{8}^{5}$: $(C_{2y}, 0)$, $(\sigma_x, 0)$: 5, 2: *a*.
E^x $\mathbf{G}_{16}^{8}$: $(\sigma_z, 0)$, $(\sigma_y, 0)$: 5, 3; 6, 3; 7, 3; 8, 3: *a*.
C^x $\mathbf{G}_{8}^{5}$: $(C_{2x}, 0)$, $(\sigma_y, 0)$: 5, 2: *a*.

67 *Cmma* D_{2h}^{21}

($F1$; $K7$; $M5$; $T3$; $Z1$.)

Γ $\mathbf{G}_{16}^{11}$: $\{C_{2z}\,|\,000\}$, $\{C_{2y}\,|\,000\}$, $\{I\,|\,\frac{1}{2}\frac{1}{2}0\}$: 5, 2; 10, 2: *c*.
Y $\mathbf{G}_{16}^{11}\otimes\mathbf{T}_2$: $\{C_{2z}\,|\,000\}$, $\{C_{2y}\,|\,000\}$, $\{I\,|\,\frac{1}{2}\frac{1}{2}0\}$; $\mathbf{t}_1$ or $\mathbf{t}_2$: 5, 2; 10, 2: *c*.
Z $\mathbf{G}_{16}^{11}\otimes\mathbf{T}_2$: $\{C_{2z}\,|\,000\}$, $\{C_{2y}\,|\,000\}$, $\{I\,|\,\frac{1}{2}\frac{1}{2}0\}$; $\mathbf{t}_3$: 5, 2; 10, 2: *c*.
T $\mathbf{G}_{16}^{11}\otimes\mathbf{T}_2$: $\{C_{2z}\,|\,000\}$, $\{C_{2y}\,|\,000\}$, $\{I\,|\,\frac{1}{2}\frac{1}{2}0\}$; $\mathbf{t}_1$ or $\mathbf{t}_2$ or $\mathbf{t}_3$: 5, 2; 10, 2: *c*.
S $\mathbf{G}_{16}^{10}$: $\{\sigma_z\,|\,\frac{1}{2}\frac{1}{2}0\}$, $\{E\,|\,010\}$, $\{I\,|\,\frac{1}{2}\frac{1}{2}0\}$: 9, 1: *a*.
R $\mathbf{G}_{16}^{10}$: $\{\sigma_z\,|\,\frac{1}{2}\frac{1}{2}0\}$, $\{E\,|\,010\}$, $\{I\,|\,\frac{1}{2}\frac{1}{2}0\}$: 9, 1: *a*.

Λ^x $\mathbf{G}_{8}^{5}$: $(C_{2z}, 0)$, $(\sigma_y, 0)$: 5, 2: *a*.
H^x $\mathbf{G}_{8}^{5}$: $(C_{2z}, 0)$, $(\sigma_y, 0)$: 5, 2: *a*.
D^x $\mathbf{G}_{4}^{1}$: $(C_{2z}, 0)$: 2, 1; 4, 1: *b*.
A^x $\mathbf{G}_{8}^{5}$: $(C_{2x}, 0)$, $(\sigma_y, 0)$: 5, 2: *a*.
Σ^x $\mathbf{G}_{8}^{5}$: $(C_{2x}, 0)$, $(\sigma_y, 0)$: 5, 2: *a*.
Δ^x $\mathbf{G}_{8}^{5}$: $(C_{2y}, 0)$, $(\sigma_x, 0)$: 5, 2: *a*.
B^x $\mathbf{G}_{8}^{5}$: $(C_{2y}, 0)$, $(\sigma_x, 0)$: 5, 2: *a*.
G^x $\mathbf{G}_{8}^{5}$: $(C_{2y}, 0)$, $(\sigma_x, 0)$: 5, 2: *a*.
F^x $\mathbf{G}_{8}^{5}$: $(C_{2y}, 0)$, $(\sigma_x, 0)$: 5, 2: *a*.
E^x $\mathbf{G}_{8}^{5}$: $(C_{2x}, 0)$, $(\sigma_y, 0)$: 5, 2: *a*.
C^x $\mathbf{G}_{8}^{5}$: $(C_{2x}, 0)$, $(\sigma_y, 0)$: 5, 2: *a*.

68 *Ccca* D_{2h}^{22}

(*F*1; *K*7; *M*5; *T*3; *Z*1.)

Γ $\mathbf{G}_{16}^{11}$: $\{C_{2z}|000\}$, $\{C_{2y}|000\}$, $\{I|\frac{1}{2}\frac{1}{2}\frac{1}{2}\}$: 5, 2; 10, 2: *c*.
Y $\mathbf{G}_{16}^{11} \otimes \mathbf{T}_2$: $\{C_{2z}|000\}$, $\{C_{2y}|000\}$, $\{I|\frac{1}{2}\frac{1}{2}\frac{1}{2}\}$; $\mathbf{t}_1$ or $\mathbf{t}_2$: 5, 2; 10, 2: *c*.
Z $\mathbf{G}_{32}^{7}$: $\{\sigma_x|\frac{1}{2}\frac{1}{2}\frac{1}{2}\}$, $\{\sigma_z|\frac{1}{2}\frac{1}{2}\frac{1}{2}\}$, $\{I|\frac{1}{2}\frac{1}{2}\frac{1}{2}\}$: 13, 3; 14, 3: *a*.
T $\mathbf{G}_{32}^{7}$: $\{\sigma_x|\frac{1}{2}\frac{1}{2}\frac{1}{2}\}$, $\{\sigma_z|\frac{1}{2}\frac{1}{2}\frac{1}{2}\}$, $\{I|\frac{1}{2}\frac{1}{2}\frac{1}{2}\}$: 13, 3; 14, 3: *a*.
S $\mathbf{G}_{16}^{10}$: $\{\sigma_z|\frac{1}{2}\frac{1}{2}\frac{1}{2}\}$, $\{E|010\}$, $\{I|\frac{1}{2}\frac{1}{2}\frac{1}{2}\}$: 9, 1: *a*.
R $\mathbf{G}_{16}^{10}$: $\{\sigma_z|\frac{1}{2}\frac{1}{2}\frac{1}{2}\}$, $\{E|010\}$, $\{I|\frac{1}{2}\frac{1}{2}\frac{1}{2}\}$: 9, 1: *a*.

Λ^x $\mathbf{G}_8^5$: $(C_{2z}, 0)$, $(\sigma_y, 0)$: 5, 2: *a*.
H^x $\mathbf{G}_8^5$: $(C_{2z}, 0)$, $(\sigma_y, 0)$: 5, 2: *a*.
D^x $\mathbf{G}_4^1$: $(C_{2z}, 0)$: 2, 1; 4, 1: *b*.
A^x $\mathbf{G}_{16}^8$: $(\sigma_z, 0)$, $(\sigma_y, 0)$: 5, 3; 6, 3; 7, 3; 8, 3: *a*.
Σ^x $\mathbf{G}_8^5$: $(C_{2x}, 0)$, $(\sigma_y, 0)$: 5, 2: *a*.
Δ^x $\mathbf{G}_8^5$: $(C_{2y}, 0)$, $(\sigma_x, 0)$: 5, 2: *a*.
B^x $\mathbf{G}_{16}^8$: $(\sigma_z, 0)$, $(C_{2y}, 0)$: 5, 3; 6, 3; 7, 3; 8, 3: *a*.
G^x $\mathbf{G}_{16}^8$: $(\sigma_z, 0)$, $(C_{2y}, 0)$: 5, 3; 6, 3; 7, 3; 8, 3: *a*.
F^x $\mathbf{G}_8^5$: $(C_{2y}, 0)$, $(\sigma_x, 0)$: 5, 2: *a*.
E^x $\mathbf{G}_{16}^8$: $(\sigma_z, 0)$, $(\sigma_y, 0)$: 5, 3; 6, 3; 7, 3; 8, 3: *a*.
C^x $\mathbf{G}_8^5$: $(C_{2x}, 0)$, $(\sigma_y, 0)$: 5, 2: *a*.

69 *Fmmm* D_{2h}^{23}

(*F*1; *K*7; *M*5; *S*14; *Z*1.)

Γ $\mathbf{G}_{16}^{11}$: $\{C_{2z}|000\}$, $\{C_{2y}|000\}$, $\{I|000\}$: 5, 2; 10, 2: *c*.
Y $\mathbf{G}_{16}^{11} \otimes \mathbf{T}_2$: $\{C_{2z}|000\}$, $\{C_{2y}|000\}$, $\{I|000\}$; $\mathbf{t}_2$ or $\mathbf{t}_3$: 5, 2; 10, 2: *c*.
X $\mathbf{G}_{16}^{11} \otimes \mathbf{T}_2$: $\{C_{2z}|000\}$, $\{C_{2y}|000\}$, $\{I|000\}$; $\mathbf{t}_1$ or $\mathbf{t}_3$: 5, 2; 10, 2: *c*.
Z $\mathbf{G}_{16}^{11} \otimes \mathbf{T}_2$: $\{C_{2z}|000\}$, $\{C_{2y}|000\}$, $\{I|000\}$; $\mathbf{t}_1$ or $\mathbf{t}_2$: 5, 2; 10, 2: *c*.
L $\mathbf{G}_4^2 \otimes \mathbf{T}_2$: $\{I|000\}$, $\{\bar{E}|000\}$; $\mathbf{t}_1$: 2, 1; 4, 1: *a*.

Λ^x $\mathbf{G}_8^5$: $(C_{2z}, 0)$, $(\sigma_y, 0)$: 5, 2: *a*.
G^x $\mathbf{G}_8^5$: $(C_{2z}, 0)$, $(\sigma_y, 0)$: 5, 2: *a*.
H^x $\mathbf{G}_8^5$: $(C_{2z}, 0)$, $(\sigma_y, 0)$: 5, 2: *a*.
Q^x $\mathbf{G}_8^5$: $(C_{2z}, 0)$, $(\sigma_y, 0)$: 5, 2: *a*.
Σ^x $\mathbf{G}_8^5$: $(C_{2x}, 0)$, $(\sigma_z, 0)$: 5, 2: *a*.
C^x $\mathbf{G}_8^5$: $(C_{2x}, 0)$, $(\sigma_z, 0)$: 5, 2: *a*..
A^x $\mathbf{G}_8^5$: $(C_{2x}, 0)$, $(\sigma_z, 0)$: 5, 2: *a*.
U^x $\mathbf{G}_8^5$: $(C_{2x}, 0)$, $(\sigma_z, 0)$: 5, 2: *a*.
Δ^x $\mathbf{G}_8^5$: $(C_{2y}, 0)$, $(\sigma_x, 0)$: 5, 2: *a*.
D^x $\mathbf{G}_8^5$: $(C_{2y}, 0)$, $(\sigma_x, 0)$: 5, 2: *a*.
B^x $\mathbf{G}_8^5$: $(C_{2y}, 0)$, $(\sigma_x, 0)$: 5, 2: *a*.
R^x $\mathbf{G}_8^5$: $(C_{2y}, 0)$, $(\sigma_x, 0)$: 5, 2: *a*.

70 *Fddd* D_{2h}^{24}

(*F*1; *K*7; *M*5; *S*14; *Z*1.)

Γ $\mathbf{G}_{16}^{11}$: $\{C_{2z}|000\}$, $\{C_{2y}|000\}$, $\{I|\frac{1}{4}\frac{1}{4}\frac{1}{4}\}$: 5, 2; 10, 2: *c*.
Y $\mathbf{G}_{32}^{7}$: $\{\sigma_x|\frac{1}{4}\frac{1}{4}\frac{1}{4}\}$, $\{\sigma_y|\frac{1}{4}\frac{1}{4}\frac{1}{4}\}$, $\{I|\frac{1}{4}\frac{1}{4}\frac{1}{4}\}$: 13, 3; 14, 3: *a*.
X $\mathbf{G}_{32}^{7}$: $\{\sigma_z|\frac{1}{4}\frac{1}{4}\frac{1}{4}\}$, $\{\sigma_x|\frac{1}{4}\frac{1}{4}\frac{1}{4}\}$, $\{I|\frac{1}{4}\frac{1}{4}\frac{1}{4}\}$: 13, 3; 14, 3: *a*.
Z $\mathbf{G}_{32}^{7}$: $\{\sigma_y|\frac{1}{4}\frac{1}{4}\frac{1}{4}\}$, $\{\sigma_z|\frac{1}{4}\frac{1}{4}\frac{1}{4}\}$, $\{I|\frac{1}{4}\frac{1}{4}\frac{1}{4}\}$: 13, 3; 14, 3: *a*.
L $\mathbf{G}_{4}^{2}\otimes\mathbf{T}_2$: $\{I|\frac{1}{4}\frac{1}{4}\frac{1}{4}\}$, $\{\bar{E}|000\}$; $\mathbf{t}_1$: 2, 1; 4, 1: *a*.

Λ^x $\mathbf{G}_{8}^{5}$: $(C_{2z}, 0)$, $(\sigma_y, 0)$: 5, 2: *a*.
G^x $\mathbf{G}_{16}^{8}$: $(\sigma_x, 0)$, $(C_{2z}, 0)$: 5, 3; 6, 3; 7, 3; 8, 3: *a*.
H^x $\mathbf{G}_{16}^{8}$: $(\sigma_y, 0)$, $(C_{2z}, 0)$: 5, 3; 6, 3; 7, 3; 8, 3: *a*.
Q^x $\mathbf{G}_{16}^{8}$: $(C_{2z}, 0)$, $(\sigma_x, 0)$: 9, 2: *b*.
Σ^x $\mathbf{G}_{8}^{5}$: $(C_{2x}, 0)$, $(\sigma_z, 0)$: 5, 2: *a*.
C^x $\mathbf{G}_{16}^{8}$: $(\sigma_y, 0)$, $(C_{2x}, 0)$: 5, 3; 6, 3; 7, 3; 8, 3: *a*.
A^x $\mathbf{G}_{16}^{8}$: $(\sigma_z, 0)$, $(\sigma_y, 0)$: 5, 3; 6, 3; 7, 3; 8, 3: *a*.
U^x $\mathbf{G}_{16}^{8}$: $(C_{2x}, 0)$, $(\sigma_y, 0)$: 9, 2: *b*.
Δ^x $\mathbf{G}_{8}^{5}$: $(C_{2y}, 0)$, $(\sigma_x, 0)$: 5, 2: *a*.
D^x $\mathbf{G}_{16}^{8}$: $(\sigma_x, 0)$, $(C_{2y}, 0)$: 5, 3; 6, 3; 7, 3; 8, 3: *a*.
B^x $\mathbf{G}_{16}^{8}$: $(\sigma_z, 0)$, $(C_{2y}, 0)$: 5, 3; 6, 3; 7, 3; 8, 3: *a*.
R^x $\mathbf{G}_{16}^{8}$: $(C_{2y}, 0)$, $(\sigma_z, 0)$: 9, 2: *b*.

71 *Immm* D_{2h}^{25}

(*F*1; *K*7; *M*5; *Z*1.)

Γ $\mathbf{G}_{16}^{11}$: $\{C_{2z}|000\}$, $\{C_{2y}|000\}$, $\{I|000\}$: 5, 2; 10, 2: *c*.
X $\mathbf{G}_{16}^{11}\otimes\mathbf{T}_2$: $\{C_{2z}|000\}$, $\{C_{2y}|000\}$, $\{I|000\}$; $\mathbf{t}_1$ or $\mathbf{t}_2$ or $\mathbf{t}_3$: 5, 2; 10, 2: *c*.
R $\mathbf{G}_{8}^{2}\otimes\mathbf{T}_2$: $\{C_{2y}|000\}$, $\{I|000\}$; $\mathbf{t}_1$: 2, 3; 4, 3; 6, 3; 8, 3: *b*.
S $\mathbf{G}_{8}^{2}\otimes\mathbf{T}_2$: $\{C_{2x}|000\}$, $\{I|000\}$; $\mathbf{t}_1$ or $\mathbf{t}_3$: 2, 3; 4, 3; 6, 3; 8, 3: *b*.
T $\mathbf{G}_{8}^{2}\otimes\mathbf{T}_2$: $\{C_{2z}|000\}$, $\{I|000\}$; $\mathbf{t}_1$ or $\mathbf{t}_2$: 2, 3; 4, 3; 6, 3; 8, 3: *b*.
W $\mathbf{G}_{8}^{5}\otimes\mathbf{T}_4$: $\{C_{2z}|000\}$, $\{C_{2y}|000\}$; $\mathbf{t}_1$ or $\mathbf{t}_2$ or $\mathbf{t}_3$: 5, 2: *a*.

Λ^x $\mathbf{G}_{8}^{5}$: $(C_{2z}, 0)$, $(\sigma_y, 0)$: 5, 2: *a*.
G^x $\mathbf{G}_{8}^{5}$: $(C_{2z}, 0)$, $(\sigma_y, 0)$: 5, 2: *a*.
P^x $\mathbf{G}_{4}^{1}$: $(C_{2z}, 0)$: 2, 3; 4, 3: *b*.
Σ^x $\mathbf{G}_{8}^{5}$: $(C_{2x}, 0)$, $(\sigma_z, 0)$: 5, 2: *a*.
F^x $\mathbf{G}_{8}^{5}$: $(C_{2x}, 0)$, $(\sigma_z, 0)$: 5, 2: *a*.
D^x $\mathbf{G}_{4}^{1}$: $(C_{2x}, 0)$: 2, 3; 4, 3: *b*.
Δ^x $\mathbf{G}_{8}^{5}$: $(C_{2y}, 0)$, $(\sigma_x, 0)$: 5, 2: *a*.
U^x $\mathbf{G}_{8}^{5}$: $(C_{2y}, 0)$, $(\sigma_x, 0)$: 5, 2: *a*.
Q^x $\mathbf{G}_{4}^{1}$: $(C_{2y}, 0)$: 2, 3; 4, 3: *b*.

72 *Ibam* D_{2h}^{26}

($F1$; $K7$; $M5$; $Z1$.)

Γ $\mathbf{G}_{16}^{11}$: $\{C_{2z}|000\}$, $\{C_{2y}|000\}$, $\{I|\frac{1}{2}\frac{1}{2}0\}$: 5, 2; 10, 2: *c*.
X $\mathbf{G}_{16}^{11}\otimes\mathbf{T}_2$: $\{C_{2z}|000\}$, $\{C_{2y}|000\}$, $\{I|\frac{1}{2}\frac{1}{2}0\}$; $\mathbf{t}_1$ or $\mathbf{t}_2$ or $\mathbf{t}_3$: 5, 2; 10, 2: *c*.
R $\mathbf{G}_{16}^{10}$: $\{\sigma_y|\frac{1}{2}\frac{1}{2}0\}$, $\{E|100\}$, $\{I|\frac{1}{2}\frac{1}{2}0\}$: 9, 1: *a*.
S $\mathbf{G}_{16}^{10}$: $\{\sigma_x|\frac{1}{2}\frac{1}{2}0\}$, $\{E|100\}$, $\{I|\frac{1}{2}\frac{1}{2}0\}$: 9, 1: *a*.
T $\mathbf{G}_8^2\otimes\mathbf{T}_2$: $\{C_{2z}|000\}$, $\{I|\frac{1}{2}\frac{1}{2}0\}$; $\mathbf{t}_1$ or $\mathbf{t}_2$: 2, 3; 4, 3; 6, 3; 8, 3: *b*.
W $\mathbf{G}_8^5\otimes\mathbf{T}_4$: $\{C_{2z}|000\}$, $\{C_{2y}|000\}$; $\mathbf{t}_1$ or $\mathbf{t}_2$ or $\mathbf{t}_3$: 5, 1: *a*.

Λ^x $\mathbf{G}_8^5$: $(C_{2z}, 0)$, $(\sigma_y, 0)$: 5, 2: *a*.
G^x $\mathbf{G}_8^5$: $(C_{2z}, 0)$, $(\sigma_y, 0)$: 5, 2: *a*.
P^x $\mathbf{G}_4^1$: $(C_{2z}, 0)$: 2, 3; 4, 3: *b*.
Σ^x $\mathbf{G}_8^5$: $(C_{2x}, 0)$, $(\sigma_z, 0)$: 5, 2: *a*.
F^x $\mathbf{G}_8^5$: $(C_{2x}, 0)$, $(\sigma_z, 0)$: 5, 2: *a*.
D^x $\mathbf{G}_4^1$: $(C_{2x}, 0)$: 2, 1; 4, 1: *b*.
Δ^x $\mathbf{G}_8^5$: $(C_{2y}, 0)$, $(\sigma_x, 0)$: 5, 2: *a*.
U^x $\mathbf{G}_8^5$: $(C_{2y}, 0)$, $(\sigma_x, 0)$: 5, 2: *a*.
Q^x $\mathbf{G}_4^1$: $(C_{2y}, 0)$: 2, 1; 4, 1: *b*.

73 *Ibca* D_{2h}^{27}

($F1$; $K7$; $M5$; $Z1$.)

Γ $\mathbf{G}_{16}^{11}$: $\{C_{2z}|\frac{1}{2}0\frac{1}{2}\}$, $\{C_{2y}|\frac{1}{2}\frac{1}{2}0\}$, $\{I|000\}$: 5, 2; 10, 2: *c*.
X $\mathbf{G}_{16}^{11}\otimes\mathbf{T}_2$: $\{C_{2z}|\frac{1}{2}0\frac{1}{2}\}$, $\{C_{2y}|\frac{1}{2}\frac{1}{2}0\}$, $\{I|000\}$: $\mathbf{t}_1$ or $\mathbf{t}_2$ or $\mathbf{t}_3$: 5, 2; 10, 2: *c*.
R $\mathbf{G}_{16}^{10}$: $\{\sigma_y|\frac{1}{2}\frac{1}{2}0\}$, $\{E|100\}$, $\{I|000\}$: 9, 1: *a*.
S $\mathbf{G}_{16}^{10}$: $\{\sigma_x|0\frac{1}{2}\frac{1}{2}\}$, $\{E|100\}$, $\{I|000\}$: 9, 1: *a*.
T $\mathbf{G}_{16}^{10}$: $\{\sigma_z|\frac{1}{2}0\frac{1}{2}\}$, $\{E|100\}$, $\{I|000\}$: 9, 1: *a*.
W $\mathbf{G}_8^5\otimes\mathbf{T}_4$: $\{C_{2x}|0-\frac{1}{2}\frac{1}{2}\}$, $\{C_{2y}|\frac{1}{2}-\frac{1}{2}0\}$; $\mathbf{t}_1$ or $\mathbf{t}_2$ or $\mathbf{t}_3$: 1, 1; 2, 1; 3, 1; 4, 1: *b*.

Λ^x $\mathbf{G}_8^5$: $(C_{2z}, 0)$, $(\sigma_y, 0)$: 5, 2: *a*.
G^x $\mathbf{G}_8^5$: $(C_{2z}, 0)$, $(\sigma_y, 0)$: 5, 2: *a*.
P^x $\mathbf{G}_8^2$: $(C_{2z}, 0)$, $(\bar{E}, 0)$: 5, 1; 7, 1: *a*.
Σ^x $\mathbf{G}_8^5$: $(C_{2x}, 0)$, $(\sigma_z, 0)$: 5, 2: *a*.
F^x $\mathbf{G}_8^5$: $(C_{2x}, 0)$, $(\sigma_z, 0)$: 5, 2: *a*.
D^x $\mathbf{G}_8^2$: $(C_{2x}, 0)$, $(\bar{E}, 0)$: 5, 1; 7, 1: *a*.
Δ^x $\mathbf{G}_8^5$: $(C_{2y}, 0)$, $(\sigma_x, 0)$: 5, 2: *a*.
U^x $\mathbf{G}_8^5$: $(C_{2y}, 0)$, $(\sigma_x, 0)$: 5, 2: *a*.
Q^x $\mathbf{G}_8^2$: $(C_{2y}, 0)$, $(\bar{E}, 0)$: 5, 1; 7, 1: *a*.

74 *Imma* D_{2h}^{28}

($F1$; $K7$; $M5$; $Z1$.)

Γ $\mathbf{G}_{16}^{11}$: $\{C_{2z} \mid \frac{1}{2}0\frac{1}{2}\}$, $\{C_{2y} \mid \frac{1}{2}\frac{1}{2}0\}$, $\{I \mid \frac{1}{2}\frac{1}{2}0\}$: 5, 2; 10, 2: *c*.

X $\mathbf{G}_{16}^{11} \otimes \mathbf{T}_2$: $\{C_{2z} \mid \frac{1}{2}0\frac{1}{2}\}$, $\{C_{2y} \mid \frac{1}{2}\frac{1}{2}0\}$, $\{I \mid \frac{1}{2}\frac{1}{2}0\}$; $\mathbf{t}_1$ or $\mathbf{t}_2$ or $\mathbf{t}_3$: 5, 2; 10, 2: *c*.

R $\mathbf{G}_8^2 \otimes \mathbf{T}_2$: $\{C_{2y} \mid \frac{1}{2}\frac{1}{2}0\}$, $\{I \mid \frac{1}{2}\frac{1}{2}0\}$; $\mathbf{t}_1$: 2, 3; 4, 3; 6, 3; 8, 3: *b*.

S $\mathbf{G}_8^2 \otimes \mathbf{T}_2$: $\{C_{2x} \mid 0\frac{1}{2}\frac{1}{2}\}$, $\{I \mid \frac{1}{2}\frac{1}{2}0\}$; $\mathbf{t}_1$ or $\mathbf{t}_3$: 2, 3; 4, 3; 6, 3; 8, 3: *b*.

T $\mathbf{G}_{16}^{10}$: $\{\sigma_z \mid 0\frac{1}{2}\frac{1}{2}\}$, $\{E \mid 100\}$, $\{I \mid \frac{1}{2}\frac{1}{2}0\}$: 9, 1: *a*.

W $\mathbf{G}_8^5 \otimes \mathbf{T}_4$: $\{C_{2x} \mid 0 -\frac{1}{2}\frac{1}{2}\}$, $\{C_{2y} \mid \frac{1}{2} -\frac{1}{2}0\}$; $\mathbf{t}_1$ or $\mathbf{t}_2$ or $\mathbf{t}_3$: 1, 3; 2, 3; 3, 3; 4, 3: *b*.

Λ^x $\mathbf{G}_8^5$: $(C_{2z}, 0)$, $(\sigma_y, 0)$: 5, 2: *a*.

G^x $\mathbf{G}_8^5$: $(C_{2z}, 0)$, $(\sigma_y, 0)$: 5, 2: *a*.

P^x $\mathbf{G}_8^2$: $(C_{2z}, 0)$, $(\bar{E}, 0)$: 5, 1; 7, 1: *a*.

Σ^x $\mathbf{G}_8^5$: $(C_{2x}, 0)$, $(\sigma_z, 0)$: 5, 2: *a*.

F^x $\mathbf{G}_8^5$: $(C_{2x}, 0)$, $(\sigma_z, 0)$: 5, 2: *a*.

D^x $\mathbf{G}_8^2$: $(C_{2x}, 0)$, $(\bar{E}, 0)$: 5, 3; 7, 3: *a*.

Δ^x $\mathbf{G}_8^5$: $(C_{2y}, 0)$, $(\sigma_x, 0)$: 5, 2: *a*.

U^x $\mathbf{G}_8^5$: $(C_{2y}, 0)$, $(\sigma_x, 0)$: 5, 2: *a*.

Q^x $\mathbf{G}_8^2$: $(C_{2y}, 0)$, $(\bar{E}, 0)$: 5, 3; 7, 3: *a*.

75 *P*4 C_4^1

($F1$; $K7$; $M5$; $Z1$.)

Γ $\mathbf{G}_8^1$: $\{C_{4z}^+ \mid 000\}$: 2, 3; 4, 3; 6, 3; 8, 3: *a*.

M $\mathbf{G}_8^1 \otimes \mathbf{T}_2$: $\{C_{4z}^+ \mid 000\}$; $\mathbf{t}_1$ or $\mathbf{t}_2$: 2, 3; 4, 3; 6, 3; 8, 3: *a*.

Z $\mathbf{G}_8^1 \otimes \mathbf{T}_2$: $\{C_{4z}^+ \mid 000\}$; $\mathbf{t}_3$: 2, 3; 4, 3; 6, 3; 8, 3: *a*.

A $\mathbf{G}_8^1 \otimes \mathbf{T}_2$: $\{C_{4z}^+ \mid 000\}$; $\mathbf{t}_1$ or $\mathbf{t}_2$ or $\mathbf{t}_3$: 2, 3; 4, 3; 6, 3; 8, 3: *a*.

R $\mathbf{G}_4^1 \otimes \mathbf{T}_2$: $\{C_{2z} \mid 000\}$; $\mathbf{t}_2$ or $\mathbf{t}_3$: 2, 3; 4, 3: *b*.

X $\mathbf{G}_4^1 \otimes \mathbf{T}_2$: $\{C_{2z} \mid 000\}$; $\mathbf{t}_2$: 2, 3; 4, 3: *b*.

Δ^x $\mathbf{G}_2^1$: $(\bar{E}, 0)$: 2, 2: *a*.

U^x $\mathbf{G}_2^1$: $(\bar{E}, 0)$: 2, 2: *a*.

Λ^x $\mathbf{G}_8^1$: $(C_{4z}^+, 0)$: 2, x; 4, x; 6, x; 8, x: *a*.

V^x $\mathbf{G}_8^1$: $(C_{4z}^+, 0)$: 2, x; 4, x; 6, x; 8, x: *a*.

Σ^x $\mathbf{G}_2^1$: $(\bar{E}, 0)$: 2, 2: *a*.

S^x $\mathbf{G}_2^1$: $(\bar{E}, 0)$: 2, 2: *a*.

Y^x $\mathbf{G}_2^1$: $(\bar{E}, 0)$: 2, 2: *a*.

T^x $\mathbf{G}_2^1$: $(\bar{E}, 0)$: 2, 2: *a*.

W^x $\mathbf{G}_4^1$: $(C_{2z}, 0)$: 2, x; 4, x: *b*.

76 $P4_1$ C_4^2

($F1$; $K7$; $M5$; $Z1$.)

Γ $\mathbf{G}_8^1$: $\{C_{4z}^+ \mid 00\frac{1}{4}\}$: 2, 3; 4, 3; 6, 3; 8, 3: a.
M $\mathbf{G}_8^1 \otimes \mathbf{T}_2$: $\{C_{4z}^+ \mid 00\frac{1}{4}\}$; $\mathbf{t}_1$ or $\mathbf{t}_2$: 2, 3; 4, 3; 6, 3; 8, 3: a.
Z $\mathbf{G}_8^1 \otimes \mathbf{T}_2$: $\{C_{4z}^+ \mid 00\frac{1}{4}\}$; $\mathbf{t}_3$: 1, 1; 3, 3; 5, 1; 7, 3: b.
A $\mathbf{G}_8^1 \otimes \mathbf{T}_2$: $\{C_{4z}^+ \mid 00\frac{1}{4}\}$; $\mathbf{t}_1$ or $\mathbf{t}_2$ or $\mathbf{t}_3$: 1, 1; 3, 3; 5, 1; 7, 3: b.
R $\mathbf{G}_4^1 \otimes \mathbf{T}_2$: $\{C_{2z} \mid 00\frac{1}{2}\}$; $\mathbf{t}_2$ or $\mathbf{t}_3$: 1, 1; 3, 1: c.
X $\mathbf{G}_4^1 \otimes \mathbf{T}_2$: $\{C_{2z} \mid 00\frac{1}{2}\}$; $\mathbf{t}_2$: 2, 3; 4, 3: b.

Δ^x $\mathbf{G}_2^1$: $(\bar{E}, 0)$: 2, 2: a.
U^x $\mathbf{G}_2^1$: $(\bar{E}, 0)$: 2, 1: a.
Λ^x $\mathbf{G}_8^1$: $(C_{4z}^+, 0)$: 2, x; 4, x; 6, x; 8, x: a.
V^x $\mathbf{G}_8^1$: $(C_{4z}^+, 0)$: 2, x; 4, x; 6, x; 8, x: a.
Σ^x $\mathbf{G}_2^1$: $(\bar{E}, 0)$: 2, 2: a.
S^x $\mathbf{G}_2^1$: $(\bar{E}, 0)$: 2, 1: a.
Y^x $\mathbf{G}_2^1$: $(\bar{E}, 0)$: 2, 2: a.
T^x $\mathbf{G}_2^1$: $(\bar{E}, 0)$: 2, 1: a.
W^x $\mathbf{G}_4^1$: $(C_{2z}, 0)$: 2, x; 4, x: b.

77 $P4_2$ C_4^3

($F1$; $K7$; $M5$; $Z1$.)

Γ $\mathbf{G}_8^1$: $\{C_{4z}^+ \mid 00\frac{1}{2}\}$: 2, 3; 4, 3; 6, 3; 8, 3: a.
M $\mathbf{G}_8^1 \otimes \mathbf{T}_2$: $\{C_{4z}^+ \mid 00\frac{1}{2}\}$; $\mathbf{t}_1$ or $\mathbf{t}_2$: 2, 3; 4, 3; 6, 3; 8, 3: a.
Z $\mathbf{G}_8^1 \otimes \mathbf{T}_2$: $\{C_{4z}^+ \mid 00\frac{1}{2}\}$; $\mathbf{t}_3$: 2, 3; 4, 3; 6, 3; 8, 3: a.
A $\mathbf{G}_8^1 \otimes \mathbf{T}_2$: $\{C_{4z}^+ \mid 00\frac{1}{2}\}$; $\mathbf{t}_1$ or $\mathbf{t}_2$ or $\mathbf{t}_3$: 2, 3; 4, 3; 6, 3; 8, 3: a.
R $\mathbf{G}_4^1 \otimes \mathbf{T}_2$: $\{C_{2z} \mid 000\}$; $\mathbf{t}_2$ or $\mathbf{t}_3$: 2, 3; 4, 3: b.
X $\mathbf{G}_4^1 \otimes \mathbf{T}_2$: $\{C_{2z} \mid 000\}$; $\mathbf{t}_2$: 2, 3; 4, 3: b.

Δ^x $\mathbf{G}_2^1$: $(\bar{E}, 0)$: 2, 2: a.
U^x $\mathbf{G}_2^1$: $(\bar{E}, 0)$: 2, 2: a.
Λ^x $\mathbf{G}_8^1$: $(C_{4z}^+, 0)$: 2, x; 4, x; 6, x; 8, x: a.
V^x $\mathbf{G}_8^1$: $(C_{4z}^+, 0)$: 2, x; 4, x; 6, x; 8, x: a.
Σ^x $\mathbf{G}_2^1$: $(\bar{E}, 0)$: 2, 2: a.
S^x $\mathbf{G}_2^1$: $(\bar{E}, 0)$: 2, 2: a.
Y^x $\mathbf{G}_2^1$: $(\bar{E}, 0)$: 2, 2: a.
T^x $\mathbf{G}_2^1$: $(\bar{E}, 0)$: 2, 2: a.
W^x $\mathbf{G}_4^1$: $(C_{2z}, 0)$: 2, x; 4, x: b.

78 $P4_3$ C_4^4

($F1$; $K7$; $M5$; $Z1$.)

Γ $\mathbf{G}_8^1$: $\{C_{4z}^+ \mid 00\frac{3}{4}\}$: 2, 3; 4, 3; 6, 3; 8, 3: a.
M $\mathbf{G}_8^1 \otimes \mathbf{T}_2$: $\{C_{4z}^+ \mid 00\frac{3}{4}\}$; $\mathbf{t}_1$ or $\mathbf{t}_2$: 2, 3; 4, 3; 6, 3; 8, 3: a.
Z $\mathbf{G}_8^1 \otimes \mathbf{T}_2$: $\{C_{4z}^+ \mid 00\frac{3}{4}\}$; $\mathbf{t}_3$: 1, 1; 3, 3; 5, 1; 7, 3: b.
A $\mathbf{G}_8^1 \otimes \mathbf{T}_2$: $\{C_{4z}^+ \mid 00\frac{3}{4}\}$; $\mathbf{t}_1$ or $\mathbf{t}_2$ or $\mathbf{t}_3$: 1, 1; 3, 3; 5, 1; 7, 3: b.
R $\mathbf{G}_4^1 \otimes \mathbf{T}_2$: $\{C_{2z} \mid 00\frac{1}{2}\}$; $\mathbf{t}_2$ or $\mathbf{t}_3$: 1, 1; 3, 1: c.
X $\mathbf{G}_4^1 \otimes \mathbf{T}_2$: $\{C_{2z} \mid 00\frac{1}{2}\}$; $\mathbf{t}_2$: 2, 3; 4, 3: b.

Δ^x $\mathbf{G}_2^1$: $(\bar{E}, 0)$: 2, 2: a.
U^x $\mathbf{G}_2^1$: $(\bar{E}, 0)$: 2, 1: a.
Λ^x $\mathbf{G}_8^1$: $(C_{4z}^+, 0)$: 2, x; 4, x; 6, x; 8, x: a.
V^x $\mathbf{G}_8^1$: $(C_{4z}^+, 0)$: 2, x; 4, x; 6, x; 8, x: a.
Σ^x $\mathbf{G}_2^1$: $(\bar{E}, 0)$: 2, 2: a.
S^x $\mathbf{G}_2^1$: $(\bar{E}, 0)$: 2, 1: a.
Y^x $\mathbf{G}_2^1$: $(\bar{E}, 0)$: 2, 2: a.
T^x $\mathbf{G}_2^1$: $(\bar{E}, 0)$: 2, 1: a.
W^x $\mathbf{G}_4^1$: $(C_{2z}, 0)$: 2, x; 4, x: b.

79 $I4$ C_4^5

($F1$; $K7$; $M5$; $Z1$.)

Γ $\mathbf{G}_8^1$: $\{C_{4z}^+ \mid 000\}$: 2, 3; 4, 3; 6, 3; 8, 3: a.
N $\mathbf{G}_2^1 \otimes \mathbf{T}_2$: $\{\bar{E} \mid 000\}$; $\mathbf{t}_2$: 2, 1: a.
X $\mathbf{G}_4^1 \otimes \mathbf{T}_2$: $\{C_{2z} \mid 000\}$; $\mathbf{t}_3$: 2, 3; 4, 3: b.
Z $\mathbf{G}_8^1 \otimes \mathbf{T}_2$: $\{C_{4z}^+ \mid 000\}$; $\mathbf{t}_1$ or $\mathbf{t}_2$ or $\mathbf{t}_3$: 2, 3; 4, 3; 6, 3; 8, 3: a.
P $\mathbf{G}_4^1 \otimes \mathbf{T}_4$: $\{C_{2z} \mid 000\}$; $\mathbf{t}_1$ or $\mathbf{t}_2$ or $\mathbf{t}_3$: 2, 3; 4, 3: b.

Λ^x $\mathbf{G}_8^1$: $(C_{4z}^+, 0)$: 2, x; 4, x; 6, x; 8, x: a.
V^x $\mathbf{G}_8^1$: $(C_{4z}^+, 0)$: 2, x; 4, x; 6, x; 8, x: a.
W^x $\mathbf{G}_4^1$: $(C_{2z}, 0)$: 2, x; 4, x: b.
Σ^x $\mathbf{G}_2^1$: $(\bar{E}, 0)$: 2, 2: a.
F^x $\mathbf{G}_2^1$: $(\bar{E}, 0)$: 2, 2: a.
Q^x $\mathbf{G}_2^1$: $(\bar{E}, 0)$: 2, x: a.
Δ^x $\mathbf{G}_2^1$: $(\bar{E}, 0)$: 2, 2: a.
U^x $\mathbf{G}_2^1$: $(\bar{E}, 0)$: 2, 2: a.
Y^x $\mathbf{G}_2^1$: $(\bar{E}, 0)$: 2, 2: a.

80 $I4_1$ C_4^6

($F1$; $K7$; $M5$; $Z1$.)

Γ $\mathbf{G}_8^1$: $\{C_{4z}^+ \mid \frac{3}{4}\frac{1}{4}\frac{1}{2}\}$: 2, 3; 4, 3; 6, 3; 8, 3: a.
N $\mathbf{G}_2^1 \otimes \mathbf{T}_2$: $\{\bar{E} \mid 000\}$; $\mathbf{t}_2$: 2, 1: a.
X $\mathbf{G}_4^1 \otimes \mathbf{T}_2$: $\{C_{2z} \mid 000\}$; $\mathbf{t}_3$: 2, 3; 4, 3: b.
Z $\mathbf{G}_8^1 \otimes \mathbf{T}_2$: $\{C_{4z}^+ \mid \frac{3}{4}\frac{1}{4}\frac{1}{2}\}$; $\mathbf{t}_1$ or $\mathbf{t}_2$ or $\mathbf{t}_3$: 2, 3; 4, 3; 6, 3; 8, 3: a.
P $\mathbf{G}_4^1 \otimes \mathbf{T}_4$: $\{C_{2z} \mid 000\}$; $\mathbf{t}_1$ or $\mathbf{t}_2$ or $\mathbf{t}_3$: 2, 2; 4, 1: b.

Λ^x $\mathbf{G}_8^1$: $(C_{4z}^+, 0)$: 2, x; 4, x; 6, x; 8, x: a.
V^x $\mathbf{G}_{16}^2$: $(C_{4z}^+, 0)$, $(\bar{E}, 1)$: 2, x; 4, x; 6, x; 8, x: f.
W^x $\mathbf{G}_4^1$: $(C_{2z}, 0)$: 2, x; 4, x: b.
Σ^x $\mathbf{G}_2^1$: $(\bar{E}, 0)$: 2, 2: a.
F^x $\mathbf{G}_2^1$: $(\bar{E}, 0)$: 2, 2: a.
Q^x $\mathbf{G}_2^1$: $(\bar{E}, 0)$: 2, x: a.
Δ^x $\mathbf{G}_2^1$: $(\bar{E}, 0)$: 2, 2: a.
U^x $\mathbf{G}_2^1$: $(\bar{E}, 0)$: 2, 2: a.
Y^x $\mathbf{G}_2^1$: $(\bar{E}, 0)$: 2, 2: a.

81 $P\bar{4}$ S_4^1

($F1$; $K7$; $M5$; $Z1$.)

Γ $\mathbf{G}_8^1$: $\{S_{4z}^+ \mid 000\}$: 2, 3; 4, 3; 6, 3; 8, 3: a.
M $\mathbf{G}_8^1 \otimes \mathbf{T}_2$: $\{S_{4z}^+ \mid 000\}$; $\mathbf{t}_1$ or $\mathbf{t}_2$: 2, 3; 4, 3; 6, 3; 8, 3: a.
Z $\mathbf{G}_8^1 \otimes \mathbf{T}_2$: $\{S_{4z}^+ \mid 000\}$; $\mathbf{t}_3$: 2, 3; 4, 3; 6, 3; 8, 3: a.
A $\mathbf{G}_8^1 \otimes \mathbf{T}_2$: $\{S_{4z}^+ \mid 000\}$; $\mathbf{t}_1$ or $\mathbf{t}_2$ or $\mathbf{t}_3$: 2, 3; 4, 3; 6, 3; 8, 3: a.
R $\mathbf{G}_4^1 \otimes \mathbf{T}_2$: $\{C_{2z} \mid 000\}$; $\mathbf{t}_2$ or $\mathbf{t}_3$: 2, 3; 4, 3: b.
X $\mathbf{G}_4^1 \otimes \mathbf{T}_2$: $\{C_{2z} \mid 000\}$; $\mathbf{t}_2$: 2, 3; 4, 3: b.

Δ^x $\mathbf{G}_2^1$: $(\bar{E}, 0)$: 2, 2: a.
U^x $\mathbf{G}_2^1$: $(\bar{E}, 0)$: 2, 2: a.
Λ^x $\mathbf{G}_4^1$: $(C_{2z}, 0)$: 2, 3; 4, 3: b.
V^x $\mathbf{G}_4^1$: $(C_{2z}, 0)$: 2, 3; 4, 3: b.
Σ^x $\mathbf{G}_2^1$: $(\bar{E}, 0)$: 2, 2: a.
S^x $\mathbf{G}_2^1$: $(\bar{E}, 0)$: 2, 2: a.
Y^x $\mathbf{G}_2^1$: $(\bar{E}, 0)$: 2, 2: a.
T^x $\mathbf{G}_2^1$: $(\bar{E}, 0)$: 2, 2: a.
W^x $\mathbf{G}_4^1$: $(C_{2z}, 0)$: 2, x; 4, x: b.

82 $I\bar{4}$ S_4^2

($F1$; $K7$; $M5$; $Z1$.)

Γ $\mathbf{G}_8^1$: $\{S_{4z}^+ \mid 000\}$: 2, 3; 4, 3; 6, 3; 8, 3: a.
N $\mathbf{G}_2^1 \otimes \mathbf{T}_2$: $\{\bar{E} \mid 000\}$; $\mathbf{t}_2$: 2, 1: a.
X $\mathbf{G}_4^1 \otimes \mathbf{T}_2$: $\{C_{2z} \mid 000\}$; $\mathbf{t}_3$: 2, 3; 4, 3: b.
Z $\mathbf{G}_8^1 \otimes \mathbf{T}_2$: $\{S_{4z}^+ \mid 000\}$; $\mathbf{t}_1$ or $\mathbf{t}_2$ or $\mathbf{t}_3$: 2, 3; 4, 3; 6, 3; 8, 3: a.
P $\mathbf{G}_8^1 \otimes \mathbf{T}_4$: $\{S_{4z}^+ \mid 000\}$; $\mathbf{t}_1$ or $\mathbf{t}_2$ or $\mathbf{t}_3$: 2, x; 4, x; 6, x; 8, x: a.

Λ^x $\mathbf{G}_4^1$: $(C_{2z}, 0)$: 2, 3; 4, 3: b.
V^x $\mathbf{G}_4^1$: $(C_{2z}, 0)$: 2, 3; 4, 3: b.
W^x $\mathbf{G}_4^1$: $(C_{2z}, 0)$: 2, x; 4, x: b.
Σ^x $\mathbf{G}_2^1$: $(\bar{E}, 0)$: 2, 2: a.
F^x $\mathbf{G}_2^1$: $(\bar{E}, 0)$: 2, 2: a.
Q^x $\mathbf{G}_2^1$: $(\bar{E}, 0)$: 2, x: a.
Δ^x $\mathbf{G}_2^1$: $(\bar{E}, 0)$: 2, 2: a.
U^x $\mathbf{G}_2^1$: $(\bar{E}, 0)$: 2, 2: a.
Y^x $\mathbf{G}_2^1$: $(\bar{E}, 0)$: 2, 2: a.

83 $P4/m$ C_{4h}^1

($F1$; $K7$; $M5$; $Z1$.)

Γ $\mathbf{G}_{16}^2$: $\{C_{4z}^+ \mid 000\}$, $\{I \mid 000\}$: 2, 3; 4, 3; 6, 3; 8, 3; 10, 3; 12, 3; 14, 3; 16, 3: d.
M $\mathbf{G}_{16}^2 \otimes \mathbf{T}_2$: $\{C_{4z}^+ \mid 000\}$, $\{I \mid 000\}$; $\mathbf{t}_1$ or $\mathbf{t}_2$: 2, 3; 4, 3; 6, 3; 8, 3; 10, 3; 12, 3; 14, 3; 16, 3: d.
Z $\mathbf{G}_{16}^2 \otimes \mathbf{T}_2$: $\{C_{4z}^+ \mid 000\}$, $\{I \mid 000\}$; $\mathbf{t}_3$: 2, 3; 4, 3; 6, 3; 8, 3; 10, 3; 12, 3; 14, 3; 16, 3: d.
A $\mathbf{G}_{16}^2 \otimes \mathbf{T}_2$: $\{C_{4z}^+ \mid 000\}$, $\{I \mid 000\}$; $\mathbf{t}_1$ or $\mathbf{t}_2$ or $\mathbf{t}_3$: 2, 3; 4, 3; 6, 3; 8, 3; 10, 3; 12, 3; 14, 3; 16, 3: d.
R $\mathbf{G}_8^2 \otimes \mathbf{T}_2$: $\{C_{2z} \mid 000\}$, $\{I \mid 000\}$; $\mathbf{t}_2$ or $\mathbf{t}_3$: 2, 3; 4, 3; 6, 3; 8, 3: b.
X $\mathbf{G}_8^2 \otimes \mathbf{T}_2$: $\{C_{2z} \mid 000\}$, $\{I \mid 000\}$; $\mathbf{t}_2$: 2, 3; 4, 3; 6, 3; 8, 3: b.

Δ^x $\mathbf{G}_4^1$: $(\sigma_z, 0)$: 2, 3; 4, 3: b.
U^x $\mathbf{G}_4^1$: $(\sigma_z, 0)$: 2, 3; 4, 3: b.
Λ^x $\mathbf{G}_8^1$: $(C_{4z}^+, 0)$: 2, 3; 4, 3; 6, 3; 8, 3: a.
V^x $\mathbf{G}_8^1$: $(C_{4z}^+, 0)$: 2, 3; 4, 3; 6, 3; 8, 3: a.
Σ^x $\mathbf{G}_4^1$: $(\sigma_z, 0)$: 2, 3; 4, 3: b.
S^x $\mathbf{G}_4^1$: $(\sigma_z, 0)$: 2, 3; 4, 3: b.
Y^x $\mathbf{G}_4^1$: $(\sigma_z, 0)$: 2, 3; 4, 3: b.
T^x $\mathbf{G}_4^1$: $(\sigma_z, 0)$: 2, 3; 4, 3: b.
W^x $\mathbf{G}_4^1$: $(C_{2z}, 0)$: 2, 3; 4, 3: b.

84 $P4_2/m$ C_{4h}^2

($F1$; $K7$; $M5$; $Z1$.)

Γ $\mathbf{G}_{16}^2$: $\{C_{4z}^+|00\frac{1}{2}\}$, $\{I|00\frac{1}{2}\}$: 2, 3; 4, 3; 6, 3; 8, 3; 10, 3; 12, 3; 14, 3; 16, 3: *d*.
M $\mathbf{G}_{16}^2 \otimes \mathbf{T}_2$: $\{C_{4z}^+|00\frac{1}{2}\}$, $\{I|00\frac{1}{2}\}$; $\mathbf{t}_1$ or $\mathbf{t}_2$: 2, 3; 4, 3; 6, 3; 8, 3; 10, 3; 12, 3; 14, 3; 16, 3: *d*.
Z $\mathbf{G}_{32}^{12}$: $\{C_{4z}^+|00\frac{1}{2}\}$, $\{E|001\}$, $\{I|00\frac{1}{2}\}$: 19, 3; 20, 3: *a*.
A $\mathbf{G}_{32}^{12}$: $\{C_{4z}^+|00\frac{1}{2}\}$, $\{E|001\}$, $\{I|00\frac{1}{2}\}$: 19, 3; 20, 3: *a*.
R $\mathbf{G}_8^2 \otimes \mathbf{T}_2$: $\{C_{2z}|000\}$, $\{I|00\frac{1}{2}\}$; $\mathbf{t}_2$ or $\mathbf{t}_3$: 2, 3; 4, 3; 6, 3; 8, 3: *b*.
X $\mathbf{G}_8^2 \otimes \mathbf{T}_2$: $\{C_{2z}|000\}$, $\{I|00\frac{1}{2}\}$; $\mathbf{t}_2$: 2, 3; 4, 3; 6, 3; 8, 3: *b*.

Δ^x $\mathbf{G}_4^1$: $(\sigma_z, 0)$: 2, 3; 4, 3: *b*.
U^x $\mathbf{G}_8^2$: $(\sigma_z, 0)$, $(E, 1)$: 5, 3; 7, 3: *a*.
Λ^x $\mathbf{G}_8^1$: $(C_{4z}^+, 0)$: 2, 3; 4, 3; 6, 3; 8, 3: *a*.
V^x $\mathbf{G}_8^1$: $(C_{4z}^+, 0)$: 2, 3; 4, 3; 6, 3; 8, 3: *a*.
Σ^x $\mathbf{G}_4^1$: $(\sigma_z, 0)$: 2, 3; 4, 3: *b*.
S^x $\mathbf{G}_8^2$: $(\sigma_z, 0)$, $(E, 1)$: 5, 3; 7, 3: *a*.
Y^x $\mathbf{G}_4^1$: $(\sigma_z, 0)$: 2, 3; 4, 3: *b*.
T^x $\mathbf{G}_8^2$: $(\sigma_z, 0)$, $(E, 1)$: 5, 3; 7, 3: *a*.
W^x $\mathbf{G}_4^1$: $(C_{2z}, 0)$: 2, 3; 4, 3: *b*.

85 $P4/n$ C_{4h}^3

($F1$; $K7$; $M5$; $Z1$.)

Γ $\mathbf{G}_{16}^2$: $\{C_{4z}^+|\frac{1}{2}\frac{1}{2}0\}$, $\{I|\frac{1}{2}\frac{1}{2}0\}$: 2, 3; 4, 3; 6, 3; 8, 3; 10, 3; 12, 3; 14, 3; 16, 3: *d*.
M $\mathbf{G}_{32}^{12}$: $\{C_{4z}^+|\frac{1}{2}\frac{1}{2}0\}$, $\{E|010\}$, $\{I|\frac{1}{2}\frac{1}{2}0\}$: 19, 3; 20, 3: *a*.
Z $\mathbf{G}_{16}^2 \otimes \mathbf{T}_2$: $\{C_{4z}^+|\frac{1}{2}\frac{1}{2}0\}$, $\{I|\frac{1}{2}\frac{1}{2}0\}$; $\mathbf{t}_3$: 2, 3; 4, 3; 6, 3; 8, 3; 10, 3; 12, 3; 14, 3; 16, 3: *d*.
A $\mathbf{G}_{32}^{12}$: $\{C_{4z}^+|\frac{1}{2}\frac{1}{2}0\}$, $\{E|010\}$, $\{I|\frac{1}{2}\frac{1}{2}0\}$: 19, 3; 20, 3: *a*.
R $\mathbf{G}_{16}^{10}$: $\{\sigma_z|\frac{1}{2}\frac{1}{2}0\}$, $\{E|001\}$, $\{I|\frac{1}{2}\frac{1}{2}0\}$: 9, 1: *a*.
X $\mathbf{G}_{16}^{10}$: $\{\sigma_z|\frac{1}{2}\frac{1}{2}0\}$, $\{E|010\}$, $\{I|\frac{1}{2}\frac{1}{2}0\}$: 9, 1: *a*.

Δ^x $\mathbf{G}_4^1$: $(\sigma_z, 0)$: 2, 3; 4, 3: *b*.
U^x $\mathbf{G}_4^1$: $(\sigma_z, 0)$: 2, 3; 4, 3: *b*.
Λ^x $\mathbf{G}_8^1$: $(C_{4z}^+, 0)$: 2, 3; 4, 3; 6, 3; 8, 3: *a*.
V^x $\mathbf{G}_{16}^2$: $(C_{4z}^+, 0)$, $(E, 1)$: 10, 3; 12, 3; 14, 3; 16, 3: *b*.
Σ^x $\mathbf{G}_4^1$: $(\sigma_z, 0)$: 2, 3; 4, 3: *b*.
S^x $\mathbf{G}_4^1$: $(\sigma_z, 0)$: 2, 3; 4, 3: *b*.
Y^x $\mathbf{G}_4^1$: $(\sigma_z, 0)$: 2, 3; 4, 3: *b*.
T^x $\mathbf{G}_4^1$: $(\sigma_z, 0)$: 2, 3; 4, 3: *b*.
W^x $\mathbf{G}_4^1$: $(C_{2z}, 0)$: 2, 1; 4, 1: *b*.

86 $P4_2/n$ C_{4h}^4

($F1$; $K7$; $M5$; $Z1$.)

Γ $\mathbf{G}_{16}^2$: $\{C_{4z}^+|\frac{1}{2}\frac{1}{2}\frac{1}{2}\}$, $\{I|\frac{1}{2}\frac{1}{2}\frac{1}{2}\}$: 2, 3; 4, 3; 6, 3; 8, 3; 10, 3; 12, 3; 14, 3; 16, 3: *d*.
M $\mathbf{G}_{32}^{12}$: $\{C_{4z}^+|\frac{1}{2}\frac{1}{2}\frac{1}{2}\}$, $\{E|010\}$, $\{I|\frac{1}{2}\frac{1}{2}\frac{1}{2}\}$: 19, 3; 20, 3: *a*.
Z $\mathbf{G}_{32}^{12}$: $\{C_{4z}^+|\frac{1}{2}\frac{1}{2}\frac{1}{2}\}$, $\{E|001\}$, $\{I|\frac{1}{2}\frac{1}{2}\frac{1}{2}\}$: 19, 3; 20, 3: *a*.
A $\mathbf{G}_{16}^2 \otimes \mathbf{T}_2$: $\{C_{4z}^+|\frac{1}{2}\frac{1}{2}\frac{1}{2}\}$, $\{I|\frac{1}{2}\frac{1}{2}\frac{1}{2}\}$; $\mathbf{t}_1$ or $\mathbf{t}_2$ or $\mathbf{t}_3$: 2, 3; 4, 3; 6, 3; 8, 3; 10, 3; 12, 3; 14, 3; 16, 3: *d*.
R $\mathbf{G}_{16}^{10}$: $\{\sigma_z|\frac{1}{2}\frac{1}{2}\frac{1}{2}\}$, $\{E|001\}$, $\{I|\frac{1}{2}\frac{1}{2}\frac{1}{2}\}$: 9, 1: *a*.
X $\mathbf{G}_{16}^{10}$: $\{\sigma_z|\frac{1}{2}\frac{1}{2}\frac{1}{2}\}$, $\{E|010\}$, $\{I|\frac{1}{2}\frac{1}{2}\frac{1}{2}\}$: 9, 1: *a*.

Δ^x $\mathbf{G}_4^1$: $(\sigma_z, 0)$: 2, 3; 4, 3: *b*.
U^x $\mathbf{G}_8^2$: $(\sigma_z, 0)$, $(E, 1)$: 5, 3; 7, 3: *a*.
Λ^x $\mathbf{G}_8^1$: $(C_{4z}^+, 0)$: 2, 3; 4, 3; 6, 3; 8, 3: *a*.
V^x $\mathbf{G}_{16}^2$: $(C_{4z}^+, 0)$, $(E, 1)$: 10, 3; 12, 3; 14, 3; 16, 3: *b*.
Σ^x $\mathbf{G}_4^1$: $(\sigma_z, 0)$: 2, 3; 4, 3: *b*.
S^x $\mathbf{G}_8^2$: $(\sigma_z, 0)$, $(E, 1)$: 5, 3; 7, 3: *a*.
Y^x $\mathbf{G}_4^1$: $(\sigma_z, 0)$: 2, 3; 4, 3: *b*.
T^x $\mathbf{G}_8^2$: $(\sigma_z, 0)$, $(E, 1)$: 5, 3; 7, 3: *a*.
W^x $\mathbf{G}_4^1$: $(C_{2z}, 0)$: 2, 1; 4, 1: *b*.

87 $I4/m$ C_{4h}^5

($F1$; $K7$; $M5$; $Z1$.)

Γ $\mathbf{G}_{16}^2$: $\{C_{4z}^+ | 000\}, \{I | 000\}$: 2, 3; 4, 3; 6, 3; 8, 3; 10, 3; 12, 3; 14, 3; 16, 3: *d*.

N $\mathbf{G}_4^2 \otimes \mathbf{T}_2$: $\{I | 000\}, \{\bar{E} | 000\}$; $\mathbf{t}_2$: 2, 1; 4, 1: *a*.

X $\mathbf{G}_8^2 \otimes \mathbf{T}_2$: $\{C_{2z} | 000\}, \{I | 000\}$; $\mathbf{t}_3$: 2, 3; 4, 3; 6, 3; 8, 3: *b*.

Z $\mathbf{G}_{16}^2 \otimes \mathbf{T}_2$: $\{C_{4z}^+ | 000\}, \{I | 000\}$; $\mathbf{t}_1$ or $\mathbf{t}_2$ or $\mathbf{t}_3$: 2, 3; 4, 3; 6, 3; 8, 3; 10, 3; 12, 3; 14, 3; 16, 3: *d*.

P $\mathbf{G}_8^1 \otimes \mathbf{T}_4$: $\{S_{4z}^+ | 000\}$; $\mathbf{t}_1$ or $\mathbf{t}_2$ or $\mathbf{t}_3$: 2, 3; 4, 3; 6, 3 ; 8, 3: *a*.

Λ^x $\mathbf{G}_8^1$: $(C_{4z}^+, 0)$: 2, 3; 4, 3; 6, 3; 8, 3: *a*.

V^x $\mathbf{G}_8^1$: $(C_{4z}^+, 0)$: 2, 3; 4, 3; 6, 3; 8, 3: *a*.

W^x $\mathbf{G}_4^1$: $(C_{2z}, 0)$: 2, 3; 4, 3: *b*.

Σ^x $\mathbf{G}_4^1$: $(\sigma_z, 0)$: 2, 3; 4, 3: *b*.

F^x $\mathbf{G}_4^1$: $(\sigma_z, 0)$: 2, 3; 4, 3: *b*.

Q^x $\mathbf{G}_2^1$: $(\bar{E}, 0)$: 2, 1: *a*.

Δ^x $\mathbf{G}_4^1$: $(\sigma_z, 0)$: 2, 3; 4, 3: *b*.

U^x $\mathbf{G}_4^1$: $(\sigma_z, 0)$: 2, 3; 4, 3: *b*.

Y^x $\mathbf{G}_4^1$: $(\sigma_z, 0)$: 2, 3; 4, 3: *b*.

88 $I4_1/a$ C_{4h}^6

($F1$; $K7$; $M5$; $Z1$.)

Γ $\mathbf{G}_{16}^2$: $\{C_{4z}^+ | \frac{3}{4}\frac{1}{4}\frac{1}{2}\}, \{I | \frac{3}{4}\frac{1}{4}\frac{1}{2}\}$: 2, 3; 4, 3; 6, 3; 8, 3; 10, 3; 12, 3; 14, 3; 16, 3: *d*.

N $\mathbf{G}_4^2 \otimes \mathbf{T}_2$: $\{I | \frac{3}{4}\frac{1}{4}\frac{1}{2}\}, \{\bar{E} | 000\}$; $\mathbf{t}_2$: 2, 1; 4, 1: *a*.

X $\mathbf{G}_{16}^{10}$: $\{\sigma_z | \frac{3}{4}\frac{1}{4}\frac{1}{2}\}, \{E | 001\}, \{I | \frac{3}{4}\frac{1}{4}\frac{1}{2}\}$: 9, 1: *a*.

Z $\mathbf{G}_{32}^{12}$: $\{C_{4z}^+ | \frac{3}{4}\frac{1}{4}\frac{1}{2}\}, \{E | 001\}, \{I | \frac{3}{4}\frac{1}{4}\frac{1}{2}\}$: 19, 3; 20, 3: *a*.

P $\mathbf{G}_8^1 \otimes \mathbf{T}_4$: $\{S_{4z}^+ | 000\}$; $\mathbf{t}_1$ or $\mathbf{t}_2$ or $\mathbf{t}_3$: 2, 3; 4, 1; 6, 3; 8, 1: *a*.

Λ^x $\mathbf{G}_8^1$: $(C_{4z}^+, 0)$: 2, 3; 4, 3; 6, 3; 8, 3: *a*.

V^x $\mathbf{G}_{16}^2$: $(C_{4z}^+, 0), (E, 1)$: 10, 3; 12, 3; 14, 3; 16, 3: *b*.

W^x $\mathbf{G}_4^1$: $(C_{2z}, 0)$: 2, 1; 4, 1: *b*.

Σ^x $\mathbf{G}_4^1$: $(\sigma_z, 0)$: 2, 3; 4, 3: *b*.

F^x $\mathbf{G}_8^2$: $(\sigma_z, 0), (E, 1)$: 5, 3; 7, 3: *a*.

Q^x $\mathbf{G}_2^1$: $(\bar{E}, 0)$: 2, 1: *a*.

Δ^x $\mathbf{G}_4^1$: $(\sigma_z, 0)$: 2, 3; 4, 3: *b*.

U^x $\mathbf{G}_8^2$: $(\sigma_z, 0), (E, 1)$: 5, 3; 7, 3: *a*.

Y^x $\mathbf{G}_4^1$: $(\sigma_z, 0)$: 2, 3; 4, 3: *b*.

89 $P422$ D_4^1

($F1$; $K7$; $M5$; $Z1$.)

Γ $\mathbf{G}_{16}^{14}$: $\{C_{4z}^+ | 000\}, \{C_{2x} | 000\}$: 6, 2; 7, 2: *a*.

M $\mathbf{G}_{16}^{14} \otimes \mathbf{T}_2$: $\{C_{4z}^+ | 000\}, \{C_{2x} | 000\}$; $\mathbf{t}_1$ or $\mathbf{t}_2$: 6, 2; 7, 2: *a*.

Z $\mathbf{G}_{16}^{14} \otimes \mathbf{T}_2$: $\{C_{4z}^+ | 000\}, \{C_{2x} | 000\}$; $\mathbf{t}_3$: 6, 2; 7, 2: *a*.

A $\mathbf{G}_{16}^{14} \otimes \mathbf{T}_2$: $\{C_{4z}^+ | 000\}, \{C_{2x} | 000\}$; $\mathbf{t}_1$ or $\mathbf{t}_2$ or $\mathbf{t}_3$: 6, 2; 7, 2: *a*.

R $\mathbf{G}_8^5 \otimes \mathbf{T}_2$: $\{C_{2z} | 000\}, \{C_{2y} | 000\}$; $\mathbf{t}_2$ or $\mathbf{t}_3$: 5, 2: *a*.

X $\mathbf{G}_8^5 \otimes \mathbf{T}_2$: $\{C_{2z} | 000\}, \{C_{2y} | 000\}$; $\mathbf{t}_2$: 5, 2: *a*.

Δ^x $\mathbf{G}_4^1$: $(C_{2y}, 0)$: 2, 2; 4, 2: *b*.

U^x $\mathbf{G}_4^1$: $(C_{2y}, 0)$: 2, 2; 4, 2: *b*.

Λ^x $\mathbf{G}_8^1$: $(C_{4z}^+, 0)$: 2, 2; 4, 2; 6, 2; 8, 2: *a*.

V^x $\mathbf{G}_8^1$: $(C_{4z}^+, 0)$: 2, 2; 4, 2; 6, 2; 8, 2: *a*.

Σ^x $\mathbf{G}_4^1$: $(C_{2a}, 0)$: 2, 2; 4, 2: *b*.

S^x $\mathbf{G}_4^1$: $(C_{2a}, 0)$: 2, 2; 4, 2: *b*.

Y^x $\mathbf{G}_4^1$: $(C_{2x}, 0)$: 2, 2; 4, 2: *b*.

T^x $\mathbf{G}_4^1$: $(C_{2x}, 0)$: 2, 2; 4, 2: *b*.

W^x $\mathbf{G}_4^1$: $(C_{2z}, 0)$: 2, 2; 4, 2: *b*.

90 $P42_12$ D_4^2

($F1$; $K7$; $M5$; $Z1$.)

Γ $\mathbf{G}_{16}^{14}$: $\{C_{4z}^+ | 000\}$, $\{C_{2x} | \frac{1}{2}\frac{1}{2}0\}$: 6, 2; 7, 2: *a*.
M $\mathbf{G}_{32}^{11}$: $\{C_{4z}^+ | 000\}$, $\{C_{2x} | \frac{1}{2}\frac{1}{2}0\}$: 6, 3; 7, 3: *a*.
Z $\mathbf{G}_{16}^{14} \otimes \mathbf{T}_2$: $\{C_{4z}^+ | 000\}$, $\{C_{2x} | \frac{1}{2}\frac{1}{2}0\}$; $\mathbf{t}_3$: 6, 2; 7, 2: *a*.
A $\mathbf{G}_{32}^{11}$: $\{C_{4z}^+ | 000\}$, $\{C_{2x} | \frac{1}{2}\frac{1}{2}0\}$: 6, 3; 7, 3: *a*.
R $\mathbf{G}_{16}^{8}$: $\{C_{2y} | \frac{1}{2}\frac{1}{2}0\}$, $\{C_{2x} | \frac{1}{2}\frac{1}{2}0\}$: 5, 3; 6, 3; 7, 3; 8, 3: *a*.
X $\mathbf{G}_{16}^{8}$: $\{C_{2y} | \frac{1}{2}\frac{1}{2}0\}$, $\{C_{2x} | \frac{1}{2}\frac{1}{2}0\}$: 5, 3; 6, 3; 7, 3; 8, 3: *a*.

Δ^x $\mathbf{G}_4^1$: $(C_{2y}, 0)$: 2, 2; 4, 2: *b*.
U^x $\mathbf{G}_4^1$: $(C_{2y}, 0)$: 2, 2; 4, 2: *b*.
Λ^x $\mathbf{G}_8^1$: $(C_{4z}^+, 0)$: 2, 2; 4, 2; 6, 2; 8, 2: *a*.
V^x $\mathbf{G}_8^1$: $(C_{4z}^+, 0)$: 2, 3; 4, 3; 6, 3; 8, 3: *a*.
Σ^x $\mathbf{G}_4^1$: $(C_{2a}, 0)$: 2, 2; 4, 2: *b*.
S^x $\mathbf{G}_4^1$: $(C_{2a}, 0)$: 2, 2; 4, 2: *b*.
Y^x $\mathbf{G}_8^2$: $(C_{2x}, 0)$, $(E, 1)$: 5, 3; 7, 3: *a*.
T^x $\mathbf{G}_8^2$: $(C_{2x}, 0)$, $(E, 1)$: 5, 3; 7, 3: *a*.
W^x $\mathbf{G}_4^1$: $(C_{2z}, 0)$: 2, 3; 4, 3: *b*.

91 $P4_122$ D_4^3

($F1$; $K7$; $M5$; $Z1$.)

Γ $\mathbf{G}_{16}^{14}$: $\{C_{4z}^+ | 00\frac{1}{4}\}$, $\{C_{2x} | 000\}$: 6, 2; 7, 2: *a*.
M $\mathbf{G}_{16}^{14} \otimes \mathbf{T}_2$: $\{C_{4z}^+ | 00\frac{1}{4}\}$, $\{C_{2x} | 000\}$; $\mathbf{t}_1$ or $\mathbf{t}_2$: 6, 2; 7, 2: *a*.
Z $\mathbf{G}_{32}^{10}$: $\{C_{4z}^+ | 00\frac{1}{4}\}$, $\{C_{2x} | 000\}$: 8, 3; 9, 3; 10, 3; 11, 3; 14, 2: *a*.
A $\mathbf{G}_{32}^{10}$: $\{C_{4z}^+ | 00\frac{1}{4}\}$, $\{C_{2x} | 000\}$: 8, 3; 9, 3; 10, 3; 11, 3; 14, 2: *a*.
R $\mathbf{G}_{16}^{8}$: $\{C_{2z} | 00\frac{1}{2}\}$, $\{C_{2y} | 00\frac{1}{2}\}$: 5, 3; 6, 3; 7, 3; 8, 3: *a*.
X $\mathbf{G}_8^5 \otimes \mathbf{T}_2$: $\{C_{2z} | 00\frac{1}{2}\}$, $\{C_{2y} | 00\frac{1}{2}\}$; $\mathbf{t}_2$: 5, 2: *a*.

Δ^x $\mathbf{G}_4^1$: $(C_{2y}, 0)$: 2, 2; 4, 2: *b*.
U^x $\mathbf{G}_8^2$: $(C_{2y}, 0)$, $(E, 1)$: 5, 3; 7, 3: *a*.
Λ^x $\mathbf{G}_8^1$: $(C_{4z}^+, 0)$: 2, 2; 4, 2; 6, 2; 8, 2: *a*.
V^x $\mathbf{G}_8^1$: $(C_{4z}^+, 0)$: 2, 2; 4, 2; 6, 2; 8, 2: *a*.
Σ^x $\mathbf{G}_4^1$: $(C_{2a}, 0)$: 2, 2; 4, 2: *b*.
S^x $\mathbf{G}_{16}^2$: $(C_{2a}, 0)$, $(\bar{E}, 0)$: 12, 3; 16, 3: *a*.
Y^x $\mathbf{G}_4^1$: $(C_{2x}, 0)$: 2, 2; 4, 2: *b*.
T^x $\mathbf{G}_4^1$: $(C_{2x}, 0)$: 2, 3; 4, 3: *b*.
W^x $\mathbf{G}_4^1$: $(C_{2z}, 0)$: 2, 2; 4, 2: *b*.

92 $P4_12_12$ D_4^4

($F1$; $K7$; $M5$; $Z1$.)

Γ $\mathbf{G}_{16}^{14}$: $\{C_{4z}^+ | 00\frac{1}{4}\}$, $\{C_{2x} | \frac{1}{2}\frac{1}{2}0\}$: 6, 2; 7, 2: *a*.
M $\mathbf{G}_{32}^{11}$: $\{C_{4z}^+ | 00\frac{1}{4}\}$, $\{C_{2x} | \frac{1}{2}\frac{1}{2}0\}$: 6, 3; 7, 3: *a*.
Z $\mathbf{G}_{32}^{10}$: $\{C_{4z}^+ | 00\frac{1}{4}\}$, $\{C_{2x} | \frac{1}{2}\frac{1}{2}0\}$: 8, 3; 9, 3; 10, 3; 11, 3; 14, 2: *a*.
A $\mathbf{G}_{32}^{11}$: $\{C_{4z}^+ | 00\frac{1}{4}\}$, $\{C_{2b} | \frac{1}{2}\frac{1}{2}\frac{3}{4}\}$: 8, 3; 9, 3; 10, 3; 11, 3; 12, 1: *b*.
R $\mathbf{G}_{16}^{8}$: $\{C_{2x} | \frac{1}{2}\frac{1}{2}0\}$, $\{C_{2y} | \frac{1}{2}\frac{1}{2}\frac{1}{2}\}$: 9, 1: *b*.
X $\mathbf{G}_{16}^{8}$: $\{C_{2y} | \frac{1}{2}\frac{1}{2}\frac{1}{2}\}$, $\{C_{2x} | \frac{1}{2}\frac{1}{2}0\}$: 5, 3; 6, 3; 7, 3; 8, 3: *a*.

Δ^x $\mathbf{G}_4^1$: $(C_{2y}, 0)$: 2, 2; 4, 2: *b*.
U^x $\mathbf{G}_8^2$: $(C_{2y}, 0)$, $(E, 1)$: 5, 3; 7, 3: *a*.
Λ^x $\mathbf{G}_8^1$: $(C_{4z}^+, 0)$: 2, 2; 4, 2; 6, 2; 8, 2: *a*.
V^x $\mathbf{G}_8^1$: $(C_{4z}^+, 0)$: 2, 3; 4, 3; 6, 3; 8, 3: *a*.
Σ^x $\mathbf{G}_4^1$: $(C_{2a}, 0)$: 2, 2; 4, 2: *b*.
S^x $\mathbf{G}_{16}^2$: $(C_{2a}, 0)$, $(\bar{E}, 0)$: 12, 3; 16, 3: *a*.
Y^x $\mathbf{G}_8^2$: $(C_{2x}, 0)$, $(E, 1)$: 5, 3; 7, 3: *a*.
T^x $\mathbf{G}_8^2$: $(C_{2x}, 0)$, $(E, 1)$: 5, 1; 7, 1: *a*.
W^x $\mathbf{G}_4^1$: $(C_{2z}, 0)$: 2, 3; 4, 3: *b*.

93 $P4_222$ D_4^5

($F1$; $K7$; $M5$; $Z1$.)

Γ $\mathbf{G}_{16}^{14}$: $\{C_{4z}^+ | 00\frac{1}{2}\}$, $\{C_{2x} | 000\}$: 6, 2; 7, 2: a,
M $\mathbf{G}_{16}^{14} \otimes \mathbf{T}_2$: $\{C_{4z}^+ | 00\frac{1}{2}\}$, $\{C_{2x} | 000\}$; $\mathbf{t}_1$ or $\mathbf{t}_2$: 6, 2; 7, 2: a.
Z $\mathbf{G}_{16}^{14} \otimes \mathbf{T}_2$: $\{C_{4z}^+ | 00\frac{1}{2}\}$, $\{C_{2x} | 000\}$; $\mathbf{t}_3$: 6, 2; 7, 2: a.
A $\mathbf{G}_{16}^{14} \otimes \mathbf{T}_2$: $\{C_{4z}^+ | 00\frac{1}{2}\}$, $\{C_{2x} | 000\}$; $\mathbf{t}_1$ or $\mathbf{t}_2$ or $\mathbf{t}_3$: 6, 2; 7, 2: a.
R $\mathbf{G}_8^5 \otimes \mathbf{T}_2$: $\{C_{2z} | 000\}$, $\{C_{2y} | 000\}$; $\mathbf{t}_2$ or $\mathbf{t}_3$: 5, 2: a.
X $\mathbf{G}_8^5 \otimes \mathbf{T}_2$: $\{C_{2z} | 000\}$, $\{C_{2y} | 000\}$; $\mathbf{t}_2$: 5, 2: a.

Δ^x $\mathbf{G}_4^1$: $(C_{2y}, 0)$: 2, 2; 4, 2: b.
U^x $\mathbf{G}_4^1$: $(C_{2y}, 0)$: 2, 2; 4, 2: b.
Λ^x $\mathbf{G}_8^1$: $(C_{4z}^+, 0)$: 2, 2; 4, 2; 6, 2; 8, 2: a.
V^x $\mathbf{G}_8^1$: $(C_{4z}^+, 0)$: 2, 2; 4, 2; 6, 2; 8, 2: a.
Σ^x $\mathbf{G}_4^1$: $(C_{2a}, 0)$: 2, 2; 4, 2: b.
S^x $\mathbf{G}_8^2$: $(C_{2a}, 0)$, $(E, 1)$: 5, 2; 7, 2: a.
Y^x $\mathbf{G}_4^1$: $(C_{2x}, 0)$: 2, 2; 4, 2: b.
T^x $\mathbf{G}_4^1$: $(C_{2x}, 0)$: 2, 2; 4, 2: b.
W^x $\mathbf{G}_4^1$: $(C_{2z}, 0)$: 2, 2; 4, 2: b.

94 $P4_22_12$ D_4^6

($F1$; $K7$; $M5$; $Z1$.)

Γ $\mathbf{G}_{16}^{14}$: $\{C_{4z}^+ | 00\frac{1}{2}\}$, $\{C_{2x} | \frac{1}{2}\frac{1}{2}0\}$: 6, 2; 7, 2: a.
M $\mathbf{G}_{32}^{11}$: $\{C_{4z}^+ | 00\frac{1}{2}\}$, $\{C_{2x} | \frac{1}{2}\frac{1}{2}0\}$: 6, 3; 7, 3: a.
Z $\mathbf{G}_{16}^{14} \otimes \mathbf{T}_2$: $\{C_{4z}^+ | 00\frac{1}{2}\}$, $\{C_{2x} | \frac{1}{2}\frac{1}{2}0\}$; $\mathbf{t}_3$: 6, 2; 7, 2: a.
A $\mathbf{G}_{32}^{11}$: $\{C_{4z}^+ | 00\frac{1}{2}\}$, $\{C_{2x} | \frac{1}{2}\frac{1}{2}0\}$: 6, 3; 7, 3: a.
R $\mathbf{G}_{16}^{8}$: $\{C_{2y} | \frac{1}{2}\frac{1}{2}0\}$, $\{C_{2z} | 000\}$: 5, 3; 6, 3; 7, 3; 8, 3: a.
X $\mathbf{G}_{16}^{8}$: $\{C_{2y} | \frac{1}{2}\frac{1}{2}0\}$, $\{C_{2z} | 000\}$: 5, 3; 6, 3; 7, 3; 8, 3: a.

Δ^x $\mathbf{G}_4^1$: $(C_{2y}, 0)$: 2, 2; 4, 2: b.
U^x $\mathbf{G}_4^1$: $(C_{2y}, 0)$: 2, 2; 4, 2: b.
Λ^x $\mathbf{G}_8^1$: $(C_{4z}^+, 0)$: 2, 2; 4, 2; 6, 2; 8, 2: a.
V^x $\mathbf{G}_8^1$: $(C_{4z}^+, 0)$: 2, 3; 4, 3; 6, 3; 8, 3: a.
Σ^x $\mathbf{G}_4^1$: $(C_{2a}, 0)$: 2, 2; 4, 2: b.
S^x $\mathbf{G}_8^2$: $(C_{2a}, 0)$, $(E, 1)$: 5, 2; 7, 2: a.
Y^x $\mathbf{G}_8^2$: $(C_{2x}, 0)$, $(E, 1)$: 5, 3; 7, 3: a.
T^x $\mathbf{G}_8^2$: $(C_{2x}, 0)$, $(E, 1)$: 5, 3; 7, 3: a.
W^x $\mathbf{G}_4^1$: $(C_{2z}, 0)$: 2, 3; 4, 3: b.

95 $P4_322$ D_4^7

($F1$; $K7$; $M5$; $Z1$.)

Γ $\mathbf{G}_{16}^{14}$: $\{C_{4z}^+ | 00\frac{3}{4}\}$, $\{C_{2x} | 000\}$: 6, 2; 7, 2: a.
M $\mathbf{G}_{16}^{14} \otimes \mathbf{T}_2$: $\{C_{4z}^+ | 00\frac{3}{4}\}$, $\{C_{2x} | 000\}$; $\mathbf{t}_1$ or $\mathbf{t}_2$: 6, 2; 7, 2: a.
Z $\mathbf{G}_{32}^{10}$: $\{C_{4z}^+ | 00\frac{3}{4}\}$, $\{C_{2x} | 000\}$: 8, 3; 9, 3; 10, 3; 11, 3; 14, 2: a.
A $\mathbf{G}_{32}^{10}$: $\{C_{4z}^+ | 00\frac{3}{4}\}$, $\{C_{2x} | 000\}$: 8, 3; 9, 3; 10, 3; 11, 3; 14, 2: a.
R $\mathbf{G}_{16}^{8}$: $\{C_{2z} | 00\frac{1}{2}\}$, $\{C_{2y} | 00\frac{1}{2}\}$: 5, 3; 6, 3; 7, 3; 8, 3: a.
X $\mathbf{G}_8^5 \otimes \mathbf{T}_2$: $\{C_{2z} | 00\frac{1}{2}\}$, $\{C_{2y} | 00\frac{1}{2}\}$; $\mathbf{t}_2$: 5, 2: a.

Δ^x $\mathbf{G}_4^1$: $(C_{2y}, 0)$: 2, 2; 4, 2: b.
U^x $\mathbf{G}_8^2$: $(C_{2y}, 0)$, $(E, 1)$: 5, 3; 7, 3: a.
Λ^x $\mathbf{G}_8^1$: $(C_{4z}^+, 0)$: 2, 2; 4, 2; 6, 2; 8, 2: a.
V^x $\mathbf{G}_8^1$: $(C_{4z}^+, 0)$: 2, 2; 4, 2; 6, 2; 8, 2: a.
Σ^x $\mathbf{G}_4^1$: $(C_{2a}, 0)$: 2, 2; 4, 2: b.
S^x $\mathbf{G}_{16}^2$: $(C_{2a}, 0)$, $(\bar{E}, 0)$: 10, 3; 14, 3: e.
Y^x $\mathbf{G}_4^1$: $(C_{2x}, 0)$: 2, 2; 4, 2: b.
T^x $\mathbf{G}_4^1$: $(C_{2x}, 0)$: 2, 3; 4, 3: b.
W^x $\mathbf{G}_4^1$: $(C_{2z}, 0)$: 2, 2; 4, 2: b.

96 $P4_32_12$ D_4^8

($F1$; $K7$; $M5$; $Z1$.)

Γ $\mathbf{G}_{16}^{14}$: $\{C_{4z}^+ \mid 00\frac{3}{4}\}$, $\{C_{2x} \mid \frac{1}{2}\frac{1}{2}0\}$: 6, 2; 7, 2: a.
M $\mathbf{G}_{32}^{11}$: $\{C_{4z}^+ \mid 00\frac{3}{4}\}$, $\{C_{2x} \mid \frac{1}{2}\frac{1}{2}0\}$: 6, 3; 7, 3: a.
Z $\mathbf{G}_{32}^{10}$: $\{C_{4z}^+ \mid 00\frac{3}{4}\}$, $\{C_{2x} \mid \frac{1}{2}\frac{1}{2}0\}$: 8, 3; 9, 3; 10, 3; 11, 3; 14, 2: a.
A $\mathbf{G}_{32}^{11}$: $\{C_{4z}^+ \mid 00\frac{3}{4}\}$, $\{C_{2b} \mid \frac{1}{2}\frac{1}{2}\frac{1}{4}\}$: 8, 3; 9, 3; 10, 3; 11, 3; 12, 1: b.
R $\mathbf{G}_{16}^{8}$: $\{C_{2x} \mid \frac{1}{2}\frac{1}{2}0\}$, $\{C_{2y} \mid \frac{1}{2}\frac{1}{2}\frac{1}{2}\}$: 9, 1: b.
X $\mathbf{G}_{16}^{8}$: $\{C_{2y} \mid \frac{1}{2}\frac{1}{2}\frac{1}{2}\}$, $\{C_{2x} \mid \frac{1}{2}\frac{1}{2}0\}$: 5, 3; 6, 3; 7, 3; 8, 3: a.

Δ^x $\mathbf{G}_4^1$: $(C_{2y}, 0)$: 2, 2; 4, 2: b.
U^x $\mathbf{G}_8^2$: $(C_{2y}, 0)$, $(E, 1)$: 5, 3; 7, 3: a.
Λ^x $\mathbf{G}_8^1$: $(C_{4z}^+, 0)$: 2, 2; 4, 2; 6, 2; 8, 2: a.
V^x $\mathbf{G}_8^1$: $(C_{4z}^+, 0)$: 2, 3; 4, 3; 6, 3; 8, 3: a.
Σ^x $\mathbf{G}_4^1$: $(C_{2a}, 0)$: 2, 2; 4, 2: b.
S^x $\mathbf{G}_{16}^2$: $(C_{2a}, 0)$, $(\bar{E}, 0)$: 10, 3; 14, 3: e.
Y^x $\mathbf{G}_8^2$: $(C_{2x}, 0)$, $(E, 1)$: 5, 3; 7, 3: a.
T^x $\mathbf{G}_8^2$: $(C_{2x}, 0)$, $(E, 1)$: 5, 1; 7, 1: a.
W^x $\mathbf{G}_4^1$: $(C_{2z}, 0)$: 2, 3; 4, 3: b.

97 $I422$ D_4^9

($F1$; $K7$; $M5$; $Z1$.)

Γ $\mathbf{G}_{16}^{14}$: $\{C_{4z}^+ \mid 000\}$, $\{C_{2x} \mid 000\}$: 6, 2; 7, 2: a.
N $\mathbf{G}_4^1 \otimes \mathbf{T}_2$: $\{C_{2y} \mid 000\}$; $\mathbf{t}_2$: 2, 3; 4, 3: b.
X $\mathbf{G}_8^5 \otimes \mathbf{T}_2$: $\{C_{2z} \mid 000\}$, $\{C_{2b} \mid 000\}$; $\mathbf{t}_3$: 5, 2: a.
Z $\mathbf{G}_{16}^{14} \otimes \mathbf{T}_2$: $\{C_{4z}^+ \mid 000\}$, $\{C_{2x} \mid 000\}$; $\mathbf{t}_1$ or $\mathbf{t}_2$ or $\mathbf{t}_3$: 6, 2; 7, 2: a.
P $\mathbf{G}_8^5 \otimes \mathbf{T}_4$: $\{C_{2z} \mid 000\}$, $\{C_{2y} \mid 000\}$; $\mathbf{t}_1$ or $\mathbf{t}_2$ or $\mathbf{t}_3$: 5, 2: a.

Λ^x $\mathbf{G}_8^1$: $(C_{4z}^+, 0)$: 2, 2; 4, 2; 6, 2; 8, 2: a.
V^x $\mathbf{G}_8^1$: $(C_{4z}^+, 0)$: 2, 2; 4, 2; 6, 2; 8, 2: a.
W^x $\mathbf{G}_4^1$: $(C_{2z}, 0)$: 2, 2; 4, 2: b.
Σ^x $\mathbf{G}_4^1$: $(C_{2x}, 0)$: 2, 2; 4, 2: b.
F^x $\mathbf{G}_4^1$: $(C_{2x}, 0)$: 2, 2; 4, 2: b.
Q^x $\mathbf{G}_4^1$: $(C_{2y}, 0)$: 2, x; 4, x: b.
Δ^x $\mathbf{G}_4^1$: $(C_{2a}, 0)$: 2, 2; 4, 2: b.
U^x $\mathbf{G}_4^1$: $(C_{2a}, 0)$: 2, 2; 4, 2: b.
Y^x $\mathbf{G}_4^1$: $(C_{2b}, 0)$: 2, 2; 4, 2: b.

98 $I4_122$ D_4^{10}

($F1$; $K7$; $M5$; $Z1$.)

Γ $\mathbf{G}_{16}^{14}$: $\{C_{4z}^+ \mid \frac{3}{4}\frac{1}{4}\frac{1}{2}\}$, $\{C_{2x} \mid 0\frac{1}{2}\frac{1}{2}\}$: 6, 2; 7, 2: a.
N $\mathbf{G}_4^1 \otimes \mathbf{T}_2$: $\{C_{2y} \mid 0\frac{1}{2}\frac{1}{2}\}$; $\mathbf{t}_2$: 2, 3; 4, 3: b.
X $\mathbf{G}_8^5 \otimes \mathbf{T}_2$: $\{C_{2z} \mid 000\}$, $\{C_{2b} \mid \frac{1}{4}\frac{1}{4}0\}$; $\mathbf{t}_3$: 5, 2: a.
Z $\mathbf{G}_{16}^{14} \otimes \mathbf{T}_2$: $\{C_{4z}^+ \mid \frac{3}{4}\frac{1}{4}\frac{1}{2}\}$, $\{C_{2x} \mid 0\frac{1}{2}\frac{1}{2}\}$; $\mathbf{t}_1$ or $\mathbf{t}_2$ or $\mathbf{t}_3$: 6, 2; 7, 2: a.
P $\mathbf{G}_8^5 \otimes \mathbf{T}_4$: $\{C_{2x} \mid 0\frac{1}{2}\frac{1}{2}\}$, $\{C_{2y} \mid 1\frac{1}{2}\frac{1}{2}\}$; $\mathbf{t}_1$ or $\mathbf{t}_2$ or $\mathbf{t}_3$: 1, 3; 2, 2; 3, 2; 4, 3: b.

Λ^x $\mathbf{G}_8^1$: $(C_{4z}^+, 0)$: 2, 2; 4, 2; 6, 2; 8, 2: a.
V^x $\mathbf{G}_{16}^2$: $(C_{4z}^+, 0)$, $(E, 1)$: 10, 2; 12, 2; 14, 2; 16, 2: b.
W^x $\mathbf{G}_4^1$: $(C_{2z}, 0)$: 2, 2; 4, 2: b.
Σ^x $\mathbf{G}_4^1$: $(C_{2x}, 0)$: 2, 2; 4, 2: b.
F^x $\mathbf{G}_4^1$: $(C_{2x}, 0)$: 2, 2; 4, 2: b.
Q^x $\mathbf{G}_8^2$: $(C_{2y}, 0)$, $(E, 1)$: 5, x; 7, x: a.
Δ^x $\mathbf{G}_4^1$: $(C_{2a}, 0)$: 2, 2; 4, 2: b.
U^x $\mathbf{G}_8^2$: $(C_{2a}, 0)$, $(E, 1)$: 5, 2; 7, 2: a.
Y^x $\mathbf{G}_4^1$: $(C_{2b}, 0)$: 2, 2; 4, 2: b.

99 *P4mm* C_{4v}^1

(*F*1; *K*7; *M*5; *Z*1.)

Γ $\mathbf{G}_{16}^{14}$: $\{C_{4z}^+ | 000\}$, $\{\sigma_y | 000\}$: 6, 2; 7, 2: *a*.
M $\mathbf{G}_{16}^{14} \otimes \mathbf{T}_2$: $\{C_{4z}^+ | 000\}$, $\{\sigma_y | 000\}$; $\mathbf{t}_1$ or $\mathbf{t}_2$: 6, 2; 7, 2: *a*.
Z $\mathbf{G}_{16}^{14} \otimes \mathbf{T}_2$: $\{C_{4z}^+ | 000\}$, $\{\sigma_y | 000\}$; $\mathbf{t}_3$: 6, 2; 7, 2: *a*.
A $\mathbf{G}_{16}^{14} \otimes \mathbf{T}_2$: $\{C_{4z}^+ | 000\}$, $\{\sigma_y | 000\}$; $\mathbf{t}_1$ or $\mathbf{t}_2$ or $\mathbf{t}_3$: 6, 2; 7, 2: *a*.
R $\mathbf{G}_{8}^{5} \otimes \mathbf{T}_2$: $\{C_{2z} | 000\}$, $\{\sigma_y | 000\}$; $\mathbf{t}_2$ or $\mathbf{t}_3$: 5, 2: *a*.
X $\mathbf{G}_{8}^{5} \otimes \mathbf{T}_2$: $\{C_{2z} | 000\}$, $\{\sigma_y | 000\}$; $\mathbf{t}_2$: 5, 2: *a*.

Δ^x $\mathbf{G}_4^1$: $(\sigma_x, 0)$: 2, 2; 4, 2: *b*.
U^x $\mathbf{G}_4^1$: $(\sigma_x, 0)$: 2, 2; 4, 2: *b*.
Λ^x $\mathbf{G}_{16}^{14}$: $(C_{4z}^+, 0)$, $(\sigma_y, 0)$: 6, *x*; 7, *x*: *a*.
V^x $\mathbf{G}_{16}^{14}$: $(C_{4z}^+, 0)$, $(\sigma_y, 0)$: 6, *x*; 7, *x*: *a*.
Σ^x $\mathbf{G}_4^1$: $(\sigma_{db}, 0)$: 2, 2; 4, 2: *b*.
S^x $\mathbf{G}_4^1$: $(\sigma_{db}, 0)$: 2, 2; 4, 2: *b*.
Y^x $\mathbf{G}_4^1$: $(\sigma_y, 0)$: 2, 2; 4, 2: *b*.
T^x $\mathbf{G}_4^1$: $(\sigma_y, 0)$: 2, 2; 4, 2: *b*.
W^x $\mathbf{G}_8^5$: $(C_{2z}, 0)$, $(\sigma_y, 0)$: 5, *x*: *a*.

100 *P4bm* C_{4v}^2

(*F*1; *K*7; *M*5; *Z*1.)

Γ $\mathbf{G}_{16}^{14}$: $\{C_{4z}^+ | 000\}$, $\{\sigma_y | \frac{1}{2}\frac{1}{2}0\}$: 6, 2; 7, 2: *a*.
M $\mathbf{G}_{32}^{11}$: $\{C_{4z}^+ | 000\}$, $\{\sigma_x | \frac{1}{2}\frac{1}{2}0\}$: 6, 3; 7, 3: *a*.
Z $\mathbf{G}_{16}^{14} \otimes \mathbf{T}_2$: $\{C_{4z}^+ | 000\}$, $\{\sigma_y | \frac{1}{2}\frac{1}{2}0\}$; $\mathbf{t}_3$: 6, 2; 7, 2: *a*.
A $\mathbf{G}_{32}^{11}$: $\{C_{4z}^+ | 000\}$, $\{\sigma_x | \frac{1}{2}\frac{1}{2}0\}$: 6, 3; 7, 3: *a*.
R $\mathbf{G}_{16}^{8}$: $\{\sigma_x | \frac{1}{2}\frac{1}{2}0\}$, $\{C_{2z} | 000\}$: 5, 3; 6, 3; 7, 3; 8, 3: *a*.
X $\mathbf{G}_{16}^{8}$: $\{\sigma_x | \frac{1}{2}\frac{1}{2}0\}$, $\{C_{2z} | 000\}$: 5, 3; 6, 3; 7, 3; 8, 3: *a*.

Δ^x $\mathbf{G}_4^1$: $(\sigma_x, 0)$: 2, 2; 4, 2: *b*.
U^x $\mathbf{G}_4^1$: $(\sigma_x, 0)$: 2, 2; 4, 2: *b*.
Λ^x $\mathbf{G}_{16}^{14}$: $(C_{4z}^+, 0)$, $(\sigma_y, 0)$: 6, *x*; 7, *x*: *a*.
V^x $\mathbf{G}_{32}^{11}$: $(C_{4z}^+, 0)$, $(\sigma_x, 0)$: 6, *x*; 7, *x*: *a*.
Σ^x $\mathbf{G}_4^1$: $(\sigma_{db}, 0)$: 2, 2; 4, 2: *b*.
S^x $\mathbf{G}_4^1$: $(\sigma_{db}, 0)$: 2, 2; 4, 2: *b*.
Y^x $\mathbf{G}_8^2$: $(\sigma_y, 0)$, $(E, 1)$: 5, 3; 7, 3: *a*.
T^x $\mathbf{G}_8^2$: $(\sigma_y, 0)$, $(E, 1)$: 5, 3; 7, 3: *a*.
W^x $\mathbf{G}_{16}^8$: $(\sigma_y, 0)$, $(C_{2z}, 0)$: 5, *x*; 6, *x*; 7, *x*; 8, *x*: *a*.

101 *P*4_2*cm* C_{4v}^3

(*F*1; *K*7; *M*5; *Z*1.)

Γ $\mathbf{G}_{16}^{14}$: $\{C_{4z}^+ | 00\frac{1}{2}\}$, $\{\sigma_y | 00\frac{1}{2}\}$: 6, 2; 7, 2: *a*.
M $\mathbf{G}_{16}^{14} \otimes \mathbf{T}_2$: $\{C_{4z}^+ | 00\frac{1}{2}\}$, $\{\sigma_y | 00\frac{1}{2}\}$; $\mathbf{t}_1$ or $\mathbf{t}_2$: 6, 2; 7, 2: *a*.
Z $\mathbf{G}_{32}^{11}$: $\{C_{4z}^+ | 00\frac{1}{2}\}$, $\{\sigma_x | 00\frac{1}{2}\}$: 6, 3; 7, 3: *a*.
A $\mathbf{G}_{32}^{11}$: $\{C_{4z}^+ | 00\frac{1}{2}\}$, $\{\sigma_x | 00\frac{1}{2}\}$: 6, 3; 7, 3: *a*.
R $\mathbf{G}_{16}^{8}$: $\{C_{2z} | 000\}$, $\{\sigma_x | 00\frac{1}{2}\}$: 9, 1: *b*.
X $\mathbf{G}_{8}^{5} \otimes \mathbf{T}_2$: $\{C_{2z} | 000\}$, $\{\sigma_y | 00\frac{1}{2}\}$; $\mathbf{t}_2$: 5, 2: *a*.

Δ^x $\mathbf{G}_4^1$: $(\sigma_x, 0)$: 2, 2; 4, 2: *b*.
U^x $\mathbf{G}_4^1$: $(\sigma_x, 0)$: 2, 3; 4, 3: *b*.
Λ^x $\mathbf{G}_{16}^{14}$: $(C_{4z}^+, 0)$, $(\sigma_y, 0)$: 6, *x*; 7, *x*: *a*.
V^x $\mathbf{G}_{16}^{14}$: $(C_{4z}^+, 0)$, $(\sigma_y, 0)$: 6, *x*; 7, *x*: *a*.
Σ^x $\mathbf{G}_4^1$: $(\sigma_{db}, 0)$: 2, 2; 4, 2: *b*.
S^x $\mathbf{G}_4^1$: $(\sigma_{db}, 0)$: 2, 2; 4, 2: *b*.
Y^x $\mathbf{G}_4^1$: $(\sigma_y, 0)$: 2, 2; 4, 2: *b*.
T^x $\mathbf{G}_4^1$: $(\sigma_y, 0)$: 2, 3; 4, 3: *b*.
W^x $\mathbf{G}_8^5$: $(C_{2z}, 0)$, $(\sigma_y, 0)$: 5, *x*: *a*.

102 $P4_2nm$ C_{4v}^4

($F1$; $K7$; $M5$; $Z1$.)

Γ $\mathbf{G}_{16}^{14}$: $\{C_{4z}^+ \mid 00\frac{1}{2}\}$, $\{\sigma_y \mid \frac{1}{2}\frac{1}{2}\frac{1}{2}\}$: 6, 2; 7, 2: a.
M $\mathbf{G}_{32}^{11}$: $\{C_{4z}^+ \mid 00\frac{1}{2}\}$, $\{\sigma_x \mid \frac{1}{2}\frac{1}{2}\frac{1}{2}\}$: 6, 3; 7, 3: a.
Z $\mathbf{G}_{32}^{11}$: $\{C_{4z}^+ \mid 00\frac{1}{2}\}$, $\{\sigma_x \mid \frac{1}{2}\frac{1}{2}\frac{1}{2}\}$: 6, 3; 7, 3: a.
A $\mathbf{G}_{16}^{14} \otimes \mathbf{T}_2$: $\{C_{4z}^+ \mid 00\frac{1}{2}\}$, $\{\sigma_x \mid \frac{1}{2}\frac{1}{2}\frac{1}{2}\}$; $\mathbf{t}_1$ or $\mathbf{t}_2$ or $\mathbf{t}_3$: 6, 2; 7, 2: a.
R $\mathbf{G}_{16}^{8}$: $\{\sigma_y \mid \frac{1}{2}\frac{1}{2}\frac{1}{2}\}$, $\{C_{2z} \mid 000\}$: 5, 3; 6, 3; 7, 3; 8, 3: a.
X $\mathbf{G}_{16}^{8}$: $\{\sigma_x \mid \frac{1}{2}\frac{1}{2}\frac{1}{2}\}$, $\{C_{2z} \mid 000\}$: 5, 3; 6, 3; 7, 3; 8, 3: a.

Δ^x $\mathbf{G}_4^1$: $(\sigma_x, 0)$: 2, 2; 4, 2: b.
U^x $\mathbf{G}_4^1$: $(\sigma_x, 0)$: 2, 3; 4, 3: b.
Λ^x $\mathbf{G}_{16}^{14}$: $(C_{4z}^+, 0)$, $(\sigma_y, 0)$: 6, x; 7, x: a.
V^x $\mathbf{G}_{32}^{11}$: $(C_{4z}^+, 0)$, $(\sigma_x, 0)$: 6, x; 7, x: a.
Σ^x $\mathbf{G}_4^1$: $(\sigma_{db}, 0)$: 2, 2; 4, 2: b.
S^x $\mathbf{G}_4^1$: $(\sigma_{db}, 0)$: 2, 2; 4, 2: b.
Y^x $\mathbf{G}_8^2$: $(\sigma_y, 0)$, $(E, 1)$: 5, 3; 7, 3: a.
T^x $\mathbf{G}_8^2$: $(\sigma_y, 0)$, $(E, 1)$: 5, 2; 7, 2: a.
W^x $\mathbf{G}_{16}^{8}$: $(\sigma_y, 0)$, $(C_{2z}, 0)$: 5, x; 6, x; 7, x; 8, x: a.

103 $P4cc$ C_{4v}^5

($F1$; $K7$; $M5$; $Z1$.)

Γ $\mathbf{G}_{16}^{14}$: $\{C_{4z}^+ \mid 000\}$, $\{\sigma_y \mid 00\frac{1}{2}\}$: 6, 2; 7, 2: a.
M $\mathbf{G}_{16}^{14} \otimes \mathbf{T}_2$: $\{C_{4z}^+ \mid 000\}$, $\{\sigma_y \mid 00\frac{1}{2}\}$; $\mathbf{t}_1$ or $\mathbf{t}_2$: 6, 2; 7, 2: a.
Z $\mathbf{G}_{32}^{10}$: $\{C_{4z}^+ \mid 000\}$, $\{\sigma_{db} \mid 00\frac{1}{2}\}$: 6, 1; 7, 1: b.
A $\mathbf{G}_{32}^{10}$: $\{C_{4z}^+ \mid 000\}$, $\{\sigma_{db} \mid 00\frac{1}{2}\}$: 6, 1; 7, 1: b.
R $\mathbf{G}_{16}^{8}$: $\{C_{2z} \mid 000\}$, $\{\sigma_x \mid 00\frac{1}{2}\}$: 9, 1: b.
X $\mathbf{G}_8^5 \otimes \mathbf{T}_2$: $\{C_{2z} \mid 000\}$, $\{\sigma_y \mid 00\frac{1}{2}\}$; $\mathbf{t}_2$: 5, 2: a.

Δ^x $\mathbf{G}_4^1$: $(\sigma_x, 0)$: 2, 2; 4, 2: b.
U^x $\mathbf{G}_4^1$: $(\sigma_x, 0)$: 2, 3; 4, 3: b.
Λ^x $\mathbf{G}_{16}^{14}$: $(C_{4z}^+, 0)$, $(\sigma_y, 0)$: 6, x; 7, x: a.
V^x $\mathbf{G}_{16}^{14}$: $(C_{4z}^+, 0)$, $(\sigma_y, 0)$: 6, x; 7, x: a.
Σ^x $\mathbf{G}_4^1$: $(\sigma_{db}, 0)$: 2, 2; 4, 2: b.
S^x $\mathbf{G}_4^1$: $(\sigma_{db}, 0)$: 2, 3; 4, 3: b.
Y^x $\mathbf{G}_4^1$: $(\sigma_y, 0)$: 2, 2; 4, 2: b.
T^x $\mathbf{G}_4^1$: $(\sigma_y, 0)$: 2, 3; 4, 3: b.
W^x $\mathbf{G}_8^5$: $(C_{2z}, 0)$, $(\sigma_y, 0)$: 5, x: a.

104 $P4nc$ C_{4v}^6

($F1$; $K7$; $M5$; $Z1$.)

Γ $\mathbf{G}_{16}^{14}$: $\{C_{4z}^+ \mid 000\}$, $\{\sigma_y \mid \frac{1}{2}\frac{1}{2}\frac{1}{2}\}$: 6, 2; 7, 2: a.
M $\mathbf{G}_{32}^{11}$: $\{C_{4z}^+ \mid 000\}$, $\{\sigma_x \mid \frac{1}{2}\frac{1}{2}\frac{1}{2}\}$: 6, 3; 7, 3: a.
Z $\mathbf{G}_{32}^{10}$: $\{C_{4z}^+ \mid 000\}$, $\{\sigma_{db} \mid \frac{1}{2}\frac{1}{2}\frac{1}{2}\}$: 6, 1; 7, 1: b.
A $\mathbf{G}_{32}^{11}$: $\{C_{4z}^+ \mid 000\}$, $\{\sigma_{db} \mid \frac{1}{2}\frac{1}{2}\frac{1}{2}\}$: 6, 3; 7, 3: a.
R $\mathbf{G}_{16}^{8}$: $\{\sigma_y \mid \frac{1}{2}\frac{1}{2}\frac{1}{2}\}$, $\{C_{2z} \mid 000\}$: 5, 3; 6, 3; 7, 3; 8, 3: a.
X $\mathbf{G}_{16}^{8}$: $\{\sigma_x \mid \frac{1}{2}\frac{1}{2}\frac{1}{2}\}$, $\{C_{2z} \mid 000\}$: 5, 3; 6, 3; 7, 3; 8, 3: a.

Δ^x $\mathbf{G}_4^1$: $(\sigma_x, 0)$: 2, 2; 4, 2: b.
U^x $\mathbf{G}_4^1$: $(\sigma_x, 0)$: 2, 3; 4, 3: b.
Λ^x $\mathbf{G}_{16}^{14}$: $(C_{4z}^+, 0)$, $(\sigma_y, 0)$: 6, x; 7, x: a.
V^x $\mathbf{G}_{32}^{11}$: $(C_{4z}^+, 0)$, $(\sigma_x, 0)$: 6, x; 7, x: a.
Σ^x $\mathbf{G}_4^1$: $(\sigma_{db}, 0)$: 2, 2; 4, 2: b.
S^x $\mathbf{G}_4^1$: $(\sigma_{db}, 0)$: 2, 3; 4, 3: b.
Y^x $\mathbf{G}_8^2$: $(\sigma_y, 0)$, $(E, 1)$: 5, 3; 7, 3: a.
T^x $\mathbf{G}_8^2$: $(\sigma_y, 0)$, $(E, 1)$: 5, 2; 7, 2: a.
W^x $\mathbf{G}_{16}^{8}$: $(\sigma_y, 0)$, $(C_{2z}, 0)$: 5, x; 6, x; 7, x; 8, x: a.

105 $P4_2mc$ C_{4v}^7

($F1$; $K7$; $M5$; $Z1$.)

Γ $\mathbf{G}_{16}^{14}$: $\{C_{4z}^+ \mid 00\frac{1}{2}\}$, $\{\sigma_y \mid 000\}$: 6, 2; 7, 2: a.
M $\mathbf{G}_{16}^{14} \otimes \mathbf{T}_2$: $\{C_{4z}^+ \mid 00\frac{1}{2}\}$, $\{\sigma_y \mid 000\}$; $\mathbf{t}_1$ or $\mathbf{t}_2$: 6, 2; 7, 2: a.
Z $\mathbf{G}_{32}^{11}$: $\{C_{4z}^+ \mid 00\frac{1}{2}\}$, $\{\sigma_{db} \mid 00\frac{1}{2}\}$: 6, 3; 7, 3: a.
A $\mathbf{G}_{32}^{11}$: $\{C_{4z}^+ \mid 00\frac{1}{2}\}$, $\{\sigma_{db} \mid 00\frac{1}{2}\}$: 6, 3; 7, 3: a.
R $\mathbf{G}_8^5 \otimes \mathbf{T}_2$: $\{C_{2z} \mid 000\}$, $\{\sigma_y \mid 000\}$; $\mathbf{t}_2$ or $\mathbf{t}_3$: 5, 2: a.
X $\mathbf{G}_8^5 \otimes \mathbf{T}_2$: $\{C_{2z} \mid 000\}$, $\{\sigma_y \mid 000\}$; $\mathbf{t}_2$: 5, 2: a.

Δ^x $\mathbf{G}_4^1$: $(\sigma_x, 0)$: 2, 2; 4, 2: b.
U^x $\mathbf{G}_4^1$: $(\sigma_x, 0)$: 2, 2; 4, 2: b.
Λ^x $\mathbf{G}_{16}^{14}$: $(C_{4z}^+, 0)$, $(\sigma_y, 0)$: 6, x; 7, x: a.
V^x $\mathbf{G}_{16}^{14}$: $(C_{4z}^+, 0)$, $(\sigma_y, 0)$: 6, x; 7, x: a.
Σ^x $\mathbf{G}_4^1$: $(\sigma_{db}, 0)$: 2, 2; 4, 2: b.
S^x $\mathbf{G}_4^1$: $(\sigma_{db}, 0)$: 2, 3; 4, 3: b.
Y^x $\mathbf{G}_4^1$: $(\sigma_y, 0)$: 2, 2; 4, 2: b.
T^x $\mathbf{G}_4^1$: $(\sigma_y, 0)$: 2, 2; 4, 2: b.
W^x $\mathbf{G}_8^5$: $(C_{2z}, 0)$, $(\sigma_y, 0)$: 5, x: a.

106 $P4_2bc$ C_{4v}^8

($F1$; $K7$; $M5$; $Z1$.)

Γ $\mathbf{G}_{16}^{14}$: $\{C_{4z}^+ \mid 00\frac{1}{2}\}$, $\{\sigma_y \mid \frac{1}{2}\frac{1}{2}0\}$: 6, 2; 7, 2: a.
M $\mathbf{G}_{32}^{11}$: $\{C_{4z}^+ \mid 00\frac{1}{2}\}$, $\{\sigma_x \mid \frac{1}{2}\frac{1}{2}0\}$: 6, 3; 7, 3: a.
Z $\mathbf{G}_{32}^{11}$: $\{C_{4z}^+ \mid 00\frac{1}{2}\}$, $\{\sigma_{db} \mid \frac{1}{2}\frac{1}{2}\frac{1}{2}\}$: 6, 3; 7, 3: a.
A $\mathbf{G}_{32}^{10}$: $\{C_{4z}^+ \mid 00\frac{1}{2}\}$, $\{\sigma_{db} \mid \frac{1}{2}\frac{1}{2}\frac{1}{2}\}$: 6, 1; 7, 1: b.
R $\mathbf{G}_{16}^8$: $\{\sigma_x \mid \frac{1}{2}\frac{1}{2}0\}$, $\{C_{2z} \mid 000\}$: 5, 3; 6, 3; 7, 3; 8, 3: a.
X $\mathbf{G}_{16}^8$: $\{\sigma_x \mid \frac{1}{2}\frac{1}{2}0\}$, $\{C_{2z} \mid 000\}$: 5, 3; 6, 3; 7, 3; 8, 3: a.

Δ^x $\mathbf{G}_4^1$: $(\sigma_x, 0)$: 2, 2; 4, 2: b.
U^x $\mathbf{G}_4^1$: $(\sigma_x, 0)$: 2, 2; 4, 2: b.
Λ^x $\mathbf{G}_{16}^{14}$: $(C_{4z}^+, 0)$, $(\sigma_y, 0)$: 6, x; 7, x: a.
V^x $\mathbf{G}_{32}^{11}$: $(C_{4z}^+, 0)$, $(\sigma_x, 0)$: 6, x; 7, x: a.
Σ^x $\mathbf{G}_4^1$: $(\sigma_{db}, 0)$: 2, 2; 4, 2: b.
S^x $\mathbf{G}_4^1$: $(\sigma_{db}, 0)$: 2, 3; 4, 3: b.
Y^x $\mathbf{G}_8^2$: $(\sigma_y, 0)$, $(E, 1)$: 5, 3; 7, 3: a.
T^x $\mathbf{G}_8^2$: $(\sigma_y, 0)$, $(E, 1)$: 5, 3; 7, 3: a.
W^x $\mathbf{G}_{16}^8$: $(\sigma_y, 0)$, $(C_{2z}, 0)$: 5, x; 6, x; 7, x; 8, x: a.

107 $I4mm$ C_{4v}^9

($F1$; $K7$; $M5$; $Z1$.)

Γ $\mathbf{G}_{16}^{14}$: $\{C_{4z}^+ \mid 000\}$, $\{\sigma_y \mid 000\}$: 6, 2; 7, 2: a.
N $\mathbf{G}_4^1 \otimes \mathbf{T}_2$: $\{\sigma_y \mid 000\}$; $\mathbf{t}_2$: 2, 3; 4, 3: b.
X $\mathbf{G}_8^5 \otimes \mathbf{T}_2$: $\{C_{2z} \mid 000\}$, $\{\sigma_{db} \mid 000\}$; $\mathbf{t}_3$: 5, 2: a.
Z $\mathbf{G}_{16}^{14} \otimes \mathbf{T}_2$: $\{C_{4z}^+ \mid 000\}$, $\{\sigma_y \mid 000\}$; $\mathbf{t}_1$ or $\mathbf{t}_2$ or $\mathbf{t}_3$: 6, 2; 7, 2: a.
P $\mathbf{G}_8^5 \otimes \mathbf{T}_4$: $\{C_{2z} \mid 000\}$, $\{\sigma_{db} \mid 000\}$; $\mathbf{t}_1$ or $\mathbf{t}_2$ or $\mathbf{t}_3$: 5, 2: a.

Λ^x $\mathbf{G}_{16}^{14}$: $(C_{4z}^+, 0)$, $(\sigma_y, 0)$: 6, x; 7, x: a.
V^x $\mathbf{G}_{16}^{14}$: $(C_{4z}^+, 0)$, $(\sigma_y, 0)$: 6, x; 7, x: a.
W^x $\mathbf{G}_8^5$: $(C_{2z}, 0)$, $(\sigma_{db}, 0)$: 5, x: a.
Σ^x $\mathbf{G}_4^1$: $(\sigma_y, 0)$: 2, 2; 4, 2: b.
F^x $\mathbf{G}_4^1$: $(\sigma_y, 0)$: 2, 2; 4, 2: b.
Q^x $\mathbf{G}_2^1$: $(\bar{E}, 0)$: 2, 2: a.
Δ^x $\mathbf{G}_4^1$: $(\sigma_{db}, 0)$: 2, 2; 4, 2: b.
U^x $\mathbf{G}_4^1$: $(\sigma_{db}, 0)$: 2, 2; 4, 2: b.
Y^x $\mathbf{G}_4^1$: $(\sigma_{da}, 0)$: 2, 2; 4, 2: b.

108 $I4cm$ C_{4v}^{10}

($F1$; $K7$; $M5$; $Z1$.)

Γ $\mathbf{G}_{16}^{14}$: $\{C_{4z}^{+} \mid 000\}$, $\{\sigma_y \mid \frac{1}{2}\frac{1}{2}0\}$: 6, 2; 7, 2: *a*.
N $\mathbf{G}_{8}^{2}$: $\{\sigma_y \mid \frac{1}{2}\frac{1}{2}0\}$, $\{\bar{E} \mid 000\}$: 5, 1; 7, 1: *a*.
X $\mathbf{G}_{8}^{5} \otimes \mathbf{T}_2$: $\{C_{2z} \mid 000\}$, $\{\sigma_{db} \mid \frac{1}{2}\frac{1}{2}0\}$; $\mathbf{t}_3$: 5, 2: *a*.
Z $\mathbf{G}_{16}^{14} \otimes \mathbf{T}_2$: $\{C_{4z}^{+} \mid 000\}$, $\{\sigma_y \mid \frac{1}{2}\frac{1}{2}0\}$; $\mathbf{t}_1$ or $\mathbf{t}_2$ or $\mathbf{t}_3$: 6, 2; 7, 2: *a*.
P $\mathbf{G}_{8}^{5} \otimes \mathbf{T}_4$: $\{C_{2z} \mid 000\}$, $\{\sigma_{db} \mid \frac{3}{2}\frac{1}{2}0\}$; $\mathbf{t}_1$ or $\mathbf{t}_2$ or $\mathbf{t}_3$: 5, 1: *a*.

Λ^x $\mathbf{G}_{16}^{14}$: $(C_{4z}^{+}, 0)$, $(\sigma_y, 0)$: 6, x; 7, x: *a*.
V^x $\mathbf{G}_{16}^{14}$: $(C_{4z}^{+}, 0)$, $(\sigma_y, 0)$: 6, x; 7, x: *a*.
W^x $\mathbf{G}_{8}^{5}$: $(C_{2z}, 0)$, $(\sigma_{db}, 0)$: 5, x: *a*.
Σ^x $\mathbf{G}_{4}^{1}$: $(\sigma_y, 0)$: 2, 2; 4, 2: *b*.
F^x $\mathbf{G}_{4}^{1}$: $(\sigma_y, 0)$: 2, 2; 4, 2: *b*.
Q^x $\mathbf{G}_{2}^{1}$: $(\bar{E}, 0)$: 2, 1: *a*.
Δ^x $\mathbf{G}_{4}^{1}$: $(\sigma_{db}, 0)$: 2, 2; 4, 2: *b*.
U^x $\mathbf{G}_{4}^{1}$: $(\sigma_{db}, 0)$: 2, 2; 4, 2: *b*.
Y^x $\mathbf{G}_{4}^{1}$: $(\sigma_{da}, 0)$: 2, 2; 4, 2: *b*.

109 $I4_1md$ C_{4v}^{11}

($F1$; $K7$; $M5$; $Z1$.)

Γ $\mathbf{G}_{16}^{14}$: $\{C_{4z}^{+} \mid \frac{3}{4}\frac{1}{4}\frac{1}{2}\}$, $\{\sigma_y \mid 000\}$: 6, 2; 7, 2: *a*.
N $\mathbf{G}_{4}^{1} \otimes \mathbf{T}_2$: $\{\sigma_y \mid 000\}$; $\mathbf{t}_2$: 2, 3; 4, 3: *b*.
X $\mathbf{G}_{16}^{8}$: $\{\sigma_{db} \mid \frac{3}{4}\frac{1}{4}\frac{1}{2}\}$, $\{C_{2z} \mid 000\}$: 5, 3; 6, 3; 7, 3; 8, 3: *a*.
Z $\mathbf{G}_{32}^{11}$: $\{C_{4z}^{+} \mid \frac{3}{4}\frac{1}{4}\frac{1}{2}\}$, $\{\sigma_{db} \mid \frac{3}{4}\frac{1}{4}\frac{1}{2}\}$: 6, 3; 7, 3: *a*.
P $\mathbf{G}_{32}^{8}$: $\{\sigma_{db} \mid \frac{3}{4}\frac{1}{4}\frac{1}{2}\}$, $\{C_{2z} \mid 000\}$: 13, 2; 14, 2; 17, 3; 18, 3: *a*.

Λ^x $\mathbf{G}_{16}^{14}$: $(C_{4z}^{+}, 0)$, $(\sigma_y, 0)$: 6, x; 7, x: *a*.
V^x $\mathbf{G}_{32}^{11}$: $(C_{4z}^{+}, 0)$, $(\sigma_{db}, 0)$: 6, x; 7, x: *a*.
W^x $\mathbf{G}_{16}^{8}$: $(\sigma_{da}, 0)$, $(C_{2z}, 0)$: 5, x; 6, x; 7, x; 8, x: *a*.
Σ^x $\mathbf{G}_{4}^{1}$: $(\sigma_y, 0)$: 2, 2; 4, 2: *b*.
F^x $\mathbf{G}_{4}^{1}$: $(\sigma_y, 0)$: 2, 2; 4, 2: *b*.
Q^x $\mathbf{G}_{2}^{1}$: $(\bar{E}, 0)$: 2, 2: *a*.
Δ^x $\mathbf{G}_{4}^{1}$: $(\sigma_{db}, 0)$: 2, 2; 4, 2: *b*.
U^x $\mathbf{G}_{4}^{1}$: $(\sigma_{db}, 0)$: 2, 3; 4, 3: *b*.
Y^x $\mathbf{G}_{8}^{2}$: $(\sigma_{da}, 0)$, $(E, 1)$: 5, 3: 7, 3: *a*.

110 $I4_1cd$ C_{4v}^{12}

($F1$; $K7$; $M5$; $Z1$.)

Γ $\mathbf{G}_{16}^{14}$: $\{C_{4z}^{+} \mid \frac{3}{4}\frac{1}{4}\frac{1}{2}\}$, $\{\sigma_y \mid \frac{1}{2}\frac{1}{2}0\}$: 6, 2; 7, 2: *a*.
N $\mathbf{G}_{8}^{2}$: $\{\sigma_y \mid \frac{1}{2}\frac{1}{2}0\}$, $\{\bar{E} \mid 000\}$: 5, 1; 7, 1: *a*.
X $\mathbf{G}_{16}^{8}$: $\{\sigma_{db} \mid \frac{1}{4}\frac{3}{4}\frac{1}{2}\}$, $\{C_{2z} \mid 000\}$: 5, 3; 6, 3; 7, 3; 8, 3: *a*.
Z $\mathbf{G}_{32}^{11}$: $\{C_{4z}^{+} \mid \frac{3}{4}\frac{1}{4}\frac{1}{2}\}$, $\{\sigma_{db} \mid \frac{1}{4}\frac{3}{4}\frac{1}{2}\}$: 6, 3; 7, 3: *a*.
P $\mathbf{G}_{32}^{8}$: $\{\sigma_{db} \mid \frac{1}{4}\frac{3}{4}\frac{1}{2}\}$, $\{C_{2z} \mid 000\}$: 13, 3; 14, 3; 17, 1; 18, 1: *a*.

Λ^x $\mathbf{G}_{16}^{14}$: $(C_{4z}^{+}, 0)$, $(\sigma_y, 0)$: 6, x; 7, x: *a*.
V^x $\mathbf{G}_{32}^{11}$: $(C_{4z}^{+}, 0)$, $(\sigma_{db}, 0)$: 6, x; 7, x: *a*.
W^x $\mathbf{G}_{16}^{8}$: $(\sigma_{da}, 0)$, $(C_{2z}, 0)$: 5, x; 6, x; 7, x; 8, x: *a*.
Σ^x $\mathbf{G}_{4}^{1}$: $(\sigma_y, 0)$: 2, 2; 4, 2: *b*.
F^x $\mathbf{G}_{4}^{1}$: $(\sigma_y, 0)$: 2, 2; 4, 2: *b*.
Q^x $\mathbf{G}_{2}^{1}$: $(\bar{E}, 0)$: 2, 1: *a*.
Δ^x $\mathbf{G}_{4}^{1}$: $(\sigma_{db}, 0)$: 2, 2; 4, 2: *b*.
U^x $\mathbf{G}_{4}^{1}$: $(\sigma_{db}, 0)$: 2, 3; 4, 3: *b*.
Y^x $\mathbf{G}_{8}^{2}$: $(\sigma_{da}, 0)$, $(E, 1)$: 5, 3; 7, 3: *a*.

111 $P\bar{4}2m$ D_{2d}^1

($F1$; $G8$; $K7$; $M5$; $Z1$.)

Γ $\mathbf{G}_{16}^{14}$: $\{S_{4z}^+ | 000\}$, $\{C_{2x} | 000\}$: 6, 2; 7, 2: a.
M $\mathbf{G}_{16}^{14} \otimes \mathbf{T}_2$: $\{S_{4z}^+ | 000\}$, $\{C_{2x} | 000\}$; $\mathbf{t}_1$ or $\mathbf{t}_2$: 6, 2; 7, 2: a.
Z $\mathbf{G}_{16}^{14} \otimes \mathbf{T}_2$: $\{S_{4z}^+ | 000\}$, $\{C_{2x} | 000\}$; $\mathbf{t}_3$: 6, 2; 7, 2: a.
A $\mathbf{G}_{16}^{14} \otimes \mathbf{T}_2$: $\{S_{4z}^+ | 000\}$, $\{C_{2x} | 000\}$; $\mathbf{t}_1$ or $\mathbf{t}_2$ or $\mathbf{t}_3$: 6, 2; 7, 2: a.
R $\mathbf{G}_8^5 \otimes \mathbf{T}_2$: $\{C_{2z} | 000\}$, $\{C_{2y} | 000\}$; $\mathbf{t}_2$ or $\mathbf{t}_3$: 5, 2: a.
X $\mathbf{G}_8^5 \otimes \mathbf{T}_2$: $\{C_{2z} | 000\}$, $\{C_{2y} | 000\}$; $\mathbf{t}_2$: 5, 2: a.

Δ^x $\mathbf{G}_4^1$: $(C_{2y}, 0)$: 2, 2; 4, 2: b.
U^x $\mathbf{G}_4^1$: $(C_{2y}, 0)$: 2, 2; 4, 2: b.
Λ^x $\mathbf{G}_8^5$: $(C_{2z}, 0)$, $(\sigma_{db}, 0)$: 5, 2: a.
V^x $\mathbf{G}_8^5$: $(C_{2z}, 0)$, $(\sigma_{db}, 0)$: 5, 2: a.
Σ^x $\mathbf{G}_4^1$: $(\sigma_{db}, 0)$: 2, 2; 4, 2: b.
S^x $\mathbf{G}_4^1$: $(\sigma_{db}, 0)$: 2, 2; 4, 2: b.
Y^x $\mathbf{G}_4^1$: $(C_{2x}, 0)$: 2, 2; 4, 2: b.
T^x $\mathbf{G}_4^1$: $(C_{2x}, 0)$: 2, 2; 4, 2: b.
W^x $\mathbf{G}_4^1$: $(C_{2z}, 0)$: 2, 2; 4, 2: b.

112 $P\bar{4}2c$ D_{2d}^2

($F1$; $K7$; $M5$; $Z1$.)

Γ $\mathbf{G}_{16}^{14}$: $\{S_{4z}^+ | 000\}$, $\{C_{2x} | 00\tfrac{1}{2}\}$: 6, 2; 7, 2: a.
M $\mathbf{G}_{16}^{14} \otimes \mathbf{T}_2$: $\{S_{4z}^+ | 000\}$, $\{C_{2x} | 00\tfrac{1}{2}\}$; $\mathbf{t}_1$ or $\mathbf{t}_2$: 6, 2; 7, 2: a.
Z $\mathbf{G}_{32}^{11}$: $\{S_{4z}^+ | 000\}$, $\{\sigma_{db} | 00\tfrac{1}{2}\}$: 6, 3; 7, 3: a.
A $\mathbf{G}_{32}^{11}$: $\{S_{4z}^+ | 000\}$, $\{\sigma_{db} | 00\tfrac{1}{2}\}$: 6, 3; 7, 3: a.
R $\mathbf{G}_8^5 \otimes \mathbf{T}_2$: $\{C_{2z} | 000\}$, $\{C_{2y} | 00\tfrac{1}{2}\}$; $\mathbf{t}_2$ or $\mathbf{t}_3$: 5, 2: a.
X $\mathbf{G}_8^5 \otimes \mathbf{T}_2$: $\{C_{2z} | 000\}$, $\{C_{2y} | 00\tfrac{1}{2}\}$; $\mathbf{t}_2$: 5, 2: a.

Δ^x $\mathbf{G}_4^1$: $(C_{2y}, 0)$: 2, 2; 4, 2: b.
U^x $\mathbf{G}_8^2$: $(C_{2y}, 0)$, $(E, 1)$: 5, 2; 7, 2: a.
Λ^x $\mathbf{G}_8^5$: $(C_{2z}, 0)$, $(\sigma_{db}, 0)$: 5, 2: a.
V^x $\mathbf{G}_8^5$: $(C_{2z}, 0)$, $(\sigma_{db}, 0)$: 5, 2: a.
Σ^x $\mathbf{G}_4^1$: $(\sigma_{db}, 0)$: 2, 2; 4, 2: b.
S^x $\mathbf{G}_4^1$: $(\sigma_{db}, 0)$: 2, 3; 4, 3: b.
Y^x $\mathbf{G}_4^1$: $(C_{2x}, 0)$: 2, 2; 4, 2: b.
T^x $\mathbf{G}_8^2$: $(C_{2x}, 0)$, $(E, 1)$: 5, 2; 7, 2: a.
W^x $\mathbf{G}_4^1$: $(C_{2z}, 0)$: 2, 2; 4, 2: b.

113 $P\bar{4}2_1m$ D_{2d}^3

($F1$; $K7$; $M5$; $Z1$.)

Γ $\mathbf{G}_{16}^{14}$: $\{S_{4z}^+ | 000\}$, $\{C_{2x} | \tfrac{1}{2}\tfrac{1}{2}0\}$: 6, 2; 7, 2: a.
M $\mathbf{G}_{32}^{11}$: $\{S_{4z}^+ | 000\}$, $\{C_{2x} | \tfrac{1}{2}\tfrac{1}{2}0\}$: 6, 3; 7, 3: a.
Z $\mathbf{G}_{16}^{14} \otimes \mathbf{T}_2$: $\{S_{4z}^+ | 000\}$, $\{C_{2x} | \tfrac{1}{2}\tfrac{1}{2}0\}$; $\mathbf{t}_3$: 6, 2; 7, 2: a.
A $\mathbf{G}_{32}^{11}$: $\{S_{4z}^+ | 000\}$, $\{C_{2x} | \tfrac{1}{2}\tfrac{1}{2}0\}$: 6, 3; 7, 3: a.
R $\mathbf{G}_{16}^8$: $\{C_{2y} | \tfrac{1}{2}\tfrac{1}{2}0\}$, $\{C_{2z} | 000\}$: 5, 3; 6, 3; 7, 3; 8, 3: a.
X $\mathbf{G}_{16}^8$: $\{C_{2y} | \tfrac{1}{2}\tfrac{1}{2}0\}$, $\{C_{2z} | 000\}$: 5, 3; 6, 3; 7, 3; 8, 3: a.

Δ^x $\mathbf{G}_4^1$: $(C_{2y}, 0)$: 2, 2; 4, 2: b.
U^x $\mathbf{G}_4^1$: $(C_{2y}, 0)$: 2, 2; 4, 2: b.
Λ^x $\mathbf{G}_8^5$: $(C_{2z}, 0)$, $(\sigma_{db}, 0)$: 5, 2: a.
V^x $\mathbf{G}_8^5$: $(C_{2z}, 0)$, $(\sigma_{db}, 0)$: 5, 1: a.
Σ^x $\mathbf{G}_4^1$: $(\sigma_{db}, 0)$: 2, 2; 4, 2: b.
S^x $\mathbf{G}_4^1$: $(\sigma_{db}, 0)$: 2, 2; 4, 2: b.
Y^x $\mathbf{G}_8^2$: $(C_{2x}, 0)$, $(E, 1)$: 5, 3; 7, 3: a.
T^x $\mathbf{G}_8^2$: $(C_{2x}, 0)$, $(E, 1)$: 5, 3; 7, 3: a.
W^x $\mathbf{G}_4^1$: $(C_{2z}, 0)$: 2, 3; 4, 3: b.

114 $P\bar{4}2_1c$ D_{2d}^4

($F1$; $K7$; $M5$; $Z1$.)

Γ $\mathbf{G}_{16}^{14}$: $\{S_{4z}^+ \mid 000\}$, $\{C_{2x} \mid \frac{1}{2}\frac{1}{2}\frac{1}{2}\}$: 6, 2; 7, 2: a.
M $\mathbf{G}_{32}^{11}$: $\{S_{4z}^+ \mid 000\}$, $\{C_{2x} \mid \frac{1}{2}\frac{1}{2}\frac{1}{2}\}$: 6, 3; 7, 3: a.
Z $\mathbf{G}_{32}^{11}$: $\{S_{4z}^+ \mid 000\}$, $\{\sigma_{db} \mid \frac{1}{2}\frac{1}{2}\frac{1}{2}\}$: 6, 3; 7, 3: a.
A $\mathbf{G}_{32}^{10}$: $\{S_{4z}^+ \mid 000\}$, $\{\sigma_{db} \mid \frac{1}{2}\frac{1}{2}\frac{1}{2}\}$: 6, 1; 7, 1: b.
R $\mathbf{G}_{16}^{8}$: $\{C_{2y} \mid \frac{1}{2}\frac{1}{2}\frac{1}{2}\}$, $\{C_{2z} \mid 000\}$: 5, 3; 6, 3; 7, 3; 8, 3: a.
X $\mathbf{G}_{16}^{8}$: $\{C_{2y} \mid \frac{1}{2}\frac{1}{2}\frac{1}{2}\}$, $\{C_{2z} \mid 000\}$: 5, 3; 6, 3; 7, 3; 8, 3: a.

Δ^x $\mathbf{G}_4^1$: $(C_{2y}, 0)$: 2, 2; 4, 2: b.
U^x $\mathbf{G}_8^2$: $(C_{2y}, 0)$, $(E, 1)$: 5, 2; 7, 2: a.
Λ^x $\mathbf{G}_8^5$: $(C_{2z}, 0)$, $(\sigma_{db}, 0)$: 5, 2: a.
V^x $\mathbf{G}_8^5$: $(C_{2z}, 0)$, $(\sigma_{db}, 0)$: 5, 1: a.
Σ^x $\mathbf{G}_4^1$: $(\sigma_{db}, 0)$: 2, 2; 4, 2: b.
S^x $\mathbf{G}_4^1$: $(\sigma_{db}, 0)$: 2, 3; 4, 3: b.
Y^x $\mathbf{G}_8^2$: $(C_{2x}, 0)$, $(E, 1)$: 5, 3; 7, 3: a.
T^x $\mathbf{G}_4^1$: $(C_{2x}, 0)$: 2, 3; 4, 3: b.
W^x $\mathbf{G}_4^1$: $(C_{2z}, 0)$: 2, 3; 4, 3: b.

115 $P\bar{4}m2$ D_{2d}^5

($F1$; $K7$; $M5$; $Z1$.)

Γ $\mathbf{G}_{16}^{14}$: $\{S_{4z}^+ \mid 000\}$, $\{C_{2a} \mid 000\}$: 6, 2; 7, 2: a.
M $\mathbf{G}_{16}^{14} \otimes \mathbf{T}_2$: $\{S_{4z}^+ \mid 000\}$, $\{C_{2a} \mid 000\}$; $\mathbf{t}_1$ or $\mathbf{t}_2$: 6, 2; 7, 2: a.
Z $\mathbf{G}_{16}^{14} \otimes \mathbf{T}_2$: $\{S_{4z}^+ \mid 000\}$, $\{C_{2a} \mid 000\}$; $\mathbf{t}_3$: 6, 2; 7, 2: a.
A $\mathbf{G}_{16}^{14} \otimes \mathbf{T}_2$: $\{S_{4z}^+ \mid 000\}$, $\{C_{2a} \mid 000\}$; $\mathbf{t}_1$ or $\mathbf{t}_2$ or $\mathbf{t}_3$: 6, 2; 7, 2: a.
R $\mathbf{G}_8^5 \otimes \mathbf{T}_2$: $\{C_{2z} \mid 000\}$, $\{\sigma_y \mid 000\}$; $\mathbf{t}_2$ or $\mathbf{t}_3$: 5, 2: a.
X $\mathbf{G}_8^5 \otimes \mathbf{T}_2$: $\{C_{2z} \mid 000\}$, $\{\sigma_y \mid 000\}$; $\mathbf{t}_2$: 5, 2: a.

Δ^x $\mathbf{G}_4^1$: $(\sigma_x, 0)$: 2, 2; 4, 2: b.
U^x $\mathbf{G}_4^1$: $(\sigma_x, 0)$: 2, 2; 4, 2: b.
Λ^x $\mathbf{G}_8^5$: $(C_{2z}, 0)$, $(\sigma_y, 0)$: 5, 2: a.
V^x $\mathbf{G}_8^5$: $(C_{2z}, 0)$, $(\sigma_y, 0)$: 5, 2: a.
Σ^x $\mathbf{G}_4^1$: $(C_{2a}, 0)$: 2, 2; 4, 2: b.
S^x $\mathbf{G}_4^1$: $(C_{2a}, 0)$: 2, 2; 4, 2: b.
Y^x $\mathbf{G}_4^1$: $(\sigma_y, 0)$: 2, 2; 4, 2: b.
T^x $\mathbf{G}_4^1$: $(\sigma_y, 0)$: 2, 2; 4, 2: b.
W^x $\mathbf{G}_8^5$: $(C_{2z}, 0)$, $(\sigma_y, 0)$: 5, x: a.

116 $P\bar{4}c2$ D_{2d}^6

($F1$; $K7$; $M5$; $Z1$.)

Γ $\mathbf{G}_{16}^{14}$: $\{S_{4z}^+ \mid 000\}$, $\{C_{2a} \mid 00\frac{1}{2}\}$: 6, 2; 7, 2: a.
M $\mathbf{G}_{16}^{14} \otimes \mathbf{T}_2$: $\{S_{4z}^+ \mid 000\}$, $\{C_{2a} \mid 00\frac{1}{2}\}$; $\mathbf{t}_1$ or $\mathbf{t}_2$: 6, 2; 7, 2: a.
Z $\mathbf{G}_{32}^{11}$: $\{S_{4z}^+ \mid 000\}$, $\{\sigma_x \mid 00\frac{1}{2}\}$: 6, 3; 7, 3: a.
A $\mathbf{G}_{32}^{11}$: $\{S_{4z}^+ \mid 000\}$, $\{\sigma_x \mid 00\frac{1}{2}\}$: 6, 3; 7, 3: a.
R $\mathbf{G}_{16}^{8}$: $\{C_{2z} \mid 000\}$, $\{\sigma_x \mid 00\frac{1}{2}\}$: 9, 1: b.
X $\mathbf{G}_8^5 \otimes \mathbf{T}_2$: $\{C_{2z} \mid 000\}$, $\{\sigma_y \mid 00\frac{1}{2}\}$; $\mathbf{t}_2$: 5, 2: a.

Δ^x $\mathbf{G}_4^1$: $(\sigma_x, 0)$: 2, 2; 4, 2: b.
U^x $\mathbf{G}_4^1$: $(\sigma_x, 0)$: 2, 3; 4, 3: b.
Λ^x $\mathbf{G}_8^5$: $(C_{2z}, 0)$, $(\sigma_y, 0)$: 5, 2: a.
V^x $\mathbf{G}_8^5$: $(C_{2z}, 0)$, $(\sigma_y, 0)$: 5, 2: a.
Σ^x $\mathbf{G}_4^1$: $(C_{2a}, 0)$: 2, 2; 4, 2: b.
S^x $\mathbf{G}_8^2$: $(C_{2a}, 0)$, $(E, 1)$: 5, 2; 7, 2: a.
Y^x $\mathbf{G}_4^1$: $(\sigma_y, 0)$: 2, 2; 4, 2: b.
T^x $\mathbf{G}_4^1$: $(\sigma_y, 0)$: 2, 3; 4, 3: b.
W^x $\mathbf{G}_8^5$: $(C_{2z}, 0)$, $(\sigma_y, 0)$: 5, x: a.

117 $P\bar{4}b2$ D_{2d}^{7}

($F1$; $K7$; $M5$; $Z1$.)

Γ $\mathbf{G}_{16}^{14}$: $\{S_{4z}^{+}\,|\,000\}$, $\{C_{2a}\,|\,\frac{1}{2}\frac{1}{2}0\}$: 6, 2; 7, 2: a.
M $\mathbf{G}_{32}^{11}$: $\{S_{4z}^{+}\,|\,000\}$, $\{\sigma_x\,|\,\frac{1}{2}\frac{1}{2}0\}$: 6, 3; 7, 3: a.
Z $\mathbf{G}_{16}^{14}\otimes\mathbf{T}_2$: $\{S_{4z}^{+}\,|\,000\}$, $\{C_{2a}\,|\,\frac{1}{2}\frac{1}{2}0\}$; $\mathbf{t}_3$: 6, 2; 7, 2: a.
A $\mathbf{G}_{32}^{11}$: $\{S_{4z}^{+}\,|\,000\}$, $\{\sigma_x\,|\,\frac{1}{2}\frac{1}{2}0\}$: 6, 3; 7, 3: a.
R $\mathbf{G}_{16}^{8}$: $\{\sigma_x\,|\,\frac{1}{2}\frac{1}{2}0\}$, $\{C_{2z}\,|\,000\}$: 5, 3; 6, 3; 7, 3; 8, 3: a.
X $\mathbf{G}_{16}^{8}$: $\{\sigma_x\,|\,\frac{1}{2}\frac{1}{2}0\}$, $\{C_{2z}\,|\,000\}$: 5, 3; 6, 3; 7, 3; 8, 3: a.

Δ^x $\mathbf{G}_{4}^{1}$: $(\sigma_x, 0)$: 2, 2; 4, 2: b.
U^x $\mathbf{G}_{4}^{1}$: $(\sigma_x, 0)$: 2, 2; 4, 2: b.
Λ^x $\mathbf{G}_{8}^{5}$: $(C_{2z}, 0)$, $(\sigma_y, 0)$: 5, 2: a.
V^x $\mathbf{G}_{16}^{8}$: $(C_{2z}, 0)$, $(\sigma_x, 0)$: 9, 2: b.
Σ^x $\mathbf{G}_{4}^{1}$: $(C_{2a}, 0)$: 2, 2; 4, 2: b.
S^x $\mathbf{G}_{4}^{1}$: $(C_{2a}, 0)$: 2, 2; 4, 2: b.
Y^x $\mathbf{G}_{8}^{2}$: $(\sigma_y, 0)$, $(E, 1)$: 5, 3; 7, 3: a.
T^x $\mathbf{G}_{8}^{2}$: $(\sigma_y, 0)$, $(E, 1)$: 5, 3; 7, 3: a.
W^x $\mathbf{G}_{16}^{8}$: $(\sigma_y, 0)$, $(C_{2z}, 0)$: 5, x; 6, x; 7, x; 8, x: a.

118 $P\bar{4}n2$ D_{2d}^{8}

($F1$; $K7$; $M5$; $Z1$.)

Γ $\mathbf{G}_{16}^{14}$: $\{S_{4z}^{+}\,|\,000\}$, $\{C_{2a}\,|\,\frac{1}{2}\frac{1}{2}\frac{1}{2}\}$: 6, 2; 7, 2: a.
M $\mathbf{G}_{32}^{11}$: $\{S_{4z}^{+}\,|\,000\}$, $\{\sigma_x\,|\,\frac{1}{2}\frac{1}{2}\frac{1}{2}\}$: 6, 3; 7, 3: a.
Z $\mathbf{G}_{32}^{11}$: $\{S_{4z}^{+}\,|\,000\}$, $\{\sigma_x\,|\,\frac{1}{2}\frac{1}{2}\frac{1}{2}\}$: 6, 3; 7, 3: a.
A $\mathbf{G}_{16}^{14}\otimes\mathbf{T}_2$: $\{S_{4z}^{+}\,|\,000\}$, $\{C_{2a}\,|\,\frac{1}{2}\frac{1}{2}\frac{1}{2}\}$; $\mathbf{t}_1$ or $\mathbf{t}_2$ or $\mathbf{t}_3$: 6, 2; 7, 2: a.
R $\mathbf{G}_{16}^{8}$: $\{\sigma_y\,|\,\frac{1}{2}\frac{1}{2}\frac{1}{2}\}$, $\{C_{2z}\,|\,000\}$: 5, 3; 6, 3; 7, 3; 8, 3: a.
X $\mathbf{G}_{16}^{8}$: $\{\sigma_x\,|\,\frac{1}{2}\frac{1}{2}\frac{1}{2}\}$, $\{C_{2z}\,|\,000\}$: 5, 3; 6, 3; 7, 3; 8, 3: a.

Δ^x $\mathbf{G}_{4}^{1}$: $(\sigma_x, 0)$: 2, 2; 4, 2: b.
U^x $\mathbf{G}_{4}^{1}$: $(\sigma_x, 0)$: 2, 3; 4, 3: b.
Λ^x $\mathbf{G}_{8}^{5}$: $(C_{2z}, 0)$, $(\sigma_y, 0)$: 5, 2: a.
V^x $\mathbf{G}_{16}^{8}$: $(C_{2z}, 0)$, $(\sigma_x, 0)$: 9, 2: b.
Σ^x $\mathbf{G}_{4}^{1}$: $(C_{2a}, 0)$: 2, 2; 4, 2: b.
S^x $\mathbf{G}_{8}^{2}$: $(C_{2a}, 0)$, $(E, 1)$: 5, 2; 7, 2: a.
Y^x $\mathbf{G}_{8}^{2}$: $(\sigma_y, 0)$, $(E, 1)$: 5, 3; 7, 3: a.
T^x $\mathbf{G}_{8}^{2}$: $(\sigma_y, 0)$, $(E, 1)$: 5, 2; 7, 2: a.
W^x $\mathbf{G}_{16}^{8}$: $(\sigma_y, 0)$, $(C_{2z}, 0)$: 5, x; 6, x; 7, x; 8, x: a.

119 $I\bar{4}m2$ D_{2d}^{9}

($F1$; $K7$; $M5$; $Z1$.)

Γ $\mathbf{G}_{16}^{14}$: $\{S_{4z}^{+}\,|\,000\}$, $\{C_{2a}\,|\,000\}$: 6, 2; 7, 2: a.
N $\mathbf{G}_{4}^{1}\otimes\mathbf{T}_2$: $\{\sigma_y\,|\,000\}$; $\mathbf{t}_2$: 2, 3; 4, 3: b.
X $\mathbf{G}_{8}^{5}\otimes\mathbf{T}_2$: $\{C_{2z}\,|\,000\}$, $\{C_{2b}\,|\,000\}$; $\mathbf{t}_3$: 5, 2: a.
Z $\mathbf{G}_{16}^{14}\otimes\mathbf{T}_2$: $\{S_{4z}^{+}\,|\,000\}$, $\{C_{2a}\,|\,000\}$; $\mathbf{t}_1$ or $\mathbf{t}_2$ or $\mathbf{t}_3$: 6, 2; 7, 2: a.
P $\mathbf{G}_{8}^{1}\otimes\mathbf{T}_4$: $\{S_{4z}^{+}\,|\,000\}$; $\mathbf{t}_1$ or $\mathbf{t}_2$ or $\mathbf{t}_3$: 2, 2; 4, 2; 6, 2; 8, 2: a.

Λ^x $\mathbf{G}_{8}^{5}$: $(C_{2z}, 0)$, $(\sigma_y, 0)$: 5, 2: a.
V^x $\mathbf{G}_{8}^{5}$: $(C_{2z}, 0)$, $(\sigma_y, 0)$: 5, 2: a.
W^x $\mathbf{G}_{4}^{1}$: $(C_{2z}, 0)$: 2, 2; 4, 2: b.
Σ^x $\mathbf{G}_{4}^{1}$: $(\sigma_y, 0)$: 2, 2; 4, 2: b.
F^x $\mathbf{G}_{4}^{1}$: $(\sigma_y, 0)$: 2, 2; 4, 2: b.
Q^x $\mathbf{G}_{2}^{1}$: $(\bar{E}, 0)$: 2, 2: a.
Δ^x $\mathbf{G}_{4}^{1}$: $(C_{2a}, 0)$: 2, 2; 4, 2: b.
U^x $\mathbf{G}_{4}^{1}$: $(C_{2a}, 0)$: 2, 2; 4, 2: b.
Y^x $\mathbf{G}_{4}^{1}$: $(C_{2b}, 0)$: 2, 2; 4, 2: b.

120 $I\bar{4}c2$ D_{2d}^{10}

($F1$; $K7$; $M5$; $Z1$.)

Γ $\mathbf{G}_{16}^{14}$: $\{S_{4z}^{+} | 000\}$, $\{C_{2a} | \frac{1}{2}\frac{1}{2}0\}$: 6, 2; 7, 2: a.
N $\mathbf{G}_{8}^{2}$: $\{\sigma_y | \frac{1}{2}\frac{1}{2}0\}$, $\{\bar{E} | 000\}$: 5, 1; 7, 1: a.
X $\mathbf{G}_{8}^{5} \otimes \mathbf{T}_2$: $\{C_{2z} | 000\}$, $\{C_{2b} | \frac{1}{2}\frac{1}{2}0\}$; $\mathbf{t}_3$: 5, 2: a.
Z $\mathbf{G}_{16}^{14} \otimes \mathbf{T}_2$: $\{S_{4z}^{+} | 000\}$, $\{C_{2a} | \frac{1}{2}\frac{1}{2}0\}$; $\mathbf{t}_1$ or $\mathbf{t}_2$ or $\mathbf{t}_3$: 6, 2; 7, 2: a.
P $\mathbf{G}_{8}^{1} \otimes \mathbf{T}_4$: $\{S_{4z}^{+} | 000\}$; $\mathbf{t}_1$ or $\mathbf{t}_2$ or $\mathbf{t}_3$: 2, 3; 4, 3; 6, 3; 8, 3: a.

Λ^x $\mathbf{G}_{8}^{5}$: $(C_{2z}, 0)$, $(\sigma_y, 0)$: 5, 2: a.
V^x $\mathbf{G}_{8}^{5}$: $(C_{2z}, 0)$, $(\sigma_y, 0)$: 5, 2: a.
W^x $\mathbf{G}_{4}^{1}$: $(C_{2z}, 0)$: 2, 2; 4, 2: b.
Σ^x $\mathbf{G}_{4}^{1}$: $(\sigma_y, 0)$: 2, 2; 4, 2: b.
F^x $\mathbf{G}_{4}^{1}$: $(\sigma_y, 0)$: 2, 2; 4, 2: b.
Q^x $\mathbf{G}_{2}^{1}$: $(\bar{E}, 0)$: 2, 1: a.
Δ^x $\mathbf{G}_{4}^{1}$: $(C_{2a}, 0)$: 2, 2; 4, 2: b.
U^x $\mathbf{G}_{4}^{1}$: $(C_{2a}, 0)$: 2, 2; 4, 2: b.
Y^x $\mathbf{G}_{4}^{1}$: $(C_{2b}, 0)$: 2, 2; 4, 2: b.

121 $I\bar{4}2m$ D_{2d}^{11}

($F1$; $K7$; $M5$; $Z1$.)

Γ $\mathbf{G}_{16}^{14}$: $\{S_{4z}^{+} | 000\}$, $\{C_{2x} | 000\}$: 6, 2; 7, 2: a.
N $\mathbf{G}_{4}^{1} \otimes \mathbf{T}_2$: $\{C_{2y} | 000\}$; $\mathbf{t}_2$: 2, 3; 4, 3: b.
X $\mathbf{G}_{8}^{5} \otimes \mathbf{T}_2$: $\{C_{2z} | 000\}$, $\{\sigma_{db} | 000\}$; $\mathbf{t}_3$: 5, 2: a.
Z $\mathbf{G}_{16}^{14} \otimes \mathbf{T}_2$: $\{S_{4z}^{+} | 000\}$, $\{C_{2x} | 000\}$; $\mathbf{t}_1$ or $\mathbf{t}_2$ or $\mathbf{t}_3$: 6, 2; 7, 2: a.
P $\mathbf{G}_{16}^{14} \otimes \mathbf{T}_4$: $\{S_{4z}^{+} | 000\}$, $\{C_{2x} | 000\}$; $\mathbf{t}_1$ or $\mathbf{t}_2$ or $\mathbf{t}_3$: 6, x; 7, x: a.

Λ^x $\mathbf{G}_{8}^{5}$: $(C_{2z}, 0)$, $(\sigma_{db}, 0)$: 5, 2: a.
V^x $\mathbf{G}_{8}^{5}$: $(C_{2z}, 0)$, $(\sigma_{db}, 0)$: 5, 2: a.
W^x $\mathbf{G}_{8}^{5}$: $(C_{2z}, 0)$, $(\sigma_{db}, 0)$: 5, x: a.
Σ^x $\mathbf{G}_{4}^{1}$: $(C_{2x}, 0)$: 2, 2; 4, 2: b.
F^x $\mathbf{G}_{4}^{1}$: $(C_{2x}, 0)$: 2, 2; 4, 2: b.
Q^x $\mathbf{G}_{4}^{1}$: $(C_{2y}, 0)$: 2, x; 4, x: b.
Δ^x $\mathbf{G}_{4}^{1}$: $(\sigma_{db}, 0)$: 2, 2; 4, 2: b.
U^x $\mathbf{G}_{4}^{1}$: $(\sigma_{db}, 0)$: 2, 2; 4, 2: b.
Y^x $\mathbf{G}_{4}^{1}$: $(\sigma_{da}, 0)$: 2, 2; 4, 2: b.

122 $I\bar{4}2d$ D_{2d}^{12}

($C2$; $F1$; $K7$; $M5$; $S1$; $S7$; $Z1$.)

Γ $\mathbf{G}_{16}^{14}$: $\{S_{4z}^{+} | 000\}$, $\{C_{2x} | \frac{1}{4}\frac{3}{4}\frac{1}{2}\}$: 6, 2; 7, 2: a.
N $\mathbf{G}_{4}^{1} \otimes \mathbf{T}_2$: $\{C_{2y} | \frac{1}{4}\frac{3}{4}\frac{1}{2}\}$; $\mathbf{t}_2$: 2, 3; 4, 3: b.
X $\mathbf{G}_{16}^{8}$: $\{\sigma_{db} | \frac{1}{4}\frac{3}{4}\frac{1}{2}\}$, $\{C_{2z} | 000\}$: 5, 3; 6, 3; 7, 3; 8, 3: a.
Z $\mathbf{G}_{32}^{11}$: $\{S_{4z}^{+} | 000\}$, $\{\sigma_{db} | \frac{1}{4}\frac{3}{4}\frac{1}{2}\}$: 6, 3; 7, 3: a.
P $\mathbf{G}_{64}^{3}$: $\{S_{4z}^{+} | 000\}$, $\{\sigma_{da} | \frac{1}{4}\frac{3}{4}\frac{1}{2}\}$: 13, x; 14, x; 15, x; 16, x; 20, x: a.

Λ^x $\mathbf{G}_{8}^{5}$: $(C_{2z}, 0)$, $(\sigma_{db}, 0)$: 5, 2: a.
V^x $\mathbf{G}_{16}^{8}$: $(C_{2z}, 0)$, $(\sigma_{db}, 0)$: 9, 2: b.
W^x $\mathbf{G}_{16}^{8}$: $(\sigma_{da}, 0)$, $(C_{2z}, 0)$: 5, x; 6, x; 7, x; 8, x: a.
Σ^x $\mathbf{G}_{4}^{1}$: $(C_{2x}, 0)$: 2, 2; 4, 2: b.
F^x $\mathbf{G}_{8}^{2}$: $(C_{2x}, 0)$, $(E, 1)$: 5, 2; 7, 2: a.
Q^x $\mathbf{G}_{16}^{2}$: $(C_{2y}, 0)$, $(\bar{E}, 0)$: 10, x; 14, x: e.
Δ^x $\mathbf{G}_{4}^{1}$: $(\sigma_{db}, 0)$: 2, 2; 4, 2: b.
U^x $\mathbf{G}_{4}^{1}$: $(\sigma_{db}, 0)$: 2, 3; 4, 3: b.
Y^x $\mathbf{G}_{8}^{2}$: $(\sigma_{da}, 0)$, $(E, 1)$: 5, 3; 7, 3: a.

123 *P4/mmm* D_{4h}^1

(*F*1; *K*7; *M*5; *O*2; *Z*1.)

Γ $\mathbf{G}_{32}^9$: $\{C_{4z}^+ \mid 000\}$, $\{C_{2x} \mid 000\}$, $\{I \mid 000\}$: 6, 2; 7, 2; 13, 2; 14, 2: *a*.
M $\mathbf{G}_{32}^9 \otimes \mathbf{T}_2$: $\{C_{4z}^+ \mid 000\}$, $\{C_{2x} \mid 000\}$, $\{I \mid 000\}$; $\mathbf{t}_1$ or $\mathbf{t}_2$: 6, 2; 7, 2; 13, 2; 14, 2: *a*.
Z $\mathbf{G}_{32}^9 \otimes \mathbf{T}_2$: $\{C_{4z}^+ \mid 000\}$, $\{C_{2x} \mid 000\}$, $\{I \mid 000\}$; $\mathbf{t}_3$: 6, 2; 7, 2; 13, 2; 14, 2: *a*.
A $\mathbf{G}_{32}^9 \otimes \mathbf{T}_2$: $\{C_{4z}^+ \mid 000\}$, $\{C_{2x} \mid 000\}$, $\{I \mid 000\}$; $\mathbf{t}_1$ or $\mathbf{t}_2$ or $\mathbf{t}_3$: 6, 2; 7, 2; 13, 2; 14, 2: *a*.
R $\mathbf{G}_{16}^{11} \otimes \mathbf{T}_2$: $\{C_{2z} \mid 000\}$, $\{C_{2y} \mid 000\}$, $\{I \mid 000\}$; $\mathbf{t}_2$ or $\mathbf{t}_3$: 5, 2; 10, 2: *c*.
X $\mathbf{G}_{16}^{11} \otimes \mathbf{T}_2$: $\{C_{2z} \mid 000\}$, $\{C_{2y} \mid 000\}$, $\{I \mid 000\}$; $\mathbf{t}_2$: 5, 2; 10, 2: *c*.

Δ^x $\mathbf{G}_8^5$: $(C_{2y}, 0)$, $(\sigma_x, 0)$: 5, 2: *a*.
U^x $\mathbf{G}_8^5$: $(C_{2y}, 0)$, $(\sigma_x, 0)$: 5, 2: *a*.
Λ^x $\mathbf{G}_{16}^{14}$: $(C_{4z}^+, 0)$, $(\sigma_y, 0)$: 6, 2; 7, 2: *a*.
V^x $\mathbf{G}_{16}^{14}$: $(C_{4z}^+, 0)$, $(\sigma_y, 0)$: 6, 2; 7, 2: *a*.
Σ^x $\mathbf{G}_8^5$: $(C_{2a}, 0)$, $(\sigma_z, 0)$: 5, 2: *a*.
S^x $\mathbf{G}_8^5$: $(C_{2a}, 0)$, $(\sigma_z, 0)$: 5, 2: *a*.
Y^x $\mathbf{G}_8^5$: $(C_{2x}, 0)$, $(\sigma_z, 0)$: 5, 2: *a*.
T^x $\mathbf{G}_8^5$: $(C_{2x}, 0)$, $(\sigma_z, 0)$: 5, 2: *a*.
W^x $\mathbf{G}_8^5$: $(C_{2z}, 0)$, $(\sigma_y, 0)$: 5, 2: *a*.

124 *P4/mcc* D_{4h}^2

(*F*1; *K*7; *M*5; *O*2; *Z*1.)

Γ $\mathbf{G}_{32}^9$: $\{C_{4z}^+ \mid 000\}$, $\{C_{2x} \mid 000\}$, $\{I \mid 00\tfrac{1}{2}\}$: 6, 2; 7, 2; 13, 2; 14, 2: *a*.
M $\mathbf{G}_{32}^9 \otimes \mathbf{T}_2$: $\{C_{4z}^+ \mid 000\}$, $\{C_{2x} \mid 000\}$, $\{I \mid 00\tfrac{1}{2}\}$; $\mathbf{t}_1$ or $\mathbf{t}_2$: 6, 2; 7, 2; 13, 2; 14, 2: *a*.
Z $\mathbf{G}_{64}^1$: $\{I \mid 00\tfrac{1}{2}\}$, $\{\sigma_{db} \mid 00\tfrac{1}{2}\}$, $\{C_{4z}^+ \mid 000\}$: 19, 3; 20, 3; 21, 3; 22, 3: *a*.
A $\mathbf{G}_{64}^1$: $\{I \mid 00\tfrac{1}{2}\}$, $\{\sigma_{db} \mid 00\tfrac{1}{2}\}$, $\{C_{4z}^+ \mid 000\}$: 19, 3; 20, 3; 21, 3; 22, 3: *a*.
R $\mathbf{G}_{32}^7$: $\{\sigma_x \mid 00\tfrac{1}{2}\}$, $\{\sigma_z \mid 00\tfrac{1}{2}\}$, $\{I \mid 00\tfrac{1}{2}\}$: 13, 3; 14, 3: *a*.
X $\mathbf{G}_{16}^{11} \otimes \mathbf{T}_2$: $\{C_{2z} \mid 000\}$, $\{C_{2y} \mid 000\}$, $\{I \mid 00\tfrac{1}{2}\}$; $\mathbf{t}_2$: 5, 2; 10, 2: *c*.

Δ^x $\mathbf{G}_8^5$: $(C_{2y}, 0)$, $(\sigma_x, 0)$: 5, 2: *a*.
U^x $\mathbf{G}_{16}^8$: $(\sigma_z, 0)$, $(C_{2y}, 0)$: 5, 3; 6, 3; 7, 3; 8, 3: *a*.
Λ^x $\mathbf{G}_{16}^{14}$: $(C_{4z}^+, 0)$, $(\sigma_y, 0)$: 6, 2; 7, 2: *a*.
V^x $\mathbf{G}_{16}^{14}$: $(C_{4z}^+, 0)$, $(\sigma_y, 0)$: 6, 2; 7, 2: *a*.
Σ^x $\mathbf{G}_8^5$: $(C_{2a}, 0)$, $(\sigma_z, 0)$: 5, 2: *a*.
S^x $\mathbf{G}_{16}^8$: $(\sigma_z, 0)$, $(C_{2a}, 0)$: 5, 3; 6, 3; 7, 3; 8, 3: *a*.
Y^x $\mathbf{G}_8^5$: $(C_{2x}, 0)$, $(\sigma_z, 0)$: 5, 2: *a*.
T^x $\mathbf{G}_{16}^8$: $(\sigma_z, 0)$, $(C_{2x}, 0)$: 5, 3; 6, 3; 7, 3; 8, 3: *a*.
W^x $\mathbf{G}_8^5$: $(C_{2z}, 0)$, $(\sigma_y, 0)$: 5, 2: *a*.

125 *P4/nbm* D_{4h}^3

(*F*1; *K*7; *M*5; *O*2; *Z*1.)

Γ $\mathbf{G}_{32}^9$: $\{C_{4z}^+ \mid \tfrac{1}{2}\tfrac{1}{2}0\}$, $\{C_{2x} \mid 000\}$, $\{I \mid \tfrac{1}{2}\tfrac{1}{2}0\}$: 6, 2; 7, 2; 13, 2; 14, 2: *a*.
M $\mathbf{G}_{64}^2$: $\{C_{4z}^+ \mid \tfrac{1}{2}\tfrac{1}{2}0\}$, $\{\sigma_x \mid \tfrac{1}{2}\tfrac{1}{2}0\}$, $\{C_{2x} \mid 000\}$: 19, 2: *a*.
Z $\mathbf{G}_{32}^9 \otimes \mathbf{T}_2$: $\{C_{4z}^+ \mid \tfrac{1}{2}\tfrac{1}{2}0\}$, $\{C_{2x} \mid 000\}$, $\{I \mid \tfrac{1}{2}\tfrac{1}{2}0\}$; $\mathbf{t}_3$: 6, 2; 7, 2; 13, 2; 14, 2: *a*.
A $\mathbf{G}_{64}^2$: $\{C_{4z}^+ \mid \tfrac{1}{2}\tfrac{1}{2}0\}$, $\{\sigma_x \mid \tfrac{1}{2}\tfrac{1}{2}0\}$, $\{C_{2x} \mid 000\}$: 19, 2: *a*.
R $\mathbf{G}_{32}^7$: $\{\sigma_x \mid \tfrac{1}{2}\tfrac{1}{2}0\}$, $\{\sigma_y \mid \tfrac{1}{2}\tfrac{1}{2}0\}$, $\{I \mid \tfrac{1}{2}\tfrac{1}{2}0\}$: 13, 3; 14, 3: *a*.
X $\mathbf{G}_{32}^7$: $\{\sigma_x \mid \tfrac{1}{2}\tfrac{1}{2}0\}$, $\{\sigma_y \mid \tfrac{1}{2}\tfrac{1}{2}0\}$, $\{I \mid \tfrac{1}{2}\tfrac{1}{2}0\}$: 13, 3; 14, 3: *a*.

Δ^x $\mathbf{G}_8^5$: $(C_{2y}, 0)$, $(\sigma_x, 0)$: 5, 2: *a*.
U^x $\mathbf{G}_8^5$: $(C_{2y}, 0)$, $(\sigma_x, 0)$: 5, 2: *a*.
Λ^x $\mathbf{G}_{16}^{14}$: $(C_{4z}^+, 0)$, $(\sigma_y, 0)$: 6, 2; 7, 2: *a*.
V^x $\mathbf{G}_{32}^{11}$: $(C_{4z}^+, 0)$, $(\sigma_x, 0)$: 6, 2; 7, 2: *a*.
Σ^x $\mathbf{G}_8^5$: $(C_{2a}, 0)$, $(\sigma_z, 0)$: 5, 2: *a*.
S^x $\mathbf{G}_8^5$: $(C_{2a}, 0)$, $(\sigma_z, 0)$: 5, 2: *a*.
Y^x $\mathbf{G}_{16}^8$: $(\sigma_y, 0)$, $(C_{2x}, 0)$: 5, 3; 6, 3; 7, 3; 8, 3: *a*.
T^x $\mathbf{G}_{16}^8$: $(\sigma_y, 0)$, $(C_{2x}, 0)$: 5, 3; 6, 3; 7, 3; 8, 3: *a*.
W^x $\mathbf{G}_{16}^8$: $(\sigma_y, 0)$, $(C_{2z}, 0)$: 5, 3; 6, 3; 7, 3; 8, 3: *a*.

126 $P4/nnc$ D_{4h}^4

($F1$; $K7$; $M5$; $O2$; $Z1$.)

Γ $\mathbf{G}_{32}^9$: $\{C_{4z}^+ \mid \frac{1}{2}\frac{1}{2}0\}$, $\{C_{2x} \mid 000\}$, $\{I \mid \frac{1}{2}\frac{1}{2}\frac{1}{2}\}$: 6, 2; 7, 2; 13, 2; 14, 2: a.
M $\mathbf{G}_{64}^2$: $\{C_{4z}^+ \mid \frac{1}{2}\frac{1}{2}0\}$, $\{\sigma_x \mid \frac{1}{2}\frac{1}{2}\frac{1}{2}\}$, $\{C_{2x} \mid 000\}$: 19, 2: a.
Z $\mathbf{G}_{64}^1$: $\{I \mid \frac{1}{2}\frac{1}{2}\frac{1}{2}\}$, $\{\sigma_{db} \mid 00\frac{1}{2}\}$, $\{C_{4z}^+ \mid \frac{1}{2}\frac{1}{2}0\}$: 19, 3; 20, 3; 21, 3; 22, 3: a.
A $\mathbf{G}_{64}^2$: $\{C_{4z}^+ \mid \frac{1}{2}\frac{1}{2}0\}$, $\{\sigma_{db} \mid 00\frac{1}{2}\}$, $\{C_{2b} \mid \frac{1}{2}\frac{1}{2}0\}$: 19, 2: a.
R $\mathbf{G}_{32}^7$: $\{\sigma_y \mid \frac{1}{2}\frac{1}{2}\frac{1}{2}\}$, $\{\sigma_x \mid \frac{1}{2}\frac{1}{2}\frac{1}{2}\}$, $\{I \mid \frac{1}{2}\frac{1}{2}\frac{1}{2}\}$: 13, 3; 14, 3: a.
X $\mathbf{G}_{32}^7$: $\{\sigma_x \mid \frac{1}{2}\frac{1}{2}\frac{1}{2}\}$, $\{\sigma_y \mid \frac{1}{2}\frac{1}{2}\frac{1}{2}\}$, $\{I \mid \frac{1}{2}\frac{1}{2}\frac{1}{2}\}$: 13, 3; 14, 3: a.

Δ^x $\mathbf{G}_8^5$: $(C_{2y}, 0)$, $(\sigma_x, 0)$: 5, 2: a.
U^x $\mathbf{G}_{16}^8$: $(\sigma_z, 0)$, $(C_{2y}, 0)$: 5, 3; 6, 3; 7, 3; 8, 3: a.
Λ^x $\mathbf{G}_{16}^{14}$: $(C_{4z}^+, 0)$, $(\sigma_y, 0)$: 6, 2; 7, 2: a.
V^x $\mathbf{G}_{32}^{11}$: $(C_{4z}^+, 0)$, $(\sigma_x, 0)$: 6, 2; 7, 2: a.
Σ^x $\mathbf{G}_8^5$: $(C_{2a}, 0)$, $(\sigma_z, 0)$: 5, 2: a.
S^x $\mathbf{G}_{16}^8$: $(\sigma_z, 0)$, $(C_{2a}, 0)$: 5, 3; 6, 3; 7, 3; 8, 3: a.
Y^x $\mathbf{G}_{16}^8$: $(\sigma_y, 0)$, $(C_{2x}, 0)$: 5, 3; 6, 3; 7, 3; 8, 3: a.
T^x $\mathbf{G}_{16}^8$: $(C_{2x}, 0)$, $(\sigma_z, 0)$: 9, 2: b.
W^x $\mathbf{G}_{16}^8$: $(\sigma_y, 0)$, $(C_{2z}, 0)$: 5, 3; 6, 3; 7, 3; 8, 3: a.

127 $P4/mbm$ D_{4h}^5

($F1$; $K4$; $K7$; $M5$; $O2$; $Z1$.)

Γ $\mathbf{G}_{32}^9$: $\{C_{4z}^+ \mid \frac{1}{2}\frac{1}{2}0\}$, $\{C_{2x} \mid \frac{1}{2}\frac{1}{2}0\}$, $\{I \mid 000\}$: 6, 2; 7, 2; 13, 2; 14, 2: a.
M $\mathbf{G}_{64}^4$: $\{C_{4z}^+ \mid \frac{1}{2}\frac{1}{2}0\}$, $\{C_{2x} \mid \frac{1}{2}\frac{1}{2}0\}$, $\{I \mid 000\}$: 6, 3; 7, 3; 20, 3; 21, 3: a.
Z $\mathbf{G}_{32}^9 \otimes \mathbf{T}_2$: $\{C_{4z}^+ \mid \frac{1}{2}\frac{1}{2}0\}$, $\{C_{2x} \mid \frac{1}{2}\frac{1}{2}0\}$, $\{I \mid 000\}$; $\mathbf{t}_3$: 6, 2; 7, 2; 13, 2; 14, 2: a.
A $\mathbf{G}_{64}^4$: $\{C_{4z}^+ \mid \frac{1}{2}\frac{1}{2}0\}$, $\{C_{2x} \mid \frac{1}{2}\frac{1}{2}0\}$, $\{I \mid 000\}$: 6, 3; 7, 3; 20, 3; 21, 3: a.
R $\mathbf{G}_{32}^7$: $\{\sigma_x \mid \frac{1}{2}\frac{1}{2}0\}$, $\{C_{2z} \mid 000\}$, $\{I \mid 000\}$: 13, 3; 14, 3: a.
X $\mathbf{G}_{32}^7$: $\{\sigma_x \mid \frac{1}{2}\frac{1}{2}0\}$, $\{C_{2z} \mid 000\}$, $\{I \mid 000\}$: 13, 3; 14, 3: a.

Δ^x $\mathbf{G}_8^5$: $(C_{2y}, 0)$, $(\sigma_x, 0)$: 5, 2: a.
U^x $\mathbf{G}_8^5$: $(C_{2y}, 0)$, $(\sigma_x, 0)$: 5, 2: a.
Λ^x $\mathbf{G}_{16}^{14}$: $(C_{4z}^+, 0)$, $(\sigma_y, 0)$: 6, 2; 7, 2: a.
V^x $\mathbf{G}_{32}^{11}$: $(C_{4z}^+, 0)$, $(\sigma_x, 0)$: 6, 3; 7, 3: a.
Σ^x $\mathbf{G}_8^5$: $(C_{2a}, 0)$, $(\sigma_z, 0)$: 5, 2: a.
S^x $\mathbf{G}_8^5$: $(C_{2a}, 0)$, $(\sigma_z, 0)$: 5, 2: a.
Y^x $\mathbf{G}_{16}^8$: $(\sigma_z, 0)$, $(C_{2x}, 0)$: 9, 1: b.
T^x $\mathbf{G}_{16}^8$: $(\sigma_z, 0)$, $(C_{2x}, 0)$: 9, 1: b.
W^x $\mathbf{G}_{16}^8$: $(\sigma_y, 0)$, $(C_{2z}, 0)$: 5, 3; 6, 3; 7, 3; 8, 3: a.

128 *P4/mnc* D_{4h}^{6}

(*F*1; *K*7; *M*5; *O*2; *Z*1.)

Γ $\mathbf{G}_{32}^{9}$: $\{C_{4z}^{+}|\frac{1}{2}\frac{1}{2}0\}$, $\{C_{2x}|\frac{1}{2}\frac{1}{2}0\}$, $\{I|00\frac{1}{2}\}$: 6, 2; 7, 2; 13, 2; 14, 2: *a*.
M $\mathbf{G}_{64}^{4}$: $\{C_{4z}^{+}|\frac{1}{2}\frac{1}{2}0\}$, $\{C_{2x}|\frac{1}{2}\frac{1}{2}0\}$, $\{I|00\frac{1}{2}\}$: 6, 3; 7, 3; 20, 3; 21, 3: *a*.
Z $\mathbf{G}_{64}^{1}$: $\{I|00\frac{1}{2}\}$, $\{\sigma_{db}|00\frac{1}{2}\}$, $\{C_{4z}^{+}|\frac{1}{2}\frac{1}{2}0\}$: 19, 3; 20, 3; 21, 3; 22, 3: *a*.
A $\mathbf{G}_{64}^{1}$: $\{I|00\frac{1}{2}\}$, $\{\sigma_{db}|00\frac{1}{2}\}$, $\{S_{4z}^{+}|\frac{1}{2}\frac{1}{2}\frac{1}{2}\}$: 19, 3; 20, 3; 21, 3; 22, 3: *a*.
R $\mathbf{G}_{32}^{14}$: $\{C_{2y}|\frac{1}{2}\frac{1}{2}0\}$, $\{C_{2z}|000\}$, $\{I|00\frac{1}{2}\}$: 5, 3; 6, 3; 7, 3; 8, 3; 15, 3; 16, 3; 17, 3; 18, 3: *a*.
X $\mathbf{G}_{32}^{7}$: $\{\sigma_{x}|\frac{1}{2}\frac{1}{2}\frac{1}{2}\}$, $\{C_{2z}|000\}$, $\{I|00\frac{1}{2}\}$: 13, 3; 14, 3: *a*.

Δ^x $\mathbf{G}_{8}^{5}$: $(C_{2y}, 0)$, $(\sigma_x, 0)$: 5, 2: *a*.
U^x $\mathbf{G}_{16}^{8}$: $(\sigma_z, 0)$, $(C_{2y}, 0)$: 5, 3; 6, 3; 7, 3; 8, 3: *a*.
Λ^x $\mathbf{G}_{16}^{14}$: $(C_{4z}^{+}, 0)$, $(\sigma_y, 0)$: 6, 2; 7, 2: *a*.
V^x $\mathbf{G}_{32}^{11}$: $(C_{4z}^{+}, 0)$, $(\sigma_x, 0)$: 6, 3; 7, 3: *a*.
Σ^x $\mathbf{G}_{8}^{5}$: $(C_{2a}, 0)$, $(\sigma_z, 0)$: 5, 2: *a*.
S^x $\mathbf{G}_{16}^{8}$: $(\sigma_z, 0)$, $(C_{2a}, 0)$: 5, 3; 6, 3; 7, 3; 8, 3: *a*.
Y^x $\mathbf{G}_{16}^{8}$: $(\sigma_z, 0)$, $(C_{2x}, 0)$: 9, 1: *b*.
T^x $\mathbf{G}_{16}^{11}$: $(C_{2x}, 0)$, $(\sigma_y, 0)$, $(E, 1)$: 6, 3; 7, 3; 8, 3; 9, 3: *b*.
W^x $\mathbf{G}_{16}^{8}$: $(\sigma_y, 0)$, $(C_{2z}, 0)$: 5, 3; 6, 3; 7, 3; 8, 3: *a*.

129 *P4/nmm* D_{4h}^{7}

(*F*1; *K*7; *M*5; *O*2; *Z*1.)

Γ $\mathbf{G}_{32}^{9}$: $\{C_{4z}^{+}|000\}$, $\{C_{2x}|\frac{1}{2}\frac{1}{2}0\}$, $\{I|\frac{1}{2}\frac{1}{2}0\}$: 6, 2; 7, 2; 13, 2; 14, 2: *a*.
M $\mathbf{G}_{64}^{2}$: $\{C_{4z}^{+}|000\}$, $\{C_{2x}|\frac{1}{2}\frac{1}{2}0\}$, $\{\sigma_x|000\}$: 19, 2: *a*.
Z $\mathbf{G}_{32}^{9}\otimes\mathbf{T}_2$: $\{C_{4z}^{+}|000\}$, $\{C_{2x}|\frac{1}{2}\frac{1}{2}0\}$, $\{I|\frac{1}{2}\frac{1}{2}0\}$; $\mathbf{t}_3$: 6, 2; 7, 2; 13, 2; 14, 2: *a*.
A $\mathbf{G}_{64}^{2}$: $\{C_{4z}^{+}|000\}$, $\{C_{2x}|\frac{1}{2}\frac{1}{2}0\}$, $\{\sigma_x|000\}$: 19, 2: *a*.
R $\mathbf{G}_{32}^{7}$: $\{C_{2y}|\frac{1}{2}\frac{1}{2}0\}$, $\{C_{2x}|\frac{1}{2}\frac{1}{2}0\}$, $\{I|\frac{1}{2}\frac{1}{2}0\}$: 13, 3; 14, 3: *a*.
X $\mathbf{G}_{32}^{7}$: $\{C_{2y}|\frac{1}{2}\frac{1}{2}0\}$, $\{C_{2x}|\frac{1}{2}\frac{1}{2}0\}$, $\{I|\frac{1}{2}\frac{1}{2}0\}$: 13, 3; 14, 3: *a*.

Δ^x $\mathbf{G}_{8}^{5}$: $(C_{2y}, 0)$, $(\sigma_x, 0)$: 5, 2: *a*.
U^x $\mathbf{G}_{8}^{5}$: $(C_{2y}, 0)$, $(\sigma_x, 0)$: 5, 2: *a*.
Λ^x $\mathbf{G}_{16}^{14}$: $(C_{4z}^{+}, 0)$, $(\sigma_y, 0)$: 6, 2; 7, 2: *a*.
V^x $\mathbf{G}_{16}^{14}$: $(C_{4z}^{+}, 0)$, $(\sigma_y, 0)$: 6, 3; 7, 3: *a*.
Σ^x $\mathbf{G}_{8}^{5}$: $(C_{2a}, 0)$, $(\sigma_z, 0)$: 5, 2: *a*.
S^x $\mathbf{G}_{8}^{5}$: $(C_{2a}, 0)$, $(\sigma_z, 0)$: 5, 2: *a*.
Y^x $\mathbf{G}_{16}^{8}$: $(C_{2x}, 0)$, $(\sigma_y, 0)$: 5, 3; 6, 3; 7, 3; 8, 3: *a*.
T^x $\mathbf{G}_{16}^{8}$: $(C_{2x}, 0)$, $(\sigma_y, 0)$: 5, 3; 6, 3; 7, 3; 8, 3: *a*.
W^x $\mathbf{G}_{8}^{5}$: $(C_{2z}, 0)$, $(\sigma_y, 0)$: 5, 1: *a*.

130 $P4/ncc$ D_{4h}^{8}

($F1$; $K7$; $M5$; $O2$; $Z1$.)

Γ $\mathbf{G}_{32}^{9}$: $\{C_{4z}^{+}\mid 000\}$, $\{C_{2x}\mid \frac{1}{2}\frac{1}{2}0\}$, $\{I\mid \frac{1}{2}\frac{1}{2}\frac{1}{2}\}$: 6, 2; 7, 2; 13, 2; 14, 2: *a*.
M $\mathbf{G}_{64}^{2}$: $\{C_{4z}^{+}\mid 000\}$, $\{C_{2x}\mid \frac{1}{2}\frac{1}{2}0\}$, $\{\sigma_{x}\mid 00\frac{1}{2}\}$: 19, 2: *a*.
Z $\mathbf{G}_{64}^{1}$: $\{I\mid \frac{1}{2}\frac{1}{2}\frac{1}{2}\}$, $\{\sigma_{db}\mid 00\frac{1}{2}\}$, $\{C_{4z}^{+}\mid 000\}$: 19, 3; 20, 3; 21, 3; 22, 3: *a*.
A $\mathbf{G}_{64}^{5}$: $\{C_{4z}^{+}\mid 000\}$, $\{\sigma_{db}\mid 00\frac{1}{2}\}$, $\{I\mid \frac{1}{2}\frac{1}{2}\frac{1}{2}\}$: 19, 1: *a*.
R $\mathbf{G}_{32}^{7}$: $\{C_{2x}\mid \frac{1}{2}\frac{1}{2}0\}$, $\{\sigma_{y}\mid 00\frac{1}{2}\}$, $\{I\mid \frac{1}{2}\frac{1}{2}\frac{1}{2}\}$: 9, 1; 10, 1: *b*.
X $\mathbf{G}_{32}^{7}$: $\{C_{2y}\mid \frac{1}{2}\frac{1}{2}0\}$, $\{C_{2x}\mid \frac{1}{2}\frac{1}{2}0\}$, $\{I\mid \frac{1}{2}\frac{1}{2}\frac{1}{2}\}$: 13, 3; 14, 3: *a*.

Δ^{x} $\mathbf{G}_{8}^{5}$: $(C_{2y}, 0)$, $(\sigma_{x}, 0)$: 5, 2: *a*.
U^{x} $\mathbf{G}_{16}^{8}$: $(\sigma_{z}, 0)$, $(C_{2y}, 0)$: 5, 3; 6, 3; 7, 3; 8, 3: *a*.
Λ^{x} $\mathbf{G}_{16}^{14}$: $(C_{4z}^{+}, 0)$, $(\sigma_{y}, 0)$: 6, 2; 7, 2: *a*.
V^{x} $\mathbf{G}_{16}^{14}$: $(C_{4z}^{+}, 0)$, $(\sigma_{y}, 0)$: 6, 3; 7, 3: *a*.
Σ^{x} $\mathbf{G}_{8}^{5}$: $(C_{2a}, 0)$, $(\sigma_{z}, 0)$: 5, 2: *a*.
S^{x} $\mathbf{G}_{16}^{8}$: $(\sigma_{z}, 0)$, $(C_{2a}, 0)$: 5, 3; 6, 3; 7, 3; 8, 3: *a*.
Y^{x} $\mathbf{G}_{16}^{8}$: $(C_{2x}, 0)$, $(\sigma_{y}, 0)$: 5, 3; 6, 3; 7, 3; 8, 3: *a*.
T^{x} $\mathbf{G}_{16}^{8}$: $(\sigma_{y}, 0)$, $(\sigma_{z}, 0)$: 9, 1: *b*.
W^{x} $\mathbf{G}_{8}^{5}$: $(C_{2z}, 0)$, $(\sigma_{y}, 0)$: 5, 1: *a*.

131 $P4_2/mmc$ D_{4h}^{9}

($F1$; $K7$; $M5$; $O2$; $S8$‡; $Z1$.)

Γ $\mathbf{G}_{32}^{9}$: $\{C_{4z}^{+}\mid 00\frac{1}{2}\}$, $\{C_{2x}\mid 000\}$, $\{I\mid 000\}$: 6, 2; 7, 2; 13, 2; 14, 2: *a*.
M $\mathbf{G}_{32}^{9}\otimes\mathbf{T}_{2}$: $\{C_{4z}^{+}\mid 00\frac{1}{2}\}$, $\{C_{2x}\mid 000\}$, $\{I\mid 000\}$; $\mathbf{t}_1$ or $\mathbf{t}_2$: 6, 2; 7, 2; 13, 2; 14, 2: *a*.
Z $\mathbf{G}_{64}^{2}$: $\{C_{4z}^{+}\mid 00\frac{1}{2}\}$, $\{\sigma_{db}\mid 00\frac{1}{2}\}$, $\{C_{2b}\mid 00\frac{1}{2}\}$: 19, 2: *a*.
A $\mathbf{G}_{64}^{2}$: $\{C_{4z}^{+}\mid 00\frac{1}{2}\}$, $\{\sigma_{db}\mid 00\frac{1}{2}\}$, $\{C_{2b}\mid 00\frac{1}{2}\}$: 19, 2: *a*.
R $\mathbf{G}_{16}^{11}\otimes\mathbf{T}_{2}$: $\{C_{2z}\mid 000\}$, $\{C_{2y}\mid 000\}$, $\{I\mid 000\}$; $\mathbf{t}_2$ or $\mathbf{t}_3$: 5, 2; 10, 2: *c*.
X $\mathbf{G}_{16}^{11}\otimes\mathbf{T}_{2}$: $\{C_{2z}\mid 000\}$, $\{C_{2y}\mid 000\}$, $\{I\mid 000\}$; $\mathbf{t}_2$: 5, 2; 10, 2: *c*.

Δ^{x} $\mathbf{G}_{8}^{5}$: $(C_{2y}, 0)$, $(\sigma_{x}, 0)$: 5, 2: *a*.
U^{x} $\mathbf{G}_{8}^{5}$: $(C_{2y}, 0)$, $(\sigma_{x}, 0)$: 5, 2: *a*.
Λ^{x} $\mathbf{G}_{16}^{14}$: $(C_{4z}^{+}, 0)$, $(\sigma_{y}, 0)$: 6, 2; 7, 2: *a*.
V^{x} $\mathbf{G}_{16}^{14}$: $(C_{4z}^{+}, 0)$, $(\sigma_{y}, 0)$: 6, 2; 7, 2: *a*.
Σ^{x} $\mathbf{G}_{8}^{5}$: $(C_{2a}, 0)$, $(\sigma_{z}, 0)$: 5, 2: *a*.
S^{x} $\mathbf{G}_{16}^{8}$: $(C_{2a}, 0)$, $(\sigma_{z}, 0)$: 5, 3; 6, 3; 7, 3; 8, 3: *a*.
Y^{x} $\mathbf{G}_{8}^{5}$: $(C_{2x}, 0)$, $(\sigma_{z}, 0)$: 5, 2: *a*.
T^{x} $\mathbf{G}_{8}^{5}$: $(C_{2x}, 0)$, $(\sigma_{z}, 0)$: 5, 2: *a*.
W^{x} $\mathbf{G}_{8}^{5}$: $(C_{2z}, 0)$, $(\sigma_{y}, 0)$: 5, 2: *a*.

‡ The tables given by Slater (1965b) for $P4_2/mnm$ (D_{4h}^{14}) in fact apply to $P4_2/mmc$ (D_{4h}^{9}) (see Gay, Albers, and Arlinghaus (1968)).

132 $P4_2/mcm$ D_{4h}^{10}

($F1$; $K7$; $M5$; $O2$; $Z1$.)

Γ $\mathbf{G}_{32}^{9}$: $\{C_{4z}^{+}\,|\,00\tfrac{1}{2}\}$, $\{C_{2x}\,|\,000\}$, $\{I\,|\,00\tfrac{1}{2}\}$: 6, 2; 7, 2; 13, 2; 14, 2: *a*.
M $\mathbf{G}_{32}^{9}\otimes\mathbf{T}_2$: $\{C_{4z}^{+}\,|\,00\tfrac{1}{2}\}$, $\{C_{2x}\,|\,000\}$, $\{I\,|\,00\tfrac{1}{2}\}$; $\mathbf{t}_1$ or $\mathbf{t}_2$: 6, 2; 7, 2; 13, 2; 14, 2: *a*.
Z $\mathbf{G}_{64}^{2}$: $\{C_{4z}^{+}\,|\,00\tfrac{1}{2}\}$, $\{\sigma_x\,|\,00\tfrac{1}{2}\}$, $\{C_{2x}\,|\,000\}$: 19, 2: *a*.
A $\mathbf{G}_{64}^{2}$: $\{C_{4z}^{+}\,|\,00\tfrac{1}{2}\}$, $\{\sigma_x\,|\,00\tfrac{1}{2}\}$, $\{C_{2x}\,|\,000\}$: 19, 2: *a*.
R $\mathbf{G}_{32}^{7}$: $\{\sigma_x\,|\,00\tfrac{1}{2}\}$, $\{\sigma_z\,|\,00\tfrac{1}{2}\}$, $\{I\,|\,00\tfrac{1}{2}\}$: 13, 3; 14, 3: *a*.
X $\mathbf{G}_{16}^{11}\otimes\mathbf{T}_2$: $\{C_{2z}\,|\,000\}$, $\{C_{2y}\,|\,000\}$, $\{I\,|\,00\tfrac{1}{2}\}$; $\mathbf{t}_2$: 5, 2; 10, 2: *c*.

Δ^x $\mathbf{G}_{8}^{5}$: $(C_{2y}, 0)$, $(\sigma_x, 0)$: 5, 2: *a*.
U^x $\mathbf{G}_{16}^{8}$: $(\sigma_z, 0)$, $(C_{2y}, 0)$: 5, 3; 6, 3; 7, 3; 8, 3: *a*.
Λ^x $\mathbf{G}_{16}^{14}$: $(C_{4z}^{+}, 0)$, $(\sigma_y, 0)$: 6, 2; 7, 2: *a*.
V^x $\mathbf{G}_{16}^{14}$: $(C_{4z}^{+}, 0)$, $(\sigma_y, 0)$: 6, 2; 7, 2: *a*.
Σ^x $\mathbf{G}_{8}^{5}$: $(C_{2a}, 0)$, $(\sigma_z, 0)$: 5, 2: *a*.
S^x $\mathbf{G}_{16}^{8}$: $(\sigma_{db}, 0)$, $(C_{2a}, 0)$: 9, 2: *b*.
Y^x $\mathbf{G}_{8}^{5}$: $(C_{2x}, 0)$, $(\sigma_z, 0)$: 5, 2: *a*.
T^x $\mathbf{G}_{16}^{8}$: $(\sigma_z, 0)$, $(C_{2x}, 0)$: 5, 3; 6, 3; 7, 3; 8, 3: *a*.
W^x $\mathbf{G}_{8}^{5}$: $(C_{2z}, 0)$, $(\sigma_y, 0)$: 5, 2: *a*.

133 $P4_2/nbc$ D_{4h}^{11}

($F1$; $K7$; $M5$; $O2$; $Z1$.)

Γ $\mathbf{G}_{32}^{9}$: $\{C_{4z}^{+}\,|\,\tfrac{1}{2}\tfrac{1}{2}\tfrac{1}{2}\}$, $\{C_{2x}\,|\,000\}$, $\{I\,|\,\tfrac{1}{2}\tfrac{1}{2}0\}$: 6, 2; 7, 2; 13, 2; 14, 2: *a*.
M $\mathbf{G}_{64}^{2}$: $\{C_{4z}^{+}\,|\,\tfrac{1}{2}\tfrac{1}{2}\tfrac{1}{2}\}$, $\{\sigma_x\,|\,\tfrac{1}{2}\tfrac{1}{2}0\}$, $\{C_{2x}\,|\,000\}$: 19, 2: *a*.
Z $\mathbf{G}_{64}^{2}$: $\{C_{4z}^{+}\,|\,\tfrac{1}{2}\tfrac{1}{2}\tfrac{1}{2}\}$, $\{\sigma_{db}\,|\,00\tfrac{1}{2}\}$, $\{C_{2b}\,|\,\tfrac{1}{2}\tfrac{1}{2}\tfrac{1}{2}\}$: 19, 2: *a*.
A $\mathbf{G}_{64}^{1}$: $\{I\,|\,\tfrac{1}{2}\tfrac{1}{2}0\}$, $\{\sigma_x\,|\,\tfrac{1}{2}\tfrac{1}{2}0\}$, $\{C_{4z}^{+}\,|\,\tfrac{1}{2}\tfrac{1}{2}\tfrac{1}{2}\}$: 19, 3; 20, 3; 21, 3; 22, 3: *a*.
R $\mathbf{G}_{32}^{7}$: $\{\sigma_x\,|\,\tfrac{1}{2}\tfrac{1}{2}0\}$, $\{\sigma_y\,|\,\tfrac{1}{2}\tfrac{1}{2}0\}$, $\{I\,|\,\tfrac{1}{2}\tfrac{1}{2}0\}$: 13, 3; 14, 3: *a*.
X $\mathbf{G}_{32}^{7}$: $\{\sigma_x\,|\,\tfrac{1}{2}\tfrac{1}{2}0\}$, $\{\sigma_y\,|\,\tfrac{1}{2}\tfrac{1}{2}0\}$, $\{I\,|\,\tfrac{1}{2}\tfrac{1}{2}0\}$: 13, 3; 14, 3: *a*.

Δ^x $\mathbf{G}_{8}^{5}$: $(C_{2y}, 0)$, $(\sigma_x, 0)$: 5, 2: *a*.
U^x $\mathbf{G}_{8}^{5}$: $(C_{2y}, 0)$, $(\sigma_x, 0)$: 5, 2: *a*.
Λ^x $\mathbf{G}_{16}^{14}$: $(C_{4z}^{+}, 0)$, $(\sigma_y, 0)$: 6, 2; 7, 2: *a*.
V^x $\mathbf{G}_{32}^{11}$: $(C_{4z}^{+}, 0)$, $(\sigma_x, 0)$: 6, 2; 7, 2: *a*.
Σ^x $\mathbf{G}_{8}^{5}$: $(C_{2a}, 0)$, $(\sigma_z, 0)$: 5, 2: *a*.
S^x $\mathbf{G}_{16}^{8}$: $(C_{2a}, 0)$, $(\sigma_z, 0)$: 5, 3; 6, 3; 7, 3; 8, 3: *a*.
Y^x $\mathbf{G}_{16}^{8}$: $(\sigma_y, 0)$, $(C_{2x}, 0)$: 5, 3; 6, 3; 7, 3; 8, 3: *a*.
T^x $\mathbf{G}_{16}^{8}$: $(\sigma_y, 0)$, $(C_{2x}, 0)$: 5, 3; 6, 3; 7, 3; 8, 3: *a*.
W^x $\mathbf{G}_{16}^{8}$: $(\sigma_y, 0)$, $(C_{2z}, 0)$: 5, 3; 6, 3; 7, 3; 8, 3: *a*.

134 $P4_2/nnm$ D_{4h}^{12}

($F1$; $K7$; $M5$; $O2$; $Z1$.)

Γ $\mathbf{G}_{32}^{9}$: $\{C_{4z}^{+}\,|\,\frac{1}{2}\frac{1}{2}\frac{1}{2}\}$, $\{C_{2x}\,|\,000\}$, $\{I\,|\,\frac{1}{2}\frac{1}{2}\frac{1}{2}\}$: 6, 2; 7, 2; 13, 2; 14, 2: *a*.
M $\mathbf{G}_{64}^{2}$: $\{C_{4z}^{+}\,|\,\frac{1}{2}\frac{1}{2}\frac{1}{2}\}$, $\{\sigma_x\,|\,\frac{1}{2}\frac{1}{2}\frac{1}{2}\}$, $\{C_{2x}\,|\,000\}$: 19, 2: *a*.
Z $\mathbf{G}_{64}^{2}$: $\{C_{4z}^{+}\,|\,\frac{1}{2}\frac{1}{2}\frac{1}{2}\}$, $\{\sigma_x\,|\,\frac{1}{2}\frac{1}{2}\frac{1}{2}\}$, $\{C_{2x}\,|\,000\}$: 19, 2: *a*.
A $\mathbf{G}_{32}^{9}\otimes\mathbf{T}_2$: $\{C_{4z}^{+}\,|\,\frac{1}{2}\frac{1}{2}\frac{1}{2}\}$, $\{C_{2x}\,|\,000\}$, $\{I\,|\,\frac{1}{2}\frac{1}{2}\frac{1}{2}\}$; $\mathbf{t}_1$ or $\mathbf{t}_2$ or $\mathbf{t}_3$: 6, 2; 7, 2; 13, 2; 14, 2: *a*.
R $\mathbf{G}_{32}^{7}$: $\{\sigma_y\,|\,\frac{1}{2}\frac{1}{2}\frac{1}{2}\}$, $\{\sigma_x\,|\,\frac{1}{2}\frac{1}{2}\frac{1}{2}\}$, $\{I\,|\,\frac{1}{2}\frac{1}{2}\frac{1}{2}\}$: 13, 3; 14, 3: *a*.
X $\mathbf{G}_{32}^{7}$: $\{\sigma_x\,|\,\frac{1}{2}\frac{1}{2}\frac{1}{2}\}$, $\{\sigma_y\,|\,\frac{1}{2}\frac{1}{2}\frac{1}{2}\}$, $\{I\,|\,\frac{1}{2}\frac{1}{2}\frac{1}{2}\}$: 13, 3; 14, 3: *a*.

Δ^x $\mathbf{G}_{8}^{5}$: $(C_{2y}, 0)$, $(\sigma_x, 0)$: 5, 2: *a*.
U^x $\mathbf{G}_{16}^{8}$: $(\sigma_z, 0)$, $(C_{2y}, 0)$: 5, 3; 6, 3; 7, 3; 8, 3: *a*.
Λ^x $\mathbf{G}_{16}^{14}$: $(C_{4z}^{+}, 0)$, $(\sigma_y, 0)$: 6, 2; 7, 2: *a*.
V^x $\mathbf{G}_{32}^{11}$: $(C_{4z}^{+}, 0)$, $(\sigma_x, 0)$: 6, 2; 7, 2: *a*.
Σ^x $\mathbf{G}_{8}^{5}$: $(C_{2a}, 0)$, $(\sigma_z, 0)$: 5, 2: *a*.
S^x $\mathbf{G}_{16}^{8}$: $(\sigma_{db}, 0)$, $(C_{2a}, 0)$: 9, 2: *b*.
Y^x $\mathbf{G}_{16}^{8}$: $(\sigma_y, 0)$, $(C_{2x}, 0)$: 5, 3; 6, 3; 7, 3; 8, 3: *a*.
T^x $\mathbf{G}_{16}^{8}$: $(C_{2x}, 0)$, $(\sigma_z, 0)$: 9, 2: *b*.
W^x $\mathbf{G}_{16}^{8}$: $(\sigma_y, 0)$, $(C_{2z}, 0)$: 5, 3; 6, 3; 7, 3; 8, 3: *a*.

135 $P4_2/mbc$ D_{4h}^{13}

($F1$; $K7$; $M5$; $O2$; $Z1$.)

Γ $\mathbf{G}_{32}^{9}$: $\{C_{4z}^{+}\,|\,\frac{1}{2}\frac{1}{2}\frac{1}{2}\}$, $\{C_{2x}\,|\,\frac{1}{2}\frac{1}{2}0\}$, $\{I\,|\,000\}$: 6, 2; 7, 2; 13, 2; 14, 2: *a*.
M $\mathbf{G}_{64}^{4}$: $\{C_{4z}^{+}\,|\,\frac{1}{2}\frac{1}{2}\frac{1}{2}\}$, $\{C_{2x}\,|\,\frac{1}{2}\frac{1}{2}0\}$, $\{I\,|\,000\}$: 6, 3; 7, 3; 20, 3; 21, 3: *a*.
Z $\mathbf{G}_{64}^{2}$: $\{C_{4z}^{+}\,|\,\frac{1}{2}\frac{1}{2}\frac{1}{2}\}$, $\{\sigma_{db}\,|\,00\frac{1}{2}\}$, $\{C_{2b}\,|\,00\frac{1}{2}\}$: 19, 2: *a*.
A $\mathbf{G}_{64}^{5}$: $\{C_{4z}^{+}\,|\,\frac{1}{2}\frac{1}{2}\frac{1}{2}\}$, $\{\sigma_{db}\,|\,00\frac{1}{2}\}$, $\{I\,|\,000\}$: 19, 1: *a*.
R $\mathbf{G}_{32}^{7}$: $\{\sigma_x\,|\,\frac{1}{2}\frac{1}{2}0\}$, $\{C_{2z}\,|\,000\}$, $\{I\,|\,000\}$: 13, 3; 14, 3: *a*.
X $\mathbf{G}_{32}^{7}$: $\{\sigma_x\,|\,\frac{1}{2}\frac{1}{2}0\}$, $\{C_{2z}\,|\,000\}$, $\{I\,|\,000\}$: 13, 3; 14, 3: *a*.

Δ^x $\mathbf{G}_{8}^{5}$: $(C_{2y}, 0)$, $(\sigma_x, 0)$: 5, 2: *a*.
U^x $\mathbf{G}_{8}^{5}$: $(C_{2y}, 0)$, $(\sigma_x, 0)$: 5, 2: *a*.
Λ^x $\mathbf{G}_{16}^{14}$: $(C_{4z}^{+}, 0)$, $(\sigma_y, 0)$: 6, 2; 7, 2: *a*.
V^x $\mathbf{G}_{32}^{11}$: $(C_{4z}^{+}, 0)$, $(\sigma_x, 0)$: 6, 3; 7, 3: *a*.
Σ^x $\mathbf{G}_{8}^{5}$: $(C_{2a}, 0)$, $(\sigma_z, 0)$: 5, 2: *a*.
S^x $\mathbf{G}_{16}^{8}$: $(C_{2a}, 0)$, $(\sigma_z, 0)$: 5, 3; 6, 3; 7, 3; 8, 3: *a*.
Y^x $\mathbf{G}_{16}^{8}$: $(\sigma_z, 0)$, $(C_{2x}, 0)$: 9, 1: *b*.
T^x $\mathbf{G}_{16}^{8}$: $(\sigma_z, 0)$, $(C_{2x}, 0)$: 9, 1: *b*.
W^x $\mathbf{G}_{16}^{8}$: $(\sigma_y, 0)$, $(C_{2z}, 0)$: 5, 3; 6, 3; 7, 3; 8, 3: *a*.

136 $P4_2/mnm$ D_{4h}^{14}

($D2$; $F1$; $G3$; $K7$; $M5$; $O1$; $O2$; $S8$‡; $Z1$.)

Γ $\mathbf{G}_{32}^{9}$: $\{C_{4z}^{+}|\frac{1}{2}\frac{1}{2}\frac{1}{2}\}, \{C_{2x}|\frac{1}{2}\frac{1}{2}0\}, \{I|00\frac{1}{2}\}$: 6, 2; 7, 2; 13, 2; 14, 2: *a*.
M $\mathbf{G}_{64}^{4}$: $\{C_{4z}^{+}|\frac{1}{2}\frac{1}{2}\frac{1}{2}\}, \{C_{2x}|\frac{1}{2}\frac{1}{2}0\}, \{I|00\frac{1}{2}\}$: 6, 3; 7, 3; 20, 3; 21, 3: *a*.
Z $\mathbf{G}_{64}^{2}$: $\{C_{4z}^{+}|\frac{1}{2}\frac{1}{2}\frac{1}{2}\}, \{\sigma_{x}|\frac{1}{2}\frac{1}{2}\frac{1}{2}\}, \{C_{2x}|\frac{1}{2}\frac{1}{2}0\}$: 19, 2: *a*.
A $\mathbf{G}_{64}^{2}$: $\{C_{4z}^{+}|\frac{1}{2}\frac{1}{2}\frac{1}{2}\}, \{C_{2x}|\frac{1}{2}\frac{1}{2}0\}, \{\sigma_{x}|\frac{1}{2}\frac{1}{2}\frac{1}{2}\}$: 19, 2: *a*.
R $\mathbf{G}_{32}^{14}$: $\{C_{2y}|\frac{1}{2}\frac{1}{2}0\}, \{C_{2z}|000\}, \{I|00\frac{1}{2}\}$: 5, 3; 6, 3; 7, 3; 8, 3; 15, 3; 16, 3; 17, 3; 18, 3: *a*.
X $\mathbf{G}_{32}^{7}$: $\{\sigma_{x}|\frac{1}{2}\frac{1}{2}\frac{1}{2}\}, \{C_{2z}|000\}, \{I|00\frac{1}{2}\}$: 13, 3; 14, 3: *a*.

Δ^{x} $\mathbf{G}_{8}^{5}$: $(C_{2y}, 0), (\sigma_{x}, 0)$: 5, 2: *a*.
U^{x} $\mathbf{G}_{16}^{8}$: $(\sigma_{z}, 0), (C_{2y}, 0)$: 5, 3; 6, 3; 7, 3; 8, 3: *a*.
Λ^{x} $\mathbf{G}_{16}^{14}$: $(C_{4z}^{+}, 0), (\sigma_{y}, 0)$: 6, 2; 7, 2: *a*.
V^{x} $\mathbf{G}_{32}^{11}$: $(C_{4z}^{+}, 0), (\sigma_{x}, 0)$: 6, 3; 7, 3: *a*.
Σ^{x} $\mathbf{G}_{8}^{5}$: $(C_{2a}, 0), (\sigma_{z}, 0)$: 5, 2: *a*.
S^{x} $\mathbf{G}_{16}^{8}$: $(\sigma_{db}, 0), (C_{2a}, 0)$: 9, 2: *b*.
Y^{x} $\mathbf{G}_{16}^{8}$: $(\sigma_{z}, 0), (C_{2x}, 0)$: 9, 1: *b*.
T^{x} $\mathbf{G}_{16}^{11}$: $(C_{2x}, 0), (\sigma_{z}, 0), (\bar{E}, 0)$: 6, 3; 7, 3; 8, 3; 9, 3: *b*.
W^{x} $\mathbf{G}_{16}^{8}$: $(\sigma_{y}, 0), (C_{2z}, 0)$: 5, 3; 6, 3; 7, 3; 8, 3: *a*.

137 $P4_2/nmc$ D_{4h}^{15}

($F1$; $K7$; $M5$; $O2$; $Z1$.)

Γ $\mathbf{G}_{32}^{9}$: $\{C_{4z}^{+}|00\frac{1}{2}\}, \{C_{2x}|\frac{1}{2}\frac{1}{2}0\}, \{I|\frac{1}{2}\frac{1}{2}0\}$: 6, 2; 7, 2; 13, 2; 14, 2: *a*.
M $\mathbf{G}_{64}^{2}$: $\{C_{4z}^{+}|00\frac{1}{2}\}, \{C_{2x}|\frac{1}{2}\frac{1}{2}0\}, \{\sigma_{x}|000\}$: 19, 2: *a*.
Z $\mathbf{G}_{64}^{2}$: $\{C_{4z}^{+}|00\frac{1}{2}\}, \{\sigma_{db}|00\frac{1}{2}\}, \{C_{2b}|\frac{1}{2}\frac{1}{2}\frac{1}{2}\}$: 19, 2: *a*.
A $\mathbf{G}_{64}^{1}$: $\{I|\frac{1}{2}\frac{1}{2}0\}, \{\sigma_{db}|00\frac{1}{2}\}, \{S_{4z}^{+}|\frac{1}{2}\frac{1}{2}\frac{1}{2}\}$: 19, 3; 20, 3; 21, 3; 22, 3: *a*.
R $\mathbf{G}_{32}^{7}$: $\{C_{2y}|\frac{1}{2}\frac{1}{2}0\}, \{C_{2x}|\frac{1}{2}\frac{1}{2}0\}, \{I|\frac{1}{2}\frac{1}{2}0\}$: 13, 3; 14, 3: *a*.
X $\mathbf{G}_{32}^{7}$: $\{C_{2y}|\frac{1}{2}\frac{1}{2}0\}, \{C_{2x}|\frac{1}{2}\frac{1}{2}0\}, \{I|\frac{1}{2}\frac{1}{2}0\}$: 13, 3; 14, 3: *a*.

Δ^{x} $\mathbf{G}_{8}^{5}$: $(C_{2y}, 0), (\sigma_{x}, 0)$: 5, 2: *a*.
U^{x} $\mathbf{G}_{8}^{5}$: $(C_{2y}, 0), (\sigma_{x}, 0)$: 5, 2: *a*.
Λ^{x} $\mathbf{G}_{16}^{14}$: $(C_{4z}^{+}, 0), (\sigma_{y}, 0)$: 6, 2; 7, 2: *a*.
V^{x} $\mathbf{G}_{16}^{14}$: $(C_{4z}^{+}, 0), (\sigma_{y}, 0)$: 6, 3; 7, 3: *a*.
Σ^{x} $\mathbf{G}_{8}^{5}$: $(C_{2a}, 0), (\sigma_{z}, 0)$: 5, 2: *a*.
S^{x} $\mathbf{G}_{16}^{8}$: $(C_{2a}, 0), (\sigma_{z}, 0)$: 5, 3; 6, 3; 7, 3; 8, 3: *a*.
Y^{x} $\mathbf{G}_{16}^{8}$: $(C_{2x}, 0), (\sigma_{z}, 0)$: 5, 3; 6, 3; 7, 3; 8, 3: *a*.
T^{x} $\mathbf{G}_{16}^{8}$: $(C_{2x}, 0), (\sigma_{z}, 0)$: 5, 3; 6, 3; 7, 3; 8, 3: *a*.
W^{x} $\mathbf{G}_{8}^{5}$: $(C_{2z}, 0), (\sigma_{y}, 0)$: 5, 1: *a*.

‡ The tables given by Slater (1965*b*) for $P4_2/mnm$ (D_{4h}^{14}) do not in fact apply to this space group but to $P4_2/mmc$ (D_{4h}^{9}) (see Gay, Albers, and Arlinghaus (1968)).

138 $P4_2/ncm$ D_{4h}^{16}

($F1$; $K7$; $M5$; $O2$; $Z1$.)

Γ $\mathbf{G}_{32}^{9}$: $\{C_{4z}^{+}\,|\,00\frac{1}{2}\}$, $\{C_{2x}\,|\,\frac{1}{2}\frac{1}{2}0\}$, $\{I\,|\,\frac{1}{2}\frac{1}{2}\frac{1}{2}\}$: 6, 2; 7, 2; 13, 2; 14, 2: *a*.
M $\mathbf{G}_{64}^{2}$: $\{C_{4z}^{+}\,|\,00\frac{1}{2}\}$, $\{C_{2x}\,|\,\frac{1}{2}\frac{1}{2}0\}$, $\{\sigma_x\,|\,00\frac{1}{2}\}$: 19, 2: *a*.
Z $\mathbf{G}_{64}^{2}$: $\{C_{4z}^{+}\,|\,00\frac{1}{2}\}$, $\{\sigma_x\,|\,00\frac{1}{2}\}$, $\{C_{2x}\,|\,\frac{1}{2}\frac{1}{2}0\}$: 19, 2: *a*.
A $\mathbf{G}_{64}^{4}$: $\{C_{4z}^{+}\,|\,00\frac{1}{2}\}$, $\{C_{2x}\,|\,\frac{1}{2}\frac{1}{2}0\}$, $\{I\,|\,\frac{1}{2}\frac{1}{2}\frac{1}{2}\}$: 6, 3; 7, 3; 20, 3; 21, 3: *a*.
R $\mathbf{G}_{32}^{7}$: $\{C_{2x}\,|\,\frac{1}{2}\frac{1}{2}0\}$, $\{\sigma_y\,|\,00\frac{1}{2}\}$, $\{I\,|\,\frac{1}{2}\frac{1}{2}\frac{1}{2}\}$: 9, 1; 10, 1: *b*.
X $\mathbf{G}_{32}^{7}$: $\{C_{2y}\,|\,\frac{1}{2}\frac{1}{2}0\}$, $\{C_{2x}\,|\,\frac{1}{2}\frac{1}{2}0\}$, $\{I\,|\,\frac{1}{2}\frac{1}{2}\frac{1}{2}\}$: 13, 3; 14, 3: *a*.

Δ^x $\mathbf{G}_{8}^{5}$: $(C_{2y}, 0)$, $(\sigma_x, 0)$: 5, 2: *a*.
U^x $\mathbf{G}_{16}^{8}$: $(\sigma_z, 0)$, $(C_{2y}, 0)$: 5, 3; 6, 3; 7, 3; 8, 3: *a*.
Λ^x $\mathbf{G}_{16}^{14}$: $(C_{4z}^{+}, 0)$, $(\sigma_y, 0)$: 6, 2; 7, 2: *a*.
V^x $\mathbf{G}_{16}^{14}$: $(C_{4z}^{+}, 0)$, $(\sigma_y, 0)$: 6, 3; 7, 3: *a*.
Σ^x $\mathbf{G}_{8}^{5}$: $(C_{2a}, 0)$, $(\sigma_z, 0)$: 5, 2: *a*.
S^x $\mathbf{G}_{16}^{8}$: $(\sigma_{db}, 0)$, $(C_{2a}, 0)$: 9, 2: *b*.
Y^x $\mathbf{G}_{16}^{8}$: $(C_{2x}, 0)$, $(\sigma_z, 0)$: 5, 3; 6, 3; 7, 3; 8, 3: *a*.
T^x $\mathbf{G}_{16}^{8}$: $(\sigma_y, 0)$, $(\sigma_z, 0)$: 9, 1: *b*.
W^x $\mathbf{G}_{8}^{5}$: $(C_{2z}, 0)$, $(\sigma_y, 0)$: 5, 1: *a*.

139 $I4/mmm$ D_{4h}^{17}

($F1$; $K7$; $M5$; $S5$; $V1$; $Z1$.)

Γ $\mathbf{G}_{32}^{9}$: $\{C_{4z}^{+}\,|\,000\}$, $\{C_{2x}\,|\,000\}$, $\{I\,|\,000\}$: 6, 2; 7, 2; 13, 2; 14, 2: *a*.
N $\mathbf{G}_{8}^{2} \otimes \mathbf{T}_2$: $\{C_{2y}\,|\,000\}$, $\{I\,|\,000\}$; $\mathbf{t}_2$: 2, 3; 4, 3; 6, 3; 8, 3: *b*.
X $\mathbf{G}_{16}^{11} \otimes \mathbf{T}_2$: $\{C_{2z}\,|\,000\}$, $\{C_{2a}\,|\,000\}$, $\{I\,|\,000\}$; $\mathbf{t}_3$: 5, 2; 10, 2: *c*.
Z $\mathbf{G}_{32}^{9} \otimes \mathbf{T}_2$: $\{C_{4z}^{+}\,|\,000\}$, $\{C_{2x}\,|\,000\}$, $\{I\,|\,000\}$; $\mathbf{t}_1$ or $\mathbf{t}_2$ or $\mathbf{t}_3$: 6, 2; 7, 2; 13, 2; 14, 2: *a*.
P $\mathbf{G}_{16}^{14} \otimes \mathbf{T}_4$: $\{S_{4z}^{+}\,|\,000\}$, $\{C_{2x}\,|\,000\}$; $\mathbf{t}_1$ or $\mathbf{t}_2$ or $\mathbf{t}_3$: 6, 2; 7, 2: *a*.

Λ^x $\mathbf{G}_{16}^{14}$: $(C_{4z}^{+}, 0)$, $(\sigma_y, 0)$: 6, 2; 7, 2: *a*.
V^x $\mathbf{G}_{16}^{14}$: $(C_{4z}^{+}, 0)$, $(\sigma_y, 0)$: 6, 2; 7, 2: *a*.
W^x $\mathbf{G}_{8}^{5}$: $(C_{2z}, 0)$, $(\sigma_{db}, 0)$: 5, 2: *a*.
Σ^x $\mathbf{G}_{8}^{5}$: $(C_{2x}, 0)$, $(\sigma_z, 0)$: 5, 2: *a*.
F^x $\mathbf{G}_{8}^{5}$: $(C_{2x}, 0)$, $(\sigma_z, 0)$: 5, 2: *a*.
Q^x $\mathbf{G}_{4}^{1}$: $(C_{2y}, 0)$: 2, 3; 4, 3: *b*.
Δ^x $\mathbf{G}_{8}^{5}$: $(C_{2a}, 0)$, $(\sigma_z, 0)$: 5, 2: *a*.
U^x $\mathbf{G}_{8}^{5}$: $(C_{2a}, 0)$, $(\sigma_z, 0)$: 5, 2: *a*.
Y^x $\mathbf{G}_{8}^{5}$: $(C_{2b}, 0)$, $(\sigma_{da}, 0)$: 5, 2: *a*.

140 $I4/mcm$ D_{4h}^{18}

($F1$; $G2$; $K7$; $M5$; $S5$; $V1$; $Z1$.)

Γ $\mathbf{G}_{32}^{9}$: $\{C_{4z}^{+}\,|\,\frac{1}{2}\frac{1}{2}0\}$, $\{C_{2x}\,|\,000\}$, $\{I\,|\,\frac{1}{2}\frac{1}{2}0\}$: 6, 2; 7, 2; 13, 2; 14, 2: *a*.
N $\mathbf{G}_{16}^{10}$: $\{\sigma_y\,|\,\frac{1}{2}\frac{1}{2}0\}$, $\{E\,|\,010\}$, $\{I\,|\,\frac{1}{2}\frac{1}{2}0\}$: 9, 1. *a*.
X $\mathbf{G}_{16}^{11} \otimes \mathbf{T}_2$: $\{C_{2z}\,|\,000\}$, $\{C_{2a}\,|\,\frac{1}{2}\frac{1}{2}0\}$, $\{I\,|\,\frac{1}{2}\frac{1}{2}0\}$; $\mathbf{t}_3$: 5, 2; 10, 2: *c*.
Z $\mathbf{G}_{32}^{9} \otimes \mathbf{T}_2$: $\{C_{4z}^{+}\,|\,\frac{1}{2}\frac{1}{2}0\}$, $\{C_{2x}\,|\,000\}$, $\{I\,|\,\frac{1}{2}\frac{1}{2}0\}$; $\mathbf{t}_1$ or $\mathbf{t}_2$ or $\mathbf{t}_3$: 6, 2; 7, 2; 13, 2; 14, 2: *a*.
P $\mathbf{G}_{16}^{14} \otimes \mathbf{T}_4$: $\{S_{4z}^{+}\,|\,000\}$, $\{C_{2x}\,|\,000\}$; $\mathbf{t}_1$ or $\mathbf{t}_2$ or $\mathbf{t}_3$: 6, 3; 7, 3: *a*.

Λ^x $\mathbf{G}_{16}^{14}$: $(C_{4z}^{+}, 0)$, $(\sigma_y, 0)$: 6, 2; 7, 2: *a*.
V^x $\mathbf{G}_{16}^{14}$: $(C_{4z}^{+}, 0)$, $(\sigma_y, 0)$: 6, 2; 7, 2: *a*.
W^x $\mathbf{G}_{8}^{5}$: $(C_{2z}, 0)$, $(\sigma_{db}, 0)$: 5, 2: *a*.
Σ^x $\mathbf{G}_{8}^{5}$: $(C_{2x}, 0)$, $(\sigma_z, 0)$: 5, 2: *a*.
F^x $\mathbf{G}_{8}^{5}$: $(C_{2x}, 0)$, $(\sigma_z, 0)$: 5, 2: *a*.
Q^x $\mathbf{G}_{4}^{1}$: $(C_{2y}, 0)$: 2, 1; 4, 1: *b*.
Δ^x $\mathbf{G}_{8}^{5}$: $(C_{2a}, 0)$, $(\sigma_z, 0)$: 5, 2: *a*.
U^x $\mathbf{G}_{8}^{5}$: $(C_{2a}, 0)$, $(\sigma_z, 0)$: 5, 2: *a*.
Y^x $\mathbf{G}_{8}^{5}$: $(C_{2b}, 0)$, $(\sigma_{da}, 0)$: 5, 2: *a*.

141 $I4_1/amd$ D_{4h}^{19}

($C3$; $F1$; $K7$; $M2$; $M4$; $M5$; $O3$; $S5$; $S13$; $V1$; $Z1$.)

Γ $\mathbf{G}_{32}^{9}$: $\{C_{4z}^{+} | 0\frac{1}{2}0\}$, $\{C_{2x} | \frac{1}{2}\frac{1}{2}0\}$, $\{I | \frac{1}{2}\frac{1}{2}0\}$: 6, 2; 7, 2; 13, 2; 14, 2: a.
N $\mathbf{G}_{8}^{2} \otimes \mathbf{T}_2$: $\{C_{2y} | 0\frac{1}{2}\frac{1}{2}\}$, $\{I | \frac{1}{2}\frac{1}{2}0\}$; $\mathbf{t}_2$: 2, 3; 4, 3; 6, 3; 8, 3: b.
X $\mathbf{G}_{32}^{7}$: $\{\sigma_{db} | 00\frac{1}{2}\}$, $\{\sigma_{da} | 0\frac{1}{2}0\}$, $\{I | \frac{1}{2}\frac{1}{2}0\}$: 13, 3; 14, 3: a.
Z $\mathbf{G}_{64}^{2}$: $\{C_{4z}^{+} | 0\frac{1}{2}0\}$, $\{\sigma_{db} | 00\frac{1}{2}\}$, $\{C_{2b} | \frac{1}{2}\frac{1}{2}\frac{1}{2}\}$: 19, 2: a.
P $\mathbf{G}_{64}^{3}$: $\{S_{4z}^{+} | \frac{1}{2}\frac{1}{2}\frac{1}{2}\}$, $\{\sigma_{db} | 10\frac{1}{2}\}$: 9, 3; 10, 3; 11, 3; 12, 3; 19, 2: b.

Λ^x $\mathbf{G}_{16}^{14}$: $(C_{4z}^{+}, 0)$, $(\sigma_y, 0)$: 6, 2; 7, 2: a.
V^x $\mathbf{G}_{32}^{9}$: $(C_{4z}^{+}, 0)$, $(\sigma_x, 0)$, $(E, 1)$: 13, 2; 14, 2: b.
W^x $\mathbf{G}_{16}^{8}$: $(C_{2z}, 0)$, $(\sigma_{db}, 0)$: 5, 3; 6, 3; 7, 3; 8, 3: a.
Σ^x $\mathbf{G}_{8}^{5}$: $(C_{2x}, 0)$, $(\sigma_z, 0)$: 5, 2: a.
F^x $\mathbf{G}_{8}^{5}$: $(C_{2x}, 0)$, $(\sigma_z, 0)$: 5, 2: a.
Q^x $\mathbf{G}_{8}^{2}$: $(C_{2y}, 0)$, $(E, 1)$: 5, 3; 7, 3: a.
Δ^x $\mathbf{G}_{8}^{5}$: $(C_{2a}, 0)$, $(\sigma_z, 0)$: 5, 2: a.
U^x $\mathbf{G}_{16}^{8}$: $(C_{2a}, 0)$, $(\sigma_z, 0)$: 5, 3; 6, 3; 7, 3; 8, 3: a.
Y^x $\mathbf{G}_{16}^{8}$: $(C_{2b}, 0)$, $(\sigma_{da}, 0)$: 5, 3; 6, 3; 7, 3; 8, 3: a.

142 $I4_1/acd$ D_{4h}^{20}

($F1$; $K7$; $M5$; $S5$; $V1$; $Z1$.)

Γ $\mathbf{G}_{32}^{9}$: $\{C_{4z}^{+} | \frac{1}{2}00\}$, $\{C_{2x} | \frac{1}{2}\frac{1}{2}0\}$, $\{I | 000\}$: 6, 2; 7, 2; 13, 2; 14, 2: a.
N $\mathbf{G}_{16}^{10}$: $\{C_{2y} | 0\frac{1}{2}\frac{1}{2}\}$, $\{E | 010\}$, $\{I | 000\}$: 10, 1: b.
X $\mathbf{G}_{32}^{7}$: $\{\sigma_z | \frac{1}{2}0\frac{1}{2}\}$, $\{\sigma_{da} | 0\frac{1}{2}0\}$, $\{I | 000\}$: 13, 3; 14, 3: a.
Z $\mathbf{G}_{64}^{2}$: $\{C_{4z}^{+} | \frac{1}{2}00\}$, $\{\sigma_{db} | 00\frac{1}{2}\}$, $\{C_{2b} | 00\frac{1}{2}\}$: 19, 2: a.
P $\mathbf{G}_{64}^{3}$: $\{S_{4z}^{+} | \frac{1}{2}\frac{1}{2}\frac{1}{2}\}$, $\{\sigma_{db} | 10\frac{1}{2}\}$: 9, 3; 10, 3; 11, 3; 12, 3; 19, 1: b.

Λ^x $\mathbf{G}_{16}^{14}$: $(C_{4z}^{+}, 0)$, $(\sigma_y, 0)$: 6, 2; 7, 2: a.
V^x $\mathbf{G}_{32}^{9}$: $(C_{4z}^{+}, 0)$, $(\sigma_x, 0)$, $(E, 1)$: 13, 2; 14, 2: b.
W^x $\mathbf{G}_{16}^{8}$: $(C_{2z}, 0)$, $(\sigma_{da}, 0)$: 5, 3; 6, 3; 7, 3; 8, 3: a.
Σ^x $\mathbf{G}_{8}^{5}$: $(C_{2x}, 0)$, $(\sigma_z, 0)$: 5, 2: a.
F^x $\mathbf{G}_{8}^{5}$: $(C_{2x}, 0)$, $(\sigma_z, 0)$: 5, 2: a.
Q^x $\mathbf{G}_{8}^{2}$: $(C_{2y}, 0)$, $(E, 1)$: 5, 1; 7, 1: a.
Δ^x $\mathbf{G}_{8}^{5}$: $(C_{2a}, 0)$, $(\sigma_z, 0)$: 5, 2: a.
U^x $\mathbf{G}_{16}^{8}$: $(C_{2a}, 0)$, $(\sigma_z, 0)$: 5, 3; 6, 3; 7, 3; 8, 3: a.
Y^x $\mathbf{G}_{16}^{8}$: $(C_{2b}, 0)$, $(\sigma_{da}, 0)$: 5, 3; 6, 3; 7, 3; 8, 3: a.

143 $P3$ C_3^1

($F1$; $K6$; $K7$; $M5$; $Z1$.)

Γ $\mathbf{G}_{6}^{1}$: $\{C_{3}^{+} | 000\}$: 2, 3; 4, 1; 6, 3: a.
M $\mathbf{G}_{2}^{1} \otimes \mathbf{T}_2$: $\{\bar{E} | 000\}$; $\mathbf{t}_2$: 2, 1: a.
A $\mathbf{G}_{6}^{1} \otimes \mathbf{T}_2$: $\{C_{3}^{+} | 000\}$; $\mathbf{t}_3$: 2, 3; 4, 1; 6, 3: a.
L $\mathbf{G}_{2}^{1} \otimes \mathbf{T}_2$: $\{\bar{E} | 000\}$; $\mathbf{t}_2$ or $\mathbf{t}_3$: 2, 1: a.
K $\mathbf{G}_{6}^{1} \otimes \mathbf{T}_3$: $\{C_{3}^{+} | 000\}$; $\mathbf{t}_1$ or $\mathbf{t}_2$: 2, x; 4, x; 6, x: a.
H $\mathbf{G}_{6}^{1} \otimes \mathbf{T}_3 \otimes \mathbf{T}_2$: $\{C_{3}^{+} | 000\}$; $\mathbf{t}_1$ or $\mathbf{t}_2$; $\mathbf{t}_3$: 2, x; 4, x; 6, x: a.

Δ^x $\mathbf{G}_{6}^{1}$: $(C_{3}^{+}, 0)$: 2, x; 4, x; 6, x: a.
U^x $\mathbf{G}_{2}^{1}$: $(\bar{E}, 0)$: 2, x: a.
P^x $\mathbf{G}_{6}^{1}$: $(C_{3}^{+}, 0)$: 2, x; 4, x; 6, x: a.
T^x $\mathbf{G}_{2}^{1}$: $(\bar{E}, 0)$: 2, x: a.
S^x $\mathbf{G}_{2}^{1}$: $(\bar{E}, 0)$: 2, x: a.
T'^x $\mathbf{G}_{2}^{1}$: $(\bar{E}, 0)$: 2, x: a.
S'^x $\mathbf{G}_{2}^{1}$: $(\bar{E}, 0)$: 2, x: a.
Σ^x $\mathbf{G}_{2}^{1}$: $(\bar{E}, 0)$: 2, x: a.
R^x $\mathbf{G}_{2}^{1}$: $(\bar{E}, 0)$: 2, x: a.

144 $P3_1$ C_3^2

($F1$; $K6$; $K7$; $M5$; $Z1$.)

Γ $\mathbf{G}_6^1$: $\{C_3^+ \mid 00\frac{1}{3}\}$: 2, 3; 4, 1; 6, 3: a.
M $\mathbf{G}_2^1 \otimes \mathbf{T}_2$: $\{\bar{E} \mid 000\}$; $\mathbf{t}_2$: 2, 1: a.
A $\mathbf{G}_6^1 \otimes \mathbf{T}_2$: $\{C_3^- \mid 00\frac{2}{3}\}$; $\mathbf{t}_3$: 2, 3; 4, 1; 6, 3: a.
L $\mathbf{G}_2^1 \otimes \mathbf{T}_2$: $\{\bar{E} \mid 000\}$; $\mathbf{t}_2$ or $\mathbf{t}_3$: 2, 1: a.
K $\mathbf{G}_6^1 \otimes \mathbf{T}_3$: $\{C_3^+ \mid 00\frac{1}{3}\}$; $\mathbf{t}_1$ or $\mathbf{t}_2$: 2, x; 4, x; 6, x: a.
H $\mathbf{G}_6^1 \otimes \mathbf{T}_3 \otimes \mathbf{T}_2$: $\{C_3^- \mid 00\frac{2}{3}\}$; $\mathbf{t}_1$ or $\mathbf{t}_2$; $\mathbf{t}_3$: 2, x; 4, x; 6, x: a.

Δ^x $\mathbf{G}_6^1$: $(C_3^+, 0)$: 2, x; 4, x; 6, x: a.
U^x $\mathbf{G}_2^1$: $(\bar{E}, 0)$: 2, x: a.
P^x $\mathbf{G}_6^1$: $(C_3^+, 0)$: 2, x; 4, x; 6, x: a.
T^x $\mathbf{G}_2^1$: $(\bar{E}, 0)$: 2, x: a.
S^x $\mathbf{G}_2^1$: $(\bar{E}, 0)$: 2, x: a.
T'^x $\mathbf{G}_2^1$: $(\bar{E}, 0)$: 2, x: a.
S'^x $\mathbf{G}_2^1$: $(\bar{E}, 0)$: 2, x: a.
Σ^x $\mathbf{G}_2^1$: $(\bar{E}, 0)$: 2, x: a.
R^x $\mathbf{G}_2^1$: $(\bar{E}, 0)$: 2, x: a.

145 $P3_2$ C_3^3

($F1$; $K6$; $K7$; $M5$; $Z1$.)

Γ $\mathbf{G}_6^1$: $\{C_3^+ \mid 00\frac{2}{3}\}$: 2, 3; 4, 1; 6, 3: a.
M $\mathbf{G}_2^1 \otimes \mathbf{T}_2$: $\{\bar{E} \mid 000\}$; $\mathbf{t}_2$: 2, 1: a.
A $\mathbf{G}_6^1 \otimes \mathbf{T}_2$: $\{C_3^+ \mid 00\frac{2}{3}\}$; $\mathbf{t}_3$: 2, 3; 4, 1; 6, 3: a.
L $\mathbf{G}_2^1 \otimes \mathbf{T}_2$: $\{\bar{E} \mid 000\}$; $\mathbf{t}_2$ or $\mathbf{t}_3$: 2, 1: a.
K $\mathbf{G}_6^1 \otimes \mathbf{T}_3$: $\{C_3^+ \mid 00\frac{2}{3}\}$; $\mathbf{t}_1$ or $\mathbf{t}_2$: 2, x; 4, x; 6, x: a.
H $\mathbf{G}_6^1 \otimes \mathbf{T}_3 \otimes \mathbf{T}_2$: $\{C_3^+ \mid 00\frac{2}{3}\}$; $\mathbf{t}_1$ or $\mathbf{t}_2$; $\mathbf{t}_3$: 2, x; 4, x; 6, x: a.

Δ^x $\mathbf{G}_6^1$: $(C_3^+, 0)$: 2, x; 4, x; 6, x: a.
U^x $\mathbf{G}_2^1$: $(\bar{E}, 0)$: 2, x: a.
P^x $\mathbf{G}_6^1$: $(C_3^+, 0)$: 2, x; 4, x; 6, x: a.
T^x $\mathbf{G}_2^1$: $(\bar{E}, 0)$: 2, x: a.
S^x $\mathbf{G}_2^1$: $(\bar{E}, 0)$: 2, x: a.
T'^x $\mathbf{G}_2^1$: $(\bar{E}, 0)$: 2, x: a.
S'^x $\mathbf{G}_2^1$: $(\bar{E}, 0)$: 2, x: a.
Σ^x $\mathbf{G}_2^1$: $(\bar{E}, 0)$: 2, x: a.
R^x $\mathbf{G}_2^1$: $(\bar{E}, 0)$: 2, x: a.

146 $R3$ C_3^4

($F1$; $K6$; $K7$; $M5$; $Z1$.)

Γ $\mathbf{G}_6^1$: $\{C_3^+ \mid 000\}$: 2, 3; 4, 1; 6, 3: a.
Z $\mathbf{G}_6^1 \otimes \mathbf{T}_2$: $\{C_3^+ \mid 000\}$; $\mathbf{t}_1$ or $\mathbf{t}_2$ or $\mathbf{t}_3$: 2, 3; 4, 1; 6, 3: a.
L $\mathbf{G}_2^1 \otimes \mathbf{T}_2$: $\{\bar{E} \mid 000\}$; $\mathbf{t}_2$: 2, 1: a.
{(a) F $\mathbf{G}_2^1 \otimes \mathbf{T}_2$: $\{\bar{E} \mid 000\}$; $\mathbf{t}_2$ or $\mathbf{t}_3$: 2, 1: a.}
{(b) F $\mathbf{G}_2^1 \otimes \mathbf{T}_2$: $\{\bar{E} \mid 000\}$; $\mathbf{t}_1$ or $\mathbf{t}_2$: 2, 1: a.}

Λ^x $\mathbf{G}_6^1$: $(C_3^+, 0)$: 2, x; 4, x; 6, x: a.
P^x $\mathbf{G}_6^1$: $(C_3^+, 0)$: 2, x; 4, x; 6, x: a.
B^x $\mathbf{G}_2^1$: $(\bar{E}, 0)$: 2, x: a.
Σ^x $\mathbf{G}_2^1$: $(\bar{E}, 0)$: 2, x: a.
Q^x $\mathbf{G}_2^1$: $(\bar{E}, 0)$: 2, x: a.
Y^x $\mathbf{G}_2^1$: $(\bar{E}, 0)$: 2, x: a.

147 $P\bar{3}$ C_{3i}^1

($F1$; $K6$; $K7$; $M5$; $Z1$.)

Γ $\mathbf{G}_{12}^6$: $\{S_6^+ \mid 000\}$, $\{I \mid 000\}$: 2, 3; 4, 1; 6, 3; 7, 1; 9, 3; 11, 3: c.
M $\mathbf{G}_4^2 \otimes \mathbf{T}_2$: $\{I \mid 000\}$, $\{\bar{E} \mid 000\}$; $\mathbf{t}_2$: 2, 1; 4, 1: a.
A $\mathbf{G}_{12}^6 \otimes \mathbf{T}_2$: $\{S_6^+ \mid 000\}$, $\{I \mid 000\}$; $\mathbf{t}_3$: 2, 3; 4, 1; 6, 3; 7, 1; 9, 3; 11, 3: c.
L $\mathbf{G}_4^2 \otimes \mathbf{T}_2$: $\{I \mid 000\}$, $\{\bar{E} \mid 000\}$; $\mathbf{t}_2$ or $\mathbf{t}_3$: 2, 1; 4, 1: a.
K $\mathbf{G}_6^1 \otimes \mathbf{T}_3$: $\{C_3^+ \mid 000\}$; $\mathbf{t}_1$ or $\mathbf{t}_2$: 2, 3; 4, 1; 6, 3: a.
H $\mathbf{G}_6^1 \otimes \mathbf{T}_3 \otimes \mathbf{T}_2$: $\{C_3^+ \mid 000\}$; $\mathbf{t}_1$ or $\mathbf{t}_2$; $\mathbf{t}_3$: 2, 3; 4, 1; 6, 3: a.

Δ^x $\mathbf{G}_6^1$: $(C_3^+, 0)$: 2, 3; 4, 1; 6, 3: a.
U^x $\mathbf{G}_2^1$: $(\bar{E}, 0)$: 2, 1: a.
P^x $\mathbf{G}_6^1$: $(C_3^+, 0)$: 2, 3; 4, 1; 6, 3: a.
T^x $\mathbf{G}_2^1$: $(\bar{E}, 0)$: 2, 1: a.
S^x $\mathbf{G}_2^1$: $(\bar{E}, 0)$: 2, 1: a.
T'^x $\mathbf{G}_2^1$: $(\bar{E}, 0)$: 2, 1: a.
S'^x $\mathbf{G}_2^1$: $(\bar{E}, 0)$: 2, 1: a.
Σ^x $\mathbf{G}_2^1$: $(\bar{E}, 0)$: 2, 1: a.
R^x $\mathbf{G}_2^1$: $(\bar{E}, 0)$: 2, 1: a.

148 $R\bar{3}$ C_{3i}^2

($F1$; $K6$; $K7$; $M5$; $S8$; $Z1$.)

Γ $\mathbf{G}_{12}^6$: $\{S_6^+ \mid 000\}$, $\{I \mid 000\}$: 2, 3; 4, 1; 6, 3; 7, 1; 9, 3; 11, 3: c.
Z $\mathbf{G}_{12}^6 \otimes \mathbf{T}_2$: $\{S_6^+ \mid 000\}$, $\{I \mid 000\}$; $\mathbf{t}_1$ or $\mathbf{t}_2$ or $\mathbf{t}_3$: 2, 3; 4, 1; 6, 3; 7, 1; 9, 3; 11, 3: c.
L $\mathbf{G}_4^2 \otimes \mathbf{T}_2$: $\{I \mid 000\}$, $\{\bar{E} \mid 000\}$; $\mathbf{t}_2$: 2, 1; 4, 1: a.
{(a) F $\mathbf{G}_4^2 \otimes \mathbf{T}_2$: $\{I \mid 000\}$, $\{\bar{E} \mid 000\}$; $\mathbf{t}_2$ or $\mathbf{t}_3$: 2, 1; 4, 1: a.}
{(b) F $\mathbf{G}_4^2 \otimes \mathbf{T}_2$: $\{I \mid 000\}$, $\{\bar{E} \mid 000\}$; $\mathbf{t}_1$ or $\mathbf{t}_2$: 2, 1; 4, 1: a.}

Λ^x $\mathbf{G}_6^1$: $(C_3^+, 0)$: 2, 3; 4, 1; 6, 3: a.
P^x $\mathbf{G}_6^1$: $(C_3^+, 0)$: 2, 3; 4, 1; 6, 3: a.
B^x $\mathbf{G}_2^1$: $(\bar{E}, 0)$: 2, 1: a.
Σ^x $\mathbf{G}_2^1$: $(\bar{E}, 0)$: 2, 1: a.
Q^x $\mathbf{G}_2^1$: $(\bar{E}, 0)$: 2, 1: a.
Y^x $\mathbf{G}_2^1$: $(\bar{E}, 0)$: 2, 1: a.

149 $P312$ D_3^1

($F1$; $K6$; $K7$; $M5$; $Z1$.)

Γ $\mathbf{G}_{12}^4$: $\{C_3^+ \mid 000\}$, $\{C'_{21} \mid 000\}$: 3, 3; 4, 3; 6, 2: a.
M $\mathbf{G}_4^1 \otimes \mathbf{T}_2$: $\{C'_{21} \mid 000\}$; $\mathbf{t}_2$: 2, 3; 4, 3: b.
A $\mathbf{G}_{12}^4 \otimes \mathbf{T}_2$: $\{C_3^+ \mid 000\}$, $\{C'_{21} \mid 000\}$; $\mathbf{t}_3$: 3, 3; 4, 3; 6, 2: a.
L $\mathbf{G}_4^1 \otimes \mathbf{T}_2$: $\{C'_{21} \mid 000\}$; $\mathbf{t}_2$ or $\mathbf{t}_3$: 2, 3; 4, 3: b.
K $\mathbf{G}_6^1 \otimes \mathbf{T}_3$: $\{C_3^+ \mid 000\}$; $\mathbf{t}_1$ or $\mathbf{t}_2$: 2, 2; 4, 2; 6, 2: a.
H $\mathbf{G}_6^1 \otimes \mathbf{T}_3 \otimes \mathbf{T}_2$: $\{C_3^+ \mid 000\}$; $\mathbf{t}_1$ or $\mathbf{t}_2$; $\mathbf{t}_3$: 2, 2; 4, 2; 6, 2: a.

Δ^x $\mathbf{G}_6^1$: $(C_3^+, 0)$: 2, 2; 4, 2; 6, 2: a.
U^x $\mathbf{G}_2^1$: $(\bar{E}, 0)$: 2, 2: a.
P^x $\mathbf{G}_6^1$: $(C_3^+, 0)$: 2, 2; 4, 2; 6, 2: a.
T^x $\mathbf{G}_2^1$: $(\bar{E}, 0)$: 2, 2: a.
S^x $\mathbf{G}_2^1$: $(\bar{E}, 0)$: 2, 2: a.
T'^x $\mathbf{G}_2^1$: $(\bar{E}, 0)$: 2, 2: a.
S'^x $\mathbf{G}_2^1$: $(\bar{E}, 0)$: 2, 2: a.
Σ^x $\mathbf{G}_4^1$: $(C'_{21}, 0)$: 2, x; 4, x: b.
R^x $\mathbf{G}_4^1$: $(C'_{21}, 0)$: 2, x; 4, x: b.

150 *P*321 D_3^2

(*F*1; *K*6; *K*7; *M*5; *Z*1)

Γ $\mathbf{G}^4_{12}$: $\{C_3^+ \mid 000\}, \{C''_{21} \mid 000\}$: 3, 3; 4, 3; 6, 2: *a*.
M $\mathbf{G}^1_4 \otimes \mathbf{T}_2$: $\{C''_{21} \mid 000\}$; $\mathbf{t}_2$: 2, 3; 4, 3: *b*.
A $\mathbf{G}^4_{12} \otimes \mathbf{T}_2$: $\{C_3^+ \mid 000\}, \{C''_{21} \mid 000\}$; $\mathbf{t}_3$: 3, 3; 4, 3; 6, 2: *a*.
L $\mathbf{G}^1_4 \otimes \mathbf{T}_2$: $\{C''_{21} \mid 000\}$; $\mathbf{t}_2$ or $\mathbf{t}_3$: 2, 3; 4, 3: *b*.
K $\mathbf{G}^4_{12} \otimes \mathbf{T}_3$: $\{C_3^+ \mid 000\}, \{C''_{21} \mid 000\}$; $\mathbf{t}_1$ or $\mathbf{t}_2$: 3, *x*; 4, *x*; 6, *x*: *a*.
H $\mathbf{G}^4_{12} \otimes \mathbf{T}_3 \otimes \mathbf{T}_2$: $\{C_3^+ \mid 000\}, \{C''_{21} \mid 000\}$; $\mathbf{t}_1$ or $\mathbf{t}_2$ or $\mathbf{t}_3$: 3, *x*; 4, *x*; 6, *x*: *a*.

Δ^x $\mathbf{G}^1_6$: $(C_3^+, 0)$: 2, 2; 4, 2; 6, 2: *a*.
U^x $\mathbf{G}^1_2$: $(\bar{E}, 0)$: 2, 2: *a*.
P^x $\mathbf{G}^1_6$: $(C_3^+, 0)$: 2, *x*; 4, *x*; 6, *x*: *a*.
T^x $\mathbf{G}^1_4$: $(C''_{22}, 0)$: 2, *x*; 4, *x*: *b*.
S^x $\mathbf{G}^1_4$: $(C''_{22}, 0)$: 2, *x*; 4, *x*: *b*.
T'^x $\mathbf{G}^1_4$: $(C''_{21}, 0)$: 2, *x*; 4, *x*: *b*.
S'^x $\mathbf{G}^1_4$: $(C''_{21}, 0)$: 2, *x*; 4, *x*: *b*.
Σ^x $\mathbf{G}^1_2$: $(\bar{E}, 0)$: 2, 2: *a*.
R^x $\mathbf{G}^1_2$: $(\bar{E}, 0)$: 2, 2: *a*.

151 $P3_112$ D_3^3

(*A*5; *F*1; *F*4; *F*5; *K*6; *K*7; *M*5; *Z*1.)

Γ $\mathbf{G}^4_{12}$: $\{C_3^+ \mid 00\frac{1}{3}\}, \{C'_{21} \mid 00\frac{2}{3}\}$: 3, 3; 4, 3; 6, 2: *a*.
M $\mathbf{G}^1_4 \otimes \mathbf{T}_2$: $\{C'_{21} \mid 00\frac{2}{3}\}$; $\mathbf{t}_2$: 2, 3; 4, 3: *b*.
A $\mathbf{G}^4_{12} \otimes \mathbf{T}_2$: $\{C_3^- \mid 00\frac{2}{3}\}, \{C'_{21} \mid 00\frac{2}{3}\}$; $\mathbf{t}_3$: 3, 3; 4, 3; 6, 2: *a*.
L $\mathbf{G}^1_4 \otimes \mathbf{T}_2$: $\{C'_{21} \mid 00\frac{2}{3}\}$; $\mathbf{t}_2$ or $\mathbf{t}_3$: 2, 3; 4, 3: *b*.
K $\mathbf{G}^1_6 \otimes \mathbf{T}_3$: $\{C_3^+ \mid 00\frac{1}{3}\}$; $\mathbf{t}_1$ or $\mathbf{t}_2$: 2, 2; 4, 2; 6, 2: *a*.
H $\mathbf{G}^1_6 \otimes \mathbf{T}_3 \otimes \mathbf{T}_2$: $\{C_3^- \mid 00\frac{2}{3}\}$; $\mathbf{t}_1$ or $\mathbf{t}_2$; $\mathbf{t}_3$: 2, 2; 4, 2; 6, 2: *a*.

Δ^x $\mathbf{G}^1_6$: $(C_3^+, 0)$: 2, 2; 4, 2; 6, 2: *a*.
U^x $\mathbf{G}^1_2$: $(\bar{E}, 0)$: 2, 2: *a*.
P^x $\mathbf{G}^1_6$: $(C_3^+, 0)$: 2, 2; 4, 2; 6, 2: *a*.
T^x $\mathbf{G}^1_2$: $(\bar{E}, 0)$: 2, 2: *a*.
S^x $\mathbf{G}^1_2$: $(\bar{E}, 0)$: 2, 2: *a*.
T'^x $\mathbf{G}^1_2$: $(\bar{E}, 0)$: 2, 2: *a*.
S'^x $\mathbf{G}^1_2$: $(\bar{E}, 0)$: 2, 2: *a*.
Σ^x $\mathbf{G}^1_4$: $(C'_{21}, 0)$: 2, *x*; 4, *x*: *b*.
R^x $\mathbf{G}^1_{12}$: $(C'_{21}, 0)$: 2, *x*; 8, *x*: *b*.

152 $P3_121$ D_3^4

(*A*6; *F*1; *F*5; *F*7; *F*8; *K*6; *K*7; *M*5; *N*2; *N*3; *P*2; *R*1; *R*4; *S*8; *T*6; *Z*1.)

Γ $\mathbf{G}^4_{12}$: $\{C_3^+ \mid 00\frac{1}{3}\}, \{C''_{21} \mid 00\frac{2}{3}\}$: 3, 3; 4, 3; 6, 2: *a*.
M $\mathbf{G}^1_4 \otimes \mathbf{T}_2$: $\{C''_{21} \mid 00\frac{2}{3}\}$; $\mathbf{t}_2$: 2, 3; 4, 3: *b*.
A $\mathbf{G}^4_{12} \otimes \mathbf{T}_2$: $\{C_3^- \mid 00\frac{2}{3}\}, \{C''_{21} \mid 00\frac{2}{3}\}$; $\mathbf{t}_3$: 3, 3; 4, 3; 6, 2: *a*.
L $\mathbf{G}^1_4 \otimes \mathbf{T}_2$: $\{C''_{21} \mid 00\frac{2}{3}\}$; $\mathbf{t}_2$ or $\mathbf{t}_3$: 2, 3; 4, 3: *b*.
K $\mathbf{G}^4_{12} \otimes \mathbf{T}_3$: $\{C_3^+ \mid 00\frac{1}{3}\}, \{C''_{21} \mid 00\frac{2}{3}\}$; $\mathbf{t}_1$ or $\mathbf{t}_2$: 3, *x*; 4, *x*; 6, *x*: *a*.
H $\mathbf{G}^4_{12} \otimes \mathbf{T}_3 \otimes \mathbf{T}_2$: $\{C_3^- \mid 00\frac{2}{3}\}, \{C''_{21} \mid 00\frac{2}{3}\}$; $\mathbf{t}_1$ or $\mathbf{t}_2$; $\mathbf{t}_3$: 3, *x*; 4, *x*; 6, *x*: *a*.

Δ^x $\mathbf{G}^1_6$: $(C_3^+, 0)$: 2, 2; 4, 2; 6, 2: *a*.
U^x $\mathbf{G}^1_2$: $(\bar{E}, 0)$: 2, 2: *a*.
P^x $\mathbf{G}^1_6$: $(C_3^+, 0)$: 2, *x*; 4, *x*; 6, *x*: *a*.
T^x $\mathbf{G}^1_4$: $(C''_{22}, 0)$: 2, *x*; 4, *x*: *b*.
S^x $\mathbf{G}^1_{12}$: $(C''_{22}, 0)$: 6, *x*; 12, *x*: *a*.
T'^x $\mathbf{G}^1_4$: $(C''_{21}, 0)$: 2, *x*; 4, *x*: *b*.
S'^x $\mathbf{G}^1_{12}$: $(C''_{21}, 0)$: 2, *x*; 8, *x*: *b*.
Σ^x $\mathbf{G}^1_2$: $(\bar{E}, 0)$: 2, 2: *a*.
R^x $\mathbf{G}^1_2$: $(\bar{E}, 0)$: 2, 2: *a*.

153 $P3_212$ D_3^5

($F1$; $K6$; $K7$; $M5$; $Z1$.)

Γ $\mathbf{G}_{12}^4$: $\{C_3^+ \mid 00\frac{2}{3}\}, \{C'_{21} \mid 00\frac{1}{3}\}$: 3, 3; 4, 3; 6, 2: a.
M $\mathbf{G}_4^1 \otimes \mathbf{T}_2$: $\{C'_{21} \mid 00\frac{1}{3}\}$; $\mathbf{t}_2$: 2, 3; 4, 3: b.
A $\mathbf{G}_{12}^4 \otimes \mathbf{T}_2$: $\{C_3^+ \mid 00\frac{2}{3}\}, \{C'_{21} \mid 00\frac{1}{3}\}$; $\mathbf{t}_3$: 3, 3; 4, 3; 6, 2: a.
L $\mathbf{G}_4^1 \otimes \mathbf{T}_2$: $\{C'_{21} \mid 00\frac{1}{3}\}$; $\mathbf{t}_2$ or $\mathbf{t}_3$: 2, 3; 4, 3: b.
K $\mathbf{G}_6^1 \otimes \mathbf{T}_3$: $\{C_3^+ \mid 00\frac{2}{3}\}$; $\mathbf{t}_1$ or $\mathbf{t}_2$: 2, 2; 4, 2; 6, 2: a.
H $\mathbf{G}_6^1 \otimes \mathbf{T}_3 \otimes \mathbf{T}_2$: $\{C_3^+ \mid 00\frac{2}{3}\}$; $\mathbf{t}_1$ or $\mathbf{t}_2$; $\mathbf{t}_3$: 2, 2; 4, 2; 6, 2: a.

Δ^x $\mathbf{G}_6^1$: $(C_3^+, 0)$: 2, 2; 4, 2; 6, 2: a.
U^x $\mathbf{G}_2^1$: $(\bar{E}, 0)$: 2, 2: a.
P^x $\mathbf{G}_6^1$: $(C_3^+, 0)$: 2, 2; 4, 2; 6, 2: a.
T^x $\mathbf{G}_2^1$: $(\bar{E}, 0)$: 2, 2: a.
S^x $\mathbf{G}_2^1$: $(\bar{E}, 0)$: 2, 2: a.
T'^x $\mathbf{G}_2^1$: $(\bar{E}, 0)$: 2, 2: a.
S'^x $\mathbf{G}_2^1$: $(\bar{E}, 0)$: 2, 2: a.
Σ^x $\mathbf{G}_4^1$: $(C'_{21}, 0)$: 2, x; 4, x: b.
R^x $\mathbf{G}_{12}^1$: $(C'_{21}, 0)$: 6, x; 12, x: a.

154 $P3_221$ D_3^6

($A6$; $F1$; $F5$; $F7$; $F8$; $K6$; $K7$; $M5$; $N2$; $N3$; $P2$; $R1$; $Z1$.)

Γ $\mathbf{G}_{12}^4$: $\{C_3^+ \mid 00\frac{2}{3}\}, \{C''_{21} \mid 00\frac{1}{3}\}$: 3, 3; 4, 3; 6, 2: a.
M $\mathbf{G}_4^1 \otimes \mathbf{T}_2$: $\{C''_{21} \mid 00\frac{1}{3}\}$; $\mathbf{t}_2$: 2, 3; 4, 3: b.
A $\mathbf{G}_{12}^4 \otimes \mathbf{T}_2$: $\{C_3^+ \mid 00\frac{2}{3}\}, \{C''_{21} \mid 00\frac{1}{3}\}$; $\mathbf{t}_3$: 3, 3; 4, 3; 6, 2: a.
L $\mathbf{G}_4^1 \otimes \mathbf{T}_2$: $\{C''_{21} \mid 00\frac{1}{3}\}$; $\mathbf{t}_2$ or $\mathbf{t}_3$: 2, 3; 4, 3: b.
K $\mathbf{G}_{12}^4 \otimes \mathbf{T}_3$: $\{C_3^+ \mid 00\frac{2}{3}\}, \{C''_{21} \mid 00\frac{1}{3}\}$; $\mathbf{t}_1$ or $\mathbf{t}_2$: 3, x; 4, x; 6, x: a.
H $\mathbf{G}_{12}^4 \otimes \mathbf{T}_3 \otimes \mathbf{T}_2$: $\{C_3^+ \mid 00\frac{2}{3}\}, \{C''_{21} \mid 00\frac{1}{3}\}$; $\mathbf{t}_1$ or $\mathbf{t}_2$; $\mathbf{t}_3$: 3, x; 4, x; 6, x: a.

Δ^x $\mathbf{G}_6^1$: $(C_3^+, 0)$: 2, 2; 4, 2; 6, 2: a.
U^x $\mathbf{G}_2^1$: $(\bar{E}, 0)$: 2, 2: a.
P^x $\mathbf{G}_6^1$: $(C_3^+, 0)$: 2, x; 4, x; 6, x: a.
T^x $\mathbf{G}_4^1$: $(C''_{22}, 0)$: 2, x; 4, x: b.
S^x $\mathbf{G}_{12}^1$: $(C''_{22}, 0)$: 2, x; 8, x: b.
T'^x $\mathbf{G}_4^1$: $(C''_{21}, 0)$: 2, x; 4, x: b.
S'^x $\mathbf{G}_{12}^1$: $(C''_{21}, 0)$: 6, x; 12, x: a.
Σ^x $\mathbf{G}_2^1$: $(\bar{E}, 0)$: 2, 2: a.
R^x $\mathbf{G}_2^1$: $(\bar{E}, 0)$: 2, 2: a.

155 $R32$ D_3^7

($F1$; $K6$; $K7$; $M5$; $S8$; $Z1$.)

Γ $\mathbf{G}_{12}^4$: $\{C_3^+ \mid 000\}, \{C'_{21} \mid 000\}$: 3, 3; 4, 3; 6, 2: a.
Z $\mathbf{G}_{12}^4 \otimes \mathbf{T}_2$: $\{C_3^+ \mid 000\}, \{C'_{21} \mid 000\}$; $\mathbf{t}_1$ or $\mathbf{t}_2$ or $\mathbf{t}_3$: 3, 3; 4, 3; 6, 2: a.
L $\mathbf{G}_4^1 \otimes \mathbf{T}_2$: $\{C'_{22} \mid 000\}$; $\mathbf{t}_2$: 2, 3; 4, 3: b.
{(a) F $\mathbf{G}_4^1 \otimes \mathbf{T}_2$: $\{C'_{21} \mid 000\}$; $\mathbf{t}_2$ or $\mathbf{t}_3$: 2, 3; 4, 3: b.}
{(b) F $\mathbf{G}_4^1 \otimes \mathbf{T}_2$: $\{C'_{23} \mid 000\}$; $\mathbf{t}_1$ or $\mathbf{t}_2$: 2, 3; 4, 3: b.}

Λ^x $\mathbf{G}_6^1$: $(C_3^+, 0)$: 2, 2; 4, 2; 6, 2: a.
P^x $\mathbf{G}_6^1$: $(C_3^+, 0)$: 2, 2; 4, 2; 6, 2: a.
B^x $\mathbf{G}_4^1$: $(C'_{21}, 0)$: 2, x; 4, x: b.
Σ^x $\mathbf{G}_4^1$: $(C'_{21}, 0)$: 2, x; 4, x: b.
Q^x $\mathbf{G}_4^1$: $(C'_{23}, 0)$: 2, x; 4, x: b.
Y^x $\mathbf{G}_4^1$: $(C'_{22}, 0)$: 2, x; 4, x: b.

156 $P3m1$ C_{3v}^1

($F1$; $K6$; $K7$; $M5$; $Z1$.)

Γ $\mathbf{G}_{12}^4$: $\{C_3^+ \mid 000\}$, $\{\sigma_{v1} \mid 000\}$: 3, 3; 4, 3; 6, 2: *a*.
M $\mathbf{G}_4^1 \otimes \mathbf{T}_2$: $\{\sigma_{v1} \mid 000\}$; $\mathbf{t}_2$: 2, 3; 4, 3: *b*.
A $\mathbf{G}_{12}^4 \otimes \mathbf{T}_2$: $\{C_3^+ \mid 000\}$, $\{\sigma_{v1} \mid 000\}$; $\mathbf{t}_3$: 3, 3; 4, 3; 6, 2: *a*.
L $\mathbf{G}_4^1 \otimes \mathbf{T}_2$: $\{\sigma_{v1} \mid 000\}$; $\mathbf{t}_2$ or $\mathbf{t}_3$: 2, 3; 4, 3: *b*.
K $\mathbf{G}_6^1 \otimes \mathbf{T}_3$: $\{C_3^+ \mid 000\}$; $\mathbf{t}_1$ or $\mathbf{t}_2$: 2, 2; 4, 2; 6, 2: *a*.
H $\mathbf{G}_6^1 \otimes \mathbf{T}_3 \otimes \mathbf{T}_2$: $\{C_3^+ \mid 000\}$; $\mathbf{t}_1$ or $\mathbf{t}_2$; $\mathbf{t}_3$: 2, 2; 4, 2; 6, 2: *a*.

Δ^x $\mathbf{G}_{12}^4$: $(C_3^+, 0)$, $(\sigma_{v1}, 0)$: 3, x; 4, x; 6, x: *a*.
U^x $\mathbf{G}_4^1$: $(\sigma_{v1}, 0)$: 2, x; 4, x: *b*.
P^x $\mathbf{G}_6^1$: $(C_3^+, 0)$: 2, x; 4, x; 6, x: *a*.
T^x $\mathbf{G}_2^1$: $(\bar{E}, 0)$: 2, 2: *a*.
S^x $\mathbf{G}_2^1$: $(\bar{E}, 0)$: 2, 2: *a*.
T'^x $\mathbf{G}_2^1$: $(\bar{E}, 0)$: 2, 2: *a*.
S'^x $\mathbf{G}_2^1$: $(\bar{E}, 0)$: 2, 2: *a*.
Σ^x $\mathbf{G}_4^1$: $(\sigma_{v1}, 0)$: 2, x; 4, x: *b*.
R^x $\mathbf{G}_4^1$: $(\sigma_{v1}, 0)$: 2, x; 4, x: *b*.

157 $P31m$ C_{3v}^2

($F1$; $K6$; $K7$; $M5$; $Z1$.)

Γ $\mathbf{G}_{12}^4$: $\{C_3^+ \mid 000\}$, $\{\sigma_{d1} \mid 000\}$: 3, 3; 4, 3; 6, 2: *a*.
M $\mathbf{G}_4^1 \otimes \mathbf{T}_2$: $\{\sigma_{d1} \mid 000\}$; $\mathbf{t}_2$: 2, 3; 4, 3: *b*.
A $\mathbf{G}_{12}^4 \otimes \mathbf{T}_2$: $\{C_3^+ \mid 000\}$, $\{\sigma_{d1} \mid 000\}$; $\mathbf{t}_3$: 3, 3; 4, 3; 6, 2: *a*.
L $\mathbf{G}_4^1 \otimes \mathbf{T}_2$: $\{\sigma_{d1} \mid 000\}$; $\mathbf{t}_2$ or $\mathbf{t}_3$: 2, 3; 4, 3: *b*.
K $\mathbf{G}_{12}^4 \otimes \mathbf{T}_3$: $\{C_3^+ \mid 000\}$, $\{\sigma_{d1} \mid 000\}$; $\mathbf{t}_1$ or $\mathbf{t}_2$: 3, x; 4, x; 6, x: *a*.
H $\mathbf{G}_{12}^4 \otimes \mathbf{T}_3 \otimes \mathbf{T}_2$: $\{C_3^+ \mid 000\}$, $\{\sigma_{d1} \mid 000\}$; $\mathbf{t}_1$ or $\mathbf{t}_2$; $\mathbf{t}_3$: 3, x; 4, x; 6, x: *a*.

Δ^x $\mathbf{G}_{12}^4$: $(C_3^+, 0)$, $(\sigma_{d1}, 0)$: 3, x; 4, x; 6, x: *a*.
U^x $\mathbf{G}_4^1$: $(\sigma_{d1}, 0)$: 2, x; 4, x: *b*.
P^x $\mathbf{G}_{12}^4$: $(C_3^+, 0)$, $(\sigma_{d1}, 0)$: 3, x; 4, x; 6, x: *a*.
T^x $\mathbf{G}_4^1$: $(\sigma_{d2}, 0)$: 2, x; 4, x: *b*.
S^x $\mathbf{G}_4^1$: $(\sigma_{d2}, 0)$: 2, x; 4, x: *b*.
T'^x $\mathbf{G}_4^1$: $(\sigma_{d1}, 0)$: 2, x; 4, x: *b*.
S'^x $\mathbf{G}_4^1$: $(\sigma_{d1}, 0)$: 2, x; 4, x: *b*.
Σ^x $\mathbf{G}_2^1$: $(\bar{E}, 0)$: 2, 2: *a*.
R^x $\mathbf{G}_2^1$: $(\bar{E}, 0)$: 2, 2: *a*.

158 $P3c1$ C_{3v}^3

($F1$; $K6$; $K7$; $M5$; $Z1$.)

Γ $\mathbf{G}_{12}^4$: $\{C_3^+ \mid 000\}$, $\{\sigma_{v1} \mid 00\frac{1}{2}\}$: 3, 3; 4, 3; 6, 2: *a*.
M $\mathbf{G}_4^1 \otimes \mathbf{T}_2$: $\{\sigma_{v1} \mid 00\frac{1}{2}\}$; $\mathbf{t}_2$: 2, 3; 4, 3: *b*.
A $\mathbf{G}_{12}^4 \otimes \mathbf{T}_2$: $\{C_3^+ \mid 001\}$, $\{\sigma_{v1} \mid 00\frac{1}{2}\}$; $\mathbf{t}_3$: 1, 1; 2, 1; 5, 1: *b*.
L $\mathbf{G}_4^1 \otimes \mathbf{T}_2$: $\{\sigma_{v1} \mid 00\frac{1}{2}\}$; $\mathbf{t}_2$ or $\mathbf{t}_3$: 1, 1; 3, 1: *c*.
K $\mathbf{G}_6^1 \otimes \mathbf{T}_3$: $\{C_3^+ \mid 000\}$; $\mathbf{t}_1$ or $\mathbf{t}_2$: 2, 2; 4, 2; 6, 2: *a*.
H $\mathbf{G}_6^1 \otimes \mathbf{T}_3 \otimes \mathbf{T}_2$: $\{C_3^+ \mid 000\}$; $\mathbf{t}_1$ or $\mathbf{t}_2$; $\mathbf{t}_3$: 2, 1; 4, 1; 6, 1: *a*.

Δ^x $\mathbf{G}_{12}^4$: $(C_3^+, 0)$, $(\sigma_{v1}, 0)$: 3, x; 4, x; 6, x: *a*.
U^x $\mathbf{G}_4^1$: $(\sigma_{v1}, 0)$: 2, x; 4, x: *b*.
P^x $\mathbf{G}_6^1$: $(C_3^+, 0)$: 2, x; 4, x; 6, x: *a*.
T^x $\mathbf{G}_2^1$: $(\bar{E}, 0)$: 2, 2: *a*.
S^x $\mathbf{G}_2^1$: $(\bar{E}, 0)$: 2, 1: *a*.
T'^x $\mathbf{G}_2^1$: $(\bar{E}, 0)$: 2, 2: *a*.
S'^x $\mathbf{G}_2^1$: $(\bar{E}, 0)$: 2, 1: *a*.
Σ^x $\mathbf{G}_4^1$: $(\sigma_{v1}, 0)$: 2, x; 4, x: *b*.
R^x $\mathbf{G}_4^1$: $(\sigma_{v1}, 0)$: 2, x; 4, x: *b*.

159 $P31c$ C_{3v}^4

($F1$; $K6$; $K7$; $M5$; $Z1$.)

Γ $\mathbf{G}_{12}^4$: $\{C_3^+ \mid 000\}$, $\{\sigma_{d1} \mid 00\frac{1}{2}\}$: 3, 3; 4, 3; 6, 2: a.
M $\mathbf{G}_4^1 \otimes \mathbf{T}_2$: $\{\sigma_{d1} \mid 00\frac{1}{2}\}$; $\mathbf{t}_2$: 2, 3; 4, 3: b.
A $\mathbf{G}_{12}^4 \otimes \mathbf{T}_2$: $\{C_3^+ \mid 001\}$, $\{\sigma_{d1} \mid 00\frac{1}{2}\}$; $\mathbf{t}_3$: 1, 1; 2, 1; 5, 1: b.
L $\mathbf{G}_4^1 \otimes \mathbf{T}_2$: $\{\sigma_{d1} \mid 00\frac{1}{2}\}$; $\mathbf{t}_2$ or $\mathbf{t}_3$: 1, 1; 3, 1: c.
K $\mathbf{G}_{12}^4 \otimes \mathbf{T}_3$: $\{C_3^+ \mid 000\}$, $\{\sigma_{d1} \mid 00\frac{1}{2}\}$; $\mathbf{t}_1$ or $\mathbf{t}_2$: 3, x; 4, x; 6, x: a.
H $\mathbf{G}_{12}^4 \otimes \mathbf{T}_3 \otimes \mathbf{T}_2$: $\{C_3^+ \mid 001\}$, $\{\sigma_{d1} \mid 00\frac{1}{2}\}$; $\mathbf{t}_1$ or $\mathbf{t}_2$; $\mathbf{t}_3$: 1, x; 2, x; 5, x: b.

Δ^x $\mathbf{G}_{12}^4$: $(C_3^+, 0)$, $(\sigma_{d1}, 0)$: 3, x; 4, x; 6, x: a.
U^x $\mathbf{G}_4^1$: $(\sigma_{d1}, 0)$: 2, x; 4, x: b.
P^x $\mathbf{G}_{12}^4$: $(C_3^+, 0)$, $(\sigma_{d1}, 0)$: 3, x; 4, x; 6, x: a.
T^x $\mathbf{G}_4^1$: $(\sigma_{d2}, 0)$: 2, x; 4, x: b.
S^x $\mathbf{G}_4^1$: $(\sigma_{d2}, 0)$: 2, x; 4, x: b.
T'^x $\mathbf{G}_4^1$: $(\sigma_{d1}, 0)$: 2, x; 4, x: b.
S'^x $\mathbf{G}_4^1$: $(\sigma_{d1}, 0)$: 2, x; 4, x: b.
Σ^x $\mathbf{G}_2^1$: $(\bar{E}, 0)$: 2, 2: a.
R^x $\mathbf{G}_2^1$: $(\bar{E}, 0)$: 2, 1: a.

160 $R3m$ C_{3v}^5

($F1$; $K6$; $K7$; $M5$; $Z1$.)

Γ $\mathbf{G}_{12}^4$: $\{C_3^+ \mid 000\}$, $\{\sigma_{d1} \mid 000\}$: 3, 3; 4, 3; 6, 2: a.
Z $\mathbf{G}_{12}^4 \otimes \mathbf{T}_2$: $\{C_3^+ \mid 000\}$, $\{\sigma_{d1} \mid 000\}$; $\mathbf{t}_1$ or $\mathbf{t}_2$ or $\mathbf{t}_3$; 3, 3; 4, 3; 6, 2: a.
L $\mathbf{G}_4^1 \otimes \mathbf{T}_2$: $\{\sigma_{d2} \mid 000\}$; $\mathbf{t}_2$: 2, 3; 4, 3: b.
(a) F $\mathbf{G}_4^1 \otimes \mathbf{T}_2$: $\{\sigma_{d1} \mid 000\}$; $\mathbf{t}_2$ or $\mathbf{t}_3$: 2, 3; 4, 3: b.
(b) F $\mathbf{G}_4^1 \otimes \mathbf{T}_2$: $\{\sigma_{d3} \mid 000\}$; $\mathbf{t}_1$ or $\mathbf{t}_2$: 2, 3; 4, 3: b.

Λ^x $\mathbf{G}_{12}^4$: $(C_3^+, 0)$, $(\sigma_{d1}, 0)$: 3, x; 4, x; 6, x: a.
P^x $\mathbf{G}_{12}^4$: $(C_3^+, 0)$, $(\sigma_{d1}, 0)$: 3, x; 4, x; 6, x: a.
B^x $\mathbf{G}_2^1$: $(\bar{E}, 0)$: 2, 2: a.
Σ^x $\mathbf{G}_2^1$: $(\bar{E}, 0)$: 2, 2: a.
Q^x $\mathbf{G}_2^1$: $(\bar{E}, 0)$: 2, 2: a.
Y^x $\mathbf{G}_2^1$: $(\bar{E}, 0)$: 2, 2: a.

161 $R3c$ C_{3v}^6

($F1$; $K6$; $K7$; $M5$; $Z1$.)

Γ $\mathbf{G}_{12}^4$: $\{C_3^+ \mid 000\}$, $\{\sigma_{d1} \mid \frac{1}{2}\frac{1}{2}\frac{1}{2}\}$: 3, 3; 4, 3; 6, 2: a.
Z $\mathbf{G}_{12}^4 \otimes \mathbf{T}_2$: $\{C_3^+ \mid 100\}$, $\{\sigma_{d1} \mid \frac{1}{2}\frac{1}{2}\frac{1}{2}\}$; $\mathbf{t}_1$ or $\mathbf{t}_2$ or $\mathbf{t}_3$: 1, 1; 2, 1; 5, 1: b.
L $\mathbf{G}_4^1 \otimes \mathbf{T}_2$: $\{\sigma_{d2} \mid \frac{1}{2}\frac{1}{2}\frac{1}{2}\}$; $\mathbf{t}_2$: 1, 1; 3, 1: c.
(a) F $\mathbf{G}_4^1 \otimes \mathbf{T}_2$: $\{\sigma_{d1} \mid \frac{1}{2}\frac{1}{2}\frac{1}{2}\}$; $\mathbf{t}_2$ or $\mathbf{t}_3$: 2, 3; 4, 3: b.
(b) F $\mathbf{G}_4^1 \otimes \mathbf{T}_2$: $\{\sigma_{d3} \mid \frac{1}{2}\frac{1}{2}\frac{1}{2}\}$; $\mathbf{t}_1$ or $\mathbf{t}_2$: 2, 3; 4, 3: b.

Λ^x $\mathbf{G}_{12}^4$: $(C_3^+, 0)$, $(\sigma_{d1}, 0)$: 3, x; 4, x; 6, x: a.
P^x $\mathbf{G}_{12}^4$: $(C_3^+, 0)$, $(\sigma_{d1}, 0)$: 3, x; 4, x; 6, x: a.
B^x $\mathbf{G}_2^1$: $(\bar{E}, 0)$: 2, 1: a.
Σ^x $\mathbf{G}_2^1$: $(\bar{E}, 0)$: 2, 2: a.
Q^x $\mathbf{G}_2^1$: $(\bar{E}, 0)$: 2, 2: a.
Y^x $\mathbf{G}_2^1$: $(\bar{E}, 0)$: 2, 1: a.

162 $P\bar{3}1m$ D_{3d}^1

($F1$; $K6$; $K7$; $M5$; $Z1$.)

Γ $\mathbf{G}_{24}^3$: $\{C_3^+ \mid 000\}$, $\{C'_{21} \mid 000\}$, $\{I \mid 000\}$: 3, 3; 4, 3; 6, 2; 9, 3; 10, 3; 12, 2: b.
M $\mathbf{G}_8^2 \otimes \mathbf{T}_2$: $\{C'_{21} \mid 000\}$, $\{I \mid 000\}$; $\mathbf{t}_2$: 2, 3; 4, 3; 6, 3; 8, 3: b.
A $\mathbf{G}_{24}^3 \otimes \mathbf{T}_2$: $\{C_3^+ \mid 000\}$, $\{C'_{21} \mid 000\}$, $\{I \mid 000\}$; $\mathbf{t}_3$: 3, 3; 4, 3; 6, 2; 9, 3; 10, 3; 12, 2: b.
L $\mathbf{G}_8^2 \otimes \mathbf{T}_2$: $\{C'_{21} \mid 000\}$, $\{I \mid 000\}$; $\mathbf{t}_2$ or $\mathbf{t}_3$: 2, 3; 4, 3; 6, 3; 8, 3: b.
K $\mathbf{G}_{12}^4 \otimes \mathbf{T}_3$: $\{C_3^+ \mid 000\}$, $\{\sigma_{d1} \mid 000\}$; $\mathbf{t}_1$ or $\mathbf{t}_2$: 3, 3; 4, 3; 6, 2: a.
H $\mathbf{G}_{12}^4 \otimes \mathbf{T}_3 \otimes \mathbf{T}_2$: $\{C_3^+ \mid 000\}$, $\{\sigma_{d1} \mid 000\}$; $\mathbf{t}_1$ or $\mathbf{t}_2$; $\mathbf{t}_3$: 3, 3; 4, 3; 6, 2: a.

Δ^x $\mathbf{G}_{12}^4$: $(C_3^+, 0)$, $(\sigma_{d1}, 0)$: 3, 3; 4, 3; 6, 2: a.
U^x $\mathbf{G}_4^1$: $(\sigma_{d1}, 0)$: 2, 3; 4, 3: b.
P^x $\mathbf{G}_{12}^4$: $(C_3^+, 0)$, $(\sigma_{d1}, 0)$: 3, 3; 4, 3; 6, 2: a.
T^x $\mathbf{G}_4^1$: $(\sigma_{d2}, 0)$: 2, 3; 4, 3: b.
S^x $\mathbf{G}_4^1$: $(\sigma_{d2}, 0)$: 2, 3; 4, 3: b.
T'^x $\mathbf{G}_4^1$: $(\sigma_{d1}, 0)$: 2, 3; 4, 3: b.
S'^x $\mathbf{G}_4^1$: $(\sigma_{d1}, 0)$: 2, 3; 4, 3: b.
Σ^x $\mathbf{G}_4^1$: $(C'_{21}, 0)$: 2, 3; 4, 3: b.
R^x $\mathbf{G}_4^1$: $(C'_{21}, 0)$: 2, 3; 4, 3: b.

163 $P\bar{3}1c$ D_{3d}^2

($F1$; $K6$; $K7$; $M5$; $Z1$.)

Γ $\mathbf{G}_{24}^3$: $\{C_3^+ \mid 000\}$, $\{C'_{21} \mid 00\frac{1}{2}\}$, $\{I \mid 000\}$: 3, 3; 4, 3; 6, 2; 9, 3; 10, 3; 12, 2: b.
M $\mathbf{G}_8^2 \otimes \mathbf{T}_2$: $\{C'_{21} \mid 00\frac{1}{2}\}$, $\{I \mid 000\}$; $\mathbf{t}_2$: 2, 3; 4, 3; 6, 3; 8, 3: b.
A $\mathbf{G}_{48}^9$: $\{\bar{C}_3^+ \mid 001\}$, $\{C'_{21} \mid 00\frac{1}{2}\}$, $\{I \mid 000\}$: 10, 1; 17, 3; 18, 3: a.
L $\mathbf{G}_{16}^{10}$: $\{C'_{21} \mid 00\frac{1}{2}\}$, $\{E \mid 001\}$, $\{I \mid 000\}$: 10, 1: b.
K $\mathbf{G}_{12}^4 \otimes \mathbf{T}_3$: $\{C_3^+ \mid 000\}$, $\{\sigma_{d1} \mid 00\frac{1}{2}\}$; $\mathbf{t}_1$ or $\mathbf{t}_2$: 3, 3; 4, 3; 6, 2: a.
H $\mathbf{G}_{12}^4 \otimes \mathbf{T}_3 \otimes \mathbf{T}_2$: $\{C_3^+ \mid 001\}$, $\{\sigma_{d1} \mid 00\frac{1}{2}\}$; $\mathbf{t}_1$ or $\mathbf{t}_2$; $\mathbf{t}_3$: 1, 3; 2, 3; 5, 2: b.

Δ^x $\mathbf{G}_{12}^4$: $(C_3^+, 0)$, $(\sigma_{d1}, 0)$: 3, 3; 4, 3; 6, 2: a.
U^x $\mathbf{G}_4^1$: $(\sigma_{d1}, 0)$: 2, 3; 4, 3: b.
P^x $\mathbf{G}_{12}^4$: $(C_3^+, 0)$, $(\sigma_{d1}, 0)$: 3, 3; 4, 3; 6, 2: a.
T^x $\mathbf{G}_4^1$: $(\sigma_{d2}, 0)$: 2, 3; 4, 3: b.
S^x $\mathbf{G}_4^1$: $(\sigma_{d2}, 0)$: 2, 3; 4, 3: b.
T'^x $\mathbf{G}_4^1$: $(\sigma_{d1}, 0)$: 2, 3; 4, 3: b.
S'^x $\mathbf{G}_4^1$: $(\sigma_{d1}, 0)$: 2, 3; 4, 3: b.
Σ^x $\mathbf{G}_4^1$: $(C'_{21}, 0)$: 2, 3; 4, 3: b.
R^x $\mathbf{G}_8^2$: $(C'_{21}, 0)$, $(E, 1)$: 5, 1; 7, 1: a.

164 $P\bar{3}m1$ D_{3d}^3

($F1$; $K6$; $K7$; $M5$; $S8$; $Z1$.)

Γ $\mathbf{G}_{24}^3$: $\{C_3^+ \mid 000\}$, $\{C''_{21} \mid 000\}$, $\{I \mid 000\}$: 3, 3; 4, 3; 6, 2; 9, 3; 10, 3; 12, 2: b.
M $\mathbf{G}_8^2 \otimes \mathbf{T}_2$: $\{C''_{21} \mid 000\}$, $\{I \mid 000\}$; $\mathbf{t}_2$: 2, 3; 4, 3; 6, 3; 8, 3: b.
A $\mathbf{G}_{24}^3 \otimes \mathbf{T}_2$: $\{C_3^+ \mid 000\}$, $\{C''_{21} \mid 000\}$, $\{I \mid 000\}$; $\mathbf{t}_3$: 3, 3; 4, 3; 6, 2; 9, 3; 10, 3; 12, 2: b.
L $\mathbf{G}_8^2 \otimes \mathbf{T}_2$: $\{C''_{21} \mid 000\}$, $\{I \mid 000\}$; $\mathbf{t}_2$ or $\mathbf{t}_3$: 2, 3; 4, 3; 6, 3; 8, 3: b.
K $\mathbf{G}_{12}^4 \otimes \mathbf{T}_3$: $\{C_3^+ \mid 000\}$, $\{C''_{21} \mid 000\}$; $\mathbf{t}_1$ or $\mathbf{t}_2$: 3, 3; 4, 3; 6, 2: a.
H $\mathbf{G}_{12}^4 \otimes \mathbf{T}_3 \otimes \mathbf{T}_2$: $\{C_3^+ \mid 000\}$, $\{C''_{21} \mid 000\}$; $\mathbf{t}_1$ or $\mathbf{t}_2$; $\mathbf{t}_3$: 3, 3; 4, 3; 6, 2: a.

Δ^x $\mathbf{G}_{12}^4$: $(C_3^+, 0)$, $(\sigma_{v1}, 0)$: 3, 3; 4, 3; 6, 2: a.
U^x $\mathbf{G}_4^1$: $(\sigma_{v1}, 0)$: 2, 3; 4, 3: b.
P^x $\mathbf{G}_6^1$: $(C_3^+, 0)$: 2, 3; 4, 1; 6, 3: a.
T^x $\mathbf{G}_4^1$: $(C''_{22}, 0)$: 2, 3; 4, 3: b.
S^x $\mathbf{G}_4^1$: $(C''_{22}, 0)$: 2, 3; 4, 3: b.
T'^x $\mathbf{G}_4^1$: $(C''_{21}, 0)$: 2, 3; 4, 3: b.
S'^x $\mathbf{G}_4^1$: $(C''_{21}, 0)$: 2, 3; 4, 3: b.
Σ^x $\mathbf{G}_4^1$: $(\sigma_{v1}, 0)$: 2, 3; 4, 3: b.
R^x $\mathbf{G}_4^1$: $(\sigma_{v1}, 0)$: 2, 3; 4, 3: b.

165 $P\bar{3}c1$ D_{3d}^{4}

($F1$; $K6$; $K7$; $M5$; $Z1$.)

Γ $\mathbf{G}_{24}^{3}$: $\{C_3^+ \mid 000\}$, $\{C''_{21} \mid 00\tfrac{1}{2}\}$, $\{I \mid 000\}$: 3, 3; 4, 3; 6, 2; 9, 3; 10, 3; 12, 2: b.
M $\mathbf{G}_{8}^{2} \otimes \mathbf{T}_2$: $\{C''_{21} \mid 00\tfrac{1}{2}\}$, $\{I \mid 000\}$; $\mathbf{t}_2$: 2, 3; 4, 3; 6, 3; 8, 3: b.
A $\mathbf{G}_{48}^{9}$: $\{\bar{C}_3^+ \mid 001\}$, $\{C''_{21} \mid 00\tfrac{1}{2}\}$, $\{I \mid 000\}$: 10, 1; 17, 3; 18, 3: a.
L $\mathbf{G}_{16}^{10}$: $\{C''_{21} \mid 00\tfrac{1}{2}\}$, $\{E \mid 001\}$, $\{I \mid 000\}$: 10, 1: b.
K $\mathbf{G}_{12}^{4} \otimes \mathbf{T}_3$: $\{C_3^+ \mid 000\}$, $\{C''_{21} \mid 00\tfrac{1}{2}\}$; $\mathbf{t}_1$ or $\mathbf{t}_2$: 3, 3; 4, 3; 6, 2: a.
H $\mathbf{G}_{12}^{4} \otimes \mathbf{T}_3 \otimes \mathbf{T}_2$: $\{C_3^+ \mid 000\}$, $\{C''_{21} \mid 00\tfrac{1}{2}\}$; $\mathbf{t}_1$ or $\mathbf{t}_2$; $\mathbf{t}_3$: 3, 1; 4, 1; 6, 1: a.

Δ^x $\mathbf{G}_{12}^{4}$: $(C_3^+, 0)$, $(\sigma_{v1}, 0)$: 3, 3; 4, 3; 6, 2: a.
U^x $\mathbf{G}_{4}^{1}$: $(\sigma_{v1}, 0)$: 2, 3; 4, 3: b.
P^x $\mathbf{G}_{6}^{1}$: $(C_3^+, 0)$: 2, 3; 4, 1; 6, 3: a.
T^x $\mathbf{G}_{4}^{1}$: $(C''_{22}, 0)$: 2, 3; 4, 3: b.
S^x $\mathbf{G}_{8}^{2}$: $(C''_{22}, 0)$, $(E, 1)$: 5, 1; 7, 1: a.
T'^x $\mathbf{G}_{4}^{1}$: $(C''_{21}, 0)$: 2, 3; 4, 3: b.
S'^x $\mathbf{G}_{8}^{2}$: $(C''_{21}, 0)$, $(E, 1)$: 5, 1; 7, 1: a.
Σ^x $\mathbf{G}_{4}^{1}$: $(\sigma_{v1}, 0)$: 2, 3; 4, 3: b.
R^x $\mathbf{G}_{4}^{1}$: $(\sigma_{v1}, 0)$: 2, 3; 4, 3: b.

166 $R\bar{3}m$ D_{3d}^{5}

($F1$; $F3$; $K6$; $K7$; $L2$; $M1$; $M5$; $S8$; $Y2$; $Z1$.)

Γ $\mathbf{G}_{24}^{3}$: $\{C_3^+ \mid 000\}$, $\{C'_{21} \mid 000\}$, $\{I \mid 000\}$: 3, 3; 4, 3; 6, 2; 9, 3; 10, 3; 12, 2: b.
Z $\mathbf{G}_{24}^{3} \otimes \mathbf{T}_2$: $\{C_3^+ \mid 000\}$, $\{C'_{21} \mid 000\}$, $\{I \mid 000\}$; $\mathbf{t}_1$ or $\mathbf{t}_2$ or $\mathbf{t}_3$: 3, 3; 4, 3; 6, 2; 9, 3; 10, 3; 12, 2: b.
L $\mathbf{G}_{8}^{2} \otimes \mathbf{T}_2$: $\{C'_{22} \mid 000\}$, $\{I \mid 000\}$; $\mathbf{t}_2$: 2, 3; 4, 3; 6, 3; 8, 3: b.
{(a) F $\mathbf{G}_{8}^{2} \otimes \mathbf{T}_2$: $\{C'_{21} \mid 000\}$, $\{I \mid 000\}$; $\mathbf{t}_2$ or $\mathbf{t}_3$: 2, 3; 4, 3; 6, 3; 8, 3: b.}
{(b) F $\mathbf{G}_{8}^{2} \otimes \mathbf{T}_2$: $\{C'_{23} \mid 000\}$, $\{I \mid 000\}$; $\mathbf{t}_1$ or $\mathbf{t}_2$: 2, 3; 4, 3; 6, 3; 8, 3: b.}

Λ^x $\mathbf{G}_{12}^{4}$: $(C_3^+, 0)$, $(\sigma_{d1}, 0)$: 3, 3; 4, 3; 6, 2: a.
P^x $\mathbf{G}_{12}^{4}$: $(C_3^+, 0)$, $(\sigma_{d1}, 0)$: 3, 3; 4, 3; 6, 2: a.
B^x $\mathbf{G}_{4}^{1}$: $(C'_{21}, 0)$: 2, 3; 4, 3: b.
Σ^x $\mathbf{G}_{4}^{1}$: $(C'_{21}, 0)$: 2, 3; 4, 3: b.
Q^x $\mathbf{G}_{4}^{1}$: $(C'_{23}, 0)$: 2, 3; 4, 3: b.
Y^x $\mathbf{G}_{4}^{1}$: $(C'_{22}, 0)$: 2, 3; 4, 3: b.

167 $R\bar{3}c$ D_{3d}^{6}

($F1$; $K6$; $K7$; $M5$; $S8$; $Z1$.)

Γ $\mathbf{G}_{24}^{3}$: $\{C_3^+ \mid 000\}$, $\{C'_{21} \mid \tfrac{1}{2}\tfrac{1}{2}\tfrac{1}{2}\}$, $\{I \mid 000\}$: 3, 3; 4, 3; 6, 2; 9, 3; 10, 3; 12, 2: b.
Z $\mathbf{G}_{48}^{9}$: $\{\bar{C}_3^+ \mid 001\}$, $\{C'_{21} \mid \tfrac{1}{2}\tfrac{1}{2}\tfrac{1}{2}\}$, $\{I \mid 000\}$: 10, 1; 17, 3; 18, 3: a.
L $\mathbf{G}_{16}^{10}$: $\{C'_{22} \mid \tfrac{1}{2}\tfrac{1}{2}\tfrac{1}{2}\}$, $\{E \mid 010\}$, $\{I \mid 000\}$: 10, 1: b.
{(a) F $\mathbf{G}_{8}^{2} \otimes \mathbf{T}_2$: $\{C'_{21} \mid \tfrac{1}{2}\tfrac{1}{2}\tfrac{1}{2}\}$, $\{I \mid 000\}$; $\mathbf{t}_2$ or $\mathbf{t}_3$: 2, 3; 4, 3; 6, 3; 8, 3: b.}
{(b) F $\mathbf{G}_{8}^{2} \otimes \mathbf{T}_2$: $\{C'_{23} \mid \tfrac{1}{2}\tfrac{1}{2}\tfrac{1}{2}\}$, $\{I \mid 000\}$; $\mathbf{t}_1$ or $\mathbf{t}_2$: 2, 3; 4, 3; 6, 3; 8, 3: b.}

Λ^x $\mathbf{G}_{12}^{4}$: $(C_3^+, 0)$, $(\sigma_{d1}, 0)$: 3, 3; 4, 3; 6, 2: a.
P^x $\mathbf{G}_{12}^{4}$: $(C_3^+, 0)$, $(\sigma_{d1}, 0)$: 3, 3; 4, 3; 6, 2: a.
B^x $\mathbf{G}_{8}^{2}$: $(C'_{21}, 0)$, $(E, 1)$: 5, 1; 7, 1: a.
Σ^x $\mathbf{G}_{4}^{1}$: $(C'_{21}, 0)$: 2, 3; 4, 3: b.
Q^x $\mathbf{G}_{4}^{1}$: $(C'_{23}, 0)$: 2, 3; 4, 3: b.
Y^x $\mathbf{G}_{8}^{2}$: $(C'_{22}, 0)$, $(E, 1)$: 5, 1; 7, 1: a.

168 $P6$ C_6^1

($F1$; $K6$; $K7$; $M5$; $Z1$.)

Γ $\mathbf{G}_{12}^1$: $\{C_6^+ \mid 000\}$: 2, 3; 4, 3; 6, 3; 8, 3; 10, 3; 12, 3: c.
M $\mathbf{G}_4^1 \otimes \mathbf{T}_2$: $\{C_2 \mid 000\}$; $\mathbf{t}_2$: 2, 3; 4, 3: b.
A $\mathbf{G}_{12}^1 \otimes \mathbf{T}_2$: $\{C_6^+ \mid 000\}$; $\mathbf{t}_3$: 2, 3; 4, 3; 6, 3; 8, 3; 10, 3; 12, 3: c.
L $\mathbf{G}_4^1 \otimes \mathbf{T}_2$: $\{C_2 \mid 000\}$; $\mathbf{t}_2$ or $\mathbf{t}_3$: 2, 3; 4, 3: b.
K $\mathbf{G}_6^1 \otimes \mathbf{T}_3$: $\{C_3^+ \mid 000\}$; $\mathbf{t}_1$ or $\mathbf{t}_2$: 2, 3; 4, 2; 6, 3: a.
H $\mathbf{G}_6^1 \otimes \mathbf{T}_3 \otimes \mathbf{T}_2$: $\{C_3^+ \mid 000\}$; $\mathbf{t}_1$ or $\mathbf{t}_2$; $\mathbf{t}_3$: 2, 3; 4, 2; 6, 3: a.

Δ^x $\mathbf{G}_{12}^1$: $(C_6^+, 0)$: 2, x; 4, x; 6, x; 8, x; 10, x; 12, x: c.
U^x $\mathbf{G}_4^1$: $(C_2, 0)$: 2, x; 4, x: b.
P^x $\mathbf{G}_6^1$: $(C_3^+, 0)$: 2, x; 4, x; 6, x: a.
T^x $\mathbf{G}_2^1$: $(\bar{E}, 0)$: 2, 2: a.
S^x $\mathbf{G}_2^1$: $(\bar{E}, 0)$: 2, 2: a.
T'^x $\mathbf{G}_2^1$: $(\bar{E}, 0)$: 2, 2: a.
S'^x $\mathbf{G}_2^1$: $(\bar{E}, 0)$: 2, 2: a.
Σ^x $\mathbf{G}_2^1$: $(\bar{E}, 0)$: 2, 2: a.
R^x $\mathbf{G}_2^1$: $(\bar{E}, 0)$: 2, 2: a.

169 $P6_1$ C_6^2

($F1$; $K6$; $K7$; $M5$; $Z1$.)

Γ $\mathbf{G}_{12}^1$: $\{C_6^+ \mid 00\frac{1}{6}\}$: 2, 3; 4, 3; 6, 3; 8, 3; 10, 3; 12, 3: c.
M $\mathbf{G}_4^1 \otimes \mathbf{T}_2$: $\{C_2 \mid 00\frac{1}{2}\}$; $\mathbf{t}_2$: 2, 3; 4, 3: b.
A $\mathbf{G}_{12}^1 \otimes \mathbf{T}_2$: $\{C_6^+ \mid 00\frac{1}{6}\}$; $\mathbf{t}_3$: 1, 1; 3, 3; 5, 3; 7, 1; 9, 3; 11, 3: d.
L $\mathbf{G}_4^1 \otimes \mathbf{T}_2$: $\{C_2 \mid 00\frac{1}{2}\}$; $\mathbf{t}_2$ or $\mathbf{t}_3$: 1, 1; 3, 1: c.
K $\mathbf{G}_6^1 \otimes \mathbf{T}_3$: $\{C_3^+ \mid 00\frac{1}{3}\}$; $\mathbf{t}_1$ or $\mathbf{t}_2$: 2, 3; 4, 2; 6, 3: a.
H $\mathbf{G}_6^1 \otimes \mathbf{T}_3 \otimes \mathbf{T}_2$: $\{C_3^- \mid 00\frac{2}{3}\}$; $\mathbf{t}_1$ or $\mathbf{t}_2$; $\mathbf{t}_3$: 2, 3; 4, 1; 6, 3: a.

Δ^x $\mathbf{G}_{12}^1$: $(C_6^+, 0)$: 2, x; 4, x; 6, x; 8, x; 10, x; 12, x: c.
U^x $\mathbf{G}_4^1$: $(C_2, 0)$: 2, x; 4, x: b.
P^x $\mathbf{G}_6^1$: $(C_3^+, 0)$: 2, x; 4, x; 6, x: a.
T^x $\mathbf{G}_2^1$: $(\bar{E}, 0)$: 2, 2: a.
S^x $\mathbf{G}_2^1$: $(\bar{E}, 0)$: 2, 1: a.
T'^x $\mathbf{G}_2^1$: $(\bar{E}, 0)$: 2, 2: a.
S'^x $\mathbf{G}_2^1$: $(\bar{E}, 0)$: 2, 1: a.
Σ^x $\mathbf{G}_2^1$: $(\bar{E}, 0)$: 2, 2: a.
R^x $\mathbf{G}_2^1$: $(\bar{E}, 0)$: 2, 1: a.

170 $P6_5$ C_6^3

($F1$; $K6$; $K7$; $M5$; $Z1$.)

Γ $\mathbf{G}_{12}^1$: $\{C_6^+ \mid 00\frac{5}{6}\}$: 2, 3; 4, 3; 6, 3; 8, 3; 10, 3; 12, 3: c.
M $\mathbf{G}_4^1 \otimes \mathbf{T}_2$: $\{C_2 \mid 00\frac{1}{2}\}$; $\mathbf{t}_2$: 2, 3; 4, 3: b.
A $\mathbf{G}_{12}^1 \otimes \mathbf{T}_2$: $\{C_6^+ \mid 00\frac{5}{6}\}$; $\mathbf{t}_3$: 1, 1; 3, 3; 5, 3; 7, 1; 9, 3; 11, 3: d.
L $\mathbf{G}_4^1 \otimes \mathbf{T}_2$: $\{C_2 \mid 00\frac{1}{2}\}$; $\mathbf{t}_2$ or $\mathbf{t}_3$: 1, 1; 3, 1: c.
K $\mathbf{G}_6^1 \otimes \mathbf{T}_3$: $\{C_3^+ \mid 00\frac{2}{3}\}$; $\mathbf{t}_1$ or $\mathbf{t}_2$: 2, 3; 4, 2; 6, 3: a.
H $\mathbf{G}_6^1 \otimes \mathbf{T}_3 \otimes \mathbf{T}_2$: $\{C_3^+ \mid 00\frac{2}{3}\}$; $\mathbf{t}_1$ or $\mathbf{t}_2$; $\mathbf{t}_3$: 2, 3; 4, 1; 6, 3: a.

Δ^x $\mathbf{G}_{12}^1$: $(C_6^+, 0)$: 2, x; 4, x; 6, x; 8, x; 10, x; 12, x: c.
U^x $\mathbf{G}_4^1$: $(C_2, 0)$: 2, x; 4, x: b.
P^x $\mathbf{G}_6^1$: $(C_3^+, 0)$: 2, x; 4, x; 6, x: a.
T^x $\mathbf{G}_2^1$: $(\bar{E}, 0)$: 2, 2: a.
S^x $\mathbf{G}_2^1$: $(\bar{E}, 0)$: 2, 1: a.
T'^x $\mathbf{G}_2^1$: $(\bar{E}, 0)$: 2, 2: a.
S'^x $\mathbf{G}_2^1$: $(\bar{E}, 0)$: 2, 1: a.
Σ^x $\mathbf{G}_2^1$: $(\bar{E}, 0)$: 2, 2: a.
R^x $\mathbf{G}_2^1$: $(\bar{E}, 0)$: 2, 1: a.

171 $P6_2$ C_6^4

($F1$; $K6$; $K7$; $M5$; $Z1$.)

Γ $\mathbf{G}^1_{12}$: $\{C_6^+ \mid 00\tfrac{1}{3}\}$: 2, 3; 4, 3; 6, 3; 8, 3; 10, 3; 12, 3: c.
M $\mathbf{G}^1_4 \otimes \mathbf{T}_2$: $\{C_2 \mid 000\}$; $\mathbf{t}_2$: 2, 3; 4, 3: b.
A $\mathbf{G}^1_{12} \otimes \mathbf{T}_2$: $\{C_6^+ \mid 00\tfrac{1}{3}\}$; $\mathbf{t}_3$: 2, 3; 4, 3; 6, 3; 8, 3; 10, 3; 12, 3: c.
L $\mathbf{G}^1_4 \otimes \mathbf{T}_2$: $\{C_2 \mid 000\}$; $\mathbf{t}_2$ or $\mathbf{t}_3$: 2, 3; 4, 3: b.
K $\mathbf{G}^1_6 \otimes \mathbf{T}_3$: $\{C_3^+ \mid 00\tfrac{2}{3}\}$; $\mathbf{t}_1$ or $\mathbf{t}_2$: 2, 3; 4, 2; 6, 3: a.
H $\mathbf{G}^1_6 \otimes \mathbf{T}_3 \otimes \mathbf{T}_2$: $\{C_3^+ \mid 00\tfrac{2}{3}\}$; $\mathbf{t}_1$ or $\mathbf{t}_2$; $\mathbf{t}_3$: 2, 3; 4, 2; 6, 3: a.

Δ^x $\mathbf{G}^1_{12}$: $(C_6^+, 0)$: 2, x; 4, x; 6, x; 8, x; 10, x; 12, x: c.
U^x $\mathbf{G}^1_4$: $(C_2, 0)$: 2, x; 4, x: b.
P^x $\mathbf{G}^1_6$: $(C_3^+, 0)$: 2, x; 4, x; 6, x: a.
T^x $\mathbf{G}^1_2$: $(\bar{E}, 0)$: 2, 2: a.
S^x $\mathbf{G}^1_2$: $(\bar{E}, 0)$: 2, 2: a.
T'^x $\mathbf{G}^1_2$: $(\bar{E}, 0)$: 2, 2: a.
S'^x $\mathbf{G}^1_2$: $(\bar{E}, 0)$: 2, 2: a.
Σ^x $\mathbf{G}^1_2$: $(\bar{E}, 0)$: 2, 2: a.
R^x $\mathbf{G}^1_2$: $(\bar{E}, 0)$: 2, 2: a.

172 $P6_4$ C_6^5

($F1$; $K6$; $K7$; $M5$; $Z1$.)

Γ $\mathbf{G}^1_{12}$: $\{C_6^+ \mid 00\tfrac{2}{3}\}$: 2, 3; 4, 3; 6, 3; 8, 3; 10, 3; 12, 3: c.
M $\mathbf{G}^1_4 \otimes \mathbf{T}_2$: $\{C_2 \mid 000\}$; $\mathbf{t}_2$: 2, 3; 4, 3: b.
A $\mathbf{G}^1_{12} \otimes \mathbf{T}_2$: $\{C_6^+ \mid 00\tfrac{2}{3}\}$; $\mathbf{t}_3$: 2, 3; 4, 3; 6, 3; 8, 3; 10, 3; 12, 3: c.
L $\mathbf{G}^1_4 \otimes \mathbf{T}_2$: $\{C_2 \mid 000\}$; $\mathbf{t}_2$ or $\mathbf{t}_3$: 2, 3; 4, 3: b.
K $\mathbf{G}^1_6 \otimes \mathbf{T}_3$: $\{C_3^+ \mid 00\tfrac{1}{3}\}$; $\mathbf{t}_1$ or $\mathbf{t}_2$: 2, 3; 4, 2; 6, 3: a.
H $\mathbf{G}^1_6 \otimes \mathbf{T}_3 \otimes \mathbf{T}_2$: $\{C_3^- \mid 00\tfrac{2}{3}\}$; $\mathbf{t}_1$ or $\mathbf{t}_2$; $\mathbf{t}_3$: 2, 3; 4, 2; 6, 3: a.

Δ^x $\mathbf{G}^1_{12}$: $(C_6^+, 0)$: 2, x; 4, x; 6, x; 8, x; 10, x; 12, x: c.
U^x $\mathbf{G}^1_4$: $(C_2, 0)$: 2, x; 4, x: b.
P^x $\mathbf{G}^1_6$: $(C_3^+, 0)$: 2, x; 4, x; 6, x: a.
T^x $\mathbf{G}^1_2$: $(\bar{E}, 0)$: 2, 2: a.
S^x $\mathbf{G}^1_2$: $(\bar{E}, 0)$: 2, 2: a.
T'^x $\mathbf{G}^1_2$: $(\bar{E}, 0)$: 2, 2: a.
S'^x $\mathbf{G}^1_2$: $(\bar{E}, 0)$: 2, 2: a.
Σ^x $\mathbf{G}^1_2$: $(\bar{E}, 0)$: 2, 2: a.
R^x $\mathbf{G}^1_2$: $(\bar{E}, 0)$: 2, 2: a.

173 $P6_3$ C_6^6

($F1$; $K6$; $K7$; $M5$; $Z1$.)

Γ $\mathbf{G}^1_{12}$: $\{C_6^+ \mid 00\tfrac{1}{2}\}$: 2, 3; 4, 3; 6, 3; 8, 3; 10, 3; 12, 3: c.
M $\mathbf{G}^1_4 \otimes \mathbf{T}_2$: $\{C_2 \mid 00\tfrac{1}{2}\}$; $\mathbf{t}_2$: 2, 3; 4, 3: b.
A $\mathbf{G}^1_{12} \otimes \mathbf{T}_2$: $\{C_6^+ \mid 00\tfrac{1}{2}\}$; $\mathbf{t}_3$: 1, 1; 3, 3; 5, 3; 7, 1; 9, 3; 11, 3: d.
L $\mathbf{G}^1_4 \otimes \mathbf{T}_2$: $\{C_2 \mid 00\tfrac{1}{2}\}$; $\mathbf{t}_2$ or $\mathbf{t}_3$: 1, 1; 3, 1: c.
K $\mathbf{G}^1_6 \otimes \mathbf{T}_3$: $\{C_3^+ \mid 000\}$; $\mathbf{t}_1$ or $\mathbf{t}_2$: 2, 3; 4, 2; 6, 3: a.
H $\mathbf{G}^1_6 \otimes \mathbf{T}_3 \otimes \mathbf{T}_2$: $\{C_3^+ \mid 000\}$; $\mathbf{t}_1$ or $\mathbf{t}_2$; $\mathbf{t}_3$: 2, 3; 4, 1; 6, 3: a.

Δ^x $\mathbf{G}^1_{12}$: $(C_6^+, 0)$: 2, x; 4, x; 6, x; 8, x; 10, x; 12, x: c.
U^x $\mathbf{G}^1_4$: $(C_2, 0)$: 2, x; 4, x: b.
P^x $\mathbf{G}^1_6$: $(C_3^+, 0)$: 2, x; 4, x; 6, x: a.
T^x $\mathbf{G}^1_2$: $(\bar{E}, 0)$: 2, 2: a.
S^x $\mathbf{G}^1_2$: $(\bar{E}, 0)$: 2, 1: a.
T'^x $\mathbf{G}^1_2$: $(\bar{E}, 0)$: 2, 2: a.
S'^x $\mathbf{G}^1_2$: $(\bar{E}, 0)$: 2, 1: a.
Σ^x $\mathbf{G}^1_2$: $(\bar{E}, 0)$: 2, 2: a.
R^x $\mathbf{G}^1_2$: $(\bar{E}, 0)$: 2, 1: a.

174 $P\bar{6}$ C_{3h}^1

($F1$; $K6$; $K7$; $M5$; $Z1$.)

Γ $\mathbf{G}_{12}^1$: $\{S_3^+ \mid 000\}$: 2, 3; 4, 3; 6, 3; 8, 3; 10, 3; 12, 3: *c*.
M $\mathbf{G}_4^1 \otimes \mathbf{T}_2$: $\{\sigma_h \mid 000\}$; $\mathbf{t}_2$: 2, 3; 4, 3: *b*.
A $\mathbf{G}_{12}^1 \otimes \mathbf{T}_2$: $\{S_3^+ \mid 000\}$; $\mathbf{t}_3$: 2, 3; 4, 3; 6, 3; 8, 3; 10, 3; 12, 3: *c*.
L $\mathbf{G}_4^1 \otimes \mathbf{T}_2$: $\{\sigma_h \mid 000\}$; $\mathbf{t}_2$ or $\mathbf{t}_3$: 2, 3; 4, 3: *b*.
K $\mathbf{G}_{12}^1 \otimes \mathbf{T}_3$: $\{S_3^+ \mid 000\}$; $\mathbf{t}_1$ or $\mathbf{t}_2$: 2, x; 4, x; 6, x; 8, x; 10, x; 12, x: *c*.
H $\mathbf{G}_{12}^1 \otimes \mathbf{T}_3 \otimes \mathbf{T}_2$: $\{S_3^+ \mid 000\}$; $\mathbf{t}_1$ or $\mathbf{t}_2$; $\mathbf{t}_3$: 2, x; 4, x; 6, x; 8, x; 10, x; 12, x: *c*.

Δ^x $\mathbf{G}_6^1$: $(C_3^+, 0)$: 2, 3; 4, 2; 6, 3: *a*.
U^x $\mathbf{G}_2^1$: $(\bar{E}, 0)$: 2, 2: *a*.
P^x $\mathbf{G}_6^1$: $(C_3^+, 0)$: 2, x; 4, x; 6, x: *a*.
T^x $\mathbf{G}_4^1$: $(\sigma_h, 0)$: 2, x; 4, x: *b*.
S^x $\mathbf{G}_4^1$: $(\sigma_h, 0)$: 2, x; 4, x: *b*.
T'^x $\mathbf{G}_4^1$: $(\sigma_h, 0)$: 2, x; 4, x: *b*.
S'^x $\mathbf{G}_4^1$: $(\sigma_h, 0)$: 2, x; 4, x: *b*.
Σ^x $\mathbf{G}_4^1$: $(\sigma_h, 0)$: 2, x; 4, x: *b*.
R^x $\mathbf{G}_4^1$: $(\sigma_h, 0)$: 2, x; 4, x: *b*.

175 $P6/m$ C_{6h}^1

($F1$; $K6$; $K7$; $M5$; $Z1$.)

Γ $\mathbf{G}_{24}^{12}$: $\{C_6^+ \mid 000\}$, $\{I \mid 000\}$: 2, 3; 4, 3; 6, 3; 8, 3; 10, 3; 12, 3; 14, 3; 16, 3; 18, 3; 20, 3; 22, 3; 24, 3: *a*.
M $\mathbf{G}_8^2 \otimes \mathbf{T}_2$: $\{C_2 \mid 000\}$, $\{I \mid 000\}$; $\mathbf{t}_2$: 2, 3; 4, 3; 6, 3; 8, 3: *b*.
A $\mathbf{G}_{24}^{12} \otimes \mathbf{T}_2$: $\{C_6^+ \mid 000\}$, $\{I \mid 000\}$; $\mathbf{t}_3$: 2, 3; 4, 3; 6, 3; 8, 3; 10, 3; 12, 3; 14, 3; 16, 3; 18, 3; 20, 3; 22, 3; 24, 3: *a*.
L $\mathbf{G}_8^2 \otimes \mathbf{T}_2$: $\{C_2 \mid 000\}$, $\{I \mid 000\}$; $\mathbf{t}_2$ or $\mathbf{t}_3$: 2, 3; 4, 3; 6, 3; 8, 3: *b*.
K $\mathbf{G}_{12}^1 \otimes \mathbf{T}_3$: $\{S_3^+ \mid 000\}$; $\mathbf{t}_1$ or $\mathbf{t}_2$: 2, 3; 4, 3; 6, 3; 8, 3; 10, 3; 12, 3: *c*.
H $\mathbf{G}_{12}^1 \otimes \mathbf{T}_3 \otimes \mathbf{T}_2$: $\{S_3^+ \mid 000\}$; $\mathbf{t}_1$ or $\mathbf{t}_2$; $\mathbf{t}_3$: 2, 3; 4, 3; 6, 3; 8, 3; 10, 3; 12, 3: *c*.

Δ^x $\mathbf{G}_{12}^1$: $(C_6^+, 0)$: 2, 3; 4, 3; 6, 3; 8, 3; 10, 3; 12, 3: *c*.
U^x $\mathbf{G}_4^1$: $(C_2, 0)$: 2, 3; 4, 3: *b*.
P^x $\mathbf{G}_6^1$: $(C_3^+, 0)$: 2, 3; 4, 1; 6, 3: *a*.
T^x $\mathbf{G}_4^1$: $(\sigma_h, 0)$: 2, 3; 4, 3: *b*.
S^x $\mathbf{G}_4^1$: $(\sigma_h, 0)$: 2, 3; 4, 3: *b*.
T'^x $\mathbf{G}_4^1$: $(\sigma_h, 0)$: 2, 3; 4, 3: *b*.
S'^x $\mathbf{G}_4^1$: $(\sigma_h, 0)$: 2, 3; 4, 3: *b*.
Σ^x $\mathbf{G}_4^1$: $(\sigma_h, 0)$: 2, 3; 4, 3: *b*.
R^x $\mathbf{G}_4^1$: $(\sigma_h, 0)$: 2, 3; 4, 3: *b*.

176 $P6_3/m$ C_{6h}^2

($F1$; $K6$; $K7$; $M5$; $M8$; $Z1$.)

Γ $\mathbf{G}_{24}^{12}$: $\{C_6^+ \mid 00\frac{1}{2}\}$, $\{I \mid 00\frac{1}{2}\}$: 2, 3; 4, 3; 6, 3; 8, 3; 10, 3; 12, 3; 14, 3; 16, 3; 18, 3; 20, 3; 22, 3; 24, 3: *a*.
M $\mathbf{G}_8^2 \otimes \mathbf{T}_2$: $\{C_2 \mid 00\frac{1}{2}\}$, $\{I \mid 00\frac{1}{2}\}$; $\mathbf{t}_2$: 2, 3; 4, 3; 6, 3; 8, 3: *b*.
A $\mathbf{G}_{48}^{11}$: $\{C_6^+ \mid 00\frac{1}{2}\}$, $\{I \mid 00\frac{1}{2}\}$, $\{\bar{E} \mid 000\}$: 28, 3; 29, 1; 30, 3: *a*.
L $\mathbf{G}_{16}^{10}$: $\{\sigma_h \mid 000\}$, $\{E \mid 001\}$, $\{I \mid 00\frac{1}{2}\}$: 10, 1: *b*.
K $\mathbf{G}_{12}^1 \otimes \mathbf{T}_3$: $\{S_3^+ \mid 000\}$; $\mathbf{t}_1$ or $\mathbf{t}_2$: 2, 3; 4, 3; 6, 3; 8, 3; 10, 3; 12, 3: *c*.
H $\mathbf{G}_{12}^1 \otimes \mathbf{T}_3 \otimes \mathbf{T}_2$: $\{S_3^+ \mid 000\}$; $\mathbf{t}_1$ or $\mathbf{t}_2$; $\mathbf{t}_3$: 2, 3; 4, 1; 6, 3; 8, 3; 10, 1; 12, 3: *c*.

Δ^x $\mathbf{G}_{12}^1$: $(C_6^+, 0)$: 2, 3; 4, 3; 6, 3; 8, 3; 10, 3; 12, 3: *c*.
U^x $\mathbf{G}_4^1$: $(C_2, 0)$: 2, 3; 4, 3: *b*.
P^x $\mathbf{G}_6^1$: $(C_3^+, 0)$: 2, 3; 4, 1; 6, 3: *a*.
T^x $\mathbf{G}_4^1$: $(\sigma_h, 0)$: 2, 3; 4, 3: *b*.
S^x $\mathbf{G}_4^1$: $(\sigma_h, 0)$: 2, 1; 4, 1: *b*.
T'^x $\mathbf{G}_4^1$: $(\sigma_h, 0)$: 2, 3; 4, 3: *b*.
S'^x $\mathbf{G}_4^1$: $(\sigma_h, 0)$: 2, 1; 4, 1: *b*.
Σ^x $\mathbf{G}_4^1$: $(\sigma_h, 0)$: 2, 3; 4, 3: *b*.
R^x $\mathbf{G}_4^1$: $(\sigma_h, 0)$: 2, 1; 4, 1: *b*.

177 $P622$ D_6^1

($F1$; $K6$; $K7$; $M5$; $Z1$.)

Γ $\mathbf{G}_{24}^{11}$: $\{C_6^+ \mid 000\}$, $\{C'_{21} \mid 000\}$: 7, 2; 8, 2; 9, 2: *a*.
M $\mathbf{G}_8^5 \otimes \mathbf{T}_2$: $\{C_2 \mid 000\}$, $\{C''_{21} \mid 000\}$; $\mathbf{t}_2$: 5, 2: *a*.
A $\mathbf{G}_{24}^{11} \otimes \mathbf{T}_2$: $\{C_6^+ \mid 000\}$, $\{C'_{21} \mid 000\}$; $\mathbf{t}_3$: 7, 2; 8, 2; 9, 2: *a*.
L $\mathbf{G}_8^5 \otimes \mathbf{T}_2$: $\{C_2 \mid 000\}$, $\{C''_{21} \mid 000\}$; $\mathbf{t}_2$ or $\mathbf{t}_3$: 5, 2: *a*.
K $\mathbf{G}_{12}^4 \otimes \mathbf{T}_3$: $\{C_3^+ \mid 000\}$, $\{C''_{21} \mid 000\}$; $\mathbf{t}_1$ or $\mathbf{t}_2$: 3, 2; 4, 2; 6, 2; *a*.
H $\mathbf{G}_{12}^4 \otimes \mathbf{T}_3 \otimes \mathbf{T}_2$: $\{C_3^+ \mid 000\}$, $\{C''_{21} \mid 000\}$; $\mathbf{t}_1$ or $\mathbf{t}_2$; $\mathbf{t}_3$: 3, 2; 4, 2; 6, 2: *a*.

Δ^x $\mathbf{G}_{12}^1$: $(C_6^+, 0)$: 2, 2; 4, 2; 6, 2; 8, 2; 10, 2; 12, 2: *c*.
U^x $\mathbf{G}_4^1$: $(C_2, 0)$: 2, 2; 4, 2: *b*.
P^x $\mathbf{G}_6^1$: $(C_3^+, 0)$: 2, 2; 4, 2; 6, 2: *a*.
T^x $\mathbf{G}_4^1$: $(C''_{22}, 0)$: 2, 2; 4, 2: *b*.
S^x $\mathbf{G}_4^1$: $(C''_{22}, 0)$: 2, 2; 4, 2: *b*.
T'^x $\mathbf{G}_4^1$: $(C''_{21}, 0)$: 2, 2; 4, 2: *b*.
S'^x $\mathbf{G}_4^1$: $(C''_{21}, 0)$: 2, 2; 4, 2: *b*.
Σ^x $\mathbf{G}_4^1$: $(C'_{21}, 0)$: 2, 2; 4, 2: *b*.
R^x $\mathbf{G}_4^1$: $(C'_{21}, 0)$: 2, 2; 4, 2: *b*.

178 $P6_122$ D_6^2

($F1$; $K6$; $K7$; $M5$; $Z1$.)

Γ $\mathbf{G}_{24}^{11}$: $\{C_6^+ \mid 00\frac{1}{6}\}$, $\{C'_{21} \mid 000\}$: 7, 2; 8, 2; 9, 2: *a*.
M $\mathbf{G}_8^5 \otimes \mathbf{T}_2$: $\{C_2 \mid 00\frac{1}{2}\}$, $\{C''_{21} \mid 00\frac{1}{2}\}$; $\mathbf{t}_2$: 5, 2: *a*.
A $\mathbf{G}_{48}^{12}$: $\{C_6^+ \mid 00\frac{1}{6}\}$, $\{C'_{21} \mid 000\}$: 10, 3; 11, 3; 12, 3; 13, 3; 14, 2; 15, 2: *a*.
L $\mathbf{G}_{16}^8$: $\{C_2 \mid 00\frac{1}{2}\}$, $\{C''_{21} \mid 00\frac{1}{2}\}$: 5, 3; 6, 3; 7, 3; 8, 3: *a*.
K $\mathbf{G}_{12}^4 \otimes \mathbf{T}_3$: $\{C_3^+ \mid 00\frac{1}{3}\}$, $\{C''_{21} \mid 00\frac{1}{2}\}$; $\mathbf{t}_1$ or $\mathbf{t}_2$: 3, 2; 4, 2; 6, 2: *a*.
H $\mathbf{G}_{12}^4 \otimes \mathbf{T}_3 \otimes \mathbf{T}_2$: $\{C_3^- \mid 00\frac{2}{3}\}$, $\{C''_{21} \mid 00\frac{1}{2}\}$; $\mathbf{t}_1$ or $\mathbf{t}_2$; $\mathbf{t}_3$: 3, 3; 4, 3; 6, 2: *a*.

Δ^x $\mathbf{G}_{12}^1$: $(C_6^+, 0)$: 2, 2; 4, 2; 6, 2; 8, 2; 10, 2; 12, 2: *c*.
U^x $\mathbf{G}_4^1$: $(C_2, 0)$: 2, 2; 4, 2: *b*.
P^x $\mathbf{G}_6^1$: $(C_3^+, 0)$: 2, 2; 4, 2; 6, 2: *a*.
T^x $\mathbf{G}_4^1$: $(C''_{22}, 0)$: 2, 2; 4, 2: *b*.
S^x $\mathbf{G}_{24}^{12}$: $(C''_{22}, 0)$, $(\bar{E}, 0)$: 17, 3; 23, 3: *b*.
T'^x $\mathbf{G}_4^1$: $(C''_{21}, 0)$: 2, 2; 4, 2: *b*.
S'^x $\mathbf{G}_8^2$: $(C''_{21}, 0)$, $(\bar{E}, 0)$: 5, 3; 7, 3: *a*.
Σ^x $\mathbf{G}_4^1$: $(C'_{21}, 0)$: 2, 2; 4, 2: *b*.
R^x $\mathbf{G}_4^1$: $(C'_{21}, 0)$: 2, 3; 4, 3: *b*.

179 $P6_522$ D_6^3

($F1$; $K6$; $K7$; $M5$; $Z1$.)

Γ $\mathbf{G}_{24}^{11}$: $\{C_6^+ \mid 00\frac{5}{6}\}$, $\{C'_{21} \mid 000\}$: 7, 2; 8, 2; 9, 2: *a*.
M $\mathbf{G}_8^5 \otimes \mathbf{T}_2$: $\{C_2 \mid 00\frac{1}{2}\}$, $\{C''_{21} \mid 00\frac{1}{2}\}$; $\mathbf{t}_2$: 5, 2: *a*.
A $\mathbf{G}_{48}^{12}$: $\{C_6^+ \mid 00\frac{5}{6}\}$, $\{C'_{21} \mid 000\}$: 10, 3; 11, 3; 12, 3; 13, 3; 14, 2; 15, 2: *a*.
L $\mathbf{G}_{16}^8$: $\{C_2 \mid 00\frac{1}{2}\}$, $\{C''_{21} \mid 00\frac{1}{2}\}$: 5, 3; 6, 3; 7, 3; 8, 3: *a*.
K $\mathbf{G}_{12}^4 \otimes \mathbf{T}_3$: $\{C_3^+ \mid 00\frac{2}{3}\}$, $\{C''_{21} \mid 00\frac{1}{2}\}$; $\mathbf{t}_1$ or $\mathbf{t}_2$: 3, 2; 4, 2; 6, 2: *a*.
H $\mathbf{G}_{12}^4 \otimes \mathbf{T}_3 \otimes \mathbf{T}_2$: $\{C_3^+ \mid 00\frac{2}{3}\}$, $\{C''_{21} \mid 00\frac{1}{2}\}$; $\mathbf{t}_1$ or $\mathbf{t}_2$; $\mathbf{t}_3$: 3, 3; 4, 3; 6, 2: *a*.

Δ^x $\mathbf{G}_{12}^1$: $(C_6^+, 0)$: 2, 2; 4, 2; 6, 2; 8, 2; 10, 2; 12, 2: *c*.
U^x $\mathbf{G}_4^1$: $(C_2, 0)$: 2, 2; 4, 2: *b*.
P^x $\mathbf{G}_6^1$: $(C_3^+, 0)$: 2, 2; 4, 2; 6, 2: *a*.
T^x $\mathbf{G}_4^1$: $(C''_{22}, 0)$: 2, 2; 4, 2: *b*.
S^x $\mathbf{G}_{24}^{12}$: $(C''_{22}, 0)$, $(\bar{E}, 0)$: 15, 3; 21, 3: *c*.
T'^x $\mathbf{G}_4^1$: $(C''_{21}, 0)$: 2, 2; 4, 2: *b*.
S'^x $\mathbf{G}_8^2$: $(C''_{21}, 0)$, $(\bar{E}, 0)$: 5, 3; 7, 3: *a*.
Σ^x $\mathbf{G}_4^1$: $(C'_{21}, 0)$: 2, 2; 4, 2: *b*.
R^x $\mathbf{G}_4^1$: $(C'_{21}, 0)$: 2, 3; 4, 3: *b*.

180 $P6_222$ D_6^4

($F1$; $F7$; $F8$; $K6$; $K7$; $M5$; $Z1$.)

Γ $\mathbf{G}_{24}^{11}$: $\{C_6^+ \mid 00\frac{1}{3}\}$, $\{C'_{21} \mid 000\}$: 7, 2; 8, 2; 9, 2: a.
M $\mathbf{G}_8^5 \otimes \mathbf{T}_2$: $\{C_2 \mid 000\}$, $\{C''_{21} \mid 000\}$; $\mathbf{t}_2$: 5, 2: a.
A $\mathbf{G}_{24}^{11} \otimes \mathbf{T}_2$: $\{C_6^+ \mid 00\frac{1}{3}\}$, $\{C'_{21} \mid 000\}$; $\mathbf{t}_3$: 7, 2; 8, 2; 9, 2: a.
L $\mathbf{G}_8^5 \otimes \mathbf{T}_2$: $\{C_2 \mid 000\}$, $\{C''_{21} \mid 000\}$; $\mathbf{t}_2$ or $\mathbf{t}_3$: 5, 2: a.
K $\mathbf{G}_{12}^4 \otimes \mathbf{T}_3$: $\{C_3^+ \mid 00\frac{2}{3}\}$, $\{C''_{21} \mid 000\}$; $\mathbf{t}_1$ or $\mathbf{t}_2$: 3, 2; 4, 2; 6, 2: a.
H $\mathbf{G}_{12}^4 \otimes \mathbf{T}_3 \otimes \mathbf{T}_2$: $\{C_3^+ \mid 00\frac{2}{3}\}$, $\{C''_{21} \mid 000\}$; $\mathbf{t}_1$ or $\mathbf{t}_2$; $\mathbf{t}_3$: 3, 2; 4, 2; 6, 2: a.

Δ^x $\mathbf{G}_{12}^1$: $(C_6^+, 0)$: 2, 2; 4, 2; 6, 2; 8, 2; 10, 2; 12, 2: c.
U^x $\mathbf{G}_4^1$: $(C_2, 0)$: 2, 2; 4, 2: b.
P^x $\mathbf{G}_6^1$: $(C_3^+, 0)$: 2, 2; 4, 2; 6, 2: a.
T^x $\mathbf{G}_4^1$: $(C''_{22}, 0)$: 2, 2; 4, 2: b.
S^x $\mathbf{G}_{12}^1$: $(C''_{22}, 0)$: 6, 2; 12, 2: a.
T'^x $\mathbf{G}_4^1$: $(C''_{21}, 0)$: 2, 2; 4, 2: b.
S'^x $\mathbf{G}_4^1$: $(C''_{21}, 0)$: 2, 2; 4, 2: b.
Σ^x $\mathbf{G}_4^1$: $(C'_{21}, 0)$: 2, 2; 4, 2: b.
R^x $\mathbf{G}_4^1$: $(C'_{21}, 0)$: 2, 2; 4, 2: b.

181 $P6_422$ D_6^5

($F1$; $F7$; $F8$; $K6$; $K7$; $M5$; $Z1$.)

Γ $\mathbf{G}_{24}^{11}$: $\{C_6^+ \mid 00\frac{2}{3}\}$, $\{C'_{21} \mid 000\}$: 7, 2; 8, 2; 9, 2: a.
M $\mathbf{G}_8^5 \otimes \mathbf{T}_2$: $\{C_2 \mid 000\}$, $\{C''_{21} \mid 000\}$; $\mathbf{t}_2$: 5, 2: a.
A $\mathbf{G}_{24}^{11} \otimes \mathbf{T}_2$: $\{C_6^+ \mid 00\frac{2}{3}\}$, $\{C'_{21} \mid 000\}$; $\mathbf{t}_3$: 7, 2; 8, 2; 9, 2: a.
L $\mathbf{G}_8^5 \otimes \mathbf{T}_2$: $\{C_2 \mid 000\}$, $\{C''_{21} \mid 000\}$; $\mathbf{t}_2$ or $\mathbf{t}_3$: 5, 2: a.
K $\mathbf{G}_{12}^4 \otimes \mathbf{T}_3$: $\{C_3^- \mid 00\frac{2}{3}\}$, $\{C''_{21} \mid 000\}$; $\mathbf{t}_1$ or $\mathbf{t}_2$: 3, 2; 4, 2; 6, 2: a.
H $\mathbf{G}_{12}^4 \otimes \mathbf{T}_3 \otimes \mathbf{T}_2$: $\{C_3^- \mid 00\frac{2}{3}\}$, $\{C''_{21} \mid 000\}$; $\mathbf{t}_1$ or $\mathbf{t}_2$; $\mathbf{t}_3$: 3, 2; 4, 2; 6, 2: a.

Δ^x $\mathbf{G}_{12}^1$: $(C_6^+, 0)$: 2, 2; 4, 2; 6, 2; 8, 2; 10, 2; 12, 2: c.
U^x $\mathbf{G}_4^1$: $(C_2, 0)$: 2, 2; 4, 2: b.
P^x $\mathbf{G}_6^1$: $(C_3^+, 0)$: 2, 2; 4, 2; 6, 2: a.
T^x $\mathbf{G}_4^1$: $(C''_{22}, 0)$: 2, 2; 4, 2: b.
S^x $\mathbf{G}_{12}^1$: $(C''_{22}, 0)$: 2, 2; 8, 2: b.
T'^x $\mathbf{G}_4^1$: $(C''_{21}, 0)$: 2, 2; 4, 2: b.
S'^x $\mathbf{G}_4^1$: $(C''_{21}, 0)$: 2, 2; 4, 2: b.
Σ^x $\mathbf{G}_4^1$: $(C'_{21}, 0)$: 2, 2; 4, 2: b.
R^x $\mathbf{G}_4^1$: $(C'_{21}, 0)$: 2, 2; 4, 2: b.

182 $P6_322$ D_6^6

($F1$; $K6$; $K7$; $M5$; $Z1$.)

Γ $\mathbf{G}_{24}^{11}$: $\{C_6^+ \mid 00\frac{1}{2}\}$, $\{C'_{21} \mid 000\}$: 7, 2; 8, 2; 9, 2: a.
M $\mathbf{G}_8^5 \otimes \mathbf{T}_2$: $\{C_2 \mid 00\frac{1}{2}\}$, $\{C''_{21} \mid 00\frac{1}{2}\}$; $\mathbf{t}_2$: 5, 2: a.
A $\mathbf{G}_{48}^{12}$: $\{C_6^+ \mid 00\frac{1}{2}\}$, $\{C'_{21} \mid 000\}$: 10, 3; 11, 3; 12, 3; 13, 3; 14, 2; 15, 2: a.
L $\mathbf{G}_{16}^8$: $\{C_2 \mid 00\frac{1}{2}\}$, $\{C''_{21} \mid 00\frac{1}{2}\}$: 5, 3; 6, 3; 7, 3; 8, 3: a.
K $\mathbf{G}_{12}^4 \otimes \mathbf{T}_3$: $\{C_3^+ \mid 000\}$, $\{C''_{21} \mid 00\frac{1}{2}\}$; $\mathbf{t}_1$ or $\mathbf{t}_2$: 3, 2; 4, 2; 6, 2: a.
H $\mathbf{G}_{12}^4 \otimes \mathbf{T}_3 \otimes \mathbf{T}_2$: $\{C_3^+ \mid 000\}$, $\{C''_{21} \mid 00\frac{1}{2}\}$; $\mathbf{t}_1$ or $\mathbf{t}_2$; $\mathbf{t}_3$: 3, 3; 4, 3; 6, 2: a.

Δ^x $\mathbf{G}_{12}^1$: $(C_6^+, 0)$: 2, 2; 4, 2; 6, 2; 8, 2; 10, 2; 12, 2: c.
U^x $\mathbf{G}_4^1$: $(C_2, 0)$: 2, 2; 4, 2: b.
P^x $\mathbf{G}_6^1$: $(C_3^+, 0)$: 2, 2; 4, 2; 6, 2: a.
T^x $\mathbf{G}_4^1$: $(C''_{22}, 0)$: 2, 2; 4, 2: b.
S^x $\mathbf{G}_8^2$: $(C''_{22}, 0)$, $(E, 1)$: 5, 3; 7, 3: a.
T'^x $\mathbf{G}_4^1$: $(C''_{21}, 0)$: 2, 2; 4, 2: b.
S'^x $\mathbf{G}_8^2$: $(C''_{21}, 0)$, $(E, 1)$: 5, 3; 7, 3: a.
Σ^x $\mathbf{G}_4^1$: $(C'_{21}, 0)$: 2, 2; 4, 2: b.
R^x $\mathbf{G}_4^1$: $(C'_{21}, 0)$: 2, 3; 4, 3: b.

183 $P6mm$ C^1_{6v}

($F1$; $K6$; $K7$; $M5$; $Z1$.)

Γ $\mathbf{G}^{11}_{24}$: $\{C_6^+ \mid 000\}, \{\sigma_{d1} \mid 000\}$: 7, 2; 8, 2; 9, 2: *a*.

M $\mathbf{G}^5_8 \otimes \mathbf{T}_2$: $\{C_2 \mid 000\}, \{\sigma_{v1} \mid 000\}$; $\mathbf{t}_2$: 5, 2: *a*.

A $\mathbf{G}^{11}_{24} \otimes \mathbf{T}_2$: $\{C_6^+ \mid 000\}, \{\sigma_{d1} \mid 000\}$; $\mathbf{t}_3$: 7, 2; 8, 2; 9, 2: *a*.

L $\mathbf{G}^5_8 \otimes \mathbf{T}_2$: $\{C_2 \mid 000\}, \{\sigma_{v1} \mid 000\}$; $\mathbf{t}_2$ or $\mathbf{t}_3$: 5, 2: *a*.

K $\mathbf{G}^4_{12} \otimes \mathbf{T}_3$: $\{C_3^+ \mid 000\}, \{\sigma_{d1} \mid 000\}$; $\mathbf{t}_1$ or $\mathbf{t}_2$: 3, 2; 4, 2; 6, 2: *a*.

H $\mathbf{G}^4_{12} \otimes \mathbf{T}_3 \otimes \mathbf{T}_2$: $\{C_3^+ \mid 000\}, \{\sigma_{d1} \mid 000\}$; $\mathbf{t}_1$ or $\mathbf{t}_2$; $\mathbf{t}_3$: 3, 2; 4, 2; 6, 2: *a*.

Δ^x $\mathbf{G}^{11}_{24}$: $(C_6^+, 0), (\sigma_{d1}, 0)$: 7, x; 8, x; 9, x: *a*.

U^x $\mathbf{G}^5_8$: $(C_2, 0), (\sigma_{v1}, 0)$: 5, x: *a*.

P^x $\mathbf{G}^4_{12}$: $(C_3^+, 0), (\sigma_{d1}, 0)$: 3, x; 4, x; 6, x: *a*.

T^x $\mathbf{G}^1_4$: $(\sigma_{d2}, 0)$: 2, 2; 4, 2: *b*.

S^x $\mathbf{G}^1_4$: $(\sigma_{d2}, 0)$: 2, 2; 4, 2: *b*.

T'^x $\mathbf{G}^1_4$: $(\sigma_{d1}, 0)$: 2, 2; 4, 2: *b*.

S'^x $\mathbf{G}^1_4$: $(\sigma_{d1}, 0)$: 2, 2; 4, 2: *b*.

Σ^x $\mathbf{G}^1_4$: $(\sigma_{v1}, 0)$: 2, 2; 4, 2: *b*.

R^x $\mathbf{G}^1_4$: $(\sigma_{v1}, 0)$: 2, 2; 4, 2: *b*.

184 $P6cc$ C^2_{6v}

($F1$; $K6$; $K7$; $M5$; $Z1$.)

Γ $\mathbf{G}^{11}_{24}$: $\{C_6^+ \mid 000\}, \{\sigma_{d1} \mid 00\tfrac{1}{2}\}$: 7, 2; 8, 2; 9, 2: *a*.

M $\mathbf{G}^5_8 \otimes \mathbf{T}_2$: $\{C_2 \mid 000\}, \{\sigma_{v1} \mid 00\tfrac{1}{2}\}$; $\mathbf{t}_2$: 5, 2: *a*.

A $\mathbf{G}^{12}_{48}$: $\{C_6^+ \mid 000\}, \{\sigma_{d1} \mid 00\tfrac{1}{2}\}$: 7, 1; 8, 1; 9, 1: *b*.

L $\mathbf{G}^8_{16}$: $\{C_2 \mid 000\}, \{\sigma_{v1} \mid 00\tfrac{1}{2}\}$: 9, 1: *b*.

K $\mathbf{G}^4_{12} \otimes \mathbf{T}_3$: $\{C_3^+ \mid 000\}, \{\sigma_{d1} \mid 00\tfrac{1}{2}\}$; $\mathbf{t}_1$ or $\mathbf{t}_2$: 3, 2; 4, 2; 6, 2: *a*.

H $\mathbf{G}^4_{12} \otimes \mathbf{T}_3 \otimes \mathbf{T}_2$: $\{C_3^+ \mid 001\}, \{\sigma_{d1} \mid 00\tfrac{1}{2}\}$; $\mathbf{t}_1$ or $\mathbf{t}_2$; $\mathbf{t}_3$: 1, 3; 2, 3; 5, 1: *b*.

Δ^x $\mathbf{G}^{11}_{24}$: $(C_6^+, 0), (\sigma_{d1}, 0)$: 7, x; 8, x; 9, x: *a*.

U^x $\mathbf{G}^5_8$: $(C_2, 0), (\sigma_{v1}, 0)$: 5, x: *a*.

P^x $\mathbf{G}^4_{12}$: $(C_3^+, 0), (\sigma_{d1}, 0)$: 3, x; 4, x; 6, x: *a*.

T^x $\mathbf{G}^1_4$: $(\sigma_{d2}, 0)$: 2, 2; 4, 2: *b*.

S^x $\mathbf{G}^1_4$: $(\sigma_{d2}, 0)$: 2, 3; 4, 3: *b*.

T'^x $\mathbf{G}^1_4$: $(\sigma_{d1}, 0)$: 2, 2; 4, 2: *b*.

S'^x $\mathbf{G}^1_4$: $(\sigma_{d1}, 0)$: 2, 3; 4, 3: *b*.

Σ^x $\mathbf{G}^1_4$: $(\sigma_{v1}, 0)$: 2, 2; 4, 2: *b*.

R^x $\mathbf{G}^1_4$: $(\sigma_{v1}, 0)$: 2, 3; 4, 3: *b*.

185 $P6_3cm$ C^3_{6v}

($F1$; $K6$; $K7$; $M5$; $Z1$.)

Γ $\mathbf{G}^{11}_{24}$: $\{C_6^+ \mid 00\tfrac{1}{2}\}, \{\sigma_{d1} \mid 000\}$: 7, 2; 8, 2; 9, 2: *a*.

M $\mathbf{G}^5_8 \otimes \mathbf{T}_2$: $\{C_2 \mid 00\tfrac{1}{2}\}, \{\sigma_{v1} \mid 00\tfrac{1}{2}\}$; $\mathbf{t}_2$: 5, 2: *a*.

A $\mathbf{G}^{13}_{48}$: $\{C_3^+ \mid 000\}, \{\sigma_{d1} \mid 000\}, \{C_2 \mid 00\tfrac{1}{2}\}$: 13, 1; 14, 3; 15, 3: *a*.

L $\mathbf{G}^8_{16}$: $\{\sigma_{d1} \mid 000\}, \{\sigma_{v1} \mid 00\tfrac{1}{2}\}$: 9, 1: *b*.

K $\mathbf{G}^4_{12} \otimes \mathbf{T}_3$: $\{C_3^+ \mid 000\}, \{\sigma_{d1} \mid 000\}$; $\mathbf{t}_1$ or $\mathbf{t}_2$: 3, 2; 4, 2; 6, 2: *a*.

H $\mathbf{G}^4_{12} \otimes \mathbf{T}_3 \otimes \mathbf{T}_2$: $\{C_3^+ \mid 000\}, \{\sigma_{d1} \mid 000\}$; $\mathbf{t}_1$ or $\mathbf{t}_2$; $\mathbf{t}_3$: 3, 1; 4, 1; 6, 1: *a*.

Δ^x $\mathbf{G}^{11}_{24}$: $(C_6^+, 0), (\sigma_{d1}, 0)$: 7, x; 8, x; 9, x: *a*.

U^x $\mathbf{G}^5_8$: $(C_2, 0), (\sigma_{v1}, 0)$: 5, x: *a*.

P^x $\mathbf{G}^4_{12}$: $(C_3^+, 0), (\sigma_{d1}, 0)$: 3, x; 4, x; 6, x: *a*.

T^x $\mathbf{G}^1_4$: $(\sigma_{d2}, 0)$: 2, 2; 4, 2: *b*.

S^x $\mathbf{G}^1_4$: $(\sigma_{d2}, 0)$: 2, 1; 4, 1: *b*.

T'^x $\mathbf{G}^1_4$: $(\sigma_{d1}, 0)$: 2, 2; 4, 2: *b*.

S'^x $\mathbf{G}^1_4$: $(\sigma_{d1}, 0)$: 2, 1; 4, 1: *b*.

Σ^x $\mathbf{G}^1_4$: $(\sigma_{v1}, 0)$: 2, 2; 4, 2: *b*.

R^x $\mathbf{G}^1_4$: $(\sigma_{v1}, 0)$: 2, 3; 4, 3: *b*.

186 $P6_3mc$ C_{6v}^4

($A1$; $B3$; $B4$; $C1$; $F1$; $G4$; $H2$; $K6$; $K7$; $M5$; $N1$; $P3$; $P4$; $R2$; $R3$; $S8$; $S10$; $Z1$.)

Γ $\mathbf{G}_{24}^{11}$: $\{C_6^+ \mid 00\frac{1}{2}\}$, $\{\sigma_{d1} \mid 00\frac{1}{2}\}$: 7, 2; 8, 2; 9, 2: *a*.
M $\mathbf{G}_8^5 \otimes \mathbf{T}_2$: $\{C_2 \mid 00\frac{1}{2}\}$, $\{\sigma_{v1} \mid 000\}$; $\mathbf{t}_2$: 5, 2: *a*.
A $\mathbf{G}_{48}^{13}$: $\{C_3^+ \mid 000\}$, $\{\sigma_{v1} \mid 000\}$, $\{C_2 \mid 00\frac{1}{2}\}$: 13, 1; 14, 3; 15, 3: *a*.
L $\mathbf{G}_{16}^8$: $\{\sigma_{v1} \mid 000\}$, $\{\sigma_{d1} \mid 00\frac{1}{2}\}$: 9, 1: *b*.
K $\mathbf{G}_{12}^4 \otimes \mathbf{T}_3$: $\{C_3^+ \mid 000\}$, $\{\sigma_{d1} \mid 00\frac{1}{2}\}$; $\mathbf{t}_1$ or $\mathbf{t}_2$: 3, 2; 4, 2; 6, 2: *a*.
H $\mathbf{G}_{12}^4 \otimes \mathbf{T}_3 \otimes \mathbf{T}_2$: $\{C_3^+ \mid 001\}$, $\{\sigma_{d1} \mid 00\frac{1}{2}\}$; $\mathbf{t}_1$ or $\mathbf{t}_2$; $\mathbf{t}_3$: 1, 3; 2, 3; 5, 2: *b*.

Δ^x $\mathbf{G}_{24}^{11}$: $(C_6^+, 0)$, $(\sigma_{d1}, 0)$: 7, *x*; 8, *x*; 9, *x*: *a*.
U^x $\mathbf{G}_8^5$: $(C_2, 0)$, $(\sigma_{v1}, 0)$: 5, *x*: *a*.
P^x $\mathbf{G}_{12}^4$: $(C_3^+, 0)$, $(\sigma_{d1}, 0)$: 3, *x*; 4, *x*; 6, *x*: *a*.
T^x $\mathbf{G}_4^1$: $(\sigma_{d2}, 0)$: 2, 2; 4, 2: *b*.
S^x $\mathbf{G}_4^1$: $(\sigma_{d2}, 0)$: 2, 3; 4, 3: *b*.
T'^x $\mathbf{G}_4^1$: $(\sigma_{d1}, 0)$: 2, 2; 4, 2: *b*.
S'^x $\mathbf{G}_4^1$: $(\sigma_{d1}, 0)$: 2, 3; 4, 3: *b*.
Σ^x $\mathbf{G}_4^1$: $(\sigma_{v1}, 0)$: 2, 2; 4, 2: *b*.
R^x $\mathbf{G}_4^1$: $(\sigma_{v1}, 0)$: 2, 1; 4, 1: *b*.

187 $P\bar{6}m2$ D_{3h}^1

($F1$; $K6$; $K7$; $M5$; $Z1$.)

Γ $\mathbf{G}_{24}^{11}$: $\{S_3^+ \mid 000\}$, $\{C'_{21} \mid 000\}$: 7, 2; 8, 2; 9, 2: *a*.
M $\mathbf{G}_8^5 \otimes \mathbf{T}_2$: $\{C'_{21} \mid 000\}$, $\{\sigma_{v1} \mid 000\}$; $\mathbf{t}_2$: 5, 2: *a*.
A $\mathbf{G}_{24}^{11} \otimes \mathbf{T}_2$: $\{S_3^+ \mid 000\}$, $\{C'_{21} \mid 000\}$; $\mathbf{t}_3$: 7, 2; 8, 2; 9, 2: *a*.
L $\mathbf{G}_8^5 \otimes \mathbf{T}_2$: $\{C'_{21} \mid 000\}$, $\{\sigma_{v1} \mid 000\}$; $\mathbf{t}_2$ or $\mathbf{t}_3$: 5, 2: *a*.
K $\mathbf{G}_{12}^1 \otimes \mathbf{T}_3$: $\{S_3^+ \mid 000\}$; $\mathbf{t}_1$ or $\mathbf{t}_2$: 2, 2; 4, 2; 6, 2; 8, 2; 10, 2; 12, 2: *c*.
H $\mathbf{G}_{12}^1 \otimes \mathbf{T}_3 \otimes \mathbf{T}_2$: $\{S_3^+ \mid 000\}$; $\mathbf{t}_1$ or $\mathbf{t}_2$; $\mathbf{t}_3$: 2, 2; 4, 2; 6, 2; 8, 2; 10, 2; 12, 2: *c*.

Δ^x $\mathbf{G}_{12}^4$: $(C_3^+, 0)$, $(\sigma_{v1}, 0)$: 3, 2; 4, 2; 6, 2: *a*.
U^x $\mathbf{G}_4^1$: $(\sigma_{v1}, 0)$: 2, 2; 4, 2: *b*.
P^x $\mathbf{G}_6^1$: $(C_3^+, 0)$: 2, 2; 4, 2; 6, 2: *a*.
T^x $\mathbf{G}_4^1$: $(\sigma_h, 0)$: 2, 2; 4, 2: *b*.
S^x $\mathbf{G}_4^1$: $(\sigma_h, 0)$: 2, 2; 4, 2: *b*.
T'^x $\mathbf{G}_4^1$: $(\sigma_h, 0)$: 2, 2; 4, 2: *b*.
S'^x $\mathbf{G}_4^1$: $(\sigma_h, 0)$: 2, 2; 4, 2: *b*.
Σ^x $\mathbf{G}_8^5$: $(C'_{21}, 0)$, $(\sigma_{v1}, 0)$: 5, *x*: *a*.
R^x $\mathbf{G}_8^5$: $(C'_{21}, 0)$, $(\sigma_{v1}, 0)$: 5, *x*: *a*.

188 $P\bar{6}c2$ D_{3h}^2

($F1$; $K6$; $K7$; $M5$; $Z1$.)

Γ $\mathbf{G}_{24}^{11}$: $\{S_3^+ \mid 000\}$, $\{C'_{21} \mid 00\frac{1}{2}\}$: 7, 2; 8, 2; 9, 2: *a*.
M $\mathbf{G}_8^5 \otimes \mathbf{T}_2$: $\{C'_{21} \mid 00\frac{1}{2}\}$, $\{\sigma_{v1} \mid 00\frac{1}{2}\}$; $\mathbf{t}_2$: 5, 2: *a*.
A $\mathbf{G}_{48}^{14}$: $\{\sigma_{v1} \mid 00\frac{1}{2}\}$, $\{C_3^+ \mid 000\}$, $\{\sigma_h \mid 000\}$: 5, 3; 6, 3; 7, 3; 8, 3; 11, 3; 12, 3: *a*.
L $\mathbf{G}_{16}^8$: $\{\sigma_{v1} \mid 00\frac{1}{2}\}$, $\{\sigma_h \mid 000\}$: 5, 3; 6, 3; 7, 3; 8, 3: *a*.
K $\mathbf{G}_{12}^1 \otimes \mathbf{T}_3$: $\{S_3^+ \mid 000\}$; $\mathbf{t}_1$ or $\mathbf{t}_2$: 2, 2; 4, 2; 6, 2; 8, 2; 10, 2; 12, 2: *c*.
H $\mathbf{G}_{12}^1 \otimes \mathbf{T}_3 \otimes \mathbf{T}_2$: $\{S_3^+ \mid 000\}$; $\mathbf{t}_1$ or $\mathbf{t}_2$; $\mathbf{t}_3$: 2, 3; 4, 3; 6, 3; 8, 3; 10, 3; 12, 3: *c*.

Δ^x $\mathbf{G}_{12}^4$: $(C_3^+, 0)$, $(\sigma_{v1}, 0)$: 3, 2; 4, 2; 6, 2: *a*.
U^x $\mathbf{G}_4^1$: $(\sigma_{v1}, 0)$: 2, 2; 4, 2: *b*.
P^x $\mathbf{G}_6^1$: $(C_3^+, 0)$: 2, 2; 4, 2; 6, 2: *a*.
T^x $\mathbf{G}_4^1$: $(\sigma_h, 0)$: 2, 2; 4, 2: *b*.
S^x $\mathbf{G}_4^1$: $(\sigma_h, 0)$: 2, 3; 4, 3: *b*.
T'^x $\mathbf{G}_4^1$: $(\sigma_h, 0)$: 2, 2; 4, 2: *b*.
S'^x $\mathbf{G}_4^1$: $(\sigma_h, 0)$: 2, 3; 4, 3: *b*.
Σ^x $\mathbf{G}_8^5$: $(C'_{21}, 0)$, $(\sigma_{v1}, 0)$: 5, *x*: *a*.
R^x $\mathbf{G}_{16}^8$: $(C'_{21}, 0)$, $(\sigma_h, 0)$: 5, *x*; 6, *x*; 7, *x*; 8, *x*: *a*.

189 $P\bar{6}2m$ D_{3h}^3

($F1$; $K6$; $K7$; $M5$; $Z1$.)

Γ $\mathbf{G}_{24}^{11}$: $\{S_3^+ \mid 000\}$, $\{C_{21}'' \mid 000\}$: 7, 2; 8, 2; 9, 2: a.
M $\mathbf{G}_8^5 \otimes \mathbf{T}_2$: $\{C_{21}'' \mid 000\}$, $\{\sigma_{d1} \mid 000\}$; $\mathbf{t}_2$: 5, 2: a.
A $\mathbf{G}_{24}^{11} \otimes \mathbf{T}_2$: $\{S_3^+ \mid 000\}$, $\{C_{21}'' \mid 000\}$; $\mathbf{t}_3$: 7, 2; 8, 2; 9, 2: a.
L $\mathbf{G}_8^5 \otimes \mathbf{T}_2$: $\{C_{21}'' \mid 000\}$, $\{\sigma_{d1} \mid 000\}$; $\mathbf{t}_2$ or $\mathbf{t}_3$: 5, 2: a.
K $\mathbf{G}_{24}^{11} \otimes \mathbf{T}_3$: $\{S_3^+ \mid 000\}$, $\{C_{21}'' \mid 000\}$; $\mathbf{t}_1$ or $\mathbf{t}_2$: 7, x; 8, x; 9, x: a.
H $\mathbf{G}_{24}^{11} \otimes \mathbf{T}_3 \otimes \mathbf{T}_2$: $\{S_3^+ \mid 000\}$, $\{C_{21}'' \mid 000\}$; $\mathbf{t}_1$ or $\mathbf{t}_2$; $\mathbf{t}_3$: 7, x; 8, x; 9, x: a.

Δ^x $\mathbf{G}_{12}^4$: $(C_3^+, 0)$, $(\sigma_{d1}, 0)$: 3, 2; 4, 2; 6, 2: a.
U^x $\mathbf{G}_4^1$: $(\sigma_{d1}, 0)$: 2, 2; 4, 2: b.
P^x $\mathbf{G}_{12}^4$: $(C_3^+, 0)$, $(\sigma_{d1}, 0)$: 3, x; 4, x; 6, x: a.
T^x $\mathbf{G}_8^5$: $(C_{22}'', 0)$, $(\sigma_{d2}, 0)$: 5, x: a.
S^x $\mathbf{G}_8^5$: $(C_{22}'', 0)$, $(\sigma_{d2}, 0)$: 5, x: a.
T'^x $\mathbf{G}_8^5$: $(C_{21}'', 0)$, $(\sigma_{d1}, 0)$: 5, x: a.
S'^x $\mathbf{G}_8^5$: $(C_{21}'', 0)$, $(\sigma_{d1}, 0)$: 5, x: a.
Σ^x $\mathbf{G}_4^1$: $(\sigma_h, 0)$: 2, 2; 4, 2: b.
R^x $\mathbf{G}_4^1$: $(\sigma_h, 0)$: 2, 2; 4, 2: b.

190 $P\bar{6}2c$ D_{3h}^4

($F1$; $K6$; $K7$; $M5$; $Z1$.)

Γ $\mathbf{G}_{24}^{11}$: $\{S_3^+ \mid 000\}$, $\{C_{21}'' \mid 00\tfrac{1}{2}\}$: 7, 2; 8, 2; 9, 2: a.
M $\mathbf{G}_8^5 \otimes \mathbf{T}_2$: $\{C_{21}'' \mid 00\tfrac{1}{2}\}$, $\{\sigma_{d1} \mid 00\tfrac{1}{2}\}$; $\mathbf{t}_2$: 5, 2: a.
A $\mathbf{G}_{48}^{14}$: $\{\sigma_{d1} \mid 00\tfrac{1}{2}\}$, $\{C_3^+ \mid 000\}$, $\{\sigma_h \mid 000\}$: 5, 3; 6, 3; 7, 3; 8, 3; 11, 3; 12, 3: a.
L $\mathbf{G}_{16}^8$: $\{\sigma_{d1} \mid 00\tfrac{1}{2}\}$, $\{\sigma_h \mid 000\}$: 5, 3; 6, 3; 7, 3; 8, 3: a.
K $\mathbf{G}_{24}^{11} \otimes \mathbf{T}_3$: $\{S_3^+ \mid 000\}$, $\{C_{21}'' \mid 00\tfrac{1}{2}\}$; $\mathbf{t}_1$ or $\mathbf{t}_2$: 7, x; 8, x; 9, x: a.
H $\mathbf{G}_{48}^{14} \otimes \mathbf{T}_3$: $\{\sigma_{d1} \mid 00\tfrac{1}{2}\}$, $\{C_3^+ \mid 000\}$, $\{\sigma_h \mid 000\}$; $\mathbf{t}_1$ or $\mathbf{t}_2$: 5, x; 6, x; 7, x; 8, x; 11, x; 12, x: a.

Δ^x $\mathbf{G}_{12}^4$: $(C_3^+, 0)$, $(\sigma_{d1}, 0)$: 3, 2; 4, 2; 6, 2: a.
U^x $\mathbf{G}_4^1$: $(\sigma_{d1}, 0)$: 2, 2; 4, 2: b.
P^x $\mathbf{G}_{12}^4$: $(C_3^+, 0)$, $(\sigma_{d1}, 0)$: 3, x; 4, x; 6, x: a.
T^x $\mathbf{G}_8^5$: $(C_{22}'', 0)$, $(\sigma_{d2}, 0)$: 5, x: a.
S^x $\mathbf{G}_{16}^8$: $(C_{22}'', 0)$, $(\sigma_h, 0)$: 5, x; 6, x; 7, x; 8, x: a.
T'^x $\mathbf{G}_8^5$: $(C_{21}'', 0)$, $(\sigma_{d1}, 0)$: 5, x: a.
S'^x $\mathbf{G}_{16}^8$: $(C_{21}'', 0)$, $(\sigma_h, 0)$: 5, x; 6, x; 7, x; 8, x: a.
Σ^x $\mathbf{G}_4^1$: $(\sigma_h, 0)$: 2, 2; 4, 2: b.
R^x $\mathbf{G}_4^1$: $(\sigma_h, 0)$: 2, 3; 4, 3: b.

191 $P6/mmm$ D_{6h}^1

($F1$; $K6$; $K7$; $M5$; $Z1$.)

Γ $\mathbf{G}_{48}^{15}$: $\{C_6^+ \mid 000\}$, $\{C_{21}' \mid 000\}$, $\{I \mid 000\}$: 7, 2; 8, 2; 9, 2; 16, 2; 17, 2; 18, 2: a.
M $\mathbf{G}_{16}^{11} \otimes \mathbf{T}_2$: $\{C_2 \mid 000\}$, $\{C_{21}'' \mid 000\}$, $\{I \mid 000\}$; $\mathbf{t}_2$: 5, 2; 10, 2: c.
A $\mathbf{G}_{48}^{15} \otimes \mathbf{T}_2$: $\{C_6^+ \mid 000\}$, $\{C_{21}' \mid 000\}$, $\{I \mid 000\}$; $\mathbf{t}_3$: 7, 2; 8, 2; 9, 2; 16, 2; 17, 2; 18, 2: a.
L $\mathbf{G}_{16}^{11} \otimes \mathbf{T}_2$: $\{C_2 \mid 000\}$, $\{C_{21}'' \mid 000\}$, $\{I \mid 000\}$; $\mathbf{t}_2$ or $\mathbf{t}_3$: 5, 2; 10, 2: c.
K $\mathbf{G}_{24}^{11} \otimes \mathbf{T}_3$: $\{S_3^+ \mid 000\}$, $\{C_{21}'' \mid 000\}$; $\mathbf{t}_1$ or $\mathbf{t}_2$: 7, 2; 8, 2; 9, 2: a.
H $\mathbf{G}_{24}^{11} \otimes \mathbf{T}_3 \otimes \mathbf{T}_2$: $\{S_3^+ \mid 000\}$, $\{C_{21}'' \mid 000\}$; $\mathbf{t}_1$ or $\mathbf{t}_2$; $\mathbf{t}_3$: 7, 2; 8, 2; 9, 2: a.

Δ^x $\mathbf{G}_{24}^{11}$: $(C_6^+, 0)$, $(\sigma_{d1}, 0)$: 7, 2; 8, 2; 9, 2: a.
U^x $\mathbf{G}_8^5$: $(C_2, 0)$, $(\sigma_{v1}, 0)$: 5, 2: a.
P^x $\mathbf{G}_{12}^4$: $(C_3^+, 0)$, $(\sigma_{d1}, 0)$: 3, 3; 4, 3; 6, 2: a.
T^x $\mathbf{G}_8^5$: $(C_{22}'', 0)$, $(\sigma_{d2}, 0)$: 5, 2: a.
S^x $\mathbf{G}_8^5$: $(C_{22}'', 0)$, $(\sigma_{d2}, 0)$: 5, 2: a.
T'^x $\mathbf{G}_8^5$: $(C_{21}'', 0)$, $(\sigma_{d1}, 0)$: 5, 2: a.
S'^x $\mathbf{G}_8^5$: $(C_{21}'', 0)$, $(\sigma_{d1}, 0)$: 5, 2: a.
Σ^x $\mathbf{G}_8^5$: $(C_{21}', 0)$, $(\sigma_{v1}, 0)$: 5, 2: a.
R^x $\mathbf{G}_8^5$: $(C_{21}', 0)$, $(\sigma_{v1}, 0)$: 5, 2: a.

192 $P6/mcc$ D^2_{6h}

($F1$; $K6$; $K7$; $M5$; $Z1$.)

Γ $\mathbf{G}^{15}_{48}$: $\{C_6^+ \mid 000\}$, $\{C'_{21} \mid 00\frac{1}{2}\}$, $\{I \mid 000\}$: 7, 2; 8, 2; 9, 2; 16, 2; 17, 2; 18, 2: *a*.
M $\mathbf{G}^{11}_{16} \otimes \mathbf{T}_2$: $\{C_2 \mid 000\}$, $\{C''_{21} \mid 00\frac{1}{2}\}$, $\{I \mid 000\}$; $\mathbf{t}_2$: 5, 2; 10, 2: *c*.
A $\mathbf{G}^{9}_{96}$: $\{\sigma_{d1} \mid 00\frac{1}{2}\}$, $\{C_3^+ \mid 000\}$, $\{\sigma_h \mid 000\}$, $\{C_2 \mid 000\}$: 19, 3; 20, 3; 21, 3; 22, 3; 23, 3; 24, 3: *a*.
L $\mathbf{G}^{7}_{32}$: $\{\sigma_{d1} \mid 00\frac{1}{2}\}$, $\{\sigma_h \mid 000\}$, $\{I \mid 000\}$: 13, 3; 14, 3: *a*.
K $\mathbf{G}^{11}_{24} \otimes \mathbf{T}_3$: $\{S_3^+ \mid 000\}$, $\{C''_{21} \mid 00\frac{1}{2}\}$; $\mathbf{t}_1$ or $\mathbf{t}_2$: 7, 2; 8, 2; 9, 2: *a*.
H $\mathbf{G}^{14}_{48} \otimes \mathbf{T}_3$: $\{\sigma_{d1} \mid 00\frac{1}{2}\}$, $\{C_3^+ \mid 000\}$, $\{\sigma_h \mid 000\}$; $\mathbf{t}_1$ or $\mathbf{t}_2$: 5, 3; 6, 3; 7, 3; 8, 3; 11, 3; 12, 3: *a*.

Δ^x $\mathbf{G}^{11}_{24}$: $(C_6^+, 0)$, $(\sigma_{d1}, 0)$: 7, 2; 8, 2; 9, 2: *a*.
U^x $\mathbf{G}^{5}_{8}$: $(C_2, 0)$, $(\sigma_{v1}, 0)$: 5, 2: *a*.
P^x $\mathbf{G}^{4}_{12}$: $(C_3^+, 0)$, $(\sigma_{d1}, 0)$: 3, 3; 4, 3; 6, 2: *a*.
T^x $\mathbf{G}^{5}_{8}$: $(C''_{22}, 0)$, $(\sigma_{d2}, 0)$: 5, 2: *a*.
S^x $\mathbf{G}^{8}_{16}$: $(C''_{22}, 0)$, $(\sigma_h, 0)$: 5, 3; 6, 3; 7, 3; 8, 3: *a*.
T'^x $\mathbf{G}^{5}_{8}$: $(C''_{21}, 0)$, $(\sigma_{d1}, 0)$: 5, 2: *a*.
S'^x $\mathbf{G}^{8}_{16}$: $(C''_{21}, 0)$, $(\sigma_h, 0)$: 5, 3; 6, 3; 7, 3; 8, 3: *a*.
Σ^x $\mathbf{G}^{5}_{8}$: $(C'_{21}, 0)$, $(\sigma_{v1}, 0)$: 5, 2: *a*.
R^x $\mathbf{G}^{8}_{16}$: $(C'_{21}, 0)$, $(\sigma_h, 0)$: 5, 3; 6, 3; 7, 3; 8, 3: *a*.

193 $P6_3/mcm$ D^3_{6h}

($F1$; $K6$; $K7$; $M5$; $Z1$.)

Γ $\mathbf{G}^{15}_{48}$: $\{C_6^+ \mid 00\frac{1}{2}\}$, $\{C'_{21} \mid 000\}$, $\{I \mid 000\}$: 7, 2; 8, 2; 9, 2; 16, 2; 17, 2; 18, 2: *a*.
M $\mathbf{G}^{11}_{16} \otimes \mathbf{T}_2$: $\{C_2 \mid 00\frac{1}{2}\}$, $\{C''_{21} \mid 00\frac{1}{2}\}$, $\{I \mid 000\}$; $\mathbf{t}_2$: 5, 2; 10, 2: *c*.
A $\mathbf{G}^{10}_{96}$: $\{C_6^+ \mid 00\frac{1}{2}\}$, $\{C'_{21} \mid 000\}$, $\{I \mid 000\}$: 15, 3; 16, 3; 24, 2: *a*.
L $\mathbf{G}^{7}_{32}$: $\{\sigma_{v1} \mid 00\frac{1}{2}\}$, $\{C'_{21} \mid 000\}$, $\{I \mid 000\}$: 13, 3; 14, 3: *a*.
K $\mathbf{G}^{11}_{24} \otimes \mathbf{T}_3$: $\{S_3^+ \mid 00\frac{1}{2}\}$, $\{C''_{21} \mid 00\frac{1}{2}\}$; $\mathbf{t}_1$ or $\mathbf{t}_2$: 7, 2; 8, 2; 9, 2: *a*.
H $\mathbf{G}^{11}_{24} \otimes \mathbf{T}_3 \otimes \mathbf{T}_2$: $\{S_3^+ \mid 00\frac{1}{2}\}$, $\{C''_{21} \mid 00\frac{1}{2}\}$; $\mathbf{t}_1$ or $\mathbf{t}_2$; $\mathbf{t}_3$: 7, 1; 8, 3; 9, 3: *a*.

Δ^x $\mathbf{G}^{11}_{24}$: $(C_6^+, 0)$, $(\sigma_{d1}, 0)$: 7, 2; 8, 2; 9, 2: *a*.
U^x $\mathbf{G}^{5}_{8}$: $(C_2, 0)$, $(\sigma_{v1}, 0)$: 5, 2: *a*.
P^x $\mathbf{G}^{4}_{12}$: $(C_3^+, 0)$, $(\sigma_{d1}, 0)$: 3, 3; 4, 3; 6, 2: *a*.
T^x $\mathbf{G}^{5}_{8}$: $(C''_{22}, 0)$, $(\sigma_{d2}, 0)$: 5, 2: *a*.
S^x $\mathbf{G}^{8}_{16}$: $(\sigma_{d2}, 0)$, $(C''_{22}, 0)$: 9, 1: *b*.
T'^x $\mathbf{G}^{5}_{8}$: $(C''_{21}, 0)$, $(\sigma_{d1}, 0)$: 5, 2: *a*.
S'^x $\mathbf{G}^{8}_{16}$: $(\sigma_{d1}, 0)$, $(C''_{21}, 0)$: 9, 1: *b*.
Σ^x $\mathbf{G}^{5}_{8}$: $(C'_{21}, 0)$, $(\sigma_{v1}, 0)$: 5, 2: *a*.
R^x $\mathbf{G}^{8}_{16}$: $(\sigma_h, 0)$, $(C'_{21}, 0)$: 5, 3; 6, 3; 7, 3; 8, 3: *a*.

194 $P6_3/mmc$ D_{6h}^4

($A2$; $A3$; $A5$; $B2$; $C5$; $E1$; $E3$; $F1$; $F2$; $F4$; $H1$; $H2$; $J1$; $K5$; $K6$; $K7$; $M3$; $M5$; $S2$; $S3$; $S8$; $Z1$.)

Γ $\mathbf{G}_{48}^{15}$: $\{C_6^+ \mid 00\frac{1}{2}\}$, $\{C'_{21} \mid 00\frac{1}{2}\}$, $\{I \mid 000\}$: 7, 2; 8, 2; 9, 2; 16, 2; 17, 2; 18, 2: *a*.
M $\mathbf{G}_{16}^{11} \otimes \mathbf{T}_2$: $\{C_2 \mid 00\frac{1}{2}\}$, $\{C''_{21} \mid 000\}$, $\{I \mid 000\}$; $\mathbf{t}_2$: 5, 2; 10, 2: *c*.
A $\mathbf{G}_{96}^{10}$: $\{C_6^+ \mid 00\frac{1}{2}\}$, $\{C''_{21} \mid 000\}$, $\{I \mid 000\}$: 15, 3; 16, 3; 24, 2: *a*.
L $\mathbf{G}_{32}^{7}$: $\{\sigma_{d1} \mid 00\frac{1}{2}\}$, $\{C''_{21} \mid 000\}$, $\{I \mid 000\}$: 13, 3; 14, 3: *a*.
K $\mathbf{G}_{24}^{11} \otimes \mathbf{T}_3$: $\{S_3^+ \mid 00\frac{1}{2}\}$, $\{C''_{21} \mid 000\}$; $\mathbf{t}_1$ or $\mathbf{t}_2$: 7, 2; 8, 2; 9, 2: *a*.
H $\mathbf{G}_{48}^{14} \otimes \mathbf{T}_3$: $\{\sigma_{d1} \mid 00\frac{1}{2}\}$, $\{C_3^+ \mid 000\}$, $\{\sigma_h \mid 00\frac{1}{2}\}$; $\mathbf{t}_1$ or $\mathbf{t}_2$: 5, 3; 6, 3; 7, 3; 8, 3; 11, 2; 12, 2: *a*.

Δ^x $\mathbf{G}_{24}^{11}$: $(C_6^+, 0)$, $(\sigma_{d1}, 0)$: 7, 2; 8, 2; 9, 2: *a*.
U^x $\mathbf{G}_8^5$: $(C_2, 0)$, $(\sigma_{v1}, 0)$: 5, 2: *a*.
P^x $\mathbf{G}_{12}^4$: $(C_3^+, 0)$, $(\sigma_{d1}, 0)$: 3, 3; 4, 3; 6, 2: *a*.
T^x $\mathbf{G}_8^5$: $(C''_{22}, 0)$, $(\sigma_{d2}, 0)$: 5, 2: *a*.
S^x $\mathbf{G}_{16}^8$: $(\sigma_h, 0)$, $(C''_{22}, 0)$: 5, 3; 6, 3; 7, 3; 8, 3: *a*.
T'^x $\mathbf{G}_8^5$: $(C''_{21}, 0)$, $(\sigma_{d1}, 0)$: 5, 2: *a*.
S'^x $\mathbf{G}_{16}^8$: $(\sigma_h, 0)$, $(C''_{21}, 0)$: 5, 3; 6, 3; 7, 3; 8, 3: *a*.
Σ^x $\mathbf{G}_8^5$: $(C'_{21}, 0)$, $(\sigma_{v1}, 0)$: 5, 2: *a*.
R^x $\mathbf{G}_{16}^8$: $(\sigma_{v1}, 0)$, $(\sigma_h, 0)$: 9, 1: *b*.

195 $P23$ T^1

($F1$; $K7$; $M5$; $T5$; $Z1$.)

Γ $\mathbf{G}_{24}^9$: $\{C_{31}^- \mid 000\}$, $\{C_{2x} \mid 000\}$, $\{\bar{C}_{2y} \mid 000\}$: 4, 2; 5, 3; 6, 3: *a*.
X $\mathbf{G}_8^5 \otimes \mathbf{T}_2$: $\{C_{2z} \mid 000\}$, $\{C_{2y} \mid 000\}$; $\mathbf{t}_2$: 5, 2: *a*.
M $\mathbf{G}_8^5 \otimes \mathbf{T}_2$: $\{C_{2z} \mid 000\}$, $\{C_{2y} \mid 000\}$; $\mathbf{t}_1$ or $\mathbf{t}_2$: 5, 2: *a*.
R $\mathbf{G}_{24}^9 \otimes \mathbf{T}_2$: $\{C_{31}^- \mid 000\}$, $\{C_{2x} \mid 000\}$, $\{\bar{C}_{2y} \mid 000\}$; $\mathbf{t}_1$ or $\mathbf{t}_2$ or $\mathbf{t}_3$: 4, 2; 5, 3; 6, 3: *a*.

Δ^x $\mathbf{G}_4^1$: $(C_{2y}, 0)$: 2, 2; 4, 2: *b*.
Σ^x $\mathbf{G}_2^1$: $(\bar{E}, 0)$: 2, 2: *a*.
Λ^x $\mathbf{G}_6^1$: $(C_{31}^-, 0)$: 2, *x*; 4, *x*; 6, *x*: *a*.
S^x $\mathbf{G}_2^1$: $(\bar{E}, 0)$: 2, 2: *a*.
Z^x $\mathbf{G}_4^1$: $(C_{2x}, 0)$: 2, 2; 4, 2: *b*.
T^x $\mathbf{G}_4^1$: $(C_{2z}, 0)$: 2, 2; 4, 2: *b*.

196 $F23$ T^2

($F1$; $K7$; $M5$; $T5$; $Z1$.)

Γ $\mathbf{G}_{24}^9$: $\{C_{31}^- \mid 000\}$, $\{C_{2x} \mid 000\}$, $\{\bar{C}_{2y} \mid 000\}$: 4, 2; 5, 3; 6, 3: *a*.
X $\mathbf{G}_8^5 \otimes \mathbf{T}_2$: $\{C_{2z} \mid 000\}$, $\{C_{2y} \mid 000\}$; $\mathbf{t}_1$ or $\mathbf{t}_3$: 5, 2: *a*.
L $\mathbf{G}_6^1 \otimes \mathbf{T}_2$: $\{C_{31}^+ \mid 000\}$; $\mathbf{t}_1$ or $\mathbf{t}_2$ or $\mathbf{t}_3$: 2, 3; 4, 1; 6, 3: *a*.
W $\mathbf{G}_4^1 \otimes \mathbf{T}_4$: $\{C_{2x} \mid 000\}$; $\mathbf{t}_2$ or $\mathbf{t}_3$: 2, 2; 4, 2: *b*.

Δ^x $\mathbf{G}_4^1$: $(C_{2y}, 0)$: 2, 2; 4, 2: *b*.
Λ^x $\mathbf{G}_6^1$: $(C_{31}^-, 0)$: 2, *x*; 4, *x*; 6, *x*: *a*.
Σ^x $\mathbf{G}_2^1$: $(\bar{E}, 0)$: 2, 2: *a*.
S^x $\mathbf{G}_2^1$: $(\bar{E}, 0)$: 2, 2: *a*.
Z^x $\mathbf{G}_4^1$: $(C_{2x}, 0)$: 2, 2; 4, 2: *b*.
Q^x $\mathbf{G}_2^1$: $(\bar{E}, 0)$: 2, *x*: *a*.

197 $I23$ T^3

($F1$; $K7$; $M5$; $T5$; $Z1$.)

Γ $\mathbf{G}^{9}_{24}$: $\{C^{-}_{31}\,|\,000\}$, $\{C_{2x}\,|\,000\}$, $\{\bar{C}_{2y}\,|\,000\}$: 4, 2; 5, 3; 6, 3: *a*.
H $\mathbf{G}^{9}_{24}\otimes\mathbf{T}_2$: $\{C^{-}_{31}\,|\,000\}$, $\{C_{2x}\,|\,000\}$, $\{\bar{C}_{2y}\,|\,000\}$; $\mathbf{t}_1$ or $\mathbf{t}_2$ or $\mathbf{t}_3$: 4, 2; 5, 3; 6, 3: *a*.
P $\mathbf{G}^{9}_{24}\otimes\mathbf{T}_4$: $\{C^{-}_{31}\,|\,000\}$, $\{C_{2x}\,|\,000\}$, $\{\bar{C}_{2y}\,|\,000\}$; $\mathbf{t}_1$ or $\mathbf{t}_2$ or $\mathbf{t}_3$: 4, x; 5, x; 6, x: *a*.
N $\mathbf{G}^{1}_{4}\otimes\mathbf{T}_2$: $\{C_{2z}\,|\,000\}$; $\mathbf{t}_3$: 2, 3; 4, 3: *b*.

Σ^x $\mathbf{G}^{1}_{2}$: $(\bar{E}, 0)$: 2, 2: *a*.
Δ^x $\mathbf{G}^{1}_{4}$: $(C_{2y}, 0)$: 2, 2; 4, 2: *b*.
Λ^x $\mathbf{G}^{1}_{6}$: $(C^{-}_{31}, 0)$: 2, x; 4, x; 6, x: *a*.
D^x $\mathbf{G}^{1}_{4}$: $(C_{2z}, 0)$: 2, x; 4, x: *b*.
G^x $\mathbf{G}^{1}_{2}$: $(\bar{E}, 0)$: 2, 2: *a*.
F^x $\mathbf{G}^{1}_{6}$: $(C^{+}_{34}, 0)$: 2, x; 4, x; 6, x: *a*.

198 $P2_13$ T^4

($F1$; $K7$; $M5$; $T5$; $Z1$.)

Γ $\mathbf{G}^{9}_{24}$: $\{C^{-}_{31}\,|\,000\}$, $\{C_{2x}\,|\,\tfrac12\tfrac12 0\}$, $\{\bar{C}_{2y}\,|\,0\tfrac12\tfrac12\}$: 4, 2; 5, 3; 6, 3: *a*.
X $\mathbf{G}^{8}_{16}$: $\{C_{2y}\,|\,0\tfrac12\tfrac12\}$, $\{C_{2x}\,|\,\tfrac12\tfrac12 0\}$: 5, 3; 6, 3; 7, 3; 8, 3: *a*.
M $\mathbf{G}^{8}_{16}$: $\{C_{2z}\,|\,\tfrac12 0\tfrac12\}$, $\{C_{2x}\,|\,\tfrac12\tfrac12 0\}$: 9, 1: *b*.
R $\mathbf{G}^{9}_{24}\otimes\mathbf{T}_2$: $\{C^{-}_{31}\,|\,010\}$, $\{C_{2x}\,|\,\tfrac12\tfrac32 0\}$, $\{\bar{C}_{2y}\,|\,0\tfrac32\tfrac12\}$; $\mathbf{t}_1$ or $\mathbf{t}_2$ or $\mathbf{t}_3$: 1, 1; 2, 3; 3, 3; 7, 1: *b*.

Δ^x $\mathbf{G}^{1}_{4}$: $(C_{2y}, 0)$: 2, 2; 4, 2: *b*.
Σ^x $\mathbf{G}^{1}_{2}$: $(\bar{E}, 0)$: 2, 2: *a*.
Λ^x $\mathbf{G}^{1}_{6}$: $(C^{-}_{31}, 0)$: 2, x; 4, x; 6, x: *a*.
S^x $\mathbf{G}^{1}_{2}$: $(\bar{E}, 0)$: 2, 1: *a*.
Z^x $\mathbf{G}^{2}_{8}$: $(C_{2x}, 0)$, $(\bar{E}, 0)$: 5, 3; 7, 3: *a*.
T^x $\mathbf{G}^{2}_{8}$: $(C_{2z}, 0)$, $(\bar{E}, 0)$: 5, 1; 7, 1: *a*.

199 $I2_13$ T^5

($C6$; $F1$; $K7$; $M5$; $T5$; $Z1$.)

Γ $\mathbf{G}^{9}_{24}$: $\{C^{-}_{31}\,|\,000\}$, $\{C_{2x}\,|\,\tfrac12\tfrac12 0\}$, $\{\bar{C}_{2y}\,|\,0\tfrac12\tfrac12\}$: 4, 2; 5, 3; 6, 3: *a*.
H $\mathbf{G}^{9}_{24}\otimes\mathbf{T}_2$: $\{C^{-}_{31}\,|\,000\}$, $\{\bar{C}_{2y}\,|\,0\tfrac12\tfrac32\}$, $\{\bar{C}_{2z}\,|\,\tfrac12 0\tfrac32\}$; $\mathbf{t}_1$ or $\mathbf{t}_2$ or $\mathbf{t}_3$: 4, 2; 5, 3; 6, 3: *a*.
P $\mathbf{G}^{9}_{24}\otimes\mathbf{T}_4$: $\{C^{-}_{31}\,|\,101\}$, $\{C_{2x}\,|\,-\tfrac12\tfrac12 0\}$, $\{\bar{C}_{2y}\,|\,0\tfrac12 -\tfrac12\}$; $\mathbf{t}_1$ or $\mathbf{t}_2$ or $\mathbf{t}_3$: 1, x; 2, x; 3, x; 7, x: *b*.
N $\mathbf{G}^{1}_{4}\otimes\mathbf{T}_2$: $\{C_{2z}\,|\,\tfrac12 0\tfrac12\}$; $\mathbf{t}_3$: 2, 3; 4, 3: *b*.

Σ^x $\mathbf{G}^{1}_{2}$: $(\bar{E}, 0)$: 2, 2: *a*.
Δ^x $\mathbf{G}^{1}_{4}$: $(C_{2y}, 0)$: 2, 2; 4, 2: *b*.
Λ^x $\mathbf{G}^{1}_{6}$: $(C^{-}_{31}, 0)$: 2, x; 4, x; 6, x: *a*.
D^x $\mathbf{G}^{2}_{8}$: $(C_{2z}, 0)$, $(\bar{E}, 0)$: 5, x; 7, x: *a*.
G^x $\mathbf{G}^{1}_{2}$: $(\bar{E}, 0)$: 2, 2: *a*.
F^x $\mathbf{G}^{6}_{12}$: $(C^{+}_{34}, 0)$, $(\bar{E}, 0)$: 7, x; 9, x; 11, x: *d*.

200 *Pm*3 T_h^1

(*F*1; *K*7; *M*5; *T*5; *Z*1.)

Γ $\mathbf{G}_{48}^{4}$: $\{C_{31}^{-} \mid 000\}$, $\{C_{2x} \mid 000\}$, $\{\bar{C}_{2y} \mid 000\}$, $\{I \mid 000\}$: 4, 2; 5, 3; 6, 3; 11, 2; 12, 3; 13, 3: *c*.
X $\mathbf{G}_{16}^{11} \otimes \mathbf{T}_2$: $\{C_{2x} \mid 000\}$, $\{C_{2y} \mid 000\}$, $\{I \mid 000\}$; $\mathbf{t}_2$: 5, 2; 10, 2: *c*.
M $\mathbf{G}_{16}^{11} \otimes \mathbf{T}_2$: $\{C_{2x} \mid 000\}$, $\{C_{2y} \mid 000\}$, $\{I \mid 000\}$; $\mathbf{t}_1$ or $\mathbf{t}_2$: 5, 2; 10, 2: *c*.
R $\mathbf{G}_{48}^{4} \otimes \mathbf{T}_2$: $\{C_{31}^{-} \mid 000\}$, $\{C_{2x} \mid 000\}$, $\{\bar{C}_{2y} \mid 000\}$, $\{I \mid 000\}$; $\mathbf{t}_1$ or $\mathbf{t}_2$ or $\mathbf{t}_3$: 4, 2; 5, 3; 6, 3; 11, 2; 12, 3; 13, 3: *c*.

Δ^x $\mathbf{G}_{8}^{5}$: $(C_{2y}, 0)$, $(\sigma_x, 0)$: 5, 2: *a*.
Σ^x $\mathbf{G}_{4}^{1}$: $(\sigma_z, 0)$: 2, 3; 4, 3: *b*.
Λ^x $\mathbf{G}_{6}^{1}$: $(C_{31}^{-}, 0)$: 2, 3; 4, 1; 6, 3: *a*.
S^x $\mathbf{G}_{4}^{1}$: $(\sigma_y, 0)$: 2, 3; 4, 3: *b*.
Z^x $\mathbf{G}_{8}^{5}$: $(C_{2x}, 0)$, $(\sigma_z, 0)$: 5, 2: *a*.
T^x $\mathbf{G}_{8}^{5}$: $(C_{2z}, 0)$, $(\sigma_y, 0)$: 5, 2: *a*.

201 *Pn*3 T_h^2

(*F*1; *K*7; *M*5; *T*5; *Z*1.)

Γ $\mathbf{G}_{48}^{4}$: $\{C_{31}^{-} \mid 000\}$, $\{C_{2x} \mid 000\}$, $\{\bar{C}_{2y} \mid 000\}$, $\{I \mid \frac{1}{2}\frac{1}{2}\frac{1}{2}\}$: 4, 2; 5, 3; 6, 3; 11, 2; 12, 3; 13, 3: *c*.
X $\mathbf{G}_{32}^{7}$: $\{\sigma_z \mid \frac{1}{2}\frac{1}{2}\frac{1}{2}\}$, $\{\sigma_y \mid \frac{1}{2}\frac{1}{2}\frac{1}{2}\}$, $\{I \mid \frac{1}{2}\frac{1}{2}\frac{1}{2}\}$: 13, 3; 14, 3: *a*.
M $\mathbf{G}_{32}^{7}$: $\{\sigma_x \mid \frac{1}{2}\frac{1}{2}\frac{1}{2}\}$, $\{\sigma_z \mid \frac{1}{2}\frac{1}{2}\frac{1}{2}\}$, $\{I \mid \frac{1}{2}\frac{1}{2}\frac{1}{2}\}$: 13, 3; 14, 3: *a*.
R $\mathbf{G}_{48}^{4} \otimes \mathbf{T}_2$: $\{C_{31}^{-} \mid 000\}$, $\{C_{2x} \mid 000\}$, $\{\bar{C}_{2y} \mid 000\}$, $\{I \mid \frac{1}{2}\frac{1}{2}\frac{1}{2}\}$; $\mathbf{t}_1$ or $\mathbf{t}_2$ or $\mathbf{t}_3$: 4, 2; 5, 3; 6, 3; 11, 2; 12, 3; 13, 3: *c*.

Δ^x $\mathbf{G}_{8}^{5}$: $(C_{2y}, 0)$, $(\sigma_x, 0)$: 5, 2: *a*.
Σ^x $\mathbf{G}_{4}^{1}$: $(\sigma_z, 0)$: 2, 3; 4, 3: *b*.
Λ^x $\mathbf{G}_{6}^{1}$: $(C_{31}^{-}, 0)$: 2, 3; 4, 1; 6, 3: *a*.
S^x $\mathbf{G}_{8}^{2}$: $(\sigma_y, 0)$, $(\bar{E}, 0)$: 5, 3; 7, 3: *a*.
Z^x $\mathbf{G}_{16}^{8}$: $(\sigma_y, 0)$, $(C_{2x}, 0)$: 5, 3; 6, 3; 7, 3; 8, 3: *a*.
T^x $\mathbf{G}_{16}^{8}$: $(C_{2z}, 0)$, $(\sigma_x, 0)$: 9, 2: *b*.

202 *Fm*3 T_h^3

(*F*1; *K*7; *M*5; *T*5; *Z*1.)

Γ $\mathbf{G}_{48}^{4}$: $\{C_{31}^{-} \mid 000\}$, $\{C_{2x} \mid 000\}$, $\{\bar{C}_{2y} \mid 000\}$, $\{I \mid 000\}$: 4, 2; 5, 3; 6, 3; 11, 2; 12, 3; 13, 3: *c*.
X $\mathbf{G}_{16}^{11} \otimes \mathbf{T}_2$: $\{C_{2x} \mid 000\}$, $\{C_{2y} \mid 000\}$, $\{I \mid 000\}$; $\mathbf{t}_1$ or $\mathbf{t}_3$: 5, 2; 10, 2: *c*.
L $\mathbf{G}_{12}^{6} \otimes \mathbf{T}_2$: $\{S_{61}^{+} \mid 000\}$, $\{\bar{E} \mid 000\}$; $\mathbf{t}_1$ or $\mathbf{t}_2$ or $\mathbf{t}_3$: 7, 1; 8, 3; 9, 3; 10, 1; 11, 3; 12, 3: *b*.
W $\mathbf{G}_{8}^{5} \otimes \mathbf{T}_4$: $\{C_{2x} \mid 000\}$, $\{\sigma_z \mid 000\}$; $\mathbf{t}_2$ or $\mathbf{t}_3$: 5, 2: *a*.

Δ^x $\mathbf{G}_{8}^{5}$: $(C_{2y}, 0)$, $(\sigma_x, 0)$: 5, 2: *a*.
Λ^x $\mathbf{G}_{6}^{1}$: $(C_{31}^{-}, 0)$: 2, 3; 4, 1; 6, 3: *a*.
Σ^x $\mathbf{G}_{4}^{1}$: $(\sigma_z, 0)$: 2, 3; 4, 3: *b*.
S^x $\mathbf{G}_{4}^{1}$: $(\sigma_y, 0)$: 2, 3; 4, 3: *b*.
Z^x $\mathbf{G}_{8}^{5}$: $(C_{2x}, 0)$, $(\sigma_z, 0)$: 5, 2: *a*.
Q^x $\mathbf{G}_{2}^{1}$: $(\bar{E}, 0)$: 2, 1: *a*.

203 *Fd*3 T_h^4

(*F*1; *K*7; *M*5; *T*5; *Z*1.)

Γ $\mathbf{G}_{48}^4$: $\{C_{31}^- \mid 000\}$, $\{C_{2x} \mid 000\}$, $\{\bar{C}_{2y} \mid 000\}$, $\{I \mid \frac{1}{4}\frac{1}{4}\frac{1}{4}\}$: 4, 2; 5, 3; 6, 3; 11, 2; 12, 3; 13, 3: *c*.
X $\mathbf{G}_{32}^7$: $\{\sigma_z \mid \frac{1}{4}\frac{1}{4}\frac{1}{4}\}$, $\{\sigma_y \mid \frac{1}{4}\frac{1}{4}\frac{1}{4}\}$, $\{I \mid \frac{1}{4}\frac{1}{4}\frac{1}{4}\}$: 13, 3; 14, 3: *a*.
L $\mathbf{G}_{12}^6 \otimes \mathbf{T}_2$: $\{S_{61}^+ \mid \frac{1}{4}\frac{1}{4}\frac{1}{4}\}$, $\{\bar{E} \mid 000\}$; $\mathbf{t}_1$ or $\mathbf{t}_2$ or $\mathbf{t}_3$: 7, 1; 8, 3; 9, 3; 10, 1; 11, 3; 12, 3: *b*.
W $\mathbf{G}_{32}^8$: $\{\sigma_z \mid \frac{1}{4}\frac{1}{4}\frac{1}{4}\}$, $\{C_{2x} \mid 000\}$: 13, 3; 14, 3; 17, 3; 18, 3: *a*.

Δ^x $\mathbf{G}_8^5$: $(C_{2y}, 0)$, $(\sigma_x, 0)$: 5, 2: *a*.
Λ^x $\mathbf{G}_6^1$: $(C_{31}^-, 0)$: 2, 3; 4, 1; 6, 3: *a*.
Σ^x $\mathbf{G}_4^1$: $(\sigma_z, 0)$: 2, 3; 4, 3: *b*.
S^x $\mathbf{G}_8^2$: $(\sigma_y, 0)$, $(\bar{E}, 0)$: 5, 3; 7, 3: *a*.
Z^x $\mathbf{G}_{16}^8$: $(\sigma_y, 0)$, $(C_{2x}, 0)$: 5, 3; 6, 3; 7, 3; 8, 3: *a*.
Q^x $\mathbf{G}_2^1$: $(\bar{E}, 0)$: 2, 1: *a*.

204 *Im*3 T_h^5

(*F*1; *K*7; *M*5; *T*5; *Z*1.)

Γ $\mathbf{G}_{48}^4$: $\{C_{31}^- \mid 000\}$, $\{C_{2x} \mid 000\}$, $\{\bar{C}_{2y} \mid 000\}$, $\{I \mid 000\}$: 4, 2; 5, 3; 6, 3; 11, 2; 12, 3; 13, 3: *c*.
H $\mathbf{G}_{48}^4 \otimes \mathbf{T}_2$: $\{C_{31}^- \mid 000\}$, $\{C_{2x} \mid 000\}$, $\{\bar{C}_{2y} \mid 000\}$, $\{I \mid 000\}$; $\mathbf{t}_1$ or $\mathbf{t}_2$ or $\mathbf{t}_3$: 4, 2; 5, 3; 6, 3; 11, 2; 12, 3; 13, 3: *c*.
P $\mathbf{G}_{24}^9 \otimes \mathbf{T}_4$: $\{C_{31}^- \mid 000\}$, $\{C_{2x} \mid 000\}$, $\{\bar{C}_{2y} \mid 000\}$; $\mathbf{t}_1$ or $\mathbf{t}_2$ or $\mathbf{t}_3$: 4, 2; 5, 3; 6, 3: *a*.
N $\mathbf{G}_8^2 \otimes \mathbf{T}_2$: $\{C_{2z} \mid 000\}$, $\{I \mid 000\}$; $\mathbf{t}_3$: 2, 3; 4, 3; 6, 3; 8, 3: *b*.

Σ^x $\mathbf{G}_4^1$: $(\sigma_z, 0)$: 2, 3; 4, 3: *b*.
Δ^x $\mathbf{G}_8^5$: $(C_{2y}, 0)$, $(\sigma_x, 0)$: 5, 2: *a*.
Λ^x $\mathbf{G}_6^1$: $(C_{31}^-, 0)$: 2, 3; 4, 1; 6, 3: *a*.
D^x $\mathbf{G}_4^1$: $(C_{2z}, 0)$: 2, 3; 4, 3: *b*.
G^x $\mathbf{G}_4^1$: $(\sigma_z, 0)$: 2, 3; 4, 3: *b*.
F^x $\mathbf{G}_6^1$: $(C_{34}^+, 0)$: 2, 3; 4, 1; 6, 3: *a*.

205 *Pa*3 T_h^6

(*F*1; *G*5; *K*7; *M*5; *S*8; *T*5; *Z*1.)

Γ $\mathbf{G}_{48}^4$: $\{C_{31}^- \mid 000\}$, $\{C_{2x} \mid \frac{1}{2}\frac{1}{2}0\}$, $\{\bar{C}_{2y} \mid 0\frac{1}{2}\frac{1}{2}\}$, $\{I \mid 000\}$: 4, 2; 5, 3; 6, 3; 11, 2; 12, 3; 13, 3: *c*.
X $\mathbf{G}_{32}^7$: $\{C_{2y} \mid 0\frac{1}{2}\frac{1}{2}\}$, $\{C_{2z} \mid \frac{1}{2}0\frac{1}{2}\}$, $\{I \mid 000\}$: 13, 3; 14, 3: *a*.
M $\mathbf{G}_{32}^7$: $\{C_{2z} \mid \frac{1}{2}0\frac{1}{2}\}$, $\{C_{2x} \mid \frac{1}{2}\frac{1}{2}0\}$, $\{I \mid 000\}$: 9, 1; 10, 1: *b*.
R $\mathbf{G}_{48}^4 \otimes \mathbf{T}_2$: $\{C_{31}^- \mid 010\}$, $\{C_{2x} \mid \frac{1}{2}\frac{3}{2}0\}$, $\{\bar{C}_{2y} \mid 0\frac{3}{2}\frac{1}{2}\}$, $\{I \mid 000\}$; $\mathbf{t}_1$ or $\mathbf{t}_2$ or $\mathbf{t}_3$: 1, 1; 2, 3; 3, 3; 7, 1; 8, 1; 9, 3; 10, 3; 14, 1: *d*.

Δ^x $\mathbf{G}_8^5$: $(C_{2y}, 0)$, $(\sigma_x, 0)$: 5, 2: *a*.
Σ^x $\mathbf{G}_4^1$: $(\sigma_z, 0)$: 2, 3; 4, 3: *b*.
Λ^x $\mathbf{G}_6^1$: $(C_{31}^-, 0)$: 2, 3; 4, 1; 6, 3: *a*.
S^x $\mathbf{G}_8^2$: $(\sigma_y, 0)$, $(\bar{E}, 0)$: 5, 1; 7, 1: *a*.
Z^x $\mathbf{G}_{16}^8$: $(\sigma_z, 0)$, $(C_{2x}, 0)$: 9, 1: *b*.
T^x $\mathbf{G}_{16}^{11}$: $(\sigma_x, 0)$, $(\sigma_y, 0)$, $(\bar{E}, 0)$: 6, 1; 7, 1; 8, 1; 9, 1: *a*.

206 $Ia3$ T_h^7

($C6$; $F1$; $K7$; $M5$; $S8$; $T5$; $Z1$.)

Γ $\mathbf{G}_{48}^{4}$: $\{C_{31}^{-}\mid 000\}$, $\{C_{2x}\mid \frac{1}{2}\frac{1}{2}0\}$, $\{\bar{C}_{2y}\mid 0\frac{1}{2}\frac{1}{2}\}$, $\{I\mid 000\}$: 4, 2; 5, 3; 6, 3; 11, 2; 12, 3; 13, 3: c.

H $\mathbf{G}_{48}^{4}\otimes\mathbf{T}_2$: $\{C_{31}^{-}\mid 000\}$, $\{\bar{C}_{2y}\mid 0\frac{1}{2}\frac{3}{2}\}$, $\{\bar{C}_{2z}\mid \frac{1}{2}0\frac{3}{2}\}$, $\{I\mid 000\}$; $\mathbf{t}_1$ or $\mathbf{t}_2$ or $\mathbf{t}_3$: 4, 2; 5, 3; 6, 3; 11, 2; 12, 3; 13, 3: c.

P $\mathbf{G}_{24}^{9}\otimes\mathbf{T}_4$: $\{C_{31}^{-}\mid 101\}$, $\{C_{2x}\mid -\frac{1}{2}\frac{1}{2}0\}$, $\{\bar{C}_{2y}\mid 0\frac{1}{2}-\frac{1}{2}\}$; $\mathbf{t}_1$ or $\mathbf{t}_2$ or $\mathbf{t}_3$: 1, 1; 2, 3; 3, 3; 7, 1: b.

N $\mathbf{G}_{16}^{10}$: $\{C_{2z}\mid \frac{1}{2}0\frac{1}{2}\}$, $\{E\mid 001\}$, $\{I\mid 000\}$: 10, 1: b.

Σ^x $\mathbf{G}_{4}^{1}$: $(\sigma_z, 0)$: 2, 3; 4, 3: b.

Δ^x $\mathbf{G}_{8}^{5}$: $(C_{2y}, 0)$, $(\sigma_x, 0)$: 5, 2: a.

Λ^x $\mathbf{G}_{6}^{1}$: $(C_{31}^{-}, 0)$: 2, 3; 4, 1; 6, 3: a.

D^x $\mathbf{G}_{8}^{2}$: $(C_{2z}, 0)$, $(\bar{E}, 0)$: 5, 1; 7, 1: a.

G^x $\mathbf{G}_{4}^{1}$: $(\sigma_z, 0)$: 2, 3; 4, 3: b.

F^x $\mathbf{G}_{12}^{6}$: $(C_{34}^{+}, 0)$, $(\bar{E}, 0)$: 7, 1; 9, 3; 11, 3: d.

207 $P432$ O^1

($F1$; $K7$; $M5$; $T5$; $Z1$.)

Γ $\mathbf{G}_{48}^{10}$: $\{C_{4x}^{+}\mid 000\}$, $\{\bar{C}_{31}^{-}\mid 000\}$, $\{C_{2b}\mid 000\}$: 4, 2; 5, 2; 8, 2: a.

X $\mathbf{G}_{16}^{14}\otimes\mathbf{T}_2$: $\{C_{4y}^{+}\mid 000\}$, $\{C_{2z}\mid 000\}$; $\mathbf{t}_2$: 6, 2; 7, 2: a.

M $\mathbf{G}_{16}^{14}\otimes\mathbf{T}_2$: $\{C_{4z}^{+}\mid 000\}$, $\{C_{2x}\mid 000\}$; $\mathbf{t}_1$ or $\mathbf{t}_2$: 6, 2; 7, 2: a.

R $\mathbf{G}_{48}^{10}\otimes\mathbf{T}_2$: $\{C_{4x}^{+}\mid 000\}$, $\{\bar{C}_{31}^{-}\mid 000\}$, $\{C_{2b}\mid 000\}$; $\mathbf{t}_1$ or $\mathbf{t}_2$ or $\mathbf{t}_3$: 4, 2; 5, 2; 8, 2: a.

Δ^x $\mathbf{G}_{8}^{1}$: $(C_{4y}^{+}, 0)$: 2, 2; 4, 2; 6, 2; 8, 2: a.

Σ^x $\mathbf{G}_{4}^{1}$: $(C_{2a}, 0)$: 2, 2; 4, 2: b.

Λ^x $\mathbf{G}_{6}^{1}$: $(C_{31}^{-}, 0)$: 2, 2; 4, 2; 6, 2: a.

S^x $\mathbf{G}_{4}^{1}$: $(C_{2c}, 0)$: 2, 2; 4, 2: b.

Z^x $\mathbf{G}_{4}^{1}$: $(C_{2x}, 0)$: 2, 2; 4, 2: b.

T^x $\mathbf{G}_{8}^{1}$: $(C_{4z}^{+}, 0)$: 2, 2; 4, 2; 6, 2; 8, 2: a.

208 $P4_232$ O^2

($F1$; $K7$; $M5$; $T5$; $Z1$.)

Γ $\mathbf{G}_{48}^{10}$: $\{C_{4x}^{+}\mid \frac{1}{2}\frac{1}{2}\frac{1}{2}\}$, $\{\bar{C}_{31}^{-}\mid 000\}$, $\{C_{2b}\mid \frac{1}{2}\frac{1}{2}\frac{1}{2}\}$: 4, 2; 5, 2; 8, 2: a.

X $\mathbf{G}_{16}^{14}\otimes\mathbf{T}_2$: $\{C_{4y}^{+}\mid \frac{1}{2}\frac{1}{2}\frac{1}{2}\}$, $\{C_{2x}\mid 000\}$; $\mathbf{t}_2$: 6, 2; 7, 2: a.

M $\mathbf{G}_{16}^{14}\otimes\mathbf{T}_2$: $\{C_{4z}^{+}\mid \frac{1}{2}\frac{1}{2}\frac{1}{2}\}$, $\{C_{2y}\mid 000\}$; $\mathbf{t}_1$ or $\mathbf{t}_2$: 6, 2; 7, 2: a.

R $\mathbf{G}_{48}^{10}\otimes\mathbf{T}_2$: $\{C_{4x}^{+}\mid \frac{1}{2}\frac{1}{2}\frac{1}{2}\}$, $\{\bar{C}_{31}^{-}\mid 000\}$, $\{C_{2b}\mid \frac{1}{2}\frac{1}{2}\frac{1}{2}\}$; $\mathbf{t}_1$ or $\mathbf{t}_2$ or $\mathbf{t}_3$: 4, 2; 5, 2; 8, 2: a.

Δ^x $\mathbf{G}_{8}^{1}$: $(C_{4y}^{+}, 0)$: 2, 2; 4, 2; 6, 2; 8, 2: a.

Σ^x $\mathbf{G}_{4}^{1}$: $(C_{2a}, 0)$: 2, 2; 4, 2: b.

Λ^x $\mathbf{G}_{6}^{1}$: $(C_{31}^{-}, 0)$: 2, 2; 4, 2; 6, 2: a.

S^x $\mathbf{G}_{8}^{2}$: $(C_{2c}, 0)$, $(\bar{E}, 0)$: 5, 2; 7, 2: a.

Z^x $\mathbf{G}_{4}^{1}$: $(C_{2x}, 0)$: 2, 2; 4, 2: b.

T^x $\mathbf{G}_{16}^{2}$: $(C_{4z}^{+}, 0)$, $(\bar{E}, 1)$: 2, 2; 4, 2; 6, 2; 8, 2: f.

209 $F432$ O^3

($F1$; $K7$; $M5$; $T5$; $Z1$.)

Γ $\mathbf{G}_{48}^{10}$: $\{C_{4x}^+|000\}$, $\{\bar{C}_{31}^-|000\}$, $\{C_{2b}|000\}$: 4, 2; 5, 2; 8, 2: a.
X $\mathbf{G}_{16}^{14}\otimes\mathbf{T}_2$: $\{C_{4y}^+|000\}$, $\{C_{2z}|000\}$; $\mathbf{t}_1$ or $\mathbf{t}_3$: 6, 2; 7, 2: a.
L $\mathbf{G}_{12}^{4}\otimes\mathbf{T}_2$: $\{C_{31}^+|000\}$, $\{C_{2b}|000\}$; $\mathbf{t}_1$ or $\mathbf{t}_2$ or $\mathbf{t}_3$: 3, 3; 4, 3; 6, 2: a.
W $\mathbf{G}_{8}^{5}\otimes\mathbf{T}_4$: $\{C_{2x}|000\}$, $\{C_{2d}|000\}$; $\mathbf{t}_2$ or $\mathbf{t}_3$: 5, 2: a.

Δ^x $\mathbf{G}_8^1$: $(C_{4y}^+, 0)$: 2, 2; 4, 2; 6, 2; 8, 2: a.
Λ^x $\mathbf{G}_6^1$: $(C_{31}^-, 0)$: 2, 2; 4, 2; 6, 2: a.
Σ^x $\mathbf{G}_4^1$: $(C_{2a}, 0)$: 2, 2; 4, 2: b.
S^x $\mathbf{G}_4^1$: $(C_{2c}, 0)$: 2, 2; 4, 2: b.
Z^x $\mathbf{G}_4^1$: $(C_{2x}, 0)$: 2, 2; 4, 2: b.
Q^x $\mathbf{G}_4^1$: $(C_{2f}, 0)$: 2, x; 4, x: b.

210 $F4_132$ O^4

($F1$; $K7$; $M5$; $T5$; $Z1$.)

Γ $\mathbf{G}_{48}^{10}$: $\{C_{4x}^+|\frac{1}{4}\frac{1}{4}\frac{1}{4}\}$, $\{\bar{C}_{31}^-|000\}$, $\{C_{2b}|\frac{1}{4}\frac{1}{4}\frac{1}{4}\}$: 4, 2; 5, 2; 8, 2: a.
X $\mathbf{G}_{16}^{14}\otimes\mathbf{T}_2$: $\{C_{4y}^-|\frac{1}{4}\frac{1}{4}\frac{1}{4}\}$, $\{C_{2x}|000\}$; $\mathbf{t}_1$ or $\mathbf{t}_3$: 6, 2; 7, 2: a.
L $\mathbf{G}_{12}^{4}\otimes\mathbf{T}_2$: $\{C_{31}^+|000\}$, $\{C_{2b}|\frac{1}{4}\frac{1}{4}\frac{1}{4}\}$; $\mathbf{t}_1$ or $\mathbf{t}_2$ or $\mathbf{t}_3$: 3, 3; 4, 3; 6, 2: a.
W $\mathbf{G}_{8}^{5}\otimes\mathbf{T}_4$: $\{C_{2d}|\frac{1}{4}\frac{1}{4}\frac{1}{4}\}$, $\{C_{2x}|010\}$; $\mathbf{t}_2$ or $\mathbf{t}_3$: 1, 3; 2, 2; 3, 3; 4, 2: b.

Δ^x $\mathbf{G}_8^1$: $(C_{4y}^+, 0)$: 2, 2; 4, 2; 6, 2; 8, 2: a.
Λ^x $\mathbf{G}_6^1$: $(C_{31}^-, 0)$: 2, 2; 4, 2; 6, 2: a.
Σ^x $\mathbf{G}_4^1$: $(C_{2a}, 0)$: 2, 2; 4, 2: b.
S^x $\mathbf{G}_8^2$: $(C_{2c}, 0)$, $(\bar{E}, 0)$: 5, 2; 7, 2: a.
Z^x $\mathbf{G}_4^1$: $(C_{2x}, 0)$: 2, 2; 4, 2: b.
Q^x $\mathbf{G}_{16}^2$: $(C_{2f}, 0)$, $(\bar{E}, 0)$: 10, x; 14, x: e.

211 $I432$ O^5

($F1$; $K7$; $M5$; $T5$; $Z1$.)

Γ $\mathbf{G}_{48}^{10}$: $\{C_{4x}^+|000\}$, $\{\bar{C}_{31}^-|000\}$, $\{C_{2b}|000\}$: 4, 2; 5, 2; 8, 2: a.
H $\mathbf{G}_{48}^{10}\otimes\mathbf{T}_2$: $\{C_{4x}^+|000\}$, $\{\bar{C}_{31}^-|000\}$, $\{C_{2b}|000\}$; $\mathbf{t}_1$ or $\mathbf{t}_2$ or $\mathbf{t}_3$: 4, 2; 5, 2; 8, 2: a.
P $\mathbf{G}_{24}^{9}\otimes\mathbf{T}_4$: $\{C_{31}^-|000\}$, $\{C_{2x}|000\}$, $\{\bar{C}_{2y}|000\}$; $\mathbf{t}_1$ or $\mathbf{t}_2$ or $\mathbf{t}_3$: 4, 2; 5, 2; 6, 2: a.
N $\mathbf{G}_{8}^{5}\otimes\mathbf{T}_2$: $\{C_{2z}|000\}$, $\{C_{2b}|000\}$; $\mathbf{t}_3$: 5, 2: a.

Σ^x $\mathbf{G}_4^1$: $(C_{2a}, 0)$: 2, 2; 4, 2: b.
Δ^x $\mathbf{G}_8^1$: $(C_{4y}^+, 0)$: 2, 2; 4, 2; 6, 2; 8, 2: a.
Λ^x $\mathbf{G}_6^1$: $(C_{31}^-, 0)$: 2, 2; 4, 2; 6, 2: a.
D^x $\mathbf{G}_4^1$: $(C_{2z}, 0)$: 2, 2; 4, 2: b.
G^x $\mathbf{G}_4^1$: $(C_{2b}, 0)$: 2, 2; 4, 2: b.
F^x $\mathbf{G}_6^1$: $(C_{34}^+, 0)$: 2, 2; 4, 2; 6, 2: a.

212 $P4_332$ O^6

($F1$; $K7$; $M5$; $T5$; $Z1$.)

Γ $\mathbf{G}_{48}^{10}$: $\{C_{4x}^+ \mid \tfrac{3}{4}\tfrac{3}{4}\tfrac{1}{4}\}$, $\{\bar{C}_{31}^- \mid 000\}$, $\{C_{2b} \mid \tfrac{1}{4}\tfrac{1}{4}\tfrac{1}{4}\}$: 4, 2; 5, 2; 8, 2: *a*.
X $\mathbf{G}_{32}^{10}$: $\{C_{4y}^- \mid \tfrac{3}{4}\tfrac{1}{4}\tfrac{3}{4}\}$, $\{C_{2x} \mid \tfrac{1}{2}\tfrac{1}{2}0\}$: 8, 3; 9, 3; 10, 3; 11, 3; 14, 2: *a*.
M $\mathbf{G}_{32}^{11}$: $\{C_{4z}^- \mid \tfrac{3}{4}\tfrac{3}{4}\tfrac{1}{4}\}$, $\{C_{2x} \mid \tfrac{1}{2}\tfrac{1}{2}0\}$: 6, 3; 7, 3: *a*.
R $\mathbf{G}_{96}^{7}$: $\{C_{2x} \mid \tfrac{1}{2}\tfrac{1}{2}0\}$, $\{C_{2y} \mid 0\tfrac{1}{2}\tfrac{1}{2}\}$, $\{C_{31}^- \mid 000\}$, $\{C_{2b} \mid \tfrac{1}{4}\tfrac{1}{4}\tfrac{1}{4}\}$: 3, 3; 4, 3; 8, 2; 13, 3; 14, 3: *a*.

Δ^x $\mathbf{G}_{8}^{1}$: $(C_{4y}^+, 0)$: 2, 2; 4, 2; 6, 2; 8, 2: *a*.
Σ^x $\mathbf{G}_{4}^{1}$: $(C_{2a}, 0)$: 2, 2; 4, 2: *b*.
Λ^x $\mathbf{G}_{6}^{1}$: $(C_{31}^-, 0)$: 2, 2; 4, 2; 6, 2: *a*.
S^x $\mathbf{G}_{16}^{2}$: $(C_{2c}, 0)$, $(\bar{E}, 0)$: 10, 3; 14, 3: *e*.
Z^x $\mathbf{G}_{8}^{2}$: $(C_{2x}, 0)$, $(\bar{E}, 0)$: 5, 3; 7, 3: *a*.
T^x $\mathbf{G}_{32}^{17}$: $(C_{4z}^+, 0)$, $(\bar{E}, 3)$: 10, 3; 12, 3; 14, 3; 16, 3: *a*.

213 $P4_132$ O^7

($F1$; $K7$; $M5$; $T5$; $Z1$.)

Γ $\mathbf{G}_{48}^{10}$: $\{C_{4x}^+ \mid \tfrac{1}{4}\tfrac{1}{4}\tfrac{3}{4}\}$, $\{\bar{C}_{31}^- \mid 000\}$, $\{C_{2b} \mid \tfrac{3}{4}\tfrac{3}{4}\tfrac{3}{4}\}$: 4, 2; 5, 2; 8, 2: *a*.
X $\mathbf{G}_{32}^{10}$: $\{C_{4y}^- \mid \tfrac{1}{4}\tfrac{3}{4}\tfrac{1}{4}\}$, $\{C_{2x} \mid \tfrac{1}{2}\tfrac{1}{2}0\}$: 8, 3; 9, 3; 10, 3; 11, 3; 14, 2: *a*.
M $\mathbf{G}_{32}^{11}$: $\{C_{4z}^- \mid \tfrac{1}{4}\tfrac{1}{4}\tfrac{3}{4}\}$, $\{C_{2x} \mid \tfrac{1}{2}\tfrac{1}{2}0\}$: 6, 3; 7, 3: *a*.
R $\mathbf{G}_{96}^{7}$: $\{C_{2x} \mid \tfrac{1}{2}\tfrac{1}{2}0\}$, $\{C_{2y} \mid 0\tfrac{1}{2}\tfrac{1}{2}\}$, $\{C_{31}^- \mid 000\}$, $\{C_{2b} \mid \tfrac{3}{4}\tfrac{3}{4}\tfrac{3}{4}\}$: 3, 3; 4, 3; 8, 2; 13, 3; 14, 3: *a*.

Δ^x $\mathbf{G}_{8}^{1}$: $(C_{4y}^+, 0)$: 2, 2; 4, 2; 6, 2; 8, 2: *a*.
Σ^x $\mathbf{G}_{4}^{1}$: $(C_{2a}, 0)$: 2, 2; 4, 2: *b*.
Λ^x $\mathbf{G}_{6}^{1}$: $(C_{31}^-, 0)$: 2, 2; 4, 2; 6, 2: *a*.
S^x $\mathbf{G}_{16}^{2}$: $(C_{2c}, 0)$, $(\bar{E}, 0)$: 12, 3; 16, 3: *a*.
Z^x $\mathbf{G}_{8}^{2}$: $(C_{2x}, 0)$, $(\bar{E}, 0)$: 5, 3; 7, 3: *a*.
T^x $\mathbf{G}_{32}^{17}$: $(C_{4z}^+, 0)$, $(\bar{E}, 3)$: 10, 3; 12, 3; 14, 3; 16, 3: *a*.

214 $I4_132$ O^8

($F1$; $K7$; $M5$; $T5$; $Z1$.)

Γ $\mathbf{G}_{48}^{10}$: $\{C_{4x}^+ \mid 00\tfrac{1}{2}\}$, $\{\bar{C}_{31}^- \mid 000\}$, $\{C_{2b} \mid \tfrac{1}{2}\tfrac{1}{2}\tfrac{1}{2}\}$: 4, 2; 5, 2; 8, 2: *a*.
H $\mathbf{G}_{48}^{10} \otimes \mathbf{T}_2$: $\{C_{4x}^+ \mid 00\tfrac{1}{2}\}$, $\{\bar{C}_{31}^- \mid 000\}$, $\{C_{2b} \mid \tfrac{3}{2}\tfrac{1}{2}\tfrac{1}{2}\}$; $\mathbf{t}_1$ or $\mathbf{t}_2$ or $\mathbf{t}_3$: 4, 2; 5, 2; 8, 2: *a*.
P $\mathbf{G}_{24}^{9} \otimes \mathbf{T}_4$: $\{C_{31}^- \mid 101\}$, $\{C_{2x} \mid -\tfrac{1}{2}\tfrac{1}{2}0\}$, $\{\bar{C}_{2y} \mid 0\tfrac{1}{2}-\tfrac{1}{2}\}$; $\mathbf{t}_1$ or $\mathbf{t}_2$ or $\mathbf{t}_3$: 1, 2; 2, 2; 3, 2; 7, 2: *b*.
N $\mathbf{G}_{8}^{5} \otimes \mathbf{T}_2$: $\{C_{2a} \mid \tfrac{1}{2}00\}$, $\{C_{2b} \mid \tfrac{1}{2}\tfrac{1}{2}\tfrac{1}{2}\}$; $\mathbf{t}_3$: 5, 2: *a*.

Σ^x $\mathbf{G}_{4}^{1}$: $(C_{2a}, 0)$: 2, 2; 4, 2: *b*.
Δ^x $\mathbf{G}_{8}^{1}$: $(C_{4y}^+, 0)$: 2, 2; 4, 2; 6, 2; 8, 2: *a*.
Λ^x $\mathbf{G}_{6}^{1}$: $(C_{31}^-, 0)$: 2, 2; 4, 2; 6, 2: *a*.
D^x $\mathbf{G}_{8}^{2}$: $(C_{2x}, 0)$, $(\bar{E}, 0)$: 5, 2; 7, 2: *a*.
G^x $\mathbf{G}_{8}^{2}$: $(C_{2b}, 0)$, $(\bar{E}, 0)$: 5, 2; 7, 2: *a*.
F^x $\mathbf{G}_{12}^{6}$: $(C_{34}^+, 0)$, $(\bar{E}, 0)$: 7, 2; 9, 2; 11, 2: *d*.

215 $P\bar{4}3m$ T_d^1

($F1$; $K7$; $M5$; $T5$; $Z1$.)

Γ $\mathbf{G}_{48}^{10}$: $\{S_{4x}^{-}\mid 000\}$, $\{\bar{C}_{31}^{-}\mid 000\}$, $\{\sigma_{db}\mid 000\}$: 4, 2; 5, 2; 8, 2: a.
X $\mathbf{G}_{16}^{14}\otimes\mathbf{T}_2$: $\{S_{4y}^{+}\mid 000\}$, $\{C_{2z}\mid 000\}$; $\mathbf{t}_2$: 6, 2; 7, 2: a.
M $\mathbf{G}_{16}^{14}\otimes\mathbf{T}_2$: $\{S_{4z}^{+}\mid 000\}$, $\{C_{2x}\mid 000\}$; $\mathbf{t}_1$ or $\mathbf{t}_2$: 6, 2; 7, 2: a.
R $\mathbf{G}_{48}^{10}\otimes\mathbf{T}_2$: $\{S_{4x}^{+}\mid 000\}$, $\{\bar{C}_{31}^{-}\mid 000\}$, $\{\sigma_{db}\mid 000\}$; $\mathbf{t}_1$ or $\mathbf{t}_2$ or $\mathbf{t}_3$: 4, 2; 5, 2; 8, 2: a.

Δ^x $\mathbf{G}_8^5$: $(C_{2y}, 0)$, $(\sigma_{dc}, 0)$: 5, 2: a.
Σ^x $\mathbf{G}_4^1$: $(\sigma_{db}, 0)$: 2, 2; 4, 2: b.
Λ^x $\mathbf{G}_{12}^4$: $(C_{31}^{-}, 0)$, $(\sigma_{db}, 0)$: 3, x; 4, x; 6, x: a.
S^x $\mathbf{G}_4^1$: $(\sigma_{de}, 0)$: 2, 2; 4, 2: b.
Z^x $\mathbf{G}_4^1$: $(C_{2x}, 0)$: 2, 2; 4, 2: b.
T^x $\mathbf{G}_8^5$: $(C_{2z}, 0)$, $(\sigma_{da}, 0)$: 5, 2: a.

216 $F\bar{4}3m$ T_d^2

($A1$; $B3$; $B4$; $B5$; $B6$; $C2$; $D4$; $F1$; $K5$; $K7$; $M5$; $M6$; $P1$; $S1$; $S6$; $S8$; $T5$; $Z1$.)

Γ $\mathbf{G}_{48}^{10}$: $\{S_{4x}^{-}\mid 000\}$, $\{\bar{C}_{31}^{-}\mid 000\}$, $\{\sigma_{db}\mid 000\}$: 4, 2; 5, 2; 8, 2: a.
X $\mathbf{G}_{16}^{14}\otimes\mathbf{T}_2$: $\{S_{4y}^{+}\mid 000\}$, $\{C_{2x}\mid 000\}$; $\mathbf{t}_1$ or $\mathbf{t}_3$: 6, 2; 7, 2: a.
L $\mathbf{G}_{12}^4\otimes\mathbf{T}_2$: $\{C_{31}^{+}\mid 000\}$, $\{\sigma_{db}\mid 000\}$; $\mathbf{t}_1$ or $\mathbf{t}_2$ or $\mathbf{t}_3$: 3, 3; 4, 3; 6, 2: a.
W $\mathbf{G}_8^1\otimes\mathbf{T}_4$: $\{S_{4x}^{+}\mid 000\}$; $\mathbf{t}_2$ or $\mathbf{t}_3$: 2, 2; 4, 2; 6, 2; 8, 2: a.

Δ^x $\mathbf{G}_8^5$: $(C_{2y}, 0)$, $(\sigma_{dc}, 0)$: 5, 2: a.
Λ^x $\mathbf{G}_{12}^4$: $(C_{31}^{-}, 0)$, $(\sigma_{db}, 0)$: 3, x; 4, x; 6, x: a.
Σ^x $\mathbf{G}_4^1$: $(\sigma_{db}, 0)$: 2, 2; 4, 2: b.
S^x $\mathbf{G}_4^1$: $(\sigma_{de}, 0)$: 2, 2; 4, 2: b.
Z^x $\mathbf{G}_4^1$: $(C_{2x}, 0)$: 2, 2; 4, 2: b.
Q^x $\mathbf{G}_2^1$: $(\bar{E}, 0)$: 2, 2: a.

217 $I\bar{4}3m$ T_d^3

($F1$; $J1$; $K7$; $M5$; $T5$; $Z1$.)

Γ $\mathbf{G}_{48}^{10}$: $\{S_{4x}^{-}\mid 000\}$, $\{\bar{C}_{31}^{-}\mid 000\}$, $\{\sigma_{db}\mid 000\}$: 4, 2; 5, 2; 8, 2: a.
H $\mathbf{G}_{48}^{10}\otimes\mathbf{T}_2$: $\{S_{4x}^{+}\mid 000\}$, $\{\bar{C}_{31}^{-}\mid 000\}$, $\{\sigma_{db}\mid 000\}$; $\mathbf{t}_1$ or $\mathbf{t}_2$ or $\mathbf{t}_3$: 4, 2; 5, 2; 8, 2: a.
P $\mathbf{G}_{48}^{10}\otimes\mathbf{T}_4$: $\{S_{4x}^{+}\mid 000\}$, $\{\bar{C}_{31}^{-}\mid 000\}$, $\{\sigma_{db}\mid 000\}$; $\mathbf{t}_1$ or $\mathbf{t}_2$ or $\mathbf{t}_3$: 4, x; 5, x; 8, x: a.
N $\mathbf{G}_8^5\otimes\mathbf{T}_2$: $\{C_{2z}\mid 000\}$, $\{\sigma_{db}\mid 000\}$; $\mathbf{t}_3$: 5, 2: a.

Σ^x $\mathbf{G}_4^1$: $(\sigma_{db}, 0)$: 2, 2; 4, 2: b.
Δ^x $\mathbf{G}_8^5$: $(C_{2y}, 0)$, $(\sigma_{de}, 0)$: 5, 2: a.
Λ^x $\mathbf{G}_{12}^4$: $(C_{31}^{-}, 0)$, $(\sigma_{db}, 0)$: 3, x; 4, x; 6, x: a.
D^x $\mathbf{G}_8^5$: $(C_{2z}, 0)$, $(\sigma_{db}, 0)$: 5, x: a.
G^x $\mathbf{G}_4^1$: $(\sigma_{da}, 0)$: 2, 2; 4, 2: b.
F^x $\mathbf{G}_{12}^4$: $(C_{34}^{+}, 0)$, $(\sigma_{da}, 0)$: 3, x; 4, x; 6, x: a.

218 $P\bar{4}3n$ T_d^4

($F1$; $K7$; $M5$; $T5$; $Z1$.)

Γ $\mathbf{G}_{48}^{10}$: $\{S_{4x}^- \mid \frac{1}{2}\frac{1}{2}\frac{1}{2}\}$, $\{\bar{C}_{31}^- \mid 000\}$, $\{\sigma_{db} \mid \frac{1}{2}\frac{1}{2}\frac{1}{2}\}$: 4, 2; 5, 2; 8, 2: *a*.
X $\mathbf{G}_{32}^{11}$: $\{S_{4y}^+ \mid \frac{1}{2}\frac{1}{2}\frac{1}{2}\}$, $\{\sigma_{dc} \mid \frac{1}{2}\frac{1}{2}\frac{1}{2}\}$: 6, 3; 7, 3: *a*.
M $\mathbf{G}_{16}^{14} \otimes \mathbf{T}_2$: $\{S_{4z}^+ \mid \frac{1}{2}\frac{1}{2}\frac{1}{2}\}$, $\{C_{2x} \mid 000\}$; $\mathbf{t}_1$ or $\mathbf{t}_2$: 6, 2; 7, 2: *a*.
R $\mathbf{G}_{96}^{7}$: $\{C_{2x} \mid 001\}$, $\{C_{2y} \mid 000\}$, $\{C_{31}^- \mid 001\}$, $\{\sigma_{db} \mid \frac{1}{2}\frac{1}{2}\frac{1}{2}\}$: 6, 3; 7, 3; 15, 1: *b*.

Δ^x $\mathbf{G}_8^5$: $(C_{2y}, 0)$, $(\sigma_{dc}, 0)$: 5, 2: *a*.
Σ^x $\mathbf{G}_4^1$: $(\sigma_{db}, 0)$: 2, 2; 4, 2: *b*.
Λ^x $\mathbf{G}_{12}^4$: $(C_{31}^-, 0)$, $(\sigma_{db}, 0)$: 3, *x*; 4, *x*; 6, *x*: *a*.
S^x $\mathbf{G}_4^1$: $(\sigma_{de}, 0)$: 2, 3; 4, 3: *b*.
Z^x $\mathbf{G}_4^1$: $(C_{2x}, 0)$: 2, 2; 4, 2: *b*.
T^x $\mathbf{G}_8^5$: $(C_{2z}, 0)$, $(\sigma_{da}, 0)$: 5, 2: *a*.

219 $F\bar{4}3c$ T_d^5

($F1$; $K7$; $M5$; $T5$; $Z1$.)

Γ $\mathbf{G}_{48}^{10}$: $\{S_{4x}^- \mid \frac{1}{2}\frac{1}{2}\frac{1}{2}\}$, $\{\bar{C}_{31}^- \mid 000\}$, $\{\sigma_{db} \mid \frac{1}{2}\frac{1}{2}\frac{1}{2}\}$: 4, 2; 5, 2; 8, 2: *a*.
X $\mathbf{G}_{16}^{14} \otimes \mathbf{T}_2$: $\{S_{4y}^+ \mid \frac{1}{2}\frac{1}{2}\frac{1}{2}\}$, $\{C_{2x} \mid 000\}$; $\mathbf{t}_1$ or $\mathbf{t}_3$: 6, 2; 7, 2: *a*.
L $\mathbf{G}_{12}^4 \otimes \mathbf{T}_2$: $\{C_{31}^+ \mid 100\}$, $\{\sigma_{db} \mid \frac{1}{2}\frac{1}{2}\frac{1}{2}\}$; $\mathbf{t}_1$ or $\mathbf{t}_2$ or $\mathbf{t}_3$: 1, 1; 2, 1; 5, 1: *b*.
W $\mathbf{G}_8^1 \otimes \mathbf{T}_4$: $\{S_{4x}^+ \mid \frac{1}{2}\frac{1}{2}\frac{1}{2}\}$; $\mathbf{t}_2$ or $\mathbf{t}_3$: 2, 3; 4, 3; 6, 3; 8, 3: *a*.

Δ^x $\mathbf{G}_8^5$: $(C_{2y}, 0)$, $(\sigma_{dc}, 0)$: 5, 2: *a*.
Λ^x $\mathbf{G}_{12}^4$: $(C_{31}^-, 0)$, $(\sigma_{db}, 0)$: 3, *x*; 4, *x*; 6, *x*: *a*.
Σ^x $\mathbf{G}_4^1$: $(\sigma_{db}, 0)$: 2, 2; 4, 2: *b*.
S^x $\mathbf{G}_4^1$: $(\sigma_{de}, 0)$: 2, 2; 4, 2: *b*.
Z^x $\mathbf{G}_4^1$: $(C_{2x}, 0)$: 2, 2; 4, 2: *b*.
Q^x $\mathbf{G}_2^1$: $(\bar{E}, 0)$: 2, 1: *a*.

220 $I\bar{4}3d$ T_d^6

($F1$; $K1$; $K7$; $M5$; $T5$; $Z1$.)

Γ $\mathbf{G}_{48}^{10}$: $\{S_{4x}^- \mid \frac{1}{2}00\}$, $\{\bar{C}_{31}^- \mid 000\}$, $\{\sigma_{db} \mid \frac{1}{2}\frac{1}{2}\frac{1}{2}\}$: 4, 2; 5, 2; 8, 2: *a*.
H $\mathbf{G}_{96}^{7}$: $\{C_{2x} \mid \frac{1}{2}\frac{1}{2}0\}$, $\{C_{2y} \mid 0\frac{1}{2}\frac{3}{2}\}$, $\{C_{31}^- \mid 001\}$, $\{\sigma_{db} \mid \frac{1}{2}\frac{1}{2}\frac{1}{2}\}$: 6, 3; 7, 3; 15, 1: *b*.
P $\mathbf{G}_{192}^{1}$: $\{C_{32}^+ \mid 0\frac{1}{2}\frac{1}{2}\}$, $\{C_{2y} \mid 0\frac{1}{2}\frac{1}{2}\}$, $\{C_{2x} \mid \frac{3}{2}\frac{3}{2}0\}$, $\{S_{4x}^+ \mid \frac{1}{2}11\}$: 17, *x*; 18, *x*; 21, *x*; 27, *x*; 28, *x*: *a*.
N $\mathbf{G}_{16}^{8}$: $\{\sigma_{db} \mid \frac{1}{2}\frac{1}{2}\frac{1}{2}\}$, $\{C_{2z} \mid \frac{1}{2}0\frac{1}{2}\}$: 5, 3; 6, 3; 7, 3; 8, 3: *a*.

Σ^x $\mathbf{G}_4^1$: $(\sigma_{db}, 0)$: 2, 2; 4, 2: *b*.
Δ^x $\mathbf{G}_8^5$: $(C_{2y}, 0)$, $(\sigma_{de}, 0)$: 5, 2: *a*.
Λ^x $\mathbf{G}_{12}^4$: $(C_{31}^-, 0)$, $(\sigma_{db}, 0)$: 3, *x*; 4, *x*; 6, *x*: *a*.
D^x $\mathbf{G}_{16}^8$: $(C_{2z}, 0)$, $(\sigma_{db}, 0)$: 5, *x*; 6, *x*; 7, *x*; 8, *x*: *a*.
G^x $\mathbf{G}_4^1$: $(\sigma_{da}, 0)$: 2, 3; 4, 3: *b*.
F^x $\mathbf{G}_{24}^3$: $(C_{34}^+, 1)$, $(\sigma_{da}, 0)$, $(E, 1)$: 9, *x*; 10, *x*; 12, *x*: *c*.

221 *Pm3m* O_h^1

(*A*4; *B*2; *B*8; *E*1; *F*1; *H*3; *J*1; *K*5; *K*7; *M*5; *N*2; *N*3; *S*4; *S*8; *T*4; *T*5; *Y*1; *Z*1.)

Γ $\mathbf{G}_{96}^{8}$: $\{C_{4x}^+ | 000\}$, $\{\bar{C}_{31}^- | 000\}$, $\{C_{2b} | 000\}$, $\{I | 000\}$: 4, 2; 5, 2; 8, 2; 12, 2; 13, 2; 16, 2: *b*.
X $\mathbf{G}_{32}^{9} \otimes \mathbf{T}_2$: $\{C_{4y}^+ | 000\}$, $\{C_{2z} | 000\}$, $\{I | 000\}$; $\mathbf{t}_2$: 6, 2; 7, 2; 13, 2; 14, 2: *a*.
M $\mathbf{G}_{32}^{9} \otimes \mathbf{T}_2$: $\{C_{4z}^+ | 000\}$, $\{C_{2x} | 000\}$, $\{I | 000\}$; $\mathbf{t}_1$ or $\mathbf{t}_2$: 6, 2; 7, 2; 13, 2; 14, 2: *a*.
R $\mathbf{G}_{96}^{8} \otimes \mathbf{T}_2$: $\{C_{4x}^+ | 000\}$, $\{\bar{C}_{31}^- | 000\}$, $\{C_{2b} | 000\}$, $\{I | 000\}$; $\mathbf{t}_1$ or $\mathbf{t}_2$ or $\mathbf{t}_3$: 4, 2; 5, 2; 8, 2; 12, 2; 13, 2; 16, 2: *b*.

Δ^x $\mathbf{G}_{16}^{14}$: $(C_{4y}^+, 0)$, $(\sigma_x, 0)$: 6, 2; 7, 2: *a*.
Σ^x $\mathbf{G}_{8}^{5}$: $(C_{2a}, 0)$, $(\sigma_z, 0)$: 5, 2: *a*.
Λ^x $\mathbf{G}_{12}^{4}$: $(C_{31}^-, 0)$, $(\sigma_{db}, 0)$: 3, 3; 4, 3; 6, 2: *a*.
S^x $\mathbf{G}_{8}^{5}$: $(C_{2c}, 0)$, $(\sigma_y, 0)$: 5, 2: *a*.
Z^x $\mathbf{G}_{8}^{5}$: $(C_{2x}, 0)$, $(\sigma_z, 0)$: 5, 2: *a*.
T^x $\mathbf{G}_{16}^{14}$: $(C_{4z}^+, 0)$, $(\sigma_y, 0)$: 6, 2; 7, 2: *a*.

222 *Pn3n* O_h^2

(*F*1; *H*3; *K*7; *M*5; *T*4; *T*5; *Z*1.)

Γ $\mathbf{G}_{96}^{8}$: $\{C_{4x}^+ | 000\}$, $\{\bar{C}_{31}^- | 000\}$, $\{C_{2b} | 000\}$, $\{I | \tfrac{1}{2}\tfrac{1}{2}\tfrac{1}{2}\}$: 4, 2; 5, 2; 8, 2; 12, 2; 13, 2; 16, 2: *b*.
X $\mathbf{G}_{64}^{1}$: $\{I | \tfrac{1}{2}\tfrac{1}{2}\tfrac{1}{2}\}$, $\{\sigma_x | \tfrac{1}{2}\tfrac{1}{2}\tfrac{1}{2}\}$, $\{C_{4y}^- | 000\}$: 19, 3; 20, 3; 21, 3; 22, 3: *a*.
M $\mathbf{G}_{64}^{2}$: $\{C_{4z}^- | 000\}$, $\{\sigma_x | \tfrac{1}{2}\tfrac{1}{2}\tfrac{1}{2}\}$, $\{C_{2x} | 000\}$: 19, 2: *a*.
R $\mathbf{G}_{192}^{2}$: $\{C_{4z}^- | 000\}$, $\{C_{2x} | 000\}$, $\{\bar{C}_{31}^- | 010\}$, $\{I | \tfrac{1}{2}\tfrac{1}{2}\tfrac{1}{2}\}$: 18, 2; 21, 3; 22, 3: *a*.

Δ^x $\mathbf{G}_{16}^{14}$: $(C_{4y}^+, 0)$, $(\sigma_x, 0)$: 6, 2; 7, 2: *a*.
Σ^x $\mathbf{G}_{8}^{5}$: $(C_{2a}, 0)$, $(\sigma_z, 0)$: 5, 2: *a*.
Λ^x $\mathbf{G}_{12}^{4}$: $(C_{31}^-, 0)$, $(\sigma_{db}, 0)$: 3, 3; 4, 3; 6, 2: *a*.
S^x $\mathbf{G}_{16}^{8}$: $(\sigma_y, 0)$, $(C_{2c}, 0)$: 5, 3; 6, 3; 7, 3; 8, 3: *a*.
Z^x $\mathbf{G}_{16}^{8}$: $(\sigma_y, 0)$, $(C_{2x}, 0)$: 5, 3; 6, 3; 7, 3; 8, 3: *a*.
T^x $\mathbf{G}_{32}^{11}$: $(C_{4z}^-, 0)$, $(\sigma_x, 0)$: 6, 2; 7, 2: *a*.

223 *Pm3n* O_h^3

(*F*1; *G*6; *H*3; *K*7; *M*5; *T*4; *T*5; *Z*1.)

Γ $\mathbf{G}_{96}^{8}$: $\{C_{4x}^+ | \tfrac{1}{2}\tfrac{1}{2}\tfrac{1}{2}\}$, $\{\bar{C}_{31}^- | 000\}$, $\{C_{2b} | \tfrac{1}{2}\tfrac{1}{2}\tfrac{1}{2}\}$, $\{I | 000\}$: 4, 2; 5, 2; 8, 2; 12, 2; 13, 2; 16, 2: *b*.
X $\mathbf{G}_{64}^{2}$: $\{C_{4y}^+ | \tfrac{1}{2}\tfrac{1}{2}\tfrac{1}{2}\}$, $\{\sigma_{de} | \tfrac{1}{2}\tfrac{1}{2}\tfrac{1}{2}\}$, $\{C_{2e} | \tfrac{1}{2}\tfrac{1}{2}\tfrac{1}{2}\}$: 19, 2: *a*.
M $\mathbf{G}_{32}^{9} \otimes \mathbf{T}_2$: $\{C_{4z}^+ | \tfrac{1}{2}\tfrac{1}{2}\tfrac{1}{2}\}$, $\{C_{2y} | 000\}$, $\{I | 000\}$; $\mathbf{t}_1$ or $\mathbf{t}_2$: 6, 2; 7, 2; 13, 2; 14, 2: *a*.
R $\mathbf{G}_{192}^{2}$: $\{C_{4z}^- | \tfrac{1}{2}\tfrac{1}{2}\tfrac{1}{2}\}$, $\{C_{2x} | 000\}$, $\{\bar{C}_{31}^- | 010\}$, $\{I | 000\}$: 18, 2; 21, 3; 22, 3: *a*.

Δ^x $\mathbf{G}_{16}^{14}$: $(C_{4y}^+, 0)$, $(\sigma_x, 0)$: 6, 2; 7, 2: *a*.
Σ^x $\mathbf{G}_{8}^{5}$: $(C_{2a}, 0)$, $(\sigma_z, 0)$: 5, 2: *a*.
Λ^x $\mathbf{G}_{12}^{4}$: $(C_{31}^-, 0)$, $(\sigma_{db}, 0)$: 3, 3; 4, 3; 6, 2: *a*.
S^x $\mathbf{G}_{16}^{8}$: $(C_{2c}, 0)$, $(\sigma_y, 0)$: 5, 3; 6, 3; 7, 3; 8, 3: *a*.
Z^x $\mathbf{G}_{8}^{5}$: $(C_{2x}, 0)$, $(\sigma_z, 0)$: 5, 2: *a*.
T^x $\mathbf{G}_{32}^{9}$: $(C_{4z}^+, 0)$, $(\sigma_x, 0)$, $(E, 1)$: 13, 2; 14, 2: *b*.

224 *Pn3m* O_h^4

(*B*7; *F*1; *H*3; *K*7; *M*5; *M*7; *T*4; *T*5; *Z*1.)

Γ $\mathbf{G}_{96}^{8}$: $\{C_{4x}^{+}|\frac{1}{2}\frac{1}{2}\frac{1}{2}\}$, $\{\bar{C}_{31}^{-}|000\}$, $\{C_{2b}|\frac{1}{2}\frac{1}{2}\frac{1}{2}\}$, $\{I|\frac{1}{2}\frac{1}{2}\frac{1}{2}\}$: 4, 2; 5, 2; 8, 2; 12, 2; 13, 2; 16, 2: *b*.
X $\mathbf{G}_{64}^{2}$: $\{C_{4y}^{+}|\frac{1}{2}\frac{1}{2}\frac{1}{2}\}$, $\{\sigma_x|\frac{1}{2}\frac{1}{2}\frac{1}{2}\}$, $\{C_{2x}|000\}$: 19, 2: *a*.
M $\mathbf{G}_{64}^{2}$: $\{C_{4z}^{-}|\frac{1}{2}\frac{1}{2}\frac{1}{2}\}$, $\{\sigma_x|\frac{1}{2}\frac{1}{2}\frac{1}{2}\}$, $\{C_{2x}|000\}$: 19, 2: *a*.
R $\mathbf{G}_{96}^{8}\otimes\mathbf{T}_2$: $\{C_{4x}^{+}|\frac{1}{2}\frac{1}{2}\frac{1}{2}\}$, $\{\bar{C}_{31}^{-}|000\}$, $\{C_{2b}|\frac{1}{2}\frac{1}{2}\frac{1}{2}\}$, $\{I|\frac{1}{2}\frac{1}{2}\frac{1}{2}\}$; $\mathbf{t}_1$ or $\mathbf{t}_2$ or $\mathbf{t}_3$: 4, 2; 5, 2; 8, 2; 12, 2; 13, 2; 16, 2: *b*.

Δ^x $\mathbf{G}_{16}^{14}$: $(C_{4y}^{+}, 0)$, $(\sigma_x, 0)$: 6, 2; 7, 2: *a*.
Σ^x $\mathbf{G}_{8}^{5}$: $(C_{2a}, 0)$, $(\sigma_z, 0)$: 5, 2: *a*.
Λ^x $\mathbf{G}_{12}^{4}$: $(C_{31}^{-}, 0)$, $(\sigma_{db}, 0)$: 3, 3; 4, 3; 6, 2: *a*.
S^x $\mathbf{G}_{16}^{8}$: $(\sigma_{de}, 0)$, $(C_{2c}, 0)$: 9, 2: *b*.
Z^x $\mathbf{G}_{16}^{8}$: $(\sigma_y, 0)$, $(C_{2x}, 0)$: 5, 3; 6, 3; 7, 3; 8, 3: *a*.
T^x $\mathbf{G}_{32}^{11}$: $(C_{4z}^{-}, 0)$, $(\sigma_y, 0)$: 6, 2; 7, 2: *a*.

225 *Fm*3*m* O_h^5

(See section 3.8; *A*4; *B*2; *B*8; *C*4; *C*5; *E*1; *F*1; *F*4; *F*6; *H*3; *J*1; *K*5; *K*7; *M*5; *P*1; *S*4; *S*8; *T*4; *T*5; *Y*3; *Z*1.)

Γ $\mathbf{G}_{96}^{8}$: $\{C_{4x}^{+}|000\}$, $\{\bar{C}_{31}^{-}|000\}$, $\{C_{2b}|000\}$, $\{I|000\}$: 4, 2; 5, 2; 8, 2; 12, 2; 13, 2; 16, 2: *b*.
X $\mathbf{G}_{32}^{9}\otimes\mathbf{T}_2$: $\{C_{4y}^{+}|000\}$, $\{C_{2x}|000\}$, $\{I|000\}$; $\mathbf{t}_1$ or $\mathbf{t}_3$: 6, 2; 7, 2; 13, 2; 14, 2: *a*.
L $\mathbf{G}_{24}^{3}\otimes\mathbf{T}_2$: $\{C_{31}^{-}|000\}$, $\{C_{2b}|000\}$, $\{I|000\}$; $\mathbf{t}_1$ or $\mathbf{t}_2$ or $\mathbf{t}_3$: 3, 3; 4, 3; 6, 2; 9, 3; 10, 3; 12, 2: *b*.
W $\mathbf{G}_{16}^{14}\otimes\mathbf{T}_4$: $\{S_{4x}^{+}|000\}$, $\{C_{2d}|000\}$; $\mathbf{t}_2$ or $\mathbf{t}_3$: 6, 2; 7, 2: *a*.

Δ^x $\mathbf{G}_{16}^{14}$: $(C_{4y}^{+}, 0)$, $(\sigma_x, 0)$: 6, 2; 7, 2: *a*.
Λ^x $\mathbf{G}_{12}^{4}$: $(C_{31}^{-}, 0)$, $(\sigma_{db}, 0)$: 3, 3; 4, 3; 6, 2: *a*.
Σ^x $\mathbf{G}_{8}^{5}$: $(C_{2a}, 0)$, $(\sigma_z, 0)$: 5, 2: *a*.
S^x $\mathbf{G}_{8}^{5}$: $(C_{2c}, 0)$, $(\sigma_y, 0)$: 5, 2: *a*.
Z^x $\mathbf{G}_{8}^{5}$: $(C_{2x}, 0)$, $(\sigma_z, 0)$: 5, 2: *a*.
Q^x $\mathbf{G}_{4}^{1}$: $(C_{2f}, 0)$: 2, 3; 4, 3: *b*.

226 *Fm*3*c* O_h^6

(*F*1; *H*3; *K*7; *M*5; *T*4; *T*5; *Z*1.)

Γ $\mathbf{G}_{96}^{8}$: $\{C_{4x}^{+}|000\}$, $\{\bar{C}_{31}^{-}|000\}$, $\{C_{2b}|000\}$, $\{I|\frac{1}{2}\frac{1}{2}\frac{1}{2}\}$: 4, 2; 5, 2; 8, 2; 12, 2; 13, 2; 16, 2: *b*.
X $\mathbf{G}_{32}^{9}\otimes\mathbf{T}_2$: $\{C_{4y}^{+}|000\}$, $\{C_{2x}|000\}$, $\{I|\frac{1}{2}\frac{1}{2}\frac{1}{2}\}$; $\mathbf{t}_1$ or $\mathbf{t}_3$: 6, 2; 7, 2; 13, 2; 14, 2: *a*.
L $\mathbf{G}_{48}^{9}$: $\{\bar{C}_{31}^{+}|010\}$, $\{C_{2b}|000\}$, $\{I|\frac{1}{2}\frac{1}{2}\frac{1}{2}\}$: 10, 1; 17, 3; 18, 3: *a*.
W $\mathbf{G}_{16}^{14}\otimes\mathbf{T}_4$: $\{S_{4x}^{-}|\frac{1}{2}\frac{3}{2}\frac{1}{2}\}$, $\{C_{2f}|000\}$; $\mathbf{t}_2$ or $\mathbf{t}_3$: 6, 3; 7, 3: *a*.

Δ^x $\mathbf{G}_{16}^{14}$: $(C_{4y}^{+}, 0)$, $(\sigma_x, 0)$: 6, 2; 7, 2: *a*.
Λ^x $\mathbf{G}_{12}^{4}$: $(C_{31}^{-}, 0)$, $(\sigma_{db}, 0)$: 3, 3; 4, 3; 6, 2: *a*.
Σ^x $\mathbf{G}_{8}^{5}$: $(C_{2a}, 0)$, $(\sigma_z, 0)$: 5, 2: *a*.
S^x $\mathbf{G}_{8}^{5}$: $(C_{2c}, 0)$, $(\sigma_y, 0)$: 5, 2: *a*.
Z^x $\mathbf{G}_{8}^{5}$: $(C_{2x}, 0)$, $(\sigma_z, 0)$: 5, 2: *a*.
Q^x $\mathbf{G}_{4}^{1}$: $(C_{2f}, 0)$: 2, 1; 4, 1: *b*.

227 *Fd3m* O_h^7

(See section 3.8; *B*5; *B*6; *D*3; *E*1; *E*2; *F*1; *H*1; *H*3; *J*1; *K*3; *K*5; *K*7; *L*1; *M*2; *M*5; *M*6; *N*3; *P*1; *P*4; *S*8; *S*11; *T*4; *T*5; *Y*3; *Z*1.)

Γ $\mathbf{G}_{96}^{8}$: $\{C_{4x}^{+}\,|\,\frac{1}{4}\frac{1}{4}\frac{1}{4}\}$, $\{\bar{C}_{31}^{-}\,|\,000\}$, $\{C_{2b}\,|\,\frac{1}{4}\frac{1}{4}\frac{1}{4}\}$, $\{I\,|\,\frac{1}{4}\frac{1}{4}\frac{1}{4}\}$: 4, 2; 5, 2; 8, 2; 12, 2; 13, 2; 16, 2: *b*.
X $\mathbf{G}_{64}^{2}$: $\{C_{4y}^{+}\,|\,\frac{1}{4}\frac{1}{4}\frac{1}{4}\}$, $\{\sigma_x\,|\,\frac{1}{4}\frac{1}{4}\frac{1}{4}\}$, $\{C_{2x}\,|\,000\}$: 19, 2: *a*.
L $\mathbf{G}_{24}^{3}\otimes\mathbf{T}_2$: $\{C_{31}^{-}\,|\,000\}$, $\{C_{2b}\,|\,\frac{1}{4}\frac{1}{4}\frac{1}{4}\}$, $\{I\,|\,\frac{1}{4}\frac{1}{4}\frac{1}{4}\}$; $\mathbf{t}_1$ or $\mathbf{t}_2$ or $\mathbf{t}_3$: 3, 3; 4, 3; 6, 2; 9, 3; 10, 3; 12, 2: *b*.
W $\mathbf{G}_{64}^{3}$: $\{S_{4x}^{-}\,|\,000\}$, $\{\sigma_y\,|\,\frac{1}{4}\frac{1}{4}\frac{1}{4}\}$: 13, 3; 14, 3; 15, 3; 16, 3; 20, 2: *a*.

Δ^x $\mathbf{G}_{16}^{14}$: $(C_{4y}^{+}, 0)$, $(\sigma_x, 0)$: 6, 2; 7, 2: *a*.
Λ^x $\mathbf{G}_{12}^{4}$: $(C_{31}^{-}, 0)$, $(\sigma_{db}, 0)$: 3, 3; 4, 3; 6, 2: *a*.
Σ^x $\mathbf{G}_{8}^{5}$: $(C_{2a}, 0)$, $(\sigma_z, 0)$: 5, 2: *a*.
S^x $\mathbf{G}_{16}^{8}$: $(\sigma_{de}, 0)$, $(\sigma_y, 0)$: 9, 2: *b*.
Z^x $\mathbf{G}_{16}^{8}$: $(\sigma_y, 0)$, $(\sigma_z, 0)$: 5, 3; 6, 3; 7, 3; 8, 3: *a*.
Q^x $\mathbf{G}_{16}^{2}$: $(C_{2f}, 0)$, $(\bar{E}, 0)$: 10, 3; 14, 3: *e*.

228 *Fd3c* O_h^8

(*F*1; *H*3; *K*7; *M*5; *T*4; *T*5; *Z*1.)

Γ $\mathbf{G}_{96}^{8}$: $\{C_{4x}^{+}\,|\,\frac{1}{4}\frac{1}{4}\frac{1}{4}\}$, $\{\bar{C}_{31}^{-}\,|\,000\}$, $\{C_{2b}\,|\,\frac{1}{4}\frac{1}{4}\frac{1}{4}\}$, $\{I\,|\,\frac{3}{4}\frac{3}{4}\frac{3}{4}\}$: 4, 2; 5, 2; 8, 2; 12, 2; 13, 2; 16, 2: *b*.
X $\mathbf{G}_{64}^{2}$: $\{C_{4y}^{+}\,|\,\frac{1}{4}\frac{1}{4}\frac{1}{4}\}$, $\{\sigma_x\,|\,\frac{3}{4}\frac{3}{4}\frac{3}{4}\}$, $\{C_{2x}\,|\,000\}$: 19, 2: *a*.
L $\mathbf{G}_{48}^{9}$: $\{\bar{C}_{31}^{+}\,|\,010\}$, $\{C_{2b}\,|\,\frac{1}{4}\frac{1}{4}\frac{1}{4}\}$, $\{I\,|\,\frac{3}{4}\frac{3}{4}\frac{3}{4}\}$: 10, 1; 17, 3; 18, 3: *a*.
W $\mathbf{G}_{64}^{3}$: $\{S_{4x}^{+}\,|\,\frac{1}{2}\frac{1}{2}\frac{1}{2}\}$, $\{\sigma_z\,|\,\frac{3}{4}\frac{3}{4}\frac{3}{4}\}$: 13, 3; 14, 3; 15, 3; 16, 3; 20, 1: *a*.

Δ^x $\mathbf{G}_{16}^{14}$: $(C_{4y}^{+}, 0)$, $(\sigma_x, 0)$: 6, 2; 7, 2: *a*.
Λ^x $\mathbf{G}_{12}^{4}$: $(C_{31}^{-}, 0)$, $(\sigma_{db}, 0)$: 3, 3; 4, 3; 6, 2: *a*.
Σ^x $\mathbf{G}_{8}^{5}$: $(C_{2a}, 0)$, $(\sigma_z, 0)$: 5, 2: *a*.
S^x $\mathbf{G}_{16}^{8}$: $(\sigma_{de}, 0)$, $(\sigma_y, 0)$: 9, 2: *b*.
Z^x $\mathbf{G}_{16}^{8}$: $(\sigma_y, 0)$, $(\sigma_z, 0)$: 5, 3; 6, 3; 7, 3; 8, 3: *a*.
Q^x $\mathbf{G}_{16}^{2}$: $(C_{2f}, 0)$, $(\bar{E}, 0)$: 10, 3; 14, 3: *e*.

229 *Im3m* O_h^9

(*A*4; *B*1; *B*2; *B*8; *C*5; *E*1; *F*1; *F*4; *H*3; *J*1; *K*5; *K*7; *M*5; *S*4; *S*8; *T*4; *T*5; *Z*1.)

Γ $\mathbf{G}_{96}^{8}$: $\{C_{4x}^{+}\,|\,000\}$, $\{\bar{C}_{31}^{-}\,|\,000\}$, $\{C_{2b}\,|\,000\}$, $\{I\,|\,000\}$: 4, 2; 5, 2; 8, 2; 12, 2; 13, 2; 16, 2: *b*.
H $\mathbf{G}_{96}^{8}\otimes\mathbf{T}_2$: $\{C_{4x}^{+}\,|\,000\}$, $\{\bar{C}_{31}^{-}\,|\,000\}$, $\{C_{2b}\,|\,000\}$, $\{I\,|\,000\}$; $\mathbf{t}_1$ or $\mathbf{t}_2$ or $\mathbf{t}_3$: 4, 2; 5, 2; 8, 2; 12, 2; 13, 2; 16, 2: *b*.
P $\mathbf{G}_{48}^{10}\otimes\mathbf{T}_4$: $\{S_{4x}^{+}\,|\,000\}$, $\{\bar{C}_{31}^{-}\,|\,000\}$, $\{\sigma_{db}\,|\,000\}$; $\mathbf{t}_1$ or $\mathbf{t}_2$ or $\mathbf{t}_3$: 4, 2; 5, 2; 8, 2: *a*.
N $\mathbf{G}_{16}^{11}\otimes\mathbf{T}_2$: $\{C_{2a}\,|\,000\}$, $\{C_{2b}\,|\,000\}$, $\{I\,|\,000\}$; $\mathbf{t}_3$: 5, 2; 10, 2: *c*.

Σ^x $\mathbf{G}_{8}^{5}$: $(C_{2a}, 0)$, $(\sigma_z, 0)$: 5, 2: *a*.
Δ^x $\mathbf{G}_{16}^{14}$: $(C_{4y}^{+}, 0)$, $(\sigma_x, 0)$: 6, 2; 7, 2: *a*.
Λ^x $\mathbf{G}_{12}^{4}$: $(C_{31}^{-}, 0)$, $(\sigma_{db}, 0)$: 3, 3; 4, 3; 6, 2: *a*.
D^x $\mathbf{G}_{8}^{5}$: $(C_{2z}, 0)$, $(\sigma_{db}, 0)$: 5, 2: *a*.
G^x $\mathbf{G}_{8}^{5}$: $(C_{2b}, 0)$, $(\sigma_{da}, 0)$: 5, 2: *a*.
F^x $\mathbf{G}_{12}^{4}$: $(C_{34}^{+}, 0)$, $(\sigma_{da}, 0)$: 3, 3; 4, 3; 6, 2: *a*.

230 *Ia*3*d* O_h^{10}

($F1$; $H3$; $K7$; $M5$; $S8$; $T4$; $T5$; $Z1$.)

Γ $\mathbf{G}_{96}^{8}$: $\{C_{4x}^{+} \mid 00\tfrac{1}{2}\}$, $\{\bar{C}_{31}^{-} \mid 000\}$, $\{C_{2b} \mid \tfrac{1}{2}\tfrac{1}{2}\tfrac{1}{2}\}$, $\{I \mid 000\}$: 4, 2; 5, 2; 8, 2; 12, 2; 13, 2; 16, 2: *b*.
H $\mathbf{G}_{192}^{2}$: $\{C_{4z}^{+} \mid 0\tfrac{1}{2}0\}$, $\{C_{2y} \mid 1\tfrac{1}{2}\tfrac{1}{2}\}$, $\{\bar{C}_{31}^{+} \mid 111\}$, $\{I \mid 000\}$: 18, 2; 21, 3; 22, 3: *a*.
P $\mathbf{G}_{192}^{1}$: $\{C_{32}^{+} \mid 0\tfrac{1}{2}\tfrac{1}{2}\}$, $\{C_{2y} \mid 0\tfrac{1}{2}\tfrac{1}{2}\}$, $\{C_{2x} \mid \tfrac{3}{2}\tfrac{3}{2}0\}$, $\{S_{4x}^{+} \mid \tfrac{1}{2}11\}$: 17, 3; 18, 3; 21, 2; 27, 3; 28, 3: *a*.
N $\mathbf{G}_{32}^{7}$: $\{\sigma_{db} \mid \tfrac{1}{2}\tfrac{1}{2}\tfrac{1}{2}\}$, $\{\sigma_{da} \mid \tfrac{1}{2}00\}$, $\{I \mid 000\}$: 13, 3; 14, 3: *a*.

Σ^x $\mathbf{G}_{8}^{5}$: $(C_{2a}, 0)$, $(\sigma_z, 0)$: 5, 2: *a*.
Δ^x $\mathbf{G}_{16}^{14}$: $(C_{4y}^{+}, 0)$, $(\sigma_x, 0)$: 6, 2; 7, 2: *a*.
Λ^x $\mathbf{G}_{12}^{4}$: $(C_{31}^{-}, 0)$, $(\sigma_{db}, 0)$: 3, 3; 4, 3; 6, 2: *a*.
D^x $\mathbf{G}_{16}^{8}$: $(C_{2z}, 0)$, $(\sigma_{db}, 0)$: 5, 3; 6, 3; 7, 3; 8, 3: *a*.
G^x $\mathbf{G}_{16}^{8}$: $(C_{2b}, 0)$, $(\sigma_{da}, 0)$: 5, 3; 6, 3; 7, 3; 8, 3: *a*.
F^x $\mathbf{G}_{24}^{3}$: $(C_{34}^{+}, 1)$, $(\sigma_{da}, 0)$, $(E, 1)$: 9, 3; 10, 3; 12, 2: *c*.

Notes to Table 6.13

(i) This table is arranged in a similar way to Table 5.7 for the single-valued representations of the space-groups (q.v.).

(ii) The space groups are arranged in the order and notation used in the *International tables for X-ray crystallography* (Henry and Lonsdale 1965) and in Tables 3.7 and 5.7 of this book.

(iii) The Brillouin zone for any space group can be identified from Table 3.7 and Figs. 3.2–3.15.

(iv) The first column lists the special points of symmetry and the lines of symmetry in the basic domain of the Brillouin zone (see section 3.3) and their positions in the Brillouin zone can be identified from Table 3.6. A superscript x is used to denote a line of symmetry.

Points of symmetry

(v) Following the label of a point of symmetry the little group ${}^H\mathbf{G}^{\dagger\mathbf{k}}$ in the sense of Herring (1942) is identified. This is done as follows: the abstract group $\mathbf{G}_{|G|}^{n}$ (see Table 5.1) is given and the space-group elements following $\mathbf{G}_{|G|}^{n}$ are to be identified with $P, Q, R, \ldots$, the generators of $\mathbf{G}_{|G|}^{n}$ in the order in which they occur.

(vi) In some cases for points of symmetry ${}^H\mathbf{G}^{\dagger\mathbf{k}}$ is written as $\mathbf{G}_{|G|}^{n} \otimes \mathbf{T}_m$, a direct product of some group $\mathbf{G}_{|G|}^{n}$ with $\mathbf{T}_m$, an Abelian group of small order, m, consisting only of translation operations of the form $\{E \mid \mathbf{t}\}$. In these cases a suitable generating element of the Abelian group is indicated after the generators of $\mathbf{G}_{|G|}^{n}$ by giving the translation $\mathbf{t}$ of $\{E \mid \mathbf{t}\}$. In a few cases ${}^H\mathbf{G}^{\dagger\mathbf{k}}$ takes the form $\mathbf{G}_{|G|}^{n} \otimes \mathbf{T}_m \otimes \mathbf{T}_s$ where $\mathbf{T}_m$ and $\mathbf{T}_s$ are two Abelian groups with generating elements $\{E \mid \mathbf{t}_m\}$ and $\{E \mid \mathbf{t}_s\}$ respectively; $\mathbf{t}_m$ and $\mathbf{t}_s$ are then listed following the generators of $\mathbf{G}_{|G|}^{n}$.

(vii) After the above information, and separated from it by a colon, is the identification of the labelling assigned to the reps of ${}^H\mathbf{G}^{\dagger\mathbf{k}}$ and of the reality of the induced space-group reps $(\Gamma_p^{\mathbf{k}} \uparrow \mathbf{G}^{\dagger})$ (see sections 4.6, 5.2, 6.3, and 7.6). Between one pair of semicolons is the code 'p, q' which is to be understood as meaning that the rep R_p of $\mathbf{G}_{|G|}^{n}$ corresponds to a small (allowed) rep, $\Gamma_p^{\mathbf{k}}$, of $\mathbf{G}^{\dagger\mathbf{k}}$, and that $(\Gamma_p^{\mathbf{k}} \uparrow \mathbf{G}^{\dagger})$ is of the qth kind ($q = 1, 2$, or 3). The usefulness of the reality of $(\Gamma_p^{\mathbf{k}} \uparrow \mathbf{G}^{\dagger})$ lies in determining whether or not extra degeneracies occur if the crystal described by the space group $\mathbf{G}^{\dagger}$ also possesses time-inversion symmetry. The relation between the reality of $(\Gamma_p^{\mathbf{k}} \uparrow \mathbf{G}^{\dagger})$ and these extra degeneracies described in the text of section 6.3 (see p. 459) is

Reality	Degeneracy
1	b
2	a
3	c

When there is no element in the space group that transforms $\mathbf{k}$ into $-\mathbf{k}$ the addition of time-reversal symmetry produces a type (x) degeneracy (see p. 460) and in this case the symbol x is used in the table in place of 1, 2, or 3.

(viii) In the cases where ${}^H\mathbf{G}^{\dagger\mathbf{k}}$ contains one or more elements of the form $\{E \mid \mathbf{t}\}$, only those reps of ${}^H\mathbf{G}^{\dagger\mathbf{k}}$ are included which are compatible with the requirement of eqn (3.4.3) $\Delta^{\mathbf{k}}(\{E \mid \mathbf{t}\}) = \exp(-i\mathbf{k}\,.\,\mathbf{t})\mathbf{1}$, for a representation of the translation group $\mathbf{T}$ of which $\{E \mid \mathbf{t}\}$ is a member; that is, only the reps of ${}^H\mathbf{G}^{\dagger\mathbf{k}}$ which lead to small reps of $\mathbf{G}^{\dagger\mathbf{k}}$ are included.

(ix) Only the double-valued reps of ${}^H\mathbf{G}^{\dagger\mathbf{k}}$ are included, that is, those reps for which $\Delta(\{\bar{E} \mid 0\})$ is negative. The single-valued reps of ${}^{H}\mathbf{G}^{\dagger\mathbf{k}}$ will of course also be found among the reps of $\mathbf{G}^{n}_{|G|}$ in this table but since these have been identified previously in Tables 5.7 and 5.8 we do not repeat them here.

(x) The labels used for the space-group reps derived from $R_1, R_2, R_3, \ldots$, etc., the reps of the abstract group $\mathbf{G}^{n}_{|G|}$ are indicated by the small letter $a, b, c, \ldots$, etc., at the end of the entry for the point of symmetry in question. The letters $a, b, c, \ldots$, etc. refer the reader to the relevant part of Table 6.14 (see the Notes to Table 6.14).

Lines of symmetry

(xi) Following the label of the line of symmetry the central extension $\overline{\mathbf{G}}^{\dagger\mathbf{k}^*}$ (see sections 3.7 and 3.8) is identified; the elements of $\overline{\mathbf{G}}^{\dagger\mathbf{k}^*}$ are of the form (H_j, α) where H_j is the point-group operation part of a space-group element and α is a member of the cyclic group $\mathbf{Z}_g$ (see eqn (3.7.31)). The (H_j, α) form the abstract group which is specified and the elements that are given are to be identified with $P, Q, R, \ldots$, the generators of $\mathbf{G}^{n}_{|G|}$, in the order in which they occur. Where alternative values of **k** for a line of symmetry were given in Table 3.6 we have used the first entry in deriving this table.

(xii) After the generators of $\overline{\mathbf{G}}^{\dagger\mathbf{k}^*}$ and separated from them by a colon is the labelling of those reps of $\mathbf{G}^{n}_{|G|}$ that satisfy eqn (3.7.33) and therefore lead to small reps of $\mathbf{G}^{\dagger\mathbf{k}}$; the reality of the induced space-group reps $(\Gamma_p^{\mathbf{k}} \uparrow \mathbf{G}^{\dagger})$ is also indicated. Between one pair of semicolons is the code 'p, q' which is to be understood as meaning that the double-valued rep $(\Gamma_p^{\mathbf{k}} \uparrow \mathbf{G}^{\dagger})$ derived from R_p of $\mathbf{G}^{n}_{|G|}$ is of the qth kind ($q = 1, 2, 3$) (see Note (vii) above). Reps that do not satisfy eqn (3.7.33) are not included and, again, the single-valued reps are not included. The letters $a, b, c, \ldots$, etc., refer the reader to the relevant part of Table 6.14 for the letters used to label the space-group reps (see the Notes to Table 6.14).

(xiii) References to work on the representations of individual space groups are given in the table, in coded form, under the appropriate space group according to the following key:

*A*1 Adler 1962.
*A*2 Altmann 1956.
*A*3 Altmann and Bradley 1965.
*A*4 Altmann and Cracknell 1965.
*A*5 Antončík and Trlifaj 1952.
*A*6 Asendorf 1957.
*B*1 Bates and Stevens 1961.
*B*2 Bell 1954.
*B*3 Birman 1959*a*.
*B*4 Birman 1959*b*.
*B*5 Birman 1962*a*.
*B*6 Birman 1962*b*, 1963.
*B*7 Bordure, Lecoy, and Savelli 1963*a*, *b*.
*B*8 Bouckaert, Smoluchowski, and Wigner 1936.
*C*1 Casella 1959.
*C*2 Chaldyshev and Karavaev 1963.
*C*3 Chen 1967.
*C*4 Chen, Berenson, and Birman 1968.
*C*5 Cornwell 1969.
*C*6 Cracknell 1965.
*C*7 Cracknell 1967*c*.
*C*8 Cracknell and Joshua 1969*a*.
*D*1 Daniel and Cracknell 1969.
*D*2 Dimmock and Wheeler 1962*b*.
*D*3 Döring and Zehler 1953.
*D*4 Dresselhaus 1955.
*E*1 Elliott 1954*b*.
*E*2 Elliott and Loudon 1960.
*E*3 Erdman 1960.
*F*1 Faddeyev 1964.
*F*2 Falicov and Cohen 1963.
*F*3 Falicov and Golin 1965.
*F*4 Falicov and Ruvalds 1968.
*F*5 Firsov 1957.
*F*6 Flower, March, and Murray 1960.
*F*7 Frei 1967*a*.
*F*8 Frei 1967*b*.
*G*1 Gashimzade 1960*a*.
*G*2 Gashimzade 1960*b*.
*G*3 Gay, Albers, and Arlinghaus 1968.
*G*4 Glasser 1959.
*G*5 Gorzkowski 1963*a*.
*G*6 Gorzkowski 1963*b*, 1964*b*.
*G*7 Gorzkowski 1964*a*.
*G*8 Gubanov and Gashimzade 1959.
*H*1 Herring 1942.
*H*2 Hsieh and Chen 1964.
*H*3 Hurley 1966.
*J*1 Jones 1960.
*J*2 Joshua and Cracknell 1969.
*K*1 Karavaev, Kudryavtseva, and Chaldyshev 1962.
*K*2 Karpus and Batarūnas 1961.
*K*3 Khartsiev 1962.
*K*4 Kitz 1965*a*.
*K*5 Koster 1957.
*K*6 Kovalev 1961*a*, *b*.
*K*7 Kovalev 1965.
*L*1 Lax and Hopfield 1961.
*L*2 Lee and Pincherle 1963.

*M*1 Mase 1958, 1959*a*.
*M*2 Mase 1959*b*; Miąsek 1960, 1963, 1966.
*M*3 Miąsek 1957*a*, *b*, 1958.
*M*4 Miąsek and Suffczyński 1961*a*, *b*, *c*.
*M*5 Miller and Love 1967.
*M*6 Montgomery 1969.
*M*7 Moskalenko 1960.
*M*8 Murphy, Caspers, and Buchanan 1964.
*N*1 Nusimovici 1965.
*N*2 Nussbaum 1962.
*N*3 Nussbaum 1966.
*O*1 Olbrychski 1961.
*O*2 Olbrychski 1963*b*.
*O*3 Olbrychski and Van Huong 1970.
*P*1 Parmenter 1955.
*P*2 Pikus 1961*a*.
*P*3 Pikus 1961*b*.
*P*4 Pikus 1961*c*.
*R*1 Rabotnikov 1960.
*R*2 Rashba 1959.
*R*3 Rashba and Sheka 1959.
*R*4 Rudra 1965*b*.
*S*1 Sandrock and Treusch 1964.
*S*2 Schiff 1955.
*S*3 Schiff 1956.
*S*4 Schlosser 1962.
*S*5 Sek 1963.
*S*6 Sheka 1960.
*S*7 Shur 1966.
*S*8 Slater 1965*b*.
*S*9 Slater, Koster, and Wood 1962; Koster 1962.
*S*10 Šlechta 1966.
*S*11 Streitwolf 1964.
*S*12 Suffczyński 1960.
*S*13 Suffczyński 1961.
*S*14 Sushkevich 1965.
*S*15 Sushkevich 1966.
*T*1 Tovstyuk and Bercha 1964.
*T*2 Tovstyuk and Gemus 1963.
*T*3 Tovstyuk and Sushkevich 1964.
*T*4 Tovstyuk and Tarnavskaya 1963.
*T*5 Tovstyuk and Tarnavskaya 1964.
*T*6 Treusch and Sandrock 1966.
*V*1 Van Huong and Olbrychski 1964.
*W*1 Waeber 1969.
*Y*1 Yamazaki 1957*a*.
*Y*2 Yamazaki 1957*b*.
*Y*3 Yanagawa 1953.
*Z*1 Zak, Casher, Glück, and Gur 1969.

TABLE 6.14

The labelling of the double-valued space-group reps

$\mathbf{G}_2^1$	R_2
a	$\bar{A}$
	Γ_2

$\mathbf{G}_4^1$	R_1	R_2	R_3	R_4
a	—	$\bar{A}_g$	—	$\bar{A}_u$
b	—	$^{2}\bar{E}$	—	$^{1}\bar{E}$
	—	Γ_3	—	Γ_4
c	$\bar{A}$	—	$\bar{B}$	—
	Γ_3	—	Γ_4	—

$\mathbf{G}_4^2$	R_2	R_4
a	$\bar{A}_g$	$\bar{A}_u$
	Γ_2^+	Γ_2^-

$\mathbf{G}_6^1$	R_1	R_2	R_3	R_4	R_5	R_6
a	—	$^{1}\bar{E}$	—	$\bar{A}$	—	$^{2}\bar{E}$
	—	Γ_4	—	Γ_6	—	Γ_5
b	$\bar{A}$	—	$^{1}\bar{E}$	—	$^{2}\bar{E}$	—
	Γ_6	—	Γ_4	—	Γ_5	—

G_8^1	R_1	R_2	R_3	R_4	R_5	R_6	R_7	R_8
a	— —	${}^1\bar{E}_1$ Γ_5	— —	${}^1\bar{E}_2$ Γ_6	— —	${}^2\bar{E}_2$ Γ_7	— —	${}^2\bar{E}_1$ Γ_8
b	$\bar{A}$ Γ_5	— —	${}^1\bar{E}$ Γ_6	— —	$\bar{B}$ Γ_7	— —	${}^2\bar{E}$ Γ_8	— —

G_8^2	R_1	R_2	R_3	R_4	R_5	R_6	R_7	R_8
a	— —	— —	— —	— —	$\bar{A}$ Γ_3	— —	$\bar{B}$ Γ_4	— —
b	— —	${}^2\bar{E}_g$ Γ_3^+	— —	${}^1\bar{E}_g$ Γ_4^+	— —	${}^2\bar{E}_u$ Γ_3^-	— —	${}^1\bar{E}_u$ Γ_4^-
c	$\bar{A}_g$ Γ_3^+	— —	$\bar{B}_g$ Γ_4^+	— —	$\bar{A}_u$ Γ_3^-	— —	$\bar{B}_u$ Γ_4^-	— —

G_8^5	R_1	R_2	R_3	R_4	R_5
a	— —	— —	— —	— —	$\bar{E}$ Γ_5
b	$\bar{A}$ Γ_2	$\bar{B}_1$ Γ_3	$\bar{B}_2$ Γ_4	$\bar{B}_3$ Γ_5	— —

G_{12}^1	R_1	R_2	R_3	R_4	R_5	R_6	R_7	R_8	R_9	R_{10}	R_{11}	R_{12}
a	— —	— —	— —	— —	— —	${}^1\bar{E}_3$ Γ_3	— —	— —	— —	— —	— —	${}^2\bar{E}_1$ Γ_4
b	— —	${}^1\bar{E}_1$ Γ_3	— —	— —	— —	— —	— —	${}^2\bar{E}_3$ Γ_4	— —	— —	— —	— —
c	— —	${}^1\bar{E}_1$ Γ_{11}	— —	${}^1\bar{E}_2$ Γ_{10}	— —	${}^1\bar{E}_3$ Γ_7	— —	${}^2\bar{E}_3$ Γ_8	— —	${}^2\bar{E}_2$ Γ_9	— —	${}^2\bar{E}_1$ Γ_{12}
d	$\bar{A}$ Γ_7	— —	${}^1\bar{E}_2$ Γ_8	— —	${}^1\bar{E}_1$ Γ_9	— —	$\bar{B}$ Γ_{10}	— —	${}^2\bar{E}_1$ Γ_{11}	— —	${}^2\bar{E}_2$ Γ_{12}	— —

$\mathbf{G}_{12}^{4}$	R_1	R_2	R_3	R_4	R_5	R_6
a	—	—	$^1\bar{E}$	$^2\bar{E}$	—	$\bar{E}_1$
	—	—	Γ_5	Γ_6	—	Γ_4
b	$\bar{A}_1$	$\bar{A}_2$	—	—	$\bar{E}$	—
	Γ_4	Γ_5	—	—	Γ_6	—

$\mathbf{G}_{12}^{6}$	R_2	R_4	R_6	R_7	R_8	R_9	R_{10}	R_{11}	R_{12}
a	—	—	—	—	$^1\bar{E}$	—	$\bar{A}$	—	$^2\bar{E}$
	—	—	—	—	Γ_4	—	Γ_5	—	Γ_6
b	—	—	—	$\bar{A}_u$	$^1\bar{E}_g$	$^2\bar{E}_u$	$\bar{A}_g$	$^1\bar{E}_u$	$^2\bar{E}_g$
	—	—	—	Γ_4^-	Γ_5^+	Γ_6^-	Γ_4^+	Γ_5^-	Γ_6^+
c	$^1\bar{E}_g$	$\bar{A}_g$	$^2\bar{E}_g$	$\bar{A}_u$	—	$^2\bar{E}_u$	—	$^1\bar{E}_u$	—
	Γ_4^+	Γ_6^+	Γ_5^+	Γ_6^-	—	Γ_5^-	—	Γ_4^-	—
d	—	—	—	$\bar{A}$	—	$^1\bar{E}$	—	$^2\bar{E}$	—
	—	—	—	Γ_4	—	Γ_5	—	Γ_6	—

$\mathbf{G}_{16}^{2}$	R_2	R_4	R_6	R_8	R_9	R_{10}	R_{11}	R_{12}	R_{13}	R_{14}	R_{15}	R_{16}
a	—	—	—	—	—	—	—	$^1\bar{E}_1$	—	—	—	$^1\bar{E}_2$
	—	—	—	—	—	—	—	Γ_3	—	—	—	Γ_4
b	—	—	—	—	—	$^1\bar{E}_1$	—	$^1\bar{E}_2$	—	$^2\bar{E}_2$	—	$^2\bar{E}_1$
	—	—	—	—	—	Γ_5	—	Γ_6	—	Γ_7	—	Γ_8
c	—	—	—	—	$\bar{A}$	—	$^1\bar{E}$	—	$\bar{B}$	—	$^2\bar{E}$	—
	—	—	—	—	Γ_5	—	Γ_6	—	Γ_7	—	Γ_8	—
d	$^1\bar{E}_{1g}$	$^1\bar{E}_{2g}$	$^2\bar{E}_{2g}$	$^2\bar{E}_{1g}$	—	$^1\bar{E}_{1u}$	—	$^1\bar{E}_{2u}$	—	$^2\bar{E}_{2u}$	—	$^2\bar{E}_{1u}$
	Γ_6^+	Γ_8^+	Γ_7^+	Γ_5^+	—	Γ_6^-	—	Γ_8^-	—	Γ_7^-	—	Γ_5^-
e	—	—	—	—	—	$^1\bar{E}_1$	—	—	—	$^1\bar{E}_2$	—	—
	—	—	—	—	—	Γ_3	—	—	—	Γ_4	—	—
f	$^1\bar{E}_1$	$^1\bar{E}_2$	$^2\bar{E}_2$	$^2\bar{E}_1$	—	—	—	—	—	—	—	—
	Γ_5	Γ_6	Γ_7	Γ_8	—	—	—	—	—	—	—	—

G_{16}^{8}	R_5	R_6	R_7	R_8	R_9
a	${}^2\bar{E}''$	${}^1\bar{E}''$	${}^1\bar{E}'$	${}^2\bar{E}'$	—
	Γ_2	Γ_3	Γ_4	Γ_5	—
b	—	—	—	—	$\bar{E}$
	—	—	—	—	Γ_5

G_{16}^{10}	R_9	R_{10}
a	$\bar{E}$	—
	Γ_2	—
b	—	$\bar{E}$
	—	Γ_2

G_{16}^{11}	R_5	R_6	R_7	R_8	R_9	R_{10}
a	—	$\bar{A}_1$	$\bar{B}_1$	$\bar{B}_2$	$\bar{B}_3$	—
b	—	$\bar{A}'$	$\bar{A}''$	$\bar{B}'$	$\bar{B}''$	—
	—	Γ_2	Γ_3	Γ_4	Γ_5	—
c	$\bar{E}_g$	—	—	—	—	$\bar{E}_u$
	Γ_5^+	—	—	—	—	Γ_5^-
d	—	—	—	—	—	$\bar{E}$
	—	—	—	—	—	Γ_5
e	$\bar{E}$	—	—	—	—	—
	Γ_5	—	—	—	—	—

G_{16}^{14}	R_6	R_7
a	$\bar{E}_1$	$\bar{E}_2$
	Γ_6	Γ_7

G_{24}^{3}	R_3	R_4	R_6	R_7	R_8	R_9	R_{10}	R_{11}	R_{12}
a	—	—	—	$\bar{A}_1$	$\bar{A}_2$	—	—	$\bar{E}$	—
	—	—	—	Γ_4	Γ_5	—	—	Γ_6	—
b	${}^1\bar{E}_g$	${}^2\bar{E}_g$	$\bar{E}_{1g}$	—	—	${}^1\bar{E}_u$	${}^2\bar{E}_u$	—	$\bar{E}_{1u}$
	Γ_5^+	Γ_6^+	Γ_4^+	—	—	Γ_5^-	Γ_6^-	—	Γ_4^-
c	—	—	—	—	—	$\bar{A}_1$	$\bar{A}_2$	—	$\bar{E}$
	—	—	—	—	—	Γ_4	Γ_5	—	Γ_6

G_{24}^{9}	R_1	R_2	R_3	R_4	R_5	R_6	R_7
a	—	—	—	$\bar{E}$	${}^1\bar{F}$	${}^2\bar{F}$	—
	—	—	—	Γ_5	Γ_6	Γ_7	—
b	$\bar{A}$	${}^1\bar{E}$	${}^2\bar{E}$	—	—	—	$\bar{T}$
	Γ_4	Γ_5	Γ_6	—	—	—	Γ_7

G_{24}^{11}	R_7	R_8	R_9
a	$\bar{E}_1$	$\bar{E}_2$	$\bar{E}_3$
	Γ_7	Γ_8	Γ_9

$\mathbf{G}_{24}^{12}$	R_2	R_4	R_6	R_8	R_{10}	R_{12}	R_{14}	R_{15}	R_{16}	R_{17}	R_{18}	R_{20}	R_{21}	R_{22}	R_{23}	R_{24}
a	$^1\bar{E}_{1g}$ Γ_{11}^+	$^1\bar{E}_{2g}$ Γ_{10}^+	$^1\bar{E}_{3g}$ Γ_7^+	$^2\bar{E}_{3g}$ Γ_8^+	$^2\bar{E}_{2g}$ Γ_9^+	$^2\bar{E}_{1g}$ Γ_{12}^+	$^1\bar{E}_{1u}$ Γ_{11}^-	— —	$^1\bar{E}_{2u}$ Γ_{10}^-	— —	$^1\bar{E}_{3u}$ Γ_7^-	$^2\bar{E}_{3u}$ Γ_8^-	— —	$^2\bar{E}_{2u}$ Γ_9^-	— —	$^2\bar{E}_{1u}$ Γ_{12}^-
b	— —	— —	— —	— —	— —	— —	— —	— —	— —	$^1\bar{E}_1$ Γ_3	— —	— —	— —	— —	$^1\bar{E}_2$ Γ_4	— —
c	— —	— —	— —	— —	— —	— —	— —	$^1\bar{E}_1$ Γ_3	— —	— —	— —	— —	$^1\bar{E}_2$ Γ_4	— —	— —	— —

$\mathbf{G}_{32}^{7}$	R_9	R_{10}	R_{13}	R_{14}
a	— —	— —	$^1\bar{F}$ Γ_3	$^2\bar{F}$ Γ_4
b	$\bar{E}_1$ Γ_3	$\bar{E}_2$ Γ_4	— —	— —

$\mathbf{G}_{32}^{8}$	R_{13}	R_{14}	R_{17}	R_{18}
a	$^1\bar{E}_1$ Γ_2	$^1\bar{E}_2$ Γ_3	$^1\bar{E}_3$ Γ_4	$^1\bar{E}_4$ Γ_5

$\mathbf{G}_{32}^{9}$	R_6	R_7	R_{13}	R_{14}
a	$\bar{E}_{1g}$ Γ_6^+	$\bar{E}_{2g}$ Γ_7^+	$\bar{E}_{1u}$ Γ_6^-	$\bar{E}_{2u}$ Γ_7^-
b	— —	— —	$\bar{E}_1$ Γ_6	$\bar{E}_2$ Γ_7

$\mathbf{G}_{32}^{10}$	R_6	R_7	R_8	R_9	R_{10}	R_{11}	R_{14}
a	— —	— —	$^1\bar{E}_1$ Γ_3	$^1\bar{E}_2$ Γ_4	$^2\bar{E}_1$ Γ_5	$^2\bar{E}_2$ Γ_6	$\bar{E}_3$ Γ_7
b	$\bar{E}_1$ Γ_6	$\bar{E}_2$ Γ_7	— —	— —	— —	— —	— —

$\mathbf{G}_{32}^{11}$	R_6	R_7	R_8	R_9	R_{10}	R_{11}	R_{12}
a	$^1\bar{F}$ Γ_6	$^2\bar{F}$ Γ_7	— —	— —	— —	— —	— —
b	— —	— —	$^1\bar{E}_1$ Γ_3	$^1\bar{E}_2$ Γ_4	$^2\bar{E}_2$ Γ_5	$^2\bar{E}_1$ Γ_6	$\bar{E}_3$ Γ_7

$\mathbf{G}_{32}^{12}$	R_{19}	R_{20}
a	$^1\bar{F}$ Γ_3	$^2\bar{F}$ Γ_4

$\mathbf{G}_{32}^{14}$	R_5	R_6	R_7	R_8	R_9	R_{15}	R_{16}	R_{17}	R_{18}	R_{19}
a	${}^2\bar{E}''_g$ Γ_2^+	${}^1\bar{E}''_g$ Γ_3^+	${}^1\bar{E}'_g$ Γ_4^+	${}^2\bar{E}'_g$ Γ_5^+	— —	${}^2\bar{E}''_u$ Γ_2^-	${}^1\bar{E}''_u$ Γ_3^-	${}^1\bar{E}'_u$ Γ_4^-	${}^2\bar{E}'_u$ Γ_5^-	— —
b	— —	— —	— —	— —	$\bar{E}_g$ Γ_5^+	— —	— —	— —	— —	$\bar{E}_u$ Γ_5^-
c	${}^2\bar{E}''_u$ Γ_2^-	${}^1\bar{E}''_u$ Γ_3^-	${}^1\bar{E}'_u$ Γ_5^-	${}^2\bar{E}'_u$ Γ_4^-	— —	${}^2\bar{E}'_g$ Γ_2^+	${}^1\bar{E}'_g$ Γ_3^+	${}^1\bar{E}''_g$ Γ_5^+	${}^2\bar{E}''_g$ Γ_4^+	— —
d	${}^2\bar{E}''_g$ Γ_5^+	${}^1\bar{E}''_g$ Γ_6^+	${}^1\bar{E}'_g$ Γ_7^+	${}^2\bar{E}'_g$ Γ_8^+	— —	${}^2\bar{E}''_u$ Γ_5^-	${}^1\bar{E}''_u$ Γ_6^-	${}^1\bar{E}'_u$ Γ_7^-	${}^2\bar{E}'_u$ Γ_8^-	— —

$\mathbf{G}_{32}^{15}$	R_6	R_7	R_8	R_9	R_{16}	R_{17}	R_{18}	R_{19}
a	$\bar{A}'_g$ Γ_2^+	$\bar{A}''_g$ Γ_3^+	$\bar{B}'_g$ Γ_4^+	$\bar{B}''_g$ Γ_5^+	$\bar{A}'_u$ Γ_2^-	$\bar{A}''_u$ Γ_3^-	$\bar{B}'_u$ Γ_4^-	$\bar{B}''_u$ Γ_5^-

$\mathbf{G}_{32}^{17}$	R_{10}	R_{12}	R_{14}	R_{16}
a	${}^1\bar{E}_1$ Γ_5	${}^1\bar{E}_2$ Γ_6	${}^1\bar{E}_3$ Γ_7	${}^1\bar{E}_4$ Γ_8

$\mathbf{G}_4^{48}$	R_1	R_2	R_3	R_4	R_5	R_6	R_7	R_8	R_9	R_{10}	R_{11}	R_{12}	R_{13}	R_{14}
a	— —	— —	— —	— —	— —	— —	— —	$\bar{A}$ Γ_4	${}^1\bar{E}$ Γ_5	${}^2\bar{E}$ Γ_6	— —	— —	— —	$\bar{T}$ Γ_7
b	— —	— —	— —	— —	— —	— —	— —	— —	— —	— —	$\bar{E}$ Γ_5	${}^1\bar{F}$ Γ_6	${}^2\bar{F}$ Γ_7	— —
c	— —	— —	— —	$\bar{E}_g$ Γ_5^+	${}^1\bar{F}_g$ Γ_6^+	${}^2\bar{F}_g$ Γ_7^+	— —	— —	— —	— —	$\bar{E}_u$ Γ_5^-	${}^1\bar{F}_u$ Γ_6^-	${}^2\bar{F}_u$ Γ_7^-	— —
d	$\bar{A}_g$ Γ_4^+	${}^1\bar{E}_g$ Γ_5^+	${}^2\bar{E}_g$ Γ_5^+	— —	— —	— —	$\bar{T}_g$ Γ_7^+	$\bar{A}_u$ Γ_4^-	${}^2\bar{E}_u$ Γ_5^-	${}^2\bar{E}_u$ Γ_6^-	— —	— —	— —	$\bar{T}_u$ Γ_7^-

$\mathbf{G}_{48}^{9}$	R_{10}	R_{17}	R_{18}
a	$\bar{E}$ Γ_4	${}^1\bar{F}$ Γ_5	${}^2\bar{F}$ Γ_6

$\mathbf{G}_{48}^{10}$	R_4	R_5	R_8
a	$\bar{E}_1$ Γ_6	$\bar{E}_2$ Γ_7	$\bar{F}$ Γ_8

$\mathbf{G}_{48}^{11}$	R_{28}	R_{29}	R_{30}
a	${}^1\bar{F}$ Γ_4	$\bar{E}$ Γ_5	${}^2\bar{F}$ Γ_6

$\mathbf{G}_{48}^{12}$	R_7	R_8	R_9	R_{10}	R_{11}	R_{12}	R_{13}	R_{14}	R_{15}
a	—	—	—	$^1\bar{E}_1$	$^2\bar{E}_1$	$^1\bar{E}_2$	$^2\bar{E}_2$	$\bar{E}_3$	$\bar{E}_4$
	—	—	—	Γ_4	Γ_5	Γ_6	Γ_7	Γ_8	Γ_9
b	$\bar{E}_1$	$\bar{E}_2$	$\bar{E}_3$	—	—	—	—	—	—
	Γ_7	Γ_8	Γ_9	—	—	—	—	—	—

$\mathbf{G}_{48}^{13}$	R_{13}	R_{14}	R_{15}
a	$\bar{E}$	$^1\bar{F}$	$^2\bar{F}$
	Γ_7	Γ_8	Γ_9

$\mathbf{G}_{48}^{14}$	R_5	R_6	R_7	R_8	R_{11}	R_{12}
a	$^1\bar{E}_1$	$^1\bar{E}_2$	$^2\bar{E}_1$	$^2\bar{E}_2$	$^1\bar{F}$	$^2\bar{F}$
	Γ_4	Γ_5	Γ_6	Γ_7	Γ_8	Γ_9

$\mathbf{G}_{48}^{15}$	R_7	R_8	R_9	R_{16}	R_{17}	R_{18}
a	$\bar{E}_{1g}$	$\bar{E}_{2g}$	$\bar{E}_{3g}$	$\bar{E}_{1u}$	$\bar{E}_{2u}$	$\bar{E}_{3u}$
	Γ_7^+	Γ_8^+	Γ_9^+	Γ_7^-	Γ_8^-	Γ_9^-

$\mathbf{G}_{64}^{1}$	R_{19}	R_{20}	R_{21}	R_{22}
a	$^1\bar{F}_1$	$^2\bar{F}_1$	$^1\bar{F}_2$	$^2\bar{F}_2$
	Γ_5	Γ_6	Γ_7	Γ_8

$\mathbf{G}_{64}^{2}$	R_{19}
a	$\bar{F}$
	Γ_5

$\mathbf{G}_{64}^{3}$	R_9	R_{10}	R_{11}	R_{12}	R_{13}	R_{14}	R_{15}	R_{16}	R_{19}	R_{20}
a	—	—	—	—	$^1\bar{E}_1$	$^1\bar{E}_2$	$^1\bar{E}_3$	$^1\bar{E}_4$	—	$^1\bar{F}$
	—	—	—	—	Γ_3	Γ_4	Γ_5	Γ_6	—	Γ_7
b	$^1\bar{E}_1$	$^1\bar{E}_2$	$^1\bar{E}_3$	$^1\bar{E}_4$	—	—	—	—	$^1\bar{F}$	—
	Γ_3	Γ_4	Γ_5	Γ_6	—	—	—	—	Γ_7	—

$\mathbf{G}_{64}^{4}$	R_6	R_7	R_{20}	R_{21}
a	$^1\bar{F}_g$	$^2\bar{F}_g$	$^1\bar{F}_u$	$^2\bar{F}_u$
	Γ_6^+	Γ_7^+	Γ_6^-	Γ_7^-

$\mathbf{G}_{64}^{5}$	R_{19}
a	$\bar{F}$
	Γ_5

$\mathbf{G}_{96}^{5}$	R_{22}	R_{23}	R_{24}	R_{28}
a	$^1\bar{E}$	$^1\bar{F}_1$	$^1\bar{F}_2$	$^1\bar{H}$
	Γ_4	Γ_5	Γ_6	Γ_7

$\mathbf{G}_{96}^{7}$	R_3	R_4	R_6	R_7	R_8	R_{13}	R_{14}	R_{15}
a	$^1\bar{E}_1$	$^2\bar{E}_1$	—	—	$\bar{E}_2$	$^1\bar{H}$	$^2\bar{H}$	—
	Γ_4	Γ_5	—	—	Γ_6	Γ_7	Γ_8	—
b	—	—	$^1\bar{E}$	$^2\bar{E}$	—	—	—	$\bar{F}$
	—	—	Γ_6	Γ_7	—	—	—	Γ_8

G_{96}^{8}	R_4	R_5	R_8	R_{12}	R_{13}	R_{16}
a	—	—	—	$\bar{E}_1$	$\bar{E}_2$	$\bar{F}$
	—	—	—	Γ_6	Γ_7	Γ_8
b	$\bar{E}_{1g}$	$\bar{E}_{2g}$	$\bar{F}_g$	$\bar{E}_{1u}$	$\bar{E}_{2u}$	$\bar{F}_u$
	Γ_6^+	Γ_7^+	Γ_8^+	Γ_6^-	Γ_7^-	Γ_8^-

G_{96}^{9}	R_{19}	R_{20}	R_{21}	R_{22}	R_{23}	R_{24}
a	${}^1\bar{F}_1$	${}^2\bar{F}_1$	${}^1\bar{F}_2$	${}^1\bar{F}_3$	${}^2\bar{F}_2$	${}^2\bar{F}_3$
	Γ_7	Γ_8	Γ_9	Γ_{10}	Γ_{11}	Γ_{12}

G_{96}^{10}	R_{15}	R_{16}	R_{24}
a	${}^1\bar{F}_1$	${}^2\bar{F}_1$	$\bar{F}_2$
	Γ_4	Γ_5	Γ_6

G_{192}^{1}	R_{17}	R_{18}	R_{21}	R_{27}	R_{28}
a	${}^1\bar{E}_1$	${}^1\bar{E}_2$	${}^1\bar{F}$	${}^1\bar{H}_1$	${}^1\bar{H}_2$
	Γ_4	Γ_5	Γ_6	Γ_7	Γ_8

G_{192}^{2}	R_{18}	R_{21}	R_{22}
a	$\bar{F}$	${}^1\bar{M}$	${}^2\bar{M}$
	Γ_5	Γ_6	Γ_7

Notes to Table 6.14

(i) For a point of symmetry ${}^H\mathbf{G}^{\dagger\mathbf{k}}$ was identified in Table 6.12 with one of the abstract groups $\mathbf{G}^n_{|\mathbf{G}|}$ and the labelling of the reps indicated by a small arabic letter *a*, *b*, *c*, . . ., etc. The labels given to the space-group reps can then be found by consulting the entry under $\mathbf{G}^n_{|\mathbf{G}|}$ in Table 6.13. The row of the entry corresponding to this small arabic letter gives the labels assigned to the reps R_1, R_2, R_3 . . ., etc. of the group $\mathbf{G}^n_{|\mathbf{G}|}$ as the reps of ${}^H\mathbf{G}^{\dagger\mathbf{k}}$ for this space group.

(ii) For a line of symmetry the interpretation of the labelling is similar except that the reps $R_1, R_2, R_3, \ldots$, etc. of the abstract group now correspond to the space-group reps derived from the reps of $\bar{\mathbf{G}}^{\dagger\mathbf{k}*}$.

(iii) The double-valued space-group reps are labelled in a similar way to that used in Table 5.8 for the single-valued reps; extended forms of both the notation of Mulliken (1933) and the Γ notation of Bouckaert, Smoluchowski, and Wigner (1936) are used. In the latter notation Γ is replaced by the letter denoting the relevant point or line of symmetry in the Brillouin zone, see section 5.3 and, especially, Note (iii) to Table 5.8.

TABLE 6.15

*The additional double-valued space-group reps for the representation domain of Pa*3 (T_h^6)

Z'^x	$\mathbf{G}_{16}^{8}$: $(\sigma_x, 0), (C_{2y}, 0)$: 5, 3; 6, 3; 7, 3; 8, 3: *a*.

Note to Table 6.15

The notation is that of Table 6.13 (see Notes to Table 6.13).

7

The magnetic groups and their corepresentations

7.1 The Shubnikov groups and their derivation

THE 230 space groups (Fedorov groups) were discussed in Chapter 1 and a detailed classification of them was given in section 3.5. These groups were regarded as the ultimate development in the study of the symmetry of a crystal until Shubnikov in 1951 introduced the idea of operations of anti-symmetry. By considering an extra coordinate, with only two possible values, in addition to the ordinary position coordinates in a crystal, for example colour (black or white), sign (+ or −), or direction of magnetic moment (parallel or anti-parallel to a given direction) it is possible to define the operation of anti-symmetry as the operation which changes the value of this coordinate, i.e. black to white and white to black for example. If this operation of anti-symmetry is denoted by $\mathscr{R}$ it is possible to have compound operations of anti-symmetry corresponding to the performance of both an ordinary point-group or space-group operation $\{R \mid \mathbf{v}\}$ together with the operation of anti-symmetry $\mathscr{R}$; this is analogous to an improper point-group operation that consists of the performance of a pure rotation operation together with the inversion operation. Allowing operations of this type it is possible to derive the *black and white* or *magnetic* point groups of which there are 58, and the *black and white* or *magnetic* space groups of which there are 1191. The magnetic point groups were derived by Shubnikov (1951) and have been conveniently tabulated by Tavger and Zaitsev (1956), and the magnetic space groups were derived in a thesis by Zamorzaev in 1953 and by Belov, Neronova, and Smirnova (1955). Two-dimensional black and white space groups were discussed by Cochran (1952). If we include the 'all white' groups and the 'grey' groups the inclusion of the operation of anti-symmetry leads to 58 + 32 + 32 point groups, = 122 point groups, and 1191 + 230 + 230 space groups, = 1651 space groups. In this chapter we use the term *Shubnikov groups* to include all these, it being clear from the context whether a point group or space group is meant. Much of the important early Russian literature on these groups has now been translated into English (Shubnikov and Belov 1964) and is readily available. An extensive account is given by Birss (1963, 1964) of the form of various tensor properties for those crystals that have to be described by one of the Shubnikov point groups. It is a matter for experimentation, usually using neutron diffraction techniques, to determine the structure of a given magnetic crystal and the orientation of its spins; it can then be assigned to the appropriate Shubnikov group.

The idea of anti-symmetry had been discussed by Heesch (1929*a*, *b*, 1930*a*, *b*) and Woods (1935*a*–*c*) long before Shubnikov's work but its importance was not realized at that time. Heesch's work was probably ignored because at that time it did not seem to have any immediate application to the description of the physical world. But more recently new kinds of magnetism have been discovered so that there are now known to be various kinds of magnetism distinguished: paramagnetism, diamagnetism, ferromagnetism, anti-ferromagnetism, ferrimagnetism, and more complicated kinds such as canted anti-ferromagnetism; also, the introduction of the technique of neutron diffraction (see, for example, Egelstaff (1965), Lomer (1966*a*, *b*)) has made it possible to determine the orientations of the spins of the various atoms in a magnetically ordered crystal. These developments have made it clear that the concept of anti-symmetry and the use of magnetic point groups and space groups are relevant to the proper crystallographic description of a large number of crystals.

7.2. The classification of Shubnikov groups

We consider point groups first. If **G** is one of the ordinary point groups discussed in the early sections of this book then there are three types of Shubnikov group corresponding to it which are denoted by I, II, or III as follows:

type I, the ordinary point groups (32),
type II, the *grey* point groups (32), and
type III, the *black and white* or *magnetic* point groups (58).

The numbers in brackets give the number of point groups of each type, and the total number of these Shubnikov point groups is 122.

In type I groups the operation of anti-symmetry $\mathscr{R}$ is not present; these groups, **G**, have already been listed in Table 1.3 and nothing further need be said about them here.

The extra coordinate, *s*, which we have introduced and allowed to take one of two values is assumed to take both values simultaneously in type II groups so that any operation of **G** leaves *s* unchanged and $\mathscr{R}$ times any operation of **G** changes black into white and white into black thereby also leaving *s* unchanged. Thus $\mathscr{R}$ times any operation of **G** is also an element of this type II group and the term *grey group* derives from the fact that *s* takes both values such as black and white simultaneously.

DEFINITION 7.2.1. A type II Shubnikov point group, **M**, is given by

$$\mathbf{M} = \mathbf{G} + \mathscr{R}\mathbf{G} \tag{7.2.1}$$

where **G** is any (ordinary) point group.

There are clearly 32 of these groups. The difference between the ordinary point groups and the grey point groups is that in the former the anti-symmetry operation $\mathscr{R}$ is not present at all, whereas in the latter it is an operation of the group, and has the effect of doubling the order of the group. Since $\mathscr{R}^2 = E$ and $\mathscr{R}$ commutes with all the elements of G the grey groups are therefore direct product groups of the form $\mathbf{M} = \mathbf{G} \otimes \{E + \mathscr{R}\}$.

In the type III Shubnikov point groups, the genuine *black and white* point groups, $\mathscr{R}$ on its own is not an element of the group, but, of the elements in a point group **G**, half of them are now multiplied by the operation of anti-symmetry, $\mathscr{R}$. The other half of the elements of **G** form a group on their own, i.e. form a halving subgroup of **G**. It may be possible to choose the halving subgroup of **G** in several different ways so that from the 32 point groups, substantially more than 32 black and white groups will be obtained; in fact there are 58 of them.

DEFINITION 7.2.2. A type III Shubnikov point group, **M**, is given by

$$\mathbf{M} = \mathbf{H} + \mathscr{R}(\mathbf{G} - \mathbf{H}), \tag{7.2.2}$$

where **H** is a halving subgroup of the (ordinary) point group **G**.

We illustrate the derivation of type III (black and white) Shubnikov point grpups from the ordinary point groups **G** by considering the example of the black and white point groups that are derived from $4mm$ (C_{4v}) the point group of the symmetry operations of a square. If a square is drawn on a piece of paper it has the symmetry

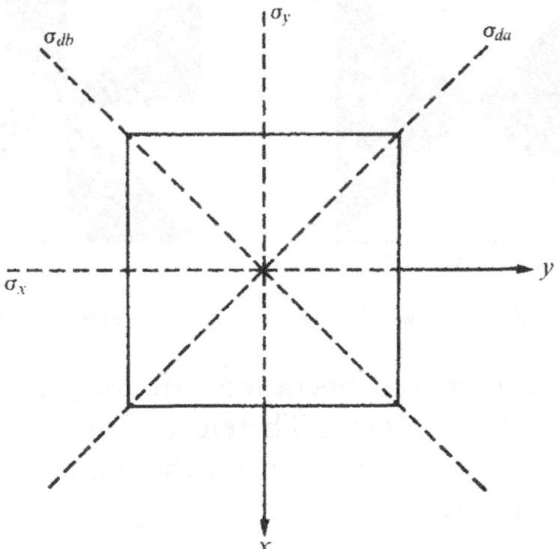

FIG. 7.1. The symmetry operations of a square.

operations E, C_{4z}^{+}, C_{4z}^{-}, C_{2z}, σ_x, σ_y, σ_{da}, and σ_{db}, see Table 1.3 and Fig. 7.1. If irregular black and white patches are drawn at random on the square these symmetry operations, other than E, will be destroyed. However, if half the square is coloured black and the other half is coloured white in some regular way, such as in Fig. 7.2(a) for example, then four of these eight symmetry operations may still survive and be symmetry operations of the coloured square. Thus E, C_{2z}, σ_x, and σ_y are still symmetry operations of the square in Fig. 7.2(a); but the other four operations are no longer symmetry operations, for instance C_{4z}^+ rotates the square by 90° and leaves a black patch where there was a white one before, and vice versa. But if each of these four operations is combined with $\mathscr{R}$, to produce $\mathscr{R}C_{4z}^+$, $\mathscr{R}C_{4z}^-$, $\mathscr{R}\sigma_{da}$, and $\mathscr{R}\sigma_{db}$, we have four colour-changing operations that are now symmetry operations of the coloured square; for example $\mathscr{R}C_{4z}^+$ rotates the square through 90° and turns black to white and white to black so that the combined effect is to leave the square indistinguishable from its initial state. Therefore instead of the original group of uncoloured operations we have four uncoloured elements and four coloured elements, so that exactly half of the original group elements remain uncoloured and exactly half of them have been combined with $\mathscr{R}$. The symmetry elements of Fig. 7.2(a) are therefore

$$E,\ C_{2z},\ \sigma_x,\ \sigma_y,\ \mathscr{R}C_{4z}^+,\ \mathscr{R}C_{4z}^-,\ \mathscr{R}\sigma_{da},\ \mathscr{R}\sigma_{db},$$

which still form a group and the group is denoted by $4'mm'$. The primes indicate the symmetry elements that are now associated with $\mathscr{R}$.‡ We could have made a different choice for the four uncoloured elements by considering the square shown in Fig. 7.2(b). For this square the four operations E, C_{4z}^+, C_{4z}^-, and C_{2z} are still symmetry

(a)

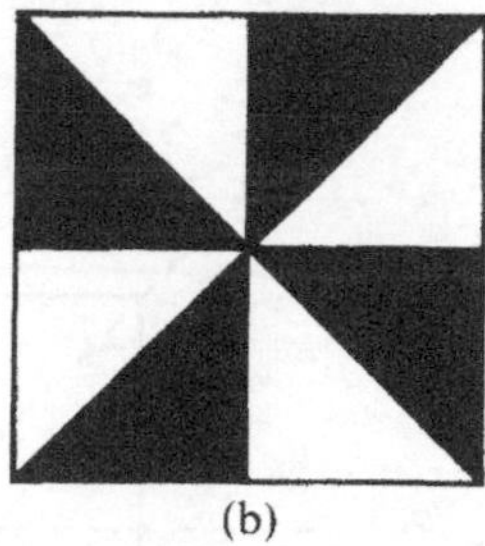

(b)

FIG. 7.2. Black and white squares to illustrate (a) $4'mm'$ and (b) $4m'm'$.

operations and the four coloured operations are $\mathscr{R}\sigma_x$, $\mathscr{R}\sigma_y$, $\mathscr{R}\sigma_{da}$, and $\mathscr{R}\sigma_{db}$. The symbol for this point group is $4m'm'$. Therefore, from the group of the eight symmetry operations of the square, that is from the point group $4mm$ (C_{4v}), we have produced two magnetic point groups, namely

$$E,\ C_{2z},\ \sigma_x,\ \sigma_y,\ \mathscr{R}C_{4z}^+,\ \mathscr{R}C_{4z}^-,\ \mathscr{R}\sigma_{da},\ \text{and}\ \mathscr{R}\sigma_{db}$$

and

$$E,\ C_{4z}^+,\ C_{4z}^-,\ C_{2z},\ \mathscr{R}\sigma_x,\ \mathscr{R}\sigma_y,\ \mathscr{R}\sigma_{da},\ \text{and}\ \mathscr{R}\sigma_{db}.$$

A third magnetic point group that contains the elements

$$E,\ C_{2z},\ \sigma_{da},\ \sigma_{db},\ \mathscr{R}C_{4z}^+,\ \mathscr{R}C_{4z}^-,\ \mathscr{R}\sigma_x,\ \text{and}\ \mathscr{R}\sigma_y$$

‡ Some authors underline the parts of the point-group symbol that denote coloured elements instead of using primes.

differs only from the first of these two groups in the orientation of the symmetry elements; the two groups are related by a 45° rotation about the z-axis.

It is possible to perform a similar analysis to the above by starting from any other of the 32 point groups and, in all, 58 different type III Shubnikov point groups can be obtained; they were listed by Tavger and Zaitsev (1956) and are reproduced in Table 7.1. Stereographic projections of the black and white point groups have been given by Koptsik (1966) and McMillan (1967). The subgroups of the black and white point groups have been investigated in detail by Ascher and Janner (1965*a*).

TABLE 7.1

The magnetic point groups

No.	**M**	**H**	**(G − H)**	*A*	
1	$\bar{1}'$	$1\,(C_1)$	I	I	*a*
2	$2'$	$1\,(C_1)$	C_{2z}	C_{2z}	*a*
3	m'	$1\,(C_1)$	σ_z	σ_z	*a*
4	$2/m'$	$2\,(C_2)$	$I;\ \sigma_z$	I	*b*
5	$2'/m$	$m\,(C_{1h})$	$I;\ C_{2z}$	I	*b*
6	$2'/m'$	$\bar{1}\,(C_i)$	$C_{2z};\ \sigma_z$	C_{2z}	*b*
7	$2'2'2$	$2\,(C_2)$	$C_{2x};\ C_{2y}$	C_{2x}	*b*
8	$m'm'2$	$2\,(C_2)$	$\sigma_x;\ \sigma_y$	σ_x	*b*
9	$m'm2'$	$m\,(C_{1h})$	$C_{2z};\ \sigma_x$	C_{2z}	*b*
10	$m'm'm'$	$222\,(D_2)$	$I;\ \sigma_x;\ \sigma_y;\ \sigma_z$	I	*c*
11	mmm'	$mm2(C_{2v})$	$C_{2x};\ C_{2y};\ I;\ \sigma_z$	I	*c*
12	$m'm'm$	$2/m\,(C_{2h})$	$C_{2x};\ C_{2y};\ \sigma_x;\ \sigma_y$	C_{2x}	*c*
13	$4'$	$2\,(C_2)$	$C_{4z}^+;\ C_{4z}^-$	C_{4z}^+	*b*
14	$\bar{4}'$	$2\,(C_2)$	$S_{4z}^-;\ S_{4z}^+$	S_{4z}^-	*b*
15	$42'2'$	$4\,(C_4)$	$C_{2x}, C_{2y};\ C_{2a}, C_{2b}$	C_{2x}	*e*
16	$4'22'$	$222\,(D_2)$	$C_{4z}^+, C_{4z}^-;\ C_{2a}, C_{2b}$	C_{2a}	*c*
17	$4/m'$	$4\,(C_4)$	$I;\ S_{4z}^-;\ \sigma_z;\ S_{4z}^+$	I	*e*
18	$4'/m'$	$\bar{4}\,(S_4)$	$I;\ C_{4z}^+;\ \sigma_z;\ C_{4z}^-$	I	*e*
19	$4'/m$	$2/m\,(C_{2h})$	$C_{4z}^+;\ C_{4z}^-;\ S_{4z}^+;\ S_{4z}^-$	C_{4z}^+	*c*
20	$4m'm'$	$4\,(C_4)$	$\sigma_x, \sigma_y;\ \sigma_{da}, \sigma_{db}$	σ_x	*e*
21	$4'mm'$	$mm2(C_{2v})$	$C_{4z}^+, C_{4z}^-;\ \sigma_{da}, \sigma_{db}$	σ_{da}	*c*
22	$\bar{4}2'm'$	$\bar{4}\,(S_4)$	$C_{2x}, C_{2y};\ \sigma_{da}, \sigma_{db}$	C_{2x}	*e*
23	$\bar{4}'2m'$	$222\,(D_2)$	$S_{4z}^-, S_{4z}^+;\ \sigma_{da}, \sigma_{db}$	σ_{da}	*c*
24	$\bar{4}'m2'$	$mm2\,(C_{2v})$	$S_{4z}^-, S_{4z}^+;\ C_{2a}, C_{2b}$	C_{2a}	*c*
25	$4/m'm'm'$	$422\,(D_4)$	$I;\ \sigma_z;\ S_{4z}^-, S_{4z}^+;\ \sigma_x, \sigma_y;\ \sigma_{da}, \sigma_{db}$	I	*g*
26	$4/m'mm$	$4mm\,(C_{4v})$	$I;\ \sigma_z;\ S_{4z}^-, S_{4z}^+;\ C_{2x}, C_{2y};\ C_{2a}, C_{2b}$	I	*g*
27	$4'/mmm'$	$mmm\,(D_{2h})$	$C_{4z}^+, C_{4z}^-;\ C_{2a}, C_{2b};\ S_{4z}^-, S_{4z}^+;\ \sigma_{da}, \sigma_{db}$	C_{2a}	*d*
28	$4'/m'm'm$	$\bar{4}2m\,(D_{2d})$	$I;\ \sigma_z;\ C_{4z}^+, C_{4z}^-;\ \sigma_x, \sigma_y;\ C_{2a}, C_{2b}$	I	*g*
29	$4/mm'm'$	$4/m\,(C_{4h})$	$C_{2x}, C_{2y};\ C_{2a}, C_{2b};\ \sigma_x, \sigma_y;\ \sigma_{da}, \sigma_{db}$	C_{2x}	*f*
30	$32'$	$3\,(C_3)$	$C'_{21}, C'_{22}, C'_{23}$	C'_{21}	*h*
31	$3m'$	$3\,(C_3)$	$\sigma_{d1}, \sigma_{d2}, \sigma_{d3}$	σ_{d1}	*h*
32	$\bar{6}'$	$3\,(C_3)$	$\sigma_h;\ S_3^-;\ S_3^+$	σ_h	*h*
33	$\bar{6}m'2'$	$\bar{6}\,(C_{3h})$	$C''_{21}, C''_{22}, C''_{23};\ \sigma_{v1}, \sigma_{v2}, \sigma_{v3}$	C'_{21}	*l*
34	$\bar{6}'m2'$	$3m\,(C_{3v})$	$\sigma_h;\ S_3^-, S_3^+;\ C'_{21}, C'_{22}, C'_{23}$	σ_h	*j*

No.	**M**	**H**	**(G − H)**	A	
35	$\bar{6}'m'2$	$32\ (D_3)$	$\sigma_h;\ S_3^-, S_3^+;\ \sigma_{v1}, \sigma_{v2}, \sigma_{v3}$	σ_h	j
36	$6'$	$3\ (C_3)$	$C_6^+;\ C_6^-;\ C_2$	C_2	h
37	$\bar{3}'$	$3\ (C_3)$	$I;\ S_6^-;\ S_6^+$	I	h
38	$\bar{3}m'$	$\bar{3}\ (C_{3i})$	$C'_{21}, C'_{22}, C'_{23};\ \sigma_{d1}, \sigma_{d2}, \sigma_{d3}$	C'_{21}	i
39	$\bar{3}'m$	$3m\ (C_{3v})$	$I;\ S_6^-, S_6^+;\ C'_{21}, C'_{22}, C'_{23}$	I	j
40	$\bar{3}'m'$	$32\ (D_3)$	$I;\ S_6^-, S_6^+;\ \sigma_{d1}, \sigma_{d2}, \sigma_{d3}$	I	j
41	$62'2'$	$6\ (C_6)$	$C'_{21}, C'_{22}, C'_{23};\ C''_{21}, C''_{22}, C''_{23}$	C'_{21}	l
42	$6'2'2$	$32\ (D_3)$	$C_2;\ C_6^+, C_6^-;\ C''_{21}, C''_{22}, C''_{23}$	C_2	j
43	$6/m'$	$6\ (C_6)$	$I;\ S_3^-;\ S_3^+;\ S_6^-;\ S_6^+;\ \sigma_h$	I	l
44	$6'/m'$	$\bar{3}\ (C_{3i})$	$C_6^+;\ C_6^-;\ C_2;\ S_3^+;\ S_3^-;\ \sigma_h$	C_2	i
45	$6'/m$	$\bar{6}\ (C_{3h})$	$I;\ S_6^-;\ S_6^+;\ C_2;\ C_6^+;\ C_6^-$	I	l
46	$6m'm'$	$6\ (C_6)$	$\sigma_{d1}, \sigma_{d2}, \sigma_{d3};\ \sigma_{v1}, \sigma_{v2}, \sigma_{v3}$	σ_{d1}	l
47	$6'm'm$	$3m\ (C_{3v})$	$C_2;\ C_6^+, C_6^-;\ \sigma_{v1}, \sigma_{v2}, \sigma_{v3}$	C_2	j
48	$6'/mmm'$	$\bar{6}2m\ (D_{3h})$	$I;\ C_2;\ S_6^+, S_6^-;\ C_6^+, C_6^-;\ \sigma_{d1}, \sigma_{d2}, \sigma_{d3};\ C''_{21}, C''_{22}, C''_{23}$	I	n
49	$6'/m'm'm$	$\bar{3}m\ (D_{3d})$	$C_2;\ \sigma_h;\ C_6^+, C_6^-;\ S_3^-, S_3^+;\ C''_{21}, C''_{22}, C''_{23};\ \sigma_{v1}, \sigma_{v2}, \sigma_{v3}$	C_2	k
50	$6/m'm'm'$	$622\ (D_6)$	$I;\ \sigma_h;\ S_6^+, S_6^-;\ S_3^+, S_3^-;\ \sigma_{d1}, \sigma_{d2}, \sigma_{d3};\ \sigma_{v1}, \sigma_{v2}, \sigma_{v3}$	I	n
51	$6/m'mm$	$6mm\ (C_{6v})$	$I;\ \sigma_h;\ S_6^+, S_6^-;\ S_3^+, S_3^-;\ C'_{21}, C'_{22}, C'_{23};\ C''_{21}, C''_{22}, C''_{23}$	I	n
52	$6/mm'm'$	$6/m\ (C_{6h})$	$C'_{21}, C'_{22}, C'_{23};\ C''_{21}, C''_{22}, C''_{23};\ \sigma_{d1}, \sigma_{d2}, \sigma_{d3};\ \sigma_{v1}, \sigma_{v2}, \sigma_{v3}$	C'_{21}	m
53	$m'3$	$23\ (T)$	$I;\ \sigma_x, \sigma_y, \sigma_z;\ S_{61}^-, S_{62}^-, S_{63}^-, S_{64}^-;\ S_{61}^+, S_{62}^+, S_{63}^+, S_{64}^+$	I	o
54	$\bar{4}'3m'$	$23\ (T)$	$\sigma_{da}, \sigma_{db}, \sigma_{dc}, \sigma_{dd}, \sigma_{de}, \sigma_{df};\ S_{4x}^+, S_{4y}^+, S_{4z}^+, S_{4x}^-, S_{4y}^-, S_{4z}^-$	σ_{da}	o
55	$4'32'$	$23\ (T)$	$C_{2a}, C_{2b}, C_{2c}, C_{2d}, C_{2e}, C_{2f};\ C_{4x}^+, C_{4y}^+, C_{4z}^+, C_{4x}^-, C_{4y}^-, C_{4z}^-$	C_{2a}	o
56	$m'3m'$	$432\ (O)$	$I;\ S_{61}^+, S_{62}^+, S_{63}^+, S_{64}^+, S_{61}^-, S_{62}^-, S_{63}^-, S_{64}^-;\ \sigma_x, \sigma_y, \sigma_z;$ $\sigma_{da}, \sigma_{db}, \sigma_{dc}, \sigma_{dd}, \sigma_{de}, \sigma_{df};\ S_{4x}^+, S_{4y}^+, S_{4z}^+, S_{4x}^-, S_{4y}^-, S_{4z}^-$	I	q
57	$m'3m$	$\bar{4}3m\ (T_d)$	$I;\ S_{61}^+, S_{62}^+, S_{63}^+, S_{64}^+, S_{61}^-, S_{62}^-, S_{63}^-, S_{64}^-;\ \sigma_x, \sigma_y, \sigma_z;$ $C_{2a}, C_{2b}, C_{2c}, C_{2d}, C_{2e}, C_{2f};\ C_{4x}^+, C_{4y}^+, C_{4z}^+, C_{4x}^-, C_{4y}^-, C_{4z}^-$	I	q
58	$m3m'$	$m3\ (T_h)$	$C_{2a}, C_{2b}, C_{2c}, C_{2d}, C_{2e}, C_{2f};\ C_{4x}^+, C_{4y}^+, C_{4z}^+, C_{4x}^-, C_{4y}^-, C_{4z}^-;$ $\sigma_{da}, \sigma_{db}, \sigma_{dc}, \sigma_{dd}, \sigma_{de}, \sigma_{df};\ S_{4x}^-, S_{4y}^-, S_{4z}^-, S_{4x}^+, S_{4y}^+, S_{4z}^+$	C_{2a}	p

Notes to Table 7.1

(i) The type III Shubnikov point groups are arranged according to the crystal systems and an arbitrary number, from 1 to 58, is assigned to each of them in column 1. The magnetic group **M** is identified in column 2 in the notation of Shubnikov and Belov (1964). In column 3 we identify **H**, the halving subgroup of uncoloured elements, in the Hermann–Mauguin notation with the Schönflies notation in brackets. The actual symmetry elements in **H** can be identified from Table 1.3, except that for $m'm2'$ (9) the elements of **H** are E and σ_y instead of the conventional E and σ_z in Table 1.3. In column 4 we identify the coloured elements, that is the elements of the coset $\mathscr{R}(\mathbf{G} - \mathbf{H})$, see eqn (7.2.2); these elements are arranged in classes with commas between elements in the same class and with semicolons separating one class from the next. The elements in column 4 are to be understood to be multiplied by $\mathscr{R}$. The information in the last two columns will be used later (see Table 7.15).

(ii) In attaching the primes to the appropriate parts of the point-group symbol in column 2 it is necessary to relate the various positions in the point-group symbols with the appropriate symmetry elements (see Table 3.3.2 of the *International tables for X-ray crystallography* (Henry and Lonsdale 1965)).

(iii) An adaptation of the Schönflies notation has sometimes been used to label the type III Shubnikov point groups; this takes the form **G(H)** so that, for example, $4'32'$ would be denoted by $O(T)$. These groups can also be labelled by using an adaptation of the Shubnikov point-group labels (see, for example, Shubnikov and Belov (1964) p. 142).

We now turn to the Shubnikov space groups. It is convenient to subdivide the Shubnikov space groups into four types rather than three. The first two types are directly analogous to the first two types of point group already discussed. The subdivision of the remaining (black and white) space groups into two types is dictated by a feature that has no direct analogy in dealing with the point groups; that is, whether or not the space group has a pure translation associated with the operation of anti-symmetry. The four types of Shubnikov space group are denoted by I, II, III, and IV as follows:

type I, the Fedorov (ordinary) space groups.(230),

type II, the *grey* space groups (230),

type III, *black and white* space groups based on ordinary Bravais lattices (674), and

type IV, *black and white* space groups based on *black and white* Bravais lattices (517).

The total number of these Shubnikov space groups is then 1651; illustrations of the symmetry operations of each of the Shubnikov space groups will be found in the book by Koptsik (1966).

The type I groups have been discussed in earlier chapters, and the generating elements of each one were listed in Table 3.7. The operation of anti-symmetry is not present in any of these groups.

Each type II group is a direct product of one of the Fedorov groups, **G**, and the group consisting of the identity operation, E, and the operation of anti-symmetry, $\mathscr{R}$. The grey group is readily given in terms of **G** by

DEFINITION 7.2.3. A type II Shubnikov space group, **M**, is given by

$$\mathbf{M} = \mathbf{G} + \mathscr{R}\mathbf{G}, \tag{7.2.3}$$

where **G** is any Fedorov group.

This should be compared with Definition 7.2.1. Nothing further need be said about the grey space groups at the moment, there are 230 of them and they are included in Table 7.4.

For a type III group there is again a direct analogy with the case of the point groups, Definition 7.2.2.

DEFINITION 7.2.4. A type III Shubnikov space group, **M**, is given by

$$\mathbf{M} = \mathbf{H} + \mathscr{R}(\mathbf{G} - \mathbf{H}), \tag{7.2.4}$$

where **H** is a halving subgroup of the Fedorov space group **G** and (**G** − **H**) contains no pure translations.

It is generally possible, as in the case of the point groups to choose a halving subgroup, **H**, in one of several ways for any given Fedorov group, **G**, so that there are substantially more than 230 of this type of Shubnikov space group. The total number of such type III space groups is 674 and they are listed in Table 7.2 where the halving subgroup **H** is also identified.

TABLE 7.2

Type III Shubnikov space groups

Fedorov group	Shubnikov group	'Coloured' generating elements	Fedorov group	Shubnikov group	'Coloured' generating elements
$P1$ (1)	—	—	$Pmc2_1$ (26)	$Pm'c2'_1$ (68)	σ_x
$P\bar{1}$ (2)	$P\bar{1}'$ (6)	I		$Pmc'2'_1$ (69)	σ_y
$P2$ (3)	$P2'$ (3)	C_{2z}		$Pm'c'2_1$ (70)	σ_x, σ_y
$P2_1$ (4)	$P2'_1$ (9)	C_{2z}	$Pcc2$ (27)	$Pc'c2'$ (80)	σ_x
$B2$ (5)	$B2'$ (15)	C_{2z}		$Pc'c'2$ (81)	σ_x, σ_y
Pm (6)	Pm' (20)	σ_z	$Pma2$ (28)	$Pm'a2'$ (89)	σ_x
Pb (7)	Pb' (26)	σ_z		$Pma'2'$ (90)	σ_y
Bm (8)	Bm' (34)	σ_z		$Pm'a'2$ (91)	σ_x, σ_y
Bb (9)	Bb' (39)	σ_z	$Pca2_1$ (29)	$Pc'a2'_1$ (101)	σ_x
$P2/m$ (10)	$P2'/m$ (44)	C_{2z}, I		$Pca'2'_1$ (102)	σ_y
	$P2/m'$ (45)	I		$Pc'a'2_1$ (103)	σ_x, σ_y
	$P2'/m'$ (46)	C_{2z}	$Pnc2$ (30)	$Pn'c2'$ (113)	σ_x
$P2_1/m$ (11)	$P2'_1/m$ (52)	C_{2z}, I		$Pnc'2'$ (114)	σ_y
	$P2_1/m'$ (53)	I		$Pn'c'2$ (115)	σ_x, σ_y
	$P2'_1/m'$ (54)	C_{2z}	$Pmn2_1$ (31)	$Pm'n2'_1$ (125)	σ_x
$B2/m$ (12)	$B2'/m$ (60)	C_{2z}, I		$Pmn'2'_1$ (126)	σ_y
	$B2/m'$ (61)	I		$Pm'n'2_1$ (127)	σ_x, σ_y
	$B2'/m'$ (62)	C_{2z}	$Pba2$ (32)	$Pb'a2'$ (137)	σ_x
$P2/b$ (13)	$P2'/b$ (67)	C_{2z}, I		$Pb'a'2$ (138)	σ_x, σ_y
	$P2/b'$ (68)	I	$Pna2_1$ (33)	$Pn'a2'_1$ (146)	σ_x
	$P2'/b'$ (69)	C_{2z}		$Pna'2'_1$ (147)	σ_y
$P2_1/b$ (14)	$P2'_1/b$ (77)	C_{2z}, I		$Pn'a'2_1$ (148)	σ_x, σ_y
	$P2_1/b$ (78)	I	$Pnn2$ (34)	$Pn'n2'$ (158)	σ_x
	$P2'_1/b'$ (79)	C_{2z}		$Pn'n'2$ (159)	σ_x, σ_y
$B2/b$ (15)	$B2'/b$ (87)	C_{2z}, I	$Cmm2$ (35)	$Cm'm2'$ (167)	σ_x
	$B2/b'$ (88)	I		$Cm'm'2$ (168)	σ_x, σ_y
	$B2'/b'$ (89)	C_{2z}	$Cmc2_1$ (36)	$Cm'c2'_1$ (174)	σ_x
$P222$ (16)	$P2'2'2$ (3)	C_{2x}, C_{2y}		$Cmc'2'_1$ (175)	σ_y
$P222_1$ (17)	$P2'2'2_1$ (9)	C_{2x}, C_{2y}		$Cm'c'2_1$ (176)	σ_x, σ_y
	$P22'2'_1$ (10)	C_{2y}	$Ccc2$ (37)	$Cc'c2'$ (182)	σ_x
$P2_12_12$ (18)	$P2'_12'_12$ (18)	C_{2x}, C_{2y}		$Cc'c'2$ (183)	σ_x, σ_y
	$P2_12'_12'$ (19)	C_{2y}	$Amm2$ (38)‡	$Am'm2'$ (189)	σ_x
$P2_12_12_1$ (19)	$P2'_12'_12_1$ (27)	C_{2x}, C_{2y}		$Amm'2'$ (190)	σ_y
$C222_1$ (20)	$C2'2'2_1$ (33)	C_{2x}, C_{2y}		$Am'm'2$ (191)	σ_x, σ_y
	$C22'2'_1$ (34)	C_{2y}	$Abm2$ (39)‡	$Ab'm2'$ (197)	σ_x
$C222$ (21)	$C2'2'2$ (40)	C_{2x}, C_{2y}		$Abm'2'$ (198)	σ_y
	$C22'2'$ (41)	C_{2y}		$Ab'm'2$ (199)	σ_x, σ_y
$F222$ (22)	$F2'2'2$ (47)	C_{2x}, C_{2y}	$Ama2$ (40)‡	$Am'a2'$ (205)	σ_x
$I222$ (23)	$I2'2'2$ (51)	C_{2x}, C_{2y}		$Ama'2'$ (206)	σ_y
$I2_12_12_1$ (24)	$I2'_12'_12_1$ (55)	C_{2x}, C_{2y}		$Am'a'2$ (207)	σ_x, σ_y
$Pmm2$ (25)	$Pm'm2'$ (59)	σ_x	$Aba2$ (41)‡	$Ab'a2'$ (213)	σ_x
	$Pm'm'2$ (60)	σ_x, σ_y		$Aba'2'$ (214)	σ_y

‡ In Table 3.1 we listed the translations suitable for C (Γ_o^b). Those for A (Γ_o^b) are $\mathbf{t}_1 = \frac{1}{2}(0, -b, c)$, $\mathbf{t}_2 = (a, 0, 0)$, $\mathbf{t}_3 = \frac{1}{2}(0, b, c)$ referred to axes $Oxyz$.

Fedorov group	Shubnikov group	'Coloured' generating elements
	Ab'a'2 (215)	σ_x, σ_y
Fmm2 (42)	*Fm'm2'* (221)	σ_x
	Fm'm'2 (222)	σ_x, σ_y
Fdd2 (43)	*Fd'd2'* (226)	σ_x
	Fd'd'2 (227)	σ_x, σ_y
Imm2 (44)	*Im'm2'* (231)	σ_x
	Im'm'2 (232)	σ_x, σ_y
Iba2 (45)	*Ib'a2'* (237)	σ_x
	Ib'a'2 (238)	σ_x, σ_y
Ima2 (46)	*Im'a2'* (243)	σ_x
	Ima'2' (244)	σ_y
	Im'a'2 (245)	σ_x, σ_y
Pmmm (47)	*Pm'mm* (251)	C_{2y}, I
	Pm'm'm (252)	C_{2x}, C_{2y}
	Pm'm'm' (253)	I
Pnnn (48)	*Pn'nn* (259)	C_{2y}, I
	Pn'n'n (260)	C_{2x}, C_{2y}
	Pn'n'n' (261)	I
Pccm (49)	*Pc'cm* (267)	C_{2y}, I
	Pccm' (268)	C_{2x}, C_{2y}, I
	Pc'c'm (269)	C_{2x}, C_{2y}
	Pc'cm' (270)	C_{2x}
	Pc'c'm' (271)	I
Pban (50)	*Pb'an* (279)	C_{2y}, I
	Pban' (280)	C_{2x}, C_{2y}, I
	Pb'a'n (281)	C_{2x}, C_{2y}
	Pb'an' (282)	C_{2x}
	Pb'a'n' (283)	I
Pmma (51)	*Pm'ma* (291)	C_{2y}, I
	Pmm'a (292)	C_{2x}, I
	Pmma' (293)	C_{2x}, C_{2y}, I
	Pm'm'a (294)	C_{2x}, C_{2y}
	Pmm'a' (295)	C_{2y}
	Pm'ma' (296)	C_{2x}
	Pm'm'a' (297)	I
Pnna (52)	*Pn'na* (307)	C_{2y}, I
	Pnn'a (308)	C_{2x}, I
	Pnna' (309)	C_{2x}, C_{2y}, I
	Pn'n'a (310)	C_{2x}, C_{2y}
	Pnn'a' (311)	C_{2y}
	Pn'na' (312)	C_{2x}
	Pn'n'a' (313)	I
Pmna (53)	*Pm'na* (323)	C_{2y}, I
	Pmn'a (324)	C_{2x}, I
	Pmna' (325)	C_{2x}, C_{2y}, I
	Pm'n'a (326)	C_{2x}, C_{2y}
	Pmn'a' (327)	C_{2y}
	Pm'na' (328)	C_{2x}
	Pm'n'a' (329)	I
Pcca (54)	*Pc'ca* (339)	C_{2y}, I
	Pcc'a (340)	C_{2x}, I
	Pcca' (341)	C_{2x}, C_{2y}, I
	Pc'c'a (342)	C_{2x}, C_{2y}
	Pcc'a' (343)	C_{2y}
	Pc'ca' (344)	C_{2x}
	Pc'c'a' (345)	I
Pbam (55)	*Pb'am* (355)	C_{2y}, I
	Pbam' (356)	C_{2x}, C_{2y}, I
	Pb'a'm (357)	C_{2x}, C_{2y}
	Pb'am' (358)	C_{2x}
	Pb'a'm' (359)	I
Pccn (56)	*Pc'cn* (367)	C_{2y}, I
	Pccn' (368)	C_{2x}, C_{2y}, I
	Pc'c'n (369)	C_{2x}, C_{2y}
	Pc'cn' (370)	C_{2x}
	Pc'c'n' (371)	I
Pbcm (57)	*Pb'cm* (379)	C_{2y}, I
	Pbc'm (380)	C_{2x}, I
	Pbcm' (381)	C_{2x}, C_{2y}, I
	Pb'c'm (382)	C_{2x}, C_{2y}
	Pbc'm' (383)	C_{2y}
	Pb'cm' (384)	C_{2x}
	Pb'c'm' (385)	I
Pnnm (58)	*Pn'nm* (395)	C_{2y}, I
	Pnnm' (396)	C_{2x}, C_{2y}, I
	Pn'n'm (397)	C_{2x}, C_{2y}
	Pnn'm' (398)	C_{2y}
	Pn'n'm' (399)	I
Pmmn (59)	*Pm'mn* (407)	C_{2y}, I
	Pmmn' (408)	C_{2x}, C_{2y}, I
	Pm'm'n (409)	C_{2x}, C_{2y}
	Pmm'n' (410)	C_{2y}
	Pm'm'n' (411)	I
Pbcn (60)	*Pb'cn* (419)	C_{2y}, I
	Pbc'n (420)	C_{2x}, I
	Pbcn' (421)	C_{2x}, C_{2y}, I
	Pb'c'n (422)	C_{2x}, C_{2y}
	Pbc'n' (423)	C_{2y}
	Pb'cn' (424)	C_{2x}
	Pb'c'n' (425)	I
Pbca (61)	*Pb'ca* (435)	C_{2y}, I
	Pb'c'a (436)	C_{2x}, C_{2y}
	Pb'c'a' (437)	I
Pnma (62)	*Pn'ma* (443)	C_{2y}, I

Fedorov group	Shubnikov group	'Coloured' generating elements	Fedorov group	Shubnikov group	'Coloured' generating elements
	$Pnm'a$ (444)	C_{2x}, I		$Im'm'm$ (536)	C_{2x}, C_{2y}
	$Pnma'$ (445)	C_{2x}, C_{2y}, I		$Im'm'm'$ (537)	I
	$Pn'm'a$ (446)	C_{2x}, C_{2y}	$Ibam$ (72)	$Ib'am$ (541)	C_{2y}, I
	$Pnm'a'$ (447)	C_{2y}		$Ibam'$ (542)	C_{2x}, C_{2y}, I
	$Pn'ma'$ (448)	C_{2x}		$Ib'a'm$ (543)	C_{2x}, C_{2y}
	$Pn'm'a'$ (449)	I		$Iba'm'$ (544)	C_{2y}
$Cmcm$ (63)	$Cm'cm$ (459)	C_{2y}, I		$Ib'a'm'$ (545)	I
	$Cmc'm$ (460)	C_{2x}, I	$Ibca$ (73)	$Ib'ca$ (550)	C_{2y}, I
	$Cmcm'$ (461)	C_{2x}, C_{2y}, I		$Ib'c'a$ (551)	C_{2x}, C_{2y}
	$Cm'c'm$ (462)	C_{2x}, C_{2y}		$Ib'c'a'$ (552)	I
	$Cmc'm'$ (463)	C_{2y}	$Imma$ (74)	$Im'ma$ (556)	C_{2y}, I
	$Cm'cm'$ (464)	C_{2x}		$Imma'$ (557)	C_{2x}, C_{2y}, I
	$Cm'c'm'$ (465)	I		$Im'm'a$ (558)	C_{2x}, C_{2y}
$Cmca$ (64)	$Cm'ca$ (471)	C_{2y}, I		$Imm'a'$ (559)	C_{2y}
	$Cmc'a$ (472)	C_{2x}, I		$Im'm'a'$ (560)	I
	$Cmca'$ (473)	C_{2x}, C_{2y}, I	$P4$ (75)	$P4'$ (3)	C_{4z}^+
	$Cm'c'a$ (474)	C_{2x}, C_{2y}	$P4_1$ (76)	$P4'_1$ (9)	C_{4z}^+
	$Cmc'a'$ (475)	C_{2y}	$P4_2$ (77)	$P4'_2$ (15)	C_{4z}^+
	$Cm'ca'$ (476)	C_{2x}	$P4_3$ (78)	$P4'_3$ (21)	C_{4z}^+
	$Cm'c'a'$ (477)	I	$I4$ (79)	$I4'$ (27)	C_{4z}^+
$Cmmm$ (65)	$Cm'mm$ (483)	C_{2y}, I	$I4_1$ (80)	$I4'_1$ (31)	C_{4z}^+
	$Cmmm'$ (484)	C_{2x}, C_{2y}, I	$P\bar{4}$ (81)	$P\bar{4}'$ (35)	S_{4z}^+
	$Cm'm'm$ (485)	C_{2x}, C_{2y}	$I\bar{4}$ (82)	$I\bar{4}'$ (41)	S_{4z}^+
	$Cmm'm'$ (486)	C_{2y}	$P4/m$ (83)	$P4'/m$ (45)	C_{4z}^+
	$Cm'm'm'$ (487)	I		$P4/m'$ (46)	I
$Cccm$ (66)	$Cc'cm$ (493)	C_{2y}, I		$P4'/m'$ (47)	C_{4z}^+, I
	$Cccm'$ (494)	C_{2x}, C_{2y}, I	$P4_2/m$ (84)	$P4'_2/m$ (53)	C_{4z}^+
	$Cc'c'm$ (495)	C_{2x}, C_{2y}		$P4_2/m'$ (54)	I
	$Ccc'm'$ (496)	C_{2y}		$P4'_2/m'$ (55)	C_{4z}^+, I
	$Cc'c'm'$ (497)	I	$P4/n$ (85)	$P4'/n$ (61)	C_{4z}^+
$Cmma$ (67)	$Cm'ma$ (503)	C_{2y}, I		$P4/n'$ (62)	I
	$Cmma'$ (504)	C_{2x}, C_{2y}, I		$P4'/n'$ (63)	C_{4z}^+, I
	$Cm'm'a$ (505)	C_{2x}, C_{2y}	$P4_2/n$ (86)	$P4'_2/n$ (69)	C_{4z}^+
	$Cmm'a'$ (506)	C_{2y}		$P4_2/n'$ (70)	I
	$Cm'm'a'$ (507)	I		$P4'_2/n'$ (71)	C_{4z}^+, I
$Ccca$ (68)	$Cc'ca$ (513)	C_{2y}, I	$I4/m$ (87)	$I4'/m$ (77)	C_{4z}^+
	$Ccca'$ (514)	C_{2x}, C_{2y}, I		$I4/m'$ (78)	I
	$Cc'c'a$ (515)	C_{2x}, C_{2y}		$I4'/m'$ (79)	C_{4z}^+, I
	$Ccc'a'$ (516)	C_{2y}	$I4_1/a$ (88)	$I4'_1/a$ (83)	C_{4z}^+
	$Cc'c'a'$ (517)	I		$I4_1/a'$ (84)	I
$Fmmm$ (69)	$Fm'mm$ (523)	C_{2y}, I		$I4'_1/a'$ (85)	C_{4z}^+, I
	$Fm'm'm$ (524)	C_{2x}, C_{2y}	$P422$ (89)	$P4'22'$ (89)	C_{4z}^+
	$Fm'm'm'$ (525)	I		$P42'2'$ (90)	C_{2x}
$Fddd$ (70)	$Fd'dd$ (529)	C_{2y}, I		$P4'2'2$ (91)	C_{4z}^+, C_{2x}
	$Fd'd'd$ (530)	C_{2x}, C_{2y}	$P42_12$ (90)	$P4'2_12'$ (97)	C_{4z}^+
	$Fd'd'd'$ (531)	I		$P42'_12'$ (98)	C_{2x}
$Immm$ (71)	$Im'mm$ (535)	C_{2y}, I		$P4'2'_12$ (99)	C_{4z}^+, C_{2x}

Fedorov group	Shubnikov group	'Coloured' generating elements	Fedorov group	Shubnikov group	'Coloured' generating elements
$P4_122$ (91)	$P4'_122'$ (105)	C^+_{4z}		$P4'_2bc'$ (222)	C^+_{4z}
	$P4_12'2'$ (106)	C_{2x}		$P4_2b'c'$ (223)	σ_x
	$P4'_12'2$ (107)	C^+_{4z}, C_{2x}	$I4mm$ (107)	$I4'm'm$ (229)	C^+_{4z}, σ_x
$P4_12_12$ (92)	$P4'_12_12'$ (113)	C^+_{4z}		$I4'mm'$ (230)	C^+_{4z}
	$P4_12'_12'$ (114)	C_{2x}		$I4m'm'$ (231)	σ_x
	$P4'_12'_12$ (115)	C^+_{4z}, C_{2x}	$I4cm$ (108)	$I4'c'm$ (235)	C^+_{4z}, σ_x
$P4_222$ (93)	$P4'_222'$ (121)	C^+_{4z}		$I4'cm'$ (236)	C^+_{4z}
	$P4_22'2'$ (122)	C_{2x}		$I4c'm'$ (237)	σ_x
	$P4'_22'2$ (123)	C^+_{4z}, C_{2x}	$I4_1md$ (109)	$I4'_1m'd$ (241)	C^+_{4z}, σ_x
$P4_22_12$ (94)	$P4'_22_12'$ (129)	C^+_{4z}		$I4'_1md'$ (242)	C^+_{4z}
	$P4_22'_12'$ (130)	C_{2x}		$I4_1m'd'$ (243)	σ_x
	$P4'_22'_12$ (131)	C^+_{4z}, C_{2x}	$I4_1cd$ (110)	$I4'_1c'd$ (247)	C^+_{4z}, σ_x
$P4_322$ (95)	$P4'_322'$ (137)	C^+_{4z}		$I4'_1cd'$ (248)	C^+_{4z}
	$P4_32'2'$ (138)	C_{2x}		$I4_1c'd'$ (249)	σ_x
	$P4'_32'2$ (139)	C^+_{4z}, C_{2x}	$P\bar{4}2m$ (111)	$P\bar{4}'2'm$ (253)	S^+_{4z}, C_{2x}
$P4_32_12$ (96)	$P4'_32_12'$ (145)	C^+_{4z}		$P\bar{4}'2m'$ (254)	S^+_{4z}
	$P4_32'_12'$ (146)	C_{2x}		$P\bar{4}2'm'$ (255)	C_{2x}
	$P4'_32'_12$ (147)	C^+_{4z}, C_{2x}	$P\bar{4}2c$ (112)	$P\bar{4}'2'c$ (261)	S^+_{4z}, C_{2x}
$I422$ (97)	$I4'22'$ (153)	C^+_{4z}		$P\bar{4}'2c'$ (262)	S^+_{4z}
	$I42'2'$ (154)	C_{2x}		$P\bar{4}2'c'$ (263)	C_{2x}
	$I4'2'2$ (155)	C^+_{4z}, C_{2x}	$P\bar{4}2_1m$ (113)	$P\bar{4}'2'_1m$ (269)	S^+_{4z}, C_{2x}
$I4_122$ (98)	$I4'_122'$ (159)	C^+_{4z}		$P\bar{4}'2_1m'$ (270)	S^+_{4z}
	$I4_12'2'$ (160)	C_{2x}		$P\bar{4}2'_1m'$ (271)	C_{2x}
	$I4'_12'2$ (161)	C^+_{4z}, C_{2x}	$P\bar{4}2_1c$ (114)	$P\bar{4}'2'_1c$ (277)	S^+_{4z}, C_{2x}
$P4mm$ (99)	$P4'm'm$ (165)	C^+_{4z}, σ_x		$P\bar{4}'2_1c'$ (278)	S^+_{4z}
	$P4'mm'$ (166)	C^+_{4z}		$P\bar{4}2'_1c'$ (279)	C_{2x}
	$P4m'm'$ (167)	σ_x	$P\bar{4}m2$ (115)	$P\bar{4}'m'2$ (285)	S^+_{4z}
$P4bm$ (100)	$P4'b'm$ (173)	C^+_{4z}, σ_x		$P\bar{4}'m2'$ (286)	S^+_{4z}, C_{2a}
	$P4'bm'$ (174)	C^+_{4z}		$P\bar{4}m'2'$ (287)	C_{2a}
	$P4b'm'$ (175)	σ_x	$P\bar{4}c2$ (116)	$P\bar{4}'c'2$ (293)	S^+_{4z}
$P4_2cm$ (101)	$P4'_2c'm$ (181)	C^+_{4z}, σ_x		$P\bar{4}'c2'$ (294)	S^+_{4z}, C_{2a}
	$P4'_2cm'$ (182)	C^+_{4z}		$P\bar{4}c'2'$ (295)	C_{2a}
	$P4_2c'm'$ (183)	σ_x	$P\bar{4}b2$ (117)	$P\bar{4}'b'2$ (301)	S^+_{4z}
$P4_2nm$ (102)	$P4'_2n'm$ (189)	C^+_{4z}, σ_x		$P\bar{4}'b2'$ (302)	S^+_{4z}, C_{2a}
	$P4'_2nm'$ (190)	C^+_{4z}		$P\bar{4}b'2'$ (303)	C_{2a}
	$P4_2n'm'$ (191)	σ_x	$P\bar{4}n2$ (118)	$P\bar{4}'n'2$ (309)	S^+_{4z}
$P4cc$ (103)	$P4'c'c$ (197)	C^+_{4z}, σ_x		$P\bar{4}'n2'$ (310)	S^+_{4z}, C_{2a}
	$P4'cc'$ (198)	C^+_{4z}		$P\bar{4}n'2'$ (311)	C_{2a}
	$P4c'c'$ (199)	σ_x	$I\bar{4}m2$ (119)	$I\bar{4}'m'2$ (317)	S^+_{4z}
$P4nc$ (104)	$P4'n'c$ (205)	C^+_{4z}, σ_x		$I\bar{4}'m2'$ (318)	S^+_{4z}, C_{2a}
	$P4'nc'$ (206)	C^+_{4z}		$I\bar{4}m'2'$ (319)	C_{2a}
	$P4n'c'$ (207)	σ_x	$I\bar{4}c2$ (120)	$I\bar{4}'c'2$ (323)	S^+_{4z}
$P4_2mc$ (105)	$P4'_2m'c$ (213)	C^+_{4z}, σ_x		$I\bar{4}'c2'$ (323)	S^+_{4z}, C_{2a}
	$P4'_2mc'$ (214)	C^+_{4z}		$I\bar{4}c'2'$ (325)	C_{2a}
	$P4_2m'c'$ (215)	σ_x	$I\bar{4}2m$ (121)	$I\bar{4}'2'm$ (329)	S^+_{4z}, C_{2x}
$P4_2bc$ (106)	$P4'_2b'c$ (221)	C^+_{4z}, σ_x		$I\bar{4}'2m'$ (330)	S^+_{4z}

Fedorov group	Shubnikov group	'Coloured' generating elements
	$I\bar{4}2'm'$ (331)	C_{2x}
$I\bar{4}2d$ (122)	$I\bar{4}'2'd$ (335)	S_{4z}^+, C_{2x}
	$I\bar{4}'2d'$ (336)	S_{4z}^+
	$I\bar{4}2'd'$ (337)	C_{2x}
$P4/mmm$ (123)	$P4/m'mm$ (341)	C_{2x}, I
	$P4'/mm'm$ (342)	C_{4z}^+, C_{2x}
	$P4'/mmm'$ (343)	C_{4z}^+
	$P4'/m'm'm$ (344)	C_{4z}^+, I
	$P4/mm'm'$ (345)	C_{2x}
	$P4'/m'mm'$ (346)	C_{4z}^+, I, C_{2x}
	$P4/m'm'm'$ (347)	I
$P4/mcc$ (124)	$P4/m'cc$ (353)	C_{2x}, I
	$P4'/mc'c$ (354)	C_{4z}^+, C_{2x}
	$P4'/mcc'$ (355)	C_{4z}^+
	$P4'/m'c'c$ (356)	C_{4z}^+, I
	$P4/mc'c'$ (357)	C_{2x}
	$P4'/m'cc'$ (358)	C_{4z}^+, I, C_{2x}
	$P4/m'c'c'$ (359)	I
$P4/nbm$ (125)	$P4/n'bm$ (365)	C_{2x}, I
	$P4'/nb'm$ (366)	C_{4z}^+, C_{2x}
	$P4'/nbm'$ (367)	C_{4z}^+
	$P4'/n'b'm$ (368)	C_{4z}^+, I
	$P4/nb'm'$ (369)	C_{2x}
	$P4'/n'bm'$ (370)	C_{4z}^+, I, C_{2x}
	$P4/n'b'm'$ (371)	I
$P4/nnc$ (126)	$P4/n'nc$ (377)	C_{2x}, I
	$P4'/nn'c$ (378)	C_{4z}^+, C_{2x}
	$P4'/nnc'$ (379)	C_{4z}^+
	$P4'/n'n'c$ (380)	C_{4z}^+, I
	$P4/nn'c'$ (381)	C_{2x}
	$P4'/n'nc'$ (382)	C_{4z}^+, I, C_{2x}
	$P4/n'n'c'$ (383)	I
$P4/mbm$ (127)	$P4/m'bm$ (389)	C_{2x}, I
	$P4'/mb'm$ (390)	C_{4z}^+, C_{2x}
	$P4'/mbm'$ (391)	C_{4z}^+
	$P4'/m'b'm$ (392)	C_{4z}^+, I
	$P4/mb'm'$ (393)	C_{2x}
	$P4'/m'bm'$ (394)	C_{4z}^+, I, C_{2x}
	$P4/m'b'm'$ (395)	I
$P4/mnc$ (128)	$P4/m'nc$ (401)	C_{2x}, I
	$P4'/mn'c$ (402)	C_{4z}^+, C_{2x}
	$P4'/mnc'$ (403)	C_{4z}^+
	$P4'/m'n'c$ (404)	C_{4z}^+, I
	$P4/mn'c'$ (405)	C_{2x}
	$P4'/m'nc'$ (406)	C_{4z}^+, I, C_{2x}
	$P4/m'n'c'$ (407)	I
$P4/nmm$ (129)	$P4/n'mm$ (413)	C_{2x}, I
	$P4'/nm'm$ (414)	C_{4z}^+, C_{2x}
	$P4'/nmm'$ (415)	C_{4z}^+
	$P4'/n'm'm$ (416)	C_{4z}^+, I
	$P4/nm'm'$ (417)	C_{2x}
	$P4'/n'mm'$ (418)	C_{4z}^+, I, C_{2x}
	$P4/n'm'm'$ (419)	I
$P4/ncc$ (130)	$P4/n'cc$ (425)	C_{2x}, I
	$P4'/nc'c$ (426)	C_{4z}^+, C_{2x}
	$P4'/ncc'$ (427)	C_{4z}^+
	$P4'/n'c'c$ (428)	C_{4z}^+, I
	$P4/nc'c'$ (429)	C_{2x}
	$P4'/n'cc'$ (430)	C_{4z}^+, I, C_{2x}
	$P4/n'c'c'$ (431)	I
$P4_2/mmc$ (131)	$P4_2/m'mc$ (437)	C_{2x}, I
	$P4'_2/mm'c$ (438)	C_{4z}^+, C_{2x}
	$P4'_2/mmc'$ (439)	C_{4z}^+
	$P4'_2/m'm'c$ (440)	C_{4z}^+, I
	$P4_2/mm'c'$ (441)	C_{2x}
	$P4'_2/m'mc'$ (442)	C_{4z}^+, I, C_{2x}
	$P4_2/m'm'c'$ (443)	I
$P4_2/mcm$ (132)	$P4_2/m'cm$ (449)	C_{2x}, I
	$P4'_2/mc'm$ (450)	C_{4z}^+, C_{2x}
	$P4'_2/mcm'$ (451)	C_{4z}^+
	$P4'_2/m'c'm$ (452)	C_{4z}^+, I
	$P4_2/mc'm'$ (453)	C_{2x}
	$P4'_2/m'cm'$ (454)	C_{4z}^+, I, C_{2x}
	$P4_2/m'c'm'$ (455)	I
$P4_2/nbc$ (133)	$P4_2/n'bc$ (461)	C_{2x}, I
	$P4'_2/nb'c$ (462)	C_{4z}^+, C_{2x}
	$P4'_2/nbc'$ (463)	C_{4z}^+
	$P4'_2/n'b'c$ (464)	C_{4z}^+, I
	$P4_2/nb'c'$ (465)	C_{2x}
	$P4'_2/n'bc'$ (466)	C_{4z}^+, I, C_{2x}
	$P4_2/n'b'c'$ (467)	I
$P4_2/nnm$ (134)	$P4_2/n'nm$ (473)	C_{2x}, I
	$P4'_2/nn'm$ (474)	C_{4z}^+, C_{2x}
	$P4'_2/nnm'$ (475)	C_{4z}^+
	$P4'_2/n'n'm$ (476)	C_{4z}^+, I
	$P4_2/nn'm'$ (477)	C_{2x}
	$P4'_2/n'nm'$ (478)	C_{4z}^+, I, C_{2x}
	$P4_2/n'n'm'$ (479)	I
$P4_2/mbc$ (135)	$P4_2/m'bc$ (485)	C_{2x}, I
	$P4'_2/mb'c$ (486)	C_{4z}^+, C_{2x}
	$P4'_2/mbc'$ (487)	C_{4z}^+
	$P4'_2/m'b'c$ (488)	C_{4z}^+, I
	$P4_2/mb'c'$ (489)	C_{2x}
	$P4'_2/m'bc'$ (490)	C_{4z}^+, I, C_{2x}

Fedorov group	Shubnikov group	'Coloured' generating elements	Fedorov group	Shubnikov group	'Coloured' generating elements
	$P4_2/m'b'c'$ (491)	I		$I4_1/ac'd'$ (567)	C_{2x}
$P4_2/mnm$ (136)	$P4_2/m'nm$ (497)	C_{2x}, I		$I4'_1/a'cd'$ (568)	C^+_{4z}, I, C_{2x}
	$P4'_2/mn'm$ (498)	C^+_{4z}, C_{2x}		$I4_1/a'c'd'$ (569)	I
	$P4'_2/mnm'$ (499)	C^+_{4z}	$P3$ (143)	—	—
	$P4'_2/m'n'm$ (500)	C^+_{4z}, I	$P3_1$ (144)	—	—
	$P4_2/mn'm'$ (501)	C_{2x}	$P3_2$ (145)	—	—
	$P4'_2/m'nm'$ (502)	C^+_{4z}, I, C_{2x}	$R3$ (146)	—	—
	$P4_2/m'n'm'$ (503)	I	$P\bar{3}$ (147)	$P\bar{3}'$ (15)	S^+_6
$P4_2/nmc$ (137)	$P4_2/n'mc$ (509)	C_{2x}, I	$R\bar{3}$ (148)	$R\bar{3}'$ (19)	S^+_6
	$P4'_2/nm'c$ (510)	C^+_{4z}, C_{2x}	$P312$ (149)	$P312'$ (23)	C'_{21}
	$P4'_2/nmc'$ (511)	C^+_{4z}	$P321$ (150)	$P32'1$ (27)	C''_{21}
	$P4'_2/n'm'c$ (512)	C^+_{4z}, I	$P3_112$ (151)	$P3_112'$ (31)	C'_{21}
	$P4_2/nm'c'$ (513)	C_{2x}	$P3_121$ (152)	$P3_12'1$ (35)	C''_{21}
	$P4'_2/n'mc'$ (514)	C^+_{4z}, I, C_{2x}	$P3_212$ (153)	$P3_212'$ (39)	C'_{21}
	$P4_2/n'm'c'$ (515)	I	$P3_221$ (154)	$P3_22'1$ (43)	C''_{21}
$P4_2/ncm$ (138)	$P4_2/n'cm$ (521)	C_{2x}, I	$R32$ (155)	$R32'$ (47)	C'_{21}
	$P4'_2/nc'm$ (522)	C^+_{4z}, C_{2x}	$P3m1$ (156)	$P3m'1$ (51)	σ_{v1}
	$P4'_2/ncm'$ (523)	C^+_{4z}	$P31m$ (157)	$P31m'$ (55)	σ_{d1}
	$P4'_2/n'c'm$ (524)	C^+_{4z}, I	$P3c1$ (158)	$P3c'1$ (59)	σ_{v1}
	$P4_2/nc'm'$ (525)	C_{2x}	$P31c$ (159)	$P31c'$ (63)	σ_{d1}
	$P4'_2/n'cm'$ (526)	C^+_{4z}, I, C_{2x}	$R3m$ (160)	$R3m'$ (67)	σ_{v1}
	$P4_2/n'c'm'$ (527)	I	$R3c$ (161)	$R3c'$ (71)	σ_{v1}
$I4/mmm$ (139)	$I4/m'mm$ (533)	C_{2x}, I	$P\bar{3}1m$ (162)	$P\bar{3}'1m$ (75)	S^+_6
	$I4'/mm'm$ (534)	C^+_{4z}, C_{2x}		$P\bar{3}'1m'$ (76)	S^+_6, σ_{d1}
	$I4'/mmm'$ (535)	C^+_{4z}		$P\bar{3}1m'$ (77)	σ_{d1}
	$I4'/m'm'm$ (536)	C^+_{4z}, I	$P\bar{3}1c$ (163)	$P\bar{3}'1c$ (81)	S^+_6
	$I4/mm'm'$ (537)	C_{2x}		$P\bar{3}'1c'$ (82)	S^+_6, σ_{d1}
	$I4'/m'mm'$ (538)	C^+_{4z}, I, C_{2x}		$P\bar{3}1c'$ (83)	σ_{d1}
	$I4/m'm'm'$ (539)	I	$P\bar{3}m1$ (164)	$P\bar{3}'m1$ (87)	S^+_6
$I4/mcm$ (140)	$I4/m'cm$ (543)	C_{2x}, I		$P\bar{3}'m'1$ (88)	S^+_6, σ_{v1}
	$I4'/mc'm$ (544)	C^+_{4z}, C_{2x}		$P\bar{3}m'1$ (89)	σ_{v1}
	$I4'/mcm'$ (545)	C^+_{4z}	$P\bar{3}c1$ (165)	$P\bar{3}'c1$ (93)	S^+_6
	$I4'/m'c'm$ (546)	C^+_{4z}, I		$P\bar{3}'c'1$ (94)	S^+_6, σ_{v1}
	$I4/mc'm'$ (547)	C_{2x}		$P\bar{3}c'1$ (95)	σ_{v1}
	$I4'/m'cm'$ (548)	C^+_{4z}, I, C_{2x}	$R\bar{3}m$ (166)	$R\bar{3}'m$ (99)	S^+_6
	$I4/m'c'm'$ (549)	I		$R\bar{3}'m'$ (100)	S^+_6, σ_{v1}
$I4_1/amd$ (141)	$I4_1/a'md$ (553)	C_{2x}, I		$R\bar{3}m'$ (101)	σ_{v1}
	$I4'_1/am'd$ (554)	C^+_{4z}, C_{2x}	$R\bar{3}c$ (167)	$R\bar{3}'c$ (105)	S^+_6
	$I4'_1/amd'$ (555)	C^+_{4z}		$R\bar{3}'c'$ (106)	S^+_6, σ_{v1}
	$I4'_1/a'm'd$ (556)	C^+_{4z}, I		$R\bar{3}c'$ (107)	σ_{v1}
	$I4_1/am'd'$ (557)	C_{2x}	$P6$ (168)	$P6'$ (111)	C^+_6
	$I4'_1/a'md'$ (558)	C^+_{4z}, I, C_{2x}	$P6_1$ (169)	$P6'_1$ (115)	C^+_6
	$I4_1/a'm'd'$ (559)	I	$P6_5$ (170)	$P6'_5$ (119)	C^+_6
$I4_1/acd$ (142)	$I4_1/a'cd$ (563)	C_{2x}, I	$P6_2$ (171)	$P6'_2$ (123)	C^+_6
	$I4'_1/ac'd$ (564)	C^+_{4z}, C_{2x}	$P6_4$ (172)	$P6'_4$ (127)	C^+_6
	$I4'_1/acd'$ (565)	C^+_{4z}	$P6_3$ (173)	$P6'_3$ (131)	C^+_6
	$I4'_1/a'c'd$ (566)	C^+_{4z}, I	$P\bar{6}$ (174)	$P\bar{6}'$ (135)	S^+_3

Fedorov group	Shubnikov group	'Coloured' generating elements
$P6/m$ (175)	$P6'/m$ (139)	C_6^+
	$P6/m'$ (140)	σ_h
	$P6'/m'$ (141)	C_6^+, σ_h
$P6_3/m$ (176)	$P6'_3/m$ (145)	C_6^+
	$P6_3/m'$ (146)	σ_h
	$P6'_3/m'$ (147)	C_6^+, σ_h
$P622$ (177)	$P6'2'2$ (151)	C_6^+, C'_{21}
	$P6'22'$ (152)	C_6^+
	$P62'2'$ (153)	C'_{21}
$P6_122$ (178)	$P6'_12'2$ (157)	C_6^+, C'_{21}
	$P6'_122'$ (158)	C_6^+
	$P6_12'2'$ (159)	C'_{21}
$P6_522$ (179)	$P6'_52'2$ (163)	C_6^+, C'_{21}
	$P6'_522'$ (164)	C_6^+
	$P6_52'2'$ (165)	C'_{21}
$P6_222$ (180)	$P6'_22'2$ (169)	C_6^+, C'_{21}
	$P6'_222'$ (170)	C_6^+
	$P6_22'2'$ (171)	C'_{21}
$P6_422$ (181)	$P6'_42'2$ (175)	C_6^+, C'_{21}
	$P6'_422'$ (176)	C_6^+
	$P6_42'2'$ (177)	C'_{21}
$P6_322$ (182)	$P6'_32'2$ (181)	C_6^+, C'_{21}
	$P6'_322'$ (182)	C_6^+
	$P6_32'2'$ (183)	C'_{21}
$P6mm$ (183)	$P6'm'm$ (187)	C_6^+
	$P6'mm'$ (188)	C_6^+, σ_{v1}
	$P6m'm'$ (189)	σ_{v1}
$P6cc$ (184)	$P6'c'c$ (193)	C_6^+
	$P6'cc'$ (194)	C_6^+, σ_{v1}
	$P6c'c'$ (195)	σ_{v1}
$P6_3cm$ (185)	$P6'_3c'm$ (199)	C_6^+
	$P6'_3cm'$ (200)	C_6^+, σ_{v1}
	$P6_3c'm'$ (201)	σ_{v1}
$P6_3mc$ (186)	$P6'_3m'c$ (205)	C_6^+
	$P6'_3mc'$ (206)	C_6^+, σ_{v1}
	$P6_3m'c'$ (207)	σ_{v1}
$P\bar{6}m2$ (187)	$P\bar{6}'m'2$ (211)	S_3^+, σ_{v1}
	$P\bar{6}'m2'$ (212)	S_3^+
	$P\bar{6}m'2'$ (213)	σ_{v1}
$P\bar{6}c2$ (188)	$P\bar{6}'c'2$ (217)	S_3^+, σ_{v1}
	$P\bar{6}'c2'$ (218)	S_3^+
	$P\bar{6}c'2'$ (219)	σ_{v1}
$P\bar{6}2m$ (189)	$P\bar{6}'2'm$ (223)	S_3^+
	$P\bar{6}'2m'$ (224)	S_3^+, σ_{d1}
	$P\bar{6}2'm'$ (225)	σ_{d1}
$P\bar{6}2c$ (190)	$P\bar{6}'2'c$ (229)	S_3^+
	$P\bar{6}'2c'$ (230)	S_3^+, σ_{d1}
	$P\bar{6}2'c'$ (231)	σ_{d1}
$P6/mmm$ (191)	$P6/m'mm$ (235)	C'_{21}, I
	$P6'/mm'm$ (236)	C_6^+, I
	$P6'/mmm'$ (237)	C_6^+, C'_{21}, I
	$P6'/m'm'm$ (238)	C_6^+, C'_{21}
	$P6'/m'mm'$ (239)	C_6^+
	$P6/mm'm'$ (240)	C'_{21}
	$P6/m'm'm'$ (241)	I
$P6/mcc$ (192)	$P6/m'cc$ (245)	C'_{21}, I
	$P6'/mc'c$ (246)	C_6^+, I
	$P6'/mcc'$ (247)	C_6^+, C'_{21}, I
	$P6'/m'c'c$ (248)	C_6^+, C'_{21}
	$P6'/m'cc'$ (249)	C_6^+
	$P6/mc'c'$ (250)	C'_{21}
	$P6/m'c'c'$ (251)	I
$P6_3/mcm$ (193)	$P6_3/m'cm$ (255)	C'_{21}, I
	$P6'_3/mc'm$ (256)	C_6^+, I
	$P6'_3/mcm'$ (257)	C_6^+, C'_{21}, I
	$P6'_3/m'c'm$ (258)	C_6^+, C'_{21}
	$P6'_3/m'cm'$ (259)	C_6^+
	$P6_3/mc'm'$ (260)	C'_{21}
	$P6_3/m'c'm'$ (261)	I
$P6_3/mmc$ (194)	$P6_3/m'mc$ (265)	C'_{21}, I
	$P6'_3/mm'c$ (266)	C_6^+, I
	$P6'_3/mmc'$ (267)	C_6^+, C'_{21}, I
	$P6'_3/m'm'c$ (268)	C_6^+, C'_{21}
	$P6'_3/m'mc'$ (269)	C_6^+
	$P6_3/mm'c'$ (270)	C'_{21}
	$P6_3/m'm'c'$ (271)	I
$P23$ (195)	—	—
$F23$ (196)	—	—
$I23$	—	—
$P2_13$ (198)	—	—
$I2_13$ (199)	—	—
$Pm3$ (200)	$Pm'3$ (16)	I
$Pn3$ (201)	$Pn'3$ (20)	I
$Fm3$ (202)	$Fm'3$ (24)	I
$Fd3$ (203)	$Fd'3$ (28)	I
$Im3$ (204)	$Im'3$ (32)	I
$Pa3$ (205)	$Pa'3$ (35)	I
$Ia3$ (206)	$Ia'3$ (39)	I
$P432$ (207)	$P4'32'$ (42)	C_{2a}
$P4_232$ (208)	$P4'_232'$ (46)	C_{2a}
$F432$ (209)	$F4'32'$ (50)	C_{2a}
$F4_132$ (210)	$F4'_132'$ (54)	C_{2a}
$I432$ (211)	$I4'32'$ (58)	C_{2a}
$P4_332$ (212)	$P4'_332'$ (61)	C_{2a}

Fedorov group	Shubnikov group	'Coloured' generating elements	Fedorov group	Shubnikov group	'Coloured' generating elements
$P4_132$ (213)	$P4_1'32'$ (65)	C_{2a}		$Pn'3m'$ (114)	I
$I4_132$ (214)	$I4_1'32'$ (69)	C_{2a}	$Fm3m$ (225)	$Fm'3m$ (118)	C_{2a}, I
$P\bar{4}3m$ (215)	$P\bar{4}'3m'$ (72)	σ_{da}		$Fm3m'$ (119)	C_{2a}
$F\bar{4}3m$ (216)	$F\bar{4}'3m'$ (76)	σ_{da}		$Fm'3m'$ (120)	I
$I\bar{4}3m$ (217)	$I\bar{4}'3m'$ (80)	σ_{da}	$Fm3c$ (226)	$Fm'3c$ (124)	C_{2a}, I
$P\bar{4}3n$ (218)	$P\bar{4}'3n'$ (83)	σ_{da}		$Fm3c'$ (125)	C_{2a}
$F\bar{4}3c$ (219)	$F\bar{4}'3c'$ (87)	σ_{da}		$Fm'3c'$ (126)	I
$I\bar{4}3d$ (220)	$I\bar{4}'3d'$ (91)	σ_{da}	$Fd3m$ (227)	$Fd'3m$ (130)	C_{2a}, I
$Pm3m$ (221)	$Pm'3m$ (94)	C_{2a}, I		$Fd3m'$ (131)	C_{2a}
	$Pm3m'$ (95)	C_{2a}		$Fd'3m'$ (132)	I
	$Pm'3m'$ (96)	I	$Fd3c$ (228)	$Fd'3c$ (136)	C_{2a}, I
$Pn3n$ (222)	$Pn'3n$ (100)	C_{2a}, I		$Fd3c'$ (137)	C_{2a}
	$Pn3n'$ (101)	C_{2a}		$Fd'3c'$ (138)	I
	$Pn'3n'$ (102)	I	$Im3m$ (229)	$Im'3m$ (142)	C_{2a}, I
$Pm3n$ (223)	$Pm'3n$ (106)	C_{2a}, I		$Im3m'$ (143)	C_{2a}
	$Pm3n'$ (107)	C_{2a}		$Im'3m'$ (144)	I
	$Pm'3n'$ (108)	I	$Ia3d$ (230)	$Ia'3d$ (147)	C_{2a}, I
$Pn3m$ (224)	$Pn'3m$ (112)	C_{2a}, I		$Ia3d'$ (148)	C_{2a}
	$Pn3m'$ (113)	C_{2a}		$Ia'3d'$ (149)	I

Notes to Table 7.2

(i) The groups are listed in the notation given by Belov, Neronova, and Smirnova (1955) as corrected by Shubnikov and Belov (1964), and are arranged according to their crystal system and, within a given crystal system, according to their isogonal point group (see also Table 7.4).

(ii) The international space group symbols of the Fedorov groups and their numbers, as given by Henry and Lonsdale (1965), are listed in column 1; the type III Shubnikov groups derived from each Fedorov group are listed in column 2, together with a somewhat arbitrary number assigned to the group by Belov, Neronova, and Smirnova (1955). Column 3 identifies those of the generating elements of the Fedorov group (see Table 3.7 and Miller and Love (1967)) which have been replaced by $\mathscr{R}$ times the element of the Fedorov group, i.e. which have become *coloured* elements; i.e. those of the generating elements which occur in $\mathscr{R}(\mathbf{G} - \mathbf{H})$ of eqn (7.2.4). A shorthand notation has been adopted in column 3; only the rotational part of the full Seitz space-group operator is given, the complete operator can be identified by reference to Table 3.7.

(iii) The invariant spin structures associated with the various magnetic space groups are given by Opechowski and Guccione (1965).

Type IV Shubnikov space groups are based on black and white Bravais lattices and just consist of all the operations of a Fedorov group **G** plus an equal number of *coloured* operations, each one being an operation in **G**, multiplied by the operation of anti-symmetry and also multiplied by an operation corresponding to a translation $\mathbf{t}_0$ of the black and white Bravais lattice. The derivation of a black and white Bravais lattice is illustrated in Fig. 7.3. If we start with an ordinary primitive cubic Bravais lattice with black lattice points and add a white lattice point to the body-centre of each unit cell we obtain the black and white cubic Bravais lattice P_I. This consists of two interpenetrating sublattices, each of which on its own would be an ordinary primitive cubic Bravais lattice, P. The black and white lattice P_I therefore has all the symmetry operations of an ordinary primitive cubic Bravais lattice, together with the product of each of those operations with $\mathscr{R}\{E \mid \mathbf{t}_0\}$ where $\mathbf{t}_0$ is the vector consisting of half the body-diagonal of the cube. The black and white Bravais lattices

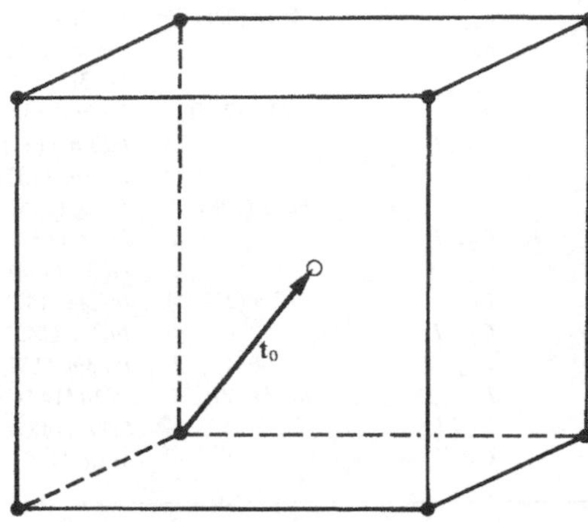

FIG. 7.3. The unit cell of the black and white (cubic) Bravais lattice, P_I.

were derived by Belov, Neronova, and Smirnova (1955), who found there to be 22 black and white lattices in addition to the 14 ordinary Bravais lattices. These black and white Bravais lattices are given in Table 7.3 together with the vector $\mathbf{t}_0$, and they are illustrated in Fig. 7.4.

DEFINITION 7.2.5. A type IV Shubnikov space group, $\mathbf{M}$, is given by

$$\mathbf{M} = \mathbf{G} + \mathscr{R}\{E \mid \mathbf{t}_0\}\mathbf{G}, \tag{7.2.5}$$

where $\mathbf{G}$ is a Fedorov group and $\mathbf{t}_0$ is the extra translation introduced when deriving the black and white Bravais lattice from the appropriate ordinary Bravais lattice.

The type IV Shubnikov space groups can be derived fairly readily from the Fedorov space groups. Each Fedorov space group, $\mathbf{G}$, is based on one of the 14 Bravais lattices, and if from this lattice there can be derived some number, x, of black and white Bravais lattices then there will be at least x type IV Shubnikov groups corresponding to the Fedorov group $\mathbf{G}$. There may be more than x type IV Shubnikov groups derived from $\mathbf{G}$ if we allow for the possibility of different orientations which can occur in the orthorhombic system. The ordinary Bravais-lattice symbol in the international space-group symbol is replaced by the symbol of the relevant black and

TABLE 7.3

The black and white Bravais lattices

Crystal system	Ordinary lattice	Black and white lattice	$\mathbf{t}_0$
Triclinic	$P(\Gamma_t)$	P_s	$\frac{1}{2}\mathbf{t}_1$
Monoclinic	$P(\Gamma_m)$	P_b	$\frac{1}{2}\mathbf{t}_3$
		P_a	$\frac{1}{2}\mathbf{t}_2$
		P_C	$\frac{1}{2}(\mathbf{t}_2 + \mathbf{t}_3)$
Monoclinic	$C(\Gamma_m^b)$	C_c	$\frac{1}{2}\mathbf{t}_1$
		C_a	$\frac{1}{2}(\mathbf{t}_2 + \mathbf{t}_3)$
Orthorhombic	$P(\Gamma_o)$	P_c	$\frac{1}{2}\mathbf{t}_3$
		$[P_a$	$\frac{1}{2}\mathbf{t}_2]$
		P_C	$\frac{1}{2}(\mathbf{t}_1 + \mathbf{t}_2)$
		$[P_A$	$\frac{1}{2}(\mathbf{t}_1 + \mathbf{t}_3)]$
		P_I	$\frac{1}{2}(\mathbf{t}_1 + \mathbf{t}_2 + \mathbf{t}_3)$
Orthorhombic	$C(\Gamma_o^b)$	C_c	$\frac{1}{2}\mathbf{t}_3$
		C_a	$\frac{1}{2}(\mathbf{t}_1 + \mathbf{t}_2)$
		C_A	$\frac{1}{2}(\mathbf{t}_1 + \mathbf{t}_2 + \mathbf{t}_3)$
	$[A(\Gamma_o^b)]$	$[A_a$	$\frac{1}{2}\mathbf{t}_2]$
		$[A_c$	$\frac{1}{2}(\mathbf{t}_1 + \mathbf{t}_3)]$
		$[A_C$	$\frac{1}{2}(\mathbf{t}_1 + \mathbf{t}_2 + \mathbf{t}_3)]$
Orthorhombic	$F(\Gamma_o^f)$	F_s	$\frac{1}{2}(\mathbf{t}_1 + \mathbf{t}_2 + \mathbf{t}_3)$
Orthorhombic	$I(\Gamma_o^v)$	I_c	$\frac{1}{2}(\mathbf{t}_1 + \mathbf{t}_2)$
Tetragonal	$P(\Gamma_q)$	P_c	$\frac{1}{2}\mathbf{t}_3$
		P_C	$\frac{1}{2}(\mathbf{t}_1 + \mathbf{t}_2)$
		P_I	$\frac{1}{2}(\mathbf{t}_1 + \mathbf{t}_2 + \mathbf{t}_3)$
Tetragonal	$I(\Gamma_q^v)$	I_c	$\frac{1}{2}(\mathbf{t}_1 + \mathbf{t}_2)$
Trigonal	$R(\Gamma_{rh})$	R_I	$\frac{1}{2}(\mathbf{t}_1 + \mathbf{t}_2 + \mathbf{t}_3)$
Hexagonal	$P(\Gamma_h)$	P_c	$\frac{1}{2}\mathbf{t}_3$
Cubic	$P(\Gamma_c)$	P_I	$\frac{1}{2}(\mathbf{t}_1 + \mathbf{t}_2 + \mathbf{t}_3)$
Cubic	$F(\Gamma_c^f)$	F_s	$\frac{1}{2}(\mathbf{t}_1 + \mathbf{t}_2 + \mathbf{t}_3)$
Cubic	$I(\Gamma_c^v)$	—	—

Notes to Table 7.3

(i) To label the black and white Bravais lattices we use the International notation adapted to the Shubnikov groups.

(ii) Column 1 lists the crystal system, column 2 the ordinary (uncoloured) Bravais lattice from which the black and white one is derived, column 3 gives the symbol of the black and white Bravais lattice and column 4 gives $\mathbf{t}_0$, in terms of the relevant $\mathbf{t}_1$, $\mathbf{t}_2$, and $\mathbf{t}_3$ of Table 3.1, except for A_a, A_c, and A_C for which $\mathbf{t}_0$ is defined in terms of the translations for $A(\Gamma_o^b)$ given in Note (iv) below.

(iii) No restrictions are understood to be imposed on the axial ratios.

(iv) In making a total count of the different Bravais lattices it must be remembered that identifications can be made between the pair (C, A) of orthorhombic ordinary Bravais lattices and between the pairs (P_c, P_a), (P_C, P_A), (C_c, A_a), (C_a, A_c) and (C_A, A_C) of orthorhombic black and white Bravais lattices, see Fig. 7.4. One member of each of these pairs is enclosed in square brackets in Table 7.3. In Table 3.1 we listed the translations suitable for $C(\Gamma_o^b)$. Those for $A(\Gamma_o^b)$ are $\mathbf{t}_1 = \frac{1}{2}(0, -b, c)$, $\mathbf{t}_2 = (a, 0, 0)$, $\mathbf{t}_3 = \frac{1}{2}(0, b, c)$ referred to axes $Oxyz$.

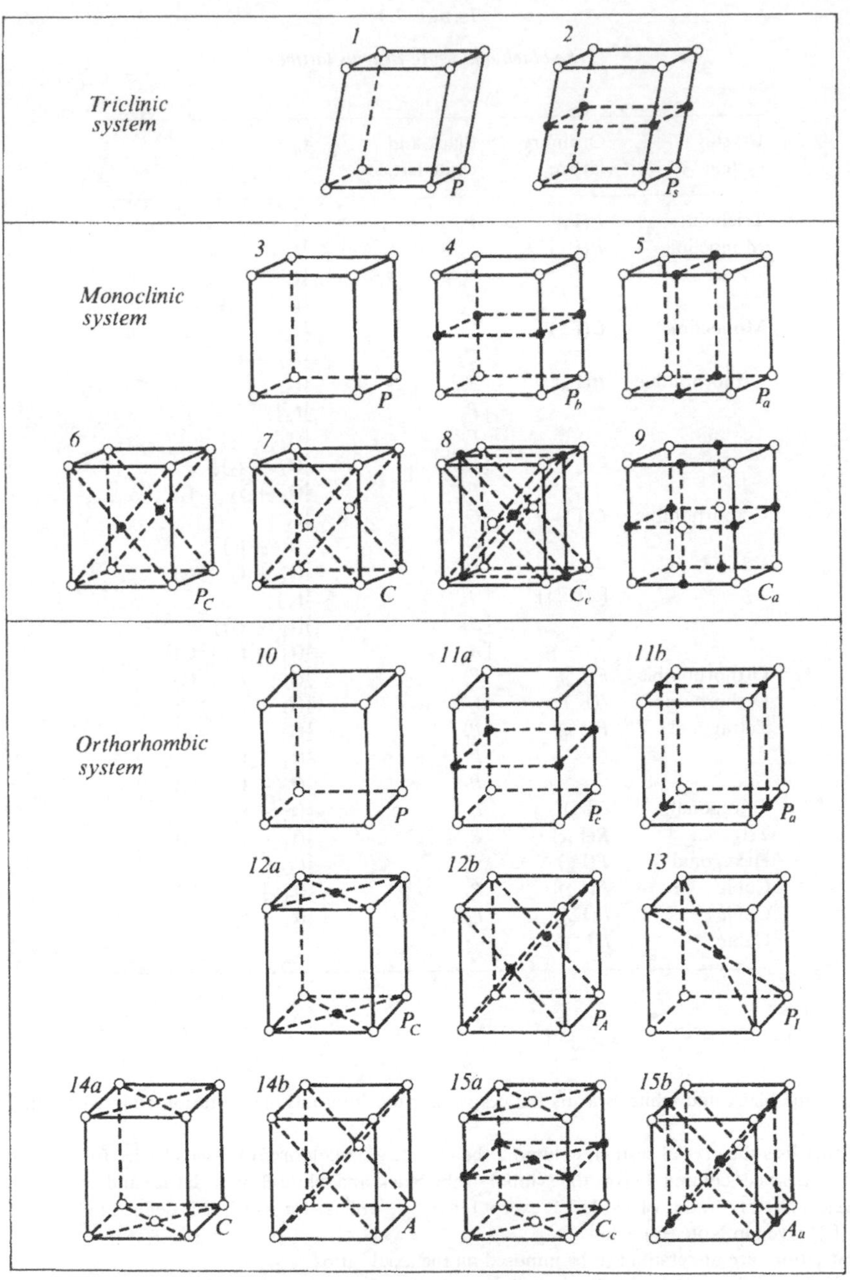

FIG. 7.4. The black and white Bravais lattices.

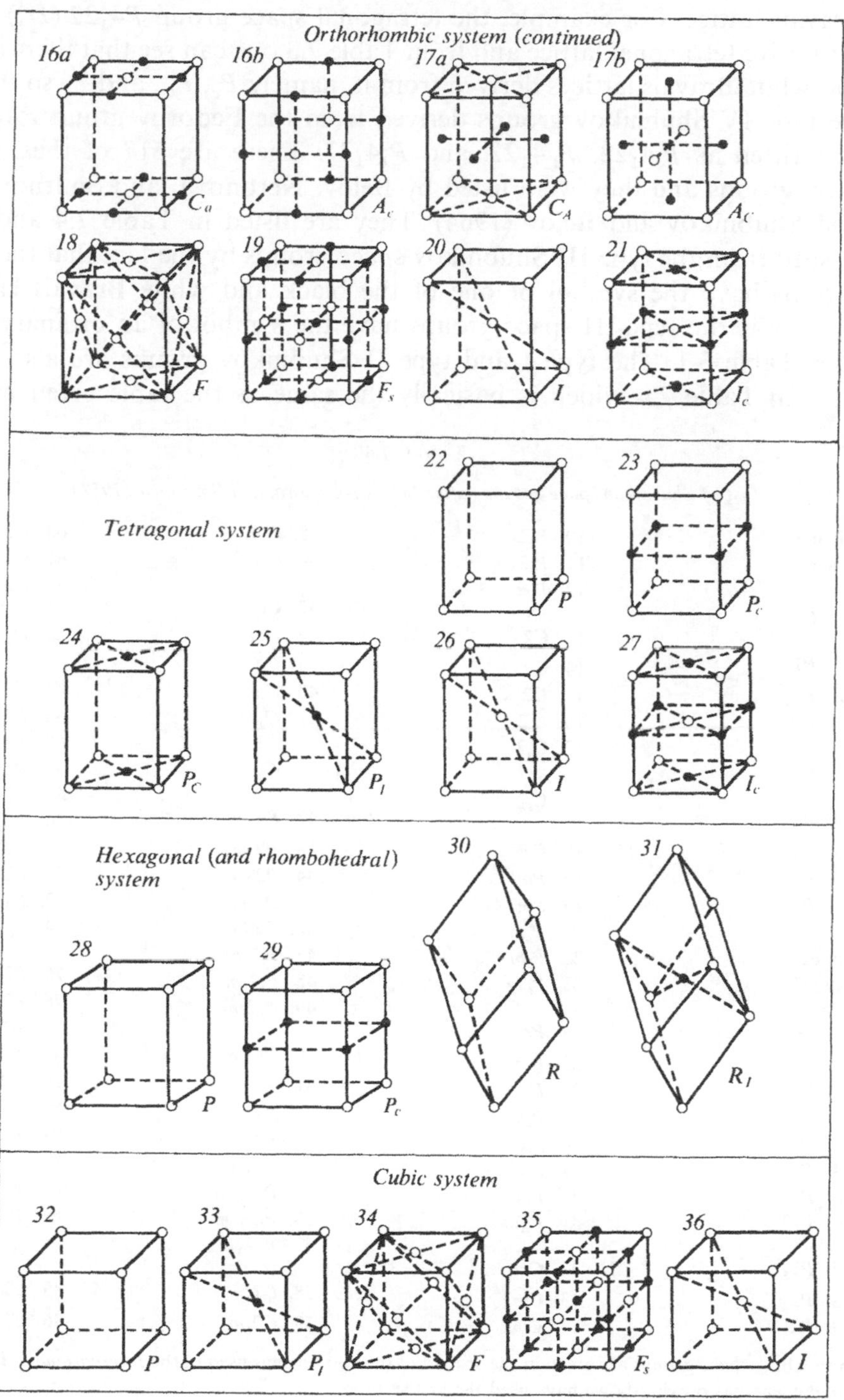
Orthorhombic system (continued)
16a
C_a
16b
A_c
17a
C_A
17b
A_C
18
F
19
F_s
20
I
21
I_c
Tetragonal system
22
P
23
P_c
24
P_C
25
P_I
26
I
27
I_c
Hexagonal (and rhombohedral) system
28
P
29
P_c
30
R
31
R_I
Cubic system
32
P
33
P_I
34
F
35
F_s
36
I

white Bravais lattice. For example, the tetragonal space group $P4_122$ (D_4^3) is based on the primitive tetragonal lattice and from Table 7.3 one can see that there are three black and white Bravais lattices derived from it, namely P_c, P_C, and P_I so that there are three type IV Shubnikov groups derived from the Fedorov group $P4_122$, and they are written as P_c4_122, P_C4_122, and P_I4_122. There are 517 of these type IV Shubnikov groups and they were listed by Belov, Neronova, and Smirnova (1955, 1957) and Shubnikov and Belov (1964). They are listed in Table 7.4 and can be distinguished from the type III Shubnikov space groups by the fact that the type IV space groups have the symbol of one of the black and white Bravais lattices in Table 7.3, wh'le the type III space groups have the symbol of an ordinary Bravais lattice, see Table 3.1. The type I and type II Shubnikov groups are also given in Table 7.4. In Table 7.4 which is basically the same as the table given by Belov,

TABLE 7.4

List of Shubnikov space groups (after Belov, Neronova, and Smirnova (1957))

Triclinic system

C_1

(1) 1. $\mathbf{P1}$
2. $P1'$
3. P_s1

C_i

(2) 4. $\mathbf{P\bar{1}}$
5. $P\bar{1}1'$
6. $P\bar{1}'$
7. $P_s\bar{1}$

Monoclinic system‡

C_2

(3) 1. $\mathbf{P2}$
2. $P21'$
3. $P2'$
4. P_a2
5. P_b2
6. P_C2

(4) 7. $\mathbf{P2_1}$
8. $P2_11'$
9. $P2'_1$
10. P_a2_1
11. P_b2_1
12. P_C2_1

(5) 13. $\mathbf{C2}$
14. $C21'$
15. $C2'$
16. C_c2
17. C_a2

C_{1h}

(6) 18. $\mathbf{Pm}$
19. $Pm1'$
20. Pm'
21. P_am
22. P_bm
23. P_Cm

(7) 24. $\mathbf{Pc}$
25. $Pc1'$
26. Pc'
27. P_ac
28. P_cc
29. P_bc
30. P_Cc
31. P_Ac

(8) 32. $\mathbf{Cm}$
33. $Cm1'$
34. Cm'
35. C_cm
36. C_am

(9) 37. $\mathbf{Cc}$
38. $Cc1'$
39. Cc'
40. C_cc
41. C_ac

C_{2h}

(10) 42. $\mathbf{P2/m}$
43. $P2/m1'$
44. $P2'/m$
45. $P2/m'$
46. $P2'/m'$
47. P_a2/m
48. P_b2/m
49. P_C2/m

(11) 50. $\mathbf{P2_1/m}$
51. $P2_1/m1'$
52. $P2'_1/m$
53. $P2_1/m'$
54. $P2'_1/m'$
55. P_a2_1/m
56. P_b2_1/m
57. P_C2_1/m

(12) 58. $\mathbf{C2/m}$
59. $C2/m1'$
60. $C2'/m$
61. $C2/m'$
62. $C2'/m'$
63. C_c2/m
64. C_a2/m

(13) 65. $\mathbf{P2/c}$
66. $P2/c1'$
67. $P2'/c$
68. $P2/c'$
69. $P2'/c'$
70. P_a2/c
71. P_b2/c
72. P_c2/c
73. P_A2/c
74. P_C2/c

(14) 75. $\mathbf{P2_1/c}$
76. $P2_1/c1'$
77. $P2'_1/c$
78. $P2_1/c'$
79. $P2'_1/c'$
80. P_a2_1/c
81. P_b2_1/c
82. P_c2_1/c
83. P_A2_1/c
84. P_C2_1/c

(15) 85. $\mathbf{C2/c}$
86. $C2/c1'$

‡ The monoclinic space groups are given in the second of the two settings used in the *International tables for X-ray crystallography* whereas previously we have used the first setting.

87. $C2'/c$
88. $C2/c'$
89. $C2'/c'$
90. C_c2/c
91. C_a2/c

Orthorhombic system

D_2

(16) 1. **$P222$**
2. $P2221'$
3. $P2'2'2$
4. P_a222
5. P_C222
6. P_I222

(17) 7. **$P222_1$**
8. $P222_11'$
9. $P2'2'2_1$
10. $P22'2'_1$
11. P_a222_1
12. P_c222_1
13. P_A222_1
14. P_C222_1
15. P_I222_1

(18) 16. **$P2_12_12$**
17. $P2_12_121'$
18. $P2'_12'_12$
19. $P2_12'_12'$
20. $P_a2_12_12$
21. $P_c2_12_12$
22. $P_A2_12_12$
23. $P_C2_12_12$
24. $P_I2_12_12$

(19) 25. **$P2_12_12_1$**
26. $P2_12_12_11'$
27. $P2'_12'_12_1$
28. $P_a2_12_12_1$
29. $P_C2_12_12_1$
30. $P_I2_12_12_1$

(20) 31. **$C222_1$**
32. $C222_11'$
33. $C2'2'2_1$
34. $C22'2'_1$
35. C_c222_1
36. C_a222_1
37. C_A222_1

(21) 38. **$C222$**
39. $C2221'$
40. $C2'2'2$
41. $C22'2'$
42. C_c222
43. C_a222
44. C_A222

(22) 45. **$F222$**
46. $F2221'$
47. $F2'2'2$
48. F_s222

(23) 49. **$I222$**
50. $I2221'$
51. $I2'2'2$
52. I_c222

(24) 53. **$I2_12_12_1$**
54. $I2_12_12_11'$
55. $I2'_12'_12_1$
56. $I_c2_12_12_1$

C_{2v}

(25) 57. **$Pmm2$**
58. $Pmm21'$
59. $Pm'm2'$
60. $Pm'm'2$
61. P_cmm2
62. P_amm2
63. P_Cmm2
64. P_Amm2
65. P_Imm2

(26) 66. **$Pmc2_1$**
67. $Pmc2_11'$
68. $Pm'c2'_1$
69. $Pmc'2'_1$
70. $Pm'c'2_1$
71. P_amc2_1
72. P_bmc2_1
73. P_cmc2_1
74. P_Amc2_1
75. P_Bmc2_1
76. P_Cmc2_1
77. P_Imc2_1

(27) 78. **$Pcc2$**
79. $Pcc21'$
80. $Pc'c2'$
81. $Pc'c'2$
82. P_ccc2
83. P_acc2
84. P_Ccc2
85. P_Acc2
86. P_Icc2

(28) 87. **$Pma2$**
88. $Pma21'$
89. $Pm'a2'$
90. $Pma'2'$
91. $Pm'a'2$
92. P_ama2
93. P_bma2
94. P_cma2
95. P_Ama2
96. P_Bma2
97. P_Cma2
98. P_Ima2

(29) 99. **$Pca2_1$**
100. $Pca2_11'$
101. $Pc'a2'_1$
102. $Pca'2'_1$
103. $Pc'a'2_1$
104. P_aca2_1
105. P_bca2_1
106. P_cca2_1
107. P_Aca2_1
108. P_Bca2_1
109. P_Cca2_1
110. P_Ica2_1

(30) 111. **$Pnc2$**
112. $Pnc21'$
113. $Pn'c2'$
114. $Pnc'2'$
115. $Pn'c'2$
116. P_anc2
117. P_bnc2
118. P_cnc2
119. P_Anc2
120. P_Bnc2
121. P_Cnc2
122. P_Inc2

(31) 123. **$Pmn2_1$**
124. $Pmn2_11'$
125. $Pm'n2'_1$
126. $Pmn'2'_1$
127. $Pm'n'2_1$
128. P_amn2_1
129. P_bmn2_1
130. P_cmn2_1
131. P_Amn2_1
132. P_Bmn2_1
133. P_Cmn2_1
134. P_Imn2_1

(32) 135. **$Pba2$**
136. $Pba21'$
137. $Pb'a2'$
138. $Pb'a'2$
139. P_cba2
140. P_aba2
141. P_Cba2
142. P_Aba2
143. P_Iba2

(33) 144. **$Pna2_1$**
145. $Pna2_11'$
146. $Pn'a2'_1$
147. $Pna'2'_1$
148. $Pn'a'2_1$
149. P_ana2_1
150. P_bna2_1
151. P_cna2_1
152. P_Ana2_1
153. P_Bna2_1
154. P_Cna2_1
155. P_Cna2_1

(34) 156. **$Pnn2$**
157. $Pnn21'$
158. $Pn'n2'$
159. $Pn'n'2$
160. P_ann2
161. P_cnn2
162. P_Ann2
163. P_Cnn2
164. P_Inn2

(35) 165. **$Cmm2$**
166. $Cmm21'$
167. $Cm'm2'$
168. $Cm'm'2$
169. C_cmm2
170. C_amm2
171. C_Amm2

(36) 172. **$Cmc2_1$**
173. $Cmc2_11'$
174. $Cm'c2'_1$
175. $Cmc'2'_1$
176. $Cm'c'2_1$
177. C_cmc2_1
178. C_amc2_1
179. C_Amc2_1

(37) 180. ***Ccc2***
181. $Ccc21'$
182. $Cc'c2'$
183. $Cc'c'2$
184. C_ccc2
185. C_acc2
186. C_Acc2

(38) 187. ***Amm2***
188. $Amm21'$
189. $Am'm2'$
190. $Amm'2'$
191. $Am'm'2$
192. A_amm2
193. A_cmm2
194. A_Cmm2

(39) 195. ***Abm2***
196. $Abm21'$
197. $Ab'm2'$
198. $Abm'2'$
199. $Ab'm'2$
200. A_abm2
201. A_cbm2
202. A_Cbm2

(40) 203. ***Ama2***
204. $Ama21'$
205. $Am'a2'$
206. $Ama'2'$
207. $Am'a'2$
208. A_ama2
209. A_cma2
210. A_Cma2

(41) 211. ***Aba2***
212. $Aba21'$
213. $Ab'a2'$
214. $Aba'2'$
215. $Ab'a'2$
216. A_aba2
217. A_cba2
218. A_Cba2

(42) 219. ***Fmm2***
220. $Fmm21'$
221. $Fm'm2'$
222. $Fm'm'2$
223. F_smm2

(43) 224. ***Fdd2***
225. $Fdd21'$
226. $Fd'd2'$
227. $Fd'd'2$
228. F_sdd2

(44) 229. ***Imm2***
230. $Imm21'$
231. $Im'm2'$
232. $Im'm'2$
233. I_cmm2
234. I_amm2

(45) 235. ***Iba2***
236. $Iba21'$
237. $Ib'a2'$
238. $Ib'a'2$
239. I_cba2
240. I_aba2

(46) 241. ***Ima2***
242. $Ima21'$
243. $Im'a2'$
244. $Ima'2'$
245. $Im'a'2$
246. I_cma2
247. I_ama2
248. I_bma2

$\boldsymbol{D_{2h}}$

(47) 249. ***Pmmm***
250. $Pmmm1'$
251. $Pm'mm$
252. $Pm'm'm$
253. $Pm'm'm'$
254. P_ammm
255. P_Cmmm
256. P_Immm

(48) 257. ***Pnnn***
258. $Pnnn1'$
259. $Pn'nn$
260. $Pn'n'n$
261. $Pn'n'n'$
262. P_annn
263. P_Cnnn
264. P_Innn

(49) 265. ***Pccm***
266. $Pccm1'$
267. $Pc'cm$
268. $Pccm'$
269. $Pc'c'm$
270. $Pc'cm'$
271. $Pc'c'm'$
272. P_accm
273. P_cccm
274. P_Accm
275. P_Cccm
276. P_Iccm

(50) 277. ***Pban***
278. $Pban1'$
279. $Pb'an$
280. $Pban'$
281. $Pb'a'n$
282. $Pb'an'$
283. $Pb'a'n'$
284. P_aban
285. P_cban
286. P_Aban
287. P_Cban
288. P_Iban

(51) 289. ***Pmma***
290. $Pmma1'$
291. $Pm'ma$
292. $Pmm'a$
293. $Pmma'$
294. $Pm'm'a$
295. $Pmm'a'$
296. $Pm'ma'$
297. $Pm'm'a'$
298. P_amma
299. P_bmma
300. P_cmma
301. P_Amma
302. P_Bmma
303. P_Cmma
304. P_Imma

(52) 305. ***Pnna***
306. $Pnna1'$
307. $Pn'na$
308. $Pnn'a$
309. $Pnna'$
310. $Pn'n'a$
311. $Pnn'a'$
312. $Pn'na'$
313. $Pn'n'a'$
314. P_anna
315. P_bnna
316. P_cnna
317. P_Anna
318. P_Bnna
319. P_Cnna
320. P_Inna

(53) 321. ***Pmna***
322. $Pmna1'$
323. $Pm'na$
324. $Pmn'a$
325. $Pmna'$
326. $Pm'n'a$
327. $Pmn'a'$
328. $Pm'na'$
329. $Pm'n'a'$
330. P_amna
331. P_bmna
332. P_cmna
333. P_Amna
334. P_Bmna
335. P_Cmna
336. P_Imna

(54) 337. ***Pcca***
338. $Pcca1'$
339. $Pc'ca$
340. $Pcc'a$
341. $Pcca'$
342. $Pc'c'a$
343. $Pcc'a'$
344. $Pc'ca'$
345. $Pc'c'a'$
346. P_acca
347. P_bcca
348. P_ccca
349. P_Acca
350. P_Bcca
351. P_Ccca
352. P_Icca

(55) 353. ***Pbam***
354. $Pbam1'$
355. $Pb'am$
356. $Pbam'$
357. $Pb'a'm$
358. $Pb'am'$
359. $Pb'a'm'$
360. P_abam
361. P_cbam
362. P_Abam
363. P_Cbam
364. P_Ibam

(56) 365. ***Pccn***
366. $Pccn1'$
367. $Pc'cn$
368. $Pccn'$
369. $Pc'c'n$
370. $Pc'cn'$

371. *Pc'c'n'*
372. $P_{a}ccn$
373. $P_{c}ccn$
374. $P_{A}ccn$
375. $P_{C}ccn$
376. $P_{I}ccn$

(57) 377. ***Pbcm***
378. *Pbcm1'*
379. *Pb'cm*
380. *Pbc'm*
381. *Pbcm'*
382. *Pb'c'm*
383. *Pbc'm'*
384. *Pb'cm'*
385. *Pb'c'm'*
386. $P_{a}bcm$
387. $P_{b}bcm$
388. $P_{c}bcm$
389. $P_{A}bcm$
390. $P_{B}bcm$
391. $P_{C}bcm$
392. $P_{I}bcm$

(58) 393. ***Pnnm***
394. *Pnnm1'*
395. *Pn'nm*
396. *Pnnm'*
397. *Pn'n'm*
398. *Pnn'm'*
399. *Pn'n'm'*
400. $P_{a}nnm$
401. $P_{c}nnm$
402. $P_{A}nnm$
403. $P_{C}nnm$
404. $P_{I}nnm$

(59) 405. ***Pmmn***
406. *Pmmn1'*
407. *Pm'mn*
408. *Pmmn'*
409. *Pm'm'n*
410. *Pmm'n'*
411. *Pm'm'n'*
412. $P_{a}mmn$
413. $P_{c}mmn$
414. $P_{A}mmn$
415. $P_{C}mmn$
416. $P_{I}mmn$

(60) 417. ***Pbcn***
418. *Pbcn1'*
419. *Pb'cn*
420. *Pbc'n*
421. *Pbcn'*
422. *Pb'c'n*
423. *Pbc'n'*
424. *Pb'cn'*
425. *Pb'c'n'*
426. $P_{a}bcn$
427. $P_{b}bcn$
428. $P_{c}bcn$
429. $P_{A}bcn$
430. $P_{B}bcn$
431. $P_{C}bcn$
432. $P_{I}bcn$

(61) 433. ***Pbca***
434. *Pbca1'*
435. *Pb'ca*
436. *Pb'c'a*
437. *Pb'c'a'*
438. $P_{a}bca$
439. $P_{C}bca$
440. $P_{I}bca$

(62) 441. ***Pnma***
442. *Pnma1'*
443. *Pn'ma*
444. *Pnm'a*
445. *Pnma'*
446. *Pn'm'a*
447. *Pnm'a'*
448. *Pn'ma'*
449. *Pn'm'a'*
450. $P_{a}nma$
451. $P_{b}nma$
452. $P_{c}nma$
453. $P_{A}nma$
454. $P_{B}nma$
455. $P_{C}nma$
456. $P_{I}nma$

(63) 457. ***Cmcm***
458. *Cmcm1'*
459. *Cm'cm*
460. *Cmc'm*
461. *Cmcm'*
462. *Cm'c'm*
463. *Cmc'm'*
464. *Cm'cm'*
465. *Cm'c'm'*
466. $C_{c}mcm$
467. $C_{a}mcm$
468. $C_{A}mcm$

(64) 469. ***Cmca***
470. *Cmca1'*
471. *Cm'ca*
472. *Cmc'a*
473. *Cmca'*
474. *Cm'c'a*
475. *Cmc'a'*
476. *Cm'ca'*
477. *Cm'c'a'*
478. $C_{c}mca$
479. $C_{a}mca$
480. $C_{A}mca$

(65) 481. ***Cmmm***
482. *Cmmm1'*
483. *Cm'mm*
484. *Cmmm'*
485. *Cm'm'm*
486. *Cmm'm'*
487. *Cm'm'm'*
488. $C_{c}mmm$
489. $C_{a}mmm$
490. $C_{A}mmm$

(66) 491. ***Cccm***
492. *Cccm1'*
493. *Cc'cm*
494. *Cccm'*
495. *Cc'c'm*
496. *Ccc'm'*
497. *Cc'c'm'*
498. $C_{c}ccm$
499. $C_{a}ccm$
500. $C_{A}ccm$

(67) 501. ***Cmma***
502. *Cmma1'*
503. *Cm'ma*
504. *Cmma'*
505. *Cm'm'a*
506. *Cmm'a'*
507. *Cm'm'a'*
508. $C_{c}mma$
509. $C_{a}mma$
510. $C_{A}mma$

(68) 511. ***Ccca***
512. *Ccca1'*
513. *Cc'ca*
514. *Ccca'*
515. *Cc'c'a*
516. *Ccc'a'*
517. *Cc'c'a'*
518. $C_{c}cca$
519. $C_{a}cca$
520. $C_{A}cca$

(69) 521. ***Fmmm***
522. *Fmmm1'*
523. *Fm'mm*
524. *Fm'm'm*
525. *Fm'm'm'*
526. $F_{s}mmm$

(70) 527. ***Fddd***
528. *Fddd1'*
529. *Fd'dd*
530. *Fd'd'd*
531. *Fd'd'd'*
532. $F_{s}ddd$

(71) 533. ***Immm***
534. *Immm1'*
535. *Im'mm*
536. *Im'm'm*
537. *Im'm'm'*
538. $I_{c}mmm$

(72) 539. ***Ibam***
540. *Ibam1'*
541. *Ib'am*
542. *Ibam'*
543. *Ib'a'm*
544. *Iba'm'*
545. *Ib'a'm'*
546. $I_{c}bam$
547. $I_{a}bam$

(73) 548. ***Ibca***
549. *Ibca1'*
550. *Ib'ca*
551. *Ib'c'a*
552. *Ib'c'a'*
553. $I_{c}bca$

(74) 554. ***Imma***
555. *Imma1'*
556. *Im'ma*
557. *Imma'*
558. *Im'm'a*
559. *Imm'a'*
560. *Im'm'a'*
561. $I_{c}mma$
562. $I_{a}mma$

Tetragonal system

C_4

(75) 1. $\mathbf{P4}$
2. $P41'$
3. $P4'$
4. P_c4
5. P_C4
6. P_I4

(76) 7. $\mathbf{P4_1}$
8. $P4_11'$
9. $P4'_1$
10. P_c4_1
11. P_C4_1
12. P_I4_1

(77) 13. $\mathbf{P4_2}$
14. $P4_21'$
15. $P4'_2$
16. P_c4_2
17. P_C4_2
18. P_I4_2

(78) 19. $\mathbf{P4_3}$
20. $P4_31'$
21. $P4'_3$
22. P_c4_3
23. P_C4_3
24. P_I4_3

(79) 25. $\mathbf{I4}$
26. $I41'$
27. $I4'$
28. I_c4

(80) 29. $\mathbf{I4_1}$
30. $I4_11'$
31. $I4'_1$
32. I_c4_1

S_4

(81) 33. $\mathbf{P\bar{4}}$
34. $P\bar{4}1'$
35. $P\bar{4}'$
36. $P_c\bar{4}$
37. $P_C\bar{4}$
38. $P_I\bar{4}$

(82) 39. $\mathbf{I\bar{4}}$
40. $I\bar{4}1'$
41. $I\bar{4}'$
42. $I_c\bar{4}$

C_{4h}

(83) 43. $\mathbf{P4/m}$
44. $P4/m1'$
45. $P4'/m$
46. $P4/m'$
47. $P4'/m'$
48. P_c4/m
49. P_C4/m
50. P_I4/m

(84) 51. $\mathbf{P4_2/m}$
52. $P4_2/m1'$
53. $P4'_2/m$
54. $P4_2/m'$
55. $P4'_2/m'$
56. P_c4_2/m
57. P_C4_2/m
58. P_I4_2/m

(85) 59. $\mathbf{P4/n}$
60. $P4/n1'$
61. $P4'/n$
62. $P4/n'$
63. $P4'/n'$
64. P_c4/n
65. P_C4/n
66. P_I4/n

(86) 67. $\mathbf{P4_2/n}$
68. $P4_2/n1'$
69. $P4'_2/n$
70. $P4_2/n'$
71. $P4'_2/n'$
72. P_c4_2/n
73. P_C4_2/n
74. P_I4_2/n

(87) 75. $\mathbf{I4/m}$
76. $I4/m1'$
77. $I4'/m$
78. $I4/m'$
79. $I4'/m'$
80. I_c4/m

(88) 81. $\mathbf{I4_1/a}$
82. $I4_1/a1'$
83. $I4'_1/a$
84. $I4_1/a'$
85. $I4'_1/a'$
86. I_c4_1/a

D_4

(89) 87. $\mathbf{P422}$
88. $P4221'$
89. $P4'22'$
90. $P42'2'$
91. $P4'2'2$
92. P_c422
93. P_C422
94. P_I422

(90) 95. $\mathbf{P42_12'}$
96. $P42_121'$
97. $P4'2_12'$
98. $P42'_12'$
99. $P4'2'_12$
100. P_c42_12
101. P_C42_12
102. P_I42_12

(91) 103. $\mathbf{P4_122}$
104. $P4_1221'$
105. $P4'_122'$
106. $P4_12'2'$
107. $P4'_12'2$
108. P_c4_122
109. P_C4_122
110. P_I4_122

(92) 111. $\mathbf{P4_12_12}$
112. $P4_12_121'$
113. $P4'_12_12'$
114. $P4_12'_12'$
115. $P4'_12'_12$
116. $P_c4_12_12$
117. $P_C4_12_12$
118. $P_I4_12_12$

(93) 119. $\mathbf{P4_222}$
120. $P4_2221'$
121. $P4'_222'$
122. $P4_22'2'$
123. $P4'_22'2$
124. P_c4_222
125. P_C4_222
126. P_I4_222

(94) 127. $\mathbf{P4_22_12}$
128. $P4_22_121'$
129. $P4'_22_12'$
130. $P4_22'_12'$
131. $P4'_22'_12$
132. $P_c4_22_12$
133. $P_C4_22_12$
134. $P_I4_22_12$

(95) 135. $\mathbf{P4_322}$
136. $P4_3221'$
137. $P4'_322'$
138. $P4_32'2'$
139. $P4'_32'2$
140. P_c4_322
141. P_C4_322
142. P_I4_322

(96) 143. $\mathbf{P4_32_12}$
144. $P4_32_121'$
145. $P4'_32_12'$
146. $P4_32'_12'$
147. $P4'_32'_12$
148. $P_c4_32_12$
149. $P_C4_32_12$
150. $P_I4_32_12$

(97) 151. $\mathbf{I422}$
152. $I4221'$
153. $I4'22'$
154. $I42'2'$
155. $I4'2'2$
156. I_c422

(98) 157. $\mathbf{I4_122}$
158. $I4_1221'$
159. $I4'_122'$
160. $I4_12'2'$
161. $I4'_12'2$
162. I_c4_122

C_{4v}

(99) 163. $\mathbf{P4mm}$
164. $P4mm1'$
165. $P4'm'm$
166. $P4'mm'$
167. $P4m'm'$
168. P_c4mm
169. P_C4mm
170. P_I4mm

(100) 171. $\mathbf{P4bm}$
172. $P4bm1'$
173. $P4'b'm$

174. $P4'bm'$
175. $P4b'm'$
176. P_c4bm
177. P_C4bm
178. P_I4bm

(101) 179. $\mathbf{P4_2cm}$
180. $P4_2cm1'$
181. $P4_2'c'm$
182. $P4_2'cm'$
183. $P4_2c'm'$
184. P_c4_2cm
185. P_C4_2cm
186. P_I4_2cm

(102) 187. $\mathbf{P4_2nm}$
188. $P4_2nm1'$
189. $P4_2'n'm$
190. $P4_2'nm'$
191. $P4_2n'm'$
192. P_c4_2nm
193. P_C4_2nm
194. P_I4_2nm

(103) 195. $\mathbf{P4cc}$
196. $P4cc1'$
197. $P4'c'c$
198. $P4'cc'$
199. $P4c'c'$
200. P_c4cc
201. P_C4cc
202. P_I4cc

(104) 203. $\mathbf{P4nc}$
204. $P4nc1'$
205. $P4'n'c$
206. $P4'nc'$
207. $P4n'c'$
208. P_c4nc
209. P_C4nc
210. P_I4nc

(105) 211. $\mathbf{P4_2mc}$
212. $P4_2mc1'$
213. $P4_2'm'c$
214. $P4_2'mc'$
215. $P4_2m'c'$
216. P_c4_2mc
217. P_C4_2mc
218. P_I4_2mc

(106) 219. $\mathbf{P4_2bc}$
220. $P4_2bc1'$
221. $P4_2'b'c$
222. $P4_2'bc'$
223. $P4_2b'c'$
224. P_c4_2bc
225. P_C4_2bc
226. P_I4_2bc

(107) 227. $\mathbf{I4mm}$
228. $I4mm1'$
229. $I4'm'm$
230. $I4'mm'$
231. $I4m'm'$
232. I_c4mm

(108) 233. $\mathbf{I4cm}$
234. $I4cm1'$
235. $I4'c'm$
236. $I4'cm'$
237. $I4c'm'$
238. I_c4cm

(109) 239. $\mathbf{I4_1md}$
240. $I4_1md1'$
241. $I4_1'm'd$
242. $I4_1'md'$
243. $I4_1m'd'$
244. I_c4_1md

(110) 245. $\mathbf{I4_1cd}$
246. $I4_1cd1'$
247. $I4_1'c'd$
248. $I4_1'cd'$
249. $I4_1c'd'$
250. I_c4_1cd

$\mathbf{D_{2d}}$

(111) 251. $\mathbf{P\bar{4}2m}$
252. $P\bar{4}2m1'$
253. $P\bar{4}'2'm$
254. $P\bar{4}'2m'$
255. $P\bar{4}2'm'$
256. $P_c\bar{4}2m$
257. $P_C\bar{4}2m$
258. $P_I\bar{4}2m$

(112) 259. $\mathbf{P\bar{4}2c}$
260. $P\bar{4}2c1'$
261. $P\bar{4}'2'c$
262. $P\bar{4}'2c'$
263. $P\bar{4}2'c'$
264. $P_c\bar{4}2c$
265. $P_C\bar{4}2c$
266. $P_I\bar{4}2c$

(113) 267. $\mathbf{P\bar{4}2_1m}$
268. $P\bar{4}2_1m1'$
269. $P\bar{4}'2_1'm$
270. $P\bar{4}'2_1m'$
271. $P\bar{4}2_1'm'$
272. $P_c\bar{4}2_1m$
273. $P_C\bar{4}2_1m$
274. $P_I\bar{4}2_1m$

(114) 275. $\mathbf{P\bar{4}2_1c}$
276. $P\bar{4}2_1c$
277. $P\bar{4}'2_1'c$
278. $P\bar{4}'2_1c$
279. $P\bar{4}2_1'c'$
280. $P_c\bar{4}2_1c$
281. $P_C\bar{4}2_1c$
282. $P_I\bar{4}2_1c$

(115) 283. $\mathbf{P\bar{4}m2}$
284. $P\bar{4}m21'$
285. $P\bar{4}'m'2$
286. $P\bar{4}'m2'$
287. $P\bar{4}m'2'$
288. $P_c\bar{4}m2$
289. $P_C\bar{4}m2$
290. $P_I\bar{4}m2$

(116) 291. $\mathbf{P\bar{4}c2}$
292. $P\bar{4}c21'$
293. $P\bar{4}'c'2$
294. $P\bar{4}'c2'$
295. $P\bar{4}c'2'$
296. $P_c\bar{4}c2$
297. $P_C\bar{4}c2$
298. $P_I\bar{4}c2$

(117) 299. $\mathbf{P\bar{4}b2}$
300. $P\bar{4}b21'$
301. $P\bar{4}'b'2$
302. $P\bar{4}'b2'$
303. $P\bar{4}b'2'$
304. $P_c\bar{4}b2$
305. $P_C\bar{4}b2$
306. $P_I\bar{4}b2$

(118) 307. $\mathbf{P\bar{4}n2}$
308. $P\bar{4}n21'$
309. $P\bar{4}'n'2$
310. $P\bar{4}'n2'$
311. $P\bar{4}n'2'$
312. $P_c\bar{4}n2$
313. $P_C\bar{4}n2$
314. $P_I\bar{4}n2$

(119) 315. $\mathbf{I\bar{4}m2}$
316. $I\bar{4}m21'$
317. $I\bar{4}'m'2$
318. $I\bar{4}'m2'$
319. $I\bar{4}m'2'$
320. $I_c\bar{4}m2$

(120) 321. $\mathbf{I\bar{4}c2}$
322. $I\bar{4}c21'$
323. $I\bar{4}'c'2$
324. $I\bar{4}'c2'$
325. $I\bar{4}c'2'$
326. $I_c\bar{4}c2$

(121) 327. $\mathbf{I\bar{4}2m}$
328. $I\bar{4}2m1'$
329. $I\bar{4}'2'm$
330. $I\bar{4}'2m'$
331. $I\bar{4}2'm'$
332. $I_c\bar{4}2m$

(122) 333. $\mathbf{I\bar{4}2d}$
334. $I\bar{4}2d1'$
335. $I\bar{4}'2'd$
336. $I\bar{4}'2d'$
337. $I\bar{4}2'd'$
338. $I_c\bar{4}2d$

$\mathbf{D_{4h}}$

(123) 339. $\mathbf{P4/mmm}$
340. $P4/mmm1'$
341. $P4/m'mm$
342. $P4'/mm'm$
343. $P4'/mmm'$
344. $P4'/m'm'm$
345. $P4/mm'm'$
346. $P4'/m'mm'$
347. $P4/m'm'm'$
348. P_c4/mmm
349. P_C4/mmm
350. P_I4/mmm

(124) 351. $\mathbf{P4/mcc}$
352. $P4/mcc1'$
353. $P4/m'cc$
354. $P4'/mc'c$
355. $P4'/mcc'$
356. $P4'/m'c'c$
357. $P4/mc'c'$
358. $P4'/m'cc'$
359. $P4/m'c'c'$
360. P_c4/mcc

361. P_C4/mcc
362. P_I4/mcc

(125) 363. **$P4/nbm$**
364. $P4/nbm1'$
365. $P4/n'bm$
366. $P4'/nb'm$
367. $P4'/nbm'$
368. $P4'/n'b'm$
369. $P4/nb'm'$
370. $P4'/n'bm'$
371. $P4/n'b'm'$
372. P_c4/nbm
373. P_C4/nbm
374. P_I4/nbm

(126) 375. **$P4/nnc$**
376. $P4/nnc1'$
377. $P4/n'nc$
378. $P4'/nn'c$
379. $P4'/nnc'$
380. $P4'/n'n'c$
381. $P4/nn'c'$
382. $P4'/n'nc'$
383. $P4/n'n'c'$
384. P_c4/nnc
385. P_C4/nnc
386. P_I4/nnc

(127) 387. **$P4/mbm$**
388. $P4/mbm1'$
389. $P4/m'bm$
390. $P4'/mb'm$
391. $P4'/mbm'$
392. $P4'/m'b'm$
393. $P4/mb'm'$
394. $P4'/m'bm'$
395. $P4/m'b'm'$
396. P_c4/mbm
397. P_C4/mbm
398. P_I4/mbm

(128) 399. **$P4/mnc$**
400. $P4/mnc1'$
401. $P4/m'nc$
402. $P4'/mn'c$
403. $P4'/mnc'$
404. $P4'/m'n'c$
405. $P4/mn'c'$
406. $P4'/m'nc'$
407. $P4/m'n'c'$
408. P_c4/mnc
409. P_C4/mnc
410. P_I4/mnc

(129) 411. **$P4/nmm$**
412. $P4/nmm1'$
413. $P4/n'mm$
414. $P4'/nm'm$
415. $P4'/nmm'$
416. $P4'/n'm'm$
417. $P4/nm'm'$
418. $P4'/n'mm'$
419. $P4/n'm'm'$
420. P_c4/nmm
421. P_C4/nmm
422. P_I4/nmm

(130) 423. **$P4/ncc$**
424. $P4/ncc1'$
425. $P4/n'cc$
426. $P4'/nc'c$
427. $P4'/ncc'$
428. $P4'/n'c'c$
429. $P4/nc'c'$
430. $P4'/n'cc'$
431. $P4/n'c'c'$
432. P_c4/ncc
433. P_C4/ncc
434. P_I4/ncc

(131) 435. **$P4_2/mmc$**
436. $P4_2/mmc1'$
437. $P4_2/m'mc$
438. $P4'_2/mm'c$
439. $P4'_2/mmc'$
440. $P4'_2/m'm'c$
441. $P4_2/mm'c'$
442. $P4'_2/m'mc'$
443. $P4_2/m'm'c'$
444. P_c4_2/mmc
445. P_C4_2/mmc
446. P_I4_2/mmc

(132) 447. **$P4_2/mcm$**
448. $P4_2/mcm1'$
449. $P4_2/m'cm$
450. $P4'_2/mc'm$
451. $P4'_2/mcm'$
452. $P4'_2/m'c'm$
453. $P4_2/mc'm'$
454. $P4'_2/m'cm'$
455. $P4_2/m'c'm'$
456. P_c4_2/mcm
457. P_C4_2/mcm
458. P_I4_2/mcm

(133) 459. **$P4_2/nbc$**
460. $P4_2/nbc1'$
461. $P4_2/n'bc$
462. $P4'_2/nb'c$
463. $P4'_2/nbc'$
464. $P4'_2/n'b'c$
465. $P4_2/nb'c'$
466. $P4'_2/n'bc'$
467. $P4_2/n'b'c'$
468. P_c4_2/nbc
469. P_C4_2/nbc
470. P_I4_2/nbc

(134) 471. **$P4_2/nnm$**
472. $P4_2/nnm1'$
473. $P4_2/n'nm$
474. $P4'_2/nn'm$
475. $P4'_2/nnm'$
476. $P4'_2/n'n'm$
477. $P4_2/nn'm'$
478. $P4'_2/n'nm'$
479. $P4_2/n'n'm'$
480. P_c4_2/nnm
481. P_C4_2/nnm
482. P_I4_2/nnm

(135) 483. **$P4_2/mbc$**
484. $P4_2/mbc1'$
485. $P4_2/m'bc$
486. $P4'_2/mb'c$
487. $P4'_2/mbc'$
488. $P4'_2/m'b'c$
489. $P4_2/mb'c'$
490. $P4'_2/m'bc'$
491. $P4_2/m'b'c'$
492. P_c4_2/mbc
493. P_C4_2/mbc
494. P_I4_2/mbc

(136) 495. **$P4_2/mnm$**
496. $P4_2/mnm1'$
497. $P4_2/m'nm$
498. $P4'_2/mn'm$
499. $P4'_2/mnm'$
500. $P4'_2/m'n'm$
501. $P4_2/mn'm'$
502. $P4'_2/m'nm'$
503. $P4_2/m'n'm'$
504. P_c4_2/mnm
505. P_C4_2/mnm
506. P_I4_2/mnm

(137) 507. **$P4_2/nmc$**
508. $P4_2/nmc1'$
509. $P4_2/n'mc$
510. $P4'_2/nm'c$
511. $P4'_2/nmc'$
512. $P4'_2/n'm'c$
513. $P4_2/nm'c'$
514. $P4'_2/n'mc'$
515. $P4_2/n'm'c'$
516. P_c4_2/nmc
517. P_C4_2/nmc
518. P_I4_2/nmc

(138) 519. **$P4_2/ncm$**
520. $P4_2/ncm1'$
521. $P4_2/n'cm$
522. $P4'_2/nc'm$
523. $P4'_2/ncm'$
524. $P4'_2/n'c'm$
525. $P4_2/nc'm'$
526. $P4'_2/n'cm'$
527. $P4_2/n'c'm'$
528. P_c4_2/ncm
529. P_C4_2/ncm
530. P_I4_2/ncm

(139) 531. **$I4/mmm$**
532. $I4/mmm1'$
533. $I4/m'mm$
534. $I4'/mm'm$
535. $I4'/mmm'$
536. $I4'/m'm'm$
537. $I4/mm'm'$
538. $I4'/m'mm'$
539. $I4/m'm'm'$
540. I_c4/mmm

(140) 541. **$I4/mcm$**
542. $I4/mcm1'$
543. $I4/m'cm$
544. $I4'/mc'm$
545. $I4'/mcm'$
546. $I4'/m'c'm$
547. $I4/mc'm'$
548. $I4'/m'cm'$
549. $I4/m'c'm'$
550. I_c4/mcm

(141) 551. **$I4_1/amd$**
552. $I4_1/amd1'$
553. $I4_1/a'md$
554. $I4'_1/am'd$
555. $I4'_1/amd'$

556. $I4'_1/a'm'd$
557. $I4_1/am'd'$
558. $I4'_1/a'md'$
559. $I4_1/a'm'd'$
560. I_c4_1/amd

(142) 561. $\mathbf{I4_1/acd}$
562. $I4_1/acd1'$
563. $I4_1/a'cd$
564. $I4'_1/ac'd$
565. $I4'_1/acd'$
566. $I4'_1/a'c'd$
567. $I4_1/ac'd'$
568. $I4'_1/a'cd'$
569. $I4_1/a'c'd'$
570. I_c4_1/acd

Hexagonal system

A. *Rhombohedral sub-system*

$\mathbf{C_3}$

(143) 1. $\mathbf{P3}$
2. $P31'$
3. P_c3

(144) 4. $\mathbf{P3_1}$
5. $P3_11'$
6. P_c3_1

(145) 7. $\mathbf{P3_2}$
8. $P3_21'$
9. P_c3_2

(146) 10. $\mathbf{R3}$
11. $R31'$
12. R_I3

$\mathbf{C_{3i}}$

(147) 13. $\mathbf{P\bar{3}}$
14. $P\bar{3}1'$
15. $P\bar{3}'$
16. $P_c\bar{3}$

(148) 17. $\mathbf{R\bar{3}}$
18. $R\bar{3}1'$
19. $R\bar{3}'$
20. $R_I\bar{3}$

$\mathbf{D_3}$

(149) 21. $\mathbf{P312}$
22. $P31'2$
23. $P312'$
24. P_c312

(150) 25. $\mathbf{P321}$
26. $P321'$
27. $P32'1$
28. P_c321

(151) 29. $\mathbf{P3_112}$
30. $P3_11'2$
31. $P3_112'$
32. P_c3_112

(152) 33. $\mathbf{P3_121}$
34. $P3_121'$
35. $P3_12'1$
36. P_c3_121

(153) 37. $\mathbf{P3_212}$
38. $P3_21'2$
39. $P3_212'$
40. P_c3_212

(154) 41. $\mathbf{P3_221}$
42. $P3_221'$
43. $P3_22'1$
44. P_c3_221

(155) 45. $\mathbf{R32}$
46. $R321'$
47. $R32'$
48. R_I32

$\mathbf{C_{3v}}$

(156) 49. $\mathbf{P3m1}$
50. $P3m1'$
51. $P3m'1$
52. P_c3m1

(157) 53. $\mathbf{P31m}$
54. $P31'm$
55. $P31m'$
56. P_c31m

(158) 57. $\mathbf{P3c1}$
58. $P3c1'$
59. $P3c'1$
60. P_c3c1

(159) 61. $\mathbf{P31c}$
62. $P31'c$
63. $P31c'$
64. P_c31c

(160) 65. $\mathbf{R3m}$
66. $R3m1'$
67. $R3m'$
68. R_I3m

(161) 69. $\mathbf{R3c}$
70. $R3c1'$
71. $R3c'$
72. R_I3c

$\mathbf{D_{3d}}$

(162) 73. $\mathbf{P\bar{3}1m}$
74. $P\bar{3}1'm$
75. $P\bar{3}'1m$
76. $P\bar{3}'1m'$
77. $P\bar{3}1m'$
78. $P_c\bar{3}1m$

(163) 79. $\mathbf{P\bar{3}1c}$
80. $P\bar{3}1'c$
81. $P\bar{3}'1c$
82. $P\bar{3}'1c'$
83. $P\bar{3}1c'$
84. $P_c\bar{3}1c$

(164) 85. $\mathbf{P\bar{3}m1}$
86. $P\bar{3}m1'$
87. $P\bar{3}'m1$
88. $P\bar{3}'m'1$
89. $P\bar{3}m'1$
90. $P_c\bar{3}m1$

(165) 91. $\mathbf{P\bar{3}c1}$
92. $P\bar{3}c1'$
93. $P\bar{3}'c1$
94. $P\bar{3}'c'1$
95. $P\bar{3}c'1$
96. $P_c\bar{3}c1$

(166) 97. $\mathbf{R\bar{3}m}$
98. $R\bar{3}m1'$
99. $R\bar{3}'m$
100. $R\bar{3}'m'$
101. $R\bar{3}m'$
102. $R_I\bar{3}m$

(167) 103. $\mathbf{R\bar{3}c}$
104. $R\bar{3}c1'$
105. $R\bar{3}'c$
106. $R\bar{3}'c'$
107. $R\bar{3}c'$
108. $R_I\bar{3}c$

B. *Hexagonal sub-system*

$\mathbf{C_6}$

(168) 109. $\mathbf{P6}$
110. $P61'$
111. $P6'$
112. P_c6

(169) 113. $\mathbf{P6_1}$
114. $P6_11'$
115. $P6'_1$
116. P_c6_1

(170) 117. $\mathbf{P6_5}$
118. $P6_51'$
119. $P6'_5$
120. P_c6_5

(171) 121. $\mathbf{P6_2}$
122. $P6_21'$
123. $P6'_2$
124. P_c6_2

(172) 125. $\mathbf{P6_4}$
126. $P6_41'$
127. $P6'_4$
128. P_c6_4

(173) 129. $\mathbf{P6_3}$
130. $P6_31'$
131. $P6'_3$
132. P_c6_3

$\mathbf{C_{3h}}$

(174) 133. $\mathbf{P\bar{6}}$
134. $P\bar{6}1'$
135. $P\bar{6}'$
136. $P_c\bar{6}$

$\mathbf{C_{6h}}$

(175) 137. $\mathbf{P6/m}$
138. $P6/m1'$

139. $P6'/m$
140. $P6/m'$
141. $P6'/m'$
142. P_c6/m

(176) 143. **$P6_3/m$**
144. $P6_3/m1'$
145. $P6'_3/m$
146. $P6_3/m'$
147. $P6'_3/m'$
148. P_c6_3/m

D_6

(177) 149. **$P622$**
150. $P6221'$
151. $P6'2'2$
152. $P6'22'$
153. $P62'2'$
154. P_c622

(178) 155. **$P6_122$**
156. $P6_1221'$
157. $P6'_12'2$
158. $P6'_122'$
159. $P6_12'2'$
160. P_c6_122

(179) 161. **$P6_522$**
162. $P6_5221'$
163. $P6'_52'2$
164. $P6'_522'$
165. $P6_52'2'$
166. P_c6_522

(180) 167. **$P6_222$**
168. $P6_2221'$
169. $P6'_22'2$
170. $P6'_222'$
171. $P6_22'2'$
172. P_c6_222

(181) 173. **$P6_422$**
174. $P6_4221'$
175. $P6'_42'2$
176. $P6'_422'$
177. $P6_42'2'$
178. P_c6_422

(182) 179. **$P6_322$**
180. $P6_3221'$
181. $P6'_32'2$
182. $P6'_322'$
183. $P6_32'2'$
184. P_c6_322

C_{6v}

(183) 185. **$P6mm$**
186. $P6mm1'$
187. $P6'm'm$
188. $P6'mm'$
189. $P6m'm'$
190. P_c6mm

(184) 191. **$P6cc$**
192. $P6cc1'$
193. $P6'c'c$
194. $P6'cc'$
195. $P6c'c'$
196. P_c6cc

(185) 197. **$P6_3cm$**
198. $P6_3cm1'$
199. $P6'_3c'm$
200. $P6'_3cm'$
201. $P6_3c'm'$
202. P_c6_3cm

(186) 203. **$P6_3mc$**
204. $P6_3mc1'$
205. $P6'_3m'c$
206. $P6'_3mc'$
207. $P6_3m'c'$
208. P_c6_3mc

D_{3h}

(187) 209. **$P\bar{6}m2$**
210. $P\bar{6}m21'$
211. $P\bar{6}'m'2$
212. $P\bar{6}'m2'$
213. $P\bar{6}m'2'$
214. $P_c\bar{6}m2$

(188) 215. **$P\bar{6}c2$**
216. $P\bar{6}c21'$
217. $P\bar{6}'c'2$
218. $P\bar{6}'c2'$
219. $P\bar{6}c'2'$
220. $P_c\bar{6}c2$

(189) 221. **$P\bar{6}2m$**
222. $P\bar{6}2m1'$
223. $P\bar{6}'2'm$
224. $P\bar{6}'2m'$
225. $P\bar{6}2'm'$
226. $P_c\bar{6}2m$

(190) 227. **$P\bar{6}2c$**
228. $P\bar{6}2c1'$
229. $P\bar{6}'2'c$
230. $P\bar{6}'2c'$
231. $P\bar{6}2'c'$
232. $P_c\bar{6}2c$

D_{6h}

(191) 233. **$P6/mmm$**
234. $P6/mmm1'$
235. $P6/m'mm$
236. $P6'/mm'm$
237. $P6'/mmm'$
238. $P6'/m'm'm$
239. $P6'/m'mm'$
240. $P6/mm'm'$
241. $P6/m'm'm'$
242. P_c6/mmm

(192) 243. **$P6/mcc$**
244. $P6/mcc1'$
245. $P6/m'cc$
246. $P6'/mc'c$
247. $P6'/mcc'$
248. $P6'/m'c'c$
249. $P6'/m'cc'$
250. $P6/mc'c'$
251. $P6/m'c'c'$
252. P_c6/mcc

(193) 253. **$P6_3/mcm$**
254. $P6_3/mcm1'$
255. $P6_3/m'cm$
256. $P6'_3/mc'm$
257. $P6'_3/mcm'$
258. $P6'_3/m'c'm$
259. $P6'_3/m'cm'$
260. $P6_3/mc'm'$
261. $P6_3/m'c'm'$
262. P_c6_3/mcm

(194) 263. **$P6_3/mmc$**
264. $P6_3/mmc1'$
265. $P6_3/m'mc$
266. $P6'_3/mm'c$
267. $P6'_3/mmc'$
268. $P6'_3/m'm'c$
269. $P6'_3/m'mc'$
270. $P6_3/mm'c'$
271. $P6_3/m'm'c'$
272. P_c6_3/mmc

Cubic system

T

(195) 1. **$P23$**
2. $P23'$
3. P_I23

(196) 4. **$F23$**
5. $F23'$
6. F_s23

(197) 7. **$I23$**
8. $I23'$

(198) 9. **$P2_13$**
10. $P2_13'$
11. P_I2_13

(199) 12. **$I2_13$**
13. $I2_13'$

T_h

(200) 14. **$Pm3$**
15. $Pm3'$
16. $Pm'3$
17. P_Im3

(201) 18. **$Pn3$**
19. $Pn3'$
20. $Pn'3$
21. P_In3

(202) 22. **$Fm3$**
23. $Fm3'$
24. $Fm'3$
25. F_sm3

(203) 26. **$Fd3$**
27. $Fd3'$
28. $Fd'3$
29. F_sd3

(204) 30. **$Im3$**
31. $Im3'$
32. $Im'3$

(205) 33. **$Pa3$**

34. *Pa3'*
35. *Pa'3*
36. P_Ia3

(206) 37. **Ia3**
38. *Ia3'*
39. *Ia'3*

O

(207) 40. **P432**
41. *P43'2*
42. *P4'32'*
43. P_I432

(208) 44. $\mathbf{P4_232}$
45. $P4_23'2$
46. $P4'_232'$
47. P_I4_232

(209) 48. **F432**
49. *F43'2*
50. *F4'32'*
51. F_s432

(210) 52. $\mathbf{F4_132}$
53. $F4_13'2$
54. $F4'_132'$
55. F_s4_132

(211) 56. **I432**
57. *I43'2*
58. *I4'32'*

(212) 59. $\mathbf{P4_332}$
60. $P4_33'2$
61. $P4'_332'$

62. P_I4_332

(213) 63. $\mathbf{P4_132}$
64. $P4_13'2$
65. $P4'_132'$
66. P_I4_132

(214) 67. $\mathbf{I4_132}$
68. $I4_13'2$
69. $I4'_132'$

$\mathbf{T_d}$

(215) 70. $\mathbf{P\bar{4}3m}$
71. $P\bar{4}3'm$
72. $P\bar{4}'3m'$
73. $P_I\bar{4}3m$

(216) 74. $\mathbf{F\bar{4}3m}$
75. $F\bar{4}3'm$
76. $F\bar{4}'3m'$
77. $F_s\bar{4}3m$

(217) 78. $\mathbf{I\bar{4}3m}$
79. $I\bar{4}3'm$
80. $I\bar{4}'3m'$

(218) 81. $\mathbf{P\bar{4}3n}$
82. $P\bar{4}3'n$
83. $P\bar{4}'3n'$
84. $P_I\bar{4}3n$

(219) 85. $\mathbf{F\bar{4}3c}$
86. $F\bar{4}3'c$
87. $F\bar{4}'3c'$
88. $F_s\bar{4}3c$

(220) 89. $\mathbf{I\bar{4}3d}$
90. $I\bar{4}3'd$
91. $I\bar{4}'3d'$

$\mathbf{O_h}$

(221) 92. **Pm3m**
93. *Pm3'm*
94. *Pm'3m*
95. *Pm3m'*
96. *Pm'3m'*
97. P_Im3m

(222) 98. **Pn3n**
99. *Pn3'n*
100. *Pn'3n*
101. *Pn3n'*
102. *Pn'3n'*
103. P_In3n

(223) 104. **Pm3n**
105. *Pm3'n*
106. *Pm'3n*
107. *Pm3n'*
108. *Pm'3n'*
109. P_Im3n

(224) 110. **Pn3m**
111. *Pn3'm*
112. *Pn'3m*
113. *Pn3m'*
114. *Pn'3m'*
115. P_In3m

(225) 116. **Fm3m**
117. *Fm3'm*
118. *Fm'3m*
119. *Fm3m'*
120. *Fm'3m'*
121. F_sm3m

(226) 122. **Fm3c**
123. *Fm3'c*
124. *Fm'3c*
125. *Fm3c'*
126. *Fm'3c'*
127. F_sm3c

(227) 128. **Fd3m**
129. *Fd3'm*
130. *Fd'3m*
131. *Fd3m'*
132. *Fd'3m'*
133. F_sd3m

(228) 134. **Fd3c**
135. *Fd3'c*
136. *Fd'3c*
137. *Fd3c'*
138. *Fd'3c'*
139. F_sd3c

(229) 140. **Im3m**
141. *Im3'm*
142. *Im'3m*
143. *Im3m'*
144. *Im'3m'*

(230) 145. **Ia3d**
146. *Ia3'd*
147. *Ia'3d*
148. *Ia3d'*
149. *Ia'3d'*

Notes to Table 7.4

(i) The 1651 Shubnikov space groups are listed in the order and notation of Belov, Neronova, and Smirnova (1957).

(ii) The Shubnikov space groups **M** that are related to a given Fedorov space group **G** (see Definitions 7.2.3–7.2.5) are listed together following **G** in the table. The 230 Fedorov space groups themselves are arranged in the order given in Volume 1 of the *International tables for X-ray crystallography* (Henry and Lonsdale 1965) which was also used in our Table 3.7. Immediately following each Fedorov space group **G** is the grey space group derived from **G** according to Definition 7.2.3. These are followed by the type III Shubnikov space groups derived according to Definition 7.2.4 which are themselves followed by the type IV Shubnikov space groups derived according to Definition 7.2.5.

(iii) The elements of any given type III Shubnikov space group can be deduced by reference to Tables 7.2 and 3.7. The elements of any given type IV Shubnikov space group can be deduced by reference to Tables 7.3 and 3.7.

(iv) Within each crystal system the Shubnikov space groups have been assigned an arbitrary number, e.g. between 1 and 91 in the monoclinic system.

Neronova, and Smirnova (1957), all the 1651 Shubnikov space groups are listed; a complete table is also given by Opechowski and Guccione (1965). Extensive graphical representations of Shubnikov space groups have been discussed by Atoji (1965). The symmetry elements and equivalent positions for each of the type III and type IV Shubnikov groups are illustrated by Koptsik (1966) with (two-coloured) diagrams similar to those used for the ordinary space groups (type I Shubnikov space groups) in the *International tables for X-ray crystallography* (Henry and Lonsdale 1965). There are alternatives to the adaptation of the international space-group labels used in Table 7.4. There is an adaptation of the Schönflies notation for type III Shubnikov space groups; this takes the form **G**(**H**) so that, for example, $P4_2'/mnm'$ would be written as $D_{4h}^{14}(D_{2h}^{12})$. For any (type I, II, III, or IV) Shubnikov space group Koptsik (1966) uses a label of the form Ш_x^y where x is the number (from 1 to 230) of the related ordinary space group and y is the number given to that Shubnikov space group in Table 7.4; in this notation, for example, $P4_2'/mnm'$ would be Ш_{136}^{499}.

It is possible to derive the type III Shubnikov groups by considering the representations of the ordinary point groups and space groups, that is, of the type I Shubnikov groups (Alexander 1962, Bertaut 1968). From the point of view of representation theory, the statements that a crystal structure has a space group **G** and that a crystal transforms according to the identity representation of the space group **G** are equivalent statements. In the same way, when a magnetic crystal structure can be described by the Shubnikov group **M** it belongs to the identity representation (strictly the identity corepresentation, see section 7.3 below) of **M**. If we consider some ordinary point group **G** then the type III Shubnikov point groups which are derived from **G** by eqn. (7.2.2) can be determined by studying the 1-dimensional representations of **G**. If the identity representation of **G** is regarded as synonymous with the type I Shubnikov group **G** itself, the remaining real 1-dimensional representations of **G** can be related to the type III Shubnikov groups **M** that are derived from **G** (Indenbom 1959). Thus (Bertaut 1968) *the number of magnetic groups is equal to the number of* 1-*dimensional representations that are distinct in the abstract sense and have characters* +1 *or* −1 *only*. Two representations are said to be 'distinct in the abstract sense' if they cannot be transformed into each other by altering the orientation of the axes.

We may illustrate this by considering again the example that was considered earlier in this section of the type III Shubnikov point groups derived from $4mm$ (C_{4v}), the group of the symmetry operations of a square. The character table of the point group $4mm$ (C_{4v}) can be found from Table 2.2 and it is shown in Table 7.5. A_1 is the identity representation and corresponds to **G**, that is, to $4mm$ (C_{4v}) itself. The representation E is degenerate and therefore is not relevant to the present discussion. In A_2 the elements E, C_{2z}, C_{4z}^+, and C_{4z}^- are represented by $+1$ while σ_x, σ_y, σ_{da}, and σ_{db} are represented by -1. The elements that are represented by $+1$ are then uncoloured in **M** while the elements represented by -1 become coloured in **M**; A_2

TABLE 7.5

Representations and magnetic groups derived from 4mm (C_{4v})

	E	$C_{4z}^{\pm}$	C_{2z}	σ_x, σ_y	σ_{da}, σ_{db}	
A_1	1	1	1	1	1	$4mm$
A_2	1	1	1	−1	−1	$4m'm'$
B_1	1	−1	1	1	−1	$4'mm'$
B_2	1	−1	1	−1	1	$4'm'm$
E	2	−2	0	0	0	—

therefore corresponds to the magnetic point group, or type III Shubnikov point group $4m'm'$ that was discussed earlier. In a similar way the representation B_1 has the elements C_{4z}^{+}, C_{4z}^{-}, σ_{da}, and σ_{db} represented by -1 so that it corresponds to the type III Shubnikov point group $4'mm'$ that was also considered earlier in this section. In a similar way B_2 leads to the magnetic point group $4'm'm$ in which the coloured elements are $\mathcal{R}C_{4z}^{+}$, $\mathcal{R}C_{4z}^{-}$, $\mathcal{R}\sigma_x$, and $\mathcal{R}\sigma_y$. However, $4'mm'$, derived from B_1 and $4'm'm$, derived from B_2 are not essentially different because one of them can be turned into the other by rotating the x and y axes by 45° about the z-axis. Therefore as described earlier it is possible to derive two type III Shubnikov point groups from the ordinary point group $4mm$ (C_{4v}). By repeating this process for each of the 32 point groups in turn all the 58 black and white point groups which were identified in Table 7.1 could be derived.

A similar argument can be applied to the determination of the type III Shubnikov space groups derived from the ordinary space group $\mathbf{G}$, by using the character table of the little group of the wave vector, $\mathbf{G}^{\mathbf{k}}$, at $\mathbf{k} = 0$. The restriction to $\mathbf{k} = 0$ has to

TABLE 7.6

Representations and magnetic groups in Pbam (D_{2h}^{9}) ($\mathbf{k} = 0$)

Representation	Characters of generating elements C_{2z}	C_{2y}	I	Characters of planes σ_x	σ_y	σ_z	Magnetic group
R_1	1	1	1	1	1	1	$Pbam$
R_2	−1	1	1	−1	1	−1	$Pb'am'$
R_3	1	−1	1	−1	−1	1	$Pb'a'm$
R_4	−1	−1	1	1	−1	−1	$Pba'm'$
R_5	1	1	−1	−1	−1	−1	$Pb'a'm'$
R_6	−1	1	−1	1	−1	1	$Pba'm$
R_7	1	−1	−1	1	1	−1	$Pbam'$
R_8	−1	−1	−1	−1	1	1	$Pb'am$

be imposed in order to ensure that the halving subgroup **H** contains all the pure translations of **G**. Each of the 1-dimensional small representations of this little group, other than the identity representation, which are distinct in the abstract sense and have real characters corresponds to one of the type III Shubnikov space groups derived from **G**. As an example we consider the space group *Pbam* (D_{2h}^9) and the type III (black and white) Shubnikov space groups derived from it. From Tables 5.1 and 5.7 the small representations can be written down and they are given in Table 7.6. The characters of the generating elements $\{C_{2z} \mid 000\}$, $\{C_{2y} \mid \frac{1}{2}\frac{1}{2}0\}$, and $\{I \mid 000\}$ are given for each representation. The three reflection planes occur in the order $\{\sigma_x \mid \frac{1}{2}\frac{1}{2}0\}$, $\{\sigma_y \mid \frac{1}{2}\frac{1}{2}0\}$, $\{\sigma_z \mid 000\}$ in the symbol of the space group *Pbam* (D_{2h}^9) so that the symbol of the magnetic space group derived from each of the representations $R_2, R_3, \ldots, R_8$ can easily be written down and it is given in the last column of Table 7.6. Not all the seven type III Shubnikov space groups are distinct. *Pba′m′* and *Pb′am′* differ only in the setting of the axes and can be transformed into each other by a rotation of 90° about the z-axis; the same is also true for *Pb′am* and *Pba′m*. We therefore have derived five type III Shubnikov space groups from the Fedorov space group *Pbam*; from Table 7.4 one can see that these are all the type III Shubnikov space groups that can be derived from *Pbam*. It is also possible to derive the type IV Shubnikov groups by repeating what we have done at $\mathbf{k} = 0$ for all the other wave vectors **k** in the Brillouin zone for which $\mathbf{G^k}$ coincides with **G**; only such **k** vectors can yield 1-dimensional reps of **G**.

Shubnikov groups and magnetic crystals

In physical terms we can now discuss the relevance of Shubnikov groups to the description of the structures of real magnetic crystals. In discussing the symmetry of magnetic crystals, the operation $\mathscr{R}$ is the operation that reverses a magnetic moment. $\mathscr{R}$ is then thought of as being the operation of time inversion. Alternatively we can think of $\mathscr{R}$ as reversing the direction of an electric current, since change of magnetic moment can be caused by a reversal of the direction of the electric current which gives rise to the magnetic moment. This reversal of the direction of the electric current is then equivalent to a reversal of the sense of the direction of the time variable since $i = \mathrm{d}q/\mathrm{d}t$. Neutron diffraction crystallographers classify magnetic crystals into the following types:

(i) *paramagnetic*, magnetic moments randomly oriented, a net moment develops in the presence of an applied magnetic field,

(ii) *ferromagnetic*, all magnetic moments parallel to one particular direction within a domain,

(iii) *anti-ferromagnetic*, all magnetic moments either parallel or antiparallel to one particular direction and arranged so that there is no net magnetic moment,

(iv) *ferrimagnetic*, all magnetic moments either parallel or anti-parallel to one particular direction, but there is a net magnetic moment because the two do not cancel completely, and

(v) *canted anti-ferromagnetic* and other complicated forms of anti-ferromagnetic structures.

We have not made any mention of diamagnetism; all the cases that we have mentioned are cases in which the orientation of a magnetic moment, which can in theory at least be either a spin or orbital magnetic moment, is under consideration and the crystal is therefore paramagnetic, ferromagnetic, or anti-ferromagnetic, etc. Diamagnetism arises from a distortion of the orbits of the electrons in the atoms of the crystal as a result of an applied magnetic field; it can therefore be regarded as being due to a current in a loop within an atom rather than to the alignment of a small permanent magnetic moment associated with the atom. But diamagnetism is weak and is always swamped by any other form of magnetism that may be present. In considerations of the symmetry of a diamagnetic crystal and its relation to the Shubnikov groups the same conditions apply as for paramagnetic crystals and so for our purposes diamagnetic crystals can be included in (i) with paramagnetic crystals.

The grey space groups, or type II Shubnikov space groups, possess the operation of time inversion, θ, itself as a symmetry operation; therefore, if a spontaneous internal magnetic field, which is described by an axial vector $\mathbf{B}$, were to develop at some point within the crystal, the presence of θ would require that an equal and opposite magnetic field should also appear at the same point. Consequently no spontaneous magnetic field can exist anywhere in the crystal. Therefore, either the individual atoms or ions within the crystal have no magnetic moments, in which case the crystal can only exhibit diamagnetism, or they do possess spontaneous magnetic moments but these spontaneous magnetic moments are randomly oriented so that they produce, on average, no magnetic field anywhere in the crystal, in which case the crystal exhibits paramagnetism.

If the operation of antisymmetry is absent, as in the case of a type I Shubnikov space group, or if it is only present in combination with rotation, reflection or translation operations of symmetry, as is the case for a type III or type IV Shubnikov space group, there are two possibilities. It may be possible for a crystal to possess a net magnetic moment, in which case it is either a ferromagnetic or a ferrimagnetic crystal; alternatively, it may be possible to have half the magnetic moments of the individual magnetic atoms or ions in the crystal arranged parallel to one particular direction and the other half anti-parallel to that direction, in which case the crystal does not possess a spontaneous magnetic moment and it is an anti-ferromagnetic crystal. It requires careful analysis of the effect of the various operations of the Shubnikov group of the crystal to determine whether a crystal with the symmetry of that group is ferromagnetic or antiferromagnetic (Neronova and Belov 1959*b*, Opechowski and Guccione 1965). The result is that a crystal described by any type I,

III, or IV Shubnikov space group is not prevented by symmetry considerations from exhibiting anti-ferromagnetism. However, of the type I and type III Shubnikov space groups only a limited number can exhibit ferromagnetism (275 in fact) and they are listed by Neronova and Belov (1959*b*) and Opechowski and Guccione (1965). A type IV Shubnikov space group cannot exhibit ferromagnetism because any spontaneous magnetic moment that may develop on one sublattice will necessarily be exactly cancelled by an equal and opposite magnetic moment that will be required by symmetry to appear on the other sublattice. Ferrimagnetic crystals satisfy exactly the same group-theoretical conditions that we have just described for ferromagnetic crystals, so that we do not need to consider ferrimagnetic crystals separately. If one investigates the point groups of all the 275 Shubnikov space groups which describe structures that may exhibit ferromagnetism they will be found to be those type I and type III Shubnikov point groups given in Table 7.7.

TABLE 7.7

Ferromagnetic Shubnikov point groups

Type I	Type III	Type I	Type III
1 (C_1)		3 (C_3)	
$\bar{1}$ (C_i)		$\bar{3}$ (C_{3i})	
2 (C_2)	$2'$		$32'$
m (C_{1h})	m'		$3m'$
$2/m$ (C_{2h})	$2'/m'$		$\bar{3}m'$
	$2'2'2$	6 (C_6)	
	$m'm2'$ $m'm'2$	$\bar{6}$ (C_{3h})	
	$m'm'm$	$6/m$ (C_{6h})	
4 (C_4)			$62'2'$
$\bar{4}$ (S_4)			$6m'm'$
$4/m$ (C_{4h})			$\bar{6}m'2'$
	$42'2'$		$6/mm'm'$
	$4m'm'$		
	$\bar{4}2'm'$		
	$4/mm'm'$		

The Brillouin zones of magnetic crystals

Let us consider some crystal that is paramagnetic above some temperature T_N, the Néel temperature, and is anti-ferromagnetic below T_N. That is, above T_N the magnetic moments of the individual atoms are randomly arranged but below T_N these magnetic moments are arranged in some regular fashion. It is assumed that no crystallographic distortion occurs at T_N although below T_N some distortion may occur as a result of the lower symmetry of the anti-ferromagnetic phase. For some crystals the Brillouin zone in the anti-ferromagnetic phase is exactly the same as the Brillouin zone in the paramagnetic phase. But for some other crystals the size and shape of the Brillouin zone will change as the crystal passes from the paramagnetic phase in which the spins are randomly oriented to the anti-ferromagnetic phase in which the spins are in an ordered array.

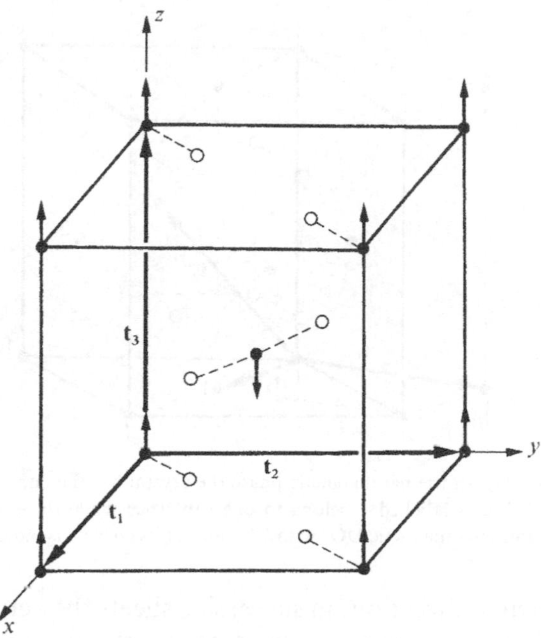

FIG. 7.5. The unit cell of anti-ferromagnetic MnF_2. ● Mn ○ F.

As an example of a crystal in which there is no change in the size of the Brillouin zone we consider MnF_2 which exhibits the rutile structure (Dimmock and Wheeler 1962*b*). The unit cell of MnF_2 is shown in Fig. 7.5. In the paramagnetic phase it belongs to one of the primitive tetragonal space groups ($P4_2/mnm$ (D_{4h}^{14})) and has the equivalent of two Mn atoms per unit cell; the basic vectors of its Bravais lattice are therefore $\mathbf{t}_1 = a\mathbf{i}$, $\mathbf{t}_2 = a\mathbf{j}$, and $\mathbf{t}_3 = c\mathbf{k}$, see Table 3.1. In the anti-ferromagnetic phase it belongs to a primitive tetragonal type III Shubnikov space group ($P4_2'/mnm'$) and has two Mn atoms per unit cell, with one Mn atom on each of the two sublattices. The basic vectors of this Bravais lattice are exactly the same as those of the crystal in its paramagnetic phase and so the Brillouin zone is also exactly the same as in the paramagnetic phase; this Brillouin zone is illustrated in Fig. 3.9.

As an example of a crystal in which a change in the size and shape of the Brillouin zone does occur as the crystal passes from the paramagnetic phase to the anti-ferromagnetic phase we consider UO_2 (Cowley and Dolling 1968). In the paramagnetic phase UO_2 has the structure shown in Fig. 7.6 in which the U^{4+} ions lie on the points of a face-centred cubic Bravais lattice and the Brillouin zone is therefore the truncated octahedron shown in Fig. 3.14. In the antiferromagnetic phase the magnetic moments of the U^{4+} ions within each horizontal sheet are all arranged

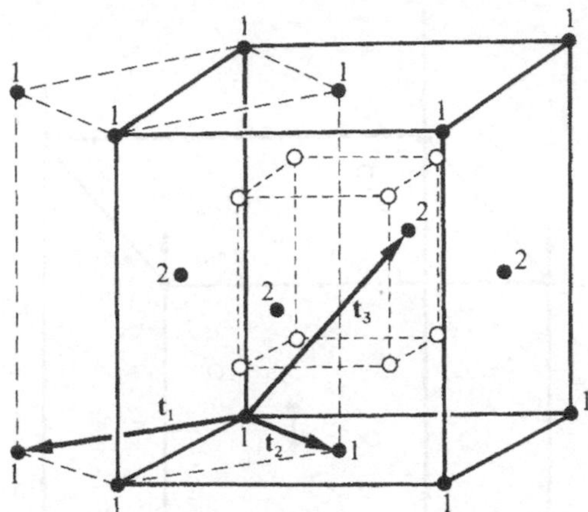

FIG. 7.6. The structure of UO_2. In the paramagnetic phase the crystal has the cubic fluorite structure. In the anti-ferromagnetic phase the U^{4+} ions labelled 1 belong to one sublattice while those labelled 2 belong to the other sublattice. The unit cell of anti-ferromagnetic UO_2 with U^{4+} ions at its corners is indicated by broken lines. ●U, ○O.

parallel to one particular direction; in successive sheets the actual direction of magnetization is reversed. Unlike the example of MnF_2 the two sublattices are therefore arranged so that the basic vectors of either sublattice are different from the basic vectors of the paramagnetic structure. In the paramagnetic structure we choose basic vectors

$$\left.\begin{aligned} \mathbf{t}_1 &= a(\tfrac{1}{2}, -\tfrac{1}{2}, 0), \\ \mathbf{t}_2 &= a(\tfrac{1}{2}, \tfrac{1}{2}, 0), \\ \text{and} \qquad \mathbf{t}_3 &= a(0, \tfrac{1}{2}, \tfrac{1}{2}) \end{aligned}\right\} \tag{7.2.6}$$

(for pedagogic convenience we have taken slightly different basic vectors from those used in Table 3.1). In the anti-ferromagnetic phase $\mathbf{t}_1$ and $\mathbf{t}_2$ are still basic vectors of sublattice 1 but $\mathbf{t}_3$ now joins a point on sublattice 1 to a point on sublattice 2 and is no longer a basic vector of the Bravais lattice of the crystal; a suitable choice of Bravais lattice vectors for the anti-ferromagnetic phase is

$$\left.\begin{aligned} \mathbf{t}'_1 &= a(\tfrac{1}{2}, -\tfrac{1}{2}, 0), \\ \mathbf{t}'_2 &= a(\tfrac{1}{2}, \tfrac{1}{2}, 0), \\ \text{and} \qquad \mathbf{t}'_3 &= a(0, 1, 1), \end{aligned}\right\} \tag{7.2.7}$$

where $\mathbf{t}_3'$ is equal to twice $\mathbf{t}_3$. These are in fact the basic vectors of a primitive tetragonal Bravais lattice, the unit cell of which is shown with broken lines in Fig. 7.6. The shape of the Brillouin zone of the anti-ferromagnetic crystal can then be found by using eqn. (3.2.2) to evaluate $\mathbf{g}_1$, $\mathbf{g}_2$, and $\mathbf{g}_3$. The volume of the fundamental anti-ferromagnetic unit cell $(=\mathbf{t}_1' \cdot (\mathbf{t}_2' \times \mathbf{t}_3'))$ is twice the volume of the paramagnetic unit cell because $\mathbf{t}_3$ has been doubled. The volume of the anti-ferromagnetic Brillouin zone is therefore equal to one-half of the volume of the Brillouin zone in the paramagnetic phase; these two Brillouin zones are illustrated in Fig. 7.7.

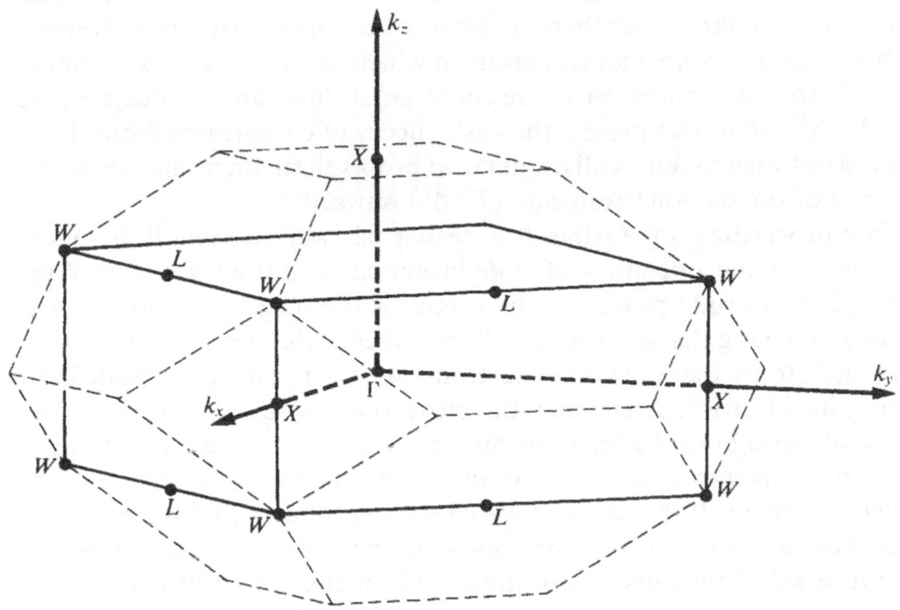

FIG. 7.7. The Brillouin zone of UO_2, continuous lines indicate the Brillouin zone of the anti-ferromagnetic structure and broken lines the paramagnetic structure.

7.3. Anti-unitary operations and the corepresentations of magnetic groups

The reps of Shubnikov groups belonging to type I were discussed in Chapters 3 and 4 and given in detail in Chapters 5 and 6. By introducing an operation of anti-symmetry, $\mathscr{R}$, we have derived a whole new collection of groups, the type II, III, and IV Shubnikov groups. Type II Shubnikov groups, defined by eqns (7.2.1) and (7.2.3) are direct product groups. If $\mathscr{R}$, the operation of anti-symmetry, were a unitary operator in the case where it is regarded as reversing the direction of a magnetic moment, there would be little difficulty in determining the representations of these groups. But in studying magnetic symmetry the operator $\mathscr{R}$ is the operation of time-inversion which we shall call θ and which is anti-unitary. It is therefore necessary to be able to extend representation theory to groups which contain anti-unitary operators. The theory was originally given by Wigner in 1932 and the most readily accessible account of this work is in the English translation of Wigner's classic book (Wigner 1959 (Chapter 26)). The development of the theory has also been discussed by Dimmock and Wheeler (1962*a*, 1962*b*) (see also Dimmock (1963*b*)) by Bradley and Davies (1968) and in an extensive series of papers by Chaldyshev, Kudryavtseva, and Karavaev (Chaldyshev 1961, Chaldyshev and Kudryavtseva 1962, Chaldyshev, Kudryavtseva, and Karavaev 1963, Karavaev, Kudryavtseva, and Chaldyshev 1962, Kudryavtseva 1965, 1967*a*, Kudryavtseva and Chaldyshev 1962*a*, *b*, 1965*a*, *b*, 1968). Whereas for a unitary group there are representations and irreducible representations, for a non-unitary group, that is, a group in which half the elements are unitary and the other half are anti-unitary, there are corepresentations and irreducible corepresentations. In this section we present the basic theory of corepresentations. For the reader who does not wish to follow all the detailed proofs the main results will be summarized at the end of the section (from eqn. (7.3.45) onwards).

Before proceeding any further it is perhaps as well to say a little more about the statement that the operation of time-inversion is anti-unitary (see Wigner (1959) Chapter 26). A crude picture of the origin of the magnetic moment of an atom is obtained by saying that according to Bohr's theory the electrons are moving in orbits within the atom; these orbiting electrons are then similar to small loops or coils carrying an electric current and therefore they produce a magnetic moment. A reversal of the sign of the time coordinate, i.e. $t \to -t$, would cause the electrons to orbit in the opposite sense and therefore would reverse the direction of the resultant magnetic moment. If the idea of *time-inversion* produces philosophical difficulties or conjures up ideas of science fiction then it may be more helpful to regard the operation as the reversal of the direction of motion of an electron in its orbit. When it comes to the inclusion of the spin angular momentum of the electrons this simple pictorial approach is not possible.

We wish to examine the effect of θ, the operation of time-inversion, on a wave function ψ. The behaviour of ψ as a function of t is determined by the time-dependent Schrödinger equation

$$H\psi = i\hbar \frac{\partial \psi}{\partial t} \tag{7.3.1}$$

where H is the Hamiltonian operator of the system. Suppose that E_k and ψ_k are the eigenvalues and eigenfunctions of H, that is, they are the solutions of the time-independent Schrödinger equation $H\psi = E\psi$. The wave function $\psi(t)$, at any time t, can be expanded in terms of the ψ_k, so that we may write $\psi(t) = \sum_k a_k(t)\psi_k$, where $a_k(0)$ are constants determined by the boundary conditions applicable to the system at $t = 0$. $a_k(t)$ can be determined by substitution into eqn (7.3.1), thus

$$H \sum_k a_k(t)\psi_k = i\hbar \frac{\partial}{\partial t}\left\{\sum_k a_k(t)\psi_k\right\}. \tag{7.3.2}$$

Therefore, using the fact that $H\psi_k = E_k\psi_k$, we have

$$\sum_k \left\{a_k(t)E_k\psi_k\right\} = i\hbar \sum_k \left\{\frac{da_k(t)}{dt}\psi_k\right\} \tag{7.3.3}$$

so that, equating coefficients of ψ_k

$$i\hbar \frac{da_k(t)}{dt} = E_k a_k(t) \tag{7.3.4}$$

and therefore

$$a_k(t) = a_k(0) \exp(-iE_k t/\hbar). \tag{7.3.5}$$

We may therefore write

$$\psi(t) = \sum_k a_k(0) \exp(-iE_k t/\hbar)\psi_k \tag{7.3.6}$$

and this is the wave function of the system at any desired time, t.

Let us now consider the effect of various operations on this wave function. If we start at $t = 0$ and perform the operation of time-inversion (1) followed by the operation of making a time displacement of $+t$ (2) the result is $+t$, see Fig. 7.8. If we start again at $t = 0$ and make a displacement of $-t$ (3) followed by the operation of time-inversion (4) the result is again $+t$, see Fig. 7.8. It is then not unreasonable to expect that the effect on $\psi(0)$ of the operations (1) and (2) should be the same as the effect of the operations (3) and (4). The effect of operation (1) is to produce $\theta\psi(0)$ and there are two possibilities; either, if θ is a linear operator,

$$\theta\psi(0) = \sum_k a_k(0)\theta\psi_k \tag{7.3.7a}$$

or, if θ is an anti-linear operator,

$$\theta\psi(0) = \sum_k a_k(0)^*\theta\psi_k. \tag{7.3.7b}$$

We now wish to investigate the effect on $\theta\psi(0)$ of the operation T which corresponds to a displacement of $+t$. We assume that the state $\theta\psi(t)$ also satisfies the time-dependent Schrödinger equation, that is eqn (7.3.1), so that the state $\theta\psi(0)$ also develops in time according to eqn (7.3.5), that is, with factors $\exp(-iE_k t/\hbar)$ rather than $\exp(+iE_k t/\hbar)$. Therefore, if θ is linear we have

$$T\theta\psi(0) = \sum_k a_k(0) \exp(-iE_k t/\hbar)\theta\psi_k, \tag{7.3.8a}$$

or, if θ is anti-linear,

$$T\theta\psi(0) = \sum_k a_k(0)^* \exp(-iE_k t/\hbar)\theta\psi_k. \tag{7.3.8b}$$

If we now start again with $\psi(0)$ and perform operation (3) we obtain

$$\psi(-t) = \sum_k a_k(0) \exp(\mathrm{i}E_k t/\hbar)\psi_k \tag{7.3.9}$$

and when operation (4) is performed on this we obtain either, if θ is linear

$$\theta\psi(-t) = \sum_k a_k(0) \exp(\mathrm{i}E_k t/\hbar)\theta\psi_k, \tag{7.3.10a}$$

or, if θ is anti-linear,

$$\theta\psi(-t) = \sum_k a_k(0)^* \exp(-\mathrm{i}E_k t/\hbar)\theta\psi_k. \tag{7.3.10b}$$

Equations (7.3.8) and (7.3.10) then must both give the wave function at the point P in Fig. 7.8 and must therefore be the same. Equations (7.3.8a) and (7.3.10a) are clearly not the same but eqns (7.3.8b) and (7.3.10b) are the same, and therefore we conclude that θ, the operation of time-inversion, must be anti-linear rather than linear. It then follows from the conservation of probability amplitudes that θ must be not only anti-linear but also anti-unitary.

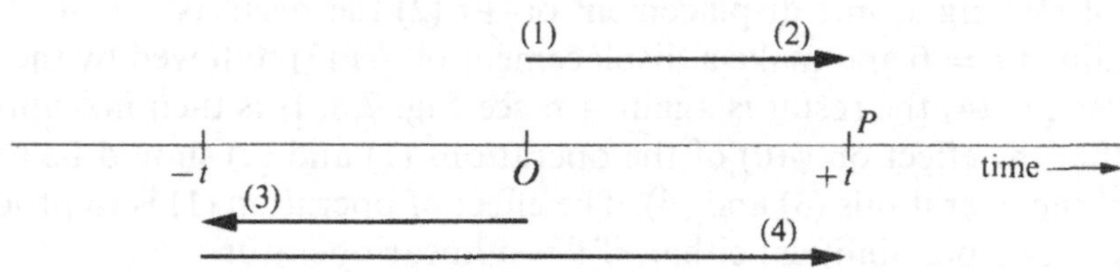

FIG. 7.8.

It is perhaps as well to emphasize that it is not an intrinsic property of the Shubnikov groups that they should be non-unitary groups; they are only non-unitary groups when they are used in the description of magnetic structures, when the colour-changing operation becomes identified with the operation of time-inversion.

Let **M** be a magnetic group and **G** its unitary subgroup of index 2 and suppose that, when we express **M** in terms of left cosets with respect to **G**,

$$\mathbf{M} = \mathbf{G} + A\mathbf{G}. \tag{7.3.11}$$

Then all the elements of the coset $A\mathbf{G}$ are anti-unitary. All magnetic groups whether they are of type II, III, or IV can be factorized in this way, so that the following analysis is suitable for any magnetic group irrespective of its type. A, of course, can be any of the anti-unitary elements of the group **M**, but once chosen we shall suppose that it is fixed once for all as coset representative. The group **G** is an invariant subgroup as indeed are all subgroups of index 2 of any group. The product of any two elements of the coset $A\mathbf{G}$ is an element of $\mathbf{G}$: this follows from the homomorphism of $\mathbf{M}$ onto the quotient group $\mathbf{M}/\mathbf{G}$ of order 2. In particular $A^2 \in \mathbf{G}$.

We now seek to determine the modifications that are necessary to the representation theory that we have already considered if it is to be appropriate to the non-unitary group $\mathbf{M}$. Let Γ be a unitary irreducible representation (i.e. a 'rep'), with dimension d, of $\mathbf{G}$ and let $\langle\psi| = \langle\psi_1, \psi_2, \ldots, \psi_d|$ be a basis for Γ. Then for any element R in the group $\mathbf{G}$

$$R\psi_i = \sum_{j=1}^{d} \psi_j \Delta(R)_{ji}, \tag{7.3.12}$$

where $\Delta(R)$ is the matrix representative of R in Γ. We shall write eqn (7.3.12) for short in the form

$$R\langle\psi| = \langle\psi|\,\Delta(R). \tag{7.3.13}$$

Let us now introduce another set of functions ϕ_i, $i = 1$ to d, where $\langle\phi|$ is produced by the action of A on $\langle\psi|$, that is,

$$A\langle\psi| = \langle\phi|. \tag{7.3.14}$$

THEOREM 7.3.1. *The vector space spanned by the functions* ψ_i, ϕ_i ($i = 1$ *to* d) *is invariant under* $\mathbf{M}$.

This can be proved fairly readily. If we consider the action of R, some element in $\mathbf{G}$, on the ϕ_i then

$$\begin{aligned} R\langle\phi| &= RA\langle\psi| \\ &= A(A^{-1}RA)\langle\psi| \\ &= A\langle\psi|\,\Delta(A^{-1}RA), && \text{since } A^{-1}RA \in \mathbf{G}, \\ &= \langle\phi|\,\Delta^*(A^{-1}RA), && \text{from eqn (7.3.14) and the fact that } A \text{ is anti-linear.} \end{aligned}$$

Hence

$$R\langle\psi, \phi| = \langle\psi, \phi| \begin{pmatrix} \Delta(R) & \mathbf{0} \\ \mathbf{0} & \Delta^*(A^{-1}RA) \end{pmatrix} \tag{7.3.15}$$

for all $R \in \mathbf{G}$. We often write $\Delta^*(A^{-1}RA) = \bar{\Delta}(R)$, which is the matrix representative of R in a rep $\bar{\Gamma}$ of $\mathbf{G}$. Now suppose we let B be any element of $A\mathbf{G}$, say AR, then

$$\begin{aligned} B\langle\psi| &= AR\langle\psi| \\ &= A\langle\psi|\,\Delta(R) && \text{from eqn (7.3.13)} \\ &= \langle\phi|\,\Delta^*(R) && \text{from eqn (7.3.14) and the fact that } A \text{ is anti-linear,} \\ &= \langle\phi|\,\Delta^*(A^{-1}B). \end{aligned}$$

Also

$$\begin{aligned} B\langle\phi| &= BA\langle\psi| && \text{from eqn (7.3.14)} \\ &= \langle\psi|\,\mathbf{\Delta}(BA) && \text{from eqn (7.3.13) since } BA \in \mathbf{G}. \end{aligned}$$

Therefore, combining these two equations we obtain

$$B\langle\psi, \phi| = \langle\psi, \phi|\begin{pmatrix} \mathbf{0} & \mathbf{\Delta}(BA) \\ \mathbf{\Delta}^*(A^{-1}B) & \mathbf{0} \end{pmatrix}. \tag{7.3.16}$$

This proves the invariance.

Using eqns (7.3.15) and (7.3.16) we can write

$$\left.\begin{aligned} \mathbf{D}(R) &= \begin{pmatrix} \mathbf{\Delta}(R) & \mathbf{0} \\ \mathbf{0} & \mathbf{\Delta}^*(A^{-1}RA) \end{pmatrix} \quad \text{all } R \in \mathbf{G}, \\ \mathbf{D}(B) &= \begin{pmatrix} \mathbf{0} & \mathbf{\Delta}(BA) \\ \mathbf{\Delta}^*(A^{-1}B) & \mathbf{0} \end{pmatrix} \quad \text{all } B \in A\mathbf{G}. \end{aligned}\right\} \tag{7.3.17}$$

This set of matrices forms what is called the corepresentation of $\mathbf{M}$ derived from Γ (Dimmock 1963*b*, Wigner 1959, 1960*a*, *b*), which we denote by $D\Gamma$. They do not obey the ordinary equation for a representation

$$\mathbf{D}(R)\mathbf{D}(S) = \mathbf{D}(RS) \tag{7.3.18}$$

where R and S are any members of the group, see section 1.3. But they do obey a more complicated set of equations,

$$\mathbf{D}(R)\mathbf{D}(S) = \mathbf{D}(RS) \tag{7.3.19a}$$

$$\mathbf{D}(R)\mathbf{D}(B) = \mathbf{D}(RB) \tag{7.3.19b}$$

$$\mathbf{D}(B)\mathbf{D}^*(R) = \mathbf{D}(BR) \tag{7.3.19c}$$

$$\mathbf{D}(B)\mathbf{D}^*(C) = \mathbf{D}(BC) \tag{7.3.19d}$$

where R and $S \in \mathbf{G}$ and B and $C \in A\mathbf{G}$, that is if the first element of the pair in the product is from $A\mathbf{G}$ then a complex conjugate appears on the second element. These equations can be checked using eqn (7.3.17). It should be noted that $\mathbf{M} \to D\Gamma$ is not a homomorphism, but that since Γ is unitary all the matrices of $D\Gamma$ are unitary.

Let us now perform a unitary transformation $\mathbf{U}$ on the basis $\langle\psi, \phi| = \langle\chi|$ such that

$$\langle\chi'| = \langle\chi|\,\mathbf{U}. \tag{7.3.20}$$

Then

$$R\langle\chi'| = \langle\chi'|\,\mathbf{D}'(R) = \langle\chi|\,\mathbf{U}\mathbf{D}'(R)$$

and

$$R\langle\chi'| = R\langle\chi|\,\mathbf{U} = \langle\chi|\,\mathbf{D}(R)\mathbf{U}$$

so that

$$\mathbf{D}'(R) = \mathbf{U}^{-1}\mathbf{D}(R)\mathbf{U}. \tag{7.3.21}$$

And

$$B\langle\chi'| = \langle\chi'|\,\mathbf{D}'(B) = \langle\chi|\,\mathbf{U}\mathbf{D}'(B)$$

but

$$B\langle\chi'| = B\langle\chi|\,\mathbf{U} = \langle\chi|\,\mathbf{D}(B)\mathbf{U}^*$$

so that

$$\mathbf{D}'(B) = \mathbf{U}^{-1}\mathbf{D}(B)\mathbf{U}^*. \tag{7.3.22}$$

This enables us to define two corepresentations of $\mathbf{M}$ to be unitarily equivalent if there exists a unitary matrix $\mathbf{U}$ such that

$$\left.\begin{aligned}\mathbf{D}'(R) &= \mathbf{U}^{-1}\mathbf{D}(R)\mathbf{U} \quad \text{all } R \in \mathbf{G}\\ \mathbf{D}'(B) &= \mathbf{U}^{-1}\mathbf{D}(B)\mathbf{U}^* \quad \text{all } B \in A\mathbf{G}.\end{aligned}\right\} \tag{7.3.23}$$

In particular, if we had chosen another anti-unitary operator $A' = TA$ where $T \in \mathbf{G}$ instead of A as coset representative the resulting corepresentation would have had as basis $\langle\psi, \phi'|$ where

$$\langle\phi'| = A'\langle\psi| = TA\langle\psi| = T\langle\phi| = \langle\phi|\,\mathbf{\Delta}^*(A^{-1}TA). \tag{7.3.24}$$

Then by choosing

$$\mathbf{U} = \begin{pmatrix}\mathbf{1} & \mathbf{0}\\ \mathbf{0} & \mathbf{\Delta}^*(A^{-1}TA)\end{pmatrix}$$

we see that $\mathbf{U}$ is unitary and the two corepresentations unitarily equivalent. This is why the choice of A in defining the corepresentation of $\mathbf{M}$ derived from Γ is immaterial: different choices lead to corepresentations which are unitarily equivalent. We can therefore speak of the corepresentation $D\Gamma$ of $\mathbf{M}$ derived from Γ without ambiguity.

Once the concept of equivalence is defined it is possible to define reducibility and irreducibility of corepresentations. This is done exactly as for ordinary representations. If the basis $\langle\chi|$ can be transformed by a unitary transformation $\mathbf{U}$ so that the resulting basis $\langle\chi'| = \langle\chi|\,\mathbf{U}$ becomes the direct sum of two spaces both invariant under $\mathbf{M}$ then $D\Gamma$ is said to be *reducible*. If not then $D\Gamma$ is said to be *irreducible*. Or in matrix language the question of reducibility is whether there is a corepresentation $D'\Gamma$ unitarily equivalent to $D\Gamma$ such that all the matrices of $D'\Gamma$ are in the same block diagonal form. In order to answer this question it is first necessary to discuss the relationship between the two representations Γ with matrices $\mathbf{\Delta}(R)$ and $\bar{\Gamma}$ with matrices $\mathbf{\Delta}^*(A^{-1}RA)$ of the subgroup $\mathbf{G}$.

First, if we write $\mathbf{\Delta}^*(A^{-1}RA) = \bar{\mathbf{\Delta}}(R)$ then $\bar{\mathbf{\Delta}}^*(A^{-1}RA) = \mathbf{\Delta}(A^{-2}RA^2) = \mathbf{\Delta}(A^{-2})\mathbf{\Delta}(R)\mathbf{\Delta}(A^2) = \mathbf{\Delta}(A^2)^{-1}\mathbf{\Delta}(R)\mathbf{\Delta}(A^2)$, which is equivalent to $\mathbf{\Delta}(R)$. From this it follows that the corepresentation of $\mathbf{M}$ derived from Γ is equivalent to the corepresentation of $\mathbf{M}$ derived from $\bar{\Gamma}$. This means that when we consider the collection of all reps of $\mathbf{G}$ some of them fall into pairs Γ and $\bar{\Gamma}$ which are mutually inequivalent and which come together to form single corepresentations of $\mathbf{M}$; as far as we know yet these corepresentations may be reducible though we shall in fact prove shortly that such corepresentations are irreducible. The remainder of the reps of $\mathbf{G}$ are such that $\Gamma \equiv \bar{\Gamma}$ and then the corepresentation of $\mathbf{M}$ contains Γ twice; we shall prove shortly that such corepresentations may or may not be reducible according to a criterion that we shall establish.

Take first the case in which Γ and $\bar{\Gamma}$ are mutually inequivalent. We suppose there exists a unitary matrix $\mathbf{U}$ which reduces $\mathbf{D}$. Since $\mathbf{D}(R)$, $R \in \mathbf{G}$ is the direct sum of irreducibles $\mathbf{\Delta}(R)$ and $\mathbf{\Delta}^*(A^{-1}RA)$ the only reduced form of $\mathbf{D}(R)$ is $\begin{pmatrix} \mathbf{X}(R) & \mathbf{0} \\ \mathbf{0} & \mathbf{Y}(R) \end{pmatrix}$ where $\mathbf{X}(R)$ is equivalent to $\mathbf{\Delta}(R)$ and $\mathbf{Y}(R)$ is equivalent to $\mathbf{\Delta}^*(A^{-1}RA)$. Then if $\mathbf{U}^{-1} = \begin{pmatrix} \mathbf{a} & \mathbf{b} \\ \mathbf{c} & \mathbf{d} \end{pmatrix}$, eqn (7.3.23) implies

$$\begin{pmatrix} \mathbf{a} & \mathbf{b} \\ \mathbf{c} & \mathbf{d} \end{pmatrix}\begin{pmatrix} \mathbf{\Delta}(R) & \mathbf{0} \\ \mathbf{0} & \mathbf{\Delta}^*(A^{-1}RA) \end{pmatrix} = \begin{pmatrix} \mathbf{X}(R) & \mathbf{0} \\ \mathbf{0} & \mathbf{Y}(R) \end{pmatrix}\begin{pmatrix} \mathbf{a} & \mathbf{b} \\ \mathbf{c} & \mathbf{d} \end{pmatrix}$$

from which

$$\mathbf{a}\mathbf{\Delta}(R) = \mathbf{X}(R)\mathbf{a}$$

$$\mathbf{d}\mathbf{\Delta}^*(A^{-1}RA) = \mathbf{Y}(R)\mathbf{d}$$

and

$$\mathbf{b}\mathbf{\Delta}^*(A^{-1}RA) = \mathbf{X}(R)\mathbf{b} = \mathbf{a}\mathbf{\Delta}(R)\mathbf{a}^{-1}\mathbf{b}$$ (**a** being non-singular since it provides the equivalence transformation between $\mathbf{\Delta}(R)$ and $\mathbf{X}(R)$)

so that

$$(\mathbf{a}^{-1}\mathbf{b})\mathbf{\Delta}^*(A^{-1}RA) = \mathbf{\Delta}(R)\mathbf{a}^{-1}\mathbf{b}.$$

But $\mathbf{\Delta}(R)$ and $\mathbf{\Delta}^*(A^{-1}RA)$ are inequivalent so that by Schur's lemma $\mathbf{a}^{-1}\mathbf{b} = \mathbf{0}$ and hence $\mathbf{b} = \mathbf{0}$. Similarly $\mathbf{c} = \mathbf{0}$, and hence $\mathbf{U}^{-1}$ is of the form $\begin{pmatrix} \mathbf{a} & \mathbf{0} \\ \mathbf{0} & \mathbf{d} \end{pmatrix}$. But no such diagonal block matrix is capable of reducing matrices of the form

$$\mathbf{D}(B) = \begin{pmatrix} \mathbf{0} & \mathbf{\Delta}(BA) \\ \mathbf{\Delta}^*(A^{-1}B) & \mathbf{0} \end{pmatrix}$$

because $\mathbf{U}^{-1}\mathbf{D}(B)\mathbf{U}^*$ will still be in off-diagonal block form. Thus if $\boldsymbol{\Delta}(R)$ and $\boldsymbol{\Delta}^*(A^{-1}RA)$ are inequivalent the corepresentation of **M** derived from Γ is irreducible.

Suppose next that $\boldsymbol{\Delta}(R)$ and $\boldsymbol{\Delta}^*(A^{-1}RA)$ are equivalent. Then there exists a unitary matrix **N** such that

$$\boldsymbol{\Delta}(R) = \mathbf{N}\boldsymbol{\Delta}^*(A^{-1}RA)\mathbf{N}^{-1} \tag{7.3.25}$$

and in particular

$$\boldsymbol{\Delta}(A^2) = \mathbf{N}\boldsymbol{\Delta}^*(A^2)\mathbf{N}^{-1}. \tag{7.3.26}$$

Now

$$\boldsymbol{\Delta}^*(R) = \mathbf{N}^*\boldsymbol{\Delta}(A^{-1}RA)\mathbf{N}^{-1*},$$

therefore

$$\begin{aligned}\boldsymbol{\Delta}^*(A^{-1}RA) &= \mathbf{N}^*\boldsymbol{\Delta}(A^{-2}RA^2)\mathbf{N}^{-1*}\\ &= \mathbf{N}^*\boldsymbol{\Delta}(A^{-2})\boldsymbol{\Delta}(R)\boldsymbol{\Delta}(A^2)\mathbf{N}^{-1*}.\end{aligned}$$

So from eqn (7.3.25) we have

$$\boldsymbol{\Delta}(R) = \mathbf{N}\mathbf{N}^*\boldsymbol{\Delta}^{-1}(A^2)\boldsymbol{\Delta}(R)\boldsymbol{\Delta}(A^2)\mathbf{N}^{-1*}\mathbf{N}^{-1}. \tag{7.3.27}$$

Equation (7.3.27) holds for all $R \in \mathbf{G}$ and since Γ is irreducible it follows from Schur's lemma that

$$\mathbf{N}\mathbf{N}^*\boldsymbol{\Delta}^{-1}(A^2) = \lambda\mathbf{1}$$

where λ is a constant. Thus $\mathbf{N}\mathbf{N}^* = \lambda\boldsymbol{\Delta}(A^2)$ or

$$\boldsymbol{\Delta}(A^2) = \frac{\mathbf{N}\mathbf{N}^*}{\lambda} \tag{7.3.28}$$

and taking the complex conjugate

$$\boldsymbol{\Delta}^*(A^2) = \frac{\mathbf{N}^*\mathbf{N}}{\lambda^*} \tag{7.3.29}$$

Hence substituting for $\boldsymbol{\Delta}(A^2)$ and $\boldsymbol{\Delta}^*(A^2)$ in eqn (7.3.26) we have

$$\frac{\mathbf{N}\mathbf{N}^*}{\lambda} = \frac{\mathbf{N}\mathbf{N}^*\mathbf{N}\mathbf{N}^{-1}}{\lambda^*} = \frac{\mathbf{N}\mathbf{N}^*}{\lambda^*}. \tag{7.3.30}$$

From which we see that λ and λ^* must be identical. Now **N** and $\boldsymbol{\Delta}(A^2)$ are unitary so that from eqn (7.3.28) $|\lambda| = 1$. Hence $\lambda = \pm 1$ are the only possible cases. It therefore follows that eqn (7.3.28) reduces to

$$\mathbf{N}\mathbf{N}^* = \pm\boldsymbol{\Delta}(A^2). \tag{7.3.31}$$

We now prove that which of these two cases holds governs whether or not the corepresentation of **M** derived from Γ is irreducible. We use the fact that $D\Gamma$ is reducible if and only if the matrices for $\mathbf{D}(R)$ and $\mathbf{D}(A)$ can be expressed simultaneously in the same block-diagonal form; every element of **M** is of the form R or RA for some $R \in \mathbf{G}$, and $\mathbf{D}(RA) = \mathbf{D}(R)\mathbf{D}(A)$ is in block-diagonal form if this is true for both $\mathbf{D}(R)$ and $\mathbf{D}(A)$.

Now $\mathbf{D}(R)$ is already in reduced form but first it is convenient to apply a transformation so that $\mathbf{D}(R)$ is in the form $\begin{pmatrix} \mathbf{\Delta}(R) & \mathbf{0} \\ \mathbf{0} & \mathbf{\Delta}(R) \end{pmatrix}$. The required unitary matrix is $\mathbf{U} = \begin{pmatrix} \mathbf{1} & \mathbf{0} \\ \mathbf{0} & \mathbf{N}^{-1} \end{pmatrix}$ for then

$$\mathbf{D}'(R) = \begin{pmatrix} \mathbf{1} & \mathbf{0} \\ \mathbf{0} & \mathbf{N} \end{pmatrix} \begin{pmatrix} \mathbf{\Delta}(R) & \mathbf{0} \\ \mathbf{0} & \mathbf{\Delta}^*(A^{-1}RA) \end{pmatrix} \begin{pmatrix} \mathbf{1} & \mathbf{0} \\ \mathbf{0} & \mathbf{N}^{-1} \end{pmatrix}$$

$$= \begin{pmatrix} \mathbf{\Delta}(R) & \mathbf{0} \\ \mathbf{0} & \mathbf{N}\mathbf{\Delta}^*(A^{-1}RA)\mathbf{N}^{-1} \end{pmatrix} = \begin{pmatrix} \mathbf{\Delta}(R) & \mathbf{0} \\ \mathbf{0} & \mathbf{\Delta}(R) \end{pmatrix}. \tag{7.3.32}$$

Also

$$\mathbf{D}'(A) = \begin{pmatrix} \mathbf{1} & \mathbf{0} \\ \mathbf{0} & \mathbf{N} \end{pmatrix} \begin{pmatrix} \mathbf{0} & \mathbf{\Delta}(A^2) \\ \mathbf{1} & \mathbf{0} \end{pmatrix} \begin{pmatrix} \mathbf{1} & \mathbf{0} \\ \mathbf{0} & \mathbf{N}^{-1*} \end{pmatrix}$$

$$= \begin{pmatrix} \mathbf{0} & \mathbf{\Delta}(A^2)\mathbf{N}^{-1*} \\ \mathbf{N} & \mathbf{0} \end{pmatrix}. \tag{7.3.33}$$

Suppose next that we try to find some unitary transformation **V** which will reduce $\mathbf{D}'(A)$ to block-diagonal form and of course leave $\mathbf{D}'(R)$ unaltered. This means that **V** must commute with $\mathbf{D}'(R)$. Let $\mathbf{V}^{-1} = \begin{pmatrix} \boldsymbol{\alpha} & \boldsymbol{\beta} \\ \boldsymbol{\gamma} & \boldsymbol{\delta} \end{pmatrix}$, then the relation $\mathbf{V}^{-1}\mathbf{D}'(R) = \mathbf{D}'(R)\mathbf{V}^{-1}$ implies, on using eqn (7.3.32), that

$$\begin{pmatrix} \boldsymbol{\alpha}\mathbf{\Delta}(R) & \boldsymbol{\beta}\mathbf{\Delta}(R) \\ \boldsymbol{\gamma}\mathbf{\Delta}(R) & \boldsymbol{\delta}\mathbf{\Delta}(R) \end{pmatrix} = \begin{pmatrix} \mathbf{\Delta}(R)\boldsymbol{\alpha} & \mathbf{\Delta}(R)\boldsymbol{\beta} \\ \mathbf{\Delta}(R)\boldsymbol{\gamma} & \mathbf{\Delta}(R)\boldsymbol{\delta} \end{pmatrix}. \tag{7.3.34}$$

Since $\mathbf{\Delta}(R)$ is irreducible it follows from Schur's lemma that $\boldsymbol{\alpha} = \lambda\mathbf{1}$, $\boldsymbol{\beta} = \mu\mathbf{1}$, $\boldsymbol{\gamma} = \nu\mathbf{1}$, and $\boldsymbol{\delta} = \rho\mathbf{1}$, where λ, μ, ν, and ρ are constants. Hence

$$\mathbf{V}^{-1} = \begin{pmatrix} \lambda\mathbf{1} & \mu\mathbf{1} \\ \nu\mathbf{1} & \rho\mathbf{1} \end{pmatrix} \tag{7.3.35}$$

where, since $\mathbf{V}^{-1}$ is unitary and hence det $\mathbf{V} \neq 0$, the constants λ, μ, ν, and ρ must satisfy

$$\lambda\rho \neq \mu\nu. \tag{7.3.36}$$

Then, if we apply the transformation with $\mathbf{V}$ to $\mathbf{D}'(A)$ we obtain

$$\mathbf{D}''(A) = \begin{pmatrix} \lambda\mathbf{1} & \mu\mathbf{1} \\ \nu\mathbf{1} & \rho\mathbf{1} \end{pmatrix} \begin{pmatrix} \mathbf{0} & \Delta(A^2)\mathbf{N}^{-1*} \\ \mathbf{N} & \mathbf{0} \end{pmatrix} \begin{pmatrix} \lambda\mathbf{1} & \nu\mathbf{1} \\ \mu\mathbf{1} & \rho\mathbf{1} \end{pmatrix} \tag{7.3.37}$$

($\mathbf{V}$ is unitary, so that $\mathbf{V}^* = \mathbf{V}^{-1T}$). On expanding the right-hand side of eqn. (7.3.37) we find that the condition for the off-diagonal terms to vanish is

$$\left.\begin{aligned} \rho\lambda\mathbf{N} + \mu\nu\Delta(A^2)\mathbf{N}^{-1*} &= \mathbf{0} \\ \mu\nu\mathbf{N} + \rho\lambda\Delta(A^2)\mathbf{N}^{-1*} &= \mathbf{0} \end{aligned}\right\}. \tag{7.3.38}$$

Equations (7.3.36) and (7.3.38) taken together imply that reduction is possible only if $\mu\nu = -\rho\lambda$. Thus of the two possibilities allowed by eqn (7.3.31) reduction is possible when $\mathbf{NN}^* = +\Delta(A^2)$ but is not possible when $\mathbf{NN}^* = -\Delta(A^2)$. When $\mathbf{NN}^* = +\Delta(A^2)$ we can choose $\lambda = \mu = \rho = 1/\sqrt{2}$ and $\nu = -1/\sqrt{2}$ and then from eqn (7.3.37)

$$\mathbf{D}''(A) = \frac{1}{2}\begin{pmatrix} \mathbf{N} + \Delta(A^2)\mathbf{N}^{-1*} & \mathbf{0} \\ \mathbf{0} & -(\mathbf{N} + \Delta(A^2)\mathbf{N}^{-1*}) \end{pmatrix}. \tag{7.3.39}$$

But $\mathbf{NN}^* = \Delta(A^2)$, so $\Delta(A^2)\mathbf{N}^{-1*} = \mathbf{N}$, giving finally

$$\mathbf{D}''(A) = \begin{pmatrix} \mathbf{N} & \mathbf{0} \\ \mathbf{0} & -\mathbf{N} \end{pmatrix}. \tag{7.3.40}$$

In the other case when $\mathbf{NN}^* = -\Delta(A^2)$ eqn (7.3.33) gives

$$\mathbf{D}'(A) = \begin{pmatrix} \mathbf{0} & -\mathbf{N} \\ \mathbf{N} & \mathbf{0} \end{pmatrix}. \tag{7.3.41}$$

If we wish to find the corepresentative of $B = RA$ then we can use eqn (7.3.19b)

$$\mathbf{D}(B) = \mathbf{D}(RA) = \mathbf{D}(R)\mathbf{D}(A) = \mathbf{D}(BA^{-1})\mathbf{D}(A). \tag{7.3.42}$$

Thus, from eqns (7.3.32) and (7.3.40), since $\mathbf{D}''(R) = \mathbf{D}'(R)$,

$$\begin{aligned} \mathbf{D}''(B) &= \begin{pmatrix} \Delta(BA^{-1}) & \mathbf{0} \\ \mathbf{0} & \Delta(BA^{-1}) \end{pmatrix} \begin{pmatrix} \mathbf{N} & \mathbf{0} \\ \mathbf{0} & -\mathbf{N} \end{pmatrix} \\ &= \begin{pmatrix} \Delta(BA^{-1})\mathbf{N} & \mathbf{0} \\ \mathbf{0} & -\Delta(BA^{-1})\mathbf{N} \end{pmatrix} \end{aligned} \tag{7.3.43}$$

and, from eqns (7.3.32) and (7.3.41),

$$\mathbf{D}'(B) = \begin{pmatrix} \mathbf{\Delta}(BA^{-1}) & \mathbf{0} \\ \mathbf{0} & \mathbf{\Delta}(BA^{-1}) \end{pmatrix} \begin{pmatrix} \mathbf{0} & -\mathbf{N} \\ \mathbf{N} & \mathbf{0} \end{pmatrix}$$

$$= \begin{pmatrix} \mathbf{0} & -\mathbf{\Delta}(BA^{-1})\mathbf{N} \\ \mathbf{\Delta}(BA^{-1})\mathbf{N} & \mathbf{0} \end{pmatrix}. \tag{7.3.44}$$

We can now summarize the three cases:

Case (a)

$$\begin{aligned} &\mathbf{\Delta}(R) = \mathbf{N}\mathbf{\Delta}^*(A^{-1}RA)\mathbf{N}^{-1}, \\ &\mathbf{N}\mathbf{N}^* = +\mathbf{\Delta}(A^2), \\ &\mathbf{D}''(R) = \mathbf{\Delta}(R), \qquad \mathbf{D}''(B) = \pm\mathbf{\Delta}(BA^{-1})\mathbf{N}. \end{aligned} \tag{7.3.45}$$

Note that the corepresentation that has the + sign is equivalent to the one that has the − sign.

Case (b)

$$\mathbf{\Delta}(R) = \mathbf{N}\mathbf{\Delta}^*(A^{-1}RA)\mathbf{N}^{-1},$$

$$\mathbf{N}\mathbf{N}^* = -\mathbf{\Delta}(A^2),$$

$$\mathbf{D}'(R) = \begin{pmatrix} \mathbf{\Delta}(R) & \mathbf{0} \\ \mathbf{0} & \mathbf{\Delta}(R) \end{pmatrix}, \qquad \mathbf{D}'(B) = \begin{pmatrix} \mathbf{0} & -\mathbf{\Delta}(BA^{-1})\mathbf{N} \\ \mathbf{\Delta}(BA^{-1})\mathbf{N} & \mathbf{0} \end{pmatrix}. \tag{7.3.46}$$

Case (c)

$\mathbf{\Delta}(R)$ is not equivalent to $\mathbf{\Delta}^*(A^{-1}RA) = \bar{\mathbf{\Delta}}(R)$.

$$\mathbf{D}(R) = \begin{pmatrix} \mathbf{\Delta}(R) & \mathbf{0} \\ \mathbf{0} & \bar{\mathbf{\Delta}}(R) \end{pmatrix}, \qquad \mathbf{D}(B) = \begin{pmatrix} \mathbf{0} & \mathbf{\Delta}(BA) \\ \bar{\mathbf{\Delta}}(BA^{-1}) & \mathbf{0} \end{pmatrix}. \tag{7.3.47}$$

Therefore starting with a rep Γ of the unitary subgroup **G** it is possible by using the appropriate one of eqns (7.3.45)–(7.3.47) to determine an irreducible corepresentation $D\Gamma$ of **M**. If one considers all the reps of **G** then one obtains a collection of irreducible corepresentations ('coreps') of **M** and, in fact, all the possible inequivalent irreducible corepresentations of **M** can be determined in this way (Wigner 1959). What is usually important is to know which of these three cases is appropriate for a given irreducible representation Γ of **G**. For this Dimmock and Wheeler (1962a) give a very simple test using the characters of Γ. If Γ^i and Γ^j are two unitary irreducible representations of **G** then from eqn (1.3.14)

$$\sum_{R \in \mathbf{G}} \mathbf{\Delta}^i(R)_{rl}\mathbf{\Delta}^{j*}(R)_{sm} = \frac{|\mathbf{G}|}{d_i}\,\delta^{ij}\delta_{rs}\delta_{lm}, \tag{7.3.48}$$

where $|\mathbf{G}|$ is the order of $\mathbf{G}$ and d_i the dimension of Γ^i. Now if $B \in A\mathbf{G}$ we have

$$\begin{aligned}\sum_{B\in A\mathbf{G}} \Delta(B^2)_{rr} &= \sum_{R\in\mathbf{G}} \Delta(ARAR)_{rr}\\ &= \sum_{R\in\mathbf{G}} \Delta(A^2)_{rs}\Delta(A^{-1}RA)_{st}\Delta(R)_{tr}.\end{aligned}$$

In cases (a) and (b) the sum is

$$\begin{aligned}\Delta(A^2)_{rs} &\sum_{R\in\mathbf{G}} \mathbf{N}^{-1*}{}_{sp}\Delta^*(R)_{pq}\mathbf{N}^*{}_{qt}\Delta(R)_{tr}\\ &= \Delta(A^2)_{rs}\mathbf{N}^{-1*}{}_{sp}\mathbf{N}^*{}_{qt}\frac{|\mathbf{G}|}{d}\delta_{pt}\delta_{qr}\\ &= \frac{|\mathbf{G}|}{d}\Delta(A^2)_{rs}\mathbf{N}^*{}_{rt}\mathbf{N}_{ts}\\ &= \pm\frac{|\mathbf{G}|}{d}\Delta(A^2)_{rs}\Delta^*(A^2)_{rs}\\ &= \pm\frac{|\mathbf{G}|}{d}\Delta(A^2)_{rs}\Delta(A^{-2})_{sr}\\ &= \pm\frac{|\mathbf{G}|}{d}\Delta(E)_{rr}\\ &= \pm|\mathbf{G}|.\end{aligned} \tag{7.3.49}$$

In the simplification we have used eqns (7.3.25) and (7.3.48) and the fact that Γ is unitary and of dimension d, so that the character of the identity is d.

In case (c) the sum is

$$\Delta(A^2)_{rs} \sum_{R\in\mathbf{G}} \bar{\Delta}^*(R)_{st}\Delta(R)_{tr} = 0. \tag{7.3.50}$$

This follows from eqn (7.3.48), Γ and $\bar{\Gamma}$ being inequivalent. Collecting these results and writing χ for the character of Γ we have

$$\sum_{B\in A\mathbf{G}} \chi(B^2) = \left.\begin{array}{r}+|\mathbf{G}| \text{ in case (a),}\\ -|\mathbf{G}| \text{ in case (b),}\\ 0 \quad \text{in case (c).}\end{array}\right\} \tag{7.3.51}$$

For the magnetic point groups it is then a relatively straightforward matter to determine their coreps (Cracknell 1966*a*, Cracknell and Wong 1967, Dimmock and Wheeler 1962*a*). The type I magnetic point groups are unitary groups and their reps, both single-valued and double-valued, have been tabulated already in earlier chapters. For a type II magnetic point group, **M**, the reps of the unitary subgroup **G** are immediately available from eqn (7.2.1). For a type III magnetic point group, **M**, the reps of the unitary subgroup may be found by consulting Table 7.1 to identify the unitary subgroup of **M**, and thus its reps. In either case, choosing a particular anti-unitary element to be A the coreps of **M** can be determined by seeing whether the reps Γ and $\bar{\Gamma}$ are equivalent. If they are not equivalent then the coreps of **M** are given by eqn (7.3.47). If Γ and $\bar{\Gamma}$ are equivalent and $\mathbf{N}\mathbf{N}^* = +\mathbf{\Delta}(A^2)$ where **N** is a unitary matrix such that, from eqn (7.3.25),

$$\mathbf{\Delta}(R) = \mathbf{N}\bar{\mathbf{\Delta}}(R)\mathbf{N}^{-1} \tag{7.3.52}$$

then the corep of **M** derived from $\mathbf{\Delta}(R)$ is given by eqn (7.3.45) and if $\mathbf{N}\mathbf{N}^* = -\mathbf{\Delta}(A^2)$ the corep of **M** derived from $\mathbf{\Delta}(R)$ is given by eqn (7.3.46). The deduction of the coreps of the magnetic space groups is complicated by the fact that they are infinite groups; we shall consider them in section 7.6. Rather than examine the matrices of Γ and $\bar{\Gamma}$ in detail one would use the test given in eqn (7.3.51) to determine whether the corep of **M** derived from Γ belongs to case (a), case (b), or case (c). Then only if it became necessary to find explicitly the matrices in eqn (7.3.45) or eqn (7.3.46) would one actually study the matrices of Γ and $\bar{\Gamma}$ and use eqn (7.3.25) to determine the matrix **N**. If the form of **N** is not immediately obvious it can be obtained (see eqn (4.5.51)) from the expression

$$\mathbf{N} = \frac{1}{|\mathbf{G}|} \sum_{R \in \mathbf{G}} \mathbf{\Delta}(R)\mathbf{X}\bar{\mathbf{\Delta}}(R^{-1}) \tag{7.3.53}$$

where **X** is a matrix conveniently chosen so that **N** is unitary; $\mathbf{X} = \mathbf{1}$ is not always a suitable choice since this choice for **X** may cause the right-hand side of eqn (7.3.53) to vanish.

7.4. The Kronecker products of corepresentations

In this section we consider the definition and reduction of the inner Kronecker product of any two irreducible corepresentations of a non-unitary group **M**. In keeping with the notation of section 7.3 we shall write Γ^i for an irreducible representation of the unitary subgroup **G** and for notational convenience later we define $\Gamma^{i'}$ where

$$\mathbf{\Delta}^{i'}(R) = \bar{\mathbf{\Delta}}^{i}(R) = \mathbf{\Delta}^{i*}(A^{-1}RA). \tag{7.4.1}$$

We denote by $D\Gamma^i$ the irreducible corepresentation of **M** derived from Γ^i.

We assume that the decompositions of inner Kronecker products within the unitary subgroup are known. That is to say, the Clebsch–Gordan coefficients $c_{ij,k}$ in the reduction

$$\Gamma^i \boxtimes \Gamma^j \equiv \sum_k c_{ij,k}\,\Gamma^k \tag{7.4.2}$$

are supposed known. Writing ψ^i for the character of Γ^i the formula for $c_{ij,k}$ is, from eqn (1.3.24),

$$c_{ij,k} = \frac{1}{|\mathbf{G}|} \sum_{R \in \mathbf{G}} \psi^i(R)\psi^j(R)\psi^k(R)^*. \tag{7.4.3}$$

A detailed analysis of how to obtain $c_{ij,k}$ from this formula in the case in which **G** is a crystal space group has already been given in section 4.7 using the method of little groups.

Suppose we define, for all elements $m \in \mathbf{M}$, and for any two irreducible corepresentations $D\Gamma^i$ and $D\Gamma^j$ the matrix $\mathbf{D}^{ij}$ whose elements are

$$\mathbf{D}^{ij}(m)_{pq,rs} = \mathbf{D}^i(m)_{pr}\mathbf{D}^j(m)_{qs} \tag{7.4.4}$$

so that the matrix $\mathbf{D}^{ij}$ is the Kronecker product of the matrices $\mathbf{D}^i$ and $\mathbf{D}^j$,

$$\mathbf{D}^{ij}(m) = \mathbf{D}^i(m) \otimes \mathbf{D}^j(m). \tag{7.4.5}$$

Then it can be fairly easily verified that the $\mathbf{D}^{ij}(m)$ form a corepresentation, which we call $D\Gamma^{ij}$, of **M**. That is to say, the matrices $\mathbf{D}^{ij}(m)$ satisfy equations of the form (7.3.19).

We know from section 7.3 that a corepresentation is uniquely determined (up to equivalence) by the characters of the elements of the unitary subgroup. Writing χ^i for the character of $D\Gamma^i$ (when restricted to **G**), then from eqn (7.4.4) we obtain, for all $R \in \mathbf{G}$,

$$\chi^{ij}(R) = \chi^i(R)\chi^j(R), \tag{7.4.6}$$

and these values characterize completely the corepresentation $D\Gamma^{ij}$. In general the corepresentation $D\Gamma^{ij}$ will be reducible. Suppose therefore that

$$D\Gamma^{ij} = D\Gamma^i \boxtimes D\Gamma^j \equiv \sum_k d_{ij,k} D\Gamma^k. \tag{7.4.7}$$

Then our problem is to determine $d_{ij,k}$. The first step is to relate $d_{ij,k}$ to the characters χ^i, χ^j, and χ^k of the unitary subgroup. This comes as a very simple generalization of eqn (7.4.3) (bearing in mind that $D\Gamma^k$ is possibly reducible under **G**), and the formula (Karavaev 1964) is

$$d_{ij,k} = \frac{\dfrac{1}{|\mathbf{G}|} \sum_{R \in \mathbf{G}} \chi^i(R)\chi^j(R)\chi^k(R)^*}{\dfrac{1}{|\mathbf{G}|} \sum_{R \in \mathbf{G}} \chi^k(R)\chi^k(R)^*}. \tag{7.4.8}$$

We now show how this formula may be simplified. If $D\Gamma^k$ is of type (a), (b), or (c) then the denominator of eqn (7.4.8) is 1, 4, or 2 respectively. This follows from eqns (7.3.45)–(7.3.47) and is a consequence of the fact that $D\Gamma^k$ contains, in case (a) just one irreducible representation of **G**, in case (b) one irreducible representation twice, and in case (c) two inequivalent irreducible representations once. That is,

$$\chi^k(R) = \left.\begin{array}{ll} \psi^k(R) & \text{in case (a),} \\ 2\psi^k(R) & \text{in case (b),} \\ \psi^k(R) + \psi^{k'}(R) & \text{in case (c).} \end{array}\right\} \tag{7.4.9}$$

Then, using the orthogonality relations for characters,

$$\frac{1}{|\mathbf{G}|}\sum_{R \in \mathbf{G}} \chi^k(R)\chi^k(R)^* = \left.\begin{array}{ll} 1 & \text{in case (a),} \\ 4 & \text{in case (b),} \\ 2 & \text{in case (c).} \end{array}\right\} \tag{7.4.10}$$

The second step is to convert eqn (7.4.8) into a relationship between $d_{ij,k}$ and the $c_{ij,k}$ of eqn (7.4.2). A laborious evaluation of the numerator of eqn (7.4.8) for each particular case is then unnecessary. Since the $c_{ij,k}$ are known the problem of the reduction of the inner Kronecker product of any two irreducible corepresentations is then solved for any non-unitary group **M**; in particular for magnetic space groups the values for $c_{ij,k}$ are first derived using the formulae of section 4.7. There are, taking into account the relation

$$d_{ij,k} = d_{ji,k} \tag{7.4.11}$$

eighteen different possibilities to consider according to whether $D\Gamma^i$, $D\Gamma^j$, and $D\Gamma^k$ are of types (a), (b), or (c). The formulae are quite straightforward to establish; one has to substitute eqns (7.4.9) and (7.4.10) into eqn (7.4.8) and to express the result in terms of $c_{ij,k}$, etc. using eqn (7.4.3). It is also necessary to use the fact that if $D\Gamma^k$ is of type (c) then if a corepresentation $D\Gamma^{ij}$ contains Γ^k it contains $\Gamma^{k'}$ an equal number of times. The values of the $d_{ij,k}$ are given in Table 7.8.

As an example of the use of Table 7.8 we consider the magnetic point group $4'mm'$. For this group **G**, the subgroup of anti-unitary elements, consists of the point group $mm2$ (C_{2v}), that is

$$\mathbf{G} = E, C_{2z}, \sigma_x, \sigma_y \tag{7.4.12}$$

and the set $A\mathbf{G}$ of anti-unitary elements consists of the elements

$$A\mathbf{G} = \theta C_{4z}^+, \theta C_{4z}^-, \theta\sigma_{da}, \theta\sigma_{db} \tag{7.4.13}$$

where θ is the operation of time inversion (see Table 7.1). The group **G** in this case therefore has four irreducible representations A_1, A_2, B_1, and B_2 (see Table 2.2). To make the notation coincide with that of this section we denote the representations A_1, A_2, B_1, and B_2 of **G** by Γ_1, Γ_2, Γ_3, and Γ_3' respectively. It can be shown by using eqn (7.3.51) that the irreducible corepresentations derived from Γ_1 and Γ_2

TABLE 7.8

Clebsch–Gordan coefficients for Kronecker products of irreducible corepresentations (Bradley and Davis 1968)

$D\Gamma^i$	$D\Gamma^j$	$D\Gamma^k$	$d_{ij,k}$
a	a	a	$c_{ij,k}$
a	a	b	$\frac{1}{2}c_{ij,k}$
a	a	c	$c_{ij,k}$
a	b	a	$2c_{ij,k}$
a	b	b	$c_{ij,k}$
a	b	c	$2c_{ij,k}$
a	c	a	$c_{ij,k} + c_{ij',k}$
a	c	b	$\frac{1}{2}c_{ij,k} + \frac{1}{2}c_{ij',k}$
a	c	c	$c_{ij,k} + c_{ij',k}$
b	b	a	$4c_{ij,k}$
b	b	b	$2c_{ij,k}$
b	b	c	$4c_{ij,k}$
b	c	a	$2c_{ij,k} + 2c_{ij',k}$
b	c	b	$c_{ij,k} + c_{ij',k}$
b	c	c	$2c_{ij,k} + 2c_{ij',k}$
c	c	a	$c_{ij,k} + c_{ij',k} + c_{i'j,k} + c_{i'j',k}$
c	c	b	$\frac{1}{2}c_{ij,k} + \frac{1}{2}c_{ij',k} + \frac{1}{2}c_{i'j,k} + \frac{1}{2}c_{i'j',k}$
c	c	c	$c_{ij,k} + c_{ij',k} + c_{i'j,k} + c_{i'j',k}$

Notes to Table 7.8

(i) The formulae are for *irreducible* corepresentations only.

(ii) Each row of the table supplies a formula for $d_{ij,k}$, as defined by eqn (7.4.7).

(iii) The appropriate row of the table to use depends on the type of the corepresentations involved. The first three entries in each row are respectively the types of $D\Gamma^i$, $D\Gamma^j$, and $D\Gamma^k$ (as defined by eqns (7.3.45)–(7.3.47).

(iv) The fourth entry in each row is the value of $d_{ij,k}$ in terms of the $c_{ij,k}$ as defined by eqn (7.4.2).

(v) Primed suffices refer to primed representations, see eqn (7.4.1). Thus $c_{ij',k}$ is the number of times that Γ^k appears in the decomposition of $\Gamma^i \boxtimes \Gamma^{j'}$

are of type (a) and the irreducible corepresentation derived from Γ_3 and Γ_3' is of type (c). We denote these corepresentations DA_1, DA_2, and DB by $D\Gamma_1$, $D\Gamma_2$, and $D\Gamma_3$ respectively.

We assume the Clebsch–Gordan coefficients for the Kronecker products of $mm2$ (C_{2v}) are known; they are given, for instance, by Koster, Dimmock, Wheeler, and Statz (1963). In fact

$$\begin{aligned} c_{11,1} = c_{12,2} = c_{13,3} = c_{13',3'} = c_{22,1} = c_{23,3'} = c_{23',3} \\ = c_{33,1} = c_{33',2} = c_{3'3',1} = 1 \end{aligned} \tag{7.4.14}$$

and all other $c_{ij,k}$ not derived from these by the relation

$$c_{ij,k} = c_{ji,k} \tag{7.4.15}$$

are zero.

We can then use the appropriate lines of Table 7.8 to obtain

$$\left.\begin{aligned} d_{11,1} &= c_{11,1} = 1, \\ d_{12,2} &= c_{12,2} = 1, \\ d_{13,3} &= c_{13,3} + c_{13',3} = 1, \\ d_{22,1} &= c_{22,1} = 1, \\ d_{23,3} &= c_{23,3} + c_{23',3} = 1, \\ d_{33,1} &= c_{33,1} + c_{33',1} + c_{3'3,1} + c_{3'3',1} = 2, \\ d_{33,2} &= c_{33,2} + c_{33',2} + c_{3'3,2} + c_{3'3',2} = 2, \end{aligned}\right\} \tag{7.4.16}$$

and all other $d_{ij,k}$ not derived from these by eqn (7.4.11) vanish. Thus the Clebsch–Gordan decomposition for $4'mm'$ is

$$\left.\begin{aligned} DA_1 \boxtimes DA_1 &= DA_1 \\ DA_1 \boxtimes DA_2 &= DA_2 \\ DA_1 \boxtimes DB &= DB \\ DA_2 \boxtimes DA_2 &= DA_1 \\ DA_2 \boxtimes DB &= DB \\ DB \boxtimes DB &= 2DA_1 + 2DA_2. \end{aligned}\right\} \tag{7.4.17}$$

The results of applying this theory to all the 58 type III magnetic point groups and the application of these results to crystal field theory have been reviewed by Cracknell (1968*a*).

7.5. The corepresentations of the magnetic point groups

The grey point groups; the consequences of invariance under time-inversion

It has already been mentioned that the effect of the operation of time inversion is to reverse the magnetic moment of an atom or ion. Thus for an ion in a paramagnetic or diamagnetic crystal, in the absence of an external magnetic field, the crystal is invariant under the operation of time inversion so that we can say that the ordinary point groups, or type I Shubnikov point groups, do not adequately describe the symmetry of the local environment of the ion and it is better described by one of the corresponding type II Shubnikov groups, the grey point groups. It is fairly straightforward to consider these groups and to derive their coreps and hence to see if any extra degeneracies are introduced by considering the invariance of the crystal under the operation of time inversion (Cracknell 1966*a*, Cracknell and Wong 1967, Dimmock and Wheeler 1962*a*, 1964).

For each of the 32 grey point groups we can choose A to be the element θ, the operation of time inversion itself; this means that $\bar{\mathbf{\Delta}}(R) = \mathbf{\Delta}(\theta^{-1}R\theta)^* = \mathbf{\Delta}(R)^*$. The reps $\boldsymbol{\Delta}(R)$ for **G** for each of the grey point groups are available from Tables 2.2 and 6.5. There are several different possibilities to consider. First, we may consider the single-valued representations; these are used in the description of particles with zero or integer spin and for these it can be shown (Wigner 1959) that $\boldsymbol{\Delta}(A^2) = \boldsymbol{\Delta}(\theta^2) = +\mathbf{1}$. For the real non-degenerate reps of **G**, $\bar{\boldsymbol{\Delta}}(R)$ is not only equivalent but also identical to $\boldsymbol{\Delta}(R)$ so that we may choose $\mathbf{N} = \mathbf{1}$. The coreps of **M** can therefore be found by using eqn (7.3.45), case (a). For complex non-degenerate reps of **G**, $\bar{\boldsymbol{\Delta}}(R) = \boldsymbol{\Delta}(R)^*$ and is inequivalent to $\boldsymbol{\Delta}(R)$, so that the coreps of **M** can be found by using eqn (7.3.47), case (c). For degenerate single-valued reps it so happens that the matrices $\boldsymbol{\Delta}(R)$ can all be chosen to have real elements, as can be seen from Table 2.3, and therefore once again $\bar{\boldsymbol{\Delta}}(R)$ is not only equivalent but also identical to $\boldsymbol{\Delta}(R)$ so that the coreps of **M** can be found using eqn (7.3.45), case (a), with $\mathbf{N} = \mathbf{1}$. This covers all possible cases for single-valued reps.

It thus remains to consider the double-valued reps; these are used in the description of particles with half-odd-integer spin and for these it can be shown (Wigner 1959) that $\boldsymbol{\Delta}(A^2) = \boldsymbol{\Delta}(\theta^2) = -\mathbf{1}$. The various possible general forms of $\boldsymbol{\Delta}(R)$ and the coreps of the grey point group **M** to which they lead can be examined one by one just as for the single-valued reps. The results for both single-valued and double-valued reps are given in Tables 7.9 and 7.11. Except in the case of degenerate reps $\boldsymbol{\Delta}(R)$ with real characters but with some matrices necessarily complex the coreps can be written down immediately using Table 7.9 and the matrices $\boldsymbol{\Delta}(R)$ themselves. For this one remaining case the characters of $\boldsymbol{\Delta}(R)$ and $\bar{\boldsymbol{\Delta}}(R)$ must be the same so that coreps derived from $\boldsymbol{\Delta}(R)$ must belong to either case (a) or case (b); which of these actually happens can be found by determining **N** either by direct inspection of $\boldsymbol{\Delta}(R)$ and

TABLE 7.9

Rules for constructing coreps of grey point groups

$\boldsymbol{\Delta}(R)$	Corep	Equation	**N**
(i) Single-valued:			
Non-degenerate, real	a	(7.3.45)	**1**
Non-degenerate, complex	c	(7.3.47)	—
Degenerate, real	a	(7.3.45)	**1**
(ii) Double-valued:			
Non-degenerate, complex	c	(7.3.47)	—
Non-degenerate, real	b	(7.3.46)	**1**
Degenerate, some characters complex	c	(7.3.47)	—
Degenerate, all characters real	a	(7.3.45)	see Table 7.11

Note to Table 7.9

The entry in column 2 indicates whether the corep derived from $\boldsymbol{\Delta}(R)$ belongs to case (a), case (b), or case (c); the entry in column 3 gives the equations to be used in determining the coreps of **M** and column 4 gives **N**.

$\bar{\mathbf{\Delta}}(R)$ or from eqn (7.3.53). Alternatively, one can use the test given in eqn (7.3.51) to determine whether the corep derived from $\mathbf{\Delta}(R)$ belongs to case (a) or case (b); however, this does not achieve very much saving of effort if one actually wishes to write down the matrices in the coreps because the matrix $\mathbf{N}$ still has to be found. In fact all coreps of this type belong to case (a).

As an example we consider the grey point group 2221′ which is derived from 222 (D_2); in Table 7.10 we list $\mathbf{\Delta}(R)$ and $\bar{\mathbf{\Delta}}(R)$ for the rep $\bar{E}$. By inspection it can easily be seen that $\mathbf{N}$ must be the matrix $\begin{pmatrix} 0 & 1 \\ -1 & 0 \end{pmatrix}$ so that $\mathbf{NN}^* = -\mathbf{1} = +\mathbf{\Delta}(A^2)$ since this is a double-valued rep of $\mathbf{G}$ and $A = \theta$. The corep of the grey point group $\mathbf{G} + \theta\mathbf{G}$ derived from the rep $\bar{E}$ of $\mathbf{G}$ can thus be written down using eqn (7.3.45) and the matrix $\mathbf{N}$ that we have just found.

TABLE 7.10

2221′, $\mathbf{\Delta}(R)$, and $\bar{\mathbf{\Delta}}(R)$ for $\bar{E}$ of 222 (D_2)

R	$\mathbf{\Delta}(R)$	$\bar{\mathbf{\Delta}}(R)$	R	$\mathbf{\Delta}(R)$	$\bar{\mathbf{\Delta}}(R)$
E	$\begin{pmatrix} 1 & 0 \\ 0 & 1 \end{pmatrix}$	$\begin{pmatrix} 1 & 0 \\ 0 & 1 \end{pmatrix}$	C_{2y}	$\begin{pmatrix} 0 & i \\ i & 0 \end{pmatrix}$	$\begin{pmatrix} 0 & -i \\ -i & 0 \end{pmatrix}$
C_{2x}	$\begin{pmatrix} 0 & 1 \\ -1 & 0 \end{pmatrix}$	$\begin{pmatrix} 0 & 1 \\ -1 & 0 \end{pmatrix}$	C_{2z}	$\begin{pmatrix} i & 0 \\ 0 & -i \end{pmatrix}$	$\begin{pmatrix} -i & 0 \\ 0 & i \end{pmatrix}$

In Table 7.11 we give the information that enables those coreps of the grey point groups that are derived from degenerate double-valued reps of the point groups to be written down. This completes the derivation of the double-valued coreps of the grey point groups; the rules for all forms of $\mathbf{\Delta}(R)$ except one have been given in Table 7.9 and for this one exception the coreps can be found using Table 7.11.

In fact one can see what extra degeneracies are present due to the presence of time-inversion symmetry without any explicit reference to the theory of corepresentations of type II Shubnikov groups. This was done by Herring (1937*a*) by considering the reality of the representations, following the earlier version of the work on time-inversion by Wigner (1932). It was to enable the reader to see at a glance whether or not invariance under the operation of time inversion would cause a particular rep of a Fedorov space group to have its degeneracy doubled, that we considered the theory of the reality of these reps in section 4.6 and stated the reality of each rep in Tables 5.7 and 6.13. The theory was illustrated in section 5.4 for the space group $F\bar{4}3c$ (T_d^5).

We shall now show that the application of the theory of the corepresentations of the type II Shubnikov groups leads to exactly the same conclusions as to the relation between the reality of the reps and the extra degeneracies to be expected in a crystal which is invariant under the operation of time inversion. The type II Shubnikov group **M** is given by eqns (7.2.1) or (7.2.3)

$$\mathbf{M} = \mathbf{G} + \theta\mathbf{G} \tag{7.5.1}$$

so that it is natural to choose as A the operation of time inversion itself.

For a system with half-odd-integer spin, $\mathbf{\Delta}(\theta^2) = -1$, and therefore

$$\mathbf{\Delta}(A^2) = -\mathbf{1}. \tag{7.5.2}$$

TABLE 7.11

The degenerate double-valued coreps of the grey point groups

$$\kappa = \begin{pmatrix} 0 & 1 \\ -1 & 0 \end{pmatrix}.$$

(Direct products are not included)

G	Rep of G	Corep of M	N	G	Rep of G	Corep of M	N
$mm2$ (C_{2v})	$\bar{E}$	a	κ	$6mm$ (C_{6v})	$\bar{E}_1, \bar{E}_2, \bar{E}_3$	a	κ
222 (D_2)	$\bar{E}$	a	κ	$\bar{6}2m$ (D_{3h})	$\bar{E}_1, \bar{E}_2, \bar{E}_3$	a	κ
32 (D_3)	$\bar{E}$	a	κ	23 (T)	$\bar{E}$	a	κ
$3m$ (C_{3v})	$\bar{E}$	a	κ		${}^1\bar{F}, {}^2\bar{F}$	c	—
422 (D_4)	$\bar{E}_1, \bar{E}_2$	a	κ	432 (O)	$\bar{E}_1, \bar{E}_2$	a	κ
$4mm$ (C_{4v})	$\bar{E}_1, \bar{E}_2$	a	κ		$\bar{F}$	a	$\begin{pmatrix} \kappa & 0 \\ 0 & \kappa \end{pmatrix}$
$\bar{4}2m$ (D_{2d})	$\bar{E}_1, \bar{E}_2$	a	κ				
622 (D_6)	$\bar{E}_1, \bar{E}_2, \bar{E}_3$	a	κ	$\bar{4}3m$ (T_d)	$\bar{E}_1, \bar{E}_2$	a	κ
					$\bar{F}$	a	$\begin{pmatrix} \kappa & 0 \\ 0 & \kappa \end{pmatrix}$

Note to Table 7.11

$\mathbf{\Delta}(R)$ is listed in column 2, in column 3 the letter a, b, or c indicates that the corep of $\mathbf{G} + \theta\mathbf{G}$ derived from that rep of **G** belongs to case (a), case (b), or case (c) respectively; the actual matrices will then be given by eqn (7.3.45), (7.3.46), or (7.3.47) respectively, and column 4 gives **N** for those cases for which it exists.

The reps $\mathbf{\Delta}(R)$ of the unitary subgroup **G** are just the ordinary reps, and there are thus three possibilities to consider:

(*i*) Γ is real,

(*ii*) Γ is equivalent to Γ^* but not to any group of real matrices,

(*iii*) Γ is not equivalent to Γ^*;

that is, in the sense of Definition 1.3.7 the rep Γ is of the first, second, or third kind respectively. In (*i*) the reps Γ and $\bar{\Gamma}$ (in which $\bar{\mathbf{\Delta}}(R) = \mathbf{\Delta}(R)^*$) are obviously not only equivalent but also identical so that we may choose $\mathbf{N} = \mathbf{1}$ and therefore $\mathbf{NN}^* = \mathbf{1}$; hence from eqn (7.5.2)

$$\mathbf{NN}^* = -\mathbf{\Delta}(A^2) \tag{7.5.3}$$

and the corep of **M** belongs to case (b), which shows that there is an extra degeneracy. In (*ii*) Γ and $\bar{\Gamma}$ are equivalent but not identical and the corep of **M** depends on **NN***. In this case Wigner (1959) has shown that $\mathbf{NN}^* = -\mathbf{1}$, so that using eqn (7.5.2)

$$\mathbf{NN}^* = -\mathbf{1} = \Delta(A^2). \tag{7.5.4}$$

The coreps of **M** therefore belong to case (a) and there is thus no extra degeneracy. Finally in (*iii*) the corep of **M** obviously belongs to case (c) and leads to an extra degeneracy. Summarizing then, for particles with half-odd-integer spin, we have

(*i*) extra degeneracy (case (b)),
(*ii*) no extra degeneracy (case (a)), and
(*iii*) extra degeneracy (case (c)).

This is the same as the results of Wigner (1932) and Herring (1937a).

For a system with zero or integer spin, eqn (7.5.2) becomes replaced by, see Wigner (1959),

$$\Delta(A^2) = +1 \tag{7.5.5}$$

and the results for cases (*i*) and (*ii*) are interchanged, while the result for case (*iii*) remains unaltered. This is also in agreement with the conclusions of Wigner and Herring, see sections 4.6 and 5.4, so that, for particles with zero or integer spin, we have

(*i*) no extra degeneracy (case (a)),
(*ii*) extra degeneracy (case (b)), and
(*iii*) extra degeneracy (case (c)).

In the above discussion, by an extra degeneracy we always mean a doubling of the degeneracy that would have resulted had **G** been the appropriate symmetry group rather than $\mathbf{M}(= \mathbf{G} + \theta\mathbf{G})$. This is because the corep $D\Gamma$ derived from Γ has twice the dimension of Γ in cases (b) and (c); on the other hand $D\Gamma$ has the same dimension as Γ in case (a) which is the case of no extra degeneracy. We have already specified the realities of the single-valued reps of the crystallographic point groups in Note (v) to Table 2.2, while the realities of the double-valued point-group reps were given in Note (vii) to Table 6.5; in both cases this gives the same results as the direct use of Table 7.9 for determining the extra degeneracies that arise if θ is added to the symmetry operations of the point group **G**.

The black and white point groups, an example m′3

We now turn to the problem of the tabulation of the coreps of the type III Shubnikov point groups, or the black and white magnetic point groups. We illustrate the deduction of these coreps with the example of the magnetic point group $m'3$ (Cracknell 1966*a*).

From Table 7.1 it can be seen that for the group $m'3$ the unitary subgroup **H** of Definition 7.2.2 is the point group 23 (T) and the character table for the single-valued representations of this group is given in Table 7.12. The fact that elements of $(\mathbf{G} - \mathbf{H})$ have been multiplied by θ does not alter the division of the group into classes, thus since $m3$ (T_h) is a direct product group $23 \otimes \bar{1}(T \otimes C_i)$, and has 8 classes, we can write

$$\begin{aligned} m'3 &= 23 + \theta I23 \\ (&= T + \theta IT). \end{aligned} \tag{7.5.6}$$

TABLE 7.12

The character table of the point group 23 (T)

H of $m'3$		E	C_{2m}	C_{3j}^{+}	C_{3j}^{-}
A	Γ_1	1	1	1	1
${}^1E$	Γ_2	1	1	ω	ω^*
${}^2E$	Γ_3	1	1	ω^*	ω
T	Γ_4	3	-1	0	0

$\omega = \exp(2\pi i/3)$.

This magnetic group is also a direct product of 23 (T) and the group $(E + \theta I)$, and it has 8 classes also:

$$\begin{array}{ll} E & \theta I \\ C_{2x}, C_{2y}, C_{2z} & \theta\sigma_x, \theta\sigma_y, \theta\sigma_z \\ C_{31}^{-}, C_{32}^{-}, C_{33}^{-}, C_{34}^{-} & \theta S_{61}^{+}, \theta S_{62}^{+}, \theta S_{63}^{+}, \theta S_{64}^{+} \\ C_{31}^{+}, C_{32}^{+}, C_{33}^{+}, C_{34}^{+} & \theta S_{61}^{-}, \theta S_{62}^{-}, \theta S_{63}^{-}, \theta S_{64}^{-}. \end{array}$$

We make the simplest possible choice of A, namely θI, and apply the results of section 7.3 in turn to each of the reps of 23 (T), the unitary subgroup $\mathbf{H}$ of the magnetic group $m'3$. For reps A and T, $\mathbf{\Delta}(R)$ is real, and since θI commutes with every element, $\bar{\mathbf{\Delta}}(R) = \mathbf{\Delta}(R)^* = \mathbf{\Delta}(R)$ and therefore $\mathbf{\Delta}(R)$ and $\bar{\mathbf{\Delta}}(R)$ are not only equivalent but also identical so that $\mathbf{NN}^* = \mathbf{1}$; therefore

$$\mathbf{NN}^* = +\mathbf{1} = +\mathbf{\Delta}(A^2) \tag{7.5.7}$$

and therefore the coreps of $\mathbf{M}$ derived from A and T belong to case (a). For the complex conjugate reps ${}^1E$ and ${}^2E$, $\bar{\Delta}(R) = \Delta(R)^*$ and thus $\Delta(R)$ and $\bar{\Delta}(R)$ are inequivalent and the coreps derived from ${}^1E$ and ${}^2E$ belong to case (c). Using eqn. (7.3.47) we find

$$\begin{aligned} \mathbf{D}(R) &= \begin{pmatrix} \mathbf{\Delta}(R) & \mathbf{0} \\ \mathbf{0} & \mathbf{\Delta}(R)^* \end{pmatrix} \\ \mathbf{D}(AR) &= \begin{pmatrix} \mathbf{0} & \mathbf{\Delta}(RA^2) \\ \mathbf{\Delta}(R)^* & \mathbf{0} \end{pmatrix} \end{aligned} \tag{7.5.8}$$

so that, using ${}^1E$, the matrices are

$$\begin{array}{ccccc} & E & C_{2m} & C_{3j}^+ & C_{3j}^- \\ \mathbf{D}(R) & \begin{pmatrix}1 & 0\\0 & 1\end{pmatrix} & \begin{pmatrix}1 & 0\\0 & 1\end{pmatrix} & \begin{pmatrix}\omega & 0\\0 & \omega^*\end{pmatrix} & \begin{pmatrix}\omega^* & 0\\0 & \omega\end{pmatrix} \\ & \theta I & \theta\sigma_m & \theta S_{6j}^- & \theta S_{6j}^+ \\ \mathbf{D}(AR) & \begin{pmatrix}0 & 1\\1 & 0\end{pmatrix} & \begin{pmatrix}0 & 1\\1 & 0\end{pmatrix} & \begin{pmatrix}0 & \omega\\\omega^* & 0\end{pmatrix} & \begin{pmatrix}0 & \omega^*\\\omega & 0\end{pmatrix}\end{array}$$

By using ${}^2E$ one would deduce a corep equivalent to this.

It remains to find the double-valued coreps of this group. The characters of the double-valued reps of **H** were given in Table 6.5 and are reproduced in Table 7.13.

TABLE 7.13

The double-valued representations of the point group 23 (T)

	E	$\bar{E}$	C_{2x}, C_{2y}, C_{2z}, $\bar{C}_{2x}, \bar{C}_{2y}, \bar{C}_{2z}$	C_{31}^-, C_{32}^-, C_{33}^-, C_{34}^-	$\bar{C}_{31}^-, \bar{C}_{32}^-$, $\bar{C}_{33}^-, \bar{C}_{34}^-$	$\bar{C}_{31}^+, \bar{C}_{32}^+$, $\bar{C}_{33}^+, \bar{C}_{34}^+$	C_{31}^+, C_{32}^+, C_{33}^+, C_{34}^+
$\bar{E}$	2	-2	0	1	-1	-1	1
${}^1\bar{F}$	2	-2	0	ω	$-\omega$	$-\omega^*$	ω^*
${}^2\bar{F}$	2	-2	0	ω^*	$-\omega^*$	$-\omega$	ω

$\omega = \exp(2\pi i/3)$.

We can still use θI as A so that once more $\bar{\mathbf{\Delta}}(R) = \mathbf{\Delta}(R)^*$. For the representation $\bar{E}$, $\bar{\mathbf{\Delta}}(R)$ must be equivalent to $\mathbf{\Delta}(R)$. To determine the coreps derived from $\bar{E}$ it is necessary to find $\mathbf{N}$; this can be done using

$$\mathbf{\Delta}(R) = \mathbf{N}\bar{\mathbf{\Delta}}(R)\mathbf{N}^{-1} \tag{7.3.25}$$

and it is fairly easy to show that $\mathbf{N} = \begin{pmatrix}0 & 1\\-1 & 0\end{pmatrix}$, so that

$$\mathbf{NN}^* = \begin{pmatrix}0 & 1\\-1 & 0\end{pmatrix}\begin{pmatrix}0 & 1\\-1 & 0\end{pmatrix} = -\mathbf{1}. \tag{7.5.9}$$

Since when we use the double-valued representations we are dealing with an odd number of electrons $\mathbf{\Delta}(A^2) = \mathbf{\Delta}((\theta I)^2) = \mathbf{\Delta}(\theta^2) = -\mathbf{1}$ and therefore $\mathbf{NN}^* = +\mathbf{\Delta}(A^2)$ and the double-valued corep of $m'3$ derived from $\bar{E}$ belongs to case (a), with

$$\mathbf{N} = \begin{pmatrix}0 & 1\\-1 & 0\end{pmatrix}. \tag{7.5.10}$$

${}^1\bar{F}$ and ${}^2\bar{F}$, however, are complex conjugate representations so that $\bar{\Delta}(R)(=\Delta(R)^*)$ and $\Delta(R)$ are inequivalent and the double-valued coreps of $m'3$ derived from ${}^1\bar{F}$ and ${}^2\bar{F}$ belong to case (c). Thus we can summarize the coreps of $m'3$ derived from the reps of 23 (T) as shown in Table 7.14. Rather than use the method of inspection which we have described it is possible to use eqn (7.3.51) to see whether a corep belongs to case (a), (b), or (c), that is

$$\sum_{B\in A\mathbf{G}} \chi(B^2) = \left.\begin{array}{l} +|\mathbf{G}| \text{ in case (a),} \\ -|\mathbf{G}| \text{ in case (b),} \\ \;\;0 \;\;\text{ in case (c).} \end{array}\right\} \tag{7.3.51}$$

The coreps in the first part of Table 7.14 are single-valued and those in the second part are double-valued. That is, the first part is relevant to the wave function of a

TABLE 7.14

The coreps of the magnetic point group m'3

Rep of **H**, 23 (T)	Case (a), (b), or (c)	**N**
A	a	$+1$
${}^1E$	c	—
${}^2E$	c	—
T	a	$\begin{pmatrix} 1 & 0 & 0 \\ 0 & 1 & 0 \\ 0 & 0 & 1 \end{pmatrix}$
$\bar{E}$	a	$\begin{pmatrix} 0 & 1 \\ -1 & 0 \end{pmatrix}$
${}^1\bar{F}$	c	—
${}^2\bar{F}$	c	—

Note to Table 7.14

For case (a) the coreps can be found using eqn (7.3.45) and the value of **N** in column 3, for case (b) they can be found using eqn (7.3.46) and the value of **N** in column 3, and for case (c) they can be found using eqn (7.3.47).

spinless particle or a particle with integer spin, while the second part is relevant to the wave function of a particle with a spin of half an odd integer. However, if one is using the common first approximation of neglecting the spin of a particle, for example as in the simple consideration of atoms neglecting spin–orbit coupling, it is the first part of the table that will be used. The fact that it was possible to choose as A an element that commutes with every element of **H** simplifies this analysis because then

$$\bar{\Delta}(R) = \Delta(R)^*, \tag{7.5.11}$$

whereas in general no such simplification of $\bar{\Delta}(R)$ is possible.

In Table 7.1 we identified for each magnetic point group **M** the unitary subgroup **H**, the classes of the set of elements $\mathscr{R}(\mathbf{G}-\mathbf{H})$ and the element A chosen for deducing the coreps of **M**. As Wigner (1959) demonstrates, the actual choice of A does not affect the coreps when they are finally derived but does involve differences in the algebra of their derivation. In Table 7.15 are listed the single-valued and double-valued reps of the unitary subgroup **H** of each magnetic group **M** together with the data which enables the coreps of **M** to be written down immediately, as in the case of $m'3$ in Table 7.14, using eqns (7.3.45)–(7.3.47). The characters, and in the case of degenerate representations the matrix representatives as well, can be found from Tables 2.2 and 2.3 for the single-valued reps and from Table 6.5 for the double-valued reps.

TABLE 7.15

The corepresentations of the magnetic point groups

(a)

M H	$\bar{1}'$ (1) 1 (C_1)		2′ (2) 1 (C_1)		m' (3) 1 (C_1)	
Rep of **H**	A	$\bar{A}$	A	$\bar{A}$	A	$\bar{A}$
Corep of **M**	a	b	a	a	a	a
N	α	α	α	α	α	α

(b)

M H	2/m' (4) 2 (C_2)				2′/m (5) m (C_{1h})				2′/m' (6) $\bar{1}$ (C_i)				2′2′2 (7) 2 (C_2)			
Rep of **H**	A	B	$^1\bar{E}$	$^2\bar{E}$	A'	A''	$^1\bar{E}$	$^2\bar{E}$	A_g	A_u	$\bar{A}_g$	$\bar{A}_u$	A	B	$^1\bar{E}$	$^2\bar{E}$
Corep of **M**	a	a	c	c	a	a	c	c	a	a	a	a	a	a	a	a
N	α	α	–	–	α	α	–	–	α	α	α	α	α	α	α	α

M H	$m'm'$2 (8) 2 (C_2)				$m'm$2′ (9) m (C_{1h})‡				4′ (13) 2 (C_2)				$\bar{4}'$ (14) 2 (C_2)			
Rep of **H**	A	B	$^1\bar{E}$	$^2\bar{E}$	A'	A''	$^1\bar{E}$	$^2\bar{E}$	A	B	$^1\bar{E}$	$^2\bar{E}$	A	B	$^1\bar{E}$	$^2\bar{E}$
Corep of **M**	a	a	a	a	a	a	a	a	a	b	c	c	a	b	c	c
N	α	α	α	α	α	α	α	α	α	α	–	–	α	α	–	–

‡ The elements of **H** are E, $\bar{E}$, σ_y, and $\bar{\sigma}_y$ instead of the conventional E, $\bar{E}$, σ_z, and $\bar{\sigma}_z$.

(c)

M **H**	$m'm'm'$ (10) 222 (D_2)					$4'22'$ (16) 222 (D_2)					$\bar{4}'2m'$ (23) 222 (D_2)				
Rep of **H**	A	B_3	B_2	B_1	$\bar{E}$	A	B_3	B_2	B_1	$\bar{E}$	A	B_3	B_2	B_1	$\bar{E}$
Corep of **M**	a	a	a	a	a	a	c	c	a	a	a	c	c	a	a
N	α	α	α	α	ρ	α	–	–	α	σ	α	–	–	α	σ

M **H**	mmm' (11) $mm2$ (C_{2v})					$4'mm'$ (21) $mm2$ (C_{2v})					$\bar{4}'m2'$ (24) $mm2$ (C_{2v})				
Rep of **H**	A_1	A_2	B_1	B_2	$\bar{E}$	A_1	A_2	B_1	B_2	$\bar{E}$	A_1	A_2	B_1	B_2	$\bar{E}$
Corep of **M**	a	a	a	a	a	a	a	c	c	a	a	a	c	c	a
N	α	α	α	α	ρ	α	α	–	–	σ	α	α	–	–	σ

M **H**	$m'm'm$ (12) $2/m$ (C_{2h})								$4'/m$ (19) $2/m$ (C_{2h})							
Rep of **H**	A_g	A_u	B_g	B_u	${}^1\bar{E}_g$	${}^2\bar{E}_g$	${}^1\bar{E}_u$	${}^2\bar{E}_u$	A_g	A_u	B_g	B_u	${}^1\bar{E}_g$	${}^2\bar{E}_g$	${}^1\bar{E}_u$	${}^2\bar{E}_u$
Corep of **M**	a	a	a	a	a	a	a	a	a	a	b	b	c	c	c	c
N	α	α	α	α	α	α	α	α	α	α	α	α	–	–	–	–

(d)

M **H**	$4'/mmm'$ (27) mmm (D_{2h})									
Rep of **H**	A_g	B_{3g}	B_{2g}	B_{1g}	A_u	B_{3u}	B_{2u}	B_{1u}	$\bar{E}_g$	$\bar{E}_u$
Corep of **M**	a	c	c	a	a	c	c	a	a	a
N	α	–	–	α	α	–	–	α	σ	σ

(e)

M **H**	$42'2'$ (15) 4 (C_4)								$4/m'$ (17) 4 (C_4)							
Rep of **H**	A	B	${}^1E$	${}^2E$	${}^1\bar{E}_1$	${}^2\bar{E}_1$	${}^1\bar{E}_2$	${}^2\bar{E}_2$	A	B	${}^1E$	${}^2E$	${}^1\bar{E}_1$	${}^2\bar{E}_1$	${}^1\bar{E}_2$	${}^2\bar{E}_2$
Corep of **M**	a	a	a	a	a	a	a	a	a	a	c	c	c	c	c	c
N	α	α	α	α	α	α	α	α	α	α	–	–	–	–	–	–

M **H**	$4m'm'$ (20) 4 (C_4)								$4'/m'$ (18) $\bar{4}$ (S_4)							
Rep of **H**	A	B	${}^1E$	${}^2E$	${}^1\bar{E}_1$	${}^2\bar{E}_1$	${}^1\bar{E}_2$	${}^2\bar{E}_2$	A	B	${}^1E$	${}^2E$	${}^1\bar{E}_1$	${}^2\bar{E}_1$	${}^1\bar{E}_2$	${}^2\bar{E}_2$
Corep of **M**	a	a	a	a	a	a	a	a	a	a	c	c	c	c	c	c
N	α	α	α	α	α	α	α	α	α	α	–	–	–	–	–	–

M H	$\bar{4}2'm'$ (22) $\bar{4}$ (S_4)							
Rep of **H**	A	B	^1E	^2E	$^1\bar{E}_1$	$^2\bar{E}_1$	$^1\bar{E}_2$	$^2\bar{E}_2$
Corep of **M**	a	a	a	a	a	a	a	a
N	α	α	α	α	α	α	α	α

(f)

M H	$4/mm'm'$ (29) $4/m$ (C_{4h})															
Rep of **H**	A_g	B_g	^1E_g	^2E_g	A_u	B_u	^1E_u	^2E_u	$^1\bar{E}_{1g}$	$^2\bar{E}_{1g}$	$^1\bar{E}_{2g}$	$^2\bar{E}_{2g}$	$^1\bar{E}_{1u}$	$^2\bar{E}_{1u}$	$^1\bar{E}_{2u}$	$^2\bar{E}_{2u}$
Corep of **M**	a	a	a	a	a	a	a	a	a	a	a	a	a	a	a	a
N	α	α	α	α	α	α	α	α	α	α	α	α	α	α	α	α

(g)

M H	$4/m'm'm'$ (25) 422 (D_4)							$4/m'mm$ (26) $4mm$ (C_{4v})						
Rep of **H**	A_1	A_2	B_1	B_2	E	$\bar{E}_1$	$\bar{E}_2$	A_1	A_2	B_1	B_2	E	$\bar{E}_1$	$\bar{E}_2$
Corep of **M**	a	a	a	a	a	a	a	a	a	a	a	a	a	a
N	α	α	α	α	ε	ρ	ρ	α	α	α	α	ε	ρ	ρ

M H	$4'/m'm'm$ (28) $\bar{4}2m$ (D_{2d})						
Rep of **H**	A_1	A_2	B_1	B_2	E	$\bar{E}_1$	$\bar{E}_2$
Corep of **M**	a	a	a	a	a	a	a
N	α	α	α	α	ε	ρ	ρ

(h)

M H	$32'$ (30) 3 (C_3)						$3m'$ (31) 3 (C_3)					
Rep of **H**	A	^1E	^2E	$\bar{A}$	$^1\bar{E}$	$^2\bar{E}$	A	^1E	^2E	$\bar{A}$	$^1\bar{E}$	$^2\bar{E}$
Corep of **M**	a	a	a	a	a	a	a	a	a	a	a	a
N	α	α	α	α	α	α	α	α	α	α	α	α

M **H**	$\bar{6}'$ (32) $3\ (C_3)$						$6'$ (36) $3\ (C_3)$					
Rep of **H**	A	^1E	^2E	$\bar{A}$	$^1\bar{E}$	$^2\bar{E}$	A	^1E	^2E	$\bar{A}$	$^1\bar{E}$	$^2\bar{E}$
Corep of **M**	a	c	c	a	c	c	a	c	c	a	c	c
N	α	–	–	α	–	–	α	–	–	α	–	–

M **H**	$\bar{3}'$ (37) $3\ (C_3)$					
Rep of **H**	A	^1E	^2E	$\bar{A}$	$^1\bar{E}$	$^2\bar{E}$
Corep of **M**	b	c	c	a	c	c
N	α	–	–	α	–	–

(i)

M **H**	$\bar{3}m'$ (38) $\bar{3}\ (C_{3i})$											
Rep of **H**	A_g	^1E_g	^2E_g	A_u	^1E_u	^2E_u	$\bar{A}_g$	$^1\bar{E}_g$	$^2\bar{E}_g$	$\bar{A}_u$	$^1\bar{E}_u$	$^2\bar{E}_u$
Corep of **M**	a	a	a	a	a	a	a	a	a	a	a	a
N	α	α	α	α	α	α	α	α	α	α	α	α

M **H**	$6'/m'$ (44) $\bar{3}\ (C_{3i})$											
Rep of **H**	A_g	^1E_g	^2E_g	A_u	^1E_u	^2E_u	$\bar{A}_g$	$^1\bar{E}_g$	$^2\bar{E}_g$	$\bar{A}_u$	$^1\bar{E}_u$	$^2\bar{E}_u$
Corep of **M**	a	c	c	a	c	c	a	c	c	a	c	c
N	α	–	–	α	–	–	α	–	–	α	–	–

(j)

M **H**	$\bar{6}'m2'$ (34) $3m\ (C_{3v})$						$\bar{3}'m$ (39) $3m\ (C_{3v})$					
Rep of **H**	A_1	A_2	E	$^1\bar{E}$	$^2\bar{E}$	$\bar{E}_1$	A_1	A_2	E	$^1\bar{E}$	$^2\bar{E}$	$\bar{E}_1$
Corep of **M**	a	a	a	a	a	a	a	a	a	c	c	a
N	α	α	ε	α	α	ε	α	α	ε	–	–	ρ

M **H**	$6'm'm$ (47) $3m\ (C_{3v})$						$\bar{6}'m'2$ (35) $32\ (D_3)$					
Rep of **H**	A_1	A_2	E	$^1\bar{E}$	$^2\bar{E}$	$\bar{E}_1$	A_1	A_2	E	$^1\bar{E}$	$^2\bar{E}$	$\bar{E}_1$
Corep of **M**	a	a	a	a	a	a	a	a	a	a	a	a
N	α	α	ε	α	α	ε	α	α	ε	α	α	ε

M H	$\bar{3}'m'$ (40) 32 (D_3)						$6'2'2$ (42) 32 (D_3)					
Rep of H	A_1	A_2	E	$^1\bar{E}$	$^2\bar{E}$	$\bar{E}_1$	A_1	A_2	E	$^1\bar{E}$	$^2\bar{E}$	$\bar{E}_1$
Corep of M	a	a	a	c	c	a	a	a	a	a	a	a
N	α	α	ε	–	–	ρ	α	α	ε	α	α	ε

(k)

M H	$6'/m'm'm$ (49) $\bar{3}m$ (D_{3d})											
Rep of H	A_{1g}	A_{2g}	E_g	A_{1u}	A_{2u}	E_u	$^1\bar{E}_g$	$^2\bar{E}_g$	$\bar{E}_{1g}$	$^1\bar{E}_u$	$^2\bar{E}_u$	$\bar{E}_{1u}$
Corep of M	a	a	a	a	a	a	a	a	a	a	a	a
N	α	α	ε	α	α	ε	α	α	ε	α	α	ε

(l)

M H	$\bar{6}m'2'$ (33) $\bar{6}$ (C_{3h})											
Rep of H	A'	$^1E''$	$^2E''$	A''	$^1E'$	$^2E'$	$^1\bar{E}_1$	$^2\bar{E}_1$	$^1\bar{E}_2$	$^2\bar{E}_2$	$^1\bar{E}_3$	$^2\bar{E}_3$
Corep of M	a	a	a	a	a	a	a	a	a	a	a	a
N	α	α	α	α	α	α	α	α	α	α	α	α

M H	$6'/m$ (45) $\bar{6}$ (C_{3h})											
Rep of H	A'	$^1E''$	$^2E''$	A''	$^1E'$	$^2E'$	$^1\bar{E}_1$	$^2\bar{E}_1$	$^1\bar{E}_2$	$^2\bar{E}_2$	$^1\bar{E}_3$	$^2\bar{E}_3$
Corep of M	a	c	c	a	c	c	c	c	c	c	c	c
N	α	–	–	α	–	–	–	–	–	–	–	–

M H	$62'2'$ (41) 6 (C_6)											
Rep of H	A	$^1E_1$	$^2E_1$	B	$^1E_2$	$^2E_2$	$^1\bar{E}_1$	$^2\bar{E}_1$	$^1\bar{E}_2$	$^2\bar{E}_2$	$^1\bar{E}_3$	$^2\bar{E}_3$
Corep of M	a	a	a	a	a	a	a	a	a	a	a	a
N	α	α	α	α	α	α	α	α	α	α	α	α

M H	$6/m'$ (43) 6 (C_6)											
Rep of H	A	$^1E_1$	$^2E_1$	B	$^1E_2$	$^2E_2$	$^1\bar{E}_1$	$^2\bar{E}_1$	$^1\bar{E}_2$	$^2\bar{E}_2$	$^1\bar{E}_3$	$^2\bar{E}_3$
Corep of M	a	c	c	a	c	c	c	c	c	c	c	c
N	α	–	–	α	–	–	–	–	–	–	–	–

M H	$6m'm'$ (46) 6 (C_6)											
Rep of **H**	A	$^1E_1$	$^2E_1$	B	$^1E_2$	$^2E_2$	$^1\bar{E}_1$	$^2\bar{E}_1$	$^1\bar{E}_2$	$^2\bar{E}_2$	$^1\bar{E}_3$	$^2\bar{E}_3$
Corep of **M**	a	a	a	a	a	a	a	a	a	a	a	a
N	α	α	α	α	α	α	α	α	α	α	α	α

(m)

M H	$6/mm'm'$ (52) $6/m$ (C_{6h})											
Rep of **H**	A_g	$^1E_{1g}$	$^2E_{1g}$	B_g	$^1E_{2g}$	$^2E_{2g}$	A_u	$^1E_{1u}$	$^2E_{1u}$	B_u	$^1E_{2u}$	$^2E_{2u}$
Corep of **M**	a	a	a	a	a	a	a	a	a	a	a	a
N	α	α	α	α	α	α	α	α	α	α	α	α

(m) cont.

M H	$6/mm'm'$ (52) $6/m$ (C_{6h})											
Rep of **H**	$^1\bar{E}_{1g}$	$^2\bar{E}_{1g}$	$^1\bar{E}_{2g}$	$^2\bar{E}_{2g}$	$^1\bar{E}_{3g}$	$^2\bar{E}_{3g}$	$^1\bar{E}_{1u}$	$^2\bar{E}_{1u}$	$^1\bar{E}_{2u}$	$^2\bar{E}_{2u}$	$^1\bar{E}_{3u}$	$^2\bar{E}_{3u}$
Corep of **M**	a	a	a	a	a	a	a	a	a	a	a	a
N	α	α	α	α	α	α	α	α	α	α	α	α

(n)

M H	$6'/mmm'$ (48) $\bar{6}2m$ (D_{3h})								
Rep of **H**	A_1'	A_2'	A_1''	A_2''	E'	E''	$\bar{E}_1$	$\bar{E}_2$	$\bar{E}_3$
Corep of **M**	a	a	a	a	a	a	a	a	a
N	α	α	α	α	ε	ε	ρ	ρ	ρ

M H	$6/m'm'm'$ (50) 622 (D_6)								
Rep of **H**	A_1	A_2	B_1	B_2	E_1	E_2	$\bar{E}_1$	$\bar{E}_2$	$\bar{E}_3$
Corep of **M**	a	a	a	a	a	a	a	a	a
N	α	α	α	α	ε	ε	ρ	ρ	ρ

M H	$6/m'mm$ (51) $6mm$ (C_{6v})								
Rep of **H**	A_1	A_2	B_1	B_2	E_1	E_2	$\bar{E}_1$	$\bar{E}_2$	$\bar{E}_3$
Corep of **M**	a	a	a	a	a	a	a	a	a
N	α	α	α	α	ε	ε	ρ	ρ	ρ

(o)

M	$m'3$ (53)							$\bar{4}'3m'$ (54)						
H	23 (T)							23 (T)						
Rep of **H**	A	^1E	^2E	T	$\bar{E}$	$^1\bar{F}$	$^2\bar{F}$	A	^1E	^2E	T	$\bar{E}$	$^1\bar{F}$	$^2\bar{F}$
Corep of **M**	a	c	c	a	a	c	c	a	a	a	a	a	a	a
N	α	–	–	E	ρ	–	–	α	α	α	F	σ	σ	σ

M	$4'32'$ (55)						
H	23 (T)						
Rep of **H**	A	^1E	^2E	T	$\bar{E}$	$^1\bar{F}$	$^2\bar{F}$
Corep of **M**	a	a	a	a	a	a	a
N	α	α	α	F	σ	σ	σ

(p)

M	$m3m'$ (58)													
H	$m3$ (T_h)													
Rep of **H**	A_g	^2E_g	^1E_g	T_g	A_u	^2E_u	^1E_u	T_u	$\bar{E}_g$	$^1\bar{F}_g$	$^2\bar{F}_g$	$\bar{E}_u$	$^1\bar{F}_u$	$^2\bar{F}_u$
Corep of **M**	a	a	a	a	a	a	a	a	a	a	a	a	a	a
N	α	α	α	F	α	α	α	F	σ	σ	σ	σ	σ	σ

(q)

M	$m'3m'$ (56)								$m'3m$ (57)							
H	432 (O)								$\bar{4}3m$ (T_d)							
Rep of **H**	A_1	A_2	E	T_1	T_2	$\bar{E}_1$	$\bar{E}_2$	$\bar{F}$	A_1	A_2	E	T_1	T_2	$\bar{E}_1$	$\bar{E}_2$	$\bar{F}$
Corep of **M**	a	a	a	a	a	a	a	a	a	a	a	a	a	a	a	a
N	α	α	ε	E	E	ρ	ρ	ψ	α	α	ε	E	E	ρ	ρ	ψ

Notes to Table 7.15

(i) The last but one row of the table specifies whether the coreps of **M** derived from the reps of **H** follow case (a), case (b), or case (c) when they are given by eqn (7.3.45), eqn (7.3.46), or eqn (7.3.47) respectively.

(ii) The matrices **N** given in the last row of each entry can be found from the following key.

$$\alpha = 1; \quad \varepsilon = \begin{pmatrix} 1 & 0 \\ 0 & 1 \end{pmatrix}; \quad \rho = \begin{pmatrix} 0 & 1 \\ -1 & 0 \end{pmatrix};$$

$$\sigma = \frac{1}{\sqrt{2}} \begin{pmatrix} 1 - \mathrm{i} & 0 \\ 0 & 1 + \mathrm{i} \end{pmatrix};$$

$$\chi = \frac{1}{2\sqrt{6}} \begin{pmatrix} (\sqrt{3} + 1) + \mathrm{i}(\sqrt{3} - 1) & 4\mathrm{i} \\ 4\mathrm{i} & (\sqrt{3} + 1) - \mathrm{i}(\sqrt{3} - 1) \end{pmatrix};$$

$$\psi = \begin{pmatrix} 0 & 1 & 0 & 0 \\ -1 & 0 & 0 & 0 \\ 0 & 0 & 0 & 1 \\ 0 & 0 & -1 & 0 \end{pmatrix}; \quad E = \begin{pmatrix} 1 & 0 & 0 \\ 0 & 1 & 0 \\ 0 & 0 & 1 \end{pmatrix}; \quad F = \begin{pmatrix} 0 & 1 & 0 \\ 1 & 0 & 0 \\ 0 & 0 & -1 \end{pmatrix}.$$

It is necessary to introduce a scheme for labelling the coreps of the magnetic point groups. The use of primes, bars, subscripts, and superscripts is already excessive so that rather than use a further modification of this nature an alternative will be adopted. We adopt the following scheme for the labelling of the coreps of the magnetic point groups (Cracknell 1968*a*). If X is the label of a given rep of the unitary subgroup (see Tables 2.2 and 6.5) then DX is used to denote the corep of **M** that is

TABLE 7.16

Notation for single-valued coreps belonging to case (c)

M	Reps of **H**	Coreps of **M**	**M**	Reps of **H**	Coreps of **M**
$4'22'$	B_2, B_3	DB	$6'$	$^1E, ^2E$	DE
$4/m'$	$^1E, ^2E$	DE	$3'$	$^1E, ^2E$	DE
$4'/m'$	$^1E, ^2E$	DE	$6/m'$	$^1E_1, ^2E_1$	DE_1
$4'mm'$	B_1, B_2	DB		$^1E_2, ^2E_2$	DE_2
$\bar{4}'2m'$	B_2, B_3	DB	$6'/m'$	$^1E_g, ^2E_g$	DE_g
$\bar{4}'m2'$	B_1, B_2	DB		$^1E_u, ^2E_u$	DE_u
$4'/mmm'$	B_{2g}, B_{3g}	DB_g	$6'/m$	$^1E', ^2E'$	DE'
	B_{2u}, B_{3u}	DB_u		$^1E'', ^2E''$	DE''
$\bar{6}'$	$^1E, ^2E$	DE	$m'3$	$^1E, ^2E$	DE

Notes to Table 7.16

(i) In column 1 the magnetic groups which have coreps belonging to case (c) are listed.

(ii) In column 3 we show the label that is used for the case (c) corep which is derived by the sticking together of the two reps of **H** which are shown in column 2.

(iii) The single-valued reps in column 2 can be identified from Tables 7.1 and 2.2.

derived from X, if the corep belongs to case (a) or case (b) (see Table 7.15). When the corep of **M** derived from X belongs to case (c) there will in fact be two reps X and Y which 'stick together' to form one corep of **M**. Starting for example with either of the reps B_1 or B_2 of the halving subgroup **H** of the magnetic group $4'mm'$ we obtain a corep of $4'mm'$ belonging to case (c) and the two coreps thus obtained are equivalent. It is therefore not necessary to distinguish between the two coreps thus derived and we denote either of them simply by DB. For all the coreps given in Table 7.15 which belong to case (c) we use the notation given in Tables 7.16 and 7.17; for all the other coreps the rule that the corep of **M** derived from the rep X of **H** is denoted by DX is perfectly adequate. The Γ notation for the labels of the reps of **H** has already been given in Tables 2.2 and 6.5 (Koster, Dimmock, Wheeler, and Statz, 1963). Once again the corep of **M** derived from the rep Γ_i of **H** can be labelled as $D\Gamma_i$. Where two reps Γ_i and Γ_j of **H** 'stick together' to form a case (c) corep of **M** this may be labelled as $D\Gamma_{i,j}$. $D\Gamma_{i,j}$ and $D\Gamma_{j,i}$ will then be equivalent, but not identical coreps of **M**. Therefore, for example, the case (c) corep of $4'22'$ derived from Γ_2 and Γ_4 would be called $D\Gamma_{2,4}$.

TABLE 7.17

Notation for double-valued coreps belonging to case (c)

M	Reps of **H**	Coreps of **M**	**M**	Reps of **H**	Coreps of **M**
$2/m'$	${}^1\bar{E}, {}^2\bar{E}$	$D\bar{E}$	$\bar{3}'$	${}^1\bar{E}, {}^2\bar{E}$	$D\bar{E}$
$2'/m$	${}^1\bar{E}, {}^2\bar{E}$	$D\bar{E}$	$\bar{3}'m$	${}^1\bar{E}, {}^2\bar{E}$	$D\bar{E}$
$4'$	${}^1\bar{E}, {}^2\bar{E}$	$D\bar{E}$	$\bar{3}'m'$	${}^1\bar{E}, {}^2\bar{E}$	$D\bar{E}$
$\bar{4}'$	${}^1\bar{E}, {}^2\bar{E}$	$D\bar{E}$	$6/m'$	${}^1\bar{E}_1, {}^2\bar{E}_1$	$D\bar{E}_1$
$4/m'$	${}^1\bar{E}_1, {}^2\bar{E}_1$	$D\bar{E}_1$		${}^1\bar{E}_2, {}^2\bar{E}_2$	$D\bar{E}_2$
	${}^1\bar{E}_2, {}^2\bar{E}_2$	$D\bar{E}_2$		${}^1\bar{E}_3, {}^2\bar{E}_3$	$D\bar{E}_3$
$4'/m'$	${}^1\bar{E}_1, {}^2\bar{E}_1$	$D\bar{E}_1$	$6'/m'$	${}^1\bar{E}_g, {}^2\bar{E}_g$	$D\bar{E}_g$
	${}^1\bar{E}_2, {}^2\bar{E}_2$	$D\bar{E}_2$		${}^1\bar{E}_u, {}^2\bar{E}_u$	$D\bar{E}_u$
$4'/m$	${}^1\bar{E}_g, {}^2\bar{E}_g$	$D\bar{E}_g$	$6'/m$	${}^1\bar{E}_1, {}^2\bar{E}_1$	$D\bar{E}_1$
	${}^1\bar{E}_u, {}^2\bar{E}_u$	$D\bar{E}_u$		${}^1\bar{E}_2, {}^2\bar{E}_2$	$D\bar{E}_2$
$\bar{6}'$	${}^1\bar{E}, {}^2\bar{E}$	$D\bar{E}$		${}^1\bar{E}_3, {}^2\bar{E}_3$	$D\bar{E}_3$
$6'$	${}^1\bar{E}, {}^2\bar{E}$	$D\bar{E}$	$m'3$	${}^1\bar{F}, {}^2\bar{F}$	$D\bar{F}$

Notes to Table 7.17

(i) In column 1 the magnetic groups which have coreps belonging to case (c) are listed.

(ii) In column 3 we show the label that is used for the case (c) corep which is derived by the sticking together of the two reps of **H** that are shown in column 2.

(iii) The double-valued reps in column 2 can be identified from Tables 7.1 and 6.5.

7.6. The corepresentations of the magnetic space groups

The relationship between the problems of determining the coreps of the magnetic space groups and determining those of the magnetic point groups is very similar to the relationship between the problems of determining the reps of the Fedorov groups and of the ordinary point groups. The theory of the determination of the reps of the Fedorov groups, that is, of the ordinary space groups, from the reps of the point groups has been considered in sections 3.7 and 3.8 and, more rigorously, in Chapter 4. It will be remembered that, using the theory of induced representations in Chapter 4, it was possible to show that the problem of determining the reps of a space group could be simplified. For each wave vector **k** in the representation domain it was found to be necessary to deduce the projective reps of the little co-group $\bar{\mathbf{G}}^{\mathbf{k}}$ having a factor system depending upon the **k** vector under consideration. Each projective rep of $\overline{\mathbf{G}}^{\mathbf{k}}$ determined a small rep of $\mathbf{G}^{\mathbf{k}}$. Each small rep of $\mathbf{G}^{\mathbf{k}}$, when induced into the space group **G**, yielded a rep of **G**. All the reps of **G** were found by allowing **k** to vary over the representation domain. For **k** a general point or on a line or plane of symmetry (see section 3.3) the projective reps of $\overline{\mathbf{G}}^{\mathbf{k}}$ were found as reps of a suitable central extension $\overline{\mathbf{G}}^{\mathbf{k}*}$. For points of symmetry (see also section 3.3) this method, though still applicable, was replaced by a more direct method due to Herring (1942), which provided the small reps of $\mathbf{G}^{\mathbf{k}}$ without first considering the little co-group. Examples of both methods were given in section 3.8. In the tabulations in Chapters 5 and 6 Herring's method was used for the points of symmetry and the central extension method for the remaining points. This theory was all for unitary groups and we now have to see how it must be modified to cover the case of the magnetic space groups that are non-unitary groups.

Suppose that the group **M** is a magnetic space group defined by

$$\mathbf{M} = \mathbf{G} + A\mathbf{G}. \tag{7.3.11}$$

If we wish to follow the procedure similar to that used in section 7.5 for magnetic point groups in order to determine the irreducible corepresentations of **M**, the first step would be to obtain the irreducible representations Δ of **G**. **G** is clearly one of the ordinary space groups, or Fedorov groups, and it is a unitary group; its single-valued and double-valued representations are available from Chapters 5 and 6. The next step would be to take characters of Δ, and to evaluate the sum $\sum_{B \in A\mathbf{G}} \chi(B^2)$; the result would determine, according to eqn (7.3.51), the type of the corepresentation of **M** derived from Δ,

$$\sum_{B \in A\mathbf{G}} \chi(B^2) = \left.\begin{array}{ll} +|\mathbf{G}| & \text{in case (a),} \\ -|\mathbf{G}| & \text{in case (b),} \\ 0 & \text{in case (c).} \end{array}\right\} \tag{7.3.51}$$

In principle given Δ, the induced representation of the whole space group **G**, this sum could be calculated, but it turns out that this would be a matter of doing more than is necessary. Indeed we have seen how Δ is determined by the small representation Γ of the little group **K**. It would obviously be better to simplify, if possible, the criterion of eqn (7.3.51) so that it involves only the characters of Γ and fewer of the group elements; in other words to determine the type of corepresentation from the small representation Γ rather than from Δ, which is a representation of the whole space group **G**.

To achieve a simplification of the left-hand side of eqn (7.3.51) we start by writing

$$\mathbf{G} = \sum_{\alpha} r_{\alpha}\mathbf{T} \tag{4.2.1}$$

where **T** is the invariant subgroup of **G** that consists of the group of lattice translations of one of the Bravais lattices. We introduce, for short, the quantity J given by

$$J = \frac{1}{|\mathbf{G}|} \sum_{B \in A\mathbf{G}} \chi(B^2) = \frac{1}{|\mathbf{G}|} \sum_{R \in \mathbf{G}} \chi(ARAR). \tag{7.6.1}$$

Using eqn (4.7.10) we obtain

$$J = \frac{1}{|\mathbf{G}|} \sum_{R \in \mathbf{G}} \sum_{\sigma}{}' \psi(r_\sigma^{-1} ARARr_\sigma), \tag{7.6.2}$$

where the r_σ are the left coset representatives in the decomposition of **G** with respect to the little group **K**, i.e. given by eqn (4.1.1)

$$\mathbf{G} = \sum_{\sigma} r_\sigma \mathbf{K} \tag{7.6.3}$$

and the prime means that for fixed R the sum over σ is restricted to those σ for which $\mathbf{K}^\sigma = r_\sigma \mathbf{K} r_\sigma^{-1}$ contains $ARAR$. Reversing the order of summation eqn (7.6.2) becomes

$$J = \frac{1}{|\mathbf{G}|} \sum_{\sigma} \sum_{R \in \mathbf{G}}{}' \psi(r_\sigma^{-1} ARARr_\sigma) \tag{7.6.4}$$

in which the prime now means that for fixed σ the sum over R is restricted to those R for which $ARAR \in \mathbf{K}^\sigma$. Now if AS is such that $ASAS \in \mathbf{K}$ then $r_\sigma AS r_\sigma^{-1} = AR$ (for some $R \in \mathbf{G}$) is such that $ARAR = r_\sigma ASAS r_\sigma^{-1} \in \mathbf{K}^\sigma$. Furthermore, for this σ, $\psi(r_\sigma^{-1} ARARr_\sigma) = \psi(ASAS)$. This means that J splits up into $|\mathbf{G}|/|\mathbf{K}|$ equal parts, one part for each σ; that is, one part for each member of the star. Thus

$$J = \frac{1}{|\mathbf{K}|} \sum_{S \in \mathbf{G}}{}' \psi(ASAS), \tag{7.6.5}$$

in which the prime now means that the sum over S is restricted to those S for which $ASAS \in \mathbf{K}$. We now make use of the fact that **T** is not only an invariant subgroup of **G** but also an invariant subgroup of **M**. We can write $Ar_\alpha = r_{\alpha'}$, for all α, so that from eqn (7.3.11)

$$\mathbf{M} = \sum_{\alpha} r_\alpha \mathbf{T} + \sum_{\alpha'} r_{\alpha'} \mathbf{T} \tag{7.6.6}$$

and further the suffices $\alpha, \beta, \ldots, \alpha', \beta', \ldots$ form a group isomorphic with $\mathbf{M}/\mathbf{T}$. With this proviso we can extend the notation of eqns (4.2.2) et seq. to include the primed suffices. Suppose we select an irreducible representation F of **T** appropriate to the little group **K** (F is, of course, 1-dimensional since **T** is Abelian); by this we mean that out of all the irreducible representations F_α of **T** defined by

$$F_\alpha(t_a) = F([t_a]_\alpha) \tag{7.6.7}$$

the set of all those unprimed suffices γ for which $F_\gamma(t_a) = F(t_a)$, for all $t_a \in \mathbf{T}$, forms a subgroup of **G**/**T** which is just the little co-group $\bar{\mathbf{K}}$. We know further that any unitary irreducible representation Γ_F^j of **K** for which

$$\Gamma_F^j \downarrow \mathbf{T} = \lambda_F^j F \tag{7.6.8}$$

(i.e. whose restriction to **T** yields an integral multiple λ_F^j of the representation F) is then a small (or allowed) representation of **K**, see Chapter 4. From eqn (7.6.8) we see that

$$\Gamma_F^j(t_a) = F(t_a)\mathbf{1}_F^j, \tag{7.6.9}$$

where $\mathbf{1}_F^j$ is the unit matrix of dimension λ_F^j. This means that we know Γ_F^j for all elements of **K** once $\Gamma_F^j(r_\gamma)$ is known for all $\gamma \in \bar{\mathbf{K}}$. We have already shown in Chapter 4 how the problem of finding all the irreducible representations of **G** can be simplified to that of finding all the irreducible projective representations of the little co-groups $\bar{\mathbf{K}}$ of all the various stars. Now if $AS = r_{\alpha'} t_a$ then $ASAS = r_{\alpha'\alpha'} t_{\alpha'\alpha'}[t_a]_{\alpha'} t_a$, so that $ASAS \in \mathbf{K}$ if $\alpha'^2 \in \bar{\mathbf{K}}$, the little co-group. Equation (7.6.4) now becomes

$$J = \frac{1}{|\mathbf{K}|} \sum_{\alpha'^2 \in \bar{\mathbf{K}}} \sum_{t_a \in \mathbf{T}} \psi(r_{\alpha'\alpha'} t_{\alpha'\alpha'}[t_a]_{\alpha'} t_a). \tag{7.6.10}$$

Using eqn (7.6.9) this yields

$$J = \frac{1}{|\mathbf{K}|} \sum_{\alpha'^2 \in \bar{\mathbf{K}}} \sum_{t_a \in \mathbf{T}} F(t_{\alpha'\alpha'}) F([t_a]_{\alpha'}) F(t_a) \psi(r_{\alpha'\alpha'}). \tag{7.6.11}$$

Now $F([t_a]_{\alpha'}) = F_{\alpha'}(t_a)$ and the sum over t_a in eqn (7.6.11) will vanish by virtue of orthogonality relations unless α' is such that $F_{\alpha'}(t_a) = F(t_a)^{-1}$. Note that all such α' satisfy $\alpha'^2 \in \bar{\mathbf{K}}$. When the sum over t_a does not vanish its value is $|\mathbf{T}|$, the order of **T**. Hence

$$J = \frac{1}{|\bar{\mathbf{K}}|} \sum_{\alpha'} \psi(r_{\alpha'}^2), \tag{7.6.12}$$

in which the sum over α' is restricted to those α' such that $F_{\alpha'} = F^{-1}$, and $|\bar{\mathbf{K}}|$ is the order of the little co-group. Of course it could be that no such α' exist in which case $J = 0$ immediately from eqn (7.6.11).

To summarize: in order to determine to which of the three cases the irreducible corepresentation of **M** derived from Δ belongs, it is necessary to determine those coset representatives $r_{\kappa'}$ in eqn (7.6.6) which satisfy $F_{\kappa'} = F^{-1}$ and evaluate the sum $\sum_{\kappa'} \psi(r_{\kappa'}^2)$, where ψ is the character of Γ in **K**. Then restricting the sum to those κ' the criterion for the type of corepresentation is

$$\sum_{\kappa'} \psi(r_{\kappa'}^2) = \left.\begin{array}{rl} +|\overline{\mathbf{K}}| & \text{in case (a),} \\ -|\overline{\mathbf{K}}| & \text{in case (b),} \\ 0 & \text{in case (c),} \end{array}\right\} \tag{7.6.13}$$

where we include the special case that if no such κ' exist then the sum is zero.

The similarity between the criteria of eqn (7.6.13) and eqn (7.3.51) gives us the clue to the definition of the magnetic little group.

DEFINITION 7.6.1. The *magnetic little co-group* $\overline{\mathbf{Q}}$ is the group of all unprimed suffices γ for which $F_\gamma = F$ together with all primed suffices κ' for which $F_{\kappa'} = F^{-1}$.

Two cases occur, either no such κ' exist at all and then $\overline{\mathbf{Q}} = \overline{\mathbf{K}}$, or there exist an equal number of primed and unprimed suffices so that $\overline{\mathbf{K}}$ is a subgroup of $\overline{\mathbf{Q}}$ of index 2. The magnetic little group $\mathbf{Q}$ is then defined to have the same relation to $\overline{\mathbf{Q}}$ as $\mathbf{K}$ has to $\overline{\mathbf{K}}$.

DEFINITION 7.6.2. If no κ' exists the *magnetic little group* is given by $\mathbf{Q} = \mathbf{K}$, but in the alternative case when both sets of suffices appear in equal numbers the *magnetic little group* $\mathbf{Q}$ is given by

$$\mathbf{Q} = \sum_{\gamma \in \overline{\mathbf{K}}} r_\gamma \mathbf{T} + \sum_{\kappa' \in \overline{\mathbf{Q}} - \overline{\mathbf{K}}} r_{\kappa'} \mathbf{T}. \tag{7.6.14}$$

In the case in which the magnetic little group coincides with the little group the significant point of interest is that we are bound to be in case (c). The appropriate corepresentation $D\Delta$ is derived from the small representation Γ by first using Theorem 4.3.1 to find Δ, and then by eqn (7.3.47) to find $D\Delta$.

The same method is of course open to us in the alternative case also where we use eqns (7.3.45)–(7.3.47) in the final step as appropriate. What is interesting is that in this case an alternative method for obtaining the irreducible corepresentation of $\mathbf{M}$ derived from Δ presents itself. Qualitatively what happens is that we can give a geometrical interpretation to the magnetic little group and the magnetic little co-group similar to the geometrical interpretation of the little group and little co-group of an ordinary space group $\mathbf{G}$ (see section 3.7). It is possible then to proceed from the small reps of the little group of $\mathbf{G}$ to the coreps of $\mathbf{M}$ via the small coreps of the magnetic little group of $\mathbf{M}$. Thus instead of proceeding by the route

$$\begin{array}{ccc} \text{small reps of little group of } \mathbf{G} & \rightarrow & \text{induced reps of } \mathbf{G} \\ & & \downarrow \\ & & \text{coreps of } \mathbf{M} \end{array}$$

we can proceed by the route:

small reps of little group of **G**

$\downarrow$

small coreps of little group of **M** $\rightarrow$ induced coreps of **M**.

The reader who is prepared to accept the geometrical interpretation of the magnetic little co-group and is not interested in justifying the steps involved in obtaining the induced coreps of **M** from the small coreps of the magnetic little group may omit the next few paragraphs and proceed to the sub-section on p. 646 entitled 'A geometrical account of magnetic little groups'.

In order to see how this method works we first note that **Q** is a magnetic group and that **K** is its unitary subgroup. Γ is an irreducible representation of **K**. So it is possible to form an irreducible corepresentation $D\Gamma$ of **Q** derived from Γ. According to eqn (7.3.51) the type of $D\Gamma$ depends on the sum $\sum_{B\in \mathbf{Q}-\mathbf{K}} \psi(B^2)$. Indeed the criterion is

$$\sum_{B \in \mathbf{Q}-\mathbf{K}} \psi(B^2) = \left.\begin{array}{ll} +|\mathbf{K}| & \text{in case (a),} \\ -|\mathbf{K}| & \text{in case (b),} \\ 0 & \text{in case (c),} \end{array}\right\} \tag{7.6.15}$$

which on performing the sum over the elements of **T** reduces to

$$\sum_{\kappa' \in \bar{\mathbf{Q}}-\mathbf{K}} \psi(r_{\kappa'}^2) = \left.\begin{array}{ll} +|\bar{\mathbf{K}}| & \text{in case (a),} \\ -|\bar{\mathbf{K}}| & \text{in case (b),} \\ 0 & \text{in case (c)} \end{array}\right\} \tag{7.6.16}$$

But this is exactly the same criterion as eqn (7.6.13) which governs the type of the irreducible corepresentation of **M** derived from Δ.

It seems appropriate to call $D\Gamma$ a small corepresentation of the magnetic little group **Q**. The alternative method for determining the irreducible corepresentation of **M** derived from Δ should now be clear: *it can also be derived by inducing the small corepresentation* $D\Gamma$ *into* **M**. It may not be entirely clear what is meant by an induced corepresentation since this is a new concept. What it means here is as follows. We are given a corepresentation $D\Gamma$ of **Q**. Since **M** is to **Q** as **G** is to **K** we can write

$$\mathbf{M} = \sum_{\sigma} r_\sigma \mathbf{Q}, \tag{7.6.17}$$

where the r_σ that appear in eqn (7.6.17) are the same as the r_σ in eqn (7.6.3) and are therefore unitary. Let the basis of $D\Gamma$ be $\langle\alpha|$; then, for all $q \in \mathbf{Q}$,

$$q\langle\alpha| = \langle\alpha|\, \mathbf{D}\Gamma(q). \tag{7.6.18}$$

Furthermore, since $D\Gamma$ is a corepresentation, we have, from eqn (7.3.19), for k_1, $k_2 \in \mathbf{K}$ and $a_1, a_2 \in \mathbf{Q} - \mathbf{K}$,

$$\mathbf{D\Gamma}(k_1)\mathbf{D\Gamma}(k_2) = \mathbf{D\Gamma}(k_1 k_2), \tag{7.6.19a}$$
$$\mathbf{D\Gamma}(k_1)\mathbf{D\Gamma}(a_2) = \mathbf{D\Gamma}(k_1 a_2), \tag{7.6.19b}$$
$$\mathbf{D\Gamma}(a_1)\mathbf{D\Gamma}^*(k_1) = \mathbf{D\Gamma}(a_1 k_1), \tag{7.6.19c}$$
$$\mathbf{D\Gamma}(a_1)\mathbf{D\Gamma}^*(a_2) = \mathbf{D\Gamma}(a_1 a_2). \tag{7.6.19d}$$

We can define for each σ in eqn (7.6.17) the set of functions

$$\langle \alpha_\sigma| = r_\sigma \langle \alpha|. \tag{7.6.20}$$

Then the set of all such sets of functions, as σ varies over the star of F, forms the basis of a vector space which is invariant under $\mathbf{M}$. To see this we suppose that $m \in \mathbf{M}$ and that

$$m r_\tau = r_\gamma q \tag{7.6.21}$$

where $q \in \mathbf{Q}$; then

$$\begin{aligned} m\langle \alpha_\tau| &= m r_\tau \langle \alpha| \\ &= r_\gamma q \langle \alpha| \\ &= r_\gamma \langle \alpha|\, \mathbf{D\Gamma}(q) \\ &= \langle \alpha_\gamma|\, \mathbf{D\Gamma}(q) \\ &= \langle \alpha_\gamma|\, \mathbf{D\Gamma}(r_\gamma^{-1} m r_\tau), \end{aligned} \tag{7.6.22}$$

in which we have made use of eqns (7.6.18), (7.6.20), and (7.6.21). In keeping with eqn (7.6.22) we define for all $m \in \mathbf{M}$ the block matrix

$$(D\Gamma \uparrow \mathbf{M})_{\gamma\tau}(m) = \mathbf{D\Gamma}(r_\gamma^{-1} m r_\tau)\delta_{\gamma, m\tau}, \tag{7.6.23}$$

where $\delta_{\gamma,m\tau} = 1$ if $m r_\tau \in r_\gamma \mathbf{Q}$ and is zero otherwise. (Notice the exact parallel between these ideas and those of induced representations as discussed in Chapter 4.) Note that m and $r_\gamma^{-1} m r_\tau$ are both unitary or both anti-unitary. Using this fact it soon follows from eqns (7.6.19) and (7.6.23) that the eqns (7.3.19) hold for $(D\Gamma \downarrow \mathbf{M})$ for all $R, S \in \mathbf{G}$ and for all $B, C \in \mathbf{M} - \mathbf{G}$. This means that $(D\Gamma \uparrow \mathbf{M})$ is a corepresentation of $\mathbf{M}$ and is said to be induced from $D\Gamma$.

We now have to establish our assertion that $(D\Gamma \uparrow \mathbf{M})$ is equivalent to the irreducible corepresentation $D\Delta$ of $\mathbf{M}$ derived from Δ.

First we remember that Δ and Γ are of the same type. Since $\Delta = (\Gamma \uparrow \mathbf{G})$ it follows at once that $(D\Gamma \uparrow \mathbf{M})$ and $D\Delta$ are of the same dimension. Secondly the induction $(D\Gamma \uparrow \mathbf{M})$ is performed using exactly the same r_σ as the induction $(\Gamma \uparrow \mathbf{G})$, see eqns (7.6.3) and (7.6.17). And since $D\Gamma$ contains Γ it follows immediately that $(D\Gamma \uparrow \mathbf{M})$ contains Δ. But we have proved that $(D\Gamma \uparrow \mathbf{M})$ is a corepresentation of $\mathbf{M}$. Furthermore it contains Δ and is of the same dimension as $D\Delta$, which is irreducible and also contains Δ. Hence $(D\Gamma \uparrow \mathbf{M})$ is also irreducible and is equivalent to $D\Delta$, the unique irreducible corepresentation of $\mathbf{M}$ derived from Δ.

Not until this point in the analysis is the definition of the magnetic little group $\mathbf{Q}$ justified. But now it is seen to be as fundamentally significant for magnetic groups, which have an invariant Abelian subgroup of unitary elements, as is the little group for non-magnetic groups with the same property. *Indeed we have proved that the irreducible corepresentations of such magnetic groups are induced from the small corepresentations of the magnetic little groups* and this can be made to include the special case in which the magnetic little group $\mathbf{Q}$ is nothing more than the little group $\mathbf{K}$, for then $\mathbf{Q}$ contains no anti-unitary elements, the small corepresentations become small representations and, since we are in case (c), the required results are obtained by inducing them straight into $\mathbf{M}$, the intermediate step to $\mathbf{G}$ being unnecessary.

Finally we study the form of the small corepresentations $D\Gamma$ of $\mathbf{Q}$ and prove a result analogous to the fact that the small representations of $\mathbf{K}$ are the irreducible projective representations of the little co-group $\overline{\mathbf{K}}$. Suppose that F is an irreducible representation of $\mathbf{T}$. In case (a), from eqns (7.3.45) and (7.6.9)

$$\mathbf{D\Gamma}(t_a) = \mathbf{\Gamma}(t_a) = F(t_a)\mathbf{1}. \tag{7.6.24}$$

In case (b), from eqns (7.3.46) and (7.6.9)

$$\mathbf{D\Gamma}(t_a) = \begin{pmatrix} \mathbf{\Gamma}(t_a) & \mathbf{0} \\ \mathbf{0} & \mathbf{\Gamma}(t_a) \end{pmatrix} = F(t_a)\mathbf{1}. \tag{7.6.25}$$

And in case (c), from eqn (7.3.47), for some fixed κ'

$$\mathbf{D\Gamma}(t_a) = \begin{pmatrix} \mathbf{\Gamma}(t_a) & \mathbf{0} \\ \mathbf{0} & \mathbf{\Gamma}^*(r_{\kappa'}^{-1}t_a r_{\kappa'}) \end{pmatrix}. \tag{7.6.26}$$

Now $\mathbf{\Gamma}^*(r_{\kappa'}^{-1}t_a r_{\kappa'}) = F^*([t_a]_{\kappa'})\mathbf{1} = F_{\kappa'}^*(t_a)\mathbf{1} = (F^{-1})^*(t_a)\mathbf{1} = F(t_a)\mathbf{1}$, where we have used the fact that, since $\kappa' \in \overline{\mathbf{Q}} - \overline{\mathbf{K}}$, $F_{\kappa'} = F^{-1}$. Hence in case (c) also

$$\mathbf{D\Gamma}(t_a) = F(t_a)\mathbf{1}. \tag{7.6.27}$$

Thus in all cases $(D\Gamma \downarrow \mathbf{T})$ is a scalar matrix. This means that we know the matrix representatives in $D\Gamma$ for all elements of $\mathbf{Q}$ once $\mathbf{D\Gamma}(r_\gamma)$ is known for all $\gamma \in \overline{\mathbf{K}}$ and $\mathbf{D\Gamma}(r_{\kappa'})$ for all $\kappa' \in \overline{\mathbf{Q}} - \overline{\mathbf{K}}$. Also from eqn (4.2.3) and the fact that $D\Gamma$ is a corepresentation it follows that for $\gamma, \delta \in \overline{\mathbf{K}}$ and $\kappa', \lambda' \in \overline{\mathbf{Q}} - \overline{\mathbf{K}}$

$$\mathbf{D\Gamma}(r_\gamma)\mathbf{D\Gamma}(r_\delta) = F(t_{\gamma\delta})\mathbf{D\Gamma}(r_{\gamma\delta}), \tag{7.6.28a}$$

$$\mathbf{D\Gamma}(r_\gamma)\mathbf{D\Gamma}(r_{\kappa'}) = F^*(t_{\gamma\kappa'})\mathbf{D\Gamma}(r_{\gamma\kappa'}), \tag{7.6.28b}$$

$$\mathbf{D\Gamma}(r_{\kappa'})\mathbf{D\Gamma}(r_\gamma)^* = F^*(t_{\kappa'\gamma})\mathbf{D\Gamma}(r_{\kappa'\gamma}), \tag{7.6.28c}$$

$$\mathbf{D\Gamma}(r_{\kappa'})\mathbf{D\Gamma}(r_{\lambda'})^* = F(t_{\kappa'\lambda'})\mathbf{D\Gamma}(r_{\kappa'\lambda'}). \tag{7.6.28d}$$

Furthermore, by virtue of eqn (4.2.11) the complex numbers F satisfy

$$F(t_{\mu\nu,\,\gamma})F(t_{\mu\nu}) = F(t_{\mu,\,\nu\gamma})F(t_{\nu\gamma}) \tag{7.6.29a}$$

for all $\mu, \nu \in \overline{\mathbf{Q}}$ and $\gamma \in \overline{\mathbf{K}}$, and

$$F(t_{\mu\nu,\,\kappa'})F^*(t_{\mu\nu}) = F(t_{\mu,\,\nu\gamma'})F(t_{\nu\kappa'}) \tag{7.6.29b}$$

for all $\mu, \nu \in \overline{\mathbf{Q}}$ and $\kappa' \in \overline{\mathbf{Q}} - \overline{\mathbf{K}}$.

The eqns (7.6.28) and (7.6.29) imply that $D\Gamma$ considered as a function on the elements of $\overline{\mathbf{Q}}$ forms an irreducible projective corepresentation of $\overline{\mathbf{Q}}$ with factor system F. The deduction of the irreducible projective corepresentations for the magnetic point groups has been made by Karavaev, Kudryavtseva, and Chaldyshev (1962), Kudryavtseva (1965), and Murthy (1966). Each such irreducible projective corepresentation of $\overline{\mathbf{Q}}$ can be derived from an irreducible projective representation of the unitary subgroup $\overline{\mathbf{K}}$, of character, say, ψ and once more there are three cases according to the criterion of eqn (7.6.13). In the present context these results follow immediately from the preceding theory; however to discuss projective corepresentations out of the present context would demand an analysis similar to that in section 7.3 but for projective representations and we do not propose to consider it here.

A geometrical account of magnetic little groups

We now interpret the above theory rather more pictorially for the case in which the group $\mathbf{M}$ is a magnetic space group or Shubnikov group. $\mathbf{G}$ is then one of the 230 crystal space groups and $\mathbf{T}$ is the subgroup of lattice translations. Elements of $\mathbf{T}$ are of the form $\{E \mid \mathbf{t}\}$, where $\mathbf{t}$ is a lattice translation. An element $r_\alpha \in \mathbf{G}$ is of the form $\{R_\alpha \mid \mathbf{v}_\alpha\}$ where R_α is a point-group operation. An element $r_{\alpha'} \in \mathbf{M} - \mathbf{G}$ is of the form $\theta\{S_{\alpha'} \mid \mathbf{w}_{\alpha'}\}$ where θ is the time-reversal operator and $S_{\alpha'}$ is a point-group operation; $\mathbf{v}_\alpha$ and $\mathbf{w}_{\alpha'}$ are either zero or are non-lattice translations. From eqn (7.6.5) it will be seen that we are writing

$$A\{R_\alpha \mid \mathbf{v}_\alpha\} = \theta\{S_{\alpha'} \mid \mathbf{w}_{\alpha'}\}. \tag{7.6.30}$$

This very general nomenclature permits us to deal simultaneously with all types of magnetic space groups; that is, no distinctions need to be drawn between types II, III, and IV Shubnikov groups.

Each irreducible representation $F^{\mathbf{k}}$ of $\mathbf{T}$ is labelled with a vector $\mathbf{k}$ from the Brillouin zone and is such that

$$F^{\mathbf{k}}(\{E \mid \mathbf{t}\}) = \exp(-i\mathbf{k}\,.\,\mathbf{t}). \tag{7.6.31}$$

Now forming the conjugate of $\{E \mid \mathbf{t}\}$ with the element $\{R_\alpha \mid \mathbf{v}_\alpha\}$ we obtain

$$\{R_\alpha \mid \mathbf{v}_\alpha\}^{-1}\{E \mid \mathbf{t}\}\{R_\alpha \mid \mathbf{v}_\alpha\} = \{E \mid R_\alpha^{-1}\mathbf{t}\}, \tag{7.6.32}$$

so that from eqn (7.6.7)

$$\begin{aligned} F_\alpha^{\mathbf{k}}(\{E \mid \mathbf{t}\}) &= F^{\mathbf{k}}(\{E \mid R_\alpha^{-1}\mathbf{t}\}) \\ &= \exp(-i\mathbf{k} \,.\, R_\alpha^{-1}\mathbf{t}) \\ &= \exp(-iR_\alpha\mathbf{k} \,.\, \mathbf{t}) \\ &= F^{R_\alpha\mathbf{k}}(\{E \mid \mathbf{t}\}). \end{aligned} \tag{7.6.33}$$

The little co-group of $F^{\mathbf{k}}$, denoted by $\overline{\mathbf{G}}^{\mathbf{k}}$ in the notation of earlier chapters, is therefore to be thought of as consisting of all those point-group operations R_γ such that $F^{R_\gamma\mathbf{k}} = F^{\mathbf{k}}$, that is which satisfy

$$R_\gamma\mathbf{k} = \mathbf{k} + \mathbf{g}, \tag{7.6.34}$$

where $\exp(-i\mathbf{g} \,.\, \mathbf{t}) = 1$ for all $\mathbf{t}$. The vectors $\mathbf{g}$ are reciprocal lattice vectors as defined in Chapter 3 and vectors of the form $(\mathbf{k} + \mathbf{g})$ are equivalent to $\mathbf{k}$. The condition (7.6.34) is therefore that $R_\gamma\mathbf{k}$ should be equivalent to $\mathbf{k}$. The little group of $F^{\mathbf{k}}$ in $\mathbf{G}$, often called the group of $\mathbf{k}$ and denoted in previous chapters by $\mathbf{G}^{\mathbf{k}}$, can then be written as

$$\mathbf{G}^{\mathbf{k}} = \sum_{R_\gamma \in \overline{\mathbf{G}}^{\mathbf{k}}} \{R_\gamma \mid \mathbf{v}_\gamma\}\mathbf{T}. \tag{7.6.35}$$

If we decompose $\mathbf{G}$ with respect to $\mathbf{G}^{\mathbf{k}}$

$$\mathbf{G} = \sum_\sigma \{R_\sigma \mid \mathbf{v}_\sigma\}\mathbf{G}^{\mathbf{k}}, \tag{7.6.36}$$

then it follows immediately that the set of representations $F^{R_\sigma\mathbf{k}}$ forms the star of $F^{\mathbf{k}}$. Geometrically one may interpret the star as consisting of all vectors $R_\sigma\mathbf{k}$ appropriate to this decomposition.

We now proceed to consider the magnetic little group, see Definition 7.6.2, and to see how to obtain a geometrical interpretation of it similar to this geometrical interpretation of the little group of an ordinary space group $\mathbf{G}$. Since θ, the operation of time-reversal, commutes with all space-group operations the conjugate of $\{E \mid \mathbf{t}\}$ formed with $\theta\{S_{\alpha'} \mid \mathbf{w}_{\alpha'}\}$ is $\{E \mid S_{\alpha'}^{-1}\mathbf{t}\}$. Hence it follows, exactly as with eqn (7.6.33), that

$$F_{\alpha'}^{\mathbf{k}}(\{E \mid \mathbf{t}\}) = F^{S_{\alpha'}\mathbf{k}}(\{E \mid \mathbf{t}\}). \tag{7.6.37}$$

In order that $S_{\alpha'}$ should belong to the magnetic little co-group $\overline{\mathbf{M}}^{\mathbf{k}}$ it is necessary for this to be equal to $(F^{\mathbf{k}})^{-1} = F^{-\mathbf{k}}$. Hence in addition to the $R_\gamma \in \mathbf{G}^{\mathbf{k}}$ the magnetic little co-group consists of all $S_{\kappa'}$ such that $S_{\kappa'}\mathbf{k}$ is equivalent to $-\mathbf{k}$. If no such $S_{\kappa'}$ exist then $\overline{\mathbf{M}}^{\mathbf{k}} = \overline{\mathbf{G}}^{\mathbf{k}}$. Finally the magnetic little group $\mathbf{M}^{\mathbf{k}}$ can be written as

$$\mathbf{M}^{\mathbf{k}} = \sum_{R_\gamma \in \overline{\mathbf{G}}^{\mathbf{k}}} \{R_\gamma \mid \mathbf{v}_\gamma\}\mathbf{T} + \sum_{S_{\kappa'} \in \overline{\mathbf{M}}^{\mathbf{k}} - \overline{\mathbf{G}}^{\mathbf{k}}} \theta\{S_{\kappa'} \mid \mathbf{w}_{\kappa'}\}\mathbf{T}. \tag{7.6.38}$$

It will be seen that we have given the definition of $\overline{\mathbf{M}}^{\mathbf{k}}$ without reference to the action of θ on the vector $\mathbf{k}$ itself. A number of authors write

$$\theta\mathbf{k} = -\mathbf{k}. \tag{7.6.39}$$

This equation has nothing at all to do with representation theory which as we have seen is concerned only with the action of the operators R_α and $S_{\alpha'}$ on $\mathbf{k}$. Indeed in the present context eqn (7.6.39) is misleading because $\theta^{-1}\{E \mid \mathbf{t}\}\theta = \{E \mid \mathbf{t}\}$ and hence $F_\theta^{\mathbf{k}}(\{E \mid \mathbf{t}\}) = F^{\mathbf{k}}(\theta^{-1}\{E \mid \mathbf{t}\}\theta) = F^{\mathbf{k}}(\{E \mid \mathbf{t}\})$ and not $F^{-\mathbf{k}}(\{E \mid \mathbf{t}\})$ as eqn. (7.6.39) might lead us to suppose. This does not mean that eqn (7.6.39) has no meaning; what it means is that

$$\theta^{-1} \exp(\mathrm{i}\mathbf{k} \,.\, \mathbf{r})\, \theta = \exp(-\mathrm{i}\mathbf{k} \,.\, \mathbf{r}), \tag{7.6.40}$$

that is, it transforms a wave function of quasi-momentum $\hbar\mathbf{k}$ into a wave function with quasi-momentum $-\hbar\mathbf{k}$. The meaning is connected with the physical interpretation of $\hbar\mathbf{k}$ as a momentum rather than the geometrical interpretation of $\mathbf{k}$ as a reciprocal length.

Finally, if $\Gamma_p^{\mathbf{k}}$ is a small representation of $\mathbf{G}^{\mathbf{k}}$ with character $\psi_p^{\mathbf{k}}$ and $\Delta_p^{\mathbf{k}}$ is the irreducible representation of $\mathbf{G}$ induced from $\Gamma_p^{\mathbf{k}}$ then the type of corepresentation of $\mathbf{M}$ derived from $\Delta_p^{\mathbf{k}}$ depends on the following criterion

$$\sum_{S_{\kappa'}} \psi_p^{\mathbf{k}}(\{S_{\kappa'} \mid \mathbf{w}_{\kappa'}\}^2) = \left.\begin{matrix} \gamma|\overline{\mathbf{K}}| & \text{in case (a),} \\ -\gamma|\overline{\mathbf{K}}| & \text{in case (b),} \\ 0 & \text{in case (c),} \end{matrix}\right\} \tag{7.6.41}$$

where, as Wigner (1932) has shown,

$$\Delta(\theta^2) = \gamma = \left.\begin{matrix} +1 & \text{for an even number of fermions} \\ -1 & \text{for an odd number of fermions} \end{matrix}\right\} \tag{7.6.42}$$

and $|\overline{\mathbf{K}}|$ is the order of $\overline{\mathbf{G}}^{\mathbf{k}}$, and in which the sum over $S_{\kappa'}$ is restricted to those point-group operations $S_{\kappa'}$ in the set $(\overline{\mathbf{M}}^{\mathbf{k}} - \overline{\mathbf{G}}^{\mathbf{k}})$ for which $S_{\kappa'}\mathbf{k}$ is equivalent to $-\mathbf{k}$, and where for purposes of evaluation one chooses for each $S_{\kappa'}$ any one element $\{S_{\kappa'} \mid \mathbf{w}_{\kappa'}\}$ such that $\theta\{S_{\kappa'} \mid \mathbf{w}_{\kappa'}\} \in \mathbf{M}$. In the case in which $\mathbf{M} = \mathbf{G} + \theta\mathbf{G}$ it reduces to the well-known criterion given originally by Herring (1937*a*) and which we have given in eqn (4.6.11) for time-reversal degeneracies in non-magnetic space groups. Thus the consideration of the theory of the corepresentations of the type II Shubnikov groups to find the degeneracies of eigenvalues is equivalent to the treatment given in section 4.6 of the reality of the representations of the ordinary space groups. Since this reality of the space-group representations was included in the tables in Chapter 5 and Chapter 6 there is little need to say much more about the corepresentations of the type II Shubnikov space groups. The basis functions of these corepresentations could be found by using the theory of section 7.3. It should perhaps be remarked that if one works with the magnetic little co-group the operators R_γ must be taken as being unitary and the operators $S_{\kappa'}$ as being anti-unitary. Alternatively one could maintain the unitarity of $S_{\kappa'}$ and write $\theta S_{\kappa'}$ for $S_{\kappa'}$ in the magnetic little co-group.

We may summarize what has been done for the Shubnikov space groups. For the Shubnikov groups of types II, III, and IV we have shown that it is possible to define a magnetic little group $\mathbf{M}^{\mathbf{k}}$, see Definition 7.6.2, which consists of all the unitary elements of $\mathbf{M}$ whose rotational parts turn $\mathbf{k}$ into $\mathbf{k} + \mathbf{g}$ and of all the anti-unitary elements of $\mathbf{M}$ whose rotational parts turn $\mathbf{k}$ into $-\mathbf{k} + \mathbf{g}$, see eqn (7.6.38). The number of anti-unitary elements in $\mathbf{M}^{\mathbf{k}}$ is equal either to the number of unitary elements or to zero. If the set of anti-unitary elements in $\mathbf{M}^{\mathbf{k}}$ is void $\mathbf{M}^{\mathbf{k}}$ is a unitary group, so that the reps of $\mathbf{M}^{\mathbf{k}}$ and therefore the reps of $\mathbf{G}$, the halving subgroup of unitary elements of $\mathbf{M}$, can be found by the methods usually employed for ordinary (Fedorov) space groups. The theory of section 7.3 can then be used to determine the induced coreps of $\mathbf{M}$ from the induced reps of $\mathbf{G}$. If the set of anti-unitary elements in $\mathbf{M}^{\mathbf{k}}$ is not void, theory similar to the theory of sections 3.7 and 3.8 and Chapter 4 shows that the corepresentations of $\mathbf{M}$ are available from the small corepresentations of $\mathbf{M}^{\mathbf{k}}$ and that, in turn, these small corepresentations can easily be found from the projective corepresentations of the magnetic little co-group $\overline{\mathbf{M}}^{\mathbf{k}}$, see eqns (7.6.28) and (7.6.29). The identification of the elements in $\overline{\mathbf{M}}^{\mathbf{k}}$ can be made by inspection of the structure of $\mathbf{M}$ and of the detail of its relationship to various Fedorov groups or ordinary space groups which have been discussed in the earlier chapters of this book. The problem of determining the required projective corepresentations DL of $\overline{\mathbf{M}}^{\mathbf{k}}$ is not a difficult task since the projective representations L of $\overline{\mathbf{G}}^{\mathbf{k}}$ from which they are derived are known (and can be obtained from the tables in Chapters 5 and 6). These corepresentations have been tabulated for all the magnetic space groups by Miller and Love (1967). An analogous theory to that given in section 7.3 for ordinary corepresentations can be developed. Alternatively, it can be shown that the projective corepresentations of $\overline{\mathbf{M}}^{\mathbf{k}}$ can be obtained as the ordinary corepresentations of a suitable central extension, $\overline{\mathbf{M}}^{\mathbf{k}*}$. It is fairly easy to see that $\overline{\mathbf{M}}^{\mathbf{k}*}$ contains $\overline{\mathbf{G}}^{\mathbf{k}*}$ as a unitary subgroup of index 2. The representations Λ of $\overline{\mathbf{G}}^{\mathbf{k}*}$ yield projective representations L of $\overline{\mathbf{G}}^{\mathbf{k}}$ (as explained in section 3.7). Given Λ it is possible to use the theory of section 7.3 to find the corepresentations $D\Lambda$ of $\overline{\mathbf{M}}^{\mathbf{k}*}$. It is these ordinary corepresentations which yield the required projective corepresentations DL of $\overline{\mathbf{M}}^{\mathbf{k}}$. Therefore, for general points in $\mathbf{k}$-space and for lines and planes of symmetry a simple development of the theory in section 3.7 is all that is required. For points of symmetry this method is still applicable, but can be replaced by a straightforward development of Herring's method, which provides the small corepresentations of $\mathbf{M}^{\mathbf{k}}$ without first considering the magnetic little co-group. All one needs to do is to use the theory of section 7.3 on the known small representations of $\mathbf{G}^{\mathbf{k}}$. We shall not consider the precise details of these developments here. However, an example of the use of the theory developed in this section to find the corepresentations of a magnetic space group will be given in section 7.7.

The grey space groups: the consequences of invariance under time-inversion

In section 5.2 we mentioned that a crystal in which there is no magnetic ordering possesses time-reversal symmetry in addition to all the spatial operations of **G**. The complete group of the crystal is then the direct product group $\mathbf{M} = \mathbf{G} \otimes (E + \theta)$, that is, the grey space group **M** derived from **G**. It was also mentioned that the addition of the extra symmetry operation θ may cause some extra degeneracies among the small reps of $\mathbf{G}^{\mathbf{k}}$. That is, in physical terms, the addition of θ may cause some extra degeneracies among the eigenvalues of a particle or quasi-particle at **k**. We also mentioned that the spectra at **k** and $-\mathbf{k}$ become identical whenever time-reversal symmetry is present. We are now in a position to justify these statements.

We use the notation that $\Gamma_p^{\mathbf{k}}$ is a small rep of $\mathbf{G}^{\mathbf{k}}$ and that $\Delta_p^{\mathbf{k}} = \Gamma_p^{\mathbf{k}} \uparrow \mathbf{G}$. For the grey space group $\mathbf{M} = \mathbf{G} + \theta\mathbf{G}$, the magnetic little group $\mathbf{M}^{\mathbf{k}}$, for a given **k**-vector, can take one of three forms

$$\mathbf{M}^{\mathbf{k}} = \mathbf{G}^{\mathbf{k}} \tag{7.6.43}$$

$$\mathbf{M}^{\mathbf{k}} = \mathbf{G}^{\mathbf{k}} + \theta\mathbf{G}^{\mathbf{k}} \tag{7.6.44}$$

$$\mathbf{M}^{\mathbf{k}} = \mathbf{G}^{\mathbf{k}} + A\mathbf{G}^{\mathbf{k}} \tag{7.6.45}$$

where A is some anti-unitary element which is a product of θ and some space-group element other than a translation (i.e. A cannot be chosen to be $\theta\,\{E \mid \mathbf{t}\}$). When we identified a rep in Tables 5.7 or 6.13 by (a), with no extra degeneracy, or (b) or (c), with an extra double-degeneracy, we were in fact prematurely using the notation, introduced in eqns (7.3.45)–(7.3.47), that coreps of $\mathbf{M}^{\mathbf{k}}$ must belong to case (a), case (b), or case (c).

If $\mathbf{M}^{\mathbf{k}}$ is given by eqn (7.6.43) there are no elements in **G** which transform **k** into $-\mathbf{k}$. This only happens for **k** vectors for which $-\mathbf{k}$ does not appear in the star of **k**. Clearly $\mathbf{M}^{\mathbf{k}}$ contains no anti-unitary operators and the degeneracies of the eigenvalues in $\mathbf{M}^{\mathbf{k}}$ are the same as in $\mathbf{G}^{\mathbf{k}}$. That is, there are no additional degeneracies at the point **k** due to the inclusion of time-reversal symmetry. However, since the set $(\mathbf{M}^{\mathbf{k}} - \mathbf{G}^{\mathbf{k}})$ is void, the induced corep $D\Delta_p^{\mathbf{k}}$ is of type (c), (since the sum in eqn (7.6.16) is trivially zero), and there is an extra degeneracy. To locate this extra degeneracy we note that $D\Delta_p^{\mathbf{k}}$ must contain both $\Delta_p^{\mathbf{k}}$ and $\bar{\Delta}_p^{\mathbf{k}}$; but since $\mathbf{M} = \mathbf{G} + \theta\mathbf{G}$ it follows that $\bar{\Delta}_p^{\mathbf{k}} = \Delta_p^{\mathbf{k}*}$ and since $\Delta_p^{\mathbf{k}} = \Gamma_p^{\mathbf{k}} \uparrow \mathbf{G}$ it follows that $\Delta_p^{\mathbf{k}*} = \Gamma_p^{\mathbf{k}*} \uparrow \mathbf{G}$. But $\Gamma_p^{\mathbf{k}*}$ is a small rep of $\mathbf{G}^{-\mathbf{k}}$. Hence the extra degeneracy produced in this case is between eigenvalues at **k** and $-\mathbf{k}$. This justifies the statement that if $+\mathbf{k}$ and $-\mathbf{k}$ are not in the same star time-reversal symmetry will nevertheless, if it is present in the crystal, cause the spectra of eigenvalues at **k** and $-\mathbf{k}$ to be identical. We call this degeneracy a degeneracy of type (x).

If $\mathbf{M}^{\mathbf{k}}$ is given by eqns (7.6.44) or (7.6.45) then there exist elements in **G** which transform **k** into $-\mathbf{k}$. So that in these cases $-\mathbf{k}$ appears in the star of **k**. If eqn (7.6.44) holds it is the elements of $\mathbf{G}^{\mathbf{k}}$ themselves which transform **k** into $-\mathbf{k}$; but by definition these elements leave **k** invariant. This implies $\mathbf{k} \equiv -\mathbf{k}$, which happens for those points of symmetry for which $\mathbf{k}$ is zero or half a reciprocal-lattice vector. If eqn (7.6.45) holds $-\mathbf{k}$ appears in the star of $\mathbf{k}$ but is not equivalent to it. In both cases it is clear that even without time-reversal symmetry the spectra of eigenvalues at $\mathbf{k}$ and $-\mathbf{k}$ are identical, as all $\mathbf{k}$ vectors in a star have identical spectra. Both these cases can be considered together as far as degeneracies are concerned. Because the spectra of eigenvalues at $\mathbf{k}$ and $-\mathbf{k}$ are already identical when only the symmetry of $\mathbf{G}$ is considered, the only additional degeneracy that can arise because of the presence of the additional elements $\theta\mathbf{G}$ in $\mathbf{M}$ will involve extra degeneracies among the various eigenvalues at $\mathbf{k}$. If the symmetry of a crystal is described by $\mathbf{G}$ the degeneracies of the eigenvalues for particles or quasi-particles with wave vector $\mathbf{k}$ will be determined by the reps $\Gamma_p^{\mathbf{k}}$ of $\mathbf{G}^{\mathbf{k}}$; if the symmetry of the crystal is described by the grey group $\mathbf{M} = \mathbf{G} + \theta\mathbf{G}$ the degeneracies of the eigenvalues will be determined by the coreps $D\Gamma_p^{\mathbf{k}}$ of $\mathbf{M}^{\mathbf{k}}$. We see, therefore, that to determine the extra degeneracies that arise due to the inclusion of time-reversal symmetry (see section 5.2) and when $-\mathbf{k}$ is in the star of $\mathbf{k}$ (i.e. eqn (7.6.44) or (7.6.45) applies), we have to determine whether the coreps of $\mathbf{M}^{\mathbf{k}}$ belong to case (a), case (b), or case (c). In case (a) there is no increase in degeneracy at $\mathbf{k}$, but in cases (b) and (c) a doubling of the degeneracy occurs. In case (b) $D\Gamma_p^{\mathbf{k}}$ contains $\Gamma_p^{\mathbf{k}}$ twice (see eqn (7.3.46)) and this implies that at $\mathbf{k}$ the extra degeneracy is between two sets of eigenvalues both belonging to the same rep $\Gamma_p^{\mathbf{k}}$. In case (c) $D\Gamma_p^{\mathbf{k}}$ contains $\Gamma_p^{\mathbf{k}}$ and $\bar{\Gamma}_p^{\mathbf{k}}$; $\bar{\Gamma}_p^{\mathbf{k}}$ is not equivalent to $\Gamma_p^{\mathbf{k}}$ and must, therefore, be equivalent to some other rep, $\Gamma_{p'}^{\mathbf{k}}$, of $\mathbf{G}^{\mathbf{k}}$ $(p' \neq p)$. Thus, in case (c) the extra degeneracy is between two sets of eigenvalues belonging to different reps $\Gamma_p^{\mathbf{k}}$ and $\Gamma_{p'}^{\mathbf{k}}$ of $\mathbf{G}^{\mathbf{k}}$ $(p' \neq p)$.

The equivalence of the criteria in eqns (7.6.13) and (7.6.16) implies that the type of the corep $D\Gamma_p^{\mathbf{k}}$ of $\mathbf{M}^{\mathbf{k}}$ is the same as the type of the induced corep $D\Delta_p^{\mathbf{k}}$ of $\mathbf{M}$. In case (b) $D\Delta_p^{\mathbf{k}}$ contains $\Delta_p^{\mathbf{k}}$ twice and in case (c) $D\Delta_p^{\mathbf{k}}$ contains $\Delta_p^{\mathbf{k}}$ and $\Delta_p^{\mathbf{k}*}$ $(=\bar{\Delta}_p^{\mathbf{k}})$. Since $\Delta_p^{\mathbf{k}*}$ is not equivalent to $\Delta_p^{\mathbf{k}}$, $\Delta_p^{\mathbf{k}*}$ cannot be induced from $\Gamma_p^{\mathbf{k}}$ but it must be induced from $\Gamma_{p'}^{\mathbf{k}}$ $(p' \neq p)$ as we would expect. In practice, because $\mathbf{M}$ is necessarily a grey group whereas $\mathbf{M}^{\mathbf{k}}$ is not necessarily a grey group (see eqn (7.6.45)), it is more convenient to determine the type of the corep $D\Delta_p^{\mathbf{k}}$ than of $D\Gamma_p^{\mathbf{k}}$. Because $\mathbf{M}$ is a grey group the condition which determines the type of the corep $D\Delta_p^{\mathbf{k}}$, which is a corep of $\mathbf{M}$ and not of $\mathbf{M}^{\mathbf{k}}$ (see eqn (7.3.51)), involves evaluating an expression which simplifies to a summation over the elements of $\mathbf{G}$:

$$\sum_{B \in \theta\mathbf{G}} \chi(B^2) = \sum_{R \in \mathbf{G}} \chi\{(\theta R)^2\} = \gamma \sum_{R \in \mathbf{G}} \chi(R^2) \tag{7.6.46}$$

where $\gamma = +1$ for particles with zero or integer spin and $\gamma = -1$ for particles with half-odd-integer spin, that is, when one uses single-valued and double-valued reps, respectively. The expression $\sum_{R \in \mathbf{G}} \chi(R^2)$ is just the expression used in Definition 1.3.7 to define the reality of the rep $\Delta_p^{\mathbf{k}}$ of $\mathbf{G}$. Thus, when $-\mathbf{k} \equiv \mathbf{k}$ (i.e. $\mathbf{M}^{\mathbf{k}}$ is given by eqn (7.6.44)) or when $-\mathbf{k} \not\equiv \mathbf{k}$ but $-\mathbf{k}$ is in the star of $\mathbf{k}$ (i.e. $M^{\mathbf{k}}$ is given by eqn

(7.6.45)) *the degeneracies at* $\mathbf{k}$ *produced by the existence of time-reversal symmetry in the crystal may be investigated by determining the realities of the induced reps* $\Delta_p^{\mathbf{k}}$ *of* $\mathbf{G}$. The determination of the reality of $\Delta_p^{\mathbf{k}}$ can be reduced to the evaluation of a summation over a finite number of terms, by means of Theorem 4.6.2; this test was first devised by Herring (1937*a*). Thus we have case (a), case (b), or case (c) according to the following pattern:

single-valued		*double-valued*	
reality of $\Delta_p^{\mathbf{k}}$	corep $D\Gamma_p^{\mathbf{k}}$	reality of $\Delta_p^{\mathbf{k}}$	corep $D\Gamma_p^{\mathbf{k}}$
1	a	1	b
2	b	2	a
3	c	3	c

where the distinction between single-valued and double-valued reps arises from the γ in eqn (7.6.46).

It must be emphasized that these extra degeneracies which occur at $\mathbf{k}$ when $-\mathbf{k}$ is in the star of $\mathbf{k}$ are governed by the reality of the full space group rep $\Delta_p^{\mathbf{k}}$ and it is these realities that are given in Tables 5.7 and 6.13. Of course, when $\mathbf{k} \equiv -\mathbf{k}$, so that $\mathbf{M}^{\mathbf{k}} = \mathbf{G}^{\mathbf{k}} + \theta\mathbf{G}^{\mathbf{k}}$ (eqn (7.6.44)), the reality of $\Delta_p^{\mathbf{k}}$ is the same as the reality of $\Gamma_p^{\mathbf{k}}$; but it is only in this situation that the reality of the small rep $\Gamma_p^{\mathbf{k}}$ has any physical significance. Finally, whenever $-\mathbf{k}$ is not in the star of $\mathbf{k}$, so that $\mathbf{M}^{\mathbf{k}} = \mathbf{G}^{\mathbf{k}}$, $\Delta_p^{\mathbf{k}}$ is trivially of reality 3 and $D\Delta_p^{\mathbf{k}}$ belongs to case (c) (see the above discussion relating to eqn (7.6.43)); we then have for both single-valued and double-valued reps a degeneracy of type (x). This is indicated in Tables 5.7 and 6.13 by using the symbol x rather than 3.

7.7. Example: the space-group corepresentations of $P4_2'/mnm'$ and their Kronecker products

As an illustration of the application of the theory of the previous sections we consider the deduction of the coreps of the type III Shubnikov space group $P4_2'/mnm'$; this is the group of antiferromagnetic MnF_2. This space group has been considered by Dimmock and Wheeler (1962*b*) and Cracknell (1967*c*). The structure is illustrated in Fig. 7.5 and the elements of the space group can be identified from Tables 7.2 and 3.7. It is clearly related to the ordinary space group $P4_2/mnm$ (D_{4h}^{14}) and from Table 7.2 we see that it has the same generating elements as $P4_2/mnm$ except that $\{C_{4z}^+|\frac{1}{2}\frac{1}{2}\frac{1}{2}\}$ has to be multiplied by θ. It is convenient in dealing with the structure of MnF_2 to use the origin and the setting of the axes given in the *International tables for X-ray crystallography* (Henry and Lonsdale 1965). The second set of entries in Table 3.7 is therefore appropriate and the generating elements are $\{I\,|\,000\}$, $\{C_{2x}\,|\,\frac{1}{2}\frac{1}{2}\frac{1}{2}\}$, and $\theta\{C_{4z}^+\,|\,\frac{1}{2}\frac{1}{2}\frac{1}{2}\}$. Coset representatives of $\mathbf{M}$ in a decomposition relative to $\mathbf{T}$, the translation group of the primitive tetragonal lattice (see Table 3.1) may therefore be taken as

$$\begin{array}{llll}
\{E\,|\,000\} & \{I\,|\,000\} & \theta\{C_{4z}^{+}\,|\,\tfrac{1}{2}\tfrac{1}{2}\tfrac{1}{2}\} & \theta\{S_{4z}^{-}\,|\,\tfrac{1}{2}\tfrac{1}{2}\tfrac{1}{2}\} \\
\{C_{2x}\,|\,\tfrac{1}{2}\tfrac{1}{2}\tfrac{1}{2}\} & \{\sigma_x\,|\,\tfrac{1}{2}\tfrac{1}{2}\tfrac{1}{2}\} & \theta\{C_{4z}^{-}\,|\,\tfrac{1}{2}\tfrac{1}{2}\tfrac{1}{2}\} & \theta\{S_{4z}^{+}\,|\,\tfrac{1}{2}\tfrac{1}{2}\tfrac{1}{2}\} \\
\{C_{2y}\,|\,\tfrac{1}{2}\tfrac{1}{2}\tfrac{1}{2}\} & \{\sigma_y\,|\,\tfrac{1}{2}\tfrac{1}{2}\tfrac{1}{2}\} & \theta\{C_{2a}\,|\,000\} & \theta\{\sigma_{da}\,|\,000\} \\
\{C_{2z}\,|\,000\} & \{\sigma_z\,|\,000\} & \theta\{C_{2b}\,|\,000\} & \theta\{\sigma_{db}\,|\,000\}.
\end{array}$$

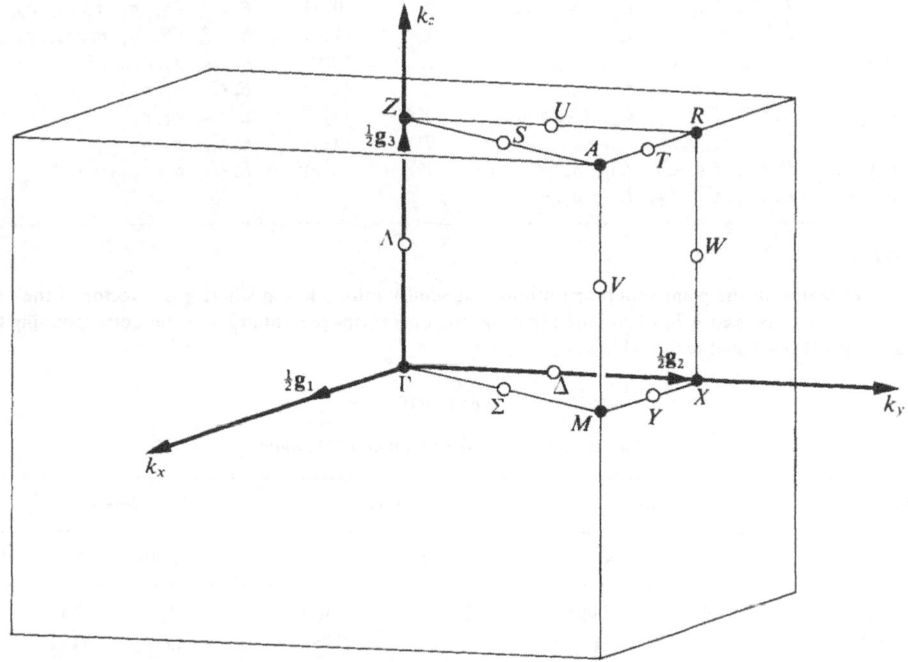

FIG. 7.9. The Brillouin zone for Γ_q. $\Gamma = (000)$; $M = (\frac{1}{2}\frac{1}{2}0)$; $Z = (00\frac{1}{2})$; $A = (\frac{1}{2}\frac{1}{2}\frac{1}{2})$; $R = (0\frac{1}{2}\frac{1}{2})$; $X = (0\frac{1}{2}0)$.

The Brillouin zone of $P4_2'/mnm'$ is illustrated in Fig. 3.9 and reproduced in Fig. 7.9. The points and lines of symmetry can be found from Table 3.6, together with **P(k)**, the set of point-group elements that transform **k** into an equivalent wave vector. The appropriate part of Table 3.6 is reproduced in Table 7.18, in which wave vectors of points and lines of symmetry are given in terms of $\mathbf{g}_1$, $\mathbf{g}_2$, and $\mathbf{g}_3$.

The first stage in determining the coreps of $P4_2'/mnm'$ lies in finding the space-group reps of the unitary subgroup, **G**. Coset representatives of **G** relative to **T** are the elements from the above list which are not associated with θ:

$$\begin{array}{ll}
\{E\,|\,000\} & \{I\,|\,000\} \\
\{C_{2x}\,|\,\tfrac{1}{2}\tfrac{1}{2}\tfrac{1}{2}\} & \{\sigma_x\,|\,\tfrac{1}{2}\tfrac{1}{2}\tfrac{1}{2}\} \\
\{C_{2y}\,|\,\tfrac{1}{2}\tfrac{1}{2}\tfrac{1}{2}\} & \{\sigma_y\,|\,\tfrac{1}{2}\tfrac{1}{2}\tfrac{1}{2}\} \\
\{C_{2z}\,|\,000\} & \{\sigma_z\,|\,000\}.
\end{array}$$

TABLE 7.18

The points and lines of symmetry of $P4'_2/mnm'$

Point or line	**k**	**P(k)**	Point or line	**k**	**P(k)**
Γ	(000)	$E, C_{4z}^{\pm}, C_{2z}, C_{2x}, C_{2y}, C_{2a}, C_{2b},$ $I, S_{4z}^{\mp}, \sigma_z, \sigma_x, \sigma_y, \sigma_{da}, \sigma_{db}.$	Δ	$(0\alpha0)$	$E, C_{2y}, \sigma_z, \sigma_x.$
			U	$(0\alpha\frac{1}{2})$	$E, C_{2y}, \sigma_z, \sigma_x.$
M	$(\frac{1}{2}\frac{1}{2}0)$	$E, C_{4z}^{\pm}, C_{2z}, C_{2x}, C_{2y}, C_{2a}, C_{2b},$ $I, S_{4z}^{\mp}, \sigma_z, \sigma_x, \sigma_y, \sigma_{da}, \sigma_{db}.$	Λ	(00α)	$E, C_{4z}^{\pm}, C_{2z}, \sigma_x, \sigma_y, \sigma_{da}, \sigma_{db}.$
			V	$(\frac{1}{2}\frac{1}{2}\alpha)$	$E, C_{4z}^{\pm}, C_{2z}, \sigma_x, \sigma_y, \sigma_{da}, \sigma_{db}.$
Z	$(00\frac{1}{2})$	$E, C_{4z}^{\pm}, C_{2z}, C_{2x}, C_{2y}, C_{2a}, C_{2b},$ $I, S_{4z}^{\mp}, \sigma_z, \sigma_x, \sigma_y, \sigma_{da}, \sigma_{db}.$	Σ	$(\alpha\alpha0)$	$E, C_{2a}, \sigma_z, \sigma_{db}.$
			S	$(\alpha\alpha\frac{1}{2})$	$E, C_{2a}, \sigma_z, \sigma_{db}.$
A	$(\frac{1}{2}\frac{1}{2}\frac{1}{2})$	$E, C_{4z}^{\pm}, C_{2z}, C_{2x}, C_{2y}, C_{2a}, C_{2b},$ $I, S_{4z}^{\mp}, \sigma_z, \sigma_x, \sigma_y, \sigma_{da}, \sigma_{db}.$	Y	$(\alpha\frac{1}{2}0)$	$E, C_{2x}, \sigma_y, \sigma_z.$
			T	$(\alpha\frac{1}{2}\frac{1}{2})$	$E, C_{2x}, \sigma_y, \sigma_z.$
R	$(0\frac{1}{2}\frac{1}{2})$	$E, C_{2x}, C_{2y}, C_{2z}, I, \sigma_x, \sigma_y, \sigma_z.$	W	$(0\frac{1}{2}\alpha)$	$E, C_{2z}, \sigma_x, \sigma_y.$
X	$(0\frac{1}{2}0)$	$E, C_{2x}, C_{2y}, C_{2z}, I, \sigma_x, \sigma_y, \sigma_z.$			

Note to Table 7.18

This table contains all the point-group operations that send **k** into $+\mathbf{k} + \mathbf{g}$ where **g** is a vector of the reciprocal lattice; there will not necessarily be uncoloured space-group operations present in $P4'_2/mnm'$ corresponding to all the point-group operations listed in the table.

TABLE 7.19

The Brillouin zones for Pnnm and $P4'_2/mnm'$

$P4'_2/mnm'$		*Pnnm*		$P4'_2/mnm'$		*Pnnm*	
Point	**k**	Point	**k**	Point	**k**	Point	**k**
Γ	(000)	Γ	(000)	Λ	(00α)	Λ	(00α)
M	$(\frac{1}{2}\frac{1}{2}0)$	S	$(\bar{\frac{1}{2}}\frac{1}{2}0)$	V	$(\frac{1}{2}\frac{1}{2}\alpha)$	Q	$(\bar{\frac{1}{2}}\frac{1}{2}\alpha)$
Z	$(00\frac{1}{2})$	Z	$(00\frac{1}{2})$	Σ	$(\alpha\alpha0)$	—	$(\bar{\alpha}\alpha0)$
A	$(\frac{1}{2}\frac{1}{2}\frac{1}{2})$	R	$(\bar{\frac{1}{2}}\frac{1}{2}\frac{1}{2})$	S	$(\alpha\alpha\frac{1}{2})$	—	$(\bar{\alpha}\alpha\frac{1}{2})$
R	$(0\frac{1}{2}\frac{1}{2})$	T	$(\bar{\frac{1}{2}}0\frac{1}{2})$	Y	$(\alpha\frac{1}{2}0)$	C	$(\bar{\frac{1}{2}}\alpha0)$
X	$(0\frac{1}{2}0)$	Y	$(\bar{\frac{1}{2}}00)$	T	$(\alpha\frac{1}{2}\frac{1}{2})$	E	$(\bar{\frac{1}{2}}\alpha\frac{1}{2})$
Δ	$(0\alpha0)$	Δ	$(\bar{\alpha}00)$	W	$(0\frac{1}{2}\alpha)$	H	$(\bar{\frac{1}{2}}0\alpha)$
U	$(0\alpha\frac{1}{2})$	B	$(\bar{\alpha}0\frac{1}{2})$				

Notes to Table 7.19

(i) For the primitive tetragonal lattice

$$\mathbf{t}_1 = (a, 0, 0);\quad \mathbf{t}_2 = (0, a, 0);\quad \mathbf{t}_3 = (0, 0, c);$$
$$\mathbf{g}_1 = (2\pi/a)(1, 0, 0);\quad \mathbf{g}_2 = (2\pi/a)(0, 1, 0);\quad \mathbf{g}_3 = (2\pi/c)(0, 0, 1);$$

while for the primitive orthorhombic lattice

$$\mathbf{t}_1 = (0, -b, 0);\quad \mathbf{t}_2 = (a, 0, 0);\quad \mathbf{t}_3 = (0, 0, c);$$
$$\mathbf{g}_1 = (2\pi/b)(0, -1, 0);\quad \mathbf{g}_2 = (2\pi/a)(1, 0, 0);\quad \mathbf{g}_3 = (2\pi/c)(0, 0, 1);$$

where in this particular case $a = b$.

(ii) $(\bar{\alpha}\alpha0)$ and $(\bar{\alpha}\alpha\frac{1}{2})$ are only planes of symmetry in *Pnnm* and are therefore not given in previous tables.

TABLE 7.20

${}^{H}\mathbf{G}^{\dagger\mathbf{k}}$ *for the points of symmetry of Pnnm*

(a) Γ ${}^{H}\mathbf{G}^{\dagger\mathbf{k}}$ for Γ of $Pnnm = \mathbf{G}^{11}_{16}$

C_1	E	$\{E\,\|\,000\}$;	C_6	R	$\{I\,\|\,000\}$;
C_2	P^2	$\{\bar{E}\,\|\,000\}$;	C_7	P^2R	$\{\bar{I}\,\|\,000\}$;
C_3	P	$\{C_{2z}\,\|\,000\}$,	C_8	PR	$\{\sigma_z\,\|\,000\}$,
	P^3	$\{\bar{C}_{2z}\,\|\,000\}$;		P^3R	$\{\bar{\sigma}_z\,\|\,000\}$;
C_4	Q	$\{C_{2y}\,\|\,\frac{1}{2}\frac{1}{2}\frac{1}{2}\}$,	C_9	QR	$\{\sigma_y\,\|\,\frac{1}{2}\frac{1}{2}\frac{1}{2}\}$,
	P^2Q	$\{\bar{C}_{2y}\,\|\,\frac{1}{2}\frac{1}{2}\frac{1}{2}\}$;		P^2QR	$\{\bar{\sigma}_y\,\|\,\frac{1}{2}\frac{1}{2}\frac{1}{2}\}$;
C_5	PQ	$\{\bar{C}_{2x}\,\|\,\frac{1}{2}\frac{1}{2}\frac{1}{2}\}$,	C_{10}	PQR	$\{\bar{\sigma}_x\,\|\,\frac{1}{2}\frac{1}{2}\frac{1}{2}\}$,
	P^3Q	$\{C_{2x}\,\|\,\frac{1}{2}\frac{1}{2}\frac{1}{2}\}$;		P^3QR	$\{\sigma_x\,\|\,\frac{1}{2}\frac{1}{2}\frac{1}{2}\}$.

(b) Y, Z, R ${}^{H}\mathbf{G}^{\dagger\mathbf{k}}$ for Y, Z, and R of $Pnnm = \mathbf{G}^{7}_{32}$

		Y	Z	R
C_1	E	$\{E\,\|\,000\}$;	$\{E\,\|\,000\}$;	$\{E\,\|\,000\}$;
C_2	P^2	$\{\bar{E}\,\|\,100\}$;	$\{\bar{E}\,\|\,001\}$;	$\{\bar{E}\,\|\,100\}$;
C_3	Q^2	$\{\bar{E}\,\|\,000\}$;	$\{\bar{E}\,\|\,000\}$;	$\{\bar{E}\,\|\,000\}$;
C_4	P^2Q^2	$\{E\,\|\,100\}$;	$\{E\,\|\,001\}$;	$\{E\,\|\,100\}$;
C_5	P	$\{\sigma_x\,\|\,\frac{1}{2}\frac{1}{2}\frac{1}{2}\}$,	$\{\sigma_x\,\|\,\frac{1}{2}\frac{1}{2}\frac{1}{2}\}$,	$\{C_{2y}\,\|\,\frac{1}{2}\frac{1}{2}\frac{1}{2}\}$,
	P^3	$\{\bar{\sigma}_x\,\|\,-\frac{1}{2}\frac{1}{2}\frac{1}{2}\}$,	$\{\bar{\sigma}_x\,\|\,\frac{1}{2}\frac{1}{2}\,-\frac{1}{2}\}$,	$\{\bar{C}_{2y}\,\|\,-\frac{1}{2}\frac{1}{2}\frac{1}{2}\}$,
	PQ^2	$\{\bar{\sigma}_x\,\|\,\frac{1}{2}\frac{1}{2}\frac{1}{2}\}$,	$\{\bar{\sigma}_x\,\|\,\frac{1}{2}\frac{1}{2}\frac{1}{2}\}$,	$\{\bar{C}_{2y}\,\|\,\frac{1}{2}\frac{1}{2}\frac{1}{2}\}$,
	P^3Q^2	$\{\sigma_x\,\|\,-\frac{1}{2}\frac{1}{2}\frac{1}{2}\}$,	$\{\sigma_x\,\|\,\frac{1}{2}\frac{1}{2}\,-\frac{1}{2}\}$;	$\{C_{2y}\,\|\,-\frac{1}{2}\frac{1}{2}\frac{1}{2}\}$;
C_6	Q	$\{C_{2z}\,\|\,000\}$,	$\{\sigma_z\,\|\,001\}$,	$\{\sigma_z\,\|\,001\}$,
	P^2Q	$\{\bar{C}_{2z}\,\|\,100\}$;	$\{\bar{\sigma}_z\,\|\,000\}$;	$\{\bar{\sigma}_z\,\|\,000\}$;
C_7	Q^3	$\{\bar{C}_{2z}\,\|\,000\}$,	$\{\bar{\sigma}_z\,\|\,001\}$,	$\{\bar{\sigma}_z\,\|\,001\}$,
	P^2Q^3	$\{C_{2z}\,\|\,100\}$;	$\{\sigma_z\,\|\,000\}$;	$\{\sigma_z\,\|\,000\}$;
C_8	PQ	$\{\bar{\sigma}_y\,\|\,\frac{1}{2}\frac{1}{2}\frac{1}{2}\}$,	$\{\bar{C}_{2y}\,\|\,\frac{1}{2}\frac{1}{2}\,-\frac{1}{2}\}$,	$\{\sigma_x\,\|\,\frac{1}{2}\frac{1}{2}\,-\frac{1}{2}\}$,
	P^3Q	$\{\sigma_y\,\|\,-\frac{1}{2}\frac{1}{2}\frac{1}{2}\}$,	$\{C_{2y}\,\|\,\frac{1}{2}\frac{1}{2}\frac{1}{2}\}$,	$\{\bar{\sigma}_x\,\|\,\frac{1}{2}\frac{1}{2}\frac{1}{2}\}$,
	PQ^3	$\{\sigma_y\,\|\,\frac{1}{2}\frac{1}{2}\frac{1}{2}\}$,	$\{C_{2y}\,\|\,\frac{1}{2}\frac{1}{2}\,-\frac{1}{2}\}$,	$\{\bar{\sigma}_x\,\|\,\frac{1}{2}\frac{1}{2}\,-\frac{1}{2}\}$,
	P^3Q^3	$\{\bar{\sigma}_y\,\|\,-\frac{1}{2}\frac{1}{2}\frac{1}{2}\}$;	$\{\bar{C}_{2y}\,\|\,\frac{1}{2}\frac{1}{2}\frac{1}{2}\}$;	$\{\sigma_x\,\|\,\frac{1}{2}\frac{1}{2}\frac{1}{2}\}$;
C_9	R	$\{I\,\|\,000\}$,	$\{I\,\|\,001\}$,	$\{I\,\|\,001\}$,
	P^2Q^2R	$\{I\,\|\,100\}$;	$\{I\,\|\,000\}$;	$\{I\,\|\,000\}$;
C_{10}	P^2R	$\{\bar{I}\,\|\,100\}$,	$\{\bar{I}\,\|\,000\}$,	$\{\bar{I}\,\|\,000\}$,
	Q^2R	$\{\bar{I}\,\|\,000\}$;	$\{\bar{I}\,\|\,001\}$;	$\{\bar{I}\,\|\,001\}$;
C_{11}	PR	$\{C_{2x}\,\|\,\frac{1}{2}\frac{1}{2}\frac{1}{2}\}$,	$\{C_{2x}\,\|\,\frac{1}{2}\frac{1}{2}\,-\frac{1}{2}\}$,	$\{\sigma_y\,\|\,\frac{1}{2}\frac{1}{2}\,-\frac{1}{2}\}$,
	P^3R	$\{\bar{C}_{2x}\,\|\,-\frac{1}{2}\frac{1}{2}\frac{1}{2}\}$,	$\{\bar{C}_{2x}\,\|\,\frac{1}{2}\frac{1}{2}\frac{1}{2}\}$,	$\{\bar{\sigma}_y\,\|\,\frac{1}{2}\frac{1}{2}\frac{1}{2}\}$,
	PQ^2R	$\{\bar{C}_{2x}\,\|\,\frac{1}{2}\frac{1}{2}\frac{1}{2}\}$,	$\{\bar{C}_{2x}\,\|\,\frac{1}{2}\frac{1}{2}\,-\frac{1}{2}\}$,	$\{\bar{\sigma}_y\,\|\,\frac{1}{2}\frac{1}{2}\,-\frac{1}{2}\}$,
	P^3Q^2R	$\{C_{2x}\,\|\,-\frac{1}{2}\frac{1}{2}\frac{1}{2}\}$;	$\{C_{2x}\,\|\,\frac{1}{2}\frac{1}{2}\frac{1}{2}\}$;	$\{\sigma_y\,\|\,\frac{1}{2}\frac{1}{2}\frac{1}{2}\}$;
C_{12}	QR	$\{\sigma_z\,\|\,000\}$,	$\{C_{2z}\,\|\,000\}$,	$\{C_{2z}\,\|\,000\}$,
	Q^3R	$\{\bar{\sigma}_z\,\|\,000\}$;	$\{\bar{C}_{2z}\,\|\,000\}$;	$\{\bar{C}_{2z}\,\|\,000\}$;
C_{13}	P^2QR	$\{\bar{\sigma}_z\,\|\,100\}$,	$\{\bar{C}_{2z}\,\|\,001\}$,	$\{\bar{C}_{2z}\,\|\,001\}$,
	P^2Q^3R	$\{\sigma_z\,\|\,100\}$;	$\{C_{2z}\,\|\,001\}$;	$\{C_{2z}\,\|\,001\}$;
C_{14}	PQR	$\{\bar{C}_{2y}\,\|\,\frac{1}{2}\frac{1}{2}\frac{1}{2}\}$,	$\{\bar{\sigma}_y\,\|\,\frac{1}{2}\frac{1}{2}\frac{1}{2}\}$,	$\{C_{2x}\,\|\,\frac{1}{2}\frac{1}{2}\frac{1}{2}\}$,
	P^3QR	$\{C_{2y}\,\|\,-\frac{1}{2}\frac{1}{2}\frac{1}{2}\}$,	$\{\sigma_y\,\|\,\frac{1}{2}\frac{1}{2}\,-\frac{1}{2}\}$,	$\{\bar{C}_{2x}\,\|\,\frac{1}{2}\frac{1}{2}\,-\frac{1}{2}\}$,
	PQ^3R	$\{C_{2y}\,\|\,\frac{1}{2}\frac{1}{2}\frac{1}{2}\}$,	$\{\sigma_y\,\|\,\frac{1}{2}\frac{1}{2}\frac{1}{2}\}$,	$\{\bar{C}_{2x}\,\|\,\frac{1}{2}\frac{1}{2}\frac{1}{2}\}$,
	P^3Q^3R	$\{\bar{C}_{2y}\,\|\,-\frac{1}{2}\frac{1}{2}\frac{1}{2}\}$;	$\{\bar{\sigma}_y\,\|\,\frac{1}{2}\frac{1}{2}\,-\frac{1}{2}\}$;	$\{C_{2x}\,\|\,\frac{1}{2}\frac{1}{2}\,-\frac{1}{2}\}$.

(c) S, T ${}^{H}\mathbf{G}^{\dagger\mathbf{k}}$ for S and T of $Pnnm$ = $\mathbf{G}_{32}^{14}$

		S	T			S	T
C_1	E	$\{E\mid 000\}$;	$\{E\mid 000\}$;	C_{11}	R	$\{I\mid 000\}$;	$\{I\mid 001\}$;
C_2	P^2	$\{\bar{E}\mid 000\}$;	$\{\bar{E}\mid 001\}$;	C_{12}	P^2R	$\{\bar{I}\mid 000\}$;	$\{\bar{I}\mid 000\}$;
C_3	Q^2	$\{\bar{E}\mid 100\}$;	$\{\bar{E}\mid 000\}$;	C_{13}	Q^2R	$\{\bar{I}\mid 100\}$;	$\{\bar{I}\mid 001\}$;
C_4	P^2Q^2	$\{E\mid 100\}$;	$\{E\mid 001\}$;	C_{14}	P^2Q^2R	$\{I\mid 100\}$;	$\{I\mid 000\}$;
C_5	P	$\{C_{2z}\mid 000\}$,	$\{\sigma_y\mid \frac{1}{2}\frac{1}{2}\frac{1}{2}\}$,	C_{15}	PR	$\{\sigma_z\mid 000\}$,	$\{C_{2y}\mid -\frac{1}{2}\frac{1}{2}\frac{1}{2}\}$,
	P^3	$\{\bar{C}_{2z}\mid 000\}$;	$\{\bar{\sigma}_y\mid \frac{1}{2}\frac{1}{2}-\frac{1}{2}\}$;		P^3R	$\{\bar{\sigma}_z\mid 000\}$;	$\{\bar{C}_{2y}\mid \frac{1}{2}\frac{1}{2}\frac{1}{2}\}$;
C_6	PQ^2	$\{\bar{C}_{2z}\mid 100\}$,	$\{\bar{\sigma}_y\mid \frac{1}{2}\frac{1}{2}\frac{1}{2}\}$,	C_{16}	PQ^2R	$\{\bar{\sigma}_z\mid 100\}$,	$\{\bar{C}_{2y}\mid -\frac{1}{2}\frac{1}{2}\frac{1}{2}\}$,
	P^3Q^2	$\{C_{2z}\mid 100\}$;	$\{\sigma_y\mid \frac{1}{2}\frac{1}{2}-\frac{1}{2}\}$;		P^3Q^2R	$\{\sigma_z\mid 100\}$;	$\{C_{2y}\mid \frac{1}{2}\frac{1}{2}\frac{1}{2}\}$;
C_7	Q	$\{\sigma_x\mid \frac{1}{2}\frac{1}{2}\frac{1}{2}\}$,	$\{C_{2z}\mid 000\}$,	C_{17}	QR	$\{C_{2x}\mid \frac{1}{2}\frac{1}{2}\frac{1}{2}\}$,	$\{\sigma_z\mid 001\}$,
	P^2Q	$\{\bar{\sigma}_x\mid \frac{1}{2}\frac{1}{2}\frac{1}{2}\}$;	$\{\bar{C}_{2z}\mid 001\}$;		P^2QR	$\{\bar{C}_{2x}\mid \frac{1}{2}\frac{1}{2}\frac{1}{2}\}$;	$\{\bar{\sigma}_z\mid 000\}$;
C_8	PQ	$\{\sigma_y\mid \frac{1}{2}\frac{1}{2}\frac{1}{2}\}$,	$\{\sigma_x\mid \frac{1}{2}\frac{1}{2}\frac{1}{2}\}$,	C_{18}	PQR	$\{C_{2y}\mid \frac{1}{2}\frac{1}{2}\frac{1}{2}\}$,	$\{C_{2x}\mid -\frac{1}{2}\frac{1}{2}\frac{1}{2}\}$,
	P^3Q	$\{\bar{\sigma}_y\mid \frac{1}{2}\frac{1}{2}\frac{1}{2}\}$;	$\{\bar{\sigma}_x\mid \frac{1}{2}\frac{1}{2}-\frac{1}{2}\}$;		P^3QR	$\{\bar{C}_{2y}\mid \frac{1}{2}\frac{1}{2}\frac{1}{2}\}$;	$\{\bar{C}_{2x}\mid \frac{1}{2}\frac{1}{2}\frac{1}{2}\}$;
C_9	Q^3	$\{\bar{\sigma}_x\mid -\frac{1}{2}\frac{1}{2}\frac{1}{2}\}$,	$\{\bar{C}_{2z}\mid 000\}$,	C_{19}	Q^3R	$\{\bar{C}_{2x}\mid -\frac{1}{2}\frac{1}{2}\frac{1}{2}\}$,	$\{\bar{\sigma}_z\mid 001\}$,
	P^2Q^3	$\{\sigma_x\mid -\frac{1}{2}\frac{1}{2}\frac{1}{2}\}$;	$\{C_{2z}\mid 001\}$;		P^2Q^3R	$\{C_{2x}\mid -\frac{1}{2}\frac{1}{2}\frac{1}{2}\}$;	$\{\sigma_z\mid 000\}$;
C_{10}	PQ^3	$\{\bar{\sigma}_y\mid -\frac{1}{2}\frac{1}{2}\frac{1}{2}\}$,	$\{\bar{\sigma}_x\mid \frac{1}{2}\frac{1}{2}\frac{1}{2}\}$,	C_{20}	PQ^3R	$\{\bar{C}_{2y}\mid -\frac{1}{2}\frac{1}{2}\frac{1}{2}\}$,	$\{\bar{C}_{2x}\mid -\frac{1}{2}\frac{1}{2}\frac{1}{2}\}$,
	P^3Q^3	$\{\sigma_y\mid -\frac{1}{2}\frac{1}{2}\frac{1}{2}\}$;	$\{\sigma_x\mid \frac{1}{2}\frac{1}{2}-\frac{1}{2}\}$;		P^3Q^3R	$\{C_{2y}\mid -\frac{1}{2}\frac{1}{2}\frac{1}{2}\}$;	$\{C_{2x}\mid \frac{1}{2}\frac{1}{2}\frac{1}{2}\}$.

Notes to Table 7.20

(i) The groups have been determined from Table 6.7. In making the correspondence a change of origin of co-ordinates has been made, using eqns (3.5.9) and (3.5.10), so that the elements correspond to the second setting given in Table 3.7; this is the more natural setting when dealing with the structure of MnF_2.

(ii) The translational parts of the Seitz space-group symbols are expressed in terms of $\mathbf{t}_1$, $\mathbf{t}_2$, and $\mathbf{t}_3$ for the primitive orthorhombic Bravais lattice, Γ_o, see Table 3.1.

These elements can be recognized from Table 3.7 as constituting the primitive orthorhombic space group *Pnnm* (D_{2h}^{12}) (using once again the second origin and setting of the axes given in Table 3.7); the generating elements of the group *Pnnm* are then $\{C_{2x}\mid \frac{1}{2}\frac{1}{2}\frac{1}{2}\}$, $\{C_{2y}\mid \frac{1}{2}\frac{1}{2}\frac{1}{2}\}$, and $\{I\mid 000\}$. We can now find the reps of the unitary subgroup of $P4_2'/mnm'$ very easily since they are just the reps of the space group *Pnnm* given in the tables in Chapters 5 and 6. The Brillouin zone of *Pnnm* is the Brillouin zone of the primitive orthorhombic Bravais lattice (see Fig. 3.5) whose notation differs somewhat from that of the primitive tetragonal Bravais lattice (see Fig. 7.9). It is therefore necessary to establish the relationship between the notations used for these two Brillouin zones before we can use the representations of *Pnnm* given in Chapters 5 and 6 as representations of the unitary subgroup of $P4_2'/mnm'$; the correspondence between these two Brillouin zones is given in Table 7.19. We can, therefore, use Tables 5.7 and 6.13 to write down the character tables of the representations of *Pnnm* at Γ, S, Z, R, T, and Y. We can use Table 6.13 to give both the single-valued and double-valued reps of *Pnnm*; ${}^{H}\mathbf{G}^{\dagger\mathbf{k}}$ (since we are using the double group this replaces ${}^{H}\mathbf{G}^{\mathbf{k}}$) is to be identified with the abstract group $\mathbf{G}_{16}^{11}$ for Γ, with $\mathbf{G}_{32}^{7}$ for Y, Z, and R and with $\mathbf{G}_{32}^{14}$ for S and T. The identification of these abstract groups with ${}^{H}\mathbf{G}^{\dagger\mathbf{k}}$ proceeds in the way that we illustrated in the example in section 5.4.

It is, therefore, not necessary to include all the details here. We simply give in Table 7.20 the identification of all the elements in ${}^{H}\mathbf{G}^{\dagger\mathbf{k}}$ in terms of the elements of the abstract groups and we list in Table 7.21 the single-valued and double-valued reps of ${}^{H}\mathbf{G}^{\dagger\mathbf{k}}$ from Tables 5.7 and 6.13 together with the space-group rep labels from Tables 5,8 and 6.14. We are of course, as always in this book, using Herring's method for the tabulation of the small reps of little groups of points of symmetry. A point that must be emphasized is that the representation domain of $P4_2'/mnm'$ is only half the size of the representation domain of its unitary subgroup (since it contains twice as many elements). Hence in making the correspondence between their Brillouin zones we are at liberty to choose one for the tetragonal lattice which contains either of two halves of the representation domain suitable for the unitary subgroup based on the orthorhombic lattice. That is, it may contain Γ, Z, R, S, Y, and T or Γ, Z, R, S, X, and U. However, in this example of $P4_2'/mnm'$ we are only concerned with the orthorhombic space group *Pnnm* as a subgroup of a tetragonal space group so that X and U are not essentially different from Y and T respectively. We may therefore identify the points of symmetry in our notation with that used by Dimmock and Wheeler (1962*b*) as follows,

This book	Γ	Z	R	S	T	Y
Dimmock and Wheeler (1962*b*)	Γ	Z	A	M	R	X.

The labels given to the representations by Dimmock and Wheeler (1962*b*) are included in Table 7.21; the translation element $\{E \mid 100\}$ that we have in ${}^{H}\mathbf{G}^{\dagger\mathbf{k}}$ at T and Y is replaced by $\{E \mid 010\}$ at X and U in the work of Dimmock and Wheeler because of the different choice of representation domain of the Brillouin zone that was made.

It now remains to use the space-group representations of *Pnnm* to determine the corepresentations of the magnetic space group $P4_2'/mnm'$. For all the points of symmetry that we have considered, $\mathbf{k}$ is equivalent to $-\mathbf{k}$ so that the magnetic little co-group simply consists of all the elements of $P4_2'/mnm'$ that send $\mathbf{k}$ into itself. From Tables 7.18 and 7.19 we see that for Y and T there will be no anti-unitary elements in $\overline{\mathbf{M}}^{\dagger\mathbf{k}}$; $\overline{\mathbf{M}}^{\dagger\mathbf{k}}$ is therefore unitary and is just equal to $\overline{\mathbf{G}}^{\dagger\mathbf{k}}$. There are, therefore, no small coreps to be determined for these points. The coreps of $P4_2'/mnm'$ obtained by inducing the small reps of $\mathbf{G}^{\dagger Y}$ and $\mathbf{G}^{\dagger T}$ into $P4_2'/mnm'$ can be seen from the application of the test in eqn (7.6.41) to belong trivially to case (c). Each of these case (c) coreps is formed by the sticking together of two small reps with different $\mathbf{k}$ vectors and this is therefore similar to the type (x) degeneracy in grey space groups discussed at the end of the previous section. For Γ, Z, R, and S the magnetic little group contains all the elements of $P4_2'/mnm'$, so that ${}^{H}\mathbf{M}^{\dagger\mathbf{k}}$ consists in each case of all the elements of ${}^{H}\mathbf{G}^{\dagger\mathbf{k}}$ given for that point in Table 7.20, together with an equal number of anti-unitary elements which may be found by multiplying ${}^{H}\mathbf{G}^{\dagger\mathbf{k}}$ by, for example, $\theta\{C_{4z}^{+} \mid \frac{1}{2}\frac{1}{2}\frac{1}{2}\}$. The coreps of ${}^{H}\mathbf{M}^{\dagger\mathbf{k}}$ which derive from each of the small reps of ${}^{H}\mathbf{G}^{\dagger\mathbf{k}}$ can then be found for each of the points Γ, Z, R, and S by using eqn (7.3.45), (7.3.46), or (7.3.47) as appropriate from the test given by eqn (7.6.41). The results of applying this test are given in Table 7.22. Since the magnetic little group is (for each of the four points) the entire magnetic group, these small coreps are already coreps of $P4_2'/mnm'$ and no further induction is necessary.

TABLE 7.21

The character tables for the points of symmetry of Pnnm

(a) Γ

	R_1	R_2	R_3	R_4	R_5	R_6	R_7	R_8	R_9	R_{10}
	A_{1g}	B_{1g}	B_{2g}	B_{3g}	$\bar{E}_g$	A_{1u}	B_{1u}	B_{2u}	B_{3u}	$\bar{E}_u$
	Γ_1^+	Γ_2^+	Γ_4^+	Γ_3^+	Γ_5^+	Γ_1^-	Γ_2^-	Γ_4^-	Γ_3^-	Γ_5^-
C_1	1	1	1	1	2	1	1	1	1	2
C_2	1	1	1	1	−2	1	1	1	1	−2
C_3	1	1	−1	−1	0	1	1	−1	−1	0
C_4	1	−1	1	−1	0	1	−1	1	−1	0
C_5	1	−1	−1	1	0	1	−1	−1	1	0
C_6	1	1	1	1	2	−1	−1	−1	−1	−2
C_7	1	1	1	1	−2	−1	−1	−1	−1	2
C_8	1	1	−1	−1	0	−1	−1	1	1	0
C_9	1	−1	1	−1	0	−1	1	−1	1	0
C_{10}	1	−1	−1	1	0	−1	1	1	−1	0

(b) Y, Z, R

	R_9	R_{10}	R_{13}	R_{14}		R_9	R_{10}	R_{13}	R_{14}
Y	E'	E''	${}^1\bar{F}$	${}^2\bar{F}$	Y	E'	E''	${}^1\bar{F}$	${}^2\bar{F}$
	Y_1	Y_2	Y_3	Y_4		Y_1	Y_2	Y_3	Y_4
	X_1	X_2	X_3	X_4		X_1	X_2	X_3	X_4
Z	E_1	E_2	${}^1\bar{F}$	${}^2\bar{F}$	Z	E_1	E_2	${}^1\bar{F}$	${}^2\bar{F}$
	Z_1	Z_2	Z_3	Z_4		Z_1	Z_2	Z_3	Z_4
	Z_1	Z_2	Z_4	Z_3		Z_1	Z_2	Z_4	Z_3
R	E_1	E_2	${}^1\bar{F}$	${}^2\bar{F}$	R	E_1	E_2	${}^1\bar{F}$	${}^2\bar{F}$
	R_1	R_2	R_3	R_4		R_1	R_2	R_3	R_4
	A_1	A_2	A_4	A_3		A_1	A_2	A_4	A_3
C_1	2	2	2	2	C_8	0	0	0	0
C_2	−2	−2	2	2	C_9	0	0	0	0
C_3	2	2	−2	−2	C_{10}	0	0	0	0
C_4	−2	−2	−2	−2	C_{11}	0	0	0	0
C_5	0	0	0	0	C_{12}	2	−2	0	0
C_6	0	0	2i	−2i	C_{13}	−2	2	0	0
C_7	0	0	−2i	2i	C_{14}	0	0	0	0

(c) S, T

	R_5	R_6	R_7	R_8	R_9	R_{15}	R_{16}	R_{17}	R_{18}	R_{19}
S	$^2E_{2g}$	$^1E_{2g}$	$^1E_{1g}$	$^2E_{1g}$	$\bar{E}_g$	$^1E_{2u}$	$^2E_{2u}$	$^2E_{1u}$	$^1E_{1u}$	$\bar{E}_u$
	S_3^+	S_4^+	S_2^+	S_1^+	S_5^+	S_4^-	S_3^-	S_1^-	S_2^-	S_5^-
	M_3^+	M_4^+	M_2^+	M_1^+	M_5^+	M_4^-	M_3^-	M_1^-	M_2^-	M_5^-
T	$^2\bar{E}_u''$	$^1\bar{E}_u''$	$^1\bar{E}_u'$	$^2\bar{E}_u'$	E_u	$^2\bar{E}_g'$	$^1\bar{E}_g'$	$^1\bar{E}_g''$	$^2\bar{E}_g''$	E_g
	T_2^+	T_3^+	T_4^+	T_5^+	T_1^+	T_2^-	T_3^-	T_4^-	T_5^-	T_1^-
	R_2^-	R_3^-	R_5^-	R_4^-	R_1^-	R_4^+	R_5^+	R_3^+	R_2^+	R_1^+
C_1	1	1	1	1	2	1	1	1	1	2
C_2	1	1	1	1	−2	1	1	1	1	−2
C_3	−1	−1	−1	−1	2	−1	−1	−1	−1	2
C_4	−1	−1	−1	−1	−2	−1	−1	−1	−1	−2
C_5	−1	−1	1	1	0	−1	−1	1	1	0
C_6	1	1	−1	−1	0	1	1	−1	−1	0
C_7	i	−i	−i	i	0	i	−i	−i	i	0
C_8	−i	i	−i	i	0	−i	i	−i	i	0
C_9	−i	i	i	−i	0	−i	i	i	−i	0
C_{10}	i	−i	i	−i	0	i	−i	i	−i	0
C_{11}	1	1	1	1	2	−1	−1	−1	−1	−2
C_{12}	1	1	1	1	−2	−1	−1	−1	−1	2
C_{13}	−1	−1	−1	−1	2	1	1	1	1	−2
C_{14}	−1	−1	−1	−1	−2	1	1	1	1	2
C_{15}	−1	−1	1	1	0	1	1	−1	−1	0
C_{16}	1	1	−1	−1	0	−1	−1	1	1	0
C_{17}	i	−i	−i	i	0	−i	i	i	−i	0
C_{18}	−i	i	−i	i	0	i	−i	i	−i	0
C_{19}	−i	i	i	−i	0	i	−i	−i	i	0
C_{20}	i	−i	i	−i	0	−i	i	−i	i	0

Notes to Table 7.21

The representations are labelled as follows. The first row gives the label used in the abstract group in Table 5.1; the second and third rows give the space-group representation labels obtained from Table 5.8 (single-valued) and Table 6.14 (double-valued); the fourth row gives the space-group representation labels used by Dimmock and Wheeler (1962*b*) (at Γ the fourth row is omitted since it is the same as the third row). Where the same abstract group occurs at several points of symmetry the abstract-group representation labels are not repeated.

TABLE 7.22

The coreps of $P4_2'/mnm'$

Γ:

	A_{1g}	B_{1g}	B_{2g} B_{3g}	$\bar{E}_g$	A_{1u}	B_{1u}	B_{2u} B_{3u}	$\bar{E}_u$
	Γ_1^+	Γ_2^+	Γ_4^+ Γ_3^+	Γ_5^+	Γ_1^-	Γ_2^-	Γ_4^- Γ_3^-	Γ_5^-
Coreps	a	a	c	a	a	a	c	a
	DA_{1g}	DB_{1g}	DB_g	$D\bar{E}_g$	DA_{1u}	DB_{1u}	DB_u	$D\bar{E}_u$
	$D\Gamma_1^+$	$D\Gamma_2^+$	$D\Gamma_{4,3}^+$	$D\Gamma_5^+$	$D\Gamma_1^-$	$D\Gamma_2^-$	$D\Gamma_{4,3}^-$	$D\Gamma_5^-$

Z:

	E_1	E_2	$^1\bar{F}$	$^2\bar{F}$
	Z_1	Z_2	Z_3	Z_4
Coreps	a	a	a	a
	DE_1	DE_2	$D^1\bar{F}$	$D^2\bar{F}$
	DZ_1	DZ_2	DZ_3	DZ_4

R:

	E_1	E_2	$^1\bar{F}$	$^2\bar{F}$
	R_1	R_2	R_3	R_4
Coreps	a	a	a	a
	DE_1	DE_2	$D^1\bar{F}$	$D^2\bar{F}$
	DR_1	DR_2	DR_3	DR_4

S:

	$^2E_{2g}$	$^1E_{2g}$	$^1E_{1g}$ $^2E_{1g}$	$\bar{E}_g$	$^1E_{2u}$	$^2E_{2u}$	$^1E_{1u}$ $^2E_{1u}$	$\bar{E}_u$
	S_3^+	S_4^+	S_2^+ S_1^+	S_5^+	S_4^-	S_3^-	S_2^- S_1^-	S_5^-
Coreps	a	a	c	a	a	a	c	a
	D^2E_{2g}	D^1E_{2g}	DE_{1g}	$D\bar{E}_g$	D^1E_{2u}	D^2E_{2u}	DE_{1u}	$D\bar{E}_u$
	DS_3^+	DS_4^+	$DS_{2,1}^+$	DS_5^+	DS_4^-	DS_3^-	$DS_{2,1}^-$	DS_5^-

Notes to Table 7.22

The representations can be identified from Table 7.21 and the third row of each entry indicates whether the corep derived from the representation above it belongs to case (a), case (b), or case (c). The matrices of the corepresentations can be written down using eqns (7.3.45), (7.3.46), and (7.3.47) for cases (a), (b), and (c) respectively.

We use the example of $P4_2'/mnm'$ to show how the theory of the Kronecker products of corepresentations given in section 7.4 applies to space-group corepresentations. This is relevant, for example, to the determination of selection rules for phonon or magnon processes in crystals belonging to one of the magnetic space groups, where it is necessary to be able to reduce the Kronecker products of the corepresentations of non-unitary groups. We require to bring together the theory of the reduction of the direct, or Kronecker, products of induced representations described in section 4.7 and the theory of the reduction of the direct products of corepresentations described in section 7.4 (Cracknell 1967*c*).

Suppose that $\mathbf{M}$ is a magnetic space group, then it can be written as

$$\mathbf{M} = \mathbf{G} + A\mathbf{G}, \tag{7.3.11}$$

where $\mathbf{G}$ is the halving subgroup of unitary elements and A is some anti-unitary element. If Γ^i, Γ^j, and Γ^k are reps induced in $\mathbf{G}$ by various small reps of little groups $\mathbf{G}^{\mathbf{k}}$, then the Clebsch–Gordan coefficients $c_{ij,k}$ in the reduction of the inner Kronecker product

$$\Gamma^i \boxtimes \Gamma^j \equiv \sum_k c_{ij,k}\Gamma^k \tag{7.7.1}$$

can be evaluated, see eqn (4.7.29). It is then a relatively simple matter to obtain the Clebsch–Gordan coefficients in the reduction of the inner Kronecker product of the coreps of $\mathbf{M}$ derived from Γ^i and Γ^j. If the coreps derived from Γ^i, Γ^j, and Γ^k are denoted by $D\Gamma^i$, $D\Gamma^j$, and $D\Gamma^k$ respectively, then we may write

$$D\Gamma^i \boxtimes D\Gamma^j \equiv \sum_k d_{ij,k} D\Gamma^k, \tag{7.7.2}$$

where the coefficients $d_{ij,k}$ can be expressed in terms of the known Clebsch–Gordan coefficients $c_{ij,k}$. The actual relationship between the $d_{ij,k}$ and the $c_{ij,k}$ depends in detail on the properties of the coreps $D\Gamma^i$, $D\Gamma^j$, and $D\Gamma^k$, that is, on whether these coreps belong to case (a), case (b), or case (c). Each of these coreps may belong to either case (a), case (b), or case (c) and expressions for $d_{ij,k}$ for all the various possible combinations which can arise were given in Table 7.8.

The reduction of the Kronecker products of the irreducible corepresentations of **M** can be divided into four stages, which we may call A, B, C, and D; we shall discuss briefly each of these in turn and then we shall illustrate them for MnF_2.

(A) *The determination of the reps of* **G**

G is one of the 230 Fedorov space groups and the deduction of the reps $(\Gamma_p^{\mathbf{k}} \uparrow \mathbf{G})$ of these space groups is a well-established process, see Chapters 3–6.

(B) *The reduction of the Kronecker products of the reps of* **G**

This process, the determination of the Clebsch–Gordan coefficients in equations of the type of eqn (7.7.1), where the reps Γ^i, Γ^j, and Γ^k are induced reps of **G** of the form $(\Gamma_p^{\mathbf{k}} \uparrow \mathbf{G})$, is also well established; it has been adequately described in section 4.7.

(C) *The determination of the coreps of* **M**

The coreps of **M** can all be determined from the reps of **G** and the theory has been described in section 7.6. As shown in section 7.6, it is enough to determine the small coreps of the magnetic little group $\mathbf{M}^{\dagger\mathbf{k}}$ for each **k** vector in the Brillouin zone to be able to characterize completely all the (induced) coreps of the magnetic space group **M**. For a magnetic space group the magnetic little group $\mathbf{M}^{\dagger\mathbf{k}}$ consists of

(i) those unitary elements of the magnetic space group whose rotational parts R transform **k** into $\mathbf{k} + \mathbf{g}$, where **g** is any vector of the reciprocal lattice, and

(ii) those anti-unitary elements of the magnetic space group whose rotational parts S transform **k** into $-\mathbf{k} + \mathbf{g}$.

The actual identification of the elements in $\mathbf{M}^{\dagger\mathbf{k}}$ can be made by inspection of the structure of **M** and of the detail of its relationship to various ordinary space groups. The magnetic little group can therefore be written as

$$\mathbf{M}^{\dagger\mathbf{k}} = \sum \{R \mid \mathbf{v}\}\mathbf{T} + \sum \theta\{S \mid \mathbf{w}\}\mathbf{T}, \tag{7.7.3}$$

where the first summation is over the elements in (i) and the second summation is over the elements in (ii), $\{R \mid \mathbf{v}\}$ and $\{S \mid \mathbf{w}\}$ are Seitz space-group symbols, and **T** is the group of unitary translation operations of the Bravais lattice on which the magnetic space group is based, see eqn (7.6.38). It is only at the points of symmetry in the Brillouin zone that one can use Herring's method (see section 3.8) and factor out from $\mathbf{M}^{\dagger\mathbf{k}}$ the group $\mathbf{T}^{\mathbf{k}}$ of all those translations **t** for which $\exp(-i\mathbf{k} \cdot \mathbf{t}) = 1$. The quotient group $\mathbf{M}^{\dagger\mathbf{k}}/\mathbf{T}^{\mathbf{k}}$ then forms a finite group which, by a typical process of identification, may be thought of as consisting of the coset representatives $\{R \mid \mathbf{v}\}$ and $\theta\{S \mid \mathbf{w}\}$ together with their products with a finite number of translations. Since elements of $\mathbf{T}^{\mathbf{k}}$ are always represented by the identity the coreps of $\mathbf{M}^{\dagger\mathbf{k}}/\mathbf{T}^{\mathbf{k}}$ lead immediately to the small coreps of $\mathbf{M}^{\dagger\mathbf{k}}$. For other **k**-vectors Herring's method is not suitable and as explained at the end of section 7.6 it is necessary, in order to obtain the small coreps of $\mathbf{M}^{\dagger\mathbf{k}}$, to consider the projective coreps of a suitable extension $\overline{\mathbf{M}}^{\dagger\mathbf{k}*}$ of the magnetic little co-group $\overline{\mathbf{M}}^{\dagger\mathbf{k}}$. However, we shall be concerned here only with points of symmetry and we shall denote $\mathbf{M}^{\dagger\mathbf{k}}/\mathbf{T}^{\mathbf{k}}$ by ${}^{H}\mathbf{M}^{\dagger\mathbf{k}}$ and similarly for other quotients relative to $\mathbf{T}^{\mathbf{k}}$.

When $\mathbf{T}^{\mathbf{k}}$ has been factored out of eqn (7.7.3) we obtain the following equation:

$$ {}^{H}\mathbf{M}^{\dagger\mathbf{k}} = {}^{H}\mathbf{G}^{\dagger\mathbf{k}} + \theta^{H}\mathbf{G}_{I}^{\dagger\mathbf{k}} \tag{7.7.4} $$

for a type II or type III magnetic space group, and

$$ {}^{H}\mathbf{M}^{\dagger\mathbf{k}} = {}^{H}\mathbf{G}^{\dagger\mathbf{k}} + \{E \mid t_0\}\theta^{H}\mathbf{G}_{I}^{\dagger\mathbf{k}} \tag{7.7.5} $$

where $\mathbf{t}_0$ is some translation operation, for a type IV magnetic space group. The identification of the elements in the unitary subgroup ${}^{H}\mathbf{G}^{\dagger\mathbf{k}}$ and in the set of anti-unitary elements $\theta^{H}\mathbf{G}_{I}^{\dagger\mathbf{k}}$ can be performed by inspection. The problems of determining the small coreps of the (infinite) little group $\mathbf{M}^{\dagger\mathbf{k}}$ therefore reduces, for the points of symmetry in the Brillouin zone, to the determination of the coreps of the (finite) group ${}^{H}\mathbf{M}^{\dagger\mathbf{k}}$; specific examples have been considered before (Bradley and Cracknell 1966; Cracknell 1965; Daniel and Cracknell 1969; Dimmock and Wheeler 1962*b*; Joshua and Cracknell 1969; Karavaev, Kudryavtseva, and Chaldyshev 1962).

(D) *The reduction of the Kronecker products of the coreps of* M

We assume that the Clebsch–Gordan coefficients $c_{ij,k}$ for the reps of the unitary subgroup **G** have been determined by the method just described under B. The Clebsch–Gordan coefficient $d_{ij,k}$ can be found from $c_{ij,k}$ by using Table 7.8 and the knowledge of whether each of the coreps $D\Gamma^{i}$, $D\Gamma^{j}$, and $D\Gamma^{k}$ belongs to case (a), case (b), or case (c), which can be determined by the process described under (C).

We now illustrate the procedure for determining the Clebsch–Gordan coefficients $d_{ij,k}$ in eqn (7.7.2) for the reduction of the Kronecker products of the coreps of the magnetic space group $P4_2'/mnm'$ of anti-ferromagnetic MnF_2 (Cracknell 1967*c*). The coreps themselves have already been identified in Tables 7.18–7.22; the unit cell and Brillouin zone are illustrated in Figs. 7.5 and 7.9.

(A) *The reps of Pnnm*

The reps of the Herring little group ${}^{H}\mathbf{G}^{\dagger\mathbf{k}}$ for each of the points of symmetry of the Brillouin zone for the space group *Pnnm* were given in Tables 7.20 and 7.21. These tables do not explicitly give the reps of ${}^{H}\mathbf{G}^{\dagger\mathbf{k}}$ for the two points X ($=\frac{1}{2}\mathbf{g}_2$) and U ($=\frac{1}{2}\mathbf{g}_2 + \frac{1}{2}\mathbf{g}_3$) because they are related by symmetry considerations to Y and T respectively in tetragonal space groups, but not in general orthorhombic space groups. Because reps of ${}^{H}\mathbf{G}^{\dagger\mathbf{k}}$ at these two points appear in the reduction of the Kronecker products of reps of *Pnnm* we would therefore need the character tables of these two points for a complete discussion of the Kronecker products of reps of *Pnnm*; these additional character tables can also be found from Tables 5.7 and 6.13.

(B) *Kronecker products for Pnnm*

We have just considered the reps of ${}^{H}\mathbf{G}^{\dagger\mathbf{k}}$ for the points of symmetry in the Brillouin zone of *Pnnm*. We now utilize the general theory given in section 4.7 to determine the reduction of the Kronecker products for this space group. That is, we require to determine the coefficients $C^{\mathbf{km},\mathbf{h}}_{pq,r}$ in

$$\Delta^{\mathbf{k}}_{p} \boxtimes \Delta^{\mathbf{m}}_{q} = \sum_{\mathbf{h},r} C^{\mathbf{km},\mathbf{h}}_{pq,r}\Delta^{\mathbf{h}}_{r}, \qquad (7.7.6)$$

where $\mathbf{k}$, $\mathbf{m}$, and $\mathbf{h}$ are the wave vectors of three points in the Brillouin zone.

The first step is to identify the star of each of the points of symmetry; the star of $\mathbf{k}$ can be found by writing $\mathbf{G}$, the complete space group, in terms of left cosets with respect to $\mathbf{G}^{\dagger\mathbf{k}}$, the group of $\mathbf{k}$. Thus from eqn (7.6.36)

$$\mathbf{G} = \sum_{i=1}^{d} \{R_i \mid \mathbf{v}_i\}\mathbf{G}^{\dagger\mathbf{k}} \qquad (7.7.7)$$

and the vectors $R_i\mathbf{k}$ ($i = 1$ to d) form the *star* of $\mathbf{k}$. We can therefore easily determine the star of each of the points of symmetry in *Pnnm*; in fact it can easily be shown that for each of the eight points of symmetry Γ, R, S, T, U, X, Y, and Z the star of $\mathbf{k}$ is just $\mathbf{k}$ itself.

Equation (4.7.19) shows how to determine the costar of a wave vector $\mathbf{k}$ with respect to the little group $\mathbf{G}^{\dagger\mathbf{h}}$; $\mathbf{G}$ is expanded into double cosets

$$\mathbf{G} = \sum_{\alpha} \mathbf{G}^{\dagger\mathbf{h}} \{R_\alpha \mid \mathbf{v}_\alpha\}\mathbf{G}^{\dagger\mathbf{k}}. \qquad (7.7.8)$$

The vectors $\mathbf{k}_\alpha = R_\alpha\mathbf{k}$ make up the *costar* of $\mathbf{k}$ with respect to $\mathbf{G}^{\dagger\mathbf{h}}$. In our particular example of *Pnnm* it can easily be shown that for each $\mathbf{k}$ and $\mathbf{h}$ this costar is just equal to $\mathbf{k}$ itself. The rules given in section 4.7 for determining the coefficients $C^{\mathbf{km},\mathbf{h}}_{pq,r}$ in eqn (7.7.6) also require that we determine the costar of $\mathbf{m}$ with respect to $\mathbf{G}^{\dagger\mathbf{h}} \cap \mathbf{G}^{\dagger\mathbf{k}}_{\alpha}$. $\mathbf{G}^{\dagger\mathbf{k}}_{\alpha}$ is the group of $\mathbf{k}_\alpha$. In this example of *Pnnm* $\mathbf{k}_\alpha$ is always just equal to $\mathbf{k}$ so that

$$\mathbf{G}^{\dagger\mathbf{h}} \cap \mathbf{G}^{\dagger\mathbf{k}}_{\alpha} = \mathbf{G}^{\dagger\mathbf{h}} \cap \mathbf{G}^{\dagger\mathbf{k}} = \mathbf{G}^{\dagger\mathbf{h}} \quad (\text{or } \mathbf{G}^{\dagger\mathbf{k}}) \qquad (7.7.9)$$

since both little groups contain the same elements. The costar of $\mathbf{m}$ with respect to

$\mathbf{G}^{\dagger \mathbf{h}} \cap \mathbf{G}_{\alpha}^{\dagger \mathbf{k}}$ is therefore simply $\mathbf{m}$ and we can summarize:

(i) the costar of $\mathbf{k}$ with respect to $\mathbf{G}^{\dagger \mathbf{h}}$ is just $\mathbf{k}$, and

(ii) the costar of $\mathbf{m}$ with respect to $\mathbf{G}^{\dagger \mathbf{h}} \cap \mathbf{G}_{\alpha}^{\dagger \mathbf{k}}$ is also just $\mathbf{m}$.

The condition given in section 4.7 (see eqns (4.7.27) and (4.7.28)) becomes simply that the terms in the summation expression for $C_{pq,r}^{\mathbf{km},\mathbf{h}}$ vanish unless

$$\mathbf{k} + \mathbf{m} \equiv \mathbf{h}. \tag{7.7.10}$$

The sets of values of $\mathbf{k}$, $\mathbf{m}$, and $\mathbf{h}$ which satisfy this condition are given in Table 7.23.

TABLE 7.23

Triples of vectors $\mathbf{k}$, $\mathbf{m}$, *and* $\mathbf{h}$

k	m	h	k	m	h	k	m	h	k	m	h	k	m	h	k	m	h	k	m	h	k	m	h
Γ	Γ	Γ	Y	Γ	Y	R	Γ	R	T	Γ	T	X	Γ	X	Z	Γ	Z	U	Γ	U	S	Γ	S
	X	X		X	S		X	T		X	R		X	Γ		X	U		X	Z		X	Y
	Y	Y		Y	Γ		Y	U		Y	Z		Y	S		Y	T		Y	R		Y	X
	Z	Z		Z	T		Z	S		Z	Y		Z	U		Z	Γ		Z	X		Z	R
	R	R		R	U		R	Γ		R	X		R	T		R	S		R	Y		R	Z
	U	U		U	R		U	Y		U	S		U	Z		U	X		U	Γ		U	T
	T	T		T	Z		T	X		T	Γ		T	R		T	Y		T	S		T	U
	S	S		S	X		S	Z		S	U		S	Y		S	R		S	T		S	Γ

The actual formula for $C_{pq,r}^{\mathbf{km},\mathbf{h}}$ is given by eqn (4.7.27) and it is now clear that for our example of *Pnnm* the summations over α and β in that expression are reduced to one term and all the Clebsch–Gordan coefficients vanish except for the triples of vectors given in Table 7.23. We can therefore simplify eqns (4.7.27) and (4.7.29) to give

$$C_{pq,r}^{\mathbf{km},\mathbf{h}} = (1/16) \sum_{\{S|\mathbf{w}\}} \chi_p^{\mathbf{k}}(\{S \mid \mathbf{w}\})\chi_q^{\mathbf{m}}(\{S \mid \mathbf{w}\})\chi_r^{\mathbf{h}}(\{S \mid \mathbf{w}\})^*, \tag{7.7.11}$$

where the summation over $\{S \mid \mathbf{w}\}$ ranges over the coset representatives of $\mathbf{T}$ (the group of translations of the Bravais lattice) in $\mathbf{G}^{\dagger \mathbf{k}}$.

It would be very lengthy and probably not very profitable to obtain the reduction of every possible Kronecker product and so we simply consider one or two examples that will illustrate how the reductions of the others may also be obtained. We consider the product of a rep of Y with a rep of Z, when from Table 7.23 we can see that $C_{pq,r}^{\mathbf{km},\mathbf{h}}$ is only non-zero if $\mathbf{h}$ is the wave vector of T. That is, the product of one rep of ${}^H\mathbf{G}^{\dagger Y}$ and one rep of ${}^H\mathbf{G}^{\dagger Z}$ can only lead to reps of ${}^H\mathbf{G}^{\dagger T}$, and which reps actually occur will be determined by the coefficients $C_{pq,r}^{\mathbf{km},\mathbf{h}}$. There are four space-group reps for Y and four space-group reps for Z so that there are 16 possible product reps. Suppose we take the reps Y_1 and Z_1 then we have to determine the coefficients

$C_{11,r}^{YZ,T}$ for all the space-group reps of T, i.e. for T_1^+, T_2^+, T_3^+, T_4^+, T_5^+, T_1^-, T_2^-, T_3^-, T_4^-, and T_5^- by using eqn (7.7.11). This is most easily done by tabulation and is illustrated for T_1^+ in Table 7.24. The Clebsch–Gordan coefficient is therefore equal to 1 for T_1^+ and it is straightforward to show that it is also equal to 1 for T_1^- and zero for all the other reps of T. We can therefore conclude that

$$Y_1 \boxtimes Z_1 = T_1^+ + T_1^-. \tag{7.7.12}$$

TABLE 7.24

Evaluation of $C_{11,r}^{YZ,T}$ for T_1^+

$\{S\mid\mathbf{w}\}$	Y_1	Z_1	T_1^+	c‡	$\{S\mid\mathbf{w}\}$	Y_1	Z_1	T_1^+	c‡
$\{E\mid\mathbf{0}\}$	2	2	2	8	$\{\bar{E}\mid\mathbf{0}\}$	2	2	2	8
$\{C_{2x}\mid\boldsymbol{\tau}\}$	0	0	0	0	$\{\bar{C}_{2x}\mid\boldsymbol{\tau}\}$	0	0	0	0
$\{C_{2y}\mid\boldsymbol{\tau}\}$	0	0	0	0	$\{\bar{C}_{2y}\mid\boldsymbol{\tau}\}$	0	0	0	0
$\{C_{2z}\mid\mathbf{0}\}$	0	2	0	0	$\{\bar{C}_{2z}\mid\mathbf{0}\}$	0	2	0	0
$\{I\mid\mathbf{0}\}$	0	0	−2	0	$\{\bar{I}\mid\mathbf{0}\}$	0	0	−2	0
$\{\sigma_x\mid\boldsymbol{\tau}\}$	0	0	0	0	$\{\bar{\sigma}_x\mid\boldsymbol{\tau}\}$	0	0	0	0
$\{\sigma_y\mid\boldsymbol{\tau}\}$	0	0	0	0	$\{\bar{\sigma}_y\mid\boldsymbol{\tau}\}$	0	0	0	0
$\{\sigma_z\mid\mathbf{0}\}$	2	0	0	0	$\{\bar{\sigma}_z\mid\mathbf{0}\}$	2	0	0	0
Total									16

‡ $c = \chi_1^Y(\{S\mid\mathbf{w}\})\chi_1^Z(\{S\mid\mathbf{w}\})\chi_r^T(\{S\mid\mathbf{w}\})^*$.

It is then possible to evaluate the coefficients $C_{pq,r}^{YZ,T}$ for all the other choices of p and q and the result of doing this is given in Table 7.25.

TABLE 7.25

Kronecker products for Y and Z in Pnnm

$Y_1 \boxtimes Z_1 = T_1^+ + T_1^-$	$Y_3 \boxtimes Z_1 = T_2^- + T_5^- + T_2^+ + T_5^+$
$Y_1 \boxtimes Z_2 = T_1^+ + T_1^-$	$Y_3 \boxtimes Z_2 = T_3^- + T_4^- + T_3^+ + T_4^+$
$Y_1 \boxtimes Z_3 = T_3^- + T_4^- + T_2^+ + T_5^+$	$Y_3 \boxtimes Z_3 = T_1^+ + T_1^-$
$Y_1 \boxtimes Z_4 = T_2^- + T_5^- + T_3^+ + T_4^+$	$Y_3 \boxtimes Z_4 = T_1^+ + T_1^-$
$Y_2 \boxtimes Z_1 = T_1^+ + T_1^-$	$Y_4 \boxtimes Z_1 = T_3^- + T_4^- + T_3^+ + T_4^+$
$Y_2 \boxtimes Z_2 = T_1^+ + T_1^-$	$Y_4 \boxtimes Z_2 = T_2^- + T_5^- + T_2^+ + T_5^+$
$Y_2 \boxtimes Z_3 = T_2^- + T_5^- + T_3^+ + T_4^+$	$Y_4 \boxtimes Z_3 = T_1^+ + T_1^-$
$Y_2 \boxtimes Z_4 = T_3^- + T_4^- + T_2^+ + T_5^+$	$Y_4 \boxtimes Z_4 = T_1^+ + T_1^-$

As a second example we consider the products of reps of $^H\mathbf{G}^{\dagger R}$ and reps of $^H\mathbf{G}^{\dagger Z}$. From Table 7.23 it can be seen that $\mathbf{h}$ must correspond to the point S for the coefficient $C_{pq,r}^{\mathbf{km},\mathbf{h}}$ not to vanish. The values of these coefficients can then be determined in a similar way to the previous example. The resulting reductions of the Kronecker products of the reps of $^H\mathbf{G}^{\dagger R}$ and $^H\mathbf{G}^{\dagger Z}$ are given in Table 7.26.

TABLE 7.26

Kronecker products for R and Z in Pnnm

$R_1 \boxtimes Z_1 = S_1^+ + S_2^+ + S_1^- + S_2^-$	$R_3 \boxtimes Z_1 = S_5^+ + S_5^-$
$R_1 \boxtimes Z_2 = S_3^+ + S_4^+ + S_3^- + S_4^-$	$R_3 \boxtimes Z_2 = S_5^+ + S_5^-$
$R_1 \boxtimes Z_3 = S_5^+ + S_5^-$	$R_3 \boxtimes Z_3 = S_1^+ + S_2^+ + S_3^- + S_4^-$
$R_1 \boxtimes Z_4 = S_5^+ + S_5^-$	$R_3 \boxtimes Z_4 = S_3^+ + S_4^+ + S_1^- + S_2^-$
$R_2 \boxtimes Z_1 = S_3^+ + S_4^+ + S_3^- + S_4^-$	$R_4 \boxtimes Z_1 = S_5^+ + S_5^-$
$R_2 \boxtimes Z_2 = S_1^+ + S_2^+ + S_1^- + S_2^-$	$R_4 \boxtimes Z_2 = S_5^+ + S_5^-$
$R_2 \boxtimes Z_3 = S_5^+ + S_5^-$	$R_4 \boxtimes Z_3 = S_3^+ + S_4^+ + S_1^- + S_2^-$
$R_2 \boxtimes Z_4 = S_5^+ + S_5^-$	$R_4 \boxtimes Z_4 = S_1^+ + S_2^+ + S_3^- + S_4^-$

(C) *The coreps of $P4_2'/mnm'$*

The determination of the corep of **M** which is derived from the rep $\mathbf{\Delta}(P)$ of **G** is made by considering the relationship between the reps $\mathbf{\Delta}(P)$ and $\overline{\mathbf{\Delta}}(P) = \mathbf{\Delta}^*(A^{-1}PA)$ of **G**. We have already mentioned that for some points in the Brillouin zone the magnetic little group is a unitary group (i.e. ${}^H\mathbf{G}_l^{\dagger\mathbf{k}} = 0$). In these cases $\mathbf{M}^{\dagger\mathbf{k}}$ is the same as $\mathbf{G}^{\dagger\mathbf{k}}$ for the unitary subgroup **G**. In our particular example $\mathbf{M}^{\dagger\mathbf{k}}$ is unitary for the points Y and T in $P4_2'/mnm'$ and it is just equal to $\mathbf{G}^{\dagger\mathbf{k}}$ for these points in *Pnnm*. The coreps of **M** induced by the small reps of $\mathbf{G}^{\dagger\mathbf{k}}$ belong to case (c) and each corep arises by the sticking together of two small reps with different **k** vectors; this is similar to the type (x) degeneracy in the grey space groups which was discussed in the previous section. However, unlike the grey space groups, the two different **k** vectors are now not necessarily equal and opposite; in fact induced coreps of $P4_2'/mnm'$ are formed by the sticking together of one small rep of $\mathbf{G}^{\dagger\mathbf{k}}$ at X and one small rep of $\mathbf{G}^{\dagger\mathbf{k}}$ at Y and by the sticking together of one small rep of $\mathbf{G}^{\dagger\mathbf{k}}$ at T and one small rep of $\mathbf{G}^{\dagger\mathbf{k}}$ at U.

For the remaining points, Γ, S, Z, and R, ${}^H\mathbf{M}^{\dagger\mathbf{k}}$ is a non-unitary group. Each of the reps of ${}^H\mathbf{G}^{\dagger\mathbf{k}}$ for these points in *Pnnm*, which were determined above, leads to a corep of ${}^H\mathbf{M}^{\dagger\mathbf{k}}$ in $P4_2'/mnm'$ and all the coreps can be obtained in this way. For each rep of *Pnnm* it is therefore necessary to determine whether the corep of $P4_2'/mnm'$ derived from it belongs to case (a), case (b), or case (c); this has already been done and the results were given in Table 7.22. A scheme for labelling the coreps of $\mathbf{M}^{\dagger\mathbf{k}}$, following the scheme we used previously for the point groups, has also been used to label the coreps in Table 7.22. If one wishes to study the transformation properties of wave functions then one actually needs to determine the matrices of the coreps of **M**. This has to be done for case (a) and case (b) by a detailed inspection of $\mathbf{\Delta}(P)$ and $\overline{\mathbf{\Delta}}(P)$, in order to find the unitary transformation relating $\mathbf{\Delta}(P)$ and $\overline{\mathbf{\Delta}}(P)$, i.e. to find a matrix **N** such that

$$\mathbf{\Delta}(P) = \mathbf{N}\mathbf{\Delta}^*(A^{-1}PA)\mathbf{N}^{-1}. \quad (7.7.13)$$

Since this determination of **N** is not strictly necessary for the present problem of the reduction of the Kronecker products of the coreps of **M** it is not included in Table 7.22.

(D) *The Kronecker products of the coreps of* $P4'_2/mnm'$

We have already noted that for certain **k** vectors the magnetic little group happens to be unitary. In this situation one can still make use of Table 7.8 because the corresponding induced corep of **M** belongs to case (c), there being a type (x) degeneracy produced between two reps, one at X and one at Y, or one at T and one at U.

It would need a great amount of space to list all the possible Kronecker products of coreps of **M** and it would also not be very profitable to do so. We therefore only consider the examples that were produced in Tables 7.25 and 7.26 for *Pnnm* and show how the reduction of the Kronecker products of coreps of $P4'_2/mnm'$ can be determined from the Kronecker products of reps of *Pnnm*.

In Table 7.25 we listed the reductions of the Kronecker products of the reps of Y and Z in *Pnnm*, such as

$$Y_1 \boxtimes Z_1 = T_1^+ + T_1^-. \tag{7.7.12}$$

We noted that the magnetic little group at Y or T in $P4'_2/mnm'$ is unitary and therefore is equal to ${}^H\mathbf{G}^{\dagger\mathbf{k}}$ for that point in *Pnnm*. The induced coreps of **M** derived from small coreps of ${}^H\mathbf{G}^{\dagger Y}$ or ${}^H\mathbf{G}^{\dagger T}$ therefore belong to case (c) as explained before (type (x) degeneracy). At Z, ${}^H\mathbf{M}^{\dagger Z}$ is a non-unitary group and therefore, using the fact that DZ_1 $(= D\Gamma^j)$ belongs to case (a), we have

$$d_{ij,k} = c_{ij,k} + c_{i'j,k}. \tag{7.7.14}$$

If $c_{ij,k}$ refers to a rep at Y then $c_{i'j,k}$ refers to the rep at X which becomes degenerate with that rep at Y to form a case (c) corep of **M**. In this example $c_{i'j,k}$ is therefore equal to 0 (see Table 7.23) and so

$$\underset{(4)}{DY_1} \boxtimes \underset{(2)}{DZ_1} = \underset{(4)}{DT_1^+} + \underset{(4)}{DT_1^-}, \tag{7.7.15}$$

where the numbers in brackets indicate the degeneracies of the coreps. Since all the coreps of ${}^H\mathbf{M}^{\dagger Z}$, i.e. DZ_1, DZ_2, DZ_3, and DZ_4, belong to case (a), eqn (7.7.14) holds for all of them and the Kronecker products for the coreps of $P4'_2/mnm'$ induced by the coreps of ${}^H\mathbf{M}^{\dagger Y}$ and ${}^H\mathbf{M}^{\dagger Z}$ can be written down very easily; this is done in Table 7.27.

TABLE 7.27

Kronecker products for Y and Z in $P4'_2/mnm'$

$DY_1 \boxtimes DZ_1 = DT_1^+ + DT_1^-$	$DY_3 \boxtimes DZ_1 = DT_2^- + DT_5^- + DT_2^+ + DT_5^+$
$DY_1 \boxtimes DZ_2 = DT_1^+ + DT_1^-$	$DY_3 \boxtimes DZ_2 = DT_3^- + DT_4^- + DT_3^+ + DT_4^+$
$DY_1 \boxtimes DZ_3 = DT_3^- + DT_4^- + DT_2^+ + DT_5^+$	$DY_3 \boxtimes DZ_3 = DT_1^+ + DT_1^-$
$DY_1 \boxtimes DZ_4 = DT_2^- + DT_5^- + DT_3^+ + DT_4^+$	$DY_3 \boxtimes DZ_4 = DT_1^+ + DT_1^-$
$DY_2 \boxtimes DZ_1 = DT_1^+ + DT_1^-$	$DY_4 \boxtimes DZ_1 = DT_3^- + DT_4^- + DT_3^+ + DT_4^+$
$DY_2 \boxtimes DZ_2 = DT_1^+ + DT_1^-$	$DY_4 \boxtimes DZ_2 = DT_2^- + DT_5^- + DT_2^+ + DT_5^+$
$DY_2 \boxtimes DZ_3 = DT_2^- + DT_5^- + DT_3^+ + DT_4^+$	$DY_4 \boxtimes DZ_3 = DT_1^+ + DT_1^-$
$DY_2 \boxtimes DZ_4 = DT_3^- + DT_4^- + DT_2^+ + DT_5^+$	$DY_4 \boxtimes DZ_4 = DT_1^+ + DT_1^-$

We now consider the second example. In Table 7.26 we listed the reduction of the Kronecker products of reps of *Pnnm* induced by the reps of ${}^{H}\mathbf{G}^{\dagger R}$ and ${}^{H}\mathbf{G}^{\dagger Z}$. R, Z, and S are all points in the Brillouin zone at which ${}^{H}\mathbf{M}^{\dagger \mathbf{k}}$ is a non-unitary group and the coreps of ${}^{H}\mathbf{M}^{\dagger \mathbf{k}}$ were identified in Table 7.22. An example of such a product is

$$R_3 \boxtimes Z_3 = S_1^+ + S_2^+ + S_3^- + S_4^-. \tag{7.7.16}$$

This is like eqn (7.7.1) and we seek to obtain the coefficients $d_{ij,k}$ in the reduction (eqn (7.7.2)) of the Kronecker products of the coreps of $P4'_2/mnm'$ induced by the coreps of ${}^{H}\mathbf{M}^{\dagger R}$ and ${}^{H}\mathbf{M}^{\dagger Z}$. From Table 7.22 the coreps DR_3, DZ_3, DS_3^-, and DS_4^- derived from R_3, Z_3, S_3^-, and S_4^- respectively belong to case (a) while the two reps S_1^+ and S_2^+ stick together to form one corep $DS_{2,1}^+$ of $P4'_2/mnm'$ which belongs to case (c). Then for S_3^- and S_4^- we have to use line 1 of Table 7.8 and for S_1^+ and S_2^+ we have to use line 3 of that table. These all have

$$d_{ij,k} = c_{ij,k} \tag{7.7.17}$$

and therefore eqn (7.7.2) becomes

$$\underset{(2)}{DR_3} \boxtimes \underset{(2)}{DZ_3} = \underset{(2)}{DS_{2,1}^+} + \underset{(1)}{DS_3^-} + \underset{(1)}{DS_4^-}, \tag{7.7.18}$$

where the numbers in brackets indicate the degeneracies of the coreps. In Table 7.28 we give the reduction of the Kronecker products of all the coreps of $P4'_2/mnm'$ induced by the coreps of ${}^{H}\mathbf{M}^{\dagger R}$ and ${}^{H}\mathbf{M}^{\dagger Z}$.

TABLE 7.28

Kronecker products for R and Z in $P4'_2/mnm'$

$DR_1 \boxtimes DZ_1 = DS_{2,1}^+ + DS_{2,1}^-$	$DR_3 \boxtimes DZ_1 = DS_5^+ + DS_5^-$
$DR_1 \boxtimes DZ_2 = DS_3^+ + DS_4^+ + DS_3^- + DS_4^-$	$DR_3 \boxtimes DZ_2 = DS_5^+ + DS_5^-$
$DR_1 \boxtimes DZ_3 = DS_5^+ + DS_5^-$	$DR_3 \boxtimes DZ_3 = DS_{2,1}^+ + DS_3^- + DS_4^-$
$DR_1 \boxtimes DZ_4 = DS_5^+ + DS_5^-$	$DR_3 \boxtimes DZ_4 = DS_3^+ + DS_4^+ + DS_{2,1}^-$
$DR_2 \boxtimes DZ_1 = DS_3^+ + DS_4^+ + DS_3^- + DS_4^-$	$DR_4 \boxtimes DZ_1 = DS_5^+ + DS_5^-$
$DR_2 \boxtimes DZ_2 = DS_{2,1}^+ + DS_{2,1}^-$	$DR_4 \boxtimes DZ_2 = DS_5^+ + DS_5^-$
$DR_2 \boxtimes DZ_3 = DS_5^+ + DS_5^-$	$DR_4 \boxtimes DZ_3 = DS_3^+ + DS_4^+ + DS_{2,1}^-$
$DR_2 \boxtimes DZ_4 = DS_5^+ + DS_5^-$	$DR_4 \boxtimes DZ_4 = DS_{2,1}^+ + DS_3^- + DS_4^-$

We have been concerned with the reduction of the Kronecker products of single-valued and double-valued coreps of magnetic space groups. One application of this theory lies in extending the well-established group-theoretical treatment of selection rules for electron or phonon processes in ordinary space groups to cover similar processes in crystals belonging to magnetic space groups. In dealing with selection rules for processes in a crystal of a particular space group we are concerned with investigating whether it is possible to show group-theoretically that some matrix element $\langle \psi_r^i, W\psi_t^k \rangle$ must vanish, where ψ_r^i and ψ_t^k are the wave functions of the final and initial states and W is some operator. In general, ψ_r^i, ψ_t^k, and W belong to coreps of the magnetic space group **M** but can be regarded, for convenience, as belonging to coreps of (usually) different magnetic little groups. The theory of selection rules in ordinary space groups has been discussed earlier, see section 4.7. Because each rep of the unitary subgroup, **G**, leads to a corep of the magnetic group **M** and all the coreps of **M** can be obtained in this way, it is possible to do all the complicated part of the reduction of the Kronecker products of coreps by considering the products of the reps of little groups of **G** in the conventional way. The reduction of the Kronecker products of the coreps of **M** can then be determined by using the theory of section 7.4. As a result of this, if one only requires a yes or no type answer to the question of whether a particular transition is allowed or not it is only really necessary to determine the selection rules for electron, phonon or magnon processes in **G**, the halving subgroup of unitary elements, and the selection rules in **M** follow immediately. For simple selection-rule questions it is then not necessary to perform the detailed analysis of the type considered above. However, if one wishes to evaluate for non-zero matrix elements the actual transition probabilities then it is necessary to consider the detailed analysis of the relationship between the decompositions of Kronecker products of the coreps of **M** and the decompositions of Kronecker products of the reps of **G**.

We have said before, in section 4.7, that to all intents and purposes one can consider the small reps $\Gamma_p^{\mathbf{k}}$ of the little group $\mathbf{G}^{\dagger\mathbf{k}}$ rather than consider the induced reps $(\Gamma_p^{\mathbf{k}} \uparrow \mathbf{G})$ of the full space group **G**. There is, however, one important point arising from the fact that the set $\theta^H \mathbf{G}_l^{\dagger\mathbf{k}}$ in eqn (7.7.4) or (7.7.5) may contain no elements at all. If this is so then the magnetic little group $\mathbf{M}^{\dagger\mathbf{k}}$ contains no anti-unitary elements at all. We have already encountered this possibility in connection with the grey space groups in section 7.6 and it means that the presence of the anti-unitary elements in **M** causes a type (x) degeneracy, that is, the spectrum of the eigenvalues of the Hamiltonian of a crystal is the same at two inequivalent **k** vectors which are unrelated in the unitary subgroup. In this situation, when $\mathbf{M}^{\dagger\mathbf{k}}$ contains no anti-unitary elements, it is necessary to recall that, strictly speaking, it is to the corep of **M** induced by one of the reps of $\mathbf{M}^{\dagger\mathbf{k}}$ that ψ_r^i, W, or ψ_t^k belongs; therefore, in using Table 7.8 it must be remembered that such a corep belongs to case (c).

7.8 Spin–space groups

The concept of a magnetic point group or magnetic space group as it has been described in this chapter is based on the assumption that an operator acts simultaneously on the space coordinates and the spin coordinate of a particle. The result of such an operation, if it is to be a symmetry operation of the crystal, leaves the atomic positions and spins indistinguishable from their starting values. However, if one makes certain simplifying assumptions concerning the form of the spin Hamiltonian of a magnetic crystal it may possess more symmetry than is described by the magnetic space groups. It is then possible to define and use a *spin–space group*, $\mathbf{G}_S$, which has considerably more symmetry than the magnetic space group of the crystal (Brinkman 1967, Brinkman and Elliott 1966*a*, *b*). The details of the exact form of the Hamiltonian, together with a consideration of the physical meaning of the approximations involved is given by Brinkman and Elliott. The gross features of, for example, the spin-wave dispersion relations can then be determined from $\mathbf{G}_S$ while the finer details may be studied by considering the appropriate magnetic space group.

For the purpose of defining a spin–space group of, for instance, a two-sublattice anti-ferromagnetic crystal we assume that the symmetry operations act separately on the space coordinates and on the spins. Atoms of the same element but with different spins are considered to be distinct while the operations that change the atomic positions do not affect the spins. We define first a group $\mathbf{G}'$ that consists of all those operations which, acting on the atomic positions but not affecting the spins, take each sublattice into itself. The spin–space group, $\mathbf{G}_S$, of the crystal then consists of the sum of all those products of the elements of $\mathbf{G}'$ with the elements of a set of symmetry operations $\mathbf{S}$, where $\mathbf{S}$ consists of all those operations which, acting on the spins alone, are symmetry operations of the spin system. It is possible for some of the elements of $\mathbf{S}$ to contain a space operation that is not in $\mathbf{G}'$. Therefore we may write

$$\mathbf{G}_S = S_1\mathbf{G}' + S_2\mathbf{G}' + S_3\mathbf{G}' + \cdots + S_n\mathbf{G}' \tag{7.8.1}$$

where $S_1, S_2, S_3, \cdots, S_n$ are the members of $\mathbf{S}$.

In the tables of the space-group representations given in Chapter 5 the existence of spin was virtually ignored. If one is considering the motions of electrons in crystals in the approximation of neglecting spin–orbit coupling this is a valid approach; it amounts to saying that, in the absence of an external magnetic field, there is no restriction on the orientations of the electrons' spins. The spin parts of the wave functions of such electrons then must belong to the representation $\mathscr{D}^{\frac{1}{2}}\{R(\alpha, \beta, \gamma)\}$ of the full rotation group, see section 6.1. The full group of this situation, which we may call $\mathbf{\Gamma}$, is then the direct product of the ordinary space group for the position co-ordinates and the full rotation group for the spins. If $R_i(\mathbf{r})$ represents a point-group operation R_i applied to the space coordinates x, y, and z only and $R_p(s)$ represents a point-group operation R_p applied to the spin coordinate only, we may then denote a complete space-group operation by the modified Seitz symbol $\{R(x, y, z, s) \mid \mathbf{v}\}$ where

$$\{R(x, y, z, s) \mid \mathbf{v}\} = \{R_i(\mathbf{r})R_p(s) \mid \mathbf{v}_{ip}\}. \tag{7.8.2}$$

It is possible for the space-group element denoted by these symbols to include the operation of time-inversion. The group $\mathbf{\Gamma}$ may then be written as

$$\mathbf{\Gamma} = \sum_i \sum_p \{R_i(\mathbf{r})R_p(s) \mid \mathbf{v}_{ip}\}\mathbf{T} \tag{7.8.3}$$

by analogy with eqn (3.5.2) for an ordinary space group, where **T** is the translation group of the Bravais lattice of the crystal. If spin–orbit coupling is introduced we obtain the double-valued space-group representations which were discussed in Chapter 6; in this case the spins are no longer entirely free and the operations $R_p(s)$ do not comprise all the operations of the 3-dimensional rotation group but are restricted to the same operations $R_i(\mathbf{r})$ as act on the space coordinates. This group we may call **G** and it can be written as

$$\mathbf{G} = \sum_i \{R_i(\mathbf{r})R_i(s) \mid \mathbf{v}_i\}\mathbf{T} \tag{7.8.4}$$

where, again, **T** is the translation group of the Bravais lattice of the crystal. If the crystal that we are considering now has some magnetic ordering these two groups **Γ** and **G** are replaced by $\mathbf{G}_S$, a spin–space group, and **M**, a magnetic space group as defined in section 7.2. The spin–space group $\mathbf{G}_S$ may be written as

$$\mathbf{G}_S = \sum_i \sum_p \{R_i'(\mathbf{r})R_p'(s) \mid \mathbf{v}_{ip}'\}\mathbf{T}_M, \tag{7.8.5}$$

where the $R_i'(\mathbf{r})$ are the symmetry operations of one magnetic sublattice and $\mathbf{T}_M$ is the translation group of the Bravais lattice of the crystal. $\mathbf{T}_M$ may be an ordinary Bravais lattice or it may be one of the black and white Bravais lattices described in section 7.2. The operations $R_p'(s)$ in eqn (7.8.5) of course do not now consist of all the elements of the 3-dimensional rotation group but only the actual symmetry operations of the ordered spin system. The presence of the second subscript p on the $\mathbf{v}_{ip}$ arises because the set of operations **S** used to derive $\mathbf{G}_S$ from **G**′ may contain some non-zero translation operation. It is possible to write $\mathbf{v}_{ip}$ in the form $\mathbf{v}_i + \boldsymbol{\tau}_p$ where the $\mathbf{v}_i$ can be regarded as originating from $R_i(\mathbf{r})$ and $\boldsymbol{\tau}_p$ can be regarded as being associated with $R_p(s)$. In **M** the addition of anisotropic exchange and other interactions means that the total effect of a symmetry operation on the space coordinates and the spins has to be considered, so that we may write **M** as

$$\mathbf{M} = \sum_i \{R_i''(\mathbf{r})R_i''(s) \mid \mathbf{v}_i''\}\mathbf{T}_M. \tag{7.8.6}$$

A diagram may best illustrate the inter-relationship that exists between **Γ**, **G**, $\mathbf{G}_S$, and **M**:

$$\begin{array}{ccc}
\boldsymbol{\Gamma} & \text{magnetic order} & \mathbf{G}_S \\
\sum_i \sum_p \{R_i(\mathbf{r})R_p(s) \mid \mathbf{v}_{ip}\}\mathbf{T} & \longrightarrow & \sum_i \sum_p \{R_i'(\mathbf{r})R_p'(s) \mid \mathbf{v}_{ip}'\}\mathbf{T}_M \\
\big\downarrow \text{spin-lattice coupling} & & \big\downarrow \\
\mathbf{G} & & \mathbf{M} \\
\sum_i \{R_i(\mathbf{r})R_i(s) \mid \mathbf{v}_i\}\mathbf{T} & \longrightarrow & \sum_i \{R_i''(\mathbf{r})R_i''(s) \mid \mathbf{v}_i''\}\mathbf{T}_M.
\end{array}$$

These subgroup relations mean that it is possible to deduce compatibility relations between the representations (or corepresentations) of $\mathbf{G}_S$ and $\mathbf{M}$. The details of these compatibility relations vary considerably from one magnetic crystal to another and some examples are given by Brinkman and Elliott.

We have already in this book considered the problem of determining the irreducible representations of $\mathbf{\Gamma}$ and $\mathbf{G}$ and the irreducible corepresentations of $\mathbf{M}$. Therefore we now give some little consideration to the modification of representation theory appropriate to $\mathbf{G}_S$ as well as the compatibility relations which must exist between $\mathbf{G}_S$ and $\mathbf{M}$. The equation

$$\mathbf{G} = \{R_1 \mid \mathbf{v}_1\}\mathbf{T} + \{R_2 \mid \mathbf{v}_2\}\mathbf{T} + \cdots + \{R_h \mid \mathbf{v}_h\}\mathbf{T}, \tag{3.5.2}$$

which expresses an ordinary space group $\mathbf{G}$ in terms of left coset representatives of $\mathbf{T}$ and was much used in the development of the theory of the representations of space groups, can be rewritten as

$$\mathbf{G} = \sum_i \{R_i \mid \mathbf{v}_i\}\mathbf{T}. \tag{7.8.7}$$

This has been replaced by a similar expansion in terms of left coset representatives of $\mathbf{T}_M$,

$$\mathbf{G}_S = \sum_i \sum_p \{R_i'(\mathbf{r})R_p'(s) \mid \mathbf{v}_{ip}'\}\mathbf{T}_M \tag{7.8.5}$$

for a spin–space group $\mathbf{G}_S$. The multiplication rule for the coset representatives that appear in eqn (7.8.7) is, of course, from section 1.5,

$$\{R_2 \mid \mathbf{v}_2\}\{R_1 \mid \mathbf{v}_1\} = \{R_2R_1 \mid \mathbf{v}_2 + R_2\mathbf{v}_1\}. \tag{1.5.3}$$

The corresponding multiplication rule for the symbols for the elements of the spin-space group $\mathbf{G}_S$ is by analogy

$$\{R_2(\mathbf{r})R_q(s) \mid \mathbf{v}_{2q}\}\{R_1(\mathbf{r})R_p(s) \mid \mathbf{v}_{1p}\} = \{R_2(\mathbf{r})R_1(\mathbf{r})R_q(s)R_p(s) \mid \mathbf{v}_{2q} + R_2(\mathbf{r})\mathbf{v}_{1p}\}, \tag{7.8.8}$$

where $R_p(s)$ and $R_q(s)$ act only on the spin coordinates and do not affect $\mathbf{v}_{1p}$ and $\mathbf{v}_{2q}$ and where the suffices p and q correspond to any pair of elements in the set $\mathbf{S}$. It is instructive to write down the effect of the operator $\{R_i(\mathbf{r})R_p(\mathbf{s}) \mid \mathbf{v}_{ip}\}$ on the spinor wave function $\psi_j(\mathbf{r})$, where j takes one of two values corresponding to $s = \pm\frac{1}{2}$; by analogy with eqns (1.5.23) and (6.2.39) we have

$$\{R_i(\mathbf{r})R_p(s) \mid \mathbf{v}_{ip}\}\psi_j(\mathbf{r}) = \sum_k \mathscr{D}^{\frac{1}{2}}\{R_p(s)\}_{jk}\psi_k[\{R_i(\mathbf{r})\}^{-1}(\mathbf{r} - \mathbf{v}_{ip})]. \tag{7.8.9}$$

There is a considerable similarity between the multiplication rule in eqn (7.8.8) and that in eqn (1.5.3); it is therefore not surprising that most of the formal theory of space-group representations should still apply to $\mathbf{G}_S$. The theory of the deduction of the corepresentations of $\mathbf{G}_S$, assuming that it contains some anti-unitary elements, via the deduction of the representations of its halving subgroup of unitary elements is formally very similar to the theory that we have considered in Chapters 3 and 4. The inclusion of the extra set of anti-unitary operations each of which contains the operation of time-inversion and the determination of the corepresentations of $\mathbf{G}_S$ from the representations of its halving subgroup is again formally similar to the theory that we have been considering earlier in this chapter. But rather than work completely *ab initio* in determining the corepresentations of the spin–space group of any given magnetic crystal it is often preferable to determine the relationships that exist between the unitary subgroup of $\mathbf{G}_S$ and one of the ordinary space groups whose little groups and their representations have been identified and tabulated in Chapters 5 and 6. The details of such relationships vary considerably according to the actual nature of the crystal in question, particularly that is on the elements in the set $\mathbf{S}$. We shall simply note a few general points and then consider an example by way of illustration.

Equation (7.8.5) is of a similar form to eqn (3.5.2) and as a result the theory of little groups, see section 3.7, can be applied to $\mathbf{G}_S$. If $\mathbf{T}_M$ corresponds to an ordinary Bravais lattice then its representations are available from section 3.4. But if $\mathbf{T}_M$ corresponds to a black and white Bravais lattice then it is a non-unitary group; however there is always a simple relationship between a black and white Bravais lattice and one of the ordinary Bravais lattices, Table 7.3, so, if they are required, the coreps of $\mathbf{T}_M$ can be written down without difficulty. It then only remains at points of symmetry to determine the irreducible representations or irreducible corepresentations of the magnetic little co-group $\overline{\mathbf{M}}^{\mathbf{k}}$, see section 7.6; along lines of symmetry or on planes one has to determine the projective representations or projective corepresentations of the central extension $\overline{\mathbf{M}}^{\mathbf{k}*}$. In $\mathbf{\Gamma}$, since the operations on the spin coordinate do not affect $\mathbf{k}$, the little group $\overline{\mathbf{G}}^{\mathbf{k}}$ at a general point in the Brillouin zone which contains no space operations $R_i(x, y, z)$ will still contain all the spin operations $R_j(s)$ of the rotation group in 3 dimensions. In $\mathbf{\Gamma}$ this leads to a double degeneracy everywhere in the Brillouin zone, as we have already mentioned, so long as spin–orbit coupling is neglected. In $\mathbf{G}_S$ the spin operations $R_j(s)$ are more restricted in number since they are just the members of $\mathbf{S}$, the set of the symmetry operations of the ordered spin system. For a general point in the Brillouin zone in a space group lacking the inversion operation, I, the little group $\mathbf{M}^{\mathbf{k}}$ can be written as

$$\mathbf{M}^{\mathbf{k}} = \sum_{p} \{E(\mathbf{r})R_p(s) \mid \boldsymbol{\tau}_p\}\mathbf{T}_M \tag{7.8.10}$$

since the only space-group operation $\{R_i(\mathbf{r}) \mid \mathbf{v}_i\}$ that leaves $\mathbf{k}$ invariant is the identity operation while the spin operations $R_p(s)$ do not affect $\mathbf{k}$ at all. The relationship between the matrix representatives of $\{E(\mathbf{r})R_p(s) \mid \boldsymbol{\tau}_p\}$ and $\{E(\mathbf{r})R_p(s) \mid \mathbf{0}\}$ is, by analogy with eqn (3.8.8),

$$\mathbf{D}\{E(\mathbf{r})R_p(s) \mid \boldsymbol{\tau}_p\} = \exp(-i\mathbf{k}\,.\,\boldsymbol{\tau}_p\}\mathbf{D}\{E(\mathbf{r})R_p(s) \mid \mathbf{0}\} \tag{7.8.11}$$

and $\mathbf{D}\{E(\mathbf{r})R_p(s) \mid \mathbf{0}\}$ is just $\mathbf{D}\{R_p(s)\}$, the matrix representative of one of the members of the point group of the spin rotations $R_p(s)$. The problem of determining the space-group representations or corepresentations of $\mathbf{G}_S$ at a general point in the Brillouin zone therefore becomes simply a matter of determining the representations of the point group $R_p(s)$. The situation is only a little more complicated if the space group contains the inversion operation, I. For lines of symmetry and planes of symmetry that are completely within the Brillouin zone the space-group representations of $\mathbf{G}_S$ can be found by relating them to the representations of some ordinary space group.

To illustrate the derivation of the representations of $\mathbf{G}_S$ at a point of symmetry suppose that we consider a simple anti-ferromagnet which can be thought of as having two interpenetrating sublattices and in which the directions of sublattice magnetization are parallel and antiparallel, respectively, to the z axis. The group $\mathbf{G}'$ of one of the sublattices is one of the ordinary (Fedorov) space groups whose representations are available from the tables in Chapters 5 and 6. The set $\mathbf{S}$ then contains the operations that leave the position of a spin unchanged, that is the point group ∞ which consists of all rotations about the spin axis. In addition $\mathbf{S}$ contains all the operations that consist of a rotation through π about any axis at right angles to the spin axis together with some operation that involves a translation $\boldsymbol{\tau}$ which sends one sublattice into the other. One example of such an operation is $\{R_i(\mathbf{r})C_{2y}(s) \mid \boldsymbol{\tau}\}$, where, in the paramagnetic phase, $\{R_i(\mathbf{r})R_p(s) \mid \boldsymbol{\tau}\}$ is an element of $\boldsymbol{\Gamma}$. In the case where there is a change of Bravais lattice when magnetic ordering occurs so that $\{E \mid \boldsymbol{\tau}\}$ is one of the elements of $\mathbf{T}$, then we can choose $R_i = E$. Finally $\mathbf{S}$ contains the product of each of the above elements with $\{E(\mathbf{r})I(s) \mid \mathbf{0}\}$. The space group of one sublattice is $\mathbf{G}'$ and therefore $\mathbf{G}_S$ contains a subgroup of operations which leave one of the spins unmoved, that is the black and white non-crystallographic point group $\infty/m'm (= \infty 2' \otimes (E + I))$. That is

$$\mathbf{G}_S = \infty/m'm \,.\, \mathbf{G}' + \{R_i(\mathbf{r})C_{2y}(s) \mid \boldsymbol{\tau}\} \,.\, \infty/m'm \,.\, \mathbf{G}'. \tag{7.8.12}$$

$\mathbf{G}_S^{\mathbf{k}}$, the little group of $\mathbf{G}_S$ at the point of symmetry $\mathbf{k}$, therefore contains the subgroup $\infty/m.\mathbf{G}'^{\mathbf{k}}$ whose reps can therefore be determined from the appropriate parts of Chapters 5 and 6 for $\mathbf{G}'^{\mathbf{k}}$ and from the reps of ∞/m which, being the axial rotation group (with the inversion I), are characterized by an m value. The small reps of the unitary subgroup $\mathbf{H}_S^{\mathbf{k}}$ of $\mathbf{G}_S^{\mathbf{k}}$ are then completely determined by considering the effect of $\{R_i(\mathbf{r})C_{2y}(s) \mid \boldsymbol{\tau}\}$ which reverses the spin and exchanges the sublattices,

$$\mathbf{H}_S^{\mathbf{k}} = \infty/m.\mathbf{G}'^{\mathbf{k}} + \{R_i(\mathbf{r})C_{2y}(s) \mid \boldsymbol{\tau}\} \,.\, \infty/m' \,.\, \mathbf{G}'^{\mathbf{k}}. \tag{7.8.13}$$

The reps of $\mathbf{H}_S^{\mathbf{k}}$ are then most easily determined by using the theory of induced representations as described and illustrated in section 4.5. The addition of the anti-unitary elements to $\mathbf{H}_S^{\mathbf{k}}$ produces the little group $\mathbf{G}_S^{\mathbf{k}}$ and the corepresentations of $\mathbf{G}_S^{\mathbf{k}}$ can then be determined from the representations of $\mathbf{H}_S^{\mathbf{k}}$ in exactly the way that we have described already in sections 7.6 and 7.7 for determining the corepresentations of $\mathbf{M}^{\mathbf{k}}$ from the representations of $\mathbf{G}^{\mathbf{k}}$, the unitary subgroup of $\mathbf{M}^{\mathbf{k}}$.

As an illustration of a spin–space group we consider the example of the structure which is possessed by anti-ferromagnetic MnF_2. The unit cell of this structure was illustrated in Fig. 7.5. The magnetic space group of this structure is $P4'_2/mnm'$ and we have already considered the determination of the irreducible corepresentations of this magnetic space group in section 7.7. The application of the theory of spin–space groups to MnF_2 was considered by Brinkman and Elliott (1966*b*). The space group of the paramagnetic structure of MnF_2, i.e. the structure shown in Fig. 7.5 without the ordering of the spins, is $P4_2/mnm$. The generating elements of $P4_2/mnm$ are, from Table 3.7, $\{C_{4z}^+ \mid \frac{1}{2}\frac{1}{2}\frac{1}{2}\}$, $\{C_{2x} \mid \frac{1}{2}\frac{1}{2}0\}$ and $\{I \mid 00\frac{1}{2}\}$. However, it is more convenient if we move the origin of the coordinate system so that it is at one of the corners of the unit cell, i.e. by $+\frac{1}{4}\mathbf{t}_3$, so that the generating elements become $\{C_{4z}^+ \mid \frac{1}{2}\frac{1}{2}\frac{1}{2}\}$, $\{C_{2x} \mid \frac{1}{2}\frac{1}{2}\frac{1}{2}\}$, and $\{I \mid 000\}$ and the origin is now at one of the Mn^{2+} ions. If we use $\boldsymbol{\tau}$ to denote the vector $\frac{1}{2}\mathbf{t}_1 + \frac{1}{2}\mathbf{t}_2 + \frac{1}{2}\mathbf{t}_3$ we may write the elements of $P4_2/mnm$ as

$$\begin{aligned}
&X_1\colon \{E \mid \mathbf{0}\}, \{C_{2z} \mid \mathbf{0}\}, \{I \mid \mathbf{0}\}, \{\sigma_z \mid \mathbf{0}\};\\
&X_2\colon \{C_{2a} \mid \mathbf{0}\}, \{C_{2b} \mid \mathbf{0}\}, \{\sigma_{da} \mid \mathbf{0}\}, \{\sigma_{db} \mid \mathbf{0}\} = X_1 . \{C_{2a} \mid \mathbf{0}\};\\
&X_3\colon \{C_{4z}^+ \mid \boldsymbol{\tau}\}, \{C_{4z}^- \mid \boldsymbol{\tau}\}, \{S_{4z}^- \mid \boldsymbol{\tau}\}, \{S_{4z}^+ \mid \boldsymbol{\tau}\} = X_1 . \{C_{4z}^+ \mid \boldsymbol{\tau}\};\\
&X_4\colon \{C_{2x} \mid \boldsymbol{\tau}\}, \{C_{2y} \mid \boldsymbol{\tau}\}, \{\sigma_x \mid \boldsymbol{\tau}\}, \{\sigma_y \mid \boldsymbol{\tau}\} \quad = X_1 . \{C_{2x} \mid \boldsymbol{\tau}\}.
\end{aligned}$$

In considering the spin–space group of MnF_2 we need first of all to identify $\mathbf{G}'$, the space group of one sublattice. The elements of $\mathbf{G}$ are of the form $\{R_i(\mathbf{r})E(s) \mid \mathbf{t}_i\}$ where $\{R_i \mid \mathbf{0}\}$ is one of the elements of $(X_1 + X_2)$ and $\mathbf{t}_i$ is a translation of the primitive tetragonal Bravais lattice, P. Therefore we may write

$$\mathbf{G}' = \sum_i \{R_i(\mathbf{r})E(s) \mid \mathbf{v}_i\}\mathbf{T} \tag{7.8.14}$$

where $\{R_i \mid \mathbf{v}_i\} \in (X_1 + X_2)$ and $\mathbf{T}$ is the group of the translational symmetry operations of P. The set $\mathbf{S}$ of the symmetry operations of the spin system contains rotations through any angle about the spin axis, products of the operation of time-inversion with a rotation through π about any axis normal to the spin axis, products of $\{C_{2y}(\mathbf{r})C_{2y}(s) \mid \boldsymbol{\tau}\}$ with a rotation through any angle about the spin axis, and products of $\{C_{2y}(\mathbf{r})\theta C_{2y}(s) \mid \boldsymbol{\tau}\}$ with a rotation through π about any axis normal to the spin axis. $\mathbf{G}_S$ may therefore be written as

$$\begin{aligned}
\mathbf{G}_S &= \mathbf{G}' . \infty/m'm + \mathbf{G}' . \{C_{2y}(\mathbf{r})C_{2y}(s) \mid \boldsymbol{\tau}\} . \infty/m'm\\
&= \mathbf{G}' . (\{E(\mathbf{r})E(s) \mid \mathbf{0}\} + \{C_{2y}(\mathbf{r})C_{2y}(s) \mid \boldsymbol{\tau}\}) . \infty/m'm,
\end{aligned} \tag{7.8.15}$$

where $\infty/m'm$ is the black and white non-crystallographic point group with elements of the form $\{E(\mathbf{r})R_p(s) \mid \mathbf{0}\}$ where the $R_p(s)$ may be either an uncoloured rotation through any angle about the z-axis or a coloured rotation through π about any axis normal to the z-axis, or the product of I with either of these two types of operation. It will be noted that there is an isomorphism between $\mathbf{G}'$ and one of the symmorphic space groups so that the representations at the points of symmetry in $\mathbf{G}'$ can easily be found. At each point of symmetry $\mathbf{G}'^{\mathbf{k}}$ is simply related, therefore, to one of the point groups and its representations are immediately available from Table 2.2. In deducing the representations of $\mathbf{H}_S^{\mathbf{k}}$ from those of $\mathbf{G}'^{\mathbf{k}}$ we study the properties of the reps of $\mathbf{G}'^{\mathbf{k}}$ under conjugation with $\{C_{2y}(\mathbf{r})C_{2y}(s) \mid \boldsymbol{\tau}\}$ since

$$\mathbf{H}_S^{\mathbf{k}} = \infty/m.\mathbf{G}'^{\mathbf{k}} + \{C_{2y}(\mathbf{r})C_{2y}(s) \mid \boldsymbol{\tau}\}.\infty/m.\mathbf{G}'^{\mathbf{k}}. \tag{7.8.16}$$

If two reps Γ_i and Γ_j of $\mathbf{G}'^{\mathbf{k}}$ are conjugate then they stick together to form one irreducible representation of $\mathbf{H}_S^{\mathbf{k}}$ but if a rep Γ_i of $\mathbf{G}'^{\mathbf{k}}$ is self-conjugate then it leads to two reps of $\mathbf{H}_S^{\mathbf{k}}$ (see section 4.5). At the point Γ each rep of $\mathbf{G}'^{\mathbf{k}}$ is self-conjugate so that each rep of $\mathbf{G}'^{\mathbf{k}}$ leads to two reps of $\mathbf{H}_S^{\mathbf{k}}$ in which the character of $\{C_{2y}(\mathbf{r})C_{2y}(s) \mid \boldsymbol{\tau}\}$ is $+n$ or $-n$. At other points the situation is a little more complicated. For example, at Z, $\mathbf{G}'^{\mathbf{k}}$ contains the elements

$$\begin{array}{ll} \{E(\mathbf{r})E(s) \mid \mathbf{0}\} & \{I(\mathbf{r})E(s) \mid \mathbf{0}\} \\ \{C_{2a}(\mathbf{r})E(s) \mid \mathbf{0}\} & \{\sigma_{da}(\mathbf{r})E(s) \mid \mathbf{0}\} \\ \{C_{2b}(\mathbf{r})E(s) \mid \mathbf{0}\} & \{\sigma_{db}(\mathbf{r})E(s) \mid \mathbf{0}\} \\ \{C_{2z}(\mathbf{r})E(s) \mid \mathbf{0}\} & \{\sigma_{z}(\mathbf{r})E(s) \mid \mathbf{0}\}. \end{array}$$

TABLE 7.29

The character table at Z in $\mathbf{G}'^{\mathbf{k}}$ in the spin–space group of MnF_2

	E	C_{2a}	C_{2b}	C_{2z}	I	σ_{da}	σ_{db}	σ_z
Z_1^+	1	1	1	1	1	1	1	1
Z_2^+	1	1	−1	−1	1	1	−1	−1
Z_3^+	1	−1	1	−1	1	−1	1	−1
Z_4^+	1	−1	−1	1	1	−1	−1	1
Z_1^-	1	1	1	1	−1	−1	−1	−1
Z_2^-	1	1	−1	−1	−1	−1	1	1
Z_3^-	1	−1	1	−1	−1	1	−1	1
Z_4^-	1	−1	−1	1	−1	1	1	−1

$$\chi\{E \mid \mathbf{t}_3\} = -1.$$

Note to Table 7.29

The elements of $\mathbf{G}'^{\mathbf{k}}$ are of the form $\{R_i(\mathbf{r})E(s) \mid \mathbf{v}_i\}$ where the R_i are given in the first row of the table and $\mathbf{v}_i = \mathbf{0}$.

The representations of $\mathbf{G}'^{\mathbf{k}}$ are therefore as shown in Table 7.29 where only those representations in which $\chi\{E \mid \mathbf{t}_3\} = -1$ are considered. The result of the conjugation of these elements with $\{C_{2y}(\mathbf{r})C_{2y}(s) \mid \boldsymbol{\tau}\}$ is to produce

$$\begin{array}{ll} \{E(\mathbf{r})E(s) \mid \mathbf{0}\} & \{I(\mathbf{r})E(s) \mid \mathbf{t}_3\} \\ \{C_{2b}(\mathbf{r})E(s) \mid \mathbf{t}_3\} & \{\sigma_{db}(\mathbf{r})E(s) \mid \mathbf{0}\} \\ \{C_{2a}(\mathbf{r})E(s) \mid \mathbf{t}_3\} & \{\sigma_{da}(\mathbf{r})E(s) \mid \mathbf{0}\} \\ \{C_{2z}(\mathbf{r})E(s) \mid \mathbf{0}\} & \{\sigma_{z}(\mathbf{r})E(s) \mid \mathbf{t}_3\} \end{array}$$

respectively, which means that the representations are conjugate in pairs: Z_1^+, Z_4^-; Z_2^+, Z_2^-; Z_3^+, Z_3^-; and Z_4^+, Z_1^-; the character table is given in Table 7.30. Each rep of $\mathbf{H}_S^{\mathbf{k}}$ at Z is therefore characterized by one of the reps in Table 7.30 together with a value of m, denoting one of the reps of the axial rotation group (with I) ∞/m. $\mathbf{G}_S^{\mathbf{k}}$ is constructed by the addition of the anti-unitary element $\{E(\mathbf{r})\theta C_{2y}(s) \mid \mathbf{0}\}$ to $\mathbf{H}_S^{\mathbf{k}}$. The corepresentations of $\mathbf{G}_S^{\mathbf{k}}$ can be determined by using eqns (7.3.45)–(7.3.47) and the test given in eqn (7.3.51) or eqn (7.6.16). The argument that we have outlined for Z can also be applied to the other special points of symmetry in the Brillouin zone.

TABLE 7.30

The character table at Z in $\mathbf{H}_S^{\mathbf{k}}$ in the spin-space group of MnF_2

	E	C_{2a}	C_{2b}	C_{2z}	I	σ_{da}	σ_{db}	σ_z
Z_1^+, Z_4^-	2	0	0	2	0	2	2	0
Z_2^+, Z_2^-	2	2	−2	−2	0	0	0	0
Z_3^+, Z_3^-	2	−2	2	−2	0	0	0	0
Z_4^+, Z_1^-	2	0	0	2	0	−2	−2	0

$$\chi\{E \mid \mathbf{t}_3\} = -2.$$

Note to Table 7.30

The characters of the remaining elements are zero. The elements given above are of the same form as those in Table 7.29.

In practice the spin Hamiltonian that is used for MnF_2 is unlikely to contain any terms that involve the F^- ions. If these F^- ions are neglected the order of the spin–space group, $\mathbf{G}_S$, will be larger than if they are included (Brinkman and Elliott 1966*a*, 1966*b*); for example $\mathbf{G}_S$ will also contain elements of the form $\{R_i(\mathbf{r})E(s) \mid \mathbf{0}\}$ where $\{R_i \mid \boldsymbol{\tau}\} \in [X_3 + X_4]$. However, if the F^- ions are neglected it must be remembered that the orders of the other groups $\boldsymbol{\Gamma}$, $\mathbf{G}$, and $\mathbf{M}$ are also likely to be increased; unless this fact is remembered the subgroup relations between $\boldsymbol{\Gamma}$, $\mathbf{G}$, $\mathbf{G}_S$, and $\mathbf{M}$ which were indicated just below eqn (7.8.6) might appear to be invalid.

7.9. Polychromatic point groups and space groups

The Shubnikov groups, or black and white groups, are derived by adding to the position coordinates x, y, and z an extra coordinate s, which takes either of two possible values. This can be extended further if the extra coordinate, s, is not now two-valued but three-valued, four-valued, or six-valued. In the same way that the two-coloured point groups and space groups were derived it is possible to derive the three-coloured, four-coloured, or six-coloured point groups or space groups. Consider the case generally of the p-coloured point group with $p = 2, 3, 4$, or 6. If we suppose that we have p colours arranged in some order, e.g. three colours, red, green, and yellow, then it is possible to define a colour-changing operation, $\mathscr{R}$, which permutes these three colours cyclically, e.g. $\mathscr{R}$ turns red into green, green into yellow, and yellow into red (see Fig. 7.10). There will be three types of p-coloured point group. A type I p-coloured point group (or space group) is simply an ordinary point group (or space group) **G**. A type II, or grey, point group can be defined by (see Definition 7.2.1):

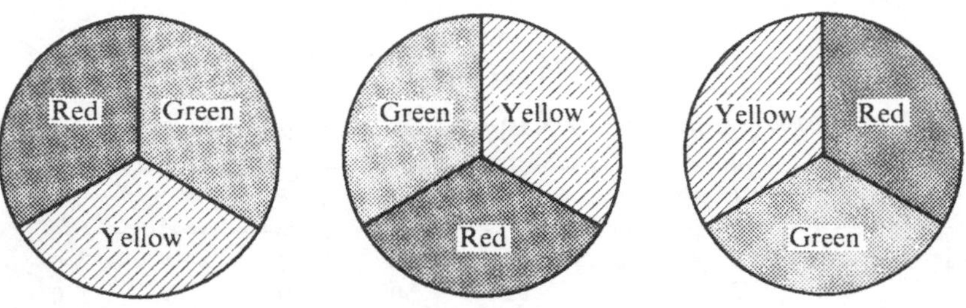

FIG. 7.10.

DEFINITION 7.9.1. A type II polychromatic point group (or space group), **C**, is given by

$$\mathbf{C} = \mathbf{G} + \mathscr{R}_p\mathbf{G} + \mathscr{R}_p^2\mathbf{G} + \cdots + \mathscr{R}_p^{p-1}\mathbf{G} \tag{7.9.1}$$

where **G** is any ordinary point group (or space group).

There are thus 32 type II polychromatic point groups and 230 type II polychromatic space groups for each value of p (= 2, 3, 4, or 6). In a similar way Definition 7.2.2 can be extended to define a type III p-coloured point group or space group.

DEFINITION 7.9.2. A type III polychromatic group, **C**, is given by

$$\mathbf{C} = \alpha_1\mathbf{H} + \alpha_2\mathscr{R}_p\mathbf{H} + \alpha_3\mathscr{R}_p^2\mathbf{H} + \cdots + \alpha_p\mathscr{R}_p^{p-1}\mathbf{H}, \tag{7.9.2}$$

where **H** is a subgroup of some ordinary point group or space group **G** and is of order $(1/p) \times$ the order of **G**, and $\alpha_1, \alpha_2, \ldots, \alpha_p$ are the left coset representatives in the expansion of **G**,

$$\mathbf{G} = \alpha_1\mathbf{H} + \alpha_2\mathbf{H} + \cdots + \alpha_p\mathbf{H}, \tag{7.9.3}$$

where α_1 is the identity element.

By deriving polychromatic Bravais lattices (p = 3, 4, or 6) Definition 7.2.5 could be extended to define p-coloured polychromatic space groups of type IV.

The procedure involved in a detailed derivation of the coloured groups with p = 3, 4, or 6 should be fairly clear from what has been done already with the black and white groups; they can readily be determined by inspection from the ordinary point groups or space groups by using Definition 7.9.2. Alternatively, in section 7.2 we noted that the 58 black and white point groups could be derived by studying the real 1-dimensional reps of the ordinary point groups and multiplying any element whose character is -1 by the colour-changing operation. In a similar way it is possible to use the complex 1-dimensional reps of the ordinary point groups to derive the three-coloured, four-coloured, and six-coloured point groups (Indenbom, Belov, and Neronova 1960). The complex numbers ω ($= e^{2\pi i/3}$), i ($= e^{2\pi i/4}$) and $-\omega^*$ ($= e^{2\pi i/6}$) can then be associated with colour-changing operations for three-coloured, four-coloured, and six-coloured point groups respectively. The two members of a pair of complex conjugate representations of an ordinary point group both lead to the same polychromatic point group. From Table 2.2 one can see that there are, in all, 18 pairs of complex conjugate representations so that there are 18 polychromatic point groups.

TABLE 7.31

The complex reps of $4/m$ (C_{4h})

	E	C_{4z}^+	C_{2z}	C_{4z}^-	I	S_{4z}^-	σ_z	S_{4z}^+	
^1E_g	1	$-i$	-1	i	1	$-i$	-1	i	$4^{(4)}/m'$
^2E_g	1	i	-1	$-i$	1	i	-1	$-i$	
^1E_u	1	$-i$	-1	i	-1	i	1	$-i$	$4^{(4)}/m$
^2E_u	1	i	-1	$-i$	-1	$-i$	1	i	

We illustrate this derivation by considering the point group $4/m$ (C_{4h}). The complex conjugate 1-dimensional reps of this point group can be obtained from Table 2.2 and are reproduced in Table 7.31. Each of the two reps ^1E_g and ^2E_g leads to a four-coloured point group in which C_{4z}^+ is associated with $\mathscr{R}_4$, the four-colour-changing operator, and I is uncoloured; this point group is denoted by $4^{(4)}/m'$. Each of the two reps ^1E_u and ^2E_u leads to a four-coloured point group in which C_{4z}^+ is associated with $\mathscr{R}_4$ and I is associated with $\mathscr{R}_2$, the two-colour-changing operator; this point group is called $4^{(4)}/m$. The elements of these two point groups are therefore

$4^{(4)}/m'$		$4^{(4)}/m$	
E	I	E	$\mathscr{R}_4^2 I$
$\mathscr{R}_4 C_{4z}^+$	$\mathscr{R}_4 S_{4z}^-$	$\mathscr{R}_4 C_{4z}^+$	$\mathscr{R}_4^3 S_{4z}^-$
$\mathscr{R}_4^2 C_{2z}$	$\mathscr{R}_4^2 \sigma_z$	$\mathscr{R}_4^2 C_{2z}$	σ_z
$\mathscr{R}_4^3 C_{4z}^-$	$\mathscr{R}_4^3 S_{4z}^+$	$\mathscr{R}_4^3 C_{4z}^-$	$\mathscr{R}_4 S_{4z}^+$

where we have used the fact that $\mathscr{R}_4^2 = \mathscr{R}_2$; these two point groups are represented diagrammatically in Fig. 7.11. The remaining polychromatic point groups can be derived in a similar way and they are also included in Fig. 7.11; their elements are identified in Table 7.32.

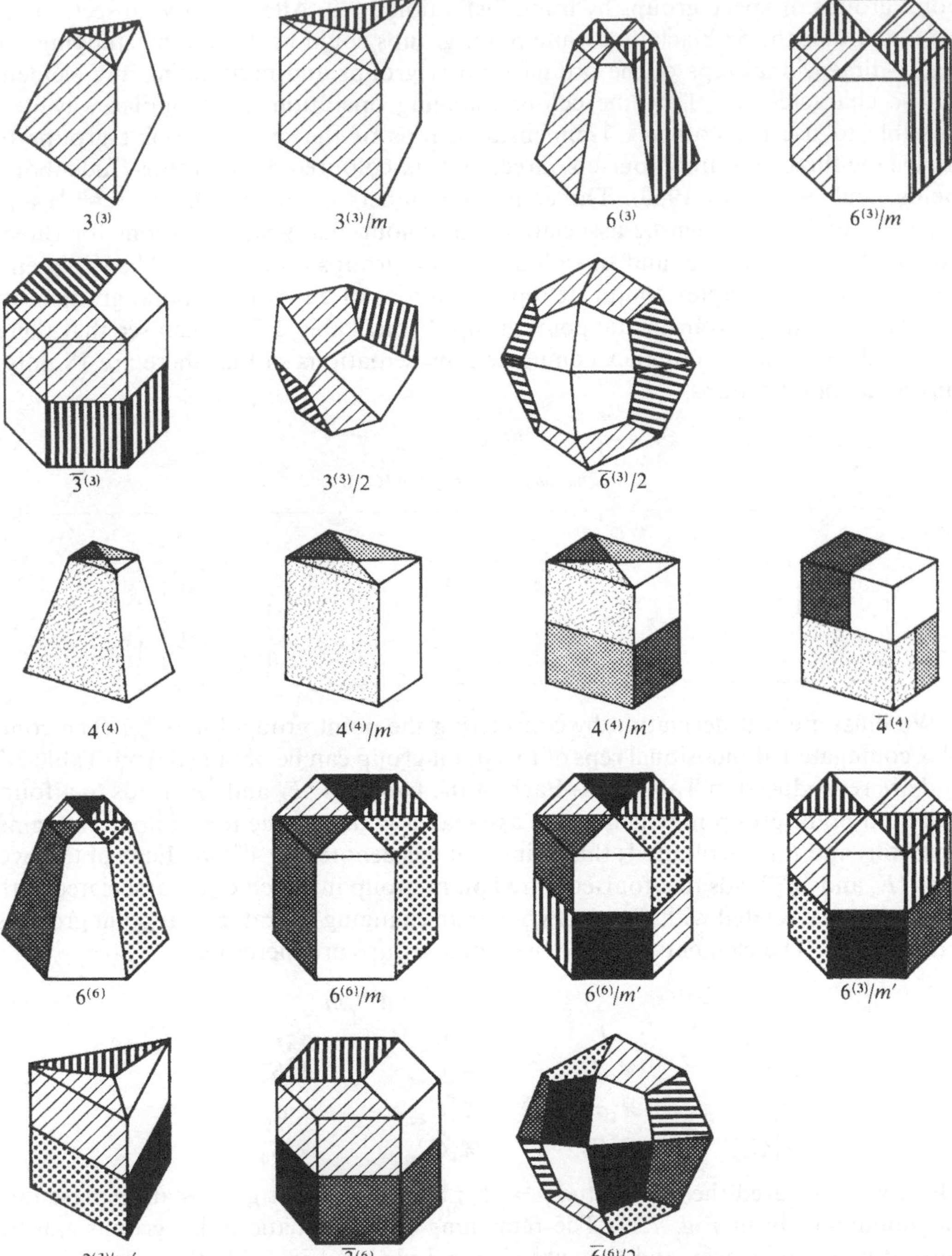

FIG. 7.11. Polychromatic point groups (Indenbom, Belov, and Neronova 1960).

TABLE 7.32

The 3-coloured, 4-coloured, and 6-coloured point groups

	C	G	Elements
1	$6^{(6)}$	6	$E;\ \mathscr{R}_6C_6^+;\ \mathscr{R}_6^2C_3^+;\ \mathscr{R}_6^3C_2;\ \mathscr{R}_6^4C_3^-;\ \mathscr{R}_6^5C_6^-$
2	$\bar{3}^{(6)}$	$\bar{3}$	$E;\ \mathscr{R}_6S_6^-;\ \mathscr{R}_6^2C_3^-;\ \mathscr{R}_6^3I;\ \mathscr{R}_6^4C_3^+;\ \mathscr{R}_6^5S_6^+$
3	$3^{(3)}/m'\ (=\bar{6}^{(6)})$	$3/m\ (=\bar{6})$	$E;\ \mathscr{R}_6S_3^-;\ \mathscr{R}_6^2C_3^+;\ \mathscr{R}_6^3\sigma_h;\ \mathscr{R}_6^4C_3^-;\ \mathscr{R}_6^5S_3^+$
4	$6^{(3)}$	6	$E;\ \mathscr{R}_3C_6^+;\ \mathscr{R}_3^2C_3^+;\ C_2;\ \mathscr{R}_3C_3^-;\ \mathscr{R}_3^2C_6^-$
5	$\bar{3}^{(3)}$	$\bar{3}$	$E;\ \mathscr{R}_3S_6^-;\ \mathscr{R}_3^2C_3^-;\ I;\ \mathscr{R}_3C_3^+;\ \mathscr{R}_3^2S_6^+$
6	$3^{(3)}/m\ (=\bar{6}^{(3)})$	$3/m\ (=\bar{6})$	$E;\ \mathscr{R}_3S_3^-;\ \mathscr{R}_3^2C_3^+;\ \sigma_h;\ \mathscr{R}_3C_3^-;\ \mathscr{R}_3^2S_3^+$
7	$6^{(6)}/m$	$6/m$	$E;\ \mathscr{R}_6C_6^+;\ \mathscr{R}_6^2C_3^+;\ \mathscr{R}_6^3C_2;\ \mathscr{R}_6^4C_3^-;\ \mathscr{R}_6^5C_6^-;\ \mathscr{R}_6^3I;\ \mathscr{R}_6^4S_3^-;\ \mathscr{R}_6^5S_6^-;\ \sigma_h;$ $\mathscr{R}_6S_6^+;\ \mathscr{R}_6^2S_3^+$
8	$6^{(3)}/m$	$6/m$	$E;\ \mathscr{R}_3C_6^+;\ \mathscr{R}_3^2C_3^+;\ C_2;\ \mathscr{R}_3C_3^-;\ \mathscr{R}_3^2C_6^-;\ I;\ \mathscr{R}_3S_3^-;\ \mathscr{R}_3^2S_6^-;\ \sigma_h;$ $\mathscr{R}_3S_6^+;\ \mathscr{R}_3^2S_3^+$
9	$6^{(6)}/m'$	$6/m$	$E;\ \mathscr{R}_6C_6^+;\ \mathscr{R}_6^2C_3^+;\ \mathscr{R}_6^3C_2;\ \mathscr{R}_6^4C_3^-;\ \mathscr{R}_6^5C_6^-;\ I;\ \mathscr{R}_6S_3^-;\ \mathscr{R}_6^2S_6^-;\ \mathscr{R}_6^3\sigma_h;$ $\mathscr{R}_6^4S_6^+;\ \mathscr{R}_6^5S_3^+$
10	$6^{(3)}/m'$	$6/m$	$E;\ \mathscr{R}_6^2C_6^+;\ \mathscr{R}_6^4C_3^+;\ C_2;\ \mathscr{R}_6^2C_3^-;\ \mathscr{R}_6^4C_6^-;\ \mathscr{R}_6^3I;\ \mathscr{R}_6^5S_3^-;\ \mathscr{R}_6S_6^-;\ \mathscr{R}_6^3\sigma_h;$ $\mathscr{R}_6^5S_6^+;\ \mathscr{R}_6S_3^+$
11	$3^{(3)}$	3	$E;\ \mathscr{R}_3C_3^+;\ \mathscr{R}_3^2C_3^-$
12	$4^{(4)}$	4	$E;\ \mathscr{R}_4C_{4z}^+;\ \mathscr{R}_4^2C_{2z};\ \mathscr{R}_4^3C_{4z}^-$
13	$\bar{4}^{(4)}$	$\bar{4}$	$E;\ \mathscr{R}_4S_{4z}^-;\ \mathscr{R}_4^2C_{2z};\ \mathscr{R}_4^3S_{4z}^+$
14	$4^{(4)}/m$	$4/m$	$E;\ \mathscr{R}_4C_{4z}^+;\ \mathscr{R}_4^2C_{2z};\ \mathscr{R}_4^3C_{4z}^-;\ \mathscr{R}_4^2I;\ \mathscr{R}_4^3S_{4z}^-;\ \sigma_z;\ \mathscr{R}_4S_{4z}^+$
15	$4^{(4)}/m'$	$4/m$	$E;\ \mathscr{R}_4C_{4z}^+;\ \mathscr{R}_4^2C_{2z};\ \mathscr{R}_4^3C_{4z}^-;\ I;\ \mathscr{R}_4S_{4z}^-;\ \mathscr{R}_4^2\sigma_z;\ \mathscr{R}_4^3S_{4z}^+$
16	$3^{(3)}/2$	$m3$	$E;\ C_{2m};\ \mathscr{R}_3C_{3j}^+;\ \mathscr{R}_3^2C_{3j}^-$
17	$\bar{6}^{(3)}/2$	$m3$	$E;\ C_{2m};\ \mathscr{R}_3C_{3j}^+;\ \mathscr{R}_3^2C_{3j}^-;\ I;\ \sigma_m;\ \mathscr{R}_3S_{6j}^-;\ \mathscr{R}_3^2S_{6j}^+$
18	$\bar{6}^{(6)}/2$	$m3$	$E;\ C_{2m};\ \mathscr{R}_6^2C_{3j}^+;\ \mathscr{R}_6^4C_{3j}^-;\ \mathscr{R}_6^3I;\ \mathscr{R}_6^3\sigma_m;\ \mathscr{R}_6^5S_{6j}^-;\ \mathscr{R}_6S_{6j}^+$

Notes to Table 7.32

(i) The numbers 1 to 18 are those of Indenbom, Belov, and Neronova (1960).

(ii) For numbers 16, 17, and 18 the subscript m takes the values x, y, and z and the subscript j takes the values 1, 2, 3, and 4.

As yet there is no definite indication of there being any use for the polychromatic groups and so we do not intend to consider their representations here. The polychromatic plane groups (2-dimensional space groups) were considered by Belov and Belova (1957), Belov, Belova, and Tarkhova (1958), and Belov and Tarkhova (1956*a*, *c*) and a corrected, translated, and edited derivation of the three-coloured, four-coloured, and six-coloured plane groups, i.e. with $p = 3$, 4, and 6, is given by Shubnikov and Belov (1964); a general discussion of polychromatic groups is given by Mackay (1957).

Appendix

After the notation that we have used for the point-group operations had been developed and after the tables contained in this book had been derived, the books by Miller and Love (1967) and Zak, Casher, Glück, and Gur (1969) were published. For the benefit of readers we therefore give the correspondence between our notation for the point-group operations and the notations used by these authors as well as that used by Kovalev (1965).

(i) $m3m(O_h)$

This book	Kovalev (1965)	Miller and Love (1967)	Zak, Casher, Glück, and Gur (1969)
E	h_1	1	E
C_{2x}	h_2	2	U^x
C_{2y}	h_3	3	U^y
C_{2z}	h_4	4	U^z
C_{31}^+	h_9	9	C_3^{xyz}
C_{32}^+	h_{11}	11	$C_3^{xy\bar{z}}$
C_{33}^+	h_{12}	12	$C_3^{\bar{x}yz}$
C_{34}^+	h_{10}	10	$C_3^{x\bar{y}z}$
C_{31}^-	h_5	5	C_3^{2xyz}
C_{32}^-	h_6	6	$C_3^{2xy\bar{z}}$
C_{33}^-	h_7	7	$C_3^{2\bar{x}yz}$
C_{34}^-	h_8	8	$C_3^{2x\bar{y}z}$
C_{4x}^+	h_{19}	19	C_4^x
C_{4y}^+	h_{24}	24	C_4^y
C_{4z}^+	h_{14}	14	C_4^z
C_{4x}^-	h_{20}	20	C_4^{3x}
C_{4y}^-	h_{22}	22	C_4^{3y}
C_{4z}^-	h_{15}	15	C_4^{3z}
C_{2a}	h_{16}	16	U^{xy}
C_{2b}	h_{13}	13	$U^{\bar{x}y}$
C_{2c}	h_{23}	23	U^{xz}
C_{2d}	h_{18}	18	U^{yz}
C_{2e}	h_{21}	21	$U^{\bar{x}z}$
C_{2f}	h_{17}	17	$U^{\bar{y}z}$

This book	Kovalev (1965)	Miller and Love (1967)	Zak, Casher, Glück, and Gur (1969)
I	h_{25}	25	I
σ_x	h_{26}	26	σ^x
σ_y	h_{27}	27	σ^y
σ_z	h_{28}	28	σ^z
S_{61}^-	h_{33}	33	S_6^{5xyz}
S_{62}^-	h_{35}	35	$S_6^{5xy\bar{z}}$
S_{63}^-	h_{36}	36	$S_6^{5\bar{x}yz}$
S_{64}^-	h_{34}	34	$S_6^{5x\bar{y}z}$
S_{61}^+	h_{29}	29	S_6^{xyz}
S_{62}^+	h_{30}	30	$S_6^{xy\bar{z}}$
S_{63}^+	h_{31}	31	$S_6^{\bar{x}yz}$
S_{64}^+	h_{32}	32	$S_6^{x\bar{y}z}$
S_{4x}^-	h_{43}	43	S_4^{3x}
S_{4y}^-	h_{48}	48	S_4^{3y}
S_{4z}^-	h_{38}	38	S_4^{3z}
S_{4x}^+	h_{44}	44	S_4^x
S_{4y}^+	h_{46}	46	S_4^y
S_{4z}^+	h_{39}	39	S_4^z
σ_{da}	h_{40}	40	σ^{xy}
σ_{db}	h_{37}	37	$\sigma^{\bar{x}y}$
σ_{dc}	h_{47}	47	σ^{xz}
σ_{dd}	h_{42}	42	σ^{yz}
σ_{de}	h_{45}	45	$\sigma^{\bar{x}z}$
σ_{df}	h_{41}	41	$\sigma^{\bar{y}z}$

(ii) $6/mmm$ (D_{6h})

This book	Kovalev (1965)	Miller and Love (1967)	Zak, Casher, Glück, and Gur (1969)	This book	Kovalev (1965)	Miller and Love (1967)	Zak, Casher, Glück, and Gur (1969)
E	h_1	1	E	I	h_{13}	13	I
C_6^+	h_2	2	C_6^z	S_3^-	h_{14}	14	S_3^{2z}
C_3^+	h_3	3	C_6^{2z}	S_6^-	h_{15}	15	S_6^{5z}
C_2	h_4	4	C_2	σ_h	h_{16}	16	σ^z
C_3^-	h_5	5	C_6^{4z}	S_6^+	h_{17}	17	S_6^z
C_6^-	h_6	6	C_6^{5z}	S_3^+	h_{18}	18	S_3^z
C'_{21}	h_{12}	10	U^2	σ_{d1}	h_{24}	22	σ^2
C'_{22}	h_{10}	8	U^1	σ_{d2}	h_{22}	20	σ^1
C'_{23}	h_8	12	U^3	σ_{d3}	h_{20}	24	σ^3
C''_{21}	h_9	7	U^x	σ_{v1}	h_{21}	19	σ^x
C''_{22}	h_7	11	U^y	σ_{v2}	h_{19}	23	σ^y
C''_{23}	h_{11}	9	U^{xy}	σ_{v3}	h_{23}	21	σ^{xy}

Bibliography

In constructing the bibliography we have adopted the following criteria. In topics that are already well-documented we have only given references to comprehensive review articles or treatises and to a few basic research papers: the mathematical crystallography of point groups and space groups (including the history of the derivation of these groups); non-crystallographic point groups; the applications to physical problems (e.g. symmetry properties of tensors, band structure, phonon and magnon dispersion relations, crystal field theory, space-group selection rules, etc.); the basic algebra underlying the theory of groups and their representations; and the determination of the space groups of individual crystalline materials by X-ray or neutron diffraction. On other topics more closely related to our central theme we have attempted to give a complete set of references: point-group representations and basis functions for these representations; the general theory of space-group representations; the determination of the representations of individual space groups or sets of space groups; the general theory of the corepresentations of magnetic groups; and the determination of the corepresentations of individual magnetic space groups. In general, we have not included references to theses, dissertations, and other unpublished reports that are relatively unobtainable.

Journal titles are abbreviated according to the *World list of scientific periodicals*, 4th edn. 1965, which gives abbreviations for journals published between 1900 and 1960. Names of journals first published after 1960, and those whose names have been altered since that date, are abbreviated in the style of the *World List*. If a journal has changed its title the abbreviation used is that which applied at the time of publication of the volume referred to. Where the language of a book or article is not English, the language is indicated.

Abelès, F. (1966) *Optical properties and electronic structure of metals and alloys*. North-Holland, Amsterdam.

Adams, E. N. (1953) The crystal momentum as a quantum mechanical operator. *J. chem. Phys.* **21**, 2013–7.

Adler, S. L. (1962) Theory of the valence band splittings at $\mathbf{k} = 0$ in zinc–blende and wurtzite structures. *Phys. Rev.* **126**, 118–22.

Aizu, K. (1964*a*) Symmetries in the paraelectric phase-transformations of ferroelectric crystals. *J. phys. Soc. Japan* **19**, 918–23.

——— (1964*b*) Ferroelectric transformations of tensorial properties in regular ferroelectrics. *Phys. Rev.* **133**, A1350–9.

Akulov, N. S. and Feldshtein, Ia. I. (1950) An application of group theory to the analysis of the anisotropy of crystals. *Dokl. Akad. Nauk SSSR* **70**, 593–6 (in Russian).

Alexander, E. (1929) Systematics of the one-dimensional space groups. *Z. Kristallogr. Kristallgeom.* **70**, 367–82 (in German); errata, *Z. Kristallogr. Kristallgeom.* **89**, 606–7 (in German).

ALEXANDER, E. and HERRMANN, K. (1929) The 80 two-dimensional space groups. *Z. Kristallogr. Kristallgeom.* **70**, 328–45 (in German).

ALEXANDER, S. (1962) Symmetry considerations in the determination of magnetic structures. *Phys. Rev.* **127**, 420–32.

ALLEN, L. C. and KARO, A. M. (1960) Basis functions for *ab initio* calculations. *Rev. mod. Phys.* **32**, 275–85.

ALTMANN, S. L. (1956) Spherical harmonics with the symmetry of the close-packed hexagonal lattice. *Proc. phys. Soc.* **A69**, 184–5.

——— (1957) On the symmetries of spherical harmonics. *Proc. Camb. phil. Soc. math. phys. Sci.* **53**, 343–67.

——— (1958) The cellular method for a close-packed hexagonal lattice. I. Theory. *Proc. R. Soc.* **A244**, 141–52.

——— (1962) Group theory. In *Quantum theory*, Vol. 2, *Aggregates of particles* (edited by D. R. Bates). Academic Press, New York.

——— (1963*a*) The crystallographic point groups as semi-direct products. *Phil. Trans. R. Soc.* **A255**, 216–40.

——— (1963*b*) Semidirect products and point groups. *Rev. mod. Phys.* **35**, 641–5.

——— (1967) The symmetry of nonrigid molecules: the Schrödinger supergroup. *Proc. R. Soc.* **A298**, 184–203.

——— and BRADLEY, C. J. (1963*a*) A note on the calculation of the matrix elements of the rotation group. *Phil. Trans. R. Soc.* **A255**, 193–8 (see also Bradley (1961)).

——— (1963*b*) On the symmetries of spherical harmonics. *Phil. Trans. R. Soc.* **A255**, 199–215.

——— (1965) Lattice harmonics. II. Hexagonal close-packed lattice. *Rev. mod. Phys.* **37**, 33–45.

——— and CRACKNELL, A. P. (1965) Lattice harmonics. I. Cubic groups. *Rev. mod. Phys.* **37**, 19–32.

ANON. (1949) Standards on piezoelectric crystals, 1949. *Proc. Inst. Radio Engrs* **37**, 1378–95.

——— (1955) Report on notation for the spectra of polyatomic molecules. *J. chem. Phys.* **23**, 1997–2011.

ANTONČÍK, E. (1959) Approximate formulation of the orthogonalized plane-wave method. *Physics Chem. Solids* **10**, 314–20.

——— (1960) On the approximate formulation of the orthogonalized plane-wave method. *Czech. J. Phys.* **10**, 22–7.

——— and TRLIFAJ, M. (1952) A note on the group analysis of the wave functions of valency electrons in a crystal. *Czech. J. Phys.* **1**, 97–109.

APPEL, J. (1964) Effect of spin–orbit coupling and other relativistic corrections on donor states in Ge and Si. *Phys. Rev.* **133**, A280–7.

ASCHER, E. (1966*a*) Some properties of spontaneous currents. *Helv. phys. Acta* **39**, 40–8.

——— (1966*b*) Magnetic anisotropy. A reformulation and its consequences. *Helv. phys. Acta* **39**, 466–76.

——— (1966*c*) Role of particular maximal subgroups in continuous phase transitions. *Phys. Lett.* **20**, 352–4.

——— and Janner, A. (1965*a*) Subgroups of black–white point groups. *Acta crystallogr.* **18**, 325–30.

ASCHER, E. and JANNER, A. (1965*b*) Algebraic aspects of crystallography, space groups as extensions. *Helv. phys. Acta* **38**, 551–72.

——— (1968) Algebraic aspects of crystallography. II. Non-primitive translations in space groups. *Communs math. Phys.* **11**, 138–67.

ASENDORF, R. H. (1957) Space group of tellurium and selenium. *J. chem. Phys.* **27**, 11–16.

ASHBY, N. and MILLER, S. C. (1965) Electric and magnetic translation group. *Phys. Rev.* **139**, A428–36.

ASTBURY, W. T. and YARDLEY, K. (1924) Tabulated data for the examination of the 230 space groups by homogeneous X-rays. *Phil. Trans. R. Soc.* **A224**, 221–57.

ATOJI, M. (1965) Graphical representations of magnetic space groups. *Am. J. Phys.* **33**, 212–19.

BÄBLER, G. (1947) On a law on the theory of crystal classes. *Comment. math. helvet.* **20**, 65–7 (in German).

BACKHOUSE, N. B. and BRADLEY, C. J. (1970) Projective representations of space groups, I: Translation groups. *Q. Jl Math.* **21**, 203–22.

BACON, G. E. (1962) *Neutron diffraction*. Clarendon Press, Oxford.

BADE, W. L. and JEHLE, H. (1953) An introduction to spinors. *Rev. mod. Phys.* **25**, 714–28.

BAGLEY, B. G. (1970) On the origin of pseudosymmetry. *J. Cryst. Growth* **6**, 323–6.

BALKANSKI, M. and DES CLOIZEAUX, J. (1960) Band structure of wurtzite-type crystals; intrinsic optical transitions in CdS. *J. Phys. Radium, Paris* **21**, 825–34 (in French).

——— and NUSIMOVICI, M. (1964) Interaction of a radiation field with the lattice vibrations at the critical points of the Brillouin zone of silicon. *Phys. Status Solidi* **5**, 635–47 (in French).

BALLHAUSEN, C. J. (1962) *Introduction to ligand field theory*. McGraw-Hill, New York.

BANDER, M. and ITZYKSON, C. (1966*a*) Group theory and the hydrogen atom (I). *Rev. mod. Phys.* **38**, 330–45.

——— ——— (1966*b*) Group theory and the hydrogen atom (II). *Rev. mod. Phys.* **38**, 346–58.

BARGMANN, V. (1954) On unitary ray representations of continuous groups. *Ann. Math.* **59**, 1–46.

——— (1962) On the representations of the rotation group. *Rev. mod. Phys.* **34**, 829–45.

BARLOW, W. (1883) Probable nature of the internal symmetry of crystals. *Nature, Lond.* **29**, 186–8, 205–7.

——— (1884) Probable nature of the internal symmetry of crystals. *Nature, Lond.* **29**, 404.

——— (1894) On the geometrical properties of homogeneous rigid structures and their application to crystals. *Z. Kristallogr. Miner.* **23**, 1–63 (in German).

BARNES, R. B., BRATTAIN, R. R., and SEITZ, F. (1935) On the structure and interpretation of the infrared absorption spectra of crystals. *Phys. Rev.* **48**, 582–602.

BARRON, T. H. K. and DOMB, C. (1955) On the cubic and hexagonal close-packed lattices. *Proc. R. Soc.* **A227**, 447–65.

——— and FISCHER, G. (1959) Brillouin zones and crystal structure factors. *Phil. Mag.* **4**, 826–9.

BASHKIROV, N. M. (1959) A generalization of Fedorov's stereohedra method. *Kristallografiya* **4**, 466–72 (in Russian); *Soviet Phys. Crystallogr.* (English Transl.) **4**, 442–7 (1960).

BASSANI, F. and CELLI, V. (1961) Energy-band structure of solids from a perturbation on the 'empty lattice'. *Physics Chem. Solids* **20**, 64–75.

——— and PARRAVICINI, G. P. (1967) Band structure and optical properties of graphite and of the layer compounds GaS and GaSe. *Nuovo Cim.* **B50**, 95–128.

BATES, C. A. and STEVENS, K. W. H. (1961) The band structure of body-centred cubic transition metals. *Proc. phys. Soc.* **78**, 1321–39.

BAXTER, R. J. (1970) Colorings of a hexagonal lattice. *J. math. Phys.* **11**, 784–9.

BAYLISS, N. S. (1949) Brillouin zones and the Mathieu equation. *Aust. J. Sci.* **12**, 12–14.

BAZAROV, I. P. (1962) On the theory of the crystal. *Physica, 's Grav.* **28**, 479–84.

BELL, D. G. (1954) Group theory and crystal lattices. *Rev. mod. Phys.* **26**, 311–20.

BELOV, K. P. (1961) *Magnetic transitions*. Consultants Bureau, New York.

BELOV, N. V. (1951) Classroom method for the derivation of space groups of symmetry. *Trudȳ Inst. Kristall.* **6**, 25–62 (in Russian); English translation, *Proc. Leeds phil. lit. Soc.*, Sci. Sect. **8**, 1–46 (1957).

——— (1954) Space groups of cubic symmetry. *Trudȳ Inst. Kristall.* **9**, 21–34 (in Russian).

——— (1956*a*) The one-dimensional infinite crystallographic groups. *Kristallografiya* **1**, 474–6 (in Russian); *Soviet Phys. Crystallogr.* (English Transl.) **1**, 372–4 (1966); also translated by Shubnikov and Belov (1964).

——— (1956*b*) Moorish patterns of the middle ages and the symmetry groups. *Kristallografiya* **1**, 610–13 (in Russian); *Soviet Phys. Crystallogr.* (English Transl.) **1**, 482–3 (1966).

——— (1956*c*) Three-dimensional mosaics with colored symmetry. *Kristallografiya* **1**, 621–5 (in Russian); *Soviet Phys. Crystallogr.* (English Transl.) **1**, 489–92 (1966); also translated by Shubnikov and Belov (1964).

——— (1956*d*) Methods of representing the cubic space groups. *Kristallografiya* **1**, 764–8 (in Russian); *Soviet Phys. Crystallogr.* (English Transl.) **1**, 596–9 (1966).

——— (1957*a*) Concerning the course in geometric crystallography for physicists. *Kristallografiya* **2**, 678–85 (in Russian); *Soviet Phys. Crystallogr.* (English Transl.) **2**, 667–73 (1958).

——— (1957*b*) Concerning the tetrahedral ($T = 23$) and gyrohedral ($O = 432$) groups. *Kristallografiya* **2**, 722–4 (in Russian); *Soviet Phys. Crystallogr.* (English Transl.) **2**, 712–14 (1958).

——— (1958) Project for a university course on Fedorov groups. *Kristallografiya* **3**, 765–72 (in Russian); *Soviet Phys. Crystallogr.* (English Transl.) **3**, 777–85 (1959).

——— (1959*a*) The hexagonal space groups. *Kristallografiya* **4**, 268–76 (in Russian); *Soviet Phys. Crystallogr.* (English Transl.) **4**, 251–8 (1960).

——— (1959*b*) On the nomenclature of the 80 plane groups in three dimensions. *Kristallografiya* **4**, 775–8 (in Russian); *Soviet Phys. Crystallogr.* (English Transl.) **4**, 730–3 (1960).

——— (1959*c*) On the basic theorem of the close-packed spherical structure. *Kristallografiya* **4**, 918–19 (in Russian); *Soviet Phys. Crystallogr.* (English Transl.) **4**, 873–4 (1960).

——— (1964) Nomenclature of the Bravais lattices. *Kristallografiya* **9**, 396–7 (in Russian); *Soviet Phys. Crystallogr.* (English Transl.) **9**, 315 (1964).

——— and BELOVA, E. N. (1957) Mosaics for 46 plane (Shubnikov) antisymmetry and for 15 (Fedorov) color groups. *Kristallografiya* **2**, 21–2 (in Russian); *Soviet Phys. Crystallogr.* (English Transl.) **2**, 16–18 (1957); also translated by Shubnikov and Belov (1964).

Belov, N. V., Belova, E. N., and Tarkhova, T. N. (1958) More about the color symmetry groups. *Kristallografiya* **3**, 618–20 (in Russian); *Soviet Phys. Crystallogr.* (English Transl.) **3**, 625–6 (1959); also translated by Shubnikov and Belov (1964).

——— and Klevtsova, R. F. (1959*a*) The simplest method of deriving the Fedorov (space) groups. *Kristallografiya* **4**, 289–92 (in Russian); *Soviet Phys. Crystallogr.* (English Transl.) **4**, 270–3 (1960).

——— ——— (1959*b*) A very simply way of deducing the space groups. *Kristallografiya* **4**, 473–6 (in Russian); *Soviet Phys. Crystallogr.* (English Transl.) **4**, 448–51 (1960).

——— Kuntsevich, T. S., and Neronova, N. N. (1962) Shubnikov (antisymmetry) groups for infinite double-sided ribbons. *Kristallografiya* **7**, 805–8 (in Russian); *Soviet Phys. Crystallogr.* (English Transl.) **7**, 651–4 (1963).

——— and Neronova, N. N. (1962) Ferromagnetic groups. *J. phys. Soc. Japan*, Suppl. B-II, 330–1.

——— ——— Donnay, J. D. H. and Donnay, G. (1962) Tables of magnetic space groups. II. Special positions. *J. phys. Soc. Japan*, Suppl. B-II, 332–5.

——— ——— and Kuntsevich, T. S. (1964) Crystal structure illustrations of Shubnikov antisymmetry groups. *Kristallografiya* **9**, 147–54 (in Russian); *Soviet Phys. Crystallogr.* (English Transl.) **9**, 117–22 (1964).

——— ——— and Smirnova, T. S. (1955) The 1651 Shubnikov groups (dichromatic space groups). *Trudŷ Inst. Kristall.* **11**, 33–67 (in Russian); translated by Shubnikov and Belov (1964).

——— ——— ——— (1957) Shubnikov groups. *Kristallografiya* **2**, 315–25 (in Russian); *Soviet Phys. Crystallogr.* (English Transl.) **2**, 311–22 (1957).

——— and Tarkhova, T. N. (1956*a*) Color symmetry groups. *Kristallografiya* **1**, 4–13 (in Russian); *Soviet Phys. Crystallogr.* (English Transl.) **1**, 5–11 (1966); also translated by Shubnikov and Belov (1964).

——— ——— (1956*b*) The 48-hedron group. *Kristallografiya* **1**, 360–1 (in Russian); *Soviet Phys. Crystallogr.* (English Transl.) **1**, 278–9 (1966).

——— ——— (1956*c*) Color symmetry groups. *Kristallografiya* **1**, 619–20 (in Russian); *Soviet Phys. Crystallogr.* (English Transl.) **1**, 487–8 (1966).

——— ——— (1960) Cayley squares for the cubic groups. *Kristallografiya* **5**, 126–34 (in Russian); *Soviet Phys. Crystallogr.* (English Transl.) **5**, 119–24 (1960).

Belova, E. N., Belov, N. V., and Shubnikov, A. V. (1948) On the number and the character of the abstract groups corresponding to the 32 crystal classes. *Dokl. Akad. Nauk SSSR* **63**, 669–72 (in Russian).

Bernal, M. J. M. and Boys, S. F. (1953) Electronic wave functions. VII. Methods of evaluating the fundamental coefficients for the expansion of vector-coupled Schrödinger integrals and some values of these. *Phil. Trans. R. Soc.* **A245**, 116–38.

Berry, R. L., Waldron, M. B., and Raynor, G. V. (1950) Outer Brillouin zones for the face-centred cubic, body-centred cubic and close-packed hexagonal structures. *Research, Lond.* **3**, 195–6.

BERTAUT, E. F. (1961*a*) Magnetic configurations. The direct method. *Physics Chem. Solids* **21**, 295–305 (in French).

——— (1961*b*) Spin configurations and group theory. *J. Phys. Radium, Paris* **22**, 321–2 (in French).

——— (1962) Lattice theory of spin configurations. *J. appl. Phys.* **33**, 1138–43.

——— (1963) Spin configurations of ionic structures: theory and practice. In *Magnetism* (edited by G. T. Rado and H. Suhl), Vol. 3, Chap. 4. Academic Press, New York.

——— (1968) Representation analysis of magnetic structures. *Acta crystallogr.* **A24**, 217–31.

BETHE, H. A. (1929) Splitting of terms in crystals. *Annln Phys.* **3**, 133–208 (in German); translated by Consultants Bureau, New York; also translated by Cracknell (1968*b*).

——— (1931) Theory of metals. Part I. Eigenvalues and eigenfunctions of the linear atomic chain. *Z. Phys.* **71**, 205–26 (in German).

——— (1934) Quantitative calculation of characteristic function of electrons in metals. *Helv. phys. Acta* **7** (Suppl. 2) 18–23 (in German).

BETTS, D. D. (1959) Solid harmonics as basis functions for cubic crystals. *Can. J. Phys.* **37**, 350–61.

——— BHATIA, A. B., and WYMAN, M. (1956) Houston's method and its application to the calculation of characteristic temperatures of cubic crystals. *Phys. Rev.* **104**, 37–42.

BHAGAVANTAM, S. (1942) Photo-elastic effect in crystals. *Proc. Indian Acad. Sci.*, **A16**, 359–65.

——— (1965) Crystal symmetry and magnetic properties. *Curr. Sci.* **34**, 72–4.

——— (1966) *Crystal symmetry and physical properties*. Academic Press, New York.

——— and PANTULU, P. V. (1963*a*) Force constants in the potential energy of a vibrating molecule (determination of the maximum number by a group theoretical method). *Proc. Indian Acad. Sci.* **A58**, 1–5.

——— ——— (1963*b*) Point defects and relaxation phenomena in crystals. *Proc. Indian Acad. Sci.* **A58**, 183–96.

——— ——— (1964*a*) Magnetic symmetry and physical properties of crystals. *Proc. Indian Acad. Sci.* **A59**, 1–13.

——— ——— (1964*b*) Crystal symmetry and physical properties: application of group theory. *Proc. Indian Acad. Sci.* **A60**, 1–10.

——— ——— (1966) Magnetic symmetry and crystal lattices. *Proc. Indian Acad. Sci.* **A63**, 391–9.

——— ——— (1967) Generalized symmetry and Neumann's principle. *Proc. Indian Acad. Sci.* **A66**, 33–9.

——— and SURYANARAYANA, D. (1949) Crystal symmetry and physical properties: application of group theory. *Acta crystallogr.* **2**, 21–6.

——— and VENKATARAYUDU, T. (1962) *Theory of groups and its application to physical problems*. Andhra University Press, Waltair, India.

BIEBERBACH, L. (1910) On the motion groups of Euclidean space. *Math. Annln* **70**, 297–336 (in German).

——— (1912) On the motion groups of Euclidean space. (Part 2) The groups having a finite fundamental domain. *Math. Annln* **72**, 400–12 (in German).

——— (1939) On the equivalence of Brillouin zones. *Mh. Math. Phys.* **48**, 509–15 (in German).

BIEDENHARN, L. C., BLATT, J. M., and ROSE, M. E. (1952) Some properties of the Racah and associated coefficients. *Rev. mod. Phys.* **24**, 249–57.

——— and ROSE, M. E. (1953) Theory of angular correlation of nuclear radiations. *Rev. mod. Phys.* **25**, 729–77.

——— and VAN DAM, H. (1965) *Quantum theory of angular momentum*. Academic Press, New York.

BIENENSTOCK, A. and EWALD, P. P. (1961) Structure theories in physical and in Fourier space. *Kristallografiya* **6**, 820–4 (in Russian); *Soviet Phys. Crystallogr.* (English Transl.) **6**, 665–7 (1962).

——— (1962) Symmetry of Fourier space. *Acta crystallogr.* **15**, 1253–61.

BIRMAN, J. L. (1958) Electronic energy bands in ZnS: potential in zincblende and wurtzite. *Phys. Rev.* **109**, 810–17.

——— (1959*a*) Some selection rules for band-band transitions in wurtzite structure. *Phys. Rev.* **114**, 1490–2.

——— (1959*b*) Simplified LCAO method for zincblende, wurtzite, and mixed crystal structures. *Phys. Rev.* **115**, 1493–505.

——— (1962*a*) Configurational instability in solids: diamond and zinc blende. *Phys. Rev.* **125**, 1959–61.

——— (1962*b*) Space group selection rules: diamond and zinc blende. *Phys. Rev.* **127**, 1093–1106.

——— (1963) Theory of infrared and Raman processes in crystals: selection rules in diamond and zincblende. *Phys. Rev.* **131**, 1489–96.

——— (1965) Space group theory and lattice dynamics. In *Lattice dynamics. Proceedings of the international conference held at Copenhagen, Denmark, August* 5–9, 1963 (edited by R. F. Wallis), pp. 669–80. Pergamon Press, Oxford.

——— (1966*a*) Full-group and subgroup methods in crystal physics. *Phys. Rev.* **150**, 771–82.

——— (1966*b*) Simplified theory of symmetry change in second-order phase transitions: application to V_3Si. *Phys. Rev. Lett.* **17**, 1216–19.

——— (1967) Predicted tetragonal modifications of V_3Si type compounds. *Chem. Phys. Lett.* **1**, 343–4.

——— LAX, M., and LOUDON, R. (1966) Intervalley-scattering selection rules in III-V semiconductors. *Phys. Rev.* **145**, 620–2.

BIRSS, R. R. (1962) Property tensors in magnetic crystal classes. *Proc. phys. Soc.* **79**, 946–53.

——— (1963) Macroscopic symmetry in space-time. *Rep. Prog. Phys.* **26**, 307–60.

——— (1964) *Symmetry and magnetism*. North-Holland, Amsterdam.

BLEANEY, B. (1959) The spin Hamiltonian of a Γ_8 quartet. *Proc. phys. Soc.* **73**, 939–42.

——— and STEVENS, K. W. H. (1953) Paramagnetic resonance. *Rep. Prog. Phys.* **16**, 108–59.

BLOCH, F. (1928) Quantum mechanics of electrons in crystal lattices. *Z. Phys.* **52**, 555–600 (in German).

BLOUNT, E. E. (1962) Formalisms of band theory. *Solid St. Phys.* **13**, 305–73.

BOERNER, H. (1955) *Representations of groups, with special consideration for the needs of modern physics*. Springer, Berlin (in German); translated, North-Holland, Amsterdam (1963).

BOHM, J. and DORNBERGER-SCHIFF, K. (1966) The nomenclature of crystallographic symmetry groups. *Acta crystallogr.* **21**, 1004–7.

BOHM, J. and DORNBERGER-SCHIFF, K. (1967) Geometrical symbols for all crystallographic symmetry groups up to three dimensions. *Acta crystallogr.* **23**, 913–33.

BOND, W. L. (1943) The mathematics of the physical properties of crystals. *Bell Syst. tech. J.* **22**, 1–72.

BORDURE, G., LECOY, G., and SAVELLI, M. (1963a) Symmetries of the Hamiltonian of a single electron in the crystalline potential. I. Generalities of the Schrödinger Hamiltonian. *Cah. Phys.* **17**, 309–18 (in French).

——— ——— ——— (1963*b*) Symmetries of the Hamiltonian of a single electron in the crystalline potential. II. Effects of spin–orbit coupling and of time reversal. *Cah. Phys.* **17**, 382–9 (in French).

BORN, M. (1915) *Dynamics of crystal lattices*. Teubner, Leipzig (in German).

——— and HEISENBERG, W. (1924) On the quantum theory of molecules. *Annln Phys.* **74**, 1–31 (in German).

——— ——— and JORDAN, P. (1926) On quantum mechanics. Part II. *Z. Phys.* **35**, 557–615 (in German) (see also Born and Jordan (1925)).

——— and HUANG, K. (1954) *Dynamical theory of crystal lattices*. Clarendon Press, Oxford.

——— and JORDAN, P. (1925) On quantum mechanics. *Z. Phys.* **34**, 858–88 (in German) (see also Born, Heisenberg, and Jordan (1926)).

——— and OPPENHEIMER, J. R. (1927) Quantum theory of molecules. *Annln Phys.* **84**, 457–84 (in German).

——— and VON KÁRMÁN, TH. (1913*a*) Theory of specific heat. *Phys. Z.* **14**, 15–19 (in German).

——— ——— (1913*b*) On the distribution of normal modes of point lattices. *Phys. Z.* **14**, 65–71 (in German).

BOUCKAERT, L. P., SMOLUCHOWSKI, R., and WIGNER, E. (1936) Theory of Brillouin zones and symmetry properties of wave functions in crystals. *Phys. Rev.* **50**, 58–67; reprinted: Knox and Gold (1964), Meijer (1964), Cracknell (1968*b*).

BOYLE, L. L. (1969) Non-crystallographic Shubnikov groups. *Acta crystallogr.* **A25**, 455–9.

BOYS, S. F. (1952) Electronic wave functions. VI. Some theorems facilitating the evaluation of Schrödinger integrals of vector-coupled functions. *Phil. Trans. R. Soc.* **A245**, 95–115.

BRADLEY, C. J. (1961) The matrix representatives $d^{l}_{m'm}(\pi/2)$ for the rotation group. Royal Society depository of unpublished mathematical tables, File No. 79; also Washington: Library of Congress collection, No. 61-18944.

——— (1966) Space groups and selection rules. *J. math. Phys.* **7**, 1145–52.

——— and CRACKNELL, A. P. (1966) Corepresentations of magnetic space groups. *Prog. theor. Phys., Kyoto* **36**, 648–9.

——— ——— (1970) Some comments on the theory of space-group representations. I. *J. Phys. C. (Solid St. Phys.)* **3**, 610–18.

——— and DAVIES, B. L. (1968) Magnetic groups and their corepresentations. *Rev. mod. Phys.* **40**, 359–79.

——— ——— (1970) Kronecker products and symmetrized squares of irreducible representations of space groups. *J. math. Phys.* **11**, 1536–52.

——— WALLIS, D. E., and CRACKNELL, A. P. (1970) Some comments on the theory of space-group representations. II. Basic domain and representation domain. *J. Phys. C. (Solid St. Phys.)* **3**, 619–26.

BRAGG, W. H., VON LAUE, M., and HERMANN, C. (1935*a*) *International tables for the determination of crystal structures*. Vol. 1. *Tables on the theory of groups*. Borntraeger, Berlin (in German, English, and French).

—— —— (1935*b*) *International tables for the determination of crystal structures*. Vol. 2. *Mathematical and Physical tables*. Borntraeger, Berlin (in German, English, and French).

BRAGG, W. L. (1920) The arrangement of atoms in crystals. *Phil. Mag.* **40**, 169–89.

BRANDENBERGER, E. (1928) Systematic presentation of the selection rules important for the crystal structures of triclinic, monoclinic and rhombic space groups. *Z. Kristallogr. Kristallgeom.* **68**, 330–62 (in German).

—— and NIGGLI, P. (1928) The systematic presentation of the selection rules important for crystal structure. *Z. Kristallogr., Kristallgeom.* **68**, 301–29.

BRAUER, E. (1933) *Introduction to the theory of groups and its applications to quantum physics*. Presses Universitaires de France, Paris (in French).

BRAUER, R. and NESBITT, C. (1937) *On the modular representations of groups of finite order*. University of Toronto Press, Toronto.

BRAVAIS, A. (1849*a*) Note on the symmetrical polyhedra of geometry. *J. Math. pures appl.* **14**, 137–40 (in French).

—— (1849*b*) Memoir on the polyhedra of symmetrical form. *J. Math. pures appl.* **14**, 141–180 (in French).

—— (1850) On the systems formed by points regularly distributed in a plane or in space. *J. Éc. polytech.* **19**, 1–128 (in French); English translation, Crystallographic Society of America, *Memoir No.* 1 (1949).

BRILLOUIN, L. (1930*a*) Electrons in metals and the rôle of Bragg's selective reflection condition. *C. r. hebd. Séanc. Acad. Sci., Paris* **191**, 198–200 (in French).

—— (1930*b*) Electrons in metals and classification of the corresponding de Broglie waves. *C. r. hebd. Séanc. Acad. Sci., Paris* **191**, 292–4 (in French).

—— (1930*c*) Free electrons in metals and the rôle of Bragg reflections. *J. Phys. Radium, Paris* **1**, 377–400 (in French).

—— (1931) *Quantum statistics and its application to the electron theory of metals*. Springer, Berlin (in German).

—— (1934) Basis of electronic theory of metals and the method of self-consistent fields. *Helv. phys. Acta* **7** (Suppl. 2), 33–46 (in French).

—— (1953) *Wave propagation in periodic structures*. Dover, New York.

BRINK, D. M. and SATCHLER, G. R. (1962) *Angular momentum*. Clarendon Press, Oxford.

BRINKMAN, H. C. (1956) *Application of spinor invariants in atomic physics*. North-Holland, Amsterdam.

BRINKMAN, W. F. (1967) Magnetic symmetry and spin waves. *J. appl. Phys.* **38**, 939–43.

—— and ELLIOTT, R. J. (1966*a*) Space group theory for spin waves. *J. appl. Phys.* **37**, 1457–9.

—— —— (1966*b*) Theory of spin–space groups. *Proc. R. Soc.* **A294**, 343–58.

BROERMAN, J. G. (1968) Double group full-zone **k.p** expansions for diamond and zincblende structures. *J. Phys. Chem. Solids* **29**, 1147–66.

BROOKS, H. and HAM, F. S. (1958) Energy bands in solids—the quantum defect method. *Phys. Rev.* **112**, 344–61.

BROWN, E. (1970) A simple alternative to double groups. *Am. J. Phys.* **38**, 704–15.

BRUSSARD, P. J. and TOLHOEK, H. A. (1957) Classical limits of Clebsch–Gordan coefficients, Racah coefficients and $D^l_{mn}(\theta, \phi, \psi)$-functions. *Physica, 's Grav.* **23**, 955–71.

BUERGER, M. J. (1935) Application of plane groups to the interpretation of Weissenberg photographs. *Z. Kristallogr., Kristallgeom.* **91**, 255–89.

——— (1942) *X-ray crystallography*. Wiley, New York.

——— (1947) Derivative crystal structures. *J. chem. Phys.* **15**, 1–16.

——— (1949) Crystallographic symmetry in reciprocal space. *Proc. natn. Acad. Sci. U.S.A.* **35**, 198–201.

——— (1950*a*) The crystallographic symmetries determinable by X-ray diffraction. *Proc. natn. Acad. Sci. U.S.A.* **36**, 324–9.

——— (1950*b*) Vector sets. *Acta crystallogr.* **3**, 87–97.

——— (1950*c*) Tables of the characteristics of the vector representations of the 230 space groups. *Acta crystallogr.* **3**, 465–71.

——— (1956) *Elementary crystallography*. Wiley, New York.

——— (1967) Some desirable modifications of the international symmetry symbols. *Proc. natn. Acad. Sci. U.S.A.* **58**, 1768–73.

BUNN, C. W. (1961) *Chemical crystallography, an introduction to optical and X-ray methods*. Clarendon Press, Oxford.

BURCKHARDT, J. J. (1930) Remarks on the arithmetic calculation of space groups. *Comment. Math. helvet.* **2**, 91–8 (in German).

——— (1934) On the theory of space groups. *Comment. math. helvet.* **6**, 159–84 (in German).

——— (1937) Space groups in many-dimensional spaces. *Comment. math. helvet.* **9**, 284–302 (in German).

——— (1966) *The motion groups of crystallography*. Birkhauser, Basel (in German).

——— (1967) The history of the discovery of the 230 space groups. *Archs Hist. exact Sci.* **4**, 235–46 (in German).

BURNSIDE, W. (1901) On group characteristics. *Proc. Lond. math. Soc.* **33**, 146–62.

——— (1903) On the representation of a group of finite order as an irreducible group of linear substitutions and the direct establishment of the relations between the group characteristics. *Proc. Lond. math. Soc.* **1**, 117–23.

——— (1911) *Theory of groups of finite order*. University Press, Cambridge; reprinted: Dover, New York (1955).

BURSTEIN, E., JOHNSON, F. A., and LOUDON, R. (1965) Selection rules for second-order infrared and Raman processes in the rocksalt structure and interpretation of the Raman spectra of NaCl, KBr, and NaI. *Phys. Rev.* **139**, A1239–45.

BURZLAFF, H., FISCHER, W., and HELLNER, E. (1968) Lattice complexes of equivalent groups. *Acta crystallogr.* **A24**, 57–67 (in German).

CALLAWAY, J. (1958). Electron energy bands in solids. *Solid St. Phys.* **7**, 99–212.

——— (1964*a*) *Energy band theory*. Academic Press, New York.

——— (1964*b*) Theory of scattering in solids. *J. math. Phys.* **5**, 783–98.

CALLEN, E. R. and CALLEN, H. B. (1963) Static magnetoelastic coupling in cubic crystals. *Phys. Rev.* **129**, 578–93.

CALLEN, H. B. (1968) Crystal symmetry and macroscopic laws. *Am. J. Phys.* **36**, 735–48.

——— CALLEN, E., and ZALVA, Z. (1970) Crystal symmetry and macroscopic laws, II. *Am. J. Phys.* **38**, 1278–84.

CANNON, J. J. (1969) Computers in group theory: a survey. *Communs Ass. comput. Mach.* **12**, 3–12.

CARDONA, M. and HARBEKE, G. (1965) Optical properties and band structure of wurtzite-type crystals and rutile. *Phys. Rev.* **137**, A1467–76.

CARMELI, M. (1968) Representations of the three-dimensional rotation group in terms of direction and angle of rotation. *J. math. Phys.* **9**, 1987–92.

CARMICHAEL, R. D. (1937) *Introduction to the theory of groups of finite order*. Ginn, Boston, Mass.

CASELLA, R. C. (1959) Symmetry of wurtzite. *Phys. Rev.* **114**, 1514–18.

CHALDYSHEV, V. A. (1958) Application of the network model to investigation of energy spectrum of electrons and (positive) holes in crystals with zincblende type lattice. *Izv. vÿssh. ucheb. Zaved.*, Fiz. (5) 65–69 (in Russian).

——— (1961) Energy spectra of electrons in crystals. I. *Izv. vÿssh. ucheb. Zaved.*, Fiz. (6) 48–51 (in Russian).

——— and KARAVAEV, G. F. (1963) The valence band structure of chalcopyrites-type compounds. *Izv. vÿssh. ucheb. Zaved.*, Fiz. (5), 103–12 (in Russian).

——— and KUDRYAVTSEVA, N. V. (1962) Energy spectra of electrons in crystals. II. *Izv. vÿssh. ucheb. Zaved.*, Fiz. (2) 104–7 (in Russian).

——— ——— and KARAVAEV, G. F. (1963) Energy spectra of electrons in crystals. V. Weighted corepresentations. *Izv. vÿssh. ucheb. Zaved.* Fiz. (2), 46–52 (in Russian) (see also Kudryavtseva and Chaldyshev (1962*a*, *b*; 1965*a*, *b*; 1966*a*, *b*)).

CHEN, H. S. and HSIEH, H. T. (1965) Reductions of the symmetrized and the antisymmetrized powers of the irreducible space group representations. *Acta phys. sin.* **21**, 519–25 (in Chinese); *Chin. J. Phys.* (English Transl.) **21**, 653–61 (1966).

CHEN, L. C., BERENSON, R., and BIRMAN, J. L. (1968) Space-group selection rules: 'rocksalt' $O_h^5 - Fm3m$. *Phys. Rev.* **170**, 639–48.

CHEN, S. H. (1967) Group-theoretical analysis of lattice vibrations in metallic β-Sn. *Phys. Rev.* **163**, 532–46.

——— and DVORAK, V. (1968) Group-theoretical analysis of lattice vibrations in molecular crystals. *J. chem. Phys.* **48**, 4060–3 (see also Pawley (1969)).

CHIU, Y. N. (1964) Irreducible tensor expansion of solid spherical harmonic-type operators in quantum mechanics. *J. math. Phys.* **5**, 283–8.

CLIFFORD, A. H. (1937) Representations induced in an invariant subgroup. *Ann. Math.* **38**, 533–50.

COCHRAN, W. (1952) The symmetry of real periodic two-dimensional functions. *Acta crystallogr.* **5**, 630–3.

COHAN, N. V. (1958) The spherical harmonics with the symmetry of the icosahedral group. *Proc. Camb. phil. Soc. math. phys. Sci.* **54**, 28–38.

COLEMAN, A. J. (1968) Induced and subduced representations. In: *Group theory and its applications* (edited by E. M. Loebl), pp. 57–118. Academic Press, New York.

CONDON, E. U. and SHORTLEY, G. H. (1935) *The theory of atomic spectra*. Cambridge University Press.

CORLISS, L. M. and HASTINGS, J. M. (1959) Symmetry of magnetic structures. *J. appl. Phys.* **30**, 279S.

CORNWELL, J. F. (1966) Direct optical absorption selection rules for the hexagonal close-packed lattice. *Phys. kondens. Materie* **4**, 327–9.

——— (1969) *Group theory and electronic energy bands in solids*. North-Holland, Amsterdam.

——— (1970) Symmetry properties of the Clebsch–Gordan coefficients of space groups. *Phys. Status Solidi* **37**, 225–35.

CORSON, E. M. (1951) *Perturbation methods in the quantum mechanics of n-electron systems*. Blackie, London.

——— (1953) *Introduction to tensors, spinors, and relativistic wave equations (relation structure)*. Blackie, London.

COTTON, F. A. (1963) *Chemical applications of group theory*. Wiley, New York.

COULSON, C. A. (1961) *Valence*. Clarendon Press, Oxford.

COUTURE, L. (1947) Selection rules for ionic crystals. *J. chem. Phys.* **15**, 153.

COWLEY, E. R. (1969) Symmetry properties of the normal modes of vibration of calcite and α-corundum. *Can. J. Phys.* **47**, 1381–91.

COWLEY, R. A. and DOLLING, G. (1968) Magnetic excitations in uranium dioxide. *Phys. Rev.* **167**, 464–77.

COXETER, H. S. M. (1948) *Regular polytopes*. Methuen, London.

——— and MOSER, W. O. J. (1965) *Generators and relations for discrete groups*. Springer, Berlin.

CRACKNELL, A. P. (1965) Corepresentations of magnetic cubic space groups. *Prog. theor. Phys., Kyoto* **33**, 812–27 (see also Bradley and Cracknell (1966)).

——— (1966*a*) Corepresentations of magnetic point groups. *Prog. theor. Phys., Kyoto* **35**, 196–213.

——— (1966*b*) Crystal field theory and magnetic point groups. *Aust. J. Phys.* **19**, 519–29.

——— (1967*a*) Crystal field theory and magnetic point groups: spin–orbit coupling. *Aust. J. Phys.* **20**, 189–92.

——— (1967*b*) Magnetic groups. *Contemp. Phys.* **8**, 459–73.

——— (1967*c*) Space-group selection rules for magnetic crystals. *Prog. theor. Phys., Kyoto* **38**, 1252–69 (see also Cracknell and Davies (1970)).

——— (1968*a*) Crystal field theory and the Shubnikov point groups. *Adv. Phys.* **17**, 367–420.

——— (1968*b*) *Applied group theory*. Pergamon Press, Oxford.

——— (1969*a*) Symmetry adapted functions for double point groups. I. Non-cubic point groups. *Proc. Camb. phil. Soc. math. phys. Sci.* **65**, 567–78 (see also Cracknell and Joshua (1970)).

——— (1969*b*) Scattering matrices for the Raman effect in magnetic crystals. *J. Phys. C. (Solid St. Phys.)* **2**, 500–11.

——— (1969*c*) Time reversal degeneracy in the electronic band structure of a magnetic metal. *J. Phys. C. (Solid St. Phys.)* **2**, 1425–33.

——— (1969*d*) The determination of the symmetries of paramagnons with applications to MnF_2 and NiF_2. *J. Phys. C. (Solid St. Phys.)* **2**, 1764–76.

——— (1969*e*) Group theory and magnetic phenomena in solids. *Rep. Prog. Phys.* **32**, 633–707.

——— (1970*a*) Time-reversal degeneracies in the band structure of a ferromagnetic metal. *Phys. Rev.* **B1**, 1261–3.

——— (1970*b*) The symmetry properties of the spin-wave dispersion relations of a ferromagnetic metal. *J. Phys. C. (Solid St. Phys.)* **3**, Metal Phys. Suppl. S175–88.

——— CRACKNELL, M. F. and DAVIES, B. L. (1970) Landau's theory of second-order phase transitions and the magnetic phase transitions in some antiferromagnetic oxides. *Phys. Status Solidi* **39**, 463–9.

——— and DAVIES, B. L. (1970) Space group selection rules for magnetic crystals. *Progr. theor. Phys., Kyoto* **43**, 1613–4.

——— and JOSHUA, S. J. (1968) The Landau theory of second-order phase transitions: $[\Gamma]^3$ and $\{\Gamma\}^2$ for crystallographic point groups. *J. Phys. A. (Proc. phys. Soc.)* **1**, 40–2.

——— ——— (1969*a*) The space group corepresentations of antiferromagnetic NiO. *Proc. Camb. phil. Soc. math. phys. Sci.* **66**, 493–504.

——— ——— (1969*b*) The spin-wave dispersion relations and the spin-wave contribution to the specific heat of antiferromagnetic UO_2. *Phys. Status Solidi* **36**, 737–45.

——— ——— (1970) Symmetry adapted functions for double point groups. II. Cubic point groups. *Proc. Camb. phil. Soc. math. phys. Sci.* **67**, 647–56

——— and WONG, K. C. (1967) Double-valued corepresentations of magnetic point groups. *Aust. J. Phys.* **20**, 173–88.

CURIE, P. (1884) On symmetry. *Bull. Soc. fr. Minér. Cristallogr.* **7**, 418–57 (in French).

——— (1894) On the symmetry in physical phenomena, symmetry of an electric field and of a magnetic field. *J. Phys., Paris* **3**, 393–415 (in French).

CURIEN, H. and DONNAY, J. D. H. (1959) The symmetry of the complete twin. *Am. Miner.* **44**, 1067–70.

——— and LE CORRE, Y. (1958) Notations of twins to help the symbolism of the coloured groups of Shubnikov. *Bull. Soc. fr. Minér. Cristallogr.* **81**, 126–32 (in French).

CURTIS, C. W. and REINER, I. (1962) *Representation theory of finite groups and associative algebras*. Wiley, New York.

DALTABUIT, E. and MCINTOSH, H. V. (1967) Representations of the magnetic symmetry groups (of crystal lattice). *Revta Mex. Fis.* **16**, 105–14 (in Spanish).

DANIEL, M. R. and CRACKNELL, A. P. (1969) Magnetic symmetry and antiferromagnetic resonance in CoO. *Phys. Rev.* **177**, 932–41.

DARWIN, C. G. (1927) The Zeeman effect and spherical harmonics. *Proc. R. Soc.* **A115**, 1–19.

DAVYDOV, A. S. (1965) *Quantum mechanics*. Pergamon Press, Oxford.

DEKKER, A. J. (1958) *Solid state physics*. Macmillan, London.

DELAUNAY, B. (1932) New representation of geometrical crystallography. *Z. Kristallogr. Kristallgeom.* **84**, 109–49 (in German).

DES CLOIZEAUX, J. (1963) Orthogonal orbitals and generalized Wannier functions. *Phys. Rev.* **129**, 554–66.

DEVONSHIRE, A. F. (1936) The rotation of molecules in fields of octahedral symmetry. *Proc. R. Soc.* **A153**, 601–21.

DIMMOCK, J. O. (1963*a*) Use of symmetry in the determination of magnetic structures. *Phys. Rev.* **130**, 1337–44.

—— (1963*b*) Representation theory for nonunitary groups. *J. math. Phys.* **4**, 1307–11.

—— (1964) On the symmetry of spin configurations in magnetic crystals. *Proc. int. Conf. Magnetism*, Nottingham, 1964, pp. 489–92.

—— and WHEELER, R. G. (1962*a*) Irreducible representations of magnetic groups. *J. Phys. Chem. Solids* **23**, 729–41.

—— —— (1962*b*) Symmetry properties of wave functions in magnetic crystals. *Phys. Rev.* **127**, 391–404.

—— —— (1964) Symmetry properties of magnetic crystals. In *The mathematics of physics and chemistry*, Vol. 2, (edited by H. Margenau and G. M. Murphy) Chap. 12. Van Nostrand, New York.

DIRAC, P. A. M. (1947) *The principles of quantum mechanics.* Clarendon Press, Oxford.

DONNAY, G., BELOV, N. V., NERONOVA, N. N., and SMIRNOVA, T. S. (1958) The Shubnikov groups. *Kristallografiya* **3**, 635–6 (in Russian); *Soviet Phys. Crystallogr.* **3**, 642–4 (1959).

—— CORLISS, L. M., DONNAY, J. D. H., ELLIOTT, N., and HASTINGS, J. M. (1958) Symmetry of magnetic structures: magnetic structure of chalcopyrite. *Phys. Rev.* **112**, 1917–23.

DONNAY, J. D. H. (1960) Early contributions to the theory of symmetry groups. *Acta crystallogr.* **13**, 1083.

—— (1969) Symbolism of rhombohedral space groups in Miller axes. *Acta crystallogr.* **A25**, 715–6.

—— and DONNAY, G. (1959) Tables of magnetic space groups. *C. r. hebd. Séanc. Acad. Sci., Paris* **248**, 3317–19 (in French).

—— —— COX, E. G., KENNARD, O., and KING, M. V. (1963) *Crystal data, determinative tables.* American Crystallographic Association, New York.

DÖRING, W. (1959) The ray-representation of crystallographic groups. *Z. Naturf.* **14a**, 343–50 (in German).

—— and ZEHLER, V. (1953) Group-theoretic investigation of the electron bands in the diamond lattice. *Annln Phys.* **13**, 214–28 (in German).

DORNBERGER-SCHIFF, K. (1956) On order–disorder structures (OD-structures). *Acta crystallogr.* **9**, 593–601.

—— (1959) On the nomenclature of the 80 plane groups in three dimensions. *Acta crystallogr.* **12**, 173.

DRESSELHAUS, G. (1955) Spin–orbit coupling effects in zinc blende structures. *Phys. Rev.* **100**, 580–6.

—— (1957) Optical absorption band edge in anisotropic crystals. *Phys. Rev.* **105**, 135–8.

DZYALOSHINSKIĬ, I. E. (1959) On the magneto-electrical effect in antiferromagnets. *Zh. éksp. teor. Fiz.* **37**, 881–2 (in Russian); *Soviet Phys. JETP* **10**, 628–9 (1960).

ECKHART, C. (1930) The application of group theory to the quantum dynamics of monatomic systems. *Rev. mod. Phys.* **2**, 305–80.

EDMONDS, A. R. (1957) *Angular momentum in quantum mechanics.* University Press, Princeton.

EGELSTAFF, P. A. (1965) *Thermal neutron scattering*. Academic Press, London.

ELERT, W. (1928) On the rotation spectrum and the vibration spectrum of a molecule of type CH_4. *Z. Phys.* **51**, 6–33 (in German).

ELLIOTT, R. J. (1954*a*) Theory of the effect of spin–orbit coupling on magnetic resonance in some semiconductors. *Phys. Rev.* **96**, 266–79.

——— (1954*b*) Spin–orbit coupling in band theory—character tables for some 'double' space groups. *Phys. Rev.* **96**, 280–7; reprinted: Knox and Gold (1964), Meijer (1964).

——— (1961) Symmetry of excitons in Cu_2O. *Phys. Rev.* **124**, 340–5.

——— and LOUDON, R. (1960) Group theory of scattering processes in crystals. *Physics Chem. Solids* **15**, 146–51.

——— ——— (1963) The possible observation of electronic Raman transitions in crystals. *Phys. Lett.* **3**, 189–91.

——— and STEVENS, K. W. H. (1952) The theory of the magnetic properties of rare earth salts: cerium ethyl sulphate. *Proc. R. Soc.* **A215**, 437–53.

——— ——— (1953*a*) The theory of magnetic resonance experiments on salts of the rare earths. *Proc. R. Soc.* **A218**, 553–66.

——— ——— (1953*b*) The magnetic properties of certain rare-earth ethyl sulphates. *Proc. R. Soc.* **A219**, 387–404.

——— and THORPE, M. F. (1967) Group theoretical selection rules in inelastic neutron scattering. *Proc. phys. Soc.* **91**, 903–16.

ENGLERT, F. (1957) Application of group theory to calculation of spin–orbit coupling in crystals. *Bull. Acad. r. Belg. Cl. Sci.* **43**, 273–83 (in French).

ERDÖS, P. (1964) The determination of the components of a tensor characterizing a crystal. *Helv. phys. Acta* **37**, 493–504.

ERDMANN, J. (1960) The symmetry of wave functions in hexagonal close-packed crystals. *Z. Naturf.* **15a**, 524–31 (in German).

EVSEEV, Z. YA. (1960) Band structure of the polaron energy spectrum. *Fizika tverd. Tela* **2**, 1222–3 (in Russian); *Soviet Phys. solid St.* (English Transl.) **2**, 1107–8 (1960).

EWALD, P. P. (1921) The 'reciprocal lattice' in structure theory. *Z. Kristallogr. Kristallgeom.* **56**, 129–56 (in German).

——— (1936) History and systematics of the use of the 'reciprocal lattice' in the study of crystal structure. *Z. Kristallogr. Kristallgeom.* **93**, 396–8 (in German).

EYRING, H., FROST, A. A., and TURKEVICH, J. (1933) Molecular symmetry and the reduction of the secular equation. *J. chem. Phys.* **1**, 777–83.

——— WALTER, J., and KIMBALL, G. E. (1944) *Quantum chemistry*. Wiley, New York.

FADDEYEV, D. K. (1964) *Tables of the principal unitary representations of Fedorov groups*. Pergamon Press, Oxford.

FALICOV, L. M. (1966) *Group theory and its physical applications*. University Press, Chicago.

——— and COHEN, M. H. (1963) Spin–orbit coupling in the band structure of magnesium and other hexagonal-close-packed metals. *Phys. Rev.* **130**, 92–7.

——— and GOLIN, S. (1965) Electronic band structure of arsenic. I. Pseudopotential approach. *Phys. Rev.* **137**, A871–82.

FALICOV, L. M. and RUVALDS, J. (1968) Symmetry of the wave functions in the band theory of ferromagnetic metals. *Phys. Rev.* **172**, 498–507.

FANO, U. (1960) Real representations of coordinate rotations. *J. math. Phys.* **1**, 417–23.

FARRELL, R. A. and MEIJER, P. H. E. (1965) First order transitions in simple magnetic systems. *Physica, 's Grav.* **31**, 725–48.

FEDOROV, E. S. (1885) Elements of the study of form. *Verh. russ. min. Ges.* **21**, 1–278 (in Russian).

——— (1891*a*) Symmetry of regular systems of figures. *Vseross. min. Obshch., Zap.* (II) **28**, 1–146 (in Russian).

——— (1891*b*) Symmetry in a plane. *Vseross. min. Obshch., Zap.* (II), **28**, 345–90 (in Russian).

——— (1892) List for the comparison of the crystallographic results of Schönflies and my own results. *Z. Kristallogr. Miner.* **20**, 25–75 (in German).

FEDOROV, F. I. (1958) The method of invariants in the optics of transparent non-magnetic crystals. *Kristallografiya* **3**, 49–56 (in Russian); *Soviet Phys. Crystallogr.* **3**, 46–52 (1958).

FICK, E. (1957) Term splitting in an electrostatic crystal field. *Z. Phys.* **147**, 307–16 (in German).

FIESCHI, R. (1957) Matter tensors in the crystallographic groups of cartesian symmetry. *Physica, 's Grav.* **23**, 972–6.

——— and FUMI, F. G. (1953) High-order matter tensors in symmetrical systems. *Nuovo Cim.* **10**, 865–82.

FIRSOV, IU. A. (1957) On the structure of the electron spectrum in lattices of the tellurium type. *Zh. éksp. teor. Fiz.* **32**, 1350–67 (in Russian); *Soviet Phys. JETP* (English Transl.) **5**, 1101–14; erratum, *Zh. éksp. teor. Fiz.* **34**, 240–2 (1958) (in Russian); *Soviet Phys. JETP* (English Transl.) **7**, 166–8 (1958).

FISCHER, J., NIEDERLE, J., and RACZKA, R. (1966) Generalized spherical functions for the noncompact rotation groups. *J. math. Phys.* **7**, 816–21.

FLODMARK, S. (1959) Electron distribution and energy bands in crystals of metal borides of the type MB_6. *Ark. Fys.* **14**, 513–50.

——— (1961) Symmetry reduction of secular matrices for crystals. *Ark. Fys.* **21**, 89–96.

——— (1963) Theory of symmetry projections in applied quantum mechanics. *Phys. Rev.* **132**, 1343–8.

FLOWER, M., MARCH, N. H., and MURRAY, A. M. (1960) Metallic transitions in ionic crystals: some group theoretical results. *Phys. Rev.* **119**, 1885–8.

FOKKER, A. D. (1933*a*) Crystal symmetry and lattice vibrations. *Physica, Eindhoven* **13**, 1–30 (in German).

——— (1933*b*) Symmetrical vibrations. *Physica, Eindhoven* **13**, 65–76 (in German).

——— (1940) The characteristic phenomena of crystalline environments. I. *Physica, 's Grav.* **7**, 385–412 (in French).

FOLDY, L. L. (1956) Synthesis of covariant particle equations. *Phys. Rev.* **102**, 568–81.

FOLLAND, N. O. and BASSANI, F. (1968) Selection rules and Mg_2Si, Mg_2Ge and Mg_2Sn. *J. Phys. Chem. Solids* **29**, 281–90.

FOX, K. and OZIER, I. (1970) Construction of tetrahedral harmonics. *J. chem. Phys.* **52**, 5044–56.

FRANK-KAMENETSKII, V. A. and SHAFRANOVSKII, I. I. (1961) The law of crystallographic limits and the principle of closest packing. *Kristallografiya* **6**, 892–900 (in Russian); *Soviet Phys. Crustallogr.* (English Transl.) **6**, 720–6 (1962).

FREI, V. (1966) Time reversal in crystals of low point symmetry. *Czech. J. Phys.* **16**, 207–27.

——— (1967*a*) The symmetry properties of the energy bands of α- and β-quartz. I. Representations of the space groups D_3^4 and D_3^6 (α-quartz) and D_6^4 and D_6^5 (β-quartz). *Czech. J. Phys.* **17**, 147–66.

——— (1967*b*) The symmetry properties of the energy bands of α- and β-quartz. II. The band spectra of α- and β-quartz and their mutual relation. *Czech. J. Phys.* **17**, 233–48.

——— (1967*c*) On energy bands of crystals with lowered point symmetry. *Phys. Status Solidi* **22**, 381–90.

——— and VELICKÝ, B. (1965) On the band structure of CdSb. *Czech. J. Phys.* **15**, 43–58.

FROBENIUS, F. G. (1896*a*) On group characters. *Sber. preuss. Akad. Wiss.* 985–1021 (in German).

——— (1896*b*) On the prime factors of a group determinant. *Sber. preuss. Akad. Wiss.* 1343–82 (in German).

——— (1898) On the relations between the characters of groups and of their subgroups. *Sber. preuss. Akad. Wiss.* 501–15 (in German).

——— (1911*a*) On the undecomposable discrete motion groups. *Sber. preuss. Akad. Wiss.* 654–65 (in German).

——— (1911*b*) Group theoretic derivation of the 32 crystal classes. *Sber. preuss. Akad. Wiss.* 681–91 (in German).

——— and SCHUR, I. (1906*a*) On the real representations of finite groups. *Sber. preuss. Akad. Wiss.* 186–208 (in German).

——— ——— (1906*b*) On the equivalence of groups of linear substitutions. *Sber. preuss. Akad. Wiss.* 209–17 (in German).

(See also Sierre (1968) for the collected papers of F. G. Frobenius.)

FUMI, F. (1947*a*) On the matrix operators of macroscopic symmetry. *Atti Accad. naz. Lincei Mem.* Classe Sci. Fis. Mat. Nat. **3**, 101–9 (in Italian).

——— (1947*b*) Axes of composite symmetry and matrix operators of improper rotations. *Atti Accad. naz. Lincei Mem.* Classe Sci. Fis. Mat. Nat. **3**, 109–114 (in Italian).

——— (1947*c*) Analytic representation of the crystal lattices. *Atti Accad. naz. Lincei Mem.* Classe Sci. Fis. Mat. Nat. **3**, 370–5 (in Italian).

——— (1947*d*) Elementary Bravais cells and Seitz' primitive translations. *Atti. Accad. naz. Lincei Mem.* Classe Sci. Fis. Mat. Nat. **3**, 376–80 (in Italian).

——— (1952*a*) Matter tensors in symmetrical systems. *Nuovo Cim.* **9**, 739–56.

——— (1952*b*) Physical properties of crystals: the direct-inspection method. *Acta crystallogr.* **5**, 44–8.

——— (1952*c*) The direct-inspection method in systems with a principal axis of symmetry. *Acta crystallogr.* **5**, 691–4 (see also Wondratschek (1953)).

GABRIEL, J. R. (1964) New methods for reduction of group representations using an extension of Schur's lemma. *J. math. Phys.* **5**, 494–504.

——— (1965) Reduction of group representations. *J. chem. Phys.* **43**, S265–7.

——— (1968) New methods for reduction of group representations. II. *J. math. Phys.* **9**, 973–6.

GABRIEL, J. R. (1969*a*) New methods for reduction of group representations. III. *J. math. Phys.* **10**, 1789–95.

——— (1969*b*) New methods for reduction of group representations. IV. *J. math. Phys.* **10**, 1932–4.

GADOLIN, A. (1869) Derivation of all crystallographic systems and their subdivision by a single general principle. *Vseross. min. Obshch., Zap.* **4**, 112–94 (in Russian).

——— (1871) Memoir on the deduction from a single principle of all the crystal systems with their subdivisions. *Acta Soc. Sci. fenn.* **9**, 1–71 (in French).

GALYARSKII, É. I. (1967) Conical groups of similarity symmetry and of antisymmetry of a different kind. *Kristallografiya* **12**, 202–7 (in Russian); *Soviet Phys. Crystallogr.* (English Transl.) **12**, 169–74 (1967).

——— and ZAMORZAEV, A. M. (1963*a*) Point symmetry groups and a different kind of antisymmetry. *Kristallografiya* **8**, 94–101 (in Russian); *Soviet Phys. Crystallogr.* (English Transl.) **8**, 68–75 (1963).

——— ——— (1963*b*) Similarity symmetric and antisymmetric groups. *Kristallografiya* **8**, 691–8 (in Russian); *Soviet Phys. Crystallogr.* (English Transl.) **8**, 553–8 (1964).

——— ——— (1965) A complete derivation of crystallographic stem groups of symmetry and different types of antisymmetry. *Kristallografiya* **10**, 147–54 (in Russian); *Soviet Phys. Crystallogr.* (English Transl.) **10**, 109–15 (1965).

GAMBA, A. (1952) Remarks on the applications of group theory to quantum physics. *Revue scient., Paris* **90**, 11–24 (in French).

——— (1953) The number of independent components of tensors in symmetrical systems. *Nuovo Cim.* **10**, 1343–4.

GANZHORN, K. (1952) Quantum mechanics of the body-centred cubic transition metals. *Z. Naturf.* **7a**, 291–2 (in German).

GARRIDO, J. (1945) On the classification of crystalline forms. *Anais Fac. Ciênc. Porto* **30**, 22–44 (in French).

GASHIMZADE, F. M. (1960*a*) Investigation of the symmetry properties of the energy bands of crystals of the SnSe and Sb_2S_3 type. *Fizika tverd. Tela* **2**, 2070–6 (in Russian); *Soviet Phys. solid St.* (English Transl.) **2**, 1856–62 (1961).

——— (1960*b*) Symmetry of energy bands in TlSe type crystals. *Fizika tverd. Tela* **2**, 3040–4 (in Russian); *Soviet Phys. solid St.* (English Transl.) **2**, 2700–4 (1961); errata, *Fizika tverd. Tela* **4**, 2282–3 (1962) (in Russian); *Soviet Phys. solid St.* (English Transl.) **4**, 1671 (1963).

GAUNT, J. A. (1928) The triplets of helium. *Phil. Trans. R. Soc.* **A228**, 151–96.

GAY, J. G., ALBERS, W. A., and ARLINGHAUS, K. J. (1968) Irreducible representations of the little groups of D_{4h}^{14}. *J. Phys. Chem. Solids* **29**, 1449–59.

GEL'FAND, I. M. and GRAEV, M. I. (1953) Unitary representations of the real unimodular group (principal non-degenerate series). *Izv. Akad. Nauk SSSR* Ser. Mat. **17**, 189–248 (in Russian); *Am. math. Soc. Transl.* **2**, 147–205 (1956).

——— MINLOS, R. A., and SHAPIRO, Z. YA. (1963) *Representations of the rotation and Lorentz groups and their applications*. Pergamon Press, Oxford.

——— and SHAPIRO, Z. YA. (1952) Representations of the group of rotations in three-dimensional space and their applications. *Usp. mat. Nauk* **7**, 3–117 (in Russian); *Am. math. Soc. Transl.* **2**, 207–316 (1956).

GHATE, P. B. (1965) Fourth-order elastic coefficients for some crystal classes. *Indian J. Phys.* **39**, 257–64.

GLASSER, M. L. (1959) Symmetry properties of the wurtzite structure. *Physics Chem. Solids* **10**, 229–41.

GLÜCK, M., GUR, Y., and ZAK, J. (1967) Double representations of space groups. *J. math. Phys.* **8**, 787–90.

GOODENOUGH, J. B. (1958) Suggestion concerning the role of wave-function symmetry in transition metals and their alloys. *J. appl. Phys.* **29**, 513–15.

GORELICK, V. S. (1968) Classification of fundamental vibrations in crystals according to the irreducible representations of their space groups. *Kristallografiya* **13**, 696–8 (in Russian); *Soviet Phys. Crystallogr.* (English Transl.) **13**, 591–3.

GORZKOWSKI, W. (1963*a*) Space group of pyrite T_h^6 (*Pa*3). *Phys. Status Solidi* **3**, 599–614.

——— (1963*b*) Representation of space group O_h^3. *Phys. Status Solidi* **3**, 910–21.

——— (1964*a*) Space group of marcasite D_{2h}^{12} (*Pnnm*). *Acta phys. pol.* **25**, 527–42.

——— (1964*b*) Compatibility relations and selection rules for the dipole electric transitions in crystals with symmetry O_h^3. *Phys. Status Solidi* **6**, 521–8.

GRANZOW, K. D. (1963) N-dimensional total orbital angular-momentum operator. *J. math. Phys.* **4**, 897–900.

——— (1964) *N*-dimensional total orbital angular-momentum operator. II. Explicit representations. *J. math. Phys.* **5**, 1474–7.

GREENAWAY, D. L. and HARBEKE, G. (1965) Band structure of bismuth telluride, bismuth selenide and their respective alloys. *J. Phys. Chem. Solids* **26**, 1585–604.

GRIFFITH, J. S. (1961) *The theory of transition metal ions.* Cambridge University Press.

——— (1962) *The irreducible tensor method for molecular symmetry groups.* Prentice-Hall, London.

GRIMLEY, T. B. (1958) The electronic structure of crystals having the sodium chloride type of lattice. *Proc. phys. Soc.* **71**, 749–57.

GUBANOV, A. I. and GASHIMZADE, F. M. (1959) Investigation of the symmetry of the electron energy zones in $CdIn_2Se_4$-type crystals. *Fizika tverd. Tela* **1**, 1411–16 (in Russian); *Soviet Phys. solid St.* (English Transl.) **1**, 1294–8 (1960).

——— ——— (1960) Structure of the energy bands in semiconductors of the $CdIn_2Se_4$ type. *Fizika tverd. Tela* **2**, 255–60 (in Russian); *Soviet Phys. solid St.* (English Transl.) **2**, 236–41 (1960).

——— and SHUR, M.S. (1965) On the dynamics of crystals having the rutile structure. *Fizika tverd. Tela* **7**, 2626–33 (in Russian); *Soviet Phys. solid St.* (English Transl.) **7**, 2124–30 (1966).

GUCCIONE, R. (1963) On the construction of the magnetic space groups. *Phys. Lett.* **5**, 105–7.

GÜNZBURG, A. M. (1929) The fundamentals of the study of the symmetry of lines and in planes. *Z. Kristallogr. Kristallgeom.* **71**, 81–94 (in German).

HAAS, C. (1965) Phase transitions in ferroelectric and antiferroelectric crystals. *Phys. Rev.* **140**, A863–8.

HAGEDORN, R. (1959) Note on symmetry operations in quantum mechanics. *Nuovo Cim.* **12** (Suppl.), 73–86.

HAHN, H. and BIEM, W. (1963) Lattice vibrations in molecular crystals. *Phys. Status Solidi* **3**, 1911–26 (in German).

HALFORD, R. S. (1946) Motions of molecules in condensed systems. I. Selection rules, relative intensities, and orientation effects for Raman and infrared spectra. *J. chem. Phys.* **14**, 8–15 (see also Couture (1947)).

HALL, G. G. (1950) The molecular orbital theory of chemical valency. VI. Properties of equivalent orbitals. *Proc. R. Soc.* **A202**, 336–44.

—— (1967) *Applied group theory*. Longmans, London.

HALL, M. (1959) *The theory of groups*. Macmillan, New York.

—— and SENIOR, J. K. (1964) *The groups of order* 2^n ($n \leqslant 6$). Macmillan, New York.

HAM, F. S. (1955) The quantum defect method. *Solid St. Phys.* **1**, 127–92.

—— and SEGALL, B. (1961) Energy bands in periodic lattices—Green's function method. *Phys. Rev.* **124**, 1786–96.

HAMERMESH, M. (1960) Galilean invariance and the Schrödinger equation. *Ann. Phys.* **9**, 518–21.

—— (1962) *Group theory and its application to physical problems*. Addison-Wesley, Reading, Mass.

HARKER, D. (1969) Diamorphism: A new phenomenon connected with colored symmetry. *Acta crystallogr.* **A25**, Suppl., S4.

HARTER, W. G. (1969) Algebraic theory of ray representations of finite groups. *J. math. Phys.* **10**, 739–52.

HAUPTMAN, H. and KARLE, J. (1956) Structure invariants and semiinvariants for non-centrosymmetric space groups. *Acta crystallogr.* **9**, 45–55.

HAÜY, R. J. (1815*a*) Memoir on a law of crystallisation called law of symmetry. *J. Mines* **37**, 215–35 (in French).

—— (1815*b*) First continuation of the memoir on the law of symmetry. *J. Mines* **37**, 347–68 (in French).

—— (1815*c*) Second continuation of the memoir on the law of symmetry. *J. Mines* **38**, 5–34 (in French).

—— (1815*d*) Third continuation of the memoir on the law of symmetry. *J. Mines* **38**, 161–74 (in French).

HAVEN, Y. and VAN SANTEN, J. H. (1954) Electrically and elastically active relaxation modes. *J. chem. Phys.* **22**, 1146–7.

HAYES, W. D. (1946) Transformation groups of the thermodynamic variables. *Q. appl. Math.* **4**, 227–32.

HEADING, J. (1958) *Matrix theory for physicists*. Longmans, London.

HEARMON, R. F. S. (1946) The elastic constants of anisotropic materials. *Rev. mod. Phys.* **18**, 409–40.

—— (1956) The elastic constants of anisotropic materials. II. *Adv. Phys.* **5**, 323–82.

HEESCH, H. (1929*a*) Structure theory of plane symmetry groups. *Z. Kristallogr. Kristallgeom.* **71**, 95–102 (in German).

HEESCH, H. (1929*b*) Systematic structure theory. II. *Z. Kristallogr. Kristallgeom.* **72**, 177–201 (in German).

——— (1930*a*) Systematic structure theory. III. On the four-dimensional groups of the three-dimensional space. *Z. Kristallogr. Kristallgeom.* **73**, 325–45 (in German).

——— (1930*b*) Systematic structure theory. IV. On the symmetry of the second kind in continua and semicontinua. *Z. Kristallogr. Kristallgeom.* **73**, 346–56 (in German).

HEINE, V. (1960) *Group theory in quantum mechanics*. Pergamon Press, Oxford.

HEISENBERG, W. (1925) On the quantum interpretation of kinematical and mechanical relationships. *Z. Phys.* **33**, 879–93 (in German).

HEJDA, B. (1967) The symmetrized combinations of plane waves for wurtzite. *Phys. Status Solidi* **24**, 659–67.

HELLWEGE, K. H. (1949*a*) Electron terms and radiation from atoms in crystals. I. Term splitting and electric-dipole radiation. *Annln Phys.* **4**, 95–126 (in German).

——— (1949*b*) Electron terms and radiation from atoms in crystals. V. Cubic crystals. *Annln Phys.* **4**, 150–60 (in German).

HENRY, N. F. M. and LONSDALE, K. (1965) *International tables for X-ray crystallography*. Vol. I. *Symmetry groups*. Kynoch, Birmingham (see also Kasper and Lonsdale (1959); MacGillavry, Rieck and Lonsdale (1962)).

HERBUT, I. F. (1967) Basic algebra of antilinear operators and some applications. *J. math. Phys.* **8**, 1345–54.

HERMAN, F. (1958) Theoretical investigation of the electronic energy band structure of solids. *Rev. mod. Phys.* **30**, 102–21.

HERMANN, C. (1928*a*) Systematic structure theory. I. A new space-group symbolism. *Z. Kristallogr. Kristallgeom.* **68**, 257–87 (in German).

——— (1928*b*) Systematic structure theory. II. Derivation of the 230 space groups from the basic vectors. *Z. Kristallogr. Kristallgeom.* **69**, 226–49 (in German).

——— (1928*c*) Systematic structure theory. III. Chain- and net-groups. *Z. Kristallogr. Kristallgeom.* **69**, 250–70 (in German).

——— (1929) Systematic structure theory. IV. Subgroups. *Z. Kristallogr. Kristallgeom.* **69**, 533–55 (in German).

——— (1931) Remarks on the previous paper by Ch. Mauguin. *Z. Kristallogr. Kristallgeom.* **76**, 559–61 (in German) (see Mauguin (1931)).

——— (1934) Tensors and crystal symmetry. *Z. Kristallogr. Kristallgeom.* **89**, 32–48 (in German).

——— (1949) Crystallography in spaces with arbitrary numbers of dimensions. I. The symmetry operations. *Acta crystallogr.* **2**, 139–45 (in German).

——— (1960) The interpretation and nomenclature of coloured space groups. *Acta crystallogr.* **13**, 1084–5.

HERRING, C. (1937*a*) Effect of time-reversal symmetry on energy bands of crystals. *Phys. Rev.* **52**, 361–5; reprinted, Meijer (1964).

——— (1937*b*) Accidental degeneracy in the energy bands of crystals. *Phys. Rev.* **52**, 365–73.

——— (1940) A new method for calculating wave functions in crystals. *Phys. Rev.* **57**, 1169–77.

——— (1942) Character tables for two space groups. *J. Franklin Inst.* **233**, 525–43; reprinted: Knox and Gold (1964).

HERZBERG, G. (1945) *Molecular spectra and molecular structure*. I. *Spectra of diatomic molecules*. Van Nostrand, Princeton, N.J.

——— (1950) *Molecular spectra and molecular structure*. II. *Infrared and Raman spectra of polyatomic molecules*. Van Nostrand, Princeton, N.J.

HERZFELD, C. M. and MEIJER, P. H. E. (1961) Group theory and crystal field theory. *Solid St. Phys*. **12**, 1–91

HESS, E. (1896) The celebration of the centenary of the birth of J. F. C. Hessel (27 April 1796). *Neues Jb. Miner., Geol. Paläont*. 107–22 (in German).

HESSEL, J. F. C. (1830) Crystal, in *Physical Dictionary* (edited by J. S. T. Gehler), Vol. 5, pp. 1023–340. Schwickert, Leipzig (in German); article reprinted, Engelmann, Leipzig (1897).

HIGMAN, B. (1955) *Applied group-theoretic and matrix methods*. Clarendon Press, Oxford; reprinted: Dover, New York (1964).

HILBERT, D. and COHN-VOSSEN, S. (1952) *Geometry and the imagination*. Chelsea, New York.

HILL, E. L. (1954) The theory of vector spherical harmonics. *Am. J. Phys*. **22**, 211–14.

HILTON, H. (1902) A comparison of various notations employed in 'Theories of crystal-structure' and a revision of the 230 groups of movements. *Phil. Mag*. **3**, 203–12.

——— (1903) *Mathematical crystallography and the theory of groups of movements*. Clarendon Press, Oxford; reprinted: Dover, New York (1963).

——— (1908) *Introduction to the theory of groups of finite order*. Clarendon Press, Oxford.

HOBSON, E. W. (1931) *The theory of spherical and ellipsoidal harmonics*. Cambridge University Press.

HODGKINSON, J. (1935) Harmonic functions with polyhedral symmetry. *J. Lond. math. Soc*. **10**, 221–6.

HOERNI, J. A. (1961) Application of the free-electron theory to three-dimensional networks. *J. chem. Phys*. **34**, 508–13.

HOLSER, W. T. (1958) Point groups and plane groups in a two-sided plane and their subgroups. *Z. Kristallogr. Kristallgeom*. **110**, 266–81.

——— (1961) Classification of symmetry groups. *Acta crystallogr*. **14**, 1236–42.

HOPFIELD, J. J. (1960) Fine structure in the optical absorption edge of anisotropic crystals. *Physics Chem. Solids* **15**, 97–107.

HORNIG, D. F. (1948) The vibrational spectra of molecules and complex ions in crystals. I. General theory. *J. chem. Phys*. **16**, 1063–76.

HOUGEN, J. T. (1962) Classification of rotational energy levels for symmetric-top molecules. *J. chem. Phys*. **37**, 1433–41.

——— (1963) Classification of rotational energy levels. II. *J. chem. Phys*. **39**, 358–65.

HOUSTON, W. V. (1948) Normal vibrations of a crystal lattice. *Rev. mod. Phys*. **20**, 161–5.

HOWARTH, D. J. and JONES, H. (1952) The cellular method of determining electronic wave functions and eigenvalues in crystals, with applications to sodium. *Proc. phys. Soc*. **A65**, 355–68.

HSIEH, H. T. and CHEN, H. S. (1964) Space group selection rules. *Acta phys. sin*. **20**, 970–90 (in Chinese).

HSÜ, C. C. and HSIEH, H. T. (1965) The matrix elements of space group operators. *Acta phys. sin*. **21**, 802–16 (in Chinese); *Chin. J. Phys*. (English Transl.) **21**, 1019–35 (1966).

HUND, F. (1936) Symmetry of a crystal lattice and the states of its electrons. *Z. Phys.* **99**, 119–36 (in German).

HURLEY, A. C. (1951) Finite rotation groups and crystal classes in four dimensions. *Proc. Camb. phil. Soc. math. phys. Sci.* **47**, 650–61.

——— (1966*a*) Ray representations of point groups and the irreducible representations of space groups and double space groups. *Phil. Trans. R. Soc.* **A260**, 1–36.

——— (1966*b*) Finite rotation groups and crystal classes in four dimensions. II. Revised tables and projection of groups of antisymmetry in three dimensions. In *Quantum theory of atoms, molecules, and the solid state. A tribute to John C. Slater* (edited by P.-O. Löwdin), pp. 571–86. Academic Press, New York.

——— NEUBÜSER, J. and WONDRATSCHEK, H. (1967) Crystal classes of four-dimensional space *R*4. *Acta crystallogr.* **22**, 605.

HUTCHINGS, M. T. (1964) Point-charge calculations of energy levels of magnetic ions in crystalline electric fields. *Solid St. Phys.* **16**, 227–73.

INDENBOM, V. L. (1959) Relation of the antisymmetry and color symmetry groups to one-dimensional representations of the ordinary symmetry groups. Isomorphisms of the Shubnikov and space groups. *Kristallografiya* **4**, 619–21 (in Russian); *Soviet Phys. Crystallogr.* (English Transl.) **4**, 578–80 (1960).

——— (1960*a*) Phase transitions without change in the number of atoms in the unit cell of the crystal. *Kristallografiya* **5**, 115–25 (in Russian); *Soviet Phys. Crystallogr.* (English Transl.) **5**, 106–15 (1960).

——— (1960*b*) Irreducible representations of the magnetic groups and allowance for magnetic symmetry. *Kristallografiya* **5**, 513–16 (in Russian); *Soviet Phys. Crystallogr.* (English Transl.) **5**, 493–6 (1961).

——— BELOV, N. V. and NERONOVA, N. N. (1960) The color-symmetry point groups (color classes). *Kristallografiya* **5**, 497–500 (in Russian); *Soviet Phys. Crystallogr.* (English Transl.) **5**, 477–81 (1961).

JAEGER, F. M. (1917) *Lectures on the principle of symmetry and its applications in all natural sciences*. Elsevier, Amsterdam.

JAFFÉ, H. H. and ORCHIN, M. (1965) *Symmetry in chemistry*. Wiley, New York.

JAHN, H. A. (1935) Rotation and vibration of the methane molecule. *Annln Phys.* **23**, 529–56 (in German).

——— (1938) Stability of polyatomic molecules in degenerate electronic states. II. Spin degeneracy. *Proc. R. Soc.* **A164**, 117–31.

——— (1949) Note on the Bhagavantam–Suryanarayana method of enumerating the physical constants of crystals. *Acta crystallogr.* **2**, 30–3 (see also Venkataryudu and Krishnamurty (1952)).

——— and TELLER, E. (1937) Stability of polyatomic molecules in degenerate electronic states. I. Orbital degeneracy. *Proc. R. Soc.* **A161**, 220–35; reprinted, Knox and Gold (1964), Cracknell (1968*b*) (see also Jahn (1938)).

JAMES, H. M. (1949*a*) Energy bands and wave functions in periodic potentials. *Phys. Rev.* **76**, 1602–10.

——— (1949*b*) Electronic states in perturbed periodic systems. *Phys. Rev.* **76**, 1611–24.

JANNER, A. (1966) On Bravais classes of magnetic lattices. *Helv. phys. Acta* **39**, 665–82.

——— and ASCHER, E. (1969*a*) Bravais classes of two-dimensional relativistic lattices. *Physica, 's Grav.* **45**, 33–66.

——— ——— (1969*b*) Relativistic crystallographic point groups in two dimensions. *Physica, 's Grav.* **45**, 67–85.

JANSEN, L. and BOON, M. (1967) *Theory of finite groups. Applications in physics.* North-Holland, Amsterdam.

JANSSEN, T. (1969) Crystallographic groups in space and time. III. Four-dimensional Euclidean crystal classes corresponding to generalized magnetic point groups. *Physica, 's Grav.* **42**, 71–92.

——— JANNER, A. and ASCHER, E. (1969*a*) Crystallographic groups in space and time. I. General definitions and basic properties. *Physica, 's Grav.* **41**, 541–65.

——— ——— ——— (1969*b*) Crystallographic groups in space and time. II. Central extensions. *Physica, 's Grav.* **42**, 41–70.

JASWON, M. A. (1959) *Studies in crystal physics.* Butterworths, London.

——— (1965) *An introduction to mathematical crystallography.* Longmans, London.

JOHNSON, F. A. and LOUDON, R. (1964) Critical-point analysis of the phonon spectra of diamond, silicon and germanium. *Proc. R. Soc.* **A281**, 274–90.

JOHNSTON, D. F. (1958) Space-group operations and time-reversal for a Dirac electron in a crystal field. *Proc. R. Soc.* **A243**, 546–54.

——— (1960) Group theory in solid state physics. *Rep. Prog. Phys.* **23**, 66–153.

JONES, H. (1934*a*) The theory of alloys in the γ-phase. *Proc. R. Soc.* **A144**, 225–34.

——— (1934*b*) Applications of the Bloch theory to the study of alloys and of the properties of bismuth. *Proc. R. Soc.* **A147**, 396–417.

——— (1960) *The theory of Brillouin zones and electronic states in crystals.* North-Holland, Amsterdam.

JORDAN, C. (1868) Memoir on the groups of movements. *Annali Mat. pura appl.* **2**, 167–215 (in French).

——— (1869) Memoir on the groups of movements. *Annali Mat. pura appl.* **2**, 322–45 (in French).

JØRGENSEN, C. K. (1962*a*) Relevant and irrelevant symmetry components. Are the Bloch energy bands the best one-electron functions? *Phys. Status Solidi* **2**, 1146–50.

——— (1962*b*) *Orbitals in atoms and molecules.* Academic Press, London.

JOSHUA, S. J. (1970) Dispersion relations and selection rules for magnons in antiferromagnetic NiF_2. *Phys. Status Solidi* **41**, 309–15.

——— and CRACKNELL, A. P. (1969) Space-group corepresentations and magnon dispersion relations in NiF_2. *J. Phys. C.* (*Solid St. Phys.*) **2**, 24–36.

JUDD, B. R. (1963) *Operator techniques in atomic spectroscopy.* McGraw-Hill, New York.

JURETSCHKE, H. J. (1952) Transformation properties of the physical constants of crystals. *Acta crystallogr.* **5**, 148–9.

——— (1955) Symmetry of galvanomagnetic effects in antimony. *Acta crystallogr.* **8**, 716–22.

KAHAN, T. (1965) *Theory of groups in classical and quantum physics.* Oliver and Boyd, Edinburgh.

KAMBER, F. and STRAUMANN, N. (1964) Group extensions in quantum theory. *Helv. phys. Acta* **37**, 563–84 (in German).

KARAVAEV, G. F. (1964) Selection rules for indirect transitions in crystals. *Fizika tverd Tela* **6**, 3678–83 (in Russian); *Soviet Phys. solid St.* (English Transl.) **6**, 2943–8 (1965).

——— (1966) Selection rules for indirect transitions in crystals with a zinc blende structure. *Izv. vyssh. ucheb. Zaved.*, Fiz. (3) 64–70 (in Russian); *Soviet Phys. J.* (English Transl.) **9**, (3) 37–42 (1966).

——— KUDRYAVTSEVA, N. V. and CHALDYSHEV, V. A. (1962) Structure of the electron energy spectrum in Th_3P_4-type crystals. *Fizika tverd. Tela* **4**, 3471–81 (in Russian); *Soviet Phys. solid St.* (English Transl.) **4**, 2540–9 (1963).

KARPUS, A. S. and BATARŪNAS, I. B. (1961) On complex band structure of semiconductors of Sb_2S_3 type. *Liet. Fiz. Rink.* **1**, 315–28 (in Russian).

KASPER, J. S. and LONSDALE, K. (1959) *International tables for X-ray crystallography*. Vol. II. *Mathematical tables*. Kynoch, Birmingham (see also Henry and Lonsdale (1965), MacGillavry, Rieck, and Lonsdale (1962)).

KATIYAR, R. S. (1970) Dynamics of the rutile structure. I. Space group representations and the normal mode analysis. *J. Phys. C (Solid St. Phys.)* **3**, 1087–96.

——— KRISHNAMURTHY, N. and KRISHNAN, R. S. (1969) Group theoretical analysis of lattice vibrations in caesium bromide. *J. Indian Inst. Sci.* **51**, 13–20.

KATZ, E. (1952) Splitting of bands in solids. *Phys. Rev.* **85**, 495–6.

KERNER, E. H. (1956) The band structure of mixed linear lattices. *Proc. phys. Soc.* **A69**, 234–44.

KHARTSIEV, V. E. (1962) An investigation of the symmetry of energy zones for CdSb and ZnSb alloys. *Fizika tverd. Tela* **4**, 983–91 (in Russian); *Soviet Phys. solid St.* (English Transl.) **4**, 721–8 (1962).

KILLINGBECK, J. (1969) An application of Löwdin projection operators to spin–orbit coupling theory. *Phys. Status Solidi* **36**, K67–9.

KIMBALL, G. E. (1940) Directed valence. *J. chem. Phys.* **8**, 188–98.

KITTEL, C. (1956) *An introduction to solid state physics*. Wiley, New York.

——— (1963) *Quantum theory of solids*. Wiley, New York.

KITZ, A. (1965*a*) The irreducible representations of space groups and the ray representations of crystallographic point groups. *Phys. Status Solidi* **8**, 813–29 (in German).

——— (1965*b*) Symmetry groups of spin distributions. *Phys. Status Solidi* **10**, 455–66 (in (German).

KLAUDER, L. T. and GAY, J. G. (1968) Note on Zak's method for constructing representations of space groups. *J. math. Phys.* **9**, 1488–9.

KLEIN, M. J. (1952) On a degeneracy theorem of Kramers. *Am. J. Phys.* **20**, 65–71.

KLEINER, W. H. (1966) Space–time symmetry of transport coefficients. *Phys. Rev.* **142**, 318–26.

——— (1967) Space–time symmetry restrictions on transport coefficients. II. Two theories compared. *Phys. Rev.* **153**, 726–7.

——— (1969) Space–time symmetry restrictions on transport coefficients. III. Thermogalvanomagnetic coefficients. *Phys. Rev.* **182**, 705–9.

KNOX, R. S. and GOLD, A. (1964) *Symmetry in the solid state*. Benjamin, New York.

KOHN, W. (1952) Variational methods for periodic lattices. *Phys. Rev.* **87**, 472–81.

KOHN, W. (1959) Analytic properties of Bloch waves and Wannier functions. *Phys. Rev.* **115**, 809–21.

——— and ROSTOCKER, N. (1954) Solution of Schrödinger's equation in periodic lattices with an application to metallic lithium. *Phys. Rev.* **94**, 1111–20.

KOOPMANS, T. (1934) Ordering of wave functions and eigenenergies to the individual electrons of an atom. *Physica, 's Grav.* **1**, 104–13 (in German).

KOPTSIK, V. A. (1957) On the superposition of symmetry groups in crystal physics. *Kristallografiya* **2**, 99–107 (in Russian); *Soviet Phys. Crystallogr.* (English Transl.) **2**, 95–103 (1957).

——— (1960) Polymorphic phase transitions and symmetry. *Kristallografiya* **5**, 932–43 (in Russian); *Soviet Phys. Crystallogr.* (English Transl.) **5**, 889–98 (1961).

——— (1963) Polar-polar and polar-neutral crystal structures *Kristallografiya* **8**, 319–27 (in Russian); *Soviet Phys. Crystallogr.* (English Transl.) **8**, 249–54 (1963).

——— (1966) *Shubnikov groups. Handbook on the symmetry and physical properties of crystal structures.* University Press, Moscow (in Russian).

——— (1967*a*) A general sketch of the development of the theory of symmetry and its applications in physical crystallography over the last 50 years. *Kristallografiya* **12**, 755–74 (in Russian); *Soviet Phys. Crystallogr.* (English Transl.) **12**, 667–83 (1968).

——— (1967*b*) Describing three-dimensional periodic magnetic structures by Shubnikov groups. *Kristallografiya* **12**, 826–30 (in Russian); *Soviet Phys. Crystallogr.* (English Transl.) **12**, 723–7 (1968).

——— and SIROTIN, YU. I. (1961) The symmetry of piezoelectric and elastic tensors and the symmetry of the physical properties of crystals. *Kristallografiya* **6**, 766–8 (in Russian); *Soviet Phys. Crystallogr.* **6**, 612–14 (1962).

KORRINGA, J. (1947) On the calculation of the energy of a Bloch wave in a metal. *Physica, 's Grav.* **13**, 392–400.

KOSTER, G. F. (1953) Localized functions in molecules and crystals. *Phys. Rev.* **89**, 67–77.

——— (1955) Extension of Hund's rule. *Phys. Rev.* **98**, 514–15.

——— (1957) Space groups and their representations. *Solid St. Phys.* **5**, 173–256.

——— (1958) Matrix elements of symmetric operators. *Phys. Rev.* **109**, 227–31.

——— (1962) Symmetry properties of the gallium energy bands. Effect of spin–orbit interaction. *Phys. Rev.* **127**, 2044–5.

——— DIMMOCK, J. O., WHEELER, R. G., and STATZ, H. (1963) *Properties of the thirty-two point groups.* M.I.T. Press, Cambridge, Mass.

——— and STATZ, H. (1959) Method of treating Zeeman splittings of paramagnetic ions in crystalline fields. *Phys. Rev.* **113**, 445–54 (see also Statz and Koster (1959)).

KOTANI, M. (1937) Note on the theory of electronic states of polyatomic molecules. *Proc. phys.-math. Soc. Japan* **19**, 460–70.

KOVALEV, O. V. (1960*a*) Possible changes of the symmetry O_h^7 during a second-order phase transition. *Fizika tverd. Tela* **2**, 1220–1 (in Russian); *Soviet Phys. solid St.* (English Transl.) **2**, 1105–6 (1960).

———(1960*b*) Concerning the degeneracy of energy levels in crystals. *Fizika tverd. Tela* **2**, 2557–66 (in Russian); *Soviet Phys. solid St.* (English Transl.) **2**, 2279–88 (1961).

——— (1961*a*) Characters of single-valued irreducible representations of space groups of a hexagonal system. I. *Ukr. fiz. Zh.* **6**, 353–65 (in Russian).

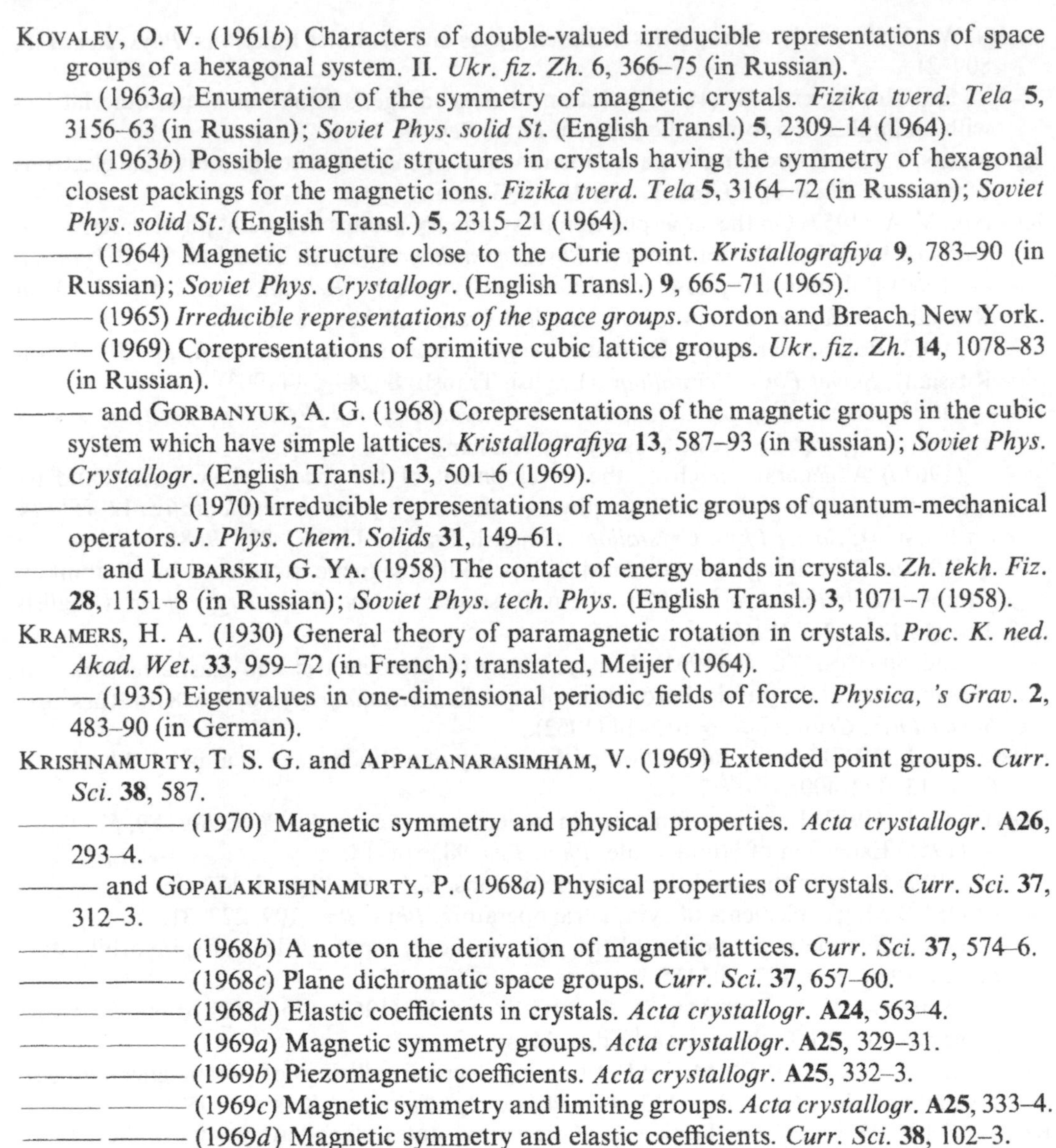

Kovalev, O. V. (1961*b*) Characters of double-valued irreducible representations of space groups of a hexagonal system. II. *Ukr. fiz. Zh.* 6, 366–75 (in Russian).

——— (1963*a*) Enumeration of the symmetry of magnetic crystals. *Fizika tverd. Tela* **5**, 3156–63 (in Russian); *Soviet Phys. solid St.* (English Transl.) **5**, 2309–14 (1964).

——— (1963*b*) Possible magnetic structures in crystals having the symmetry of hexagonal closest packings for the magnetic ions. *Fizika tverd. Tela* **5**, 3164–72 (in Russian); *Soviet Phys. solid St.* (English Transl.) **5**, 2315–21 (1964).

——— (1964) Magnetic structure close to the Curie point. *Kristallografiya* **9**, 783–90 (in Russian); *Soviet Phys. Crystallogr.* (English Transl.) **9**, 665–71 (1965).

——— (1965) *Irreducible representations of the space groups.* Gordon and Breach, New York.

——— (1969) Corepresentations of primitive cubic lattice groups. *Ukr. fiz. Zh.* **14**, 1078–83 (in Russian).

——— and Gorbanyuk, A. G. (1968) Corepresentations of the magnetic groups in the cubic system which have simple lattices. *Kristallografiya* **13**, 587–93 (in Russian); *Soviet Phys. Crystallogr.* (English Transl.) **13**, 501–6 (1969).

——— ——— (1970) Irreducible representations of magnetic groups of quantum-mechanical operators. *J. Phys. Chem. Solids* **31**, 149–61.

——— and Liubarskii, G. Ya. (1958) The contact of energy bands in crystals. *Zh. tekh. Fiz.* **28**, 1151–8 (in Russian); *Soviet Phys. tech. Phys.* (English Transl.) **3**, 1071–7 (1958).

Kramers, H. A. (1930) General theory of paramagnetic rotation in crystals. *Proc. K. ned. Akad. Wet.* **33**, 959–72 (in French); translated, Meijer (1964).

——— (1935) Eigenvalues in one-dimensional periodic fields of force. *Physica, 's Grav.* **2**, 483–90 (in German).

Krishnamurty, T. S. G. and Appalanarasimham, V. (1969) Extended point groups. *Curr. Sci.* **38**, 587.

——— ——— (1970) Magnetic symmetry and physical properties. *Acta crystallogr.* **A26**, 293–4.

——— and Gopalakrishnamurty, P. (1968*a*) Physical properties of crystals. *Curr. Sci.* **37**, 312–3.

——— ——— (1968*b*) A note on the derivation of magnetic lattices. *Curr. Sci.* **37**, 574–6.

——— ——— (1968*c*) Plane dichromatic space groups. *Curr. Sci.* **37**, 657–60.

——— ——— (1968*d*) Elastic coefficients in crystals. *Acta crystallogr.* **A24**, 563–4.

——— ——— (1969*a*) Magnetic symmetry groups. *Acta crystallogr.* **A25**, 329–31.

——— ——— (1969*b*) Piezomagnetic coefficients. *Acta crystallogr.* **A25**, 332–3.

——— ——— (1969*c*) Magnetic symmetry and limiting groups. *Acta crystallogr.* **A25**, 333–4.

——— ——— (1969*d*) Magnetic symmetry and elastic coefficients. *Curr. Sci.* **38**, 102–3.

Krishnan, R. S. (1965) Raman spectra of cubic crystals. *Indian J. pure appl. Phys.* **3**, 424–30.

Kronig, R. de L. and Penney, W. G. (1931) Quantum mechanics of electrons in crystal lattices. *Proc. R. Soc.* **A130**, 499–513.

Kucher, T. I. (1962) Eigenfrequencies of the lattice vibrations of silicon and diamond. *Fizika tverd. Tela* **4**, 992–8 (in Russian); *Soviet Phys. solid St.* (English Transl.) **4**, 729–34 (1962).

Kudryavtseva, N. V. (1965) Possible structure of the energy spectrum of electrons in crystals, taking into account the operation of time reversal. *Fizika tverd. Tela* 7, 998–1007 (in Russian); *Soviet Phys. solid St.* (English Transl.) 7, 803–10 (1965).

KUDRYAVTSEVA, N. V. (1967*a*) The question of investigating the energy spectrum of electrons in a crystal classification of space–time groups. *Izv. vȳssh. ucheb. Zaved.* Fiz. (5) 135–8 (in Russian); *Soviet Phys. J.* (English Transl.) **10**, (5) 74–6 (1967).

——— (1967*b*) Determination of zero slope points. *Fizika tverd. Tela* **9**, 2364–8 (in Russian); *Soviet Phys. solid St.* (English Transl.) **9**, 1850–3 (1968).

——— (1968) Zero-slope points in crystals. *Fizika tverd. Tela* **10**, 1616–21 (in Russian); *Soviet Phys. solid St.* (English transl.) **10**, 1280–3 (1968).

——— (1969) Zero-slope points in crystals. *Fizika tverd. Tela* **11**, 1031–2 (in Russian); *Soviet Phys. solid St.* (English Transl.) **11**, 840 (1969).

——— and CHALDYSHEV, V. A. (1962*a*) Energy spectra of electrons in a single crystal. III. Some properties of the weighted corepresentation. *Izv. vȳssh. ucheb. Zaved.* Fiz. (3) 133–9 (in Russian).

——— ——— (1962*b*) Energy spectra of electrons in a single crystal. IV. *Izv. vȳssh. ucheb. Zaved.* Fiz. (4) 98–106 (in Russian).

——— ——— (1965*a*) Energy spectra of electrons in crystals. VI. Associated corepresentations. *Izv. vȳssh. ucheb. Zaved.* Fiz. (2) 57–64 (in Russian); *Soviet Phys. J.* (English Transl.) (2) 37–42 (1965).

——— ——— (1965*b*) Energy spectra of electrons in crystals. VII. Classification of weighted corepresentations. *Izv. vȳssh. ucheb. Zaved.* Fiz. (3) 104–11 (in Russian); *Soviet Phys. J.* (English Transl.) (3), 71–5 (1965).

——— ——— (1966*a*) Energy spectrum of electrons in a crystal. VIII. Characters of weighted corepresentations. *Izv. vȳssh. ucheb. Zaved.* Fiz. (1) 93–100 (in Russian); *Soviet Phys. J.* (English Transl.) **9** (1), 57–60 (1966).

——— ——— (1966*b*) Energy spectra of electrons in a crystal. IX. Characters of weighted corepresentations of the point groups D_{2h}, D_{4h} and D_{6h}. *Izv. vȳssh. ucheb. Zaved.* Fiz. (4) 108–9 (in Russian); *Soviet Phys. J.* (English Transl.) **9** (4), 66–70 (1966).

——— ——— (1968) The characters of the irreducible non-equivalent weighted corepresentations of some point groups. *Izv. vȳssh. ucheb. Zaved.* Fiz. (11) 23 (in Russian); *Soviet Phys. J.* (English Transl.) **11** in press. (See also Chaldyshev (1961), Chaldyshev and Kudryavtseva (1962), and Chaldyshev, Kudryavtseva, and Karavaev (1963)).

KUNTSEVICH, T. S. and BELOV, N. V. (1968) Geometrical interpretation of point structure symmetry and details of a Bravais lattice in {4}. *Acta crystallogr.* **A24**, 42–51 (in Russian).

——— ——— (1970) Four-dimensional Bravais lattices. *Kristallografiya* **15**, 215–29 (in (Russian); *Soviet Phys. Crystallogr.* (English Transl.) **15**, 180–92.

KYNCH, G. J. (1937) Multiplet structure in a crystalline electric field of cubic symmetry. *Trans. Faraday Soc.* **33**, 1402–18.

LANDAU, L. D. and LIFSHITZ, E. M. (1958*a*) *Quantum mechanics. Non-relativistic theory.* Pergamon Press, Oxford.

——— ———(1958*b*) *Statistical Physics.* Pergamon Press, Oxford.

——— ——— (1960) *Electrodynamics of continuous media.* Pergamon Press, Oxford.

LAX, M. (1962) Influence of time reversal on selection rules connecting different points in the Brillouin zone. *Proc. int. Conf. Phys. Semicond.*, Exeter, pp. 396–402; reprinted: Knox and Gold (1964).

——— (1965) Subgroup techniques in crystal and molecular physics. *Phys. Rev.* **138**, A793–802.

——— and BURSTEIN, E. (1955) Infrared lattice absorption in ionic and homopolar crystals. *Phys. Rev.* **97**, 39–52.

——— and HOPFIELD, J. J. (1961) Selection rules connecting different points in the Brillouin zone. *Phys. Rev.* **124**, 115–23.

LE CORRE, Y. (1958*a*) The magnetic crystallographic groups and their properties. *J. Phys. Radium, Paris* **19**, 750–64 (in French).

——— (1958*b*) The two-coloured symmetry groups and their applications. *Bull. Soc. fr. Minér. Cristallogr.* **81**, 120–5 (in French).

LEDINEGG, E. and URBAN, P. (1952) On the group-theoretical treatment of the linear atomic chain. *Acta phys. austriaca* **6**, 7–29 (in German).

LEE, P. M. and PINCHERLE, L. (1963) The electronic band structure of bismuth telluride. *Proc. phys. Soc.* **81**, 461–9.

LEECH, J. W. and NEWMAN, D. J. (1969) *How to use groups*. Methuen, London.

LEGGETT, A. J. and TER HAAR, D. (1965) Finite linewidths and 'forbidden' three-phonon interactions. *Phys. Rev.* **139**, A779–88.

LEHMANN, A. M. (1969*a*) Determination of magnetic structures in h.c.p. crystals. I. *Acta phys. pol.* **36**, 245–58.

——— (1969*b*) Determination of magnetic structures in h.c.p. crystals. II. *Acta phys. pol.* **36**, 529–39.

LEHRER-ILAMED, Y. (1964) On the direct calculations of the representations of the three-dimensional pure rotation group. *Proc. Camb. phil. Soc. math. phys. Sci.* **60**, 61–6.

LIETZ, M. and RÖSSLER, U. (1964) Determination of the energy band structures of crystals with the chalcopyrites structure by **k.p** perturbation theory. *Z. Naturf.* **19a**, 850–6 (in German).

LIFSHITZ, I. M. and KAGANOV, M. I. (1959) Some problems of the electron theory of metals. I. Classical and quantum mechanics of electrons in metals. *Usp. fiz. Nauk.* **69**, 419–58 (in Russian); *Soviet Phys. Usp.* (English Transl.) **2**, 831–55 (1960).

LITTLEWOOD, D. E. (1950) *The theory of group characters and matrix representations of groups*. Clarendon Press, Oxford.

LIU, L. (1962) Effects of spin–orbit coupling in Si and Ge. *Phys. Rev.* **126**, 1317–28.

LITVIN, D. B. and ZAK, J. (1968) Clebsch–Gordan coefficients for space groups. *J. math. Phys.* **9**, 212–21.

LOEB, A. L. (1962) A modular algebra for the description of crystal structures. *Acta crystallogr.* **15**, 219–26.

LOMER, W. M. (1955) The valence bands in two-dimensional graphite. *Proc. R. Soc.* **A227**, 330–49.

——— (1966*a*) Neutron spectroscopy of solids. Part I. *Contemp. Phys.* **7**, 278–93.

——— (1966*b*) Neutron spectroscopy of solids. Part II. *Contemp. Phys.* **7**, 401–18.

LOMONT, J. S. (1959) *Applications of finite groups*. Academic Press, New York.

LONGUET-HIGGINS, H. C. (1963) The symmetry of non-rigid molecules. *Molec. Phys.* **6**, 445–60.

LOUDON, R. (1964) The Raman effect in crystals. *Adv. Phys.* **13**, 423–82; erratum, *Adv. Phys.* **14**, 621 (1965).

——— (1965) General space-group selection rules for two-phonon processes. *Phys. Rev.* **137**, A1784–7.

——— (1968) Theory of infrared and optical spectra of antiferromagnets. *Adv. Phys.* **17**, 243–80.

LOW, W. (1960) *Paramagnetic resonance in solids*. Academic Press, New York (*Solid St. Phys.* Suppl. 2).

LÖWDIN, P.-O. (1964) Angular momentum wave functions constructed by projector operators. *Rev. mod. Phys.* **36**, 966–76.

——— (1966) *Quantum theory of atoms, molecules and the solid state. A tribute to John C. Slater*. Academic Press, New York.

LUEHRMANN, A. W. (1968) Crystal symmetries of plane-wave-like functions. I. The symmorphic space groups. *Adv. Phys.* **17**, 1–77.

LUKESH, J. S. (1950) On the symmetry of graphite. *Phys. Rev.* **80**, 226–9.

LUTTINGER, J. M. (1956) Quantum theory of cyclotron resonance in semiconductors: general theory. *Phys. Rev.* **102**, 1030–41.

LYUBARSKII, G. YA. (1960) *The application of group theory in physics*. Pergamon Press, Oxford.

LYUBIMOV, V. N. (1963*a*) Space symmetry of electric and magnetic dipole structures. *Fizika tverd. Tela* **5**, 951–3 (in Russian); *Soviet Phys. solid St.* (English Transl.) **5**, 697–8 (1963).

——— (1963*b*) Space symmetry of electric and magnetic dipole structures. *Kristallografiya* **8**, 699–705 (in Russian); *Soviet Phys. Crystallogr.* (English Transl.) **8**, 559–64 (1964).

——— (1965*a*) The symmetry of multipolar lattices. *Kristallografiya* **10**, 405–7 (in Russian); *Soviet Phys. Crystallogr.* (English Transl.) **10**, 325–7 (1965).

——— (1965*b*) The interaction of polarization and magnetization in crystals. *Kristallografiya* **10**, 520–4 (in Russian); *Soviet Phys. Crystallogr.* (English Transl.) **10**, 433–6 (1966).

——— (1967) Derivation of symmetry and antisymmetry groups. *Kristallografiya* **12**, 348–9 (in Russian); *Soviet Phys. Crystallogr.* (English Transl.) **12**, 290–1 (1967).

——— and ZHELUDEV, I. S. (1963) Crystal structures with and without dipoles. *Kristallografiya* **8**, 313–18 (in Russian); *Soviet Phys. Crystallogr.* (English Transl.) **8**, 246–8.

MCCLURE, D. S. (1959*a*) Electronic spectra of molecules and ions in crystals. Part I. Molecular crystals. *Solid St. Phys.* **8**, 1–47.

——— (1959*b*) Electronic spectra of molecules and ions in crystals. Part II. Spectra of ions in crystals. *Solid St. Phys.* **9**, 399–525.

MCCUBBIN, W. L. (1966) Symmetry orbitals and UV selection rules for the polymethylene chain. *Phys. Status Solidi* **16**, 289–93.

MACGILLAVRY, C. G., RIECK, G. D., and LONSDALE, K. (1962) *International tables for X-ray crystallography*. Vol. III. *Physical and chemical tables*. Kynoch, Birmingham (see also Henry and Lonsdale (1965), Kasper and Lonsdale (1959)).

McIntosh, H. V. (1960*a*) Towards a theory of the crystallographic point groups. *J. molec. Spectrosc.* **5**, 269–83.

——— (1960*b*) Symmetry-adapted functions belonging to the symmetric groups. *J. math. Phys.* **1**, 453–60.

——— (1962) On matrices which anticommute with a Hamiltonian. *J. molec. Spectrosc.* **8**, 169–92.

——— (1963) Symmetry adapted functions belonging to the crystallographic groups. *J. molec. Spectrosc.* **10**, 51–74.

Mackay, A. L. (1957) Extensions of space-group theory. *Acta crystallogr.* **10**, 543–8.

——— and Pawley, G. S. (1963) Bravais lattices in four-dimensional space. *Acta crystallogr.* **16**, 11–19 (see also Kuntsevich and Belov (1968)).

Mackay, G. W. (1949) Imprimitivity for representations of locally compact groups. I. *Proc. natn. Acad. Sci. U.S.A.* **35**, 537–45.

——— (1950*a*) Imprimitivity for representations of locally compact groups. II. Intertwining numbers for the representations. *C. r. hebd. Séanc. Acad. Sci., Paris* **230**, 808–9 (in French).

——— (1950*b*) Imprimitivity for representations of locally compact groups. III. Kronecker products and intertwining numbers. *C. r. hebd. Séanc. Acad. Sci., Paris* **230**, 908–9 (in French).

——— (1951) On induced representations of groups. *Am. J. Math.* **73**, 576–92.

——— (1952) Induced representations of locally compact groups. I. *Ann. Math.* **55**, 101–39.

——— (1953*a*) Induced representations of locally compact groups. II. The Frobenius reciprocity theorem. *Ann. Math.* **58**, 193–221.

——— (1953*b*) Symmetric and antisymmetric Kronecker squares and intertwining numbers of induced representations of finite groups. *Am. J. Math.* **75**, 387–405.

——— (1958) Unitary representations of group extensions. *Acta math., Stockh.* **99**, 265–311.

——— (1968) *Induced representations of groups and quantum mechanics.* Benjamin, New York.

McLachlan, D. (1956) Symmetry in reciprocal space. *Acta crystallogr.* **9**, 318.

McMillan, J. A. (1967) Stereographic projections of the colored crystallographic point groups. *Am. J. Phys.* **35**, 1049–55.

——— (1969) Symmetry and properties of crystals: theorem of group intersection. *Am. J. Phys.* **37**, 793–9.

McWeeny, R. (1955) On the basis of orbital theories. *Proc. R. Soc.* **A232**, 114–35.

——— (1963) *Symmetry. An introduction to group theory and its applications.* Pergamon Press, Oxford.

Malíšek, V. (1968) A survey of symmetry point groups and characters of its irreducible representations. *Czech. J. Phys.* **18**, 965–9.

Maradudin, A. A., Montroll, E. W., and Weiss, G. H. (1963) *Theory of lattice dynamics in the harmonic approximation.* Academic Press, New York (*Solid St. Phys.* Suppl. 3).

——— and Vosko, S. H. (1968) Symmetry properties of the normal vibrations of a crystal. *Rev. mod. Phys.* **40**, 1–37 (see also Warren (1968)).

Margenau, H. and Murphy, G. M. (1956) *The mathematics of physics and chemistry.* Van Nostrand, New York.

MARGENAU, H. and MURPHY, G. M. (1964) *The mathematics of physics and chemistry*, Vol. II. Van Nostrand, New York.

MARIOT, L. (1957) Finite groups of symmetry and determination of solutions of the Schrödinger equation. *J. Phys. Radium, Paris* **18**, 345–56 (in French).

——— (1962) *Group theory and solid state physics*. Prentice Hall, Englewood Cliffs, N.J.

MASE, S. (1958) Electronic structure of bismuth type crystals. *J. phys. Soc. Japan* **13**, 434–45.

——— (1959*a*) Electronic structure of bismuth type crystals. II. *J. phys. Soc. Japan* **14**, 584–9.

——— (1959*b*) Algebraic method to obtain irreducible representations of space groups with an application to white tin. *J. phys. Soc. Japan* **14**, 1538–50.

MATOSSI, F. (1951) Irreducible representations of cubic groups. *J. chem. Phys.* **19**, 1612–13.

——— (1961) *Group theory of normal modes of point systems*. Springer, Berlin (in German).

MAUGUIN, CH. (1931) On the symbolism of the repetition groups or the symmetry of crystalline assemblages. *Z. Kristallogr. Kristallgeom.* **76**, 542–58 (in French) (see also Hermann (1931)).

MAUTNER, F. I. (1950*a*) Unitary representations of locally compact groups. I. *Ann. Math.* **51**, 1–25.

——— (1950*b*) Unitary representations of locally compact groups. II. *Ann. Math.* **52**, 528–56.

——— (1951) A generalization of the Frobenius reciprocity theorem. *Proc. natn. Acad. Sci. U.S.A.* **37**, 431–5.

——— (1952) Induced representations. *Am. J. Math.* **74**, 737–58.

MEHTA, M. L. and SRIVASTAVA, P. K. (1968) Irreducible corepresentations of groups having a compact simple Lie group as a subgroup of index 2. *J. math. Phys.* **9**, 1375–85.

MEIJER, P. H. E. (1954) Calculation of wave functions in a symmetrical crystalline field. *Phys. Rev.* **95**, 1443–9.

——— (1964) *Group theory and solid state physics. A selection of papers*, Vol. 1. Gordon and Breach, New York.

——— (1969) *Group theory and solid state physics. A selection of papers*. Vol. 2. Gordon and Breach, New York.

——— and BAUER, E. (1962) *Group theory: the application to quantum mechanics*. North-Holland, Amsterdam.

MEISTER, A. G., CLEVELAND, F. F., and MURRAY, M. J. (1943) Interpretation of the spectra of polyatomic molecules by use of group theory. *Am. J. Phys.* **11**, 239–47.

MELVIN, M. A. (1956) Simplification in finding symmetry-adapted eigenfunctions. *Rev. mod. Phys.* **28**, 18–44.

MENZIES, A. C. (1953) Raman effect in solids. *Rep. Prog. Phys.* **16**, 83–107.

MEYER, B. (1954) On the symmetries of spherical harmonics. *Can. J. Math.* **6**, 135–57.

MIĄSEK, M. (1957*a*) The application of the tight binding method to the investigation of energy bands in hexagonal close-packed structure. I. *Acta phys. pol.* **16**, 343–68.

——— (1957*b*) The application of the tight binding method to the investigation of energy bands in hexagonal close-packed structure. II. *Acta phys. pol.* **16**, 447–69.

——— (1958) The application of the tight binding method to the investigation of energy bands in hexagonal close-packed structure. III. *Acta phys. pol.* **17**, 371–87.

MIĄSEK, M. (1960) Tight binding method for white tin. *Bull. Acad. pol. Sci., Sér. Sci. math. astr. phys.* **8**, 89–93.

——— (1963) Band structure of white tin. *Phys. Rev.* **130**, 11–16.

——— (1966) The symmetrized plane waves for white tin structure. *Acta phys. pol.* **29**, 141–61.

——— and SUFFCZYŃSKI, M. (1961*a*) Space group of white tin. I. Symmetry points. *Bull. Acad. pol. Sci., Sér. Sci. math. astr. phys.* **9**, 477–82.

——— (1961*b*) Space group of white tin. II. Symmetry lines and planes. *Bull. Acad. pol. Sci., Sér. Sci. math. astr. phys.* **9**, 483–7 (see also Suffczyński (1961)).

——— (1961*c*) Space group of white tin. IV. Basis functions for the irreducible representations at symmetry points. *Bull. Acad. pol. Sci., Sér. Sci. math. astr. phys.* **9**, 609–15.

MIHEEV, V. I. (1950) The number of homology classes of crystals. *Dokl. Akad. Nauk SSSR* **71**, 667–70 (in Russian).

MILLER, G. A. (1894, 1946). Miller's papers on abstract groups are so numerous that we do not attempt to list any of them, but refer the reader to *The collected works of George Abram Miller*. University of Illinois, Urbana, Vol. I, 1935; Vol. II, 1938; Vol. III, 1946; Vol. IV, 1955; Vol. V, 1959.

MILLER, S. C. and LOVE, W. F. (1967) *Tables of irreducible representations of space groups and co-representations of magnetic space groups*. Pruett, Boulder, Col.

MINDEN, H. T. (1952) The complete symmetry group for internal rotation in CH_3CF_3 and like molecules. *J. chem. Phys.* **20**, 1964–5.

MITRA, S. S. (1962) Vibration spectra of solids. *Solid St. Phys.* **13**, 1–80.

——— (1964) Space group selection rules and infrared active phonon processes in GaAs. *Phys. Lett.* **11**, 119–21.

MOFFITT, W. E. and BALLHAUSEN, C. J. (1956) Quantum theory. *A. Rev. phys. Chem.* **7**, 107–36.

MOKIEVSKII, V. A. and SHAFRANOVSKII, I. I. (1957) Symmetry, antisymmetry, and pseudosymmetry of nucleation surfaces. *Kristallografiya* **2**, 23–8 (in Russian); *Soviet Phys. Crystallogr.* (English Transl.) **2**, 19–23 (1957).

——— ——— (1961) A graphic method for describing crystal forms. *Kristallografiya* **6**, 944–8 (in Russian); *Soviet Phys. Crystallogr.* (English Transl.) **6**, 761–4 (1962).

MOLCHANOV, A. G. (1969) Selection rules for longitudinal scattering of excitons by phonons in optically active crystals. *Fizika tverd. Tela.* **11**, 1184–7 (in Russian); *Soviet Phys. solid St.* (English Transl.) **11**, 963–5.

MONTGOMERY, H. (1969) The symmetry of lattice vibrations in the zincblende and diamond structures. *Proc. R. Soc.* **A309**, 521–49.

MONTROLL, E. W. (1954) Frequency spectrum of vibrations of a crystal lattice. *Am. math. Mon.* **61**, 46–73.

MORGAN, T. N. (1968) Symmetry of electron states in GaP. *Phys. Rev. Lett.* **21**, 819–23.

MORSE, P. M. (1930) The quantum mechanics of electrons in crystals. *Phys. Rev.* **35**, 1310–24.

MOSES, H. E. (1965) Irreducible representations of the rotation group in terms of Euler's theorem. *Nuovo Cim.* **A40**, 1120–38.

——— (1966) Irreducible representations of the rotation group in terms of the axis and angle of rotation. *Ann. Phys.* **37**, 224–6.

——— (1967) Irreducible representations of the rotation group in terms of the axis and angle of rotation. II. Orthogonality relations between matrix elements and representations of rotations in the parameter space. *Ann. Phys.* **42**, 343–6.

MOSKALENKO, S. A. (1960) Exciton absorption of light in a Cu_2O crystal. 1. Absence of constant external fields. *Fizika tverd. Tela* **2**, 1755–65 (in Russian); *Soviet Phys. solid St.* (English Transl.) **2**, 1587–96.

MOTT, N. F. and JONES, H. (1936) *The theory of the properties of metals and alloys.* Oxford University Press; reprinted: Dover, New York (1958).

MOTZOK, D. (1929*a*) Substitution theory as an analysis of symmetry study. *Z. Kristallogr. Kristallgeom.* **71**, 406–11 (in German).

——— (1929*b*) Substitution theory as an analysis of symmetry study. *Z. Kristallogr. Kristallgeom.* **72**, 249–71 (in German).

——— (1930*a*) Composition and transformation in the study of symmetry. *Z. Kristallogr. Kristallgeom.* **73**, 434–41 (in German).

——— (1930*b*) Multi-dimensional symmetry and substitution theory. *Z. Kristallogr. Kristallgeom.* **75**, 345–62 (in German).

MUELLER, E. (1946) The study of ornaments as an application of the theory of finite groups. *Euclides, Madr.* **6**, 42–52 (in Spanish).

MUELLER, F. M. and PRIESTLEY, M. G. (1966) Inversion of cubic de Haas-van Alphen data, with an application to palladium. *Phys. Rev.* **148**, 638–43.

MÜLLER, K. (1957*a*) The rational determination of the irreducible parts of a given tensor representation in the 32 crystal classes. *Abh. braunschw. wiss. Ges.* **9**, 116–26 (in German).

——— (1957*b*) The construction of the character of eigensymmetrical tensor relations. *Abh. braunschw. wiss. Ges.* **9**, 127–34 (in German).

MULLIKEN, R. S. (1933) Electronic states of polyatomic molecules and valence. IV. Electronic states, quantum theory of the double bond. *Phys. Rev.* **43**, 279–302.

MURNAGHAN, F. D. (1938) *The theory of group representations.* Johns Hopkins Press, Baltimore.

——— (1952*a*) On the invariant theory of the classical groups. *Proc. natn. Acad. Sci. U.S.A.* **38**, 966–73.

——— (1952*b*) On the decomposition of tensors by contraction. *Proc. natn. Acad. Sci. U.S.A.* **38**, 973–9.

——— (1962) *The unitary and rotation groups.* Spartan, Washington, D.C.

MURPHY, J., CASPERS, H. H., and BUCHANAN, R. A. (1964) Symmetry coordinates and lattice vibrations of $LaCl_3$. *J. chem. Phys.* **40**, 743–53.

MURTHY, M. V. (1966) Ray representations of finite non-unitary groups. *J. math. Phys.* **7**, 853–7.

MUTO, Y. and YANO, K. (1937) On the theory of spinors. *Proc. phys.-math. Soc. Japan* **19**, 413–35 (in French).

NERONOVA, N. N. (1966) Classification principles for symmetry groups and groups of a different kind of antisymmetry. I. Scheme of the crystallographic symmetry groups and groups of a different kind of antisymmetry. *Kristallografiya* **11**, 495–504 (in Russian); *Soviet Phys. Crystallogr.* (English Transl.) **11**, 445–52 (1967).

NERONOVA, N. N. (1967*a*) Classification principles for the symmetry and antisymmetry groups. II. Special spatial elements as the basis of classification for symmetry groups. *Kristallografiya* **13**, 3–10 (in Russian); *Soviet Phys. Crystallogr.* (English Transl.) **12**, 3–8 (1967).

——— (1967*b*) Classification principles for symmetry groups and groups of different kinds of antisymmetry. III. Brief survey of 'crystallographic' symmetry groups of three-dimensional continua, discontinua, semicontinua and aperiodic objects. *Kristallografiya* **12**, 191–3 (in Russian); *Soviet Phys. Crystallogr.* (English Transl.) **12**, 159–61 (1967).

——— and BELOV, N. V. (1959*a*) The symmetry of ferroelectrics. *Dokl. Akad. Nauk SSSR* **129**, 556–7 (in Russian); *Soviet Phys. Dokl.* (English Transl.) **4**, 1179–80 (1959).

——— ——— (1959*b*) Ferromagnetic and ferroelectric space groups. *Kristallografiya* **4**, 807–12 (in Russian); *Soviet Phys. Crystallogr.* (English Transl.) **4**, 769–74 (1960).

——— ——— (1961*a*) A single scheme for the classical and black-and-white crystallographic symmetry groups. *Kristallografiya* **6**, 3–12 (in Russian); *Soviet Phys. Crystallogr.* (English Transl.) **6**, 1–9 (1961).

——— ——— (1961*b*) Color antisymmetry mosaics. *Kristallografiya* **6**, 831–9 (in Russian); *Soviet Phys. Crystallogr.* (English Transl.) **6**, 672–8 (1962).

NESBET, R. K. (1958) The use of projection operators in the configuration interaction problem. *Ann. Phys.* **3**, 397–407.

——— (1961) Construction of symmetry-adapted functions in the many-particle problem. *J. math. Phys.* **2**, 701–9.

NEUSTADT, R. J., CAGLE, F. W., and WASER, J. (1968) Vector algebra and the relations between direct and reciprocal lattice quantities. *Acta crystallogr.* **A24**, 247–8.

NIELSON, J. R. and BERRYMAN, L. H. (1949) A general method of obtaining molecular symmetry coordinates. *J. chem. Phys.* **17**, 659–62.

NIGGLI, A. (1953) Character tables as an expression of the symmetry properties of molecules and crystals. *Schweiz. miner. petrogr. Mitt.* **33**, 21–113 (in German).

——— (1959) The systematic and group-theoretical derivations of symmetry-, antisymmetry-, and degeneration-symmetry groups. *Z. Kristallogr. Kristallgeom.* **111**, 288–300 (in German).

——— and WONDRATSCHEK, H. (1960) A generalization of point groups. I. The simple cryptosymmetry. *Z. Kristallogr. Kristallgeom.* **114**, 215–31 (in German).

NIGGLI, P. (1924) The surface symmetries of homogeneous discontinua. *Z. Kristallogr. Kristallgeom.* **60**, 283–98 (in German).

——— (1926) The regular point distribution along a line in a plane. *Z. Kristallogr. Kristallgeom.* **63**, 255–74 (in German).

——— (1927) The topological structure analysis. I. *Z. Kristallogr. Kristallgeom.* **65**, 391–415 (in German).

——— (1928) The topological structure analysis. II. *Z. Kristallogr. Kristallgeom.* **68**, 404–66 (in German).

——— (1946) New formulation of crystallography. *Experientia* **2**, 336–49 (in German).

——— (1949*a*) The geometrical foundations of selection rules for normal modes and term-splitting in molecule- and crystal-bondings. 1. The derivation and representation of the single symmetry operation of normal-mode systems. *Helv. chim. Acta* **32**, 770–83 (in German).

NIGGLI, P. (1949*b*) The geometrical foundations of selection rules for normal modes and term-splitting in molecule- and crystal-bondings. 2. The derivation and representation of the single symmetry operation of normal-mode systems. *Helv. chim. Acta* **32**, 913–24 (in German).

——— (1949*c*) The geometrical foundations of selection rules for normal modes and term-splitting in molecule- and crystal-bondings. 3. The determination of the degrees of freedom of the modes. *Helv. chim. Acta* **32**, 1453–69 (in German).

——— (1949*d*) The complete and unique identification of the space groups by tables of characters. I. *Acta crystallogr.* **2**, 263–70 (in German).

——— (1950) The complete and unambiguous designation of space groups by tables of characters. II. *Acta crystallogr.* **3**, 429–33 (in German).

——— (1951) Vector presentation of 230 space groups. *Acta crystallogr.* **4**, 190 (in German).

——— and NOWACKI, W. (1935) The arithmetical concept of crystal classes and the derivation of the space groups based upon this concept. *Z. Kristallogr. Kristallgeom.* **91**, 321–35 (in German).

NOVOSADOV, B. K., SAULEVICH, L. K., SVIRIDOV, D. T., and SMIRNOV, YU. F. (1969) The decomposition of direct products of irreducible representations of space groups. *Dokl. Akad. Nauk SSSR* **184**, 82–4 (in Russian); *Soviet Phys. Dokl.* (English Transl.) **14**, 50–2.

NOWACKI, W. (1933) The non-crystallographic point groups. *Z. Kristallogr. Kristallgeom.* **86**, 19–31 (in German).

——— (1954) On the number of different space groups. *Schweiz. miner. petrogr. Mitt.* **34**, 160–8 (in German).

——— EDENHARTER, A. and MATSUMOTO, T. (1967) *Crystal data, systematic tables.* American Crystallographic Association, New York.

NUSIMOVICI, M. (1965) Selection rules for Raman scattering and infrared absorption in wurtzit. *J. Phys. Radium, Paris* **26**, 689–96 (in French).

NUSSBAUM, A. (1962) Group theory and the energy band structure of semiconductors. *Proc. Inst. Radio Engrs.* **50**, 1762–80.

——— (1966) Crystal symmetry, group theory and band structure calculations. *Solid St. Phys.* **18**, 165–272.

——— (1968) Group theory and normal modes. *Am. J. Phys.* **36**, 529–39.

NYE, J. F. (1957) *Physical properties of crystals.* Clarendon Press, Oxford.

OEHLER, O. and GÜNTHARD, HS. H. (1968) Symmetry coordinates and factorization of vibrational problems for crystals with symmorphic space groups. *J. chem. Phys.* **48**, 2032–5.

OKADA, S. (1958) Symmetrical property of eigenfunctions of an electron in a one-dimensional periodic potential field. *Sci. Rep. Res. Insts Tôhoku Univ.* **A42**, 130–7.

OKUBO, S. (1965) Algebra of currents and symmetry. *Phys. Lett.* **17**, 172–4.

OLBRYCHSKI, K. (1961) Representations of the space group of rutile (TiO_2). *Bull. Acad. pol. Sci., Sér. Sci. math. astr. phys.* **9**, 537–42.

OLBRYCHSKI, K. (1963*a*) Abstract definition and representations of a space group. *Phys. Status Solidi* **3**, 1868–75.

——— (1963*b*) Representations of 16 space groups from the class D_{4h}. *Phys. Status Solidi* **3**, 2143–54.

——— and VAN HUONG, N. (1970) Selection rules for the space group of white tin. *Acta phys. pol.* **A37**, 369–89.

ONODERA, Y. and OKAZAKI, M. (1966) Tables of basis functions for double point groups. *J. phys. Soc. Japan* **21**, 2400–8.

OPECHOWSKI, W. (1940) On the 'double' crystallographic groups. *Physica, 's Grav.* **7**, 552–62 (in French); translated by Knox and Gold (1964), Meijer (1964), Cracknell (1968*b*).

——— and GUCCIONE, R. (1965) Magnetic symmetry. In *Magnetism* Vol. 2A, (edited by G. T. Rado and H. Suhl), Chap. 3. Academic Press, New York.

ORGEL, L. E. (1966) *An introduction to transition-metal chemistry ligand-field theory*. Methuen, London.

OSIMA, M. (1941) Note on the Kronecker product of representations of a group. *Proc. imp. Acad. Japan* **17**, 411–13.

OVANDER, L. N. (1960) The form of the Raman tensor. *Optika Spektrosk.* **9**, 571–5 (in Russian); *Optics Spectrosc., N.Y.* (English Transl.) **9**, 302–4 (1960).

——— (1964) Raman scattering tensor in crystals. *Fizika tverd. Tela* **6**, 361–7 (in Russian); *Soviet Phys. solid St.* (English Transl.) **6**, 290–4 (1964).

OVERHAUSER, A. W. (1956) Multiplet structure of excitons in ionic crystals. *Phys. Rev.* **101**, 1702–12.

OVERHOF, H. and RÖSSLER, U. (1968) Magnetic space groups and their irreducible representations. *Phys. Status Solidi* **26**, 461–8.

OZEROV, R. P. (1967) Systematic extinctions of reflections in the coherent magnetic scattering of slow neutrons, due to the presence of symmetry and antisymmetry elements in collinear magnetic materials. *Kristallografiya* **12**, 239–51 (in Russian); *Soviet Phys. Crystallogr.* (English Transl.) **12**, 199–208 (1967).

PABST, A. (1962) The 179 two-sided, two-colored band groups and their relations. *Z. Kristallogr. Kristallgeom.* **117**, 128–34.

PALISTRANT, A. F. (1963) Groups of layer symmetry and antisymmetry of different types. *Kristallografiya* **8**, 783–5 (in Russian); *Soviet Phys. Crystallogr.* (English Transl.) **8**, 624–6 (1964).

——— (1965) Planar point groups of symmetry and different types of antisymmetry. *Kristallografiya* **10**, 3–9 (in Russian); *Soviet Phys. Crystallogr.* (English Transl.) **10**, 1–5 (1965).

——— (1966) Two-dimensional groups of color symmetry and color antisymmetry of different kinds. *Kristallografiya* **11**, 707–13 (in Russian); *Soviet Phys. Crystallogr.* (English Transl.) **11**, 609–13 (1967).

——— (1967) Groups of colored symmetry and of different kinds of antisymmetry of layers. *Kristallografiya* **12**, 194–201 (in Russian); *Soviet Phys. Crystallogr.* (English Transl.) **12**, 162–8 (1967).

——— (1968) Planar point groups with color symmetry and different classes of antisymmetry.

Kristallografiya **13**, 955–9 (in Russian); *Soviet Phys. Crystallogr.* (English Transl.) **13**, 833–6 (1969).

PALISTRANT, A. F. and ZAMORZAEV, A. M. (1963) The symmetry and antisymmetry of groups of layers. *Kristallografiya* **8**, 166–73 (in Russian); *Soviet Phys. Crystallogr.* (English Transl.) **8**, 120–6 (1963).

——— ——— (1964) Groups of symmetry and of different types of antisymmetry of borders and ribbons. *Kristallografiya* **9**, 155–61 (in Russian); *Soviet Phys. Crystallogr.* (English Transl.) **9**, 123–8 (1964).

PANDRES, D. (1965) Schrödinger basis for spinor representations of the three-dimensional rotation group. *J. math. Phys.* **6**, 1098–102.

——— and JACOBSON, D. A. (1968) Scalar product for harmonic functions of the group SU(2). *J. math. Phys.* **9**, 1401–3.

PANTULU, P. V. and RADHAKRISHNA, S. (1967) A method of deriving Shubnikov groups. *Proc. Indian Acad. Sci.* **A66**, 107–11.

PARMENTER, R. H. (1955) Symmetry properties of the energy bands of the zinc blende structure. *Phys. Rev.* **100**, 573–9.

PARODI, O. (1965) Group theory methods for the study of crystal fields. *J. Phys. Paris* **26**, 531–6 (in French).

PARZEN, G. (1953) Electronic energy bands in metals. *Phys. Rev.* **89**, 237–43.

PAULI, W. (1927) Quantum mechanics of the magnetic electron. *Z. Phys.* **43**, 601–23 (in German).

PAWLEY, G. S. (1961*a*) Mosaics for color antisymmetry groups. *Kristallografiya* **6**, 109–11 (in Russian); *Soviet Phys. Crystallogr.* (English Transl.) **6**, 87–9 (1961).

——— (1961*b*) Symmetry and antisymmetry of irregular tetrahedra in a non-lattice array. *Z. Kristallogr. Kristallgeom.* **116**, 1–12.

——— (1961*c*) Coloured polyhedra. *Acta crystallogr.* **14**, 319–20.

——— (1962) Plane groups on polyhedra. *Acta crystallogr.* **15**, 49–53.

——— (1969) Comment on group-theoretical analysis of lattice vibrations in molecular crystals. *J. chem. Phys.* **50**, 2785–6.

PEIERLS, R. E. (1930) Theories of electrical and thermal conductivity of metals. *Annln. Phys.* **4**, 121–48 (in German).

——— (1955) *Quantum theory of solids*. Clarendon Press, Oxford.

PEISER, H. S. and WACHTMAN, J. B. (1965) Reduction of crystallographic point groups to subgroups by homogeneous stress. *J. Res. natn. Bur. Stand.* **A69**, 309–23.

——— ——— and DICKSON, R. W. (1963) Reduction of space groups to subgroups by homogeneous strain. *J. Res. natn. Bur. Stand.* **A67**, 395–401.

——— ——— MUNLEY, F. A., and MCCLEARY, L. C. (1965) Splitting of equivalent points in noncentrosymmetric space groups into subsets under homogeneous stress. *J. Res. natn. Bur. Stand.* **A69**, 461–79 (see also Wachtman and Peiser (1965)).

PERCUS, J. K. and ROTENBERG, A. (1962) Exact eigenfunctions of angular momentum by rotational projection. *J. math. Phys.* **3**, 928–32.

PHILLIPS, F. C. (1963*a*) *An introduction to crystallography*. Longmans, London.

PHILLIPS, J. C. (1956) Critical points and lattice vibration spectra. *Phys. Rev.* **104**, 1263–77; erratum, *Phys. Rev.* **105**, 1933 (1957).

——— (1961) Generalized Koopmans' theorem. *Phys. Rev.* **123**, 420–4.

PHILLIPS, J. C. (1963*b*) Exciton-induced images of phonon spectra in ultra-violet reflectance edges. *Phys. Rev. Lett.* **10**, 329–32.

——— and KLEINMAN, L. (1959) New method for calculating wave functions in crystals and molecules. *Phys. Rev.* **116**, 287–94.

PHILLIPS, T. G. and ROSENBERG, H. M. (1966) Spin waves in ferromagnets. *Rep. Prog. Phys.* **29**, 285–332.

PIKUS, G. E. (1961*a*) The energy spectrum in crystals with a tellurium lattice allowing for spin–orbital reaction. *Fizika tverd. Tela* **3**, 2809–12 (in Russian); *Soviet Phys. solid St.* (English Transl.) **3**, 2050–2 (1962).

——— (1961*b*) A new method of calculating the energy spectrum of carriers in semiconductors. I. Neglecting spin–orbit interaction. *Zh. éksp. teor. Fiz.* **41**, 1258–73 (in Russian); *Soviet Phys. JETP* (English Transl.) **14**, 898–907 (1962).

——— (1961*c*) A new method of calculating the energy spectrum of carriers in semiconductors. II. Account of spin–orbit interaction. *Zh. éksp. teor. Fiz.* **41**, 1507–21 (in Russian); *Soviet Phys. JETP* (English Transl.) **14**, 1075–85 (1962).

PINCHERLE, L. (1960) Band structure calculations in solids. *Rep. Prog. Phys.* **23**, 355–94.

POBEDIMSKAYA, E. A. and BELOV, N. V. (1965) Body-centred packing. *Kristallografiya* **10**, 908–11 (in Russian); *Soviet Phys. Crystallogr.* (English Transl.) **10**, 753–5 (1966).

POLLACK, S. A. and SATTEN, R. A. (1962) Electron-optical phonon interaction for paramagnetic ions in crystalline fields. *J. chem. Phys.* **36**, 804–16.

PÓLYA, G. (1924) About the analogy of the crystal symmetry in the plane. *Z. Kristallogr. Kristallgeom.* **60**, 278–82 (in German).

——— and MEYER, B. (1949*a*) On the symmetries of the Laplace spherical functions. *C. r. hebd. Séanc. Acad. Sci., Paris* **228**, 28–30 (in French).

——— ——— (1949*b*) On the Laplace spherical functions of a given crystallographic symmetry. *C. r. hebd. Séanc. Acad. Sci., Paris* **228**, 1083–4 (in French).

POOLE, E. G. C. (1932) Spherical harmonics having polyhedral symmetry. *Proc. Lond. math. Soc.* **33**, 435–56.

PRINS, J. A. (1947) A thermodynamical substitution group. *Physica, 's Grav.* **13**, 417–21.

PRZYSTAWA, J. (1968) Magnetic symmetry of uranium compounds of UX_2 type. *Phys. Status Solidi* **30**, K115–17.

PU, F. C. and YAN, C. W. (1964) The magnetic structures of crystals with magnetic unit cell un-identical with its chemical unit cell. *Acta phys. sin.* **20**, 825–45 (in Chinese).

PUFF, H. (1970) Contribution to the theory of cubic harmonics. *Phys. Status Solidi* **41**, 11–22.

RABOTNIKOV, YU. L. (1960) Application of group-theory methods to study the lattice oscillations of tellurium. *Fizika tverd. Tela* **2**, 2095–108 (in Russian); *Soviet Phys. solid St.* (English Transl.) **2**, 1879–90 (1961).

RACAH, G. (1942*a*) Theory of complex spectra. I. *Phys. Rev.* **61**, 186–97.

——— (1942*b*) Theory of complex spectra. II. *Phys. Rev.* **62**, 438–62.

——— (1943) Theory of complex spectra. III. *Phys. Rev.* **63**, 367–82.

——— (1949) Theory of complex spectra. IV. *Phys. Rev.* **76**, 1352–65.

RAGHAVACHARYULU, I. V. V. (1961*a*) Representations of space groups. *Can. J. Phys.* **39**, 830–40.

RAGHAVACHARYULU, I. V. V. (1961*b*) Normal modes of vibrations of crystals. *Can. J. Phys.* **39**, 1704–20.

——— and BHAVANACHARYULU, I. (1962) Simplifications in finding the allowable representations of space groups. *Can. J. Phys.* **40**, 1490–5.

——— and SHRESTHA, C. B. (1961*a*) Irreducible representations of line groups. *J. molec. Spectrosc.* 7, 46–52.

——— ——— (1961*b*) Irreducible representations of plane groups. *J. molec. Spectrosc.* **7**, 341–61.

——— ——— (1966) A new compatibility relation of space-group representations. *Can. J. Phys.* **44**, 444–5.

RAHMAN, A. (1953) Enumeration of physical constants of crystals. *Acta crystallogr.* **6**, 426–7.

RAIMES, S. (1961) *The wave mechanics of electrons in metals.* North-Holland, Amsterdam.

RASHBA, É. I. (1959) Symmetry of energy-bands in crystals of wurtzite type. I. Symmetry of bands disregarding spin–orbit interaction. *Fizika tverd. Tela* **1**, 407–21 (in Russian); *Soviet Phys. solid St.* (English Transl.) **1**, 368–80.

——— and SHEKA, V. I. (1959) The energy band symmetry in wurtzite-type crystals. II. Band symmetry with allowance for spin interactions. *Fizika tverd. Tela* Suppl. II, 162–76.

RAYNOR, G. V. (1947) *An introduction to the electron theory of metals.* Institute of Metals, London.

ROOS, M. (1964) On the construction of a unitary matrix with elements of given moduli. *J. math. Phys.* **5**, 1609–11.

ROSE, M. E. (1954) Spherical tensors in physics. *Proc. phys. Soc.* **A67**, 239–47.

——— (1957) *Elementary theory of angular momentum.* Wiley, New York.

ROSENTHAL, J. E. and MURPHY, G. M. (1936) Group theory and the vibrations of polyatomic molecules. *Rev. mod. Phys.* **8**, 317–46.

ROSENWALD, D. (1967) Cubic symmetry effects in solid He^3. *Phys. Rev.* **154**, 160–70.

ROTENBERG, M., BIVINS, R., METROPOLIS, N., and WOOTEN, J. K. (1959) *The 3-j and 6-j symbols.* Technology Press, M. I. T., Cambridge, Mass.

ROTH, L. M. (1962) Theory of Bloch electrons in a magnetic field. *J. Phys. Chem. Solids* **23**, 433–46.

ROY, C. L. (1967) Electronic energy-states of one-dimensional mixed crystals. *Indian J. Phys.* **40**, 639–47.

RUDRA, P. (1965*a*) On projective representations of finite groups. *J. math. Phys.* **6**, 1273–7.

——— (1965*b*) On irreducible representations of space groups. *J. math. Phys.* **6**, 1278–82.

——— (1966) On irreducible tensor operators for finite groups. *J. math. Phys.* **7**, 935–7.

SACHS, M. (1957) Selection rules for the absorption of polarized electromagnetic radiation by mobile electrons in crystals. *Phys. Rev.* **107**, 437–45.

——— (1963) *Solid state theory.* McGraw-Hill, New York.

SACK, R. A. (1964*a*) Generalization of Laplace's expansion to arbitrary powers and functions of the distance between two points. *J. math. Phys.* **5**, 245–51.

——— (1964*b*) Three-dimensional addition theorem for arbitrary functions involving expansions in spherical harmonics. *J. math. Phys.* **5**, 252–9.

——— (1964*c*) Two-center expansion for the powers of the distance between two points. *J. math. Phys.* **5**, 260–8.

SAFFREN, M. M. and SLATER, J. C. (1953) An augmented plane-wave method for the periodic potential problem. II. *Phys. Rev.* **92**, 1126–8.

SAHNI, V. C. and VENKATARAMAN, G. (1970*a*) On a method of constructing irreducible multiplier representations. *Phys. kondens. Materie* **11**, 199–211.

——— ——— (1970*b*) Irreducible multiplier representations for some points in the Brillouin zone of D_{6h}^4. *Phys. kondens. Materie* **11**, 212–9.

SANDROCK, R. and TREUSCH, J. (1964) Symmetry properties of the energy bands of chalcopyrites structures. *Z. Naturf.* **19a**, 844–50 (in German).

SATTEN, R. A. (1957) Crystalline field selection rules: the effect of vibration-electronic interaction. *J. chem. Phys.* **27**, 286–99.

——— (1958) On the theory of electronic-vibrational interaction for rare earth and actinide series ions in crystals. II. Octahedral complexes. *J. chem. Phys.* **29**, 658–64.

SAULEVICH, L. K., SVIRIDOV, D. T., and SMIRNOV, YU. F. (1970*a*) Decomposition of the direct products of irreducible representations of the nonsymmorphic groups using weighted representations. *Kristallografiya* **15**, 410–14 (in Russian); *Soviet Phys. Crystallogr.* (English Transl.) **15**, 351–4.

——— ——— (1970*b*) Clebsch–Gordan coefficients for the nonsymmorphic space groups. *Kristallografiya* **15**, 415–20 (in Russian); *Soviet Phys. Crystallogr.* (English Transl.) **15**, 355–9.

SCHIEBOLD, E. (1929) On a new derivation of the nomenclature of the 230 crystallographic space groups. *Abh. sächs. Akad. Wiss. Math.-phys. Kl.* **40**, 9–204 (in German).

SCHIFF, B. (1955) The cellular method for a close-packed hexagonal lattice, with applications to titanium. *Proc. phys. Soc.* **A68**, 686–95.

——— (1956) The cellular method for close-packed hexagonal titanium. *Proc. phys. Soc.* **A69**, 185–6 (see also Altmann (1956)).

SCHIFF, L. I. (1954) Paper representations of the non-cubic crystal classes. *Am. J. Phys.* **22**, 621–2.

——— (1955) *Quantum mechanics.* McGraw-Hill, New York.

SCHLOSSER, H. (1962) Symmetrized combinations of plane waves and matrix elements of the Hamiltonian for cubic lattices. *J. Phys. Chem. Solids* **23**, 963–9.

SCHÖNFLIES, A. (1887*a*) On the groups of motion. *Math. Annln* **28**, 319–42 (in German).

——— (1887*b*) On the groups of motion (second part). *Math. Annln* **29**, 50–80 (in German).

——— (1889) On groups of transformations of space into itself. *Math. Annln* **34**, 172–203 (in German).

——— (1891) *Theory of crystal structure.* Teubner, Leipzig (in German).

SCHOUTEN, J. A. (1951) *Tensor analysis for physicists.* Oxford University Press.

SCHUR, I. (1904) On the representation of finite groups by fractional linear substitutions. *J. reine angew. Math.* **127**, 20–50 (in German).

——— (1907) Investigations on the representation of finite groups by fractional linear substitutions. *J. reine angew. Math.* **132**, 85–137 (in German).

——— (1911) Concerning groups of linear substitutions with coefficients from an algebraic set of numbers. *Math. Annln* **71**, 355–67 (in German).

SEGAL, I. E. (1950) The two-sided regular representation of a unimodular locally compact group. *Ann. Math.* **51**, 293–8.

SEITZ, F. (1934) Matrix-algebraic development of the crystallographic groups. *Z. Kristallogr. Kristallgeom.* **88**, 433–59; reprinted: Meijer (1964).

——— (1935*a*) Matrix-algebraic development of the crystallographic groups. Part II. *Z. Kristallogr. Kristallgeom.* **90**, 289–313; reprinted: Meijer (1964).

——— (1935*b*) Matrix-algebraic development of the crystallographic groups. Part III. *Z. Kristallogr. Kristallgeom.* **91**, 336–66; reprinted: Meijer (1964).

——— (1936*a*) Matrix-algebraic development of the crystallographic groups. Part IV. *Z. Kristallogr. Kristallgeom.* **94**, 100–30; reprinted: Meijer (1964).

——— (1936*b*) On the reduction of space groups. *Ann. Math.* **37**, 17–28; reprinted: Knox and Gold (1964), Cracknell (1968*b*).

SĘK, Z. (1963) Space groups from the class D_{4h} based on body-centred tetragonal lattice. *Phys. Status Solidi* **3**, 2155–65.

SERBER, R. (1934) The solution of problems involving permutation degeneracy. *J. chem. Phys.* **2**, 697–710.

SERRET, J. A. (1866) *Course of higher algebra.* Villars, Paris (in French).

SHAFRANOVSKII, I. I. (1957) Development of the study of crystal forms. *Kristallografiya* **2**, 326–33 (in Russian); *Soviet Phys. Crystallogr.* (English Transl.) **2**, 323–9 (1958).

——— (1959) Geometrical varieties of face forms for crystals falling in the classes of low symmetry. *Kristallografiya* **4**, 293–301 (in Russian); *Soviet Phys. Crystallogr.* (English Transl.) **4**, 274–80 (1960).

——— (1960*a*) An extended study of crystal forms and the crystal morphology of twins. *Kristallografiya* **5**, 525–9 (in Russian); *Soviet Phys. Crystallogr.* (English Transl.) **5**, 504–8 (1961).

——— (1961) Extension of the theory of crystalline forms and of the morphology of twins. *Bull. Soc. fr. Minér. Cristallogr.* **84**, 20–4 (in French).

——— (1962) Varieties of the cube in crystallography. *Kristallografiya* **7**, 664–70 (in Russian); *Soviet Phys. Crystallogr.* (English Transl.) **7**, 539–43 (1963).

——— (1963) *Evgraf Stepanovich Fedorov.* USSR Academy of Sciences, Moscow (in Russian).

——— and PIS'MENNYI, V. A. (1961) General forms of twinned structures. *Kristallografiya* **6**, 31–42 (in Russian); *Soviet Phys. Crystallogr.* (English Transl.) **6**, 27–33 (1961).

SHAM, L. J. and ZIMAN, J. M. (1963) The electron-phonon interaction. *Solid St. Phys.* **15**, 221–98.

SHEKA, V. I. (1960) The symmetry of the energy zones for a spinning electron. *Fizika tverd. Tela* **2**, 1211–19 (in Russian); *Soviet Phys. solid St.* (English Transl.) **2**, 1096–104 (1960).

SHOCKLEY, W. (1937) The empty lattice test of the cellular method in solids. *Phys. Rev.* **52**, 866–72.

SHOEMAKER, D. P. (1949) A method for calculating the energy of a Bloch wave in a metal. *Physica, 's Grav.* **15**, 34–9.

SHUBNIKOV, A. V. (1930) On the symmetry of semicontinua. *Z. Kristallogr. Kristallgeom.* **73**, 430–3 (in German).

——— (1951) *Symmetry and antisymmetry of finite figures.* USSR Academy of Sciences, Moscow (in Russian); translated in Shubnikov and Belov (1964).

——— (1954*a*) Correction to the book *Symmetry and antisymmetry of finite figures. Trudȳ Inst. Kristall.* **9**, 383 (in Russian).

SHUBNIKOV, A. V. (1954*b*) Antisymmetry of finite figures. *Trudȳ Inst. Kristall.* **10**, 10–12 (in French).

——— (1958) Antisymmetry of textures. *Kristallografiya* **3**, 263–8 (in Russian); *Soviet Phys. Crystallogr.* (English Transl.) **3**, 269–73 (1958).

——— (1959*a*) Symmetry and antisymmetry of rods and semicontinua with principal axis of infinite order and finite transfers along it. *Kristallografiya* **4**, 279–85 (in Russian); *Soviet Phys. Crystallogr.* (English Transl.) **4**, 261–6 (1960).

——— (1959*b*) Complete systematics of point groups of symmetry. *Kristallografiya* **4**, 286–8 (in Russian); *Soviet Phys. Crystallogr.* (English Transl.) **4**, 267–9 (1960).

——— (1960*a*) Time reversal as an operation of antisymmetry. *Kristallografiya* **5**, 328–33 (in Russian); *Soviet Phys. Crystallogr.* (English Transl.) **5**, 309–14 (1960).

——— (1960*b*) Symmetry of similarity (preliminary communication). *Kristallografiya* **5**, 489–96 (in Russian); *Soviet Phys. Crystallogr.* (English Transl.) **5**, 469–76 (1961).

——— (1961) Complete systematics of black-white point groups. *Kristallografiya* **6**, 490–5 (in Russian); *Soviet Phys. Crystallogr.* (English Transl.) **6**, 394–8 (1962).

——— (1962*a*) Symmetry and antisymmetry groups (classes) of finite strips. *Kristallografiya* **7**, 3–6 (in Russian); *Soviet Phys. Crystallogr.* (English Transl.) **7**, 1–4 (1962).

——— (1962*b*) Black-white groups of infinite ribbons. *Kristallografiya* **7**, 186–91 (in Russian); *Soviet Phys. Crystallogr.* (English Transl.) **7**, 145–9 (1962).

——— (1962*c*) On the attribution of all crystallographic symmetry groups to three-dimensional groups. *Kristallografiya* **7**, 490–5 (in Russian); *Soviet Phys. Crystallogr.* (English Transl.) **7**, 394–8 (1962).

——— (1963*a*) Incompleteness of 'A unified scheme of crystallographic groups'. *Kristallografiya* **8**, 131–2 (in Russian); *Soviet Phys. Crystallogr.* (English Transl.) **8**, 101–2 (1963).

——— (1963*b*) The information contained in a regular system of points. *Kristallografiya* **8**, 943–4 (in Russian); *Soviet Phys. Crystallogr.* (English Transl.) **8**, 760–1 (1964).

——— (1963*c*) The so-called homology of crystals. *Dokl. Akad. Nauk SSR* **88**, 453–6 (in Russian).

——— (1964) On so-called complete symmetry. *Kristallografiya* **9**, 587–8 (in Russian); *Soviet Phys. Crystallogr.* (English Transl.) **9**, 502–3 (1965).

——— (1965) Thirty-two crystallographic groups containing only rotations and antirotations. *Kristallografiya* **10**, 775–8 (in Russian); *Soviet Phys. Crystallogr.* (English Transl.) **10**, 655–7 (1966).

——— (1966) Thirty-two crystallographic groups containing only rotations and mirror antirotations. *Kristallografiya* **11**, 365–7 (in Russian); *Soviet Phys. Crystallogr.* (English Transl.) **11**, 331–2 (1966).

——— and BELOV, N. V. (1964) *Colored symmetry*. Pergamon Press, Oxford.

SHULL, C. G. and WOLLAN, E. O. (1956) Applications of neutron diffraction to solid state problems. *Solid St. Phys.* **2**, 137–217.

SHUR, M. S. (1966) Theorem about third order invariants for irreducible representations of space groups. *Zh. éksp. teor. Fiz.* **51**, 1260–2 (in Russian); *Soviet Phys. JETP* (English Transl.) **24**, 845–6 (1967).

——— (1967) Projection operators for the irreducible representations of space groups.

Kristallografiya **12**, 981–8 (in Russian); *Soviet Phys. Crystallogr.* (English Transl.) **12**, 857–62 (1968).

SHUVALOV, L. A. (1957) Dielectric and piezoelectric properties of polarized ceramic $BaTiO_3$ in its various ferroelectric phases. *Kristallografiya* **2**, 119–29 (in Russian); *Soviet Phys. Crystallogr.* (English Transl.) **2**, 115–24 (1957).

——— (1959) Ferromagnetic phase transitions and the symmetry of crystals. *Kristallografiya* **4**, 399–409 (in Russian); *Soviet Phys. Crystallogr.* (English Transl.) **4**, 371–80 (1960).

——— (1962*a*) Antisymmetry and its concrete modifications. *Kristallografiya* **7**, 520–5 (in Russian); *Soviet Phys. Crystallogr.* (English Transl.) **7**, 418–22 (1963).

——— (1962*b*) Limit groups of double antisymmetry. *Kristallografiya* **7**, 822–5 (in Russian); *Soviet Phys. Crystallogr.* (English Transl.) **7**, 669–72 (1963).

——— (1963) A crystallographic classification of ferroelectrics, ferroelectric phase transformations and peculiarities of domain structure, and some physical properties of ferroelectrics of the different classes. *Kristallografiya* **8**, 617–24 (in Russian); *Soviet Phys. Crystallogr.* (English Transl.) **8**, 495–500 (1964).

——— and BELOV, N. V. (1962) The symmetry of crystals in which ferromagnetic and ferroelectric properties appear simultaneously. *Kristallografiya* **7**, 192–4 (in Russian); *Soviet Phys. Crystallogr.* (English Transl.) **7**, 150–1 (1962); errata, *Kristallografiya* **8**, 304 (1962) (in Russian); *Soviet Phys. Crystallogr.* (English Transl.) **8**, 237 (1962).

——— and SONIN, A. S. (1961) The crystallography of antiferroelectrics. *Kristallografiya* **6**, 323–30 (in Russian); *Soviet Phys. Crystallogr.* (English Transl.) **6**, 258–62 (1961).

——— and TAVGER, B. A. (1958) Symmetry of magnetostrictive properties of crystals. *Kristallografiya* **3**, 756–9 (in Russian); *Soviet Phys. Crystallogr.* (English Transl.) **3**, 765–8 (1958).

SIERRE, J.-P. (1968) *Ferdinand Georg Frobenius: collected papers.* Three volumes. Springer, Berlin (in German).

SILBERMAN-ROMAN, E. (1936) *On the Moebius strips for crystalline symmetries.* Les Presses Univ. de France, Paris (in French).

SINHA, K. P. and UPADHYAYA, U. N. (1962) Phonon-magnon interaction in magnetic crystals. *Phys. Rev.* **127**, 432–9.

SIROTIN, YU. I. (1960) Group tensor spaces. *Kristallografiya* **5**, 171–9 (in Russian); *Soviet Phys. Crystallogr.* (English Transl.) **5**, 157–65 (1960).

——— (1961) Plotting tensors of a given symmetry. *Kristallografiya* **6**, 331–40 (in Russian); *Soviet Phys. Crystallogr.* (English Transl.) **6**, 263–71 (1961).

——— (1962) Magnetic symmetry of a tensor and the energy of magnetic anisotropy. *Kristallografiya* **7**, 89–96 (in Russian); *Soviet Phys. Crystallogr.* (English Transl.) **7**, 71–77 (1962).

——— (1963) Possible changes in the point-group magnetic symmetry in second-order ferromagnetic transitions. *Kristallografiya* **8**, 259–60 (in Russian); *Soviet Phys. Crystallogr.* (English Transl.) **8**, 195–6 (1963).

——— and KOPTSIK, V. A. (1963) Magnetic spatial symmetry of tensors. *Dokl. Akad. Nauk SSSR* **151**, 328–31 (in Russian).

SIVARDIÈRE, J. (1968) Analogy between the representations of double and space groups. *C. r. hebd. Séanc. Acad. Sci., Paris* **B266**, 453–5 (in French).

SIVARDIÈRE, J. (1969*a*) Properties of space groups of a given class and lattice. Construction of space groups and magnetic groups. *Acta crystallogr.* **A25**, 363–7 (in French).

——— (1969*b*) New method for the construction of magnetic groups. *Acta crystallogr.* **A25**, 658–65 (in French).

——— (1969*c*) A simple method for determining crystal, magnetic and many-coloured classes. *Bull. Soc. fr. Minér. Cryst.* **92**, 3–8 (in French).

——— (1969*d*) Structure of the point groups in three and four dimensions and generalization of the method of Heesch. *Bull. Soc. fr. Minér. Cryst.* **92**, 405–8 (in French).

——— (1970*a*) Study of the symmetry of some magnetic configurations by Bertaut's method. *Acta crystallogr.* **A26**, 101–5.

——— (1970*b*) Direct construction of the space groups and magnetic groups. *Bull. Soc. fr. Minér. Cryst.* **93**, 146–52 (in French).

SLATER, J. C. (1937) Wave functions in a periodic potential. *Phys. Rev.* **51**, 846–51.

——— (1951) Note on superlattices and Brillouin zones. *Phys. Rev.* **84**, 179–81.

——— (1953) An augmented plane wave method for the periodic potential problem. *Phys. Rev.* **92**, 603–8 (see also Saffren and Slater (1953)).

——— (1963) *Quantum theory of molecules and solids.* Vol. 1. *Electron structure of molecules.* McGraw-Hill, New York.

——— (1965*a*) Space groups and wave-function symmetry in crystals. *Rev. mod. Phys.* **37**, 68–83.

——— (1965*b*) *Quantum theory of molecules and solids.* Vol. 2. *Symmetry and energy bands in crystals.* McGraw-Hill, New York.

——— (1967) *Quantum theory of molecules and solids.* Vol. 3. *Insulators, semiconductors and metals.* McGraw-Hill, New York.

——— and KOSTER, G. F. (1954) Simplified LCAO method for the periodic potential problem. *Phys. Rev.* **94**, 1498–524.

——— ——— and WOOD, J. H. (1962) Symmetry and free electron properties of the gallium energy bands. *Phys. Rev.* **126**, 1307–17.

ŠLECHTA, J. (1966) Application of theory of weighted representations to group analysis of wave functions of AlN crystals. *Česk. Časopis Fys.* **16**, 1–5 (in Czech).

SMITH, C. S. (1958) Macroscopic symmetry and properties of crystals. *Solid St. Phys.* **6**, 175–249.

SMITH, D. and THORNLEY, J. H. M. (1966) The use of 'operator equivalents'. *Proc. phys. Soc.* **89**, 779–81.

SMITH, G. F. (1964) On the generation of anisotropic tensors. *J. math. Phys.* **5**, 1612–4.

SOHNCKE, L. (1879) *Development of a theory of crystal structure.* Teubner, Leipzig (in German).

——— (1884) Probable nature of the internal symmetry of crystals. *Nature, Lond.* **29**, 383–4 (see also Barlow (1883, 1884)).

——— (1892) Two theories of crystal structure. *Z. Kristallogr. Kristallgeom.* **20**, 445–67 (in German).

SOKOLOV, A. V. and SHIROKOVSKII, V. P. (1956) Group theory method in the quantum physics of a solid body (point symmetry). *Usp. fiz. Nauk.* **60**, 617–88 (in Russian).

——— ——— (1957) Description of electronic states in a cubic crystal. *Fizika Metall.* **4**, 3–8 (in Russian); *Physics Metals Metallogr., N.Y.* (English Transl.) **4**, 1–5 (1957).

SOKOLOV, A. V. and SHIROKOVSKII, V. P. (1960) Group theoretical methods in the quantum physics of solids (spatial symmetry). *Usp. fiz. Nauk* **71**, 485–513 (in Russian); *Soviet Phys. Usp.* (English Transl.) **3**, 551–66 (1961).

SONIN, A. S. and ZHELUDEV, I. S. (1959) Space groups and ferroelectric phase transitions. *Kristallografiya* **4**, 487–97 (in Russian); *Soviet Phys. Crystallogr.* (English Transl.) **4**, 460–9 (1960).

SPEISER, A. (1923) *The theory of groups of finite order.* Springer, Berlin (in German).

SPONER, H. and TELLER, E. (1941) Electronic spectra of polyatomic molecules. *Rev. mod. Phys.* **13**, 75–170.

STATZ, H. and KOSTER, G. F. (1959) Zeeman splittings of paramagnetic atoms in crystalline fields. *Phys. Rev.* **115**, 1568–77 (see also Koster and Statz (1959)).

STENO, N. (1669) *Dissertation concerning a solid naturally contained within a solid.* Stella, Florence (in Latin); translated: Winter (1916).

STEVENS, K. W. H. (1952) Matrix elements and operator equivalents connected with the magnetic properties of rare earth ions. *Proc. phys. Soc.* **A65**, 209–15.

STIEFEL, E. (1952) Two applications of group characters to the solution of boundary-value problems. *J. Res. natn. Bur. Stand.* **48**, 424–7.

STREITWOLF, H. W. (1964) Symmetries in the lattice vibrations problem. *Phys. Status Solidi* **5**, 383–90 (in German).

——— (1969*a*) On space group selection rules. *Phys. Status Solidi* **33**, 217–23.

——— (1969*b*) Selection rules for the space group C_{6v}^4 (wurtzite). *Phys. Status Solidi* **33**, 225–33.

STRUTT, M. J. O. (1928*a*) Vibration of string with sinc distribution of mass. *Annln Phys.* **85**, 129–36 (in German).

——— (1928b) Wave mechanics of the atomic lattice. *Annln Phys.* **86**, 319–24 (in German).

SUFFCZYŃSKI, M. (1960) The group theoretical analysis of α-uranium. *Physics Chem. Solids* **16**, 174–6.

——— (1961) Space group of white tin. III. Double group. *Bull. Acad. pol. Sci., Sér. Sci. math. astr. phys.* **9**, 489–95 (see also Miąsek and Suffczyński (1961*a*, *b*, *c*)).

SUSHKEVICH, T. N. (1965) Symmetry of the energetic structure of C_{2v}^{18}, C_{2v}^{19}, D_2^7, D_{2h}^{23}, D_{2h}^{24} group crystals. *Ukr. fiz. Zh.* **10**, 861–6 (in Ukrainian).

——— (1966) Symmetry of carrier energy zones in crystals of monoclinal syngony. *Ukr. fiz. Zh.* **11**, 739–44 (in Ukrainian).

SYLOW, L. (1872) Theorems on the groups of substitutions. *Math. Annln* **5**, 584–94 (in French).

TAM, W. G. and OPECHOWSKI, W. (1966) Magnetic space groups of an electron in a crystal. *Phys. Lett.* **23**, 212–13.

TANABE, Y. and KAMIMURA, H. (1958) On the absorption spectra of complex ions. IV. The effect of the spin–orbit interaction and the field of lower symmetry on *d*-electrons in cubic field. *J. phys. Soc. Japan* **13**, 394–411.

——— and SUGANO, S. (1954*a*) On the absorption spectra of complex ions. I. *J. phys. Soc. Japan* **9**, 753–66.

——— ——— (1954*b*) On the absorption spectra of complex ions. II. *J. phys. Soc. Japan* **9**, 766–79.

TANABE, Y. and KAMIMURA, H. (1956) On the absorption spectra of complex ions. III. The calculation of the crystalline field strength. *J. phys. Soc. Japan* **11**, 864–77.

TAVGER, B. A. (1958*a*) The symmetry of ferromagnetics and antiferromagnetics. *Kristallografiya* **3**, 339–41 (in Russian); *Soviet Phys. Crystallogr.* (English Transl.) **3**, 341–3 (1958).

——— (1958*b*) Symmetry of piezomagnetism of antiferromagnetics. *Kristallografiya* **3**, 342–5 (in Russian); *Soviet Phys. Crystallogr.* (English Transl.) **3**, 344–7 (1958).

——— (1960) Ultimate magnetic symmetry of physical systems. *Kristallografiya* **5**, 677–80 (in Russian); *Soviet Phys. Crystallogr.* (English Transl.) **5**, 646–9 (1961).

——— and ZAITSEV, V. M. (1956) Magnetic symmetry of crystals. *Zh. éksp. teor. Fiz.* **30**, 564–8 (in Russian); *Soviet Phys. JETP.* (English Transl.) **3**, 430–7 (1956); reprinted: Meijer (1964), Cracknell (1968*b*).

TELEMAN, E. and GLODEANU, A. (1967) Basis functions for the double groups O_h and D_{3h}. *Revue roum. Sci. Tech. Electrotech. Energet.* **12**, 551–60.

TEMPLE, G. (1960) *Cartesian tensors, an introduction.* Methuen, London.

TERPSTRA, P. (1955) *Introduction to the space groups.* Wolters, Groningen.

THEIMER, O. (1956) Selection rules and temperature dependence of the first-order Raman effect in crystals. *Can. J. Phys.* **34**, 312–38.

TINKHAM, M. (1964) *Group theory and quantum mechanics.* McGraw-Hill, New York.

TOBIN, M. C. (1960) Irreducible representations, symmetry coordinates and the secular equation for line groups. *J. molec. Spectrosc.* **4**, 349–58.

——— (1961) Symmetry-adapted LCAO MO's for long chain polyenes. *J. chem. Phys.* **35**, 1905–6.

TOVSTYUK, K. D. and BERCHA, D. M. (1964) Band symmetry in crystals belonging to $D_{2h}^1 - D_{2h}^{16}$, $D_2^1 - D_2^4$ and $C_{2v}^1 - C_{2v}^{10}$, *Fizika tverd. Tela* **6**, 662–79 (in Russian); *Soviet Phys. solid St.* (English Transl.) **6**, 517–32 (1964).

——— and GEMUS, D. M. (1963) Structure of the spectrum in CdSb-type crystals. *Fizika tverd. Tela* **5**, 142–6 (in Russian); *Soviet Phys. solid St.* (English Transl.) **5**, 100–3 (1963).

——— and SUSHKEVICH, T. N. (1964) Symmetry of bands in crystals of groups $C_{2v}^{11} - C_{2v}^{17}$, D_2^5, D_2^6, $D_{2h}^{17} - D_{2h}^{22}$. *Ukr. fiz. Zh.* **9**, 933–42 (in Ukrainian).

——— and TARNAVSKAYA, M. V. (1963) Investigation of the energy spectrum in crystals with $O_h^1 - O_h^{10}$ structures. *Fizika tverd. Tela* **5**, 819–38 (in Russian); *Soviet Phys. solid St.* (English Transl.) **5**, 597–613 (1963).

——— ——— (1964) Symmetry of the energy bands of current carriers in crystals of cubic syngony. *Ukr. fiz. Zh.* **9**, 629–41 (in Ukrainian).

TREUSCH, J. and SANDROCK, R. (1966) Energy band structures of selenium and tellurium (Kohn–Rostoker method). *Phys. Status Solidi* **16**, 487–97.

TURNER, R. E. and GOODINGS, D. A. (1965) Phase factors of Bloch functions in the localized approximation of the Koster–Slater theory. *Proc. phys. Soc.* **86**, 87–92.

USOV, O. A. and KOTOV, B. A. (1970) Selection rules for the inelastic scattering of neutrons in crystals with the würtzite structure. *Fizika I. Tekhnika Paluprovodnikov* **4**, 830–6 (in Russian); *Soviet Phys., Semicond.* (English Transl.) **4**, 703–8.

VAINSHTEIN, B. K. (1960) Antisymmetry of transformation of Fourier forms. *Kristallografiya* **5**, 341–5 (in Russian); *Soviet Phys. Crystallogr.* (English Transl.) **5**, 323–7 (1960).

VAINSHTEIN, B. K. and ZVYAGIN, B. B. (1963) Representation of crystal-lattice symmetry in reciprocal space. *Kristallografiya* **8**, 147–57 (in Russian); *Soviet Phys. Crystallogr.* (English Transl.) **8**, 107–14.

VAN DER WAERDEN, B. L. and BURCKHARDT, J. J. (1961) Coloured groups. *Z. Kristallogr. Kristallgeom.* **115**, 231–4 (in German).

VAN HOVE, L. (1953) The occurrence of singularities in the elastic frequency distribution of a crystal. *Phys. Rev.* **89**, 1189–93.

VAN HUONG, N. and OLBRYCHSKI, K. (1964) Symmetrized plane waves for the groups $D_{4h}^{17} - D_{4h}^{20}$ at four symmetry points of the Brillouin zone. *Phys. Status Solidi* **7**, 57–65.

VAN KRANENDONK, J. and VAN VLECK, J. H. (1958) Spin waves. *Rev. mod. Phys.* **30**, 1–23.

VAN VLECK, J. H. (1932) *Theory of electric and magnetic susceptibilities.* Clarendon Press, Oxford.

——— (1951) The coupling of angular momentum vectors in molecules. *Rev. mod. Phys.* **23**, 213–27.

VAN WAGENINGEN, R. (1964) Explicit polynomial expressions for finite rotation operators. *Nucl. Phys.* **60**, 250–63.

VENKATARAYUDU, T. and KRISHNAMURTY, T. S. G. (1949*a*) Symmetry properties of wave functions in crystals. Part I. Two-dimensional lattices. *Proc. Indian Acad. Sci.* **A29**, 137–47.

——— ——— (1949*b*) Symmetry properties of wave functions in crystals. Part II. Three-dimensional lattices. *Proc. Indian Acad. Sci.* **A29**, 148–61.

——— ——— (1952) Physical constants of isotropic solids. *Acta crystallogr.* **5**, 287.

——— and RAGHAVACHARYULU, I. V. V. (1959) Reduction of solvable groups. *Proc. Indian Acad. Sci.* **A50**, 196–201.

VON DER LAGE, F. C. and BETHE, H. A. (1947) A method of obtaining electronic eigenfunctions and eigenvalues in solids with an application to sodium. *Phys. Rev.* **71**, 612–22; reprinted, Meijer (1964).

VON NEUMANN, J. (1961, 1963) *John von Neumann: collected works* (edited by A. H. Taub), Vol. I, 1961; Vol. II, 1961; Vol. III, 1961; Vol. IV, 1962; Vol. V, 1963; Vol. VI, 1963. Pergamon Press, Oxford.

WACHTMAN, J. B. and PEISER, H. S. (1962) Symmetry considerations for internal friction caused by jumping of point defects in crystals. *Appl. Phys. Lett.* **1**, 20–2.

——— ——— (1965) Splitting of a set of equivalent sites in centrosymmetric space groups into subsets under homogeneous stress. *J. Res. natn. Bur. Stand.* **A69**, 193–207 (see also Peiser and Wachtman (1965) and Peiser, Wachtman, Munley, and McCleary (1965)).

WAEBER, W. B. (1969) Lattice vibrations of gallium metal. I. Group-theoretical analysis. *J. Phys. C (Solid St. Phys.)* **2**, 882–902.

WAINWRIGHT, T. and PARZEN, G. (1953) Electronic energy bands in crystals. *Phys. Rev.* **92**, 1129–34 (see also Parzen (1953)).

WALLIS, R. F. (1965) *Lattice dynamics. Proceedings of the international conference at Copen-*

hagen, Denmark, August 5–9, 1963. Pergamon Press, Oxford (*J. Phys. Chem. Solids Suppl.* 1).

WANG, G. V. (1966) Determinations of the symmetries of single-electronic levels and phonons in a crystal. *Acta phys. sin.* **22**, 197–204 (in Chinese), *Chin. J. Phys.* (English Transl.) **22**, 155–60 (1966).

WANNIER, G. H. (1959) *Elements of solid state theory.* Cambridge University Press.

WARREN, J. L. (1968) Further considerations on the symmetry properties of the normal vibrations of a crystal. *Rev. mod. Phys.* **40**, 38–76 (see also Maradudin and Vosko (1968)).

WATSON, J. K. G. (1965) The correlation of the symmetry classifications of states of nonrigid molecules. *Can. J. Phys.* **43**, 1996–2007.

WEBER, L. (1929) The symmetry of homogeneous plane point systems. *Z. Kristallogr. Kristallgeom.* **70**, 309–27.

WEISS, R. J. (1963) *Solid state physics for metallurgists.* Pergamon Press, Oxford.

WELLS, A. F. (1962) *Structural inorganic chemistry.* Clarendon Press, Oxford.

WEYL, H. (1931) *The theory of groups and quantum mechanics.* Methuen, London.

——— (1939) *The classical groups, their invariants and representations.* University Press, Princeton, N.J.

——— (1952) *Symmetry.* University Press, Princeton, N. J.

WIGNER, E. P. (1930) On the characteristic elastic vibrations of symmetrical systems. *Nachr. Akad. Wiss. Goettingen* Math-phys. K1. 133–46 (in German); translated in Knox and Gold (1964), Meijer (1964), Cracknell (1968*b*).

——— (1932) On the operation of time reversal in quantum mechanics. *Nachr. Akad. Wiss. Goettingen* Math.-Phys. K1. 546–59 (in German); translated in Meijer (1964).

——— (1941) On representations of certain finite groups. *Am. J. Math.* **63**, 57–63.

——— (1959) *Group theory and its application to the quantum mechanics of atomic spectra.* Academic Press, New York.

——— (1960*a*) Normal form of antiunitary operators. *J. math. Phys.* **1**, 409–13.

——— (1960*b*) Phenomenological distinction between unitary and antiunitary symmetry operators. *J. math. Phys.* **1**, 414–16.

——— and SEITZ, F. (1933) On the constitution of metallic sodium. *Phys. Rev.* **43**, 804–10.

——— ——— (1934) On the constitution of metallic sodium. II. *Phys. Rev.* **46**, 509–24.

(A list of E. P. Wigner's papers is given in *Rev. mod. Phys.* **34**, 587–91 (1962).)

WILSON, A. H. (1953) *The theory of metals.* Cambridge University Press.

WILSON, E. B. (1934) The degeneracy, selection rules, and other properties of the normal vibrations of certain polyatomic molecules. *J. chem. Phys.* **2**, 432–9; reprinted: Cracknell (1968*b*).

——— DECIUS, J. C. and CROSS, P. C. (1955) *Molecular vibrations.* McGraw-Hill, New York.

WINSTON, H. (1954) Studies on localized orbitals. *Phys. Rev.* **94**, 328–36.

——— and HALFORD, R. S. (1949) Motions of molecules in condensed systems. V. Classification of motions and selection rules for spectra according to space symmetry. *J. chem. Phys.* **17**, 607–16.

WINTER, J. G. (1916) *The Prodromus of Nicolaus Steno's Dissertation concerning a solid body enclosed by process of nature within a solid.* Macmillan, New York.

WINTGEN, G. (1941) On the representation theory of space groups. *Math. Annln* **118**, 195–215 (in German).

WITTKE, O. (1962) The colour-symmetry groups and cryptosymmetry groups associated with the 32 crystallographic point groups. *Z. Kristallogr. Kristallgeom.* **117**, 153–65.

——— and GARRIDO, J. (1959) Symmetry of the polychromatic polyhedra. *Bull. Soc. fr. Minér. Cristallogr.* **82**, 223–30 (in French).

WOHLFARTH, E. P. (1953) The energy band structure of a linear metal. *Proc. phys. Soc.* **A66**, 889–93.

WOLF, K. L. (1949) Symmetry and polarity. *Studium gen.* **2**, 213–24 (in German).

——— and WOLFF, R. (1956) *Symmetry*. Böhlau, Cologne.

WONDRATSCHEK, H. (1953) Remarks on the problem of tensor symmetries. *Acta crystallogr.* **6**, 107 (in German).

——— and NEUBÜSER, J. (1967) Determination of the symmetry elements of a space group from the 'general positions' listed in *International tables for X-ray crystallography*, Vol. I. *Acta crystallogr.* **23**, 349–52.

——— and NIGGLI, A. (1961) A generalization of point groups. II. The multiple cryptosymmetry. *Z. Kristallogr. Kristallgeom.* **115**, 1–20 (in German).

WONG, E. Y. (1969) Note on the time-reversal operator. *J. chem. Phys.* **50**, 4120–1.

WOODRUFF, T. O. (1957) The orthogonalized plane-wave method. *Solid St. Phys.* **4**, 367–411.

WOODS, H. J. (1935*a*) The geometrical basis of pattern design. Part I. Point and line symmetry in simple figures and borders. *J. Textile Inst.* **26**, T197–210.

——— (1935*b*) The geometrical basis of pattern design. Part II. Nets and sateens. *J. Textile Inst.* **26**, T293–308.

——— (1935*c*) The geometrical basis of pattern design. Part III. Geometrical symmetry in plane patterns. *J. Textile Inst.* **26**, T341–57.

WOOSTER, W. A. (1938) *Crystal physics*. Cambridge University Press.

WYBOURNE, B. G. (1970) *Symmetry principles and atomic spectroscopy*. Wiley, New York.

WYCKOFF, R. W. G. (1930) *The analytical expression of the results of the theory of space groups*. Carnegie Institution, Washington.

——— (1963) *Crystal structures*, Vol. 1. Interscience, New York.

——— (1964) *Crystal structures*, Vol. 2. *Inorganic compounds* RX_n, R_nMX_2, R_nMX_3. Interscience, New York.

——— (1965) *Crystal structures*, Vol. 3. *Inorganic compounds* $R_x(MX_4)_y$, $R_x(MX_p)_y$, *hydrates and ammoniates*. Interscience, New York.

——— (1966) *Crystal structures*, Vol. 5. *The structures of aliphatic compounds*. Interscience, New York.

——— (1968) *Crystal structures*, Vol. 4. *Miscellaneous inorganic compounds, silicates and basic structural information*. Interscience, New York.

YAGUCHI, K. (1967) Crystallographic and magnetic properties of Gd_3S_4 and Gd_3Se_4. *J. phys. Soc. Japan* **22**, 673–4.

YAMASHITA, J., WAKOH, S., and ASANO, S. (1966) Band theory of super-lattice CoFe. *J. phys. Soc. Japan* **21**, 53–61.

YAMAZAKI, M. (1957*a*) Group-theoretical treatment of the energy bands in metal borides MeB_6. *J. phys. Soc. Japan* **12**, 1–6.

——— (1957*b*) Electronic band structure of boron carbide. *J. chem. Phys.* **27**, 746–51.

YANAGAWA, S. (1953) Theory of the normal modes of vibration in crystals. *Prog. theor. Phys., Kyoto* **10**, 83–94.

YI, P. N., OZIER, I. and ANDERSON, C. H. (1968) Theory of nuclear hyperfine interactions in spherical top molecules. *Phys. Rev.* **165**, 92–109.

YOSHIMORI, A. (1959) A new type of antiferromagnetic structure in the rutile type crystal. *J. phys. Soc. Japan* **14**, 807–21.

YUTSIS, A. P., LEVINSON, I. B., and VANAGAS, V. V. (1962) *Mathematical apparatus of the theory of angular momentum*. Israel Program for Scientific Translations, Jerusalem.

ZAK, J. (1960) Method to obtain the character tables of nonsymmorphic space groups. *J. math. Phys.* **1**, 165–71 (see also Klauder and Gay (1968)).

——— (1962) Selection rules for integrals of Bloch functions. *J. math. Phys.* **3**, 1278–9.

——— (1964*a*) Magnetic translation group. *Phys. Rev.* **134**, A1602–6.

——— (1964*b*) Magnetic translation group. II. Irreducible representations. *Phys. Rev.* **134**, A1607–11.

——— (1964*c*) Group-theoretical consideration of Landau level broadening in crystals. *Phys. Rev.* **136**, A776–80.

——— (1964*d*) One-dimensional equation for a two-dimensional Bloch electron in a magnetic field. *Phys. Rev.* **136**, A1647–9.

——— (1966) Selection rules for scattering processes in crystals. *Phys. Rev.* **151**, 464–7.

——— CASHER, A., GLÜCK, M. and GUR, Y. (1969) *The irreducible representations of space groups*. Benjamin, New York.

ZAMORZAEV, A. M. (1957) Generalization of Fedorov groups. *Kristallografiya* **2**, 15–20 (in Russian); *Soviet Phys. Crystallogr.* (English Transl.) **2**, 10–15 (1957).

——— (1958) Derivation of new Shubnikov groups. *Kristallografiya* **3**, 399–404 (in Russian); *Soviet Phys. Crystallogr.* (English Transl.) **3**, 401–6 (1958).

——— (1962) On the 1651 Shubnikov groups. *Kristallografiya* **7**, 813–21 (in Russian); *Soviet Phys. Crystallogr.* (English Transl.) **7**, 661–8 (1963).

——— (1963) Symmetry groups and groups of the various kinds of antisymmetry. *Kristallografiya* **8**, 307–12 (in Russian); *Soviet Phys. Crystallogr.* (English Transl.) **8**, 241–5 (1963).

——— (1966) Similarity symmetry space groups. *Dokl. Akad. Nauk SSSR* **167**, 334–7 (in Russian); *Soviet Phys. Dokl.* (English Transl.) **11**, 192–4 (1966).

——— (1967) Quasisymmetry (p-symmetry) groups. *Kristallografiya* **12**, 819–25 (in Russian); *Soviet Phys. Crystallogr.* (English Transl.) **12**, 717–22.

——— (1969) Space groups of colour symmetry. *Kristallografiya* **14**, 195–200 (in Russian); *Soviet Phys. Crystallogr.* (English Transl.) **14**, 155–9 (1969).

——— and PALISTRANT, A. F. (1960) The two-dimensional Shubnikov groups. *Kristallografiya* **5**, 517–24 (in Russian); *Soviet Phys. Crystallogr.* (English Transl.) **5**, 497–503 (1961).

——— ——— (1961) Mosaics for 167 two-dimensional Shubnikov groups (three lowest orders). *Kristallografiya* **6**, 163–76 (in Russian); *Soviet Phys. Crystallogr.* (English Transl.) **6**, 127–40 (1961).

——— ——— (1964) The number of generalized Shubnikov space groups. *Kristallografiya* **9**, 778–82 (in Russian); *Soviet Phys. Crystallogr.* (English Transl.) **9**, 660–4 (1965).

——— and SOKOLOV, E. I. (1957) Symmetry and various kinds of antisymmetry of finite bodies. *Kristallografiya* **2**, 9–14 (in Russian); *Soviet Phys. Crystallogr.* (English Transl.) **2**, 5–9 (1957).

ZAMORZAEV, A. M. and TSEKINOVSKII, B. V. (1968) Four-dimensional Bravais lattices. *Kristallografiya* **13**, 211–14 (in Russian); *Soviet Phys. Crystallogr.* (English Transl.) **13**, 165–8 (1968).

ZASSENHAUS, M. (1948) On an algorithm for the determination of space groups. *Comment. math. helvet.* **21**, 117–41 (in German).

ZAWADSKI, W. and LAX, B. (1966) Two-band model for Bloch electrons in crossed electric and magnetic fields. *Phys. Rev. Lett.* **16**, 1001–3.

ZEHLER, V. (1953) The calculation of the energy bands in the diamond crystal. *Annln Phys.* **13**, 229–52 (in German).

ZHELUDEV, I. S. (1957*a*) Symmetry and piezoelectric properties of crystals and 'textures'. *Kristallografiya* **2**, 89–98 (in Russian); *Soviet Phys. Crystallogr.* (English Transl.) **2**, 86–94 (1958).

——— (1957*b*) Symmetry of scalars, vectors and tensors of second rank. *Kristallografiya* **2**, 207–16 (in Russian); *Soviet Phys. Crystallogr.* (English Transl.) **2**, 202–10 (1958).

——— (1957*c*) The point groups of symmetry of crystals and their physical interpretation. *Kristallografiya* **2**, 728–33 (in Russian); *Soviet Phys. Crystallogr.* (English Transl.) **2**, 718–22 (1959).

——— (1960*a*) Similar and dissimilar features of antisymmetry, magnetic symmetry and complete symmetry. *Izv. Akad. Nauk SSSR* **24**, 1436–9 (in Russian); *Bull. Acad. Sci. USSR phys. Ser.* (English Transl.) **24**, 1425–8 (1960).

——— (1960*b*) Complete symmetry of scalars, vectors and tensors of rank two. *Kristallografiya* **5**, 346–53 (in Russian); *Soviet Phys. Crystallogr.* (English Transl.) **5**, 328–34 (1960).

——— (1960*c*) Complete ultimate symmetry groups of scalars, vectors, and second-rank tensors and their combinations. *Kristallografiya* **5**, 508–12 (in Russian); *Soviet Phys. Crystallogr.* (English Transl.) **5**, 489–92 (1961).

——— (1964*a*) The change of complete symmetry of the crystals during the phase transitions in ferroelectrics and ferromagnets. *Proc. Indian Acad. Sci.* **A59**, 168–84.

——— (1964*b*) Cubic and ultimate groups of complete symmetry. *Proc. Indian Acad. Sci.* **A59**, 191–202.

——— and SHUVALOV, L. A. (1956) Ferroelectric phase transitions and crystal symmetry. *Kristallografiya* **1**, 681–8 (in Russian); *Soviet Phys. Crystallogr.* (English Transl.) **1**, 537–42 (1956).

——— ——— (1957) Orientation of domains and macrosymmetry of the characteristics of ferroelectric monocrystals. *Izv. Akad. Nauk SSSR* **21**, 264–74 (in Russian); *Bull. Acad. Sci. USSR phys. Ser.* (English Transl.) **21**, 266–76 (1957).

ZIMAN, J. M. (1960) *Electrons and phonons.* Clarendon Press, Oxford.

ZOCHER, H. and TÖRÖK, C. (1953) About space–time asymmetry in the realm of classical general and crystal physics. *Proc. natn. Acad. Sci. U.S.A.* **39**, 681–6.

Subject index

Note

(i) Where reference is made to the definition of an entry the page number is in bold type.
(ii) Where reference is made to tabular material or figures relevant to an entry the page numbers are in italic type.

Printed and bound by CPI Group (UK) Ltd, Croydon, CR0 4YY